5G Mobile Communications

Wei Xiang • Kan Zheng
Xuemin (Sherman) Shen
Editors

5G Mobile Communications

Springer

Editors
Wei Xiang
James Cook University
Cairns, QLD, Australia

Xuemin (Sherman) Shen
University of Waterloo
Waterloo, ON, Canada

Kan Zheng
Beijing University of Posts
and Telecommunications
Beijing, China

ISBN 978-3-319-34206-1 ISBN 978-3-319-34208-5 (eBook)
DOI 10.1007/978-3-319-34208-5

Library of Congress Control Number: 2016950231

Printed on acid-free paper

This Springer imprint is published by Springer Nature
The registered company is Springer International Publishing AG Switzerland

Preface

Mobile communications have been instrumental in transforming our contemporary societies in the past decades. From the first-generation (1G) of analogue mobile phone system to the newest commercial fourth-generation (4G) long-term evolution (LTE) networks deployed widely across the global, mobile communications have fundamentally changed the ways as to how humans in the modern society access, exchange, and share information with each other. Currently, we are at an era of mobile Internet with explosive big data. The growing demand for mobile data traffic and the proliferation of applications requiring high data rates have raised a significant interest in the definition of new standards in the mobile market. This calls for next-generation mobile communication systems, which should have to respond to an avalanche of traffic, an explosion in the number of connected devices, and the large diversity of use cases and requirements.

Against the above backdrop, the fifth-generation (5G) mobile communications are fast emerging to tackle the challenges brought by an exponential increase in wireless data traffic. On top of the massive increase in data volumes and rates, a formidable challenge for the 5G networks to deal with is how to connect billions of smart devices such as surveillance cameras, smart-home/grid devices, connected sensors, etc. The primary goals of 5G networks are to support a 1000-fold gain in capacity, connections for at least 100 billion devices, and 10 Gb/s delivered to individual users. Furthermore, new 5G networks will be able to provide mass low-latency and ultrareliable connectivity between people, machines, and devices, which will ultimately usher in the era of the Internet of Things (IoT). To meet these enormous challenges, disruptive innovations and drastic improvements need to be made in the mobile network architecture design in both the physical and upper layers.

The International Telecommunications Union (ITU) has stipulated 2020 to be the target year of standardising future 5G mobile networks. Although the detailed technical approaches to implementing 5G mobile networks remain uncertain at the time of this writing, several breakthrough 5G techniques stand out such as massive MIMO and millimetre-wave (mmWave) communications. This book aims to be one of the first comprehensive books to reveal the enabling techniques underpinning next-generation 5G networks and to address the challenges and opportunities brought by 5G mobile communications. Specifically, the book is divided into three major parts: Part I Overview of 5G Networks, Part II Transmission and Design Techniques for 5G Networks, and Part III Networking Techniques and Applications for 5G Networks.

Part I of this book provides a comprehensive introduction to and overview of 5G networks. It consists of three chapters.

The chapter "An Overview of 5G Requirements" presents an overview of next-generation 5G mobile networks. To facilitate the study of 5G requirements and to provide guidance to 5G technical design, this chapter discusses several typical deployment scenarios including indoor hotspot, dense urban, urban macro, rural, and high-speed scenarios. It also presents high-level key capabilities and detailed technical requirements for 5G networks. Some technical performance metrics of 5G networks are also discussed.

The chapter "Spectrum Analysis and Regulations for 5G" discusses various aspects of 5G spectrum issues. It is expected that 5G requires much more bandwidth as well as more flexibility in spectrum usage and management. The suitable frequency ranges of 5G will include those bands below 6 GHz such as re-farmed 2G/3G spectrum, identified frequency bands for IMT, and also WRC-15 candidate bands. However, due to the scarcity of spectrum below 6 GHz, it is imperative to seek potential frequency ranges above 6 GHz. Controlled spectrum sharing is an important way of reusing spectrum to complement current licensed dedicated spectrum, which is still the foundation for the operation of 5G systems.

The chapter "Spectrum Sharing for 5G" first introduces spectrum sharing for 5G systems, which consists of multiple spectrum types with different scenarios. Then, spectrum sharing techniques mapped into different scenarios are introduced, i.e. coordination protocol, GLDB support, cognitive/DSA, and MAC-based coexistence. Besides, current applications of these techniques in real systems are described. Finally, spectrum sharing directions for 5G systems are analysed for different spectrum sharing techniques. It is concluded that licensed dedicated spectrum will continue to be the dominant spectrum usage method for 5G systems due to the possibility to control interference and guarantee coverage, while other spectrum sharing scenarios will act as complementary spectrum usage methods when beneficial.

Part II of this book presents new transmission and design techniques for 5G networks with a focus on physical-layer enabling techniques. It contains 11 chapters.

The chapter "Massive MIMO Communications" argues that every new network generation needs to make a leap in area data throughput, to manage the growing wireless data traffic. Massive MIMO technology can bring at least tenfold improvements in area throughput by increasing the spectral efficiency (bit/s/Hz/cell) while using the same bandwidth and density of base stations as in current networks. These extraordinary gains are achieved by equipping the base stations with arrays of a hundred antennas to enable spatial multiplexing of tens of user terminals.

This chapter explains the basic motivations and communication theory behind the massive MIMO technology and provides implementation-related design guidelines.

The chapter "Millimeter-Wave Mobile Communications" introduces key technologies of mmWave communications. Channel measurements show that mmWave signals suffer from much larger propagation losses and are suitable for small cell coverage. A hybrid network is presented where mmWave is used for capacity enhancement in hotspots, and a low-frequency network is applied for seamless coverage. Uniform air interface is a consequence to simplify the design between mmWave bands and low-frequency bands. Unified access and backhaul technique not only reduces the cost of backhaul but also can meet the requirement of 1000 times capacity enhancement over LTE systems.

The chapter "Non-Orthogonal Multiple Access (NOMA) for Cellular Future Radio Access" introduces state-of-the-art NOMA techniques and evaluates the low density spreading (LDS)-based system, which is a strong candidate for the next generation of mobile networks due to its well-known advantages compared to state-of-the-art techniques based on orthogonal frequency division multiple access (OFDMA). Furthermore, the effect of LDS parameters such as density factor and maximum number of users at each time instance on the sum rate is evaluated. The effect of irregularity on the complexity is also discussed. Moreover, it is shown that the loss of achievable rates which is caused by modulation can be compensated by using a suitable channel coding scheme.

The chapter "New Multicarrier Modulations for 5G" presents recent advances in filter bank multicarrier (FBMC) techniques and compares them with the conventional cyclic prefix (CP)-OFDM approach, in the context of 5G. After a brief description of some adaptations of CP-OFDM, FBMC combined with offset-QAM is considered, pointing out the crucial issue of subchannel equalisation. Then, an alternative approach is proposed, FBMC combined with pulse amplitude modulation (PAM). FBMC-PAM is an attractive option whenever asynchronous access and high level of out-of-band rejection are required. Finally, the case of nonoverlapping emitted symbols is considered, and a CP-less OFDM scheme with frequency domain equaliser in the receiver is included in the performance comparison.

The chapter "Fundamentals of Faster-than-Nyquist Signaling" presents the fundamentals of Faster-than-Nyquist (FTN) signalling. As originally introduced, FTN increases the bit rate in the signalling bandwidth by packing symbols closer in time, at the cost of introducing intersymbol interference (ISI). The chapter begins with the Euclidean distance properties of bandwidth-efficient pulses at FTN rates and describes receivers that mitigate the severe ISI. The FTN achievable information rate is compared with the Nyquist information rate for practical pulses. It then discusses the FTN extension to multicarrier systems with not only time packing but also subcarrier, optimising both the time and frequency packing.

The chapter "Generalized Frequency Division Multiplexing: A Flexible Multi-Carrier Waveform for 5G" aims to develop a unified air interface that can be configured on-the-fly to address emerging 5G applications. Apart from an ever-increasing demand for data rates, 5G is facing new applications such as Tactile

Internet and the Internet of Things. Being aligned with the whole concept of software-defined networking, this chapter introduces the multicarrier waveform termed generalised frequency division multiplexing (GFDM) as the basis for realising such a flexible physical design.

The chapter "Spectrally Efficient Frequency Division Multiplexing for 5G" focuses on novel multicarrier communication techniques, which share the common goal of increasing spectrum efficiency in future communication systems. In particular, a technology termed spectrally efficient frequency division multiplexing (SEFDM) is described in detail outlining its benefits, challenges, and trade-offs when compared to the current state of the art. A decade of research has been devoted to examining SEFDM from different angles: mathematical modelling, algorithm optimisation, hardware implementation, and system experimentation. The aim of this chapter is to therefore give a taste of this technology, and in doing so, the chapter concludes by outlining a number of experimental test beds which have been developed for the purpose of evaluating the performance of SEFDM in practical scenarios.

The chapter "Full-Duplex Wireless Communications for 5G" introduces full-duplex (FD) wireless communications for 5G, which enables simultaneous transmission and reception over the same frequency band. In this way, the spectral efficiency can be improved significantly compared with half-duplex (HD). However, there exists severe self-interference (SI), signal leakage from the local transmitter to its own receiver. Three different classes of SI mitigation techniques are presented in this chapter, i.e. propagation-domain SI suppression, analogue-domain SI cancellation, and digital-domain SI cancellation. Furthermore, the system performance of several FD schemes in several different application scenarios is presented. Theoretically, the spectral efficiency of FD bidirectional and cooperative communications can be doubled, while for cognitive radio networks, the FD-based protocol can achieve much better sensing performance than the traditional HD-based cognitive radio schemes.

The chapter "Device-to-Device Communications over 5G Systems: Standardization, Challenges and Open Issues" introduces one of the key enabling technologies at the heart of the 5G systems, namely, device-to-device (D2D) communications. The potential of D2D communication paradigm holding the promise to overcome the limitations of conventional cellular systems with very high bit rates, low delay, and low power consumption is illustrated. Starting from an overview of D2D communication technology, this chapter will browse through the main aspects that characterise the proximity services, with a view on the standardisation process, the challenges, and the open issues.

The chapter "M2M Communications in 5G" provides an overview of machine-type communications (MTC) within the context of 5G networks. The Internet of Everything foresees a hyperconnected World where humans, things, and machines will need to coexist together. They will be interconnected and Internet-connected via communication networks. In specific, the authors review the key novel challenges of MTC: what is new with regard to human-type traffic (HTC). They then analyse existing communication technologies and how suitable they are for MTC. Finally, the authors identify key technology enablers being considered for the design of 5G networks and provide an outlook for the future.

The chapter "Design Techniques of 5G Mobile Devices in the Dark Silicon Era" is concerned with the design of the prospected 5G mobile communication system, which needs wide skills in wireless communication, analogue circuit design, embedded system, microwave technology, and so forth. System-level analyses, design space exploration, and performance trade-offs are some key steps that enable the design of low-cost, energy-efficient, ubiquitous, and flexible transceiver. This chapter provides comprehensive design techniques for 5G mobile communication in the dark silicon era using More than Moore technology (MtM).

Part III of this book focuses primarily on the networking and application layer techniques for 5G networks, which includes 12 chapters.

The chapter "Ultra-Dense Network Architecture and Technologies for 5G" presents the ultra-density network (UDN), which is the most promising way to meet the ultrahigh area capacity requirement for 5G. The content of this chapter includes characters of UDN scenarios, network architecture design and key technologies like flexible networking, wireless backhauling, multi-RAT coordination, mobility management, interference management, and radio resource management.

The chapter "5G RAN Architecture: C-RAN with NGFI" describes cloud radio access networks (C-RAN), which are viewed as one of the key RAN architectures for 5G networks, with evolved architecture based on a newly designed fronthaul interface, dubbed the next-generation fronthaul interface (NGFI). The design principles and the challenges of NGFI are introduced. A prototype is further developed to verify the applicability of NGFI-based C-RAN.

The chapter "User-Centric Wireless Network for 5G" addresses the concept of user-centric wireless network for 5G from the perspective of fulfilling multiple user experience requirements in 5G. Four key technical directions are studied based on a gap analysis between LTE technology and 5G requirements, i.e. user-centric 5G access network architecture design, flexible functionality and deployment, smart user and traffic awareness and management, and high-efficient low-cost network operation. These key technologies work together with cross-layer and end-to-end solutions to provide the user-centric 5G ecosystem.

The chapter "Energy Harvesting Based Green Heterogeneous Wireless Access for 5G" is concerned with the issues of energy harvesting for future 5G cellular systems. A feasible and efficient method to tackle this issue is to let the communication systems harvest energy from renewable energy sources instead of fossil fuels. However, by employing the energy harvesting (EH) technique, the instability of renewable energy resources introduces new challenges on the design of the upcoming 5G systems. This chapter focuses on uplink access schemes and power allocations for EH-based heterogeneous networks. First, a heterogeneous access model incorporating EH-based mobile users is proposed and followed by a throughput maximisation framework. Then, by classifying transmission policies into two main categories (i.e. single-channel vs. multichannel scenarios), the proposed framework is concretised under various practical conditions, including the availability of central control, causality of harvested energy, channel state information, and others.

The chapter "Resource Management in Sustainable Green HetNets with Renewable Energy Sources" investigates the energy sustainability performance of a green HetNet where the small cell base stations (SBSs) are powered by green energy. Specifically, we first develop an analytical framework to study the energy sustainability of each SBS. Then, we propose a distributed admission control strategy at SBSs striking a balance between resource utilisation and energy sustainability. Extensive simulations validate the analytical framework and demonstrate that relaxing the admission control criteria can improve resource utilisation when the energy is abundant, but may significantly degrade resource utilisation instead when the energy comes short due to poor sustainability performance.

The chapter "Resource Allocation for Cooperative D2D Communication Networks" studies various resource allocation policies for cooperative device-to-device (D2D) communications in systems operating under OFDMA or cognitive radio architectures. A variety of system models are explored, wherein additional features, such as, packet storage, energy harvesting, and cognitive radio capabilities, are incorporated at the user devices so as to enable cooperative D2D communications. Computationally efficient solutions are provided for multiple resource optimisation problems including power allocation, subcarrier allocation, subcarrier pairing, and relay selection. Simulation results demonstrate that the sum-throughput performance can be improved whenever the user devices are equipped with cooperative D2D capabilities.

The chapter "Fog Computing and Its Applications in 5G" explains the emergence of fog computing as a promising, practical, and efficient solution tailored to serving mobile traffics. Fog computing deploys highly virtualised computing and communication facilities at the proximity of mobile users. Dedicated to serving the mobile users, fog computing explores the predictable service demand patterns of mobile users and typically provides desirable localised services accordingly. It can provide mobile users with the demanded services via low-latency and short-distance local connections. The authors introduce the main features of fog computing and describe its concept, architecture, and design goals. Lastly, they discuss the potential research issues from the perspective of 5G networking.

The chapter "A Conceptual 5G Vehicular Networking Architecture" shows how 5G communication systems will help to enable connected future cars to implement automated functions in short term and fully autonomous operation in long term. The authors review the well-known existing communication technologies for connected cars and analyse their shortcomings. Towards this end, they outline the innovation areas that 5G aims to address in order to mitigate the limitations of the current technologies.

The chapter "Communications Protocol Design for 5G Vehicular Networks" provides an overview on existing standards in vehicular networking and highlights new emerging trends towards an integrated infrastructure based on the interworking of heterogeneous technologies. Next-generation mobile vehicular networks are first characterised by providing an insight on relevant stable standards in wireless communication technologies, with a special focus on heterogeneous vehicular networks. Furthermore, the chapter discusses a general framework supporting opportunistic networking scheme and outlines novel application and use cases based on social- and context-awareness paradigms.

The chapter "Next-Generation High-Efficiency WLAN" centres around the topic of next-generation high-efficiency WLAN technology. With the increasing demands for WLAN and the deployment of carrier-WiFi networks, the number of WiFi public hotspots worldwide is expected to increase dramatically. To face this huge increase in the number of densely deployed WiFi networks, and the massive amount of data to be supported by these networks in indoor and outdoor environments, it is necessary to improve the current WiFi standard and define specifications for high-efficiency wireless local area networks (HEWs). This chapter introduces emerging HEW technology, including its typical use cases, environments, and potential techniques that can be applied for HEWs. The typical HEW use cases are first given, followed by an analysis of the main requirements from these use cases and environments. Then, potential techniques, including enhanced medium access and spatial frequency reuse, are presented and discussed.

The chapter "Shaping 5G for the Tactile Internet" investigates the topic of the Tactile Internet, which is expected to have a massive impact on business and society. It has the potential to revolutionise almost every segment of the society by enabling wireless control and remote operation in a range of scenarios. The next-generation (5G) mobile communication networks will play an important role in realising the Tactile Internet. This chapter investigates the interesting area of 5G and Tactile Internet intersection. Key requirements for the Tactile Internet, from a networking perspective, have been identified, after introducing exciting Tactile Internet applications. The chapter covers several technical issues and challenges in shaping 5G networks for realising the vision of the Tactile Internet. The most important challenge would be to ensure tight and scalable integration of various technological solutions into a single network.

We would like to thank all the authors who submitted their research work to this book. We would also like to acknowledge the contribution of many experts who have participated in the review process and offered comments and suggestions to the authors to improve their works. We would also like to express our sincere appreciation to the editors from Springer for their support and assistance during the development of this book.

Cairns, QLD, Australia — Wei Xiang
Beijing, China — Kan Zheng
Waterloo, ON, Canada — Xuemin (Sherman) Shen

Contents

Part I
Overview of 5G Networks

An Overview of 5G Requirements

Dajie Jiang and Guangyi Liu

Abstract Mobile Internet and IoT (Internet of Things) are the two main market drivers for 5G. There will be a massive number of use cases for Mobile Internet and IoT, such as augmented reality, virtual reality, remote computing, eHealth services, automotive driving and so on. All these use cases can be grouped into three usage scenarios, i.e., eMBB (Enhanced mobile broadband), mMTC (Massive machine type communications) and URLLC (Ultra-reliable and low latency communications). Eight key capabilities including peak data rate, latency and connection density, etc., are defined to meet the requirements of usage scenarios. Based on the usage scenarios, several typical deployment scenarios including indoor hotspots, dense urban, urban macro, rural and high-speed scenarios are specified, together with the detailed technical requirements for 5G. Both the deployment scenarios and technical requirements are essential guidance for 5G technical design.

1 Introduction

A mobile and connected society is emerging in the near future, which is characterized by a tremendous amount of growth in connectivity, traffic volume and a much broader range of use scenarios [1]. Some typical trends are summarized as follows:

- Explosive growth of data traffic: There will be an explosive growth in traffic. The global data traffic will increase by more than 200 times from 2010 to 2020, and about 20,000 times from 2010 to 2030;
- Great increase in connected devices: While smart phones are expected to remain as the main personal devices, the number of other kinds of devices, including wearable devices and MTC devices will continue to increase;
- Continuous emergence of new services: Different kinds of services, e.g., services from enterprises, from vertical industries and Internet companies, etc. will be exploited.

D. Jiang (✉) • G. Liu
China Mobile Research Institute, Beijing, China
e-mail: jiangdajie@chinamobile.com

W. Xiang et al. (eds.), *5G Mobile Communications*,
DOI 10.1007/978-3-319-34208-5_1

The fifth-generation (5G) mobile communications system will emerge to meet new and unprecedented demands beyond the capability of previous generations of systems.

There are two phases of 5G requirements research by different organizations. Phase 1 focuses on 5G use cases and high-level key capabilities of 5G networks, and can be regarded as the 5G vision stage. In Phase 1, ITU has released the vision recommendation [2] and defined the key capabilities of 5G. 3GPP started the smarter program [4] and studied 5G use cases and requirements. NGMN completed a 5G whitepaper and defined a large number of 5G use cases and requirements [5]. IMT-2020 (5G) Promotion Group released the 5G vision and requirements whitepaper in May 2014 [1], which aims to contribute to the ITU-R work in Phase 1. Phase 2 focuses on 5G deployment scenarios and detailed technical requirements. There are two important reports in Phase 2. One is the IMT-2020 technical performance requirements from ITU-R which will be completed by February 2017 [3], while the other is the scenarios and requirements technical report from 3GPP which will be completed in March 2017 [6]. NGMN has started the relevant work at the beginning of 2015 and drafted several liaisons to 3GPP and ITU by March 2016 [7, 8]. IMT-2020 (5G) PG plans to complete the evaluation scenarios and the KPI report in the first half of 2016, and will have an impact on the work of ITU and 3GPP in Phase 2 (Fig. 1).

The rest of this chapter is organized as follows. The outcomes of Phase 1, i.e., 5G use cases and high-level key capabilities are introduced in Sects. 2 and 3, respectively. The latest status of Phase 2 including deployment scenarios

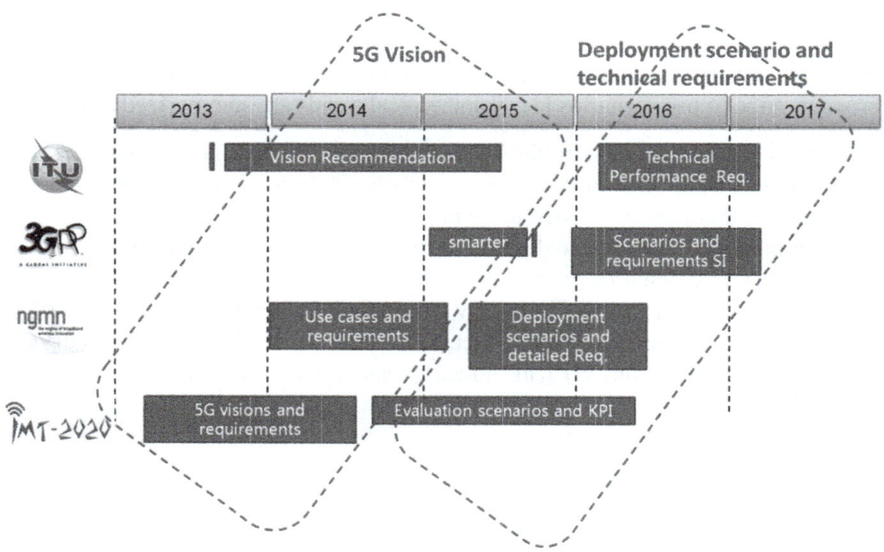

Fig. 1 Overview of 5G requirements research by different organizations

and detailed technical requirements are presented in Sects. 4 and 5, respectively. Section 6 presents the operational requirements, while Sect. 7 draws concluding remarks.

2 Use Cases and Challenges

It is foreseen that there will be a huge number of use cases in the upcoming 5G era. In [1], an overall vision for 5G life is illustrated in Fig. 2. 5G will penetrate into every single element of our future society and create an all-dimensional, user-centered information ecosystem. 5G will break the limitation of time and space to enable an immersive and interactive user experience. 5G will also shorten the distance between human and things, and implement a seamless integration to achieve easy and smart interconnection between people and all things. 5G will enable us to realize the vision—“Information is a finger away, and everything will be kept in touch”.

There are a great many use cases proposed by different organizations [1, 9–11]. Mobile Internet and the Internet of Things (IoT) are the two main market drivers in the future development of mobile communications [1], and they will trigger a large range of use cases.

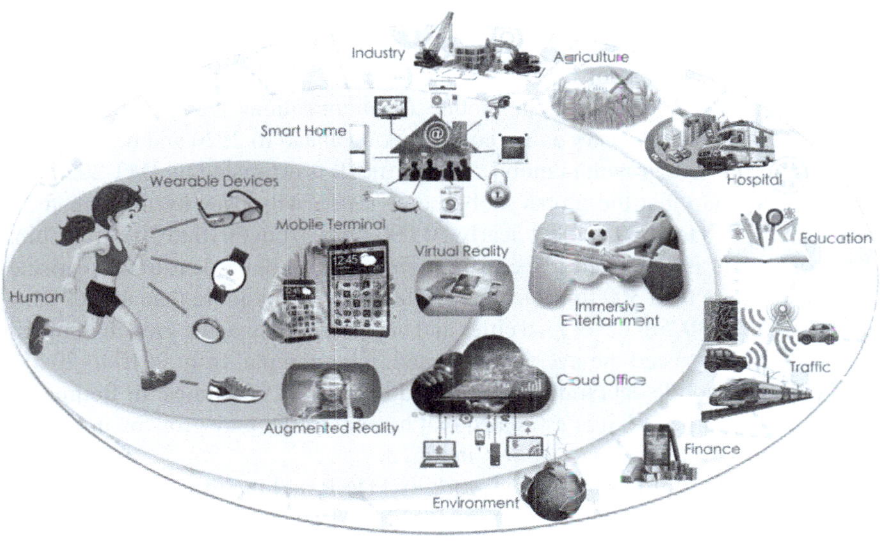

Fig. 2 Overall vision of 5G

2.1 Use Cases of Mobile Internet

Mobile Internet is disrupting the traditional business model of mobile communications, enabling unprecedented user experiences and making a profound impact on every aspect of people's work and life. Looking ahead to 2020 and beyond, mobile Internet will promote the continued evolution of the way human interacting information, and provide users with ultimate experience through more immersive services including but not limited to:

- Video services, such as immersive Ultra High Definition (UHD) and three-dimensional (3D) video
- Augmented reality
- Virtual reality
- Video/photo sharing in stadium/open air gathering
- Online gaming applications
- Mobile cloud/desktop cloud
- Tactile Internet
- Remote computing
- 3D connectivity: aircraft
- 3D connectivity: drones
- Collaborative robots
- Broadcast-like services, like local, regional and national news and information
- Smart office

The future development of Mobile Internet will trigger the growth of mobile traffic by a magnitude of thousands in the future, and promote a new wave of upgrades and a revolution in mobile communications technologies and the telecommunications industry as a whole. Looking ahead to 2020 and beyond, there will be an explosive growth in mobile data traffic. It is estimated by IMT-2020 (5G) Promotion Group that the global mobile data traffic will grow by more than 200 times from 2010 to 2020, and by nearly 20,000 times from 2010 to 2030. In China, the growth rate is projected to be even higher, with mobile data traffic expected to grow by more than 300 times from 2010 to 2020 and by more than 40,000 times from 2010 to 2030. For developed cities and hotspots in China, the growth of mobile data traffic will exceed the average projected. For example, from 2010 to 2020 in Shanghai, the mobile data traffic is projected to grow by 600 times. In Beijing and during this same period, it is estimated that hotspot traffic may grow by up to 1000 times. The above estimation is shown in Fig. 3.

There is some traffic anticipation work by ITU-R and the results are detailed in Report ITU-R M.[IMT.2020BEYOND TRAFFIC] [12]. This report contains global IMT traffic estimates beyond 2020 from several sources. These estimates anticipate that global IMT traffic will grow by 10–100 times from 2020 to 2030. The main drivers behind the anticipated traffic growth include increased video usage, device proliferation and application uptake. These are expected to evolve over time, and this evolution will differ between countries due to social and economic differences.

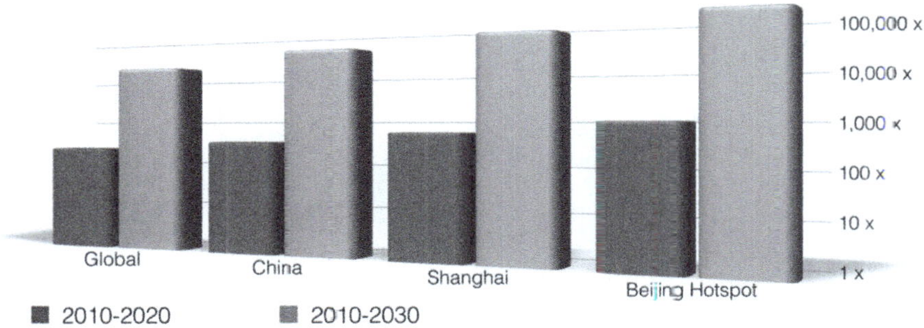

Fig. 3 2010–2030 growth of mobile data traffic

Traffic asymmetry aspects for this period are also presented by ITU-R. It is observed that the current average traffic asymmetry ratio of mobile broadband is in favor of the downlink, and this is expected to increase due to a growing demand for audio-visual content.

Mobile Internet is aiming at people-oriented communications with a focus on the user experience. Towards 2020 and beyond, the increasing popularity of ultra-high definition (UHD), 3D and video immersion will significantly drive up the data rates. For example, with a hundredfold compression, the transmission of 8K (3D) video will require a transmission rate close to 1 Gbps. Services, such as augmented reality, desktop cloud, and online gaming will not only pose a challenge to uplink and downlink data transmission rates but also generate a stringent requirement for the so-called "imperceptible latency". In the future, vast amounts of individuals and office data will be stored in the cloud. Such massive data activity will require transmission rates to be comparable to optical fiber communications, which will lead to enormous traffic challenges for mobile communications networks particularly in hotspot areas. Over-the-top (OTT) services, such as social networking, will be counted among leading applications going forward, and the associated frequently-occurring small packets will devour signaling resources. At the same time, consumers will continue to demand better experiences on mobile communications wherever they are. A consistent service experience is expected in all scenarios, including ultra-dense scenarios such as stadiums, open-air gatherings and concerts, and high-speed moving scenarios such as high-speed trains, vehicles and subways.

The total number of devices connected by global mobile communications networks will reach 100 billion in the future. By 2020, it is predicted that the number of mobile terminals around the world will surpass ten billion, of which China will contribute over two billion, as shown in Fig. 4.

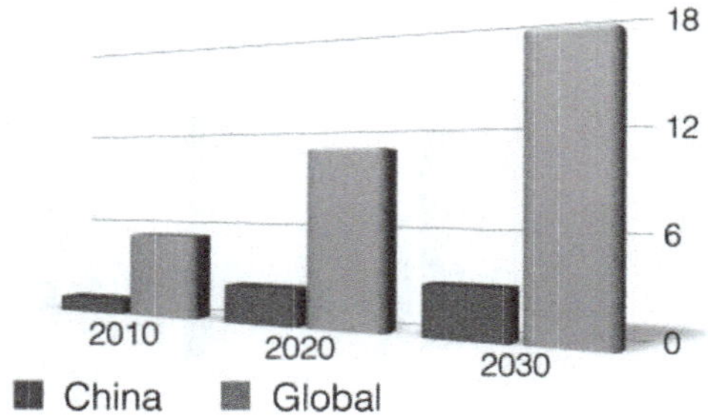

Fig. 4 2010–2030 growth of mobile device

2.2 *Use Cases of Internet of Things*

The IoT has extended the scope of mobile communications services from interpersonal communications to interconnection between tings (smart devices), and between people and things, allowing mobile communications technologies to penetrate into broader industries and fields. By 2020 and beyond, applications such as mobile health, Internet of Vehicles (IoV), smart home, industrial control, and environmental monitoring will drive the explosive growth of IoT applications, facilitating hundreds of billions of devices to connect to a network creating a true "Internet of Everything". This will give rise to emerging industries of a unprecedented scale and instill infinite vitality to mobile communications. Meanwhile, the massive number of interconnected devices and the diversified IoT services will also pose new challenges to mobile communications. The potential IoT use cases include:

- Smart Grid and critical infrastructure monitoring
- Environmental monitoring
- Smart agriculture
- Smart metering
- eHealth services
- Remote object manipulation like remote surgery
- Automotive driving/Internet of vehicles
- Smart wearables, like sports and fitness
- Sensor networks
- Mobile video surveillance
- Smart cities
- Smart transportation
- Smart home
- Industrial control

IMT-2020 (5G) Promotion Group has estimated the number of IoT devices in future years as shown in Fig. 5. It is projected that the total number of devices connected by the global mobile communications network will reach 100 billion in the future. By 2020, the number of mobile terminals around the world will surpass ten billion, of which China will contribute over two billion. The number of IoT

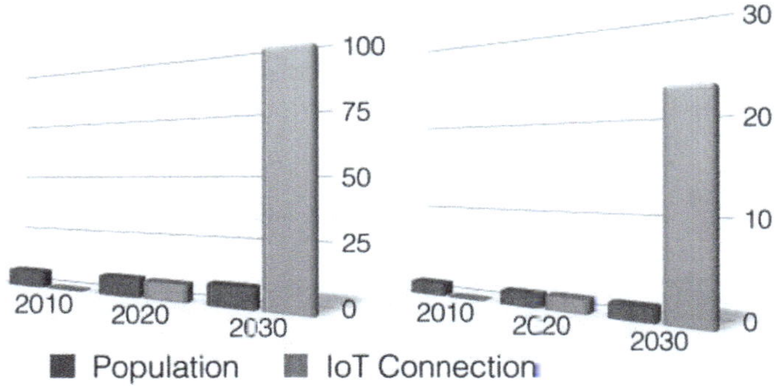

Fig. 5 2010–2030 growth of IoT connections

connections will also expand rapidly, reaching the size of global population of seven billion by 2020. By 2030, the number of global IoT connections will reach 100 billion. Among all types of terminals, smart phones will contribute most traffic, and IoT terminals will contribute less, even though the number of devices is much larger.

IoT is focused on communications between things and between things and people, involving not only individual users, but also a large number of various vertical industrial customers. IoT services types and relevant requirements of IoT services are very diverse. For services such as the smart home, smart grid, environmental monitoring, smart agriculture, and smart metering, the network will be required to support a massive amount of device connections and frequently-occurring small data packets. Services like video surveillance and mobile health will have strict requirements on transmission rates, while services such as IOV and industrial control will demand millisecond-level latency and nearly 100 % reliability. In addition, many IoT devices may be deployed in remote, or in areas where transmission losses can be a problem, such as indoor corners, basements and tunnels. Therefore, the coverage of mobile communications networks need to be further enhanced. In order to penetrate into more IoT services, 5G should be more flexible and more scalable, to support massive device connections and meet diverse user requirements.

Users expect better and yet more cost-effective services and experiences with mobile Internet and IoT. In addition to satisfying cost and experience demands, 5G will also need to meet extremely high security requirements, particularly for services such as e-banking, security monitoring, safe driving, and mobile health. 5G will also be able to support lower power consumption to build greener mobile communication networks and to enable much longer terminal battery life, especially for some IoT devices.

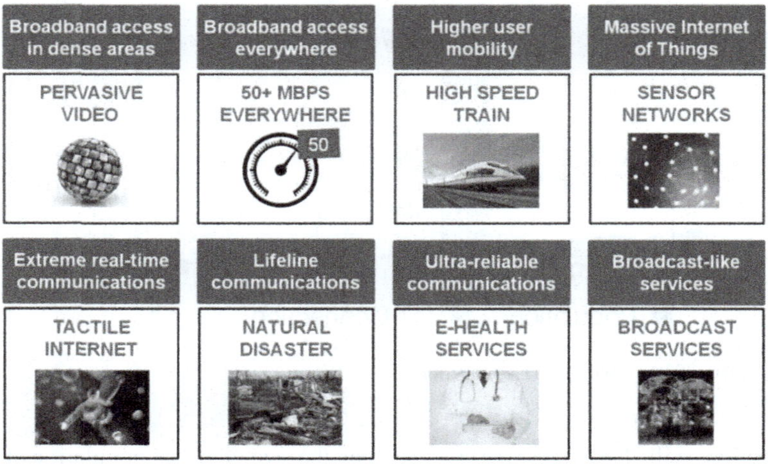

Fig. 6 5G use case families and related use case examples

2.3 *Classification of 5G Use Cases*

5G will support a large variety of use cases which are emerging now or will emerge in the future. Different use cases have varying characteristic and requirements. It is helpful to group countless emerging use cases into several use case families. Use cases in each use case family share similar characteristic and requirements.

NGMN has developed 25 use cases for 5G as representative examples, which are grouped into eight use case families. The following diagram [5] illustrates the eight use case families with one example use case given for each family, and these families and their corresponding use case examples are described in Fig. 6.

ITU-R has concluded three usage scenarios (use case groups) addressing different use case characteristics in Fig. 7 [2]:

- Enhanced mobile broadband: Mobile broadband addresses human-centric use cases for access to multi-media content, services and data. The demand for mobile broadband will continue to increase, leading to enhanced mobile broadband. The enhanced mobile broadband usage scenario will come with new application areas and requirements in addition to existing mobile broadband applications for improved performance and an increasingly seamless user experience.
- Ultra-reliable and low latency communications: This use case has stringent requirements for capabilities such as throughput, latency and availability. Some examples include wireless control of industrial manufacturing or production processes, remote medical surgery, distribution automation in a smart grid, transportation safety, etc.

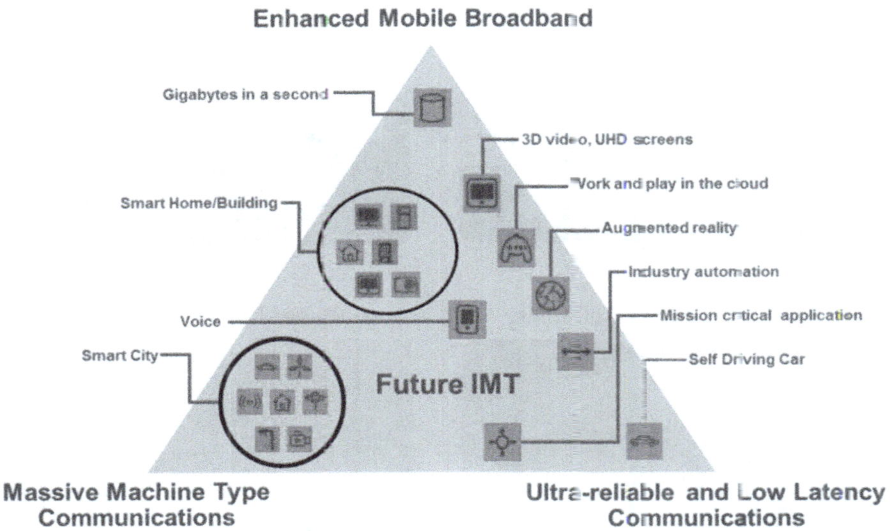

Fig. 7 5G usage scenarios

- Massive machine type communications: This use case is characterized by a very large number of connected devices typically transmitting a relatively low volume of non-delay-sensitive data. Devices are required to be of low cost, and have a very long battery life.

Similarly, IMT-2020 (5G) Promotion Group has proposed four technical scenarios for 5G in Fig. 8 which are well in line with the three usage scenarios from ITU-R. The main difference is that the eMBB scenario from ITU-R is divided into two technical scenarios, i.e., the seamless wide-area coverage scenario and high-capacity hot-spot scenario.

For the seamless wide-area coverage scenario, seamless coverage and medium to high mobility are desired, with much improved user data rate compared to existing data rates. However, the data rate requirement may be relaxed compared to the hotspot scenario. For the high-capacity hot-spot scenario, i.e., for an area with high user density, very high traffic capacity is needed, while the requirement for mobility is low and the user data rate is higher than that of wide area coverage. For the wide area coverage case, seamless coverage and medium to high mobility are desired, with much improved user data rate compared to existing data rates. However, the data rate requirement may be relaxed compared with the hotspot.

Figure 9 shows the mapping from the eight use case families proposed by NGMN to the three usage scenarios defined by ITU-R. The eMBB usage scenario consists of broadband access in dense areas, broadband access everywhere, higher user mobility and broadcast-like services. URLLC consists of extreme real-time communications, lifeline communications and ultra-reliable communications. mMTC corresponds to massive Internet of Things.

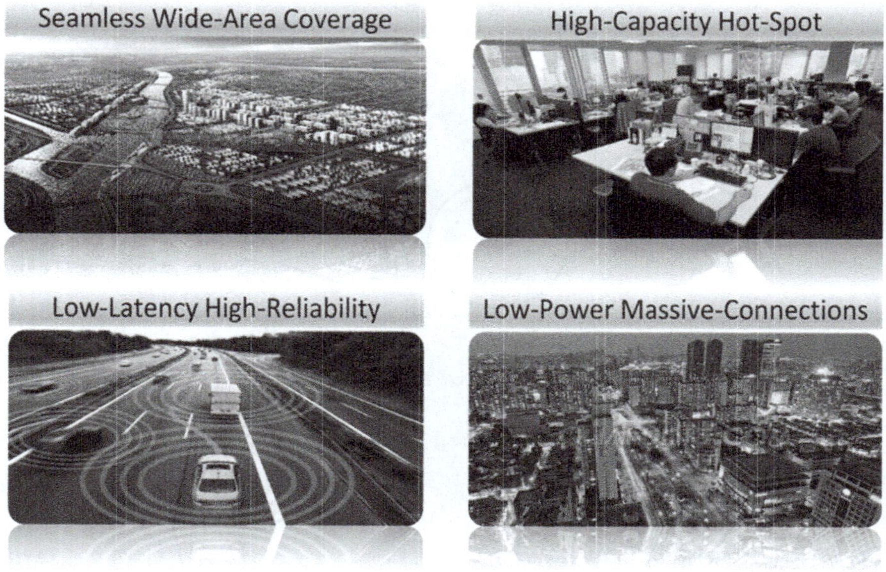

Fig. 8 Technical scenarios for 5G

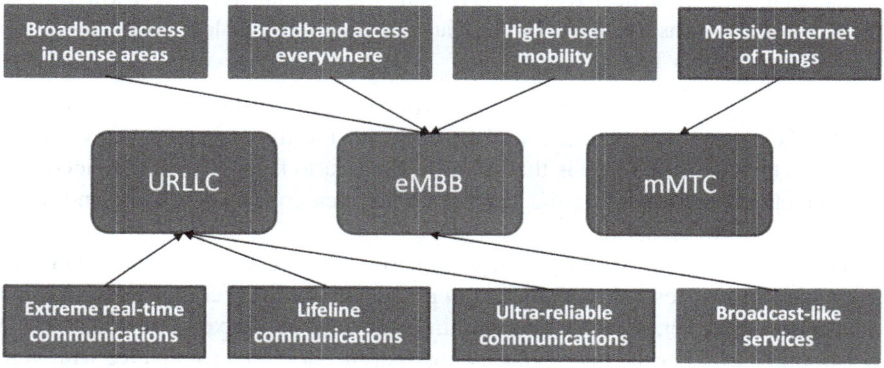

Fig. 9 Mapping eight use case families to three use scenarios

3 High-Level Key Capabilities

IMT-2020 (5G) PG groups the 5G high-level requirements into several performance indicators and efficiency indicators. The key performance indicators for 5G include the user experienced data rate, connection density, end-to-end delay, traffic volume density, mobility, and peak date rate. Their definitions are listed in Table 1. The value for each performance indicator and the relevant scenario are illustrated in Fig. 10.

Table 1 5G Performance indicators

Performance indicators	Definition
User experienced data rate (bps)	Minimum achievable data rate for a user in real network environment
Connection density (/km^2)	Total number of connected devices per unit area
End-to-end latency (ms)	Duration between the transmission of a data packet from the source node and the successful reception at destination node
Traffic volume density (bps/km^2)	Total data rate of all users per unit area
Mobility (km/h)	Relative speed between receiver and transmitter under certain performance requirement
Peak date rate (bps)	Maximum achievable data rate per user

Several problems are anticipated if today's networks are used to handle the explosive development of mobile Internet and IoT:

- The energy efficiency level, overall cost per bit and complexity of network deployment and maintenance cannot effectively handle 1000 times traffic growth and the massive number of connected devices in the next decade;
- Co-existence of multiple radio access technologies (RAT) causes increased complexity and degraded user experience;
- Existing networks can not realize accurate monitoring of network resources and effective awareness of services, and therefore they cannot intelligently fulfill the diversified requirements of future users and services;
- Widely distributed and fragmented spectrum will cause interference and co-existence complexity.

To solve these problems, 5G should have the following capabilities to achieve sustainability. In terms of network construction and deployment, 5G networks need to:

- Provide higher network capacity and better coverage, while decreasing the complexity and cost of network deployment, especially the deployment of ultra-dense networks.
- Have a flexible and scalable architecture to adapt to the diverse needs of users and services.
- Make flexible and efficient use of various spectrum resources, including paired and unpaired spectrum, re-farmed spectrum and new spectrum, low-frequency and high-frequency bands, and licensed and unlicensed bands.
- Have stronger device-connection capabilities to deal with the access requirements of huge amounts of IoT devices.

In terms of operation and maintenance (O&M), 5G needs to:

- Improve network energy efficiency and the O&M cost-per-bit to cope with data traffic growth and the diverse needs of various services and applications.

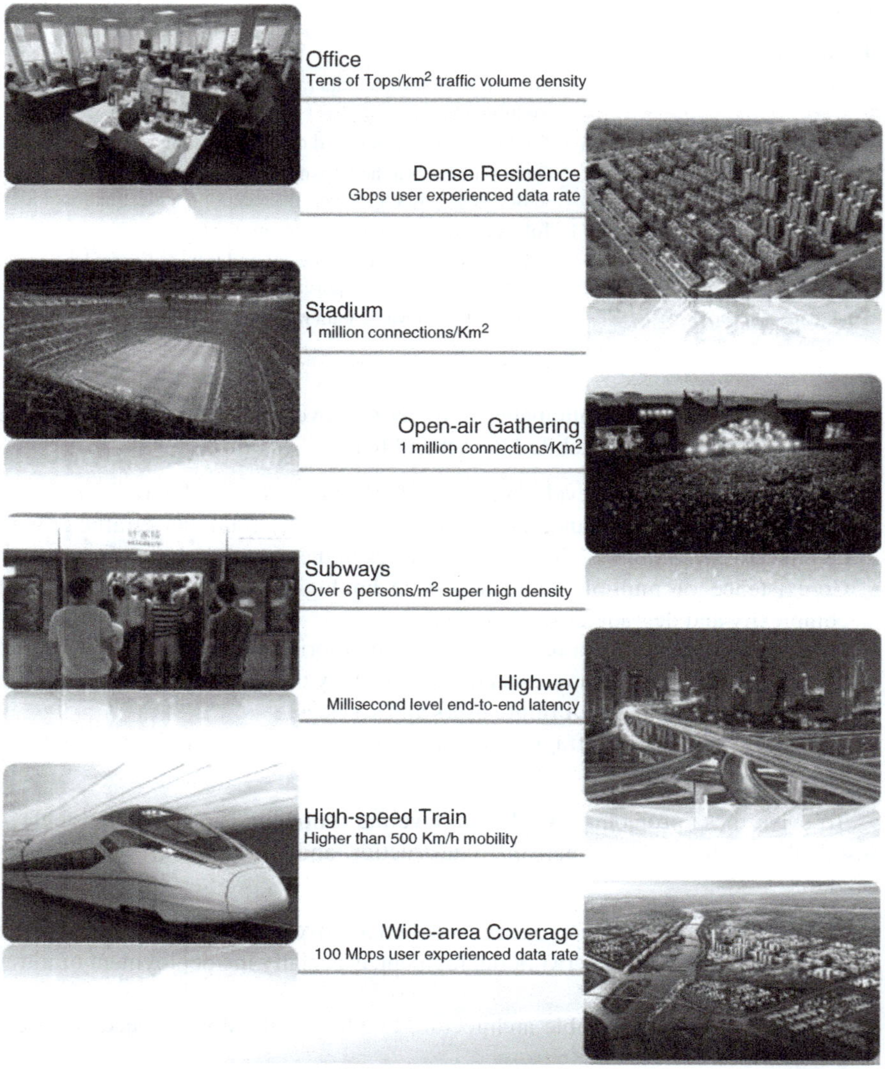

Fig. 10 Challenging scenarios and performance indicators

- Reduce the complexity caused by the co-existence of multiple radio access technologies, network upgrades, and the introduction of new features and functions, to improve users' experience.
- Make intelligent optimization based on awareness of users behaviors and services contents
- Provide a variety of network security solutions to meet the needs of all types of devices and services of mobile internet and IoT.

Table 2 5G Key efficiency indicators

Efficiency indicators	Definition
Spectrum efficiency (bps/Hz/cell or bps/Hz/km^2)	Data throughput per unit of spectrum resource per cell (or per unit area)
Energy efficiency (bit/J)	Number of bits that can be transmitted per joule of energy
Cost efficiency (bit/¥)	Number of bits that can be transmitted per unit cost

Spectrum utilization, energy consumption and cost are the three key factors which must be addressed in sustainable mobile communication networks. In order to achieve sustainability, 5G needs to make significant improvements in the following aspects (Table 2):

- Spectrum efficiency: 3–5 times
- Energy efficiency: 100+ times
- Cost efficiency: 100+ times

5G systems must dramatically outperform previous generation systems. 5G should support

- User experienced data rate: 0.1–1 Gbps
- Connection density: one million connections per square kilometer
- End-to-end latency: millisecond level
- Traffic volume density: tens of Gbps per square kilometer
- Mobility: higher than 500 km per hour
- Peak data rate: tens of Gbps

Among these requirements, the user experienced data rate, connection density and end-to-end latency are the three most fundamental ones. Meanwhile, 5G needs to significantly improve the efficiency of network deployment and operations. Compared with 4G, 5G should have 3–5 times improvement on spectrum efficiency and more than 100 times improvement on energy and cost efficiency.

The performance requirements and efficiency requirements define the key capabilities of 5G, which can be illustrated as a “blooming flower” depicted in Fig. 11. The petals and leaves rely on each other. The petals represent the six key capabilities in terms of performance and can fulfill the diverse requirements of future services and scenarios. The leaves represent the three key capabilities in terms of efficiency, and can guarantee the sustainable development of 5G. The top of each petal means the maximum value of the corresponding capability.

The key capabilities of IMT-2020 defined by ITU-R are shown in Fig. 12, compared with those of IMT-Advanced. The values for each key capability are shown in Table 3.

All key capabilities may to some extent be important for most use cases, and the relevance of certain key capabilities may be significantly different, depending on the use cases/scenario. The importance of each key capability for the usage scenarios

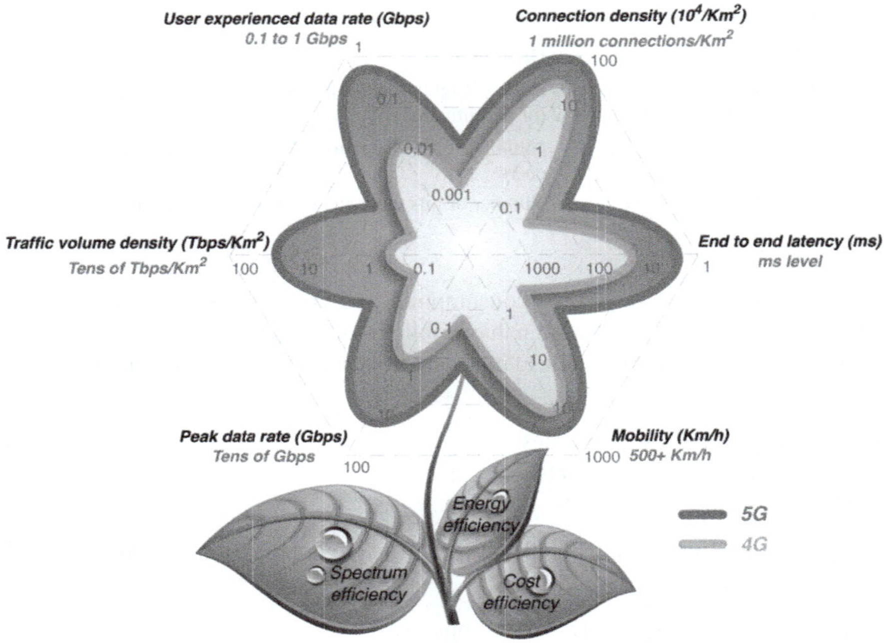

Fig. 11 5G key capabilities

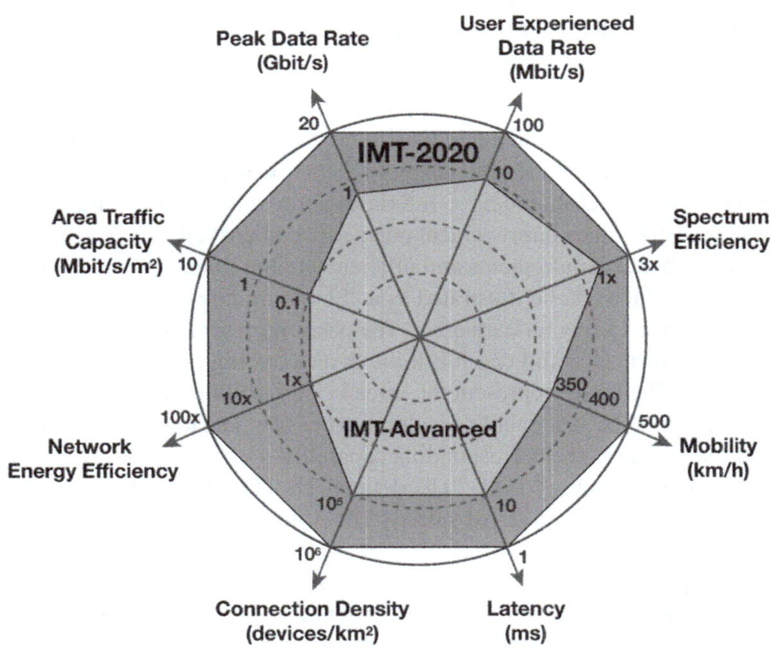

Fig. 12 Enhancement of key capabilities from IMT-Advanced to IMT-2020

Table 3 5G Key capabilities and values from ITU-R

Key capabilities	Values
Peak data rate	20 Gbps
User experienced data rate	0.1–1 Gbps
Latency	1 ms over-the-air
Mobility	500 km/h
Connection density	10^6/km^2
Energy efficiency	100 times compared with IMT-Advanced
Spectrum efficiency	3–5 times compared with IMT-Advanced
Area traffic capacity	10 Mbit/s/m^2

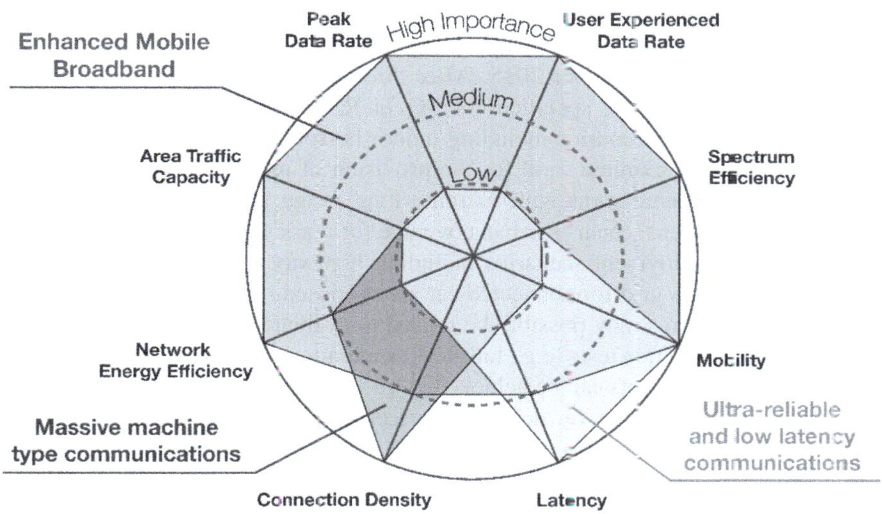

Fig. 13 Importance of key capabilities in different usage scenarios

enhanced mobile broadband, ultra-reliable and low latency communication and massive machine-type communication is illustrated in Fig. 13. This is done using an indicative scaling in three steps as "high", "medium" and "low".

In the enhanced mobile broadband scenario, the user experienced data rate, area traffic capacity, peak data rate, mobility, energy efficiency and spectrum efficiency all have high importance, but mobility and the user experienced data rate would not have equal importance simultaneously in all use cases. For example, in hotspots, a higher user experienced data rate, but a lower mobility, would be required than in the wide area coverage case.

In the ultra-reliable and low latency communications scenarios, low latency is of highest importance, e.g., safety critical applications. Such capability would be required in some high mobility cases as well, e.g., transportation safety, while high data rates could be less important.

In the massive machine type communication scenario, high connection density is needed to support a tremendous number of devices in the network that may transmit only occasionally, at low bit rate and with zero/very low mobility. A low cost device with long operational lifetime is vital for this usage scenario.

4 Deployments Scenarios

Use cases will be delivered across a wide range of environments. To facilitate the study of 5G requirements and provide guidance to 5G technical design, several typical deployment scenarios need to be specified.

3GPP has started study on scenarios and requirements for next generation access technologies [6] in December 2015. After three months' study, there are totally 10 deployment scenarios specified for 5G in RAN plenary meeting in March 2016. 7 deployment scenarios including indoor hotspot, dense urban, rural, urban macro, high speed, extreme rural for the provision of minimal services over long distances and extreme rural with extreme long range are proposed mainly for eMBB, 1 deployment scenario, urban coverage for massive connection is proposed for mMTC, 2 deployment scenarios including highway scenario (for Internet of vehicles) and urban grid for connected car are proposed for URLLC. Some eMBB deployment scenarios may possibly be reused to evaluate mMTC and URLLC, or some specific evaluation tests (e.g., link-level simulation) can be developed to check whether the requirements can be achieved.

High-level descriptions on deployment scenarios including carrier frequency, aggregated system bandwidth, network layout including Inter-Site-Distance (ISD), UE distribution, UE moving speed and service profile are proposed in the following sections.

- Indoor hotspots

The indoor hotspot deployment scenario focuses on small coverage per cell and high user throughput or user density in buildings. The key characteristics of this deployment scenario are high capacity, high user density and consistent user experience indoor.

This scenario represents indoor offices with a total area of 120 m × 50 m. 12 small cells are deployed with an ISD of 20 m. The BS antenna height is 3 m. The carrier frequency options include 4 GHz, around 30 and 70 GHz. The bandwidth for 4 GHz is up to 200 MHz. The bandwidth for around both 30 and 70 GHz is up to 1 GHz. 10 users per cell are distributed uniformly and all users are indoors with 3 km per hour velocity. Full buffer and/or burst traffic model is assumed.

- Dense urban

The dense urban heterogeneous deployment scenario focuses on macro cells with micro cells and high user densities and traffic loads in city centres and dense urban areas. The key characteristics of this deployment scenario are high traffic loads, outdoor and outdoor-to-indoor coverage. This scenario will be interference-limited, using macro TRPs with or without micro cells. A continuous cellular layout and the associated interference shall be assumed.

The ISD for the macro cells is 200 m. There are 3 micro cells per macro cell. The macro BS antenna height is 25 m and micro BS antenna height is 10 m. The carrier frequency for macro cell is 4 GHz. The carrier frequencies for micro cell include 4 GHz, around 30 GHz and around 70 GHz. The bandwidth for 4 GHz is up to 200 MHz and bandwidth for both around 30 GHz and around 70 GHz is up to 1 GHz. Full buffer and/or burst traffic model is assumed. 10 UEs are distributed per micro sector. 80 % users are indoor with a moving speed of 3 km per hour and 20 % is in cars with a velocity of 30 km per hour.

- Urban macro

The urban macro homogeneous deployment scenario focuses on large cells and continuous coverage. The key characteristics of this scenario are continuous and ubiquitous coverage in urban areas. This scenario will be interference-limited, using macro TRPs (i.e. radio access points above rooftop level).

The ISD in this scenario is 500 m. The BS antenna height is 35 m. The carrier frequency is 4 GHz, 2 GHz, and around 30 GHz. The bandwidth for 4 GHz is up to 200 MHz and bandwidth for 2 GHz is up to 100 MHz. Full buffer and/or burst traffic model is assumed. 10 UEs are distributed per cell. 80 % users are indoor with 3 km per hour velocity and 20 % is in cars with 30 km per hour velocity.

- Rural

The rural deployment scenario focuses on larger and continuous coverage. The key characteristics of this scenario are continuous wide area coverage supporting high speed vehicles. This scenario will be noise-limited and/or interference-limited, using macro TRPs.

The ISD in this scenario is 1732 m or 5000 m. The BS antenna height is 35 m. The carrier frequency is 700 MHz. The bandwidth is up to 20 MHz. Full buffer and/or burst traffic model is assumed. 10 UEs are distributed per cell. 50 % users are indoors with 3 km per hour velocity and 50 % is in cars with 120 km per hour velocity.

4 GHz and 2 GHz frequency are also considered in this scenario.

- High speed

The high speed deployment scenario focuses on continuous coverage along track in high speed trains. The key characteristics of this scenario are consistent user experience with very high mobility. In this deployment scenario, dedicated linear deployment along railway line is considered and UEs are located in train carriages. If the antenna of relay node for eNB-to-Relay is located at top of one carriage of the train, the antenna of relay node for Relay-to-UE could be distributed to all carriages.

- Extreme rural for the provision of minimal services over long distances

The extreme rural deployment scenario is defined to allow the provision of minimal services over long distances for low average revenue per user (ARPU) and low density areas including both humans and machines. The key characteristics of this scenario are macro cells with very large area coverage supporting basic data and voice services, with low to moderate user throughput and low user density.

- Extreme rural with extreme long range

The extreme rural long range deployment scenario is defined to allow for the provision of services for very large areas such as wilderness or areas with only highways. The key characteristics of this scenario are macro cells with very large area coverage supporting basic data speeds and voice services, with low to moderate user throughput and low user density.

- Urban coverage for massive connection

The urban coverage for massive connection scenario focuses on large cells and continuous coverage to provide mMTC. The key characteristics of this scenario are continuous and ubiquitous coverage in urban areas, with very high connection density of mMTC devices. This deployment scenario is for the evaluation of the KPI of connection density.

- Highway scenario

The highway deployment scenario focuses on scenario of vehicles on highways with high speeds. The main KPIs evaluated under this scenario include reliability/availability under high speeds/mobility (and thus frequent handover operations).

- Urban grid for connected car

The urban macro deployment scenario focuses on the scenario of highly dense deployed vehicles placed in urban areas. It can cover the scenario where freeways lead through an urban grid. The main KPI evaluated under this scenario are reliability/availability/latency in high network load and high UE density scenarios.

5 Detailed Technical Requirements

- Peak data rate

Peak data rate is the highest theoretical data rate which is the received data rate assuming error-free conditions assignable to a single mobile station, when all assignable radio resources for the corresponding link direction are utilized (i.e., excluding radio resources that are used for physical layer synchronization, reference signals or pilots, guard bands and guard times).

The target for peak data rates are 20 Gbps for downlink and 10 Gbps for uplink.

- Peak spectral efficiency

Peak spectral efficiency is the highest theoretical data rate (normalized by bandwidth), which is the received data rate assuming error-free conditions assignable to a single mobile station, when all assignable radio resources for the corresponding link direction are utilized (i.e., excluding radio resources that are used for physical layer synchronization, reference signals or pilots, guard bands and guard times).

The targets for peak spectral efficiency are 30 bps/Hz for downlink and 15 bps/Hz for uplink.

Higher frequency bands imply higher bandwidths but lower spectral efficiency, while lower frequency bands results in lower bandwidths but higher spectral efficiency. Thus, the peak data rate cannot be directly derived from the peak spectral efficiency and bandwidth multiplication.

- Bandwidth

Bandwidth means the maximal aggregated total system bandwidth. It may be supported by a single or multiple RF carriers.

The value for this KPI is for further study.

- Control plane latency

Control plane latency refers to the time to move from a battery efficient state (e.g., IDLE) to the start of continuous data transfer (e.g., ACTIVE).

The target for control plane latency should be 10 ms.

- User plane latency

The time taken to successfully deliver an application layer packet/message from the radio protocol layer 2/3 SDU ingress point to the radio protocol layer 2/3 SDU egress point via the radio interface in both the uplink and downlink directions, where neither device nor base station reception is restricted by DRX.

For URLLC the target for user plane latency should be 0.5 ms for UL and 0.5 ms for DL. Furthermore, if possible, the latency should also be low enough to support the use of next-generation access technologies as a wireless transport technology that can be used within the next-generation access architecture.

For other cases, the target for user plane latency should be 4 ms for UL and 4 ms for DL.

- Latency for infrequent small packets

For infrequent application layer small packet/message transfer, the time it takes to successfully deliver an application layer packet/message from the radio protocol layer 2/3 SDU ingress point at the mobile device to the radio protocol layer 2/3 SDU egress point in the RAN, when the mobile device starts from its most "battery efficient" state.

- Mobility interruption time

Mobility interruption time means the shortest time duration supported by the system during which a user terminal cannot exchange user plane packets with any base station during transition.

The target for mobility interruption time should be zero.

This KPI is for intra-system mobility.

- Inter-system mobility

Inter-system mobility refers to the ability to support mobility between the IMT-2020 system and at least one IMT system.

- Reliability

Reliability can be evaluated by the success probability of transmitting e.g., 20 bytes within 1 ms, which is the time it takes to deliver a small data packet from the radio protocol layer 2/3 SDU ingress point to the radio protocol layer 2/3 SDU egress point of the radio interface, at a certain channel quality (e.g., coverage-edge).

The target for reliability should be $1–10^{-5}$ within 1 ms.

- Coverage

"Maximum coupling loss" (MCL) on the uplink and downlink between device and Base Station site (antenna connector(s)) for a data rate of 160 bps, where the data rate is observed at the egress/ingress point of the radio protocol stack in uplink and downlink.

The target for coverage should be 164 dB.

- UE battery life

The UE battery life is determined by the battery life of the UE without recharge. For mMTC, the UE battery life in extreme coverage depends on the activity of mobile originated data transfer consisting of 200 bytes UL per day followed by 20 bytes DL from MCL of 164 dB, assuming a stored energy capacity of 5 Wh.

The target for UE battery life is beyond 10 years.

- UE energy efficiency

UE energy efficiency means the capability of a UE to sustain much better mobile broadband data rate while minimizing the UE modem energy consumption.

- Cell/Transmission Point/TRP spectral efficiency

TRP spectral efficiency is defined as the aggregate throughput of all users (the number of correctly received bits, i.e., the number of bits contained in the service data units (SDUs) delivered to Layer 3, over a certain period of time) divided by the channel bandwidth divided by the number of TRPs. A three-sector site consists of 3 TRPs. In case of multiple discontinuous "carriers" (one carrier refers to a continuous block of spectrum), this KPI should be calculated per carrier. In this case, the aggregate throughput, channel bandwidth, and the number of TRPs on the specific carrier are employed.

The target considered as a starting point for eMBB deployment scenarios is in the order of 3× IMT-Advanced requirements for full buffer.

- Area traffic capacity

Area traffic capacity means total traffic throughput served per geographic area (in Mbit/s/m^2). This metric can be evaluated by two different traffic models, i.e., full buffer model and non-full buffer model.

Full buffer model:

The total traffic throughput served per geographic area (in Mbit/s/m^2). The computation of this metric is based on full buffer traffic.

Non-full buffer model:

The total traffic throughput served per geographic area (in Mbit/s/m^2). Both the user experienced data rate and the area traffic capacity need to be evaluated at the same time using the same traffic model.

The area traffic capacity is a measure of how much traffic a network can carry per unit area. It depends on the site density, bandwidth and spectrum efficiency. In the special case of a single layer single band system, it may be expressed as:

$$\begin{aligned}\text{area capacity }\left(\text{bps}/\text{m}^2\right) &= \text{site density }\left(\text{site}/\text{m}^2\right) \times \text{bandwidth (Hz)}\\ &\quad \times \text{spectrum efficiency (bps/Hz/site)}\,.\end{aligned}$$

Based on the above, it is proposed to use the spectrum efficiency results together with the assumptions on the available bandwidth and site density in order to derive a quantitative area traffic capacity KPI for information.

- User experienced data rate

User experienced data rate can be evaluated for non-full buffer traffic and for full buffer traffic. Non-full buffer simulations are preferred for the evaluation of this KPI.

For non-full buffer traffic, the user experienced data rate is 5 % of the user throughput. User throughput (during active time) is defined as the size of a burst divided by the time between the arrival of the first packet of the burst and the reception of the last packet of the burst.

The target values for the user experienced data rate are associated with non-full buffer evaluation. The non-full buffer user experienced data rate target is applicable at the non-full buffer area traffic capacity traffic level.

For full buffer traffic, the user experienced data rate is calculated as:

$$\text{user experienced data rate} = 5\%\ \text{user spectrum efficiency} \times \text{bandwidth}$$

To improve user experienced data rates, 3GPP can develop standards with means for high 5 % user spectrum efficiency. To this end, 5 % user spectrum efficiency gains in the order of three times that IMT-Advanced are proposed. Furthermore, 3GPP can develop standards with means for large bandwidth support. Towards this end, it is proposed that at least 1 GHz aggregated bandwidth shall be supported.

The available bandwidth and site density, which both have a strong impact on the available user experienced data rates, are however beyond the control of 3GPP.

Based on this, the full buffer experienced user data rate is evaluated for information without numerical requirements.

- 5th percentile user spectrum efficiency

5th percentile user spectrum efficiency means the 5 % point of the cumulative distribution function (CDF) of the normalized user throughput. The (normalized) user throughput is defined as the average user throughput (the number of correctly received bits by users, i.e., the number of bits contained in the SDU delivered to Layer 3, over a certain period of time, divided by the channel bandwidth and is measured in bit/s/Hz. The channel bandwidth for this purpose is defined as the effective bandwidth multiplies the frequency reuse factor, where the effective bandwidth is the operating bandwidth normalized appropriately considering the uplink/downlink ratio. In the case of multiple discontinuous "carriers" (one carrier refers to a continuous block of spectrum), this KPI should be calculated per carrier. In this case, the user throughput and channel bandwidth on the specific carrier are employed.

The target considered as a starting point for eMBB deployment scenarios is in the order of 3× IMT-Advanced requirements for full buffer.

- Connection density

Connection density refers to the total number of devices fulfilling specific QoS per unit area (per km^2). QoS definition should take into account the amount of data or access request generated within a time X (to be studied in future) that can be sent or received within a given time, Y (to be studied in future), with Z% (to be studied in future) probability.

The target for connection density should be 1,000,000 device/km^2 in the urban environment.

- Mobility

Mobility means the maximum user speed (km/h) at which a defined QoS can be achieved.

The target for mobility target is 500 km/h.

- Network energy efficiency

The capability is to minimize the RAN energy consumption while providing much better area traffic capacity.

Qualitative KPIs as the baseline and quantitative KPI are for further study.

6 Requirements for RAN Architecture

The RAN design for the Next Generation Radio Access Technologies shall be designed to fulfill the following requirements:

- The RAN architecture shall support tight interworking between the new RAT and LTE.
- Considering high performing inter-RAT mobility and aggregation of data flows via at least dual connectivity between LTE and new RAT. This shall be supported for both collocated and non-collocated site deployments;
- The RAN architecture shall support connectivity through multiple transmission points, either collocated or non-collocated;
- The RAN architecture shall enable the separation of control plane signaling and user plane data from different sites;
- The RAN architecture shall support interfaces supporting effective inter-site scheduling coordination;
- Different options and flexibility for splitting the RAN architecture shall be allowed;
- The RAN architecture shall allow for deployment flexibility e.g., to host relevant RAN, CN and application functions close together at the edges of the network, when needed, e.g., to enable context aware service delivery, low latency services, etc;
- The RAN architecture shall allow for C-plane/U-plane separation;
- The RAN architecture shall allow deployments using Network Function Virtualization;
- The RAN architecture shall allow for the RAN and the CN to evolve independently;
- The RAN architecture shall allow for the operation of Network Slicing;
- The RAN architecture shall support sharing of the RAN between multiple operators;
- The design of the RAN architecture shall allow for rapid and efficient deployment of new services;
- The design of the RAN architecture shall allow the support of 3GPP defined service classes (e.g. interactive, background, streaming and conversational);
- The design of the RAN architecture shall enable lower CAPEX/OPEX with respect to current networks to achieve the same level of services;
- RAN-CN interfaces and RAN internal interfaces (both between new RAT logical nodes/functions and between new RAT and LTE logical nodes/functions) shall be open for multi-vendor interoperability;
- The RAN architecture shall support operator-controlled sidelink (device-to-device) operation, both in coverage and out of coverage

7 Chapter Summary

Mobile Internet and IoT are the two main market drivers in the future development of mobile communications, and they will provide a broad range of prospects for 5G. There will be a massive number of use cases in 5G era, such as augmented reality, virtual reality, remote computing, eHealth services, automotive driving and so on. All these use cases can be grouped into three categories, i.e., eMBB, mMTC and URLLC. To facilitate the study of 5G requirements and provide guidance to 5G technical design, several typical deployment scenarios are specified including indoor hotspots, dense urban, urban macro, rural and high-speed scenarios.

This chapter also presents high-level key capabilities and detailed technical requirements for 5G. 5G will be able to sustainably satisfy the requirement of thousands of times mobile data traffic growth. 5G will provide users with fiber-like access data rate and "zero" latency user experience. It will be capable of connecting 100 billion devices. 5G will be able to deliver a consistent experience across a variety of scenarios including the cases of ultra-high traffic volume density, ultra-high connection density and ultra-high mobility. 5G will also be able to provide intelligent optimization based on services and user awareness, and will improve energy and cost efficiency by over a hundred times, enabling us all to realize the vision of 5G—"Information a finger away, everything in touch".

References

1. Whitepaper, 5G vision and requirements, IMT-2020 (5G) Promotion Group
2. Recommendation M.2083, Framework and overall objectives of the future development of IMT for 2020 and beyond, ITU-R
3. Workplan, timeline, process and deliverables for the future development of IMT, ITU-R WP5D
4. New WID on New Services and Markets Technology Enablers – SmarTer, Vodafone, February 2015
5. 5G white paper, NGMN, March 2015
6. RP-160689, Study on Scenarios and Requirements for Next Generation Access Technologies, 3GPP TR 38.913 V0.3.0, March 2016
7. Further Elaboration on NGMN Requirement Metrics and Deployment Scenarios for IMT-2020 in Response to Liaison Statement from ITU-R WP5D from 4th February 2015, NGMN Alliance, February 2016
8. Further elaboration on NGMN KPIs and requirements for 5G, NGMN Alliance, March 2016
9. Deliverable D1.1, Scenarios, requirements and KPIs for 5G mobile and wireless system, METIS
10. Whitepaper, 5G NEW WAVE TOWARDS FUTURE SOCIETIES IN THE 2020S, 5G forum
11. Whitepaper, 4G Americas Recommendations on 5G Requirements and Solutions, 4G Americas
12. Report, IMT Traffic estimates for the years 2020 to 2030, ITU-R

Spectrum Analysis and Regulations for 5G

Tan Wang, Gen Li, Biao Huang, Qingyu Miao, Jian Fang, Pengpeng Li, Haifeng Tan, Wei Li, Jiaxin Ding, Jingchun Li, and Ying Wang

Abstract In 5G vision, the spectrum issue is one of the most important parts. Governments, agencies, standardization organizations and research institutions from many countries pay high attention to the 5G spectrum strategies. The appeals for international harmonized spectrum and full band spectrum access are intense, the range of which are possibly from 0-100GHz. This chapter addresses the current spectrum for mobile communications, the future spectrum demand, possible candidate frequency bands and spectrum management considerations. Information from international and regional telecommunications such as ITU, CEPT, APT, and also from different countries are collected and analyzed. Some academic views are also provided for future work.

1 Introduction

Mobile communication has gone deep into people's lives. For the year 2020 and beyond, all kinds of new services will continue to emerge, mobile data traffic will show explosive growth, and connections from terminals will increase exponentially. With the rapid development of Mobile Internet and Internet of Things, the 5G will be integrated into each field of the society in the future, and construct a full range of information ecosystem, providing a broad prospect.

Spectrum is the one of the most valuable resource for mobile communications. Among the 5G studies and pre-standardization works, the spectrum issue is one of the most important parts. In this context, the global 5G research institutions pay high attention to the research on spectrum, such as EU FP7 METIS project [1] and China's IMT-2020 Promotion Group [http://www.imt-2020.org.cn/zh]. In addition, some countries put the spectrum for next generation of mobile communication in

T. Wang (✉) • B. Huang • J. Fang • P. Li • H. Tan • W. Li • J. Ding • J. Li
State Radio Monitoring Center/State Radio Spectrum Management Center, MIIT, Beijing, China
e-mail: wangtan@srrc.org.cn

G. Li • Q. Miao
Ericsson, China Radio Research Lab, Beijing, China

Y. Wang
Beijing University of Posts and Telecommunications (BUPT), Beijing, China

W. Xiang et al. (eds.), *5G Mobile Communications*,
DOI 10.1007/978-3-319-34208-5_2

Fig. 1 General consideration of spectrum for 5G

a prominent position [2, 3]. Furthermore, in the framework of ITU, the ITU-R study groups (e.g., WP 5D) and related regional telecommunication organizations (including ASMG, APT, CEPT, CITEL, RCC, ATU and so on) are actively conducting studies for 5G spectrum.

According to current research, generally, spectrum below 6 GHz is the best resource in the near future [4]. These bands mainly include re-farmed 2G/3G spectrum, identified frequency band for IMT in Radio Regulations and also WRC-15 candidate bands. However, due to its scarcity and increasingly difficulty to realize international harmonization after WRC-15, it is the time to seek spectrum above 6 GHz, as shown in Fig. 1. The 5G spectrum solution will be a comprehensive outcome, indicating combinations of different frequency ranges for different scenarios. It is expected that 5G requires more in the total amount, wider with respect to individual bandwidths, greater in the range and more flexible in the usage and management pattern.

In this chapter, 5G spectrum related issues are investigated, discussed and summarized, taking typical regions and countries as examples around the world including Asia-Pacific, Europe and Americas. In Sect. 2, the current used spectrum for mobile communication is described. Section 3 focuses on the 5G spectrum demand, followed by the 5G potential candidate frequency bands in Sect. 4. Some new ideas in 5G spectrum management are investigated in Sect. 5 to give some food for thought. We conclude this chapter in Sect. 6.

2 Current Spectrum for Mobile Communications

The sharing of information on frequency allocation and spectrum usages will be beneficial to reach harmonization on spectrum usages among all countries around the world, especially for mobile communications. Moreover, as described in Sect. 4, the current allocated spectrum for International Mobile Telecommunication (IMT) is an important part of 5G candidate spectrum. In this regard, this chapter starts with the usage status of mobile communication spectrum. In different countries, due to the different development stages of mobile communication industry, the total amount of spectrum, different frequency ranges for 2G/3G/4G vary widely, but also reflects a certain degree of consistency.

Take the situation of Asia as the first step. In Asia, it was agreed to develop an APT Report on information of mobile operators' frequencies, technologies and license durations in Asia Pacific countries [5]. The APT report is a compilation

Table 1 Operators, frequency bands, used technology and license duration of Japan

Japan				
	Frequency band (MHz)			
Operator	Uplink	Downlink	Technology	License duration
NTT DoCoMo, Inc.	728–738	783–793	LTE	Period: 5 years (Note)
	830–845	875–890	WCDMA/LTE	
	1447.9–1462.9	1495.9–1510.9	LTE	
	1764.9–1784.9	1859.9–1879.9	WCDMA/LTE	
	1940–1960	2130–2150	WCDMA/LTE	
	3480–3520 (TDD)		LTE	
KDDI Corporation	718–728	773–783	LTE	
	815–830	860–875	CDMA2000/LTE	
	1437.9–1447.9	1485.9–1495.9	LTE	
	1920–1940	2110–2130	CDMA2000/LTE	
	3520–3560 (TDD)		LTE	
Softbank Mobile Corp.	900–915	945–960	WCDMA/LTE	
	1427.9–1437.9	1475.9–1485.9	WCDMA	
	1960–1980	2150–2170	WCDMA/LTE	
	3560–3600 (TDD)		LTE	
Ymobile Corp.	738–748	793–803	LTE	
	1749.9–1764.9	1844.9–1859.9	WCDMA/LTE	
Ymobile Corp.	1884.5–1915.7 (TDD)		PHS	
Wireless City Planning Inc.	2545–2575 (TDD)		AXGP (Advanced eXtended Global Platform)	
UQ Communications Inc.	2595–2645 (TDD)		WiMAX/WiMAX2+	

Note: License is granted to Base Station each. Expiration date of license for each station is different

of responses to the Questionnaire and up to now includes information from 26 members such as Australia, China, India, Iran, Japan, Korea, and Thailand, etc. One example with relatively new information is shown in Table 1.

For US, the public mobile communications spectrum is mainly reflected in Title 47 Telecommunication [http://reboot.fcc.gov/reform/systems/spectrum-dashboard]. As for Europe, the ECC is in response of developing European common positions and proposals for efficient use of radio spectrum in the framework of international and regional bodies. A European Table of Frequency Allocations and Applications (ECA Table) [6] is provided by CEPT.

In general, the frequency table for mobile communication in countries can be found in national official publications or output documents in regional organiza-tions. It is shown that almost all the frequency ranges are focus on the spectrum below 3 GHz. Another observation from these data is that generally the amount of spectrum currently used for mobile communications is several hundred megahertz.

3 5G Spectrum Demand

Every few years, ITU-R sets up agenda items in advance to study the future spectrum demand for IMT, and to support the consideration of additional spectrum allocations. Currently, ITU-R has already finished the study on IMT spectrum demand towards the year 2020. Some countries have also started the study on spectrum demand beyond 2020.

3.1 Demand in the Year 2020

How to calculate IMT spectrum demands? Generally, a methodology starts with an analysis of future market and traffic volume, moves on to calculate and distribute the traffic on different RATs, and then calculates the required capacity, before concluding the estimation. The actual calculation process can be very complicated, when there are a variety of traffic types, different environments and multiple cell types of different RATs. For example, imagine how to estimate the data rates of a high quality video streaming user located in indoor offices, connecting with future 5G small cells, in the year 2025.

Many countries have made contributions to the calculation methodology and output results. Table 2 summarizes national spectrum requirements as provided by some countries and organizations during the study of ITU-R, in the form of total amount for all the operators in one country [7]. It should be noted that these national spectrum requirements have differences in the methodology used and assumptions made (e.g., differences in traffic/radio-aspects related parameters, differences in estimation year, differences in estimates based on whether the spectrum requirements are total or additional, etc.).

Some of them, such as GSMA (GSM Association) and UK, focused on the improvements of existing ITU-R method specified in Recommendation M.1768-1.

The methodology of Recommendation ITU-R M.1768-1 is developed by ITU-R Study Group, which is used in WRC-07 to calculate the spectrum demand in the future. The methodology provides the spectrum requirements of IMT as a whole, and divided between two radio access technique groups (RATGs):

- RATG 1: Pre-IMT systems, IMT-2000 and its enhancements;
- RATG 2: IMT-Advanced.

The methodology reflects certain recent advances in IMT technologies and the deployment of IMT networks such as the introduction of spectrum sharing between the macro and micro cell layers in IMT Advanced, and the introduction of a new spectrum granularity parameter for IMT systems.

Figure 2 summarizes the steps of the calculation algorithm employed in Recommendation ITU-R M.1768-1 in conjunction with relevant input parameters (detailed descriptions of these parameters are provided in Sect. 4) [7]. The methodology starts

Table 2 Summary of national spectrum requirements in some countries [7]

Source	US	Australia	Russia	China	GSMA	India	UK
Estimation year	Until 2014	Until 2020	2020	2015, 2020	2020	2017, 2020	2020
Spectrum requirements	Additional 275 MHz by 2014	1081 MHz in total	1065 MHz in total	570–690 MHz in total by 20151490–1810 MHz in total by 2020	1600–1800 MHz in total for some countries	Additional 300 MHz by 2017Additional another 200 MHz by 2020	775–1080 MHz in total for low2230–2770 in total MHz for high
Methodology	Original	Original	Original	Rec. ITU-R M.1768-1	Complementary to Rec. ITU-R M.1768-1	Original	Rec. ITU-R M.1768-1

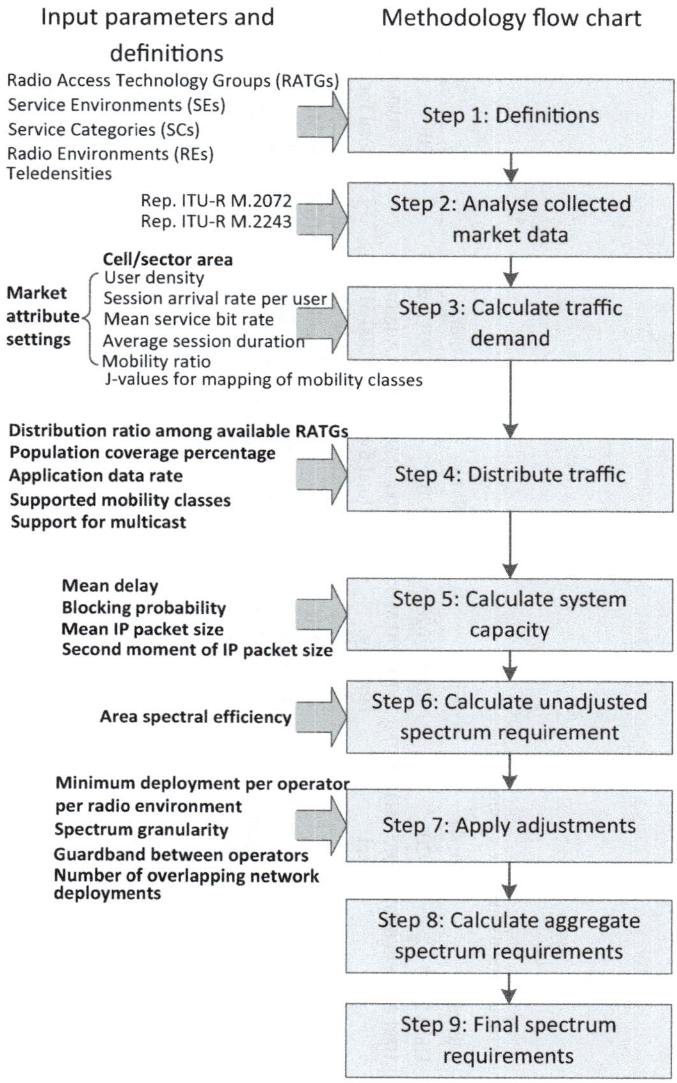

Fig. 2 Steps of calculation algorithm and relevant input parameters [3]

from market studies that characterize all of the traffic carried by IMT and other mobile systems, corresponding to Steps 2–3. In Step 4, the total traffic obtained from the market studies is distributed among different radio environments (cell layers) and RATGs according to factors such as traffic characteristics (required data rates and user mobility), RATG capabilities (supported data rates, available cell types and their coverages), etc. The system capacity required to carry the offered traffic is then calculated in Step 5 using separate capacity calculation algorithms

Table 3 Total spectrum requirements for both RATG 1 and RATG 2 in the year 2020 [7]

Total spectrum requirements	RATG 1	RATG 2	RATGs 1 and 2
Lower user density settings	440 MHz	900 MHz	1340 MHz
Higher user density settings	540 MHz	1420 MHz	1960 MHz

for reservation-based traffic and packet-based traffic respectively. Initial spectrum estimates are obtained from the capacity requirements by dividing the latter by the spectral efficiencies (Step 6). Then, adjustments are made to take into account network deployments with the spectrum requirements being aggregated over the relevant deployments (Steps 7–8). Finally, the methodology outputs the overall spectrum requirements of RATG 1 and RATG 2, which collectively denote IMT systems (Step 9).

There are differences in the markets and deployments and timings of the mobile data growth in different countries. Therefore, two settings are developed to characterize lower and higher user density settings. These two sets of market study input parameter values are considered in the calculations to characterize differences in the user densities in different countries. Table 3 shows the calculated spectrum requirements for both RATGs 1 and 2.

In some countries, national spectrum requirement can be lower than the estimate derived by lower user density settings and in some other countries, national spectrum requirement can be higher than the estimate derived by higher user density settings.

Besides using existing methodologies, there are also original methodologies proposed by US [8], Australia [9], Russia [10] and so on. These methodologies are reflected as case studies based on national considerations. In the following, a brief example is shown from China, which is conducted from the IMT-2020 Promotion Group.

The main point of the new method is to estimate the upper bound of demand. In fact, it is proved by the operational data that the area which has the largest spectrum demand is always in metropolises. When requirements of such a scenario are satisfied, the solution might be applicable for others as well. In the calculation, a typical hotspot zone in Beijing is selected as the research area. To gather the historical data within, the government issued an investigation letter to all network operators. The data survey is based on the operator's network management system in China, which is a major difference between the proposed method and other international methods. Supported by these first-hand data, analyses are made to estimate the traffic increase, traffic distribution and Base Station (BS) deployments in the future. The general flow chart of the methodology is shown in Fig. 3. According to the calculation, the total IMT spectrum demand in China is 1350–1810 MHz in 2020 [11], which are a little different from but similar with the result from the calculation by M.1768-1 (1490–1810 MHz).

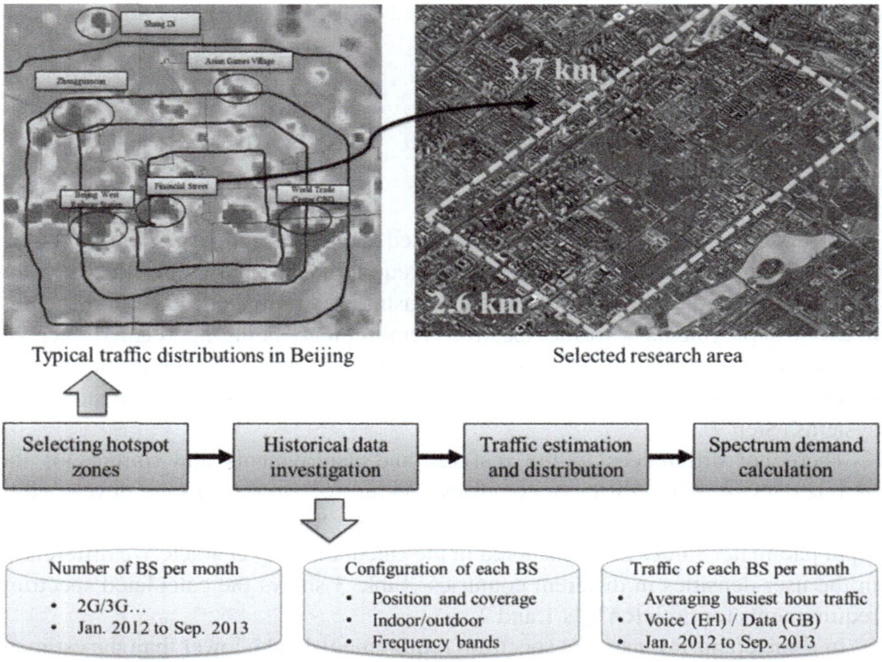

Fig. 3 Flow chart of the spectrum demand methodology proposed by China, where a 3.7 km × 2.6 km square area around Beijing Financial Street is selected as the typical dense traffic area [4]

The obtained estimation results regardless from different methods are huge. Compared with current identified spectrum for IMT in different countries, there are still several hundred to even 1000 MHz deficit. Moreover, in the period of 5G commercialization after 2020, the total spectrum demand may continue to increase.

3.2 *Demand Beyond the Year 2020*

The above result is, to some extent, a total amount of spectrum and the range depends on the assumptions made in the estimation process. However, in view of 5G beyond 2020, it is probably difficult to measure the new demand only in a total number. For example, in 5G indoor high traffic scenarios, in order to achieve high peak data rates, the demand for frequency bandwidth may be up to several GHz. This can be solved by using higher frequencies and denser deployments [12]. But these solutions may not be the best choice for outdoor wide area scenarios. Therefore, it will be useful to separately estimate how much spectrum is required for coverage, capacity, performance and connections for each 5G scenario as Fig. 4 shows, to

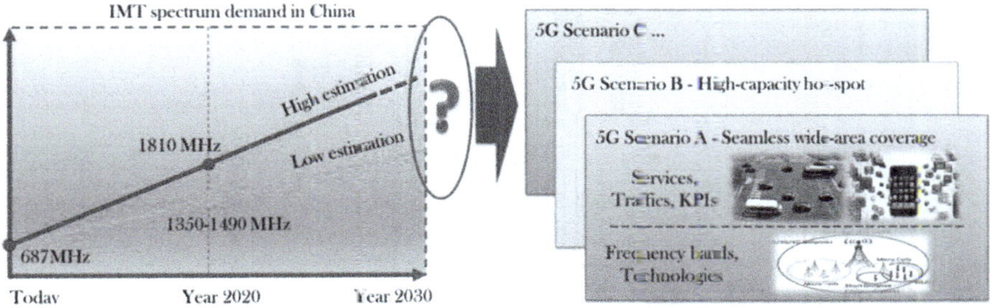

Fig. 4 A scenario-based solution for spectrum demand in 2030 [http://www.imt-2020.org.cn/zh]

perform a mapping onto different frequency bands, such as bands below 1 GHz, between 1 and 6 GHz and above 6 GHz. It may be believed that the results from different scenarios will vary a lot.

Generally, the 5G spectrum demand estimation will be a comprehensive outcome, indicating spectrum solutions for different scenarios. It is expected that 5G requires more in the total amount, wider with respect to individual bandwidths, greater in the range and more flexible in the usage pattern.

4 5G Candidate Frequency Bands

Is the world ready to provide sufficient frequency bands for 5G? From the study point of view, the potential bands can be divided into two parts: bands below 6 GHz and above [4].

4.1 Spectrum Below 6 GHz

For wireless communications, lower frequencies provide better coverage. Currently, almost all countries are using spectrum below 6 GHz for IMT systems. Besides achieving high data rates, it is also necessary to guarantee wide-area coverage and outdoor-to-indoor coverage in 5G. Therefore, spectrum below 6 GHz forms a very important part for the 5G spectrum solution. Potential 5G spectrum below 6 GHz includes the following aspects:

- Spectrum re-farming

In order to realize the benefits of new generation of mobile communication systems, regulators need to deliver efforts to enable re-farming of spectrum in the frequency band occupied by the old ones.

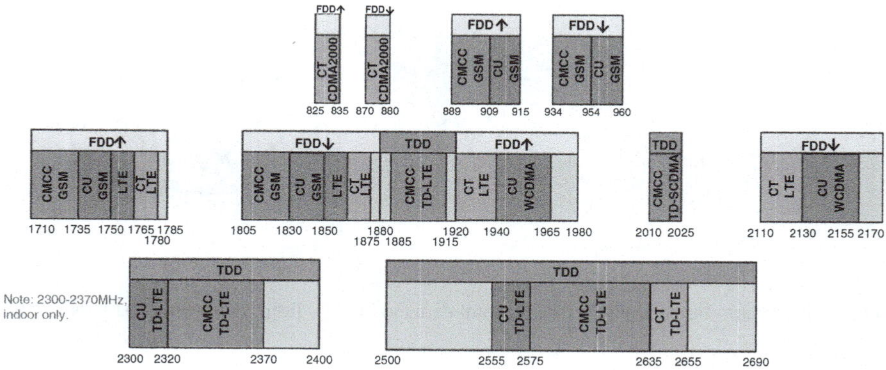

Fig. 5 Frequency bands used for mobile technology in China (CT: China Telecom, CM: China Mobile, CU: China Unicom)

Taking China as an example, up to Feb. 2016, 517 MHz been allotted to three operators for eight networks, as shown in Fig. 5. For the two operators authorized for LTE hybrid network, China Telecom has 2×15 MHz spectrum for LTE FDD, while China Unicom only has 2×10 MHz in the beginning. Market competition has prompted Chinese Unicom to accelerate the pace of 2G spectrum re-farming for 3G and 4G networks. On the other hand, owing the world largest TD-LTE network, China Mobile has also expressed willingness to re-farm 2G spectrum for LTE FDD.

When 5G is put into use, it is expected that some spectrum from the old generation could be re-farmed for it as well. However, re-farming does not increase the total spectrum amount.

- Identified spectrum

In the past ITU-R studies, sometimes it is difficult to reach global or even regional agreement for the usage of some frequency bands. Therefore some Resolutions from WRC will not modify the Table of Frequency Allocations uniformly, but usually take effect in the form of adding footnotes for some countries. There are many cases for IMT frequency bands planning in this way.

For example, footnote **5.432A** specifies that "In Korea (Rep. of), Japan and Pakistan, the band 3400–3500 MHz is identified for International Mobile Telecommunications (IMT). This identification does not preclude the use of this band by any application of the services to which it is allocated and does not establish priority in the Radio Regulations…"

However, some identified IMT spectrum is not for immediate use. For example, in Regulations of Radio Frequency Division of China, there are several frequency bands identified for IMT specified in footnote CHN28 such as 2300–2400 MHz and 3400–3600 MHz [13], but only when the compatibility studies are finished. Currently in China, LTE-Hi (LTE Hotspot/indoor) is proposed to meet the coverage requirement of hotspot and indoor scenarios. Compared with current IMT, LTE-Hi is targeting higher frequency, including 3400–3600 MHz.

For 2300–2400 MHz, after careful studies, IMT systems have proven the capability of coexisting with radio location services, but only limited to indoor use. This band has been assigned in Dec. 2013, for deploying TD-LTE systems.

Nevertheless, other frequency bands are still to be studied with respect to compatibility before official use. For example, 3400–3600 MHz is already used as extended C band for satellite services in China, since it can provide better propagation characteristics against the rain attenuation than higher frequency bands. Therefore, up to now, allowing this band for IMT systems needs further demonstration and coordination.

- New frequency bands after WRC-15

In accordance with resolves to invite ITU-R 2 of Resolution 233 (WRC-12), there are 19 candidate frequency bands proposed for IMT from different administrations, in the framework of WRC-15 Agenda Item (AI) 1.1 [14], as shown in Fig. 6.

These frequency bands are considered as potential candidate frequency bands from among the suitable frequency ranges provided by the ITU-R. These frequency bands were proposed by one or more administrations and were studied by the ITU-R. For a frequency band to be included, the ITU-R JTG 4-5-6-7 developed the following criteria: it must have been proposed by an administration and have been studied.

In studying AI 1.1, inputs were received reflecting views on certain frequency ranges/bands. Some of the views are listed in Table 4. Studies related to various frequency bands (studies initiated, carried out, not completed, completed), as well as an overview of which services were and were not studied for each band, are addressed in [14].

Different administrations or regions may have different considerations in supporting these frequency bands. The final results could be found in Provisional Final Acts of WRC-15 [15]. It shows that, to realize global harmonization, it is necessary to achieve further regional convergence.

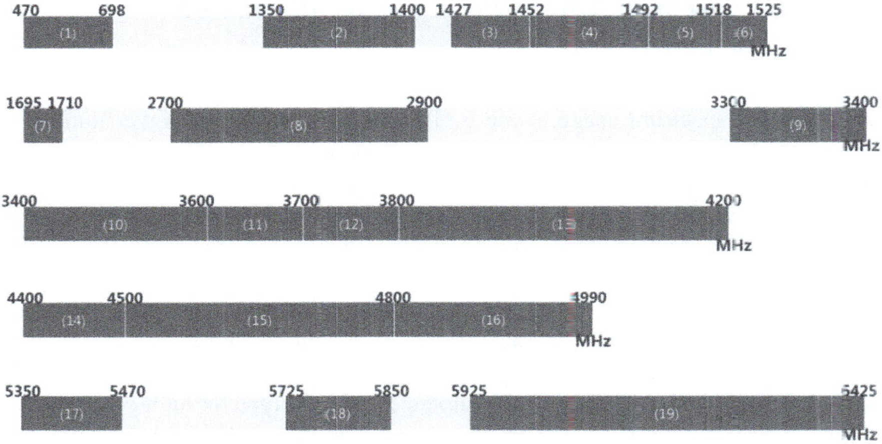

Fig. 6 Possible candidate bands in WRC-15 Agenda Item 1.1 before the conference

Table 4 Views on suitable frequency ranges/bands

Frequency ranges (MHz)		Views[a]					
		EU	US	Russia	Japan	Korea	China
1	470–694/698	s	√	×	×	√	
2	1350–1400	×		×			
3	1427–1452	√	×	×	√		
4	1452–1492	√	×	×	√	√	
5	1492–1518	√	×	×	√		
6	1518–1525	×	×	×	√		
7	1695–1710	×	√				×
8	2700–2900	×	s				×
9	3300–3400	×					√
10	3400–3600	√	s	×	√	√	×
11	3600–3700	√	s	×	√	√	×
12	3700–3800	√	s	×	√	√	×
13	3800–4200	×	s	×	√	√	×
14	4400–4500	×			√		√
15	4500–4800	×			√		×
16	4800–4990	×			√	√	√
17	5350–5470	×	√				
18	5725–5850	s					
19	5925–6425	s		√			×

[a] √: support, ×: oppose, s: study

Spectrum below 6 GHz are the best resources for IMT in the near future. However, due to its scarcity and increasingly difficulty to realize international harmonization, it is the time to seek spectrum above 6 GHz.

4.2 New Spectrum After WRC-15, Especially the Bands Above 6 GHz

Traditionally, spectrum usage above 6 GHz such as millimeter wave (mmWave) often applies for point-to-point communications in large power systems such as satellite systems and microwave systems.

From the perspective of public mobile communication, mmWave communication technology also attracts peoples' attention. Samsung from Korea conducted studies and tests for the above 6 GHz, especially at 13.4–14 GHz, 18.1–18.6 GHz, 27.0–29.5 GHz and 38.0–39.5 GHz. At 28 GHz, deploying 64 antennas, Samsung has done the experiments of the adaptive beam-forming technology to implement 1 Gbps peak rate within 200 m, and its error rate is less than 0.01 % [16]. Moreover, at 11 GHz, NTT DoCoMo implemented prototype for high-speed mobile vehicle communications. In addition, manufacturers such as Ericsson, Huawei, and

NSN show their vision for mmWave communications on various bands such as 14 GHz, E-band (71–76 GHz/81–86 GHz) and so on. It is obvious that mmWave communication is drawing attentions all over the world as one of the potentially essential technologies of the next generation of mobile communication systems.

The 60 GHz International Broadband Wireless Access Systems (BWAS) are drawing peoples' attention. It mainly focuses on short-distance, high-rate communications. The transmission distance is usually within 10 m, and TDD mode is applied. The specific operating frequency is 56–66 GHz (59–64 GHz in China). The three related international standards are ECMA 387, IEEE 802.15.3c and IEEE 802.11ad. In addition, IEEE establishes IEEE 802.11aj to formulate the specific criteria of 60 and 45 GHz usage in China.

In 2001, FCC allotted a continuous section of spectrum (57–64 GHz) for wireless communication in unlicensed mode [17]. In Europe, ECC also allotted 57–66 GHz unlicensed spectrum. Specifically, 62–63 GHz and 65–66 GHz were for Mobile Broadband System, and 59–62 GHz were for Wireless Local Area Network (WLAN) [18]. In 2006, 59–64 GHz was planned in China for wireless short-distance technologies. Similarly, Japan and Australia conducted the related spectrum planning in this frequency band. The detail procedure is shown in Fig. 7. In 2013, China allotted 40.5–42.3 GHz band and 48.4–50.2 GHz band for fixed point-to-point wireless access system (light license management), 42.3–47 GHz band and 47.2–48.4 GHz band for mobile point-to-point wireless access system (unlicensed management).

However, due to the characteristics of high frequency, the key technologies applied for low frequency are hard to directly put into use. How to utilize the advantage and overcome the disadvantage are both opportunities and challenges [19].

- Principles for candidate bands selection [20]

In the Table of Frequency Allocations, spectrum above 6 GHz are mainly allocated to fixed service, mobile services, radio location service, radio navigation service, fixed satellite service, inter-satellite services and broadcasting satellite service, etc. In practical use, for example, there are digital microwave relay systems, navigation

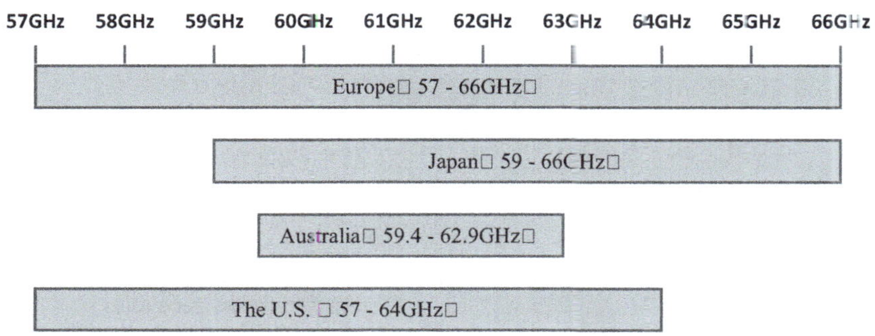

Fig. 7 International spectrum allocation of 60 GHz

and air traffic control radar system and satellite communication systems. High frequency radio communication has different characteristics in radio propagation antenna and RF, compared with lower frequency communication. For candidate bands selection for mobile communications, the following principle could be considered.

Compliance Bands allocated to Mobile Services (or in footnotes identified) are preferred.

Consistency To select potential candidates for international harmonization, to take into account the future promotion in WRC.

Security Fully consideration on the electromagnetic compatibility between systems, to ensure the protection of mobile systems as well as other systems.

Continuous To take the advantage of broadband (e.g., >500 MHz) from high frequency to guarantee that the system can obtain a higher efficiency.

Effective Taking into account the propagation characteristics of high frequency and industrial hardware manufacturing capability, to select the appropriate frequency to ensure the effective design and realization of the system, terminal, instrument, etc.

- Some potential frequency ranges for 5G above 6–100 GHz

Preliminary studies from different administrations or organizations show that a lot of potential suitable frequency ranges could be found from 6 to 100 GHz. This sub-section summaries some considerations from research institutes, countries and regional organizations.

4.2.1 METIS

The METIS project has delivered study results on potential suitable frequency ranges [21] in terms of a prioritization of the bands as shown in Figs. 8 and 9,

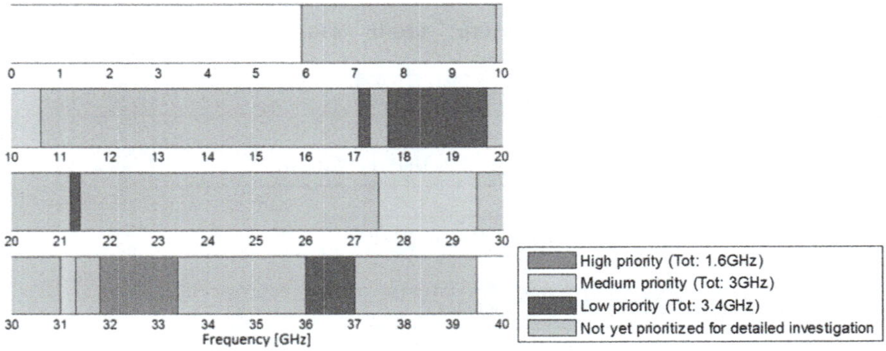

Fig. 8 Opportunities of potential sharing with current incumbent within 5.925–40.5 GHz [21]

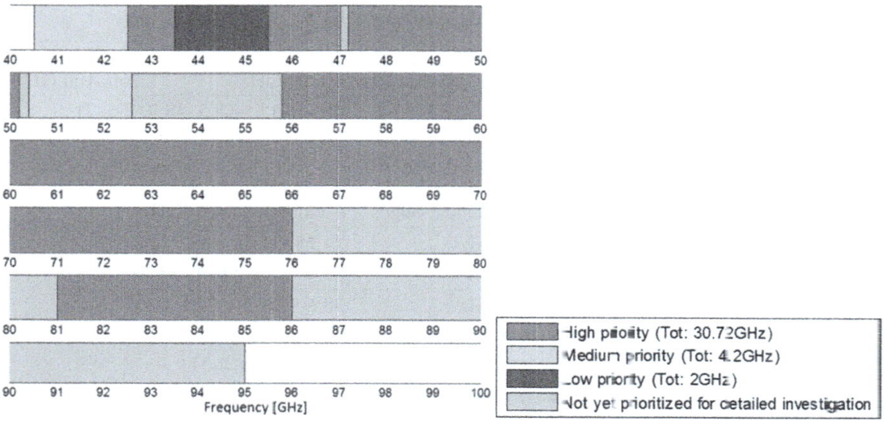

Fig. 9 Opportunities of potential sharing with current incumbent within 40.5–100 GHz [21]

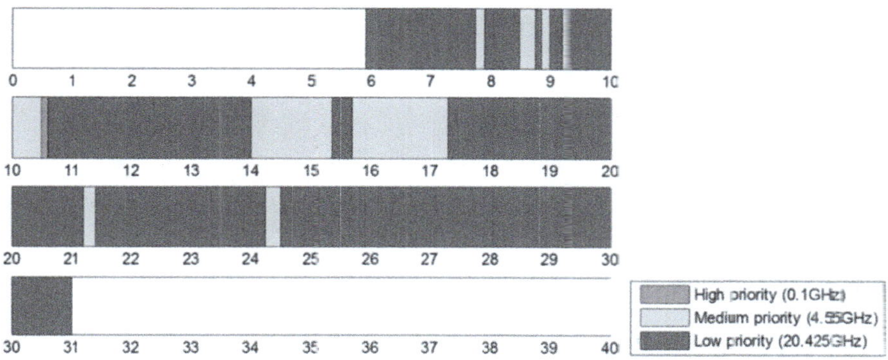

Fig. 10 Opportunities of potential sharing with current incumbent for outdoor scenario [21]

indicating the most relevant bands for initial consideration in regulation when starting the procedure to find additional bands for 5G systems considering only current allocations.

To further shed light on the opportunities in the 5.925–31 GHz range, an additional band assessment with modified assessment criteria has been performed, as shown in Fig. 10. In this assessment, a reduced target minimal contiguous bandwidth of 60 MHz was used, and a larger focus was given to outdoor deployments.

4.2.2 Ofcom [22, 23]

For spectrum above 6 GHz, the Ofcom started with a national Call For Input (CFI) for the potential candidate bands and summarized the response as shown in Table 5.

Table 5 Ofcom 5G candidate frequency bands above 6 GHz

Frequency ranges	6–20 GHz	20–40 GHz	40–60 GHz		60–100 GHz
Candidate bands	10.125–10.22510.475–10.575	31.8–33.4	40.5–43.5	45.5–48.9	66–71

Table 6 FCC 5G candidate frequency bands above 6 GHz

Frequency ranges	Candidate bands (GHz)
24 GHz	24.25–24.45, and 25.05–25.25
LMDS	27.5–28.35, 29.1–29.25 and 31–31.3
39 GHz	38.6–40
37/42 GHz	37–38.6 and 42–42.5
60 GHz	57–64 and 64–71
70/80 GHz	71–76 and 81–86

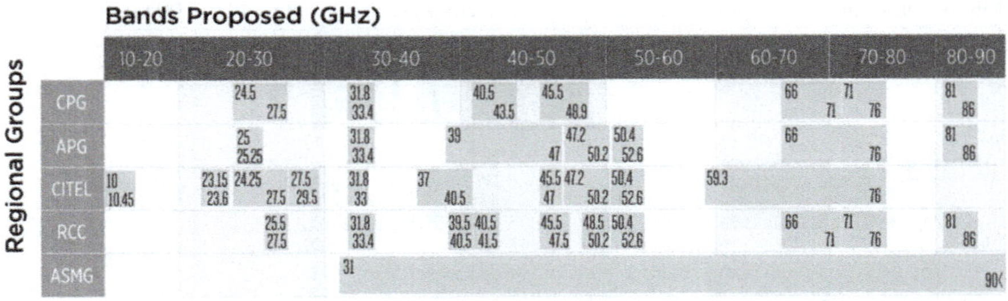

Fig. 11 Views from different regional organizations on identification of frequency bands for IMT in portion(s) of the frequency range between 6 and 100 GHz

4.2.3 FCC [24]

Similarly, FCC started with a Notice Of Inquiry (NOI) the potential candidate bands and only consider bands above 24 GHz as shown in Table 6.

4.2.4 Regional Organizations

Up to Oct. 2015, several regional organizations have made preliminary views for WRC-15 Agenda Item 10. Almost all the organizations have agreed to establish a new agenda item for WRC-19, to consider identification of frequency bands for IMT in portion(s) of the frequency range between 6 and 100 GHz. These views are shown in Fig. 11.

From the regulatory perspective, it is vital to perform solid research of these bands, including channel measurements, system modeling and detailed compatibility studies with currently used services [25].

4.2.5 Provisional Final Acts of WRC-15

After WRC-15, a new Agenda Item 1.13 is established in WRC-19, to consider identification of frequency bands for the future development of IMT, including possible additional allocations to the mobile service on a primary basis [15]. The related frequency bands are:

- 24.25–27.5 GHz2, 37–40.5 GHz, 42.5–43.5 GHz, 45.5–47 GHz, 47.2–50.2 GHz, 50.4–52.6 GHz, 66–76 GHz and 81–86 GHz, which have allocations to the mobile service on a primary basis; and
- 31.8–33.4 GHz, 40.5–42.5 GHz and 47–47.2 GHz, which may require additional allocations to the mobile service on a primary basis.

4.3 *Spectrum for IoT*

Compared with previous generations of mobile communications, 5G needs to meet extremely high performance requirements in more diverse scenarios. Besides scenarios such as seamless wide-area coverage and high-capacity hot-spot, there is also a need of low-power massive-connections including machine-to-machine and man-to-machine, driven by internet of things (IoT).

Low-power massive-connection scenario mainly targets sensing and data collecting use cases, such as environmental monitoring and intelligent agriculture. This scenario is characterized by small data packets, low power consumption, low cost, and massive connections. Specifically, probably at least one million connections per squared kilometer need to be supported [26].

The suitable frequency ranges of such 5G usage scenarios will be wide and rather different with those for communication between people. An example is shown in Table 7, which is the list the frequency bands used for smart metering application in Power Grid Management Systems in some parts of the world [27].

When considering certain spectrum for Short Range Devices (SRDs) as one part of 5G, e.g., RFID, the 5G potential frequency ranges will be much wider.

5 Spectrum Management Considerations

New scenarios, service types, spectrum demand and usage methods of 5G will definitely bring new challenges to current spectrum management. In this section, some new policies and policy related studies on spectrum management for mobile communications from a number of countries are introduced, reflecting the future 5G spectrum management development trends.

Table 7 Example of frequency bands in wireless metering in Power Grid Management Systems [27]

Frequency (MHz)	Area/region
40–230 (part of), 470–694/698	North America, UK, Europe, Africa, and Japan
169.4–169.8125	Europe
220–222	Some parts of ITU Region 2
223–235	China
410–430	Parts of Europe
450–470	North America, parts of Europe
470–510	China
470–698	North America and Europe
779–787	China
868–870	Europe
870–876	Parts of Europe
896–901	North America
901–902	North America
902–928	North America, South America, Australia
915–921	Parts of Europe
917–923.5	Korea
920–928	Japan
928–960	North America
950–958	Japan

To be clear, some new ideas mentioned in this chapter do not represent the current policy and are considered as research work aiming to give some food for thought. But from various aspects of the related work, they could be a possible trend.

5.1 US

Foreseeing the rapid development of mobile communications, US pay high attention to the related spectrum resources management, with carefully planned steps to carry out a series of related work.

In June 2010, the Presidential Memorandum entitled of Unleashing the Wireless Broadband Revolution was released. Accordingly, the National Telecommunications and Information Administration (NTIA) collaborating with the Federal Communications Commission (FCC) are planned to make 500 MHz of Federal and nonfederal spectrum available for wireless broadband use within 10 years.

Furthermore, in July 2012, the report from PCAST (President of the Science and Technology Advisory Committee) to the US President propose to released 1000 MHz band to "create of spectrum super-highway" [2].

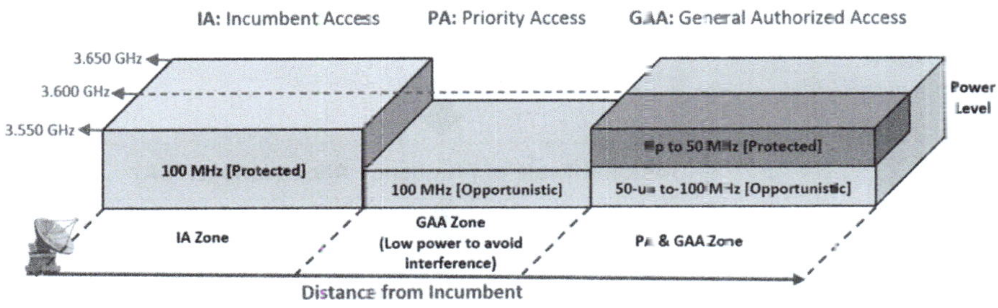

Fig. 12 Proposed Licensing Model for 3.5 GHz (source: http://www.ischool.pitt.edu/)

In June 2013, based on the above report, the Presidential Memorandum entitled of Expanding America's Leadership in Wireless Innovation was released. A Spectrum Policy Team was set up to seek how to bring spectrum sharing measures into current spectrum management of NTIA and FCC for a study period of 1 year.

July 11, 2014, NTIA and the FCC issued a joint notice for comments on Model City for demonstrating and evaluating advanced spectrum sharing technologies recommended by PCAST. More than a dozen replies are received from AT&T, Wireless Innovation Forum, Dynamic Spectrum Alliance, Telecommunications Industry Association, etc.

When seeking the candidate bands to release 1000 MHz the 3550–3700 MHz band are selected as a possible one. On April 2014, FCC proposed Rules to make up this 150 MHz of spectrum available for new citizens broadband radio service. This innovative rule significantly promotes the intensive shared use of spectrum in the 3.5 GHz Band.

On April 2015, the FCC finally adopted rules for the Citizens Broadband Radio Service, creating a new spectrum band and taking a major step forward in spectrum policy by authorizing advanced spectrum sharing among commercial and federal operators, as shown in Fig. 12.

5.2 *EU*

Recently, ECC states that License Shared Access (LSA) is a complementary spectrum management tool that fits under an "individual licensing regime" [28] and the national legislation framework is shown in Fig. 13. National Tables of Frequency Allocations (NTFAs) primarily specify the radio services authorized by an individual administration in frequency bands and the entities which have access to them. National frequency assignments, as derived from the ITU concept, allows the fine management of frequency bands in accordance with the rules set in NTFAs, particularly in bands shared by different type of users and also in respect of coexistence issues in adjacent bands. They may contain sensible data and their management requires confidentiality procedures. Under this, two different ways

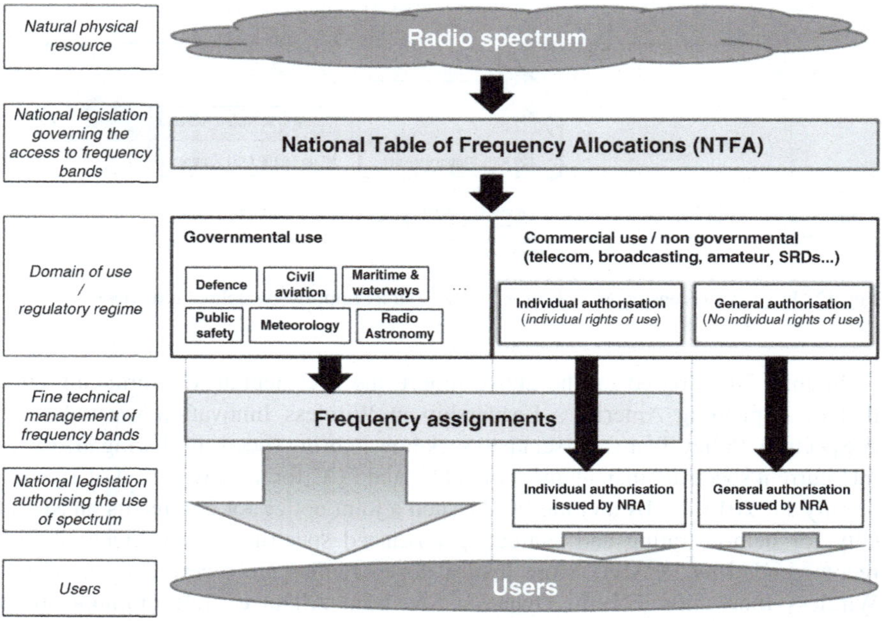

Fig. 13 National legislation from the radio spectrum to users [28]

of using spectrum are introduced: Government use covering various domains (e.g. defense, civil aviation, maritime & waterways, public safety, meteorology, science), and commercial use understood as the public legal act issued by NRAs for the purpose of delivering spectrum usage rights to private entities or citizens. The following two terminologies should also be distinguished for commercial use:

- Individual authorization (Individual rights of use);
- General authorization (No individual rights of use).

Individual rights of use are given for limited duration and do not constitute property act of the frequencies by the operator as frequencies are part of the national domain.

5.3 *China*

At present, the spectrum license pattern in China mainly includes both licensed and unlicensed use. In addition, there is a light license mode, in which the user only needs to report but not apply for setting up a new radio station to the government for recording.

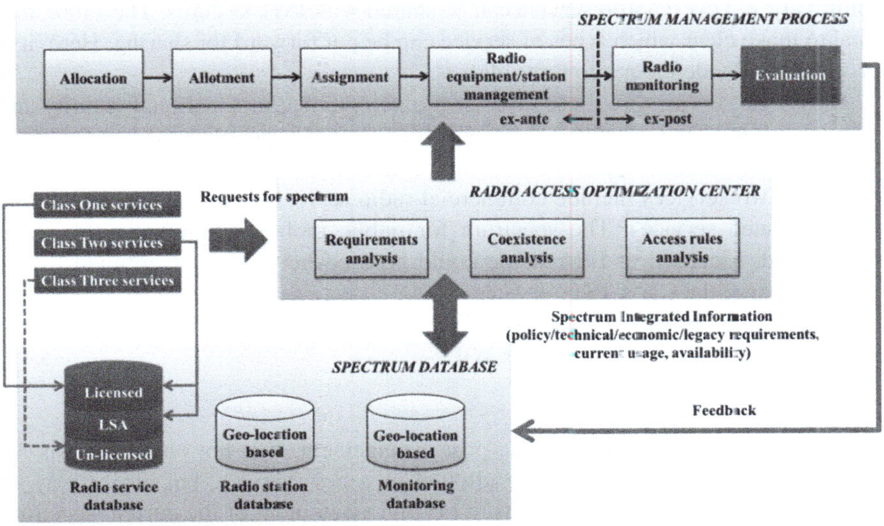

Fig. 14 New elements studied in current spectrum management framework in China [4]

In such a management framework, spectrum sharing is not well supported. As 5G brings much more challenges in spectrum demand and also new spectrum usage methods, the spectrum management has to keep pace with such innovation. Currently based on the current framework of spectrum management in China shown in Fig. 14, two new elements are being studied, namely services classification and spectrum assessment.

The *basic spectrum management process* is shown at the top of Fig. 14. The spectrum allocation (to radio services), allotment (to users or systems) and assignment (to specific radio stations) as well as the radio equipment and station management belong to ex-ante management. The radio monitoring belongs to ex-post management, including radio occupancy measurements, signal parameter and transmitter inspection, illegal transmitter detection and finding and so on [http://www.srrc.org.cn/english/]. The *radio access optimization center*, composed of relevant organizations and institutions, is in charge of spectrum demand calculation, coexistence and access analysis. The *spectrum database*, established in State Radio Monitoring Center and State Radio Spectrum Management Center (SRMC/SRSMC), stores all the information of radio services, stations and monitoring reports, which indicate the current spectrum usage and availability. When a new service access request arrives, the *center* checks the *database*, performs integrated analysis and outputs the solution to support the *process*, which conversely inputs the information to the *database* for further use, establishing a closed-loop management.

- Services classification

Intuitively, not every radio service can be shared with IMT systems. Therefore, it is vital to make clear which types of service can be put forward for sharing. Here, it is proposed to divide them into three classes.

Class One services are those involving government affairs, national security and people's safety. They usually occupy specified licensed spectrum for free and need strict protection. Their spectrum cannot be shared with others.

Class Two services include commercial radio services, and also general public and dedicated services. The spectrum for public mobile communication in 5G is included. Licensed or LSA mode could be used here. The government would flexibly charge the Class Two services.

Class Three services are dedicated for public free use. The access process might also need spectrum sensing, such as WiFi.

- Spectrum assessment

Spectrum assessment belongs to ex-post management. It is not only the basis for whether a frequency band can be allotted to a new service, but also a way of supervising its utilization efficiency. In [4], the assessment of the current spectrum usage is proposed in China for the first time. Based on assessment results, the government can adjust or even retrieve the spectrum allotment.

To make fair assessment, a scientific and effective Key Performance Indicator (KPI) system is required to be established. The KPI currently under studied mainly includes three aspects.

- Radio monitoring related information, such as noise and radio occupancy measurement results in different scenarios and locations.
- Radio stations related information, such as location, RF parameters and also the related statistics.
- Service related information, e.g., different Quality of Service (QoS) and protection requirements.

The KPI system is necessary for making reasonable conclusions in spectrum management. It is important to study the feasibility of providing such information in a technology- and service-neutral way, and whether it allows for a practical and efficient spectrum evaluation process.

6 Summary

Different aspects of 5G spectrum issues are discussed in this chapter. Generally, it is expected that 5G requires more in the total amount, wider with respect to individual bandwidths, greater in the range and more flexible in the usage and management pattern.

The suitable frequency ranges of 5G will include those bands below 6 GHz such as re-farmed 2G/3G spectrum, identified frequency bands for IMT and also WRC-15 candidate bands. However, due to the scarcity of spectrum below 6 GHz, which has almost been fully utilized and it will be even more difficult to find internationally harmonized spectrum after WRC-15, it is necessary to seek potential frequency ranges above 6 GHz.

The controlled spectrum sharing is an important way of re-using spectrum to complement current licensed dedicated spectrum which is still the basis for operation of 5G systems.

Acknowledgments This chapter is supported by Chinese National Key Project under Grant No. 2014ZX03001027 and No. 2015ZX03002008, National High Technology Research and Development Program ("863" Program) of China under Grant No. 2014AA01A707, and Beijing Natural Science Foundation(4132050).

References

1. ICT-317669 METIS project, Intermediate description of the spectrum needs and usage principles. Deliverable D5.1, April 2013, https://www.metis2020.com/documents/deliverables
2. G. Documents, *Realizing the Full Potential of Government-Held Spectrum to Spur Economic Growth - PCAST - July 2012* (IEEE USA Books & eBooks, 2012)
3. Ofcom, Spectrum management strategy - Ofcom's strategic direction and priorities for managing spectrum over the next 10 years, April, 2014
4. T. Wang, G. Li, J. Ding, Q. Miao, J. Li, Y. Wang, 5g spectrum: is China ready? Commun. Mag. IEEE **53**, 58–65 (2015)
5. AWG-18/TMP-18, A Draft Revision of APT Report on Information of Mobile Operators' Frequencies, Technologies and License Durations in Asia Pacific Countries, Kyoto, Japan, March 2015
6. Electronic Communications Committee (ECC), The European Table of Frequency Allocations and Applications in the Frequency Range 8.3 kHz to 3000 GHz (ECA Table), May 2014
7. Report ITU-R M.2290-0, Future spectrum requirements estimate for terrestrial IMT, Dec. 2013
8. Federal Communications Commission, Mobile Broadband: The Benefits of Additional Spectrum, Oct. 2010
9. ITU-R Document 5D/66, Relevant Work Undertaken by Australia Regarding the Potential Identification of Additional Spectrum for IMT, July 2012
10. ITU-R Document 5D/118, Future IMT Spectrum Requirements Assessment for the Russian Federation, Sep. 2012
11. H. Biao, W. Tan, Spectrum requirements calculation based on Urban Hotspot in China for mobile communication in the year 2020. ZTE Technol. J. **20**(2), 5–10 (2014)
12. S. Chen, J. Zhao, The requirements, challenges and technologies for 5G of terrestrial mobile telecommunication. IEEE Commun. Mag. **52**(5), 36–43 (2014)
13. MIIT of China, Regulations of Radio Frequency Division of People's Republic of China, Feb. 2014
14. ITU-R, Report of the CPM on operational and regulatory/procedural matters to the World Radio Communication Conference, 2015
15. Provisional Final Acts WRC-15, Geneva, http://www.itu.int/pub/R-ACT-WRC.11-2015
16. W. Roh et al., Millimeter-wave beamforming as an enabling technology for 5G cellular communications: theoretical feasibility and prototype results. IEEE Commun. Mag. (2014)
17. FCC, Code of Federal Regulation, Title 47 Telecommunication, Chapter 1, Part 15.255
18. ETSI DTR/ERM-RM-049, Electromagnetic compatibility and radio spectrum matters (ERM); system reference document; technical characteristics of multiple gigabit wireless systems in the 60 GHz range. Mar. 2006

19. S. Chen, J. Zhao, M. Ai, D. Liu, Y. Peng, Virtual RATs and a flexible and tailored radio access network evolving to 5G. IEEE Commun. Mag. **53**(6), 52–58 (2015)
20. APG, APG15-5/OUT-40: Preliminary APT Common Proposals on WRC-15 Agenda Item 10, July 31st 2015
21. METIS deliverable D5.3, Description of the spectrum needs and usage principles, Aug. 2014
22. Ofcom, Spectrum above 6 GHz for future mobile communications: Call for Input, Jan. 2015
23. Ofcom, Laying the foundations for next generation mobile services: Update on bands above 6 GHz, April 2015
24. FCC, Notice of Inquiry, Oct. 2014
25. Y. Wang, J. Xu, L. Jiang, Challenges of system-level simulations and performance evaluation for 5G wireless networks. IEEE Access **2**, 1553–1561 (12/2014)
26. IMT-2020 (5G) Promotion Group, 5G Wireless Technology Architecture White Paper, May 2015
27. ITU-R Working Party 1A, REPORT ITU-R SM.2351-0 Smart grid utility management systems, June 2015
28. ECC Report 205, licensed shared access, Feb. 2014

Spectrum Sharing for 5G

Gen Li, Tan Wang, Qingyu Miao, Ying Wang, and Biao Huang

Abstract Spectrum sharing for 5G is motivated by the fact that significantly more spectrum and much wider bandwidths than what is available today will be needed in order to realize the performance targets of 5G. First, spectrum sharing scenarios, i.e. vertical sharing and horizontal sharing, are summarized for different type of spectrum. Second, a thorough review of current spectrum sharing techniques are provided including coordination protocol, GLDB, DSA and MAC-based collision avoidance. Then spectrum sharing is analyzed specific to 5G design and one general architecture to enable 5G spectrum sharing is proposed. Finally, it is concluded that spectrum sharing becomes more and more important for 5G systems as a complementary way of using spectrum.

1 Introduction

Currently, traditional 2G, 3G and 4G cellular networks have so far been designed for dedicated, licensed spectrum. With that, cellular technology can only operate in a subset of the spectrum that is in principle available. More recently, other forms of shared spectrum including unlicensed spectrum is close to application in near future for 4G systems, i.e. licensed assisted access (LAA) LTE which is ongoing work in 3GPP. Significantly more spectrum and much wider bandwidths than what is available today will be needed in order to realize the performance targets of 5G, i.e. more system capacity and better user experience. One way is to try the best to fill the spectrum gap by finding new spectrum, especially going to higher frequency band which is less crowded than what is being used today. The other way is to improve

G. Li (✉) • Q. Miao
Ericsson, China Radio Research Lab, Beijing, China
e-mail: gen.li@ericsson.com

T. Wang • B. Huang
State Radio Monitoring Center, MIIT, Beijing, China

Y. Wang
Beijing University of Posts and Telecommunications (BUPT), Beijing, China

W. Xiang et al. (eds.), *5G Mobile Communications*,
DOI 10.1007/978-3-319-34208-5_3

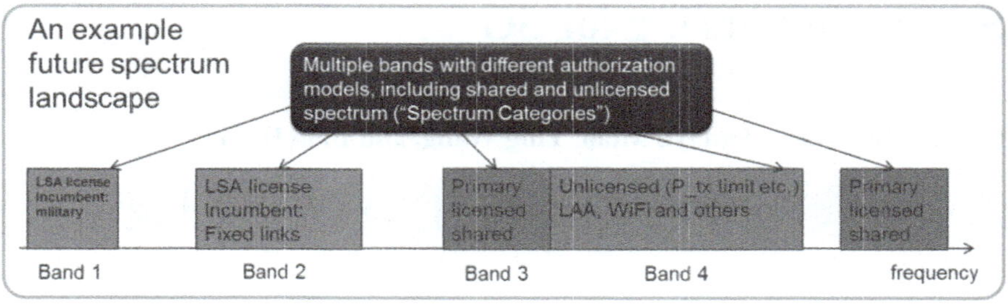

Fig. 1 Example future spectrum landscape [1]

total spectrum efficiency by spectrum sharing. Generally speaking, it is expected that the spectrum for 5G will, at least partly, be available under spectrum sharing regimes.

As an example future spectrum landscape one can envision the situation depicted in Fig. 1. Therein the 5G system is able to access primary and licensed bands as well as other bands in an LSA fashion or unlicensed bands. The typical choice would be to go with the primary band for reliability, but local interference situations could make it more beneficial for some nodes to use other bands with other regulatory frameworks.

This chapter first introduces various spectrum sharing scenarios corresponding to different type of spectrum in Sect. 2. Then Sect. 3 presents a thorough review of current spectrum sharing techniques, including coordination protocol, GLDB, DSA and MAC-based collision avoidance. Next in Sect. 4, it also analyzes the limitations of the aforementioned techniques and proposed general architecture and the corresponding research directions in 5G. Finally Sect. 5 concludes this chapter.

2 Spectrum Sharing Scenario

It is well known that current 2G, 3G and 4G systems only use licensed dedicated spectrum. However, according to the analysis in this chapter, 5G systems will need significantly more spectrum than today and it is hard to find enough bands. Thus using only licensed mode may not meet 5G requirement very well. This section provides a complete summary of spectrum sharing scenarios as the following categories [1, 2]:

- *Vertical sharing* refers to spectrum sharing between users of different priority (e.g. primary and secondary), i.e., unequal rights of spectrum access.
- *Horizontal sharing* is sharing between systems that have the same priorities in the spectrum, i.e. different users have fair access rights to the spectrum. If the sharing users in the spectrum adopt the same technology, it is called *homogenous horizontal sharing*; otherwise *heterogeneous horizontal sharing*.

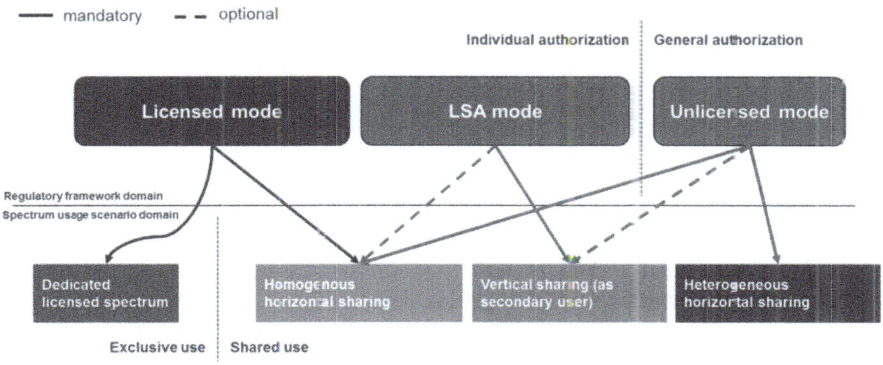

Fig. 2 Illustration of spectrum type and related usage scenario [2]

First, as shown in Fig. 2, the above-mentioned spectrum sharing scenarios are mapped into different types of spectrum, i.e. licensed, licensed shared access (LSA) and unlicensed as follows:

2.1 Licensed Mode

In licensed mode, spectrum is allotted to wireless systems as primary use. The only relevant spectrum sharing scenario is homogenous horizontal sharing, i.e. sharing spectrum with other operators using the same RAT, which is also called co-primary or inter-operator spectrum sharing. There are two different ways to achieve horizontal sharing, i.e. mutual renting and limited spectrum tool.

- In the *mutual renting* sub-scenario the spectrum resources in a band are subdivided into several blocks and each block is licensed to one operator. Operators mutually allow other operators to "rent" parts of their licensed resources. An operator can rent resources from multiple other operators simultaneously. The actual "owner" of a resource has always strict priority in accessing its licensed part of the spectrum, including the possibility of preemption at any time. This approach could be applied to bands that have initially been used in a dedicated licensed way where there is a need to increase peak data rate beyond what is possible within one licensed block.
- The *limited spectrum pool* scenario allows an operator to obtain an authorization, usually a license, to use up to the whole band on a shared basis with a limited number of other known authorized users. This setup does not provide guarantee for instantaneous access to a minimum amount of spectrum, but it is envisioned that mutual agreements between licensees are such that the long term share of an individual operator has a predictable minimum value. This is similar to

the sharing situation in unlicensed bands but it is a priori known how many authorizations a regulator will give out and what the rules for the band will be, hence providing the necessary certainty for investments into large networks.

2.2 LSA Mode

In LSA mode, a licensee has the authorization to access spectrum that is unused by an incumbent user at certain locations and/or times. This *vertical sharing* is based on well-defined conditions which are parts of a sharing license. In the initially developed LSA concept, the licenses are expected to be long term and exclusive for one system or operator only. As it evolves, the LSA concept may in the longer term be combined with *homogenous horizontal sharing* so that the issued licenses could be non-exclusive and require several licensees to coexist.

LSA is currently being developed through the European Commission's Radio Spectrum Policy Group (RSPG), which issued a report on LSA in November 2013 with recommendations to member countries. Now ECC FM 52 has essentially concludes this LSA regulation work and only the final round of ECC approval remains. CEPT ECC has started a study to use the 3.6–3.8 GHz band for LSA in order to enable sharing between FSS and mobile systems. Standardization activities on LSA technical conditions and geo-location databases are ongoing at ETSI (the European Telecommunications Standards Institute), where they are being carried out by the Technical Committee on Reconfigurable Radio Systems (RRS).

2.3 Unlicensed Mode

In unlicensed mode, a wireless system has to share spectrum with other unlicensed systems. For *heterogeneous horizontal sharing* in an unlicensed band a system must be prepared for coexistence with any other technology that may be present in the band [3]. Meanwhile, it is unavoidable to have *homogeneous horizontal sharing* between different systems with the same RAT. *Vertical sharing* functionality may also be required for a system operating in unlicensed mode if a primary user exists in the band. Therefore the complexity to implement spectrum sharing is more complicated and seems non-reliable compared to others since it almost involves all kinds of sharing scenario.

Furthermore, due to the usual restrictions on maximum transmit power in unlicensed bands; building standalone contiguous coverage areas for mobility in unlicensed spectrum requires a very large number of sites. Unrestricted access to the bands also creates unpredictable interference conditions that typically lead to a decrease in accessible capacity when traffic levels are high. The result is high costs and intense operational complexity. License-exempt spectrum creates a low barrier to market entry. This has market advantages but also causes issues with controlling the numbers of users and operators, and in turn with controlling the quality of access and service. It is also difficult to repurpose license-exempt spectrum to adapt to changing regulatory needs, causing trouble in terms of long term spectrum management.

3 Spectrum Sharing Techniques

There are several different ways to technically realize the above spectrum sharing and to implement corresponding regulatory requirements. As illustrated in Fig. 3, four different kinds of spectrum sharing techniques are mapping to the most possible application scenarios. In this section we outline a few of the most common or promising techniques.

3.1 Coordination Protocol

Coordination protocol means explicit information exchange between the sharing systems via some well-defined interface to avoid interference in sharing spectrum. The protocol defines the behavior of the nodes when receiving certain messages or when certain events take place. This includes the possibility for defining behavior depending on what relation nodes have with each other, e.g. interference relation. Both centralized and distributed structure can be applied to coordination as illustrated in Fig. 4. Here the connections between networks or between network and central manager can be backbone wired backhaul or over the air. The advantage of this approach is that it can achieve more efficient spectrum sharing with enough information on the interference between different networks by employing smart resource allocation. One obvious drawback is the mandatory need of connection between different networks or to central manager for multiple networks, which may not be easy to have in most scenarios.

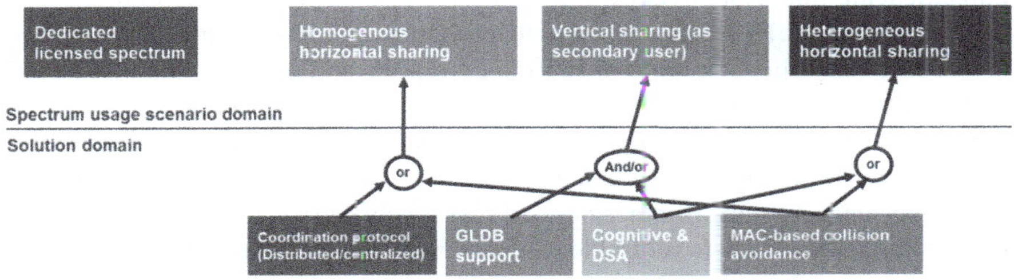

Fig. 3 Summary of mapping from spectrum sharing techniques to scenarios

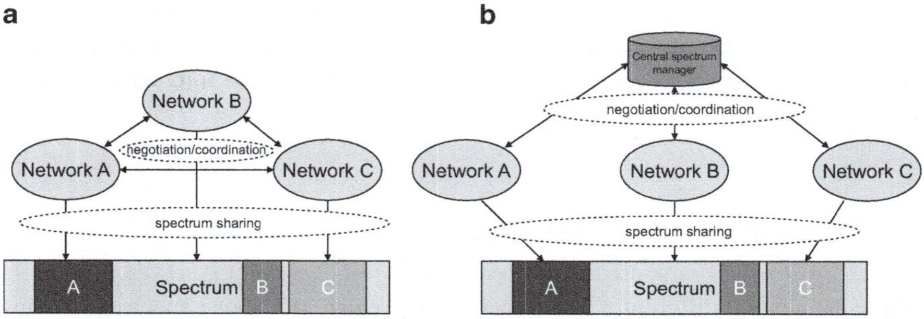

Fig. 4 Coordination structure: (**a**) distributed coordination; (**b**) centralized coordination

An example for such an approach is the Coordinated Coexistence Protocol of IEEE 802.16h [4]. 802.16h employs distributed structure with over-the-air connection. When a base station (BS) needs to send coordination messages to its neighbor BS, it may select one or more user equipments (UE) as forwarders and send the message over the air. The detailed procedures are listed as follows:

- System selects the forwarding UE according to the Radio Signal Strength Indicator (RSSI) of neighbor BS detected by the interference victim SSs, i.e. the SSs with higher neighbor BS RSSI is better;
- After identifying the forwarding UE, the serving BS requests access to the neighbor BS through the forwarding UE;
- Upon receiving the above access neighbor BS request from serving BS, the forwarding UE will try to access the neighbor BS following a procedure similar to the normal network entry process;
- If the forwarding UE accesses the neighbor BS successfully, it shall respond to its serving BS with a message that includes the parameter "UE has successfully accessed the requested neighbor BS." The serving BS and its neighbor BS can communicate with each other via the forwarding UE.

After connection establishment, resource is coordinated between different systems. IEEE 802.16h specifically defines a coexistence frame structure used to implement multiple systems coexist as shown in Fig. 5. A coexistence frame consists of four IEEE 802.16 MAC frame (MAC frame is called subframe here), and is divided into the following types of subframes: master subframe, slave subframe and shared subframe. Each of the coexistence system is required to obtain a subframe as its master frame. Other coexisting systems claim slave frame on this subframe. It is worth mentioning that each subframe can only be the master frame by only one coexistence system. In other networks, if a subframe is announced as master frame for one system, this subframe can only be slave subframes of other systems. Systems can use the maximum power for transmission in the period of master frame. Within slave subframe, systems cannot transmit at all or only be allowed to transmit following restrictions by master systems.

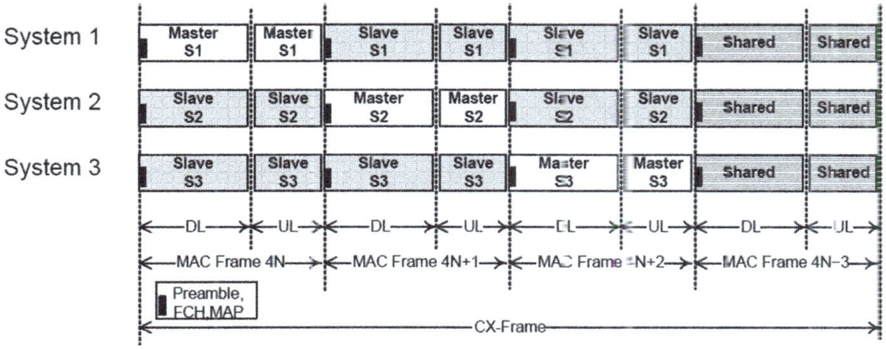

Fig. 5 Coexistence frame structure in 802.16h network [4]

3.2 Geolocation Database (GLDB) Support

In this solution the system queries a data base to acquire information on what resources are available (i.e., unused by other systems of typically higher priority) for operation in its current location. The geo-location data base provides rules for operating in the geographical region of the system. A rudimentary geo-location data base would provide a list of available channels and possibly associated transmits power (EIRP) limitations to a querying system for a specified geographical region.

The GLDB approach has been considered as an appropriate solution to manage unlicensed access to TV white spaces while ensuring that the bands may, in the future, be vacated of unlicensed transmissions. In this case this solution does not provide exclusive use of the spectrum to any system. The presence of a centrally controlled resource manager opens up for the possibility of future repurposing of the bands, since the behavior of the radios the database controls can be influenced, including the possibility to prevent transmissions completely. In this way a band could be opened up for usage not compatible with the database-controlled devices. For example, 802.22 use such approach to enable coexistence with TV systems. It is assumed that all devices must be equipped with satellite technology (e.g. GPS) which provide positioning and also gives time reference for inter-BS synchronization. Before communication, UE needs to inform BS the location and possibly antenna pattern, height and etc. Then BS queries database for allowed channels and powers. Finally BS will communicate with corresponding UE with received configuration limit.

GLDB is also seemed as a good solution to achieve vertical sharing in LSA regime [5]. Typical examples are government and military spectrum subject to sparse use, either in terms of geographic coverage or temporal characteristics. One simple LSA GLDB structure is illustrated in Fig. 6. The LSA controller in LSA licensee retrieves the information about spectrum made available under the LSA scheme from the LSA GLDB of the incumbent spectrum user and therefore has information on the available spectrum in any given location where its base stations are sited. Under this scheme, the base stations are allowed to transmit in the LSA

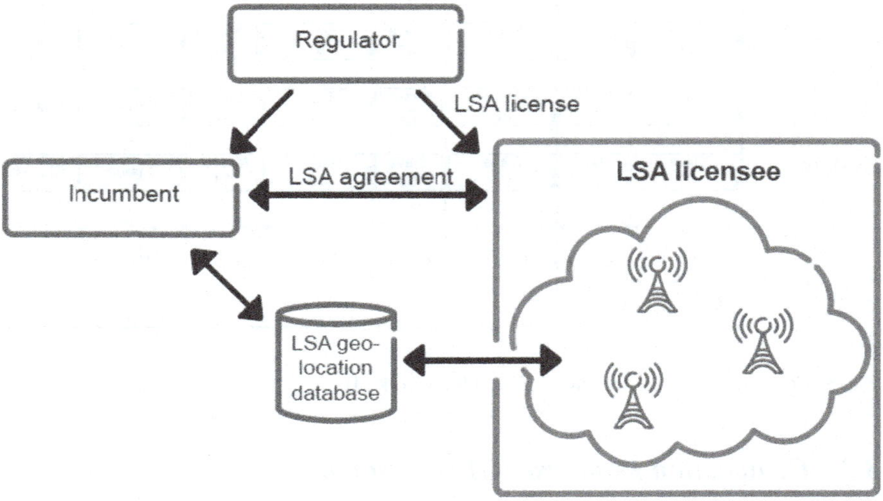

Fig. 6 GLDB architecture for LSA [5]

band only where the LSA controller indicates that the spectrum is available and no harmful interference is caused to the incumbent. This management of access to LSA spectrum is completely transparent to the mobile device which simply roams between spectrum bands according to the Radio Resource Management (RRM) messages it receives from the network.

3.3 Cognitive/Dynamic Spectrum Access (DSA)

Cognitive radio is an automatic machine that uses software defined radio to change frequency band to adapt and sense the surrounding environment. They recognize radio spectrum by sensing when it is unused by the incumbent radio system and use this spectrum in an intelligent way [6].

The idea of cognitive radio (CR) was first presented officially in an article by Joseph Mitola III and Gerald Q. Maguire, Jr in 1999. It was a new approach in wireless communications that Mitola described as: "The point in which wireless personal digital assistants (PDAs) and the related networks are sufficiently computationally intelligent about radio resources and related computer-to-computer communications to detect user communications needs as a function of use context, and to provide radio resources and wireless services most appropriate to those needs" [7]. Later, other researchers or organizations have evolving definitions. For example, the U.S. Federal Communications Commission (FCC) defines cognitive radio as an aware adaptive radio that senses the radio environment and adapts to it to enable more efficient use of the available spectrum [8]. In general, all definitions of cognitive radio have the common features that the cognitive radio should, to some extent, be aware of its radio frequency environment and be able to adapt to optimize its operation.

There are two different types of systems present in situations in which cognitive radios are used to dynamically access the spectrum: (1) A *primary user* on a portion of the spectrum is a user who owns a license that legally permits the user to use the spectrum for a purpose specified by the license; (2) A *secondary user* is a user that access spectrum that is licensed by a primary user. This access has typically to be regulated in such a way that the secondary user does not cause harmful interference to the license holder.

There are two important functions to achieve good spectrum sharing between primary user and secondary user:

- *Spectrum sensing* function available in secondary systems to sense status of primary systems to find spectrum opportunity;
- *Dynamic spectrum access (DSA)* function available in secondary system to control spectrum access dynamically according to varying environment of primary systems.

3.3.1 Spectrum Sensing

The main challenge of spectrum sensing is the strict requirement of sensitivity. In order to incorporate the noise uncertainty and multi-path fading/shadowing uncertainty into the specification, the sensing node requires identifying the presence of primary signal in the low signal to noise ratio (SNR) region. In Table 1, several typical spectrum sensing algorithms are summarized including basic energy detection and other feature detection methods that improve the sensitivity by exploiting a priori knowledge embedded in the primary signal.

In general, the above schemes are oriented for local spectrum sensing in one node. However, if the primary signal is in the deep fading or obstructed by a large obstacle, the received power at secondary node is too weak to detect. Thus hidden node problem occurs in this situation. This uncertainty can be effectively mitigated by the spatial diversity in the cooperative sensing. Additionally, the cooperative sensing can relax the sensitivity requirement of local sensing and increase the agility of making sensing decision. The procedure of cooperative sensing is as follows: all cognitive radios that have agreed to participate in the cooperative sensing process sense the spectrum and communicate their results to the other cognitive radios. Based on the information type sent by the individual CR, the cooperative sensing can be categorized by soft decision fusion and hard decision fusion.

- *Hard decision fusion* is the merge of the information to produce a common final decision for the whole system if the information that it has received consists of individual preliminary decisions;

Table 1 Summary of spectrum sensing technique

Technique	Description	Requirement	Remark
Energy detection [9, 10]	Measure the radio frequency energy or the received signal strength indicator over certain observation time and compare the accumulated metric to a predefined threshold	No	(+) Easily implemented without a priori knowledge (−) Sensitive to the noise and interference uncertainty (−) Can't distinguish primary signal with other interference
Coherent detection [11, 12]	Detect known pilot signal of primary systems with matched filter and compare obtained metric to a predefined threshold	Requires the knowledge of primary signal at both PHY layer (e.g. pulse shaping, modulation) and MAC layer	(+) can distinguish primary signal from interference and noise (+) Work well in low SNR (+) Shorter observation time (−) More implementation complexity by performing timing, carrier sync and equalization
Covariance-based detection [13]	Estimate sample covariance matrix of the received signal and make spectrum whiteness tests from eigenvalue-based or covariance statistic-based statistics	Primary signal correlated due to dispersive channel, correlation of multiple receiver antennas or over-sampling effect	(+) Exploit more general feature of primary signal without signal-specific knowledge (−) Performance highly depends on correlation of primary signal
Cyclostationary detection [14, 15]	Analyze the cyclic autocorrelation function or its equivalent Fourier transformation spectrum correlation to seek cyclostationary feature	Cyclostationary feature in primary signal caused by modulation and coding, hopping sequence, cyclic prefix	(+) Capability of differentiating the primary signal from the interference and noise (−) More implementation complexity with wide scan of the cyclic frequencies

- *Soft decision fusion* means what it receives is channel statistics or raw data the procedure of creating the final decision.

The soft decision fusion outperforms the hard decision fusion at the expense of large signaling overhead to forward the sensing data. As an alternative, the quantized decision fusion is desirable to achieve the tradeoff between these two issues. It is shown that even a simple cooperative scheme, such as the OR-rule, may improve the performance of the cognitive secondary system significantly. The probability of false detection can be made arbitrarily small while maintaining the level of interference protection of the primary system. This is achieved by adding more cooperating cognitive users that are experiencing uncorrelated fading or shadowing. Considering only few cooperating uncorrelated users give considerable improvement in the sensitivity, compared to the case without cooperation.

3.3.2 Dynamic Spectrum Access

After sensing the status of primary systems, secondary systems need to determine how to access the spectrum according to dynamic sensing results. There are three modes of operation dealing with the spectrum access of the cognitive radio (or secondary user) in fundamentally different ways and it is thus reasonable to use them to classify the cognitive radio behavior.

Interweave Operation

The basic idea of interweave operation is that the secondary systems scan the radio environment to find unused parts of the spectrum, so called spectrum holes. Once such holes are identified the secondary users start to communicate using these unused frequencies. Since the radio frequency environment is constantly changing the available spectrum holes change with time and the secondary users must constantly scan the spectrum to keep track of the spectrum holes and possibly predict where they may appear. To enable communication using spectrum to be possible, the secondary transmitter needs to communicate to the intended receiver with the chosen transmission parameters. This might be performed over a dedicated pilot channel, possibly located in an unlicensed band. An alternative way to initialize communication in a secondary system is to send a trigger on the frequency intended to be used. The trigger may be followed by a standardized pilot signal containing the necessary transmission parameters and the intended receiver identity. This possibility arises since the secondary receiver constantly scans the entire frequency range and hence, may be able to detect the trigger signal and start to receive using the parameters specified by the pilot following the trigger signal.

Underlay Operation

Underlay access means operating below the noise floor of the primary systems, involving an undercurrent of secondary system communications without primary systems being aware of. In details, a secondary system that operates in the underlay mode transmits at the same time as the primary systems of the spectrum under consideration. However, the secondary system uses a spectrum mask to make sure that the interference caused to the primary users is kept below the noise floor, or below the interference limit. In order for the secondary systems to obtain a sufficient signal-to-noise ratio (SNR), a wide bandwidth might be used for the secondary signal.

There are obvious difficulties of ensuring a sufficiently low level of interference at a primary receiver that is most likely hidden from the secondary transmitter. For this reason underlay operation is most likely only applicable for short and low power communication between secondary users. One possible implementation of underlay spectrum access is via ultra-wide band (UWB) techniques at low power.

Overlay Operation

For overlay operation, the level of interference at a primary receiver is kept below a certain tolerable level. The idea in this channel model is that the secondary transmitter (cognitive radio) obtains the message to be transmitted in the primary system in a non-causal way and uses dirty paper coding technique, or possibly linear interference cancellation techniques, to eliminate the interference at the secondary receiver due to the message sent in the primary system. This approach does not try to limit the interference to the primary system and hence operation according to this scheme violates the idea of a cognitive radio acting as a secondary system not causing harmful interference to the primary system. For this reason the cognitive radio channel is sometimes denoted a selfish approach to overlay operation.

3.3.3 Application System Examples

The most simple application example is Dynamic Frequency Selection (DFS) in 5 GHz spectrum for Wi-Fi coexistence with Radar systems. Radar systems are given priority over all other devices operating in the DFS bands, i.e. 5250–5350 MHz and 5470–5725 MHz. As a result, all devices operating at these frequencies must stop transmitting when radar is detected on the current channel. So this is a typical vertical sharing scenario. Secondary system needs to detect if there is Radar systems all the time. It adopts interweave operation mode, i.e. secondary systems must not transmit on that channel for 30 min (non-occupancy period) from the time of detection.

Another example is IEEE 802.22 protocol operating on TV band. It is a wireless communication standard for operating as a non-interfering secondary user in TV whitespace with fixed communication nodes, aimed at bringing broadband internet access to rural environments worldwide. Along with GLDB support, spectrum sensing is also adopted to mitigate the interference to TV and wireless microphones. Spectrum sensing is mandatory in both BSs and user terminals. Three types of incumbents are needed to sense, i.e. analog TV, digital TV and low power licensed device (e.g. wireless microphones). There are several sensing requirements specified as sensing receiver sensitivity (0 dBi in all directions), channel detection time (2 s from start of incumbent transmission), probability of detection (0.9 @ sensitivity), probability of false alarm (0.1 @ sensitivity) and etc. To detect the incumbent systems, network-wide quiet periods are scheduled by BS for sensing time of both BS and UEs. Besides 802.22 protocol, there are several other ones aiming to work on TV white space, i.e. 802.11af, ECMA and etc. All their spectrum sensing parts are similar standardized according to regulation rules.

3.4 *MAC-Based Coexistence Mechanism*

The MAC protocol of a system can contain functionality that enables spectrum sharing especially. The key essence is to avoid possible collision which may occur between different systems or operators but can't guarantee that there is no interference at all. The MAC is designed to alleviate the collision as much as possible, i.e. 'best-effort' collision avoidance. So this tool is more suitable for horizontal sharing. Examples of this approach include listen before talk (LBT)-based CSMA/CA protocol in IEEE 802.11 and LAA-LTE, frequency hopping in Bluetooth and etc.

A particularly interesting instance of MAC behavior for coexistence is the LBT-based CSMA/CA. One example using this protocol is IEEE 802.11 WiFi systems. As shown in Fig. 7, the basic protocol is applied by all stations, including the access point, i.e. in both downlink and uplink. A station that wishes to transmit a packet first senses the medium. If the medium is sensed idle for a certain minimum time, a so-called Distributed Inter Frame Space (DIFS, 34 μs for 11n), the packet is transmitted. If the medium is busy, the station first defers until the medium is sensed idle. When this occurs, the station does not transmit immediately, which would lead to collisions if more than one station was deferring. Instead, the station sets a backoff timer to a random number, and does not transmit until this timer has expired. The backoff timer is only decreased when the medium is sensed idle, whereas whenever the medium is sensed busy, a deferral state is entered where the backoff timer is not decreased. When the backoff timer expires, the packet is transmitted. Another example is LAA-LTE in unlicensed band which is using LBT-based CSMA/CA to deal with coexistence with WiFi and between different operators as well. The listen-before-talk mechanism is similar with WiFi and the

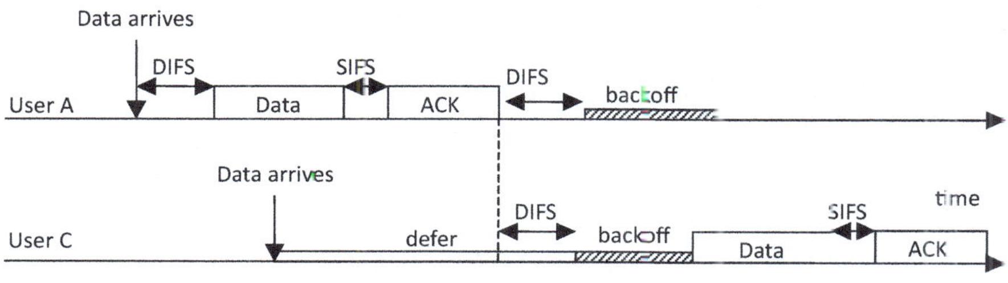

Fig. 7 Illustration of IEEE 802.11 MAC

most difference is that LAA-LTE systems can't transmit the data any time since the data should be scheduled by downlink control channel in certain resource part. One solution is to insert redundancy signal until control channel boundary when the channel is determined to be idle. Both the simulation and demo results show that LAA-LTE is a good neighbor to WiFi systems and can make both systems working better than before [16].

Bluetooth operates at 2.4 GHz short-range radio frequency band as WiFi does. But it uses a different radio technology called frequency-hopping spread spectrum to deal with coexistence problem. Bluetooth divides transmitted data into packets, and transmits each packet on one of 79 designated Bluetooth channels. Each channel has a bandwidth of 1 MHz. Bluetooth 4.0 uses 2 MHz spacing, which accommodates 40 channels. The first channel starts at 2402 MHz and continues up to 2480 MHz in 1 MHz steps. It usually performs 1600 hops per second, with Adaptive Frequency-Hopping (AFH) enabled.

3.5 Conclusion

It is clearly seen that various spectrum sharing techniques are already used in existing wireless systems but not for cellular systems. However, the application of them is different according to the features. Coordination protocol is more feasible for horizontal sharing between the same technologies since connection between different systems is very difficult to be established. GLDB and cognitive radio are two important techniques to enable vertical sharing between different systems. GLDB is more reliable to guarantee the protection to primary systems but centralized structure and database information collection of other systems are needed which is a lot of overhead. Cognitive radio is flexible as a distributed way for coexistence with primary systems although it may not be reliable when spectrum sensing. MAC-based coexistence mechanism such as CSMA is very promising for sharing between systems in unlicensed band. To conclude, these spectrum sharing techniques in existing wireless systems will provide a good reference for the application in 5G cellular systems.

4 5G Spectrum Sharing Considerations

4.1 5G Spectrum Sharing Enablers

5G systems have some natural characteristics to facilitate the above-mentioned spectrum sharing technique, which is explained in detail as follows:

- *Vertical sharing* can be a good way to unlock many bands for 5G systems as soon as possible. As mentioned in this chapter, the candidate bands for 5G systems are not "clean" even for frequency above 6 GHz. Almost all candidate bands have other existing allocations and already in use, e.g. fixed services or radar systems. Coexistence with them may not in every case be easy and thus one way to use these bands for 5G systems is to repurpose them for licensed cellular usage. In some cases this approach may require a lot of effort and time to succeed. In order to make them available for 5G use in relatively short time, one possible compromise is to introduce LSA. 5G systems would then be operating under the constraint of protecting the incumbent service. To allow an efficient, spectrum database or spectrum sensing technique may be employed and details will be described in later sections.

- *Homogenous horizontal sharing* could be feasible for 5G systems particularly in high frequency band, i.e. above 6 GHz. Using these bands might not target to provide ubiquitous coverage but non-continual coverage islands. Further different operators may be required to serve different traffic volumes at the same instant. In such case, spectrum sharing techniques can bring potential gain from statistical multiplexing. Besides, propagation conditions at high frequencies combined with high-gain beamforming for 5G systems is a promising venue for making good use of such frequencies.
- *Heterogeneous horizontal sharing* will likely only be employed in unlicensed spectrum and can be used to boost capacity for 5G systems. The use of unlicensed spectrum in 5 GHz for 4G systems is being discussed in 3GPP in combination of licensed spectrum for critical control signaling. It is very likely that unlicensed mode will continue to be a complementary method of using spectrum for 5G systems, especially for indoor enterprise solutions.

4.2 5G General Architecture for Spectrum Sharing

The impact of spectrum sharing on 5G system architecture discussed in this section aims at providing basic logical architecture support for implementation of 5G spectrum sharing in all scenarios foreseen by all spectrum sharing techniques. For specific scenario, only parts of functions are enabled according to the needs.

In order to meet the above functionality requirements, new logical entities need to be introduced in the overall logical 5G spectrum sharing architecture. The general view of additional logical entities overlaid to the physical structure is given in Fig. 8.

4.2.1 Spectrum Management Client (SMaC)

As shown in Fig. 8, SMaC is implemented in each Access Node (AN) to handle spectrum-related functions. Basically, there will be one spectrum information database which provides a list of supported frequency bands from hardware point of view, comprising of frequency range, band type (i.e. primary user mode, unlicensed

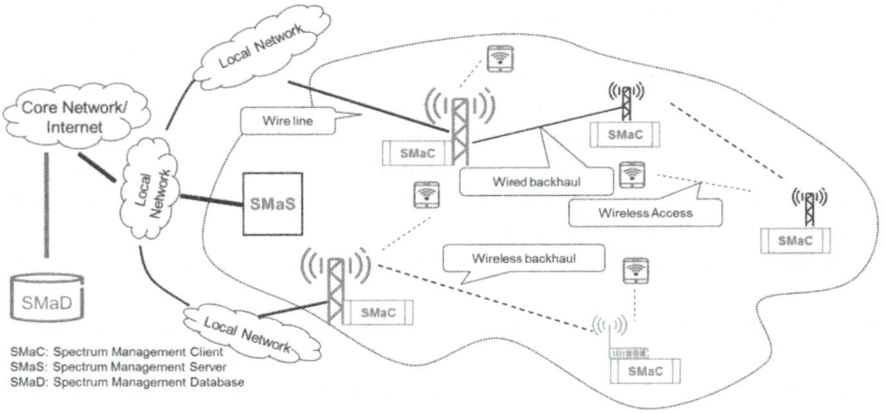

Fig. 8 Logical spectrum management entities (*blue*) over physical 5G architecture

and LSA) and potential coexisting system type (e.g. Radar, Wi-Fi, other 5G systems and etc.). The basic function for SMaC is to check the band information from spectrum information database which may be stored in memory part of AN. In other words, when AN is willing to operate or operating in certain band, the implemented SMaC inside is able to know which spectrum sharing scenarios are needed by checking the stored band information in its database.

For example, if 5G system is operating on TVWS, it will know that this is unlicensed band with primary users. Thus the following spectrum sharing scenarios listed in Fig. 2 are needed in this band: *vertical sharing* as secondary user and *horizontal sharing* including heterogeneous and homogeneous. Additionally, it may be also known that there will be TV systems as primary user and WiFi system as coexisting system. Then to enable coexistence with different systems in different spectrum sharing scenario, different tool box should be selected (e.g. GLDB for coexistence with TV systems in the scenario of vertical coexistence as secondary user). SMaC is an import entity introduced to support implementation of spectrum sharing tools, which comprises of the following possible support functions (SF):

- **SF for GLDB**: request spectrum availability information and receive spectrum information directly from GLDB or indirectly from the Spectrum Management Server (SMaS introduced below);
- **SF for Cognitive/DSA and MAC-based coexistence**: local spectrum sensing (energy, feature detection or virtual sensing) for primary systems (e.g. Radar etc.) or horizontal systems (e.g. Wi-Fi systems and etc.);
- **SF for Cognitive/DSA**: send or receive the spectrum sensing related information to other logical functionalities (e.g. SMaS or other SMaCs) for cooperative spectrum sensing;
- **SF for Coordination Protocol**: local interference measurement information collection from other 5G systems in the same operating channel and exchange information with other SMaCs located in neighbor ANs if needed;

- **SF for all**: make its own decision on which part of the spectrum is available for use based on collected information or receive centralized decision or recommendation information for band availability from other logical functionalities (e.g. SMaS).

Note that not all the SFs listed above should be implemented in SMaC. It depends on which spectrum sharing tool is selected and how it is implemented.

4.2.2 Spectrum Management Server (SMaS)

The spectrum management server (SMaS) is an operational assisting logical entity for centralized solution in part of 5G networks. Here "centralized" means it can collect information from other distributed entities (i.e. SMaC) and may provide a centralized decision or recommendation for spectrum availability based on gathered information (e.g. sensing). Having a SMaS as a centralized logical entity is optional, i.e. depending on whether the spectrum sharing tool is implemented in a centralized manner. One benefit of this is to provide more accurate estimation on the spectrum availability. For example, SMaS can gather sensing information from multiple SMaCs for cooperative spectrum sensing, which will avoid hidden node problem and improve spectrum discovery performance. The SMaS can be implemented in a centralized control node if existing. The SMaS may have one or more of the following functions:

- **SF for GLDB support**: If applicable (i.e.in scenarios where this is required), send request for spectrum status and receive UDN-wide spectrum information directly from GLDB/spectrum broker (i.e. SMaS can act as a proxy for the SMaCs);
- **SF for Cognitive/DSA and MAC-based coexistence**: Receive the spectrum sensing related information to other logical functionalities (e.g. SMaS or other SMaCs) for cooperative spectrum sensing;
- **SF for all**: Make centralized recommendation on spectrum availability information for its connected SMaCs and Send centralized recommendation on spectrum availability information to its connected SMaCs.

Note that SMaS is a purely logical entity to form the logical spectrum sharing architecture. Even when there is a SMaS in one 5G systems, only part of the listed functions may be selected to be implemented for a specific solution.

4.2.3 Spectrum Management Database (SMaD)

A SMaD is the global centralized logical functionality to provide spectrum information about primary systems. It is typically located outside the 5G networks. Obviously GLDB is a typical SMaD. The detailed functions are listed as follows:

- Keep track of the spectrum usage status of the primary system and estimate the spectrum availability and corresponding requirements for different areas and keep the information updated;
- Receive spectrum requests and related information from SMaS or SMaCs;
- Send spectrum information (availability status and constraint or grant) to other logical functionalities (e.g. SMaS or SMaC).

Obviously most of the functions listed in above section require logical connections between the different logical entities. For example, the implementation of the first listed function for the SMaC needs the connection between SMaC and SMaD or SMaS. In addition, a protocol running over those interfaces needs to be defined as well to ensure that the information transmitted on the connection can be understood by the communicating logical entities. Therefore, a logical architecture can be established by including logical entities, connections and protocols for spectrum sharing. Given different spectrum sharing requirements, different logical architectures may be designed. In order to provide a complete view for spectrum sharing support, our proposed architecture here is aiming to support the implementation of all spectrum sharing toolbox components in the different spectrum sharing scenarios from Fig. 2.

From Fig. 9, it can be seen that for different kinds of logical connections, there will be different kinds of spectrum sharing information exchange between the end points of the logical connection. Therefore, different protocols should be developed to make the communication understandable. In Fig. 10, it is exemplified by detailing some behavior by the various flavors of the Spectrum Management Protocol (SMP):

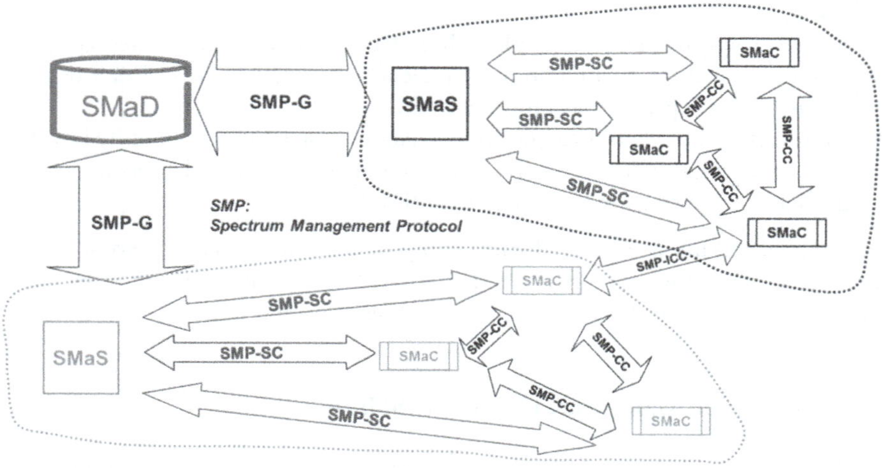

Fig. 9 Logical 5G spectrum sharing architecture

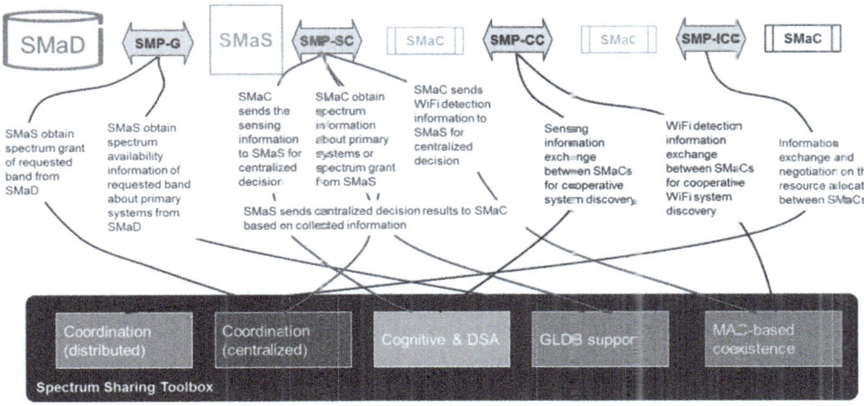

Fig. 10 Illustration of spectrum management protocols

- SMP-G (SMP-Global): Protocol used to communicate with external services such as a geo-location database or a spectrum coordinator or broker entity
- SMP-SC (SMP-Server Client): Protocol used for communication between the SMaS and the SMaC functionalities
- SMP-CC (SMP-Client Client): Protocol used for communication between SMaC functional entities residing in different nodes.
- SMP-ICC (SMP-Inter Cluster Coordination): Protocol used for information exchange and negotiating between different node clusters.

4.3 Spectrum Sharing Techniques for 5G

For 5G, one main difference is that the operating frequency will go even higher especially beyond 6 GHz. However, current application spectrum sharing techniques are employed in low frequency band, i.e. below 6 GHz. So sharing high frequency band for 5G systems will bring significant impact to current existing spectrum sharing techniques described in Sect. 3. The detailed impacts are analyzed below as a starting point for 5G spectrum sharing applications.

4.3.1 Coordination Protocol for 5G

For 5G systems, coordination protocol can be one feasible candidate tool to achieve homogenous horizontal sharing between different sharing networks or operators. It is expected that antenna directivity with massive MIMO, which is one important feature for 5G, could alleviate inter-network interference which enables aggressive spatial resource reuse between different networks. At the same time, it brings the difference for interference coordination design from original node pair-based in omni-mode to future link pair-based in high-gain beamforming mode. A ‘coordination context’ may be introduced for interfering links that require coordination for optimized performance. The objective is a solution that coordinates radio resource usage on a “per link” basis, i.e. only transmission links that actually create non-negligible interference (beyond a given threshold) to particular other links should be subject to coordination. Here, coordination means that transmissions of an inter-network interference link pair (one link in one 5G system and one in the other) are subject to scheduling constraints so that they cannot be scheduled on the same (time/frequency) radio resources. The two interfering networks negotiate an agreement on such a resource partitioning and record it in the form of a Coordination Context (CC) which is stored in a Coordination Context Database (CCDB) in each network. The CC thus represents a constraint that needs to be considered by the scheduler in each network. For example as shown in Fig. 11, A1 and B1 is an interfering link pair between different 5G systems. As a result of coordination, they will be coordinated to that the blue resources are reserved for A1 and the red resources for B1. In general, the resource division for one interference link pair can be in time, frequency or both.

To enable such interference coordination, reliable connections between different networks or operators are needed. This may be more practical for LSA spectrum since one centralized controller may be available for providing spectrum information to achieve vertical sharing. This LSA controller will be a good bridge to handle such kind of coordination between different networks if horizontal sharing is valid.

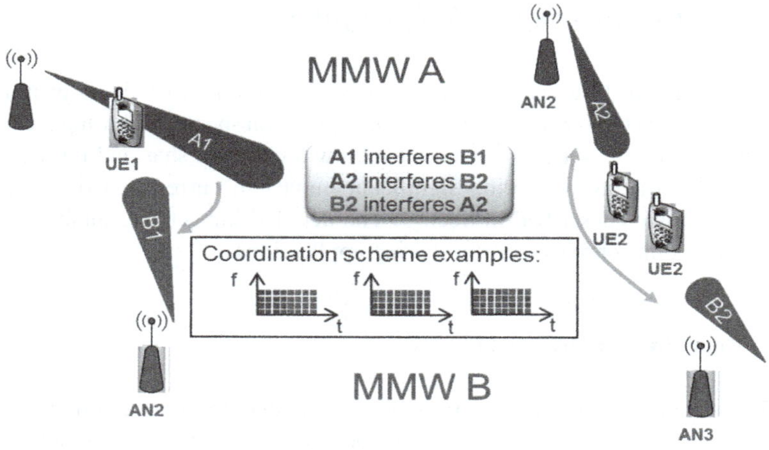

Fig. 11 Illustration of coordination context concept

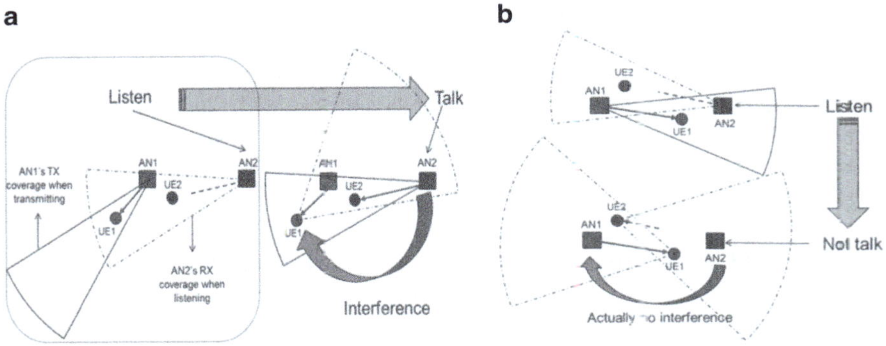

Fig. 12 Problem illustration of directional LBT: (**a**) hidden node problem; (**b**) exposed node problem

4.3.2 MAC-Based Coexistence for 5G

In high frequency band, high-gain beamforming is available for 5G system design, which will bring severe hidden node problem and exposed node problem for traditional LBT. On one aspect, hidden node problem is that transmitter of one link can't hear the current link transmission but there will be interference when they are transmitting simultaneously. One example of hidden node problem is given in Fig. 12a. In left side, AN1 is transmitting data to UE1 and AN2 is listening. But AN2 is not in TX coverage of AN1 and thus considers the channel is available. Then in right side, AN2 starts to transmit data to UE2. Obviously UE1 is interfered by AN2's transmission. On the other aspect, exposed node problem is that transmitter of one link hears the current link transmission but actually there will be not interference when they are transmitting simultaneously. As shown in Fig. 12b, AN1 is transmitting to UE1 and AN2 hear it. So AN2 determines the carrier/channel is busy and will not transmit. But actually there will be no interference between these two links because AN1 is not in RX coverage of UE2 and AN2 is not in RX coverage of UE1.

Based on our simple evaluation and analysis, it can be concluded that traditional 'listen', i.e., physical carrier sensing, is not effective to avoid transmission collision in directional transmission case especially with narrow TX&RX beam width, Particularly there exists severe hidden node problem that almost 100 % real interference case can't be detected and exposed node problem that almost 100 % detected interference is not real interference. The key reason is that transmitter listen to transmitter doesn't represent interference in directional transmission case. Solution directions to this are to involve receiver into LBT process and two different schemes are proposed. One is enhanced LBT with smart listening and notification, i.e. transmitter listens to directional CTS and receiver listens to directional RTS. The other is 'listen after talk', whose key idea is 'talk' first anyway and then collision solving. It is well known that the default mode of LBT for transmitter is 'not to send' and data is sent only when it is confirmed that the channel is idle by listening. Listen after talk concept adopts opposite logic as follows: the default mode for transmitter is 'to send' and data is not sent only when it is confirmed that channel is occupied by interfering transmissions. This is motivated that low interference situation exists for high frequency band case with high gain beamforming. In such situation, listen after talk will reduce overhead compared to LBT due to no waste of backoff time in most cases.

4.3.3 GLDB & Cognitive Radio for Vertical Sharing for 5G

As mentioned in Sects. 2 and 3, current vertical sharing primary systems are TV systems or Radar systems in low frequency band. But from spectrum allocation given by the ITU-R shows that mobile allocation above 6 GHz is almost always allocated together with fixed link services. Above 6 GHz, fixed link services are currently utilizing many spectrum bands. If 5G system intend to use that spectrum, then mechanism to protect the fixed services will be required. Fixed link applications are usually point to point, with very directive/high gain antennas. GLDB and cognitive radio may be employed but additional features should be developed. For GLDB support, the antenna pattern, transmission direction and etc. may also be included in the information database besides location and power information. Furthermore, smart exclusive zone generation should be performed according to coexistence evaluation between fixed link system and 5G systems. For spectrum sensing-based cognitive radio, the hidden node problem and exposed node problem also exists, cf. Fig. 12. More advanced spectrum sensing algorithms are needed to solve these problems of high-gain beamforming case in high frequency band.

5 Summary

First, spectrum sharing becomes more and more important for 5G systems, which will comprise of multiple spectrum type with different scenarios, i.e., vertical sharing and horizontal sharing. Next, spectrum sharing techniques mapping into different scenarios are introduced, i.e. coordination protocol, GLDB support, cognitive/DSA and MAC-based coexistence. Besides, current applications of these techniques in real systems are described. Finally spectrum sharing directions for 5G systems are analyzed for different spectrum sharing techniques.

In general, it should be clear that applying these spectrum sharing techniques to practice need additional functions. These may involve signaling overhead and implementation limits which means more work is needed. Licensed dedicated spectrum will continue to be the dominant spectrum usage method for 5G systems due to the possibility to control interference and guarantee coverage, while other spectrum sharing scenarios will act as complementary spectrum usage methods when beneficial.

References

1. T. Irnich, J. Kronander, Y. Selén, G. Li, Spectrum sharing scenarios and resulting technical requirements for 5G systems, in *Workshop on Spectrum Sharing Strategies for Wireless Broadband Services, PIMRC 2013*, London, UK, 8–11 September 2013
2. T. Wang, G. Li et al., 5G Spectrum, is China Ready?, published on IEEE Communication Magazine 5G Spectrum: Enabling the Future Mobile Landscape, July 2015
3. M. Xing, Y. Peng, T. Xia, H. Long, K. Zheng, Adaptive spectrum sharing of LTE co-existing with WLAN in unlicensed frequency bands, in *IEEE 81st Vehicular Technology Conference 2015-Spring*, Glasgow, Scotland, 11–14 May 2015
4. IEEE Standard for Local and metropolitan area networks Part 16. Air Interface for Broadband Wireless Access Systems Amendment 2: Improved Coexistence Mechanisms for License-Exempt Operation. IEEE Std 802.16h-2010 (Amendment to IEEE Std 802.16-2009), July 30, 2010. doi:10.1109/IEEESTD.2010.5538195
5. Ericsson White Paper on Spectrum Sharing. http://www.ericsson.com/res/docs/whitepapers/wp-spectrum-sharing.pdf, Oct. 2013
6. Cognitive Radio and Dynamic Spectrum Access, Lars Berlemann Deutsche Telekom, Germany and Stefan Mangold Swisscom, Switzerland
7. Cognitive Radio An Integrated Agent Architecture for Software Defined Radio-Dissertation, Doctor of Technology Joseph Mitola III, Royal Institute of Technology (KTH), ISSN 1403-5286, ISRN KTH/IT/AVH-00/01-SE, 8 May 2000
8. Federal Communications Commission, Cognitive Radio Technologies Proceeding (CRTP) ET Docket No. 03-108, March 2005. http://www.fcc.gov/oet/cognitiveradio/
9. F.F. Digham, M.-S. Alouini, M.K. Simon, On the energy detection of unknown signals over fading channels, in *IEEE ICC 2003*, Washington, May 2003
10. Y.C. Liang, Y. Zeng, E.C.Y. Peh, A.T. Hoang, Sensing-throughput tradeoff for cognitive radio networks. IEEE Trans. Wireless Commun. **7**(4), 1326–1337 (2008)
11. M.K. Steveny, *Fundamentals of Statistical Signal Processing, Volume 2: Detection Theory* (Prentice Hall PTR, 1998)
12. H.-S. Chen, W. Gao, D.G. Daut, Signature based spectrum sensing algorithms for IEEE 802.22 WRAN, in *IEEE International Conference on Communications (ICC)*, June 2007, pp. 6487–6492
13. Y. Zeng, Y.C. Liang, Spectrum-sensing algorithms for cognitive radio based on statistical covariances. IEEE Trans. Veh. Technol. **58**(4), 1804–1815 (2009)
14. A. Dandawate, G.B. Giannakis, Statistical tests for presence of cyclostationarity. IEEE Trans. Sig. Process. **42**, 2355–2369 (1994)
15. D. Birru, K. Challapali, B. Dong, Detection of the presence of television signals embedded in noise using cyclostationary toolbox. U.S. Patent 2010/0157066, 2010
16. A. Mukherjee, J.F. Cheng et al., System architecture and coexistence evaluation of licensed-assisted access LTE with IEEE 802.11, in *IEEE ICC 2015*, London, June 2015

Part II
Transmission and Design Techniques for 5G Networks

Massive MIMO Communications

Trinh Van Chien and Emil Björnson

Abstract Every new network generation needs to make a leap in area data throughput, to manage the growing wireless data traffic. The Massive MIMO technology can bring at least ten-fold improvements in area throughput by increasing the spectral efficiency (bit/s/Hz/cell), while using the same bandwidth and density of base stations as in current networks. These extraordinary gains are achieved by equipping the base stations with arrays of a hundred antennas to enable spatial multiplexing of tens of user terminals. This chapter explains the basic motivations and communication theory behind the Massive MIMO technology, and provides implementation-related design guidelines.

1 Introduction

Much higher area data throughput is required in future cellular networks, since the global demand for wireless data traffic is continuously growing. This goal can be achieved without the need for more bandwidth or additional base stations if the spectral efficiency (measured in bit/s/Hz/cell) is improved. This chapter explains why the Massive MIMO (multiple-input multiple-output) communication technology, where multi-antenna base stations spatially multiplex a multitude of user terminals over the entire bandwidth, is well-suited for this purpose. The rationale behind the Massive MIMO concept and its transmission protocol is explained from a historical perspective in Sect. 2. Next, Sect. 3 provides a basic communication theoretic performance analysis. Closed-form spectral efficiency expressions are derived and the key properties and performance limitations of Massive MIMO are highlighted. The chapter is concluded by Sect. 4 where implementation-related design guidelines are given, particularly regarding power allocation and the reuse of pilot sequences for efficient channel estimation. Multi-cell simulations are provided to showcase that the Massive MIMO technology can provide tenfold or even 50-fold improvements in spectral efficiency over contemporary technology, without

T. Van Chien • E. Björnson (✉)
Department of Electrical Engineering (ISY), Linköping University,
SE-581 83 Linköping, Sweden
e-mail: trinh.van.chien@liu.se; emil.bjornson@liu.se

W. Xiang et al. (eds.), *5G Mobile Communications*,
DOI 10.1007/978-3-319-34208-5_4

the need for advanced signal processing or network coordination. Finally, the full mathematical details are provided in Appendix at the end of this chapter.

2 Importance of Improving the Spectral Efficiency

The wireless information traffic has doubled every two and a half years since the beginning of wireless communications, as observed by Martin Cooper at ArrayComm in the nineties. Different technologies and use cases have dominated in different periods, but the exponential increase is currently driven by wireless data traffic in cellular and local area networks. There are no indications that this trend will break anytime soon; in fact, a slightly faster traffic growth is predicted in the well-reputed Cisco Visual Networking Index and Ericsson Mobility Report.

To keep up with the rapid traffic growth, a key goal of the 5G technologies is to improve the area throughput by orders of magnitude; 100× and even 1000× higher throughput are regularly mentioned as 5G design goals. The area throughput of a wireless network is measured in bit/s/km^2 and can be modeled as follows:

$$\text{Area throughput (bit/s/km}^2) =$$
$$\text{Bandwidth (Hz)} \times \text{Cell density (cells/km}^2) \times \text{Spectral efficiency (bit/s/Hz/cell)}.$$

This simple formula reveals that there are three main components that can be improved to yield higher area throughput: (1) more bandwidth can be allocated for 5G services; (2) the network can be densified by adding more cells with independently operating access points; and (3) the efficiency of the data transmissions (per cell and for a given amount of bandwidth) can be improved.

The improvements in area throughput in previous network generations have greatly resulted from cell densification and allocation of more bandwidth. In urban environments, where contemporary networks are facing the highest traffic demands, cellular networks are nowadays deployed with a few hundred meters inter-site distances and wireless local area networks (WLANs) are available almost everywhere. Further cell densification is certainly possible, but it appears that we are reaching a saturation point. Moreover, the most valuable frequency bands are below 6 GHz because these frequencies can provide good network coverage and service quality, while higher bands might only work well under short-range line-of-sight conditions. In a typical country like Sweden, the cellular and WLAN technologies have in total been allocated more than 1 GHz of bandwidth in the interval below 6 GHz and thus we cannot expect any major bandwidth improvements either.

In contrast, the spectral efficiency (SE) has not seen any major improvements in previous network generations. Hence, it might be a factor that can be greatly improved in the future and possibly become the primary way to achieve high area throughput in 5G networks. In this chapter, we describe the rationale and background of the physical-layer technology Massive multiple-input multiple-output (MIMO), which provides the means to improve the SE of future networks by one or two orders of magnitude.

2.1 Multi-User MIMO Communication

The SE of a single-input single-output (SISO) communication channel, from a single-antenna transmitter to a single-antenna receiver, is upper bounded by the Shannon capacity, which has the form $\log_2(1 + \overline{SNR})$ bits/Hz for additive white Gaussian noise (AWGN) channels. The SISO capacity is thus a logarithmic function of the signal-to-noise ratio (SNR), denoted here as $\overline{SNR}$. To improve the SE we need to increase the SNR, which corresponds to increasing the power of the transmitted signal. For example, suppose we have a system that operates at 2 bit/s/Hz and we would like to double its SE to 4 bit/s/Hz, then this corresponds to improving the SNR by a factor 5, from 3 to 15. The next doubling of the SE, from 4 to 8 bit/s/Hz, requires another 17 times more power. In other words, the logarithm of the SE expression forces us to increase the transmit power exponentially fast to achieve a linear increase in the SE of the SISO channel. This is clearly a very inefficient and non-scalable way to improve the SE, and the approach also breaks down when there are interfering transmissions in other cells that scale their transmit powers in the same manner. We therefore need to identify another way to improve the SE of cellular networks.

Each base station (BS) in a cellular network serves a multitude of user terminals. Traditionally, the time/frequency resources have been divided into resource blocks and only one of the user terminals was active per block. This terminal can then receive a single data stream with an SE quantified as $\log_2(1 + \overline{SNR})$. The efficient way to increase the SE of a cellular network is to have multiple parallel transmissions. If there are $\mathbb{G}$ parallel and independent transmissions, the sum SE becomes $\mathbb{G} \log_2(1 + \overline{SNR})$ where $\mathbb{G}$ acts as a multiplicative pre-log factor. Parallel transmissions can be realized by having multiple transmit antennas and multiple receive antennas. There are two distinct cases:

1. Point-to-point MIMO [39], where a BS with multiple antennas communicates with a single user terminal having multiple antennas.
2. Multi-user MIMO [34], where a BS with multiple antennas communicates with multiple user terminals, each having one or multiple antennas.

There are many reasons why multi-user MIMO is the most scalable and attractive solution [17]. Firstly, the wavelength is 5–30 cm in the frequency range of cellular communication (1–6 GHz). This limits the number of antennas that can be deployed in a compact user terminal for point-to-point MIMO, while one can have almost any number of spatially separated single-antenna terminals in multi-user MIMO. This is an important distinction since the number of simultaneous data streams that can be separated by MIMO processing equals the minimum of the number of transmit and receive antennas. Secondly, the wireless propagation channel to a user terminal is likely to have only a few dominating paths, which limits the ability to convey multiple parallel data streams to a terminal in point-to-point MIMO. The corresponding restriction on multi-user MIMO is that the users need to be, say, a few meters apart to have sufficiently different channel characteristics, which is a very loose restriction that is true in most practical scenarios. Thirdly, advanced signal processing is needed at the terminals in point-to-point MIMO to detect the multiple data streams, while each terminal in multi-user MIMO only needs to detect a single data stream.

The canonical multi-user MIMO system consists of a BS with M antennas that serves K single-antenna terminals; see Fig. 1 for a schematic illustration. The BS multiplexes one data stream per user in the downlink and receives one stream per user in the uplink. Simply speaking, the BS uses its antennas to direct each signal towards its desired receiver in the downlink, and to separate the multiple signals received in the uplink. If the terminal is equipped with multiple antennas, it is often beneficial to use these extra antennas to mitigate interference and improve the SNR rather than sending multiple data streams [6]. For the ease of exposition, this chapter concentrates on single-antenna terminals. In this case, $\min(M, K)$ represents the maximal number of data streams that can be simultaneously transmitted in the cell, while still being separable in the spatial domain. The number $\min(M, K)$ is referred to as the *multiplexing gain* of a multi-user MIMO system.

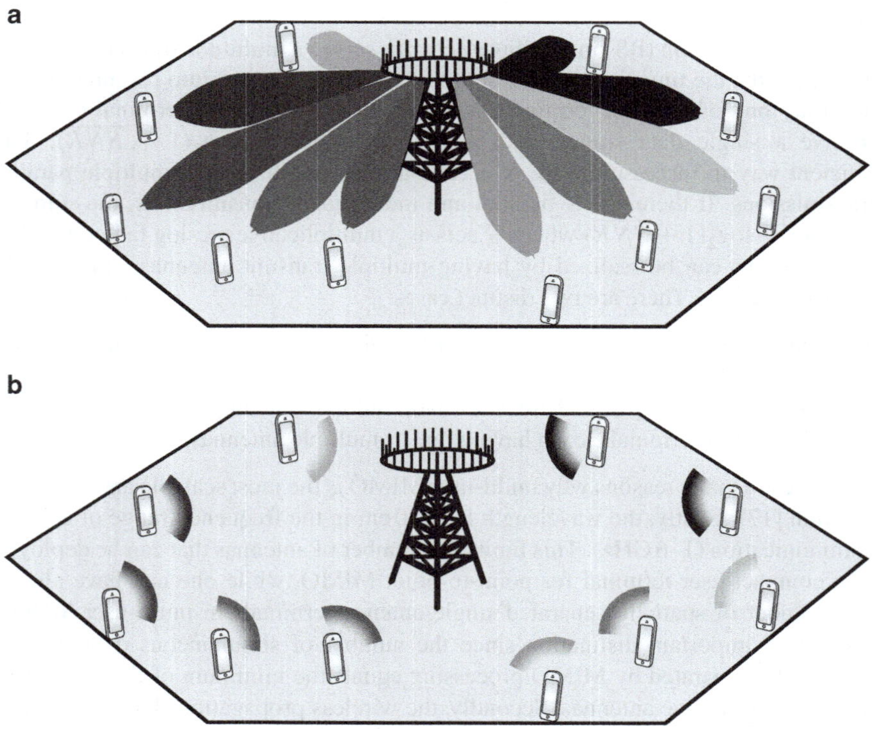

Fig. 1 Illustration of the downlink and uplink transmission in a multi-user MIMO system, where the BS is equipped with M antennas and serves K user terminals simultaneously. This illustration focuses on line-of-sight propagation where the downlink signals can be viewed as angular beams, but multi-user MIMO works equally well in non-line-of-sight conditions. (**a**) Downlink in multi-user MIMO. (**b**) Uplink in multi-user MIMO

2.2 Lessons Learned

The research on multi-user MIMO, particularly with multi-antenna BSs, has been going on for decades. Some notable early works are the array processing papers [1, 38, 44, 47], the patent [36] on spatial division multiple access (SDMA), and the seminal information-theoretic works [11, 18, 42, 43, 46] that characterized the achievable multi-user capacity regions, assuming that perfect channel state information (CSI) is available in the system. In this section, we summarize some of the main design insights that have been obtained over the years.

Capacity-achieving transmission schemes for multi-user MIMO are based upon non-linear signal processing; for example, the dirty-paper coding (DPC) scheme that achieves the downlink capacity and the successive interference cancelation (SIC) scheme that achieves the uplink capacity. The intuition behind these schemes is that the inter-user interference needs to be suppressed, by interference-aware transmit processing or iterative interference-aware receive processing, to achieve the optimal performance. These non-linear schemes naturally require extensive computations and accurate CSI, because otherwise the attempts to subtract interference cause more harm than good.

How large are the gains of optimal non-linear processing (e.g., DPC and SIC) over simplified linear processing schemes where each user terminal is treated separately? To investigate this, let us provide a numerical example where $K = 10$ user terminals are simultaneously served by a BS with M antennas. For simplicity, each user is assumed to have an average SNR of 5 dB, there is perfect CSI available everywhere, and the channels are modeled as uncorrelated Rayleigh fading (this is defined in detail in Sect. 3). Figure 2 shows the average sum SE, as a function of M, achieved by sum capacity-achieving non-linear processing and a simplified linear processing scheme called zero-forcing (ZF), which attempts to suppress all interference. The results are representative for both uplink and downlink transmissions.

This simulation shows that the non-linear processing greatly outperforms linear ZF when $M \approx K$. The operating point $M = K$ makes particular sense from a multiplexing perspective since the multiplexing gain $\min(M, K)$ does not improve if we let M increase for a fixed K. Nevertheless, Fig. 2 shows that there are other reasons to consider $M > K$; the capacity increases and the performance with linear ZF processing approaches the capacity. Already at $M = 20$ (i.e., $M/K = 2$) there is only a small gap between optimal non-linear processing and linear ZF. In fact, both schemes also approach the upper curve in Fig. 2 which represents the upper bound where the interference between the users is neglected. This shows that we can basically serve all the K users as if each one of them was alone in the cell.

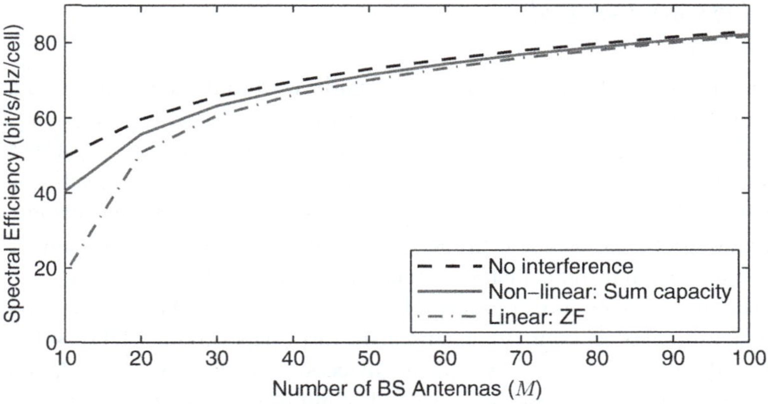

Fig. 2 Average spectral efficiency in a multi-user MIMO system with $K = 10$ users and varying number of BS antennas. Each user has an average SNR of 5 dB and the channels are Rayleigh fading. The sum capacity is compared with the performance of linear ZF processing and the upper bound when neglecting all interference. The results are representative for both uplink and downlink

First lesson learned: Linear processing, such as ZF, provides a sum spectral efficiency close to the sum capacity when $M \gg K$.

The performance analysis and optimization of linear processing schemes have received much attention from academic researchers. While non-linear schemes are hard to implement but relatively easy to analyze and optimize, linear processing schemes have proved to have the opposite characteristics. In particular, computing the optimal downlink linear precoding is an NP-hard problem in many cases [27], which requires monotonic optimization tools to solve; see for example [5]. Nevertheless, the suboptimal ZF curve in Fig. 2 was generated without any complicated optimization, thus showing that the optimal linear processing obtained in [5] can only bring noticeable gains over simple ZF for $M \approx K$, which is the regime where we have learnt not to operate.

As mentioned earlier, the BS needs CSI in multi-user MIMO systems to separate the signals associated with the different users. Perfect CSI can typically not be achieved in practice, since the channels are changing over time and frequency, and thus must be estimated using limited resources. The channel estimation of a frequency-selective channel can be handled by splitting the frequency resources into multiple independent frequency-flat subchannels that can be estimated separately. A known pilot sequence is transmitted over each such subchannel and the received signal is used to estimate the channel response. In order to explore all spatial channel dimensions, this sequence must at least have the same length as the number of transmit antennas [4]. This means that a pilot sequence sent by the BS needs to have the length M, while the combined pilot sequence sent by the single-antenna user terminals needs to have the length K.

There are two ways of implementing the downlink and uplink transmission over a given frequency band. In frequency division duplex (FDD) mode the bandwidth is split into two separate parts: one for the uplink and one for the downlink. Pilot sequences are needed in both the downlink and the uplink due to the frequency-selective fading, giving an average pilot length of $(M-K)/2$ per subchannel. There is an alternative time-division duplex (TDD) mode where the whole bandwidth is used for both downlink and uplink transmission, but separated in time. If the system switches between downlink and uplink faster than the channels are changing, then it is sufficient to learn the channels in only one of the directions. This leads to an average pilot length of $\min(M, K)$ per subchannel, if we send pilots only in the most efficient direction. In the preferable operating regime of $M \gg K$, we note that TDD systems should send pilots only in the uplink and the pilot length becomes $\min(M, K) = K$. We conclude that TDD is the preferable mode since it not only requires shorter pilots than FDD, but is also highly scalable since the pilot length is independent of the number of BS antennas.

We give a concrete numerical example in Fig. 3 for downlink transmission with $K = 10$ users, an SNR of 5 dB, and uncorrelated Rayleigh fading channels. Two linear precoding schemes are considered; (a) maximum ratio (MR) and (b) zero-forcing (ZF). These schemes are later defined mathematically in Sect. 3. This simulation compares the SE obtained when having perfect CSI with the performance when having CSI estimated with pilot sequences of length τ_p. The SE is shown as a function of the number of BS antennas, M, and we compare TDD mode using $\tau_p = K = 10$ with FDD mode using either $\tau_p = 10$, $\tau_p = M$, or $\tau_p = \min(M, 50)$, where the latter models an arbitrarily chosen maximum pilot length of 50 (e.g., motivated by pilot overhead constraints).

In TDD mode there is a visible performance loss in Fig. 3 as compared to having perfect CSI. The loss with MR precoding is very small, which shows that it is robust to estimation errors. The performance loss is larger for ZF precoding, since estimation errors make it harder to suppress interference, but we notice that ZF anyway provide higher performance than MR for all considered M. We notice that the performance losses are substantially constant irrespective of the number of BS antennas, thus TDD systems always benefit from adding more antennas. In contrast, FDD systems only benefits from adding more antennas if the pilot sequences are also made longer, as in the case $\tau_p = M$. With $\tau_p = 10$ there is no benefit from having more than ten antennas, while the performance saturates at 50 antennas when $\tau_p = \min(M, 50)$. In summary, TDD operation is fully scalable with respect to the number of BS antennas, while FDD operation can only handle more antennas by also increasing the pilot overhead. It is practically feasible to deploy FDD systems with many antennas, particularly for slowly varying channels where we can accept a large pilot overhead, but TDD is always the better choice in this respect.

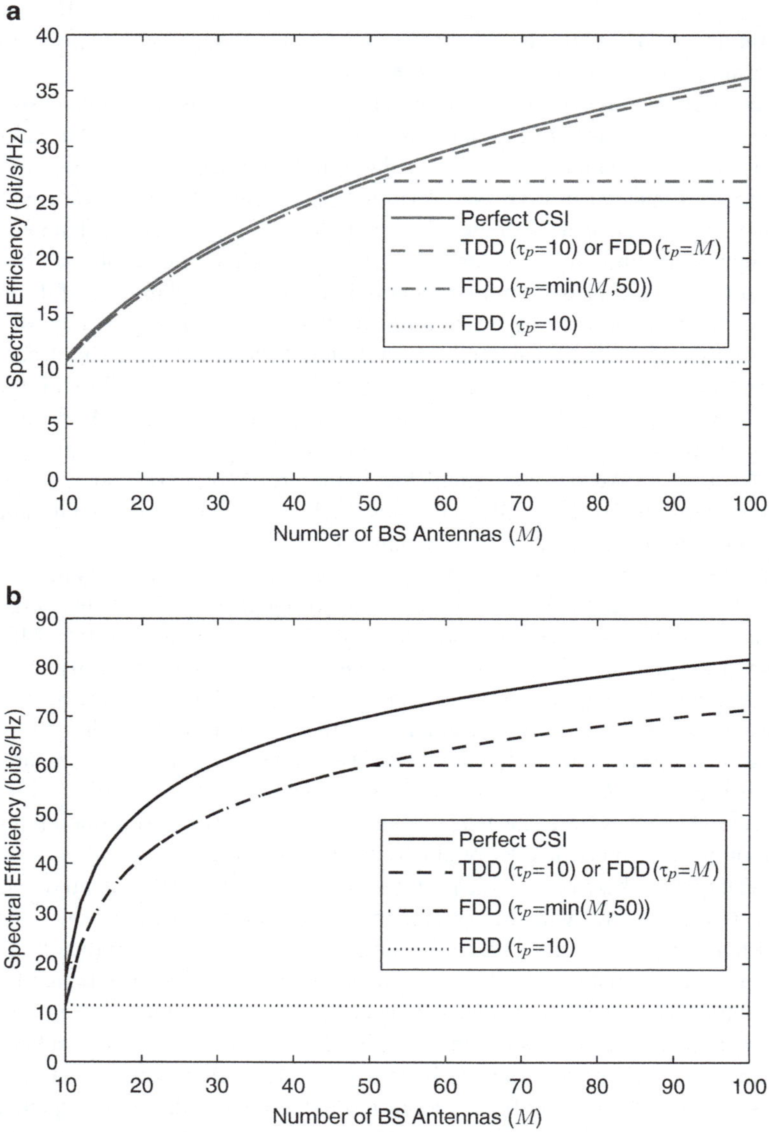

Fig. 3 Average downlink spectral efficiency, as a function of the number of BS antennas, with different processing schemes and different types of CSI available at the BS. (**a**) Downlink simulation with maximum ratio precoding. (**b**) Downlink simulation with zero-forcing precoding

Second lesson learned: The channel estimation is simplified when operating in TDD mode, since the pilot sequences only need to be of length K irrespective of the number of BS antennas M.

Note that the uplink works in the same way in the TDD and FDD modes, while the distinct benefit of TDD in terms of scalability appears in the downlink.

2.2.1 Favorable Propagation

Recall from Fig. 2 that by adding more BS antennas, both the sum capacity-achieving non-linear processing and the simplified linear ZF processing approached the case without interference. This is not a coincidence but a fundamental property that is referred to as *favorable propagation*.

Let $\mathbf{h}_1, \mathbf{h}_2 \in \mathbb{C}^M$ represent the channel responses between a BS and two different user terminals. If these vectors are non-zero and orthogonal in the sense that

$$\mathbf{h}_1^H \mathbf{h}_2 = 0, \tag{1}$$

where $(\cdot)^H$ denotes the conjugate transpose, then the BS can completely separate the signals s_1, s_2 transmitted by the users when it observes $\mathbf{y} = \mathbf{h}_1 s_1 + \mathbf{h}_2 s_2$. By simply computing the inner product between $\mathbf{y}$ and $\mathbf{h}_1$, the BS obtains

$$\mathbf{h}_1^H \mathbf{y} = \mathbf{h}_1^H \mathbf{h}_1 s_1 + \mathbf{h}_1^H \mathbf{h}_2 s_2 = \|\mathbf{h}_1\|^2 s_1 \tag{2}$$

where the inter-user interference disappeared due to (1). The same thing can be done for the second user: $\mathbf{h}_2^H \mathbf{y} = \|\mathbf{h}_2\|^2 s_2$. Note that the BS needs perfect knowledge of $\mathbf{h}_1$ and $\mathbf{h}_2$ to compute these inner products. The channel orthogonality in (1) is called favorable propagation, since the two users can communicate with the BS without affecting each other.

Is there a chance that practical channels offer favorable propagation? Probably not according to the strict definition that $\mathbf{h}_1^H \mathbf{h}_2 = 0$ but an approximate form of favorable propagation is achieved in non-line-of-sight scenarios with rich scattering:

Lemma 1. *Suppose that $\mathbf{h}_1 \in \mathbb{C}^M$ and $\mathbf{h}_2 \in \mathbb{C}^M$ have independent random entries with zero mean, identical distribution, and bounded fourth-order moments, then*

$$\frac{\mathbf{h}_1^H \mathbf{h}_2}{M} \to 0 \tag{3}$$

almost surely as $M \to \infty$.

Proof. This is a consequence of the law of large numbers. A direct proof is provided along with Theorem 3.7 in [14]. ∎

This lemma shows that the inner product between $\mathbf{h}_1$ and $\mathbf{h}_2$, if normalized with the number of BS antennas, goes asymptotically to zero as M increases. We refer to this as *asymptotic favorable propagation* and note that this phenomenon explains the behaviors in Fig. 2; the difference between having no inter-user interference and suppressing the interference by ZF becomes smaller and smaller as the number of antennas increases, because the loss in desired signal gain when using ZF reduces when the user channels become more orthogonal.

One special case in which Lemma 1 holds is $\mathbf{h}_1, \mathbf{h}_2 \sim \mathcal{CN}(\mathbf{0}, \mathbf{I}_M)$, where $\mathcal{CN}(\cdot, \cdot)$ denotes a multi-variate circularly symmetric complex Gaussian distribution and $\mathbf{I}_M$ is the $M \times M$ identity matrix. This is known as *uncorrelated Rayleigh fading* and in this case one can even prove that the variance of the inner product in (3) is $1/M$ and thus decreases linearly with the number of antennas [32]. Many academic works on Massive MIMO systems consider Rayleigh fading channels, due to the analytic tractability of Gaussian distributions. Nevertheless, Lemma 1 shows that asymptotic favorable propagation holds for other random channel distributions as well. This mathematical result can be extended to also include correlation between the elements in a channel vector. One can also derive similar analytic results for line-of-sight propagation [32] and behaviors that resemble asymptotic favorable propagation have been observed also in the real-world multi-user MIMO channel measurements presented in [16, 19].

Third lesson learned: Most wireless channels seem to provide asymptotic favorable propagation.

This lesson is yet another reason to design multi-user MIMO systems with $M \gg K$. It is, however, important to note that $(\mathbf{h}_1^H \mathbf{h}_2)/M \to 0$, as $M \to \infty$, does not imply that $\mathbf{h}_1^H \mathbf{h}_2 \to 0$. Strict favorable propagation is unlikely to appear in practical or theoretical channels. In fact, the inner product $\mathbf{h}_1^H \mathbf{h}_2$ grows roughly as $\sqrt{M}$ for Rayleigh fading channels. The key point is that this correlation has a negligible impact, since the SE depends on $(\mathbf{h}_1^H \mathbf{h}_2)/M$ which goes to zero roughly as $1/\sqrt{M}$. Moreover, the main suppression of inter-user interference appears already at relatively small number of antennas due to the square root.

2.3 Massive MIMO Concept

The Massive MIMO concept was proposed in the seminal paper [28] and described in the patent [29], both of which have received numerous scientific awards. Massive MIMO takes multi-user MIMO communications to a new level by designing a highly scalable communication protocol that utilizes the three lessons described in Sect. 2.2. The basic information and communication theoretic limits of this 5G technology were established in early works such as [7, 20, 21, 23, 30]. In this chapter we define Massive MIMO as follows:

> **Massive MIMO** is a multi-user MIMO system with M antennas and K users per BS. The system is characterized by $M \gg K$ and operates in TDD mode using linear uplink and downlink processing.

This definition does not manifest any particular ratio between M and K, or any particular orders of magnitude that these parameters should have. One attractive example is a system with M in the range of 100–200 antennas, serving between $K = 1$ and $K = 40$ users depending on the data traffic variations. The first public real-time implementation of Massive MIMO is the LuMaMi testbed described in [41], which features $M = 100$ and $K = 10$. We stress that other definitions of Massive MIMO are available in other works and can both be more restrictive (e.g., require certain dimensionality of M and K) and looser (e.g., also include FDD mode), but in this chapter we only consider the definition above.

The BS antenna array typically consists of M dipole antennas, each having an effective size $\lambda/2 \times \lambda/2$, where λ is the wavelength. This means that an array area of 1 m^2 can fit 100 antennas at a 1.5 GHz carrier frequency and 400 antennas at 3 GHz. Each antenna is attached to a separate transceiver chain, so that the system can access the individual received signal at each antenna and select the individual signals to be transmitted from each antenna. The array can have any geometry; linear, rectangular, cylindrical, and distributed arrays are described in [25]. It is important to note that no model of the array geometry is exploited in the Massive MIMO processing, thus the antennas can be deployed arbitrarily without any geometrical array calibration.

The basic Massive MIMO transmission protocol is illustrated in Fig. 4. The time–frequency resources are divided into blocks of size B_c Hz and T_c s, with the purpose of making each user channel approximately frequency-flat and static within a block. Hence, the bandwidth B_c is selected to be smaller or equal to the anticipated channel coherence bandwidth among the users, while T_c is smaller or equal to the anticipated channel coherence time of the users. For this particular reason, each

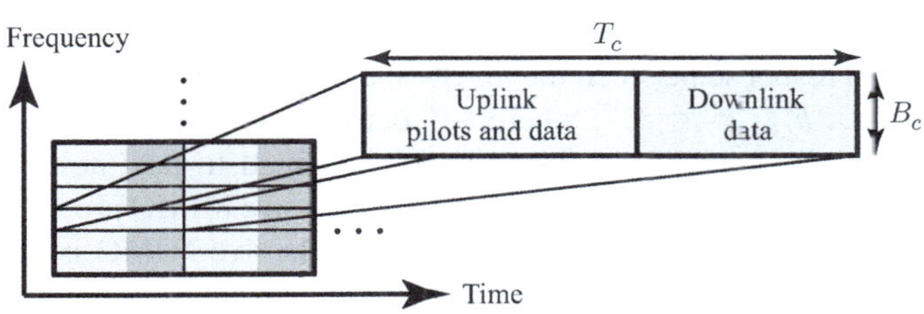

Fig. 4 Illustration of the basic Massive MIMO transmission protocol, where the time–frequency resources are divided into coherence intervals, each containing $\tau_c = B_c T_c$ transmission symbols. Each coherence interval contains uplink pilot sequences and can be used for both uplink and downlink payload data transmission based on TDD operation

block is referred to as a *coherence interval*. The number of transmission symbols that fit into a coherence interval is given by $\tau_c = B_c T_c$, due to the Nyquist-Shannon sampling theorem. The dimensionality of the coherence interval depends greatly on the anticipated system application. For example, a coherence interval of $\tau_c = 200$ symbols can be obtained with $B_c = 200\,\text{kHz}$ and $T_c = 1\,\text{ms}$, which supports highway user velocities in urban environments at 2 GHz carrier frequencies. Much larger coherence intervals (e.g., τ_c at the order of 10^3 or 10^4) can be obtained by limiting the application to scenarios with low user mobility and short delay spread.

Each coherence interval is operated in TDD mode and can contain both downlink and uplink payload transmissions. To enable channel estimation at the BS, τ_p of the symbols in each coherence interval are allocated for uplink transmission of pilot sequences (where $\tau_p \geq K$), while the remaining $\tau_c - \tau_p$ symbols can be allocated arbitrarily between uplink and downlink payload data transmissions.

We let γ^{UL} and γ^{DL} denote the fractions of uplink and downlink payload transmission, respectively. This means that the uplink contains $\gamma^{\text{UL}}(\tau_c - \tau_p)$ data symbols and the downlink contains $\gamma^{\text{DL}}(\tau_c - \tau_p)$ data symbols per coherence interval. Naturally, these fractions satisfies $\gamma^{\text{UL}} + \gamma^{\text{DL}} = 1$ and $\gamma^{\text{UL}}, \gamma^{\text{DL}} \geq 0$. Notice that no downlink pilots are assumed in this protocol, since the effective precoded channels converge to their mean values when the BS has many antennas (due to the law of large numbers). It is certainly possible to also send a small amount of downlink pilots, particularly for estimating the small fading variations of the effective precoded channels, but the additional gains from doing this appears to be small in many relevant Massive MIMO cases [31].

Based on this definition of Massive MIMO, the next sections analyze how large SEs that the transmission protocol can offer in 5G cellular networks.

3 Performance Analysis

In this section, we describe the uplink detection and downlink precoding of a Massive MIMO network, and analyze the achievable system performance. We consider a basic Massive MIMO network comprising L cells, each consisting of a BS with M antennas and K single-antenna user terminals.

The channel response between the lth BS and user k in the ith cell is denoted by $\mathbf{h}_{i,k}^{l} = [h_{i,k,1}^{l} \; \dots \; h_{i,k,M}^{l}]^T \in \mathbb{C}^M$, where $(\cdot)^T$ denotes the transpose. These channel vectors are ergodic random variables that are assumed to take new independent realizations in each coherence interval; recall the Massive MIMO protocol described in Sect. 2.3. To show that the general concept of Massive MIMO is applicable in any propagation environment, we keep the performance analysis general by only defining the basic statistical channel properties: the mean value and variance of each channel coefficient $h_{i,k,m}^{l}$ (note that m stands for the mth antenna at BS l, for $m = 1, \dots, M$). We let

$$\bar{\mathbf{h}}^l_{i,k} = \mathbb{E}\{\mathbf{h}^l_{i,k}\} = [\bar{h}^l_{i,k,1} \, \dots \, \bar{h}^l_{i,k,M}]^T \tag{4}$$

denote the vector of mean values. The variance of the mth coefficient of $\mathbf{h}^l_{i,k}$ is denoted by

$$\beta^l_{i,k} = \mathbb{V}\{h^l_{i,k,m}\}, \tag{5}$$

which is independent of the antenna index m (assuming that the large-scale fading is stationary over the BS array). We also assume that each BS and user can keep perfect track of these long-term statistical properties, and that the user channels are statistically independent.

Using these channel properties, we now analyze the uplink and the downlink.

3.1 Uplink with Linear Detection

For each uplink symbol, the received baseband signal $\mathbf{y}_l \in \mathbb{C}^M$ at the lth BS is modeled as

$$\mathbf{y}_l = \sum_{i=1}^{L} \sum_{k=1}^{K} \mathbf{h}^l_{i,k} \sqrt{p_{i,k}} x_{i,k} + \mathbf{n}_l, \tag{6}$$

where $x_{i,k}$ is the normalized transmission symbol (with $\mathbb{E}\{|x_{i,k}|^2\} = 1$) and $p_{i,k}$ is the transmit power of user k in cell i. The receiver hardware at the BS is contaminated by additive white noise, as modeled by the vector $\mathbf{n}_l \in \mathbb{C}^M$ which is zero-mean circularly symmetric complex Gaussian distributed with variance σ^2_{UL}; that is, $\mathbf{n}_l \sim \mathcal{CN}(\mathbf{0}, \sigma^2_{\mathrm{UL}} \mathbf{I}_M)$.

The matrix notations $\mathbf{H}^l_i = [\mathbf{h}^l_{i,1} \, \dots \, \mathbf{h}^l_{i,K}] \in \mathbb{C}^{M \times K}$, $\mathbf{P}_i = \mathrm{diag}(p_{i,1}, \dots, p_{i,K}) \in \mathbb{C}^{K \times K}$, and $\mathbf{x}_i = [x_{i,1} \, \dots \, x_{i,K}]^T \in \mathbb{C}^K$ can be used to write the multi-cell multi-user MIMO system model from (6) in a compact matrix form:

$$\mathbf{y}_l = \sum_{i=1}^{L} \mathbf{H}^l_i \mathbf{P}^{1/2}_i \mathbf{x}_i + \mathbf{n}_l. \tag{7}$$

The channels $\mathbf{h}^l_{i,k}$ need to be estimated at BS l to perform good detection and this is done in the uplink by letting each user transmit a sequence of τ_p pilot symbols; see Fig. 4. We let $\tau_p = fK$ for some positive integer f (e.g., $1, 2, \dots$) which is called the *pilot reuse factor*. This allows for linear independence between a total of τ_p different pilot sequences. This is, by design, sufficient to allocate independent pilot sequence to the K users in each cell and to also divide the L cells into f disjoint cell groups having fully independent pilot sequences. The benefit of having multiple cell groups is reduced interference during the pilot transmission and the corresponding gains in estimation quality are quantified below.

The uplink received signal $\mathbf{Y}_l^{\text{pilot}} \in \mathbb{C}^{M \times \tau_p}$ at the lth BS during pilot transmission is

$$\mathbf{Y}_l^{\text{pilot}} = \sum_{i=1}^{L} \mathbf{H}_i^l \mathbf{P}_i^{1/2} \boldsymbol{\Phi}_i^H + \mathbf{N}_l \tag{8}$$

and collects the received signal from (7) over the τ_p pilot symbols. Here, $\boldsymbol{\Phi}_i = [\boldsymbol{\phi}_{i,1} \dots \boldsymbol{\phi}_{i,K}] \in \mathbb{C}^{\tau_p \times K}$ denotes the pilot matrix used by the K users in the ith cell, where $\boldsymbol{\phi}_{i,k} \in \mathbb{C}^{\tau_p}$ is the pilot sequence used by the kth user in that cell. The pilot matrix satisfies $\boldsymbol{\Phi}_i^H \boldsymbol{\Phi}_i = \tau_p \mathbf{I}_K$. Moreover, $\boldsymbol{\Phi}_l^H \boldsymbol{\Phi}_i = \tau_p \mathbf{I}_K$ if cell l and cell j belong to the same cell group (i.e., use the same set of pilots), while $\boldsymbol{\Phi}_l^H \boldsymbol{\Phi}_i = \mathbf{0}$ if the two cells belong to different cell groups. For notational convenience, we let $\mathscr{P}_l \subset \{1, \dots, L\}$ denote the set of cell indices that belong to the same cell group as cell l, including l itself. Some particular examples are given later in Fig. 7.

By using the channel mean and variances, defined in the beginning of Sect. 3, we can use the linear minimum mean square error (LMMSE) estimator to separately acquire each element of $\mathbf{h}_{i,k}^l$ from the received pilot signal (8), which was proposed in [37] as a low-complexity estimation scheme. The channel estimate $\hat{\mathbf{h}}_{i,k}^l$ related to the true channel response $\mathbf{h}_{i,k}^l$ is given by the following lemma.

Lemma 2. *Suppose that BS l estimates each channel coefficient separately from its received signal* (8) *using an LMMSE estimator. BS l can then estimate the channel to the kth user in the jth cell as*

$$\hat{\mathbf{h}}_{j,k}^l = \bar{\mathbf{h}}_{j,k}^l + \frac{\sqrt{p_{j,k}}\beta_{j,k}^l}{\sum_{i \in \mathscr{P}_j} p_{i,k}\tau_p \beta_{i,k}^l + \sigma_{\text{UL}}^2} \left(\mathbf{Y}_l^{\text{pilot}} \boldsymbol{\phi}_{j,k} - \sum_{i \in \mathscr{P}_j} \sqrt{p_{i,k}} \tau_p \bar{\mathbf{h}}_{i,k}^l \right). \tag{9}$$

Each element of the uncorrelated estimation error $\mathbf{e}_{j,k}^l = \mathbf{h}_{j,k}^l - \hat{\mathbf{h}}_{j,k}^l$ has zero mean and the variance

$$\text{MSE}_{j,k}^l = \beta_{j,k}^l \left(1 - \frac{p_{j,k}\tau_p \beta_{j,k}^l}{\sum_{i \in \mathscr{P}_j} p_{i,k}\tau_p \beta_{i,k}^l + \sigma_{\text{UL}}^2} \right). \tag{10}$$

Proof. The proof is available in Appendix at the end of this chapter. ■

It is worth emphasizing that the estimation error variance in (10) is independent of M, thus the estimation quality per channel coefficient is not affected by adding more antennas at the BS. Note that Lemma 2 holds for any correlation between the channel coefficients, since each coefficient is estimated separately. If the channel coefficients are correlated, with a known correlation structure and distribution, the estimation quality would improve with the number of antennas if the estimator is modified appropriately [4]. We also stress that the estimation error is only affected by noise and interference from the users in the same cell group that are allocated the same pilot sequence. In addition, we notice that the estimate in (9) can be computed using elementary linear algebra operations, with low computational complexity.

Using the channel estimates derived in Lemma 2, in this chapter we analyze the performance of a Massive MIMO network with non-cooperative BSs. During uplink payload data transmission this means that the BS in cell l only utilizes its own received signal $\mathbf{y}_l$ in (6) and only targets to detect the signals sent by its own K users. Signals coming from users in other cells are perceived as inter-cell interference and eventually treated as additional noise. The BS in cell l discriminates the signal transmitted by its kth user from the interference by multiplying the received signal in (6) with a linear detection vector $\mathbf{v}_{l,k} \in \mathbb{C}^M$ as follows:

$$\begin{aligned} \mathbf{v}_{l,k}^H \mathbf{y}_l &= \sum_{i=1}^{L} \sum_{t=1}^{K} \mathbf{v}_{l,k}^H \mathbf{h}_{i,t}^l \sqrt{p_{i,t}} x_{i,t} + \mathbf{v}_{l,k}^H \mathbf{n}_l \\ &= \underbrace{\mathbf{v}_{l,k}^H \mathbf{h}_{l,k}^l \sqrt{p_{l,k}} x_{l,k}}_{\text{Desired signal}} + \underbrace{\sum_{\substack{t=1 \\ t\neq k}}^{K} \mathbf{v}_{l,k}^H \mathbf{h}_{l,t}^l \sqrt{p_{l,t}} x_{l,t}}_{\text{Intra-cell interference}} + \underbrace{\sum_{\substack{i=1 \\ i\neq l}}^{L} \sum_{t=1}^{K} \mathbf{v}_{l,k}^H \mathbf{h}_{i,t}^l \sqrt{p_{i,t}} x_{i,t}}_{\text{Inter-cell interference}} + \underbrace{\mathbf{v}_{l,k}^H \mathbf{n}_l}_{\text{Residual noise}} \end{aligned} \tag{11}$$

where $x_{i,t}$ is the transmitted data symbol from user t in cell i. As seen from (11), the processed received signal is the superposition of four parts: the desired signal, intra-cell interference, inter-cell interference, and residual noise. Since the linear detection vector $\mathbf{v}_{l,k}$ appears in all these terms, it can be used to amplify the desired signal, suppress the interference, and/or suppress the noise. More precisely, by gathering the detection vectors at BS l in matrix form as $\mathbf{V}_l = [\mathbf{v}_{l,1} \dots \mathbf{v}_{l,K}] \in \mathbb{C}^{M\times K}$, there are two main schemes being considered in the Massive MIMO literature: maximum ratio (MR) and zero-forcing (ZF). These are given by

$$\mathbf{V}_l = \begin{cases} \hat{\mathbf{H}}_l^l, & \text{for MR,} \\ \hat{\mathbf{H}}_l^l \left((\hat{\mathbf{H}}_l^l)^H \hat{\mathbf{H}}_l^l \right)^{-1}, & \text{for ZF.} \end{cases} \tag{12}$$

MR detection exploits the M observations in $\mathbf{y}_l$ to maximize the ratio between the average signal gain in (11) and the norm of the detection vector:

$$\mathbb{E}\left\{ \frac{\mathbf{v}_{l,k}^H \mathbf{h}_{l,k}^l}{\|\mathbf{v}_{l,k}\|} \right\} = \frac{\mathbf{v}_{l,k}^H \hat{\mathbf{h}}_{l,k}^l}{\|\mathbf{v}_{l,k}\|} \leq \|\hat{\mathbf{h}}_{l,k}^l\| \tag{13}$$

where the expectation is computed with respect to the zero-mean channel estimation error. The inequality in (13) is satisfied with equality by $\mathbf{v}_{l,k} = \hat{\mathbf{h}}_{l,k}^l$ (leading to MR detection with $\mathbf{V}_l = \hat{\mathbf{H}}_l^l$). In contrast, the ZF detection matrix utilizes the M observations over the antennas to minimize the average intra-cell interference, while retaining the desired signals:

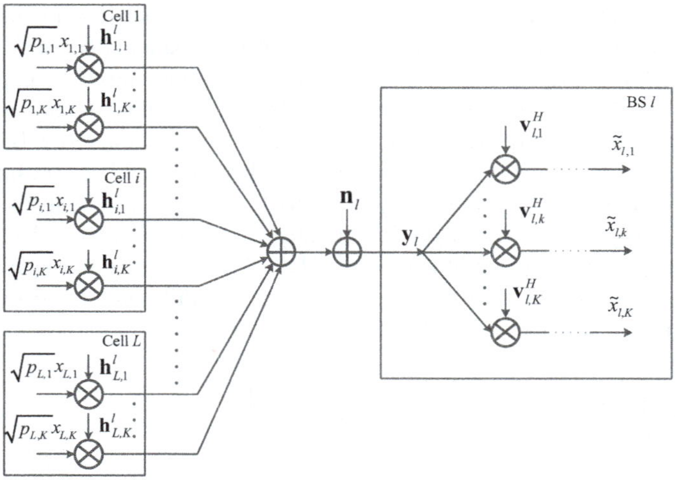

Fig. 5 Block diagram of the uplink transmission with linear detection in a multi-cell multi-user MIMO network, where BS l receives a linear combination of the signals transmitted from all K users in all L cells

$$\mathbb{E}\{\mathbf{V}_l^H \mathbf{H}_l^l \mathbf{P}_l^{1/2} \mathbf{x}_l\} = \mathbf{V}_l^H \hat{\mathbf{H}}_l^l \mathbf{P}_l^{1/2} \mathbf{x}_l = \left((\hat{\mathbf{H}}_l^l)^H \hat{\mathbf{H}}_l^l \right)^{-1} \left((\hat{\mathbf{H}}_l^l)^H \hat{\mathbf{H}}_l^l \right) \mathbf{P}_l^{1/2} \mathbf{x}_l = \mathbf{P}_l^{1/2} \mathbf{x}_l \quad (14)$$

where the expectation is computed with respect to the zero-mean channel estimation error and the second equality follows from the ZF detection matrix definition. The average processed signal becomes $\mathbf{P}_l^{1/2}\mathbf{x}_l = [\sqrt{p_{l,1}} x_{l,1} \ \dots \ \sqrt{p_{l,K}} x_{l,K}]^T$, which contains no intra-cell interference. Note that the inverse of the $K \times K$ matrix $(\hat{\mathbf{H}}_l^l)^H \hat{\mathbf{H}}_l^l$ only exists if $M \geq K$. There are also multi-cell variants of ZF detection that can be used to cancel out inter-cell interference; see for example [2] and [9].

A block diagram of the uplink transmission with linear detection is provided in Fig. 5. The purpose of the detection is to make the detected signal $\tilde{x}_{l,k}$ at BS l equal to the true signal $x_{l,k}$, at least up to a scaling factor. Due to noise and estimation errors, there is always a mismatch between the signals which is why the communication link has a limited capacity. If the true signal $x_{l,k}$ originates from a discrete constellation set $\mathscr{X}$ (e.g., a quadrature amplitude modulation (QAM)), $\tilde{x}_{l,k}$ is selected based on $\mathbf{v}_{l,k}^H \mathbf{y}_l$ by finding the minimum distance over all the candidates $x \in \mathscr{X}$:

$$\tilde{x}_{l,k} = \min_{x \in \mathscr{X}} \left| \mathbf{v}_{l,k}^H \mathbf{y}_l - \mathbf{v}_{l,k}^H \hat{\mathbf{h}}_{l,k}^l \sqrt{p_{l,k}}\, x \right|^2. \quad (15)$$

This expression can be utilized to compute bit error rates and similar uncoded performance metrics. Since modern communication systems apply channel coding over relatively long data blocks, to protect against errors, the ergodic channel capacity is a more appropriate performance metric in 5G networks. It merits to note that the ergodic capacities of the individual communication links are hard to characterize exactly, particularly under imperfect channel knowledge, but tractable lower bounds are obtained by the following theorem.

Theorem 1. *In the uplink, a lower bound on the ergodic capacity of an arbitrary user k in cell l is*

$$R_{l,k}^{\mathrm{UL}} = \gamma^{\mathrm{UL}} \left(1 - \frac{\tau_p}{\tau_c}\right) \log_2\left(1 + \mathrm{SINR}_{l,k}^{\mathrm{UL}}\right), \tag{16}$$

where the signal-to-interference-and-noise ratio (SINR) is

$$\mathrm{SINR}_{l,k}^{\mathrm{UL}} = \frac{p_{l,k} \left|\mathbb{E}\left\{\mathbf{v}_{l,k}^H \mathbf{h}_{l,k}^l\right\}\right|^2}{\sum_{i=1}^{L}\sum_{t=1}^{K} p_{i,t}\mathbb{E}\left\{|\mathbf{v}_{l,k}^H \mathbf{h}_{i,t}^l|^2\right\} - p_{l,k}\left|\mathbb{E}\left\{\mathbf{v}_{l,k}^H \mathbf{h}_{l,k}^l\right\}\right|^2 + \sigma_{\mathrm{UL}}^2 \mathbb{E}\left\{\|\mathbf{v}_{l,k}\|^2\right\}}. \tag{17}$$

Proof. The proof is available in Appendix at the end of this chapter. ■

Theorem 1 demonstrates that the achievable SE of an arbitrary user k in cell l in a Massive MIMO network can be described by an SINR term $\mathrm{SINR}_{l,k}^{\mathrm{UL}}$ that contains expectations with respect to the small-scale channel fading. The numerator contains the gain of the desired signal, while the denominator contains three different terms. The first term is the average power of all the signals, including both multi-user interference and the desired signal, while the second term subtracts the part of the desired signal power that is usable for decoding. The third term is the effective noise power. The pre-log factor $(1 - \frac{\tau_p}{\tau_c})$ compensates for the fact that τ_p/τ_c of the transmission symbols contain pilots instead of payload data. The SE is also multiplied by γ^{UL}, which was defined earlier as the fraction of uplink data. Clearly, MR detection aims at maximizing the numerator of $\mathrm{SINR}_{l,k}^{\mathrm{UL}}$, while ZF detection tries to minimize the intra-cell interference.

The expectations in Theorem 1 can be computed numerically for any channel distribution and any detection scheme. In the case of MR detection, the desired signal gain $|\mathbb{E}\{\mathbf{v}_{l,k}^H \mathbf{h}_{l,k}^l\}|^2$ grows as M^2 for most channel distributions, while the noise term $\sigma_{\mathrm{UL}}^2 \mathbb{E}\{\|\mathbf{v}_{l,k}\|^2\}$ only grows as M and thus becomes less significant the more antennas are deployed at the BS. This property is known as the *array gain* from coherent detection. The behavior of the multi-user interference terms greatly depends on the channel distribution, but typically these terms will also have the slower scaling of M [32], except for users that interfered with each other during pilot transmission (i.e., appeared in each other's expressions (10) for the estimation error variance). The latter is a phenomenon called *pilot contamination* and is further discussed later in this chapter.

To demonstrate these properties in detail, we now consider the special case in which the channel between BS l and user k in cell i is uncorrelated Rayleigh fading:

$$\mathbf{h}_{i,k}^l \sim \mathbb{C}\mathcal{N}\left(\mathbf{0}, \beta_{i,k}^l \mathbf{I}_M\right). \tag{18}$$

Hence, $\bar{\mathbf{h}}_{i,k}^{l} = \mathbb{E}\{\mathbf{h}_{i,k}^{l}\} = \mathbf{0}$, which means that there is no line-of-sight channel component. This special case is relevant in rich-scattering environments where the channel does not have any statistically dominating directivity.

Subsequently, the LMMSE estimate in Lemma 2 simplifies to

$$\hat{\mathbf{h}}_{j,k}^{l} = \frac{\sqrt{p_{j,k}}\beta_{j,k}^{l}}{\sum\limits_{i\in\mathscr{P}_j} p_{i,k}\tau_p\beta_{i,k}^{l} + \sigma_{\mathrm{UL}}^2}\mathbf{Y}_l^{\mathrm{pilot}}\boldsymbol{\phi}_{j,k} \tag{19}$$

and becomes circularly-symmetric complex Gaussian distributed:

$$\hat{\mathbf{h}}_{j,k}^{l} \sim \mathbb{C}\mathscr{N}\left(\mathbf{0}, (\beta_{j,k}^{l} - \mathrm{MSE}_{j,k}^{l})\mathbf{I}_M\right). \tag{20}$$

There is an important relationship between the two estimated channels $\hat{\mathbf{h}}_{l,k}^{l}$ and $\hat{\mathbf{h}}_{i,k}^{l}$ for cell indices i and l such as $i \in \mathscr{P}_l$, expressed by

$$\hat{\mathbf{h}}_{i,k}^{l} = \frac{\sqrt{p_{i,k}}\beta_{i,k}^{l}}{\sqrt{p_{l,k}}\beta_{l,k}^{l}}\hat{\mathbf{h}}_{l,k}^{l}. \tag{21}$$

This equation shows that BS l cannot tell apart the channels of users that send the same pilot sequence; the estimates are the same up to a scaling factor. This fact is the cause of pilot contamination and will have a key impact on the performance, as shown later.

Moreover, the LMMSE estimator in (19) is also the MMSE estimator in the special case of Rayleigh fading, since the channels are Gaussian distributed [24]. By using these key properties, the ergodic SE in Theorem 1 can be computed in closed form for MR and ZF detection, as shown by the following corollary.

Corollary 1. *In the uplink, if all channels are uncorrelated Rayleigh fading, the lower bound on the ergodic capacity of user k in cell l stated in Theorem 1 becomes*

$$R_{l,k}^{\mathrm{UL}} = \gamma^{\mathrm{UL}}\left(1 - \frac{\tau_p}{\tau_c}\right)\log_2\left(1 + \mathrm{SINR}_{l,k}^{\mathrm{UL}}\right), \tag{22}$$

where the SINR is

$$\mathrm{SINR}_{l,k}^{\mathrm{UL}} = \frac{Gp_{l,k}\beta_{l,k}^{l}\frac{p_{l,k}\tau_p\beta_{l,k}^{l}}{\sum_{i'\in\mathscr{P}_l}p_{i',k}\tau_p\beta_{i',k}^{l}+\sigma_{\mathrm{UL}}^2}}{G\sum\limits_{i\in\mathscr{P}_l\setminus\{l\}}p_{i,k}\beta_{i,k}^{l}\frac{p_{i,k}\tau_p\beta_{i,k}^{l}}{\sum_{i'\in\mathscr{P}_l}p_{i',k}\tau_p\beta_{i',k}^{l}+\sigma_{\mathrm{UL}}^2} + \sum\limits_{i=1}^{L}\sum\limits_{t=1}^{K}p_{i,t}z_{i,t}^{l} + \sigma_{\mathrm{UL}}^2} \tag{23}$$

and the parameters G and $z_{i,t}^{l}$ depend on the choice of detection scheme. MR gives $G = M$ and $z_{i,t}^{l} = \beta_{i,t}^{l}$, while ZF gives $G = M - K$ and

$$z_{i,t}^{l} = \begin{cases} \mathrm{MSE}_{i,t}^{l}, & \text{for } i \in \mathscr{P}_l, \\ \beta_{i,t}^{l}, & \text{otherwise}. \end{cases}$$

Proof. The proof is available in Appendix at the end of this chapter. ■

The closed-form achievable SE expressions in Corollary 1 provide many insights on the advantages of spatial multi-user multiplexing and the effects of channel estimation. Firstly, the desired signal term in the numerator of (23) scales with the number of BS antennas, proportionally to M and $M - K$ with MR and ZF, respectively. This array gain is multiplied with the average received signal power per antenna, $p_{l,k}\beta^l_{l,k}$, and the relative channel estimation quality

$$\frac{p_{l,k}\tau_p\beta^l_{l,k}}{\sum_{i'\in\mathscr{P}_l} p_{i',k}\tau_p\beta^l_{i',k} + \sigma^2_{\text{UL}}}, \tag{24}$$

which is a number between 0 and 1 (where 1 is perfect CSI and 0 is no CSI).

Secondly, we notice that the first term of the denominator in (23) has a similar structure as the desired signal and represents the coherent pilot contamination—interference that is amplified along with the desired signals due the BS's inability to tell apart users that use the same pilot sequence. The pilot contamination degrades the SINR by adding additional interference that scales as M or $M - K$, depending on the detection scheme. However, since pilot contamination only arises at BS l from the interfering user in cell i in $\mathscr{P}_l$, the network can suppress pilot contamination by increasing the pilot reuse factor f and by designing the cell groups appropriately. To understand how to suppress pilot contamination, we have a look at the ratio between the pilot contamination term and the signal term in (23):

$$\frac{G \sum\limits_{i\in\mathscr{P}_l\setminus\{l\}} p_{i,k}\beta^l_{i,k} \frac{p_{i,k}\tau_p\beta^l_{i,k}}{\sum_{i'\in\mathscr{P}_l} p_{i',k}\tau_p\beta^l_{i',k}+\sigma^2_{\text{UL}}}}{G p_{l,k}\beta^l_{l,k} \frac{p_{l,k}\tau_p\beta^l_{l,k}}{\sum_{i'\in\mathscr{P}_l} p_{i',k}\tau_p\beta^l_{i',k}+\sigma^2_{\text{UL}}}} = \sum_{i\in\mathscr{P}_l\setminus\{l\}} \left(\frac{p_{i,k}\beta^l_{i,k}}{p_{l,k}\beta^l_{l,k}}\right)^2. \tag{25}$$

This ratio represents the relative strength of the pilot contamination and (25) should preferably be small. The pilot contamination caused by UE k in cell i is small whenever $\beta^l_{i,k}/\beta^l_{l,k}$ is small, which occurs when either $\beta^l_{l,k}$ is large (i.e., the desired user is close to its serving BS) or $\beta^l_{i,k}$ is small (i.e., the interfering cell is far away). The cell groups should be designed based on these properties, and this issue is further discussed in Sect. 4.2.

Thirdly, the performance in Corollary 1 is also affected by classical noise and interference. Since MR focuses only on maximizing the SNR, the interference term $\sum_{i=1}^{L}\sum_{t=1}^{K} p_{i,t}\beta^l_{i,t}$ is simply the average signal power received at any antenna of BS l. In contrast, ZF pays attention to the intra-cell interference and takes no notice of the noise. The interference suppression replaces the full channel variance $\beta^l_{i,t}$ in the aforementioned interference summation with the estimation error variance $\text{MSE}^l_{i,t}$ for cells $i \in \mathscr{P}_l$. Due to the imperfect CSI (i.e., $\text{MSE}^l_{i,t} > 0$) not all intra-cell interference can be removed by ZF. However, the pilot contamination also has the positive effect that not only intra-cell interference is suppressed, but also the inter-cell interference coming from other users in the same cell group (which use the same pilots as in cell l). The fact that the interference and noise terms are independent of M, while the desired signal scales with M, is a consequence of the asymptotic favorable propagation that was described in Sect. 2.2.1.

If we limit the scope to a single-cell network, achievable SE expressions can be obtained directly from Corollary 1 by simply setting $\mathscr{P}_l = \{l\}$ and removing the interference from all other cells $j \in \{1, \ldots, L\} \setminus \{l\}$. For simplicity of exposition, we leave out the cell index l in this special case.

Corollary 2. *In the single-cell uplink, if all channels are uncorrelated Rayleigh fading, a lower bound on the ergodic SE of an arbitrary user k is given by*

$$R_k^{\mathrm{UL}} = \gamma^{\mathrm{UL}} \left(1 - \frac{\tau_p}{\tau_c}\right) \log_2 \left(1 + \frac{G p_k^2 \tau_p \beta_k^2}{\left(p_k \tau_p \beta_k + \sigma_{\mathrm{UL}}^2\right) \left(\sum\limits_{t=1}^{K} p_t z_t + \sigma_{\mathrm{UL}}^2 \right)} \right). \tag{26}$$

Here, the parameters G and z_t depend on the detection scheme. MR gives $G = M$ and $z_t = \beta_t$, while ZF gives $G = M - K$ and $z_t = \frac{\beta_t \sigma_{\mathrm{UL}}^2}{p_t \tau_p \beta_t + \sigma_{\mathrm{UL}}^2}$.

This corollary shows that the spatial multi-user multiplexing capability is even greater in isolated single-cell networks. The most notable difference compared to a multi-cell network is the lack of inter-cell interference, both during data and pilot transmission. In other words, the interference only originates from users within the own cell, while pilot contamination vanishes thanks to the orthogonality of all pilot sequences in the cell. The SE per cell is therefore higher in single-cell networks than in the multi-cell networks—at least if the cell geometry is the same and we only neglect inter-cell interference. The motivation of having multiple cells is, of course, to cover a larger area and thereby achieve much higher total SE. The scenarios when the interference suppression of ZF is beneficial as compared to MR can be identified from Corollary 2 as the cases when

$$\frac{M - K}{\sum\limits_{t=1}^{K} \frac{p_t \beta_t \sigma_{\mathrm{UL}}^2}{p_t \tau_p \beta_t + \sigma_{\mathrm{UL}}^2} + \sigma_{\mathrm{UL}}^2} > \frac{M}{\sum\limits_{t=1}^{K} p_t \beta_t + \sigma_{\mathrm{UL}}^2}. \tag{27}$$

To summarize, we have derived uplink SE expressions for Massive MIMO networks, for general channel distributions in Theorem 1 and for Rayleigh fading in Corollary 1. In the latter case, the expressions are in closed form and can thus be computed and analyzed directly, without having to simulate any channel fading realizations. These expressions are used in Sect. 4 to illustrate the anticipated performance of Massive MIMO networks.

3.2 Downlink with Linear Precoding

Next, we consider the downlink of a Massive MIMO network where the BSs are transmitting signals to their users. For an arbitrary BS l, we let $\mathbf{x}_l \in \mathbb{C}^M$ denote the transmitted signal vector intended for its K users. We consider linear precoding where this vector is computed as

$$\mathbf{x}_l = \sum_{t=1}^{K} \sqrt{\rho_{l,t}} \mathbf{w}_{l,t} s_{l,t}, \tag{28}$$

where the payload symbol $s_{l,t}$ intended for user t in cell l has unit transmit power $\mathbb{E}\{|s_{l,t}|^2\} = 1$ and $\rho_{l,t}$ represents the transmit power allocated to this particular user. Moreover, $\mathbf{w}_{l,t} \in \mathbb{C}^M$, for $t = 1, \ldots, K$, are the corresponding linear precoding vectors that determine the spatial directivity of the signal sent to each user.

The received signal $y_{l,k} \in \mathbb{C}$ at user k in cell l is modeled as

$$y_{l,k} = \sum_{i=1}^{L} (\mathbf{h}_{l,k}^{i})^H \mathbf{x}_i + n_{l,k} \tag{29}$$

where $n_{l,k} \sim \mathbb{C}\mathcal{N}(0, \sigma_{\mathrm{DL}}^2)$ is the additive white noise with variance σ_{DL}^2. Notice that $\mathbf{h}_{l,k}^{i}$ is the same channel response as in the uplink, due to the reciprocity of physical propagation channels (within a coherence interval).[1] A block diagram of the downlink transmission is provided in Fig. 6. Since there are no downlink pilots in the Massive MIMO protocol described in Sect. 2.3, the users are assumed to only know the channel statistics. The lack of instantaneous CSI would greatly reduce the performance of small MIMO systems, but Massive MIMO works well without it since the effective precoded channels quickly approach their mean as more antennas are added. Hence, coherent downlink reception is possible using only statistical CSI. This leads to a low-complexity communication solution where all the intelligence is placed at the BS. Since the ergodic capacity is hard to characterize in this case, the following theorem derives a lower bound on the capacity between user k in cell l and its serving BS.

Theorem 2. *In the downlink, a lower bound on the ergodic rate an arbitrary user k in cell l is*

$$R_{l,k}^{\mathrm{DL}} = \gamma^{\mathrm{DL}} \left(1 - \frac{\tau_p}{\tau_c}\right) \log_2 \left(1 + \mathrm{SINR}_{l,k}^{\mathrm{DL}}\right), \tag{30}$$

[1] In fact, the reciprocal channel is $(\mathbf{h}_{l,k}^{i})^T$, using the regular transpose instead of the conjugate transpose as in (29), but since the only difference is a complex conjugation we can characterize the performance using (29) without loss of generality. The reason to use the conjugate transpose is that the notation becomes easier and the relation to the uplink is clearer.

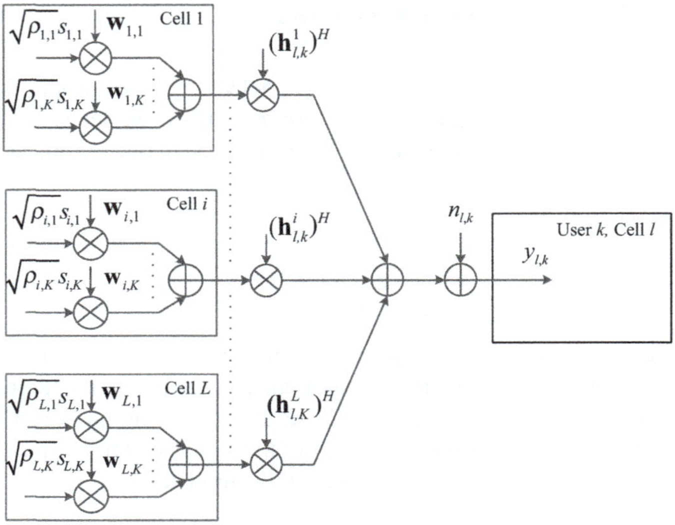

Fig. 6 Block diagram of the downlink transmission with linear precoding in a multi-cell MIMO system, where BSs equipped with M antennas are transmitting signals that reach user k in cell l

where the SINR is

$$\mathrm{SINR}_{l,k}^{\mathrm{DL}} = \frac{\rho_{l,k} \left| \mathbb{E}\left\{ (\mathbf{h}_{l,k}^{l})^{H} \mathbf{w}_{l,k} \right\} \right|^2}{\sum_{i=1}^{L} \sum_{t=1}^{K} \rho_{i,t} \mathbb{E}\left\{ |(\mathbf{h}_{l,k}^{i})^{H} \mathbf{w}_{i,t}|^2 \right\} - \rho_{l,k} \left| \mathbb{E}\left\{ (\mathbf{h}_{l,k}^{l})^{H} \mathbf{w}_{l,k}^{l} \right\} \right|^2 + \sigma_{\mathrm{DL}}^2}. \quad (31)$$

Proof. The proof is available in Appendix at the end of this chapter. ■

The downlink achievable SE provided in Theorem 2 holds for any channel distributions and choice of precoding vectors. Since the uplink and downlink channels are reciprocal, it would make sense if the uplink and downlink performance were also somehow connected. The downlink achievable SE in Theorem 2 indeed bears much similarity with the corresponding uplink expression in Theorem 1. The desired signal terms are the same, except for the potentially different transmit power parameters and the fact that the detection vector is replaced by the corresponding precoding vector. The interference terms have a similar structure, but the indices are swapped between the channel vector and the processing vector. This is because the uplink interference arrives through different channels for different users while all the downlink interference from a particular cell comes through the same channel from the BS. These observations lead to the following *uplink–downlink duality* [9, 10]:

Lemma 3. *Suppose that the downlink precoding vectors are selected as*

$$\mathbf{w}_{l,k} = \frac{\mathbf{v}_{l,k}}{\sqrt{\mathbb{E}\{\|\mathbf{v}_{l,k}\|^2\}}} \tag{32}$$

based on the uplink detection vectors $\mathbf{v}_{l,k}$, for all l and k. For any given uplink powers $p_{i,t}$ (for $i = 1, \ldots, L$ and $t = 1, \ldots, K$), there exist a corresponding set of downlink powers $\rho_{i,t}$ (for $i = 1, \ldots, L$ and $t = 1, \ldots, K$) such that

$$\mathrm{SINR}_{l,k}^{\mathrm{UL}} = \mathrm{SINR}_{l,k}^{\mathrm{DL}} \tag{33}$$

for all l and k, and

$$\frac{\sum_{i=1}^{L}\sum_{t=1}^{K} p_{i,t}}{\sigma_{\mathrm{UL}}^2} = \frac{\sum_{i=1}^{L}\sum_{t=1}^{K} \rho_{i,t}}{\sigma_{\mathrm{DL}}^2}. \tag{34}$$

Proof. The proof is available in Appendix at the end of this chapter. ■

This lemma shows that the same performance can be achieved in both the uplink and the downlink, if the downlink power is allocated in a particular way based on the uplink powers and the precoding vectors are selected based on the detection vectors as in (32). The downlink powers are computed according to (72), which is given in Appendix at the end of this chapter since the important thing for now is that there exist a collection of downlink powers that give exactly the same performance in both directions. If $\sigma_{\mathrm{UL}}^2 = \sigma_{\mathrm{DL}}^2$, then the same total transmit power is used in both directions of the Massive MIMO network; however the power will generally be distributed differently over the users.

Motivated by the uplink–downlink duality, it makes sense to consider MR and ZF precoding as the main downlink precoding schemes. These are defined as

$$\mathbf{w}_{l,k} = \begin{cases} \dfrac{\hat{\mathbf{h}}_{l,k}^{l}}{\sqrt{\mathbb{E}\left\{\|\hat{\mathbf{h}}_{l,k}^{l}\|^2\right\}}}, & \text{for MR,} \\ \dfrac{\hat{\mathbf{H}}_{l}^{l}\mathbf{r}_{l,k}}{\sqrt{\mathbb{E}\left\{\|\hat{\mathbf{H}}_{l}^{l}\mathbf{r}_{l,k}\|^2\right\}}}, & \text{for ZF,} \end{cases} \tag{35}$$

where $\mathbf{r}_{l,k}$ denotes the kth column of $((\hat{\mathbf{H}}_{l}^{l})^{H}\hat{\mathbf{H}}_{l}^{l})^{-1}$.

Similar to the uplink performance analysis, we now compute the downlink SE in closed form for uncorrelated Rayleigh fading channels, as defined in (18). Because of the channel reciprocity, the channel estimates obtained at the BSs in the uplink can also be used in the downlink. In particular, the channel estimates $\hat{\mathbf{h}}_{i,k}^{i}$ and $\hat{\mathbf{h}}_{i,k}^{l}$ for cell indices i and l with $l \in \mathscr{P}_i$ are still related as

$$\hat{\mathbf{h}}^{i}_{l,k} = \frac{\sqrt{p_{l,k}}\beta^{i}_{l,k}}{\sqrt{p_{i,k}}\beta^{i}_{i,k}}\hat{\mathbf{h}}^{i}_{i,k}, \tag{36}$$

thus showing that pilot contamination exists also in the downlink; that is, BS i cannot precode signals toward its user k without also precode the signal towards user k in cell $i \in \mathscr{P}_l$. The next corollary specializes Theorem 2 for Rayleigh fading channels.

Corollary 3. *In the downlink, if all channels are uncorrelated Rayleigh fading, the lower bound on the ergodic capacity of user k stated in Theorem 2 becomes*

$$R^{\mathrm{DL}}_{l,k} = \gamma^{\mathrm{DL}}\left(1 - \frac{\tau_p}{\tau_c}\right)\log_2\left(1 + \mathrm{SINR}^{\mathrm{DL}}_{l,k}\right), \tag{37}$$

where the SINR is

$$\mathrm{SINR}^{\mathrm{DL}}_{l,k} = \frac{G\rho_{l,k}\beta^{l}_{l,k}\frac{p_{l,k}\tau_p\beta^{l}_{l,k}}{\sum_{i'\in\mathscr{P}_l} p_{i',k}\tau_p\beta^{l}_{i',k}+\sigma^2_{\mathrm{UL}}}}{G\sum\limits_{i\in\mathscr{P}_l\setminus\{l\}}\rho_{i,k}\beta^{i}_{l,k}\frac{p_{l,k}\tau_p\beta^{i}_{l,k}}{\sum_{i'\in\mathscr{P}_l} p_{i',k}\tau_p\beta^{i}_{i',k}+\sigma^2_{\mathrm{UL}}} + \sum\limits_{i=1}^{L}\sum\limits_{t=1}^{K}\rho_{i,t}z^{i}_{l,k} + \sigma^2_{\mathrm{DL}}}. \tag{38}$$

The parameters G and $z^i_{l,k}$ are specified by the precoding scheme. MR precoding gives $G = M$ and $z^i_{l,k} = \beta^i_{l,k}$, while ZF precoding gives $G = M - K$ and

$$z^{i}_{l,k} = \begin{cases} \mathrm{MSE}^{i}_{l,k}, & \text{for } i \in \mathscr{P}_l, \\ \beta^{i}_{l,k}, & \text{otherwise.} \end{cases}$$

Proof. The proof is available in Appendix at the end of this chapter. ■

For Rayleigh fading channels, Corollary 3 shows that the array gain, pilot contamination, and all other attributes of MR and ZF precoding are very similar to the uplink counterparts. Hence, the same kind of observations can be made from Corollary 3 as previously done for Corollary 1.

In the single-cell scenario, the SE expression in Corollary 3 simplifies to the following result.

Corollary 4. *In the single-cell downlink, if all channels are uncorrelated Rayleigh fading, a lower bound on the ergodic SE of an arbitrary user k is given by*

$$R^{\mathrm{DL}}_{k} = \gamma^{\mathrm{DL}}\left(1 - \frac{\tau_p}{\tau_c}\right)\log_2\left(1 + \frac{G\rho_k p_k\tau_p\beta_k^2}{\left(p_k\tau_p\beta_k + \sigma^2_{\mathrm{UL}}\right)\left(z_k\sum_{t=1}^{K}\rho_t + \sigma^2_{\mathrm{DL}}\right)}\right). \tag{39}$$

The parameters G and z_k depend on the precoding scheme. MR gives $G = M$ and $z_k = \beta_k$, while ZF obtains $G = M - K$ and $z_k = \frac{\beta_k\sigma^2_{\mathrm{UL}}}{p_k\tau_p\beta_k+\sigma^2_{\mathrm{UL}}}$.

We conclude the analytical part of this chapter by recalling that the uplink and downlink spectral efficiencies with Massive MIMO can be easily computed from Theorems 1 and 2 for any channel distributions and processing schemes. In the uncorrelated Rayleigh fading case there are even closed-form expressions. The same SINR performance can be achieved in the uplink and downlink, based on what is known as uplink–downlink duality. The intuition is that the downlink precoding and uplink detection vectors should be the same, but that the power allocation needs to be adapted differently in the two cases.

4 Design Guidelines and Anticipated Spectral Efficiency Gains

In this section, we provide some basic design guidelines for Massive MIMO networks and showcase the SEs that the technology can deliver to 5G networks according to the theory developed in Sect. 3. For illustrative purposes, we consider a classic cellular network topology with hexagonal cells, where each cell can be illustrated as in Fig. 1. In other words, the BS is deployed in the center of the cell, while the K users are distributed over the cell area. When many cells of this type are placed next to each other, the cellular network has the shape showed in Fig. 7. While conventional cellular networks use sectorization to split each cell into, say, three static sectors, this is not assumed here. This is because the spatial transceiver processing at the BS in Massive MIMO basically creates K virtual sectors, adapted dynamically to the positions of the current set of users.

4.1 Power Allocation

The average transmit power of user k in cell j is denoted by $p_{j,k}$ in the uplink and by $\rho_{j,k}$ in the downlink. These are important design parameters that determine the SEs of the users; see Theorem 1 (for the uplink) and Theorem 2 (for the downlink). Since inter-user interference is an important factor in any multi-user MIMO system, each transmit power coefficient affects not only the strength of the desired signal at the desired user, but also the amount of interference caused to all the other users in the network (although the interference is most severe within a cell and between neighboring cells). The selection of these transmit power coefficients is referred to as *power allocation* and needs to be addressed properly.

A key property of Massive MIMO is that the small-scale fading in time and the frequency-selective fading variations are negligible, since they essentially average out over the many antennas at each BS. For example, the SE expressions for Rayleigh fading channels in Corollaries 1–4 only depend on the channel variances $\beta_{i,k}^{l}$ and not on the instantaneous realizations of the corresponding channel vectors $\mathbf{h}^{l}_{i,k}$. Therefore, there is no need to change the power allocation between each coherence interval, but only over the longer time frame where the channel variances change, due to modifications in the large-scale propagation behaviors (e.g., caused by user mobility). This is a substantial increase of the time frame in which power allocation decisions are to be made, from milliseconds to seconds. This fact makes it possible to optimize and coordinate the power allocation across cells, in ways that have not been possible in the past due computational or delay limitations.

A structured approach to power allocation is to find the transmit powers that jointly maximize the network utility functions $U^{\text{UL}}(\{R^{\text{UL}}_{l,k}\})$ and $U^{\text{DL}}(\{R^{\text{DL}}_{l,k}\})$ in the uplink and downlink, respectively. These utilities are increasing functions of the users' SEs, where $\{R^{\text{UL}}_{l,k}\}$ and $\{R^{\text{DL}}_{l,k}\}$ denote the sets of all SEs. Some particular examples of network utility functions are [8]

$$U(\{R_{l,k}\}) = \begin{cases} \sum_{l=1}^{L} \sum_{k=1}^{K} R_{l,k}, & \text{Sum utility,} \\ \prod_{l=1}^{L} \prod_{k=1}^{K} R_{l,k}, & \text{Proportional fairness,} \\ \min_{l\in\{1,\dots,L\},\, k\in\{1,\dots,K\}} R_{l,k}, & \text{Max-min fairness,} \end{cases} \tag{40}$$

where we have omitted the uplink/downlink superscripts since the same type of utility function can be utilized in both cases. These utilities are often maximized with respect to a given power budget per user (in the uplink) and per BS (in the downlink). For brevity, we will not provide any further mathematical details, but briefly outline what is known around power allocation for Massive MIMO.

Maximization of the sum utility (SU) provides high SEs to users with good average channel conditions, at the expense of low SE for users with bad average channel conditions. In contrast, max-min fairness (MMF) enforces that each user should get equal SE, which effectively means that users with good channels reduce their SEs to cause less interference to the users with bad channels. Proportional fairness (PF) can be shown to lie in between these extremes. The SU achieves the highest sum SE, since this is really what is optimized by this utility function, while MMF trades some of the sum SE to obtain a uniform user experience. The choice of network utility function is a matter of subjective taste, since there is no objectively optimal utility function [8]. Nevertheless, there seems to be a trend towards more fairness-emphasizing utilities in the Massive MIMO literature [9, 33, 45], motivated by the fact that contemporary networks are designed to provide high peak rates, while the cell edge performance is modest and needs to be improved in 5G. In the uplink, another important aspect to consider in the power allocation is the fact that a BS cannot simultaneously receive desired user signals of very different power levels, since then the weak signals will then drown in the quantization noise caused by the analog-to-digital conversion. Hence, even if the channel attenuation might differ by 50 dB within a cell, these variations need to be brought down to, say, 10 dB by the uplink power allocation.

From a numerical optimization perspective, the downlink power allocation problem (for fixed uplink power allocation) has the same mathematical structure as the seemingly different scenario of single-antenna multi-cell communications with perfect CSI. The downlink utility optimization can therefore be solved using the techniques described in [3, 27, 35] and references therein. In general, the PF and MMF utilities give rise to convex optimization problems that can be solved efficiently with guaranteed convergence to the global optimum. These algorithms can also be implemented in a distributed fashion [3]. The SU problem is, in contrast, provable non-convex and hard to solve [27], which means that the optimal solution cannot be found under any practical constraints on complexity.

The uplink power allocation is more complicated than power allocation in the downlink; for example, because the SE expression in Corollary 1 contains both $p_{j,k}$ and $p_{j,k}^2$ (while the downlink SE expressions only contain the linear term $\rho_{j,k}$). Nevertheless, there are several efficient algorithms that maximize the MMF utility [12, 13, 45], and the approach in [13] can also maximize the SU and PF utilities with MR and ZF detection. The work [26] provides an alternative methodology to maximize an approximation of the SU metric for other detection methods.

In summary, power allocation is used in Massive MIMO to distribute the sum SE over the individual users. There are plenty of algorithms that can be used to optimize the power allocation, depending on the utility function that is used in the system.

4.2 Non-Universal Pilot Reuse

An important insight from the theoretical analysis in Sect. 3 is that the SE of a particular cell j is influenced by the pilot signaling carried out in other cells. The degradations in CSI estimation quality and pilot contaminated interference are caused only by the interfering cells in $\mathscr{P}_j$ that use the same pilot sequences as cell j. Since the channel attenuation of the interference increases with distance, one would like these interfering cells to be as far away from cell j as possible—and the same is desirable for all cells in $\mathscr{P}_j$.

Recall that the pilot reuse factor $f = \tau_p/K$ was assumed to be an integer in Sect. 3, which leads to a division of the L cells into f disjoint cell groups. The case $f = 1$ is known as universal pilot reuse and $f > 1$ is called non-universal pilot reuse. Since the hexagonal cell topology has a multiple of six cells in each tier of interfering cells, the smallest pilot reuse factors that give rise to symmetric pilot reuse patterns are $f = 1, f = 3$, and $f = 4$ [15]. Examples of such reuse patterns are given in Fig. 7, where cells with different colors use different subsets of the pilot sequences. The cells with the same color use exactly the same subset of pilots and therefore cause pilot contamination to each other, while there is no contamination between cells with different colors. If the center cell in Fig. 7 has index j, then $\mathscr{P}_j$ is the set of all cells having the same color. By increasing the pilot reuse factor, there are more colors and therefore fewer interfering cells in each group. We note that with a pilot reuse factor of $f = 4$, one can either divide the cells into four different disjoint groups (as in the lower left example in Fig. 7) or divide each cell into two subcells: cell edge and cell center (as in the lower right example in Fig. 7). The latter

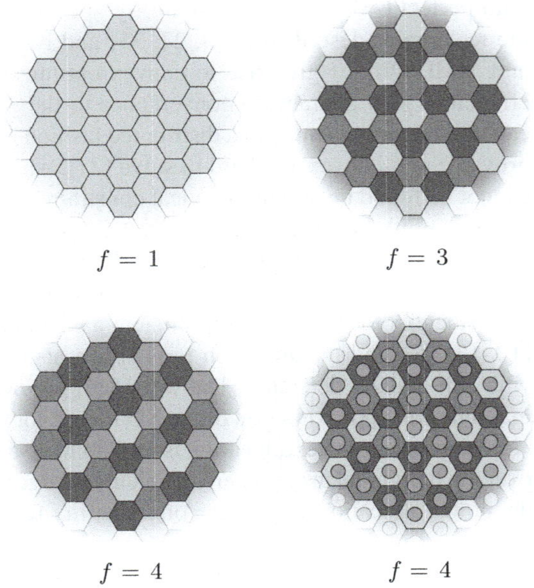

Fig. 7 Illustration of potential symmetric reuse patterns created by three different pilot reuse factors, f, in a cellular network with hexagonal cells. In the *lower right* case, each cell is divided into two sub-cells with different sets of pilots. If j is the index of a particular cell, then $\mathscr{P}_j$ is the index set of all cells having the same color. Only the cells with the same *color* use the same pilot sequences, and thereby degrade each other's CSI estimation quality and cause pilot contaminated interference

is known as *fractional pilot reuse* and can be used to have less frequent pilot reuse at the cell edges than in the cell centers [2], because it is users at the cell edges that are most sensitive to pilot contamination.

To give a concrete example, consider a Massive MIMO scenario with $M = 200$ BS antennas and a coherence interval of $\tau_c = 400$ symbols. The users are assumed to be uniformly distributed in the cell, except for the 10 % cell center, and the channels are modeled as uncorrelated Rayleigh fading with a distant-dependent channel attenuation with pathloss exponent 3.7. We consider the pilot reuse factors $f \in \{1, 3, 4\}$, but not the fractional pilot reuse case. Recall from the uplink–downlink duality in Lemma 3 that the same SE is achievable in the uplink and the downlink, thus it is sufficient to study the uplink. We assume a simple power allocation policy

$$p_{j,k} = \frac{\delta}{\beta_{j,k}^{j}} \quad j = 1, \ldots, L, \quad k = 1, \ldots, K, \tag{41}$$

where $\delta \geq 0$ is a design parameter that determines the SNR achieved at each BS antenna: $p_{j,k}\beta_{j,k}^{j}/\sigma_{\text{BS}}^2 = \delta/\sigma_{\text{BS}}^2$.[2] This is called statistical channel inversion power allocation.

[2]This explicit uplink power allocation policy is very similar to what the uplink MMF utility would give [3], but has the benefit of not requiring any numerical optimization.

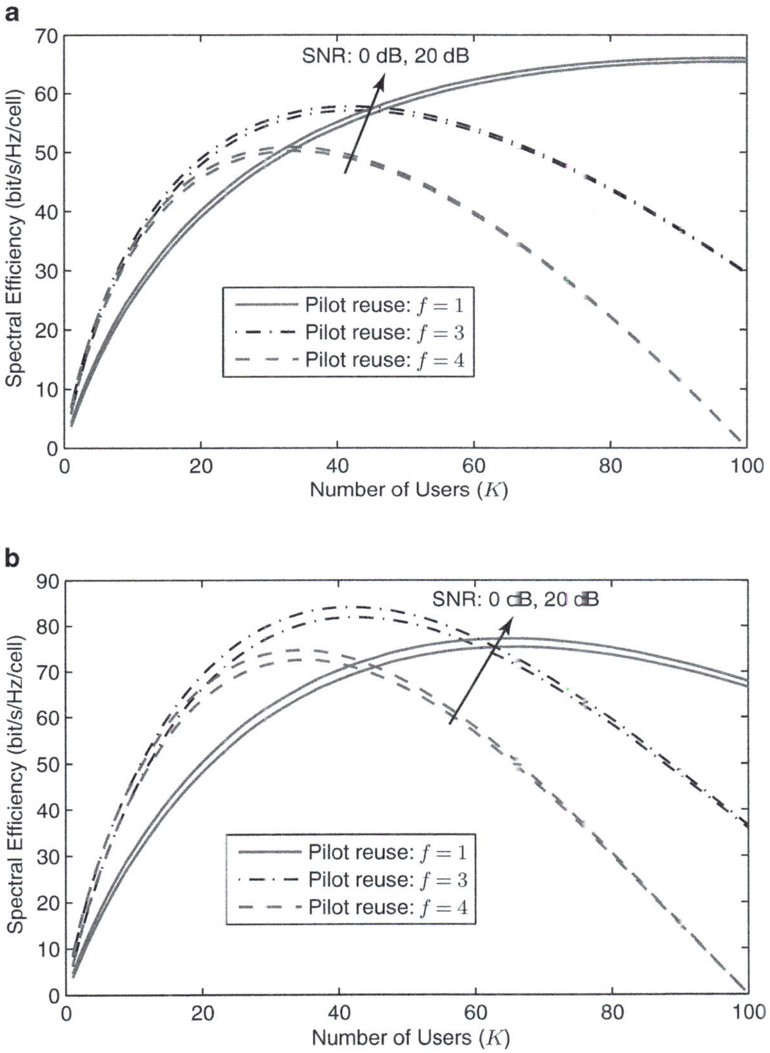

Fig. 8 Average spectral efficiency, as a function of the number of users, with different processing schemes and pilot reuse factors. Two different SNR levels are considered: $\delta/\sigma_{\text{BS}}^2 = 0\,\text{dB}$ or $20\,\text{dB}$. (**a**) Maximum ratio detection. (**b**) Zero-forcing detection

Figure 8 shows the average SE for different number of users, for both MR and ZF detection. The first observation from Fig. 8 is that the two SNR levels, $\delta/\sigma_{\text{BS}}^2 = 0\,\text{dB}$ and $20\,\text{dB}$, give essentially the same performance. This shows that Massive MIMO works equally well at high and low SNRs, since the array gain makes the SE interference-limited and not noise-limited. Next, we notice that different pilot reuse factors are desirable at different user loads (i.e., number of users K). A pilot reuse of $f = 3$ is desired at low load, while $f = 1$ is needed to reduce the prelog factor $(1 - fK/\tau_c)$ when K is large. By selecting f properly, one can always operate on the top curve in Fig. 8 and then Massive MIMO can provide a high SE over a wide range of different number of users. In fact, the technology provides a relatively stable SE for any $K > 10$. This removes the need for intricate scheduling in Massive MIMO networks, because all active users can basically be served simultaneously in every coherence interval (or at least up to $\tau_c/2$ users, to leave half of the coherence interval for data, which is a number that is typically more than a hundred [9]); the high sum SE is then shared between all the users.

Another observation is that the difference in SE between ZF and MR is relatively small in Fig. 8; ZF only gives a performance gain of between 3 and 45 %, depending on the number of users. This should be compared with the single-cell simulation in Fig. 3, where ZF provided more than twice the SEs as MR. The reason for the more modest performance gap is that also ZF suffers from interference in the multi-cell case, since the pilot contamination and many inter-cell interferers make it impossible to cancel all interference.

In summary, the pilot reuse factor is an important design parameter in Massive MIMO networks and the best choice depends on the user load. As shown in [3], it also depends on the propagation environment and the number of BS antennas.

4.3 How High Spectral Efficiency can Massive MIMO Deliver?

We conclude this chapter by showcasing the SE that the Massive MIMO technology can deliver in the uplink and downlink of 5G networks—which is the same due to the uplink–downlink duality. We continue the previous simulation example from Fig. 8, but focus on ZF processing with pilot reuse $f = 3$ and a power allocation policy that gives an SNR of 0 dB to everyone. Note that these design choices are motivated by the previous simulation results.

Figure 9 shows the SE as a function of the number of BS antennas M. The number of active users is optimized for each M to get the highest SE, and the optimal user numbers are also shown in the figure. A reasonable performance baseline is the IMT-Advanced requirements for 4G networks, provided in [22]. This document specifies spectral efficiencies in the range of 2–3 bit/s/Hz/cell, depending on the simulation scenario. In comparison, the Massive MIMO network considered in Fig. 9 achieves 52 bit/s/Hz/cell using $M = 100$ antennas, which is a 17× to 26× improvement over IMT-Advanced. With $M = 400$ antennas the Massive MIMO system achieves 114 bit/s/Hz/cell, which is an incredible 38× to 57× improvement over IMT-Advanced. These improvements are between one and two orders of magnitude!

It is important to notice that the number of active users increase alongside the SE in Fig. 9. If one divides the top curve with the bottom curve, this gives the SE per user. Interestingly, the SE per user lies in the modest range of 1–2.5 bit/s/Hz.

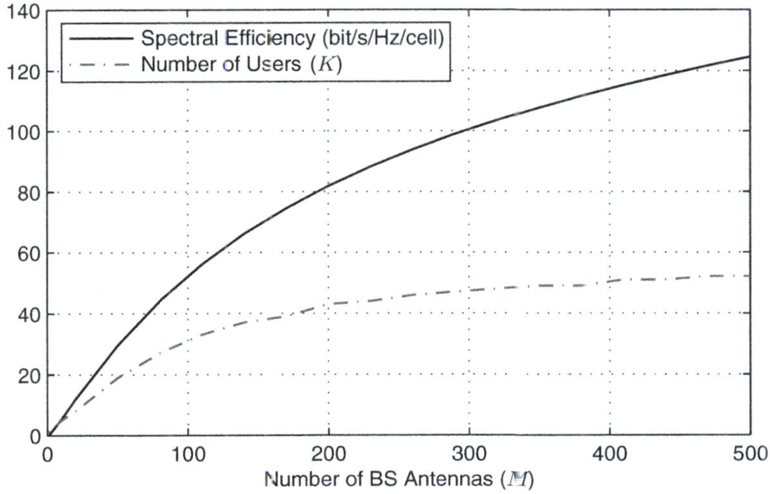

Fig. 9 Average spectral efficiency, as a function of the number of BS antennas, with ZF processing, a pilot reuse factor $f = 3$, and an SNR of 0 dB. The number of users is optimized for each antenna number to yield the highest SE, and the corresponding number of users is also shown

Such spectral efficiencies can be achieved in practice using conventional modulation schemes, such as 16-QAM with appropriate channel coding.

In conclusion, Massive MIMO can theoretically provide tenfold or even 50-fold improvements in SE over IMT-Advanced. These huge improvements are mainly achieved by serving many users simultaneously, while the SE per user is in the conventional range. Huge gains are achieved already at $M = 100$ or fewer BS antennas. These are indeed very encouraging results indicating that the Massive MIMO technology is key to not only improve the SE, but can also be the driving force towards achieving orders of magnitude higher area throughput in 5G technologies.

Acknowledgements The authors of this chapter have received funding from the European Union's Horizon 2020 research and innovation programme under grant agreement No 641985 (5Gwireless). The authors are also supported by ELLIIT and CENIIT. We would also like to thank Daniel Verenzuela and Meysam Sadeghi for helping out with the proof-reading.

Appendix

Proof of Lemma 2

Recall that $\boldsymbol{\phi}_{j,k} \in \mathbb{C}^{\tau_p}$ is the pilot sequence used by the kth user in the jth cell, where $\boldsymbol{\Phi}_j = [\boldsymbol{\phi}_{j,1} \ldots \boldsymbol{\phi}_{j,K}]$. Since the desired channel $\mathbf{h}_{j,k}^l$ only appears as $\mathbf{h}_{j,k}^l \boldsymbol{\phi}_{j,k}^H$ in (8), a sufficient statistic for estimating this channel is given by

$$\begin{aligned} \mathbf{Y}_l^{\text{pilot}} \boldsymbol{\phi}_{j,k} &= \sum_{i=1}^{L} \mathbf{H}_i^l \mathbf{P}_i \boldsymbol{\Phi}_i^H \boldsymbol{\phi}_{j,k} + \mathbf{N}_l \boldsymbol{\phi}_{j,k} \\ &= \sum_{i \in \mathscr{P}_j} \sqrt{p_{i,k}} \tau_p \mathbf{h}_{i,k}^l + \tilde{\mathbf{n}}_{l,j,k} \end{aligned} \tag{42}$$

where $\tilde{\mathbf{n}}_{l,j,k} = \mathbf{N}_l \boldsymbol{\phi}_{j,k} = [\tilde{n}_{l,j,k,1} \ldots \tilde{n}_{l,j,k,M}]^T \sim \mathbb{C}\mathscr{N}(\mathbf{0}, \tau_p \sigma_{\text{UL}}^2 \mathbf{I}_M)$. The second equality follows from the assumed orthogonality of the pilot sequences.

Based on (42), we compute a separate LMMSE estimate of each element of $\mathbf{h}_{j,k}^l$. If $y_{l,j,k,m} \in \mathbb{C}$ denotes the mth row of the vector in (42), then

$$y_{l,j,k,m} = \sum_{i \in \mathscr{P}_j} \sqrt{p_{i,k}} \tau_p h_{i,k,m}^l + \tilde{n}_{l,j,k,m}. \tag{43}$$

By the definition of LMMSE estimation [24], the LMMSE estimate of $h_{j,k,m}^l$ is given by

$$\hat{h}_{j,k,m}^l = \mathbb{E}\left\{h_{j,k,m}^l\right\} + \frac{\operatorname{Cov}\left\{h_{j,k,m}^l, y_{l,j,k,m}\right\}}{\mathbb{V}\left\{y_{l,j,k,m}\right\}} \left(y_{l,j,k,m} - \mathbb{E}\left\{y_{l,j,k,m}\right\}\right) \tag{44}$$

where we recall that $\mathbb{E}\left\{h_{j,k,m}^l\right\} = \bar{h}_{j,k,m}^l$ by definition and $\operatorname{Cov}\{\cdot,\cdot\}$ stands for covariance. Moreover, we have

$$\mathbb{E}\left\{y_{l,j,k,m}\right\} = \sum_{i \in \mathscr{P}_j} \sqrt{p_{i,k}} \tau_p \bar{h}_{i,k,m}^l, \tag{45}$$

$$\operatorname{Cov}\left\{h_{j,k,m}^l, y_{l,j,k,m}\right\} = \sqrt{p_{j,k}} \tau_p \beta_{j,k}^l, \tag{46}$$

$$\mathbb{V}\left\{y_{l,j,k,m}\right\} = \sum_{i \in \mathscr{P}_j} p_{i,k} \tau_p^2 \beta_{i,k}^l + \tau_p \sigma_{\text{UL}}^2. \tag{47}$$

The estimation expression in (9) is obtained by substituting (45)–(47) into (44) and writing the result in vector form. The variance of the estimate is then given by

$$\mathbb{V}\left\{\hat{h}_{j,k,m}^l\right\} = \frac{\left|\operatorname{Cov}\left\{h_{j,k,m}^l, y_{l,j,k,m}\right\}\right|^2}{\mathbb{V}\left\{y_{l,j,k,m}\right\}} = \frac{p_{j,k} \tau_p (\beta_{j,k}^l)^2}{\sum_{i \in \mathscr{P}_j} p_{i,k} \tau_p \beta_{i,k}^l + \sigma_{\text{UL}}^2}, \tag{48}$$

while the estimation error variance in (10) is obtained as $\beta_{j,k}^l - \mathbb{V}\left\{\hat{h}_{j,k,m}^l\right\}$ since the LMMSE estimate and its error are uncorrelated [24].

Proof of Theorem 1

The ergodic capacity $C_{l,k}^{\mathrm{UL}}$ with linear detection and pilot-based channel estimation is the supremum of the mutual information between the input signal $x_{l,k}$ and the output signal $\mathbf{v}_{l,k}^{H}\mathbf{y}_l$ in (11). The supremum is taken over the distribution of the unit-variance input signal $x_{l,k}$, thus a lower bound is obtained by assuming that $x_{l,k} \sim \mathbb{C}\mathscr{N}(0,1)$. Let $\widehat{\mathscr{H}}$ denote the channel estimates available as side-information at the receiver. We then have that

$$\begin{aligned} C_{l,k}^{\mathrm{UL}} &\geq \gamma^{\mathrm{UL}}\left(1-\frac{\tau_p}{\tau_c}\right) I(x_{l,k}; \mathbf{v}_{l,k}^{H}\mathbf{y}_l, \widehat{\mathscr{H}}) \\ &= \gamma^{\mathrm{UL}}\left(1-\frac{\tau_p}{\tau_c}\right)\left(h(x_{l,k}) - h(x_{l,k}|\mathbf{v}_{l,k}^{H}\mathbf{y}_l, \widehat{\mathscr{H}})\right) \\ &= \gamma^{\mathrm{UL}}\left(1-\frac{\tau_p}{\tau_c}\right)\left(\log_2(\pi e) - h(x_{l,k}|\mathbf{v}_{l,k}^{H}\mathbf{y}_l, \widehat{\mathscr{H}})\right) \end{aligned} \tag{49}$$

where $I(\cdot;\cdot)$ denotes the mutual information under the suboptimally assumed Gaussian signal distribution and $h(\cdot)$ is the differential entropy function. The first equality follows from the definition of mutual information and the second equality uses the entropy expression for complex Gaussian random variables. The factor $\gamma^{\mathrm{UL}}(1-\frac{\tau_p}{\tau_c})$ is the fraction of transmission symbols used for uplink data in each coherence interval. It remains to characterize $h(x_{l,k}|\mathbf{v}_{l,k}^{H}\mathbf{y}_l, \widehat{\mathscr{H}})$, which is done by finding a tractable upper bound on this term:

$$\begin{aligned} h(x_{l,k}|\mathbf{v}_{l,k}^{H}\mathbf{y}_l, \widehat{\mathscr{H}}) &= h(x_{l,k} - \alpha\mathbf{v}_{l,k}^{H}\mathbf{y}_l|\mathbf{v}_{l,k}^{H}\mathbf{y}_l, \widehat{\mathscr{H}}) \\ &\leq h(x_{l,k} - \alpha\mathbf{v}_{l,k}^{H}\mathbf{y}_l) \\ &\leq \log_2\left(\pi e \mathbb{E}\{|x_{l,k} - \alpha\mathbf{v}_{l,k}^{H}\mathbf{y}_l|^2\}\right) \end{aligned} \tag{50}$$

where the equality follows from the fact that subtracting a known variable $\alpha\mathbf{v}_{l,k}^{H}\mathbf{y}_l$, for some deterministic scalar α, does not change the entropy. The first inequality follows from dropping the knowledge of $\mathbf{v}_{l,k}^{H}\mathbf{y}_l$ and $\widehat{\mathscr{H}}$ which increases the entropy, while the second inequality follows from exploiting the fact that the highest entropy is obtained when $x_{l,k} - \alpha\mathbf{v}_{l,k}^{H}\mathbf{y}_l$ is a zero-mean complex Gaussian random variable with the same second-order moment as the original variable has.

The last step of the proof is to select α to get the tightest upper bound in (50), which corresponds to the minimization problem

$$\min_{\alpha} \mathbb{E}\left\{|x_{l,k} - \alpha\mathbf{v}_{l,k}^{H}\mathbf{y}_l|^2\right\} = \frac{1}{1+\mathrm{SINR}_{l,k}^{\mathrm{UL}}}, \tag{51}$$

which is solved by first computing the expectation with respect to the independent signals $x_{i,t}$, for all i and t, then finding the optimal α by equating the first derivative of the expression (with respect to α) to zero, and substituting it back into the expression. From (49) to (51) we now have

$$\begin{aligned} C_{l,k}^{\mathrm{UL}} &\geq \gamma^{\mathrm{UL}}\left(1-\frac{\tau_p}{\tau_c}\right)\left(\log_2(\pi e) - \log_2\left(\pi e \frac{1}{1+\mathrm{SINR}_{l,k}^{\mathrm{UL}}}\right)\right) \\ &= \gamma^{\mathrm{UL}}\left(1-\frac{\tau_p}{\tau_c}\right)\log_2\left(1+\mathrm{SINR}_{l,k}^{\mathrm{UL}}\right), \end{aligned} \tag{52}$$

which is the result stated in the theorem.

Proof of Corollary 1

Before computing the SINR expression in (23) for Rayleigh fading channels, we recall that

$$\mathbb{V}\left\{h_{i,t,m}^{l}\right\} = \beta_{i,t}^{l}, \tag{53}$$

$$\mathbb{V}\left\{\hat{h}_{i,t,m}^{l}\right\} = \frac{p_{i,t}\tau_p(\beta_{i,t}^{l})^2}{\sum_{i' \in \mathscr{P}_l} p_{i',t}\tau_p\beta_{i',t}^{l} + \sigma_{\mathrm{UL}}^2}, \tag{54}$$

$$\mathrm{MSE}_{i,t}^{l} = \beta_{i,t}^{l}\left(1 - \frac{p_{i,t}\tau_p\beta_{i,t}^{l}}{\sum_{i' \in \mathscr{P}_l} p_{i',t}\tau_p\beta_{i',t}^{l} + \sigma_{\mathrm{UL}}^2}\right), \tag{55}$$

for the channel between an arbitrary user t in cell i ($i = 1, \ldots, L$) and BS l. Note that m is used as an arbitrary antenna index since the channel variance is the same for all antennas. The corollary is first proved in the case of MR detection, where $\mathbf{v}_{l,k} = \hat{\mathbf{h}}_{l,k}^{l}$, in which case $\mathrm{SINR}_{l,k}^{\mathrm{UL}}$ in Theorem 1 becomes

$$\mathrm{SINR}_{l,k}^{\mathrm{MR,UL}} = \frac{p_{l,k}\left|\mathbb{E}\left\{(\hat{\mathbf{h}}_{l,k}^{l})^H\mathbf{h}_{l,k}^{l}\right\}\right|^2}{\sum\limits_{i=1}^{L}\sum\limits_{t=1}^{K} p_{i,t}\mathbb{E}\left\{|(\hat{\mathbf{h}}_{l,k}^{l})^H\mathbf{h}_{i,t}^{l}|^2\right\} - p_{l,k}\left|\mathbb{E}\left\{(\hat{\mathbf{h}}_{l,k}^{l})^H\mathbf{h}_{l,k}^{l}\right\}\right|^2 + \sigma_{\mathrm{UL}}^2\mathbb{E}\left\{\|\hat{\mathbf{h}}_{l,k}^{l}\|^2\right\}}. \tag{56}$$

It remains to compute the expectations in the numerator and denominator of (56). Since $\mathbf{h}_{l,k}^{l} = \hat{\mathbf{h}}_{l,k}^{l} + \mathbf{e}_{l,k}^{l}$, as stated in Lemma 2, the numerator is computed as

$$p_{l,k}\left|\mathbb{E}\left\{(\hat{\mathbf{h}}_{l,k}^{l})^H\mathbf{h}_{l,k}^{l}\right\}\right|^2 = M^2 p_{l,k}\left(\mathbb{V}\left\{\hat{h}_{l,k,m}^{l}\right\}\right)^2. \tag{57}$$

When computing the denominator, we decompose its first term into three parts based on the pilot reuse; the first two parts contain the cells that use the same pilot sequences as cell l (i.e., all cells in $\mathscr{P}_l$) and the third part contains the remaining cells. We then observe that

$$\begin{aligned}
&\sum_{i=1}^{L}\sum_{t=1}^{K} p_{i,t}\mathbb{E}\left\{|(\hat{\mathbf{h}}_{l,k}^{l})^H\mathbf{h}_{i,t}^{l}|^2\right\} - p_{l,k}\left|\mathbb{E}\left\{(\hat{\mathbf{h}}_{l,k}^{l})^H\mathbf{h}_{l,k}^{l}\right\}\right|^2 + \sigma_{\mathrm{UL}}^2\mathbb{E}\left\{\|\hat{\mathbf{h}}_{l,k}^{l}\|^2\right\} \\
&= \sum_{i \in \mathscr{P}_l} p_{i,k}\mathbb{E}\left\{|(\hat{\mathbf{h}}_{l,k}^{l})^H\mathbf{h}_{i,k}^{l}|^2\right\} + \sum_{i \in \mathscr{P}_l}\sum_{\substack{t=1 \\ t \neq k}}^{K} p_{i,t}\mathbb{E}\left\{|(\hat{\mathbf{h}}_{l,k}^{l})^H\mathbf{h}_{i,t}^{l}|^2\right\} \\
&\quad + \sum_{i \notin \mathscr{P}_l}\sum_{t=1}^{K} p_{i,t}\mathbb{E}\left\{|(\hat{\mathbf{h}}_{l,k}^{l})^H\mathbf{h}_{i,t}^{l}|^2\right\} - p_{l,k}\left|\mathbb{E}\left\{(\hat{\mathbf{h}}_{l,k}^{l})^H\mathbf{h}_{l,k}^{l}\right\}\right|^2 + \sigma_{\mathrm{UL}}^2\mathbb{E}\left\{\|\hat{\mathbf{h}}_{l,k}^{l}\|^2\right\} \qquad (58) \\
&= M^2\mathbb{V}\left\{\hat{h}_{l,k,m}^{l}\right\}\sum_{i \in \mathscr{P}_l \setminus \{l\}} p_{i,k}\mathbb{V}\left\{\hat{h}_{i,k,m}^{l}\right\} + M\mathbb{V}\left\{\hat{h}_{l,k,m}^{l}\right\}\sum_{i \in \mathscr{P}_l}\sum_{t=1}^{K} p_{i,t}\mathbb{V}\left\{h_{i,t,m}^{l}\right\} \\
&\quad + M\mathbb{V}\left\{\hat{h}_{l,k,m}^{l}\right\}\sum_{i \notin \mathscr{P}_l}\sum_{t=1}^{K} p_{i,t}\mathbb{V}\left\{h_{i,t,m}^{l}\right\} + M\mathbb{V}\left\{\hat{h}_{l,k,m}^{l}\right\}\sigma_{\mathrm{UL}}^2.
\end{aligned}$$

The first term in the second expression of (58) demonstrates the effect of pilot contamination and is computed by using (21) and the independence between the MMSE estimate and its estimation error.

Besides, we handle the expectation $\mathbb{E}\{\|\hat{\mathbf{h}}^l_{l,k}\|^4\}$ by virtue of Lemma 2.9 in [40], since $\hat{\mathbf{h}}^l_{l,k}(\hat{\mathbf{h}}^l_{l,k})^H$ is an $M \times M$ central complex Wishart matrix with one degree of freedom:

$$\mathbb{E}\left\{\|\hat{\mathbf{h}}^l_{l,k}\|^4\right\} = \mathbb{E}\left\{\text{tr}^2\left(\hat{\mathbf{h}}^l_{l,k}(\hat{\mathbf{h}}^l_{l,k})^H\right)\right\} = M(M+1)\left(\mathbb{V}\left\{\hat{h}^l_{l,k,m}\right\}\right)^2, \tag{59}$$

where tr(·) stands for the trace of a matrix. In contrast, the second term of the middle expression of (58) is computed by the fact that the remaining users in $\mathscr{P}_l$ use pilot sequences that are orthogonal to the pilot sequence of user k. The third term in (58) is computed based on the independence between the channel estimates in cell l and the channels in other cells not belong to $\mathscr{P}_l$, while the last term follows from the fact that $\mathbb{E}\left\{\|\hat{\mathbf{h}}^l_{l,k}\|^2\right\} = M\mathbb{V}\left\{\hat{h}^l_{l,k,m}\right\}$.

Substituting (57) and (58) into (56), the SINR expression with MR detection becomes

$$\text{SINR}^{\text{MR,UL}}_{l,k} = \frac{Mp_{l,k}\mathbb{V}\left\{\hat{h}^l_{l,k,m}\right\}}{M\sum\limits_{i\in\mathscr{P}_l\setminus\{l\}} p_{i,k}\mathbb{V}\left\{\hat{h}^l_{i,k,m}\right\} + \sum\limits_{i\in\mathscr{P}_l}\sum\limits_{t=1}^{K} p_{i,t}\mathbb{V}\left\{h^l_{i,t,m}\right\} + \sigma^2_{\text{UL}}} \tag{60}$$

which equals the expression in the corollary by further substituting (53)–(55) into (60).

In case of ZF detection, the channel inversion structure yields the property

$$\mathbb{E}\left\{\mathbf{v}^H_{l,k}\mathbf{h}^l_{l,k}\right\} = 1. \tag{61}$$

Additionally, the noise term in (17) is computed as

$$\sigma^2_{\text{UL}}\mathbb{E}\left\{\|\mathbf{v}_{l,k}\|^2\right\} = \sigma^2_{\text{UL}}\mathbb{E}\left\{\text{tr}\left[\left((\hat{\mathbf{H}}^l_l)^H\hat{\mathbf{H}}^l_l\right)^{-1}\right]_{k,k}\right\} = \frac{\sigma^2_{\text{UL}}}{(M-K)\mathbb{V}\left\{\hat{h}^l_{l,k,m}\right\}} \tag{62}$$

by utilizing the fact that $(\hat{\mathbf{H}}^l_l)^H\hat{\mathbf{H}}^l_l$ is a $K \times K$ central complex Wishart matrix with M degrees of freedom and applying Lemma 2.10 in [40] to compute the trace of the inverse. Note that $[\cdot]_{k,k}$ is used here to denote the kth diagonal element of a matrix.

Substituting (61) and (62) into (17), we achieve the SINR expression

$$\text{SINR}^{\text{ZF,UL}}_{l,k} = \frac{p_{l,k}}{\sum\limits_{i=1}^{L}\sum\limits_{t=1}^{K} p_{i,t}\mathbb{E}\left\{|\mathbf{v}^H_{l,k}\mathbf{h}^l_{i,t}|^2\right\} - p_{l,k} + \frac{\sigma^2_{\text{UL}}}{(M-K)\mathbb{V}\{\hat{h}^l_{l,k,m}\}}}. \tag{63}$$

To compute the remaining expectations, we utilize the pilot reuse patterns together with the ZF properties to decompose the expectation term in (63) into three terms:

$$\begin{aligned}
&\sum_{i=1}^{L}\sum_{t=1}^{K} p_{i,t}\mathbb{E}\left\{|\mathbf{v}^H_{l,k}\mathbf{h}^l_{i,t}|^2\right\} \\
&= \sum_{i\in\mathscr{P}_l} p_{i,k}\mathbb{E}\left\{|\mathbf{v}^H_{l,k}\hat{\mathbf{h}}^l_{i,k}|^2\right\} + \sum_{i\in\mathscr{P}_l}\sum_{t=1}^{K} p_{i,t}\mathbb{E}\left\{|\mathbf{v}^H_{l,k}\mathbf{e}^l_{i,t}|^2\right\} + \sum_{i\notin\mathscr{P}_l}\sum_{t=1}^{K} p_{i,t}\mathbb{E}\left\{|\mathbf{v}^H_{l,k}\mathbf{h}^l_{i,t}|^2\right\} \\
&= \sum_{i\in\mathscr{P}_l} \frac{p^2_{i,k}(\beta^l_{i,k})^2}{p_{l,k}(\beta^l_{l,k})^2} + \sum_{i\in\mathscr{P}_l}\sum_{t=1}^{K} \frac{p_{i,t}\text{MSE}^l_{i,t}}{(M-K)\mathbb{V}\left\{\hat{h}^l_{l,k,m}\right\}} + \sum_{i\notin\mathscr{P}_l}\sum_{t=1}^{K} \frac{p_{i,t}\mathbb{V}\left\{h^l_{i,t,m}\right\}}{(M-K)\mathbb{V}\left\{\hat{h}^l_{l,k,m}\right\}}.
\end{aligned} \tag{64}$$

In the last equality of (64), the first term is obtained by utilizing the relationship between user channels for cells in $\mathscr{P}_l$ as stated in (21). The second and third terms follow directly from the independence between the ZF detection vector, the estimation errors for channels in $\mathscr{P}_l$ and the complete channels for cells not in $\mathscr{P}_l$. Moreover, Lemma 2.10 in [40] is again used to compute the expectation of the inverse of the central complex Wishart matrix $(\hat{\mathbf{H}}_l^l)^H \hat{\mathbf{H}}_l^l$. Substituting (64) back into (63) and utilizing the properties in (53)–(55), the final SINR expression for ZF is obtained.

Proof of Theorem 2

Substituting (28) into (29), the received signal at user k in cell l is

$$\begin{aligned} y_{l,k} &= \sum_{i=1}^{L} (\mathbf{h}_{l,k}^i)^H \sum_{t=1}^{K} \sqrt{\rho_{i,t}} \mathbf{w}_{i,t} s_{i,t} + n_{l,k} \\ &= \underbrace{\sqrt{\rho_{l,k}} (\mathbf{h}_{l,k}^l)^H \mathbf{w}_{l,k} s_{l,k}}_{\text{Desired signal}} + \underbrace{\sum_{\substack{t=1 \\ t \neq k}}^{K} \sqrt{\rho_{l,t}} (\mathbf{h}_{l,k}^l)^H \mathbf{w}_{l,t} s_{l,t}}_{\text{Intra-cell interference}} + \underbrace{\sum_{\substack{i=1 \\ i \neq l}}^{L} \sum_{t=1}^{K} \sqrt{\rho_{i,t}} (\mathbf{h}_{l,k}^i)^H \mathbf{w}_{i,t} s_{i,t}}_{\text{Inter-cell interference}} + \underbrace{n_{l,k}}_{\text{Noise}} . \end{aligned} \tag{65}$$

The last row of (65) shows that $s_{l,k}$ is the desired signal that we want to detect, under additive noise, intra-cell, and inter-cell interference. Similar to Theorem 1, if $C_{l,k}^{\text{DL}}$ is the ergodic capacity with linear precoding, then we compute a lower bound on the mutual information between $s_{l,k}$ and $y_{l,k}$ by considering a potentially suboptimal Gaussian signal distribution, $s_{l,k} \sim \mathscr{CN}(0,1)$, and bounding the corresponding conditional mutual information $I(s_{l,k}; y_{l,k})$ as follows:

$$\begin{aligned} C_{l,k}^{\text{DL}} &\geq \gamma^{\text{DL}} \left(1 - \frac{\tau_p}{\tau_c}\right) I(s_{l,k}; y_{l,k}) \\ &\geq \gamma^{\text{DL}} \left(1 - \frac{\tau_p}{\tau_c}\right) \left(\log_2(\pi e) - h(s_{l,k}|y_{l,k})\right) \\ &\geq \log_2\left(1 + \text{SINR}_{l,k}^{\text{DL}}\right) \end{aligned} \tag{66}$$

where the inequalities follow from the same procedures as in (50)–(52). The lower bound on the ergodic capacity $R_{l,k}^{\text{DL}}$ in (30) is then obtained. Note that in contrast to the proof of Theorem 1, the receiver does not have any side-information with channel estimates in the downlink.

Proof of Lemma 3

Let $\xi_{l,1}, \ldots, \xi_{l,K}$ be the uplink SINRs of the K users in cell l that are achieved by Theorem 1 for the given detection vectors and uplink power coefficients, such that the equations $\text{SINR}_{l,k}^{\text{UL}} = \xi_{l,k}$ hold for $l = 1, \ldots, L$ and $k = 1, \ldots, K$. From this condition we get

$$\xi_{l,k} \frac{\mathbb{E}\left\{\|\mathbf{v}_{l,k}\|^2\right\}}{\left|\mathbb{E}\left\{\mathbf{v}_{l,k}^H \mathbf{h}_{l,k}^l\right\}\right|^2} = \frac{p_{l,k}}{\sum_{i=1}^{L} \sum_{t=1}^{K} p_{i,t} \frac{\mathbb{E}\{|\mathbf{v}_{l,k}^H \mathbf{h}_{i,t}^l|^2\}}{\mathbb{E}\{\|\mathbf{v}_{l,k}\|^2\}} - p_{l,k} \frac{|\mathbb{E}\{\mathbf{v}_{l,k}^H \mathbf{h}_{l,k}^l\}|^2}{\mathbb{E}\{\|\mathbf{v}_{l,k}\|^2\}} + \sigma_{\text{UL}}^2}, \tag{67}$$

by multiplying each side of the equation $\text{SINR}_{l,k}^{\text{UL}} = \xi_{l,k}$ with $\mathbb{E}\left\{\|\mathbf{v}_{l,k}\|^2\right\} / |\mathbb{E}\{\mathbf{v}_{l,k}^H \mathbf{h}_{l,k}^l\}|^2$.

The goal is to prove that also $\text{SINR}_{l,k}^{\text{DL}} = \xi_{l,k}$ holds if the downlink precoding vectors in (32) are used and the downlink transmit power coefficients are selected appropriately. According to the definition of the downlink precoding vectors, the equation $\text{SINR}_{l,k}^{\text{DL}} = \xi_{l,k}$ can be written as

$$\xi_{l,k}\frac{\mathbb{E}\left\{\|\mathbf{v}_{l,k}\|^2\right\}}{\left|\mathbb{E}\left\{(\mathbf{h}_{l,k}^{l})^H\mathbf{v}_{l,k}\right\}\right|^2} = \frac{\rho_{l,k}}{\sum\limits_{i=1}^{L}\sum\limits_{t=1}^{K}\rho_{i,t}\frac{\mathbb{E}\{|(\mathbf{h}_{l,k}^{i})^H\mathbf{v}_{i,t}|^2\}}{\mathbb{E}\{\|\mathbf{v}_{i,t}\|^2\}} - \rho_{l,k}\frac{|\mathbb{E}\{(\mathbf{h}_{l,k}^{l})^H\mathbf{v}_{l,k}\}|^2}{\mathbb{E}\{\|\mathbf{v}_{l,k}\|^2\}} + \sigma_{\text{DL}}^2}, \tag{68}$$

by multiplying each side of the equation with $\mathbb{E}\left\{\|\mathbf{v}_{l,k}\|^2\right\}/|\mathbb{E}\{(\mathbf{h}_{l,k}^{l})^H\mathbf{v}_{l,k}\}|^2$.

Let us define a diagonal matrix $\mathbf{D}$ and a matrix $\boldsymbol{\Psi}$, both of size $KL \times KL$. Let $\mathbf{D}_l \in \mathbb{C}^{K\times K}$ be the lth diagonal block of $\mathbf{D}$ and let $\boldsymbol{\Psi}_{l,i} \in \mathbb{C}^{K\times K}$ be the (l,i)th block of $\boldsymbol{\Psi}$. The elements of these blocks are defined as

$$[\mathbf{D}_l]_{k,k} = \frac{\xi_{l,k}\mathbb{E}\left\{\|\mathbf{v}_{l,k}\|^2\right\}}{\left|\mathbb{E}\left\{(\mathbf{h}_{l,k}^{l})^H\mathbf{v}_{l,k}\right\}\right|^2}, \tag{69}$$

$$[\boldsymbol{\Psi}_{l,i}]_{k,t} = \begin{cases} \frac{\mathbb{E}\{|(\mathbf{h}_{l,k}^{i})^H\mathbf{v}_{i,t}|^2\}}{\mathbb{E}\{\|\mathbf{v}_{i,t}\|^2\}} - \frac{|\mathbb{E}\{(\mathbf{h}_{l,k}^{l})^H\mathbf{v}_{l,k}\}|^2}{\mathbb{E}\{\|\mathbf{v}_{l,k}\|^2\}}, & \text{for } l=i \text{ and } t=k, \\ \frac{\mathbb{E}\{|(\mathbf{h}_{l,k}^{i})^H\mathbf{v}_{i,t}|^2\}}{\mathbb{E}\{\|\mathbf{v}_{i,t}\|^2\}}, & \text{otherwise.} \end{cases} \tag{70}$$

Using this notation, the KL equations in (67) and (68) respectively become

$$\frac{\mathbf{p}}{\sigma_{\text{UL}}^2} = \left(\mathbf{I}_{KL} - \mathbf{D}\boldsymbol{\Psi}^T\right)^{-1}\mathbf{D}\mathbf{1}_{KL} \tag{71}$$

$$\frac{\boldsymbol{\rho}}{\sigma_{\text{DL}}^2} = \left(\mathbf{I}_{KL} - \mathbf{D}\boldsymbol{\Psi}\right)^{-1}\mathbf{D}\mathbf{1}_{KL}, \tag{72}$$

where $\mathbf{p} = [\mathbf{p}_1^T \dots \mathbf{p}_L^T]^T$ and $\mathbf{p}_i = [p_{i,1} \dots p_{i,K}]^T$ contain the uplink transmit powers, $\boldsymbol{\rho} = [\boldsymbol{\rho}_1^T \dots \boldsymbol{\rho}_L^T]^T$ and $\boldsymbol{\rho}_i = [\rho_{i,1} \dots \rho_{i,K}]^T$ contain the downlink transmit powers, $\mathbf{1}_{KL}$ is a $KL \times 1$ vector with only ones, and $\mathbf{I}_{KL}$ is the $KL \times KL$ identity matrix. These equations give the uplink and downlink transmit powers that provide the SINRs $\xi_{l,1},\dots,\xi_{l,K}$ in cell l, but only if the inverses $(\mathbf{I}_{KL} - \mathbf{D}\boldsymbol{\Psi}^T)^{-1}$ and $(\mathbf{I}_{KL} - \mathbf{D}\boldsymbol{\Psi})^{-1}$ exist.

Since $\mathbf{I}_{KL} - \mathbf{D}\boldsymbol{\Psi}^T$ and $\mathbf{I}_{KL} - \mathbf{D}\boldsymbol{\Psi}$ have the same eigenvalues, either both or none of the inverses exist. Recall that we have selected $\xi_{l,k}$ (for $l = 1,\dots,L$, $k = 1,\dots,K$) as the SINRs that were actually achieved in the uplink, thus the inverses must exist and (72) gives the downlink transmit powers that achieves the same SINRs in the downlink as in the uplink. It is also straightforward to show that

$$\frac{\mathbf{1}_{KL}^T\mathbf{p}}{\sigma_{\text{UL}}^2} = \frac{\boldsymbol{\rho}^T\mathbf{1}_{KL}}{\sigma_{\text{DL}}^2}, \tag{73}$$

which corresponds to the relationship between the total transmit power in the uplink and downlink stated in the lemma.

Proof of Corollary 3

The proof follows along the same lines as the proof of Corollary 1, because the same expectations are involved, thus the variances summarized in (53)–(55) are still useful. We briefly summarize the proof of Corollary 3 as follows.

We need to compute all the expectations in (31). MR precoding gives the desired signal power

$$\rho_{l,k}\left|\mathbb{E}\left\{(\mathbf{h}_{l,k}^{l})^{H}\mathbf{w}_{l,k}\right\}\right|^{2}=\frac{\rho_{l,k}}{M\mathbb{V}\left\{\hat{h}_{l,k,m}^{l}\right\}}\left|\mathbb{E}\left\{\|\hat{\mathbf{h}}_{l,k}^{l}\|^{2}\right\}\right|^{2}=M\rho_{l,k}\mathbb{V}\left\{\hat{h}_{l,k,m}^{l}\right\} \tag{74}$$

and the denominator is computed as

$$\begin{aligned}
&\sum_{i=1}^{L}\sum_{t=1}^{K}\rho_{i,t}\mathbb{E}\left\{|(\mathbf{h}_{l,k}^{i})^{H}\mathbf{w}_{i,t}|^{2}\right\}-\rho_{l,k}\left|\mathbb{E}\left\{(\mathbf{h}_{l,k}^{l})^{H}\mathbf{w}_{l,k}^{l}\right\}\right|^{2}+\sigma_{\mathrm{DL}}^{2}\\
&=\sum_{i\in\mathscr{P}_{l}}\rho_{i,k}\mathbb{E}\left\{|(\mathbf{h}_{l,k}^{i})^{H}\mathbf{w}_{i,k}|^{2}\right\}+\sum_{i\in\mathscr{P}_{l}}\sum_{\substack{t=1\\t\neq k}}^{K}\rho_{i,t}\mathbb{E}\left\{|(\mathbf{h}_{l,k}^{i})^{H}\mathbf{w}_{i,t}|^{2}\right\}\\
&\quad+\sum_{i\notin\mathscr{P}_{l}}\sum_{t=1}^{K}\rho_{i,t}\mathbb{E}\left\{|(\mathbf{h}_{l,k}^{i})^{H}\mathbf{w}_{i,t}|^{2}\right\}-\rho_{l,k}\left|\mathbb{E}\left\{(\mathbf{h}_{l,k}^{l})^{H}\mathbf{w}_{l,k}^{l}\right\}\right|^{2}+\sigma_{\mathrm{DL}}^{2}\\
&=M\sum_{i\in\mathscr{P}_{l}\setminus\{l\}}\rho_{i,k}\mathbb{V}\left\{\hat{h}_{i,k,m}^{i}\right\}+\sum_{i=1}^{L}\sum_{t=1}^{K}\rho_{i,t}\mathbb{V}\left\{h_{l,k,m}^{i}\right\}+\sigma_{\mathrm{DL}}^{2}.
\end{aligned} \tag{75}$$

Substituting (74) and (75) into (31), yields the SINR expression stated for MR in the corollary.

Next, we consider ZF precoding for which we notice that

$$\mathbb{E}\left\{\left\|\hat{\mathbf{H}}_{i}^{i}\mathbf{r}_{i,t}\right\|^{2}\right\}=\mathbb{E}\left\{\left[\left((\hat{\mathbf{H}}_{i}^{i})^{H}\hat{\mathbf{H}}_{i}^{i}\right)^{-1}\right]_{t,t}\right\}=\frac{1}{(M-K)\mathbb{V}\left\{\hat{h}_{i,t,m}^{i}\right\}} \tag{76}$$

by utilizing the fact that $(\hat{\mathbf{H}}_{i}^{i})^{H}\hat{\mathbf{H}}_{i}^{i}$ is a central complex Wishart matrix with M degrees of freedom. Hence, the ZF precoding vector becomes

$$\mathbf{w}_{i,t}=\sqrt{(M-K)\mathbb{V}\left\{\hat{h}_{i,t,m}^{i}\right\}}\hat{\mathbf{H}}_{i}^{i}\mathbf{r}_{i,t}. \tag{77}$$

Using this precoding vector, we compute the numerator and denominator of (31) as follows:

$$\rho_{l,k}\left|\mathbb{E}\left\{(\mathbf{h}_{l,k}^{l})^{H}\mathbf{w}_{l,k}^{l}\right\}\right|^{2}=(M-K)\rho_{l,k}\mathbb{V}\left\{\hat{h}_{l,k,m}^{l}\right\}, \tag{78}$$

$$\begin{aligned}
&\sum_{i=1}^{L}\sum_{t=1}^{K}\rho_{i,t}\mathbb{E}\left\{|(\mathbf{h}_{l,k}^{i})^{H}\mathbf{w}_{i,t}|^{2}\right\}-\rho_{l,k}\left|\mathbb{E}\left\{(\mathbf{h}_{l,k}^{l})^{H}\mathbf{w}_{l,k}\right\}\right|^{2}+\sigma_{\mathrm{DL}}^{2}\\
&=(M-K)\sum_{i\in\mathscr{P}_{l}\setminus\{l\}}\rho_{i,k}\mathbb{V}\left\{\hat{h}_{l,k,m}^{i}\right\}+\sum_{i\in\mathscr{P}_{l}}\sum_{t=1}^{K}\rho_{i,t}\mathrm{MSE}_{l,k}^{i}+\sum_{i\notin\mathscr{P}_{l}}\sum_{t=1}^{K}\rho_{i,t}\mathbb{V}\{h_{i,t,m}\}+\sigma_{\mathrm{DL}}^{2}.
\end{aligned} \tag{79}$$

Substituting (78) and (79) into (31), yields the SINR expression stated for ZF in the corollary.

References

1. S. Anderson, M. Millnert, M. Viberg, B. Wahlberg, An adaptive array for mobile communication systems. IEEE Trans. Veh. Technol. **40**(1), 230–236 (1991)
2. I. Atzeni, J. Arnau, M. Debbah, Fractional pilot reuse in massive MIMO systems, in *Proceedings of the IEEE ICC*, 2015
3. E. Björnson, E. Jorswieck, Optimal resource allocation in coordinated multi-cell systems. Found. Trends Commun. Inf. Theor. **9**(2–3), 113–381 (2013)
4. E. Björnson, B. Ottersten, A framework for training-based estimation in arbitrarily correlated Rician MIMO channels with Rician disturbance. IEEE Trans. Signal Process. **58**(3), 1807–1820 (2010)
5. E. Björnson, G. Zheng, M. Bengtsson, B. Ottersten, Robust monotonic optimization framework for multicell MISO systems. IEEE Trans. Signal Process. **60**(5), 2508–2523 (2012)
6. E. Björnson, M. Kountouris, M. Bengtsson, B. Ottersten, Receive combining vs. multi-stream multiplexing in downlink systems with multi-antenna users. IEEE Trans. Signal Process. **61**(13), 3431–3446 (2013)
7. E. Björnson, J. Hoydis, M. Kountouris, M. Debbah, Massive MIMO systems with non-ideal hardware: Energy efficiency, estimation, and capacity limits. IEEE Trans. Inf. Theory **60**(11), 7112–7139 (2014)
8. E. Björnson, E. Jorswieck, M. Debbah, B. Ottersten, Multi-objective signal processing optimization: the way to balance conflicting metrics in 5G systems. IEEE Signal Process. Mag. **31**(6), 14–23 (2014)
9. E. Björnson, E. Larsson, M. Debbah, Massive MIMO for maximal spectral efficiency: how many users and pilots should be allocated? IEEE Trans. Wireless Commun. **15**(2), 1293–1308 (2016)
10. H. Boche, M. Schubert, A general duality theory for uplink and downlink beamforming, in *Proceedings of the IEEE VTC-Fall* (2002), pp. 87–91
11. G. Caire, S. Shamai, On the achievable throughput of a multiantenna Gaussian broadcast channel. IEEE Trans. Inf. Theory **49**(7), 1691–1706 (2003
12. H.V. Cheng, E. Björnson, E.G. Larsson, Uplink pilot and data power control for single cell massive MIMO systems with MRC, in *Proceedings of the IEEE ISWCS*, 2015
13. H.V. Cheng, E. Björnson, E.G. Larsson, Optimal pilot and payload power control in single-cell massive MIMO systems. IEEE Trans. Signal Process (submitted)
14. R. Couillet, M. Debbah, *Random Matrix Methods for Wireless Communications* (Cambridge University Press, Cambridge, 2011)
15. D. Cox, Cochannel interference considerations in frequency reuse small-coverage-area radio systems. IEEE Trans. Commun. **30**(1), 135–142 (1982)
16. X. Gao, O. Edfors, F. Rusek, F. Tufvesson, Linear pre-coding performance in measured very-large MIMO channels, in *Proceedings of the IEEE VTC Fall*, 2011
17. D. Gesbert, M. Kountouris, R. Heath, C.B. Chae, T. Sälzer, Shifting the MIMO paradigm. IEEE Signal Process. Mag. **24**(5), 36–46 (2007)
18. A. Goldsmith, S. Jafar, N. Jindal, S. Vishwanath, Capacity limits of MIMO channels. IEEE J. Sel. Areas Commun. **21**(5), 684–702 (2003)
19. J. Hoydis, C. Hoek, T. Wild, S. ten Brink, Channel measurements for large antenna arrays, in *International Symposium on Wireless Communications Systems (ISWCS)*, 2012
20. J. Hoydis, S. ten Brink, M. Debbah, Massive MIMO in the UL/DL of cellular networks: How many antennas do we need? IEEE J. Sel. Areas Commun. **31**(2), 160–171 (2013)
21. H. Huh, G. Caire, H. Papadopoulos, S. Ramprashad, Achieving "massive MIMO" spectral efficiency with a not-so-large number of antennas. IEEE Trans. Wireless Commun. **11**(9), 3226–3239 (2012)
22. ITU: Requirements related to technical performance for IMT-advanced radio interface(s). Tech. rep., ITU-R M.2134 (2008)

23. J. Jose, A. Ashikhmin, T.L. Marzetta, S. Vishwanath, Pilot contamination and precoding in multi-cell TDD systems. IEEE Trans. Commun. **10**(8), 2640–2651 (2011)
24. S. Kay, *Fundamentals of Statistical Signal Processing: Estimation Theory* (Prentice Hall, Englewood Cliffs, NJ, 1993)
25. E.G. Larsson, F. Tufvesson, O. Edfors, T.L. Marzetta, Massive MIMO for next generation wireless systems. IEEE Commun. Mag. **52**(2), 186–195 (2014)
26. X. Li, E. Björnson, E.G. Larsson, S. Zhou, J. Wang, Massive MIMO with multi-cell MMSE processing: exploiting all pilots for interference suppression. IEEE Trans. Wireless Commun. (2015, submitted). Available: http://arxiv.org/abs/1505.03682
27. Y.F. Liu, Y.H. Dai, Z.Q. Luo, Coordinated beamforming for MISO interference channel: complexity analysis and efficient algorithms. IEEE Trans. Signal Process. **59**(3), 1142–1157 (2011)
28. T.L. Marzetta, Noncooperative cellular wireless with unlimited numbers of base station antennas. IEEE Trans. Wireless Commun. **9**(11), 3590–3600 (2010)
29. T.L. Marzetta, A. Ashikhmin, MIMO system having a plurality of service antennas for data transmission and reception and method thereof, US Patent 8,594,215, 2011
30. H. Ngo, E. Larsson, T. Marzetta, Energy and spectral efficiency of very large multiuser MIMO systems. IEEE Trans. Commun. **61**(4), 1436–1449 (2013)
31. H.Q. Ngo, E.G. Larsson, T.L. Marzetta, Massive MU-MIMO downlink TDD systems with linear precoding and downlink pilots. IEEE Trans. Commun. **61**(4), 1436–1449 (2013)
32. H. Ngo, E. Larsson, T. Marzetta, Aspects of favorable propagation in massive MIMO, in *Proceedings of the EUSIPCO*, 2014
33. H.Q. Ngo, A.E. Ashikhmin, H. Yang, E.G. Larsson, T.L. Marzetta, Cell-free massive MIMO: uniformly great service for everyone, in *Proceedings of the IEEE SPAWC*, 2015
34. A. Paulraj, C. Papadias, Space-time processing for wireless communications. IEEE Signal Process. Mag. **14**(6), 49–83 (1997)
35. L. Qian, Y. Zhang, J. Huang, MAPEL: achieving global optimality for a non-convex wireless power control problem. IEEE Trans. Wireless Commun. **8**(3), 1553–1563 (2009)
36. R. Roy, B. Ottersten, Spatial division multiple access wireless communication systems. US Patent 5,515,378, 1991
37. N. Shariati, E. Björnson, M. Bengtsson, M. Debbah, Low-complexity polynomial channel estimation in large-scale MIMO with arbitrary statistics. IEEE J. Sel. Top. Signal Process. **8**(5), 815–830 (2014)
38. S.C. Swales, M.A. Beach, D.J. Edwards, J.P. McGeehan, The performance enhancement of multibeam adaptive base-station antennas for cellular land mobile radio systems. IEEE Trans. Veh. Technol. **39**(1), 56–67 (1990)
39. E. Telatar, Capacity of multi-antenna Gaussian channels. Eur. Trans. Telecom. **10**(6), 585–595 (1999)
40. A.M. Tulino, S. Verdú, Random matrix theory and wireless communications. Found. Trends Commun. Inf. Theor. **1**(1), 1–182 (2004)
41. J. Vieira, S. Malkowsky, K. Nieman, Z. Miers, N. Kundargi, L. Liu, I.C. Wong, V. Öwall, O. Edfors, F. Tufvesson, A flexible 100-antenna testbed for massive MIMO, in *Proceedings of the IEEE Globecom Workshop - Massive MIMO: From Theory to Practice*, 2014
42. P. Viswanath, D. Tse, Sum capacity of the vector Gaussian broadcast channel and uplink-downlink duality. IEEE Trans. Inf. Theor. **49**(8), 1912–1921 (2003)
43. H. Weingarten, Y. Steinberg, S. Shamai, The capacity region of the Gaussian multiple-input multiple-output broadcast channel. IEEE Trans. Inf. Theory **52**(9), 3936–3964 (2006)
44. J. Winters, Optimum combining for indoor radio systems with multiple users. IEEE Trans. Commun. **35**(11), 1222–1230 (1987)
45. H. Yang, T.L. Marzetta, A macro cellular wireless network with uniformly high user throughputs, in *Proceedings of the IEEE VTC-Fall*, 2014
46. W. Yu, Uplink-downlink duality via minimax duality. IEEE Trans. Inf. Theory **52**(2), 361–374 (2006)
47. P. Zetterberg, B. Ottersten, The spectrum efficiency of a base station antenna array system for spatially selective transmission. IEEE Trans. Veh. Technol. **44**(3), 651–660 (1995)

Millimeter-Wave Mobile Communications

Yi Wang and Zhenyu Shi

Abstract Mobile communications in millimeter wave (mmWave) bands have recently attained a wide range of research due to the available ultra-broad spectrum bands. In this chapter, we introduce the key technologies of mmWave communications based on pioneering researches. Channel measurement and modeling as a fundamental issue is presented in Sect. 2. Beam-tracking technique based on large-scale antenna array is studied in Sect. 3. Network architecture, particularly considering unified access and backhaul, is presented in Sect. 4. Current prototypes are introduced in Sect. 5. Finally we summarize the chapter in Sect. 6.

1 Background

The development of wireless cellular networks has been growing rapidly during the past decades. There are still exploding demands for high quality data communications in applications such as smart phones, tablets and video streaming. The capacity of cellular networks in 2020 is envisioned to be as much as 1000 times compared to current fourth-generation (4G) technology. Meanwhile, the cellular network is expected to be capable of connecting 50 billion devices for wireless services.

Recently, mmWave communication has been investigated because of its ultra broad spectrum band [1, 2]. Typical mmWave frequencies range from 6 GHz to 100 GHz including 6 GHz, 15 GHz, 28 GHz [3, 4], 38 GHz [3], 60 GHz [5], and E-band (71–76 GHz, 81–86 GHz) [1, 4]. From channel measurements, the propagation loss for mmWave transmission is quite large. Its transmission range is thus limited. Therefore, mmWave transmission is more suitable for small cells for data rate and dense user scenarios. Owing to these features, mmWave transmission is tailored for a hotspot scenario and has become a promising candidate for the fifth generation (5G) [7–9, 16, 18, 22].

Y. Wang (✉) • Z. Shi
Huawei Technologies Co., Ltd., Jinsu Road 200, Shanghai, P. R. China
e-mail: yi.wang@huawei.com; zhenyu.shi@huawei.com

W. Xiang et al. (eds.), *5G Mobile Communications*,
DOI 10.1007/978-3-319-34208-5_5

2 mmWave Channel Measurements and Modeling

Channel propagation property is a fundamental topic of mmWave communications [17]. The channel model will affect the spectrum allocation in the International Telecomm Union (ITU) and the World Radio Conference (WRC) as well as system design and performance evaluation in 3rd Generation Partnership Project (3GPP) standards.

2.1 Ultra-Wide Band Channel Sounder

An ultra wide band channel measurement sounder for transmitter and receiver has been developed by Huawei Chengdu branch, see Fig. 1. The front-end can be changed in order to match different carrier frequencies. The transmitter has a signal generator for signal generation, and the receiver uses a signal analyzer for measurement data acquisition. A transmitter (TX) and receiver (RX) synchronized spectrum sweeping solution is adopted. In this way, 2 GHz or even larger bandwidth can be measured. The signal generator at TX sites and the signal analyzer at RX sites are controlled remotely by a personal computer through network.

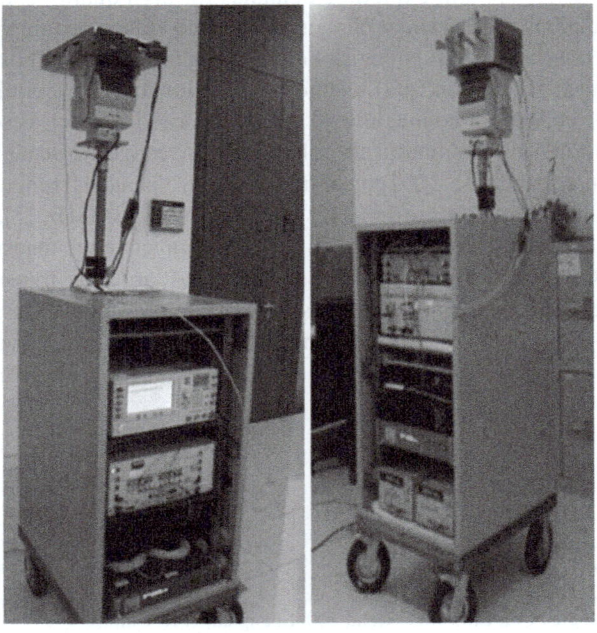

Fig. 1 Channel sounder at E-band, transmitter (*left*) and receiver (*right*)

Frequency bands from 71 to 73 GHz were measured. Vertically polarized horn antennas with 25 dBi gain and 10° half power beam width (HPBW) are used for both outdoor and indoor measurements. At the receiver, the channel information from four different neighboring angles of arrival can be caught by a four-channel RF-front together with four horn antennas.

2.2 Channel Measurement

In this section we present current measurements in E-band. The typical scenarios of mmWave communications are indoor hotspots and outdoor hotspots, these two scenarios have been measured in Huawei Chengdu branch.

The outdoor scenario is selected in dense office buildings, see Fig. 2. Both transmitter and receiver are set at the height of 2–4 m. In non line-of-sight (NLOS) channel, there are building and trees for blocking but there are reflected signals from the opposite building where glass window and concrete wall provide reflections. Occasionally there are a few people going across the road which causes body penetration loss. A dining room is selected as the indoor hotspot scenario, see Fig. 3. The dining room is 70 m long and 25 m wide. Food lines are located at both ends. Most walls are made of glass and the rest are made of concrete.

For the line-of-sight (LOS) TX-RX location, the TX and RX antennas were aligned by means of optical. Then the TX antenna fixed pointing at the aligned angle. The RX antenna adjusts elevation angle of the aligned line to −10°, 0° and 10°, respectively, with the antenna scanning around the entire 360° azimuth plane in 10° (i.e. antenna half power beam width) step.

For the NLOS scenario, there are two steps to complete the measurement procedure. First, searching the strongest path in each TX-RX location. In this step, both the TX and the RX antennas were automatically rotated to find the strongest

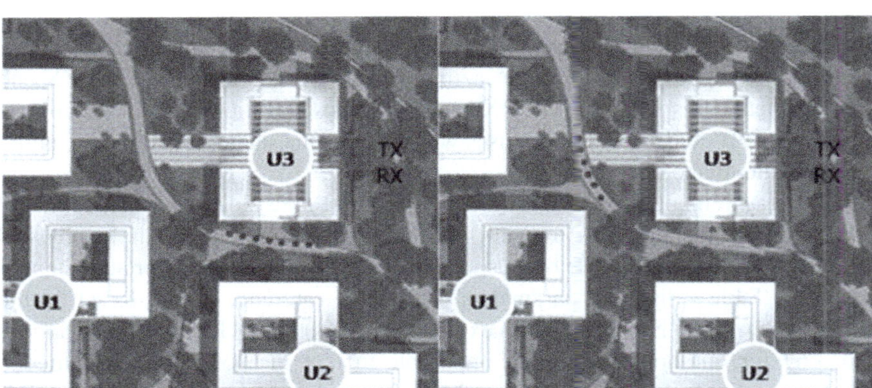

Fig. 2 UMi outdoor scenario, LOS (*left*) and NLOS (*right*)

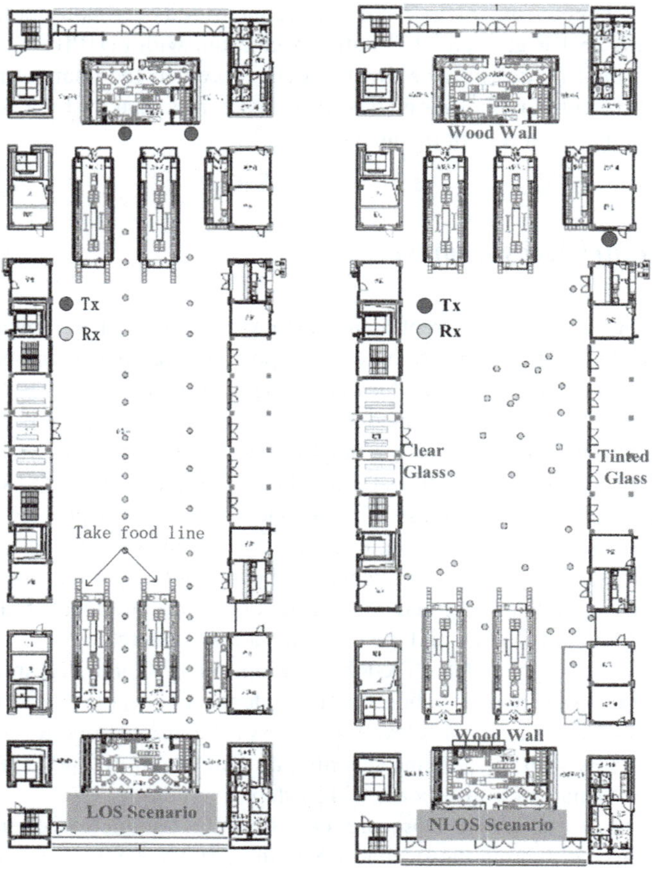

Fig. 3 Indoor hotspot scenario, LOS (*left*) and NLOS (*right*)

received power. Second, Setting the pointing angle of the TX and RX antenna for maximum received power as the 0° azimuth and elevation angles for the TX and the RX antenna, respectively. The RX antenna adjusts elevation angle of the aligned line to −10°, 0° and 10°, three statuses with the antenna scanning around the entire 360° azimuth plane in 10° step.

The measurement results are shown in Fig. 4. We take close-in reference model with reference distance 1 m for path loss since all the parameters have physical meanings [1]. It is reasonable because the measurements cannot cover all distances and sites. Table 1 lists the parameters of close-in reference model based on the measurements. The NLOS path loss exponents (PLEs) of 72 GHz is 4.08 for indoor and 3.67 for outdoor which are comparable to that of 2 GHz in 3GPP model. However, a more accurate path loss channel model and more measurement data are required before making any conclusion.

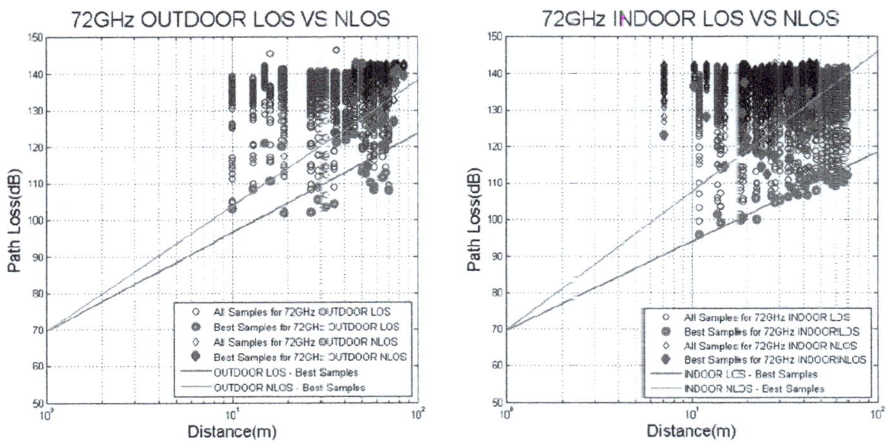

Fig. 4 E-band path loss in UMi outdoor and indoor

Table 1 The path loss parameters in 72 GHz, close-in reference model with reference distance 1 m is applied

Scenario	LOS/NLOS	PLE(β)	α	σ
Indoor	LOS	2.58	69.47	2.38
	NLOS	4.08	69.47	10.63
Outdoor	LOS	2.86	69.47	9.76
	NLOS	3.67	69.47	8.34

2.3 *Channel Modeling*

There are a number of organizations and researchers working on channel modeling in mmWave bands (ref. the survey in [1, 6]). Mobile and Wireless Communications Enablers for the Twenty-twenty Information Society (METIS) project in Europe proposed a map-based channel model which is essentially a simplified ray-tracing method to model channels and fully depends on the defined map. Institute for Electrical and Electronic Engineers (IEEE) 802.11ay standard is making efforts to extend the Saleh-Valenzuela (S-V) model in 60 GHz to larger distance up to 100 m for both outdoor and indoor scenarios. In June 2015, 3GPP Radio Access Network (RAN) agreed to initiate a study item to study channel model for above-6 GHz frequency bands. Meanwhile, International Telecommunication Union Radiocommunication Section (ITU-R) Working Party 5D (WP5D) announced to standardize channel model in above-6 GHz, which is expected to reach a "standard" channel model in mid-2017. A fact is that the channel models in ITU and 3GPP are spatial channel models (SCM) for sub-6 GHz bands, which mainly come from WINNER project researches. 3GPP further extended SCM to three-dimension (3D) cases. Both 3G/4G cellular systems are based on the common SCM model to do system-level evaluation. It is promising that the channel model for 5G systems will be an extension of the SCM model.

A new feature for mmWave communications is that high-gain beams are required in base-station (BS) and potentially in user terminal (UE) in order to compensate the large propagation loss of mmWave signals. Large-scale antenna array is widely used to form high-gain beams. For example, a size of 66 mm by 66 mm may accommodate 1024 antenna elements at E-band, and can form HPBW as narrow as 3°. Thus mmWave transmission link will be sensitive to angle changes in departure and arrival. It is different from the system in sub-6 GHz bands where wide beams are used and the angle variation can be ignored in a drop. Therefore, it is necessary to introduce variant angles for mmWave channel models.

In the framework of 3GPP SCM, WINNER project studied the variant angles and variant delay spreads in a drop. An accurate formula for variant angles is derived in LOS channel, and random angles are approximated for NLOS channel in the local coordination systems (LCS). Here, an independent study on variant angles are investigated and the contributions include: (1) derivation of variant angles in the global coordinate system (GCS); (2) approximation of variant angles with linear method; (3) extension of variant angles to 3D SCM model. Please find more detailed derivations and ray-tracing based simulations in [10, 11].

The idea of SCM model with variant angles (VA-SCM) is to update the angles at each time when the receiver is moving within the duration of one drop. A drop duration denoted by T_{max} is around 1000 transmission time intervals (TTIs) in 3GPP SCM where one TTI corresponds to 1 s. Following 3GPP SCM framework, large-scale parameters (e.g. path loss, delay spread, number of clusters, angular spread etc.) are fixed in a drop. The channel impulse response consists of multiple clusters, and each cluster has different time delays and average receive power which are randomly generated in terms of a given probability distribution function. Each cluster is the superposition of a number of rays. For NLOS cluster, the corresponding scatters are randomly generated for each drop, and the scatters for every ray in a cluster are distributed in a small range depending on angular spread in configuration.

Variant angles are introduced for each ray including azimuth angle of departure and arrival (AoD, AoA) and zenith angle of departure and arrival (ZoD, ZoA). Receiver location at each time can be accurately calculated based on the moving speed of UE and the moving direction. Accordingly the angles can be updated with transmitter and receiver information in the global coordination system (GCS). However, an accurate method to calculate variant angles has high computational complexity, particularly in link-level simulation where channel impulse responses are calculated for each sample or each data symbol. Linear approximation is an efficient way to reduce complexity with acceptable errors. A linear model for variant angles of the pth ray of the qth cluster is given by

$$\begin{aligned}
\theta_{p,q,ZoA}(t) &= \theta_{p,q,ZoA}(t_0) + S_{p,q,ZoA} \cdot (t - t_0)\,, \quad t \in [t_0, t_0 + T_{max}] \\
\theta_{p,q,ZoD}(t) &= \theta_{p,q,ZoD}(t_0) + S_{p,q,ZoD} \cdot (t - t_0)\,, \quad t \in [t_0, t_0 + T_{max}] \\
\phi_{p,q,AoA}(t) &= \phi_{p,q,AoA}(t_0) + S_{p,q,AoA} \cdot (t - t_0)\,, \quad t \in [t_0, t_0 + T_{max}] \\
\phi_{p,q,AoD}(t) &= \phi_{p,q,AoD}(t_0) + S_{p,q,AoD} \cdot (t - t_0)\,, \quad t \in [t_0, t_0 + T_{max}]
\end{aligned} \tag{1}$$

where $S_{p,q,ZoA}$ and $S_{p,q,ZoD}$ are the slopes of variant angles in vertical direction and $S_{p,q,AoA}$ and $S_{p,q,AoD}$ are the slopes of variant angles in horizontal direction. Notice the four slopes are fixed in a drop period T_{max} although they can be extended to time-varying version but at the cost of higher computational complexity.

For easy of description, p and q is omitted in the rest of the section. Assuming receiver is moving at velocity v with angle ϕ_v. h_{BS} and h_{UE} are the height of BS and UE. For LOS cluster, the expression of AoD and ZoD slopes are given by

$$\begin{aligned} S_{ZoD} &= -S_{ZoA} = \frac{v\cos(\phi_v - \phi_{AoD}(t_0))}{(h_{BS} - h_{UE})/\cos(\theta_{ZoD}(t_0))}, \\ S_{AoD} &= S_{AoA} = -\frac{v\sin(\phi_v - \phi_{AoD}(t_0))}{(h_{BS} - h_{UE})\tan(\theta_{ZoD}(t_0))}. \end{aligned} \tag{2}$$

For NLOS cluster with one reflection ray, the model can be simplified by introducing a virtual UE which is the mirror image of UE based on the reflection surface, as is shown in Fig. 5. The simplified slopes in NLOS channel are given by

$$\begin{aligned} S_{ZoD} &= -S_{ZoA} = -\frac{v\cos(\phi_v + \phi_{AoD}(t_0) - \phi_{RS})}{(h_{BS} - h_{UE})/\cos(\theta_{ZoD}(t_0))}, \\ S_{AoD} &= S_{AoA} = -\frac{v\sin(\phi_v + \phi_{AoD}(t_0) - \phi_{RS})}{(h_{BS} - h_{UE})\tan(\theta_{ZoD}(t_0))}, \end{aligned} \tag{3}$$

where ϕ_{RS} is the angle of the reflection surface and it can be deduced from the initial ϕ_{AoD} and ϕ_{AoA}.

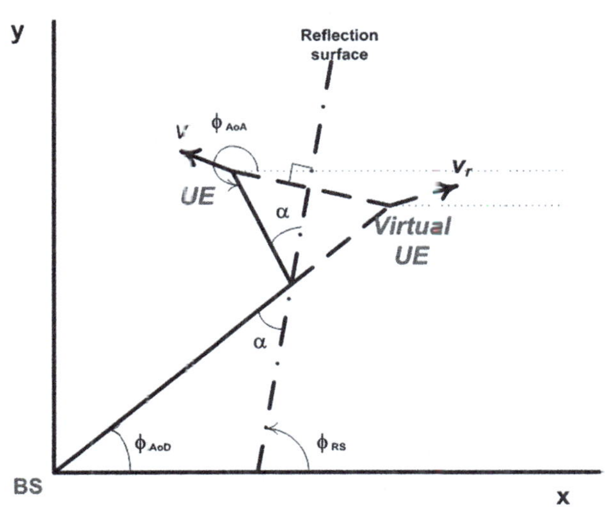

Fig. 5 Global coordination system of an NLOS ray

For NLOS clusters with multiple reflections, it is hard to obtain the accurate location of reflectors. A solution is to take multiple reflections as one-order reflection and randomly set the reflection surface. The channel coefficients of each cluster can be generated following the steps of SCM by considering time variant angles.

3 Beamforming and Beam Tracking in mmWave

The air interface of mmWave communications is featured with antenna-array based beamforming and tracking. Both mmWave base station (mB) and user equipment (UE) use antenna array to compensate the large path loss of mmWave propagation. Such scheme with high-gain narrow beams brings challenges for air interface design. In this section, the concept of uniform air interface and multi-mode beamforming are presented.

3.1 Uniform Air Interface

In all spectrum access system of 5G, the mmWave communication is not an independent air interface but subscribes to several constraints from lower frequency air interface and unified access and backhaul (UAB) architecture. Hence the concept of uniform air interface is expected to be realized with parameter configuration and to flexibly meet different requirements.

Firstly, there might be multiple mmWave bands allocated to 5G, and air interface should be uniform among different carrier frequencies, such as frame and physical numerologies. It is recommended that mmWave could reuse the 10 ms frame structure and 1 ms subframe. Slot can be redesigned to be 125 μs for all mmWave bands. The continuous spectrum bands may have similar channel propagation properties, for example, 6–20 GHz, 20–50 GHz, and 50–90 GHz. Different physical numerologies can be used for different frequency band groups.

Secondly, UAB architecture requests a uniform air interface design between radio access and backhaul. It is better that backhaul works like a special UE of mB where backhaul and UEs share the same radio resource. The difference is that backhaul will use special link configuration, e.g. beams, coding and modulation, to meet its quality of service (QoS) requirements. This will be further discussed in multi-mode beamforming.

Thirdly, mmWave communication allows efficient waveform design to meet different QoS requirements in different scenarios, such as power efficiency, latency etc. Both orthogonal frequency division multiplexing (OFDM) and single carrier OFDM (SC-OFDM) waveforms used in long term evolution (LTE) can be used for mmWave communications. Other candidate waveforms might be filtered OFDM (F-OFDM) and sparse code multiple access (SCMA) currently proposed for 5G low frequency systems. A good combination of them is to use software defined adaptive air interface to select different waveforms for different scenario usages. However, a common constraint is that the waveform is robust to phase noise which may dominate other distortions in front-end. Peak to average power (PAPR) is also an important issue to waveform selection. It is expected to deploy efficient PAPR reduction algorithms to improve power amplifier efficiency particularly when using multiple carrier waveforms.

3.2 *Multi-Mode Beamforming*

Multi-mode beamforming (MM-BF) scheme is proposed in this section. It allocates antenna elements, intermedium and radio frequency (IRF) channels and baseband units to the beams of backhaul links and radio access adaptively. Antenna arrays are divided into sub-arrays, and the beamforming processing is carried out in not only digital baseband but also IRF part. User data streams and backhaul streams are first put into adaptive multiple-input multiple-output (MIMO)mode selection (AMMS), and AMMS will do MIMO processing of each stream based on the current channel environment, where MIMO mode can be spatial multiplexing (SM), space time coding (STC), space time beamforming, etc. After AMMS, the data streams are precoded in digital domain based on beam requirements and downlink channel estimation results, then the signal streams passes through digital-to-analogue converter, infrared (IR) and radio frequency (RF) processing. Before the signals are fed to the antenna elements, the phase of signals should be shifted for each antenna element based beams' requirements and channel estimation results. Such structure can flexibly realize multiple antennas techniques including MIMO, beamforming, diversity and their hybrid forms.

In a mobile scenario with multiple users accessing, the challenge is to design algorithms to align the narrow beams between mB and UEs. Overhead cost, complexity and tracking ability are the key merits of figure to evaluate performance. Two beam phases are proposed to finish the beam alignment: beam training and beam tracking. Beam training performs a rough beam alignment where both quasi-omni-directional beams and wide beams can be used for training. Since exhaustive beam search might bring a high cost in design pilots, there are potential methods to shorten beam training period and overhead. Hierarchical beam training method [12] is an efficient way which firstly uses sector-level beam to do training and then uses narrow beams for searching. Frame design for beam training, for example, centralized training can also shorten training time.

Beam tracking performs channel information update during the time when there is no beam training. The updated channel information is typically AoA and AoD for transmitter and receiver to do beamforming. There are two methods to solve the problem. One is to use reference signal and old channel information to predict the AoD and AoA for next data transmission. The authors in [13] propose a new channel tracking technique based on sequentially updating the beamforming at transmitter and receiver. Numerical results demonstrate the superior channel tracking ability of the proposed technique over various baselines in the literature. The second method is to estimate the AoA and AoD based on reference signal. The challenge is that the reference signal cost is increasing with the antenna size in BS and UE which is prohibitive in case of antenna size larger than 16. An efficient method to overcome the problem is to use compression sensing (CS) technique which well exploit the sparse property of mmWave channel, and can significantly reduce the overhead. The authors in [14] propose a joint channel estimation and beamforming method where CS is used to estimate AoA and AoD jointly. Results show that the overhead can be saved up to 75 % compared to traditional non-CS estimation method, e.g., least square algorithm, under practical scenarios. Furthermore, the proposed method in [14] has only 2–3 dB loss compared to the method with perfect channel information.

4 Unified Access and Backhaul in mmWave

In the design of 5G cellular networks, the mmWave band is expected to be served as a complementary to the low frequency band. The basic structure is a hybrid network where the low frequency and high frequency communication coexist. The conventional LTE is responsible for seamless coverage due to its longer range. In the meanwhile, the mmWave communications are served as multiple hotspots scattered inside the range of the network. Based on these rules, a prototype of unified access and backhaul network has been introduced. The UAB network utilizes a C/U split configuration. The control-plane (C-plane) is managed by macro base stations (MBs) through low frequency bands while the user plane (U-plane) is processed via mmWave base stations (mBs) through high frequency bands. A large number of UEs associated with mBs will bring significant challenges to the backhaul link from the mBs to the MBs. The cost of fiber link to the small cells will be prohibitively expensive. Hence, the wireless backhaul can be applied to reduce the cost. The conventional LTE adopt out-of-band backhaul since the frequency band in radio access is already quite limited. In mmWave networks, an in-band backhaul is suitable thanks to the abundant available spectrum of high frequency bands. In this section, three methods are proposed to determine the resource allocation in the in-band backhaul scheme.

4.1 Network Topology

The UAB topology is illustrated in Fig. 6. As the figure shows, there are mainly three kinds of entities in this network: MBs, mBs and UEs. MBs are responsible for transmitting control information operating in 2 GHz low frequency bands. In Fig. 6, each MB will cover three macro cells. Thus, the whole network is covered

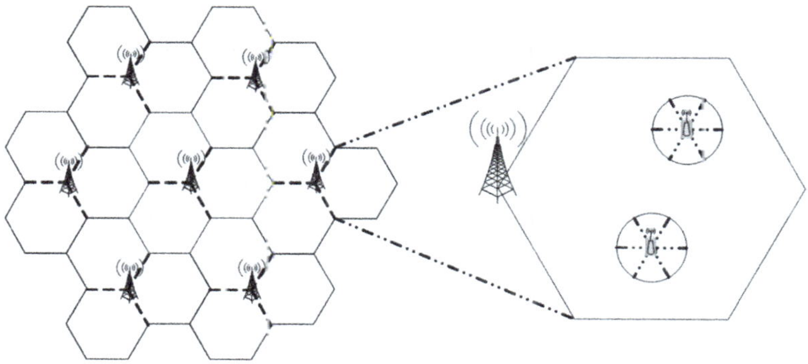

Fig. 6 UAB network topology with seven MBs and twenty-one macro cells

by the macro cells. Inside each macro cell, mBs are deployed. The mBs will cover a relatively smaller range, functioning as small cells. Each mB can support six pico cells with 60° each. Inside each pico cell, there will be a large number of UEs connecting to the mBs. These UEs communicate with their associated mBs via high frequency bands. For those UEs that are unable to connect with mBs, they can choose to communicate with the MBs via low frequency bands.

In mmWave UAB networks, both the radio access and the backhaul share a continuous bandwidth in mmWave frequency band. In order to mitigate interference between radio access and backhaul, it is necessary to separate them either in time or in frequency resource blocks. All mBs share a same resource partition ratio so as not to interfere with each other. The ratios can also be dynamically adapted over time.

4.2 Dynamic Resource Allocation

The resource can be divided either in time or in frequency. Frequency division is taken as an example here. The extension to time division is straightforward. Assuming that the overall bandwidth is B, where backhaul (BH) is allocated bandwidth B^{BH} and radio access (RA) is allocated bandwidth B^{RA}. Since the overall bandwidth is a fixed value, $B = B^{BH} + B^{RA}$ is always fulfilled.

Consider the nth mB, the backhaul throughput can be written as a function of B^{BH} by

$$T_n^{BH} = R_n^{BH} B^{BH} \tag{4}$$

and for radio access the function is

$$T_n^{RA} = R_n^{RA} B^{RA} = R_n^{RA} (B - B^{BH}) , \tag{5}$$

where R_n^{BH} and R_n^{RA} are the corresponding data rates in bit/s/Hz. R_n^{BH} is determined by the link quality between the mB and MB. It can be treated as the backhaul bandwidth efficiency. R_n^{RA} is determined by UEs inside the mB's range. Hence, this value is computed by summing up all the active UEs' bandwidth efficiency at a certain time. Both R_n^{BH} and R_n^{RA} can be pre-calculated by measuring the throughputs during a certain time period and then dividing the current bandwidth.

For any mB, its throughput T_n should be limited by the minimum value of the two throughput values:

$$T_n = \min(T_n^{BH}, T_n^{RA}) \ . \tag{6}$$

By taking (4) and (5) into (6), and after some derivations, T_n can be written as a function of B^{BH}, provided by

$$T_n(B^{BH}) = \begin{cases} R_n^{BH} B^{BH} \text{ if} B^{BH} < B_{n,max}^{BH} \\ \frac{R_n^{BH} R_n^{RA}}{R_n^{BH} + R_n^{RA}} B \text{ if} B^{BH} = B_{n,max}^{BH} \\ R_n^{RA} (B - B^{BH}) \text{ if} B^{BH} > B_{n,max}^{BH} \end{cases} . \tag{7}$$

This function is a triangular function. $T_n(B^{BH})$ is first linearly increasing with a slope of R_n^{BH}. In this range, the backhaul throughput is lower than the radio access throughput. The overall throughput is "BH-limited". The value reaches its maximum when

$$B_{n,max}^{BH} = \frac{R_n^{RA}}{R_n^{BH} + R_n^{RA}} B \ . \tag{8}$$

and the peak value is

$$T_{n,max} = \frac{R_n^{BH} R_n^{RA}}{R_n^{BH} + R_n^{RA}} B. \tag{9}$$

Afterwards, it is reducing linearly with a slope of R_n^{RA}. At this range, the radio access throughput is lower than the backhaul throughput, so the overall throughput is "RA-limited".

For multiple mB, each mB will have different peak values and corresponding backhaul bandwidth. Under the assumption that the backhaul bandwidth is the same for every mB. In order to satisfy different requirements, three dynamic resource allocation methods are introduced and listed as follows:

- Max-Min: Maximization of the minimum throughput
- Max-Sum: Maximization of the sum throughput
- Quasi-PF: Maximization of the satisfactory factor

4.2.1 Max-Min

According to the definition, the minimum throughput of all mBs can be expressed by

$$T_{min}(B^{BH}) = \min(T_1, T_2, \cdots, T_N) . \tag{10}$$

The goal is to find a suitable backhaul bandwidth, where $T_{min}(B^{BH})$ is maximum:

$$B_m^{BH} = \arg\max\{T_{min}(B^{BH})\} . \tag{11}$$

In fact, the minimum throughput is determined by the minimum R_{min}^{RA} and the minimum R_{min}^{BH} of all mBs, with

$$R_{min}^{BH} = \min(R_1^{BH}, R_2^{BH}, \ldots, R_N^{BH}) ,$$

and

$$R_{min}^{RA} = \min(R_1^{RA}, R_2^{RA}, \ldots, R_N^{RA}) .$$

Thus, the minimum throughput is again a triangular function which is of the form

$$T_{min}(B^{BH}) = \begin{cases} R_{min}^{BH} B^{BH} \text{ if} B^{BH} < B_m^{BH} \\ \frac{R_{min}^{BH} R_{min}^{RA}}{R_{min}^{BH} + R_{min}^{RA}} B \text{ if} B^{BH} = B_m^{BH} \\ R_{min}^{RA}(B - B^{BH}) \text{ if} B^{BH} > B_m^{BH} \end{cases} . \tag{12}$$

To achieve the maximum value for $T_{min}(B^{BH})$, the chosen backhaul bandwidth for all mBs should be

$$B_m^{BH} = \frac{R_{min}^{RA}}{R_{min}^{BH} + R_{min}^{RA}} B . \tag{13}$$

and the radio access bandwidth is

$$B_m^{RA} = B - B_m^{BH} . \tag{14}$$

4.2.2 Max-Sum

The purpose of this algorithm is to maximize the summation of all mBs' throughputs $\sum T_n$. For different mBs, the triangular functions are different in the ascending slope, the descending slope and the peak point. When the backhaul bandwidth starts from zero, all mBs are BH limited. With all the throughputs increasing, the sum throughput is increasing at the maximum speed. When the backhaul bandwidth reaches the first peak of a certain mB, this mB's throughput begins to decrease

Algorithm 1 Max-Sum Algorithm

1: Computing the chosen BH bandwidth where T_n is maximum for each mB, reordering them from minimum to maximum, with the resulting sequence $[B'^{BH}_1, B'^{BH}_2, \cdots, B'^{BH}_i, \cdots B'^{BH}_N]$;
2: Based on the above mentioned sequence, listing the corresponding BH and RA rates sequence as $[R'^{BH}_1, R'^{BH}_2, \cdots, R'^{BH}_i, \cdots R'^{BH}_N]$ and $[R'^{RA}_1, R'^{RA}_2, \cdots, R'^{RA}_i, \cdots R'^{RA}_N]$;
3: Writing a general form of $\sum T_n = \sum_{i=l+1}^{N} R'^{BH}_i B^{BH} + \sum_{i=1}^{l} R'^{RA}_i (B - B^{BH})$, $B^{BH} \in [B'^{BH}_l, B'^{BH}_{l+1}]$;
4: Finding $[B'^{BH}_l, B'^{BH}_{l+1}]$ where $\frac{\partial(\sum T_n)}{\partial B^{BH}}$ is minimum positive;
5: **return** $B^{BH}_m = B'^{BH}_{l+1}$ as the uniform backhaul bandwidth for all mBs;

while the other mBs' throughputs continue increasing, but the ascending rate of the sum throughput is lower than before. The ascending rate continues decreasing after each mB's peak reaches until the sum throughput begins to decrease. At that point, the sum throughput is maximum. Afterwards, the descending rate speeds up until all mBs are RA limited. The function of $\sum T_n$ is a piecewise function with multiple sections.

Basically, the main goal of this method is to find the peak point of the sum throughput given the BH and RA rates of all mBs:

$$B^{BH}_m = \arg\max\{\sum T_n(B^{BH})\} . \tag{15}$$

The detailed derivations of this procedure can be found in [15], where the major steps are summarized in Algorithm 1.

4.2.3 Quasi-PF

The previous two algorithms only focus on the instantaneous bandwidth allocation where the history of the allocation is not considered. Both Max-Min and Max-Sum allocate resources based on the current radio access and backhaul rates. However, it may cause unfairness among mBs. For instance, if one mB has better channel conditions, the Max-Sum resource allocation will always provide it with the maximum data rates while other mBs' will always transmit at lower rates. In the long run, the mBs with high throughput will become extremely higher, and mBs with low throughputs will become extremely lower and UEs of those low throughput mBs will not be satisfied. To satisfy as many UEs as possible, the current resource allocation should also take the accumulated historical throughputs into account. Hence, the idea of the well-known proportional fairness (PF) algorithm [19] can be borrowed herein.

To evaluate among mBs, a satisfactory factor is introduced for each mB and each time index. The mB's satisfactory factor at the tth TTI $\eta_n(t)$ is defined as

$$\eta_n(t) = \frac{T_{n,max}(t)}{\overline{T}_{n,pre}(t)} , \tag{16}$$

where $T_{n,max}(t)$ is the maximum achievable throughput of the nth mB at current time index. This value is determined by the current R_n^{RA} and R_n^{BH} using (9). $\overline{T}_{n,pre}(t)$ is the average throughput consumed by this mB during the previous $t-1$ TTIs. The larger the nominator, the higher maximum throughput can achieve currently. The smaller the denominator, the less average throughput has been transmitted previously. Therefore, to compromise a balance between the achievable maximum throughput and the satisfactory of UEs, the mB with the largest satisfactory factor should be chosen:

$$n = \arg\max(\eta_n(t)) \ . \tag{17}$$

In case the mB with the largest $\eta_n(t)$ has been found, the chosen overall system's BH bandwidth is the peak point of this mB's triangular function.

To evaluate the proposed algorithms, simulations are carried out in the UAB networks. The network topology follows Fig. 6, operating in E-band. Radio access and backhaul share a total bandwidth of 5 GHz. To compare the algorithms in various kinds of scenarios, three different UE densities are considered, which are

- Case 1: 20 % mBs have high UE loads, 80 % mBs have low UE loads;
- Case 2: 50 % mBs have high UE loads, 50 % mBs have low UE loads;
- Case 3: all mBs have the same UE loads.

Figure 7 compares the system's performance of three cases. The mean throughput is computed by summing up the throughputs of all mB and then dividing the mB

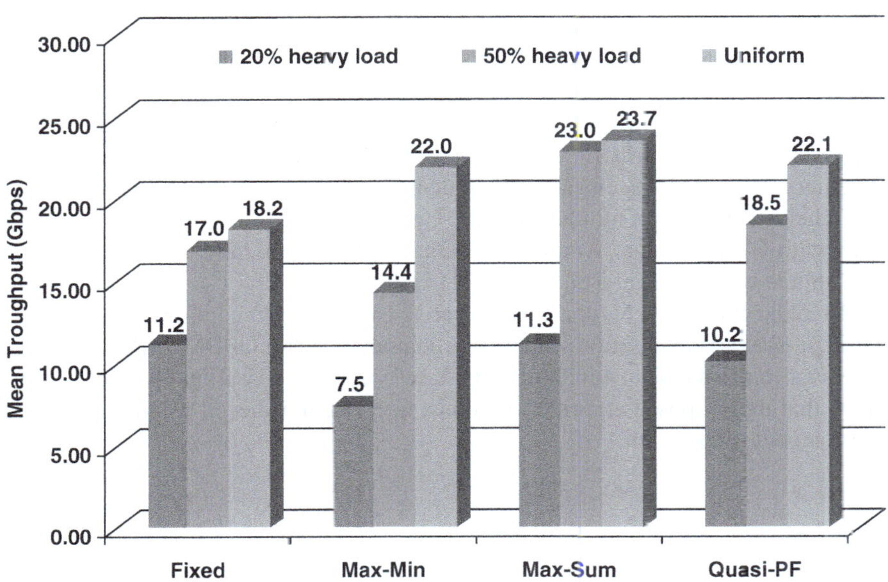

Fig. 7 Mean throughputs for different cases

number and the TTI number. It can be found that the Max-Min algorithm provides the lowest mean throughput for all three cases. This is reasonable as this way of allocation only satisfies the worst case. The Max-Sum algorithm in all three cases delivers the best performance. The throughput of Quasi-PF algorithm is in between the fixed case and the Max-Sum algorithm in case 2 and case 3. There is 35.3 and 30.2 % performance improvement compared to the fixed one in case 2 and case 3, respectively. Hence, it can be concluded that in terms of the whole system throughput, the Max-Sum algorithm provides the best solution. The Quasi-PF algorithm aims at providing a better trade off between overall capacity and mB fairness.

5 Prototype Verification

Several companies have announced their prototype verification for mmWave communications. Samsung realized a peak data rate of 7.5 Gbit/s at 28 GHz frequency; DoCoMo realized a peak data rate of 10 Gbit/s at 11 GHz frequency [20]. Ericsson realized a peak data rate of 5 Gbit/s at 15 GHz. Huawei and Nokia demonstrated a peak data rate of 115 Gbit/s [4] and 10 Gbit/s [21] in 72 GHz bands, respectively. For the waveform of prototypes, Nokia applies single carrier waveform (called NCP-SC) which is essentially discrete fourier transform (DFT)-precoded OFDM waveform but cyclic prefix is null. Samsung, DoCoMo, Ericsson and Huawei are using OFDM waveforms in prototype.

In this section, a 115 Gbit/s prototype working in E-band (72 GHz) is presented and a mmWave platform supporting multiple bands. The 115 Gbit/s prototype, see Fig. 8, was designed to demonstrate the peak data rate by using a total available bandwidth of 10 GHz in E-band. Point-to-point transmission is demonstrated in indoor LOS scenario. MIMO is configured with 2 polarized antennas in mB and 2 polarized antennas in UE. With LOS MIMO, two data streams are transmitted together. An OFDM waveform and low-density parity-check (LDPC) channel coding scheme is used for link transmission. Considering 20 % overhead, two streams, each with 64 Quadrature Amplitude Modulation (QAM), have been successfully transmitted which deliver a data rate of 115 Gbit/s.

To reduce the severe phase noise at E-band, a low complexity two stage estimation and compensation scheme was demonstrated for OFDM-MIMO systems. This is a joint channel estimation and phase noise compensation method. Results show that the proposed algorithm is robust to phase noise even when the detailed noise model is uncertain.

Fig. 8 115 Gbps prototype E-band transmitter and receiver (*left*), real-time receiver performance (*right*)

6 Chapter Summary

The key technologies of mmWave mobile communications are investigated in this chapter. Channel measurements show that mmWave signals suffer from much larger propagation loss and are suitable for small cell coverage. A hybrid network is presented where mmWave is used for capacity enhancement in hotspots and low frequency network is applied for seamless coverage. Uniform air interface is a consequence to simplify the design between mmWave bands and low frequency bands. Unified access and backhaul technique not only reduces the cost of backhaul, but also can meet the requirement of 1000 times capacity enhancement than LTE system.

References

1. T. Rappaport, R.W. Heath, R.C. Daniels, J.N. Murdock, *Millimeter Wave Wireless Communications* (Prentice Hall, Englewood Cliffs, NJ, 2014)
2. Y. Wang, J. Li, L. Huang, Y. Jing, A. Georgakopoulos, P. Demestichas, Spectrum broadening to higher-frequency bands to support high data rates. IEEE Veh. Technol. Mag. **9**, 39–46 (2014)
3. A.I. Sulyman, A.T. Nassar, M. Samimi, G.R. MacCartney, T.S. Rappaport, A. Alsanie, Radio propagation path loss models for 5G cellular networks in the 28 GHz and 38 GHz millimeter-wave bands. IEEE Commun. Mag. **52**, 78–86 (2014)

4. Y. Wang, L. Huang, Z. Shi, H. Huang, D. Steer, J. Li, G. Wang, W. Tong, An introduction to 5G mmWave communications, in *Proceedings of IEEE Globecom 2015 Workshop*, San Diego, 2015
5. Channel models for 60 GHz WLAN systems. IEEE Std. 802.11-09/00 334r8 (2010)
6. P. Almers, E. Bonek, A. Burr et al., Survey of channel and radio propagation models for wireless MIMO systems. EURASIP J. Wirel. Commun. Netw. **2007**, 19pp (2007). Article ID 19070
7. W. Roh, J.Y. Seol, J.H. Park, B. Lee, J. Lee, Y. Kim, J. Cho, K. Cheun, Millimeter-wave beamforming as an enabling technology for 5G cellular communications: theoretical feasibility and prototype results. IEEE Commun. Mag. **52**, 106–113 (2014)
8. Preliminary draft new recommendation ITU-R M.[IMT.VISION], ITU-R WP 5D std., Document 5D/TEMP/548(Rev.3) (2015)
9. Getting ready for 5G. 3GPP std. TSG RAN 67, Shanghai, 9–12 March 2015
10. Y. Wang, L. Huang, Z. Shi, K. Liu, X. Zou, A millimeter wave channel model with variant angles under 3GPP SCM framework, in *Proceedings of IEEE PIMRC 2015 Workshop*, Hong Kong, 2015
11. Y. Wang, Z. Shi, M. Du, W. Tong, A millimeter wave spatial channel model with variant angles and variant path loss, in *Proceedings of IEEE WCNC 2016*, Doha, 2016
12. Wireless LAN medium access control (MAC) and physical layer (PHY) specifications. IEEE std 802.11ad. (2012)
13. J. He, T. Kim, H. Ghauch, K. Liu, G. Wang, Millimeter wave MIMO channel tracking systems, in *Proceedings of IEEE Globecom '14 Workshop*, Austin (2014), pp. 414–419
14. H. Huang, K. Liu, R. Wen, Y. Wang, G. Wang, Joint channel estimation and beamforming for millimeter wave cellular system, in *Proceedings of IEEE Globecom '15*, San Diego, 2015
15. Z. Shi, Y. Wang, L. Huang, T. Wang, Dynamic resource allocation in mmWave unified access and backhaul network, in *Proceedings of IEEE PIMRC '15 Workshop*, Hongkong, 2015
16. H. Li, L. Huang, Y. Wang, Scheduling schemes for interference suppression in millimeter-wave cellular network, in *Proceedings of IEEE PIMRC '15 Workshop*, Hongkong, 2015
17. G. MacCartney, J. Zhang, S. Nie, T.S. Rappaport, Path loss models for 5G millimeter wave propagation channels in urban microcells, in *Proceedings of IEEE Globecom '13* (2013), pp. 3948–3953
18. J. Zhang, A. Beletchi, Y. Yi, H. Zhuang, Capacity performance of millimeter wave heterogeneous networks at 28 GHz/73 GHz, in *Proceedings of IEEE Globecom '14 Workshop*, Austin, 2014
19. P. Viswanath, D. Tse, R. Laroia, Opportunistic beamforming using dumb antennas. IEEE Trans. Inf. Theory **48**(6), 1277–1294 (2002)
20. S. Suyama, J. Shen, Y. Oda, H. Suzuki, K. Fukawa, DoCoMo and Tokyo Institute of Technology achieve world's first 10 Gbps packet transmission in outdoor experiment, in *Proceedings of IEEE PIMRC '13*, 2013
21. Nokia, Nokia networks showcases 5G speed of 10Gbps with NI at the Brooklyn 5G summit, 2015. http://networks.nokia.com/news-events/press-room/press-releases/nokia-networks-showcases-5g-speed-of-10gbps-with-ni-at-the-brooklyn-5g-summit
22. 3GPP TSG RAN E-UTRA, Further advancements for E-UTRA physical layer aspects (Release 9). 3GPP std. TR 36.814 v9.0.0 (2010)

Non-Orthogonal Multiple Access (NOMA) for Future Radio Access

Razieh Razavi, Mehrdad Dianati, and Muhammad Ali Imran

Abstract Radio access technologies for cellular mobile communications are typically characterized by multiple access schemes, e.g., frequency division multiple access (FDMA), time division multiple access (TDMA), code division multiple access (CDMA), and OFDMA. In the 4th generation (4G) mobile communication systems such as Long-Term Evolution (LTE) (Au et al., Uplink contention based SCMA for 5G radio access. Globecom Workshops (GC Wkshps), 2014. doi:10.1109/GLOCOMW.2014.7063547) and LTE-Advanced (Baracca et al., IEEE Trans. Commun., 2011. doi:10.1109/TCOMM.2011.121410.090252; Barry et al., Digital Communication, Kluwer, Dordrecht, 2004), standardized by the 3rd Generation Partnership Project (3GPP), orthogonal multiple access based on OFDMA or single carrier (SC)-FDMA is adopted. Orthogonal multiple access was a reasonable choice for achieving good system-level throughput performance with simple single-user detection. However, considering the trend in 5G, achieving significant gains in capacity and system throughput performance is a high priority requirement in view of the recent exponential increase in the volume of mobile traffic. In addition the proposed system should be able to support enhanced delay-sensitive high-volume services such as video streaming and cloud computing. Another high-level target of 5G is reduced cost, higher energy efficiency and robustness against emergencies.

1 Introduction

Radio access technologies for cellular mobile communications are typically characterized by multiple access schemes, e.g., frequency division multiple access (FDMA), time division multiple access (TDMA), code division multiple

R. Razavi (✉)
Group Research and Development, Vodafone Group Services Limited, Emerald House, Newbury Business Park, London Road, Newbury, Berks., RG14 2PZ, UK
e-mail: Razieh.Razavi@vodafone.com

M. Dianati • M.A. Imran
Institute for Communication Systems (ICS), Home of the 5G Innovation Centre, University of Surrey, Surrey GU2 7XH, UK
e-mail: m.dianati@surrey.ac.uk; m.Imran@surrey.ac.uk

W. Xiang et al. (eds.), *5G Mobile Communications*,
DOI 10.1007/978-3-319-34208-5_6

access (CDMA), and OFDMA. In the 4th generation (4G) mobile communication systems such as Long-Term Evolution (LTE) and LTE-Advanced, orthogonal multiple access based on OFDMA or single carrier (SC)-FDMA is adopted. Orthogonal multiple access was a reasonable choice for achieving good system-level throughput performance with simple single-user detection. However, considering the trend in 5G, achieving significant gains in capacity and system throughput performance is a high priority requirement in view of the recent exponential increase in the volume of mobile traffic. In addition the proposed system should be able to support enhanced delay-sensitive high-volume services such as video streaming and cloud computing. Another high-level target of 5G is reduced cost, higher energy efficiency and robustness against emergencies. 5G has to provide increased capacity per unit network cost and be energy efficient and resilient to natural disasters. Figure 1 shows how NOMA fits in the system architecture of 5G.

To accommodate such demands, novel radio access technologies are needed. In this chapter, we introduce non-orthogonal multiple access techniques and evaluate the low density spreading based system which is a strong candidate for the next generation of mobile networks. we investigate the achievable rate region of the low density signature (LDS) schemes for flat and frequency selective channels. The main objective is to define an effective channel matrix taking into account sparse allocation of chips to the users that forces certain elements of the effective channel matrix be zero. In other words, the effective channel matrix becomes sparse since each user is allowed to use a certain set of chips. We take into account practical considerations such as finite size modulation and convolutional coding to improve the accuracy of the proposed analysis.

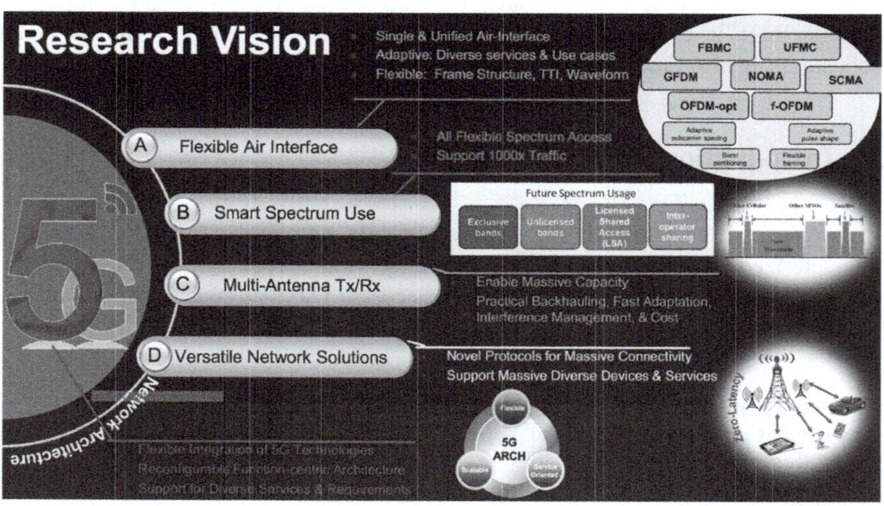

Fig. 1 5G system architecture [33]

2 State of the Art Non-Orthogonal Multiple Access Techniques

Multiple access techniques enable users to share the available resources, in other words, using these techniques multiple users can effectively share radio resources. These techniques provide each user: access to system resources with conflict avoidance; fair resource allocation or different quality of service (QoS) satisfaction; as well as system capacity enhancement. Although there are many different multiple access techniques in the literature, this section focuses only on recently proposed state of the art techniques and those which are adopted in 3G and LTE-Advanced.

2.1 Code Division Multiple Access (CDMA)

CDMA is a single carrier technique that enables multiple users to share the same spectrum using their unique code sequence (spreading code). The code sequence, applied to an information-bearing, narrow-band signal, spreads the signal across the operating frequency band. Every additional user increases interference level (background noise) in the channel because all users use the entire bandwidth. Therefore, the interference level depends on the number of users and their transmission power. Power control is mandatory in order to overcome the near-far effect and to increase system capacity. Code sequences have to be chosen very carefully because well designed codes can provide orthogonality between users. Good cross-correlation features can decrease multiple access interference (MAI) whereas good auto-correlation features can decrease inter symbol interference (ISI)

2.2 Large Area Synchronised CDMA

Considerable research efforts have been invested in designing spreading sequences for CDMA systems, which benefit from zero correlation values, when the relative delay-induced code offset is in the so-called zero correlation zone (ZCZ) or interference free window (IFW) of the spreading code [12]. Combining the large area codes introduced in [22] with loosely synchronised codes [35] resulted in large area synchronised (LAS) CDMA spreading sequences. The resultant LAS codes exhibit an IFW, where both the off-peak aperiodic auto-correlation values and the aperiodic cross-correlation values become zero. Therefore this system has zero ISI and zero MAI, as long as the time-offset of the codes is within the IFW. One disadvantage of LAS codes is that the number of codes having an IFW is limited. Furthermore, the auto-correlation and cross-correlation function of LAS codes have a higher value outside the IFW compared to that of traditional random codes. Therefor, LAS-CDMA faces more serious MAI and multipath interference compared to conventional CDMA when it operates in an asynchronous manner [43].

2.3 Multi-Carrier CDMA (MC-CDMA)

In MC-CDMA systems a data symbol is sent on multiple subcarriers by using a spreading code, which is different for the multiple access users [26]. Multiple user signals overlap in the time and frequency domain but they can be detected at the receiver by using the knowledge of spreading codes. Therefore, MC-CDMA can be considered as a combination of OFDM and CDMA schemes, combining the benefits of both schemes. Thus, this system is considered to be a suitable approach to cope with challenging service demands due to its ability of exploiting both time and frequency resources [8, 13, 28, 45]. In the uplink channel using MC-CDMA, non-orthogonality of received effective signatures causes multi-user interference (MUI). Multiuser detection can be used to moderate the detrimental effects of MUI similar to CDMA systems.

One of the advantages of MC-CDMA systems is that since this system is a combination of OFDM and CDMA, it can decrease the symbol rate in each subcarrier so that a longer symbol duration makes it easier to quasi-synchronise the transmission [18]. However, implementation of optimum multiuser detector (MUD) is computationally complex for MC-CDMA systems.

In order to reduce the complexity of MUD for MC-CDMA systems, Group Orthogonal MC-CDMA (GO-MC-CDMA) [44] has been proposed. The basic idea of this scheme is to partition the available subcarriers into groups and distribute users among the groups. The users that are assigned subcarriers of the same group are separated using spreading codes. The users in each group are immune to interference from other groups. The main advantage of GO-MC-CDMA is that each group behaves as an independent MC-CDMA system with a smaller number of users, making the optimum MUD computationally feasible within each group. It has been shown that by dividing users into subgroups, GO-MC-CDMA is able to achieve a performance very close to the single-user bound [44], with a reasonable computational complexity.

2.4 Interleave Division Multiple Access (IDMA)

An interleaver-based multiple access scheme has been studied in [23, 29] for high spectral efficiency, improved performance and low receiver complexity. This scheme relies on interleaving as the only means to distinguish the signals from different users, and hence it has been called interleave-division multiple-access. IDMA inherits many properties of CDMA based systems, in particular, diversity against fading and mitigation of the worst-case other-cell user interference problem. Furthermore, it allows a very simple chip-by-chip iterative MUD strategy [23, 29]. In this system, the normalized MUD cost (per user) is independent of the number of users. Furthermore, since user separation can be done by using interleaving, the whole bandwidth can be devoted for coding and, thus, paving the way for the information theoretic view regarding the coding-spreading saga, IDMA, in essence, may achieve high spectral efficiency. However, in order to achieve optimality, IDMA requires a large number of observations. The observation may come from a large number of spreading chips or a large frame size. It has been shown in [24] that there exists a trade-off between performance and the length of the interleaver.

2.5 Low Density Signature CDMA (LDS-CDMA)

In order to reduce the complexity of MUD for CDMA systems, a novel spreading signature structure was introduced. By introducing the new structure, termed as LDS structure, higher loading became possible for those systems [20]. Such CDMA systems with LDS structure are called LDS-CDMA. In this system, the interference is managed intelligently at the transmitters side. The main idea of this new technique is to switch off a large number of chips for each symbol so the signature matrix will be a sparse matrix. In other words, each user will spread its data over a small number of chips (or effective processing gain $v_k \ll N$). Consequently at each received chip, a user will have only maximum of $d_c - 1$ interferers, where $d_c \ll K$ is the number of users that are allowed to interfere to each other at each chip. The complexity order of a maximum a posteriori (MAP) receiver turns out to be $\mathcal{O}(|\mathbb{X}|^{d_c})$ instead of $\mathcal{O}(|\mathbb{X}|^K)$ of an optimum receiver for a conventional CDMA system.

Further reduction in complexity of multiuser detection for LDS-CDMA system is possible using grouped-based technique introduced in [42]. This is done by arranging the interfering users of a chip into two groups and approximating the log-likelihood ratio (LLR) of the interference coming from other group as a single symmetric Gaussian distributed variable.

The LDS-CDMA has been reported to approach near single-user performance even under a load of 200 % [20]. However one drawback of LDS-CDMA is that in wideband channels, its performance degrades due to multipath fading. This happens because the ISI introduced by multipath fading causes different chips interfere to each other. Therefore, more number of users will interfere in one chip; this increases the density of LDS's graph by adding edges to the graph representation of LDS. To overcome this issue LDS-OFDM was proposed [21].

To improve the performance in high order modulations, sparse code multiple access (SCMA) is introduced in [1, 27, 37]. This scheme is a multi-dimensional codebook-based non-orthogonal spreading technique. In SCMA, the procedure of bit to QAM symbol mapping and spreading are combined together and incoming bits are directly mapped to multi-dimensional codewords of SCMA codebook sets. Each layer has its dedicated codebook. Shaping gain of a multi-dimensional constellation is one of the main sources of the performance improvement in comparison to the simple repetition of QAM symbols in LDS.

2.6 Space-Division Multiple-Access (SDMA)

Antennas arrays, sometimes called smart multiple access antennas, have been introduced in mobile radio systems to exploit the spatial domain. In these systems, signals are received and sent only from and into a limited angular range, following the directional nature of multipath. This improves coverage or link quality in noise-limited situations and enhances capacity in interference-limited situations.

SDMA systems enable two or more users, associated to the same base station, to use the same time and frequency and code resources on the grounds of their physical location or spatial separation [36]. Therefore, we can discriminate amongst the users by exploiting the fact that different users impinge different spatial signatures on the receive antenna array. This is generally done with directional antennas. SDMA techniques can be used for increasing cell coverage areas. Also, they are useful in reducing the interference for improved service quality [25]. We showed a summary of the discussed NOMA schemes in Table 1.

3 Performance Evaluation of LDS Based Systems

In this section, we will evaluate LDS based systems since it is a strong candidate for 5th generation (5G) of mobile networks. LDS was initially proposed for CDMA systems, known as LDS-CDMA in [20]. The main idea is to spread information of each user on a limited number of chips, in contrast to conventional CDMA which spreads information of each user on all chips. Consequently, for a system with K users, only a limited number of users are allowed to interfere at each chip.

Table 1 Summary of different NOMA schemes

CDMA	Several transmitters can send information simultaneously over a single communication channel. CDMA employs spread-spectrum technology and a special coding scheme.
LAS-CDMA	Applies a new set of spreading codes that exhibit interference free window to separate users in the wireless channel.
MC-CDMA	Spreads each user symbol in the frequency domain. That is, each user symbol is carried over multiple orthogonal subcarriers.
IDMA	Interleavers are used as the only means for user separation. IDMA inherits many advantages from CDMA, such as diversity against fading and mitigation of the worst-case other-cell user interference problem.
LDS-CDMA	Uses spreading sequences of low density for spreading the data symbols in time domain. This technique benefits from a less complex MUD compared to conventional CDMA with optimum MUD.
SCMA	Enables two or more users, associated to the same base station, to use the same time and frequency and code resources on the grounds of their physical location or spatial separation.

The effective number of interfering users at chip n will be reduced to $\eta_n \ll K$. Thus, the complexity order of the MUD for LDS-CDMA receiver will reduce to $\mathcal{O}\left(|\mathbb{X}|^{\eta_n}\right)$ compared to $\mathcal{O}\left(|\mathbb{X}|^{K}\right)$ of CDMA.

Recently, most of the modern communication systems adopt orthogonal frequency division multiplexing (OFDM) transmission techniques due to its resilience against inter symbol interference (ISI) for wideband channels. To this end, LDS for MC-CDMA (LDS-OFDM) has been proposed in [21, 31]. In this scheme, the information bits of each user are spread over a subset of sub-carriers. Therefore, transmission of each user will only be interfered by a small number of users. Similar to LDS-CDMA, detection of LDS-OFDM can be based on the message passing algorithm (MPA). An LDS-OFDM system can be considered as a system that uses LDS as multiple access technique and OFDM for multi-carrier transmission. In [21], we have shown that LDS-OFDM outperforms OFDMA in terms of bit error rate (BER). Figure 2 shows the BER results of 200 % loaded BPSK LDS-OFDMA and QPSK OFDMA when both are using the same error correcting convolutional code with rate 1/2. As we can see from this Figure, even for system with channel coding still LDS-OFDMA maintains its superiority to OFDMA systems by approximately 3.4 dB and 5.54 dB at a BER of 10^{-3} and 10^{-5}, respectively.

One of the non-trivial and open issues for LDS schemes is the analysis of its achievable rate region. Analysis of the conventional MAC for fading channels has already been studied in the literature in [5, 39], where two notions of capacity for the fading channel are considered. The first is the classic notion of Shannon capacity with a constraint on BER, which is directly applied to fading channels. The second analysis defines an alternative notion of capacity by setting a constraint on coding delay for fading channels.

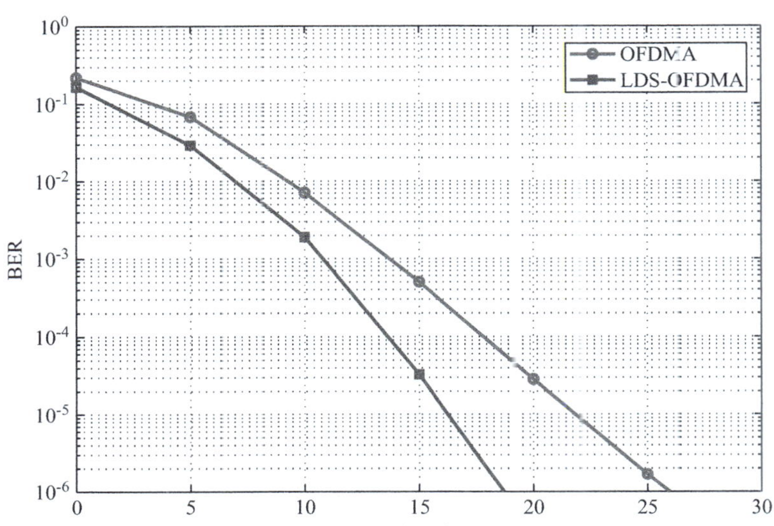

Fig. 2 QPSK OFDMA vs. 200 % overloaded LDS-OFDMA BER performance for coded case

3.1 System Model

We consider a multiple access channel (MAC) which typically corresponds to the uplink communications from multiple users to a single base station in practical system. The following sub-sections explain the important aspects of our system model, including the block diagram of LDS schemes, the corresponding LDS design, and the signal model for LDS schemes.

3.1.1 LDS Schemes

The block diagram of a general LDS based system is shown in Fig. 3. In this scheme, unique low density spreading sequences (signatures) are assigned to each user, and the receiver has full knowledge of these assignments. Low density spreading sequences are designed in a way that allow each user to spread its data only over a small number of chips. Then, zero padding is performed such that the processing gain remains the same as that of a conventional CDMA system. The spreading sequences are then uniquely permuted for each user such that the resultant signature matrix become highly sparse. As shown in Fig. 3, for both LDS-CDMA and LDS-OFDM systems the encoded users' data will be first spread using low density signature. Then, for LDS-CDMA, the sub-carrier mapper allocate one wideband carrier to all users, while for LDS-OFDM, the sub-carrier mapper assign orthogonal sub-carriers to different users and generates an OFDM signal. In other words, for LDS-OFDM the spreading is followed by OFDM modulation, which means that each chip is mapped to a single sub-carrier. Thus, the combination of the spread users' symbols plus the AWGN form the received OFDM symbol. As shown in Fig. 4, an LDS system with K users and N chips can be represented by a factor graph $\mathcal{G}(\mathcal{U}, \mathcal{C})$, where the symbols of users are represented by variable nodes $u \in \mathcal{U}$ and chips are represented by function nodes $c \in \mathcal{C}$. In this graph edges

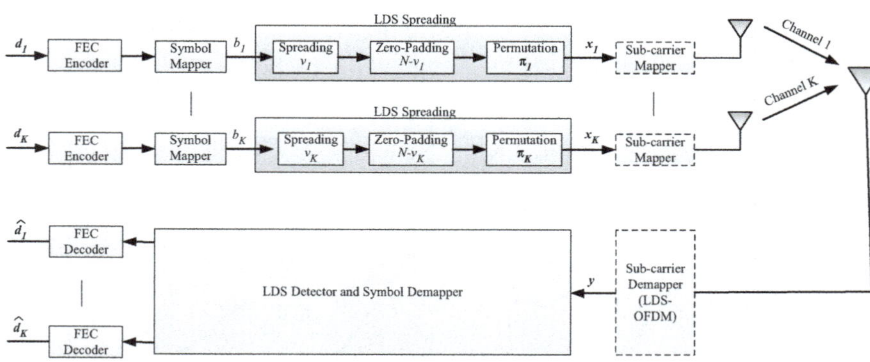

Fig. 3 Block diagram of LDS scheme

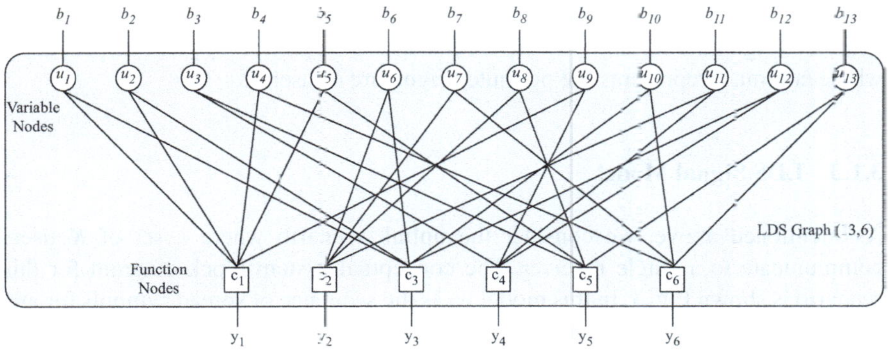

Fig. 4 Factor graph representation of LDS scheme

represent the connection between the received chip and the corresponding users. At the receiver, chip level iterated (CLi) MUD is used to detect different users' symbols [20]. In this iterative MUD, using MPA in an iterative manner, messages containing the reliability values of the corresponding variable nodes are exchanged between the variable nodes and function nodes. At the end, after appropriate number of iterations the soft output which is the calculated log likelihood ratio at each variable node will be sent to the channel decoder.

3.1.2 Low Density Signature Design

In this section, we explain how low density signatures, i.e., the sequences which are mainly populated with zeros, are created. This structure helps switch off a large number of chips in each user's signature. We consider an LDS system with K users with user indices $k = 1, \ldots, K$. LDS with density factor $0 < \zeta_k \leq 1$, which represents the ratio of the number of chips used by kth user to the total number of chips N, is generated in a process with the following steps:

1. A random signature (vector) with length $\nu_k = \zeta_k N$ for user k is generated. Each element of this vector is a random variable with a Gaussian distribution. This means that the effective processing gain for user k is equal to ν_k.
2. Zero padding is carried out by adding $N - \nu_k$ zeros at the end of the signature for user k such that the processing gain of the system becomes N.
3. The permutation pattern $\boldsymbol{\pi}_k = \{\pi_k(1), \pi_k(2), \ldots, \pi_k(N)\}$ is chosen randomly and applied to the zero-padded signatures of user k. Thus, the none-zero components of the permuted signature for user k will appear in $\mathscr{C}_k = \{\pi_k(1), \pi_k(2), \ldots, \pi_k(\nu_k)\}$ positions. Furthermore, we define $\mathscr{P} = \{\boldsymbol{\pi}_1, \boldsymbol{\pi}_2, \ldots, \boldsymbol{\pi}_K\}$ as the set of all users' permutation patterns.

Let $\boldsymbol{s}_k = [s_{k,1}, \ldots, s_{k,N}]^T$ be the generated signature for user k. By arranging the generated signature for all users the LDS matrix $\boldsymbol{S} = [\boldsymbol{s}_1, \ldots, \boldsymbol{s}_K]$ will be produced, where column k represents the permuted signature of user k.

3.1.3 LDS Signal Model

As mentioned above, we consider the uplink scenario where a set of K users communicate to a single receiver. The conceptual system block diagram for this scenario is shown Fig. 3. In this model, $\boldsymbol{x}_k$ is the sequence of spread symbols for any single modulated symbol of user k, namely, b_k at the transmitter side:

$$\boldsymbol{x}_k = b_k \boldsymbol{s}_k, \tag{1}$$

where $\boldsymbol{s}_k$ is the signature for user k. Denote $\boldsymbol{y} = [y_1, \ldots, y_N]^T$ as the corresponding received sequence at the receiver side after match filtering and equalization. Considering a perfect symbol-synchronous conventional CDMA system under flat fading channel at chip level, the received signal sampled at ith time instance (chip) can be written as follows [41]:

$$y_i = \sum_{k=1}^{K} b_k s_{k,i} h_{k,i} + v_i, \tag{2}$$

where v_i is the complex white Gaussian noise with variance per real symbol $N_0/2$ and $h_{k,i}$ is the fading channel gain of the kth user at time instance i. User k is also subject to an average transmitter power constraint of $\bar{P}_k$. This system model is provided for a time varying channel, therefore, according to [30] the coherence time of channel for a vehicle traveling 60 mph using a 900 MHz carrier is around 2.22 ms while according to Interim Standard 95 (IS-95), the symbol length is around 10 ms which is more than the coherence time. Therefore, in our system model a unique fading coefficient is not multiplied by all chips.

In LDS schemes, for a given time instance $i = 1, \ldots, N$; let $\mathscr{K}_i$ denote the set of users that are active at time instance i, i.e., $\mathscr{K}_i = \{k | i \in \mathscr{C}_k\}$. Considering the LDS scheme explained earlier, for LDS-CDMA, (2) can be given by

$$y_i = \sum_{k \in \mathscr{K}_i} b_k s_{k,i} h_{k,i} + v_i. \tag{3}$$

This equation demonstrates that the number of interfering users on time instance i is significantly reduced in an LDS system. Note that in this model, we consider fading effects which are frequency nonselective. For LDS-OFDM, for both selective and nonselective frequency fading, the received signal is the same as (3) if i is equal to the frequency index.

3.2 Achievable Rate Region of LDS-CDMA

Capacity of conventional MAC is analysed in [7, 10, 11, 19, 32]. In multi-user wireless communication, the channel capacity is characterized by achievable rate region, where each point in the region represents a vector of simultaneously achievable rates for the users, constrained by probabilities of bit or symbol error rate for each user [10]. The set of all such achievable rate vectors is known as the capacity region, and dubbed as the achievable rate region for LDS based multiple access in this chapter.

We shall now focus on the scenario of interest in this section, where the channels are time-varying. It is assumed that the receiver has full channel state information (CSI). The capacity region of time varying fading MAC is well known [16, 34]; it is the set of all rate vectors $\boldsymbol{R}$ calculated as follows:

$$\mathscr{C}_{MAC}(\boldsymbol{h}, \boldsymbol{P}) = \left\{ \boldsymbol{R} : \boldsymbol{R}(\mathscr{S}) \leq W\mathbb{E}_{\boldsymbol{h}}\left[\log\left(1 + \frac{\sum_{k\in\mathscr{S}} |h_k|^2 P_k}{WN_0}\right)\right], \forall \mathscr{S} \subseteq \{1, \ldots, K\} \right\}, \tag{4}$$

where $\boldsymbol{P} = [P_1, \ldots, P_K]$ and $\boldsymbol{h} = [h_1, \ldots, h_K]$ is a random vector with elements which have the stationary distribution of the joint fading processes and W is the bandwidth of the channel. In this chapter, for any vector $\boldsymbol{v}$, the notation $\boldsymbol{v}(\mathscr{S})$ is used to denote $\sum_{i\in\mathscr{S}} \boldsymbol{v}(i)$. A rigorous proof of this result can be found in [34]. An intuitive understanding of this result can be obtained by viewing capacities in terms of time averages of mutual information [16].

The achievability of the upper bound on the rate region is subject to the fact that received messages are decoded to a message in the codebook which is uniquely jointly typical [10]. For a fixed density factor ζ_k, as the overall signature length approaches infinity, the length of the full density part will also approach to infinity. We can extend the joint typicality definition to low density signatures by applying the original definition. A detailed proof is given in the Appendix.

As mentioned earlier, the spreading sequence for user k contains many zeros and some non-zero elements. Thus, the achievable rate region of LDS-CDMA under time varying fading channel can be derived by defining an equivalent LDS channel: $\tilde{h}_{k,i} = a_{k,i}h_{k,i}$ where $a_{i,k}$ denotes the activity of user k at time i, specifically $a_{k,i}$ will be equal to one if $k \in \mathscr{K}_i$ otherwise it will be zero. For notation brevity, all users are assumed to have the same density factor, i.e., $\zeta = \zeta_k, \forall k = 1, \ldots, K$. Thus, in order to derive the achievable rate region of fading LDS-CDMA we must replace the original fading MAC channel with its LDS equivalent channel in (4) as shown bellow

$$\mathscr{C}_{LDS}(\tilde{\boldsymbol{h}}, \tilde{\boldsymbol{P}}) = \left\{ \boldsymbol{R} : \boldsymbol{R}(S) \leq W\mathbb{E}_{\tilde{\boldsymbol{a}}}\left[\log\left(1 + \frac{\sum_{k\in\mathscr{S}} \left|\tilde{h}_k\right|^2 \tilde{P}_k}{WN_0}\right)\right], \forall \mathscr{S} \subseteq \{1, \ldots, K\} \right\}. \tag{5}$$

where $\tilde{P}_k : P_k/\zeta$, as for a proportion of time the users don't send data, thus, their transmit power denoted by $\tilde{P}_k$ will be larger than their original transmit power.

More specifically, for a fixed permutation pattern and thus a fixed low density structure, the derived achievable rate region will be

$$\mathscr{C}_{LDS}(\boldsymbol{h},\tilde{\boldsymbol{P}}) = \left\{\boldsymbol{R}:\boldsymbol{R}(\mathscr{S}) \le \frac{1}{N}\sum_{i=1}^{N} W\mathbb{E}_{\boldsymbol{h}}\left[\log\left(1+\frac{\sum_{k\in\mathscr{S}} a_{k,i}\,|h_k|^2\,\tilde{P}_k}{WN_0}\right)\right], \forall\mathscr{S}\subseteq\{1,\ldots,K\}\right\}. \tag{6}$$

Now if we take into account the ensemble of LDS scheme over all possible permutation patterns, the activity parameters will be independent and follow Bernoulli distribution with probability of being 1, given as ζ. Therefore, the achievable rate region will be

$$\mathscr{C}_{LDS}(\boldsymbol{a},\boldsymbol{h},\boldsymbol{P}) = \left\{\boldsymbol{R}:\boldsymbol{R}(\mathscr{S}) \le W\mathbb{E}_{\boldsymbol{a},\boldsymbol{h}}\left[\log\left(1+\frac{\sum_{k\in\mathscr{S}} a_k\,|h_k|^2\,P_k/\zeta}{WN_0}\right)\right], \forall\mathscr{S}\subseteq\{1,\ldots,K\}\right\}, \tag{7}$$

where $\boldsymbol{a} = [a_1,\ldots,a_K] \in \{0,1\}^K$ is a vector that contains Bernoulli components with probability ζ. It can be noted that the derived model allows any number of users to be active in a given symbol because of the Bernoulli distribution of the activity of the users. In this regard, we call the ensemble of the produced low density signatures as $\mathrm{LDS}_K(\zeta,K)$. If the number of active users is limited, it becomes less than a specific number $\xi \le K$, the generated LDS will be called $\mathrm{LDS}_K(\zeta,\xi)$. In this case, the achievable rate region still will have the same form as (15), but the distribution for the elements of activity vector $\boldsymbol{a}$ must be modified. The original independent and identically distributed Bernoulli distribution of the element of the activity vector imply

$$p(\boldsymbol{a};\zeta,K) = \prod_{k=1}^{K} \zeta^{a_k}(1-\zeta)^{1-a_k}. \tag{8}$$

On the other hand, the distribution for $\mathrm{LDS_K}(\zeta,\xi)$ can be expressed in the following form:

$$p(\boldsymbol{a};\zeta,\xi) = \begin{cases}\beta\prod_{k=1}^{K}\zeta^{a_k}(1-\zeta)^{1-a_k}, & \sum_{k=1}^{K} a_k \le \xi\\ 0, & \text{otherwise}\end{cases} \tag{9}$$

where β is a normalizing factor and is expressed as follows:

$$\beta = \left(\sum_{d=0}^{\xi} \binom{K}{d} \zeta^d (1-\zeta)^{K-d} \right)^{-1}. \tag{10}$$

It should be noted that the channel resources will be unused if there exist a chip with no active user in our signature. Therefore, it is necessary to further restrict the LDS ensemble to those that have at least one user active at a given time instance. Inspired by the schemes used for expurgating the codes and the union bound for error probability [4, 6], we expurgated the condition where no user is active at a chip or time instance. This expurgated scheme will be called $\mathrm{LDS}_K(\zeta, \xi^+)$ with the activity vector distribution

$$p(\boldsymbol{a}; \zeta, \xi^+) = \begin{cases} \beta^+ \prod_{k=1}^{K} \zeta^{a_k} (1-\zeta)^{1-a_k}, & 0 < \sum_{k=1}^{K} a_k \leq \xi \\ 0, & \text{otherwise} \end{cases} \tag{11}$$

where

$$\beta^+ = \left(\sum_{d=1}^{\xi} \binom{K}{d} \zeta^d (1-\zeta)^{K-d} \right)^{-1}. \tag{12}$$

By introducing the constraint of maximum number of active users at a specific time instance, the probability of having value 1 in each element of activity vector, $\hat{\zeta}$, will diverge from its initial value, ζ, according to number of active users d. $\hat{\zeta}$ can be calculated as follows:

$$\hat{\zeta} = \sum_{d=1}^{\xi} p(d)\, \hat{\zeta}(d), \tag{13}$$

$$p(d) = \binom{K}{d} (1-\zeta)^{K-d} \zeta^d \beta^+ \quad \& \quad \hat{\zeta}(d) = \frac{d}{K}, \tag{14}$$

$$\Rightarrow \quad \hat{\zeta} = \sum_{d=1}^{\xi} \binom{K}{d} \zeta^d (1-\zeta)^{K-d} \beta^+ \frac{d}{K}. \tag{15}$$

By plugging $\hat{\zeta}$ from (15) into (7) we obtain the achievable rate region of LDS-CDMA. Monte Carlo simulations can be used to compute and evaluate achievable rate of LDS-CDMA given by (7). Here, we present the simulation results for a representative scenario with 20 users. The channel bandwidth is considered to be 5 MHz. For comparison, achievable sum rate (SR) of LDS-CDMA is compared

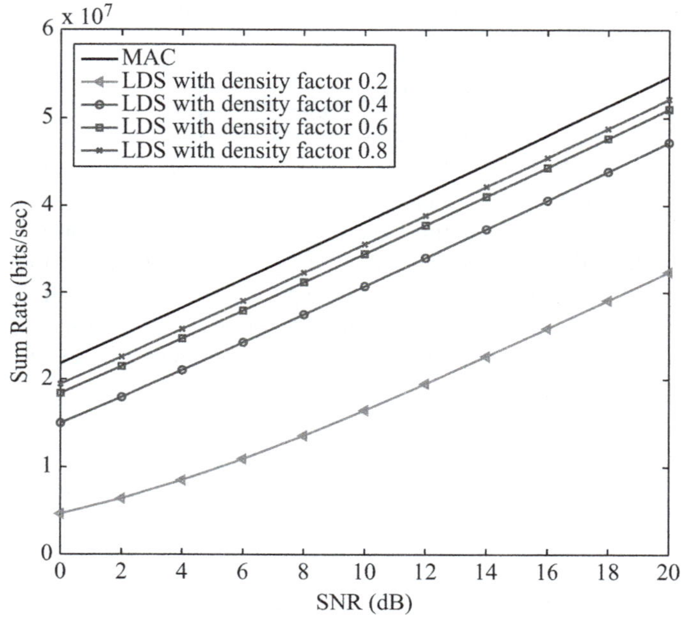

Fig. 5 Sum rate for different density factors with maximum number of allowed users per chip equal to 8

to that of conventional MAC in a similar scenario. Figures 5 and 6 show the comparison between the SR of LDS-CDMA and conventional MAC for various density factors and maximum number of users allowed per chip, respectively. As it can be seen from Fig. 5, conventional MAC outperforms LDS-CDMA in terms of achieved sum rate. This performance loss is traded for significant reduction of the complexity as it will be discussed later. These figures show that for LDS systems, increasing ζ will increase the SR and reducing the maximum number of allowed users at each time instance will also result in higher SR. Furthermore, Fig. 5 implies that there is a slightly large gap between the SR curves for density factor 0.4 and 0.2, while the gap decrease dramatically for density factors larger than 0.4. Thus, density factor 0.4 would be an optimal choice as we would loos less capacity while reducing the complexity enormously.

As shown in Fig. 7, by applying LDS structure to MAC, the complexity order of the MUD will reduce from total number of users, $\mathscr{O}\left(|\mathbb{X}|^{K}\right)$, to the maximum number of allowed users per chip, $\mathscr{O}\left(|\mathbb{X}|^{\eta_n}\right)$. Figure 7 shows that by decreasing the density factor, the computational complexity of MUD for LDS systems reduces significantly.

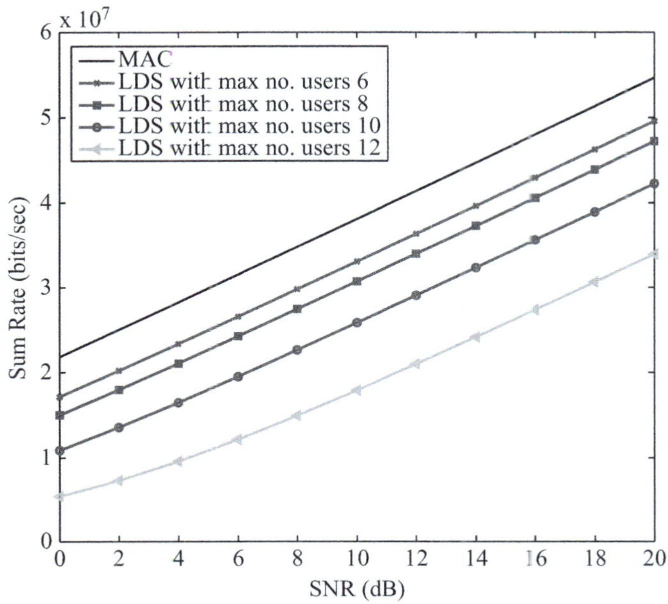

Fig. 6 Sum rate for different maximum number of allowed users per chip and density factor equal to 0.4

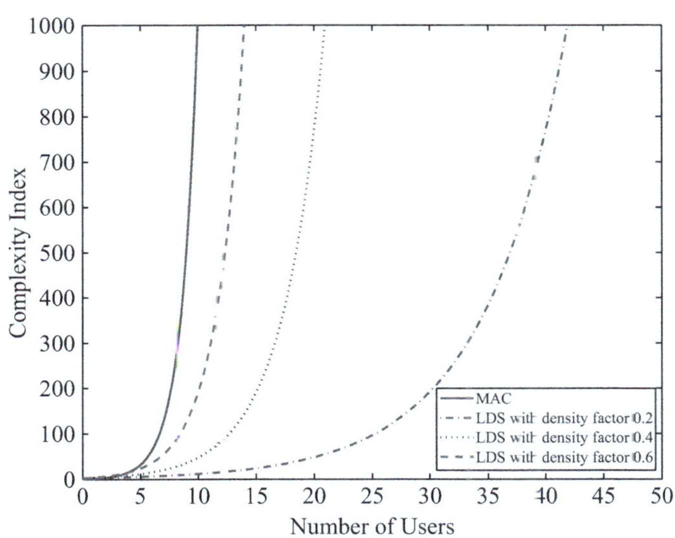

Fig. 7 Complexity of MUD for MAC in comparison with LDS with different density factors

3.3 Achievable Rate Region for LDS-OFDM

OFDM is widely used in current wireless communications systems and is a core technique for the fourth generation (4G) cellular telephony. OFDM provides a high spectral efficiency over a frequency selective channel [2]. In this regard under wideband channels applying LDS scheme to OFDM would help us avoid multi-path fading effect on the density of graph and have a better performance. In LDS-OFDM, each chip will play the role of an OFDM sub-carrier to make the system more robust in frequency selective environment.

In this section, we aim to find the achievable rate region for LDS-OFDM. To this end, we will firstly explain how LDS-OFDM differs from LDS-CDMA conceptually.

In LDS-OFDM, instead of sending data on different chips or time instances, the data is spread on orthogonal sub-carriers, thus, in LDS-OFDM we switch from time domain to frequency domain. Figure 8 shows how the two schemes are different in variables in our analysis and different textures show different users that share the same chip or frequency. This figure also shows that power allocation is not necessarily uniform; thus, a data symbol might be transmitted over its assigned sub-carriers with different power levels and with different complex phases. The analysis in previous section can be extended to LDS-OFDM by replacing time instance parameters with frequency, thus, we will have

$$\tilde{h}_k(f) = a_{k,f} h_k(f), \tag{16}$$

where $a_{k,f}$ denotes the activity of user k at sub-carrier f, in other words, $a_{k,f}$ is equal to one if user k spread its data over sub-carrier f otherwise it will be zero. Thus, the achievable rate region for LDS-OFDM is calculated by finding the expected value of the channel on frequency domain

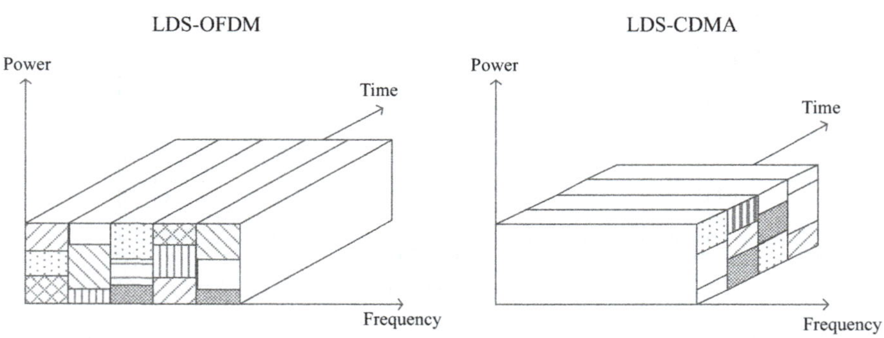

Fig. 8 LDS-OFDM vs. LDS-CDMA

$$\mathscr{C}_{LDS-OFDM}(\boldsymbol{a},\boldsymbol{h},\boldsymbol{P}) = \left\{\boldsymbol{R} : \boldsymbol{R}(\mathscr{S}) \leq W\mathbb{E}_{\boldsymbol{a},\boldsymbol{h}}\left[\log\left(1 + \frac{\sum_{k\in\mathscr{S}} a_k |h_k|^2 P_k/\zeta}{WN_0}\right)\right], \forall\mathscr{S} \subseteq \{1,\ldots,K\}\right\}. \quad (17)$$

This shows the mentioned analysis is valid for LDS-OFDM as well.

3.4 Effect of Irregularity on LDS Schemes

The uplink BER performance of a multi-user system such as LDS-CDMA can be improved significantly by careful design of low density signature. As discussed in Sect. II, an LDS system can be represented by a factor graph, where the symbols of users and chips are presented by variable nodes and function nodes respectively. The graph edges represent the connection between the received chip and the corresponding user. Therefore, this graph can be called regular if all of its variable nodes and function nodes have the same number of edges. A main factor in design of low density signature is its regularity. This analysis will help us evaluate the effect of irregular LDS structure on performance improvement and its cost in terms of complexity. Throughout this section, for ease of analysis, the LDS structures are classified as:

- Regular: When $\nu_k = \nu$, $\forall k = 1,\ldots,K$, and $\eta_n = \eta$, $\forall n = 1,\ldots N$
- Irregular: Otherwise

In order to analyse the effect of irregularity, initially the effect of number of active users on sum rate is shown in Fig. 9. The evaluation is done for a system with 20 users with SNR=20 dB and density factor equal to 0.4. The difference between the SR for different number of active users is the penalty which is paid for reducing the complexity in LDS. For irregular LDS, different number of users are active at different chips. We define

$$\alpha = W\log\left(1 + \frac{\sum_{k\in\mathscr{S}} a_k |h_k|^2 P_k/\zeta}{WN_0}\right). \quad (18)$$

Considering that SR is related to $\mathbb{E}_{\boldsymbol{a},\boldsymbol{h}}[\alpha]$, we can say that the expectation is taken over different combinations of active users. Having $\boldsymbol{a} \in \mathscr{A}$, where $\mathscr{A}$ is the set of possible combinations of active users,

$$\mathscr{A} : \left\{\boldsymbol{a} \middle| \sum_{k=1}^{K} a_k = d_1 \text{ or } \sum_{k=1}^{K} a_k = d_2 \ldots \text{ or } \sum_{k=}^{K} a_k = d_\xi\right\}. \quad (19)$$

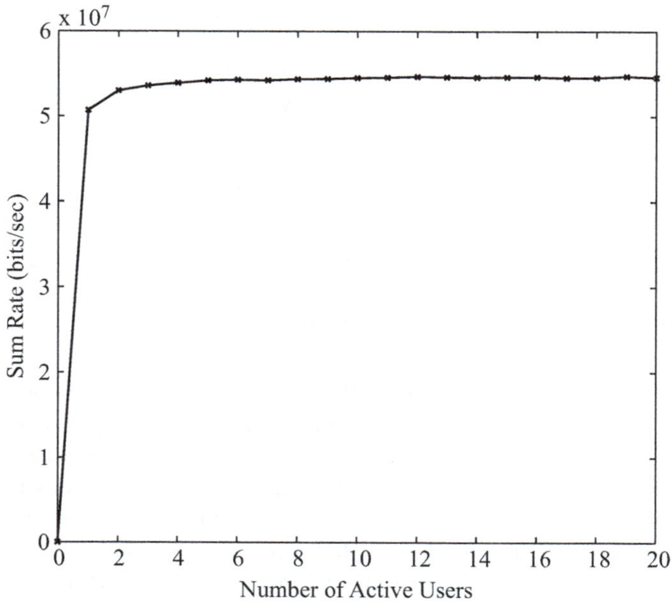

Fig. 9 Effect of number of active users on sum rate

By dividing the set $\mathscr{A}$ in ξ subsets $\mathscr{A}_1, \ldots, \mathscr{A}_\xi$ we will have

$$\mathbb{E}_{\boldsymbol{a},\boldsymbol{h}}[\alpha] = \mathbb{E}_{\boldsymbol{a}|\mathscr{A}_1,\boldsymbol{h}}[\alpha]\, p(\mathscr{A}_1) + \mathbb{E}_{\boldsymbol{a}|\mathscr{A}_2,\boldsymbol{h}}[\alpha]\, p(\mathscr{A}_2) + \ldots + \mathbb{E}_{\boldsymbol{a}|\mathscr{A}_\xi,\boldsymbol{h}}[\alpha]\, p\left(\mathscr{A}_\xi\right), \quad (20)$$

where

$$\mathscr{A}_j : \left\{\boldsymbol{a} | \sum_{k=1}^{K} a_k = d_j\right\}. \quad (21)$$

This shows that $\mathbb{E}_{\boldsymbol{a},\boldsymbol{h}}[\alpha]$ is linear with respect to $\mathbb{E}_{\boldsymbol{a}|A_j,\boldsymbol{h}}[\alpha]$. Thus, the rate of LDS scheme with different number of active users is linear with respect to the rate of LDS scheme with fixed number of active users at all chips. On the other hand, the complexity increases exponentially by the number of active users. Therefore, by combining various number of active users (irregular LDS) we will have more complexity for a fixed SR, because the SR changes linearly while complexity changes exponentially. For instance if we consider a system with number of active users 4 and 6, the SR will be similar to a regular system with 5 active users (due to linearity) while the complexity is more as it is shown in Fig. 10. Therefore, regular LDS design introduces less complexity while keeping the SR the same.

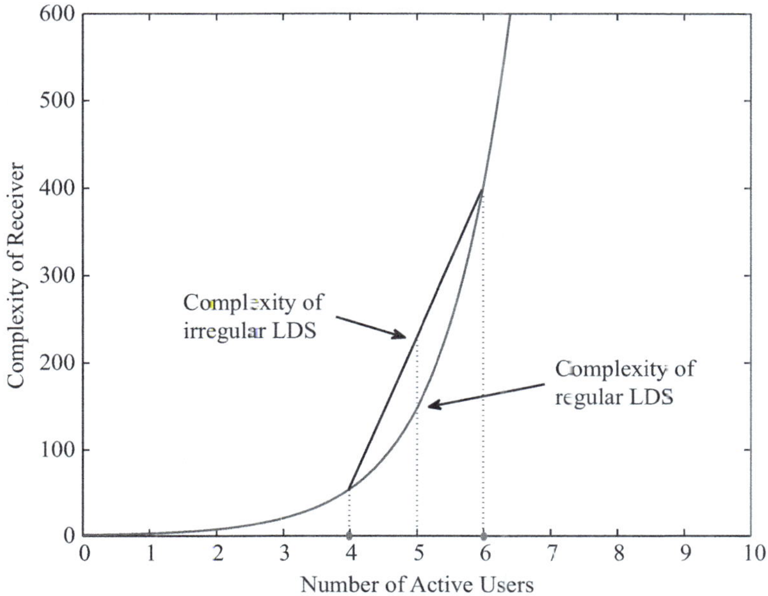

Fig. 10 Complexity of irregular LDS in comparison with similar regular LDS

3.5 *Analysis of Achievable Region of LDS System with Practical Modulation and Coding Schemes*

We analysed a system with infinite constellation space size. However, practical systems employ a finite constellation size. To analyse a practical system with finite modulation size, in this section we extend our analysis to a system with Quadrature Amplitude Modulation (QAM) modulation and convolutional channel coding.

3.5.1 Achievable Rate Region of QAM LDS System

The SNR gap approximation, first introduces in [9], is an effective tool to model the achievable rates when a suboptimal transmission scheme is used for a given probability of symbol error, P_e. Thus, using the gap approximation the rate achieved by a discrete constellation over an AWGN channel can be expressed in a form similar to Shannon's capacity formula. In this chapter, for finding the effect of modulation on the achievable rate of LDS schemes we introduce the SNR gap concept. To achieve a certain data rate for a given probability of error with modulation, a value of SNR greater than that suggested by Shannon's capacity is required. This extra amount of power is called the SNR gap, denoted in this chapter by Γ_{mod}. The SNR gap is widely studied for single user case under AWGN channel for different modulation types by considering the error probability relation with modulation order and SNR. Since LDS scheme is a multi-user scenario, we are interested in the analysis of the SNR gap for multi-user case, provided in [15]. On the other hand for the case of fading channel with non-adaptive transmission (constant transmit power and rate), the average BER is obtained by integrating the BER in AWGN over the fading distribution [17]. Considering that in previous section, for calculating the achievable rate the average was taken over fading process, the calculation of achievable rate of modulated LDS system can be performed for fading channel straightforwardly. The sum rate for the LDS system where users apply the same modulation constellation, $\mathscr{C}_{M-LDS}$, can be expressed as follows:

$$\mathscr{C}_{M-LDS}(\boldsymbol{a}, \boldsymbol{h}, \boldsymbol{P}) = \left\{ \boldsymbol{R} : \boldsymbol{R}(S) \leq W\mathbb{E}_{\boldsymbol{a},\boldsymbol{h}}\left[\log\left(1 + \frac{\sum_{k\in S} a_k \left|h_k\right|^2 P_k/\zeta}{\Gamma_{\text{mod}}\, W N_0}\right)\right], \forall S \subseteq \{1, \ldots, K\} \right\}. \tag{22}$$

For a system with M-QAM modulation Γ_{mod} is calculated as follows [38]:

$$\Gamma_{\text{mod}}\left(P_e, M\right) = \left[Q^{-1}\left(\frac{M P_e}{2\left(M-1\right)}\right)\right]^2 /3. \tag{23}$$

In order to have a fair comparison for different modulation orders we need to consider the E_b/N_0 gap, $\Gamma^{'}_{\text{mod}}$, which is calculated as follows [3]:

$$\Gamma^{'}_{\text{mod}}\left(P_b, M\right) = \left[Q^{-1}\left(\frac{P_b \log_2 M}{4\left(1 - 2^{\frac{\log_2 M}{2}}\right)}\right)\right]^2 /3. \tag{24}$$

By plugging $\Gamma^{'}$ from (32) into (30) we obtain the achievable rate region of modulated LDS MAC. The relation between SR and E_b/N_0 is presented in Fig. 11 for an LDS system with ζ equal to 0.8 and maximum of 8 users allowed per chip. Furthermore, this figure shows that the comparison of different modulation orders; 64-QAM, 16-QAM, and QPSK (quadrature phase shift keying) for $P_b = 10^{-6}$. Figure 11 shows the SR is reduced 10 Mbits/sec at $E_b/N_0 = 14$ for QPSK modulation, thus, even at reasonably good E_b/N_0, the modulated scheme is about 6 dB, short of achievable rate. Also it is shown higher order modulations leads to better SR which matches the results given in [14].

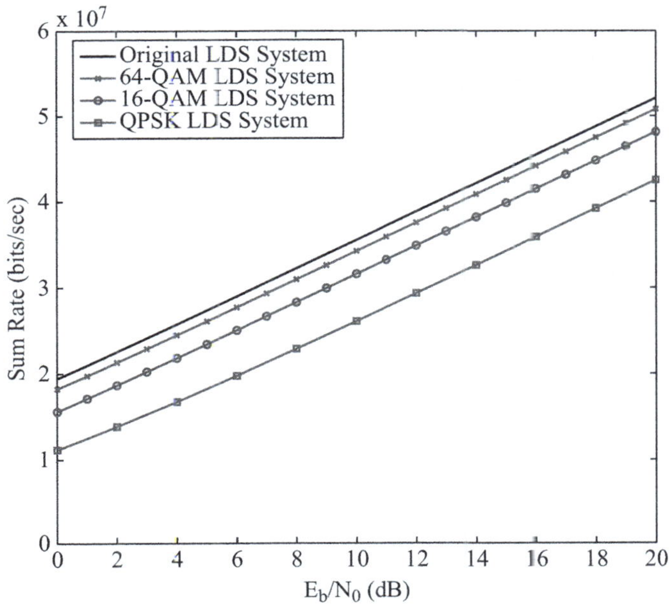

Fig. 11 Effect of modulation order on sum rate of LDS system

3.5.2 Achievable Rate Region of Convolutional Coded LDS System

Having introduced commonly used signal constellations for data transmission of LDS scheme, several performance measures are used to compare the coded systems based on these constellations. In order to take into account the effect of coding in the achievable rate region, coding gain is defined [9]. Coding gain, γ, specifies the improvement of one constellation over another when used to transmit the same information. Once the gain with respect to un-coded transmission for a given coding method is known, the gap becomes

$$\Gamma_{\text{cod}} = \Gamma_{\text{mod}} - 10\log_2 \gamma_{eff}, \tag{25}$$

where γ_{eff} is the effective coding gain. For a rate k/n binary linear convolutional code C with free distance d, the nominal coding gain can be calculated as follows [14]:

$$\gamma(C) = \left(\frac{k}{n}\right) d. \tag{26}$$

To estimate from $\gamma(C)$ the effective coding gain $\gamma_{eff}(C)$, we use the normalized error coefficient $K(C) = K_d/k$, and apply the rule introduced in [14]. Here, K_d is the number of code sequences with minimum weight d in a given k-input

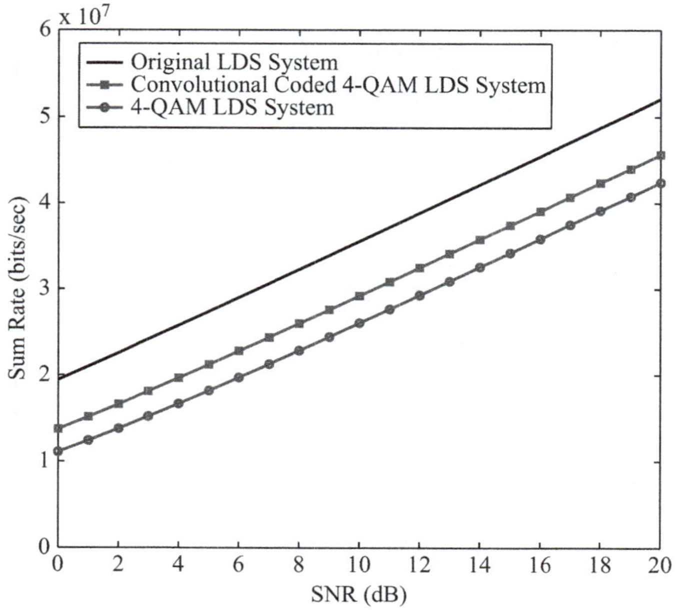

Fig. 12 Effect of coding on sum rate of LDS system

n-output block. We should emphasis that estimated effective coding gain assume soft decisions, maximum likelihood decoding.

Figure 12 shows the effect of coding on the sum rate of LDS with density factor 0.8 and maximum number of allowed user equal to 8. The code type is a half rate convolutional code with $d = 5$ and K_d equal to 1, thus, the nominal coding gain is $\gamma(C) = 5/2$ (4 dB), and the effective coding gain is also 4 dB. This figure illustrates that this simple convolutional code was able to increase the SR curve 4 Mbits/sec at SNR equal to 16 dB.

3.6 Frequency-Selective Fading Channels

In the previous sections we have analysed a flat fading model which is appropriate if the Nyquist sampling period is large compared to the delay spread of the multi-paths in the received signal, so that the individual paths are not resolvable in the sampled system. For wideband applications, the multi-paths can be resolved, and hence channel has memory. The appropriate model in this case is the time-varying frequency-selective fading channel. In this section, we will extend our previous results for LDS-CDMA to this model. For LDS-OFDM, frequency-selective fading channel is converted to parallel flat-fading channels in frequency domain, thus, the model remains the same for frequency selective fading channels.

First the signal model for LDS-CDMA under frequency selective fading channel must be driven. The baseband signal model for conventional MAC is well known [17]. Calling the notations given in Sect. 3.1.3, the signal model for LDS-CDMA is shown for a channel with l paths (which results in l taps in the impulse respond for fading multi-path channel),

$$y_i = \sum_{k \in \mathscr{K}_i} \sum_{l} h_{k,i,l} x_{k,i-l} + v_i. \tag{27}$$

We denote $h_{k,i,l}$ as the lth channel filter tap for user k at time instance i and $x_{k,i} = b_k s_{k,i}$ similar to the expression given in Sect. 3.1.3. In order to analyse the achievable rate region of this channel, when both the transmitters and the receiver can track the channel, we will first focus on the case where the channel is time-invariant. Therefore, we will have

$$y_i = \sum_{k \in \mathscr{K}_i} \sum_{l} h_{k,l} x_{k,i-l} + v_i. \tag{28}$$

Considering that this channel has the characteristics of a Gaussian multiple access channel with ISI, the achievable rate region of this channel for conventional MAC is given in [7]. For LDS-CDMA as discussed earlier the channel is multiplied by the activity parameters which are independent and follow Bernoulli distribution. Considering that the distribution of activity parameters will remain Bernoulli after the Fourier transform, for LDS-CDMA the achievable rate region is

$$\mathscr{C}_{LDS}(\boldsymbol{a}, \boldsymbol{P}) = \Bigg\{ \boldsymbol{R} : \boldsymbol{R}(\mathscr{S}) \leqslant$$
$$\mathbb{E}_{\boldsymbol{a}} \left[\int_{-\frac{W}{2}}^{\frac{W}{2}} \log \left(1 + \frac{\sum_{k \in \mathscr{S}} a_k \left| \hat{h}_k(f) \right|^2 P_k / \zeta}{W N_0} \right) df \right] \quad \forall \mathscr{S} \subset \{1, \ldots, K\} \Bigg\}. \tag{29}$$

where $\hat{h}_k(f)$ is the Fourier transform of the channel. Next the general case where the channel is time-varying is analysed. To do so we assume that the time variations of the channel are random and ergodic and the channel varies very slowly compared to the multi-path delay spread. The achievable rate region related to such channel for a convectional MAC is given in [16]. For LDS-CDMA after applying the concept of activity parameters the achievable rate region will be

$$\mathscr{C}_{LDS}(\boldsymbol{a}, \hat{\boldsymbol{h}}, \boldsymbol{P}) = \Bigg\{ \boldsymbol{R} : \boldsymbol{R}(\mathscr{S}) \leqslant$$
$$\mathbb{E}_{\boldsymbol{a}, \hat{\boldsymbol{h}}} \left[\int_{-\frac{W}{2}}^{\frac{W}{2}} \log \left(1 + \frac{\sum_{k \in \mathscr{S}} a_k \left| \hat{h}_k(f) \right|^2 P_k / \zeta}{W N_0} \right) df \right] \forall \mathscr{S} \subset \{1, \ldots, K\} \Bigg\}, \tag{30}$$

where $\hat{\boldsymbol{h}} = [\hat{h}_1(f), \ldots, \hat{h}_K(f)]$, thus the expectation is taken over fading states and the activity elements. For each time instance t, $\hat{h}_k(f,t)$ is the frequency response of user k at fading state t.

We use a numerical example to evaluate the validity of our assumption of slow variations of the channel in time domain compared to multi-path delay spread in practice. Consider a typical mobile scenario where the vehicle is moving at 60 km/h and the carrier frequency is 1 GHz. Time variation of the channel becomes uncorrelated every half carrier wavelength over distance [17]. Thus, the time constant associated with the fading effect is in the order of time taken for the mobile to travel one wavelength at the transmitter frequency which is 0.0185 s for this example. In indoor channels delay spread typically ranges from 10 to 1000 ns, in suburbs it ranges from 200 to 2000 ns, and in urban areas it ranges from 1 to 30 μs [17]. Hence, even for the urban areas, the time variation of the channel is significantly slower than the delay spread which indicate that the assumption is quite reasonable for typical wireless systems.

Figure 13 shows the SR for both LDS-OFDM and LDS-CDMA under frequency-selective fading channel. For simulations, ITU Vehicular B channel model is considered and all the 20 users share a total bandwidth of 60 MHz. As expected LDS-OFDM showed to have a better sum rate compared to LDS-CDMA, this is because LDS-OFDM is able to combat the negative effect of frequency-selective fading channel by dividing it into a number of flat fading sub-channels.

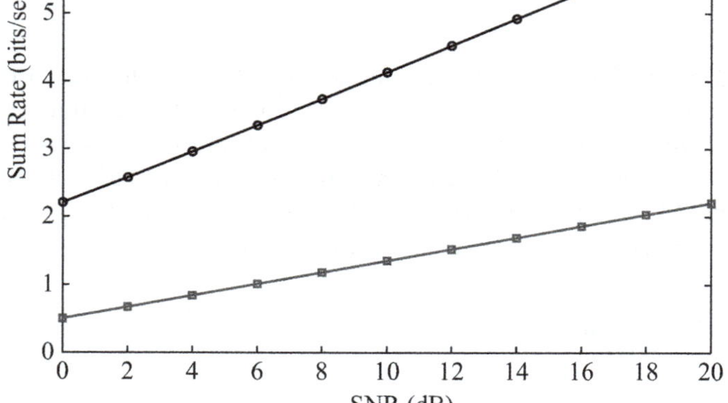

Fig. 13 Sum rate for LDS-CDMA and LDS-OFDM with maximum number of allowed users per chip/sub-carrier equal to 8

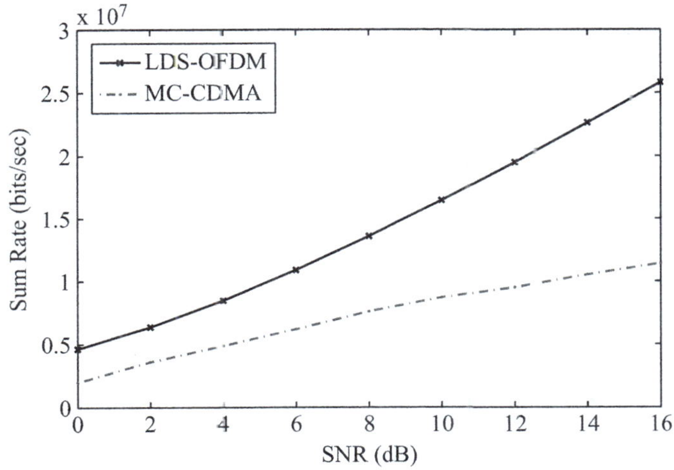

Fig. 14 Sum rate for LDS-OFDM system in comparison with MC-CDMA system with MMSE MUD

Finally, in order to provide more motivation for using an LDS-OFDM system in practice we compare the sum-rate of the LDS-OFDM system with a low-complexity multiuser detection MAC system. Therefore, Fig. 14 shows the sum rate for an LDS-OFDM system with density factor equal to 0.2 and an MC-CDMA system with minimum mean-square error (MMSE) multiuser detector. The simulation results for the MC-CDMA system with load 2.5 are produced using the derivations carried out in [40]. As can be seen LDS-OFDMA is able to achieve higher sum rate under frequency selective fading channel, and this superiority is even more at higher SNRs.

4 Summary

In this chapter, we first reviewed different state of the art non orthogonal multiple access techniques. We then rigorously analysed the performance of LDS based non-orthogonal multiple access techniques that are considered to be among promising solutions for the next generation of cellular access networks due to their well-known advantages compared to state of the art techniques based on OFDMA. In this regard the achievable rate region of LDS schemes is calculated using information theoretic analysis by introducing an effective channel gain, based on the activity of users. It is shown that LDS significantly reduces complexity of multi-user detection at the price of moderate loss of the system performance in terms of achievable sum rate. Furthermore, the effect of LDS parameters such as density factor and maximum number of users at each time instance on the sum rate is evaluated. The results show that the achievable rate approaches to that of MAC when increasing the density factor towards one. The effect of irregularity on the complexity has also been evaluated, which leads to the conclusion that regular signatures are better in terms of complexity. Furthermore, we have extended our model by taking into account practical system with a finite constellation size; the effect of finite size modulation was introduced into the our model using SNR gap. Furthermore, it was shown that the loss of achievable rate which had been caused by modulation can be compensated by using a suitable channel coding scheme. Finally, comparing with the MC-CDMA system with an MMSE MUD we showed that LDS-OFDM is able to achieve higher sum rate under frequency selective fading channel, and this superiority is even more at higher SNRs.

Appendix: Joint Typicality for Low Density Signatures

Here we explain how the joint typicality definition is extended to low density signatures (codewords). The signatures can be consider as codewords because in theory coding is a scheme that maps a block of symbols to another block of variables which can be generated randomly. Therefore, considering that LDS scheme also maps a symbol to a block of variables which are generated randomly we can say that the signature matrix in LDS scheme plays the role of a codebook (please note that term coding here doesn't mean channel coding). First we assume that low density codebook is generated randomly using a probability distribution with high probability of random variable being equal to zero. Therefore, $(X_1, X_2, \ldots, X_k)$ denote a finite collection of discrete random variables with some fixed joint distribution. Let $\mathscr{B}$ denote an ordered subset of these random variables and consider N independent copies of $\mathscr{B}$ denoted by $\mathscr{B}_1, \mathscr{B}_2, \ldots, \mathscr{B}_N$. As before, let $\mathscr{C}_k$ denote the set of time instances over which user k spreads its data, i.e., $\mathscr{C}_k = \{i|k \in \mathscr{K}_i\}$. We further define $\mathscr{D}(\mathscr{B})$ as the ordered set of time indices over which users related to the members of $\mathscr{B}$ are active, i.e., $\mathscr{D}(\mathscr{B}) = \bigcup_{k\in\hat{\mathscr{B}}} \mathscr{C}_k$, where $\hat{\mathscr{B}}$ is the ordered set of user indices that their random variables are a member of $\mathscr{B}$. Considering that each of the copies is independent of the other subsets, the joint probability mass function of the subsets will be

$$p(\mathscr{B}_1, \mathscr{B}_2, \ldots, \mathscr{B}_N) = \prod_{i\in\mathscr{D}(\mathscr{B})} p(\mathscr{B}_i), \tag{31}$$

where $\mathscr{D}(\mathscr{B})$ has at least ν_k members, let N' be the size of $\mathscr{D}(\mathscr{B})$. Considering that $\zeta_k = \frac{\nu_k}{N}$, for a fixed density factor, ζ_k, as the overall signature length, N, approaches to infinity, the length of full density portion, ν_k, will also approach to infinity. Thus, for sufficiently large N, N' will approach infinity as well. Therefore, by the law of large numbers for sufficiently large N, for any subset $\mathscr{B}$ of random variables,

$$-\frac{1}{N'} \log p(\mathscr{B}_1, \mathscr{B}_2, \ldots, \mathscr{B}_N) = -\frac{1}{N'} \sum_{i\in\mathscr{D}(\mathscr{B})} \log p(\mathscr{B}_i) \to H(\mathscr{B}), \tag{32}$$

where the convergence takes place with probability 1 for all 2^k subsets $\mathscr{B} \subseteq \{X^{(1)}, \ldots, X^{(k)}\}$.

Definition. The set $A_\epsilon^{(N)}$ of jointly typical N-sequences $(\mathbf{x}_1 \ldots, \mathbf{x}_k)$ with empirical entropies ϵ-close to the true entropies is defined by [10]

$$A_\epsilon^{(N)}(X^{(1)}, \ldots, X^{(k)}) = A_\epsilon^{(N)} =$$
$$\left\{ (\mathbf{x}_1, \ldots, \mathbf{x}_k) : \left| -\frac{1}{N} \log p(\boldsymbol{b}) - H(\mathscr{B}) \right| < \epsilon, \forall \mathscr{B} \subseteq \{X^{(1)}, \ldots, X^{(k)}\} \right\}.$$

Furthermore, let $A_\epsilon^{(N)}(\mathscr{B})$ denote the restriction of $A_\epsilon^{(N)}$ to the coordinates of $\mathscr{B}$. For low density codes we can write

$$A_\epsilon^{(N)} = \left\{ (\mathbf{x}_1, \ldots, \mathbf{x}_k) : \left| -\frac{1}{N'} \log p(\boldsymbol{b}) - H(\mathscr{B}) \right| < \epsilon, \forall \mathscr{B} \subseteq \{X^{(1)}, \ldots, X^{(k)}\} \right\}. \tag{33}$$

Lemma. *For any $\epsilon > 0$, for sufficiently large N, the probability of having jointly typical sequences is*

$$\Pr(A_\epsilon^{(N)}(\mathscr{B})) \geq 1 - \epsilon, \qquad \forall \mathscr{B} \subseteq \{X^{(1)}, \ldots, X^{(k)}\} \tag{34}$$

Proof. This follows from the law of large numbers for random variables in the definition of $A_\epsilon^{(N)}(\mathscr{B})$. Calling (6), given $\epsilon > 0$, there exist N_1 such that for all $N > N_1$,

$$\Pr\left(\left| -\frac{1}{N'} \log p(\mathscr{B}_1, \mathscr{B}_2, \ldots, \mathscr{B}_N) - H(\mathscr{B}) \right| \geq \epsilon \right) < \epsilon, \qquad \forall \mathscr{B} \subseteq \{X^{(1)}, \ldots, X^{(k)}\} \tag{35}$$

$$\Rightarrow \Pr\left(\left| -\frac{1}{N'} \log p(\boldsymbol{b}) - H(\mathscr{B}) \right| \geq \epsilon \right) < \epsilon, \qquad \forall \mathscr{B} \subseteq \{X^{(1)}, \ldots, X^{(k)}\} \tag{36}$$

Thus, for sufficiently large N for low density codes we will have,

$$\Pr(A_\epsilon^{(N)}(\mathscr{B})) \geq 1 - \epsilon, \qquad \forall \mathscr{B} \subseteq \{X^{(1)}, \ldots, X^{(k)}\} \tag{37}$$

Therefore, we could validate the joint typicality concept for low density signatures.

References

1. K. Au, Zh. Liqing, H. Nikopour et al., Uplink contention based SCMA for 5G radio access, in *Globecom Workshops (GC Wkshps)*, 2014. doi:10.1109/GLOCOMW.2014.7063547
2. P. Baracca, S. Tomasin, L. Vangelista et al., Per sub-block equalization of very long OFDM blocks in mobile communications. IEEE Trans. Commun. (2011). doi:10.1109/TCOMM.2011.121410.090252
3. J.R. Barry, E.A. Lee, D.E. Messerschmit, *Digital Communication* (Kluwer, Dordrecht, 2004)
4. E. Biglieri, G. Caire, G. Taricco, Expurgating the union bound to error probability: a generalization of the Verdu-Shields theorem, in *Proceedings of 1997 IEEE International Symposium on Information Theory*, 1997. doi:10.1109/ISIT.1997.613310
5. E. Biglieri, J. Proakis, S. Shamai, Fading channels: information-theoretic and communications aspects. IEEE Trans. Inf. Theory (1998). doi:10.1109/18.720551
6. G. Caire, G. Taricco, E. Biglieri, Bit-interleaved coded modulation. IEEE Trans. Inf. Theory (1998). doi:10.1109/18.669123
7. R.S. Cheng, S. Verdu, Gaussian multi-access channels with ISI: capacity region and multi-user water-filling. IEEE Trans. Inf. Theory (1993). doi:10.1109/18.256487
8. A. Chouly, A. Brajal, S. Jourdan, Orthogonal multicarrier techniques applied to direct sequence spread spectrum CDMA systems, in *Proceedings of IEEE Global Telecommunications Conference, including a Communications Theory Mini-Conference. IEEE Technical Program Conference Record, Houston. GLOBECOM '93*, 1993. doi:10.1109/GLOCOM.1993.318364
9. J. Cioffi, *Course notes for Digital Communication: Signal Processing* (Stanford Bookstore Custom Publishing, Stanford, CA, 2007)
10. T.M. Cover, *Elements of Information Theory* (Wiley, New York, 2006)
11. T. Cover, R. McEliece, E. Posner, Asynchronous multiple-access channel capacity. IEEE Trans. Inf. Theory (1981). doi:10.1109/TIT.1981.1056382
12. P. Fan, L. Hao, Generalized orthogonal sequences and their applications in synchronous CDMA systems. IEICE Trans. Fundam. **E83–A**, 2054–2069 (2000)
13. K. Fazel, Performance of CDMA/OFDM for mobile communication system, in *Proceedings of 2nd International Conference on Universal Personal Communications: Gateway to the 21st Century*, 1993. doi:10.1109/ICUPC.1993.528524
14. J. Forney, G. Ungerboeck, Modulation and coding for linear Gaussian channels. IEEE Trans. Inf. Theory (1998). doi:10.1109/18.720542
15. C.H.F. Fung, W. Yu, T.J. Lim, Precoding for the multiantenna downlink: multiuser SNR gap and optimal user ordering. IEEE Trans. Commun. (2007). doi:10.1109/TCOMM.2006.885095
16. R. Gallager, An inequality on the capacity region of multi-access fading, in *Channels Communications and Cryptography—Two Sides of One Tapestry* (1994), pp. 129–139
17. A. Goldsmith, *Wireless Communication* (Cambridge University Press, Cambridge, 2005)
18. S. Hara, R. Prasad, Overview of multicarrier CDMA. IEEE Commun. Mag. (1997). doi:10.1109/ISSSTA.1996.563752
19. Ch. Hon-Fah, M. Motani, Capacity region of the asynchronous gaussian vector multiple-access channel. IEEE Trans. Inf. Theory (2013). doi:10.1109/TIT.2013.2272012
20. R. Hoshyar, F.P. Wathan, R. Tafazolli, Novel low-density signature for synchronous CDMA systems over AWGN channel. IEEE Trans. Signal Process. (2008). doi:10.1109/TSP.2007.909320
21. R. Hoshyar, R. Razavi, M. Al-Imari, LDS-OFDM an efficient multiple access technique, in *Proceedings of IEEE 71st Vehicular Technology Conference (VTC 2010-Spring)*, 2010. doi:10.1109/VETECS.2010.5493941
22. D. Li, A high spectrum efficient multiple access code. Communications (1999). APCC/OECC '99. doi:10.1109/APCC.1999.824954
23. L. Liu, L.K. Leung, L. Ping, Simple iterative chip-by-chip multiuser detection for CDMA systems, in *Proceedings of VTC 2003-Spring Vehicular Technology Conference. The 57th IEEE Semiannual*, 2003. doi:10.1109/VETECS.2003.1207209

24. L. Liu, J. Tong, L. Ping, Analysis and optimization of CDMA systems with chip-level interleavers. IEEE J. Sel. Areas Commun. (2006). doi:10.1109/JSAC.2005.858896
25. M.P. Lotter, P. Van Rooyen, An overview of space division multiple access techniques in cellular systems, in *Proceedings of South African Symposium Communications and Signal Processing COMSIG '98*, 1998. doi:10.1109/COMSIG.1998.736941
26. R.V. Nee, R. Prasad, *OFDM for Wireless Multimedia Communications* (Artech House, Boston, 2000)
27. H. Nikopour, E. Yi, A. Bayesteh et al., SCMA for downlink multiple access of 5G wireless networks, in *IEEE Global Communications Conference (GLOBECOM)* (2014). doi:10.1109/GLOCOM.2014.7037423
28. A. Persson, T. Ottosson, E. Strom, Time-frequency localized CDMA for downlink multi-carrier systems, in *Proceedings of IEEE Seventh International Spread Spectrum Techniques and Applications Symposium*, 2002. doi:10.1109/ISSSTA.2002.1049298
29. L. Ping, L. Liu, K. Wu, W.K. Leung, Interleave division multiple-access. IEEE Trans. Wireless Commun. (2006). doi:10.1109/TWC.2006.1618943
30. T.S. Rappaport, *Wireless Communications: Principles and Practice* (Prentice Hall, Englewood Cliffs, NJ, 2007)
31. R. Razavi, M. Al-Imari, M.A. Imran, On receiver design for uplink low density signature OFDM (LDS-OFDM). IEEE Trans. Commun. (2012). doi:10.1109/TCOMM.2012.082812.110284
32. Z. Rezki, M.S. Alouini, On the capacity of multiple access and broadcast fading channels with full channel state information at low SNR. IEEE Trans. Wireless Commun. (2014). doi:10.1109/TWC.2013.120113.130895
33. E. Schulz, Forward 2020 5G (2015). http://cwbackoffice.co.uk/Presentation/RTSS
34. S. Shamai, A.D. Wyner, Information-theoretic considerations for symmetric, cellular, multiple-access fading channels. ii. IEEE Trans. Inf. Theory (1997). doi:10.1109/18.641554
35. S. Stanczak, H. Boche, M. Haardt, Are LAS-codes a miracle? in *Proceedings of IEEE Global Telecommunications Conference GLOBECOM '01*, 2001. doi:10.1109/GLOCOM.2001.965185
36. B. Suard, G. Xu, H. Liu, T. Kailath, Uplink channel capacity of space-division-multiple-access schemes. IEEE Trans. Inf. Theory (1998). doi:10.1109/18.681322
37. M. Taherzadeh, H. Nikopour, A. Bayesteh, H. Baligh, SCMA codebook design, in *IEEE 80th Vehicular Technology Conference (VTC Fall)*, 2014. doi:10.1109/VTCFall.2014.6966170
38. D. Toumpakaris, J. Lee, The gap approximation for Gaussian multiple access channels, in *Proceedings of IEEE Global Telecommunications Conference GLOBECOM*, 2009. doi:10.1109/GLOCOM.2009.5425889
39. D.N.C. Tse, S.V. Hanly, Multiaccess fading channels. i. Polymatroid structure, optimal resource allocation and throughput capacities. IEEE Trans. Inf. Theory (1998). doi:10.1109/18.737513
40. A.M. Tulino, L. Li, S. Verdu, Spectral efficiency of multicarrier CDMA. IEEE Trans. Inf. Theory (2005). doi:10.1109/TIT.2004.840875
41. S. Verdu, *Multiuser Detection* (Cambridge University Press, Cambridge, 1998)
42. F. Wathan, R. Hoshyar, R. Tafazolli, Dynamic grouped chip-level iterated multiuser detection based on Gaussian forcing technique. IEEE Commun. Lett. (2008). doi:10.1109/LCOMM.2008.071931
43. H. Wei, L. Hanzo, On the uplink performance of LAS-CDMA. IEEE Trans. Wireless Commun. (2006). doi:10.1109/TWC.2006.1633372
44. C. Xiaodong, Z. Shengli, G.B. Giannakis, Group-orthogonal multicarrier CDMA. IEEE Trans. Commun. (2004). doi:10.1109/TCOMM.2003.822174
45. N. Yee, J.P. Linnartz, BER of multi-carrier CDMA in an indoor Rician fading channel, in *Proceedings of Conference Signals, Systems and Computers Record of The Twenty-Seventh Asilomar Conference*, 1993. doi:10.1109/ACSSC.1993.342548

New Multicarrier Modulations for 5G

Davide Mattera, Mario Tanda, and Maurice Bellanger

Abstract Recent advances in filter bank multicarrier (FBMC) techniques are presented and compared with the conventional cyclic prefix (CP)-OFDM approach, in the context of 5G. After a brief description of some adaptations of CP-OFDM, FBMC combined with offset-QAM is considered, pointing out the crucial issue of subchannel equalization to compensate for the absence of cyclic prefix. Implementation has to be carried out in the frequency domain, in order to reach high performance without further increasing latency and allow for per user equalization and carrier frequency offset compensation. However, due to the delay objectives and the computational complexity, the prototype filter length, and overlapping factor, must be small, which limits the performance. Then, an alternative approach is proposed, FBMC combined with pulse amplitude modulation (PAM). Like OFDM, FBMC-PAM possesses the perfect reconstruction property and, equipped with the sine prototype filter, it yields good performance with the overlapping factor $K = 2$, which makes it an attractive option whenever asynchronous access and high level of out-of-band rejection are required. Finally, the case of nonoverlapping emitted symbols is considered and a CP-less OFDM scheme with frequency domain equalizer in the receiver is included in the performance comparison.

1 Introduction

Many specific requirements for 5G are linked to the multi-user context, such as asynchronous access, coexistence issues and per-user channel equalization and carrier frequency offset compensation. In addition, latency has to be kept to a minimum. At the system level, it is necessary to maintain coherence between the

D. Mattera • M. Tanda (✉)
Dipartimento di Ingegneria Elettrica e delle Tecnologie dell'Informazione, Università degli Studi di Napoli Federico II, Via Claudio, 21, 80125 Napoli, Italy
e-mail: mattera@unina.it; tanda@unina.it

M. Bellanger
CNAM-Electronique, 292 rue Saint-Martin, 75141, Paris Cedex 03, France
e-mail: maurice.bellanger@cnam.fr

W. Xiang et al. (eds.), *5G Mobile Communications*,
DOI 10.1007/978-3-319-34208-5_7

performance of the signal processing techniques and the existing or foreseeable capabilities of radiofrequency interfaces. Obviously, these requirements have a critical impact on the selection of the physical layer techniques.

Multicarrier modulations have proved successful in cable and wireless communication systems because of their spectral efficiency and their operational flexibility and they are strong candidates for the physical layer in 5G and beyond. So far, the most widely deployed technique is cyclic-prefix orthogonal frequency division multiplexing (CP-OFDM), which is simple in concept and easy to implement. Basically, it consists of an inverse fast Fourier transform (IFFT) in the transmitter and an FFT in the receiver, whose size M defines the number of subcarriers in the system. When the blocks of M time samples generated by the IFFT are fed to the FFT, the subcarriers are orthogonal and the input data samples are perfectly reconstructed. In the presence of channel distortion, a cyclic prefix has to be introduced at the transmitter side so that orthogonality can be restored at the receiver side, through an appropriate time shift, under the condition that the length of the cyclic prefix exceeds the duration of the channel impulse response. Then, the demodulation process is completed by per subcarrier frequency domain equalization, which is the key to the spectral efficiency of the scheme.

The CP-OFDM approach is well suited to the transmission of single user over a bandwidth interval much larger than the channel coherence bandwidth. In the presence of several simultaneous users exploiting the same OFDM frame, it is necessary to make sure that time and frequency alignment is obtained with sufficient accuracy, so that the orthogonality of the signals transmitted by all the users is secured. The cost of such an alignment depends on the number of users involved in the process. In case of independent adjacent users, the global spectral efficiency is impacted by the filtering effect of the multicarrier modulations involved. The basic OFDM scheme offers limited spectral separation and a number of requirements defined by the use case families considered for the definition of the 5G standards [6] tend to make the approach unfit to the considered scenarios. Specifically, the limited filtering effect of OFDM leads to losses in spectral efficiency that can be severe in case of asynchronous access and coexistence issues, as well as in the presence of fragmented spectrum. Then, in order to keep spectral efficiency in such conditions, the filter bank multicarrier (FBMC) modulation has been proposed as an alternative to OFDM. In a filter bank, a low-pass prototype filter with KM coefficients is shifted on the frequency axis by integer multiples of the subcarrier spacing, leading to a set of M subchannels. A key difference with OFDM is that K emitted multicarrier symbols overlap, and K, the so-called overlapping factor, determines the filter attenuation characteristics and, thus, the spectral separation between adjacent users. With this approach, the cyclic-prefix concept can be abandoned and the channel distortion must be mitigated by subchannel equalization. In fact, FBMC without multitap or frequency-domain subchannel equalization can be considered similar to OFDM with a very short cyclic prefix.

Regarding implementation, two options exist, as shown in Fig. 1, for the receiver. In the time domain scheme, a set of M digital filters, having K coefficients each and called polyphase network (PPN), is cascaded with the FFT. In the frequency domain

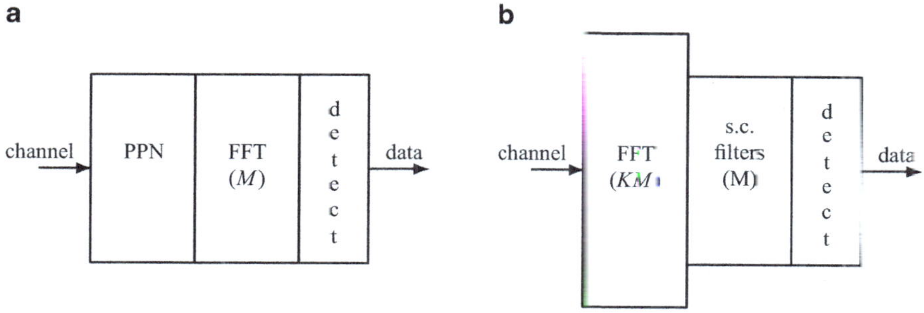

Fig. 1 Structures of filter banks in FBMC receivers: (**a**) PPN-FFT scheme, (**b**) Frequency domain scheme

scheme, an FFT of size KM is followed by a set of per-subchannel decimation filters whose coefficients are derived from the time domain coefficients through discrete Fourier transform. Of course, the frequency domain scheme requires more computations and it is practical only for small overlapping factors, typically 2 or 3. However, it has two key benefits

- channel equalization is performed per subchannel at the FFT output, which leads to high performance with no additional delay, contrary to the PPN approach which requires time domain equalization with the corresponding delay. The same applies to per-user carrier frequency offset (CFO) compensation.
- although it is performed per subchannel, the equalization at the FFT output before filtering is equivalent to the global equalization of the received signal, which means that perfect reconstruction techniques can be used and orthogonality is possible, as in OFDM.

Now, to complete the system description, data processing is considered. In OFDM systems, the data sequence is mapped into quadrature amplitude modulation (QAM) samples at the input of the transmitter IFFT. This mapping cannot be applied to FBMC, because of the filter overlapping.

The signals transmitted on the subcarriers $k-1$ and $k+1$ generate interference in the subcarrier k and there are two approaches to achieve full rate transmission in spite of this interference

- Offset-QAM modulation leading to FBMC-OQAM systems. The principle relies on the impulse responses of interference filters that cross the zero axis at multiples of the symbol period M, but with a shift of $M/2$ between the real part and the imaginary part. Then, full rate is obtained if the symbol rate is doubled and real and imaginary data samples alternate in the time domain and in the frequency domain;
- Pulse amplitude modulation (PAM) leading to FBMC-PAM systems. The phase shift $\pi/2$ is applied to the subcarriers of adjacent sub-channels and the transmitter input is a real sequence of data samples.

The poor spectral containment and the synchronization requirements in multiple access are the weak points of the classical OFDM system that make it unfit to the next generation communication systems; its simpler evolutions improve on such issues but introduce a further loss in terms of spectral efficiency. The FBMC-OQAM system achieves significant improvements over classical OFDM system in terms of spectral containment and robustness to the imperfections of the timing and CFO estimation procedures with the same spectral efficiency; such improvements are more significant for larger K. Its main drawbacks are the increase of the computational complexity and the inefficiency in short-burst transmissions, which is also an important requirement for next generation systems; both the complexity and the inefficiency increase with K. The FBMC-PAM, instead, is superior to the classical OFDM system in terms of spectral containment and synchronization requirement but it also exhibits smaller complexity than FBMC- OQAM; moreover, the inefficiency of FBMC-PAM in short-burst transmissions is marginal. Therefore, in many use cases, FBMC-PAM represents a good compromise; only in extreme scenarios, characterized by severe requirements in terms of spectral containment and spectrum fragmentation, FBMC-OQAM with $K = 3$ (or also with $K = 4$) may become necessary.

Following this brief overview of the multicarrier transmission field, the rest of the chapter is organized as follows. In Sect. 2, the classical OFDM system is presented and some of the evolutions which have been proposed to improve filtering are discussed. Section 3 is dedicated to the FBMC-OQAM system, with emphasis on the frequency domain implementation option and the performance of channel equalization and CFO compensation. Then, in Sect. 4, the FBMC-PAM system is considered in combination with perfect reconstruction filter design, so that the orthogonality property is obtained, as in OFDM. Finally, a CP-less OFDM scheme with frequency domain equalization in the receiver is discussed in Sect. 5 and, in Sect. 6, the considered systems are compared by simulation experiments.

2 The Classical OFDM and Its Evolutions

For the sake of clarity, we shortly resume the characteristics of the standard CP-OFDM; the classical rectangular pulse $g^{(O)}[n]$ is used to transmit a complex-valued symbol:

$$g^{(O)}[n] = u_{M+M_p}[n] \triangleq \begin{cases} 1 & n \in \{0, 1, \ldots, M + M_p - 1\} \\ 0 & n \notin \{0, 1, \ldots, M + M_p - 1\} \end{cases}. \quad (1)$$

The pulse transmitted in the first symbol period $\{0, 1, \ldots, M + M_p - 1\}$ on the kth subcarrier is written as

$$g_k^{(O)}[n] \triangleq g^{(O)}[n]e^{j2\pi \frac{k}{M}(n-M_p)}. \quad (2)$$

The problem of intersymbol interference is solved by using a symbol period $M + M_p$ larger than the pulse duration M; the length of the cyclic prefix M_p has to be larger than the channel delay spread M_h in order to eliminate the interference effects from previous symbol intervals on the current one; for such a reason, the first samples of the received signal, which are affected by pulse transmitting the previous symbols, are not utilized at the receiver. The interferences among the signals transmitted on the different subcarriers are determined by the coefficients:

$$I_k^{(O)}[n_k, \varepsilon_k] \triangleq \left| \sum_{n=0}^{M-1} g^{(O)}[n + M_p - n_k] \left(g^{(O)}[n + M_p]\right)^* e^{j2\pi \frac{k+\varepsilon_k}{M} n} \right|$$
$$= \left| \int_{-1/2}^{1/2} U_M(F) U_M^* \left(F + \frac{k + \varepsilon_k}{M}\right) e^{-j2\pi \bar{n}_k F} \mathrm{d}F \right| \qquad (3)$$

where $n_k = \tilde{n}_k + n_0$ and $\varepsilon_k = \tilde{\varepsilon}_k + \epsilon_0$ with $\tilde{n}_k$ and $\tilde{\varepsilon}_k$ denoting the relative timing offset and the relative normalized CFO, respectively, of the user transmitting at a distance of k subcarriers from that of interest and with n_0 and ϵ_0 denoting the estimation error of the timing offset and of the normalized CFO, respectively, of the user of interest. Moreover, for $|n_k| \leq M/2$, we have defined

$$\bar{n}_k \triangleq \begin{cases} 0 & 0 \leq n_k \leq M_p \\ n_k - M_p & n_k > M_p \\ n_k & n_k < 0 \end{cases} \qquad (4)$$

and

$$U_M(F) \triangleq \mathscr{F}\{u_M[n]\} = e^{-j\pi F(M-1)} \frac{\sin(\pi FM)}{\sin(\pi F)}. \qquad (5)$$

We here consider the case where the subcarrier signal, which is transmitted on kth subcarrier adjacent to that of interest, is possibly used by a different user (in such a case $\tilde{n}_k \neq 0$ unless a proper timing alignment procedure is employed). The well-known property $I_k^{(O)}[0, 0] = 0 \ \forall k \neq 0$ guarantees that there is no interference between two subcarriers employed by the same user (i.e., $\tilde{n}_k = \tilde{\varepsilon}_k = 0$) provided that $M_h \leq M_p$ and the timing offset and CFO of the user of interest can be perfectly estimated (i.e., $n_0 = \varepsilon_0 = 0$). This property, which holds for $0 \leq n_k \leq M_p$ (since $\bar{n}_k = 0$ for such values of n_k), has allowed the great diffusion of CP-OFDM in classical downlink scenarios. At the same time, the fact that the same property does not hold when $\bar{n}_k \neq 0$ or $\varepsilon_k \neq 0$ represents one of the main drawbacks that complicated the use of CP-OFDM in the uplink scenarios.

In order to reduce $I_k^{(O)}[n_k, \varepsilon_k]$, the spectral behavior of the chosen pulse has to be modified. In fact, (3) clearly shows that, for $|k| \geq 2$ (i.e., for all subcarriers except the two adjacent ones), the first lobes of the two pulses are separated. Such a separation would imply that $I_k^{(O)}[n_k, \varepsilon_k]$ is negligible also when $(\bar{n}_k, \varepsilon_k) \neq (0, 0)$

provided that the secondary lobes of the chosen pulse $U_M(F)$ were sufficiently small. In such a case, the relative timing offset n_k and the relative normalized CFO ε_k could be also very large without generating significant interferences from non-adjacent subcarriers. This would imply that the devices of the radio access network would not need to be precisely synchronized in time and frequency with each other as it is currently required in the classical OFDM systems. Such a requirement will tend to become more cumbersome (and, in perspective, also unfeasible) as the support of multiple traffic types, which is going to be required by the wireless radio interface of the next 5G standard, will imply the need to maintain synchronized a massive number of low-rate machine-type communication devices, along with the usual high-rate hand-held devices [19, 20].

The main disadvantage of the classical OFDM system lies, therefore, in the poor spectral containment of $U_M(F)$ outside its first lobe, i.e., outside the bandwidth $(-\frac{1}{M}, \frac{1}{M})$. The same spectral behavior of $U_M(F)$ is responsible for the other limit of OFDM: the difficulty to efficiently transmit over a small bandwidth fragment. In fact, the power spectral density (PSD) $P_{s_k}^{(\mathrm{O})}(F)$ of the signal $s_k^{(\mathrm{O})}[n]$ transmitted on the kth subcarrier

$$s_k^{(\mathrm{O})}[n] \triangleq \sqrt{\frac{E_s}{M+M_p}} \sum_{i=-\infty}^{+\infty} c_k[i] g_k^{(\mathrm{O})}[n - i\left(M + M_p\right)] \tag{6}$$

for a complex-valued symbol sequence $c_k[i]$ satisfying the condition

$$E\left[c_k[i]c_k^*[i-m]\right] = \delta[m], \tag{7}$$

where $\delta[\cdot]$ is the Kronecker delta, can be written as

$$P_{s_k}^{(\mathrm{O})}(F) = \frac{E_s}{\left(M+M_p\right)^2} \left| U_{M+M_p}\left(F - \frac{k}{M}\right)\right|^2 = E_s \frac{\sin^2\left(\pi\left(F-\frac{k}{M}\right)\left(M+M_p\right)\right)}{\left(M+M_p\right)^2 \sin^2\left(\pi\left(F-\frac{k}{M}\right)\right)} \tag{8}$$

and, assuming the independence of the symbol sequences used on the different subcarriers, the PSD of the signal transmitted by each user can be written as $\sum_{k\in\mathscr{A}} P_{s_k}^{(\mathrm{O})}(F)$ where $\mathscr{A}$ is the set of subcarriers exploited by the considered user. Here we have denoted with E_s the energy transmitted for each complex-valued symbol.

Let us show in detail with an example the problem created by the poor spectral containment of the pulse used in CP-OFDM in the transmission over a bandwidth fragment. Assume that the bandwidth available for transmission is the numerical bandwidth $[0.2, 0.4]$ and let us choose the following parameters $M = 256$ and $M_p = M/8$; if there were no constraint on the PSD outside the available bandwidth, the set $\{53, 54, \ldots, 101\}$ of subcarriers could be used by the transmitter and the resulting PSD is depicted in Fig. 2. However, if we introduce the constraint that the PSD outside the chosen bandwidth has to be -30 dB below the maximum value of the PSD, which is a realistic constraint when we resort to a fragmented usage of the spectrum, we have to note that the standard CP-OFDM cannot be employed.

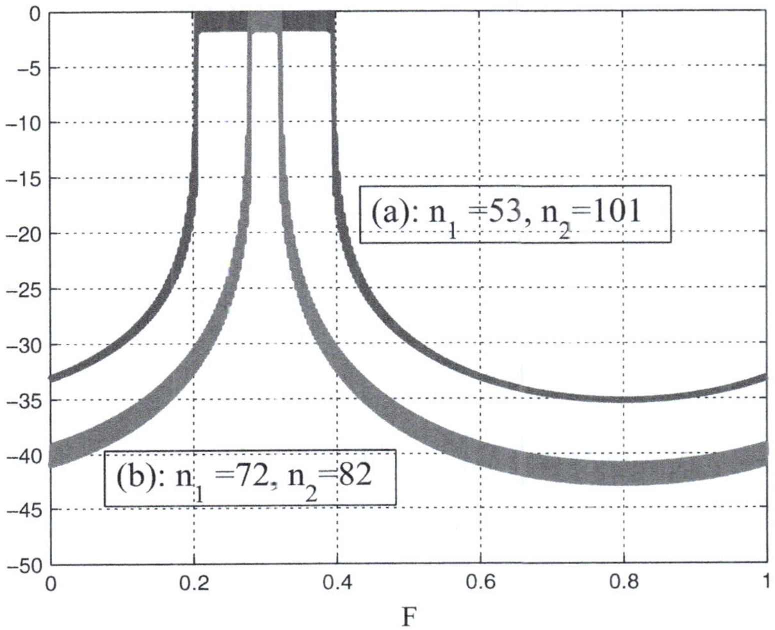

Fig. 2 The PSD of the transmitted signal, defined as $\frac{1}{E_s}\sum_{k=n_1}^{n_2} F_{s_k}^{(O)}(F)$, when *(a)* all the subcarriers inside the bandwidth fragment (0.2, 0.4) are employed and *(b)* when the minimum number of guard subcarriers needed to satisfy the constraint on PSD outside the considered fragment at -30 dB are used

The first basic line to circumvent the problem of spectral containment is the well-known practice of using the null guard subcarriers. With reference to the previous example, we can transmit only on the subcarriers in the following set $\{72, 73, \ldots, 82\}$; the resulting PSD is also reported in Fig. 2. If we had specified the slightly stronger constraint of -35 dB outside the chosen bandwidth, the set of the subcarriers that could be used would be further reduced to the following set: $\{78, 79\}$; more selective constraints cannot be satisfied by CP-OFDM by resorting to the null subcarriers. It is clear that the strong reduction of the number of active subcarriers is caused by the slow decrease of the pulse in (1); we pass from 49 to 11 (or possibly to 2) subcarriers. Therefore, using the null subcarriers can solve the problem of spectral containment, but only if the constraint introduced on the out-of-band PSD is not more severe than 35–40 dB; in any case, also if the problem can be solved, this implies a significant loss in the achieved bit-rate, also when the CP-OFDM has to be employed in the downlink scenario on a large fragment such as the considered numerical bandwidth (0.2, 0.4) including in this example $N_f = 49$ subcarriers. In some scenarios, also smaller values of N_f could be available, making the spectral inefficiency of the classical OFDM system using guard subcarriers also more severe.

2.1 The Time Separation Approach for Evolving Classical OFDM

With regard to the theoretical proposals of a multicarrier system for data transmission, the classical OFDM technique is not the oldest one [7]. The main difficulties that delayed the adoption of the multicarrier techniques mainly lie in their computational complexity; the first version commonly adopted in various standards, the classical OFDM system, was therefore that minimizing the computational complexity. The overall network constraints were designed in order to circumvent the main drawback of the OFDM system, the poor spectral containment. However, many researchers have tried to propose different variants of the classical OFDM with smaller out-of-band emissions. In the last years, many contributions appeared in the literature (e.g., see [2, 3] and references therein) are proposing evolved multicarrier systems since the overall network scenario is making appear much probable that, for the next 5G standard, the classical OFDM system may be abandoned, at least in some use cases.

Solving the problem of the spectral containment requires to adopt a basic pulse whose spectral secondary lobes can be neglected. As it is well-known, this is however associated with the increase of the width of the main lobe. However, the subcarrier bandwidth is usually imposed by external constraints; therefore, adopting the same subcarrier bandwidth, the new basic pulse needs to become longer in time; let us denote with $(M + M_p)/\rho$ the length of the new pulse with the amplitude of the secondary lobes sufficiently small, i.e., let us denote with ρ the loss of spectral efficiency due to the increased pulse length.

This imposes a basic choice whether to maintain or to abandon the approach followed in classical OFDM about the issue of intersymbol interference. A first line of evolution, named here *time separation*, motivated by the greater compatibility with the current technology, proposes to limit the changes to the implementation platform, and in particular to the equalization stage. Therefore, two time-shifted versions of the basic pulse are still maintained separated in the time domain; such a separation is imposed at the transmitter output and is maintained also at the receiver input by resorting to an adequate cyclic prefix. In other terms, the improvement of the spectral containment is mainly paid in terms of the bit-rate achieved on the given bandwidth, i.e., in terms of spectral efficiency, while the computational complexity of the classic OFDM transceiver is maintained.

The line of *time-separation* includes the so-called *pulse-shaping* OFDM where the rectangular pulse in (1) is replaced by a properly chosen pulse shape; the choice of the new pulse has to take into account both the issue of spectral containment and that of the intercarrier interference (ICI). In particular, one of the first proposals considered in [8] for pulse-shaping OFDM is the well-known Hanning pulse. An important choice in the category of pulse-shaping is given by the time-window approach where the central rectangular shape of the pulse in (1) is maintained but a tail is appended on the two sides of the pulse in order to improve the spectral containment; typically, the tail is given a cosine shape, so obtaining that the pulse is shaped as a raised-cosine (RC) in the time domain:

$$g^{(RC)}[n] \triangleq \begin{cases} \frac{1-\cos\left(\pi \frac{n+1}{M_r+1}\right)}{2} & 0 \le n \le M_r - 1 \\ 1 & M_r \le n \le M_r + M + M_p - 1 \\ \frac{1+\cos\left(\pi \frac{n+1-M-M_p-M_r}{M_r+1}\right)}{2} & M + M_p + M_r \le n \le M + M_p + 2M_r - 1 \\ 0 & n \notin \{0, 1, \dots, M + M_p + 2M_r - 1\} \end{cases} \tag{9}$$

where $M_r = (M + M_p)(\frac{1}{\rho} - 1)$. Such a choice provides three important advantages: the first advantage is that the receiver adaptation to the new pulse-shape can be avoided by selecting at the receiver only the central rectangular part of the transmitted raised-cosine pulse. The second advantage, connected to the first one, is that the tail on the right side of the previous pulse can be time overlapped to that on the left side of the next pulse, consequently reducing the loss in spectral efficiency. The third advantage is given by the fact that using only the central rectangular part of the pulse also guarantees the absence of ICI.

The principal analysis about such a line of evolution for OFDM concerns the quantitative values of the loss of spectral efficiency due to the need to improve the spectral containment. It can be written as follows:

$$\rho\left(1 - \frac{N_g}{N_f}\right)\left(\frac{M}{M + M_p}\right) \tag{10}$$

where we have denoted with ρ the loss of spectral efficiency due to the pulse length increase, with N_g the number of guard subcarriers and with N_f the maximum number of subcarriers that could be employed in the available spectral fragment. Note that the last two terms are already present in the classical OFDM while the loss factor ρ, which is unit in classical OFDM, is peculiar of the considered line of the *time separation*.

In order to quantify such a loss, in Figs. 3 and 4 the PSD of the transmitted signal at the edge of the spectral bandwidth is reported; the various curves correspond to different values of ρ, which determines the pulse length; they are plotted by assuming that the PSD of the signal transmitted on each subcarrier is unit at the subcarrier center. The figures refer to the two usually considered pulses in pulse shaping [18] (the Hanning pulse and the RC one) for $M+M_p = 256(1+1/8) = 288$.

With reference to the Hanning pulse, we obtain that the achieved out-of-band attenuation depends on the number of guard subcarriers and on the loss of spectral efficiency. Since also the use of guard subcarriers causes a loss of spectral efficiency, the choice of the pulse length should take into account the two factors ρ and $(1 - N_g/N_f)$. For example, if the out-of-band attenuation is set to -40 dB, we cannot

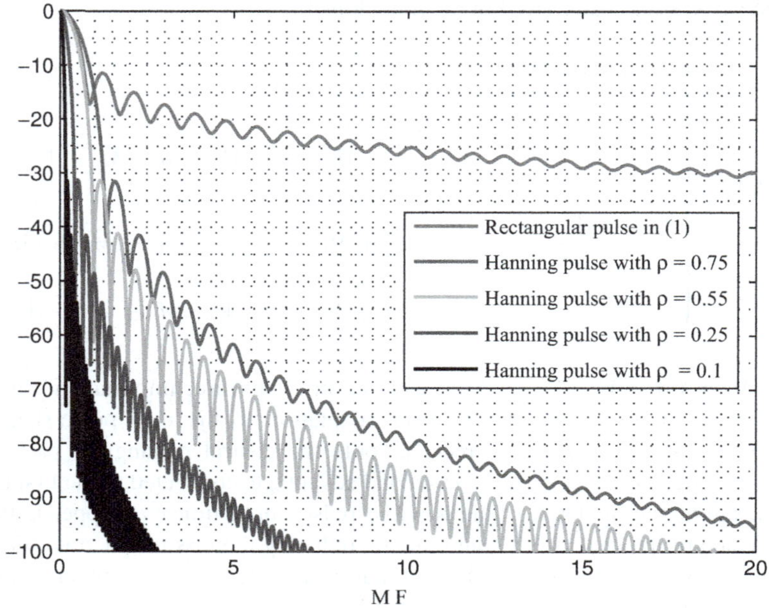

Fig. 3 PSD of the transmitted signal at the edge of the spectral bandwidth with the Hanning pulse for $\rho \in \{0.75, 0.55, 0.25, 0.1\}$ and with the rectangular pulse in (1)

achieve it with $\rho = 1$, but we can achieve it with $\rho = 0.75$ and 2 guard subcarriers, or with $\rho = 0.55$ and $N_g = 1$ or with $\rho = 0.1$ and $N_g = 0$. It is evident that often the first choice is the most convenient.

When we have to achieve important out-of-band attenuation, it can be useful to resort to longer pulses obtained with smaller values of ρ. For example, if we have to achieve an out-of-band attenuation of -80 dB, 3 guard subcarriers are sufficient with $\rho = 0.25$ while 10 subcarriers are needed when $\rho = 0.75$. The overall efficiencies in the two cases are

$$(1 - 6/N_f)0.25 \qquad (1 - 20/N_f)0.75 \tag{11}$$

where N_f is the number of subcarriers in the assigned bandwidth fragment. Therefore, the best choice between the two considered alternatives depends on the condition $N_f > 27$, and, obviously for $7 \le N_f \le 27$, only a single choice would be possible. Similar reasoning applies also with the RC pulse, whose spectral behavior is reported in Fig. 4 again for $M + M_p = 256(1 + 1/8) = 288$, and the results can be also worst. To achieve an out-of-band attenuation of -80 dB, 2 guard subcarriers are needed for $\rho = 0.25$ while 16 subcarriers are needed for $\rho = 0.75$ and the new condition becomes $N_f > 46$. To achieve, instead, an out-of-band attenuation of -40 dB, no guard subcarrier is needed for $\rho = 0.25$ and 3 guard subcarriers are needed for $\rho = 0.75$, the relative condition becomes $N_f > 9$.

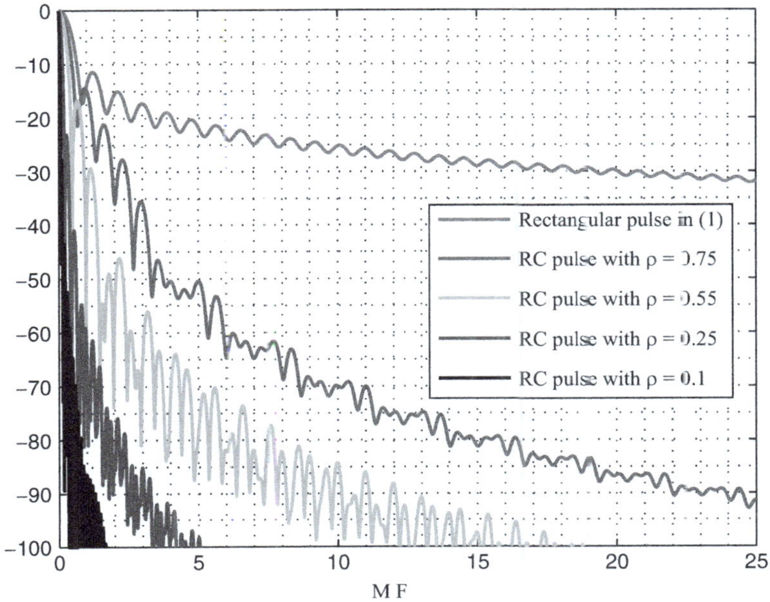

Fig. 4 PSD of the transmitted signal at the edge of the spectral bandwidth with the RC pulse in (9) for $\rho \in \{1, 0.75, 0.55, 0.25, 0.1\}$ and with the rectangular pulse in (1)

We can deduce from the above considered example the following general result about the analysis of the pulse shaping approach. When the fragment is composed by a small number of subcarriers and a strong constraint is imposed on the out-of-band attenuation, the best choice is obtained with small values of ρ, and therefore the pulse-shaped OFDM achieves the spectral containment by accepting a significant loss in terms of overall spectral efficiency. Otherwise, when the required out-of-band attenuation is moderate or low, and the fragment under consideration is composed of a large number of subcarriers, the pulse-shaping OFDM can use values of ρ around 0.75 and, therefore, pulse-shaped OFDM suffers from a moderate loss in terms of spectral efficiency for achieving its target of spectral containment. Both factors affect the loss of spectral efficiency: the required out-of-band attenuation and the size of the bandwidth fragment (in terms of number of subcarriers). It is important to note that the imperfections of the analog amplifiers can make appear not interesting the capability to achieve severe values of out-of-band attenuation; on the other hand, the possibility to allocate small bandwidth fragments to the machine-type communications can make appear interesting the case where N_f is particularly small.

The network scenarios considered in 5G standardization include the extreme cases. There are use cases where the loss of spectral efficiency related to the use, with minor changes, of the classical OFDM transceiver can be considered contained and therefore still acceptable, in the meantime that the technology improvements become available. Note however that the time will render less and less attractive solutions that have their strong advantage in the limited computational complexity; this is especially true as the available bandwidth is going to become more important. Such use cases concern the availability of bandwidth fragments assigned to the transceiver whose size is relatively large and the spectral containment required is moderate. On the other hand, there are use cases where such conditions are not satisfied and to maintain the line of *time separation* requires a significant reduction of the bit-rate, specified by the parameter ρ.

The same reasoning applies to the other approaches that fall within the line of *time separation* to simplify the standard approach to equalization. In particular, filtered OFDM [1] follows the same general reasoning already developed for pulse-shaping OFDM and, therefore, accepts similar losses in spectral efficiency in order to simply improve the standard OFDM system.

In such an approach, in fact, the final filter with $M_f + 1$ taps operates on the sum of the signals transmitted on the different subcarriers by the same user, which is equivalent to use filtered versions of the rectangular pulse in (1). There are two main differences:

- the first one lies in the fact that each subcarrier signal is constructed with a different equivalent pulse;
- the second difference lies in the fact that the transmitted signal is constructed by properly filtering the sum of the signals generated using the same standard pulse on each subcarrier; this implies a complexity cost associated with such a final filtering.

The length of the cyclic prefix introduced before the final filtering has to be increased from M_p to $M_f + M_p$, i.e., the cyclic prefix must be increased by M_f in order to maintain the line of the time separation. The choice $M_f = (M + M_p)\left(\frac{1}{\rho} - 1\right)$ implies that the resulting spectral efficiency can be still expressed as in (10) provided that the first M_f samples of a time interval overlaps with the final M_f samples of the previous interval. By selecting (for the FFT processing) the central block of M samples, not affected by the filtering effects of the transmission filter and of the multipath channel, the orthogonality of the different pulses is preserved and, therefore, the classical OFDM receiver can be employed also in filtered OFDM. The fraction of wasted energy is equal to $\frac{M_f + M_p}{M + M_f + M_p}$; the effects of violating the usual condition $M_p \geq M_h$ can be alleviated by the fact that the final M_f samples of the pulse transmitted in the previous time interval (overlapped on the first M_f samples of the pulse transmitted in the present time interval) exhibits smaller power in comparison with that of the useful samples; following such a perspective, schemes resorting to the extreme choice $M_p = 0$ have been proposed (e.g., [17]) in the literature in order to reduce the loss in spectral efficiency; this implies the abandon of the line of *time separation* and is discussed in the next section. As in pulse-shaping OFDM, also in filtered OFDM the choice of ρ has to trade-off (achieving similar results) the reduction of the spectral efficiency in (10) with the increase of the filter length M_f, and therefore with the consequent advantage in terms of spectral containment.

To complete the scenario followed in the line of *time separation*, we have to finally discuss the approach to achieving the spectral containment that makes use of the coefficients transmitted on the guard subcarriers (consequently, renamed cancellation subcarriers) in each symbol period as a function of the data symbol transmitted on the active subcarriers. Such coefficients are usually set to zero; different algorithms have been proposed to achieve the spectral containment. The target of large spectral efficiency, as well as the power lost on the guard subcarriers, should be limited by the small percentage of such subcarriers but this also limits the degrees of freedom required to achieve spectral containment on a large bandwidth.

The use of cancellation subcarriers does not appear useful to obtain significant increase of the value of ρ achieved through pulse-shaping with *time separation* because the obtained spectral containment is often important only in limited bandwidths outside the useful fragment.

The use of cancellation subcarriers can be interesting when we have already accepted to operate along the line of *time separation* with reasonable values of ρ; in such a case, we can reduce the number of guard and cancellation subcarriers, further slightly improving the overall spectral efficiency, by paying a limited price in terms of computational complexity and wasted transmitted power (see, for example, [11]).

3 Frequency Domain FBMC-OQAM

The choice of a basic pulse longer than M is necessary for solving the problem of the spectral containment, as already discussed in Sect. 2.1. The requirement to not pay such a solution in terms of spectral efficiency imposes to abandon the line of the *time separation* and, therefore, to admit the time overlapping of the signals associated with the successive symbols transmitted on the same subcarrier (as well as with those transmitted on the other subcarriers). The resulting transceiver design has therefore to consider not only the problem of ICI, as already done in the classical OFDM and in its variants introduced in Sect. 2, but also the problem of ISI, both on the flat and on the dispersive channel.

On a flat channel the two problems are solved by a careful transceiver design; the transceiver has also to be adapted in order to operate on a dispersive channel and in presence of timing and carrier offsets of the signal received by different transmitters. We first discuss how to design the transceiver on a flat channel, then we discuss its adaptation to the channel imperfections through proper procedures for equalization and synchronization.

3.1 Transceiver Design on a Flat Channel

The length of the basic pulse, usually called the prototype filter, is denoted with KM where K is the overlap parameter; we do not use ρ to denote it, as done in Sect. 2.1, because the increase of pulse length does not affect the spectral efficiency.

Once abandoned the line of the *time separation*, also the cyclic prefix can be avoided; however, the pulse length KM should be roughly dimensioned as the length $(M + M_p)/\rho$ of the pulses considered in Sect. 2.1 in order to obtain the same level of spectral containment.

On time-dispersive channels, we aim to solve the problem of the spectral containment without any loss in terms of spectral efficiency; we need to use the designed prototype filter with an intercarrier spacing of $1/M$, which generates interferences between adjacent subcarriers. The presence of non-negligible secondary lobes in the spectral shape $U_M(F)$ of the rectangular pulse (1) used in the classical OFDM system is crucial to guarantee the orthogonality of adjacent subcarriers. Figure 5 shows how the secondary lobes are crucial to make null the real part of the integral in (3) for $\overline{n}_k = \varepsilon_k = 0$.

When the secondary lobes can be neglected, the orthogonality does not hold any more. Let us call $g[n]$ the prototype filter designed in the interval $I_K \triangleq \{-(KM/2-1), \ldots, -1, 0, 1, \ldots, KM/2-1\}$ (i.e., $g[n] = 0$ for $n \notin I_K$) in order to achieve the required spectral containment and assume that the design has imposed that the filter be real valued and symmetric (i.e, with $g[n] = g[-n]$). Consequently, its Fourier transform $G(F)$ is real and even. The orthogonality between adjacent subcarriers requires to consider the product $G^{\text{prod}}(F) \triangleq G(F)G(F - 1/M)$. If the

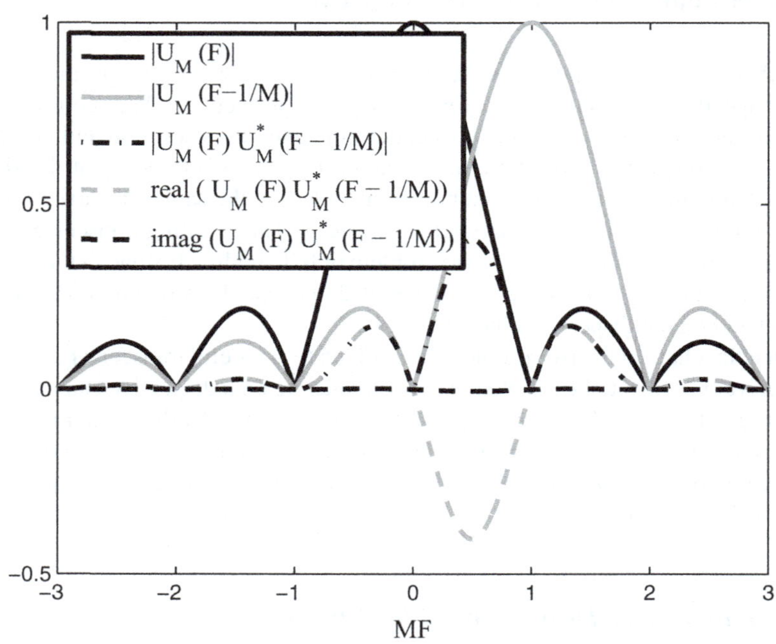

Fig. 5 A plot of $|U_M(F)|$, $|U_M(F-1/M)|$, $|U_M(F)U_M^*(F-1/M)|$, $\Re\left\{U_M(F)U_M^*(F-1/M)\right\}$, and $\Im\left\{U_M(F)U_M^*(F-1/M)\right\}$ versus MF for $M = 256$

secondary lobes are negligible, the energy of $G^{\text{prod}}(F)$ is mainly concentrated in the interval $[0, 1/M]$; moreover, in such interval, the symmetry of $G(F)$ implies that $G^{\text{prod}}(F)$ is symmetric around the point $F = 1/(2M)$. Consequently, $G^{\text{prod}}(F)$ can be reasonably approximated, as far as the secondary lobes are negligible, with the function $G^{\text{symm}}(F - \frac{1}{2M})$ where:

$$G^{\text{symm}}(F) \triangleq \begin{cases} G^{\text{prod}}(F + \frac{1}{2M}) & |F| \leq \frac{1}{2M} \\ 0 & \text{otherwise} \end{cases}. \tag{12}$$

Consequently, $G^{\text{symm}}(F)$ is real and even and, also its inverse Fourier transform $g^{\text{symm}}[n]$ is real (and also symmetric). Then, the inverse Fourier of $G^{\text{prod}}(F)$ can be written as $g^{\text{symm}}[n]e^{j2\pi\frac{1}{2M}n}$. The symmetry property imposed on the designed filter and the spectral containment of the pulse energy within its main lobe implies, therefore, that the scalar product r_i between $g[n - iM/2]$ and $g[n]e^{j2\pi\frac{1}{M}n}$ can be written as follows:

$$\begin{aligned} r_i &\triangleq \sum_{n=-\infty}^{+\infty} g\left[n - i\frac{M}{2}\right]\left(g[n]e^{j2\pi\frac{1}{M}n}\right)^* = \int_{-\frac{1}{2}}^{\frac{1}{2}} G(F)G(F - 1/M)e^{-j2\pi i\frac{M}{2}F}\mathrm{d}F \\ &\simeq \left[g^{\text{symm}}[n]e^{j2\pi\frac{1}{2M}n}\right]_{|n=-iM/2} = j^{-i}g^{\text{symm}}\left[i\frac{M}{2}\right]. \end{aligned} \tag{13}$$

The relation (13) implies that the scalar product between adjacent subcarriers is purely real for even i and purely imaginary for odd i because we have already shown that $g^{\text{symm}}[n]$ is real valued. Therefore, we can still recover the transmitted symbol provided that a purely imaginary symbol is transmitted on the adjacent interfering subcarrier for even i and a purely real one for odd i when a purely real symbol is transmitted on the useful one and, *vice versa*, a purely real one for even i and a purely imaginary one for odd i when a purely imaginary symbol is transmitted on the useful one; here $iM/2$ represents the misalignment between the pulses transmitted on the adjacent subcarriers, as shown in the left-hand side of (13). More specifically, the signal $s_k[n]$ transmitted on the k subcarrier is written as follows

$$s_k[n] \triangleq \sum_{i=-\infty}^{+\infty} j^i a_k[i] g_k\left[n - i\frac{M}{2}\right] \tag{14}$$

where the pulse $g_k[n]$ used for transmissions on the kth subcarrier is defined as follows:

$$g_k[n] \triangleq j^k g[n] e^{j2\pi\frac{k}{M}n} \tag{15}$$

and $a_k[i]$ denotes the real-valued symbol transmitted on the kth subcarrier with the ith time-shifted pulse. Relation (13) implies that the scalar product between the interfering pulse $j^{i+m}g_{k+1}[n-(i+m)M/2]$ on the adjacent subcarrier and the useful one $j^i g_k[n - iM/2]$ is purely imaginary. Therefore, the adjacent subcarrier does not interfere with the detection of the useful symbol $a_k[i]$ provided that it is real-valued. Moreover, the PSD of the signal $s_k[n]$ can be written as $\frac{P_a}{M}\left|G\left(F - \frac{k}{M}\right)\right|^2$ where $Pa \triangleq E[|a_k[i]|^2]$; furthermore, the PSDs of the signal transmitted on each subcarrier are summed when the symbols transmitted on the different subcarriers are assumed to be uncorrelated.

Let us also note that the basic property used here (i.e., $j^i r_i$ real-valued) does not hold for relative delays different from $iM/2$ (see relation (13)); therefore, a relative timing offset between the signals received on adjacent subcarriers generates intercarrier interference. This represents a problem when a timing estimation error is present while, also in an uplink channel that does not use a procedure for timing alignment among users, no interference is present in practice when the timing offset of the useful signal is perfectly estimated, provided that a single guard subcarrier among the different users is introduced.

No ICI problem from non-adjacent subcarriers is present because the secondary lobes of the designed prototype filter $g[n]$ are negligible. The spectral separation of the signals transmitted on non-adjacent subcarriers also implies that the presence of a relative timing offset and of a relative CFO among different users does not affect the receiver performance. It is only necessary that the receiver correctly estimates the timing offset and the CFO of each user; no timing and CFO alignment procedure among the different users is necessary, which is a crucial advantage when the number of different users is large.

The signal design in (14) and (15) is also able to solve the ISI problem in the considered transceiver. In fact, the scalar product between the interfering term $j^{i+m} g_k[n - (i+m)M/2]$ and the useful pulse $j^i g_k[n - iM/2]$ is purely imaginary for odd m; this property straightforwardly follows from the fact that

$$\begin{aligned} g_k[n-(i+m)M/2]g_k^*[n-iM/2] &= \left[j^k g[n-(i+m)M/2]e^{j2\pi\frac{k}{M}(n-(i+m)M/2)}\right] \\ &\quad \times \left[j^k g[n-iM/2]e^{j2\pi\frac{k}{M}(n-iM/2)}\right]^* \\ &= g[n-(i+m)M/2]e^{-j2\pi\frac{k}{M}mM/2}g[n-iM/2] \quad (16) \end{aligned}$$

is real-valued. Such a scalar product, for even $m \neq 0$, is equivalent to the scalar product between the interfering pulse $g_k[n - iM/2 - mM]$ and the useful one $g_k[n - iM/2]$, or also between the pulses $g_k[n - mM]$ and $g_k[n]$. It is well-known that the interference between the last two pulses is null provided that $g_k[n]$ satisfies the Nyquist condition, i.e., provided that the following quantity

$$\sum_{\ell=0}^{M-1}\left|G_k\left(F - \frac{\ell}{M}\right)\right|^2 = \sum_{\ell=0}^{M-1}\left|G\left(F - \frac{\ell}{M}\right)\right|^2 \quad (17)$$

Table 1 Coefficients of the prototype filter in (18) for $K \in \{2, 3, 4\}$

$K = 2$	$K = 3$	$K = 4$
$G_0 = 1$	$G_0 = 1$	$G_0 = 1$
$G_1 = 1/\sqrt{2}$	$G_1 = 0.911438$	$G_1 = 0.97196$
–	$G_2 = 0.411438$	$G_2 = 1/\sqrt{2}$
–	–	$G_3 = 0.235147$

is constant with F; here we have denoted with $G_k(\cdot)$ and $G(\cdot)$ the Fourier transform of $g_k[\cdot]$ and $g[\cdot]$, respectively. The equality in (17), which straightforwardly follows from the definition (15), implies that the ISI problem for even m can be solved by introducing a further constraint in the prototype filter design; moreover, it implies that the transmitted signal is white when all the subcarriers are active. The prototype filter, therefore, has to be real-valued, symmetric, with the right-hand side of (17) constant with F, and with secondary lobes sufficiently small, of course.

A method for filter design often employed in the literature determines [4] the filter according to the following expression:

$$g[n] = G_0 + 2\sum_{q=1}^{K-1} G_q \cos\left(2\pi \frac{q}{KM} n\right) \qquad n \in I_K \tag{18}$$

and null otherwise. The coefficients optimized to achieve the design requirements are specified in Table 1; the behavior of the obtained prototype filters is reported in Fig. 6. We can notice the strong improvement in the spectral selectivity for $K = 3$ and for $K = 4$; however, such improvement corresponds to a complexity increase of the transceiver. We have also to note a poorer spectral containment for $K = 2$, which is a cheaper choice from the computational point of view. Therefore, the method considered in the present section is appropriate when severe spectral containment is required and we are ready to pay for the corresponding increase of computational complexity. A better solution for the case $K = 2$ will be presented in Sect. 4. Let us also note that the orthogonality condition is only achieved approximately both because the secondary lobes are approximately null and because the condition (17) is only approximated. However, large signal-to- interference ratios are achieved with the considered filters.

3.2 *Transmitter Structure*

When the transmitted signal can be written as

$$s[n] \triangleq \sum_{k\in\mathscr{A}} s_k[n] \tag{19}$$

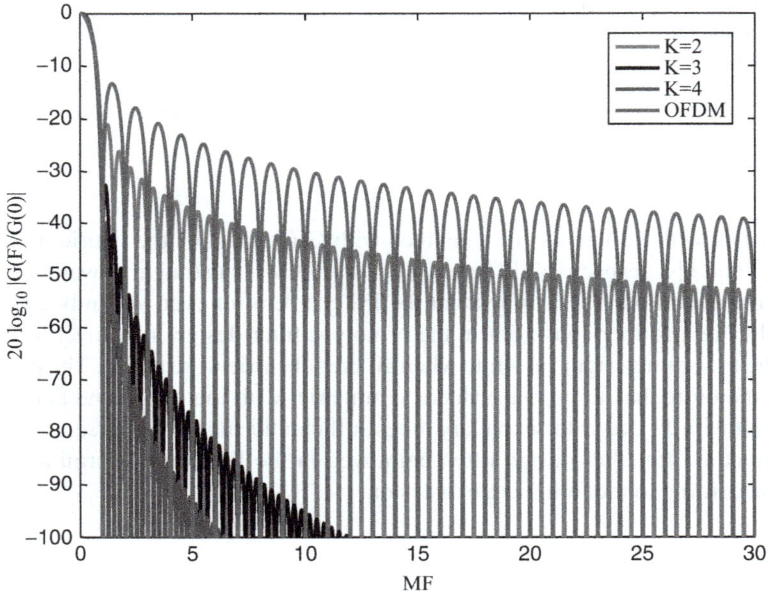

Fig. 6 The spectral shape of the different prototype filters proposed in [4] for $K \in \{2, 3, 4\}$

with $s_k[n]$ defined in (14) and $\mathscr{A}$ denoting the set of the active subcarriers, we can construct [13] the transmitter using a traditional polyphase structure and the receiver according to the recently proposed frequency-despreading structure.

The standard structure for the implementation of the FBMC-OQAM transmitter has been first proposed in [9]; it first separates the overall signal as $s[n] = s^R[n] + js^I[n - M/2]$, we call $a^R_{i,k}$ the real-valued subsequence of the symbols $a_k[i]$ with halved bit-rate transmitted by the signal $s^R[n]$ and with $a^I_{i,k}$ the real-valued subsequence of the symbols $a_k[i]$ with halved bit-rate transmitted by the signal $s^I[n]$. A first part of the structure constructs the signal $s^R[n]$ as follows while the second part constructs the signal $s^I[n]$.

The generation of the sequence $s^R[n]$ is equivalent to the generation of the sequence of $M \times 1$ vectors $\mathbf{d}_i^{(R)}$ whose kth component $d_{i,k}^{(R)}$ is equal to $s^R[iM + k]$ for $k \in \{0, 1, \ldots, M - 1\}$. Such a vector can be written [13] as follows:

$$\mathbf{d}_i^{(R)} = \sum_{\ell=-\infty}^{+\infty} \mathbf{b}_\ell^{(R)} \odot \mathbf{g}_{i-\ell} \tag{20}$$

where $\odot$ denotes the Hadamard product, the vector $\mathbf{g}_i$ is defined so that its kth component is $g[k + iM - KM/2]$ and $\mathbf{b}_\ell^{(R)}$ is the IDFT of the vector with kth component equal to $j^k a^R_{\ell,k}$ ($k = 0, 1, \ldots, M - 1$). The finite length property of the prototype filter implies that the sum in (20) can be limited to $\ell \in \{i-K+1, i-K+2, \ldots, i\}$. Consequently, (20) can be rewritten as

$$\mathbf{d}_i^{(R)} = \mathbf{g}_0 \odot \mathbf{b}_i^{(R)} + \mathbf{g}_1 \odot \mathbf{b}_{i-1}^{(R)} + \ldots + \mathbf{g}_{K-1} \odot \mathbf{b}_{i-(K-1)}^{(R)} \quad (21)$$

The name PPN, often used for the standard structure, is related to the use of the polyphase components of the prototype filter in (21). Analogously, the vector $\mathbf{d}_i^{(I)}$ whose kth ($k = 0, 1, \ldots, M-1$) component is equal to $s^I[iM + k]$ can be evaluated according to the following relation:

$$\mathbf{d}_i^{(I)} = \mathbf{g}_0 \odot \mathbf{b}_i^{(I)} + \mathbf{g}_1 \odot \mathbf{b}_{i-1}^{(I)} + \ldots + \mathbf{g}_{K-1} \odot \mathbf{b}_{i-(K-1)}^{(I)} \quad (22)$$

where $\mathbf{b}_\ell^{(I)}$ is the IDFT of the vector with kth component equal to $j^k a_{\ell,k}^I$ ($k = 0, 1, \ldots, M-1$).

3.3 Receiver Structure

The previously described orthogonality property implies that the optimum receiver (in the maximum likelihood (ML) sense) in AWGN channel requires to evaluate the real part of the scalar product between the received signal $r[n]$ and the useful pulse $j^i g_k\left[n - i\frac{M}{2}\right]$ in order to estimate $a_k[i]$:

$$D_k^i \triangleq \Re\left\{\sum_{n=-\infty}^{+\infty} r[n] j^{-i} g_k^*\left[n - i\frac{M}{2}\right]\right\} = (-1)^{kK} \sum_{q=-(K-1)}^{K-1} (-1)^{qK} G_q \Re\left\{R_{i,kK+q} j^{-i-k}\right\} \quad (23)$$

where

$$r_i[n] \triangleq r\left[n + i\frac{M}{2}\right] \quad (24)$$

and with $r_{i,\ell}$ the ℓth component ($\ell = 0, 1, \ldots, KM-1$) of the vector

$$\mathbf{r}_i \triangleq \left[r_i\left[-\frac{KM}{2}\right], r_i\left[-\frac{KM}{2}+1\right], \ldots, r_i\left[\frac{KM}{2}-1\right]\right] \quad (25)$$

whose DFT has been denoted with $\mathbf{R}_i$. The last equality in (23) follows from the use of the expression (18) and specifies a simple structure, called frequency-despreading, for implementing the receiver. The most complex stage lies in the implementation of the DFT of $\mathbf{r}_i$ over KM points every $M/2$ time steps and by the further M (one for each value of the index k) linear combinations of $2K-1$ elements of the DFT outputs with the same coefficients $(-1)^{qK} G_q$.

In the presence of the normalized carrier frequency offset ε on the received signal, the previous expression can be straightforwardly extended. In fact, in such a case, the ML receiver in AWGN channel can be written as

$$D_k^i \triangleq \Re\left\{\sum_{n=-\infty}^{+\infty} r[n]e^{-j2\pi\frac{\varepsilon}{M}n} j^{-i} g_k^*\left[n - i\frac{M}{2}\right]\right\}$$
$$\simeq (-1)^{kK} \sum_{q=-(K_\varepsilon-1)}^{K_\varepsilon-1} (-1)^{qK_\varepsilon} \Re\left\{G_{q,\varepsilon}^* R_{i,kK+q} j^{-i-k}\right\}. \qquad (26)$$

The parameters K_ε and $G_{q,\varepsilon}$ are defined by the following approximation, used for deriving (26):

$$g[n]e^{j2\pi\frac{\varepsilon}{M}n} \simeq \sum_{q=-(K_\varepsilon-1)}^{K_\varepsilon-1} G_{q,\varepsilon} e^{j2\pi\frac{q}{KM}n} \qquad (27)$$

and, therefore, $MG_{q,\varepsilon}$ is the value of $G(F)$ for $F = \frac{q-K\varepsilon}{KM}$. Since it can be simply evaluated by Dirichlet relation with minor complexity increase [14], the structure (26) can easily compensate in the frequency domain the CFO of the received signal. The complexity increase is usually very small since the value K_ε is much close to K because the spectral containment of the prototype filter makes negligible the increase $K_\varepsilon - K$ necessary to achieve a reasonable approximation in (27).

The proposed scheme exhibits an increase in terms of computational complexity with respect to the classical OFDM. In fact, the main computational weight is given by the DFT stage. In the transmitter, implemented according to the polyphase structure, we need an FFT over M points to be performed with a rate $2/M$, i.e., once every $M/2$ samples. This requires to roughly double the complexity of the OFDM transceiver where an FFT over M points is required every $M + M_p$ samples. The receiver, implemented according to the frequency-despreading structure, roughly requires to calculate an FFT over KM samples with rate $2/M$ and, therefore, exhibits a complexity increase roughly quantified by the factor $2K$; we could simply spend only a factor 2 of receiver complexity increase by resorting to the polyphase structure also at the receiver.

When the complexity counting includes also the computational complexity required for frequency-domain CFO compensation, the complexity-increase factor becomes smaller because classical OFDM (and polyphase receiver for FBMC-OQAM) is not able to perform it with the same efficiency [5, 14]. This becomes particularly important when the receiver has to operate with signals transmitted by different users, each with its own value of CFO. While, in the case of the single user transmitting over all the subcarriers, the upper-bound for an inefficient CFO compensation in the classical OFDM system is the complexity doubling required by the time-domain CFO compensation, when the number of users becomes larger, such upper bound can become much larger than the complexity of the proposed structure. A detailed comparison of the computational complexity of the classical

OFDM receiver including CFO compensation procedures and that of the frequency-despreading structure is not available yet; however, the traditional conclusion of a complexity-increase factor equal to $2K$ (of the proposed FBMC-OQAM in comparison to classical OFDM system), which is certainly a good approximation when the complexity of the CFO compensation is not included in the counting, seems much far from the practice, especially when a large number of different users is considered at the receiver. An important reduction of the complexity-increase factor is further encountered when the procedures for timing and CFO alignment among the different users in an uplink scenario, only needed in the classical OFDM system but not in the FBMC-OQAM system, are included in the counting. For such a reason, we do not consider the complexity increase a significant reason to not adopt the FBMC-OQAM system.

3.4 Equalization Stage

The equivalent channel is usually modeled as a linear filter with impulse response $h_c[n]$, assumed not null in the interval $[0, 1, \ldots, M_h]$; then, the received signal $r_t[n]$ can be written as

$$r_t[n] = \sum_{\ell=0}^{M_h} h_c[\ell]s[n-\ell] \tag{28}$$

where $s[n]$ is specified in (19). The multipath channel introduces new delayed versions of the useful transmitted signal; consequently, new delayed versions of the interfering signals, whose delay (once set to zero the delay of the useful component) is not multiple of $M/2$, are now present at the receiver and affect the basic transceiver considered in the previous stage; therefore, the receiver needs a proper equalization stage.

The equalization stage is present also in the classical OFDM receiver; however, the presence of cyclic prefix much simplifies its design: a single-tap equalizer per each FFT output is sufficient for implementing the optimum equalizer. In fact, the main purpose of introducing the cyclic prefix at the transmitter is the simplification of the equalizer stage at the receiver.

The abandon of the line of the *time separation* makes appear insufficient the simple single-tap equalization stage. Nevertheless, the standard solution for equalization stage in FBMC-OQAM closely resembles the equalization stage in OFDM; in fact, the receiver structure (26) is usually modified by replacing the FFT output $R_{i,m}$ with $R_{i,m}c_m$, i.e., a single-tap equalizer is commonly used at the receiver output; the design of the coefficient c_m can follow a zero-forcing approach or an MMSE approach. When we follow a zero-forcing approach, the coefficient c_m is set as $H_c^{-1}(m/(KM))$ where $H_c(F)$ is the Fourier transform of the channel impulse response. As already done with reference to the previously considered CFO compensation stage, here we are neglecting the effect of the estimation error.

It is well known in the literature (e.g., see [13]) that the single-tap equalizer is only an approximate solution, i.e., the receiver equipped with such an equalizer stage shows a performance loss with respect to that achieved on the flat channel, with the same fading statistics, unless the value of M is sufficiently large. Determining the value of M for which such approximate solution shows limited performance loss is often neglected in the literature; however, it is usually considered that the use of the single-tap structure as equalization stage of the FBMC-OQAM receiver provides minor performance loss only if M is sufficiently large. In other terms, there may be values of M for which it makes sense to consider alternative structures.

An interesting equalizer structure suited to the considered receiver scenario where an FFT bank is already present at the receiver is the frequency-domain implementation of a linear equalizer [10]. Such a linear equalizer can be designed by zero-forcing approach; then, its form can be written as $H_c^{-1}(F)$ when we resort to the zero-forcing design but we can also use a different expression provided that the noise power (or better the noise PSD) has been correctly estimated.

Since the output of such a linear equalizer represents the input of the receiver structure described in (26), we need its output values $\tilde{r}[n]$ in the interval $I_i \triangleq \{(i-K)M/2, \ldots, (i+K)M/2-1\}$ in order to be able to construct the vector $\mathbf{r}_i$ in (25) and evaluate its DFT $\mathbf{R}_i$. When the equalization stage is perfect, the useful component of $\tilde{r}[n]$ coincides with $s[n]$ while the noise may be amplified in correspondence of the weaker subcarriers.

To evaluate the output samples of the signal $\tilde{r}[n]$ in the interval I_i, we need to extract from the received signal $r_h[n]$ the samples in the interval $\overline{I}_i \triangleq \{(i-K)M/2-n_c, \ldots, (i+K)M/2-1+n_a\}$. Such an interval includes (in its central part) the interval I_i and is extended using n_c samples on the left side and n_a samples on the right side. We denote its length with $\overline{KM} \triangleq KM + n_c + n_a$ and, consequently, define the input signal $\overline{r}_h^{(i)}[n]$ as follows:

$$\overline{r}_h^{(i)}[n] = \begin{cases} r_h[n] & n \in \overline{I}_i \\ 0 & n \notin \overline{I}_i \end{cases} . \tag{29}$$

Let us assume that we design the linear filter according to the zero-forcing criterion as the filter with impulse response $H_c^{-1}(F)$ evaluated as follows:

$$H_c^{-1}(F) \simeq \sum_{n=-M_a}^{M_c} h_{\text{inv}}[n] e^{-j2\pi nF} \tag{30}$$

where M_c denotes the causal memory and M_a the anticausal memory of the inverse filter to be implemented with impulse response $h_{\text{inv}}[n]$. Since the channel is modeled as an FIR filter, its inverse will have an infinite length; we assume that the length of the memories M_a and M_c is sufficient to reasonably neglect the unavoidable differences between the FIR structure and the actual IIR structure so that the equality holds for any F with good approximation. Note that the values of M_a and M_c mainly depend on the zeros of the channel closest to the unit circle on the internal and external side, respectively. Once accepted that we are trying to implement a noncausal filter with finite memory specified in (30), we first consider the implementation of the causal FIR filter with impulse response $h_{\text{inv}}[n-M_a]$, whose output can be therefore written as $\tilde{r}[n-M_a]$. To evaluate such an output, let us first collect the nonnull values of the signal $\overline{r}_h^{(i)}[n]$ in the vector $\overline{\mathbf{r}}_i$ of length $\overline{K}M$ and let us define the vector $\mathbf{h}_{\text{inv}}^{(a)} \triangleq \mathbf{h}_{\text{inv}} \odot \mathbf{w}$ of the same size where the ℓth component ($\ell = 0, 1, \ldots, \overline{K}M - 1$) of $\mathbf{h}_{\text{inv}}$ is equal to the Fourier transform of $h_{\text{inv}}[n]$ (i.e., $H_c^{-1}(F)$) and the ℓth component of the vector $\mathbf{w}$ is equal to $e^{-j2\pi M_a F}$, in both cases for $F = \ell/(\overline{K}M)$. Consequently, it is well known that the vector $\widetilde{\mathbf{r}}_i^{\text{w}}$ of the same size defined as follows:

$$\widetilde{\mathbf{r}}_i^{\text{w}} = \text{IDFT}\left[\text{DFT}\left[\overline{\mathbf{r}}_i\right] \odot \mathbf{h}_{\text{inv}}^{(a)}\right] = \text{IDFT}\left[\text{DFT}\left[\overline{\mathbf{r}}_i\right] \odot \mathbf{h}_{\text{inv}} \odot \mathbf{w}\right] \tag{31}$$

is such that its ℓth component $\tilde{r}_{i,\ell}^{\text{w}}$ is equal to

$$\tilde{r}_{i,\ell}^{\text{w}} = \tilde{r}[n - M_a]_{|n=(i-K)M/2-n_c+\ell} \qquad \ell \geq M_a + M_c \tag{32}$$

with obviously $\ell \leq \overline{K}M - 1$; the number of points of the DFT and IDFT in (31) is equal to $\overline{K}M$. Moreover, the vector

$$\widetilde{\mathbf{r}}_i = \text{IDFT}\left[\text{DFT}\left[\overline{\mathbf{r}}_i\right] \odot \mathbf{h}_{\text{inv}}\right] \tag{33}$$

is the circularly shifted version of $\tilde{r}_{i,\ell}^{\text{w}}$ of M_a samples on the left. Therefore, its ℓth component satisfies the following relation:

$$\tilde{r}_{i,\text{mod}_{\overline{K}M}(\ell-M_a)} = \tilde{r}_{i,\ell}^{\text{w}} = \tilde{r}[(i-K)M/2 - n_c + \ell - M_a] \qquad \ell \geq M_a + M_c \tag{34}$$

or, equivalently,

$$\tilde{r}_{i,\ell-M_a} = \tilde{r}[(i-K)M/2 - n_c + \ell - M_a] \qquad \ell \geq M_a + M_c \tag{35}$$

and, also,

$$\tilde{r}_{i,\ell} = \tilde{r}[(i-K)M/2 - n_c + \ell], \qquad M_c \leq \ell \leq \overline{K}M - 1 - M_a. \tag{36}$$

The relation (36) shows that we can extract the desired values of $\tilde{r}[n]$ for $n \in I_i$, provided that $n_c \geq M_c$ and $n_a \geq M_a$. In such a case, the desired vector of length KM is obtained from the values of $\tilde{r}_{i,\ell}$ from $\ell = n_c$ to $\ell = n_c + KM - 1$. Therefore, the frequency domain equalizer (FDE) requires

- to select a block of length KM samples of the received signal, centered on the sample $iM/2$;

- to further increase its length by including at least M_c samples on the left of the block and M_a samples at its right side, so determining the vector $\overline{\mathbf{r}}_i$ of length $\overline{K}M$;
- to determine the vector $\widetilde{\mathbf{r}}_i$ according to (33); here, we use the channel response since the ℓth component of $\mathbf{h}_{\text{inv}}$ is equal to $H_c^{-1}(F)$ for $F = \ell/(\overline{K}M)$;
- to extract from it the central part constituted by the desired KM components starting from the component with index n_c.

The choice of the parameters n_c and n_a is critical because they have to be larger than M_c and M_a, which depend on the specific channel realization; therefore, they have to be chosen sufficiently larger than the values usually encountered in practice; moreover, we have to consider that the value $\overline{K}M$ should be chosen to be a power of two. For example, one could choose $K = 3$ and $n_c = n_a = M/2$ or one could choose $K = 2$ and $n_c = n_a = M$ or, also, when $K = 1$, $n_c = n_a = M/2$.

The increase of the computational complexity is due to the DFT and IDFT stages described in (33), whose size is also slightly larger than the DFT already present in the receiver. Therefore, the considered receiver structure with its computational cost could be also used in the classical OFDM system to avoid the loss of spectral efficiency due to the cyclic prefix, as well as with reference to the transceiver described in the next section. In fact, it operates on the received signal trying to restore the transmitted signal independently of the transmitter and the receiver structure.

An adaptive behavior of such equalizer could allow to reduce the computational complexity of the structure. In fact, as soon as the channel impulse response is estimated, the values of M_c and M_a needed to satisfy (30) can be evaluated and the FFT and IFFT size $\overline{K}M$ modified accordingly; this would minimize the number of FFT (and IFFT) points required in correspondence of each channel realization.

The selected vector represents the input of the standard receiver structure, called $\mathbf{r}_i$ in (25); therefore, we have to perform again the DFT of the KM vector with the selected components. Therefore, we have a series of three stages, first the DFT in (31), then the IFDT in (31), and finally the DFT in the standard receiver. The first two stages are over $\overline{K}M$ points while the last one is over KM points. In the special case where the number of points of the longer DFT is the double of the shorter one, we can perform also the IDFT on the shorter number of points provided that a few further processing is included. In particular, let us denote with N the shorter length and assume that the central part of the IDFT has to be extracted and processed according to the DFT; in such a case, we should calculate the following quantities

$$y[n] = \sum_{k=0}^{2N-1} X_k e^{j2\pi \frac{k}{2N} n} \qquad Y_k = \frac{1}{N} \sum_{n=N/2}^{3N/2-1} y[n] e^{-j2\pi \frac{k}{N}(n-N/2)} \tag{37}$$

where X_k represents the kth component of the IDFT input while Y_k can be also expressed as

$$Y_k = \frac{(-1)^k X_{2k}}{N} + \frac{j}{N} \sum_{n=0}^{N-1} \left[\sum_{q=0}^{N-1} (-1)^q X_{2q+1} e^{-j2\pi \frac{q}{N} n} \right] e^{j\pi \frac{n}{N}} e^{-j2\pi \frac{k}{N} n}. \tag{38}$$

Therefore, also the IDFT can be performed over N points. In such a case, we pass from a single FFT over N points usually employed with single-tap equalizer to an FFT over $2N$ followed by an IFFT and an FFT over N points; the complexity-increase factor due to the equalization stage is therefore roughly equal to 4.

4 FBMC-PAM

The good spectral behavior of the prototype filter designed in (18) for $K = 4$ motivated its frequent use in various contributions appeared in the literature. However, the choice $K = 3$ is particularly interesting when the choice $\overline{K} = 4$ is sufficient for the equalization stage. The choice $K = 4$, instead, is not much suited to the considered equalization stage when the FFT length is imposed to be a power of two; in fact, the choice $\overline{K} = 8$ would become mandatory, with negative effects on the transceiver latency. Let us note that the transceiver latency is given by $\overline{K}M - n_c = KM + n_a$ plus the time interval needed to implement the receiver processing; it is true that pulse lengths longer than M are needed to achieve the spectral containment but the further length introduced for the specific purpose of FBMC-OQAM implementation represents a contribution to the network latency that could be avoided; for such a reason, the choice $\overline{K} = 8$ should be avoided. Obviously, we can also employ a receiver structure where $\overline{K}M$ is not a power of two to maintain the transceiver latency at its minimum possible value. Also the choice $K = 4$ with single-tap equalizer is often employed in the literature and may represent a good compromise.

In order to decrease the network latency, the choice $K = 2$ is particularly interesting; however, as shown in Fig. 6, the spectral behavior of the prototype filter obtained in (18) for the case $K = 2$ is particularly poor, and this can give problems in many practical applications since many properties of the FBMC-OQAM system are related to the significant spectral containment of the desired filter. For such a reason, we consider a different multicarrier system, the FBMC-PAM [15], able of improving the spectral containment of the FBMC-OQAM system when the choice $K = 2$ is adopted.

Let us consider the following extension

$$T_c[k,n] \triangleq \sin\left[\frac{\pi}{2M}\left(n+\frac{1}{2}\right)\right] \exp\left[j\frac{\pi}{M}\left(k+\frac{1}{2}\right)\left(n+\frac{1}{2}+\frac{M}{2}\right)\right] \tag{39}$$

of the kth real-valued modulated lapped transform $T[k,n] = \Re\{T_c[k,n]\}$ introduced by Malvar in its seminal contribution [12]. The transform in (39), which is defined for $k, n \in \mathcal{K}_{2M} \triangleq \{0, 1, \ldots, 2M-1\}$ and is assumed null otherwise, can be used to transmit $2M$ real data symbols at rate $1/M$. The transmitted signal is

$$s[n] = \sum_{k=0}^{2M-1} s_k[n] \qquad s_k[n] = \sum_{i=-\infty}^{+\infty} a_k[i] T_c[k, n-iM]. \tag{40}$$

When the data symbols are statistically independent with zero-mean and variance P_a, it can be shown [15] that $s[n]$ is wide-sense stationary and white while the PSD $S_k(F)$ of the signal $s_k[n]$ transmitted on the kth subcarrier is

$$S_k(F) = \frac{P_a}{M} \left| \sin\left(\frac{\pi}{4M}\right) \frac{\cos\left(2\pi M\left(F - \frac{k+1/2}{2M}\right)\right)\cos\left(\pi\left(F - \frac{k+1/2}{2M}\right)\right)}{\cos\left(2\pi\left(F - \frac{k+1/2}{2M}\right)\right) - \cos\left(\frac{\pi}{2M}\right)} \right|^2 . \tag{41}$$

The study of $S_k(F)$ is useful when not all the subcarriers are active; it is centered around $F_k \triangleq \frac{k+1/2}{2M}$ and is practically separated from $S_{k+2}(F)$ and $S_{k-2}(F)$ as shown in Fig. 7, which reports the behavior of $S_0(F)$, $S_1(F)$, and $S_2(F)$ under the assumption that these PSDs at the central frequency are unit. Moreover, the property of orthogonality of the Malvar transform guarantees (see [15] for the details) that the signals transmitted on adjacent subcarriers do not interfere. Therefore, subcarrier

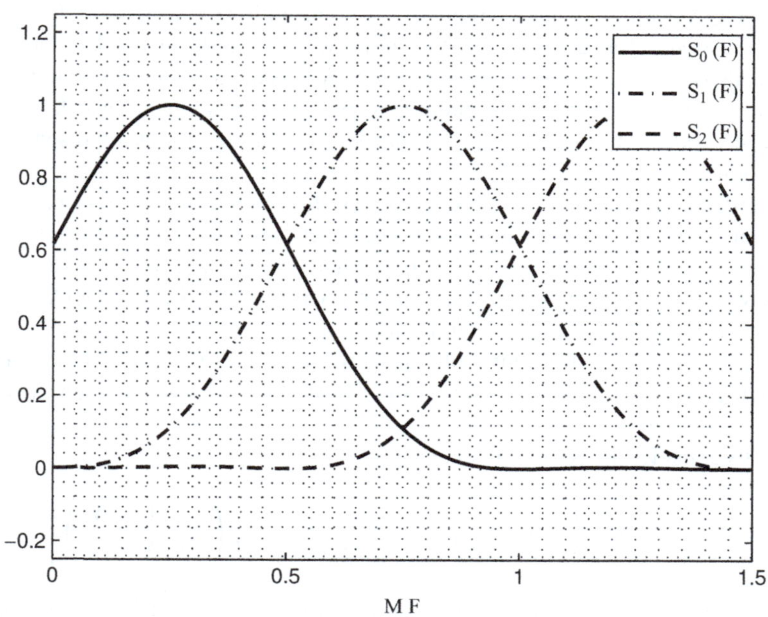

Fig. 7 PSDs of the signal $s_k[n]$ for $k = 0$, $k = 1$, and $k = 2$

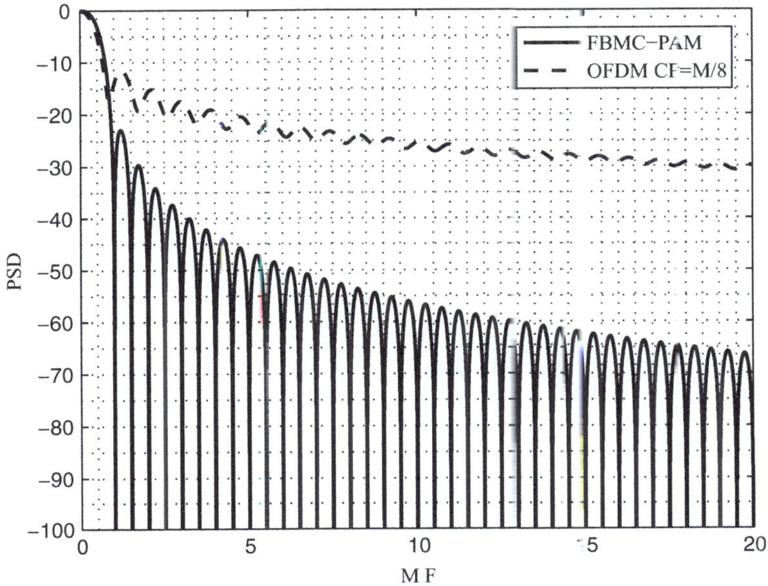

Fig. 8 PSD in dB of the transmitted signal at the edge of the spectral bandwidth (FBMC-PAM). The PSD of the classical OFDM signal with the cyclic prefix $M_p = M/8$ is inserted for comparison

spacing is set to $1/(2M)$ and the subcarriers are used by the $2M$ real data symbols to be transmitted in a multicarrier symbol period. For such a reason, the transceiver is called FBMC-PAM. Figure 8, which reports the PSD of the transmitted signal at the edge of the spectral bandwidth, shows that the achieved levels of spectral containment are similar to those obtained by the signals considered in Figs. 3 and 4: in particular, if the out-of-band attenuation is set to -40 dB, a guard band equal to $2 \times 3/M$ is sufficient. Therefore, FBMC-PAM spectral efficiency is expressed by (10) with $\rho = 1$ and $M_p = 0$.

When the signal $s[n]$ is transmitted on a flat AWGN channel with channel gain α and $r[n]$ denotes the received signal, the ML receiver of the symbol $a_k[i]$ can be written as

$$\Re\left\{\frac{1}{\alpha M}\sum_{n=0}^{2M-1} r[iM+n]\sin\left(\frac{\pi}{2M}\left(n+\frac{1}{2}\right)\right)e^{-j\frac{\pi}{M}\left(k+\frac{1}{2}\right)\left(n+\frac{1}{2}+\frac{M}{2}\right)}\right\} \quad (42)$$

provided that no synchronization error is present and the data symbols are statistically independent. The proof, which is provided in [15], is based on the properties of the Malvar transform.

4.1 Efficient Transceiver Implementation

Transceiver implementation represents a crucial system issue. The structures in (39) for the transmitter side and in (42) for the receiver side represent a basic step: they are mainly useful for understanding the behavior of the proposed transceiver but they are not an effective implementation tool. For implementation purposes, it is important to verify, as proved in [15], that efficient transceiver structures can be obtained. In particular, at the transmitter side, the ith block $[s[iM], s[iM+1], \ldots, s[iM+2M-1]]$ of the FBMC-PAM signal in (40) is the IFFT of the vector whose components, for $k \in \mathcal{K}_{2M}$, are

$$(1-j)\left[\theta\, T_i[\mathrm{mod}_{2M}[k-1]] - T_i[k]\right] \tag{43}$$

with

$$T_i[k] \triangleq \theta^k j^k \frac{a_k[i]}{2\sqrt{2}} \qquad \theta \triangleq \exp\left(j\frac{\pi}{2M}\right). \tag{44}$$

The structure is named [15] frequency-spreading transmitter. Also an efficient receiver structure can be derived; more specifically, at the receiver side, the decision variable in (42) can be obtained as the real part of $D_k[i]$ defined as follows:

$$D_k[i] = \frac{(1+j)j^{-k}}{2\sqrt{2}\alpha}\left[\theta^{-(k+1)} R_i[\mathrm{mod}_{2M}[k+1]] - \theta^{-k} R_i[k]\right] \tag{45}$$

where $R_i[k]$ represents the DFT of $r[iM+n]$:

$$R_i[k] \triangleq \frac{1}{2M}\sum_{n=0}^{2M-1} r_i[n] e^{-j\frac{2\pi}{2M}nk} \qquad r_i[n] \triangleq r[iM+n] \qquad n = 0, 1, \ldots, 2M-1. \tag{46}$$

Such a structure is named frequency-despreading; its implementation complexity is sufficiently small so that the proposed system appears a good compromise between the spectral containment described by Fig. 7 and the computational increase.

Let us observe that the obtained expression (45) for the decision variable in (42) in terms of the FFT of the vector $r_i[n]$ in (46) suggests that frequency domain equalization can be performed by exploiting the FFT of the channel impulse response. Moreover, it also suggests that a CFO compensation procedure analogous to that introduced with reference to FBMC-OQAM can be implemented.

For illustration, the block diagram of the corresponding multicarrier receiver is shown in Fig. 9.

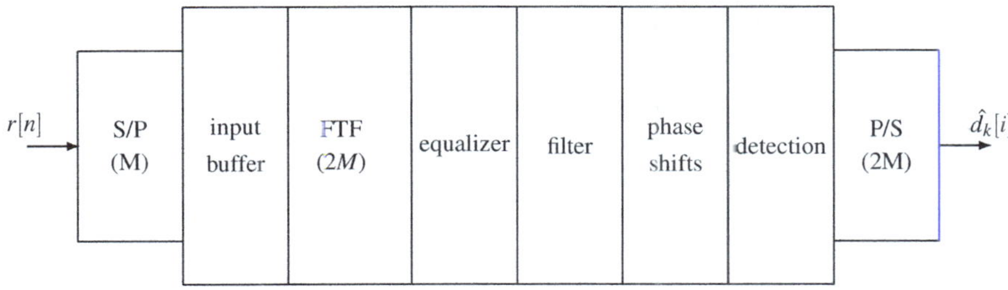

Fig. 9 Efficient implementation of the FBMC-PAM receiver

4.2 Evaluating the Computational Complexity

The computational complexity of the transceiver structure in (44) and (45) has been evaluated in [15]. The total number of real-valued flops (per discrete-time sample) required by the transmitter in (43) can be approximately written as

$$8 \log(M) + 20. \tag{47}$$

Moreover, the total number of real-valued flops (per discrete-time sample) required by the receiver in (45) can be approximately written as

$$8 \log(M) + 28. \tag{48}$$

In the case of the transmitter for classical OFDM the number of real-valued flops per discrete-time sample can be approximately written as follows:

$$4 \log(M). \tag{49}$$

The same expression approximately represents the complexity of the receiver with the single-tap equalizer for classical OFDM. We can therefore verify that the proposed system has a computational complexity more than double than that of the classical OFDM system.

Further improvements of the transmitter structure can be achieved by using a slightly modified structure proposed in [15]; in such a case, it can be shown that the complexity-reduction factor is around 35 % in comparison with that of the implementation structure in (43) and (44). This makes appear practically negligible the increase of the implementation complexity at the transmitter of the proposed structure (in comparison with the classical OFDM system). The comparison with the complexity of the FBMC-OQAM system with $K = 2$ has to take into account that such a system needs to perform a DFT of the same size at a double rate. Therefore, the complexity of the FBMC-OQAM with $K = 2$ is double than that of the FBMC-PAM here considered.

4.3 Further Stages of the FBMC-PAM Transceiver

Analogously to the FBMC-OQAM transceiver, the FBMC-PAM transceiver needs to be equipped with a stage for CFO compensation and with a stage for channel equalization at the receiver side. We do not discuss in detail such stages since they can be implemented following the lines discussed in Sect. 3. In particular, it is reasonable that the CFO compensation procedure described in Sect. 3.3 can be extended to the FBMC-PAM receiver by following an analogous derivation. Consequently, we can assume that, unlike the classical OFDM system, the FBMC-PAM can compensate the CFO at the receiver with marginal increase of its computational complexity. This implies that the doubling of the required complexity discussed in the previous section can become much more limited when the procedure for CFO compensation is taken into account.

The equalization stage can be easily implemented by adopting the FDE described in Sect. 3.4; in fact, we have already discussed its independence of the specific used transceiver. In particular, in the considered receiver, we need to extract a filter output block of length $N = 2M$, as the central part of the extended block of length $\overline{K}M$; here a reasonable choice is $n_c = n_a = M$ which corresponds to the choice $\overline{K}M = 4M$; the resulting transceiver latency amounts to $\overline{K}M - n_c = 3M$. In such a case the computational cost of the equalization stage is given by an FFT over $4M$ points, an IFFT over $2M$ points according to (38), and, finally, the complex-valued multiplications for $H_c^{-1}(\cdot)$ and for the term $e^{j\pi \frac{n}{N}}$ in (38). The number of real-valued flops per each discrete-time sample can be written as:

$$16\log(4M) - 24 + 8\log(2M) - 12 + 12 = 24\log(M) + 16. \tag{50}$$

If we add such value to the receiver complexity in (48) we arrive at about $32\log(M)$, which is eight times larger than the complexity of the classical OFDM receiver reported in (49).

5 CP-Less OFDM and Compatibility

The perfect reconstruction property obtained with the sine filter in the previous section can be extended to other kinds of filters and, particularly, a square filter having M non-zero coefficients equal to unit. Then, the emitted symbols no longer overlap, as in OFDM. Since there is no cyclic prefix, frequency domain equalization is critical and the performance of the scheme is linked to the size of the FFT window employed in the receiver.

The operation of the equalizer is illustrated in Fig. 10. A sequence of symbols generated by the transmitter is applied to the channel and, in the receiver, symbol i has to be restored. The length of the FFT window is $2M$, the length of the channel impulse response is denoted with M_h, the length of the non-negligible causal part

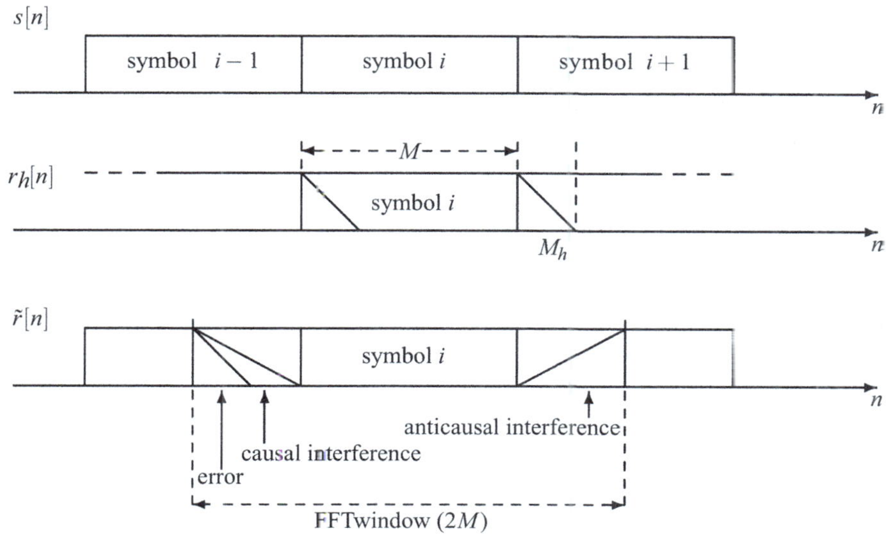

Fig. 10 Frequency domain equalizer operation in CP-less OFDM: the transmitted signal $s[n]$ is a sequence of symbol intervals of length M, not separated by a cyclic prefix. The received signal $r_h[n]$ is the output of the linear convolution where M_h is the channel memory. The error term in the equalizer input is due to the difference between the linear and the circular convolution and has length M_h. The intervals named causal interference of length $M_h + M_c$ and anticausal interference of length M_a represent the samples of the equalizer output $\tilde{r}[n]$ affected by the error at the equalizer input

of its inverse is denoted with M_c, and, finally, the length of the non-negligible anticausal part of the channel inverse is denoted with M_a. At the receiver input, the samples of the symbol of interest i are spread over $M + M_h$ samples. Then, the equalizer restores symbol i in its proper length as well as the sections of the adjacent symbols which falls in the FFT window. However, due to the circular property of the FFT, an error signal of length M_h is present at the equalizer input: it is the sum of the missing samples of symbol $i + 1$ and the overlapping samples of symbol $i - 1$. Then, the circular convolution of this error signal with the inverse channel impulse response generates a causal interference signal of length $M_h + M_c$ and an anticausal signal of length M_a; both the interference samples should not overlap with the desired symbol.

If the desired symbol is positioned in the middle of the FFT window, the condition for perfect restoration of the signal is

$$M_h + M_c < (\overline{K} - 1)M/2, \qquad M_a < (\overline{K} - 1)M/2. \tag{51}$$

In fact, the optimal position for the desired symbol is determined by the channel. A wireless transmission channel can be of the line-of-sight (LOS) type or non line-of-sight (NLOS). Moreover, some residual timing offset may be added. Therefore, the error signal can occur at both ends of the FFT window and, in such conditions, it is reasonable to center the desired symbol.

Condition (51) clearly shows that the inverse channel impulse response has a critical impact on the performance of the system. If necessary, the oversize factor $\overline{K}$ can be increased but at the cost of more computational complexity and processing delay. In practice, $\overline{K}$ is mostly limited to 2, 3 or 5, with delays $1.5M$, $2M$ and $3M$ respectively, and some performance degradation will have to be tolerated for large M_a and M_c values. An important observation is that the error signal in Fig. 10 disappears when symbols $i-1$, i and $i+1$ are identical. This can be used for channel identification with the help of a preamble. The main reason for using CP-less OFDM instead of the classical OFDM is that the scheme is robust to timing offsets, as shown in the next section, and, therefore, it is appropriate for asynchronous access.

6 Simulation Results and Comparison

The objective of the simulations is to compare the above techniques in a scenario where all the multicarrier systems can operate, but with CP-OFDM as the reference. Therefore, we are considering a scenario where the OFDM system is working with optimum performance and minimal receiver complexity and transceiver latency while we are interested to evaluate the loss of the alternative systems in terms of computational complexity, transceiver latency, and FDE equalizer performance needed to achieve the improvement in terms of spectral efficiency. In particular, we are interested in determining the values of $n_c = n_a$ needed to achieve a satisfactory performance in the case $K = 2$ because we are interested to limit the transceiver latency.

The main settings of our experiments follow:

- we have set the parameter $M = 256$; note that the subcarrier spacing is equal to $1/M$ in OFDM and FBMC-OQAM systems and $1/(2M)$ in FBMC-PAM;
- we have set the prototype filter for the FBMC-OQAM system according to (18) with $K = 2$;
- we have used as channel model the ITU-R vehicular B [16] where the different paths have the following discrete delays [0 1 25 36 48 56] and the following rms amplitudes [0.75 1 0.23 0.316 0.055 0.16];
- we have set the percentage of active subcarriers in OFDM, FBMC-OQAM, and FBMC-PAM transceivers as 82, 89, and 89 % of the overall number of subcarriers, respectively. Taking into account also the presence of the cyclic prefix in the OFDM system (we have set $M_p = M/4$), the ratio between the FBMC spectral efficiency and the OFDM spectral efficiency is equal to 1.36;
- the BER is obtained by averaging over all the active sub-carriers of a given multicarrier symbol and over 10^4 independent channel realizations;

- we have used the single-tap equalizer for OFDM and for FBMC-OQAM transceiver equipped with the standard polyphase receiver while we have used the FDE structure described in Sect. 3.4 with different values of $n_c = n_a$ for the FBMC-PAM and for the CP-less OFDM. More specifically, we have denoted the different solutions under test as follows:
  - *one-tap-FBMC-2M*: the FBMC-OQAM system with $K = 2$, using the standard polyphase receiver and the single-tap equalizer;
  - *FS-FBMC-2M*: the FBMC-OQAM system with $K = 2$, using the frequency-despreading receiver and the single-tap equalizer;
  - *OFDM-CP=0*: the CP-less OFDM system using the single-tap equalizer;
  - *FBMC-PAM-2M*: the FBMC-PAM system using the single-tap equalizer;
  - *OFDM-3M*: the CP-less OFDM system using the FDE with $n_c = n_a = M$;
  - *FBMC-PAM-2M-4M*: the FBMC-PAM using the FDE with $n_c = n_a = M$;
  - *OFDM-5M*: the CP-less OFDM system using the FDE with $n_c = n_a = 2M$;
  - *OFDM CP=M/4*: the classical OFDM system with cyclic prefix $M_p = 64 = M/4 > M_h = 56$ using the single-tap equalizer.

Figure 11 shows the performance of the considered systems when the 4QAM/2PAM symbol constellation is used. The first obvious result that can be noted from the figure is that the CP-less OFDM cannot work with the simple single-tap equalizer; the other solutions using the single-tap equalizer provide performance

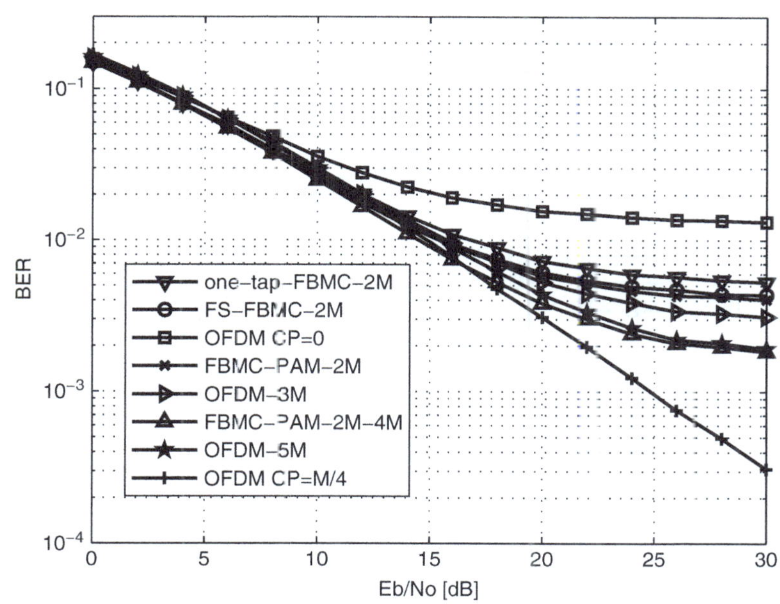

Fig. 11 Performance of the different systems under test in the considered scenario when the 4QAM/2PAM symbol constellation is used

much similar among them, except for the solution with polyphase structure, with slightly worse performance. Such solutions exhibit significant performance loss in comparison with the classical OFDM when E_b/N_0 becomes larger than 15 dB. The solutions using the FDE structures exhibit the best performance; among them, the CP-less OFDM exhibits the same performance of the FBMC-PAM only when the same transceiver latency is accepted (in fact, the solution "OFDM-5M" uses $n_a = 2M$ so arriving at a total transceiver latency of $3M$, which is the same latency of the solution "FBMC-PAM-2M-4M") while the solution "OFDM-3M" with minor latency also exhibits worse performance. It is interesting to note that the choice $n_a = M$ is sufficient to achieve good performance for the FBMC-PAM structure; in practice, it shows significant performance loss in comparison with the OFDM system, which is transmitting at a minor rate on the same bandwidth, only when $E_b/N_0 > 20$ dB. Analogous results are shown in Fig. 12: the use of the 64QAM/8PAM constellation makes notice more clearly the differences among the different structures. We can observe in particular that the polyphase structure of the FBMC-OQAM receiver with single-tap equalizer is much worse than the other structures using the single-tap equalizer. Moreover, the performance of the CP-less OFDM receiver with $n_a = M$ is similar to that achieved by the single-tap structures while also the choice $n_a = 2M$ pays about 4 dB in terms of E_b/N_0 in comparison with the classical OFDM when the required BER is set to 10^{-2}.

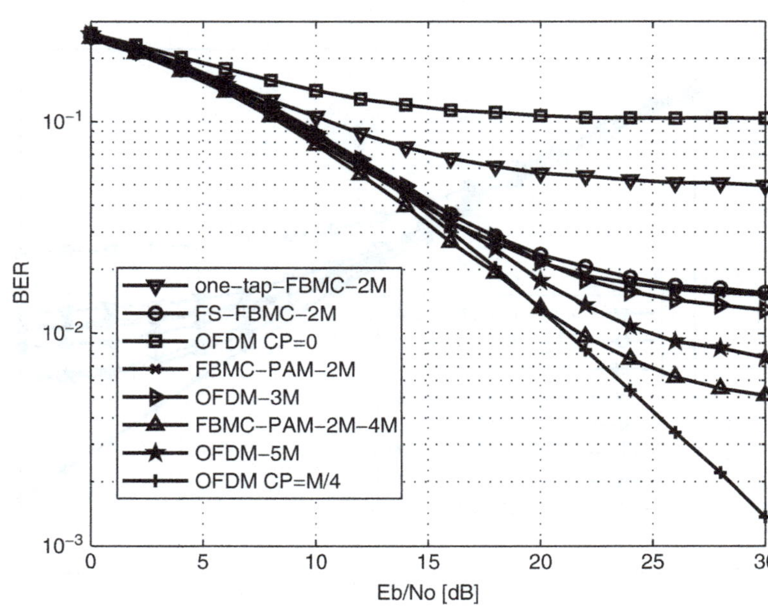

Fig. 12 Performance of the different systems under test in the considered scenario when the 64QAM/8PAM symbol constellation is used

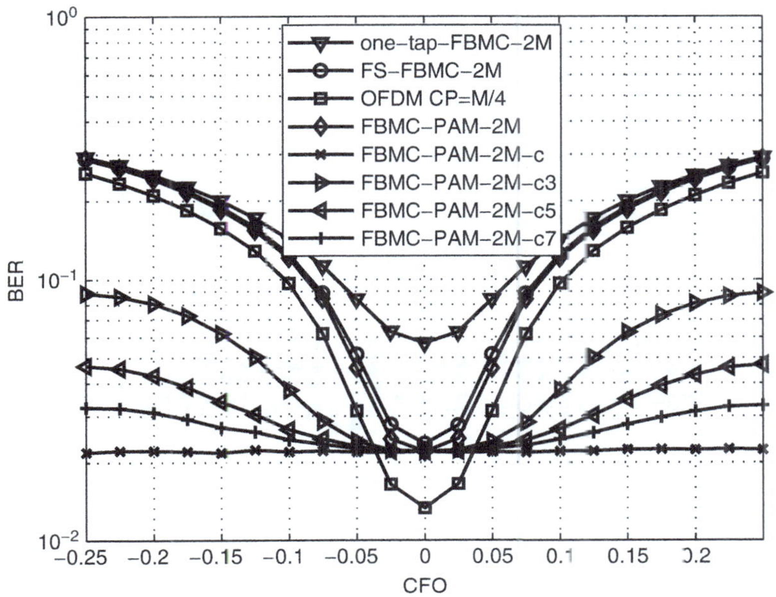

Fig. 13 Performance of the different systems under test as a function of the CFO with $E_b/N_0 =$ 20 dB and using a 64QAM/8PAM constellation

Finally, note that practically no loss in terms of E_b/N_0 is payed by the FBMC-PAM-2M-4M solution for a target BER of 10^{-2}. Therefore, such solution, which starts to pay a performance loss in comparison with the classical OFDM system only for $E_b/N_0 > 22$ dB, is able to gain its spectral-efficiency advantage in comparison with the classical OFDM system by paying with a factor three in terms of transceiver latency and with a factor eight in terms of computational complexity.

The considered complexity comparison does not take into account the cost of the CFO compensation procedure and the cost of the procedures for timing and CFO alignment among the different users, also because the procedures for CFO compensation to be employed in FBMC-PAM systems have not been defined yet in their details. However, we have used a general less efficient procedure described in [5] for the CFO compensation. We have reported in Fig. 13 the results of the simulation experiments (carried out for $E_b/N_0 = 20$ dB and using a 64QAM/8PAM constellation) aimed at verifying the effect of an uncompensated CFO on the most important of the considered solutions but also the performance achieved by applying to FBMC-PAM the CFO compensation procedure described in [5]. We denote with "FBMC-PAM-2M-c" the perfectly compensated structure with an heavy computational requirement and with "FBMC-PAM-2M-cB" the structure employing B coefficients for the compensation procedure. We can notice that all the solutions analyzed in Fig. 13 are sensitive to the uncompensated CFO but, however, a few coefficients are sufficient for CFO compensation in the FBMC-PAM

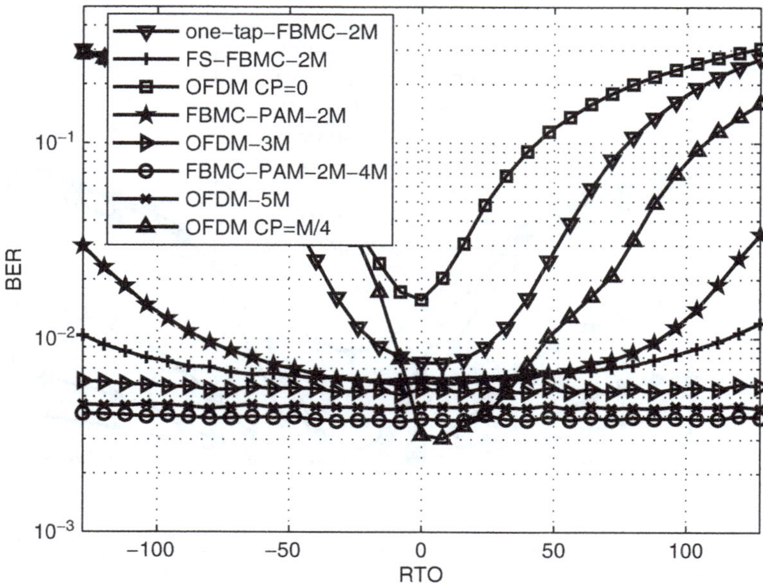

Fig. 14 Performance of the different systems under test as a function of the residual timing offset (RTO) with $E_b/N_0 = 20\,\text{dB}$ and using a 4QAM/2PAM constellation

system. When a procedure analogous to that described in (26) for FBMC-OQAM will be available, the increase of the computational complexity in comparison with OFDM systems will become smaller; the results show that there is space for such improvement in CFO compensation for FBMC-PAM systems which is able to converge also with a minor number of coefficients.

Finally, we have analyzed the effect of a timing offset on the received signal. We have considered that the residual timing offset is estimated after that the receiver FFT has been performed and we analyzed the performance of the different methods for frequency-domain timing-offset compensation. The results of the experiment carried out by setting $E_b/N_0 = 20\,\text{dB}$ and using a 4QAM/2PAM constellation, reported in Fig. 14, confirm an obvious result: the single-tap equalizer structures cannot compensate the residual timing offset while those employing the FDE can compensate almost perfectly and without significantly increasing their computational complexity. It is worth noting that, among the solutions employing single-tap equalizer, the FBMC-OQAM and the FBMC-PAM exhibit a minor performance loss due to imperfections in timing-offset compensation. Moreover, the results of the experiment carried using the 64QAM/8PAM constellation, reported in Fig. 15, confirm the previous results and show that the advantage of the structures using the FDE is also larger.

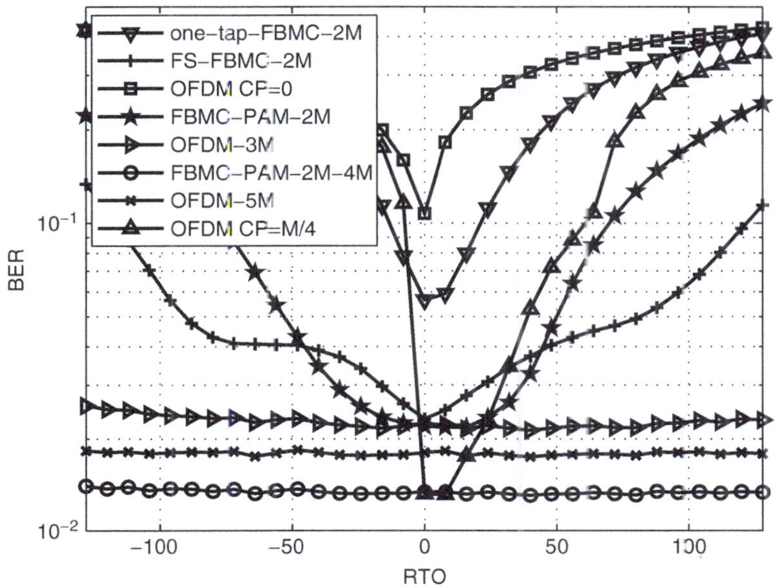

Fig. 15 Performance of the different systems under test as a function of the residual timing offset (RTO) with $E_b/N_0 = 20$ dB and using a 64QAM/8PAM constellation

7 Conclusions

Some evolutions of OFDM, particularly per-user filtering and pulse shaping, lead to significant improvements in spectral separation, with moderate loss in spectral efficiency and limited increase in complexity. Although asynchronous access may remain an issue with these techniques, they are likely to be strong candidates for 5G standardization. A different perspective, which can be complementary, has been proposed above, with different trade-offs in terms of performance and complexity. It aims at maximum spectral efficiency, with adjustable performance in spectral separation and transmission characteristics. The principle is to use filter banks with frequency domain processing, exploiting the FFT. Of the two approaches presented, namely FBMC-OQAM and FBMC-PAM, the latter seems particularly appropriate to meet the objectives of 5G in terms of spectral separation, limited latency increase with respect to CP-OFDM and asynchronous access capability. In the baseline option, the increase in computational complexity is limited If necessary, the equalizer performance can be enhanced, at the cost of more computations in the receiver, as well as more processing delay. Overall, the FBMC approach, and particularly the FBMC-PAM technique, is likely to be another strong candidate for 5G standardization.

References

1. J. Abdoli, M. Jia, J. Ma, Filtered OFDM: a new waveform for future wireless systems, in *2015 IEEE 16th International Workshop on Signal Processing Advances in Wireless Communications (SPAWC)* (2015), pp. 66–70
2. E. Bala, J. Li, R. Yang, Shaping spectral leakage: a novel low-complexity transceiver architecture for cognitive radio. IEEE Veh. Technol. Mag. **8**(3), 38–46 (2013)
3. P. Banelli, S. Buzzi, G. Colavolpe, A. Modenini, F. Rusek, A. Ugolini, Modulation formats and waveforms for 5G networks: who will be the heir of OFDM? An overview of alternative modulation schemes for improved spectral efficiency. IEEE Signal Process. Mag. **31**(6), 80–93 (2014)
4. M. Bellanger, Specification and design of a prototype filter for filter bank based multicarrier transmissions, in *2001 Proceedings of the IEEE International Conference on Acoustics, Speech, and Signal Processing (ICASSP)*, vol. IV (2001), pp. 2417–2420
5. J. Choi, C. Lee, H.W. Jung, Y.H. Lee, Carrier frequency offset compensation for uplink OFDM-OFDMA systems. IEEE Commun. Lett. **4**(12), 414–416 (2000)
6. J. Erfanian, R.E. Hattachi, B. Daly et al., NGMN 5G white paper (2015). Available at www.dieti.unina.it/dieti/nostro.pdf
7. B. Farhang-Boroujeny, C.H. Yuen, Cosine modulated and offset QAM filter bank multicarrier techniques: a continuous-time prospect. EURASIP J. Adv. Signal Process. **2010**, 16 (2010)
8. M. Gudmundson, P.O. Anderson, Adjacent channel interference in an OFDM system, in *IEEE 46th Vehicular Technology Conference Mobile Technology for the Human Race*, vol. 2 (1996), pp. 918–922
9. B. Hirosaki, An orthogonally multiplexed QAM system using the discrete fourier transform. IEEE Trans. Commun. **29**(7), 982–989 (1981)
10. J. Li, E. Bala, R. Yang, Sliding-window frequency-domain equalization for multi-mode communication systems, in *Proceedings of 9th Annual Conference on Systems Applications and technology (LISAT 2013), Long Island, NY, USA, May 3* (2013), pp. 1–6
11. A. Loulou, M. Renfors, Enhanced OFDM for fragmented spectrum use in 5G systems. Trans. Emerg. Telecommun. Technol. **26**, 31–45 (2015). doi:10.1002/ett.2898
12. H. Malvar, Modulated QMF filter banks with perfect reconstruction. Electron. Lett. **26**(13), 906–907 (1990)
13. D. Mattera, M. Tanda, M. Bellanger, Analysis of an FBMC/OQAM scheme for asynchronous access in wireless communications. EURASIP J. Adv. Signal Process. **2015**, 23 (2015)
14. D. Mattera, M. Tanda, M. Bellanger, Frequency domain CFO compensation for FBMC systems. Signal Process. **114**, 183–197 (2015)
15. D. Mattera, M. Tanda, M. Bellanger, Filter bank multicarrier with PAM modulation for future wireless systems. Signal Process. **120**, 594–606 (2016)
16. Recommendation ITU-R M. 1225, Guidelines for evaluation of radio transmission technologies for IMT-2000 (1997)
17. V. Vakilian, T. Wild, F. Schaich, S. ten Brink, J.F. Frigon, Universal-filtered multi-carrier technique for wireless systems beyond LTE, in *IEEE Globecom Workshops (GC Wkshps)* (2013), pp. 223–228
18. J.J. van de Beek, P.O. Brjesson, M.L. Boucheret, D. Landstrom, J.M. Arenas, P. Odling, S.K. Wilson, Three non-pilot based time- and frequency estimators for OFDM. Signal Process. **80**(7), 1321–1334 (2000)
19. G. Wunder, M. Kasparick, S. ten Brink, F. Schaich, T. Wild, I. Gaspar, E. Ohlmer, S. Krone, N. Michailow, A. Navarro, G. Fettweis, D. Ktenas, V. Berg, M. Dryjanski, S. Pietrzyk, B. Eged, 5GNOW: challenging the LTE design paradigms of orthogonality and synchronicity, in *2013 IEEE 77th Vehicular Technology Conference (VTC Spring)* (2013), pp. 1–5
20. G. Wunder, P. Jung, M. Kasparick, T. Wild, F. Schaich, Y. Chen, S. Brink, I. Gaspar, N. Michailow, A. Festag, L. Mendes, N. Cassiau, D. Ktenas, M. Dryjanski, S. Pietrzyk, B. Eged, P. Vago, F. Wiedmann, 5GNOW: non-orthogonal, asynchronous waveforms for future mobile applications. IEEE Commun. Mag. **52**(2), 97–105 (2014)

Fundamentals of Faster-than-Nyquist Signaling

Angelos D. Liveris and Costas N. Georghiades

Abstract This chapter presents the fundamentals of Faster-than- Nyquist (FTN) signaling. As originally introduced, FTN increases the bit-rate in the signaling bandwidth by packing symbols closer in time, at the cost of introducing intersymbol interference (ISI). We begin with the Euclidean distance properties of bandwidth efficient pulses at FTN rates and describe receivers that mitigate the severe ISI. The FTN achievable information rate is compared with the Nyquist information rate for practical pulses. We then discuss FTN extension to multicarrier systems with not only time packing but also subcarrier, optimizing both the time and frequency packing.

1 Introduction

The basic idea behind Faster-than-Nyquist signaling is based on challenging the time orthogonality assumption in a typical single carrier system shown in Fig. 1 for the binary case. In such a system the transmit signal is

$$s_{\mathbf{a}}(t) = \sum_{k} a_k g(t - k\tau), \tag{1}$$

where $\boldsymbol{a} = [a_k]$ is the sequence of transmitted pulse-amplitude modulation (PAM) symbols and $g(t)$ is the transmit pulse shaping filter. The transmit signal is then passed through the channel, such as an additive white Gaussian noise (AWGN) channel considered here with zero-mean noise $n(t)$ and power-spectral density $N_0/2$.

A.D. Liveris (✉)
Microwave Networks Inc., 4000 Greenbriar, Stafford, TX 77477
e-mail: alive@ieee.org

C.N. Georghiades
Electrical and Computer Engineering Department, Texas A&M University, College Station, TX 77843–3128
e-mail: georghiades@tamu.edu

W. Xiang et al. (eds.), *5G Mobile Communications*,
DOI 10.1007/978-3-319-34208-5_8

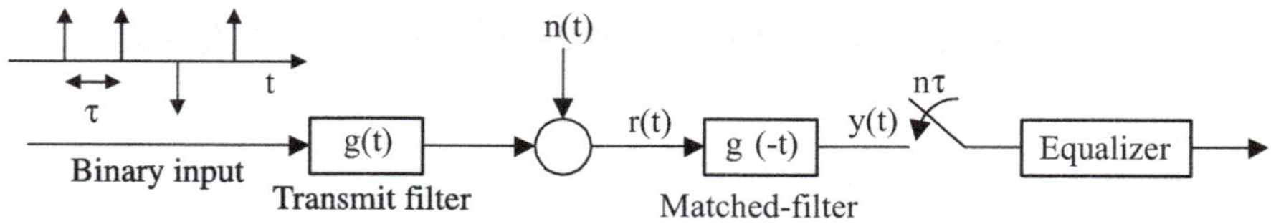

Fig. 1 Single carrier communication system for binary faster-than-Nyquist signaling

The received signal $r(t)$ is first passed through a receive matched filter $g^*(-t)$, resulting in the output signal

$$y(t) = \sum_k a_k x(t - k\tau) + n'(t), \tag{2}$$

where $x(t) = g(t) * g^*(-t)$ and $n'(t)$ is the filtered noise with autocorrelation $R_{n'n'}(\Delta t) = x(\Delta t)N_0/2$. The matched filter output signal $y(t)$ is then sampled at the signaling rate $1/\tau$

$$y(n\tau) = \sum_k a_k\, x\left((n-k)\tau\right) + n'(n\tau). \tag{3}$$

To limit the bandwidth occupied by the transmit signal, a limited bandwidth transmit filter $g(t)$ is used. The ideal filter is the sinc filter with a strictly bandlimited, square pulse shaped frequency response. This results in the end-to-end response being a sinc pulse, i.e.,

$$x(t) = E_s \operatorname{sinc}\left(\frac{t}{T}\right) = E_s \frac{\sin(\pi t/T)}{\pi t/T}, \tag{4}$$

where assuming baseband transmission $W = 1/2T$ is the bandwidth it occupies. E_s is the average PAM symbol energy.

If we use the ideal sinc pulse and the successive signal pulses are transmitted at $T = \tau$, then the matched filter output samples become

$$y(nT) = a_n\, x\,(0) + n'(nT), \tag{5}$$

i.e., each sample only depends on a single transmitted symbol and the noise samples become independent, significantly simplifying the PAM symbol detection process. This is the Nyquist constraint or time orthogonality principle and the $1/T$ signaling rate is called the Nyquist rate.

To improve bandwidth efficiency, we could signal faster with $\tau < T$ so that the signal bandwidth, e.g., using the sinc pulse, does not change, but more PAM symbols are transmitted in the same time as shown in Fig. 2. In this case the matched filter output samples in (3) will include interference from the other symbols in the

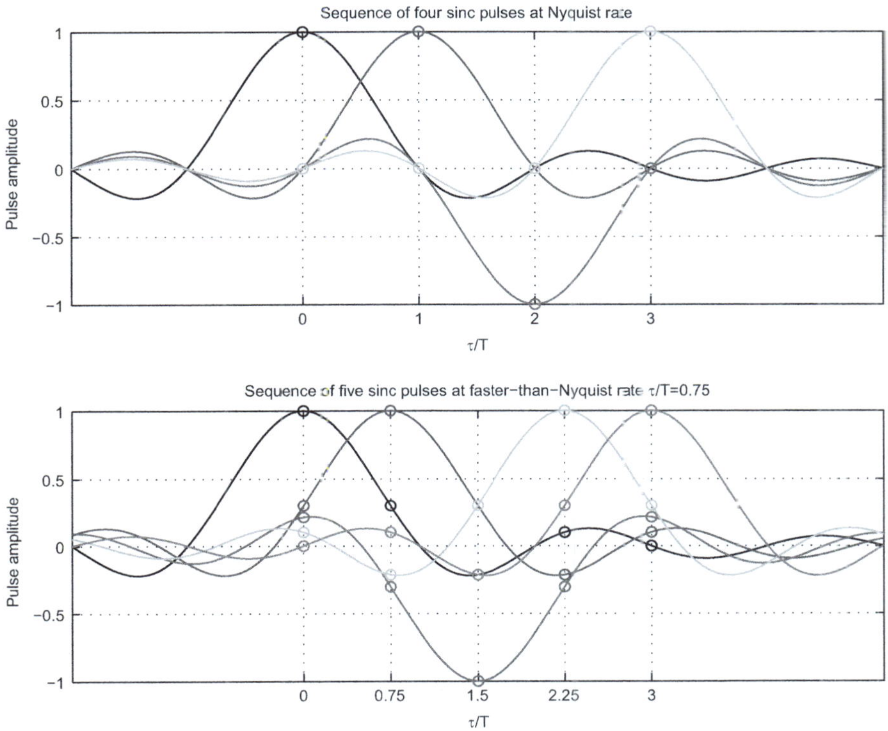

Fig. 2 Sequence of sinc pulses at the Nyquist rate (*top*) and at the faster-than-Nyquist rate $\tau/T = 0.75$ (*bottom*). The values of each pulse at the sampling instants is also shown in both cases

sequence as the Nyquist constraint no longer holds as the sampling instants in Fig. 2 no longer correspond to the zero crossings of the neighboring pulses.

Surprisingly, with the ideal sinc pulses Mazo [1] was the first to realize that the minimum Euclidean distance between binary sequences transmitted at a rate up to 25 % higher than Nyquist was the same as the minimum distance of sequences transmitted at the Nyquist rate. This minimum distance limit is referred to as the Mazo limit in the FTN literature.

Since the minimum Euclidean distance between sequences referred to maximum likelihood detection and the intersymbol interference (ISI) between symbols when signaling faster than the Nyquist rate was severe, theoretically infinite for the sinc pulse, achieving the promised minimum distance gains meant significant receiver complexity. Given that at about the same time the Viterbi algorithm was just being introduced for ISI receivers [2, 3], it seemed unrealistic to come up with a limited complexity receiver that takes advantage of Mazo's minimum distance results [4]. Mazo himself admitted that his FTN research was triggered by "curiosity" [5].

In the decades that followed several developments changed this landscape:

(1) First, reduced complexity receivers started being introduced with performance close to the maximum likelihood receivers even with severe intersymbol interference [6–10]. This development allowed handling the FTN intersymbol interference with reasonable complexity.
(2) Second, the introduction of turbo codes [11] led to applying the turbo concept to several other scenarios, such as turbo equalization [12]. This allowed achieving coding gains over channels with severe ISI with an iterative receiver at reasonable complexity. The channel capacity approaching turbo codes also directed more focus on the achievable information rates than just the maximum likelihood performance. This further helped strengthen the FTN approach as a way to take advantage of all available frequencies in the channel as we will see next.
(3) The extension of the "time-packing" approach of FTN to frequency-packing in a multicarrier system [13] as well as to combined time and frequency packing, the two-dimensional Mazo limit [14], showed even more promising minimum distance gains.
(4) Modern telecommunication systems and their constant quest for increased throughput started reaching a limit of resources, especially bandwidth, even though processing power kept growing.

All these developments led to renewed interest in FTN lately as signal processing techniques used in current telecommunication systems keep advancing. Meanwhile, resources, such as bandwidth, have become more scarce and their efficiency must improve even at the cost of increased processing complexity. So, FTN is considered for the 5G standard [15] and for satellite and optical communications [16, 17].

In the following sections we address several key aspects of faster-than-Nyquist (FTN) signaling.

2 Minimum Distance

The ideal sinc pulse discussed in the Introduction is a special case of the general family of raised-cosine pulses [18]

$$x(t) = E_s \,\mathrm{sinc}(t/T)\, \frac{\cos(\pi \alpha t/T)}{1-(2\alpha t/T)^2}, \tag{6}$$

with $0 \leq \alpha \leq 1$ being the roll-off factor and $W = (1+\alpha)/(2T)$ the bandwidth they occupy. The ideal sinc pulse has $\alpha = 0$. In practical communications systems a non-zero roll-off factor α is used, typically with α in the range of 0.1–0.5 for higher bandwidth efficiency.

ISI free transmission satisfying the Nyquist constraint for these pulses means $\tau = T = (1+\alpha)/(2W) \geq 1/(2W)$. In [19] Mazo's d_{min} observations for the binary case were numerically extended to non-zero roll-off α as shown in Fig. 3 for roll-off

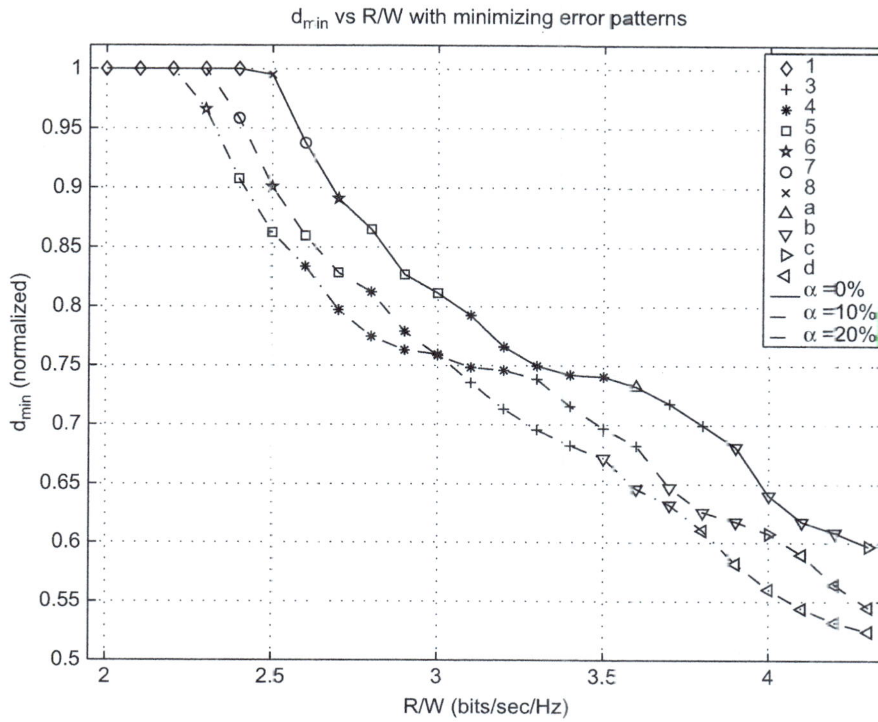

Fig. 3 Minimum Euclidean distance d_{min} as a function of the bandwidth efficiency for binary faster-than-Nyquist signaling (FNS) for the sinc pulse ($\alpha = 0\,\%$) and the raised-cosine pulses with roll-off $\alpha = 10, 20\,\%$. The error sequences $\boldsymbol{e}_{min}$ are also shown. The numbers show the length of the single alternating block $\boldsymbol{e}_{min}$ [e.g., $5 = \pm(\ldots 000 - + - + - 000 \ldots)$] and the letters stand for $\boldsymbol{e}_{min}$ containing more than one block [$a = \pm(\ldots 000 - + - 0 + - + 000 \ldots)$, $b = \pm(\ldots 000 - +0 - +0 - +000 \ldots)$, $c = \pm(\ldots 000 - +0 - +00 - +0 - +000 \ldots)$, $d = \pm(\ldots 000 - +0 - +000 \ldots)$]

factors of $\alpha = 0, 10, 20\,\%$. The minimum distance is plotted as a function of the bandwidth efficiency R/W, i.e., the ratio of the signaling rate $R = 1/\tau$ to the utilized bandwidth $W = (1+\alpha)/(2T)$ (measured in bits/s/Hz). The Nyquist (orthogonal) signaling rate is $1/T$ corresponding to a bandwidth efficiency of $2/(1+\alpha)$ bits/s/Hz. The highest Nyquist rate bandwidth efficiency is 2 bits/s/Hz, achieved by the ideal sinc pulse ($\alpha = 0$).

We define the normalized minimum Euclidean distance d_{min} as

$$d_{min} = \min_{\boldsymbol{e} \neq \boldsymbol{0}} \frac{d(\boldsymbol{e})}{2\sqrt{E_b}}, \tag{7}$$

where $\boldsymbol{e} = \boldsymbol{a} - \boldsymbol{b}$ is the difference (error) sequence between the transmitted sequences $\boldsymbol{a}$ and $\boldsymbol{b}$, with $e_k = (a_k - b_k) \in \{0, \pm 2\}$ for the binary case, and the distance $d(\boldsymbol{e})$ defined as

$$d^2(\boldsymbol{e}) = \int_{-\infty}^{\infty} [s_a(t) - s_b(t)]^2 \, dt = \sum_k \sum_l e_k e_l x\left((k-l)\tau\right). \tag{8}$$

Figure 3 shows that the d_{min} between binary sequences is not reduced below the d_{min} for the Nyquist rate case. Even for non-zero roll-off this d_{min} behavior extends to high enough bandwidth efficiency R/W that overcomes the bandwidth efficiency loss caused by the non-zero roll-off.

As long as there is no change in d_{min} compared to the ISI-free case, the error sequence $\boldsymbol{e}_{min}$ that yields d_{min} is one with only a single nonzero component; i.e., a pair of input binary sequences is at d_{min} if the two sequences differ at a single position. Surprisingly, the error patterns $\boldsymbol{e}_{min}$ that first cause degradation in d_{min} contain a single long block of alternating polarity and, more generally, they consist of one or more blocks of alternating polarity, as Fig. 3 shows. This was also observed in [1, 20, 21] for sinc pulses.

These error events $\boldsymbol{e}_{min}$ are depicted for all three roll-off factors. The numbers in the legend stand for error sequences containing a single block of alternating polarity of this length; for example, 5 represents the error pattern $\pm(\ldots 000 - + - + - 000\ldots)$, where $+$ and $-$ correspond to $e_k = 2$ and $e_k = -2$, respectively. Each letter corresponds to an error pattern consisting of more than one block. Thus, $a = \pm(\ldots 000 - + - 0 + - + 000\ldots)$, $b = \pm(\ldots 000 - +0 - +0 - +000\ldots)$, $c = \pm(\ldots 000 - +0 - +00 - +0 - +000\ldots)$, $d = \pm(\ldots 000 - +0 - +000\ldots)$. Close to the "knee", the point where d_{min} starts decreasing, the d_{min} error event is a single block of alternating polarity, while for higher rates, first its length is reduced and then it starts splitting into two or three alternating blocks.

Such patterns can only arise from a pair of binary sequences that include alternating blocks of some length with opposite polarity. What this means for the single block of alternating polarity close to the "knee" is that the probability of such sequences is rather low. Therefore, the d_{min} reductions close to the "knee" are not likely to affect the maximum likelihood performance of systems operating at this FTN rate [1]. The lower bound on the bit error rate is given by combining the error event lower bound from [22]

$$P_{event} \geq \Pr\left[\boldsymbol{e}_{min}\right] Q\left(d_{min}\sqrt{\frac{2E_b}{N_0}}\right) \tag{9}$$

with the number of bit errors resulting from $\boldsymbol{e}_{min}$, i.e.,

$$P_e \geq \sum_k |e_{min,k}| \Pr\left[\boldsymbol{e}_{min}\right] Q\left(d_{min}\sqrt{\frac{2E_b}{N_0}}\right) \tag{10}$$

where $\boldsymbol{e}_{min} = [e_{min,k}]$ and $Q(\cdot)$ is the area under the tail of the Gaussian probability density function.

For example, at the bandwidth efficiency of $R/W = 2.4$ bits/s/Hz with $\alpha = 10\,\%$, $d_{min} = 0.959$ and

$$\boldsymbol{e}_{min} = \pm[\ldots 0,\ 0,\ 0,\ -2,\ +2,\ -2,\ +2,\ -2,\ +2,\ -2,\ 0,\ 0,\ 0,\ldots]. \quad (11)$$

For this example the lower bound of the maximum likelihood performance is given by

$$P_e \geq \frac{7}{64} Q\left(0.959\sqrt{\frac{2E_b}{N_0}}\right). \quad (12)$$

Due to the low multiplicity of the d_{min} error event in this example, this lower bound is less than the no-ISI $Q\left(\sqrt{\frac{2E_b}{N_0}}\right)$ performance even for medium-to-high SNR as we will see in the next section.

These results were extended to non-binary PAM symbols in [23], where it was found that for the same roll-off factor α that the same τ caused the binary, the 4-PAM, and the 8-PAM minimum distance to start decreasing from the corresponding binary, 4-PAM, and 8-PAM pairwise (no-ISI) distance. Of course, at the same τ 8-PAM and then 4-PAM achieve better bandwidth efficiency. The target bandwidth efficiency determines the best combination of signaling alphabet and τ, which is operating close to the d_{min} curve "knee" of each signaling alphabet. For example, with $\alpha = 30\,\%$ at bandwidth efficiency of 4–5 bits/s/Hz the 4-PAM faster-than-Nyquist d_{min} is larger than the binary and 8-PAM faster-than-Nyquist d_{min} [23].

Another approach based on the structured form of error events and the associated binary sequences is to use constrained codes that do not allow the occurrence of input sequences to the transmit filter that generate such error events [19]. This keeps d_{min} constant for even higher signaling rates at the expense of rate loss and also allows the use of limited complexity receivers.

An alternative view taken in [23] is to use a short linear precoding filter with 4–8 taps in the transmitter that does not add any redundancy. Optimizing its coefficients can improve the binary and 4-PAM d_{min} without the rate loss penalty of the constrained coding approach. This approach adds additional complexity to the receiver side.

3 Reduced Complexity Receivers for Uncoded FTN

The intersymbol interference (ISI) introduced by FTN signaling in a single matched filter output sample extends to a large number of symbols. So, the maximum likelihood approach [2, 3] of considering all the interfering symbols results in significant complexity that is not possible to implement.

The first approach to limiting the complexity of a detector handling the severe ISI caused by FTN is to truncate the number of states in the Viterbi algorithm. This means that only the ISI from the neighboring symbols, which typically cause most of the ISI, as shown in Table 1, is managed by the detector while the rest of the ISI is left untreated leading to performance degradation.

Approximating this residual ISI as Gaussian noise leads to the realization that it can cause an error floor at high signal-to-noise ratio (SNR) as the residual ISI is constant and does not change with the SNR [19].

Fortunately, there are several approaches to considering the complete ISI or to mitigating the effect of the residual ISI within the detector. One of the simplest approaches is to include decision feedback cancelation of the residual ISI inside the Viterbi algorithm as proposed in [10].

Results for both these approaches, truncating the Viterbi algorithm memory without and with decision feedback, can be seen in Fig. 4. The d_{min} lower bound for FTN is based on the d_{min} discussion of the previous section.

Another detection approach presented in [19] was based on a soft-output reduced state Viterbi detection followed by linear filtering to cancel the residual ISI.

Other reduced complexity approaches are based on the M-algorithm, which can handle non-binary modulation with lower complexity than reduced state detection [23]. In [23] it was also observed that the M-algorithm does not work well with the Ungerboeck receiver model used in this section [3, 10], which operates directly on the matched filter output samples. Instead, an approximate whitened matched filter (WMF) is employed, following Forney's receiver model [2], with which the M-algorithm performs well despite the minor mismatch required to implement the WMF. Further analysis and advantages of the WMF model for FTN signaling are explored in [24].

4 Turbo Equalization for Coded FTN

The previous section presented low-complexity receivers for uncoded FTN transmission. In this section we address low-complexity receivers for coded FTN transmission. Figure 5 shows such a binary coded FTN system. The channel encoder output is interleaved before it is passed to the FTN pulse shaping. At the receiver the matched filter output samples are now processed by a turbo equalizer, which performs iterative detection and decoding [12].

When combining FTN ISI detection and decoding in the turbo equalizer, there are two aspects that need to be considered:

(1) A reduced complexity soft-input soft-output (SISO) detector is needed for the severe ISI introduced by FTN signaling. This way the soft output can be input into the channel decoder and allow iterations between the two receiver modules.

Table 1 Raised cosine ISI coefficients for the closest ten pulses from one side for $\alpha = 10\,\%$ and $\tau/T = 0.75$

$x(0)$	$x(0.75T)$	$x(1.5T)$	$x(2.25T)$	$x(3T)$	$x(3.75T)$	$x(4.5T)$	$x(5.25T)$	$x(6T)$	$x(6.75T)$	$x(7.5T)$
1.0000	0.2985	−0.2078	0.0954	0.0000	−0.0525	0.0582	−0.0328	−0.0000	0.0212	−0.0240

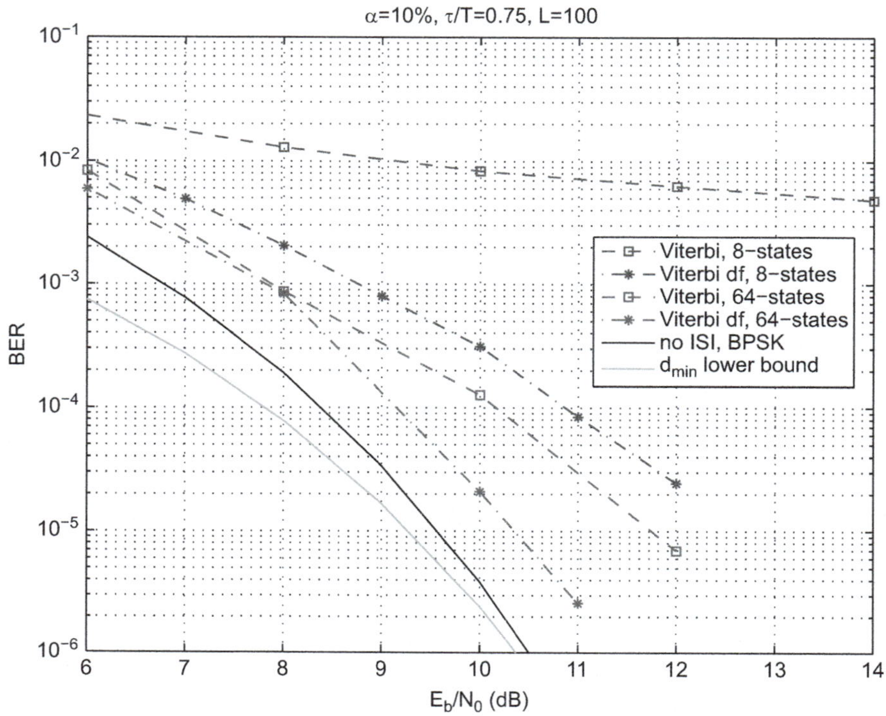

Fig. 4 Reduced state Viterbi BER for the raised-cosine pulses with roll-off $\alpha = 10\,\%$ with bandwidth efficiency 2.42 bits/s/Hz

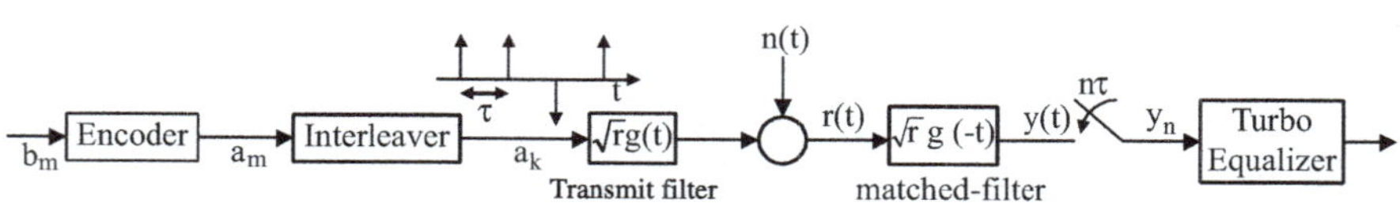

Fig. 5 Communication system for coded binary faster-than-Nyquist signaling

(2) The effect of the interleaver is important not just due to its ability to decorrelate the iterative equalizer detector and decoder input data but also due to the nature of the long error events with FTN signaling. The interleaver will allow breaking the long alternating polarity error events, so, they can more easily be corrected by the decoder.

An advantage of the turbo equalizer is that the channel decoder helps make the suboptimum ISI detection more reliable, especially, its residual ISI cancelation, which in some reduced complexity detectors, such as the decision feedback we considered in the uncoded FTN detection section, could partly be based on tentative decisions.

What the analysis of the information rates suggests (more details in the next section) is that even the use of a very low complexity ISI detector combined with a channel decoder in an iterative (turbo) equalization scheme can work as well as a more complex ISI detector. The ISI detector beyond the SISO requirement must also consider the whole length of the ISI, even if it just performs tentative-decision cancelation of the interference.

To demonstrate this, we consider here the approach from [25] simplified for the binary case and with some adjustments from [19]. The basic idea is that the mean value of the FTN received symbols is used to cancel their effect on neighboring symbols and the uncertainty in these FTN received symbols is included in the generation of the soft-output probabilities.

When the FTN symbols are received, their statistics consist only of their received values. So, rewriting (3)

$$y_n = y(n\tau) = x(0)a_n + \sum_{k\neq n} a_k\, x\left((n-k)\tau\right) + n'(n\tau). \tag{13}$$

the intersymbol interference in y_n from the neighboring symbols can be approximated as an additional source of Gaussian noise when considering only the current sample to make a decision on a_n. The channel information can be written as the following log-likelihood ratio

$$L_n^{ch} = \log\frac{\Pr[a_n = +1|y_n]}{\Pr[a_n = -1|y_n]} = 4\frac{E_s}{N_0+\rho}y_n, \tag{14}$$

where $E_s = x(0)$ and ρ is defined as the scaled variance of the residual ISI in y_n

$$\rho = \frac{2}{x(0)}E\left[\sum_{k\neq n} a_k\, x\left((n-k)\tau\right)\right]^2 = \frac{2}{x(0)}\sum_{k\neq 0} x^2\left(k\tau\right). \tag{15}$$

The channel log-likelihood ratio (LLR) is the only piece of information available initially at the receiver. This can provide an initial estimate of the probabilities for a_n as

$$\Pr[a_n = +1|y_n] = \frac{e^{L_n^{ch}}}{1+e^{L_n^{ch}}} \text{ and } \Pr[a_n = -1|y_n] = \frac{1}{1+e^{L_n^{ch}}}. \tag{16}$$

Based on these probabilities the mean value and the variance of a_n can be derived as

$$\mu_n = \frac{e^{L_n^{ch}}-1}{e^{L_n^{ch}}+1} \tag{17}$$

and

$$\sigma_n^2 = \frac{4e^{L_n^{ch}}}{\left(e^{L_n^{ch}} + 1\right)^2}. \tag{18}$$

The soft-output ISI detector can then use the mean values μ_n to cancel the interference

$$\tilde{y}_n = y_n - \sum_{k \neq n} \mu_k x\left((n-k)\tau\right) \tag{19}$$

and the variance values σ_n^2 to scale the soft-output log-likelihood ratios

$$L_n^{det} = 4\frac{E_s}{N_0 + \tilde{\rho}}\tilde{y}_n, \tag{20}$$

where

$$\tilde{\rho} = \frac{2}{x(0)} \sum_{k \neq 0} x^2(k\tau)\, \sigma_k^2. \tag{21}$$

The soft-output LLR L_n^{det} is then deinterleaved and passed to the channel decoder. The soft-output of the channel decoder can then be interleaved and the output of the interleaver L_n^{dec} can be passed back to the ISI detector to improve the ISI cancelation by adding the extrinsic part of this LLR L_n^{ap} to L_n^{ch} in (16), where L_n^{ap} is given from

$$L_n^{ap} = L_n^{dec} - L_n^{det}. \tag{22}$$

The block diagram of the turbo equalizer is shown in Fig. 6.

The performance of this turbo equalizer has been simulated in Fig. 7 using the rate 1/2 [5,7] convolutional code and an s-random 4000-bit interleaver. The convolutional code decoder is the standard BCJR decoder [26]. The lower bound for the performance without ISI shown in Fig. 7 is based on the minimum Hamming distance $d_{H,min}$ of the [5,7] convolutional code, i.e.,

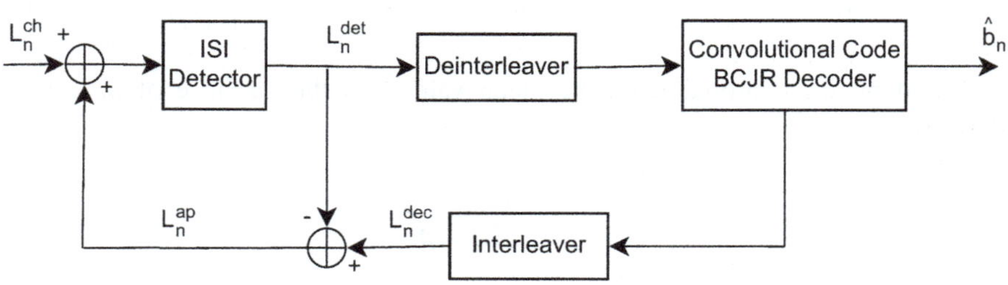

Fig. 6 Block diagram of the turbo equalizer used for coded Faster-than-Nyquist

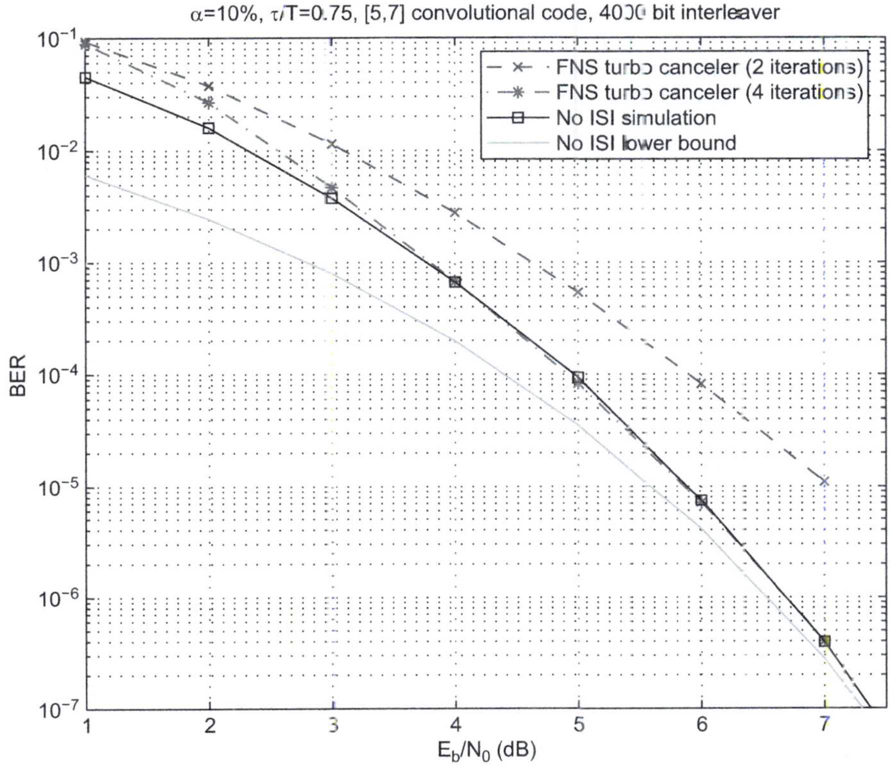

Fig. 7 Performance of convolutionally coded Faster-than-Nyquist signaling for the raised-cosine pulses with roll-off $\alpha = 10\,\%$ with bandwidth efficiency 1.21 bits/s/Hz

$$P_e \geq Q\left(\sqrt{rd_{H,min}\frac{2E_b}{N_0}}\right), \tag{23}$$

where $r = 1/2$ is the rate of the [5,7] convolutional code and $d_{H,min} = 5$.

A similar FTN interference cancelation approach with a turbo code instead of a convolutional code was used for one of the receivers in [27].

Another approach with both serial and parallel concatenation of convolutional codes [28] used the more complex ISI detector from [29] and truncated the number of states. The serial concatenation used a rate-1 precoder, whose decoding was performed in the same BCJR as the ISI detector without any increase in complexity, and resulted in improved performance compared to not including the precoder.

The reduced complexity BCJR from [29] can also be used for binary and nonbinary transmitted alphabets and was combined with LDPC codes in [30]. In [30] it is found that it is beneficial to increase spectrum efficiency by keeping the code rate constant and signaling at higher rates than by using higher code rates. In addition, for limited ISI detector complexity, which means a reduced number of states in the ISI detector, there is a spectrum efficiency level, about 1.4 bits/s/Hz for the parameters used in [30], at which LDPC-coded FTN 16QAM starts outperforming LDPC-coded FTN QPSK.

Frequency-domain equalization generating soft output for a serial concatenation of convolutional codes was proposed in [31]. The rate-1 precoder in this case was followed by a second interleaver showing significantly improved performance compared to not including the precoder and the second interleaver.

5 Achievable Information Rates

The ultimate limit of the FTN gains is determined by the channel capacity based on Shannon's capacity formula [32]

$$C = W\log_2\left(1 + \frac{P}{WN_0}\right), \tag{24}$$

where P is the transmit signal power and W is the signal bandwidth. Shannon's formula assumes that the ideal sinc pulses are used in transmission.

However, in practical systems the roll-off of the pulse bandwidth causes some of this capacity to remain unused, as for the rolloff to be contained within the same bandwidth the signaling rate needs to be reduced. For example, for Nyquist transmission with a root raised cosine pulses of roll-off α, the achievable information rate becomes:

$$I_{\text{Nyquist},\alpha} = \frac{1}{2T}\log_2\left(1 + \frac{2PT}{N_0}\right), \tag{25}$$

where T is the signaling rate and $W = (1+\alpha)/2T$.

The last two equations are compared in Fig. 8, which shows that taking into account the pulse roll-off reduces the achievable information rate when maintaining the Nyquist orthogonality constraint and the transmit signal power P. The larger the roll-off the larger this information rate loss. Note that even though this constrained information rate is affected by the pulse roll-off, it does not take into account the actual pulse shape. The information rate (25) is the same even if different pulses with the same roll-off were used, e.g., triangular.

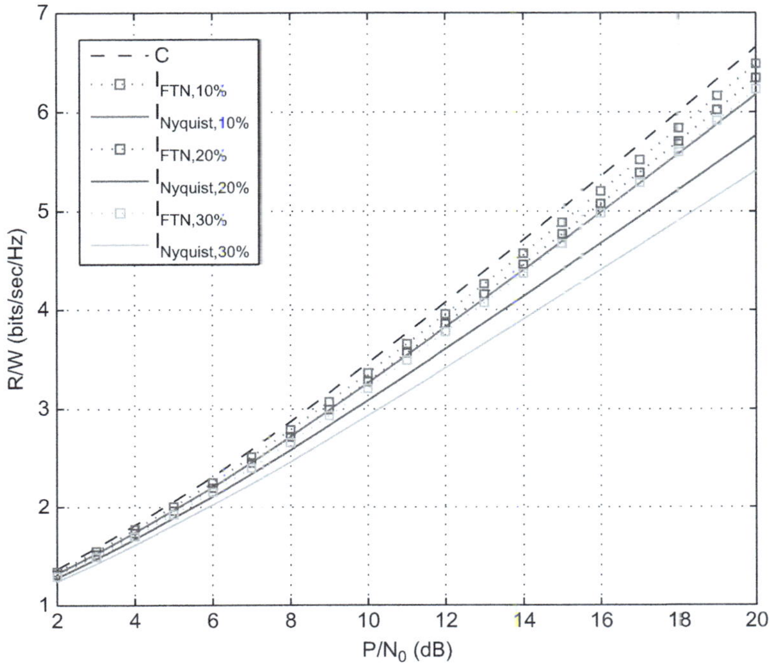

Fig. 8 Achievable information rates for faster-than-Nyquist signaling and Nyquist signaling with roll-off $\alpha = 10, 20, 30\,\%$

FTN can be viewed as a method to regain some of the lost information rate by taking advantage of the complete pulse shape of the signal. It was shown in [33] that the achievable information rate with FTN using independently and identically distributed (i.i.d.) input is

$$I_{\text{FTN},\alpha} = \int_0^W \log_2\left(1 + \frac{2P}{N_0}|G(f)|^2\right) df, \tag{26}$$

where $G(f)$ is the Fourier transform of the transmit pulse shape $g(t)$ with $\int_{-W}^{W} |G(f)|^2 df = 1$. This information rate can be achieved by FTN when $\tau \geq \frac{1}{2W} = \frac{T}{1+\alpha}$ [16, 33], i.e., by signaling fast enough to make up for the non-zero roll-off factor. The main observation from (26) is that at high SNR the weaker parts of the spectrum can also contribute noticeably to the achievable information rate.

Figure 8 shows how the FTN information rate with raised cosine pulses compares to the Nyquist information rate for different roll-off. For zero roll-off, i.e., when FTN also uses the sinc pulse, it was proven in [34] that the FTN information rate equals the Shannon capacity.

Figure 8 establishes a clear advantage of FTN signaling in practical systems, which employ non-zero roll-off transmit pulses. This comparison is based on Gaussian input signals. Since we signal faster than the Nyquist rate, the assumption that the transmitted pulses have to be orthogonal at the Nyquist rate is no longer relevant. Dropping this assumption [35] further bridged the gap between the channel capacity C in Fig. 8 and the FTN information rate at high SNR by optimizing with respect to the pulse shape.

For more practical input alphabets the constrained input information rate has so far been established with lower-bounds. These lower-bounds are based on constraining the receiver structure, such as symbol-by-symbol detection [25] or MMSE-DFE detection [36], and finding the achievable rate of this suboptimal receiver.

6 Frequency Packing

Bringing the pulses closer together in time, as described previously, is also called time packing. Another approach to improving the bandwidth efficiency is with frequency packing in the context of multicarrier modulation. In frequency packing the neighboring carriers start overlapping generating inter-carrier interference, but as in the case of time packing, less overall bandwidth is used resulting in improved bandwidth efficiency.

The transmitted multicarrier signal is expressed as follows:

$$s(t) = \sum_m \left[\sum_k a_{m,k} g(t - k\tau) \right] e^{-j2\pi mFt}, \tag{27}$$

where F is the frequency spacing of the carriers.

Starting with the filter bank multicarrier approach in [37], one can require that the frequency content of neighboring subcarriers does not overlap to guarantee that no intercarrier interference (ICI) occurs. This frequency spacing constraint together with the Nyquist time constraint are together called the generalized Nyquist constraints [37].

Following the time approach, one can start bringing the subcarriers closer together causing them to overlap and cause ICI but becoming more bandwidth efficient. Such approaches were introduced in [13, 38] as a method to increase the OFDM bandwidth efficiency as shown in Fig. 9.

It is possible to further improve bandwidth efficiency in (27) by allowing both time and frequency packing and optimizing the trade-off between them [14]. This yields additional gains in the sense that the two-dimensional time-frequency limit, where the minimum distance between transmitted signals becomes smaller than the minimum distance for the orthogonal time-frequency transmission, extends to higher spectral efficiency than in time-packing or frequency-packing alone.

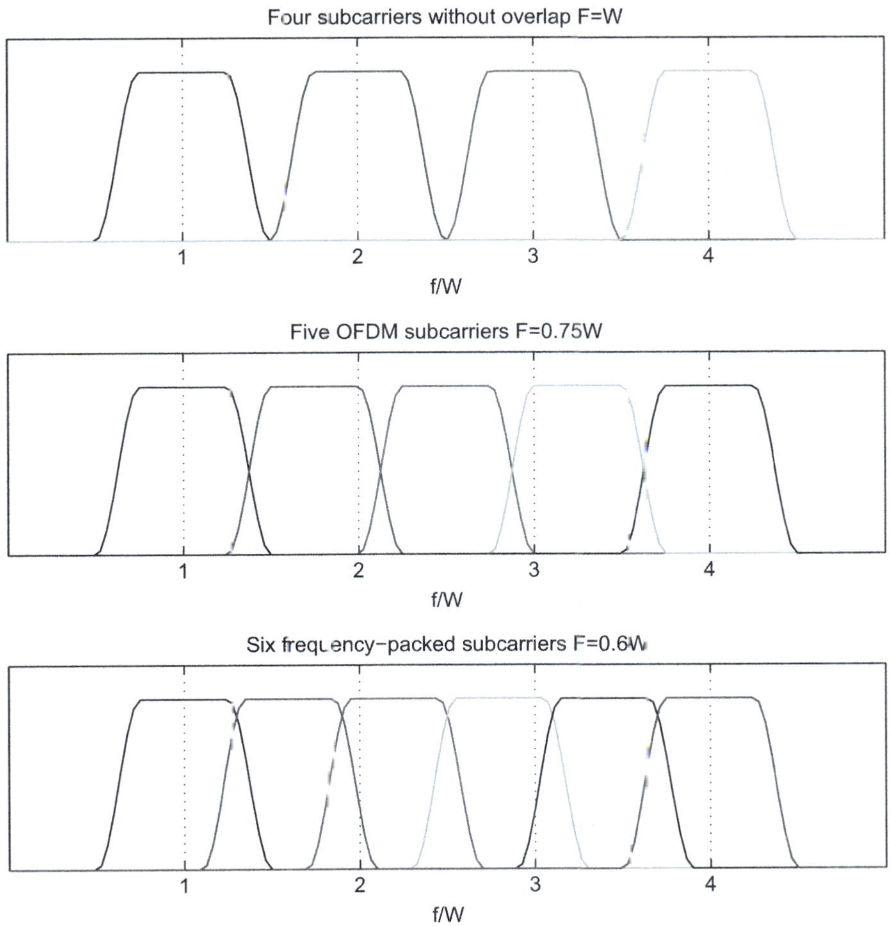

Fig. 9 Subcarriers in multicarrier transmission when the subcarrier spectrum does not overlap (*top*), in OFDM (*middle*), and in the frequency packed approach. W is the bandwidth of each subcarrier and in this example all subcarriers fit within a total bandwidth of $4W$

Orthogonal multicarrier modulation, such as OFDM, is more bandwidth efficient than single carrier modulation, because the unused roll-off is only that of the subcarriers at the edges of the whole band. Therefore, the information rate gains of single carrier modulation are more limited in multicarrier systems [15, 39], but combining both time and frequency offers other advantages, mainly additional flexibility in pulse shape design and rate adaptivity and use in the single carrier uplink [15] as well as less overhead and better latency in other 5G applications [40]. These advantages of FTN have made it an alternative to the time and frequency orthogonality in the 5G standard [15].

The same low complexity interference canceler that was described in the Turbo Equalization section earlier can be used to mitigate the effects of both the intersymbol and intercarrier interference from the neighboring symbols. Such schemes approach the coded performance without interference with both convolutional codes [39] and LDPC codes [25]. Similar interference cancelation approaches dealing with intercarrier interference when only frequency packing is used can perform about 1dB worse than coded OFDM at the same bit rate with the frequency-packed signal using 60 % of the OFDM bandwidth [41].

7 Discussion

In this section we outline some of the recent research activity in directions relevant to modern communication systems.

Rusek [42] proved that the AWGN d_{min} Mazo limit can be extended to the MIMO case and also reported results on the fading channel. Recent work on the broadcast channel suggests that FTN signaling is also an alternative approach to achieving the Gaussian input capacity for two users [43]. A turbo-coded broadcast scheme was introduced approaching this capacity [27]. An alternative uncoded scheme with more than two users and detecting all users' signals was proposed in [44]. Very promising sum capacity gains with FTN signaling were also reported for the synchronous and asynchronous multiple access channels in [45, 46] but no receiver design has been proposed yet to take advantage of these gains. This multiuser work focuses only on the classical time-packed FTN signaling.

Practical directions of faster-than-Nyquist signaling have been investigated recently with hardware implementations [5, 47] considering several aspects of the algorithms involved in an FTN system. The focus of these implementations has been on multicarrier systems as these seem to be more relevant to 5G communication systems. Other potential benefits of FTN signaling in practical systems, such as the peak-to-average power ratio (PAPR), have been explored in [48].

8 Conclusion

As the increasing amount of recently published FTN work shows, FTN signaling has been a rather active research area in the last few years because of its more efficient use of spectrum resources. There are several open questions that remain unanswered, especially with multiple transmit and receive antennas, in multiuser systems and in practical implementations, and therefore, the associated research activity is expected to intensify in the coming years.

Acknowledgements The authors would like to thankfully acknowledge support from Texas Instruments for their original work on faster-than-Nyquist signaling.

References

1. J.E. Mazo, Faster-than-Nyquist signaling. Bell Syst. Tech. J. **54**, 1451–1462 (1975)
2. G.D. Forney, Maximum-likelihood sequence estimation of digital sequences in the presence of intersymbol interference. IEEE Trans. Inf. Theory **IT-18**, 363–378 (1972)
3. G. Ungerboeck, Adaptive maximum-likelihood receiver for carrier-modulated data-transmission systems. IEEE Trans. Commun. **COM-22**, 624–636 (1974)
4. G.J. Foschini, Contrasting performance of faster binary signaling with QAM. Bell Syst. Tech. J. **63**, 1419–1445 (1984)
5. D. Dasalukunte, V. Owall, F. Rusek, J.B. Anderson, *Faster than Nyquist Signaling: Algorithms to Silicon* (Springer, Cham, 2014)
6. J. Anderson, S. Mohan, Sequential coding algorithms: a survey and cost analysis. IEEE Trans. Commun. **32**, 169–176 (1984)
7. M.V. Eyuboglu, S.U.H. Qureshi, Reduced-state sequence estimation with set partitioning and decision feedback. IEEE Trans. Commun. **COM-36**, 13–20 (1988)
8. A. Duel-Hallen, C. Heegard, Delayed decision-feedback sequence estimation. IEEE Trans. Commun. **COM-37**, 428–436 (1989)
9. K. Balachandran, J.B. Anderson, Reduced complexity sequence detection for nonminimum phase intersymbol interference channels. IEEE Trans. Inf. Theory **43**, 275–280 (1997)
10. A. Hafeez, W.E. Stark, Decision feedback sequence estimation for unwhitened ISI channels with applications to multiuser detection. IEEE J. Select. Areas Commun. **JSAC-16**, 1785–1795 (1998)
11. C. Berrou, A. Glavieux, P. Thitimajshima, Near Shannon limit error-correcting coding and decoding: turbo-codes, in *Proceedings of ICC '93*, vol. 2 (1993), pp. 1064–1070
12. C. Douillard, M. Jezequel, C. Berrou, A. Picart, P. Didier, A. Glavieux, Iterative correction of intersymbol interference: turbo-equalization. Eur. Trans. Telecommun. **6**, 507–511 (1995)
13. M. Rodrigues, I. Darwazeh, A spectrally efficient frequency division multiplexing based communication system. In: *8th International OFDM-Workshop, Hamburg, Germany*, pp. 70–74 (2003)
14. F. Rusek, J.B. Anderson, The two dimensional Mazo limit, in *Proceedings of ISIT '05* (2005), pp. 970–974
15. P. Banelli, S. Buzzi, G. Colavolpe, A. Modenini, F. Rusek, A. Ugolini, Modulation formats and waveforms for 5G networks: Who will be the heir of OFDM?: An overview of alternative modulation schemes for improved spectral efficiency. IEEE Signal Process. Mag. **31**, 80–93 (2014)
16. A. Modenini, F. Rusek, G. Colavolpe, Faster-than-Nyquist signaling for next generation communication architectures, in *2014 Proc. 22nd Europ. Signal Proc. Conf. (EUSIPCO)* (2014), pp. 1856–1860
17. M. Secondini, T. Foggi, F. Fresi, G. Meloni, F. Cavaliere, G. Colavolpe, E. Forestieri, L. Poti, R. Sabella, G. Prati, Optical time-frequency packing: principles, design, implementation, and experimental demonstration. J. Lightwave Technol. **33**(17), 3558–3570 (2015)
18. J.G. Proakis, *Digital Communications*, 4th edn. (McGraw-Hill, New York, 2000)
19. A.D. Liveris, C.N. Georghiades, Exploiting faster-than-Nyquist signaling. IEEE Trans. Commun. **51**, 1502–1511 (2003)
20. J.E. Mazo, H.J. Landau, On the minimum distance problem for faster-than-Nyquist signaling. IEEE Trans. Inf. Theory **IT-34**, 1420–1427 (1988)
21. D. Hajela, On computing the minimum distance for faster than Nyquist signaling. IEEE Trans. Inf. Theory **IT-36**, 289–295 (1990)
22. G.D. Forney, Lower bounds on error probability in the presence of large intersymbol interference. IEEE Trans. Commun. **COM-20**, 76–77 (1972)
23. F. Rusek, J.B. Anderson, Non binary and precoded faster than Nyquist signaling. IEEE Trans. Commun. **56**, 808–817 (2008)

24. A. Prlja, J.B. Anderson, F. Rusek, Receivers for faster-than-Nyquist signaling with and without turbo equalization, in *Proc. IEEE Int. Symp. Inf. Theory* (2008), pp. 464–468
25. A. Barbieri, D. Fertonani, G. Colavolpe, Time-frequency packing for linear modulations: spectral efficiency and practical detection schemes. IEEE Trans. Commun. **COM-57**, 2951–2959 (2009)
26. L.R. Bahl , J. Cocke , F. Jeinek, J. Raviv, Optimal decoding of linear codes for minimizing symbol error rate. IEEE Trans. Inf. Theory **20**, 284–287 (1974)
27. Y.J.D. Kim, J. Bajcsy, An architecture for faster than Nyquist turbo broadcasting, in *Proc. 7th Int. Symp. on Turbo Codes and Iterative Inform.* (2012), pp. 170–174
28. F. Rusek, J.B. Anderson, Serial and parallel concatenations based on faster than Nyquist signaling, in *Proceedings of ISIT '06* (2006), pp. 1993–1997
29. G. Colavolpe, A. Barbieri, On MAP symbol detection for ISI channels using the Ungerboeck observation model. IEEE Commun. Lett. **9**, 720–722 (2005)
30. J. Yu, J. Park, F. Rusek, B. Kudryashov, I. Bocharova, High order modulation in faster-than-Nyquist signaling communication systems, in *Proceedings of 2014 IEEE Vehicular Technology Conference (VTC Fall)* (2014), pp. 1–5
31. S. Sugiura, L. Hanzo, Frequency-domain-equalization-aided iterative detection of faster-than-Nyquist signaling. IEEE Trans. Veh. Technol. **64**, 2122–2128 (2015)
32. C.E. Shannon, Communication in the presence of noise. Proc. IRE **37**, 10–21 (1949)
33. F. Rusek, J.B. Anderson, Constrained capacities for faster-than-Nyquist signaling. IEEE Trans. Inf. Theory **55**, 764–775 (2009)
34. A.D. Liveris, On Distributed Coding, Quantization of Channel Measurements and Faster-than-Nyquist Signaling, Ph.D. thesis, Texas A&M University, 2004
35. J. Zhou, D. Li, X. Wang, Generalized faster-than-Nyquist signaling, in *IEEE Int. Symp. Inform. Theory* (2012), pp. 1–5
36. Y.G. Yoo, J.H. Cho, Asymptotic optimality of binary faster-than-Nyquist signaling. IEEE Commun. Lett. **14**, 788–790 (2010)
37. B. Farhang-Boroujeny, OFDM versus filter bank multicarrier. IEEE Signal Process. Mag. **28**, 92–112 (2011)
38. I. Kanaras, A. Chorti, M.R.D. Rodrigues, I. Darwazeh, Spectrally efficient FDM signals: bandwidth gain at the expense of receiver complexity, in *IEEE Int. Conf. Commun. (ICC)* (2009), pp. 1–6
39. J.B. Anderson, F. Rusek, V. Owall, Faster-than-Nyquist signaling. Proc. IEEE **101**, 1817–1830 (2013)
40. G. Wunder, P. Jung, M. Kasparick, T. Wild, F. Schaich, Y. Chen, S. Brink, I. Gaspar, M. Michailow, A. Festag, L. Mendes, N. Cassiau, D. Ktenas, M. Dryjanski, S. Pietrzyk, B. Eged, P. Vago, F. Wiedmann, 5GNOW: non-orthogonal, asynchronous waveforms for future mobile applications. IEEE Commun. Mag. **52**, 97–105 (2014)
41. T. Xu, I. Darwazeh, Spectrally efficient FDM: spectrum saving technique for 5G?, in *Int. Conf. 5G for Ubiquitous Connectivity (5GU)* (2014), pp. 273–278
42. F. Rusek, On the existence of the Mazo-limit on MIMO channels. IEEE Trans. Wirel. Commun. **8**, 1118–1121 (2009)
43. Y.J.D. Kim, J. Bajcsy, Faster than Nyquist broadcast signaling, in *Proc. 26th Bienn. Symp. Commun., Kingston* (2012), pp. 186–189
44. M.J. Abdoli, M. Jia, Trellis decoding for multi-user faster-than-Nyquist transmission, in *IEEE Veh. Tech. Conf. (VTC Fall)* (2014), pp. 1–5
45. Y. Feng, J. Bajcsy, On faster-than-Nyquist transmission over a Multiple-Access Channel, in *IEEE Mil. Commun. Conf. (MILCOM)* (2014), pp. 824–829
46. Y. Feng, J. Bajcsy, Improving throughput of faster-than-Nyquist signaling over Multiple-Access Channels, in *IEEE Veh. Tech. Conf. (VTC Spring)* (2015), pp. 1–5
47. P.N. Whatmough, M.R. Perrett, S. Isam, I. Darwazeh, VLSI architecture for a reconfigurable spectrally efficient FDM baseband transmitter. IEEE Trans. Circuits Syst. I Regul. Pap. **59**, 1107–1118 (2012)
48. C. Le, M. Schellmann, M. Fuhrwerk, J. Peissig, On the practical benefits of faster-than-Nyquist signaling, in *Int. Conf. Advanced Techn. for Commun. (ATC)* (2014), pp. 208–213

Generalized Frequency Division Multiplexing: A Flexible Multi-Carrier Waveform for 5G

Maximilian Matthé, Ivan Simões Gaspar, Luciano Leonel Mendes, Dan Zhang, Martin Danneberg, Nicola Michailow, and Gerhard Fettweis

Abstract The next generation of wireless networks will face different challenges from new scenarios. The conventional Orthogonal Frequency Division Multiplexing (OFDM) has shown difficulty in fulfilling all demanding requirements. This chapter presents Generalized Frequency Division Multiplexing (GFDM) as a strong waveform candidate for future wireless communications systems which can be combined with several techniques such as precoding or Offset Quadrature Amplitude Modulation (OQAM) and which offers the flexibility to emulate a variety of other popular waveforms as corner cases. This property suggests GFDM as a key technology to allow reconfiguration of the physical layer (PHY), enabling a fast and dynamic evolution of the infrastructure. Additionally, multicarrier transmission theory is covered in terms of Gabor theory. Details on synchronization, channel estimation algorithms and MIMO techniques for GFDM are presented and a description of a proof-of-concept demonstrator shows the suitability of GFDM for future wireless networks.

1 GFDM: A General Solution for 5G

The fifth generation (5G) of mobile cellular systems needs to deal with a wide range of services, each of which can have very different requirements. For instance, the four main scenarios illustrated in Fig. 1 respectively require loose synchronization for Internet of Things (IoT); low latency for Tactile Internet; reliable, efficient, and robust high throughput for bitpitpe communication; high coverage and dynamic spectrum allocation with low out-of-band (OOB) emission and Cognitive Radio (CR) techniques for Wireless Regional Area Network (WRAN) applications.

M. Matthé (✉) • I.S. Gaspar • D. Zhang • M. Danneberg • N. Michailow • G. Fettweis
Vodafone Chair Mobile Communication Systems, Technische Universität Dresden, Dresden, Germany
e-mail: maximilian.matthe@ifn.et.tu-dresden.de; ivan.gaspar@ifn.et.tu-dresden.de; dan.zhang@ifn.et.tu-dresden.de; martin.danneberg@ifn.et.tu-dresden.de; nicola.michailow@ifn.et.tu-dresden.de; gerhard.fettweis@ifn.et.tu-dresden.de

L.L. Mendes
Instituto Nacional de Telecomunicações, Santa Rita do Sapucaí, MG, Brazil
e-mail: luciano@inatel.br

W. Xiang et al. (eds.), *5G Mobile Communications*,
DOI 10.1007/978-3-319-34208-5_9

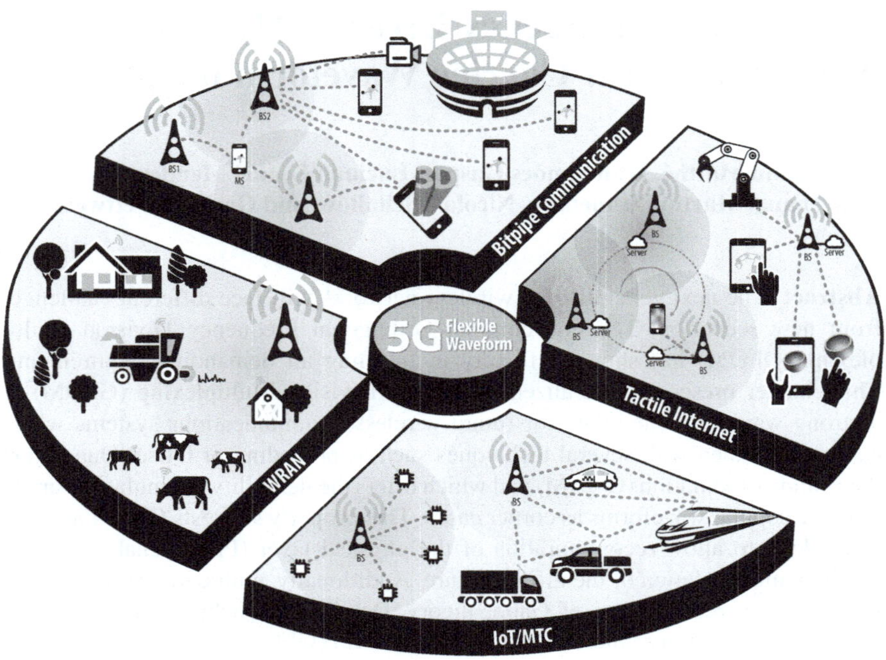

Fig. 1 Four of the main scenarios envisioned for 5G networks

A physical layer (PHY) with unprecedented flexibility will be of great benefit to users such that various applications can be efficiently and effectively addressed by proper reconfiguration. Certainly, all these scenarios will benefit from multiple-input multiple-output (MIMO)-techniques, such as increased data rate, enhanced robustness and the additional degree of freedom in space for multi-user and multi-cell interference management. In order to maximize MIMO profits, the 5G PHY must also be able to cope with multi-antenna interference and related algorithms need to be implemented with an affordable complexity. 5G related topics have been intensely researched lately and several projects were funded by the European Union [1], e.g. METIS and 5GNOW.

Instead of selecting one specific waveform for each 5G scenario, it is much more desirable to adopt a single flexible waveform that can be easily reconfigured to address a multitude of applications. More importantly, such a *software-defined waveform (SDW)* builds a foundation at PHY for preparing the paradigm shift towards software-defined virtualization. Namely, the SDW shall be generated in programmable hardware, based on Application Specific Integrated Processors (ASIP), field programmable gate array (FPGA), or software with manageable cost. By means of a cost effective approach that exposes the time-frequency resource grid and waveform engineering capabilities to software, it becomes feasible to customize PHY such that it can be seen as a virtual service for upper layers.

In this context, GFDM [2] provides a very flexible time-frequency structure that favors software exposition, controlling, and virtualization. Starting from introducing some basics of GFDM, the main goal of this section is to explore the GFDM flexibility to achieve a framework that covers all major waveform candidates for the different 5G scenarios. Furthermore, we will show that the main waveforms considered for 5G are, in fact, corner cases for GFDM. Also, the flexibility offered by GFDM is sufficient and necessary to fulfill diverse quality of service (QoS) requirements in 5G.

The remainder of this chapter is organized as follows: This section introduces the basics of the GFDM system, including modulation and demodulation equations, waveform engineering, precoding and application of OQAM modulation. The theoretic foundation for multicarrier systems in terms of Gabor theory is described in the subsequent Sect. 2. Section 3 covers development of synchronization and channel estimation algorithms for GFDM in detail whereas Sect. 4 is dedicated to a detailed description of MIMO techniques for GFDM. Finally, Sect. 5 considers implementation aspects for a real-world system demonstrator and the chapter is concluded in Sect. 6.

1.1 Basics

GFDM has a block structure where $N = KM$ complex-valued data symbols $d_{k,m}$ are transmitted on K subcarriers and M time-slots (or subsymbols). Each subcarrier is pulse-shaped by a filter impulse response with $S = RT$ samples, where T is the number of periods of the filter and R is the number of samples per period. The use of a circular pulse shaping filter is a key feature of GFDM when it was initially proposed [3]. Its advantages include (1) keeping the GFDM signal compact in time; and (2) enabling low complexity signal processing, e.g. [4, 5]. Denoting modulo S operation as $\langle\rangle_s$, the GFDM transmit sequence is given by

$$x[n] = \sum_{k=0}^{K-1} \sum_{m=0}^{M-1} d_{k,m} g\left[\langle n - mP\rangle_S\right] \exp\left(j2\pi \frac{kQ}{S} n\right), \quad n = 0, 1, \ldots, S-1, \tag{1}$$

where P is the spacing between adjacent subsymbols and Q is the distance between adjacent subcarriers. The subsymbol and subcarrier spacing factors are respectively given by

$$\begin{aligned} \nu_t &= \frac{P}{R} \\ \nu_f &= \frac{Q}{T}, \end{aligned} \tag{2}$$

Table 1 Terminology

Variable	Meaning
R	Samples per period in the filter
T	Periods in the filter
$S = RT$	Total number of samples in the signal
P	Subsymbol spacing in time domain
Q	Subcarrier spacing in frequency domain
$\nu_t = P/R$	Subsymbols distance factor
$\nu_f = Q/T$	Subcarriers distance factor
$K = RT/Q = R/\nu_f = S/Q$	Subcarriers per block
$M = TR/P = T/\nu_t = S/P$	Subsymbols per block
$N = KM$	Number of data symbols per block

which allows us to rewrite (1) as

$$x[n] = \sum_{k=0}^{K-1}\sum_{m=0}^{M-1} d_{k,m} g\left[\langle n - m\,\nu_t R\rangle_S\right] \exp\left(j2\pi \frac{k\,\nu_f T}{S} n\right), \quad n = 0, 1, \ldots, S-1. \quad (3)$$

When $\nu_t < 1$ the overlapping between the subsymbols increases, leading to a higher intersymbol interference (ISI) and better spectrum efficiency. For $\nu_t > 1$, the subsymbols are taken apart from each other, reducing the time overlapping and decreasing the spectrum efficiency. The same reasoning is valid for ν_f, but now in the frequency domain. Table 1 summarizes the parameters used to describe the GFDM signal.

An interesting configuration is achieved when $\nu_t = \nu_f = 1$, which will be called "critically sampled GFDM". In this case, the GFDM sequence can be written as

$$x[n] = \sum_{k=0}^{K-1}\sum_{m=0}^{M-1} d_{k,m} g\left[\langle n - mK\rangle_N\right] \exp\left(j2\pi \frac{k}{K} n\right), \quad n = 0, 1, \ldots, N-1 \quad (4)$$

where Fig. 2 depicts the block diagram of such GFDM modulation. A cyclic prefix (CP) is added to protect the GFDM sequence from the multipath channel. Assuming that the CP length is larger than the channel impulse response, the received signal after CP removal (see Fig. 3) is given by

$$y[n] = x[n] \circledast h[n] + w[n], \quad (5)$$

where $\circledast$ denotes the circular convolution operation, $h[n]$ is the channel impulse response and $w[n]$ is the additive white Gaussian noise (AWGN) with variance σ_w^2. After synchronization and channel estimation, a frequency-domain equalizer (FDE) is used to compensate the effects of the multipath channel, leading to a equalized

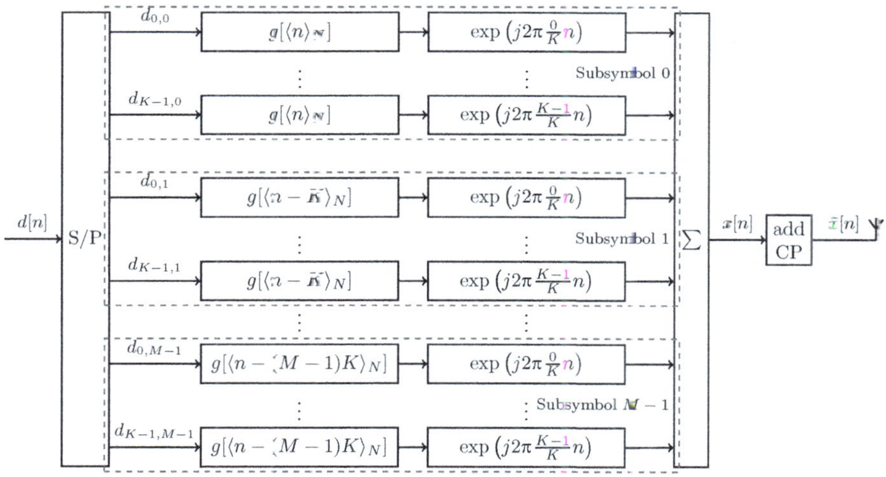

Fig. 2 Block diagram of the GFDM modulator

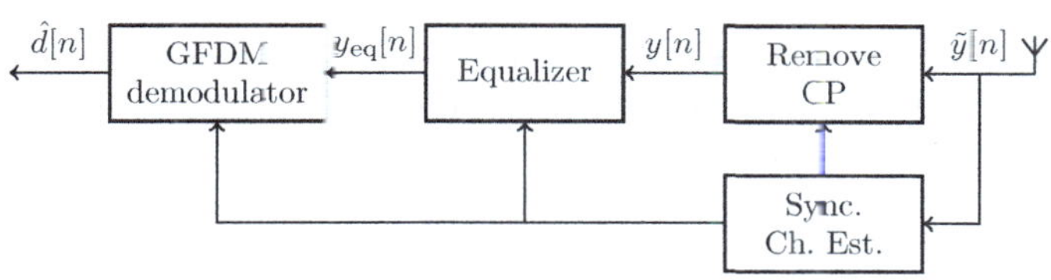

Fig. 3 Block diagram of the GFDM receiver

received sequence $y_{eq}[n]$. A set of filters is derived from a prototype receive filter $\gamma[n]$ as

$$\gamma_{k,m} = \gamma[\langle -n \rangle_N] \exp(-j2\pi \tfrac{k}{K} n), \tag{6}$$

and the recovered data symbols are given by

$$\hat{d}_{k,m} = \gamma^*_{k,m} \circledast y_{eq}[n]|_{n=mK}. \tag{7}$$

Different receive filters can be used to recover the data symbols and matrix notation can be useful to design linear receivers for GFDM. The column vectors containing the samples of the transmit filters are given by

$$[\mathbf{g}_{k,m}]_n = g_{k,m}[n], \tag{8}$$

and the modulation matrix is arranged by

$$\mathbf{A} = [\mathbf{g}_{0,0}\ \mathbf{g}_{1,0} \cdots \mathbf{g}_{K-1,0}\ \mathbf{g}_{0,1} \cdots \mathbf{g}_{K-1,M-1}]\,. \tag{9}$$

Let us further organize the *KM* data symbols of one GFDM block into a $(K \times M)$ data matrix, given by

$$\mathbf{D} = \begin{bmatrix} d_{0,0} & d_{0,1} & \dots & d_{0,M-1} \\ d_{1,0} & d_{1,1} & \dots & d_{1,M-1} \\ \vdots & \vdots & \ddots & \vdots \\ d_{K-1,0} & d_{K-1,1} & \dots & d_{K-1,M-1} \end{bmatrix} = \left[\mathbf{d}_{c,0}\ \mathbf{d}_{c,1} \dots \mathbf{d}_{c,M-1} \right] = \begin{bmatrix} \mathbf{d}_{r,0} \\ \mathbf{d}_{r,1} \\ \vdots \\ \mathbf{d}_{r,K-1} \end{bmatrix} \quad (10)$$

Then, the GFDM transmit vector is given by

$$\mathbf{x} = \mathbf{A}\mathbf{d}, \quad (11)$$

where the data vector $\mathbf{d}$ is obtained by stacking the columns of $\mathbf{D}$ one after another

$$\mathbf{d} = \left[\mathbf{d}_{c,0}^{T}\ \mathbf{d}_{c,1}^{T}\ \dots\ \mathbf{d}_{c,M-1}^{T} \right]^{T}. \quad (12)$$

The prototype filter used on the transmit side can lead to a non-orthogonal system, which means that ISI and intercarrier interference (ICI) can be introduced among the data symbols by the modulation process. The GFDM demodulator must be able to tackle this self-interference to minimize the impact on the symbol error rate (SER) performance. After FDE, the demodulation process can be performed as

$$\hat{\mathbf{d}} = \mathbf{B}\mathbf{y}_{eq}, \quad (13)$$

where $\mathbf{B}$ is the demodulation matrix. Different approaches can be used to demodulate the GFDM signal based on the transmit matrix, as presented in Table 2. The Matched filter (MF) demodulator maximizes the signal-to-noise ratio (SNR), but it is not able to remove the self-interference. The zero-forcing (ZF) demodulator can remove the interference, but it enhances the noise. The minimum mean square error (MMSE) demodulator achieves a good balance between MF and ZF by taking the SNR into account. Also, the MMSE incorporates the channel matrix $\mathbf{H}$, which means that the FDE is not necessary prior the demodulation in this case.

Figure 4 shows two possible configurations of GFDM and the corresponding OFDM counterpart. Configuration (a) concatenates several OFDM symbols, while keeping the subcarrier spacing and symbol duration. Hence, the multitude of CPs of the several OFDM symbols can shrinked to one single CP for the GFDM signal. However, as the number and distance of subcarriers is kept, the PAPR for

Table 2 Linear demodulation matrices

Type	Definition
MF	$\mathbf{B}_{MF} = \mathbf{A}^{H}$
ZF	$\mathbf{B}_{ZF} = \mathbf{A}^{-1}$
MMSE	$\mathbf{B}_{MMSE} = (\mathbf{R}_w + \mathbf{A}^{H}\mathbf{H}^{H}\mathbf{H}\mathbf{A})^{-1}\mathbf{A}^{H}\mathbf{H}^{H}$

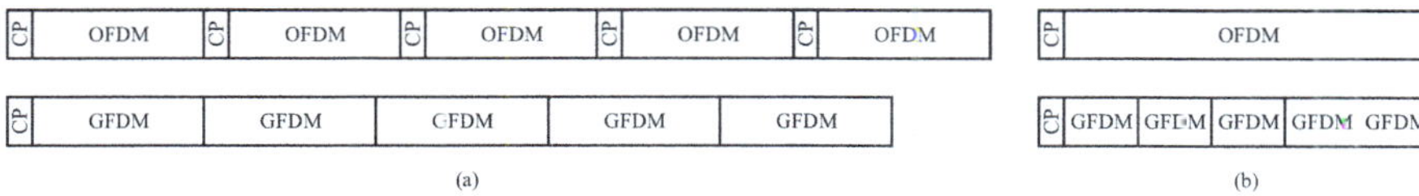

Fig. 4 Different configurations for GFDM with relation to equivalent OFDM system. (**a**) Configuration suitable for downlink, where GFDM and OFDM have equal number of subcarriers and same bandwidth. The CP overhead of OFDM is saved by GFDM. (**b**) Configuration suitable for uplink, where GFDM has fewer but wider subcarriers than OFDM. Using such configuration, GFDM can achieve lower PAPR than OFDM. One special case is to apply a Dirichlet kernel for pulse shaping, emulating SC-FDMA

Table 3 Comparison of implementation complexity for GFDM equivalent OFDM, with configurations (a) and (b) referring to Fig. 4

	Complex multiplications per TX symbol	
	Transmitter	Receiver (incl. FDE)
GFDM	$\frac{1}{KM}(\frac{KM}{2}(M+\log_2 K))$	$\frac{1}{KM}(\frac{KM}{2}(M+3\log_2 K)+KM)$
OFDM (a)	$\frac{1}{KM}\frac{KM}{2}\log_2 K$	$\frac{1}{KM}(\frac{KM}{2}\log_2 K - KM)$
OFDM (b)	$\frac{1}{KM}\frac{KM}{2}\log_2 KM$	$\frac{1}{KM}(\frac{KM}{2}\log_2 KM + KM)$

Values taken from [6]

this configuration is as high as for OFDM, making this configuration preferable for a downlink transmission from the base station. On the other hand, PAPR can be reduced by increasing the subcarrier bandwidth and reducing the number of subcarriers. This configuration is shown in Fig. 4b, where one GFDM block is designed to have the same length as one OFDM symbol. Hence, the subcarriers become wider and the block contains fewer subcarriers. Accordingly the PAPR reduces, similar to the principle of SC-FDMA, where several OFDM subcarriers are combined to a single channel for the SC-FDMA system. Hence, this configuration is suitable for uplink transmission.

In the literature, several algorithms for a reduced complexity implementation of the GFDM modulator and demodulator are given [4–8]. Table 3 shows the number of required complex multiplications for current proposals for implementation of GFDM, compared to the FFT/IFFT approach of OFDM. As shown, the additionally to FFT/IFFT operations for frequency domain conversion and FDE, the GFDM implementation requires an extra amount of roughly $M/2$ multiplications per data symbol which accounts for the overlapping of subsymbols in time or frequency domain.

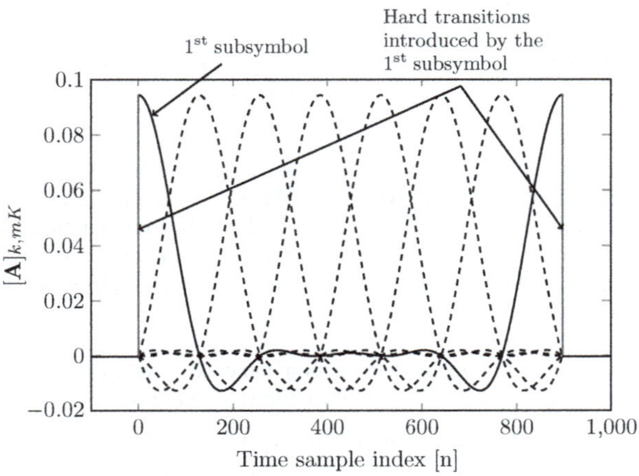

Fig. 5 The first subsymbol of the GFDM block introduces abrupt variations in the time-domain signal, which leads to high OOB emissions

1.2 Waveform Engineering

The flexibility of GFDM allows for designing a signal that has a very low OOB radiation. This section contributes with solutions for low OOB radiation.

1.2.1 GFDM with Guard Symbol

In order to reduce the OOB emission, it is necessary to smooth the transitions between the GFDM blocks. The circularity of the signal in time-domain allows a simple and elegant solution to reduce the abrupt change between GFDM blocks. As can be seen in Fig. 5, the first subsymbol wraps around the edges of the block, introducing the abrupt amplitudes discontinuities.

By erasing the first sub-symbol, a guard symbol is introduced between the GFDM blocks and the edges of the signal fade out towards zero, making the transitions between blocks smooth. This technique is called guard-symbol GFDM (GS-GFDM) and Fig. 6 shows the time-domain signal and the corresponding spectrum.

The addition of the CP would introduce hard transitions between blocks once again. One solution to avoid this problem is to make also the last subsymbol null and making $N_{\mathrm{CP}} = K$. The drawback of this approach is the throughput reduction, given by

$$R_{\mathrm{GS}} = \frac{M-2}{M} \times \frac{KM}{KM+K} = \frac{M-2}{M+1}. \tag{14}$$

From (14), it is clear that this technique becomes interesting for scenarios where M is large.

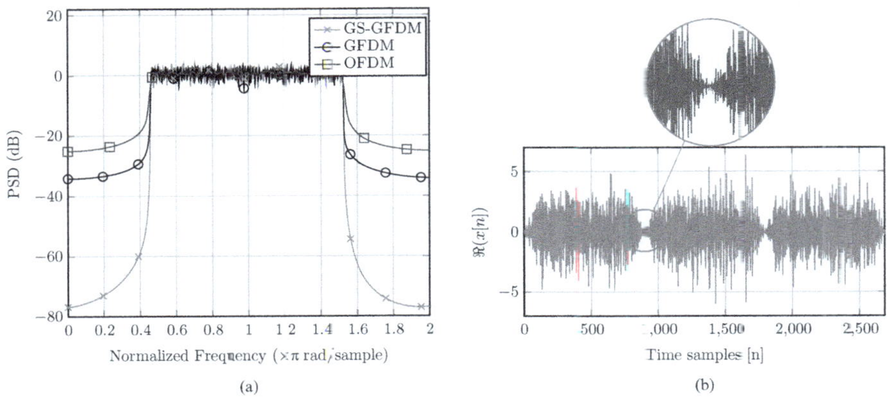

Fig. 6 GS-GFDM signal. (**a**) Comparison of the GS-GFDM, GFDM and OFDM PSDs for the following parameters: $K = 128$ with 68 active subcarriers, $M = 7$, RC with $\alpha = 0.5$. (**b**) Real part of the GS-GFDM signal, highlighting the smooth transition between blocks

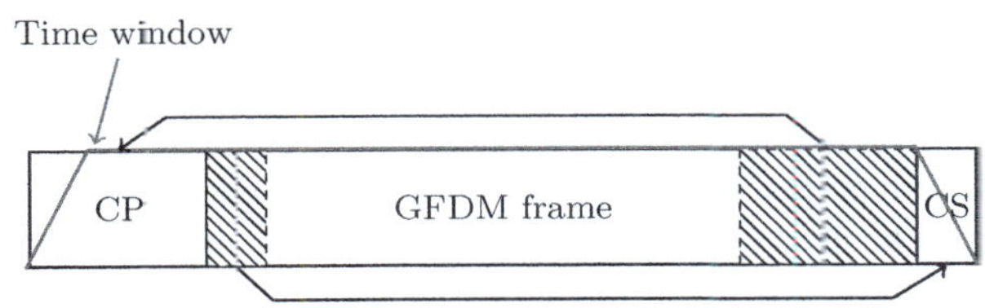

Fig. 7 W-GFDM time-domain signal. A time-window is used to smooth the transitions between GFDM blocks

1.2.2 Windowed-GFDM

For windowed GFDM (W-GFDM), a time-window is applied for smoothing the transition between GFDM blocks, as depicted in Fig. 7.

A CP with length $N_{CP} = N_{CH} + N_W$ and a cyclic suffix (CS) with length $N_{CS} = N_W$, where N_{CH} is the length of the channel impulse response and N_W is the length of the time window transition, are added to the GFDM block. Notice that the CS is just the copy of the first N_{CS} samples of the GFDM block to its end. The time window is defined as

$$w[n] = \begin{cases} w_{\text{rise}}[n] & 0 \leq n < N_W \\ 1 & N_W \leq n \leq N_{CP} + N \\ w_{\text{fall}}[n] & N_{CP} + N < n < N_{CP} + N + N_W \end{cases} \tag{15}$$

where $w_{\text{up}}[n]$ and $w_{\text{down}}[n]$ are the ramp-up and ramp-down segments of the time window, respectively. The ramp-up and ramp-down segments can assume different shapes. The most common cases are linear, RC or fourth order RC [2]. Figure 8 shows the PSD achieved by W-GFDM when linear and cosine ramp-up and

Fig. 8 PSD of the W-GFDM using linear and cosine time windows compared with OFDM

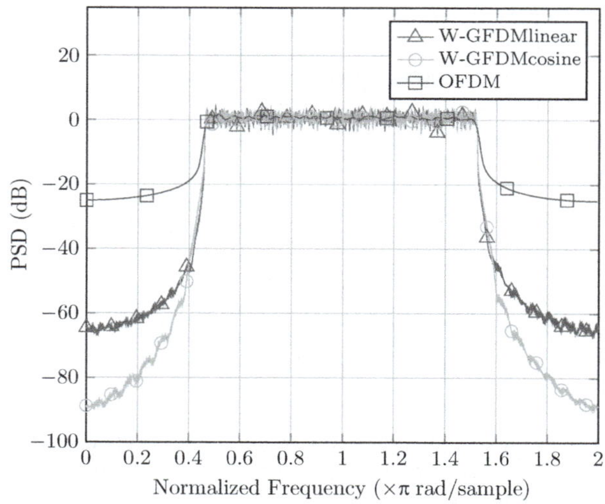

ram-down, with 32 samples each, are employed. Clearly, varying the ramp-up and ramp-down sequences impacts the OOB emission. Sequences that introduce low derivative inflexion points in the edges of the signal provide lower OOB emission.

W-GFDM can be used to achieve low OOB emission and still keep high spectral efficiency, even when M is low. The rate loss introduced by the ramp-up and ramp-down edges is given by

$$R_{\mathrm{W}} = \frac{N}{N + N_{\mathrm{CP}} + N_{\mathrm{W}}}. \tag{16}$$

Since N_{CP} is defined as a function of the channel impulse response, it equally affects the rate loss of GS-GFDM and W-GFDM. The ramp-up and the ramp-down sequences are much smaller than the GFDM block length, meaning R_{W} is typically higher than R_{GS}. From Figs. 6 and 8, it is possible to conclude that both GS-GFDM and W-GFDM can achieve similar OOB emission. Therefore, W-GFDM can be seen as a more promising solution for a high efficient PHY with low OOB emissions.

1.3 Precoding: WHT

One challenge for low-latency scenarios is to achieve reliable communication with single shot transmission over frequency selective channels. In this case, relatively small packages must be received with low probability of error since the low latency requirement does not allow for retransmissions of missed packages. Precoding using WHT can efficiently increase the GFDM robustness over multipath channels [9].

The main idea is to spread the data symbols over all subcarriers, so data symbols can be correctly detected on the receiver side even when a subset of subcarriers suffers from severe attenuation.

In order to achieve this goal, for each subsymbol, the data symbols $\mathbf{d}_{\mathrm{c},m}$ are linearly combined using the Walsh-Hadamard matrix

$$\boldsymbol{\Omega}_K = \frac{1}{\sqrt{2}} \begin{bmatrix} \boldsymbol{\Omega}_{\frac{K}{2}} & \boldsymbol{\Omega}_{\frac{K}{2}} \\ \boldsymbol{\Omega}_{\frac{K}{2}} & -\boldsymbol{\Omega}_{\frac{K}{2}} \end{bmatrix}, \tag{17}$$

with $\boldsymbol{\Omega}_1 = [1]$. Hence, the data coefficients transmitted in the mth subsymbol are given by

$$\mathbf{c}_m = \boldsymbol{\Omega}_K \mathbf{d}_{\mathrm{c},m}. \tag{18}$$

On the receiver side, after the GFDM demodulation, the data symbols can be reconstructed as

$$\hat{\mathbf{d}}_{\mathrm{c},m} = \boldsymbol{\Omega}_K^{\mathrm{H}} \hat{\mathbf{c}}_m = \boldsymbol{\Omega}_K \hat{\mathbf{c}}_m. \tag{19}$$

Figure 9 shows the WHT-GFDM SER performance assuming the parameters presented in Table 4 and the channel delay profiles shown in Table 5. The figure shows that the gain introduced by the WHT highly depends on the channel delay profile. Higher gain is expected when the channel presents narrow and deep notches in the frequency response, while smaller gain is obtained under channel with mild frequency responses. Therefore, WHT-GFDM is an interesting scheme to be used in scenarios where low OOB emission and robustness must be combined.

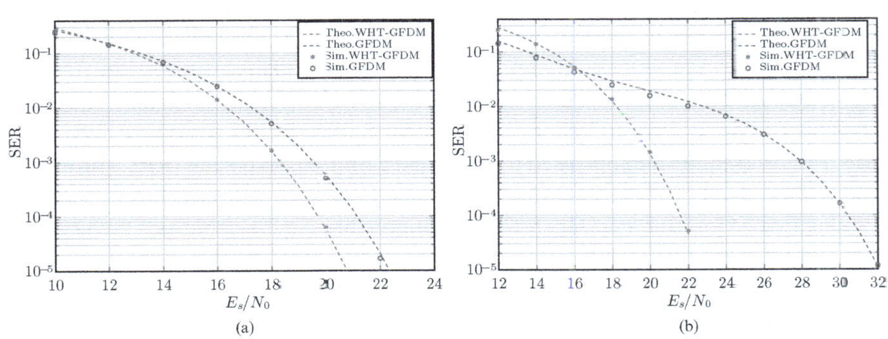

Fig. 9 SER performance of WHT-GFDM over FSC using the ZF demodulator. (**a**) SER performance over channel A. (**b**) SER performance over channel B

Table 4 Simulation parameters

Parameter	Value
Mapping	16-QAM
Transmit filter	RC
Roll-off (α)	0.25
Number of subcarriers (K)	64
Number of subsymbols (M)	7
GFDM block duration	256 μs
CP duration	32 μs
Windowing	Not used

Table 5 Channel delay profiles

Ch. A	Gain (dB)	0	−8	−14	–	–	–	–
	Delay (μs)	0	4.57	9.14	–	–	–	–
Ch. B	Gain (dB)	0	−10	−12	−13	−16	−20	−22
	Delay (μs)	0	2.85	4.57	6.28	9.71	15.43	20

1.4 OQAM

In general, GFDM is a non-orthogonal waveform. However, it is possible to make it orthogonal by choosing some special pulse shaping filters, such as Dirichlet pulses. Another solution is to combine GFDM with OQAM [10].

Instead of direct modulation of the in-phase (I) and quadrature-phase (Q) symbol input, OQAM can be seen as two independent Gabor expansions where the symbol input is restricted to be either only the real or only the imaginary component that achieves the orthogonality condition.

For square root Nyquist prototype filters with bandwidth limited to two subcarriers, the I-input introduces only real interference in the adjacent subcarriers, while the Q-input produces only imaginary interference. This aspect is illustrated in Fig. 10 using the ambiguity function $\mathscr{X}(m, k)$ to calculate the contribution from a transmitted symbol to every other position of the half-symbol spaced time-frequency grid. In Fig. 10, the I-input is used to modulate a root raised cosine (RRC) prototype filter with roll-off 0.5. The magnitude of real and imaginary part of $\mathscr{X}(m, k)$ reveals regions free of interference. Therefore, orthogonality can be achieved if every even and odd subcarrier is modulated with I and Q inputs, respectively.

Figure 10 also shows that I-inputs produce zero real interference in the adjacent subcarriers at every half subsymbol shift while Q-inputs produce zero imaginary interference. Hence, a second Gabor expansion with a time-shift of half subsymbol duration can be used to transmit data free of interference.

The use of OQAM considering pulse shaping filters with arbitrary length and overlapping factors greatly increases the flexibility of the waveform. Recently, frequency-shift OQAM has also being introduced as an alternative for using shorter

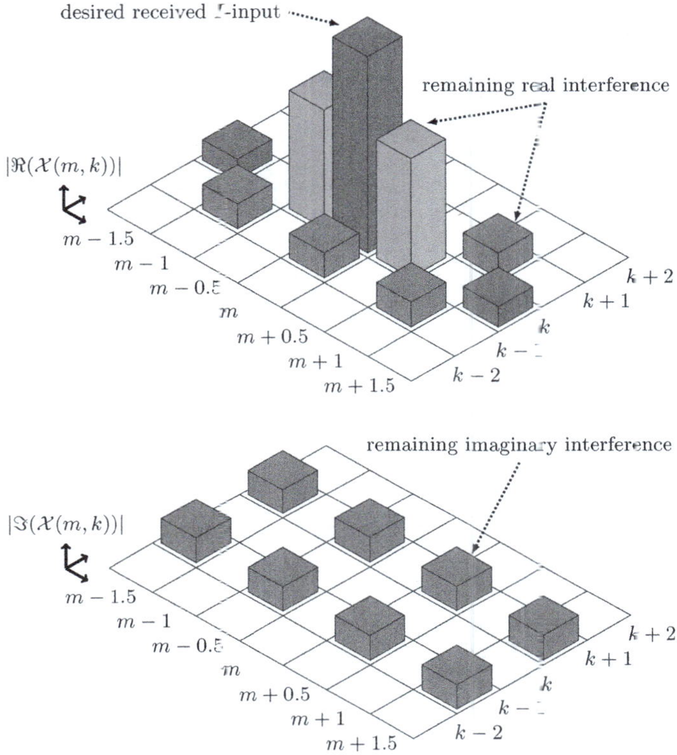

Fig. 10 Illustration of the ambiguity function $\mathscr{X}(m,k)$, with the magnitude of the residual real (*top*) and imaginary (*bottom*) values along the neighbors subsymbols and subcarriers, for a GFDM transmitted *I*-input using RRC prototype filter with roll-off = 0.5

pulses in time [11]. Also, low complexity implementation that fully explores the cyclic principles has been addressed in [2] and allows to harvest the benefits of OQAM.

1.5 Waveform Framework

This section illustrates how GFDM can be used as a framework for multicarrier waveforms. To this end, the parameters and properties of GFDM necessary to emulate prominent waveforms are listed in Table 6. All contestants have common roots in the filtered multicarrier systems proposed by Chang [12] and Saltzberg [13]. Although this framework covers the most popular waveforms, note that some candidates may not be covered. For example, Universal Filtered Multicarrier (UFMC) [14] applies linear filtering to a set of subcarriers and does not use CP, making it incompatible with GFDM.

Table 6 GFDM parameters for emulating 5G waveform candidates

		Block				FBMC	FBMC	FBMC			
Design space	GFDM	OFDM	OFDM	SC-FDE	SC-FDM	OQAM	FMT	COQAM	CB-FMT	FTN	SEFDM
# subcarriers	K	K	K	1	K	K	K	K	K	K	K
# subsymbols	M	1	M	M	M	M	M	M	M	M	1
Scaling freq.	ν_f	1	1	1	1	1	> 1	1	> 1	1	< 1
Scaling time	ν_t	1	1	1	1	1	1	1	1	< 1	1
Silent subsym.	M_s	–	–	–	–	M_p	M_p	–	–	M_p	–
Filter imp. resp.	Cyclic	Rect	Rect	Dirichlet	Dirichlet	$\sqrt{\text{Nyquist}}$	$\sqrt{\text{Nyquist}}$	Cyclic	Cyclic	IOTA	Rect
Offset mod.	(Yes)	(Yes)	No	No	No	Yes	No	Yes	No	Yes	No
Cyclic prefix	Yes	Yes	Yes	Yes	Yes	No	No	Yes	Yes	No	Yes
Orthogonal	(Yes)	Yes	Yes	Yes	Yes	Yes	Yes	Yes	Yes	No	No
Application scenarios	All	Legacy syst.	Bitpipe	IoT, MTC	IoT, MTC	WRAN, bitpipe	WRAN	Tactile internet	Tactile internet	Bitpipe	Bitpipe
Beneficial features	Flex.	Orth.	Small CP overhead	Low PAPR	Low PAPR	Low OOB	Low OOB	No filter tail	No filter tail	Spectral eff.	Spectral eff.

In the context of this framework, the different waveforms are characterized by two aspects. First, parameters related to the dimensions of the underlying resource grid are explored. This includes the number of subcarriers K and subsymbols M in the system. The scaling factor in time ν_t and frequency ν_f can theoretically take values of any rational number larger than zero, while numbers close to one are meaningful because they relate to critically sampled Gabor frames. Additionally, the option to have guard symbols, i.e., the presence of silent subsymbols, is relevant for some candidates. The second set of features is related to the properties of the signal. Here, the choice of the pulse shaping filter is a significant attribute and the presence or absence of circularity constitutes a characteristic feature. Moreover, the use of OQAM is needed for some waveforms, aiming to achieve real orthogonality. Further, some waveforms rely on a CP to allow transmission of a block based frame structure in a time dispersive channel, while others don't use CP in order to achieve higher spectrum efficiency.

The family of *classical waveforms* includes OFDM, block OFDM, Single Carrier Frequency Domain Equalization (SC-FDE) and single carrier frequency division multiplexing (SC-FDM). Particularly OFDM and SC-FDM have been relevant for the development of the fourth generation (4G) cellular standard Long-Term Evolution (LTE). All four waveforms in this category have in common that $\nu_f = 1$ and $\nu_t = 1$, which allows to meet the Nyquist criterion. Silent subsymbols are not employed, the CP and regular quadrature amplitude modulation (QAM) are used in the default configuration. OFDM and block OFDM are corner cases of GFDM, where a rectangular pulse is used. Additionally, OFDM is restricted to one subsymbol, while block OFDM constitutes the concatenation of multiple OFDM symbols in time to create a block with a single common CP. Similarly, SC-FDE and SC-FDM can also be considered as corner cases of GFDM. However, here a Dirichlet pulse is used and analogously, the number of subcarriers in SC-FDE is $K = 1$, while SC-FDM is a concatenation in frequency of multiple SC-FDE signals. All waveforms in this category share property of orthogonality, but with different sensitivities towards various radio frequency (RF) imperfections, for instance SC-FDE is well known for its low peak-to-average power ratio (PAPR), which greatly benefits the mobile stations (MS) in terms of transmit power efficiency and reduced cost of the power amplifier.

The family of *filter bank waveforms* revolves around filtering the subcarriers in the system and still retaining orthogonality. As the names suggest, Filterbank multicarrier (FBMC)-OQAM [15] and its cyclic extension FBMC-COQAM [16] rely on offset modulation, while in FBMC-FMT and cyclic block filtered multitone (CB-FMT) [17] the spacing between the subcarriers is increased such that they do not overlap, i.e. $\nu_f > 1$. Also, a separation between cyclic and non-cyclic prototype filters can be made. In this context, silent subsymbols become relevant. The best spectral efficiency is achieved with no silent subsymbols $M_s = 0$, while $M_s > 0$ helps to improve the spectral properties of the signal. Using a sufficiently large number of silent subsymbols at the beginning and the end of a block allows to emulate non-cyclic filters from a cyclic prototype filter response, in order to generate

FBMC-OQAM and FBMC-FMT bursts. More precisely, M_p is the length of the prototype filter and $M_s = M_p$. Lastly, the CP is only compatible with cyclic filters.

Generally, the waveform can become non-orthogonal depending on the use of specific filters and for a given value of ν_f and ν_t. This is addressed in the final category that consists of the *non-orthogonal multicarrier techniques* Faster than Nyquist (FTN) [15] and spectrally efficient frequency division multiplexing (SEFDM) [18]. The key property of FTN is $\nu_t < 1$, which results in an increment of the subsymbol data rate. The isotropic orthogonal transform algorithm (IOTA) pulse, in combination with OQAM, has been proposed in order to avoid the use of CP. Since the impulse response of the filter is not cyclic, M_p subsymbols are silent. Analogously, the idea of SEFDM is to increase the density of subcarriers in the available bandwidth, i.e. $\nu_f < 1$. Here, $M = 1$ because each block consists of a single subsymbol that is filtered with a rectangular pulse and a CP is prepended to combat multipath propagation. In this case, regular QAM is employed. Clearly, the amount of squeezing without severely impacting the error rate performance is limited. The Mazo limit states that this threshold is around 25 % for both schemes.

1.6 Multi-Service and Multi-User Systems

In [14], a unified framework was proposed for delivering various services. GFDM has good time and frequency properties to fit into that framework. Namely, one subcarrier as a guard band is sufficient to divide the resource grid for serving different applications in an interference free manner. Benefiting from this feature, a single resource grid can convey various GFDM settings, including frame structure and pilot pattern, for target applications. This concept is particularly of interest for base station (BS) which can support programmable transceivers. On contrary, a MS does not need to have all possible features of GFDM implemented on its PHY, but only the features that are requested for the services supported by this equipment.

The GFDM time and frequency properties are also beneficial to mitigate the multi-user interference coming from asynchronous users sharing the resource grid [19]. With a single subcarrier as the guard band, the spectrum of the different users does not overlap. As such, the time misalignment among users does not lead to multi-user interference. Since the side lobes of the subcarriers spectrum are negligible, only coarse frequency synchronization among the devices is required to avoid spectrum overlapping due to frequency misalignments.

2 Gabor Transform: Theory Behind GFDM

In order to better understand the common principles of multicarrier modulation, consider a signal s. When looking at its time-domain representation $s(t)$, exact information about the behavior at any time instant is revealed. However, this representation contains no information at all about frequency components at these positions. Instead, the Fourier transform (FT) of the signal provides exact information about frequency components, but on the other hand, no information on time-domain behaviour is obtained. The short-time Fourier transform (STFT) provides time-localized information about frequency components of a given signal by looking at the FT of the signal multiplied with a time window. However, the output of the STFT is highly redundant, since time and frequency sampling is kept independent.

To remove this redundancy, Dennis Gabor proposed in 1947 to represent a signal as a linear combination of Gaussian functions that are shifted in time and frequency to positions in a regular grid given by

$$s(t) = \sum_{k\in\mathbb{Z}} \sum_{m\in\mathbb{Z}} a_{k,m} g(t - m\Delta T) \exp(j2\pi \Delta F k t). \tag{20}$$

where ΔT, ΔF describes grid distance in time and frequency, respectively and $g(t)$ describes the Gaussian bell function. Gabor proposed $\Delta F \Delta T = 1$. This technique is illustrated in Fig. 11. Gabor selected the Gaussian function as the basis function due to its best simultaneous localization in time and frequency, so that local behavior of the signal is most accurately described. According to Gabor, the original signal is fully characterized by the coefficients $a_{k,m}$ multiplying the Gaussian functions, establishing the foundation of time-frequency analysis [20]. More precisely, the uniqueness and existence of such an expansion was later analyzed and it was shown to critically depend on the density of the grid of time-frequency shifts, which is defined as the product $\rho \triangleq \Delta T \Delta F$ of spacing in time ΔT and frequency ΔF. Oversampled densities $\rho > 1$ imply non-unique expansions whereas with undersampled densities $\rho < 1$, the expansion coefficients only exist for certain signals. For critical densities $\rho = 1$, the expansion is unique and only exists for certain windows.

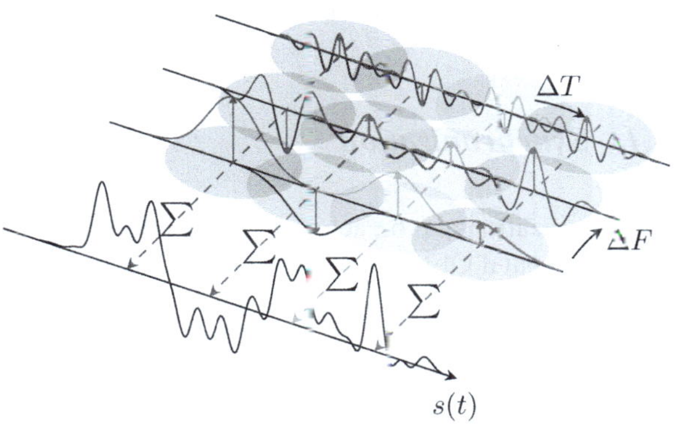

Fig. 11 Illustration of Gabor expansion. The expanded signal is the sum of scaled time-frequency shifts of a prototype window. The scaling factors are given by the Gabor expansion coefficients

Nowadays, the linear combination of time-frequency shifted windows (20) is known as a Gabor expansion and the calculation of the STFT with a certain window at a regular grid is known as a Gabor transform with window $\gamma(t)$ given by [21]

$$a_{k,m} = \int_{\mathbb{R}} s(t)\gamma^*(t - m\Delta T)\exp(-j2\pi\Delta Fkt)dt \tag{21}$$

$$= \langle s(t), \gamma(t - m\Delta t)\exp(j2\pi\Delta Fkt)\rangle_{\mathbb{C}}\,. \tag{22}$$

The expansion and transform windows are in a dual relation. Accordingly, the coefficients used to expand to a certain signal with a given window $g(t)$ are provided by the Gabor transform of that signal with the dual window $\gamma(t)$. In case the window and its dual are equal, i.e. $\gamma(t) = g(t)$, the window is said to be orthogonal and expansion and transform reduce to well-known orthogonal expansion series

$$s(t) = \sum_{k\in\mathbb{Z}}\sum_{m\in\mathbb{Z}} \langle s(t), g_{k,m}(t)\rangle_{\mathbb{C}}\, g_{k,m}(t), \tag{23}$$

with $g_{k,m}(t) = g(t - m\Delta T)\exp(j2\pi\Delta Fkt)$.

A prominent example for Gabor expansion and transform is OFDM, which performs a Gabor expansion using a finite discrete set of rectangular window functions with length $T_S + T_{CP}$ in time and shifts of $1/T_S$ in the frequency grid. At the receiver, the dual receiver window are rectangular windows with length T_S in order to achieve orthogonality between subcarriers.

3 Synchronization and Channel Estimation

As depicted in Fig. 2, synchronization and channel estimation are required for coherent equalization and demodulation. In this section, we introduce some basics of synchronization and channel estimation for GFDM systems.

3.1 Synchronization

Synchronization can be achieved in GFDM on a block basis, allowing adaptation of fundamental OFDM solutions to estimate symbol time offset (STO) and carrier frequency offset (CFO) [22, 23], but low OOB emission is a factor to be specifically considered in GFDM.

Algorithms originally proposed to OFDM [23–25] can achieve one-shot synchronization using a straightforward proposal of a separated preamble. For instance, this concept can be exploited with low OOB in the GFDM case by using a windowed preamble (Fig. 12) for best isolation of data and training sequence, or by embedding

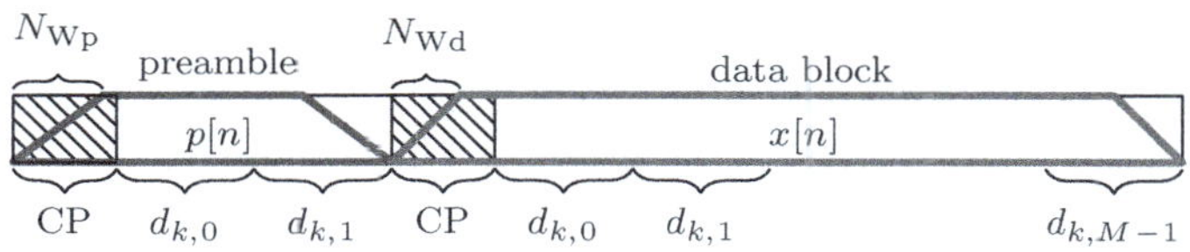

Fig. 12 Windowed preamble preceding a GFDM data block

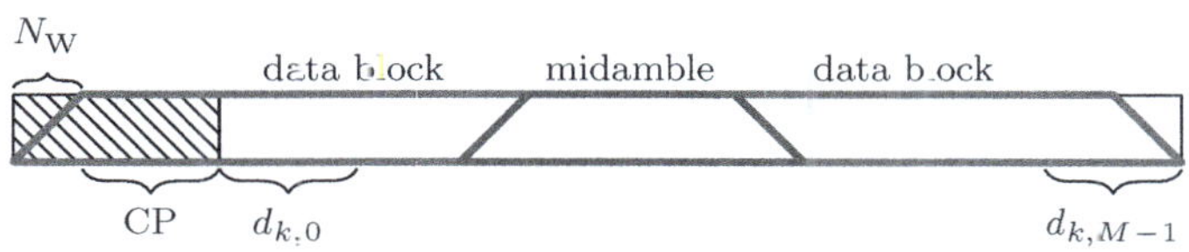

Fig. 13 Embedded midamble in a GFDM data block

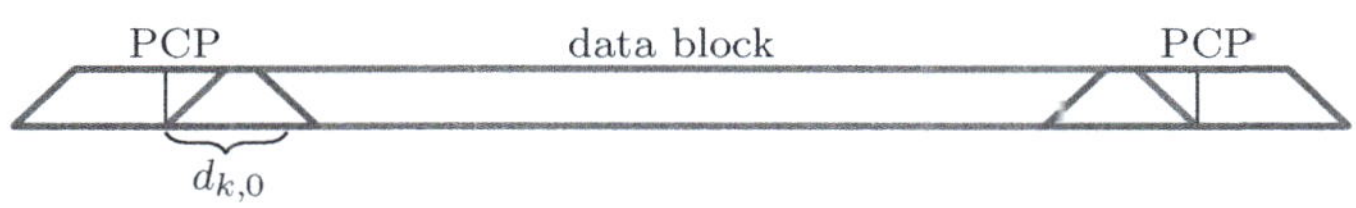

Fig. 14 PCP in a GFDM data block

the training sequence as a midamble [26] (Fig. 13) or by a pseudo circular preamble (PCP) approach [27] (Fig. 14) for a self containing block transmission.

Classically, consecutive transmission of identical signals allows the exploration of robust coarse autocorrelation metrics [23, 24]. This can be achieved with subsymbols configured to carry the same pseudo-noise (PN) sequence data in all subcarriers in the first two time slots, $\mathbf{d}_{c,0}$ and $\mathbf{d}_{c,1}$, in the preamble case, or two consecutive time slots in the center of the block for the midamble case, or by the use of only even or odd subcarriers in the first time slot for the PCP scheme. In the latter one, the training information can be used to replace the traditional use of the CP with known cyclic data-aided information and easily achieve soft concatenation of blocks in continuous transmissions.

In the isolated preamble case, low OOB emission is obtained by pinching the block boundary, where a windowed preamble precedes a data block and forms a burst with a double pinching pattern. Different pinching lengths N_{Wp} and N_{Wd} can be applied respectively to the preamble and data blocks in order to achieve a desired emission mask. Regarding the midamble design, low OOB emission can be obtained by using guard subsymbols to smooth the block boundary, in this case the midamble and data forms a single burst with one pinching pattern. The midamble is placed in the center of the block, which minimizes its time difference to any subsymbol and can later also be used for equalization purposes. And with respect to the PCP approach, low OOB emission can be obtained by concatenating consecutive blocks, once in this case the training sequence forms a continuous transmission without transitions between blocks.

In the following the isolated preamble case will be presented in details.

Given that $r[n]$ is a set of received samples containing at least one complete windowed preamble, the two identical halves are identified with an autocorrelation metric, which is integrated along the CP and CS length to remove plateau effects [23, 24], leading to the metric

$$\mu[n] = \sum_{\iota=-N_{\mathrm{CP}}}^{0} \sum_{k=0}^{N-1} r[n+\iota+k]^* r\left[n+\iota+k+\frac{N}{2}\right], \tag{24}$$

where $N = MK$ denotes the length of the GFDM preamble. The argument that maximizes the absolute value of the metric $\mu[n]$ is taken as a coarse STO

$$\hat{n}_{\mathrm{c}} = \underset{n}{\operatorname{argmax}} |\mu[n]|, \tag{25}$$

while the angle of $\mu[\hat{n}_{\mathrm{c}}]$ is used to estimate the CFO as

$$\hat{\epsilon} = \frac{\angle \mu[\hat{n}_{\mathrm{c}}]}{\pi}. \tag{26}$$

The value $\hat{\epsilon}$ is employed to correct the CFO in the received sequence and a cross-correlation operation is then performed as

$$c[n] = \frac{1}{N} \sum_{k=0}^{N-1} r[n+k]^* e^{j2\pi \frac{\hat{\epsilon}}{N}(n+k)} p[k], \tag{27}$$

where $p[k]$ represents the known GFDM preamble.

To suppress side peaks that arise from the two halves and from the CP and CS parts, $c[n]$ is combined with $\mu[n]$ and an optimized estimation of the STO is obtained by searching the peak value in the range $\mathscr{S} = \{\hat{n}_{\mathrm{c}} - \frac{N}{2}, \ldots, \hat{n}_{\mathrm{c}} + \frac{N}{2}\}$ around the coarse STO estimation by

$$\hat{n}_{\mathrm{o}} = \underset{n \in \mathscr{S}}{\operatorname{argmax}} \left(|c[n]||\mu[n]|\right). \tag{28}$$

Depending on SNR and channel conditions, the search range $\mathscr{S}$ can be reduced in order to decrease implementation complexity. This synchronization procedure is robust for single path channels, but in a time-variant FSC, the primary echo can be lower than other echoes and the strongest peak will not represent the correct STO. Thus, an additional search before $\hat{n}_{\mathrm{o}}$ can reveal if there is another yet undetected peak to be considered as the primary one. For samples that do not belong to the preamble, the output of the cross-correlation can be interpreted as a complex Gaussian random sequence and a threshold criteria, depending on an acceptable probability of false alarm p_{FA}, can reveal the presence of multipaths before $\hat{n}_{\mathrm{o}}$.

Considering that the first peak of (27) is within the range $(\hat{n}_{\mathrm{o}} - \lambda, \hat{n}_{\mathrm{o}}]$, where $\lambda < N_{\mathrm{CP}}$ is an adjustable parameter, the threshold is defined as

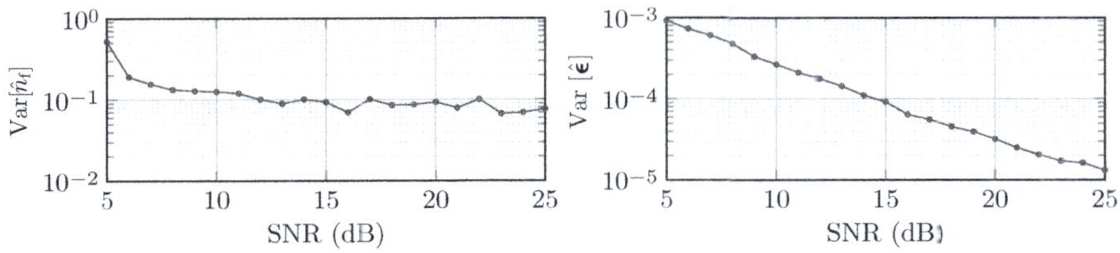

Fig. 15 Performance evaluation of variance of the fine STO estimation $\hat{n}_f$ and CFO estimation $\hat{\epsilon}$ for the GFDM preamble in a time-variant multipath channel

Table 7 Isolated preamble parameters

Parameter	Value
Mapping	QPSK
Transmit filter	Rect
Transmit window	RC
Roll-off (α)	0
Number of subcarriers (K)	64
Number of subsymbols (M)	2
CP length (N_{CP})	16 samples
CS length (N_{CS})	0 samples

$$T_{\mathrm{Th}} = \sqrt{-\frac{4}{\pi}\ln\left(p_{\mathrm{FA}}\right)}\left(\frac{1}{\frac{N}{2}-2\lambda}\sum_{k=-\frac{N}{2}+\lambda}^{-\lambda}\left|c[\hat{n}_c+k]\right|\right) \tag{29}$$

and the fine STO estimation for the first multipath is finally given by

$$\hat{n}_f = \underset{\hat{n}_o-\lambda<n\leq\hat{n}_o}{\mathrm{argfirst}}\left(|c[n]| > T_{\mathrm{Th}}\right). \tag{30}$$

The performance evaluation in terms of variance of normalized STO and CFO estimations is presented in Fig. 15 for the W-GFDM preamble following the parameters presented in Table 7 but with $M = 2$.

A $w_{RC}[n]$ window function with $N_W = 16$ is used, $\lambda = 16$, $p_{FA} = 10^{-4}$ and the time-variant FSC with exponential decaying taps from 0 to -10 dB. For an SNR range higher than 5 dB the variance of the STO estimation stabilizes within tenths of a sample due the time variant fading effect in the multipath channel. The variance of the CFO estimation starts from thousandths of the subcarrier bandwidth and gets linearly better (in log scale) with increasing SNR.

The results obtained with the double pinching configuration show that burst synchronization can be achieved without penalties to the performance when compared with the results presented in [25]. The smooth block boundaries at the edges of the preamble and the data portion of the signal are particularly important as keeping OOB emission low is an important feature of GFDM.

3.2 Channel Estimation

Coherent detection used in modern mobile wireless communication standards requires knowledge of the wireless channel between transmit and receive terminals. Such channel state information (CSI) is usually obtained by transmitting known data (a.k.a. pilots) between the transmitted payload. At the receiver the effect of the wireless channel onto the known data is analyzed and the underlying CSI is extracted. Wireless channels are usually modelled as time-variant multipath channels, which are again modelled with the tapped-delay line principle, where the delays and gains represent the corresponding channel impulse response (CIR). Because of the variability of the wireless channel due to mobility or changes in the environment, the CIR changes over time. In this chapter, we assume a block-fading channel, which implies that the CIR is assumed constant for the duration of one GFDM block.

Generally, two methods for channel estimation exist: Preamble-based and pilot-based methods. In the former, a dedicated signal portion is only allocated with known data. Payload is subsequently transmitted independently of the preamble. The approach has the advantage that the preamble can be processed separately, hence reducing estimation complexity. The estimated CIR is then used for the payload block. In pilot-based methods, known data and payload are multiplexed and transmitted within the same resource block. In this case, channel estimation needs to consider if pilots and payload are orthogonal and if not, the payload needs to be additionally considered in the estimation, increasing estimation complexity. However, with the multiplexed approach, faster-changing wireless channels can be utilized, as the payload is virtually transmitted at the same time as the pilots, compared to the preamble-based approach.

In this chapter, we analyze the pilot-based channel estimation techniques for GFDM. Consider the GFDM data block vector $\mathbf{d}$ to be shared between payload and pilots such that

$$\mathbf{d} = \mathbf{d}_d + \mathbf{d}_p, \tag{31}$$

where $\mathbf{d}_d$ contains the payload and $\mathbf{d}_p$ describes the pilot symbols. Note that a single time-frequency resource can only be allocated to either pilot or payload, hence $\mathbf{d}_d$ and $\mathbf{d}_p$ cannot be non-zero at the same index and $\mathbf{d}_d \circ \mathbf{d}_p = \mathbf{0}$, where $\circ$ denotes the Hadamard (elementwise) product. Accordingly, assuming perfect synchronization, the received signal $\mathbf{y}$ in the time domain is given by

$$\mathbf{y} = \widetilde{\mathbf{H}}\mathbf{A}(\mathbf{d}_d + \mathbf{d}_p) + \mathbf{n}, \tag{32}$$

where $\mathbf{n}$ is the AWGN at the receiver and $\widetilde{\mathbf{H}} = \mathrm{circ}(\mathbf{h})$ is the channel matrix performing circular convolution with the CIR $\mathbf{h}$. Transforming the received signal into the frequency domain yields

$$\mathbf{Y} = \mathbf{F}\mathbf{y} = \mathbf{F}\widetilde{\mathbf{H}}\mathbf{F}^{\mathrm{H}} \underbrace{\mathbf{F}\mathbf{A}\mathbf{d}_p}_{\mathbf{x}_p} + \mathbf{F}\widetilde{\mathbf{H}}\mathbf{F}^{\mathrm{H}} \underbrace{\mathbf{F}\mathbf{A}\mathbf{d}_d}_{\mathbf{x}_d} - \mathbf{F}\mathbf{n}, \tag{33}$$

with $\mathbf{F}$ being the unitary Fourier transform matrix. With denoting the channel frequency response by $\mathbf{H} = \mathbf{F}\mathbf{h}$, this is transformed into

$$\mathbf{Y} = \mathrm{diag}(\mathbf{H})\mathbf{x}_p + \mathrm{diag}(\mathbf{H})\mathbf{x}_d + \mathbf{N} \tag{34}$$

$$= \underbrace{\mathrm{diag}(\mathbf{x}_p)}_{\mathbf{X}_p}\mathbf{H} + \underbrace{\mathrm{diag}(\mathbf{x}_d)}_{\mathbf{X}_d}\mathbf{H} + \mathbf{N}. \tag{35}$$

With the introduced notation, the final system model used for deriving the estimation algorithms is hence given by

$$\mathbf{Y} = \mathbf{X}_p\mathbf{H} + \mathbf{X}_d\mathbf{H} + \mathbf{N}, \tag{36}$$

where $\mathbf{X}_p$ is a diagonal matrix with the known pilot signal in the frequency domain (FD) on the diagonal. $\mathbf{X}_d$ is a random diagonal matrix with the payload signal in the frequency domain on the diagonal. Hence, the linear model in (36) describes the received signal $\mathbf{Y}$ to be constructed of a deterministic part $\mathbf{X}_p\mathbf{H}$, a payload interference part $\mathbf{X}_d\mathbf{H}$ and a noise part $\mathbf{N}$. The influence of the payload interference is a significant difference from channel estimation for OFDM systems, where pilots and payload are separated with the Fourier transform. This on the one hand increases estimation complexity, but on the other hand enables to achieve a higher resolution in frequency domain for the estimation. From the model (36) we can now derive several basic estimators.

3.2.1 Least-Squares Channel Estimation

Direct application of Least Squares (LS) or ZF estimation results in

$$\widehat{\mathbf{H}}_{LS} = \mathbf{X}_p^{+}\mathbf{Y} \tag{37}$$

$$= \mathbf{H} + \mathbf{X}_p^{+}\mathbf{X}_d\mathbf{H} + \mathbf{X}_p^{+}\mathbf{N}, \tag{38}$$

where $\mathbf{X}_p^{+}$ denotes the pseudoinverse of $\mathbf{X}_p$. This estimator is unbiased, however it does not use any a-priori knowledge of the channel behaviour. Apparently, the noise is directly added to the estimation and noise enhancement occurs due to the multiplication with the pseudo inverse. Let $\chi_p = \mathbf{X}_p^{+}$, the mean-squared error (MSE) of this estimator is given by

$$\mathrm{MSE}(\widehat{\mathbf{H}}_{LS}) = E[(\widehat{\mathbf{H}}_{LS} - \mathbf{H})^{\mathrm{H}}(\widehat{\mathbf{H}}_{LS} - \mathbf{H})] \tag{39}$$

$$= E[(\mathbf{H}^{\mathrm{H}}\mathbf{X}_d^{\mathrm{H}}\chi_p^{\mathrm{H}} + \mathbf{N}^{\mathrm{H}}\chi_p^{\mathrm{H}})(\chi_p\mathbf{X}_d\mathbf{H} + \chi_p\mathbf{N})] \tag{40}$$

$$= E[\mathbf{H}^{\mathrm{H}}\mathbf{X}_d^{\mathrm{H}}\chi_p^{\mathrm{H}}\chi_p\mathbf{X}_d\mathbf{H}] + E[\mathbf{N}^{\mathrm{H}}\chi_p^{\mathrm{H}}\chi_p\mathbf{N}], \tag{41}$$

where the last step follows since $\mathbf{H}$ and $\mathbf{N}$ are uncorrelated. In the first term, both $\mathbf{H}$ and $\mathbf{X}_d$ are random variables, however as $\mathbf{X}_d$ is diagonal, only element-wise multiplication of the random variables appears. Accordingly, the MSE is expressed as [28, Chap. 5.0.5]

$$\mathrm{MSE}(\widehat{\mathbf{H}}_{LS}) = \mathrm{tr}(\boldsymbol{\chi}_p^{\mathrm{H}}\boldsymbol{\chi}_p(\mathbf{R}_{\mathbf{HH}} \circ \mathbf{R}_{\mathbf{X}_d\mathbf{X}_d})) + \sigma_n^2\mathrm{tr}(\boldsymbol{\chi}_p^{\mathrm{H}}\boldsymbol{\chi}_p), \tag{42}$$

where $\mathbf{R}_{\mathbf{HH}}$ and $\mathbf{R}_{\mathbf{X}_d\mathbf{X}_d}$ denote the covariance matrices of the channel $\mathbf{H}$ and the data $\mathbf{X}_d$, respectively. Apparently, the MSE consists of two terms, where only one depends on the channel noise. Hence, an error floor due to interference is expected, coming from overlapping payload and pilot signals. In order to improve the estimation performance one can apply the knowledge of the maximum channel length, which is given by the CP of the used system. Hence, the CIR is obtained by IDFT of the estimated frequency response, truncated after the N_{CP}th sample and transformed back into the frequency domain. The improved estimate $\widehat{\mathbf{H}}_{LS,t}$ is then given by

$$\widehat{\mathbf{H}}_{LS,t} = \mathbf{FTF}^{\mathrm{H}}\boldsymbol{\chi}_p\mathbf{Y}, \tag{43}$$

where $\mathbf{T}$ performs the truncation, being a diagonal matrix where only the first N_{CP} elements of the diagonal are 1. Let $\mathbf{U} = \mathbf{FTF}^{\mathrm{H}}$, the resulting MSE is calculated by

$$\mathrm{MSE}(\widehat{\mathbf{H}}_{LS,t}) = \mathrm{tr}(\boldsymbol{\chi}_p^{\mathrm{H}}\mathbf{U}^{\mathrm{H}}\mathbf{U}\boldsymbol{\chi}_p(\mathbf{R}_{\mathbf{HH}} \circ \mathbf{R}_{\mathbf{X}_d\mathbf{X}_d})) + \sigma_n^2\mathrm{tr}(\boldsymbol{\chi}_p^{\mathrm{H}}\mathbf{U}^{\mathrm{H}}\mathbf{U}\boldsymbol{\chi}_p) \tag{44}$$

3.2.2 Linear MMSE Channel Estimation

Starting from the linear model (36), a linear MMSE (LMMSE) estimator can be directly derived, given by

$$\widehat{\mathbf{H}}_{LM} = \mathbf{R}_{\mathbf{HY}}\mathbf{R}_{\mathbf{YY}}^{-1}\mathbf{Y}, \tag{45}$$

where $\mathbf{R}_{\mathbf{YY}}$ and $\mathbf{R}_{\mathbf{HY}}$ are given by

$$\mathbf{R}_{\mathbf{HY}} = \mathbf{R}_{\mathbf{HH}}\mathbf{X}_p^{\mathrm{H}} \tag{46}$$

$$\mathbf{R}_{\mathbf{YY}} = \mathbf{X}_p\mathbf{R}_{\mathbf{HH}}\mathbf{X}_p^{\mathrm{H}} + \mathbf{R}_{\mathbf{HH}} \circ \mathbf{R}_{\mathbf{X}_d\mathbf{X}_d} + \sigma_n^2\mathbf{I}. \tag{47}$$

The calculation of the LMMSE estimator requires a-priori knowledge of the autocorrelation of the channel, i.e. the power delay profile needs to be known in advance. For example, this can be accomplished averaging over LS-estimates. The expected MSE for the LMMSE estimator is given by

$$\mathrm{MSE}(\widehat{\mathbf{H}}_{LM}) = \mathrm{tr}(\mathbf{R}_{\mathbf{HH}} - \mathbf{R}_{\widehat{\mathbf{H}}_{LM}\widehat{\mathbf{H}}_{LM}}) \tag{48}$$

$$= \mathrm{tr}(\mathbf{R}_{\mathbf{HH}} - \mathbf{R}_{\mathbf{HY}}\mathbf{R}_{\mathbf{YY}}^{-1}\mathbf{R}_{\mathbf{HY}}^{\mathrm{H}}) \tag{49}$$

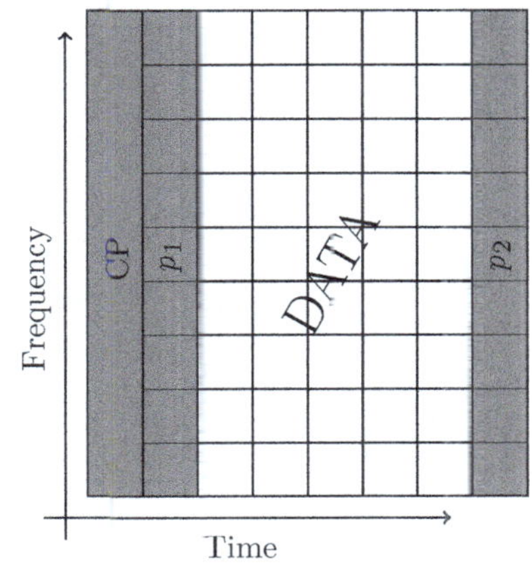

Fig. 16 Proposed pilot arrangement

Table 8 System parameters used for simulation

Parameter	Symbol	Value
Number of subsymbols	M	7
Number of subcarriers	K	64
Cyclic prefix length	N_{CP}	12
Prototype filter	$g[n]$	RC
Rolloff factor	α	0.5
Channel power delay profile		Exponential

3.2.3 Pilot Structure and Simulation Results

In this section, we compare the estimation accuracy of the three presented channel estimation algorithms. We use the pilot structure shown in Fig. 16. The system parameters for the presented evaluation are given in Table 8.

For the simulation either two (p_1, p_2) or only one (p_1) pilot subsymbols were used. Considering (42) and (44), the noise enhancement depends on the pseudo-inverse of the pilot signal. Accordingly, care needs to be taken when designing this signal. Values close to zero in the frequency domain of the pilot signal need to be avoided in order to not increase the noise. In the present simulation, Zadoff-Chu sequences are used for pilots on each subsymbol, where the sequence on the second pilot subsymbol is a circular shift of the first pilot subsymbol. This way it can be assured that both pilot sequences do not add up to zero in the frequency domain.

The estimation performance for all proposed estimators is shown in Fig. 17. Apparently, the naive LS estimation shows the worst estimation performance. The application of a-priori knowledge of the channel length significantly reduces the estimation MSE. Both LS estimators exhibit high error floors resulting from the interference part of the GFDM block, which is not accounted for in the estimation. On the other hand, the LMMSE estimator provides a significantly improved estimation performance, since it also takes self-interference into account. With this,

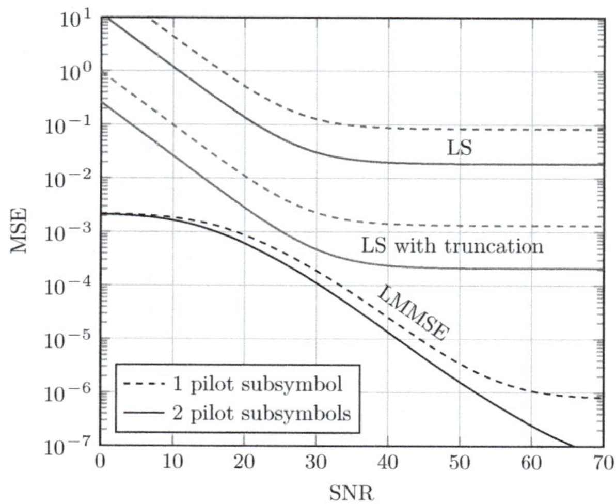

Fig. 17 Mean squared errors of proposed estimators

the error floor is significantly reduced. In overall, the estimation performance is far superior compared to LS estimators. However, the LMMSE estimator requires knowledge of channel statistics.

In Fig. 17 the estimation performance is presented for both one and two used pilot subsymbols. According to the simulation, for LS estimation, the addition of the second pilot subsymbol reduces the error floor by roughly one order of magnitude and improves estimation by 3 dB SNR. With the LMMSE estimator, performance improves by roughly 1 dB, and the error floor is also reduced. However, for the LMMSE estimation, the improvement from addition of the second subsymbol is marginal compared to the improvement of LS estimators.

4 Multi-Antenna Systems

A 5G waveform must be compatible with MIMO techniques in order to achieve the necessary robustness and throughput in mobile environments. In this section, we introduce space time block code for GFDM to achieve transmit diversity. Furthermore, near-maximum likelihood (ML) detection algorithms are presented for MIMO systems using GFDM based spatial multiplexing (SM).

4.1 Space Time Block Code

This section describes a simple approach to achieve transmit diversity with GFDM, using time-reverse space time code (TR-STC), which has been proposed by Al-Dhahir [29] as a space time code (STC) for single carrier transmission over frequency-selective channels. The proposal takes two subsequent data blocks $\mathbf{x}_i$ and $\mathbf{x}_{i+1}$ of length N which are separated by a CP. Let their corresponding discrete Fourier transforms be $\mathbf{X}_{(\cdot)} = \mathbf{F}\mathbf{x}_{(\cdot)}$, and $\mathbf{F}$ denotes the unitary Fourier transform matrix. Then, for two subsequent time slots the transmit signals on both antennas are given by

	Antenna 1	Antenna 2
Block i	$\mathbf{F}^{\mathrm{H}}\mathbf{X}_i$	$-\mathbf{F}^{\mathrm{H}}\mathbf{X}^*_{i+1}$
Block $i+1$	$\mathbf{F}^{\mathrm{H}}\mathbf{X}_{i+1}$	$\mathbf{F}^{\mathrm{H}}\mathbf{X}^*_i$,

(50)

where i is an even number and $(\cdot)^{\mathrm{H}}$ is the transpose-conjugate operator. Note that the name "time-reversal space-time coding" is reasoned by the property

$$(\mathbf{F}^{\mathrm{H}}\mathbf{X}^*_i)_n = x^*_i[\langle -n\rangle_N] \tag{51}$$

of the discrete Fourier transform.

At the receiver, after removing the CP, the CIR $\mathbf{h}_{j,l}$ between the jth transmit and lth receiving antenna circularly convolves the transmitted signals. Then, their frequency domain representation is calculated. Accordingly, assuming a block-fading channel during the transmission of two subsequent blocks, the frequency domain of the received blocks are given by

$$\mathbf{Y}_{i,l} = \mathbf{H}_{1,l}\mathbf{X}_i - \mathbf{H}_{2,l}\mathbf{X}^*_{i+1} + \mathbf{W}_{1,l} \tag{52}$$

$$\mathbf{Y}_{i+1,l} = \mathbf{H}_{1,l}\mathbf{X}_{i+1} + \mathbf{H}_{2,l}\mathbf{X}^*_i + \mathbf{W}_{2,l}, \tag{53}$$

where $\mathbf{H}_{j,l} = \mathrm{diag}(\mathbf{F}\mathbf{h}_{j,i})$. STC combining can then be carried out in the frequency domain according to

$$\widehat{\mathbf{X}}_i = \mathbf{H}_{eq}^{-1}\sum_{l=1}^{L}\mathbf{H}^*_{1,l}\mathbf{Y}_{i,l} + \mathbf{H}_{2,l}\mathbf{Y}^*_{i+1,l}$$
$$\widehat{\mathbf{X}}_{i+1} = \mathbf{H}_{eq}^{-1}\sum_{l=1}^{L}\mathbf{H}^*_{1,l}\mathbf{Y}_{i+1,l} - \mathbf{H}_{2,l}\mathbf{Y}^*_{i,l}, \tag{54}$$

$$\text{where } \mathbf{H}_{eq} = \sum_{j=1}^{2}\sum_{l=1}^{L}\mathbf{H}^*_{j,l}\mathbf{H}_{j,l} \tag{55}$$

and L is the number of receiving antennas. Finally, going back to the time domain with the inverse Fourier transform yields the estimates of the transmitted blocks

$$\hat{\mathbf{x}}_i = \mathbf{F}^{\mathrm{H}} \widehat{\mathbf{X}}_i. \tag{56}$$

When now identifying $\{\mathbf{x}_i\}_{i=0,1}$ with two subsequent GFDM blocks, we can apply the proposed STC technique to achieve transmit diversity. At the transmitter the GFDM modulated data blocks are space-time encoded with the technique above. At the receiver, the GFDM data blocks are recovered from the STC, yielding transmit diversity, and subsequently demodulated by the GFDM demodulator.

4.2 *Spatial Multiplexing*

Consider a $R \times T$ SM MIMO GFDM system with no transmit CSI and perfect CSI at the receiver side. The transmit signal of the tth transmit antenna is given by

$$\mathbf{x}_t = \mathbf{A}\mathbf{d}_t, \tag{57}$$

where $\mathbf{A}$ is the GFDM modulation matrix and $\mathbf{d}_t$ contains the transmit data of the tth antenna, taken from the set $\mathcal{S}$ of constellation symbols. The signal at the receive antennas can then be compactly written as

$$\underbrace{\begin{bmatrix} \mathbf{y}_1 \\ \vdots \\ \mathbf{y}_R \end{bmatrix}}_{\mathbf{y}} = \underbrace{\begin{bmatrix} \mathbf{H}_{1,1}\mathbf{A} & \dots & \mathbf{H}_{1,T}\mathbf{A} \\ \vdots & \ddots & \vdots \\ \mathbf{H}_{R,1}\mathbf{A} & \dots & \mathbf{H}_{R,T}\mathbf{A} \end{bmatrix}}_{\widetilde{\mathbf{H}}} \underbrace{\begin{bmatrix} \mathbf{d}_1 \\ \vdots \\ \mathbf{d}_T \end{bmatrix}}_{\mathbf{d}} + \mathbf{n}, \tag{58}$$

where $\widetilde{\mathbf{H}}$ represents the equivalent channel of the MIMO system. $\mathbf{H}_{r,t}$ denotes the circulant channel matrix between the tth transmit and the rth receive antenna, $\mathbf{d}$ contains all $\mathbf{d}_t$ and $\mathbf{n}$ describes AWGN with covariance $N_0\mathbf{I}_{RN}$. The transmit symbols are energy-normalized such that $E[\mathbf{d}\mathbf{d}^{\mathrm{H}}] = \mathbf{I}$.

4.2.1 Near-ML Detection

At the receiver, the ML solution $\hat{\mathbf{d}}$ for (58) is

$$\hat{\mathbf{d}} = \arg \max_{\mathbf{d} \in \mathcal{S}^U} \|\mathbf{y} - \widetilde{\mathbf{H}}\mathbf{d}\|^2, \tag{59}$$

where $U = TM|\mathcal{K}|$ denotes the number of transmitted symbols per block and $\mathcal{K}$ is the set of allocated subcarriers on each antenna. Obviously, applying a brute-force solution to (59) is prohibitively complex and hence, solutions with reduced complexity are of interest. In the orthogonal system OFDM $\widetilde{\mathbf{H}}$ can be block-diagonalized by the discrete Fourier transform (DFT), so that (59) reduces into $|\mathscr{K}|$ problems of dimension T. For each of these problems, a sphere decoder (SD) [30] can be applied to find the optimal ML solution. However, since the $\mathbf{H}_{r,t}\mathbf{A}$ cannot be simultaneously diagonalized, FD decoupling cannot be achieved in GFDM. To overcome the complexity issue, a combination of successive interference cancellation (SIC) with SD which achieves near-ML performance in high-SNR regimes is presented in the following.

Let the MMSE sorted QR-decomposition (SQRD) [31] of $\widetilde{\mathbf{H}}$ be

$$\begin{bmatrix} \widetilde{\mathbf{H}} \\ \sqrt{N_0}\mathbf{I} \end{bmatrix} = \begin{bmatrix} \mathbf{Q}_1 \\ \mathbf{Q}_2 \end{bmatrix} \mathbf{R}\mathbf{P}^{\mathrm{T}}. \tag{60}$$

There, the matrix $\mathbf{P}$ permutes the columns of $\widetilde{\mathbf{H}}$ prior to decomposition. The received signal is multiplied with $\mathbf{Q}_1^{\mathrm{H}}$, yielding [31]

$$\tilde{\mathbf{y}} = \mathbf{Q}_1^{\mathrm{H}}\mathbf{y} = \mathbf{R}\mathbf{P}^{\mathrm{T}}\mathbf{d} - \sqrt{N_0}\mathbf{Q}_2^{\mathrm{H}}\mathbf{P}^{\mathrm{T}}\mathbf{d} + \mathbf{Q}_1^{\mathrm{H}}\mathbf{n}, \tag{61}$$

where the first, second and third term represent the useful signal, interference traded off for reduced noise enhancement and channel noise, respectively. The interference is ignored to yield

$$\tilde{\mathbf{y}} = \mathbf{R}\tilde{\mathbf{d}} + \tilde{\mathbf{n}} \tag{62}$$

with $\tilde{\mathbf{d}} = \mathbf{P}^{\mathrm{T}}\mathbf{d}$ and $\tilde{\mathbf{n}} = \mathbf{Q}_1^{\mathrm{H}}\mathbf{n}$. Still, the problem in (62) is of dimension $U \times U$ and direct application of SD requires prohibitive complexity. Instead, we introduce a factor S that divides the problem of size U into a chain of U/S problems of size S each. These subproblems are then subsequently solved. The proposed algorithm is given by

while Dim($\tilde{\mathbf{y}}$) > 0 **do**
  $\hat{\tilde{\mathbf{d}}}_S = \mathrm{SD}(\tilde{\mathbf{y}}_S, \mathbf{R}_{S,S})$ ▷ Jointly detect last S streams.
  $\tilde{\mathbf{y}} \leftarrow \tilde{\mathbf{y}} - \mathbf{R}_{:,S}\hat{\tilde{\mathbf{d}}}_S$ ▷ Cancel interference.
  $\tilde{\mathbf{y}} \leftarrow \tilde{\mathbf{y}}_{\bar{S}}, \mathbf{R} \leftarrow \mathbf{R}_{\bar{S},\bar{S}}$ ▷ Reduce system size.
end while.

There, the subscripts $S, \bar{S}, :$ denote the last S, all but the last S, and all elements of the subscripted object, respectively. For matrices, two subscripts are used, where the first operates on rows and the second operates on columns. The proposed algorithm first performs joint detection of a subset of size S of the complete problem. Afterwards, this detection is considered to be correct and the interference is cancelled from the remaining problems. Subsequently, it proceeds to the next subset until all subsets are processed. With the parameter S, detection complexity

Table 9 System parameters used for simulation

Parameter		Symbol	Value
GFDM	# subsymbols, # subcarriers	M, K	$5, 32$
	# allocated subcarriers	$\lvert\mathcal{K}\rvert$	16
	Prototype filter	$g[n]$	RC, $\alpha = 1$
OFDM	Block length	N	$160 = MK$
	# allocated subcarriers		$80 = M\lvert\mathcal{K}\rvert$
General	CP length	N_{CP}	16
	Modulation	$\mathcal{S}$	64-QAM
	# transmit, receive antennas	T, R	2×2
Channel PDP	Exponential, 0 to −10 dB, N_{CP} taps		

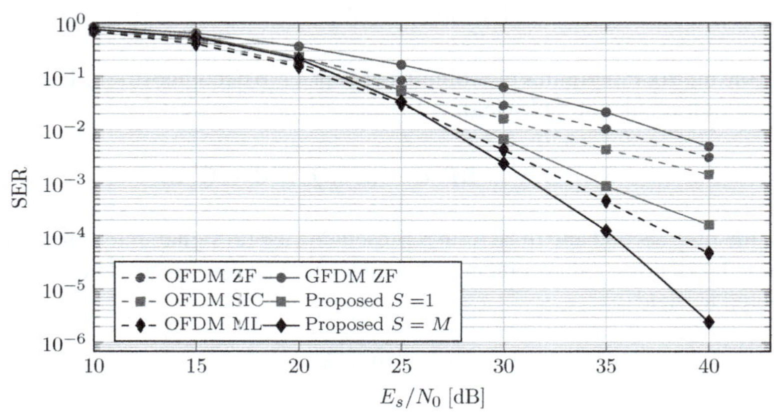

Fig. 18 SER performance of proposed the detection algorithm in a fading multipath channel

and performance can be controlled, since with $S = 1$, standard SIC detection [31] with low complexity and with $S = U$, full ML detection with prohibitive complexity is performed.

In Table 9 the system parameters used for performance simulation of the proposed algorithm in independent Rayleigh-fading multipath channels are given. In Fig. 18, we compare the uncoded SER performance of the proposed algorithm with common detection methods for OFDM. For reference, also the SER performance for ZF equalization is provided, which shows that the SER of GFDM is 3 dB worse than OFDM due to the self-interference of the GFDM system. However, with the proposed algorithm, the SER of GFDM is found to be below that of OFDM both for pure SIC ($S = 1$) and SIC+ML ($S = M$) detection already at a SER of $\approx 5 \cdot 10^{-2}$. Note that for $S = M$, all subsymbols on one GFDM subcarrier are detected simultaneously. Since one subsymbol is transmitted on several frequency samples in GFDM, frequency diversity can be inherently exploited. This property cannot be achieved with OFDM, since for OFDM only one frequency sample is available for each symbol. Accordingly, GFDM can achieve a lower SER than OFDM for higher SNR values.

4.2.2 Soft MIMO Detection

After the invention of turbo codes, the concept of iterative decoding has been extended and has found tremendous applications in the transceiver design for wireless communications systems. Particularly, the detector as one key functional unit at the receiver shall not only estimate the transmitted symbols, i.e., hard output, but also provide the reliability information of them, i.e., soft output. In short, soft detection algorithms are necessary for iterative receiver designs.

Assume the data symbol vector $\mathbf{d} \in \mathscr{S}^U$ is generated from the codeword $\mathbf{c} \in \{0,1\}^L$ with $L \triangleq U \log_2 |\mathscr{S}|$. The optimal soft detector computes the a-posteriori log-likelihood ratio (LLR) of the lth code bit c_l in the code bit vector $\mathbf{c}$ according to [32]

$$\lambda_l^{[\mathrm{p}]} = \log\left[\frac{\sum_{\mathbf{c}:c_l=1} P(\mathbf{c}|\mathbf{y},\widetilde{\mathbf{H}})}{\sum_{\mathbf{c}:c_l=0} P(\mathbf{c}|\mathbf{y},\widetilde{\mathbf{H}})}\right] = \log\left[\frac{\sum_{\mathbf{c}:c_l=1} P(\mathbf{y}|\mathbf{c},\widetilde{\mathbf{H}})P(\mathbf{c};\lambda_l^{[\mathrm{a}]})}{\sum_{\mathbf{c}:c_l=0} P(\mathbf{y}|\mathbf{c},\widetilde{\mathbf{H}})P(\mathbf{c};\lambda_l^{[\mathrm{a}]})}\right], \tag{63}$$

where the a-priori distribution of the codeword, i.e., $P(\mathbf{c};\lambda_l^{[\mathrm{a}]})$, is parameterized by the a-priori LLRs $\{\lambda_l^{[\mathrm{a}]}\}$ fed back from the decoder

$$P(\mathbf{c};\lambda_l^{[\mathrm{a}]}) = \prod_{l=0}^{L-1} P(c_l) = \prod_{l=0}^{L-1} \frac{e^{c_l \lambda_l^{[\mathrm{a}]}}}{1+e^{\lambda_l^{[\mathrm{a}]}}} \tag{64}$$

In the case of MIMO-GFDM, the presence of ICI and ISI requires us to marginalize the distribution $P(\mathbf{c}|\mathbf{y},\widetilde{\mathbf{H}})$ over the complete codeword $\mathbf{c}$ except c_l for obtaining $\lambda_l^{[\mathrm{p}]}$. For practical codeword lengths, approximations are evidently inevitable.

One possible approximation is to extend the above presented SIC-ML detection algorithm by applying the soft-input soft-output SD [30] instead of hard-input hard-output SD. As low complexity alternatives to SD for near-optimum soft MIMO detection, Markov chain Monte Carlo (MCMC) methods have been widely researched in the literature, e.g. [33–36]. Instead of using tree search to locate the most likely transmitted data symbols, they use Gibbs samplers to generate promising candidates based on a proposal distribution. While preserving near-optimum detection performance, it is possible to reduce exponential detection complexity to polynomial complexity. One issue with the MCMC technique discovered in the literature is the *stalling problem*, typically occurring at high signal-to-noise ratios (SNRs). The produced candidate set only consists of local optimums. To tackle this problem, we can activate multiple Gibbs sampling process, each of which generates samples independently [34]. We can also control the Gibbs sampler by a temperature parameter, e.g., in [35] and [37]. For more details, we refer readers to [38].

5 Proof-of-Concept: MIMO-GFDM Demonstrator

The development of a proof-of-concept GFDM transceiver with Software-Defined Radios is an important step from theory towards practice. Therefore, the described GFDM MIMO System is implemented on the National Instruments (NI) USRP-RIO 2953R platform. This section briefly describes the development of a 2×2 MIMO-GFDM demonstrator including the employment of TR-STC for achieving diversity gain, and preamble-based synchronization and channel estimation.

5.1 Implementation Model

An overview of the processing blocks used to implement the transmitter and receiver can be found in Fig. 19.

At the transmitter, the inputs $\mathbf{d_1}$ and $\mathbf{d_2}$ are two data vectors that carry N QAM symbols each. They are respectively filtered by a set of circular filters, yielding the modulated waveforms $\mathbf{x}_{\mathrm{d},1}$ and $\mathbf{x}_{\mathrm{d},2}$. Those two signals are fed to the TR-STC block, which applies the space-time block coding as described in Sect. 4.1. Next, the two predefined preamble signals $\mathbf{x}_{\mathrm{p},1}$ and $\mathbf{x}_{\mathrm{p},2}$ are acquired from a look-up table (LUT). From this point, all six signals undergo the same processing steps of CP appending and windowing. Lastly, the frame is assembled according to the structure displayed in Fig. 20. The samples of the two channels passed to the hardware are collected in $\mathbf{x}_1$ and $\mathbf{x}_2$.

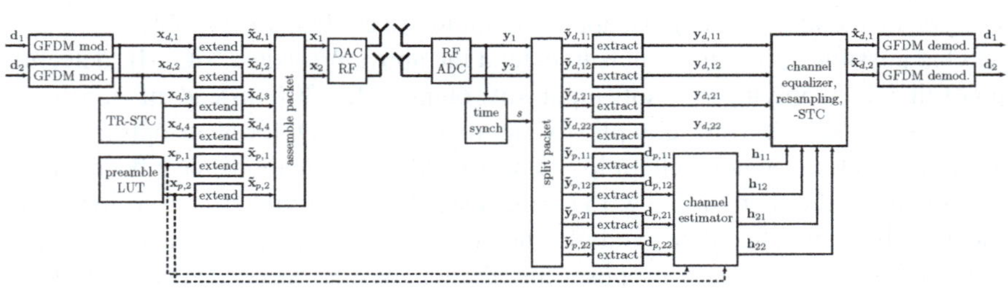

Fig. 19 Transceiver block diagram

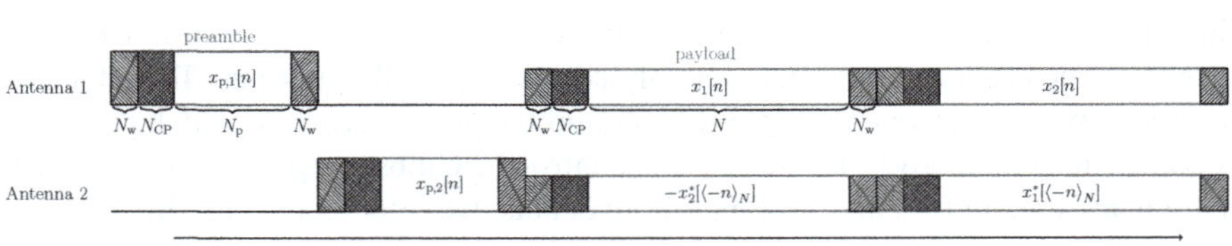

Fig. 20 The TR-STC GFDM frame structure

On the receiver side, the software-defined radio (SDR) hardware provides the received signals $\mathbf{y}_1$ and $\mathbf{y}_2$, which each contain twice as many samples as in one transmitted frame. The signal on the first channel is used to obtain the start index of the frame s through cross-correlation with the known preamble. In the subsequent extraction blocks, the received signals $\mathbf{y}_1$ and $\mathbf{y}_2$ are split up into four data blocks $\mathbf{y}_{d,11}, \mathbf{y}_{d,12}$ and $\mathbf{y}_{d,21}, \mathbf{y}_{d,22}$ and four preamble blocks $\mathbf{y}_{p,11}, \mathbf{y}_{p,21}, \mathbf{y}_{p,22}$ and $\mathbf{y}_{p,12}$, where the first index indicates the receive antenna and the second index corresponds to the time slot. The CP is removed in the same step. The channel estimation block uses the reference preamble signals $\mathbf{x}_{p,1}, \mathbf{x}_{p,2}$ to calculate the channel impulse responses $\mathbf{h}_{11}$, $\mathbf{h}_{12}, \mathbf{h}_{21}, \mathbf{h}_{22}$ based on the LS criterion. In the following block, the channel responses are resampled and then used by the space-time decoder to combine the received samples according to (54). The results are the estimated receive samples $\hat{\mathbf{x}}_{d,1}$ and $\hat{\mathbf{x}}_{d,2}$. These are then passed to the GFDM demodulator, which implements a linear receiver, i.e., ZF, to detect the data symbols $\hat{\mathbf{d}}_{d,1}$ and $\hat{\mathbf{d}}_{d,2}$. From the constellation points the error vector magnitude (EVM) and instantaneous SER can be used to evaluate the system performance later on.

5.2 Demonstrator Setup

The demonstrator consists of two terminals, one configured as transmitter and another set as receiver, as presented in Fig. 21. Each terminal is connected to a controlling PC and to a GPS reference clock to eliminate frequency offsets. On the transmitter host PC, data is generated, modulated with the GFDM waveform and TR-STC encoded. Table 10 presents the relevant parameters used in this experiment. The digital in phase and quadrature (IQ) samples are transferred via a PCI-Express connection to the transmitter front-end, which then performs digital to analog conversion and sends the signal over the air. At the receiver side, the IQ samples are captured and passed to the receiver host PC via the PCI-Express connection, where the waveform is demodulated.

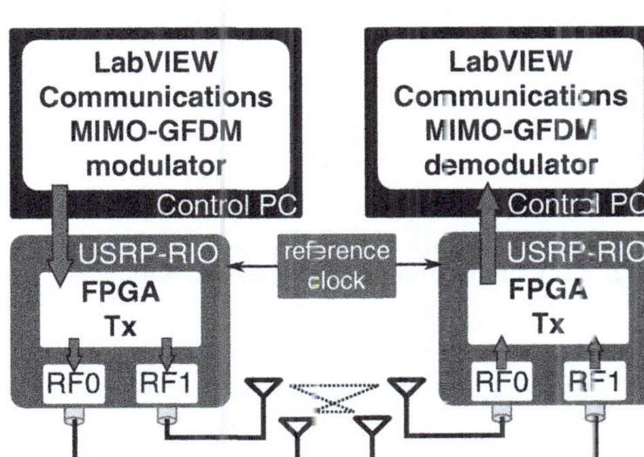

Fig. 21 Schematic overview of the hardware setup

Table 10 Parameters used in the demonstrator

Parameter	Variable	Value
Number of subcarriers (data)	K, K_p	128
Number of subcarriers (preamble)	K_p	128
Number of subsymbols (data)	M	15
Number of subsymbols (preamble)	M_p	5
Number of transmit samples (data)	$N = KM$	1920
Number of transmit samples (preamble)	$N_p = K_p M_p$	1920
Set of active subcarriers (data)	K_{set}	11 to 54
Set of active subcarriers (preamble)	$K_{\text{p, set}}$	10 to 55
Set of active subsymbols (data)	M_{set}	1 to 15
Set of active subsymbols (preamble)	$M_{\text{p, set}}$	1 to 5
Prototype subcarrier filter	$\mathbf{g}$	RRC
Prototype filter roll-off	α	0.4767
Window coefficients	$\mathbf{w}$	
Window function roll-off	β	0.5
Length of window flanks	$N_{\text{w}} = \beta K$	32
Maximum channel delay spread	N_{ch}	
Length of cyclic prefix	$N_{\text{CP}} = N_{\text{ch}}$	32
Modulation order	M_{u}	4
Number of channels	–	2
Center frequency	–	1.99 GHz
Sampling frequency	–	10 MHz
Transmitter gain	–	0 dB
Receiver gain	–	0 dB

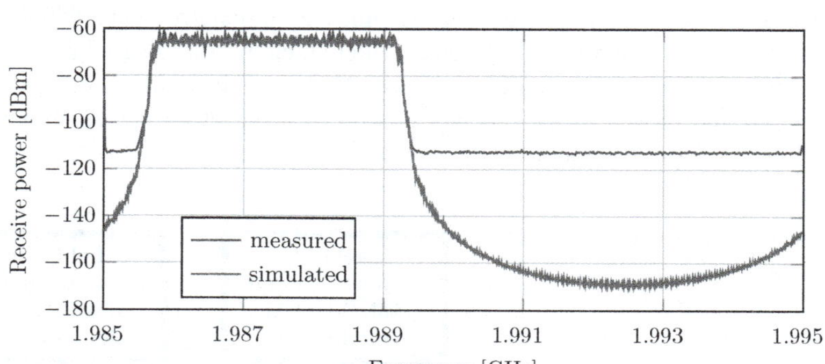

Fig. 22 Comparison of measured and simulated spectrum

Figure 22 shows the measured spectrum from a spectrum analyzer and compares it with simulation results. The measured OOB radiation is at around −48 dB. This result is achieved due to the pulse shaping filter applied at each subcarrier. Therefore, GFDM is favorable in communication scenarios with highly fragmented spectrum or in CRs. In [39], the authors have shown that GFDM has less influence and disturbance to legacy systems than OFDM. The simulated OOB level is considerably lower than the measured OOB level because it does not take into account the spectrum analyzer sensitivity and RF impairments, such as IQ-imbalance and nonlinearities, introduced by the USRP front-end.

6 Summary

The various scenarios and applications that must be covered by 5G networks will demand an unprecedented flexibility of the PHY layer. Although using a specific waveform to address the requirements of each scenario is one option, it is more desirable to have a single waveform that can be shaped to deal with the challenges imposed by the different applications. This chapter has proved GFDM to be a flexible multicarrier modulation that can be tuned to cover the major 4G and 5G waveforms. Namely, it has an efficient time-frequency resource grid for conveying information. This has been well exemplified in the context of Gabor transform. On top of it, OQAM can be applied in both time and frequency domains to obtain an orthogonal system. Precoding can be used to broaden the flexibility even more, for instance, precoding matrices can be used to achieve higher performance over frequency selective channels or lower OOB emissions. Roughly synchronized devices can share the time-frequency grid by using a single subcarrier as guard-band, while an efficient use of the CP can deal with long channel delay profiles and time-misalignments among the users. The overall GFDM performance can be further enhanced by MIMO techniques. All these features make GFDM a strong candidate for the 5G PHY layer.

This chapter has also addressed the receiver design for GFDM-based systems. It covers synchronization, channel estimation and MIMO detection. As an important step from theory to practice, the development of a MIMO-GFDM demonstrator has been presented at the end of this chapter.

References

1. P. Pirinen, A brief overview of 5G research activities, in *1st International Conference on 5G for Ubiquitous Connectivity* (2014), pp. 17–22
2. N. Michailow, M. Matthé, I. Gaspar, A. Caldevilla, L. Mendes, A. Festag, G. Fettweis, Generalized frequency division multiplexing for 5th generation cellular networks. IEEE Trans. Commun. **62**(9), 3045–3061 (2014)
3. G. Fettweis, M. Krondorf, S. Bittner, GFDM - generalized frequency division multiplexing, in *VTC Spring 2009 - IEEE 69th Vehicular Technology Conference* (IEEE, Barcelona, 2009), pp. 1–4
4. M. Matthé, I.S. Gaspar, G. Fettweis, Short paper: reduced complexity calculation of LMMSE filter coefficients, in *Proceedings IEEE 82nd Vehicular Technology Conference*, Boston (2015)

5. I.S. Gaspar, N. Michailow, A.N. Caldevilla, E. Ohlmer, S. Krone, G. Fettweis, Low complexity GFDM receiver based on sparse frequency domain processing, in *IEEE 77th Vehicular Technology Conference, 2013. VTC Spring* (2013)
6. A. Farhang, N. Marchetti, L.E. Doyle, Low complexity modem design for GFDM. IEEE Trans. Signal Process. **99**, 1–1 (2015). Available: http://ieeexplore.ieee.org/lpdocs/epic03/wrapper.htm?arnumber=7332798
7. N. Michailow, I. Gaspar, Generalized frequency division multiplexing: analysis of an alternative multi-carrier technique for next generation cellular systems, in … *Systems (ISWCS), …* (2012). Available: http://ieeexplore.ieee.org/xpls/abs_all.jsp?arnumber=6328352
8. M. Matthé, L. Mendes, N. Michailow, D. Zhang, G. Fettweis, Widely linear estimation for space-time-coded GFDM in low-latency applications. IEEE Trans. Commun. **63**(11), 4501–4509 (2015)
9. N. Michailow, L. Mendes, M. Matthé, I. Gaspar, A. Festag, G. Fettweis, Robust WHT-GFDM for the next generation of wireless networks. IEEE Commun. Lett. **19**(1), 106–109 (2015)
10. H. Bölcskei, Orthogonal frequency division multiplexing based on offset QAM, in *Advances in Gabor Analysis*, ed. by H. Feichtinger, T. Strohmer. Applied and Numerical Harmonic Analysis (Birkhäuser, Boston, 2003), pp. 321–352
11. I. Gaspar, M. Matthé, N. Michailow, L. Mendes, D. Zhang, G. Fettweis, Frequency-shift Offset-QAM for GFDM. IEEE Commun. Lett. **99**, 1–1 (2015)
12. R.W. Chang, Synthesis of band-limited orthogonal signals for multichannel data transmission. Bell Syst. Tech. J. **45**(10), 1775–1796 (1966)
13. B. Saltzberg, Performance of an efficient parallel data transmission system. IEEE Trans. Commun. Technol. **15**(6), 805–811 (1967)
14. G. Wunder, P. Jung, M. Kasparick, T. Wild, F. Schaich, Y. Chen, S. Brink, I. Gaspar, N. Michailow, A. Festag, L. Mendes, N. Cassiau, D. Ktenas, M. Dryjanski, S. Pietrzyk, B. Eged, P. Vago, F. Wiedmann, 5GNOW: non-orthogonal, asynchronous waveforms for future mobile applications. IEEE Commun. Mag. **52**(2), 97–105 (2014)
15. P. Banelli, S. Buzzi, G. Colavolpe, A. Modenini, F. Rusek, A. Ugolini, Modulation formats and waveforms for 5G networks: Who will be the heir of OFDM?: an overview of alternative modulation schemes for improved spectral efficiency. IEEE Signal Process. Mag. **31**(6), 80–93 (2014)
16. H. Lin, P. Siohan, Multi-carrier modulation analysis and WCP-COQAM proposal. EURASIP J. Adv. Signal Process. **2014**(1), 79 (2014). Available: http://asp.eurasipjournals.com/content/2014/1/79
17. A.M. Tonello, A novel multi-carrier scheme: cyclic block filtered multitone modulation, in *2013 IEEE International Conference on Communications (ICC)* (IEEE, Budapest, 2013), pp. 5263–5267
18. I. Kanaras, A. Chorti, M.R. Rodrigues, I. Darwazeh, Spectrally efficient fdm signals: bandwidth gain at the expense of receiver complexity, in *IEEE International Conference on Communications, 2009. ICC'09* (IEEE, Dresden, 2009)
19. M. Matthé, L. Mendes, I. Gaspar, N. Michailow, G. Fettweis, Multi-user time-reversal STC-GFDM for 5G networks. IEEE Trans. Wirel. Commun. (submitted, 2014). Available: http://jwcn.eurasipjournals.springeropen.com/articles/10.1186/s13638-015-0366-6
20. K. Gröchenig, *Foundations of Time-Frequency Analysis* (Birkhäuser, Boston, 2001)
21. J.J. Benedetto, C. Heil, D.F. Walnut, Gabor systems and the Balian-Low theorem, in *Gabor Analysis and Algorithms*, ed. by H.G. Feichtinger, T. Strohmer (Birkhäuser, Boston, 1998), Chap. 2, pp. 85–122
22. J. van de Beek, M. Sandell, P. Borjesson, ML Estimation of time and frequency offset in OFDM systems. IEEE Trans. Signal Process. **45**(7), 1800–1805 (1997)
23. T. Schmidl, D. Cox, Robust frequency and timing synchronization for OFDM. IEEE Trans. Commun. **45**(12), 1613–1621 (1997)
24. H. Minn, M. Zeng, V. Bhargava, On Timing Offset Estimation for OFDM Systems. IEEE Commun. Lett. **4**(7), 242–244 (2000)

25. A. Awoseyila, C. Kasparis, B.G. Evans, Improved preamble-aided timing estimation for OFDM systems. IEEE Commun. Lett. **12**(11), 825–827 (2008)
26. I. Gaspar, G. Fettweis, An embedded midamble synchronization approach for generalized frequency division multiplexing, in *IEEE Global Communications Conference (GLOBECOM)* (2015)
27. I. Gaspar, A. Festag, G. Fettweis, Synchronization using a pseudo-circular preamble for generalized frequency division multiplexing in vehicular communication, in *IEEE 82nd Vehicular Technology Conference (VTC Spring)* (2015)
28. R.A. Horn, C.R. Johnson, *Topics in Matrix Analysis* (Cambridge University Press, Cambridge, 1994). Available: https://books.google.com/books?id=LeuNXB2bl5EC&pgis=1
29. N. Al-Dhahir, Single-carrier frequency-domain equalization for space-time block-coded transmissions over frequency-selective fading channels. IEEE Commun. Lett. **5**(7), 304–306 (2001). Available: http://ieeexplore.ieee.org/lpdocs/epic03/wrapper.htm?arnumber=935750
30. C. Studer, H. Bölcskei, Soft-input soft-output single tree-search sphere decoding. IEEE Trans. Inf. Theory **56**(10), 4827–4842 (2010)
31. D. Wubben, R. Bohnke, V. Kühn, K.-D. Kammeyer, MMSE extension of V-BLAST based on sorted QR decomposition, in *2003 IEEE 58th Vehicular Technology Conference, 2003. VTC 2003-Fall*, vol. 1 (2003), pp. 508–512
32. A. Chindapol, J. Ritcey, Design, analysis, and performance evaluation for BICM-ID with square QAM constellations in Rayleigh fading channels. IEEE J. Sel. Areas Commun. **19**(5), 944–957 (2001)
33. B. Farhang-Boroujeny, H. Zhu, Z. Shi, Markov chain Monte Carlo algorithms for CDMA and MIMO communication systems. IEEE Trans. Signal Process. **54**(5), 1896–1909 (2006)
34. X. Mao, P. Amini, B. Farhang-Boroujeny, Markov chain Monte Carlo MIMO detection methods for high signal-to-noise ratio regimes, in *Proceedings of IEEE Global Telecommunications Conference (GLOBECOM)*, Washington, DC (2007)
35. M. Senst, G. Ascheid, A Rao-Blackwellized Markov chain Monte Carlo algorithm for efficient MIMO detection, in *Proc. IEEE Int. Conf. on Commun. (ICC)*, Tokyo (2011)
36. R. Datta, N. Michailow, M. Lentmaier, G. Fettweis, GFDM interference cancellation for flexible cognitive radio PHY design, in *2012 IEEE Vehicular Technology Conference (VTC Fall)* (IEEE, Quebec City, 2012), pp. 1–5
37. T. Datta, N. Ashok Kumar, A. Chockalingam, B. Rajan, A novel MCMC algorithm for near-optimal detection in large-scale uplink mulituser MIMO systems, in *Proc. IEEE Inf. Theory and Appl. Workshop (ITA)*, San Diego (2012), pp. 69–77
38. D. Zhang, M. Matthé, L. Mendes, G. Fettweis, A Markov Chain Monte Carlo algorithm for near-optimum detection of MIMO-GFDM signals, in *Proc. IEEE Int. Symp. on Personal, Indoor and Mobile Radio Commun. (PIMRC)*, Hong Kong (2015)
39. M. Danneberg, R. Datta, A. Festag, G. Fettweis, Experimental testbed for 5g cognitive radio access in 4g lte cellular systems, in *2014 IEEE 8th Sensor Array and Multichannel Signal Processing Workshop (SAM)* (2014), pp. 321–324

Spectrally Efficient Frequency Division Multiplexing for 5G

Izzat Darwazeh, Ryan C. Grammenos, and Tongyang Xu

Abstract The focus of this chapter is on novel multi-carrier communication techniques, which share the common goal of increasing spectrum efficiency in future communication systems. In particular, a technology termed Spectrally Efficient Frequency Division Multiplexing (SEFDM) is described in detail outlining its benefits, challenges and trade-offs when compared to the current state-of-the-art. A decade of research has been devoted to examining SEFDM from different angles; mathematical modelling, algorithm optimisation, hardware implementation and system experimentation. The aim of this chapter is to therefore give a taste of this technology and in doing so, the chapter is organised as follows; first, it is explained how SEFDM fits within the remit of future 5th Generation (5G) communication systems; second, the design principles and implementation trade-offs associated with SEFDM systems are described; third, a number of linear and more sophisticated polynomial detection schemes are compared in terms of performance and complexity; finally, the chapter concludes by outlining a number of experimental testbeds which have been developed for the purpose of evaluating the performance of SEFDM in practical scenarios.

1 Introduction

1.1 Transition from 4G to 5G

The beginning of the twenty-first century has been marked by a pressing need for more spectrum [9]. Networks are confronted with an ever-growing number of users who run bandwidth-hungry applications using a wide range of devices. Yet, spectrum is a scarce and an expensive resource. This fact coupled with users' demand for higher data rates, better Quality of Service (QoS) and lower costs, has led telecommunication operators and engineers to recognise that the best means for increasing capacity in future communication systems is via improved spectrum efficiency.

I. Darwazeh (✉) • R.C. Grammenos • T. Xu
University College London, London, UK
e-mail: i.darwazeh@ucl.ac.uk; r.grammenos@ucl.ac.uk; tongyang.xu.11@ucl.ac.uk

W. Xiang et al. (eds.), *5G Mobile Communications*,
DOI 10.1007/978-3-319-34208-5_10

To this end, academia and industry are working together to define the characteristics of the next communication standard, commonly referred to nowadays as 5G systems [4]. Emerging technologies and disruptive ecosystems, such as machine-to-machine (M2M) communications and the Internet of Things (IoT), are also considered to be important pieces of the 5G puzzle and come under the umbrella of the recently coined term *tactile* Internet [18]. Two major European projects addressing 5G are METIS [20] and 5GNOW [76]. In particular, 5GNOW explores novel physical layer schemes which use non-orthogonal waveforms to overcome the strict orthogonality and synchronisation criteria specified in existing systems.

In line with these innovative projects, this chapter describes a novel Multi-carrier Modulation (MCM) scheme termed SEFDM, which promises to deliver bandwidth savings at the expense of receiver complexity [50].

1.2 Multi-Carrier Schemes for 5G

Modulation schemes for high-speed data transmission have been of continuing interest for well over half a century [34]. The advancement of modern communication systems, however, can be attributed to the use of MCM techniques, which provide better immunity against multipath fading compared to single carrier schemes.

While Orthogonal Frequency Division Multiplexing (OFDM) is regarded as one of the key technology enablers in 4th Generation (4G) mobile telecommunication networks [73], it has certain disadvantages which make it inapt for spectrum optimisation, as depicted in Fig. 1. Thereby, MCM techniques aimed at improving spectrum utilisation are gaining increasingly more attention by the academic community and the business world across the globe [68]. These techniques propose the packing of more data within a conventional time-frequency plane by:

- Deliberately and counter-intuitively violating the orthogonality principle defined for OFDM systems, hence reducing the frequency spacing between the sub-carriers to save bandwidth or increasing the data rate within the same bandwidth.
- Transmitting non-orthogonal pulses with special properties.
- Applying advanced detection algorithms at the receiver, such as Sphere Decoding (SD), which has been made feasible due to advances in silicon technology.

One of the first systems designed to increase the data rate relative to an OFDM system was OFDM/Offset Quadrature Amplitude Modulation (OQAM) [33]. This was achieved by discarding the guard interval. As a result, alternative prototype functions providing good localisation in the time domain had to be employed to mitigate the effects of Inter-symbol Interference (ISI). The use of non-orthogonal functions offered new degrees of freedom favouring the design of improved pulses which were better-suited for minimising the Inter-carrier Interference (ICI) and ISI effects present in frequency-selective and time-dispersive channels [52]. An example of a prototype function offering quasi-optimum localisation is the Isotropic Orthogonal Transform Algorithm (IOTA) function which is generated by applying an IOTA filter to a Gaussian function [43].

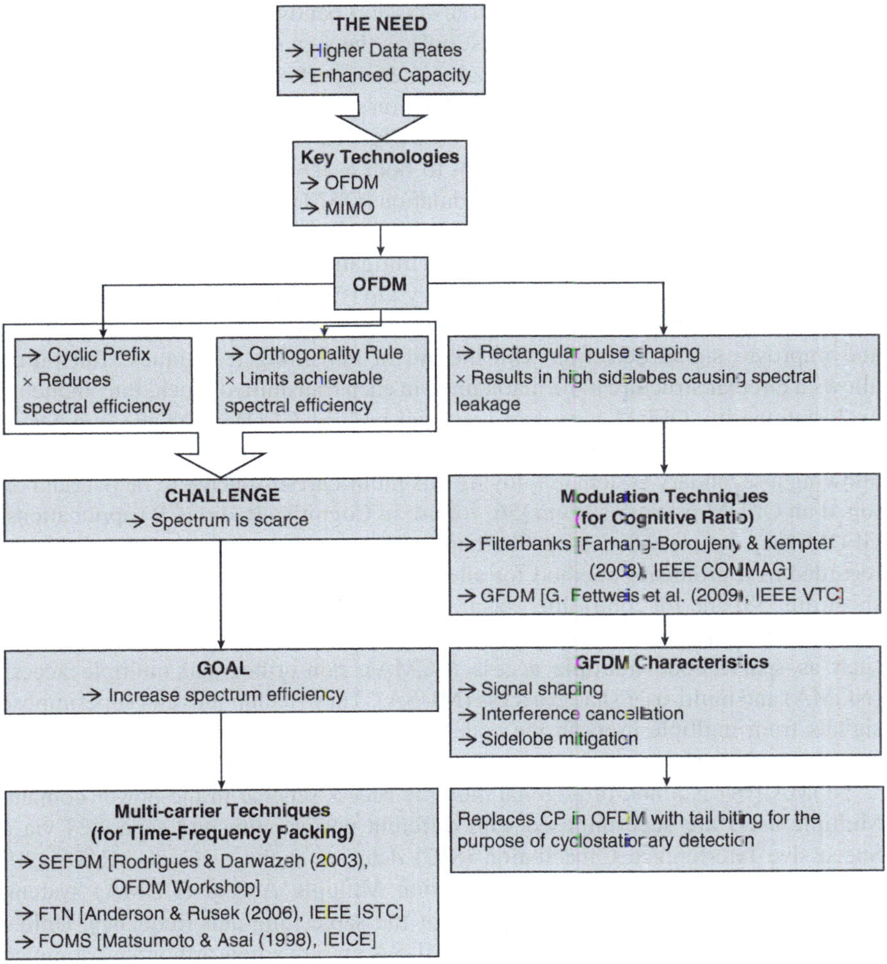

Fig. 1 Transition from 4G to 5G: challenges and promising solutions

In retrospect, a number of systems have been explored with the aim of increasing spectrum efficiency through bandwidth savings and/or flexible frequency allocation [70]. Examples of systems which reduce spectrum utilisation by half but constrained to one-dimensional (1D) modulation schemes include Fast Orthogonal Frequency Division Multiplexing (FOFDM) [65] and M-ary Amplitude Shift Keying (ASK) OFDM (MASK-OFDM) [77]. Systems designed to operate with two-dimensional (2D) modulation methods include High Compaction MCM (HC-MCM) [29], Overlapped FDM (Ov-FDM) [44], Faster than Nyquist (FTN) signalling [3], Frequency Overlapped Multi-carrier System (FOMS) [54], as well as SEFDM [66].

FTN systems increase the data rate within the Nyquist signal bandwidth at the expense of introducing additional ISI [53]. This was proved in 1975 by Mazo [55], who showed that for a reduction in transmission bandwidth by a factor $\rho = 0.8$ with reference to conventional Nyquist pulses, the signalling rate can be increased by $(\frac{1-\rho}{\rho}) \cdot 100 = 25\,\%$ without a penalty in error performance [55]. This concept, however, was validated only for 1D modulation schemes, such as Pulse Amplitude Modulation (PAM), ASK and Binary Phase Shift Keying (BPSK). In 2005, Rusek and Anderson [3] extended the concept to both a 2D and multi-carrier space, for example 16-Quadrature Amplitude Modulation (QAM) modulation [67] and multi-stream FTN signalling [3].

A different system that offers sidelobe mitigation, pulse shaping and interference cancellation, is Generalized Frequency Division Multiplexing (GFDM) [19]. GFDM reduces out-of-band radiation, thanks to Raised Cosine (RC) pulse shaping, and improves signal detection with the aid of tail biting techniques. Tail biting allows a circular structure to be maintained in each transmitted block, thus inducing cyclostationarity. OFDM is in fact a special case of GFDM with the number of blocks set equal to one [61]. It has been demonstrated that GFDM is well-suited for allowing a secondary system employing this multi-carrier scheme to be overlaid on top of an OFDM primary system [56] for use in Cognitive Radio (CR) applications. GFDM may be classified as a filterbank multi-carrier technique [17], the latter regarded as a promising method for allowing *agile waveforming* over fragmented spectrum [58] and for controlling spectral leakage.

The non-orthogonal concept may also be applied in multiple access scenarios such as sparse code multiple access (SCMA), non-orthogonal multiple access (NOMA) and multi-user shared access (MUSA). These techniques can superimpose signals from multiple users in the code-domain or the power-domain to enhance system access performance.

NOMA [69] is a non-orthogonal multiple access scheme in the power domain. Multiple users are superimposed with different power gains and separated via a Successive Interference Cancellation (SIC) detector at the receiver. MUSA [30] is an advanced multi-carrier Code Division Multiple Access (CDMA) system, which allows multiple users to transmit at the same time and frequency. Unlike the typical CDMA spreading sequence, MUSA employs non-orthogonal complex spreading sequences at the transmitter to modulate signals of each user leading to a denser deployment of existing resources. At the receiver, an advanced SIC detection algorithm is operated to remove the interference and recover each user's data. In SCMA [59], symbols from multiple users are sparsely spread in multiple layers which overlap in the frequency domain to enable non-orthogonal multiple access. All these non-orthogonal multiple access schemes aim to non-orthogonally superimpose multiple existing orthogonal waveforms from different users.

1.3 Spectrally Efficient Frequency Division Multiplexing (SEFDM)

SEFDM is a non-orthogonal multi-carrier technique which plays an important role in the area of non-orthogonal waveform design. It has similar aims and characteristics to other non-orthogonal multi-carrier systems, for example FTN,

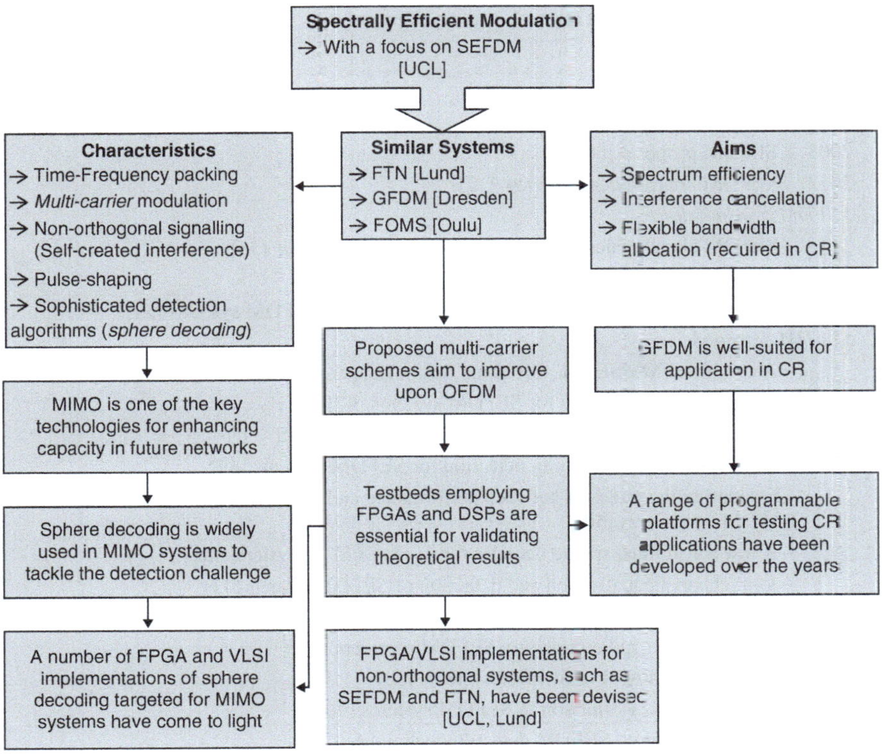

Fig. 2 Aims and developments of non-orthogonal multi-carrier schemes

GFDM and FOMS, as illustrated in Fig. 2. The legend keys correspond to the following descriptions:

- **UCL**: This refers to the team working on SEFDM. The team is led by Izzat Darwazeh who is affiliated with University College London (UCL), United Kingdom. Notable researchers (current and alumni) working on SEFDM include: M.R.D. Rodrigues, I. Kanaras, A. Chorti, R.G. Clegg, S. Isam, M.R Perrett, R.C. Grammenos, P.N. Whatmough and T. Xu.
- **Lund**: This refers to the team working on FTN. The team is led by John B. Anderson who is affiliated with Lund University, Sweden. Notable researchers include: F. Rusek, V. Öwall and D. Dasalukunte.
- **Dresden**: This refers to the team working on GFDM. The team is led by Gerhard Fettweis who is affiliated with Dresden University, Germany.
- **Oulu**: This refers to the team working on FOMS. The team is led by Tadashi Matsumoto who is affiliated with Oulu University, Finland.

The history of SEFDM can be traced back to 2002 when an OFDM based technique termed FOFDM [65] which doubled the data rate was proposed. Subsequently,

Table 1 Timeline of SEFDM research and development

Year	Literature
SEFDM principles	
2002	– Fast OFDM proposal [65]
2003	– SEFDM proposal [66]
2012	– SEFDM: a near-decade review [12]
SEFDM transmission	
2010	– SEFDM signal generation using IDFT based techniques [36] – A precoded SEFDM system [35]
2012	– Characterisation of the self-created interference in SEFDM systems [39]
SEFDM detection	
2008	– A combined MMSE-ML technique for the detection of SEFDM signals [46] – Near optimum detection for SEFDM systems [47]
2009	– A pruned SD algorithm for use in SEFDM systems [48] – Semi definite programming detection for SEFDM systems [49] – Evaluating the trade-off between performance and complexity for different SEFDM receivers [50]
2010	– A fast constrained sphere decoder for use in SEFDM systems [51] – Joint channel equalisation and detection of SEFDM signals [11]
2011	– Detection of SEFDM signals using a truncated SVD approach [41] – Evaluation of fixed complexity SEFDM receivers [37] – Reducing the PAPR in SEFDM systems [38]
2012	– Robust channel estimation in SEFDM systems [40]
2013	– DSP evaluation of SEFDM detection techniques [26] – Improving the detection of SEFDM signals through soft decision [84]
2014	– Detection of SEFDM signals for higher order modulation schemes [79] – Block SEFDM: Reducing the complexity of SEFDM using a multi-band architecture [80] – Bandwidth saving in a multi-path fading channel using turbo equalisation [78]
SEFDM implementation	
2011	– FPGA implementation of a reconfigurable SEFDM transmitter [62] – VLSI architecture for a reconfigurable SEFDM transmitter [74] – FPGA design of a truncated SVD based SEFDM receiver [27]
2012	– Verification and evaluation of an FPGA based SEFDM transmitter [63] – Hardware implementation of a reconfigurable SEFDM receiver [25]
2013	– FPGA implementation of a real-time FSD detector for use in SEFDM systems [83]
Other SEFDM systems	
2009	– Application of SEFDM in physical layer security [10]
2014	– The first experimental optical SEFDM system [13]
2015	– Practical implementation of a beyond LTE-Advanced SEFDM system [82] – First demonstration of a radio-over-fibre SEFDM system [57] – Transmission of a 24 Gb/s Dual Polarization Coherent Optical SEFDM signal [60]

the principle of SEFDM was proposed in 2003 [66] to overcome the limitations encountered with FOFDM, as discussed in Sect. 2.1. Since then, a number of researchers have worked on different aspects of SEFDM signals ranging from novel algorithms and techniques to generate and detect such signals, through to their practical implementation using state-of-the-art hardware devices. Table 1 gives an account of the key milestones achieved during the study of SEFDM signals over the last decade.

2 SEFDM Design Issues

The ever-increasing demand for higher data rates and improved coverage has called for a change of regime with the ultimate goal of optimising spectrum utilisation. SEFDM was developed to address the limits imposed upon spectral efficiency in OFDM systems and offers trade-offs in complexity and performance for diverse bandwidth compression rates.

2.1 Preliminary Concepts

Taking into account that the Cyclic Prefix (CP) in OFDM systems reduces bandwidth efficiency by a factor of $\frac{N}{N+N_g}$, where N is the number of sub-carriers and N_g is the length of the CP in terms of discrete time samples, the first attempts to improve spectral efficiency focused on removing the CP and replacing rectangular pulses by pulses which were better localised in the time and frequency domains to avoid ISI. This alternative scheme termed OFDM/OQAM, which offsets the data timing by half a symbol, can guarantee orthogonality between sub-carriers only for real-valued symbols [43].

One of the key publications examining the relationship between time, frequency, orthogonality and bandwidth efficiency appeared in 1945 by Gabor [21]. This work along with the work by Mazo [55] inspired the pursuit of systems like Multi-stream FTN [3], GFDM [19], Filterbank Multi-carrier (FBMC) [16] and SEFDM. Furthermore, the use of FTN and SEFDM has been extended beyond wireless systems to optical links, as they appear in the work by Colavolpe [6] and Darwazeh [13], respectively. These techniques may be subsumed under one term as *time-frequency packing*. A timeline of the fundamental concepts, methods and systems developed over the years for time-frequency packing is illustrated in Fig. 3.

In a conventional Frequency Division Multiplexing (FDM) system, the frequency spacing between the sub-carriers is such that it obeys Nyquist's first criterion, also known as the Nyquist ISI criterion, which is the condition for zero ISI between transmitted symbols [64]. As the name suggests, FTN results in the transmission of more data within the minimum bandwidth specified by the Nyquist criterion.

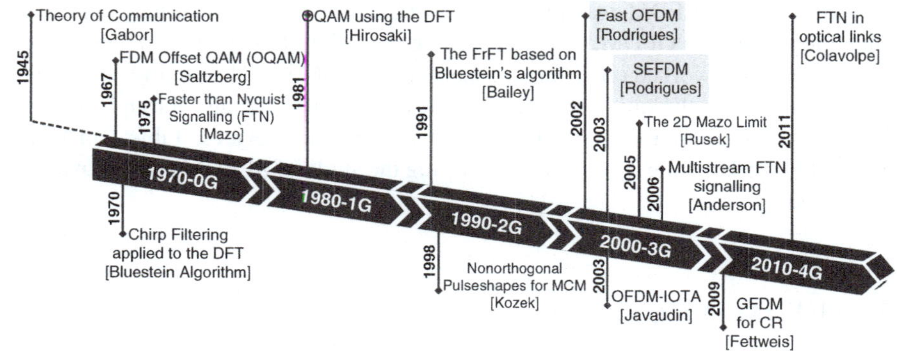

Fig. 3 Timeline of spectrally efficient concepts and non-orthogonal communication techniques

This causes ISI which for a multi-carrier system also leads to ICI. Pulse shaping filters are therefore employed whose bandwidth and spectral shape determine how closely adjacent sub-carriers overlap with each other. Achievable information rates for FTN have been investigated recently [85]. While the focus here is on SEFDM, as proposed by Rodrigues and Darwazeh in 2003 [66], FTN, GFDM and SEFDM have a number of characteristics in common, such as:

- Aiming to increase spectral efficiency or favouring flexible bandwidth allocation;
- Using multi-carrier modulation similar to OFDM;
- Exploiting non-orthogonal signalling at the expense of self-created interference;
- Making use of pulse shaping filters to mitigate ISI (in FTN and GFDM);
- Employing iterative, tree-search and soft-decoding detection algorithms to recover the transmitted data successfully.

FOFDM [65] and MASK-OFDM [77] are special cases of spectrally efficient multi-carrier systems. These techniques operate only for non-complex modulation schemes, in other words 1D signals, such as BPSK and ASK, allowing the spacing between the sub-carriers to be reduced by half relative to a conventional OFDM system. This limitation motivated the invention of SEFDM [66], which allows the use of complex modulation schemes, such as M-QAM with $M > 2$. SEFDM offers an additional degree of freedom over OFDM by relaxing the stringent orthogonality constraint defined for OFDM. The attractiveness of SEFDM compared to other spectrally efficient techniques is the fact that it does not require special pulse shaping functions. Instead, it uses rectangular pulse shaping as in OFDM.

2.2 *SEFDM Signals*

SEFDM is based on similar key principles to OFDM in that a high data rate serial input stream is split into N parallel, overlapping sub-carriers. Consequently, the conventional transceiver for SEFDM, as originally proposed in [66], resembles

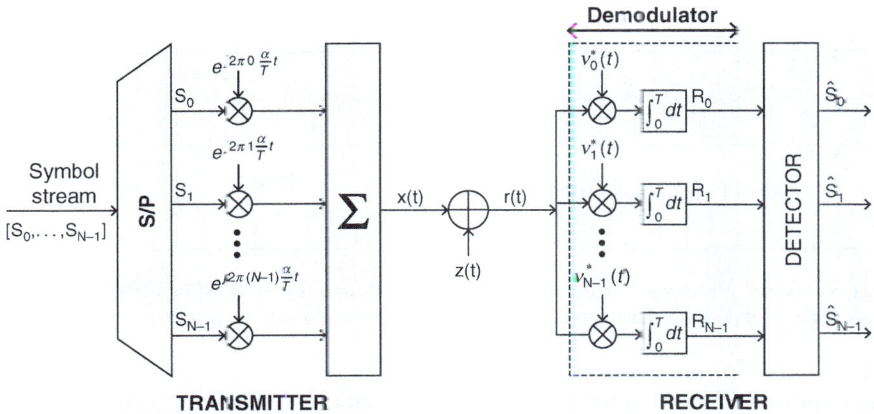

Fig. 4 Conceptual architecture for an SEFDM transceiver using a bank of modems

that for OFDM, as illustrated in Fig. 4. Figure 4 shows that the SEFDM receiver is different from that of OFDM in that it incorporates two stages. The first stage involves a demodulation of the incoming received signal whose output is then fed to a second stage being a detector to generate estimates of the transmitted data symbols. It is assumed that no pulse shaping filters are employed, in other words, rectangular pulses are used, such as in OFDM.

As depicted in Fig. 4, N complex input symbols, denoted by $s_n = s_{n\Re e} + js_{n\Im m}$, modulate N sub-carriers generating a continuous time domain SEFDM signal expressed as[1]

$$x(t) = \sum_{l=-\infty}^{+\infty} \sum_{n=0}^{N-1} s_{l,n} e^{j2\pi n \Delta f \alpha t}, \tag{1}$$

where α is a key system parameter in SEFDM systems and dictates the level of bandwidth compression. Contrary to OFDM, SEFDM deliberately and counter-intuitively violates the orthogonality rule defined for OFDM by reducing the spacing between the sub-carriers. Hence, in SEFDM the sub-carrier spacing is equal to $\Delta f = \frac{\alpha}{T}$, where T is the SEFDM symbol duration with $\alpha < 1$ corresponding to an SEFDM signal and $\alpha = 1$ to an OFDM signal.

A general comparison between SEFDM and OFDM is illustrated in Fig. 5 to demonstrate the benefit of bandwidth saving in SEFDM. Figure 5a shows a typical OFDM spectrum, where the sub-carrier spacing Δf_1 equals $\frac{1}{T}$. Figure 5b depicts an SEFDM system with sub-carrier spacing $\Delta f_2 = \frac{\alpha}{T}$. Figure 5c presents

[1]The notations $\Re e$ and $\Im m$ represent the real and imaginary parts of a complex number, respectively, l denotes the SEFDM symbol index, n denotes the sub-carrier index and $s_{l,n}$ represents the information symbol conveyed by sub-carrier with index n during the time stamp with index l.

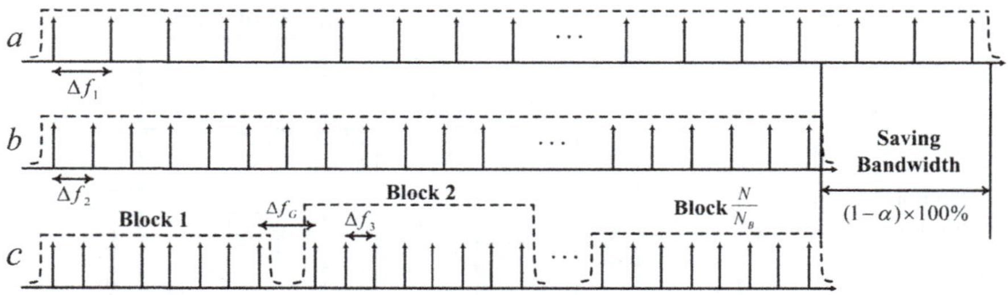

Fig. 5 Spectra illustration for (**a**) OFDM, (**b**) SEFDM, and (**c**) Block-SEFDM. Each impulse represents a single sub-carrier and there are N sub-carriers overall for each system, respectively

an improved SEFDM system termed block-spectrally efficient frequency division multiplexing (B-SEFDM). It is evident that both the conventional SEFDM and B-SEFDM systems can save $(1 - \alpha) \times 100\,\%$ of bandwidth compared to an OFDM system.

In OFDM, the frequency spacing Δf between the overlapping sub-carriers has to equal the reciprocal of the OFDM symbol period T to comply with the orthogonality rule. Mathematically, the orthogonality rule guarantees that the sub-carriers are mutually independent, in other words their cross-correlation is zero. In SEFDM, however, this is not the case, since the sub-carriers are packed closer together in a non-orthogonal manner. This gives rise to cross-talk between the sub-carriers, in other words self-created ICI, which presents a key challenge in SEFDM systems.

The self-created ICI present in SEFDM systems stems from the fact that the cross-correlation between the sub-carriers is no longer zero. This non-zero cross-correlation may in turn be quantified using the so-called correlation matrix **C**, which describes the self-created ICI. The magnitude of the non-zero cross-correlation determines the ill-conditioning of the matrix **C**. Ill-conditioning means that small changes in the entries of the input matrix yield very large changes in the solution matrix. These small changes are typically caused by round-off errors during matrix computations. As a result of this ill-conditioning, the correlation matrix **C** may tend to singularity,[2] thus becoming non-invertible and complicating the design of receiver architectures.

The elements of the correlation matrix **C** are given by

$$\begin{aligned} c_{m,n} &= \frac{1}{Q}\sum_{k=0}^{Q-1} e^{\frac{j2\pi mk\alpha}{Q}} e^{-\frac{j2\pi nk\alpha}{Q}} \\ &= \frac{1}{Q} \times \begin{Bmatrix} Q & , m = n \\ \frac{1-e^{j2\pi\alpha(m-n)}}{1-e^{\frac{j2\pi\alpha(m-n)}{Q}}} & , m \neq n \end{Bmatrix}. \end{aligned} \tag{2}$$

[2] A matrix is singular if its determinant is equal to zero [71].

where m,n are indices of any two sub-carriers in the system with $m, n = 1, \ldots, N$. The diagonal terms ($m = n$) correspond to the auto-correlation of each sub-carrier with itself, which according to Eq. (2) is equal to one. The off-diagonal terms ($m \neq n$) measure the cross-correlation between any two overlapping sub-carriers, which for an SEFDM system would be non-zero. From this description and with reference to Eq. (2), it should be evident that for an OFDM system, the correlation matrix **C** is the identity matrix of size $N \times N$.

In SEFDM, the conditioning of the correlation matrix **C** depends on the system size, in terms of the number of sub-carriers N, and the level of bandwidth compression α. The matrix becomes more ill-conditioned as N is increased and/or the value of α is decreased leading to performance degradation. These concepts, as well as solutions to address these challenges are examined in further detail in Sect. 3.

3 Generation and Detection of SEFDM Signals

This section examines the transceiver architecture for an SEFDM system assuming an Additive White Gaussian Noise (AWGN) channel. Different detection techniques are compared and contrasted in terms of their error performance, complexity and ease of implementation.

3.1 SEFDM Signal Transmission

The conventional method of generating SEFDM signals using a bank of analogue modulators does not scale well with the increase in the number of sub-carriers, taking into account that the complexity of this method is of the order $\mathcal{O}(N^2)$. Hence, the potential of employing Discrete Fourier Transform (DFT) operations to address this issue appears particularly attractive.

Since SEFDM is inherently a non-orthogonal system this means that standard DFT operations cannot be used directly. To this end, a first example of an algorithm employing the DFT in a modified configuration to generate non-orthogonal signals is the Inverse Fractional Fourier Transform (IFrFT) as proposed by Bailey [5], which considers fractional roots of unity. In analogy to an Inverse Discrete Fourier Transform (IDFT) operation, the output of the IFrFT may be expressed as

$$x_{Fr}(m) = \frac{1}{N} \sum_{n=0}^{N-1} s_n e^{j2\pi m\alpha n/N}, \tag{3}$$

which converges to the IDFT formula for $\alpha = 1$. As noted in [5], the IFrFT can be computed in a fast and efficient manner using the algorithm proposed by

Bluestein [7]. Using this latter algorithm the computation of the IFrFT requires $20N \log_2 N$ floating-point operations. While this results in the algorithm being 20 times more computationally intensive compared to the conventional IDFT algorithm, this technique still yields a significantly lower complexity compared to the analogue counterpart.

Notwithstanding, the ultimate goal is to generate non-orthogonal signals using standard DFT operations. The first attempts date back to the 1980s when Hirosaki used the DFT for an OQAM OFDM system [33]. A similar approach was adopted later on in FOFDM systems where the value of α is fixed at 0.5. More recently, however, standard DFT operations have been employed in non-orthogonal systems where the value of α is arbitrary. A first example appears in the HC-MCM system proposed by Hamamura [29]. Similar techniques were developed later on by Isam and Darwazeh [36] for use in SEFDM systems. These latter methods allow SEFDM signals to be generated using IDFT operations, thus offering the same advantages in terms of complexity as OFDM.

The equation for the continuous time SEFDM signal was given in Eq. (1). Let T_s be the sampling period introduced to discretise the time and let ρ be an oversampling factor such that $Q = \rho N$ with $\rho \geqslant 1$. Then the discrete time axis will be given by mT_s while the symbol period will be given by $T = QT_s$. Consequently, the discrete time expression for a single SEFDM symbol may be expressed as

$$x(m) = \frac{1}{\sqrt{Q}} \sum_{n=0}^{N-1} s_n e^{j2\pi m\alpha n/Q}, \tag{4}$$

where $x(m)$ represents the time samples indexed by the integer m, Q represents the total number of time samples contained in a single SEFDM symbol, $\mathbf{s} = [s_0, \ldots, s_{N-1}]'$ is a vector of complex data symbols modulated using Phase Shift Keying (PSK) or QAM modulation and $[\cdot]'$ denotes a vector or matrix transpose operation. Specifically, $\mathbf{s}$ represents the input data vector with $\mathbf{s} \in \mathfrak{M}^N$ where $\mathfrak{M}$ represents a discrete alphabet, in other words the set of constellation points. The size of the alphabet, known as the constellation cardinality, is denoted by M. Where QAM is employed, it is assumed that the constellation is rectangular with M equal to even powers of two and that the data symbols are encoded using Gray encoding.

Three different, nevertheless equivalent, SEFDM transmitter types were proposed by Isam, which are illustrated in Fig. 6 with details provided in [36]. In these architectures, the bandwidth compression α is expressed as a ratio of integers, in other words $\alpha = \frac{b}{c}$ where $b, c \in \mathbb{Z}_{>0}$ with $b < c$. To simplify the forthcoming discussion and without loss of generality, let it be assumed that no oversampling is used ($\rho = 1$) yielding $Q = N$. The transmitters shown in Fig. 6 are then designed to offer diverse trade-offs.

Type 1, known as the *proportional inputs* transmitter, is the simplest method in that it employs a standard IDFT block of length $\frac{N}{\alpha}$. This means that the size of this block compared to a conventional OFDM system is $\frac{1}{\alpha}$ times larger. The number of input data symbols remains the same, in other words N with the remaining $\frac{N}{\alpha} - N$

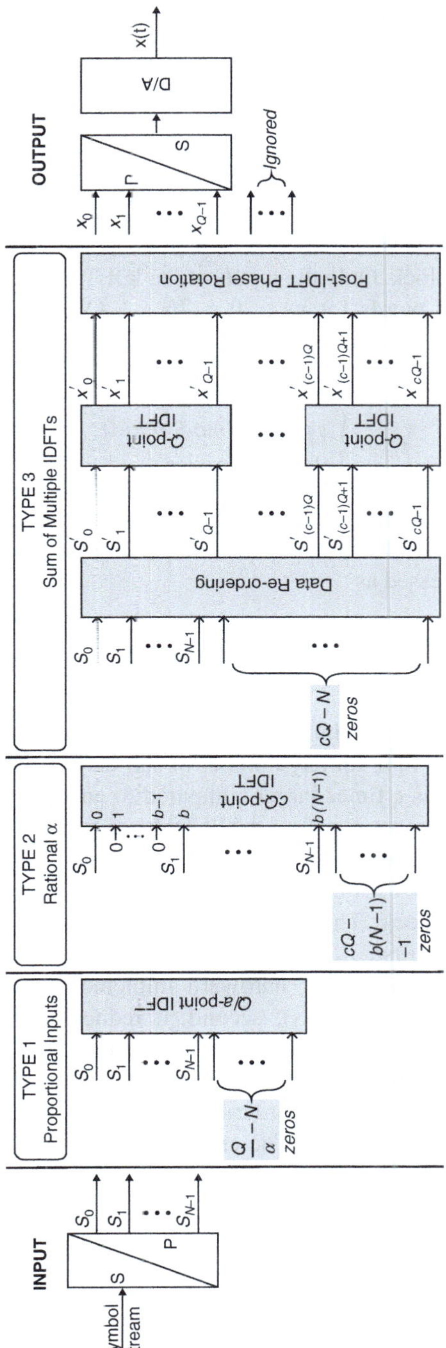

Fig. 6 DFT based SEFDM transmitter types

IDFT inputs padded with zeros. The output of the IDFT is truncated, hence only N samples are taken forward to construct the SEFDM signal while the remaining outputs are ignored. This results in the SEFDM signal transmitting the same amount of information over the same number of sub-carriers as OFDM but with the benefit of reduced bandwidth utilisation. The key drawback of the Type 1 transmitter is that it requires the ratio $\frac{N}{\alpha}$ to be an integer number to prevent performance degradation. Consequently, this limits the flexibility in choosing the value of α.

For this reason, two alternative transmitter schemes were also proposed, namely Type 2 and Type 3, shown in Fig. 6. Type 2, known as the *rational* α transmitter, uses a cN length IDFT block. Only the inputs of the IDFT whose indices are integer multiples of b, in other words for $n = 0, b, 2b, \ldots, (N-1)b$ are linked to the input data symbols while the remaining IDFT inputs are padded with zeros. This is achieved via the following condition

$$s_n = \begin{cases} s_{\frac{n}{b}} & \text{if } n \; mod \;\; b = 0 \\ 0 & \text{otherwise} \end{cases} \tag{5}$$

where *mod* is defined as the modulus of the remainder after division.

Complying with the condition of Eq. (5), the generation process for the Type 2 transmitter may be expressed as

$$x(m) = \frac{1}{\sqrt{N}} \sum_{n=0}^{cN-1} s_n e^{j2\pi \frac{mn}{cN}}. \tag{6}$$

The benefit of this transmitter technique is that b and c can be chosen independently to give any desired value of α for any number of sub-carriers N. The length of the IDFT block in this case is c times larger compared to an equivalent OFDM block and b times larger compared to the Type 1 SEFDM transmitter.

Thereby, the Type 3 transmitter, known as the *sum of multiple IDFTs*, was proposed. This transmitter uses c separate IDFT blocks each of length N instead of one large cN IDFT block. The advantage of this transmitter over the Type 2 architecture is two-fold; first, it allows multiple IDFT blocks to be configured in parallel, a feature attractive for hardware implementation employing Field Programmable Gate Arrays (FPGAs); second, it reduces the computation time required for the IDFT processing stage provided that all c IDFT blocks can operate concurrently in parallel. It is important to note that the Type 3 transmitter requires two additional operations, namely a reordering stage before the IDFT block, as well as a post-IDFT phase rotation stage. The expression for the Type 3 transmitter is therefore given by

$$x(m) = \frac{1}{\sqrt{N}} \sum_{u=0}^{c-1} e^{j2\pi \frac{um}{cN}} \sum_{n=0}^{N-1} s_{u+nc} e^{j2\pi \frac{mn}{N}}. \tag{7}$$

3.2 SEFDM Signal Demodulation

As illustrated in Fig. 4, the SEFDM signal $x(t)$ is transmitted through an AWGN channel and arrives at the receiver as

$$r(t) = x(t) + z(t), \tag{8}$$

where $z(t)$ represents the noise impairment introduced by the channel. The output of the receiver's demodulator is acquired by correlating the received signal with a conjugate set of correlation functions $v_n(t)$, thus giving

$$R_n = \int_0^T r(t)v_n^*(t)dt, n = 0, \ldots, N-1. \tag{9}$$

The receiver model expressed mathematically by Eq. (9) is valid given two key assumptions; first, the demodulator is a linear filter to prevent noise colouring; second, the noise is Gaussian, in other words random and not burst, the latter being the case in a multipath propagation scenario due to fading and Doppler effects.

Assuming ideal sampling and ideal timing, the linear statistical model of Eq. (9) may be represented in matrix form as

$$\mathbf{R} = (\mathbf{X} + \mathbf{Z})\mathbf{V}^H, \tag{10}$$

where $\mathbf{R} = [R_n]$ is the $N \times 1$ vector of observation statistics. Recalling from Eq. (4), each discrete SEFDM symbol consists of Q time samples, hence the transmitted signal may be expressed in matrix form as

$$\mathbf{X} = \mathbf{\Phi s} \tag{11}$$

where $\mathbf{X} = [X_m]$ is a Q-dimensional vector of time samples, $\mathbf{s} = [s_n]$ is an N-dimensional vector of complex transmitted data symbols and $\mathbf{\Phi}$ is a $Q \times N$ sub-carrier matrix with elements equal to $e^{\frac{j2\pi mn\alpha}{Q}}$. The vector $\mathbf{Z} = [Z_m]$ represents Q noise samples, which are complex independent and identically distributed (i.i.d.) Gaussian samples with mean $\mu = 0$ and variance $\sigma^2 = \frac{\mathcal{N}_0}{2}$, where $\mathcal{N}_0$ is the noise spectral density. Finally, $\mathbf{V}$ is the matrix representation of the basis functions where $\mathbf{V}^H$ denotes the Hermitian, or otherwise complex conjugate transpose, of matrix $\mathbf{V}$.

The functions $v_n(t)$ must be orthonormal or orthogonal [45] to satisfy the linear filtering requirement pertaining to Eq. (9). Two methods have been proposed to generate the basis functions $v_n(t)$. The first method uses *orthonormalisation procedures*, such as Gram-Schmidt (GS) or Löwdin processes, to generate an orthonormal base from a set of linearly independent but non-orthogonal sub-carriers [23, 45]. The second method uses *matched filtering*, as defined in [1]. The two methods are equivalent, as established in [1].

In matched filtering, the conjugate basis functions $v_n^*(t)$ equal the complex conjugate sub-carriers $\mathbf{\Phi}^H$, in other words

$$R_n = \int_0^T r(t)v_n^*(t)dt = \int_0^T r(t)e^{-j2\pi n\alpha\Delta ft}dt, n = 0, \ldots, N-1. \tag{12}$$

Subsequently, Eq. (10) becomes

$$\mathbf{R} = \mathbf{\Phi}^H\mathbf{X} + \mathbf{\Phi}^H\mathbf{Z} = \mathbf{\Phi}^H\mathbf{\Phi}\mathbf{s} + \mathbf{\Phi}^H\mathbf{Z} = \mathbf{C}\mathbf{s} + \mathbf{Z}_{\mathbf{\Phi}^H} \tag{13}$$

where $\mathbf{C}$ is the sub-carriers correlation matrix and describes the self-created interference between adjacent sub-carriers, as explained in Sect. 2.2, while $Z_{\mathbf{\Phi}^H}$ is the AWGN correlated with the conjugate sub-carriers.

If instead of matched filtering we used orthonormalisation procedures, the sub-carriers correlation matrix $\mathbf{C}$ would be replaced with the projections matrix $\mathbf{D}$, where $\mathbf{D}$ is given by

$$\mathbf{D} = \mathbf{B}^H\mathbf{\Phi}, \tag{14}$$

and $\mathbf{B}$ is the matrix representation of the orthonormal basis functions.

In conclusion, the linear statistical model for the receiver may be expressed in general form as

$$\mathbf{R} = \mathbf{H}\mathbf{s} + \mathbf{Z}, \tag{15}$$

where $\mathbf{H}$ is termed the interference matrix and corresponds to either the projections matrix $\mathbf{D}$ or the sub-carriers correlation matrix $\mathbf{C}$ depending on the basis functions $v_n(t)$ employed at the receiver.

3.3 SEFDM Signal Detection

Figure 7 depicts a practical transceiver comprising three different stages for the demodulation and detection of SEFDM signals. The first stage is a demodulator which, as described in Sect. 3.2, may be implemented using orthonormalisation processes or matched filtering. Matched filtering is preferred since it is the optimum receiving filter and can be realised using DFT operations. The second stage involves the use of linear or iterative techniques while the third stage employs more sophisticated algorithms, such as Maximum Likelihood (ML) or SD. A snapshot of the demodulation and detection methods employed in SEFDM is depicted in Fig. 8.

Section 3.2 explained that the SEFDM demodulator is responsible for generating the statistics vector $\mathbf{R}$. Each element of this vector $\mathbf{R}$ is merely a discrete sample which has a value proportional to the energy of the received symbol with the added

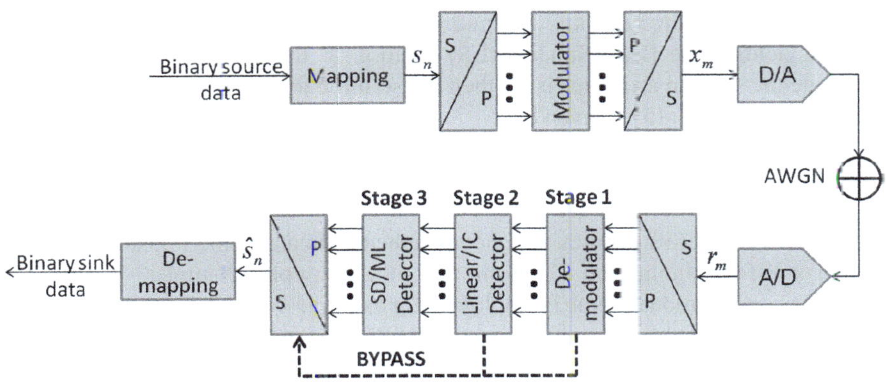

Fig. 7 Practical SEFDM transceiver

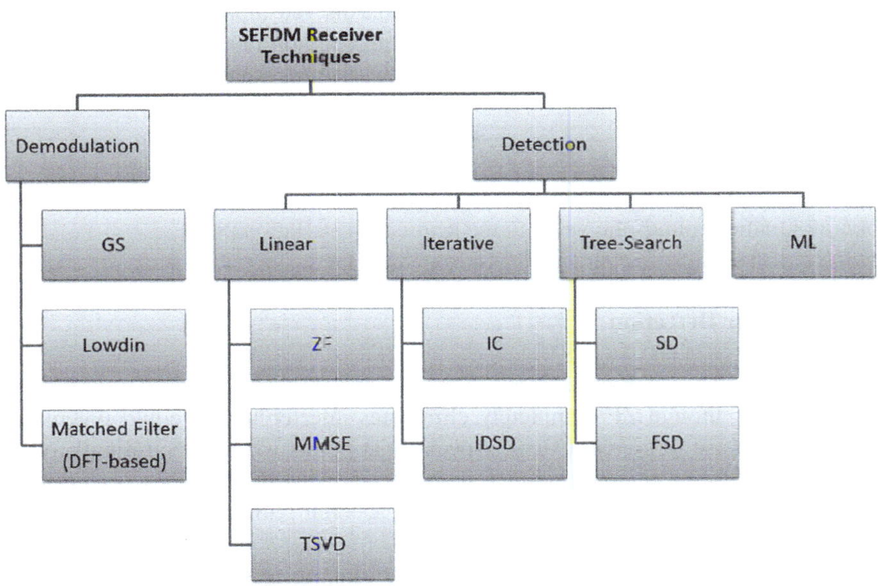

Fig. 8 SEFDM demodulation and detection techniques

noise. The task of the detector is to obtain the best possible estimate $\hat{\mathbf{s}}$ of each transmitted data symbol, where $\hat{\mathbf{s}}$ denotes the estimate of the transmitted signal vector $\mathbf{s}$ after slicing, based on the values of $\mathbf{R}$ and the properties of the sub-carriers correlation matrix $\mathbf{C}$.

Recalling from Eq. (15), the linear statistical model for the received SEFDM signal is given by $\mathbf{R} = \mathbf{Hs} + \mathbf{Z}$. Linear detection methods treat this relation as an unconstrained linear estimation problem and thus aim to recover the transmitted symbols by cancelling out the self-created interference. The most popular methods include Zero Forcing (ZF) and Minimum Mean Squared Error (MMSE) with the former yielding the Least Squares (LS) solution. If the estimator matrix used to generate the symbol estimates is **G**, then the solution to the unconstrained problem with respect to Eq. (15) is given by

$$\tilde{\mathbf{s}} = \mathbf{GR}, \tag{16}$$

with $\tilde{\mathbf{s}} \in \mathbb{C}^N$, where $\tilde{\mathbf{s}}$ denotes the unconstrained estimate of the transmitted signal vector $\mathbf{s}$ before slicing and $\mathbb{C}^N$ is the set of all complex N-tuples[3] assuming a complex constellation scheme is employed, for example 4-QAM. If the constellation scheme contains only real values, such as BPSK, the set $\mathbb{C}^N$ may be reduced to the set $\mathbb{R}^N$.

The solution of Eq. (16) is referred to as the unconstrained estimate since it does not take into account that the elements of **s** can in fact take only the values pertaining to a limited set of constellation points defined in a discrete alphabet $\mathfrak{M}^N$. To this end, the problem is constrained by applying a slicing operation to each of the entries of $\tilde{\mathbf{s}}$ yielding

$$\hat{\mathbf{s}} = \lfloor \tilde{\mathbf{s}} \rceil , \tag{17}$$

with $\hat{\mathbf{s}} \in \mathbb{Z}^N$ where $\mathbb{Z}^N$ denotes the set of integer N-tuples and $\lfloor \cdot \rceil$ denotes a slicing operator, which rounds the value to the nearest constellation point.

3.3.1 Linear Detection

The ZF detector generates symbol estimates by forcing the interference terms to zero. Even though ZF completely eliminates interference, it could potentially amplify the noise to a great extent, especially at points where spectral nulls occur [22], leading to severe performance degradation. MMSE improves performance over ZF by taking the presence of noise into account with the aim of minimising the total expected error.

Previous work has shown that the ill-conditioning of the interference matrix significantly degrades the estimates generated using the ZF or MMSE detectors [1, 45]. Moreover, when the interference matrix is singular it becomes non-invertible, thus the ZF technique cannot be employed. A technique termed Truncated Singular Value Decomposition (TSVD) was therefore proposed to address this challenge [41].

The TSVD approach generates an approximate pseudoinverse $\mathbf{H}_\xi$ of the interference matrix **H**. The truncation of the small singular values renders $\mathbf{H}_\xi$ less sensitive to perturbations yielding a solution with higher confidence levels.

[3] An N-tuple is a sequence or ordered set of N elements.

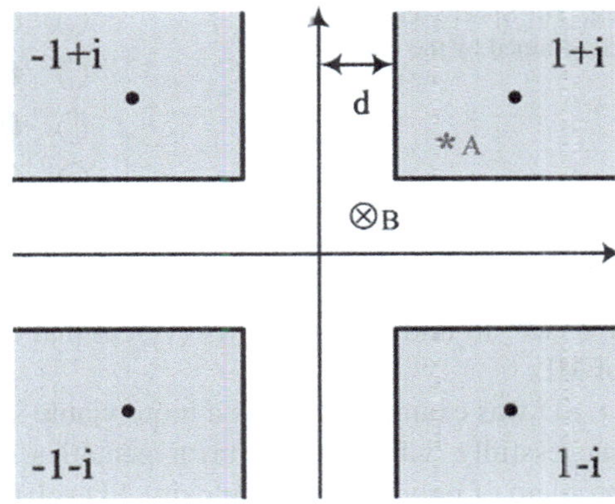

Fig. 9 Soft mapping principle

3.3.2 Iterative Detection

Iterative Cancellation (IC) techniques were applied by Kanaras in [45] and by Grammenos in [24] to the projections matrix **D** and to the sub-carriers correlation matrix **C**, respectively, to generate symbol estimates. Having generated an upper triangular matrix, the symbol estimates are obtained via back-substitution. IC results in improved Bit Error Rate (BER) performance over ZF and TSVD. This is due to the fact that IC constrains the symbol estimate on each iteration, in other words it takes into account that each entry of **s** pertains to a value from a finite set of constellation points $\mathscr{M}$. For this reason, such IC techniques are classified as *hard mapping* detectors.

An improved iterative detector which uses *soft mapping* was applied to SEFDM first in [32] and subsequently in [84]. The soft mapping principle is illustrated in Fig. 9. The white uncertainty interval is defined by $d = 1 - \frac{m}{\upsilon}$, where m is the m_{th} iteration and υ denotes the number of iterations. Only points that fall in the grey area can be mapped to the corresponding constellation point. Other points remain unchanged and left to the next iteration. It should be noted that the uncertainty area is reduced after each iteration. Hence, the algorithm always converges to a solution so long as a sufficient number of iterations takes place.

3.3.3 Maximum Likelihood and Sphere Decoding

It is well known that ML is the optimum detector for achieving the theoretical error performance. The detection problem in this case may be expressed as follows

$$\hat{\mathbf{s}}_{ML} = \arg\min_{\mathbf{s}\in\mathscr{M}^N} \|\mathbf{R} - \mathbf{H}\mathbf{s}\|^2 , \quad (18)$$

which translates into finding the vector symbol $\hat{\mathbf{s}}$ from all possible transmitted vector symbols $\mathbf{s} \in \mathscr{M}^N$ that minimises the Euclidean distance, in other words the vector symbol that has most likely been transmitted. Assuming that complex Gaussian noise has been added to each symbol and that all transmitted vector symbols

Fig. 10 Sphere search constrained by the radius g

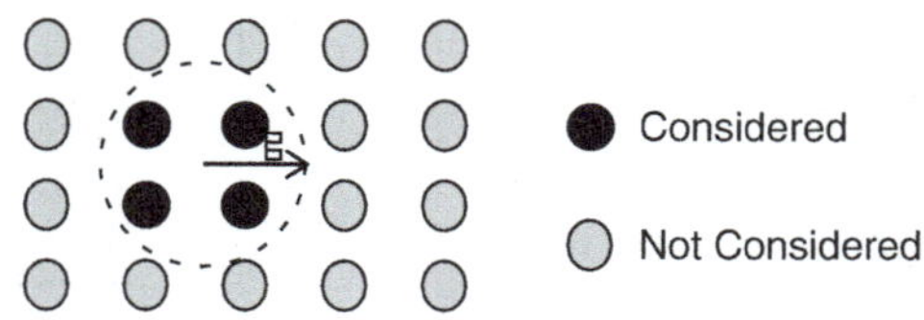

have equal probability of occurring, the ML algorithm searches over the entire set $\mathbf{s} \in \mathscr{M}^N$ to find $\hat{\mathbf{s}}$. Hence, it is evident that this hinders a practical implementation of ML.

SD was examined [45] as a more viable solution for recovering SEFDM signals successfully with tolerable error penalties since it reduces the ML solution to a problem of polynomial complexity. SD solves the same problem as ML, yet in this case, the search plane is limited to an N-dimensional hypersphere, whereby only the vector symbols that lie within this hypersphere are considered as possible solutions. Thereby, Eq. (18) becomes

$$\hat{\mathbf{s}}_{SD} = \arg \min_{\mathbf{s} \in \mathscr{M}^N} \|\mathbf{R} - \mathbf{H}\mathbf{s}\|^2 \leqslant g, \tag{19}$$

where g is a scalar denoting the radius of the sphere which is centred around the statistics vector $\mathbf{R}$, as illustrated in Fig. 10. SD is classified as a tree-search algorithm because the constellation points refer to the nodes of the tree and the algorithm is executed by traversing the levels and branches of the tree until a complete path is found.

The conventional SD algorithm has a variable computational complexity which changes according to the noise in the channel and the properties of the interference matrix. For this reason, the Fixed Sphere Decoding (FSD) was proposed in [41] and its algorithm is outlined in Fig. 11. The FSD guarantees a constant throughput by fixing the complexity of the conventional SD algorithm. This complexity is independent of the noise present in the system, as well as the ill-conditioning of the interference matrix.

The complexity of the algorithm is user-defined by setting the parameter ζ, which defines the complexity order of the algorithm, with $W = (\sqrt{M}^{\zeta})$ where W denotes the tree width. During the initialisation phase, the first ζ levels are fully expanded. In the subsequent levels, only the best W candidate nodes are expanded. The tree width determines the complexity of the FSD, as it defines the number of node visits per level. The larger the tree width, the better the performance at the expense of increased complexity. From Fig. 11, it should be evident that the update interval centres θ_{ij} at each level can be computed independently in parallel rendering the FSD algorithm an attractive choice for hardware implementation using FPGAs.

The sorting stage which performs comparisons amongst all candidate nodes B_i at each level constitutes one of the main bottlenecks. The larger the tree width, the higher the number of comparisons. This leads to an increase in the overall latency of the algorithm and subsequently reduction in throughput. On the other hand, reducing

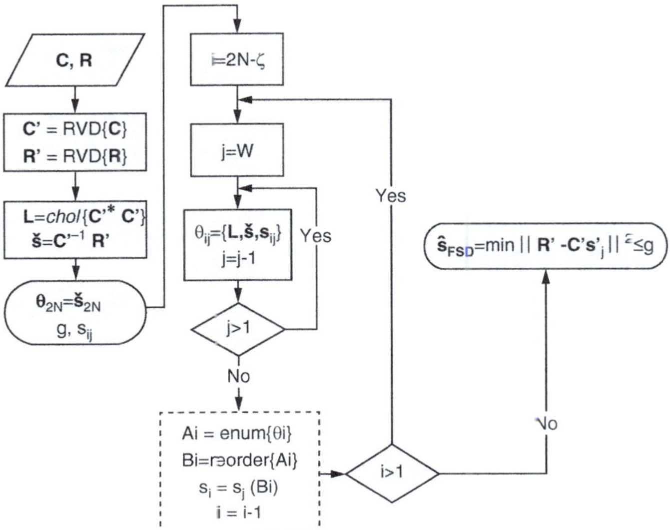

Fig. 11 Flowchart of the FSD algorithm for the detection of SEFDM signals

the tree width or removing the sorting stage affects the order in which a parent node branches out to a child node, thus degrading BER performance. Hence, to strike a balance between the conflicting objectives of good BER performance and tangible computational complexity, two techniques were applied; the first, uses Sort Free (SF) detection which reduces latency thanks to the simplified sorting stage implemented in the algorithm; the second technique, involves the utilisation of Modified Real Valued Decomposition (MRVD), which offers improved error rates. These techniques have been shown to deliver performance gains in Multiple Input Multiple Output (MIMO) systems [2].

Figure 12 compares the spanning tree for two variants of the FSD with different sorting strategies, namely the Shortest Path (SP), entailed in the original FSD algorithm, and the novel SF technique, demonstrating the simplified structure of the latter version. On the other hand, the improved performance offered by MRVD lies in the fact that, at each level, the algorithm flips between the real and imaginary parts of the next symbol in line. Therefore, MRVD can be regarded as providing 'more complete' solutions compared to the conventional Real-valued Decomposition (RVD).

3.3.4 Numerical Results

In this section, we compare different detection schemes in terms of error performance for varying system sizes and levels of bandwidth compression. Table 2 summarises the system parameters employed in the simulations.

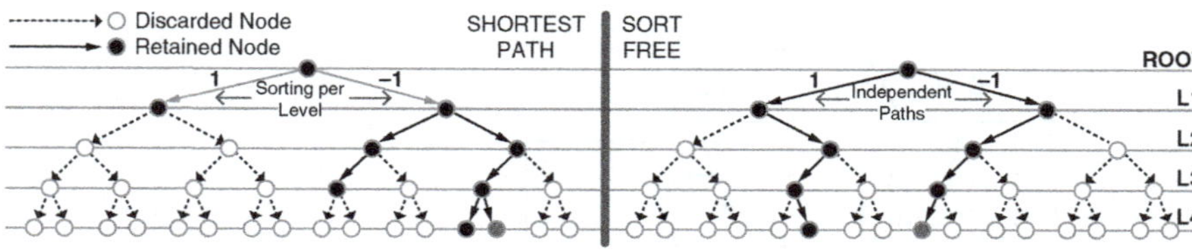

Fig. 12 Tree of Shortest-Path FSD (*left*) vs. Sort-Free FSD (*right*) ($N = 2$, $W = 2$)

Table 2 Simulation parameters for different detection schemes

Parameter	Value
Total number of sub-carriers N_{IFFT}	32 / 64
Used sub-carriers N	16
Bandwidth compression factor α	0.7 / 0.8
Modulation scheme	BPSK / 4-QAM
Oversampling factor ρ	1 / 2
E_b/N_o range	1–20 dB
Number of runs	10,000

The simulation methodology adopted has the following characteristics:

- The source data is generated using a uniform random number generator.
- All results are averaged over 10,000 randomly generated SEFDM symbols unless otherwise stated.
- The bits are mapped to symbols using either BPSK or 4-QAM unless otherwise stated. In the case of 4-QAM, it is assumed that Gray coding is employed.
- The number of bits processed is approximately equal to $\frac{10}{DesiredBER}$.
- The channel is linear, discrete-time and time-invariant.
- The only source of error is due to AWGN.[4]
- It is assumed that the receiver has perfect knowledge of the interference matrix **H** at all times under all conditions and that decisions yielding symbol estimates are made strictly based on the received amplitude of the noisy signal.

The legend keys that appear in the figures correspond to the following descriptions:

- *Matched Filtering (MF)*: The symbol estimates are given by $\hat{\mathbf{s}}_{MF} = \lfloor \mathbf{R} \rceil = \lfloor \mathbf{\Phi}^H \mathbf{r} \rceil$ where $\mathbf{r}$ is a vector representing discrete time samples of the received signal having propagated through an AWGN channel.
- *Matched Filtering followed by Zero Forcing (MF-ZF)*: The symbol estimates are given by $\hat{\mathbf{s}}_{ZF} = \lfloor \mathbf{C}^{-1}\mathbf{R} \rceil$.

[4]Results showing the performance of SEFDM in fading channels have been published in the work of Chorti et al. [11] and Isam et al. [40].

- *Demodulation using Iterative Modified Gram Schmidt (IMGS) processes [45] followed by Zero Forcing (IMGS-ZF)*: The symbol estimates are given by $\hat{\mathbf{s}}_{ZF_{IMGS}} = \lfloor \mathbf{D}^{-1}\mathbf{R} \rceil$.
- *Matched Filtering followed by Truncated Singular Value Decomposition (MF-TSVD)*: The symbol estimates are given by $\hat{\mathbf{s}}_{TSVD} = \lfloor \mathbf{C}_{\blacksquare}\mathbf{R} \rceil$.
- *Matched Filtering followed by Iterative Cancellation (MF-IC)*: The symbol estimates are given by $\hat{\mathbf{s}}_{IC} = \hat{s}_m$ with $m = 1, \ldots, N$ and $\hat{s}_m = \left\lfloor \frac{1}{c_{m,m}}(R_m - \sum_{n=m+1}^{N} c_{m,n}\hat{s}_n) \right\rceil$.
- *Demodulation using Iterative Modified Gram Schmidt processes followed by Iterative Cancellation (IMGS-IC)*: The symbol estimates are given by $\hat{\mathbf{s}}_{IC_{IMGS}} = \hat{s}_m$ with $m = 1, \ldots, N$ and $\hat{s}_m = \left\lfloor \frac{1}{d_{m,m}}(R_m - \sum_{n=m+1}^{N} d_{m,n}\hat{s}_n) \right\rceil$.
- *Demodulation using Löwdin processes [45] followed by Iterative Cancellation (Lowdin-IC)*: The symbol estimates are given by $\hat{\mathbf{s}}_{IC_{Lowdin}} = \hat{s}_m$ where $\hat{s}_m = \left\lfloor \frac{1}{l_{m,m}}(R_m - \sum_{n=m+1}^{N} l_{m,n}\hat{s}_n) \right\rceil$, with $m = 1, \ldots, N$ and where $l_{m,n}$ denotes the elements of the Hermitian matrix generated using the Löwdin orthonormalisation process.
- *Matched Filtering followed by Minimum Mean Squared Error (MF-MMSE)*: The symbol estimates are given by $\hat{\mathbf{s}}_{MMSE} = \left\lfloor \mathbf{C}\left(\mathbf{C}\mathbf{C}^H + \frac{\sigma^2}{\sigma_s^2}\mathbf{I}\right)^{-1}\mathbf{R} \right\rceil$.
- *Iterative Detection with soft decision (ID)*: The symbol estimates are given by $\hat{\mathbf{s}}_{ID} = S_n$ with $S_n = \lambda\mathbf{R} + (e - \lambda\mathbf{C})S_{n-1}$ where S_n is an N-dimensional vector of recovered symbols after n iterations, S_{n-1} is an N-dimensional vector of estimated symbols after $n-1$ iterations, e is an $N \times N$ identity matrix and λ is a convergence factor that determines how many iterations are used in the detection.

Figure 13 shows that for BPSK, the IC schemes using MF or Löwdin orthonormalisation offer the best and commensurate error performance. For Signal-to-Noise Ratio (SNR) regimes where $\frac{E_b}{N_0} \geqslant 14\,\text{dB}$, the IC method employing IMGS approaches the performance of TSVD and MMSE. A key observation from Fig. 13 is that MF on its own can provide the same BER performance than both MF-TSVD and MF-MMSE. This is an important result as it allows the linear stage in Fig. 7 to be bypassed, should one require the lowest complexity receiver (a straightforward demodulator) at the expense of error performance when compared to an IC receiver.

Figure 14 compares the BER of the original FSD detector against its variants. It is clear that TSVD and MRVD when separately applied to the standalone SF yield significant performance gains, especially in the higher SNR regimes. When both techniques are combined and applied to SF, the resultant TSVD-MRVD-SF FSD detector approaches the performance of the more optimum, shortest-path TSVD-FSD detector, yet with the benefit of reduced computational complexity.

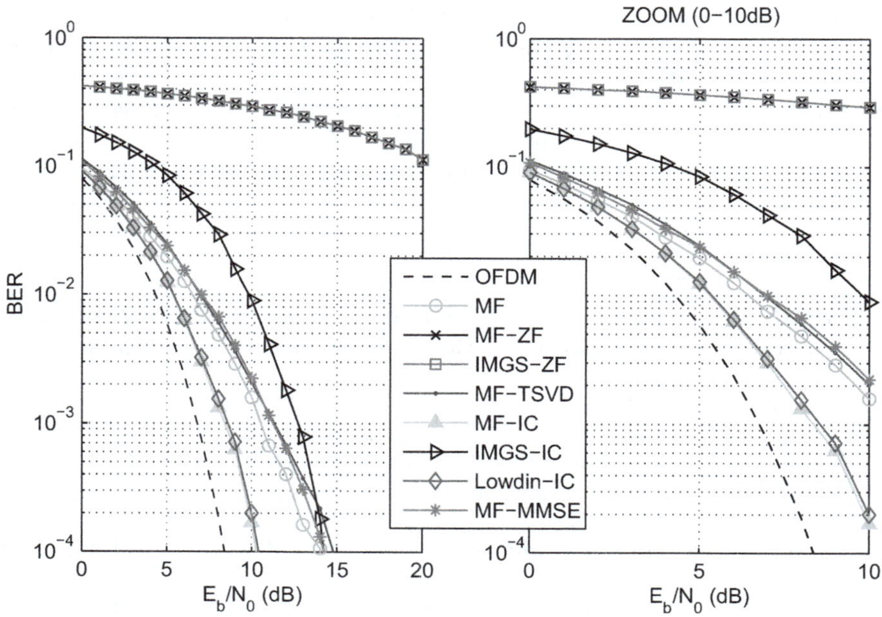

Fig. 13 BER of different detection schemes for BPSK with $N = 16$, $\alpha = 0.8$ and $\rho = 2$

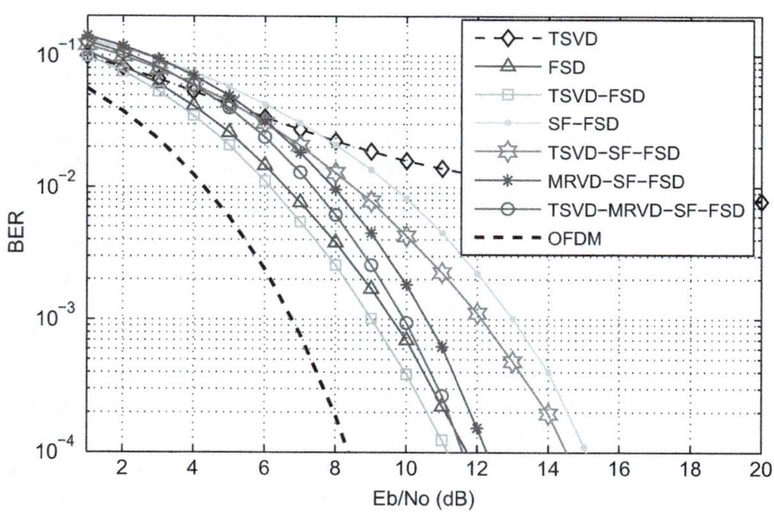

Fig. 14 BER performance for the different FSD variants (4-QAM, $\alpha = 0.8$, $N = 16$ and $W = 16$)

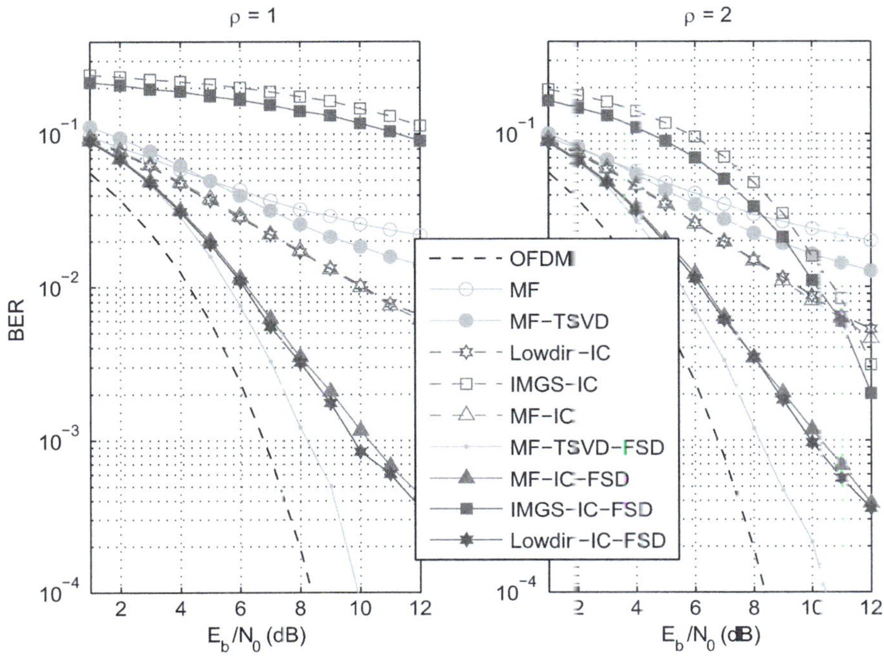

Fig. 15 BER of hybrid and non-hybrid detection schemes for $\rho = 1$ (*left sub-plot*) and $\rho = 2$ (*right sub-plot*) ($N = 16$, $\alpha = 0.8$, FSD tree width $W = 16$)

Figure 15 shows that the hybrid TSVD-FSD detector outperforms all other hybrid and non-hybrid detectors (which employ hard mapping). Figure 16, however, demonstrates that the soft iterative detector combined with FSD (ID-FSD) is superior to TSVD-FSD for different bandwidth compression factors. More importantly, when $\alpha = 0.8$, ID-FSD offers near-optimum performance.

3.4 Coded SEFDM

Section 3.3 demonstrated that iterative techniques, such as iterative cancellation and iterative detection with soft decision, offer good BER performance. As the system size increases, however, giving rise to higher interference power, even these methods fail to recover SEFDM signals successfully.

To address this challenge, Turbo equalisation was introduced in the SEFDM receiver [78], as an improved means for mitigating the self-created interference. A detailed description of the Turbo equalisation principle may be found in [28]. A functional block diagram of the Turbo-SEFDM system is illustrated in Fig. 17. This system architecture was conceptually described in [78, 31] and experimentally

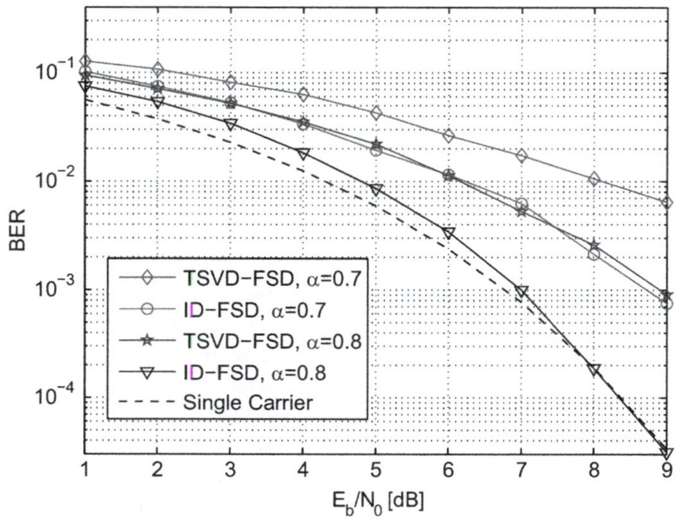

Fig. 16 BER performance for ID-FSD and TSVD-FSD detectors carrying 4QAM symbols for $N = 16$, $\upsilon = 20$ and $W = 16$

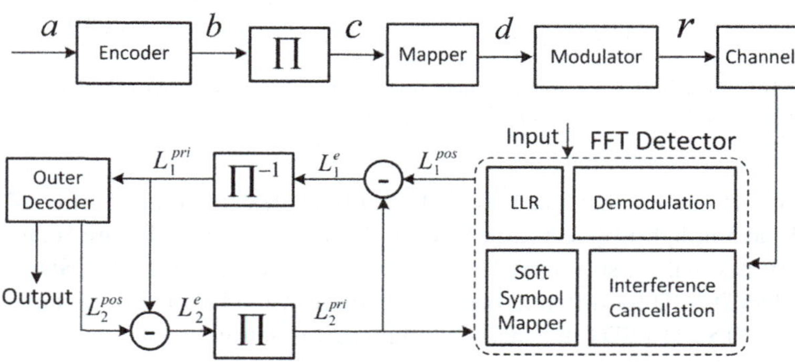

Fig. 17 Functional block diagram of the Turbo-SEFDM system

evaluated in [82]. Here, we present a summary of the proposed system and interested readers may resort to the aforementioned references for a detailed description.

At the transmitter side, the bit stream labelled **a** is encoded using either convolutional codes or Turbo codes. Subsequently, the coded bit stream labelled **b** is permuted within the interleaver module **Π**. The interleaved bit stream **c** is then mapped to complex symbols **d** and finally modulated **r** using an SEFDM modulator [75].

At the receiver side, the signal is demodulated using an SEFDM FFT based demodulator [75]. The log-likelihood ratio (LLR) unit generates soft information, referred to as *a posteriori* information and denoted by L_1^{pos}, for each bit. The sign of the LLR value reflects the sign of the bit while its magnitude determines

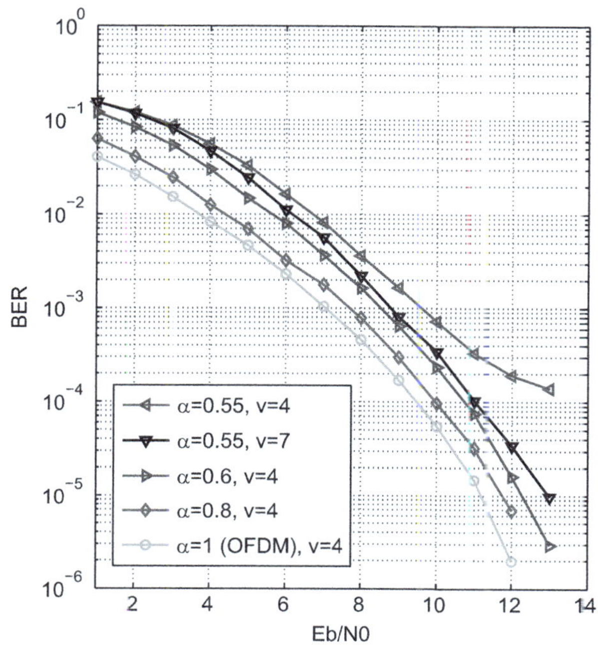

Fig. 18 Performance of Turbo-SEFDM in a multipath fading channel with $N = 1024$ at various α

the reliability of the sign of the bit. This soft information is then transformed to extrinsic information, L_1^e by subtracting the *a priori* information, L_2^{pri}. This extrinsic information is exchanged between the detector and the outer decoder in an iterative fashion. On each iteration, the interference represented by the updated soft symbols, is subtracted from the original demodulated symbols. Hence, the receiver maximises the *a posteriori* probability (APP) for a given bit.

Figures 18 and 19 demonstrate the performance of Turbo-SEFDM in a frequency selective fading channel where $h(t) = 0.8765\delta(t) - 0.2279\delta(t - T_s) + 0.1315\delta(t - 4T_s) - 0.4032e^{\frac{j\pi}{2}}\delta(t - 7T_s)$. It is assumed that perfect channel state information (CSI) is known at the receiver. Figure 18 illustrates that the bandwidth can be compressed up to 45 % with an error penalty of up to 2 dB using only four to seven iterations. Figure 19 shows that the best performance can be achieved after just four iterations for a bandwidth saving of 40 %.

4 SEFDM Implementation Aspects

The dimension of the SEFDM problem presents significant challenges compared to the application of similar detection methods in MIMO systems. In this section, we discuss the complexity and performance trade-offs for the different detectors examined in the previous sections. We also explore testbeds that were devised in

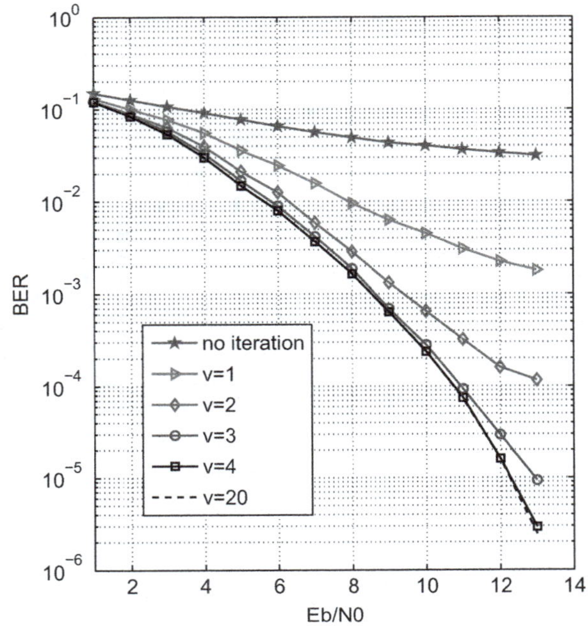

Fig. 19 Performance of Turbo-SEFDM in a multipath fading channel with $\alpha = 0.6$ for a different number of iterations

recent years for the implementation of SEFDM, as well as the recent application and experimental demonstration of SEFDM in other systems, notably in Long Term Evolution (LTE), optical communications and radio-over-fibre (RoF).

4.1 Performance and Complexity Trade-offs

SEFDM increases multiplexing gain with respect to an OFDM system by permitting the use of more sub-carriers in the default OFDM bandwidth. This gain, however, comes at the expense of complex signal processing at the receiver which is required in order to separate the sub-carriers suffering from self-created ICI. A linear increase in the number of sub-carriers, however, results in a more than linear increase in the complexity order of both the demodulation and detection stages.

4.1.1 In Demodulation, Linear and Iterative Detection

A demodulator employing GS or Löwdin orthonormalisation has the same complexity order as a conventional analogue multi-carrier system equal to $\mathcal{O}(N^2)$. The matched filter technique addressed this issue by substituting these processes with DFT operations, thus allowing the use of the Fast Fourier Transform (FFT) algorithm reducing the complexity order to $\mathcal{O}(Nlog_2N)$. The linear detection methods explored in Sect. 3.3 involve matrix inversion or decomposition. Analytical tools like MATLAB, generally apply Gaussian elimination to compute the inverse of a matrix, such as the computation required in the LS solution of the ZF estimate. The TSVD detector uses the standard Singular Value Decomposition (SVD) matrix decomposition method to generate an approximate inverse. In both cases, the order of complexity is assumed to be cubic $\mathcal{O}(N^3)$ [8, 45]. The inverted matrix is then multiplied by the statistics vector **R**, which implies that the total number of multiplications amounts to N^2. The slicing operator acts as a simple threshold detector, therefore its complexity is considered to be negligible compared to the complexity of the prerequisite matrix operations.

Conversely, iterative detection techniques operate on upper triangular matrices, where the total number of multiplications equals $\frac{N}{2}(N+1) \approx \frac{N^2}{2}$, which is less than the number of multiplications required for the ZF or TSVD techniques. Iterative detection requires the inversion of only the diagonal elements of the matrix.

In summary, the benefits and drawbacks of using linear or iterative detection are as follows:

- **Linear detection**: The fundamental difficulty in this case lies in the inversion of the sub-carriers correlation matrix **C** which has a large dimension equal to $N \times N$ and is also complex-valued which instantly doubles the dimension of the problem to $2N \times N$. The computational complexity associated with the inversion or decomposition of matrices is often prohibitive in wireless systems [22]. **C** in SEFDM systems is deterministic and has constant values for a given configuration of N and α. Thereby, the elements of $\mathbf{C}^{-1}$ only need to be generated once, a process which can take place offline, and subsequently stored in block memories. The computation of the symbol estimates is then a straightforward complex-valued matrix multiplication between **R** and $\mathbf{C}^{-1}$. The default number of words required for the storage of the elements pertaining to $\mathbf{C}^{-1}$ is N^2. Matrix multiplication can be implemented efficiently through the concurrent execution of multiple multiply-accumulate operations using FPGAs.
- **Iterative detection**: Iterative detection alleviates the need for matrix inversion which means that the results suffer less from round-off errors. The performance of iterative detection largely depends on the correctness of the first symbol decoded since any errors during the initial stages of the algorithm's execution will propagate through the iterative process and could potentially lead to severe BER degradation. It does, however, require a cascaded architecture which hinders parallel processing. One possible solution to this issue is to decompose the entire iteration process into a single step. In FPGA terms, this decomposition would necessitate combinatorial logic, which is generally avoided as it increases the load on the clock severely limiting the maximum achievable clock frequency.

Hence, based on the above, it is evident that linear and iterative schemes present different challenges with varying trade-offs. The choice depends on the specific technology adopted to implement these algorithms in practice. FPGAs offer significant performance gains over conventional Digital Signal Processors (DSPs) provided that their parallel processing capabilities are leveraged.

4.1.2 In Sphere Decoding

The complexity of an algorithm is determined by two aspects; first, the number of arithmetic operations required to execute the algorithm, such as multiplications, additions and subtractions; second, the number of compare-select operations that take place during the sorting stage.

While the conventional SD provides a significant reduction in complexity compared to ML, its execution is highly sensitive to noise and the interference properties of SEFDM systems. An authoritative account of the complexity associated with SD is given by Hassibi and Vikalo [31], as well as Jalden and Ottersten [42]. There is also rich literature discussing these issues in MIMO systems, a notable example being the work presented by Viterbo and Boutros [72].

This variable complexity associated with the conventional SD renders it inapt for hardware realisation. In addition, the sequential nature of the algorithm does not fit well with the parallel processing capabilities of FPGAs. For this reason, the FSD was explored which fixes the algorithm's execution time and has an architecture which is better tailored for FPGA implementation. A key enhancement of the FSD was the simplification of the sorting stage, which was found to be the main bottleneck in the algorithm's execution.

Table 3 compares the total number of operations required at each tree level for the SP and SF detection schemes. The depth of the tree, commensurate with the system dimension, is denoted by D which is equal to $2N$ as a result of using RVD. From this table, it is evident that the number of comparisons at each level is fixed and independent of the system dimension. For the SF method, a single comparator is required for each node to compute the minimum distances to the centre point. The number of comparisons is equal to the tree width. On the contrary, the SP technique requires multiple cascaded minimum finders to perform the reordering of the candidate nodes with respect to the minimum distance from the centre point. In both cases, the number of comparisons is dictated by the tree width.

Table 3 Total number of operations per level for the Shortest-Path and Sort-Free FSD detectors

Metric	Shortest-path (SP)	Sort-free (SF)
Multiplications	$W(D-k)$	$W(D-k)$
Additions	$W(2(D-k)+1+\sqrt{M})$	$W(2(D-k)+1)$
Comparisons	$\frac{1}{2}(MW^2+\sqrt{M}W+4W)$	W

4.2 Experimental Testbeds

Much work has been done with respect to the hardware development of an SEFDM transceiver. Independent FPGA realisations of a reconfigurable SEFDM transmitter and ZF\TSVD receiver have appeared in [62] and [25], respectively. In [62], the Type 3 transmitter, discussed in Sect. 3.1, was implemented using multiple Inverse Fast Fourier Transform (IFFT) blocks configured in parallel, thus achieving the lowest latency implementation possible.

Subsequently, a route to a Very Large Scale Integration (VLSI) implementation of a reconfigurable SEFDM transmitter in 32-nm CMOS was published in [75] in which a pruned IFFT architecture was introduced. Running at a clock frequency of 100 MHz, the throughput can reach 17.4 Mbps with QPSK modulation, and up to 52.2 Mbps with 64QAM modulation, with a bandwidth saving of up to 50 % relative to OFDM. Here, it is also worth mentioning that VLSI implementations of FTN transceivers have also come to light [15].

A joint hardware implementation of the hybrid TSVD-FSD receiver with the aid of FPGAs and DSPs was presented in [26]. A high-level representation of the complete hardware system is depicted in Fig. 20. The linear detector, details of which are published in [25], was realised on an FPGA device with a flexible and reconfigurable design supporting different system sizes, oversampling factors, modulation orders and levels of bandwidth compression while providing a theoretical data rate of up to 136.8 Mbps. On the DSP side, a commercially available DSP development environment was used to implement the SP and SF FSD detectors on the device and to handle the data transfer between the DSP and the computer terminal.

The SEFDM concept has been expanded and evaluated in other experimental systems. In [81], a multi-band architecture termed B-SEFDM was tested in a practical, LTE like environment. A new optical SEFDM (O-SEFDM) system was proposed in [14] and subsequently demonstrated in [13]. Results [13] showed that 4-QAM O-SEFDM can outperform 8-QAM optical OFDM by 1.6 dB. This latter work constitutes the first experimental verification of 25 % optical faster than Nyquist rate. More recently, a dual polarisation coherent optical fibre system employing O-SEFDM was developed [60]. This system is capable of operating at 24 Gb/s while occupying a total bandwidth ranging from 6 GHz (for $\alpha = 1$) to 4.02 GHz (for $\alpha = 0.67$).

Finally, the use of SEFDM in RoF was also demonstrated, for the first time, for beyond 4G cellular networking [57]. In this work, a proof-of-concept SEFDM-LTE type radio over multi-mode fibre transmission was shown to operate successfully at 36 Mb/s.

5 Conclusions

This chapter considered the design, implementation and performance evaluation of a novel multi-carrier modulation technique termed SEFDM, which improves spectrum efficiency at the expense of receiver complexity. The key challenge that

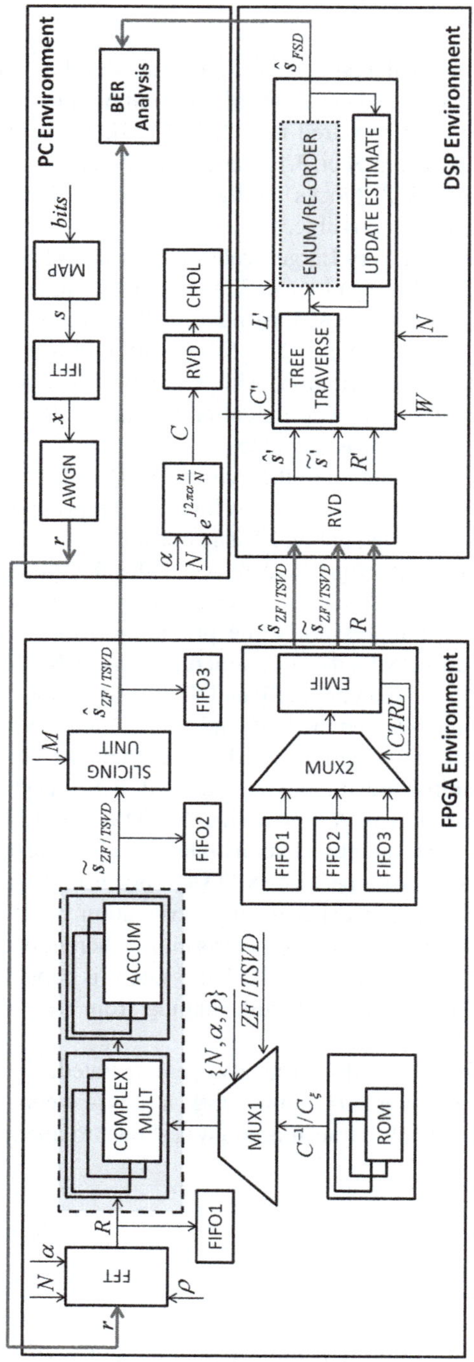

Fig. 20 System block diagram for the hybrid TSVD-FSD hardware detector implementation

arises in SEFDM systems is the self-created interference which complicates the design of receiver architectures. To this end, sophisticated detection algorithms have been developed to address this challenge. Subsequently, these algorithms are optimised to make them better-suited for application in the real world. Finally, a number of hardware prototypes and experimental testbeds have been realised to assess the feasibility of employing SEFDM in practical systems including wireless, optical and RoF systems.

References

1. S.I. Ahmed, Spectrally Efficient FDM Communication Signals and Transceivers : Design, Mathematical Modeling and System Optimization, PhD thesis, University College London (UCL), 2011
2. K. Amiri, C. Dick, R. Rao, J.R. Cavallaro, Flex-Sphere: an FPGA configurable sort-free sphere detector for multi-user MIMO wireless system, in *Proceedings of the Software Defined Radio Technology Conference*, Washington, DC, SDR, 2008
3. J.B. Anderson, F. Rusek, Improving OFDM : multistream faster-than-Nyquist signaling, in *4th International Symposium Turbo Codes and Related Topics, ISTC*, Munich, 2006
4. J.G. Andrews, S. Buzzi, W. Choi, S.V. Hanly, A. Lozano, A.C.K. Soong, J.C. Zhang, What will 5G be? IEEE J. Sel. Areas Commun. **32**(6), 1065–1082 (2014)
5. D.H. Bailey, The fractional Fourier transform and applications. SIAM Rev. **33**(3), 389–404 (1991)
6. A Barbieri, D. Fertonani, G. Colavolpe, Improving the spectral efficiency of linear modulations through time-frequency packing, in *IEEE International Symposium on Information Theory, ISIT*, Toronto, vol. 2, 2008, pp. 2742–2746
7. L. Bluestein, A linear filtering approach to the computation of discrete Fourier transform. IEEE Trans. Audio Electroacoust. **18**(4), 451–455 (1970)
8. A. Burg, VLSI Circuits for MIMO Communication Systems, PhD thesis, Swiss Federal Institute of Technology (ETH), Zurich, 2006
9. J.W. Burns, Measuring spectrum efficiency - The art of spectrum utilisation metrics. Tech. rep., Aegis Systems Ltd., 2002
10. A. Chorti, I. Kanaras, Masked M-QAM OFDM : a simple approach for enhancing the security of OFDM systems, in *IEEE 20th International Symposium Personal, Indoor and Mobile Radio Communication, PIMRC*, Tokyo, 2009, pp. 1682–1686
11. A. Chorti, I. Kanaras, M.R.D. Rodrigues, I. Darwazeh, Joint channel equalization and detection of Spectrally Efficient {FDM} signals, in *2010 IEEE 21st International Symposium on Personal Indoor and Mobile Radio Communications (PIMRC)*, 2010, pp. 177–182
12. R.G. Clegg, S. Isam, I. Kanaras, I. Darwazeh, A practical system for improved efficiency in frequency division multiplexed wireless networks. IET Commun. **6**(4), 449 (2012)
13. I. Darwazeh, T. Xu, T. Gui, Y. Bao, Z. Li, Optical SEFDM system; bandwidth saving using non-orthogonal sub-carriers. Photon. Technol. Lett. IEEE **26**(4), 352–355 (2014)
14. I. Darwazeh, T. Xu, T. Gui, Y. Bao, Z. Li, Optical spectrally efficient FDM system for electrical and optical bandwidth saving, in *2014 IEEE International Conference on Communications (ICC)*, 2014, pp. 3432–3437
15. D. Dasalukunte, F. Rusek, V. Öwall, Multicarrier faster-than-Nyquist transceivers: hardware architecture and performance analysis. IEEE Trans. Circ. Syst. I: Reg. Papers **58**, 1–12 (2010)
16. B. Farhang-Boroujeny, Filter bank multicarrier modulation : a waveform candidate for 5G and beyond. Adv. Electr. Eng. **2014**, 25 (2014)

17. B. Farhang-Boroujeny, R. Kempter, Multicarrier communication techniques for spectrum sensing and communication in cognitive radios. IEEE Commun. Mag. **46**(4), 80–85 (2008)
18. G.P. Fettweis, The tactile internet: applications and challenges. IEEE Veh. Technol. Mag. **9**(1), 64–70 (2014)
19. G. Fettweis, M. Krondorf, S. Bittner, GFDM - Generalized Frequency Division Multiplexing, in *IEEE 69th Vehicular Technology Conference, VTC*, Barcelona, 2009
20. FP 7 European Project 317669, METIS (Mobile and wireless communications Enablers for Twenty-twenty (2020) Information Society) (2012). https://www.metis2020.com/
21. D. Gabor, Theory of communication. J IEE: Part III: Radio Commun. Eng. **93**(26), 429–441 (1946)
22. A. Goldsmith, *Wireless Communications* (Cambridge University Press, Cambridge, 2005)
23. G.H. Golub, C.F.V.V. Loan, *Matrix Computations*, 3rd edn. (The Johns Hopkins University Press, Baltimore, MD, 1996)
24. R.C. Grammenos, Spectrum Optimisation in Wireless Communication Systems: Technology Evaluation, System Design and Practical Implementation. PhD thesis, University College London - Department of Electronic and Electrical Engineering, 2013
25. R.C. Grammenos, I. Darwazeh, Hardware implementation of a practical complexity spectrally efficient FDM reconfigurable receiver, in *IEEE 23rd International Symposium Personal, Indoor and Mobile Radio Communication, PIMRC*, Sydney, 2012
26. R.C. Grammenos, I. Darwazeh, Performance trade-offs and DSP evaluation of spectrally efficient FDM detection techniques, in *IEEE International Conference Communication, ICC*, Budapest (2013, to appear)
27. R.C. Grammenos, S. Isam, I. Darwazeh, FPGA design of a truncated SVD based receiver for the detection of SEFDM signals, in *IEEE 22nd International Symposium Personal, Indoor and Mobile Radio Communication, PIMRC*, Toronto, 2011
28. J. Hagenauer, The turbo principle: tutorial introduction and state of the art, in *Proceedings of the International Symposium Turbo Codes*, 1997, pp. 1–11
29. M. Hamamura, S. Tachikawa Bandwidth efficiency improvement for multi-carrier systems, in *IEEE 15th International Symposium Personal, Indoor and Mobile Radio Communication, PIMRC*, Barcelona, 2004, pp. 48–52
30. S. Hara, R. Prasad, Overview of multicarrier {CDMA}. Commun. Mag. IEEE **35**(12), 126–133 (1997)
31. B. Hassibi, H. Vikalo, On the sphere-decoding algorithm I. Expected complexity. IEEE Trans. Signal Process **53**(8), 2806–2818 (2005)
32. S.J. Heydari, M. Ferdosizadeh, F.A. Marvasti, Iterative detection with soft decision in spectrally efficient FDM systems. Tech. Rep. (2013). arXiv:1304.4003v1
33. B. Hirosaki, An orthogonally multiplexed QAM system using the discrete Fourier transform. IEEE Trans. Commun. **29**(7), 982–989 (1981)
34. E. Hopner, An experimental modulation-demodulation scheme for high-speed data transmission. IBM J. Res. Dev. **3**, 74–84 (1959)
35. S. Isam, I. Darwazeh, Precoded spectrally efficient FDM system, in *IEEE 21st International Symposium Personal Indoor and Mobile Radio Communication, PIMRC*, Istanbul, 2010, pp. 99–104
36. S. Isam, I. Darwazeh, Simple DSP-IDFT techniques for generating spectrally efficient FDM system signals, in *7th IEEE, IET International Symposium Communication Systems Networks and Digital Signal Processing, CSNDSP*, Newcastle, 2010
37. S. Isam, I. Darwazeh, Design and performance assessment of fixed complexity spectrally efficient FDM receivers, in *IEEE 73rd Vehicular Technology Conference, VTC*, Budapest, 2011
38. S. Isam, I. Darwazeh, Peak to average power ratio reduction in spectrally efficient FDM systems, in *IEEE International Conference Telecommunication, ICT*, Ayia Napa, 2011
39. S. Isam, I. Darwazeh, Characterizing the intercarrier interference of non-orthogonal spectrally efficient FDM system, in *8th IEEE, IET International Symposium Communication Systems Networks and Digital Signal Processing, CSNDSP*, Poznan, 2012

40. S. Isam, I. Darwazeh, Robust channel estimation for spectrally efficient FDM system, in *IEEE International Conference Telecommunication, ICT*, Jounieh, 2012
41. S. Isam, I. Kanaras, I. Darwazeh A truncated SVD approach for fixed complexity spectrally efficient FDM receivers, in *IEEE Wireless Communication and Networking Conference, WCNC*, Cancun, 2011
42. J. Jalden, B. Ottersten, On the complexity of sphere decoding in digital communications. IEEE Trans. Signal Process. **53**(4), 1474–1484 (2005)
43. J.P. Javaudin, D. Lacroix, Technical description of the OFDM/IOTA modulation. Tech. rep., France Telecom R&D, Tech. Rep. R1-03-168, 2003
44. W. Jian, Y. Xun, Z. Xi-lin, D. Li, The prefix design and performance analysis of DFT-based overlapped frequency division multiplexing (OvFDM-DFT) system, in *3rd International Workshop on Signal Design and Its Applications in Communication, IWSDA*, Chengdu, 2007, pp. 361–364
45. I. Kanaras, Spectrally Efficient Multicarrier Communication Systems: Signal Detection, Mathematical Modelling and Optimisation. PhD thesis, University College London (UCL), 2010
46. I. Kanaras, A. Chorti, M.R.D. Rodrigues, I. Darwazeh, A combined MMSE-ML detection for a spectrally efficient non orthogonal FDM signal, in *5th International Conference on Broadband Communications, Networks, and Systems, BROADNETS*, London, 2008
47. I. Kanaras, A. Chorti, M.R.D. Rodrigues, I. Darwazeh, An optimum detection for a spectrally efficient non-orthogonal FDM system, in *Proceedings of the 13th International OFDM-Workshop*, Hamburg, 2008
48. I. Kanaras, A. Chorti, M.R.D. Rodrigues, I. Darwazeh, A new quasi-optimal detection algorithm for a non orthogonal Spectrally Efficient FDM, in *9th International Symposium Communications and Information Technologies, ISCIT*, Incheon, 2009
49. I. Kanaras, A. Chorti, M.R.D. Rodrigues, I. Darwazeh, Investigation of a semidefinite programming detection for a spectrally efficient FDM system, in *IEEE 20th International Symposium Personal, Indoor and Mobile Radio Communication, PIMRC*, Tokyo, 2009
50. I. Kanaras, A. Chorti, M.R.D. Rodrigues, I. Darwazeh, Spectrally efficient FDM signals: bandwidth gain at the expense of receiver complexity, in *IEEE International Conference on Communications, ICC*, Dresden, 2009
51. I. Kanaras, A. Chorti, M.R.D. Rodrigues, A fast constrained sphere decoder for ill conditioned communication systems. IEEE Commun. Lett. **14**(11), 999–1001 (2010)
52. W. Kozek, A.F. Molisch, Nonorthogonal pulseshapes for multicarrier communications in doubly dispersive channels. IEEE J. Sel. Areas Commun. **16**(8), 1579–1589 (1998)
53. A.D. Liveris, C.N. Georghiades, Exploiting faster-than-Nyquist signaling. IEEE Trans. Commun. **51**(9), 1502–1511 (2003)
54. T. Matsumoto, T. Asai Frequency-overlapped multicarrier signaling with inter-carrier interference cancellation (FOLMSIC). IEICE SAT **97**(486), 69–76 (1998)
55. J.E. Mazo Faster-than-Nyquist signaling. Bell Syst. Tech. J. **54**(8), 1451–1462 (1975)
56. N. Michailow, M. Lentmaier Integration of a GFDM secondary system in an OFDM primary system, in *Future Network and Mobile Summit*, Warsaw, 2011
57. S. Mikroulis, T. Xu, J.E. Mitchell, I. Darwazeh, First demonstration of a spectrally efficient {FDM} radio over fiber system topology for beyond {4G} cellular networking, in *European Conference on Network and Optical Communications, NOC 2015*, 2015
58. M. Nekovee, Cognitive radio access to TV white spaces: spectrum opportunities, commercial applications and remaining technology challenges, in *IEEE International Symposium on New Frontiers Dynamic Spectrum Access Networks, DySPAN* , Singapore, 2010
59. H. Nikopour, H. Baligh, Sparse code multiple access, in *2013 IEEE 24th International Symposium on Personal Indoor and Mobile Radio Communications (PIMRC)*, 2013, pp. 332–336
60. D. Nopchinda, T. Xu, R. Maher, B. C. Thomsen, I. Darwazeh, Dual Polarization Coherent Optical Spectrally Efficient Frequency Division Multiplexing. IEEE Photon. Technol. Lett. **28**(1), pp. 83–86, (2016)

61. D. Panaitopol, R. Datta, G. Fettweis, Cyclostationary detection of cognitive radio systems using GFDM modulation, in *IEEE Wireless Communication and Networking Conference, WCNC*, Paris, 2012, pp. 930–934
62. M.R. Perrett, I. Darwazeh, Flexible hardware architecture of SEFDM transmitters with real-time non-orthogonal adjustment, in *IEEE International Conference on Telecommunications, ICT*, Ayia Napa, 2011
63. M.R. Perrett, R.C. Grammenos, I. Darwazeh, A verification methodology for the detection of spectrally efficient FDM signals generated using reconfigurable hardware, in *IEEE International Conference on Communications, ICC*, Ottawa, 2012
64. J. Proakis, *Digital Communications*, 4th edn. (McGraw-Hill, New York, 2000)
65. M.R.D. Rodrigues, I. Darwazeh, Fast OFDM: a proposal for doubling the data rate of OFDM schemes, in *IEEE International Conference on Telecommunication, ICT*, Beijing, 2002, pp. 484–487
66. M.R.D. Rodrigues, I. Darwazeh, A spectrally efficient frequency division multiplexing based communications system, in *Proceedings of the 8th International OFDM Workshop*, Hamburg, 2003, pp. 48–49
67. F. Rusek, J.B. Anderson, The two dimensional Mazo limit, in *Proceedings of the International Symposium Information Theory, ISIT*, Adelaide, 2005, pp. 970–974
68. A. Sahin, I. Guvenc, H. Arslan, A survey on multicarrier communications: prototype filters, lattice structures, and implementation aspects. *IEEE Commun Surv. Tutor.* **16**(3), 1312–1338 (2014)
69. Y. Saito, A. Benjebbour, Y. Kishiyama, T. Nakamura, System-level performance evaluation of downlink non-orthogonal multiple access {(NOMA)}, in *2013 IEEE 24th International Symposium on Personal Indoor and Mobile Radio Communications (PIMRC)*, 2013, pp. 611–615
70. F. Schaich, T. Wild, Waveform contenders for 5G - OFDM vs. FBMC vs. UFMC, in *6th International Symposium Communication, Control, and Signal Processing*, 2014, pp. 457–460
71. G. Strang, *Introduction to Linear Algebra*, 4th edn. (Wellesley Cambridge Press, Wellesley, 2009)
72. E. Viterbo, J. Boutros A universal lattice code decoder for fading channels. IEEE Trans. Inf. Theory **45**(5), 1639–1642 (1999)
73. R. Westwick, Mobile WiMAX versus LTE: a comparison of next generation mobile broadband technologies. J. Telecommun. Manage. **1**(1), 79–85 (2008)
74. P.N. Whatmough, M.R. Perrett, S. Isam, I. Darwazeh, VLSI architecture for a reconfigurable spectrally efficient FDM baseband transmitter. IEEE Trans. Circuits Syst. I: Reg. Papers **59**(5), 1–12 (2012)
75. P.N. Whatmough, M.R. Perrett, S. Isam, I. Darwazeh, {VLSI} architecture for a reconfigurable spectrally efficient {FDM} baseband transmitter. IEEE Trans. Circuits Syst. I: Reg. Papers **59**(5), 1107–1118 (2012)
76. G. Wunder, P. Jung, M. Kasparick, T. Wild, F. Schaich, Y. Chen, S. Brink, I. Gaspar, N. Michailow, A. Festag, L. Mendes, N. Cassiau, D. Ktenas, M. Dryjanski, S. Pietrzyk, B. Eged, P. Vago, F. Wiedmann, {5GNOW}: non-orthogonal, asynchronous waveforms for future mobile applications. Commun. Mag. IEEE **52**(2), 97–105 (2014)
77. F. Xiong, M-ary amplitude shift keying OFDM system. IEEE Trans. Commun. **51**(10), 1638–1642 (2003)
78. T. Xu, I. Darwazeh, A soft detector for spectrally efficient systems with non-orthogonal overlapped sub-carriers. Commun. Lett. IEEE **18**(10), 1847–1850 (2014)
79. T. Xu, I. Darwazeh, M-QAM signal detection for a non-orthogonal system using an improved fixed sphere decoder, in *9th IEEE/IET International Symposium on Communication Systems, Networks & Digital Signal Processing 2014 (CSNDSP14)*, Manchester, 2014, pp. 623–627
80. T. Xu, I. Darwazeh, Multi-band reduced complexity spectrally efficient FDM systems, in *9th IEEE/IET International Symposium on Communication Systems, Networks & Digital Signal Processing 2014 (CSNDSP14)*, Manchester, 2014, pp. 904–909
81. T. Xu, I. Darwazeh, Spectrally efficient FDM: spectrum saving technique for 5G?, in *2014 1st International Conference on 5G for Ubiquitous Connectivity (5GU)*, 2014, pp. 273–278

82. T. Xu, I. Darwazeh, Bandwidth compressed carrier aggregation, in *IEEE ICC 2015 - Workshop on 5G & Beyond - Enabling Technologies and Applications (ICC'15 - Workshops 23)*, London, 2015
83. T. Xu, R.C. Grammenos, I. Darwazeh, FPGA implementations of real-time detectors for a spectrally efficient FDM system, in *IEEE International Conference on Telecommunications, ICT*, Casablanca, 2013
84. T. Xu, R.C. Grammenos, F. Marvasti, I. Darwazeh, An improved fixed sphere decoder employing soft decision for the detection of non-orthogonal signals. IEEE Commun. Lett. **17**(10), 1964–1967 (2013)
85. Y.G. Yoo, J.H. Cho, Asymptotic optimality of binary faster-than-Nyquist signaling. IEEE Commun. Lett. **14**(9), 788–790 (2010)

Full-Duplex Wireless Communications for 5G

Mingxin Zhou, Yun Liao, and Lingyang Song

Abstract In this chapter, we will introduce Full-Duplex (FD) wireless communications for 5G, which enables simultaneous transmission and reception over the same frequency band. In this way, the spectral efficiency can be improved significantly compared with half-duplex (HD). However, there exists severe self interference (SI), signal leakage from the local transmitter to its own receiver. Three different classes of SI mitigation techniques are presented in this chapter: propagation-domain SI suppression, analog-domain SI cancelation, and digital-domain SI cancelation. Furthermore, the system performance of several FD schemes in several different application scenarios is presented. Theoretically, the spectral efficiency of FD bidirectional and cooperative communications can be doubled, while for cognitive radio networks, the FD-based protocol can achieve much better sensing performance than the traditional HD-based cognitive radio schemes.

1 Introduction

The enormous increase in demand for data traffic drives the quest for efficient use of the available and valuable resources. However, most of the existing wireless communication devices can only operate in half-duplex, which dissipates the precious resources by exploiting either time-division or frequency-division. A full-duplex (FD) system simultaneously transmits and receives signals over the same frequency band [1], providing the potential of doubling the capacity compared to conventional half-duplex systems [2] and sustaining the evolution of the fifth generation (5G) technologies within limited spectrum. Beyond improvement of spectral efficiency, FD can also be beneficial to high layers, such as medium access (MAC) layer. From the perspective of MAC layer, enabling frame level FD transmissions, where a terminal is able to reliably receive an incoming frame while simultaneously transmitting an outgoing frame, can provide terminals with

M. Zhou • Y. Liao • L. Song (✉)
School of Electronics Engineering and Computer Science, Peking University,
Beijing 100871, China
e-mail: lingyang.song@gmail.com

W. Xiang et al. (eds.), *5G Mobile Communications*,
DOI 10.1007/978-3-319-34208-5_11

new capabilities. For example, in FD cognitive radio networks, secondary users can sense and access the vacant spectrum simultaneously, leading to significant sensing performance and data transmission efficiency enhancement [2, 3].

Despite the attractive features of FD, it is greatly affected by the signal leakage from the local output to input, referred to as self-interference (SI). Because it is locally generated, the power of the SI is much stronger than that of the desired signal. Apart from the direct SI, reflected interference signals from nearby obstacles also exist. Recent years, there has been a large number of SI cancelation techniques [2] making FD communications more and more feasible, which will be introduce in the next subsection.

Overall, FD is a promising technology that can be employed in the next-generation wireless networks. There are two reasons as follows: First, there is a widespread consensus that most conventional approaches to increasing spectral efficiency, such as advances in modulation, coding and MIMO techniques, have now been exhausted, and the FD system tends to be a non-conventional approach that can further increase the spectral efficiency. In addition, Indoor dense networks to be deployed in 5G, such as small-cell systems, are making the SI cancelation problem more manageable than previous long-coverage macro-cell networks. This is because in small-cell systems, the transmit power is smaller, and the path loss is reduced due to its short coverage.

1.1 Self-interference Cancelation Techniques

First, an example of small-cell cellular systems is provided to demonstrate the impact of SI [2]. The maximum allowable transmit powers of the base stations and mobile handsets are 24 and 21 dBm in small-cell systems, respectively, and the noise floors are both -100 dBm. If we assume 15 dB isolation between the base station's transmit and receive signal paths, then the SI of the base station is $24 - 15 - (-100) = 109$ dB above the noise floor. This means a FD base station must cancel self interference by about 109 dB to achieve the link SINR equivalent to that of a half duplex system.

Why is self-interference hard to subtract? After all, the FD transceiver knows the signal being transmitted, it can just subtract it from the received signal. However, this is not the whole picture. Although the transceiver knows the original version of the clean transmitted digital baseband signal, once it is converted to analog and up-converted to the carrier frequency and then transmitted through a fading channel, the received version is quite different from the original one. Actually, after all these processes, the received version is a complicated distorted function of the original clean signal along with noise.

Thus, sophisticated approaches to SI cancelation is needed to obtain the desired signal. In the following, these approaches are described, which can be typically classified into three types, namely propagation-domain, analog-circuit-domain, and digital-domain approaches [2].

1.1.1 Propagation-Domain Self-interference Suppression

Wireless-propagation-domain isolation techniques aim to electromagnetically isolate the transmit chain from the receive chain, i.e., to suppress the self-interference before it manifests in the receive chain circuitry, so that the downstream receiver hardware does not need to accurately process signals with a huge dynamic range.

In separate-antenna systems, the path loss between the FD terminals' transmit and receive antennas (or antenna arrays) can be increased by spacing them apart and/or by placing absorptive shielding between them, as quantified in [4–6]. Although the simplicity makes the pass-loss-based approaches attractive, their effectiveness is greatly limited by the size of device: the smaller the device, the less room there is to implement such techniques. Cross-polarization technique offers an additional mechanism to electromagnetically isolate the transmit and receive antennas. For example, an FD terminal can be sophistically designed so that it transmits only horizontally polarized signals and receives only vertically polarized signals, and the interference between them can be avoided [5, 7]. Similarly, with directional transmit and/or receive antennas (i.e., antennas with non-uniform radiation/sensing patterns), one may align their null directions to achieve the same goal [8]. In fact, by carefully placing a single receive antenna at precisely a location where the carrier waveforms are exactly 180° out of phase, one can near-perfectly null the received signal at the receive antenna, and the self-interference can be near-perfectly canceled [1, 7].

1.1.2 Analog-Circuit-Domain Self-interference Cancellation

Analog-circuit-domain cancellation focuses on cancelation of SI in the analog receive-chain circuitry before the ADC. The basic idea is to model and predict the distortions caused by SI and compensate the received signal accordingly in the analog domain. This cancelation may occur either before or after the downconverter and the (low-noise amplifier) LNA. In these techniques, a signal that resembles the SI at the receive-chain is generated, and utilized to cancel SI by signal subtraction. The generated signal can either be tapped at the transmit antenna, or obtained from the digital domain, and apply the necessary gain/phase/delay adjustments digitally (where it is much easier to do so), and then convert it to the analog-circuit domain for use in self-interference cancellation [9, 10].

For the cancelation schemes before the downconverter, i.e., at RF, the canceling signal also needs to be upconverted to RF. The analog cancelers can be classified based on whether the canceling signal is generated by processing the self-interference signal prior to or after upconversion. Those cancelers where the canceling signal is generated by processing prior to upconversion are called pre-mixer cancelers, whereas cancelers where the canceling signal is generated by processing afterwards are called post-mixer cancelers [11]. Figure 1a shows the structure of a pre-mixer canceler with processing function $f(\cdot)$, in which $x[n]$ is the

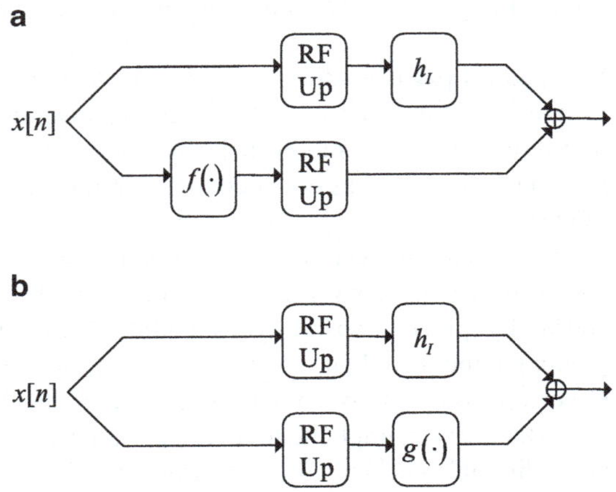

Fig. 1 Schematics of the pre-mixer (**a**) and post-mixer (**b**) analog cancelers

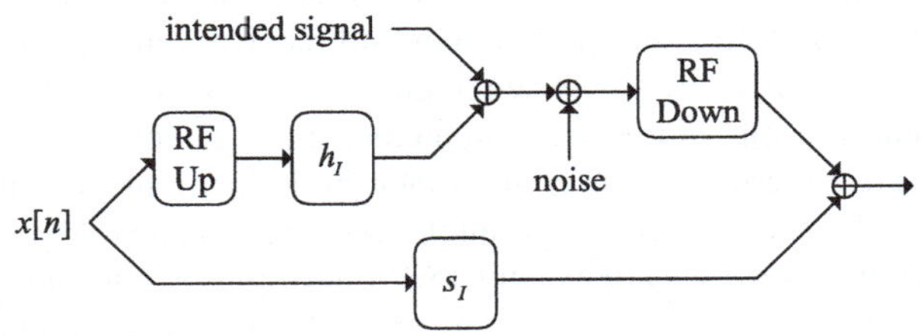

Fig. 2 Structure of a baseband analog canceler

original signal, and h_I is the self-interference channel from the transmit-chain to the receive-chain. Figure 1b is the schematic of a post-mixer canceler with processing function $g(\cdot)$. Functions $f(\cdot)$ and $g(\cdot)$ are ideal if they can completely eliminate self-interference from the received signal.

An analog canceler where the canceling signal is generated in baseband and the cancelation occurs in the analog baseband is called a baseband analog canceler. Figure 2 shows a representation of the baseband analog canceler. In baseband analog cancelers, transmitted signal $x[n]$ is processed by function $s(\cdot)$, and added directly to the received signal to perform the cancelation.

1.1.3 Digital-Domain Self-interference Cancellation

Digital-domain SI cancelation works in the digital domain after the received signal has been quantized by the ADC by applying sophisticated DSP techniques to the received signal [2]. The advantage of using digital-domain SI cancellation techniques is the reduction of circuit complexity and power consumption. However, the dynamic range of the ADC imposes strict limit on the maximum amount of SI suppression, which means that to implement digital-domain methods, a sufficient amount of the SI suppression must be done before the ADC using other techniques like the propagation-domain and/or analog-circuit-domain methods described above. In this sense, the digital-domain cancellation can be regarded as the last step of defense against self-interference, where the goal is to cancel the self-interference left over from the propagation-domain and analog-circuit-domain approaches.

1.2 Key Application Scenarios

FD technology, with its key feature of enabling simultaneous transmission and reception and its potential of doubling the spectral efficiency, has ignited great interest in both academia and industry. We introduce the following three key application scenarios: FD bidirectional communications, FD cooperative communications, FD cognitive radio networks.

Figure 3 depicts FD bidirectional communication between a pair of FD transceivers. The two FD nodes transmit and receive data to and from each other over the same frequency band simultaneously. The two nodes A and B are equipped with N_A and N_B antennas, respectively. Let $link_{AB}$ denote the transmission link from node A to B, and $link_{BA}$ to denote the link from B to A. Each node receives a combination of its desired signal from the other node and SI from its own transmit antennas. In this scenario, if the SI can be reduced completely, then the spectral efficiency will be doubled by using FD rather than half duplex (HD) communications.

The next application of FD to wireless communication came in the context of cooperative communications illustrated in Fig. 4, which generally consist of three types of nodes, a source node, a relay node, and a destination node. The source

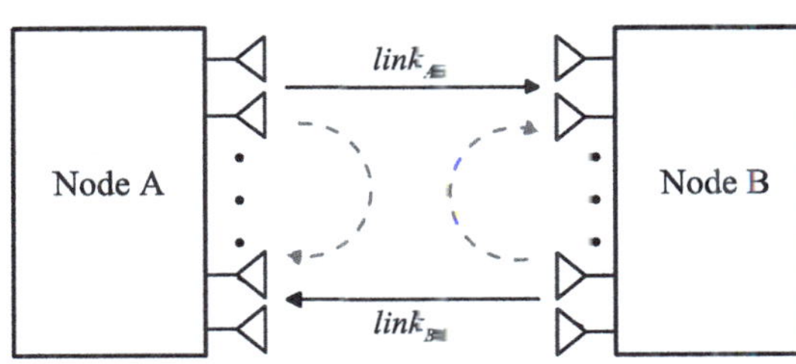

Fig. 3 Full-duplex bidirectional communications

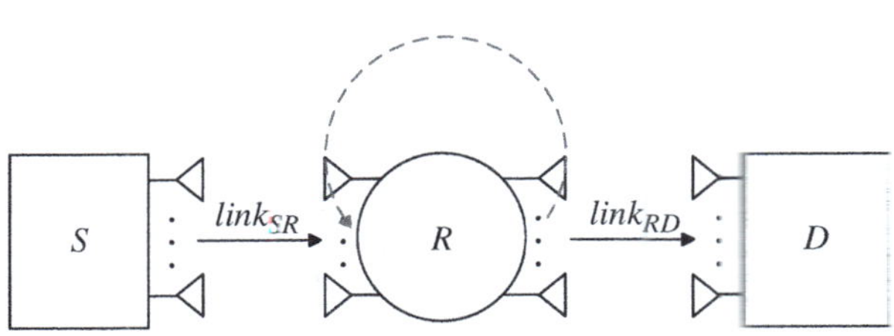

Fig. 4 Full-duplex cooperative communications

node wants to transmit its data signal to the destination node. The relay node is used for increase the coverage of the source node, i.e., it receives, amplifies and retransmits the wireless signal. Let us denote the transmission links from S to R as $link_{SR}$ and R to D as $link_{RD}$, respectively. When FD transmission is applied to the relay node, however, the source and relay nodes can simultaneously transmit their signals via $link_{SR}$ and $link_{RD}$ in the same frequency band. Similar to the FD bidirectional scenario, the FD cooperative communications can theoretically double the spectral efficiency.

Finally, the FD cognitive radio networks (CRNs) is considered. In CRNs, the secondary (unlicensed) users (SUs) are allowed to share the licensed spectrum with the primary (licensed) users (PUs) as long as SUs do not cause harmful interference at the PUs. Spectrum sensing is essential in identifying spectrum holes that can be utilized by SUs. Conventionally, SUs need to periodically suspend their ongoing transmissions to perform spectrum sensing, which leads to spectrum waste and tardy reaction to the PU's state change. When SUs have FD capability, they can perform simultaneous sensing and transmission, as shown in Fig. 5, and thus to bypass the intrinsic problems in conventional HD CRNs.

Apart from the aforementioned application scenarios, full-duplex wireless can be applied to more communication scenarios, such as cellular networks, ad-hoc networks, local area networks, satellite communications, vehicle communications, etc. Taking FD cellular network as an example, the base station can operate in FD, then it can support simultaneous uplink and downlink communications, potentially doubling the spectral efficiency.

The rest of the chapter is organised as follows: Section 2 introduces FD bidirectional communications, including the FD/HD mode switching and transmit-receive antenna pair selection schemes. In Sect. 3, the FD cooperative communications are described. The one-way joint relay and antenna mode selection and two-way

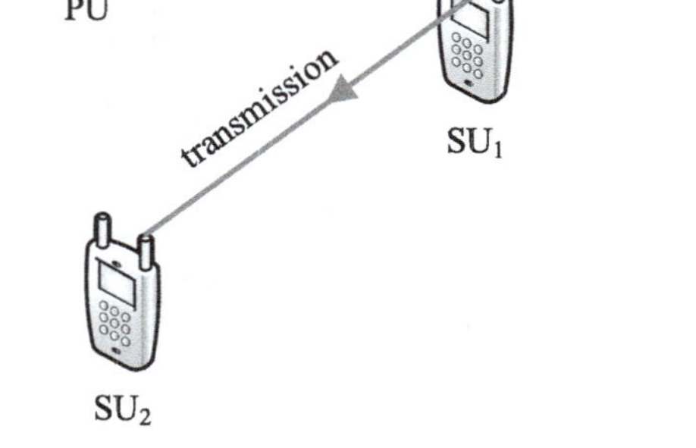

Fig. 5 Full-duplex cognitive radio network

relay selection schemes are briefly introduced in this section, and the performance analysis of the schemes is provided. Finally, the "listen-and-talk" protocol of FD CRNs is presented in the last section.

2 Full-Duplex Bidirectional Communications

Thanks to the various state-of-the-art approaches for self-IC schemes, SI is no longer a critical bottleneck to implementing a practical FD system. In this section, a few signal processing techniques of FD bidirectional communications are elaborated, and the analysis of their performance is given.

2.1 *Mode Switching Between Full-Duplex and Half-Duplex*

Typically, there are two kinds of baseline for FD and HD mode switching: fixed number of antennas, and fixed number of RFs. When the number of antennas is fixed, the system can be adaptively switched between FD and half-duplex spatial-multiplexing (HD-SM), where both modes have the same degree of freedom. On the other hand, when the number of transmit and receive RFs is fixed, the number of active RFs in FD can be twice of that in HD,[1] thus the degree of freedom of FD is potentially doubled. In this section, only the first kind is considered, and details of the second kind can be found in [12].

The system model is illustrated in Fig. 6, where each node is equipped with two antennas. In FD, one antenna is used for transmission and the other is used for reception at each node, while, in HD-SM, both antennas at one node are used for transmission and those at the other node are used for reception. In FD, the SI

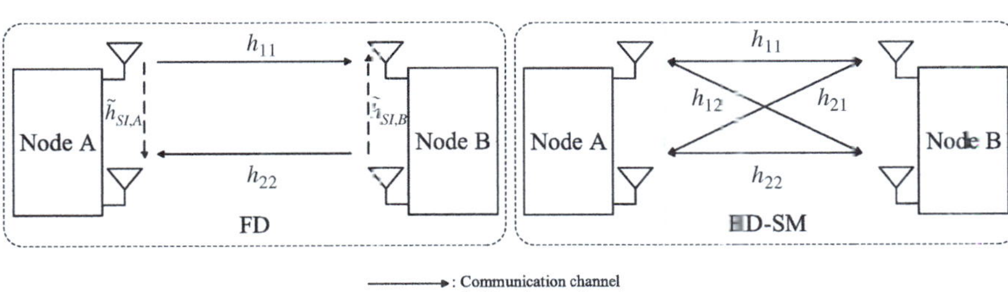

Fig. 6 Mode switching between FD and HD-SM

[1]In FD, all of the RFs are active, while only transmit RFs at the transmitter and receive RFs at the receiver in HD are active.

can be partially cancelled by the aforementioned SI cancelation techniques, and the residual interference can generally be modeled to follow Rayleigh distribution. On the other hand, the HD-SM system can use all of the communication links and avoid the self-interference problem, however, it may suffer from the sum-rate loss caused by the spatial correlation. As depicted in Fig. 6, there are overall 6 links, containing 4 communication links and 2 self-interference links. The communication links between the two nodes can be denoted as $\mathbf{H} = \left[h_{ij}\right]$, $i, j = 1, 2$, the entry of which follows independent identically distributed (i.i.d.) Rayleigh distribution with zero mean and unit variance, while the residual self-interference channel is $\tilde{h}_{SIk}$, $k = A, B$, with zero mean and unit variance.

In FD, the two nodes employ 2 communication links to exchange information. After self interference cancellation, both nodes receive a combination of the signal transmitted by the other node, the residual self interference, and noise

$$\begin{aligned} y_A &= \sqrt{\rho} h_{11} x_B + \sqrt{\eta} \tilde{h}_{SI,A} x_A + n_A \\ y_B &= \sqrt{\rho} h_{22} x_A + \sqrt{\eta} \tilde{h}_{SI,B} x_B + n_B, \end{aligned} \tag{1}$$

where ρ and η denote the average SNR and INR, respectively. x_A, x_B denote the signal transmitted from nodes A and B, and n_A, n_B denote the AWGN with zero mean and unit variance at the corresponding nodes.

At the same time, when the HD-SM system is employed, assuming time division duplex in this paper, each node processes one half of the time resources for transmission through a 2×2 channel $\mathbf{H}$. We first consider received signal at node B, which is given by

$$\mathbf{y}_{B,SM} = \sqrt{\rho} \mathbf{H} \mathbf{x}_A + \mathbf{n}_B, \tag{2}$$

where $\mathbf{n}_B$ is an AWGN vector with unit variance. $\mathbf{x}_A$ is the signal vector transmitted from node A.

The HD-SM system can avoid the self-interference problem, however, it suffers from the sum-rate loss caused by the spatial correlation. Under the separable correlation model [13], it is possible to express the channel matrix as $\mathbf{H} = \mathbf{\Phi}_R^{1/2} \mathbf{H}_0 \mathbf{\Phi}_T^{1/2}$, where $\mathbf{H}_0$ is an i.i.d. zero mean complex Gaussian random matrix with unit variance, and $\mathbf{\Phi}_R$ and $\mathbf{\Phi}_T$ represent the normalized correlation matrixes at receiver and transmitter, respectively. Assuming a block-fading reciprocal channel, where channel varies independently at different time slots, and the forward channel is reciprocal with the reverse channel. Because the two nodes process one half of time resource, transmit/receive correlation matrices at a node are receive/transmit correlation matrices at the other node. Thus, the received signal at node A is symmetric, and the average achievable rates for the two nodes are equal. For simplicity, in the following sections, only node B is considered as the receiver.

2.1.1 Average Sum Rates

Due to symmetry, the average rates of the two directions are equal, thus the average sum rate is a doubled rate of one direction

$$\bar{R}_{FD} = 2\mathrm{E}\left[\log_2\left(1 + \frac{\rho|h_{22}|^2}{\eta\left|\tilde{h}_{SI,A}\right|^2 + 1}\right)\right], \tag{3}$$

After some manipulations, the closed form expression can be expressed as [14]

$$\bar{R}_{FD}\frac{2\exp\left(\frac{1}{\rho}\right)\log_2 e}{\left(\frac{\eta}{\rho} - 1\right)}\left[\exp\left(\frac{1}{\eta} - \frac{1}{\rho}\right)\mathrm{E}_1\left(\frac{1}{\eta}\right) - \mathrm{E}_1\left(\frac{1}{\rho}\right)\right]. \tag{4}$$

where $\mathrm{E}_1(\cdot)$ is the exponential integral function of the first order.

Next, the performance of HD-SM is given. The case of HD-SM with power constraint ρ and equal power allocation at the transmitter is considered. The eigenvalues of the transmit and receive correlation matrixes, $\boldsymbol{\Phi}_R$ and $\boldsymbol{\Phi}_T$ are represented as $\phi_{r,1} > \phi_{r,2}$ and $\phi_{t,1} > \phi_{t,2}$, respectively. Therefore, the sum rate of the correlated HD-SM system can be represented as [13]

$$\begin{aligned}\bar{R}_{SM} &= \mathrm{E}\left\{\log_2\det\left(\mathbf{I} + \frac{\rho}{2}\mathbf{H}_0^H\boldsymbol{\Phi}_R^H\mathbf{H}_0\boldsymbol{\Phi}_T\right)\right\} \\ &= \frac{2\sum\limits_{l=1}^{2}\det\left(\Xi\left(l\right)\right)}{\ln(2)\rho(\phi_{t,2} - \phi_{t,1})(\phi_{r,2} - \phi_{r,1})},\end{aligned} \tag{5}$$

where $\Xi(l)$, $l = 1, 2$, are 2×2 matrixes with entries given by

$$\{\Xi(l)\}_{i,j} = \begin{cases}\frac{\rho}{2}\phi_{t,i}\phi_{r,j}\exp\left(\frac{2}{\rho\phi_{t,i}\phi_{r,j}}\right)\mathrm{E}_1\left(\frac{2}{\rho\phi_{t,i}\phi_{r,j}}\right), & i = l\\ \frac{\rho}{2}\phi_{t,i}\phi_{r,j} + 1, & i \neq l\end{cases} \tag{6}$$

As a result, combining the exact sum rates of the FD mode (4) and the HD-SM mode (5), we can define

$$\begin{aligned}\Delta R^{EX} &\triangleq \bar{R}_{FD} - \bar{R}_{SM}. \\ &= \frac{2\exp\left(\frac{1}{\rho}\right)\log_2 e}{\left(\frac{\eta}{\rho} - 1\right)}\left[\exp\left(\frac{1}{\eta} - \frac{1}{\rho}\right)\mathrm{E}_1\left(\frac{1}{\eta}\right) - \mathrm{E}_1\left(\frac{1}{\rho}\right)\right] \\ &\quad - \frac{2\sum\limits_{l=1}^{2}\det\left(\Xi\left(l\right)\right)}{\ln(2)\rho(\phi_{t,2} - \phi_{t,1})(\phi_{r,2} - \phi_{r,1})}.\end{aligned} \tag{7}$$

The system can adaptively select between FD and HD-SM modes depending on ΔR^{EX}. That is, the system selects HD-SM, if $\Delta R^{EX} < 0$; otherwise, it selects FD. The criterion is determined by three statistical factors: the average INR η in FD, the spatial correlation $\mathbf{\Phi}_T$, $\mathbf{\Phi}_R$ in HD-SM, and the average SNR ρ. However, it is rather complicated to determine the exact criterion through these three factors. In the following, the approximate results are provided to facilitate the mode switching.

2.1.2 Approximate Results

First the approximate average sum rate of the FD system is calculated at high SNR. Using the approximation $\mathrm{E}_1(\varepsilon) \approx -\gamma - \ln(\varepsilon)$ and $\exp(\varepsilon) \approx 1 + \varepsilon$ for $\varepsilon = \frac{1}{\rho} \to 0$ [15], the approximate sum rate of the FD system in (4) can be rewritten as

$$\begin{aligned} \bar{R}_{FD}^{AP} =& 2\frac{\rho+1}{\rho-\eta}\log_2\rho - 2\gamma\frac{\rho+1}{\rho-\eta}\log_2 e \\ &- \frac{2\left(\rho^2-1\right)\log_2 e}{(\rho-\eta)\,\rho}\exp\left(\frac{1}{\eta}\right)\mathrm{E}_1\left(\frac{1}{\eta}\right), \end{aligned} \tag{8}$$

where $\gamma \approx 0.5772$ is Euler's constant [15]. Furthermore, at low or moderate INR, i.e., $\rho \gg \eta$, (8) can be well approximated by

$$\bar{R}_{FD}^{AP} = [2\log_2\rho - 2\gamma\log_2 e] - \exp\left(\frac{1}{\eta}\right)\mathrm{E}_1\left(\frac{1}{\eta}\right)\log_2 e. \tag{9}$$

The first term is the high-SNR sum rate of the FD system without residual self interference, and the second term denotes the rate degradation caused by the residual self interference.

On the other hand, at high SNR, the average sum rate of the HD-SM system can be well approximated by Paulraj et al. [16]

$$\begin{aligned} \bar{R}_{SM}^{AP} &= \mathrm{E}\left\{\log_2\det\left(\frac{\rho}{2}\mathbf{H}_0\mathbf{H}_0^H\right)\right\} + \log_2\det\left(\mathbf{\Phi}_T\right) \\ &\qquad + \log_2\det\left(\mathbf{\Phi}_R\right) \\ &\approx \left[2\log_2\left(\frac{\rho}{2}\right) + (1-2\gamma)\log_2 e\right] + \log_2\det\left(\mathbf{\Phi}_T\right) \\ &\qquad + \log_2\det\left(\mathbf{\Phi}_R\right). \end{aligned} \tag{10}$$

The first term is the high-SNR sum rate for independent fading HD-SM. The second and third terms represent the rate degradation due to the spatial correlations at transmit and receive sides, respectively. It is worth noting that $\det\left(\mathbf{\Phi}_T\right)$ and $\det\left(\mathbf{\Phi}_R\right)$ are less than one, and a larger correlation $\mathbf{\Phi}$ implies a smaller $\det(\mathbf{\Phi})$.

Similarly, an approximate selection criterion can be derived by calculating the difference between the approximate average sum rates in (10) and (23)

$$\begin{aligned}\Delta R^{AP} &\triangleq \bar{R}_{FD}^{AP} - \bar{R}_{SM}^{AP} \\ &= [2 - \log_2 e] - \log_2\left[\det\left(\boldsymbol{\Phi}_T\right)\det\left(\boldsymbol{\Phi}_R\right)\right] \\ &\quad - 2\exp\left(\frac{1}{\eta}\right)\mathrm{E}_1\left(\frac{1}{\eta}\right)\log_2 e,\end{aligned} \tag{11}$$

Given the high-SNR assumption, ΔR^{AP} is just determined by two factors: monotonically increasing with spatial correlation in the HD-SM system, and monotonically decreasing with the average INR η, but independent of SNR ρ.

Let $\Delta R^{AP} = 0$, and the approximate selection threshold can be calculated. We consider the exponential correlation model [13] with equal correlation coefficients at both sides

$$\boldsymbol{\Phi}_T = \boldsymbol{\Phi}_R = \begin{bmatrix} 1 & \beta^* \\ \beta & 1 \end{bmatrix}, \tag{12}$$

Then, the selection threshold of spatial correlation βth can be given by

$$|\beta \mathrm{th}| = \begin{cases} \sqrt{1 - \frac{2}{\exp\left[0.5 + \exp\left(\frac{1}{\eta}\right)\mathrm{E}_1\left(\frac{1}{\eta}\right)\right]}}, & \eta > \eta_0 \\ 0, & \eta \le \eta_0 \end{cases}. \tag{13}$$

Given $\eta \ge \eta_0$, the system selects HD-SM, if $|\beta| > |\beta \mathrm{th}|$; otherwise, it selects FD. If $\eta \le \eta_0$, ΔR_{AP} is always positive, i.e., FD is always superior to HD-SM even if the MIMO channel is independent. η_0 is the value that satisfies

$$\begin{aligned} 1 - \frac{2}{\exp\left[0.5 + \exp\left(\frac{1}{\eta_0}\right)\mathrm{E}_1\left(\frac{1}{\eta_0}\right)\right]} &= 0 \\ \exp\left(\frac{1}{\eta_0}\right)\mathrm{E}_1\left(\frac{1}{\eta_0}\right) &= \ln 2 - 0.5 \end{aligned} \tag{14}$$

By using $\frac{1}{2}\ln\left(1 + \frac{2}{x}\right) < \exp(x)\mathrm{E}_1(x) < \ln\left(1 + \frac{1}{x}\right)$ [15] for $x = \frac{1}{\eta_0}$, we can obtain

$$\frac{2}{\sqrt{e}} - 1 < \eta_0 < \frac{2}{e} - \frac{1}{2}. \tag{15}$$

Or equivalently, we have $-6.71\ \mathrm{dB} < \eta_0 < -6.28\ \mathrm{dB}$.

The derivative of $|\beta\text{th}|$ can be written as

$$\frac{d|\beta\text{th}|}{d\eta} = \frac{d|\beta\text{th}|}{dt}\frac{dt}{d\eta}, \tag{16}$$

where $t = \exp\left(\frac{1}{\eta}\right)\mathrm{E}_1\left(\frac{1}{\eta}\right)$. It is obvious that $\frac{d|\beta\text{th}|}{dt} > 0$, and $\frac{dt}{d\eta}$ can be given by

$$\begin{aligned}\frac{dt}{d\eta} &= \frac{1}{\eta} - \frac{1}{\eta^2}\exp\left(\frac{1}{\eta}\right)\mathrm{E}_1\left(\frac{1}{\eta}\right) \\ &> \frac{1}{\eta} - \frac{1}{\eta}\frac{\eta+1}{2\eta+1} \\ &= \frac{1}{2\eta+1} > 0\end{aligned} \tag{17}$$

where the first inequality is obtained by using $x\exp(x)\mathrm{E}_1(x) < \frac{x+1}{x+2}$ [15] for $x = \frac{1}{\eta}$. Thus, we can obtain $\frac{d|\beta\text{th}|}{d\eta} > 0$, which implies that the spatial correlation threshold in (13) monotonically increases with self interference.

Figure 7 depicts the exact and approximate switching thresholds between FD and HD-SM, where we assume $\beta_r = \beta_t = \beta$ without loss of generality. The exact threshold is obtained by comparing the exact sum rates of the FD and HD-SM systems in (4) and (5). The approximate threshold is given in (13). Above the threshold is the area where the FD system outperforms the HD-SM system, whereas the HD-SM system achieves a better performance in the region below the curve.

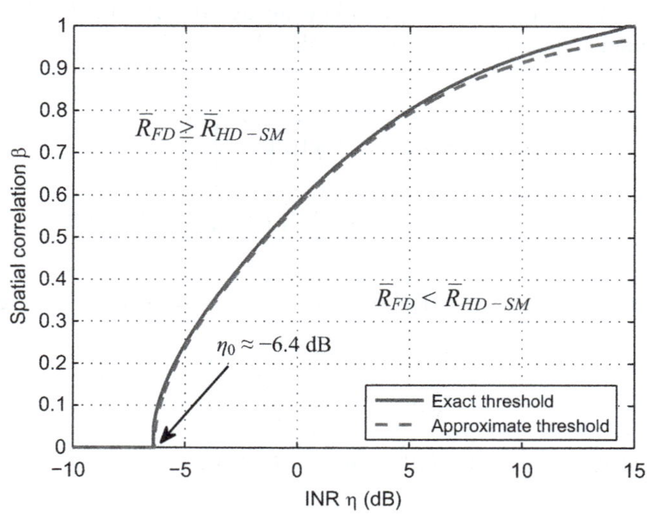

Fig. 7 Exact and approximate switching thresholds, with SNR = 30 dB

Moreover, through simulation, we can obtain $\eta_0 \approx -6.4$ dB, which accords with the aforementioned analysis $-6.71\,\text{dB} < \eta_0 < -6.28\,\text{dB}$. Through this figure, it can be shown that at high SNR, the approximate threshold is effective to determine whether the FD system or the HD-SM system should be employed in a specific condition.

2.2 Transmit-Receive Antenna Pair Selection

In this subsection, a transmit-receive antenna pair selection (TRAPS) strategy [17, 18] is elaborated, where the functions of the antennas are not predefined, and the system can adaptively configure the antennas for transmission and reception based on the given criteria. In the following, the maximum sum rate and minimum symbol error rate (SER) criteria are introduced, and their performance is analyzed.

As shown in Fig. 8, a bidirectional communication scenario with two nodes, N_1 and N_2, is considered , where each node is equipped with two antennas, which can be dynamically configured for transmission and reception. There exist four kinds of possible combinations (patterns), as depicted in Fig. 8. The channels between the two nodes exhibit independent Rayleigh fading. Let $h_{i\bar{i}}^{(j)}$ denote the channel of pattern (j) from node $N_{\bar{i}}$ to N_i, $i = 1, 2$, $j = 1, 2, 3, 4$, $\bar{i} = 3 - i$, in Fig. 8.

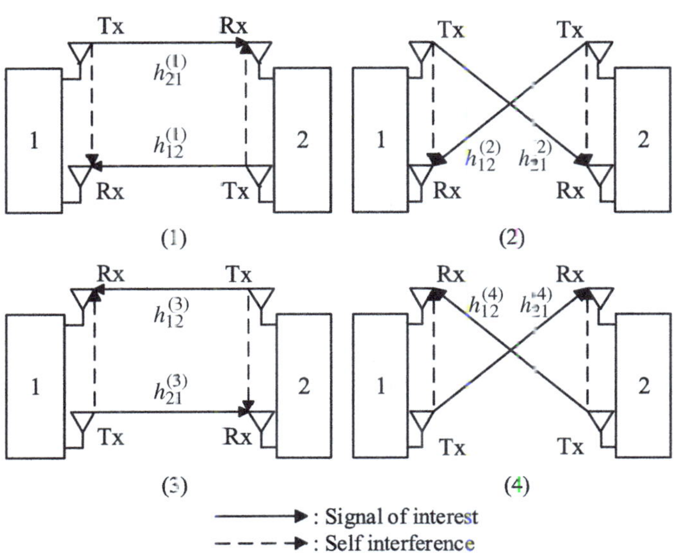

Fig. 8 System model of TRAPS scheme

We assume that they are all complex-Gaussian random variables with zero mean and unit variance, i.e., $h_{i\bar{i}}^{(j)} \sim \mathcal{CN}(0,1)$. The transmit powers of the two nodes are equal, denoted by P_s.

The received signal at node N_i can be expressed as

$$y_i^{(j)} = \sqrt{P_s} h_{i\bar{i}}^{(j)} x_{\bar{i}}^{(j)} + \sqrt{\eta P_s} w_i^{(j)} + n_i^{(j)}, \tag{18}$$

where $x_{\bar{i}}^{(j)}$ denotes the transmitted signal from node $\bar{i}$ in pattern j. The second term represents the residual interference after cancelation. w_i follows $\mathcal{CN}(0,1)$, and η denotes the interference cancelation factor, indicating the cancelation capability. A small η indicates a good performance of interference cancelation. n_i denotes the additive white Gaussian noise (AWGN) at node i, with zero mean and unit variance. Therefore, the instantaneous received signal-to-interference-plus-noise ratio (SINR) of node N_i in pattern j can be expressed as

$$\rho_i^{(j)} = \frac{P_s}{\eta P_s + 1} |h_{i\bar{i}}^{(j)}|^2. \tag{19}$$

The probability density function (PDF) of $\rho_i^{(j)}$ is given by

$$f_{\rho_i^{(j)}}(x) = \frac{\eta P_s + 1}{P_s} e^{-\frac{\eta P_s + 1}{P_s} x}. \tag{20}$$

Furthermore, we have $|h_{i\bar{i}}^{(1)}|^2 = |h_{\bar{i}i}^{(3)}|^2$, $|h_{i\bar{i}}^{(2)}|^2 = |h_{\bar{i}i}^{(4)}|^2$. Thus, pattern (1), (3) and pattern (2), (4) have the same instantaneous SINR at each node, respectively. For simplicity, only patterns (1) and (2) are considered for the rest of this subsection.

2.2.1 Maximum Sum Rate Criterion

In the Max-SR approach, where the pattern is selected with a higher sum rate of the two nodes

$$\mathcal{P} = \arg\max_j \left(\sum_{i=1}^{2} R_i^{(j)} \right), \tag{21}$$

where $R_i^{(j)} = \log_2 \left(1 + \frac{P_s}{\eta P_s + 1} |h_{i\bar{i}}^{(j)}|^2 \right)$ represents the rate of node N_i in pattern (j). Thus, the sum rate of the Max-SR approach can be given by

$$\bar{R}_{max} = \mathbb{E}\left[\max_j \left(\sum_{i=1}^{2} R_i^{(j)} \right) \right], \tag{22}$$

By using $\max(A+B, C+D) \le \max(A, C) + \max(B, D)$, for $A, B, C, D \ge 0$,[2] An analytical upper bound is derived as

$$\bar{R}_{ub} = \mathbb{E}\left[\sum_{i=1}^{2} \max_{j}\left(R_i^{(j)}\right)\right]. \tag{23}$$

After some manipulations, (23) can be rewritten as

$$\bar{R}_{ub} = 2\left[2e^{\frac{\eta P_s+1}{P_s}}\mathrm{E}_1\left(\frac{\eta P_s+1}{P_s}\right) - e^{\frac{2\eta P_s+2}{P_s}}\mathrm{E}_1\left(\frac{2\eta P_s+2}{P_s}\right)\right]\log_2 e. \tag{24}$$

In the high-SNR regime, the average sum rate is obtained as

$$\lim_{P_s\to\infty} \bar{R}_{ub} = 2e^{\eta}\mathrm{E}_1(\eta)\log_2 e + 2\left[e^{\eta}\mathrm{E}_1(\eta) - e^{2\eta}\mathrm{E}_1(2\eta)\right]\log_2 e. \tag{25}$$

The first term of (25) is the sum rate of traditional FD in the high-SNR regime, while the second term, which is easily proved to be positive, represents the improvement of the sum rate ceiling by using Max-SR approach.

We define $\Delta R = e^{\eta}\mathrm{E}_1(\eta) - e^{2\eta}\mathrm{E}_1(2\eta)$, and we have $d\Delta R/d\eta < 0$. It indicates that when the residual self interference gets smaller, the improvement of the sum rate increases. Furthermore, when the self interference is canceled properly, i.e., $\eta \to 0$, an upper bound of ΔR can be obtained, by using $\mathrm{E}_1(\eta) \approx -\gamma - \ln\eta$, $e^{\eta} \approx 1-\eta$

$$\Delta R \le \Delta R_{\eta\to 0} \approx (1+\eta)\ln 2 + \eta(\gamma + \ln 2\eta) \approx \ln 2. \tag{26}$$

Substituting (26) into (25), it can be shown that the improvement of the sum rate increases with the improvement of the interference cancelation ability, i.e., the decrease of η, and it is upper bounded by 2.

Figure 9 plots the sum rate versus the transmit power P_s for the FD with or without TRAPS. The exact sum rate curves are given by Monte Carlo simulations with the interference cancelation factor $\eta = \{-3, -10, -20\ \mathrm{dB}\}$. The upper bound of the Max-SR is depicted by (24). Figure 9 shows that the Max-SR achieves a sum rate gain, and converges to a higher sum rate ceiling than that of the traditional FD. In addition, with the improvement of interference cancelation, i.e., the decrease of η, the sum rate gain increases correspondingly.

[2]This inequality can be proved by classified discussion on the different relationships of A, B, C and D.

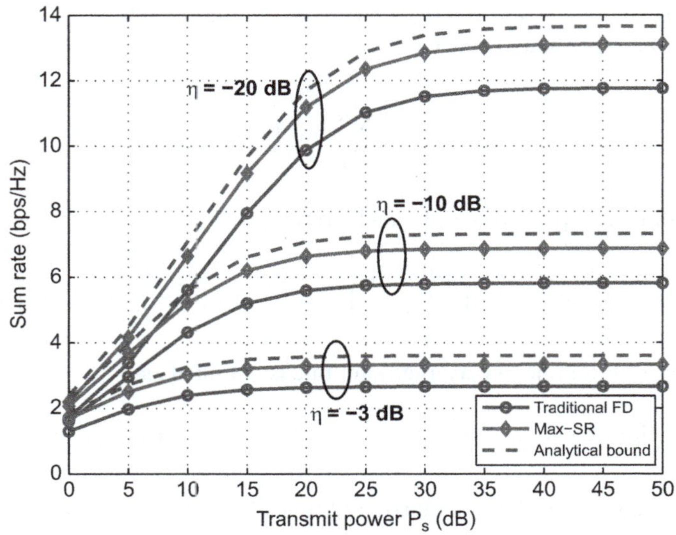

Fig. 9 Analytical and simulated sum rate versus transmit power P_s of Max-SR and traditional FD, with $\eta = \{-3, -10, -20\,\text{dB}\}$

2.2.2 Minimum SER Antenna Criterion

In the Min-SER criterion, the pattern is selected with a lower sum SER of the two nodes

$$\mathcal{P} = \arg\min_{j} \left(\frac{1}{2} \sum_{i=1}^{2} \text{SER}_i^{(j)} \right), \tag{27}$$

where $\text{SER}_i^{(j)}$ denotes the SER at N_i from the other node in pattern (j). The average SER of the Min-SER approach is conditioned on the instantaneous SINR and can be written as

$$\overline{\text{SER}}_{min} = \frac{1}{2} \text{E} \left[\min_{j} \left(\sum_{i=1}^{2} Q\left(\sqrt{b\rho_i^{(j)}} \right) \right) \right], \tag{28}$$

where $Q\{\cdot\}$ is the Gaussian-Q function, and b is a constant determined by the modulation format, e.g., $b = 2$ for binary phase-shift keying (BPSK).

In most cases, the sum SER of the two nodes is dominated by the larger one, and thus, a simplified selection criterion can be considered for which the larger instantaneous SER is minimized so as to minimize the average sum SER in a suboptimal way, referred to as Min-Max selection criterion [19]. The selected pattern is denoted by $\mathcal{P}$, and we have

$$\mathcal{P} = \arg\min_{j} \left(\frac{1}{2} \max_{i} \left(\mathrm{SER}_i^{(j)} \right) \right). \tag{29}$$

Since $Q(\cdot)$ is a monotonic decreasing function, the Min-Max selection criterion in (29) can be further written as

$$\mathcal{P} = \arg\max_{j} \left(\frac{1}{2} \min_{i} \left(\rho_i^{(j)} \right) \right). \tag{30}$$

As the performance of the Min-Max criterion is almost the same as the optimal criterion, which will be verified through the simulations, then the average SER based on the Min-Max criterion is analyzed. Let SER_i^{mm} denote the instantaneous SER at node N_i, based on the Min-Max criterion. The average SER based on the Min-Max criterion can be written as

$$\overline{\mathrm{SER}}_{mm} = \frac{1}{2} \mathbb{E} \left[\mathrm{SER}_1^{mm} + \mathrm{SER}_2^{mm} \right]. \tag{31}$$

Furthermore, as earlier discussed, the sum SER (31) can be lower bounded by using the larger SER

$$\begin{aligned} \overline{\mathrm{SER}}_{lb} &= \frac{1}{2} \mathbb{E} \left[\max \left(\mathrm{SER}_1^{mm}, \mathrm{SER}_2^{mm} \right) \right] \\ &= \frac{1}{2} \mathbb{E} \left[Q \left(\sqrt{b z_{\max}} \right) \right], \end{aligned} \tag{32}$$

where we define $z_{\max} = \max_{j} \left(\min_{i} \left(\rho_i^{(j)} \right) \right)$, and based on (20), the pdf of $z_{\max}$ is given by the order statistics [20]

$$f_{z_{\max}}(z) = \frac{4\eta P_s + 4}{P_s} \left(e^{-\frac{2\eta P_s + 2}{P_s} z} - e^{-\frac{4\eta P_s + 4}{P_s} z} \right). \tag{33}$$

Its corresponding cdf can be written as

$$F_{z_{\max}}(z) = 1 - 2e^{-\frac{2\eta P_s + 2}{P_s} z} + e^{-\frac{4\eta P_s + 4}{P_s} z}. \tag{34}$$

By substituting (34) into (32), the lower bound of the SER can be given by

$$\overline{\mathrm{SER}}_{lb} = \frac{1}{2} + \frac{1}{2} \sqrt{\frac{bP_s}{(b + 8\eta)P_s + 8}} - \sqrt{\frac{bP_s}{(b + 4\eta)P_s + 4}}. \tag{35}$$

At high SNR, (35) can be rewritten as

$$\lim_{P_s \to \infty} \overline{SER_{lb}} = \frac{1}{2} + \frac{1}{2}\sqrt{\frac{b}{b+8\eta}} - \sqrt{\frac{b}{b+4\eta}}. \quad (36)$$

From (36), it indicates that the SER of the Min-Max criterion also converges to an error floor.

The average SER of traditional FD without TRAPS is also provided for comparison

$$\overline{\mathrm{SER}_t} = \frac{1}{2} - \frac{1}{2}\sqrt{\frac{bP_s}{(b+2\eta)P_s + 2}}. \quad (37)$$

At high SNR, the SER of the traditional FD without TRAPS can be rewritten as

$$\lim_{P_s \to \infty} \overline{\mathrm{SER}_t} = \frac{1}{2} - \frac{1}{2}\sqrt{\frac{b}{b+2\eta}}. \quad (38)$$

By using Taylor expansion, the SER lower bound of the Min-Max criterion (36) and the average SER of traditional FD (38) can be rewritten as

$$\begin{aligned} \overline{\mathrm{SER}_t} &\approx \tfrac{\eta}{b} - \tfrac{3}{2}\left(\tfrac{\eta}{b}\right)^2 + \cdots, \\ \overline{\mathrm{SER}_{lb}} &\approx 6\left(\tfrac{\eta}{b}\right)^2 + \cdots. \end{aligned} \quad (39)$$

It clearly indicates in (39) that the error floor of TRAPS based on the Min-Max criterion is limited by η^2, while that of traditional FD is limited by η. Assuming $\eta \to 0$, we have $\overline{\mathrm{SER}_{lb}} \ll \overline{\mathrm{SER}_t}$, and the improvement of the average SER significantly increases as η gets smaller.

Figure 10 compares the Min-SER approach with the traditional FD. The simulations are performed for a BPSK modulation, with $\eta = \{-3, -10, -20\}$ dB. The simulated SER based on the Min-Max criterion is also included, showing that the Min-Max criterion achieves almost the same performance as the optimal criterion. The analytical approximation using only the larger node's SER (35) is included, which is tight at medium to high SINR. Another important observation is that the SER curves converge to different error floors. The Min-SER approach has much lower error floor than the traditional FD, and the average SER gain gets larger with the decrease of the residual self interference.

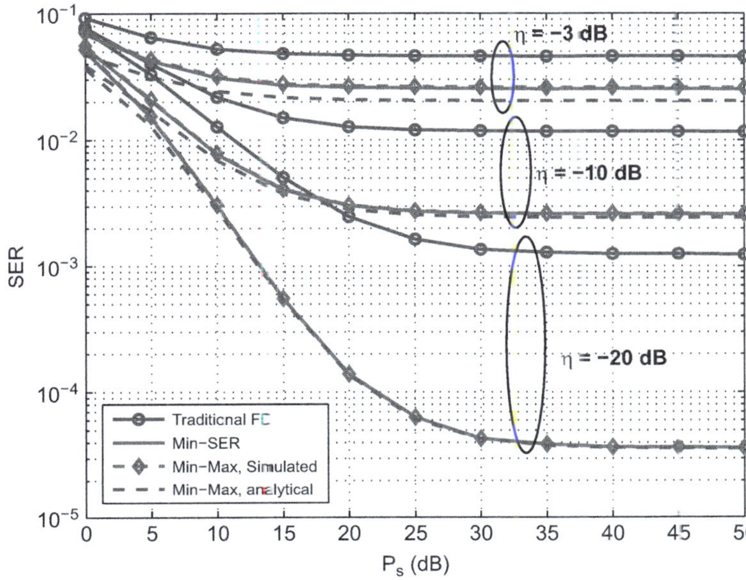

Fig. 10 Analytical and simulated SER versus transmit power P_s of Min-SER and traditional FD, with $\eta = \{-3, -10, -20\,\text{dB}\}$

3 Full-Duplex Cooperative Communications

Cooperative communications, or relaying, is a promising technology that can improve the throughput and coverage of the access points significantly. The traditional relay operates in the HD mode due to its implementation simplicity, leading to inefficient use of the limited spectral efficiency. To integrate the benefits of FD and relaying, it is natural to make relays operate in the FD mode [21–24]. In this section, two schemes in one-way and two-way FD relay communications will be introduced.

3.1 One-Way Joint Relay and Tx/Rx Antenna Mode Selection

In this subsection, the joint relay and Tx/Rx antenna mode selection scheme (JRAMS) [23] is described, where there are one source (S), one destination (D), and N AF relay nodes, as shown in Fig. 11. We assume that there is no direct channel link between the source and the destination due to the strong path loss and attenuation in this link. Both source and relay nodes use the same time-frequency resource, and

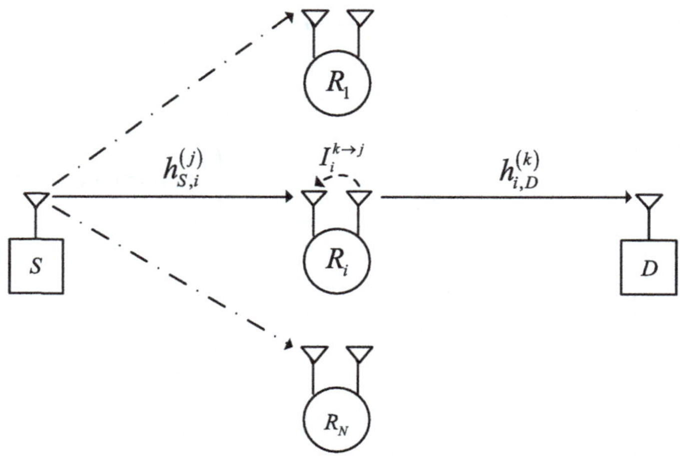

Fig. 11 System model of the multiple AF relay system, $j, k \in \{1, 2\}$ and $j \neq k$

the relay nodes work in the full-duplex mode with two antennas (one for receiving and one for transmitting), denoted by T_1 and T_2. In the JRAMS scheme, only the optimal relay with the optimal configuration of FD Tx and Rx antenna mode is selected to forward the signals from the source to the destination. Note that for each relay, the Tx/Rx mode of the two antennas is adaptively determined based on the instantaneous channel conditions between the source/destination and the two antennas of the FD relay nodes.

Assuming that the ith relay node, denoted by R_i, is selected to forward the signal, and the relay node R_i configures its antenna T_j/T_k to receive/transmit the signal, $i \in \{1, 2, \ldots, N\}, j, k \in \{1, 2\}$ and $j \neq k$.

The signal received by antenna T_j of the relay R_i is given as

$$y_i^{(j)}[t] = h_{S,i}^{(j)}\sqrt{P_S}x[t] + I_i^{k\to j}\sqrt{P_R}x_i[t] + n_i^{(j)}[t], \tag{40}$$

where $h_{S,i}^{(j)}$ is the channel link between the source and the antenna T_j of the relay R_i, and $I_i^{k\to j}$ is the residual loop interference from the transmit antenna T_k to the receive antenna T_j of the relay R_i. $x[t]$ and $x_i[t]$ are the tth signal with the unit power, transmitted from the source and the relay R_i, respectively. P_S and P_R are the transmit power of the source and relay node. $n_i^{(j)}$ is the additive white Gaussian noise at the receive antenna T_j with the power σ^2.

Upon receiving the signal from the source, the relay uses the AF protocol to forward the following signal.

$$x_i[t] = \beta_i^{k\to j}y_i^{(j)}[t], \tag{41}$$

where $\beta_i^{k\to j}$ is the power amplification factor to ensure that the average power of signal $x_{R_i}[t]$ satisfies the following power constraint,

$$\mathbb{E}\left[|x_i[t]|^2\right] = \left(\beta_i^{k\to j}\right)^2\left(|h_{S,i}^{(j)}|^2 P_S + |I_i^{k\to j}|^2 P_R + \sigma^2\right) \leq 1. \tag{42}$$

From Eq. (42), we have

$$\beta_i^{k\to j} = \sqrt{\frac{1}{|h_{S,i}^{(j)}|^2 P_S + |I_i^{k\to j}|^2 P_R + \sigma^2}}. \tag{43}$$

The received signal at the destination is given by

$$y_D[t] = h_{i,D}^{(k)}\sqrt{P_R}x_i[t] + n_D[t], \tag{44}$$

where $h_{i,D}^{(k)}$ is the channel link from the transmit antenna T_k of the relay R_i to destination, and n_D is the additive white Gaussian noise with mean zero and variance σ^2 at the destination.

The end-to-end SINR can be obtained as

$$\gamma_i^{k\to j} = \frac{P_S P_R\left|h_{S,i}^{(j)}h_{i,D}^{(k)}\beta_i^{k\to j}\right|^2}{P_R^2\left|h_{i,D}^{(k)}I_i^{k\to j}\beta_i^{k\to j}\right|^2 + P_R\left|h_{i,D}^{(k)}\beta_i^{k\to j}\right|^2\sigma^2 + \sigma^2}. \tag{45}$$

Substituting (43) into (45), the end-to-end SINR via the relay R_i can be calculated as

$$\gamma_i^{k\to j} = \frac{P_S P_R \gamma_{S,i}^{(j)}\gamma_{i,D}^{(k)}}{P_S\gamma_{S,i}^{(j)} + (P_R\gamma_{i,D}^{(k)} + 1)(P_R\gamma_{LI,i}^{k\to j} + 1)}, \tag{46}$$

where $\gamma_{S,i}^{(j)} = |h_{S,i}^{(j)}|^2/\sigma^2$, $\gamma_{i,D}^{(k)} = |h_{i,D}^{(k)}|^2/\sigma^2$, and $\gamma_{LI,i}^{k\to j} = |I_i^{k\to j}|^2/\sigma^2$, representing the link gains between the source/destination and the relay R_i, and the interference gain from the transmit antenna to the receive antenna at the relay R_i respectively.

The block Rayleigh fading are assumed for all links, where the channel remains unchanged during a transmit slot, and changes independently from one slot to another. The mean value of channel gains between the source/destination and the Tx/Rx antenna of the FD relay nodes, $\gamma_{S,i}^{(j)}$ and $\gamma_{i,D}^{(k)}$, are denoted by $\lambda_{S,i}^{(j)}$ and $\lambda_{i,D}^{(k)}$ respectively. We also assume that the average channel gains between the source/destination and the two antennas of the relay R_i is same, $\lambda_{S,i}^{(1)} = \lambda_{S,i}^{(2)} = \lambda_{S,i}$ and $\lambda_{i,D}^{(1)} = \lambda_{i,D}^{(2)} = \lambda_{i,D}$. The residual loop interference between Tx and Rx antenna at relay is modeled as a block Rayleigh fading channel. The mutual interferences between Tx and Rx antenna at each relay node, $\gamma_{LI,i}^{1\to 2}$ and $\gamma_{LI,i}^{2\to 1}$ at the relay node R_i, are assumed to be the same due to the same loop interference cancellation method with the same configurations. Therefore, the interference gain of the relay R_i, $\gamma_{LI,i}^{1\to 2} = \gamma_{LI,i}^{1\to 2} = \gamma_{LI,i}$, follows the exponential distribution with the expectation of $\lambda_{LI,i}$. The interferences at different relay nodes are independent.

In the system under consideration, each relay node is equipped with two antennas. One antenna is used to receive the signal from the source node, and the other is to forward the signal at the same time. There are two possible channel link pairs between one relay and source/destination nodes, depending on which antenna of T_1 and T_2 is used as Tx and Rx antenna at relay. Those are $mode^{2\rightarrow 1}$ where relay R_i configures the antenna T_1 as the receive antenna and T_2 as the transmit antenna, and $mode^{1\rightarrow 2}$ where relay R_i configures T_2 as the receive antenna and T_1 as the transmit antenna. The optimal Tx/Rx antenna mode for relay R_i is determined based on the maximal end-to-end SINR criterion as follows,

$$mode_i = \arg\max\{\gamma_i^{1\rightarrow 2}, \gamma_i^{2\rightarrow 1}\}, \tag{47}$$

where $\gamma_i^{2\rightarrow 1}$ denotes the end-to-end SINR when the relay R_i works under $mode^{2\rightarrow 1}$, and can be computed via Eq. (46). Similarly, $\gamma_i^{1\rightarrow 2}$ is the end-to-end SINR via the relay R_i under $mode^{1\rightarrow 2}$. The CSI of $h_{S,i}^{(j)}$ and $h_{i,D}^{(k)}$ can be measured by the standard pilot-based channel estimation. The CSIs of the loop interference channel $I_i^{j\rightarrow k}$ can be obtained through sufficient training. It is also assumed that the very reliable feedback channels are deployed, and CSIs can be estimated and transmitted to the decision node accurately.

In the multiple relay system, the optimal relay with the optimal Tx/Rx antenna mode configuration is selected to optimize the end-to-end SINR performance. Therefore, JRAMS scheme can be formulated as

$$\{R_{opt}, mode_{opt}\} = \arg\max_i\{\gamma_i\}, \tag{48}$$

where $\gamma_i = \max\{\gamma_i^{1\rightarrow 2}, \gamma_i^{2\rightarrow 1}\}$. At the beginning of each transmission, all the CSIs are measured and collected at a decision node (say source node). The system computes the end-to-end SINR of the $2N$ candidate relaying modes, and selects the optimal relay R_{opt} with the optimal antenna mode configuration $mode_{opt}$ based on the end-to-end SINR maximization criterion in the JRAMS scheme as shown in equation above. The selected relay node R_{opt} will be used to help to forward the signal from the source to the destination under the optimal mode $mode_{opt}$.

Then, the performance analysis of the JRAMS scheme is provided. The approximate CDF expression of the end-to-end SINR via the ith FD relay node, denoted by γ_i, can be calculated as [23]

$$F_i(x) = 1 - \frac{2}{1+\eta_i x} I_i(x) + \frac{1}{1+2\eta_i x} I_i^2(x), \tag{49}$$

where

$$I_i(x) = 2\sqrt{\frac{x^2+x}{P_S P_R \lambda_{S,i}\lambda_{i,D}}} K_1\left(2\sqrt{\frac{x^2+x}{P_S P_R \lambda_{S,i}\lambda_{i,D}}}\right) e^{-\left(\frac{1}{P_S\lambda_{S,i}}+\frac{1}{P_R\lambda_{i,D}}\right)x}. \tag{50}$$

and $K_1(\cdot)$ is the first order modified Bessel function of the second kind [15], and $\eta_i = P_R\lambda_{LI,i}/P_S\lambda_{S,i}$.

When the average link SNRs, $P_S\lambda_{S,i}$ and $P_R\lambda_{i,D}$, go to infinity, $2\sqrt{\frac{x^2-x}{P_SP_R\lambda_{S,i}\lambda_{i,D}}}$ $K_1\left(2\sqrt{\frac{x^2+x}{P_SP_R\lambda_{S,i}\lambda_{i,D}}}\right)$ approaches one, and thus $I_i(x) \to 1$. The distribution of the received SINR via the ith relay in the high transmit power scenarios will approach

$$F_i^{\infty}(x) = 1 - \frac{2}{1+\eta_i x} + \frac{1}{1+2\eta_i x}. \tag{51}$$

In the high link SNR conditions, the distribution of the end-to-end SINR approaches a special distribution which is only determined by the loop interference level η_i at the relay node. This indicates that the performance of the FD relay system is similar in the high link SNR scenarios, and increasing the transmit power at the transmitter will be helpless to improving the system performance.

In this scheme, the optimal relay with the largest end-to-end SINR is selected to forward the signal. Therefore, the outage probability of JRAMS scheme of the N relay networks can be calculated as

$$P_{out}(x) = \prod_{i=1}^{N}\left[1 - \frac{2}{1+\eta_i x}I_i(x) + \frac{1}{1+2\eta_i x}I_i^2(x)\right]. \tag{52}$$

The transmit power of the source and the relay is assumed same, $P_S = P_R = P_t$. In the high transmit power scenario, the function $I_i(x)$ approaches $e^{-(\frac{1}{P_t\lambda_{S,i}}+\frac{1}{P_t\lambda_{i,D}})x}$, the outage probability can be written as

$$P_{out}(x) \approx \prod_{i=1}^{N}\left(1 - \frac{2}{1+\eta_i x}e^{-\frac{1}{P_t}(\frac{1}{\lambda_{S,i}}+\frac{1}{\lambda_{i,D}})x} + \frac{1}{1+2\eta_i x}e^{-\frac{2}{P_t}(\frac{1}{\lambda_{S,i}}+\frac{1}{\lambda_{i,D}})x}\right). \tag{53}$$

Based on the finite-SNR diversity order defined in [25], the diversity order d_{JRAMS} of the JRAMS scheme is obtained as

$$d_{JRAMS}(\lambda) = \sum_{i=1}^{N}\frac{2x}{P_t}\left(\frac{1}{\lambda_{S,i}}+\frac{1}{\lambda_{i,D}}\right)\frac{\frac{1}{1+\eta_i x}e^{-\frac{1}{P_t}(\frac{1}{\lambda_{S,i}}+\frac{1}{\lambda_{i,D}})x} - \frac{1}{1+2\eta_i x}e^{-\frac{2}{P_t}(\frac{1}{\lambda_{S,i}}+\frac{1}{\lambda_{i,D}})x}}{1 - \frac{2}{1+\eta_i x}e^{-\frac{1}{P_t}(\frac{1}{\lambda_{S,i}}+\frac{1}{\lambda_{i,D}})x} + \frac{1}{1+2\eta_i x}e^{-\frac{2}{P_t}(\frac{1}{\lambda_{S,i}}+\frac{1}{\lambda_{i,D}})x}} \tag{54}$$

The diversity order can be well estimated as by using the Taylor expansion

$$d_{JRAMS}(\lambda) \approx \sum_{i=1}^{N} 2\frac{1+A_i}{1+2A_i+2A_i^2}, \tag{55}$$

where $A_i = \frac{P_t \eta_i \lambda_{S,i} \lambda_{i,D}}{\lambda_{S,i} + \lambda_{i,D}}$. It can be observed that when the transmit power P_t increases to infinity, the diversity order approaches zero, representing the performance floor of the FD relaying system. It also shows that in the small loop interference condition, $A_i \to 0$, the diversity order of JRAMS scheme can reach almost $2N$ before the reaching the performance floor.

The performance floor of the outage probability are expressed as

$$P_{out}^{\infty}(x) = \prod_{i=1}^{N} \left(1 - \frac{2}{1 + \eta_i x} + \frac{1}{1 + 2\eta_i x} \right). \tag{56}$$

The performance floor of the proposed scheme is determined only by the loop interference level η_i of the N relay nodes, and is lower than that without the Tx/Rx antenna mode selection.

Figure 12 plots the outage performance versus the transmit power for the FD relay system with $N = 1, 2, 3$ and the loop interference level $\eta = 0.05$. The performance of the conventional optimal relay selection (ORS) scheme, proposed in [22], is also illustrated in Fig. 12 for comparison. It can be observed from the figure that there is a performance floor which coincides with the result computed by (56). In the same loop interference environment, the proposed JRAMS scheme outperforms that with the fixed antenna configuration obviously.

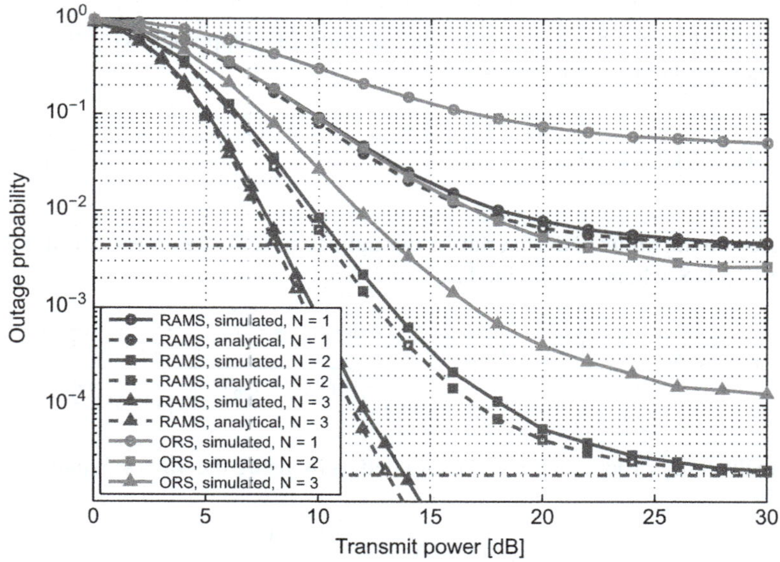

Fig. 12 Outage probability of JRAMS scheme versus the transmit power, $\eta = 0.05$

3.2 Two-Way FD Relay Selection Scheme

In this subsection, a two-way FD relay selection (FRS) scheme with AF protocol [24] is introduced, as illustrated in Fig. 13. The system consists of two sources, denoted by S_j, $j = 1, 2$, and N relays, denoted by R_i, $i = 1, \ldots, N$. The transmit powers of each source and each relay are denoted by p_s and p_r, respectively. The direct link between sources does not exist due to the shadowing effect. The block-fading channel coefficient between S_j and R_i is reciprocal and denoted by h_{ji}, which follows independent complex-Gaussian distribution with zero mean and normalized variance.

In contrast to the system model of traditional two-way HD relay [19], each communication node in Fig. 13 operates in the FD mode. We assume that each node is equipped with one FD anttena [26, 27]. The residual self-interference channels at S_j and R_i are denoted by $\tilde{h}_{S_j}$ and $\tilde{h}_{R_i}$. Considering the transmission by the relay R_i at the time slot κ, the sources S_j, $j = 1, 2$, transmit their respective symbols $x_j[\kappa]$ to R_i. Simultaneously, R_i broadcasts another symbol $t_i[\kappa]$ to the sources. Therefore, the received signal at R_i is

$$r_i[\kappa] = \sqrt{p_s} x_1[\kappa] h_{1i} + \sqrt{p_s} x_2[\kappa] h_{2i} + \tilde{h}_{R_i} t_i[\kappa] + n_{r_i}[\kappa] \tag{57}$$

where $\mathbb{E}\left\{x_j[\kappa]^2\right\} = 1$, $n_{r_i}[\kappa]$ is the additive white Gaussian noise (AWGN) at R_i with zero-mean and variance of σ_n^2, and the residual self-interference at the relay $e_{r_i} = \tilde{h}_{R_i} t_i[\kappa]$ is modeled by AWGN with zero-mean and variance of $\sigma_{e,r}^2$.

The relay R_i adopts the AF protocol, and thus $t_i[\kappa]$ is the amplification of the prior received signal, i.e.,

$$t_i[\kappa] = \sqrt{p_r} \beta_{r_i} r_i[\kappa - 1] \tag{58}$$

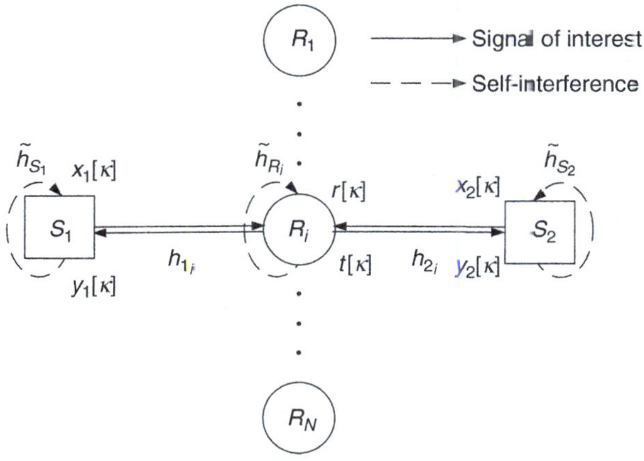

Fig. 13 System model of two-way FRS scheme

where β_{r_i} is the amplification factor of R_i. From (57) and (58), $t_i[\kappa]$ can be expanded as

$$t_i[\kappa] = \beta_{r_i}\sqrt{p_r}\sum_{m=1}^{\infty}\left(\tilde{h}_{r_i}\beta_{r_i}\sqrt{p_r}\right)^{m-1}\{h_{1i}\sqrt{p_s}x_1[\kappa-m]+h_{2i}\sqrt{p_s}x_2[\kappa-m]+n_{r_i}[\kappa-m]\}. \tag{59}$$

To prevent the oscillation at R_i, the amplification factor β_{r_i} should be

$$\beta_{r_i} = \frac{1}{\sqrt{p_s|h_{1i}|^2 + p_s|h_{2i}|^2 + p_r|\tilde{h}_{R_i}|^2 + \sigma_n^2}} \tag{60}$$

Therefore, according to (59) and (60), the variance of $t_i[\kappa]$ is finite, i.e.,

$$\mathbb{E}\{t_i[\kappa]^2\} = \left(|h_{1i}|^2 p_s + |h_{2i}|^2 p_s + \sigma_n^2\right)\frac{\beta_{r_i}^2 p_r}{1 - |\tilde{h}_{R_i}|^2\beta_{r_i}^2 p_r} = p_r. \tag{61}$$

The received signal at $S_j, j = 1, 2$, is expressed as

$$y_j[\kappa] = h_{ji}t_i[\kappa] + n_j[\kappa] + \tilde{h}_{S_j}\sqrt{p_s}x_j[\kappa] \tag{62}$$

where $n_j[t]$ is the AWGN at S_j, and the residual self-interference at the source $e_{s_j} = \tilde{h}_{S_j}\sqrt{p_s}x_j[\kappa]$ is modeled by AWGN with zero-mean and variance of $\sigma_{e,s}^2$. Note that the self-interference at sources and that at relays have different impact on the performance. Specifically, the self-interference at the relay is iterative because of (57) and (58), and might make the relay oscillation, whereas the self-interference at the source is non-iterative.

According to (57), (58), and (62), the received signal at S_j can be expanded as

$$y_j[\kappa] = h_{ji}h_{ji}\beta_{r_i}\sqrt{p_r p_s}x_j[\kappa-1] + h_{ji}h_{\bar{j}i}\beta_{r_i}\sqrt{p_r p_s}x_{\bar{j}}[\kappa-1] \tag{63}$$

$$+ h_{ji}\beta_{r_i}\sqrt{p_r}\tilde{h}_{R_i}t_i[\kappa-1] + \tilde{h}_{S_j}\sqrt{p_s}x_j[\kappa] \tag{64}$$

$$+ h_{ji}\beta_{r_i}\sqrt{p_r}n_{r_i}[\kappa-1] + n_j[\kappa] \tag{65}$$

where $\{j, \bar{j}\} = \{1, 2\}$ or $\{2, 1\}$. The first term in (63) can be subtracted totally because of network coding technique [28], and the second term in (63) is the useful signal from $S_{\bar{j}}$; the first term and the second term in (64) are the residual self-interference from R_i and S_j, respectively; the first term and the second term in (65) are the noise from R_i and S_j, respectively.

After some manipulation, the instantaneous received SINR at S_j via R_i is written as

$$\gamma_{ji} = \frac{\tilde{\psi}_s|h_{ji}|^2\tilde{\psi}_r|h_{\bar{j}i}|^2}{\left(\tilde{\psi}_s + \tilde{\psi}_r\right)|h_{ji}|^2 + \tilde{\psi}_s|h_{\bar{j}i}|^2 + 1}, \tag{66}$$

where $\tilde{\psi}_s = \frac{p_s}{(\sigma_{e,r}^2+1)\sigma_n^2}$ and $\tilde{\psi}_r = \frac{p_r}{(\sigma_{e,s}^2+1)\sigma_n^2}$, and $\sigma_{e,s}^2$ and $\sigma_{e,r}^2$ are the variances of residual self-interference at the sources and the relays, respectively.

According to [24], in the two-way relay, the optimal FRS scheme is

$$k = \max_{i=1,\ldots,N} \min\left(|h_{1i}|^2, |h_{2i}|^2\right) \tag{67}$$

where k is the index of the selected relay. The RS scheme (67) is optimal in minimizing the outage probability and symbol error rate, and also maximizing the ergodic capacity.

In the following, only the bit error rate (BER) analysis is provided for the two-way FRS scheme, other performance analysis can be found in [24]. Prior to that, the distribution function of γ_{jk} is calculated, which is determined by the distribution of $|h_{jk}|^2$ according to (66). According to the RS scheme (67), the CDF of $|h_{jk}|^2$ is expressed as

$$F_{|h_{jk}|^2}(z) = 1 - N\sum_{m=0}^{N-1}\binom{N-1}{m}\frac{(-1)^m}{2m+1}\left\{\exp(-z) + \frac{m}{m+1}\exp\left[-2(m+1)z\right]\right\} \tag{68}$$

where $j = 1, 2$.

With the help of the CDF of $|h_{jk}|^2$, a tight lower bound of BER for S_j can be obtained

$$\overline{BER}_j \geq \frac{\alpha}{2} - \frac{3\sqrt{2}\pi\alpha\sqrt{\beta}}{2\tilde{\psi}_r\tilde{\psi}_h}N^2\sum_{m=0}^{N-1}\sum_{n=0}^{N-1}\binom{N-1}{m}\binom{N-1}{n}\frac{(-1)^m}{2m+1}\frac{(-1)^n}{2n+1}\left[f_{11}+f_{12}+f_{21}+f_{22}\right], \tag{69}$$

where

$$f_{11} = \left[\left(\sqrt{\frac{1}{\tilde{\psi}_r}} + \sqrt{\frac{1}{\tilde{\psi}_h}}\right)^2 + \frac{\beta}{2}\right]^{-\frac{5}{2}} F\left(\frac{5}{2}, \frac{3}{2}; 2; \frac{\left(\sqrt{1/\tilde{\psi}_r} - \sqrt{1/\tilde{\psi}_h}\right)^2 + \beta/2}{\left(\sqrt{1/\tilde{\psi}_r} + \sqrt{1/\tilde{\psi}_h}\right)^2 + \beta/2}\right), \tag{70}$$

$$f_{12} = 2n\left[\left(\sqrt{\frac{1}{\tilde{\psi}_r}} + \sqrt{\frac{\xi_n}{\tilde{\psi}_h}}\right)^2 + \frac{\beta}{2}\right]^{-\frac{5}{2}} F\left(\frac{5}{2}, \frac{3}{2}; 2; \frac{\left(\sqrt{1/\tilde{\psi}_r} - \sqrt{\xi_n/\tilde{\psi}_h}\right)^2 + \beta/2}{\left(\sqrt{1/\tilde{\psi}_r} + \sqrt{\xi_n/\tilde{\psi}_h}\right)^2 + \beta/2}\right), \tag{71}$$

$$f_{21} = 2m\left[\left(\sqrt{\frac{\xi_m}{\tilde{\psi}_r}} + \sqrt{\frac{1}{\tilde{\psi}_h}}\right)^2 + \frac{\beta}{2}\right]^{-\frac{5}{2}} F\left(\frac{5}{2}, \frac{3}{2}; 2; \frac{\left(\sqrt{\xi_m/\tilde{\psi}_r} - \sqrt{1/\tilde{\psi}_h}\right)^2 + \beta/2}{\left(\sqrt{\xi_m/\tilde{\psi}_r} + \sqrt{1/\tilde{\psi}_h}\right)^2 + \beta/2}\right), \quad (72)$$

and

$$f_{22} = 4mn\left[\left(\sqrt{\frac{\xi_m}{\tilde{\psi}_r}} + \sqrt{\frac{\xi_n}{\tilde{\psi}_h}}\right)^2 + \frac{\beta}{2}\right]^{-\frac{5}{2}} F\left(\frac{5}{2}, \frac{3}{2}; 2; \frac{\left(\sqrt{\xi_m/\tilde{\psi}_r} - \sqrt{\xi_n/\tilde{\psi}_h}\right)^2 + \beta/2}{\left(\sqrt{\xi_m/\tilde{\psi}_r} + \sqrt{\xi_n/\tilde{\psi}_h}\right)^2 + \beta/2}\right). \quad (73)$$

In addition, $\xi_m = 2m + 2$, $\xi_n = 2n + 2$, $\tilde{\psi}_h = \frac{\tilde{\psi}_s\tilde{\psi}_r}{\tilde{\psi}_s+\tilde{\psi}_r}$, and $F(a, b; c; z)$ is the Confluent Hypergeometric function [15].

Figure 14 depicts the BER performance versus SNR $= p/\sigma_n^2$, when the variance of residual self-interference $\sigma_e^2 = 0.1,\ 0.5,$ and 2, and the number of relay $N = 4$. The figure reveals that the two-way FD relay has significant gain than the two-way HD relay all over the SNR regime, when the residual self-interference is small. For instance, the performance gain is about 1 dB in high SNR when $\sigma_e^2 = 0.5$, and the

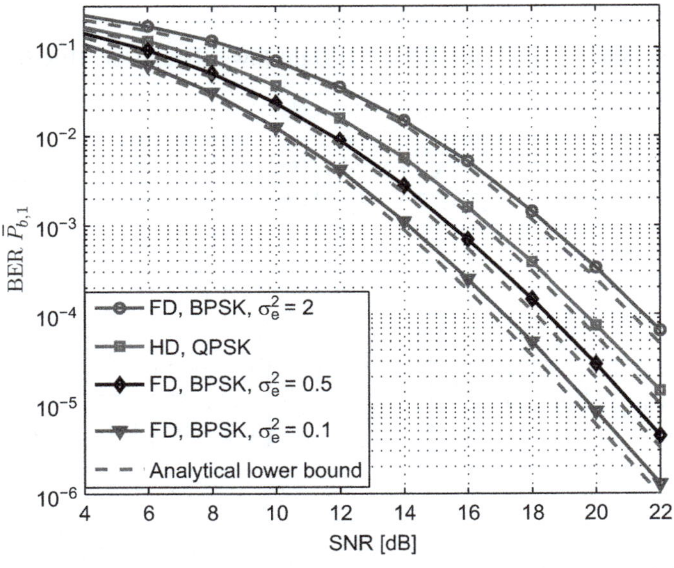

Fig. 14 BER of the two-way FD relay selection and two-way HD relay selection, when the number of relay $N = 4$

gain increases to 2 dB when the variance of the residual self-interference decreases to $\sigma_e^2 = 0.1$. However, when the residual self-interference is large, the two-way FD relay has worse performance than the two-way HD relay.

4 Full-Duplex Cognitive Radio Networks

Cognitive radio, which enables spectrum sharing between licensed (PUs) and unlicensed (SUs) users, has been regarded as a promising solution to spectrum shortage, and caused wide attention for more than a decade [29, 30]. Conventionally, the so-called "listen-before-talk" strategy in which SUs sense the target channel before transmission has been extensively studied [31]. In the listen-before-talk scheme, the sensing and transmission are separated in time domain, which leads to two main problems: (1) transmission time reduction due to sensing, and (2) sensing accuracy impairment due to data transmission. With the help of FD techniques, SUs no longer need to stop their transmissions periodically for sensing. Instead, they can perform simultaneous spectrum sensing and data transmission, which can naturally bypass the above problems. In this section, a "listen-and-talk" (LAT) protocol for FD CR users based on the idea of simultaneous sensing and transmission [32, 33] is introduced. The protocol design as well as detailed performance analysis is provided.

4.1 Listen-and-Talk Protocol

A CRN is considered, consisting of one PU and one SU pair, where SU_1 needs to transmit data to SU_2. Each SU is equipped with two antennas Ant_1 and Ant_2. The spectrum band occupancy by the PU can be modeled as an alternating ON/OFF random process where the PU can access the spectrum at any time. We assume that the probabilities of the PU's arrival and departure remain the same across the time, and the holding time of either state is distributed as the exponential distribution [34]. We denote the variables of the idle period and busy period of the PU as t_0 and t_1, respectively. And let $\tau_0 = \mathbb{E}[t_0]$ and $\tau_1 = \mathbb{E}[t_1]$ represent the average idle and busy duration.

Figure 15 shows the sensing and spectrum access procedure of the LAT protocol. SU_1 performs sensing and transmission simultaneously by using the FD technique: Ant_1 senses the spectrum continuously while Ant_2 transmits data when a spectrum hole is detected. Specifically, SU_1 keeps sensing the spectrum with Ant_1 with sampling frequency f_s, which is shown in the line with down arrows. At the end of each slot with duration T, SU_1 combines all samples in the slot and makes the decision of the PU's presence. The decisions are represented by the small circles, in which the higher ones denote that the PU is judged active, while the lower ones denote otherwise. The activity of SU_1 is instructed by the sensing decisions,

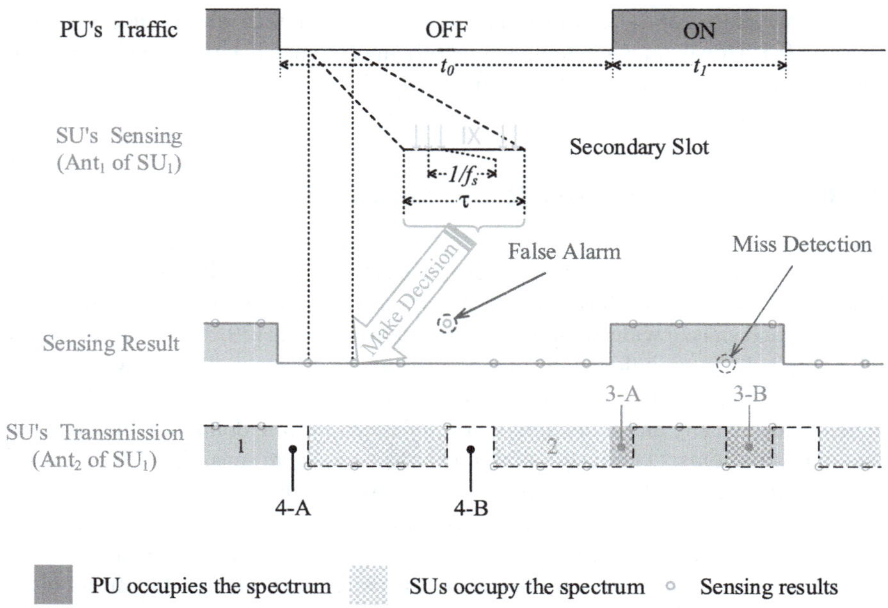

Fig. 15 The LAT protocol

i.e., SU_1 can access the spectrum in the following slot when the PU is judged absent, and it needs to backoff otherwise.

On the implementation of the LAT protocol, the residual self-interference becomes a major impediment to accurate sensing, i.e., when SU_1 is transmitting, part of the transmitted signal will be received by the sensing antenna Ant_1, which obscures the potential signal from the PU. Specifically, the received signal at Ant_1 can be expressed as

$$y = \begin{cases} h_s s_p + \delta_1 w + u, & \mathcal{H}_1, \\ \delta_1 w + u, & \mathcal{H}_0, \end{cases} \tag{74}$$

where $\mathcal{H}_1$ and $\mathcal{H}_0$ are the hypotheses under which the PU is busy or idle, respectively; s_p denotes the signal of the PU, h_s is the channel from the PU to Ant_1 of SU_1, and $u \sim \mathcal{CN}\left(0, \sigma_u^2\right)$ denotes the complex-valued Gaussian noise. Without loss of generality, we assume that s_p is PSK modulated with variance σ_p^2, and h_s is a Rayleigh channel with zero mean and variance σ_h^2. Variable w denotes the residual self-interference at Ant_1, which is modeled as Rayleigh distribution with zero mean and variance $\chi^2\sigma_s^2$, with σ_s^2 denoting the secondary transmit power and χ^2 representing the degree of self-interference suppression, which is defined as

$$\chi^2 := \frac{\text{Power of the RSI}}{\text{Transmit power}}.$$

The parameter χ^2 is commonly expressed in dBs, and indicates how well can the self-interference be suppressed. The Boolean variable δ_1 represents SU_1's activity. As shown above, the distribution of the received signal for sensing is related to the activity of SU_1, indicating that the sensing strategy needs to be adjusted according to SU_1's own activity.

For simplicity and without loss of generality, energy detection is adopted as the sensing scheme, and the test statistics O can be given as:

$$O = \frac{1}{N_s} \sum_{n=1}^{N_s} |y(n)|^2, \tag{75}$$

where $y(n)$ denotes the nth sample received by Ant_1 in a slot. With a chosen threshold ϵ, the spectrum is judged occupied when $O \geq \epsilon$, otherwise the spectrum is idle, and the false alarm and miss detection probabilities can be defined as, respectively,

$$\begin{aligned} P_f(\epsilon) &= \Pr(O > \epsilon | \mathcal{H}_0), \\ P_m(\epsilon) &= \Pr(O < \epsilon | \mathcal{H}_1). \end{aligned} \tag{76}$$

As mentioned above, the sensing threshold needs to be changed according to SU_1's activity, i.e., when SU_1 is silent, it chooses ϵ_0 as the sensing threshold, and when it is active, it chooses a different ϵ_1. In the design of the sensing threshold, one of the most important constraints is that the interference to the PU must be under a certain level. This constraint is considered as the collision ratio between SUs and the PU, defined as

$$P_c = \lim_{t\to\infty} \frac{\text{Collision duration}}{\text{PU's transmission time during } [0, t]}.$$

Combining the PU's state change and SU_1's activity, the collision ratio can be derived following the solid arrows in Fig. 16, in which the "state transition" block refers to the Markov chain among the four states of the system: (1) only the PU uses the spectrum; (2) only the SUs use the spectrum; (3) collision; and (4) spectrum waste. Further, for simplicity in threshold design, we artificially add a constraint $P_m^0 = P_m^1 = P_m$, i.e., the miss detection probability remains the same when SU_1 is either active or silent. $P_c - \left(1 - e^{-\frac{T}{\tau_1}}\right)/2$ is a good approximation of P_m in common cases where the PU changes its activity sufficiently slowly.

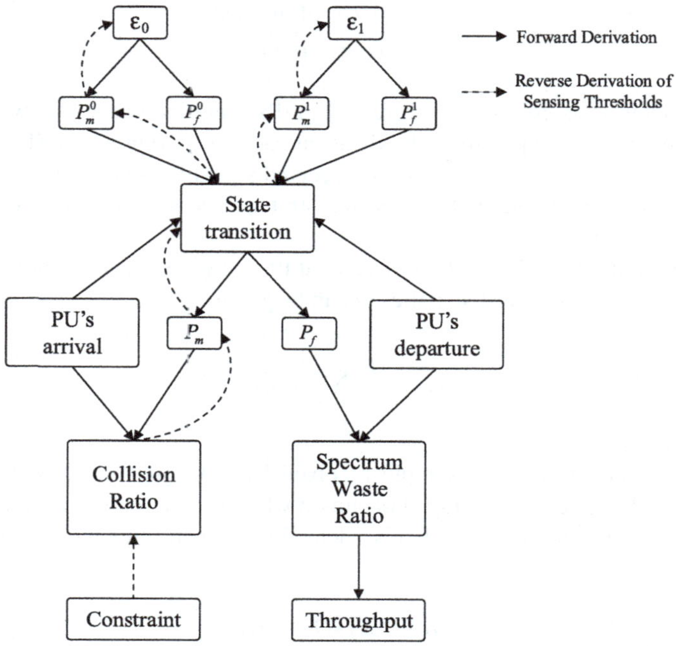

Fig. 16 The schematic design procedure of the sensing thresholds of the LAT protocol

Then, the sensing thresholds are given by

$$\begin{aligned}\epsilon_{\delta_1} &= \epsilon_{\delta_1}(P_c) = \left(\frac{\mathcal{Q}^{-1}(1-P_m)}{\sqrt{N_s}}+1\right)(1+\gamma_s+\gamma_i\delta_1)\,\sigma_u^2 \\ &= \left(\frac{\mathcal{Q}^{-1}\left(\frac{3}{2}-P_c-\frac{1}{2}e^{-T/\tau_1}\right)}{\sqrt{N_s}}+1\right)(1+\gamma_s+\gamma_i\delta_1)\,\sigma_u^2,\ \delta_1\in\{0,1\}\end{aligned} \tag{77}$$

in which $\gamma_s = \frac{\sigma_p \sigma_h^2}{\sigma_u^2}$ denotes the signal-to-noise ratio (SNR) in sensing, and $\gamma_i = \frac{\chi^2 \sigma_s^2}{\sigma_u^2}$ is the interference-to-noise ratio (INR).

4.2 *Spectrum Utilization Efficiency*

With the thresholds in (77), the spectrum utilization efficiency can be derived by following the procedure in Fig. 16. Specifically, the spectrum waste ratio, defined as

$$P_w = \lim_{t\to\infty}\frac{\text{Duration of unused spectrum hole}}{\text{Total length of spectrum holes during } [0,t]},$$

can be calculated as

$$P_w = \frac{\mu}{2} + \frac{\left(\frac{1}{\mu} - 1\right) P_f^1 + 1 - P_m}{1 + \left(\frac{1}{\mu} - 1\right)\left(1 - P_f^0 + P_f^1\right)}\Bigg|_{\mu = 1 - e^{-\frac{T}{\tau_0}}}$$
$$= \frac{1 - e^{-T/\tau_0}}{2} + \frac{\frac{e^{-T/\tau_0}}{1-e^{-T/\tau_0}} P_f^1 + 1 - P_m}{1 + \frac{e^{-T/\tau_0}}{1-e^{-T/\tau_0}}\left(1 - P_f^0 + P_f^1\right)}, \tag{78}$$

where P_f^0 and P_f^1 denote the false alarm probability when SU_1 is silent and active, respectively, which can be expressed as

$$P_f^{\delta_1}(P_m) = \mathcal{Q}\left(\mathcal{Q}^{-1}(1 - P_m)\left(1 + \frac{\gamma_s}{1 + \gamma_i \delta_1}\right) + \frac{\gamma_s}{1 + \gamma_i \delta_1}\sqrt{N_s}\right), \tag{79}$$

where $\mathcal{Q}(\cdot)$ is the complementary distribution function of the standard Gaussian distribution.

In Fig. 17, the receiver operating characteristic curves (ROCs) is used to present the sensing performance. Both the ROCs of LAT and conventional "listen-before-talk" (LBT) protocols are presented. For comparison fairness, in the calculation and

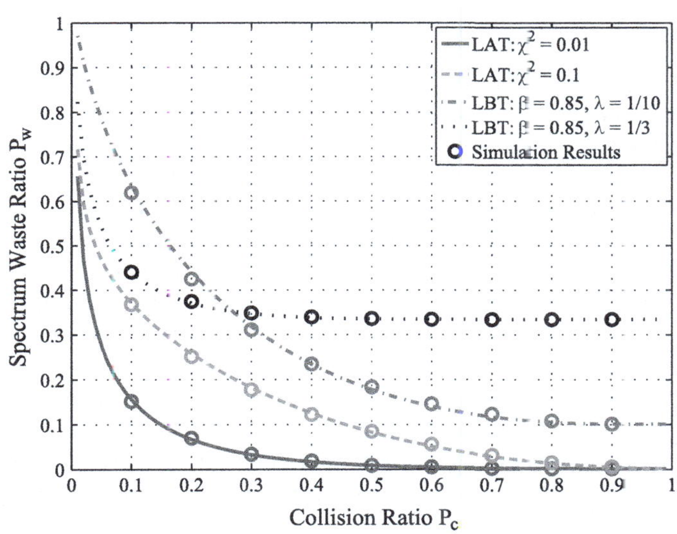

Fig. 17 ROCs in sensing. In this figure, the probability of the PU's arrival $\mu = 1/500$, departure $\nu = 6/500$, the sample number of a slot N_s is 300, normalized secondary transmit power $\sigma_s^2/\sigma_u^2 =$ 10 dB, sensing SNR $\gamma_s = -8$ dB, the RSI factor χ^2 in the LAT varies between 0.1 and 0.01, the spatial correlation coefficient β is 0.85, and the ratio of sensing duration changes between 1/3 and 1/10

simulation of the LBT protocol, we also assume that each SU is equipped with two antennas, which means that SU_1 can use two antennas for sensing. The spatial correlation of the two antennas is represented by β, and the percentage of sensing duration in a slot is denoted by λ. In Fig. 17, smaller area under a curve denotes better sensing performance, i.e., the sensing performance of the LAT is much better than the LBT. It can be seen that for the curves in the LAT protocol, the solid red line is lower than the dashed pink one, which indicates that smaller residual self-interference leads to better sensing performance. It is noteworthy that the ratio of spectrum waste of the LBT protocol can never be lower than λ, while that of the LAT can be quite close to zero if the PU's state change is sufficiently slow. Also, in the left side of Fig. 17 where the allowable P_c is small, P_w in the LAT decreases much more sharply with the increase of P_c than P'_w in the LBT, which also implies better sensing performance of the LAT in the real case when P_c is strictly constrained.

4.3 *Power-Throughput Tradeoff*

The secondary throughput can be obtained from (78):

$$C = R \cdot (1 - P_w) = \log_2(1 + \gamma_t) \cdot \left(\frac{1 + e^{-T/\tau_0}}{2} - \frac{\frac{e^{-T/\tau_0}}{1-e^{-T/\tau_0}} P_f^1 + 1 - P_m}{1 + \frac{e^{-T/\tau_0}}{1-e^{-T/\tau_0}} \left(1 - P_f^0 + P_f^1\right)} \right), \tag{80}$$

where $\gamma_t = \frac{\sigma_s^2 \sigma_t^2}{\sigma_u^2}$ represents the SNR in transmission, with σ_t^2 denotes the pass loss of the transmit channel from SU_1 to SU_2.

In the expression of SU_1's throughput in (80), there are two factors: R and $(1 - P_w)$. On one hand, R is positively proportional to SU_1's transmit power σ_s^2. On the other hand, however, it can be proved that the spectrum waste ratio P_w increases with the secondary transmit power σ_s^2. Thus, there may exist a tradeoff between secondary transmit power and secondary throughput in this protocol: when the secondary transmit power is low, the residual self-interference is negligible, the spectrum is used more fully with small P_w, yet the ceiling throughput is limited by R; when the transmit power increases, the sensing performance get deteriorated, while at the same time, SU_1 can transmit more data in a single slot.

Figure 18 shows the throughput performance of the LAT protocol in terms of secondary transmit power. The solid and dotted lines represent the analytical performance of the LAT protocol, and the asterisks (*) denote the analytical local optimal transmit power. The small circles are the simulated results, which match the analytical performance well. The thin solid line depicts the ideal case with perfect self-interference cancelation. Without residual self-interference, the sensing performance is no longer affected by transmit power, and the throughput always goes up with the power. This line is also the upperbound of the LAT performance. The thick dash-dotted, dotted and dash lines in the middle are the typical cases, in

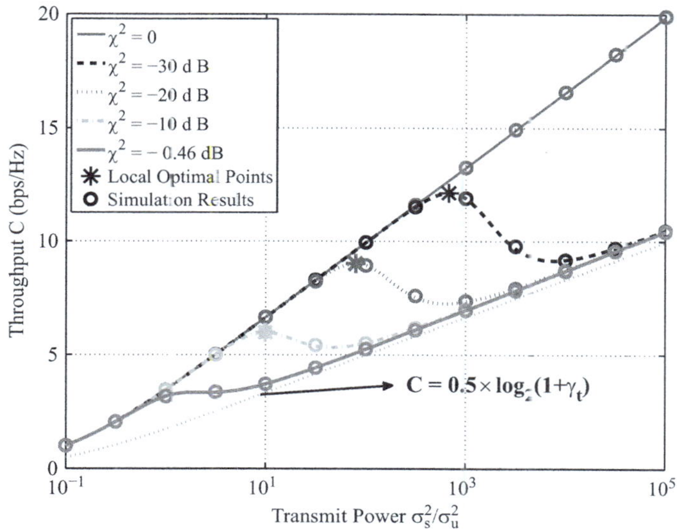

Fig. 18 Power-Throughput curves in terms of different degree of self-interference cancelation factor χ^2, where the probability of the PU's arrival is 0.002/slot, departure is 0.012/slot, the collision ratio $P_c = 0.1$, the sample number of a slot N_s is 300, sensing SNR $\gamma_s = -5$ dB

which we can clearly observe the power-throughput tradeoff and identify the local optimal power. With the improvement of self-interference cancelation (χ^2 from 0.1 to 0.01 to 0.001), the local optimal transmit power increases, and the corresponding throughput goes to a higher level. This makes sense since the smaller the residual self-interference is, the better it approaches the ideal case, and the deterioration cause by self-interference becomes dominant under a higher power. Also, when χ^2 is sufficiently large, i.e., $\chi^2 = 0.9$ in the figure, there exists no power-throughput tradeoff.

One noticeable feature of Fig. 18 is that when self-interference exists, all curves approach the thin dotted line $C = 0.5 \log_2 (1 + \gamma_t)$ when the power goes up. This line indicates the case that the spectrum waste is 0.5. When the transmit power is too large, severe self-interference largely degrades the performance of spectrum sensing, and the false alarm probability becomes unbearably high. It is likely that whenever SU_1 begins transmission, the spectrum sensing result falsely indicates that the PU has arrived due to false alarm, and SU_1 stops transmission in the next slot. Once SU_1 becomes silent, it can clearly detect the PU's absence, and begins transmission in the next slot again. And the state of SU_1 changes every slot even when the PU does not arrive at all. In this case, the utility efficiency of the spectrum hole is approximately 0.5, which is clearly shown in Fig. 18. Also, it can be seen that the larger χ^2 is, the earlier the sensing gets unbearable and the throughput approaches the orange line.

5 Conclusions and Future Work

In this chapter, FD wireless communications for 5G have been presented, which allows simultaneous transmission and reception over the same frequency band, and potentially increases the spectral efficiency. The most fundamental challenge of FD implementation comes from the performance degradation induced by SI. To this end, we described several SI cancelation techniques in propagation-domain, analog-circuit-domain and digital-domain, which enable FD communications in different application scenarios, such as bidirectional communications, cooperative networks, and cognitive networks. In addition, the performance of several schemes was analyzed, including HD/FD mode switching and TRAPS scheme in FD bidirectional communications, one-way JRAMS and two way RS scheme in FD cooperative networks, and LAT in FD CRN. The existing work mainly focused on the analysis of physical layer, however, much work remains to be done for analysis of higher layers, such as MAC layer and network layer. These are great opportunities for inter-discipline wireless researchers to develop more efficient wireless networks in the future.

References

1. J.I. Choi, M. Jain, K. Srinivasan, P. Levis, S. Katti, Achieving single channel, full duplex wireless communication, in *Proceedings of the Sixteenth Annual International Conference on Mobile Computing and Networking* (ACM, New York, 2010), pp. 1–12
2. A. Sabharwal, P. Schniter, D. Guo, D.W. Bliss, S. Rangarajan, R. Wichman, In-band full-duplex wireless: challenges and opportunities. IEEE J. Sel. Areas Commun. **32**(9), 1637–1652 (2014)
3. Y. Liao, K. Bian, L. Song, Z. Han, Full-duplex mac protocol design and analysis. IEEE Commun. Lett. **19**(7), 1185–1188 (2015)
4. M. Duarte, A. Sabharwal, Full-duplex wireless communications using off-the-shelf radios: feasibility and first results, in *2010 Conference Record of the Forty Fourth Asilomar Conference on Signals, Systems and Computers (ASILOMAR)* (IEEE, New York, 2010), pp. 1558–1562
5. E. Everett, A. Sahai, A. Sabharwal, Passive self-interference suppression for full-duplex infrastructure nodes. IEEE Trans. Wirel. Commun. **13**(2), 680–694 (2014)
6. A. Sahai, G. Patel, A. Sabharwal, Pushing the limits of full-duplex: design and real-time implementation, arXiv preprint arXiv:1107.0607, 2011
7. E. Aryafar, M.A. Khojastepour, K. Sundaresan, S. Rangarajan, M. Chiang, MIDU: enabling MIMO full duplex, in *Proceedings of the 18th Annual International Conference on Mobile Computing and Networking* (ACM, New York, 2012), pp. 257–268
8. E. Everett, M. Duarte, C. Dick, A. Sabharwal, Empowering full-duplex wireless communication by exploiting directional diversity, in *2011 Conference Record of the Forty Fifth Asilomar Conference on Signals, Systems and Computers (ASILOMAR)* (IEEE, New York, 2011), pp. 2002–2006
9. M. Duarte, C. Dick, A. Sabharwal, Experiment-driven characterization of full-duplex wireless systems. IEEE Trans. Wirel. Commun. **11**(12), 4296–4307 (2012)
10. M. Duarte, A. Sabharwal, V. Aggarwal, R. Jana, K. Ramakrishnan, C.W. Rice, N. Shankaranarayanan, Design and characterization of a full-duplex multiantenna system for WIFI networks. IEEE Trans. Veh. Technol. **63**(3), 1160–1177 (2014)

11. A. Sahai, G. Patel, C. Dick, A. Sabharwal, On the impact of phase noise on active cancelation in wireless full-duplex. IEEE Trans. Veh. Technol. **62**(9), 4494–4510 (2013)
12. B.P. Day, A.R. Margetts, D.W. Bliss, P. Schniter, Full-duplex bidirectional MIMO: achievable rates under limited dynamic range. IEEE Trans. Signal Process. **60**(7), 3702–3713 (2012)
13. M. Kiessling, J. Speidel, Mutual information of MIMO channels in correlated Rayleigh fading environments-a general solution, in *IEEE International Conference on Communications. 2004*, vol. 2 (IEEE, New York, 2004), pp. 814–818
14. M. Zhou, H. Cui, L. Song, Y. Li, Is full duplex configuration better than mimo spatial multiplexing? in *2015 IEEE International Conference on Communications in China* (IEEE, New York, 2015)
15. M. Abramowitz, I.A. Stegun, *Handbook of Mathematical Functions: With Formulas, Graphs, and Mathematical Tables* (Courier Corporation, New York, 2012)
16. A. Paulraj, R. Nabar, D. Gore, *Introduction to Space-Time Wireless Communications* (Cambridge University Press, Cambridge, 2003)
17. M. Zhou, H. Cui, L. Song, B. Jiao, Transmit-receive antenna pair selection in full duplex systems. IEEE Trans. Wirel. Commun. **3**(1), 34–37 (2014)
18. M. Zhou, L. Song, Y. Li, X. Li, Simultaneous bidirectional link selection in full duplex MIMO systems. IEEE Trans. Wirel. Commun. **14**(7), 4052–4062 (2015)
19. L. Song, Relay selection for two-way relaying with amplify-and-forward protocols. IEEE Trans. Veh. Technol. **60**(4), 1954–1959 (2011)
20. H.A. David, H.N. Nagaraja, *Order Statistics*. Wiley Online Library, 1970
21. G.-P. Liu, R. Yu, H. Ji, V. Leung, X. Li, In-band full-duplex relaying: a survey, research issues and challenges. IEEE Commun. Surv. Tutorials **17**(2), 500–524 (2015)
22. I. Krikidis, H. Suraweera, P.J. Smith, C. Yuen et al., Full-duplex relay selection for amplify-and-forward cooperative networks. IEEE Trans. Wirel. Commun. **11**(12), 4381–4393 (2012)
23. K. Yang, H. Cui, L. Song, Y. Li, Efficient full-duplex relaying with joint antenna-relay selection and self-interference suppression. IEEE Trans. Wirel. Commun. **14**, 3991–4005 (2015)
24. H. Cui, M. Ma, L. Song, B. Jiao, Relay selection for two-way full duplex relay networks with amplify-and-forward protocol. IEEE Trans. Wirel. Commun. **13**(7), 3768–3777 (2014)
25. R. Narasimhan, A. Ekbal, J.M. Cioffi, Finite-SNR diversity-multiplexing tradeoff of space-time codes, in *2005 IEEE International Conference on Communications, 2005. ICC 2005*, vol. 1 (IEEE, New York, 2005), pp. 458–462
26. Y. Hua, P. Liang, Y. Ma, A.C. Cirik, Q. Gao, A method for broadband full-duplex MIMO radio. IEEE Signal Process. Lett. **19**(12), 793–796 (2012)
27. D. Bharadia, E. McMilin, S. Katti, Full duplex radios, in *ACM SIGCOMM Computer Communication Review*, vol. 43 (ACM, New York, 2013), pp. 375–386
28. R.H. Louie, Y. Li, B. Vucetic, Practical physical layer network coding for two-way relay channels: performance analysis and comparison. IEEE Trans. Wirel. Commun. **9**(2), 764–777 (2010)
29. J. Mitola III, G.Q. Maguire Jr., Cognitive radio: making software radios more personal. IEEE Pers. Commun. **6**(4), 13–18 (1999)
30. J. Mitola, Cognitive radio - an integrated agent architecture for software defined radio, 2000
31. T. Yücek, H. Arslan, A survey of spectrum sensing algorithms for cognitive radio applications. IEEE Commun. Surv. Tutorials **11**(1), 116–130 (2009)
32. Y. Liao, T. Wang, L. Song, Z. Han, Listen-and-talk: full-duplex cognitive radio networks, in *IEEE Global Communications Conference (GLOBECOM), 2014* (IEEE, New York, 2014), pp. 3068–3073
33. Y. Liao, L. Song, Z. Han, Y. Li, Full duplex cognitive radio: a new design paradigm for enhancing spectrum usage. IEEE Commun. Mag. **53**(5), 138–145 (2015)
34. S. Huang, X. Liu, Z. Ding, Opportunistic spectrum access in cognitive radio networks, in *IEEE INFOCOM 2008. The 27th Conference on Computer Communications* (IEEE, New York, 2008)

Device-to-Device Communications over 5G Systems: Standardization, Challenges and Open Issues

G. Araniti, A. Raschellà, A. Orsino, L. Militano, and M. Condoluci

Abstract This chapter deals with a comprehensive analysis of the challenges behind the introduction of Device-to-Device (D2D) communications in the context of 5th Generation (5G) communications. In fact, although D2D communications have emerged as an efficient solution for local traffic between mobile User Equipments (UEs) in proximity in cellular environments, has recently also attracted attention as a key enabling technology for 5G wireless networks. In details, this chapter addresses an overview of the current standardization, its integration in the cellular system architecture, future challenges and open issues. This chapter also provides a performance analysis, which illustrates the important achievements in terms of data rate in a scenario where the proximity communications between devices in an LTE-A system has been introduced for multicast downloading services.

1 Introduction

Recently, telco operators have been facing an increasing demand for high data rate as a consequence of a proliferation of innovative applications and services emerging in daily routines of mobile users. In this context, Device-to-Device (D2D) communications have attracted significant attention as a key enabling technology of 5th generation (5G) wireless networks [1] as witnessed by the recent Long Term Evolution (LTE) release 12 [2] from 3rd Generation Partnership Project (3GPP). Direct D2D communications are indeed a promising technological solution for local traffic between mobile User Equipments (UEs) in proximity in cellular

G. Araniti (✉) • A. Orsino • L. Militano
University Mediterranea of Reggio Calabria, Reggio Calabria, Italy
e-mail: araniti@unirc.it; antonino.orsino@unirc.it; leonardo.militano@unirc.it

A. Raschellà
Liverpool John Moores University, Liverpool, UK
e-mail: a.raschella@ljmu.ac.uk

M. Condoluci
King's College London, London, UK
e-mail: massimo.condoluci@kcl.ac.uk

W. Xiang et al. (eds.), *5G Mobile Communications*,
DOI 10.1007/978-3-319-34208-5_12

environments. In particular, devices being close to each other can activate direct links and bypass the base station (BS) or access point (AP) by either using cellular communications resources or using alternative radio technologies such as Wi-Fi.

The conventional cellular communications paradigm involving the BS remains still appropriate when users are not usually close enough to make possible direct communications and for traditional low data rate mobile services such as voice calls and text messages. However, this communication scheme can become inadequate with the modern high data rate services (e.g., video conferencing, interactive gaming, and mobile social networking). When the UEs are in range for D2D communications, this innovative communication paradigm is undoubtedly an advantageous solution to be considered to optimize the spectral efficiency of the network. Noteworthy, the advantages of D2D communications are not only limited to an improved spectral efficiency. In fact, also link coverage, throughput, energy consumption can benefit from this communication technology. Moreover, robust public safety communications can be supported where devices provide at least local connectivity in case of damage to the radio infrastructure.

D2D communications can be enabled through either cellular network transmitting on cellular spectrum, known as *inband communication*, or existing technologies that consider unlicensed bands such as Wi-Fi access points, known as *outband communication*. Concerning the inband approach, the cellular spectrum may be shared between D2D and cellular communications, which is knows as the *underlay inband D2D mode* [3]. In this case a main field of investigation is the mitigation of the interference between D2D and cellular communication [4]. As alternative, other works such as [5] propose to assign a portion of the cellular resources to D2D communications to avoid interference problems, which is known as *overlay inband D2D mode*. With this approach, resource allocation becomes the most important aspect to be investigated in order to avoid wasting precious spectrum resources [6]. The coordination between radio interfaces is either controlled by the BS (i.e, *controlled outband mode*) or by the users (i.e., *autonomous outband mode*). However, in both outband modalities baseline rules are needed to coordinate the communication over two different bands where also a second radio interface is involved (e.g., Wi-Fi Direct). Hence, the studies on outband D2D involve aspects such as power consumption and inter-technology architectural design [7–9].

Despite of the promising features of the D2D communications, there are some important challenges and issues still waiting for a solution before a widespread use of D2D communications in next 5G systems is granted. These include: (1) device discovery procedures with the aim to detect the presence of other UEs in the neighbourhood; (2) link setup strategies in order to properly select the spectrum to be used in the D2D radio links between interested UEs; (3) interference avoidance mechanisms that make possible the coexistence among D2D UEs with cellular network; (4) new devices design. Moreover, some research works has been conducted on as mobility management [10], multicast and broadcast communications and radio access procedures for D2D received little attention so far.

The interest for D2D communications as a key technology in the context of 5G mobile systems is witnessed by the initiatives of the mobile and wireless communications enablers for the twenty-twenty information society (e.g., METIS European project [11], 5G-PPP association [12], Networld2020 platform [13]). The main objective of these new initiatives is to handle the foundation of 5G, the next generation mobile and the wireless communications systems. In particular, for what concerns the D2D communications, the METIS project is currently evaluating its role in various scenarios such as vehicle-to-vehicle communications, national security and public safety, cellular network offloading, or service advertisement.

In the remaining of the chapter a general overview of D2D communications will be provided firstly introducing its possible role in the context of 5G systems. Then, an overview of the standardization, the system architecture, and some examples of scenarios and use cases will be given. Particular attention will then devoted to D2D communications over cellular networks and possible services exploiting this novel technology. Then a particular scenario will be presented, showing the potentialities offered by this technologies in enhancing the main performance features for the UEs in a downlink multicast service. In the last part of the chapter challenges and open issues, such as mobility and full-duplex communications will be discussed.

2 The Role of D2D Communications in Emerging 5G Systems

The D2D communications technology has been addressed in 3GPP LTE release 12 system [14]; notwithstanding, it is expected to have a complete standardization of proximity services in next 3GPP releases 13 and 14 [15]. As mentioned in the introduction to this chapter, the exploitation of D2D communications between UEs in proximity is expected to achieve improvements in terms of spectrum utilization, overall throughput, energy consumption, and to guarantee better public safety networks management. In what is presented next, a general overview of the current D2D standardization process is provided together with the system architecture proposed to integrate this new technology in the current cellular systems, and a number of possible applications in different scenarios and use cases.

2.1 Standardization Overview

The standardization process is an aspect of utmost importance to be considered for the commercial feasibility and future deployment of new technologies. In the particular case of D2D communications, although direct communications are already provided by the use of unlicensed Industrial, Scientific and Medical (ISM) bands (e.g., Wi-Fi Direct), its standardization in the context of the cellular system is currently still ongoing. A first example of the introduction of D2D communications into the LTE-Advanced (LTE-A) network is provided by Qualcomm Company, which developed a mobile communication system called

FlashLinq [16]. In particular, FlashLinq is a PHY/MAC network architecture, which allows cellular devices automatically and continuously discovering thousands of other FlashLinq enabled devices within 1 km and communicating peer-to-peer, at broadband speeds and without the need of intermediary infrastructures. Moreover, peer-to-peer communications enabled through Qualcomm's FlashLinq can share connectivity with a cellular network technology unlike Wi-Fi Direct's-based peer-to-peer. FlashLinq discovery procedure is carried out by broadcasting public/private expressions mapped into tiny 128-bit packages of data, which represent basic information of either devices or users.

From a standardization point of view, 3GPP is focusing its efforts on D2D communications (recently begun in release 12 [14]) for public safety Proximity Services (ProSe) [17]. This strategy has been initially targeted to allow LTE becoming a competitive broadband communication technology for public safety networks used by first responders. However, from a technical perspective point of view, the exploitation of the proximity nature of the communicating devices will provide the further performance benefits: (1) D2D UEs will be able to exploit high data rate with a low delay due to the short range; (2) compared to traditional downlink/uplink cellular communication, D2D will enable energy savings and improve radio resource utilization; (3) cellular data traffic offloading and, consequently, lower overload in the network. In detail, the 3GPP Radio Access Network (RAN) working group has proposed in TR 36.843 Rel. 12 [17] two basic functions for supporting *ProSe discovery* and *ProSe communications* over the LTE radio interface. ProSe discovery allows an UE using the LTE air interface to identify other UEs in proximity. Two kinds of ProSe discovery exist, namely *restricted* and *open*; the difference consists in whether the permission is necessary or not for the discovery for a UE. ProSe communication instead, is the data communication between two UEs in proximity using the LTE air interface. 3GPP Services working group (SA1) has defined in specification TR 22.803 [18] the use cases and scenarios for ProSe. In the document, conditions for service flows and potential requirements for different use cases are analyzed in order to provide a support for D2D systems design. Some examples of use cases and scenarios identified for general commercial/social use and network offloading are summarized below.

The following terms are defined by 3GPP in the description of D2D use cases:

- *ProSe Discovery*: it is a process that identifies a UE in proximity of another, using EvolvedUMTS Terrestrial Radio Access Network (E-UTRAN).
- *ProSe Communication*: it is a communication between two UEs in proximity through an E-UTRAN communication path established between the UEs. The communication path can for example be established directly between the UEs or routed via local evolved-NodeB (eNB).
- *ProSe-enabled UE*: it is a UE that supports ProSe Discovery and/or ProSe Communication.
- *LTE D2D*: it is a series of technologies characterized by ProSe capability.

2.2 Uses Cases and Scenarios Presented in 3GPP Rel. 12

Some examples of use cases for ProSe Discovery and ProSe Communication scenarios defined by 3GPP SA1 in specification TR 22.803 [17] are given below.

Restricted/Open ProSe Discovery: these are use cases for a basic ProSe Discovery scenario that can be exploited for any kind of application. In case of restricted ProSe Discovery, a ProSe-enabled UE discovers another UE in proximity only if it has previously achieved the permission; while, in case of open ProSe Discovery, a ProSe-enabled UE is able to discover neighbor devices without the necessity of a permission. An example of restricted use case is the friend discovery in a social network where the discovery is constrained by the UE's privacy settings. While a shop/restaurant advertisement is an example of open use case because shops and restaurants are open to be discovered by all the possible ProSe-enabled UEs in proximity, being free of privacy issues.

Network ProSe Discovery: it is a use case for ProSe Discovery scenarios where the Mobile Network Operator (MNO) verifies if a UE has the permission to discover another UE and the proximity. Therefore, in this case the network should be able to determine and provide the ProSe-enabled UEs with their proximity.

Service Continuity Between Infrastructure and E-UTRA ProSe Communication Paths: this is a use case for a ProSe Communication scenario where the operator is able to switch user traffic from the initial infrastructure communication path to the ProSe communication one. Then, the traffic can be addressed again towards an infrastructure path, without being perceived by the users. Hence, the operator should be able to dynamically control the proximity criteria (e.g., range, channel conditions, achievable QoS) for switching between the two communication paths.

ProSe-Assisted WLAN Direct Communications: WLAN direct communication is a use case available between ProSe-enabled UEs with WLAN capability when they are in Wi-Fi Direct communications range. It is based on the ProSe Discovery and the WLAN configuration information from the 3GPP Evolved Packet Core (EPC). In this case the operator is able to switch data session between infrastructure path and WLAN ProSe communication path.

ProSe Application Provided by the Third-Party Application Developer: in this case the operator can provide ProSe capability features in a series of APIs to third-party application developers. Through this cooperation between the operator and third-party application developers, the user can download and use a wide variety of new ProSe applications created by third-party developers. In this case the operator's network and the ProSe enabled UE should provide a mechanism that enables to identify, authenticate and authorize the third-party application to use ProSe capability features.

In Table 1 the available specifications together with the corresponding main topics provided by the 3GPP working groups is summarized. It can be noticed the presence of the mentioned SA1 and RAN working group handling, respectively, feasibility study for ProSe and LTE radio interface issues. Other examples of topics

Table 1 Available documents for D2D

SA1	TR 22.803	Feasibility study for Proximity Services (ProSe)
	TS 22.278	Service requirements for the Evolved Packet System (EPS)
	TS 22.115	Service aspects; changing and billing
	TS 21.905	Vocabulary for 3GPP specifications
SA2	TR 23.703	Study on architecture enhancements to support Proximity-based Services (ProSe)
	TS 23.303	Proximity-based Services (ProSe); Stage 2
SA3	TR 33.833	Study on security issues to support Proximity Services
RAN 1 & RAN 2	TR 36.843	Study on LTE device to device proximity services—Radio Aspects
CT1	TS 24.333	Proximity-services management object (MO)
	TS 24.334	Proximity-services (ProSe) user equipment (UE) to Proximity-services function aspects; Stage 3

under investigation supporting ProSe are the study of the architecture, security issues and Management Objects (MOs) representing parameters that handle the configuration of ProSe-enabled UEs.

2.3 System Architecture

In order to support the scenarios illustrated earlier in this chapter, the enhancements in the LTE architecture illustrated in Fig. 1 have been proposed. In details, this architecture aims at meeting the following requirements introduced by the 3GPP specifications to:

- Allow the operator to control the ProSe discovery feature in its network and authorizing the functionalities required for the ProSe discovery of each UE.
- Allow the ProSe communication or ProSe-assisted WLAN Direct communication and seamless service continuity when switching user traffic between an infrastructure path and a ProSe communication of the ProSe-enabled UEs.
- Allow an authorized third party ProSe application interacting with the 3GPP network in order to use the ProSe services provided by the network.
- Be able to control ProSe communication between ProSe-enabled UEs in case the UEs are served by either the same or different eNBs.
- Handle the ProSe-related security functions that correspond to privacy, support for regulatory functions including Lawful Interception, and authentication upon ProSe discovery and ProSe communication.
- Allow the operator's authorization and authentication of the third-party applications before making use of the ProSe features.

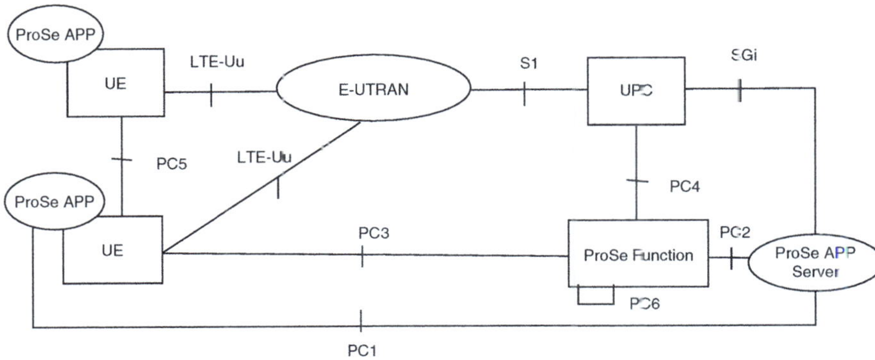

Fig. 1 3GPP D2D proposed architecture

As depicted in Fig. 1, in addition to the entities of the conventional LTE architecture (i.e., Evolved UMTS Terrestrial Radio Access Network, E-UTRAN and Evolved Packet Core, EPC), several new entities are required. These new entities are illustrated in the following.

- *Application servers (ProSe App Server)*: include the ProSe capability in order to build the application functionalities, such as specific agencies (PSAP) in case of Public Safety or social media in the commercial cases. These applications are proposed outside the 3GPP architecture; notwithstanding there may be reference points toward 3GPP entities. Moreover, the Application server can communicate with an application in the UE.
- *Applications in the UE (ProSe UEs App)*: exploit the ProSe capability in order to build the application functionality. An example may be the communication between members of Public Safety groups as well as social media application that requires finding other UEs in proximity.
- *ProSe Functions*: are a reference point toward the ProSe App Server, the EPC, and the UE. The functionalities may comprise (but not limited to): (1) inter-networking via a reference point toward the third-party applications; (2) authorization and configuration of UEs for discovery and direct communication, (3) allowing the functionality of the EPC-level ProSe discovery, and charging. Notice that for the interconnection of the new entities and the connection with the conventional LTE ones, seven new interfaces/reference points are illustrated in the figure as PC1, PC2, PC3, PC4, PC5, PC6, and SGi (Fig. 1).

2.4 Application Scenarios

Applications of 5G D2D communications include local service, emergency communication, and the Internet of Things (IoT) enhancement. A brief description of these applications is provided in the following.

2.4.1 Local Service

In this scenario, user data is directly transmitted between terminals without being routed through the network side. Local service is usually utilized for social apps that are a basic D2D application based on the proximity feature. Through the D2D discovery and communication functions, a user can find other close users in order to share data or play games with them.

Another basic application of local service is the local data transmission, which exploits the proximity and direct data transmission characteristics of D2D to extend mobile applications while saving spectrum resources and then, making possible a new source of revenue for operators. In fact, local advertising service based on proximity can accurately target people in order to improve its benefits. Some examples of local transmissions conceived to improve commercial benefits are: a shopping mall where discounts and commercial promotions are sent to people walking into or around the mall; a cinema where information about movies and showtimes can be sent to people close by.

A third application of local service is the cellular traffic offloading that can reduce network overloading problems. In fact, consider that nowadays media services are becoming more and more popular; their massive traffic flows cause an extensive pressure on core networks and spectrum resources. In this context, D2D-based local media services allow operators to save spectrum resources in their core networks. In hotspot areas, operators or content providers can exploit media servers storing popular media services and sending them in D2D modality to the users. Alternatively, users can utilize D2D communications to obtain the media content from close terminals which have obtained media services. This enables to optimize the downlink transmission pressure of operator cellular networks. Furthermore, the cellular communication between short-distance users can be switched to the D2D modality in order to offload cellular traffic.

2.4.2 Emergency Communications

Natural disasters such as earthquakes can damage traditional communication network infrastructures making networks not available and causing enormous rescue efforts. This problem could be overcome through the introduction of D2D communications. In fact, although the communication network infrastructures may be irremediably affected, a wireless network can still be created between terminals based on the D2D connections. This means that an ad hoc network can be set up based on multi-hop D2D to guarantee smooth wireless communication between users. Moreover, a wireless network affected by terrain or buildings can have blind spots. With single-hop or multi-hop D2D communication, users may be connected in the blind spots to other users, which are in coverage areas and then, be connected to the wireless network.

2.4.3 IoT Enhancement

One of the main aims of designing new mobile communication technologies is to create an extensive interconnection among different networks involving various types of terminals. This is the motivation, which has pushed forward the development of the Internet of Things (IoT) in the cellular communication framework. The industry forecast says that by 2020 there will be 50 billion cellular access terminals on a global scale and most of them will be devices with the IoT feature. In this context, the connection between D2D with IoT will drive towards a truly interconnected wireless network.

A common application of D2D-based IoT challenge is vehicle-to-vehicle (V2V) communication in the Internet of Vehicles (IoV). For instance, when a vehicle runs at high speeds, it can warn close vehicles in D2D mode before it changes lanes or slows down. According to the received messages, close vehicles warn drivers or even automatically handle the driving in an emergency situation; hence, thanks to this application drivers can react more quickly to diminish the number of traffic accidents. D2D communications provide inherent advantages when they are considered in the context of IoV security issues also thanks to their favorable features in terms of communication delay and neighbor discovery.

As there exist many IoT devices in a 5G network, access load is becoming a serious issue to be taken into account. Nevertheless, D2D-based network access is expected to improve this problem. In a scenario characterized by many terminals, low-cost terminals can access close special terminals in D2D modality instead of direct connections with BSs. Moreover, if multiple special terminals are isolated, the wireless resources for accessing low-cost terminals may be reutilized by these special terminals. Notice that this not only improves access pressure on BSs, but also optimizes the spectrum efficiency. Furthermore, the D2D-based access modality is more flexible and costs less than the small cell structure of the existing 4G networks.

In a smart home application, a smart terminal may be considered as a special terminal. Wireless appliances in the smart home access the smart terminal in D2D modality; while, the smart terminal may access the BS in a traditional cellular mode. The cellular-based D2D communication can represent a real breakthrough for the development of the smart home industry.

2.4.4 Other Applications

D2D communications may also be considered in other potential scenarios, such as multiuser MIMO enhancement, cooperative relaying, and virtual MIMO. In the context of the traditional multiuser MIMO, BSs find precoding weights based on the feedback received by the terminals in the respective channel in order to create nulls and delete interference between users. Through the introduction of D2D communications, paired users may directly exchange information about channel status. Hence, terminals can put together channel status information to be sent to the BSs improving the performance of multi-user MIMO.

D2D communications may also contribute to solve problems in new wireless communication scenarios. For instance, in the indoor positioning terminals may not achieve satellite signals if they are indoors. In this case, the traditional satellite-based positioning cannot work efficiently. In case of D2D-based indoor positioning, either pre-deployed terminals with given location information, or usual outdoor terminals with given position can detect the location of terminals to be localized, and support indoor positioning at a low cost in 5G networks.

3 State-of-the-art on D2D Communications Over Cellular Networks

D2D communications are expected to play a key role in the ecosystem of future 5G cellular networks. This is motivated by two aspects: (1) the amount of data traffic exchanged over radio mobile systems is exponentially increasing and this dictates novel communications paradigms for radio mobile networks; (2) use cases for D2D communications presented above represent key 5G services. As a consequence, the natively support of D2D communications becomes crucial in 5G systems.

D2D communication was initially proposed in cellular networks as a new paradigm to enhance network performance. Several studies in the literature have already discussed the improvements in terms of spectral efficiency and reduced communication delay that D2D communication can provide in cellular networks [19–24]. On the other hand, this new paradigm presents several aspects to be investigated in terms for instance of interference control overhead and network protocols. Therefore, the feasibility of D2D communications in the context of LTE-A is currently a fascinating topic under investigation by academia, industry, and the standardization bodies. A general overview of state-of-the-art applications based on D2D communications for future 5G wireless systems is given next in both, uplink and downlink scenarios. Then, some examples of services where D2D communications have been efficiently exploited in LTE-A networks will be illustrated and assessed through exhaustive performance evaluation.

Several studies addressing D2D communications for downlink services can be found in the literature, covering several aspects and applications as for instance mobile data offloading [25], cell coverage extension [26] or content sharing [27, 28]. Recently, D2D communications have been taken into account also for downloading multicast services with focus on direct device communications over short links of a different technology than the cellular one. To cite some of them, in [29] a subset of mobile devices are considered as anchor points in a cell to forward the multicast data received from the BS to other devices in proximity through multihop ad-hoc Wi-Fi links. In [30] cellular users directly communicate to carry out cooperative retransmissions using generic short-range communication capabilities. However, the use of heterogeneous wireless interfaces introduces several issues in terms of content synchronization that become essential in case of multicast video streaming applications. Moreover, as also stated in [31], the use of cellular D2D links provides several benefits compared to *outband* D2D links, like Wi-Fi, in terms of improved user throughput. Although, the focus of the literature has been mainly on technical issues for downlink services, uplink direction scenarios are of undoubted interest as also witnessed by recent publications, such as [32] where relaying on smartphones is proposed to transmit emergency messages from disconnected areas. Multihop D2D communications have been also investigated in a very few recent works. In [33, 34] network-assisted D2D communication is addressed with an analysis on power control and mode selection on the direct links. However, the analysis refers to a more traditional two-hop scenario, with a UE or the eNodeB as the last hop node. Similarly, multihop D2D communication is considered in [35, 36] for end-to-end Machine-to-Machine and human-traffic connectivity.

As an example of D2D communications over cellular LTE-A links a downlink scenario for multicast transmission is considered in order to efficiently overcome the limitations identified in [29, 30]. In details, in the scenario proposed in this study a portion of multicast users, which sense poor channel qualities is split into clusters. The members of these clusters are served through cellular D2D transmissions, while the remaining users (i.e., those with better channel quality) are served over cellular transmission from the BS.

All the solutions illustrated in this section exploit D2D communications relying on LTE-A network infrastructure. In LTE-A, Orthogonal Frequency Division Multiple Access (OFDMA) and Single Carrier Frequency Division Multiple Access (SC-FDMA) are considered, respectively, in case of downlink and uplink. The eNodeB handles the spectrum resources by providing the appropriate number of RBs to each scheduled user and by selecting the Modulation and Coding Scheme (MCS) for each RB. Scheduling solutions are based on the Channel Quality Indicator (CQI) feedback, which is sent by a UE to the eNodeB over dedicated control channels. Each CQI value correspond to a given maximum supported MCS as specified in [37]. The MCS parameters can be adapted at every CQI Feedback Cycle (CFC), which can last one or several Transmission Time Intervals (TTIs) where one TTI is 1 ms.

3.1 A Case Study: D2D-Enhanced Conventional Multicast Scheme for Video Delivery

Nowadays, several research and industrial organizations are considering solutions to best manage the increased traffic and the dissimilar channel quality sensed by users in the same multicast group and efficiently handle the available resources [38]. A basic approach to this issues is the Conventional Multicast Scheme (CMS) [38], which serves all multicast users in a cell at every TTI, by constraining the data rate to the user with the worst channel conditions. However, this approach translates into poor performance in terms of data rate and low satisfaction levels for users with good channel situations. An alternative solution is the Opportunistic Multicast Scheme (OMS) [39], which only serves users with the best channel conditions in each TTI. This enables to enhance the network data rate, but short-term fairness is no more assured to users and this could affect the delivery of time sensitive services such as video to some users. In this study it is demonstrated how multicast delivery strategies in LTE-A systems can benefit from D2D communications. For instance, neighboring UEs that belong to the same multicast group can exploit direct links by utilizing cellular radio resources [21, 40], to cope with adverse cell-edge effects. This study exploits the D2D communications in the multicast communication framework to complement CMS in order to address some of the limitations of the latter. In details, it is proposed a D2D-enhanced Conventional Multicast Scheme (CMS) with Single Frequency (D2D-SF) by combining in an innovative way CMS-based and D2D content delivery in order to improve the aggregate data rate of the cell, while also keeping the short-term fairness between devices. D2D-SF solution makes use of one or more mobile subscriber devices as forwarding devices (FD) to retransmit the data received from the BS over direct local links to other members of the multicast group. It is considered that the D2D links exploit uplink frequencies, as suggested in [22] and all the FDs in the same cell simultaneously use *the same frequency* to transmit multicast data over the D2D links, as illustrated in [41, 42]. The receivers take into account these retransmissions as multipath components of the same signal. D2D-SF is defined in conjunction with a policy for selecting FDs based on the devices clustering. Simulation results demonstrate how significant enhancements of the overall performance when delivering video content can be achieved compared to both state-of-the-art multicast strategies and a D2D novel solution that does not consider the single-frequency paradigm.

3.1.1 System Model and Problem Formulation

The LTE/MBMS architecture [42] is composed as follows (i.e., please refers to Fig. 2 to have a general overview). The access network is made of the eNodeB and the MultiCell/Multicast Coordination Entity (MCE), in charge of transmission parameters configuration in single- and multi-cell mode, respectively. The core network involves: Mobility Management Entity (MME) that is in charge for

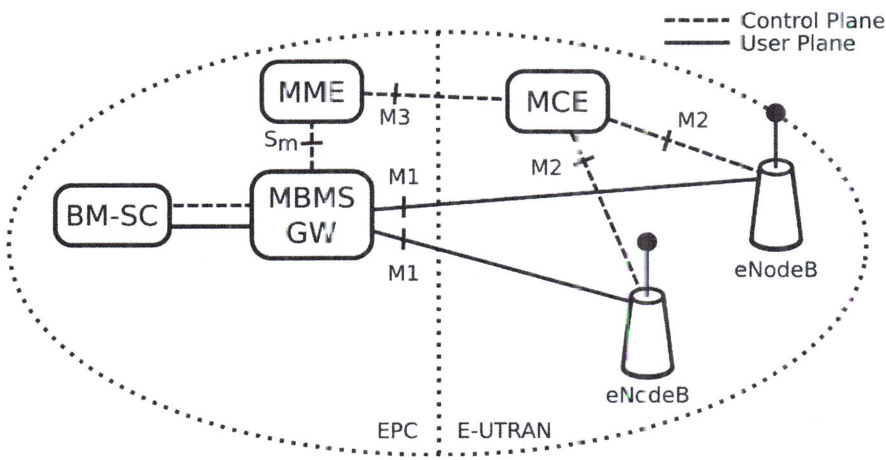

Fig. 2 LTE/MBMS architecture

authentication, security, and mobility management procedures; MBMS Gateway (MBMS-GW), a logical entity whose principal function is data packets forwarding to eNodeBs; Broadcast Multicast-Service Center (BM-SC), which is the MBMS traffic source accomplishing also service announcement and group membership functions. MBMS entities rely on the LTE-A system illustrated in Sect. 3 considering TDD, by referring to the frame structure *type 2* foreseen by 3GPP [41] and *configuration 1* which guarantees an equal number of downlink and uplink slots over the frame.

Let us consider a group of UEs that is interested in the same multicast content provided by a single LTE-A cell. Under this condition a multicast video delivery can be provided exploiting the enhancements offered by D2D communications among the involved devices. In the considered LTE-A cell, a set of users, denoted by $\mathscr{K}$, is associated to the same multicast group (MG). The eNodeB performs link adaptation procedures on both cellular and D2D links by managing N available RBs according to the CQI feedbacks collected from each user. Let C be the number of available CQI levels and let $c_k \in \{1, 2, \ldots, C\}$ be the CQI reported by multicast member k, with $k \in \mathscr{K}$. Furthermore, let $c_{k,j}$ be the CQI value for each D2D link between nodes $k, j \in \mathscr{K}$, $k \neq j$. For a given MCS value m, the available data rate depends on the number of assigned RBs and on the spectral efficiency for the given MCS, b_m expressed in bit/s/Hz as explained in [31]. Hence, the spectral efficiency is denoted with b_m^{dl} and b_m^{ul} (where $m = 1, \ldots, C$) respectively in downlink and uplink transmissions. Moreover, $f^{dl}(m, n_m)$ and $f^{ul}(m, n_m)$ represent the data rate respectively in downlink and uplink transmissions adopting the MCS associated to the CQI m, as a function of m and the assigned RBs. The proposed radio resource management (RRM) scheme is responsible of deciding which multicast configuration to allow, by this meaning: (1) the set of UEs directly served by the eNodeB in downlink, (2) the MCS for the downlink transmission, (3) the cluster configuration for D2D relaying, and (4) the resource allocation and the MCS assignment for the transmissions of each activated FD. The eNodeB carries out the following steps when the service delivery starts. In details, a single execution of the listed steps is carried out. However, when significant variations in the channel conditions are sensed, these steps should be repeated to update the service configuration.

1. *Service registration*: The eNodeB advertises the multicast service and all interested UEs within the cell join this service to create a single MG.
2. *CQI collection*: The eNodeB collects the CQI feedbacks from all UEs belonging to the MG, i.e., $c_k \; \forall k \in \mathcal{K}$.
3. *D2D CQI collection*: The eNodeB collects the $c_{k,j}$ values from all UEs $k, j \in \mathcal{K}$, $k \neq j$ belonging to the MG. This information will be utilized to discover the UEs that can be reachable through D2D links by selected FDs in the MG.
4. *FD selection and cluster formation*: Being $\tilde{\mathcal{C}} \subseteq \{1, 2, \ldots, C\}$ the set of CQI levels in downlink for the UEs in the MG, for each $m \in \tilde{\mathcal{C}}$ the eNodeB computes: (1) the set of UEs that can correctly decode data if served by the BS, i.e., $\mathcal{K}_m^{dl} = \{k \in \mathcal{K} | c_k \geq m\}$; *(ii)* the subset of served UEs $\mathcal{R}_m \subseteq \mathcal{K}_m^{dl}$, that can act as FDs; (3) the remaining UEs that are not served by the eNodeB, but can be served by a FD through D2D connections. Hence, the eNodeB determines a D2D CQI matrix (DCM) based on the $c_{k,j}$ values (where $k \in \mathcal{K}_m^{dl}$ and $j \in \mathcal{K} \setminus \mathcal{K}_m^{dl}$) for all the links between the potential FDs (the matrix rows) and the remaining nodes (the DCM columns). A $c_{k,j} = 0$ value in the DCM represents that a D2D link cannot be activated between nodes k and j. Following the values in the DCM, the eNodeB will select the subset of UEs $\mathcal{D}_{m,r} \subseteq \mathcal{K} \setminus \mathcal{K}_m^{dl}$ to be associated to each allowed FD $r \in \mathcal{R}_m$.
5. *D2D link configuration*: For each CQI level $m \in \tilde{\mathcal{C}}$ evaluated for downlink transmissions, the eNodeB determines the resource $N_{m,r}^{ul}$ and the MCS level $l_{m,r}$, to be utilized on the D2D link for each FD $r \in \mathcal{R}_m$. D2D links can be either unicast or multicast. A conservative approach is considered in the multicast case; hence, the FD serves all UEs in the D2D cluster in a single transmission by using the MCS corresponding to the worst CQI value in the DCM, i.e., $l_{m,r} = \min_{k \in \mathcal{D}_{m,r}} \{c_{r,k}\}$ for FD r. This study takes into account two alternative policies according to which the FDs manage the uplink frequencies to send data in their own D2D cluster. The first policy associates different resources to the different FDs; the second one implements the novel single-frequency-based D2D paradigm, i.e., all the FDs considers the same RBs. In the former case, disjoint sets of RBs are assigned to the D2D links. In the latter case the amount of resources assigned to the D2D links are constrained by the cluster with the lowest activated MCS. In general, devices connected on a D2D link are expected to be at a short distance and with good channel conditions, thus they need a lower amount of resources compared to those needed for a direct cellular communication.
6. *Multicast service activation and resource allocation*: Finally, the eNodeB chooses the solution to activate, which is the one that optimizes the system data rate under the constraint that all the UEs in a MG are served, either through

direct cellular links or through D2D links. In details, after the selection of the MCS level m^* to activate in downlink and of the corresponding $\mathcal{K}_{m^*}^{dl}$, $\mathcal{R}_{m^*}$, $\mathcal{D}_{m^*,r}$, $N_{m^*,r}^{ul}$, and $l_{m^*,r}$ values, the eNodeB assigns the available resources.

All values $m \in \tilde{\mathcal{C}}$ are potential CQI levels to activate in downlink. For each of the CQI levels a cluster formation algorithm is implemented to provide a *configuration* of FDs and corresponding D2D clusters. If a given tested level is *eligible*, then the corresponding data rate Ω_m is determined. A cluster configuration is taken into account *eligible* if the FDs are able to forward the total amount of bits received from the eNodeB over the D2D links to all users not served by the cellular link. This needs the following conditions: (1) the enabled FDs can successfully serve all the nodes belonging to $\mathcal{K} \setminus \mathcal{K}_m^{dl}$ via D2D links, and (2) the N available resources are enough to relay all data to the D2D receivers. If instead, no cluster configuration for the tested CQI level m can be found, thus the iteration on the $m \in \tilde{\mathcal{C}}$ value is stopped and the final selection is carried out. In details, the iteration can be stopped as the tested CQI levels follow an order from the minimum to the maximum CQI value, and with higher values for the CQI level in downlink the probability of getting an *eligible* configuration is reduced. Fundamental steps in the implementation of the proposed RRM previously discussed, are the FD selection and cluster formation and the D2D link configuration with the radio resource allocation. Concerning the first strategy, let us consider the generic iteration where the m-th CQI level is tested for downlink transmission. Given $\mathcal{K}_m^{dl}$, the set of UEs that can correctly decode the data according to the considered CQI, and based on the DCM, the eNodeB determines which nodes can potentially act as FDs for the remaining $\mathcal{K} \setminus \mathcal{K}_m^{dl}$ nodes. Based on this information the eNodeB can assign the resources to each D2D link. Let us focus now on the BSC policy, which is based on the idea that the eNodeB chooses "the best" FD for each UE not served in downlink. In particular, the best FD for each node j belonging to $\mathcal{K} \setminus \mathcal{K}_m^{dl}$ is considered as the node $r \in \mathcal{K}_m^{dl}$ which assures the best D2D link conditions. In those cases where more than one FD can assure the same CQIs, the eNodeB selects the FD serving more users in order to limit the number of FDs.

3.1.2 Performance Results

Main simulation parameters are listed in Table 2. $R = 100$ RBs are available in the LTE system on a 20 MHz channel bandwidth. Channel conditions for the UEs are evaluated in terms of signal to interference and noise ratio (SINR) experienced over each sub-carrier [43] when path loss and fading phenomena affect the signal reception. The effective SINR is mapped onto the CQI level that guarantees a block error rate (BLER) smaller than 1 % [43, 44].

The following metrics have been considered to assess the performance of the proposed solutions with respect to CMS [38] and OMS [39]:

- *mean data rate* is the mean data rate value experienced by the multicast members;

Table 2 Main simulation parameters

Parameter	Value
Cell radius	500 m
Frame Structure	Type 2 (TDD)
TTI	1 ms (11 OFDM data symbols plus 3 control symbols)
Cyclic prefix/Useful signal frame length	16.67 μs / 66.67 μs
TDD configuration	1
Carrier Frequency	2.5 GHz
eNodeB Tx power	46 dBm
D2D node Tx power	23 dBm
Noise power	−174 dBm/Hz
Path loss (cell link)	128.1 + 37.6 log(d), d[km]
Path loss (D2D link, NLOS)	40 log(d) + 30 log(f) + 49, d[km], f[Hz]
Path loss (D2D link, LOS)	16.9 log(d) + 20 log (f/5) + 46.8, d[m], f[GHz]
Shadowing standard deviation	10 dB (cell mode); 12 dB (D2D mode)
RB size	12 sub-carriers, 0.5 ms
Sub-carrier spacing	15 kHz
BLER target	1 % [44]

- *aggregate data rate (ADR)* is the sum of the data rates experienced by the multicast users;

The performance analysis in terms of the metrics indicated focuses on the video streaming towards multicast users lasting 1 s; within this interval the BS adapts the transmission parameters every scheduling frame (i.e., 10 ms) [45]. In particular, for this analysis, two different study cases are considered:

- **Case A:** This case analyses the impact that the channel bandwidth has on the considered strategies. In this case the multicast group size $|\mathscr{K}|$ is set to 200, whereas a variable number of resources R (ranging from 10 to 100 RBs) is dedicated to the service. UEs are distributed within an area of 100 m × 100 m located near the cell-edge;
- **Case B:** This case analyses the impact a varying multicast group size has on the considered strategies. The number of available resources R is set to 100 RBs, whereas the number of UEs $|\mathscr{K}|$ ranges from 20 to 200. The same cell-edge distribution of UEs as in case A is considered.

Video parameters are set in accordance to [46], where adaptive video coding [47] is considered to be performed at the BS. Video parameters are tuned such that the video stream has an average bit rate between 256 kbps and a maximum value, which depends on the channel quality experienced by multicast users.

Case A The results achieved are plotted in Fig. 3. As expected, both mean data rate and ADR improve with the number of available RBs for all solutions. The proposed D2D-based strategies outperform CMS, with a better performance for *D2D-SF*. The

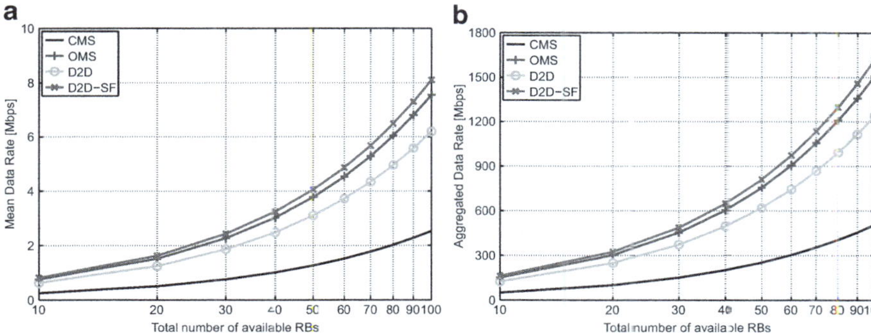

Fig. 3 Performance analysis for video streaming analysis: study case A. (**a**) Mean UE data rate, (**b**) aggregate data rate

single-frequency solution also outperforms OMS. The benefit compared to CMS is constant with the number of available RBs and equal to 144 % for the *D2D* and *D2D-SF* solutions. The mean data rate for CMS is lower than *D2D* (left-hand in Fig. 3). The data rate of the OMS solution gets higher values than the *D2D* policy, but lower values compared to the solution based on the single-frequency paradigm. The price to pay when adopting OMS is the decrease in the number of served users and in short-term fairness. A similar behavior is achieved for ADR (right-hand in Fig. 3).

Case B The performance obtained by varying the multicast group size is shown in Fig. 4. Also in this case, the novel scheme considering the single-frequency paradigm, i.e., *D2D-SF*, outperforms the others. It is worth noting that for all the solutions, the mean data rate slightly diminishes when the number of users in the cell increases, as illustrated in left-hand plot in Fig. 4. In details, the CMS shows a performance reduction of about 29 % when changing from 20 to 200 UEs. When considering *D2D* there is a 25 % of reduction. This is an expected result as greater the number of users in the group, higher the risk of having users with very low channel conditions, which limit the overall performance. Also OMS senses a reduction in its offered mean data rate with the increase in number of UEs. Nevertheless, this phenomenon is less evident, as the mean data rate diminishes from 7.9 Mbps (with 20 UEs) to 7.5 Mbps (with 200 UEs). As expected, the right-hand in Fig. 4 illustrates how the ADR value increases with the number of multicast members in the cell. Furthermore, the gain introduced by *D2D* with respect to the CMS policy is larger when the number of multicast users increases.

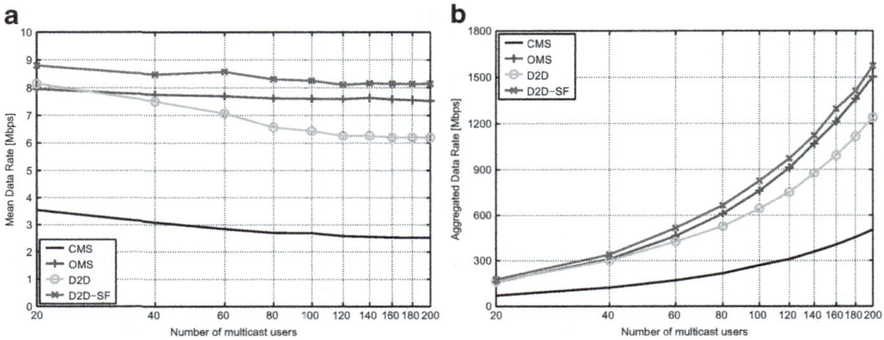

Fig. 4 Performance analysis for video streaming analysis: study case B. (**a**) Mean UE data rate, (**b**) aggregate data rate

4 Challenges and Ongoing Research

Future 5G cellular networks are attracting the interest of the research community and engineers from the entire world. In particular, 5G technologies are expected to attain 1000 times higher mobile data volume per unit area, 10–100 times higher number of connecting devices and user data rate, 10 times longer battery life, and five times reduced latency [48]. However, 5G is currently without an official definition. Notwithstanding, the research community has already highlighted how the above objectives can be potentially reached through a multi-tier heterogeneous network architecture together with several technologies such as spatial modulation, millimeter wave (mmWave), visible light communication (VLC), and massive MIMO. In the following new challenges in the context of D2D communications as innovative technology supporting 5G will be illustrated.

4.1 Mobility

The development of wireless networking technologies has brought enormous changes to people's everyday life. In particular, the vision of 5G networks empowers users to communicate through different wireless access technologies enabling new applications in modern communication paradigms, such as vehicular ad hoc networks (VANETs) and mobile cloud computing (MCC). These scenarios, however, introduce to one of the major challenging features related to D2D communications, namely user mobility. In particular, it affects network connectivity and the design of network protocols. On the other hand, another important aspect to be considered is the complex traffic flow related to a dynamic network topology and the double communication opportunities (i.e., infrastructure or infrastructureless wireless networks). The interest for D2D communications in Vehicle-to-Vehicle

(V2V) communication is related to the strict delay requirements in traffic safety use cases [49]. In particular, for collision avoidance systems it will be essential to have very low latency. Hence, it can obtain more complete and reliable information, which would be useful to avoid accidents. By using D2D communications rather than relying on infrastructure, it is possible to both diminish the latency and to develop a solution, which works without cellular network coverage so that a car can get more complete and reliable information to avoid accidents.

4.2 Heterogeneous Networks

D2D communications may be also considered as an efficient solution for Multiuser cooperative communication (MUCC). For instance, a benefited user (BU) is in an area experiencing a poor cellular signal, but there is another user in the proximity sensing a good cellular signal. This user may be considered for helping the BU to enhance its experienced signal and then, acting as a so-called supporting user (SU). There exist two radio links for the BU: BU towards eNB (i.e., link directly) and SU directly towards picocells. Moreover, both BU and SU communicate with each other through LTE-A D2D communication. This kind of communication enhances the throughput because the system will always be able to schedule the best user experiencing the most favorable channel quality. There exist different channel/pipes and the probability that all channels are degraded at the same time is quite small, and this will improve reliability. Moreover, any single pipe improvement may be also considered at the same time (e.g., MIMO). Furthermore, MUCC improves the performance of this scheme through the context information (e.g., location information) of the users.

4.3 mmWave

Millimeter wave (mmWave) communication operating in the 10–300 GHz band, appear to be an innovative candidate for next-generation cellular systems by which multiple gigabit-per-second data rates can be supported. On the other hand, enabling mmWave cellular systems in practice requires properly dealing with the channel impairments and propagation characteristics of the high frequency bands. In fact, the main propagation-related obstacles in realizing mmWave cellular are that free-space path loss is much larger in mmWave due to the higher carrier frequency, scattering is less significant which reduce the available diversity, and no line-of-sight (LOS) links are weaker making blockage and coverage holes more pronounced. Moreover, the noise power is larger due to the use of larger bandwidth channels. Furthermore, it is difficult for mmWave signals to penetrate through solid materials that confines outdoor mmWave signals to streets and other outdoor structures, although some signal power may get inside the buildings through glass windows and wood doors.

The mentioned propagation features give opportunities and advantages in allowing D2D communications over directional mmWave networks. In fact, mmWave can be considered in D2D-enabled wireless devices for direct short range communications amongst users or machines in near proximity. Moreover, D2D communication can also provide a path between two wireless devices and between wireless devices and mmWave BSs by relay if LOS paths are no available. Furthermore, mmWave communication with highly directional antennas gives wireless connections characterized by a high data rate for BS to BS (B2B) communications, in contrast with 4G cellular networks where fiber links communications between BSs are considered. Finally, interference management is one of the most important issue that may be enhanced through the exploitation of D2D communications. Due to the directional antennas and large propagation loss, mmWave communication has relatively low multi-user interference (MUI) that may support simultaneous communications over the same radio spectrum. Enabling multiple concurrent D2D links, the network capacity and spectrum efficiency may be efficiently enhanced. Hence, new network architectures and resource sharing strategies that will account for the directional interference sources are needed in mmWave 5G cellular networks to fully exploit the enhancements of both technologies.

4.4 Massive-MIMO

Massive MIMO is a technology whose features can be efficiently exploited in the context of D2D communications in uplink scenarios of cellular networks. In details, in a massive MIMO system, each BS utilizes a very large antenna array to serve multiple users in each time-frequency resource block [50]. If the number of antennas at a BS is considerably greater than the number of served users, the channel of each user towards the BS is closely orthogonal to that of any other user. This aspect enables to process techniques for very simple transmit or receive such as matched filtering to be nearly optimal with enough antennas even in the presence of interference. This means that, a large antenna array at a BS may drive D2D communications towards a very low interference at the uplink massive MIMO BS, making D2D very simple and appealing in massive MIMO systems. However, although D2D-to-cellular interference can be effectively managed by the large antenna array at a BS, cellular interference provoked in D2D communications holds and can be worse in a massive MIMO system. In details, massive MIMO is a multi-user transmission strategy developed in order to support multiple users in each time-frequency block; the number of users simultaneously active in an uplink scenario is scalable with the number of antennas at the BS. Considering this increased number of uplink transmitters, the D2D connections reutilizing uplink radio resources will sense increased interference. In order to protect D2D connections against such increased interference, the number of simultaneously active uplink users might have to be limited.

4.5 Full-Duplex

Duplex mode is a recent innovative research direction related to D2D communications development. In particular, when considering FDD, two receivers and two antennas are needed. The Tx antenna and Rx one should have enough isolation distance. It is always challenging implementing into UE with small physical size. While TDD has the advantage that uplink downlink transmissions/receptions are performed in different time slots and in the same frequency band. This is basically why most of D2D researchers are currently focusing on TDD modality [51]. It is predicted that in 5G D2D will be supported by both TDD and FDD. Notwithstanding, D2D TDD mode will be the principal type considered thanks to its lower cost, lower complexity, and the possible benefit of better frequency availability.

5 Conclusion

In this chapter an overview of the D2D-communications has been presented as one of the promising technological features for the definition of future 5G wireless networks. The details related to the standardization process, the possible applications and technological issues, have been browsed together with the important benefits such a technology can offer. In the second part of the chapter a practical application of D2D communications has been discussed, with reference to a scenario where the proximity communications between devices in an LTE-A system, introduces benefits to a multicast downloading service. Important improvements in terms of data rate increase for the involved users has been demonstrated through simulative results. Finally, in the last part of this chapter open challenges and research field directions have been introduced for the design of novel technological solutions integrating also mobility issues, mmWave communications, massive MIMO and Full-Duplex.

References

1. X. Lin, J.G. Andrews, A. Ghosh, R. Ratasuk, An overview on 3GPP device-to-device proximity services. IEEE Commun. Mag. **52**(4), 40–48 (2014)
2. 3GPP, 3rd Generation Partnership Project; Technical Specification Group SA; Feasibility Study for Proximity Services (ProSe) (Release 12). TR 22.803 V1.0.0, August 2012
3. C.-H. Yu, K. Doppler, C. Ribeiro, O. Tirkkonen, Performance impact of fading interference to device-to-device communication underlaying cellular networks, in *Proceedings of IEEE PIMRC* (IEEE, Piscataway, 2009), pp. 858–862
4. L. Lei, Y. Zhang, X. Shen, C. Lin, Z. Zhong, Performance analysis of device-to-device communications with dynamic interference using stochastic petri nets. IEEE Trans. Wirel. Commun. **12**(12), 6121–6141 (2013)

5. Y. Pei, Y.-C. Liang, Resource allocation for device-to-device communication overlaying two-way cellular networks. IEEE Trans. Wirel. Commun. **12**(7), 3611–3621 (2013)
6. L. Lei, Y. Kuang, X. Shen, C. Lin, Z. Zhong, Resource control in network assisted device-to-device communications: solutions and challenges. IEEE Commun. Mag. **52**(6), 108–117 (2014)
7. A. Asadi, V. Mancuso, Energy efficient opportunistic uplink packet forwarding in hybrid wireless networks, in *Proceedings of the Fourth International Conference on Future Energy Systems* (2013), pp. 261–262
8. A. Asadi, V. Mancuso, On the compound impact of opportunistic scheduling and D2D communications in cellular networks, in *Proceedings of the 16th ACM International Conference on Modeling, Analysis and Simulation of Wireless and Mobile Systems (MSWiM '13)* (ACM, New York, 2013), pp. 279–288
9. M. Condoluci, L. Militano, A. Orsino, J. Alonso-Zarate, G. Araniti, LTE-direct vs. WiFi-direct for machine-type communications over LTE-A systems, in *2015 IEEE 26th Annual International Symposium on Personal, Indoor, and Mobile Radio Communications (PIMRC)*, 30 August–2 September 2015, pp. 2298–2302
10. A. Orsino, M. Gapeyenko, L. Militano, D. Moltchanov, S. Andreev, Y. Koucheryavy, G. Araniti, Assisted handover based on device-to-device communications in 3GPP LTE systems, in *IEEE Globecom Workshop on Emerging Technologies for 5G Wireless Cellular Networks*, December 2015
11. EU Project METIS, https://www.metis2020.com
12. The 5G Infrastructure Public Private Partnership (5GPPP), https://5g-ppp.eu
13. NetWorld2020, http://networld2020.eu
14. D. Astely, E. Dahlman, G. Fodor, S. Parkvall, J. Sachs, LTE release 12 and beyond [Accepted From Open Call]. IEEE Commun. Mag. **51**(7), 154–160 (2013)
15. Ericsson white paper: LTE Release 13, http://www.ericsson.com/res/docs/whitepapers/150417-wp-lte-release-13.pdf
16. W. Xinzhou, S. Tavildar, S. Shakkottai, T. Richardson, L. Junyi, R. Laroia, A. Jovicic, FlashLinQ: a synchronous distributed scheduler for peer-to-peer ad hoc networks. IEEE/ACM Trans. Networking **21**(4), 1215–1228 (2013)
17. 3GPP TR 36.843, Study on LTE device to device proximity services; Radio aspects, v12.0.1, March 2014
18. 3GPP TR 22.803, Feasibility study for proximity services (ProSe), v12.2.0, June 2013
19. IEEE Standard 802.16e-2005, Air interface for fixed and mobile broadband wireless access systems amendment for physical and medium access control layers for combined fixed and mobile operation in licensed band (2005)
20. B. Kaufman, B. Aazhang, Cellular networks with an overlaid device to device network, in *Proceedings of Asilomar Conference on Signals, Systems and Computers* (2008), pp. 1537–1541
21. K. Doppler, M. Rinne, C. Wijting, C. Ribeiro, K. Hugl, Device-to-device communication as an underlay to LTE-advanced networks, *IEEE Commun. Mag.* **47**(12), 42–49 (2009)
22. K. Doppler, M.P. Rinne, P. Janis, C. Ribeiro, K. Hugl, Device-to-device communications; functional prospects for LTE-Advanced networks, in *Proceedings of IEEE ICC Workshops* (2009), pp. 1–6
23. A. Osseiran, K. Doppler, C. Ribeiro, M. Xiao, M. Skoglund, J. Manssour, Advances in device-to-device communications and network coding for IMT-Advanced, in *ICT Mobile Summit* (2009)
24. T. Peng, Q. Lu, H. Wang, S. Xu, W. Wang, Interference avoidance mechanisms in the hybrid cellular and device-to-device systems, in *Proceedings of IEEE PIMRC* (2009), pp. 617–621
25. S. Andreev, O. Galinina, A. Pyattaev, K. Johnsson, Y. Koucheryavy, Analyzing assisted offloading of cellular user sessions onto d2d links in unlicensed bands. IEEE J. Sel. Areas Commun. **33**, 67–80 (2015)
26. G. Fodor, E. Dahlman, G. Mildh, S. Parkvall, N. Reider, G. Miklos, Z. Turanyi, Design aspects of network assisted device-to-device communications. IEEE Commun. Mag. **50**, 170–177 (2012)

27. J. Seo, T. Kwon, V. Leung, Social groupcasting algorithm for wireless cellular multicast services. IEEE Commun. Lett. **17**, 47–50 (2013)
28. L. Militano, M. Condoluci, G. Araniti, A. Molinaro, A. Iera, G.-M. Muntean, Single frequency-based device-to-device enhanced video delivery for evolved multimedia broadcast and multicast services. IEEE Trans. Broadcast. **61**(2), 263–278 (2015)
29. S.C. Spinella, G. Araniti, A. Iera, A. Molinaro, Integration of ad-hoc networks with infrastructured systems for multicast services provisioning, in *International Conference on Ultra Modern Telecommunications and Workshops, 2009. ICUMT '09*, 12–14 October 2009, pp. 1–6
30. Q. Zhang, F.H.P. Fitzek, V.B. Iversen, Design and performance evaluation of cooperative retransmission scheme for reliable multicast services in cellular controlled P2P networks, in *IEEE 18th International Symposium on Personal, Indoor and Mobile Radio Communications, (PIMRC)*, Athens, Greece, September 2007
31. L. Militano, M. Condoluci, G. Araniti, A. Molinaro, A. Iera, F.H.P. Fitzek, Wi-Fi cooperation or D2D-based multicast content distribution in LTE-A: a comparative analysis, in *IEEE International Conference on Communications Workshops (ICC)*, June 2014, pp. 296–301
32. H. Nishiyama, M. Ito, N. Kato, Relay-by-smartphone: realizing multihop device-to-device communications. IEEE Commun. Mag. **52**(4), 56–65 (2014)
33. J. da Silva, G. Fodor, T. Maciel, Performance analysis of network-assisted two-hop d2d communications, in *Globecom Workshops (GC Wkshps)*, December 2014, pp. 1050–1056
34. L. Lei, X. Shen, M. Dohler, C. Lin, Z. Zhong, Queuing models with applications to mode selection in device-to-device communications underlaying cellular networks. IEEE Trans. Wirel. Commun. **13**(12), 6697–6715 (2014)
35. G. Rigazzi, F. Chiti, R. Fantacci, C. Carlini, Multi-hop d2d networking and resource management scheme for m2m communications over lte-a systems, in *International Wireless Communications and Mobile Computing Conference (IWCMC)* (IEEE, 2014), pp. 973–978
36. L. Militano, A. Orsino, G. Araniti, A. Molinaro, A. Iera, A constrained coalition formation game for multihop D2D content uploading. IEEE Trans. Wirel. Commun., **15**(3), 2012–2024 (2015)
37. X. Lu, P. Wang, D. Niyato, A layered coalitional game framework of wireless relay network. IEEE Trans. Veh. Technol. (63), 472–478 (2014)
38. A. Richard, A. Dadlani, K. Kim, Multicast scheduling and resource allocation algorithms for OFDMA-based systems: a survey. IEEE Commun. Surv. Tutor. **15**, 240–256 (2013)
39. T.P. Low, M.O. Pun, Y.W.P. Hong, C.C.J. Kuo, Optimized opportunistic multicast scheduling (OMS) over wireless cellular networks. IEEE Trans. Wirel. Commun. **9**, 791–801 (2009)
40. L. Lei, Z. Zhong, C. Lin, X. Shen, Operator controlled device-to-device communications in LTE-advanced networks. IEEE Wirel. Commun. **19**, 96–104 (2012)
41. 3GPP TS 36.300, Evolved Universal Terrestrial Radio Access (E-UTRA) and Evolved Universal Terrestrial Radio Access Network (E-UTRAN), Rel. 11, September 2012
42. 3GPP TS 36.440, General aspects and principles for interfaces supporting Multimedia Broadcast Multicast Service (MBMS) within E-UTRAN, Rel. 11, September 2012
43. C. Mehlfuhrer, M. Wrulich, J.C. Ikuno, D. Bosanska, M. Rupp, Simulating the long term evolution physical layer, in *17th European Signal Processing Conference (EUSIPCO)*, Glasgow, Scotland, August 2009, pp. 1471–1478
44. A. Urie, A. Rudrapatna, C. Raman, J.M. Hanriot, Evolved multimedia broadcast multicast service in LTE: an assessment of system performance under realistic radio network engineering conditions. Bell Labs Tech. J. **18**, 57–76 (2013)
45. S. Deb, S. Jaiswal, K. Nagaraj, Real-time video multicast in WiMAX networks, in *IEEE INFOCOM 2008. The 27th Conference on Computer Communications*, April 2008, pp. 1579–1587
46. S. Sharangi, R. Krishnamurti, M. Hefeeda, Energy-efficient multicasting of scalable video streams over WiMAX networks. IEEE Trans. Multimedia **13**, 102–115 (2011)
47. Y. Wang, L.P. Chau, K.H. Yap, Bit-rate allocation for broadcasting of scalable video over wireless networks. IEEE Trans. Broadcast. **56**, 288–295 (2010)
48. J.G. Andrews, S. Buzzi, C. Wan, S.V. Hanly, A. Lozano, A.C.K. Soong, J.C. Zhang, What will 5G be. IEEE J. Select. Areas Commun. **32**(6), 1065–1082 (2014)

49. G. Piro, A. Orsino, C. Campolo, G. Araniti, G. Boggia, A. Molinaro, D2D in LTE vehicular networking: system model and upper bound performance, in *2015 7th International Congress on Ultra Modern Telecommunications and Control Systems and Workshops (ICUMT)*, 6–8 October 2015, pp. 281–286
50. X. Lin, R.W. Heath, J.G. Andrews, The interplay between massive MIMO and underlaid D2D networking. IEEE Trans. Wirel. Commun. **14**(6), 3337–3351 (2015)
51. L. Wang, T. Fei, T. Svensson, D. Feng, M. Song, S. Li, Exploiting full duplex for device-to-device communications in heterogeneous networks. IEEE Commun. Mag. **53**(5), 146–152 (2015)

M2M Communications in 5G

Jesus Alonso-Zarate and Mischa Dohler

Abstract This chapter provides a summary of the State-of-the-Art and future trends related to wireless connectivity solutions for Machine-to-Machine (M2M) Communications and Machine-Type Communications (MTC) that will be part of 5G networks, and technologies beyond 5G.

1 Introduction

In 2009, the Wireless World Research Forum (WWRF) envisioned a total of 7 Trillion (7,000,000,000,000) of connected devices by 2017 all around the world [1]. Such expectations led the magazine Forbes to claim in August 2014 that the Internet of Things (IoT) was the most hyped technology, overtaking Big Data (Fig. 1).

Today, it is possible to assure that the figure envisioned by the WWRF back in 2009 was maybe a bit too optimistic, according to the lower figures brought down to billions or even millions as stated in various market research reports. However, even though the figures may differ by some orders of magnitude, there is a common agreement on the fact the Internet of Things (IoT), or Internet of Everything (IoE), is here to stay, grow, and change the world as we know it. The reality is that, according to Machina Research [2], the number of sensor and actuator shipments in 2014 reached 23.64 billion (which mainly includes the dozen sensors found in a typical smart phone), the number of total connected devices was of more than 14.4 billion, and the number of M2M connections were above 4 billion.

Such promising market figures are motivated by the fact that the concept of having all objects interconnected and Internet-connected is very appealing and can foster the creation of innovative applications and business models, and boost societal

J. Alonso-Zarate (✉)
Centre Tecnològic de Telecomunicacions de Catalunya (CTTC), Castelldefels, Barcelona, Spain
e-mail: jesus.alonso@cttc.es

M. Dohler
King's College London, London, UK
e-mail: mischa.dohler@kcl.ac.uk

W. Xiang et al. (eds.), *5G Mobile Communications*,
DOI 10.1007/978-3-319-34208-5_13

Fig. 1 IoT: hype or ripe?

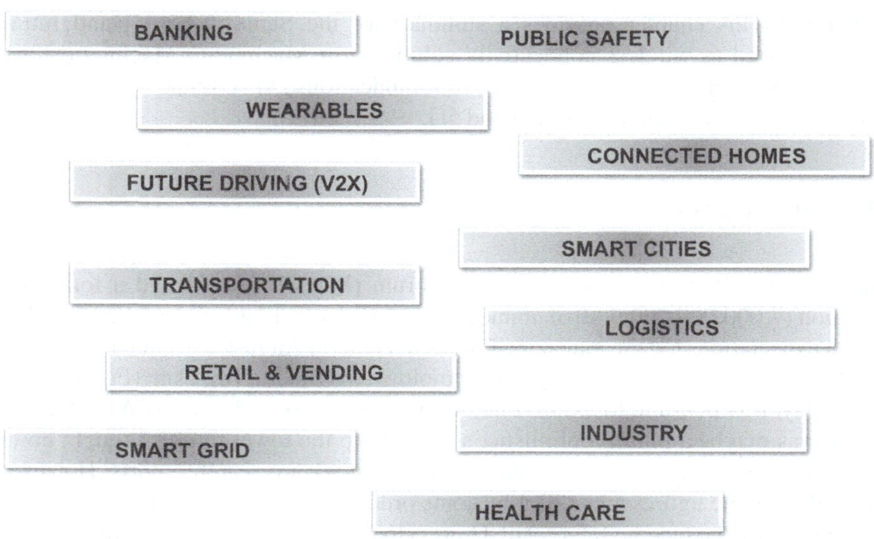

Fig. 2 Key IoT verticals

well-being altogether. The wide range of IoT applications is overwhelmingly wide, including smart grids, smart cities, industries 4.0, smart automotive driving, eHealth, and wearable technology, among many others (Fig. 2).

Smart grids constitute one of the key markets, and they have received a lot of attention from both academia and industry. The vision is to improve the efficiency of the electric grid so that the energy demands and waste can be reduced by making a smarter use of energy, the dependency on natural resources can be softened, and the distribution of electricity can become more efficient. The use of automated communicating devices monitoring the state of the grid and the instantaneous use of the resources is essential to attain the objective of achieving a truly smart grid. Very much related to smart grids is the concept of automatic smart meter infrastructure (AMI), which may pose a wider view of the application scenario. Besides electricity, the measurement of water or gas consumption can improve the efficiency in which these resources are dispatched and consumed. A smart and automated reading of meters is beneficial for all involved players: consumers can accurately track their consumption and thus learn to be more efficient; on the other side, facilities, besides avoiding the manual reading and reporting of meters, can better dimension the production of resources, thus leading to cost savings.

Smart cities constitute one of the most appealing applications and markets. Automated parking search, public lights, container levels, notification in case of incident, warning in case of vandalism of security threat, are some of the example of applications that could make cities more efficient by enabling communications between devices. Indeed, the Smart City concept has triggered the attention of many cities, which are struggling to become the worldwide references in applying technology to their city operations. Barcelona, London, New York, or Sydney are just few examples of emerging technology hubs devoted to the Smart City concept.

The concept of the **Industrial Internet**, as introduced by General Electric (GE) a few years ago, promises improvement of efficiencies in industries, driving down the costs of manufacturing, and creating a revolution in a sector which has an impact in many verticals. According to the numbers presented by GE in Fig. 3, improving the efficiency of industrial processes by a factor 1 % can lead to huge economic gains and savings, thus becoming a primary application domain for the IoT.

The **automotive** sector has become a key vertical market for the IoT. Enabling remote monitoring, in-vehicle diagnosis, or car-to-car communication to improve

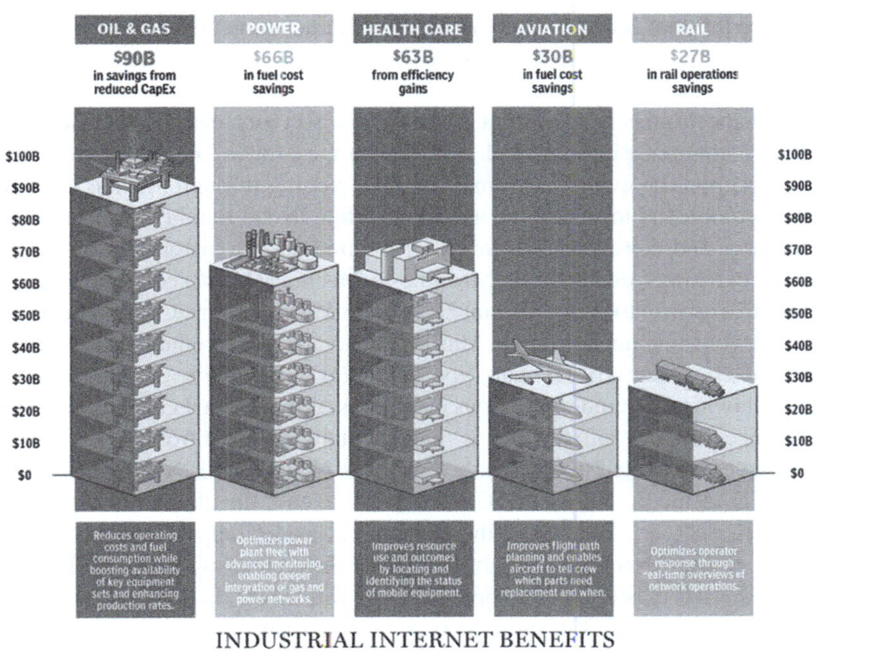

Fig. 3 Industrial Internet: the power of 1 % (benefits over 15 years when scaled up across the economic system)

safety on-road emerge as key domains to drive the innovation of the IoT. The European eCall initiative, that targets the installation of SIM cards in each vehicle for automatic notification in case of accident, will enable more sophisticated vehicle-based applications in the near future. This is just the beginning.

Today, we also see the advent of **wearable technology**. Bringing technology down to the individual domain, having sensors and actuators interacting with individuals becomes the latest trend in the development of the IoT. Smart clothes, watches, earrings, or glasses, will completely change the world as we know it. **eHealth** raises as a big subdomain of wearable technology. Remote diagnosis and even drug dosing are just two simple examples of applications that can benefit from having automated communications between in- or on-body sensors, actuators located at the patient home, and the centralized servers installed at hospitals.

All in all, the use of automated communications among machines, devices, and people can lead to a smarter and more efficient world. A world which is different from what we now today. However, before this new world becomes a reality, there are still many challenges that need to be faced. Some are related to the new economics triggered by the IoT, or even the societal implications that such hyper-connectivity will impose. However, in this chapter we focus on the technological challenges, many of which are still pertinent because prior technology attempts failed. In particular, the focus will be on the challenges that the definition of the next generation of wireless communications will need to cope with. This is the **5G**.

5G will be the next generation of wireless communications, and its deployment is expected to start by 2020. The Wireless Radio communications Conference (WRC) was held in November 2015, officially starting-off discussions about what 5G will be and what spectrum is needed for it. Over the last 30 years, every new generation of wireless communications has had to deal with the growing need for more data bits and capacity. The evolution from 2G to 4G, passing through 3G, has seen an increase in complexity to satisfy the needs of Mobile Broad Band (MBB) services, requiring more and more data. However, for the first time in history, the arrival of the Internet of Things, dealing with Machine-to-Machine (M2M) communications, also referred to as with the more special 3GPP term Machine-Type Communications (MTC), calls for something different. This new generation of communications has to deal with the coexistence of Human-Type Traffic (HTC), possibly calling for more capacity following the trend of previous evolutions of the various generations, with the emergence of MTC traffic, which is fundamentally different from HTC.

This chapter is devoted to discuss the emerging MTC wireless communications and their role in the definition of 5G. Towards this end, the chapter is organized as follows. In Sect. 2, we describe the elements of an IoT application and define the concept of MTC and M2M, highlighting the differences with HTC traffic. In Sect. 3, we discuss existing technologies to cope with MTC. In Sect. 4, we set the objectives to be met by 5G in order to cope with the various requirements posed by MTC applications. Towards the consecution of these objectives, key enabling technologies are identified and discussed in Sect. 5. Finally, in Sect. 6, an outlook and future perspective of MTC in 5G is provided to conclude the chapter.

2 M2M Communications: What Is New?

M2M or MTC constitute a fundamental part of any IoT application. Making things simple for the sake of understanding, the elements of an IoT application are (Fig. 4):

- **End M2M devices**: electronic platforms consisting of at least some kind of energy source (battery or energy harvester), a microprocessor to execute "smart" instructions, a radio transceiver and antenna to communicate with other devices, and sensors and actuators to interact with the physical world, hence the frequently coined term cyber-physical system (see Fig. 5).
- **M2M gateways**: devices in charge of coordinating and concentrating data from a number of end devices in an M2M area network and providing them with connectivity to the Internet.
- **Communications Networks**: backhaul and core networks to connect either the end devices or M2M gateways to the Internet or other external networks. These networks can be either static of virtualized dynamically.
- **M2M Platforms**: hardware/software entities (e.g. data centers, computers, data bases in the cloud, etc.) in charge of providing data ***storage*** and (big) data ***processing*** capabilities to process the data obtained from the sensors and take some actions (*business intelligence*).
- **End User devices**: any kind of device which enables the interaction between the cyber-physical system and humans. This could be, for example, a mobile app running on a smartphone or tablet, or a web service browsed from a laptop.

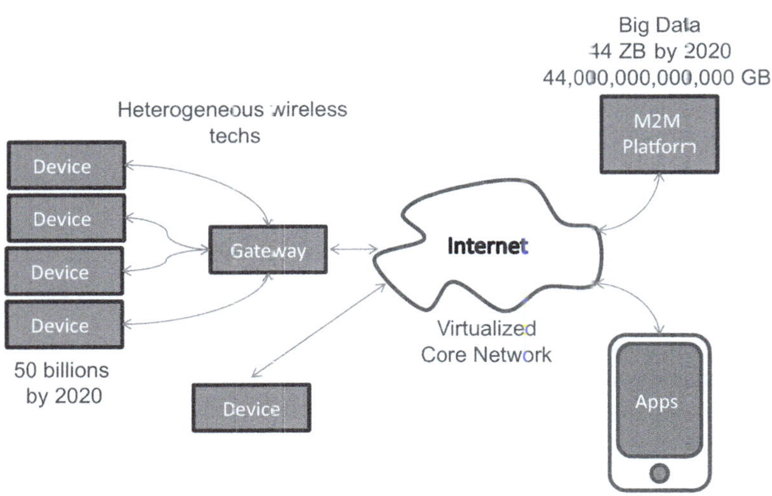

Fig. 4 Elements of an IoT application

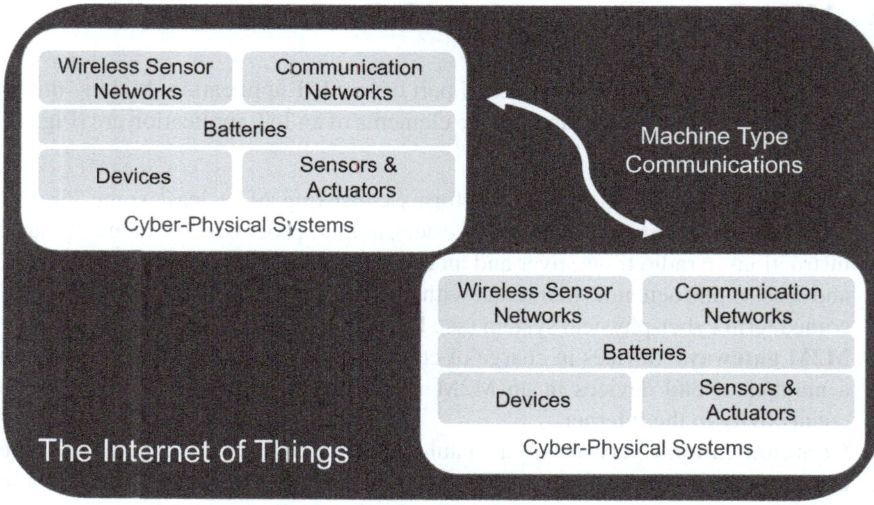

Fig. 5 The Internet of Things and Machine-Type Communications

The term M2M or MTC is typically used to refer to any exchange of information between machines into this ecosystem. Note that MTC traffic can travel in the access network from end M2M devices to other end M2M devices or to M2M gateways, from the M2M gateways to the core communication networks, within the communication networks themselves, from these networks to/from M2M platforms, and in data exchanges to/from the end user devices. Essentially, M2M or MTC refer to any communication where at least one of the ends is a machine.

These communications between machines, or between machines and people, MTC, are conceptually different from HTC [3]. This implies that the design of future communication networks has to take these differences into account. Some of the main differences that the specification of 5G will have to deal with are [4]:

- **Small Data Transmission**: the data exchanges of MTC are typically small. Think for example of the transmission of a "flag" to indicate whether a light is on or off, or the case of a reading from a smart meter. Few bytes will suffice for that purpose. Compared to HTC traffic, where the payloads are typically very large, MTC deals with small data.
- **Uplink transmissions**: the weight of the uplink of MTC is comparable, or even higher, than downlink. This implies a design shift from traditional HTC communications systems, primarily designed for the downlink. Despite the emergence of interactive applications, still most applications rely on the downlink. The growth of video transmissions and broadcast of multimedia material points at an even higher asynchrony in the data flows for HTC.
- **Energy efficiency**: yes, all communications must be as efficient as possible from the energy point of view to avoid the need for constant battery recharges. However. In MTC, this becomes particularly critical, since in many cases, it will

be not feasible to recharge the batteries of all deployed devices due to scalability (size of the network) or reachability. Note that in some cases, the devices will be installed in wide spread areas, and thus it will turn unfeasible to recharge or replace the batteries individually. The energy efficiency needs to be boosted by a factor 10 at least, so that the lifetime of devices can be measured in tens of years.
- **Amount of devices**: it is expected that the amount of devices connected per aggregation point will increase by a factor 1000 with the IoT. Therefore, MTC devices will be deployed in massive networks.
- **Infrequent transmissions**: exploiting "deep" duty cycles will be key to ensure scalability and energy efficiency of MTC applications. Therefore, the infrequent transmission of data is expected from MTC devices. This has a great impact into the design of communication networks, since the systems need to be designed for a huge number of devices, knowing that just a small proportion of them will be active at any given time.
- **Low Mobility**: many MTC applications are characterized by having very low mobility patterns; in many cases, the devices will be completely immobile. This has strong implications in those systems that integrate control signaling to allow mobility and roaming.
- **Heterogeneous requirements**: the wide variety of applications related to the IoT pose diverse, and sometimes counterpoised requirements. For example, some MTC applications will be delay-constrained, while other will be highly delay tolerant. Therefore, the design of communications networks for MTC will have a to deal with a great variety of different requirements involving capacity, reliability, scalability, delay, Quality of Service (QoS), etc.

The new generation of wireless communications will need to cope with all these requirements of MTC, while ensuring that all HTC services are not jeopardized.

3 Existing Communication Technologies for MTC

As it has been discussed in the previous section, MTC are fundamentally different from HTC. Therefore, new emerging communications systems will need to handle with the coexistence of both types of traffic. However, before starting to define a new technology in 5G to cope with MTC, it is necessary to understand which technologies are available today and how suitable (or not) they are to handle such complex scenario that is being formed today.

Figure 6 shows some of the key available wireless communications technologies used to today to power the IoT.

Before continuing with the discussion, they reader may have noted that there are no wired solutions represented in the figure. A wired solution, understood as a solution where sensors or actuators are physically connected to a gateway that provides connectivity to the backbone network, provides an extremely reliable communication channel, with high rates, low delays, and extremely high security.

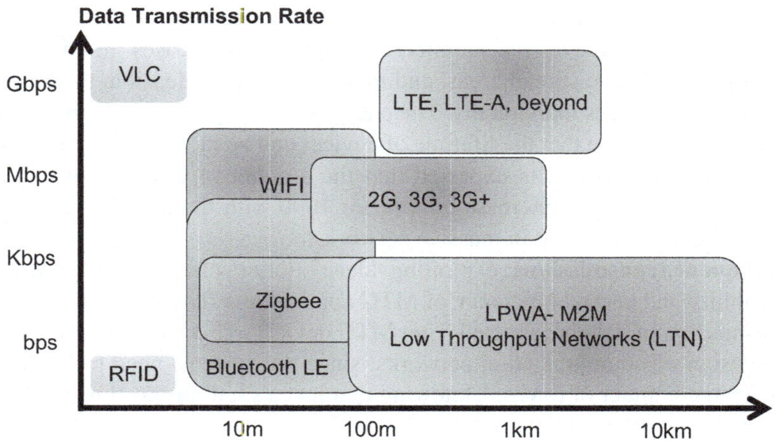

Fig. 6 Existing wireless technologies

Information cannot be overheard by a third party unless cables are physically intercepted. However, although wired solutions are cheap to maintain, their deployment might be an important entrance barrier for some IoT applications. The main drawback of wired solutions is the lack of mobility and scalability, as the addition of new devices to an M2M network requires the cabling of the new devices.

Therefore, the IoT will be mainly driven by wireless solutions. Connecting billions of devices through wires does not seem a promising approach. Therefore, even though some applications will still rely on cables, it is expected that the majority of M2M applications running the show of the IoT will be based on wireless technologies.

Looking at Fig. 6, it is possible to see how available wireless communication technologies offer different solutions in terms of range and transmission rate. The ones shown in the figure and which will be discussed later on in this chapter are:

- **Visible Light Communications** (high rate, short range).
- **Radio Frequency Identification** (low rate, short range).
- **WiFi** (high rate, medium range).
- **Zigbee-like** (low rate, shot single-hop range and long range via multi-hop).
- **Bluetooth Low Energy** (low rate, short range).
- **2G, 3G, 3G+** (medium rate, long range).
- **LTE, LTE-A**, and beyond (high rate, long range).
- **Low Power Wide Area (LPWA)**—(low rate, very long range).

It is important to note that when it comes to selecting the most suitable technology, other performance indicators may become even more relevant that range or rate; note for example the case of eHealth solutions where security and privacy may be the key performance indicators, or automotive applications, where latency and reliability may become the critical parameters to evaluate. An example

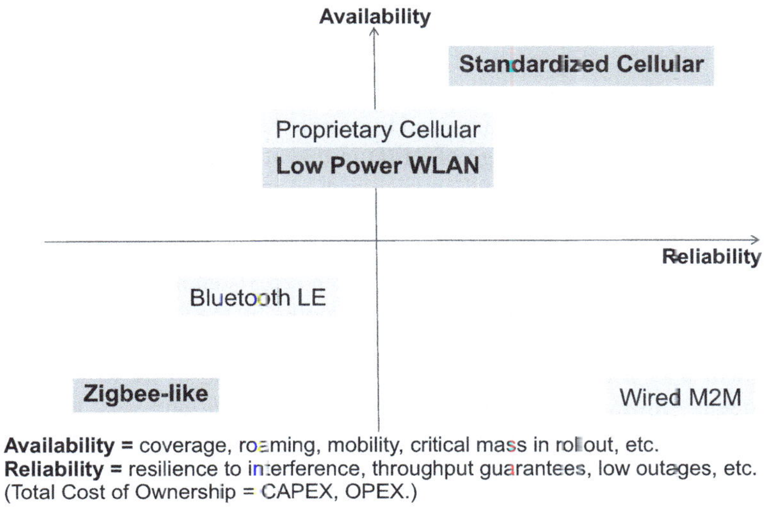

Fig. 7 Business criteria for technology selection

of business criteria for selecting various technologies for M2M is shown in Fig. 7 where availability and reliability are traded for the purpose of selecting the most suitable technology for each IoT application.

In the following sections, each of the technologies shown in Fig. 6 are briefly described and their suitability for MTC is quantitatively assessed.

3.1 Visible Light Communication (VLC)

The use of modulated LED (Light Emitting Diode) to transmit data is not a mainstream technology yet, but it has a huge potential to become ubiquitous in the coming years [5]. The reasons are pretty simple:

- It is an extremely secure technology; since the light cannot go through walls, only receivers in line of sight of the transmitter can receive data.
- It has no associated radio emissions, thus avoiding limitations of use in "sensitive" areas such as hospitals, primary schools, or chemical plants, among others.
- It has high bandwidth compared to radio technologies, thus leading to high data transmission rates.

On the counter-side, VLC suffers from not being a mainstream technology, thus not mature enough as compared to other radio solutions. Therefore, its cost is still too high to be widely accepted to deploy billions of devices. Also, the limited range and need for line of sight, while providing strong security, it constitutes a limitation of its usage for certain applications.

3.2 *Radio Frequency Identification (RFID)*

Many runners all around the world use the technology based on the ISO 18000 Standard to measure the time they need to complete 10k, 21k or marathons. This is probably the most popular usage of RFID in the world. However, this technology is used in a plethora of other applications, such as proximity access cards, tracking pets or goods for logistics, among many others. By the year 2014, the RFID market was quantified in being close to a 9 billion US dollar market, showing its strength as a tagging technology. Some market research reports claim that the potential of RFID will make it grow up to a 27 billion US dollar by the year 2024.

The relationship between RFID and the IoT is undisputed, and its usage to tag things will play a key role in the future, especially due to its very low cost of deployment. However, the limited data transmission rate and very low coverage limits its applicability to very specific applications, where, by the way, it may see no competition at all in the coming years.

3.3 *Wireless Local Area Networks (IEEE 802.11)*

The IEEE 802.11 Standard is the most successful commercialized standard for Wireless Local Area Networks WLANs. The WiFi alliance was very successful in maximizing the market penetration of WiFi devices that today can be found almost everywhere. This technology operates in license-free bands, the so-called Industrial, Scientific and Medical (ISM) bands.

The ubiquity of WiFi is becoming undisputed. The number of Access Points (Aps) available in urban cities is growing, and the footprint of WiFi is becoming denser. Today's data plans to not allow sharing these WiFi APs, but, being the infrastructure already there, it is just a matter of finding the business case to make all these connecting points available to the public.

In addition, the IEEE 802.11 standards family is extremely dynamic, and new amendments are constantly being developed and evolved to increase capacity (e.g., 802.11ac and ad) or to enable new uses cases such as the IoE, home automation, smart grid or wearable consumers (e.g., 802.11ah).

The latter, i.e., the novel IEEE 802.11ah project, is particularly interesting since it can potentially become an accelerator for IoT uptake. The standardization work of the corresponding ah task group had commenced in November 2010 where the prospective technology is generally based on a variation of IEEE 802.11ac standard but down-clocked by a factor of 10. It is currently being developed to enable low-cost long-range (up to 1 km) connectivity across massive M2M deployments with high spectral and energy efficiencies. Today, thousands of M2M devices may already be found in dense urban areas, which required providing support for up to 6 K machines connecting to a single access point.

Fortunately, IEEE 802.11ah technology does not need to maintain backward compatibility with the other representatives of the IEEE 802.11 family. Operating over different frequencies, 802.11ah could thus afford defining novel compact frame formats, as well as offering more efficient mechanisms to support a large population of devices, advanced channel access schemes, as well as important power saving and throughput enhancements. As the result, 802.11ah is believed to significantly enrich the family of 802.11 protocols, which already receive increasing attention from mobile network operators willing to introduce low-cost connectivity in unlicensed bands.

All these amendments are being already developed with security measures in place, thus making WiFi a very trustful and secure technology. For all these reasons, the cost of WiFi chips is dramatically going down, thus becoming very competitive to become a major driver of the IoT [6].

On the downside, all IEEE 802.11-based Standards operate in an increasingly crowded license-free ISM band, thus suffering from inter-system interference, suffering from limitations in the transmit power, and suffering from a lack of network planning. The overlapping of access point's coverage can lead to very low performance in highly dense environments.

3.4 IEEE 802.15.4 (Zigbee-Like Solutions)

The IEEE 802.15.4 Standard, promoted by the Zigbee Alliance, was originally defined as the most energy efficient solution for short-range communications also operating in ISM bands. Unfortunately, some recent research works have shown that WiFi used in Low Power mode can be more efficient than an IEEE 802.15.4 Network [7]. WiFi in Low Power mode is nothing else than using smart duty cycling mechanisms to let devices go to sleep whenever they have no data to be transmitted or received. Since energy consumption is the result of power multiplied by time, it has been shown that is it more energy efficient to transmit for short periods of time at high power (also reaching long distances and reducing the complexity of multi-hopping), rather than transmitting for long periods of time at low power.

Further, Zigbee-like solutions suffer from many drawbacks that limit their applications to very specific and delimited applications, such as industrial environments. Zigbee has not followed the market success story of WiFi and the infrastructure is barely found. In addition, the need for multi-hop transmissions to cover medium and large distances makes its design, implementation, and operation very costly and complex.

3.5 Bluetooth Low Energy (BLE)

Bluetooth also operates in ISM bands. Therefore, it also suffers from the effects of operating in an increasingly crowded band. The last generation of Bluetooth products, based on the specification Bluetooth 4.0 and featuring very low energy consumption have become very popular for the IoT. An example can be seen with the iBeacon technology made popular by Apple, which lets BLE receivers to detect "beacons" transmitted from a base station at close proximity (<10 m) and retrieve data of interest from the Internet based on the geo-location information. This has a huge potential for delivering personalized services, and this is the reason why a number of companies exploiting the iBeacon technology is emerging. BLE has something on favor; the low cost and extremely high ubiquity it has achieved. Almost all smartphones today are shipped with a BLE transceiver, and thus can communicate via BLE with other devices and access points. In this case, the low transmission rate and short range, play on favor of a technology that is increasingly reaching the IoT market for very specific applications.

3.6 3GPP Solutions

3.6.1 The Power of Standardized Cellular Technologies

All the technologies discussed so far in previous sections operate in ISM bands; this means that they are frequency bands free of license which can be used by anyone under certain regulations (transmission power and transmission time). The inter- and intra-system interference thus limits the reliability of communications conducted in these bands. If a Service Level Agreement (SLA) has to be committed and some guarantees of QoS are to be provided, communication in license-bands becomes the only alternative.

3GPP Standards for cellular communications operate in license bands. Mobile Network Operator (MNOs) pay big amounts of money to bid for chunks of frequency bands where they can operate without (legal) interference from other systems. For this reason, they invest in deploying a vast infrastructure which can provide customers with reliable communications. Today, the coverage of standardized cellular systems is enormous, and almost very populated inch in the planet has some kind of cellular coverage.

In addition, these networks can provide mobility and roaming, something that cannot be offered by any other radio system today.

Unfortunately, as it is discussed below, the evolution of standardized cellular systems has followed the increasing demand for higher data rates, and not the specific needs of MTC described in Sect. 2 of this chapter. However, from release 12 on of 3GPP, new emerging releases for cellular systems are starting to deal with both HTC and MTC.

3.6.2 From 2G to 4G

Global System for Mobile Communications (GSM) broke into the market in 1990, providing a cellular network infrastructure designed for the transmission of voice calls. This technology showed an extraordinary market penetration and today provides connectivity worldwide. With the irruption of the Internet by the mid-1990s, the transmission of data and the sense of ubiquitous connectivity has become almost a commodity for humans, who demand more and more data transmission capability and some degree of ensured QoS. Unfortunately, GSM was designed for the transmission of voice, and its capability to provide efficient data transmission was very limited. The response of the standard bodies towards this need was the continuous evolution of the cellular technology.

General Packet Radio Service (GPRS) was the first evolution of GSM to provide increased data transmission capabilities, raising the concept of the 2.5 generation of mobile communications (2.5G). Then came the third generation networks (3G), with technologies such as EDGE (Enhanced Data Rates for GSM Evolution), UMTS (Universal Mobile Telecommunications Systems) and CDMA200. High Data Packet Access (HDPA) is called the 3.5G, and still offers increased data transmission capabilities. Even though the data transmission capabilities with 3G and 3.5G are considerably higher than the capability of GSM, the non-top increasing demand for multimedia contents and interactive applications such as social networking, still feeds the need for increased data rates and higher performance. This is the motivation for the fourth generation (4G) of cellular mobile communications, with transmission rates up to 1 Gbps are envisioned in the downlink. This is the Long Term Evolution (LTE)—Release 8 and LTE Advanced—Release 10.

From 2G to 4G, the evolution of cellular networks has been driven by the need for higher transmission rate and increased QoS to meet the increasing demands of humans to use multimedia and interactive applications. However, now is time to evolve and design communication networks so that they can efficiently cope with HTC and MTC. Indeed, the Release 12 of 3GPP already included, for the first time, some improvements towards MTC traffic [8]. One was, for example, the inclusion of a power saving mode for a newly defined Category-0 of user terminals, with reduced complexity for the sake of lower cost and complexity of MTC devices. Further progress will be achieved in Release 13 bringing more complexity reduction (75 % complexity reduction compared to Cat-1 modems), reduced bandwidth, down to 1,4MHx, 10+ years of battery, and 15–20 dB coverage extension

3.6.3 Release 13 and Road to 5G

3GPP Standards continue their evolution to meet the requirements of next generation networks and enable new use cases, new services, and new applications. For this reason, the next releases of 3GPP will continue aiming at improving the capacity of networks for Mobile Broad Band, but also for MTC services.

For the MBB, the next efforts aim at the so-called LTE-Unlicensed. The idea is that LTE can operate also in unlicensed bands via coexistence with WiFi networks, but also via usage of LTE radio in unlicensed bands. While the former option has been progressively integrated into the releases of 3GPP, the concept of LTE-U has been a brand fresh approach for 3GPPP standards to operate in the 5 GHz band. The design of Licensed Assisted Access (LAA) will be the key to ensure the right operation of LTE-U. Essentially, this consists in using unlicensed under the control and cooperation of the main cellular network. Critical data would travel through the main site using licensed spectrum, while best effort traffic can be offloaded via the unlicensed links. Other techniques being included in next releases of 3GPP aim at including enhanced LTE carrier aggregation, exploitation of full-dimension MIMO, and downlink multi-user transmission using superposition coding.

As for the IoT, the activities of the 3GPP towards 5G can be divided into two big groups:

(1) **Further enhancements of LTE for Machine-Type Communications**, building on the work initiated in the release 12 (which introduced the Category-0 of terminals and new power saving states), reducing the bandwidth channel operation down to 1.4 MHz (in contrast with the 20 MHz of release 12), enhancing coverage with an increased 15 dB budget for extended coverage, simplifying PHY/MAC functions, and further improving power usage.
(2) Design of a brand new narrow-band radio optimized for the IoT; on June 2016, the 3GPP completed the standardization of NB-IOT. This is part of Release 13 and includes optimizations for MTC. NB-IOT can operate in-band or out-of-band of licensed frequencies, thus offering very high flexibility. NB-IOT features very narrowband operation with a bandwidth below 200 kHz (thus compatible with GSM bands). NB-IOT focuses on very low data rate transmissions (below 100 bps), with very limited mobility support (to reduce signaling), and ensures very large coverage and both extremely low cost and power consumption.

Transversal to both HTC and MTC, 3GPP is also working on innovative Device-to-Device communications, enabling the by-pass of the communication through the eNodeB (base stations) to dramatically reduce end-to-end latency.

3.7 Low Power Wide Area Networks (LPWA)

While 3GPP and IEEE do their homework standardizing technologies for MTC, some proprietary solutions have started to gain the market. These are the so-called Low Power Wide Area (LPWA) networks, which offer **very large coverage**, being able to give service to an entire city with a very little number of connection points, **offer extremely low cost**, since they have been optimized for MTC and cannot provide services for HTC, **offer a SLA to the customer**, thus becoming very interesting option for the end users, and what is more important: **they are already in the market** [9].

Examples of such solutions are those proposed by Neul, Sigfox, and the LoRa Alliance (out of Cycleo/Semtech), each of them offering different business models and technical solutions.

4 Key Objectives for MTC in 5G

Future communication networks must be designed to ensure that the following objectives can be meet to satisfy the new needs posed by emerging MTC services [10]:

1. *Reduce* **latency** of wireless communications, enabling end-to-end ***latencies below 5 ms*** for mission-critical applications where real-time transmissions are required [11, 12]. Today's wireless technology is unable to provide such delay figures, thus blocking some potential applications, which are expected to have a huge impact on economy, industry, and society in various vertical sectors such as automotive, industrial automation, or smart grid, among others.
2. *Improve* **reliability** of wireless communications, defined as the probability that a certain amount of data to/from an end user is successfully transmitted to another peer within a predefined time frame, i.e., before a certain deadline expires. The amount of data to be transmitted and the deadline depend on the service characteristics of the underlying use case. In this context, a reliability of at least 99.9999 % is targeted within the project, i.e., the probability that a packet is not delivered within the specified deadline must be below 10^{-6} (or equivalently, at most one in a million packets does not arrive within the specified deadline, on average). For certain types of applications (e.g., industrial automation), an even higher reliability of up to 10^{-9} packet loss rate (outside latency) is sought.
3. *Ensure* **availability** of communication networks by extending network coverage, ensuring network access, and providing roaming support to make sure that critical applications are not in outage when needed. An availability of 99.999 % is sought (i.e., less than ~300 s of accumulated outage per annum), aiming at equating availability of wireless and wired solutions.
4. *Guarantee* **interoperability** of heterogeneous communication technologies; in a hyper-connected world with humans, things, and machines connected, it is necessary to make sure that 5G can cope with a great variety of heterogeneous applications, devices, and radio technologies.
5. *Increase* the **number of connected devices** by a factor of up to 100 with respect to currently deployed LTE networks (typically up to 250 simultaneous users per cell); regardless of the particular numbers of the many publicly available market forecasts, there is no doubt that 5G networks will need to cope with an unprecedented number of connected devices [13].

6. *Boost* **energy efficiency** by a factor of 10 with respect to currently deployed MTC technology, to significantly reduce energy consumption and operational costs of the network infrastructure, and allow battery-driven devices to operate without the need to recharge or replace batteries frequently, which would constrain the number of potential applications [14].
7. ***Secure*** communications [15]; many MTC applications will involve personal or confidential data that must not reach the public domain, nor be modified or replayed by unauthorized parties. This becomes a severe challenge when the amount of data to be transmitted is very small and the acceptable securing overhead must be kept to the minimum.

Note that not all specific objectives above are equally relevant to all applications. For example, ultra-low latency is essential for certain future driving applications, whereas energy efficiency is essential for ultra-low power MTC. Furthermore, achieving certain objectives may entail sacrificing others (e.g., very high energy efficiency can usually only be achieved with long sleep modes, which increases latency and has an impact on reliability and availability). All these trade-offs must be well understood in the long run.

5 Key Enabling Technologies for MTC in 5G

In order to achieve the objectives defined in the previous sections, many solutions have to be explored at various domains. Considering the particular focus of this chapter towards available wireless communication networks, the following non-comprehensive list of key enabling technologies are identified:

(1) In order to ensure scalability of networks, it is necessary to seek for innovative ways of managing the access to shared resources. Looking at the radio access, still today most of wireless technologies are based on ALOHA-based protocols, which suffer from congestion as the number of users increase. Solutions such as those proposed in [16, 17], based on **DQ technologies** constitute innovative ways of sharing resources that must be further understood. Beyond the radio access, such sharing techniques could be applied to any domain where common resources need to be shared, such as available resources in a cloud.

(2) Collisions in the wireless medium must be further exploited to attain useful information. **Advanced coding techniques** from the signal processing domain can be brought to the protocol design domain to turn collisions into useful realizations of data transmissions [18]. Success Interference Cancellation (**SIC**) techniques constitute a promising technology to be further studied and whose implementation must be better matured [19].

(3) Enabling **Device-to-Device** communications, avoiding the two-hop path from devices to infrastructure and from infrastructure to devices, can dramatically reduce the energy of devices, offload main cellular networks, and achieve end-to-end latencies which cannot be attained today [20]. The use of D2D has

started to be integrated into 3GPPP releases, but still many functionalities need to be further studied and understood to make them viable in future generation networks. Notably, inter-operator D2D functionalities are needed to make D2D a truly scalable solution; i.e. a Siemens sensor on a T-Mobile data plan should be able to communicate directly with an ABB actuator on a Telefonica data plan. The usage of LTE in unlicensed bands (**LTE-U**) is also emerging as an alternative to the D2D facilitated by other radios such as WiFi.

(4) **Heterogeneous networks**, combining different cell-sizes and technologies to create a unique and ubiquitous communication infrastructure. As it was pointed in [21], the greatest increases in capacity over the last 35 years have been achieved thanks to making cells smaller. Future 5G networks will be comprised of a gamut of technologies and sites that need to coexist together in harmony.

(5) Network Function Virtualization (**NFV**) and Software Defined Networks (**SDN**) are promising technologies to make the backhaul and core networks flexible and capable to managing a variable and possibly unexpectedly dynamic network conditions [22]. Moving processing power and functions along the network, closer to the edge, can improve the overall QoS perceived by the users, enabling applications which today cannot be realized due to the limitations of an extremely static core network configuration.

6 Summary and Outlook

The definition of 5G must take the emergence of the IoT and its need to coexist with Mobile Broad Band services into account all along the way of the specification process. Involving communication between machines (either M2M or MTC), communication networks need to deal with traffic patterns which are fundamentally different from HTC. These have been discussed in Sect. 2 of this chapter. Existing wireless communication technologies presented and discussed in Sect. 3 were originally designed for HTC, and thus still further improvements for the future are to be sought. Standardization bodies all around the world are conducting great efforts to standardized new technical solutions that can cope with the particular requirements of MTC. In the meanwhile, some proprietary solutions are already available in the market to cover the needs to today's application. In the long run, it is difficult to ensure that the IoT will become mainstream unless real standards are defined and widely accepted. This was the example of WiFi, and this must be the example for the IoT.

The definition of requirements for the IoT, due to the great variety of possible applications and markets, needs to be agreed with vertical players which are the final end users of the technology. Once these requirements are set, new emerging technology enablers need to be explored to make sure that the necessities are covered, as it has been discussed in Sect. 5. At the air interface, massive MIMO, disruptive access methods based on something different to ALOHA, such as DQ technologies, exploitation of Successive Interference Cancellation, or Device-to-Device communications have been identified as key enablers, identifying the use of LTE in unlicensed bands (LTE-U) as an emerging area for research. At the architectural level, both Network Function Virtualization (NFV) and Software Defined Networks (SDN) will be core enablers for the core networks of the future.

Global initiatives are working together to make the realization of 5G a reality so that it can efficiently deal with both traffic generated by humans and machines in good harmony.

References

1. User Scenarios 2020 – A Worldwide Wireless Future. WWRF Outlook, July 2009. Available online
2. The need for low cost, high reach, wide area connectivity for the Internet of Things: A Mobile Network Operator's Perspective. White Paper, February 2014
3. A. Laya, K. Wang, L. Alonso, J. Alonso-Zarate, J. Markendahl, Future challenges in efficiently supporting M2M in the long term evolution (LTE) standards, in *Machine-to-Machine (M2M) Communications, Architecture, Performance and Applications* (Woodhead Publishing, 12 January 2015)
4. K. Zheng, S. Ou, J. Alonso-Zarate, M. Dohler, F. Liu, H. Zhu, Challenges of massive access in highly dense LTE-advanced networks with machine-to-machine communications. IEEE Wirel. Commun. **21**(3), 12–18 (2014)
5. A. Jovicic, Li. Junyi, T. Richardson, Visible light communication: opportunities, challenges and the path to market. IEEE Commun. Mag. **51**(12) (2013)
6. S. Tozlu, M. Senel, Wi Mao, A. Keshavarzian, Wi-Fi enabled sensors for the Internet of Things: a practical approach. IEEE Commun. Mag. **50**(6) (2012)
7. S. Tozlu, Feasibility of Wi-Fi enabled sensors for the Internet of Things, in *Proceedings of the Wireless Communications and Mobile Computing Conference (IWCMC)*, 2011
8. 3GPP Technical Report 37.869, "Study on Enhancements to Machine-Type Communications (MTC) and other Mobile Data Applications", V12.0.0, Sep. 2013
9. X. Xiong, K. Zheng, X. Rongtao, X. Wei, P. Chatzimisios, Low power wide area machine-to-machine networks: key techniques and prototype. IEEE Commun. Mag. **53**(9), 64–71 (2015)
10. E. Dahlman, G. Mildh, S. Parkvall, J. Peisa, J. Sachs, Y. Selén, J. Sköld, 5G wireless access: requirements and realization. IEEE Commun. Mag. (2014)
11. N.A. Johansson, Y.-P.E. Wang, E. Eriksson, M. Hessler, Radio access for ultra-reliable and low-latency 5G communications, in *IEEE International Conference on Communication (ICC)*, London, June 2015
12. O.N.C. Yilmaz, Y.-P.E. Wang, N.A. Johansson, N. Brahmi, S.A. Ashraf, J. Sachs, Analysis of ultra-reliable and low-latency 5G communication for a factory automation use case, in *IEEE International Conference on Communication (ICC)*, London, June 2015
13. A. Laya, L. Alonso, J. Alonso-Zarate, Is the random access channel of LTE and LTE-A suitable for M2M communications? A survey of alternatives. IEEE Commun. Surv. Tutorials, Special Issue on Machine-to-Machine Communications, **16**(1) (2014)
14. T. Tirronen, A. Larmo, J. Sachs, B. Lindoff, N. Wiberg, Machine-to-machine communication with long-term evolution with reduced device energy consumption. Trans. Emerg. Telecommun. Technol. **24**(4), 413–426 (2013)
15. A. Grau, Can you trust your fridge? IEEE Spectrum **52**(3) (2015)
16. J. Alonso-Zárate, E. Kartsakli, A. Cateura, C. Verikoukis, L. Alonso, A near-optimum cross-layered distributed queuing protocol for wireless LAN. IEEE Wirel. Commun. Mag., Special Issue on MAC protocols for WLAN, **15**(1), 48–55 (February 2008), IF. 3.180, ISSN 1536-1284

17. A. Laya, C. Kalalas, F. Vazquez-Gallego, L. Alonso, J. Alonso-Zarate, Goodbye, ALOHA!, IEEE Access, April 2016, Special Section: The Plethora of Research in Internet of Things (IoT)
18. N.K. Pratas, H. Thomsen, C. Stefanovic, P. Popovski, Code-expanded random access for machine-type communications, in *2012 IEEE Global Communications Conference (Globecom) Workshops*, Dec. 2012
19. E. Paolini, C. Stefanovic, G. Liva, P. Popovski, Coded random access: applying codes on graphs to design random access protocols. IEEE Commun. Mag. **53**(6), 144–150 (2015)
20. A. Laya, K. Wang, A. Widaa, J. Alonso-Zarate, J. Markendahl, L. Alonso, Device-to-device communications and small cells: enabling spectrum reuse for dense networks. IEEE Wirel. Commun. Mag. **21**(4), 98–105 (2014)
21. M. Dohler, R.W. Heath, A. Lozano, C.B. Papadis, R.A. Valenzuela, Is the PHY layer dead? IEEE Commun. Mag. **49**(4) (2011)
22. M. Condoluci, M. Dohler, G. Araniti, A. Molinaro, K. Zheng, Toward 5G Densenets: architectural advances for effective machine-type communications over femtocells. IEEE Commun. Mag. **53**(1), 134–141 (2015)

Design Techniques of 5G Mobile Devices in the Dark Silicon Era

Imed Ben Dhaou and Hannu Tenhunen

Abstract In the internet of things age, future communication technologies should provide the necessary bandwidth and latency for the connection of billion devices and the development of ubiquitous applications to improve the quality of life. The design of the prospected mobile communication system needs wide skills in wireless communication, analog circuit design, embedded system, microwave technology, and so forth. System level analyses, design space exploration, performance tradeoffs are some key steps that enable the design of low-cost, energy efficient, ubiquitous and flexible transceiver. This chapter provides comprehensive design techniques for 5G mobile communication in the dark silicon era and using More than Moore technology (MtM).

1 Introduction

Wireless sensor networks have enabled the development of a plethora of smart and ubiquitous applications (ambient intelligence, smart-home, intelligent transportation system, smart-city, etc.) Fig. 1 shows an electronic system that aims to combat traffic noise in urban areas [1]. The system uses an acoustic sensor (microphone) combined with a DSP system, a communication block, and a CCTV camera. The communication block is used to access the Internet wirelessly. The Long-term evaluation system (4G LTE) offers the required data-rate and latency to the anti-traffic noise system.

The growing importance of the Internet of Things (IoT) and the ever increasing demands for smart and ubiquitous systems has led to major technical challenges beyond the capabilities of the 4G system.

To address the growing need for higher-bandwidth, low-latency, and the connection of billion devices, the international communities have started to look for a replacement to the current 4G system.

I. Ben Dhaou (✉)
College of Engineering, Qassim University, Saudi Arabia
e-mail: phd.imed.bendhaou@ieee.org

H. Tenhunen
School of Information and Communication Technology Royal Institute of Technology, Sweden

W. Xiang et al. (eds.), *5G Mobile Communications*,
DOI 10.1007/978-3-319-34208-5_14

Fig. 1 An anti-traffic noise system proposed in [1]

As discussed in [2], there are five disruptive technologies for the 5G. These technologies are: (1) a device-centric architecture, (2) a millimeter wave communication, (3) a massive MIMO system, (4) a smarter device, and (5) a native support for machine-to-machine communication.

It is expected that the 5G system will enable the integration of the various wireless communication technologies (GSM, WCDMA, 3G, 4G, DVB-T, WLAN, RFID, Zigbee, to name a few) [3].

To quench the future wireless communication requirements, the authors of [4] suggest that the 5G communication network should offer for example a round-trip latency less than 1 ms, 1000× aggregate data-rate compared to 4G and guarantees 100 Mbps for 95 % of users.

The success of the 5G terminals will be contingent to the fabrication technology. CMOS is the technology of choice for the manufacturing of the base-band signal processing blocks. For the RF front-end, CMOS has also been used to fabricate the various RF front-end blocks (mixers, amplifiers, oscillators, down-converters, filters, and so forth) [5].

The growing number of wireless standards, the need for a single radio transceiver and the advances in CMOS technology are the main factors that have contributed to the birth of the software defined radio (SDR). The SDR aims to design a flexible and reconfigurable radio architecture [6]. The reconfiguration is done using a specialized software. The SDR has brought into light various design techniques for the RF front-end and the base-band signal processing. Among the emerging paradigm is the *dirty RF* [7]. The design of the SDR architecture using the dirty-RF is accomplished using design trade-offs at the system level. The analog front-end is designed using a direct-conversion receiver architecture. The advantages of the direct-conversion receiver (DCR) compared to the superheterodyne are low-cost, reduced complexity and energy efficiency [8]. However, the DCR architecture suffers from various impairments such as even-order distortion, dc-offset, I/Q mismatches, and 1/f noise [9].

The dirty-RF paradigm allows to design imperfect analog front-end. To mitigate the noisy RF front-end, the receiver should use effective numeric algorithms. These algorithms seek to minimize the bit-error rate or symbol error rate for a wide-range of signal to noise ratios.

Among the popular techniques in the dirty-RF paradigm are: (1) scrambling algorithms to reduce the peak-to-average power ratio for OFDM signal [10], and (2) a frequency offset compensation scheme combined with channel estimation for OFDM DCR architecture [11, 12].

In recent years, the mobile telephony systems (devices and network) have migrated from the voice-oriented applications to ubiquitous computing. Current smart phones offer the user a plethora of pervasive applications some of which are: location based services, navigation [13], augmented reality, and vehicle to pedestrian communication [14].

The development of the smart phone is accredited to the progress in sensor technology, the development of the system-in-package, the progress in the embedded system design, and so forth. Figure 2 shows a generic smart-phone architecture.

A smart-phone contains the following blocks: (1) a heterogeneous multi-core processor (base-band, protocol, application and GPU processors), (2) a RF front-end (amplifiers, mixers, filers, oscillator, ADC, DAC, and so forth), (3) sensors (GPS, gyroscope, micro-phone, cameras, accelerometer, etc.), (4) a machine-to-machine communication platform (Bluetooth, NFC, V2X communication, and so on), and (5) a user interface (keyboard, loudspeaker, tactile screen, and so forth).

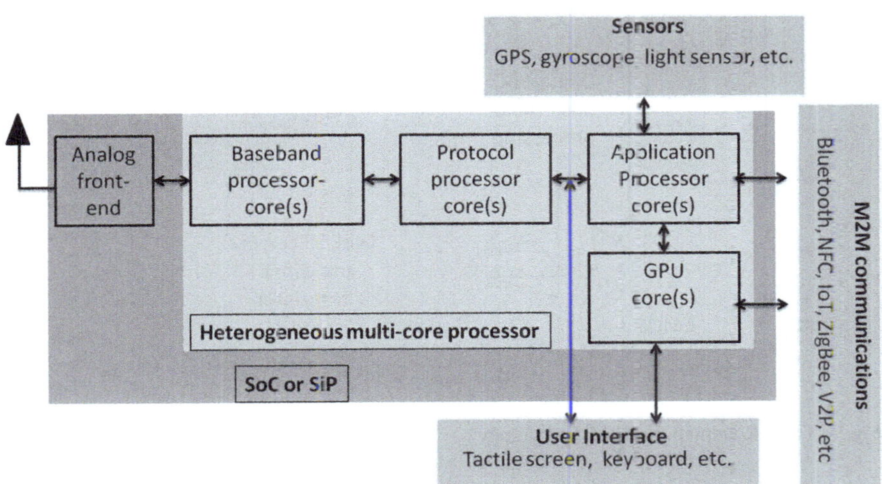

Fig. 2 Smart-phone architecture

For years, Moore's law has been the vehicle for aggressive transistor scaling. The technology scaling paradigm allows to design faster, cheaper and energy-efficient systems. However the non-digital blocks do not benefit equally from the scaling. Moreover, in the nano-meter regime, the transistor feature sizes and interconnects are hitting the borderline. Hence, many research efforts have started to look for a replacement to the CMOS technology beyond Moore's law [15–18]. The international roadmap for the semiconductor has provided a dual trends for the technology roadmap. The first trend is the continuous scaling of the transistor feature size (More Moore). The second trend is the continuous integration of more non-digital blocks (sensors, RF, actuators, and so forth). Figure 3 depicts these dual trends.

The rest of the chapter is organized as follows. Section 2 describes the problems associated with the interconnect centric design, reports the NoC as a new methodologies to address the complexity associated with on-chip communication and elaborates several techniques to design baseband circuitry in the dark silicon age. Section 3 presents the multi-gate MOSFET technology for the design of heterogeneous multi-core process and the RF front-end for the feature size smaller than 22 nm. Section 5 overviews the various technology and methodologies to build sensors for 5G ubiquitous devices. Finally, Sect. 7 summarizes the chapter.

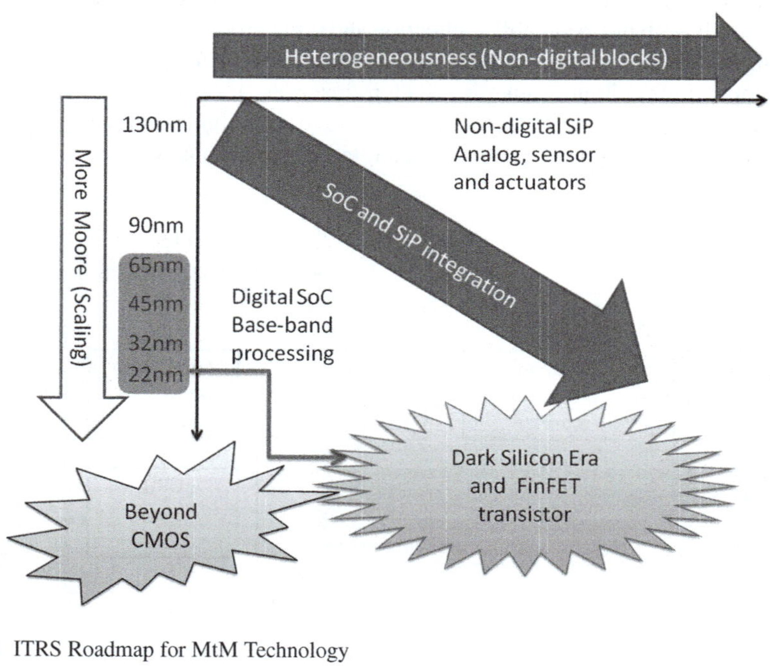

Fig. 3 ITRS Roadmap for MtM Technology

2 Signal Processor Design in the Dark Silicon Era

CMOS is the mainframe technology for the design of digital integrated circuitry. Its popularity is attributed to a number of factors such as: (1) low cost, (2) high yield, (3) low power consumption, (4) high-noise immunity, and (5) the scaling capabilities.

In 1965, Gordan Moore postulated his famous prediction, stating that the number of transistor will double every 18 months. Ever since its postulation, Moore's law has been the guideline for aggressive transistor scaling and large scale integration. Figure 4 plots the transistor density (number of transistor per square millimeter) for the generations of Intel microprocessor. The figure clearly shows that technology scaling spurs an exponential increase in the transistor density.

Technology scaling enables the implementation of faster, smaller and energy efficient transistor. To understand the underlying mechanisms behind technology scaling, we will revisit the device characteristics in the nano-meter regime (short-channel effect) [19]. The NMOS transistor (shown in Fig. 5) is built using specially selected layers of metal and insulator. The transistor is a four terminal device (body, gate, drain and source). The body of the NMOS transistor is the p-type semiconductor.

The NMOS device operates in linear, saturated or cutoff mode. Using the long-channel model, the expression for the drain-source current of an NMOS transistor of length L and width W is shown in (1).

$$I_{ds} = \begin{cases} 0 & \text{if } V_{gs} < V_t \\ \beta(V_{gs} - V_t - \frac{V_{ds}}{2})V_{ds} & \text{if } V_{ds} < V_{gs} - V_t \\ \frac{\beta}{2}(V_{gs} - V_t)^2 & \text{if } V_{ds} > V_{gs} - V_t \end{cases} \quad (1)$$

where $\beta = \mu C_{ox}\frac{W}{L}$, C_{ox} is the gate oxide capacitance, μ is the electron mobility, V_t is the threshold voltage, V_{ds} is the drain-source voltage, and V_{gs} is the gate-source voltage.

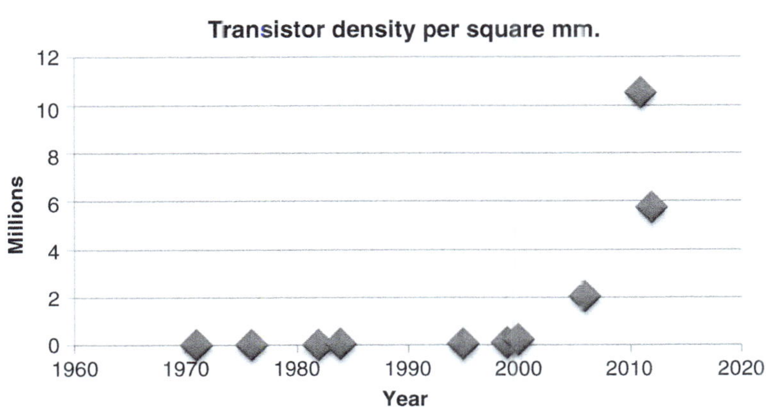

Fig. 4 Evolution of transistor density for Intel microprocessor

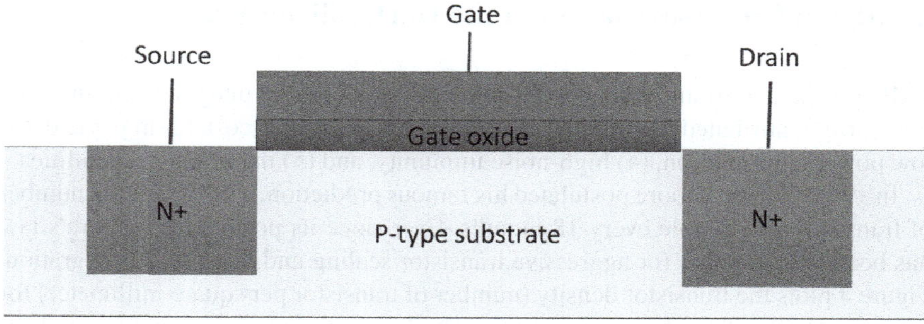

Fig. 5 Layout of NMOS transistor

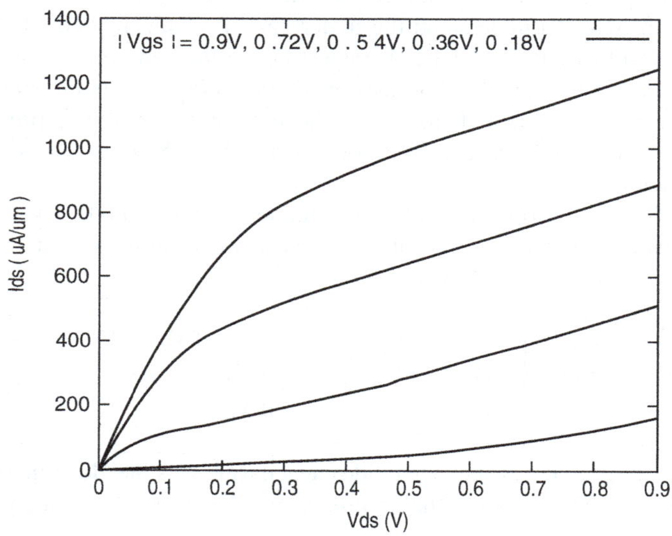

Fig. 6 IV characteristic for a 32 nm NMOS transistor

The expression for the IV characteristics of the NMOS transistor has been calculated under the assumption that the lateral electric field between the source and the drain is constant.

In sub-nanometer technology (below 90 nm), the secondary effects neglected in long-channel model have become significant. The prominent ramifications are for instance mobility degradation, velocity saturation, channel length modulation, drain-induced barrier lowering, and sub-threshold leakage [19].

To illustrate the IV characteristic of NMOS transistor in sub-nanometer technology, we used the predictive technology model described in [20] to obtain the IV characteristics of the transistor. Figure 6 depicts the relationship between I_{ds} and V_{ds} in the three operating regions for the 32 nm NMOS transistor.

2.1 Low-Power Design

The power consumption of the CMOS circuits is composed of the dynamic and static components. The dynamic power consumption is the sum of the short-circuit power consumption and the power consumed due to the switched capacitance. The latter is the dominant. As a result, the widely accepted expression for the dynamic power consumption is shown in (2).

$$P_{dynamic} \approx \alpha C V_{dd}^{2} f, \qquad (2)$$

where f is the clock frequency, α is the switching activity, C is the circuit capacitance, and V_{dd} is the supply voltage.

In the idle status, the CMOS circuitry consumes static power which is due to leakage currents (sub threshold, gate, junction) and contention current.

In nano-meter design, the sub-threshold leakage current has started to represent over 40 % of the total power consumption [21]. Several techniques have been proposed to bring down the leakage power consumption. Those techniques have been developed for the various abstraction levels (system, algorithm, architecture, and circuit). The power gating technique has been reported as an effective way to cancel the leakage current [19]. There are many schemes and circuit styles for power gating (fine-grain, coarse-grain, and so forth). Figure 7 exemplifies circuits for power gating [19].

Over the last decades, several approaches have been proposed to reduce the dynamic power consumption [22]. It has been widely accepted that higher abstraction levels have larger savings than lower level. The system level power optimization can lead to a sturdy power savings. The dynamic voltage combined with the

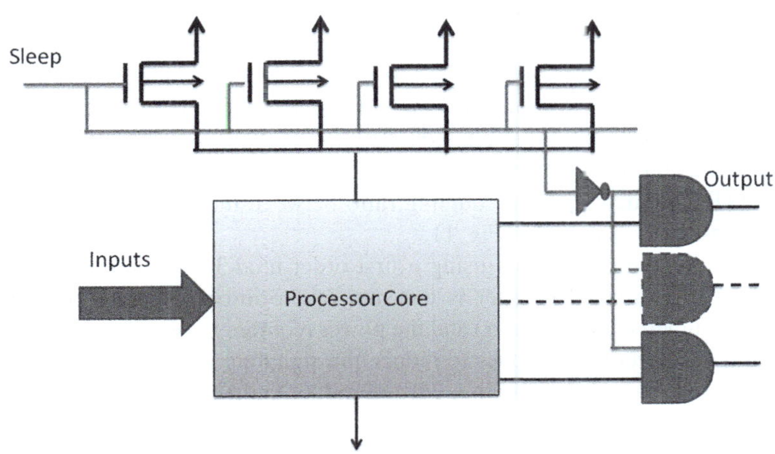

Fig. 7 Circuit for power gating

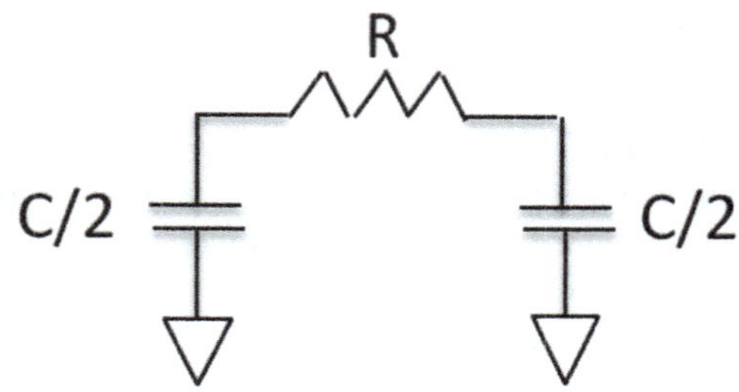

Fig. 8 π RC model for on-chip interconnect

frequency scaling (DVFS) is a well established technique and has been shown very effective to reduce the power consumed by a modern processor [23].

2.2 *Interconnect Centric Design: The NoC Era*

The increasing number of devices has created routability and electromigration problems. To cope with the problem, the semiconductor foundries have increased the number of metalization levels, proposed new interconnect architectures and developed interconnect prediction models.

In nanometer regime, interconnect capacitances have started to be several orders of magnitude larger than the intrinsic capacitances of the CMOS transistors. There exist several ways to model the on-chip wires [24]. The most widely used model is the RC wire depicted in Fig. 8. The model can be used to estimate the propagation delay, power consumption, and so forth. It can also be used to optimize the wire under a given delay budget (buffer insertion, wire resizing, and so forth).

Using the RC wire model shown in Fig. 8, ITRS predicted that delay for intermediate wire will increase as shown in Fig. 9.

Both the interconnection delay and the coupling capacitance for the intermediate wire is experiencing an increase. The interconnect model used by ITRS to predict the coupling coefficient is depicted in Fig. 10. A 2-D static field solver has been used to determine the values for the self and coupling capacitances. The model takes into account the fringing capacitance.

The ITRS predicts that the coupling ratio between the intermediate wires will remain constant as depicted in Fig. 11.

In [25], the authors showed using a first-order model that the dynamic power dissipation for a loaded inverter is the sum of the intrinsic power consumption (power consumed by the inverter) and the power required to send the voltage across a global wire. Several techniques to reduce the signaling power consumption have been proposed (low-voltage swing, current mode, multi-level signaling, etc.) [24, 26–28].

The coupling capacitance has created a new design problems. The widely reported issues associated with the capacitive couplings are timing closure, crosstalk noise, glitches, and power consumption. The impact of the crosstalk on the physical parameters of the circuit depends on the alignment of the aggressors and the

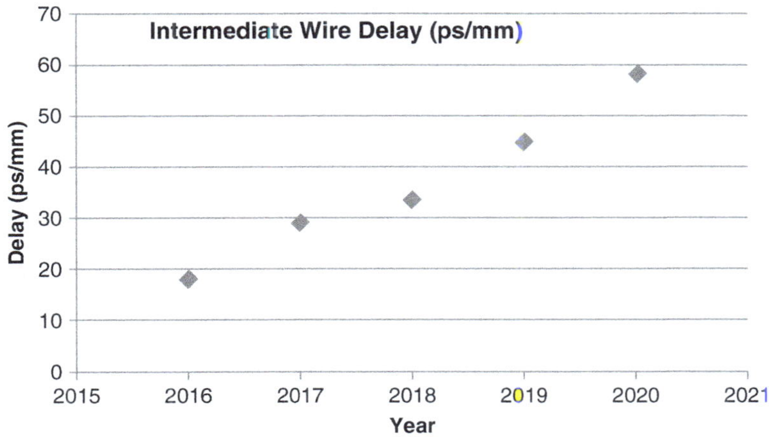

Fig. 9 Projected intermediate wire delay (ps/mm)

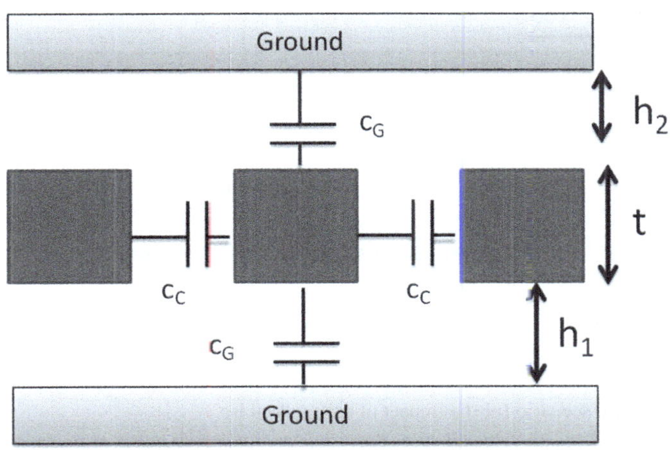

Fig. 10 Interconnect model used by ITRS to compute self and coupling capacitance

victims [24]. The worst case occurs when the aggressor and the victim switch in the opposite direction as exemplified in Fig. 12.

In [29], the author presented a time-dependent power estimation model for capacitively coupled circuits. Figure 13 quantifies the impact of the crosstalk capacitance on the dynamic power consumption. The dynamic power has been obtained through spice simulation. The curves clearly show that the maximum power is consumed when the aggressor and the victim switch in the opposite direction.

To overcome the bottleneck caused by wires and to sustain the performance improvements, Networks-On-Chip (NoC) has emerged as a new design paradigm for SoC. NoC motivations come from many design corners such as the migration towards multi-core implementation (many core architecture), the growing silicon

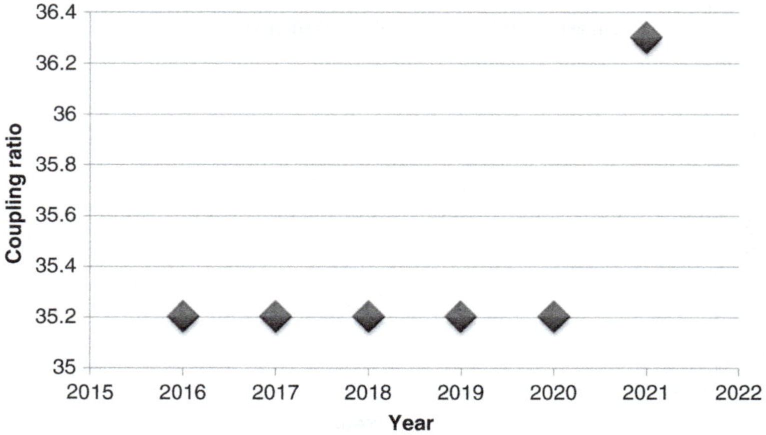

Fig. 11 Coupling ratio for intermediate and global wires

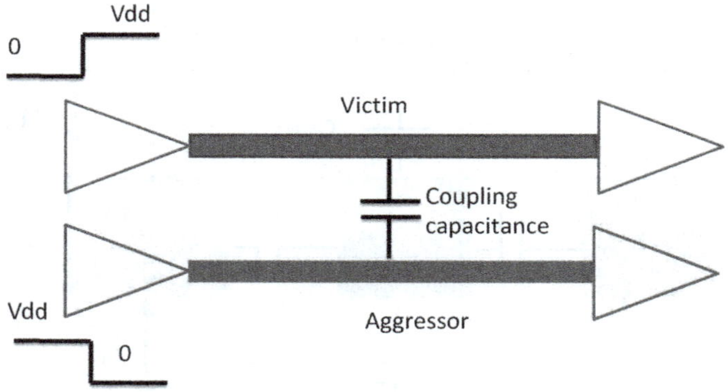

Fig. 12 Worst case scenario for the impact of crosstalk noise

area devoted to routings, and the difficulties to implement point-to-point communication among cores or Processing Elements (PEs) [30]. Figure 14 depicts a 3x3 2D NoC architecture. Its worth pointing out that there are plenty of NoC architectures beyond the 2D mesh network.

The growing number of cores has created design bottlenecks for the NoC architecture. The widely reported design challenges are the high communication latency for distant cores and the extra power consumed by the switches [31]. Several network and communication technologies have been reported in the literature. Millimeter-wave NoC architecture using a hierarchical small-world networks proposed in [32] is a promising solution that solves some the issues related to the NoC design using multi-hope network.

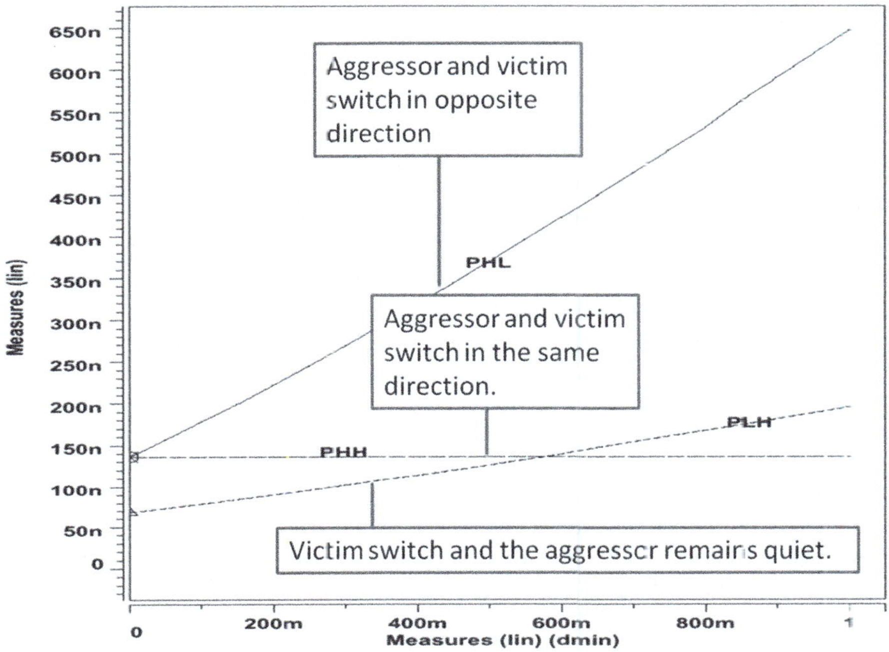

Fig. 13 Impact of the coupling capacitance on the dynamic power consumption

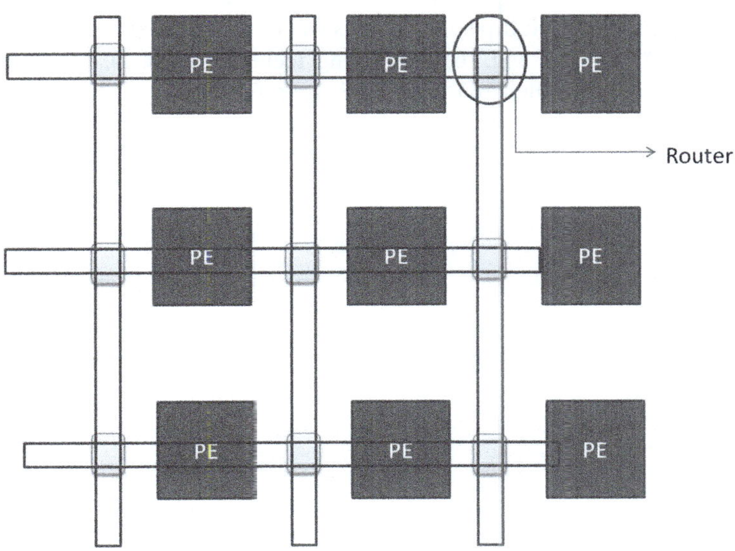

Fig. 14 3x3 2D mesh network

2.3 Dark Silicon Era

As a result of the aggressive transistor scaling and the ever increasing number of transistor density, voltage scaling can no more follow the Dennard's scaling law. This phenomena has led to a steady increase in power density. To prevent the processor from overheating, it has been observed that the number of powered transistors need to be shrunk using for example power gating. As a result, for a given period of time a percentage of the transistors is dark. This effect is referred to as *the dark silicon* [33]. The amount of the dark silicon is expected to escalate as the transistor feature sizes decrease (cf. Fig. 15.)

The dark silicon has emerged as a new design challenges that prevent circuit designers from increasing processor performance. As a results, new design methodologies and solutions should be proposed to reduce the dark silicon area.

In [34], the authors developed scaling models for device, core, and multicore. The models have been used to (1) study the percentage of dark silicon for future nodes, (2) develop the Pareto frontiers, and (3) investigate the impact of various architecture parameters on the processor's performance in the dark silicon era. The authors have considered the following parameters: (1) memory bandwidth, cache size, simultaneous multi-threading (SMT), and parallelism. Table 1 summarizes the reported results.

The table clearly shows that the parallelism is one important design consideration that can alleviate the impact of the dark silicon. For the 5G mobile communication system, most of the DSP algorithms can easily be parallelized using the MIMO system (Multiple Input Multiple Output) instead of the SISO system (Single Input Single Output) [35] (cf. Fig. 16). Unfolding techniques can also be applied to increase the level of parallelism for the DSP algorithms.

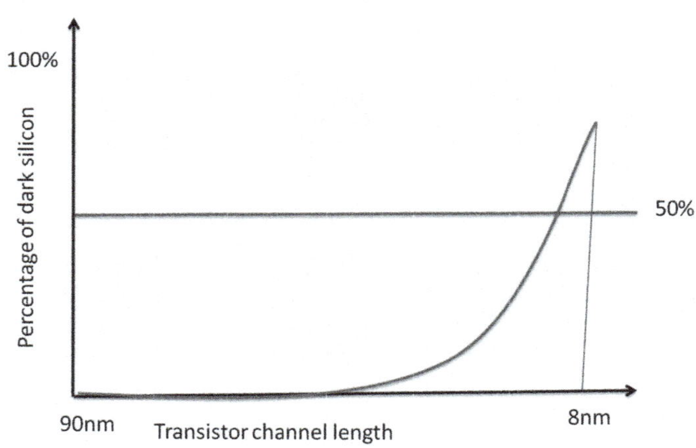

Fig. 15 Growing problem of dark silicon area [34]

Table 1 Impact of architecture parameters on processor performance

Parameter	Maximum improvements
Cache area	20 %
Memory bandwidth	2×
2-way SMT	1.5×
8-way SMT	0.3–2.5×
Parallelism	6.3×

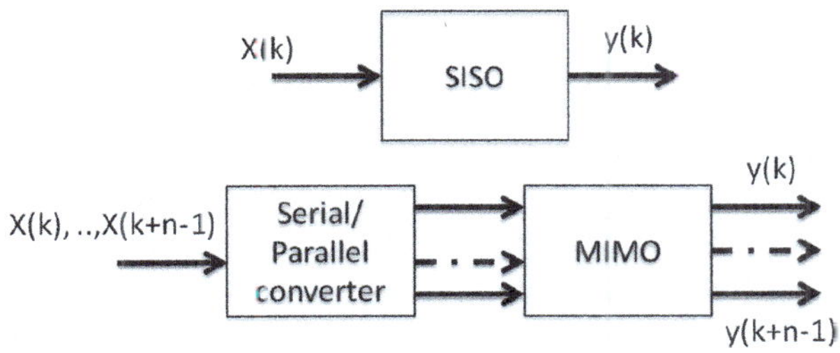

Fig. 16 SISO to MIMO transformations for DSP algorithms

The systolic architecture is yet another design technique that can be used to implement the DSP algorithm using concurrent and identical processing elements (PEs). The dependence graph is used to determine the number of PEs [35, 36]. Contrary to the MPSOC, the Amdahl's law is inapplicable to the systolic processor. This suggest that the prediction reported in [34] can be too pessimistic for the DSP systems implemented using systolic array.

In [37], the authors firstly analyzed the source of energy dissipation in CPU. Secondly, they defined the utilization wall. Finally they proposed the deployment of energy saving cores (c-cores) to reduce the ratio energy/instruction.

In [38], the authors showed the advantage of the heterogeneous cores (processor) over homogeneous ones. They devised an algorithm to determine the optimal number of the heterogeneous cores for a given power budget and dark silicon constraints. The authors showed that the synthesized heterogeneous multi-cores can lead to up-to 60 % improvement in performance compared to the homogeneous design.

As the NoC has become a design solution to address the growing problem of on-chip communication, it has been observed that the NoC circuitry can consume a significant portion of the energy. The highest reported value is 33 % which is consumed by Intel's cloud computing processor [23]. To bring this energy down, many published reports considered schemes based on dynamic voltage and frequency scaling (DVFS) [39, 40]. However, in the dark silicon era, DVFS scheme has become very ineffective because of the various factors such as the growing leakage power dissipation. In [41], the authors devised multi-layer NoC routers. Each layer has a unique voltage and frequency pair, referred to as a VF level. The levels have been defined for a target CMOS technology node. This scheme sacrifices silicon area to reduce the percentage of the dark silicon.

3 FinFeT for Future SoC Designs

It is well documented that the scaling process for the bulk CMOS is reaching a dead end beyond 22 nm [42, 43]. This fact is attributed primarily to the short-channel effect and leakage current. In 70 nm node technology, it has been reported that the active mode leakage current is responsible for over 40 % of the total power consumption [21].

To address the shortcomings of the bulk CMOS technology and to enable the progress of technology scaling, the FinFET has been proposed as a substitute for the bulk CMOS transistor. It is widely reported that the FinFET overcomes the problem associated with short-channel effect, offers higher current drive, and dissipates less standby current compared to the bulk MOSFET technology [44].

The FinFET transistor is a MOSFET with double gates. There exist several competing realizations for the double-gated transistor. In [45] double-gated transistor on SOI, independent double-gate FinFET and planar double-gate SOI were reported. Figure 17 shows these three structures.

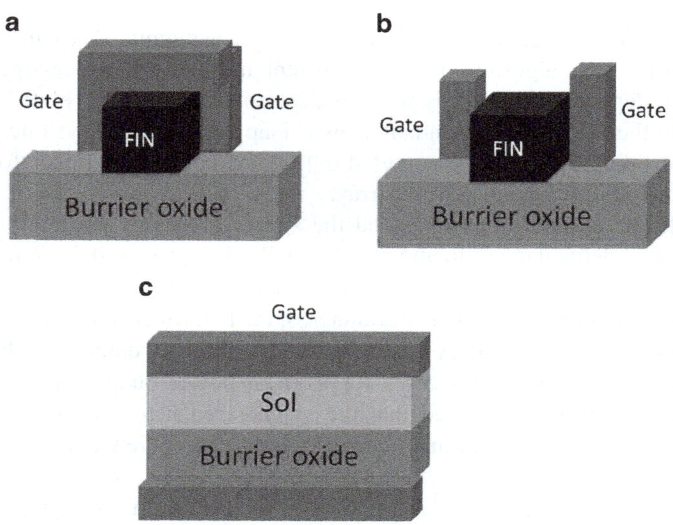

Fig. 17 FinFET structure:(**a**) shorted-gate FinFET, (**b**) independent double-gate FinFET, (**c**) planar double-gate FinFET on SOI

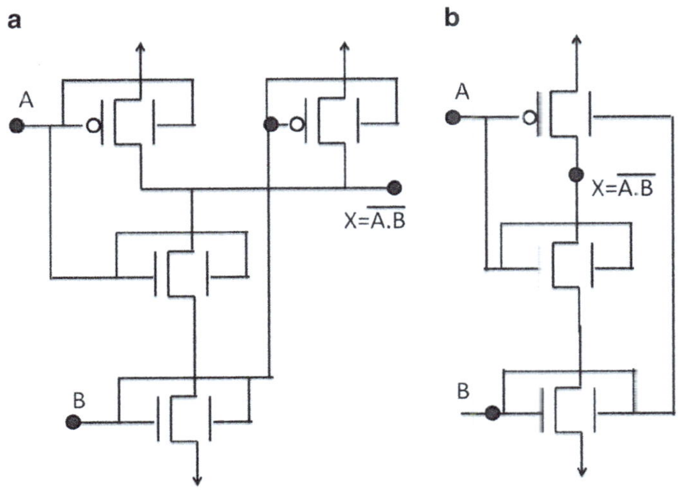

Fig. 18 A 2-input NAND logic gate built using: (**a**) a shorted-gate FinFET, and (**b**) independent gate FinFET transistor

3.1 Design Techniques Using FinFET Transistor

The primitive gates built using multi-gate MOSFET is fundamentally different from the bulk CMOS. Figure 18 shows a 2-input NAND gate built using, respectively, shorted-gate and independent-gate FinFET transistor [46].

From Fig. 18, we can see that (1) the IG-FinFET consumes less transistor than the SG-FinFET, and (2) the SG-FinFET transistor can be fabricated easily using the bulk CMOS process.

The results on the ISCAS benchmark reported in [47] show that the IG-FinFET consumes less power but slower than the SG-FinFET. Figure 19 shows a low-power NAND gate built using FinFET and IG-FinFET [47]. The supply voltages V_L and V_H are used to reverse-bias the IG-FinFET transistor.

The FinFET transistor did not completely solved the leakage current problem. In [47], it is reported that the leakage power consumption for the ISCAS'85 benchmark represents on average 33.1 % of the total power consumption. The authors presented several schemes to reduce the leakage power dissipation under delay or relaxed delay constraints.

The emerging FinFET technology for the manufacturing of future integrated circuits raised a deep concern for the projection made for the dark silicon area reported in [33]. The authors of [48] studied the dark-silicon problem using the FinFET transistor. The study was performed for the 7 nm FinFET technology. Two types of processor cores were used in the study. These processor cores are the Nehalem out-of-order and LEON3. The reported results show that the 50 % dark silicon prediction reported in [33] is very pessimistic. They further demonstrated that the near-threshold voltage regime is a powerful way to reduce the percentage of dark-silicon.

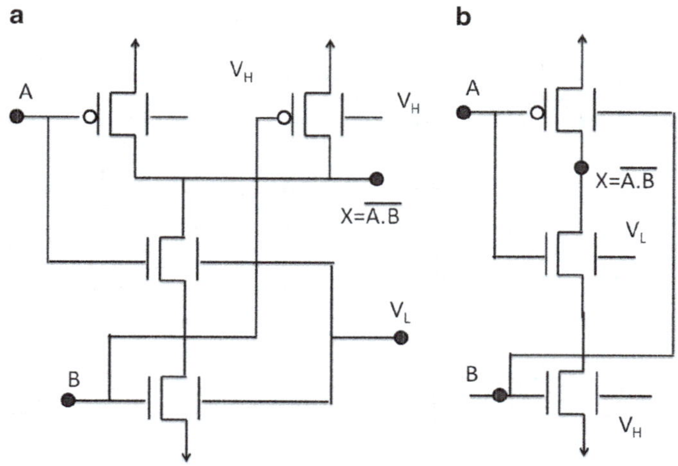

Fig. 19 A low-power NAND gate built using: (**a**) Low-power FinFET, (**b**) Low-power IG-FinFET [47]

4 RF Design Using FinFET Transistor

For the RF design, it has been reported in [49] that the FinFET outperforms the bulk CMOS for frequency below 5 GHz. However, for high frequency (millimeter wave band) the authors found that the FinFET suffers from low drive current. The authors of [50], concluded that for the RF frequency higher than 10 GHz, the bulk CMOS is better fitted to design RF blocks.

In recent years, several reports showed the potential of FinFET for the construction of RF circuits in millimeter wave band [51, 52].

5 Sensors for 5G Device

In modern days, thanks to the fast development of the integration technology, sensors have become widespread, ubiquitous and low-cost. In addition, the sensor technology has been subjected to conspicuous development over time. There exist several technologies to fabricate sensors. To determine the preponderant technology, we conducted extensive web search for the up-to-date commercial sensors. The results are presented in Table 2. The table shows that the MEMS is the prevalent technology for sensor manufacturing.

Table 2 Sensor technology and applications

Sensor type	Application	Technology
Mechanical	Pressure sensing	MEMS
Thermal	Air conditioning	MEMS
Biological	Blood glucose	MEMS
Chemical	Environmental monitoring	MEMS
Optic	Ambient light detection	Silicon
magnetic	Gyroscope	Hall IC

6 Microelectromechanical System

Microsystem or Microelectromechanical System or MEMS for short is a system that has both electronic and non-electronic devices [53]. MEMS contains the following blocks:

1. A Transducer to convert physical quantities into electrical signal and vice-versa.
2. An actuator and sensor for action.
3. A controller which is an integrated circuit devised to take decision.

As opposed to the integrated circuits, MEMS is diverse in fabrication, functionalities, and applications. In other words, MEMS fabrication is not confined to a certain material and steps. What makes MEMS attractive for sensor is it is size, fidelity and yield. Furthermore, the current success of MEMS is credited to abundant of factors, such as its similarity with the integrated circuit, low-cost, and the existing of CAD tools.

Contemporary smart phones and tablets contain a bunch of sensors fabricated using MEMS (gyroscope, accelerometer, and microphone). There is also a growing interest to build RF front-end using MEMS (RF-MEMS). To develop smart applications, future smart phones will integrate electronic nose. The nose will be responsible for detecting volatile compound.

Using the mobile Internet device as a driving force for the development of MEMS, a road map has been determined for the following devices: accelerometers, gyroscopes, integration path for inertial measurement units, microphone, RF MEMS, and electronic nose.

7 Summary

Future wireless transceiver will be implemented using MtM technology. In this technology, the number of sensors is expected to increase at a rate proportional to the Moore scaling paradigm. For the base-band processor, the new design challenge is the increasing portion of the dark silicon area. In this chapter, we discussed potential solutions to reduce the impact of dark silicon and we proposed a multitude of architecture solutions. These solutions are spread across all the design abstraction levels (system, algorithm, circuit and device). Multi-gate FET transistor has plenty of advantages compared to the bulk CMOS for the design of both RF front-end and base-band processor in the technology below 20 nm.

Acknowledgements The authors deeply acknowledge Dr. Mehrdad Dianati for his support and interest that has shown to the chapter.

References

1. I.B. Dhaou, An electronic system to combat drifting and traffic noises on Saudi roads, in *Proceeding of IV* (2012), pp. 217–222
2. F. Boccardi, R.W. Heath Jr., A. Lozano, T.L. Marzetta, P. Popovski, Five disruptive technology directions for 5G. IEEE Commun. Mag. **2**(52), 74–80 (2014)
3. K.E. Skouby, P. Lynggaard, Smart home and smart city solutions enabled by 5G, IoT, AAI and CoT services, in *Proceedings of IC3I* (2014), pp. 874–878
4. J. Andrews, S. Buzzi, W. Choi, S. Hanly, A. Lozano, A. Soong, J. Zhang, What will 5G be? IEEE J. Select. Areas Commun. **32**(6), 1065–1082 (2014)
5. T.H. Lee, *The Design of CMOS Radio-Frequency Integrated Circuits*, 2nd edn. (Cambridge University Press, Cambridge, 2003)
6. P.B. Kenington, *RF and Baseband Techniques for Software Defined Radio* (Artech House, Boston, 2005)
7. G. Fettweis, M. Lohning, D. Petrovic, M. Windisch, P. Zillmann, W. Rave, Dirty RF: a new paradigm, in *Proceeding of PIMRC*, vol. 4 (2005), pp. 2347–2355
8. W. Namgoong, T.H. Meng, Direct-conversion RF receiver design. IEEE Trans. Commun. **49**(3), 518–529 (2002)
9. L. Noor, A. Anpalagan, Direct conversion receiver for radio communication systems. IEEE Potentials **24**(5), 32–35 (2005)
10. T. Jiang, Y. Wu, An overview: peak-to-average power ratio reduction techniques for OFDM signals. IEEE Trans. Broadcast. **54**(2), 257–268 (2008)
11. I.B. Dhaou, C.C. Logothetis, H. Tehnunen, On the robustness and performance tradeoffs for OFDM channel estimation, in *Proceedings of the 13th International Conference On Wireless Communication*, vol. 1 (2001), pp. 41–48
12. D. Tandur, M. Moonen, Compensation of RF impairments in MIMO OFDM systems, in *Proceeding of ICASSP* (2008), pp. 3097–3100
13. I.B. Dhaou, Client-server network architecture for safe pilgrim journey in the Kingdom of Saudi Arabia, in *Proceeding of IV* (2010), pp. 1043–1048
14. W.Z. Khan, X. Yang , M.Y. Aalsalem, Q. Arshad, Mobile phone sensing systems: a survey. Commun. Surv. Tutor. **15**(1), 402–427 (2013)
15. R.K. Cavin, P. Lugli, V.V. Zhirnov, Science and engineering beyond moore's law. Proc. IEEE **100**, 1720–1749 (2012)
16. K. Roy, J. Byunghoo, A. Raghunathan, Integrated systems in the more-than-moore era: designing low-cost energy-efficient systems using heterogeneous components, in *Proceeding of VLSID* (2010), pp. 464–469
17. W. Arden, M. Brillout, P. Cogez, M. Graef, B. Huizing, R. Mahnkopf More-than-moore white paper, International Roadmap for Semiconductor, ITRS, 2010
18. G.Q. Zhang, A.J. van Roosmalen (eds.), *More than Moore Creating High Value Micro/Nanoelectronics Systems* (Springer, New York, 2009)
19. N. Weste, D. Harris, *Principles of CMOS VLSI Design*, 4th edn. (Pearson, Boston, 2011)
20. [Online]. Available: http://ptm.asu.edu

21. J. Kao, S. Narendra, A. Chandrakasan, Subthreshold, leakage modeling and reduction techniques, in *Proceeding of ICCAD* (2002), pp. 141–148
22. J. Rabaey, *Low Power Design Essentials* (Springer, New York, 2009)
23. J. Howard et al., A 48-Core IA-32 processor in 45 nm CMOS using on-die message-passing and DVFS for performance and power scaling. IEEE J. Solid-State Circuits **46**(1), 173–183 (2011)
24. I.B. Dhaou, Efficient interconnect modeling for SoC systems with application to interconnection delay estimation, Technical Report, TUCS 2003
25. I.B. Dhaou, H. Tenhunen, V. Sundararajan, K.K. Parhi, Energy efficient signaling in deep-submicron technology, in *Proceeding of ISCAS*, vol. 5 (2001), pp. 411–414
26. I.B. Dhaou, Low-power design techniques in deep-submicron with application to wireless transceiver design, Ph.D. Dissertation, Royal Institute of Technology, Sweden 2002
27. E. Mensink, D. Schinkel, E.A.M. Klumperink, E. van Tuijl, B. Nauta, Power efficient gigabit communication over capacitively driven RC-limited on-chip interconnects. IEEE J. Solid-State Circuits **45**(2), 447–457 (2010)
28. J. Nurmi, H. Tenhunen, J. Isoaho, A. Jantsch (eds.), *Interconnect-Centric Design for Advanced SoC and NoC* (Springer, New York, 2005)
29. D. Khalil, D. Sinha, H. Zhou, Y. Ismail, A timing-dependent power estimation framework considering coupling. IEEE Trans. Very Large Scale Integr. VLSI Syst. **17**(6), 843–847 (2009)
30. H. Tenhunen, A. Jantsch (eds.), *Networks on Chip* (Springer, New York, 2003)
31. R. Marculescu, U. Ogras, L. Peh, N. Jerger, Y. Hoskote, Outstanding research problems in NoC design: system, microarchitecture, and circuit Perspectives. IEEE Trans. Comput. Aided Des. Integr. Circuits Syst. **28**(1), 3–21 (2009)
32. S. Deb, K. Chang, X. Yu, S.P. Sah, M. Cosic, A. Ganguly, P.P Pande, B. Belzer, D. Heo, Design of an energy-efficient CMOS-compatible NoC architecture with millimeter-wave wireless interconnects. IEEE Trans. Comput. **62**(12), 2382–2396 (2013)
33. H. Esmaeilzadeh et al., Dark silicon and the end of multicore scaling, in *Proceeding of 38th International Symposium on Computer Architecture (ISCA)* (2011), pp. 365–376
34. H. Esameilzadeh, E. Blem, R.St. Amant, K. Sankaralingam, D. Burger, Power limitation and dark silicon challenge the future of multicore. ACM Trans. Comput. Syst. **30**(3), 11:1–11:27, August 2012
35. K.K. Parhi, *VLSI Digital Signal Processing Systems: Design and Implementation* (Wiley Interscience, New York, 1999)
36. N. Petkov, *Systolic Parallel Processing* (North Holland Publishing, New York, 1992)
37. N.G. Hotta, J. Sampson, Q. Zheng, V. Bhatt, J. Auricchop, GreenDroid: an architecture for the dark silicon age, in *Proceeding of ASP-DAC* (2012), pp. 100–105
38. Y. Turkhia, B. Raghunathan, S. Garg, D. Marculescu, HaDeS: architectural synthesis for heterogeneous dark silicon chip multi-processors, in *Proceeding of DAC* (2013), pp. 1–7
39. R. David, P. Bogdan, R. Marculescu, U. Ogras, Dynamic power management of voltage-frequency island partitioned networks-on-chip using intel's single-chip cloud computer, in *Proceeding of NoCS* (2011), pp. 257–258
40. M.K. Yadav, M.R. Casu, M. Zamboni, A simple DVFS controller for a NoC switch, in *Proceeding of PRIME* (2012), pp. 131–134
41. H. Bokhari, H. Javaid, M. Shafique, J. Henkel, darkNoC: designing energy-efficient network-on-chip with multi-vt cells for dark silicon, in *Proceeding of DAC* (2014), pp. 1–6
42. D. Frank, R. Dennard, E. Nowak, P. Solomon, Y. Taur, H.-S. P. Wong, Device scaling limits of Si MOSFETs and their application dependencies. Proc. IEEE **89**(3), 259–288 (2001)
43. Y. Bin et al., FinFET scaling to 10 nm gate length, in *Proceeding of IEDM* (2002), pp. 251–254
44. T.-J. King, FinFETs for nanoscale CMOS digital integrated circuits, in *Proceeding of ICCAD* (2005), pp. 207–210
45. Y.S. Chauhan, D.D. Lu, V. Sriramkumar, S. Khandelwal, J.P. Duarte, N. Payvadosi, A. Niknejad, Ch. Hu, *FinFET Modeling for IC Simulation and Design: Using the BSIM-CMG Standard* (Academic, London, 2015)
46. D. Chen, N.K. Jha (eds.), *Nanoelectronic Circuit Design* (Springer, New York, 2011)

47. A. Mutterja, N. Agarwal, N.K. Jha, CMOS logic design with independent gate FinFETs, in *Proceeding of ICCAD* (2007), pp. 560–567
48. A. Shafaei, Y. Wang, S. Ramadurgam, Y. Xue, P. Bogdan, M. Pedram, Analyzing the dark silicon phenomenon in a many-core chip multi-processor under deeply-scaled process technologies, in *Proceeding of GLSVLSI* (2015), pp. 127–132
49. V. Subramanian, B. Parvais, J. Borremans, A. Mercha, D. Linten, P. Wambacq, J. Loo, M. Dehan, C. Gustin, N. Collaert, S. Kubicek, R. Lander, J. Hooker, F. Cubaynes, S. Donnay, M. Jurczak, G. Groeseneken, W. Sansen, S. Decoutere, Planar bulk MOSFET versus FinFETs: an analog/RF perspective, IEEE Trans. Electron Devices **53**(12), 3071–3079 (2006)
50. M. Dehan et al., Perspectives of (sub-) 32 nm CMOS for analog/RF and mm-wave applications, in *Proceeding of EuMIC*, 27–28 October 2008, pp. 103–106
51. S.K. Lin, J.L. Kuo, W. Huei, A 60 GHz sub-harmonic resistive FET mixer using 0.13 μm CMOS technology. IEEE Microw. Compon. Lett. **21**(10), 562–564 (2011)
52. S.C. Woo et al., Device design guidelines for nanoscale FinFETs in RF/analog applications. IEEE Electron Device Lett. **33**(9), 1234–1236 (2012)
53. S.E. Lyshevski, *MEMS and NEMS: Systems, Devices, and Structures* (CRC Press, Boca Raton, 2001)

Part III
Networking Techniques and Applications for 5G Networks

Ultra-Dense Network Architecture and Technologies for 5G

Shanzhi Chen, Fei Qin, Bo Hu, Xi Li, and Jiamin Liu

Abstract In recent years, with the growing popularity of smart device, our daily life has come to revolve around with spectacularly successful mobile Internet services, which lead to the explosion of data traffic in mobile communication networks. The requirement on communication networks has become a critical issue. By 2020, the global mobile traffic volume will have about 1000 times growth compared to that of 2010. Recent research on 5G requirements indicates that the traffic density in crowded city or hotspot area will reach 20 Tbps/Km^2. Ultra dense network(UDN) has been introduced to meet the traffic capacity requirement of 5G. as a most promising method. Challenges, network architectures, key technologies will be discussed in this section.

1 Introduction

In recent years, with the growing popularity of smart device, our daily life has come to revolve around with spectacularly successful mobile Internet services, which lead to the explosion of data traffic in mobile communication networks. The requirement on communication networks has become a critical issue. By 2020, the global mobile traffic volume will have about 1000 times growth compared to that of 2010 [1, 2]. According to the history of 4G and the former generations, the next generation of mobile network, i.e. 5G which is towards 2020 and beyond would reach 10 Gbps, which is 10 times compared to the peak data rate of 4G. Recent research on 5G requirements indicates that the traffic density in crowded city or hotspot area will

This work was supported in parts by the National Natural Science Foundation of China for Distinguished Young Scholar under Grant 61425012.

S. Chen (✉) • F. Qin • J. Liu
China Academy of Telecommunications Technology (CATT) and Datang Telecom Technology & Industry Group, Beijing, China
e-mail: chensz@datanggroup.cn

B. Hu • X. Li
Beijing University of Posts and Telecommunications (BUPT), Beijing, China

W. Xiang et al. (eds.), *5G Mobile Communications*,
DOI 10.1007/978-3-319-34208-5_15

reach 20 Tbps/km^2 [3]. In addition, other requirements such as lower latency, higher spectral efficiency and energy efficiency are also included.

There are usually three ways to improve throughput in wireless system: (1) Improving spectral efficiency through new coding and modulation technologies; (2) Increasing spectrum bandwidth; (3) cell splitting to improve spectrum reuse density. In [4], it is observed that, by 2008, the wireless capacity has increased 1 million times from 1957. Among these gains, 25 times improvements came from wider spectrum and 25 times improvements were contributed by advanced air interface design, and 1600 times improvements were due to the reduction in cell sizes and transmit distance. From technique evolving viewpoint, the potential gain derived from the advance of wireless modulation and multi-antenna technologies, is approaching to the upper bound. In LTE-A, the theoretical peak spectral efficiency has reached 30 bps/Hz through 8-layer spatial multiplexing. This value nearly reaches the maximum of typical wireless transmitting technology. Secondly, the spectrum resource is limited. The continuous development of radio services such as satellite, broadcasting, scientist, fixed and mobile terminal result that spectrum become very rare resources. The International Mobile Telecommunication (IMT) forecast that requirement for spectrum in year 2020 will reach 1490–1810 MHz in China [5]. It is a big challenge to identify and allocate enough spectrum resource to IMT system.

Based upon the above analysis, it is hard to meet 5G traffic capacity requirement through increasing spectral efficiency and allocating more spectrum bands. Therefore, we can conclude that the increasing of the wireless access point (AP) density with smaller coverage is the most efficient way to improve the system capacity, especially in hotspot scenarios [6]. China promoted LTE-Hi [7] to increase the system throughput for hotspot areas. LTE-Hi is a Long Term Evolution (LTE) based hotspot solution, which aims to provide higher performance in hotspot scenarios by exploiting wider bandwidth in higher frequency. In ITU-R Report M.2320 [8], Ultra Dense Network (UDN) is promoted as one of the technology trends to meet the high throughput requirements of 5G. And METIS [9] also take the Ultra Density Network as one of the most important topics towards the mobile system for 2020 and beyond. On the other hand, Since Bell Lab proposed the concept of cellular network in 1970s, network architecture has been adopted in the successive generations of mobile networks. For architecture evolution perspective, authors in [10] point out that the macro cellular and local small cell coexistence architecture in 5G will replace the macro cellular-dominated architecture.

Typical scenarios of UDN include: office, apartment, open-air gathering, stadium, subways and railway station [3]. The common requirements in these scenarios are huge numbers of connections, high density network traffic and high data rate. To meet these requirements, the AP(s) need to be deployed densely, with minimum Inter-Site Distances (ISDs) of tens meters or below, i.e. one or more APs per room for indoor scenarios, and one AP on each lamp post for outdoor scenarios.

2 Concept and Challenges of UDN

2.1 *What Is UDN*

The main difference between UDN and traditional cellular network is the APs density. In UDN, the radius of AP coverage is about 10 m and there are thousands of APs in 1 km^2. But in traditional cellular network, cell range is more than 500 m and usually less than 3–5 base stations (BSs) in 1 km^2. Correspondingly, only one or several terminals are connected to one UDN AP, whereas hundreds or even thousands of active users are resident in one macro cell. Table 1 gives the comparison between the UDN and traditional cellular network.

Another key point is that the type of APs in UDN is diversified. Small cell station, relay station, distributed Remote Radio Head (RRH) and user equipment (UE) itself can act as AP in UDN. However the macro BS in traditional cellular network is the dominate access point.

Besides the above features, higher spectrum and wider band, heterogeneous and irregular deployment, flexible backhaul, lower user mobility are also differences between UDN and traditional cellular networks.

In [9], METIS defines UDN as a stand-alone system that will be optimized for hotspot areas where the highest traffic increase will be observed. The concept of UDN includes radio access technologies, small cell integration/interaction and wireless backhauling. The authors in [11] suggested that UDN is a major technology to meet the requirements of ultra-high traffic volume density.

In general, UDN is a new wireless network solution for hotspot scenarios, to provide higher throughput and better user experience. In UDN, the AP density is comparable to or even higher than the user density. Different types of APs will tightly cooperate to achieve higher spectrum efficiency, lower power consumption and seamless mobility.

Table 1 Comparison between UDN and the traditional cellular network

ITEM	UDN	Traditional cellular network
Deployment Scenarios	Indoor, Hotspot	Wide coverage
AP density	More than 1000/km^2	3–5/km^2
AP Coverage	Around 10 m	Several hundred meters and more
AP types	Small-Cell, Pico, Femto, UE relay, Relay	Macro/Micro BS
AP Backhaul	Ideal/Non-ideal, Wired/Wireless	Ideal, Wired
User Density	High	Low/Medium
User Mobility	Low mobility	High mobility
Traffic density	High	Low/Medium
Deployment	Heterogeneous, irregular	Single layer, regular cell
System Bandwidth	Hundreds of MHz	Tens of MHz
Spectrum	>3 GHz (up to mm Wave)	<3 GHz

2.2 Challenges and Technical Directions of UDN

In order to provide very high traffic density and better user experience in UDN, there are many new challenges including network architecture, mobility management, interference management and others.

2.2.1 Challenges of Network Architecture

Traditional cellular network architectures are designed for wide area seamless coverage. Due to the differences of UDN and traditional cellular network discussed in Table 1, there are many problems to utilize traditional cellular network architecture (e.g. 4G architecture) in UDN.

- Signaling overhead and lengthy data path
  Too many functions such as service control and mobility control are centralized in Core Network (CN), which include Mobility Management Entity (MME) and Packet Data Network Gateway (PGW)/Serving Gateway (SGW). It is not efficient for UDN with high traffic throughput and ultra-dense APs deployment, and it leads to signaling overhead and lengthy data transmission path between APs and CN.
- Frequent handover
  The tight coupling of user plane and control plane over one air interface will result in frequent handover when the AP coverage is very small. It is not efficient and flexible in heterogeneous networking within both macro and UDN AP coverage.
- All distributed functions
  Higher layer process, Radio Resource Management (RRM), mobility management functions are distributed on each AP independently. In order to better support advanced interference management and resource management for UDN, the functions on each distributed AP needs to be centralized.
- Better user experience
  UDN is target to provide smooth handover and very high data rate for each user with ultra dense AP. The simple data gathering and transmission function of Local GateWay (LGW) cannot support better user experience. More functions are needed in LGW.

Therefore, a new architecture of UDN is needed to support high density AP deployment and flexible network management. In this new architecture, a local centralized user service center is necessary for knowing and measuring user's radio environment. Besides, the RRM and user service control center much closer to the user are required to provide better joint processing and Quality of Service (QoS) control, and a lower mobility anchor is also needed. Meanwhile, the CN functions should be simplified to provide only high level service to user.

2.2.2 Challenges of Mobility Management

With very small coverage and irregular network topology, mobility management in UDN is quite different from traditional cellular network.

In the scenarios that ultra-dense APs are deployed, the coverage radius of these APs is extremely small, i.e. only a few meters to tens meters. According to traditional handover judgment method, frequent handover may occur and the interruption probability of the user's experience of high data rate may increase. Moreover, from the perspective of the network side, it also means a high handover signaling overhead. How to solve the problem is essential for UDN [6, 12].

2.2.3 Challenges of Interference Management

Huge amount of APs may bring much higher throughput and better user experience, but may also lead problems. Interference management may directly impact the system performance [13]. Along with the multiplexing of resources for more access opportunities, interference increases as well and becomes more complex related to traditional cellular network. We need to solve the following problems:

- The ultra-density environment results in more interference sources. For example, in the crowd subway trains, lots of terminals and APs exist, and signals may have more reflected and scattering paths. Then transmission power should increase, which turns out more interference than before, however.
- Decreasing interference and increasing the resource utility become two parts in the contradiction and we need to find the proper tradeoff point.
- The existing parameters to evaluate interference impact such as interference threshold may not reflect the overall performance of the networks, and then more suitable parameters should be used to give a better indication between interference managing results and throughput, energy efficiency and other system level parameters.

Therefore, we should set up suitable interference models, analyze the typical wireless transmission scenario, and then propose effective interference control approach for UDN.

2.2.4 Challenges of Flexible Networking

With extreme densification and complex heterogeneous deployment, network planning and optimization will become difficult for UDN. It is critical to further enhance Self Organization Network (SON) to support flexible networking.

Huge amounts of APs in UDN make it more complex to realize self-configuration, self-optimization and self-healing.

Ultra high throughput, ultra low latency, ultra high reliability, massive connections are required to be provided in UDN. On another side, UDN is a very complex network to cover both indoor and outdoor scenarios, with ideal backhaul AP deployment and non-ideal backhaul AP deployment. From view of the access technologies, there are 5G new access technologies, LTE access technologies, WLAN access technology working together. So, flexible network architecture with intelligent network sensing and automotive network optimization is very important to flexible networking and increasing spectrum efficiency in UDN.

Based on the above analysis, new network architecture for UDN is an important direction. Key technologies towards flexible networking, self-backhauling, multi-RAT coordination, advanced interference management, advanced mobility management and radio resource management are also essential for UDN.

In addition, due to different radio scenarios and higher spectrum and wider bandwidth, new transmission technologies and new air interfaces design are also need to further enhance the system performance for UDN. Millimeter wave communication is a very attractive direction for UDN [14], which can provide Gbps user experience and tens of Gbps throughput per AP. And it's also very suitable for wireless backhauling for UDN APs. Details of millimeter wave communication are discussed in other chapters of this book.

3 Network Architecture of UDN

3.1 Overview

The topology of UDN toward 5G can be summarized with following key characteristics: (1) high density of mobile APs distributed randomly without network planning [15]; (2) various kinds of APs with different backhaul or fronthaul capabilities [16]; (3) heterogeneous networks with different overage ranges, different spectrum band and multiple radio access technologies (RATs) [14, 17].

Based on above characteristics, in order to provide high throughput and better user experience, some new principles and methods for network architecture design are introduced:

3.1.1 Localization and Flatter

It is a promising cost effective method to offloading the 1000 times of traffic within local UDN area. A flatter architecture is needed to support localization and decrease the cost of transmission. Localization of data path and controlling function are introduced by METIS and NGMN [18, 19] toward 5G UDN.

3.1.2 U/C Separation

New principle in [7] is the user plane and control plane separation. It's an effective way to provide user high data rate through density small cells but without mobility and connectivity experience degradation. This feature is supported in 3GPP Release 12 through dual/multi-node connections.

3.1.3 User-Centric

Toward 5G, UDN is quite different from traditional cellular network. It is very important to introduce user-centric principle [6], the network will organize the dense APs dynamically to serve the user to make it feel like always in the centre of a cell. User-centric ultra-dense network (UUDN) [6], including user-centric mobility management, user-centric clustering for data plane [20] and user-specific network controlling, can provide better user experience and higher spectrum efficiency.

3.1.4 C-RAN, Distributed-RAN and Flexible-RAN

Centralized Radio Access Network (RAN) and distributed RAN architectures for UDN are discussed in [15] For centralized architecture, it can provide better joint processing and get higher spectrum efficiency, but need very high backhaul/fronthaul capability. For distributed architecture, it is more flexible for network deployment, but interference management is difficult and has lower spectrum efficiency. So a more flexible architecture is preferred to adaptively connect APs with different backhaul/front haul capability [18].

3.1.5 SDN & NFV

The new trends of Software Defined Network (SDN) and Network Function Virtualization (NFV) are also impacts the design of the network architecture for 5G [18, 21, 22]. With the decoupling of controlling and forwarding, software and hardware, the network architecture can be redesigned to define the logical functions and their interfaces but not the entities and their interfaces. This will also enabling even more flexible architecture towards UDN 5G.

In summary, the trends of UDN architecture are becoming more flatter, localized, U/C separation, user-centric, flexible and intelligent. Some examples toward small coverage density network are given in following subsections.

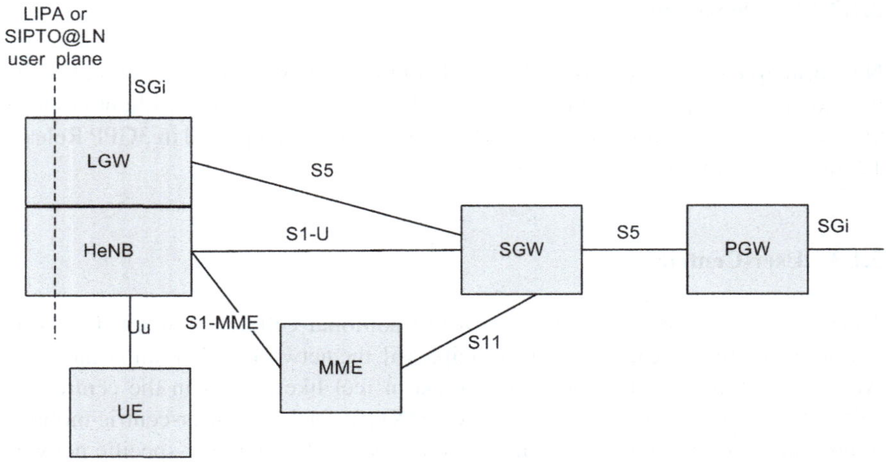

Fig. 1 4G LTE/SAE network architecture

3.2 GPP HeNB Architecture

Home enhanced Node B (HeNB) is target for dense small coverage for hotspot area. A typical 3GPP architecture for HeNB under the 4G LTE/SAE framework, is shown in Fig. 1 [23, 24]:

In this architecture, the network side mainly includes five entities: HeNB, LGW MME, SGW and PGW. The function of each entity can be found in [15]. The function of HeNB is almost same as traditional macro BS except smaller coverage. And the user side control plane and data plane are managed by CN. This architecture also derives the localized features, which enable the function of Local Internet Packet Access (LIPA) and Selected IP Traffic Offload (SIPTO) for data offloading.

The HeNB architecture is easy for equipment development at early stage deployment with lower density of HeNBs. But when APs density increased, challenges of the architecture are as following:

(1) Many functions such as service control and mobility management are centralized in CN. It is not efficient with high traffic throughput and dense APs deployment. This will lead to extra signaling overhead and longer data transmission latency between APs and CN.
(2) The tight coupling of user plane and control plane at air interface will result in frequency handover for ultra dense network.
(3) The data transmission function of LGW cannot support user-centric service experience. More functions are needed to be collected and supported in LGW.

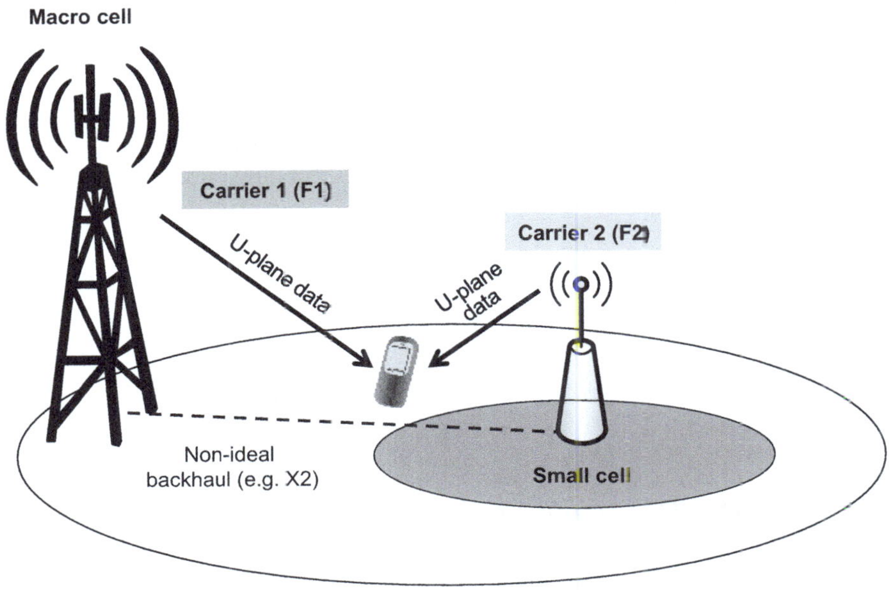

Fig. 2 Inter-node radio resource aggregation

3.3 Small Cell Enhancement (SCE) Architecture

To further support the heterogeneous network deployment scenarios, 3GPP developed the SCE architecture in release 12 [25].

A “dual connectivity” is used to refer to operation where a given UE consumes radio resources provided by at least two different network points connected with non-ideal backhaul. Dual connection can be done by aggregating radio resources in more than one eNB for user plane data transmission as illustrated in Fig. 2. By this solution, signaling overhead towards the CN can be potentially saved by keeping the mobility anchor in the macro cell.

In dual connectivity operation, a UE has only one RRC connection. With this principle, only the Master eNB (MeNB) generates the final RRC messages to be sent towards the UE. The MeNB and Secondary eNB (SeNB) will tightly coordinated to provide a UE dual User plane connections as illustrated in Fig. 3. The UE RRC entity sees all messages coming only from one entity (in the MeNB) and the UE only replies back to that entity.

Within these characteristics, the mobile anchor is moved to macro station, handover signaling overhead is reduced when the user is moving among small cells. But there are still many RRC signaling between high speed UEs and macro station during the movement for the SeNB updating procedure.

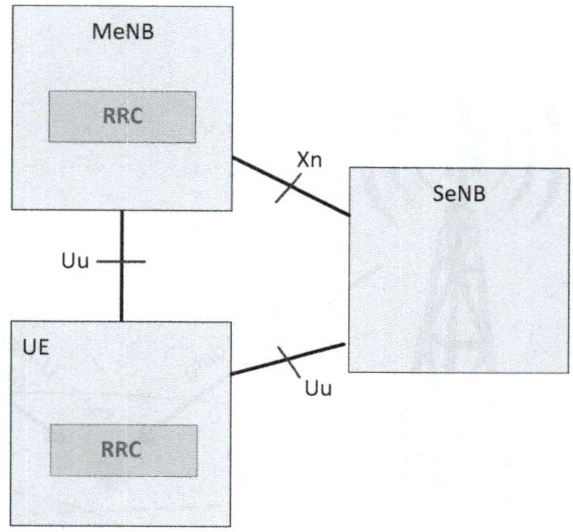

Fig. 3 Radio interface C-plane architecture for dual connectivity

3.4 METIS UDN Architecture

UDN is identified as one of the five Horizon Topics (HTs) of METIS project toward 5G, and also one of the four main technical enablers [7, 18]. Considering the trends of SDN and NFV, in order to support wide range of scenarios and use cases for 5G, METIS designed a general 5G architecture with flexible manners as shown in Fig. 4-1 in [18]. The architecture can be used for UDN by different function selection at different network domain as shown in Fig. 4-3 in [18].

Based on the separation of user plane and control plane, six functions are defined in this architecture. Synchronize radio control plane functions (SRC) and synchronize radio user plane functions (SRU) are deployed very close to end users. It supports Multi-APs and RRHs with ideal backhaul with centralized PHY/MAC, like C-RAN technologies.

Asynchronous radio control plane functions (ARC) and asynchronous radio user plane (ARU) together with CN user plane (Core-U) are located RAN side, which is much closer to user than traditional 3GPP SAE architecture. They can provide higher layer control functions like mobility management, QoS control, inter-RAT coordination, and also user plane function including local breakout.

CN control plane function (Core-C) and CN user plane functions (Core-U) are located in aggregation and CN side to provide high level management like Authentication, Authorization and Accounting (AAA) and roaming.

METIS UDN architecture based on the C-RAN, SDN, NFV technologies with principle of localized, U/C split concept, it's a flexible and efficient 5G UDN architecture.

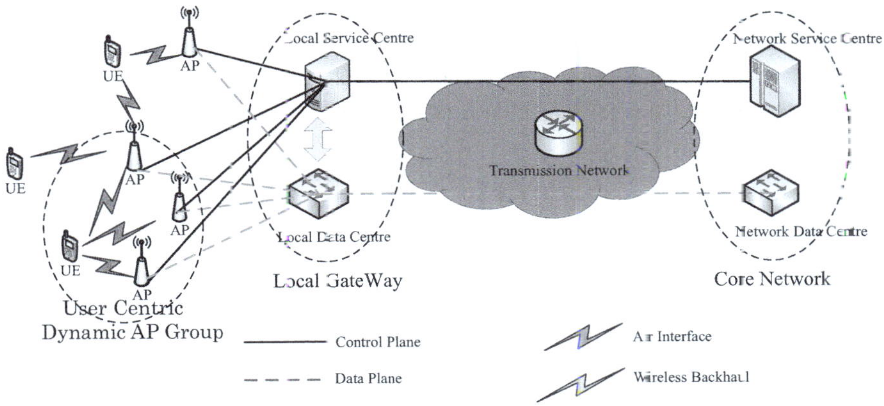

Fig. 4 User centric architecture

3.5 *User-Centric Architecture for UDN*

In [6], user-centric UDN (UUDN) architecture with the principles of user-centric, localized, U/C separation. It can be easily and flexible deployed with SDN and NFV manner as Fig. 4.

In this architecture, there are no more "cells" logically and physically from user's perspective. Dense APs in one area will be organized intelligently to follow user's movement and provide data transmission on-demand.

Four functional entities are introduced to provide user-centric services. At the radio side, local service center (LSC) and local data center (LDC) are introduced to provide the logical decoupling of control plane and user plane. All kinds of APs are connected to the LSC and LDC by various backhaul (idea/no-ideal, wire/wireless). At the core network side, network service center (NSC) and network data center (NDC) are introduced to provide the control and transmit functions. The LSC and LDC can be integrated into one Local Gateway as a physical entity, and NSC and NDC may also be integrated into one CN entity.

The functions and interfaces of each entity in UUDN are as following [6]:

(1) AP is the radio access channel for UE including the data plane and the control plane. APs can be built with RF, PHY, MAC and IP layer functions or combinations of them based on the backhaul capacity. If the AP has only RF, the PHY to IP layers will be centralized into LDC. With this architecture, LDC can provide joint processing in PHY layer. The advanced signal process can thus be used to avoid the inter-AP interference.

(2) LSC is the control service center to organize a dynamic APs Group (APG) to serve one user. It will have the new functions of user-centric RRM, multi-RAT coordination, effective QoS control, user-centric mobility management, and local radio link control.

Table 2 Characteristics of the network architectures toward UDN

	3GPP HeNB architecture	3GPP SCE architecture	METIS UDN architecture [18]	UUDN architecture [5]
Localization and flatter	LIPA & SIPTO	LIPA & SIPTO	Localized (User Plane and Part of Control Plane)	Localized (User Plane and Most of the Control Plane)
U/C separation	No	Yes	Yes	Yes
SDN & NFV based	No	No	Yes	Yes
Flexible Backhauling	No	No	Yes	Yes
User-Centric	No	No	No	Yes

(3) LDC is the local data center to dealing with user data transmission. It will provide the user plane functions including higher layer process, dynamic AP channel processing. It also has the functions of Multi-AP coordination based and multi radio bear convergence for users.
(4) NSC is the network data center to provide the functions of user policy control, AAA, and high level mobility (roaming, inter NSC handover) etc.
(5) NDC is worked as packet data gateway at the network side.

With this architecture, LSC and LDC are very close to the location of APs, so that it is easy to provide user-centric service functions, advanced resource management, and interference management. It is more flexible for UUDN deployment with the decoupling of user plane and control plane, and decentralizing the CN functions to LSC and LDC. And signaling overhead and backhaul overhead can be greatly reduced for UDN scenarios.

3.6 Summary of Architecture

The trends of UDN architecture are becoming more flatter, localized, U/C separation, user-centric, flexible and intelligent. A good UDN architecture will better support very dense APs work efficiently at complicated deployment scenarios. The characteristics of some typical architectures toward UDN are summarized in the following Table 2.

4 Key Technologies of UDN

Based on the new architecture, many key technologies can be introduced to provide high QoE, high area spectrum efficiency and low cost. The promising technologies are summarized below.

4.1 Flexible Networking

Flexible networking technologies, as SON in 4G [26], are promising technologies to enable UDN in real deployment. With extreme densification, UDN will face huge challenges on self-installation, self-configuration of physical layer Identifier (ID), Automatic Neighbour Relations (ANR), Mobile Load Balancing (MLB) and etc.

4.1.1 Self-Installation

In super dense deployment of the network, the number of APs is very large. It is difficult for operators to configure and initialize each device manually, and then require that each AP can be plug and play [26]. Using the mechanism, the BSs automatically install in LTE, APs automatically start and stop initialization, including automatically get IP address, automatically establish the connection with the Operation Administration and Maintenance (OAM) server, automatically download the software, initial the configuration of wireless and transmission.

4.1.2 Physical Layer Cell ID and Self-Configuration

In UDN, hundreds or even thousands of APs will be deployed in the same area. The amount of AP IDs may not be enough. The types of AP are diverse, i.e. including mobile AP. Such AP needs to support self-configuration ID. In order to reduce the collision and confusion probability of AP IDs, there are two methods [26]. One is to provide sufficient number of IDs in the design of physical layer; the other is that adjacent AP can automatically avoid using the same ID through signal detection and coordination.

Method 1 is to provide enough number of the physical IDs (for example, thousands of IDs) in 5G system design. But increasing the number of physical IDs is proportionate to the complexity of the physical layer design of the system. It will lead to higher request numbers to the signal decoding ability of the terminal and increases equipment cost. Therefore, to design physical layer IDs for UDN system, trade off between ID amount and system complexity need to be considered.

When the number of physical ID increased, it is more difficult to allocate ID to each AP Manually. Therefore, physical ID self-configuration and self-adjustment mechanism are needed to avoid potential collision or confusion. Method 2 requires that each AP can listen to the signal on the radio to obtain physical layer IDs of neighbour APs to avoid the configuration of these IDs.

Due to the limitation of the transmitting power and coverage of AP, signal listening could not detect neighbour AP which is out the coverage range. Then there may be physical layer ID confusion. This can be coordinated through the signaling between APs, similar to the mechanism of LTE.

4.1.3 Automatic Neighbour Relations

In UDN, AP number is huge and changing and the neighbor relationship is more complex compared to LTE, which makes it unimaginable to establish and maintain AP's neighbour list manually. Therefore, it is essential to set up automatic self-configuration function to support neighbouring topological relations.

- Centralized mode
  In scenario with the macro BS coverage, the AP can adopt LTE ANR mechanism. APs can establish the neighbour relations and network interfaces through reading macro BS broadcast information, For APs under the coverage of the macro BS, the macro BS can use beam forming technology to receive and measure the specific pilot signal of each AP, which can calculate the direction angle and the approximate distance with respect to macro BS. Based on this information, macro BS can draw the network topology map of APs in its coverage region. Then based on the network topology map, macro BS can specify the neighbor list for each AP automatically.
- Distributed mode
  In distributed mode, each AP can receive the wireless signal of the neighbour APs directly, try to read the system broadcast information of them, get their global identity information and automatically create a neighbour relations between APs. In order to discovery more neighbor cells, new discovery signal design is need.

In other semi-distributed mode, dynamic cluster structure between APs is adopted. Each AP cluster will have a cluster head node. Neighbour relation map is maintained by the cluster head node.

4.1.4 Load Balancing

In UDN, the coverage of AP is smaller than traditional eNB. UE's serving AP will change fast. Load balancing among APs, for example resource load balancing, is very important in order to achieve the purpose of improving the system efficiency and user experience.

- Scenario with macro BS coverage
  In UDN with dual-connectivity mechanism, the control plane of the UE is carried by macro BS. According to the load condition of the APs, the RRM algorithms of macro BS can timely update UE's serving APs and achieve load balancing between APs.
- Scenario without macro BS coverage
  In localized structure, a centralized RRM can gather the users' measurements information offload conditions of each AP. And then it will select the appropriate AP resources to provide services for the UE and achieve load balancing between APs globally.

4.2 Backhauling

Deployment scenario that all the APs have ideal wired backhaul linked to the network will not always exist. Hence the APs with wireless backhaul cannot be avoided in UDN [27].

In 3GPP Release 10, solutions related to relay node has been standardized [28]. But the solution of relay's wireless backhaul has some limitations. Firstly, 3GPP relay can only support one hop transmission, which means one AP can only select one BS with wired backhaul as donor eNB (DeNB). But in UDN scenarios, with the AP density increasing and coverage decreasing, one point is that it is not easy to find donor AP with idea backhaul in one hop coverage range [29]. Hence **multi-hop backhaul** can further enhance the flexibility of self-backhauling. Another point is that one AP can discover more than one UDN AP, sometimes macro BS, around it can also provide relay links. So **multi-connectivity backhauling** can provide much higher throughput to relay links. Secondly, 3GPP relay can only serve as transparent backhaul links for APs without wired link.

Meanwhile, more tight coordination between APs is preferred to provide joint transmission and joint beamforming. This requires more frequent signaling interaction and data forwarding among coordination APs group. **Wireless interfaces between APs** is a effective way to directly exchange information on the air and reduce interface delay from current delay (i.e. 20 ms for X2) to 1 ms level in UDN scenario [30].

4.2.1 Multi-Hop Backhaul Mechanisms

Enabling multi-hop makes the wireless backhaul mechanism more complexity. It is not easy to assign the donor APs for each wireless backhauling AP manually. Donor AP discovery and adaptive backhaul path detection and updating in UDN are the key issues.

The process of selecting the DeNB is usually completed in the OAM configuration with the process of measurement and access by the relay as a UE. The above pre-configuration process is not suitable for the APs in flexible deployment, and donor selection method of the current relay is so simple that it cannot cover all the wireless APs in UDN.

For the wireless path selection after the wireless APs are switched on, two kinds of the modes are included from the overall structures that are centralized mode and distributed mode.

Furthermore, due to the need of the nomadic deployment, the demand for energy saving in UDN, complex network environment and mobility of access node, the APs may be frequently switched on and off or removed, which results that the established path can no longer continue to provide services. Due to the changing of the traffic demand and link status, it is possible that the original wireless path cannot meet the requirements. So it is necessary to consider and design the process of backhaul path detection and updating.

- Donor AP discovery
  When an AP without wired link switch on, it need to firstly work as UE to discovery it's neighbor AP which can be served as it's donor AP. Discovery signal should be properly designed and assigned on the time and spectrum domain. It need to support enough coverage range but do not take too much overhead. After discover the AP, more system information and measurement information are also needed to help the wireless AP to select donor AP. These information including Reference Signal Received Power (RSRP), backhaul capacity, system load and so on can be broadcasted on the air.
- Wireless backhaul path detection
  Wireless backhaul path detection refers to the measurement of the link and transmission conditions for the current wireless backhaul path, and the gathering and processing of the measurement results, in order to form the assessment of the state of the current wireless link. When a wireless backhaul link cannot meet the requirements of the transmission or cannot provide services, it should be reported and feedback as soon as possible and then start the subsequent updating process.
- Wireless backhaul path updating
  The basic process of the wireless backhaul path updating is to destroy the old path and establish a new path, in which the method of the new path establishment is basically same as the process of the path establishment.

4.2.2 Multi-Connectivity Backhaul Mechanisms

Compared to legacy network, multi-connectivity backhaul can provide more abundant routing path selection [31]. When the quality of one path become poor, user data and signaling can be transmitted throughput other path. Furthermore, the total amount of traffic rate is the sum of all the paths when multiple paths are used at the same time. Multi-connectivity backhaul can enhance the reliability and increase the data rate of backhaul. In order to facilitate the management and coordination in a number of connections, it needs to have a primary connection, and the others are the secondary connections. The method of the primary connection decision is usually that the macro BS serves for the primary connection and the rest of the APs serve for the secondary connections, or the first access AP serves for the primary connection and the rest of the APs serve for the secondary connections. The primary connection will have most of the controlling functions and be easy for centralized control and management. For convenience, the secondary connections can also retain some control functions in order to control itself related transmission.

In the multi-path backhaul connections of the wireless APs, the air interface backhaul bearer of the wireless APs are complex, which are a aggregation of bearers of all UEs served by the wireless AP, i.e., the multiple UE bearers with the same characteristics converge into an air interface backhaul bearer of the wireless AP. Based on such complex mapping relations, when to split the wireless APs bearers, it needs to consider the unified processing in access network and to avoid the complexity in CN.

4.2.3 Wireless Interfaces Between APs

In the solution of 3GPP relay, relay node is equivalent to a cell belonging to its DeNB from the view of the neighbour nodes. The DeNB acts as a proxy gateway of the relay node and processes all kinds of network interface as an agent This idea can still be used in UDN wireless backhaul scenarios with multi-connection/multi-hop backhaul link. But it needs to enhance the design due to the change of the network architecture and the increasing of the wireless backhaul connections. Furthermore, the reduced distances between APs in UDN also make the AP can transfer some of dynamic coordination messages each other through the air interface to enhance the whole system transmission efficiency.

4.3 Multi-RATs Coordination

Along with the development of UDN, multi-access technologies coexistence is a long-term trend, especially 4G, 5G and WLAN. How to run and maintain multiple networks effectively, reduce maintenance costs and save energy, is a key issue to be resolved. Multi-RAT tight integration will become an important direction in the future wireless communication standardization fields.

How to coordinately use the various wireless access technologies and how to further improve the overall operational efficiency and user experience are a problem that need to be solved by multi-RAT integration technology [18]. Shown below for multi-RAT integration logic function schematic figure, multi-RAT, i.e. 5G/4G/WiFi, is converged through a centralized wireless network control entity.

In 3GPP Release-12 and Release-13, one of network integration technologies has been studied and standardized, which are heterogeneous network integration based on WLAN-LTE interworking [32]. The aim of WLAN-LTE interworking is to solve the problem of UE's mobility and how a UE to simultaneously use LTE cell's resource and WLAN resource. As shown in the following figure of a possible multi-layer coverage network environment, UE can work simultaneously in eNB and WLAN AP. In this scenario, eNB is responsible for the mobility management and control of UE. WLAN APs only participate in data transmission

In addition, the existing technology can also support the close integration between LTE and WLAN. As shown in Fig. 5 of a joint deployment scenario of WLAN AP(s) and LTE eNB(s), similarly, UE can work simultaneously in 4G eNB and WLAN AP while a new Xw interface is used between AP and eNB. eNB can transfer part or all of UE's data to WLAN AP to obtain the services provided by WiFi and realize resources aggregation and users' throughput improvement.

Future potential research areas include:

(1) Access technology intelligently selection: Constructing a user's non aware access technology selection mechanism. According to the real-time network state, the wireless environment, combining with the intelligent service sensing

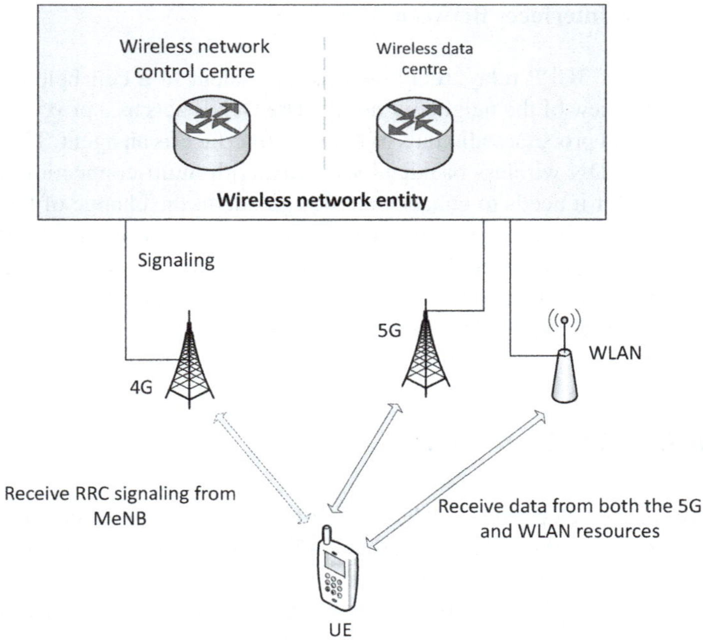

Fig. 5 Multi-RAT coordination architecture

technology, the different services will be mapped to the most suitable access technology.

(2) Multi-RAT multi-connection technology: Terminals can simultaneously establish connections with different RAT network nodes and realize multiple flow parallel transmission to improve the throughput, enhance the user experience, achieve services dynamic splitting and convergence between the different access network technologies.

4.4 Mobility Management

In UDN, mobility support and association decision is important because the small cell size causes frequent handovers, which introduce frequent short term service disruptions to AP and then significantly affect the user experience [33]. Meanwhile, the mobility management should be joint-optimized with the resource management and interference coordination in UDN. As a consequence, an effective mobile management mechanism can provide not only higher mobility performance but also an opportunity to meet higher throughput requirements for users [34].

The UDN is a multi-layer and multi-RAT network, network functions are built with these multiple levels of hierarchy. The centralized mobility anchors naturally match with the hierarchical architecture, which will lead to more signaling load to the core network and suboptimal paths transmission. The advantage of centralized mobility management (CMM) mechanism resides in its simplicity, because the central anchor can follow user movements by simply rerouting the packets over tunnels created with the access router where the UE is currently connected. However, the single-handed mobility anchor thus easily becomes a single point of failure. Therefore, more flat and flexible mobility architectures and mobility management approaches are expected to be proposed [35].

4.4.1 Mobility Management Architecture in UDN

Today's trend points to the deployment of UDN in order to provide ubiquitous connectivity at high data rates. However, this is hard to couple with the current mobile networks' architecture, which is heavily centralized, posing difficult challenges when coping with high-frequent measurement, intolerable handover failure. Additionally, future 5G networks will exhibit disparate types of services, posing different connectivity requirement. The authors in [36] propose that distributed mobility management (DMM) is emerging as valid framework to design future mobile network architectures, taking into account the requirements for high-frequent handover and the rise of extremely dense wireless access networks.

In extremely dense wireless access networks, it is important to design a more flat and flexible mobility management architecture. The authors in [12] propose a systematic and unified methodology for designing mobility management architecture. The design philosophy is applied in the proposed function reference model and protocol reference model of Mobility-Driven Networks (MDN) to abstract the mobility support functions, entities, and principles.

Consider a Wireless Local Area Network consisting of Access Points that can provide only a limited cone of coverage, which leads to frequent changes of serving APs. The network as a whole, is, however, required to provide full coverage of a relatively large area, compared with the coverage of a single Network AP. At this point, the authors in [37] illustrate the structure of a Blanket Wireless Local Area Network (BWLAN), which serves a number of End Nodes (ENs). The BWLAN structure also contains centralized services and databases, as certain network operations, e.g., admission control, are best handled via a centralized gateway. In this architecture, each EN communicates with a set of NAPs through the wireless medium, and these NAPs will be referred to as the EN's "blanket." The notion of a blanket allows the network to locate each EN and transmit to the EN only from the NAPs that can reach it.

Existing cellular networks are insufficient in meeting future mobile users' demand, in part due to inflexible and expensive equipment as well as complex and non-agile control plane. Software defined networking is emerging as a natural solution for next generation cellular networks as it enables further network programmability. An all-SDN network is described in [38] and followed by a unified approach to mobility, handover and connectivity management. It has proposed an alternative realization is to keep the handover control at the BS and RAN controllers but in close coordination with the routing control that is realized at the network controller, thus decreasing the signaling load to CN.

4.4.2 User-Centric Mobility Management

UDN is envisioned as a key deployment to address coverage and capacity issues in emerging wireless systems. In this case, the number of APs is likely comparable to the number of users.

Location areas are statically configured in existing network, but the boundary of a traditional location area in UDN becomes unclear. Thus, the location management mode will be redesigned, from "static AP planning" to "dynamic AP cooperation". In [39], an overview of challenges and requirements for mobility management in user-centric networks is given, and a new distributed and dynamic per-application mobility management solution is presented. Therefore, the authors in [6] propose a Dynamical APs Grouping method (DAPGing), which is considered a user-centered mobility management approach. With this method, every registered user has a unique APG with a unique APG-ID. APG context will be stored in a local serve center (LSC), and most of the DAPGing processes will be executed by LSC. Some high level processes such as authentication, handover are managed by network serve center (NSC). While a user is moving, its APG will be dynamically adjusted to support its movement, which is quite different from traditional mobility management and handover process. In traditional cellular network, users are handover from one cell to another. While in their proposed de-cellular ultra-dense network structure, the network will follow user's movement. The DAPGing makes mobility management functionality different. There are three following mobility scenarios and related approaches of user-centric mobility management:

(1) For Intra LSC or Inter LSC mobility, only APG members may perform refreshment for the terminal that has radio bearers, the APG-ID is not changed.
(2) For Inter NSC mobility, the APG should be possibly moved from one LSC to another LSC that connected to different NSC, assuming that the APG-ID is unique within one NSC.
(3) For Inter NSC mobility, or the mobility between UUDN and the traditional cellular network, the APG handover procedure is applied, e.g. from on one NSC to another NSC, or from one NSC to another traditional cellular network.

Furthermore, mobility management in UDN needs to be considered with interference coordination and resource management.

4.5 Interference Management

Interference management is very important in UDN to meet the high throughput requirement with short inter site distance. The resource multiplexing is adopted to increase the access opportunities, and then also brings more challenges for interference control. Traditional methods include interference randomization, interference cancellation, and inter-cell interference coordination (ICIC), interference alignment and etc. It needs further discussion to use these techniques in UDN. Meanwhile, new technologies such as millimeter-wave introduced into UDN may also present new challenges [14].

The research on interference management may include, but not limited to, the following:

(1) Channel model and capacity analysis. The wireless transmission environments in UDN become very complex with densified APs in multi-layer and multi-RATs condition. Therefore, the effective channel models need to be setup for various scenarios. Meanwhile, the channel capacity should be studied. The research in UDN just begins and needs further investigation. It is worth to note that this research direction relates to not only interference evaluation, but also other key technologies, such as coding method, antenna technique, etc.
(2) Interference model based on the effective evaluation methods. The ultra-dense environment results in more interference sources. For example, in crowd subway trains, lots of terminals and APs exist, therefore, signals may have more reflecting and scattering paths. Proper model should be set up to describe the interference level. Meanwhile, the existing parameters to measure and evaluate the impact of interference, such as interference temperature and interference threshold, may not reflect the overall interference measure and performance control of the networks. More suitable parameters should be discussed to give a better indication between interference managing results and throughput, associated with energy efficiency and other system level parameters [5].
(3) Interference management technologies with proper complexity. Interference management technologies have been studied with many research results [13] in traditional cellular networks. Considering the constraint in UDN, these techniques should be optimized with proper modification.

   - Interference cancellation. It provides methods to regenerate the interfering signals through various coding methods and then subtract them from the desired signal [40]. It requires for other interfering users' information. Considering the complexity, it is usually used in the BSs. In UDN, APs' functions may be simplified. Therefore, interference cancellation should be modified.
   - Inter-cell interference coordination. Fractional frequency reuse (FFR) and soft frequency reuse (SFR) have been proposed to control the neighbour cells interference from spectrum planning [41, 42]. Dynamic ICIC (D-ICIC) has the advantage of flexibility with many existing works [43]. The frequency

and time domain radio resources could be reused under certain limitation in multihop cellular networks [44]. These methods need coordination among BSs, therefore the signaling overhead generated in the process of exchanging information may impact the network performance.

- Coordinated multipoint transmission and reception (CoMP). One classic scheme to achieve CoMP is joint processing (JP)/joint transmission (JT), which is regarded as an advanced downlink solution and mainly focuses on achieving spectral efficiency in LTE-A. It is also possible to use the idea in UND, under the effective cooperation among APs.
- Interference alignment (IA). Each user is capable of reaching higher Degrees of Freedom (DoF) by using IA. DoF known as capacity prolong or multiplexing gain [45], is a good characterization of the approximate capacity because it becomes increasingly accurate in the high signal noise ratio (SNR) regime. IA regulates the interference falling into a particular signal subspace and leaves the residual subspace interference free. In UDN, appropriate pre-encoder could be designed at the transmitter with the purposes of preprocessing signals.

(4) Interference problems brought by new technologies. Advanced technologies have been introduced into UDN, to provide wider usable spectrum, higher throughput and better user experience. Millimeter-wave and non-orthogonal multiple access technology are considered as some of the most promising candidates and will be used in wireless access and backhauling. In millimeter-wave communication, high-gain beam forming is adopted to mitigate path loss and ensure low interference [14]. Besides, other traditional interference management methods should also be used to improve the overall performance [46].

The interference management needs to be jointly considered with resource management, mobility management and network deployment [5]. During the design of relative algorithms, interference control should be taken into consideration. Therefore, the high throughput requirements may be fulfilled in UDN.

4.6 Radio Resource Management

As mentioned before, RRM in UDN faces new challenges from the complex communication environment and skyrocketing throughput requirements. More services will be developed with various QoS demands, requiring shorter setup time and delay, as well as reduced signaling overhead and energy consumption [47]. This means more flexible and efficient resource management schemes. Furthermore, the heterogeneous networks coexisting topology brings multi-RATs and multi-layer scenarios. How to jointly allocate resources in different networks to maximize the overall utility efficiency and optimize the system performance is an important and interesting problem for UDN.

The research on RRM of UDN may include, but not limited to, the following directions:

(1) Terminal association problem and corresponding resource allocation method. The neighbour APs usually have very short ISD with each other. This means the user may be under several APs' coverage, and then enjoy accessing services from one or more APs. Depending on the association condition, the resource assigned to serve the user may vary. The impacting factors may include the connected APs number of the terminal, the available resources, the required data rate and other QoS parameters, etc. Some researchers point out that the possible frequency band for UDN may extend from GHz to millimeter-wave, with wider bandwidths, higher flexibility in use of uplink/downlink resources based on time division duplexing (TDD) access, and physical layer waveform based on the modified METIS proposals [48]. This indicates wider choices for the scarce spectrum resources, as well as more complicated allocating algorithms.
(2) Cooperation and load balance among APs. When multiple APs serve for a terminal, these APs need close cooperation with each other to provide better services. The cooperation may come from PHY layer and MAC layer. Then the assigned resources for each AP should comply with the cooperation method. Meanwhile, considering the different APs' capability and access condition, the load balance and fairness among the APs should be considered. Besides, the necessary signaling overhead for the cooperation and load balance also need to be considered. In traditional cellular networks, the interaction among cells may be impacted in practical constraints due to backhaul delay, backhaul capacity and user mobility [49]. Whereas in UDN, these signaling messages may be dealt with in the local control unit and then save the processing cost.
(3) Power allocation and energy-saving scheme. In UDN densified Aps indicate the possibility of higher overall energy consumption. Therefore, the green communication technologies are essential for the network. The power allocation should be considered to secure the overlapping of the wireless signals and QoS requirements, under effective interference control methods. Also, in the future mobile networks, the traffic load distribution will not only be impacted by the geographic area, but also relate with time, such as the night-day human behavior and daily commuting between offices and residential areas [44]. In order to save the energy consumption of idle APs, many researchers have proposed switch on/off methods for UDN. This will directly impact the resource allocation and access algorithms.
(4) Scenario of heterogeneous networks. Multiple networks coexisting with each other will become a typical deployment condition in the future communication networks. It means that UDN should also consider the coexistence and cooperation with other traditional cellular networks. Under this heterogeneous manner and multi-RATs interfaces, RRM has more challenges. The operations of different RATs have been independently defined by respective standards, but may lead to suboptimal usage of the wireless resources [50]. Multiple networks should jointly manage the radio resource and then improve the overall system

performance. The access control in heterogeneous networks needs further discussion when considering the joint optimization of RRM. The load balance among multiple networks, cooperation transmission and wireless backhaul will also have more flexible and complex schemes.

It has been pointed out that the RRM in UDN should be jointly considered, not only within the aforementioned research directions, but also with interference control, mobility management, multi-RATs, backhaul and other respects [5]. Although many researchers have paid attention to the relative fields, there is no well recognized resolution yet. Based on the cognition of transmission environment, the limited resources should be flexibly allocated and adjusted. From AP side or network side, two different approaches may be considered [5].

(1) When there are sufficient available resources and necessary function from the terminal, RRM may be fully distributed. The AP side will be responsible for the resource allocation process. It requires the APs should be capable of environment sensing on idle channels, interference levels, and neighbour APs' conditions. The advantages of this approach include the high efficiency of local resource managing and real-time fulfilling for the users requirements. Yet its disadvantages are also obvious. Negotiation process should be introduced to settle the conflict among APs. Then the processing complexity will increase. Also the requirements will be very high for environment cognitive ability of APs.
(2) The second approach needs the network side to participate and to offer center control in certain degree. A local control unit maintains the resource list in the neighbour areas, such as an available resource pool. It also controls the resource allocation process for the local APs. The APs may get the resource information by periodical broadcast or dedicated request. Once there is need for transmission, AP would then select the proper resource from the pool, the resource list kept by the local control unit may change accordingly. By this approach, RRM would achieve optimization by avoiding resource contention and decreasing interferences. The disadvantage is that it is hard to determine the neighbour area of the resource list in real cases.

The design of RRM algorithms would become hot research fields for UDN. Feasible and effective schemes will greatly improve the overall performance of the network and fulfill the users' requirements.

5 Conclusion

UDN is a technology to meet the mobile traffic requirements of 5G. It can provide better user experience even in extremely user density scenarios. Ultra density of APs is the key feature of UDN. Higher spectrum efficiency and energy efficiency, flexible networking, lower cost are the most important targets of UDN. User-centric and localization are the new concept for UDN architecture design. Based on the new architecture, key technologies such as mobility management, interference management, SON and multi-RAT coordination need to be further enhanced.

Acknowledgment The authors would like to give special thanks to Dr. Ming Ai and Mr. Zhonglin Chen of CATT for their kind reviews and revisions.

References

1. ITU-R M.2243, Assessment of the global mobile broadband deployments and forecasts for International Mobile Telecommunications, March 2011
2. Cisco, Visual Networking Index: Global Mobile Data Traffic Forecast Update 2013–2018, February 2014
3. IMT-2020 (5G) Promotion Group, 5G Vision and Requirement, May 2014
4. V. Chandrasekhar, J.G. Andrews et al., Femtocell networks: a survey. IEEE Commun. Mag. **46**(9), 59–67 (2008)
5. ITU-R M.2290, Future spectrum requirements estimate for terrestrial IMT, 2014
6. S. Chen, F. Qin et al., User-centric ultra-dense networks (UUDN) for 5G: challenges, methodologies and directions. IEEE Wirel. Commun. Mag. **23**(2), 78–85 (2016)
7. S. Chen, Y. Wang et al., LTE-Hi: a new solution to future wireless mobile broadband challenges and requirements. IEEE Wirel. Commun. Mag. **21**(3), 70–78 (2014)
8. ITU-R M.2320, Future technology trends of terrestrial IMT systems, November 2014
9. P. Popovski, V. Braun et al., Initial report on horizontal topics, first results and 5G system concept, METIS Deliverable D6.2, 2014
10. S. Chen, J. Zhao, The requirements, challenges and technologies for 5G of terrestrial mobile telecommunication. IEEE Commun. Mag. **52**(5), 36–43 (2014)
11. IMT-2020 (5G) Promotion Group, 5G Wireless Technology Architecture, May 2015
12. S. Chen, Y. Shi et al., Mobility-driven networks (MDN): from evolutions to visions of mobility management. IEEE Netw. Mag. **28**(4), 66–73 (2014)
13. H. Zhang, S. Chen et al., Interference management for heterogeneous network with spectral efficiency improvement. IEEE Wirel. Commun. Mag. **22**(2), 101–107 (2015)
14. R. Baldemair, T. Irnich et al., Ultra-dense networks in millimeter-wave frequencies. IEEE Commun. Mag. **53**(1), 202–208 (2015)
15. H. Wang, X. Zhou et al., Coverage and throughput analysis with a non-uniform small cell deployment. IEEE Trans. Wirel. Commun. **13**(4), 2047–2059 (2014)
16. P. Rost, C.J. Bernardos et al., Cloud technologies for flexible 5G radio access networks. IEEE Commun. Mag. **52**(5), 68–76 (2014)
17. I. Hwang, B. Song et al., A holistic view on hyper-dense heterogeneous and small cell networks. IEEE Commun. Mag. **51**(6), 20–27 (2013)
18. U. Doetsch, N. Bayer et al., Final Report on Architecture, METIS Deliverable D6.4, 2015
19. R. El Hattachi, J. Erfanian et al., NGMN 5G Initiative White Paper, February 2015
20. V. Garcia, Y.Q. Zhou et al., Coordinated multipoint transmission in dense cellular networks with user-centric adaptive clustering. IEEE Trans. Wirel. Commun. **13**(8), 4297–4308 (2014)
21. H. Ali-Ahmad, C. Cicconetti et al., CROWD: an SDN approach for DenseNets, in *2013 Second European Workshop on Software Defined Networks (EWSDN)*, pp. 25–31, October 2013
22. H. Wang, S. Chen et al., SoftNet: a software defined decentralized mobile network architecture toward 5G. IEEE Netw. Mag. **29**(2), 16–22 (2015)
23. 3GPP TS 23.401 v12.3.0, GPRS Enhancements for E-UTRAN Access, December 2013
24. 3GPP TS 33.401, 3GPP System Architecture Evolution (SAE); Security architecture (Release 12), October 2014

25. 3GPP TR36.842 v12.0.0, Study on Small Cell enhancements for E-UTRA and E-UTRAN: Higher layer aspects, January 2014
26. 3GPP TR 36.902, Evolved Universal Terrestrial Radio Access Network (E-UTRAN): Self-configuring and self-optimizing network (SON) use cases and solutions, 2009
27. S. Tombaz, P. Monti et al., Is backhaul becoming a bottleneck for green wireless access networks? in *2014 IEEE International Conference on Communications (ICC2014)*, pp. 4029–4035, June 2014
28. 3GPP TS 36.216, Evolved Universal Terrestrial Radio Access (E-UTRA): Physical layer for relaying operation, 2014
29. X. Ge, H. Cheng et al., 5G wireless backhaul networks: challenges and research advances. IEEE Netw. Mag. **28**(6), 6–11 (2014)
30. X. Su, K. Hi Chang, A comparative study on wireless backhaul solutions for beyond 4G network, in *2013 International Conference on Information Networking (ICOIN2013)*, pp. 505–510, January 2013
31. J. Núñez Martínez, Self-organized backpressure routing for the wireless mesh backhaul of small cells, Universitat Politècnica De Catalunya, 2014
32. 3GPP TS 36.331, Evolved Universal Terrestrial Radio Access (E-UTRA): Radio Resource Control (RRC), 2014
33. J. Zhang, J. Feng et al., Mobility enhancement and performance evaluation for 5G ultra dense networks, in *2015 IEEE Wireless Communications and Networking Conference (WCNC)*, pp. 1793–1798, March 2015
34. P. Kela, J. Turkka et al., Borderless Mobility in 5G Outdoor Ultra-Dense Networks, the journal for rapid open access publishing, pp. 1462–1476, August 2015
35. F. Giust, L. Cominardi, C.J. Bernardos, Distributed mobility management for future 5G networks: overview and analysis of existing approaches. IEEE Commun. Mag. **53**, 141–148 (2015)
36. J.C. Zuniga, C.J. Bernardos et al., Distributed mobility management: a standards landscape. IEEE Commun. Mag. **51**(3), 80–87 (2013)
37. A. Reznik, C. Ye et al., Mobility Management for Dense Networks, vol. 34, pp. 1–4, May 2011
38. V. Yazici, U.C. Kozat et al., A new control plane for 5G network architecture with a case study on unified handoff, mobility, and routing management. IEEE Commun. Mag. **52**(11), 76–85 (2014)
39. F. Sivrikaya, S. Salsano, M. Bonola, M. Trenca, *Mobility Support in User-Centric Networks, Part IV of User-Centric Networking* (Springer International Publishing, 2014), pp. 269–288
40. B. Kaufman, E. Erkip et al., Femtocells in cellular radio networks with successive interference cancellation, in *2011 IEEE International Conference on Communications Workshops (ICC)*, pp. 1–5, June 2011
41. Z. Xu, L. Geoffrey Ye et al., Throughput and optimal threshold for FFR schemes in OFDMA cellular networks. IEEE Trans. Wirel. Commun. **11**(8), 2776–2785 (2012)
42. S.-P. Chung, Y.-W. Chen, Performance analysis of call admission control in SFR-based LTE systems. IEEE Commun. Lett. **16**(7), 1014–1017 (2012)
43. K. Zheng, Y. Wang et al., Graph-based interference coordination scheme in orthogonal frequency-division multiplexing access Femtocell networks. IET Commun. **5**(7), 2533–2541 (2011)
44. K. Zheng, B. Fan, J. Liu, Y. Lin, W. Wang, Interference coordination for OFDM-based multihop LTE-advanced networks. IEEE Wirel. Commun. **18**(7), 54–63 (2011)
45. H.-M. Anders, N. Aria, The multiplexing gain of wireless networks, in *2005 International Symposium on Information Theory (ISIT2005)*, pp. 2065–2069, September 2005
46. K. Zheng, Y. Wang, W. Wang, M. Dohler, J. Wang, Energy-efficient wireless in-home: the need for interference-controlled Femtocells. IEEE Wirel. Commun. **18**(6), 36–44 (2011)
47. R. Baldemair, E. Dahlman et al., Evolving wireless communications: addressing the challenges and expectations of the future. IEEE Veh. Technol. Mag. **8**(1), 24–30 (2013)
48. H. Tullberg, Z. Li et al., Towards the METIS 5G concept: first view on horizontal topics concepts, in *2014 European Conference on Networks and Communications (EuCNC2014)*, pp. 1–5, October 2014

49. R. Wang, H. Hu et al., Potentials and challenges of C-RAN supporting multi-RATs toward 5G mobile networks. IEEE Access 2, 1187–1195 (2014)
50. Y.-N. Li, J. Li et al., Energy efficient small cell operation under ultra dense cloud radio access networks, in *2014 Globecom Workshops (GC Workshops)*, pp. 1120–1125, December 2014

5G RAN Architecture: C-RAN with NGFI

Chih-Lin I, Jinri Huang, Yannan Yuan, and Shijia Ma

Abstract This chapter discusses Cloud Radio Access Networks (C-RAN), which has been viewed as one of the key RAN architectures for 5G networks. First, the basic concept and the initially defined architecture of C-RAN are recalled, including the major benefits in terms of acceleration of network deployment, cost reduction and facilitation of 5G technologies. Then the major challenges of C-RAN realization are analyzed. One of the key challenges lies in fronthaul (FH) transportation which may limit C-RAN implementation in 5G. To address the issue, a new FH interface called Next Generation Fronthaul Interface (NGFI) is proposed. The C-RAN architecture itself evolves with the NGFI interface. The design principles for NGFI are presented including decoupling the FH bandwidth from the number of antennas, decoupling cell and user equipment processing and focusing on high-performance-gain collaborative technologies. NGFI claims the advantages of reduced bandwidth as well as improved transmission efficiency by exploiting the tidal wave effect of mobile network traffic. The transmission of NGFI is Ethernet-based to enjoy the benefits of flexibility and reliability. The major impact, challenges and potential solutions of Ethernet-based FH network are also analyzed. In addition, a prototype is developed and presented to verify NGFI.

1 Introduction

C-RAN which stands for Centralized, Collaborative, Cloud and Clean RAN was first proposed by China Mobile Research Institute in 2009 [1, 2]. The original motivation for the proposal of the new concept comes from practical needs in network deployment. Traditionally, network deployment requires a separate room per site with supporting facilities such as air conditioning to accommodate the base station (BS) or baseband unit (BBU). Every site consumes an equipment room. As the number of sites increases greatly, which is particularly the case for LTE, the number of equipment rooms required increases too. Traditional methods of deployment are

Chih-Lin I • J. Huang (✉) • Y. Yuan • S. Ma
Green Communication Technology Research Center, China Mobile Research Institute, Beijing, China
e-mail: huangjinri@chinamobile.com

W. Xiang et al. (eds.), *5G Mobile Communications*,
DOI 10.1007/978-3-319-34208-5_16

becoming increasingly difficult since available real estate is becoming scarcer and rental costs are increasing. In this case, centralization of different BSs or BBUs into one location seems to be a natural solution for network deployment.

However, C-RAN does not stop at the centralization stage. As 5G recently became the research focus, it was realized that C-RAN serves as an ideal framework in support of 5G technologies. C-RAN is supposed to be able to accommodate and facilitate several 5G technologies such as Large Scale Antenna Systems (LSAS), full duplex, ultra-dense networks cooperative communication, mainly thanks to its inherent centralization nature as well as the flexibility and scalability of a cloud-based implementation [1]. In this sense C-RAN has become an essential element for future 5G systems.

In the following of this chapter, first the C-RAN concept will be recalled briefly and then an evolved C-RAN architecture which is based on a newly defined FH interface called NGFI will be proposed. The definition and design principles of NGFI will be described in detail with the advantages and challenges analyzed. Furthermore, different function split solutions will be analyzed, using LTE as an example which is critical to NGFI realization. The FH network design will also be analyzed from the perspectives of synchronization, jitter and delay. In addition a Proof-of-Concept to verify the NGFI will be described, followed by the conclusion.

2 C-RAN Basics

This section gives a basic introduction of the C-RAN concept, features and advantages.

2.1 The Concept of C-RAN

With distributed BSs as the basic component, a C-RAN system centralizes different baseband processing resources together to form a pool in which the resources could be managed and dynamically allocated on demand on a pool level. Figure 1 shows the C-RAN architecture, which consists of three parts [2].

- Baseband resource pool: A baseband pool is a pool of resources from which resources can be instantiated to a "soft" BBU. In C-RAN the BBU is soft in the sense that the processing resources and capabilities are dynamically allocated and could be reconfigured based on real-time conditions (e.g. traffic status).
- Remote Radio Head (RRH) networks: RRHs are the same as in traditional systems to provide basic wireless signal coverage.
- Transport networks: A transport network provides a connection between a BBU instance in a pool and the remote RRHs. It could be of different forms depending on the scenario. Some examples include direct fiber connection via dark fiber, microwave transmission and fiber transport networks.

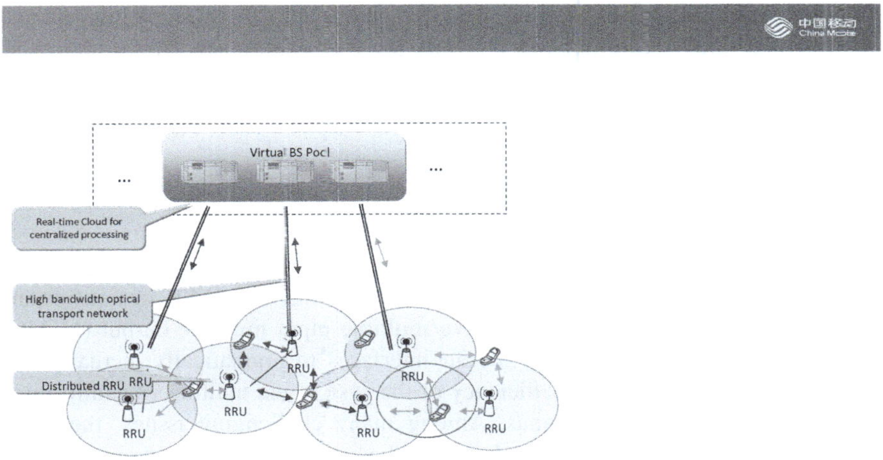

Fig. 1 C-RAN architecture

2.2 *Features and Advantages of C-RAN*

Aggregating the computing resources into one central location, centralization is the foundational feature and the first step toward a complete C-RAN. Centralization allows the aggregated computation nodes to share the same facilities, for instance air-conditioning, which greatly reduces the power consumption and, therefore, operating expense (OPEX). In a trial in China Mobile's commercial networks, it was demonstrated that power consumption could be reduced by up to 70 % with C-RAN deployment compared with traditional deployment methods [2]. In addition, centralization makes it easier to find a smaller number of central offices to accommodate the equipment, which in turn can speed up network construction.

Generalization of platform is another essential feature of C-RANs. The use of general purpose platforms (GPPs) not only reduces the procurement cost for operators but more importantly lays down the basis for the implementation of virtualization technology.

Based on the GPP, virtualization technology can be implemented to bring various benefits including improved resource utility efficiency, flexibility, scalability and energy efficiency, as they have already achieved and demonstrated in modern data centers in IT industry. In particular, with virtualization, the resource can be allocated on demand for different scenarios, traffic patterns, applications and radio access technologies (RATs). C-RAN in this sense can be called both "soft" and "green".

C-RAN is also "open" in the sense that C-RAN is designed to provide a set of standard application programming interfaces (APIs) to outside parties to encourage new service development on the edge and to provide interoperability. For example, dozens of radio parameters on user or cell level, e.g. the channel quality of a user or the number of users in a cell can be exposed so that the edge applications can adjust parameters such as uplink/downlink (UL/DL) audio/video bit rates, picture and text compression rates accordingly. In this way the user experience can be improved. With its openness, C-RAN aims to push forward the prosperity of the ecosystem.

2.3 Support of 5G Technologies

C-RAN may serve as an ideal framework for many potential technologies in various 5G aspects, such as duplex mode, network optimization, multi-RAT coexistence and spectrum sharing. For example, massive multiple-input multiple-output (MIMO) is identified as a key enabler for 5G air interfaces to dramatically increase both energy efficiency and spectral efficiency of the system via multiuser beamforming with a large number of antennas. Among many challenging issues, the inter-cell pilot contamination severely degrades the performance of massive MIMO, which mandates joint design of UL pilots and joint optimization of multiuser beamforming in adjacent cells. In addition, from the aspect of multiple access control, non-orthogonal multiple access (NOMA) supports more than two users on the same resource, thereby requiring more efficient and flexible means of system optimization. As to 5G duplex mode design, full duplex is regarded as an important direction of evolution. Full duplex base stations are capable of transmitting and receiving simultaneously on the same time and frequency resource, thus promising a maximum 100 % capacity improvement with well-scheduled DL and UL transmissions. The major issue for both NOMA and full duplex lies on the much more severe intra- and inter-cell interference. Especially in full duplex systems, on top of the conventional DL to DL and UL to UL interferences, there exist many complicated interferences between DL and UL. These interferences must be mitigated efficiently. In addition, one important network-level consideration of 5G is the extremely high base station density required to meet the ever-increasing mobile traffic requirements. Naturally, the high-density network design involves joint consideration of many issues, like the interference mitigation in a heterogeneous network, control and data decoupling, DL and UL decoupling, carrier aggregation and Coordinated Multi-point (CoMP) etc. As to the multi-RAT coordination, 5G networks need to coexist harmoniously all the existing 3G, 4G and WiFi networks. The information sharing between these different networks and corresponding joint scheduling are essential to its success. Finally, due to the scarcity of wireless spectrum, spectrum sharing among all wireless networks (and operators) will fundamentally enhance resource utilization efficiency.

The performance improvements promised by the above technologies, however, require systematic information sharing and joint optimization, which is difficult to achieve with traditional distributed architecture. Fortunately, thanks to the inherent capability of central processing, such information sharing becomes feasible in the C-RAN context and therefore C-RAN is deemed to be able to facilitate the realization of those 5G key technologies.

2.4 Fronthaul Challenges and Existing Major Solutions

There are two major challenges for C-RAN realization. One is the FH challenge to enable efficient and scalable centralization and the other is resource cloudification. The ultimate goal of C-RAN is to realize the features of cloudification, i.e. processing resources can be dynamically allocated to form soft BBU entities. The solution to this is to implement virtualization technology which has been pervasive in IT industry for cloud applications. Thanks to the extremely strict real-time requirements for signal processing in wireless networks, the implementation of virtualization may require special design and optimization. Some challenges for this include optimization of operating systems and the hypervisor, optimization of virtualization management functions, I/O virtualization to improve the virtual machines (VM)'s I/O performance and its compatibility with live migration. Readers could find more information in [2, 3]. In this chapter we will mainly focus on the challenges on the FH side since centralization is the basis to realize other features of C-RAN.

FH is defined as the link between a BBU and a RRH. Typical examples of FH interfaces include Common Public Radio Interface (CPRI) and the Open Base Station Architecture Initiative (OBSAI) [4].

Centralization is the critical first step required in order to realize all the other features of C-RAN. Centralization aggregates different BBUs into one central office with shared facilities. The key challenge for centralization is that it requires a large number of fiber resources if using direct fiber connections. The issue can be illustrated by the following example. In a TD-LTE system with 20 MHz bandwidth and RRHs of 8 antennas, which is the most common scenario in CMCC's networks, the CPRI data rate between the BBU and the RRH for one TD-LTE carrier transmission is as high as 9.8 Gbps. When considering both UL and DL, 2 fiber connections would be required with 10 Gb/s optical modules. Since usually one site consists of three sectors with each supporting at least one carrier, the number of fiber connections for one site can be as high as 6, which is difficult to achieve for most operators due to limited fiber resources.

In order to overcome the fiber disadvantage in a centralized implementation, various solutions have been developed. Some are mature enough while others are still in the early stage.

One of the solutions is CPRI compression. There are various kinds of compression techniques such as non-linear quantization and IQ data compression [5]. So far, major vendors have been able to support 2:1 compression with lossless system performance. In [3], the authors have verified the functions in their commercial LTE C-RAN networks. Another solution is Single Fiber Bi-Direction (SFBD) which allows simultaneous transmission of UL and DL on the same fiber. SFBD could further reduce the usage of fiber by another 50 %. As a result, when combining SFBD with compression, fiber resources can be reduced 4-fold with lossless performance.

The most common and effective FH solution today is to use wavelength-division multiplexing (WDM) technology. With WDM dozens of carriers could be carried on a single fiber, which greatly conserve fiber resources. Although the introduction of new transport nodes in the network may also cause additional delay and noise jitter, it turns out that the expense is affordable. In [3] the authors have tested the mainstream WDM FH equipment in the commercial LTE C-RAN network and found that the additional delay and jitter contributed by the WDM equipment are small enough to be negligible. Even with the optical transport network (OTN) technology which is supposed to have much higher delay than WDM, the delay is still around 20–30 μs, small enough to have little impact on the system performance.

However there is one common disadvantage for all the solutions today, no matter whether they are WDM, CPRI compression, or SFBD, which is that the object that these transport solutions carry is CPRI data. Due to the inherent shortcomings of CPRI (see next section), the transport efficiency is low.

In order to better support C-RAN large-scale deployment and key 5G technologies, a new FH interface is proposed in the following sections.

3 NGFI: Next Generation Fronthaul Interface

In this section, the design of the fronthaul interface in order to better support 5G technologies, especially C-RAN will be introduced.

3.1 Motivation

As mentioned in previous sections, on the road of C-RAN realization, the FH issue has been one of the biggest challenges. Due to the high FH data rate centralization requires consumption of a high number of fiber cores which are scarce and not easy to afford. Although other transport technologies such as WDM and OTN could save fiber consumption, the cost on the introduction of additional transport equipment makes the economic viability a concern to operators. Some operators are still not very convinced on the merits of C-RAN deployment. Therefore in 5G to enable large-scale C-RAN deployment it is necessary to reduce FH bandwidth.

Current FH interfaces could also raise new issues for C-RAN virtualization. C-RAN is supposed to run on a general-purpose processor (GPP) platform consisting of standard IT servers, storage and switches. GPPs do not provide a FH interface for telecom applications. To support FH, either a new interface should be created on the GPP platform or an adapter card is needed, both complicating the system and introducing additional costs. It would be desirable to base the 5G FH interface on existing GPP interfaces which could maximize efficiency and save costs.

In addition, scalability issues exist for today's FH technology to support widely discussed 5G technologies, including LSAS, CoMP processing etc. For example, it is possible that a 5G LSAS RRH could be equipped with 64 or even 128 antennas. With LTE, the FH bandwidth will rise to the order of 100 Gb/s, which makes it unaffordable. It is clear that the impact of the number of antennas on FH should be minimized to the greatest extent possible.

In order to address the issues above, one possible idea is to change the FH interface fundamentally.

In the following chapter, CPRI will be used to represent traditional FH solutions since it is the most widely adopted in the current industry.

3.2 *CPRI Re-Visiting*

The CPRI interface helps to separate the BBU and the RRH to enable the deployment of distributed base stations. Although in traditional networks the CPRI is mainly deployed over short distances, usually in the order of several meters or several hundred meters, it can support up to 40 km between the BBU and the RRH. CPRI has been working well for traditional mobile networks including 2G, 3G and 4G. There are three traits for the CPRI interface.

- The CPRI line rate is constant regardless of traffic;
- The mapping between the BBU and the RRH is fixed one-to-one correspondence and not flexible;
- The sampling I/Q data rate is dependent on the number of antennas.

With networks evolving to 5G, CPRI is becoming more and more unsuitable to accommodate the evolution. First, it is well known that mobile traffic varies in the temporal dimension, which is called tidal wave effect. For example, the data traffic in an office area is high in the daytime and plummets at midnight. For dense urban areas, the tidal wave effect is noticeable. However, the CPRI data stream is synchronous digital hierarchy-alike (SDH-alike), which means that it is constant regardless of the changes in traffic [4]. Even when there is no user traffic in the network, there are still CPRI streams running between the BBU and the RRH. This is a waste of bandwidth and leads to low utilization efficiency.

Second, with CPRI an RRH is in one-to-one correspondence with a BBU. The relationship is configured offline. It may cause a concern in the context of C-RAN. In C-RAN, the baseband units are centralized and virtualized in a pool. Reliability becomes extremely important as each pool takes care of thousands of users. Therefore for the sake of protection, it would be desirable if in C-RAN, one RRH could be automatically switched to another BBU pool. Current CPRI however, does not support such flexible and automatic re-mapping.

Finally, the CPRI bandwidth is dependent on the number of antennas. As the number of antennas increases, the CPRI data rate increases in proportion. This could become a major hindrance for CPRI's applicability in 5G as far as multiple antenna technologies are concerned.

Based on the analysis above, it is believed that the CPRI interface that is universally applied throughout the industry is primarily applicable to peer-to-peer connections. Due to the disadvantages such as low transmission efficiency, poor flexibility and poor scalability, and particularly due to the high cost of centralized deployment, it is difficult for current FH interfaces to meet the evolving needs for 5G-oriented FH networking.

To increase the efficient use of network resources and better support the evolution of wireless networks toward 5G systems, as well as to better support C-RAN-centralized deployment, a new BBU-RRH interface is proposed in the following sections.

3.3 Definition and Design Principles of NGFI

The motivation for designing a new FH interface is to facilitate C-RAN deployment, to make it compatible with GPP platforms and to be scalable enough to support the evolution to 5G.

The new FH is defined as the Next Generation Fronthaul Interface (NGFI) between the BBU and the RRH with the following four features [6, 7]:

- Its data rate should be traffic-dependent and therefore support statistical multiplexing.
- The mapping between BBU and RRH should be of one-to-many correspondence and flexible;
- It should be independent of the number of antennas.
- It should be packet-based, i.e. the FH data could be packetized and transported via packet-switched networks.

With NGFI, several advantages are expected, including:

- The FH transmission efficiency could be improved. This is because NGFI is traffic-aware. Therefore it could fully leverage the tidal effect of mobile networks so that the FH data rate is variable with traffic change.
- With reduced average data rates, it could further reduce the pressure on FH transport networks for the sake of C-RAN deployment.
- NGFI could support 5G antenna technologies well thanks to the antenna-independent feature.
- The packet-based feature makes it possible to use Ethernet to transport FH data. The benefits would be several-fold. First, the Ethernet interface is the most common interface on standard IT servers and the use of Ethernet makes C-RAN virtualization easier and cheaper. The operation and maintenance are therefore straightforward. The Ethernet can fully make use of the dynamic nature of NGFI to realize statistical multiplexing. The flexible routing capability could also be used to realize multiple paths between the BBU pool and the RRH and to enable FH resources sharing.

The fundamental way to realize NGFI is to re-design the function partitioning between the BBU and the RRH. Furthermore, the NGFI interface will have a great impact on the design of the underlined transport networks for C-RAN deployment.

In the next section, the design and analysis of BBU-RRH function split as well as the packet-switched FH networks will be introduced.

3.4 NGFI-Based C-RAN Architecture

NGFI requires the baseband (BB) function to split between the BBU and the RRH, i.e. moving partial BB functions from the local BBU to the remote RRH. C-RAN architecture evolved with NGFI. The NGFI-based C-RAN architecture is shown in Fig. 2 which consists of the following components:

- Radio Aggregation Unit (RAU): The moved partial BB functions formed a new entity which is called radio aggregation unit. RAU is a logical concept and its realization depends on implementation solutions. For example, RAU could be integrated to the RRH to form a new-type RRH. Alternatively, it could also be an independent hardware entity.
- Remote Radio Systems (RRS): A RRS consists of an RAU and multiple RRHs. It is expected that collaboration could happen among different RRHs via the RAU within the same area coverage of a RRS. There could be multiple RRS in a C-RAN network.

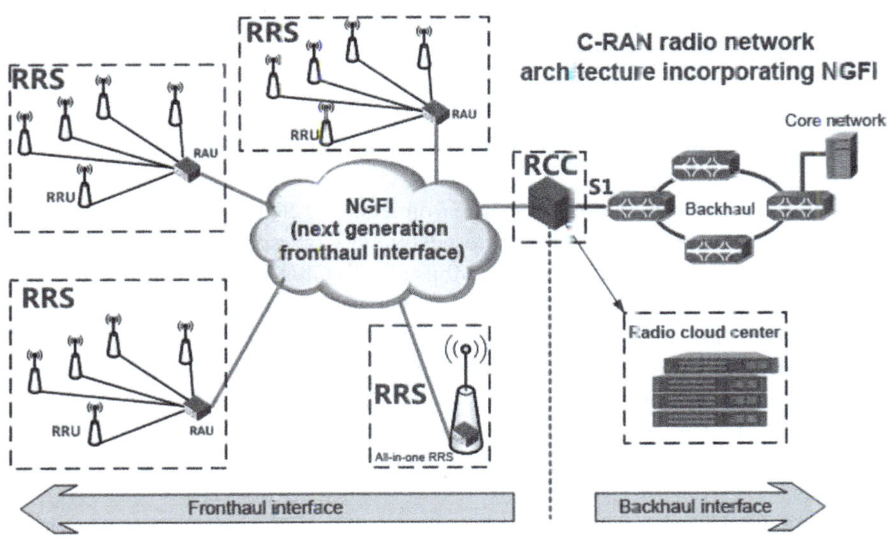

Fig. 2 NGFI-based C-RAN architecture

- Radio Cloud Center (RCC): The remaining BB functions, together with high-layer functionalities constitute a RCC. RCC is the place where all the processing resources are pooled into a cloud with virtualization technology.

4 Analysis of RCC-RRS Function Split

Traditionally, the baseband-related functions are processed by the BBU while the RRH processes radio frequency-related functions. It is this simple partitioning that leads to the shortcomings of CPRI, as mentioned above. Therefore, the NGFI design should start with a paradigm shift by rethinking and redesigning the function split between the RCC and the RRS. Moreover, the function split between RCC and RRS may be different for different scenarios. For example, generally the number of the antennas of each radio head is 2 in indoor distributed systems, in which fewer functions may be moved from RCC to RRS. On the contrary, for outdoor scenarios, the number of the antennas of a massive MIMO radio head could be 128 or even larger. In this case more functions need be moved from RCC to RRS to reduce the FH bandwidth.

4.1 RCC-RRS Function Split Principles

Function split is the key way to achieve NGFI and different split options would lead to different features for fronthaul interface. In this section the principles for desirable function split schemes will be discussed.

4.1.1 Decoupling the FH Bandwidth from the Number of Antennas

The air interface bandwidth per carrier on 2G, TD-SCDMA and TD-LTE are 0.2 MHz, 1.6 MHz and 20 MHz, respectively. Correspondingly, FH transport bandwidth per carrier is 30 Mbps, 400 Mbps and 10 Gb/s, respectively. At the same time, FH is facing a bandwidth explosion, considering the rapid traffic growth in 5G (potentially 1000× by 2020) and C-RAN-centralized deployment. Compared with the air interface bandwidth, the existing FH interface transportation efficiency is low. One of the most important reasons is that FH bandwidth is proportional to the number of antennas. In order to increase transport efficiency, a RCC/RRS function split principle should enable NGFI to decouple FH bandwidth from the number of antennas.

Taking TD-LTE as an example, an 8-antenna TD-LTE carrier FH bandwidth is 10 Gb/s based on the current BBU/RRH function split. For a C-RAN baseband pool with 100 TD-LTE carriers, the overall FH bandwidth would be as high as 1 Tb/s. If it is a 128-antenna LSAS system, a TD-LTE carrier FH bandwidth will increase to 160 Gb/s. Then the overall FH bandwidth of the C-RAN baseband pool will increase to 16 Tb/s. Thus, the existing FH blocks the C-RAN centralized deployment and is facing a big challenge with the increase of antenna numbers. In order to reduce the bandwidth, one potential idea is to redesign the RCC/RRS function split. In particular, in order to phase out the effect of antenna numbers, it is proposed that antenna related functions should be moved from the RCC to the RRS. For LTE, the antenna related functions include DL antenna mapping, FFT/iFFT, channel estimation and equalization. It is shown that an LTE carrier FH bandwidth may decrease in the order of 100 Mb/s no matter how many antennas are used. Therefore, FH bandwidth will decrease significantly, if the RCC/RRS function split can decouple non-antenna related processing and antenna related processing.

4.1.2 Decoupling Cell/UE Processing

Dynamic variation is a major feature of wireless traffic. The tidal wave effect is obvious in many wireless deployment scenarios such as residential, office and commercial districts. Moreover, the traffic load of most areas is usually in the valley between late night and early morning. However, the current constant-rate FH does not match the mobile traffic features, which results in a waste of resources.

In order to quantify wireless traffic features, a TD-LTE traffic investigation on a commercial LTE network was done via network monitoring systems. There were six base station sites in the investigation, each having at least three carriers. Two of them are indoor distributed systems while the others are outdoor macro base station sites. The investigation period was 7 days during which the traffic load was sampled and collected by network monitoring systems every 15 min.

In Figs. 3 and 4, the DL and UL traffic load distribution of the above six base station sites are shown. Site 1, 2, 3 and 4 are outdoor sites. Site 5 and 6 are indoor sites. Based on the statistical data, several traffic load features were observed as follows:

- The tidal wave effect is obvious in the test districts, where the traffic load is almost zero for 12 h.
- Even when the site is at "peak" status the resource block (RB) usage is not high. The duration of a site at peak status is short, usually not exceeding 30 min.
- The probability that different sites are simultaneously in peak status is almost zero.
- When the site is at "valley" status the RB usage is low and the duration is long, usually exceeding 2 h.
- Frequently, different sites are in valley status at the same time. Moreover, the overlapping time is long among different sites.
- Most of time the traffic load stays at the average level, which is low.

Based on the above observations, it is clear that constant-rate FH transport does not match the mobile traffic features, which results in a waste of re-sources. To address this issue, it is first observed that the existing base band processing can be divided into cell processing and user equipment (UE) processing. Cell processing

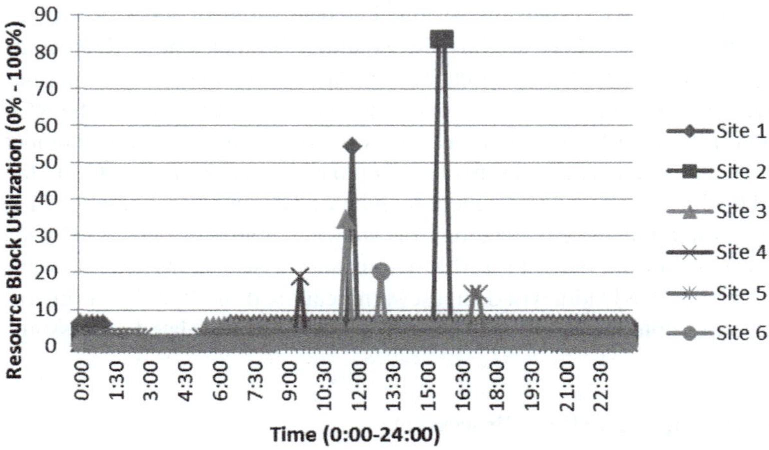

Fig. 3 DL traffic load distribution

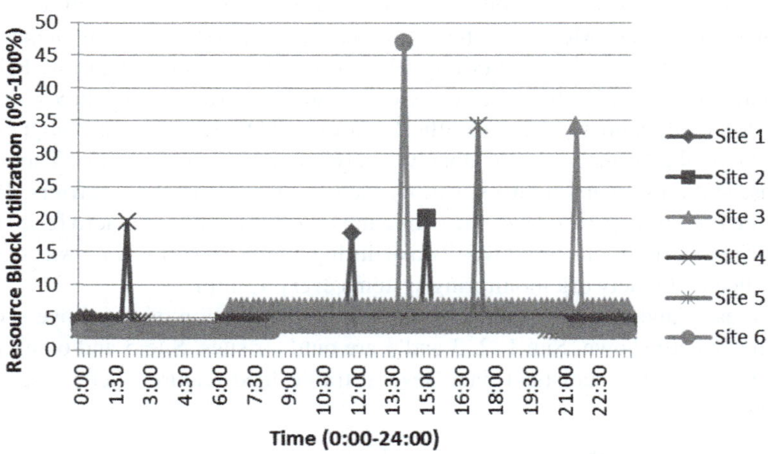

Fig. 4 UL traffic load distribution

is irrelevant to traffic load and is fixed no matter how many UEs are active. Some examples of such processing units in LTE include inverse fast Fourier transform/fast Fourier transform (iFFT/FFT), cyclic prefix (CP) addition/removal, cell-specific reference signal/primary synchronization signal/secondary synchronization signal (CRS/PSS/SSS) generation and physical broadcast channel (PBCH) processing. It is therefore proposed to move these cell processing functions from the RCC to the RRS, i.e., decoupling the cell and UE processing.

If cell processing is moved from the RCC to the RRS, the FH bandwidth will be lower and load dependent. The load-dependent feature gives an opportunity to exploit the statistical multiplexing gain when it comes to FH transport network design for C-RAN deployment. Although the peak throughput requirements for radio resources are very high and rising with the evolution of RAN, the statistical result of our commercial network shows that the duration of a site at peak status is short, usually not exceeding 30 min. Moreover, the peak throughput occurrence time is usually different among different sites. Therefore, every site can also reach peak throughput under FH network statistic multiplexing model. In order to not decrease the air interface performance, FH network statistic multiplexing must be designed suitably according to the business model of different scenarios. Thanks to statistical multiplexing, the bandwidth needed for transport of a number of FH links in C-RAN could be reduced greatly and therefore the cost is diminished.

Cell/UE processing decoupling can further help reduce power consumption and enhance the network reliability. This is because cell basic cover-age signal processing is a kind of cell processing. Therefore, cell basic coverage will be provided by the RRS if cell processing functions are moved from the RCC to the RRS. On the one hand, RCC software can be switched to a dormant state to save power when there is no active UE. On the other hand, RRS is able to provide continuous air interface coverage, even when RCC breaks down. This way, it provides sufficient time for RCC fault processing.

4.1.3 Focusing on High-Performance-Gain Collaborative Technologies

CoMP has been viewed as one of the important 5G technology candidates to improve system performance, which can be divided into two classes: MAC layer coordination and physical layer coordination. For example, collaborative scheduling (CS) is one of the MAC layer coordinated mechanisms. Joint reception (JR) and joint transmission (JT) are physical layer coordinated technologies. The design of NGFI should take into account support for CoMP. The above two principles lead to a low bandwidth and traffic dependent FH. In the meantime, some physical-layer-coordinated technologies are difficult to implement in RCC since some collaborative information has be processed and terminated by RRS. Therefore, it is proposed that data exchange among RRS can be supported on FH network. However, data exchange among RRS is not necessary because it is found that the performance gain of JR/JT decreases significantly as the number of antennas increases [8]. Moreover, it is also found that MAC-level collaborative technologies can bring comparable performance gains with lower complexity, easier implementation and fewer constraints.

In order to verify this, a CoMP field trial was conducted in 2014 in which two CoMP schemes including JT and CS were examined and compared. The testing zone was a central business district including around 7000 active UEs, which is a typical CoMP test scenario. It was covered by 35 base stations with different antenna heights in which inter-cell interference is serious because of the high ratio of overlap area.

Test results show that cell edge UE throughput increased by 127.45 % $\sim$ 173.65 % when the serving cell reference signal received power varied from −88 dBm to

Table 1 Utilization factor of JT and CS

Network load	JT/CoMP	CS/CoMP	CoMP/Total
20 %	20.04 %	79.96 %	42.02 %
50 %	18.09 %	81.91 %	34.43 %
70 %	20.89 %	79.11 %	54.33 %
100 %	24.89 %	75.11 %	47.31 %

−106 dBm. Table 1 shows the utilization factor of JT, CS and the CoMP application. In Table 1, the "CoMP/Total" means the utilization factor of CoMP which is defined as the ratio of the number of CoMP transmission time intervals (TTIs) to the total number of test TTIs. Similarly, the "JT/CoMP" and "CS/CoMP" are defined as the ratio of the number of JT and CS TTIs to that of CoMP TTIs, respectively.

From Table 1, it is found that the network load growth results in a small reduction of CS usage. For example, when the network load increases to 100 %, the usage ratio of CS is still around 75 %, similar to other cases. The usage of JT is only around 25 %, much lower than CS. It is therefore fair to say that most of the performance gain is contributed by CS. Compared with JT, CS does not need complex matrix computing. It is easier to implement CS with contemporary base station equipment. On the contrary, JT performance is influenced by antenna calibration accuracy, channel estimation accuracy and channel variation speed, all requiring high FH bandwidth.

The test results demonstrated that MAC-level collaborative technologies could mitigate most network interference. Therefore, RCC should focus on high performance gain collaborative technologies rather than all the collaborative technologies. This principle provides guidance on how to make a tradeoff between wireless and FH performance.

4.2 Analysis and Comparison of LTE Function Split Solutions

As shown in Fig. 5, taking LTE as an example, potential function split solutions for the RCC-RRS interface are discussed in this section. Baseband processing functions can be divided into UE processing (related to traffic level) and cell processing (unrelated to traffic level). The green blocks in the figure are traffic-dependent UE processing modules, and the yellow blocks are traffic-independent cell processing modules. The blue block, channel estimation and equalization, is also UE processing. The complexity of channel estimation and equalization is correlated with the number of receiving antennas.

In order to estimate the FH bandwidth demands of different RCC-RRS function split solutions, the following assumptions are made:

1) Single 20 MHz LTE carrier
2) 2 antenna ports
3) 8 antennas

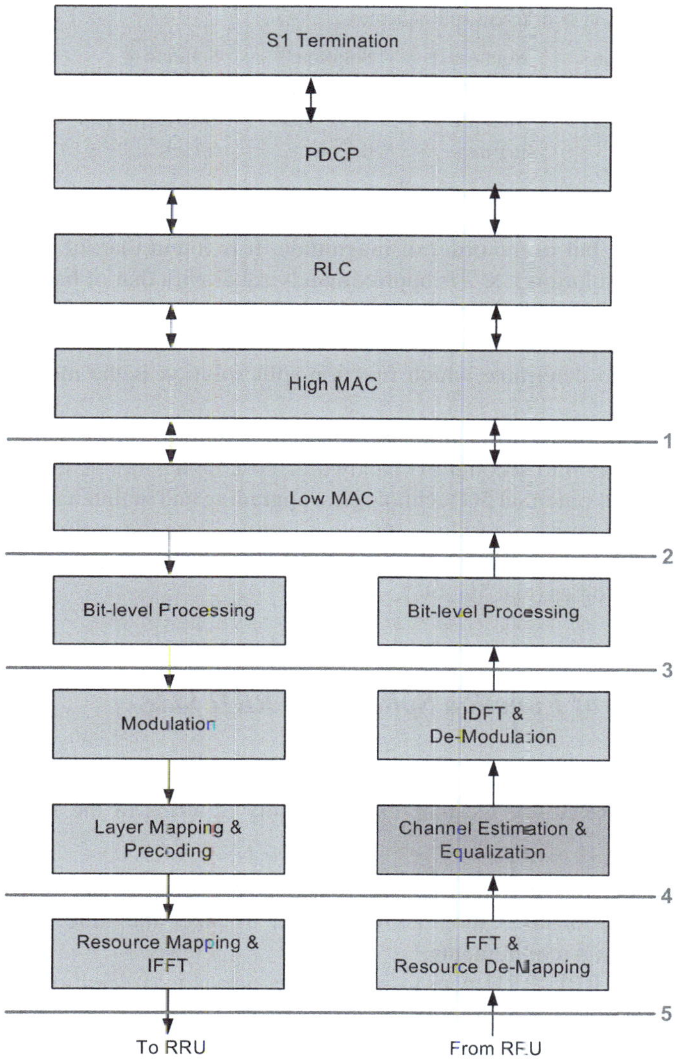

Fig. 5 Potential RCC-RRS function split solutions in LTE

4) The highest modulation type of DL is 64 quadrature amplitude modulation (QAM). The highest modulation type of UL is 16 QAM.
5) Maximum number of users: 100

As shown in Table 2, "DL/UL" means DL/UL bandwidth requirement per carrier which does not include synchronization data or Ethernet headers. In the table, the bandwidth of solution 1, 2, 3, and 4 is the maximum bandwidth, and the bandwidth of existing solution 5 is constant. Delay is shown not in the exact

Table 2 Comparison of different split schemes

	Solution 1	Solution 2	Solution 3	Solution 4	Solution 5
DL	174 Mb/s	179.2 Mb/s	125.2 Mb/s	498 Mb/s	9830.4 M b/s
UL	99 Mb/s	78.6 Mb/s	464.6 Mb/s	2689.2 Mb/s	9830.4 M b/s
Delay	1 ms	0.1 ms	0.1 ms	0.1 ms	0.1 ms

numerical value but in the order of magnitude. It is found that the FH bandwidth requirement of solution 1 & 2 is approximately equal with that of backhaul. All the function split solution 1, 2 and 3 meet the RCC-RRS function split principles, as described in the previous subsections. Further research on each potential solution is still needed to determine which function split solution is the most suitable for NGFI.

Based on the analysis of the above potential solutions, NGFI is a compromise solution that takes into account FH bandwidth, cost, delay, mobile traffic characteristics, support for potential 5G technologies, upgrading and maintenance challenges, support for cooperative algorithms and other factors. With the evolution of C-RAN, NGFI also should be evolving to meet the FH requirements of next-generation wireless technologies (e.g. 5G, 6G).

4.3 Impact of Function Split on the RRH Side

Compared with the traditional RRH, RCC-RRS function split increases the complexity on the RRH side to some extent, mainly in terms of the following three aspects:

- Partial baseband function processing is added on the RRH side.
- Clock synchronization modules are added to meet the time and frequency synchronization requirements.
- The existing BBU-RRH peer-to-peer connection is expanded to multipoint-to-multipoint connection. Therefore, RRS should be viewed as an independent network element, and RRS management functions should be also added accordingly.

Moreover, there are two ways to replace the partial baseband functions to be moved from the BBU side. One is to integrate them into the RRH, leading to a new type of RRH. The other is to add dedicated equipment to deal with the partial baseband function processing, which could be co-located with the original RRH. For the integrated RRH, adding partial baseband function processing will increase the power dissipation, which brings some impact on the volume and weight. The volume, weight, power dissipation and transmission power level of the RRH are closely interrelated. According to the estimate of the current RRH cooling capacity, at the upper limit of temperature, the heat volume is approximately 15 W/L.

Compared with the existing RRH, the volume of the integrated RRH will increase by 1 L for each additional 15 W of power consumption, at the same time with a corresponding 1 kg weight increase. Currently, based on the commercial network power dissipation test, it was shown that the power dissipation of the RRH is in the order of 100 W and that of BBU is in the order of 10 W. Therefore, the power dissipation of partial baseband processing should be in the order of 1 W. Preliminary predictions are that adding partial baseband processing on the RRH side will only bring a slight increase in power dissipation. However, the exact impact of the power dissipation and cost needs to be evaluated quantitatively in more details in the future.

5 NGFI FH Network Design

The principle of NGFI points out that a desirable FH interface should be packet-based, which makes it easily transmitted by packet-switched networks, especially Ethernet. This could make full use of the advantages of Ethernet to achieve multi-point to multi-point connection, statistical multiplexing, and flexible routing. In the meantime, the adoption of Ethernet also brings new challenges. In this section, such challenges will be analyzed and potential solutions will be given based on wireless data packetization.

5.1 *Wireless Data Packetization*

Wireless data packetization refers to using Ethernet packets to transmit wireless data between the RCC and the RRS. Compared with time-division-multiplexing (TDM) packaging in traditional CPRI interface, data packetization could provide greater flexibility and scalability, meet the need for high-layer wireless data transmission, be capable of identifying user traffic and achieve efficient transmission of wireless user loads and related control loads in packet-switched networks. To meet the high requirements of wireless data transmission and retain the advantages of packetization, achieving packetized transmission based on wireless data characteristics is one of the problems that must be considered in realizing NGFI interfaces in the transport field.

Given that wireless payloads are very sensitive to synchronization, jitter and delay, it may be necessary to introduce specific packet headers to deliver in-band control information such as timestamps or sequential numbers to facilitate coordination between the receiving and transmitting ends. The process, i.e. introduction of packet headers into wireless loads is called encapsulation. Encapsulation could be "structure aware" or "structure agnostic". The design of encapsulation should take into account several factors such as the application scenarios, overheads and other requirements. In addition, synchronization technologies necessary for wireless data transportation are another research focus. The encapsulation process could be irrelevant to either the baseband function split schemes or the specific radio technologies in order to have better forward compatibility and technological neutrality.

Packetized wireless payloads should be transparent with respect to the transportation networks and could be carried by different transmission technologies, including Ethernet, Multiprotocol Label Switching-Transport Profile (MPLS-TP), IP routing and L2/L3 MPLS switching. Different transportation technologies carry the wireless payloads in different ways based on their own characteristics and capabilities. At the same time, to distinguish such loads from other existing protocol or packet types, it may be necessary to introduce new encapsulation or protocol identifiers for the wireless loads. Different transportation technologies vary in terms of the carrying and forwarding efficiency, network/traffic scalability and adaptability, operation and maintenance management capabilities and ecosystem. Appropriate choices should be made based on the overall strategy of the transmission networks, together with the technical characteristics of each transmission technology.

In Fig. 6, an example of FH topology of a C-RAN system is shown. In the example function split between the RCC and the RRS is not considered. Therefore the RCC is the same as BBU pool while the RRS is the same as the RRH. As shown in the figure, all the RRHs are connected to a BBU pool through a ring Ethernet

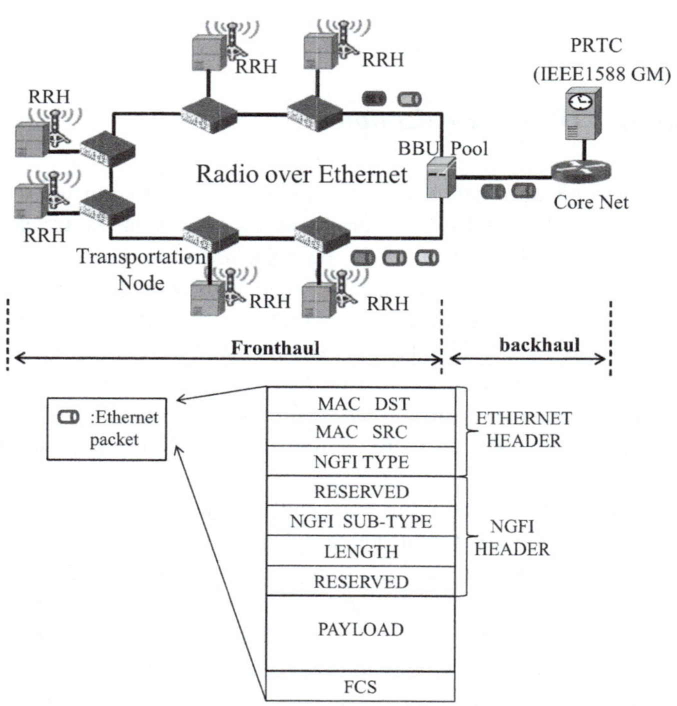

Fig. 6 A FH topology example and illustrative Ethernet packet format to support NGFI

network. There are multiple routes between BBU and RRH to help to enhance network reliability. When one of routes fails, FH packets can be transported through another route.

FH DL data is encapsulated with Ethernet headers in the BBU and de-encapsulated from Ethernet packets in the RRH and vice versa for UL. In Fig. 6, the structure of an NGFI-supporting Ethernet packet is proposed. It includes traditional Ethernet header, NGFI header and payload. Source Media Access Control (MAC) address, destination MAC address and packet type are fill in the traditional Ethernet header. The packet type here is the NGFI type which is new to distinguish NGFI packets from other packets. The NGFI header consists of NGFI packet sub-type, packet length and reserved field for protocol extension. There could be at least two NGFI packet sub-types. One is for wireless data while the other is for control & management data. For control and management type, it may include link delay test, link status monitoring, RRH configuration and RRH status reporting.

5.2 Main Challenges of NGFI Transmission

5.2.1 NGFI Synchronization Issues and Potential Solutions

TDM systems require strict synchronization which includes two aspects—frequency and time (or phase). For Time Division-Synchronous Code Division Multiple Access (TD-SDMA) and TD-LTE, the accuracy of frequency synchronization should be in the range of ± 0.05 ppm while the accuracy of time synchronization should be in the range of ± 1.5 μs [9].

In CPRI three types of data including wireless protocol data, synchronization data and control & management data are packaged together and transmitted in TDM mode. Upon receiving the CPRI frames, the Clock and Data Recovery (CDR) circuit of the RRH can extract frequency information to achieve frequency synchronization. Meanwhile, CPRI transport time is nearly constant and can be measured by the BBU. Based on the measurement, timing between the BBU and RRH can be configured in advance. With timing information extracted from CPRI frames, time synchronization at the RRH could be achieved.

For Ethernet-based NGFI, contrary to CPRI, the transportation time of FH data is no longer constant due to the packet-switched nature of Ethernet. As a result, frequency and time synchronization between the RCC and the RRS becomes potentially difficult.

The potential NGFI synchronization scheme includes the introduction of Global Position System (GPS) or the BeiDou system on the RAU side or synchronization via FH networks. If the synchronization scheme adopts GPS or the BeiDou system on the RAU side as the synchronization source, the impact of the NGFI will be relatively small. If it adopts clock synchronization from the FH network, it is necessary to carefully consider the requirements of the synchronization clock with respect to the FH network. Synchronization accuracy is related to the scale of the

FH network and performance of the transmission equipment. Synchronous Ethernet (Sync-E) and 1588v2 are potential mechanisms to achieve synchronization by FH networks.

Sync-E could be a promising solution in NGFI to achieve frequency synchronization. In fact, traditional backhaul network have already adopted Sync-E technology to achieve frequency synchronization. In LTE, the macro station frequency synchronization index is defined as 0.05 ppm to meet the frequency synchronization error requirements of wireless air interfaces. The working principle of SYNC-E is similar with CPRI as both of them use an 8B/10B encoder in the physical layer. The performance of Sync-E could be found in ITU-T G.8262 [10]. In this reference document, it is shown that under free-running conditions, the EEC output frequency accuracy should not be greater than 4.6 ppm with regard to a reference traceable to a clock define in ITU-T G.811 [11]. What's more, the document of G.811 shows if the slave clock has tracked the Primary Reference Clock (PRC), then the maximum allowable fractional frequency offset for observation times greater than one week is 1 part in 10^{11}, over all applicable operational conditions. Based on these observations, Sync-E can meet demands for frequency synchronization. Therefore, in the NGFI environment, RRSs can acquire frequency synchronization via Sync-E solution in the FH network.

As for the time synchronization, backhaul network with traditional BBUs must fulfill the need for air interface synchronization of the Time-Division-Duplex (TDD) system. According to the 3GPP definition, air interface accuracy should be within $\pm 1.5\ \mu s$. To achieve time synchronization, a potential solution is to use 1588v2 which is a high-accuracy time synchronization protocol based on packetized networks. To implement 1588v2, 1588v2 modules should be added in both the RAU and the RRS. The performance of 1588v2 could be found in the document of G.8273.2 [12], which states that the maximum absolute time error for the clock of class A is 100 ns and for T-TSC Class B is 70 ns. So the accuracy requirement of time synchronization for $\pm 1.5\ \mu s$ can be achieved with 1588v2. In fact, the performance of 1588v2 has been verified in China Mobile's commercial networks. Considering the additional accuracy loss by BBU data processing, China Mobile even requires much stricter time synchronization accuracy requirement which is that the time error introduced by Packet Transport Network (PTN) equipment after 30 hops cannot exceed $1\ \mu s$ [13].

However, when it comes to the time synchronization in the FH network, there are two major issues needed to be solved. The first issue is time hopping. When the RRS obtains time offset information from 1588v2, it needs to compensate this offset. The usual way is to adjust this offset directly, but it brings a vertical step change, which is called time hopping. Time hopping can result in discontinuous transmission in the RRS, which is intolerable for mobile communication. One potential solution is to use frequency-adjustable oscillators to calibrate time in RRS, which can adjust the oscillator frequency gradually to ensure continuous time variation.

Another issue is that cooperative technologies on the PHY level in LTE require higher demands on the time synchronization. For example, for CoMP, synchronization errors between different antennas of different RRH (RRH in the same site or RRH in different sites) must be less than 130 ns in order to support DL joint delivery functionality between multiple RRHs [14]. This imposes a big challenge for Ethernet-based NGFI even when 1588v2 is leveraged. It cannot be achieved by 1588v2 currently. Therefore, more efforts are needed to figure out how to meet the stricter time synchronization requirement such as considering GPS solutions on the BBU, improving the accuracy of timestamps, increasing clock frequency and optimizing the deviation-adjusting algorithms.

5.2.2 NGFI Transmission Jitter Issues and Potential Solutions

NGFI interface transmission jitter is defined as the fluctuation in the time taken for data to be transmitted from the RCC network interface to the RRS network interface. Transmission jitter is mainly reflected in two aspects. First, the wireless equipment on both sides needs to implement data caching to compensate for jitter. The greater the jitter is, the bigger the data buffer and the more complex the equipment will be. Second, the baseband's actual processing time sequence must be designed according to the latest data arrival time. The greater the jitter is, the shorter the available baseband processing time will be.

In CPRI, jitter is negligible while it is common and unavoidable in Ethernet-based NGFI since all the packets are processed in every network node based on the store-and-forward pattern and the presence of transit nodes further increases traffic message jitter, which leads to transportation latency fluctuation. To guarantee wireless performance, transmission devices must be carefully designed or optimized to reduce the jitter and meet the jitter requirements of the NGFI. For example, an appropriate circular buffering may be needed in RRS. On the one hand, data packets can be sorted in order. On the other hand, data is continuously sent to air interface because transportation jitter can be isolated by the buffer.

5.2.3 NGFI Transport Delay Issues and Potential Solutions

NGFI interface RCC-RRS data transmission delay in NGFI is defined as the total time required for transmission and switching of data from RCC network interfaces to RRS network interfaces. Compared with traditional CPRI interfaces, new NGFI interfaces will create additional delay losses, mainly due to the additional switching delay in the switched networks. Taking China mobile's current PTN as an example, for PTN equipment, the processing time of one hop is 50 μs. A typical PTN ring consists of 6 nodes (hops) and has 20 km length. The transmission time of fiber for 20 km is 100 μs. In the case of 6 hops, the total delay is $100 + 50*6 = 400$ μs [13]. The maximum transportation delay is another big factor which influences FH transportation performance. This is because that current LTE protocols require the strict interaction time of HARQ on both the UE side and the system side. It is specified by 3GPP that in LTE when the RRH receives a frame from UE, it must respond by transmitting the responding DL frame within 3 ms. The 3 ms time budget is consumed by the BBU processing, the RRH processing and FH transportation which includes the transmission latency on fiber and processing latency by FH nodes. The more time that the BBU and RRH processing occupy, the lower the allocable budget for FH transportation. Therefore, if RCC-RRS functional partitioning points are placed within the HARQ process, this will influence both data transmission delay and data processing delay. If the transmission delay is extended, it will then be necessary to shorten the processing delay, which will place higher demands on RCC chip processing capacity. If RCC-RRS functional partitioning points are placed outside the HARQ process, there will be too many functions preceding the remote location, which will affect multi-carrier cooperative performance. Therefore, transmission delay is a key factor in NGFI design. And the maximum allowable transport delay for FH networks requires co-design from both wireless and transportation perspectives.

6 NGFI PoC

A Proof-of-Concept (PoC) was developed in order to verify the NGFI concept and the feasibility, as shown in the Fig. 7. In this demo packet-based FH networks are targeted, i.e. Ethernet FH networks. The verification of RCC-RRS function split schemes are not carried out here. In this demo two specific Field Programmable Gate Array (FPGA) cards were developed on the BBU and the RRH respectively, which are used to transform Ethernet packets to and from CPRI frames. Taking the DL transmission as an example, the FPGA board on the BBU side encapsulates the

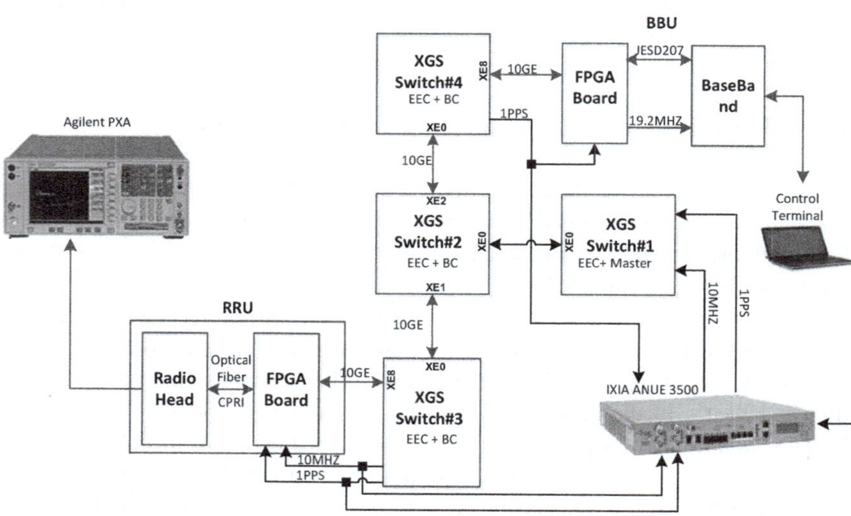

Fig. 7 NGFI demo architecture

baseband frames into Ethernet packets, and then transmits them to the FH networks. As for the NGFI networks, it consists of three Ethernet switches connected directly. All of them support 1588v2 and Sync-E mechanisms to provide the necessary time and frequency synchronization information between the BBU and the RRH.

In this demo, a synchronization tester was used with a GPS signal as the source input for timing. Next, the time and frequency information is injected to the Ethernet FH networks by a switch device, which served as the master time and frequency server for the NGFI networks. On the RRH site, the FPGA card can recover the time and frequency information and use this to extract and assemble the CPRI frames from the Ethernet packets. In addition, the FPGA card could feed the time and frequency information back to the tester to calculate the difference.

To evaluate the synchronization performance, the Maximum Time Interval Error (MTIE) which is a key indicator for time and frequency synchronization performance, was tested. It was found that the MTIE for time synchronization in this demo is around 15 ns through 3 hops. The frequency deviation is around 0.01 ppb, much lower than the CPRI requirement of 2 ppb. From the results, it can be seen that the Ethernet-based FH networks with specific design could meet the synchronization requirements of CPRI cases.

The Error Vector Magnitude (EVM) which is a key parameter to indicate the wireless performance was also tested. For comparison, a distributed BS with CPRI connection was set up, as shown in Fig. 8. E-UTRA Test Model (ETM) is used to test the wireless performance. The results are shown in Table 3. It is found that the EVM of this PoC is only slightly higher than that of traditional systems. For example, with ETM2 (64QAM) configuration, the EVM of the NGFI demo was 1.17 %, a bit higher than that of CPRI-based distributed BS which was 1.07 %. Both of them meet 3GPP requirements. Therefore, it could be concluded that the adoption of packet-based FH networks has no impact on wireless system performance. The feasibility of the NGFI concept was thus initially verified.

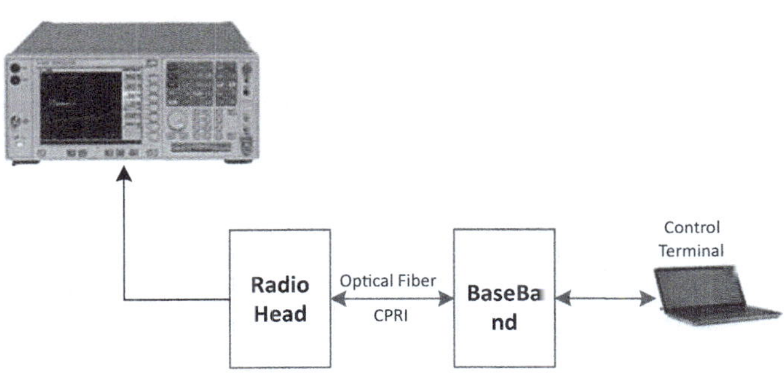

Fig. 8 Distributed BS with direct CPRI connection

Table 3 Wireless performance in terms of ETM

	Traditional BS with CPRI	NGFI with 3 hops	3GPP requirement
ETM2 (64QAM)	1.07 %	1.17 %	9 %
ETM3.1 (64QAM)	2.13 %	2.19 %	9 %
ETM3.2 (16QAM)	3.14 %	3.21 %	13.5 %
ETM3.2 (QPSK)	4.50 %	4.79 %	18.5 %

7 Conclusions

C-RAN has been viewed as an essential technology for 5G. In this chapter, first the basic concepts, features, benefits and challenges of C-RAN are briefly recalled. One of the major challenges for C-RAN lies on the fronthaul transportation. To address the challenge, a new FH interface called Next Generation Fronthaul Interface is proposed. The proposal of NGFI aims to address the shortcomings of traditional FH interfaces such as low transmission efficiency, poor scalability and ultimately to facilitate large-scale deployment of C-RAN and to support other 5G technologies. The design principles are described and the major impact, challenges and potential solutions on FH transport networks are analyzed. NGFI requires re-design of the BBU-RRH function split and packetization of FH data. By decoupling the FH bandwidth from the antenna number, NGFI can better support large antenna technologies. In addition, the cell-processing functions should be decoupled from the UE-processing functions to make NGFI traffic-aware, which can enable exploiting the statistical multiplexing gain to improve efficiency and further reduce power consumption. It is also suggested that the function split schemes for DL and UL could be different to improve the flexibility and efficiency. The use of Ethernet for NGFI transmission brings the benefits of improved reliability and flexibility due to the packet-switching nature of Ethernet. While Sync-E and 1588v2 could be introduced to address the time and frequency synchronization issues, they still need careful design in order to support collaborative technologies. In the meantime, jitter and latency remain the other key difficulties to overcome to finally realize NGFI. Finally a PoC was introduced with the conclusion that the feasibility of adopting Ethernet as FH networks was initially verified.

In addition to the challenges analyzed in this chapter, in the future there remains a lot of work to do to deeply understand NGFI. For example, the control, data and management channels that are transported via NGFI should be analyzed and carefully designed to make NGFI a better fit in different 5G architectures. In addition, the coexistence of NGFI data and other traffic on the same FH networks needs further examination.

Acknowledgments We would like to express our sincere gratitude to our partner, Broadcom for development of the NGFI PoC. We also owe special thanks to all C-RAN team members in China Mobile for helpful discussion and valuable comments.

References

1. I. Chih-Lin, C. Rowell, S. Han, Z. Xu, G. Li, Z. Pan, Toward green and soft: a 5G perspective. IEEE Commun. Mag. **52**(2), 65–73 (2014)
2. C. M. R. Institute, C-RAN White Paper 3.0: the road towards green ran. [Online]. Available: labs.chinamobile.com/cran, 2014
3. I. Chih-Lin, J. Huang, R. Duan, C. Cui, J.X. Jiang, L. Li, Recent progress on C-RAN centralization and cloudification. IEEE Access **2**, 1030–1039 (2014)
4. CPRI, Common Public Radio Interface (CPRI) Specification (V6.0), Tech. Rep. Aug. 2013. Online: http://www.cpri.info
5. Light Radio Portfolio: Technical Overview. Alcatel Lucent, Feb. 2011
6. Chih-Lin I, Y. Yuan, J. Huang, S. Ma, R. Duan, C. Cui, Rethink Fronthaul for Soft RAN. IEEE Commun. Mag. **53**(9), 82–88 (2015)
7. C. M. R. Institute, White Paper of Next Generation Fronthaul Interface. [Online]. Available: labs.chinamobile.com/cran, 2015
8. A. Davydov, G. Morozov, I. Bolotin, A. Papathanassiou, Evaluation of joint transmission CoMP in C-RAN based LTE-A HetNets with large coordination areas, in *IEEE Globecom Workshops (GC Wkshps)*, Dec. 2013
9. 3GPP TS 36.101, User Equipment (UE) radio transmission and reception (Release 9), Version 8.9.0, Mar. 2009
10. ITU-T Recommendation G.8262/Y.1362, Timing characteristics of synchronous Ethernet equipment slave clock (EEC), Aug. 2007
11. ITU-T Recommendation G.811, Timing characteristics of primary reference clocks, Sep. 2009
12. ITU-T Recommendation G.8273.2/Y.1368.2, Timing characteristics of telecom boundary clocks and telecom time slave clocks, Apr. 2014
13. YD/T 1704-2007, General Technical Requirement for Intelligentized PSTN Network, 2007
14. 3GPP TS 36.104, Base Station (BS) radio transmission and reception (Release 11), Version 11.4.0, Mar. 2013

User-Centric Wireless Network for 5G

Yunlu Liu and Guangyi Liu

Abstract In order to better meet the future requirements of mobile Internet and Internet of Things, to fulfill multiple user experience requirements, such as low latency, high data rate, high reliability, low energy consumption, is taken as a key goal of 5G system. Therefore, flatter network architecture, flexible functionality and topology, smart user and traffic awareness, high efficient network operation with lower cost, etc. to facilitate the user-centric wireless network becomes the elements of 5G eco-system access part design.

1 Introduction

In order to better meet the future requirements of mobile Internet and Internet of Things, efficiency and agility are crucial to 5G. Specifically, Mobile Internet is to provide advanced user experience, such as Augmented Reality, Virtual Reality, ultra high definition video, mobile cloud. The requirements of Mobile Internet include Gbit data rate, *ms*-level delay, ultra-high traffic density, ultra-high connection density, and user experience guarantee in the high mobility scenarios, such as high speed train, express way. While Internet of Things broadens the objective of mobile communication from people and people to things and things, such as mobile hospital, vehicle to vehicle communication, smart home, industry control. The requirements of Internet of Things include high volume connections, ultra-low delay, low cost, etc.

Based on the study of 5G requirements, the gap analysis between LTE technology and 5G requirements is studied as shown in Table 1. Here we list the most related requirements to wireless network design, i.e., challenging scenarios including connectivity density, ultra-high volume density, diverse services referring to huge difference between the requirements, "0" latency and QoE (Quality of Experience), easy operation including diverse deployment scenarios and low cost.

Connectivity density may be fulfilled by Long Term Evolution (LTE) via massive machine type communication (MTC) technology, as well as increasing the number

Y. Liu (✉) • G. Liu
Wireless Department, China Mobile Research Institute, Beijing, China
e-mail: lyunlu@gmail.com

W. Xiang et al. (eds.), *5G Mobile Communications*,
DOI 10.1007/978-3-319-34208-5_17

Table 1 Gap analysis and 5G guidelines

Challenges	Requirements	LTE capabilities	5G Guidelines
Challenging scenarios	Connectivity density (one million connections/km^2)	Might be fulfilled, but not economic	Cost-efficient operation
Diverse services	Ultra high volume density (tens of Tbps/km^2)	No core network becomes a bottleneck	Flatter architecture and cost-efficient operation
	Huge difference between the requirements	No similar processing	Flexible topology and smart RAN
Easy operation	"0" latency	No long routing path through core network	Flatter architecture
	QoE+	No blind access network	Smart RAN
	Diverse deployment scenarios	No lack of flexibility of network topology	Flexible topology
	Lower cost (USD/bit)	Need to be improved	Cost-efficient operation

of base stations supported by the technologies such as inter cell interference coordination (ICIC) and coordinated multi-point transmission (CoMP) to improve performance [1, 2]. However, due to the limitation of LTE architecture and air interface design, it is inefficient to realize one million connections/km^2 requirement in LTE. Thus, the cost-efficient operation is needed.

ICIC and CoMP technologies can help enhance volume density to some extent, while it is hard for the tens of Tbps/km^2 volume density 5G requirement. Besides, under the 4th generation network (4G) architecture, all data transferred through the core network, which makes core network a bottleneck to fulfill ultra high volume density requirement. It is possible to be solved by new flatter architecture design and cost-efficient operation.

There are different 5G requirements, which are hardly to be fulfilled by unified procedure as in 4G, e.g., low latency vs. high reliability, enhanced mobile broadband (eMBB) vs. massive MTC. In this circumstance, smart Radio Access Network (RAN) is required to provide traffic awareness and distinction to improve the user experience, which cannot be well provided in LTE RAN. Flexible topology supports self-adaptive topology and functionality to fulfill diverse requirements. Flatter architecture can facilitate ms-level latency.

Due to the fixed network topology, it is hard to realize diverse deployment scenarios. Moreover, the low cost is always one of the most important requirements of communication systems. Therefore, flexible topology and cost-efficient operation are important elements of 5G design.

Therefore, the flatter network architecture, flexible functionality and topology, smart user and traffic awareness, and highly efficient low cost network operation are expected to be key elements of user-centric 5G access network design.

In this chapter, several potential technologies are addressed in order to fulfill the requirements as analyzed above.

User centric 5G access network architecture design: a flatter network architecture, with some functions and entities shifted down to RAN side, can help meet the "0" latency requirement.

Flexible functionality and deployment: to match diverse scenarios in 5G, it is necessary to provide flexible access point deployment and functionality. It may include supporting diverse access points, plug and play access points, and functionality slicing, among other things.

Smart user and traffic awareness and management: to better meet diverse user requirements of the mobile Internet, the 5G system needs to adapt services and user requirements based on user behavior awareness, such as multi-dimension QCI design, personalized local content provisioning, user behavior awareness.

High efficient low cost network operation: in addition to a higher data rate and lower delay, the operation cost efficiency requires flexible and self-organized operations.

2 User Centric 5G Access Network Architecture Design

To improve end-to-end (E2E) delay and user experience, the architecture is considered to be much flatter than the earlier systems such as LTE, where some functions are to be shifted down to RAN side, such as multi-RAT coordination, service awareness, localized caching, data processing and forwarding. It helps shorten the traffic path, provide customized service and flexible core-agnostic multi-RAT access.

Besides, flexible spectrum deployment and multi-connectivity have to be taken into architecture design account, to fully use broad range of spectrum, support flexible ultra-dense deployment, and ultimately bring high throughput.

Figure 1 shows the E2E 5G architecture proposed in [3]. It comprises three layers including the infrastructure resource layer, the business enablement layer and the business application layer, an E2E management, as well as an orchestration entity.

The infrastructure resource layer consists of the physical resources of a fixed-mobile converged network, comprising 5G devices (in the form of (smart) phones, wearable devices, machine type modules and others), access nodes, cloud nodes for processing and storing resources [4, 5], networking nodes and associated links. 5G devices may have multiple configurable capabilities and act as a relay/hub or a computing/storage resource, depending on the context [6, 7]. Hence, 5G devices are also considered as part of the configurable infrastructure resource. The resources are exposed not only to higher layers but also to the end-to-end management and orchestration entity through relevant access points. Performance

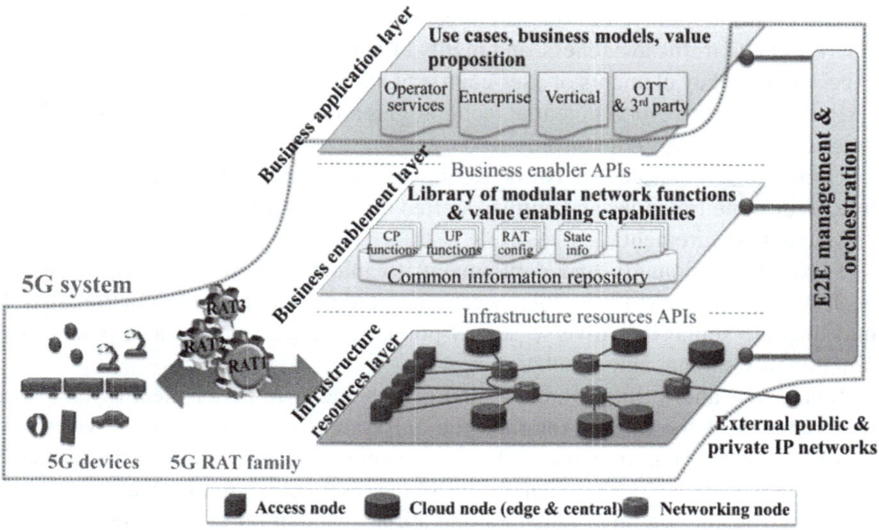

Fig. 1 5G Architecture [3]

and status monitoring as well as configurations are intrinsic part of such an access point. The business enablement layer is a library of all functions required within a converged network in the form of modular architecture building blocks. It includes functions realized by software modules that can be retrieved from the repository to the desired location, and a set of configuration parameters for certain parts of the network, e.g., radio access. The functions and capabilities are called upon request by the orchestration entity, through relevant access point. For certain functions, multiple variants might exist, e.g., different implementations of the same functionality which have different performance or characteristics. The different levels of performance and capabilities offered could be utilized to differentiate the network functionality much more than in today's networks (e.g., to offer as mobility function nomadic mobility, vehicular mobility, or aviation mobility, depending on specific needs). The business application layer contains specific applications and services of the operator, enterprise, verticals or third parties that utilize the 5G network. The interface to the end-to-end management and orchestration entity allows, for example, to build dedicated network slices for an application, or to map an application to existing network slices [3].

Based on the E2E architecture design, a paradigm of 5G access network is shown in Fig. 2. In the architecture, one logical module called "General Network Controller" is introduced to fulfill localized control function. The module works together with "Radio data centric" model as a black box to provide unified core-agnostic multi-RAT and diverse access style, and the shifted down functions.

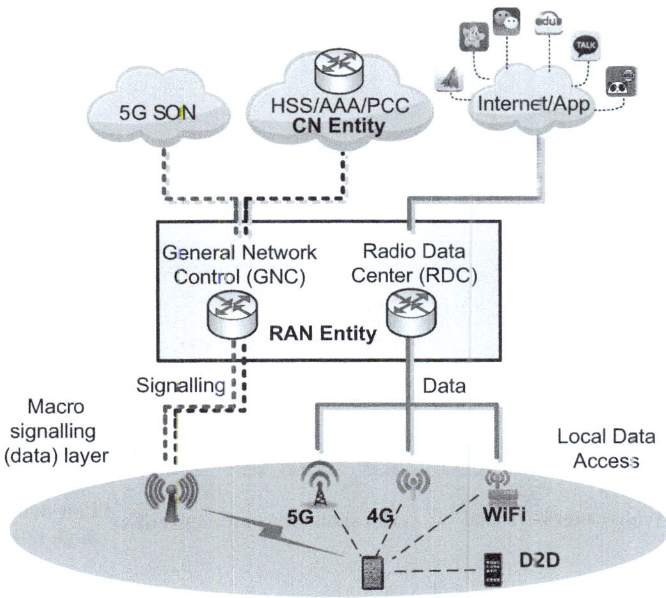

Fig. 2 A paradigm of 5G access network

- Signaling and data decoupling
  To meet 5G "0" latency requirement, the traffic path is needed to be shorten. While from the point of control plane, it is better to keep a relative high anchor for seamless mobility compared to data plane. It is hard to meet both data and control plane requirements. Therefore, control and data plane decoupling is used to tackle this issue. A method to support control and data plane decoupling is to provide Macro signaling access point and data access point, i.e.,

  - Macro signaling access point
    The macro signaling access point is responsible for initial access and system information for data access point.
  - Data access point
    Data access point is activated as required. The transmission efficiency can be supported by several technologies such as: massive MIMO, beam forming, positioning.

    Usually, there are two evolutionary ways from LTE to 5G, i.e.,

    Option 1: At the first stage, signaling access point is upgraded from eNB, data access point is from 5G as shown in Fig. 2. Then, 5G system can work standalone at the second stage,, which means both signaling and data access points are based on 5G.

    Option 2: 5G system is standalone without co-existence with LTE.

    Although the technical routine may be different of the two ways, the objective of 5G design is kept the same, i.e., a flatter, flexible and user centric system.

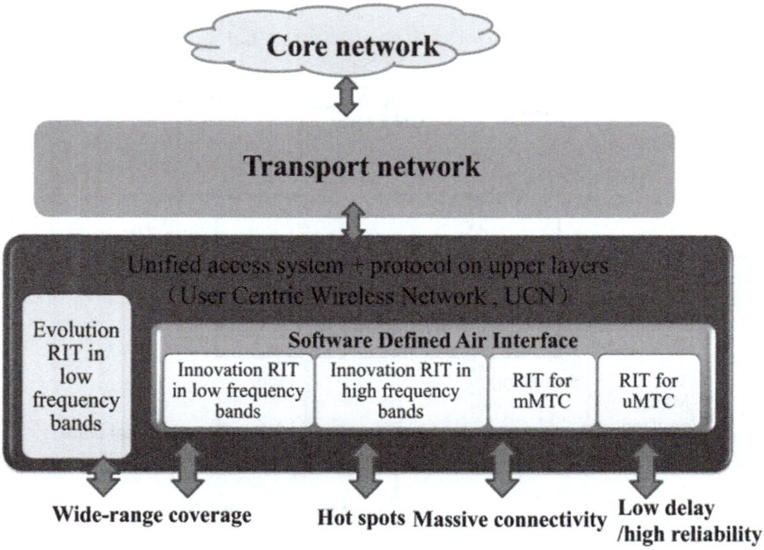

Fig. 3 An example of Multi-RIT access networks

- Data plane functionality decentralization
  To meet the low latency and diverse 5G requirements, some data plane functionalities are to be put on the edge of the network to facilitate local breakout/switch/data provision.
- Multi-RIT (Radio Interface Technology) access
  In 5G, it is necessary to provide multiple RITs to support diverse application requirements. An example is shown in Fig. 3. The evolution RIT in low frequency bands, innovation RIT in low frequency bands, innovation RIT in high frequency bands, RIT for mMTC (massive machine type communication), and RIT for uMTC (ultra-reliable machine type communication) should be provided. Evolution RIT in low frequency bands and innovation RIT in low frequency bands are to support wide-range coverage. Innovation RIT in high frequency bands, RIT for mMTC is to support hot spot scenario. Massive connectivity is supported by RIT for mMTC. Moreover, RIT for uMTC provides low latency and high reliability access. Although multiple RITs are necessary to meet diverse requirements, unified access system via software defined air interface is to be supported to decrease system/UE (User Equipment) complexity and enhance user experience.

3 Flexible Functionality and Deployment

Network/device functions and RAT configuration have to be tailored for each use case, leveraging the Network Function Virtualization (NFV) and Software Defined Network (SDN) concepts. Thus, the network should support flexible composition of network functions, as well as. their flexible allocation and location. The network functions need to be scalable so that capacity is provided when and where needed. Even when particular functions or nodes become unavailable, e.g., due to disaster events, the system should support graceful degradation instead of service interruption. To improve such robustness, the state information should be split from functions and nodes so that contexts could be easily relocated and restored even in failure events. 5G aims to virtualize as many functions as possible, including the radio baseband processing. Although some functions may still run on non-virtualized platforms, e.g., to meet state-of-the-art performance targets, they should be programmable and configurable using C-plane functions according to Software Defined Network (SDN) principles [8].

3.1 *Flexible Deployment*

As shown in Fig. 4, diverse low cost access points are to be supported in 5G. Based on the classification according to functionality, there are coverage access point and hotspot access point. The functionality of coverage access point is to guarantee the basic coverage and uniform access. It can be the anchor of mobility and also the signaling access point due to the relative larger coverage.

Hotspot access point is to fulfill the high throughput requirements in certain area. It needs to be plug and in equipment deployed in hotspot areas as required.

According to level support of protocol stack, there are full stack access point and semi-stack access point.

The former includes PHY, MAC, RLC, PDCP and RRC for control plane or IP for user plane. The latter inherits the protocol stack of eNB (evolved Node Base-station) in LTE to provide complete protocol treatment including IP, PDCP, RLC, MAC and PHY.

Semi-stack access point is proposed to simplify the design of access point and reduce cost. It is expected to be low cost and easy deployed.

3.2 *Flexible Functionality*

Software and hardware decoupling of network equipment is the fundamental of flexible functionality. Thus, diverse network functionalities can be flexibly implemented as required. In this circumstance, the network provides multiple

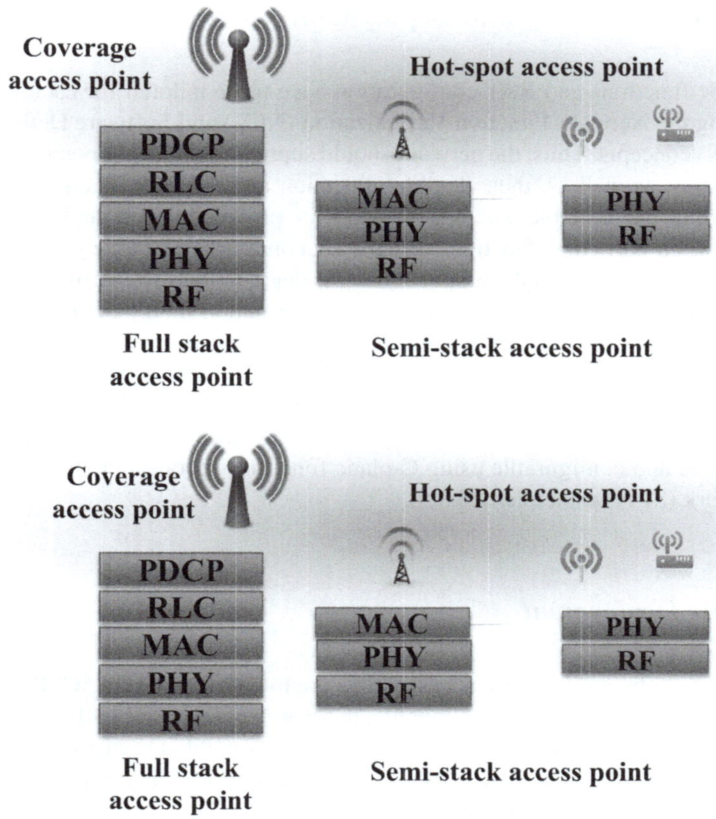

Fig. 4 A paradigm of flexible network deployment

functionality modules to fulfill diverse requirements. A group of functionalities to meet certain requirements constitute a slice. Figure 5 presents an example, which has three different slices, i.e., eMBB (enhanced mobile broadband) slice for mobile Internet service, mMTC slice for massive machine type access, uMTC slice for traffic with low latency and high reliability requirement. The functionality modules for different slices may be different. As shown in Fig. 5, eMBB slice may use content and service module implemented at RAN side for content awareness/analysis/local caching for hot content, and forwarding function module at RAN side for local breakout for low latency service. For example, the regional user and service requirements are analyzed via big data algorithms with user and traffic information. Then, the corresponding network functionality module can be deployed according to the analysis e.g., to deploy Multimedia Broadcast Multicast Service (MBMS) functionality module in areas with the broadcasting requirement, to deploy device to device (D2D) functionality module in areas with the group communication requirement, and to deploy safety functionality module in areas with

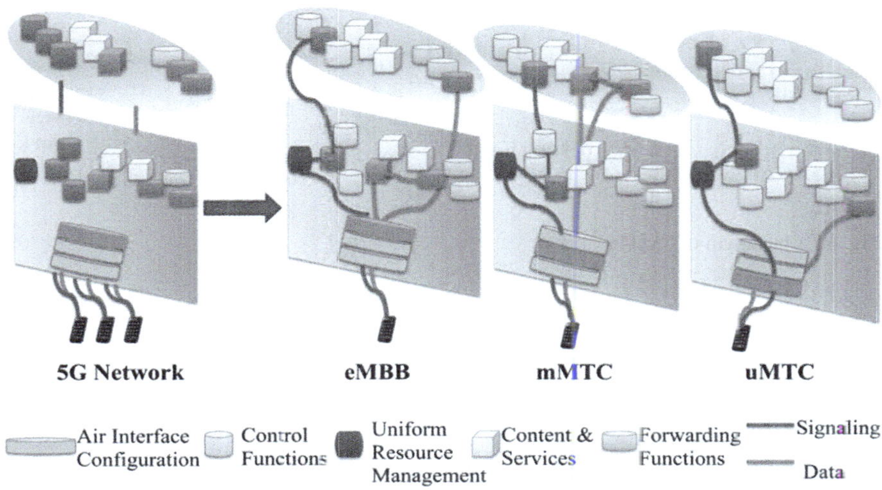

Fig. 5 A paradigm of network slicing

the high requirement of security. mMTC slice can use content and service module implemented at core network side for data processing, and forwarding function module at core network side for latency insensitive traffic. uMTC slice directly uses forwarding function module at RAN side for local breakout to meet low latency requirement.

Regarding the control plane, all slices go through the uniform resource management module to provide unified radio access regardless of air interface technology.

3.3 *Local Breakout*

As described in the flexible functionality section above, the traffic can be transmitted by a short path via local breakout in order to meet the low latency requirement. Figure 6 shows the architecture of local breakout. The data can be directly steered to local network or Internet on the edge. Signaling traffic can continue to traverse the mobile operator network. It can not only shorten the transmission time, but also release the traffic pressure at the core network. There are several potential solutions, e.g., LIPA (Local IP Access) and SIPTO (Selected IP Traffic Offload) are proposed in 3GPP [9]. The main idea of LIPA is to access local network through local gateway (L-GW). SIPTO is to steer part of the traffic to Internet directly through L-GW. There are usually two deployment manners, i.e., collocated L-GW which may have mobility issue among base stations, and non-collocated L-GW with mobility issue among L-GWs, each of which covers several base stations. There are still several open issues before commercial ready, such as the mobility issue described above, as well as UE and core network complexity. Therefore, studies on local breakout are

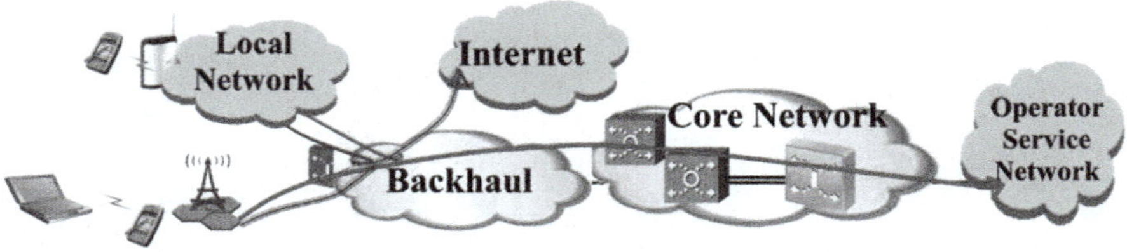

Fig. 6 A paradigm of local breakout

not limited to this and go on broadly. No matter which solutions are to be put to use, local breakout is an inevitable branch of user-centric 5G network.

4 Smart User and Traffic Awareness and Management

Smart user and traffic awareness and management, together with user-centric 5G access network architecture and flexible deployment and functionality provide the foundation of user centric network.

4.1 *Multi-Dimension QCI [QoS (Quality of Service) Class Indicator] Design*

QCI is introduced in LTE to meet the requirements of diverse traffic. Radio access network schedules and allocates wireless resource according to the indication from the core network via QCI. As shown in Fig. 7, the QoS management in LTE is a hierarchical E2E system, which can be divided into two parts, i.e., EPS bearer traffic and outer bearer traffic. The traffic of the same bearer has the same QoS guarantee.

There are several reasons that it can also not fulfill the requirements of diverse user and traffic addressed above.

1. There are only nine kinds of QCI in current standardization. It is not enough for much more diverse traffic requirements of 5G network.
2. The current Access network is "blind", and knows nothing about users and traffic. The strategy of wireless resource schedule and allocation is from the QCI indication of the core network. It is relatively complicate and inefficient. 5G wireless network should have the capability of awareness and adaptation of the user and the traffic.
3. The difference of user requirements and behavior is not taken into consideration of network design before 5G. To better meet the diverse user requirements and improve user experience, the user requirements and behavior, e.g., daily mobility track, preferred traffic content, have to be considered in 5G network design.

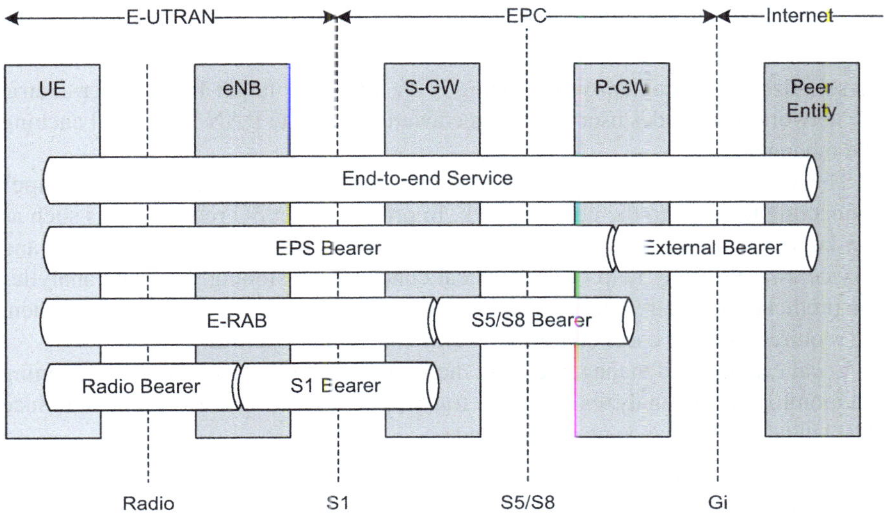

Fig. 7 EPC bearer architecture in LTE systems

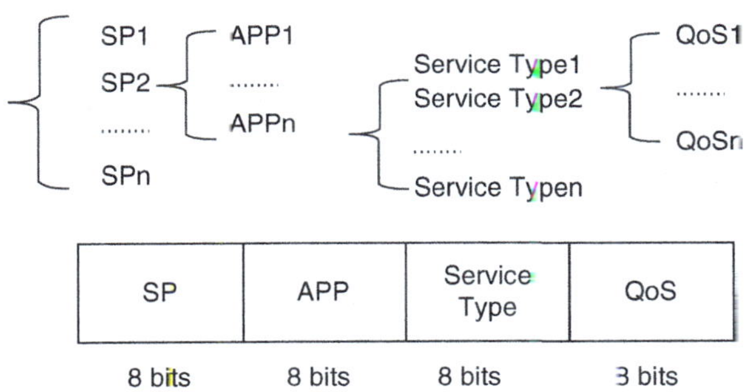

Fig. 8 Hierarchical QoS system

4. The consistency of user experience is an important requirement of 5G network, especially for dense or high speed scenarios, such as stadium, high speed railway.

To better meet the requirements of diverse traffic in 5G network, new QoS guarantee mechanism is to be designed. A paradigm of new QoS guarantee mechanism is addressed in this chapter.

The mechanism is designed based on hierarchical bit-mapping. The information includes service provider, access point name, traffic type, QoS requirements, and so on. Let us take 32 bits QoS parameter as an example. As shown in Fig. 8, the first 8 bits represents service provider, the consequent 8 bits are used for application name, the next 8 bits is for traffic type, and the last 8 bits indicates QoS parameter. The information is helpful for wireless resource scheduling and allocation.

4.2 Personalized Local Content Provisioning

Personalized local content provisioning is an important target for the user-centric 5G network. It includes user and service awareness at the RAN side, local caching and management.

The access network used to be user and service agnostic, which is a "blind pipe" connecting the user to the core network. In order to meet 5G requirements such as low latency and user-centric requirements, the access network needs to be user and service aware. It may help facilitate local content provisioning. Big data analytics can predict user requirements with mining the data from the network [10–14]. Then, the requirements are checked locally if they can be met via local content.

Local caching and management are the basis to fulfill local content provisioning via monitoring and analyzes of source traffic, providing a local copy. It can reduce E2E latency and enhance user experience.

- User and traffic information acquisition
  The first step is to collect traffic attribute via packet analysis, which may include the application type (FTP, HTTP, etc.), server address, port number, traffic content, etc. Then, this data is analyzed comprehensively via clustering algorithm to label the traffic, e.g., as sports news, entertainment video, or romance/action movie.
- User requirement analysis and prediction
  The big data analysis algorithm, especially recommendation algorithms can recommend contents that the user may be interested in by considering traffic labels, user attributes, terminal types, etc.
- Local caching and content management
  The popular content is to be copied locally, as in the case of a downloaded movie, or to be downloaded from the application server and cached locally, as in the case of sports news, by matching the traffic labels with the content users may be interested in.
- Content provisioning
  When the user initiates an application request, the system will check whether the corresponding content is already locally cached and thus can be sent directly. On the other hand, according to the content recommendation via big data analysis, the system can check if it is cached locally and push to the user directly.

4.3 User Behavior Awareness

Wireless resources need to be optimized according to user and service requirements in 5G to improve efficiency and the user experience.

It is found that a 93 % potential predictability in user mobility (route and time) across the whole user base by studying the mobility patterns of anonymized mobile phone users [15]. As shown in Fig. 9, some passengers often get on the train

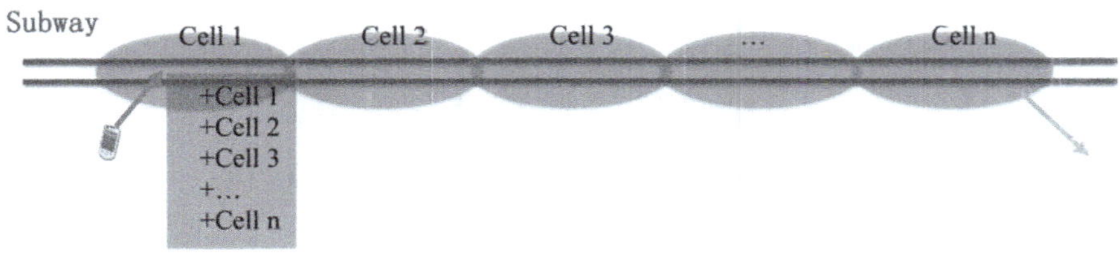

Fig. 9 A paradigm of user behavior awareness

at a certain station via subway or high-speed rail, and get off at another station. The motion trajectory can be predicted via big data analysis to pre-configure the network.

The access points can record the historical access point list of each user, and directly upload it to a central processing module or to a target access point when the serving access point is changed. Then, the collected user motion data is analyzed to predict the motion trajectory via the big data algorithms. As shown in Fig. 9, when access point 1 (Cell 1) get the predication that the user motion trajectory is from Cell 1 to Cell n via Cell 2, 3, and n-1, it will send the configuration of Cell 2, 3, ..., n to the corresponding terminal to facilitate pre-configuration of these cells.

5 High Efficiency Network Operation

Operation efficiency is an important requirement in 5G network design. Self operation network (SON) techniques have been studied in LTE, including PCI optimization, self-configuration of neighbor cell, minimization of drive test (MDT), mobile load balancing, energy saving, etc. However, there is no mature technique to let the SON work well in practical systems so far due to the limitation of LTE network architecture. Therefore, SON should be taken into consideration at the stage of 5G network design.

The high efficiency network operation system includes traffic sensing, user capability sensing, centralized and personalized functionality management and distribution, unified wireless platform, among other modules.

The system is divided into two parts, i.e., the decision making domain and the implementation domain. The decision making domain is mainly responsible for the collection and management of user information, as well as service and terminal state, network upgrade, selection, configuration and unified management, etc. The implementation domain is responsible for the user, terminal and network status reporting, network configuration and dynamic deployment, etc.

Figure 10 shows a paradigm of big data based network operation. It includes user information collection and strategy making which belong to decision making domain, as well as UE context management and function configuration which compose implementation domain.

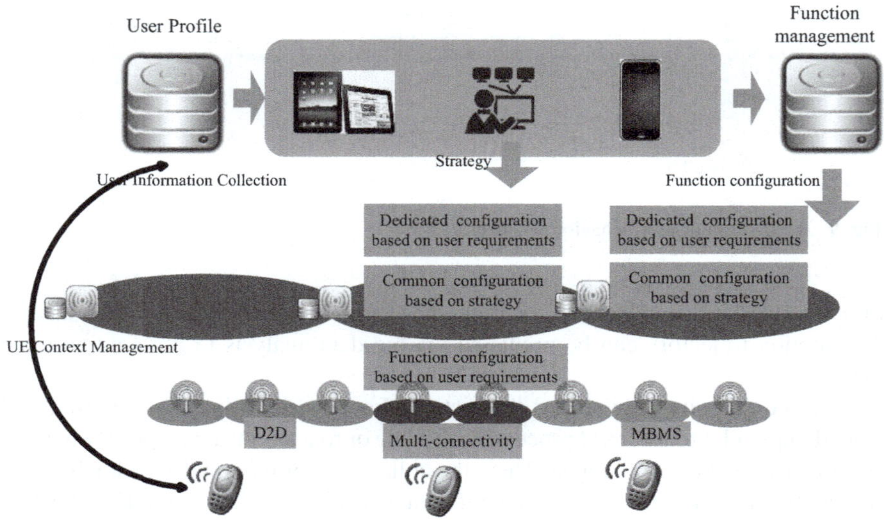

Fig. 10 A paradigm of big data based network operation

The decision making domain provides the basic configuration to initialize the network. Then, it chooses and deploys the functionalities and access points which meet the user requirements. The implementation domain builds multi-connectivity bearers with terminals via dynamic access points, functionality and configuration according to the requirements after receiving the personalized configuration.

The efficiency of the big data based network operation can be maximized, through optimal resource allocation (frequency, time, antenna, power, etc) to each access point and to each user, such that the required service is provided from the optimal set of access points with optimally allocated resources. For example, in the 5G ultra dense deployment scenario, it may be a problem for users to have too many neighbor cells, which may be redundant and consume too much unnecessary power, and sometimes interfere with each other. To resolve this issue, big data algorithms can analyze the user and service requirements with the collection of network status. Then, the system can choose the corresponding configuration of neighboring cell according to the analysis. For example, it is possible that not all the access points support downlink 256QAM. If a service needs the downlink 256QAM functionality, only those access points with this functionality need to be configured as the neighboring cells, with remaining neighboring cells turned off for power saving. The decision making domain needs to have the whole picture of functionality distribution in the network and the requirements of users and services. The implementation domain configures the terminal according to the command from the decision making domain to fulfill user-centric neighboring cell configuration.

6 Chapter Summary

In this chapter, the concept of user-centric wireless network for 5G is addressed to fulfill multiple user experience requirements in 5G. Four key technical directions are studied based on the gap analysis between LTE technology and 5G requirements, i.e., user centric 5G access network architecture design, flexible functionality and deployment, smart user and traffic awareness and management, and high efficient low cost network operation. These key technologies work together with cross-layer and E2E solutions [16–19] to provide the user-centric 5G eco-system.

A.1 Annex-User Experience Requirements Analysis [3]

To design a user-centric 5G network, it is better to study carefully of 5G user experience requirements. In this part, the analysis of user experience requirements, including consistent user experience, user experience data rate, latency, mobility, is provided from.

User experience requirements address the end users perception of technology when consuming one or more services. User experience will have to be managed in highly heterogeneous environment and under different user scenarios/contexts.

A.2 Consistent User Experience

The 5G system should be able to deliver a consistent user experience over time for a given service everywhere the service is offered.

Consistent user experience is defined by service-dependent minimum KPIs (e.g. data rate, latency) being met over the service coverage area, with a level of variation configurable by the operator. These service-dependent KPIs are for further study. A consistent user experience across time and space depends obviously on the technology performance and capabilities, and on the operator deployment. The requirements address only the technology performance and capabilities.

A.3 User Experienced Data Rate

Data rate requirements are expressed in terms of user experienced data rate, measured in bit/s at the application layer. The required user experienced data rate should be available to at least 95 % of the location and time in the considered environment (i.e. at the "cell-edge"). The user experienced data rate requirement depends on the targeted application/use case. It is set as the minimum user experienced data rate required for the user to get a good experience of the targeted application/use case.

Use case specific user experienced data rates up to 1 Gb/s should be supported in some specific environments, like indoor offices, while at least 50 Mb/s shall be available everywhere at a reasonable cost.

A.4 Latency

When considering latency requirements, the following metrics are considered:

- E2E Latency: Measures the duration between the transmission of a small data packet from the application layer at the source node and the successful reception at the application layer at the destination node and back.
- User Plane Latency: Measures the time it takes for a small data packet from user terminal to the L2/L3 interface of the 5G system destination node and back.

The E2E latency is the latency perceived by the end user. It accounts for the time needed for the data packet to cross all the nodes up to the application server and back, which includes nodes of the 5G system and nodes potentially outside the 5G system. In contrast, the user plane latency is limited to the 5G system only. Both latency metrics approximately coincide when the application server is located within the 5G system. In the latter case the latency is minimized when the application server is co-located with a radio node, e.g., the radio base station or another user terminal (for the case of device-to-device communication). As a result, the requirements on minimum latency are expressed in terms of E2E latency.

The 5G system should be able to provide 10 ms E2E latency in general and 1 ms E2E latency for the use cases which require extremely low latency. Note these latency targets assume the application layer processing time is negligible to the delay introduced by transport and switching.

The 5G system should also give the end user the perception of being always connected. The establishment of the initial access to the network (or status change from idle state to connected) should then be instantaneous from the human user perspective.

A.5 Mobility

Mobility refers to the system's ability to provide seamless service experience to users that are moving. In addition to mobile users, 5G use cases identify that 5G networks will support an increasingly large segment of static and nomadic users/devices. 5G solutions therefore should not assume mobility support for all devices and services but rather provide mobility on demand only to those devices and services that need it. In other words, mobility on-demand should be supported, ranging from very high mobility, such as high-speed trains/airplanes, to low mobility or stationary devices such as smart meters.

The mobility requirements are expressed in terms of the relative speed between receiver and transmitter at which consistent user experience should be ensured.

References

1. K. Zheng, Y. Wang, W. Wang, M. Dohler, J. Wang, Energy-efficient wireless in-home: the need for interference-controlled femtocells. IEEE Wirel. Commun. **18**(6), 36–44 (2011)
2. M. Mukherjee, L. Shu, V. Kumar, P. Kumar, R. Matam, Reduced out-of-band radiation-based filter optimization for UFMC systems in 5G, in *Proceedings of the 11th International Wireless Communications and Mobile Computing Conference (IWCMC 2015)*, Dubrovnik, Croatia, 24–27 August 2015
3. NGMN 5G White Paper, NGMN Alliance, March 2015
4. W. Fang, Y. Li, H. Zhang, N. Xiong, J. Lai, A.V. Vasilakos, On the throughput-energy tradeoff for data transmission between cloud and mobile devices. Inform. Sci. **283**, 79–93 (2014)
5. W. Fang, X. Yin, Y. An, N. Xiong, Q. Guo, J. Li, Optimal scheduling for data transmission between mobile devices and cloud. Inform. Sci. **301**, 169–180 (2015)
6. K. Zheng, B. Fan, Z. Ma, G. Liu, X. Shen, W. Wang, Multihop cellular networks toward LTE-advanced. IEEE Veh. Technol. Mag. **4**(3), 40–47 (2009)
7. M. Mukherjee, L. Shu, Y. Zhang, Z. Zhou, K. Wang, Joint power and reduced spectral leakage based resource allocation for D2D communications in 5G, in *15th International Conference on Algorithms and Architectures for Parallel Processing (ICA3PP)*, Zhangjiajie, China, 18–20 November 2015
8. IMT-2020(5G) PG-white paper on 5G vision and requirements, China IMT-2020 Promotion Group, May 2014
9. 3GPP TR23.859 V12.0.1. Local IP access (LIPA) mobility and Selected IP Traffic Offload (SIPTO) at the local network (Release 12)
10. K. Zheng, Z. Yang, K. Zhang, P. Chatzimisios, W. Xiang, K. Yang, Big data driven optimization for mobile networks towards 5G. IEEE Netw. Mag. **30**(1), 44–51 (2016)
11. C.-L. I, Y. Liu, S. Han, S. Wang, G. Liu, On big data analytics for greener and softer RAN. IEEE Access **3**, 3068–3075 (2015)
12. J. Wu, H. Liu, H. Xiong, J. Cao, J. Chen, K-means-based consensus clustering: a unified view. IEEE Trans. Knowl. Data Eng. **27**(1), 155–169 (2015)
13. Y. Chen, L. Shu, A.M. Ortiz, N. Crespi, L. Lv, Locating in crowdsourcing-based dataspace: wireless indoor localization without special devices. Mob. Netw. Appl. **19**(4), 534–542 (2014)
14. Y. Chen, A.M. Ortiz, N. Crespi, L. Shu, L. Lv, Reality mining: digging the impact of friendship and location on crowd behavior, in *Mobile and Ubiquitous Systems: Computing, Networking (MobiQuitous), and Services*, 2013, pp. 142–154
15. C. Song, Z. Qu, N. Blumm et al., Limits of predictability in human mobility. Science. **327**(5968), 1018–1021 (2010)
16. J. Wang, J. Wen, Y. Han, J. Zhang, C. Li, Z. Xiong, CUBIC-FIT: a high performance and TCP CUBIC friendly congestion control algorithm. IEEE Commun. Lett. **17**(8), 1664–1667 (2013)
17. J. Wang, J. Wen, Y. Han, J. Zhang, C. Li, Z. Xiong, DC-Vegas: a delay-based TCP congestion control algorithm for datacenter applications. J. Netw. Comput. Appl. **53**, 103–114 (2015)
18. J. Wang, J. Wen, Y. Han, J. Zhang, C. Li, Z. Xiong, Achieving high throughput and TCP Reno fairness in delay-based TCP over large networks. Front. Comput. Sci. **8**(3), 426–439 (2014)
19. H. Wang, S. Chen, H. Xu, M. Ai, Y. Shi, SoftNet: a software defined decentralized mobile network architecture toward 5G. IEEE Netw. **29**(2), 16–22 (2015)

References

Energy Harvesting Based Green Heterogeneous Wireless Access for 5G

Hang Li, Chuan Huang, Fuad E. Alsaadi, Abdullah M. Dobaie, and Shuguang Cui

Abstract A worldwide challenge for the design of future cellular systems is to meet the increasing energy demand, while, on the other hand, to lower the emission of greenhouse gases for achieving the environment sustainability. A feasible and efficient method to tackle this issue is to let the communication systems harvest energy from renewable energy sources instead of fossil fuels. However, by employing the energy harvesting (EH) technique, the instability of renewable energy resources introduce new challenges on the design of the upcoming 5G systems. In this chapter, we focus on uplink access schemes and power allocations for EH based heterogeneous networks. First, a heterogeneous access model incorporating EH based mobile users is proposed and followed by a throughput maximization framework. Then, by classifying transmission policies into two main categories (i.e., single-channel vs. multi-channel scenarios), the proposed framework is concretized under various practical conditions, including the availability of central control, causality of harvested energy, channel state information, and others. Finally, future research directions and open problems are discussed.

H. Li
Department of Electrical and Computer Engineering, Texas A&M University, College Station, USA
e-mail: david_lihang@tamu.edu

C. Huang
National Key Laboratory of Science and Technology on Communications, University of Electronic Science and Technology of China, Chengdu, China
e-mail: huangch@uestc.edu.cn

F.E. Alsaadi • A.M. Dobaie
Department of Electrical and Computer Engineering, King Abdulaziz University, Jeddah, Saudi Arabia
e-mail: fuad.alsaadi@gmail.com; adobaie@kau.edu.sa

S. Cui (✉)
Department of Electrical and Computer Engineering, University of California, Davis, USA
e-mail: sgcui@ucdavis.edu

W. Xiang et al. (eds.), *5G Mobile Communications*,
DOI 10.1007/978-3-319-34208-5_18

1 Introduction

As data service is prioritized by service operators, it now enjoys an unprecedented popularity across the world. Compared to the voice and text services, data service incurs more intensive communication usages at mobile users (MUs), like Internet surfing, picture uploading, video streaming, and many others, which could quickly deplete the energy stored in the battery. It is not surprising that the energy cost of the cellular system keeps growing [11]. Taking the electrical energy consumption of all cell phones as an example: a regular cell phone consumes about 4 kWh [27] annually, and the number of cell phones worldwide is about seven billion in 2015 [13], which approximately requires us to burn 15 million tons of coal at each year [35]. Although significant efforts have been made such as developing energy efficient transmission technologies to save energy and prolong battery lifetimes, the high energy cost and high carbon-oxygen emission issues have not been fundamentally solved. When further considering the energy consumption at the base stations (BSs) and data centers, such an energy challenge will inevitably draw both environmental and financial concerns for service operators and mobile users [14].

The 5G system, as a new generation of the cellular system, is expected to mitigate the above issues by becoming "green": lowering both the energy cost and greenhouse gas emission. The most effective way to achieve this is to shift the paradigm of energy sources from the conventional fossil fuels to renewable energy sources. Recently, with the rapid development of hardware technologies, energy harvesting (EH) techniques were developed, capable of efficiently converting the environmental energy (e.g., solar, wind, thermal, vibration, etc.) or man-made energy (e.g., vibration, RF) into electric energy. So far, many prototypes of EH based wireless sensors have been proposed [32] and various EH products have been commercialized [25]. As such, EH based wireless devices are expected to enjoy large-scale applications in 5G systems.

One important characteristic of 5G wireless systems is to achieve an aggregate data rate (characterized by bits/s per unit area) more than 1000× higher than that in 4G. The heterogeneous network (HetNet) architecture, which incorporates multiple radio access technologies (RATs), is a promising solution to realize such a capacity boosting. As such, the 5G system could support not only new standards (e.g., mmWave based ones with device-to-device (D2D) communications [2, 36]), but also the conventional 3G, 4G, and WiFi standards.

For the conventional battery or grid powered HetNet, the design of access and power allocation policies only accounts the limitations on the average and/or peak power consumptions, which are usually fixed over the whole operation period. When the transmitter is powered by energy harvesters, however, the design of the transmission policies faces a new type of constraints, named *EH constraints*, i.e., the power consumption is limited by the accumulated energy up to the current time. Besides, the ambient energy sources are usually random and intermittent. Thus, the amount of harvested energy per unit time may vary, which brings extra unreliability to the system performance. In addition, hardware related limitations also need to be accounted into the transmission policy design for practical purposes, which include the battery storage efficiency, the full/half energy-duplex consideration, and the supporting circuit power overhead.

In this chapter, we consider the design of the transmission policy (i.e., the control of the channel access and transmission power) for EH-based heterogeneous networks, where the network provides either single-channel or multi-channel access for MUs, and each MU is equipped with an EH-based power supply. We propose an optimization framework to determine throughput-optimal transmission policies, where such a framework incorporates various practical considerations, such as finite time horizon, causality of the state information, and others. Following this framework, we also introduce the state-of-art transmission policies by describing their application scenarios and how they achieve efficient operations.

The rest of this chapter is organized as follows. In Sect. 2, we first present a heterogeneous wireless access model, and then consider a general optimization framework to maximize the average sum rate. In Sects. 3 and 4, we introduce transmission policies in single-channel and multi-channel scenarios, respectively. Finally, in Sect. 5, we summarize the existing access and power allocation policies, and discuss future research problems in EH based heterogeneous wireless access.

2 System Model and Optimization Framework

Given a targeted area, there are multiple EH-based mobile users (MUs) that are ready to transmit their data. To meet MUs' transmission requirements, different types of base stations (BSs) serve as wireless access points for the MUs. To fully and efficiently coordinate the BSs, a cloud-based radio access network (C-RAN) architecture has been proposed in [7], where a virtual BS pool performs the centralized management, including joint signal processing, inter-cell interference (ICI) mitigating, radio management, and MU scheduling. Here we adopt C-RAN in our framework for the heterogeneous access as shown in Fig. 1.

A BS can be a marco cell station, small cell station, or WiFi access point, and thus multiple RATs can be supported by multiple BSs handling the signals transmitted from MUs over certain radio frequencies. The signals propagate through a certain wireless medium, which could be divided into multiple parallel channels via various schemes [41], e.g., frequency division multiple access (FDMA), code division multiple access (CDMA), or orthogonal frequency division multiplexing (OFDM). In Fig. 1, there are M parallel channels for N MUs. A particular channel could be reused over space by different BSs. Note that if two BSs are too close, like BS 1 and BS 2, and both use channel 1, serious interference may be caused. Although parallel channels could be created in many ways, the total number of independent channel usages (the degree of freedom) is finite and usually smaller than the number of MUs in most practical multiple access scenarios. This requires

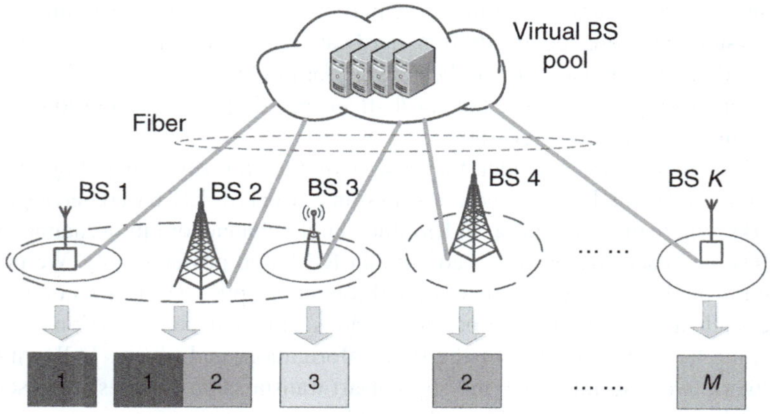

Fig. 1 Network architecture of the C-RAN [7] with heterogeneous wireless access, where K BSs are connected to the virtual BS pool via fiber links, and different BSs may use the different or the same wireless channels

the system designer to develop smart algorithms that can efficiently manage the channel resources so as to maximize the desired utility. Here, we only focus on the uplink wireless accesses of multiple MUs and assume that the C-RAN can perfectly deal with all the jobs at the receiver side.

2.1 System Model

For the purpose of exposition, we tailor the system model of C-RAN and consider a simplified heterogeneous wireless access model as shown in Fig. 2, involving M parallel synchronized channels for N EH-based MUs. Each MU can each access more than one channels. The system is slotted over time with unit slot length. The considered time horizon is T, i.e., $1 \leq t \leq T$, where t is the slot index. Moreover, T can be either deterministic or random, which depends on the utility of interests. The deterministic case is commonly used in most studies, and T is either finite or infinite. In some applications, T can also be treated as a random variable, which will be shown later.

For the nth MU, the EH rate (i.e., energy harvested per slot) at time slot t is denoted by $E_n(t)$, and the battery energy level at the beginning of time slot t is denoted by $B_n(t)$. In general, the power consumption at the MU can be roughly divided into two parts: circuit power and transmission power. Note that the circuit power can be usually modeled as a constant value (see Sect. IX in [34]) that can be approximately treated as a constant offset in the total power consumption. Thus, to investigate the limit of the system performance, it is reasonable to first ignore

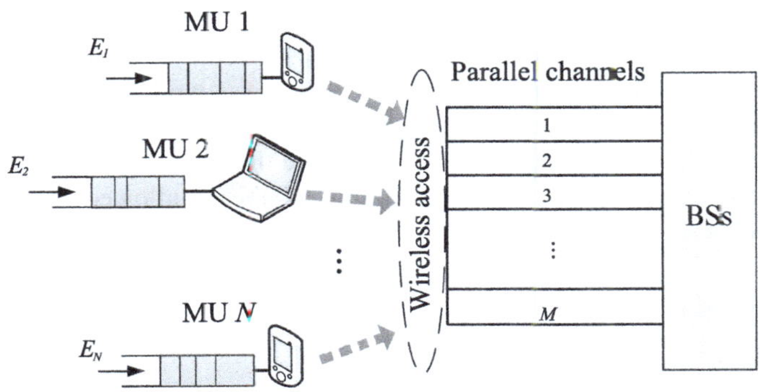

Fig. 2 Access model for the heterogeneous network with EH-based MUs. Multiple BSs provide in total of M parallel channels for the N EH-based MUs, and each MU can access more than one channels

the circuit power consumption and purely focus on the transmission power control. Suppose that the total transmission power of MU n during time slot t is $P_n(t)$ (which is also the transmission energy given the unit slot length); the battery evolution can be expressed as

$$B_n(t+1) = \max\{0, \min\{B_n(t) + E_n(t) - P_n(t), B_{n,max}\}\}, \tag{1}$$

where $B_{n,max} > 0$ is the battery capacity.[1] We call $\{E_n(t), B_n(t)\}_{1\leq n\leq N}$ as energy state information (ESI) at time t. In general, for the nth MU, $\{E_n(t)\}_{t\geq 1}$ is a random process whose statistics is often viewed as a prior knowledge.

In our considered system, each MU could access more than one channels simultaneously. We let $s_{n,m}(t)$ be an indicator function such that

$$s_{n,m}(t) = \begin{cases} 1, & \text{the } n\text{th MU accesses channel } m \text{ in time slot } t; \\ 0, & \text{otherwise.} \end{cases} \tag{2}$$

We want to point out that $s_{n,m}(t)$ can be either deterministic or random depending on the adopted access scheme. For example, if a central controller was in charge of the access control and directly assigned channel m to the nth MU at time t, there would be $s_{n,m}(t) = 1$, which is deterministic. However, if the controller randomly assigned a channel to the nth MU, $s_{n,m}(t)$ would become a random variable depending on the channel selection policy. A typical selection is to assign the channel with the highest channel power gain at time t to the nth MU such that the instantaneous transmission

[1]Sometimes $B_{n,max}$ could be treated as positive infinity when the EH rate is relatively small compared to the batter capacity. For example, an AA-sized NiMH battery has a capacity of 7.7 kJ, which may require a couple of hours to be fully charged by some commercial solar panels (1 W ~ 400 W) [21, 30], which generate relatively small $E_n(t)$.

rate is maximized. On the other hand, in a distributed access scenario, if the nth MU randomly accessed channel m, the indicator $s_{n,m}(t)$ would be a random variable. In short, access variables $\{s_{n,m}(t)\}$ are crucial in transmission policy design and we will further specify their characteristics later.

In general, the interference is caused by more than one MUs transmitting over the same channel. Suppose that the maximum number of allowed collisions in the mth channel is set as N_m, i.e.,

$$0 \leq \sum_{n=1}^{N} s_{n,m}(t) \leq N_m, \text{ for all } m = 1, 2, \ldots, M. \tag{3}$$

When $N_m \geq 1$, i.e., more than one MUs access the same channel and transmit simultaneously, their signals are mixed together, which may cause errors when the C-RAN decodes the information for each MU. Since here we do not focus on how C-RAN reconstructs information from the collided signals, we simply treat N_m as a fixed system parameter. In some scenarios, only one MU is allowed to access one particular channel at each time slot, i.e., $N_m = 1$, in order to fully mitigate the interference, where many access schemes have been proposed under this setup [3, 9, 22].

From the MU point of view, we have

$$0 \leq \sum_{m=1}^{M} s_{n,m}(t) \leq A_n, \text{ for all } n = 1, 2, \ldots, N, \tag{4}$$

where A_n is the number of channels that the nth MU could use, and is determined by hardware limitations and resource management policies. Since the total number of channels is M, it follows that $1 \leq A_n \leq M$. Thus, (3) and (4) together capture the characteristic of the proposed heterogeneous wireless access. Again, it is worth noticing that $\{s_{n,m}(t)\}$ could be determined in either a centralized or distributed manner.

Let $x_n(t)$ be the transmitted signal by MU n. Then, the received signal at the BSs (processed at C-RAN) over channel m is given by

$$y_m(t) = \sum_{n=1}^{N} h_{n,m}(t) s_{n,m}(t) \sqrt{P_{n,m}(t)} x_n(t) + z_m(t), \tag{5}$$

where $h_{n,m}(t)$ is the channel coefficient treated as a constant during one time slot, $P_{n,m}(t)$ is the transmission power of the nth MU over channel m, and $z_m(t)$ is the circularly symmetric complex Gaussian (CSCG) noise with zero mean and variance σ_m^2. Note that the total transmission power at each MU is $\sum_{m=1}^{M} s_{n,m}(t) P_{n,m}(t) = P_n(t)$. For notation simplicity, we let $g_{n,m}(t) = \frac{|h_{n,m}(t)|^2}{\sigma_m^2}$ be the effective channel power gain, and call $\{g_{n,m}(t)\}$ as the channel state information (CSI). By applying

Table 1 Key features for the system setup

System features	Conditions
Time horizon	Finite ($T < \infty$) or infinite ($T \to \infty$)
Central controller	Available (centralized) or unavailable (distributed)
CSI	Non-causal or causal
ESI	Non-causal or causal
Channel accessibility	Single-channel or multi-channel

the Shannon capacity formula over a multiple access channel [33], the sum rate of all N MUs over the mth channel is given by $\log\left(1 + \sum_{n=1}^{N} s_{n,m}(t)g_{n,m}(t)P_{n,m}(t)\right)^2$ for $1 \le m \le M$. Thus, the total instantaneous rate of the uplink is given by

$$\sum_{m=1}^{M} \log\left(1 + \sum_{n=1}^{N} s_{n,m}(t)g_{n,m}(t)P_{n,m}(t)\right). \tag{6}$$

So far, we have presented the basic system parameters including: the environment variables, i.e., time horizon T, ESI and CSI; and the controlling variables, i.e., the access variables $\{s_{n,m}(t)\}$ and power allocations $\{P_{n,m}(t)\}$ for all n, m and t. Note that these variables can be further specified according to the interest of system design. We list some further system specifications in Table 1, which will be discussed in details when we introduce transmission polices in Sects. 3 and 4.

2.2 *Throughput Maximization*

In this subsection, we present a general formulation framework for throughput maximization and introduce some useful techniques to find the solution structure of design variables $\{s_{n,m}(t)\}$ and $\{P_{n,m}(t)\}$.

For the considered system, the instantaneous rate of the uplink is given by (6). However, we do not directly consider the sum rate over the desired time horizon T as the objective function. Instead, we introduce an auxiliary variable λ as the objective function such that

$$\mathbf{E}\left[\sum_{t=1}^{T}\sum_{m=1}^{M} \log\left(1 + \sum_{n=1}^{N} s_{n,m}(t)g_{n,m}(t)P_{n,m}(t)\right) - \lambda T\right] = 0, \tag{7}$$

[2]Note that the notation "log" could represent either the logarithm with base 2 or base e according to the applications. In this chapter, we use this unified notation.

where $\mathbf{E}[\cdot]$ is the expectation operator over some dynamic statistics (if any). The initial conditions are $\{g_{n,m}(1)\}$ and $\{E_{n,m}(1), B_{n,m}(1)\}$. We employ λ since condition (7) is able to provide us certain flexibility to define the utility of interest. If there is no specific regulations on T, an intuitive observation is that the variable λ can represent the expected value of the average uplink throughput, i.e., condition (7) is equivalent to

$$\mathbf{E}\left[\frac{1}{T}\sum_{t=1}^{T}\sum_{m=1}^{M}\log\left(1+\sum_{n=1}^{N}s_{n,m}(t)g_{n,m}(t)P_{n,m}(t)\right)\right]=\lambda. \tag{8}$$

In fact, as it will be shown later, λ can also represent the average rate per transmission, like "rate-of-return," if we consider some particular dynamic optimization techniques.

Besides condition (7), the access variable $\{s_{n,m}(t)\}$ should satisfy constraints (3) and (4) for $1 \le t \le T$. Moreover, the total transmission power $P_n(t) = \sum_{m=1}^{M} s_{n,m}(t)P_{n,m}(t)$ is bounded and satisfies a new type of EH constraint, i.e.,

$$0 \le P_n(t) \le B_n(t), \text{ for all } n = 1, 2, \ldots, N, \tag{9}$$

where $B_n(t)$ is the battery energy level satisfying (1).

Thus, the throughput maximization problem can be formulated as

$$\begin{aligned} &\text{maximize} \quad \lambda \\ &\text{subject to} \ \ (7), \\ &\qquad\qquad (3),\ (4) \text{ for } 1 \le t \le T, \\ &\qquad\qquad (9) \text{ for } 1 \le t \le T. \end{aligned} \tag{10}$$

Note that when $T \to \infty$ (or $T \to \infty$ almost surely when T is random), all constraints in this framework still need to hold, and λ in this case can represent the long term average throughput.

The solution to problem (10) is highly dependent on the system setups listed in Table 1. One such major condition is the causality of the system information (CSI or ESI). When both the CSI and ESI could be non-causally obtained, there is no randomness in problem (10) and the expectation operator could be dropped off. Then, problem (10) is solvable by applying some offline optimization tools, e.g., convex or mix-integer optimization.

When the knowledge about CSI or ESI is obtained from statistics, problem (10) could be solved by some dynamic optimization techniques, which usually require more definitions beyond the formulations described above. There are four main quantities in a dynamic optimization framework as follows:

1. **The space of system states**. In problem (10), the system state is represented by both CSI and ESI, i.e., $\{g_{n,m}(t)\}$ and $\{E_{n,m}(t), B_{n,m}(t)\}$. Thus, the space of the system states consists of all possible values of CSI and ESI.
2. **The space of actions**. This space is composed by all possible values of the design variables, i.e., the access control $\{s_{n,m}(t)\}$ and the transmission power levels $\{P_{n,m}(t)\}$. In each time slot t, the decision maker chooses a set of actions $\{s_{n,m}(t), P_{n,m}(t)\}_{1 \le n \le N, 1 \le m \le M}$ that must satisfy all constraints in problem (10). Note that in most existing literatures on EH-based wireless access design [19, 34], the spaces of system states and actions are assumed to be fixed over time.
3. **The reward**. For a given system state and action at the tth time slot, the corresponding instantaneous reward for problem (10) is instantaneous throughput (6).
4. **The transition of system states**. The transition of the system states is dependent on the system state and the actions at the current time slot. Particularly, the battery states $\{B_{n,m}(t)\}$ are influenced by both the actions and $\{E_{n,m}(t)\}$ according to the battery evolution in (1). The transition can be either deterministic or dynamic. In the dynamic case, the transition is often described by the state transition probability.

Now, solving problem (10) can be achieved by identifying the sequence of actions $\{s_{n,m}(t)\}$ and $\{P_{n,m}(t)\}$.

To derive the optimal solutions, popular dynamic frameworks have been applied in studying EH-based wireless communications, including the Markov decision process (MDP) [24, 34] (and the references therein), partially observable MDP (POMDP) [3, 12] (and the references therein), restless multi-arm bandit (RMAB) [3], and optimal stopping [20, 21]. The characteristics of these frameworks that reflect their applications are briefly discussed below.

MDP and POMDP: MDP is a controlled discrete-time stochastic process, where the decision maker can *perfectly know* the current state and the state transition probability of the system. Moreover, the system state at the next time slot is *conditionally independent* of all previous states and actions. The decision maker will choose an action from an action space according to the current reward, and this action will influence the evolution of the system state.

The POMDP is a generalization of an MDP, which permits certain *uncertainty* of the state information. Such uncertainty may be resulted from the indirect or partial observations over the system state Thus, an additional observation model needs to be incorporated into the POMDP. For example, if the MU cannot perfectly obtain its CSI, a hypothesis testing process is commonly used to identify the channel status.

RMAB and Optimal stopping: RMAB makes decisions for multiple stochastic processes simultaneously, which fits the scenario of multiple MUs where the state of each MU is a stochastic process. The decision maker needs to deal with the tradeoff between allocating resources for a high current reward (*exploitation*) and acquiring new knowledge with a prospect of a better future reward (*exploration*). Moreover, the considered multiple stochastic processes keep the state evolution even if the decision maker does not make any actions, while in classic MAP, the processes are "frozen" if there are no actions. When the state transition of the system has Markov properties, RMAB is also an MDP (or POMDP if the state information cannot be obtained directly).

If the decision maker is required to decide when to stop the decision process (a binary decision), it becomes an optimal stopping problem. The main tradeoff is also the exploration vs. exploitation, while in optimal stopping, the exploitation also implies that the decision process is finished.

Despite the different formulations for the dynamic case, the solution under these frameworks, however, usually turns out to be nothing but dynamic programming. When the knowledge about the system (e.g., the reward or state transition probability) is unknown, the solution may turn to machine learning algorithms (e.g., reinforcement learning, Q-learning) [4, 28].

It should be pointed out that for either case of the causality over CSI/ESI, problem (10) has not been solved in general. Next, we only focus on several typical scenarios and discuss the state-of-art transmission polices. To highlight the flexibility of channel management, we introduce transmission policies in two main scenarios:

- Single-channel scenario, i.e., $M = 1$: This is a typical access model when the channel resources are limited, which is commonly used in ad hoc, sensor, and cellular network models.
- Multi-channel scenario, i.e., $M \geq 2$: In this scenario, there are more than one channels for MUs to access at each given time, while for data transmissions, an MU can use either only one channel, or more than one channels simultaneously. When the hardware of the MU supports multi-channel transmissions, possible multiple RATs, it can fulfil a transmission over multiple channels simultaneously.

Note that the channel accessibility sets limitations on channel access indicators $\{s_{n,m}(t)\}$, and thus constraints (3) and (4) need to be modified accordingly, which will be specified later.

3 Single-Channel Scenario

Consider a scenario where multiple MUs are sharing limited channel resources, and this can be modeled as the single-channel scenario, i.e., $M = 1$. For this case, transmission policies that solve problem (10) can be further classified according to their ability of mitigating the interference among different MUs:

1. Without collision: The access control is designed to remove the interference by guaranteeing that each MU solely uses the channel during its data transmission.
2. With collision: Note that the interference could be mitigated in different levels. If the receiver could handle interferences among MUs by adopting some advanced interference management techniques [2], the access control can allow multiple MUs to access one channel at the same time. In this case, the single-channel scenario degrades to the conventional multiple access channel.

3.1 Without Collision

In this subsection, the access policy is required to manage the data transmissions of MUs one by one over time such that no transmission collisions happen. Hence, TDMA is one feasible scheduling scheme, where in each time slot the channel can only be assigned to one MU. CDMA with orthogonal codes could also support multiple MUs.

3.1.1 Centralized Access and Power Control

When TDMA can be implemented in a centralized manner, it is optimal for the controller to pick the MU with the highest effective transmission rate according to its CSI and ESI in each time slot. That is, the selected MU should jointly have a good channel condition and sufficient energy in the battery.

The authors in [16] adopted the above idea and studied the case when all MUs have the same mean EH rate. The selection process in their proposed policy depends only on the channel condition. Specifically, the region of the channel power gain is divided into multiple interval units, based on some discrete-rate transmission schemes (e.g., M-QAM). Each channel state has an one-to-one mapping to a transmission rate, which determines the required transmission power. In order to meet the constraint of the average power consumption (i.e., the mean of EH rates), the length of interval unit must be designed accordingly. Then, at each time slot, the controller picks the MU with the highest ratio between the channel power gain and the length of the interval unit. The intuition is that the one with a higher ratio should attain a higher rate level. Note that if the channel statistics are identical, the length of the unit interval of the channel state will be the same, which is equivalent to pick the MU with the highest channel power gain. Due to the EH constraint, the selected MU may fail to transmit at the desired rate. In this case, the selected MU can just keep silent. It is shown in [16] that by using such a policy, the average uplink throughput can achieve the similar performance as the system with conventional power supplies. In other words, the effect of EH constraints can approximately vanish.

The above centralized policy requires the knowledge of instantaneous CSI of all MUs, which may induce difficulties in implementations. Without the need of CSI, another TDMA-based transmission policy was investigated in [22], which lets the MUs access the channel and transmit one by one in a fixed order. Such a fixed TDMA scheme enjoys a lower computation complexity at the central controller compared to some aggressive selection algorithms that require the instantaneous status from each MU, like the one introduced in the previous paragraph. One transmission period is the total number of time slots, during which all MUs transmit for one time slot. As such, the total number of time slots for one transmission period equals N since each MU only uses the channel once in each period. The channel access indicators $\{s_{n,1}(t)\}$ in our framework are preassigned by the controller. Suppose that the nth MU is assigned in the tth slot, and let $d_n(t)$ be the duration of the time slot t. Then, the total time of one transmission period is $\sum_{n=1}^{N}\sum_{t=1}^{N} d_n(t)$.

The controller manages the transmission time for each MU in order to maximize the expected sum rate under an additional constraint that the average transmission period cannot exceed a fixed number. The authors in [22] showed that the optimal transmission policy is the equal-power scheme, i.e., each MU should set its transmission power at the same level which is dependent on the energy information of all MUs. The optimal transmission time, on the other hand, is proportional to the mean EH rate such that a higher EH rate implies a longer transmission time. A simpler policy, equal-time scheme, was also studied in [22]. Unlike the equal-power scheme, the equal-time scheme simply divides one transmission period equally and gives all MUs the same transmission time. The power allocation uses the greedy algorithm. Comparing these two policies, the equal-power scheme outperforms the equal-time one when the battery has an infinite capacity. Whereas, they are similar when the battery has a finite capacity.

3.1.2 Distributed Access and Power Control

Centralized algorithms usually lead to a better throughput performance, while extensive computation complexity can be incurred at the central controller, especially in a large-scale system case. To relax the complexity, or even remove the necessity of the central controller, the investigation of distributed algorithms has drawn great attentions over the past years.

Commonly used distributed random access schemes, such as ALOHA and CSMA, are all applicable for problem (10). In this case, access variables $\{s_n(t)\}$

are determined locally[3] according to the chosen schemes. For example, if all N MUs equally contend for the channel use with probability $1/N$ at each time slot, we obtain

$$s_n(t) = \begin{cases} 1, \text{ with probability (w.p.) } \frac{1}{N}, \\ 0, \text{ w.p. } 1 - \frac{1}{N}. \end{cases} \quad (11)$$

In this example, a successful channel contention happens when only one MU accesses the channel, i.e., $s_n(t) = 1$ and $s_i(t) = 0$ for $i \neq n$, where $1 \leq n, i \leq N$.

For the power control, an intuitive and feasible scheme is the best-effort delivery method, which is basically the same as the greedy algorithm such that whenever an MU occupies the channel, it will use up all available energy for data transmission without considering the channel condition and the battery status. This scheme is easy to implement, while the throughput performance may not be optimal.

An improved best-effort method, named distributed opportunistic framework, was designed in [21], which maximizes the long term average throughput of the network by jointly taking the advantages of the causal CSI and ESI. The proposed opportunistic framework operates in two stages: *channel probing* (CP) and *energy probing* (EP). In the first stage, all MUs keep contending for the channel use independently until there is only one occupying the channel. The successful MU obtains the CSI through some handshaking mechanism with the BS. This stage is called CP. After CP, the successful MU can possibly transmit over multiple time slots and then release the channel. The total number of time slots over which MU could transmit is called a transmission block. Moreover, the MU can chooses one of the following actions based on its local CSI and ESI right after CP:

(a) releases the channel (if CSI and ESI indicate that the transmission rate is lower than a given threshold) and let all links re-contend; or
(b) directly transmits until the end of the transmission block; or
(c) holds the channel, and starts the second stage EP.

It is worth noting that each MU keeps harvesting energy until it starts a transmission, and after each round of CP, only the successful MU makes a choice among three actions as listed above. When the successful MU decides not to take action (a) or (b) defined above, it starts the second stage EP, i.e., action (c), to obtain more energy. During this stage, the MU chooses to continue harvesting energy slot by slot, and then ends EP by action (a) or (b), i.e., either releasing the channel or transmitting over the rest of the transmission block.

Our proposed framework described in Sect. 2.2 can be used to optimize the two-stage probing strategy. Besides the constraints described in (10), we need to further specify $\{s_n(t)\}$ in the two-stage probing scenario. Note that $\{s_n(t)\}_{1 \leq n \leq N}$ are all random during the period of channel contentions. If some MU, say the nth MU, occupied the channel and transmitted over multiple time slots, $\{s_n(t)\}_{1 \leq n \leq N}$

[3]Here, we remove the channel index m since we only consider the single-channel scenario.

would become determined during the period of the transmission, i.e., there must be $s_n(t) = 1$ and $s_i(t) = 0$ for $i \neq n$ during this time. Here, we let Q denote the number of time slots of the transmission block.

The transmission power is determined by the remaining time within a transmission block (since the MU may perform EP) and the energy level when the MU starts a transmission. Here, we adopt a unform power allocation scheme. For example, if the number of remaining time slots was $\widehat{Q}$ and the transmission started at t_0, the transmission power would be $P_n(t) = \frac{B_n(t_0)}{\widehat{Q}}$ and keep the same during $\widehat{Q}$ slots.

We let L_C and L_E be the stopping rules of CP and EP, respectively. Stopping rule L_C indicates the change point of $\{s_n(t)\}$ becoming deterministic, and stopping rule L_E regulates the transmission power within the transmission block. Since each CP may take a random number of slots, we let K_i be the number of slots for the ith CP. Then, the total number of slots to fulfil one transmission is given by

$$S(L_C, L_E) = \sum_{i=1}^{L_C-1} L_{E,i} + \sum_{i=1}^{L_C} K_i + Q. \tag{12}$$

If we keep using the stopping rules L_C and L_E for J times, the time horizon is given as

$$T = \sum_{j=1}^{J} S_j(L_C, L_E). \tag{13}$$

The infinite time horizon can be realized by letting $J \to \infty$, which leads to $T \to \infty$ a.s. In Fig. 3, we illustrate one realization of a successful transmission, which costs L_C rounds of CP, and also L_E slots of EP in the last CP.

Then, we can find the optimal stopping rules L_C^* and L_E^* by plugging (13) into the framework given in (10). More precisely, condition (7) becomes

$$\mathbf{E}\left[\sum_{j=1}^{J}\sum_{t=1}^{S_j(L_C,L_E)} \log\left(1 + s_n(t)g_n(t)P_n(t)\right) - \lambda \sum_{j=1}^{J} S_j(L_C, L_E)\right] = 0. \tag{14}$$

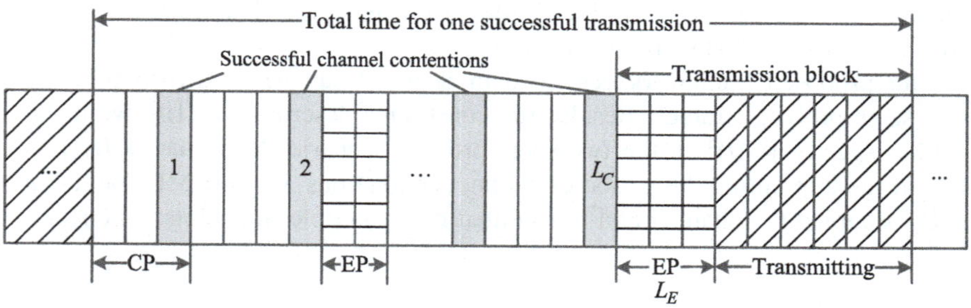

Fig. 3 One realization of a successful transmission by applying the distributed opportunistic framework [21]

By letting $J \to \infty$, we obtain

$$\mathbf{E}\left[\sum_{t=1}^{S(L_C,L_E)} \log\left(1 + s_n(t)g_n(t)P_n(t)\right) - \lambda S(L_C, L_E)\right] = 0, \tag{15}$$

which is equivalent to

$$\lambda = \frac{\mathbf{E}\left[\sum_{t=1}^{S(L_C,L_E)} \log\left(1 + s_n(t)g_n(t)P_n(t)\right)\right]}{\mathbf{E}\left[S(L_C, L_E)\right]}. \tag{16}$$

Here, λ can also be viewed as the "rate-of-return" in the sense that Eq (16) describes the average rate per transmission.

According to the optimal stopping rules L_C^* and L_E^* derived in [21], the successful MU will transmit only when the CSI and ESI jointly indicate a higher transmission rate than a given threshold, which can be found via a simple one dimension searching algorithm. However, such an opportunistic framework does not guarantee the fairness among MUs. When applying this scheme, the transmission will be more likely to be fulfilled by those MUs with the better channel conditions and the higher EH rates.

Setting the performance of the best-effort delivery as the benchmark. In Fig. 4, it is shown how the throughput gains change as the number of MUs increases under the condition that the battery has a finite capacity. It is observed that the gain from EP decreases when the number of MUs increases. This is due to the fact that the more

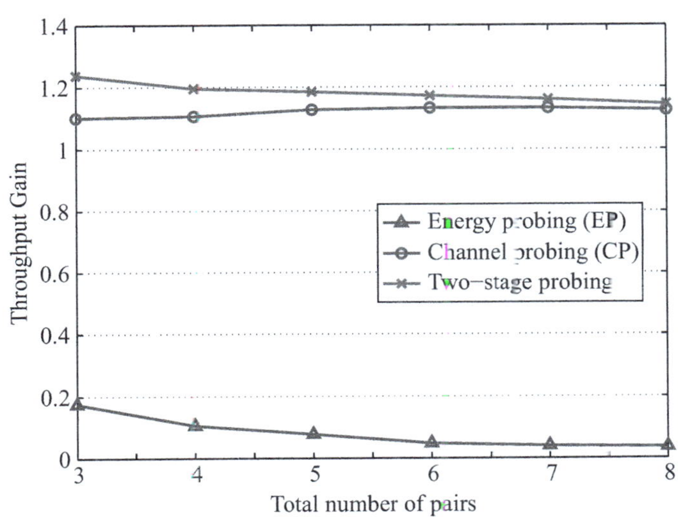

Fig. 4 The throughput gain over the best-effort delivery method

MUs are present, the higher probability that CP will last a longer time. Therefore, more MUs would maintain a high energy level and EP is rarely triggered after CP. It is also observed that the throughput gain from CP increases with the increase in the number of MUs since CP could utilize the multi-user diversity of both channel gains and energy levels. The CP can offer a much higher gain than EP, and thus the gain from the two-stage probing is comparable to CP as the size of the network increases.

3.2 With Collision

In this case, multiple MUs can transmit over the same channel simultaneously, which is just the *multiple access channel* under EH constraints [17, 37, 39]. Existing literatures have studied the sum rate maximization problem in some "ideal" scenarios. The algorithms introduced next are all centralized in the sense that controller needs to regulate the sum transmission power in order to achieve the optimality.

A typical case is Gaussian multiple access scenario with two MUs and the ESI is non-causally known [39]. One key observation is that if the goal is to maximize the sum rate only, the optimal sum power has a stair-case structure, which is the same as the single MU case (see Sect. III in [34]). Furthermore, if the fairness of MUs is accounted, the sequence of the optimal sum power allocation needs to be split into two subsequences in order to achieve the largest throughput region of two MUs. The optimal power allocation for each MU can be computed by applying the generalized iterative water-filling algorithm [39]. For the fading channel with non-causal CSI, authors in [37] proposed an efficient power allocation scheme by considering a finite battery capacity and an upper limit on the instant transmission power.

When ESI is causally known at MUs, an online algorithm is proposed for Gaussian multiple access in [17] with $T \to \infty$. One important observation is that the optimal power allocation should make the energy level of the battery stationary for a given EH rate process. Then, the optimal power allocation can be expressed by the stationary distribution of the battery and the distribution of the EH rate. For the infinite battery capacity case, the solution is the "constant" power strategy. Specifically, the transmission power level is equal to the mean of the EH rates. If the current energy is not sufficient enough, the MU just uses up the available energy. This is similar to the algorithm mentioned in [29] for point-to-point channels. For the finite battery capacity, it is shown that the optimal power allocation does not have a close-form expression and it can be solved numerically.

4 Multi-Channel Scenario

In this section, we consider the scenario when there are multiple channels for MUs to access, i.e., $M \geq 2$. Moreover, there is an additional condition in this case: whether or not the MU can transmit over more than one channels simultaneously.

- Transmit over one channel, i.e., $A_n = 1$ in constraint (4): This condition implies that although there are more than one channels to access, the MU is restricted to use at most one channel at each time. Such a restriction could equalize the opportunity of channel usages for different MUs to some extent, and also simplifies the design of access algorithms.
- Transmit over multiple channels, i.e., $A_n \geq 2$: Under this condition, the MU should first decide or be assigned with the available channels, and then apply the water-filling algorithm for the power allocation according to the current CSI and ESI. Obviously, the MU throughput could be potentially enhanced compared to the case with $A_n = 1$; but the hardware complexity and the total power consumption would be increased as well.

Next, we will introduce the transmission polices for problem (10) for two special scenarios according to the number of MUs: $N \geq 2$ and $N = 1$. In each subsection, we will further consider the realization of problem (10) for cases with $A_n = 1$ and $A_n \geq 2$.

4.1 Network Throughput Maximization with $N \geq 2$

In this subsection, we mainly introduce the centralized transmission schemes. We also modify constraint (3) by letting $N_m = 1$ for all $m = 1, 2, \ldots, M$, i.e., each channel is only allowed to be accessed by at most one MU in each time slot. With this modified constraint, the designed access policy can avoid the collisions of data transmissions and improve the system reliability. Similar to the single-channel scenario, TDMA is one feasible scheduling scheme. The main difference is that for each time slot, a group of MUs need to be scheduled to multiple channels. Obviously, when the number of MUs exceeds the number of available channels, some MUs cannot be scheduled for a given time slot.

First, we consider the case $A_n = 1$ for all $1 \leq n \leq N$, i.e., the MU can only transmit over one channel. The average throughput of the network can be maximized by wisely choosing the set of scheduled MUs and controlling the transmission power according to the CSI and ESI of all MUs. However, the controller can hardly obtain the information of all MUs. To facilitate this, all MUs need to send their own ESI to the BSs, and also the BSs need to feedback the CSI to each MU. This process will introduce extra scheduling issues and may cost a long period when the number of MUs is large. As such, a reasonable condition is that the CSI and ESI of all MUs is causally and partially known by the controller. That is, the controller only obtains the new CSI and ESI of the MUs who are scheduled in the current slot. Then, the controller can use all available CSI and ESI (may include the history) to optimize the current uplink throughput or to predict the future status of MUs.

Under this condition, the centralized schemes for problem (10) can be derived by applying some dynamic frameworks (e.g., MDP, PMDP, or RMAB). Here, we only introduce two centralized schemes: uniformizing random ordered policy (UROP) and myopic policy. Note that they may not be throughput-optimal since their throughput formulations are not exactly the same as (6). However, both are feasible and do not involve extensive computations.

- *UROP*: The authors in [9] studied the access policy for the packet level under the assumption of a constant energy consumption for delivering a packet. Since the battery status of each MU is random and is unknown at the receiver side, the problem here is how to schedule the accesses of MUs such that the expected average throughput is maximized. The solution proposed in [9] is a UROP: briefly, one round of scheduling starts with randomly ordering all N MUs. Then, the controller picks the first M MUs in the order list and let them access M parallel channels, where each user can only access one channel for each time slot. In the next time slot, the MUs who did not transmit in the previous time slot will be replaced by those who have not been scheduled. The controller selects the unscheduled MUs according the order list. After such a replacement, the newly scheduled M MUs will transmit for one time slot. This procedure will keep running until all MUs transmit at least once. Then, the controller will start a new round of scheduling by creating a new order list. It is shown in [9] that from the network point of view, when the average packet rate is smaller than the average delivering rate (i.e., the total number of channel usages divided by N), UROP leads to minimum waste energy.
- *Myopic Policy*: Given that the current states of MUs are unknown at the controller side in each time slot, the authors in [3] proposed an "myopic" policy that only maximizes the expected instantaneous sum rate. The power allocation in this policy follows the greedy algorithm that uses all available energy for data transmission. The scheduling in this policy is similar to UROP. The difference is that the order list is not created randomly. Instead, the controller orders MUs according to the length of the idle time after the previous transmission of each MU. Clearly, the longer the idle time, the more energy that can be possibly harvested. In each time slot, the first M MUs in the order list can access channels. If the scheduled MU transmits, it will be added to the end of the order list in the next time slot since it uses up all energy. The MUs who have not transmitted (either scheduled or not) will go up in the order list. This myopic policy is proved in [3] to be throughput-optimal over the considered time horizon.

Next, we consider $A_n \geq 2$. In this case, the centralized scheduling is a mixed integer programming problem, which is non-convex and hard to solve. However, if we consider the centralized instantaneous scheduling (i.e., optimizing the throughput only in a given time slot), maximizing (6) is similar to the throughput maximization problem for a centralized uplink OFDM wireless system, which has already been extensively investigated [18, 38]. It is worth noting that in OFDM systems, another key problem is the frequency bandwidth allocation, while in our considered system model, the frequency band is fixed and we only focus on the problem of channel assignment and power allocation Thus, those algorithms concerning channel assignments and power allocations proposed for OFDM systems can be directly modified and applied in maximizing the throughput in our considered system.

The transmission polices designed for the case of $N_m = 1$ can be extended to the general case of $N_m \geq 2$ by reassigning the channels to the unscheduled MUs. Then, there are more MUs scheduled in each time slot. However, the general case also means that the access control does not avoid the interference among different MUs, which exerts more difficulties on the design of the optimal power control.

4.2 *User Throughput Maximization with* $N = 1$

In this subsection, we focus on the scenario when the channel resources are relatively sufficient, e.g., late at night in the business area or working hours in the resident area. Such a scenario can be modeled as the single-MU scenario $N = 1$ with multiple channels. For the case $A_n = 1$, an intuitive access strategy for the MU is to pick the channel with the highest channel power gain among all the available channels. To obtain such information, the MU needs to probe channels with the cooperation of the BS. The access schemes implemented at MUs have been studied [15, 31], especially in cognitive radio networks. In general, the MU can perform one of the three main access mechanisms:

1. Parallel channel probing: Parallel probing can be realized by broadcasting a *query* message and receiving *reply* messages that contains the CSI of the probed channels. Then, the MU can access any one and start the transmission.
2. Sequential channel probing with recall: The MU probes channels one by one according to a certain order. When the MU stops probing, it can access any channel that has been probed.
3. Sequential channel probing without recall: The MU probes channels one by one according to a certain order. When the MU stops probing, it can only access the last probed channel.

Although the first two mechanisms may provide more options when the MU chooses which channel to access, both have apparent drawbacks that prevent from being implemented. Parallel channel probing requires a compatible hardware of broadcasting and also causes relatively high instantaneous power consumption. Sequential channel probing with recall needs the MU to memorize the state of the probed channels. However, the CSI may change when the MU accesses the channel once again.

On the contrary, sequential channel probing without recall does not carry the above shortcomings and enjoys much more attentions than the other two. Such a probing mechanism has been extensively studied for cognitive radio networks [6, 31, 41]. However, for the EH-based heterogeneous access problem, the specific transmission polices accounting the energy information have not been well studied. As we discussed in Sect. 2.2, two dynamic frameworks, POMDP and optimal stopping, are ideal tools to formulate the sequential decision making process and can be used to derive the optimal transmission policies.

For the case $A_n \geq 2$, we apply a save-then-transmit protocol as the basic structure of the transmission policy. As its name implies, this protocol lets the MU accumulate energy in the first step, and then lets the MU use all available energy for data transmission. Such a protocol enjoys two major characteristics:

1. It captures the fundamental operation of EH-based systems that the energy must be accumulated in the first place, even in an energy full-duplex fashion. However, the time required for harvesting energy will induce some penalties.
2. It is simple, tractable, and applicable for any EH-based systems.

It is easy to see that the main design issue of the save-then-transmit protocol is to determine when to start data transmission such that the average throughput is maximized.[4] In the following, we first investigate the case when this protocol is used over a finite time horizon; and then, we extend the protocol to the infinite time horizon case.

4.2.1 Finite Time Horizon

The save-then-transmit protocol applied over a finite time horizon is depicted in Fig. 5, where the time horizon is normalized as one time slot. The MU spends a certain portion of time on harvesting energy, and then starts channel probing for CSI. Such a probing process is similar to sensing with recall [31]. After channel probing, the MU transmits over the selected channel(s).

Clearly, spending time on harvesting energy and channel probing will increase the instantaneous rate, while the effective transmission time will be decreased. To deal with this tradeoff, two main design variables need to be optimized for this protocol: the saving ratio (indicating the portion within one time slot for harvesting energy) and the number of channels to probe. Note that the first design variable is a continuous number, while the second one is an integer. Thus, maximizing the average throughput is also a mixed integer programming problem. An algorithm proposed in [40] can be used to solve this problem, which randomly selects many pairs of feasible solutions and chooses the best one.

[4] Note that this protocol can also be optimized to minimize the outage probability for a given time slot [23].

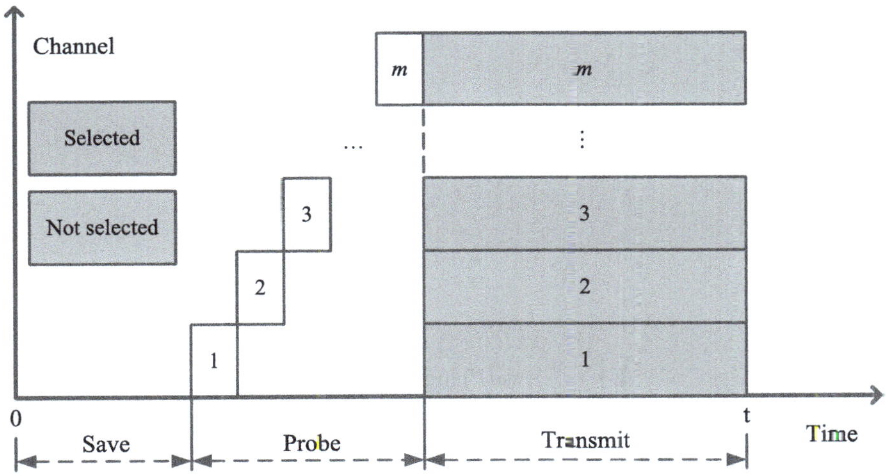

Fig. 5 A realization of the save-then-transmit protocol within one time slot [40]

4.2.2 Infinite Time Horizon

Next, we extend the protocol in Fig. 5 to the infinite time horizon case (e.g., multiple time slots). In this case, the MU can save the energy for more than one time slot and then start the transmission. Besides the two design variables (saving ratio and the number of channels to probe) mentioned above, another important variable is the stopping rule: when to stop saving and to start the transmission. The stopping rule determines the action after the MU probes channels. The MU has the choice to transmit or skip the transmission in the current time slot according to its current CSI and ESI. The skipping is preferred when the CSI and ESI jointly indicate a low transmission rate.

The stopping rule can be reflected in the indicators $\{s_n(t)\}$, where the index n for MUs is omitted since $N = 1$. Particularly, if we let L be the stopping rule such that the MU transmits only at the time slot L, there must be $\{s_m(t) = 0\}_{1 \leq m \leq M}$ for $1 \leq t < L$. At time slot L, $\{s_m(t)\}_{1 \leq m \leq M}$ are determined by the channel selection scheme. Thus, by applying the stopping rule L, the total transmission power $P(L)$ satisfies

$$0 \leq P(L) \leq B(L) = \min\left\{\sum_{t=1}^{L-1} E(t), B_{max}\right\}, \tag{17}$$

If this stopping rule is used for J times, condition (7) can be written as

$$\mathbf{E}\left[\sum_{j=1}^{J}\sum_{t=1}^{L_j}\sum_{m=1}^{M} \log\left(1 + s_m(t)g_m(t)P_m(t)\right) - \lambda \sum_{j=1}^{J} L_j\right] = 0, \tag{18}$$

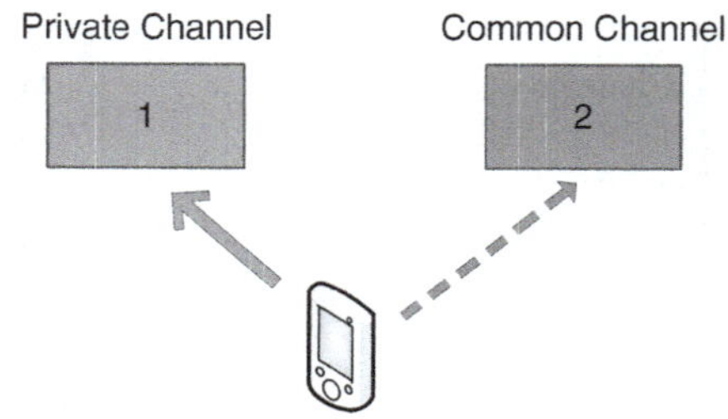

Fig. 6 The hybrid access model

which leads to

$$\lambda = \frac{\mathbf{E}\left[\sum_{m=1}^{M} s_m(L) \log\left(1 + g_m(L)P_m(L)\right)\right]}{\mathbf{E}[L]}, \tag{19}$$

by letting $J \to \infty$, where $\sum_{m=1}^{M} s_m(L)P_m(L) = P(L)$. Then, we can use framework (10) to find the optimal stopping rule. Such a save-then-transmit protocol with optimal stopping rule can be named as a opportunistic transmission policy since it explores the state information over both time and space.

It is worth noticing that the term "opportunistic spectrum access" [41] in cognitive radio networks is quite different from the opportunistic transmission discussed above. The opportunistic spectrum access describes the spectrum access of the secondary users in a hierarchy network model. In such a network, the secondary user detects the availability of the licensed channel and then decides whether transmit or not. The term "opportunistic" is used in the sense that the secondary user makes a decision that can only achieve certain optimality in probability. Although the term "opportunistic" may represent different philosophies, some techniques proposed for spectrum access in EH-based cognitive radio networks [26] are also applicable in EH-based heterogeneous access.

To better illustrate the save-then-transmit protocol with optimal stopping rule, we set the saving ratio as zero and assume that the time for the MU to learn the CSI is negligible compared to the time slot length. That is, we only focus on the design of the stopping rule L. Particularly, we consider an interesting case of a hybrid access scenario as shown in Fig. 6, where the MU has both deterministic access via one private channel, and dynamic access with certain probability via one common channel. Such a common channel usually can be offered by a local BS, e.g., a WiFi router. The MU can only access the common channel with certain probability given that the common channel may be close temporarily. In this hybrid access scenario, the total number of parallel channels is $M = 2$. Following the notations in Sect. 2, we let $\{g_1(t)\}_{t\geq 1}$ and $\{g_2(t)\}_{t\geq 1}$ be the sequence of channel power gains for the private channel and the common channel, respectively. Then, the instantaneous rate is written as

$$R(t) = \sum_{m=1}^{2} \log\left(1 + s_m(t) g_m(t) P_m(t)\right).$$

We assume that $\{g_2(t)\}_{t\geq 1}$ are independent and identically distributed (i.i.d.) over time, given that the distance from the MU to the local BS is short and thus the common channel often experiences fast channel fading. Note that the private channel can always be accessed, i.e., we have $s_1(t) = 1$ for all t. The common channel can only be accessed randomly, and the MU can transmit via the common channel only when the MU secures it. Let p_s denote the probability that the MU can secure the common channel, termed *securing probability*. Note that the securing probability may not necessarily equal the access probability, but their relations can be derived according to the system setups. Here, for notation simplicity, we assume that whenever the MU accesses the common channel, it can also secure the channel. Then, $s_2(t)$ is a random variable such that

$$s_2(t) = \begin{cases} 1, \text{ w.p. } p_s; \\ 0, \text{ w.p. } 1 - p_s. \end{cases}$$

If we let L be the stopping rule, λ in (19) can be rewritten as

$$\lambda = \frac{\mathbf{E}\left[R(L)\right]}{\mathbf{E}[L]}. \tag{20}$$

Thus, the design of the save-then-transmit protocol is now to find the optimal stopping rule L^* that maximizes (20).

When $\{g_1(t), E(t)\}_{t\geq 1}$ are modeled as Markovian processes, the optimal stopping rule L^* was proved to exist in [20] and have a state-dependent threshold-based structure under both finite and infinite battery capacity assumptions. The optimal throughput is proved to be strictly increasing over the securing probability.

When $\{g_1(t), E(t)\}_{t\geq 1}$ are modeled as i.i.d. processes, respectively, the corresponding optimal stopping rule is proved to be a pure-threshold policy, i.e., the threshold does not change over time, which could be found via one-dimension search. We plot the variation of the average throughput as the threshold changes in Fig. 7, and observe that the average throughput could be optimized by adjusting the threshold.

In Fig. 8, we compare the throughput performance attained by the opportunistic transmission policy (with the optimal stopping rule L^*) and the best-effort delivery method. Note that the optimal power allocation with traditional power supply was investigated and shown to outperform the opportunistic policy [20]. Such observation may be predictable given that the EH constraint sets a hard cap on the instant transmission power, while the conventional average power constraint does not.

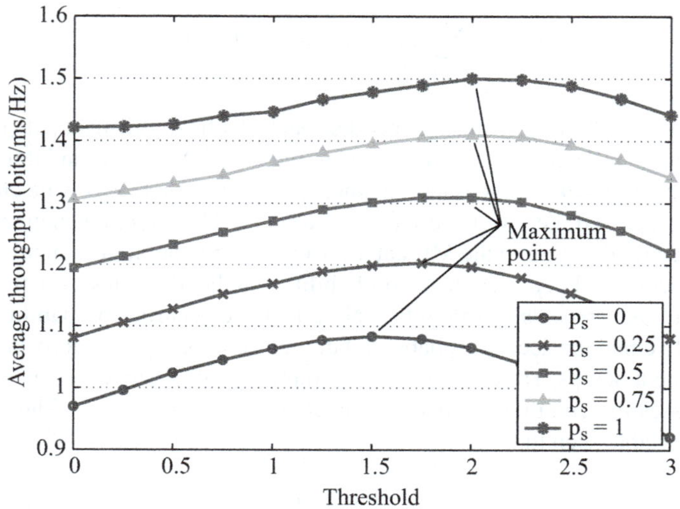

Fig. 7 The average throughput over the threshold

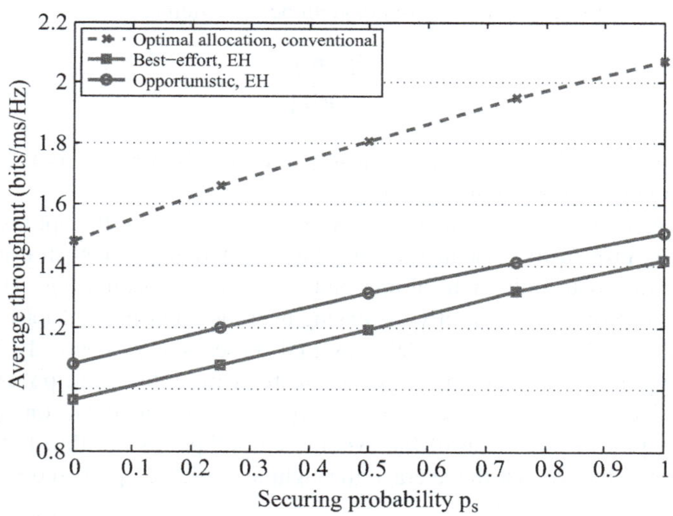

Fig. 8 The throughput performance over different transmission policies

So far, we have discussed the multiple-channel scenario for both $N \geq 2$ and $N = 1$ cases. Again, we want to point out that accessing multiple channels at the same time may be too "luxury" for MUs. Some major issues include:

- Complexity at MU: The hardware at the MU should support multi-channel access, e.g., adopting multiple RF front ends. Meanwhile, power allocations over

multiple parallel channels may result in a higher power consumption on circuit overhead, which exerts a heavy burden on EH power supply.
- Interference: When the MU transmits via multiple channels, it means that other MUs cannot use these channels or may cause interference if they access. From the network operator point of view, allowing one MU to access multiple channels at the same time may decrease the whole benefits of the system.
- Needs of BS coordinations: The received signals from the same MU over multiple channels should be jointly processed. This requires the C-RAN to perfectly coordinate multiple BSs in a real-time fashion.

5 Summary and Discussion

EH-based power supply can drive wireless systems in a green fashion, while it also introduces both EH constraints and energy unreliability. These drawbacks should be mitigated as much as possible in different layers of the system design. In this chapter, we considered the EH-based heterogeneous wireless access, and proposed a throughput maximization framework across both physical and medium access layers. Transmission policies, including access and transmission power controls, were discussed in single and multi-channel scenarios.

EH-based heterogeneous wireless access is still under development, and many issues remain to be resolved. As the end of this chapter, some potential directions for future works are presented below.

Learning Approach Most of current studies adopt the dynamic programming technique to find the optimal transmission policies under the assumptions that the rewards and state transition probabilities of the system are known. However, given that the number of MUs keeps increasing, dynamic programming based solutions are no longer efficient when the state space of CSI/ESI of MUs is large and the statistics of CSI/ESI may not be fully accessible. Thus, it is worth to investigate learning based approaches, e.g., reinforcement learning, which are more feasible in the large-scale scenario since the learning approach does not require the statistics of the system and is extendable to deal with the large environments (see Chap. 3 in [5]).

QoS vs. QoE In general, the throughput-optimal transmission policies may not satisfy the quality of service (QoS) and the quality of experience (QoE) requirement at the same time. Sometimes there may exist a tradeoff between QoS and QoE. For example, too high QoS may decrease QoE since high QoS would consume more resources on both the user (e.g., energy stored in the battery) and operator (e.g., network resources, BS power) sides [1].

Effect of EH-Based BSs In this chapter, we only consider the heterogeneous access with EH-based MUs. Furthermore, BSs can also be driven or partially driven by EH-based power supplies to lower the energy cost on fuels and carbon-oxygen emission. The employment of EH at BSs has drawn a great attention in recent years [10, 14, 42]. Particularly, the availability of EH-based BSs [8] has been analyzed. By accounting EH constraints at both MU and BS sides, the design of the channel access and power control becomes much more challenging.

MU-to-MU Communications The proposed heterogeneous wireless access model in this chapter captures the characteristic of the "MU-to-BS" communications, which belong to the medium-to-long range communications. As the short range communication being developed rapidly in recent years, 5G systems are also expected to incorporate device-to-device (D2D) communications where more and more EH-based devices could be involved. Thus, the wireless access control will mix multiple layers, and more tractable models need to be developed.

References

1. P.K. Agyapong, M. Iwamura, D. Staehle et al., Design considerations for a 5G network architecture. IEEE. Commun. Mag. **52**(11), 65–75 (2014)
2. J.G. Andrews, S. Buzzi, W. Choi et al., What will 5G be? IEEE J. Sel. Areas Commun. **32**(6), 1065–1082 (2014)
3. P. Blasco, D. Gündüz, Multi-access communications with energy harvesting: a multi-armed bandit model and the optimality of the myopic policy. IEEE J. Sel. Areas Commun. **33**(3), 585–597 (2015)
4. P. Blasco, D. Gündüz, M. Dohler, A learning theoretic approach to energy harvesting communication system optimization. IEEE Trans. Wirel. Commun. **12**(4), 1872–1882 (2013)
5. L. Busoniu, R. Babuska, B. De Schutter et al., *Reinforcement Learning and Dynamic Programming Using Function Approximators*, vol. 39 (CRC press, Boca Raton, 2010)
6. Y. Chen, Q. Zhao, A. Swami, Distributed spectrum sensing and access in cognitive radio networks with energy constraint. IEEE Trans. Signal Process. **57**(2), 783–797 (2009)
7. China mobile, C-RAN the road towards green RAN. White paper, ver. 2.5 (2011)
8. H.S. Dhillon, Y. Li, P. Nuggehalli et al., Fundamentals of heterogeneous cellular networks with energy harvesting. IEEE Trans. Wirel. Commun. **13**(5), 2782–2797 (2014)
9. O.M. Gul, E. Uysal-Biyikoglu, A randomized scheduling algorithm for energy harvesting wireless sensor networks achieving nearly 100% throughput, in *IEEE WCNC 14: Topics in Mobile and Wireless Networks. IEEE Wireless Communications and Networking Conference*, Istanbul (2014)
10. H.A.H. Hassan, L. Nuaymi, A. Pelov, Renewable energy in cellular networks: a survey, in *IEEE OnlineGreenComm 2013: Topics in Green Wireless Networks. IEEE Online Conference on Green Communications* (2013)
11. A. He, A. Amanna, T. Tsou et al., Green communications: a call for power efficient wireless systems. J. Commun. **6**(4), 340–351 (2011)
12. X. Huang, T. Han, N. Ansari, On green energy powered cognitive radio networks. IEEE Commun. Surv. Tutorials **17**(2), 827–842 (2015)
13. International Telecommunication Union, The key 2005–2015 ICT data for the world. http://www.itu.int/en/ITU-D/Statistics/Pages/stat/default.aspx. Accessed 26 May 2015
14. M. Ismail, W.H. Zhuang, E. Serpedin et al., A survey on green mobile networking: from the perspectives of network operators and mobile users. IEEE Commun. Surv. Tutorials. **17**(3), 1535–1556 (2015)
15. Z. Ji, Y. Yang, J. Zhou et al., Exploiting medium access diversity in rate adaptive wireless LANs, in *MobiCom 04. The 10th Annual International Conference on Mobile Computing and Networking*, Philadelphia (2004)

16. P.S. Khairnar, N.B. Mehta, Discrete-rate adaptation and selection in energy harvesting wireless systems. IEEE Trans. Wirel. Commun. **14**(1), 219–229 (2015)
17. M.B. Khuzani, P. Mitran, On online energy harvesting in multiple access communication systems. IEEE Trans. Inf. Theory **60**(3), 1883–1898 (2014)
18. K. Kim, Y. Han, S.L. Kim, Joint subcarrier and power allocation in uplink OFDMA systems. IEEE Commun. Lett. **9**(6), 526–528 (2005)
19. M.L. Ku, W. Li, Y. Chen et al., Advances in energy harvesting communications: past, present, and future challenges. IEEE Commun. Surv. Tutorials. **18**(2), 1384–1412 (2015)
20. H. Li, C. Huang, F.E. Alsaadi et al., Opportunistic multi-channel access in heterogeneous 5G network with renewable energy supplies (2015). arXiv:1412.4458. Accessed 21 May 2015
21. H. Li, C. Huang, P. Zhang et al., Distributed opportunistic scheduling for energy harvesting based wireless networks: a two-stage probing approach. IEEE/ACM Trans. Netw. **24**(3), 1618–1631 (2015)
22. J. Liu, H. Dai, W. Chen, On throughput maximization of time division multiple access with energy harvesting users. IEEE Trans. Veh. Technol. **65**(4), 2457–2470 (2015)
23. S. Luo, R. Zhang, T.J. Lim, Optimal save-then-transmit protocol for energy harvesting wireless transmitters. IEEE Trans. Wirel. Commun. **12**(3), 1196–1207 (2013)
24. N. Michelusi, M. Zorzi, Optimal adaptive random multiaccess in energy harvesting wireless sensor networks. IEEE Trans. Commun. **63**(4), 1355–1372 (2015)
25. R. Moghe, Y. Yang, F. Lambert et al., A scoping study of electric and magnetic field energy harvesting for wireless sensor networks in power system applications, in *ECCE. IEEE Energy Conversion Congress and Exposition*, San Jose (2009)
26. S. Park, D. Hong, Achievable throughput of energy harvesting cognitive radio networks. IEEE Trans. Wirel. Commun. **13**(2), 1010–1022 (2014)
27. A. Pasternack, The environmental costs (and benefits) of our cell phones (2009). http://www.treehugger.com/clean-technology/the-environmental-costs-and-benefits-of-our-cell-phones.html. Accessed 13 May 2015
28. K.J. Prabuchandran, S.K. Meena, S. Bhatnagar, Q-learning based energy management policies for a single sensor node with finite buffer. IEEE Wirel. Commun. Lett. **2**(1), 82–85 (2013)
29. R. Rajesh, V. Sharma, P. Viswanath, Capacity of Gaussian channels with energy harvesting and processing cost. IEEE Trans. Inf. Theory **60**(5), 2563–2575 (2014)
30. V. Sharma, U. Mukherji, V. Joseph, Optimal energy management policies for energy harvesting sensor nodes. IEEE Trans. Wirel. Commun. **9**(4), 1326–1336 (2010)
31. T. Shu, M. Krunz, Sequential opportunistic spectrum access with imperfect channel sensing. Ad Hoc Netw. **11**(3), 778–797 (2012)
32. S. Sudevalayam, P. Kulkarni, Energy harvesting sensor nodes: survey and implications. IEEE Commun. Surv. Tutorials **13**(3), 443–461 (2011)
33. D.N.C. Tse, S. Hanly, Multiple-access fading channels: Part I: polymatroid structure, optimal resource allocation, and throughput capacities. IEEE Trans. Inf. Theory **44**(7), 2796–2815 (1998)
34. S. Ulukus, A. Yener, E. Erkip et al., Energy harvesting wireless communications: a review of recent advances. IEEE J. Sel. Areas Commun. **33**(3), 360–381 (2015)
35. U.S. Energy Information Administration, How much coal, natural gas, or petroleum is used to generate a kilowatthour of electricity? (2015). http://www.eia.gov/tools/faqs/. Accessed 12 June 2015
36. L. Wang, F. Tian, T. Svensson et al., Exploiting full duplex for device-to-device communications in heterogeneous networks. IEEE Commun. Mag. **53**(5), 146–152 (2015)
37. Z. Wang, V. Aggarwal, X. Wang, Iterative dynamic water-filling for fading multiple-access channels with energy harvesting. IEEE J. Sel. Areas Commun. **33**(3), 382–395 (2015)
38. E. Yaacoub, Z. Dawy, A survey on uplink resource allocation in OFDMA wireless networks. IEEE Commun. Surv. Tutorials **14**(2), 322–337 (2012)
39. J. Yang, S. Ulukus, Optimal packet scheduling in multiple access channel with energy harvesting transmitters. J. Commun. Netw. **14**(2), 140–149 (2012)

40. S. Yin, E. Zhang, L. Yin et al., Optimal saving-sensing-transmitting structure in self-powered cognitive radio systems with wireless energy harvesting. In IEEE ICC 2013: topics in cognitive radio and networks symposium, in *IEEE International Conference on Communications*, Budapest (2013)
41. Q. Zhao, B.M. Sadler, A survey of dynamic spectrum access. IEEE Signal Process. Mag. **24**(3), 79–89 (2007)
42. S. Zhou, J. Gong, Z. Zhou et al., GreenDelivery: proactive content caching and push with energy-harvesting-based small cells. IEEE Commun. Mag. **53**(4), 142–149 (2015)

Resource Management in Sustainable Green HetNets with Renewable Energy Sources

Ran Zhang, Miao Wang, Lin X. Cai, Yu Cheng, Xuemin (Sherman) Shen, and Liang-Liang Xie

Abstract Green energy has become a promising alternative energy source for powering wireless cellular networks by effectively reducing the network operational expenditure (OPEX) and carbon footprints. However, green energy sources such as solar and wind are harvested from the environment and their availability and capacity are by nature unstable, which poses great challenges to achieve sustainable network operation. In this chapter, we study the energy sustainable performance of a green HetNet where the small cell base stations (SBSs) are powered by green energy sources. Specifically, we first develop an analytical framework to study the energy sustainability of each SBS. The energy buffer at each SBS is modeled as a G/G/1 queue with arbitrary patterns of energy charging and discharging. We apply the diffusion approximation to analyze the transient evolution of the energy buffer, and derive the probability distribution of the queue length and the energy depletion time for a given initial energy level. Based on the energy sustainability analysis, we propose a distributed admission control strategy at SBSs striking a balance between high resource utilization and energy sustainability in the green HetNet. Extensive simulations are conducted to validate the analytical framework and evaluate the sustainability performance of the green HetNets using the proposed distributed admission control scheme. The simulation results demonstrate that relaxing the admission control criteria for the SBSs can improve the resource utilization (i.e, power and spectrum) of the system when the energy is abundant but can significantly degrade the resource utilization instead when the energy comes short due to poor sustainability performance (i.e., frequent depletion of the SBSs).

R. Zhang (✉) • M. Wang • X. (Sherman) Shen • L.-L. Xie
Department of Electrical and Computer Engineering, University of Waterloo, Waterloo, ON, Canada, K2B8E3
e-mail: r62zhang@uwaterloo.ca

L.X. Cai • Y. Cheng
Department of Electrical and Computer Engineering, Illinois Institute of Technology, Chicago, IL, USA, 60616-3793

W. Xiang et al. (eds.), *5G Mobile Communications*,
DOI 10.1007/978-3-319-34208-5_19

1 Introduction

The wireless communication industry is facing a great challenge with the expanding demand of mobile subscribers for broadband multimedia services. A recent study forecasted that the global mobile data traffic will increase tenfold between 2014 and 2019, and mobile video will account for more than 70 % of the total data by the end of 2019 [1]. As such, the main objective of the next generation mobile wireless networks is to further increase the network capacity and provision quality of service (QoS) guarantee for high densities of mobile users at the minimum cost. Dense deployment of heterogeneous networks (HetNets) has been recognized as a desirable and feasible solution for increasing the spatial network capacity and QoS provisioning in the fourth generation wireless networks [2]. However, a large number of small cell base stations (BSs) account for high capital expenditure (CAPEX) and operational expenditure (OPEX). In addition, a high volume of energy is required to power multiple small cell BSs (SBSs) in the cellular network, which also results in more greenhouse gases and carbon footprints, besides the increased energy costs. Therefore, one of the most critical issues in the next generation 5G wireless network is to design and develop new green technologies to reduce the energy consumption and costs of HetNets.

To achieve a green and sustainable wireless network, several research projects have been launched worldwide. OPERANET project aims to improve the power efficiency of mobile radio networks by monitoring and controlling the energy use and optimize the energy consumption at the BSs [3]. Mobile VCEs Green Radio programme examined the green techniques across the protocol stack for power reductio [4]. Most existing green communication networking solutions mainly focus on minimizing the energy consumption of the cellular network infrastructure, i.e., BSs, by improving hardware design, network planning [5, 6], energy efficient communications [7–9], and radio resource management [10–12]. With the recent advances in green energy techniques, it becomes a promising enabling technology to power SBSs with renewable energy sources such as solar and wind. By analyzing both the CAPEX and OPEX, it is shown in [13] that deploying green energy powered HetNets, especially the SBSs, is an economically convenient and sustainable solution for mobile operators, and it is possible to use harvested energy of BSs to serve plausible traffic loads in a dense HetNet [14].

Unlike traditional energy supply, green energy sources are by nature sustainable in the long term yet are variable and unstable in the short term, as the energy harvesting process is highly dependent on the charging environment, i.e., the weather, the geo-location, and the time. For example, a wind turbine provides intermittent power while a solar panel can supply relatively continuous power with varying output over a day and through the seasons. To fully utilize the harvested green energy, usually a rechargeable battery of large capacity is needed to store the charged energy and to provide a reliable and sustainable power supply for future use. In such a green energy powered wireless HetNet, the main performance metric has been shifted from energy efficiency to energy sustainability [15], i.e., the dynamically harvested energy at the BSs can sustain the traffic demands of mobile users in a multi-tier HetNet. For instance, a green BS powered by renewable energy sources can be on and maintain normal operation when the charged energy is sufficient to provision services to its associated mobile users. However, a green BS will become off and out of service if the charged energy is used up and cannot sustain the service demands of mobile users. When this happens, all users associated with the out-of-power BS will experience excessive service delays or even service interruptions. Thus, it is indispensable to deliberately design the network resource management and admission control strategies to minimize the probability that an SBS depletes its energy and becomes out of service.

In this article, we first present a literature survey on existing green communication and networking techniques, and identify the relevant challenging issues in the design and development of a green HetNet powered by renewable energy sources. We then study resource management issues in a multi-tier HetNet where a macro cell BS (MBS) and multiple SBSs co-exist to provide services for mobile users in the network area. Specifically, we first develop an analytical model to analyze the energy buffer evolution at SBSs, characterizing the dynamics in the energy charging and discharging processes. Based on the analytical model, we propose an distributed admission control strategy to ensure the quality of service (QoS) requirements of admitted users, while maximizing the resource utilization and ensuring the sustainable performance of the green HetNets.

The main contributions of the chapter are four folds.

- Analytical model of energy buffer: A generic analytical model is developed to study the transient evolution of energy buffer at a HetNet BS. Basically, the rechargeable battery at each BS is used as an energy buffer to store the harvested energy, which can be modeled as a G/G/1 queue with general energy charging and discharging processes. Based on the first two statistical moments, i.e., the mean and variance, of the energy charging and discharging intervals, the diffusion approximation is applied to derive closed-form distributions of the transient queue length, i.e., the available energy level.
- Energy sustainability analysis: Using the developed analytical model, we then analyze the energy sustainability performance of the green HetNets. Specifically, two main metrics are derived to evaluate the sustainable performance of BSs: (1) the energy depletion probability: the probability that a BS depletes the energy and becomes unavailable for service; and (2) the duration for an BS to deplete its energy: given the current available energy level, the time duration that the BS can continue to serve the users until it depletes the energy and becomes unavailable.
- Distributed admission control: Based on the energy sustainability analysis, a distributed admission control strategy is deployed at BSs to provision QoS guarantee for admitted users and ensure desirable energy sustainability performance of the network while achieving high resource utilization.
- Performance evaluation: We conduct extensive simulations to validate the analytical model of energy buffer and examine the sustainable performance of the proposed distributed admission control scheme. We study the network stainability under different admission control parameters, and further investigate

the impacts of the energy depletion at one or multiple BSs on the overall performance of the green HetNets.

The remainder of the chapter is organized as follows. In Sect. 2, we give a broad overview of the existing works on green radio communication networks in the literature. In Sect. 3, the system model is presented including the network scenario and the energy model. An analytical framework is developed to study the energy sustainability performance of the BSs in Sect. 4, followed by an adaptive admission control strategy in Sect. 5. Numerical results are presented in Sect. 6. Finally, we conclude the chapter and introduce our future work in Sect. 7.

2 Understanding the Green HetNets

In this section, the role of the green HetNets in 5G wireless communications is first described. Then, the state-of-the-art of the research works related to green HetNets is introduced in details. Specifically, we classify the related research works into the following two categories: (1) energy efficient green HetNets; and (2) green energy powered wireless networks.

2.1 Green HetNets in 5G

The wireless communications in 5G era are expected to fulfill a wide variety of requirements from both the operator and the user perspectives. On one hand, in order to meet the explosive growth of the mobile traffic, the wireless industry needs to significantly boost the system capacity by both excavating more spectrum resources and increasing the spectrum efficiency. On the other hand, the unprecedented expansion and densification of the wireless networks will cause huge energy consumption and cost, which has become a major challenge in commercializing the 5G systems. Green networking, as an energy-saving and eco-friendly approach to solve the energy-related challenges, has won ever-increasing attention in both academia and industry by its foreseeable potentials. The meaning of "green" has two folds: energy efficiency and renewable energy sources. Developing energy-efficient mechanisms for the radio access part of the system improves the system service quality without increasing much energy cost; and using renewable energy sources to power the wireless networks can effectively reduce the carbon footprint and relieve the greenhouse effects. Both folds have gained wide consensus and been incorporated as the main issues of the 5G development. Figure 1 illustrates the connections among different research issues covered by 5G, where rectangles with red rims show how the subject investigated in this chapter is connected to 5G.

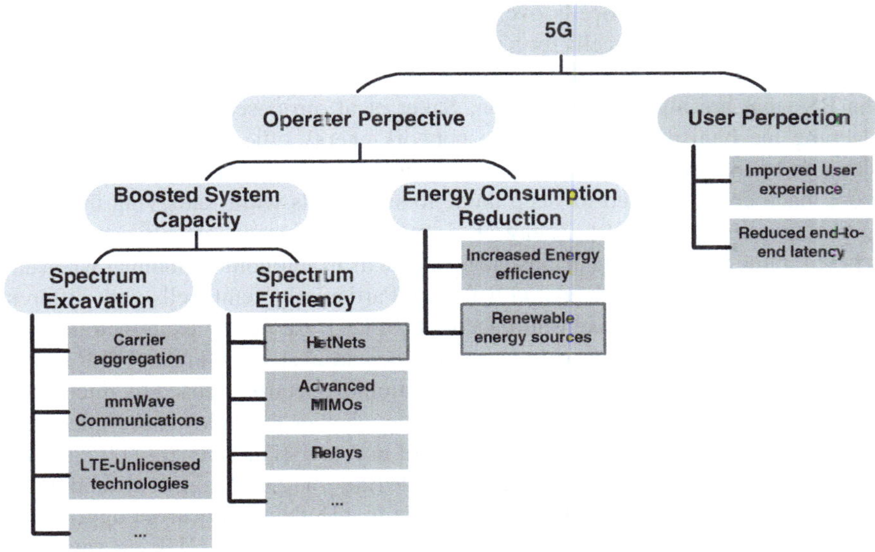

Fig. 1 Connections among different research issues covered by 5G

2.2 *Energy Efficient Green HetNets*

It is reported that the major part of the energy consumption (as much as 70 %) for typical cellular networks comes from the radio access part [16]. Therefore, improving the energy efficiency of the cellular access network is of crucial importance in reducing the OPEX as well as cutting down the carbon footprints. To this end, the HetNet becomes a promising solution in significantly enhancing the network energy efficiency as densely deployed small cells can provide high-speed services at a low transmission power. There have been a massive number of existing works studying energy efficiency improvement in HetNets [7, 8], which can be generally divided into two categories according to different pointcuts, namely network topology control and radio resource management.

Network topology control is an effective methodology for improving energy efficiency of HetNets, by optimizing the energy efficiency through deliberate network planning and user association strategies subject to a set of system and user constraints, e.g., the maximum power constraints of the BSs and the user QoS requirements. As for the networking planning, the optimal network deployment in terms of the minimal number of BSs and the deployment locations are determined based on different metrics in different scenarios [17–21]. For instance, in [17], Zhang et al. studied the optimal energy efficiency in a HetNet consisting of a macrocell and multiple small cells. Exploiting the stochastic geometry theory [22], the optimal small cell density was derived to maximize the energy efficiency under coverage constraints. In [18], Klessig et al. investigated a HetNet composed of a random number of microcells overlaid with multiple hexagonal macrocells. The optimal number of microcells for a reference cell area was determined to maximize the energy efficiency under different network loads. In [19], instead of dealing with the BSs that are always powered on, Saker et al. proposed to utilize macrocell sleeping mechanisms to improve the network energy efficiency. Specifically, the sleep and wake-up of the macrocells are carefully controlled to offload the users from macrocells switched to sleep mode to small cells while satisfying the users' QoS requirements. In [20], Zheng et al. put forward an energy-efficient scheme where femtocells cooperate in clusters to decide in an autonomous manner when and which femtocells should sleep and wake up. Particularly, femtocells with different user access policies are taken into account. In [21], Cao et al. examined how the energy efficiency of HetNets is impacted by the partial spectrum reuse (PSR) factor of the microcells, which is defined as the portion of spectrum reused by microcells. Optimal PSR factor was derived to determine whether to deploy more microcells or switch off less macro BSs. In the aspect of user association, energy efficiency is studied on both user side [23] and network side [24]. In [23], Zhang et al. studied the energy-saving performance of user equipment in a HetNet with carrier aggregation (CA) [25] and demonstrated that CA can be leveraged to save UE energy while satisfying the user demand. In [24], an user association algorithm was proposed to strike an energy-saving balance between the radio access and the backhaul links in HetNets.

Improving the energy efficiency of HetNets via resource management mainly aims at optimizing network resource allocation (e.g., bandwidth and power) to maximize energy efficiency, subject to the intra- and inter-tier interference in HetNets [10–12, 26, 27]. For instance, [10] studied how to optimize the transmission power of the femtocells to mitigate inter-cell interference while maximizing the energy efficiency of the system. The results revealed that a much higher energy efficiency could be achieved by sacrificing a relatively low loss in the spectral efficiency. In [11], the orthogonal frequency division multiple access (OFDMA) based HetNets were studied, where the joint subchannel allocation and power control problem was formulated to maximize the sum energy efficiency of all BSs subject to the BS power constraints and subchannel exclusivity constraints. In [12], the energy efficiency was investigated in a cognitive radio cellular network where the resource allocation problem was modelled as a Stackelberg game between cognitive macrocells and femtocells. In [26], energy-efficient pre-coding scheme for coordinated multi-point (CoMP) transmission in HetNets was studied, where the MBS and SBS can cooperatively transmit to a same user to provide better services. The optimal precoder structure was obtained to minimize the energy consumption with considering the inter-cell interference. In [27], the backhaul link power consumption of the CoMP HetNet was considered, and the resource allocation problem was modelled into an energy efficiency maximization problem subject to the spectral efficiency and user QoS constraints.

2.3 Green Energy Powered Wireless Networks

Recent advances in green energy technologies make it a cost-effective solution to power wireless networks compared with the traditional energy supplies. The optimal rate adaptation in a single user energy harvesting system was studied in [28] to minimize the packet delivery time. In [29], the optimum transmission policies was presented for energy harvesting nodes with limited batteries. In [30], a directional water-filling algorithm was introduced to achieve optimal transmission of energy harvesting nodes over a wireless fading channel. Optimal power adaptation schemes were studied in [31] to maximize the achievable rate of a single-user channel in a device-to-device communication underlaying green cellular network, considering both total power constraint and the BS power constraint, and further extended in [15] to a single user relay channel for cooperative communications.

Besides the studies in a single user channel, the system performance of green energy powered network were examined in different contexts. Solar energy powered sensor network was first studied in [32], and experiments showed that near-perpetual operation of a sensor node was possible with the developed prototypes of sustainable sensors. The solar and wind powered wireless local area networks (WLANs) were studied in [33, 34] and a solar battery configuration methodology along with an outage control algorithm were proposed to minimize the energy outage at wireless nodes. In [35], an optimal green AP placement problem was formulated to find a minimal number of green access points (APs) powered by renewable energy to be deployed in constrained locations and to provision QoS to mobile users. The network throughput and energy sustainability in the green-energy-powered maritime wireless network was studied in [36]. Resource allocation and energy management in sustainable mesh networks have also been studied. In [37], the fundamental design criterion of a sustainable wireless network was elaborated. It explained that under the green network paradigm powered by renewable energy, communication and networking protocols across the protocol stack should be re-visited, as the design criteria has been shifted from energy efficiency to energy sustainability. Different routing schemes are examined and compared in the context of sustainable networks in [38], aiming to distribute the traffic loads evenly over the network to improve the network sustainability. It was shown in that traditional routing strategies which did not consider environmental energy supply achieve poor performance, and it was crucial to adapt the routing decisions to the time varying power supply conditions in a sustainable network environment. However, how to accurately capture and model the power conditions was not discussed. In our previous work [39], a generic energy model was developed to characterize the dynamic energy charging and discharging processes, and an adaptive resource management scheme was proposed for a green mesh network using renewable energy. The objective was to maximize the energy sustainability of the network, or equivalently, to minimize the failure probability that the mesh APs depleted their energy and went out of service by distributing the traffic loads appropriately across the mesh network.

With the dense deployment of small cells to improve the capacity of a multi-tiered cellular network, it is well recognized that small cells can benefit from green energy supplies while macro cells usually require fairly high power and less likely to use harvested energy to serve mobile users. Generally, green energy powered BSs have a higher CAPEX but a lower OPEX, and vice versa for traditional BSs. By careful analyzing the CAPEX and OPEX of the two types of BSs, it was shown in [13] that gradually deploying green energy powered small cell BSs was an economically convenient solution for mobile operators to provide sustainable network services. The spatial and temporal dynamics of energy harvesting in a cellular network was studied in [14, 40–42]. The availability region for a set of self-powered BSs was investigated in [14], considering the uncoordinated on and off operations of green BSs based on the available energy levels. A new network architecture that overlays an uplink cellular network was proposed in [40] to deploy power beacons which can charge mobile devices by microwave radiation. An energy field model was developed to analyze the coverage of self-powered green cellular network in [41], where the energy harvesters are distributively deployed in clusters and the generated energy from the same cluster was aggregated and then redistributed to power BSs. In [42], a joint network deployment and resource management problem was formulated for a two-tier cellular network such that the minimal number of small cell BSs were deployed with optimal subcarrier allocation to ensure full network connectivity and users' QoS requirements can be fulfilled with the harvested energy under the cost constraint.

The aforementioned works mainly studied network deployment or power management schemes to achieve a sustainable green cellular network. To the best of our technology, the resource management issues especially the admission control strategy for a green HetNet is not well investigated yet. In addition, due to the dynamic energy charging and limited battery size, the energy outage may occur stochastically. No existing work has extensively studied the impacts of energy depletion of one or multiple SBSs on the overall sustainable network performance.

3 System Model

3.1 Network Model

We consider a heterogeneous network composed of different types of base stations serving different coverage areas. As shown in Fig. 2, there is one macrocell and multiple small cells. Due to the high power consumption and critical always-on operation requirements, the MBS is powered by both renewable energy and traditional energy from the electricity grid[1]; while the small cell base stations (SBSs) are powered by renewable energy supplies. In such a HetNet, the MBS uses the whole spectrum band for communications and each SBS uses only a

[1]For the MBS, renewable energy source is the first choice. The electricity grid provides supplemental energy source to guarantee the power supply of the MBS at all times.

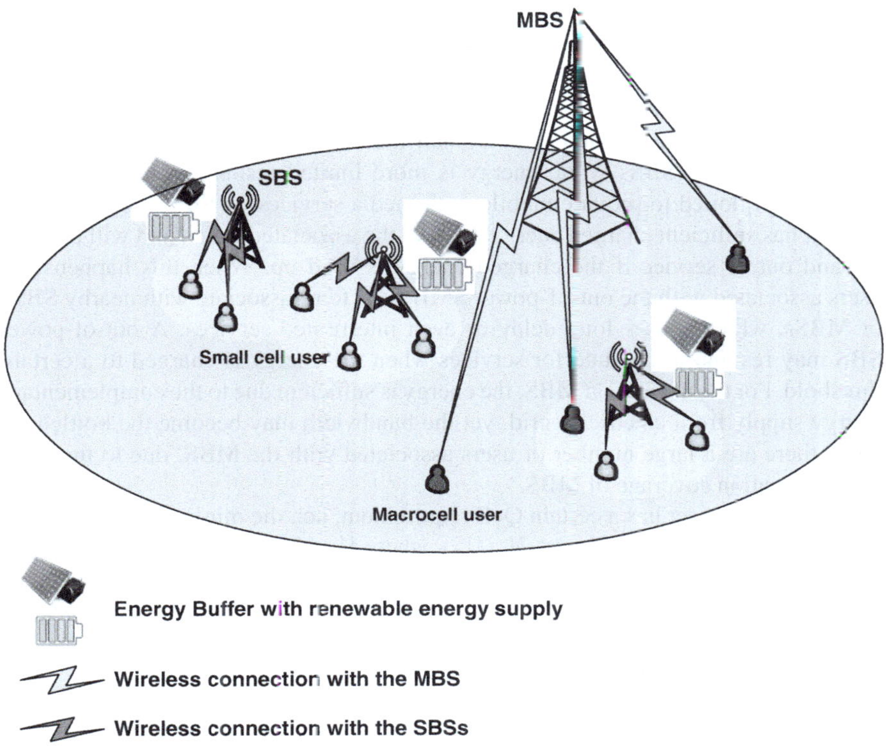

Fig. 2 Green HetNets with renewable energy supplies

subset of the spectrum bands for communications. Generally, orthogonal frequency division multiple access (OFDMA) is adopted as the radio access technology for both macrocells and small cells, and small cells will experience co-channel interference from the macrocell and the neighboring small cells when they use the same frequency subcarriers for communication. Denote the MBS as m and the set of SBSs as $\mathbb{S}$. Each small cell BS can provide high-rate multimedia services to the users within its coverage range. One typical network scenario is a household or an office building where a green SBSs with solar panels or wind turbines can be installed on the roof of the building to harvest energy from the environment, e.g., sunlight or wind. When one user arrives in the system, it checks for the available BSs in the neighborhood. In a high density HetNet, it is possible that a user is within the coverage areas of multiple SBSs, and the user will first choose the closest SBS to associate with, provided that it can be accommodated by the target SBS based on certain admission control criteria. If the association request is rejected, the user will choose the next closest SBS to associate with. The process is repeated until the user is successfully admitted by a BS, either a small cell BS in its neighborhood or the MBS if no nearby small cell has sufficient network resources to serve the user.

As the initial step, only downlink transmissions are considered for both macrocells and small cells. For small cells, bandwidth is usually sufficient and the renewable energy supply is the bottleneck. This is acceptable since bandwidth can be rich with advanced wireless technologies (e.g., carrier aggregation in both licensed and unlicensed band[2] and a small number of users due to the relatively small coverage of SBSs while energy is more limited resource when renewable energy is exploited to provide mobile multimedia services. A SBS is on operation when it has sufficient charged energy to serve the associated users, and will become off and out of service if the charged energy is used up. When this happens, all users associated with the out-of-power SMB need to re-associate with nearby SBSs or MBSs, which causes long delay or even interrupted services. A out-of-power SBS may resume to on state for services when the energy is charged to a certain threshold. For the always-on MBS, the energy is sufficient due to the complementary energy supply from electricity grid, yet the bandwidth may become the bottleneck when there are a large number of users associated with the MBS, due to the large communication coverage of MBS.

Each mobile user has a certain QoS requirement, i.e., the minimum transmission rate of user i is denoted as r_i, $i \in \mathbb{U}_s$ ($\mathbb{U}_m$), where $\mathbb{U}_s$ ($\mathbb{U}_m$) is the set of users that are associated with SBS s (MBS m). According to the Shannon theory, r_i in small cells can be derived as

$$r_i = W_{si} \log(1 + \frac{P_t {d_{si}}^{-\alpha}}{n_0 W_{si} + \sum\limits_{j \in \{m\} \cup S \backslash s} I_{ji}}), \quad \forall i \in \mathbb{U}_s, \ \forall s \in \mathbb{S} \tag{1}$$

where W_{si} is the bandwidth assigned by SBS s to its associated user i, P_t is the transmission power of the SBS to the associated user i, d_{si} is the distance between the SBS s and the associated user i, α is the pass loss exponent, n_0 is the noise power spectrum density, and I_{ji} is the perceived interference at user i from base station (BS) j (either MBS or SBS). The perceived interference I_{ji} is a stochastic variable which is closely related to the distance between user i and neighboring BS j, the transmission power of BS j and the sub-carrier allocation at each BS. The perceived interference can be effectively estimated leveraging standard estimation methods [43]. Since bandwidth is rich for small cells in the considered scenario, W_{si} is adaptively adjusted to combat the perceived interference to satisfy the transmission rate requirement (i.e., r_i) of each user. Similarly, if user i is served by the MBS m, the transmission rate of user i is

$$r_i = W_{mi} \log(1 + \frac{P'_t {d_{mi}}^{-\alpha}}{n_0 W_{mi} + \sum\limits_{j \in \{m\} \cup S \backslash m} I_{ji}}), \quad \forall i \in \mathbb{U}_m. \tag{2}$$

where P'_t denotes the transmission power of MBS to each macrocell user.

[2]FCC opened 100 MHz spectrum in 5 GHz band for unlicensed use in 2014, and plans to open an additional 195 MHz spectrum in the near future. The abundant unlicensed spectrum band provides great potential for cellular operators to deploy high density small cells that operates in unlicensed 5 GHz band.

3.2 Energy Model

Each green BS is equipped with an energy buffer that stores the harvested renewable energy for future use. Denote the buffer energy level at time t by $E_s(t)$, $s \in \mathbb{S} \cup \{m\}$. Let $E_s(0)$ be the initial energy buffer level, and $C_s(t)$ be the amount of charged energy over the time interval $[t-1, t]$. Due to different environmental conditions (e.g., various sunlight intensities or wind strengths at different geographical positions and time periods), the charging rate of an SBS may vary uniquely with each other. Thus, in this paper, the buffer charging process is modelled as a continuous-time stochastic event arrival process with an arbitrary but stationary distribution; the charged energy units are considered as the arrival events in the buffer. Denote the mean and variance of the inter-arrivals of the energy charging as μ_c and σ_c^2, respectively. For SBS s, the energy stored in the buffer at time t is subject to the boundary conditions, i.e., $0 < E_s^{min} \leq E_s(t) \leq E_s^{max}$. E_s^{min} is the minimum required energy in the buffer to ensure normal BS operations such as signal processing and cooling system; E_s^{max} is the maximum capacity of the energy buffer. Without loss of generality, this energy buffer model can be further simplified by letting $0 \leq E_s(t) \leq B_s$ where $B_s = E_s^{max} - E_s^{min}$. In the following section, we study the case that the buffer capacity is sufficiently large to store the harvested energy, i.e., $B_s \to \infty$. The analysis for finite buffer capacity will be included in the future work.

Denote the energy used for of a packet transmission as e^u. We normalized that $e^u = 1$ energy unit for presentation convenience. The packet arrival of each user is modelled as a stochastic process with an arbitrary but stationary distribution. The mean and variance of the inter-packet arrival are μ_0 and $\sigma_0{}^2$, respectively. Given the traffic model for a single user, the mean and variance of the aggregated packet arrivals for one SBS s can also be measured, which are equivalent to the mean (denoted as $\mu_{s,d}$) and variance (denoted as $\sigma_{s,d}^2$) of the discharging interval of the energy buffer in SBS s, respectively.

Given the above energy buffer model, the energy queue evolves as follows. The total amount of energy of SBS s discharged during $[t-1, t]$, denoted as $D_s(t)$, is given by

$$D_s(t) = \sum_{i \in \mathbb{U}_s} N_i(t) e^u, \tag{3}$$

where $N_i(t)$ denotes the number of packets of user i transmitted by SBS s within $[t-1, t]$. Therefore, the energy level of SBS s at time t is calculated as

$$E_s(t) = \min\{\max\{E_s(t-1) + C_s(t) - D_s(t), 0\}, B_s\}, \forall s \tag{4}$$

The evolution of the energy buffer of an SBS is illustrated in Fig. 3. Generally, the energy charging and discharging are random processes, and the rates of which vary over time yet their first and second moments can be measured. The energy buffer increases when the energy charging rate is higher than the energy discharging rate, and decreases otherwise. An energy buffer may deplete temporally when the energy level reaches 0, i.e., $E_s(t) = 0$.

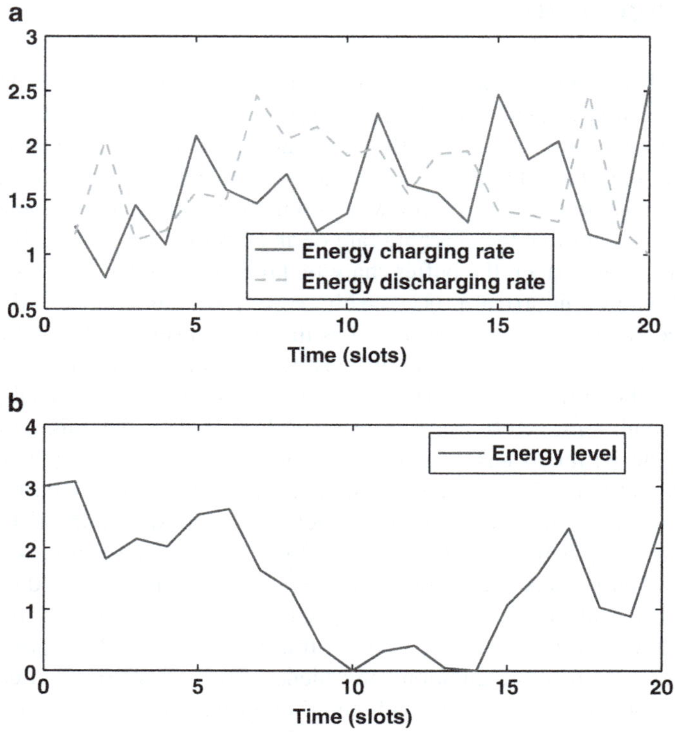

Fig. 3 Energy buffer evolution under random charging and discharging processes. (**a**) Energy charging and discharging processes. (**b**) Energy buffer evolution

4 Energy Sustainability Analysis

In this section, a generic analytical framework is developed to analyze the energy buffer. The diffusion approximation approach is applied to study the energy buffer evolution in the infinite buffer size case. For notational simplicity, we omit the SBS index s in the notations of the previous section (e.g., $E_s(t)$ to $E(t)$). Given the initial energy level $E(0)$, the buffer may be depleted when $E(t)$ reaches 0 due to energy discharging for downlink transmissions. When one SBS is depleted, it becomes temporally unaccessible and all the associated users need to find alternative BSs (other nearby SBSs or the MBS) to connect with based on the association policy described in Sect. 3.1. Therefore, it is critical to ensure energy sustainability of SBSs. To address this issue, we need to answer the following questions: (1) how to derive the depletion probability for a given initial energy level in the buffer and cell load conditions, (2) how to design effective admission control policy based on the depletion probability, and (3) to what extend the depletion of one or multiple SBSs will degrade the overall network performance.

With infinite buffer capacity, all the harvested energy can be stored in the buffer. According to the system model, both the charging and discharging processes are random processes with arbitrary but stationary distributions. Thus, the energy buffer can be modelled into a G/G/1 queue. The mean and variance of the inter-charging (inter-discharging) intervals are μ_c (μ_d) and σ_c^2 (σ_d^2), respectively. To make the G/G/1 queue tractable for analysis, the diffusion approximation method is exploited by approximating the discrete-event queue via a continuous random process [44]. Specifically, the discrete energy queue length $E(t)$ is approximated to a continuous random process $Y(t)$ in the way that the increase of $Y(t)$ over a small period dt follows a normal distribution, i.e.,

$$\begin{aligned} &dY(t) = Y(t+dt) - Y(t) = \mu dt + n(t)\sigma\sqrt{dt}, \\ &\text{where } \mu = \frac{1}{\mu_c} - \frac{1}{\mu_d},\ \ \sigma^2 = \frac{\sigma_c^2}{\mu_c^3} + \frac{\sigma_d^2}{\mu_d^3}, \end{aligned} \tag{5}$$

In Eq. (5) μ and σ are the mean and standard variance of $dY(t)$, respectively, and $n(t)$ is the standard Gaussian process (i.e., with zero mean and unit variance). Given the initial energy level $E(0) = e_0$, the conditional probability density function (pdf) of $Y(t)$, which also refers to the energy buffer level at time t $E(t)$, can be expressed as

$$f_Y(y, t|e_0) = \frac{\mathrm{P}(y < Y(t) \le y + dy|e_0)}{dy}. \tag{6}$$

According to [44], $f_Y(y, t|e_0)$ satisfies the forward diffusion equation

$$\frac{\partial f_Y(y, t|e_0)}{\partial t} = \frac{\sigma^2}{2}\frac{\partial^2 f_Y(y, t|e_0)}{\partial y^2} - \mu\frac{\partial f_Y(y, t|e_0)}{\partial y}, \quad \text{for} \quad Y(t) \ge 0 \text{ and } t > 0. \tag{7}$$

To solve the above forward diffusion equations, the method of images [45] is utilized to obtain the expression of the conditional p.d.f. as

$$f_Y(y, t|e_0) = \frac{\partial}{\partial y}\left\{Z\left(\frac{y - e_0 - \mu t}{\sigma\sqrt{t}}\right) - e^{2\mu y/\sigma^2} Z\left(-\frac{y - e_0 + \mu t}{\sigma\sqrt{t}}\right)\right\}, \tag{8}$$

where $Z(y)$ is the standard normal integration, $Z(y) = \int_{-\infty}^{y} \frac{1}{\sqrt{2\pi}} e^{-x^2/2} dx$.

After obtaining the conditional p.d.f. of the energy buffer size at time t, i.e., $f_Y(y, t|e_0)$, two main performance metrics can be derived: (1) the depletion probability given initial energy level e_0; and (2) the distribution of the energy depletion time, denoted by $T(e_0)$, i.e., from the initial time with the energy level of e_0 till the time when the energy buffer full depletes and reaches zero for the first time, $E[t] = 0$. Thus, we have

$$T(e_0) = \inf(t \ge 0|Y(t) = 0, Y(0) = e_0). \tag{9}$$

The energy depletion time is an important sustainability metric for green HetNets which indicates the maximum time duration that the SBS can sustain the demands of the small cell operation before it becomes out of service, for a given initial energy level. In stochastic theory, $T(e_0)$ is also referred to as the first passage time of $Y(t)$ from e_0 to 0. The p.d.f. of $T(e_0)$, denoted as $f_T(t|e_0)$, can be derived similarly to $f_Y(y, t|e_0)$ based on the diffusion equations with the absorbing barrier at 0. According to [44], the diffusion equations are given by

$$f_T(t|e_0) = \lim_{x\to 0}\Big\{\frac{\sigma^2}{2}\frac{\partial f_Y(y,t|e_0)}{\partial y} - \mu f_Y(y,t|e_0)\Big\}. \tag{10}$$

Based on Eq. (10), the p.d.f. of $T(e_0)$ can be derived as

$$f_T(t|e_0) = \frac{e_0}{\sqrt{2\pi\sigma^2 t^3}}\exp\Big[-\frac{(e_0+\mu t)^2}{2\sigma^2 t}\Big]. \tag{11}$$

Then, the moment generation function of $T(e_0)$ can be expressed as

$$\Phi_T(s|e_0) = \int_0^{+\infty}\exp(st)f_T(t|e_0)dt = \exp\Big[-\frac{e_0}{\sigma^2}(\mu + \sqrt{\mu^2 - 2\sigma^2 s})\Big]. \tag{12}$$

The moment generation function can be utilized to derive the first and second moments of a random process by calculating the first-order and second-order differential equations of Eq. (12) with respective to s, respectively. Thus, the mean and variance of $T(e_0)$ when $\mu \neq 0$ are calculated as

$$E[T(e_0)] = \frac{\partial\Phi_T(s|e_0)}{\partial s}|_{s=0} = \begin{cases}\exp(-2e_0\mu/\sigma^2)e_0/\mu, & \text{if } \mu > 0,\\ -e_0/\mu, & \text{if } \mu < 0;\end{cases} \tag{13}$$

and

$$\begin{aligned} V[T(e_0)] &= E[T(e_0)^2] - E^2[T(e_0)] = \frac{\partial^2\Phi_T(s|e_0)}{\partial s^2}|_{s=0} - E^2[T(e_0)] \\ &= \begin{cases}\exp(\frac{-2\mu e_0}{\sigma^2})(\frac{e_0^2}{\mu^2} + \frac{e_0\sigma^2}{\mu^3}) - \exp(\frac{-4e_0\mu}{\sigma^2})\frac{e_0^2}{\mu^2}, & \text{if } \mu > 0,\\ -e_0\sigma^2/\mu^3, & \text{if } \mu < 0.\end{cases}\end{aligned} \tag{14}$$

If $s \to 0$, $\Phi_T(s|e_0)$ will result in the conditional probability of approaching to the absorbing barrier (i.e., $Y(t) = 0$), given the initial energy level of e_0. Thus, we have

$$\mathrm{P}(Y = 0|e_0) = \lim_{s\to 0}\Phi_T(s|e_0) = \begin{cases}\exp(-2e_0\mu/\sigma^2), & \text{if } \mu \geq 0\\ 1, & \text{if } \mu < 0.\end{cases} \tag{15}$$

It can be seen from Eq. (15) that when the energy charging rate is smaller than the energy discharging rate, i.e., $\mu < 0$ (or equivalently, $1/\mu_a < 1/\mu_d$), the energy buffer will deplete eventually and $Y(t|e_0)$ will become 0 with probability 1. When $\mu \geq 0$, although the energy charging rate is no less than the energy discharging rate, it is still possible that the energy buffer may deplete with a certain probability due to the variances of the charging and discharging rates. In such case, the depletion probability is dependent on the initial energy level e_0, and the first and second moments of the energy charging and discharging processes.

5 Adaptive Admission Control Strategy for Small Cells

In this section, a distributed energy-aware user admission control strategy is proposed for small cells based on the queueing analysis of the energy buffer. Generally in a green HetNet, as a user will always attempt to associate with a nearby SBS first for achieving a higher transmission rate at a shorter distance and with a lower transmission energy cost, compared with the MBS; and a user will associate with the MBS only when no nearby SBS can accommodate it. Besides the bandwidth and energy efficiency of small cell communications, it is desirable because the MBS may need to use the energy from the electricity grid to serve wireless users, which involves extra cost. Thus, our admission control strategy at SBSs aims at protecting each SBS from being depleted its energy and becoming out of service due to admitting too many users while helping relieve the traffic burden from the MBS as much as possible, taking into consideration the transient energy level, cell load conditions, and charging capabilities. The strategy is implemented in each SBS and is conducted when the network conditions change (e.g., when a user arrives at or leaves a small cell, energy buffer of a small cell depletes, and a small cell recovers from depletion).

Basically, the admission control strategy tackles the tradeoff between the sustainability of the small cells and the traffic offloading from the macrocell. On one hand, according to Eq. (15), when the energy charging rate is higher than the energy discharging rate at one SBS (i.e., $\mu > 0$), the SBS depletion probability reduces as the discharging rate reduces. In other words, admitting a fewer number of users to one SBS can help improve the SBS sustainability. On the other hand, if a fewer number users are admitted to small cells, the MBS has to provision QoS to more users, probably resulting in bandwidth exhaustion after the network capacity is reached when the user density is high. When the bandwidth exhaustion occurs, no more users should be admitted, otherwise the QoS requirements of some macrocell users cannot be guaranteed, leading to significant user dissatisfaction. Therefore, the energy-aware admission control strategy should deliberately adjust the admission criteria to well balance the tradeoff.

In a small cell, each admitted user should be provided with guaranteed service; that is, once the user is admitted by a small cell, its transmission rate requirement is guaranteed by 100 %. When a new user sends an association request to a SBS at time t, the SBS will check the traffic demands of the user, admit it only if its depletion probability condition is under a desired threshold ϵ after admitting the user, based on its current energy level (i.e., $E(t)$) and the energy charging capability, i.e.,

$$\begin{aligned} &\mathrm{P}(Y = 0|E(t)) \leq \epsilon \ll 1. \\ &\Rightarrow \mathrm{P}(Y = 0|E(t)) = \exp\left[-2E(t)\mu/\sigma^2\right] < \epsilon, \ \mu > 0. \end{aligned} \tag{16}$$

The threshold ϵ is an adjustable parameter but should be kept far less than 1, $\epsilon << 1$, to guarantee a sufficiently small risk of energy depletion, which leads to a necessary condition that the charging rate of the energy buffer is required to be larger than the discharging rate. Smaller ϵ indicates more strict criteria in admitting a user. Based on the admission criteria of small cells, the admission control strategy is outlined as follows.

- When a user arrives in the network, it first creates a BS set (denoted as $\mathbb{S}_c$) within whose coverage it is located, and sort the BSs (SBSs and the MBS) in the descending order of the signal strength. The user will first send a network association request to the BS with the highest signal strength, usually the closest SBS. Upon receiving the request, the SBS will examine whether the admission criteria is satisfied. If so, the user will be admitted and associated with the SBS; if the condition cannot be satisfied, the user will be rejected and have to attempt network association with the next BS in the list of $\mathbb{S}_c$. If no small cells can accommodate the user, the user will be associated to the MBS for communications, given the network capacity is not reached or the bandwidth is not exhausted.
- When a user leaves a small cell so that the SBS can accommodate more users, the SBS will broadcast this information periodically. If any user within the coverage receives the message, it will send an association request to the SBS if the SBS has a better channel condition compared with the currently associated BS (MBS or SBS). The SBS processes the requests based on the first-come-first-serve principle and accommodate the requesting user(s) if the admission criteria (16) is satisfied in order to offload the traffic burden from the MBS and provide more bandwidth and energy-efficient communications.
- In addition, if by any chance an SBS depletes its energy (with a probability smaller than ϵ), the SBS will be temporally unaccessible, and all the associated users may need to find an available SBS or the MBS for network access.
- After an SBS depletes, no users can be served, and the energy discharging rate becomes zero. The SBS starts the recovery process where the energy is accumulated in the battery or energy buffer. When the energy queue accumulates to e_0, the SBS starts to broadcast the information periodically. Users receiving the messages update the list of available SBSs and sort them in the descending order of the signal strength. A user served by other BSs may choose to associate with the resumed SBS if it provides better services, i.e., a higher signal strength, for energy-efficient communications.

To summarize, flow charts are presented in the following to illustrate the admission control and user association procedures for all situations (Fig. 4).

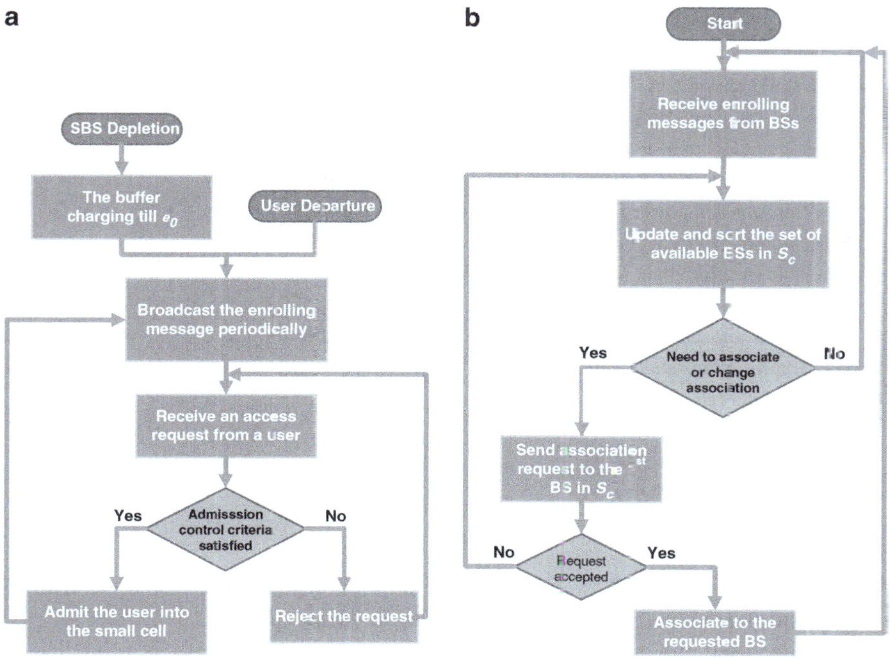

Fig. 4 Flow chart for admission control and user association procedures for all situations. (**a**) Admission control procedure in SBSs. (**b**) User association procedure

6 Performance Evaluation

In this section, simulations are conducted to validate the analytical framework of energy buffer and evaluate the impact of admission control criteria on the resource (i.e., transmission power and bandwidth) utilization of the green HetNet powered by the renewable energy sources.

6.1 Simulation Setup

In the simulation, we have a dense HetNet composed of one macrocell and N_s ($N_s = 30$ if not otherwise specified) overlaid small cells, as shown in Fig. 2. The N_s small cells (with a radius of r meters) are uniformly distributed within the macrocell coverage (with a radius of R meters) and may overlap with each other. The user arrivals and departures are modelled with Poisson processes with equal rates so that the number of users in the network can be kept at the same level in long term. For the user spatial distributions, one user may stay in a small cell with probability 5/6 and outside of the small cells with probability 1/6. Users are uniformly distributed

Table 1 Simulation parameters

Parameters	Values
Time slot duration	1 ms
Coverage radius of the macrocell, R	500 m
Coverage radius of a small cell, r	30 m
MBS power spectrum density, P_m	−23.5 dBm/Hz
SBS transmission power for a user packet, P_t	−6.49 dBm
Noise power spectrum density, n_0	−174 dBm/Hz
Available spectrum bandwidth for the MBS	10 MHz
Subcarrier bandwidth	15 kHz
Subcarriers allocation strategy	Random selection
User transmission rate requirement, r_i	300 kbps
User packet size	150 bits
Path loss component, α	2
Initial energy level of SBS, E_0	Uniform distributed on [5,25] energy units
Admission control criteria, ϵ	[0.001,0.5]

both inside and outside the small cells. The simulation starts with 360 users with 10 users inside each small cell and 60 users outside the small cells.

For any user, the packet arrival process at the associated BS is modelled as Poisson process with an average inter-packet interval of $u_0 = 20$ time slots. According to the system model, the transmission energy of an SBS for one packet is $e^u = 1$ energy unit via adjusting the allocated bandwidth. For the energy charging process at a SBS, the inter-charging interval is randomly selected from $\{1, 2, 3, 4\}$ time slots with a certain probability mass function. For instance, if the probability for choosing $\{1, 2, 3, 4\}$ is $\{0.4, 0.4, 0.1, 0.1\}$, respectively, the mean and variance for the inter-charging interval will be $\mu_c = 1.9$ and $\sigma_c^2 = 0.89$, respectively. Other main parameters used in the simulations are summarized in Table 1. The simulation data is obtained via averaging over 1000 simulation runs.

6.2 Simulation Results

We first examine the validity of the analytical model of energy buffer by investigating the evolution process of an energy buffer. The depletion probability of an energy buffer with different initial energy levels are shown in Fig. 5. We let $\mu > 0$ in the corresponding simulations since the depletion probability is 1 when $\mu < 0$. To reflect the relationship between the energy buffer and the cell load conditions, ten users are connected to an SBS without leaving the network. Two values of the average inter-packet interval are simulated (i.e., $\mu_0 = 20$ and $\mu_0 = 21$), respectively. It can be seen that the depletion probability descends with the increase of initial energy level e_0, and the exponential trends of the analytical

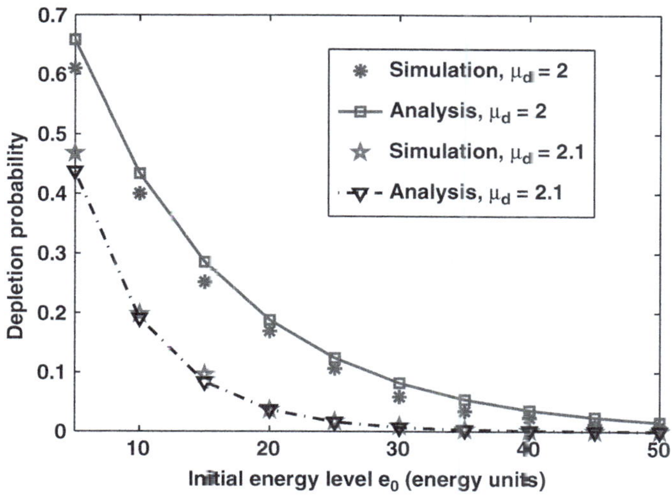

Fig. 5 The depletion probability for an energy buffer $P(Y(0) = 0|e_0)$ vs. initial energy level e_0. Default values: $\mu_c = 1.9$, $\sigma_c^2 = 0.39$, $\mu_d = 2$, and $\sigma_d^2 = 4$

results approach well with the simulation ones. With a higher initial energy level or an energy charging rate, it is less likely that the energy buffer will depletes, and thus the depletion probability is smaller. Besides that, the gap between the analytical and simulation results is slightly smaller when $\mu_d = 2.1$ than that when $\mu_d = 2$. This is as expected since the diffusion approximation approach is proven to be more accurate when the queue length is larger [44], and the average queue size when $\mu_d = 2.1$ is larger since the corresponding discharging rate is smaller than that when $\mu_d = 2$. This explanation also applies to the observation that the gap is slightly larger when the initial energy level e_0 is smaller.

The cumulative distribution function (CDF) of the first passage time from the initial energy level to 0 (i.e., $T(e_0)$) is presented in Fig. 6. Three e_0 values are simulated. It can be observed that the analytical results match well with the simulation results. In addition, for a given $T(e_0)$ value, the cumulative probability is larger for a smaller e_0. This is because when the energy charging rate is smaller than the energy discharging rate (i.e., $1/\mu_c < 1/\mu_d$), the queue depletes faster with a smaller initial energy e_0, leading to a smaller first passage time from e_0 to 0, i.e., $T(e_0)$. As a result, when e_0 is smaller, it is less likely that the SBS can sustain the traffic demands of its cell for a certain time, or equivalently the probability that $T(e_0)$ is smaller than a given value is higher.

After verifying the energy model, we then simulate the HetNet to investigate the impact of the admission control criteria ϵ on the resource utilization of the network, especially the MBS. For different ϵ values, two metrics are examined, i.e., the bandwidth exhaustion probability when the network capacity is reached, and the overall transmission power of the MBS. The results are shown in Fig. 7. It can

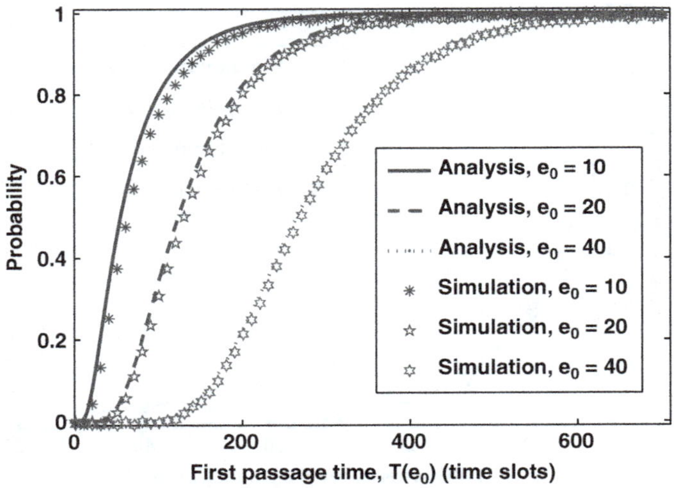

Fig. 6 The CDF of the first passage time from $Y(0) = e_0$ to $Y(t) = 0$, i.e., $T(e_0)$. Default values: $\mu_c = 1.9$, $\sigma_c^2 = 0.89$, $\mu_d = 1.5$, and $\sigma_d^2 = 2.25$

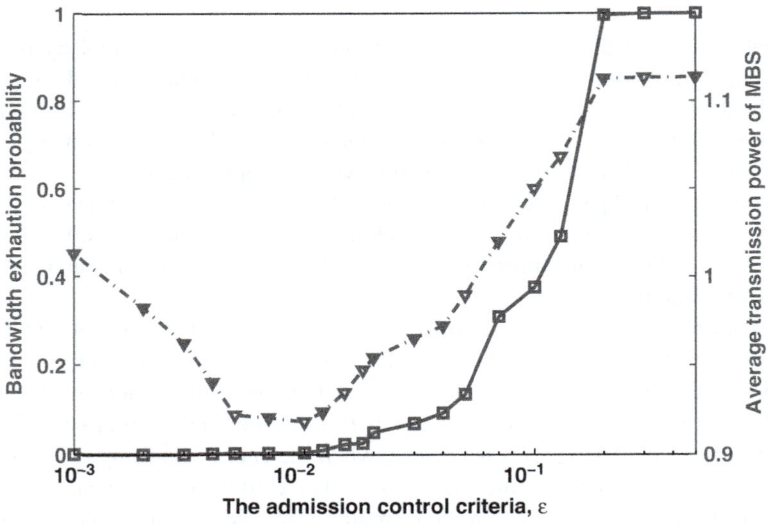

Fig. 7 The impact of admission control criteria ϵ on bandwidth and power consumption of the MBS. Default values: $\mu_c = 1.9$, $\sigma_c^2 = 0.89$, $\mu_0 = 20$, $\sigma_0^2 = 400$, and $N_s = 30$

be seen that when ϵ increases from 0.001 to 0.5, the average overall transmission power of the MBS first decreases and then increases. This can be explained as follows. When ϵ is small, increasing ϵ will increase the number of admitted users in each small cell. As a result, fewer users are connected with the MBS, thus the overall transmission power of the MBS decreases. When ϵ is large, increasing ϵ

will lead to a larger depletion probability for small cells. Consequently, although more users can be admitted with a larger ϵ for small cells, more small cells may deplete its energy and become unavailable for services; and the associated users with the unavailable SBSs will then connect to the MBS instead. In such a case, more transmission powers are used at the MBSs to provide services to the users whose associated SBSs are depleted.

The bandwidth exhaustion probability represents the probability that the MBS runs out of bandwidth or reaches its network capacity in providing QoS guaranteed service to the associated users. When bandwidth exhaustion occurs, the following user requests will be rejected, otherwise the QoS requirements of some associated users cannot be satisfied, which result in degraded service dissatisfaction of the users. It can be seen from Fig. 7 that when ϵ increases, the bandwidth exhaustion probability first stays 0, and then monotonically increases to 1 when the capacity is reached and no more user can be admitted. The monotonic trend indicates the dominative role that the small cell depletion probability plays in the bandwidth utilization of the MBS. When ϵ is very small, although a fewer number of users can be admitted to the small cells, the macrocell users do not used up the bandwidth. When ϵ becomes large, although more users can be admitted in the small cells, small cell energy depletion occurs more as well due to the loose admission control criteria. Depleted SBSs will results in more users associated with the MBS and thus more bandwidth is needed at the MBS. As a result, the bandwidth exhaustion probability increases. When bandwidth is exhausted, the average transmission power of the MBS remains unchanged when the network capacity is reached.

A more intuitive illustration for bandwidth exhaustion can be found in Fig. 8, which exhibits in one simulation run the total required bandwidth from all the macrocell users as time changes. It can be seen that when $\epsilon = 0.001$, the small cell depletion probability is very small, and the impulse caused by the depletion only occurs occasionally. The bandwidth exhaustion does not happen in the whole horizon. When ϵ increases, the depletion probability increases, resulting in that more than 1 small cells deplete their energy at the same time. Consequently, bandwidth

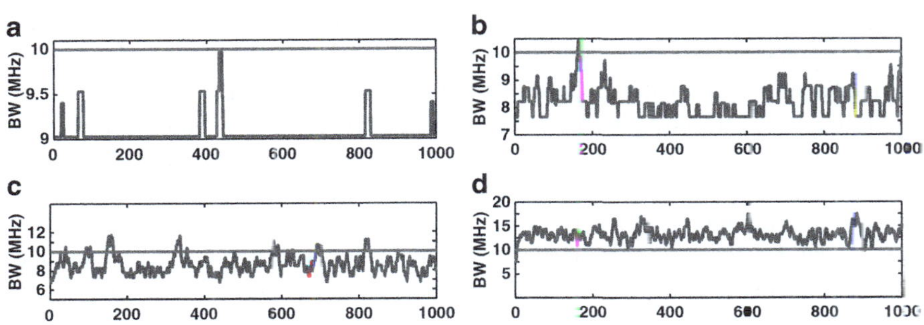

Fig. 8 The bandwidth exhaustion conditions over time. The notation *BW* denotes the aggregate required bandwidth from all macrocell users in each time slot. (**a**) $\epsilon = 0.001$. (**b**) $\epsilon = 0.01$. (**c**) $\epsilon = 0.1$. (**d**) $\epsilon = 0.2$

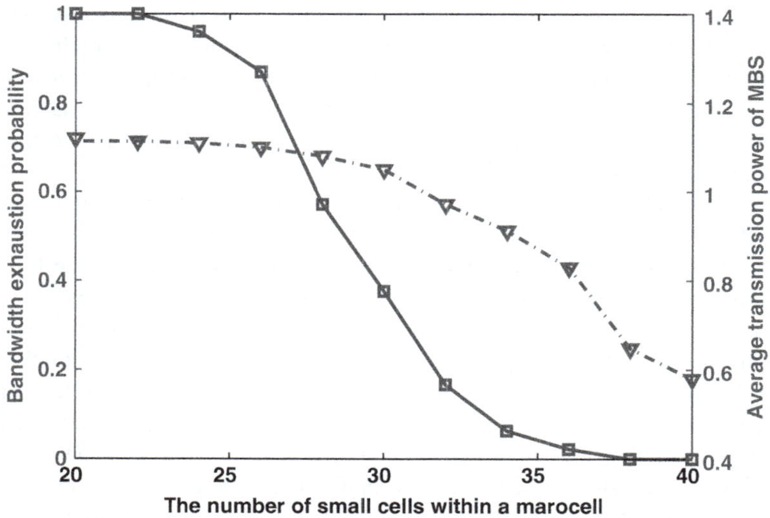

Fig. 9 The bandwidth exhaustion probability and average transmission power of the MBS vs. the small cell density. Default values: $\mu_c = 1.9$, $\sigma_c^2 = 0.89$, $\mu_0 = 20$, $\sigma_0^2 = 400$, and $\epsilon = 0.1$

exhaustion may occur occasionally. When $\epsilon = 0.1$, bandwidth exhaustion becomes more frequent due to s higher depletion probability. When ϵ is large (e.g., 0.2), bandwidth is exhausted in most of the time slots except the very beginning. Therefore, it can be concluded from Figs. 7 and 8 that optimal admission control criteria ϵ exists for small cells in striking the tradeoff between small cell sustainability and macrocell resource utilization.

Figure 9 presents how the bandwidth exhaustion probability and average transmission power of the MBS are affected by the small cell density when the admission control criteria ϵ is fixed. It can be seen that with increased number of small cells in one macrocell, both the bandwidth exhaustion probability and the average MBS transmission power decrease. The results are intuitive since given ϵ, more small cells can accommodate more users and reduce the number of macrocell users without increasing the depletion probability. Consequently, more MBS transmission power is saved and the bandwidth exhaustion probability is brought down.

Figure 10 shows how the bandwidth exhaustion probability and average transmission power of the MBS change with the average initial energy level of the SBSs. In each simulation run, the initial energy of each SBS is uniformly and randomly chosen from the interval $[e_0 - 3, e_0 + 3]$. The average value is obtained over 1000 runs. It can be observed that starting from the average value of 9, both examined metrics decrease as the average initial energy level of the SBSs increase. The reason is similar with that for Fig. 9, i.e., higher average initial energy level allows one SBS to accommodate more users and reduce the number of users connected to the MBS. In addition, one interesting observation may be noted that a trough occurs when the average initial energy level increases from 5 to 9 energy units. This phenomenon is

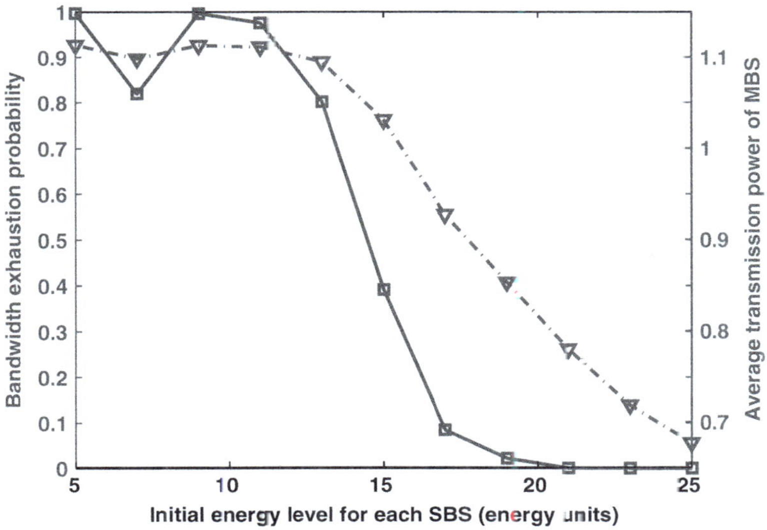

Fig. 10 The bandwidth exhaustion probability and average transmission power of the MBS vs. the average initial energy level e_0 of the SBSs. Default values: $\mu_c = 1.9$, $\sigma_c^2 = 0.89$, $\mu_0 = 20$, $\sigma_0^2 = 400$, $\epsilon = 0.1$, and $N_s = 30$

counter-intuitive at first glance but reasonable after a second thought. When $e_0 = 5$ and $e_0 = 7$, the number of users that can be admitted into a small cell satisfying $\epsilon = 0.1$ is the same, but the number of admitted users gradually increases when $e_0 = 9$ and $e_0 = 11$, etc. This is because when the initial energy level is low, the energy depletion probability is relatively high, and users should be admitted conservatively. When the initial energy is abundant, more users can be admitted. This explains why the number of admitted users do not change when e_0 increases from 5 to 7, but changes afterwards when e_0 is large. We also validate the number of admitted users in the simulations by checking the intermediate results. Given the same number of admitted users, the SBS depletion probability when $e_0 = 7$ is considerably smaller than that when $e_0 = 5$ according to Eq. (15). As a result, the SBSs when $e_0 = 7$ are less likely to be depleted, leading to a smaller average number of macrocell users. When $e_0 = 9$, the number of users that can be admitted into a small cell increases (i.e., the discharging rate is increased), resulting into a larger SBS depletion probability and larger average number of macrocell users.

Figure 11 illustrates how the average small cell throughput changes with different ϵ values when $\mu_0 = 20$ and $\mu_0 = 21$, respectively. Different μ_0 values indicate different energy discharging rates from the energy buffer. It can be found that for either curve, the average throughput of small cells first increases slightly and then drops dramatically as ϵ increases. The reason is as follows. There are two major factors that affect the average small cell throughput: SBS depletion probability and ϵ. When ϵ plays a dominant role, i.e., the performance gain of more admitted users

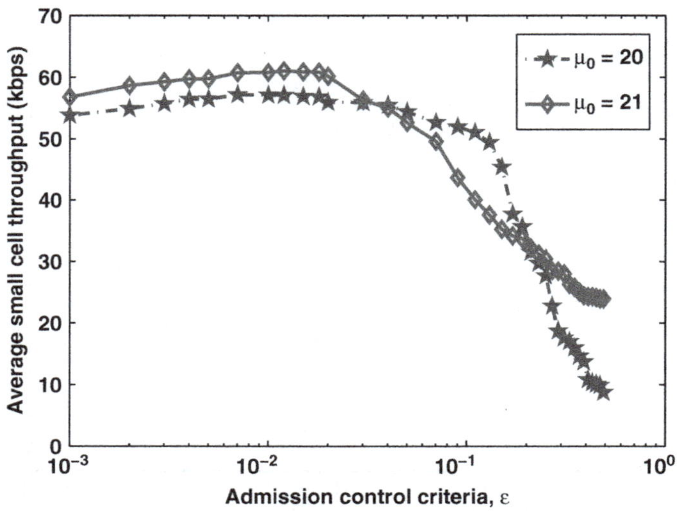

Fig. 11 The impact of admission control criteria ϵ on the average small cell throughput. Default values: $\mu_c = 1.9$, $\sigma_c^2 = 0.89$, and $N_s = 30$

due to looser admission control criteria surpasses the performance loss due to higher depletion probability, the average small cell throughput increases. The situation is reverse when the negative impacts of energy depletion becomes the dominant factor that surpasses the performance gain of more admitted users. In addition, the dominant factors vary at different charging rates and admission control criteria, and we observe that there are two cross points between the two curves in Fig. 11.

7 Conclusion

In this chapter, we have developed a generic analytical framework to study the energy sustainability of a green HetNet powered by renewable energy sources, by characterizing the dynamics in the energy charging and discharging processes. We have derived the closed-form solutions of the energy buffer at SBSs, i.e., the energy depletion probability and average energy depletion time. Based on the analysis, we have further proposed a distributed admission control scheme to strike the balance between high resource utilization of the green HetNet and the energy sustainability of SBSs, by effectively controlling the volume of the offloaded traffic from the macrocell to small cells.

In our future work, we plan to jointly consider bandwidth allocation and power management in a green HetNet powered by sustainable energy sources, considering the inter- and intra-tier interference among different cells, and finite energy buffers installed at BSs. How to optimally deploy the minimal number of green BSs to ensure network coverage and fully sustainable network operation of a green HetNet will also be under investigation.

References

1. Cisco White Paper, Cisco Visual Networking Index: Global Mobile Data Traffic Forecast Update 2014–2019 (2015)
2. R. Zhang, M. Wang, L.X. Cai, Z. Zheng, X. Shen, L.L. Xie, LTE-unlicensed: The future of spectrum aggregation for cellular networks. IEEE Wirel. Commun. Mag., **22**(3), 150–159 (2015).
3. OPERA-Net I/II, Optimising Power Efficiency in Mobile Radio Networks (2015). Available: http://projects.celticplus.eu/opera-net2/index.html
4. Mobile VCE Project, Green radio (2013). Available: http://www.mobilevce.com/green-radio
5. L.X. Cai, L. Cai, X. Shen, J.W. Mark, Resource management and QoS provisioning for IPTV over mmWave-based WPANs with directional antenna. Mob. Netw. Appl. **14**(2), 210–219 (2009)
6. P. Grant, S. Fletcher, Mobile basestations: reducing energy. Eng. Technol. Mag. **6**(2), (2011)
7. S. Navaratnarajah, A. Saeed, M. Dianati, M.A. Imran, Energy efficiency in heterogeneous wireless access networks. IEEE Wirel. Commun. Mag. **20**(5), 37–43 (2013)
8. J.B. Rao, A.O. Fapojuwo, A survey of energy efficient resource management techniques for multicell cellular networks. IEEE Commun. Surv. Tutorials **16**(1), First Quarter (2014)
9. R. Zhang, M. Wang, Z. Zheng, X. Shen, L.L. Xie, Cross-layer carrier selection and power control for LTE-A uplink with carrier aggregation, in *Proceedings of IEEE Globecom'13* (2013)
10. S. Mclaughlin, P.M. Grant, J.S. Thompson, H. Haas, D.I. Laurenson, C. Khirallah, Y. Hou, R. Wang, Techniques for improving cellular radio base station energy efficiency. IEEE Wirel. Commun. Mag. **18**(5), 10–17 (2011)
11. X. Xiao, X. Tao, Y. Jia, J. Lu, An energy-efficient hybrid structure with resource allocation in OFDMA networks, in *Proceedings of IEEE WCNC'11* (2011)
12. R. Xie, F.R. Yu, H. Ji, Y. Li, Energy-efficient resource allocation for heterogeneous cognitive radio networks with femtocells. IEEE Trans. Wirel. Commun. **11**(11), 3910–3920 (2012)
13. G. Piro, M. Miozzo, G. Forte, N. Baldo, L.A. Grieco, G. Boggia, P. Dini, HetNets powered by renewable energy sources: sustainable next-generation cellular networks. IEEE Internet Comput. **17**(1), 32–39 (2013)
14. H.S. Dhillon, Y. Li, P. Nuggehalli, Z. Pi, J.G. Andrews, Fundamentals of heterogeneous cellular networks with energy harvesting. IEEE Trans. Wirel. Commun. **13**(5), 2782–2797 (2014)
15. Z. Zheng, X. Zhang, L.X. Cai, R. Zhang, X. Shen, Sustainable communication and networking in two-tier green cellular networks. IEEE Wirel. Commun. Mag. **21**(4), 47–53 (2014)
16. T. Edler, S. Lundberg, Energy efficiency enhancements in radio access networks, in *Ericsson Review* (2004). Available: http://www.ericsson.com/ericsson/corpinfo/publications/review/2004_01/files/2004015.pdf
17. X. Zhang, Z. Su, Z. Yan, W. Wang, Energy-efficiency study for two-tier heterogeneous networks (HetNets) under coverage performance constraints. Mob. Netw. Appl. **18**(4), 567–577 (2013)
18. H. Klessig, A.J. Fehske, G.P. Fettweis, Energy efficiency gains in interference-limited heterogeneous cellular mobile radio networks with random micro site deployment, in *IEEE 34th Sarnoff Symposium* (2011), pp. 1–6
19. L. Saker, S.E. Elayoubi, R. Combes, T. Chahed, Optimal control of wake up mechanisms of femtocells in heterogeneous networks. IEEE J. Sel. Areas Commun. **30**(3), 664–672 (2012)

20. W. Zheng, W. Li, Y. Xie, X. Wen, Hybrid BS-cooperative power management scheme with self-organized sleep mode in virtual cell-based femto networks. Int. J. Distrib. Sens. Netw. **2012**, 12 pp. (2012)
21. D. Cao, S. Zhou, Z. Niu, Improving the energy efficiency of two-tier heterogeneous cellular networks through partial spectrum reuse. IEEE Trans. Wirel. Commun. **12**(8), 4129–4141 (2013)
22. R. Zhang, M. Wang, Z. Zheng, X. Shen, L.L. Xie, Stochastic geometric performance analysis for carrier aggregation in LTE-A systems, in *Proceedings of IEEE ICC'14* (2014)
23. M. Lauridsen, H. Wang, P. Mogensen, LTE UE energy saving by applying carrier aggregation in a HetNet scenario, in *Proceedings of IEEE VTC Spring'13* (2013)
24. A. Mesodiakaki, F. Adelantado, L. Alonso, C. Verikoukis, Energy-efficient user association in cognitive heterogeneous networks. IEEE Commun. Mag. **52**(7), 22–29 (2014)
25. R. Zhang, Z. Zheng, M. Wang, X. Shen, L.L. Xie, Equivalent capacity in carrier aggregation-based LTE-A systems: a probabilistic analysis. IEEE Trans. Wirel. Commun. **13**(11), 6444–6460 (2014)
26. Z. Xu, C. Yang, G.Y. Li, Y. Liu, S. Xu, Energy-efficient CoMP precoding in heterogeneous networks. IEEE Trans. Signal Process. **62**(4), 1005–1017 (2014)
27. K.M.S. Huq, S. Mumtaz, J. Bachmatiuk, J. Rodriguez, X. Wang, R.L. Aguiar, Green HetNet CoMP: energy efficiency analysis and optimization. IEEE Trans. Veh. Technol. **64**, 4670–4683 (2014)
28. J. Yang, S. Ulukus, Optimal packet scheduling in an energy harvesting communication systems. IEEE Trans. Commun. **60**(1), 220–230 (2012)
29. K. Tutuncuoglu, A. Yener, Optimum transmission policies for battery limited energy harvesting nodes. IEEE Trans. Wirel. Commun. **11**(3), 1180–1189 (2012)
30. O. Ozel, K. Tutuncuoglu, J. Yang, S. Ulukus, A. Yener, Transmission with energy harvesting nodes in fading wireless channels: optimal policies. IEEE J. Sel. Areas Commun. **29**(8), 1732–1743 (2011)
31. X. Zhang, X. Shen, L.L. Xie, Joint subcarrier and power allocation for cooperative communications in LTE-advanced networks. IEEE Trans. Wirel. Commun. **13**(2), 658–668 (2014)
32. J. Taneja, J. Jeong, D. Culler, Design, modeling, and capacity planning for micro-solar power sensor networks, *Proceedings of IEEE ICPN'08* (2008)
33. A. Sayegh, T.D. Todd, M.N. Smadi, Resource allocation and cost in hybrid solar/wind powered WLAN mesh nodes, in *Wireless Mesh Networks: Architectures and Protocols* (Springer, New York, 2008), pp. 167–189
34. A. Farbod, T.D. Todd, Resource Allocation and Outage Control for Solar-Powered WLAN Mesh Networks. IEEE Trans. Mob. Comput. **6**(8), 960–970 (2007)
35. Z. Zheng, L.X. Cai, M. Dong, X. Shen, H.V. Poor, Constrained AP placement with rate adaptation in green WLAN mesh networks, in *Proceedings of IEEE GLOBECOM'11* (2011)
36. T. Yang, Z. Zheng, H. Liang, R. Deng, N. Cheng, X. Shen, Green energy and content aware data transmission in maritime wireless communication network. IEEE Trans. Intell. Transp. Syst. **16**(2), 751–762 (2015)
37. L.X. Cai, Y. Liu, T.H. Luan, X. Shen, J.W. Mark, H.V. Poor, Dimensioning network deployment and resource management in green mesh networks. IEEE Wirel. Commun. Mag. **18**(5), 58–65 (2011)
38. E. Lattanzi, E. Regini, A. Acquaviva, A. Bogliolo, Energetic sustainability of routing algorithms for energy-harvesting wireless sensor networks. Comput. Commun. **30**, 2976–2986 (2007). Available: http://dx.doi.org/10.1016/j.comcom.2007.05.035
39. L.X. Cai, Y. Liu, T.H. Luan, X. Shen, J.W. Mark, H.V. Poor, Sustainability analysis and resource management for wireless mesh networks with renewable energy supplies. IEEE J. Sel. Areas Commun. **32**(2), 345–355 (2014)
40. K. Huang, V.K.N. Lau, Enabling wireless power transfer in cellular networks: architecture, modelling and deployment. IEEE Trans. Wirel. Commun. **13**, 902–912 (2014)
41. K. Huang, V.O.K. Li, Renewables powered cellular networks: energy field and network coverage. IEEE Trans. Wirel. Commun., to appear

42. Z. Zheng, L.X. Cai, R. Zhang, X. Shen, H.V. Poor, RNP-SA: joint relay placement and sub-carrier allocation in wireless communication networks with sustainable energy. IEEE Trans Wirel. Commun. **11**(10), 3818–3828 (2012)
43. S. Wang, W. Guo, M.D. McDonnell, Downlink interference estimation without feedback for heterogeneous network interference avoidance, in *Proc. IEEE ICT'14* (2014)
44. D.R. Cox, H.D. Miller, *The Theory of Stochastic Processes* (Chapman and Hall, London, 1965)
45. H. Kobayashi, Application of the diffusion approximation to queueing networks I: equilibrium queue distributions. J. ACM **21**(2), 316–328 (1974)

Resource Allocation for Cooperative D2D Communication Networks

Shankhanaad Mallick, Roya Arab Loodaricheh, K.N.R. Surya Vara Prasad, and Vijay Bhargava

Abstract Device-to-device (D2D) communications technology is currently being investigated as a potential enabler for the fifth generation (5G) communication networks. Significant performance gains are achievable in a cooperative D2D framework, wherein the user equipments (UEs) cooperate with each other to enable a variety of low-latency proximity-based services or to establish indirect communication links with the Base Station (BS) whenever direct service coverage is not possible. This chapter is focused on the throughput gains achievable in the latter scenario, i.e., when few UEs perform relaying operations to provide indirect service coverage to other UEs. In this direction, resource allocation problems are formulated for a variety of system models operating under the orthogonal frequency division multiple access (OFDMA) cellular or cognitive radio (CR) access architectures. The performance of mobile D2D relaying under different scenarios is evaluated. The system models are designed to study the benefits of incorporating additional capabilities at the devices, such as packet storage (using buffers), energy-harvesting, and cognitive spectrum access within the cooperative D2D framework. Depending on the system model, efficient algorithms are proposed to obtain optimal power allocation, subcarrier assignment, subcarrier pairing, and relay-UE selection policies which maximize the system throughput under a variety of system-dependant constraints. Simulation results demonstrate the effectiveness of our proposed algorithms and the performance improvement of mobile D2D-relaying networks over conventional networks.

1 Introduction

Device-to-device (D2D) communication is a radio technology which allows nearby devices (such as mobile phones and computers) to establish local communication links with each other. The local communication links, established either on the licensed or the unlicensed spectrum, allow traffic to flow directly between them without traversing the conventional path via the base stations (BSs) [1]. Consider,

S. Mallick (✉) • R. Arab Loodaricheh • K.N.R. Surya Vara Prasad • V. Bhargava
University of British Columbia, Vancouver, BC, Canada
e-mail: shankha@ece.ubc.ca; royaa@ece.ubc.ca; surya@ece.ubc.ca; vijayb@ece.ubc.ca

W. Xiang et al. (eds.), *5G Mobile Communications*,
DOI 10.1007/978-3-319-34208-5_20

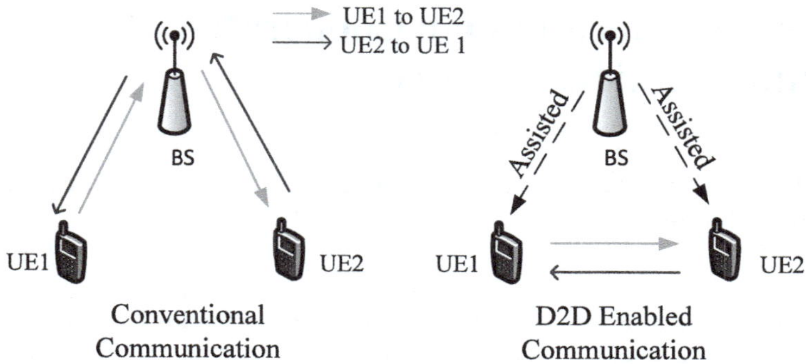

Fig. 1 In conventional systems, UE1 and UE2 communicate via the BS, irrespective of their proximity to each other. However, in D2D enabled communication, network latencies and throughput rates can be significantly improved by establishing local communications with the help of signalling assistance from the BS

for example, a scenario where two user equipments (UEs), located within close proximity to each other, should share gigabytes of data on a regular basis through a wireless medium. If D2D communications are enabled, as shown in Fig. 1, the BS coordinating these devices is no longer a traffic bottleneck between the source and the destination.

D2D communications can potentially improve the end-user experience by reducing network latency, reducing power consumption, increasing peak data rates, and facilitating several proximity-based services [2]. Moreover, since D2D is a short-distance based communication technology, sufficiently separated D2D links can simultaneously operate on the same frequency band, thereby facilitating dense spectral reuse across the network. Due to these added benefits, D2D has been specified by 3GPP in LTE Release 12, as a technology with focus on the applications of proximity-based services and public safety communication [3, 4].

In proximity-based services, devices detect their proximity and subsequently trigger a variety of services including, for example, mobile social networking, mobile marketing, proximal multi-player gaming, advertising, local exchange of information, and smart vehicle-to-vehicle communications [5]. In public safety support, D2D can provide local connectivity to the UEs when the central radio infrastructure is damaged. Therefore, UEs can benefit significantly from a number of public safety services with the help of D2D technology. However, D2D brings a unique set of research challenges to the network designer. The most important research challenges include the design of smart devices, information security, interference management, and mobility management.

1.1 Cooperative D2D as a Technology Enabler for 5G

D2D communication is also being considered as one of the state-of-the-art technologies for the 5G network architecture [6, 7]. This is because D2D-enabled 5G networks can support a plethora of services such as, cellular network offloading, low-latency tactile applications, vehicle-to-vehicle communication, national security and service advertisement. Recent research shows that D2D communications within the shared cellular spectrum controlled by cellular infrastructure achieve the following potential gains:

1. Spectral or network capacity gain: by sharing radio resources efficiently between conventional and D2D-enabled UEs.
2. User data rate gain: high peak rates may be achieved when UEs are close to each other with strong propagation links between them.
3. Latency gain: direct communication between UEs bypassing the network infrastructure reduces the end-to-end latency.

Additionally, D2D connectivity makes it possible for mobile devices in the multi-tier 5G network architecture to function as transmission relays for each other. Such cooperative mobile relaying methods have recently attracted significant attention because they can greatly improve the network performance without deploying new BSs [8]. By allowing relay-based communication, it is possible to improve network coverage and quality of service and to achieve similar performance benefits as in multiple-input-multiple-output (MIMO) systems. The idea of relaying is particularly attractive for the cell-edge users and for the users in coverage holes. In conventional relay systems, fixed relay stations are installed on a per-demand basis. Fixed relay stations not only result in significant infrastructure and maintenance costs for the operators, but also consume significant amounts of power. In this regard, mobile UEs with strong communication links to the BS can help as relays, via D2D communication, to other UEs with heavy blockage and shadowing at no additional cost to the mobile operators. In other words, mobile relaying via D2D communication can enhance the performance of cellular networks by taking the advantage of "shadowing diversity".

Consider the D2D relaying example illustrated in Fig. 2. The BS has to communicate some information with the destination UE. However, due to network blockage, direct communication with the destination UE cannot be established. The Relay UE, which is closer to the BS and has strong communication links with it, can act as a relay for the destination UE to enable indirect communication between the BS and the destination UE. In other words, if the cellular link of a mobile UE is weak, strong D2D links can establish a cellular connection via relay-UEs. Additionally, as opposed to the high power consumption in fixed relay stations, D2D relays operate at significantly lower power consumption levels and can therefore, potentially act as an energy-efficient replacement for the fixed relay stations.

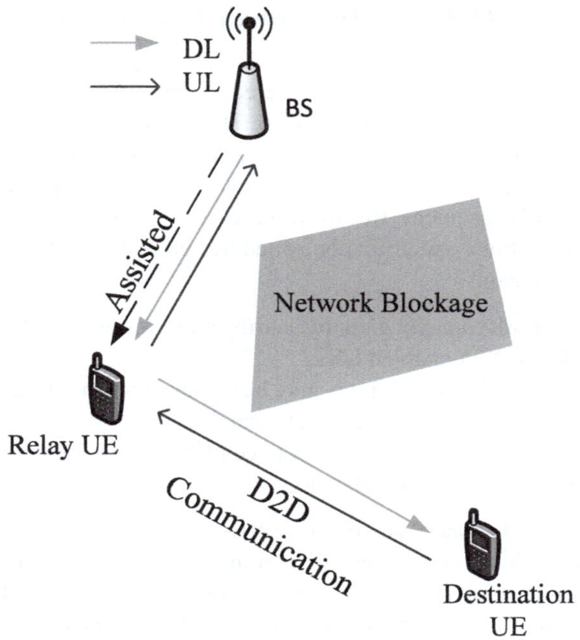

Fig. 2 System model for cellular wireless systems with cooperative D2D communication. UEs with strong communication links from the BS can send/receive data directly from the BS. However, UEs with bad communications links from the BS (for example, due to network blockage) can benefit from D2D communication with relay UEs to send/receive the data from the BS

1.2 Chapter Organization

The concept of D2D relaying has motivated researchers to investigate design methods which jointly exploit the advantages of both D2D and relaying technologies. Prominent research areas include opportunistic relay-UE selection, relay-UE power allocation, interference management, minimization of relay battery power, and maximizing network energy efficiency, improving coverage, and so on. In this chapter, different resource allocation problems associated with cooperative D2D networks operating under a variety of architectural settings are discussed, including OFDMA and cognitive radio (CR) networks. In Sect. 2, resource allocation problems for downlink transmission in OFDMA-based buffer-aided D2D relaying networks are investigated. Efficient algorithms are developed to solve the problem of power and subcarrier allocation such that the system capacity is maximized. In Sect. 3, energy-harvesting technology is considered in OFDMA systems where the UEs are capable of harvesting energy using the "wireless power transfer (WPT)" technology [9] and only the harvested energy is used for cooperative D2D-relaying. The joint optimization problem of power allocation, subcarrier allocation and relay-UE selection is studied. In Sect. 4, a cognitive radio network (CRN) is considered, where the secondary users (SUs) from different small cell primary user (PU) networks attempt communication with each other. Cooperative D2D-relaying is utilized to enable communication among the SUs. Efficient relay-CR selection and power allocation schemes are developed for realistic networks under channel uncertainty. Finally, in Sect. 5, we provide few concluding remarks and directions for future work.

2 Resource Allocation for OFDMA-Based Buffer-Aided D2D Relaying Networks with Frequency Reuse

In this section, we consider the scenario illustrated earlier in Fig. 2, in which BS cannot communicate with the destination UE due to heavy obstructions or shadowing. Our goal is to establish indirect communication links between the BS and the UEs which are blocked from direct communication with the BS. In this regard, some other UEs with fair connection to the BS can assist as relays. As discussed earlier in Sect. 1.1, D2D relaying enhances the system performance by taking advantage of "shadowing diversity".

Specifically, we investigate D2D relaying for downlink transmission in OFDMA cellular networks. To analyze the performance gains achievable via D2D-relaying in this OFDMA system, we first classify the D2D pairs[1] based on the level of proximity with each other. Firstly, we consider a scenario where all the D2D pairs are very close to each other and their coverage areas overlap. In this case, since reusing the subcarriers among the D2D pairs will introduce significant interference to the D2D links, we do not attempt subcarrier reuse among the D2D pairs. Next, we consider a scenario where the D2D pairs are far enough from each other, and therefore their coverage areas are disjoint. In this case, full subcarrier reuse can be attempted with the help of buffer-aided relaying in all of the D2D pairs. We discuss relevant resource allocation problems and analyze the throughput gains achievable in both of the above scenarios. Later, we provide a general motivation on how frequency reuse can be attempted for a general scenario where D2D pairs are randomly distributed in a cell, i.e, the two aforementioned scenarios co-exist simultaneously within the cell.

2.1 System Model

Let us assume that all the mobile UEs which can directly communicate with the BS are interested in relaying. However we opportunistically select few UEs for relaying based on the performance gains they can offer to the system. The important fact

[1]We refer to a relay-UE and destination-UE combination as a D2D pair if a D2D communication link can be established between them.

in the D2D communication is that the UEs can only communicate within short distances. Therefore, the relay-UEs (also referred to as relays interchangeably) and the UEs in coverage holes (also referred to as destination-UEs, or d-UEs, interchangeably) must be in the coverage area of each other. Also, to make the problem general, we assume that the d-UEs can receive assistance from multiple relays. Let N orthogonal subcarriers, each with bandwidth W be available for resource allocation in the system. Let K be the number of d-UEs, i.e., UEs requiring relay assistance. Further, let us denote the BS, relay-UEs and d-UEs as B, R, and U, respectively. The set of the relays which can assist the kth UE is denoted by R_k. The channel gain and the transmission power from the BS to relay $m_k \in R_k$ over subcarrier i at time slot t are denoted by $g^{B,R}_{k,m_k,i,t}$ and $P^{B,R}_{k,m_k,i,t}$, respectively. Similarly, the channel gain from relay $m_k \in R_k$ to the kth UE over subcarrier j at time slot t is denoted by $g^{R,U}_{k,m_k,j,t}$. The transmission power of any relay in the candidate set, i.e, R_k, $\forall k$, over any subcarrier, i.e, $\forall i \in \{1, 2, \ldots N\}$, is the same fixed value denoted by P_R. We implement the popular amplify-and-forward (AF) protocol [10] for relaying.

2.2 D2D Relaying Without Frequency Reuse

We start with the first case in which all the D2D pairs are close to each other and their coverage areas overlap. As a result, frequency reuse may introduce excessive interference into the system. Note that, due to indirect communication via mobile relaying, downlink transmission from the BS to a given d-UE happens over two-time-slot intervals. During the first time slot, the BS transmits information to a relay-UE. This information is relayed by the relay-UE to the d-UE during the second time-slot. We also assume that each relay-UE has a unit buffer size, i.e, if all the BS-to-relay transmissions occur in the tth time slot, all the relay-to-d-UE transmissions occur in the $(t + 1)$th time sot. Under the AF relaying protocol, the achievable throughput $R_{k,m_k,i}$ for downlink transmission from the BS to the kth UE via relay m_k over subcarrier i is given by

$$R_{k,m_k,i} = \frac{1}{2}\log_2\left(1 + \frac{P^{B,R}_{k,m_k,i} P_R g^{B,R}_{k,m_k,i} g^{R,U}_{k,m_k,i}}{P^{B,R}_{k,m_k,i} g^{B,R}_{k,m_k,i} + P_R g^{R,U}_{k,m_k,i}} \cdot \frac{1}{N_0 W}\right), \tag{1}$$

where N_0 is the noise variance assuming additive white Gaussian noise (AWGN) channels. In this orthogonal transmission case, the time index is omitted, i.e, $g^{B,R}_{k,m_k,i}$ is used in place of $g^{B,R}_{k,m_k,i,t}$, because we assume the unit buffer size at all the relays. Using (1), we formulate the following resource allocation problem to maximize the sum throughput in the system

$$\begin{aligned}
&\underset{P,S}{\text{maximize}} \quad \sum_{k=1}^{K}\sum_{m_k\in R_k}\sum_{i=1}^{N} s_{k,m_k,i}R_{k,m_k,i}\\
&\text{subject to :}\\
&C_1 : \sum_{k=1}^{K}\sum_{m_k\in R_k}\sum_{i=1}^{N} s_{k,m_k,i}P_{k,m_k,i}^{B,R} \leq P_{max},\\
&C_2 : s_{k,m_k,i} \in \{0,1\}, \quad \forall i,k,m_k,\\
&C_3 : \sum_{k=1}^{K}\sum_{m_k\in R_k} s_{k,m_k,i} \leq 1, \quad \forall i,\\
&C_4 : P_{k,m_k,i}^{B,R} \geq 0, \quad \forall i,k,m_k,
\end{aligned} \tag{2}$$

where $P = \{P_{k,m_k,i}^{B,R}\}$ and $S = \{s_{k,m_k,i}\}$ denote the power allocation and subcarrier allocation policies, respectively.

C_1 is the power budget constraint with P_{max} being the power budget for the BS. C_2 shows that the subcarrier assignment indicators $s_{k,m_k,i}$ are binary integer variables. $s_{k,m_k,i} = 1$ if BS is transmitting to the kth UE via relay m_k over subcarrier i and $s_{k,m_k,i} = 0$ otherwise. C_3 ensures that the subcarriers are not reused among the D2D pairs to avoid interference. C_4 ensures that the power allocation variables are non-negative.

In simple terms, our goal in (2) is to allocate the BS transmission power and the N subcarriers such that the sum-throughput in the system is maximized. From an optimization perspective, the problem in (2) is a mixed integer non-linear program (MINLP) which, in its original form, is difficult to solve in polynomial time due to the coupled integer and continuous variables [11–13]. Therefore, to obtain a real-time solution, we attempt a series of mathematical modifications and simplifications as shown in Fig. 3.

Firstly, we relax the integer binary variables, i.e., the subcarrier assignment indicators $s_{k,m_k,i}$ into continuous variables in the interval $[0,1]$. Using this relaxation, we introduce auxiliary power variables as

$$\tilde{P}_{k,m_k,i}^{B,R} = s_{k,m_k,i}P_{k,m_k,i}^{B,R}. \tag{3}$$

When the original power variables in (2) are replaced by the auxiliary power variables defined in (3), the optimization problem (2) is relaxed into the following convex optimization problem

$$\min_{\lambda\geq 0}\max_{\tilde{P},S} L(\lambda,\tilde{P},S), \tag{4}$$

where L is a Lagrangian function and λ is the associated Lagrangian multiplier. Therefore, efficient convex optimization algorithms can be used to solve the relaxed problem. Consider, for example, the popular dual decomposition technique wherein a master problem and few sub-problems are solved iteratively until convergence is achieved. The dual decomposition technique can be adopted to solve (4) by allowing the master problem to update λs and the sub-problems to find optimal power and subcarrier assignment for a given λ.

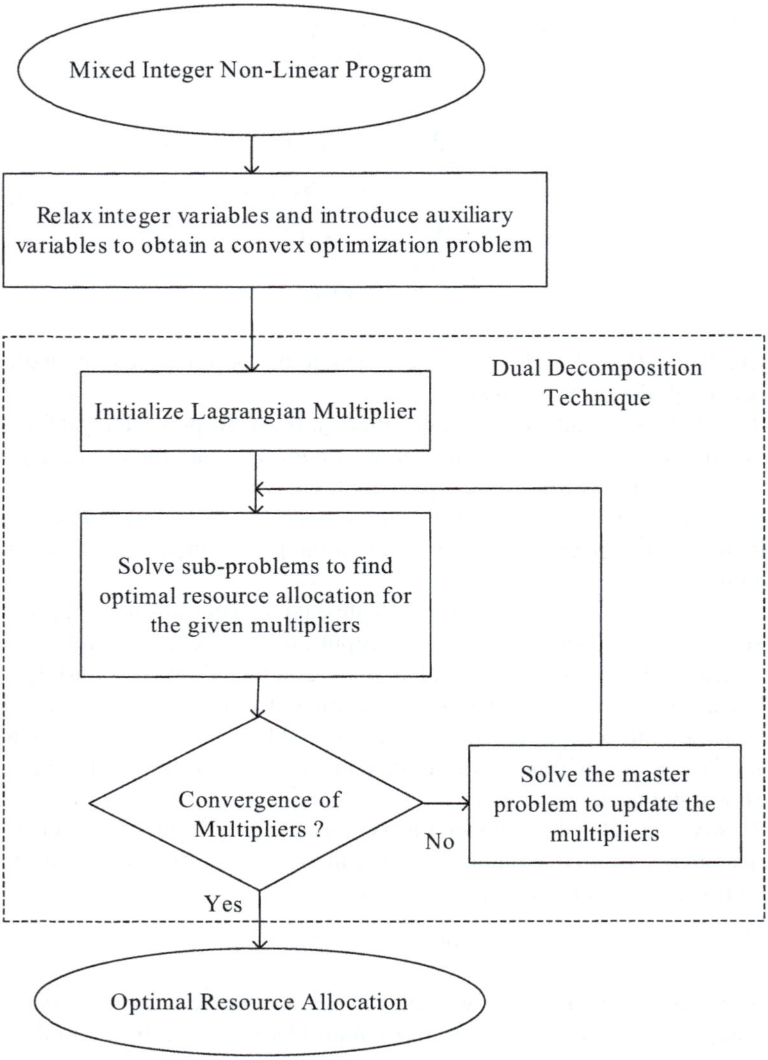

Fig. 3 Optimization framework to solve (2)

We begin with the following definition of the Lagrangian function

$$L(\lambda, \tilde{P}, S) = \sum_{k=1}^{K} \sum_{m_k \in R_k} \sum_{i=1}^{N} s_{k,m_k,i} \frac{1}{2} \log_2 \left(1 + \frac{\frac{\tilde{P}^{B,R}_{k,m_k,i} g^{B,R}_{k,m_k,i}}{s_{k,m_k,i} N_0 W} P_R g^{R,U}_{k,m_k,i}}{\frac{\tilde{P}^{B,R}_{k,m_k,i} g^{B,R}_{k,m_k,i}}{s_{k,m_k,i}} + P_R g^{R,U}_{k,m_k,i}} \right) + \lambda \Big(P_{max} - \sum_{k=1}^{K} \sum_{m_k \in R_k} \sum_{i=1}^{N} \tilde{P}^{B,R}_{k,m_k,i} \Big), \tag{5}$$

where we consider the power constraint C_1 only and initialize the Lagrangian multiplier λ with a non-negative value. Constraints C_2, C_3, C_4 are not considered in (5) but will be satisfied in subsequent steps. By applying KKT conditions to (5), we can derive optimal auxiliary power variables $\tilde{P^*}^{B,R}_{k,m_k,i}$ and the corresponding throughput $\tilde{R^*}_{k,m_k,i}$ in terms of the subcarrier assignment indicators $s_{k,m_k,i}$ and the Lagrangian multiplier λ. In the derivation process, constraint C_4 can be satisfied by offsetting negative $\tilde{P^*}^{B,R}_{k,m_k,i}$ values to zero. Plugging back $\tilde{P^*}^{B,R}_{k,m_k,i}$ and $\tilde{R^*}_{k,m_k,i}$ into the sub-problem (5) and re-arranging the equations, we obtain the simplified problem

$$\begin{aligned}&\underset{S}{\text{maximize}} \quad \sum_{k=1}^{K}\sum_{m_k\in R_k}\sum_{i=1}^{N} s_{k,m_k,i}H_{k,m_k,i} + \lambda P_{max}\\ &\text{subject to} \ : \ C_3,\end{aligned} \tag{6}$$

where we have

$$H_{k,m_k,i} = \tilde{R^*}_{k,m_k,i} - \lambda \tilde{P^*}^{B,R}_{k,m_k,i}. \tag{7}$$

Upon inspection, one may observe that (6) is a linear programming problem [14] with respect to $s_{k,m_k,i}$, i.e., it is a linear subcarrier assignment problem. Applying totally unimodularity theorem [15, 16] shows that the problem (6) has integer optimal values even if the optimization variables $S = \{s_{k,m_k,i}\}$ span over the continuous interval [0,1]. Optimal solution to the linear subcarrier assignment problem (6) can be obtained using simple algorithms such as Algorithm 1.

Algorithm 1 Subcarrier allocation algorithm I

1: Find the best user-relay pair for each subcarrier as
$(k^*, m_k^*) = \underset{k,m_k}{\text{maximize}}\ H_{k,m_k,i}\ \forall i.$

2: **return** S^*

Note that, constraint C_3 can be satisfied while implementing Algorithm 1. Also, since Algorithm 1 yields optimal solutions, constraint C_2 is satisfied inherently due to totally unimodularity theorem. Since the optimal power allocation and subcarrier assignment values are now available, the master problem of updating the Lagrange multiplier λ can be solved using standard sub-gradient methods [17] as follows

$$\lambda(l+1) = \Big[\lambda(l) - \eta\big(P_{max} - \sum_{k=1}^{K}\sum_{m_k\in R_k}\sum_{i=1}^{N} \tilde{P}^{B,R}_{k,m_k,i}\big)\Big]^+, \tag{8}$$

where η is a constant step size parameter, l is the iteration index, and $[x]^+ = \max(0, x)$.

The master problem and the sub-problems are solved iteratively until we observe a convergence in the Lagrangian multiplier, i.e, λ, values. The complexity of the proposed algorithm is polynomial in time and is significantly lower than the brute-force exhaustive search method.

2.3 *Buffer-Aided D2D Relaying with Frequency Reuse*

We now consider the second scenario wherein the D2D pairs are far enough from each other such that their coverage areas do not overlap. Therefore, frequency reuse can now be attempted to fully utilize the available spectrum. The resource allocation scheme proposed in the previous section, i.e, for orthogonal transmission, may not be spectrally efficient for this system because frequency reuse was not considered. Note that, frequency reuse is not possible in the first time slot, as the BS is broadcasting the messages. However, in the second time slot, frequency reuse can be attempted to establish D2D communication between relay-UEs and d-UEs. To exploit the benefits of frequency reuse, we propose a communication protocol using buffer-aided relaying.

As a motivational example, let us consider a system with two D2D pairs and two subcarriers. During the first time-slot, the BS broadcasts two packets, i.e, one to each relay-UE, over the two subcarriers. During the second time-slot, the BS broadcasts another two packets over the two subcarriers. Therefore, at the end of the second time-slot, we have four packets at the relay-UE buffers, i.e, two packets at each relay-UE. Since the coverage areas of the two D2D pairs do not overlap, both the relay-UEs can use the two subcarriers simultaneously for D2D communications. That is, during the third time slot, each relay-UE can forward its two packets over the two subcarriers without causing any interference to other relay-UEs. Therefore, using buffer-aided relaying with full frequency reuse, four packets are transmitted to two d-UEs in three time slots. On the other hand, the orthogonal transmission, i.e, without frequency reuse, scheme discussed previously in Sect. 2.2 requires four time slots to transfer the same four packets from the BS to the two d-UEs.

To understand the benefits of frequency reuse, we now formulate a resource allocation problem similar to (2) wherein the objective is to maximize the sum throughput in the system. Similar to the previous scenario, we have K destination UEs and N subcarriers. We assume that the BS broadcasts messages to the relay-UEs over T time slots and that the relay-UEs forward these messages to d-UEs during time slot $T + 1$. Thereby, the following resource allocation problem can be formulated as

$$
\begin{aligned}
&\underset{P,S}{\text{maximize}} \quad \sum_{k=1}^{K} \sum_{m_k \in R_k} \sum_{i=1}^{N} \sum_{j=1}^{N} \sum_{t=1}^{T} s_{k,m_k,i,j,t} R_{k,m_k,i,j,t} \\
&\text{subject to}: \\
&C_1: \sum_{k=1}^{K} \sum_{m_k \in R_k} \sum_{i=1}^{N} \sum_{j=1}^{N} s_{k,m_k,i,j,t} P^{B,R}_{k,m_k,i,t} \le P_{max}, \ \forall t, \\
&C_2: s_{k,m_k,i,j,t} \in \{0,1\}, \quad \forall i,j,k,m_k,t, \\
&C_3: \sum_{k=1}^{K} \sum_{m_k \in R_k} \sum_{j=1}^{N} s_{k,m_k,i,j,t} \le 1, \quad \forall i,t, \\
&C_4: \sum_{t=1}^{T} \sum_{m_k \in R_k} \sum_{i=1}^{N} s_{k,m_k,i,j,t} \le 1, \quad \forall k,j, \\
&C_5: \sum_{t=1}^{T} \sum_{m_k \in R_k} \sum_{i=1}^{N} \sum_{j=1}^{N} s_{k,m_k,i,j,t} \le N, \quad \forall k, \\
&C_6: P^{B,R}_{k,m_k,i,t} \ge 0, \quad \forall i,k,m_k,t,
\end{aligned}
\tag{9}
$$

where $P = \{P^{B,R}_{k,m_k,i,t}\}$ and $S = \{s_{k,m_k,i,j,t}\}$ denote the power allocation and subcarrier allocation policies, respectively. Here, C_1 is the power budget constraint for the BS and P_{max} is the maximum power budget for the BS. C_2 shows that the subcarrier assignment indicators $s_{k,m_k,i,j,t}$ are binary integer variables. $s_{k,m_k,i,j,t} = 1$ if the BS is transmitting messages to a relay m_k over subcarrier i during time slot t, and the relay m_k forwards these messages to the kth UE over subcarrier j during time slot $(T+1)$. $s_{k,m_k,i,j,t} = 0$ otherwise. C_3 implies that the BS cannot reuse any subcarrier within a given time-slot for broadcasting, i.e, to avoid interference. C_4 implies that the subcarriers cannot be reused during a given time-slot for communication within a D2D pair, i.e, to avoid interference. C_5 implies that each relay-UE can have a buffer length upto N because a maximum of N messages, i.e, using all the N subcarriers, can be forwarded by a given relay-UE in a given time slot. C_6 ensures that the power variables are non-negative.

In simple terms, our goal in (10) is to allocate the BS power and assign the N subcarriers such that the total system throughput is maximized. The achievable throughput $R_{k,m_k,i,j,t}$ for downlink transmission from BS to UE k, via the relay m_k, over subcarrier i and j in the tth and $(T+1)$th time slots , respectively, is given by

$$
R_{k,m_k,i,j,t} = \frac{1}{2}\log_2\left(1 + \frac{P^{B,R}_{k,m_k,i,t} P_R g^{B,R}_{k,m_k,i,t} g^{R,U}_{k,m_k,j,T+1}}{P^{B,R}_{k,m_k,i,t} g^{B,R}_{k,m_k,i,t} + P_R g^{R,U}_{k,m_k,j,T+1}} \cdot \frac{1}{N_0 W}\right). \tag{10}
$$

Similar to the scenario in the previous section, the original problem is MINLP. Therefore, we follow a similar solution approach to solve (10). We begin by relaxing the integer variables into continuous intervals and defining auxiliary power variables similar to (3). Plugging these changes into the original problem makes it convex and the dual decomposition technique described in Fig. 3 can be applied. The corresponding dual Lagrangian function is given by

$$
\begin{aligned}
L(\boldsymbol{\lambda}, \tilde{P}, S) = & \sum_{k=1}^{K} \sum_{m_k \in R_k} \sum_{i=1}^{N} \sum_{j=1}^{N} \sum_{t=1}^{T} s_{k,m_k,i,j,t} \\
& \frac{1}{2}\log_2 \left(1 + \frac{\frac{\tilde{P}^{B,R}_{k,m_k,i,t} g^{B,R}_{k,m_k,i,t}}{s_{k,m_k,i,j,t} N_0 W} P_R g^{R,U}_{k,m_k j,T+1}}{\frac{\tilde{P}^{B,R}_{k,m_k,i,t} g^{B,R}_{k,m_k,i,t}}{s_{k,m_k,i,j,t}} + P_R g^{R,U}_{k,m_k j,T+1}} \right) + \\
& \sum_{t=1}^{T} \lambda_t \Big(P_{max} - \sum_{k=1}^{K} \sum_{m_k \in R_k} \sum_{i=1}^{N} \sum_{j=1}^{N} \tilde{P}^{B,R}_{k,m_k,i,t} \Big),
\end{aligned}
\tag{11}
$$

where $\boldsymbol{\lambda} = \{\lambda_t\}$ is the vector of non-negative Lagrangian multipliers associated with the power constraints in C_1. The constraints C_2, C_3, C_4, C_5, C_6 are not considered in (11) but will be satisfied in subsequent steps. As in the previous solution approach, for a given Lagrangian multiplier λ, optimal auxiliary power variables $\tilde{P*}^{B,R}_{k,m_k,i,t}$ and corresponding throughput rates $\tilde{R*}_{k,m_k,i,j,t}$ can be obtained by applying KKT conditions to the sub-problems. While deriving $\tilde{P*}^{B,R}_{k,m_k,i,t}$, constraint C_6 can be satisfied by offsetting negative values to zero. Plugging back $\tilde{P*}^{B,R}_{k,m_k,i,t}$ and $\tilde{R*}_{k,m_k,i,j,t}$ into the sub-problem and re-arranging the equations results in the following simplified problem

$$
\begin{aligned}
& \underset{S}{\text{maximize}} \quad \sum_{k=1}^{K} \sum_{m_k \in R_k} \sum_{i=1}^{N} \sum_{j=1}^{N} \sum_{t=1}^{T} s_{k,m_k,i,j,t} H_{k,m_k,i,j,t} + \sum_{t=1}^{T} \lambda_t P_{max} \\
& \text{subject to} : \; C_3, C_4, C_5,
\end{aligned}
\tag{12}
$$

where

$$
H_{k,m_k,i,j,t} = \tilde{R*}_{k,m_k,i,j,t} - \lambda_t \tilde{P*}^{B,R}_{k,m_k,i,t}. \tag{13}
$$

That is, as encountered in the previous solution approach, we are required to solve a linear integer programming problem (12) so as to obtain optimal subcarrier assignment for a given Lagrangian multiplier λ. Therefore, we propose Algorithm 2 to solve the linear integer programming problem (12) in polynomial time.

Algorithm 2 begins with finding the best relay-UE for all the subcarriers i used by the BS for broadcasting messages and all the subcarriers j used by the relay-UEs for forwarding those messages. Here, the subcarriers are guaranteed to be used at most once in each broadcasting time-slot, i.e, constraint C_3 is satisfied. However, subcarrier reuse is permitted in the second time slot because the D2D pairs are far apart and their coverage areas do not overlap, i.e, constraint C_4 is satisfied. We also ensure that each relay receives and transmits over the same number of subcarriers. In the process, constraint C_5 is also satisfied. Towards the end, we use ordered subcarrier pairing [18] such that the subcarrier with the greatest channel gain among the BS-relay communication links is assigned to the subcarrier with the greatest channel gain among the relay to d-UE communication links. Similarly, subcarriers with the second largest gains are paired, and so on. Such a subcarrier pairing is known to maximize the system throughput [18]. Note, however, that an

Algorithm 2 Subcarrier allocation algorithm II

1: For each subcarrier, find the relay-UE with the best channel gain to each destination-UE:
$m_k^* = \underset{m_k}{\text{maximize}}\, g_{k,m_k,j,T+1}^{R,U}, \forall k, j.$
2: Count the number of subcarriers assigned to each relay-UE and denote it as $a(m_k)$.
3: **for** i=1:N
4: **for** t=1:T
5: For each subcarrier i used by the BS to broadcast messages to the relay-UEs, find the D2D pair with the best channel gain at time slot t:
$(k^*, m_k^*) = \underset{k,m_k}{\text{maximize}}\, g_{k,m_k,i,t}^{B,R}.$
6: $a(m_k^*) = a(m_k^*) - 1;$
7: **if** $a(m_k^*) \geq 0$, assignment (i, t, k^*, m_k^*) is valid.
else assignment (i, t, k^*, m_k^*) is not valid and relay-UE m_k^* is omitted from the relay selection pool.
8: **end**
9: **end**
10: For each relay m_k, implement an ordered subcarrier pairing [18].
11: **return** S^*

optimal subcarrier assignment can not be found in polynomial time and that the discussed subcarrier assignment algorithm, although sub-optimal, provides a real-time solution. With subcarrier assignment, the solution process for sub-problems is complete.

The Lagrangian multiplier set $\{\lambda_t\}$ is updated in the master problem using sub-gradient methods [14] such as follows

$$\lambda_t(l+1) = \left[\lambda_t(l) - \eta\Big(P_{max} - \sum_{k=1}^{K} \sum_{m_k \in R_k} \sum_{i=1}^{N} \sum_{j=1}^{N} \tilde{P}_{k,m_k,i,t}^{B,R}\Big)\right]^+, \forall t, \tag{14}$$

where η is a constant step size parameter. The master problem and the sub-problems are iteratively solved until a convergence in the Lagrangian multiplier values, i.e, λ_t, is observed. This completes our discussion on resource allocation for buffer-aided D2D relaying systems.

2.4 Simulation Results

In this section, we present two numerical results to illustrate and compare the performance gains achieved by the proposed resource allocation methods. The simulation setup comprises three candidate relays for each UE. The transmission power for each relay P_R is set to 0.1 W. Figure 4 shows average system capacity versus P_{max} for different number of UEs K considering $N = 16$ subcarriers. We compare the achievable performance gains of a scenario where frequency reuse is possible, i.e, when D2D pairs are far enough from each other with disjoint coverage area, against a scenario where frequency reuse is not possible, i.e, D2D pairs have

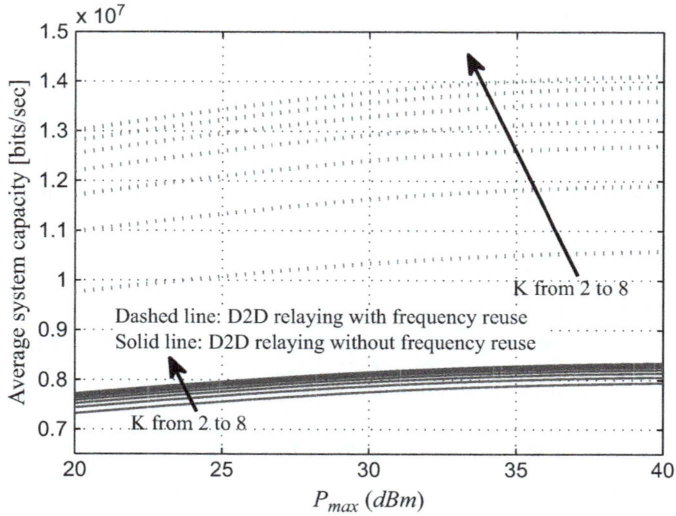

Fig. 4 Average system capacity versus BS power budget P_{max} for different number of UEs K. Result shows significant improvement of system capacity when frequency is reused in D2D links

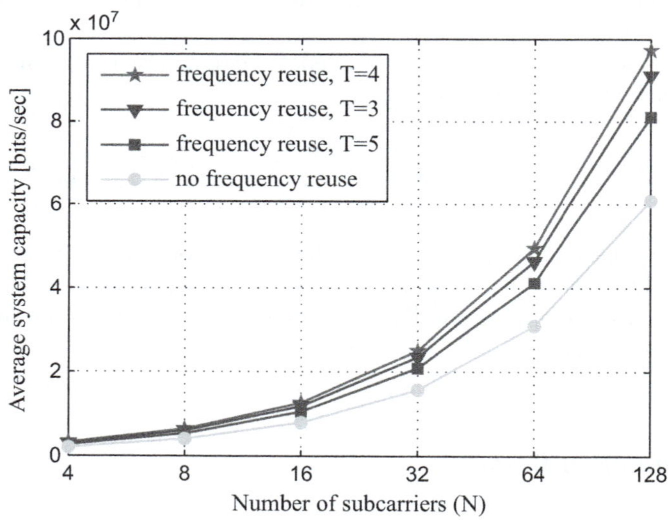

Fig. 5 Average system capacity versus number of subcarriers N for $P_{max} = 35$ dBm and $K = 4$

overlapping coverage area. Significant system capacity improvement is observed for resource allocation schemes with frequency reuse. With the increase of number of UEs, i.e, K, the frequency reusing factor improves and as a result, the capacity gain increases.

In Fig. 5, we show average system capacity versus number of subcarriers N for $P_{max} = 35\,\text{dBm}$ and $K = 4$. As expected the system capacity increases with the increase of the number of subcarriers. Here, we show the performance of the resource allocation scheme with frequency reuse for different values of T, where T is the number of time slots for broadcasting. We notice that by increasing T from 3 to 4, the system capacity increases. However, the system capacity decreases when we change T from 4 to 5. The reason is when T is greater than the number of D2D pairs, the number of messages are greater than the size of the buffer at the relays. Therefore, some of the messages are discarded which will result in having lower system capacity than expected.

2.5 D2D Relaying: Motivation for a General Scenario

In the discussions so far, we have considered two theoretically extreme scenarios in the context of D2D relaying. In the first scenario, the D2D pairs are close to each other and their coverage areas overlap. Therefore, since frequency reuse cannot be attempted, we proposed resource allocation algorithms considering the orthogonality of the subcarriers. In the second scenario, the D2D pairs are far enough from each other and their coverage areas are disjoint. Therefore, we proposed resource allocation which fully reuse the subcarriers for D2D communication

However, in a practical setup, D2D pairs would be distributed randomly in the cell. As a result, both the scenarios discussed in the previous sections can co-exist simultaneously. That is, few D2D pairs would experience coverage overlap with each other, while few other D2D pairs would be disjoint from each other. Resource allocation for such a general scenario can be attempted by developing an analogy based on the conventional frequency reuse patterns used in cellular networks.

In conventional cellular systems, the available frequency spectrum is generally reused in sufficiently separated cells. While adjacent cells use different frequencies to reduce inter-cell interference, the available set of frequencies are usually exhausted within a cluster of cells due to limited spectrum. However, the same frequencies can be reused in adjacent cell-clusters. A more aggressive and effective approach from an interference perspective is the fractional frequency reuse (FFR) technique, where the cells are in-turn partitioned into smaller spatial regions and the available frequency spectrum is reused over these regions.

Similar frequency reuse ideas can be applied in the context of D2D relaying networks. The cellular operator can reserve a dedicated in-band spectrum for D2D communications. Since D2D communications occur over short distances, ultra-dense frequency reuse can be attempted. In other words, several clusters of D2D pairs, with each cluster fully exhausting the available frequency resources, can co-exist within each cell. The D2D pairs within in the same cluster should use orthogonal frequencies, i.e, resource allocation as per the first scenario (c.f. Sect. 2.2) should be attempted. On the other hand, D2D pairs in different clusters can reuse the same set of available frequencies, i.e, resource allocation as per the second scenario (c.f. Sect. 2.3) should be attempted. On a slightly advanced note, intelligent FFR techniques can also be adopted for D2D communications (for example, see [19]) to minimize interference in the system.

2.6 Summary

In this section, we proposed resource allocation algorithms for OFDMA cellular systems, in which BS communicates with destination-UEs via mobile relay-UEs using D2D links. We considered two scenarios based on the proximity of the D2D pairs. Firstly, we considered that D2D pairs are very close to each other and their coverage areas overlap. In this case, we proposed subcarrier and power allocation without frequency reuse in order to avoid the interferences among the D2D links. Next, we considered that the D2D pairs are far enough and their coverage areas are disjoint. In this case, we proposed subcarrier allocation policy with full frequency reuse with the help of buffer-aided relaying. We also provided motivation for resource allocation in a general scenario where D2D pairs are randomly distributed in a cell, i.e., the two aforementioned scenarios co-exist within the cell. Numerical results show that buffer-aided D2D relaying with frequency reuse can significantly improve the performance of the system compared to the conventional D2D relaying scheme with no frequency reuse.

3 Resource Allocation for Energy Harvesting-Based OFDMA Systems with Selective D2D Relaying

In the previous section, we assumed all the relay-UEs are interested in relaying information. However, this assumption is too optimistic for UEs due to their capacity-limited batteries. Without any incentive, the UEs may not be interested in relaying information using their own battery power. In this regard, energy harvesting technology can offer energy incentives for the relay-UEs.

D2D relaying improves the performance of the system if energy harvesting (EH) capabilities are incorporated at the relay-UEs. This is because the relay-UEs can utilize the energy harvested from external resources for relaying operations. Typically, the relays can harvest energy from environmental resources such as wind, solar, and thermal energy or through wireless power transfer from the BS. Environmental energy resources are often not reliable for wireless communication because they are subject to environmental fluctuations and therefore, may not be available anytime and anywhere. On the other hand, in the wireless power transfer (WPT) technology [20], the BS transfers both information and power to the relays using radio frequency (RF) signals. The relays can harvest the energy transmitted by the BS and utilize the harvested energy to establish D2D links with the UEs in the coverage holes. Thereby, relays facilitate indirect communication between the BS and the UEs in coverage holes. Note that, WPT also offers reliability because the downlink RF transmissions to the relays can be controlled at the BS.

In this section, we study the performance gains achieved by incorporating energy harvesting capabilities into a D2D-relaying OFDMA system. Specifically, the relays harvest energy transferred via WPT by the BS and utilize the harvested energy for relaying operations. Among the various kinds of energy-harvesting receivers proposed in the literature, we focus on relays with energy splitting receivers [21]. Basically, using an energy splitting receiver, the relays split a received signal into two streams: one of them is used for decoding information and the other is used for energy harvesting. Thereby, we develop resource allocation algorithms for maximizing the system throughput. Specifically, we discuss the joint optimization problem of power allocation, subcarrier assignment, and relay-UE selection considering selective cooperation of UE devices. The optimization problem is again an MINLP, which is in general very difficult to solve in its original form due to the coupled integer and continuous variables [11–13]. Similar to the solution approach discussed in Sect. 2.2, we propose a computationally efficient solution approach based on the dual decomposition method. The proposed scheme is more effective and energy-efficient in comparison to the other existing schemes in the literature.

3.1 System Model

Consider a downlink single-cell network where the BS is placed in the center of the cell. We consider an OFDMA system with N orthogonal subcarriers, each of them with a bandwidth of W. There are two groups of mobile UEs and the number of UEs in the first and second group is M and K, respectively. The first group UEs are close to the BS and have strong direct link channels. The second group UEs are far apart from the BS and have weak direct link channels due to heavy blockage. When cooperation is beneficial for the second group, UEs in the first group act as relays for them via D2D cooperation. In this section, we consider the downlink transmission from the BS to the second group UEs only via D2D-relay assistance from the first group of UEs. We also assume that the first group UEs use separate bands for their own communication with the BS. In the remainder of the section, we will consider second group as users and the first group as relays. The popular DF relaying protocol [22] is used whenever D2D cooperation is beneficial. The relay UEs decide whether to cooperate or not based on the channel conditions.

We denote the relays, i.e., UEs in the first group, by m, and the users, i.e. UEs in the second group, by k. The transmission modes corresponding to cooperative and non-cooperative are superscripted with (C) and (NC), respectively. The channel gains from BS B to relay m, and from relay m to user k are defined as $g_{B,m}^{i,(C,1)}$ and $g_{m,k}^{i,(C,2)}$, respectively. The superscript $i,(C,1)$ and $i,(C,2)$ indicate the cooperative transmission mode over subcarrier i in time slot one and two, respectively. Similarly, the channel gain for the non-cooperative, i.e, direct, link is defined as $g_{B,k}^{i,(NC)}$. In addition, we define $\alpha_{B,m}^{i,(C,1)} = \frac{g_{B,m}^{i,(C,1)}}{N_0 W}$, $\alpha_{m,k}^{i,(C,2)} = \frac{g_{m,k}^{i,(C,2)}}{N_0 W}$ and $\alpha_{B,k}^{i,(NC)} = \frac{g_{B,k}^{i,(NC)}}{N_0 W}$, where N_0 is the variance of the AWGN received by each receiver.

We assume that each UE has two sources of power, one from the battery and the other from the harvested energy. The harvested energy is received by the power splitting operation. Particularly, in the case of cooperative transmission, relays split the received signal from the BS into two streams: one for harvesting the energy (E), and the other for decoding the information (I) with the ratios of $\rho_{m,k}^{i,(E)}$ and $\rho_{m,k}^{i,(I)}$, respectively. The harvested energy is used by the relays for D2D transmission, i.e., only to forward the data for the users. Therefore, UEs do not consume their own batteries for the purpose of relaying. Since we consider downlink transmissions only, the role of energy harvesting at the UEs in the second group is ignored in our discussions.

3.2 Resource Optimization Problem Formulation

We now formulate a resource allocation problem, where the objective is to maximize the total capacity of the system. The capacity of the cooperative link can be calculated as

$$R_{m,k}^{i,(C)} = \frac{1}{2}\min\Big\{\log_2(1 + \alpha_{B,m}^{i,(C,1)}\rho_{m,k}^{i,(I)}P_{B,m}^{i,(C)}), \\ \log_2(1 + \alpha_{m,k}^{i,(C,2)}\alpha_{B,m}^{i,(C,1)}\rho_{m,k}^{i,(E)}P_{B,m}^{i,(C)})\Big\}, \qquad (15)$$

where $P_{B,m}^{i,(C)}$ is the transmission power of BS B on subcarrier i over the cooperative link to send the data for relay m. $\alpha_{B,m}^{i,(C,1)}P_{B,m}^{i,(C)}$ is the power received by relay m after attenuation by the fading channel in the first hop. The relay splits the received signal into two power streams with the ratios of $\rho_{m,k}^{i,(I)}$ and $\rho_{m,k}^{i,(E)}$, where $\rho_{m,k}^{i,(I)}$ and $\rho_{m,k}^{i,(E)}$ are used for decoding the information and for harvesting, respectively. In the second time slot, the relay will use the harvested power given by $\alpha_{B,m}^{i,(C,1)}\rho_{m,k}^{i,(E)}P_{B,m}^{i,(C)}$ to forward the data for user k. It is assumed that all the energy harvested in first time slot in the relay will be completely utilized for forwarding the information in the second time slot. The remaining harvested energy, if any, at the end of each transmission frame is assumed to be discarded. This assumption simplifies the problem as resources can be optimized over each transmission frame. On the other hand, if we assume that the UEs are capable of storing the harvested energy for future use, dynamic programming methods [23] can be implemented to efficiently utilize the harvested energy over multiple transmission frames. For simplicity, we do not consider the second scenario where dynamic programming is required.

If the direct link is better than the cooperative link, the BS directly transmits data to the user. The capacity of the direct link is calculated as

$$R_k^{i,(NC)} = \log_2(1 + \alpha_{B,k}^{i,(NC)} P_{B,k}^{i,(NC)}), \tag{16}$$

where $P_{B,k}^{i,(NC)}$ is the transmission power of BS B on subcarrier i over the direct link to user k.

Our objective is to maximize the total capacity of the system subject to the power constraint of the BS. The joint optimization problem of subcarrier assignment, power allocation, transmission mode, i.e, cooperative or non-cooperative, and relay selection is formulated as

$$\begin{aligned}
&\underset{P,S,\rho}{\text{maximize}} \quad \sum_{k=1}^{K}\sum_{m=1}^{M}\sum_{i=1}^{N} s_{m,k}^{i,(C)} R_{m,k}^{i,(C)} + \sum_{k=1}^{K}\sum_{i=1}^{N} s_k^{i,(NC)} R_k^{i,(NC)} \\
&\text{subject to :} \\
&C_1 : \sum_{k=1}^{K}\sum_{m=1}^{M}\sum_{i=1}^{N} s_{m,k}^{i,(C)} P_{B,m}^{i,(C)} + \\
&\qquad \sum_{k=1}^{K}\sum_{i=1}^{N} s_k^{i,(NC)} P_{B,k}^{i,(NC)} \le P_{max}, \\
&C_2 : s_{m,k}^{i,(C)}, s_k^{i,(NC)} \in \{0,1\}, \quad \forall i,k,m, \\
&C_3 : \sum_{k=1}^{K}\sum_{m=1}^{M} s_{m,k}^{i,(C)} + \sum_{k=1}^{K} s_k^{i,(NC)} \le 1, \quad \forall i, \\
&C_4 : \rho_{m,k}^{i,(I)} + \rho_{m,k}^{i,(E)} = s_{m,k}^{i,(C)}, \quad \forall i,k,m, \\
&C_5 : P_{B,m}^{i,(C)}, P_{B,k}^{i,(NC)}, \rho_{m,k}^{i,(I)}, \rho_{m,k}^{i,(E)} \ge 0, \quad \forall i,k,m, \\
&C_6 : \sum_{k=1}^{K}\sum_{i=1}^{N} s_{m,k}^{i,(C)} \le n_m, \quad \forall m, \\
&C_7 : \alpha_{B,m}^{i,(C,1)} \rho_{m,k}^{i,(I)} = \alpha_{B,m}^{i,(C,1)} \alpha_{m,k}^{i,(C,2)} \rho_{m,k}^{i,(E)}, \quad \forall i,k,m,
\end{aligned} \tag{17}$$

where P, ρ and S denote the power allocation policy, power splitting factors and subcarrier allocation policy, respectively. C_1 is the power budget constraint, where P_{max} is the power budget of the BS. C_2 shows that $s_{m,k}^{i,(C)}$ and $s_k^{i,(NC)}$ are binary integer variables, i.e, indicators. If BS is transmitting to user k with the assistance of relay m over subcarrier i, $s_{m,k}^{i,(C)}$ is one, otherwise it is zero. Similarly $s_k^{i,(NC)}$ is defined for non-cooperative transmission. C_3 implies that each subcarrier can be used only once to avoid interference. It also states that over each subcarrier either cooperative or non-cooperative mode can be used. C_4 indicates that the sum of the power splitting ratio in each subcarrier should be equal to one if that subcarrier is selected for cooperation, i.e., the splitter is not producing any energy and power wastage is ignored. C_5 states that the power and splitting variables are non-negative. C_6 indicates that each relay can only assist over n_m subcarriers due to the limitation of the power splitter and capacity of the energy harvester, where n_m is known a priori. The constraint C_7 is incorporated to avoid information loss and wastage of power at the relays. The problem stated above is a MINLP which is very difficult to solve in general. An efficient approach which uses the dual decomposition method to obtain an optimal solution for this problem is given in [24].

3.3 Solution Approach

In order to solve the problem (17), first we obtain the power splitting ratios for each cooperative link from C_4 and C_7 as

$$\begin{aligned} \rho_{m,k}^{i,(I)*} &= \frac{h_{m,k}^{i,(C,2)}}{1 + h_{m,k}^{i,(C,2)}}, \\ \rho_{m,k}^{i,(E)*} &= \frac{1}{1 + h_{m,k}^{i,(C,2)}}. \end{aligned} \tag{18}$$

Since we have obtained the optimal power-splitting factors in (18), the variables of the optimization problem in (17) are reduced to (P, S), i.e., transmission power and subcarrier allocation policy.

We follow a solution strategy similar to our previous discussions on MINLP problems (c.f. Sect. 2.2). An overview of the solution approach is provided here. The major concern for solving problem (17) is the presence of integer subcarrier assignment variables and constraints. Therefore, we relax the indicators S to the continuous [0,1] interval and define auxiliary power variables as $\tilde{P}_{A,m}^{i,(C)} = s_{m,k}^{i,(C)} P_{A,m}^{i,(C)}$ and $\tilde{P}_{A,k}^{i,(NC)} = s_k^{i,(NC)} P_{A,k}^{i,(NC)}$ to make the optimization problem tractable. The resulting optimization problem becomes convex with respect to the auxiliary and relaxed variables given by $\tilde{P}$ and S, respectively. Moreover, strong duality holds, i.e., the duality gap is zero. Therefore, we can solve the Lagrangian dual problem and still obtain the optimal solution of the relaxed problem [24]. An optimal subcarrier allocation policy can be obtained using the popular Hungarian algorithm [25] (c.f. [24] for details) . It can be proved that the solution of the relaxed problem is optimal and has integer values and hence, we can obtain the optimal solution of the original problem (17). The proof is based on the concept of totally unimodularity [15]. The proof of convexity and optimality of the solution for integer variables can be derived similar to [16].

3.4 Simulation Results

Let us evaluate the performance of the proposed D2D relaying system for two scenarios: (1) with energy harvesting via WPT and (2) without energy harvesting capabilities. We assume that in both cases P_{max} is the limit on the total power budget of the system. The noise power spectral density N_0 and bandwidth of each subcarrier W are assumed to be 5×10^5 W/Hz and 20 kHz, respectively. The number of users in each group is 8, i.e., $M = K = 8$ and number of subcarriers for each relay n_m is 4.

Figure 6 depicts the average system capacity in terms of the power budget of the BS P_{max} for different number of subcarriers. As expected, the total capacity of the system grows both with the increase of P_{max} and the number of subcarriers. Figure 7

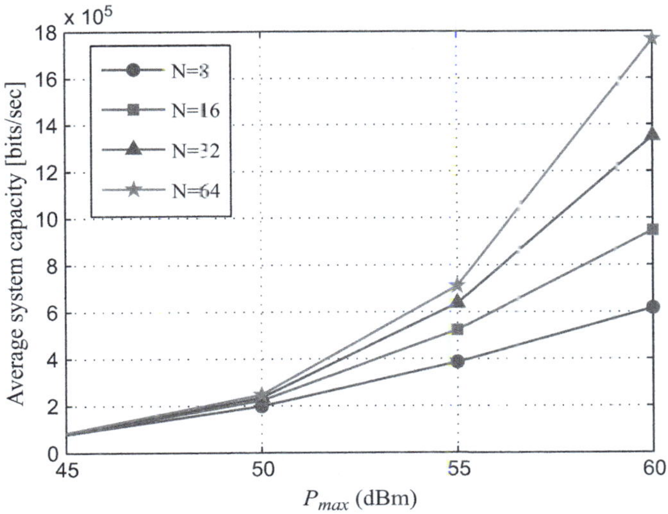

Fig. 6 Average System Throughput vs. P_{max} for different number of subcarriers

shows the energy efficiency in terms of P_{max} for different number of subcarriers. The energy efficiency is defined as R_T/P_T, where R_T is the total system capacity and P_T is the total power consumed in the system. P_T contains both the transmission and the constant power of all the nodes. We compare our proposed scheme with a scenario where there is no energy harvesting. As shown in the figure, our proposed scheme performs better than "No harvesting" scenarios in terms of energy efficiency. It is worth mentioning that in our proposed energy harvesting scheme, the relay UEs do not consume energy from their batteries for the purpose of relaying. Therefore, in terms of battery fairness, our scheme outperforms the "No harvesting" scenario.

3.5 *Summary*

In this section, we proposed resource allocation algorithm for energy harvesting based OFDMA systems with selective D2D relaying We assumed that the UEs are capable of energy harvesting and do not use their own power for relaying purpose. The harvested energy works as an incentive for the UEs to participate in relaying. A novel method based on dual decomposition technique is proposed to solve the joint optimization problem of power and subcarrier allocation, and relay-UE and transmit mode selection, optimally. Numerical results show the performance improvement of our proposed system over a system without energy harvesting.

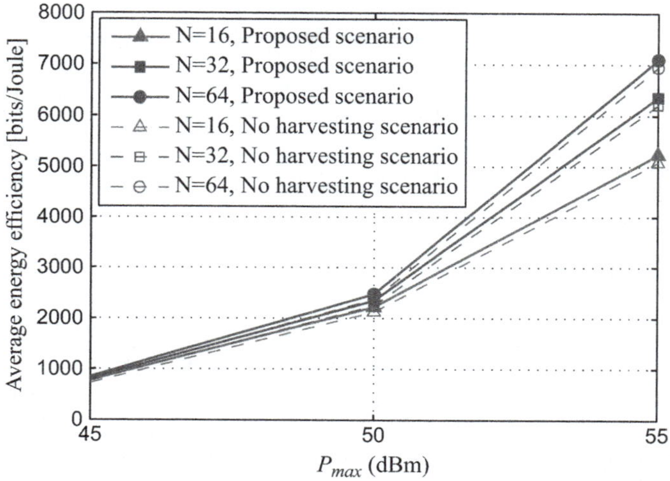

Fig. 7 Average energy efficiency vs. P_{max} for different number of subcarriers

4 Robust Resource Allocation for Cognitive Small Cell Networks with D2D (CR to CR) Cooperation

D2D cooperation can also be utilized to hold in-band communications between devices in the network. Consider, for example, a CR network where an unlicensed UE, i.e., secondary UE, can transmit over the licensed, i.e., primary UE (PU), band, as long as the interference received in the PU band is within the limit set by the regulatory bodies [26]. In a dense CR network with several closely-spaced small cells, the secondary UEs (SUs) operating in different small cells might be interested in communicating with each other for reasons such as data offloading and low-latency application sharing. However, due to interference constraints in the PU bands or due to poor direct communication links, these SUs may not be able to hold direct D2D communications. In such scenarios, D2D cooperation techniques can establish indirect D2D communication links [26–28]. An example scenario is illustrated in Fig. 8.

Figure 8 describes a D2D communication scenario between a source SU s and a destination SU d located in two different small cell networks operating in PU sub-bands B_s and B_d within the system bandwidth, respectively. The two SUs s and d are reasonably far apart but are within the D2D communication range of each other. In such a scenario, any transmission between the SUs in their respective PU bands can add significant interference to the system because the PU receivers can be much closer than the SU receivers. For example, if the source s transmits information directly to the destination d, the PU receivers close to the source s experience significant interference. This may not be permissible because the SUs, being unlicensed UEs, should only opportunistically utilize the available spectrum

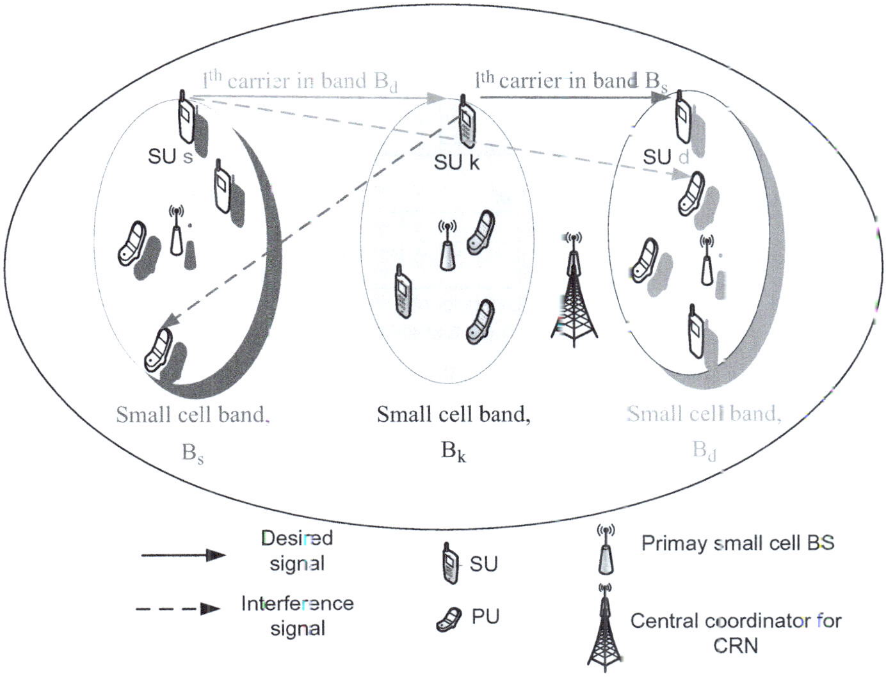

Fig. 8 SUs s and d operating in two different small cells cannot establish direct D2D communication links with each other due to poor communication links or due to PU band interference constraints. Another SU k operating in a different small cell can relay information between s and d by establishing separate D2D links with s and d

while maintaining the interference levels to PUs within an acceptable limit. In such a scenario with PU band interference constraints, another SU k operating in a different PU small cell can relay information between SU s and SU d. For example, the source can transmit a message to the relay k over the PU band of the destination, i.e., B_d, without exceeding the interference limit. The relay k can then forward that message to the destination d over the PU band of the source or the destination depending on the received interference level. With the help of such D2D cooperation, low-powered transmissions are enabled over the small cell PU bands while ensuring that the PU interference levels are acceptable. A similar argument is also valid for scenarios where the SUs s and d are not within the D2D communication range of each other. The relay SU k can facilitate a low-powered indirect communication link between SUs s and d.

In this section, we focus on such CRN scenarios wherein D2D relays establish indirect communication links between SUs located in different small cell networks. We study optimal resource allocation in such D2D relaying systems to improve the CRN performance while keeping the PU band interference below acceptable limits. Specifically, we focus on solving the joint relay-CR selection and power allocation

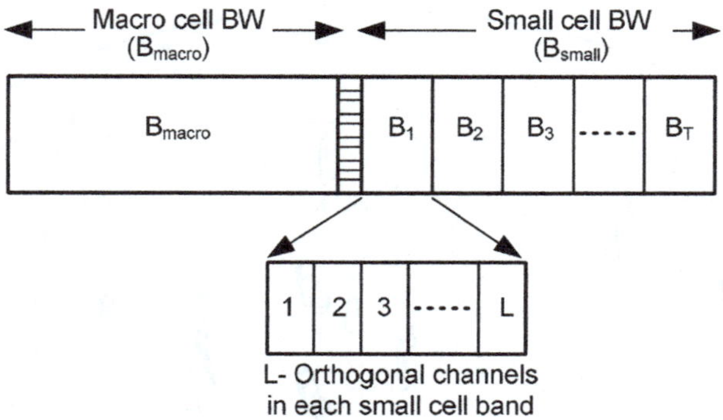

Fig. 9 Dedicated spectrum allocation to avoid macro cell interferences

problem. Note that, unlike with the systems discussed in previous sections, the performance of CR networks depends heavily on channel uncertainties for the SU-PU channels and channel estimation errors for the SU-SU channels. Therefore, our goal is to propose robust solution to the joint optimization problem by taking channel uncertainties and channel estimation errors into account.

4.1 System Model

Consider a CR network where the SUs are operating from small cell (for example, micro, pico, or femto-cell) PU networks. We also assume that the small cell networks operate in conjunction with a macro cell and that a dedicated spectrum allocation technique (see [29, 30] for example techniques) is employed to avoid interference in the small cells due to macro cell. Basically, as shown in Fig. 9, the total licensed spectrum bandwidth BW is divided into two non-overlapping parts B_{macro} and B_{small}, where B_{macro} is used by the macrocell layer and B_{small} is used by the small cell layer. The total small cell BW B_{small} is further divided into T non-overlapping parts and each small cell BW contains L orthogonal channels to avoid the interference among the licensed small cell UEs.

Under such a deployment scenario, as shown in Fig. 8, a set of SU sources $\mathscr{S} = \{1, 2, \ldots, s, \ldots, S\}$ needs to communicate with a set of SU destinations $\mathscr{D} = \{1, 2, \ldots, d, \ldots, D\}$, where the source and destination nodes are far apart and operating from two different small cell PU networks with frequency bands B_s and B_d, respectively. We consider that SUs can use the spectrum bands of PUs such that the maximum amount of interference generated at the PU receivers is below a specified limit. As PU receivers are much closer than the SU receivers to the SU transmitter, any transmission of the SUs in their respective PU bands creates significant interference to the PU receivers. For example, a SU source s may not be able to communicate in band B_s and keep the interference to the PU receivers within the limits. Since destination node d is far apart, a direct long distance transmission in band B_d requires high transmission power, which may not be available at the SUs. High power SU transmission may cause excessive interference to the other intermediate small cell PU receivers operating in the same frequency band. As a result, SU s may not be able to transmit directly to SU d in band B_d and therefore, we look for low-powered CR transmission, which does not create significant interference to the PU networks.

Consider another set of SUs $\mathscr{R} = \{1, 2, \ldots, k, \ldots, K\}$, which is located in between the source and destination nodes and operates from a different PU network with frequency band B_k. These intermediate group of users can assist in establishing a communication between the source and the destination nodes via D2D relaying. An SU source group $\mathscr{S}$ can transmit their messages in band B_d, which can be decoded by the intermediate relay group $\mathscr{R}$. This transmission over band B_d keeps the interference received at the PUs operating in band B_d within limits. The relay group $\mathscr{R}$ decodes the messages sent by $\mathscr{S}$ and forwards them to the destination over band B_s or B_d depending on their location and the interference received over the PU band. However, the frequency band switching from B_d to B_s is beneficial as it facilitates full-duplex communication at the relay SUs [26]. Similarly, the destination group $\mathscr{D}$ can transmit their messages in band B_s and the relay group can forward them over band B_d, and a dual hop D2D communication is possible with the help of the relays. We assume that the SUs in different small cell networks can operate in different frequency bands [26].

The CR network is assumed to have dedicated control channels. During the signaling period, which is a small portion of a transmission frame, SU-SU channels are estimated from a training sequence and resource allocation is performed. Each transmission frame consists of a block of messages and the channel impulse response is assumed to be time invariant, i.e, slow fading, within a frame. A central coordinator for the CR network performs the resource allocation and scheduling. We further assume, for simplicity, that each relay SU can help only one SU during one transmission frame of the CR network. This is a reasonable assumption as relays are small SU devices and supporting multiple users require higher power, which is only possible for the fixed dedicated relay stations mainly used for cell extension purposes in a cooperative or heterogeneous network. Since each relay SU can support a single SU at a time, the problem of choosing a suitable relay from the candidate relay set for each SU arises. In this direction, a simple relay selection strategy, where the relay with the *best* channel conditions is selected, is known to be close to optimal. Moreover, such a simple relay selection strategy has low computational complexity and introduces low overhead into the system [31].

The dual hop communication between a user s and destination d via relay k consists of two phases. In the first phase, user s transmits the message x_s to the relays in carrier l of band B_d with power p_s. Without loss of generality, we assume that the l-th carrier of the PU band B_d and B_s is assigned to user s and relay k, respectively, before the data transmission phase. The received message at relay k is given by

$$y_k = \sqrt{p_s} h_{s,k} x_s + z_k, \tag{19}$$

where $h_{s,k}$ is the channel coefficient for the user s to relay k link. The channel coefficients are statistically modeled as independent and identically distributed (i.i.d) circularly symmetric complex Gaussian (CSCG) random variables (RVs) with zero-mean. z_k denotes the i.i.d. AWGN at the receiver with zero-mean and unit variance.

Relay k decodes the received message and forwards it in the second phase to destination d in carrier l of band B_s with power p_k. The received message at destination d forwarded by relay k is given by

$$y_d = \sqrt{p_k} h_{k,d} x_k + z_d, \tag{20}$$

where $h_{k,d}$ is the channel coefficient between relay k and destination d. z_d denotes AWGN at the receiver with zero-mean and unit variance.

PUs $pu(d)$ and $pu(s)$ that use the carrier l in band B_d and B_s, respectively, will also receive the transmission from the SUs as interference during the first and second phase of the transmission. The interference received at the PUs can be expressed as

$$\begin{aligned} y_{pu(d)} &= \sqrt{p_s} h_{s,pu(d)} x_s, \quad \forall pu(d) \in \mathscr{B}_d^{(l)}, \\ y_{pu(s)} &= \sqrt{p_k} h_{k,pu(s)} x_k, \quad \forall pu(s) \in \mathscr{B}_s^{(l)}, \end{aligned} \tag{21}$$

respectively. $h_{s,pu(d)}$ and $h_{k,pu(s)}$ are the channel coefficients between SU s and PU receiver $pu(d)$ and SU k and PU receiver $pu(s)$, respectively. We denote $\mathscr{B}_d^{(l)}$ and $\mathscr{B}_s^{(l)}$ as the set of PUs using the l-th carrier of band B_d and B_s in different small cells, respectively.

The received signal from the CR network, $y_{pu(d)}$ and $y_{pu(s)}$ will result in added interference to the PU receivers in band B_d and B_s. To protect the rights of the PUs as licensed owners of the bandwidth, interference generated at the PU receivers should be kept below a specified limit. There are two types of interference constraints (1) per-channel or carrier interference constraint and (2) total interference constraint. Since per-carrier interference constraint is more restrictive than that of the total interference constraint [32], we consider per-carrier interference constraint for resource allocation in our research, which is given by

$$\begin{aligned} g_{s,pu(d)} p_s &\le I_{th}^{B_d(l)}, \quad \forall pu(d) \in \mathscr{B}_d^{(l)}, \\ g_{k,pu(s)} p_k &\le I_{th}^{B_s(l)}, \quad \forall pu(s) \in \mathscr{B}_s^{(l)}, \end{aligned} \tag{22}$$

in which $I_{th}^{B_d(l)}$ and $I_{th}^{B_s(l)}$ are the maximum allowed per-carrier interference limit for each CR transmission in carrier l of band B_d and B_s, respectively. $g_{s,pu(d)} = |h_{s,pu(d)}|^2$ and $g_{k,pu(s)} = |h_{k,pu(s)}|^2$ are the channel power gains from SU transmitters to PU receivers.

Considering half-duplex decode-and-forward (DF) transmission [33], the mutual information of the links between user s and relay k in the first time slot and between relay k and destination d in the second time slot are given by

$$C_{s,k}^{(1)} = \frac{1}{2}\log_2(1 + |h_{s,k}|^2 p_s),$$
$$C_{k,d}^{(2)} = \frac{1}{2}\log_2(1 + |h_{k,d}|^2 p_k), \tag{23}$$

respectively, where the factor $\frac{1}{2}$ comes from two time slot transmission. $|h_{s,k}|^2$ and $|h_{k,d}|^2$ are the channel power gains of the links between user s and relay k, and relay k and destination d, respectively. We assume that there is no data buffer at the UEs of the CRN and the maximum achievable data rate of the dual hop link between user s and destination d is limited by the minimum data rate of the two hops.

For our considered cooperative CRN, estimation errors and outdated CSI make the SU-SU channels imperfect [12, 34]. To satisfy the interference constraints in the PU bands, the channels between SU-PU need to be estimated. However, it is extremely difficult to estimate these channels perfectly. Keeping this in mind, we formulate a resource allocation problem by considering two types of channel estimation error models.

4.2 SU to SU Channel Model with Imperfectness

For cooperative D2D-based CR transmission, we consider that training sequences are used to obtain the CSI of SU-SU channels and the resulting channel uncertainty is unbounded Gaussian [35]. SU to SU transmission channel coefficients are modeled as $h = \hat{h} + e$, where $\hat{h}$ is the estimated channel and e denotes the error in estimation. The error e is modeled as i.i.d and distributed as complex Gaussian with zero mean and variance given by σ_e^2. We also assume that the estimated channel $\hat{h}$ and the error e are statistically independent.

Using $h = \hat{h} + e$, the mutual information expressions in (23) can be re-written as

$$C_{s,k}^{(1)} = \tfrac{1}{2}\log_2(1 + \left|\hat{h}_{s,k} + e_{s,k}\right|^2 p_s),$$
$$C_{k,d}^{(2)} = \tfrac{1}{2}\log_2(1 + \left|\hat{h}_{k,d} + e_{k,d}\right|^2 p_k). \tag{24}$$

Depending on the signal detection methods employed, lower bounds on the ergodic capacities can be derived. Note that, with CSI error, a channel outage occurs whenever the actual instantaneous mutual information drops below the target data rate r_s. Such a condition is mathematically given by

$$\min(C_{s,k}^{(1)}, C_{k,d}^{(2)}) < r_s, \tag{25}$$

even when channel capacity achieving coding is applied for error protection [12]. This is because the instantaneous mutual information in (24) is a random variable due to the error in estimation. Note that the lower bounds on ergodic capacities do not capture such outage events. Therefore, we should define a new system performance measure which captures the effect of outage events.

In this direction, we adopt the system goodput [12, 36] as the performance measure and weighted system goodput as the objective function of the problem. In the following, we first define the instantaneous goodputs $U_{s,k}^{inst}$ and $U_{k,d}^{inst}$, i.e., bits per second per Hz successfully delivered to relay k and destination d, respectively as

$$\begin{aligned} U_{s,k}^{inst} &= r_s \times 1(r_s \leq C_{s,k}^{(1)}), \\ U_{k,d}^{inst} &= r_s \times 1(r_s \leq C_{k,d}^{(2)}), \end{aligned} \tag{26}$$

where $1(\cdot)$ is an indicator function, i.e., the function value is 1 if $(\cdot)$ is true and zero otherwise.

The average goodput $\mathrm{U}_{s,k}^{goodput}$ for each (s, k) pair can be defined as minimum of the total average bits per second per Hz successfully delivered to relay k and destination d, and is given by

$$\begin{aligned} \mathrm{U}_{s,k}^{goodput} &= E\left\{\min(U_{s,k}^{inst}, U_{k,d}^{inst})\right\} \\ &= E_{\hat{h}_{s,k},\hat{h}_{k,d}} \times \left\{\min\left(r_s E_{\hat{h}_{s,k}} \times \left[1\left(r_s \leq C_{s,k}^{(1)}\right)\right],\right.\right. \\ &\left.\left. r_s E_{\hat{h}_{k,d}} \times \left[1\left(r_s \leq C_{k,d}^{(2)}\right)\right]\right)\right\} \\ &= E_{\hat{h}_{s,k},\hat{h}_{k,d}} \times \left\{\min\left[r_s \times \Pr.\left(r_s \leq C_{s,k}^{(1)} | \hat{h}_{s,k}\right),\right.\right. \\ &\left.\left. r_s \times \Pr.\left(r_s \leq C_{k,d}^{(2)} | \hat{h}_{k,d}\right)\right]\right\}. \end{aligned} \tag{27}$$

$E_{\hat{h}}(\cdot)$ denotes the statistical expectation with respect to the RV $\hat{h}$. Finally, by incorporating the channel outage events, we define a new objective function of the problem as the average weighted system goodput,

$$\mathrm{U}_{goodput} = \sum_{s=1}^{S} \sum_{k=1}^{K} w_s \, \mathrm{U}_{s,k}^{goodput}. \tag{28}$$

4.3 SU to PU Channel Models with Uncertainty

(a) Model-1: Probabilistic

We assume that CR transmitters cannot estimate the instantaneous channels and have information only about the distribution type and the corresponding distribution parameters of the channel gains among the CR transmitter to the PU receivers. We consider a probabilistic model, which requires only the knowledge of channel gain statistics instead of instantaneous CSI. We assume that the CR transmitters have some knowledge of the transmit power of the PUs and hence can obtain the mean channel gains from PUs to themselves by exploiting the pilot signals from the PUs [37]. Using the reciprocal property of the wireless channels, some statistics of the SU-PU channels can be obtained [37, 38].

For example, suppose the SU to PU channels $h_{s,pu(d)}$ and $h_{k,pu(s)}$ are statistically modeled as i.i.d circularly symmetric complex Gaussian RVs with zero mean and variances $\sigma^2_{s,pu(d)}$, and $\sigma^2_{k,pu(s)}$, respectively. We assume that only the mean and the variance of the channels are known to the CRN. Since the channel coefficients are distributed as CSCG, their channel power gains $g_{s,pu(d)}$ and $g_{k,pu(s)}$ are exponentially distributed with parameters $\lambda_{s,pu(d)} = 1/(\sigma^2_{s,pu(d)})$ and $\lambda_{k,pu(s)} = 1/(\sigma^2_{k,pu(s)})$, respectively.

(b) Model-2: Bounded Uncertainty

In this case, we assume that the uncertainty among the SU-PU channels is bounded and no statistical knowledge is available about it. Without restricting ourselves to any specific estimation schemes, we model the channel as $h = \hat{h} + \epsilon$, where ϵ is the estimation error which determines how far the actual channel h can extend in both real and imaginary parts from the estimated value $\hat{h}$. The value of ϵ depends on the type of the error and the accuracy of the estimation method used. We use the bounded uncertainty model presented in [39]. This model can be viewed as deterministic by considering the worst case error, i.e., $|\epsilon| = \epsilon_{max}$ [40]. The interference constraints C_3 and C_4 are satisfied for the worst case and hence guaranteed for all channels.

The channel gain with uncertainty can be written as

$$g = |h|^2 = (\hat{h} + \epsilon)(\hat{h} + \epsilon)^* = \hat{g} + \delta, \tag{29}$$

where $\hat{g} = \hat{h}\hat{h}^*$ is the estimated channel gain, $\delta = \epsilon\epsilon^* + 2\Re(\hat{h}\epsilon^*)$ is the uncertainty in estimation and $\Re(\cdot)$ is the real part of $(\cdot)$. For each h that lies in some bounded region, the channel gain g, lies on a set of line segment [39], given by

$$g \in \mathscr{L} = \{\hat{g} + \delta v \mid |v| \leq 1\}. \tag{30}$$

To determine the worst possible case, we consider the maximum error given by $|\epsilon| = \epsilon_{max}$. Note that the line segment, $\mathscr{L}$, is generally asymmetric around the estimated gain $\hat{g}$ for $\epsilon \neq 0$. For the worst case scenario, we take the maximum absolute uncertainty, $\delta^{max} = \epsilon_{max}\epsilon^*_{max} + 2|\Re(\hat{h}\epsilon^*_{max})|$, as the uncertainty on both sides of $\hat{g}$, and re-write the channel gain as

$$g \in \mathscr{L} = \{\hat{g} + \delta^{max} v \mid |v| \leq 1\}. \tag{31}$$

where the estimated channel gain $\hat{g}$ is the mid point of the line segment $\mathscr{L}$. Therefore, the channel gains to the PU receiver in the l-th carrier of band B_d and B_s from the CR transmitter s and k can be written as

$$g_{s,pu(d)} \in \mathcal{L}_{s,pu(d)} = \{\hat{g}_{s,pu(d)} + \delta^{max}_{s,pu(d)} v \mid |v| \leq 1\},$$
$$g_{k,pu(s)} \in \mathcal{L}_{k,pu(s)} = \{\hat{g}_{k,pu(s)} + \delta^{max}_{k,pu(s)} v \mid |v| \leq 1\}. \quad (32)$$

4.4 Problem Formulation with Imperfect CSI

In this section, we formulate the robust resource allocation problem taking into account the estimation errors in the SU-SU channels and the uncertainties among the SU-PU channels. The optimization objective is to maximize the weighted system goodput of the CRN. To achieve that, we need to optimally select the relays and allocate power for CR transmission. In addition, we need to satisfy the QoS of the SUs such that the interference received at the PU receivers remain within limits. In order to solve the optimization problem, we first obtain the power allocation solution for each feasible (s, k) pair and then find the optimal assignment pairs (s^*, k^*) [41].

The robust power allocation problem is probabilistic and can be written as

$$\begin{aligned}
\forall (s,k) \in S \times K : \quad & \underset{\{p_s, p_k\}}{\text{argmax}} \ \mathrm{U}^{goodput}_{s,k}, \\
& \text{subject to :,} \\
& C_1 : \Pr.\left(r_s > C^{(1)}_{s,k} | \hat{h}_{s,k}\right) \leq p_{out}, \\
& C_2 : \Pr.\left(r_s > C^{(2)}_{k,d} | \hat{h}_{k,d}\right) \leq p_{out}, \\
& C_3 : \Pr.\left(g_{s,pu(d)} p_s \leq I^{B_d(l)}_{th}\right) \geq c, \quad \forall pu(d) \in \mathcal{B}^{(l)}_d, \\
& C_4 : \Pr.\left(g_{k,pu(s)} p_k \leq I^{B_s(l)}_{th}\right) \geq c, \quad \forall pu(s) \in \mathcal{B}^{(l)}_s, \\
& C_5 : 0 \leq p_s, \ p_k \leq P_{max},
\end{aligned} \quad (33)$$

where C_1 and C_2 are the QoS constraints, which guarantee the target data rate r_s of user s with a probability of $1 - p_{out}$. Constraints C_3 and C_4 incorporate *Model-1* of SU-PU channel uncertainty. These constraints satisfy the interference to PU bands with a probability c, which is a predefined term known to the CR coordinator. C_5 is the non-negative individual power budget constraint for the UEs and limits the power to a maximum of P_{max}.

For *Model-2* that considers bounded uncertainty, the optimization problem given in (33) would remain the same, except for the constraints C_3 and C_4 which depend on the channel uncertainty model. New constraints corresponding to C_3 and C_4 can be written from (32) considering the worst case scenario as

$$\begin{aligned}
& C_3 : \sup\{g_{s,pu(d)} p_s \mid g_{s,pu(d)} \in \mathcal{L}_{s,pu(d)}\} \leq I^{B_d(l)}_{th}, \\
& \qquad\qquad \forall pu(d) \in \mathcal{B}^{(l)}_d \\
& C_4 : \sup\{g_{k,pu(s)} p_k \mid g_{k,pu(s)} \in \mathcal{L}_{k,pu(s)}\} \leq I^{B_s(l)}_{th}, \\
& \qquad\qquad \forall pu(s) \in \mathcal{B}^{(l)}_s.
\end{aligned} \quad (34)$$

Note that supremum of a set, $\sup\{\cdot\}$, captures the essence of worst case optimization.

4.5 Optimization Framework for the Proposed Scheme

In this section, we briefly describe the optimization framework for the formulated probabilistic optimization problem (33). We first transform the probabilistic problem to deterministic and develop a power allocation algorithm. Once the power allocation solution is obtained, we can determine the optimal relay assignment for the transformed problem based on the Hungarian algorithm [25].

In order to solve the problem efficiently, we incorporate the outage-probability (or, QoS) constraints C_1 and C_2 of (33) into the objective function. This is possible if the constraints are fulfilled with equality at the optimal point, which is generally the case for the low outage probabilities required in practical applications [12]. Thus, the resulting optimization problem contains stronger constraints than the original problem. Accordingly, the first QoS constraint becomes

$$C_1 : \Pr.\left(r_s > C^{(1)}_{s,k}|\hat{h}_{s,k}\right) = p_{out}. \tag{35}$$

Using (24), C_1 can be expressed as

$$\Pr.\left(\left|\hat{h}_{s,k} + e_{s,k}\right|^2 \leq \frac{(2^{2r_s}-1)}{p_s}|\hat{h}_{s,k}\right) = p_{out}, \tag{36}$$

where $\chi^2_{2_{s,k}} = \left|\hat{h}_{s,k} + e_{s,k}\right|^2$ is a non-central chi-square distributed random variable with non-centrality parameter, $\lambda^2_{s,k} = \dfrac{|\hat{h}_{s,k}|^2}{\sigma^2_{e_{s,k}}/2}$ and degrees of freedom $n = 2$.

To obtain a closed-form result, we approximate the non-central chi-square distribution by the central chi-square $\chi^2_{2_{s,k}}(0)$ [42, 43] and evaluate as

$$\Pr.\left(\chi^2_{2_{s,k}} \leq \frac{2^{2r_s}-1}{p_s}\right) \approx \Pr.\left(\chi^2_{2_{s,k}}(0) \leq \frac{(2^{2r_s}-1)/p_s}{1+\lambda^2_{s,k}/2}\right) = 1 - e^{-\dfrac{(2^{2r_s}-1)/p_s}{2\left(1+\lambda^2_{s,k}/2\right)}}. \tag{37}$$

Using the expression of (37) in (36), we can obtain the guaranteed QoS considering channel estimation error in SU source to relay links as

$$r_s = \frac{1}{2}\log_2(1 + p_s\alpha_{s,k}), \tag{38}$$

where $\alpha_{s,k} = [-\ln(1-p_{out})] \times 2\left(1+\lambda_{s,k}^2/2\right)$.

Similarly, C_2 can be transformed into an equality constraint and a similar expression to (38) can be derived.

Since the power and interference constraints are instantaneous in (33), as far as the optimization problem is concerned, the average goodput $\mathrm{U}_{s,k}^{goodput}$ maximization is equivalent to the maximization of the instantaneous goodput for each set of channel gains, although both criteria are different in general. Therefore, we can write

$$\begin{aligned}
&\underset{\{p_s,p_k\}}{\text{argmax}} \quad \mathrm{U}_{s,k}^{goodput} \\
&= E_{\hat{h}_{s,k},\hat{h}_{k,d}} \times \left\{\min\left(r_s \times \Pr.\left(r_s \le C_{s,k}^{(1)}|\hat{h}_{s,k}\right), r_s \times \Pr.\left(r_s \le C_{k,d}^{(2)}|\hat{h}_{k,d}\right)\right)\right\} \\
&\equiv \underset{\{p_s,p_k\}}{\text{argmax}} \quad \left\{\min\left(r_s \times \Pr.\left(r_s \le C_{s,k}^{(1)}|\hat{h}_{s,k}\right), r_s \times \Pr.\left(r_s \le C_{k,d}^{(2)}|\hat{h}_{k,d}\right)\right)\right\} \\
&= \underset{\{p_s,p_k\}}{\text{argmax}} \quad \min\left\{(1-p_{out})\tfrac{1}{2}\log_2(1+p_s\alpha_{s,k}), (1-p_{out})\tfrac{1}{2}\log_2(1+p_k\alpha_{k,d})\right\},
\end{aligned} \tag{39}$$

where "≡" denotes the equivalence relation.

4.6 Optimization Framework for Model-1: Probabilistic Constraints

Since $g_{s,pu(d)}$ is exponentially distributed with parameter $1/(\sigma_{s,pu(d)}^2)$, the left hand side of the first constraint C_1 in (33) can be evaluated as

$$\Pr.\left(g_{s,pu(d)}p_s \le I_{th}^{B_d(l)}\right) = 1 - e^{-\left(\frac{I_{th}^{B_d(l)}}{\sigma_{s,pu(d)}^2 p_s}\right)}. \tag{40}$$

Therefore, the problem (33) can be written as a deterministic optimization problem given by

$$\begin{aligned}
\forall(s,k) \in S \times K: \quad & \underset{\{p_s,p_k\}}{\text{argmax}} \quad \min\left\{\frac{(1-p_{out})}{2}\log_2(1+p_s\alpha_{s,k}), \frac{(1-p_{out})}{2}\log_2(1+p_k\alpha_{k,d})\right\} \\
& \text{subject to}: \\
& C_1: p_s \le \frac{I_{th}^{B_d(l)}}{\sigma_{s,pu(d)}^2\{-\ln(1-c)\}}, \quad \forall pu(d) \in \mathscr{B}_d^{(l)}, \\
& C_2: p_k \le \frac{I_{th}^{B_s(l)}}{\sigma_{k,pu(s)}^2\{-\ln(1-c)\}}, \quad \forall pu(s) \in \mathscr{B}_s^{(l)}, \\
& C_3: 0 \le p_s, p_k \le P_{max}.
\end{aligned} \tag{41}$$

4.7 *Optimization Framework for Model-2: Bounded Uncertainty Constraints*

For model-2 with bounded uncertainty region given in (34), we propose worst case optimization approach in which the interference constraints are satisfied for all possible values of the channel gains that lie in some bounded region. Therefore the constraint C_3 in (34) can now be expressed as

$$\begin{aligned} &\sup\{g_{s,pu(d)}p_s \mid g_{s,pu(d)} \in \mathscr{L}_{s,pu(d)}\} \leq I_{th}^{B_d(l)}, \quad \forall pu(d) \in \mathscr{B}_d^{(l)} \\ \Rightarrow\ &\sup\{(\hat{g}_{s,pu(d)} + g_{s,pu(d)}^{max} v)p_s \mid |v| \leq 1\} \leq I_{th}^{B_d(l)}, \quad \forall pu(d) \in \mathscr{B}_d^{(l)} \\ \Rightarrow\ &\hat{g}_{s,pu(d)}p_s + g_{s,pu(d)}^{max}p_s \leq I_{th}^{B_d(l)}, \quad \forall pu(d) \in \mathscr{B}_d^{(l)}. \end{aligned} \tag{42}$$

Following a similar procedure, a similar expression to (42) can be derived for the second interference constraint C_4 of (34). Finally, replacing C_1 and C_2 in (41), respectively with the constraints derived here, we obtain an equivalent deterministic optimization problem for Model-2.

4.8 *Solution Approach*

It can be proved that the robust power optimization problem in (41) (for both the channel uncertainty models) is convex and can be solved efficiently. The algorithm to obtain the solution for problem (41) is given in Algorithm 3.

In Algorithm 3, we obtain the closed form analytical expression for optimal p_s^* and p_k^* for user-relay (s,k) pair for problem (41).

The optimal relay assignment can be obtained by forming a $S \times K$ matrix $\mathbf{B} = [b(s,k)]$, such that

$$b(s,k) = w_s\left[(1-p_{out})\frac{1}{2}\log_2(1+p_s^*\alpha_{s,k})\right] = w_s\left[(1-p_{out})\frac{1}{2}\log_2(1+p_k^*\alpha_{k,d})\right].$$

By applying Hungarian algorithm [25] on matrix $\mathbf{B}$, we obtain the optimal (s^*, k^*) pairs that maximize the weighted goodput of the CRN.

4.9 *Simulation Results*

We conduct simulations to show the effectiveness of our proposed solution and compare the performance with the baseline scheme that ignores the imperfectness in CSI. Different system parameters are considered to analyze the influence of different channel estimation errors on the system performance. Frequency-flat Rayleigh fading channels are assumed in all simulations. Each of the channel coefficients is

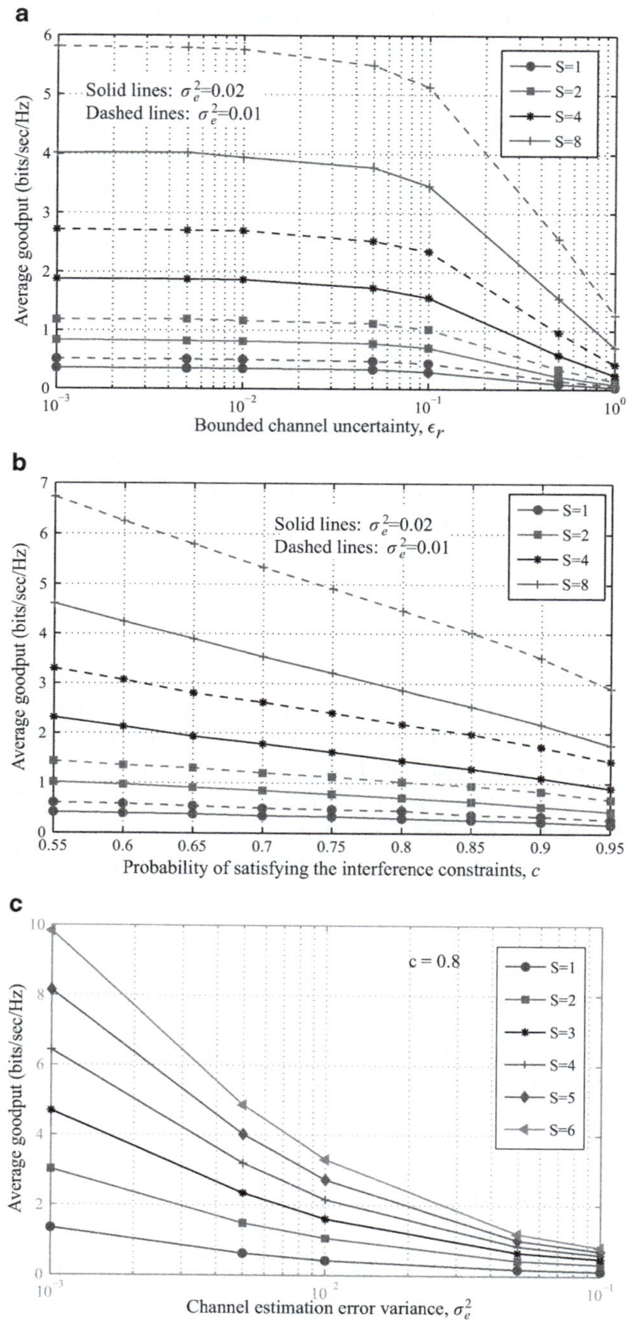

Fig. 10 Average goodput versus different uncertainty models among the SU-PU channels and SU-SU channels. (**a**) Average goodput versus uncertainty distance, ϵ_r among the channels of SU transmitters to PU receivers. (**b**) Average goodput versus probability of satisfying the interference constraints c. (**c**) Average goodput versus channel estimation error variance σ_e^2 of SU-SU channels

Algorithm 3 Power allocation for problem (41)

Require: $\alpha_{s,k}$, $\alpha_{k,d}$, $I_{th}^{B_d(l)}$, $I_{th}^{B_s(l)}$, c, P_{max}, $\sigma^2_{s,pu(d)}$, $\sigma^2_{k,pu(s)}$, $\hat{g}_{s,pu(d)}$, $\hat{g}_{k,pu(s)}$, $\delta^{max}_{s,pu(d)}$, $\delta^{max}_{k,pu(s)}$.

1: Calculate Pow_s and Pow_k given by

$$\text{Pow}_s = \min\left(\frac{I_{th}^{B_d(l)}}{\sigma^2_{s,pu(d)}\{-\ln(1-c)\}}, P_{max}\right) \text{; for model-1}$$

$$\text{Pow}_s = \min\left(\frac{I_{th}^{B_d(l)}}{\hat{g}_{s,pu(d)} + \delta^{max}_{s,pu(d)}}, P_{max}\right) \text{; for model-2}$$

$$\text{Pow}_k = \min\left(\frac{I_{th}^{B_s(l)}}{\sigma^2_{k,pu(s)}\{-\ln(1-c)\}}, P_{max}\right) \text{; for model-1}$$

$$\text{Pow}_k = \min\left(\frac{I_{th}^{B_s(l)}}{\hat{g}_{k,pu(s)} + \delta^{max}_{k,pu(s)}}, P_{max}\right) \text{; for model-2}$$

2: **if** $\left(\frac{\text{Pow}_s \alpha_{s,k}}{\alpha_{k,d}} \le \text{Pow}_k\right)$ **then**

3: $\quad p_s^* = \text{Pow}_s$, $p_k^* = \frac{\text{Pow}_s \alpha_{s,k}}{\alpha_{k,d}}$

4: **else**

5: $\quad p_k^* = \text{Pow}_k$, $p_s^* = \frac{\text{Pow}_k \alpha_{k,d}}{\alpha_{s,k}}$

6: **end if**

an i.i.d. circularly symmetric complex Gaussian random variable and distributed as $\mathscr{CN}\left(0, \frac{1}{(1+d)^\alpha}\right)$ [44], where d is the normalized distance among various nodes in the network and $\alpha = 4$ is the path loss coefficient.

The bounded uncertainty region for model-2, as discussed in Sect. 4.7, is considered to be rectangular with four equal sides. The uncertainty distance, i.e., the distance from the center of the rectangle to the center of each side (or, equivalently half length of a side) is given by ϵ_r. Same channel estimation error variance σ_e^2 is assumed for all SU to SU channels. We also consider that the maximum per-carrier interference limit in both band B_d and B_s is the same, i.e., $I_{th}^{B_d(l)} = I_{th}^{B_s(l)} = I_{th}$. Since both the source and the relays are UEs, the maximum power available are assumed to be the same. Unless stated otherwise, probability of outage, p_{out}, is assumed to be 0.1, equal priority is given to all the users by setting $w_s = 1$, maximum power, P_{max} and interference threshold limit, I_{th}, are set at 1 and 0.1 W, respectively. The results are averaged over 10,000 independent channel realizations.

Figure 10a–c show the average system goodput versus the uncertainty distance ϵ_r, probability of satisfying the interference constraints c, and the channel estimation error variance σ_e^2 of the SU channels, respectively, for different number of users S. We see that average goodput decreases with the increase of ϵ_r, c, and σ_e^2 and increases with the increase of the number of users S. The effect of error variance σ_e^2 of the SU channels is more significant at the lower values of ϵ_r and c and at higher number of users. This is because low ϵ_r or c values indicate better estimation of the channels between SU transmitters and PU receivers. Low ϵ_r or c values allow higher transmission capability of the CRN as the interference threshold

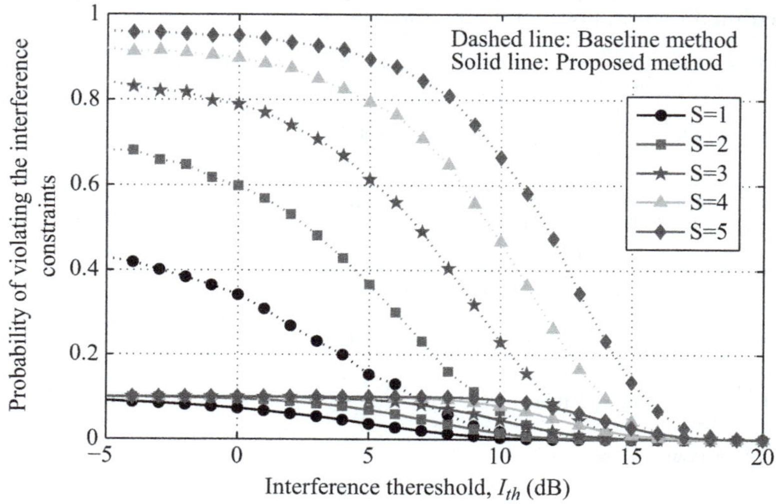

Fig. 11 Probability of violating the interference constraints versus interference threshold I_{th}

margin becomes more flexible. Similarly, with increasing number of users, more power is transmitted by the CRN. Since the average goodput is proportional to the transmission power and inversely proportional to σ_e^2, the effect of estimation error in the SU-SU channels becomes more detrimental at high power regimes.

Figure 11 shows the probability of violating the interference constraints given in (33) versus interference threshold I_{th} for different number of users. In this simulation, we consider that some uncertainty is associated with the channels of SU transmitters to PU receivers and compare the performance of our proposed scheme with the baseline scheme which assumes the estimated channel to be perfect. We observe frequent violations of the constraints for the baseline scheme in the active region of I_{th} for ignoring the channel uncertainty in the design process. Since increasing the number of users use more power to improve the system goodput, the probability of violation increases with the increase of the number of users. Similar trends of interference violation can be observed for the resource allocation schemes proposed in [45, 46]. For our proposed probabilistic interference scheme, i.e., model-1, we set $c = 0.9$. That is, we allow the interference constraints to be violated with a probability of 0.1 when the constraints are active. As a result, we see that the maximum value of interference violation is $1 - c = 0.1$, which is much lower than the baseline scheme. Model-2, i.e., the bounded uncertainty model is designed to satisfy the interference constraints for the worst case scenario. As a result, no violation is observed for our proposed scheme when the uncertainty lies in some bounded region.

4.10 Summary

We have developed robust power allocation and relay-CR selection schemes for a CR network where the SUs from different small cell PU networks communicate with each other. In such a network with tight interference constraints in the PU band, close proximity of the PUs and limited power budgets of SUs, the communication among the SUs is established via D2D cooperation. We have developed efficient resource allocation schemes with the provision of QoS to each SU considering different channel uncertainty models in the network. Channel outage events, which have resulted from the imperfect CSI under slow fading channels, have been taken into account in our algorithm. System goodput of the CR network has been maximized while satisfying the interference constraints for the PU bands both probabilistically and for the worst case scenario. The joint optimization problem is probabilistic and non-convex, which is very difficult to solve in its original form. The problem has been approximated and transformed into a convex deterministic form and a closed form analytical solution for power allocation has been derived The closed form power allocation solution helped us to develop an efficient relay-CR selection scheme based on Hungarian algorithm. Our results reveal that the implications of ignoring the imperfectness among different channels are violations of interference constraints and failure to provide the QoS, which ultimately result in retransmissions and wastage of power.

5 Conclusion and Future Research Directions

In this chapter, we discussed several resource allocation problems which attempt throughput maximization in cooperative D2D networks operating under OFDMA or cognitive radio architectures A variety of system models were considered to understand how cooperative D2D communications can benefit from incorporating packet storage using buffers, energy-harvesting, and cognitive radio capabilities at UEs in the network. For each system, we provided efficient solutions to joint resource optimization problems including power allocation, subcarrier allocation, subcarrier pairing, and relay-UE selection. Significant throughput gains are observed in each system, thus proving that the system performance can be significantly improved whenever the UEs are enabled with cooperative D2D capabilities.

Several interesting research works can be pursued to further investigate the potential performance gains offered by cooperative D2D communications. Energy Efficiency is becoming an important performance metric for 5G networks because there are growing concerns about the global carbon footprint of telecommunication systems. Energy Efficiency, which is the ratio of the total achieved throughput to the total power consumption in the system, is a fractional and nonlinear function and is different from the conventional performance measures. Therefore, potential future research works can be to develop energy efficient resource allocation schemes for the cooperative D2D systems discussed in this chapter. Another extension of this research can be to develop an intermix of the presented system models to develop a comprehensive understanding of the benefits offered by cooperative D2D. For example, resource allocation problems can be formulated to understand the benefits of incorporating both packet-storage and energy-harvesting capabilities at the UEs in cooperative D2D networks. Also, investigating the role of strict and soft fractional frequency reuse patterns in enabling dense cooperative D2D networks with spectrum availability constraints can be a possible future research work.

Another research direction in the field of cooperative D2D and mobile relays can be investigating radio resource management problems such as opportunistic D2D mode selection, i.e, switch opportunistically between overlay D2D and underlay D2D, user association, link adaptation, multi-casting, and delay-tolerant data offloading. Appropriate interference mitigation, load-balancing, and scheduling algorithms can be derived when cooperative D2D capabilities are incorporated in ultra-dense multi-tier heterogenous networks (HetNets). Also, the benefits of cooperative D2D communications can be evaluated when operated in conjunction with other 5G enabling technologies, such as massive MIMO, millimeter wave spectrum, and full-duplex radio technologies. Wherever applicable, distributed resource allocation algorithms can be proposed based on suitable trade-offs between computational complexities and expected delays.

References

1. S. Mumtaz, J. Rodriguez, *Smart Device to Smart Device Communication* (Springer, Cham, 2014)
2. L. Song, D. Niyato, Z. Han, E. Hossain, *Wireless Device-to-Device Communications and Networks* (Cambridge University Press, New York, NY, 2015)
3. K.W. Choi, Z. Han, Device-to-device discovery for proximity-based service in LTE-advanced system IEEE J. Sel. Areas Commun. **33**(1), 55–66 (2015)
4. L. Lei, Z. Zhong, C. Lin, X. Shen, Operator controlled device-to-device communications in LTE-advanced networks. IEEE Wirel. Commun. **19**(3), 96–104 (2012)
5. L. Wei, R. Q. Hu, Y. Qian, G. Wu, Enable device-to-device communications underlaying cellular networks: challenges and research aspects. IEEE Commun. Mag. **52**(6) 90–96 (2014)
6. L. Song, D. Niyato, Z. Han, E. Hossain, Game-theoretic resource allocation methods for device-to-device (D2D) communication. IEEE Wirel. Commun. **21**(3), 136–144 (2014)
7. M.N. Tehrani, M. Uysal, H. Yanikomeroglu, Device-to-device communication in 5G cellular networks: challenges, solutions, and future directions. IEEE Commun. Mag. **52**(5), 86–92 (2014)
8. A. Asadi, Q. Wang, V. Mancuso, A survey on device-to-device communication in cellular networks. IEEE Commun. Surv. Tutorials **16**(4), 1801–1819 (2014)
9. S. Bi, C. Ho, R. Zhang, Wireless powered communication: opportunities and challenges. IEEE Commun. Mag. **53**(4), 117–125 (2015)
10. S. Berger, M. Kuhn, A. Wittneben et al., Recent advances in amplify-and-forward two-hop relaying. IEEE Commun. Mag. **47**(7), 50–56 (2009)
11. K.T.K Cheung, S. Yang, L. Hanzo, Achieving maximum energy-efficiency in multi-relay OFDMA cellular networks: a fractional programming approach. IEEE Trans. Commun. **61**(7), 2746–2757 (2013)

12. D.W.K. Ng, R. Schober, Cross-layer scheduling for OFDMA amplify-and-forward relay networks. IEEE Trans. Veh. Technol. **59**(3), 1443–1458 (2010)
13. M.S. Alam, J.W. Mark, X. Shen, Relay selection and resource allocation for multi-user cooperative OFDMA networks. IEEE Trans. Wirel. Commun. **12**(5), 2193–2205 (2013)
14. S. Boyd, L. Vandenberghe, *Convex Optimization* (Cambridge University Press, New York, NY, 2004)
15. S.O. Krumke, *Integer Programming. Polyhedra and Algorithms*. Draft, January 4, 2006
16. R. Arab Loodaricheh, S. Mallick, V.K. Bhargava, Energy efficient resource allocation for OFDMA cellular networks with user cooperation and QoS provisioning. IEEE Trans. Wirel. Commun. **13**(11), 6132–6146 (2014)
17. S. Boyd, L. Xiao, A. Mutapcic, Subgradient methods, in *Notes for EE392o Stanford University Autumn*, 2003–2004
18. W. Dang, M. Tao, H. Mu, J. Huang, Subcarrier-pair based resource allocation for cooperative multi-relay OFDM systems. IEEE Trans. Wirel. Commun. **9**(5), 1640–1649 (2010)
19. H. Zhu, J. Wang, Device-to-device communication in cellular networks with fractional frequency reuse, in *Proceedings of 2014 IEEE ICC*, pp. 5503–5507
20. L. Liu, R. Zhang, K.C. Chua, Wireless information transfer with opportunistic energy harvesting. IEEE Trans. Wirel. Commun. **12**(1), 288–300 (2013)
21. D.W.K. Ng, E.S. Lo, R. Schober, Wireless information and power transfer: energy efficiency optimization in OFDMA systems. IEEE Trans. Wirel. Commun. **12**, 6352–6370 (2013)
22. T. Wang, A. Cano, B. Giannakis et al., High-performance cooperative demodulation with decode-and-forward relays. IEEE Trans. Commun. **55**(7), 1427–1438 (2007)
23. D.P. Bertsekas, *Dynamic Programming and Optimal Control*. vol 1, no. 2 (Athena Scientific, Belmont, MA, 1995)
24. R. Arab Loodaricheh, S. Mallick, V.K. Bhargava, Resource allocation for OFDMA systems with selective relaying and energy harvesting, in *Proceedings of 2014 IEEE VTC (Fall)*, pp. 1–5
25. H.W. Kuhn, The Hungarian method for the assignment problem. Naval Res. Logistic Q. **2**, 83–97 (1955)
26. G. Zhao, C. Yang, G.Y. Li, D. Li, A. Soong, Power and channel allocation for cooperative relay in cognitive radio networks. IEEE J. Sel. Top. Sign. Proces. **5**(1), 151–159 (2011)
27. X. Gong, W. Yuan, W. Liu, W. Cheng, S. Wang, A cooperative relay scheme for secondary communication in cognitive radio networks, in *Proceedings of 2008 IEEE Globecom*, pp. 1–6
28. C. Sun, K.B. Letaief, User cooperation in heterogeneous cognitive radio networks with interference reduction, in *Proceedings of 2008 IEEE ICC*, pp. 3193–3197
29. D. López-Pérez, A. Valcarce, G. Roche, J. Zhang, OFDMA femtocells: a roadmap on interference avoidance. IEEE Commun. Mag. **47**(9), 41–48 (2009)
30. I. Demirdogen, I. Guvenc, H. Arslan, A simulation study of performance trade-offs in open access femtocell networks, in *Proceedings of 2010 IEEE PIMRC*, pp. 151–156
31. S. Kadloor, R. Adve, Relay selection and power allocation in cooperative cellular networks. IEEE Trans. Wirel. Commun. **9**(5), 1676–1685 (2010)
32. T. Al-Khasib, M. Shenouda, L. Lampe, Dynamic spectrum management for multiple-antenna cognitive radio systems: design with imperfect CSI. IEEE Trans. Wirel. Commun. **10**(9), 2850–2859 (2011)
33. J.N. Laneman, D.N.C. Tse, G.W. Wornell, Cooperative diversity in wireless networks: efficient protocols and outage behavior. IEEE Trans. Inf. Theory **50**(12), 3062–3080 (2004)
34. M. Medard, The effect upon channel capacity in wireless communications of perfect and imperfect knowledge of the channel. IEEE Trans. Inf. Theory **46**(3), 933–946 (2000)
35. G. Zheng, S. Ma, K.-K. Wong, T.-S. Ng, Robust beamforming in cognitive radio. IEEE Trans. Wirel. Commun. **9**(2), 570–576 (2010)
36. Z.K.M. Ho, V.K.N. Lau, R.S.K. Cheng, Closed loop cross layer scheduling for goodput maximization in frequency selective environment with no CSIT. in *Proceedings of 2007 IEEE WCNC*, 299–303
37. D. I. Kim, L. B. Le, E. Hossain, Joint rate and power allocation for cognitive radios in dynamic spectrum access environment. IEEE Trans. Wirel. Commun. **7**(12), 5517–5527 (2008)

38. L. Zhang, Y.-C. Liang, Y. Xin, H. Poor, Robust cognitive beamforming with partial channel state information. IEEE Trans. Wirel. Commun. **8**(8), 4143–4153 (2009)
39. S. Mallick, M.M. Rashid, V.K. Bhargava, Joint relay selection and power allocation for decode-and-forward cellular relay network with channel uncertainty. IEEE Trans. Wirel. Commun. **11**(10), 3496–3508 (2012)
40. G. Zheng, K.-K. Wong, B. Otterston, Robust cognitive beamforming with bounded channel uncertainties. IEEE Trans. Signal Process. **57**(12), 4871–4881 (2009)
41. S. Mallick, R. Devarajan, R. Arab Loodaricheh, V.K. Bhargava, Robust resource optimization for cooperative cognitive radio networks with imperfect CSI. IEEE Trans. Wirel. Commun. **14**(2), 907–920 (2015)
42. P.-J. Chung, H. Du, J. Gondzio, A probabilistic constraint approach for robust transmit beamforming with imperfect channel information. IEEE Trans. Signal Process. **59**(6), 2773–2782 (2011)
43. D.R. Cox, N. Reid, Approximations to non-central distributions. Can. J. Stat. **15**(2), 105–114 (1987)
44. I. Hammerstrom, A. Wittneben, On the optimal power allocation for nonregenerative OFDM relay links, in *Proc. 2006 IEEE ICC*, pp. 4463–4468
45. M. Choi, J. Park, S. Choi, Simplified power allocation scheme for cognitive multi-node relay networks. IEEE Trans. Wirel. Commun. **11**(6), 2008–2012 (2012)
46. M. Shaat, F. Bader, Asymptotically optimal resource allocation in OFDM-based cognitive networks with multiple relays. IEEE Trans. Wirel. Commun. **11**(3), 892–897 (2012)

Fog Computing and Its Applications in 5G

Longxiang Gao, Tom H. Luan, Bo Liu, Wanlei Zhou, and Shui Yu

Abstract With smartphones becoming our everyday companions high-quality mobile applications have become an important integral of people's lives. The intensive and ubiquitous use of mobile applications have led to explosive growth of mobile data traffics. To accommodate the surge mobile traffic yet providing the guaranteed service quality to mobile users represent a key issue of 5G mobile networks. This motivates the emergence of Fog computing as a promising, practical and efficient solution tailored to serving mobile traffics. Fog computing deploys highly virtualized computing and communication facilities at the proximity of mobile users. Dedicated to serving the mobile users, Fog computing explores the predictable service demand patterns of mobile users and typically provides desirable localized services accordingly. Stitching above features, Fog computing can provide mobile users with the demanded services through low-latency and short-distance local connections. In this chapter, we introduce the main features of Fog computing and describe its concept, architecture and design goals. Lastly, we discuss on the potential research issues from the perspective of 5G networking.

1 Introduction

Smartphones have already become our everyday companions. In 2011, the smartphone shipment worldwide overtook that of PCs for the first time in history, and now the smartphone penetration has reached 75 % in US. It is envisioned by Cisco that the average number of connected mobile devices per person will hit 6.56 in 2020, due to the proliferating use of "Internet of Things" applications, e.g., smart home, smart community, and emerging mobile electronics, e.g., wearable devices.

Smart devices have brought rich computing and communication capability to the palm of our hand. As a result, rich mobile applications are enabled to enhance our day-to-day experiences by enabling productivity, connectivity and achievement

L. Gao • T.H. Luan (✉) • B. Liu • W. Zhou • S. Yu
School of Information Technology, Deakin University, Burwood, Victoria, Australia
e-mail: longx@deakin.edu.au; tom.luan@deakin.edu.au; bo.liu@deakin.edu.au; wanlei.zhou@deakin.edu.au; shui.yu@deakin.edu.au

W. Xiang et al. (eds.), *5G Mobile Communications*,
DOI 10.1007/978-3-319-34208-5_21

of our goals. For example, we may be already addicted to mobile applications everyday for social connectivity and to fulfill our multimedia messaging needs. On each passing day, mobile applications connect with wearable devices to read our heartbeat and track our health conditions, adjust the temperature and light of our room by communicating with smart home facility, coerce us to take that morning walk, offer a different route that will help us avoid the rush hour traffic, and become increasingly intelligent by understanding our mobilities, gestures and social activities. Apparently, from changing the way we communicate to revolutionizing the way we work and live, mobile electronics and applications pervade our daily lives everywhere.

The proliferation and pervasive use of mobile applications inevitably leads to the explosive growth of the mobile data traffic. To accommodate the surge mobile traffic and in the meantime provide guaranteed service quality to mobile users represent the key issue of next generation mobile networks. This motivates the emergence of Fog computing as a promising, practical and efficient solution that extends cloud computing to better serving mobile traffics. The term "Fog computing" was first proposed by Cisco in 2012 [1]. Similar systems typically known as edge computing, such as Cyber Foraging [2], Cloudlets [3] can date back to early 2000.

In this chapter, we introduce why the Fog computing is promising, the main features of Fog computing and describe its concept, architecture and design goals. Then we demonstrate a case study on how the Fog computing can improve the network performance in 5G environment, followed by a discussion on the potential research issues from the perspective of 5G networking.

2 Fog Computing Architecture

Fog computing extends the cloud-based Internet by introducing an intermediate layer between mobile devices and cloud, aiming at the smooth, low-latency service delivery from the cloud to mobile. This accordingly leads to a three hierarchy Mobile-Fog-Cloud architecture as depicted in Fig. 1.

The intermediate Fog layer is composed of geo-distributed Fog servers which are deployed at the edge of networks, e.g., parks, bus terminals, shopping centers, etc. Each Fog server is a highly virtualized computing system, similar to a light-weight cloud server, and is equipped with the on-board large-volume data storage, compute and wireless communication facility. The role of Fog servers is to bridge the mobile users and cloud. On one hand, Fog servers directly communicate with the mobile users through single-hop wireless connections using the off-the-shelf wireless interfaces, such as WiFi and Bluetooth. With the on-board compute facility and pre-cached contents, they can independently provide pre-defined service applications to mobile users without assistances from cloud or Internet. On the other hand, the Fog servers can be connected to the cloud so as to leverage the rich functions and application tools of the cloud. The next section describes some typical examples of Fog computing in details.

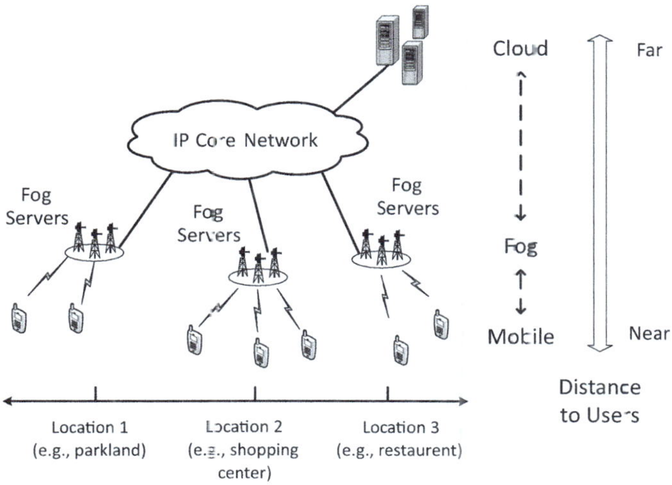

Fig. 1 Fog computing architecture

To summarize, the purpose of Fog computing is to place a handful of compute, storage and communication resources in the proximity of mobile users, and therefore to serve mobile users with the local short-distance high-rate connections. This overcomes the drawback of cloud which is far to mobile users with elongated service delays. Therefore, the fog is interpreted as "the cloud close to the ground" [1].

3 Why Fog Computing?

Nowadays, the evolving of Internet has shown two obvious trends. First, the could-based architecture is adopted to host major applications and storage. As predicted by Cisco Global Cloud Index, the global cloud traffic will account for more than three-fourths of total data center traffic by 2018. Second, the Internet users have shifted predominantly from using desktop computers to smartphones and tablets. With cloud becoming the overarching approach for service delivery and information retrieval, and mobile users becoming the major service consumers, the seamless interconnection of cloud computing and mobile applications therefore represents a key issue in the 5G mobile networks and motivates the emergence of Fog computing.

To show the rationale of Fog computing, in what follows we take a retrospect study by revisiting the design of cloud-based Internet and service requirements of mobile users.

3.1 Global and Local Information

The cloud computing represents an efficient and scalable centralized solution for information management and distribution. It is efficient to serve the information requests from the traditional desktop users. To be specific, the desktop users, typically accessing Internet at homes and offices are often interested in the information which is irrelevant to their locations, such as the world news, stock market at different cities or countries, to name a few. We refer to such information as the *global information*. As a contrast, we refer to the location-based service information related to the location of users as the *local information*. Cloud computing favors desktop users with an optimized approach for serving the global information services. With a scalable and efficient approach to store and manage the information originated at different locations of the world and using a static public IP, cloud computing conveniently distribute the cached global information from a remote central server to desktops worldwide.

The mobile users, however, have distinguished service requirements from the desktop users. This requires the current cloud-based Internet to be modified accordingly to cater to the specific service requirements of mobile users. In specific, different from desktop users, mobile users, particularly smartphones, are typically in the outdoor environments. This makes their service requirements closely related to their current locations. In other words, mobile users are more interested in the *local information* around them. For example, a mobile user in a shopping center tends to be interested in the sales, open hour, restaurants and events inside the attended shopping center; such information become useless once he/she leaves the shopping center. In another example, a traveller to a city would seek for information on the places of interest, local news and weather conditions of the specific city, while such information of other places is useless. The massive demand of location-based mobile services is also reported in [4].

3.2 Physical and Communication Distance

The cloud-based Internet can be inefficient to serve the local information desired by mobile users. As a motivating example shown in Fig. 2a, assuming that a mobile user inside a shopping center intends to retrieve flyers of stores within the shopping center. To do this using the cloud-based Internet, the stores may need to first upload their flyers to a remote cloud server over Internet, and then direct mobile users to retrieve the desired information from the remote cloud server. In other words, although the *physical distance* between the mobile user (destination) and stores (original source) is short, using the remote cloud as the information depot, the actual *communication distance* can be far, e.g., from the cloud server to mobile user in this example.

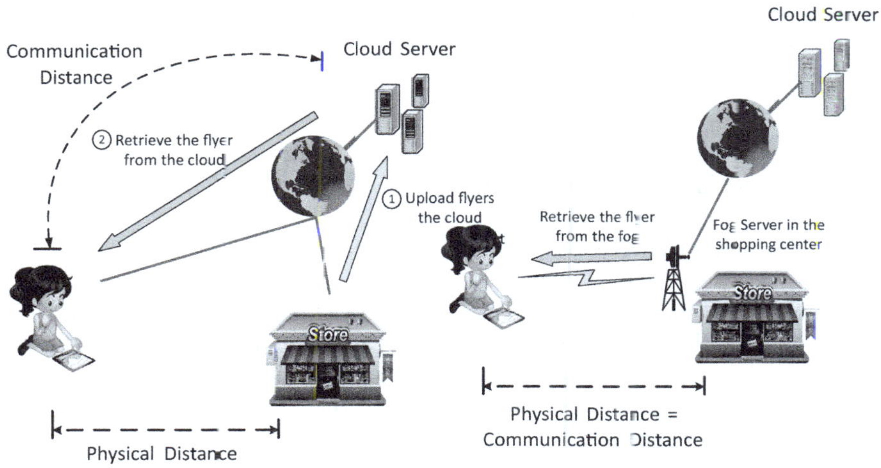

Fig. 2 Example: download the flyer of a nearby store. (**a**) Retrieving the flyer from the cloud. (**b**) Retrieving the flyer from the fog

The Fog computing paradigm represents a practical and efficient solution to resolve the mismatch between physical and communication distances. As a remedy shown in Fig. 2b, a Fog server can be deployed inside the shopping center and to distribute the local store flyers to mobile users. As such, the physical distance is equal to the communication distance and users can acquire low-latency desirable services.

By minimizing the communication distance, the Fog computing therefore brings the following two advantages:

- **To mobile users**: compared to cloud, the Fog computing can provide enhanced service quality with much increased data rate and reduced latency and response time. Moreover, by reducing the bandwidth cost of data transmission in the backbone, the users can also be benefited from the reduced service cost.
- **To network**: by avoiding the duplicated back and forth traffic between cloud and mobile user, not only the backbone bandwidth can be significantly saved, the energy consumption of core networks can also be greatly reduced, which contributes to the sustainable development of networking.

4 Components of Fog Computing

The Fog thus behaves as a surrogate of Cloud or a private Cloud at the user's premises. This enables Fog servers to be more efficient to handle the localized computation requests. Therefore, Fog computing targets to deliver the localized and location-based service applications to mobile users. In what follows, we showcase some examples of Fog computing implementation from this perspective, and discuss on the features of a Fog server as a comparison to Cloud server.

4.1 Exemplary Implementations

4.1.1 Shopping Center

Assuming that a number of Fog servers are deployed inside a multi-floor shopping center, which collectively form an integrated localized information system. The Fog servers at different floors can pre-cache floor-related contents, such as the layout and ads of stores on a particular floor. The Fog servers can deliver engaged services including indoor navigation, ads distribution and feedback collections to mobile users through WiFi.

4.1.2 Parkland

The Fog computing system can be deployed in the parkland to provide localized travel services. For instance, Fog servers can be deployed at the entrance and other important locations of the park. The Fog server at the entrance can pre-cache information including park map, travel guide and local accommodations; other Fog servers at different locations inside the park can be incorporated with sensor networks for environment monitoring and provide navigation to travellers. By connecting the Fog servers to the park administration office and cloud, the Fog servers can be used as an information gateway to send timely alerts and notifications to travellers.

4.1.3 Inter-State Bus

Greyhound has launched "BLUE" [5], an on-board Fog computing system over inter-state buses for entertainment services. As an example illustrated in Fig. 3, a Fog server can be deployed inside the bus and provides on-board video streaming, gaming and social networking services to travellers using WiFi. The on-board Fog server connects to the Cloud through cellular networks to refresh the pre-catched contents and update application services. Using its computing facility, the Fog server can also collect and process user's data, such as number of travellers and their feedbacks, and report to cloud.

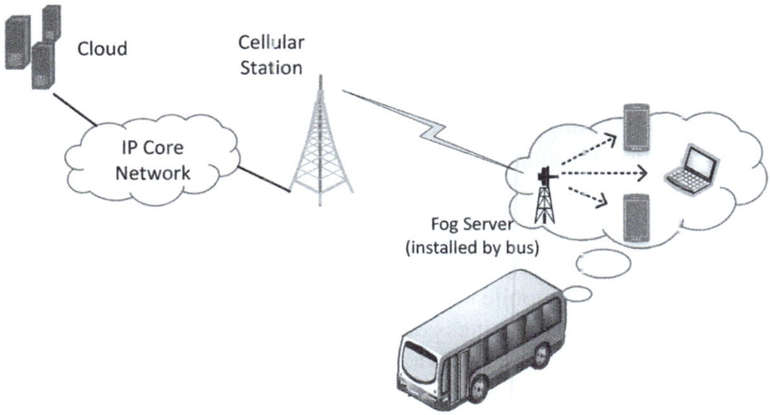

Fig. 3 On-board fog computing system

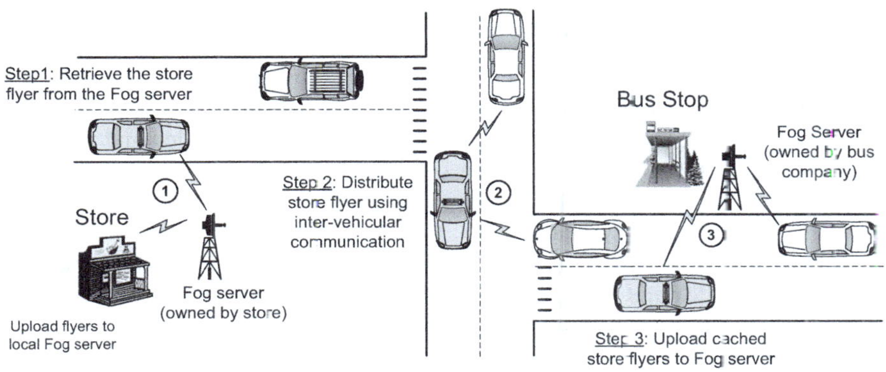

Fig. 4 Fog computing for content distribution in vehicular ad hoc networks

4.1.4 Vehicular Fog Computing Networks

Luan et al. [6] present the application of Fog computing as an integrated large-scale network for localized content disseminations. Figure 4 shows a motivating scenario. Assuming that a store installs a Fog server at its parking lot with the purpose to distribute the store flyer. In step 1, the store uploads flyers to the Fog server via wireless connections, and the Fog server distributes the flyers wirelessly to vehicles driving through its coverage using wireless communications. With the vehicle moving to different locations, it can further disseminate the cached flyers to other vehicles using wireless communications, as depicted in Step 2. In Step 3, the flyers can also be retrieved and cached at other Fog servers deployed at different locations, e.g., bus stop, and further propagated in the network.

4.2 Comparison to Cloud Computing

Fog computing is featured by the dedication to serving the localized and location-based applications. To this end, a Fog server manages its on-board resources to fully explore the location information and predictable user demand with the following functions.

- **Wireless**: Fog computing is dedicated to serving the mobile users. Each Fog server typically has limited wireless coverage, e.g., 200 m using WiFi, and directly interacts with mobile users using the single-hop wireless connections.
- **Local Services**: Fog computing is dedicated to serving the localized information and providing location-based service applications. For example, the Fog computing system deployed in a specific park only provide the navigation services within the park.
- **Distributed Management**: A Fog computing system may typically be deployed and managed by the local business, with the purpose to deliver designated contents and services to specific user groups.

Using the example in Fig. 4 for illustration, the Fog servers provide localized content distribution using wireless communications, which matches the first two features. The Fog server deployed nearby the store may be installed and managed by the store owner for the distribution of store flyers; the Fog server at the bus stop may be managed by the bus company for the distribution of bus information, e.g., bus schedules, safety manual, etc., to mobile users waiting for the bus. The Fog computing system in [6] is therefore distributedly constructed with Fog servers distributed installed and managed by different entities to serve their own purposes, which matches the third feature.

Table 1 summarizes the differences between Fog computing and Cloud computing.

By targeting to different user groups at different locations, Fog computing extends Cloud computing to better serve local mobile traffics. As the system architecture shown in Fig. 1, the Fog servers deployed at different locations would be used to deliver engaged services specified by their owners. The Fog servers at different locations can connect to the same cloud and form an integrated Fog computing system in a wide region.

4.3 Components of Fog Computing

4.3.1 Storage

In a predefined service area, a Fog server predicts the mobile user's demand on information and pre-cache the desirable information accordingly using a proactive way in its storage. Such information can be either retrieved from the Cloud or uploaded by its owner. For example, the Fog server installed at a restaurant can

Table 1 Comparison of fog computing and cloud computing

	Fog computing	Cloud computing
Target user	Mobile users	General Internet users
Service type	Limited localized information services related to specific deployment locations	Global information collected from worldwide
Hardware	Limited storage, compute power and wireless interface	Ample and scalable storage space and compute power
Distance to users	In the physical proximity and communicate through single-hop wireless connection	Faraway from users and communicate through IP networks
Working environment	Outdoor (streets, parklands, etc.) or indoor (restaurants, shopping malls, etc.)	Warehouse-size building with air conditioning systems
Deployment	Centralized or distributed in reginal areas by local business (local telecommunication vendor, shopping mall retailer, etc.)	Centralized and maintained by Amazon, Google, etc.

pre-cache the menu of the restaurant and dish recipes to serve the mobile users inside the restaurant. In another example, the Fog servers deployed in the airport can pre-cache the flight and local transportation information which is desirable to travellers in the airport. Therefore, the key design issue of Fog computing is to predict the user's demand and proactively select the contents to cache in the geo-distributed Fog servers based on the specific locations.

The Content Delivery Network (CDN) [7] represents the most mature catch networks and extensively pursued in both academic and industry. CDN is the Internet-based cache network by deploying cache servers at the edge of Internet to reduce the download delay of contents from remote sites. CDN mainly targets to serve traditional desktop Internet users, which have much broader interests and blur service demands that are more difficult to predict than those of mobile users. With precise service region, Fog computing has more clear target users of specific service demand. It is thus key for Fog servers to explore this feature to fully utilize its storage and computing resources to provide the most desirable services to mobile users.

Similar to Fog computing, the Information Centric Network (ICN) [8] is also a wireless cache infrastructure which provides content distribution services to mobile users with distributed cache servers. Different from the cache servers in ICN, the Fog servers are intelligent computing unit. Therefore, the Fog servers are not only used for caching, but also as a computing infrastructure to interact with mobile users and devices for real-time data processing. The Fog servers can be connected to the cloud and accordingly utilize the extensive computing power and big data tools for rich applications other than content distribution, such as internet of things, vehicular communications and smart grid applications [1].

Baştuğ et al. [9] also show that the information demand patterns of mobile users are predictable to an extent and propose to proactively pre-cache the desirable information before users request it. The social relations and device to device communications are leveraged. Unlike Fog computing, the proactive caching scheme in [9] is not explicitly used to serve local information services. As a more broad and generic paradigm, Fog computing can incorporate the proactive caching framework as described in [9].

4.3.2 Compute

A salient feature that differentiates Fog computing from the traditional cache networks is that Fog servers are intelligent compute system. This allow a Fog server to autonomously and independently serve local computation and data processing requests from mobile users. Satyanarayanan et al. [10] shows the applications of Fog computing in the cognitive applications. In another example, a Fog server inside the shopping mall or parkland can maintain an on-board geographic information system, and provide the real-time navigation and video streaming to connect mobile users.

Bridging the mobile and Cloud, a Fog server can also be conveniently used to collect the environmental data or demographic data from mobile users at the deployed spot, and transport the collected big data to Cloud for in-depth data analysis; the results can be provided to third party for strategic and valuable insights on business and government event planning, execution and measurement.

Despite of the high computing power, the Cloud is faraway from mobile users and can hardly support real-time computing intensive applications due to the bandwidth-constrained IP networks. The demand of real-time resource-intensive mobile applications, e.g., cognitive and internet-of-things applications, motivates the design of ubiquitous edge computing system [10, 11]. Cloudlets [3, 10] adopt the same framework of Fog computing, in which a Cloudlet server, similar to the Fog server, is deployed in the proximity of mobile users and processes the computing requests of mobile devices at real-time for video streaming and data processing. A comparison of processing delays using Cloudlets and Amazon Clouds is shown in http://elijah.cs.cmu.edu/demo.html. Transparent computing [11] is a highly virtualized system, which targets to develop the computing system transparent to users with cross-platform and cross-application support.

The Fog computing is a generic platform for edge computing and focuses on the localized service applications and computation requests. The prototype and techniques in [10, 11] can be incorporated in Fog computing framework.

4.3.3 Communication

Fog server can equip with different wireless interfaces, e.g., WiFi, Bluetooth and visible light communications [12] according to the specific application scenarios. The Fog computing differs from traditional radio access networks, e.g., WiFi and Femtocell networks, in two important ways.

Cross-Layer Design Unlike traditional WiFi access points, the Fog server manages an autonomous, all-inclusive network by providing both service applications and wireless communications to mobile users in the coverage. Therefore, a Fog server can work without Internet connections as that in [6]. Note that the Fog computing tailors its applications based on the specific deployment location and environment, and therefore is highly service-oriented. To this end, a Fog server can manage all the communication layers and effectively enables the cross-layer design [13] to provide the best service quality to users. For example, as in "BLUE" [5], a Fog server can cache a number of videos and deliver Youtube-like video streaming services to mobile users in the proximity. In this case, based on the context, wireless channel and video popularity information, the video services can be conveniently adapted towards the optimal performance via cross-layer adjustifications.

Predictable Location-Based Service The key of Fog computing is to provide the localized network and information applications to mobile users, whereas the traditional radio access networks focus on the provision of Internet applications and global information. With this distinguished feature, the design of Fog computing communications needs to consider the specific deployment environment and the features of mobile users in the considered scenario. For example, a Fog computing system deployed in the shopping mall needs to address the diverse mobilities of users, whereas the similar system deployed in the inter-state bus [5] only needs to consider static on-board passengers.

5 Case Study: Hybrid Data Dissemination in Fog Computing

In this section, we demonstrate a case study based on Fog computing and show how Fog computing can be incorporated into the 5G network towards improved performance to mobile users.

5G technique will make streaming applications becoming more and more popular, however the long latency may severely affect the user experience and is not tolerable. To address this issue, Fog computing can move Cloud services from remote Internet to the edge of networks and makes streaming content much closer to mobile users, which significantly decreases the streaming latency. On the contrary, the data dissemination from Cloud to every Fog servers can be expensive, which may takes the hudge 5G bandwidth resource. In addition, note that since the majority of streaming applications are video based, such as movies, teleplay and product advertisement, such contents are not always necessary to be strictly up-to-date and 1 or 2 days latency is affordable. Therefore, the Fog computing and delay tolerant network (DTN) techniques can be combined together to improve the performance of 5G network.

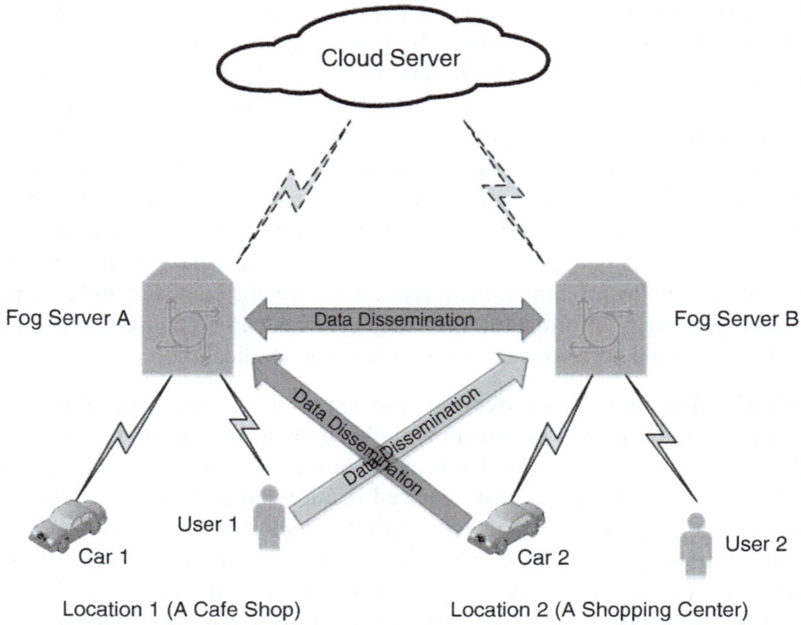

Fig. 5 Data dissemination in fog computing based on delay tolerant network technique

5.1 System Model

To have an efficient data dissemination in Fog computing, DTN is used to offload data among Fog servers. For example, in Fig. 5, if Fog server A has some small content, such as a new video ads, but Fog server B does not. In previous, Fog server B needs to get the update from a Cloud server directly. With DTN technique, user 1 could download this content when he has a coffee in this cafe shop. After a couple of hours, he goes to shopping center for shopping. When he moves into the transmission range of Fog server B, the content storied in user 1's mobile device is automatically upload to Fog server B and this store-carry-forward process is completed. For a large content, e.g. a high definition movie, transmitted from Fog server B to Fog server A, vehicle based DTN is used. In this example, if car 2 is parked in shopping center and within the Fog server B's transmission range, it downloads this movie into its local storage. Once this vehicle moves to the transmission range of Fog server A, this movie is uploaded to it and this data dissemination is completed.

In addition of the above DTN based data dissemination between mobile user and Fog server, direct data dissemination from Fog server to Fog server based on DTN technique is also available. As shown in Fig. 6, there is an on-board Fog server on a tourism bus, where all passengers on this bus can access its Fog server to watch movies or play games. If a passenger wants to find some new stuff, such as "just in"

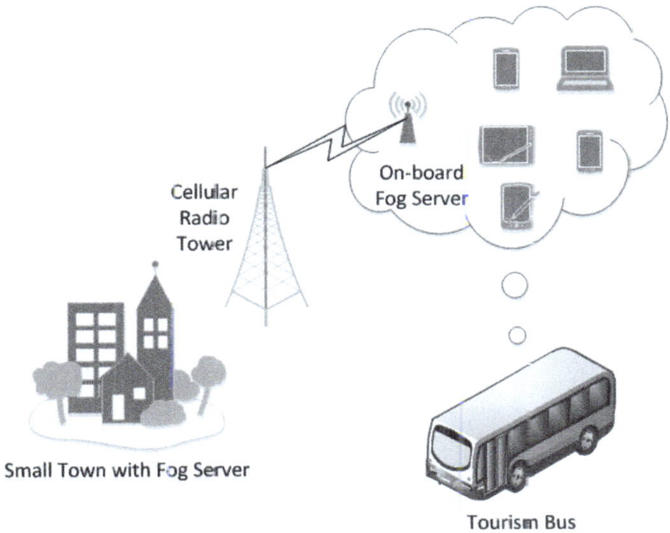

Fig. 6 Data dissemination between fog servers

news, Fog server can use the cellular network to retrieve this content immediately. When this bus travel to a small town along its route, it can synchronize its content with the Fog server located in this small town to update both servers' content list. If the Fog server in this small town has some content while other small towns along the route do not have, this tourism bus could download these contents into its local storage and carry to those small towns where they need these contents.

With the above DTN based data dissemination techniques, we propose a hybrid data dissemination model, as shown in Fig. 7. This hybrid model not only includes the normal data dissemination between Cloud servers and Fog servers, but also involves large amount low-cost DTN based data disseminations, which can be used between Fog servers and mobile users and among Fog servers. To organize these data dissemination, we re-identify the function of Cloud servers. In this model, the main function of Cloud server is to act as the "control plane" to determine the Fog server needed to be updated with the required content and control data dissemination process, as shown in Fig. 8. Fog servers and part of Cloud servers are treated as "data plane" to provide data dissemination service.

This data dissemination model has three components, namely as *Data Structures*, *Protocol Messages* and *Algorithms*. Data structures use tables to store key information which is used to determine the path of data dissemination. Protocol messages use various tapes of messages to discover content and mobile devices associated with a Fog server, exchange content list, and other tasks to learn and maintain accurate information about the network. Algorithms are used to calculate the best data dissemination path.

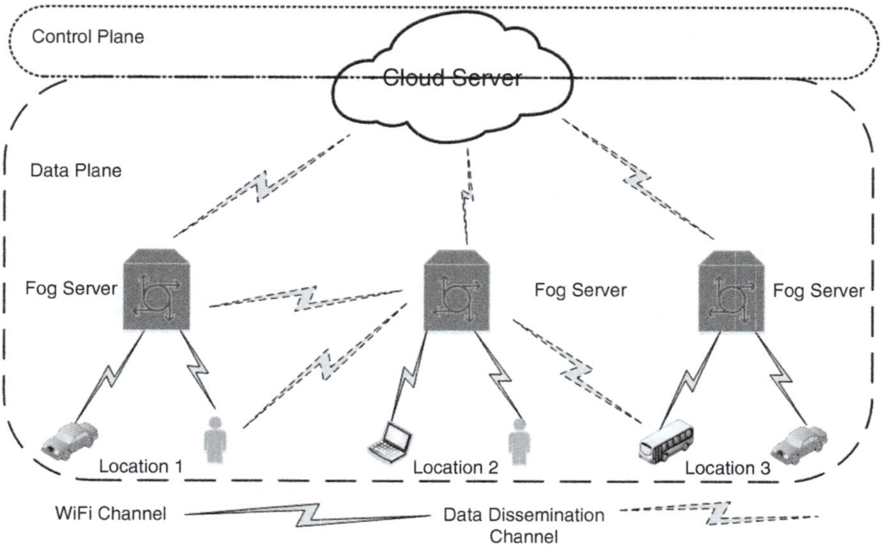

Fig. 7 Hybrid data dissemination model in fog computing

Cloud server in this model needs to have an overall information and its data structures include the following tables:

- Fog Server List Table: a table to record all Fog servers, which are managed by Cloud server. This table includes Fog server's ID, content ID in each of Fog servers, mobile device ID associated with each of Fog servers.
- Global Content List Table: a table to record all public contents (not include these private content created by Fog server owner) in Fog servers or supposed to be in Fog servers. This table includes content ID, the size of each contents, Fog server's ID (for these Fog servers who have this content), date of update, validation time.
- Mobile Devices's Movement Pattern Table: a table to record mobile device's ID, Fog server ID (whom mobile device linked before), linked time, social attribute, geographic movement pattern.

A Fog server needs to have a table to record content ID, the size of this content, mobile device's ID which linked with this Fog server, the linked duration of this mobile device. For mobile devices, they need to record the content ID which they carry on, their movement path, time duration with a Fog server and its ID.

In order to collect and exchange the above information, several data messages are used in this model.

- Hello Message between Fog Servers and Cloud Server: an update message from a Fog server to its Cloud provider, which includes its content ID and associated mobile devices ID. This is a triggered message.

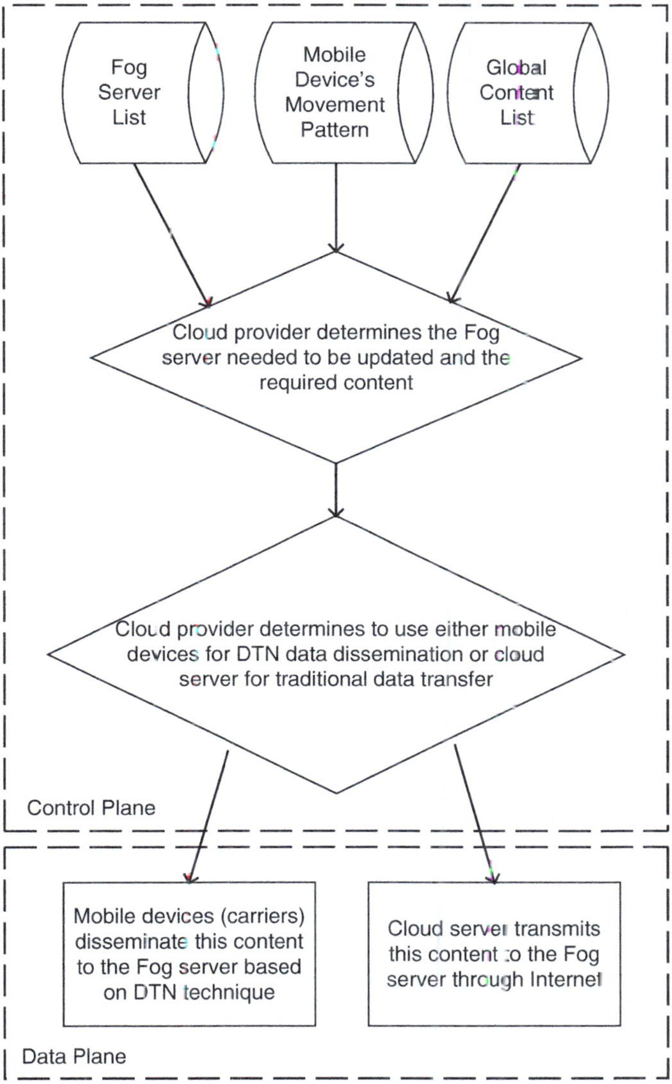

Fig. 8 Control plane and data plane in this model

- DTN Data Dissemination Request Message: a message sent from Cloud provider to a Fog server. When a Cloud provider determines there is a content need to be updated for Fog server A, if Fog server B has this content and its associated mobile device has potential move to Fog server A, a DTN data dissemination request message is sent to Fog server B to ask it to disseminate the content through that mobile device.

- DTN Data Dissemination Accept Message: a message sent from Fog server to its Cloud provider. Once a Fog server received a DTN data dissemination request message and it sends the content to the corresponding mobile device, it sends this DTN data dissemination accept message to its Cloud provider to confirm that this content has been sent out.
- DTN Data Dissemination Decline Message: a message sent from Fog server to its Cloud provider. When a Fog server received a DTN data dissemination request message, it's associated carrier (mobile device), for some reason, does not receive the complete content before this carrier leave the current Fog server. In this case, the DTN data dissemination decline message is sent back to the Cloud provider.
- DTN Data Dissemination Acknowledgement Message: a message sent from Fog server to its Cloud provider. When a Fog server receives the assigned content from carrier, it sends this acknowledgement message to Cloud provider to confirm that this DTN data dissemination is completed.

Algorithms together with other operations in this hybrid data dissemination model are described in the following sub-section.

5.2 Data Dissemination

Hybrid data dissemination is determined by control plane, as shown in Fig. 8, where Cloud provider in the control plane has the global information to control the data plane. The main data flow control algorithm conducted by Cloud provider is illustrated in Algorithm 1, where Cloud provider checks its global Fog server and content lists to determine the Fog server that needed to be updated, and the required content. It also checks which Fog server has this content. If none of Fog server has this content, Cloud provider sends updated content to that Fog server directly by using traditional Cloud based techniques, such as broadband and cellular networks [14]. Otherwise, the DTN based data dissemination is applied by using Algorithm 2 to choose mobile devices, which are connecting with these selected Fog servers, as carriers to provider DTN based data dissemination services.

The carriers selection is based on their delivery time and delivery probability (Algorithm 3). A pre-determined content delay threshold, T_{delay}, which is an attribute of this content and can also be treated as content delivery priority, is provided by the Cloud provider. Only those mobile devices (carriers) with a shorter delivery time compared with the pre-determined delay time, and a higher delivery probability are selected as potential DTN based data dissemination carriers.

Once these carriers are chosen, Cloud provider sends DTN data dissemination request message to each of the selected Fog servers to ask them send the required content to the Fog server which is needed to be updated. If the content is transmitted to the carried successfully, a DTN data dissemination accept message is sent back

Algorithm 1 Data Flow Control Algorithm

Step 1: Cloud provider compares its "Global Content List" table with "Fog Server List" table to determine which Fog server is needed to be updated. In this case, Fog sever F_d is determined and content C is needed to be updated.
if There is no other Fog server has this content **then**
  Cloud provider sends this content to the Fog server directly, which is the same as traditional Cloud service and this dissemination process is finished.
else
  Move to the next step
end if
Step 2: Cloud provider determines a list of Fog servers, $< F_{c1}, F_{c2}, F_{c3}, \cdots >$, which have this content.
Step 3: Algorithm 2 is used to select n most suitable carrier, $Carrier_n$, to provide this DTN dissemination service.
if n greater than 0 **then**
  Move to the next step
else
  Cloud provider sends this content to the Fog server directly, which is the same as traditional Cloud service and this dissemination process is finished.
end if
Step 4: Cloud provider sends "DTN Data Dissemination Request" message to each of selected Fog servers to ask them send the content C to F_d by using the carrier (mobile device) determined in **Step 3**.
Step 5: Once a Fog server receives "DTN Data Dissemination Request" message, it sends the content C along with the destination, F_d, to the selected carrier.
if Content C is transferred to the selected carrier completely **then**
  This Fog server sends the "DTN Data Dissemination Accept" message to its Cloud provider and move to the next step.
else
  This Fog server send the "DTN Data Dissemination Decline" message to its Cloud provider. When Cloud provider receives this message, it repeats *Step 3* to get the "n+1" Fog server, if it has, and continue from *Step 4*.
end if
Step 6: Once F_d receives the content C, it sends "DTN Data Dissemination Acknowledgement" message to Cloud provider.
Step 7:
if Cloud provider receives the "DTN Data Dissemination Acknowledgement" message within a pre-defined period, T_{delay}, in Algorithm 2 **then**
  It updates "Fog Server List" and "Global Content List" tables, and this data dissemination is finished
else
  It repeats from the *Step 1*
end if

to Cloud provider confirming this content has been sent out. Otherwise, a DTN data dissemination decline message is sent out. For example, a mobile device (carrier) left the Fog server's coverage area.

When the Fog server, who needs this content, receives the content, it sends a DTN data dissemination acknowledgement message to Cloud provider to confirm it has received the content and this DTN based data dissemination process is finished.

Algorithm 2 DTN Data Dissemination Carrier Selection Algorithm

Step 1: Cloud provider determines the affordable delay time, T_{delay}, of this content. T_{delay} Cloud be treated as the priority of this content.

Step 2: Cloud provider checks its "Fog Server List" and "Mobile Devices's Movement Pattern" tables to find the list of mobile users accessed F_d before, $MobList_d(i)$, and their average connection time with F_d, $Time_d(i)$, where i is the ID of connected mobile device.

Step 3: These mobile devices with a short connection time are filtered out, as they are not able to upload the content to the Fog server:

for each of mobile device in $MobList_d(i)$ **do**

if

$$\frac{Size_C}{Speed_i} > Time_d(i)$$

then

This mobile device is filtered out from $MobList_d(i)$ and a new list $MobList'_d(r)$ is formed, where r is the number of mobile device meeting the above condition

end if

end for

Step 4: $MobList'_d(r)$ is further classified into two categories, *scheduled* and *non-scheduled* visit lists. Scheduled visit list stores these mobile devices which are pre-determined to visit a particular Fog server, such as airport shuttle bus. The rest of filtered mobile devices are classified into the non-scheduled visit list.

Step 5: For scheduled mobile devices, Si, as long as its delivery time, which is the time from now to its next scheduled visit time, is within the T_{delay}, it is added into the DTN data dissemination carrier list, $< Carrier_{S1}, Carrier_{S2}, \cdots, Carrier_{Sx} >$, where x the total number of carriers selected to add into the carrier list.

Step 6: Non-scheduled mobile device list, NSi, is re-ordered by mobile devices' delivery probability to F_d based on Algorithm 3. Cloud provider select the top y mobile devices according their delivery probabilities to add them into the DTN dissemination carrier list, where the number of y is the largest number to satisfy the following condition:

$$\frac{\sum_{i=1}^{x} DeliTime_{Si} + \sum_{i=1}^{y} DeliTime_{NSi}}{x+y} < T_{delay}$$

End: Now $x + y$ mobile devices are selected as carriers to provide DTN data dissemination service.

Otherwise, Cloud provider needs to re-select mobile nodes as carriers or directly sends the content using traditional method. Detailed hybrid data dissemination processed are illustrated in Algorithms 1–3, and notations used in these three algorithm are explained in Table 2.

6 Future Research Topics of Fog Computing in 5G

Based on the Mobile-Fog-Cloud hierarchy shown in Fig. 9, we envision potential research directions from the communication efficiency's viewpoint as follows.

Algorithm 3 Mobile Device Delivery Probability

Step 1: For each of mobile devices, m, Cloud provider collects its contact frequency with F_d, $ConFre_m$, geographic locations and visit times of the three most recently visited Fog servers, $Loc_m < Lan, Lon, T >$.

Step 2: Based on the three most recent visited history, $Loc_m < Lan, Lon, T >$, and real distance from the map, average movement speed and direction of mobile device m could be generated as $Speed_m$ and $Direction_m$.

Step 3: The expected delivery time from mobile device m to Fog server F_d, $DeliTime_m$, is calculated by using $Speed_m$, $Direction_m$ and the geographic distance between both them.

if $DeliTime_m > T_{delay}$ **then**

The delivery probability of this mobile device, $DeliProb_m$, is marked as 0 and this algorithm is finished

else

Move to next step

end if

Step 4: Assume there are n suitable mobile devices left in this step. The overall delivery probability of mobile device m is calculated as:

$$DeliProb_m = \frac{ConFre_m}{\sum_{i=1}^{n} ConFre_i} \times \left(1 - \frac{DeliTime_m}{\sum_{i=1}^{n} DeliTime_i}\right)$$

and each of them is added into the delivery probability list, $DeliProbList[n]$.

Step 5: Sort $DeliProbList[i]$ in ascending order

Set $u = 1, j = n$

while $u \leq n$ **do**

  while $j > u$ **do**

    if $DeliProbList[j-1] > DeliProbList[j]$ **then**

      swap($DeliProbList[j-1]$, $DeliProbList[j]$)

    end if

    $j--$

  end while

  $u++$

end while

This delivery probability list is ready to be used for Algorithm 2.

6.1 Communications Between Mobile and Fog

Note that a Fog server manages 3-D resources including storage, computing and communication. The service quality acquired by users relies on the collective performance of resource utilization from all the three dimensions. Moreover, as Fog computing typically provides pre-defined application services and targets to specific user groups, the service-oriented resource allocation customized to the specific deployment environments is thus necessary. For example, considering the on-board Fog computing system inside the inter-state bus as in Fig. 3, three types of traffics may coexist including video streaming, gaming and web surfing delivered through the same Fog server. As such, a cross-layer MAC design at the Fog server can be devised based on the application's information. Considering that Fog servers have limited storage and deliver limited localized services only, another key design issue is how to optimally select the desirable information contents to cache at each Fog server and determine the appropriate service applications which cause the least service failure rates to mobile users. The solution needs to consider the predictable pattern of mobile service requests, available storage and compute power of a Fog server.

The Fog computing can also be incorporated with the 5G cellular networks. In this case, by making the cellular base station a Fog server with on-board storage and compute facility, the entire Fog system can provide greater coverage and dedicated services to cellular users.

6.2 Communications Between Fog and Cloud

The cloud performs two roles in the integrated Fog computing system. First, the cloud is the central controller of Fog servers deployed at different locations. With each Fog server focusing on the service delivery to mobile users at specific locations, the cloud manages and coordinates the geo-distributed Fog server clusters at different regions. Second, the cloud is the central information depot. The Fog servers at different locations select the information contents from the cloud and then deliver the copied contents from its cache to the mobile users. With above two roles, the design goal of the communications between fog and cloud can be twofold: (1) how to enable the reliable and scalable control of Fog servers at the cloud; and (2) how to develop the scalable data routing scheme from cloud to Fog server for content updates.

Note that the dual functions of cloud as stated above well match the architecture of a software-defined networking (SDN) [15, 16], which decouples the traffic routing to the control plane and data plane. It is thus promising to apply the SDN scheme for the control of Fog computing.

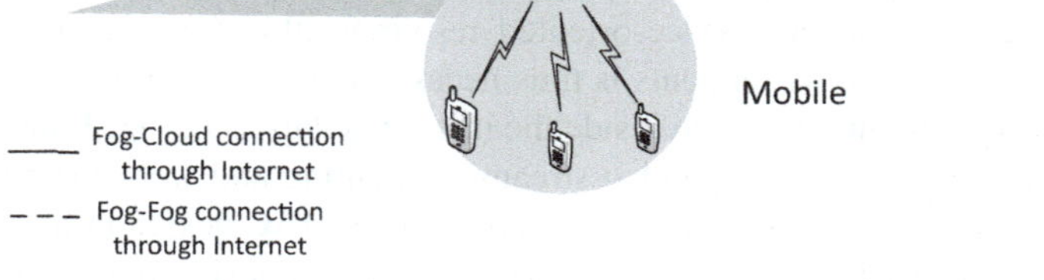

Fig. 9 Mobile-fog-cloud architecture of fog computing

Table 2 Notation used in Sect. 5.2

Notation	Definition
F_d	A fog server needed to be updated (the destination of data dissemination)
C	A content needed to be updated and disseminated
F_i	The fog server i
$Carrier_i$	A content carrier (mobile device) i to provide DTN based data dissemination service
T_{delay}	The maximum affordable delay time of a content. It is a time period from now to its must updated time, such as a shopping center promotion video must be released on every Wednesday
$MobList_d(i)$	The ith mobile device (carrier) in the mobile list attached with fog server d
$Time_d(i)$	An average connection period between mobile device i and fog server (d). It is a maximum time window used to upload/download a content to a fog server
$Size_C$	The size of content C
$Speed_i$	The wireless transmission speed of mobile device i
$Loc_i < Lan, Lon, T >$	The geographic location vector of mobile device i visited at latitude Lan and longitude Lon on time T
$Direction_i$	The expected movement direction of mobile device i
$DeliTime_i$	An average content delivery time of Carrier i to the destined fog server
$ConFre_m$	A contact frequency between the mobile device m and its destined fog server
$DeliProbList[n]$	A list to store all carriers (mobile devices) delivery probability to the destined fog server

6.3 Internet-of-Things Applications

As Fog servers are deployed at the physical spot close to mobile users and can be equipped with sensors, it is convenient to incorporate the Fog computing with the Internet-of-things applications. Bonomi [1] and Stojmenovic and Wen [17] present the examples of adopting Fog computing in the applications of smartgrid, vehicular networks and sensor networks.

7 Conclusion

This book chapter introduced the Fog computing under 5G environment.

This article presents Fog computing, a new networking frontier dedicated to serving mobile users. By deploying reserved compute and communication resources at the edge, Fog computing absorbs the intensive mobile traffic using local fast-rate connections and relieves the long back and forth data transmissions among cloud and mobile devices. This significantly improves the service quality perceived by mobile users and, more importantly, greatly save both the bandwidth cost and energy consumptions inside the Internet backbone. Therefore, Fog computing represents a scalable, sustainable and efficient solution to enable the convergence of cloud-based Internet and the mobile computing. The purpose of this article is to investigate on the major motivation and design goals of Fog computing from the networking's perspective. We emphasis that the emergence of Fog computing is motivated by the predictable service demands of mobile users, and Fog computing is thus mainly used to fulfill the service requests on localized information. As a Fog server possesses hardware resources in three-dimensions (storage, compute and communications), the 3-D service-oriented resource allocations are therefore the key of Fog computing. Moreover, with the three-tier Mobile-Fog-Cloud architecture and rich potential applications in both mobile networking and Internet-of-things, the Fog computing also opens broad research issues on network management, traffic engineering, big data and novel service delivery. Therefore, we envision a bright future of Fog computing.

References

1. F. Bonomi, R. Milito, J. Zhu, S. Addepalli, Fog computing and its role in the internet of things, in *Proceedings of ACM MCC* (2012), pp. 13–16
2. R. Balan, J. Flinn, M. Satyanarayanan, S. Sinnamohideen, H.I. Yang, The case for cyber foraging, in *Proceedings of ACM SIGOPS* (2002), pp. 87–92
3. M. Satyanarayanan, P. Bahl, R. Caceres, N. Davies, The case for vm-based cloudlets in mobile computing, No. 99 (2011)
4. Location-Based Mobile Services Have Huge Untapped Potential Worldwide (2012)
5. G.M. Relations, Greyhound launches BLUE, an exclusive Wi-Fi enabled onboard entertainment system. https://www.greyhound.com/en/newsroom/viewrelease.aspx?id=528&year=2013. Accessed on Dec 2014
6. T.H. Luan, L.X. Cai, J. Chen, X. Shen, F. Bai, Vtube: towards the media rich city life with autonomous vehicular content distribution, in *Proceedings of IEEE SECON* (2011), pp. 359–367
7. G. Peng, CDN: content distribution network (2004). arXiv preprint cs/0411069
8. B. Ahlgren, C. Dannewitz, C. Imbrenda, D. Kutscher, B. Ohlman, A survey of information-centric networking. IEEE Commun. Mag. **50**(7), 26–36 (2012)
9. E. Baştuğ, M. Bennis, M. Debbah, Living on the edge: the role of proactive caching in 5G wireless networks. IEEE Commun. Mag. **52**, 82–89 (2014)
10. M. Satyanarayanan, Z. Chen, K. Ha, W. Hu, W. Richter, P. Pillai, Cloudlets: at the leading edge of mobile-cloud convergence, in *Proceedings of MobiCASE* (2014)
11. Y. Zhang, Y. Zhou, Transparent computing: a new paradigm for pervasive computing, in *Ubiquitous Intelligence and Computing: Third International Conference* (Springer, Berlin/Heidelberg, 2006), pp. 1–11
12. T. Komine, M. Nakagawa, Fundamental analysis for visible-light communication system using LED lights. IEEE Trans. Consum. Electron. **50**(1), 100–107 (2004)

13. C.X. Lin, X. Shen, J.W. Mark, L. Cai, Y. Xiao, Voice capacity analysis of WLAN with unbalanced traffic. IEEE Trans. Veh. Technol. **55**(3), 752–761 (2005)
14. K. Zheng, F. Hu, W. Wang, W. Xiang, M. Dohler, Radio resource allocation in lte-advanced cellular networks with m2m communications. IEEE Commun. Mag. **50**, 184–192 (2012)
15. H. Kim, N. Feamster, Improving network management with software defined networking. IEEE Commun. Mag. **51**, 114–119 (2013)
16. K. Zheng, L. Hou, H. Meng, Q. Zheng, N. Lu, L. Lei, Soft-defined heterogeneous vehicular network: architecture and challenges. CoRR (2015). abs/1510.06579
17. I. Stojmenovic, S. Wen, The fog computing paradigm: scenarios and security issues, in *Proceedings of FedCSIS* (2014)

A Conceptual 5G Vehicular Networking Architecture

The Driving Force to Connected Autonomy

Konstantinos Katsaros and Mehrdad Dianati

Abstract This chapter presents a thorough investigation on current vehicular networking architectures (access technologies and overlay networks) and their (r)evolution towards the 5G era. The main driving force behind vehicular networking is to increase safety, with several other applications exploiting this ecosystem for traffic efficiency and infotainment provision. The most prominent existing candidates for vehicular networking are based on dedicated short range communications (DSRC) and cellular (4G) communications. In addition, the maturity of cloud computing has accommodated the invasion of vehicular space with cloud-based services. Nevertheless, current architectures can not meet the latency requirements of Intelligent Transport Systems (ITS) applications in highly congested and mobile environments. The future trend of autonomous driving pushes current networking architectures further to their limits with hard real-time requirements. Vehicular networks in 5G have to address five major challenges that affect current architectures: congestion, mobility management, backhaul networking, air interface and security. As networking transforms from simple connectivity provision, to service and content provision, fog computing approaches with caching and pre-fetching improve significantly the performance of the networks. The cloudification of network resources through software defined networking (SDN)/network function virtualization (NFV) principles, is another promising enabler for efficient vehicular networking in 5G. Finally, new wireless access mechanisms combined with current DSRC and 4G will enable to bring *the vehicles in the cloud*.

1 Introduction

Connected vehicles enable a broad range of applications for safety, traffic efficiency and infotainment, and in the future facilitate connected autonomous vehicles. However, it is shown that the current communication systems are not sufficiently

K. Katsaros (✉) • M. Dianati
Institute for Communication Systems, University of Surrey, Guildford GU2 7XH, UK
e-mail: k.katsaros@surrey.ac.uk; m.dianati@surrey.ac.uk

W. Xiang et al. (eds.), *5G Mobile Communications*,
DOI 10.1007/978-3-319-34208-5_22

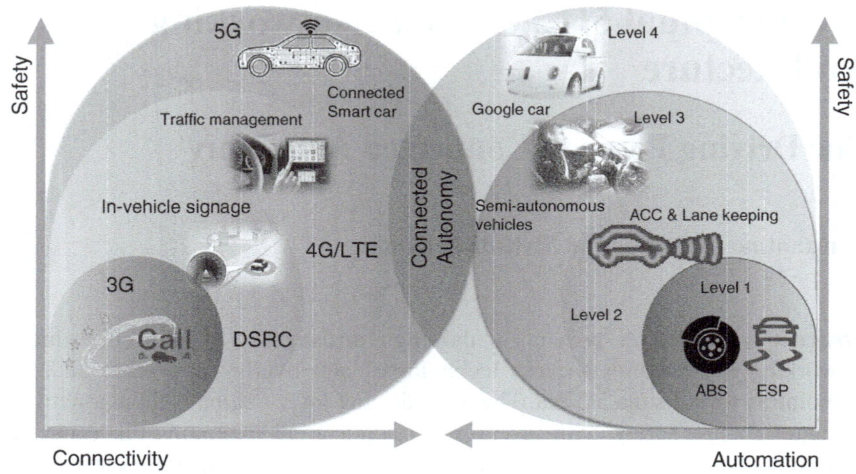

Fig. 1 Evolution of vehicular safety systems with connectivity and automation (NHTSA classification [50]: Level 1: Function specific automation; Level 2: Combined function automation; Level 3: Limited self driving automation; Level 4: Full self-driving automation)

reliable for the market [59] and something that raises concerns to the public when vehicles are travelling through traffic automatically with no driver behind the wheel. On the other hand, academia and industry are capitalizing on the application requirements and weaknesses of current communication systems to research and design the fifth generation (5G) networks.

There are presently two parallel paths in automotive industry that 5G will seek to converge; *connected* and *autonomous* vehicles (Fig. 1). Safety has always been one of the driving forces of innovation in automotive industry. From passive and active safety systems such as seatbelts and sensors, which alleviate the damage of an accident and alert drivers, we are currently at what is known as *connected mobility*. Intelligent Transportation Systems (ITS) are the building block for connected mobility. They are the outcome of coupling transportation systems with wireless communications offering users several innovative and effective services. It is envisaged that the future vehicles will be able to form ad-hoc networks between them, or connect to infrastructure in order to exchange important traffic and safety related information in highway and urban environments. There is a strong demand from the automotive industry, driven by market forces and consumer needs, to transform future cars into smart, connected systems, *a smartphone on wheels*, that offer personalised services and have the capability to adapt to the users' preferences and needs. Several communication technologies have been proposed and investigated for vehicular network applications from academia, industry and governmental bodies. Each of these technologies have certain properties that make them suitable for particular type of applications, mostly dictated by the end-to-end delay requirements, and the dissemination mode i.e. broadcast, geocast, or unicast.

However two technologies, namely, dedicated short range communication (DSRC), e.g., IEEE WAVE, ETSI ITS G5, and cellular, e.g., 3GPP LTE, are the most promising potential candidates.

What is more, state-of-the-art cars in the market currently provide some simple autonomous features such as emergency braking function, adaptive cruise control, and lane assist system. In addition, advanced navigation and infotainment systems take advantage of the Internet connectivity via cellular networks to provide enhanced and dynamic navigation services. Research in robotics, inspired by unmanned space vehicles, has given birth to driverless concept cars such as Google robotic car[1] empowered by proximity sensing and on-board autonomous processing.

Given that by 2030–2050 nearly all cars are expected to be autonomous [42], it is emerging that a combination of connected vehicles and autonomy can significantly boost the performance, safety, and reliability of new generations of autonomous transportation system. The binding component to achieve this, is 5G communications. Connected autonomous vehicles will lead to vastly improved safety and fewer collisions. KPMG estimates that by 2030 connected autonomous cars could save over 2500 lives and prevent more than 25,000 serious accidents in the UK. Production and dissemination of traveller information will be improved with the exploitation of Internet of Things (IoT) and Big Data analytics facilitated by low-latency, fully connected vehicles and cloud infrastructure. Travelling experience will be enriched for autonomous vehicles' occupants with augmented-reality infotainment systems. In this direction, Ericsson has proposed the Connected Traffic Cloud—a managed cloud concept that enables the sharing of real-time traffic and road condition data between connected vehicles and road traffic authorities. Future cars will be highly autonomous, smart systems that relieve the drivers from tedious tasks and reduce the negative impacts of human errors on safety of passengers and pedestrians. They will be connected to the internet and their surrounding environment (other cars and infrastructures), having access to a massive amount of information and processing power from the cloud and awareness about their surroundings environment. This will improve safety, efficiency, environmental friendliness, and comfort level of future transport systems. Such significant increase in digital capabilities of future cars is expected to have enormous positive economic and societal impacts.

Following this brief introduction to the vehicular networking ecosystems and their potential benefits, Sect. 2 provides a more detailed description of the vehicular applications and the specific requirements they have from a networking perspective. Then, in Sect. 3 current vehicular communication systems and their limitations are presented, demonstrating different access technologies as well as networking architectures. Based on the requirements for current and future vehicular application, as well as the limitations of current vehicular communication systems, a conceptual vehicular networking architecture towards 5G is proposed in Sect. 4. Finally, Sect. 5 concludes this chapter.

[1]Google Self-Driving Car Project [online]: https://plus.google.com/+GoogleSelfDrivingCars.

2 Vehicular Applications and Requirements

In this section, current and futuristic use cases of vehicular networks and their specific communication requirements are reviewed. These include latency limits, reliability, as well as security and privacy requirements, focusing only on off-board communications.[2] The basic requirements and use cases for 5G networking have been presented in Chap. "Spectrum Analysis and Regulations for 5G". and can be summarized in the following list compared to current 4G:

- Increase in mobile data volume (bps/km^2) and peak data rates by a factor of x100 and x1000, respectively;
- Reduce the end-to-end delay to 1 ms;
- Increase reliability and support higher mobility; and
- Increase energy efficiency by 10 %.

Latency requirements for the different categories of ITS applications are specified by ETSI [66]. Ultra low latency (e.g., 1 ms for the round trip delay) will be required in order to ensure the exchange of kinematic information for distributed control algorithms [23] for future ITS. Furthermore, the architecture of ITS systems is shifting towards a cloud-based system in order to off-load heavy calculations and storage to the cloud, facilitate vehicle maintenance and deliver new services to customers with a short time-to-market [16].

There is a long list of desirable applications considered for the short-term deployment of ITS [66] that can be grouped into four categories:

- Safety applications focusing on reducing accidents and their impact, e.g., road hazard warning [19];
- Traffic efficiency applications aiming at more efficiency and thus greener traffic, e.g., green light optimal speed advisory (GLOSA) [38];
- Comfort and infotainment applications, e.g., automatic map updates or video streaming [64]; and
- Manufacturer oriented applications, e.g., automatic software updates [27].

However, new application types have been foreseen for future deployments of automated and connected vehicles, e.g. augmented reality navigation, co-operative electronic stability control [23].

The following subsections analyse some of the use cases and their requirements, e.g. for safety applications, and general requirements for vehicular networking, e.g. security.

[2]Vehicle to Vehicle and Vehicle to Infrastructure.

2.1 Safety Applications

Safety-related applications aim to enhance driver's road perspective by analysing received messages such as Cooperative Awareness Message (CAM), Basic Safety Message (BSM) or Decentralized Environment Notification Message (DENM) broadcast by other vehicles and road side units (RSUs) in the communication range. These messages carry the status of the transmitting vehicle like kinematic information, type of vehicle etc, or information related to specific event, e.g. an accident. They are transmitted with frequencies ranging from 1 to 10 Hz. ETSI specifies the latency limit for the reception of these messages to 100 ms [66]. Currently, these applications are realised by in-vehicle systems that display warnings or emit sound to inform the driver. In the future, these warnings can be displayed on augmented dashboards as overlay information on top of what a driver is seeing through the front window, e.g. MINI's augmented reality.[3] It is also very important that cars can communicate not only with each other but also with other participants in the city, such as pedestrians or cyclists.

2.2 Comfort and Infotainment Applications

Comfort and infotainment applications are mainly constrained by the available network capacity. For example, the trend in video streaming is to support "4K" content. Netflix, a content provider, currently encodes 4K streams at ∼15 Mbps and with projection to get it down to 10–12 Mbps for 30 fps streams. Even though such download speeds may be available with 4G technology, the challenge here is how to scale to multiple users with high mobility, as future users will expect to continue their good quality connection independent of their location and speed. In terms of latency requirements, ETSI has given a higher limit for this type of applications at 500 ms [66], but some types of applications, such as VoIP, are also constraint by jitter.

2.3 IoT/Monitoring Cloud-Assisted Applications

Internet of Things (IoT) is gaining momentum and it is foreseen to reach more than 50 millions connected devices by 2020 [12], with 11 millions being embedded cellular-based telematics [4]. A recent survey[4] showed that application related to

[3]MINI's augmented-reality glasses allow drivers to see through the body of their car [Online]:http://goo.gl/oQUwal.

[4]The 10 most popular Internet of Things applications right now [online]: http://iot-analytics.com/10-internet-of-things-applications/.

connected vehicles, like smart parking and remote-controlled car, are in the 10 most popular applications. In addition, the electrification of vehicles and its interconnection with smart-grid has spurred new use-cases that will require very good scalability and ability to efficiently manage heterogeneous networks [9]. Currently, each vehicle has an average of 60–100 sensors on board. This is projected to reach 200 by 2020 as vehicles become "smarter". All these sensors travel together with the vehicle like a swarm, which has lead to the development of several applications. One of the most futuristic and highly computational expensive use cased of IoT has been proposed in the work of Jain et al. [35]. It exploits several inputs from vehicle sensors to pre-assess the driver behaviour and anticipate future manoeuvres (3.5 s before they occur) only within ∼3.6 ms computation time. In the future, individual driver input shall be fuzzed with the aid of cloud infrastructure and big-data analytics with surrounding drivers to provide co-operative assessment of manoeuvres with the same computational time limit.

2.4 Connected Automated Driving

There is a plethora of automated functions currently implemented on-board, like active suspension systems, that interconnect sensors and actuators around the vehicle through a ultra-low latency wired network. However, future co-operative automated functions should support equivalent characteristics while fusing sensor information from other vehicles. The computational capabilities of the vehicles are currently restricted, which limits the supported functionalities. Recent literature has revealed that the performance of the control systems can be significantly enhanced if the on-board information and processing is augmented by the information and processing power from the cloud [73]. In the industry, Electrobit[5] has collaborated with Infineon and NVIDIA to develop an automated driving platform that will increase the computing power and intelligence of vehicles required for automated driving. On the other hand, ZF Friedrichshafen AG has launched a cloud-based smart urban electric vehicle[6] With two semi-automated driver assistance functions they aim to increase comfort and safety of the passengers. Concretely, Smart Parking Assist maneuvers the vehicle into virtually any small parking space—remote-controlled at the push of a button using mobile devices such as a smartphone or smartwatch. Comfortable, efficient motoring is possible with the concept vehicle thanks to the cloud-based PreVision Cloud Assist assistance function. Where necessary, the system reduces the drive torque, for instance, in good time before entering the bend and thus throttles back the speed without any mechanical braking. Most

[5]Elektrobit (EB) collaborates with Infineon and NVIDIA to deliver first-of-its-kind automated driving platform—Press Release [online] https://goo.gl/LxxaLk.

[6]ZF Press Release "Collective Knowledge: PreVision Cloud Assist in the Advanced Urban Vehicle from ZF", July 2015.

of the research on automated functions is focused on light vehicles. However, trucks and buses have also been enhanced with autonomous functions. For example Daimler's research on autonomous trucks has produced functions such as the new anti-jackknifing system, which Daimler engineers have developed to ensure that extra-long articulated buses remain controllable in difficult situations. All these characteristics of automated driving make it a challenging topic as it requires ultra-reliable, fault-tolerant networks, low end-to-end delay among vehicle and with cloud-based infrastructure, and it must work everywhere. In addition, driver-less vehicles should be able to communicate with other "vulnerable" road users such as bicycles and pedestrians.

2.5 Security and Privacy

This section provides an overview of the security and privacy requirements for vehicular networking. Achieving high level of security with low communication cost is challenging. In addition, concerns regarding privacy breaches introduces another challenge to bring this type of technology to the masses. As cars become smarter, more and more vehicle components interconnect through the onboard vehicular network. This trend has been reflected in the number of lines of code running on all the electronic control units (ECUs). A modern high-end car features around 100 million lines of code, and this number is planned to grow to 200–300 millions in the near future. This creates new threats as external adversaries can potentially intrude and take control of critical vehicle functions [36]. A classification of the threats in modern vehicles can be found in [46]. The functionalities related to security and privacy that have to by provided by the communication system include the followings.

- *Data origin authentication* ensures the validity of the origin of a message. Vehicles commonly use digital certificates to perform authentication. A digital signature typically contains information such as name of a vehicle or its role, a public key, validity, and signature.
- *Data Integrity* ensures that the contents of messages are not altered on transit by an adversary, which could undermine the safety of users of the system and lead to fatalities. This can be useful in multi-hop message relay scenarios where messages are passed on from one node to the other. Integrity checks can also help prevent replay attacks in which location and time stamp information can be altered to deceive message recipients.
- *Network Availability* is a key requirement of vehicular communication system for implementation of critical applications such as emergency message dissemination. Although provisioning of total and unconditional availability is not feasible in practice, it is crucial for network engineers and application developers to have a clear understanding of the availability constraints and implement proper

Table 1 Vehicular application requirements

Application type	Latency limit (ms)	Bandwidth	Security requirements	Examples use cases
Infotainment	500	✓✓✓	✓	Video streaming
Traffic efficiency	200	✓	✓✓	e-Toll collection
Safety	100	✓	✓	Collision warning
Reliable M2M	20	✓	✓✓	Auto-cruise
Real-time virtual reality	5	✓✓	✓✓	Augmented reality navigation
Control	1	✓✓	✓✓✓	Co-operative electronic stability control

mechanisms to combat potential Denial of Service (DoS) attacks and mitigate the risk associated with the availability of the network.
- *Privacy* in vehicular communication systems refers to the capability that ensures personal identifying information of the drivers is not disclosed to third parties. Hence, the network often and by default is required to ensure that activities of drivers cannot be traced by the adversaries. However, conditional privacy stipulates that it should be possible to reveal the identity of offending drivers for revocation and criminal prosecution. *Anonymity* is a technique of hiding the physical identity of a vehicle such as IP address and electronic number plates and thus providing the required privacy. The Vehicular Public Key Infrastructure (VPKI) should ensure a means to offer conditional anonymity to drivers by separating vehicular information from personal information about drivers.
- *Confidentiality* provision of the VPKI ensures that messages cannot be easily read or understood by third parties even if they are intercepted. However, for the reasons of safety, message confidentiality should not be implemented for emergency messages. Hence, broadcast applications such as DENM and CAM do not typically require confidentiality.

A summary of application requirements in terms of latency limits, bandwidth and security is presented in Table 1. Applications are grouped in five types and example use cases are given for each type.

3 Current Vehicular Communication Systems

This section presents the state-of-the-art in vehicular networking architectures and their ability to meet the application requirements specified in Sect. 2. To this end, the most prominent existing wireless access technology candidates for vehicular communications are dedicated short-range communications (DSRC) based on IEEE 802.11p standard and cellular communications with 3GPP LTE. In addition, hybrid solutions which integrate DSRC and LTE have emerged that aim to benefit from advantages of both technologies. There are also higher layer (overlay) networking architectures alternatives like information centric networking (ICN), which decouples locator and identifier of contents, and vehicular cloud computing. Content-based applications have promoted the introduction of ICN, with its most popular candidate the Named-Data Networking (NDN), in the vehicular networking sphere. Vehicular cloud extends the concepts of mobile cloud computing in the vehicular ecosystem taking intro account the characteristics of the vehicular applications These five vehicular networking architectures will be analysed in the remainder of the section.

3.1 *Dedicated Short Range Communication (DSRC)*

Dedicated short range communication (DSRC) is based on a two-way wireless system operating in the 5.9 GHz licensed spectrum band. In the US, Federal Communication Commission has allocated 75 MHz spectrum, whereas in Europe, ETSI allocated 30 MHz in the same band. Both in Europe and the US, the principal physical and MAC layers are based on the IEEE 802.11p standard. In Europe, that has been implemented within the ETSI ITS G5 [21], with ITS-G5A being dedicated to safety related applications and ITS-G5B to non-safety applications. On the other hand, in US the implementation follows the IEEE WAVE standard [32]. The main differences of the two implementations are located on upper layers and particularly the network as shown in Fig. 2. Both implementations promote a dual stack architecture, splitting traffic to ITS-specific applications and generic IP-based applications. The later are served by TCP/UDP or other transport protocol over an IPv6 network. These are non-safety applications and are carried by ITS G5 B or UMTS in the European stack. It is possible to translate IPv6 to GeoAddress and use the GeoNetworking layer [68]. For ITS-specific applications, ETSI has proposed the Basic Transport Protocol (BTP), a UDP-like transport protocol, operating on top of the GeoNetworking layer. GeoNetworking provides different methods of packet forwarding (broadcast, geocast and unicast) based on geographic routing. On the other hand, IEEE WAVE architecture proposed the Wave Short Message Protocol (WSMP) as the main network layer for ITS-specific traffic. Contrary to ETSI GeoNetworking, WSMP is based purely on single-hop broadcast, hence no routing protocol is employed as no application requiring forwarding has been designed in WAVE architecture.

3.1.1 DSRC Performance in Vehicular Environments

There have been several performance evaluations of DSRC networks either for benchmarking or for comparison with optimised protocol proposals. In Fig. 3, we present a baseline evaluation of the performance of a standard DSRC (IEEE 802.11p-based) network in highway and urban scenarios with respect to packet delivery ratio and delay. All vehicles periodically broadcast single-hop packets

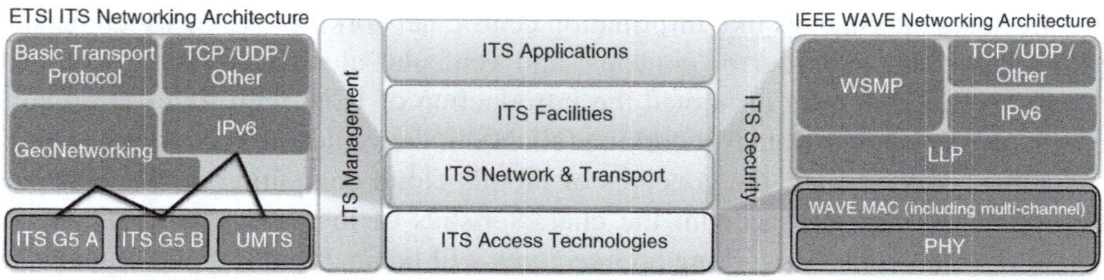

Fig. 2 DSRC protocol stack in Europe (*left*) [20] and US (*right*) [32]

of same type (e.g. CAMs) and same size (300 bytes) with frequency 10 Hz. For our evaluations we have assumed a vehicle density of 1400 and 1800 veh/h/lane with three lanes per direction following a Poisson arrival and vehicles travel with speeds varying from 60–120 km/h[7] in highway scenarios. In urban scenario, the vehicle density is 900 veh/h/lane with two lanes per direction again following a Poisson arrival. Vehicles travel in a Manhattan grid road topology with speeds from 50–60 km/h. The path loss model in both scenarios considers the large scale and shadowing effects and in the urban scenarios buildings obstruct line-of-sight communications.

Evaluations from state-of-the-art works like Wu et al. [74] and Mir [48] conform with the baseline results in Fig. 3. In principle DSRC can meet the strict latency constrains for active safety in most cases, but as the vehicle density and vehicle speed increase the percentage of packets received over the delay limits increase. For DSRC networks, the delay increases when the transmission frequency and number of vehicle increase. Sparse networks exhibit better performance, however, as the network load increases, denser networks with higher transmission frequencies result in higher end-to-end delays. The end-to-end delay increases as the vehicle speed increases. Vehicles spend more time in contending for the common channel as the result of larger neighborhood densities and because of the fact that they need to access the channel more frequently for higher transmission frequencies.

The previous evaluation only covered the single-hop broadcast scenarios supporting safety and traffic efficiency applications. For infotainment and cloud-based applications, unicast multi-hop communications are required. By design DSRC cannot meet the data rate requirements for high throughput infotainment applications such as 4K TV, as the high data rate of 27 Mbps configuration of IEEE 802.11p is only achievable with a very good channel quality. Such situation is obtained only with static nodes and line-of-sight communication, while lower rates (6–12 Mbs) that are more robust, are selected for highly mobile cases. Further, connection to RSUs and from there to the internet is intermittent with an average duration of a few seconds depending on vehicle speed. Handing-off between RSUs introduces significant delay for address re-configuration and authentication in the new access point [83]. Clearly the ultra-low latency of future real-time cloud-based control applications can not be met in current DSRC systems.

---

[7] The geography of transport systems—Highway Speed, Flow and Density [online] http://goo.gl/biXc6d.

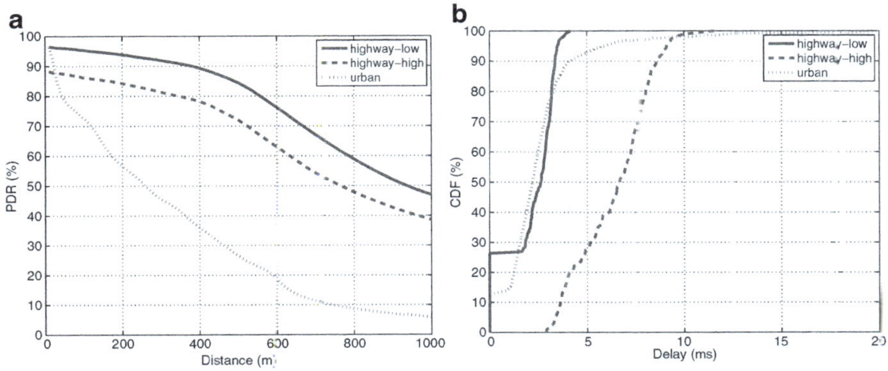

Fig. 3 DSRC Performance in highway and urban scenarios. (**a**) Packet delivery ratio. (**b**) CDF of single-hop delay

3.2 *Cellular Network Communications*

Cellular communications have long been used in vehicular networking, from GSM up to the most recent LTE networks, due to the wide penetration of the technology. One representative use case of cellular networking for vehicular applications is the European initiative *e-Call* that will be mandatory for all new vehicles shipped after April 2018. In case of a crash, e-Call equipped vehicles will automatically call the nearest emergency centre using Europe's single emergency number (112), as well as send a 'minimum set of data', which includes the location of the vehicle. ETSI has proposed a framework to integrate mobile networks in cooperative ITS (C-ITS) architectures [67] as depicted in Fig. 4. The framework identifies related functional requirements on the ITS architecture, amendments and modifications of existing standards on C-ITS in order to enable usage of public mobile cellular networks. One of the most important requirements in the architecture is a network server[8] responsible to the re-broadcasting of messages as LTE design does not allow Vehicle-to-Vehicle broadcast. A recently emerged technology that enables direct vehicle-to-vehicle communications when they are in proximity is Device-to-Device (D2D). A feasibility study of D2D in a vehicular network context has been performed in [11] that demonstrated promising results in boosting spectrum utilization. LTE has already been trialled by manufacturers for emulating non-safety critical V2V applications, such as exchanging non-time critical information including hyper-local weather, road conditions, and traffic data directly between vehicles.

[8]Reflector as dubbed in CoCarX project [15].

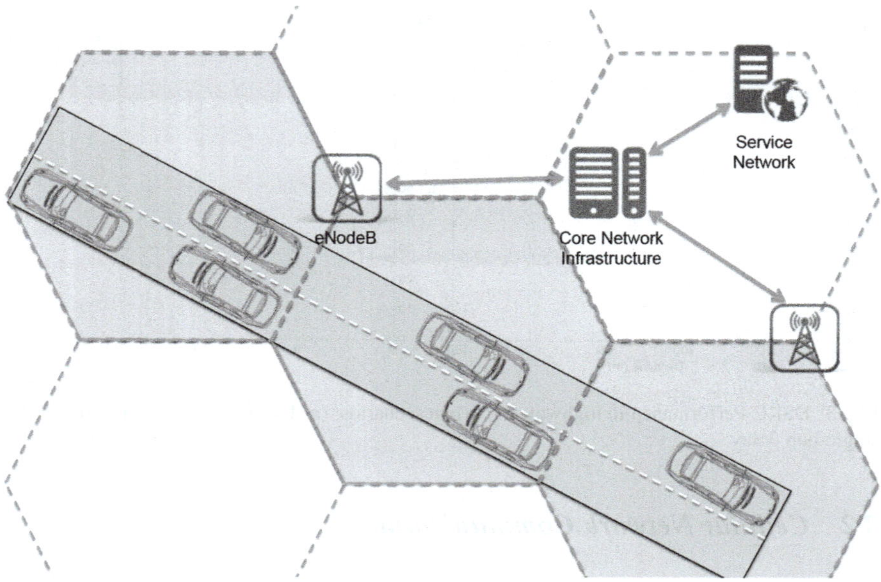

Fig. 4 Architecture for mobile networks in cooperative ITS

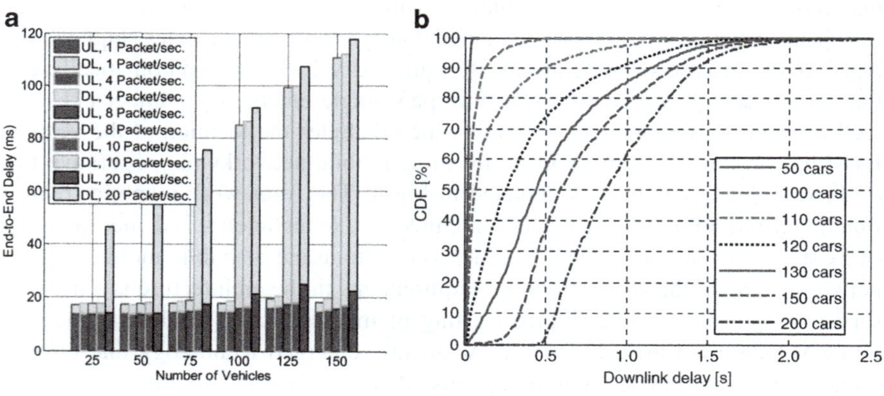

Fig. 5 LTE Performance. (a) End-to-End delay [48]. (b) CDF Downlink delay [67]

3.2.1 Cellular Networks Performance in Vehicular Environments

The design of LTE networks targets a user plane latency of 20 ms. These figures exclude the latency introduced by the backhaul network (i.e., RAN edge-Core edge, including security gateway), which is currently about 10 ms [65]. Unfortunately, the actual end-to-end delay is much greater than the target specifications as reported by OpenSignal for US mobile network providers [24].

Performance evaluations of LTE networks in the vehicular context exist in the literature [6, 48, 55, 67] and samples are presented in Fig. 5. One common assumption followed by all these works is that only vehicular traffic exists on the network, hence their results are generally over optimistic. Their results can be summarised in the following. LTE can meet the latency requirements for co-operative and active safety applications in limited scenarios (e.g., low loads, low number of vehicles). However, it can provide better service to infotainment applications due to better mobility support and higher available bandwidth. The end-to-end delay for the LTE network with vehicular traffic application only (i.e., no background traffic flows) is analysed as follows (Fig. 5a). As the network size and transmission frequency increase, the end-to-end delay increases. For higher network load, the total delay however gradually increases. While the uplink delays largely remain intact, the downlink delays increase significantly as the number of vehicles and transmission frequency increases. This can be also seen in Fig. 5b, where the CDF of downlink delays is analysed. Further, the increase of participating vehicles and the frequency of broadcast messages has a significant effect on the latency. The evaluation of LTE for CAM and DENM dissemination in [55] exhibits such behavior. The use of eMBMS to broadcast messages can potentially reduce the delay and increase the number of supported nodes/cell [6]. Moreover, even though LTE provides some service differentiation through the Evolved Packet switched System (EPS) Bearer mechanism, none of them can guarantee the ultra-low delay requirements of real-time control applications.

3.3 *Hybrid Networking Architecture*

Hybrid solutions that integrate DSRC with cellular networks, in loose or tight coupling, can alleviate some of the burdens and extend the range of supported scenarios for both LTE and DSRC usage, but still there are limitations. For example, the works in [79, 82] propose mechanisms that assist routing in DSRC systems with dissemination of data through cellular network, looking at the tradeoff between higher delivery ratio and lower latency. However, these techniques still cannot meet the latency requirements in Table 1. Another use of cellular systems in conjunction with ad-hoc networks is to facilitate efficient clustering mechanisms [57, 63]. In these works, only the cluster heads are connected to the cellular networks and data is disseminated using DSRC within the clusters. Clustering mechanisms are useful for reducing effective network size and potentially reduce latency within each cluster however the highly dynamic nature of vehicular networks requires very efficient and low overhead mechanisms. Further, tight coupled hybrid networking architectures facilitate with off-loading traffic from cellular networks to DSRC either for capacity improvements or for economic reasons [10, 39, 40]. Vehicular traffic could select either networks depending on the type of traffic (e.g. signalling or data traffic as in [39]), or based on a utility function taking into account the cost, signal quality and other indicators. However, several challenges arise in such architectures that relate to the offloading strategies, the practicality of DSRC infrastructure deployments that need to be tackled in order to fully benefit from these networks.

3.4 Information Centric Networking

There is an increasing trend in networking research to decouple locator and identifier of content, such as the IP addresses, from the actual content. This networking architecture is known as Information Centric Networking (ICN) and the most notable implementation is the Named-Data Networking (NDN) architecture [80]. Other ICN architectures include Data Oriented Network Architecture (DONA) [41] with a clean-slate redesign of Internet naming and name resolution, and Content Centric Networking (CCNx) [34] addressing scalability, security and flexibility challenges of modern Internet. In principle, ICN architectures are based on the publish-subscribe paradigm, where intermediate nodes/routers can cache the information and serve future consumers. This is depicted in a "narrow waist" approach Fig. 6, where the IP-based networking architecture on the left is compared with ICN-based architecture on the right. As mentioned above, ICN decouples content from the identifiers, and the whole network architecture is now constructed around the content itself, rather than the IPs. The characteristics of vehicular content, such as geography—and time-dependency of certain events, and the fact that it targets a large set of vehicles independent of their IPs, makes ICN a suitable candidate networking architecture. In addition, ICN provides support of content caching, which is beneficial for the intermittent DSRC-based networks. Recent works on ICN related to the vehicular networks ecosystem include [8, 29, 71, 75].

3.5 Vehicular Cloud Networking

Several automotive companies are promoting cloud connectivity services through cellular systems in their latest high-end models such as the Ford Sync[9] or the Apple CarPlay.[10] Such services are designed to deliver applications in a car for information, entertainment, making service appointments and other needs. In the future, remote diagnostics and operating system upgrades could be feasible. Model Predictive Control (MPC) techniques that are foreseen in future autonomous vehicles and require significant computing resources could be provided as a cloud service as well [25]. Content-based networking presented in Sect. 3.4 has promoted further the use of cloud-based architectures. Vehicular cloud networking shares the principles of mobile cloud networking where each component of the network is considered *a service* [28, 30, 44, 62, 73, 81]. A basic architecture for vehicular cloud networking is presented in Fig. 7. Cloud-based services, ranging from ITS-specific to generic content services, can be accessible either through cellular access networks

[9]Ford SYNC technology [Online]:www.ford.com/technology/sync.

[10]Apple CarPlay [Online]:www.apple.com/ios/carplay.

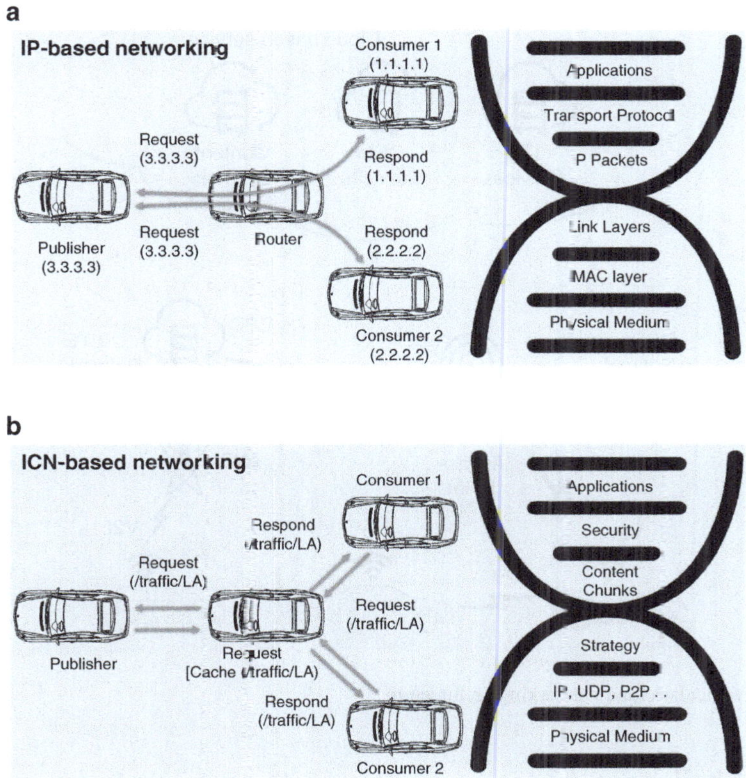

Fig. 6 ICN-based compared to IP-based vehicular architecture. (**a**) IP-based networking. (**b**) ICN-based networking

or DSRC-based. However, such architectures introduce significant delays as current cloud-based services are quite centralised with small number of spatially distributed servers.

4 Vehicular Networking in 5G

This section aims to analyse the challenges and candidate 5G technologies that have been recently under discussion and their implications on vehicular networking. More advanced networked-car visions will rely on capabilities that come with the 5G technology and perhaps will augment today's DSRC and 4G networks. For example, BMW anticipates that 5G mobile networks could be vital in providing the mission-critical reliability as it seeks to deploy self-driving cars onto city streets.

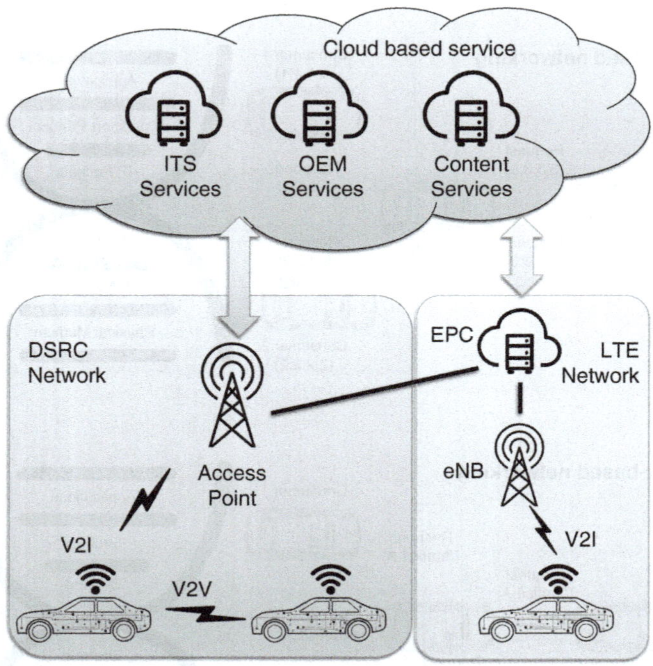

Fig. 7 Vehicular cloud networking architecture

4.1 Challenges

In Sect. 3, current vehicular networking architectures were reviewed, determining their limitations to meet applications' requirements, due to congestion, mobility and other bottlenecks. Existing research reveals that nominal, congestion free, access network latency in the range of 1–2 ms is feasible [23] for communications between cars and the edge of the cloud. But, progress toward this objective requires novel research into wireless transmission, network architectures and functions/protocols to reduce serialisation, queuing, and network processing latencies. For reliability and security, additional degrees of freedom in 5G communication systems (eg, spatial diversity in dense base station deployments) as well as physical layer encryption techniques have great potentials. These challenges and existing solutions are identified in the remainder of the section.

4.1.1 Congestion

The performance analysis both of DSRC and LTE systems, in Sects. 3.1.1 and 3.2.1 respectively, has outlined a common limitation on both architectures due to congestion. The effective network size and traffic load play a significant role to the performance of latency. Due to the limited available resources, competition among entities to access the shared radio resource introduces latency. In order to avoid congestion caused due to large network size or heavy load, multiple solutions have been proposed.

First, congestion can be avoided by introducing clustering which groups the large amount of users into clusters with fewer users and competition. Clustering-based algorithms [53] aim to seek opportunities to minimize the consumption of radio resources via integrating information, working as one station, and reusing radio resources in different clusters. Second, reuse of radio resources via smart allocation of multiple non-overlapping channels helps reduce the contention level, e.g., multichannel MAC design [78]. It is critical to have a technology that works even when the network operator is not there, as they are out of range for example. Device-to-device (D2D) communication is a potential solution for such situations. D2D communications [11, 56] also target to tackle the problem by reducing contention level in the system via encouraging local p2p communications. Meanwhile, D2D communications easily avoid redundancy in terms of signalling for controlling and maintenance as in the traditional cellular networks. Coordination (e.g., among Base Stations) on top of D2D communication systems expedites radio resource utilization.

For DSRC systems, ETSI has proposed several decentralized congestion control (DCC) mechanisms [69]. For example, the transmit power control or the DCC sensitivity control mechanisms work on the same principle as clustering, multi-channel MAC and D2D that aim to reduce the effective network size, thus reducing the congestion levels. On the other hand, transmit data rate control, transmit access control mechanisms and the proposed TCP-like congestion control [60] tackle the congestion from another perspective, that of the offered load. For the future vehicular networking, it is unlikely to use a single radio resource management scheme to satisfy the stringent requirements of reliability and latency but more like a combination of efficient MAC schemes. A novel design of the integration of the above mentioned solutions potentially helps reduce the latency caused by congestion.

4.1.2 Mobility Management

Vehicular networks are characterised by nodes that are constantly moving on semi-predictive but predefined trajectories over the underlying road topology, contrary to the stationary IoT networks or the low random mobility of pedestrians. This level of node mobility results also to the need of network mobility as vehicles travel under the coverage of different RSUs. It has already been mentioned in Sect. 3.1 that the mobility and sparse RSU deployments result in intermittent connections for DSRC networks and increased delays due to hand-off procedures. However, similar challenge is envisioned in future 5G networks, where the densification of the network with the use of small cells will increase the probability of hand-off. Therefore, optimising hand-off procedure is a crucial task in the process of reducing delay and having more reliable connections.

Solutions based on Network Mobility (NEMO) and Mobile IP (MIPv6) are aiming to reduce the handoff time utilizing predictive techniques to perform registration in advance [3, 43, 61]. However, the centralized architecture of MIPv6 is inefficient for highly dynamic vehicular networks and a distributed mobility management approach would be more appropriate [45]. Other solutions include Location Identity Separation Protocol (LISP) [22] that natively supports mobility with route optimization, multi-homing dual stack and network mobility. LISP is also compatible with the ICN principles of locator/identifier separation. Further, the utilization of Stream Control Transmission Protocol's (SCTP) inherent multi-homing support and Media Independent Handover (IEEE 802.21 standard) have also been investigated for efficient handover [52]. Efficient and simultaneous use of all available access networks combined with minimal signalling and predictive techniques ("make before brake") would help reduce the delay introduced by the mobility of the user.

4.1.3 Backhaul Network

Mane services have been moved to a cloud based architecture from data storages e.g. dropbox, up to complete operating systems e.g. Chrome OS. Particularly for ITS services, the use of cloud has been demonstrated in Sect. 3.5 with current infotainment services and futuristic control. However, the location of the server would significantly impact the performance of the service, as control techniques can tolerate very low latency and content has to be available seamlessly.

Currently, cloud-based services are centralised in proprietary data centres and the connection of the network provider with that data centre will impact the delay. The locality of the remote host with respect to the end-user (vehicle) is a major source of latency for applications that require such connections. The latency will depend on the type of links and distance of the network gateway and the remote host, as well as the routing rules, which may contradict with the Network Neutrality that FCC recently vote in favor. Thus, bringing the remote host closer to the end-user, what is dubbed as fog computing [5, 14], reduces the backhaul delay. Fog computing within a vehicular network architecture entitles spatially distributing service components and caching content within a flat network provider core and the vehicles [37, 72].

4.1.4 Air Interface

As mentioned in the beginning of this section, 1 ms end-to-end delay is feasible in congestion free networks and a proposed division of that delay in different components is demonstrated in Fig. 8. Such architecture will require 0.2 ms for the

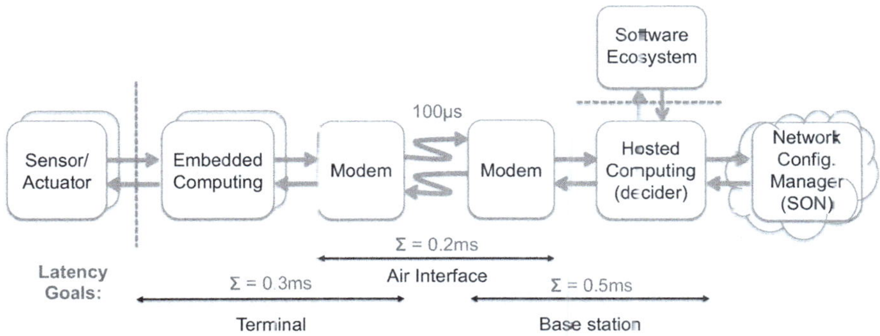

Fig. 8 Proposed end-to-end delay budget in 5G [23]

air interface, that accounts for ~15 timeslots[11] the current IEEE 802.11p standard. Current access categories (ACs) with low priorities in the IEEE 802.11p can not satisfy this delay limit and only AC0 and AC1 can fulfil this requirement in theory, under the assumptions of very low contention level and no collisions. Moreover, in current LTE systems, the duration of a single frame is 10 ms consisting of 10 sub-frames with two slots per sub-frame. Large packets would require segmentation to over multiple sub-frames, which would increase the transmission delay of those packets. Further, from the 10 sub-frames, maximum 6 can be used for MBMS, while the rest are left to unicast traffic. This impacts the delay of broadcast packets, which for ITS applications is the majority of traffic.

It is clear that in order to support 0.2 ms air-interface delay in DSRC-based networks, the backoff procedure and relevant mechanisms has to be redesigned. Potential solutions include a mixture of TDMA and CSMA access in a super-frame cycle. Ultra-high AC would have deterministic delay through the TDMA part of the frame cycle and lower AC's would content on the CSMA cycle. Such access mechanisms are employed for on-board networks based on FlexRay [26]. Several air interface solutions specifically designed for reliable vehicular communications have been proposed and analysed in METIS 2020 project [47]. Most notable, the coded slotted Aloha (CSA) MAC technique enables reliable ad-hoc communications that operate in nearly double the network sizes of standard IEEE 802.11p networks. In terms of cellular technologies, the current LTE frame structure is not able to support the 0.2 ms limit. There is a proposal for flexible TDD frame structure [51] that targets to provide the overall latency of 1 ms able to support future ITS use cases such as cooperative autonomous driving, tactile Internet and real time control. Mechanisms to reduce transmission time interval (TTI) and retransmission processes as well as provision of better QoS differentiation have enabled the reduction of latency nearly five times [47].

[11] Assuming each timeslot duration is 13 μs.

4.1.5 Security

Security in vehicular communication systems is vital as explained in Sect. 2.5. However, maintaining a secure network (e.g. exchanging keys, encryption) and authenticating source of a message introduces significant latency [54], both in terms of communication and processing. The processing time increases remarkably in dense networks [7]. High beacon rate (e.g., 10 Hz for CAMs) also leads to significant computational overhead in terms of authentication, apart from the congestion reported in Sect. 4.1.1.

The majority of the delay related to security is spent on the processing/checking the legitimation of messages e.g., 43.2 % of the CPU time is devoted to the message check [58]. In order to reduce this delay, the authentication/checking mechanism has to be redesigned. Lightweight filter security schemes have the potential to reduce overall processing time. Such techniques reduce the amount of information to be checked each time or allow fewer checks according to the "credit" for each user. Other solutions aim to reduce the overhead introduced by security mechanisms. They appropriately design effective security and privacy-enhancing schemes such as group signature schemes, hybrid pseudonym schemes that can protect legitimate users as effectively as unsecured systems [7]. However, even those techniques are not fast enough to reach the 1 ms delay budget for the most demanding applications. In order to perform faster computations for security, physical layer security performed by system on chip (SoC) components or exploiting propagation randomness to establish secret keys [49]. Finally, the VPKI which is currently employed to manage security key and pseudonyms ([70]) has to be redesigned as it is developed around DSRC systems since most applications are using V2V broadcast messages. However, in the 5G architecture, DSRC and cellular are expected to be tightly coupled, thus the security protocols used should be shared or at least compatible so that vehicles can communicate seamlessly through any technology.

4.2 Conceptual Vehicular Architecture in 5G

Having considered the requirements and potential solutions to realise secure, ultra-low latency, high-reliability vehicular networks, a *conceptual* vehicular networking architecture in 5G is presented in Fig. 9. This follows a revolutionary approach, rather than simply evolution from the current 4G architecture, empowered by intelligence of the cloud and cooperation among cars and infrastructures. The architectural design is divided into four layers: (a) the generic cloud/internet service layer, (b) the core network cloud layer, (c) the radio access network layer and finally (d) the vehicle and RSU space. Cloud computing technologies are mature enough to be adopted by mobile network providers, hence each of these layers is represented as a cloud. This ecosystem shall be able to support current and future ITS applications in the most demanding scenarios. Following a top-down approach, these layers and functionalities within each of them are described in more details in the remainder of the section.

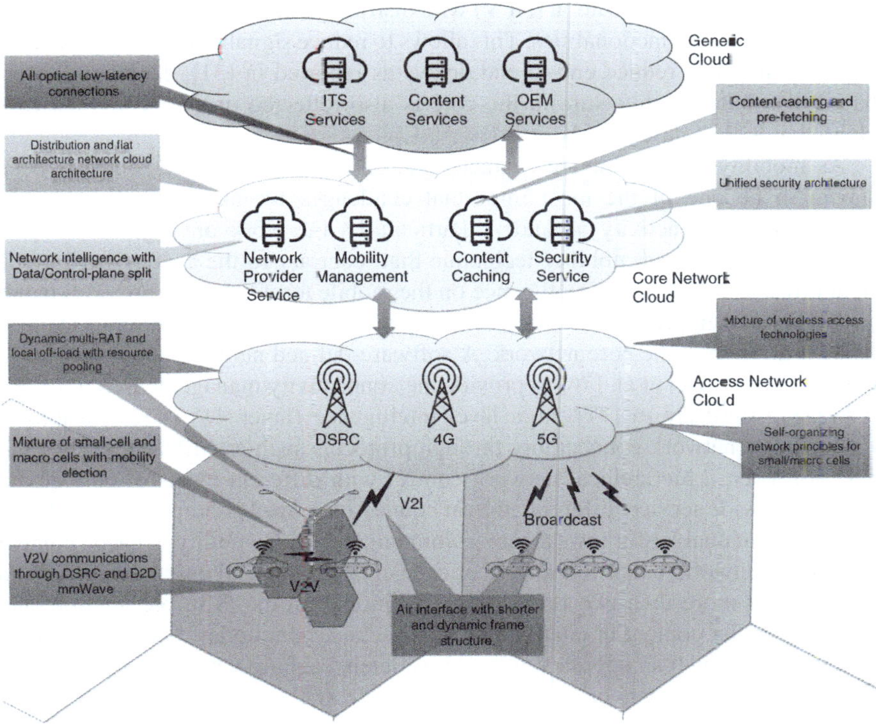

Fig. 9 Conceptual 5G architecture

4.2.1 Generic Cloud

The generic cloud represents the current cloud infrastructure where public authorities and OEMs provide their ITS-specific services. In addition, generic content providers are included in this layer. Following the basic design considerations for 5G architecture, the services have to be spatially distributed even in the generic cloud. The connections of these data-centres with the core network have to be all-optical in order to provide low-latency and high reliability.

4.2.2 Core Network Cloud

The core network architecture should be completely revamped. Even though state-of-art LTE-A evolved packet core (EPC) architecture is simpler than the previous 3G core architecture, by reducing network entities, supporting an all-IP network and smart off-loading functions, it still introduces significant delays. In the conceptual 5G architecture of Fig. 9, most of the current EPC functionalities are represented by the *Network Provider Services*. Their implementation is based on the principles of network function virtualization (NFV) with distributed and flat architecture forming clusters of service functionalities. This allows to reduce signalling traffic by as much as 70 %, and thus reduce end-to-end delays as reported in [31]. This flat cluster-based distributed architecture in the core is also reflected on the access network cloud that will be described later.

As mobility is a distinctive characteristic of vehicular networks and has been shown to be one of the most important challenges for 5G networks, there is a lot of research activity around it. Particularly, the work of Yegin et al. [77] proposes a distributed mobility technique that, contrary to the centralised Mobile IP approaches, pushes the intelligence on the mobile terminal. This provides better scalability and improves efficiency as the terminal is in better position to identify its own flows than the core network. A software-defined network approach is also employed by Yazici et al. [76] to provide the connectivity management as a service (CMaaS). Similarly to [77], high level intelligence (layer 4+) is responsible to maintain the network connectivity through protocols such as MP-TCP, SCTP and SIP. However, a hierarchical network control with different levels of complexity that can provide service differentiation at connectivity level is proposed in CMaaS.

One major enabler for low-latency communication is the ability to cache content within the network, either in the core or even closer to the end user. As multimedia accounts for more than two thirds of global mobile traffic [13], pre-fetching and caching popular content in intermediate nodes can accommodate users of the same content faster. Such an approach of mobile content caching within the EPC as well as the radio access network is described in [72]. It is shown that this technique can reduce the delay by 80 % depending on the amount of cached content in addition to increase the economic benefits for the network provider. Apart from content caching, functionality distribution throughout the core and access network can resolve problems related to congestion and latency. This is described as *fog computing* [14] and provides compute, storage, and network services at the network edge. It also adds intelligence in the network platform itself to orchestrate the underlying resources in real-time.

Last important component of the core network is the security management service. It has been mentioned previously that in order to meet the delay constraints physical layer security mechanisms have to be employed. However, the core network should implement a sophisticated distributed VPKI. With the densification of the radio access network and the use of multiple radio access technologies, the probability of a vehicle being required to handover and re-authenticate increases. Similar to the network mobility solutions, a SDN-based approach for authentication is proposed in [18]. In this technique user-related security context including identity, location, direction and physical layer characteristics, is shared by the SDN controller to the predicted next cell access point. This accelerates the authentication process and increases the levels of tolerance to network failures as the security context is pre-shared among access points.

4.2.3 Access Network Cloud

Significant redesigns in the wireless access network will appear in the future 5G with the enabling technologies described in Part II "Transmission and Design Techniques for 5G Networks" of this book. These will characterise the 5G access, however current technologies such as 4G and DSRC will not eclipse. Within the 5G access network cloud, several technologies have to work in synergy in a multi-tier architecture from macro cells providing a ubiquitous service coverage to D2D communications through mm-Wave and DSRC for inter-vehicle connections.

Using the principles of software defined radio (SDR), hardware and software are decoupled. This enables network intelligence to flexible control the available radio resources either from the network or the user perspective. Operators can dynamically configure the network capacity faster, which increases the flexibility for network expansion and bottleneck elimination. However, the resulting increase in operational complexity calls for advanced self-organizing techniques for the underlying networks. Exploitation of big data analytics of control, signalling and contextual data within the core network provides unprecedented amounts of system-level intelligence. This network intelligence can be fed into proactive self-organised network (SON) engines to enable better resource orchestration as reported in [33].

The dynamic multi-RAT access will also stimulate further off-load from macro cells to small cells or DSRC in order to increase performance and reduce the cost of communications. There are a lot of research works on how and when a user shall switch from one technology to another. At the moment, SNR-based measurements are the preferred method. However Andreev et al. [2] have shown that load-aware user-centric schemes, which augment SNR measurements with additional information about network loading, can improve the performance compared to only SNR-based selection.

4.2.4 Vehicle Network Cloud

The last layer in this conceptual architectural design is the vehicle itself and the RSUs. The principles of fog computing can also be employed at this layer, which is the edge of the access network as depicted in Fig. 10. Such a technique is presented in [62], where RSUs form a cloud employing SDN to dynamically instantiate, replicate, and/or migrate services. This improves the reconfigurations costs and infrastructure delay when compared to purist service installations.

It has been already mentioned that the number of electronic control units on a vehicle has increased. These interconnected ECUs along with the assisting sensors and actuators are generally classified as Cyber-Physical System (CPS). Within the CPS a vehicle can be interpreted as a resource for sensing, data storage, computing, data relaying, and a means for locating other objects [1]. However, it is a trade-off between sophisticated sensors and computational on-board and the QoS communications capabilities. Less sophisticated on-board systems would require more cloud-based assistance hence higher QoS communications and vice versa. The focus

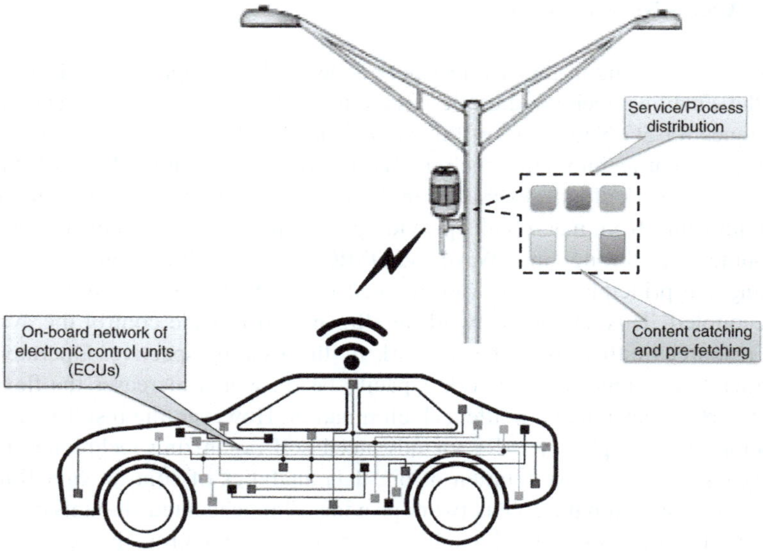

Fig. 10 Vehicle and RSU cyber-physical ecosystem

should be shifted from bringing "the internet in the vehicle" to bringing "the vehicle in the internet", converting the vehicles from a simple content consumers to a content generators. This will also increase the need for high-speed links in uplink transmission for both vehicle-to-vehicle and vehicle-to-infrastructure. In the absence of infrastructure, either due to early-state of deployment or due to network malfunction, vehicles can also form a vehicular cloud [17]. For example, parked vehicles can be used as a temporary network and storage infrastructure which can increase the resilience and reliability of the vehicular network.

5 Summary

The fifth generation of mobile communications systems is evangelising intelligent systems that work together to give the impression of unlimited data capacity, impressive speeds and impeccable efficiency compared to current 4G networks. In addition, these kind of services would also require guaranteed performance all the way from the cloud that runs the service to the vehicle, and back again. These characteristics will enable new types of applications such as connected autonomous vehicles, which are at the moment infeasible due to constraints such as latency or capacity limitations of current vehicular communication systems. The future connected autonomous vehicles will lead to fewer accidents and improve fuel efficiency as huge volumes of data can be fuzzed in the cloud and provide real-time information to large amount of vehicles. Wideband connectivity will enable operating system updates, driving software, and support next-generation applications.

Key enablers of the future vehicular networking in 5G are analysed in this chapter. Cloud computing and SDN are two of the driving forces for the 5G core network where resources can be dynamically orchestrated and re-configured fast. In the access network current DSRC and 4G technologies should be optimised and cooperate with the new 5G radios. Vehicles and RSUs will also play a significant role in the network operations with service distribution and content caching, apart from being just the end-user of those services and contents. Finally, what is required for 5G is a series of trade-offs to enable a balance between the IP approach and the ICN/NDN approach possibly as a regionalised NDN approach with classical IP-based policies between local domains.

Acknowledgements The authors would like to thank Jerry Foster, 5G systems architect at 5GIC, for sharing his wisdom during the conceptual 5G architecture definition.

References

1. S. Abdelhamid, H. Hassanein, G. Takahara, Vehicle as a resource (VaaR). IEEE Netw. **29**(1), 12–17 (2015). doi:10.1109/MNET.2015.7018198
2. S. Andreev, M. Gerasimenko, O. Galinina, Y. Koucheryavy, N. Himayat, S.P. Yeh, S. Talwar, Intelligent access network selection in converged multi-radio heterogeneous networks. IEEE Wirel. Commun. **21**(6), 86–96 (2014). doi:10.1109/MWC.2014.7000976
3. R. Baldessari, A. Festag, J. Abeille, Nemo meets vanet: a deployability analysis of network mobility in vehicular communication, in *7th International Conference on ITS Telecommunications, 2007. ITST '07* (2007), pp. 1–6. doi:10.1109/ITST.2007.4295897
4. S. Bayless, A. Guan, *Connected Vehicle Technical Insigts - Vehicle Applications and Wireless Interoperability*. Technology Scan Series 2011–2015 (ITS America, Washington, DC, 2015)
5. F. Bonomi, R. Milito, J. Zhu, S. Addepalli, Fog computing and its role in the internet of things, in *Proceedings of the First Edition of the MCC Workshop on Mobile Cloud Computing. MCC '12* (2012), pp. 13–16. doi:10.1145/2342509.2342513
6. J. Calabuig, J. Monserrat, D. Gozalvez, O. Klemp, Safety on the roads: LTE alternatives for sending ITS messages. IEEE Veh. Technol. Mag. **9**(4), 61–70 (2014). doi:10.1109/MVT.2014.2362272
7. G. Calandriello, P. Papadimitratos, J.P. Hubaux, A. Lioy, On the performance of secure vehicular communication systems. IEEE Trans. Dependable Secure Comput. **8**(6), 898–912 (2011). doi:10.1109/TDSC.2010.58
8. M. Chen, D.O. Mau, Y. T. Zhang, T. Taleb, V.C. Leung, VENDNET: VEhicular Named Data NETwork. Veh. Commun. **1**(4), 208–213 (2014). doi:10.1016/j.vehcom.2014.002
9. X. Cheng, X. Hu, L.L. Yang, I. Husain, K. Inoue, P. Krein, R. Lefevre, Y. Li, H. Nishi, J. Taiber, F.Y. Wang, Y. Zha, W. Gao, Z. Li, Electrified vehicles and the smart grid: the ITS perspective. IEEE Trans. Intell. Transp. Syst. **15**(4), 1388–1404 (2014). doi:10.1109/TITS.2014.2332472
10. N. Cheng, N. Lu, N. X.S. Zhang, Shen, J.W. Mark, Vehicular wifi offloading: challenges and solutions. Veh. Commun. **1**(1), 13–21 (2014)
11. X. Cheng, L. Yang, X. Shen, D2D for intelligent transportation systems: a feasibility study. IEEE Trans. Intell. Transp. Syst. **PP**, 1–10 (2015). doi:10.1109/TITS.2014.2377074

12. Cisco Systems Inc., The internet of things - how the next evolution of the internet is changing everything. White paper (2011)
13. Cisco Systems Inc., Cisco visual networking index: global mobile data traffic forecast update, 2012–2017. Technical Report (2013)
14. Cisco Systems Inc., Fog computing. Cisco Technology Radar (2014)
15. CoCar Consortium, CoCarX Coperative Cars eXtended ITS services and communication architecture. Deliverable D3 (2011)
16. COST IC 1004, Scientific challenges towards 5G mobile communications. White paper (2013)
17. F. Dressler, P. Handle, C. Sommer, Towards a vehicular cloud - using parked vehicles as a temporary network and storage infrastructure, in *International Workshop on Wireless and Mobile Technologies for Smart Cities, WiMobCity '14* (2014), pp. 11–18. doi:10.1145/2633661.2633671
18. X. Duan, X. Wang, Authentication handover and privacy protection in 5g hetnets using software-defined networking. IEEE Commun. Mag. **53**(4), 28–35 (2015). doi:10.1109/MCOM.2015.7081072
19. T. ElBatt, S.K. Goel, G. Holland, H. Krishnan, J. Parikh, Cooperative collision warning using dedicated short range wireless communications, in *Proceedings of the 3rd International Workshop on Vehicular Ad Hoc Networks* (2006), pp. 1–9. doi:10.1145/1161064.1161066
20. EN 302 665, Intelligent Transport Systems (ITS) - Communications Architecture. European standard, ETSI (2010)
21. ES 202 663, Intelligent Transport Systems (ITS) - European profile standard for the physical and medium access control layer of Intelligent Transport Systems operating in the 5GHz frequency band. European standard, ETSI (2009)
22. D. Farinacci, V. Fuller, D. Meyer, D. Lewis, The Locator/ID Separation Protocol (LISP). RFC 6830 (Experimental) (2013). http://www.ietf.org/rfc/rfc6830.txt
23. G. Fettweis, S. Alamouti, 5G: personal mobile internet beyond what cellular did to telephony. IEEE Commun. Mag. **52**(2), 140–145 (2014). doi:10.1109/MCOM.2014.6736754
24. Fierce Wireless, 3G/4G wireless network latency: comparing Verizon, AT&T, Sprint and T-Mobile in February 2014 (2014). http://goo.gl/OkiHl6
25. D. Filev, J. Lu, D. Hrovat, Future mobility: integrated vehicle control with cloud computing. ASME Dyn. Syst. Control Mag. **1**(1), 18–24 (2013)
26. FlexRay Consortium, Flexray communications system-protocol specification. Technical Report (2005)
27. S. Gaudin, Ford uses Microsoft cloud to seamlessly update cars (2015), http://goo.gl/1dWhrL. Accessed: 15 May 2015
28. M. Gerla, E.K. Lee, G. Pau, U. Lee, Internet of vehicles: from intelligent grid to autonomous cars and vehicular clouds, in *2014 IEEE World Forum on Internet of Things (WF-IoT)* (2014), pp. 241–246. doi:10.1109/WF-IoT.2014.6803166
29. G. Grassi, D. Pesavento, G. Pau, L. Zhang, S. Fdida, Navigo: interest forwarding by geolocations in vehicular named data networking. ArXiv e-prints (2015)
30. L. Gu, D. Zeng, S. Guo, Vehicular cloud computing: a survey, in *2013 IEEE Globecom Workshops (GC Wkshps)* (2013), pp. 403–407. doi:10.1109/GLOCOMW.2013.6825021
31. H. Hawilo, A. Shami, M. Mirahmadi, R. Asal, NFV: state of the art, challenges, and implementation in next generation mobile networks (vepc). IEEE Netw. **28**(6), 18–26 (2014). doi:10.1109/MNET.2014.6963800
32. IEEE, IEEE Guide for Wireless Access in Vehicular Environments (WAVE) - Architecture. IEEE Std. 1609.0-2013 (2014), pp. 1–78. doi:10.1109/IEEESTD.2014.6755433
33. A. Imran, A. Zoha, Challenges in 5g: how to empower son with big data for enabling 5g. IEEE Netw., **28**(6), 27–33 (2014). doi:10.1109/MNET.2014.6963801
34. V. Jacobson, D.K. Smetters, J.D. Thornton, M.F. Plass, N.H. Briggs, R.L. Braynard, Networking named content, in *Proceedings of the 5th International Conference on Emerging Networking Experiments and Technologies, CoNEXT '09* (2009), pp. 1–12. doi:10.1145/1658939.1658941

35. A. Jain, H.S. Koppula, B. Raghavan, A. Saxena, Know before you do: anticipating maneuvers via learning temporal driving models. arXiv preprint arXiv:1504.02789 (2015)
36. W. Jones, BMW fixes software flaw that would've let hackers unlock doors. IEEE Spectrum (2015). http://goo.gl/Io9Yht
37. K. Kanai, T. Muto, H. Kisara, J. Katto, T. Tsuda, W. Kameyama, Y.J. Park, T. Sato, Proactive content caching utilizing transportation systems and its evaluation by field experiment, in *2014 IEEE Global Communications Conference (GLOBECOM)* (2014), pp. 1382–1387. doi:10.1109/GLOCOM.2014.7037001
38. K. Katsaros, R. Kernchen, M. Dianati, D. Rieck, C. Zinoviou, Application of vehicular communications for improving the efficiency of traffic in urban areas. Wirel. Commun. Mob. Comput. **11**(12), 1657–1667 (2011)
39. K. Katsaros, M. Dianati, L. Le, Effective implementation of location services for VANETs in hybrid network infrastructures, in *International Conference on Communications (ICC) Workshops* (2013), pp. 521–525. doi:10.1109/ICCW.2013.6649289
40. K. Katsaros, M. Dianati, R. Tafazolli, G. Xiaolong, End-to-End delay bound analysis for location-based routing in hybrid vehicular networks. IEEE Trans. Veh. Technol. **PP**(99), 1–1 (2015). doi: 10.1109/TVT.2015.2482362
41. T. Koponen, M. Chawla, B.G. Chun, A. Ermolinskiy, K.H. Kim, S. Shenker, I. Stoica, A data-oriented (and beyond) network architecture. SIGCOMM Comput. Commun. Rev. **37**(4), 181–192 (2007). doi:10.1145/1282427.1282402
42. KPMG, Connected and autonomous vehicles—the uk economic opportunity. Technical Report (2015)
43. J.H. Lee, T. Ernst, N. Chilamkurti, Performance analysis of PMIPv6-based network mobility for intelligent transportation systems. IEEE Trans. Veh. Technol. **61**(1), 74–85 (2012). doi:10.1109/TVT.2011.2157949
44. E. Lee, E.K. Lee, M. Gerla, S. Oh, Vehicular cloud networking: architecture and design principles. IEEE Commun. Mag. **52**(2), 148–155 (2014). doi:10.1109/MCOM.2014.6736755
45. P. Matzakos, J. Härri, B. Villeforceix, C. Bonnet, An IPv6 architecture for cloud-to-vehicle smart mobility services over heterogeneous vehicular networks, in *3rd International Conference on Connected Vehicles & Expo ICCVE 2014* (2014)
46. C. McCarthy, K. Harnett, A. Carter, Characterization of potential security threats in modern automobiles - a composite modeling approach. DOT HS 812 074, NHTSA (2014)
47. METIS Consortium, Proposed solutions for new radio access. Deliverable D2.4 (2015)
48. Z.H. Mir, F. Filali, LTE and IEEE 802.11p for vehicular networking: a performance evaluation. EURASIP J. Wirel. Commun. Netw. **2014**(89) (2014). doi:10.1186/1687-1499-2014-89
49. A. Mukherjee, S. Fakoorian, J. Huang, A. Swindlehurst, Principles of physical layer security in multiuser wireless networks: a survey. IEEE Commun. Surv. Tutorials **16**(3), 1550–1573 (2014). doi:10.1109/SURV.2014.012314.00178
50. NHTSA, Preliminary statement of policy concerning automated vehicles (2013)
51. Nokia, 5g radio access system design aspects. White paper (2015)
52. N. Omheni, F. Zarai, M.S. Obaidat, K.F. Hsiao, L. Kamoun, A novel media independent handover-based approach for vertical handover over heterogeneous wireless networks. Int. J. Commun. Syst. **27**(5), 811–824 (2014). doi:10.1002/dac.2628
53. B. Peng, T. Peng, Z. Liu, Y. Yang, C. Hu, Cluster-based multicast transmission for device-to-device (d2d) communication, in *2013 IEEE 78th Vehicular Technology Conference (VTC Fall)*, pp. 1–5 (2013). doi:10.1109/VTCFall.2013.6692134
54. J. Petit, Z. Mammeri, Authentication and consensus overhead in vehicular ad hoc networks. Telecommun. Syst. **52**(4), 2699–2712 (2013). doi:10.1007/s11235-011-9589-y
55. M. Phan, R. Rembarz, S. Sories, A capacity analysis for the transmission of event and cooperative awareness messages in lte networks, in *ITS World Congress* (2011)
56. J. Qiao, X. Shen, J. Mark, Q. Shen, Y. He, L. Lei, Enabling device-to-device communications in millimeter-wave 5g cellular networks. IEEE Commun. Mag. **53**(1), 209–215 (2015). doi:10.1109/MCOM.2015.7010536

57. G. Remy, S. Senouci, F. Jan, Y. Gourhant, Lte4v2x: Lte for a centralized vanet organization, in *2011 IEEE Global Telecommunications Conference (GLOBECOM 2011)*, pp. 1–6 (2011). doi:10.1109/GLOCOM.2011.6133884
58. N. Ristanovic, P. Papadimitratos, G. Theodorakopoulos, J.P. Hubaux, J.Y. Le Boudec, Adaptive message authentication for multi-hop networks, in *2011 Eighth International Conference on Wireless On-Demand Network Systems and Services (WONS)* (2011), pp. 96–103. doi:10.1109/WONS.2011.5720206
59. P.E. Ross, Cars that talk need wireless that works. IEEE Spectrum (2014). http://goo.gl/juFrIA
60. S. Ruehrup, P. Fuxjaeger, D. Smely, TCP-like congestion control for broadcast channel access in VANETs, in *International Conference on Connected Vehicles and Expo (ICCVE)* (2014)
61. S. Ryu, K.J. Park, J.W. Choi, Enhanced fast handover for network mobility in intelligent transportation systems. IEEE Trans. Veh. Technol. **63**(1), 357–371 (2014). doi:10.1109/TVT.2013.2272059
62. M. Salahuddin, A. Al-Fuqaha, M. Guizani, Software-defined networking for rsu clouds in support of the internet of vehicles. IEEE Internet Things J. **2**(2), 133–144 (2015). doi:10.1109/JIOT.2014.2368356
63. R. Sivaraj, A. Gopalakrishna, M. Chandra, P. Balamuralidhar, Qos-enabled group communication in integrated vanet-lte heterogeneous wireless networks, in *2011 IEEE 7th International Conference on Wireless and Mobile Computing, Networking and Communications (WiMob)* (2011), pp. 17–24. doi:10.1109/WiMOB.2011.6085417
64. F. Soldo, C. Casetti, C. Chiasserini, P. Chaparro, Video streaming distribution in vanets. IEEE Trans. Parallel Distrib. Syst. **22**(7), 1085–1091 (2011). doi:10.1109/TPDS.2010.173
65. STOKE, Latency considerations in LTE, Implications to security gateway. White Paper (2014)
66. TR 102 638, Intelligent Transport Systems (ITS) - Vehicular communications - Vehicular communications; basic set of applications; definitions. Technical Report, ETSI (2009)
67. TR 102 692, Intelligent Transport Systems (ITS) - Framework for public mobile networks in Cooperative ITS (C-ITS). Technical report, ETSI (2012)
68. TS 102 636-6-1, Intelligent Transport Systems (ITS) - Vehicular communications - GeoNetworking - Part 6: internet integration - Sub-part 1: Transmission of IPv6 packets over GeoNetworking protocols. Technical specification, ETSI (2011)
69. TS 102 687, Intelligent Transport Systems - Decentralized congestion control mechanisms for intelligent transport systems operating in the 5GHz range; access layer part. Technical specification, ETSI (2011)
70. TS 102 940, Intelligent Transport Systems - Security - ITS communications security architecture and security management. Technical specification, ETSI (2012)
71. L. Wang, R. Wakikawa, R. Kuntz, R. Vuyyuru, L. Zhang, Data naming in vehicle-to-vehicle communications, in *2012 IEEE Conference on Computer Communications Workshops (INFOCOM WKSHPS)* (2012), pp. 328–333. doi:10.1109/INFCOMW.2012.6193515
72. X. Wang, M. Chen, T. Taleb, A. Ksentini, V. Leung, Cache in the air: exploiting content caching and delivery techniques for 5G systems. IEEE Commun. Mag. **52**(2), 131–139 (2014). doi:10.1109/MCOM.2014.6736753
73. M. Whaiduzzaman, M. Sookhak, A. Gani, R. Buyya, A survey on vehicular cloud computing. J. Netw. Comput. Appl. **40**, 325–344 (2014). doi:10.1016/j.jnca.2013.08.004
74. X. Wu, S. Subramanian, R. Guha, R.G. White, J. Li, K.W. Lu, A. Bucceri, T. Zhang, Vehicular communications using DSRC: challenges, enhancements, and evolution. IEEE J. Sel. Areas Commun. **31**(9), 399–408 (2013). doi:10.1109/JSAC.2013.SUP.0513036
75. Z. Yan, S. Zeadally, Y.J. Park, A novel vehicular information network architecture based on named data networking (ndn). IEEE Internet Things J. **1**(6), 525–532 (2014). doi:10.1109/JIOT.2014.2354294

76. V. Yazici, U. Kozat, M.O. Sunay, A new control plane for 5g network architecture with a case study on unified handoff, mobility, and routing management. IEEE Commun. Mag. **52**(11), 76–85 (2014). doi:10.1109/MCOM.2014.6957146
77. A. Yegin, J. Park, K. Kweon, J. Lee, Terminal-centric distribution and orchestration of ip mobility for 5g networks. IEEE Commun. Mag. **52**(11), 86–92 (2014). doi:10.1109/MCOM.2014.6957147
78. X. Yin, X. Ma, K. Trivedi, A. Vinel, Performance and reliability evaluation of BSM broadcasting in DSRC with multi-channel schemes. IEEE Trans. Commun. **63**(12), 3101–3113 (2014). doi:10.1109/TC.2013.175
79. L. Zeng, Y. Zhu, 3g-assisted routing in vehicular networks, in *2012 IEEE Sensors* (2012), pp. 1–4. doi:10.1109/ICSENS.2012.6411481
80. L. Zhang, A. Afanasyev, J. Burke, V. Jacobson, K. Claffy, P. Crowley, C. Papadopoulos, L. Wang, B. Zhang, Named data networking. SIGCOMM Comput. Commun. Rev. **44**(3), 66–73 (2014). doi:10.1145/2656877.2656887
81. N. Zhang, N. Cheng, A. T. Gamage, K. Zhang, J.W. Mark, X.S. Shen, Cloud assisted HetNets toward 5G wireless networks. IEEE Commun. Mag. **53**(6), 59–65 (2015). doi:10.1109/MCOM.2015.7120046
82. Q. Zhao, Y. Zhu, C. Chen, H. Zhu, B. Li, When 3g meets vanet: 3g-assisted data delivery in vanets. IEEE Sensors J. **13**(10), 3575–3584 (2013). doi:10.1109/JSEN.2013.2265304
83. K. Zhu, D. Niyato, P. Wang, E. Hossain, D.I. Kim, Mobility and handoff management in vehicular networks: a survey. Wirel. Commun. Mob. Comput. **11**(4), 459–476 (2011). doi:10.1002/wcm.853

Communications Protocol Design for 5G Vehicular Networks

Francesco Chiti, Romano Fantacci, Dino Giuli, Federica Paganelli, and Giovanni Rigazzi

Abstract This chapter provides an overview on existing standards in vehicular networking and highlights new emerging trends towards an integrated infrastructure based on the interworking of heterogeneous technologies. Next-Generation Mobile Vehicular Networks are first characterized by providing an insight on relevant stable standards in wireless communications technologies, with a special focus on Heterogeneous Vehicular Networks. Furthermore, the chapter discusses a general framework supporting opportunistic networking scheme and outlines novel application and use cases based on social- and context-awareness paradigms.

1 Introduction

The automotive industry is a challenging domain for the integration and evolution of heterogeneous Information and Communication Technologies (ICTs). Indeed, in order to provide drivers with a more pleasant and safer driving experience, vehicles are gradually endowed with sensing, actuation, computing, user-interaction and communications capabilities. Boosted by marketing and commercial competition, vehicles are thus becoming *intelligent nodes* capable of interconnecting, cooperating and even autonomously adapting to the surrounding environment. Since drivers usually spend a non-negligible fraction of time in vehicles and their behaviour may have a strong impact at a collective scale (e.g., pollution and noise in a city), the adoption of advanced ICT approaches for achieving a more safe, green and effective mobility management is still an open research issue. Moreover, the familiarity with the *social networking* paradigm makes people used to share information on-line with (in)direct *contacts* and use it in their everyday life depending on the degree of trust.

F. Chiti • R. Fantacci (✉) • D. Giuli • G. Rigazzi
Department of Information Engineering, University of Florence, Florence, Italy
e-mail: francesco.chiti@unifi.it; romano.fantacci@unifi.it; dino.giuli@unifi.it; giovanni.rigazzi@unifi.it

F. Paganelli
National Interuniversity Consortium for Telecommunications, Research Unit at the University of Florence, Florence, Italy
e-mail: federica.paganelli@unifi.it

W. Xiang et al. (eds.), *5G Mobile Communications*,
DOI 10.1007/978-3-319-34208-5_23

It is worth noticing that this vision inherently deals with multimedia information, as it aggregates, process and disseminates sensed data (originating both on board and on the road side) integrated with warning/alarms, navigation suggestions, and context details that come from a specific area or even the Internet.

The widespread adoption of existing social-based traffic and navigation applications, paves the way for introducing novel vehicular services and applications, that are typically location-based as well as community-driven. To this purpose, wireless communications may enable the effective exchange of *mobile* information among vehicles in a participatory way to enhance open information sharing and knowledge exchange processes. The actual trends in mobile networking are: (1) to extend services offered by traditional providers with unlimited peer-to-peer capabilities, and (2) to dynamically integrate available communications segments to exploit all communication opportunities. In this context, the crucial point is how to guarantee high-capacity and reliable connectivity without affecting the system scalability and cost-effectiveness. Due to the attractive benefits in terms of enhanced data rate and ultra-low latency, 5G represents the most promising solution to enable novel and efficient Intelligent Transport System (ITS) applications and to support the internetworking with existing technologies.

In this chapter, we provide an overview on existing standards in vehicular networking and highlight new emerging trends towards an integrated infrastructure based on the interworking of heterogeneous technologies. Section 2 introduces Next-Generation Mobile Vehicular Networks by providing an insight on relevant stable standards in wireless communications technologies, and reviewing emerging 5G vehicular communication solutions, with a special focus on Heterogeneous Vehicular NETworks (HetVNET). As a preliminary result towards this integrated networking scheme, we propose a general framework supporting opportunistic networking scheme as overlay of an LTE-A system. In Sect. 3 a review on game theory inspired approaches for clustering schemes is provided. Finally, Sect. 4 provides an overview on novel application and use cases based on social- and context-awareness that are enabled by the emerging technological paradigms in Vehicular Networking.

2 Next-Generation Mobile Vehicular Networks

Enabling inter-vehicle communications paves the way to a plethora of novel applications for Intelligent Transportation Systems (ITS), where the main goal is not only improving safety and efficiency of transportation systems, but also providing multimedia contents and information to the mobile users without the support of cellular networks. Based on the application area, vehicular applications can be classified as follows:

1. *Traffic management applications*. To address traffic and environmental related issues, as bottlenecks and fuel consumption, traffic management applications aim at improving the vehicle traffic flow, traffic coordination and traffic assistance, and at providing updated local information, maps and information of relevance

bounded in space and time. This type of time-to-live traffic application is usually used in vehicle-to-backoffice or vehicle-to-roadside scenarios. They may use beacons or alerts, a multihop position-based communication regime, and their latency cannot be higher than 400 ms, whereas the packet delivery ratio cannot be lower than 95 %.

2. *Safety applications* Minimizing the probability of traffic accidents and avoiding collisions situations that most frequently occur between vehicles and other objects such as animals, trees, and pedestrians, are the end goals of safety applications. This type of applications relies on real-time information and uses a vehicle-to-vehicle communication scheme. They provide information and assistance to drivers to avoid traffic accidents, which are mainly caused by human errors [32]. Vehicles and roadside units share information, which is then used to predict a dangerous situation. Moreover, this information is used to locate dangerous locations on roads. They use beacon messages, a single-hop position-based or fast-bidirectional communication regime, and their latency cannot exceed 100 ms, whereas the packet delivery ratio cannot be lower than 99 %. Some examples of safety applications include: intersection collision warning, lane change assistance, overtaking vehicle warning, head on collision warning, and emergency vehicle warning.
3. *Driver assistance and infotainment applications* These applications provide services such as comfort and driving assistance. This class of applications attempts to support all features needed by drivers and passengers for a convenient travel. Driver assistance applications provide information about repair notifications, remote diagnostics, context information, navigation information, and alerts. These applications usually use vehicle-to-backoffice or vehicle-to-roadside communication. They utilize normal messages and bidirectional communication; their latency cannot be higher than 400 ms, whereas the packet delivery ratio cannot be lower than 95 % [48]. Infotainment applications also are known as in-car comfort entertainment, and they usually do not use inter-vehicular communications. These applications are usually found inside vehicles or at vehicle-to-roadside settings. They use alerts, a multihop position-based communication scheme, and their latency cannot be higher than 400 ms, whereas the packet delivery ratio cannot be lower than 95 % [48]. Applications in this category include cooperative local services and global Internet services.

2.1 *Standardization Trends and Perspectives*

Wireless communications are experiencing an impressive evolutionary trend, making users able to independently collect and share information through heterogeneous devices and networks. Specifically, the simultaneous availability of multi-radio technologies compels their integration and management towards enhanced *social inspired* networking opportunities.

An interesting case study is represented by the so called Vehicular Ad-hoc Networks (VANETs), which aim at revolutionising the travelling experience especially within a smart city [34]. The underlying concept is indeed to provide each vehicle with a *wireless identity* which is not simply a *global* identifier, as it happens in IPv6, but it is strictly related with user profiles (drivers and passengers) and the context as well. This allows to increase the interoperability and to *extend the horizon*, thus increasing the safety and optimizing the resources.

However, this vision strictly depends upon the presence of a communications infrastructure providing connectivity to all the vehicles. In the following we briefly overview the candidate technologies [24], spanning from available approaches towards an enhanced paradigm of *ad hoc* interaction that relies upon next generation networks.

2.1.1 DSRC-WAVE

The available technology for supporting the Intelligent Transportation Systems (ITS) applications in the short-range communications is represented by Wireless Access in Vehicular Environments (WAVE), also referred to as Dedicated Short Range Communication (DSRC). It is an approved amendment to the IEEE 802.11 standard, whose protocol suite results from the combination of IEEE 802.11p and IEEE 1609. The former focuses on physical and MAC layers for vehicular environments, while resource management, security, networking and multichannel operation are handled according to the IEEE 1609 protocol suite.

IEEE 802.11p is based on the popular IEEE 802.11 WiFi, exploiting the same medium access scheme—based on the exchange of Request to Send/Clear to Send packets (RTS/CTS)—to alleviate packet collisions in case of unicast communications. However, this technique might not be enabled when a packet is sent in broadcast and the packet collision probability is consequently increased. Furthermore, the quality of service (QoS) support is provided by using the enhanced distributed channel access (EDCA) scheme, which cannot guarantee high packet delivery probability due to the huge number of collisions generated by high priority packets, characterized by a small and fixed Contention Window (CW) size. The entire 75-MHz bandwidth is divided into seven channels: one channel, i.e., Control Channel (CCH), is only used to convey emergency and safety traffic, while six Service Channels (SCH) are occupied by infotainment and multimedia traffic transmission. Further, a channel switching scheme allows the devices to tune to CCH during the CCH time interval and then switch to a specific service channel during the SCH interval.

2.1.2 ISO CALM

Continuous Air-interface Long and Medium range (CALM) is the protocol suite standardized by ISO/TC 204 WG for Europe [30]. The CALM CI (Communication Interface) provides all the functionalities to manage the physical and link layers and can support diverse wireless technologies, ranging from infrared communications to cellular and satellite communications. Moreover, the CALM networking layer coordinates network and transport layers and can either employ IPv6 mobility support protocols or the CALM FAST protocol to perform unicast and broadcast transmissions and store and forward operations.

2.1.3 C2C-CC

Car2Car Communication Consortium (C2C-CC) represents an European open industrial standard mostly focused on (non) safety applications development. It proposed the C2CNet architecture which is not completely IPv6 oriented, while supporting multi-hop communications through ad hoc designed routing protocols and vehicles beaconing [30]. Besides, it is inherently conceived for a multi-radio environments supporting both IEEE 802.11a/b/g/n, IEEE 802.11p and cellular technologies.

2.1.4 B3G Cellular Networks

Wide-area cellular networks, especially 4G Long Term Evolution-Advanced (LTE-A), are expected to directly support vehicular applications by means of self-organizing femto-cells, traffic shaping and *vertical* roaming across different radio access technologies (RATs). Despite limitations on the radio access flexibility and on the resource allocation efficiency, especially in the case of small data transmission, LTE-A networks offer benefits in terms of spectral efficiency, cost reduction and capacity. Furthermore, 4G systems intrinsically guarantee long operative life, scalability and higher performance, thus representing a remarkable solution to support VANET-like applications. As a consequence, cellular systems are being adopted by numerous car manufacturers with the aim of offering diverse services, e.g., remote vehicle monitoring, infotainment or assisted driving experience.

Along with the Machine Type Communication (MTC) definition, the Third Generation Partnership Project (3GPP) Release 12 introduces in LTE-A the support to Device-to-Device (D2D) or direct-mode communications, enabling P2P transmission between devices in proximity [5]. This functionality fosters a new generation of devices able to communicate with each other, without any kind of human intervention and without involving the cellular infrastructure for the user plane. On the other hand, D2D communications pose also several challenges. First of all, direct mode communications should be established without affecting traditional communications via base station, i.e., by avoiding possible *interferences* with other devices. Secondly, peer and service discovery functionalities need to be introduced, since a device is typically not aware of other terminals in the proximity. Following the paradigm of Multi-hop Cellular Networks (MCN) [28, 33, 35], D2D also provides more flexibility in a cellular network, as direct links enable the communication among the terminals using multiple hops.

The evolutionary trend, presented in Fig. 1, points out a convergence of the existing VANET standards, i.e., WAVE, CALM and C2C-CC, towards an integrated

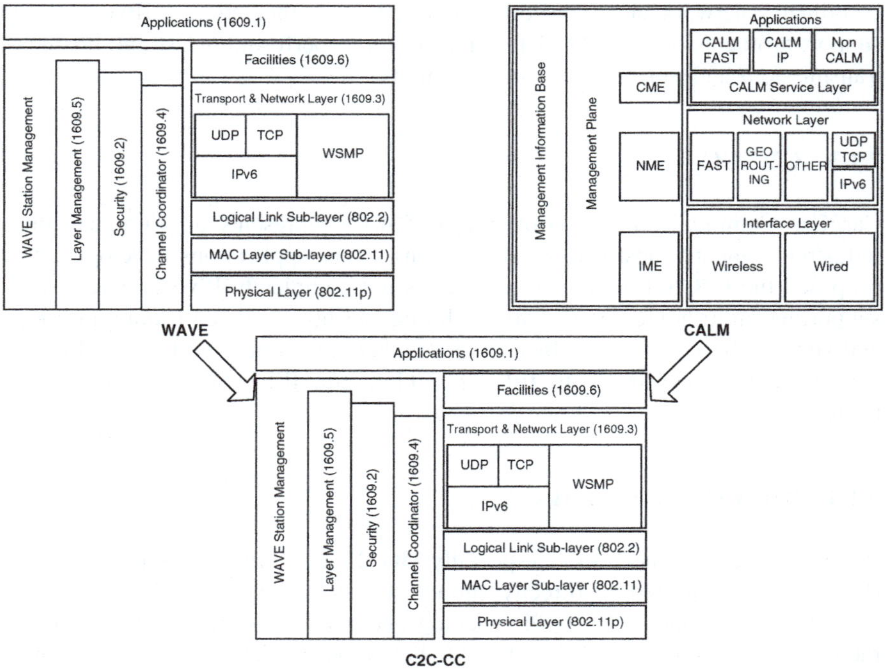

Fig. 1 Convergence of the existing VANET standards towards an integrated protocol stack, as pointed out in [30]

protocol stack, as it is anticipated in [30]. In particular, it is evident an increasing degree of interoperability with different air interfaces both at the Physical and MAC layers, while the higher layers are expected to be unified by the adoption of IPv6 oriented protocols, according to the Internet of Things (IoT) paradigm.

Mobile network evolution is expected to converge by 2020 into the novel 5G system effectively integrating both wired and wireless segments to accomplish expanded use cases and applications [7, 44]. In particular, new services are expected such as the massive sensor and vehicular-to-anything communications. The latter aspect is involved with the so called safety critical applications, requiring extremely low set-up times and transmission delays.

5G is going to be conceived as an unplanned *ultra dense* system, where nodes are able to self organize and to improve cell performance. This trend is clearly pointed out by the emphasis 3GPP working group is progressively posing on *small* cells, whose limit is represented by the D2D technology, provided an optimal coordination/cooperation among devices/cells. As the system is assumed to be inherently heterogeneous, a higher degree of flexibility is required for both radio access and backhaul networks. At the light of the above considerations, it can be expected that VANET will represent a relevant case study for 5G architecture, effectively integrating existing WAVE, CALM and C2C-CC functional stacks, as highlighted in Fig. 1.

2.2 *Internetworking with 5G Systems*

Next 5G communication systems are expected not only to accommodate the massive growth of the mobile data demand generated by the huge number of future connected devices, but also to support diverse wireless applications and satisfy a new user expectation of the network. Accordingly, 5G will lead to higher data rate and network capacity necessary to cope with the extreme network densification, and will guarantee lower latency (i.e., 1 ms), higher reliability (i.e., 99.999 %) and better energy efficiency [22]. To satisfy all the requirements, close integration of vehicles into cellular networks will be of paramount importance and novel heterogeneous networking solutions will assume vehicles to be able to help the network deliver the traffic as well as generate/consume data.

2.2.1 Communication Modes

Due to significant evolution in vehicle technology, future cars will represent the integration of a plethora of emerging information processing, communication, control and electronic systems. Vehicles will be equipped with diverse technologies, including light detection and ranging (LIDAR), radar, advanced cameras and GPS. Intelligent cars will then collect the information about the surrounding environment generated by each sensor and elaborate decisions to make the driving experience more secure and comfortable [47]. In this context, 5G will enable cooperative behavior applications by providing low-latency and high reliable V2X communication, thus disseminating relevant information related to traffic and road condition or sending warning messages in the event of accidents [2]. As described in the next section, vehicles will leverage multi-hop D2D communications, where short-range direct connectivity allows to exchange critical data with minimum delay, while every mobile terminal is also in charge of forwarding the packets in order to communicate with farther terminals without relying on cellular base stations. Moreover, D2D communications will be established on the licensed spectrum, thus guaranteeing high reliability, improved security and easy integration with the cellular network. Fig. 2a shows the D2D links established among close vehicles exchanging information generated by on-board sensors.

Due to the reduced battery power constraint and the easy multiple antenna deployment, vehicular communications also represent an attractive solutions to extend cellular coverage and enhance network capacity for cellular users in the proximity [11]. As shown in Fig. 2b, vehicles can act as mobile Small Cells (mSC) providing better quality of experience for passengers and pedestrians and helping small cells to accommodate the traffic. As a result, drastic network densification can be obtained if effective interference management techniques are employed and vehicle density is sufficient to guarantee adequate data offloading.

Vehicle Content Distribution Networks (VCDN) aim at enabling high-bandwidth content distribution in vehicular networks by opportunistically communicating with

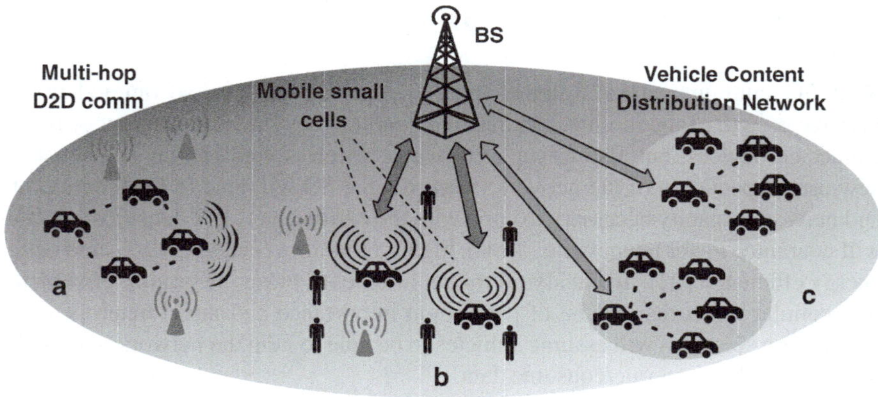

Fig. 2 5G vehicular communication solutions: multi-hop D2D communication (**a**), mobile small cell network (**b**) and vehicle content distribution networking (**c**)

close Access Points (AP) to download multimedia content [38]. Fig. 2c illustrates a VCDN within a cellular network. A clustering approach can be applied to select a cluster-head vehicle, which downloads data from the BS and disseminates into the cluster via short-range D2D connectivity. In turn, this approach leads to a reduction in the number of connections vehicle-BS, even though clusters are largely affected by the traffic conditions and vehicle density.

2.2.2 Heterogeneous Vehicular Networking

Heterogeneous Networks (HetNet) are a popular and attractive solution to increase the network capacity and enable high user data rates in cellular systems [15]. Usually, HetNets refer to the coexistence of various cells with different coverage, namely, macro, small and femto-cells, which require sophisticated interference cancellation techniques to combat the interference originated from nearby cells. In the context of vehicular networking, Heterogeneous Vehicular NETworks (Het-VNET) identify complex systems integrating diverse radio technologies, spanning from IEEE 802.11p DSRC to 3G/4G cellular networks [16]. As an example, an LTE network can guarantee global connectivity and ubiquitous coverage and can help to send multimedia traffic demanding high bandwidth, while DSRC is mostly useful to deliver data generated by safety and early warning applications due to the reduced latency. However, the major challenges associated with the adoption of multi-radio technologies lie in the need for (1) a dynamic support of multiple network configurations, (2) a flexible and effective radio resource management strategy and (3) an efficient mechanism to satisfy the QoS requirements depending on the vehicular application. Inspired by the solutions presented in [48], we envision a 5G vehicular communication system able to perform the joint management of multiple radio technologies and the resource allocation according to the specific

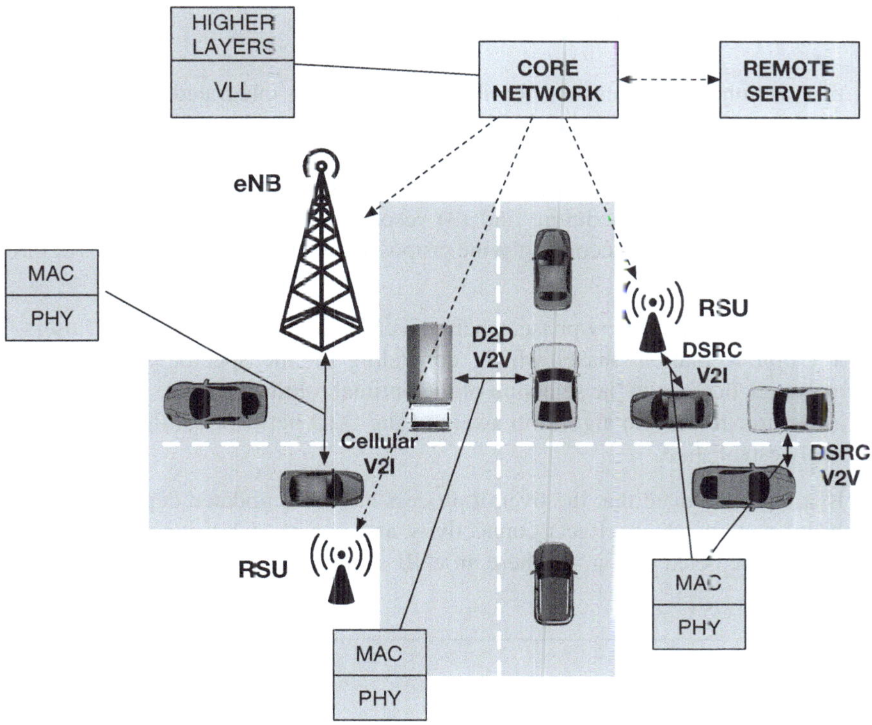

Fig. 3 Proposed 5G architecture for the integration DSRC-LTE in vehicular applications

application. The proposed 5G architecture is illustrated in Fig. 3. Specifically, we suppose that the core network handles key functionalities, ranging from security functions to packet aggregation, being directly connected to remote servers offering diverse services for vehicular users. Furthermore, the HetVNET RAN consists of both cellular LTE eNBs and DSRC RSUs, which guarantee the communication vehicle-to-infrastructure. We also assumed two possible communication modes:

1. V2I, which can be performed through either the cellular link toward a BS or the DSRC interface by exploiting RSUs deployed along the road.
2. V2V, enabling the direct communication between vehicles via LTE D2D connections or DSRC links.

The radio access technology adaption is then accomplished by including a Virtual Link Layer (VLL) into the core network protocol stack. This layer encompasses all the functionalities required to handle different radio interfaces and allows the load balancing among the different systems involved. Moreover, efficient radio interface coordination helps to guarantee the QoS support for safety and multimedia applications. To this end, we assume that network virtualization methods are conveniently applied to obtain virtual resources [9], which abstract the radio resources and the physical layers of each wireless system within the HetVNET.

2.3 Multihop D2D Paradigm

In this section, we address the feasibility issues of an integrated framework able to support opportunistic networking scheme as *overlay* of an LTE-A system as a first step toward the integrated networking envisaged by 5G systems. In particular, we adopted a network-assisted solution to manage (1) network and services discovery, (2) users scheduling and (3) resources allocation, depending on the network configuration. Accordingly, the proposed integrated design consists of three subsequent phases:

1. a D2D *cluster* discovery phase, where the eNB identifies *potential* D2D users,
2. a *graph* formation phase, which, depending on the specific services to be activated, deals with the selection of the optimal routing scheme,
3. a resource allocation algorithm aware of the D2D network configuration previously established.

It could be noticed that the overall process has to be updated depending on the mobility pattern which effects connectivity and service definition. The reference scenario is depicted in Fig. 4, where an eNB serves both pedestrian and vehicular

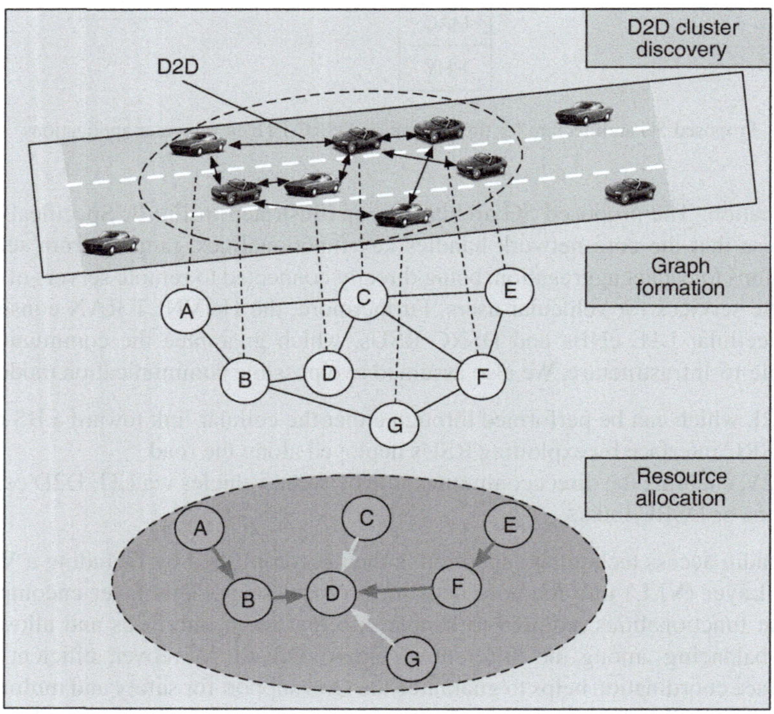

Fig. 4 Proposed integrated network management frameworks across the specific phases

users. We assume all the devices capable of operating both in *basic* cellular mode, as well as in *direct* mode (MD2D), i.e., by directly exchanging data. We focus on a specific use case, according to which all the node send traffic information to a *unique* collector[1] in charge of data aggregating and transmitting to eNB.[2] Furthermore, due to transmit power constraints, we assume no direct communication between sensing nodes and the collector node, thus implying a multi-hop communication scheme and an optimized routing scheme. For the sake of simplicity, we consider an *overlay* resource allocation, where resources dedicated to cellular and D2D users are mutually orthogonal.

2.3.1 D2D Cluster Discovery

As previously mentioned, a network assisted approach, i.e., via the eNB, is a valid solution to support the cluster and services discovery phase since the selection of the D2D candidates can be performed by the network entity. The eNB can decide how clusters are formed by taking into account the context information provided by incoming vehicles (e.g. location, direction, speed, final destination). For instance, vehicles moving towards opposite directions can be partitioned into two different clusters, even though they are close. Moreover, the eNB could also maintain a registry of service and resources made available by cluster members in order to facilitate the dynamic instantiation of *composite* services within a cluster through the steps of service discovery and provider selection. Once a D2D cluster is discovered by collecting the availability of each UE to cooperate and share particular services, an authentication procedure is applied in order to allow direct communication between devices in LTE-A networks. Besides, a direct beaconing is necessary to measure and disseminate the channel state information (CSI) about *all* the *potential* links between cluster members. The eNB is also in charge of assisting a vehicle joining an already formed cluster by taking into account the context information delivered by the vehicle: when a vehicle is approaching an ENB it can, indeed, advertise its position, resource and service capabilities, application and user interests and availability to join D2D clusters. Based on this information, the eNB can select the active clusters that best match the context information and extend these D2D clusters accordingly.[3]

---

[1]The collector selection can be performed by the eNB or by the D2D users, according to a centralized or distributed approach, respectively.

[2]We assumed that D2D communications can be performed only during the *uplink* frame.

[3]It can be noticed that this procedure might require the cooperation of the vehicle itself and additional eNBs due to roaming.

2.3.2 Graph Construction

Once the information regarding the cooperating nodes has been collected by the eNB, the network topology can be selected, depending on the specific use case, as explained din Sect. 4.3. Then this topology is optimized according to a particular target function. It is worth noting that in VSNs all the possible topologies are inherently data centric, since the kind of information affects the message passing scheme and the *logical* connectivity as well. For instance, data aggregation could be effectively performed via a *tree* topology where a single node is in charge of collecting and refining data sensed by other ones and to eventually deliver to the eNB. To this end, Routing Protocol for Low-power and Lossy networks (RPL) is a widely adopted solution to establish a convenient loop free route toward a given destination [1].

2.3.3 Network-Aware Resource Allocation Algorithm

Once the topology has been established, a serious drawback is represented by potential bottlenecks, which for a tree topology involve nodes with lower rank value, i.e., nodes closer to the root (collector).[4]

An effective resource allocation strategy should take into account the rank of each node, providing more resources to the nodes closest to the collector. A heuristic strategy could assign resources according to a *proportional fair* scheduling approach, by setting the scheduling weights w_i for the i-th node taking into account the its estimated rate r_i[5] as well as the rate of the sub-tree $\mathscr{T}_i$ rooted at the i-th node ($\mathscr{T}$ represents the overall tree). In particular, we have:

$$w_i \doteq \frac{r_i + \sum_{j \in \mathscr{T}_i} r_j}{\sum_{k \in \mathscr{T}} r_k + \sum_{j \in \mathscr{T}_k} r_j} \tag{1}$$

2.3.4 Performance Analysis

We investigated the performance achievable by the proposed approach by focusing on a low-to-moderate mobility scenario, where only *one* Resource Block (RB) is reserved for VANET traffics with a minimum power transmission and a robust Modulation and Coding Scheme (MCS).[6] We referred to an anycast data dissemination, as explained in Fig. 8, where, depending on the mobility patterns, several topologies are possible.

[4]This problem is commonly addressed by means of the *min-cut graph analysis*.

[5]This information is collected by the eNB during the services discovery phase.

[6]In particular, we adopted the MCS 5, according to LTE standard.

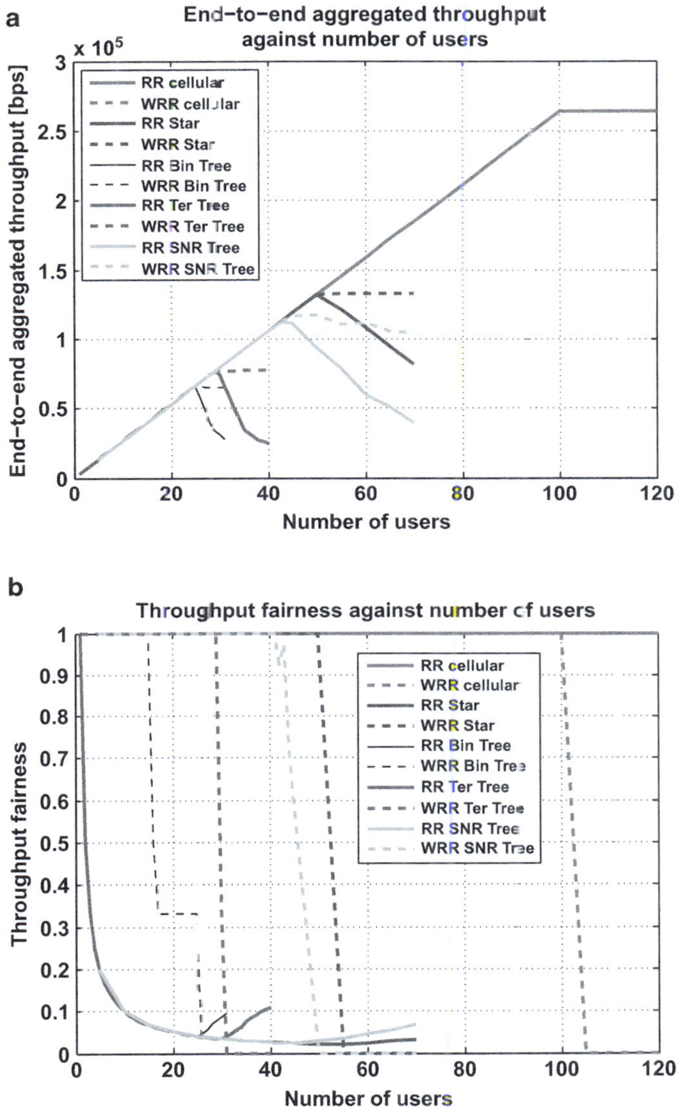

Fig. 5 End-to-end aggregated throughput and fairness comparisons for a practical case study

In Fig. 5 the performance in terms of throughput and fairness, respectively, are investigated for different network topologies, including basic cellular mode, single hop D2D (star), multihop D2D with binary, ternary or stochastic trees. Besides, the network unaware (RR) and aware (WRR) scheduling approaches have been compared. It can be noticed a *linear* throughput increasing for all the topologies, while in saturation conditions WRR scheduling scheme achieves better performance

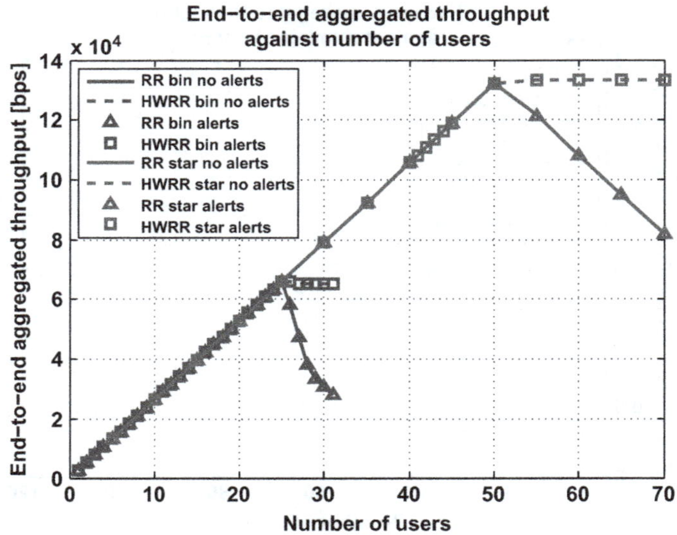

Fig. 6 Throughput performance in the case of a star and a binary tree topologies with and without alarm flow for WRR and HWRR algorithms

than RR. In addition, it is worth pointing out that the availability of one RB allows the effective management of at most 100 vehicles arranged in a star topology, while the *effective* cluster size decrease for tree topology. Further, WRR is more fair than RR, even though in saturation WRR implies the *starvation* effect.

We investigated also a different scenario where two traffic flows are considered, a background *data* and a sporadic *alarm* flows, with different rates (specifically data flows is 100 times higher than the other one) and priorities. In Fig. 6, the throughput performance are investigated in the case of a star a binary tree topologies with and without alarm flow for WRR and HWRR algorithms, pointing out an unchanged network capacity. However, in Fig. 7, the delivery delay performance are also evaluated for WRR and HWRR algorithms, respectively, highlighting that the latter approach is more capable of handling flow priority and matching time constraints.

3 Advanced Communications Protocol Design

3.1 Small-World Networking

The topic of D2D communications has been recently investigated in several possible applications, as data offloading [3], cell coverage extension [18], content sharing [37], and disaster area alerting [33] with a specific focus mainly on downlink

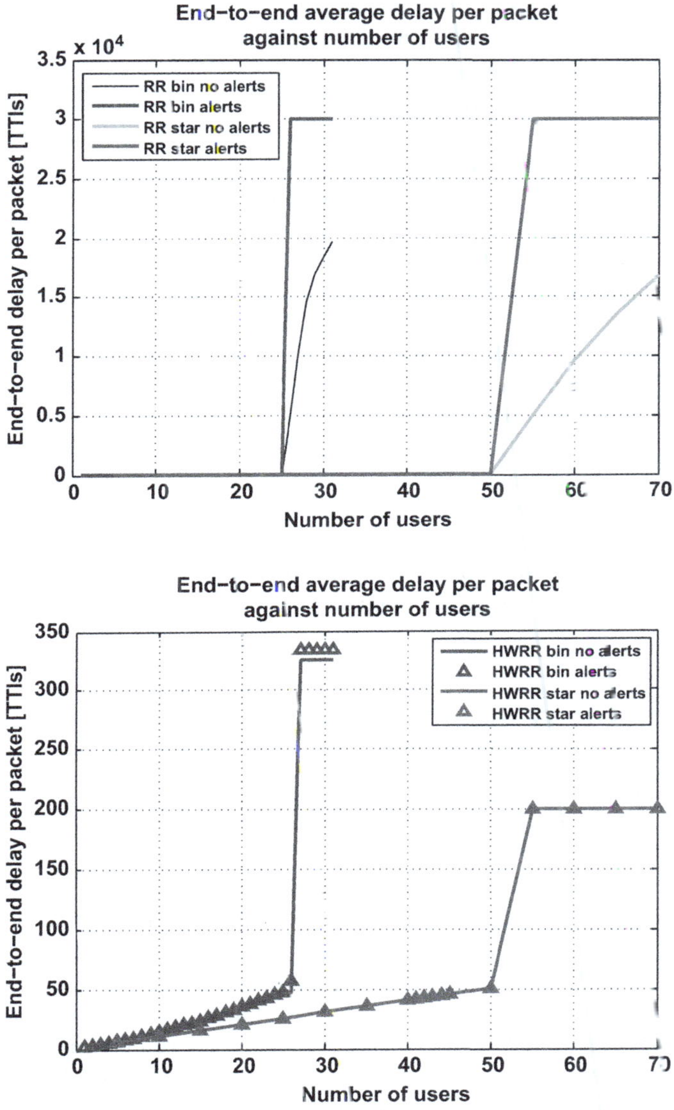

Fig. 7 End-to-end delay per packet performance in the case of a star and a binary tree topologies with and without alarm flow for WRR and HWRR algorithms

issues. However, these contributions are limited to the case of single hop scheme, while the more general case of multihop D2D communications still deserves more efforts. In [12] network-assisted D2D communication is addressed with an analysis on power control limited to a more traditional two-hop scenario. Conversely, multihop D2D communication is considered in [35] for end-to-end Machine-to-Machine connectivity, deriving a topology aware resource allocation scheme for overlay networks and avoiding co-channel interference.

Despite these contributions, the development of suitable protocols which effectively support information sharing is still an open issue. Recent research on large scale self-organized distributed networks, as well as in social communities, suggests an alternative perspective for routing protocol design [25]. It relies on the evidence of scale-free properties for typical VANET deployments [31], which implies a *small-world* network structure. The analysis of the so called *network clustering coefficient*[7] quantifies the clustering properties of a specific community of devices and motivates the introduction of clustering protocols.

This is particularly relevant for the design of a joint clustering and routing scheme, which dynamically organises nodes into groups, referred to as clusters, where each cluster is coordinated by one of the vehicles, referred to as the cluster head (CH) and the rest of vehicles are referred to as the ordinary nodes (ONs) [45]. The optimal CH selection relies on a combination of several parameters, such as the ID, degree, energy level, position, speed and direction, and is typically NP-hard to solve. Several heuristic policies [10, 42] have been proposed in ad hoc networks, such as the lowest-ID, highest-degree and node-weight heuristics. However heuristics are always not optimal.

3.2 Game Theory Inspired Approaches

Generally speaking, Game theory (GT) has been widely considered as a promising approach to model the interaction among independent nodes in self-organizing, decentralized and autonomic networks. In particular, non-cooperative models address the interaction among individual rational decision makers, called players. In such models players are assumed selfish, and select their strategies in order to maximize their own performance without caring about the global network performance, leading to steady-state equilibrium that are socially undesirable. As examples of application the case of a medium access control based on the CSMA/CA is considered in [8], while the packet forwarding problem is analyzed in [17, 41]. Likewise, Cooperative GT, focuses on the study of coalition of players, that act like a single entity. In particular, [36] proposes a novel classification of coalitional games, whose main class is that of the canonical games. In this case the superadditivity property holds, meaning that the formation of large coalitions is never detrimental to any of the involved players. The objectives of a canonical game are to study the stability and fairness of the grand coalition, i.e., the coalition of all the players, where stability means finding a payoff allocation which guarantees that

[7]In particular, it represents the probability that two nodes connected to a common node are also connected between themselves.

no group of players have an incentive to leave the grand coalition. In [26] the authors use the canonical games model to tackle the problem of how to fairly allocate the transmission rates between a numbers of users accessing a wireless Gaussian MAC channel, guaranteeing the stability under the envy-free fairness criterion. Finally in [21], canonical coalition games are used to solve an inherent problem in packet forwarding ad hoc networks.

Only recently, GT and coalitional games have been successfully applied also to a few D2D communications, by first analysing open issues [40]. In [27] the problem of uplink radio resource allocation is studied when multiple D2D pairs and cellular users share the available resources. In addition, GT is applied to ensure energy-efficient D2D resource allocation [49], and a simple coalitional game is studied for energy-efficient D2D communications in Long Term Evolution (LTE) public safety networks [46].

The aforementioned limitations of centralized optimization and heuristic solutions, especially for clustering schemes, have led to an interesting body of literature that deals with the use of coalitional graph games, where a specific graph interconnects the players to maximize their individual payoffs [14]. The coalitional graph games principle has been applied to several wireless-oriented applications, such as wireless physical layer security improvement [6], network formation for maximizing data exchange [4], and decentralized joint relay selection and power allocation [43]. Despite their potential, such game-theoretical approaches have their own shortcomings, such as requiring the knowledge of other players' actions, the solution equilibrium is evaluated from each single player's point of view, which doesn't fit the setting of two distinct sets of players (CHs and ONs) in the VANET clustering problem. Recently, matching theory has becoming a promising technique which can overcome some limitations of game theory and optimization [20].

4 Novel ITS Applications Based on Context and Social Awareness

Vehicular Social Networks (VSNs) [29] are an emerging type of network which allows the exchange of information among drivers, passengers, as well as vehicles. The exploitation of social relationships can improve and extend the added value provided by applications of vehicular networks (e.g. navigation safety applications, navigation efficiency, entertainment, participatory and urban sensing, emergency [19]). Indeed, leveraging the social dimension in vehicular networking can bring several advantages:

- ease the dissemination of information by exploiting users common interests and preferences thus avoiding to deliver information to uninterested parties;
- ease the on-time and continuous production of up-to-date and capillary information on the road conditions and traffic status by turning users and vehicles into producers of information;

- enhance the exchange of information, media content and recommendations for entertainment and tourism;
- support new mechanisms for reputation building and establishment of trust among nodes in Vehicular Ad-Hoc Networks (VANETs) [24], (for instance the relationship degree in online social networks could contribute to define the degree of trust with an unknown node entering the network).

With respect to on-line and mobile social networks, VSNs pose several challenges:

- the topology of a VSN changes very often and it is formed by peers that move at high speed;
- contact duration of vehicles on the move is short (in the order of tens of seconds [29]);
- the technological infrastructure is made by heterogeneous components, such as On-Board Units (OBU), vehicle sensors and actuators, mobile phones, Road Side Units (RSU) and remote services.

The nature of VSNs is inherently dynamic and dependent on context changes, as discussed hereafter.

4.1 Context-Awareness in VSNs

One of the first and more widely adopted definition of context was provided by Dey [13]: "Context is any information that can be used to characterize the situation of an entity. An entity is a person, place or object that is considered relevant to the interaction between a user and an application, including the user and applications themselves".

Context-awareness was defined as the capability of a system to use context for providing relevant information and/or services to the user, depending on the user's task. In the last decade, a lot of research has been conducted in the field of context-aware systems to propose novel context representation and reasoning models and context management frameworks and also to widen the scope of application of context-aware adaptation mechanisms, beyond information provisioning to end users, to include several application domains, including the automotive one.

Context-awareness plays a relevant role in VSNs for several reasons:

- context information can be used to dynamically detect and build communities. Relations in VSNs can be built by taking into account highly variable information such as common geographical location, final destination and/or part of the route, but also rarely changing information, such as habits (e.g., daily routes to work), friendship, family relationships, common interests and trust;
- members of VSNs can exchange context updates, such as change of speed, occurrence of congestion events or accidents, PoI recommendations;

- provided services can adapt to current context, e.g., navigation services can suggest a new route, alerts are propagated only to nodes that can be affected by the alerted event.

It is expected [19] that the evolution of Vehicular Networks will intersect that of smart phones and Wireless Sensor Networks (WSN), thus extending the range of context sources and the type of context information that can be gathered in the vehicular environment and opening up the opportunity of novel context-aware services. Ad-hoc networking mechanisms provided by wireless and cellular technologies are key enablers for VSNs since they enable the establishment of opportunistic networks relaying on the cooperation of dynamic groups of peers (i.e., smart phones, vehicle OBUs, smart objects along the road infrastructure).

Moreover, these opportunistic networks can be promptly put in place or reconfigured when the situation requires a boost in the degree of social interactions in order to face unexpected events. For instance, as soon as an incident occurs, a network can be setup to alert the vehicles approaching that location.

Context in vehicular/automotive environments can be represented by referring to the following main categories:

- *User*
- *Vehicle*
- *Social Context*
- *Physical context*
- *Application*
- *Network*
- *Device.*

This context information can be used to trigger appropriate adaptation at different scopes (in-vehicle, vehicle-to-vehicle, vehicle-to-remote services) in order to enhance the quality of travel for drivers and passengers and assure their safety. Table 1 provides some main examples of context-aware adaptation actions.

4.2 Service Overlay

The appropriate use of context-aware adaptation paradigms in the operation of application, network and communication services is considered a key enabler not only for VSN but also, on a wider scope, for 5G technology [39]. Indeed, 5G infrastructure is envisioned as a *Neural bearer* where *Everything* (applications, processing resources, network, data, etc.) can be provided as a service and most of intelligence, including context-aware adaptation logic, will be distributed at the edge, i.e., in the aggregation and access segments up to the end user premises [39]. As a consequence, it is possible to envisage an overlay of heterogeneous service capabilities pervasively distributed and interconnected and deeply integrated through the 5G network infrastructure that can be dynamically composed and

Table 1 Context-aware adaptation in VANETS

Context		Context-aware adaptation									
Context Category	*Description*	Warning (collision, security distance, etc.)	Prevent the propagation of irrelevant information	Suggest new route to avoid delay due to incident/congestions	Location and direction-aware delivery of information	Keep information "alive" in certain locations and over certain intervals of time	Adapt information delivery and input from user to current user activity and level of attention	Efficient message routing	Select of the optimal transmission rate,	Dynamic building g of groups	Adaptive users scheduling and resource allocation
User	User's profile and preferences, user habits, user activity, driver behavior.	X	X	X			X				
Vehicle	Location, direction, speed.	X		X	X	X				X	
Social	People/Vehicles nearby, distance among vehicles, short- and long-term social relationships.	X	X			X		X		X	X
Physical context	Time, weather information, type of road and traffic density, congestion, incident.	X		X	X	X			X	X	X
Application	Navigation applications, tourist guides, Driving assistance systems, PoI recommendation	X	X	X	X	X	X	X		X	
Network	Information characterizing VANETs, such as topology, bandwidth, traffic performance and link quality							X	X		X
Device	Information characterizing devices with user interfaces (e.g. smart phones, OBU, etc.), such as device hardware and software configuration, battery power, memory consumption, received signal strength of available access networks, etc.						X	X			

adapted to context changes. In this perspective, it is worth mentioning the IEEE NGSON standard [23] since it specifies a service-overlay network over IP-based infrastructure that provides context-aware and dynamically adaptive service control and delivery functions. The NGSON functional architecture specifies a set of functional entities (FEs) for main NGSON capabilities: service composition, service discovery and negotiation, service routing, context information management, content delivery, and service policy decision to negotiate and enforce Quality of Service (QoS) at service and transport levels with underlying networks.

The envisioned 5G-enabled VANET scenario poses new challenges for the physical deployment of the NGSON distributed overlay. Indeed, to accommodate the requirement of pervasive intelligence and adaptation capabilities, we envision a NGSON physical deployment where FEs would be distributed at least across three levels: locally (on the vehicle and in a proximity-based group of vehicles), at the edge (access/aggregation network) and remotely (network operators and service providers data centers). Thanks to opportunistic communication, hw and sw resources hosted by nodes, such as smart-phones and vehicle OBUs, can be exposed as a service and dynamically composed across different nodes for providing added-value services. For instance a video captured by a cam on a vehicle can be processed by a video compressor provided by a smart-phone and then diffused to the other peers of the network.

4.3 Application Scenarios and Communications Use Cases

Cooperation among vehicles can enable a wide range of application services for the benefit of several categories of end users: drivers of private vehicles, drivers of public or collective transport vehicles, transportation authorities, public

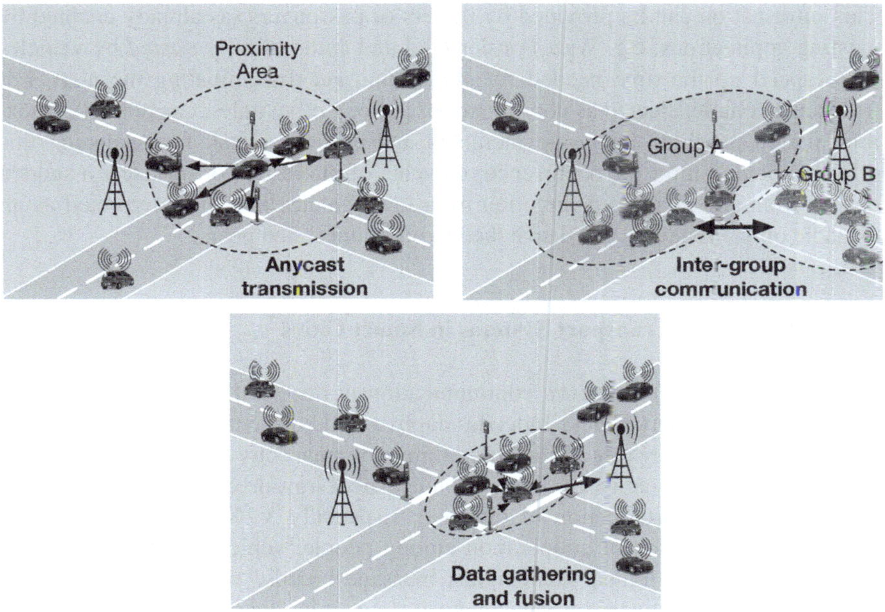

Fig. 8 Potential use cases from a data exchanging and networking perspective

administrations, etc. Hereafter we provide some scenarios to illustrate how context-aware applications can be empowered by multi-hop D2D communication. To this purpose, we preliminary focus on three different networking use cases, starting from (1) critical information diffusion performed by a single vehicle towards a potential group of neighbours, up to (2) data exchange among convoys when a communication opportunity arises, and (3) data aggregation and fusion performed by a vehicle acting as gateway to the Internet, as depicted in Fig. 8. As soon as communications become less sporadic and context aware, an ad hoc networking scheme is required: in particular we indicated (1) a basic flooding scheme or more advanced solutions relying on (2) disruption-tolerant networking (DTN) or classical mobile ad hoc networking (MANET).

4.3.1 Safety and Early Warning Applications

Safety and early warning applications are one of the most investigated types of applications in vehicular systems. Leveraging the dynamic exchange of context information among OBUs and RSUs, dynamic communities can be built that group vehicles that are in the same geographical area and eventually share a significant part of their route or have the final destination in common. For example, events of interest are: an abrupt deceleration of a vehicle or a group of vehicles, an obstacle on the road, a sudden change of weather conditions, a breakdown of a vehicle.

This information can be provided by drivers or passengers (as already enabled by existing applications, e.g. Waze) or detected and automatically *posted* by vehicles so to speed up the time needed for generating and disseminating the alert. For instance a vehicle, elected as coordinator of a group of vehicles, can collect context information provided by group members and execute a task for analysing this information and infer the occurrence of events of interests. For instance, a sudden deceleration and change of direction by a few vehicles may be interpreted as an obstacle on the roadway and dispatched to other interested peers.

4.3.2 Autonomic Transport Systems in Smart Cities

The paradigm of *opportunistic* communication is one of the main ICT enablers in the fields of Smart Cities. Indeed, adaptive and resilient communications are expected to play a key role in the achievement of smart city goals (i.e. sustainable economic growth, quality of life, rational use of resources through the active involvement and participation of citizens). Especially V2X communication can enable novel scenarios of cooperation among people, vehicles and urban infrastructure (e.g. traffic lights, street lights, message boards, etc.) so to achieve the best compromise between private interests (e.g. minimize travel time to work) and public goals (e.g. reduce CO2 and noise levels). People, vehicles and smart urban objects can act as sensors and exchange context information. Distributed or centralized decision approaches can be implemented that based on this information can dynamically adapt the use of urban resources (e.g., change traffic light policies, allow or avoid access to a road). For instance, priority at a crossing road could be dynamically decided according to current situation: e.g. number of vehicles on a lane, a congestion after the crossroad, an approaching ambulance, etc. Drivers can be appropriately alerted through road messages and in vehicle alerts. Furthermore, with the introduction of autonomous vehicle control systems, vehicles could cooperatively agree on a traffic flow policy and autonomously enforce the decision (e.g. a group of vehicles approaching a traffic light can smoothly decelerate in order not to stop at the traffic light).

5 Summary

In this chapter we have analyzed main current trends and perspective in Vehicular Networking in 5G. We first provided an overview on existing standards and highlighted expected benefits enabled by the interworking with 5G systems. We proposed a 5G vehicular communication systems able to perform the joint management of multiple radio technologies and the resource allocation according to specific application requirements. We pointed out main open issues regarding the protocol design and we discussed the adoption of game theory inspired methodologies in the VANET clustering problem. Finally, we provided an insight on recent trends in context- and social-aware vehicular applications enabled by the envisaged evolution in Vehicular Networking.

References

1. A. Aijaz, A. Aghvami, Cognitive machine-to-machine communications for internet-of-things: a protocol stack perspective. IEEE Internet Things J. **2**(2), 103–112 (2015)
2. M. Amoozadeh, A. Raghuramu, C. Chen-Nee, D. Ghosal, H.M. Zhang, J. Rowe, K. Levitt, Security vulnerabilities of connected vehicle streams and their impact on cooperative driving. IEEE Commun. Mag. **53**(6), 126–132 (2015)
3. S. Andreev, O. Galinina, A. Pyattaev, K. Johnsson, Y. Koucheryavy, Analyzing assisted offloading of cellular user sessions onto D2D links in unlicensed bands. IEEE J. Sel. Areas Commun. **33**, 67–80 (2015)
4. E. Arcaute, R. Johari, S. Mannor, Network formation: bilateral contracting and myopic dynamics. IEEE Trans. Autom. Control **54**(8), 1765–1778 (2009)
5. D. Astely, E. Dahlman, G. Fodor, S. Parkvall, J. Sachs, LTE release 12 and beyond. IEEE Commun. Mag. **51**(7), 154–160 (2013)
6. S. Bayat, R.H.Y. Louie, Z. Han, B. Vucetic, Y. Li, Physical-layer security in distributed wireless networks using matching theory. IEEE Trans. Inf. Forensics Secur. **8**(5), 717–732 (2013)
7. F. Boccardi, R. Heath, A. Lozano, T. Marzetta, P. Popovski, Five disruptive technology directions for 5g. IEEE Commun. Mag. **52**(2), 74–80 (2014)
8. M. Cagalj, S. Ganeriwal, I. Aad, J.-P. Hubaux, On selfish behavior in CSMA/CA networks, in *IEEE INFOCOM* (2005)
9. L. Chengchao, F.R. Yu, Wireless network virtualization: a survey, some research issues and challenges. IEEE Commun. Surv. Tutorials **17**(1), 358,380 (2015)
10. F. Chiti, R. Fantacci, R. Mastandrea, G. Rigazzi, A.S. Sarmiento, E. López, A distributed clustering scheme with self nomination: proposal and application to critical monitoring. Springer Wirel. Netw. **21**(1), 329–345 (2015)
11. COST IC1004, White Paper on scientific challenges towards 5G mobile communications (2013)
12. J. da Silva, G. Fodor, T. Maciel, Performance analysis of network-assisted two-hop d2d communications, in *IEEE Globecom 2014 Workshops* (2014), pp. 1050–1056
13. A.K. Dey, Understanding and using context. Pers. Ubiquit. Comput. **5**(1), 4–7 (2001)
14. L. Dong, Z. Han, A.P. Petropulu, H.V. Poor, Improving wireless physical layer security via cooperating relays. IEEE Trans. Signal Process. **3**(58), 1875–1888 (2010)
15. B. Dongwoon, P. Checlhee, L. Jungwon, N. Hoang, J. Singh, A. Gupta, P. Zhouyue, K. Taeyoon, L. Chaiman, K. Min-Goo, K. Inyup, LTE-advanced modem design: challenges and perspectives. IEEE Commun. Mag. **50**(2), 178–186 (2012)
16. F. Dressler, H. Hartenstein, O. Altintas, O. Tonguz, Inter-vehicle communication: quo vadis. IEEE Commun. Mag. **52**(6), 170–177 (2014)
17. M. Felegyhazi, L. Buttyan, J.-P. Hubaux, Ash equilibria of packet forwarding strategies in wireless ad hoc networks. IEEE Trans. Mob. Commun. **5**(5), 463–476 (2006)
18. G. Fodor, E. Dahlman, G. Mildh, S. Parkvall, N. Reider, G. Miklos, Z. Turanyi, Design aspects of network assisted device-to-device communications. IEEE Commun. Mag. **50**, 170–177 (2012)
19. M. Gerla, L. Kleinrock, Vehicular networks and the future of the mobile internet. Comput. Netw. **55**(2), 457–469 (2011)
20. Y. Gu, W. Saad, M. Bennis, M. Debbah, Z. Han, Matching theory for future wireless networks: fundamentals and applications. IEEE Commun. Mag. **53**(5), 52–59 (May 2015)

21. Z. Han, V. Poor, Coalitional games with cooperative transmission: a cure for the curse of boundary nodes in selfish packet-forwarding wireless networks. IEEE Trans. Commun. **57**, 203–213 (2009)
22. ICT-317669 METIS Project, Requirements and General Design Principles for New Air Interface. Deliverable D2.1, August 2013, https://www.metis2020.com/documents/deliverables
23. IEEE Standard 1903, The functional architecture of next generation service overlay networks (2011)
24. G. Karagiannis, O. Altintas, E. Ekici, G. Heijenk, B. Jarupan, K. Lin, T. Weil, Vehicular networking: A survey and tutorial on requirements, architectures, challenges, standards and solutions. IEEE Commun. Surv. Tutorials **13**(4), 584–616 (2011)
25. J.M. Kleinberg, Navigation in a small world. Nature **406**, 845 (2000)
26. R. La, V. Anantharam, A game-theoretic look at the gaussian multi-access channel, in *DIMACS Series in Discrete Mathematics and Theoretical Computer Science*, vol. 66, pp. 87–106 (2003)
27. Y. Li, D. Jin, J. Yuan, Z. Han, Coalitional games for resource allocation in the device-to-device uplink underlaying cellular networks. IEEE Trans. Wirel. Commun. **13**, 3965–3977 (2014)
28. Y.-D. Lin, Y.-C. Hsu, Multihop cellular: a new architecture for wireless communications, in *INFOCOM 2000. Proceedings of the Nineteenth Annual Joint Conference of the IEEE Computer and Communications Societies*, vol. 3 (IEEE, 2000), pp. 1273–1282
29. F. Mezghani, R. Dhaou, M. Nogueira, A.-L. Beylot, Content dissemination in vehicular social networks: taxonomy and user satisfaction. IEEE Commun. Mag. **52**(12), 34–40 (2014)
30. S. Mohammad, A. Rasheed, A. Qayyum, Vanet architectures and protocol stacks: a survey, in *Communication Technologies for Vehicles*, ed. by T. Strang, A. Festag, A. Vinel, R. Mehmood, C. Rico Garcia, M. Röckl. Lecture Notes in Computer Science, Springer Berlin Heidelberg, vol. 6596 (2011), pp. 95–105
31. R. Monteiro, S. Sargento, W. Viriyasitavat, O.K. Tonguz, Improving VANET protocols via network science, in *IEEE Vehicular Networking Conference (VNC)* (2012)
32. National Highway Traffic Safety Administration (NHTSA), National Motor Vehicle Crash Causation Survey. U.S. Department of Transportation, Report DOT HS 811 059 (2008)
33. H. Nishiyama, M. Ito, N. Kato, Relay-by-smartphone: realizing multihop device-to-device communications. IEEE Commun. Mag. **52**(4), 56–65 (2014)
34. A. Rasheed, H. Zia, F. Hashmi, U. Hadi, W. Naim, S. Ajmal, Fleet and convoy management using VANET. J. Comput. Netw. **1**(1), 1–9 (2013)
35. G. Rigazzi, F. Chiti, R. Fantacci, C. Carlini, Multi-hop d2d networking and resource management scheme for m2m communications over LTE-a systems, in *2014 International Wireless Communications and Mobile Computing Conference (IWCMC)* (2014), pp. 973–978
36. W. Saad et al., Coalitional game theory for communication networks. IEEE Signal Process. Mag. **26**(5), 77–97 (2009)
37. J. Seo, T. Kwon, V. Leung, Social groupcasting algorithm for wireless cellular multicast services. IEEE Commun. Lett. **17**, 47–50 (2013)
38. U. Shevade, Y.C. Chen, L. Qiu, Y. Zhang, V. Chandar, M.K. Han, H.H. Song, Y. Seung, Enabling high-bandwidth vehicular content distribution, in *Proceedings of the 6th International Conference* (ACM, 2010), p. 23
39. D. Soldani, A. Manzalini, Horizon 2020 and beyond: On the 5g operating system for a true digital society. IEEE Veh. Technol. Mag. **10**(1), 32–42 (2015)
40. L. Song, D. Niyato, Z. Han, E. Hossain, Game-theoretic resource allocation methods for device-to-device communication. IEEE Wirel. Commun. **21**(3), 136–144 (2014)
41. V. Srinivasan et al., Cooperation in wireless ad hoc networks, in *IEEE INFOCOM* (2003)
42. D. Tian, Y. Wang, H. Xia, F. Cai, Clustering multi-hop information dissemination method in vehicular ad hoc networks. Intell. Transp. Syst. **7**(4), 464–472 (2013)
43. B. Wang, Z. Han, K.J.R. Liu, Distributed relay selection and power control for multiuser cooperative communication networks using stackelberg game. IEEE Trans. Mob. Comput. **7**(8), 975–990 (2009)

44. C.-X. Wang, F. Haider, X. Gao, X.-H. You, Y. Yang, D. Yuan, H. Aggoune, H. Haas, S. Fletcher, E. Hepsaydir, Cellular architecture and key technologies for 5g wireless communication networks. IEEE Commun. Mag. **52**(2), 122–130 (2014)
45. T.L. Willke, P. Tientrakool, N.F. Maxemchuk, A survey of inter-vehicle communication protocols and their applications. IEEE Commun. Surv. Tutorials **11**(2), 3–20 (2009)
46. E. Yaacoub, O. Kubbar, Energy-efficient device-to-device communications in LTE public safety networks, in *IEEE Globecom 2012 Workshops* (IEEE, 2012), pp. 391–395
47. S.H. Yu, O. Shih, H.M. Tsai, N. Wisitpongphan, R. Roberts, Smart automotive lighting for vehicle safety IEEE Commun. Mag. **51**(12), 50–59 (2013)
48. K. Zheng, Q. Zheng, P. Chatzimisios, W. Xiang, Y. Zhou, Heterogeneous Vehicular Networking: A Survey on Architecture, Challenges, and Solutions. IEEE Commun. Surv. Tutorials, **17**(4), 2377–2396, (2015) Fourthquarter
49. Z. Zhou, M. Dong, K. Ota, R. Shi, Z. Liu, T. Sato, Game-theoretic approach to energy-efficient resource allocation in device-to-device underlay communications. IET Commun. **9**(3), 375–385, (2015)

Next-Generation High-Efficiency WLAN

Nan Cheng and Xuemin (Sherman) Shen

Abstract Currently, the exponential growth of mobile data traffic has put an increasingly heavy burden on the cellular network, and results in severe overload problem. As a cost-effective Internet access solution, WiFi networks consume a major portion of the global Internet traffic, and greatly offload the cellular network. However, with the increasing demands for WLAN and the deployment of carrier-WiFi networks, the number of WiFi public hotspots worldwide is expected to increase dramatically. To face this huge increase in the number of densely deployed WiFi networks, and the massive amount of data to be supported by these networks in indoor and outdoor environments, it is necessary to improve the current WiFi standard and define specifications for high-efficiency wireless local area networks (HEWs). In this chapter, the emerging HEW technology is introduced and discussed, including typical use cases, environments, and potential techniques that can be applied for HEWs. We first give the typical HEW use cases, and analyze the main requirements from these use cases and environments. Then, potential techniques, including enhanced medium access, and spatial frequency reuse, are presented and discussed.

1 Introduction

1.1 High-Efficiency WLAN

As one of the most successful wireless access technologies, IEEE 802.11 wireless local area networks (WLANs) provide cost-effective Internet access that can satisfy the communication requirements of most current wireless services and applications. Due to the high data rate, low deployment complexity, and low price, WLANs have been widely deployed around the world and experienced tremendous growth

N. Cheng (✉) • X. (Sherman) Shen
Department of Electrical and Computer Engineering, University of Waterloo, Waterloo, ON, Canada N2L 3G1
e-mail: wmchengnan@gmail.com

W. Xiang et al. (eds.), *5G Mobile Communications*,
DOI 10.1007/978-3-319-34208-5_24

in the recent years. Like many other wireless technologies, since the first standard IEEE 802.11a-1999, WLANs or WiFi technologies have evolved and integrated the latest technologies, in order to continue to improve the spectrum utilization and network performance. In IEEE 802.11n-2009, single-user multiple-input multiple-output antennas (SU-MIMO) is added to support a maximum data rate from 54 to 600 Mbps. In addition, frame aggregation is employed to reduce the protocol overhead. IEEE 802.11ac-2013 further extends the features, including multi-user MIMO (MU-MIMO), extending the channel bandwidth to 160 MHz, and high-density modulation (up to 256-QAM). With these feature, IEEE-802.11ac based WLANs can support maximum 866 Mbps data rate. IEEE 802.11af and 802.11ah extend the WLAN spectrum bands to the TV white spaces and sub 1G bands with cognitive radio technologies, so as to support long-range and power saving applications such as machine-to-machine (M2M) communications. In IEEE 802.11p is designed on 5.9 GHz to support Intelligent Transportation Systems (ITS) applications, such as toll collection, vehicle safety applications, and high-speed Internet access. The most popular IEEE 802.11 amendments are shown in Table 1.

The emerging 5G network is envisioned as a heterogeneous network, with macrocells, microcells, small cells and relays working cooperatively [36], as shown in Fig. 1. Different networks may have different features and satisfy different demands. For example, the macrocells provide the largest coverage, and offer high-rate data service to time-intensive applications, however with a relatively high price. On the other hand, small cells can provide local high-speed Internet access with low price, and can greatly enhance the spectrum efficiency. As shown in Fig. 1, in 5G heterogeneous networks, WiFi networks play an important role in delivering local mobile data in a cost-effective way. WiFi is recognized as one of the primary offloading technologies [1]. By delivering data originally targeted for cellular networks by WiFi, which is referred to as WiFi offloading, the congestion of cellular networks can be alleviated. WiFi offloading has been extensively studied for stationary or slow moving users [1, 6, 28, 35]. It is shown that around 65 % of the cellular traffic can be offloaded by merely using the most straightforward way of simply switching the IP connection from the cellular network to WiFi when the WiFi connectivity is available. In addition, significant amount (above 80 %) of data can be offloaded by delaying the data application [28].

In spite of the high data rates supported, the next-generation WLANs face three challenges. First, the popularity of WLANs has led and will continue to lead to very high WLAN deployment, which is a primary issue to address. Second, the emerging data-craving applications, such as real-time high-quality video streaming will significantly increase the users' throughput requirements. Third, the outdoor WLANs are getting more popular, which has a more complex communication environment. For the first challenge, legacy WLANs will find its difficulty to well behave in the scenario of very dense deployment of access points (APs). Based on the carrier sense multiple access with collision avoidance (CSMA/CA) mechanism WLANs employ, in the dense scenario, the contention problem will be severe and the channel utilization is poor. In addition, in the dense scenario, WLANs are not noise-limited system any more, and the signal from nearby co-channel basic

Table 1 Popular IEEE 802.11 amendments

Amendment	Frequency (GHz)	Bandwidth (MHz)	Maximum data rate (Mbps)	Responsibility
a	5	20	54	Enable up to 54 Mbps data transmission in 5 GHz unlicensed band by utilizing orthogonal frequency division multiplexing (OFDM)
b	2.4	22	11	Enable up to 11 Mbps data transmission in 2.4 GHz unlicensed band by utilizing Direct-sequence spread spectrum (DSSS)
g	2.4	20	54	Enable up to 54 Mbps data transmission in 2.4 GHz unlicensed band by utilizing OFDM
n	2.4/5	20/40	150	Utilize MIMO to support higher data rates
ac	2.4/5	20/40/80/160	866.7	Support MU-MIMO to further enhance data rate

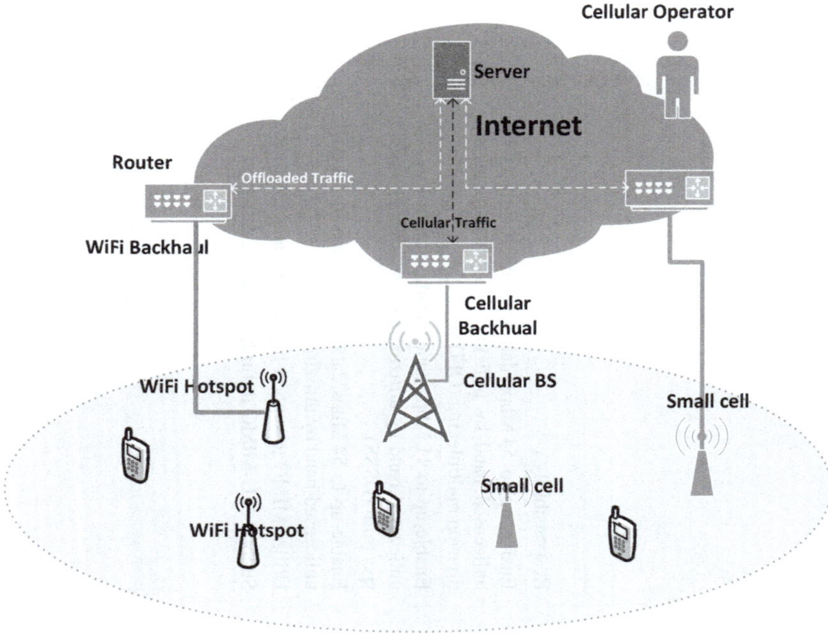

Fig. 1 Heterogeneous 5G networks with small cells and WLANs

service sets (BSSs) may significantly interfere with the received signal, leading to low signal-to-interference-plus-noise (SINR), and thus degrade the performance. The overlapping BSS (OBSS) problem should be well addressed in next-generation WLAN. For the second challenge, since the current WiFi amendments are aimed at improving the network throughput performance, advanced mechanisms should be investigated to improve the single user's throughput. For the third challenge, the channel condition in outdoor environments is more unreliable, since the delay spread can be large due to the multi-path transmission. In addition, the non-LOS channel makes it difficult to maintain links, even with a good SINR. Due to the possible long transmission range, the power of user devices can be a limiting factor. Also, the interference can be severe, caused by unmanaged networks, and home gateways signal leaking outdoors.

To address these challenges, the High-Efficiency WLAN (HEW) Study Group [13] is working on a new 802.11 amendment called IEEE 802.11ax-2019. A typical dense deployment of HEW is depicted in Fig. 2, where there is a large number of access points (APs) and a large number of stations (STAs) associated with each AP. The scope of 802.11ax is stated in [15]:

> This amendment defines standardized modifications to both the IEEE 802.11 physical layers (PHY) and the IEEE 802.11 Medium Access Control layer (MAC) that enable at least one mode of operation capable of supporting at least four times improvement in the average throughput per station (measured at the MAC data service access point) in a dense

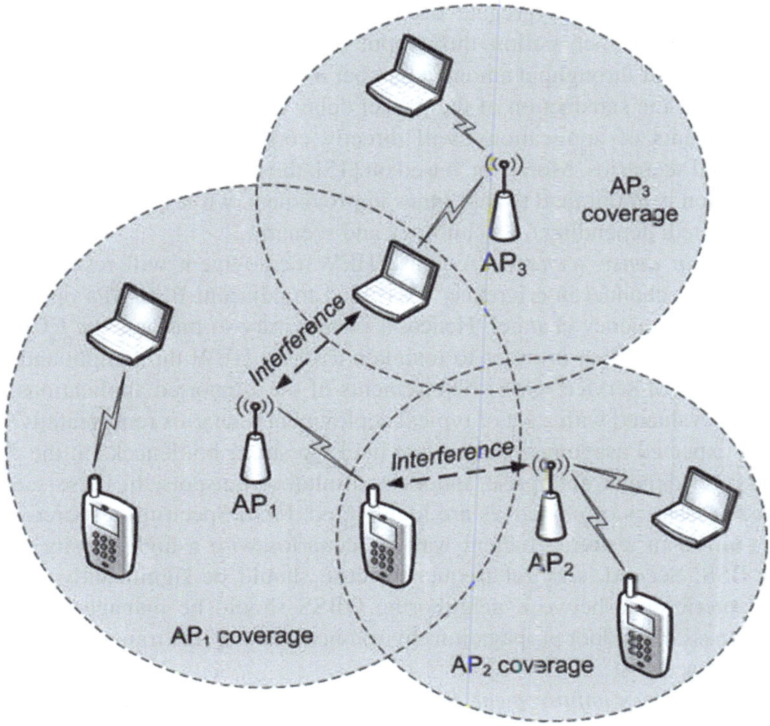

Fig. 2 A dense HEW deployment

> deployment scenario, while maintaining or improving the power efficiency per station. This amendment defines operations in frequency bands between 1 and 6 GHz. The new amendment shall enable backward compatibility and coexistence with legacy IEEE 802.11 devices operating in the same band.

In this chapter, we mainly present the features of the IEEE 802.11ax-2019 HEW, and discuss the potential techniques.

1.2 Requirements and Performance Metrics of HEW

From the above statement, it can be seen that the new IEEE 802.11ax HEW, as one of the next-generation WiFi technologies, has the following requirements:

- *Throughput enhancement per station.* Due to the varied throughput requirements of individual users and emerging applications, the throughput performance should be enhanced from the perspective of not only the network, but also individual users. Average throughput per station is directly proportional to both area throughput and aggregate BSS throughput. Therefore, the 5th percentile measure of the per station throughput can be used to evaluate the desired distribution of throughput among a number of stations in an area. These metrics, along with the satisfaction of the packet delay and the packet error ratio (PER) requirements of applications, will directly correspond to user experience in identified scenarios. Moreover, based on [15], the increase in average throughput per station is not limited to four times improvement, while a range of 5–10 times are targeted, depending on technology and scenario.

- *Efficient in dense scenario*. A dense HEW deployment will result in a high level of co-channel interference (CCI), due to adjacent BSS APs operating on the same frequency channel. Hence, it is necessary to mitigate the CCI among neighbouring BSSs, in order to maintain a stable HEW throughput and satisfy the quality of service (QoS) requirements of the supported applications. HEW will be evaluated with a set of typical deployment scenarios representative of the main expected usage models that are likely to suffer bottlenecks in the coming years: residential, enterprise, indoor and outdoor hotspots. In these scenarios, three categories of objectives are highlighted. First, Spectrum resources should be utilized in a more efficient way in scenarios with a high density of STAs per BSS. Second, spectral frequency reuse should be significantly enhanced, and interference between neighboring OBSS should be managed. Third, the robustness in outdoor propagation environments and uplink transmissions should be increased.
- *Power efficiency improvement*. Battery is one bottleneck of the mobile devices. Applications and WLAN infrastructure assume that the devices are constantly connected, which results in a large battery drain. In addition, current WLAN devices spend 90 % of time listening, which also consumes much energy. Several works in the literature have proposed mechanisms to enhance the STA power efficiency by using sleeping mechanisms, however, none has considered a dense deployment of WLANs in an outdoor environment. Power efficiency is intended to measure consumption of devices which can reasonably be assumed to be powered by batteries and will take into account average power consumption for a given scenario.
- *Backward compatibility*. The HEW should be able to provide backward compatibility, such that legacy devices can work in HEW, and in turn HEW devices can work well in legacy networks.

To evaluate the performance of HEW, performance metrics should be properly defined. In [15], it is stated that HEW aims to improve the metrics that reflect user experience, including per station throughput, the 5th percentile of per station throughput, and area throughput, in the scenarios of indoor/outdoor dense deployment. The detailed performance metrics of HEW are suggested and discussed in [12]. Among them, several important metrics are listed as follows.

- *Self-interference suppression*. This is a measurement of the ability of a device to receive signal when it is transmitting signal on the same frequency channel. Self-interference suppression is critical for full-duplex transmission, which is one of the key technologies of HEW.
- *MAC efficiency*. Defined by MAC throughput divided by PHY throughput, MAC efficiency is a proper metric to evaluate the performance of HEW MAC protocols.
- *Spectrum Efficiency*. Spectrum Efficiency is measured in bps/Hz, which should be maximized to enhance the network performance.
- *Throughput*. The 5 percentile and 95 percentile user throughput should be measured to evaluate the throughput at the center and edge of the cell. In dense deployment scenario, improving the throughput performance is important yet challenging, due to the presence of severe collision and interference.
- *Robustness*. Robustness evaluates the reliability and strategy of wireless connections in different scenarios. Association strategies include always connected, connected with minimum rate, and blocked when the admitted data rate is not met
- *Availability (or outage)*. Given the link received signal is good, the communication link can still be unavailable due to high congestion, overloading, or severe interference.
- *User experience*. User experience may include network selection and association latency, association successful ratio, data transmission latency, and handoff latency and successful ratio.

In the following of this chapter, we first study the HEW use cases, and then discuss several key enabling technologies, including enhanced 802.11 distributed coordination function (DCF) and spectrum spatial reuse. At last, the chapter is concluded.

2 HEW Use Cases

In this section, we list and discuss the usage models of IEEE 802.11ax HEW. Different from legacy WLAN, most of the use cases for HEW are characterized by a high density of STAs and/or BSSs. Users need WiFi access anytime and everywhere, which pushes its deployment/usage characterized by a high density of STAs and BSSs, such as hotspots in airport/train stations, malls, stadium, parks, streets, and campus. In addition, the increased usage of WiFi also leads to a higher density in “traditional” WiFi environment. Therefore, clearly defining the usage models can understand the specific requirements of each individual usage model, which can in turn facilitate the investigation of HEW key technologies.

2.1 Environment

Environment is the type of place in which the network of the use case is deployed. In HEW, seven environments are defined, which are enterprise, small office, hotspot in public places, outdoor hotspots, home, campus, and airplane/bus/train/ship, respectively.

- *Enterprise*. The enterprise environment can be on desk and in cubicle characterized by short range and line of sight (LOS) transmission, and conference room characterized by medium range and mostly line of sight transmission.
- *Small office*. The small office environment is characterized by single BSS with limited number of users and unmanageable interferences.
- *Hotspot in public places*. WLAN will be widely deployed in public areas, such as transportation hubs (airport, train station, and bus station), exhibition hall, and shopping malls. This environment is characterized by many STAs in a BSS.
- *Outdoor hotspots*. The typical examples include park, streets, stadium, and areas with special crowded events. The outdoor WLANs can be co-located with cellular base stations such as small cells, and user equipments such as private mobile APs.
- *Home*. The typical examples include dense apartment buildings and town houses, where WLANs can interfere each other.
- *Campus (Educational space, Hospital)*. The typical examples include auditorium/lecture halls/classrooms in the educational space, video conferencing/telepresence, and hospitals where remote medical assistance for operations is via wireless networks.
- *Airplane/bus/train/ship* WLANs can provide Internet access in large vehicles, such as in airplane and train. This environment is characterized by many STAs in a BSS.

Most environments are characterized by the overlap of multiple WLANs in the dense area that need to cohabit efficiently and fairly. There are different types of network overlap: (1) one or multiple cluster of APs (extended service set (ESS)), where each ESS is managed by a controller; (2) one or multiple stand-alone APs, each with its own management entity; (3) one or multiple single-link networks for P2P communications; and (4) the overlapping networks can be legacy networks (11a/b/g/n/ac).

2.2 Applications

In HEW, application is defined as a source and/or sink of wireless data that relates to a particular type of user activity. Several new/enhanced applications are considered, which include wireless docking, unified communications, display sharing, cloud computing, video distribution at home, progressive streaming, user generated content (UGC) upload and sharing, interactive multimedia and gaming, real-time video analytics and augmented reality, wearable devices, geo-location services, etc.

2.3 *Usage Models*

In the current discussion of HEW, five main usage models are defined, each of which include several different environments. A list of usage models, along with the environments, are given in Table 2, as listed in [16].

Comparing to the use cases defined in previous IEEE 802.11 amendments, such as 802.11ac, there are substantial differences with respect to the environments, applications, objectives, and metrics. In terms of environments, hotspots for cellular offload and public access in dense areas for indoor and outdoor are unique, and for general home and enterprise scenarios, it is more challenging towards high density deployment. For applications, HEW usage models consider a mix of traffic and clients, and focus on the quality of experience (QoE) of all clients in these more dense scenarios. For objectives, unlike previous IEEE 802.11 amendments which aim to improve the peak throughput, HEW targets to improve the WLAN efficiency by improving user experience and overall capacity per area. For metrics, the candidate metric that are under discussion include average per user throughput, 5th percentile of per user throughput cumulative density function (CDF), and area throughput. In the following, each use case is introduced.

2.3.1 Use Case 1

Use case 1 focuses on the scenarios with both high density of APs and high density of STAs per AP. This is the most important and challenging use case of HEW, where the high density of APs can cause severe interference, and the high density of STAs may congest the network, leading to packet collisions and retransmission. The typical environments in use case 1 are shown in Fig. 3.

(1a) Stadium. The stadium environment is an open area with few obstacles and multiple APs deployed by single/multiple operators. Most of the transmissions are LOS and the layout of APs are frequently changed. Interference can happen between APs belonging to the same managed ESS, and between unmanaged APs due to very high density deployment. In addition, interference may also caused by cellular networks and bluetooth devices. The traffic is characterized by bursty in time and is uneven due to users' participation in physical space. Example applications may include many users requesting high-rate best effort recreational content, receiving VHD video feed highly compressed, and following ESPN event or Twitter for supplemental event content.

Table 2 Usage models classified by requirement characteristics

Use case no.	Use case	Env. no.	Environment
1	High density of APs and high density of STAs per AP	a	Stadium
		b	Airport/train stations
		c	Exhibition hall
		d	Shopping malls
		e	E-Education
		f	Multi-media Mesh backhaul
2	High density of STAs—indoor	a	Dense wireless office
		b	Public transportation
		c	Lecture hall
		d	Manufacturing floor automation
3	High density of APs (low/medium density of STAs per AP)—indoor	a	Dense apartment building
		b	Community WiFi
4	High density of APs and high density of STAs per AP—outdoor	a	Super dense urban street
		b	Pico-cell street deployment
		c	Macro-cell street deployment
5	Throughput-demanding applications	a	Surgery/health care
		b	Production in stadium
		c	Smart car

Fig. 3 Typical environments of use case 1. (**a**) Stadium. (**b**) Airport (**c**) Exhibition hall. (**d**) Shopping mall

(1b) Airports and train stations. In airports and train stations, travelers are using the networks deployed by many service providers to check emails, watch movies, play online games, and access cloud services. The environment can be very complex and WLAN data transmissions can suffer severe interference, which may come from APs belonging to the same managed ESS, APs belonging to different managed ESS, and unmanaged networks.

(1c) Exhibition hall. Exhibition halls are an typical example of high density use cases. There may be many WLANs deployed by the exhibitors and event coordinator, through which people access the digital contents like demonstration and promotion videos of exhibitors. People also require Internet access to visit exhibitors' web sites, to send video and pictures to the companies, etc. In the indoor open spaces, there will be many APs and STAs and may suffer severe interference from APs belonging to the same managed ESS, stand-alone APs in the exhibition booths, and unmanaged networks.

(1d) Shopping mall. In shopping malls, shoppers can use the WLANs to localize themselves in the mall and receive coupons (potentially along with augmented reality). For classical public access, users can also use the network to visit websites, watch videos from YouTube, play online games, and access cloud services.

2.3.2 Use Case 2

Use case 2 is characterized by high density of STAs in indoor environments. In this category, the density of APs may not be extremely high, but the density of STAs per AP is very high, resulting in the congested networks, and degraded performance. The typical environments in use case 2 are shown in Fig. 4.

(2a) Dense wireless office. In a dense office, devices are operating in close proximity, with the distance between two devices smaller than 50 meters. There are multiple APs per floor, and 20–30 STAs per AP. The computers of an office can wirelessly associate to an external server and display with user configuration. This requires a reliable WLAN transmission so that users do not feel a delay indicating the wireless communication. In addition, in

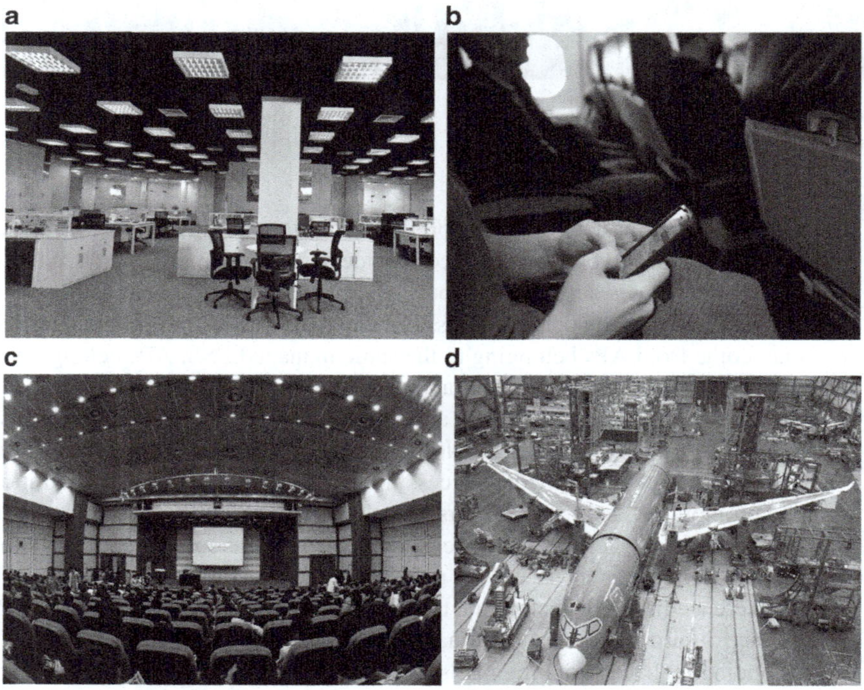

Fig. 4 Typical environments of use case 2. (**a**) Dense wireless office. (**b**) Public transportation. (**c**) Lecture hall. (**d**) Manufacturing floor automation

office users may have their own fixed or mobile devices, which also requires Internet access with high data rates. The interference in this environment can be caused by peer-to-peer networks, between APs belonging to the same managed ESS, and between APs belonging to different managed ESS due to the presence of multiple enterprise in the same building.

(2b) Public transportation. In the public transportation vehicles, such as trains, buses, and planes, crowded users may have operational WLAN network for Internet access, especially during the peak hours, such as commuting time. The WLAN can provide contents such as video, audio, and interactive gaming through onboard entertainment system, and can also offer high-rate Internet access for in-vehicle users. Interference in this scenario can be caused by unmanaged networks, as well as by surrounding WLANs for public transportation systems.

(2c) Campus network—lecture hall. In a lecture hall, a professor can remotely deliver a lecture to hundreds of students. The lecture comprises of both real-time video of the professor and supplemental content. The professor can receive real-time video of the lecture hall, and can zoom the video to view the audience, while the students can use wireless devices to watch the video and access the supplemental content simultaneously.

(2d) Manufacturing floor automation. On a manufacturing floor, there are thousands of tasks happening each minute, and many of the tasks requirement communications. In such as scenario, a mix of applications may have very different QoS requirements. For example, streaming of live or CAD video requires high throughput, very low delay, and high reliability. Machine-machine communications and robotic material handling requires high reliability but less time sensitivity and lower data rate.

2.4 Key Technologies

HEW considers MAC and PHY technologies that can significantly improve WLAN efficiency and robustness. Specifically, it should make more efficient use of spectrum resources in dense deployment scenarios, increase robustness in outdoor propagation environments, and improve the power efficiency. To achieve these goals, many new and efficient technologies should be employed to work together. We list the potential technologies in Table 3, and discuss two of them, i.e., enhanced DCF and spatial spectrum reuse, in detail in the rest of the chapter.

3 Enhanced 802.11 DCF

The distributed coordination function (DCF) is the fundamental access method of CSMA/CA technique, which has not been enhanced since the release of the legacy IEEE 802.11 standard. However, CSMA/CA techniques have poor performance

Table 3 Potential technologies for HEW

Category	Technology
Edge throughput enhancement	HARQ
	Larger contention period
MAC enhancement	Basic access mechanism enhancements
	Dynamic sensitivity control
	Traffic prioritization, QoE
	Multicast transmissions
MIMO/Beamforming	Massive MIMO, MIMO precoding
	DL/UL MU-MIMO
	Beamforming for OBSS
	Beamforming for interference handling
Multiplexing schemes	OFDMA, SDMA, OFDM-IDMA, FFR
	TD-uCSMA
	Channel bonding
Overlapping BSS handling	Interference management, antenna pattern nulling
	Efficient resource utilization
	Control frame transmission reduction
Full-duplex (FD)	MAC/PHY mechanisms for enabling In-band FD
	Enhancements for enabling out-band FD

in very dense scenario due to collision and interference. IEEE 802.11ax HEW aims to enhance or change the underlying CSMA/CA scheme to minimize the collisions. There are two ways to do this: changing to a centralized MAC scheme, or enhancing the CSMA/CA based protocol. Centralized solutions, such as the hybrid coordination function controlled channel access (HCCA), have never been employed in WLAN. Therefore, to maintain backward compatibility, HEW sticks to the enhancement of CSMA/CA protocols. Potential development of the DCF operation is required in order to improve its performance in a highly dense scenario, i.e., when a BSS consists of a large number of STAs. To achieve this goal, two techniques can be employed. First, **full duplex technology** allows the simultaneous signal transmission and reception, which can increasing the number of simultaneous transmissions within a BSS and reduce transmission collision probability. Second, the **enhanced access category (AC)** can decrease the channel time for control information transmissions in order to improve the channel utilization.

3.1 Full Duplex Technology

In IEEE 802.11 DCF, it is assumed that each STA is equipped with a half-duplex physical layer, i.e., either transmits or receives signal at one time. However, recently, the development of a full duplex technology has significantly advanced, which

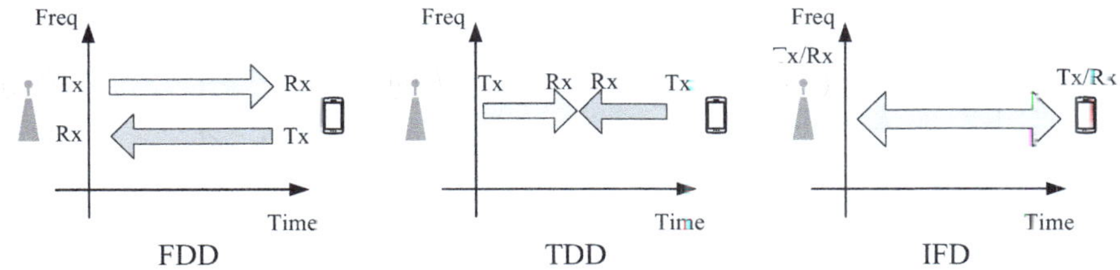

Fig. 5 In-band full-duplex

allows a transceiver to simultaneously transmit and receive signals on the same frequency [2, 5, 24]. The concept of full duplex technique is shown in Fig. 5. With frequency division duplexing (FDD) or out-band full-duplxe (OFD), a device can transmit and receive at the same time, however, on different frequency bands, which wastes frequency resource. With time division duplexing (TDD) or in-band half-duplex (IHD), a device can transmit and receive on the same frequency band, but at different times, which wastes time resource. Differently, in-band full-duplex (IFD) radio can transmit and receive simultaneously on the same frequency channel, and thus does not waste frequency and time resources. The merit of IFD is that it can achieve high spectral efficiency. Given a link between two nodes a and b, we can calculate the theoretical ergodic capacities of ideal IFD as

$$C_{IFD} = C_{ab} + C_{ba}, \tag{1}$$

where $C_{xy} = \log_2(1 + SINR_{xy})$, while the theoretical ergodic capacities of OFD and IHD can be calculated by

$$C_{IFD} = \frac{1}{2}C_{ab} + \frac{1}{2}C_{ba}. \tag{2}$$

Therefore, theoretically, IFD can achieve twice spectra efficiency than OFD and IHD. However, there are two main challenges of implementing, self-interference (SI) and CCI. Figure 6 shows how SI is generated and influence the IFD operation. Since the average transmit power of WLAN is 20 dBm, and the noise floor is around −90 dBm, the SI should be canceled by 110 dB to a negligible level [2]. CCI may happen when an IFD capable AP supports the IHD capable users, where user A can transmit to the AP while AP is transmitting to user B. Thus, the transmission of user A causes interference at user B. Self-interference cancellation technologies are proposed to reduce the SI, and achieve more IFD gain. There are mainly three categories of SIC technologies, which are propagation SIC (PSIC) in antenna domain, analog SIC (ASIC) in circuit domain, and digital SIC (DSIC) in baseband domain. However, for PSIC, the merit of IFD in terms of spectral efficiency disappears since PSIC mainly employs different antennas for Tx and Rx, and cancel SI by physical isolation between Tx and Rx antennas. In addition, the

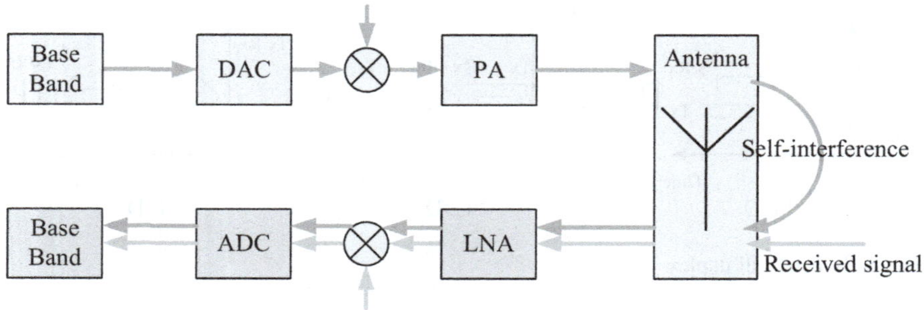

Fig. 6 Self-interference

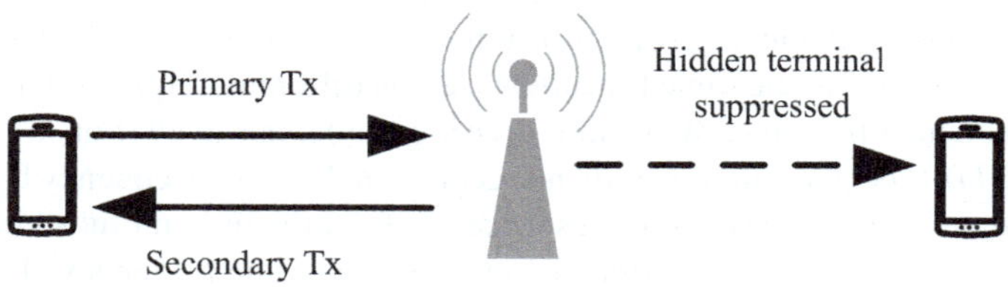

Fig. 7 IFD solves hidden terminal problem

size of IFD transceiver gets larger. Therefore, SIC mainly rely on ASIC and DSIC, which operates with single antenna. With ASIC, The comparison of state-of-the-art SIC performance is shown in Table 4, which is summarized in [17].

Besides up to 2 times spectral efficiency, IFD can also facilitate advanced MAC protocols to solve many problems using half-duplex transceiver. It can reduce the DCF overhead by eliminating request-to-send (RTS)/clear-to-send (CTS) handshaking information exchange which precedes the actual data transmission [11]. This is because the receiver can immediately start to transmit, which suppresses nearby nodes, as shown in Fig. 7. In addition, when a WLAN is saturated, the bottleneck of the network is the AP, since it has the same chance to transmit as each of its serving clients. With IFD, the problem can be solved since the AP can transmit when STAs are sending packets to it. To date, a few studies have considered the development of MAC schemes based on a full duplex physical transceiver. Some of these studies propose techniques that are not based on the CSMA/CA mechanism [27]. In [27], AP behaves as a central node, scheduling the transmissions of all STAs under control. The AP first sends a probe packet, and all STAs registered under the AP will send the information of transmission length and current interference level to the AP. Based on the information, a schedule called Janus determines the packets to be transmitted concurrently, and the mode (half- or full-duplex) and data rate using a heuristic algorithm to achieve close to the best throughput performance. Simulation results show that Janus MAC can achieve nearly twice throughput comparing to the legacy MAC schemes. However, it may not be practical since centralized MAC is not favored by WLAN implementations.

Table 4 Comparison of state-of-the-art SIC performance

Institute	Year (Standard)	Freq. (MHz)	BW (MHz)	PSIC (dB)	ASIC (dB)	DSIC (dB)	Total SIC (dB)
NEC (Japan)	2011	5	10	55	None	20	75
Rice University (US)	2011	2.4	10	57	24	None	81
	2012	2.4	20	65	20	85	
	2012	2.4	20	71	24	95	
Stanford University (US)	2010	2.48	5	30	25	15	70
	2011	2.4	10		45	28	73
	2013	2.45	80	None	60	50	110
DUPLO	2014	2.45	6	None	50	50	110
RF Window (Korea)	LTE WCDMA	2	20	60	None	10	70
WITHUS (Korea)	LTE WCDMA	2	10	35	None	35	70
AirPoint (Korea)	LTE TDD	2.2		55	None	35	90
SOLiD (Korea)		2		65	None	35	100

On the other hand, other research has developed full-duplex MAC schemes based on CSMA/CA [24, 33]. There is one case the hidden terminal cannot be addressed by IFD, where the primary and secondary packets may have different lengths. For example, in Fig. 7, if the secondary transmission is shorter than the primary transmission, the hidden terminal is no more suppressed. In [24], a busy tone mechanism is employed to address this problem. When a nodes finish transmitting before finish receiving, it will transmit a predefined busy tone, until it finishes receiving; when a node receives a packet but has no packet to transmit, it will also transmit the busy tone. Through the hardware implementation of a simple case where two full duplex nodes transmit packet to each other, it is shown that the proposed full-duplex MAC scheme can achieve better performance in packet reception ratio, throughput and fairness than half-duplex MAC schemes. In [33], a full-duplex physical layer (FD-PHY) is proposed and implemented, based on which a full-duplex MAC (FD-MAC) scheme is developed and implemented for infrastructure-based WLANs. To maximize full-duplex capability and guarantee fairness, queue information needs to be shared with neighboring nodes, and all nodes should be allowed to access the medium. To do this, three mechanisms are employed in FD-MAC. First mechanism is called shared random backoff, the backoff counters of two communicating nodes are synchronized so that they can transmit at the same time while other nodes are allowed to access channel during the backoff of the communicating nodes. The second mechanism is snooping on the header of all transmission within the radio range, even when the node is not transmitting during the network allocation vector (NAV). This enables the nodes to estimate their local topology, and discover the clique or hidden node with themselves. Lastly, the FD-MAC employs virtual contention resolution mechanisms to further balance the objectives between maximize full-duplex opportunities and fairness.

Current research on IFD-based MAC is still in its infant stage. The target MAC scheme should be able to maximize the benefit of full-duplex capabilities (i.e., the opportunity of simultaneous transmission and receiving by a STA) in infrastructure and independent BSSs, support half duplex operation (for backward compatibility), and maintain fairness among all half duplex and full duplex STAs. In addition, the employment of such technology will also facilitate developing more efficient transmit power control and interference mitigation schemes.

3.2 Enhanced Access Category

The aim of HEWs is to satisfy the QoS requirements of various applications, in both dense indoor and dense outdoor environments, while reducing the power consumption. In HEWs, due to the DCF scheme, a STA's average throughput decreases with the increasing number of STAs that are being served by the AP, and the worse channel condition: the former is due to increase of transmission collisions, and the latter is due to that the STA with weak channel take a long time for its data transmission [3, 29]. To satisfy STAs' QoS requirements, the pervious WLAN standards provide a basic AC scheme and an enhanced DCF scheme called Enhanced distributed channel access (EDCA) [9]. With the AC scheme, a STA is accepted or rejected based on medium time which is calculated from the minimum physical layer rate of the STA [8, 11, 26]. However, Such AC scheme is inefficient because calculating the medium time results in an underutilized channel. In addition, AC-based EDCA cannot guarantee the strict QoS requirements of real-time voice and video services [4]. Therefore, other AC schemes (e.g., [8, 11, 26, 30]) and further improvements for EDCA (e.g., [4]) have been proposed. However, none of these schemes has been designed for a dense WLAN deployment in an outdoor environment. In such deployment scenario, multiple APs coexist, wireless channels rapidly varys, and all the WLAN frequency channels are utilized by adjacent APs. Therefore, to achieve robust and optimal throughput and QoS performance, advanced AC and dynamic handover schemes based on these characteristics should be developed.

4 Spatial Frequency Reuse

OBSS is one of the main problems that HEW aims to solve, since high density deployment of WiFi networks is one targeting scenario of HEW. Intuitively, improving spatial reuse can efficiently mitigate OBSS. Spatial reuse has been extensively studied in the literature. In this section, we present a literature survey of two kinds of spatial reuse improvement methods, namely enhanced clear channel assessment (CCA) and transmit power control (TPC).

4.1 Enhanced Clear Channel Assessment

CCA is a function for WiFi to determine the current usage state of the wireless medium, which is defined in IEEE 802.11-2007 standards [10]. There are two functions in CCA, namely carrier sense (CCA-CS) and energy detection (CCA-ED). The former is used for a receiver to detect and decode WiFi preamble, while the latter is used to detect non-WiFi energy. CCA functions in the way that if a WiFi signal is detected above the CCA-CS level (or CCA-CS threshold), or a non-WiFi signal is detected above the CCA-ED level (or CCA-ED threshold) at a STA, it indicates that the wireless medium is currently captured by other STAs or non-WiFi devices, and thus the STA has to stay off the air in order to avoid collisions. In [10], the CCA-CS and CCA-ED levels are defined, which are summarized in Table 5.

In IEEE 802.11-2007 standards, the CCA levels are defined relatively low values, which can guarantee a large coverage of the WLAN, minimize the interference (signals received below the CCA level is considered as noise), and mitigate the hidden terminal problem. This can work well in low density WiFi with proper AP

Table 5 CCA levels defined in [10]

Bandwidth (MHz)	CCA-CS level (dBm)	CCA-ED level (dBm)
5	−88	−78
10	−85	−75
20	−82	−72

channel selection algorithms. However, in HEW, BSSs are deployed densely, and OBSSs are very common. Therefore, in HEW, with a low CCA level, an ongoing transmission can prevent many STAs of nearby BSSs from transmitting, which in turn degrades the overall performance of HEW. In order to improve the performance of HEW (mean and 5 % throughput), an intuitive solution is to allow concurrent transmissions with high data rates. To this end, enhanced CCA mechanisms are required to improve the spatial reuse of HEW.

4.1.1 Fixed CCA Enhancements

A straightforward solution is to increase the CCA threshold (CCAT) to a certain value such that HEW STAs can transmit in spite of the presence of signals from other OBSS STAs. The advantages of this kind of CCA schemes are easy-implementing, and in some cases, they can achieve high efficiency. In [14], the throughput performance of several CCATs in the scenario of indoor small BSSs are evaluated through simulations. It is shown that the average throughput increases with the CCAT, while the 5 % point in the throughput decreases with the CCAT. This is because with a higher CCAT, more concurrent transmissions in the same area are allowed, but the cell edge users suffer low SINR due to low signal from the AP and high interference from OBSS STAs. Theoretically, with a higher CCAT, the interference will become severer, and more collisions may happen since more concurrent transmissions are permitted. Therefore, the joint impact of concurrent transmissions, interference, and collision on the HEW performance should be investigated. In [25], the performance of varied CCATs is studied through a simulation of a dense cell-like topology of co-channel BSSs. It is demonstrated that the positive impact of spatial reuse dominates over the negative impacts of collision and interference in the dense scenario due to the short distances between APs and associated STAs, and the throughput performance is increased by 190 % when choosing optimal CCATs. In [38], the throughput performance of OBSSs with and without CCA adaption is evaluated. The simulation results show that with CCA adaption, the throughput of HEW STAs and legacy STAs can be increased to 59.6 and 12 Mbps from 14 Mbps and 2 kbps without CCA adaption, respectively. The fairness issue between HEW STAs and legacy STAs (no CCA adaption used) is one critical issue in CCA adaption, which will be discussed later. A simulation study of CCA performance is presented in [21], where the impact of different CCATs in HEW simulation scenario 1 (residential), 2 (wireless office), and 3 (indoor small

BSS) is investigated. It is shown that when the CCAT varies from −82 to −30 dBm, both mean and 5 % throughput increase with the CCAT in scenario 2, while in scenario 1 and 3, there are specific values of CCAT to achieve maximum throughput, and the optimal value for mean and 5 % throughput may be different. This is in accordance with the observations found in [22].

4.1.2 Dynamic CCA Enhancements

Fix CCA enhancement methods can be efficient in the scenarios where the network is deployed regularly, and the mobility of STAs is low. However, with a more complicated scenario where a huge number of BSSs are deployed densely and irregularly, and the STAs may move fast and frequently, fix and global CCA enhancement methods may not be suitable to achieve satisfying performance. Therefore, dynamic CCA schemes are required to adapt to the complicated scenarios and the STA mobility. Since the dynamic CCA is to adjust the radio sensitivity, they are usually referred to as dynamic sensitivity control (DSC).

In [32], a centralized joint CCA and channel bandwidth control scheme is proposed to adaptively control the CCAT of APs, and optimize the channel bandwidth allocated to STAs, in order to maximize the transmission opportunity of APs, and guarantee the required SINR of STAs. Simulation results show that the proposed joint CCA and channel bandwidth control scheme can achieve approximate twice in average throughput comparing to only CCA adaption is employed. In [31], a centralized CCA adaption scheme is proposed, in which CCATs of APs are set according to the current positions of STAs. A central controller uses a camera to obtain the positions of STAs, and CCATs for APs are then calculated based on the obtained positions in order to achieve a high SINR for STAs.

In [23], a simple distributed DSC scheme is proposed, where the CCATs of STAs are dynamically adapted to the received signal from the AP. STAs measure the received signal strength indication (RSSI) of the AP beacon, and adapts its CCAT S by

$$S = \min(R, L) - M, \tag{3}$$

where R denotes the RSSI of the AP beacon, L sets the upper limit of R, and M denotes a margin value. The upper limit L of AP beacon RSSI is to prevent STAs close to AP having too high CCAT which may result in severe hidden terminal problem. The value of margin M should be large enough to guarantee a large enough SNR and to deal with sudden changes in reception of beacon signals. With the proposed DSC scheme, the coverage of a BSS greatly reduces, and therefore the spatial reuse is improved. Through an analysis to a cell-like topology, it is shown that the network capacity can be improved by 7.58 times. In [7], a distributed physical carrier sensing tuning scheme is proposed for dense multiple access point architecture where a STA can associate with multiple APs to exploit the spatial diversity. To enhance the spatial reuse efficiency, the CCATs of STAs are adjusted based on the packet loss rate and busy ratio. Simulation results show that the proposed scheme can greatly increase the system throughput comparing with the basic DCF scheme. In [18], a distributed CCA adaption scheme is proposed for each PLCP protocol data unit (PPDU). The transmitting device announces the CCAT it permits other devices to use for each PPDU, and dynamically adjusts the value. The aim of the adjustment is that if a STA successfully transmits packets successively, the surrounding STAs can use a higher CCAT to improve the spatial reuse; if a transmission fails, the STA will reset the CCAT of surrounding STAs to a low level in order for the successful transmission of following packets.

4.1.3 Fairness Among CCA STAs and Legacy STAs

HEW should be backward compatible, which means HEW should able to coexist with legacy networks and STAs with a certain level of performance preservation of the legacy network. However, with CCA enhancement implemented, HEW devices may have great advantages over legacy devices, which causes severe fairness issues. Consider a room where CCA-enhanced HEW and legacy WiFi network coexist densely, and HEW STAs have higher CCATs than legacy STAs. Due to the CSMA scheme, HEW STAs may have much higher transmission opportunity than legacy STAs since they can transmit when legacy STAs are transmitting, but legacy STAs cannot transmit when HEW STAs are transmitting since the wireless medium is sensed to be busy. In [38], it shows that the throughput of legacy STAs is about only 2 kbps while HEW STAs have 14 Mbps. In [19], through a simulation in residential scenario, it is shown that by changing CCAT from −82 to −62 dBm, the throughput of HEW STAs increases by 36 %, but the throughput of legacy STAs reduces by 48 %.

In [20], the reasons for the fairness issues are explained, and solutions are proposed accordingly. The first reason is that if a CCA-enabled transmission happens during the TXOP of the legacy device, the successive legacy transmission will be delayed until the end of the TXOP of CCA-enabled devices. To address this problem, the TXOP of CCA-enabled devices can be reduced such that the two TXOPs end at the same time. Another fairness issue is that if the CCA-enabled devices transmit in the middle of legacy devices' TXOP, collisions at legacy STAs may be caused, resulting higher power consumption due to retransmissions and reduced throughput. This issue can be addressed by carefully adjusting the transmit power of HEW devices. In [14], a joint TPC and DSC scheme is proposed to enhance the fairness between HEW and legacy devices. The TPC schemes will be discussed in detail in the following part.

4.2 CCA with TPC

TPC schemes can also be used to enhance the spatial reuse in HEW. By properly adjusting the transmit power, the interference among nearby APs and STAs can be mitigated, collisions can be reduced, and more transmission opportunities will arise. In [37], a joint channel allocation and power control algorithm is proposed to adjust the transmit power of APs to minimize the coverage overlap among neighboring BSSs. First, channels are allocated to APs such that neighboring APs have large channel spacing. Then, the transmit power of APs is adapted to STAs' distance to the AP, in order to keep the SINR at STAs a constant. A minimum transmit power is set to keep the AP-STA link from breaking. Simulation results show that the proposed scheme can guarantee a 20 Mbps throughput even when the network density is high (200 APs in the simulated area), while with random channel allocation the throughput is only around 12 Mbps. In addition, interference can also be reduced by more than three times. In [34], a TPC scheme called distributed spatial reuse is devised for each STA to transmit at an exact power in order to minimize the interference and maximize the communication pair. A greedy maximum independent set algorithm is designed to select maximum number of non-interfering communication pair. Through evaluation, it is shown that the proposed scheme can improve the performance of both throughput and delay.

CCA schemes can reduce the coverage of BSSs, and increase concurrent transmission, but have the fairness issue with legacy networks. TPC schemes can decrease the interference among STAs, and control the SINR as required. However, TPC is not effective enough to reduce the coverage and increase the spatial reuse. Therefore, a combination of the two schemes is envisioned to increase the spatial reuse and guarantee the fairness to legacy networks at the same time. In [39], the relation between CCAT, transmit power, and data rate is theoretically analyzed, and based on the results, a distributed joint tuning scheme is proposed to enhance the spatial reuse and guarantee the fairness. The maximum achievable data rate is first estimated based on the RSSI of received AP frame. Then, the transmit power is reduced to save power and reduce interference, and CCAT is increased to improve spatial reuse accordingly. Simulation results show that the aggregate throughput increases from 27 to 55 Mbps by employing the proposed scheme. In [14], a joint CCA and TPC scheme is proposed in which the receiver requests the transmitter to transmit at power equal to CCAT plus margin. In this way, HEW STAs can have higher throughput according to higher CCAT, while legacy STAs have the RSSI of margin value, which guarantees the SINR (and data rate) to a certain level.

5 Conclusion

In summary, the next-generation WLAN can employ advanced PHY and MAC schemes to better utilize the spectrum resource, increase the network throughput, while enhance the per-user throughput to satisfy the varying QoS requirements. WLAN will continue to be one of the most popular access technologies, providing high data rate, low price, and more reliable services to mobile users.

References

1. A. Aijaz, H. Aghvami, M. Amani, A survey on mobile data offloading: technical and business perspectives. IEEE Wirel. Commun. **20**(2), 104–112 (2013)
2. D. Bharadia, E. McMilin, S. Katti, Full duplex radios. Proc. ACM SIGCOMM **43**(4), 375–386 (2013)
3. G. Bianchi, Performance analysis of the ieee 802.11 distributed coordination function. IEEE J. Sel. Areas Commun. **18**(3), 535–547 (2000)
4. X. Chen, H. Zhai, X. Tian, Y. Fang, Supporting QOS in IEEE 802.11 e wireless lans. IEEE Trans. Wirel. Commun. **5**(8), 2217–2227 (2006)
5. J.I. Choi, M. Jain, K. Srinivasan, P. Levis, S. Katti, Achieving single channel, full duplex wireless communication, in *Proceedings of ACM MobiCom* (2010)
6. S. Dimatteo, P. Hui, B. Han, V.O. Li, Cellular traffic offloading through WiFi networks, in *Proceedings of the IEEE MASS* (Spain, 2011)
7. Y. Hua, Q. Zhang, Z. Niu, Distributed physical carrier sensing adaptation scheme in cooperative map WLAN, in *Proceedings of IEEE GLOBECOM, Hawaii, USA* (2009)
8. M.L. Huang, S. Lee, S.-C. Park, An efficient admission control algorithm for ieee 802.11 e WLAN, in *Proceedings of IEEE VTC-Fall* (2008)
9. IEEE, Part 11: wireless LAN medium access control (MAC) and physical layer (PHY) specifications - amendment 8: medium access control (MAC) quality of service enhancements. Technical Report (2005)
10. IEEE, Especificacion 802.11-2007. Technical Report (2007)
11. IEEE, Part 11: wireless LAN medium access control (MAC) and physical layer (PHY) specifications. Technical Report (2012)
12. IEEE, 802.11-13/1054: HEW evaluation metrics suggestion, Technical Report (2013)
13. IEEE, IEEE 802.11 study group. Status of IEEE 802.11 HEW study group, Technical Report (2014) [Online]. Available: http://www.ieee802.org/11/Reports/hewupdate.htm
14. IEEE, 802.11-14/1207: OBSS reuse mechanism which preserves fairness. Technical Report (2014)
15. IEEE, 802.11-14/0165: 802.11 HEW SG proposed par, Technical Report (2014)
16. IEEE, 802.11-13/0657: usage models for IEEE 802.11 high efficiency WLAN study group (HEW SG) Liaison with WFA, Technical Report (2014)
17. IEEE, 802.11-15-0043-01-00ax: in-band full duplex radios and system performance, Technical Report (2014)
18. IEEE, 802.11-14/0872: a protocol framework for dynamic CCA. Technical Report (2014)
19. IEEE, 802.11-14/0372: HEW system level simulations on increased spatial reuse. Technical Report (2014)
20. IEEE, 802.11-14/0637: spatial reuse and coexistence with legacy devices. Technical Report (2014)
21. IEEE, 802.11-14/0889: performance gains from CCA optimisation. Technical Report (2014)
22. IEEE, 802.11-14/0846: CCA study in residential scenario. Technical Report (2014)
23. IEEE, 802.11-13/1012: dynamic sensitivity control. Technical Report (2014)
24. M. Jain, J.I. Choi, T. Kim, D. Bharadia, S. Seth, K. Srinivasan, P. Levis, S. Katti, P. Sinha, Practical, real-time, full duplex wireless, in *Proceedings of ACM MobiCom* (2011)
25. I. Jamil, L. Cariou, J.-F. Helard, Improving the capacity of future IEEE 802.11 high efficiency WLANs, in *Proceedings of IEEE ICT, Lisbon* (2014)

26. T. Kim, S. Lee, S.-C. Park, Call admission control based on adaptive physical rate for EDCA in IEEE 802.11 e WLAN system, in *Proceedings of IEEE CCNC* (2008), pp. 59–61
27. J.Y. Kim, O. Mashayekhi, H. Qu, M. Kazadiieva, P. Levis, Janus: a novel MAC protocol for full duplex radio. Stanford CS Tech reports in http://hci.stanford.edu/cstr/ CSTR **2**(7), 23 (2013)
28. K. Lee, J. Lee, Y. Yi, I. Rhee, S. Chong, Mobile data offloading: how much can WiFi deliver? IEEE/ACM Trans. Netw. **21**(2), 536–550 (2013)
29. P. Liu, Z. Tao, S. Narayanan, T. Korakis, S.S. Panwar, Coopmac: a cooperative MAC for wireless lans. IEEE J. Sel. Areas Commun. **25**(2), 340–354 (2007)
30. K. Medepalli, F. Tobagi, D. Famolari, T. Kodama et al., On optimization of csma/ca based wireless lans: Part ii-mitigating efficiency loss, in *Proceedings of IEEE ICC* (2006)
31. K. Murakami, T. Ito, S. Ishihara, Improving the spatial reuse of IEEE 802.11 WLAN by adaptive carrier sense threshold of access points based on node positions, in *Proceedings of IEEE ICMU, Hakodate City* (2015)
32. T. Nakahira, K. Ishihara, Y. Asai, Y. Takatori, R. Kudo, M. Mizoguchi Centralized control of carrier sense threshold and channel bandwidth in high-density WLANs, in *Proceedings of IEEE APMC, Sendai, Japan* (2014)
33. A. Sahai, G. Patel, A. Sabharwal, Pushing the limits of full-duplex: design and real-time implementation (2011). arXiv:1107.0607
34. K.-P. Shih, C.-M. Chou, M.-Y. Lu, S.-M. Chen, A distributed spatial reuse (DSR) MAC protocol for IEEE 802.11 ad-hoc wireless LANs, in *Proceedings of IEEE ISCC, Cartagena* (2005)
35. S. Singh, H. Dhillon, J. Andrews, Offloading in heterogeneous networks: modeling, analysis, and design insights. IEEE Trans. Wirel. Commun. **12**(5), 2484–2497 (2012)
36. C.-X. Wang, F. Haider, X. Gao, X.-H. You, Y. Yang, D. Yuan, H. Aggoune, H. Haas, S. Fletcher, E. Hepsaydir, Cellular architecture and key technologies for 5g wireless communication networks. IEEE Commun. Mag. **52**(2), 122–130 (2014)
37. Y. Wu, Y. Sun, Y. Ji, J. Mao, Y. Liu, A joint channel allocation and power control scheme for interference mitigation in high-density WLANs, in *Proceedings of IEEE ICCT, Guilin* (2013)
38. A.E. Xhafa, A. Batra, A. Zaks, On the coexistence of overlapping BSSs in WLANs, in *Proceedings of IEEE VTC, Baltimore* (2007)
39. Z. Zhou, Y. Zhu, Z. Niu, J. Zhu, Joint tuning of physical carrier sensing, power and rate in high-density WLAN, in *Proceedings of IEEE APCC, Bangkok* (2007)

Shaping 5G for the Tactile Internet

Adnan Aijaz, Meryem Simsek, Mischa Dohler, and Gerhard Fettweis

Abstract The Tactile Internet is envisioned to transport touch and actuation in real-time. It will enable unprecedented applications and revolutionize almost every segment of the society. It is expected that the next generation 5G mobile communications networks will enable the Tactile Internet at the wireless edge. The tactile internet creates daunting new requirements for 5G network design. This chapter focuses on the key technological concepts which lay at the intersection of 5G and the Tactile Internet. The chapter outlines the key application areas of the Tactile Internet. This is followed by an end-to-end architecture of the Tactile Internet. The chapter also presents the key technical requirements of the Tactile Internet along with some potential solutions to meet these requirements. Such solutions revolve around protocol-level and system-level innovations.

1 Introduction

The Internet is unarguably the most important invention of the twentieth century. What originally started as a network of a handful of nodes, went on to define the economies of the late twentieth century. The wireless revolution in general and the unprecedented development of mobile/cellular technologies in particular, has totally transformed the way we perceive the Internet today by creating what could be termed as the Mobile Internet. The focus of mobile communications is now moving towards providing ubiquitous connectivity for machines and devices which will enable the Internet-of-Things. In a natural evolution to these Internet embodiments, the notion of Tactile Internet [1, 2] is emerging, in which

A. Aijaz (✉)
Toshiba Research Europe Ltd., Cambridge, UK
e-mail: adnan.aijaz@toshiba-trel.com

M. Simsek • G. Fettweis
Technical University Dresden, Dresden, Germany
e-mail: meryem.simsek@tu-dresden.de; fettweis@ifn.et.tu-dresden.de

M. Dohler
King's College London, London, UK
e-mail: mischa.dohler@kcl.ac.uk

W. Xiang et al. (eds.), *5G Mobile Communications*,
DOI 10.1007/978-3-319-34208-5_25

ultra-reliable, ultra-responsible, and sufficiently intelligent network connectivity will enable the delivery of real-time control and physical haptic experience, remotely. Unlike the conventional Internet, which provides the medium for voice and data communications, the Tactile Internet will provide the ability to transport touch and actuation through the Internet in real-time. Currently, the conventional Internet and its different embodiments are widely used for delivery content services (voice telephony, text messaging, file sharing, etc.). The Tactile Internet will provide a true paradigm shift from content delivery to remote skill-set delivery, and thereby revolutionize almost every segment of the society. Recently, standardization efforts for the Tactile Internet have started to emerge within IEEE and ETSI. State-of-the-art fourth generation (4G) mobile communications networks do not largely fulfil the technical requirements of the Tactile Internet. Therefore, it is expected that the fifth generation (5G) mobile communications networks will underpin the Tactile Internet at the wireless edge.

5G networks are expected to be commercially deployed by 2020. An early assessment of 5G scenarios and requirements has been carried out in the flagship EU METIS project [17] and more recently by the telecommunications industry alliance NGMN [18]. With Tactile Internet, daunting new requirements for 5G arise. Meeting such requirements will require revolutionary approaches to 5G network design.

Research efforts for both 5G and the Tactile Internet are in infancy. The main focus of this chapter is thus the intersection of 5G and Tactile Internet. To this end, this chapter is organized as follows. In Sect. 2 we provide a brief overview of 5G networks. In Sect. 3 we outline some of the key application areas of the Tactile Internet. Section 4 covers the end-to-end architecture for the Tactile Internet. In Sect. 5 we outline the key technical requirements for the Tactile Internet from a networking perspective. Section 6 discusses some potential solutions for 5G networks in realizing the Tactile Internet. Finally, the chapter is concluded in Sect. 7.

2 5G: A Brief Overview

The fifth generation (5G) of mobile communication networks has attracted a lot of attention from industry, research centers, and academia. 5G is expected to enable a fully mobile and connected society and to empower socio-economic transformations and to fulfill the wireless communication requirements for 2020 and beyond. The demands of a fully mobile and connected society are characterized by the tremendous growth in connectivity and density/volume of traffic, the required multi-layer densification in enabling this, and the broad range of use cases and business models expected. Overall there is a common understanding that 5G should not only support an evolution of traditional mobile communication services, such as personal mobile multimedia communication or personal mobile broadband services; 5G should in addition address novel use cases and connect a massive number of devices to the Internet of Things. The addressed novel use cases for

5G include machine-type communication in fields like, e.g. smart grids, vehicular communication and intelligent transport systems. These use cases have stringent requirements which were not in the focus of former mobile network generations. Hence, 5G is expected to cope with massive and stringent requirements in terms of throughput, ultra-low latency, ultra-high reliability, higher connectivity density, security and privacy, and higher mobility range when and where needed [4]. Hence, 5G will operate in a highly heterogeneous environment with a mix of various requirements for different use cases. This leads to the need to rethink the overall system architecture for 5G to enable flexibility for optimized network usage. The 5G architecture should include modular network functions which can be deployed and scaled on demand. The broad range of targeted 5G capabilities, i.e. stringent requirements and flexible, highly reactive network architecture, will make 5G an important enabler for the Tactile Internet—when humans will wirelessly control real and virtual objects in real-time—but not before we overcome enormous system design challenges.

3 Key Tactile Internet Applications

The Tactile Internet opens whole new areas of technology to the Internet, based on very low end-to-end latency. It enables control of real and virtual objects and adds a new dimension to human-machine interaction in real-time through the conveyance of the perception of touch through communication. The Tactile Internet has been described as a communication infrastructure combining low latency, very short transit time, high availability and high reliability with a high level of security [9]. Associated with cloud computing proximity through e.g. mobile edge-clouds and combined with the virtual or augmented reality for sensory and haptic controls, the Tactile Internet addresses areas with reaction times in the order of a millisecond. This enables the Tactile Internet to encompass mission critical applications (e.g., manufacturing, transportation, and healthcare), as well as non-critical applications (e.g., edutainment and events).

3.1 Manufacturing

Around the world, traditional manufacturing industry is over to be revolutionized by a digital transformation that is accelerated by exponentially growing technologies, e.g. intelligent robots, sensors etc. The widespread adoption by manufacturing industry around the framework of information and communication technology is now paving the way for distributive approaches to development, production, and the entire logistics chain.

Industrial countries have already witnessed three industrial revolutions, which could also be described as disruptive leaps in industrial processes resulting in significantly higher productivity [10]. In the first industrial revolution (starting in the 1780s), mechanical production facilities with the help of water and steam power was introduced. The second industrial revolution started in the 1870s and enabled the introduction of division of labor and mass production with the help of electrical energy. The use of electronic and Information Technology (IT) systems further automated the production in the third industrial revolution together with the first programmable logic controller in 1969. The upcoming fourth industrial revolution (industry 4.0) is based on real-time enabled cyber-physical systems and comes with key changes in manufacturing, engineering, material usage and supply chain and life cycle management, i.e. leading to flexible and self-organized smart factories. With increased agility and flexibility in the production process, industry 4.0 will enable to individualize any product and allow to react faster on changing market requirements. Cyber-physical systems in industry 4.0 comprise smart machines, storage systems and productions facilities capable of autonomously exchanging (wirelessly) information, triggering actions and controlling each other independently.

Industry 4.0 will naturally come together will some challenges. The services and applications provided by cyber-physical system platforms will connect people, objects, and systems to each other. This possesses novel requirements on safety, security, and reliability for everything from sensors to user interfaces in real-time [11]. Among other issues, this will involve addressing the challenges posed by the wide range of different data sources and devices. Identified critical system parameters in smart factories are latency (and jitter), safety, and energy consumption. Latency requirements of machines are in the range of several milliseconds, the sensitivity of rapidly moving devices' control circuits is significantly below 1 ms per sensor, while subsystems rely on a latency of several micro seconds. Hence, smart factories together with automation in industry is a key application field in the Tactile Internet.

3.2 Autonomous Transportation

Technological advancements are creating a continuum between conventional, fully human-driven vehicles and autonomous vehicles, which partially or fully drive themselves. Autonomous vehicle technology with self-driving features will revolutionize the driving experience, and consumers will need time to learn how to use and manage the new features. In partially self-driving vehicles, automobiles are set-up with certain features such as self-parking, emergency braking, and adaptive cruise control and are already reality in selected applications that feature controlled environments, such as mining and farming. One of the next steps in autonomous vehicles with partially self-driving capabilities will be most-likely on-highway trucks as the first vehicles to feature the autonomous driving technology on public roads.

Fully self-driving cars, on the other hand, are also no longer a futuristic idea. Various companies have already released self-driving features that give the car the ability to drive itself. Compared to partially self-driving cars, fully self-driving cars' steering wheel will no longer exist. The vehicles will drive by themselves by using the same system of sensors, radar and GPS mapping that partially self-driving vehicles employ. Besides self-evident benefits, such as easier parking, significantly reduced (zero in the optimal case) accident rates, one of the key profits of self-driving cars is time saving. Autonomous car drivers will be able to spend travelling time working, relaxing, or accessing entertainment/digital media on the road which will have a big economic impact. However, these benefits come at the cost of supplementary challenges requiring significant improvement in communication technologies and hardware for the sensor domain, to get a reliable environment model, and for the connectivity domain, to obtain information from other cars and the infrastructure.

To facilitate fully self-driving capabilities, recent advancement in 5G wireless communication technologies and automobiles have enabled the evolution of Intelligent Transport System (ITS) to wirelessly connected and communicating vehicles. ITS addresses various vehicular traffic issues like traffic congestion, information dissemination, accident avoidance etc. Vehicular Ad-hoc NETwork (VANET), in particular, which is a distinctive class of Mobile ad-hoc Network (MANET), is an integral component of ITS. Hereby, wireless communication technologies play a vital role in supporting both Vehicle to Vehicle (V2V) and Vehicle to Infrastructure (V2I) communication in VANET. Typically, a VANET consists of four major components namely: Vehicles, (GPS enabled) Devices/Sensors, road-side info-stations and Traffic Management Centre (TMC) [14, 15]. These components communicate using wireless communication standards/protocols that regulate the various aspects of communication such as data transmission range and rate, latency and security.

Enabling real-time communication and enhanced security, a considerable and sustainable reduction of road accidents and traffic jams can be realized by fully autonomous driving. The time needed for collision avoidance in today's applications for vehicle safety is below 10 ms [12]. In case of a bidirectional data exchange for automatic driving manoeuvres, a latency in the order of a millisecond will likely be needed [13]. This can technically be realized by the Tactile Internet and its 1 ms end-to-end latency. Fully autonomous driving is expected to change the traffic behavior entirely. Especially small distances between automated vehicles, in particular in platoons, potentially safety critical situations need to be detected earlier than with human drivers. This requires ultra-high reliable and proactive/predictive behavior in future wireless communication systems.

3.3 e-Health

Information Technology (IT) is a key in the field of e-health. Most of the technological advancement in wireless networking has been applied to advance healthcare services. E-health and health care services are information based and, hence, a better utilization of information can make health services more incorporated to enhance the patient safety. Consequently, wireless standards and IT need to be further improved to provide efficient, reliable, and robust real-time health services in tele-diagnosis, tele-surgery and tele-rehabilitation (i.e. tele-medicine). In [16], the impact of latency on a sergeant's precision in a robotic-assisted remote telepresence surgery has been examined. It has been shown that a non-real-time system leads to errors in the sergeant's behavior.

Tele-medicine uses (wireless) communication and information technology to overcome geographic distances, and increase access to health care services. This is particularly beneficial for emergency situations, long distances between patient and sergeant, and for rural and underserved communities in developing countries suffering from absence of access to health care.

Using advanced tele-diagnostic tools, medical expertise could be available anywhere and anytime regardless of the physician's location. Hereby, a tele-robot at the patient's location will be controlled by the physician, so that not only audio and/or visual information but also haptic feedback is provided. The same technical principle is applied to tele-surgery applications.

E-health comes together with stringent requirements on the reliability of wireless connection. Especially in tele-surgery and tele-diagnostic, reliability is of particular importance. Unreliable connectivity can lead to delayed imaging, so that poor image resolution may limit the efficacy of the sergeant's remote handling. In addition, an accurate tele-medical treatment can only be realized with haptic feedback which in turn is possible if the (physician/) human-to-machine interaction can be facilitated in real-time. This requires a deterministic real-time behavior which is not supported by recent communication systems. An end-to-end latency of a few milliseconds together with ultra-high reliability in wireless link connection and data transmission is required in e-health which can be realized by the Tactile Internet.

3.4 Edutainment and Gaming

Edutainment is a combined word of education and entertainment and describes educational content with entertainment value. Teaching of exacting manual operation demanding fine-motor skills in a verbal or classical way is challenging and require permanent repetitions and corrections by the teacher. The Tactile Internet promises a completely new way of educations based on the haptic interaction of teachers and students, i.e. while a student performs a manual operation (on a virtual or real object), the teacher can follow him and correct him in case of failures remotely. To realize this in a timely synchronous manner, an identical multi-modal, i.e. visual, auditory, and haptic human-machine-interface, which operates on an extremely low end-to-end latency, is essential. This can be applied to medical students during their first surgeries or to students learning to play a musical instrument. Furthermore, the Tactile Internet will enable the distributed interactive playing of music, e.g. joint practicing of band members. This is with today's technologies not possible due to the high end-to-end latency.

Serious gaming, on the other hand, are real-world simulations designed for the purpose of solving a problem. The end-to-end latency in the interaction between players and games is a key factor influencing the quality of players' experience and the game's usability, since the delay influences directly the perceived realism of the game. Other key requirements in gaming are the synchronicity and extremely low delays in multi-player games. If this is not achieved, each player will experience a different status of the game and a so-called lag, which leads in the worst case to unfairness among players.

3.5 *Further Application Areas*

Other Tactile Internet applications are for example virtual and augmented reality. The virtual reality is a shared, haptic virtual environment in which several users are physically coupled via a simulation tool to jointly/collaboratively perform tasks by perceiving the objects not only audio-visually but also via the touch sense. In augmented reality, on the other hand, the combination of real and computer generated content is visualized in the user's field of view. The major goal of future augmented reality applications, compared to today's static information augmentation, is the visualization of dynamic content and up-to-date information. Haptic feedback in virtual reality is a prerequisite for high-fidelity interaction. Especially, the perception of objects in virtual reality via the sense of touch leads to various applications relying on high level of precision. This precision can only be realized if the latency between the users and the virtual reality is a few milliseconds. The augmentation of additional information into a user's field of view enables the development of many assistance systems, e.g. maintenance, driver-assistance systems, education. With the Tactile Internet the content in augmented reality can be moved from static to dynamic. This enables a real-time virtual extension of a user's field of view, so that possible dangerous events can be identified and avoided.

Unmanned autonomous or remotely controlled systems are increasingly used in a large number of contexts to support humans in dangerous and difficult-to-reach environments, remotely controlled by humans, or for tasks that are too tedious or repetitive for humans. The remote control of an unmanned aircraft, for example, can be realized with high precision and without any reaction delay with a reduced end-to-end latency as a Tactile Internet application.

Besides these applications, efficient and reliable smart grids together with electro mobility, smart metering, and the automation of energy distribution and cloud computing (especially mobile edge clouds) are some other applications that can significantly benefit from the availability of the Tactile Internet.

4 End-to-End Architecture for the Tactile Internet

Unlike the conventional Internet which provides the medium for audio and visual transport, the Tactile Internet is envisioned to provide the medium for transporting touch, actuation, and control in real-time. The Tactile Internet will enable haptic communications [3] as one of the primary applications. The exchange of haptic information generally closes a global control loop with strict latency constraints due to the presence of haptic feedback. In the absence of haptic feedback, there is conventional audio/visual feedback only and no notion of a control loop is involved. Another key application of the Tactile Internet will be networked control systems, wherein sensors and actuators are connected and highly dynamic processes are controlled. The control and feedback signals are exchanged in the form of information packets through the network, closing a global control loop and leading to strict latency constraints. Unlike haptic communications, the human element is not involved in case of networked control systems.

As shown in Fig. 1, the end-to-end architecture for the Tactile Internet can be split into three distinct domains: a master/controller domain, a network domain, and a slave/controlled domain. In case of haptic communications and remote operation, the master domain consists of a human (operator) and human system interface (HSI). The HSI (usually a haptic device) converts the human input into a tactile input through various coding techniques and allows a user to manipulate objects in real and virtual environments. The master domain primarily controls the operation of the remote domain as discussed later. The master domain also has provisioning for audio and visual feedbacks which not only enable non-haptic control and remote operation but also play a critical role in enhancing the perceptual performance. In case of networked control systems, the master domain comprises a controller which gives commands to the sensor and actuator system.

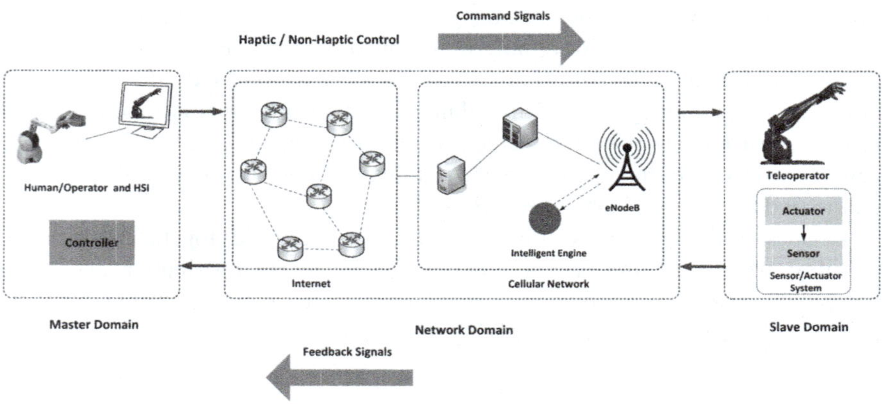

Fig. 1 An illustration of the end-to-end Tactile Internet architecture

The slave domain consists of a teleoperator (controlled robot) in case of haptic communications, which is directly controlled by the master domain through various command signals. The teleoperator interacts with the remote environment and usually sends a haptic feedback on the reverse path. Typically, no *a priori* knowledge exists about the environment. In case of networked control systems, the slave domain consists of a sensor and actuator system. The sensors measure the state of the system whereas the actuators manipulate the system based on the commands received from the controller.

The network domain provides the medium for communication between master and slave domains. The Tactile Internet requires ultra-reliable and ultra-responsive network connectivity that would enable typical reliabilities and latencies for real-time control and remote operation. The underlying 5G-driven communication architecture is expected to meet the key requirements in realizing the vision of the Tactile Internet.

5 Networking Requirements for the Tactile Internet

The Tactile Internet will not be realized without overcoming an enormous set of cross-disciplinary design challenges, in order to meet the underlying requirements. Some of the most stringent design challenges for the Tactile Internet have been recently presented in [2]. In the following, we highlight the key technical requirements, from a networking perspective, for realizing the Tactile Internet.

5.1 Ultra-Reliable Connectivity

The Tactile Internet is expected to service critical areas of the society, and therefore requires ultra-reliable network connectivity. The term ultra-reliability is quantified as seven nines reliability i.e., an outage probability of 10^{-7}, which translates to milliseconds of outage per day. Ultra-reliability is critical for keeping packet losses to a minimum especially in error prone wireless environments. Packet losses results in various types of artifacts for haptic communications that directly affect the activity of the operator. Similarly, packet losses may destabilize a networked control systems.

5.2 Ultra-Responsive Connectivity

Due to the real-time context, the Tactile Internet requires ultra-responsive network connectivity on the order of single-digit ms, to achieve truly immersive steering and control. For tactile application requiring no feedback, latency of up to 5 ms is tolerable. However, in case of feedback, the latency requirement becomes much more stringent; on the order of 1 ms.

The end-to-end latency in a communication system includes the time spent in processing and transmission of information at the transmitter end, propagation of information to the receiver end, and re-transmission from the receiver end (including processing and propagation time) in case of feedback. Considering the end-to-end latency in a communication systems, the latency budget for air-interface is reduced by an order of magnitude.

5.3 Edge-Intelligence

Edge-Intelligence would be critical in overcoming two critical challenges in realizing the Tactile Internet. The first challenge is how to overcome the physical limitation arising due to the finite speed of light. This limits the range of tactile actuation to approximately 150 km (assuming no transmission, processing, etc.). The second challenge arises due to the fundamental nature of wireless channel. Even if a communication infrastructure providing latency of 1 ms is realized, deterministic delay guarantees cannot be provided as the wireless channel is time-varying in nature. Therefore, stability of control loop in case of haptic communications and networked control systems cannot be guaranteed. To overcome these challenges, the Tactile Internet must support a hybrid composition of machine and human actuation mixing real tactile actuation with intelligence-based predictive actuation. Such predictive actuation should be in close proximity of the tactile edge. Therefore, the edge of the network (mobile edge cloud) must be equipped with intelligence to facilitate predictive caching as well as interpolation/extrapolation of human actions. This necessitates the development of novel artificial intelligence techniques for edge-cloud architectures.

5.4 Security and Privacy

Security and privacy are become particularly important requirements for the Tactile Internet as its application areas range from large-scale industrial systems to critical infrastructures and services. Security becomes equally important for wireless transmission and remotely operated entities. With stringent latency constraints, security must be embedded in the physical transmission and ideally be of low computational overhead. Novel coding techniques need to be developed for tactile applications that allow only the legitimate receivers to process a secure message. Absolute security will, hereby, be achieved if an illegitimate receiver cannot decode the date even with infinite computational power. This rises a challenge, especially in massive connectivity applications. Identification of legitimate receivers requires novel, reliable and low-delay methods. One such method could be the usage of hardware specific attributes such as biometric fingerprints.

5.5 Codecs Family

The Tactile Internet must handle the tactile information in the same way as the conventional audio/visual information. Hence, a standard codecs family, similar to the state-of-the-art audio (ITU-TH.264) and video (ISO/IEC MPEG-4) codecs is needed, which facilitate transmission of tactile data over packet-switched networks. Such a codec family would be a key enabler for scalability at the network edge and universal uptake.

6 Network Design for 5G-Enabled Tactile Internet

The Tactile Internet creates daunting new requirements for 5G networks. Such requirements can be met through advances in Physical layer techniques, novel approaches to legacy wireless access and radio resource management protocols, and innovative architectural designs. In the following, we discuss some potential solutions to meet the technical requirements for realizing the Tactile Internet.

6.1 Network Slicing

5G networks will support a range of use cases, spanning different vertical industries, not all of which share the same service requirements as the Tactile Internet. Therefore, one of the most important challenges lying at the intersection of 5G and the Tactile Internet is how one network, based on a common physical infrastructure, can be efficiently shared among different vertical applications. The research community has a general consensus that 5G networks must be designed in a flexible manner to cater for the service requirements of different vertical applications. Such flexibility is possible through network slicing [5], which will be indispensable for 5G network design. A network slice can be defined as a connectivity service, based on various customizable logical network (and associated device) functions, supporting the requirements of a particular use case. The basic idea behind network slicing is to create multiple logical networks over a common physical infrastructure, with each network tailored to the specific needs of a use case. Such envisioned network slicing would be possible through network function virtualization (NFV) [6] and software-defined networking (SDN) [7] paradigms. NFV provides the separation of network functions from the hardware infrastructure which can managed as software module in any standard cloud-computing infrastructure. SDN decouples control and data planes and enables direct programmability of network control through software-based controllers. Therefore, the end-to-end 5G network architecture must be designed to facilitate flexible network slicing.

6.2 Physical Layer

Given its attractive features, orthogonal frequency division multiple access (OFDM) is the primary Physical layer candidate for 5G networks. To overcome some of the weaknesses of OFDM, several variants have been proposed recently such as filter bank multi-carrier (FBMC) [FBMC], generalized frequency division multiplexing (GFDM) [GFDM], bi-orthogonal OFDM (BFDM) [BFDM], etc. It should be noted that OFDM, with its current adaptation for legacy 4G networks is not suitable for tactile applications. An end-to-end latency constraint of 1 ms pushes one-way air-interface transmission budget to approximately 100 μs. However, OFDM symbol duration in LTE/LTE-A is approximately 70 μs. One way to meeting this transmission budget is to change the underlying OFDM numerology in terms of symbol duration, sub-carrier spacing, FFT block size, cyclic prefix etc. Given the increasingly software nature of radios, it can be easily expected that OFDM would be implemented in a *tunable* manner to cater for requirements of different use cases.

In addition to waveform design issues, novel enhancements are required at the Physical layer for meeting the stringent Tactile Internet requirements. These include enabling high levels of diversity (for reliable signal detection and decoding), implementing fast channel estimation/equalization techniques, and using faster decoding mechanisms.

6.3 Protocol Stack Optimizations

The stringent requirements of the Tactile Internet require a revisit of legacy wireless access protocols. In order to reduce air-interface latency, optimizations are needed for both control and data planes. For example, to ensure the stability of closed-loop tactile applications, the base station must support a fast connection re-establishment feature at the radio link layer so that any ongoing tactile session is transparent to radio link failures. From a medium access control (MAC) layer perspective, the transmission time interval (TTI) needs to be reduced in order to achieve end-to-end latency on the order of 1 ms. Such reduction of TTI would be possible through spectrum aggregation techniques. To achieve the required levels of reliability, diversity becomes critically important. However, diversity must be embedded in frequency and spatial domains considering its enhanced latency if applied in the time domain. Hence, link-level techniques like hybrid ARQ (HARQ) must be replaced with suitable equivalents in frequency or spatial domains. Improving wireless connectivity to the required level for tactile applications is not an invincible task. By exploiting frequency diversity over multiple uncorrelated links, required reliability level can be approached.

6.4 Radio Resource Management

Radio resource management is a key functionality of cellular networks. It has a direct impact on QoS of a user and performance of higher layers. With the introduction of tactile applications into the 5G ecosystem resource management becomes particularly challenging as available resources are shared with other human-to-human (H2H) and machine-to-machine (M2M) applications having different and often conflicting service requirements. Due to stringent service requirements radio resource must be allocated on priority for tactile applications without any external competition. In this context, slicing of radio resources (within broader context of network slicing) becomes particularly attractive. A separate slice of radio resources can be allocated to tactile applications which remains dedicated for any ongoing operation. Such radio resource slicing would be achieved through virtualization of radio resources. In this context, a flexible radio resource slicing strategy is desirable for 5G networks which not only maximizes the utility of various applications, but also maintains high utilization of the overall radio resource pool. Another benefit of such radio resource slicing is the application specific customization of radio resources within the allocated slice. Such customization of radio resources for haptic communications has been recently investigated in [8].

6.5 Core Networking

The key functionalities of the 5G core network relevant to the Tactile Internet include dynamic application-aware QoS provisioning, edge-cloud access, and security. Overall, a *thin* core network is desirable with substantial decrease in the protocol overhead. Core network thinning would be achieved through its functional decomposition and moving some of the functionalities to the access network. This will reduce the number of nodes in the data path and hence reduce the end-to-end latency. Functional decomposition is a method to separate out tightly coupled sub-functions of a network entity and can be achieved in either horizontal or vertical manner. The software-defined networking (SDN) paradigm is particularly attractive for the 5G core networking. It will introduce the required programmability and hence the flexibility for such functional decomposition through which data flows can be tailored according to the requirements of different applications.

The existing IP security functionalities are sufficient for providing the required security in the Tactile Internet. However, the placement of IP security far from the tactile edges increases end-to-end latency. Hence, novel approaches are needed to provide adequate security for tactile applications with minimal delays.

Finally, latency in the core Internet must be reduced, which is currently variable and largely dictated by queuing delays and geographic routing policies.

7 Concluding Remarks

The Tactile Internet is expected to have a massive impact on business and society. It will revolutionize almost every segment of the society by enabling wireless control and remote operation in a range of scenarios. It will create new opportunities for vendors, operators, content providers, and other members of the service chain.

The next generation (5G) mobile communications networks will play an important role in realizing the Tactile Internet. Research for both 5G networks and Tactile Internet is at a nascent stage. This chapter investigated the interesting area of 5G and Tactile Internet intersection. After introducing exciting Tactile Internet applications, key requirements for the Tactile Internet, from a networking perspective, have been identified. The chapter covered several technical issues and challenges in shaping 5G networks for realizing the vision of the Tactile Internet. The most important challenge would be to ensure tight and scalable integration of various technological solutions into a single network.

References

1. G. Fettweis, The tactile internet: applications and challenges. IEEE Veh. Technol. Mag. **9**(1), 64–70 (2014)
2. A. Aijaz, M. Dohler, A.H. Aghvami, V. Friderikos, M. Frodigh, Realizing the tactile internet: haptic communications over next generation 5G cellular networks. IEEE Wirel. Commun. (2015), [Online]. Available: http://arxiv.org/abs/1510.02826
3. E. Steinbach et al., Haptic communications. Proc. IEEE **100**(4), 937–956 (2012)
4. J. Andrews et al., What will 5G be? IEEE J. Sel. Areas Commun. **32**(6), 1065–1082 (2014)
5. Ericsson, 5G systems: enabling industry and society transformation. *White Paper*, Jan. 2015. [Online]. Available: http://www.ericsson.com/res/docs/whitepapers/what-is-a-5g-system.pdf
6. B. Han, V. Gopalakrishnan, L. Ji, S. Lee, Network function virtualization: challenges and opportunities for innovations. IEEE Commun. Mag. **53**(2), 90–97 (2015)
7. B. Nunes, M. Mendonca, X.-N. Nguyen, K. Obraczka, T. Turletti, A survey of software-defined networking: past, present, and future of programmable networks. IEEE Commun. Surv. Tutorials **16**(3), 1617–1634 (2014)
8. A. Aijaz, Towards 5G-enabled tactile internet: radio resource allocation for haptic communications, in *Proceedings of the IEEE Wireless Communications and Networking Conference (WCNC) Workshops*, April 2016
9. ITU-T, The Tactile Internet. Technology Watch Report, 2014. [Online]. https://www.itu.int/dms_pub/itu-t/oth/23/01/T23010000230001PDFE.pdf
10. Deloitte, Industry 4.0 – challenges and solutions for the digital transformation and use of exponential technologies. *White Paper*, Oct. 2014. [Online]. http://www2.deloitte.com/content/dam/Deloitte/ch/Documents/manufacturing/ch-en-manufacturing-industry-4-0-24102014.pdf
11. N. Nikaein, S. Krea, Latency for real-time machine-to-machine communication in LTE-based system architecture, in *European Wireless (EW) Conference*, Vienna, Austria, April 2011
12. J.D. Lee, D.V. McGehee, T.L. Brown, M.L. Reyes, Collision warning timing, driver distraction, and driver response to imminent rear-end collisions in a high-fidelity driving simulator. Hum. Factors Ergon. Soc. **44**(2), 314–334 (2002)

13. S.B. Mer, Smart vehicle-to-vehicle communication with 5G technology. Int. J. Recent Innov. Trends Comput. Commun. 3(5), 3241–3244 (2015)
14. M. Faezipour, M. Nourani, A. Saeed, S. Addepalli, Progress and challenges in intelligent vehicle area networks. Commun. ACM **55**(2), 90–100 (2012)
15. T. Kosch, I. Kulp, M. Bechler, M. Strassberger, B. Weyl, R. Lasowski, Communication architecture for cooperative systems in Europe. IEEE Commun. Mag. **47**(5), 116–125 (2009)
16. M. Anvari et al., The impact of latency on surgical precision and task completion during robotic-assisted remote telepresence surgery. Comput. Aided Surg. **10**(2), 93–99 (2005)
17. A. Osseiran et al., Scenarios for 5G mobile and wireless communications: the vision of the METIS project. IEEE Commun. Mag. **52**(5), 26–35 (2014)
18. NGMN, 5G White Paper. Feb. 2015. [Online]. https://www.ngmn.org/uploads/media/NGMN_5G_White_Paper_V1_0.pdf

The manufacturer's authorised representative in the EU is Springer Nature Customer Service Centre GmbH, Europaplatz 3, 69115 Heidelberg, Germany. If you have any concerns regarding our products, please contact ProductSafety@springernature.com

Printed and bound by CPI Group (UK) Ltd, Croydon, CR0 4YY
30/09/2026
02225570-0020

Sustainable Solid Waste Management

Edited by
Jonathan W. C. Wong
Rao Y. Surampalli
Tian C. Zhang
Rajeshwar D. Tyagi
Ammaiyappan Selvam

Sponsored by the
Hazardous, Toxic, and Radioactive Waste Engineering Committee of the
Environmental Council of the
Environmental and Water Resources Institute of the
American Society of Civil Engineers

ASCE

EWRI ASCE

Published by the American Society of Civil Engineers

Library of Congress Cataloging-in-Publication Data

Names: Wong, Jonathan W. C., editor.
Title: Sustainable solid waste management / edited by Jonathan W. C. Wong, Rao Y. Surampalli, Tian C. Zhang, Rajeshwar D. Tyagi, Ammaiyappan Selvam ; sponsored by the Hazardous, Toxic, and Radioactive Waste Engineering Committee of the Environmental Council of the Environmental and Water Resources Institute of the American Society of Civil Engineers.
Other titles: Sustainable solid waste management (Reston, Va.)
Description: Reston, Virginia : American Society of Civil Engineers, [2016] | Includes bibliographical references and index.
Identifiers: LCCN 2015049311 (print) | LCCN 2016006528 (ebook) | ISBN 9780784414101 (print) | ISBN 9780784479308 (pdf)
Subjects: LCSH: Refuse and refuse disposal. | Sustainable engineering.
Classification: LCC TD791 .S875 2016 (print) | LCC TD791 (ebook) | DDC 628.4/4–dc23
LC record available at http://lccn.loc.gov/2015049311

Published by American Society of Civil Engineers
1801 Alexander Bell Drive
Reston, Virginia 20191-4382
www.asce.org/bookstore | ascelibrary.org

Errata: Errata, if any, can be found at http://dx.doi.org/10.1061/9780784414101.

ISBN 978-0-7844-1410-1 (print)
ISBN 978-0-7844-7930-8 (PDF)
Manufactured in the United States of America.

22 21 20 19 18 17 16 1 2 3 4 5

Contents

Preface

The uncontrolled consumption lifestyle is the root cause of the huge waste generation problem of our modern world. The solid waste generation rates are currently higher in industrialized countries. However, the rates will more than double over the next 20 years in lower income countries. According to the World Health Organization report, world cities alone currently generate about 1.3 billion tonnes of solid waste per year, and is predicted to be around 2.2 billion tonnes by 2025. Increases in the quantity of waste generation necessitate the budget to be apportioned to manage these wastes. By 2025, an increase of about 83% in the annual budget, when compared with 2012, is estimated to cope with the predicted increase of 69% in waste quantity. Furthermore, the cost increases will be more than 4 to 5 folds in low-income countries and lower-middle income countries. Currently the solid management practices in the low-income countries are not adequate to ensure proper disposal and environmental hygiene, justifying the escalation in the estimated budget. The impacts of solid waste are significantly influencing the human lives, contaminating the environment through greenhouse gas emission, bioaerosols, spreading of potentially pathogenic organisms and rapid use of resources. Due to the rapid urbanization and industrial development around the world, this steady increase in the waste generation has put pressure on the government in handling the diverse nature of wastes.

In fact, the waste management industry has witnessed a range of technological and sociological changes at different levels. Despite that, the problems associated with solid waste generation are persistent due to increase in the generation of wastes and its subsequent impact on the environment, and the use of compounds/materials deleterious to the ecosystems. This, together with the dwindling of resources, have made it clear that the past efforts have made the public and governments believe that "wastes are not waste anymore but indeed resources for our future." The topic of current interest is how to tap into their potential with minimal footprint on environment. Most of the developed countries have adopted 'waste minimization' as a strategy to reduce our dependence on landfill disposal as well as resource conservation. This paradigm shift has already transformed waste management into an industry of resource conservation; an effective public-private partnership is inevitable to achieve a win-win situation. Yet, the policy makers and government cannot entirely depend on the service providers/operators; they must be adequately and appropriately informed of the available technologies and their limitations from collection to the disposal.

Sustainable solid waste management involves waste minimization, utilizing waste as resources, appropriate disposal and an efficient management system supported by competent policies. The ultimate goal of such management practices is to protect the environment and allow rational use of resources leaving this planet hospitable for future generations. It is also essential to cultivate the environmental responsibility of relevant stakeholders, especially the community at large and the policy makers in sustainable solid waste management. In the last decade, new technologies and policies have been introduced and their impacts and success are yet to be appraised. Furthermore, every treatment option must be characterized for their contribution to the environmental deterioration right before the inception.

Rapidly depleting mineral and energy sources around the world necessitates serious consideration of the recycling and waste-to-energy options. Therefore, a multi-dimensional approach is required to manage solid wastes that warrant the latest developments to be shared efficiently and rapidly. Therefore, the aim of this book is to enlighten the readers on the latest developments and issues to be resolved to achieve sustainable solid waste management. This book will be a valuable tool not just for policy makers and planners, but also for municipal engineers, researchers and other professionals engaged in solid waste management.

During the last decade, there was steady developments in the (1) policies focusing on waste minimization including reduction at source and recycling, (2) new technologies for waste treatments towards waste reduction, (3) safe and efficient treatment and disposal of hazardous and other special wastes, and (4) development and maintenance of engineered landfills and landfill mining. These issues were logically divided into four sections respectively while an additional section covering the legal frameworks and the role and use of life cycle assessment as a potential tool for waste management industry were added as the fifth section.

We believe that the information provided in this book will be a guide and an eye-opener for government and local authorities to commit themselves to infuse sustainability in the waste management industry. In addition, the contents will increase the awareness in readers about the potential social and economic benefits made from the sustainable approaches, and guide the academics in their focus to further improve policies and technologies.

The editors gratefully acknowledge the hard work and patience of all the authors who have contributed to this book. The views or opinions expressed in each chapter of this book are those of the authors and should not be construed as opinions of the organizations they work for.

—JWW, RYS, TCZ, RDT, AS

About the Editors

Dr. Jonathan W. C. Wong is a professor in the Department of Biology, Hong Kong Baptist University, Hong Kong SAR. In addition he is serving as the director of Sino-Forest Applied Research Centre for Pearl River Delta Environment (ARCPE), as well as the Hong Kong Organic Resource Centre (HKORC). He received his M.Phil. degree from the Chinese University Hong Kong and Ph.D. degree from Murdoch University, Australia. He joined the Hong Kong Baptist University in 1992 and continue to serve in various capacities. Prof. Wong is serving as a Visiting Professor in China University of Agriculture, Nanjing Agricultural University and Shangdong University, China. He conducts research in organic waste management and treatment with particular emphasis on composting and bioenergy production striving towards zero-waste organic waste disposal. He chaired two international conferences (ICSWHK 2011 and 2013), and the third conference in is 2015. He has served/serving as member of International Advisory Board, Scientific committee and in other capacities of various International Conferences. Prof. Wong was bestowed with the Medal of Honor by the Government of Hong Kong Special Administrative Region in 2011 for his service and contribution to the Environment. He is serving in the Editorial Board of six journals, notably, as Regional Editor of 'Environmental Technology' and Board member of 'Bioresource Technology'. Prof. Wong has authored over 500 publications including refereed journals, conferences proceedings, four patents, more than 100 technical reports, editing 12 books/journals volumes and others.

Dr. Rao Y. Surampalli, P.E., Dist.M.ASCE, is President and Chief Executive Officer of the Global Institute for Energy, Environment and Sustainability. He was with the U.S.Environmental Protection Agency (USEPA) for 29 years and retired as an Engineer Director in October, 2014. He received his M.S. and Ph.D. degrees in Environmental Engineering from Oklahoma State University and Iowa State University, respectively. He is a Registered Professional Engineer in the branches of Civil and Environmental Engineering, and also a Board Certified Environmental Engineer (BCEE) of the American Academy of Environmental Engineers (AAEE) and a Diplomate of the American Academy of Water Resources Engineers (DWRE). He is an Adjunct Professor in seven universities and distinguished/honorary visiting professor in six universities. Currently, he serves, or has served on 66 national and international committees, review panels, or advisory boards including the ASCE National Committee on Energy, Environment and Water Policy. He also served as President of Civil Engineering Certification (CEC), Inc., that was started by ASCE for Board Certification of various specialties within civil engineering. He is a Distinguished Engineering Alumnus of both the Oklahoma State and Iowa State Universities, and is an elected Fellow of the American Association for the Advancement of Science, an elected Member of the European Academy of Sciences and Arts, an elected Member of the Russian Academy of Engineering, an elected Fellow of the Water Environment Federation and International Water Association, and a Distinguished Member of the American Society of Civil Engineers. He also is Editor-in-Chief of the ASCE Journal of Hazardous, Toxic and Radioactive Waste, past Vice-Chair of Editorial Board of Water Environment Research Journal, and serves on the editorial boards of 8 other refereed environmental journals. He has authored over 500 technical publications in journals and conference proceedings, including 12 patents, 16 books, 64 book chapters, and presented over 110 plenary/keynote or invited presentations worldwide.

Dr. Tian C. Zhang, P.E., is a Professor in the department of Civil Engineering at the University of Nebraska-Lincoln (UNL), USA. He received his Ph.D. in environmental engineering from the University of Cincinnati in 1994. He joined the UNL faculty in August 1994. Professor Zhang teaches courses related to water/wastewater treatment, remediation of hazardous wastes, and non-point pollution control. Professor Zhang's research involves fundamentals and applications of nanotechnology and conventional technology for water, wastewater, and storm-water treatment and management, remediation of contaminated environments, and detection/control of emerging contaminants in the environment. Professor Zhang has published more than 78 peer-reviewed journal papers, 45 book chapters and 9 books since 1994. Professor Zhang is a member of American Association for the Advancement of Science (AAAS), the Water Environmental Federation (WEF), and Association of Environmental Engineering and Science Professors (AEESP). Professor Zhang is a Diplomate of Water Resources Engineer (D.WRE) of the American Academy of Water Resources Engineers, and Board Certified Environmental Engineers (BCEE) of the American Academy of Environmental Engineers, Fellow of American Society of Civil Engineers (F. ASCE), and Academician of European Academy of Sciences and Arts. Professor Zhang is the Associate Editor of *Journal of Environmental Engineering* (since 2007), *Journal of Hazardous, Toxic, and Radioactive Waste* (since 2006), and *Water Environment Research* (since 2008). He has been a registered professional engineer in Nebraska, USA since 2000.

Dr. Rajeshwar D. Tyagi is an internationally recognized Professor with 'Institut national de la recherche scientifique – Eau, terre, et environment', (INRS-ETE), University of Québec, Canada. He holds Canada Research Chair on, 'Bioconversion of wastewater and wastewater sludge to value added products.' He conducts research on hazardous/solids waste management, water/wastewater treatment, sludge treatment/disposal, and bioconversion of wastewater and wastewater sludge into value added products. He has developed the novel technologies of simultaneous sewage sludge digestion and metal leaching, bioconversion of wastewater sludge (biosolids) into *Bacillus thuringiensis* based biopesticides, bioplastics, biofertilisers and biocontrol agents. He is a recipient of the ASCE State-of-the Art of Civil Engineering Award, ASCE Rudolph Hering Medal, ASCE Wesley Horner Medal and ASCE Best Practice Oriented Paper Award. He also received 2010 Global Honor Award (Applied Research) of International Water Association, 2010 Grand Prize (University research) of American Academy of Environmental Engineers, and Excellence Award of Natural Sciences and Engineering Research Council of Canada for Industry-University collaborative research. Dr. Tyagi has published/presented over 600 papers in refereed journals, conferences proceedings and is the author of eleven books, seventy book chapters, ten research reports and nine patents.

Dr. Ammaiyappan Selvam is currently working as a Research Assistant Professor in Sino-Forest Applied Research Centre for Pearl River Delta Environment, and Department of Biology, Hong Kong Baptist University, Hong Kong SAR. He received his M.Phil. and Ph.D. degrees from the University of Madras, Chennai, India. He worked as a Post-doctoral researcher at Centre for Environmental Studies, Anna University (India), Department of Microbiology and Biochemistry, National Taiwan University (Taiwan) and Department of Biology, Hong Kong Baptist University (Hong Kong SAR). His research interests are biomass and bioenergy production from organic wastes, phyto/bio-remediation; and his innovative ideas earned him four best paper/poster awards in the International conferences. He served as members of scientific/working committee in five International conferences; and serving as an editorial member of the 'Environmental Technology' and ad-hoc reviewer to a number of environmental journals. Dr. Selvam has authored over 100 publications including 43 papers in refereed journals, 40 full papers in conferences proceedings, 6 edited books/journal volumes, 2 set of genomic sequences and others.

Contributing Authors

Konstantinos Abeliotis, *Harokopio University, Athens, Greece*
Mahtab Ahmad, *Kangwon National University, Chuncheon, Korea*
Moussa-Mallave Alhadj-Mallah, *Zhejiang University, Hangzhou, P.R. China*
Mukesh K. Awasthi, *Hong Kong Baptist University, Hong Kong SAR*
Puspendu Bhunia, *Indian Institute of Technology, Bhubaneswar, India*
Sébastien Bonnet, *King Mongkut's University of Technology Thonburi, Thailand*
Xu Cai, *Zhejiang University, Hangzhou, P.R. China*
Yong Chi, *Zhejiang University, Hangzhou, P.R. China*
Moe N. Chowdhury, *Jackson State University, Mississippi, USA*
Lee Man Chu, *The Chinese University of Hong Kong, Hong Kong SAR*
Kui-Hao Chuang, *Central Taiwan University of Science and Technology, Taiwan*
Christina Chroni, *Harokopio University, Athens, Greece*
Shabbir H. Gheewala, *King Mongkut's University of Technology Thonburi, Thailand*
Sadhan K. Ghosh, *Jadavpur University, Kolkata, India*
Sannidhya K. Ghosh, *University of Colorado Boulder, Boulder, CO, USA*
Xu Han, *Zhejiang University, Hangzhou, P.R. China*
Hiroshan Hettiarachchi, *United Nations University (UNU-FLORES), Germany*
Patrick Hettiaratchi, *University of Calgary, Calgary, Alberta, Canada*
Qunxing Huang, *Zhejiang University, Hangzhou, P.R. China*
Kurian Joseph, *Anna University, Chennai, India*
Prakriti Kashyap, *Asian Institute of Technology, Pathumthani, Thailand*
Geonha Kim, *Hannam University, Daejeon, Korea*
Kye Hoon Kim, *The University of Seoul, Seoul, Korea*
Sunil Kumar, *CSIR-NEERI, Maharashtra, India*
Katia Lasaridi, *Harokopio University, Athens, Greece*
Xiaodong Li, *Zhejiang University, Hangzhou, P.R. China*
Shengxuan Lin, *Nanyang Technological University, Singapore*
Shengyong Lu, *Zhejiang University, Hangzhou, P.R. China*
Feiyan Mao, *Zhejiang University, Hangzhou, P.R. China*
Jay N. Meegoda, *New Jersey Institute of Technology, Newark, USA*
Tanaji T. More, *INRS-ETE, Université du Quebéc, Québec, QC, Canada*
Yong Sik Ok, *Kangwon National University, Chuncheon, Korea*
Sridhar Pilli, *INRS-ETE, Université du Quebéc, Québec, QC, Canada*
Anushka Upamali Rajapaksha, *Institute of Fundamental Studies, Kandy, Sri Lanka*
Anushuya Ramakrishnan, *University of Nebraska-Lincoln, Lincoln, NE, USA*
Prangya Ranjan Rout, *Indian Institute of Technology, Bhubaneswar, India*

Ammaiyappan Selvam, *Hong Kong Baptist University, Hong Kong SAR*
Kaimin Shih, *The University of Hong Kong, Hong Kong SAR*
Yoshiyuki Shinogi, *Kyushu University, Japan*
Avick Sil, *Environmental Policy & Research India, Maharashtra, India*
Xiaolong Sun, *Nanyang Technological University, Singapore*
Rao Y. Surampalli, *University of Nebraska-Lincoln, Lincoln, NE, USA*
Rajeshwar D. Tyagi, *INRS-ETE, Université du Quebéc, Québec, QC, Canada*
Marjorie Valix, *The University of Sydney, NSW, Australia*
Chettiyappan Visvanathan, *Asian Institute of Technology, Pathumthani, Thailand*
MeththikaVithanage, *Institute of Fundamental Studies, Kandy, Sri Lanka*
Jing-Yuan Wang, *Nanyang Technological University, Singapore*
Jun Wang, *Zhejiang University, Hangzhou, P.R. China*
Lin Wang, *Nanyang Technological University, Singapore*
Zhonghui Wang, *Nanyang Technological University, Singapore*
Jonathan W.C. Wong, *Hong Kong Baptist University, Hong Kong SAR*
Jianhua Yan, *Zhejiang University, Hangzhou, P.R. China*
Song Yan, *INRS-ETE, Université du Quebéc, Québec, QC, Canada*
Tian C. Zhang, *University of Nebraska-Lincoln, Lincoln, NE, USA*
Lei Zhao, *Nanyang Technological University, Singapore*
Antonis A. Zorpas, *Cyprus Open University, Cyprus*

CHAPTER 1

Waste Management and Sustainability: An Introduction

Ammaiyappan Selvam[1]
Jonathan Wong[2]

1.1 INTRODUCTION

The interaction between the human and the environment is a complex phenomenon. The Earth's capacity to support the humans is determined not just by the basic food requirements but also by our levels of resource consumption, amount of waste generation, technologies employed in varying applications. With the increase in the population and increasing resource utilization pattern, in fact we have surpassed the carry capacity of the planet. The Earth's environmental assets are now insufficient to sustain our demands and economic activities (Postel 1994). The global warming has underscored the danger of overstepping the Earth's ability to absorb our waste products. In addition, the consequences of exceeding the sustainable supply of essential resources and how far we already moved in this chain are also not well recognized, and often approached in an economic and industrial mindset.

The earth's ability to absorb our waste is a major factor influencing the adaptation of the waste treatment technologies. Landfilling is probably the oldest organized waste management technology. Until 1970s, the landfilling has been carried out as an unceremonious dumping of waste in any convenient location without considering the health, safety, environmental protection, or cost efficiency (Poulsen 2014). But now the scenario changed not because of the perception and importance of the waste treatment but on other issues. In the urban settings, availability of space for landfilling is becoming scare and a very serious issue. Influenced by the "Not In My Back Yard (NIMBY)" syndrome, the problem creates political pressure to divert the wastes to other treatment technologies. Currently, the trend in the countries having advanced waste management system is to reduce the waste end up in landfills. For example, in Hong Kong, the scarcity of the landfill space, rather than the resource utilization, was the driving force of

[1]Hong Kong Baptist Univ., Hong Kong SAR.
[2]Hong Kong Baptist Univ., Hong Kong SAR.

the initiative. Two decades ago, only about 30% of the MSW produced were recycled. Only in the last decade, the importance of waste reduction and recycling was realized and through extensive education program on the Source Separation of Domestic Waste, a recovery rate of 48% was achieved in 2012 (Wong and Selvam 2014).

In addition to recycling, the increasing waste-to energy programmes and the developments in the technology and pollution control devices further reduced the quantities reaching landfills mainly in Europe while it could be a blueprint for other countries in the future. This is especially applicable to the regions where finding new landfill space is a challenge and those regions where these technologies are not yet fully implemented. It may also be expected that improved and more sustainable product design would be available in the near future that would change the face of the energy recovery systems. Despite the oddities, landfilling is unavoidable and often the final inert fraction needs to be buried. The landfill design, operation and management is continuously being researched and new techniques are inducted to avoid the air and water pollution. For example, the landfill gas collection provide scope for reducing the greenhouse gas (GHG). But the economics of the landfill collection and energy recovery need to be demonstrated convincingly. Because, the average methane content of landfill gas is around 50% due to partial oxidation in the landfill and part of the gas produced in landfills is lost to the atmosphere even with an effective gas collection system (Poulsen 2014).This low level of methane in the landfill gas necessitate critical upgrading works jeopardizing the advantages of the landfill gas extraction.

1.2 ENVIRONMENTAL HEALTH

The organic fraction of municipal solid waste (MSW) is an important component, not just because of the quantity it is generated but also for its potential negative impacts on the environmental and human health. Upon improper management of this fraction its adverse impact will continue until it has fully decomposed or stabilized in the landfill and this takes about at least a couple of decades. Consequently, organic waste landfilling put the air, soil and water matrices at the risk of contamination which tend to aggravate the people against any waste treatment/disposal activities in the city and create a pressure for the governments. Organic waste recycling and treatment technologies are gaining attention in the recent past. A combination of anaerobic digestion and composting technologies facilitate the recovery of energy and nutrients; however, adopting such technologies is influenced by the space available for treatments and environmental performance of the technology, and the market potential and economics of the operation also pose great impact. If the technology expects subsidies from the government, then the policies will also play a crucial role in the induction of such technologies.

Majority of the human population in developing countries lives in rural areas (United Nations 2009). However, the treatment of rural MSW is not adequate and often threat the health of the environment. For example, in China, nearly 65% population lives in the rural and in 2011, about 200 million tonnes of MSW were generated and was higher than the 180 million tonnes generated from the cities. In developed nations, the waste collection, transport and treatment are practiced as an extension of the service provided in the urban setups. Therefore, an effective prevention of pollution due to the generated wastes is achieved. However, since the waste collection in the rural requires higher cost due to the low density of the rural households than the urban settings. For example, collection, transportation and treatment of 1 tonne of MSW in the urban and rural areas of Hangzhou, China costs about 175 and 293 US$/ton, respectively (He 2012). Such situation reduces the efficiency of collection, transport and treatment in the rural areas resulting in higher pollution subsequently causing environmental as well as public health issues. Therefore, waste management in the rural areas are as important as performed in the cities.

Waste reduction and recycling play a crucial role in the modern waste management hierarchy. On one hand they reduce the waste diverted to the landfills and on the other hand reduces the waste to be treated reducing the revenue required. We have entered an era in which global prosperity increasingly depends on effective resource utilization and reducing consumption levels. With the global demand and strong territorial boundaries characterized by different living standards, it is impossible to control the consumption of resources equally. However, on the other hand resources should be identified without affecting the Earth's natural resources. One potential way is to recycle the enormous quantity of waste to exploit them as resources adopting a 'cradle to cradle' approach.

Identifying the 'future' problems of waste recycling and treatment relies on the materials used in the contemporary products. Unfortunately, development of appropriate recycling and treatment technologies is often initiated only when sufficient quantities are appeared in the waste stream (Pomberger and Ragossnig 2014). By the time, the possibly hazardous waste components already exerted significant effect on the environment. Therefore it is important to identify and assess the materials and compounds of the 'future wastes' carefully. Potential and competent solid waste management programme can reduce the environmental impact and has a direct impact on public health. The public health includes human lives in the vicinity of the waste storage and disposal and the thousands of personals involved in the waste handling.

1.3 DRIVING FORCES OF SUSTAINABILITY

Increase in population and rapid consumption of resources are strong driving forces necessitating the changes in our life style that control the resource utilization pattern and effective utilization of the resources. The developments in the technology and environmental education can help to achieve these changes. The appropriate technologies should no longer consider to excavate the natural resources instead allow us to benefit from the resources we have (Postel 1994). The developments of material recovery facilities could achieve effective resource recovery and could be an alternative to the unregulated scavenging of the recyclables in the developing countries. Yet, the modern facilities did not completely reach these regions due to economic reasons. It is expected that the next two decades could witness such a development because the 'waste management' already became an industry and many companies now target the developing nations due to the high scope for development. But the policies of respective countries hold the key for such development.

Reuse and recovery of the inorganic components of the waste stream currently gained more importance; but special attention should be given to organic residues. The organic biodegradable fraction could represent about 50% of the waste (by weight) in the developing countries and energy recovery possibilities makes this fraction an attractive and alternative to the fossil fuels. An accurate knowledge of the quantity and composition of the waste is essential to the success of the resource recovery and energy recovery programmes and it is the basis for adopting successful technologies from other regions. However, adopting technologies from another country should be assessed with the existing waste management practices and networks and the robustness of the technology. Despite all these factors, steady supply of waste materials and a market for the reclaimed resource are the key drivers influencing the entire recycling and recovery program, without which the landfilling will hold its place firmly.

1.4 INTEGRATED WASTE MANAGEMENT AND SUSTAINABILITY

Integrated waste management is a concept for designing and implementing new waste management systems and for analyzing and optimizing existing systems. In this concept, both technical and non-technical components of the management system should be analyzed together. Currently, with the development of new policies, regulations and waste management business as an industry, non-technical components including public engagement and education is unavoidable and key to the successful implementation of many recycling and recovery programmes. One classical example is the public opposition of the incineration facilities worldwide mainly due the conception that the incinerators are the sources of dioxins, which often underscores effectiveness of incinerators in reducing the waste volume and the rate of waste treatment. Besides, the developments in pollution abatement systems and the gasification technologies deserve a thorough assessment in controlling the contaminants entering the environment. Most importantly, engaging the public in such assessments and educating them of the needs and challenges of the waste treatment and disposal of a particular region or country are required for successful implementation of the new waste treatment facilities. Currently, a collaboration among the public, private, and informal sectors is evident and it is ideal to organize the environmental education and public engagement through one of these channels for effective implementation.

The waste management hierarchy witnessed changes in the recent decade and currently recycling and recovery is focused more than the landfilling. For example, the hierarchy adopted by the Hong Kong Blueprint for Sustainable Use of Resources (2013–2022) is presented in Figure 1.1. This Blueprint, for the first time recognizes the waste as resources and proposes a lot of initiatives to achieve the set target. Indeed, the initiations on waste recovery of the previous framework will provide a fine template for this framework; while the waste avoidance and a firm action towards waste-to-energy technologies, waste reduction through producer responsibility scheme (PRS) and waste charging are the targeted actions [Environment Bureau (EB) 2013]. Similarly, many countries are now adopting the 4 R policies which is the basis for the sustainable resource utilization.

As mentioned before, the carrying capacity of the earth is being challenged constantly while the environmental health is paying the price for the economic activities. In addition, the resources are declining while demand is increasing with the modified lifestyle and demand. As the modern waste management hierarchy insists, public participation and changing their perception are the first step to achieve the waste reduction while recycling and reuse require the support of technology also. Energy and nutrient recovery are technology based while their acceptance could be a victim of NIMBY syndrome if not approached appropriately. Landfilling is unavoidable but the level of sustainability targeted and achieved in the other aspects would determine the quantity reaching the landfills.

Therefore, a modern integrated waste management is the need of the time while the sustainability must be infused in all components considering the supply

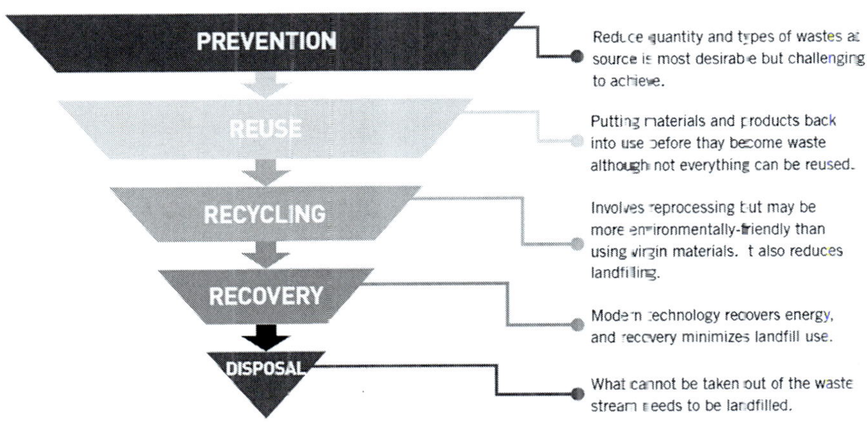

Figure 1.1. Waste management hierarchy adopted for the Hong Kong Blueprint for Sustainable Use of Resources (2013–2022)
Source: EB (2013)

and demand of the resources. It is inevitable that the wastes are now resources and it is our responsibility how we are going to use them. As evident from the past experience, it is not easy, but not impossible if we really consider the Earth as "our home".

References

EB (Environment Bureau). (2013). "Hong Kong blueprint for sustainable use of resources 2013–2022." Hong Kong Special Administrative Region, Hong Kong.

He, P. J. (2012). "Municipal solid waste in rural areas of developing country: Do we need special treatment mode?" *Waste Manage.*, **32**(7), 1289–1290.

Pomberger, R., and Ragossnig, A. (2014). "Future waste—Waste future." *Waste Manage. Res.*, **32**(2), 89–90.

Postel, S. (1994). "Carrying capacity: Earth's bottom line." State of the Word—A World-watch Institute Rep. on Progress towards a Sustainable Society, Worldwatch Institute, Washington, DC, 3–21.

Poulsen, T. G. (2014). "Landfilling, past, present and future." *Waste Manage. Res.*, **32**(3), 177–178.

United Nations. (2009). "World demographic trends report of the secretary-general: Commission on population and, development.", UN.

Wong, J. W. C., and Selvam, A. (2014). "Waste management in Hong Kong: A lesson to learn." Proc., 4th Int. Conf. on Solid Waste Management and Exhibition on Municipal Services, Urban Development & Clean Technology, Acharya NGR Agriculture Univ., Hyderabad, India.

CHAPTER 2

Waste Characteristics and Generation

Avick Sil[1]
Sunil Kumar[2]

2.1 WHAT IS WASTE?

Waste is useless, unwanted or discarded material resulting from agricultural, commercial, communal and industrial activities. According to Defra (2012), waste is any substance or object, which the holder discards or intends or is required to discard. Now the question arises, is it important to define waste? The answer is yes. Why it is important, because it can lead to formulation of waste management practices also from health aspects (Das et al. 2012). One can take the example of Television (TV). If a TV is not working and it is beyond the capability of repaired, it would be discarded and treated as waste. Now, one should understand a TV may contain many hazardous substances such as heavy metals like mercury (Hg) in its Cathode Ray Tube. This mercury can be hazardous to environment and human health. Therefore, a TV cannot be disposed like any other waste; and needs to be disposed properly. Similar example could be the batteries, which contains lead. Hence, definition of waste is needed for management, policy development, environmental and human health point of view.

Next question arises, how one can identify a waste or how one can assess or treat a particular substance as waste. This can be achieved through asking yourself/ oneself variety of questions as highlighted in Figure 2.1.

Thus, these questions needs to be raised before treating a discard or unwanted substance or object as waste. This would aid in better decision making and reduce burden of waste into the environment; and also useful for decision makers and regulatory authorities.

Waste management is one of the most challenging tasks in urban and rural areas (Sil et al. 2011). Waste management must begin at source itself. In some developed countries segregation of waste into different components like biodegradable, non-biodegradable, recyclable and inert is compulsory (Bulkeley et al.

[1]Environmental Policy & Research India (EPRI), Maharashtra, India.
[2]CSIR-NEERI, Maharashtra, India.

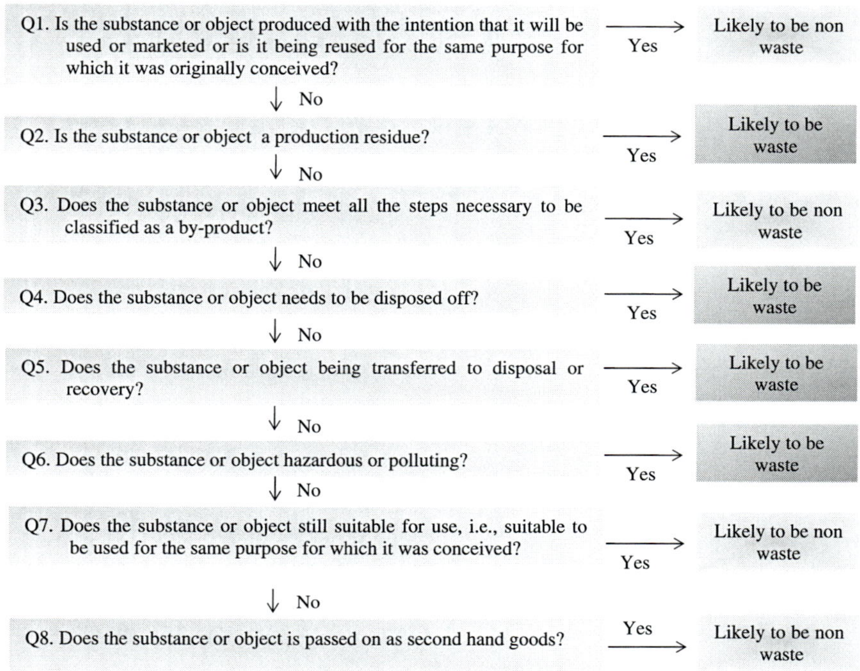

Figure 2.1. Questionnaire based assessment of waste
Source: Defra (2012)

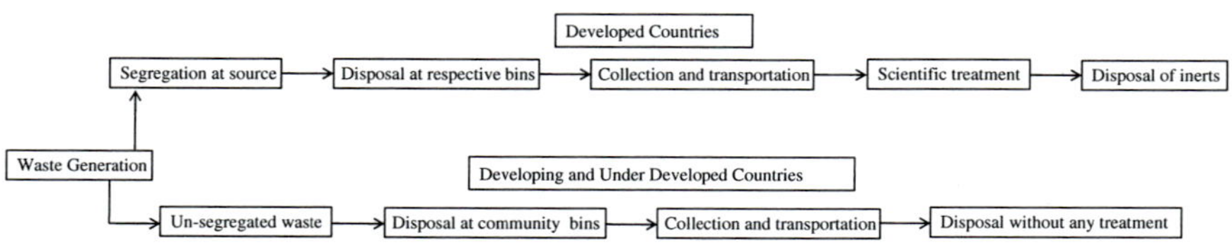

Figure 2.2. Waste management strategies
Source: Bulkeley et al. (2005); Sil et al. (2011)

2005). It is non-existence or negligible in developing and under developed countries (Kumar et al. 2004). The current management strategies of MSW are highlighted in Figure 2.2.

Thus there is a tremendous scope of improvement in terms of MSW in developing and under developed countries. They could follow and adopt the measures and policies from developed nations. This could be done with joint venture assignments leading to knowledge sharing and improving the existing MSW management system. Sil et al. (2011) have reported that most of the transportation vehicles operating in developing countries are outdated and the need to be retrofitted with clean technologies. This again provides scope of

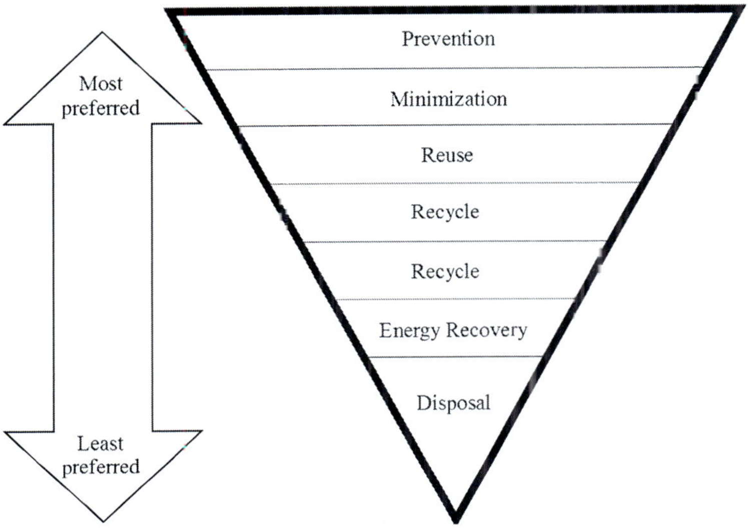

Figure 2.3. Waste management hierarchy

research and development, in developing nations. Waste management is an integrated process of knowledge, segregation, transportation, treatment and management. If the wastes are managed properly, it can lead to production of energy and CDM potentials (Kumar et al. 2004). Sustainable waste management takes into account various aspects, which are being highlighted in Figure 2.3.

Prevention and minimization of waste could be possible at the source level, for e.g. industries level (Phillips et al. 2006). This can be achieved using clean technology mechanism. One can opt for reuse and recycle of waste. This can be carried out for industrial, hazardous and MSW (Kumar et al. 2007). Biodegradable component of MSW has tremendous energy recovery potential. Hence this waste should be treated to produce energy like electricity, biogas, CNG, etc (Kumar et al. 2004; Wakadikar et al. 2012). This can be achieved using waste to energy methodologies, incineration, gasification, biomethanation and many more. More research and development is needed in this sector for optimization of particular technology at different climatic conditions of different parts of the world. If all these measures are taken care, very less quantity of waste would be disposed off in landfill sites.

2.2 TYPES, SOURCES, AND CHARACTERISTICS OF WASTE

Waste can be classified depending on its source. Classification of waste is also determined by the socio-economic sector generating the waste (Li et al. 2011). The types of wastes include:

- Municipal solid waste (MSW)
- Industrial waste
- Hazardous waste
- Agricultural waste
- Medical waste
- Sludge
- Chemical waste
- Biological waste
- E-waste

The flow of waste in environment is cyclic and a continuous process as presented in Figure 2.4. The raw materials like agricultural products, metals, steels, etc. are used for manufacturing various products. Agricultural waste could be generated at this stage. The next step of manufacturing products could generate industrial scraps (Vesilind et al. 2002), which can be reused. However, industrial waste could be sent back to the environment. The finished products are used by individuals and this can lead to generation of domestic waste, often termed as MSW. All these wastes are sent back to environment for their treatment and disposals. After treatment of waste, energy produced can be used for production process as renewable source of energy. The sectors are described in detail in the following sections.

2.2.1 Municipal Solid Waste (MSW)

MSW includes durable goods, non-durable goods, containers and packaging, food wastes and yard trimmings, and miscellaneous inorganic wastes. Sometime it is also referred as garbage. Garbage (e.g., milk cartons and coffee grounds), refuse (e.g., metal scrap, wallboard, and empty containers), sludge from waste treatment plants, water supply treatment plants, or air pollution control facilities (e.g., scrubber sludge). Other discarded material, including solid, semi-solid, liquid, or contained gaseous material resulting from industrial, commercial, mining,

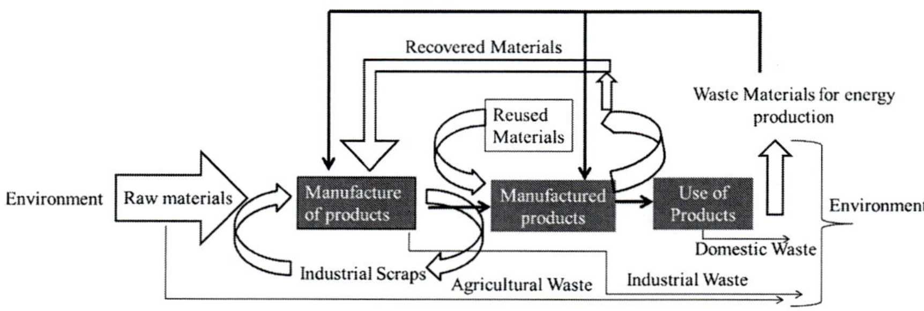

Figure 2.4. Cyclic distributions of products and raw materials in environment
Source: Adapted from Vesilind et al. (2002)

Table 2.1. Sources of MSW

Source	*Activities, location of waste generated*	*Types of waste*
Residential	Single-family and multi-family dwellings, low, medium and high-rise apartments, etc.	Food wastes, rubbish, ashes, special wastes
Commercial and Institutional	Stores, restaurants, markets,office buildings, hotels, motels, schools, print shops, auto repair shops, medical facilities and institutions	Food wastes, rubbish, ashes, demolition and constructionwastes, special wastes, occasionally hazardous wastes
Open areas	Streets, alleys, parks, vacant lots, playgrounds, beaches, highways, recreational areas, etc.	Street sweepings, roadside litter, rubbish, and other special wastes
Treatment plant sites	Water, sewage and industrialwaste water treatment processes	Sludges

Source: Buenrostro et al. (2001); Sethi et al. (2012)

agricultural, and community activities (e.g., boiler slag or fly ash). MSW is mainly generated from households and wastes of similar character from shops, market and offices, open areas, and treatment plant sites (Gómez et al. 2009). Each of these sources of municipal waste is elaborated in Table 2.1

Depending upon the sources, there could be various types of MSW as highlighted in Table 2.2. MSW mainly consists of biodegradable, non-biodegradable and inert components (Wakadikar et al. 2012). The physiological composition of MSW from different continent is presented in Table 2.3.

As observed from Table 2.3, it can be concluded that organic waste component is one the challenging wastes needing proper treatment and disposal. MSW also comprise of non-biodegradable and inert components. The heavy metal composition of biodegradable waste is given in Table 2.4. Similarly, the heavy metal characterization of non-biodegradable is presented in Table 2.5. Hence, treatment of both biodegradable and non-biodegradable components of waste is very much essential for both environmental and human health point of view.

As per COWI (2004), the standard composition of MSW in European, Commonwealth of Independent States (CIS) and Central and Eastern Europe (CEE) countries are given in Table 2.6.

Table 2.2. Types of Solid waste

Component	*Descriptions*
Food wastes	The animal, fruit or vegetable residues resulting from the handling, preparation, cooking and post-consumption leftovers of foods. Food wastes are putrescible, so they will decompose rapidly, especially in warm weather
Rubbish	Combustible and non-combustible solid wastes, excluding food wastes or putrescible materials. Typically, combustible rubbish consists of materials such as paper, cardboard, plastics, textiles, rubber, leather, wood, furniture, and garden trimmings. Non- combustible rubbish consists of glass, crockery, tin cans, aluminium cans, ferrous and non-ferrous metals, dirt and construction material
Ashes and Residues	Materials remaining from the burning wood, coal, coke, and other combustible wastes. Ashes and residues are normally composed of fine, powdery materials, cinders, clinkers, and small amounts of burned and partially burned materials
Demolition and construction wastes	Wastes from razed buildings and other structures are classified as demolition wastes. Wastes from the construction, remodelling, and repairing of residential, commercial, and industrial buildings and similar structures are classified as construction wastes. These wastes may include dirt, stones, concrete, bricks, plaster, lumber, shingles, and plumbing, heating, and electrical parts. They are usually of inert in nature. The main exception is asbestos, for which special disposal is required
Special wastes	Wastes such as street sweepings, roadside litter, catch-basin debris, dead animals, trash like abandoned vehicles, electrical appliances are classifies as special wastes
Treatment plant wastes and dredged soil	The solid and semisolid wastes from water, sewage and industrial wastewater treatment facilities are included in this classification. Sewage sludge is slurry of fine organic-rich particles with a highly variable chemical composition depending on the sources of the effluent and the type and efficiency of the treatment processes. Sewage sludge may have heavy metals and water-soluble synthetic organic compounds, but they may also contain greases, oils and microbial biomass such as bacteria. Dredged materials are excavated from river estuaries, harbours and other waterways to aid navigation

Source: Sethi et al. (2012); Borgatto et al. (2008)

Table 2.3. Composition of MSW from different continents

Continent	*Organic (%)*	*Paper (%)*	*Plastic (%)*	*Glass (%)*	*Metal (%)*	*Others (%)*
Europe	35	21	10	6	4	23
Asia	51	14	11	4	4	16
South America	51	15	11	3	3	18
North America	42	22	13	3	6	14
Australia	52	18	8	5	4	13
Africa	56	9	10	3	3	20

Source: World Bank (2010)

Thus, from the above Table 2.6, one could conclude that MSW mainly consist of food waste (organic waste) which can be treated to generate energy (Dolgen et al. 2005). There are various treatment options available. In order to select appropriate treatment option, segregation and proper analysis of MSW is needed.

2.2.2 Industrial Waste

Industrial waste is defined as waste generated by manufacturing or industrial processes. It includes biodegradable waste, dirt and gravel, masonry and concrete, scrap metals, trash, oil, solvents, chemicals, weed grass and trees, wood and scrap lumber, and similar wastes. Industrial solid waste may be solid, liquid or gases (Musee et al. 2008). It could be further classified into hazardous and non-hazardous waste. Non-hazardous industrial wastes are those that do not meet the EPA's definition of hazardous waste and are not municipal waste (Das et al. 2012). Thus, it could be stated that non-hazardous industrial wastes are in between the characteristics of MSW and hazardous waste. These wastes may be toxic, ignitable, corrosive or reactive. If improperly managed, this waste can pose dangerous health and environmental consequences.

2.2.3 Hazardous Waste

U.S. EPA (2005) defined hazardous waste as a waste, or combination of wastes, which because of its quantity, concentration, or physical, chemical, or infectious characteristics may (a) cause, or significantly contribute to, an increase in mortality or an increase in serious irreversible, or incapacitating reversible, illness; or (b) pose a substantial threat or potential hazard to human health or the environment when improperly treated, stored, transported, or disposed of, or otherwise managed. In order words, hazardous wastes are certain commercial products such as cleaning fluids, paints or pesticides discarded by commercial establishments or individuals, which may be harmful to human health and environment.

USEPA, as per the characteristics of hazardous wastes, identifies it as a waste, which can cause death or illness in humans or ecological damage (Li et al. 2012).

Table 2.4. Heavy metal composition in of biodegradable waste (mg/kg dry weight basis)

Category	*As*	*Cd*	*Cr*	*Sb*	*Hg*	*Mn*	*Pb*	*P*	*Zn*	*Cu*	*Ni*	*Source*
Paper/ cardboard	2	1	20		0.1		42					a
Paper/ cardboard	2	11	67		<0.1		127					b
Paper/ cardboard	<5	1	10		0.2		23					c
Paper/ cardboard (range)		0.03–0.3	20–30				72–228		80–180			d
Paper/ cardboard		5.26	10.9				44.3		393			e
Paper/ cardboard		0.4					23		424			f
Wood	6	12	231		1		277					b
Wood	<5	<1	5		<0.1		21					c
Food waste			7	8					17			g
Meat waste								8200 ± 2400	32 ± 21			i
Fish waste						6 ± 7		20400 ± 6400	62 ± 12	1 ± 1		i
Fish and vegetable waste						8 ± 2		1700 ± 400	11 ± 3	4 ± 1		i
Restaurant waste							11 ± 16	14400 ± 6300	17 ± 19	1 ± 1		i

Household waste		2 ± 5	30 ± 23		55 ± 71	206 ± 360	2000 ± 800	160 ± 223	289 ± 764		i
Organic waste	5	1	25	0.1		76					a
Organic waste	37	5	310	<0.1		188					b
Organic waste	30	1	55	0.5		90					c
Organic waste		1	607			407		1010			e
Organic waste		0.4				37		241			f
Organic waste (range)		0.06–0.45	10–15			20–156		50–100			c
Compost: (mean)		0.69	21	0.2		105		186	53	15	h
Compost: (range)		0.47–1.12	11-33	0.08–0.34		48–219		100–268	32–92	11–21	h

Sources: a) Burnley (2007); b) Otte (1995); c) White et al. (1999); e) Chandler et al. (1997); f) Rotter et al. (2004); g) Jung et al. (2006); h) Dimambro et al. (2006)

Table 2.5. Chemical composition of non-biodegradable waste (mg/kg dry weight basis)

Category	*As*	*Cd*	*Cr*	*Sb*	*Se*	*Hg*	*Pb*	*Zn*	*Cu*	*Ni*	*Sn*	*P*	*Source*
Total plastics	3	388	853			<0.1	302						a
Total plastics (range)		0.39–2.1	20–22.5				32–108	30–80					b
Total plastics		40.4	106				602	284					c
Polymer market		40	50	2600			1900	300	100	30	20	1200	e
Dense plastics	2	32	181			0.1	879						f
Film plastics	4	4	195			0.1	1595						f
Film plastics		1.3					219	730					g
PVC	<0.5	66	29			0.2	50						h
Plastics		67.4					500	4380					g
Packaging plastics		0.9					109	768					g
Packaging plastics			420	26	0.4		11	50					h
Textiles	2	1	25			0.1	35						f
Textiles	2	16	934			<0.1	30						a
Textiles	<5	1	17			<0.1	35						h
Textiles (range)		0.06–0.3	7.5–10				4–12	10–20					b
Textiles		32.6	3160				773	4000					c
Textiles		1					35	520					g

Sources: a) Otte (1995); b) White et al. (1999); c) Chandler et al. (1997); d) Morf et al. (2007); e) Vehlow and Mark (1997); f) Burnley (2007); g) Rotter et al. (2004); h) Jung et al. (2006)

Table 2.6. Standard composition of MSW in EU, CIS, and CEE countries

Sources	*Fractions*	*EU (%)*	*CEE (%)*	*CIS (%)*
Multi storey houses in urban areas	Paper	23.3	7.1	8.5
	Cardboard	5.3	4.5	6.4
	Cardboard, bulky	0.4	0.4	0.4
	Plastics	5.6	5.4	6.4
	Glass	4.3	7.2	8.5
	Metal	1.8	2.7	3.4
	Metal, bulky	2.1	0.9	0.9
	Food	28.7	44.6	46.7
	Garden	3.1	2.2	2.1
	Non-combustible, residual	1.6	1.8	4.2
	Non-combustible, residual bulky	10.2	2.2	2.1
	Hazardous	0.3	0.2	0.2
	WEEE	1.2	0.4	0.4
Household, rural areas	Paper	24.5	13.0	14.2
	Cardboard	5.9	10.8	11.8
	Cardboard, bulky	1.0	-	-
	Plastics	5.9	10.8	11.8
	Glass	4.2	13.0	14.2
	Metal	2.0	2.2	2.4
	Metal, bulky	3.3	-	-
	Food	26.1	25.8	28.3
	Garden	-	-	-
	Non-combustible, residual	4.2	2.2	2.4
	Non-combustible, residual bulky	3.3	-	-
	Combustible, residual	8.2	21.6	14.2
	Combustible, residual, bulky	9.8	-	-
	Hazardous	0.3	0.2	0.2
	WEEE	1.3	0.4	0.5
Commerce and services	Paper	38.0	30.0	25.0
	Cardboard	10.0	14.0	15.0
	Plastics	10.0	15.0	20.0
	Glass	2.0	2.0	1.0
	Metal	3.0	2.0	1.0
	Food	15.0	16.0	17.0
	Garden	5.0	5.0	5.0
	Non-combustible, residual	5.0	5.0	5.0
	Combustible, residual	10.0	10.0	10.0
	WEEE	2.0	1.0	1.0

(Continued)

Table 2.6. Standard composition of MSW in EU, CIS, and CEE countries (Continued)

Sources	Fractions	EU (%)	CEE (%)	CIS (%)
Industries	Paper	20.0	25.0	33.0
	Cardboard	25.0	25.0	24.0
	Plastics	4.0	6.0	7.0
	Glass	1.0	3.0	5.0
	Metal	25.0	15.0	4.0
	Food	6.0	4.0	1.0
	Non-combustible, residual	8.0	10.0	12.0
	Combustible, residual	11.0	12.0	14.0

Source: COWI (2004)

USEPA has recommended four characteristics as the means for identifying hazardous waste (U.S. EPA 2005):

- Ignitability,
- Corrosivity,
- Reactivity, and
- Toxicity. USEPA has listed all the substances as per their characteristics and given them a unique code for their identification. Ignitable, corrosive, and reactive wastes carry the waste codes D001, D002, and D003, respectively; while wastes displaying toxicity carry waste codes D004 to D043.

2.2.3.1 Ignitability

Ignitable wastes can readily catch fire and lead to combustion. Paints, cleaners, and other industrial wastes pose such fire hazard. Most ignitable wastes are liquid in physical form. Flash point test as the method is used for determining whether a liquid waste is combustible enough to be treated as hazardous (Talini et al. 2005). The flash point test determines the lowest temperature at which a chemical ignites when exposed to flame. Many wastes in solid or non-liquid physical form (e.g., wood, paper) can also readily catch fire and sustain combustion, but EPA did not intend to regulate most of these non-liquid materials as ignitable wastes (U.S. EPA 2005). A non-liquid waste is only hazardous due to ignitability if it can spontaneously catch fire under normal handling conditions and can burn so vigorously that it creates a hazard. Certain compressed gases and chemicals called oxidizers can also be ignitable (Lu et al. 2012). Ignitable wastes carry the waste code D001.

2.2.3.2 Corrosivity

Corrosive wastes are acidic or alkaline (basic) wastes, which can readily corrode or dissolve flesh, metal, or other materials (Supplit et al. 2007). They are also among the most common hazardous wastes. Waste sulphuric acid from automotive batteries is an example of a corrosive waste. There are two criteria to identify corrosive hazardous wastes. The first is a pH test. Aqueous wastes with a pH greater than or equal to 12.5, or less than or equal to 2 are corrosive under EPA's rules. A waste may also be corrosive if it has the ability to corrode steel in a specific EPA-approved test protocol (U.S. EPA 2005). Corrosive wastes carry the waste code D002.

2.2.3.3 *Reactivity*

A reactive waste is one that readily explodes or undergoes violent reactions (Weltens et al. 2012). Common examples are discarded munitions or explosives. In many cases, there is no reliable test method to evaluate a waste's potential to explode or react violently under common handling conditions. Therefore, EPA uses narrative criteria to define most reactive wastes and allows waste handlers to use their best judgment in determining if a waste is sufficiently reactive to be regulated (U.S. EPA 2005). This is possible because reactive hazardous wastes are relatively uncommon and the dangers they pose are well known to the few waste handlers who deal with them. Wastes exhibiting the characteristic of reactivity are assigned the waste code D003; and considered reactive if it meets any of the following criteria (Weltens et al. 2012):

- it can explode or violently react when exposed to water, when heated, or under normal handling conditions;
- it can create toxic fumes or gases when exposed to water or under normal handling conditions;
- it generates toxic levels of sulfide or cyanide gas when exposed to a pH range of 2 through 12.5.

2.2.3.4 *Toxicity Characteristics*

The leaching of toxic compounds or elements into groundwater or drinking water supplies from wastes disposed off in landfills is one of the most common ways to identify toxicity of the chemicals found in industrial wastes (U.S. EPA 2005). USEPA developed a characteristic design to identify wastes likely to leach dangerous concentrations of certain known toxic chemicals into groundwater. In order to predict whether any particular waste is likely to leach chemicals into groundwater in the absence of special restrictions on its handling, EPA first designed a lab procedure that replicates the leaching process and other effects that occur when wastes are buried in a typical municipal landfill. This laboratory procedure is known as the Toxicity Characteristic Leaching Procedure (TCLP) (Sloot and Kosson 2012). Using the TCLP on a waste sample creates a liquid leachate that is similar to the liquid expected to be found in the ground near a landfill containing the same waste. Once the leachate is created in the laboratory, a waste handler must determine whether it contains any of 39 different toxic chemicals above specified regulatory levels. If the leachate sample contains a sufficient concentration of one of the specified chemicals, the waste exhibits the toxicity characteristic (TC). EPA used groundwater modelling studies and toxicity data for a number of common toxic compounds and elements to set these threshold concentration levels (U.S. EPA 2005). The list of 39 chemicals and their toxicity level is given in Table 2.7.

Industries are the main sources of hazardous waste. Their process activities lead to generation of hazardous waste. Table 2.8 provides list of hazardous waste and their industrial sources. In order to identify hazardous waste, one can refer the listing of USEPA, which provides codes for hazardous waste as listed below:

- Ignitable waste
- Corrosive waste
- Reactive waste
- Toxicity characteristic waste
- Acute hazardous waste
- Toxic waste

As per the listing, USEPA list hazardous wastes in F, K, P and U. The description and types of waste included in these categories are given in Table 2.9.

Based on different criteria mentioned above, identification strategy for hazardous waste is developed. This is given in Figure 2.5.

As per Figure 2.5, one can access or identify a waste into hazardous waste by a questionnaire survey. This could be treated as preliminary assessment. If waste fall on hazardous waste category, through preliminary assessment, confirmatory tests are carried out to confirm the waste as hazardous waste. Once confirmed, the waste has to be managed and treated as hazardous waste. If the waste fails or gives negative result from preliminary and confirmatory test, it should be treated and managed as solid waste. Thus, identification of hazardous waste is a complex process so is its treatment and management.

2.2.4 Agricultural Waste

Agricultural waste is any substance or object from premises used for agriculture or horticulture, which the holder discards, intends to discard or is required to discard. It is waste specifically generated by agricultural activities (Buenrostro et al. 2001). For example, waste which came from a farm shop or a vegetable packing plant would not be considered as agricultural waste. Some examples of agricultural waste include empty pesticide containers, old silage wrap, out of date medicines and wormers, used tyres, surplus milk, packaging waste, silage plastics, scrap metals, batteries, pesticides, etc. Agricultural waste consist of almond, baggase, straw, corn, cotton, peanut, rice hulls, straw and wheat straw. They have very high heating or calorific value. Other chemical component includes ash content, carbon, oxygen and nitrogen (Zhang et al. 2011).

Table 2.7. Toxicity characteristic level

Code	*Contaminants*	*mg/l*
D004	Arsenic	5.0
D005	Barium	100.0
D018	Benzene	0.5
D006	Cadmium	1.0
D019	Carbon tetrachloride	0.5
D020	Chlordane	0.03
D021	Chlorobenzene	100.0
D022	Chloroform	6.0
D007	Chromium	5.0
D023	o-cresol	200.0
D024	m-cresol	200.0
D025	p-cresol	200.0
D026	Total cresol	200.0
D016	2,4-D	10.0
D027	1,4-Dichlorobenzene	7.5
D028	1,2-Dicholoethane	0.5
D029	1,1-Dichloroethylene	0.7
D030	2,4-Dinitrotoluene	0.13
D012	Endrin	0.02
D031	Heptachlor (and its epoxide)	0.003
D032	Hexachlorobenzene	0.13
D033	Hexachlorobutadiene	0.5
D034	Hexachloroethane	3.0
D008	Lead	5.0
D013	Lindane	0.4
D009	Mercury	0.2
D014	Methoxychlor	10.0
D035	Methyl ethyl ketone	200.0
D036	Nitrobenzene	2.0
D037	Pentachlorophenol	100.0
D038	Pyridine	5.0
D010	Selenium	1.0
D011	Silver	5.0
D039	Tetrachloroethylene	0.7
D015	Toxaphene	0.5
D040	Trichloroethylene	0.5
D041	2,4,5-Trichlorophenol	400.0
D042	2,4,6-Trichlorophenol	2.0
D017	2,4,5-TP (Silvex)	1.0
D043	Vinyl Chloride	0.2

Source: U.S. EPA (2005)

Table 2.8. Examples of hazardous waste from different sources

Industries	*Waste type*	*Sources*
Chemical manufacturers	Acids and bases Spent solvents Reactive waste Wastewater containing organic constituent	Manassaram et al. (2003); Musee et al. (2008)
Painting	Heavy metal solution Waste ink Solvents Ink sludge containing heavy metals	Andrade et al. (2012)
Petroleum refining	Wastewater containing benzene and other hydrocarbons Sludge	Alidi (1996); Das et al. (2012)
Leather product manufacturing	Toluene and benzene	Kolomaznik et al. (2008)
Paper	Heavy metals Ignitable solvents	Duan et al. (2008)
Construction industries	Ignitable paint waste Spent solvents Strong acid and bases	Duan et al. (2008); Zhao et al. (2010)
Metal manufacturing	Sludge containing heavy metals Cyanide waste Paint waste	Duan et al. (2008); Zhao et al. (2010)

2.2.5 Medical Waste

Medical waste is also known as infectious waste or biomedical waste. It is defined as waste generated during diagnosis, testing, treatment, research or production of biological products for humans or animals. It includes syringes, live vaccines, laboratory samples, body parts, bodily fluids and waste, sharp needles, cultures and lancets. The main sources are hospitals, medical clinics and laboratories (Marinković et al. 2008). This waste can be detrimental to human health hence public and individuals must follow procedures not to come in contact with this kind of waste. The categories and types of medical waste includes human anatomical waste, animal waste, microbiological and biotechnology waste, waste sharps, discarded medicines, liquid wastes, etc. (Patwary et al. 2011).

Table 2.9. Listing of hazardous waste

List	*Description*	*Types of Waste*
F	Non industry specific process waste	• Spent solvent • Electroplating and other metal finishing wastes • Dioxin containing wastes • Chlorinated aliphatic hydrocarbons • Wood preserving wastes • Petroleum refinery wastewater treatment sludge • Multi source leachate
K	Industry specific process waste	• Wood preservation • Organic chemicals • Pesticides • Petroleum refining • Ink formulation • Coking • Veterinary pharmaceuticals • Inorganic pigments • Inorganic chemicals • Explosives • Iron and steel • Primary aluminium • Secondary lead
P and U*	Discarded unused chemicals	• Commercial chemical products • Chemical intermediates • Contaminated residues, soil and debris • Containers and inner liners • Technical grade solutions • Sole grade chemicals • Active ingredients

*P list contain acute hazardous waste while U list contain toxic constituents
Source: U.S. EPA (2005)

2.2.6 E Waste

E-waste is the term, which is commonly used to describe "Waste Electrical and Electronic Equipment (WEEE)". Informal processing of E-waste in developing countries may cause serious health and pollution problems, though these countries are also most likely to reuse and repair electronics (Tasaki et al. 2006). All electronic scrap components, such as CRTs, may contain contaminants such as lead, cadmium, beryllium, or brominated flame-retardants. Even in developed countries recycling and disposal of E-waste may involve significant

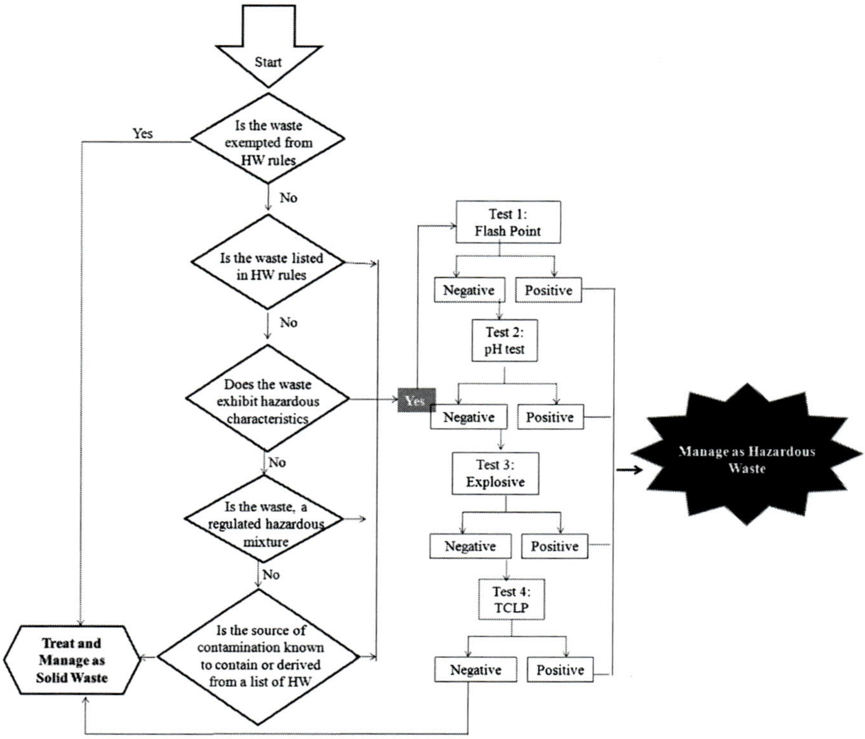

Figure 2.5. Identification criteria for hazardous waste
Source: Adapted from U.S. EPA (2005)

risk to workers and communities and great care must be taken to avoid unsafe exposure in recycling operations and leaching of material such as heavy metals from landfills and incinerator ashes. The types and sources of E waste are given in Table 2.10.

E-waste comprises 70% of overall toxic waste in USA (Lim and Schoenung 2010). It is also estimated that about 50 million cell phones are replaced every month worldwide. E-waste contains various types of hazardous chemical compounds. Some of the compounds present in E-wastes are polychlorinated biphenyls (PCB), tetrabromo-bisphenol A (TBBA), polybrominated biphenyls (PBB), polybrominated diphenyl ethers (PBDE), chlorofluorocarbon (CFC), polyvinyl chloride (PVC), arsenic, barium, beryllium, cadmium, chromium VI, lead, lithium, mercury, nickel, selenium, zinc sulphide, toner dust and other radioactive substance (Bouvier and Wagner 2011).

All these compounds need to be treated before their disposal. Otherwise, they could cause both environmental and human health hazards. Some the diseases that could be caused are liver damage, cancer, bronchitis, etc. Environmental damage could be much more dangerous as these heavy meats could contaminate soil, ground and surface water and thereby affecting the human population.

Table 2.10. Sources and types of E-waste

Sources	*Types*
Household appliances	Washing machines, dryers, refrigerators, irons, TV, monitors, toaster, coffee machines, air conditioners
Offices, information and communication centres	PCs, laptops, mobiles, telephones, fax machines, copiers, printers, Xerox machines
Entertainment and consumer electronics	Televisions, VCR, DVD players, CD players, DVD, CDs, radio, cassettes, Hi-Fi sets
Lighting equipments	Fluorescent tubes, sodium lamps, used bulbs, led light, CFL, tube lights
Electric and electronic tools	Drills, electric saws, sewing machines, lawn movers, electric motors, fans
Toys and sports materials	Electric toy trains, cars, bikes, remote controls, coin slot machines, treadmills, etc
Medical instruments and equipments	
Surveillance and control equipments	CCTV cameras, monitors
Automatic issuing machines	

Source: Lim and Schoenung (2010); Bouvier and Wagner (2011)

2.3 GENERATION OF WASTE

Generation of waste is directly proportional to population growth (Wakadikar et al. 2012). Generation of MSW is increasing with leaps and bounds (Lu et al. 2006). Generation of industrial waste and hazardous waste depend on production rate and demand in the market. If the demand is more, production will be more and this lead to more generation of industrial and hazardous waste. Some countries are enforcing rules and regulation towards use of clean technologies to reduce the generation of industrial and hazardous waste from processing plants and units. As per the World Bank (2010), the developed countries are the main producers of waste. They generate about 46% of total global waste. This is mainly due to the behavioural and cultural aspect of the people residing in developed countries (Bogner et al. 2007). Figure 2.6 highlights the contribution of waste generation from different countries as per their economic status.

It could also be estimated that most of the waste generated in high income countries are treated properly. This does not hamper the progress of the nation as compared with other low, middle and upper middle income countries. MSW generation also depends on culture and behavioural aspect of individual as well. As

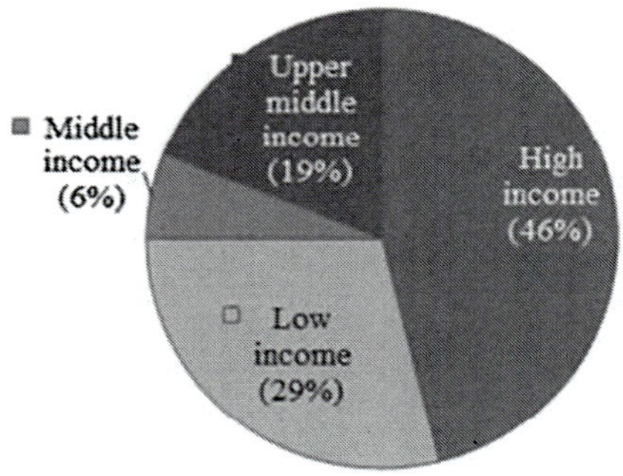

Figure 2.6. Waste generation from different countries

per Bogner et al. (2007), the generation rate of MSW from low, high and middle income country is 0.1–0.6, 0.2–0.5, and 0.3–0.8 tonnes/capita/year, respectively (Bonger et al. 2007).

In the low income and middle income group countries, the population is divided as per urban and rural areas (Alsan et al. 2006). The people residing in these areas differ in their cultural values, behaviour, habits, etc., thereby per capita generation of waste also differs. If one wishes to compare the per capita waste generation rate, they should consider the urban population of low income countries and population of high income countries. Comparing waste generation rate from urban population of low and high income countries show that waste generation higher for high income countries as compared with low income (Bogner et al. 2007).

Indian Government has carried out a research and reported the MSW generation rate in Indian cities as per the population range (Sharholy et al. 2007); which are presented in Table 2.11. This data indicates that waste generation rate increases with population. This is mainly due to life style, social, cultural and economic activities of people.

Thus, all these data could be useful in formulating, strategizing and predicting MSW generation from different parts of the world including low income, high income and middle income countries. Similarly, E-waste generation

Table 2.11. MSW generation rate as per population range

Population range (millions)	*MSW generation (kg/capita/day)*
0.1–0.5	0.21
0.5–1.0	0.25
1.0–2.0	0.27
2.0–5.0	0.35
>5.0	0.50

Source: MoUD (2000)

is also increasing with leaps and bounds. E-waste is also transported from developed countries to India and China for their disposal and treatment (Hicks et al. 2005). This trans-boundary trade of E-waste is of great concern. E-waste must be treated and disposed in scientific manner to avoid environmental hazards. Per capita collection of E-waste ranges from 1.1–28.1 kg/person/year in European countries (Ongondo et al. 2011). Finally, it can be stated that waste generation depends on social, cultural and behavioural activities of the nation. There are many sources of wastes and they contain various contaminants like heavy metals. Hence, all the materials need to be collected, processed and disposed in a scientific manner to avoid environmental and human hazards. If they are not disposed properly it could lead to various challenges like ground water contamination, social nuisance, disposal of waste on lands leading to green House Gas emissions, etc.

Finally, it can be stated that waste generation depends on social, cultural and behavioural activities of the nation. There are many sources of wastes and they contain various contaminants like heavy metals. Hence, all the materials need to be collected, processed and disposed in a scientific manner to avoid environmental and human hazards. If they are not disposed properly it could lead to various challenges like ground water contamination, social nuisance, greenhouse gas emissions, etc.

2.4 WASTE TESTING AND ANALYSIS

Testing and analysis of waste are important to understand and determine the type of waste. The identification of waste (Figure 2.1) and hazardous waste (Figure 2.5) are already mentioned earlier. Hazardous wastes can be confirmed with assessment of their ignitability, corrosivity, reactivity and toxicity. One could follow methodologies of standard procedures like ASTM standards or APHA. Based on the assessment, level of treatment and disposal criteria for the hazardous and industrial waste could be chalked out. But MSW has mixed wastes. Hence proper segregation is necessary to chalk out its treatment and disposal strategies. Based on the analysis report one could select appropriate technology for treatment of waste (Kumar et al. 2004). MSW must be investigated for its physical and chemical properties. The physical parameters include density (specific weight), moisture content, particle size distribution, field capacity and permeability. Density is defined as weight of a material per unit volume (kg/m^3 or lb/ft^3); and it varies with geographic locations, season and time (D'Alisa et al. 2012). Sometime density is important in order to chalk out number and volume of transportation vehicles and size of disposal facility (Chandrappa and Das 2012). The moisture content of the waste is expressed as percentage of the wet waste present in the MSW. It can be determined by the Eq. 2.1.

$$M = (w - d)/w \times 100 \qquad \text{(Eq. 2.1)}$$

where, M is moisture content (%); w is weight of wet waste (g); d is weight of dry waste (g).

Table 2.12. Typical density and moisture content of MSW

	Density (kg/m^3)		Moisture content (%)	
Component	Range	Typical	Range	Typical
Food Waste	130–480	290	50–80	70
Paper	40–130	89	4–10	6
Plastics	40–130	64	1–4	2
Glass	160–480	194	1–4	2

Source: D'Alisa et al. (2012)

Typical density and moisture content of waste are highlighted in Table 2.12.

The size and distribution of components of wastes are important for recovery of materials (von Blottnitz et al. 2002). Mechanical separation process like screens and magnetic separators are used for recovery of various materials. The particle size can be determined by the Eqs. 2.2 to 2.4

$$Sc = L \qquad \text{(Eq. 2.2)}$$

$$Sc = ((L + W))/2 \qquad \text{(Eq. 2.3)}$$

$$Sc = ((L + W + h))/3 \qquad \text{(Eq. 2.4)}$$

where, Sc is size of component (mm); L is length (mm); W is width (mm); and h is height (mm). Field capacity is the amount of moisture that can be retained by the wastes. It is important for determining the formation of leachate in the landfills (Wu et al. 2012). It varies with pressure and decomposition rate of wastes. However, typically its value ranges from 50–60% for MSW (Wu et al. 2012). Permeability, sometime referred to as hydraulic conductivity, of wastes governs movement of liquids and gases in landfill. It depends on pore size distribution, surface area and porosity (Wu et al. 2012). Other physical properties, which are important for decision-making, are colour, optical, magnetic and electric properties of waste (Chandrappa and Das 2012).

Chemical analysis of MSW is important to evaluate alternate processing and recovery options from wastes. It is mainly carried out through proximate analysis, fusion through ash, ultimate analysis (major elements) and energy content (Jansen et al. 2004). Proximate analysis is mainly carried out for combustible components of MSW (Thipkhunthod et al. 2005). It includes following assessments:

- Volatile solids (ignition at 950 °C in absence of oxygen)
- Fixed carbon (combustible residue left after ignition)
- Ash (weight of residue after combustion)

Chandrappa and Das (2012) estimated that volatiles component of mixed food waste, mixed paper, plastic and mixed MSW was 21.4%, 75.9%, 95.8%, and 52.0% respectively, while their carbon content was found to be 3.6%, 8.4%, 2.0% and 7.0% respectively, and finally ash content was 5.0%, 5.4%, 2.0% and 20.0% respectively. Chandrappa and Das (2012) also reported ash content of glass material in the range of 96–99%.

Fusion point of ash is the temperature at which the ash formed from burning of waste would form a solid (clinker) by fusion and agglomeration. Typical fusing temperatures of wastes are 1100–1200°C. Ultimate analysis involves determination of major elements like carbon, hydrogen, oxygen, nitrogen, sulfur, ash and sometime few halogen compounds (Parikh et al. 2007; Jansen et al. 2004). It is essential for predicting the chemical composition of organic component of MSW, especially C/N ratio. It is important to maintain proper and suitable C/N ratio for biological transformations. Analysis of waste for carbon, nitrogen, hydrogen and sulphur can be carried out using CHNS analysers.

The energy content of the waste is determined using bomb calorimeter tests. The energy is mainly recovered from the organic component of MSW. The average energy content of typical MSW is approximately 10,000 kJ/kg (Chandrappa and Das 2012).

Biological properties of MSW depend on water soluble constituents like sugar, starch, amino acids, and various organic acids; hemicelluloses; cellulose; fats, oil and waxes; lignin; lignocelluloses; proteins, etc. present in waste (Tejada et al. 2009). Solid waste contains bacteria, protozoa, fungi etc. (Chandrappa and Das 2012). Numerous human and vector borne diseases are caused by these microorganisms. Fungi can cause infection to skin, hair, nail and lungs (Ellis et al. 2007). Solid waste can host an array of insects (cockroaches, dung beetles, ants, termites, mosquitoes, honey bees and house flies), arthropods (spider and scorpions) and annelids (centipede, millipede and earthworms) (Chandrappa and Das 2012).

Microorganisms play an important role in the decomposition of decomposable fraction of solid waste. Thermophilic bacteria would break down proteins and other easily biodegradable material. Fungi and actinomycetes would degrade complex organic matter like cellulose and lignin. Breakdown and decomposition of MSW leads to generation of methane (CH_4). A number of studies has been carriedout to determine the CH_4 potential of solid organic substrates and MSW. Different assays were described as ISO standards, which can be categorized in two groups; one group deals with the definition of anaerobic biodegradability (ISO 14853–999; ASTM D 5511–1994; ASTM 5210–1992; ASTM E 2170–2001; ISO 15473–2002 and others) while other deals with the ultimate biodegradability (ISO 11734–1995; ISO/DIS 14853–1999; UK Environment Agency 2005) (Wakadikar et al. 2012). These assays include the biochemical methane potential (BMP), dynamic respiration rate, and the COD test (Owen et al. 1979). Of these, the BMP assay is the most popular method to estimate CH_4 potential of wastes. BMP assay requires minimal labor, cost for set-up, monitoring and provides accurate results (Wakadikar et al. 2012). Laboratory scale experimental set up for BMP assay is highlighted in Figure 2.7. The methane potential of different wastes is given in Table 2.13.

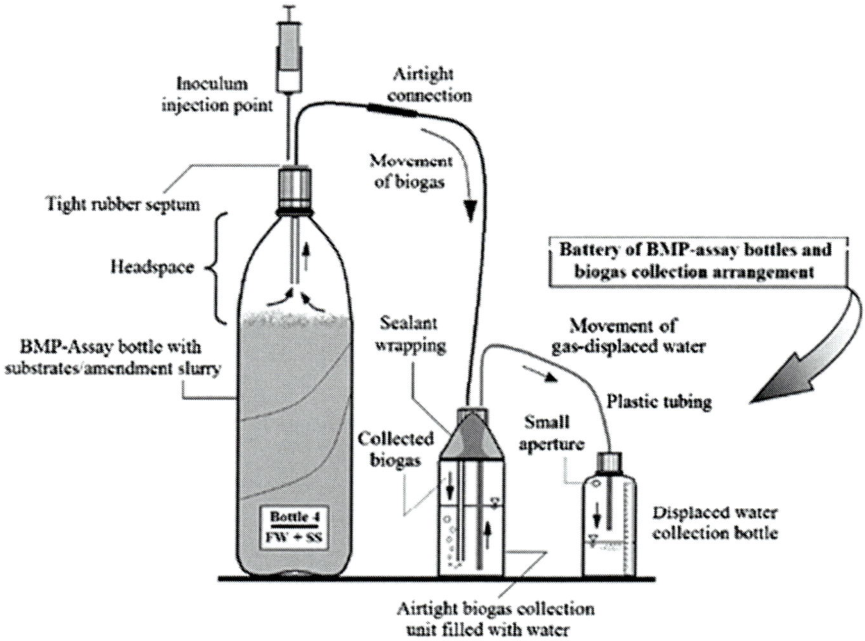

Figure 2.7. Experimental set up of BMP assay (Wakadikar et al. 2012)

Table 2.13. BMP of different kind of wastes

Types	*BMP ($l\,CH_4/g$ VS)*
Cellulose/paper	0.93–3.50
Food waste	0.97–1.81
Vegetable waste	0.24–1.05
Mixed MSW	0.10–0.57

Note: Leachate was used as VS; Source: Wakadikar et al. (2012)

BMP assays could be used as medium to understand the methane and energy potential of wastes. BMP assay depend on type of volatile solids (VS) and their rate of application. But results of BMP assay must be interpreted with utmost care (Wakadikar et al. 2012).

2.5 CONCLUSION

There are various kinds of waste including MSW, industrial, hazardous, biomedical and E-wastes. Each kind of waste has its own characteristics and must be treated properly to avoid environmental damage and health impacts. Waste generation also increases with time, population explosions and habits that influence the waste management, which is becoming one of the challenging tasks of the modern society. Reduction of waste is very challenging and its generation may increase with time. But waste can be treated as resource and many valuable products can be obtained from them that would lead to green jobs and betterment of society.

References

Alidi, A. S. (1996). "A multi objective optimization model for the waste management of the petrochemical industry." *Appl. Math. Modell.*, **20**(12), 925–933.

Alsan, M., Bloom, D. E., and Canning, D. (2006). "The effect of population health on foreign direct investment inflows to low- and middle-income countries." *World Dev.*, **34**(4), 613–630.

Andrade, L. C., Míguez, G. C., Gómez, M. C. T., and Bugallo, P. M. B (2012). "Management strategy for hazardous waste from atomised SME: application to the printing industry." *J. Cleaner Prod.*, **35**, 214–229.

Bogner, J., et al. (2007). "Waste management." *Climate Change 2007: Mitigation. Contribution of Working Group III to the Fourth Assessment Report of the Intergovernmental Panel on Climate Change*, B. Metz, O. R. Davidson, P. R. Bosch, R. Dave, and L. A. Meyer, eds., Cambridge University Press, Cambridge, U.K.

Borgatto, A. V. A., Izzo, R. L. S., and Mahler, C. F. (2008). "Application of a morphologic classification of Brazilian MSW." *Proc., GeoCongress*, ASCE, Reston, VA.

Bouvier, R., and Wagner, T. (2011). "The influence of collection facility attributes on household collection rates of electronic waste: The case of televisions and computer monitors." *Resour. Conserv. Recycl.*, **55**(11), 1051–1059.

Buenrostro, O., Bocco, G., and Cram, S. (2001). "Classification of sources of municipal solid wastes in developing countries." *Resour. Conserv. Recycl.*, **32**(1), 29–41.

Bulkeley, H., Watson, M., Hudson, R., and Weaver, P. (2005). "Governing municipal waste: Towards a new analytical framework." *J. Environ. Policy Plann.*, 7(1), 1–23.

Burnley, S. J. (2007). "The use of chemical composition data in waste management planning—A case study." *Waste Manage.*, **27**(3), 327–336.

Chandler, A. J., et al. (1997). *Municipal solid waste incinerator residues*, International Ash Working Group (IAWG), Amsterdam, the Netherlands.

Chandrappa, R., and Das, D. B. (2012). *Solid waste management: Principles and practice*, Springer, Berlin.

Chang, Y. F., Lin, C. J., Chyan, J. M., Chen, I. M., and Chang, J. E. (2007). "Multiple regression models for the lower heating value of municipal solid waste in Taiwan." *J. Environ. Manage.*, **85**(4), 891–899.

COWI. (2004). *The feasible model, version 2, user manual and documentation*, Kongens Lyngby, Denmark.

D'Alisa, G., Di Nola, M. F., and Giampietro, M. (2012). "A multi-scale analysis of urban waste metabolism: Density of waste disposed in Campania." *J. Cleaner Prod.*, **35**, 59–70.

Das, A., Gupta, A. K., and Mazumder, T. N. (2012). "Vulnerability assessment using hazard potency for regions generating industrial hazardous waste." *J. Hazard. Mater.*, **209**, 308–317.

Defra. (2012). "Guidance on the legal definition of waste and its application." ⟨www.defra.gov.uk/ . . . /files/pb13813-waste-legal-def-guide.pdf⟩ (Jul. 25, 2014).

Dimambro, M. E., Lillywhite, R. D., and Rahn, C. (2006). *Biodegradable municipal waste composts: Analysis and application to agriculture*, Warwick HRI, Univ. of Warwick, U.K.

Dolgen, D., Sarptas, H., Alpaslan, N., and Kucukgul, O. (2005). "Energy potential of municipal solid wastes." *Energy Sources*, **27**(15), 1483–1492.

Duan, H., Huang, Q., Wang, Q., Zhou, B., and Li, J. (2008). "Hazardous waste generation and management in China: A review." *J. Hazard. Mater.*, **158**(2-3), 221–227.

Ellis, J. G., Dodds, P. N., and Lawrence, G. J. (2007). "The role of secreted proteins in diseases of plants caused by rust, powdery mildew and smut fungi." *Curr. Opin. Microbiol.*, **10**(4), 326–331.

Gómez, G., Meneses, M., Ballinas, L., and Castells, F. (2009). "Seasonal characterization of municipal solid waste (MSW) in the city of Chihuahua, Mexico." *Waste Manage.*, **29**(7), 2018–2024.

Hicks, C., Dietmar, R., and Eugster, M. (2005). "The recycling and disposal of electrical and electronic waste in China—Legislative and market responses." *Environ. Impact Assess. Rev.*, **25**(5), 459–471.

Jansen, J. C., Spliid, H., Hansen, T. L., Svärd, Å., and Christensen, T. H. (2004). "Assessment of sampling and chemical analysis of source-separated organic household waste." *Waste Manage.*, **24**(6), 541–549.

Jung, C. H., Matsuto, T., and Tanaka, N. (2006). "Flow analysis of metals in a municipal solid waste management system." *Waste Manage.*, **26**(12), 1337–1348.

Kolomaznik, K., Adamek, M., Andel, I., and Uhlirova, M. (2008). "Leather waste—Potential threat to human health, and a new technology of its treatment." *J. Hazard. Mater.*, **160**(2-3), 514–520.

Kumar, S., Gawaikar, V., Gaikwad, S. A., and Mukherjee, S. (2004). "Cost-benefit analysis of landfill system with gas recovery for municipal solid waste management: A case study." *Int. J. Environ. Stud.*, **61**(6), 637–650.

Kumar, S., Mukherjee, S., Chakrabarti, T., and Devotta, S. (2007). "Hazardous waste management system in India: an overview." *Crit. Rev. Env. Sci. Technol.*, **38**(1), 43–71.

Li, Y., Li, J., Chen, S., and Diao, W. (2012). "Establishing indices for groundwater contamination risk assessment in the vicinity of hazardous waste landfills in China." *Environ. Pollut.*, **165**, 77–90.

Li, Z., Fu, H., and Qu, X. (2011). "Estimating municipal solid waste generation by different activities and various resident groups: A case study of Beijing." *Sci. Total Environ.*, **409**(20), 4406–4414.

Lim, S. R., and Schoenung, J. M. (2010). "Human health and ecological toxicity potentials due to heavy metal content in waste electronic devices with flat panel displays." *J. Hazard. Mater.*, **177**(1-3), 251–259.

Lu, L. T., Hsiao, T. Y., Shang, N. C., Yu, Y. H., and Ma, H. W. (2006). "MSW management for waste minimization in Taiwan: The last two decades." *Waste Manage.*, **26**(6), 661–667.

Lu, W., Rankin, J. G., Bondra, A., Trader, C., Heeren, A., and Harrington, P. B. (2012). "Ignitable liquid identification using gas chromatography/mass spectrometry data by projected difference resolution mapping and fuzzy rule-building expert system classification." *Forensic Sci. Int.*, **220**(1-3), 210–218.

Manassaram, D. M., Orr, M. F., and Kaye, W. E. (2003). "Hazardous substances events associated with the manufacturing of chemicals and allied products." *J. Hazard. Mater.*, **104**(1-3), 123–135.

Marinković, N., Vitale, K., Holcer, N. J., Džakula, A., and Pavić, T. (2008). "Management of hazardous medical waste in Croatia." *Waste Manage.*, **28**(6), 1049–1056.

Morf, L. S., Tremp, J., Gloor, R., Schuppisser, F., Stengele, M., and Taverna, R. (2007). "Metals, non-metals and PCB in electrical and electronic waste—Actual levels in Switzerland." *Waste Manage.*, **27**(10), 1306–1316.

MoUD (Ministry of Urban Development). (2000). "Government of India, Solid waste management manual." Government of India, India.

Musee, N., Lorenzen, L., and Aldrich, C. (2008). "New methodology for hazardous waste classification using fuzzy set theory: Part I. Knowledge acquisition." *J. Hazard. Mater.*, **154**(1–3), 1040–1051.

Ongondo, F. O., Williams, I. D., and Cherrett, T. J. (2011). "How are WEEE doing? A global review of the management of electrical and electronic wastes." *Waste Manage.*, **31**(4), 714–730.

Otte, P. (1995). "Analysis of metals and calorific value in components from household waste 1988–1992." Rep. No. 776201024, National Institute of Public Health and Environmental Protection, Bilthoven, Netherlands.

Owen, W. F., Stuckey, D. C., Healy, J. B., Jr., Young, L. Y., and McCarty, P. L. (1979). "Bioassay for monitoring biochemical methane potential and anaerobic toxicity." *Water Res.*, **13**(6), 485–492.

Parikh, J., Channiwala, S. A., and Ghosal, G. K. (2007). "A correlation for calculating elemental composition from proximate analysis of biomass materials." *Fuel*, **86**(12-13), 1710–1719.

Patwary, M. A., O'Hare, W. T., and Sarker, M. H. (2011). "Assessment of occupational and environmental safety associated with medical waste disposal in developing countries: A qualitative approach." *Saf. Sci.*, **49**(8-9), 1200–1207.

Phillips, P. S., Barnes, R., Bates, M. P., and Coskeran, T. (2006). "A critical appraisal of an UK county waste minimisation programme: The requirement for regional facilitated development of industrial symbiosis/ecology." *Resour. Conserv. Recycl.*, **46**(3), 242–264.

Rotter, V. S., Kost, T., Winkler, J., and Bilitewski, B. (2004). "Material flow analysis of RDF-production processes." *Waste Manage.*, **24**(10), 1005–1021.

Sethi, S., Kothiyal, N. C., Nema, A. K., and Kaushik, M. K. (2012). "Characterization of Municipal Solid Waste of Jalandhar City, Punjab, India." *J. Hazard. Toxic Radioact. Waste*, 10.1061/(ASCE)HZ.2153-5515.0000156, 97–106.

Sharholy, M., Ahmad, K., Vaishya, R. C., and Gupta, R. D. (2007). "Municipal solid waste characteristics and management in Allahabad, India." *Waste Manage.*, **27**(4), 490–496.

Sil, A., Wakadikar, K., Kumar, S., and Kumar, R. (2011). "Driving characteristics of waste transportation vehicles and their effect on emission lLoad: A case study of Mumbai, India." *Waste Manage.*, **31**(7), 1661–1665.

Sloot, H. A., and Kosson, D. S. (2012). "Use of characterisation leaching tests and associated modelling tools in assessing the hazardous nature of wastes." *J. Hazard. Mater.*, **207-208**, 36–43.

Supplit, R., Koch, T., and Schubert, U. (2007). "Evaluation of the anti-corrosive effect of acid pickling and sol-gel coating on magnesium AZ31 alloy." *Corros. Sci.*, **49**(7), 3015–3023.

Talini, I., Yamantürk, R., Aydın, E., and Başakçılardan-Kabakçi, S. (2005). "A rating system for determination of hazardous wastes." *J. Hazard. Mater.*, **126**(1-3), 23–30.

Tasaki, T., Hashimoto, S., and Moriguchi, Y. (2006). "A quantitative method to evaluate the level of material use in lease/reuse systems of electrical and electronic equipment." *J. Cleaner Prod.*, **14**(17), 1519–1528.

Tejada, M., García-Martínez, A. M., and Parrado, J. (2009). "Relationships between biological and chemical parameters on the composting of a municipal solid waste." *Bioresour. Technol.*, **100**(17), 4062–4065.

Thipkhunthod, P., Meeyoo, V., Rangsunvigit, P., Kitiyanan, B., Siemanond, K., and Rirksomboon, T. (2005). "Predicting the heating value of sewage sludges in Thailand from proximate and ultimate analyses." *Fuel*, **84**(7-8), 849–857.

U.S. EPA. (2005). "Solid waste and emergency response (5305W)." EPA530-K-05-012, Washington, DC.

Vehlow, J., and Mark, F. E. (1997). "Electrical and electronic plastic waste co-combustion with municipal solid waste for energy recovery." Association of Plastic Manufacturers in Europe, Brussels, Belgium.

Vesilind, P. A., Worrell, W., and Reinhart, D. (2002). *Solid waste engineering*, Brooks/Cole, Pacific Grove, CA.

von Blottnitz, H., Pehlken, A., and Pretz, T. (2002). "The description of solid wastes by particle mass instead of particle size distributions." *Resour. Conserv. Recycl.*, **34**(3), 193–207.

Wakadikar, K., Sil, A., Kumar, S., Kumar, R., and Mudhoo, A. (2012). "Influence of sewage sludge and leachate on biochemical methane potential of waste biomass." *J. Bioremed. Biodeg.*, S8:002.

Weltens, R., Vanermen, G., Tirez, K., Robbens, J., Deprez, K., and Michiels, L. (2012). "Screening tests for hazard classification of complex waste materials—Selection of methods." *Waste Manage.*, **32**(12), 2208–2217.

White, P. R., Franke, M., and Hindle, P. (1999). *Integrated solid waste management, a lifecycle inventory*, Aspen Publication, New York.

World Bank. (2010). "Waste generation—Urban development series." ⟨http://siteresources.worldbank.org/INTURBANDEVELOPMENT/Resources/336387-1334852610766/Chap3.pdf⟩ (Jul. 25, 2014).

Wu, H., Wang, H., Zhao, Y., Chen, T., and Lu, W. (2012). "Evolution of unsaturated hydraulic properties of municipal solid waste with landfill depth and age." *Waste Manage.*, **32**(3), 463–470.

Zhang, J., et al. (2011). "Effects of physico-chemical parameters on the bacterial and fungal communities during agricultural waste composting." *Bioresour. Technol.*, **102**(3), 2950–2956.

Zhao, W., Leeftink, R. B., and Rotter, V. S. (2010). "Evaluation of the economic feasibility for the recycling of construction and demolition waste in China-The case of Chongqing." *Resour. Conserv. Recycl.*, **54**(6), 377–389.

CHAPTER 3

Waste Storage, Segregation, and Collection

Kurian Joseph[1]
Ammaiyappan Selvam[2]
Jonathan W. C. Wong[3]

3.1 INTRODUCTION

Waste management includes collection, transport, processing, recycling and disposal of wastes. Of these the collection influences the rest of the operations thus significantly influences the overall waste management. Furthermore, collection lies at the hub of an integrated waste management and a costly process of the waste management because of the labor intensity of the work and the massive use of trucks in the collection process. Factors including environmental economic, technical, legislational, institutional, and political issues play crucial role in shaping the waste management system and the collection systems also (Beliën et al. 2014; Kogler 2007). Collection systems becoming complex than in the past due to the massive urbanization, which increases the importance of the waste collection in the modern society.

Street sweeping, waste collection and transport are the most publicly visible aspect of waste management to avoid nuisance and public health problems. When waste is putout from the household and left uncontained on the street results in street sweepers having to manually pick up the waste. In many low income countries, waste is often manually handled several times before final disposal; at the point of arising, collection points and transfer stations. Waste collection activity accounts for about 80% of all costs associated with waste management (Beliën et al. 2011; Tavares et al. 2009). Investment, operation and maintenance costs of vehicles, and salaries of collection workers comprise the vast majority of the costs of collection services.

The high public visibility and high costs of waste collection and associated services means that it is an important area for the municipality to improve its

[1]Anna Univ., Chennai, India.
[2]Hong Kong Baptist Univ., Hong Kong SAR.
[3]Hong Kong Baptist Univ., Hong Kong SAR.

performance. Implementing a collection and transfer system involves the finalization of the collection routes, purchasing and managing collection and transfer equipment, hiring and training of personnel, developing and managing contracts with private collection companies, providing information to the public and maintenance of facilities. The most effective waste management systems are those that effectively combine high service standards with high rates of materials recovery and recycling. This requires consideration of several issues such as the

- waste generation and characteristics (rates, density)
- number of households and of inhabitants per household
- existing storage facilities at households
- distance to collection/transfer point
- estimated number of required staff and vehicles
- existing facilities and infrastructure (storage facilities, roads, drains)
- existing practises (waste pickers and waste recycling)
- decision-making structure (area representatives, opinion leaders)
- communication channels for information to households
- ability and willingness to contribute for waste management
- required equipment (containers, bins)
- frequency of collection (storage capacity and odour emissions)
- point of secondary collection (communal storage; house-to-house) and
- integration of waste pickers

Significant improvements to service performance can be achieved when multiple manual handling of waste is minimized and waste components are separated with high purity. In many countries an informal materials recovery and recycling system operates alongside the waste collection service. Identifying goals, objectives, and constraints can help guide the planning process. This chapter presents an overview of the critical issues associated with waste collection and waste separation systems.

3.2 SOURCE SEGREGATION

The separation of recyclable materials into individual components, either by the generator or at curbside by the collector, in defined, different waste groups according to the specific treatment and disposal requirements is known as source segregation. Source separation of MSW becomes an integral components of planning and executing an Integrated Waste Management system. Besides, its adoption is likely to increase in the coming years due to increasing preference for the recycling. It requires a high degree of homeowner involvement, and has high collection costs but low processing costs. It is a prerequisite for efficient waste recycling and is critical in reducing waste quantities, returning resources back to use, and optimizing the financial and environmental burden of waste management. Separately collected MSW fractions from the household can greatly assist recycling and resource recovery at a local level by providing a clean and reliable feedstock. Segregation should be carried out by the waste generator and it must take place at source. Source separated collection of wastes has several options, including selective collection of paper, glass, metal, plastics or the organic fraction as well as their multiple combinations. These options focus on the convenience of collecting recyclables separately in order to reduce recycling costs.

In its simplest form, the wet materials (food scraps, yard trimmings, disposable diapers, soiled paper and animal waste) are separated from other materials for collection. This wet stream is usually composted. Other materials, including recyclables, form the dry portion. Some communities collect all of their dry waste mixed and separate recyclables during processing. Others require further separation of dry materials into recyclable and non-recyclable fractions. At source, households separate recyclables from the domestic waste and sell them to itinerant waste buyers on the basis of the weight of each individual material. Itinerant waste buyers move around streets with a basket or a push cart, on foot or on bicycle and buy recyclables collected at source and/or from street waste pickers. In high-income areas, domestic servants do it and at primary collection stage, sweepers employed by local government collect waste from different sources.

The quality of the segregation system will depend on the quality of the segregation concept and quality of the implementation of the concept. While considering to implement a source separate collection system, it should be kept in mind that the overall waste collection costs will be higher than that for mixed waste collection. Moreover, the success depends greatly on public participation which in turn will need sustained education and communication programmes. Segregated waste should not be mixed during transport and storage. The sustenance of public participation requires the subsequent processing of wastes in segregated manner. Further issues associated with segregated collection of waste are:

- special vehicles might be needed for separate collection.
- encouraging the establishment of formal or informal collection cooperatives.
- public information and education schemes focused on encouraging householders, communities and businesses to segregate their wastes (either for the informal or formal sector).

The primary benefit of source segregated collection is the revenue generation potential from selling recyclables, which also contribute to natural resources preservation in terms of raw materials and energy. It will also reduce the requirements for waste disposal. The compost produced from source segregated organics will be of very good quality.

Local bodies may optimize voluntary optimizing, Non-Governmental Organizations (NGOs) or co-operatives to take up the work of organizing street rag-pickers and elevate them to door step "waste collectors" by motivating them to stop picking up soiled and contaminated solid waste from streets, bins or disposal sites and instead improve their lot by collecting recyclable clean materials from the doorstep at regular intervals of time. Local Bodies may actively associate resident associations, trade & Industry associations, Community Based Organizations (CBOs) and NGOs in creating awareness among the people to segregate recyclable material at source and hand it over to a designated waste collector.

3.3 WASTE STORAGE

Storage of waste at the source of waste generation till it is collected is the first essential step of waste management. In many developing countries, in the absence of system of waste storage at source, the waste is thrown on the streets, treating streets as receptacle of waste. Even if local bodies make arrangements to remove all the waste disposed of by the citizens on the street on day to day basis, the city will remain clean only for two to three hours and not beyond till the habit of throwing waste on the streets is not changed. There is, therefore, a need to educate the people to store waste at source, dispose of the waste as per the directions of the local bodies and effectively participate in the activities of the local bodies to keep the cities clean.

In residential areas where refuse is collected manually, either plastic bags or standard-sized metal or plastic containers are typically required for waste storage. When automatic or semiautomatic collection systems are used, solid waste containers must be specifically designed to fit the truck-mounted loading mechanisms. The factors that must be considered in the on-site storage of solid wastes include the type of container to be used, container location, collection method and public health and aesthetic aspects. To a large extent, the types and capacities of containers depend on the characteristics of wastes to be stored, collection frequency and the space available for placement of the containers. Following good practices (UNEP IETC 1996) may be considered while choosing appropriate waste container systems.

- containers shall be adequate capacity, corrosion resistant and made of local, recycled, or readily available materials.
- containers that are easy to identify, due to either shape, colour or special markings are to be preferred
- containers shall be sturdy and/or easy to repair or replace.
- containers shall prevent access by waste pickers and appropriate to the terrain.

If community storage sites are being used, the storage points should be at intervals convenient to the generators and should be designed so that waste is deposited into containers at that site to facilitate secondary collection. With a view to ensure that streets and public places are not littered with waste materials generated while on a move, litter bins may be provided on important streets, markets, public places, tourist spots, bus and railway stations, large commercial complexes, etc. at a distance ranging from 25 m to 250 m depending on the local condition.

3.4 WASTE COLLECTION

Waste Collection, often divided into 'primary' and 'secondary' services, is the process of picking up waste from residences or collection points, loading into vehicles and transporting them to locations for processing, transfer or disposal. Primary collection is the means by which waste is collected from its source (dwellings and commercial premises) and transported to community storage, transfer points or even disposal sites. Secondary collection is the collection of solid waste for the second time, for example from community collection points, prior to its transport (often as part of a collection round by larger vehicles) to a transfer station, treatment facility or disposal site. Waste collection occupies the central position of an integrated waste management system. The way that waste materials are collected determines which waste management options can subsequently be used. Moreover, the selected collection method will significantly influence the quality of recovered materials. In many instances, the collection point will be an interface where waste generators and waste collectors must be carefully managed if the system is to be effective. Waste generators require waste collection with minimal inconvenience, while collectors must be able to collect waste in a way that is compatible with the planned treatment and processing methods if the waste management system is to be sustainable (McDougall et al. 2001).

Waste collection and transfer are often the most difficult components to design due to the involvement of many factors. To simply the design of collection systems, the following components must be considered and these components also help to middy an existing collection system (Pferdehirt 1995):

- Define community goals and constraints.
- Characterize waste generation and service area.
- Determine public and private collection and transfer options.
- Determine system funding structure.
- Identify waste preparation and collection procedures.
- Identify collection equipment and crew size requirements.
- Evaluate transfer needs and options.
- Evaluate collection and transfer alternatives.
- Develop collection routes and schedules.
- Implement the collection system.
- Monitor system performance; adjust as necessary.

The management of collection is most difficult and complex in an urban environment because the generation of residential and commercial-industrial solid waste and recyclables takes place in every home, every apartment building, and every commercial and industrial facility, as well as in the streets, parks, and even vacant areas. As the patterns of waste generation become more diffuse and the total quantity of waste increases, the logistics of collection become more complex. A comparison of the different collection options are summarized in Table 3.1.

The door to door collection facilitates maximum household separation and avoids low participation behaviour, because greater control is exercised. In low-income countries, waste is often manually handled several times before final disposal: at the point of arising, collection points and transfer stations. When waste is putout from the household, it is very often left uncontained on the street. This results in street sweepers or collection workers having to manually pick and scrape up waste. Where communal containers are used, collection workers often have to manually handle waste out of the containers. Where vehicles do not have tipping gear, further manual handling is required to unload them at transfer and/or disposal sites.

Kerbside collection is perhaps the most commonly used method in high-income areas of the industrialized countries, having replaced back door collection to reduce the labour costs. Furthermore, the productivity of the collection can be enhanced by using standardized containers and providing a less frequent collection service. Climatic conditions generally dictate that waste in developing countries should not be stored for periods in excess of four days, so if household generation rates are low this method of collection is rarely economical in low- and medium-income areas. Backyard or back door collection, requiring very minimal effort from the residents/waste generators, is becoming less common in industrialized countries due to its expensive nature while is rarely practiced in developing countries (UNEP 2010).

Key decisions of collection system design are the management and administration structure for the service standard of service to be provided, the agency undertaking collection (municipality, municipal enterprise, waste pickers, householders), the point of collection (from dwelling, from kerb, from communal station), the types of vehicle to be used, storage system and type/size/material of waste storage containers to be used and whether separation at source of recyclable materials is economically viable and must be allowed for and the frequency of collection.

The collection rate is influenced by the type of collection systems adopted, which in turn is affected by the national income. As presented in Figure 3.1, low income countries tend to have low collection rates, around 41%; whereas, an average of 98% was observed in high-income countries (Hoornweg and Bhada-Tata 2012).

The rate of waste generation has a significant influence on the choice of collection system and vehicle also. Countrywide average rates of waste generation in most industrialized countries lie between 0.8 and 1.4 kg/capita/day. In developing

Table 3.1. Comparison of waste collection options

Sl. No	Collection option	Operation	Comments
1	Door to door collection	Waste collector knocks on each door or rings doorbell and waits for waste to be brought out by resident.	Convenient for resident. Little waste on street. Residents must be available to hand waste over. Not suitable for apartment buildings because of the amount of walking required.
2	Block collection	Collector sounds horn or rings bell and waits at specified locations for residents to bring waste to the collection vehicle.	Economical. Less waste on streets. No permanent container or storage to cause complaints. If all family members are out when collector comes, waste must be left outside for collection. It may be scattered by wind, animals and waste pickers.
3	Kerbside collection	Waste is left outside property in a container and picked up by passing vehicle, or swept up and collected by sweeper.	Convenient. No permanent public storage. Waste that is left out may be scattered by wind, animals, children or waste pickers. If collection service is delayed, waste may not be collected or some time, causing considerable nuisance.
4	Back Yard collection	Collection labour enters property to remove waste.	Very convenient for residents. No waste in street. The most expensive system. Cultural beliefs, security considerations or architectural styles may prevent labourers from entering properties.

(Continued)

Table 3-1. Comparison of waste collection options (Continued)

Sl. No	*Collection option*	*Operation*	*Comments*
5	Dumping at designated location (Drop off centres)	Residents and other generators are required to dump their waste at a specified location or in a masonry enclosure.	Low capital costs. Loading the waste into trucks is slow and unhygienic. Waste is scattered around the collection point. Adjacent residents and shopkeepers protest about the smell and appearance.
6	Shared container	Residents and other generators put their waste inside a container which is emptied or removed.	Low operating costs. If containers are not maintained they quickly corrode or are damaged. Adjacent residents complain about the smell and appearance.

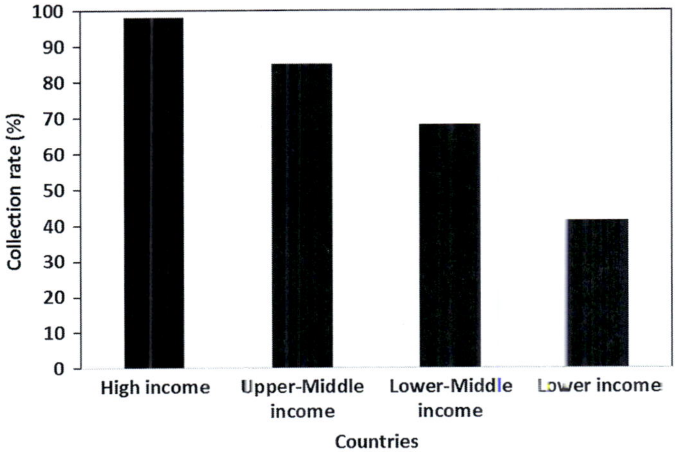

Figure. 3.1. Waste collection rate of countries by income

countries, the average generation rate is generally within the range of 0.2 to 0.4 kg/capita/day, but the actual generation rate varies greatly between different parts of any city, and may be as low as 0.1 kg/capita/day in the low-income urban areas, the generation rates in such areas being as little as one half the citywide average (UNEP 2010).

3.4.1 Frequency of Collection

Frequency of collection is a fundamental parameter of any waste collection system. Waste quantities arising, coupled with waste container capacities and climate conditions, customer expectations and cost will affect the frequency of collection. In hot and humid climates waste decomposes rapidly, posing an aesthetic and health risk. Denser communities may require more frequent waste collection services. Seasonal changes in waste generation will also require flexibility in the collection service. Some communities are accustomed to a collection seven days a week, whilst other collection agencies are striving for just once each week. If fly breeding is to be controlled, the waste should be collected twice a week in hot climates. Furthermore, the concentration of putrescible matter in wastes generated in economically developing countries usually is high.

Putrescible materials serve as breeding media for flies and are potential sources of foul doors. The eggs of the housefly can hatch in as little as one day. However, the larvae feed for about five days before pupation, which takes an additional three days. The total period to reach maturation may be as little as seven days in tropical countries. A weekly collection, therefore, prevents the production of adult flies in the stored wastes, provided that the larvae are unable to migrate from the container. On the other hand, decomposition of the wastes becomes noticeable during the first two to three days of storage. Consequently, aesthetic standards may be a more critical factor than the lifecycle of the fly (UNEP 2005).

Other factors to consider are the odours caused by decomposition and the accumulated quantities. If residents are accustomed to daily collection it may not be politically feasible to reduce the frequency to twice a week. In some cities waste is collected on the day of rest (Sunday or Friday). Some collect waste at night, perhaps for cultural reasons or because of the weather or traffic congestion. Changing collection frequency (USEPA 1999) include lowering the operating costs by improving operational productivity, savings in labor, capital, and maintenance costs.

3.4.2 Collection Vehicles

Many technical factors have a direct bearing on the selection of a collection system and vehicle for any particular situation while the choices of vehicle and storage system are interrelated. The choice of waste collection vehicle is influenced by a range of factors including the waste generation rate, loading heights, stage of collection (ie, primary or secondary), the method of solid waste collection adopted, the nature of the waste (density, volume, and composition), the topography of the city, the budget available for transport and the size (range) of the collection schedule, distance to disposal site, traffic restrictions and road conditions. The common types of primary collection vehicles in cities of developing countries are the

- handcart which is pushed by the operator as he walks along.
- pedal tricycle with a tray or box in front of or behind the operator.
- tractor-trailer units operated by a collector driver.

The handcart operated by one person is very cheap but with limited capacity (0.25 to 1 m^3). A labourer can push 150 kg of waste in a well-designed and maintained cart. If the cart is used for street sweepings, it is appropriate to use less capacity because it takes more time to collect the waste and there may be community storage containers at frequent intervals so that it is not necessary to carry the waste over a long distance. The decision over human, bicycle or machine-powered vehicles is influenced by haul distance, road width and condition, traffic density and the total quantity of waste to be removed. If the cart has only two wheels, the operator is always required to apply a lifting force, adding to fatigue and possible back injuries. At least three wheels are suggested. But carts with three wheels may roll by themselves if left on sloping ground. Four wheels make the cart more stable allowing easier crossing of a hole in the road.

The optimum vehicle choice may be a trade-off between vehicle size and labour productivity. The important features to remember when selecting the collection vehicle are

- to optimize the payload while avoiding the need for manual lifting to high levels;
- to use compaction vehicles only when they are proved to be compatible with the type of waste requiring collection, and more efficient and effective.
- preferably to use vehicles that incorporate a tipping mechanism.

It is recommended to select vehicles that use the minimum amount of energy and technical complexity necessary to collect the targeted materials. Compaction vehicles are the standard vehicle for MSW collection in industrialized countries. They tend to have a high capital and operating cost, require extensive and regular maintenance and work best on wide paved streets and with waste of a low density and moisture content. Their application in developing countries should take optimizing of these issues. However, when the waste densities are low, as in industrialized countries, it is economical to compact the refuse to increase its density and reduce its volume and maximize payloads. This is the reason that compaction vehicles are mainly used in the industrialized countries. Compactor trucks are typically designed to compact light refuse, having a density of around 100 kg/m^3, to a compressed density of up to 500 kg/m^3, so that the truck can carry a full load. However, it is interesting to note that the density of the wastes in the developing countries is usually higher than 300 kg/m^3 and even much higher at some specific instances, therefore the use of compacter trucks must be carefully investigated before inducting such fleets (UNEP 2010). Locally made equipment with traditional vehicle design and local expertise are to be preferred whenever possible. Non-compactor trucks, dump trucks or vans may be used where population is dispersed, or waste is already dense.

The practice of labors tipping the contents of their carts onto the ground and then scooping the waste up into another container for transfer is inefficient (wasting time), unhealthy (forcing the labor to touch or have close contact with waste) and polluting (often some of the waste is left on the ground or is scattered by the wind). This has to be avoided by use of split-level sites (waste can be tipped directly from the cart into the bulk container) or by containerizing the refuse in a number of bins that are small enough so that they can be lifted and tipped into the bulk container.

The size of the bins should be large enough that big items of waste cannot bridge across the rim and prevent the efficient utilization of the bin's capacity. Bins that are rectangular may be difficult to pack if the dimensions of large items are larger than the smaller horizontal dimension of the bin. Square and rectangular bins may be easily deformed and develop dangerous jagged projections at the corners. Bins that are circular do not fit together well, considerable space is lost between bins.

The tricycle has comparatively high productivity in terms of higher holding capacity (1.5 m^3) and higher speed with little impact on environment. Simple, motor powered vehicles should be considered where longer distances require larger payloads and higher speeds. A wide variety of trucks are used for collection. Factors considered in selecting such vehicles include

- the weight of waste that the truck can actually carry
- cost of purchase and operation, including fuel and maintenance
- suitability of the vehicle for the local roads considering width, congestion, and surface conditions, ease of loading and unloading and getting spares for repairs

It is important to design and specify all items of equipment carefully. Even simple equipment such as brooms and carts are used in large numbers and so the impact or a more efficient or durable design can be very significant in economic terms. In many instances vehicles that have been developed for waste collection in high-income countries have proven unsuitable for use in a developing country context. Thus, in many cases, improvements to solid waste collection services can be best promoted through the development of local capability for manufacturing vehicles and equipment. Where the import of vehicles is essential, a great deal of attention needs to be placed on ensuring that they are appropriate to local conditions, and can be maintained and operated effectively. In areas with narrow streets smaller collection vehicles often prove to be more appropriate. In areas with high traffic flows, careful selection of vehicles and timing of collection in off-peak hours, can allow services to be improved.

Although larger collection vehicles offer significant advantages in terms of economies of scale, care needs to be taken to ensure that they can be loaded easily. Manual loading of waste into vehicles, and lack of covering during waste transport often leads to a high degree of spillage and littering. Systematic implementation of waste covering during transport is a simple management measure that can greatly reduce littering. Many of the larger vehicles have side walls which are more than 2 m above the ground, and as a consequence require a two-person loading process where one worker has to stand on the vehicle and receive the containers passed up to them from personnel at ground level. This can cause collection workers significant physical strain, and greatly reduce the efficiency of the collection service.

Good, preventative, maintenance of vehicles is essential to an efficient and effective collection service. Efficient vehicle maintenance, fleet management and financial planning can significantly improve a municipality's ability to continuously upgrade its fleet. Often, municipalities strive to optimize their collection vehicles assuming that similarity will result in cost-efficient operation and maintenance. This optimization may result in the exclusion of large areas of cities from the collection service. These considerations need to be balanced in developing the collection vehicle fleet. Vehicles that are designed to operate in low-density urban areas with wide and well-paved roads do not often perform well in locations with narrow, steep and poor-quality roads and/or with a high-density population. The continued use of vehicles beyond their economic life has a serious negative impact on the efficiency and reliability of the collection system.

Collection vehicle routes can be planned to maximize the productivity of vehicles through efficient routing, correct frequency and optimization of crew frequency. Collection routes must be planned to optimal efficiency in order to minimize transport distances and ensure an equitable distribution of workloads amongst crews. Vehicle productivity may be increased by selecting and planning routes so that the quantity of waste collected daily is optimum. The time and motion study helps here by first measuring and then optimizing workloads in detail. In many cases several small vehicles will be found to be more productive than a single large vehicle carrying the equivalent combined load because of their relative ease of loading and maneuverability. The load of waste should be thoroughly covered during transport; this is particularly important for motor vehicles travelling at 30 km/h or more.

3.5 WASTE SEPARATION

In case the waste is collected in mixed form, the recyclables with the general waste stream are recovered by manually or mechanically sorting at a suitable site such as Material Recycling Facility (MRF) (Table 3.2). Such sorting by the municipality has the advantage of eliminating the dependence on the public and ensuring that the recycling does occur. The disadvantage however, is that the value of the recyclable materials is reduced since being mixed in and compacted with other garbage can have adverse effects on the quality of the recyclable material.

3.6 HEALTH AND SAFETY ISSUES

Health and safety are especially important because waste collection employees encounter many hazards during each working day due to

- multiple manual handling of MSW
- handling of unsegregated and uncontained clinical and special wastes
- busy roads and heavy traffic
- rough and sharp edged containers that can cause cuts and infections
- exposure to injury from vehicle movements and mechanisms
- heavy containers and high vehicle loading that can cause back injuries.

The overall effect of multiple manual handling of waste is to expose workers to significant health and safety risks that significantly impair the efficiency of the entire waste collection service. Where communal containers are used, collection workers often have to manually handle waste out of the containers. Where vehicles do not have tipping gear, further manual handling is required to unload them at transfer and/or disposal sites. Health and safety considerations that are particularly relevant to waste collection and transfer in developing countries include the following:

- climatic conditions such as tropical heat and humidity often dictate that waste is collected several times per week, and daily in many places.
- manual methods of collection often result in both formal waste workers and informal waste pickers coming in direct contact with wastes that sometimes contain faecal matter (human and animal) and infected medical wastes.
- waste is often thrown out (uncontained) onto the street before collection, in low-income countries. This leads to greater exposure to insects and vermin, scavenging by animals, and increased public health risks.

Table 3.2. Waste Separation Techniques

Sl. No.	*Method of sorting*	*Brief description*
1	Manual sorting	Manual sorting of waste is still very much a technique that is used in the world today. Furthermore it is often a preliminary step in MRFs to remove the heavy object which tend to affect the delicate instruments downstream.
2	Trommel separators/drum screens	The separation is based on the size of the material is achieved using trommel and drum screens. Waste is fed into a large rotating drum which is perforated with holes of a certain size. Materials smaller than the diameter of the holes will be able to drop through, but larger particles will remain in the drum.
3	Air classification	Separation of light and heavy fractions based on density using air as the media
3	Magnetic separator	Ferrous (magnetic) materials are separated from nonmagnetic materials.
4	Eddy current separator	An 'eddy current' occurs when a conductor is exposed to a changing magnetic field. Put simply, it is an electromagnetic way of dividing ferrous and non-ferrous metals. This method is applied specifically to separate the metals.
5	Induction sorting	Material is sent along a conveyor belt with a series of sensors underneath. These sensors locate different types of metal which are then separated by a system of fast air jets which are linked to the sensors.
6	Near infrared sensors (NIR)	When materials are illuminated they mostly reflect light in the near infrared wavelength spectrum. The NIR sensor can distinguish between different materials based on the way they reflect light.
7	X-ray technology	X-rays can be used to distinguish between different types of waste based on their density (x-ray detection of PVC)

- uncollected waste can block drains and create stagnant water, with consequent risks of mosquitoes breeding.
- municipalities typically have very limited financial resources available for hiring of additional staff or purchasing extra vehicles.
- public awareness of health implications of waste management is often limited.

Collection workers perform an extremely important and valuable role in society and should be regarded accordingly. Low workforce discipline can seriously hamper service performance and the reliability and regularity of services. Measures that can be employed to boost morale include improving overall working conditions, providing uniforms, incentives for good performance, and training and team building. Community-based services are particularly crucial where waste needs to be collected frequently and municipal resources to provide necessary services are highly constrained. Counterbalancing the above, however, communities often show considerable willingness to participate in cleaning up their areas, and improving their overall health and safety conditions. People do not want to live in an unclean environment, and have in many cases proven extremely willing to participate in good initiatives that they feel will benefit them directly.

3.7 CONCLUSION

Efficient, sanitary, and customer-responsive collection of solid wastes is at the heart of a well-run waste management system. The following factors are worth considering when planning a new collection system or evaluating waste collection and transfer alternatives.

- containment of wastes at the point of primary and secondary collection to ease handling and loading, and reduce health and safety risks.
- development of adequate supporting facilities, such as collection points.
- providing the right vehicles for the right job
- costs of required new equipment and ability of community to obtain financing for it
- costs to operate collection system and transfer facilities
- compatibility of total costs with budget available for solid waste services
- differences in levels of service provided by alternative systems
- ability of system to meet public's demands or expectations for service
- methods for financing system costs and public acceptability of those methods
- system's effects on efforts to meet the municipality's waste reduction and management goals

- compatibility of roles for public and private sectors with political support for them
- public's interest or disinterest in changing present arrangements for collecting solid waste and recyclables
- consideration of vehicle modifications to assist loading and maximizing vehicle loads.
- involving the private sector in vehicle maintenance for example via leasing of vehicles or outsourcing vehicle maintenance.
- getting the correct balance between labor and mechanization.
- efficient use and maintenance of motor vehicles in order to achieve high productivity.
- improving management performance by training supervisors
- planning for seasonal variations in waste arising and collection frequency.
- ongoing and constructive dialogue between neighboring municipalities and between informal and municipal collection sectors.
- single handling of waste through more efficient management, introduction of appropriate containers and vehicle modifications.
- improve the moral and public image of workers

It is very important that each community should carefully evaluate which type of collection system, or combination of systems, will best meet their needs. Minimum physical infrastructure for waste collection includes a central garage, with parking space for motor vehicles, and district depots for assembling and controlling collectors, handcarts, and animal carts. The locations of the depots should minimize travel time between depot and working area. Systems that provide for the direct emptying of portable storage containers into a vehicle offer the highest productivity and the lowest health risk to workers.

References

Beliën, J., De Boeck, L., and van Ackere, J. (2014). "Municipal solid waste collection and management problems: A literature review." *Transp. Sci.*, **48**(1), 78–102.

Beliën, J., De Boeck, L., and van Ackere, J. (2011). "Municipal solid waste collection problems: A literature review." *HUB research papers 2011/34: Economics & Management*, K.U. Leuven Associates, Leuven, Belgium, 35.

Hoornweg, D., and Bhada-Tata, P. (2012). "What a waste: A global review of solid waste management." Rep. No. 15, World Bank, Washington.

Kogler, T. (2007). "Waste collection, ISWA working group on collection and transportation technology." ⟨http://www.iswa.org/uploads/tx_iswaknowledgebase/ctt_2007_2.pdf⟩ (May 2, 2014).

McDougall, F. R., White, P. R., Franke, M., and Hindle, P. (2001). *Integrated solid waste management: A life cycle inventory*, 2nd Ed., Blackwell Publications, Oxford, 544.

Pferdehirt, W. (1995). "Collection and transfer." *Decision makers' guide to solid waste management*, Vol. **2**, U.S. EPA, Washington, DC.

Tavares, G., Zsigraiova, Z., Semiao, V., and Carvalho, M. G. (2009). "Optimisation of MSW collection routes for minimum fuel consumption using 3D GIS modeling." *Waste Manage.*, **29**(3), 1176–1185.

UNEP IETC (United Nations Environment Programme International Environmental Technology Centre). (1996). "International source book on environmentally sound technologies for municipal solid waste management." Washington, DC.

UNEP (United Nations Environment Programme). (2005). "Solid waste management." CalRecovery, Washington, DC.

UNEP (United Nations Environment Programme). (2010). "Collection of municipal solid waste in developing Countries." *United Nations Human Settlements Programme (UN-HABITAT)*, Gutenberg Press, Malta, 198.

U.S. EPA. (1999). "Collection efficiency strategies for success.", Washington, DC.

CHAPTER 4
Waste Prevention

Katia Lasaridi[1]
Christina Chroni[2]
Antonis A. Zorpas[3]
Konstantinos Abeliotis[4]

4.1 INTRODUCTION

Waste generationis the other side of the coin of resources exploitation and potential scarcity. Therefore, waste generationis interwoven with global environmental security and governance and poses a problem that has grave environmental, social and economic repercussions for all nations and future generations. In the context of sustainability, waste is considered a measure of inefficiency, on the contrary "waste avoidance would further sustainable conduct" (Pongrácz 2009). Hence, waste prevention is becoming a priority in many waste management policies worldwide (Zorpas and Lasaridi 2013).

The definition of waste prevention, as well as the development of prevention actions, is an extremely broad and sometimes controversial topic. A wide range of definitions have been drafted and critically analysed in the literature, national and international waste policies specifically addressing waste prevention have been developed. In addition, methods to implement waste prevention measures in all steps of a product's life cycle have been examined, both by the industry and the academia. Moreover, the barriers – economic, administrative, social, psychological, etc - for the implementation of waste prevention have been investigated in several reports and papers. However, the understanding of the basic principle governing the implementation of waste prevention into practice requires a multi – disciplinary approach and hence, has many knowledge gaps.

In this chapter, the key points on the different aspects of waste prevention are discussed in a multidisciplinary context, in an attempt to contribute to the better understanding of waste prevention to a wide audience with specialisation and/or

[1]Harokopio Univ., Athens, Greece.
[2]Harokopio Univ., Athens, Greece.
[3]Cyprus Open University, Cyprus.
[4]Harokopio Univ., Athens, Greece.

interest in waste management. An effort has been made to clearly define the term "waste prevention" and its importance towards sustainability is discussed. The policy context on waste prevention in selected countries and supranational entities that have pioneered the efforts for waste prevention has been presented. An analysis of waste prevention actions and initiatives that can be applied to production supply, consumption and the household for different key waste streams is also presented. Finally, the major barriers affecting waste prevention are examined, and successful efforts undertakenby central and local authorities, businesses, consumers and social enterprises by adopting a choice of best practices to mitigate negative notion of the barriers, have been presented.

4.2 WHAT IS WASTE PREVENTION?

According to the revised European Union (EU) Directive on Waste (2008/98/EC), widely known as Waste Framework Directive (WFD - Article 3, clause 12 & 13), waste prevention encompasses complex actions and measures, and a range of policy options in a broader sense, taken before a material or a product is characterised as waste. These actions and measures ought to reduce:

- the amount of waste, through a Paradigm shift in the production and consumption patterns, including the re-design of products, their re-use or the extension of their life span;
- the adverse impact of the generated waste on both the environment and human health; and
- the content of harmful substances (i.e. the toxicity) in materials and products.

A common point of dispute in the interpretation of the aforementioned (and any other widely accepted) definition is the word "before", as it raises boundary issues concerning the useful life of a product, material or substance. Although the WFD clarifies the distinction between waste and by-products (EC 2012), in practice several uncertainties exist throughout the whole supply and consumption chain. On the consumer level, quite often the characterisation of a product as waste rather than as a repairable/reusable/second-hand item is determined by the economic and social status, the convenience of disposability or the "latest" technological or commercial trends. On the waste management level, the distinction between "reuse", which is considered a waste prevention policy and the "preparing for reuse", which is a waste management activity (more specifically, the second step of the waste hierarchy – see §4.5.1), seems to be a matter of perception and legal definition, with no technical orenvironmental justification, as it "is fully dependent on the distinction between a waste and a non-waste"(EC 2010). Overall, it is widely recognised that there is a difficulty defining which actions should be labelled as furthering waste prevention (Lilja 2009).

Based on the above definition, waste prevention can be achieved through three different, but often intersecting pathways (Figure 4.1):

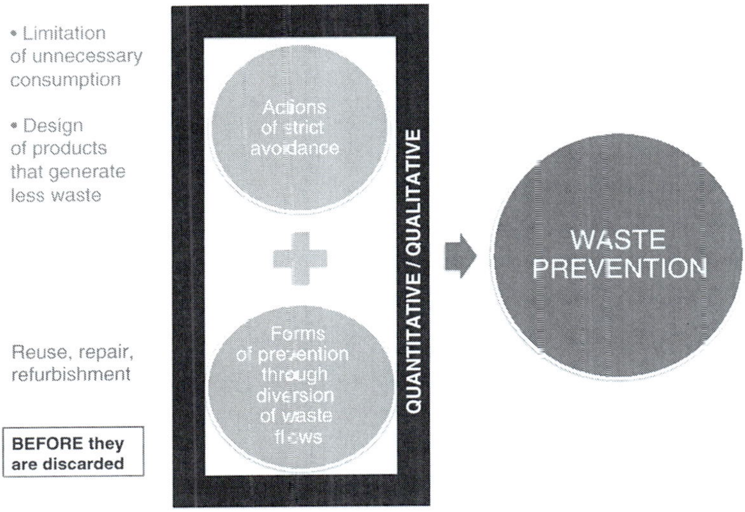

Figure 4.1. The three pathways of waste prevention: strict avoidance of waste, diversion of material and product flows from disposal, and elimination of waste toxicity

- the strict avoidance of waste;
- the diversion of material and product flows from disposal; and
- the elimination of waste toxicity.

The first two paths concern the quantitative side of waste prevention, while the third the qualitative. In the first pathway, waste prevention targets waste generation at its source, exploiting policies that limit unnecessary consumption and leading to the design and/or the consumption of products that generate less waste (EC 2012). The second approach, refers to materials and products that approach or have already reached their end-of-life, but are not yet discarded. It includes actions that extend the useful life or encourage the re-use of products. The third path, the elimination of waste toxicity, touches on the qualitative side of waste prevention. It is widely known, that the contemporary production chain uses a large variety of synthetic materials that are not found in natural ecosystems and posean enormous negative impact on the environment. The reduction of the hazardous content of waste will limit the human and environmental exposure to hazardous substances (EC 2012).

A relatively different approach to the concept of waste prevention has been introduced by the Organisation for Economic Co-operation and Development (OECD), seeing it as part of waste minimisation (Figure 4.2). The latter embraces all the material saving and recovery options, which can be applied before and after a material/product has been labelled as waste (OECD 2002). Referring to a report of EEA (2002), Salhofer et al. (2008) indicated that waste minimisation includes preventive measures (such as *prevention, reduction at source* and *reuse of products*)

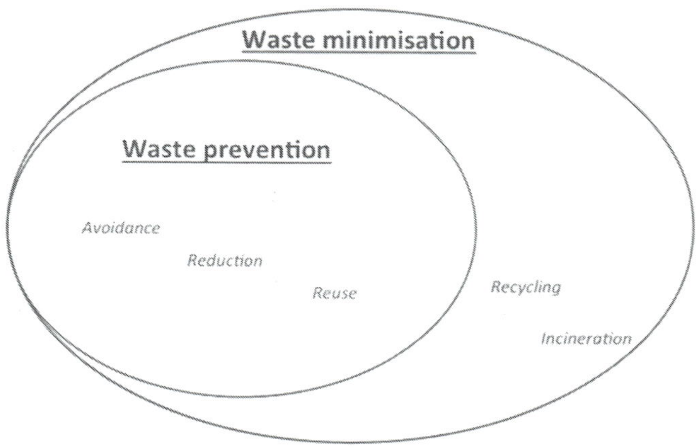

Figure 4.2. An illustration of waste prevention in the context of integrated waste management, based on the definition of the OECD (2002)

as well as some other waste management procedures aiming at *quality improvements* and *recycling*.

Yet, approaches from both the OECD and the WFD have two important characteristics in common:

- the use of a "life-cycle thinking" on waste management policies, and
- the placement of waste prevention within the framework of waste management.

Lately, a transition of the concept of waste prevention, from a waste management approach to an integrated context of improved and efficient use of resources (OECD 2005) and a shift of the global production and consumption Paradigm is being proposed. The International Solid Waste Association (ISWA) underlined the current need to integrate waste management into material and energy flows and depicted the relation of waste management and waste prevention to the whole resources management cycle, in the context of the so sought transition to the Circular Economy (ISWA 2012). A new perception of use of resources might hold the key solution for the waste issue.

4.2.1 Core Concepts

A basic understanding of two key concepts is essential for approaching waste prevention. The first is to realise that actions for waste prevention can be taken at all the steps in a product's life cycle, from production and supply to consumption and discard (Figure 4.3). Various factors play a key role in each step and different policy tools are best suited to promote prevention. In all the steps, policies and actions for waste prevention typically focus on *priority waste streams*, i.e. those waste streams for which, due to their quantity and environmental footprint, prevention can most effectively reduce environmental impacts and achieve quantifiable environmental benefits. For some waste streams large differences

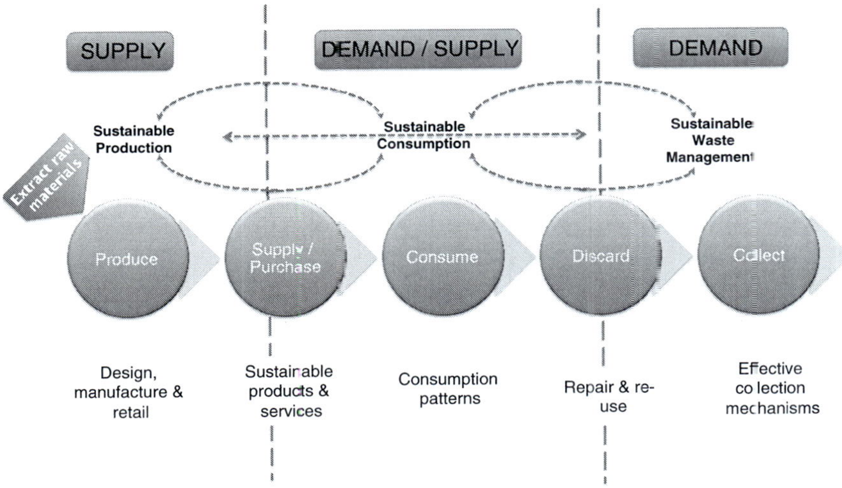

Figure 4.3. Waste prevention refers to all the steps in a product's life cycle
Source: Adapted from DEFRA (2009)

can be achieved at the production and supply stage, while others are better suited for actions at the consumption and discard level. For most waste streams, however, significant prevention potential may be identified at all levels of the product's life cycle.

In sections 4.6 and 4.7 of this chapter the prevention potential for the main priority waste streams is discussed, i.e. food waste, packaging waste, waste from electrical and electronic equipment (WEEE), construction and demolition waste, bulky and other waste, at the level of production and supply; and for food waste, packaging waste, non-packaging paper, WEEE and green waste, at the level of consumption/discard. In addition, at the level of production, the issue of planned obsolescence and the need to lessen it are tackled (section 4.5.4).

The second concept concerns the barriers waste prevention implementation. Although waste prevention has been placed at the top of the waste management hierarchy in many national and international policy frameworks, waste prevention policies have not brought the expected significant results so far. Mainly, the cause rest on measurement limitations and therefore lack of specific quantitative targets, conflict of interests, inconsistent definitions, the lack of comprehensive strategies (Salhofer et al. 2008) and the human attitudes and environmental awareness (Pongrácz 2009). Both concepts are thoroughly discussed in this chapter.

4.3 THE GROWING BURDEN OF WASTE

Whether one refers to an industrialised or a developing country, the increasing waste amounts are indisputably a major issue of concern and ought to be treated seriously. The accurate estimation of the overall waste generated per capita is difficult, mainly due to shortage of data, methodological uncertainties and definition differences (OECD 2000; Tonjes and Greene 2012; Wilson et al. 2012); however, the overall trend is steadily on the rise. According to the best available estimates, in the OECD countries, waste generation increased by 18% between 1995 and 2007 (OECD 2010).

The difficulty in accurate estimation of waste generation is outlined by Tonjes and Greene (2012), when reporting a quite wide range of the waste quantities in the United States of America (USA). More specifically, they state that,in 2008, the generated municipal solid waste (MSW) in the USA was estimated to be between 226.4 and 353.4 million tonnes, which amounts to 745–1161 kg capita^{-1} year^{-1} (2.0–3.2 kg capita^{-1} year^{-1}). The broad variations in estimated quantities highlights the methodological differences and the disagreement of the "two best known assessments of the quantity of US MSW", delivered by the US Environmental Protection Agency and the team of the BioCycle magazine, respectively, on the statistics and the basic waste infrastructure facts (Tonjes and Greene 2012). However, even the more modest of the above two estimates indicates the magnitude of the waste problem. Furthermore, it should benoted that the "hidden flows", i.e. the overall material requirements supporting an economy and the displacement of environmental assetswhich are never actually absorbed into the economic sphere (e.g. overburden from mining operations) (OECD 2000) are not included in the aforementioned figures.

In EU-27 alone, three billion tonnes of waste was generatedin 2006, corresponding to an average of 6 tonnes per person (EEA 2010). Although the generation of MSW in the EU-27 appears to be stabilising (at approximately 520 kg/capita) since 2000, there are substantial differences among states; some Member States (MS) report stabilisation or even a reduction in waste generation, while others show constant increase (EEA 2011), with the average per capita quantity still presenting a wide divergence. In addition, most EU officials and academics agree that further investigation is required to clarify the causes of the aforementioned decline in those Member States. The extent of the influence of the economic recession, potential changes in the reporting methodology or the progress of waste prevention implementation may all have their part in the reported figures.

The factors driving waste generation are not always easily identified, for the existing evidence is often inadequate to draw sound conclusions (DEFRA 2007). Notwithstanding this, the drivers that have been largely indicated in the relevant literature (OECD 2000; DEFRA 2007; WRAP 2012), are illustrated in Figure 4.4 and can be classified in the following six categories:

1. world population;
2. affluence;
3. production and consumption patterns (strongly dependent on lifestyle, social class, behavioural traits and shopping habits);
4. technology;

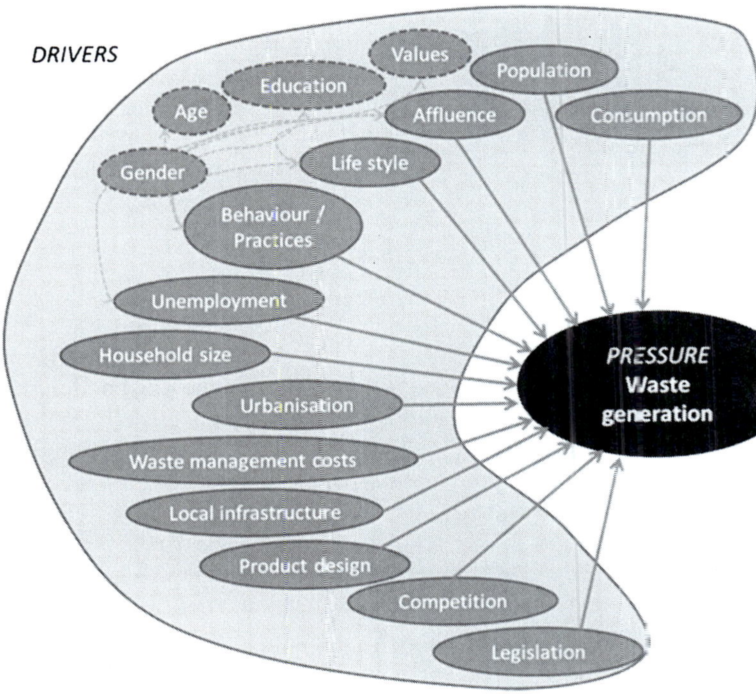

Figure 4.4. Overview of drivers that effect waste generation
Source: WRAP (2012). Reproduced with permission from WRAP

5. households' characteristics (size, age, presence of children, pets, etc.); and
6. seasonal variations (affecting food waste, garden waste and paper – the first especially in low-income countries, where there is a strong dependence on local seasonal products, as well as in areas economically reliant on seasonal tourism).

Among these factors, of special interest is the strong association of the unit waste production (i.e. waste produced per capita) with economic growth, as expressed by Gross Domestic Product (GDP)/capita as presented in Figure 4.5 (UNEP 2011). In a world of increasing population and – hopefully – increasing affluence, the challenge for the waste and resources sector cannot be ignored.

Waste prevention measures typically focus on the last four of the aforementioned categories, with special emphasis on the production and consumption patterns and reinforcing the belief that waste management is also a social discipline. The world population and affluence are usually - if not always - placed beyond the objectives of waste prevention policies.

Against the backdrop of these factors, the issue of decoupling economic growth from waste generation and the related adverse environmental impacts is crucial (Mazzanti and Zoboli 2008; Fell et al. 2010; Sjöström and Östblom 2010)

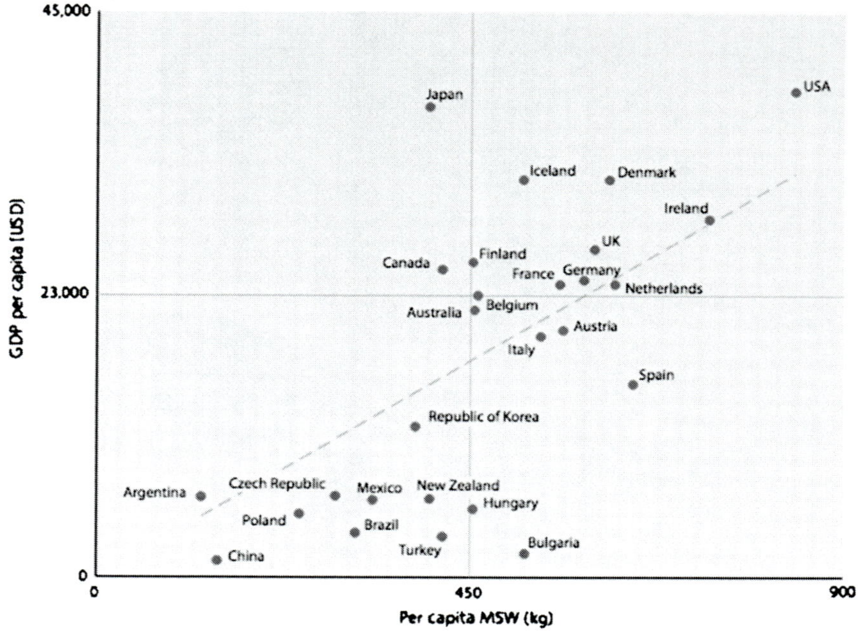

Figure 4.5. GDP per capita versus MSW per capita for different countries, (different years per country, all in the period 2002–2006)
Source: UNEP (2011). Reproduced with permission

and forms the cornerstone of the EU waste policy, as expressed in the "Thematic Strategy on the Prevention and Recycling of Waste"and the WFD. On the positive side, based on waste generation and the GDP data, the EC (2012) reported recently that "the link between economic growth and waste generation is beginning to weaken in the EU" (Figure 4.6). However, an absolute decoupling has not been observed (Mazzanti and Zoboli 2008; Gentil et al. 2011). According to EEA (2011), the causes of the decoupling might differ from one country to another. More specifically, in the EU-12 group (Bulgaria, Cyprus, Czech Republic, Estonia, Hungary, Latvia, Lithuania, Malta, Poland, Romania, Slovakia and Slovenia), decoupling might occur due to changes in measurement methods, while in the EU-15 group (Austria, Belgium, Denmark, Finland, France, Germany, Greece, Ireland, Italy, Luxembourg, the Netherlands, Portugal, Spain, Sweden, and the United Kingdom), it could be partly attributed to the consumption patterns, the waste collection methods or the implementation of waste prevention initiatives (EEA 2011).

Comparing the solid waste management in twenty cities around the world, Wilson et al. (2012) highlighted that waste generation per capita increases with both the development level and the income level, as they are reflected, respectively, by the UNDP's Human Development Index and the Gross National Income per capita, in agreement with data in Figure 4.5. However, it remains questionable

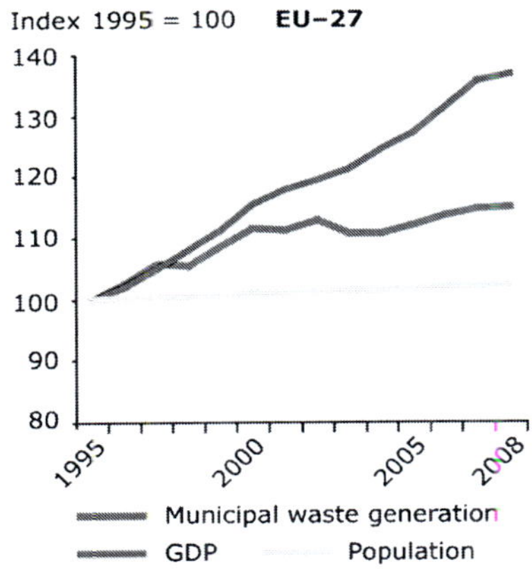

Figure 4.6. Trends in the municipal waste generation in EU-27 in comparison with GDP
Source: EEA (2010). Reproduced with permission

whether decoupling is a realistic concept (Fell et al. 2010), especially without concerted action encompassing suitable financial instruments. In this context, there is a strong argument that waste generation would only be constrained if producers and consumers were obligated to pay the full social and environmental cost of their waste (OECD 2000).

4.4 WASTE PREVENTION IN THE CONTEXT OF SUSTAINABILITY

Waste management influences a range of broader concerns such as energy and material security, climate change, environmental protection, economy, justice and equity, which are all directly related to the three pillars of sustainability, i.e. environmental protection, economic development and social justice. Waste prevention is a useful multipurpose tool for the consolidation of sustainability, as it brings about, direct or indirect waysof tackling:

- the growing – quantitative and qualitative - waste burden;
- the resource-intensive production and consumption patterns; and
- the waste management costs.

In the short and long term, waste prevention furthers the efficient use of resources and can be perceived as a substantial element for fostering sustainable environmental management. Therefore, it is not surprising that quite often waste prevention is integrated in national sustainable development programmes or policies. Waste prevention opportunities occur at every stage of the life cycle of a material or product; therefore, it may play a significant role in relation to the reduction of waste generation andthe overall waste management planning, as well as to other upstream steps of the production line. It may also contribute to the restriction of the (over)exploitation of resources by leveraging their use.

The contribution of waste prevention actions to the abatement of global climate change could be valuable, since the greenhouse gases (GHG) emissions of the waste sector account for approximately 5% of the global greenhouse budget (IPCC 2006). According to Yusuf et al. (2012), the main sources of anthropogenic methane emissions belong to three sectors, namely agriculture (53%), energy (28%) and waste (19%). However, if the overall impact of resources extraction and use is considered, the impact of inefficient use of resources / waste production on climate change is considerably larger, albeit difficult to estimate at an acceptable level of accuracy.

Considering what is environmentally desirable, is not always the socio-economically preferable solution. Certain waste prevention actions, such as efforts to extend product's life span, the banning of goods with hazardous materials and the promotion of service provision in lieu of purchasing equipment, might exercise a negative pressure on market economies. This rather short-sighted view may be responsible for reluctance by the industry to adopt waste prevention promoting models, in spite of the warm relevant rhetoric. However, waste prevention is not competing with economic development, both in the short and long term, as it enhances resources security while allowing for new, less resource-intensive, profit-making models. By evolving and implementing a suitable waste prevention plan, households, industry, waste management companies can reduce the amount of waste, without distorting the contemporary market economy.

In terms of domestic household economics, the linking of waste prevention with a more reasonable consumptive pattern results in saving considerable amounts of the family budget. On a larger scale, the efforts of the industry to prevent their waste are undoubtedly motivated and/or stipulated by the potential margin for profit (to a lesser degree, it is also dictated by legal and environmental imperatives compulsions). At waste management level, the integration of the waste prevention concept into the system would lead to the reduction of costs and creation of new job opportunities, mainly through the enhancement of repair and reuse networks. Overall, the waste prevention concept could have a tremendous impact on development and economic growth, promoting a more sustainable model for the exploitation of valuable resources(e.g. rare earth, phosphorus). Conversely, the lack of resources would pose serious threats to development, economic growth and interstate relations.

An important aspect of waste prevention refers to food waste, especially their avoidable fraction. Hunger still remains a serious problem in the 21st century, and food waste turns is a significant ethical and environmental issue. In EU-MS alone, 79 million citizens live below the poverty line, while approximately 89 Mt (amounting to 179 kg per capita) of food waste is disposed annually (EC 2011a).

Ethical issue related more or less directly to waste prevention, is the concern about environmental equity when addressing legal and illegal shipments of waste in Less Developed Countries (LDC). Even in the legal shipment, additional measures must be taken to ensure that waste is treated in a safe way for the wellbeing of workers and the inhabitants. Ongondo et al. (2011) cited legal but ethically questionable cases of using child labour in dismantling WEEE exported from Europe to LDC.

4.5 THE POLICY CONTEXT

In order to counteract the growing waste generation, policies providing robust guidelines on waste prevention and a suitable regulating framework are deemed a necessity. Waste prevention is now a key element of waste management and environmental policy frameworks, on regional, national and supranational (i.e. EU) level. To comply with the provisions of these policies, several countries have attempted to institutionalise waste prevention. In the following sections the policy framework for the EU, the US and selected countries, which have taken important steps towards the promotion of waste prevention, is briefly presented.

4.5.1 The EU Policy Framework

At the EU level, the first request to the MS to support actions for the prevention or the recycling of certain waste streams was issued through the Council Directive on Waste (75/442/EEC) in 1975, when the Union was a Community of nine states. With time, the request went through a "maturation" stage, being exposed to a series of programmes, strategies and directives (such as the Community Strategy for Waste Management – SEC (89) 934 Final 1989). Eventually, it became a priority within the Thematic Strategy on Prevention and Recycling of Waste (COM (2005) 666 Final) and a requirement with the WFD. Today the core of the current EU policy framework consists of the Thematic Strategy on Prevention and Recycling of Waste coupled with two framework legislation instruments – namely the WFD and the Waste Shipment Regulation and a range of stream specific Directives. The Thematic Strategy sets the principles for waste policy makers, while the legislation binds the MS to take action within their own legal and administrative system.

The legislation framework of the EU waste policy opts for a major transformation of the waste management landscape for new and older MS (Lasaridi 2009). More specifically, the WFD requires MS to prepare and introduce Waste Prevention Plans within a certain time frame. Attempt to clarify the end of waste criteria, stimulates MS to take measures for separate collection of biowaste and enhances the extended producer responsibility (EPR) policies. An efficient waste prevention plan has to comprise all materials and products flows, from cradle to grave. Therefore, it has to be linked to the waste management sector, mining sector, industry, product designers, service providers, retailers and consumers.

The EU waste policy is based on the idea of the *waste hierarchy*, in other words various waste management options are ranked in terms of their environmental impact. It was introduced in the Community Strategy for Waste Management (SEC(89)934 Final 1989), but reaffirmed and established in the revised WFD

(2008/98/EC). Waste prevention was placed at the top of the ladder as the most favoured option, followed by (in descending order) preparing for reuse, recycling, energy/material recovery and disposal in properly engineered landfills – the latter "being the terminal option" (EC 2011c). The European Commission (EC 2011c) has clarified that the hierarchical system of waste management has not necessarily imperative in character, especially "since different waste treatment methods can have different environmental and health outcomes"; a deviation from the hierarchy is acceptable, if it is justified by life-cycle thinking.

Of particular interest is the main regulation on waste management in EU consists of setting fixed targets for the reduction of specific waste streams (e.g. Directives on Landfilling - 1999/31/EC, Packaging and Packaging Waste - 94/62/EEC, End-of-Life Vehicles - 2000/53/EC,WEEE - 2002/96/EC, Batteries -2006/66/EC, etc). By issuing the WFD, the EU attempted to create a planning and institutional framework for the proper implementation of waste sectorial regulatory instruments, making a shift from specific waste streams regulation to integrated principles and planning.

4.5.2 US Policy

Since 1976, the waste management in US has been run under the federal Resource Conservation and Recovery Act (RCRA) statute, regulations and programs (U.S. EPA 2002). The RCRA amended a former Act, namely the Solid Waste Disposal Act (SWDA), which was passed in 1965. It sets the requirements that were intended to streamline the waste management practices in different States.

According to the US Congress, the RCRA was passed to ensure that wastes were treated in a manner that protects both human health and the environment. It was set up to be a pollution prevention measure (U.S. EPA 2002), in a broader concept that also embraced waste prevention. Since the RCRA was signed, a range of improvements in waste management practices have been achieved across the USA: open dumping was banned; uncontrolled dumping of hazardous industrial waste has dwindled, while MSW landfills have been upgraded. Still, critics argue about how RCRA program decides when a material has turned into waste; in addition, the program persisted in focusing on "end of the pipe" controls instead of focusing on upstream pollution prevention measures (U.S. EPA 2003).

4.5.3 Specific National Initiatives

4.5.3.1 United Kingdom

Although England and the devolved Assemblies or Parliaments of Wales, Northern Ireland and Scotland have developed different waste policies and are independently responsible for their delivery, they do work closely together within the same EU legislative framework (DEFRA 2011). The waste policy of England, developed since 2000, has brought significant progress in the national waste management system. It has been shaped by the UK Department for Environment, Food and Rural Affairs (DEFRA) through the Waste Strategy 2000 and Waste

Strategy 2007, both of which were aimed to decouple waste generation from economic growth. England's waste policy has been employed towards a "zero waste economy", considering (from the government's point of view) that a zero waste economy does not mean cessation of waste generation. A significant progress has been achieved (but not exclusively), is attributed to the introduction of the landfill tax escalator, the Landfill Allowance Trading Scheme (known as LATS), the Waste Implementation Programme, the Waste and Resources Action Programme (WRAP) and the Business Resource Efficiency and Waste (BREW) programme (Phillips et al. 2011).

In December 2013, the Waste Prevention Programme for England, entitled "Prevention is Better than Cure" was published after extensive public consultation. It also aims at a zero waste economy by enabling better resource efficiency and waste prevention, in compliance with the revised WFD, and contains plans for (DEFRA 2011; HM Government 2013):

- additional EPR deals;
- engaging households to waste prevention and reuse through rewards; and
- working with business on both the "promotion of the economic case for new business models which help to prevent waste" and "on the design of products enable easier upgrades, repair or recycling at end of life".

To date, good progress has been reported towards the initial goals. In the March 2012 progress report, DEFRA stated that "evidence to support waste prevention measures is being developed, together with stakeholders and WRAP" (DEFRA 2012).

In Ireland, the National Waste Prevention Programme was introduced in 2004, aimed at delivering significant results on waste prevention, as well as on the integration of "initiatives addressing awareness-raising, technical and financial assistance, training and incentive mechanisms" (U.S. EPA 2012). It is being led by the Environmental Protection Agency with its implementation managed by the National Waste Prevention Committee which provides strategic directions for the EPA.Within the National Waste Prevention Programme of Ireland, programmes such as the Local Authority Prevention Network, the Green Business Initiative, the Green Hospitality Award, the Stop Food Waste and the Packaging Waste Prevention Programme, have been developed and are running.

In Scotland, the first Zero Waste Plan was launched in June 2010, envisaging a nation "where all waste is seen as a resource; Waste is minimised; valuable resources are not disposed of in landfills, and most waste is sorted, leaving only limited amounts to be treated". The waste strategy document (Towards Zero Waste) for Waleswas developed by the Welsh Assembly Government, mirroring a vision for achieving more sustainable waste management. It builds upon the previous waste strategy document, entitled "Wise about Waste" issued in 2002.

4.5.3.2 Finland

Finland seems to be a step ahead of the requirements of EU waste policy and legislation and therefore can provide some useful lessons to other EU-MS. Waste prevention was encompassed in Finnish waste legislation through the promulgation of its National Waste Act in 1993 - that is a couple of years before Finland joined the EU. At the time when the WFD required the EU-MS to develop a waste prevention plan, Finland had already endorsed one, in the ongoing Finnish National Waste Plan for 2007–2016.

Despite the progress in the adoption of policy instruments and the improvement of waste management infrastructures, by the early 2000s the Finnish waste policy had not proved effective in promoting waste prevention (Melanen et al. 2002). After analysing the effects and effectiveness of the Finnish waste policy, Melanen et al. (2002) recommended that economic instruments should be used more extensively, supporting the efficient use of natural resources and materials. Similarly, Lilja (2009) recommended a shift from waste prevention to the promotion of material efficiency. The authors discuss the difficulties that arose in preparing the Finish Waste Prevention Plan, and that waste prevention should not be taken as a strategy for waste management butshould be regardedas an integrated environmental policy (i.e. resources policy).

4.5.3.3 *Japan*

Driven by the need for economic development in a resource restricted country, intensive efforts have been taken by the government of Japan to promote the policy of "3R‘s" (notably, Reduction, Reuse and Recycling) in their waste management and sustainable consumption and production policies (Kurisu and Bortoleto 2011). This was embodied in legislation setting targets for waste prevention and avoidance of final disposal. In 2008, the Cabinet of Japan approved the Second Fundamental Plan for Establishing a Sound Material-Cycle Society, which aims at the optimisation of natural resources exploitation. The Plan introduces targets for three material flow indicators, namely resource productivity, cyclical use rate and the final disposal quantity (Kawai and Osako 2012). Comparing the per capita waste generation rate of Japan with that of other countries with similar GDP per capita, the Japanese waste prevention campaigns and the related attitudes may be considered as a success story (see Figure 4.5).

4.5.3.4 *The Netherlands*

Since the ‘60s, the waste management system in Netherlands has been transformed "from uncontrolled landfilling to a differentiated waste-handling system of recycling, incineration with energy use and controlled landfilling" (Kemp 2007). A milestone for this transition was the centralisation of the responsibility of waste management at central government authorities. The centralisation was proposed by the Commission Epema in 1996 (Kemp 2007), but obtained legal basis through the amendment of the Environmental Management Act on 8 May 2002 (ETC/SPC 2012). The latter stipulated the Ministry for Housing, Spatial Planning and the Environment to deliver a Waste Management Plan every six years, covering the needs of the national authorities, provinces and municipalities, imposing the implementation of international regulations and setting the rules for the management of specific waste streams (ETC/SPC 2012). The first Dutch Waste Management Plan was introduced in 2003, aiming mainly at increasing the proportion of waste utilisation (including Waste-to-Energy incineration) from 81% in 2000 to 86% in 2012. The second Plan was delivered in 2009 with a view to 2021, focusing on the environmental policy and sustainability. The Plan lays emphasis on waste prevention as a part of waste management, given that its objectives are (ETC/SPC 2012):

- the tackling of waste generation through the decoupling from economic growth;
- the reduction the environmental impact through recovery and reuse; and
- the minimisation of the environmental impacts of the different product chains.

4.6 WASTE PREVENTION AT THE LEVEL OF PRODUCTION AND SUPPLY

As discussed in section 4.2.1, actions for waste prevention can be taken at all the steps in a product's life cycle, from production and supply to consumption and discard (Figure 4.3). In the following sub-sections the prevention potential and suitable policies, at the level of production and supply, for the main priority waste streams are discussed, namely for food waste, packaging, WEEE, construction and demolition waste, bulky and other waste.

4.6.1 Food Waste

By definition, *food waste* is "composed of raw or cooked food materials and includes food loss, before, during or after meal preparation in the household, as well as food discarded in the process of manufacturing, distribution, retail and food service activities" (EC 2011a), while *food loss* "refers to a decrease in mass (dry matter) or nutritional value (quality) of food that was originally intended" but is no more suitable "for human consumption" (FAO 2013). *Food wastage* comprises both food waste and loss.

Food and Agricultural Organization (FAO 2011) estimated that approximately one third of food produced for human consumption is either lost or wasted throughout the food supply chain, from agricultural production and post-harvest handling and storage to processing, transportation, distribution and consumption (Moomaw et al. 2012). According to a (preliminary) study for food waste in the EU-27, the retail-wholesale sector produces approximately 5% of the total amount of food waste (4.4 Mt per year), the manufacturing sector 39% (35 Mt per year), households 42% (38 Mt per year), while food service - catering 14% (12.3 Mt per year) (EC 2011a). However, from a worldwide point of view, the contribution of each sector differs substantially due to a range of reasons dependent on the

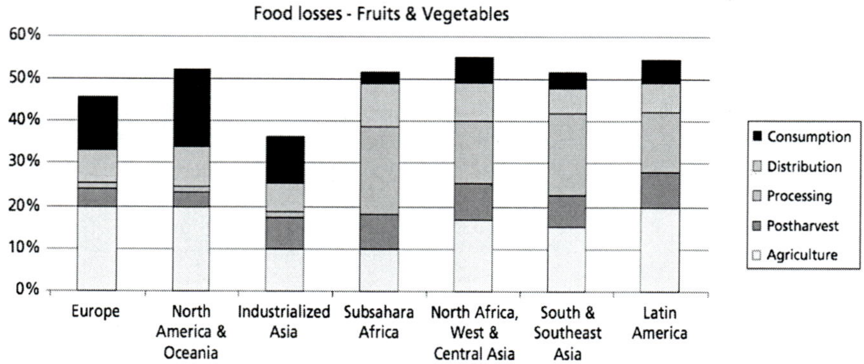

Figure 4.7. Fruit and vegetables lost or wasted at different sectors of the food supply chain in different regions
Source: FAO (2011); reproduced with permission from Springer

socio-economic conditions of each country (Figure 4.7). Parfitt et al. (2010) stated that in less developed countries food losses are higher at the immediate post-harvest stages, while in industrialised and developing economies they are higher for perishable foods after they have been prepared for market, or, in fact, sold to the consumer. A recent analysis of the food supply sectors by region, showed that (FAO 2013):

- upstream losses occurring at agricultural production phase appear homogenous across regions, representing about one third of each region's food wastage; and
- downstream wastage occurring at consumption level is much more variable, with wastage in middle- and high-income regions at 31-39%, but much lower in low-income regions, at 4–16%.

Although food production and supply chains are optimised for the minimisation of (mainly economical) costs, still they constitute a significant source of wastage, due to a multitude of sector specific causes. In the retail and wholesale sectors, the causes for waste generation can be objective or subjective, attributed to (IVL Swedish Environmental Research Institute 2011; Mena et al. 2011):

- the social norms, which form consumers' behaviour and attitudes, knowledge, status and food culture;
- technical issues such as storage conditions, proper handling, etc.; and
- the operation and management of the stores including stock management, management of full shelves, marketing strategies and standards.

In the manufacturing sector, a large amount of food wastage is unavoidable, as it comprises bones, internal organs etc. The rest is mainly attributed to technical malfunctions (shortcomings in anticipating demand, packaging damage – EC

2011a, b). In respect to the distribution sector, the main causes for food waste are related to supply chain inefficiencies, high product specificity, and last but not least, temperature sensitivity.

Based on Figure. 4.7, the production and supply sectors alike, could reduce food waste through the following indicative considerations:

- *Donations:* Both sectors can donate food that otherwise would be discarded due to unprofitable prices, surpluses and weakness to cover the marketable criteria. Food donation can be exploited as a means to prevent food wastage, while meeting social needs. Industries and retailers could benefit from the cost savings from landfill/disposal diversion. In the USA, the national network of food banks (Second Harvest Food Banks) coordinates the distribution of surplus food to 25 million people (Schneider 2008).
- *Development of synergies:* By establishing industrial symbiosis and eco-industrial parks, one firm's waste can serve as a resource to others (Lehtoranta et al. 2011). For instance, one's firm food waste can be used as animal food.
- *Clarification of current food data labels:* Research on date labelling showed that 45–49% of consumers in the UK misunderstood the meaning of the date labels "best before" and "use by" (Cox et al. 2010). In Greece, during severe economic crisis, 42% of consumers would discard perfectly edible food due to label misunderstanding (Abeliotis et al. 2014).
- *Updating the production chain:* By adopting more efficient energy and material technologies, the production chain can be improved to generate less food waste.
- *Appropriate storage conditions:* The edible life of a product can be prolonged byensuring optimal storage conditions, thus reducing wastage. Technical improvements focused on proper storage of food either in the retail or the wholesale sector (e.g. shelves packing, packaging, etc.) could lead to the extension of life span of perishable food.
- *Education of personnel in the retail-wholesale sectors:* Training the personnel on the appropriate handling of food products is a good way to reduce damage. Research has shown that the personnel adopt waste prevention measures more easily when information on the environmental and economic impact of food waste is provided to them (IVL Swedish Environmental Research Institute 2011).
- *Management of orders in the retail sector:* The stock management and the purchases should be optimised as not to lead to surplus amounts of unsalable food.

4.6.2 Packaging Waste

According to the EU packaging Directive (94/62/EC) packaging waste is defined as "all products made by any material of any nature to be used for the containment, protection, handling, delivery and presentation of goods, from raw materials to processed goods, from the producer to the user or the consumer".

Since packaging is a means for ensuring hygienic conditions or for avoiding products' damage, packaging waste is to a certain point unavoidable. However, over- packaging can pose a significant environmental problem. The generation of packaging waste can be influenced by the materials used, the geometry of the package, the content and the information on the package (Langley et al. 2011). The primarywaste materials from packaging waste, are glass, aluminium, steel, plastic and paper/cardboard.

Apparently, one of the most promising and favoured packaging waste prevention strategies is *dematerialisation*. The term was introduced by van der Voet et al. (2004), in order to describe the "process of fulfilling society's functions with a decreasing use of material resources". For instance, for packaging waste prevention, dematerialisation could be translated into actions aimed at substituting one type of packaging with a lighter one (light-weighting) or the provision of more concentrated products, which would be available in smaller packaging for the same functional units (e.g. detergents). The actions mentioned below could enable both the manufacturing and distribution/retail sectors to reduce packaging waste (Cleary 2010; Weidema et al. 2004):

- *Inclusion of a deposit return system:* When adopting such a system, the manufacturer returns an amount to the consumer/user in exchange for the returned packaging. This has been most frequently implemented with beverage and beer bottles and in many countries it is based on automated reverse vending systems with excellent results (e.g. PET bottles in the Netherlands).
- *Promotion of refillable /returnable packaging:* Refillable packages (bottles, jars, metal boxes) containing products, such as dishwashing liquids and liquid soap, which are accompanied by product recharges from lighter packages that are widely available and promoted in the market.
- *Promotion of reusable bags:* It can be achieved by a series of actions, such as banning or charging for plastic shopping bags, encouraging retailers to reduce their use of plastic bags, etc.In this front, the recent legislative initiative of the European Commission to amend the Packaging Directive to reduce the use of plastic carrier bags with a thickness below 50 microns (EC 2013) is of particular interest,several countries have achieved remarkable results through the use of economic instruments (e.g. Ireland, Denmark, Italy).
- *Light-weighting a package:* Although this is an obvious choice that has been successfully adopted in several cases, a careful approach is need that may otherwise restrict its full potential. Given that the properties of a product are a function of many variables (functionality, technical characteristics, aesthetics, additional service provision, etc.), there is a possibility that the candidate product/ substitute is not equivalent to the previous one.

4.6.3 Waste from Electrical and Electronic Equipment

The rapid growth of electrical and electronic equipment (EEE) consumptionhas led to an explosion of obsolete appliances and devices, with low or null resale value, and the corresponding e-waste, estimated worldwide at approximately 40 million tonnes per year (Serranti et al. 2013). Waste from electrical and electronic equipment, referred to as WEEE or e-waste, include a wide range of different appliances (e.g. consumer equipment, large and small home appliances, monitoring and control instruments, electrical and electronic tools, automatic dispensers, etc), and broad range of materials. Therefore WEEE, apart from being a growing environmental problem and a source of toxic substances, can also be considered as a "secondary resource" of the base and precious metals, as well as rare earths (Hobohm et al. 2013; Serranti et al. 2013).

The skyrocketing amount of WEEE is not only a problem of the affluent countries, asless developed and developing countries often become receivers of used EEE or (legal and illegal) WEEE shipments. Exporting WEEE to those countries can prove cost-effective for the exporter, although it poses great risks for the environment and human health. According to Schwarzer et al. (2005), sometimes illegal waste exports are even "hidden under the umbrella of charity".

A prerequisite for the successful implementation of waste prevention measures targeting WEEE is the active involvement of the public and all stakeholders (Ongondo et al. 2011). However, this is not as straightforward as it might seem, because in EEE manufacturing sector, the producer is a broader notion. Preventing WEEE at the production line and the distribution/retail sector and reducing their impacts on the environment and human health, could be achieved by the following actions:

- *Extension of the life of EEE.* It could be achieved through the development of proper design, hardware and software that would enable easy upgrade of devices.
- *Avoidance of the use of substances mentioned in RoHS Directive:* The RoHS (Restriction of the use of Hazardous Substances) EU Directive prohibits (in the EU) the presence of EEE containing lead, hexavalent chromium, cadmium, mercury and polybrominated biphenyl and diphenyletherflame retardants. This might not lead to the reduction of waste amounts, but it will cushion the impact on the environment and on human health, through toxicity reduction. This is particularly important in cases where WEEE'sare exported to countries with low waste treatment standards (EC 2011c).
- *Inclusion of EPR based instruments:* The complete connection of the production chain with the end-of-life products and their environmental burden is considered necessary to promote waste prevention, recycling and overall resources efficiency (Ongondo et al. 2011). Indicative examples of EPR based instruments for WEEE are the take-back obligations and deposit-refund systems. It is important to note that there are voices doubting the effectiveness

of EPR strategies, at least within the frame of WEEE Directive. Lauridsen and Jorgensen (2010), consider that using EPR strategies "as a guiding design principle" has resulted in a "situation where only the waste management objectives of the directives have been realised, whereas the vision of designing for waste reduction has been left out".

- *Adoption of Product Service Systems:* For details, please refer to section 4.6.5.
- *Development of compatible EEE parts:* For example, in 2009, the major mobile phone manufacturers committed to providing harmonised chargers in the EU, through a Memorandum of Understanding.

4.6.4 Reducing Planned Obsolescence

Planned obsolescence can be defined as the "deliberate curtailment of a product's life span" (Cooper 2005). It stipulates designing a product such that it shortens the product's useful life, by making it functional or fashionable just for a limited period of time. This is a noticeable trend of the modern consumerist era: before the middle of the 20th century, the purchase of durable products was deemed an investment. The tendency for designing products with a set lifespan was born in USA London (1932), as a way to overcome the economic depression of the 1930s. The concept has since thrived in the market economies and became an essential factor for the economic growth, the reduction of unemployment and the increase of taxes and GDP.

The main mechanisms used to shorten the usable life of a product are (Guiltinan 2009):

- the limited functional life design (known as "death dating");
- the design for limited repair (e.g. the disposable single-use cameras or components that are soldered instead of being screwed); and
- the design aesthetics that lead to reduced satisfaction as fashion changes.
- On the less known, bright side of planned obsolescence, two points are highlighted:
- Without planned obsolescence, energyinefficient appliances would have been stocked and still be in use, increasing energy consumption and GHG emissions (Fishman et al. 1993; Cooper 2005).
- Planned obsolescence led - in a way - to disposable products that are necessary for hygienic and health reasons (e.g. medical gloves, syringes).

Currently, however, the mounting amounts of functioning durable goods ending up in landfills have led to renewed criticism of planned product obsolescence (Guiltinan 2009). On the dark side of it, planned obsolescence leads people (consumers) to throw away devices that are not failed, goods that are not torn. The parameters which influence the policy of planned obsolescence are schematically illustrated in Figure 4.8.

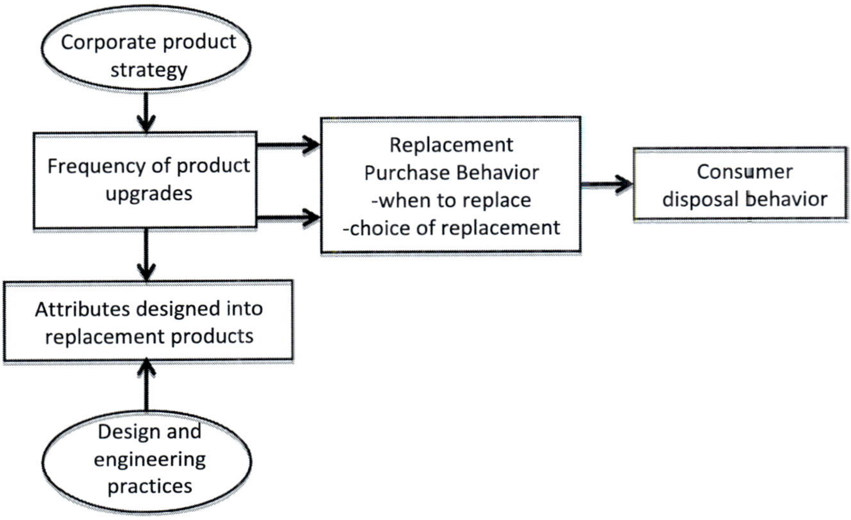

Figure 4.8. Decisions and influences linking planned obsolescence and the environment
(Source: Guiltinan 2009)

Lately, the need for a Paradigm shift for the elimination, or at least the curbing, of planned obsolescence is recognised and reinforced by the broader need to address resources exploitation and waste generation. This shift constitutes one of the central pillars of eco-design and life-cycle design. Both of these design movements aim at reducing the adverse impacts of products on the environment and human health by introducing life-cycle thinking in product design (Braungart and McDonough 2009). Increasing the longevity of a product requires improvements in both its durability and maintenance processes and capacity (appropriate use, repair, upgrading if applicable, and reuse). In this respect special attention should be paid to both good practices and theoretical models encompassing product efficiency and sufficiency, to show how the longer-lasting products may contribute to sustainable development (Cooper 2005).

4.6.5 Transition from Products to Products Service Systems

In the route to sustainable production and supply, the promotion of service provisions instead of goods seems increasingly interesting. Although in most cases this modern trend is stimulated by financial profit, it may indisputably provide cleaner product concepts, satisfying market demand with the use of less resources (Heiskanen and Jalas 2003). A range of firms are investigating the option of service provision, by using some of their products as platforms for extra service offerings (Gottberg et al. 2010). Bartolomeo et al. (2003) mention that under certain circumstances both the manufacturer and its customer consider the service component more significant than the production of goods.

A particular type of service provision is the Product Service Systems (PSS), a concept that historically emerged from a case study on the modern trends in service provisions (Gottberg et al. 2010). A PSS consists of products and services combined to work jointly and satisfy the consumer's demands in an integrated, convenient, economic and resource-efficient manner. It can have many forms, such as machinery leasing, laundry services or e-shopping.

4.6.6 Construction and Demolition Waste

In 2006, approximately 970 million tons of waste were generated in the construction sector in EU-27 (EC 2010), with nearly 90% of them being mineral. In 2008, a decrease was noted, with the generation being curtailed to 860 million tons but it is not clear whether this reduction was due to the economic recession that had affected some of the EU-MS, the paucity of data and reporting changes (EC 2011b) or waste prevention measures (for example crushed concrete being reused in construction).

When developing waste prevention measures for construction and demolition (C&D) waste, policy makers have to confront:

- materials with hazardous substances, calling for special treatment (i.e. PAH's, ozone depleting substances);
- limitations to the transportation of mineral materials, due to escalated costs; sometimes, this problem is compounded by the scarcity and therefore high costs of raw materials;
- exposure of the employees to hazardous materials during de-construction; and
- the fact that the effects of C&D waste prevention measures are often not seen in the short-term, due to the relative long lifespan of a building.

It is widely accepted that the enhancement of C&D waste management systems through the development of preventive measures during all stages of the building design and construction processes is needed (del Río Merino et al. 2011). The inclusion of a life-thinking approach to the training of all stakeholders intervening in the construction process, coupled with a turn of mind-set that C&D waste should be viewed not as a burden but as a potential source for profit, are prerequisites for the curtailment of C&D waste (del Río Merino et al. 2011; Llatas 2011).

A series of measures (under the control of different personnel e.g. the construction manager and the site manager) for the improvement of C&D waste management is explicitly reported in the paper of del Río Merino et al. (2011). Waste prevention measures should aim at:

- *Encapsulating the waste prevention principles and introducing a life-thinking approach to building design:* In particular, the design must increase the flexibility of the building to changes (i.e. refurbishment) and enable the user to adapt it according to his/her needs with the minimum cost (Morgan and

Stevenson 2005). It should also cover the deconstruction of the building in the long-term. The Scottish Ecological Design Association (SEDA) has publicised a useful guide about design for deconstruction in this regard (Morgan and Stevenson 2005).

- *Utilisation of eco-efficient or reclaimed materials:* A good example of C&D waste prevention strategy is the Austrian programme "Nachhaltig Wirtschaften" (Sustainable Housekeeping), which financially supports pilot and demonstration projects on the innovative utilisation of renewable and eco-efficient construction materials for low energy consuming buildings (EC 2010).
- *Developing a network for contractors, builders and architects:* The network will be employed as a platform for purchasing/offering materials from C&D processes (e.g. the excavated soil during the construction of a building).
- Setting compulsory limits to the content of hazardous materials and labelling eco-designed and ecological constructions.

4.6.7 Bulky Waste

According to the definition of the European Environmental Agency (EEA), the term *bulky* labels all the sizable items of waste, such as furniture and big appliances, which are not usually accepted by the conventional waste collection system. In this chapter, appliances are excluded from the discussion of bulky waste (as they are better classified in the WEEE stream, which was discussed in section 4.6.3), while mattresses, textiles (clothes, curtain, accessories) and shoes are included.

Bulky wastes are very heterogeneous in nature and comprise of many composite materials. Still, the dominant materials are wood, textiles and metals. Their collection varies widely from country to country, from street collection to central collection points and collection on request and can be free of charge for the citizen (i.e. included in the standard municipal taxes/waste management charges) or paid upon use of the service (Ministry of Life, Austria–Lebensministerium.at 2011). In all cases, bulky wastemay pose significant challenges on the waste collection and management systems and their fly-tipping can be a common problem, even in developed countries.

The following actions could enable both the manufacturing and distribution/retail sectors to lessen their bulky waste:

- *Design furniture that can be easily renovated and/or repaired:* The design and the manufacturing of modular furniture could ensure the easy repair of a part and thus enhance this practice.
- Organising after-sales departments for the repair/service offurniture.
- *Development of a network for the reconditioning and marketing of used furniture:* An indicative example of the implementation of this concept is the network "Ecomoebel", which is supported by the German Federal

Ministry of Education and Research (http://www.ecomoebel.de). The network is responsible for the reconditioning/repair of used furniture and provides a certificate which guarantees low environmental impact and high quality standards.

- Development and promotion of second hand markets.
- *Production of guides (leaflets/manuals):* For example, guides (leaflet/manuals) for the proper cleaning of clothes and shoes and for the proper use and maintenance of furniture.

4.6.8 Other Waste

Waste generated during children infancy is also of major concern. Disposable diapers, toys, baby and nursery products raise great environmental issues, due to their amounts, composition and their short useful life-span. Their amount and environmental impacts could be curtailed by:

- *Shift to reusable diapers:* Salhofer et al. (2008) demonstrated that reusable diapers are preferable to disposable ones, with respect to MSW generation. However, they also indicated that when taking into consideration the whole life cycle of them, the outcomes depend on the geographical region.
- *Organising leasing services for baby and nursery products:* Cradles, baby carriages and bathinettes are samples of the large range of items, which seem absolutely necessary during infancy and useless when children reach the age of four. Although parents tend to gift them to other families, still there is a significant amount that is discarded. The development of leasing services for theseitems could offer an effective solution towards waste prevention in this products' stream.
- *Design and development of flexible baby and nursery furniture and accessories:* Improvements in the design and construction of these products, in a way that they may be easily repaired or renovated in order to satisfy a child's needs over time, could also reduce the quantity of the relevant wastes.
- *Design toys with extended life-span:* Increasing durability for many items and eliminating planned obsolescence of toys is a good basis for the prevention of their waste. Additionally, leasing centres could be developed.

4.7 WASTE PREVENTION AT THE LEVEL OF CONSUMPTION AND THE HOUSEHOLD

In this section the prevention potential for the main priority waste streams at the level of consumption and the household, as well as suitable prevention policies and measures are discussed. Food waste, packaging waste, non-packaging paper waste, WEEE and green waste are specifically tackled in the following sub-sections.

4.7.1 Food Waste

Since the publication of the findings of WRAP (2009) that consumers in the UK throw away 31% of the food that they buy, food waste at the consumption and household level is becoming an increasingly significant global issue. Even though the generation of food waste from households is only one part of the whole picture, there is growing evidence that their contribution to the problem is particularly significant (Sharp et al. 2010; EC 2011a). It is roughly estimated that, in EU-27, households are responsible for 42% of the total amount of food wastegenerated (EC 2011a).

In order to tackle food waste generation in households, a number of socio-demographic, behavioural and attitudinal factors that could have a clear impact on food waste reduction have been investigated (Evans 2012; Koivupuro et al. 2012; Stefan et al. 2013). Koivupuro et al. (2012) showed that, in Finland, the amount of food waste is statistically correlated with the size and the type of household, the gender of the person who has the main responsibility for shopping, people's view on the waste prevention potential, the provision of food products in low prices, and the household's view on the impact of purchasing different package sizes. Understanding such behaviours is crucial for planning effective campaigns for food waste reduction: which groups of people to focus on, channels of engagement and design of campaign material (WRAP 2009).

Among the different waste prevention initiatives that can be implemented by waste-stream at the household level, food waste ranks at the top, with the highest prevention potential (Cox et al. 2010). Besides the obvious environmental impacts of food waste (i.e. increasing stress on the waste management system, significant contribution to climate change and exploitation of limited resources, valuable for sustaining life on the Earth system, such as soil and water), they are also associated with economic impacts, such as business and consumers' savings (Nahman et al. 2012), social and ethical issues, such as global food and water security (Koivupuro et al. 2012) and the expanding dimensions of malnutrition, in both hunger/undernourishment (Nahman et al. 2012) and obesity terms (Parfitt et al. 2010).

At the household level, food waste is associated with certain attitudes and behaviours, such as meal planning, cupboard checking and list making before going for shopping, proper storage of food items, use of food leftovers, cooking the right amount of food and good understanding of expiration date labels (WRAP 2009; Abeliotis et al. 2012a, b; Koivupuro et al. 2012; Stefan et al. 2013; Quested et al. 2013; Abeliotis et al. 2014). The study of the behavioural aspects of waste prevention has only recently appeared in the literature and is still very limited. Findings so far indicate that waste prevention behaviour has different dimensions from recycling behaviour and therefore requires different approaches to increase people's engagement to relevant policies (Bortoleto et al. 2012). These authors developed and tested a structural equation model, as a variant of the Theory of Planned Behaviour (TPB) for the prediction of household waste prevention. The model indicated that personal norms and perceived behavioural control are the main predictors for waste prevention behaviourin Portugal. Focusing on food waste, Stefan et al. (2013) applied the TPBto the study of food waste prevention behaviour of consumers in Romania. According to their findings, for the Romanian consumers, food waste generation is mainly driven by routines related to food provisioning rather than by a conscious intention not to waste food.

In spite of the indisputable value of these pioneering studies on waste prevention behaviour, as well as the studies of WRAP (2009, 2012), Koivupuro et al. (2012) and Schneider and Lebersorger (2009) on the patterns and causal factors of food waste generation,more research is needed, in order to guide efficient campaigns to promote households engagement in food waste prevention initiatives.

4.7.2 Packaging Waste

The contribution of individuals to the prevention of packaging waste could be catalytic, as the successful implementation of any action at the consumption/household level depends on its integration into daily routines and consequently on individuals' behaviour and attitudes. Consumers and/or users can participate in packaging waste prevention initiatives both in direct and indirect ways. In particular, they can exercise pressure on the market by favouring products with lighter or more environmentally friendly packaging or by preferring refillable, returnable and/or compacted products. In addition, they could devise alternative uses (reuse) for several package items such as jars, wrapping paper, ribbons and gift boxes.

Consumers and/or users could be considered as the final "checkpoint" for dematerialisation strategies (for the definition, please refer to 4.6.2). Although these strategies are rather promising, they strongly hinge on the consumers' and/or users' acceptance. More specifically, a dematerialised action will not be adopted by the public, unless it does not affect negatively the standard of living, its perception or even the commonly acceptable life-style trends (Cleary 2010). The consumer should be convinced that there is a reasonable motive for the substitution of a product that he/she is happy to use, with another one that is supposed to be more environmental friendly. This point should always be considered when assessing the potential of waste prevention actions within an LCA approach (Cleary 2010).

4.7.3 Paper Waste (Non-Packaging)

Municipal paper waste is defined as waste originating from printing, packaging and sanitary paper. In this section, packaging paper is excluded as it is collectively discussed in the section on packaging waste (4.7.2). Here, the discussion on paper waste refers to newsprint, magazines, leaflets, archived paper, catalogues, telephone directories, unprinted sheets and cartons, as well as handkerchief, table napkins, paper towels and toilet paper.

The actions that could enable households and consumers to reduce their paper waste include:

- *efforts to eliminate junk mail*: several studies have shown that posting a "No junk mail" sticker on the door may be an effective manner to avoid unsolicited advertising;
- *requiring e-statements*, instead of printed paper format statements;
- *the use of cloth napkins*; and, to a different level,
- switch to e-publications and the paperless office.

4.7.4 Waste from Electrical and Electronic Equipment

Whereas the production and supply sectors can contribute to the curtailment of WEEE through ever-improving technologies and services, directing users' behaviour towards WEEE prevention is not easy to achieve (Ongondo et al. 2011). The fact that electrical and electronic equipment (EEE) are assimilated into the mainstream with alarming speed, coupled with the improved environmental and functional efficiency of newer models, boil down to the conclusion that the quantities of WEEE will keep increasing. Taking under consideration the dominating "throw-away" consumption patterns that dominate contemporary affluent market economies, Ongondo et al. (2011) postulated that WEEE are more expected "to be generated than prevented at the consumer end of the spectrum".

Re-use of EEE could be an effective waste prevention measure (Ongondo et al. 2013). However, a number of barriers have to be surpassed for re-use to be successfully implemented. A significant barrier lies on environmental grounds and restricts the life-cycle benefits of EEE re-use: the limitations on EEE re-use, which are set by the newer, more efficient and less energy-consuming products have been widely discussed in the literature (Kissling et al. 2013). Other barriers relate to economic and consumer trust grounds. The tumbling resale values and the lack of trained personnel for repairs, in many developed economies result into high repair prices that cannot compete with the new products market (Ongondo et al. 2011). Moreover, consumers doubt the quality of refurbished EEE, as these often lack any guarantee. Ongondo et al. (2013) suggested that this can be surpassed through a standardisation approach (such the novel UK Publicly Available Specification – PAS 141), which will ensure quality and enhance consumer confidence.

Another aspect that needs to be considered when developing WEEE prevention plans at the consumer/household level, is the phenomenon of electronic waste stockpiling. Owners in many cases do not discard useless or damaged electronic items, but they store them in attics, basements etc., adopting in this way a type of end-of-life (EoL) management at home (Wagner 2009; Ongondo et al. 2011). This phenomenon is quite prominent in both North America and China, though for different reasons. In the former, the possible reasons are EoL fees, convenience, perceived minimum risk presented by intact electronic items, sufficient storage capacity and perceived value of the items, meaning that the owners feel they may use them in the future (Ongondo et al. 2011; Wagner 2009). In the latter, stockpiling is mainly attributed to the belief that EEE might prove useful in the near future or that they are valuable and their owners expect a kind of compensation for their WEEE (Li et al. 2006; Ongondo et al. 2011).

Despite the aforementioned difficulties, the following WEEE prevention activities may be implemented at the consumer/household level:

- *donation of unwanted EEE* to friends, charities, schools;
- *repairing* an appliance or upgrading an electronic device rather than discarding it and replacing it by a new model;
- sharing electric and/or electronic tools with neighbours;
- *setting criteria that promote waste prevention when purchasing EEE.* For instance, favouring environmentally efficient equipment, select equipment with the potential to be upgraded with low cost, or select EEE with compatible parts.

4.7.5 Green Waste

Green waste consist of biodegradable materials, such as grass, leaves and branches, derived from private and public gardens and parks. The quantity of green waste and its environmental impact at the household level could be reduced by:

- *Selecting native or easily adaptable plant species,* as they withstand the local weather and soil conditions. Perennial plants that do not need frequent cutting should be favoured, as annual and biennial plants have a short life span and end up into waste within one or two years.
- *Taking into consideration the growing potential of plants.* By choosing the right species for the right place, clippings and trimmings will be reduced.
- Composting green waste.

4.8 BARRIERS TO WASTE PREVENTION

Waste prevention measurement limitations, conflict of interests, inconsistent definitions and the lack of a comprehensive strategy often feature as the main obstacles for the effective implementation of waste prevention policies (Salhofer et al. 2008). Moreover, apart from technological and potential legislation barriers, human attitudes and environmental awareness (or the lack of it) are the most important drivers for waste prevention, defining the success or failure of any prevention initiative. In the following sub-sections, the most important barriers to waste prevention are discussed.

4.8.1 Barriers with Respect to Consumer Behaviour

On a conceptual basis, the engagement of the public in waste prevention actions is of major importance for the successful implementation and the progress of any relevant policy. Hence, understanding public behaviour and attitudes has attracted the interest of many researchers (Cooper 2005; Barr 2007; Peattie 2010). Barr (2007) demonstrated that "people with ecocentric values and those with strong environmental concerns based on the concept of sustainable human development, are more likely to reduce their waste".

In practice, there is a discrepancy between the expected and the actual participation of the public to waste prevention activities (Salhofer et al. 2008). The authors claim that conservative measures, which would not influence the consumption rates, could result to a waste reduction of 1–3% of MSW. However, the accurate assessment of both the potential and the actual effect of the different waste prevention practices faces plenty of methodological issues. Cox et al. (2010) suggest that there is an inability to estimate to what extent waste prevention behaviour is practiced, because there is not just one, but many different behaviours. The motivations for and the barriers to engaging consumers in waste prevention behaviours are as many as the diverse activities (Cox et al. 2010). The most common barriers could be broadly categorised into the following four groups, which are further discussed below:

- the elements that dictate how easily an action can be adopted;
- the shortcomings in the distinction of waste prevention from recycling, especially for the general public;
- the contemporary consumption patterns; and
- the persons' self-image.

4.8.1.1 Drivers for Adopting a Waste Prevention Action

What is environmentally desirable cannot always be translated into an adoptable action or behaviour (DEFRA 2009). Its adoption or rejection depends on variables such as the social norms and habits, information diffusion, economic cost and inconvenience. A discrepancy between the expected and the actual engagement of people in waste prevention actions stems from established ethnic and social norms and habits – in particular when these norms and habits enhance and/or impose behaviours that does not favour waste prevention principles. For instance, in Greece, the preparation of lunch or dinner is seen as an action that strengthens family ties. Therefore, the majority of people do not prefer pre-prepared food (Abeliotis et al. 2012a) although it may produce less waste than preparing food at home (at least, at the household level).

A variable that always influences the degree of acceptance of a new behaviour or action is the way that the latter is promoted to the public. Quite often, the rejection derives from misleading and distorted data and shortcomings in the dissemination of relevant, coherent information. The strong negative reaction of Greeks (by the general public, the press, official organisations and many opinion leaders) to the initiative of the Ministry of Development to allow the sales of non-perishable food products that have passed their "best use before" date at lower prices, in the autumn of 2012, is an indicative example. Greek people, under the pressure of the economic recession, equated products past their "best use before" date to expired perishable food that is not safe to be consumed and considered this action as an imposed reduction in their quality of living and the food safety standards that they had conquered.

Engaging people can be easily obstructed by lack of knowledge and skills about how to carry out the proposed waste prevention actions, especially when coupled with lack of infrastructure or organising structures (DEFRA 2009). Another significant variable is the cost. It can constitute both a motivation and a barrier for the engagement of consumers in different waste prevention behaviours (DEFRA 2009). The cost savings from using refillable products (e.g. beverage packaging) is an indicative motivation example. On the other hand, encouraging consumers to purchase more goods than they need by offering "buy two, get one free" packs, can be a major barrier to waste prevention. As another example, Salhofer et al. (2008) mention the elevated cost of (certain) goods made of biodegradable materials as well as the costs incurred by industry for endorsing waste prevention measures in the production chainas barriers to waste prevention.

The degree to which an action is (in) convenient dictates whether it will be adopted by the consumers. Inconvenience has been reported as a barrier to the adoption of reusable nappies and certain PSS. However, in many cases the inconvenience is quite illusive; consumers believe that an action is inconvenient because they have never tried to implement it.

4.8.1.2 Confusion about Waste Prevention and Recycling

Once a consumer's intention to engage in waste prevention behaviour is established, another issue needs to be considered: research has demonstrated that the consumers often confuse prevention with recycling, for the recycling norm has been endorsed in their routine (DEFRA 2010). Cox et al. (2010) stated that there are hints that kerbside recycling may inhibit the development of waste prevention habits and therefore concluded that "it cannot be assumed that prevention is the next natural step from recycling". According to an older study (Barr 2007), waste reduction and reuse behaviour is predicted by underlying environmental values, knowledge and concern-based variables, while recycling is proved to be a highly normative behaviour. Therefore, prior to the implementation of any waste prevention programme, it should be stressed that waste prevention occurs before recycling.

Recycling, undoubtably, contributes to the reduction of resources' exploitation; however, it requires waste, meaning that materials and energy should be consumed for waste generation and treatment (collection, transportation/shipping, recycling). Furthermore, during the recycling process, an amount of waste is inevitably produced. Waste prevention calls for no energy consumption or material transportation. At this point it might be important to stress that waste prevention is not meant to be a substitute for recycling, but something to run in parallel with it.

4.8.1.3 Contemporary Consumption Patterns

Contemporary consumption patterns act as a barrier to waste prevention behaviours for the same reasons for which they persists to provoke the ever-growing waste burden. They counteract waste prevention, because consumption is not only linked to the satisfaction of the basic needs for subsistence or survival (food, shelter, clothing) but it is also extended to other fundamental needs, i.e. affection, participation, leisure and identity. Given that for most people in modern societies their consumption "footprint" provides self-esteem and personal value, it seems insurmountable for an individual to curtail consumption. The effect of the contemporary consumer behaviours and patterns is well mirrored in the term "shopping therapy", which may be very difficult to counteract through environmental awareness campaigns.

4.8.1.4 Self-Image

In many cases, the adoption of waste prevention behaviours is inhibited by the way that people think of themselves as individuals or a part of the society. This barrier may have two aspects. The first concerns people who feel that someone else – usually, retailers, business or the industry - is responsible for waste generation (DEFRA 2009) and therefore they themselves do not have to do something for waste prevention. According to Cox et al. (2010), this side worsens the attitude of apathy that some people might have. The second is that many people consider that their "small" contribution to the waste problem is so tiny that any effort to reduce it would not be worthwhile (Tucker and Douglas 2007).

4.8.2 Barriers to Waste Prevention Assessment

Although waste prevention has been placed on top of the waste management hierarchy, it is a common understanding that waste prevention policies have not brought significant results, so far. To some, the rise of this discrepancy is attributed to factors such as the difficulty to measure prevention, conflict of interests and/or inconsistent definitions (Salhofer et al. 2008). Overall, the above mentioned barriers can be considered as shortcomings of different elements of waste prevention assessment. Conversely, there is a wide consensus that the understanding of "the potential environmental benefits of waste prevention ... through appropriate assessment, can lead to more efficient and sustainable waste management infrastructures" (Gentil et al. 2011). Apart from deciding which instrument/practice fits best in each case, the efficient assessment can also (OECD 2000):

- provide information for the improvement of pre-existing policies, relevant to waste prevention,
- provide evidence on the functioning of policy and management processes to ensure that intentions are translated into practice expeditiously, and
- support the prompt communication with target-groups.

In this section, the focus is on the measurement limitations and the conflict of interests, while the definitions have been discussed already in section 4.1.

4.8.2.1 Measurement Limitations

Although the assessment of waste prevention activities is imperative for the development of an effective strategic plan, it still poses large difficulties regarding the methodology for monitoring and measuring, i.e. for the quantification of prevention and the selection of the most suitable methodology for the assessment. As Zorpas and Lasaridi (2013) state "as waste prevention is becoming increasingly important... it is crucial to develop reliable methods to monitor, measure and evaluate its benefits". Waste prevention includes a variety of different options, such as: (i) reducing or even eliminating the consumption and therefore the production of certain goods; (ii) substitution of certain products by others, with smaller environmental footprint; and (iii) extending the utilisation phase of products and/or materials. Thus, there is no general method which can be applied in all cases to evaluate the effectiveness of different specific activities.

The measurement limitations have confronted researchers and policy makers with lack of data or poor quality data. Sharp et al. (2010) elucidated this problem, by setting a single simple question: "How can you measure something that isn't there?" They also highlighted three important aspects of the problem:

- the participation in waste prevention actions cannot be visually recorded;
- the problem is compounded by the difficulty to discern whether an observed reduction in waste quantities is attributed to prevention or other factors (e.g.economic crisis); and
- even in the case that some changes can be classified as waste prevention, it is difficult or impossible to decide whether they are accidental or maintainable.

The main methods that have been employed in order to monitor waste prevention actions are: self-weighing, monitoring or reporting; use of collection round data to accurately measure waste increase;use of control and pilot groups; attitude and behaviour surveys including metrics, interviews and focus groups; participation surveys; compositional analysis; conversion factors; and Point of Sales Data (Sharp et al. 2010; Zorpas and Lasaridi 2013). In one of the largest studies of waste prevention worldwide, which took place inthe UK under the WRAP programme, waste prevention was monitored and evaluated through the estimation of waste mass reduction and behavioural studies (Sharp et al. 2010).

Another point of concern emerges when one has to evaluate the overall environmental effectiveness of different waste prevention actions. In this case, setting objective benchmarks is a clear necessity (Salhofer et al. 2008). An extensive review of the international literature reveals that, in spite of its shortcomings, the use of life cycle assessment (LCA) is a well-established methodology for carrying out the environmental assessment of waste prevention activities (Cleary 2010; Gentil et al. 2011; Nessi et al. 2012).

4.8.2.2 *Conflict of Interests*

A hurdle that is quite difficult to overcome concerns conflicts of interest. The issue is two fold: the implementation and success of waste prevention actions clashes with both the recycling and/or incineration-oriented waste management systems, and the production chains.

There is inevitably a conflict of interest between the implementation of waste prevention policies and the compatible waste management systems, because a reduction in waste generation is translated into a reduction of the inputs to that system. Moreover, the direct environmental benefits of prevention, when limiting assessment to the waste management system, are rather minor, at least for the non-hazardous waste streams, such as paper, packaging and food waste (Gentil et al. 2011). Especially for advanced waste management systems, based on recycling and energy recovery, prevention does not offer any significant further savings. However, when the avoided production due to waste prevention is added to the assessment, significant environmental benefits are demonstrated for all types of waste management systems. The benefits are expected to be even more significant for waste streams containing hazardous compounds, such as batteries and WEEE (Gentil et al. 2011). Therefore, in a conflict of interests between prevention and the established waste management systems, prevention should have a priority on environmental and sustainability grounds. Moreover, it should be clear that realistic waste prevention – at least for the foreseeable future - could not be expected to exceed 10–15% of the waste quantities produced, therefore its impact on the viability of the existing waste management systems and infrastructure would be limited.

When evaluating the effectiveness of a waste prevention action versus a compatible waste treatment, two aspects should be taken into consideration:

- the overall economic, social and environmental impacts of both of them, since waste policies must reflect political decisions, not only cost/benefit calculations; and
- the boundaries of each action.

Concerning the production chain, significant conflict of interests may arise if prevention measures dictated the banning or replacement of products that sell in large numbers. For instance, the swap to reusable diapers may be hindered by the manufacturers of disposable diapers. However, for most non-hazardous products and/or materials, waste prevention policies do not rely on direct bans but rather on attitude and behaviour changes, as well as measures for the provision of alternatives. As the effect of such policies and measures is usually gradual, alternative production chains and business models can be developed, restricting the intensity of such conflicts.

4.9 BEST PRACTICES IN THE EU

The Paradigm shift in the waste and resources management sectors, that will actually achieve significant and measurable prevention and ultimately contribute to sustainability, has as prerequisite the implementation of effective strategies. A range of waste prevention strategies, with varying degrees of success, have been applied in several EU countries. The strategies applied fall into three categories with different levels of engagement of central or local authorities: i) diffusion of information; ii) promotional campaigns; and iii) setting regulations. According to the European Commission (EC 2012), "these strategies are complementary and can be integrated into other relevant existing policy areas or can compose a stand-alone national waste prevention programme". A range of best practices (per waste stream) is presented and briefly analysed below, to feature good examples and stimulate waste prevention.

4.9.1 Biodegradable Waste – Food Waste

The "Love Food Hate Waste" Campaign (launched by WRAP in 2007, UK): This has proved to be one of the most successful food waste related awareness campaigns. Over the first two years of its application (from its launch in 2007 to 2009) in the UK, it contributed to the prevention of approximately 137,000 tonnes of food waste. The campaign aims at raising awareness, personalising the issue of food waste prevention, and developing "encourage and enable" actions. Consumers are its target group and are approached in both direct and indirect ways (via strategic partners, i.e. local authorities, institutes and retailers). The focus point of the campaign is to provide simple, easily adoptable solutions, which will cause small changes in consumers' daily routine and lead to waste prevention (http://lovefoodhatewaste.com).

Approved Food & Drink Company (Private sector 2009, UK): This is a food redistribution programme, which provides through its website dry food products that are near or past their "best before" date at a lower price (http://www.approvedfood.co.uk).

Buon Samaritano (a range of stakeholders 2005, Italy): The "Buon Samaritano" (Good Samaritan) project was launched in 2005 by the municipality of Torino and Amiat. It is a food redistribution programme, which ensures that uneaten but still edible meals from school canteens and super markets are collected and distributed to charity organisations. It is estimated that approximately 150 kg of bread and 50 kg of fruit are recovered every day from school canteens. In 2008, the project is said to have recovered over 81,000 kg of food (http://ec.europa. eu/food/food/sustainability/gp_food_redistribution_en.print.htm#4-2).

Réduisonsnos Déchets (National Authority 2005, France): The awareness campaign "Réduisonsnos Déchets" (Reducing our Waste) was developed by the ADEME, the French Environmental and Energy Management Agency, in order to provide information to households about waste generation and prevention. This campaign also participates and promotes the European Week for Waste Reduction (EWWR) initiative (http://www.reduisonsnosdechets.fr).

Eurest restaurant food waste campaign (Private sector 2010, Sweden): The chain of Eurest restaurants has developed a scheme in order to quantify food waste and publicise the measurements to staff and customers, explaining in parallel the impact of food waste on the environment and providing information about waste prevention. The 150 restaurants of the Eurest publicise a spreadsheet with a graph showing how much waste they produce every single day. The scheme has resulted in 23% reduction of waste generation in the chain. (http://ec.europa.eu/food/food/sustainability/good_practices_en.htm).

Anti-waste workshops (Local Authority 2009, Belgium): Cooking workshops with the aim of reducing food waste are offered by the Bruxelles Environment (with no fees), to the local community. In 2009, approximately 1,000 citizens were trained in these workshops (http://ec.europa.eu/food/food/sustainability/gp_awareness_information_education_en.print.htm#2-1).

4.9.2 Junk Mail

Stop Pub (National Authority 2004, France): The "Stop-pub" operation is a key feature of the "Réduisonsnos Déchets" (Reducing our Waste) awareness campaign, with the aim to reduce junk mail using a post box sticker and to stipulate public engagement in waste prevention actions. Over the first year of its implementation, 2.6 million stickers were requested. Seventy percent of the participants claimed that they received less junk mail (http://ec.europa.eu/environment/waste/prevention/pdf/Stop_Pub_Factsheet.pdf).

Fighting advertising paper (Local Authority 2006, Italy): In 2006 the municipality of Dogliani, Italy, commenced a public awareness campaign on reducing unwanted advertising mail and paper. Among other actions, the campaign promoted a "no junk mail" sticker, which was deemed successful in reducing unwanted mail.

Junk the Junk Mail (Local Authority 2001, Scotland): The "Junk the Junk Mail" campaign was launched in 2001 by the Sandwell Metropolitan Borough Council (Scotland) and offered a good opportunity to test the effectiveness of the Mail Preference Service (MPS) which is offered for free in UK. The MPS enables citizens to have their addresses removed from or added to the lists of the direct mail industry. The campaign contributed to the reduction of total unwanted mail by approximately 30%.

4.9.3 Green Business

Green Business Initiative (National Authority 2008, Ireland): The Green Business Initiative was developed and organised by the Irish Environmental Protection Agency. It commenced in 2008 with a flagship project: the development of the "greenbusiness.ie" website and the "Green Hospitality" project (U.S. EPA 2012). The website features interactive waste auditing tools, allowing small- and medium-sized business to gather data on resource use. By 2011, 479 active members and up to 6.1 million euros of potential savings due to resource efficiency activities were recorded (U.S. EPA 2012). Members may request a Resource Efficiency Assessment (REA). Up to 2011 forty five REAs have been completed (http://greenbusiness.ie).

Green Hospitality (National Authority 2008, Ireland): The "Green Hospitality" scheme is the only Irish environmental certification programme recognized internationally. It consists of a step-by-step guide to environmental management within the sectors of hospitality and catering. The efforts of its members have led to a 50% reduction of waste going to landfill (which includes waste prevention, recycling and energy recovery), 50% reduction of water consumption and 30% reduction of energy consumption. The scheme offers the "Green Hospitality Awards" (over the first year of its implementation, approximately 200 hotels had signed up - 80 of these hotels were given the award), the Green Hospitality Eco-Label, Green Hospitality Workshops, networking, conferences etc (http://greenhospitality.ie).

4.10 CONCLUSIONS

Since waste prevention was placed at the top of the waste management hierarchy and, therefore, brought to the forefront of discussions on environmental security and sustainable development, its relation to the whole resources management chain has been intensively investigated.

The relevant discourse and a multitude of studies indicate that waste prevention activities can – and should – be taken at all steps of the product's life chain. Waste prevention should be considered a shared priority of Central and Local Authorities, the private sector and individuals / consumers. In each step of a product's life cycle, attention is paid on certain priority waste streams, in order to minimise the negative environmental impacts while maximising the possible environmental, economic and social benefits.

Effective actions are underway by international organisations, central and local authorities in various countries, as well as the private sector, setting precedents for worldwide adaptation and replication. Among them, some initiatives focus on restraining unsustainable consumption patterns. Yet, much more work needs to be done, engaging the main principles of sustainable production, supply, consumption and discard patterns (e.g. resource efficiency, dematerialisation, eco-design, lifestyle) and waste management. Moreover, there are still several barriers and challenges, such methodological limitations, conflicts of interest and established human attitudes and behaviours that need to be resolved. An effort to approach waste prevention targets, considering social and market patterns, might hold a key to the design of effective waste prevention measures and the move to the–so needed–resource efficient societies.

4.11 ACKNOWLEDGMENTS

The preparation of this chapter was supported by the LIFE10ENV/GR/622 project "WASP Tool", which is co-funded by the European Union LIFE+ programme.

References

Abeliotis, K., Lasaridi, K., and Chroni, C. (2011a). "Attitudes and behaviour of Greek households towards food waste generation." *Proc., Int. Conf. on Solid Waste 2011—Moving Towards Sustainable Resource Management*, J.W.C. Wong, K. Fricke, R. Y. Surampalli, and A. Selvam, eds., Hong Kong Baptist Univ., Hong Kong SAR, P.R. China, 93.

Abeliotis, K., Lasaridi, K., Chroni, C., and Papakosta, A. (2012b). "Preliminary assessment of food waste generation in households in Greece." *Proc., 8th Int. Conf. ORBIT 2012—Global Assessment for Organic Resources and Waste Management*, Rennes, France, 15.

Abeliotis, K., Lasaridi, K., and Chroni, C. (2014). "Attitudes and behaviour of Greek households regarding food waste prevention." *Waste Manage. Res.*, **32**(3), 237–240.

Barr, S. (2007). "Factors influencing environmental attitudes and behaviors: A U.K. case study of household waste management."*Environ. Behav.*, **39**(4), 435–473.

Bartolomeo, M., et al. (2003). "Eco-efficient producer services—What are they, how do they benefit customers and the environment and how likely are they to develop and be extensively utilized?" *J. Cleaner Prod.*, **11**(8), 829–837.

Bortoleto, A. P., Kurisu, K. H., and Hanaki, K. (2012). "Model development for household waste prevention behaviour." *Waste Manage.*, **32**(12), 2195–2207

Braungart, M., and McDonough, W. (2009). *Cradle to cradle: Remaking the way we makethings*, Vintage, London.

Cleary, J. (2010). "The incorporation of waste prevention activities into life cycle assessments of municipal solid waste management systems: methodological issues." *Int. J. Life Cycle Assess.*, **15**(6), 579–589.

Cooper, T. (2005). "Slower consumption. reflections on product life spans and the" throwaway society"." *J. Ind. Ecol.*, **9**(1-2), 51–67.

Cox, J., Giorgi, S., Sharp, V., Strange, K., Wilson, D. C., and Blakey, N. (2010). "Household waste prevention—A review of evidence." *Waste Manage. Res.*, **28**(3), 193–219.

DEFRA. (2007). "Lifestyle scenarios & waste composition." Final Rep. WR0104, Brook Lyndhurst, London.

DEFRA. (2009). "Household waste prevention evidence review." Final Rep. WR1204, Brook Lyndhurst, London.

DEFRA. (2010). "Waste prevention in Hampshire: Small changes big challenges." Final Rep. WR0117, Brook Lyndhurst, London.

DEFRA. (2011). "Government review of waste policy in England 2011." London.

DEFRA. (2012). "Progress with delivery of commitments fromthe government's review of waste policy in England (2011)." London.

del Río Merino, M., Garcia Navarro, J., and Villoria Sáez, P. (2011). "Legal aspects which implement good practice measures in the management of construction and demolition waste." *TOBCTJ*, **5**(1), 124–130.

EC (European Commission). (2010). "Analysis of the evolution of waste reduction and the scope of waste prevention." Final Rep. NV.G.4/FRA/2008/0112, ARCADIS for the European Commission DG Environment, Brussels, Belgium.

EC (European Commission). (2011a). "Preparatory study on food waste across EU-27." Technical Rep. 2010-054, Brussels, Belgium.

EC (European Commission). (2011b). "Evolution of (bio-) waste generation/prevention and (bio-) waste prevention indicators." Final Rep. ENV.G.4/FRA/2008/0112, Brussels, Belgium.

EC (European Commission). (2011c). "Report from the commission to the european parliament, the council, the european economic and social committee and the committee of the regions on the thematic strategy on the prevention and recycling of waste." Brussels.

EC (European Commission). (2012). "Preparing a waste prevention programme, guidance document." European Commission DG Environment, Brussels, Belgium.

EC (European Commission). (2013). "Proposal for a Directive of the European Parliament and of the Council amending Directive 94/62/EC on Packaging and Packaging Waste to reduce the consumption of lightweight plastic carrier bags." Brussels, Belgium.

EEA (European Environment Agency). (2002). "Case studies on waste minimisation practices in Europe." Topic Rep. 2/2002, Copenhagen.

EEA (European Environment Agency). (2010). "The European environment—State and outlook 2010: Synthesis." Copenhagen.

EEA (European Environment Agency). (2011). "Municipal waste generation (CSI 016/waste 001)." ⟨http://www.eea.europa.eu/data-and-maps/indicators/municipal-waste-generation/municipal-waste-generation-assessment-published-4⟩ (Nov. 17, 2013).

ETC/SCP (European Topic Centre on Sustainable Consumption and Production). (2012). "Factsheets for Netherlands." ⟨http://scp.eionet.europa.eu/facts/factsheets_waste/2009_edition/factsheet?country=NL⟩ (Nov. 17, 2013).

Evans, D. (2012). "Binning, gifting and recovery: The conduits of disposal in household food consumption." *Environ. Plann. D Soc. Space*, **30**(6), 1123–1137.

FAO. (2011). "Global food losses and waste: Extent, causes and prevention." ⟨http://www.fao.org/docrep/014/mb060e/mb060e00.pdf⟩ (Mar. 14, 2013).

FAO. (2013). "Food wastage footprint, Impacts on Natural Resources." ⟨http://www.fao.org/docrep/018/i3347e/i3347e.pdf⟩ (Feb. 17, 2016).

Fell, D., Cox, J., and Wilson, D. (2010). "Future waste growth, modelling and decoupling." *Waste Manage. Res.*, **28**(3), 281–286.

Fishman, A., Gandal, N., and Shy, O. (1993). "Planned obsolescence as an engine of technological progress." *J. Ind. Econ.*, **41**(4), 361–370.

Gentil, E. C., Gallo, D., and Christensen, T. H. (2011). "Environmental evaluation of municipal waste prevention." *Waste Manage.*, **31**(12), 2371–2379.

Gottberg, A., Longhurst, P. J., and Cook, M. B. (2010). "Exploring the potential of product service systems to achieve household waste prevention on new housing developments in the UK." *Waste Manage. Res.*, **28**(3), 228–235.

Guiltinan, J. (2009). "Creative destruction and destructive creations: Environmental ethics and planned obsolescence." *J. Bus. Ethics*, **89**, 19–28.

Heiskanen, E., and Jalas, M. (2003). "Can services lead to radical eco-efficiency improvements? A review of the debate and evidence." *CSR Environ. Manage.*, **10**(4), 186–198.

HM Government. (2013). "Prevention is better than cure: The role of waste prevention in moving to a resource efficient economy." ⟨http://www.gov.uk/government/publications⟩ (Dec. 20, 2013).

Hobohm, J., Szczepaniak, K., and Kuchta, K. (2013). "Rare earth metals—Recycling and pre-treatment." *Proc., 14th Int. Waste Management and Landfill Symp.*, CISA, Padova, Italy.

IPCC (Intergovernmental Panel on Climate Change). (2006). "IPCC guidelines for national greenhouse gas inventories." ⟨http://www.ipcc-nggip.iges.or.jp/public/2006gl/index.htm⟩ (Apr. 20, 2008).

ISWA. (2012). "ISWA key issue paper on waste prevention, waste minimization and resource management." *Proc., ISWA World Congress, ISWA Working Group Recycling and Waste Minimization, 2011*, Florence, Italy.

IVL Swedish Environmental Research Institute. (2011). "Initiatives on prevention of food waste in the retail and wholesale trades." ⟨http://www.ivl.se/download/18.7df4c4e812d2-da6a416800089028/B1988.pdf⟩ (Mar. 14, 2013).

Kawai, K., and Osako, M. (2012). "Reduction of natural resource consumption in cement production in Japan by waste utilization." *J. Mater. Cycles Waste Manage.*, **14**(2), 94–101.

Kemp, R. (2007). "An example of a "managed transition": The transformation of the waste management subsystem in the Netherlands (1960-2000)." *Innovations towards sustainability*, M. Lehmann-Waffenschmidt, ed., Springer, Berlin.

Kissling, R., et al. (2013). "Success factors and barriers in re-use of electrical and electronic equipment." *Resour. Conserv. Recycl.*, **80**, 21–31.

Koivupuro, H.-K., et al. (2012). "Influence of socio-demographical, behavioural and attitudinal factors on the amount of avoidable food waste generated in Finnish households." *Int. J. Consum. Stud.*, **36**(2), 183–191.

Kurishu, K. H., and Bortoleto, A. P. (2011). "Comparison of waste prevention behaviours among three Japanese megacity regions in the context of local measures and socio-demographics."*Waste Manage.*, **31**(7), 1441–1449.

Langley, J., Turnet, N., and Yoxall, A. (2011). "Attributes of packaging and influences on waste." *Packag. Technol. Sci.*, **24**(3), 161–175.

Lasaridi, K. (2009). "Implementing the landfill directive in Greece: Problems, perspectives and lessons to be learned." *Geogr. J.*, **175**(4), 261–273.

Lauridsen, E. H., and Jorgensen, U. (2010). "Sustainable transition of electronic products through waste policy." *Res. Policy*, **39**(4), 486–494.

Lehtoranta, S., Nissinen, A., Mattila, T., and Melanen, M. (2011). "Industrial symbiosis and the policy instruments of sustainable consumption and production." *J. Cleaner Prod.*, **19** (16), 1865–1875.10.1016/j.jclepro.2011.04.002

Li, J., Tian, B., Liu, T., Liu, H., Wen, X., and Honda, S. (2006). "Status-quo of e-waste management in mainland China." *J. Mater. Cycles Waste Manage.*, **8**(1), 13–20.

Lilja, R. (2009). "From waste prevention to promotion of material efficiency: Change of discourse in the waste policy of Finland." *J. Cleaner Prod.*, **17**(2), 129–136.

Llatas, C. (2011). "A model for quantifying construction waste in projects according to the European waste list." *Waste Manage.*, **31**(6), 1261–1276.

London, B. (1932). "Ending the depression through planned obsolescence." ⟨http//: en.wikipedia.org/wiki/File:London_(1932)_Ending_the_depression_through_planned_obsolescence.pdf⟩ (Mar. 14, 2013).

Mazzanti, M., and Zoboli, R. (2008). "Waste generation, waste disposal and policy effectiveness: Evidence on decoupling from the European Union." *Resour. Conserv. Recycl.*, **52**(10), 1221–1234.

Melanen, M., Kautto, P., Saarikoski, H., Ilomaki, M., and Yli-Kauppila, H. (2002). "Finnish waste policy—Effects and effectiveness." *Resour. Conserv. Recycl.*, **35**(1-2), 1–15.

Mena, C., Adenso-Diaz, B., and Yurt, O. (2011). "The causes of food waste in the supplier-retailer interface: Evidences from the UK and Spain." *Resour. Conserv. Recycl.*, **55**(6), 648–658.

Moomaw, W., Griffin, T., Kurczak, K., and Lomax, J. (2012). "The critical role of global food consumption patterns in achieving sustainable food systems and food for all." United Nations Environment Programme, Division of Technology, Industry and Economics, Paris, France.

Morgan, C., and Stevenson, F. (2005). "Design for deconstruction. Design Guides for Scotland: No 1." Scottish Ecological Design Association, Edinburgh.

Nahman, A., de Lange, W., Oelofse, S., and Godfrey, L. (2012). "The costs of household food waste in South Africa." *Waste Manage.*, **32**(11), 2147–2153.

Nessi, S., Rigamonti, L., and Grosso, M. (2012). "LCA of waste prevention activities: A case study for drinking water in Italy." *J. Environ. Manage.*, **108**, 73–83.

OECD. (2000). "Working group on waste prevention and recycling: OECD workshop on waste prevention—Toward performance indicators." ⟨http://www.oesc.org⟩ (Mar. 14, 2013).

OECD. (2002). "Working group on waste prevention and recycling—Working group on environmental information and outlooks: OECD workshop on waste prevention—Toward Performance Indicators." ⟨http://www.oesc.org⟩ (Mar. 14, 2013).

OECD. (2005). "Working group on waste prevention and recycling: Outcome of the first OECD workshop on sustainable materials management." ⟨http://www.oecd.org⟩ (Mar. 14, 2013).

OECD. (2010). "Total amount generated of municipal waste." *OECD Factbook 2010: Economic, environmental and social statistics*, OECD Publishing, Paris.

Ongondo, F. O., Williams, I. D., and Cherrett, T. J. (2011). "How are WEEE doing? A global review of the management of electrical and electronic wastes." *Waste Manage.*, **31**(4), 714–730.

Ongondo, F. O., Williams, I. D., Dietrich, J., and Carroll, C. (2013). "ICT reuse in socio-economic enterprises." *Waste Manage.*, **33**(12), 2600–2606.

Parfitt, J., Barthel, M., and Macnaughton, S. (2010). "Food waste within food supply chains: Quantification and potential for change to 2050." *Philos. Trans. R. Soc. London Ser. B*, **365**(1554), 3065–3081.

Peattie, K. (2010). "Green consumption: Behavior and norms." *Annu. Rev. Environ. Resour.*, **35**(1), 195–228.

Phillips, P. S., Tudor, T., Bird, H., and Bates, M. (2011). "A critical review of a key waste strategy initiative in England: Zero waste places 2008–2009." *Resour. Conserv. Recycl.*, **55**(3), 335–343.

Pongrácz, E. (2009). "Through waste prevention towards corporate sustainability: Analysis of the concept of waste and a review of attitudes towards waste prevention." *Sustainable Dev.*, **17**(2), 92–101.

Quested, T. E., Marsh, E., Stunell, D., and Parry, A. D. (2013). "Spaghetti soup: The complex world of food waste behaviours." *Resour. Conserv. Recycl.*, **79**, 43–51.

Salhofer, S., Obersteiner, G., Schneider, F., and Lebersorger, S. (2008). "Potentials for the prevention of municipal solid waste." *Waste Manage.*, **28**(2), 245–259.

Schneider, F. (2008). "Wasting food – An insistent behaviour." *Proc., Waste – The Social Context 2008 Conf.*, Alberta, Canada, X-1.

Schneider, F., and Lebersorger, S. (2009). "Households attitudes and behaviour towards wasting food—A case study." *Proc., 12th Sardinia Symp.*, Margherita di Pula, Cagliari, Italy.

Schwarzer, S., De Bono, A., Giulani, G., Kluser, S., and Peduzzi, P. (2005). "E-waste, the hidden side of IT equipment's manufacturing and use." ⟨http://www.grid.unep.ch/products/3_Reports/ew_ewaste.en.pdf⟩ (Mar. 14, 2013).

Serranti, S., Palmieri, R., and Bonifazi, G. (2013). "Opto-electronic combined characterization of end-of-life mobile phone wastes for a sustainable processing." *Proc., 14th Int. Waste Management and Landfill Symp.*, S. Margherita di Pula, Cagliari, Sardinia, Italy.

Sharp, V., Giorgi, S., and Wilson, D. C. (2010). "Methods to monitor and evaluate household waste prevention." *Waste Manage. Res.*, **28**(3), 269–280.

Sjöström, M., and Östblom, G. (2010). "Decoupling waste generation from economic growth—A CGE analysis of the Swedish case." *Ecol. Econ.*, **69**(7), 1545–1552.

Stefan, V., van Herpen, E., Tudoran, A., and Lähteenmäki, L. (2013). "Avoiding food waste by Romanian consumers: The importance of planning and shopping routines." *Food Qual. Preference*, **28**(1), 375–381.

Tonjes, D. J., and Greene, K. L. (2012). "A review of national municipal solid waste generation assessments in the USA." *Waste Manage. Res.*, **30**(8), 758–771.

Tucker, P., and Douglas, P. (2007). "Understanding household waste prevention behaviour." Final Rep. WR 0112, DEFRA, Scotland.

UNEP. (2011). "Waste, Investing in energy and resource efficiency." ⟨http://www.unep.org/greeneconomy/Portals/88/documents/ger/GER_8_Waste.pdf⟩ (Feb 17, 2016).

U.S. EPA. (2002). "25 years of RCRA: Building on our past to protect our future." Office of Solid Waste and Emergency Response, Washington, DC.

U.S. EPA. (2003). "Beyond RCRA. Waste and materials management in the year 2020." Office of Solid Waste, Washington, DC.

U.S. EPA. (2012). "National waste prevention programme: Seventh annual report–for year 2011." Washington, DC.

van DerVoet, E., van Oers, L., and Nikolic, I. (2004). "Dematerialization: not just a matter of weight." *J. Ind. Ecol.*, **8**(4), 121–137.

Wagner, T. P. (2009). "Shared responsibility for managing electronic waste: a case study of Maine, USA." *Waste Manage.*, **29**(12), 3014–3021.

Weidema, B., Wenzel, H., Petersen, C., and Hansen, K. (2004). "The product, functional unit and reference flows in LCA." Danish Environmental Protection Agency.

Wilson, D. C., Rodic, L., Scheinberg, A., Velis, C., and Alabaster, G. (2012). "Comparative analysis of solid waste management in 20 cities." *Waste Manage. Res.*, **30**(3), 237–254.

WRAP. (2009). "Household food and drink waste in UK." Banbury.

WRAP. (2012). "Decoupling of waste and economic indicators." Final Rep. SAP142-002, BIO Intelligence Service, Metroeconomica, Resource Futures, Banbury, Oxon.

Yusuf, R. O., Noor, Z. Z., Abba, A. H., Abu Hassan, M. A., and Mohd Din, M. F. (2012). "Methane emission by sectors: A comprehensive review of emission sources and mitigation methods." *Renew. Sustainable Energy Rev.*, **16**(7), 5059–5070.

Zorpas, A. A., and Lasaridi, K. (2013). "Measuring waste prevention." *Waste Manage.*, **33**(5), 1047–1056.

CHAPTER 5

Material Recycling and Resource Recovery

Jing-Yuan Wang[1]
Lei Zhao[2]
Xiaolong Sun[3]
Zhonghui Wang[4]
Lin Wang[5]
Shengxuan Lin[6]

5.1 INTRODUCTION

Recycling is refers to the conversion of waste into valuable resources. For example, plastic is shredded and remoulded into new plastic products, and newspapers are recovered as the feedstock for cardboard or newspaper again. In an integrated waste management system, recycling is always listed near the top of a hierarchy, which was designed to achieve maximum economic, social and environmental returns (Schmidt et al. 2007). However, recycling alone is not enough to completely utilize the waste as a valuable resource. Therefore, resource recovery is also an important component and a prerequisite that will be addressed in this chapter.

Resource recovery refers to the process of acquiring energy or materials from wastes (Bartone 1990). It is a methodology constituting reuse, recycling, and waste-to-energy. Through resource recovery, the products are utilized optimally, virgin natural resources are saved, and the waste to be disposed is minimized. The Table 5.1 summarises some of the virgin resources saved by recovering the resources.

[1]Nanyang Technological Univ., Singapore.
[2]Nanyang Technological Univ., Singapore.
[3]Nanyang Technological Univ., Singapore.
[4]Nanyang Technological Univ., Singapore.
[5]Nanyang Technological Univ., Singapore.
[6]Nanyang Technological Univ., Singapore.

Table 5.1. Virgin resources saved by recovering the resource

Resource recovered	*Virgin resources saved*
Plastics	Recycling of plastics reduces the amount of crude oil for producing new plastics.
Metals	Recycling of metal waste like aluminium drink cans will reduce the amount of bauxite (aluminium ore) required.
Paper	Recycling waste paper can reduce the amount of trees needed to produce new paper.
Demolition waste	Recycling of demolition waste can substitute the consumption of natural sand and other building materials.
Waste oil	Recycling of waste oil can substitute fuel oil or diesel as it can be converted into biodiesel.

People have been involved in resource recovery long before they are aware of it. Animal manure and human waste have been collected and used in land applications for centuries. By the late 1980s, the modern environmental awareness has gradually been triggered by global warming, atmosphere ozone depletion, scarcity of landfill space and exhaustion of nature resources. This has spurred the booming development of resource recovery technologies and legislations. For example the European Union directive says that a minimum of 22.5% by weight for plastics needs to be recycled before 2008 (European Parliament and Council Directive 1994). On the other hand in Asia, China has set specific laws to recycle and recover solid waste products since 1 April 2005 (He et al. 2006).

In developed countries, a variety of national or interstate legislations have been established in these years. Article 191 (ex. art. 174 EC) describes the objectives and principles of the EU environmental policy, which indicates that preventative action should be taken to preferentially reduce the environmental impact at source, and the polluter should pay. Under the 'polluter pays' principle, the producer should take the underlying responsibility for environmental pollution, and have the incentive to design and manufacture of their products with the longest possible life span and to find the best way for recovery and disposal at the end of product life (Duke 2009; Raunio 2009).

On the other hand, in developing countries like China, Colombia, Mexico, Philippines and Thailand, non-professional collectors (garbage pickers) contribute to the majority of recycling practices. These practices includes (Wilson et al. 2006):

1. Purchasing of waste
    - Door to door collection of dry recyclables.
    - Selling to recycling shops.

2. Picking of street waste
   - Recovering secondary waste materials from streets or garbage bins.
3. Municipal waste collection crew
   - Recovering of secondary waste materials from vehicles transporting municipal solid waste to disposal sites.
4. Picking of waste from dumps
   - Scavenging waste in landfills before the site is covered.

These garbage pickers, who make a living out of recycling, provide an important resource recovery service where there is no official waste recycling system in place. In Beijing alone, there are around 160,000 garbage pickers collect a large portion of the plastic sheeting, office printouts, bottles, radiators, and scraps of cardboard from municipal solid waste (Branigan 2009).

To achieve efficient resource recovery from waste, three consecutive steps are involved. The first step is the separation of resources from waste stream. There are two major separation approaches for solid waste: source separation and material recovery facility (MRF). The second step is processing of the resources to materials or energy. For example, glass and ash can be processed, and used as construction materials. Finally, the third step is the utilization of the materials or energy by consumers or industries (Pichtel 2010)

However, there are some issues concerning resource recovery practices. Firstly, lack of motivation to recycle. The lack of motivation is due to the disposal habits of consumers. Several solutions were devised to tackle this problem, one of the solutions is to promote recycling by "pay-as-you-throw" schemes (Neo 2010). Some companies like NTUC Fairprice have introduced a 10 cents rebate to customers who bring their own bags (NTUC FairPrice 2011). Secondly, excessive costs and efforts are required to retrofit existing processes and facilities to suit resource recovery systems. Therefore, both technological and legislative measures should be addressed to promote a more robust resource recovery system. Table 5.2 gives an overview of the current challenges, processes and possible resource recovery solutions.

To sum up, waste is not waste, but a misplaced resource from our daily life and industrial manufacturing activities. With proper management and suitable technologies, waste can be recovered as reusable/new materials energy, and other products with values. This resource recovery approach is a key component of modern solid waste management, and it provides both environmental and economic benefits.

5.2 RESOURCE RECOVERY PROGRESS AND STATISTICS

The world has experienced a steady change in waste management, which relies less on landfill and increasingly involved in resource recovery currently. This movement is driven by national legislation, increasing people's awareness of environmental protection and economic incentives. For instance, many countries have established

Table 5.2. Challenges, current processes, and solutions of resource recovery practices

Challenges	*Current process*	*Solution*
Disposal habits	• Small scale collection of waste by rag-and-bone man • Excessive use of disposal materials, for example, plastic bags from supermarkets.	• Disincentives and incentives to cultivate recycling habits. • Peer motivation and education
Technical and cost barriers	• Incineration of non-recyclable waste. • Landfilling waste which cannot be incinerated.	• Legislative measures to lower investment cost. • Research new technologies to retrofit new technologies into existing infrastructures.

stepwise targets and guidelines for their national resource recovery from waste. The guidelines mainly aimed at waste generators including both individual citizens and business sectors. Since 1990, most states in US have set specific guidelines for reducing the waste generation through a combination of source reduction, recycling and composting (Pichtel 2010). Moreover, the resource recovery activities attracted increasing investment and support due to the rising prices of raw materials and energy supply, which is coupled with the land scarcity for landfill and restrictions on certain types of wastes to landfill. However, there is still a considerable space for improvement in waste management, as disposal is dominant on an average level and the practices of resource recovery are unevenly developed among countries with different economic and social status. Even in developed regions like EU, 49% of waste was disposed, around 46% was recovered and recycled, and 5% was sent for energy recovery in 2008. The resource recovery rate varied from less than 10% in Bulgaria, Greece, Malta and Romania to more than 85% in Belgium and Denmark (Eurostat 2015). Resource recovery is emphasised by land constraint counties like Singapore, which has an overall recycling rate up to 60% (National Environmental Agency 2012). Currently, countries such as Ireland, Italy, Portugal and United Kingdom, have steady increase of their recovery rate by nearly 1% per year since 2000 (EEA 2013). Looking into the specific waste streams, in Singapore, high recycling rate was achieved in paper/cardboard, construction debris, wood/timber, metals, used slag and scrap tyres (Table 5.3).

In the EU, resource recovery has been successfully employed on paper and cardboard, bio-waste, glass, plastic, and metals (Fischer and Werge 2009), and this

Table 5.3. Waste Statistics and Recycling Rate for 2012 in Singapore

Waste Type	*Waste disposed of (tonne)*	*Total waste Recycled (tonne)*	*Total waste output (tonne)*	*Recycling rate (%)*
Construction debris	9,100	1,282,700	1,291,800	99
Used slag	9,800	358,000	367,800	97
Ferrous metals	54,800	1,331,200	1,386,000	96
Scrap tyres	2,300	16,100	18,400	88
Non-ferrous metals	25,200	97,300	122,500	79
Wood/timber*	107,800	236,000	343,800	69
Paper/cardboard	555,300	696,700	1,252,000	56
Horticultural waste*	139,800	108,000	247,800	44
Glass	55,400	19,500	74,900	26
Food waste	618,100	85,100	703,200	12
Textile/leather	141,000	18,300	159,300	11
Plastics	721,300	82,100	803,400	10
Others (stones, ceramics & rubber)	333,800	4,600	338,400	1
Sludge	160,200	0	160,200	0
Total	2,933,900	4,335,600	7,269,500	60

*Includes 110,300 tonnes used as fuel in biomass power plants

Source: NEA (2012). Reproduced with permission from National Environment Agency of Singapore

has continuously contribute to substitution of certain raw materials, especially paper and cardboard, and iron and steel (Figure 5.1; EEA 2013). The success of EU's practice lies in a combination of policy measures and recycling technologies. In the EU, the following policy measures contributed positively in achieving a higher level of recycling in their MSW management system (Tojo and Fischer 2011)

1. Ban on landfilling of specific parts of MSW, such as mixed waste with a considerable content of biodegradable organic components, bio waste and combustible waste;
2. Taxation on incineration and landfill;
3. Setting recycling targets specific for MSW;
4. Setting specific mandates for paper waste such as source separation, collection and recycling targets.

According to a outlook model developed by the European Environment Agency, by 2020, resource recovery (including the recycled, composed and incinerated) would increase from 60% in 2008 to 72%, while the rest for landfill would stay at around 28% in the EU 27 (excluding Cyprus), plus Norway and

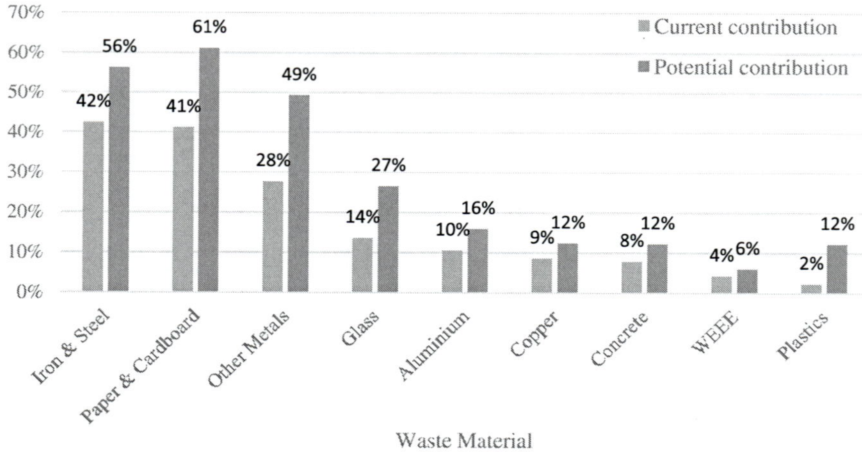

Figure 5.1. Current and potential contribution of resource recovery to meet EU demand for various materials, 2006
Source: EEA 2013; reproduced with permission from EEA

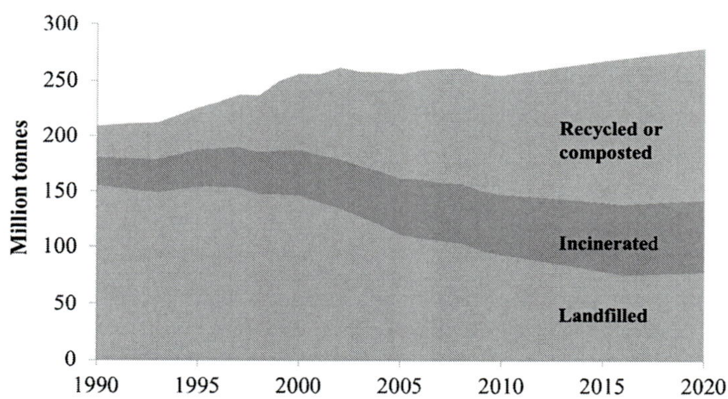

Figure 5.2. Trends and outlook for management of municipal waste in the EU-27 plus Norway and Switzerland (without Cyprus), baseline scenario

Switzerland (Figure 5.2), based on the historic trends in the development of rate of recycling, composting, incineration and landfill (Bakas et al. 2011).

5.3 PLASTIC AND RUBBER

5.3.1 Plastic

Waste plastics in MSW are mainly from post-consumer household products, such as plastic bags, soft-drink bottles, and shampoo bottles, etc. There are three major

Table 5.4. Features of municipal plastic wastes recycling methods (Al-Salem et al. 2009; Pichtel 2010)

Recycling methods	*Purpose*	*Features*	*Methodology*
Mechanical recycling	To recycle the plastic waste as the feedstock for new plastic product.	It is performed on separated plastics with high purity. Pre-treatments such as separation and cleaning are required.	Extrusion moulding, injection moulding, blow moulding, vacuum moulding, and inflation moulding.
Chemical recycling	To recover monomer or fuel from plastic wastes.	It can be used for both separated and mixed plastic wastes recycling.	Pyrolysis, gasification, and hydrolysis.
Energy recovery	To burn the plastic wastes for heat or power generation.	It is suitable for treating contaminated and mixed plastic waste.	Incineration.

categories of waste plastics recovery: mechanical recycling, chemical recycling, and energy recovery. The features of the recycling methods of municipal plastic wastes are summarized and listed in Table 5.4.

5.3.1.1 Mechanical Recycling of Waste Plastics

Plastics are synthetic polymers which fall in two categories, thermoplastics and thermosets. Thermoplastics can be melted and moulded repeatedly without chemical reactions, while thermosets can only be melted and moulded once, as irreversible chemical reaction occurs. Thus mechanical recycling is only appreciable for thermoplastics. Different plastics should not be mixed due to their different polymer structures. Even for those products constituted by multi-polymer, the polymers should be fed in separated flows (Giles et al. 2005). Therefore, presorting and separating are essential for waste plastics mechanical recycling. Resin identification code was introduced by the Society of the Plastic Industry (SPI) in 1988 to distinguish different types of plastic products by numbers (Table 5.5).

Table 5.5. Society of Plastics Industry (SPI) resin identification codes

Resin codes	*Descriptions*
1	Polyethylene Terephthalate (PET)
2	High-density polyethylene (HDPE)
3	Polyvinyl chloride (PVC)
4	Low-density polyethylene (LDPE)
5	Polypropylene (PP)
6	Polystyrene (PS)
7	Others

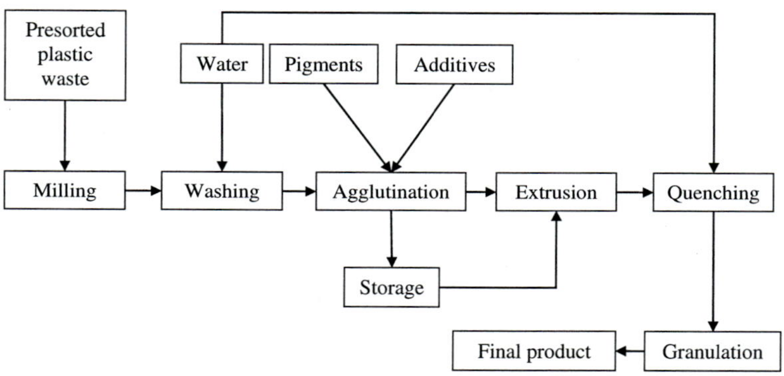

Figure 5.3. General scheme for mechanical recycling of waste plastic
Source: Aznar et al. 2006; reproduced with permission from Elsevier

The scheme of mechanical recycling of waste plastic is illustrated in Figure 5.3. The steps involved are:

- Milling: In this step, the presorted single-polymer plastic wastes are milled into small particles.
- Washing: The milled plastic is washed by either water or chemical agent (like caustic soda and surfactants) or both depending on the conditions of the feedstock.
- Agglutination: Pigments and additives will be added into the raw material if necessary.
- Extrusion: Recycled plastics are extruded into strands and then pellets.
- Quenching: The plastic pellets go through water-cooling and become granulates, which can be sold as feedstock in plastic manufacture industry.

The recycling plastic granulate can fully or partially replace the virgin plastic resin in moulding. Taking PET as an example, recycled PET was widely used to manufacture textile, carpet, apparel, and bottles. The quality of such products is comparable to those produced using virgin resin. Another practice to use recycled plastic is re-bonding (Zia et al. 2007). The mixture of recycled plastic powders and binders are fed to the compressor. The recycled granules appeared as filler in polyester moulding compounds. Products like carpet underlay and athletic mats can be manufactured from this process.

5.3.1.2 Chemical Recycling of Waste Plastics

Chemical recycling refers to converting waste plastics into either monomer or petrochemical via thermo-chemical means. The main advantage of chemical recycling is its flexibility of treating multi-composited and contaminated plastic waste with limited use of pre-treatment. In addition, recycling of both thermoplastics and thermo-sets are feasible. The technologies used in chemical recycling are similar to those employed in petrochemical industry, e.g. pyrolysis; and the products can be utilised either as fuel or feedstock in chemical industry. By their nature, the condensation polymers such as PET and polyamides are more likely to undergo degradation to produce monomer, while the vinyl polymers like PE and PS produce numerous components which can be used as fuel. Since the fossil fuel is getting scarce, the plastic-to-fuel technologies are receiving much attention. There are many research and industrial cases about converting plastic waste to fuel oil or other value added product like lubricant (Demirbas 2008; Filip et al. 2011; Horvat 1996; Lin and Yang 2007; McCaffrey et al. 1995).

Pyrolysis is one of the most popular methods for oil recovery from plastic wastes. It refers to a thermal process that the plastic wastes are heated up in an inert atmosphere. It can be conducted at various levels of temperature, retention times, pressures, and in the presence or absence of catalysts. Through pyrolysis, the plastic wastes are converted into three phases of products liquid oil, gas, and wax, respectively. Table 5.6 lists the advantages of fuel recovery from plastic waste by pyrolysis.

Pyrolysis was successfully industrialised in Germany, Japan, UK, etc (Scheirs and Kaminsky 2006). One popular process was BP polymer cracking process established in Scotland in 1999 (Tukker et al. 1999) and the process flowchart is shown in Figure 5.4. The plant has a capacity of 25,000 tonnes/year. The plastic wastes are shredded first then fed into the fluidised bed reactor operating at 500°C. Plastics are cracked in the oxygen-free environment into hydrocarbons. A lime absorber is set up to neutralise the HCl released from PVC pyrolysis. Fuel oil and wax are collected in condensers.

The trends of this technology are multi-step pyrolysis technology (Aguado et al. 2007), catalyst synthesis (Arandes et al. 2004), and hydrothermal process for converting PET to oil (Grause et al. 2011) etc.

5.3.1.3 Energy Recovery from Waste Plastics

For mechanical and chemical recycling of plastic waste, a crucial issue prior to its recycling is pretreatment such as separation (Zhao et al. 2012). Incineration

Table 5.6. Advantages of fuel recovery from plastic waste by pyrolysis

Advantages	*Descriptions*
Operational	• Lower temperature; • No air pollution control requirements; • Not affected by types of plastics; • Flexible product choices.
Environmental	• Reduction of landfilling; • Less carbon emission; • Less pollutant formation.
Economic	• Relatively lower cost; • Valuable products; • Growing market for recovered products.

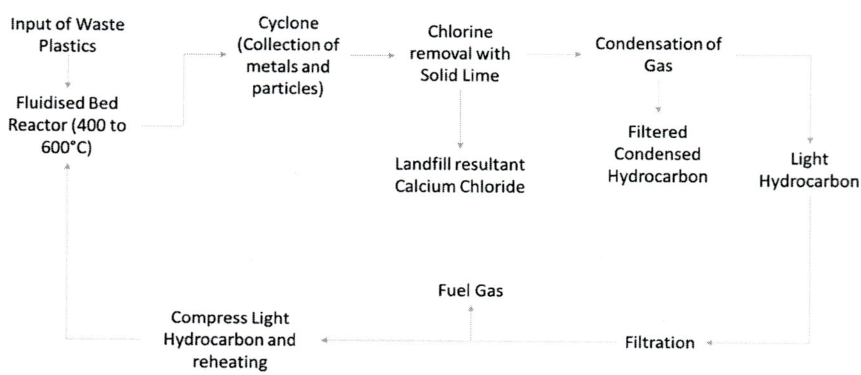

Figure 5.4. Flowchart of BP polymer cracking process
Source: Data from Tukker et al. 1999

recovers heat or power from plastic waste by burning them. The presence of waste plastic increases the energy value of MSW due to its high calorific value of about 35 MJ/kg (Mastellone 1999). Both grate furnace and fluidised bed incinerator can be applied for energy recovery. Comparing to pyrolysis, incineration requires a simple or no pretreatment of feedstock, however it emits signification amount of pollutants which must be treated properly. The pros and cons of mechanical recycling, chemical recycling, and energy recovery of plastic waste are summarized in Table 5.7.

5.3.2 Rubber

Post-consumer rubber tires are the most representative rubber wastes in MSW. Tire recycling refers to diverse technologies in order to convert the waste rubber into a broad range of materials or products which can be categorised into four

Table 5.7. Summary of the technologies for plastic recycling

Recycling methods	*Products*	*Advantage*	*Disadvantage*	*Remark*
Mechanical recycling	Plastic products	Simple and low cost.	Stringent quality requirement: clean and well separated.	Best option for leftover bits and pieces recycling. Good option for PET bottle recycling.
Chemical recycling	Fuel oil and chemicals	High-value products, flexible, low pollution.	Immature yet for treating mixed plastic waste.	Highly potential option for plastic waste recycling in the future. Good option for the cities which don't have WTE facilities.
Energy recovery	Power and/ or heat	High volume reduction through treatment	High cost, pollution, relatively low valued product	Best option for mixed plastic waste treatment currently. Only applicable for the cities which are operating WTE facilities.

levels of treatments (Figure 5.5). Through rubber recycling, virgin rubber resources and construction materials can be partially substituted (Shulman 2004). These recycling technologies include retreading and splitting, pyrolysis, rubber-derived fuel, and rubberised asphalt and concrete.

5.3.2.1 Retreading and Splitting

Waste tires are processed in a recapping system to introduce new tread to the tire (Pal et al. 2008). After the retreading, the waste tires can be reused as commercial products. Splitting involves removal of the metal parts and tread first by cutting, stamping and peeling. Then the rest are recycled in rubber manufacture industry

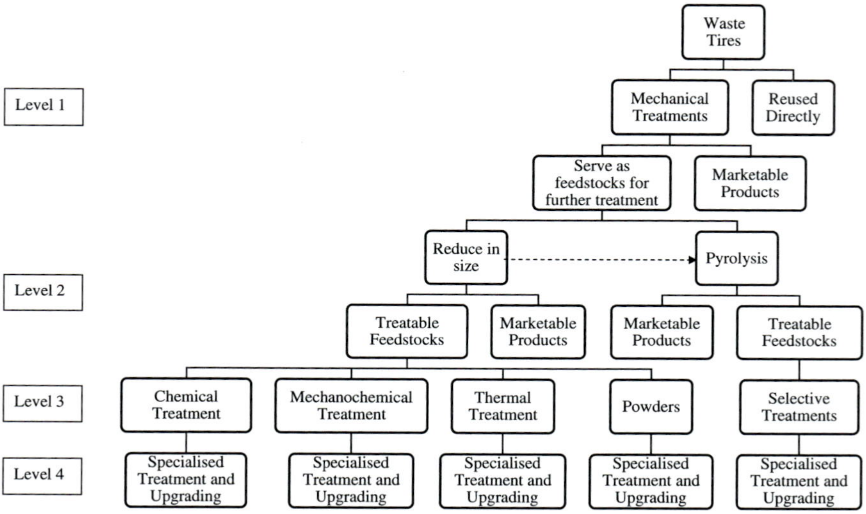

Figure 5.5. Four levels of waste tire treatments beginning with mechanical treatments and slowly progressing up the hierarchy to more sophisticated means of treatment to produce outputs with distinctive characteristics and features

where they can be die cast into various products such as gaskets, seals, and insulators (Shulman 2004).

5.3.2.2 Pyrolysis

Pyrolysis is a flexible way to recycle waste tires and other rubber products, which can recover oil, carbon black and many monomer chemicals by heating the rubber wastes in an oxygen-free atmosphere. Various technologies of rubber pyrolysis have been developed, such as fluidized bed, rotary kiln, or ablative system, etc. These technologies result in different efficiency of conversion, among which continuous rotary kiln is considered as an optimal option (Karthikeyan et al. 2012). Typical operating conditions of rubber pyrolysis are 450–550°C and atmosphere pressure, though vacuum system is developed as well. Oil with high calorific value of 43 MJ/kg and residual materials, i.e. carbon black can be harvested from such processes (Kaminsky and Mennerich 2001). Other newly developed technologies like microwave pyrolysis, plasma pyrolysis, and supercritical pyrolysis were also brought to practice (Bashir et al. 2012; Huang et al. 2003; Piskorz et al. 1999). Pyrolysis attracts growing attention for its application in feedstock recycling of rubber waste. However, further improvements are required in the area of handling of the addictives in rubber products such as carbon black and zinc oxide and upgrading the final products (Scheirs and Kaminsky 2006).

5.3.2.3 Rubber-Derived Fuel

A major recovery route for post-consumer tires is the use of shredded tires in the cement manufactory as supplementary fuel (Scheirs and Kaminsky 2006). The waste tires have approximately three times the energy content of MSW (De et al. 2005); and the rubber-derived fuel (RDF) has a better quality than the refuse-derived fuel. It was reported that 25% of coal consumption in cement industry can be substituted by RDF. Alternatively, rubber waste can serve as a good energy source for power generation. It was reported that 25% of waste rubber tires in the EU and 45% in the USA were recycled as rubber-derived fuel (Shulman 2004). Nowadays stringent emission limitation and safety concerns, such as NOx and CO emission, soil and water contamination caused by combustion residue, are hindering the approach of RDF. It was reported that the environmental impact of RDF combustion was at least 2 orders of magnitude higher than coal alone (Alvarez et al. 2004). However, this recycling method is still one of the most effective ways to face the ever increasing amount of rubber wastes.

5.3.2.4 Rubberised Asphalt and Concrete

The addition of waste rubber to asphalt and concrete has drawn great attentions (Baek et al. 2009; Bissada and Anani 1984; Kluitenberg et al. 1991; Xiao et al. 2007; Xiao et al. 2011). Two different, wet and dry, processes are applied for the manufacturing of rubberised asphalt (McQuillen et al. 1988). In the dry process, the recycled rubbers are premixed with aggregate before blending with asphalts. The PlusRide system and the TAK system are two popular systems utilised in dry process. In the wet process, recycled rubber is pre-blended with asphalt cement prior to introducing into the hot mix. The efficient mixing of rubber and asphalt results in an improvement of uniform dispersion of the additive in the asphalt mixture. Rubberised asphalt is widely applied to road construction or surface of ground, e.g. tennis courts or athletic grounds (Scheirs and Kaminsky 2006). In rubberised concrete manufactory, recycled rubber is used to substitute the aggregate even cement (Chung and Hong 2002). Rubberised concrete has many advantages compared to the conventional one, which includes increased crack resistance and shock wave absorption, reduced heat conductivity, reduced noise levels, lighter weight, and increased resistance to acid rain (De et al. 2005). Compare to other recycling methods, employing waste rubber in asphalt or concrete is more viable because the market is broader while technical barriers are less.

5.4 METALS

Metal is a limited resource and it plays an important role in our daily life and development of the economy. Therefore, there is a need to recycle metals to prevent depletion of metal ore supplies and reducing environmental impact. According to (Gitlitz 2003) from 1960s to 2000, world output of primary aluminium alone has increased 5 times. Furthermore, recycling of metals provides several advantages. Firstly, recycling of metals turns useless waste back into precious resources and relieves pressure on landfills. Secondly, recycling of waste metals substitutes mining industry. The substitution of mining prevents the use of toxic materials like mercury for gold mining. Lastly, recycling metals could reduce energy consumption. For example recycling of aluminium by remelting only requires about 5–10% of the amount of energy required to extract aluminium from bauxite (Quinkertz et al. 2001).

With the advance of metal recycling technologies like pyrometallurgy (separation of metals at molten state), hydrometallurgy (separation by controlled reaction of metals with various chemicals), and biotechnology (utilising micro-organisms to feed on other metals leaving targeted metals behind), scrap metal can be reprocessed into new materials with minimal loss of quality. Moreover, lower energy is required and lower CO_2 is generated than metal ore extraction (Cumbul Altay et al. 2011). Up to date, the most recycled metals are briefly categorised into two types: a) ferrous metal, e.g. iron and steel; and b) non-ferrous metals, e.g. Al, Pb, and Cu.

5.4.1 Ferrous Metals

Ferrous metal recycling market is highly developed all over the world. It is by far the most recyclable and profitable waste material, and the applications of recycled ferrous metals have been well accepted (Tam and Tam 2006). Scrap steel contains nearly 100% metal and is almost totally recycled and allows repeated recycling. It comes from municipalities, demolition operations, automobile dismantlers, ship yards, and industrial plants (Pichtel 2010). Ferrous scraps from municipalities are often commingled with non-ferrous metals, and should be separated by magnetic process. One of the most seen ferrous metals waste from municipalities is large, bulky household appliance such as washing machines, refrigerators, and stoves. Other recyclable and hazardous materials in such waste are firstly removed, and the metal is shredded or baled. Preferably, steel from demolition site should be reused directly. If it is unsuitable for direct reuse, it will be remelted to produce new steel.

Urbanisation results in an increase in the number of vehicles, and thus the end-of-life vehicle (ELV) has become a major waste stream of ferrous metals. The recycling of ELV is usually carried out by the car manufacturers and recycling companies (Kim et al. 2004). In the recycling process, ELVs are first dismantled to recycle any re-usable parts, and hazardous materials, including batteries, refrigerants, used oil, and antifreeze are removed. Then the residues are compressed and sent to the shredding facility. The shredded residues, of which ferrous metal consisted of around 70% of the mass, will be sent for remelting (Cumbul Altay et al. 2011; Kim et al. 2004).

5.4.2 Non-Ferrous Metals

Aluminium is one of the most widely applied non-ferrous metals in our life. Numerous products from different sectors are made of aluminium alloy, such as cans, vehicles, sports equipment, and window frames. Aluminium can, one of the most seen aluminium products, contains more than 90% aluminium (Schlesinger 2007). Its large quantity and high metal content makes it a valuable waste material. Generally, old aluminium cans collected are flattened by small, low-volume processer and sold to a nearby wholesaler, where larger operation will take place to bale, densify, or shred the cans. The baled or shredded cans are shipped by trucks, rail cars, or ships to regional mills or reclamation plants. Other types of aluminium wastes collected may first need to be shredded to separate the aluminium from other metals and attachments. During the process, the waste is reduced to a size convenient for transportation to reclamation plants and further processing. At the reclamation plant the baled aluminium cans are conveyed to a delacquering oven to remove coating and moisture. The hot shredded aluminium is passed through a screen to remove dirt and other contaminants, and then fed directly into a reverberatory furnace to be heated to 650°C (Pichtel 2010). In the furnace, the metal is melted and blended with other alloy elements as needed. The melted aluminium is then removed from the furnace and casted into ingots, which will be shipped to can manufacturer to product new cans.

Another non-ferrous metal usually found in waste is Pb scrap from spent automobile and industrial lead-acid batteries (LAB). Processing recycling lead from waste LAB consists of two main stages: a) pre-process stage, where the waste LAB is scrapped and metal and non-metal parts are separated; and b) in-process stage, where elemental lead is recovered from lead compounds in scrapped batteries in smelting furnaces at 1200–1260°C and then purified in a post-process stage (Genaidy et al. 2009). Depending on the smelter's practices and process, the lead-containing wastes are either directly fed to the smelting furnace, or desulphurised before being fed to the furnace.

Number of non-ferrous metals, such as Cu, Zn, and Pb, can be found from construction waste and printed circuit boards (PCB). While chemical methods to recover such non-ferrous metals are limited by the emission of toxic liquid or gas causing secondary pollution to the environment (Li et al. 2007) mechanical recycling methods are widely used in the industry. Mechanical recycling of PCB mainly includes two steps, disassembly and separation. Main approach to separate non-ferrous metal is electric conductivity-based separation, which includes: a) eddy current separation, relying on the electric conductivity and density characteristics of the materials, and b) corona electrostatic separation, utilizing electric conductivity properties of the materials only. Sink-float separation in heavy media is also applied in the separation of Al and Pb (Cui and Forssberg 2003).

Metal recycling is a very mature industry today, and adequate supply is one of the most significant aspects of maintaining the metal recycling industry. Advanced recycling infrastructure and waste collection system are important too. In the recycling practice, bulky items such as castings and extrusions are much more easily recycled than thin, heavily adulterated items such as aluminium foil and wiring. Waste with high metals concentration has higher recycling value and rate. Some metal manufacturers also recycle the scrap and residues from their own processes. Challenges of the current technologies lie in secondary pollution prevention and control. The most concerned pollutions from metal recycle process are: gaseous emission from aluminium recycling (Schlesinger 2007), fine particles and toxic heavy metals from PCB recycling, and Pb and Sn-Pb alloy from ELV (Wang and Chen 2012). Table 5.8 shows a summary of technologies employed to recover waste metals.

5.5 GLASS WASTE

5.5.1 Background

Glass waste constitutes a large portion of household and industrial waste due to its weight and density. Glass waste family generally consists of glass containers, window glass, glass wall, tempered glass, lamp glass, lab glassware, etc. The composition differs among different types of glass. The main streams of glass waste, soda-lime glass for containers and windows, are composed of about 63–81% silica (SiO_2) plus sodium oxide (Na_2O) from soda ash, lime (CaO), and several minor additives. The compositions of soda-lime glass and borosilicate glass are shown in Table 5.9 and 5.10, respectively.

Nowadays, glass waste is being considered as a resource rather than a waste. As glass is a very inert and stable material, if landfilled, it will take up the space without degradation for thousands of years. Glass is generally produced from sand, lime and soda, and it consumes about 40 percent more power to produce from raw materials than it does with recycled materials. Hence, for every ton of glass that is recycled to make new glass products, 693 pounds of carbon dioxide can be saved (http://glass.recycling123.info/flow-chart-for-recycling-glass-bottles.php).

The best recycling option for glass waste is, of course, to make it into new glass products by crushing, sorting, and remelting in glass manufacturing plants. Glass that is crushed and ready to be remelted is called cullet. To ensure an acceptable product quality, glass waste should be separated by types or chemical compositions, and then by colours. The most common types used for consumer containers are clear glass, green glass, and brown/amber glass. The sorted glass waste would have higher value than unsorted ones. Today, automatic optical colour sorting machines for glass cullet are already available. This has increased the recycling rate of glass waste greatly.

The beauty of glass is that it is one of the few materials that can be recycled indefinitely. A recycled glass container is just as strong as one made from virgin materials, and it can be recycled again and again without any loss of quality.

Table 5.8. Summary of the technologies for metals

Metal	*Sources of waste metals*	*Recovering process*	*Environmental benefits of the process*	*Environmental challenges of the process*
Ferrous	1) Municipalities 2) Demolition operations 3) Automobile dismantles 4) Ship yards 5) Industrial plants	1) Shredding 2) Magnetic separation	1) No requirement of special chemicals. 2) Relieving stress on landfills.	1) Large amount of electricity to run the machineries.
Lead	1) Batteries 2) Construction waste 3) PCB	1) Sink float 2) Eddy Current Separation 3) Corona Electrostatic Separation	1) Relieving stress on landfills. 2) Reduces amount of toxic waste entering the ecosystem.	1) Release of fine particles and toxic metals. 2) Gas emissions
Aluminium	1) Soft drink cans 2) Sports equipment (bicycles, badminton rackets) 3) Cooking foils	1) Sink float 2) Corona Electrostatic Separation	1) Lowers energy requirement in comparison to primary aluminium production. 2) Relieving stress on landfills.	1) Release of fine particles and toxic metals. 2) Gas emissions (furans, dioxides and hydrochloric acid)

(Continued)

Table 5.8. Summary of the technologies for metals (Continued)

Metal	*Sources of waste metals*	*Recovering process*	*Environmental benefits of the process*	*Environmental challenges of the process*
Copper	1) Electrical wires 2) Plumbing tubes 3) Heat exchangers	1) Sink float 2) Eddy Current Separation 3) Corona Electrostatic Separation	1) Relieving stress on landfills.	1) Release of fine particles and toxic metals. 2) Gas emissions
Zinc	1) Batteries 2) Construction sites 3) PCB	1) Electric arc furnace	1) Relieving stress on landfills. 2) Prevents pollution to rivers.	1) Release of fine particles and toxic metals. 2) Gas emissions

Table 5.9. Soda-lime glass composition

	Typical container glass	*Typical window glass*	*Approximate limits (mol %)*
SiO_2	74.42	71.86	63–81
Al_2O_3	0.75	0.08	0–2
MgO	0.30	5.64	0–6
CaO	11.27	9.23	7–14
Li_2O	0.00	0.00	0–2
Na_2O	12.9	13.13	9–15
K_2O	0.19	0.02	0–1.5
Fe_2O_3	0.01	0.04	0–0.6
Cr_2O_3	0.00	0.00	0–0.2
MnO_2	0.00	0.00	0–0.2
Co_3O_4	0.00	0.00	0–0.1
TiO_2	0.01	0.01	0–0.8
SO_3	0.16	0.00	0–0.2
Se	0.00	0.00	0–0.1

Source: http://glassproperties.com/glasses/. Reproduced with permission from Alexander Fluegel

Table 5.10. Borosilicate glass compositions

	Typical E glass	*Approximate limits*
SiO_2	56.99	43–74
B_2O_3	6.12	0–8.5
Al_2O_3	8.78	6–10
MgO	6.50	0.5–9
CaO	19.64	15–28
Na_2O	0.61	0–2.5
K_2O	0.00	0–0.5
Fe_2O_3	0.13	0–0.3
TiO_2	0.44	0–1
F	0.70	0–2

Source: http://glassproperties.com/glasses/. Reproduced with permission from Alexander Fluegel

However, not all the waste glass is recyclable. Some waste glass cannot be recycled through conventional way due to some special additives used in glass making. Moreover, mixed colour waste glass with contaminants and glass with coatings or laminations that cannot be removed easily are not possible to be sorted and recycled even with the modern sorting technologies.

5.5.2 Secondary Recycling of Glass Waste

As a complement of the conventional glass recycling process by remelting, i.e. primary recycling method, a number of secondary recycling methods have been developed and practiced. These include use as construction materials, industrial fillers and abrasives, water treatment media, gardening and landscaping, and so on.

5.5.2.1 Cement & Concrete

Use of glass waste in cement and concrete is one of the most studied topics for glass recycling because of the large demand and low quality requirements. The main application is the partial replacement for cement or aggregates in concrete and mortar (Byars et al. 2004; Shayan and Xu 2004; Topçu and Canbaz 2004). Most of the research found that the use of waste glasses as concrete aggregates has slight negative effects on the workability, strength and freezing-thawing resistance of the concrete (Federico and Chidiac 2009). However, the main concern is expansion and cracking of the concrete containing glass aggregates, which may result from the potential corrosion of glass aggregates. This may be overcome by the replacement of Portland cement with pozzolanic materials such as fly ash, silica fume and metakaolin.

On the other hand, being amorphous and containing relatively large quantities of silicon and calcium, glass is, in theory, pozzolanic or even cementitious in nature when it is finely ground. Ground glass powders exhibit very good pozzolanic reactivity and can be used as cement replacement (Shayan and Xu 2006; Shi et al. 2004). Its pozzolanic reactivity increases with increasing fineness. Nevertheless, alkalis in the glass powder can cause alkali-aggregate reaction and expansion if aggregates are alkali-reactive. The alkali-aggregate reaction expansion decreases as glass proportion increases, and will be under the deleterious limit if the glass replacement is 50% or more. The combined use of other supplementary cementing materials such as coal fly ash, ground blast furnace slag and metakaolin can also decrease the expansion from alkali-aggregate reaction. Lithium salt can be a very effective additive to prevent the alkali-aggregate reaction expansion of concrete containing glass powders.

Some research shows that the presence of ground glass powder has little influence on the kinetics of cement hydration. However, the production of portlandite is enhanced relative to the control, and calcium silicate hydrate (CSH) levels were found to be higher in the pastes containing glass (Dyer and Dhir 2001). Dyer and Dhir (2001) also measured the compressive strength development of cement pastes containing different colour of glass, e.g. white, green, and amber cullet. Some differences in the strength development of mortars containing different-coloured finely ground glass cullets were found: white and green result in a slight increase in 28-day compressive strengths relative to the glass-free control at replacement levels of around 10%, whereas finely ground amber glass cullet merely achieves similar strengths to the control. Besides, the rate of strength gain in mortars containing finely ground glass cullet is much higher between 7 and 28 days compared to the control.

5.5.2.2 Abrasive

Blasting is a requirement throughout the lifespan of a ship, for maintenance of the hull and preparation for recoating at regular intervals. The marine industry makes extensive use of grit or copper slag blasting to clean and prepare steel for production processes and then coating of the final product. Waste glass powder, in fact, could be a good abrasive medium for this application. UK WRAP has conducted a study in 2006 to evaluate the feasibility of using waste glass powder as a substitute for sand and copper slag for blasting from a technical, commercial, environmental, market and operational point of view (Shipbuilders and Shipprepairers Association 2006). The benefits of using waste glass abrasives for blasting lie in three aspects, including performance, environmental impact, and health & safety. The waste glass abrasive performs as well, if not better in some cases, than conventional abrasives (Shipbuilders and Shipprepairers Association 2006). As glass particles are angular to subangular, they have the ability to cut many coatings exceptionally well. Unlike other abrasives, glass does not contain significant chlorides or other salts which can accelerate the corrosion of cleaned surfaces. In addition, glass abrasives may be reused more than some other lower cost abrasives before becoming "spent".

Unlike copper slag and other metal slags, recycled glass is non-toxic, inert, and contains little heavy metal residues. This has major benefits especially when working in and around environmentally sensitive areas such as watercourses. Also, disposal of "spent" glass abrasives will have less negative environmental impacts. Glass has the potential for lower dust generation than other abrasives during use, therefore being relatively safer to use in public areas. Unlike many other mineral based abrasives, glass has no detectable "free" or "crystalline" silica. This greatly reduces the potential for worker health hazards from silicosis, a lung damaging disease which can be fatal.

5.5.2.3 Water Filtration Media

Filtration is a common process in both clean and wastewater treatment, facilitating the removal of suspended matter and associated contaminants in the influent stream. One of the most common techniques for the mechanical filtration of water is sand filtration. However, a common problem where filtration is applied in biological treatment processes is 'biofouling', which involves the media becoming clogged with bacterial matter. As a consequence, increased backwashing is required to dislodge the matter, along with increased use of energy and chemicals in linked treatment processes due to the reduced efficiency of the filter.

Research has shown that, by controlling the surface chemistry of certain types of processed waste glass materials, it is possible to achieve self-sterilization as well as surface catalytic and adsorption properties, useful attributes for filtration media

(Shipbuilders and Shipprepairers Association 2006). WRAP study has shown that crushed and screened waste glass particles has the potential to reduce these problems and thus to provide a more resilient and efficient media in a wide range of pressure and rapid gravity filter applications. The use of waste glass filtration media can improve the treatment efficiency by up to 30% due to media resistance to biofouling and contamination. Lower backwash rates can also be achieved by up to 50%. Moreover, the process stability and longevity of the filter is improved.

However, another research project (Rutledge and Gagnon 2002) reported that in general, the particle removal capabilities of the crushed-glass filter were slightly poorer than those of a sand filter, as quantified in a field application in the community of Orangedale, Nova Scotia. The crushed glass used in this project was initially sieved and washed, with no other pretreatment. The glass particles had a higher angularity and slightly higher uniformity coefficient than sand. Compared to sand, the initial start-up performance of the crushed glass filter was more variable and appeared to improve along the testing period. These show that further studies are required to determine the optimal particle size gradation of the glass media and their configuration in the filtration system.

Although the market potential for recycled glass filter media in a range of industry sectors and treatment applications is considerable, to open up these markets is dependent on, other than technical issues, the progress towards securing regulatory approval for the product, full end-user acceptance and the development of stable supply-chain relationships where feedstock material of a required quality and quantity can be supplied.

5.5.2.4 Foam Glass

Foam glass, also referred to as cellular glass, has been commercially available since the 1930's. Originally, it was manufactured from virgin glass. Today, the main foam glass producers in China, Europe and North America use a high percentage of processed post-consumer glass in their products. The basic principle of foam glass manufacturing is to generate a gas in glass melt at a temperature above 700°C. The gas expands thus producing a structure of cells to form a porous body. The foam glass can either be produced in aggregate form or block and shape form, depending on the applications. Foam glass is best suited as a rigid thermal and sound insulation material in roofs, walls, pipes, and traffic areas, where other insulation products may be compressed resulting in an uneven surface and the loss of insulating properties. Foam glass has excellent fire resistant properties and it is biologically inert.

Foam glass is a high-value product which also needs a high capital investment. It was reported by WRAP study (Hurley 2003) that the energy consumption is not a major operating cost as the technology with reduced energy consumption has been well developed. It is the payback of capital and plant labour costs that contribute to the largest portion of the costs. Another key economic consideration as well as an environmental consideration for the foam glass production is the collection and transportation of the glass cullet. It is recommended that the plant should be situated near to the source of waste glass. To further reduce the cost, the use of low-grade glass waste e.g. contaminated glass and CRT glass that cannot be accepted by primary recycling plant should be explored.

5.5.2.5 *Other Applications*

Waste glass can be pulverised and employed as industrial mineral fillers ranging in size from coarse grit to micrometre scale powders. In products such as paint and elastomeric coatings, finely sized fillers can provide volume and increase viscosity. Some of the common materials currently used as industrial mineral fillers, such as flint, talc, and various dry clays, contain crystalline silica in varying amounts. It has been recognised that crystalline silica-containing dust can cause serious lung disorders such as silicosis. Thus, concerns for worker health have prompted increased interest in identifying substitute filler materials. Waste glass is in amorphous state with less than 1% crystalline silica, which could be an excellent substitute for this. These applications typically require to pulverise waste glass to a mesh size finer than 150 mesh for coatings, and finer than 325 mesh for smooth surface paints.

A study in Taiwan making asphalt concrete road with pulverized waste glass with sizes smaller than 4.75 mm showed low permeability and higher frictional coefficients that are beneficial to the roadway safety (Chen et al. 2002). The waste glass also enhanced light reflection for night visibility. On the other hand, the stability values of asphalt concrete with waste glass are lower than that of without glass addition, but are well within the limit specified in the related code requirement. Another study in Australia showed crushed recycled glass with sizes similar or smaller than fine gravels exhibit geotechnical behaviour similar to natural aggregates (Disfani et al. 2011). Shear strength tests indicate that the glass fillers encompass shear strength parameters similar to that of natural sand and gravel mixtures comprising of angular particles. However, coarser recycled glass was found to be unsuitable for geotechnical engineering applications.

Waste glass has also been widely used as a raw material or a flux agent in the production of porcelain (Rambaldi et al. 2007), glass-ceramics (Bernardo et al 2010), and bricks (Loryuenyong et al. 2009). Results indicated that glass addition can both improve the performance of the products and save the production cost by reducing the material and energy consumption.

5.6 PAPER

Paper production from cellulose fibre can be traced back to ancient China at the early 2nd century. Nowadays, thousands of paper products are being used in our daily life, for writing, printing, packages and cleaning etc. However, behind the success of paper industry, paper waste is becoming a prominent issue in waste management. In 2010, the global consumption of paper has reached 394 million tonnes and kept rising year-by-year (Confederation of European Paper Industries 2011). For example, wastepaper (WP), which accounts for up to 55% of total waste

Table 5.11. Summarised general grades of paper according to 2001 European List of Standard Grades of Recovered Paper and Board

	Grade	Examples
1	Ordinary	Mixed paper and board, supermarket corrugated paper and board, unsold magazines, telephone books, mixed newspaper and magazines
2	Medium	Newspapers, unsold newspapers, sorted office paper, white wood-free books, coloured wood-free magazines, carbonless copy paper
3	High	Mixed lightly coloured printer shavings, white business forms, white wood-free computer print-out, printed bleached sulphate board, white newsprint and shavings
4	Kraft	New shavings of corrugated board, used corrugated Kraft and Kraft sacks
5	Special	Mixed recovered paper and board, mixed packaging

stream, is considered as the largest group in municipal solid waste (MSW) (Sangkharak 2011). Towards this, recycling is a better solution than landfill or incineration. Wastepaper utilisation can effectively prevent the rising price and concomitantly save the limited sources such as woods for paper production. In addition, recycling helps to minimise the pollution from burning wastepaper and the need for new landfill space. It is worth noting that approximately 19% of all paper is non-recyclable, such as toilet tissue, construction paper materials, archival paper and paper contaminated by organic waste (Williams 2005).

Wastepaper can be graded into different categories according to its quality, usage and origin. The criteria vary among countries. More than 50 recovered paper grades have been listed by the Institute of Scrap Recycling Industries (ISRI). But broadly speaking, there are five common grades of paper (Table 5.11).

Compared to the lower grade waste paper, the higher grade paper require less pretreatment and can be used as paper pulp substitute in application, such as printing paper and tissues; while the lower quality wastepaper is better treated as packing materials. Medium grade wastepaper requires further processing to de-ink and therefore is suitable for reuse as newspaper raw material. However, wastepaper cannot be recycled forever. Paper fibres are eventually broken down and become shorter during recycling processes. Technically speaking, the fibres can be recycled about four times. Once the used paper is not able to be recycled, it can be incinerated.

Wastepaper requires several processes to separate the fibres from contaminants which include inks, wood, plastics, adhesives, staples, sand, coating and inorganic fillers, including a starting stage of pulping paper products into a system, followed by screening and cleaning to remove contaminants, and then other

Table 5.12. Summary on general unit processes of recycled wastepaper

Processes	*Purpose*	*Operation*
Pulping	To defibre the paper and degrade contaminants	By wetting the paper, circulation to separate fibres, attrition to tear paper into flakes
Screening	To separate a mixed pulp flow	Including contaminant screening, fractionation, washing and dewatering
Cleaning	To remove contaminants from pulp	Using cleaners/hydrocyclones to separate contaminants based on the density differences
Deinking	To remove ink from the pulp	Using ink-removing system and deinking chemicals, followed by washing, flotation, kneading and dispersion processes
Bleaching	To bleach fibres	Including colour stripping, delignification, and brightening processes and using bleaching agents (chemicals or enzymes)
Water clarification	To clarify and reuse water	Recent techniques include sedimentation, dissolved-air flotation and filtration

further processing (deinking, bleaching, etc.) to clean and thicken the pulp. Table 5.12 shows the common unit processes (Doshi and Dyer 2001) that depends on the requirements for cleanliness and brightness of the final products.

These recycled technologies are successful to recover the cellulosic fibre and provide variable applications of wastepaper. Recycled fibres can be used alone or combined with other materials in making paper products. Recent applications of recycled paper mainly include:

- *Packaging*: Packing is the major application of recycled wastepaper. Compared to petroleum-based materials, paper is more environmental-friendly and biodegradable. However, when using recycled paper as food packager, safety issue is a concern that is being one of the major focuses of recent researches (Biedermann and Grob 2010).
- *News printing*: About 25% of recycled paper can be used for newspapers and magazines (Williams 2005), however the quality may be a little compromised.
- *Biomass*: Recovered paper as feedstock in the pulp and paper industry for paper production will lead lower energy input and less CO_2 emissions (Laurijssen et al. 2010). Some pretreated wastepaper, such as office paper (87%) and newspaper (80.34%), provide high amount of cellulose and can

thus be used as cost-effective substrates for the production of reducing sugar and ethanol (Sangkharak 2011), as well as methane (Baba et al. 2013).

- *Building materials*: Wastepaper can be mixed in concrete to improve flexural behaviour, ductility, toughness, and shrinkage characteristics of the cement (Sujivorakul et al. 2012).

Recycling rate of wastepaper varies around the world. The numbers are considerable in most developed countries, such as Europe, the Unite States, Japan, Hong Kong and Taiwan. For example in 2011, the European paper recycling rate reached 71.7% (European Recovered Paper Council 2012) showing global leadership in paper recycling. Some countries such as Germany and UK export recycled paper. A major problem in the paper recovery is the large amount of paper and paper products in household waste, which is difficult for recycling and add burden to the existing landfills. In addition, the paper market control policy will affect the use of wastepaper fibres. Equilibrium should be found when raw and recycled fibres increase with the greater consumption of paper and paper products.

5.7 OTHER WASTES

5.7.1 Waste Oil

Lubricant oil is used to reduce the friction and wear in machines. Basically, it is classified into two major groups: vehicle lubricants and industrial lubricants. Lubricating oil consists of 90% base oils (petroleum fraction) and 10% chemical additives and other components. After certain service time, the chemical additives in the oil will exhaust. Moreover, the oil will be contaminated by particles from metallic materials, such as Fe, Cd, Cr, and Pb (Al-Ghouti and Al-Atoum 2009). Used oil also has more water, sediment and oxidation products. Such degradation makes the oil physically and chemically unsuitable for further use and must be replaced. The used lubricant oil can be reused as fuel oils or auxiliary fuels in the cogeneration system (Tsai 2011).The quality of used oil can be improved by removal of metals and oxidation products by polar solvents mixtures (Rincon et al. 2005a, b), removal of metal particles and dusts by hollow fibre membranes (Cao et al. 2009), catalytic desulphurisation (Bhaskar et al. 2004) and catalytic pyrolysis (Demirbas 2008)

Waste cooking oil (WCO) is the residue from kitchens, restaurants, and food factories. WCO might be directly reused as fuel oil, yet the high level of free fatty acids and water contents will cause damage to the engine. Hence, technologies to convert WCO into biodiesel are applied. Transesterification is the most widely used process in the production of biodiesel from WCO (Zhang et al. 2003). It is a catalysed reaction of vegetable oil in the presence of alcohol to yield biodiesel and glycerol. Methanol is the most applied alcohol in this process due to its low price, and the product from transesterification using methanol has low molecular weight, low viscosity, and low pour point (Yaakob et al. 2013). Sometimes pre-treatment for WCO is required, such as filtration to remove solid residues and dehydration to remove water (Maddikeri et al. 2012). So far, the price of biodiesel from WCO is higher than diesel from fossil fuels. Thus, more research is needed to reduce the production cost and improve the yield and quality of the biodiesel.

5.7.2 Construction Waste

From pure monetary considerations, recycling of construction waste is only attractive when the recycled product is competitive with natural resources in terms of cost and quantity. Recycled construction materials will be more competitive in regions when both raw materials and landfill sites are lacking. According to Coelho and de Brito (2013), recycling of construction materials can be a lucrative business. It is possible to break even within a short period of 2 years. However, there are many barriers to enter this business, including high initial investments and technical challenge in the technology.

Materials that can be recycled from construction waste include asphalt, bricks, concrete, and masonry. Plastics, rubber, ferrous metals, non-ferrous metals, glass, paper and cardboard could also be found in construction waste and their recycling practices were discussed in previous sections. Asphalt waste can be used for the production of new asphalt. The new asphalt can contain up to 10–15% recycled asphalt (Tam and Tam 2006). Used asphalt concrete, which is a mixture of bitumen and mineral aggregates, can be recycled as construction aggregates after simple processing e.g. crushing and screening; and can be used for road surface and parking lot.

Bricks from demolition may be contaminated with mortar, rendering and plaster, and are often mixed with other materials such as timber and concrete. Separation of the recyclable bricks with its contaminants is usually difficult and requires hand sorting, thus the sorting and cleaning of bricks is labour-intensive and costly.

The most common way to recycle concrete rubble is to use it as road base. Another way is to use it as replacement for natural aggregates (Soutsos et al. 2011, 2012). The challenges of the process lie in the performance of recycled aggregates, as they often contain contaminants, such as soft soil and old concrete rubbles, which may affect the properties of the products (Kou et al. 2012).

Timber waste is produced in large quantity from construction all over the world. It can be reused in construction after cleaning, de-nailing and sizing. Undamaged wood can be reused as plank, beam, door, floorboard, rafter, panel, pallet, and pile. Timber waste can also be recycled for energy, such as making fuel or charcoal for power generation.

References

Aguado, J., Serrano, D., San Miguel, G., Castro, M., and Madrid, S. (2007). "Feedstock recycling of polyethylene in a two-step thermo-catalytic reaction system." *J. Anal. Appl. Pyrol.*, **79**(1), 415–423.

Al-Ghouti, M. A., and Al-Atoum, L. (2009). "Virgin and recycled engine oil differentiation: A spectroscopic study." *J. Environ. Manage.*, **90**(1), 187–195.

Al-Salem, S., Lettieri, P., and Baeyens, J. (2009). "Recycling and recovery routes of plastic solid waste (PSW): A review." *Waste Manage.*, **29**(10), 2625–2643.

Alvarez, R., et al. (2004). "Soil, water, and air environmental impact from tire rubber/coal fluidized-bed cocombustion." *Energy Fuels*, **18**(6), 1633–1639.

Arandes, J. M., Ereña, J., Azkoiti, M. J., López-Valerio, D., and Bilbao, J. (2004). "Valorization by thermal cracking over silica of polyolefins dissolved in LCO." *Fuel Process. Technol.*, **85**(2), 125–140.

Association, S. A. S. (2006). "Full scale operational trials involving the use of recycled glass as a blasting medium in the marine sector." WRAP.

Aznar, M. P., Caballero, M. A., Sancho, J. A., and Francés, E. (2006). "Plastic waste elimination by co-gasification with coal and biomass in fluidized bed with air in pilot plant." *Fuel Process. Technol.*, **87**(5), 409–420.

Baba, Y., Tada, C., Fukuda, Y., and Nakai, Y. (2013). "Improvement of methane production from waste paper by pretreatment with rumen fluid." *Bioresour. Technol.*, **128**, 94–99.

Baek, S., Kim, H., Doh, Y., and Kim, K. (2009). "Estimation of high-temperature properties of rubberized asphalt using chromatograph." *KSCE J. Civ. Eng.*, **13**(3), 161–167.

Bakas, I., Sieck, M., Hermann, T., Andersen, F. M., and Larsen, H. (2011). "Projections of municipal waste management and greenhouse gases." ⟨http://scp.eionet.europa.eu/wp/2011wp4⟩ (Jan. 14, 2014).

Bartone, C. R. (1990). "Economic and policy issues in resource recovery from municipal solid wastes." *Resour. Conserv. Recycl.*, **4**(1–2), 7–23.

Bashir, N., Hussain, K., Khan, K. M., Hussain, Z., and Perveen, S. (2012). "A new method for the co-liquefaction of coal and waste tyre rubber into useful products using microwave metal interaction pyrolysis." *J. Chem. Soc. Pak.*, **34**(1), 162–167.

Bernardo, E., Bonomo, E., and Dattoli, A. (2010). "Optimisation of sintered glass–ceramics from an industrial waste glass." *Ceram. Int.*, **36**(5), 1675–1680.

Bhaskar, T., et al. (2004). "Recycling of waste lubricant oil into chemical feedstock or fuel oil over supported iron oxide catalysts." *Fuel*, **83**(1), 9–15.

Biedermann, M., and Grob, K. (2010). "Is recycled newspaper suitable for food contact materials? Technical grade mineral oils from printing inks." *Eur. Food Res. Technol.*, **230**(5), 785–796.

Bissada, A. F., and Anani, A. A. (1984). "Evaluation of rubberized limestone filler in asphalt paving mixtures." *Rubber Chem. Technol.*, **57**(1), 1–18.

Branigan, T. (2009). "From east to west, a chain collapses." *Guardian*, 8.

Byars, E. A., Morales-Hernandezz, B., and HuiYing, Z. (2004). "Waste glass as concrete aggregate and Pozzolan: Laboratory and industrial projects." *Concrete*, **38**(1), 41–44.

Cao, Y., Yan, F., Li, J., Liang, X., and He, B. (2009). "Used lubricating oil recycling using a membrane filtration: Analysis of efficiency, structural and composing." *Desalin. Water Treat.*, **11**(1-3), 73–80.

Chen, G., et al. (2002). "Glass recycling in cement production—An innovative approach." *Waste Manage.*, **22**(7), 747–753.

Chung, K. H., and Hong, Y. K. (2002). "Scrap tire/aggregate composite: Composition and primary characterizations for pavement material." *Polym. Compos.*, **23**(5), 852–857.

Coelho, A., and de Brito, J. (2013). "Economic viability analysis of a construction and demolition waste recycling plant in Portugal—Part I: Location, materials, technology and economic analysis." *J. Cleaner Prod.*, **39**, 338–352.

Confederation of European Paper Industries. (2011). "CEPI key statistics 2011." Brussels, Belgium.

Cui, J., and Forssberg, E. (2003). "Mechanical recycling of waste electric and electronic equipment: A review." *J. Hazard. Mater.*, **99**(3), 243–263.
Cumbul Altay, M., Sivri, N., Onat, B., Şahin, Ü., Zorağa, M., and Fatih Altay, H. (2011). "Recycle of metals for end-of-life vehicles (ELVs) and relation to Kyoto protocol." *Renew. Sustainable Energy Rev.*, **15**(5), 2447–2451.
Demirbas, A. (2008). "Gasoline-like fuel from waste engine oil via catalytic pyrolysis." *Energy Sources Part A*, **30**(16), 1433–1441.
De, S. K., Avraam, I., and Khait, K. (2005). *Rubber recycling*, Taylor & Francis/CRC Press, Boca Raton, FL.
Disfani, M., Arulrajah, A., Bo, M., and Hankour, R. (2011). "Recycled crushed glass in road work applications." *Waste Manage.*, **31**(11), 2341–2351.
Doshi, M. R., and Dyer, J. M. (2001). "Paper: Recycling and recycled materials." *Encyclopedia of materials: Science and technology*, K. H. J. Buschow, R. W. Cahn, M. C. Flemings, B. Ilschner, E. J. Kramer, S. Mahajan, and P. Veyssière, eds., 2nd Ed., Elsevier Oxford, 6711–6720.
Duke, S. (2009). "Providing for european-level diplomacy after Lisbon: The case of the european external action service." *Hague J. Diplomacy*, **4**(2), 211–233.
Dyer, T. D., and Dhir, R. K. (2001). "Chemical reactions of glass cullet used as cement component." *J. Mater. Civ. Eng.*, 10.1061/(ASCE)0899-1561(2001)13:6(412), 412–417.
European Environment Agency. (2013). "Recycling's current and potential contribution (*) to meeting EU demand for various materials, 2006." R.C.A.P. Contribution, Copenhagen, Denmark.
European Environment Agency. (2013). *Managing municipal solid waste*, Copenhagen, Denmark.
European Recovered Paper Council. (2012). "Paper recycling monitoring report 2012." Brussels, Belgium.
European Parliment and Council Directive. (1994). "European parliament and council directive 94/62/EC of 20 December 1994 on packaging and packaging waste." *OJ L*, **365**, 10.
Eurostat. (2015). "Waste statistics." Copenhagen, Denmark.
FairPrice, N. (2011). "Sustainable environment, NTUC FairPrice." ⟨http://www.csr.fairprice.com.sg/sustainable-environment.html⟩ (Jan. 14, 2014).
Federico, L. M., and Chidiac, S. E. (2009). "Waste glass as a supplementary cementitious material in concrete–critical review of treatment methods." *Cement and concrete composites*, **31**(8), 606–610.
Filip, M., Pop, A., and Rusu, T. (2011). "GS-MS and FTIR studies of the liquid products obtained by thermal and catalytic degradation of polystyrene waste." *Stud. Univ. Babes-Bolyai Chem.*, **56**(4), 27–40.
Fischer, C., and Werge, M. (2009). "EU as a recycling society present recycling levels of municipal waste and construction & demolition waste in the EU." ⟨http://scp.eionet.europa.eu/publications/wp2009_2/wp/WP2009_2⟩ (Jan. 14, 2014).
Genaidy, A., Sequeira, R., Tolaymat, T., Kohler, J., and Rinder, M. (2009). "Evidence-based integrated environmental solutions for secondary lead smelters: Pollution prevention and waste minimization technologies and practices." *Sci. Total Environ.*, **407**(10), 3239–3268.
Giles, H., Jr., Wagner, J., Jr., and Mount III, E., (2005). *Extrusion-The definitive processing guide and handbook*, William Andrew Publishing, Norwich, New York.
Gitlitz, J. (2003). "The role of the consumer in reducing primary aluminum demand." International Strategic Roundtable on the Aluminum Industry, Culver City, CA.
Grause, G., Handa, T., Kameda, T., Mizoguchi, T., and Yoshioka, T. (2011). "Hydrolytic degradation of poly (ethylene terephthalate) in a pyrolytic two step process to obtain benzene rich oil." *J. Appl. Polym. Sci.*, **120**(6), 3687–3694.

He, W., et al. (2006). "WEEE recovery strategies and the WEEE treatment status in China." *J. Hazard. Mater.*, **136**(3), 502–512.

Horvat, N. (1996). *Tertiary polymer recycling: study of polyethylene thermolysis and polyethylene oil hydrogenation*, Univ. of Waterloo, Waterloo, Ontario.

Huang, H., Tang, L., and Wu, C. (2003). "Characterization of gaseous and solid product from thermal plasma pyrolysis of waste rubber."*Environ. Sci. Technol.*, **37**(19), 4463–4467.

Hurley, J. (2003). "A UK market survey for foam glass." , WRAP, Oxon, U.K.

Kaminsky, W., and Mennerich, C. (2001). "Pyrolysis of synthetic tire rubber in a fluidised-bed reactor to yield 1, 3-butadiene, styrene and carbon black." *J. Anal. Appl. Pyrol.*, **58**, 803–811.

Karthikeyan, S., Sathiskumar, C., and Moorthy, R. S. (2012). "Effect of process parameters on tire pyrolysis: A review." *J. Sci. Ind. Res.*, **71**(5), 309–315.

Kim, K. H., et al. (2004). "Management status of end-of-life vehicles and characteristics of automobile shredder residues in Korea." *Waste Manage.*, **24**(6), 533–540.

Kluitenberg, G., Bilskie, J., and Horton, R. (1991). "Rubberized asphalt for sealing cores of shrinking soil." *Soil Sci. Soc. Am. J.*, **55**(5), 1504–1507.

Kou, S. C., Poon, C. S., and Wan, H. W. (2012). "Properties of concrete prepared with low-grade recycled aggregates." *Constr. Build. Mater.*, **36**, 881–889.

Laurijssen, J., Marsidi, M., Westenbroek, A., Worrell, E., and Faaij, A. (2010). "Paper and biomass for energy? The impact of paper recycling on energy and CO2 emissions." *Resour. Conserv. Recycl.*, **54**(12), 1208–1218.

Li, J., Lu, H., Guo, J., Xu, Z., and Zhou, Y. (2007). "Recycle technology for recovering resources and products from waste printed circuit boards." *Environ. Sci. Technol.*, **41**(6), 1995–2000.

Lin, Y. H., and Yang, M. H. (2007). "Catalytic pyrolysis of polyolefin waste into valuable hydrocarbons over reused catalyst from refinery FCC units." *Appl. Catal. A*, **328**(2), 132–139.

Loryuenyong, V., Panyachai, T., Kaewsimork, K., and Siritai, C. (2009). "Effects of recycled glass substitution on the physical and mechanical properties of clay bricks." *Waste Manage.*, **29**(10), 2717–2721.

Maddikeri, G. L., Pandit, A. B., and Gogate, P. R. (2012). "Intensification approaches for biodiesel synthesis from waste cooking oil: a review." *Ind. Eng. Chem. Res.*, **51**(45), 14610–14628.

Mastellone, M. (1999). "Thermal treatments of plastic wastes by means of fluidized bed reactors." Ph.D. thesis, Second Univ. of Naples, Caserta, Italy.

McCaffrey, W., Kamal, M., and Cooper, D. (1995). "Thermolysis of polyethylene." *Polym. Degrad. Stab.*, **47**(1), 133–139.

McQuillen, J. L., Jr., Takallou, H. B., Hicks, R. G., and Esch, D. (1988). "Economic analysis of rubber-modified asphalt mixes." *J. Transp. Eng.*, 10.1061/(ASCE)0733-947X(1988)114:3(259), 259–277.

NEA (National Environmental Agency). (2012). "Environmental protection division annual reports." Singapore.

Neo, H. (2010). "The potential of large-scale urban waste recycling: A case study of the National Recycling Programme in Singapore." *Soc. Nat. Resour.*, **23**(9), 872–887.

Pal, K., Das, T., Pal, S. K., and Das, C. K. (2008). "Use of carboxylated nitrile rubber and natural rubber blends as retreading compound for OTR tires." *Polym. Eng. Sci.*, **48**(12), 2410–2417.

Pichtel, J. (2010). *Waste management practices: Municipal, hazardous, and industrial*, CRC Press, Boca Raton, FL.

Piskorz, J., Majerski, P., Radlein, D., Wik, T., and Scott, D. (1999). "Recovery of carbon black from scrap rubber." *Energy Fuel*, **13**(3), 544–551.

Quinkertz, R., Rombach, G., and Liebig, D. (2001). "A scenario to optimise the energy demand of aluminium production depending on the recycling quota." *Resour. Conserv. Recycl.*, **33**(3), 217–234.

Rambaldi, E., Carty, W., Tucci, A., and Esposito, L. (2007). "Using waste glass as a partial flux substitution and pyroplastic deformation of a porcelain stoneware tile body." *Ceram. Int.*, **33**(5), 727–733.

Raunio, T. (2009). "National parliaments and European integration: What we know and agenda for future research." *J. Legal Stud.*, **15**(4), 317–334.

Rincon, J., Canizares, P., and Garcia, M. T. (2005a). "Regeneration of used lubricant oil by polar solvent extraction." *Ind. Eng. Chem. Res.*, **44**(12), 4373–4379.

Rincon, J., Canizares, P., and Garcia, M. T. (2005b). "Waste oil recycling using mixtures of polar solvents." *Ind. Eng. Chem. Res.*, **44**(20), 7854–7859.

Rutledge, S. O., and Gagnon, G. A. (2002). "Comparing crushed recycled glass to silica sand for dual media filtration." *J. Environ. Eng. Sci.*, **1**(5), 349–358.

Sangkharak, K. (2011). "Optimization of enzymatic hydrolysis for ethanol production by simultaneous saccharification and fermentation of wastepaper." *Waste Manage. Res.*, **29**(11), 1134–1144.

Scheirs, J., and Kaminsky, W. (2006). *Feedstock recycling and pyrolysis of waste plastics*, Wiley, Chichester, West Sussex.

Schlesinger, M. E. (2007). *Aluminum recycling*, 2nd Ed., CRC/Taylor & Francis, Boca Raton, FL, 225.

Schmidt, J. H., Holm, P., Merrild, A., and Christensen, P. (2007). "Life cycle assessment of the waste hierarchy—A Danish case study on waste paper." *Waste Manage.*, **27**(11), 1519–1530.

Shayan, A., and Xu, A. (2004). "Value-added utilisation of waste glass in concrete." *Cem. Concr. Res.*, **34**(1), 81–89.

Shayan, A., and Xu, A. (2006). "Performance of glass powder as a pozzolanic material in concrete: A field trial on concrete slabs." *Cem. Concr. Res.*, **36**(3), 457–468.

Shi, C., Wu, Y., Shao, Y., and Riefler, C. (2004). "Alkali-aggregate reaction of concrete containing ground glass powder." *Proc., 12th Int. Conf. on AAR in Concrete*, International Academic Publishers, Sonning Common, U.K.

Shulman, V. L., Rapra Technology, and ebrary. (2004). "Tyre recycling Rapra review reports." ⟨http://site.ebrary.com/lib/rochester/Doc?id=10236872⟩ (Dec. 1, 2012).

Soutsos, M. N., Tang, K., and Millard, S. G. (2011). "Use of recycled demolition aggregate in precast products, phase II: Concrete paving blocks." *Constr. Build. Mater.*, **25**(7), 3131–3143.

Soutsos, M. N., Tang, K., and Millard, S. G. (2012). "The use of recycled demolition aggregate in precast concrete products–Phase III: Concrete pavement flags." *Constr. Build. Mater.*, **36**, 674–680.

Sujivorakul, C., Muhummud, T., and Dokkhan, N. (2012). *High performance light-weight cement composite plates using wastepaper fibers and wire mesh High performance fiber reinforced cement composites 6*, Springer, Dordrecht, Netherlands, 495–502.

Tam, V. W. Y., and Tam, C. M. (2006). "A review on the viable technology for construction waste recycling." *Resour. Conserv. Recycl.*, **47**(3), 209–221.

Tojo, N., and Fischer, C. (2011). "Europe as a recycling society, European recycling policies in relation to the actual recycling achieved." European Environment Agency, Copenhagen, Denmark.

Topçu, I. B., and Canbaz, M. (2004). "Properties of concrete containing waste glass." *Cem. Concr. Res.*, **34**(2), 267–274.

Tsai, W. T. (2011). "An analysis of used lubricant recycling, energy utilization and its environmental benefit in Taiwan." *Energy*, **36**(7), 4333–4339.

Tukker, A., de Groot, H., Simons, L., and Wiegersma, S. (1999). "Chemical recycling of plastic waste: PVC and other resins." European Commission, Delft, Netherlands.

Wang, J., and Chen, M. (2012). "Management status of end-of-life vehicles and development strategies of used automotive electronic control components recycling industry in China." *Waste Manage. Res.*, **30**(11), 1198–1207.

Williams, P. T. (2005). *Waste recycling waste treatment and disposal*, Wiley, Chichester, U.K., 127–170.

Wilson, D. C., Velis, C., and Cheeseman, C. (2006). "Role of informal sector recycling in waste management in developing countries." *Habitat Int.*, **30**(4), 797–808.

Xiao, F., Amirkhanian, S., and Juang, C. H. (2007). "Rutting resistance of rubberized asphalt concrete pavements containing reclaimed asphalt pavement mixtures." *J. Mater. Civ. Eng.*, 10.1061/(ASCE)0899-1561(2007)19:6(475), 475–483.

Xiao, F., Amirkhanian, S. N., and Zhao, W. (2010). "Laboratory investigation of fatigue characteristics of rubberized asphalt mixtures containing warm asphalt additives at a low temperature." *J. Test. Eval.*, **39**(2), 1–6.

Yaakob, Z., Mohammad, M., Alherbawi, M., Alam, Z., and Sopian, K. (2013). "Overview of the production of biodiesel from waste cooking oil." *Renewable Sustainable Energy Rev.*, **18**(0), 184–193.

Zhang, Y., Dubé, M. A., McLean, D. D., and Kates, M. (2003). "Biodiesel production from waste cooking oil: 1. Process design and technological assessment." *Bioresour. Technol.*, **89**(1), 1–16.

Zhao, L., Wang, Z. H., Chen, D. Z., Ma, X. B., and Luan, J. (2012). "Influence of impurities on waste plastics pyrolysis: Products and emissions." *Environ. Sci.*, **33**(1), 329–336.

Zia, K. M., Bhatti, H. N., and Ahmad Bhatti, I. (2007). "Methods for polyurethane and polyurethane composites, recycling and recovery: A review." *React. Funct. Polym.*, **67**(8), 675–692.

CHAPTER 6

Public Engagement for Implementation of Waste Reduction and Recycling Policies

Chettiyappan Visvanathan[1]
Prakriti Kashyap[2]

6.1 INTRODUCTION

Solid waste generation is a result of peoples' lifestyles and activities; hence public participation becomes an integral requirement of waste management. Reduce, Reuse and Recycle (3R) of waste are the best approach to sustainable waste management. Translating these 3R policies into the ground level implementation requires enthusiastic public engagement; otherwise even the best of these policies fizzle out if not well received by the public. Public participation ranges from passive reception of services to the active participation in decision-making and shaping up the plans and policies.

Public involvement in waste management activities has gradually expanded in terms of the level and effectiveness of the participation. In earlier days, public participation was understood as equivalent to 'informing or educating.' With time and experiences, it is now understood that public participation is more than just information. An extended understanding of participation is 'engaging' public into 3R activities and in environmental decision making. Engaging the public effectively is a way to bring a sense of public ownership, hence attributing to continual public support and sustainability of the project activities. Public contribution in decision-making requires 'empowering' the public and empowerment is certainly more than just informing.

Usually public participation in 3R activities occurs at many areas and levels namely; waste reduction activities, source segregation of waste, participation in community recycling activities, or buying "green", and gathering public consensus

[1]Asian Institute of Technology, Pathumthani, Thailand.
[2]Asian Institute of Technology, Pathumthani, Thailand.

for the waste treatment options and infrastructures and evaluation of waste management services and programs.

Private sector participation (PSP) has come across as an important policy component in some of the progressive waste management frameworks; however, public participation is still conducted mostly at scattered 'activity' level rather than embedded as a 'policy' requirement. It is thus very important to address the issues of public engagement in the policy framework as the legal requirement.

This chapter builds with the background on the meaning and various levels of public participation in environmental matters. Examples of solid waste management from Asia are picked to showcase why public participation is important in waste reduction and recycling activities. These examples also dissect the possible strategic drivers and methodologies used for public engagement. The chapter concludes with the need for framing 3R policy that 'embeds-engage-and-empower' the public in waste management governance hence sharing the responsibilities of the local authorities (LAs) on waste management.

6.2 DEFINING PUBLIC PARTICIPATION

Participation is defined as an active process where intended beneficiaries influence the program outcomes and gain personal growth (Oakley 1989). Community participation is approached with the understanding that people know it better what they want, if involved in programs they will make a contribution of resources, become committed to the activities, and develop skills, knowledge and experience (Rifkin and Kangere 2002). Public participation also cultivates a feeling of ownership in projects, ensures transparency and accountability hence contributes to better governance. Promoting community participation in decision making increases the legitimacy and wide acceptance of the decisions.

Public participation occurs at various stages of policy and program formulation and implementation. It can range in the increasing spectrum of – 'to inform', 'to consult', 'to collaborate and partner', and 'to empower' (Arnstein 1969; IDEA 2003; IAP2 2004). The tools used at each stage of participation are also different. Each of these stages illustrates the differences in the involvement and power diffusion. The degree of participation is usually based on the objective of participation. Effective public participation may use one or a combination of these stages and tools. Whatever may be the level of participation, the active relationship between the government (and other private third parties) and communities are effective to strengthen public trust, improve transparency, unbiased monitoring of the programs, and also to enhance the citizens' capacity to make the projects and policies successful.

LAs and the local community are the two important stakeholders in the public participation process. For ensuring the effective public participation, both the parties are required to fulfill their entitled roles. LAs usually play the role of an initiator for participation, because they are the ones who formulate policies and programs to the public. Their role can be of a convener, mediator, catalyst, funder, technical assistance provider and capacity builder. The community may also initiate the participation and perform the roles of a debater/discussant, contributor, implementer and evaluator (IDEA 2003). Non-Governmental Organizations (NGOs) and Community Based Organizations (CBOs) usually facilitate (and empower the community) the participation process between the LAs and the community. If all of these stakeholders fulfill their respective roles, an effective participation and partnership evolves that secures the division of labor, combining resources, mutual support and shared obligations.

6.3 PUBLIC PARTICIPATION IN WASTE MANAGEMENT SYSTEMS

Against the backdrop of economic prosperity and over consumption of commodities, increased quantity and complex composition of waste is a direct resultant of public lifestyles. Such increment in waste quantity and complexity makes it exceedingly difficult for the LAs alone to perform solid waste management activities. Public participation hence becomes a critical factor of the waste management system. Effective public participation can materialize a reduction in waste amount, less/no littering of public spaces, high level segregation of waste, as well as ensuring proper monitoring of waste management services. The benefits of public participation are consensus, commitment and legitimacy over the waste management programs. Public participation can occur in various community groups- household level involvement, school level activities on waste reduction and recycling, to community level participation in decision making.

Increasing responsibilities but tightly constrained human resources, capacities, technology and budget makes the LAs to look out for the partnership with various stakeholders. With public private partnership (PPP), the LAs shift their roles from a 'service provider' to the 'facilitator of service.' LAs then focus on planning and management, while a private company or a community takes up the day to day operation related to waste management. Figure 6.1 below illustrates different actors of PPP.

In the context of participation and partnership, the private sector participation is given comparatively more weightage in LA's practices. This makes sense as involving private sector means sharing the technology and the financial burden of the waste management responsibilities of the LA. Community participation on the other hand may not directly contribute to the technology and funding resources, but indirectly does so by waste reduction and segregation at source and thus lowering the waste management cost for the LAs. Without public participation even the best laid policies and the private sector's participation may meet a rejection. Similarly, NGOs/CBOs are other important stakeholders as they facilitate the public participation process or take part in waste management activities as social entrepreneurs. No matter whether LA does the job itself or hires private

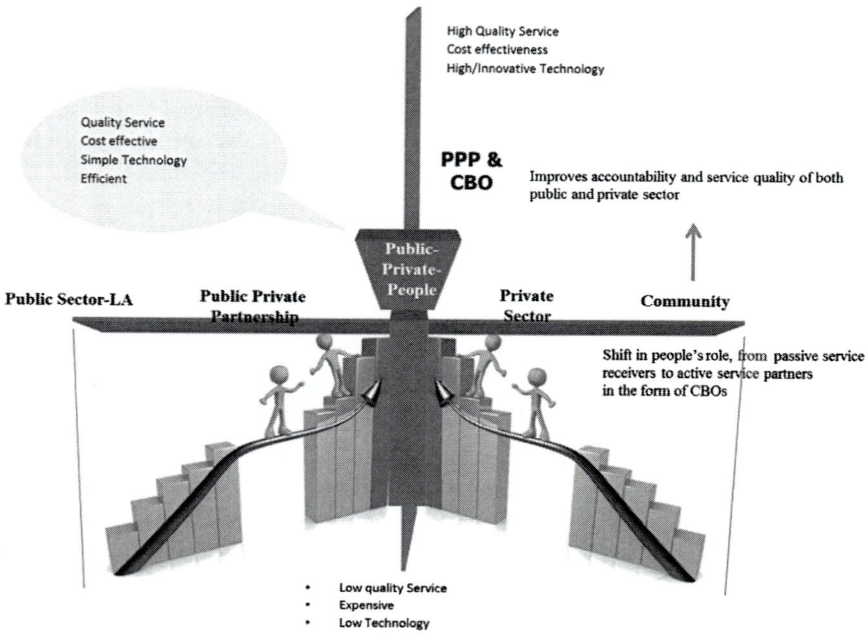

Figure 6.1. Actors and their roles in sustainable solid waste management system

sector or NGOs to perform the waste management activities, local community remains the central.

Community engagement is normally sought when the municipalities introduce new waste management policies and programs at the neighborhood. A typical example of the PPP could be observed, when implementing household waste source separation rules (biodegradable and non-biodegradables). It requires direct and active participation of each individual from the community. The usual approach of community participation used in such case are the public education and awareness building through mass media campaigns, advertisements etc, and engaging NGOs/CBOs. Tie-up (sub-contracting) with the private waste collector company is a typical case of PPP.

Therefore, PPP is a balancing act, where both the private sector and local communities should be given equal importance. The public participation has to be inclusive. Heavy privatization should not drive off the informal sector as well as the community. There are cases where solid waste management privatization such as in Delhi, India, has met strong public resistance as the informal sector was completely warded off from collection of recyclable hence affecting their livelihoods. It is therefore a farsighted decision of the LAs to invest time, resources and efforts to equally address public awareness and community participation in the solid waste management policy along with other policy components-legal and institutional framework, finance and cost recovery, and private sector participation.

6.4 PUBLIC PARTICIPATION POLICY IN GLOBAL CONTEXT

At a global level, the importance of community participation in environmental matters has been explicitly expressed in the Principle 10 of the Agenda 21 adopted at the Rio Conference. However, the actual translation of the principle has not been that extensively and equally done in many of the national/regional policy frameworks. Policy making structures and legal systems are formed within social and cultural contexts. The extent of public involvement varies considerably among countries as well as the degree by which the public participation instruments are used in different actors and groups. Organization for Economic Co-operation and Development (OECD) countries has long-standing traditions of extensive citizen engagement in public policy development and the design and delivery of services. With the EU Directives and international conventions a broad spectrum of instruments Europe stands strong in facilitating public participation in decision making processes. Developing Asian, African and South American countries are also giving importance to more transparent and public centric participation. In these countries, community based waste management activities are part of the livelihood opportunities.

Building in the backdrop of famous international treaty Principle 10 of the Rio Declaration, Europe has accrued public participation in environmental decision making through 1998's Aarhus Convention. The convention was agreed by the UN Economic Commission for Europe (UNECE) on the 25th of June 1998 in the Danish city of Århus at the Fourth Ministerial Conference in the "Environment for Europe" process. The Convention entered into force in October, 2001. The Public Participation Directive of the EU that introduced the requirements of the Aarhus Convention in the EU law guarantees the rights to the public to participate in the environmental decision-making procedures (J&E 2011). The convention has three main pillars: *Access to Information*, *Public Participation in Decision-making*, and *Access to Justice in environmental matters*. The list where public participation in decision-making is obligatory includes; activities in the fields of energy, metals, minerals and chemical industries, waste management, major wastewater treatment plants etc. The Waste Framework Directive, the overarching waste regulation directive in the EU also relatively pays attention to public participation. It guarantees the public participation through:

- Opportunity for public participation in the elaboration of the waste management plans and waste prevention programs, and have access to them once elaborated; and
- The plans and programs shall be placed on a publicly available website.

The Environmental Impact Assessment (EIA) also sets the public participation in environmental decision-making as a fundamental principle. The inclusion of the views of the affected and interested public helps to ensure the decision making process equitable and fair, and leads to more informed choice and better environmental outcomes. The inclusion of public participation issue into directives and policy framework works as legal instruments guarantees the public participation process and raise the capacity of the public.

Most of the public participation issues are associated with siting the waste treatment facilities (specially the hazardous waste facility, as this is one area that often meets public opposition. The US EPA developed the "RCRA Expanded Public Participation Rule" to empower communities to become more actively involved in local environmental decision making, especially in permitting hazardous waste facilities. Prior to this rule, RCRA provided opportunities for formal public involvement at two key points in the permitting process: 1) when the permitting agency announced its intent to grant or deny a permit and 2) when a facility requested a modification of a permit that had already been granted. With the "RCRA expanded Public Participation, public can access to information about facilities before permitting process, and eliminates confusion or delays in the permitting process (U.S. EPA 1996).

Through the Canadian Environmental Protection Act, 1999 the Government of Canada encourages and supports public participation in environmental decision-making, including solid waste management. One Part of the Act is specifically dedicated to address the opportunities for public participation, including the obligation to create the Environmental Registry. The Environment Registry, required under section 12 of CEPA 1999, provides timely and comprehensive access to public documents, and is a key instrument for fulfilling the Government's commitment to encourage public participation in decision-making. It also invites industries, individuals, interest groups and others to participate in the public consultations and decision-making processes that take place under the Act (CEPA 1999).

Admittedly, unlike Europe, Asia is still behind in terms of putting the public participation issue as a policy/legislative framework. Nevertheless, most of the developing Asian countries have voluntary public participation as a component of waste management plans/programs implementation, and as a source of livelihood rather than the policy requirement.

6.5 TYPICAL AREAS OF PUBLIC PARTICIPATION IN WASTE MANAGEMENT SYSTEMS

Following are some of the specific areas of public engagement in waste management system:

6.5.1 Waste Collection and Anti-Littering

Usually municipalities are not well equipped with finance and infrastructures to cover the primary collection of domestic refuse for entire locality. In such cases the municipalities usually depend on third parties to collect household waste and deposit the collected waste at a central location for municipal pickup. This kind of arrangement is done with a contractual hiring of private or local community waste collectors. On a voluntary basis, the strong community organizations may get involved in solid waste collection services through self-help approaches (neighborhood groups). If a private collector is hired, the community collects money from each household and pays the waste collection fee. Regular neighborhood clean up campaigns such as street sweeping and cleaning public spaces, parks may also be solely organized on a voluntary and rotational basis among the local residents. This does not involve monetary contribution but contribution in-kind, the labor.

Corporation of Chennai, India is a typical example of participatory municipal solid waste management, where all the three actors (LA, Community, Private Company, and local NGOs/CBOs) are involved. Since 2003, the Corporation has started a door-to-door collection of domestic waste, covering 70% of the city. While the 30% (Zone VI, VIII, X) collection is handled by a Singapore based private company ONYX. The overall daily waste collection reaches 95% (Corporation of Chennai 2012). The private company's role is of street sweeping, mechanized waste collection and treatment. Chennai is the first in India to establish the PPP. Besides partnership with one big private company, Corporation of Chennai also had engaged many local CBOs. Civic Exnora is one such CBO that works for the rehabilitation of waste scavengers. Its activities include community motivation and mobilization for source segregation through awareness activities and door-to-door waste collection. Similarly, '*Kumari Nagar*' another CBO run by local housewives, is involved in waste collection and small scale recycling activities. Such engagement of NGO/CBO and the private sector has empowered the local community to actively participate in the waste management services than just be a passive receiver of services.

6.5.2 Public Education and Awareness on Using New Waste Management System

This is the most essential aspects of public participation in waste management systems. When new rules, programs are launched by LAs the public is informed about the new system and even educated on using the new facilities and tools. For instance, if a local authority announces the new rule of household waste separation and distributes color coded waste bins for the same, the households should be notified of the new rules and provided with specific information. Some of the common information includes- the date effective from, which waste goes into which color bin, rewards and punishments involved, and collection schedule for waste types etc. Lack of proper information at an early stage may hamper the segregation process and thus the whole essence of the source separation scheme. Another typical example is the demonstration and training on the use of household compost bins, if the service providers are introducing household composting as waste reduction and recycling activity.

6.5.3 Developing an Integrated Solid Waste Management Plan

Effective planning and sustainable municipal solid waste management systems require an understanding of the needs and preferences of a wide range of stakeholders that have roles in solid waste generation, collection, re-use, transportation, and disposal, and in the service delivery, costs, and corresponding environmental and social impacts (Bernstein 2004). It is thus equally plausible to engage community in bottom-up approach to planning at an early stage of waste management project formulation. Such inclusive participation can bring public feedbacks concerning financial arrangements, including fee setting for waste collection, and selection of waste treatment options.

6.5.4 Siting and Technology Choice for Waste Treatment and Disposal

Choosing an appropriate technology and siting the facility (recycling unit, landfill site, waste to energy plants etc.) involves complex processes, and are often performed by the experts and technicians, leaving behind the public consultations. But, the choice of technology and siting of the facility is of concern to the public too. Experiences suggest that usually siting a waste treatment or recycling or disposal facilities suffers Not in My Back Yard (NIMBY) syndrome. Siting such facilities without prior consultation and consensus with the public can backfire the investment, time and efforts, as the public opposition may halt the functioning of the facility. Examples of public opposition to landfill siting are common in Asia.

Not only the siting of the treatment facility, but also the technology choice also requires the careful consideration and stakeholders' consensus. Normally technologies that are obsolete and can have environmental and public health consequences are usually opposed by the public as well as NGOs/CBOs. The classic example is the worldwide campaign against incineration run by the Greenpeace. Beside technical stuffs, even simple managerial decisions such as collection vehicle routing, placing the central pick up points, waste collection scheduling, etc., can also enter into public consultation for consensus decision making ensuring the program's sustainability. Late or non-inclusion of public consultation on technology design and siting of the facility may cause additional resources to un-do the design features and respective decisions made.

6.5.5 Effective Monitoring and Evaluation

Public participation can also significantly contribute towards effective and unbiased monitoring of both public and private sectors' waste management services. A typical example could be the monitoring of the waste collection and disposal services either by a LA or a private company- whether they are regular or not, or whether or not the landfill activities are performed with safety measures etc. Another example includes, the community acting as watchdogs for its own neighborhood's waste management activities. Catching the foul play by fellow residents in terms of littering or non-segregation of waste is a typical example.

6.6 KEY INGREDIENTS OF PUBLIC PARTICIPATION

Public participation offers many benefits but the participation process is often time-consuming and complex. Successful public participation requires certain essential ingredients that drives and sustains the participation.

6.6.1 Education and Awareness

Waste management involves the behavior of an individual to a great extent. Therefore, changing behavior and attitude is the first step towards effective public engagement. When people understand the connection between their behaviors and environmental harm, they are more likely to make their behavior environment friendly (Babcock 2009). Behavioral changes are generally attenuated with increased awareness though dissemination of information; awareness brings conscience and hence changes in the attitude. Awareness raising activities involve education, training, information sharing and mass media advertisements. Correcting the NIMBY element is also possible through education and awareness, public consultations, debates, and demonstrations of the technologies by the technical experts to the public representatives. Public education and awareness are usually seen and performed as a one-time attempt, usually at the beginning of a project. But, continued awareness and information is important to monitor and keep a track on the sustained public participation.

6.6.2 Rewards and Recognition of Public Efforts

Proper evaluation and recognition is important to keep the spirit of public efforts high. Rewards, certification for the activities that demonstrate notable source reduction and recycling behavior of individuals, households, organization, company or a community as a whole are some of the typical forms of felicitation and direct recognition for public's efforts. Similarly, at household level, the volume based fee also works as an indirect monetary reward for practicing waste reduction and segregation, and paying less fees.

6.6.3 Accessible Alternatives, Facilities, and Infrastructures

Public coordinates and participates in waste reduction and segregation activities if only there are suitable, convenient and affordable alternatives and infrastructural supports. Typically, municipalities often announce some rules such as ban of plastic carry bag or provision of 3-4 color coded separate bins for waste segregation. Both the efforts are praiseworthy attempts to improve the waste management system; however, the LAs often fail to provide suitable alternative solutions. Provision of alternative to plastic bags, such as paper, jute bags, etc., has to be easily available and affordable to the public. People may not willing to pay expensive price to limited alternative options, or people though want to follow the rule may be unable to do so if paper, jute bags are not easily available. In the second example of providing separate bins, not all households can accommodate 3-4 wheelie bins at their premises, hence alternative must be thought of. Such practical consideration must be well thought to assure effective public participation.

6.6.4 Strict Enforcement of the Policy and Program

Stricter and mandatory enforcement of rules with a system of punishment for the non-compliance is one of the measures that ensure higher participation rates (Kipperberg 2007). But, stricter enforcement and compliance isn't a one way business. For the sustained public participation, the LAs should also comply with its own rules and targets sincerely. For instance, when LA orders some rules for the public and provides supporting infrastructures, performs its duty effectively, public builds trust on LA's effort and continue their support. Whereas if a LA enforces waste separation rule at source but in reality LA has no investments made in the recycling of segregated waste, public will be disappointed and hence disinterested to support waste segregation efforts. Similarly, good performance of municipal staffs and private sector such as timely collection of source separated waste and proper maintenance and operation of disposal sites, recycling centers, etc., will encourage the public to continue their participation.

6.6.5 Institutional Arrangement in LAs for Public Participation

Some local authorities have a special department or units to facilitate community participation for waste management activities. Though having a special unit is not mandatory, however, it shows commitment towards flourishing the public participation. For example, Kathmandu Metropolitan City (KMC) Office's Environment Division has a Community Mobilization Unit (CMU) under the Solid Waste Management Department. The main role of the CMU is to educate local communities and involve them in waste management activities of KMC. CMU works with local community groups, youth volunteers, school children, and other relevant stakeholders with awareness raising and capacity building programs (SWMRMC 2004). Though the success of each community activity requires independent evaluation, it is worth mentioning the following community outreach programs under CMU:

Children and Environment Program: CMU understands children are the important part of the community. When educated at a young age, values and lessons learnt are ingrained as lifelong habits. Children can also expand the outreach of the message to their friends and families. CMU has created 50 Nature clubs in local schools, whose main function is to encourage waste reduction and recycling practices in schools. Typical activities include waste separation, school level composting, and revenue generation for the club's activity by selling the recyclables thus collected. While educating children about better waste management programs, such clubs also build capacity of club members' leadership skills, report writing and other event management skills. Such clubs in association with CMU organize neighborhood cleaning programs, various competitions on environmental occasions, mass rally and awareness campaigns too.

City Volunteer (CV): City volunteers are a group of young people, normally the college students who are enthusiastic to work for clean and green environment. Started in 2002 with 25 volunteers, about 100 youth have been trained to become "City Volunteers". These CVs work as a link between the municipality and the community to promote household composting and recycling, and community involvement in keeping their neighborhood clean. These CVs work with students of Nature club to strategize clubs' activity. As a motivational factor to CVs, CMU provides some vocational training, sends CVs for participation in environmental workshops and conferences, and also involve in surveys and research activities of the municipality.

Community Participation and Training: CMU also provides the community groups with technical trainings, and at times with the financial assistance of "seed" money to initiate the recycling activity. It has supported Ward Environment Committees in several wards (administrative units).

Demonstration of Environmental Technologies: CMU is promoting technologies such as compost bins and vermicomposting kits. Designed and prepared by the municipality, these household composting techniques are promoted to manage kitchen waste. Besides, CMU also provides demonstration and training on the use of these composting kits. At a community scale, the municipality also assisted in establishing several community compost plants.

Community Recycling Centers: Community Recycling Center is equivalent to any other junk shops, but established and managed by municipalities. Such recycling centers in various administrative wards encourage people for waste segregation as it buys recyclable at an attractive price. Household composts produced using the compost kits are bought at these centers.

Mass Education: CMU utilizes radio and television media, message boards and regular exhibitions to reach out to the general public. KMC also provides annual Kathmandu Environment Award to individuals and organizations that are actively involved in environmental activities.

6.7 SELECTED EXAMPLES OF PUBLIC PARTICIPATION IN WASTE REDUCTION AND RECYCLING FROM ASIA

Waste reduction and recycling has an important role in overall waste management strategy. Ground realities illustrate that the success of waste reduction and recycling activities are dependent on the degree of public participation; otherwise even the best laid plans are projects cannot meet the practical success. Here are some of those examples from selected Asian countries, where public participation had a greater role to play in achieving waste reduction and recycling targets.

6.7.1 Examples of Public Participation in Waste Reduction

6.7.1.1 Public Participation in Waste Reduction and Recycling in Hong Kong

Policy Framework for the Management of Municipal Solid Waste (2005–2014) is the principal waste management policy in Hong Kong. Under this framework arrangement, the Government has been promoting waste reduction and recycling through the territory-wide Program on Source Separation of Waste (SSW) in Domestic and Commercial/industrial buildings since 2005. The SSW program was later given a legislative status in 2008. The SSW Program aims to enhance recycling by encouraging building owners and managers to set up waste separation facilities at locations as close as possible to the sources of waste generation. Residents in rural villages are encouraged to use recycling bins placed in communal refuse collection points. Funding support to place recycling bins in buildings is provided through Environment and Conservation Fund (ECF). Following the SSW, about half of Hong Kong's Municipal Solid Waste (MSW) is recovered for re-use or recycling. SSW began as a trial in 13 housing estates in 2004, at the end of June 2012, 1875 housing estates/buildings had signed up to join the SSW Program (HKEPD 2012).

The success of the SSW is very much attributed to the much concerted participation from the community. Without a strong local community network, the SSW implementation will not be satisfactory. To attract public participation toward SSW, the Government is supporting the SSW Program with many publicity campaigns at various strata of the community (Wong et al. 2011). Mass media campaigns from TV and radio announcements, advertisements in newspapers and on public transport, talks, exhibitions and road shows, as well as specific promotional programs are targeted at individual housing estates, businesses and schools. In order to motivate participation for SSW, not only the housing estates, but also the business sectors are encouraged to practice waste reduction and recycling.

As a motivation factor, the Environmental Protection Department (EPD), the Hong Kong Association of Property Management Companies and the Hong Kong Productivity Council jointly started Hong Kong Awards for Environmental Excellence since 2011. The forum provides certificates to the award winning housing estates and to other participating commercial and industrial buildings. The monetary value of the recyclable selling also motivated the housing estates to enroll into this program. To broaden the participation in waste reduction and recycling, the Government continues to expand the SSW Program to some older districts as well as rural areas. For this expansion, the EPD calls for partnership with NGOs who have a strong tie with local resident groups through the subsidy of ECF.

With all these efforts in place to secure effective public participation, in an Action Plan announced in 2011, Hong Kong targets to raise MSW recovery (for reuse or recycling) rate to 55% by 2015. The government continues the public engagement process with the aim of encouraging greater public awareness of and participation in this process. In addition to SSW, the EPD has also introduced many other household level recycling programs, such as:

- *Fluorescent Lamp Recycling Program (FLRP)*, a voluntary Producer Responsibility Scheme that is funded by 15 members of the lighting trade and supported by the EPD, and three green groups (Friends of the Earth, Green Power and Tai Po Environmental Association). It is also supported by the Hong Kong Association of Property Management Companies, Housing Authority, Housing Society, and three companies (Sun Hung Kai Properties Ltd, Pricerite Stores Ltd and Japan Home Centre (HK) Ltd), which help to set up public collection points at shopping malls and retail outlets. The initial goal is to collect 400,000 spent lamps from households each year. The FLRP provides households with free collection and treatment for all used mercury-containing lamps, including compact fluorescent lamps, fluorescent tubes and high intensity discharge lamps. The objective is to reduce the environmental risk from improper disposal of these lamps and to recover resources.
- *Computer Recycling Program (CRP)* is another example of voluntary producer responsibility, launched in January 2008. Along with take back and recycling of the old computers, CRP also raises public awareness on the reduction, reuse of computer waste.

Similarly, there are other programs on recovery of waste electrical and electronic equipment, recycling of rechargeable battery, reduction on plastic shopping bags etc. Experiences from Hong Kong show that public participation is essential for the success of waste reduction and recovery program. Hong Kong Government realizes this and hence considers public participation as vital part of the SSW program. Public education and awareness is a large part of the SSW. The ECF sets aside HK$10 million to promote environmental initiatives under the MSW Policy Framework, including public education. Further another HK$1 billion was added into the ECF in 2008 for educational, research and technology demonstration projects, including waste reduction and recovery programs (HKEPD 2012).

6.7.1.2 Public Participation in Waste Reduction and Recycling in Taiwan

Taiwan's waste reduction, reuse, and recycling and green consumption dates back to 2003, when the Taiwan Environmental Protection Administration's (TEPA) formulated the policy on municipal solid waste (MSW). The zero waste objectives set in the policy framework are based on the volumes of MSW collected in 2001. The goals of volume reduction are set at 25%, 40% and 75% by the end of 2007, 2011 and 2020. TEPA's major initiatives to recycle MSW are presented in Figure 6.2.

"Per Bag Trash Collection Fee" is the key policy adopted by Taipei City, targeted for waste reduction and recycling. Under this rule, all residents are required to purchase special garbage bags designed by the Department of Environmental Protection (DEP). The disposal bags are available in eight different

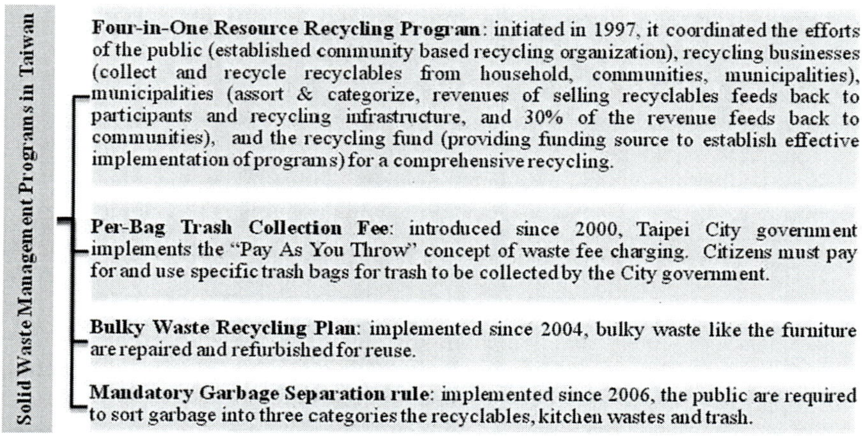

Figure 6.2. Waste reduction and recycling programs in Taiwan

size specifications: 5 L, 14 L, 25 L, 33 L, 45 L, 76 L, 92 L, and 120 L. These bags are provided with a logo to avoid counterfeit; and are available at 24-hour convenience stores and superstores. This is a volume based waste collection fee charging system that implements the principle of "Pay As You Throw." The collection fee is 0.45 NT$/L. This policy acquired public support and has successfully reduced the waste volume by 67% and boosted the recycling volume by 44%. As of 2009, daily garbage generation rate was 0.5 kg per capita, dropped from the highest record in 1998 (1.14 kg/c/d) (Houng 2010). Before this rule, the waste fee was based on the amount of tap water used per household, which had no correlation, and hence no motivation and chance to reduce the trash fee.

Stricter enforcement with respective reward and punishment systems was the major factor that garnered public support. Non-use of DEP bag is charged between 1200–4500 NT$. Similarly, the trash collection has to be followed as per the specific rules such as no bags should be filled beyond the dotted line or tied improperly, or there should not be any leaks and pricks from overstuffing and from sharps. Similarly the counterfeiting of the DEP trash bag, if found would be penalized between 30,000 to 100,000 NT$. Another notable effort for public participation process used here was the "watchdog" of the neighbors. If anyone found of illicit dumping of trash, photographs will be taken and posted over the internet and also fined with 1800-6000 NT$. Also as a motivational factor, anyone reporting illicit dump or counterfeiters is awarded with 20% of the fine collected.

Taiwan has made a great deal of progress in reducing waste and expanding recycling programs. Other Waste Minimization efforts introduced in Taiwan include:

- Restriction on the Use of Plastic Shopping Bags and Disposable Tableware (July 2002): Yearly usage of plastic disposable tableware was reduced by 2 billion pieces, a reduction of 86%.

- Restriction of Excessive Packaging (July 2006.07): Yearly usage of plastic shopping bags was reduced by 2 billion pieces, a reduction of 58%.
- Green Mark Program launched (1992): Consumers are encouraged to purchase items bearing the Green Mark logo, which denotes recyclability. As of May 2010, around 5,500 products carried the Green Mark (Ministry of Foreign Affairs 2012).

Example from Taipei is a behavioral evidence of active public participation through mandatory and stricter enforcement of regulations. Significant source reduction was achievable with the command and control instrument of “Pay As You Throw” based waste collection and disposal programs, as well as reward and punishment system as motivational factors.

6.7.1.3 Public Participation in Waste Reduction and Recycling in Thailand

School Garbage Bank (SGB) stands a long success of community participation in recycling activities in Thailand. It was flamed by Wongpanich Company, a private recycling company, in 1999, with collaboration from local authorities. The first SGB was launched at Panpi temple municipality school in Pitsanulok city municipality to support the students and others in the community to participate in sorting trash, recycle, and sell recyclable.

The garbage bank is where members bring the recyclable and receive an earning in exchange, depending on the volume and the type of the recovered material. Youth in each family (especially school student) can join the membership of the bank. Adults can register for members too. The garbage bank was operated by the setting up of the youth committee that comprised of 12 members with 2 adult advisors by the selection of the community committee. The management committee map is responsible for administrative management of the bank, public relation, activity promotion, accountancy, and collaboration of the buyers to sell. With the success of the garbage bank, in 2001 it was expanded to more than 30 provinces, with more than 500 such garbage banks. Each one was estimated to decrease the garbage for 3–5 tons per month and the overall could help decreasing the garbage in each year around 18,000–30,000 tons that could save the country’s budget for garbage management (Sasaki 2006).

Similar approach to school and community based waste collection and recycling activities have been started by the Thai Beverage Carton Group (TBCG) since 2004. TBGC is a collaborative initiative of major beverage carton manufacturers; Tetra Pak (Thailand) and SIG Combibloc Company. The objective of the group is to encourage and support consumers to segregate used food and beverage carton systemically. The Beverage carton group runs continuous campaigns to educate people to segregate used food and beverage carton waste. At present, around 1,200 schools, 500 households, and 300 organizations and factories are participating in the campaign The used beverage cartons (specially the milk cartons from School Milk Program) are collected separately in these participating entities, and are then sent to the recycling facility via TBGC.

School garbage bank is an example of smart strategy of helping the local authorities in waste management and recycling activities through targeting various sections of the community- starting with the school children and expanding to the families and the entire community. This program also creates a collaborative platform for NGO/CBOs and private sector participation.

6.7.1.4 Public Participation in Waste Reduction and Recycling in Sri Lanka

An Example from Negombo Municipal Council (MC) is an illustration of a monitoring of the municipal waste collection service of the active citizens. In 2003 the Negombo MC started the waste collection by municipal staffs with the bell system to notify the residents on the arrival of waste collection service. This system helped citizens to make the garbage ready to be handed to the municipal collection vehicle. People appreciated this initiative. However, after a few months, drivers forgot to play the music-bell, the speakers on the waste collection tractor also had problems, and hence people could not be aware/ready for the time the municipal collection vehicles arrive for waste collection in the neighborhood. As a result, Negombo MC failed to comply with the garbage collection schedule. But when many street committee members and citizens raised their complaints to the Mayor, the drivers resumed the bell system while routing for waste collection in the community (NSWMSC 2008).

Similarly, during the construction of the Moon Plain Sanitary Landfill in Nuwara Eliya, a monitoring committee was formed to get public acceptance of the landfill site. The steering committee members included the representative (2 Buddhist Priests) of the neighborhood, whose role was to ensure the landfill operation was in compliance with the method stated in the landfill operation manual, and provide transparency of the operation of the landfill (NSWMSC 2008). The monitoring committee held meetings periodically and visited the site for assessment and made the monitoring results available to the public.

Above examples illustrate that public participation can in fact perform the role of evaluator or watchdog to the services provided by LAs. Such effective monitoring usually improves the LA's capacity for waste management. In the second example, the choice of religious leaders as the committee member is worth mentioning strategy, because they are the highly regarded members of the community and their participation in the steering committee entrusted the local communities that the landfill construction is in good accordance with the manual. Therefore, who we choose as the participants can also positively influence the overall community participation and performance of the project.

6.7.2 Cases of Failed Waste Reduction and Recycling Programs Due to Lack of Continued Public Participation

Experiences from above cases show public participation are vital for the success of any waste reduction and recovery programs. It has also to be noticed that public participation is not a one-shot requirement. The successes of the programs rely on continuous and effective public participation, throughout the project. In an ideal case, it is also desirable for the public to continue the waste reduction and recycling efforts even after the withdrawal of LA or the project proponent's support. For the long term sustainability of the programs the community needs to be empowered to take over the programs. Empowerment would be required in transferring management skills and addressing the financial needs too. Turning waste management into community level green entrepreneurship is one of the potential areas of empowerment.

There are various examples of failure of waste recovery plans and programs due to lack of, and discontinued public participation. Although less in number, there are also the cases of absolute halt or closure of the waste management programs and treatment facilities.

Vilappilsala Solid Waste Processing Plant is one recent example that met a strong public resistance. Thiruvananthapuram city in Kerala, India, is estimated to generate about 265 tons of MSW every day of which 150 MT has been estimated to be bio-degradable in nature. Thiruvananthapuram Municipal Corporation (TMC) in order to implement an integrated approach to solid waste management practices in line with the Municipal Solid Waste (Management & Handling) Rules, 2000, established a Solid Waste Processing Facility at Vilappilsala, 13 km away from the city. The waste processing plant received around fifty garbage trucks daily for composting. Composting caused leachate issues, which the local residents blamed for the flawed functioning of the plant. People in the vicinity got affected by this and resisted the operation of the plant demanding an immediate and permanent shutdown. Local government was against the closure of the plant. The public protest and outcry went up to the court. The Kerala High Court ordered the State government to provide police protection for the Kerala Corporation for smooth functioning of the plant. Supreme Court too declined to interfere with the High Court's decision. Despite this court order, the public resistance took the shape of violent street protests. State Government had to withdraw the operation of the plant. After nearly eight months of closing down of the plant, the Council tried to apply corrective measures through leachate treatment but the locals did not allow shifting the equipments required to set up a low-cost treatment at the Vilappilsala waste processing plant in a recent protest of August 2012. Public demands the total shut down of the plant. With no hope of resumption of the waste processing plant at Vilappilsala, the closed factory is turning out to be a huge financial liability for the corporation putting the 7 billion Indian Rupee of investment to the processing plant at risk (Kumar 2012). Though unfortunate, such incidents clearly showcase the importance of public acceptance over the decisions of waste facility siting. And the only way to gather public acceptance is through public engagement at the planning phase.

Recycling industries failing to acquire enough waste feed materials for their recycling activities can also be linked to lack of public participation. Examples below illustrate how lack of segregated waste causes the recycling unit to manually separate the waste, which increase the cost but decreases efficiency and sometimes a monetary loss because the recycling unit cannot run at its full design capacity. Poorly separated waste also compromises the quality of the recycled products, hence affecting the recycling business.

Aerobic Composting, Nonthaburi, Thailand can utilize about 20 tonnes of organic waste a day. With the ineffective waste segregation at point source; the composting process needs high energy for aeration and takes about 90 days to produce mature compost; and the compost produced is of low nutrient content and value. During implementation of the project, various combinations of wastes were tried out to get the optimum compost output. The municipality does not have the capacity to invest in waste segregation units. It is planning to add manure to increase the nutrient content of the compost so that it will have a better market value. Similarly another waste to energy and fertilizer project in the Rayong Municipality in Thailand treats organic components of MSW using anaerobic digestion technology producing energy-biogas for electricity and the organic fertilizer as a byproduct. The plant has a capacity of handling 25,500 tons of biowaste/year producing 5,800 tons of soil conditioner and 3,826 MWh surplus of electricity annually. But due to improper source segregation of plastic bags, the plant is not able to operate continuously and produce the desired output (AIT 2011).

A case of Fluorescent tube recycling by Philips Electronics (Thailand) Ltd., is one of the many cases that highlights the needs for continued public education and awareness. In 2006, the Pollution Control Department (PCD) of Thailand initiated a partnership project with Philips and Toshiba. These companies started their own recycling facilities, but LAs helped in collecting waste fluorescent lamps. During 2007–08, the PCD did an effective campaign for collecting back the fluorescent tubes from households via municipalities for its new household hazardous waste management scheme. This initially helped the recycling facility to acquire enough feed material, but later on the PCD did not keep up with the awareness and information campaigns on reminding the households to segregate the waste fluorescent lamps. It was one of the reasons quoted for the fluorescent lamp recycling facility currently running below its design capacity (AIT 2011).

These examples highlight the need for the municipality and the private investor to ensure continuous and effective segregation at source in order to ensure quality waste feed suitable for the processing and continuous operation of the plant.

6.7.3 Examples of Public Participation in Waste Recycling Program

Successful waste reduction and segregation is the basis of a recycling program. Effective public participation in waste reduction activities thus indirectly supports waste recycling. But a direct public interaction in waste recycling comes through the integration of informal recyclers, who are generally the scavengers and small scale scattered recycling enterprises. Recycling in developing countries cannot be decoupled from such informal recyclers. Hence in order to increase the recycling rate and efficiency, LAs need to acknowledge the participation of these informal recyclers and empower and strengthen them. Some of the activities that LAs can take to strengthen the informal recycler's networks are discussed below.

6.7.3.1 Formalization of Informal Waste Scavengers

Formalizing the informal waste scavengers into a group and integrating into the formal systems of waste management is a way forward to 'gaining' and 'greening' the waste recycling efforts in developing countries. The scattered waste scavengers are usually institutionalized into cooperatives, association, forum or even micro-enterprises (with facilitation from NGO/CBOs and sometimes the government authorities). Formalization of the informal sector provides a legal identity of the informal sector making it easy for the LAs to coordinate with them as legal stakeholders. NGOs/CBOs can act as an intermediary to flourish an environment of trust, and investment between local governments and the (organized) informal sector. The LA can ensure the meaningful participation of the (organized) informal sector through contracting and sub-letting the waste management services, mainly the household waste collection. Such arrangements avoid conflicts of displacing the informal sector from their livelihood opportunities. Such kind of engagement can be termed as pro-poor public participation, as the participation positively influences the overall livelihood of the informal sector.

One such example of institutionalization of the informal sector is from Mumbai, India, where the waste pickers organized themselves and formed a NGO. Three organizations *Aakar, Stree Mukti Sanghatna* and Forum of Recyclers Communities and Environment (FORCE) are registered as NGO and are recognized by the Municipal Corporation of Greater Mumbai (MCGM). The MCGM issues identity cards to these waste pickers, and the NGOs have entered into contractual arrangements for waste collection with the MCGM. In addition, *Stree Mukti Sanghathan* and FORCE are also provided with collection vehicles for recyclable collection and a space for running scrap shops (Chikarmane et al. 2008).

6.7.3.2 Operation of Recycling Centers

Participation of informal recyclers can be boosted also through providing the marketplace to sell the recyclables collected by these scavengers. LA operated recycling centers can provide them a stable and fair price for their recyclables often gaining the independence from middleman brokerage. Providing job opportunities for the informal sector in such recycling centers are also a way of integration.

Sri Lankan municipal corporations have such recycling shops in operation in many urban councils. Termed as *Sampath Kendralaya or Piyasa* in local language, is the urban council managed and operated junk shops that buy recyclable. A point card system was introduced as an extra incentive mechanism for continued action of the citizens. A person who sells the non-biodegradable waste gets points for each type of the recyclables sold on a card system, and the amount gets accumulated, which he can withdraw as in a bulk amount. The Balangoda

Municipal Council established 5 small recyclable waste collection centers *Sampath Keendralaya*, in schools in the year 2009. The number has doubled to 10 in 2011. There are larger collection centers *Smapath Piyasa* at city center. Thus far 15 metric tons of recyclable waste from each of the 10 centers is collected and the council paid US$ 480 to students (Friend of the Human Race 2012). This program has significantly changed the behavior of students towards waste segregation, motivated by the prospect of additional income.

6.8 CONCLUSIONS

In solid waste management system, it is critical to consult and engage the public. Public participation in waste management can range from public education and information to planning a waste management policy and program, designing a project involving technology selection and siting of the facilities, and even independent monitoring and evaluation of the services provided by both LA and/or private sector.

Public participation offers a consensus decision, commitment, transparency and ownership. People are more likely to recycle when they observe others in their vicinity recycling. Social norms can be activated by individuals and communities in publicly demonstrating their commitment to sustainable waste practices. Degree of community participation generally decides the success of a waste management policy/program. LAs may provide a wide range of policies and recycling facilities, but failing to influence, inform, consult, collaborate and engage the public may produce no results. Degree of public participation and power sharing depends on the objectives of the participation, tools and approaches of participation can thus be chosen accordingly.

Public participation is promoted partly by legislations, and partly by the work of the industry and voluntary initiatives. The voluntary initiatives come from governmental &/or non-governmental organizations as well as from other independent actors, industry, and the research community. Sometime the 'compulsory' regulations such as recycling or pay as you throw rules are required to shape people's behavior and legitimize public engagement, while the 'voluntary' commitments from the public can be garnered through many awareness raising campaigns, communicating the personal and community benefits of participating in waste management activities, and motivational incentives such as rewards and recognition.

The proper inclusion of public participation issues into policy layout is essential to make the public participation an integral and mandatory component than an optional choice. All in all, waste reduction and segregation is inevitably linked to people's behavior or attitudinal corrections of reckless consumption of goods. Hence, waste management has a social dimension along with technology needs, and the inclusive public engagement is the only way to engender this collective social responsibility.

References

AIT (Asian Institute of Technology). (2011). "Policy study on 3R technology transfer." Institute of Global Environmental Strategies (IGES), Asia Resource Circulation Policy Research, Pathumthani, Thailand.

Arnstein, S. R. (1969). "A ladder of citizen participation." *J. Am Plann. Assoc.*, **35**(4), 216–224.

Babcock, H. M. (2009). "Assuming personal responsibility for improving the environment: Moving toward a new environmental norm." *Harv. Envtl. L. Rev.*, **33**(1), 117–175.

Bernstein, J. (2004). "Toolkit- social assessment and public participation in municipal solid waste management." ECSSD Urban Environment Thematic Group, World Bank, Washington, DC.

CEPA (Canadian Environmental Protection Act). (1999). "Canadian environmental protection act, 1999. Government of Canada." ⟨http://laws-lois.justice.gc.ca/eng/acts/C-15.31⟩ (Jan. 9, 2014).

Chikarmane, P., Narayan, L., Chaturvedi, B. (2008). "Recycling livelihoods—Integration of the informal recycling sector in solid waste management in India." SNDT Women's Univ., Chintan Environmental Research and Action Group, India.

Corporation of Chennai. (2012). "Solid waste management." ⟨http://www.chennaicorporation.gov.in/departments/solid-wastemanagement/index.htm⟩ (Sep. 1, 2012).

Friend of the Human Race. (2012). "3R society. News/Ezines." ⟨http://www.friend-of-the-human-race.org/3r-society⟩ (Sep. 5, 2012).

HKEPD. (2012). "Hong Kong waste reduction, environmental protection department, the government of the Hong Kong special administrative region." ⟨https://www.wastereduction.gov.hk/en/index.htm⟩ (Aug. 29, 2012).

Houng, H. J. (2010). "Chinese Taipei report, OECD global forum on environment: sustainable materials management." ⟨http://www.oecd.org/environment/waste/46326190.pdf⟩ (Aug. 29, 2012).

IAP 2. (2004). "Public participation spectrum." International Association for Public Participation Australasia, Louisville, CO.

IDEA (International Institute for Democracy and Electoral Assistance). (2003). "Democracy at the local level- A guide for the south Caucasus. Part 2: Participatory democracy." Stockholm, Sweden.

J&E (Justice and Environment). (2011). "Public participation in waste management related decision-making." Dvorakova, Czech Republic.

Kipperberg, G. (2007). "A comparison of household recycling behaviors in Norway and the United States." *Environ. Resour. Econ.*, **36**(2), 215–235.

Kumar, A. J. (2012). "Corporation's 70-crore investment at Vilappilsala goes waste. Newsprint Times of India." ⟨http://articles.timesofindia.indiatimes.com/2012-08-08/thiruvananthapuram/33099877_1_vilappilsala-treatment-plant-leachate⟩ (Sep. 10, 2012).

Ministry of Foreign Affairs. (2012). "Recycling and waste management." ⟨http://www.taiwan.gov.tw/ct.asp?xItem=45011&ctNode=1927&mp=1001⟩ (May 23, 2011).

NSWMSC (National Solid Waste Management Support Centre) (2008). "Encouraging public participation in solid waste management—Promotion manual for public private partnership in solid waste management (Version 1, 2008)." Sri Lanka.

Oakley, P. (1989). *Community involvement in health development*. WHO, Geneva.

Rifkin, S. B., and Kangere, M. (2002). "What is participation?" CBR A Participatory Strategy in Africa, Centre for International Child Health, Univ. College London, London.

Sasaki, S. (2006). "Better practice of waste management in Thailand." Asia-Pacific Forum for Environment and Development (APFED), Kanagawa, Japan. ⟨http://wasteinthai.cocolog-nifty.com/blog/files/UNEP.pdf⟩ (Sep. 6, 2012).

SWMRMC (Solid Waste Management and Resource Mobilization Center). (2004). "Solid waste management in Nepalese municipalities- Kathmandu Metropolitan City." Clean Energy Nepal and Environment and Public Health Organization, Kathmandu, Nepal.

U.S. EPA. (1996). "RCRA expanded public participation rule." Washington, DC.

Wong, L. T. K., Wong, W. K. W., Wong, W. Y., and Chui, F. Y. P. (2011). "Waste reduction and recycling in Hong Kong." *Proc., Int. Conf. on Solid Waste 2011- Moving Towards Sustainable Resource Management*, Hong Kong Baptist Univ., Hong Kong SAR, P.R. China, 57–59.

CHAPTER 7

Thermal Treatment Techniques: Incineration, Gasification, and Pyrolysis

Jianhua Yan[1]
Qunxing Huang[2]
Shengyong Lu[3]
Xiaodong Li[4]
Yong Chi[5]

7.1 INTRODUCTION

7.1.1 State of the Art

The disposal of solid waste has become increasingly intricate and costly, because of the decrease in space available for landfills and the growing concern about the living environment. Thermal treatment has the merit to reduce the volume of waste and to render it innocuous, by destruction of pathogens and toxic chemicals. The composition and structure of solid waste are destroyed thoroughly through the high temperature, with the principles of thermal stability.

The thermal technologies (shown in Figure 7.1) are essential components of integrated solid waste management systems, and are also already operating successfully as confirmed by several studies (Brunner et al. 2004; Porteous 2005; Psomopoulos et al. 2009). They are characterized by higher temperatures and conversion rates than most others, i.e. the biochemical and physicochemical processes, and they can dispose different types of solid waste, in particular the unsorted residual Municipal Solid Waste (MSW). Their main benefits are: (i) significant reduction of waste volume (typically, about 80 ~ 90%) and mass (typically, about 70 ~ 80%) (Consonni et al. 2005); (ii) short processing time, from few minutes (fluidized bed

[1]Zhejiang Univ., Hangzhou, P.R. China.
[2]Zhejiang Univ., Hangzhou, P.R. China.
[3]Zhejiang Univ., Hangzhou, P.R. China.
[4]Zhejiang Univ., Hangzhou, P.R. China.
[5]Zhejiang Univ., Hangzhou, P.R. China.

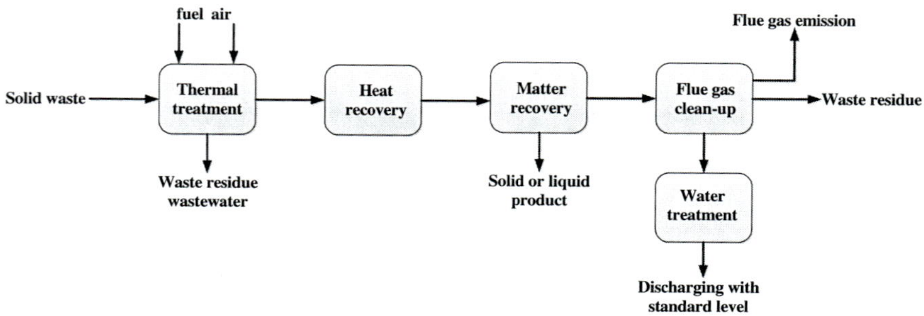

Figure 7.1. Basic flow chart of thermal treatment processes

units) to more than one hour (mechanical grate or stoker units) or even a large part of a day (some pyrolysis units). (McKay 2002); (iii) destruction of organic contaminants, such as halogenated hydrocarbons (dioxins and furans), polycyclic aromatic hydrocarbons (PAHs), phenols, cyanides (Buekens and Cen 2011); (iv) concentration and immobilization of inorganic contaminants, such as heavy metals; which may be usefully and safely utilized or else disposed of (ISWA 2008); (v) thorough disinfection, which is a good method for disposal of pathogenic waste and organics polluted waste; (vi) utilization of ferrous and non-ferrous metals from fly and bottom ash (Wiles 1996); (vii) avoided environmental burdens as evidenced through life cycle assessment studies (Arena et al. 2003; Azapagic et al. 2004). Waste-to-energy was shown to have less environmental impact than almost any other source of electricity (Arena 2012); (viii) utilization of the renewable energy of solid waste, such as electricity and process heat(Cheng et al. 2007); yet, the cost of the operation can be offset to some extent by energy sales. Thermal treatment technology includes three main thermal conversion processes: combustion, gasification and pyrolysis, which are compared schematically in Table 7.1. Other thermal treatments include roasting (disposal of solid waste under the melting point), dewatering and drying, thermal decomposition, and sintering.

Incineration is the process of controlled and complete combustion of solid waste in an oxidizing atmosphere. The solid waste is decomposed at high temperature and oxidized strongly during the process. It allows eventual energy recovery and destruction of toxic waste, for example, waste from hospitals. The temperature in the incinerators varies between 850°C and 1200°C. One of the most attractive advantages is to reduce the original volume of combustible solid waste by 80–90%. The premise of the incineration for solid waste should consider the characteristics and composition of the waste and heat value. Solid waste incinerators can be divided into mechanical grate furnace, fluidized bed incinerators and rotary kiln incinerator.

Pyrolysis is the chemical decomposition process of organic solid waste into solid (char), liquid (tar) and gaseous (syngas) fuel, under reducing conditions. During pyrolysis emissions are reduced by retaining volatiles, such as alkali salts and heavy metals (except mercury and cadmium), and – to some extent - sulphur and chlorine within the process residues. Especially, pyrolysis largely prevents

Table 7.1. Main characteristics of the chemical processes for thermal treatment of solid waste (Arena 2012; Arena and Mastellone 2009)

	Combustion	*Pyrolysis*	*Gasification*
Aim of the process	To maximize waste conversion to high temperature flue gases, mainly CO_2 and H_2O	To maximize thermal decomposition of solid waste to gases and condensed phases	To maximize waste conversion to high heating value fuel gases, mainly CO, H_2 and CH_4
Operating conditions (Reaction environment)	Oxidizing (oxidant amount larger than that required by stoichiometric combustion)	Total absence of any oxidant	Reducing (oxidant amount lower than that required by stoichiometric combustion)
Reactant gas	Air	None	Air, pure oxygen, oxygen-enriched air, steam
Temperature	850 °C ~ 1200 °C	500 °C ~ 800 °C	550 ~ 900 °C (in air gasification), 1000 ~ 1600 °C
Pressure	atmospheric	Slight over-pressure	atmospheric
Types of reactors	Grate furnace, fluidized bed incinerator and rotary kiln	batch, semi-batch and continuous reactors; fixed bed, fluidized bed and screw kiln reactors	fixed bed, fluidized-bed, rotary kiln bed gasifiers and other gasifiers
Process output (Produced gases)	CO_2, H_2O	CO, H_2, CH_4 and other hydrocarbons	CO, H_2, CO_2, H_2O, CH_4
Pollutants	SO_2, NO_x, HCl, PCDD/F, particulate	H_2S, HCl, NH_3, HCN, tar, particulate	H_2S, HCl, COS, NH_3, HCN, tar, alkali, particulate
Gas cleaning	Treated in air pollution control units to meet the emission limits and then sent to the stack	It is possible to clean the syngas to meet the standards of chemicals production processes or those of high efficiency energy conversion devices	It is possible to clean the syngas to meet the standards of chemicals production processes or those of high efficiency energy conversion devices

(Continued)

Table 7.1. Main characteristics of the chemical processes for thermal treatment of solid waste (Arena 2012; Arena and Mastellone 2009) (Continued)

	Combustion	*Pyrolysis*	*Gasification*
Ash	Bottom ash can be treated to recover ferrous (iron, steel) and non-ferrous metals (such as aluminium, copper and zinc) and inert materials (to be utilized as a sustainable building material). Air Pollution Control residues are generally treated and disposed as industrial waste	Often having not negligible carbon content. Treated and disposed as industrial special waste	As for combustion process. Bottom ash are often produced as vitreous slag that can be utilized as backfilling material for road construction

PCDD/F formation and reduces thermal NO_x formation due to lower temperatures and reducing conditions (Malkow 2004). According to the feeding and discharge procedures, pyrolysis reactors can be categorized into batch, semi-batch and continuous units; based on the heat transfer methods followed as well as on the flow and reaction patterns, the reactors can be classified into fixed bed, fluidized bed and screw kiln ones.

The objective of gasification has generally been to produce fuel- or synthesis gases, which would be stored and used when required, under partially oxidizing conditions. Gasification is a kind of waste-to-energy conversion scheme that often offers the most attractive solution to both waste disposal and energy problems. This technology has become one of the important thermal disposal methods for MSW. At present, the main types of MSW gasifiers include fixed bed, fluidized-bed, rotary kiln bed gasifiers, and other gasifiers.

7.1.2 Applicable Scope and Development Prospects

Solid waste can be subdivided into MSW, general industrial solid waste, and hazardous solid waste. For MSW, the main thermal treatment method is combustion. In most of the developed countries such as USA, Japan, Australia, and Britain, the MSW has been sorted before collected and combusted. However, in some developing countries, such as China, the MSW is not sorted or prepared before the combustion, so waste sorting is an important direction that China should be working on. For hazardous solid waste, rotary kiln incinerator is the most frequently applied reactor, which can dispose of most hazardous solid waste, like medical waste, chlorinated organic solvents, washing water for pesticide, biological waste, sewage sludge, and contaminated soil.

Within the waste management hierarchy, thermal disposal especially incineration with energy recovery is a desired and viable option often used in industrial nations. Incineration has the advantages of low residual waste volume, reduced pollutant emissions, and generating thermal and electric energy. Therefore, those countries where land resources are limited, such as parts of the EU and Japan, prefer incineration, whereas countries with vast territories, such as the US and Canada, continue to apply landfill disposal (Baggio et al. 2008). However, solid waste incinerators show low energy efficiencies (13 ~ 24%) mainly due to lower steam temperature and pressure required to prevent severe boiler corrosion, fouling and slagging (fireside problems). This results in energy efficiencies typically of about 15% (Malkow 2004). Moreover, pollutant emissions of SO_x, HCl, NO_x, and PAHs, polychlorinated dibenzo-p-dioxin/-furans (PCDD/Fs), polychlorinated biphenyls (PCBs), as well as heavy metals (Hg, Cd) also cause a major problem of perception. Whereas, pyrolysis and in particular gasification are able to solve these problems more effectively, provide energy more efficiently, and make thermal treatment more environmentally friendly and economical. Gasification technology of MSW is relatively mature in Japan, Europe and United States and other developed countries. The key performance figures of biomass burning with respect to cost, efficiency, and resulting cost ranges of power and heat are presented in Table 7.2. They are receiving significant

Table 7.2. Global overview of current and projected performance data for the main conversion routes of biomass to power and heat

Conversion option		*Typical capacity range*	*Net efficiency (low heat value basis)*	*Investment cost range (Euro/kW)*
Combustion	Heat	1 ~ 5 MW	70 ~ 90%	300 ~ 700/kWth
	CHP	0.1 ~ 1 MWe	60 ~ 90%	
	Co-combustion	5 ~ 20 MWe	30 ~ 40% (electrical)	2.5 ~ 1600
Gasification	Heat	Around 100's kWth	80 ~ 90% (overall)	Depending on capacity
	CHP gas engine	0.1 ~ 1 MWe	15 ~ 30%	3.000 ~ 1.000
	BIG/CC	30 ~ 100 MWe	40 ~ 50%	5.000 ~ 3.500
Pyrolysis	Bio-oil	100's kWth	60 ~ 70%	Not commercially available

Source: Faaij 2006; reproduced with permission from Springer

attention around the world and may be the main thermal treatment techniques in the future.

7.2 SOLID WASTE INCINERATION TECHNOLOGY

7.2.1 Solid Waste Combustion Characteristics

Compared with traditional waste disposal methods such as dumping or landfill, which are restricted by less land availability near metropolitan areas and much stricter sanitary requirements, incineration is more attractive and will therefore play an important role as a practical disposal method (Liang et al. 2008).

Solid waste, especially the MSW, is a low quality fuel (Liang et al. 2008). Its heat value is mainly affected by four components of the waste: moisture, volatile matter, fixed carbon and ash. The impact of moisture and ash contents on net calorific value and combustion properties is tremendous, especially the moisture content caused by weather change. Moisture content has an important impact on the combustion efficiency of fuel. High moisture content causes ignition difficulties. Based on experience, when the lower calorific value of waste exceeds 3.35 MJ/kg, auto-ignition proceeds without adding auxiliary fuel. The major concerns of waste combustion include combustion efficiency (unburned char in exit ash), emissions of harmful gases (CO, VOC, hydrocarbon, NOx, dioxin etc.), heavy metal partition between solid and gas phases and flame stability of high moisture content (Yang et al. 2004).

Jin (2002) studied the typical components of household waste separately, using a differential thermal balance. The various components of the differential thermogravimetric curve show that the weight loss of every sample obviously had different bands: biomass waste presented two bands, representing the combustion reaction of volatile matter and coke; for the plastic sample, there was only one band, related to the combustion of volatiles; PVC combustion was relatively complicated, presenting three bands: the first with mainly HCl evolution; the second showing evolution and combustion of benzene, toluene and other hydrocarbons; finally, the third band related to coke combustion.

In China, a survey shows that MSW contains high moisture (up to 50%) and ash (up to 35%); 23 main cities feature average calorific value of 4.125 MJ/kg (Jin 2002) far below the typical MSW heat value 10.25 MJ/kg from UK cities (Liang et al. 2008). It means that ignition and sustained combustion of garbage are more difficult. So specific techniques should be developed and much more challenging work should be done.

The materials to be burned experiences the following three stages: (1) drying, (2) burning, and (3) residue burnout. The process can also be divided into four successive sub-processes: evaporation of moisture from the solids, volatile release/char formation, burning of the hydrocarbon volatiles in the gaseous phase and combustion of charcoal residues (Yang et al. 2004). During the process, 4 factors should be controlled well in the actual operation to sustain the process efficiency:

- *Temperature*: Pollutants in the waste need appropriate combustion temperature and a certain residence time to be burnt out. In theory, for most organic solid waste, the temperature ranges from 800 to 1000 °C, and usually between 800 ~ 900°C.
- *Residence time* directly influences the effect and extent of waste incineration and the presence of TOC and CO of tail gas. The residence time of gas in combustion chamber tends to be 1 ~ 2 s.
- *Turbulence intensity:* To burn out the waste completely, we must make waste and combustion air fully contacted and combustion gas and combustion air fully mixed.
- *Excess air ratio:* The temperature of combustion and its effect on the combustion process is directly depending on the fuel/oxidizer ratio as well as on the O_2 concentration (Rogaume et al. 2009). Too high air ratio will increase the auxiliary fuel quantity and the blast volume. What is more treatment of more tail gas leads to higher operating cost. Conversely, low excess air ratio will result in incomplete combustion, soots and the harmful substances cannot be completely decomposed. Generally, O_2 content of the incinerator flue gas should be about 10% (Qing et al. 2005).

The tail gas from the incineration contains pollutants, which are to be removed to reduce the emissions. The products in the flue gas mainly include:

- *Products of incomplete combustion (PIC)* generally contain carbon monoxide, hydrocarbons, alkenes, aldehydes, ketones, alcohols, organic acids, soot, etc.

These materials not only pollute the environment, but also indicate the existence of problems for the combustion in the furnace that need to be resolved in time.

- *Dust* refers to the waste of inert metal salts, metal oxide or incomplete combustion material, etc. Especially, a kind of particles called $PM_{2.5}$, whose diameter is less than 2.5 μm, is considered to be one of the factors responsible for respiratory diseases.
- *Acidic gas* contains HCl, SO_2, SO_3, NOx, etc. Without effective treatment, emissions will cause air pollution. But for the waste combustion, it may be more complex, for example, S- and N-containing compounds had been proposed for their inhibition ability of PCDD/F formation during co-combustion or post combustion processes (Lu et al. 2007).
- *Heavy metal pollutants*, includes the elemental state, oxide and chloride of lead, mercury, cadmium, arsenic, chromium, etc. They can be enriched in fly ash, and have long resident time in the atmosphere. The microbial degradation is hard; and in the human body, heavy metal can be accumulated affecting the health for long time. Increasing understanding of pollutants produced during the garbage incineration, the heavy metal pollution around world has caused wide public concern and attention.
- *Dioxin (PCDDs/PCDFs):* Polychlorinated dibenzo-p-dioxins (PCDDs) and polychlorinated dibenzofurans (PCDFs) are persistent organic pollutants. These compounds are collectively termed 'dioxins'. Their chemical structure is shown in Figure 7.2. Highly chlorinated congeners exist in the solid state while the low chlorinated congeners are in the gas phase at room temperature. PCDD/Fs are generated as unwanted by-products from different processes such as municipal waste incineration (MWI), metal industries, bushfires and prescribed and uncontrolled burning (Altarawneh et al. 2009).

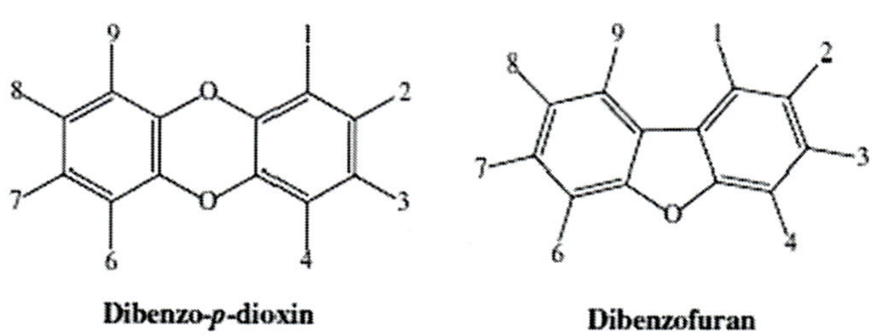

(a) (b)

Figure 7.2. General structures for (a) PCDDs and (b) PCDFs. Each numbered position can be substituted either by a hydrogen atom or by a chlorine atom, resulting in 75 different isomers in the PCDDs family and 135 in the PCDFs family

7.2.2 Incineration Techniques

Solid waste incineration started in the 1870s. The intermittent fixed- hearth incinerators were first established in UK and USA in 1874 and 1885, respectively. In the early 20th century, MSW incineration plants were built in Europe, North America and Japan. During this period, the mechanical grate incinerator was invented and developed, at first for coal. Since 1970s, solid waste incineration technologies witnessed rapid development and at present they are the main disposal methods for MSW in many developed countries such as Switzerland, Denmark and Japan (Zhang and Zhao 2000). The main solid waste incineration techniques in domestic and abroad can be divided into stoker combustion, fluidized bed combustion and rotating combustion, the corresponding incinerators are mechanical grate, fluidized bed and rotary kiln furnaces, respectively; and a comparison is presented in Table 7.3.

As the most mature and developed technology, stoker combustion has been widely applied in many countries. The mechanical grate is the most important factor for ensuring the support of MSW as well as stable combustion and there are many kinds of grates such as fixed grate, chain grate, roller grate and reciprocating grate, etc. Schematic diagrams of two typical grate furnaces are shown in Figure 7.3. The flame transfers from the ignited layer to the non-ignited layer and then the stoker combustion process forms. As presented in Figure 7.4, during combustion, solid waste pass through three zones: preheat zone, primary combustion zone and burnout zone (Pan 2007). Solid waste absorbs heat from furnace wall radiation resulting in convection of combustion gases and the heat generation at the interior of the solid waste layer. Under the effect of grate, the solid waste layer is strongly shaken and stirred, and as a consequence, it becomes loose and its permeability is also increased. This change is of great help for ignition and combustion of the solid waste.

The grate furnace has the following characteristics (Zhang 2004):

- Most MSW does not need to be classified and grinded, except for bulky waste.
- Solid wastes are not perfectly distributed in the furnace, only waste with high calorific value can achieve good combustion efficiency.
- Power requirement of the blowers is small; thus power consumption is low.
- Generation of fly ash and clinker ignition loss of slag are both low.
- As there is no coal mixing with solid waste, the dust content of the flue gas is low consequently reducing the load of dust collector and operation cost.
- Heat loss of the furnace body increases since the volume of the furnace is relatively high.
- Air-inlet channels are easy to be blocked by the lime-soil of the solid waste.

Fluidized bed incinerators consist of a refractory-lined combustion vessel partly filled with particles of quartz sand, which has high heat capacity. Combustion air is supplied through a distribution plate at the base of the vessel at a velocity sufficient to fluidize the bed (Saxena and Jotshi 1996). The structure of the air

Table 7.3. Comparison of the stoker, fluidized bed, and rotary kiln

Item	*Grate furnace*	*Fluidized bed*	*Rotary kiln*
The main areas where the technology is applied.	Europe, USA, Japan	Japan, USA	USA, Denmark, Switzerland
Disposal capacity	≥200 t/d	≤150 t/d	≥200 t/d
Furnace type	Chain grate, rolling grate and reciprocating grate	Bubbling and circulating fluidized bed	Slag type and non-slag type
Reaction rate	Medium	Medium or high	Low
Residence time of waste (h)	>1	>1	1–2
Residence time of flue gas (s)	>2	>3	Rotary kiln:1.2 Burnout chamber: 1.5
Temperature in furnace	Surface of waste layer: 800°C Flue gas: 800–1000 °C	800–900 °C	Rotary kiln: 600–800 °C Burn-out chamber: 1000–1200 °C
High temperature corrosion	High temperature corrosion is a common problem for all these technologies and there is no effective method to deal with it at present.		
Applicable fuel	Mainly for MSW and domestic waste	There is no restriction for this incinerator	Mainly for medical waste and hazardous waste
Combustion performance	Combustion of this technology is stable and slow. Burnout rate of this technology is high.	Combustion stability of this technology is not very well, but combustion velocity and burnout rate of this technology are both very high.	Combustion stability of this technology is not very well and combustion velocity is not very high, but the burnout rate is relatively high.

Disadvantages	(1) The cost of construction, operation and maintenance is high (2) The grate is easy to be damaged when it is used for combustion of waste with high calorific (3) It is not suitable for wastes with high moisture or big volume	(1) Circulating medium is necessary (2) The waste should be grinded until its diameter is smaller than 15cm before combustion (3) The flow rate of the flue gas is high, thus the hearth material is easy to be damaged (4) A large amount of fly ash is emitted	(1) The transmission device is complex (2) The dust concentration in the flue gas is high and refractory material in the inner wall of the furnace is easy to be damaged (3) The volume of the rotary kiln is too large that the site area is very huge

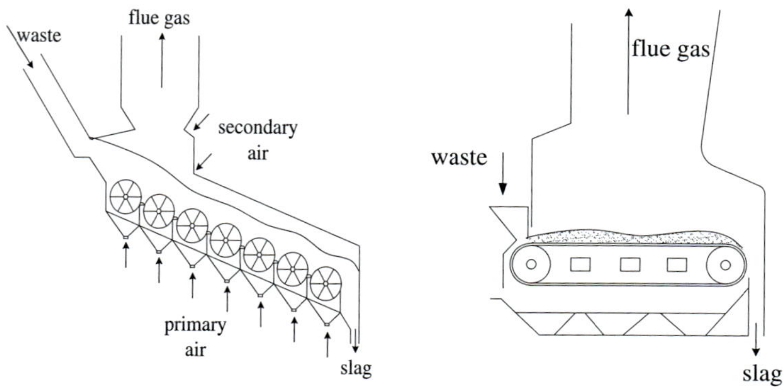

Figure 7.3. Schematic diagram of rolling grate and chain grate

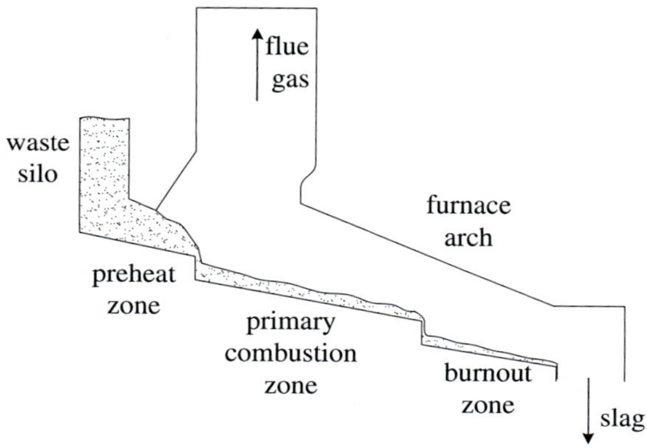

Figure 7.4. The process of solid waste combustion on grate

distribution plate and the blast cap are designed as inverted cone and L type structures, respectively. Temperature in the furnace is controlled between 800°C and 900°C. The working characteristics of fluidized bed incinerator can be described as follows: the bed material is preheated by fuel oil until the temperature in the furnace reaches 600°C, and then solid waste will be sent into the furnace for combustion. The primary air is introduced into the furnace through blast cap and air distribution plate in turn. The secondary air is introduced in from the upside of the fluidized bed (Du et al. 2003). Under the effect of strong mixing and heat transfer of bed material, solid waste can be ignited rapidly and the combustion in the furnace is homogeneous. Fluidized bed incinerators can be operated in two distinct modes, namely bubbling and circulating bed. The schematic diagram of a circulating fluidized bed unit can be seen in Figure 7.5. In the circulating bed design, air velocities are greater and the solids are blown overhead, separated in a cyclone and returned to the combustion chamber. These incinerators are primarily used for sludge or shredded solid waste (Anthony 1995; Saxena and Jotshi

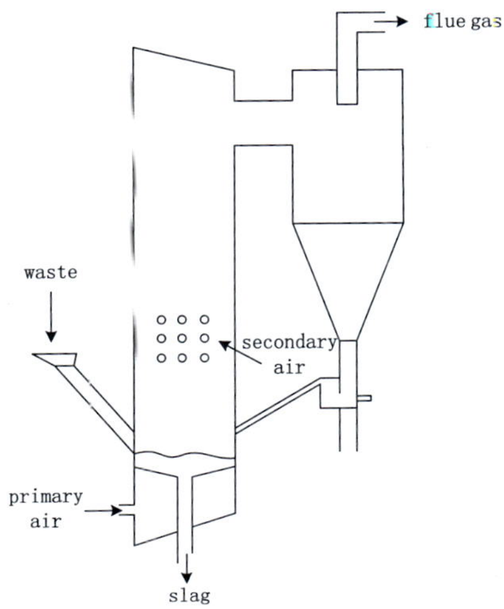

Figure 7.5. The schematic diagram of fluidized bed

1996). The fluidized bed incinerator has the following characteristics (Song et al. 2007; Ying 2010):

- Preheating, drying, ignition and combustion of waste can take place simultaneously, and the solid waste can combust steadily and continuously.
- The fuel adaptability of this incinerator is high: MSW, sewage sludge, residual oil and tar from refinery and agricultural waste etc. can all be treated by the fluidized bed. Furthermore, the burnout rate of bed material is also very high and the clinker ignition loss is usually lower than 1–2%.
- Excess air is low and air staged feeding technology is adopted for nitrogen-rich waste; as a consequence the emissions of NO_x, SO_2 are well controlled.
- As combustion intensity and heat transfer rates are high, the size of the furnace is relatively small leading to reduced investment.
- There is no rotary mechanical equipment in the fluidized bed, thus the construction cost is relatively low.
- Resistance in the furnace is high and as a result the power consumption is high.

In rotary kiln combustion, the main part is a rotary kiln, the inner wall of which is made up of refractory brick or water wall tube to protect the drum. The rotary kiln is usually under the condition of slant setting; its diameter and length are usually 4–6 m and 10–20 m, respectively. Owing to the process of rollover, the solid waste can have a sufficient contact with air and as a result the combustion is relatively complete. Characteristics of the rotary kiln are as follows (Cheng and Hu 2010; Zhang 2004):

- The rotary kiln is appropriate for the solid waste with wide range of moisture content and having substances like medical waste and hazardous waste that are difficult for combustion.

- There is no need for the solid waste pre-treatment before combustion.
- This incinerator runs flexibly and the waste is mixed thoroughly, thus the combustion is complete and the destruction rate can be 99.99%.
- Disposal capacity of this incinerator has wide accommodative amplitude and the residence time is adjustable because the rotation speed of the furnace is adjustable.
- The two ends of the rotary kiln are not well sealed; as a result, oxygen content in the furnace cannot be perfectly controlled.

7.2.3 Waste Dewatering and Drying Technology

With the continuous improvement of sewage treatment capacity, the number of sewage treatment plants is increasing. The disposal of a large majority of the residual sludge wastewater becomes a pressing problem. Since the water content of original sludge is generally about 95%~98% (Shen 2010), original sludge is inconvenient for disposal, transportation and recycling. However from 1980s, the successful use of sludge dewatering and drying technology allows them to quickly be promoted.

Dehydration includes natural and mechanical dewatering. Natural dehydration is moisture removal by infiltration, evaporation and others; and is greatly influenced by climatic conditions. Mechanical dewatering is the sludge dewatering process using mechanical equipments. After mechanical dewatering, the water content of sludge can be reduced to about 80%. But sludge retains high water content after dewatering, and therefore needs drying.

Sludge drying is conducted after mechanical dewatering, which can reduce the water content of sludge from 80% to <10%~50%. Now the common drying techniques include thermal, solar, microwave heating, bio-drying and so on. Among these, thermal drying technology is the most widely used and the most mature technology (Wang et al. 2010).

Thermal drying technology: The sludge thermal drying techniques can be divided into three categories: *direct thermal drying, indirect thermal drying and direct-indirect combined drying.* Direct thermal drying process is that heat transfers from heating medium to wet sludge through a heat exchanger; and the thermal medium and sludge are separated during the process. The thermal transfer efficiency and evaporation rate of this method are not as good as that of direct thermal drying. Common indirect drying equipment is disc thermal dryer, thin-film thermal dryer, etc. (Qiu et al. 2007). Direct-indirect joint-drying technology is the integration direct and indirect techniques. For example, high-speed film dryer designed by VOMM (Wiesman 1994), new fluidized bed dryer designed by Boyles (Boyles 1993) and belt dryer launched by Envirex (Smithmab 1994) belong to this type.

Solar drying technology: This technology uses solar energy as energy source and is combined with the traditional greenhouse drying technology. After solar drying, the moisture content of sludge can be reduced to about 20% and the sludge can be incinerated. It can save more than 80% of fuel costs (Ma and Chen 2011). The earliest commercial application of solar drying for sludge was reported in the sewage treatment plant ISTAnglagen-bau GmbH in southern Germany in 1994 (Luboschik 1999). In recent years, as sludge production rises and traditional sludge treatment methods cannot meet the environmental requirements, this technique is promoted in Europe, especially in France and Germany, such as the Solia process of Veolia and the Helantis process of Degrémont (Gu and Yang 2008); and the Thermosystem process using a robot jointly reported by University of Hohenheim in Germany and Thermo-System Industrie- und Trocknungstechnik GmbH (Bux et al. 2011; Starcevic et al. 2007). Now the research and application of this technology have been increasing in China.

Microwave drying technology: Microwave drying technology uses microwave to treat disposal. The water content of sludge after microwave drying is reduced from 80% to 60% or less. The power consumption per hour of the microwave drying is less than 400 kWh/t (Li 2007), saving a lot of time and energy. Now the microwave drying technology mainly includes microwave de-oiling of oily sludge, pre-treatment of sludge using microwave, microwave-assisted heating drying of original sludge and microwave-assisted mechanical dewatering of sludge.

Biological drying technology: Biological drying technology primarily uses microorganisms to decompose dewatered sludge under higher temperature that accelerates the loss of water at the same time resulting in sludge with low moisture content (Chen 2011). Biological drying was first proposed by Jewell et al. (1984) at Cornell University in 1994, and is now popular in European countries. Eco-Deco in Italy, Herhof in Greece, and Kehlsen Company in Germany develop this technology and build biological drying projects in many countries (Wu et al. 2010).

Although sludge dewatering and drying technology is improving continuously, the main bottleneck restricting its wide application is the high energy consumption. Nature and composition of the sludge are the critical parameters affecting the dewatering and drying; and deeper understanding on the mechanism of dewatering will enable us to develop new potential methods for drying.

7.2.4 Case Study

Sludge drying and incineration project of Zhejiang Province is presented here as an example to analyse the performance of its sludge drying device. With sewage sludge incineration as the focus, researches on key techniques; and development of development and demonstration projects were performed by Institute of Thermal Power Engineering of Zhejiang University. This project began on 2007 and built a 220 m^2 hollow paddle sludge drying machine, the largest one in China. This dryer can treat 100 t/per day sludge of 85% water content.

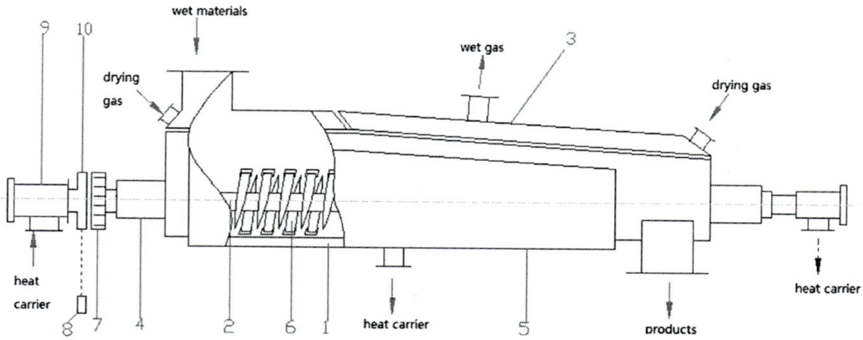

Figure 7.6. The hollow paddle sludge drying machine. 1-W-type tank; 2-hollow hot shaft; 3-upper cover; 4-bearing and stuffing box; 5-jacket; 6-wedge-shaped blade; 7-gear; 8-deceleration device; 9-rotary joint; 10-sprocket wheel

The hollow paddle sludge dryer, an indirect heat drying equipment, is mainly composed of a cylinder with a jacket, a hollow paddle shaft, a drive device and a heat source, shown in Figure 7.6. The body of dryer includes a cylinder, rotary axis, and a pair of shaft blades which set into a wedge and have auxiliary blades at the edge to enhance stirring intensity and to eliminate the heat transfer dead ends (Ma et al. 2006). The specific drying process is that the wet sludge is continuously dosed into the charging port of the dryer by a screw feeder; and is stirred by the rotation of blades to renew the heated interface and fully contact with the heated body and blades for evaporating water. Meanwhile, dry sludge is conveyed to the exit as the blades rotate, which continue stirring during the transportation so that moisture continues to evaporate. Finally, the uniform dry sludge is discharged from the exit. The water vapour evaporated enters condensing water film device, where vapour is condensed into liquid and then discharged into the sewer network. The hollow paddle sludge dryer is shown in Figure 7.6.

The characteristics of using the hollow paddle dryer are that it can use the existing steam generating equipment and use residual heat steam (0.4–0.6 MPa) from a nearby thermal power plant to dry. The drying system is shown in Figure 7.7. Its drying temperature is 180°C. The condensate steam can be reused and the wedge-shaped blades have the ability to self-purify and transfer sludge. After drying, moisture content of sludge is reduced from 85% to 40%. Dry sludge returns to the thermal power plant to be burned in circulating fluidized bed incinerator, thus sludge drying and subsequent sludge processing is solved. Heat source for this hollow paddle sludge drying machine is residual heat steam; and this method realizes the energy utilization, sludge reduction and cost savings. Although it is a traditional dewatering and drying technology, this technology has some advanced features and with continuous efforts on improvement, it is believed that more and better techniques will be developed.

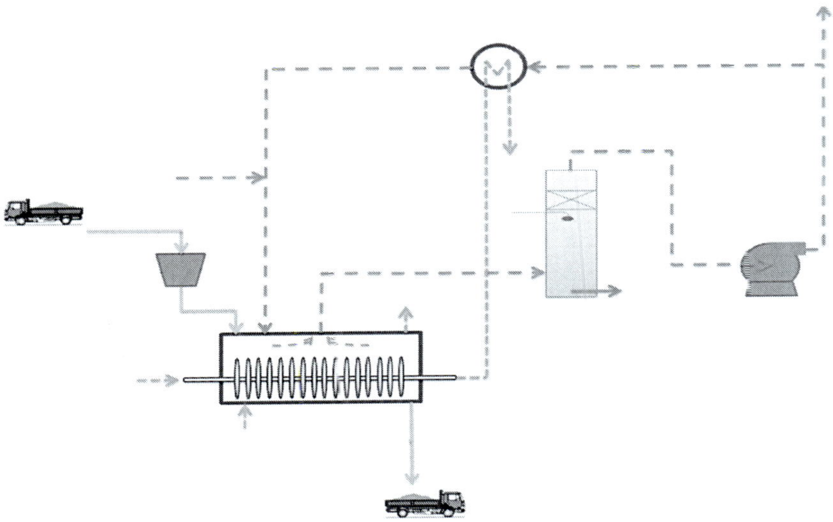

Figure 7.7. The sludge drying system

7.3 PYROLYSIS AND GASIFICATION TECHNOLOGIES

7.3.1 Pyrolysis Technology and Equipment

The concept of pyrolysis of MSW gained popularity in the 1960s as it was assumed that since MSW is typically about 60% organic matter, it would be well suited to pyrolytic treatment. By the mid-1970s, studies in Europe and the United States on pyrolysis of MSW were completed; some of them involved the construction and operation of demonstration plants. By late 1970s, however, both technical and economic difficulties surrounding the pyrolysis of MSW arose, which resulted in reduced interest and expectations from the technology. Since that time, the pyrolysis of MSW has been a subject of intensive investigation but the application continues to face technical limitations.

Pyrolysis is the thermal decomposition of feedstock at a range of temperatures in the absence of oxygen. The end product is a mixture of solids (char), liquids (oxygenated oils), and syngas (consisting of CO_2, CO, CH_4, H_2). The pyrolytic oils and syngas can be used directly as boiler fuel or refined for higher quality uses such as engine fuels, chemicals, adhesives, and other products. The solid residue is a combination of non-combustible inorganic materials and carbon.

Pyrolysis requires thermal energy that is usually applied indirectly by thermal conduction through the walls. Heat transfer from the reactor walls occurs by filling the reactor with inert gas which also provides a transport medium for the removal of gaseous products. The composition of the pyrolytic product can be modified by the temperature, speed of process, and rate of heat transfer. Liquid products (pyrolytic oils) are produced by lower temperatures while syngas is produced by higher temperatures. The syngas produced can be combusted in a separate reaction chamber to produce thermal energy which can then be used to produce steam for electricity production.

Pyrolysis generally takes place at lower temperatures than used for gasification which results in less volatilization of carbon and certain other pollutants, such as heavy metals and dioxin precursors. The relatively low temperatures allow for better metal recovery before the residual pyrolysis products enter the high temperature chamber, where they are vitrified. Pyrolysis provides a number of advantages:

1. By using less oxygen, fewer air emissions may be produced.
2. It takes trash and converts it into valuable new products, oil, carbon and gas.
3. It allows for the recovery of metals and glass either before or after the process.
4. It is suitable for a heterogeneous waste stream.
5. Issues identified in relation to the pyrolysis process include:
   - Low energy outputs
   - The requirement for a properly sealed reaction chamber for safe operation. The pyrolysis process is highly sensitive to the presence of air. Accidental incursions of air can result in process upsets and increase the risk of explosive reactions.
   - The requirement for pre-treatment of the MSW.

Now pyrolysis technology have been used for treating a number of waste streams such as scrap tires, mixed plastic waste and certain forms of biomass such as wood waste. The reactor type has significant influence on the heat transfer rate, mixing of waste with pyrolysis products, residence time and the reflux level of the primary products. Reactors can be classified into batch, semi-batch and continuous or classified based on types of reactor bed. In the batch reactor, the materials are fed into the reactor in batches either at the start of the process or after all of the fed materials are processed. In the continuous reactor, the feed materials are input from one part and the products are led out from the other part of the reactor. A semi-batch reactor removes the pyrolysis products continuously once they are generated but the feed materials are added initially before the pyrolysis process starts. Some semi-batch process uses inert carrier gas to help remove the pyrolysis products.

Based on the heat transfer methods, flow patterns of the feedstock and products, the pyrolysis reactors can be classified into fixed bed reactor, fluidized bed reactor and screw kiln reactor. In the fixed bed reactor, the pyrolysis occurs on a stationary bed, which is easy to design and operate. However, the irregular size and shape of the feedstock plastics may cause feeding problems in continuous process and the low thermal conductivity of the plastics results in large temperature gradient in batch process devices. In some systems, the fixed bed reactors are only used as the secondary pyrolysis reactor because the products from the primary pyrolysis are mainly in liquid and gaseous phase which can be easily fed into the fixed bed.

The fluidized bed reactor has been used in most commercial plants, where gaseous products or inert gas flow through an expanded bed of feedstock and other bed materials, forming bubbles or eddies. The advantages of fluidized bed reactor are the homogeneity of both temperature and composition. Heat and mass transfer rates are much higher than the fixed bed thus the low thermal conductivity is no longer a problem in fluidized bed reactors. The dimensions and the bed material are the key parameters affecting the pyrolysis and products. Loss of bed materials and separation of bed material from the gases are other issues needing consideration.

In recent years, a new reaction system named screw kiln reactor has been widely applied for pyrolysis processing. In this type of reactor, an extruder screw the feedstock from a feeder in an oxygen free environment, while the solid residues and pyrolysis products are separated and collected from the other end of the extruder. The extruder is heated by external heat sources. The high viscosity of waste is not a problem for the flow in screw kiln reactor because the flow is driven by the external motor. Melted waste or even solid waste particles can be fed into this reactor. The small diameter of the extruder and good mixing of the materials make the radial temperature gradient negligible. The process is relatively stable and does not use bed material as in the fluidized bed reactor. The feeding rate can be controlled by adjusting the rotation speed of the extruder, which also determines the residence time of waste.

Waste plastic pyrolysis. For a variety of waste plastic mixture, current research focuses on the mutual effect between various substances in the pyrolysis process. Chowlu et al. (2009) observed synergistic effect for the mixed samples of PP composition >40 wt.%; and PP: LDPE at 65:35 wt.% is the most preferable one due to its low activation energy and formation of significantly high amount of hydrocarbons of ≤C10. Ciliz et al. (2004) found that the C3–C4 formation increased with the increase of PP in the mixture of PE: PP. The addition of PP waste plastic to the waste mixtures caused an increase in the alkene/alkane ratio for both C2 and C3. Bhaskar et al. (2004) found that the use of calcium carbonate carbon composite (Ca-C) completely removed the chlorine and bromine content from the liquid products during PP/PE/PS/PVC/HIPS-Br pyrolysis, however in the presence of PET, the catalytic experiment (Ca-C, 8 g) yielded liquid products containing 310 ppm of Br and 20 ppm of Cl. In addition, the Ca-C increased the yield of liquid products about 3–6 wt. %, as well as enhanced the gaseous product evolution and decreased the yield of residue.

Waste paper &wood pyrolysis. As mentioned previously, temperature has a great effect on the pyrolysis. Wu et al.(Wu et al. 2002) found that the yields of non-hydrocarbon and hydrocarbon gases were about 10.46 and 0.49% at 623 K, 33.68 and 0.89% at 700 K, 64.52 and 1.05% at 788 K, and 79.10 and 1.63% at 938 K by thermo gravimetric analysis, respectively. For the waste paper processing in a fluidized bed pyrolysis, the best temperature was observed to be 600 ~ 700°C, at which, the final products contained 55% to 65% pyrolysis gas, 25% to 30% pyrolysis oil and 15% to 20% coke (Ni et al. 2006).

About 6% to 12% of the total mass of the MSW is wood. During the pyrolysis of milled beech wood, Demirbas (2005) found that from 60 s to 90 s period of pyrolysis process, the average heating rate is higher than 6 K/s for the samples with +0.060 mm and +0.150 mm (diameter of particles). The average increase of temperature is slow from 170 s to 350 s period. The increase of temperature is highly regular in 450–600 s periods. The increases of liquid yields from pyrolysis are considerably sharp for all the samples in 90–170 s periods. The highest increase of liquid yield and the lowest completion time (445 s) was obtained from the beech wood sample with +0.150 mm particle size.

Park et al. (2010) considered two distinct sequential thermal regimes during wood pyrolysis. First, an endothermic reaction that causes a center temperature plateau. During this period, loss of a large amount of solid mass occurs. Second, a steep temperature rise with the centre temperature peak exceeding the surface temperature. This occurs after the centre temperature plateau. A small or negligible amount of solid mass loss occurs during this exothermic reaction period. Therefore, the endothermic period corresponds to primary reactions of active volatiles generation, especially cellulose decomposition; and the exothermic period corresponds to the conversion of the intermediate solid to final char (Park et al. 2010). Luo et al. (2005) developed a model of wood pyrolysis in a fluidized bed reactor simulating the effect of main operation parameters on product distribution. The model shows that reaction temperature plays a major important role; and the particles less than 500 μm could achieve a high heating-up rate to meet pyrolysis demand.

Multicomponent MSW pyrolysis. Luo et al. (2010) observed increasing gas yield and concurrent decrease in char and tar with increasing temperature during pyrolysis of Refuse Derived Fuel (RDF) in a lab-scale fixed bed reactor. Smaller particles produce more gaseous products, less tar and char, but smaller MSW particles produce more H_2 and CO and are more favorable for gas quality and yield. Buah et al. (2007) found that thermo gravimetric analysis of MSW in the form of RDF showed two zones of weight loss; the main weight loss within the lower temperature zone (240–380°C) was due mainly due to the degradation of the cellulosic fraction of the RDF, while the second peak of weight loss, occurring between 410 and 500°C was due mainly to degradation of plastic components. The pyrolysis of MSW in the form of RDF in the fixed bed reactor showed that the yield of char fell as the pyrolysis temperature was raised from 400°C to 700°C, whereas that of oil/wax and gas products increased (Buah et al. 2007).

7.3.2 Gasification Technology and Equipment

Gasification is a process, in which thermal chemical reactions of waste and gasification agent occur under high temperature and control-oxygen; and generate combustible gases. The amount of air required for gasification is less than the amount required for combustion, so the concentration of the flue gas produced in gasification process is less than the incineration process. The gas generated by the gasification process can be used for power generation and heating. Due to good volume reduction, more conducive to environmental protection and pollution control, this technology has become one of the important methods of thermal disposal of MSW.

Due to the complex and diverse components of MSW, the gasification reaction mechanism is very complex, and includes a battery of series-parallel unit reactions. Generally, it is believed that the complete gasification process is divided into two stages. The first one is thermal decomposition, in which waste is devolatilized and gasified under 600°C, generating a variety of hydrocarbons, fixed carbon and ash. In the second stage, fixed carbon reacts with water vapor or hydrogen, meanwhile releasing hydrocarbons and other combustible gases. The major chemical reactions of gasification are as follows (Liscinsky 2002; McKendry 2002):

$$C + O_2 \rightarrow CO_2 - 393.1\ \text{kJ/mol}$$

$$2C + O_2 \rightarrow 2CO - 220.8\ \text{kJ/mol}$$

$$2CO + O_2 \rightarrow 2CO_2 - 172.3\ \text{kJ/mol}$$

$$C + O_2 \rightarrow 2CO + 171.5\ \text{kJ/mol}$$

$$C + H_2O \rightarrow CO + H_2 + 131.4\ \text{kJ/mol}$$

$$C + 2H_2O \rightarrow CO_2 + 2H_2 + 75.1\ \text{kJ/mol}$$

$$C + 2H_2 \rightarrow CH_4 - 74.0\ \text{kJ/mol}$$

$$CO + H_2O \rightarrow CO_2 + H_2 - 41.0\ \text{kJ/mol}$$

The above reactions are reversible reactions, and the content of combustible components in the product gases depends on the reaction temperature, pressure, and concentration of reactants, etc. Under normal circumstances, the heat released by the oxidation reaction of carbon with oxygen facilitates pyrolysis of the waste and gasification of fixed carbon. The ideal situation is that the heat released from the partial oxidation of the waste is equal to the heat absorbed by the process of thermal decomposition, then the entire system need no additional heat (Jiao 2009).

Based on the gasification temperature, garbage gasification technology can be divided into the low-temperature and high-temperature gasification. The low-temperature gasification technology mainly includes three techniques: direct heating, indirect heating and thermoselect gasification technology. The first two are commonly used and their schematic diagrams are presented in Figures 7.8 and 7.9, respectively. For high-temperature gasification, gasification area is divided into a drying zone, a thermal decomposition gasification zone and a slag zone. The

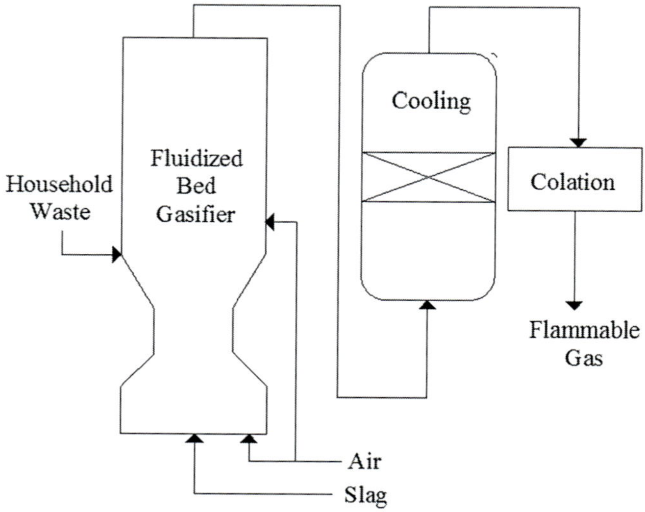

Figure 7.8. Direct heating gasification technology

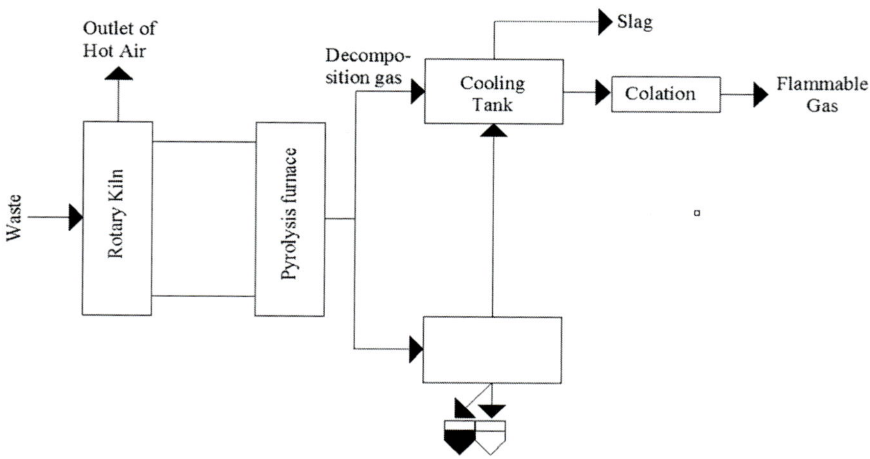

Figure 7.9. Indirect heating gasification technology

major advantage is that the technology can effectively control the generation of dioxin.

Internationally, gasification technology of MSW is relatively mature in Europe and United States and other developed countries. Switzerland SA Company has developed thermoselect gasifier and built plastic waste treatment plant in Germany. MTCI Company of USA has successfully developed multiple models of fluidized bed waste gasifier using indirect heating gasification technology and the gasification temperature is generally 700°C. The U.S. also developed the outside heating gasifier (SilvaGas TM) more suitable for gasifying waste containing a large proportion of wood. In addition, the EPA has developed Torrax technology, the UUC Company has developed Purox technology, and German has developed FLK technology and so on Japan has also done a lot of research work, and has a distinct advantage in the commercialization (Xiao et al. 2006). In China, with the economic development and government support, several garbage gasification techniques suited to China's actual situation have been successfully developed and some of the techniques have been put into commercial operation after years of experimental research. Tongji University researched gasifying MSW to produce gases in a fluidized bed in the late 1980s and proposed a program that uses the organic of waste to produce gases and the inorganic to make bricks (Wen et al. 1992; Wu et al. 1989). In recent years, Zhejiang University carried out a series of experimental study on biomass gasification and designed a circulating fluidized bed gas - steam - electrical cogeneration process systems using coal and biomass as fuel (Cen et al. 1993). Southeast University studied the low-temperature gasification technology of MSW in a circulating fluidized bed, designed and built a daily processing capacity of 3 t hot experimental system (Xiao et al. 2001).

At present, the main types of MSW gasifier include fixed bed, fluidized-bed, and the rotary kiln bed gasifier and are described in detail below.

- *Fixed Bed* gasification technology has been very mature, especially pressurized fixed bed gasification has been successfully developed. In this process, crushed waste material is sent to furnace from the top-loading port and the material is gasified hierarchically in the furnace. The flow of the gas production in the furnace relies on the induced draft fan installed on the outlet side of gas production that sends air into furnace. The reaction speed of the fixed bed gasifier is relatively slow. According to the flow direction of the waste materials and the vaporized medium in the bed, fixed bed gasifier can be divided into cocurrent, countercurrent and cross-flow gasifiers (Figure 7.10).
- *Fluidized Bed Gasifier*: The basic working principle of the fluidized bed gasifier is that the gasification agent blows fluidizing medium and waste feedstock through the air distribution plate. The strong upward flow will make the fluidizing medium and waste feedstock floating like a liquid boiling, and in the fluidized state. Fluidized bed gasifier is divided into bubbling bed gasifier, circulating fluidized bed gasifier, dual fluidized bed gasifier and entrained gasifier, and the common two are bubbling bed gasifier and circulating fluidized bed gasifier (Figure 7.11).
- *Rotary kiln bed gasifier*. According to the heating mode, the rotary kiln reactor is divided into external heat and internal heat. The reactor has a strong adaptability for materials, such as solid of different sizes and shapes and liquid and gas waste. The reactor is easy to control and operate, but not sufficient for pyrolysis; and the exit of rotary kiln is susceptible to gas leak.

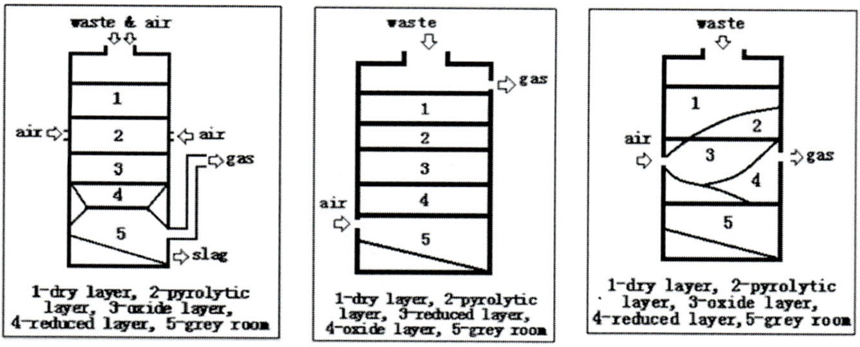

(a) concurrent gasifier (b) countercurrent gasifier (c) cross-flow gasifier

Figure 7.10. Different kinds of fixed bed gasifier

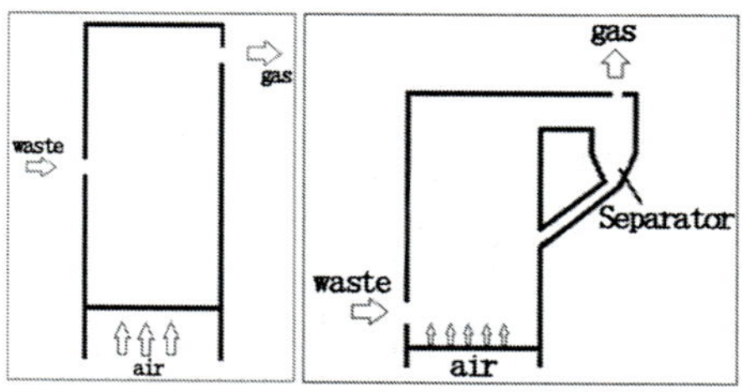

Figure 7.11. Bubbling bed (left) and circulating fluidized bed gasifier (right)

In summary, MSW gasification is a new type of waste treatment technology. Compared with the conventional method of waste disposal, gasification technology has efficient utilization of energy, good environmental properties, and has broad prospects for development.

7.3.3 Case Study (Hazardous Waste)

Hazardous waste is a subset of solid waste and is defined as any solid waste that can cause, or significantly contribute to an increase in mortality or an increase in serious irreversible, or incapacitating reversible illness; or pose a substantial or potential hazard to human health or the environment when improperly treated, stored, transported, or disposed of, or otherwise managed.

Huzhou Shiji Qing Solid Waste Disposal Center in cooperation with Zhejiang University constructed and put into operation a 10 t/d rotary-fluidizing multiple-stage pyrolysis-incineration furnace system during 2004 (Shen 2005); and the technical process is shown in Figure 7.12. In this system, the medical waste is fed by a feeding system into the main combustion chamber (rotary kiln) and moves towards the rear end of the kiln while the kiln is rotating. During the movement inside the kiln, the medical waste undergoes several stages, including preheating, drying, pyrolysis, gasification and partial combustion. The rate of

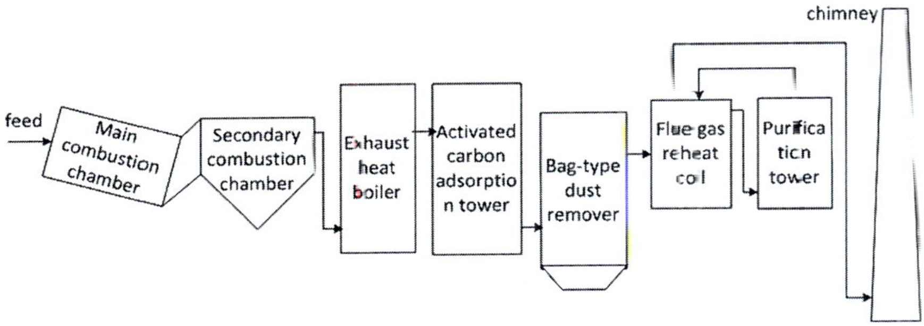

Figure 7.12. Schematic of the rotary kiln and fluidized bed multi-stage incineration

waste flow through the kiln is a function of the speed of the kiln, which is variable, and the rake or angle of the kiln to the horizontal, which is normally fixed.

The residence time of medical waste inside the kiln and the treating capacity of the rotary kiln can be regulated by adjusting the angle of inclination or the rotary speed. The large capacity of the rotary kiln can ensure that the medical waste to be treated in time during equipment maintenance. Air typically in excess of the stoichiometric requirement is provided to the kiln to help burn out the waste. A secondary combustion chamber is part of the kiln system. Off-gas from the kiln contains volatiles from the waste that have not burned out, and burnout is completed in the secondary chamber. The waste is agitated in the kiln by the rotating motion of the kiln, normally in the range of 1–3 rpm. As the waste is subjected to turbulence, it is contacted by air and thus combustion is encouraged. This turbulence, however, also increases the particulate load in the flue off-gases. The kiln system generally requires more extensive air emission control than the modular units. The kiln system is a continuous system, i.e., waste is continually fed to the unit, normally with a charging ram. By the time the charge reaches the end of the kiln, it has burned out to an ash which can be discharged dry or into water quench. The combination of a kiln plus secondary burner equals the full-load rating (usually expressed in the units of million Btu's per hour) of the incineration system.

A high temperature preheater has been designed in order to preheat the air which enters into the furnace, so as to reclaim some part of the heat. The high temperature flue gases from secondary combustion chamber are cooled down to 500°C in the air preheater. The generated air with high temperature can be used as pyrolysis gasification air of the rotary kiln. High temperature air increases the incineration temperature in the combustion chamber and makes the medical waste to be combusted without auxiliary fuel. The flue gas treatment system includes flue gas quenching system, flue gas purification system, bag filter dust collector, lime slurry preparation, delivery system, control system, etc. The dust collection system is composed of a bag-type dust separator, bag-type dust ash hooper and conveying ash apparatus. As the heat value of MW is relatively high, it is easy to burn, so it has good adaptability to use rotary kiln incinerator. However, the following things to be considered:

- The bottom of feeding device and the head of main combustion chamber should be cooled down to keep feeding continuous and equality.
- The main combustion chamber and the secondary combustion chamber should decrease the rate of temperature change, and guarantee outlet temperature of the main combustion be higher than the melting temperature of glass at the same time.
- The exhaust-heat boiler should has closed reheat device and dust removal device.

With proper design and operation, incineration techniques can meet the environmental emissions standards (Lee and Huffman 1996); and can remain be the most effective and best available treatment technology for destruction of toxic and infectious medical waste streams.

7.4 FLUE GAS CLEANING AND DIOXIN ISSUE

7.4.1 Air Pollution Control—Techniques and Devices

Incineration is conducted with a number of outputs, which include ash and flue gas emission. Before flue gas cleaning systems were introduced, the flue gas was discharged into atmosphere that led to pollution. These emissions are now typically controlled using a wide range of air pollution control techniques and devices.

Control of heavy metals. Heavy metals in fly ash from MSW incinerators are present in high concentrations. Therefore fly ash may be a potential source of heavy metals which can be relatively easily removed during the thermal treatment of fly ash. Only mercury and thallium show sizeable vapor pressure in cleaned flue gas. Obvious countermeasures are: mercury oxidation to oxides or chlorides, acid scrubbing, and adsorption on (possibly specially impregnated) activated carbon (AC).

ASH DEC technology, by ASH DEC Umwelt AG company, is viable when heavy metals need to be removed from ash. Additives (NaCl, MgO, and $NaHCO_3$) are mixed with ash, and the mixture is treated in a furnace at 900–1050°C for 20 minutes. Heavy metals evaporate during the treatment, and only the phosphorus-rich ash will remain. Subsequently, heavy metals precipitate in air emission cleaning processes as different metal hydroxides. (Boutoussov 2007)

Control of particulates. Activated carbon injection (ACI) could effectively remove air pollutants adsorbed in the particles, such as PCDD/Fs, VOCs, PAHs (Cudahy and Helsel 2000) and heavy metals. Activated carbon can generally be applied in three ways, i.e. entrained flow or called activated carbon injection (ACI), fixed bed and moving bed adsorption (Buekens and Huang 1998). In the activated carbon injection method, powder activated carbon (PAC) is injected upstream of the baghouse filter (BF) and accumulates on the filter bag surface, and flue gases are made to pass through the AC + residual dust layer, thus realize the removal of particles.

Baghouse filter (BF) operates in a manner similar to a household vacuum cleaner. Dust-laden gases pass through fabric bags, where the dry particles are captured on the fabric surface. After enough dust has built up on the filters, as indicated by a build-up in pressure across the fabric, dust is periodically removed by blowing air back through the fabric, pulsing the fabric with a blast of air, or shaking the fabric. Dust from the fabric then falls to a collection hopper where it is removed. As dust builds up on the fabric, the dust layer itself can act as a filter aid improving the removal efficiency of the device. BF can collect over 99.9% of the entering particles, even fine PM. Fabrics used in baghouses can be made of a number of different materials, selected for the particular application. Common materials for these filters include Nomex, fiberglass, Teflon, and even spun stainless steels.

ESPs are relatively large, low velocity dust collection devices that remove particles in much the same way that static electricity in clothing picks up small pieces of lint. Transformers are used to develop extremely high voltage drops between charging electrodes and collecting plates. The electrical field produced in the gas stream as it passes through the high voltage discharge introduces a charge on the particles, which is then attracted to the collecting plates. Periodically the collected dust is removed from the collecting plates by a hammer or rapper striking the top of the plates (rapping) dislodging the particulate, which falls in to a bottom hopper for removal. ESPs are often configured as a series of collecting plates to improve overall collection efficiency. Efficiencies exceeding 99% can be achieved, and ESPs are used in many of the same applications as baghouses.

Control of acid gases. Historically, several solutions have been developed to solve the acid gas problem: first of all wet scrubbing, then dry scrubbing, followed closely by semi-wet scrubbing, and finally semi-dry scrubbing. Generally, acid gases are neutralised by injecting hydrated lime, $Ca(OH)_2$, into the flue gas in one (or more) out of three modes: (1) dry, (2) semi-dry, i.e. after further moistening of the flue gas, and (3) semi-wet scrubbing, injecting lime slurries into the flue gas; and a comparison is presented in Table 7.4.

Wet scrubbing is an even more efficient solution, since the principal acid gas, HCl, is eminently water soluble; yet it is also more complex and capital-intensive, because of the necessity of maintaining a separate water circuit and treating the resulting wastewater by removing organic compounds as well as sludge and heavy metals. Wet scrubbing generally is conducted in two steps: in a first acid scrubbing (pH 0 to 2), the bulk of HCl and HF is removed whereas SO_2 only follows in a second step, operating under mild acid or at times basic conditions (pH 6 to 8). At high pH values, CO_2 also is increasingly absorbed, considerably inflating the consumption of neutralising agents (Buekens et al. 2012).

Dry scrubbing has the tremendous advantage of uttermost simplicity and almost negligible investment cost: hydrated lime is injected into the flue gas in sufficient amount and the resulting spent salts are removed during dust collection.

Table 7.4. Comparison of different FGD Techniques

FGD technology	*Wet*	*Dry*	*Semi-dry*
Desulfurization efficiency (%)	90 ~ 99	50 ~ 80	65 ~ 85
Investment percentage*	10	6 ~ 7	8
Advantages	High desulfurization efficiency, high utilization rate of limestone, low energy consumption, high capacity of desulfurization tower, high reliability	Mature technology, relatively simple process, high reliability	A combination of some advantages of the former two: high reaction speed and high desulfurization efficiency (Wet), no emission of waste water and acid, easy handling of desulfurization products (Dry)

*Investment percentage: the percentage accounted for by desulfurization equipment of all equipment

However, this simple system is strongly limited in efficiency, even when a large multiple of the stoichiometric requirements is injected (Buekens et al. 2012).

Semi-dry scrubbing increases the reactivity of hydrated lime by conditioning (i.e., moistening) the flue gas prior to its scrubbing. This combines some of the qualities of wet scrubbing with the advantage of gaining dry reaction products. A suspension of finely divided hydrated lime is contacted with the flue gas, and the acid gases are rapidly neutralised by the moist lime particles that are dried in the process and eventually separated by a baghouse filter. The latter provides further neutralization of acid gases, as well as adsorption of various pollutants, including aerosols enriched in heavy metals and semi-volatiles (PAHs and dioxins). Usual operating problems relate to inadequate dispersion by rotary slurry dispensers or lances, slurry misdistribution and impaction of lime particles on the walls, as well as clogging and corrosion by hygroscopic salts. Today, it is often regarded as a Best-Available-Technology (BAT) (European Commission 2006)

Control of nitrogen oxides. NOx can be thermally (Selective Non Catalytic Reduction, SNCR) or catalytically (Selective Catalytic Reduction, SCR) reduced by means of ammonia, urea, or amines (organic N-compounds), hydrocarbons (re-burning) or other reducing agents. Thermal reduction of NO (SNCR) proceeds when ammonia or urea is injected into the flue gas while it is still at 750–1000°C while SCR operates at medium-low temperature (200–400°C). A variety of reducing agents and catalysts may produce the desired results, yet most used are ammonia and urea and honeycomb or pellet type catalyst (sub-stoichiometric $V_2O_5 - x$ on a TiO_2 carrier with WO_3 as promoter). SNCR alone is capable of attaining the EU limit value of 200 mg NO_2/Nm^3; while SCR attains a much higher reduction, e.g. <70 mg; NO_2/Nm^3 (Buekens 2004). In the SCR reactor, ammonia and NOx react to form nitrogen and water; and >90% NOx removal is possible with these systems.

Catalytic reactors perform thermal destruction functions like incinerators, but at lower temperatures and for selected waste gases only. They incorporate beds of solid catalytic material that the unwanted gases pass through, typically for oxidation or reduction purposes. Problems with catalytic reactors may include the flow short-circuiting through the bed, poisoning of the catalysts, and excessive oxidation followed by thermal failure. Poisoning or contamination of the catalyst can cause a severe drop in performance over time.

7.4.2 Dioxin Emission and Control

Dioxins first came to light in the waste incinerator fly ash at 1977 (Olie et al. 1977), since then their formation mechanism and emission control technology have developed tremendously and always attract much public attention. A large number of scholars devoted themselves to improve waste incineration technology, strengthening environmental pollution control and guaranteeing environment health. More than 90 nations in the world, including China, signed the “Persistent organic pollutants (POPs) Stockholm convention” in 2001, which calls for controlling the emission and the prohibition of the use of 12 kinds of typical POPs (including dioxins).

Table 7.5. Dioxins release inventory of China for the reference year of 2004

	Projects	*Annual total emissions/g*				
		Atmosphere	*Water*	*Product*	*Soil*	*Total*
Emission source	Waste incineration	610.5	0	0	1147.1	1757.6
	Steel and other metal production	2486.2	13.5	0	2167.2	4666.9
	Heat and power generation	1304.4	0	0	588.1	1892.5
	Mineral products production	413.6	0	0		413.6
	Transportation	119.7	0	0		119.7
	Uncontrolled combustion process	63.5	0	0	953.2	1016.7
	Chemicals and consumption goods production and use	0.68	23.16	174.39	68.9	267.13
	Waste treatment and landfill	0	4.50	0	43.2	47.7
	Others	44.2	0	0	11	55.2
Total		5042.78	41.16	174.39	4978.7	10237.03

Source: Lv et al. 2008; reproduced with permission from Environmental Pollution and Control

According to the dioxins emission status of China in 2004, as is listed in Table 7.5 (Lv et al. 2008), waste incineration (including MSW incinerator [MSWI], hazardous waste incinerator [HWI] and medical waste incinerator [MWI]) is one of the main dioxins emission sources. Although, no exact data about dioxin emissions on a national scale is reported for recent years, we can also conjecture that the dioxins yield grows accordingly, with the rapid development of waste incineration in China. Fujian province performed dioxin source updating investigation and found that non-ferrous metal production, steel and other metal production, iron ore sintering process and waste incineration are the four main dioxin emission sources occupying 75% of the total emissions (Huang 2010); of which, dioxins discharged from waste incineration (807.26 mg I-TEQ/y) takes up 11%.

Dioxins emissions from waste incineration in Taiwan and USA (Chen 2004; Lin et al. 2007; Ryan 2001) have been listed in Table 7.6. Incineration is widely used in USA and a mass of dioxins was released, especially from MSW incineration (1100 g I-TEQ/y, much more than that in China). MSW incineration is also

Table 7.6. Dioxins emissions from waste incineration in Taiwan region and USA (g I-TEQ/y)

Nation or Region	*MSW*	*Hazardous Waste*	*Medical waste*
USA (1998)	1100	477	6
Taiwan (2000)	3.5	0.26	0.39
Taiwan (2004)	1.3	0.34	1.1

the main dioxins emission source in Taiwan region, however, dioxins released from MSW incineration in 2004 became much lower than that in 2000.

Dioxins can be generated through two different processes at least: high temperature gas phase synthesis and low temperature heterogeneous catalytic synthesis. In combustion process, dioxins are mainly synthesized through low temperature heterogeneous catalytic synthesis, which can be subdivided into two different types, depending on the source molecules taking part in the process: (1) directly synthesized from precursors, such as chlorobenzene and chlorophenol; and (2) *de novo* synthesis from simple small molecules which contain elements like carbon, hydrogen, oxygen or chlorine. Numerous researchers have focused on investigating which synthesis process dominates in the dioxins formation and they proposed different viewpoints. After analyzing the PCDD/Fs data from various combustion processes, including actual incinerator tests and simulation studies, Huang and Buekens (1995) considered *de novo* synthesis is the main reason for PCDD/Fs formation in incinerator flue gas. However, Altwicker (1996) compared *de novo* synthesis and precursor synthesis rates, discovering that the reaction rate of latter is much faster than that of the former. Moreover, chlorophenol is more likely to form PCDD/Fs than chlorobenzene does (Addink et al. 1995).

Dioxin has drawn extensive public attention. According to the Hazardous waste incineration pollution control standard GB18484-2001, the current dioxin emission standard for medical waste incinerator flue gas in China is 0.5 ng I-TEQ/Nm^3 (11 Vol.% O_2). The more stringent limit level in Europe is 0.1 ng I-TEQ/Nm^3. In order to fulfil the environmental demand and reduce the health risk to surrounding residents, the implementation and development of dioxin emission control technology are quite an emergency.

- *Activated carbon adsorption* is the simplest and most widely applied dioxin tail gas removal technology, due to its huge surface area and strong adsorption capacity. AC is able to adsorb gaseous phase as well as solid phase dioxin or even the ones attached to particles. This method has some drawbacks: (1) AC supplies as sufficient carbon source, $CuCl_2$ and $FeCl_2$ in the fabric filter and the appropriate operation temperature (150 ~ 300°C) of air pollutant control devices facilitate the dioxin formation, increase dioxins total emission amount (including those exist in fly ash and flue gas) (Chang and Lin 2001); and (2) merely transfers pollutants from the gas to solid phase, further treatment for fly ash is needed.

- *Heterogeneous catalytic decomposition.* Catalytic decomposition technology can achieve real destruction of PCDD/Fs without generating secondary pollutants. Some studies indicated that the TiO_2-based V_2O_5/WO_3 catalysts originally designed for the removal of nitrogen oxides (NOx) via SCR are effective in the decomposition of PCDD/Fs at the same temperature as that used for the DeNOx. Catalytic destruction of PCDD/Fs can be combined with catalytic deNOx-SCR processes, so that no additional system has to be built. During the last decade, some commercial SCR catalysts have been optimized for combined PCDD/Fs/NOx removal. This was achieved mainly by increasing the oxidation potential of the catalysts using higher vanadium content. Catalyst employed for de-NOx costs 1.800 ~ 2.200 Euro/ton NOx; however, the price is 10 ~ 15% higher when the catalyst is simultaneously applied for dioxin removal. Besides, extra expense is also required to maintain catalyst activity influencing the operation cost.
- *Photo decomposition and photocatalytic decomposition.* In the early 1970s, scientists have conducted research on photo-decomposition of organic compounds. To further improve decomposition efficiency, photo-catalytic decomposition, e.g., heterogeneous TiO2 or other inorganic catalysts, has been proposed.
- *Plasma assisted decomposition:* Plasma provides abundant active groups and electrons with high energy, which facilitate the decomposition of organic compounds. As for pollutants in gas or liquid phase, non-equilibrium plasma technology under atmosphere is widely applied, including corona discharge, dielectric barrier discharge, gliding arc discharge, electron beam irradiation, etc. As for organic compounds in solid and that absorbed on solid, thermal plasma (e.g. DC plasma) is generally used.
- *Inhibitor addition technology:* Directly mixing inhibitor with waste before operation or injecting into the tail of incinerator can effectively inhibit the PCDD/Fs formation. There are three kinds of inhibitors: alkaline inhibitor, S-containing compounds and N-containing compounds. Alkaline inhibitor is often added to absorb acid gases, especially HCl. Frequently-used alkaline inhibitors like CaO, $CaCO_3$, $Ca(OH)_2$, $MgCO_3$, MgO and Mg(OH)2 absorb acid gas by reacting like $Ca(OH)_2 + 2HCl \rightarrow CaCl_2 + 2H_2O$. So they are quite effective in removing acid gases (HCl, HBr, HF). The inhibition of PCDD/Fs formation by alkaline inhibitors relies on controlling HCl concentration. S-containing compounds includes S, SO_2, Na_2S, $Na_2S_2O_3$, sulfonate and sulfonamides. The inhibition mechanism can be considered as following: (1) reacting with active catalyst to form inert compounds, as $CuO + SO_2 + 1/2O_2 = CuSO_4$; (2) consuming chlorinating agent as $Cl_2 + SO_2 + H_2O = 2HCl + SO_3$; and (3) reacting with dioxin precursor. The formation of PCDD/Fs can be effectively inhibited by ammonia with an efficiency of about 80%. However, in a few investigations, ammonia showed no inhibition effect. More research is required to understand the inhibition effects and mechanism of N-containing compounds.

- *Low temperature decomposition:* This technology is usually applied to remove dioxins in fly ash. Hagenmaier et al. (1987) proposed this method, in which fly ash is heated to 300 ~ 600°C under inert atmosphere and holding for a certain time, so that the dioxin in fly ash can be effectively decomposed. Recently, some companies conducted pilot tests and applied it commercially to decompose dioxin or polychlorinated biphenyl in fly ash and solid (Troxler et al. 2010).

7.5 CONCLUSION

For solid wastes, incineration continues to be the most common method of thermal treatment as a Waste To Energy technology. It is reasonable to anticipate that this technology would be proposed for new solid waste disposal facilities. This technique becomes very mature with the advantages of simple mechanical structure, simple operation and easy maintenance besides a few decades of operational performance. But solid waste incineration emits secondary pollutions such as $PM_{2.5}$, dioxins, SO_2 and heavy metals.

Compared with incineration, other thermal treatment technologies such as gasification and pyrolysis have a higher energy efficiency and lower generation of secondary pollution, especially dioxins. But they are not so widely used due to the limitations such as complexity, and difficulty in handling making them not a viable technology. However, the industry continues to evolve and technologies continue to advance; and these limitations can be resolved in future. In Europe, Japan and America, some solid waste gasification plants have been put to run for years, and these plants satisfy the stringent regulations with energy recovery making gasification as a viable alternative solution for waste treatment. On the other hand, the technical reliability and economic convenience of gasification still needs to be proved as there are only 100 WTE plants in the world operating with this technique.

References

Addink, R., Olie, K., and Cnubben, P. (1995). "Formation of polychlorinated dibenzo-p-dioxins/dibenzofurans on fly ash from precursors and carbon model compounds." *Carbon*, **33**(10), 1463–1471.

Altarawneh, M., Dlugogorski, B. Z., Kennedy, E. M., and Mackie, J. C. (2009). "Mechanisms for formation, chlorination, dechlorination and destruction of polychlorinated dibenzo-p-dioxins and dibenzofurans (PCDD/Fs)." *Prog. Energy Combust. Sci.*, **35**(3), 245–274.

Altwicker, E. R. (1996). "Relative rates of formation of polychlorinated dioxins and furans from precursor and de novo reactions." *Chemosphere*, **33**(10), 1897–1904.

Anthony, E. J. (1995). "Fluidized bed combustion of alternative solid fuels; status, successes and problems of the technology." *Prog. Energy Combust. Sci.*, **21**(3), 239–268.

Arena, U. (2012). "Process and technological aspects of municipal solid waste gasification. A review." *Waste Manage.*, **32**(4), 625–639.

Arena, U., and Mastellone, M. L. (2009). "Fluidized bed gasification of RDF and PDF." AMRA Scientific Rep., AMRA, Napoli, Italy.

Arena, U., Mastellone, M. L., and Perugini, F. (2003). "The environmental performance of alternative solid waste management options: A life cycle assessment study." *Chem. Eng. J.*, **96**(1-3), 207–222.

Azapagic, A., Perdan, S., and Clift, R. (2004). *Sustainable development in practice*, Wiley, Chicester, U.K.

Baggio, P., Baratieri, M., Gasparella, A., and Longo, G. A. (2008). "Energy and environmental analysis of an innovative system based on municipal solid waste (MSW) pyrolysis and combined cycle." *Appl. Therm. Eng.*, **28**(2–3), 136–144.

Bhaskar, T., et al. (2004). "Pyrolysis studies of PP/PE/PS/PVC/HIPS-Br plastics mixed with PET and dehalogenation (Br, Cl) of the liquid products." *J. Anal. Appl. Pyrol.*, **72**(1), 27–33.

Boutoussov, M. (2007). "Thermal process for separating off heavy metals from ash in agglomerated form." WO Patent 2,007,124,527.

Boyles, F. (1993). "The Vommturbo dryer system." *Seminar on the drying of sewage sludge*, Wakefield, U.K.

Brunner, P. H., Morf, L., and Rechberger, H. (2004). "Thermal waste treatment–a necessary element for sustainable waste management." *Solid waste: Assessment, monitoring, remediation*, I. Twardowska, H. E. Allen, A. F. Kettrup, and W. J. Lacy, eds., Elsevier, Amsterdam, Netherlands.

Buah, W. K., Cunliffe, A. M., and Williams, P. T. (2007). "Characterization of products from the pyrolysis of municipal solid waste." *Process. Saf. Environ.*, **85**(5), 450–457.

Buekens, A. (2004). "Encyclopedia of life support systems (EOLSS)." *Auspices of the UNESCO*, Eolss Publishers, Oxford, U.K.

Buekens, A., and Cen, K. (2011). "Waste incineration, PVC, and dioxins." *J. Mater. Cycles Waste Manage.*, **13**(3), 190–197.

Buekens, A., and Huang, H. (1998). "Comparative evaluation of techniques for controlling the formation and emission of chlorinated dioxins/furans in municipal waste incineration." *J. Hazard. Mater.*, **62**(1), 1–33.

Buekens, A., Huang, J., Shengyong, L., and Xiaodong, L. (2012). *Incinerator flue gas treatment—A worldwide perspective*, Seoul/Ilsan KINTEX, Korea.

Bux, M., Baumann, R., Philipp, W., Conrad, T., and Mühlbauer, W. (2011). "Class-a by solar drying recent experiences in Europe." *Proc., WEFTEC (Water Environment Federation) Congress*, Georgia, Atlanta.

Cen, K., Fang, M., and Luo, Z. (1993). "Research on rice hull gasification gas steam cogeneration using a circulation fluidized bed." *Proc., Rubow, L.N., 12th Int. Conf. on FBC*, ASME Press, New York, 197.

Chang, M. B., and Lin, J. J. (2001). "Memory effect on the dioxin emissions from municipal waste incinerator in Taiwan." *Chemosphere*, **45**(8), 1151–1157.

Chen, C. (2004). "The emission inventory of PCDD/PCDF in Taiwan." *Chemosphere*, **54**(10), 1413–1420.

Cheng, H., and Hu, Y. (2010). "Municipal solid waste (MSW) as a renewable source of energy: Current and future practices in China." *Bioresour. Technol.*, **101**(11), 3816–3824.

Cheng, H., Zhang, Y., Meng, A., and Li, Q. (2007). "Municipal solid waste fueled power generation in China: A case study of waste-to-energy in Changchun City." *Environ. Sci. Technol.*, **41**(21), 7509–7515.

Chen, Q. (2011). "The current situation of biodrying in sewage sludge." *China Resour. Compr. Util.*, **29**(2), 35–38.

Chowlu, A. C. K., Reddy, P. K., and Ghoshal, A. K. (2009). "Pyrolytic decomposition and model-free kinetics analysis of mixture of polypropylene (PP) and low-density polyethylene (LDPE)." *Thermochim. Acta*, **485**(1), 20–25.

Ciliz, N. K., Ekinci, E., and Snape, C. E. (2004). "Pyrolysis of virgin and waste polypropylene and its mixtures with waste polyethylene and polystyrene." *Waste Manage.*, **24**(2), 173–181.

Consonni, S., Giugliano, M., and Grosso, M. (2005). "Alternative strategies for energy recovery from municipal solid waste: Part A: Mass and energy balances." *Waste Manage.*, **25**(2), 123–135.

Cudahy, J. J., and Helsel, R. W. (2000). "Removal of products of incomplete combustion with carbon." *Waste Manage.*, **20**, 339–345.

Demirbas, A. (2005). "Pyrolysis of ground beech wood in irregular heating rate conditions." *J. Anal. Appl. Pyrol.*, **73**(1), 39–43.

Du, J., Wang, H. B., and Jin, X. (2003). "An Analysis summary of urban waste incineration technology in our and foreign countries." *Ind. Boiler*, 15–19.

European Commission. (2006). "Integrated pollution prevention and control-reference document on the best available techniques for waste incineration." Brussels, Belgium.

Faaij, A. (2006). "Modern biomass conversion technologies." *Mitigation Adaptation Strategies Global Change*, **11**(2), 335–367.

Gu, Z., and Yang, D. (2008). "Application of solar drying for sludge treatment in Europe." *Sichuan Environ.*, **27**(6), 93–96.

Guan, Y. (2010). "Summary of municipal refuse burning technology." *Power Syst. Eng.*, **26**(2), 74.

Hagenmaier, H., Kraft, M., Brunner, H., and Haag, R. (1987). "Catalytic effects of fly-ash from waste incineration facilities on the formation and decomposition of polychlorinated dibenzo-para-dioxins and polychlorinated dibenzofurans." *Environ. Sci. Technol.*, **21**(11), 1080–1084.

Huang, H., and Buekens, A. (1995). "On the mechanisms of dioxin formation in combustion processes." *Chemosphere*, **31**(9), 4099–4117.

Huang, Y., and Ning, Z. (2010). "Dioxin-like pops pollution status updating survey in Fujian province." *Chem. Eng. Equip.*, **10**(3), 179–181.

ISWA. (2008). "Management of APC residues from WTE Plants." *ISWA-WG thermal treatment of waste*, 2nd Ed., Copenhagen.

Jewell, W. J., Dondero, N. C., and Van Soest, P. J. (1984). "High temperature stabilization and moisture removal from animal wastes for byproduct recovery." Final Rep. SEA/CR 616-15-168, USDA, Washington, DC.

Jiao, Z. (2009). "Experimental Study on MSW Fludized Bed Gasification and the Prediction Model." Ph.D. thesis, Zhejiang Univ., Hangzhou.

Jin, Y. (2002). "Study on MSW combustion characteristics and a new CFB incineration technology." Ph.D. thesis, Zhejiang Univ., Hangzhou.

Lee, C. C., and Huffman, G. L. (1996). "Medical waste management/incineration." *J. Hazard. Mater.*, **48**(1–3), 1–30.

Liang, L., et al. (2008). "Experimental study on effects of moisture content on combustion characteristics of simulated municipal solid wastes in a fixed bed." *Bioresour. Technol.*, **99**(15), 7238–7246.

Lin, L., Lee, W., Li, H., Wang, M., and Chang-Chien, G. (2007). "Characterization and inventory of PCDD/F emissions from coal-fired power plants and other sources in Taiwan." *Chemosphere*, **68**(9), 1642–1649.

Liscinsky, D. (2002). *Biomass gasification and power generation using advanced gas turbine system*, United Technologies Research Center, East Hartford, CT, 3–8.

Li, Z. (2007). "The microwave drying performance research of municipal wastewater treatment plant." M.S. thesis, Chongqing Univ., Chongqing.

Luboschik, U. (1999). "Solar sludge drying based on the IST process." *Renewable Energy*, **16**(1–4), 785–788.

Luo, S., Xiao, B., Hu, Z., Liu, S., Guan, Y., and Cai, L. (2010). "Influence of particle size on pyrolysis and gasification performance of municipal solid waste in a fixed bed reactor." *Bioresour. Technol.*, **101**(16), 6517–6520.

Luo, Z., Wang, S., and Cen, K. (2005). "A model of wood flash pyrolysis in fluidized bed reactor." *Renewable Energy*, **30**(3), 377–392.

Lu, S., Yan, J., Li, X., Ni, M., and Cen, K. (2007). "Laboratory-scale study of the suppression of PCDD/F emission during coal and MSW co-incineration." *J. Environ. Sci.*, **19**(6), 762–767.

Lv, Y., Huang, J., Yu, G., and Yang, X. (2008). "Comparison of the dioxins release inventories of China and other countries." *Environ. Pollut. Cont.*, **30**, 71–74.

Malkow, T. (2004). "Novel and innovative pyrolysis and gasification technologies for energy efficient and environmentally sound MSW disposal." *Waste Manage.*, **24**(1), 53–79.

Ma, X., Jiang, X., and Ma, Z. (2006). "Experimental study on drying sewage sludge with paddle dryer." *Energy Eng.*, **3**, 57–60.

Ma, W., and Chen, X. (2012). "Solar sludge drying technology applications and processes." *Sci. Technol. Info.*, **27**, 435–436.

McKay, G. (2002). "Dioxin characterisation, formation and minimisation during municipal solid waste (MSW) incineration: Review." *Chem. Eng. J.*, **86**(3), 343–368.

McKendry, P. (2002). "Energy production from biomass (part 3): gasification technologies." *Bioresour. Technol.*, **83**(1), 55–63.

Ni, M., Xiao, G., and Chi, Y. (2006). "Study on pyrolysis and gasification of wood in MSW." *J. Environ. Sci.*, **18**, 407–415.

Olie, K., Vermeulen, P., and Hutzinger, O. (1977). "Chlorodibenzo-p-dioxins and chlorodibenzofurans are trace components of fly ash and flue gas of some municipal incinerators in The Netherlands." *Chemosphere*, **6**(8), 455–459.

Pan, Z. (2007). "A comparison and selection consideration of MSW incineration technology." *Urban Manage. Sci. Technol.*, **9**(3), 44–47.

Park, W. C., Atreya, A., and Baum, H. R. (2010). "Experimental and theoretical investigation of heat and mass transfer processes during wood pyrolysis." *Combust. Flame*, **157**(3), 481–494.

Porteous, A. (2005). "Why energy from waste incineration is an essential component of environmentally responsible waste management." *Waste Manage.*, **25**(4), 451–459.

Psomopoulos, C. S., Bourka, A., and Themelis, N. J. (2009). "Waste-to-energy: A review of the status and benefits in USA." *Waste Manage.*, **29**(5), 1718–1724.

Qing, S., Wang, H., Lu, G., and Hu, J. (2005). "The study on clean production of MSWI-20 incinerator." *Ind. Heat.*, **34**(3), 18–20.

Qiu, J., Wu, Q., and Wei, Z. (2007). "Research advances in municipal sludge drying." *Ecol. Environ.*, **16**(2), 667–671.

Rogaume, T., Jabouille, F., and Torero, J. L. (2009). "Effect of excess air on grate combustion of solid wastes and on gaseous products." *Int. J. Therm. Sci.*, **48**(1), 165–173.

Ryan, S. (2001). "The formation of polychlorinated dibenzo-p-dioxins/dibenzofurans (PCDD/F) via the de novo synthesis in the catalytic extraction process (CEP) and the role of iron chloride." Ph.D. dissertation, Rensselaer Polytechinic Institute, New York.

Saxena, S. C., and Jotshi, C. K. (1996). "Management and combustion of hazardous wastes." *Prog. Energy Combust. Sci.*, **22**(5), 401–425.

Shen, Q. (2010). "Review of sludge processing technology." *Sci. Technol. Innov. Herald*, **23**, 64.

Shen, X. (2005). "Study on pyrolysis and incineration of medical waste in a rotary kiln." Ph.D. thesis, Zhejiang Univ., Hangzhou.

Smith, M. A. B. (1994). "The development of fluid beds for drying sewage sludge." *Seminar on the Drying of Sewage Sludge*, Wakefield, U.K.

Song, Z., Lu, Y., Liang, Y., and Yang, W. (2007). "Present situation of the development of municipal domestic waste incineration technology." *Environ. Sanit. Eng.*, **15**(1), 21–24.

Starcevic, N., Bux, M., Rohr, T., Lutz, K., Metodi, K., and Ritterbusch, S. (2007). "Development of a robot for biomass handling in a solar greenhouse dryer." *Agric. Eng. Int. CIGR E J.*, 9.

Troxler, W. L., Hunt, J. W., Taylor, J., and McNiven, C. (2010). "Thermal desorption treatment of dioxin-contaminated soil at the former allied feeds site, Sydney, Australia." *Environ. Eng. Sci.*, **27**(7), 613–622.

Wang, R., Liu, J., and Yu, Y. (2010). "Municipal sludge characteristics and drying technologies." *Water Wastewater Eng.*, **36**(z1), 153–158.

Wen, W., Wei, D., and Zhuang, Y. (1992). "Investigation on gasification of municipal waste garbage." *Gas Heat*, **12**(4), 4–10.

Wiesman, J. (1994). "Zurich's search for the ideal solution to sewage sludge disposal." *Water Qual. Int.*, **2**, 16–19.

Wiles, C. C. (1996). "Municipal solid waste combustion ash: State-of-the-knowledge." *J. Hazard. Mater.*, **47**(1–3), 325–344.

Wu, C., Chang, C., and Tseng, C. (2002). "Pyrolysis products of uncoated printing and writing paper of MSW." *Fuel*, **81**(6), 719–725.

Wu, J., Wen, W., Wang, B., and Chen, C. (1989). "An influence of RDF properties on carbonization and gasification." *J. Tongji Univ.*, **17**(1), 113–122.

Wu, W., Ma, Q., and Li, R. (2010). "Comparative analysis of state-of-the-art technology for waste treatment." *Municipal Admin. Technol.*, **1**, 18–21.

Xiao, B., Wang, Y., and Su, Q. (2006). "The research of treating municipal solid waste by gasification." *China Resour. Compr. Util.*, **24**(10), 18–20.

Xiao, R., Jin, B., and Lan, J. (2001). "Second generation municipal solids waste incinerator-gasification and melting furnace." *Ind. Furnace*, **4**, 39–44.

Yang, Y. B., Ryu, C., Goodfellow, J., Sharifi, V. N., and Swithenbank, J. (2004). "Modelling waste combustion in grate furnaces." *Process Saf. Environ. Prot.*, **82**(3), 208–222.

Zhang, Y. (2004). "Municipal solid waste incineration technology and control of dioxins emission." *Ind. Boiler*, **5**, 1–7.

Zhang, Y., and Zhao, Y. (2000). *Municipal solid waste incineration technology*. Chemical Industry Press, Beijing.

CHAPTER 8

Anaerobic Digestion or Co-Digestion for Sustainable Solid Waste Treatment/ Management

Sridhar Pilli[1]
Tanaji T. More[2]
Song Yan[3]
Rajeshwar D. Tyagi[4]
Rao Y. Surampalli[5]
Tian C. Zhang[6]

In this chapter, an attempt to summarise the literature on the details of the anaerobic digestion (AD) and co-digestion has been made. This chapter mainly discusses AD and co-digestion of waste, i.e., section 8.1 describes the increasing needs of AD or co-digestion for managing the solid waste in different countries. Section 8.2 describes the principles of AD and co-digestion as well as how AD evolves into co-digestion. Section 8.3 describes the types of anaerobic digester or co-digesters and the major parameters affecting their performance. Section 8.4 gives the details of the feed stocks required for AD or co-digestion. Then section 8.5 details about substrate pre-treatment technologies for AD and co-digestion. Section 8.6 discusses the design of AD and co-digestion processes, the principles of mass, electron and energy balance and evaluation of biogas production. Section 8.7 briefs the case studies on anaerobic co-digestion. Section 8.8 discusses the advantages and disadvantages of the anaerobic technology. Section 8.9 briefs the future research required for improving the AD

[1]INRS-ETE, Université du Quebéc, Québec, QC, Canada.
[2]INRS-ETE, Université du Quebéc, Québec, QC, Canada.
[3]INRS-ETE, Université du Quebéc, Québec, QC, Canada.
[4]INRS-ETE, Université du Quebéc, Québec, QC, Canada.
[5]Univ. of Nebraska-Lincoln, Lincoln, NE.
[6]Univ. of Nebraska-Lincoln, Lincoln, NE.

processes. The conclusions about AD or co-digestion are summarised in section 8.10, and the references are listed in section 8.11.

8.1 INCREASING NEEDS FOR AD OR CO-DIGESTION

Continued population growth and economic development have contributed to increasing generation of solid organic wastes/materials (such as food scraps, fats, oils, grease, manure, sludge, etc.), around the world. A decade back, the urban municipal solid waste generation around the world was 0.65 kg person per day (i.e., 2.9 billion urban residents generated 0.68 billion tonnes per year); Hoornweg and Bhada-Tata (2012) estimated that today solid wastes (these) generation has increased to about 1.2 kg per person per day (3 billion residents) and, in the future (i.e., by 2025), to 1.42 kg/capita/day of municipal solid waste (or 4.3 billion urban residents generating 2.2 billion tonnes per year). Moreover, the estimated solid waste management cost in 2010 is $205.5 billion, and will be $375 billion by 2025. With such a rapid increase in the solid waste generation rate and the corresponding cost (Khalid et al. 2011; Muhammad Nasir et al. 2012), the safe and cost-effective disposal of waste is a major concern because decomposition of these organic wastes could result in large-scale contamination of land, water, and air, and treatment/management of the solid waste would otherwise consume resources. Over the decades most of the solid waste is landfilled or incinerated which are unsustainable technologies (or not green technologies). With major concerns of cost, quantity, treatment, disposal, environment, climate and greenhouse gas (GHG) emissions, solid waste management has been subjected to major developments during the last decade. With increased rules, regulations, energy consumption and fossil fuels diminution, there is a requirement to utilize the organic materials/wastes to generate sustainable energy. Moreover, the management/policy makers should have an "ultimate goal to ensure that no solid waste is landfilled or incinerated until it has been demonstrated that energy recovery is not an economically viable option."

To utilize the solid waste for the energy recovery, researchers have studied different treatment options, and concluded that AD or co-digestion is the most successful (fully demonstrated technology) and innovative technology to generate energy from the source separated organics, or with mixed and residual waste/materials (Angelidaki et al. 2006; Nguyen et al. 2007; Alvarez et al. 2010; Khalid et al. 2011; Esposito et al. 2012). AD or co-digestion technology has been implemented at different magnitudes around the world (Lou et al. 2012) because of the bio-energy recovery from the produced methane and offsetting the usage of fossil fuels. It has been extensively used in small scales in developing countries (mostly in rural areas) to recover energy from the farm waste for lighting, cooking and heating. There are 30 million small-scale digesters, of which most of them are located in China, India, and the rest of them are in developing countries like Nepal, Sri Lanka, etc. (Chen et al. 2008; Gautam et al. 2009; Mahapartha et al. 2009; Levis et al. 2010; Ferrer et al. 2011; Levis and Barlaz 2011; Lou et al. 2012).

In the developed countries, Germany is leading on implementing the AD or co-digestion technology, having 6800 large-scale anaerobic digesters (IEA 2011). Moreover, most of the waste generated (8.6 million metric tons/year of source separated biowaste) is composted (around 90%), and the rest (900,000 metric tonnes) was anaerobically digested (Baere and Mattheeuws 2010). To minimize the GHG emissions and to increase the biogas production, a further addition of AD to the existing composting infrastructure is planned in Germany. In Europe, the implementation of AD technology for solid waste management has been increased remarkably in the first decade of the 21st century (more than 70 plants were installed during 2006–2010). There are 244 installed AD plants digesting almost 8 million tonnes of organic waste. It was estimated that AD represents 25% of biological treatment. European Union has targeted to supply 5% of the energy demands with renewable energy from AD biogas by year 2020 (Holm-Nielson et al. 2009; Kim and Oh 2011). In the UK, AD technology has been constituted for decades for treating the sewage sludge. There are 147 plants treating 13.3 million tonnes per year of wet sewage sludge, and 66 plants are treating approximately 1 million tonnes per year of wet food and agricultural waste. The electricity produced (151 million watts) from these plants will provide the electricity for approximately 300,000 homes annually (www.parliament.uk/briefing-papers/post-pn-387.pdf).

In North America, Toronto is the first major city to implement the AD technology for treating the organic waste since 2002 (Press 2008). Further, the city council of Toronto approved the "Target 70" plan to install two 55,000 metric tons/year digesters, i.e., to divert 70% of the waste to AD. In the US, AD is used on a large scale at wastewater treatment plants and also at large farms for treating the manure and waste. By the end of 2011, livestock manure was digested in 176 anaerobic digesters, which generated about 541 million kWh of electricity. By digesting the manure and utilising the energy recovered from the biogas 301,000 metric tons of equivalent CO_2 emissions had minimised by offsetting the fossil fuel usage. Moreover it had reduced 55,000 metric tonnes of direct methane emissions from manure (http://www.epa.gov/agstar/ documents/2011_digester_update. pdf).

The most prominent treatment methods of the organic waste were composting and incineration. In the US, combustion of municipal solid waste with energy recovery is getting popular gradually; there are 86 facilities processing more than 28 million tonnes waste per year with production of 2,720 megawatts of power per year (USEPA 2010). The USEPA estimated that the food waste comprises 12.4% of the municipal solid waste; if 50% of the food waste generated each year is anaerobically digested, it will produce electricity enough to power 2.5 million homes. In the North West, small-scale anaerobic digesters are established, e.g., the State of Wisconsin has 17 anaerobic digesters in operation (Press 2008), and some are under development, such as Columbia biogas, Green lane energy, etc. (Sullivan 2012) Revolution Energy Solutions have installed a patented psychropilic (low temperature) sequence-batch anaerobic digester at Lochmead Farm Oregon, to process manure generated from 750 cows into 1.5 MW of power (Sullivan 2012). Current application of AD technology on an industrial scale in various

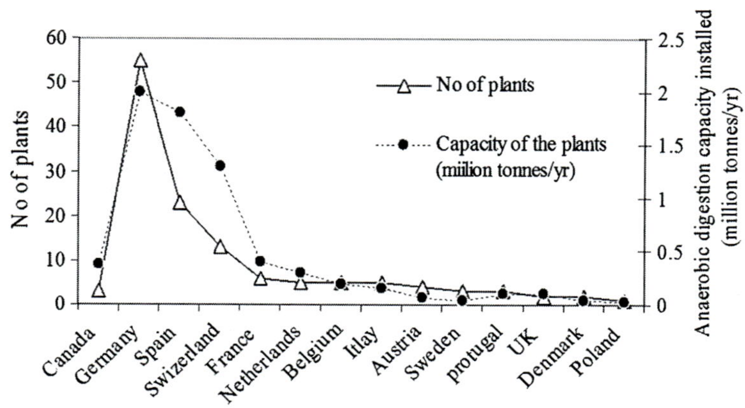

Figure 8.1. Industrial application of AD technology in selected countries

countries is given in Figure 8.1 (Press 2008; Baere and Mattheeuws 2010; Levis et al. 2010).

The common AD technologies in use include Wassa process, Dry anaerobic composting process (Dranco), Kompogas process, Valorga process, Super Blue Box Recycling (SUBBOR) process, Biopercolat process, Biocel process, Sequential Batch Anaerobic Composting (SEBAC) process, Anaerobic Phased Solids (APS) Digester process, BioConverter, and BTA process, (Angelidaki et al. 2003; IEA 2008; Rapport et al. 2012). The details of these AD technologies with advantages and disadvantages are summarized by Rapport et al. (2012). In addition, the numbers of plants with capacity and type of operation were also reported by Rapport et al. (2012) and in the report, commercial assessment on AD technology for biomass projects by Renewable East (http://www.biomassenergycentre.org.uk/). There are 107 companies around the world operating the anaerobic digesters or co-digesters. AD or co-digestion of waste is considered as solution for the waste management. During last 5 years there are more than hundreds of publications and case studies on AD or co-digestion of wastes, however, researchers are still carrying out research on different organic wastes to co-digest in order to reduce the digester volume, increase the methane production, and reduce the volume of waste.

8.2 ANAEROBIC DIGESTION AND CO-DIGESTION

Principles of AD. Traditionally, anaerobic digestion processes are used as conventional digestion processes for sludge solids degradation in wastewater treatment plants. Anaerobic waste decomposition will generate methane and biogas, which are principal GHGs. For example, decomposition of 1 metric tonne of grass waste will release 90–140 m^3 of methane and 50–110 m^3 of carbon dioxide into the atmosphere (Yu et al. 2002; Muhammad Nasir et al. 2012). For this reason, many wastewater treatment plants are equipped with facilities (e.g., internal combustion engines or gas turbines for power generation) to recover and utilise methane. However, most of these facilities are setting these instead of being used for energy production. This is due to the complexity of facilities, the cost of maintenance and operation, and the need to treat CO_2 and trace compounds (e.g., ~20,000 ppmv H_2S) in the biogas. With increased concerns for the disposal of the solid waste generated from different sectors (agriculture, industrial, etc.), AD has been used for stabilising, reducing volume and generating renewable biogas from the solid wastes, and thus, becomes a sustainable/green technology.

AD is the microbial conversion process/breakdown of organic materials/wastes (mainly biodegradable matter) into a methane and carbon dioxide mixture (called biogas), anaerobic biomass and organic residuals (that can be used as a fertilizer) in the absence of elemental oxygen. In general, AD includes at least four distinct stages (Fig. 8.2), namely hydrolysis, acidogenesis, acetogenesis and methanogenesis (Tchobanoglous et al. 2003). The 1st stage involves hydrolytic microorganisms, which hydrolyse the complex substrates (carbohydrates, lipids, proteins, etc.) to dissolved monomers (sugars, fatty acids, amino acids, etc.). The 2nd stage involves acedogenic microorganisms to convert monomers to CO_2, H_2, volatile fatty acids (VFAs) and alcohols. The 3rd stage involves hydrogen producing acetogens that convert the simple monomers and VFAs to acetate, H_2, and CO_2. The 4th stage involves methanogenic microorganisms that utilize the H_2, CO_2 and acetate to produce CH_4 and CO_2.

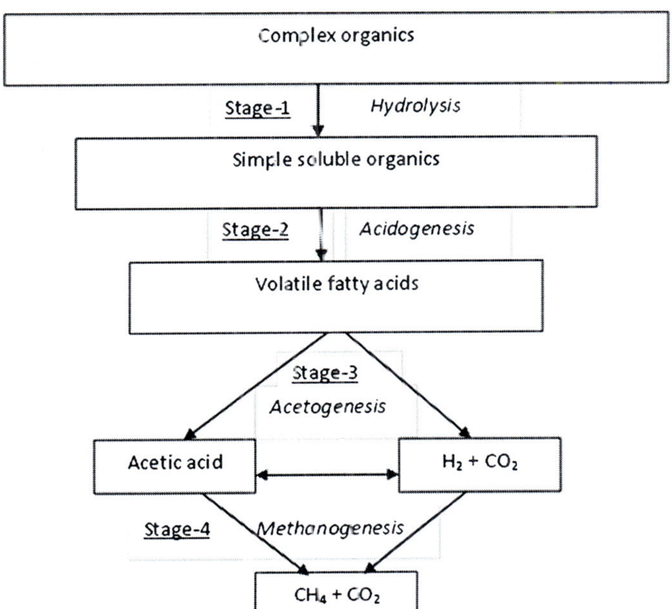

Figure 8.2. Different stages of AD or co-digestion processes

It should be pointed out that AD occurs in a series of microbiological processes with the participation of many different groups of microorganisms (Harper and Pohland 1986). Since the 1980s, substrate conversion rates and nutritional requirements of major groups of anaerobic bacteria have been explored in more detail, focusing particularly on the role of hydrogen in selecting populations and directing the nature of metabolic intermediates (Harper and Pohland 1986; Zhang and Noike 1993). It has been reported that the increase of H_2 pressure in a digester will usually cause a change in VFA distribution (e.g., the propionate accumulation) or even the type of fermentation (Boone and Xun 1987; Thiele et al. 1988), but will not inhibit the syntrophic methanogenic reactions in microbial aggregates (i.e., floc) (Thiele et al. 1988). Accumulated hydrogen strongly inhibits the degradation of volatile fatty acids, such as propionate and butyrate, resulting in a consequent deterioration of the normal operation. If hydrogen production exceeds the maximum ability of the methanogenic biomass to degrade hydrogen, there will be a rapid and large increase in the hydrogen concentration prior to digester failures. Retention time of anaerobic digesters may not affect the population levels of hydrolytic bacteria and H_2-producing acetogens, but may have a considerable effect on the population levels of methanogens, homoacetogens and sulfate-reducing bacteria and on the composition of fermentative products (Zhang and Noike 1994). So far, there have been huge numbers of publications and reports on the AD of waste. Some of the reviews on AD process include Nallathambi Gunaseelan (1997), Mata-Alvarez et al. (2000), Demirel and Yenigun (2002), Salminen and Rintala (2002), Ahring (2003), Zhou et al. (2006), Pesta (2007), Gray et al. (2008), Madsen et al. (2011), Li et al. (2011), Muhammad Nasir et al. (2012) and Rapport et al. (2012).

Co-digestion. Anaerobically digesting two or more homogeneous wastes together to produce biogas is defined as co-digestion. The anaerobic sludge digestion process to CH_4 and CO_2 is a slow process and requires a longer retention time. In particular, intracellular biopolymers solubilisation and conversion of solid degradable organics to the lower molecular weight compounds through hydrolysis is a rate limiting step (Khanal et al. 2007; Pilli et al. 2011). This results in a longer hydraulic retention time (30–60 d, Tchobanoglous et al. 2003). The reduced digestion rate (e.g., the first-order digestion rate constant = 0.15 day^{-1} for sludge, Shimizu et al. 1993) is due to non-availability of the readily biodegradable organic matter for AD. Therefore, to increase the biogas production and the rate of digestion, organic substrates are added to the anaerobic digester. Moreover, AD is sensitive to the waste loads and toxic materials. Therefore, to enhance the biogas generation and minimise the toxic effect and waste loading, co-digestion of substrates has been recognized as an effective process to overcome the problems (Esposito et al. 2012; Khalid et al. 2011; Rapport et al. 2012). The main advantages of the co-digestion are given in Table 8.1.

There is a dramatic increase in the number of papers published on the co-digestion. Some of the recent studies include using the co-digestion for treatment of hazardous tannery solid waste with primary sludge (Thangamani et al. 2010), agricultural and industrial wastes (Kacprzak et al. 2010), municipal solid organic

Table 8.1. Advantages of co-digestion

Description	*Advantages*	*Reference*
Methane production	Relatively higher methane production compared to AD. Microbial synergism effect leads to the higher methane yield	Carucci et al. (2005); El-Mashad and Zhang (2007); Esposito et al. (2012)
Nutrient balance	Some of the waste such as manure has a low C/N ratio (commonly as low as 8–15). Therefore, the addition of waste with a high C/N ratio (e.g., food waste) will increase the AD efficiency. On the other hand, if the waste contain a high C/N ratio (e.g., cellulose waste: 173/1–1000/1) can be diluted with the addition of low C/N waste. Thus, balance of the nutrients can be maintained in the digester with the addition of co-substrates	Carucci et al. (2005); Zhang et al. (2008); Siddiqi et al. (2011); Atandi and Rahman (2012); Esposito et al. (2012);
Volume of feedstocks	When resources or feedstocks are limited, co-digestion will help in increasing the volume of the feedstock (wastes) and biogas generation. For example: manure generated from a single farm may not be sufficient in large scale digesters; therefore adding others waste (co-digestion) will sustain the AD	Braun and Wellinger (2010); Atandi and Rahman (2012); Esposito et al. (2012)
Homo-genisation	Proper homogenisation of particulate matter, floating matter, or settling waste can be achieved with mixing animal manures or sewage sludge	Braun and Wellinger (2010)
Toxicity	Due to dilution of compounds in co-digestion the toxicity of the specific compounds is minimised. Further, detoxification may be possible due to co-metabolism.	Agdag and Sponza (2007); Montusiewiz and Lebiocka (2011); Kawai et al. (2012); Pastor et al. (2013)

(Continued)

Table 8.1. Advantages of co-digestion (Continued)

Description	*Advantages*	*Reference*
Economics	Co-digestion improves the economics of AD facility, i.e., tipping fee for processing the waste, and moreover, it will increase biogas production and minimises the fossil fuel usage.	Wu (2007); U.S. EPA (2011); Atandi and Rahman (2012)
GHGs	Co-digestion will minimise the GHG emissions, i.e., with increased biogas energy will offset the fossil fuel usage. Moreover it will reduce uncontrolled emissions from land fill and land disposal of solid waste	Gray et al. (2008); Atandi and Rahman (2012)
Fertiliser	The digestate produced from the co-digestion has a fertiliser value	Esposito et al. (2012); Khalid et al. (2011)

waste with melon residue (Anjum et al. 2012), dairy manure with mulched switchgrass (Frigon et al. 2012), dairy cow manure and high concentrated food processing waste (Yamashiro et al. 2013), swine manure with energy crop residues (Cuetos et al. 2011), kitchen waste and fruit–vegetable waste (Yang et al. 2013), whey and cow manure (Hublin et al. 2012), sunflower hulls and poultry manure (Demirel et al. 2013), used oils and urban landfill leachates with sewage sludge (Pastor et al. 2013), and algal biomass with lipid residue (Park and Li 2012), etc. There have been several reviews on anaerobic co-digestion, such as Salminen and Rintala (2002), Mata-alvarez et al. (2011), Atandi and Rahman (2012), Bond et al. (2012), Esposito et al. (2012), Long et al. (2012) and Kangle et al. (2012).

Evolution of AD/Co-digestion Technologies. Conventional AD requires longer retention time and produces lower biogas. In the 1960s, the conventional AD process has been extended into two-stage (in some papers, also called two-phase) AD (Ghosh and Pohland 1974). In two-stage AD, hydrolysis, acidification and liquefaction are performed in the first stage (i.e., acid-forming phase) and methane production via acetate and hydrogen/carbon dioxide in the second stage (i.e., methane-forming phase) (Appels et al. 2008). It should be noted that the two-stage AD is different from the temperature-wise two-stage AD, e.g., a mesophilic anaerobic stage (digester) followed by a thermophilic anaerobic stage (digester), which was developed to eradicate the pathogen and to obtain Class A solids. Considerable studies have been conducted since 1970s to understand the two-stage AD and its applications (Chartrain et al. 1987; Zhang and Noike 1991, 1993). The main advantages of the two-stage AD processes are that the population levels of hydrogenotrophs (e.g., H_2-utilizing methanogenic bacteria and homoacetogens) and are in the same order of magnitude as the acidogenic bacterial in the two-stage system, while the population levels of hydrogenotrophs in the conventional single stage anaerobic digester were 0.01 to 0.1 times of that of acidogenic bacterial (Zhang and Noike 1991). Phase separation does not always mean the complete separation of the acid and methane formers, rather, it strengthens the ecological relationships among the trophic groups of microorganisms in each phase (Zhang and Noike 1991, 1993). While their principles, design, process behavior as well as operation and maintenances (O/M) are well known, the two-stage AD processes have almost never been used for sludge treatment in full-scale wastewater treatment plants (WWTPs).

On the other hand, feasibility of using AD for garbage treatment was studied as early as the 1930s, with different digesters (e.g., bench-scale batch digester, standard Imhoff digester, large continuous stirred-tank digesters or even medium-scale two-stage anaerobic digesters) (CIWMB 2008). Pilot-scale tests were conducted between the 1970s and 1990s in the U.S. and Europe to test the feasibility of using AD processes for solid waste treatment, particularly for municipal solid waste (MSW) treatment in order to i) reduce the amount of material being landfilled, ii) stabilize organic material before disposal in order to reduce future environmental impacts from air and water emissions, and iii) recover energy (CIWMB 2008). For example, a pilot two-stage digester was tested in the 1990s at the Illinois Institute of Technology on sewage sludge mixed with the MSW consisted of 72% paper, food, and garden waste. The two-stage digester consisted of an inclined 76 L (20 gal) first stage followed by a vertical 230 L (60 gal) packed-bed reactor for the second stage. The pilot system was operated in various modes at a range of OLR from 2.4 to 7.5 kg/m^3-d and HRT from 7 to 13 d, resulting in methane yields of 0.138–0.222 m^3/kg VS (CIWMB 2008). Fig. 8.3 shows the evolution of solid waste anaerobic digester capacity in Europe. Today, AD and aerobic composting of MSW have been well established in Europe, largely due to waste and energy policies in Europe (e.g., The Landfill Directive). By the end of 2006, there were some 124 anaerobic digester plants with capacity greater than 3,000 MT/y (3,300 tons/y) treating feedstock composed of at least 10% MSW (CIWMB 2008).

Since the 1990s, the hydrolysis stage has been enhanced with various pre-treatment technologies (physical, chemical, physio-chemical, and biological) in order to increase methane production and reduce the retention time. Pre-treatment processes have been adopted to pre-treat feedstocks (generated from different sources). Evaluations of these pre-treatment technologies to increase the methane production are still at lab-scale studies. Similarly, co-digestion of feedstocks with sludge was studied initially to increase the cost-effectiveness of the AD process; later it was adopted for co-digesting different feedstocks. With research advancements, anaerobic co-digestion is also adopted for digesting semi-dry and dry organics wastes. Still researchers are evaluating and optimising the co-digestion process for different feedstocks. In addition, the treatment of solid wastes using AD or co-digestion adds several new challenges because of the variety in the feedstock and the space limitations where such facilities would be located. The organic fraction of MSW (OFMSW) may contain

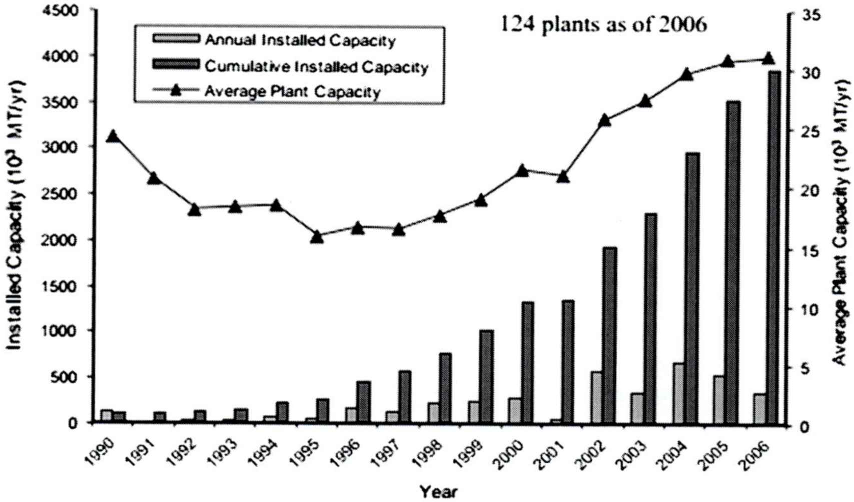

Figure 8.3. Solid waste anaerobic digester capacity in Europe. Annual installed capacity for 1990 includes all years prior to 1990; all other years include only new installations. Average plant capacity indicates the number of plants in operation
Source: CIWMB 2008; reproduced with permission from California Department of Resources Recycling and Recovery (Cal Recycle)

food, yard waste or paper in varying concentrations, sizes, and composition. Furthermore, MSW is contaminated with non-organics, such as glass and metal, and therefore requires pre-treatment to separate the feedstock. Although the ideal waste stream for an AD plant would be source-separated organics, the reality is that there is always a small degree of contamination that must be handled on site. Nevertheless, AD or co-digestion is the focal point for generation of renewable energy for biomass; the related technologies have the opportunity to be an integral part of the solution to two of the most pressing environmental concerns of urban centers: waste management and renewable energy.

In addition, there are new technology applications for the biogas upgrading (Petersson and Wellinger 2009) and the technologies used for energy production with biogas, depending upon the end use. Several biogas cleaning/upgrading technologies include pressure swing adsorption (PSA), cryogenic upgrading, adsorption, water scrubbing (WA), organic physical scrubbing, chemical scrubbing and membranes. Out of these technologies, the most widely used ones are pressure swing adsorption, water scrubbing, organic physical scrubbing and chemical scrubbing. More than 100 AD or co-digestion plants from 16 countries (Austria, Canada, France, Germany, Iceland, Japan, Norway, Netherland, South Korea, Spain, Sweden, Switzerland, USA and UK) are upgrading with new technologies for biogas purification (Petersson and Wellinger 2009). Finally, the technologies used for energy production with biogas need to improve. As shown in Table 8.2, ICE, the most used biogas energy recovery systems, presents the worst

Table 8.2. Characteristics of different ways for energy recovery of biogas[a]

Method for energy recovery	*Electric efficiency*	*Fuel consumption (MJ/kWh)*	*Emission (μg/kJ)*		*Installed power (MW)*	*Investment Cost (M$)*	*O & M ($/kW-y)*
			NO_x	*CO*			
ICE	33%	11	56.6	56.6	1	1.2	115.2
GT	28%	12.9	15	19	0.84	1.26	99.84
ORC (Geothermal)	18%	19.2	16	18.9	0.54	0.81	61.44
MCFC	50%	7.2	Trace	1.4	1.5	4.2	96
SCE	38.6%	9.4	3.11	15	–	–	–
SOFC	–	–	–	–	1.5	5.25	84

[a]Adapted from Shan et al. (2010). The hypotheses include availability of the systems (320 d/y), systems life (10 years) or fuel cell stack life (5 years), and fuel cell stack cost equal to one third of the fuel cell system cost. ICE = Internal Combustion Engines; GT = Gas Turbines; ORC = Organic Rankine Cycle; MCFC = Molten Carbonate Fuel Cells; SCE = Stirling Cycle Engine; and SOFC = Solid Oxide Fuel Cell

environmental performance of all. The emissions of Stirling cycle engine (SCE) and high temperature fuel cells (both MCFC and SOFC) are very low, which warrants more research.

8.3 DIGESTER TYPES AND MAJOR AFFECTING PARAMETERS

8.3.1 Types of Anaerobic Digesters and Co-Digesters

Different types of anaerobic digesters or co-digesters have been used for digestion of solid waste/feedstocks. The design of the anaerobic digesters or co-digesters depends upon the quantity and type of the feedstock that is going to be processed. As shown in Fig. 8.4 and Table 8.3, classification of the anaerobic digester can be done on the basis of the following parameters: i) moisture content, ii) feeding mode, iii) temperature, iv) mode of operation, and v) shape/structure. Anaerobic digesters are broadly classified as the dry and wet digester; further classifications are on the basis of feeding mode and temperature (Fig. 8.4). Initially the anaerobic digesters are operated under wet conditions (solid content <15%w/w) for sludge volume minimization and weight reductions in the WWTPs. To enhance the digestion and methane production and to manage the feedstocks or solid waste, the feedstock is added to the sludge wet digesters. Thus, dry digesters are evolved (solids content >15%w/w). Similarly, mesophilic and thermophilic digesters are evolved on the basis of the temperature range. Mostly the high solids digesters are

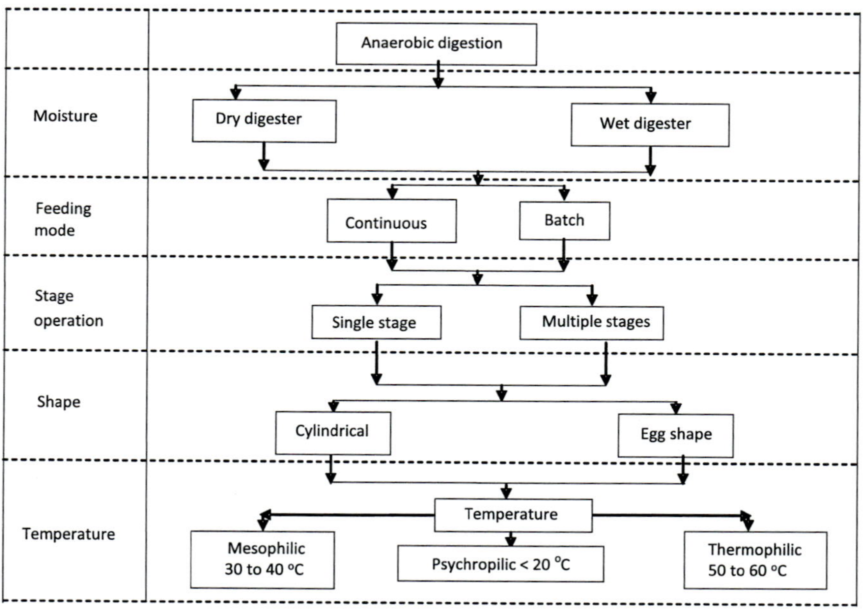

Figure 8.4. Classification of dry digester and wet digester

Table 8.3. Types of anaerobic digesters used for different feedstocks

Parameter	*Type of digester*	*Conditions*	*Advantages*	*Disadvantages*
Moisture content	Dry digester	Dry solids content >15% w/w. There is no definitive division between these two designations	Required smaller volume, very small volume is required and volume heat requirement. Produces compost.	Additional equipments for mixing and material flow.
	Wet digester	When dry solids content in the digester is ≤15% w/w	Dilution to the inhibitory substance; requires less sophisticated mechanical equipment; produces liquid/granular fertilizer.	Pre treatment is required; requires high consumption of heat and larger volume digesters
Feeding mode	Continuous	Fed continuously with digestate material; fully degraded material is continuously removed from the bottom of the reactor.	Digesters can be used as standby or storage tanks	Operational problems and requires high skill labor
	Batch	The substrate is fed into the digester until the decomposition process is near completion	Low cost; simplified material handling; Reduced pre-sorting and treatment; Higher rate and extent of digestion than landfill bioreactors	Requires increased pre-treatment to provide a suitable refined digestate material.

(Continued)

Table 8.3. Types of anaerobic digesters used for different feedstocks (Continued)

Parameter	*Type of digester*	*Conditions*	*Advantages*	*Disadvantages*
Temperature	Mesophilic	Temperature between 30 to 40°C.	More robust and tolerant process than thermophilic.	Larger digestion tanks, separate sanitization stage is required.
	Thermophilic	Temperature between 50 to 65°C.	Increased biogas production.	Higher maintenance cost; need effective control; process more sensitive to environmental variables.
Stage	Single stage	All digestion occurs in one digester vessel	Simpler designs suffer less frequent technical failures and have smaller investment costs.	Requires to proceed under the same operating conditions despite differences in growth rates and optimal pH of the microbial groups involved in each step.
	Multiple stage	Consists of several digesters, often the acid forming stage is separated from the methane forming stage	Offers optimization of the digestion conditions; increased biogas production.	High sophisticated equipments required, complicated designs, high investment cost.

Shape	Cylindrical	They are not often less than 6 m or >38 minimum diameter; the water depth should be >7.5 m	Reactor shape allows a relatively large volume for biogas storage, easy in construction, cost can be competitive.	Inefficient mixing and dead spaces, accumulation of grit and silt, digester have to shut down for removing scum and silt
	Egg shape	Similar to cylindrical digesters and consist of unconfined gas mixing, mechanical draft-tube	Efficient mixing; minimum grit accumulation; more homogeneous biomass; lower O/M cost; cleaning frequency required, less land area	External gas storage is required high–sophisticated structures, greater foundation design requirements and seismic consideration, high cost

Table 8.4. Important parameters that affect AD processes

Parameters	*Conc.*	*Effect on AD or co-digestion*	*Reference*
Ammonia	4051–5737 (mg NH_4-N/L)	At this concentration methanogens will not sustain, only *Methanosarcina barkeri, Methanobacterium thermoautotrophicum*, and *Methanobacterium formicicum* can sustain at 10 gNH_3/L. It will change in the intracellular pH, increases maintenance energy requirement, and inhibition of a specific enzyme reaction. The ammonia concentration less than 200 mg/L will enhance the AD process. Total ammonia (free ammonia plus ammonium) concentrations of 1.5–3.0 g-N/L at pH 7.4 and above will cause inhibition but at mesophilic temperature if ammonia is above 3.0 g-N/L, will cause inhibition at all pH levels.	Chen et al. (2008); Khalid et al. (2011)
pH	6.5 to 7.2	The hydrolysis and acidogenesis will occur at low pH 5.3 to 6.5. If the pH is less than 6.5 which will result in the toxicity to methanogens and if the pH is greater than 7.5 the ammonia concentration will increase and inhibit the methane production (ammonia toxicity). Methanogenic bacteria are extremely sensitive to pH with optimum between 6.5 and 7.2.	Kim et al. (2003); Turovskiy and Mathai (2006); Chen et al. (2008); Ward et al. (2008);

Temperature	30–37°C (mesophilic) 49–57°C (thermophilic)	Growth rates of microbes tend to increase with temperature, within the temperature tolerances of the microorganisms. Too high temperature, however, will cause the metabolic rate to decline due to degradation (denaturing) of enzymes. Waste with high concentration of ammonia is inhibitory and less stable at thermophilic than at mesophilic temperature during AD. Mesophilic digestion is most commonly employed with the majority of installations operating at ~35°C, and thermophilic digestion has been used on a limited basis.	Taricska et al. (2007); Chen et al. (2008); Tchobanoglous et al. (2003); Chae et al. (2008); Khalid et al. (2011);
Alkalinity and VFAs/alkalinity ratio	1.4:1 (bicarbonate/VFA)	Alkalinity is essential for pH control. Alkalinity serves as a buffer that prevents rapid change in pH. The VFAs produced during AD tend to reduce the pH. A molar ratio of at least 1.4:1 of bicarbonate/VFA should be maintained for a stable and well buffered digestion process. Methane production will be minimised.	Appels et al. (2008); Chen et al. (2008)
C/N ratio	20–35	The C/N ratio of 20–30 will provide sufficient nitrogen for the AD process. At C/N ratio higher than 30 causes slower microorganisms multiplication and structural material metabolism of microorganisms. This will result in lower substrate degradation. For fruit and vegetable waste, C/N ratio 22 to 25 is the optimum. Co-digestion of algal sludge and waste paper the optimum C/N ratio is 20–25. During the hydrolysis and acidogenesis the required C:N:P:S ratio is 500:15:5:3 and for methanogenesis 600:15:5:3 is required.	Khalid et al. (2011)

(Continued)

Table 8.4. Important parameters that affect AD processes (Continued)

Parameters	*Conc.*	*Effect on AD or co-digestion*	*Reference*
Nutrients	mg/g acetate Iron = 0.002 Cobalt = 0.004 Nickel = 0.003 Zinc = 0.02	There is no special nutrient required during the hydrolysis and acidogenesis. Nutrient levels should be at least in excess of the optimum concentrations needed by the methanogenic bacteria, since these are the most severely inhibited by slight nutrient deficiencies. For growth of methanogens bacteria the specific metals required are nickel, cobalt, and molybdenum and iron.	Tchobanoglous et al. (2003); Chen et al. (2008); Kangle et al. (2012); Khalid et al. (2011)
Retention time	3–55 days	For solid waste the solids retention time can vary from 3–55 days it varies with types of the organic material. For dry digester it can vary from 14 to 30 days and it can be as low as 3 days. Solid waste from poultry and slaughter house required 50–100 days retention time. A decrease in the SRT decreases the extent of the reactions and vice versa. Each time sludge is withdrawn a fraction of the bacterial population is removed thus implying that the cell growth must at least compensate the cell removal to ensure steady state and avoid process failure.	Appels et al. (2008); Chen et al. (2008); Khalid et al. (2011)
Mixing	2 min /hr	Proper mixing of waste is one of the most important operating parameters of the AD in achieving optimum performance. It helps in preventing stratification of waste, improves the contact between the micro-organisms and substrate, keeps solid particles in suspension, temperature gradients within the digester, releasing the biogas from the digester content and improves the bacterial population's ability to obtain nutrients	McMahon et al. (2001); Stroot et al. (2001); Kaparaju et al. (2007); Sulaiman et al. (2009)

operated under thermophilic conditions. Nevertheless, mesophilic and low-solids AD plants are most commonly employed because of their advantages thermophilic digestion has been used on a limited basis (Taricska et al. 2007).

8.3.2 Parameters Affecting AD or Co-Digestion

The growth of microorganisms and degradation of waste are of paramount importance during the AD or co-digestion process. To enhance the microbial activity and to increase the methane production the operational parameters of the anaerobic digester or co-digester have to be controlled. Chen et al. (2008), Appels et al. (2008), Khalid et al. (2011), Esposito et al. (2012) and Kangle et al. (2012) have summarised the effect of the parameters on AD or co-digestion process. As shown in Table 8.4, the parameters influencing the AD process are pH, ammonia, VFA/alkalinity ratio, C/N ratio, nutrient, retention time and mixing. Growth of microorganisms, mainly methanogens is vital for biogas production. Table 8.4 shows the operational limits and the effect of these parameters during AD or co-digestion on methane production. Increases in ammonia concentration, VFA/alkalinity, C/N ratio, nutrient concentration, mixing and pH will inhibit the growth of methanogens, and thus, decrease the methane production as the concentrations or levels of the parameters are directly related to methanogenic activity.

8.4 FEEDSTOCKS FOR AD/CO-DIGESTION

Any material which can be used as substrate for AD or co-digestion to produce biogas is the feedstock of anaerobic co-digestion. Initially, researchers studied the co-digestion of waste activated sludge and municipal solid waste (Pauss et al. 1984; Esposito et al. 2012). Over the years researchers have studied the biogas generation potential of the different waste materials. Moreover, the concept of anaerobic co-digestion was studied with different feedstocks generated from different sources. Feedstocks for AD or co-digestion can be divided into the following categories on the basis of the source of generation (Steffen et al. 1998): (i) feedstocks from agriculture sectors, i.e., algal biomass, energy crops (corn, grass, sunflower), harvest remains, manure (from cattle, pig, and poultry); (ii) feedstocks generated from communities, i.e., collected municipal organic solid waste from markets and households, food waste, grass clippings/garden waste, organic waste, etc; and (iii) feedstocks generated from industrial sectors, i.e., food/beverage processing, dairy, starch industry, sugar industry, pharmaceutical industry, cosmetic industry, biochemical industry, pulp and paper, slaughter house, etc.

Feedstock composition is a major parameter for AD or co-digestion. It will influence in deciding the type of digester, reactor configuration (design and operational considerations like mixing, etc.), and also on volume of biogas, bacterial physiology, digestate quality and economic of the digester facility (Siddiqi et al. 2011; Esposito et al. 2012). Researchers have combined different waste generated from different categories for co-digestion. The feedstock generation, collection and separation varies from continent to continent and from municipalities to municipalities. The feedstock characteristics vary from country to country, and they also differ with the season, life style of the people and other environmental conditions. The physical, chemical and biological characteristics of feedstock and their effect on AD or co-digestion are discussed below.

8.4.1 Physical Characteristics of Feedstocks

Knowing the physical characteristic of the feedstock will help in assessing the possibility for energy recovery and in the selection and operation of equipments and facilities. The specific density, moisture content and particle size distribution are the physical characteristics that will have a direct impact on bio-methane production during the AD or co-digestion (Pakmowski and Muller 2000, 2003; Chen et al. 2008; Khalid et al. 2011; Esposito et al. 2012; Menardo et al. 2012). The feedstocks generated from different sources contain different materials with different specific densities. It is important to know the density of the feedstock in order to assess the total mass and volume that should be fed in the AD or co-digestion system on the basis of the capacity of the digester. The moisture content of the feedstock varies, depending upon the source, season and humidity. Moisture content is important to assess the solids content of the waste for loading into an anaerobic digester or co-digester. The physical characteristics of the feedstock will reflect the type of pre-treatment required.

Moreover, AD or co-digestion processes can be distinguished as wet or dry digester on the basis of the moisture content (i.e., the total solids concentration of the feed) (Lissens et al. 2001). If the solids concentration is less than 15% (w/w), it is called the wet AD or co-digestion process; when the concentration reaches 20–40% (w/w), it is called dry digestion or co-digestion (Hartmann and Ahring 2006). The sludge generated from the secondary wastewater treatment with the moisture content greater than 97% (v/v) has low methane production. Therefore, to reduce the moisture content of the sludge several other organic wastes (organic feedstocks) are added to improve the methane production. A feed stock having high moisture requires high energy to maintain digestion temperature. The particle size of the feedstock has a significant role during the AD (Esposito et al. 2012). Smaller size of feedstock provides greater surface area for the enzymatic attack during hydrolysis (Palmowski and muller 2003; Hartmann and Ahring 2006). Many researchers (Mshandet et al. 2006; Tumutegyereize et al. 2011; Menardo et al. 2012) have reported that particle size will influence methane production. Mechanical mixers or grinders are used as pre-treatment devices to reduce the particle size of the feedstock/waste.

8.4.2 Chemical Characteristic of Feedstocks

The chemical compounds present in the feedstock undergo various forms and changes during the solid waste management. The rate of degradation and conversion of the feedstock during the treatment processes is assessed through chemical analysis. The chemical characteristics of the feedstock will directly influence the methane production during the AD or co-digestion. Therefore, chemical characteristics evaluation is essential for choosing the alternative digestion options. The chemical characteristics mainly include pH, carbon, hydrogen, nitrogen, phosphorus, sulphur, ash, toxicity and calorific value. pH is the operating parameter of AD or co-digestion. The methanogenesis cannot survive at low pH ≤ 6.3 and at high pH ≥ 7.5. Carbon, hydrogen, nitrogen, phosphorus, sulphur and ash are estimated by ultimate analysis. The quantification of these chemical data will be used in characterising the chemical composition of the organic matter in the feedstock/solid waste. Moreover, the data will help in achieving the proper mix of feedstocks to attain the suitable C/N ratio for AD. The effect of carbohydrates, proteins, fats, carbon, nitrogen and C/N ratios during anaerobic co-digestion is detailed in anaerobic co-digestion of organic wastes by Esposito et al. (2012).

The toxicity characteristics of the feedstock include heavy metals, pesticides, insecticides, which will influence the treatment pathway and disposal pathway. The carbohydrate, proteins, lipids, and natural fibres are biochemical compounds of the feedstock. Characterization of these compounds will help in understanding the fate in the feedstock as it moves through the AD or co-digestion process.

8.4.3 Biological Characteristics of Feedstocks

Biological characteristics of the feedstocks change from place to place. The biological characterization is important to estimate the amount of the methane produced and also to calculate the amount of feedstock degraded and volume reduced. The organic fraction is one of the important biological characteristics to estimate what are biodegraded to CO_2, CH_4, inert organic and inorganic solids (Zhang et al. 2007). The anaerobic digesters are designed on the basis of the biodegradability of the organic matter. To enhance the AD or co-digestion for methane production, organic feedstock is added to the sewage sludge digester. Moreover, characterising the putrescible nature of the organic matter and digesting or co-digesting them will minimise the production of odours and generation of flies.

8.5 PRE-TREATMENT OF FEEDSTOCKS FOR AD OR CO-DIGESTION

Pre-treatment methods have mainly evolved to increase the dewaterability of the sludge; further they have been adopted to enhance to AD efficiency of sludge. With the increase in applications of AD technology for waste management, the pre-treatment methods have also been become dominant in enhancing the biogas production. This section presents main pre-treatment methods adopted for sludge and pre-treatment methods for feedstock generated from different sources.

8.5.1 Pre-treatment of Sludge

The AD of sludge requires longer retention time for breaking the complex substrates to CH_4 and CO_2. To improve the availability of biodegradable and soluble organic matter to the microbiological process during AD, pre-treatment of sludge is necessary (Pilli et al. 2011; Khanal et al. 2007). Pre-treatment of sludge is required to rupture the cell wall and to facilitate the release of intracellular matter into the aqueous phase and thus increase biodegradability and enhance the AD rate with reduced retention time and with higher biogas production (Khanal et al. 2007; Carrere et al. 2012). Thus, using an efficient pre-treatment process, the substrate can be made better accessible for the anaerobic microorganisms, increasing the degree of degradation and increasing the methanogenic potential of the sludge.

With the advancements in technology, considerable studies have been conducted on physical (freezing & thawing and osmotic shock), chemical (ozonization, Fenton reagent, acids and bases), thermal (thermal hydrolysis and microwave), mechanical (ultrasound stirred ball mills, high pressure homogenizer, pulse power technique, rheology homogenizers, and lysate centrifuge-technique), and biological pre-treatment (enzymatic lysis) processes to substantially increase biogas production. Carrere et al. (2012), Pilli et al. (2010), Sheng et al. (2011), and Tyagi and Lo (2011), have published review papers summarizing the different pre-treatment process in enhancing the AD of sludge. Mudhoo (2012) have published a book on the pre-treatment methods used to increase the biogas production from the wastes. The pre-treatment processes can enhance the biodegradability of sludge by several orders; however, economical constraints of these technologies have limited their scale-up and lab-to-field implementation. Table 8.5 shows the pre-treatment technologies that can improve methane production and AD efficiency of the sludge.

8.5.2 Pre-treatment of Different Feedstocks

Pre-treatment of feedstocks is required to enhance the biogas production due to the poor availability of easily biodegradable substrates. Increasing the biogas production will increase economics of the AD or co-digestion. In the literature there are various pre-treatment technologies used to enhance biogas production from different substrates or feedstocks, and they are summarized in Table 8.6 in this chapter. Mosier et al. (2005) discussed the features of promising technologies for pre-treatment of lignocellulosic biomass. During the last decade, various researchers were using the pre-treatment/pre-processing methods for increasing the co-digestion efficiency of substrates/feedstocks (Esposito et al. 2012). The commonly used pre-treatment methods are presented in Table 8.6 with advantages and disadvantages. Espostio et al. (2012) summarised the physical, biological and physical–chemical pre-treatments methods to enhance the co-digestion. The remaining of this section discusses recent studies on pre-treatment and their effects on biogas production during anaerobic co-digestion.

Table 8.5. Methods used for sludge pre-treatment

Pre-treatment method	*Compared with control, increase in (%)*		*Solids degra-dation(%)*	*Advantages*	*Disadvantages*
	Biodegrade-ability	*Biogas production*			
Ultra-sonication	50–80	10–100	40–70	Increased dewaterability, no odour, easy to implement, operation stability is high, technical state-of-art is good, no clogging problems, energy recovery	Erosion in the sonotrode; high energy consumption.
Thermal pre-treatment	60–85	40–100	60–80	Increased dewaterability, best sludge disinfection, Energy recovery	Odour problems, corrosion problems, operation stability is poor, toxic and refractory compounds formation.
Ball mills	15	10	40–60	Easy in operation, no odour generation	Huge erosion in the grinding chamber, high energy losses, clogging problems, low degree of disintegration, operation stability is low.
Fenton pre-treatment	30–70	10–75	20–45	Increased dewaterability, sludge disinfection.	Erosion problems, requires low pH & high safety
Ozone pre-treatment	10–15	8	36	Increased dewaterability, sludge disinfection.	High energy consumption, low degree of disintegration.

Table 8.6. Pre-treatment or pre-processing methods used for different solid feedstocks

Methods or pre-processes	*Advantages*	*Disadvantages*	*Reference*
Physical			
Source separation Manual sorting Freezing and thawing	Will minimise the further processing of source separated organic waste, which could be directly used for the AD process. Sorting will increase the efficiency of the methane production and reduce the volume of waste to be managed. Mechanical operations will be easy. Freezing and thawing improves the AD of the kitchen waste. Removing materials may decrease the quality of the digestate	Require high energy	Mata-Alvarez (2003); Omar et al. (2009); Mata-Alvarez et al. (2011);
Chemical			
Acid, base, ozone, etc.	Accelerate the methane production and minimises the retention time of the AD	High cost for the required chemicals	Mata-Alvarez et al. (2000); Mata-Alvarez (2003)
Mechanical			
Crushing, grading, magnetic separation, eddy current separation, ballistic separation, trommelling, sorting, sonolysis and blender	Separation of the waste mixture into reusable, e.g. inorganic, organic, light and heavy fractions (RDF). Provision of optimum waste characteristics for the biological pre-treatment. Increase in the available specific surface improves the biological process	Requires high operation and maintenance cost. Replacement of the wear and tear parts	Mata-Alvarez et al. (2000); Mata-Alvarez (2003)
Physico-chemical			
Physical process + acid, alkali are combined	Minimises the solids retention time in the AD. Increases the biogas production	Requires energy and chemicals	Cheng et al. (2008); Khalid et al. (2011); Carrere et al. (2012)
Thermal			
Thermal hydrolysis	Pathogens will be killed, produces sanitised digestate, particle size will be reduced, increases the biogas production rate	Required high energy	Mata- Alvarez (2003); Sasaki et al. (2010);

Physical Pre-treatment of Waste/Feed Stocks. Physical pre-treatment/ pre-processing methods are mainly used to remove the inorganic matter (metal, glass, sand, etc.) from the solid waste, which improve the AD or co-digestion, methane production, volume of waste to handle, and ease of operations (mixing, pumping, etc.). Removing inorganic metals will reduce the wear of equipment, clogging, and increases the digester volume. Sorting, separation, and freezing and thawing are the important physical pre-treatments for the solid waste co-digestion. The anaerobic co-digestion process is dependent on these pre-treatment methods. Sorting the toxic and inert materials will increase the anaerobic co-digestion efficiency (Aragaw et al. 2013). Separation of higher density materials, metals, etc. will ease the operations and increases the digester volume. Freezing and thawing will improve the methane production, which is mainly used for source separated organic waste (Omar et al. 2009).

Mechanical Pre-treatment of Waste prior to Co-digestion. Mechanical pre-treatment processes are used in order to reduce the particle size of the feedstocks/ waste materials. Reducing the particle size of feedstocks/waste materials increases direct surface area of contact with microorganism, increases the biogas production and reduces the volume of the substrate. Particle size of the substrate influences the co-digestion operating parameters and efficiency (Cesaro et al. 2012; Espostio et al. 2012). Recent studies have reported to use ultrasonication as pre-treatment to enhance the co-digestion efficiency (Luste et al. 2009; Saifuddin and Fazlili 2009; Blank and Hoffmann 2011; Castrillón et al. 2011; Elbeshbishy et al. 2011; Saha et al. 2011; Hidalgo et al. 2012; Castrillón et al. 2013a, b).

During the last 5 years researchers were using ultrasonication (or sonolysis) to enhance the biodegradability of waste activated sludge. It is well analysed and documented that ultrasonication reduces the particles size of the sludge solids. Cesaro et al. (2012) studied the effect of sonolysis on solid waste and its effect on biogas production and concluded that sonolysis can significantly improve the solubilisation of organic solid waste and increases the biogas production, i.e., 24% higher than the control. Monou et al. (2009) studied the anaerobic co-digestion of solid fruit waste by using a food blender to blend oranges, bananas, and water melon to reduce the particle size of the wastes; they concluded that anaerobic co-digestion is a viable treatment option for industrial biowastes and livestock wastes produced.

Chemical and Physico-chemical Pre-treatment. Chemical addition (alkali, acid, base or ozone, etc.) will hydrolyse the organic matter and minimises the hydraulic retention of the organic waste in AD and enhances the biogas production. When both the physical and chemical pre-treatment methods combined together to pre-treat the feedstock, they are defined as physico-chemical processes. Espostio et al. (2012) summarised the literature results on chemical pre-treatment of waste prior to co-digestion. Taherzadeh and Karimi (2008) summarised different pre-treatment processes (physical, chemical, physic-chemical and biological processes) to pre-treat lignocellulose waste for improving the biogas production.

Biological Pre-treatment. Researchers have used biological pre-treatment to enhance the AD of sludge (Komastu et al. 2007; Espostio et al. 2012). Espostio et al. (2012) detailed the biological pre-treatment studies reported in the literature to enhance AD. During the biological pre-treatment microorganisms are used to hydrolyse the cells. Recent studies on biological pre-treatment of various substrates are reported by Komastu et al. (2007), Forgacs et al. (2013) and Kassuwi et al. (2012).

8.6 PROCESS DESIGN AND MASS/ELECTRON/ENERGY BALANCE

8.6.1 Process Design

Process design is very critical for an energy efficient process. As shown in Table 8.3 and Fig. 8.4, the type of feedstocks, organic loading rate, hydraulic retention time, temperature of the reactor, and mixing type are the important parameters that influence the AD or co-digestion process, and therefore, their design. As described above, the feed stocks available for the bio-energy production can be classified on the basis of the source of generation; they can be classified as low, medium and high solids feedstocks as well. Table 8.7 shows the type of anaerobic digesters to be designed on the basis of the feedstock classification.

Design of AD processes is on the basis of the theoretical relationships among hydraulic retention time, volumetric loading, volatile solids destruction and observed volume reduction (Tchobanoglous et al. 2003). Temperature is also an important parameter as microorganisms are very sensitive to the temperature. On the basis of temperature, digesters are designed as mesophilic and thermophilic ones. Design of mixing systems is a crucial element. Proper mixing is required in an anaerobic digester or co-digester to avoid stratification, temperature gradient, deposition of solids as well as sums and crust formation; moreover, it will also help in release of biogas. Having an efficient mixing system will maintain physical, chemical, and biological uniformity, which will directly influence the biogas production.

8.6.2 Mass/Electron/Energy Balance

Establishing a self-sustainable process is important. Therefore, it is necessary to estimate energy requirements in the anaerobic process. To do so, two tools, that is, mass and/or electron and energy balance, are always used. With mass and energy balance evaluation, a proper design and operation of an AD system can be achieved. Here, the concepts and principles of how to use these tools are described.

Mass/Electron Balance. Mass balance takes into account inputs, outputs and transformations of matter on the basis of the law of mass conservation. Fig. 8.5 shows the mass of the AD of solid waste. Fig. 8.6 shows a simplified mass balance of a sustainable MSW management system.

The microbial degradation of the substrate to CH_4 and CO_2 during anaerobic co-digestion can be calculated on the basis of electron and mass balance as given in Eq. 8.1:

Table 8.7. Classification of feedstock and types of digester used

	Feedstock		
Parameters	*Low solids*	*Medium solids*	*High solids*
Classified	<3% total solids by weight, low suspended solids, readily mixed	3 to 12% total solids by weight, slurry phase but can be mixed with traditional methods	Up to 25% total solids by weight, solid processing system requires sophisticated mixing methods.
Feed stocks	Secondary wastewater treatment, spent beverages & out of spec/expired products, hydraulic flush manure systems (swine), commercial wastewater, port-a-let pumpings, whey and cheese wastes.	Dairy manure, fats/oils/grease (FOG), paunch manure, "Scraped" swine manure, industrial DAF sludge, used restaurant cooking oil	Organic fraction of MSW, agriculture residues, food processing waste, food residuals, clarifier sludge (pulp/paper), green and yard waste
Type of Anaerobic digesters used	Completely mixed reactors, filter reactors, sludge blanket reactors, fixed-film packed bed reactors.	Plug flow reactors, complete mix slurry reactors, slurry-loop reactors, acid-phased or temperature phased, contact reactors.	Dry continuous (plug flow, complete mix), dry batch with permeate recycle, sequencing batch reactors.

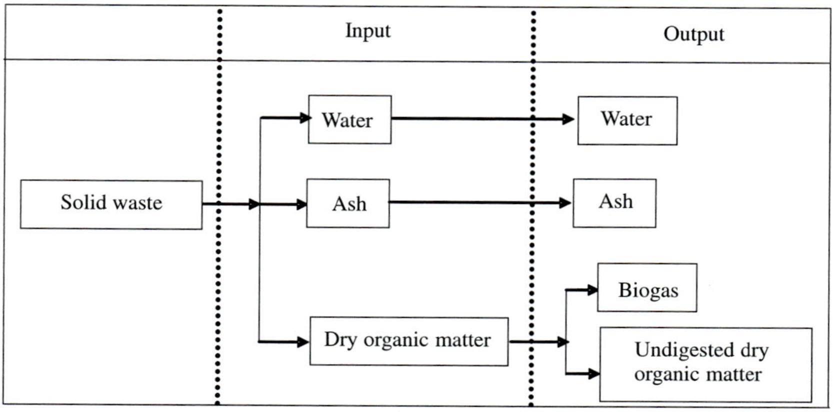

Figure 8.5. Mass balance of AD of solid waste

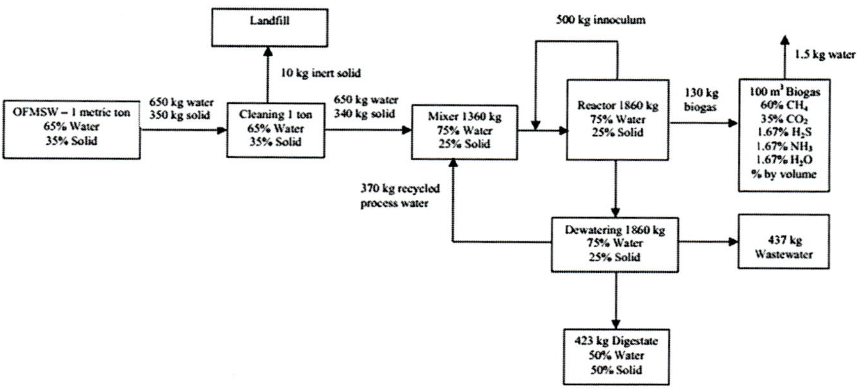

Figure 8.6. An example of simplified mass balance of a sustainable MSW management system

Source: Adapted from Ostrem 2004; reproduced with permission from Columbia University

$$\text{CnHaObNz} + \text{X}H_2O \rightarrow \text{Y}CO_2 + \text{K}CH_4 \qquad \text{(Eq. 8.1)}$$

During the degradation process of a substrate the nutrients with C, H, P, and Fe donate 4, 1, 5, and 3 electrons, respectively, and O, N, and S accepts 2, 3, and 2 electrons, respectively. Therefore, the redox conversion of the general substrate can be defined by the term 4n + a – 2b – 3z (i.e., referred to the available electrons in the organic material). When 4n + a – 2b – 3z is referred to 1 carbon atom it is described as the "degree of reducibility" of a substrate and denoted as "Dr" as given in Eq. 8.2 (Sobotka et al. 1983):

$$X=(n-a/4-b/2+3z/4); \quad Y=(n/2-a/8+b/4-3z/8);$$
$$K=(n/2-a/8+b/4-3z/8); \text{ and } Dr=4+a/n-2b/n-3z/n \qquad \text{(Eq. 8.2)}$$

For example, mass and energy balance for ethanol degradation and the degree of reducibility of each parameter are shown below:

$$CH_3CH_2OH \rightarrow 3CH_4 + CO_2 \qquad \text{(Eq. 8.3)}$$

when referred to 1 C the degree of reducibility is given as

$$C_2H_6O \quad Dr = 6; \quad [4 + (6/2) - (2 * 1/2)] \qquad \text{(Eq. 8.4)}$$

$$CH_4 \quad Dr = 8; \quad [4 + (4/1)] \qquad \text{(Eq. 8.5)}$$

$$CO_2 \quad Dr = 0; \quad [4 - (2 * 2/1)] \qquad \text{(Eq. 8.6)}$$

The maximum degree of reducibility is “8” for methane and minimum is “0” for CO_2; the degree of reducibility of any substrate will fall within the range of 0–8. The mass and energy balance provides a theoretical formula for the methane percent content in the biogas (as shown in Eq. 8.3) (Sobotka et al. 1983). Thus measuring the methane content of the biogas produced with the addition of organic waste will give the percentage degree of degradation of the organic substrate (e.g., 75% for ethanol as per Eq. 8.7). Hence it will be used as a parameter to evaluate the efficiency of the organic substrate degradation.

$$\%CH_4 = [(4 + a/n - 2b/n - 3z/n)/8] \times 100 \qquad \text{(Eq. 8.7)}$$

Energy Balance in AD or Co-digestion Processes. The energy balance links the amount of energy consumed, energy generated, and net energy produced in a system together as shown in Eq. 8.8. Furthermore, energy balance is very important to estimate the biogas production, GHG emissions and economical feasibility of the AD.

$$NE = Q - (Q1 + Q2 + Q3 + Q4 + Q5 + Q6 + Q7 + Q8) \qquad \text{(Eq. 8.8)}$$

where Q1 = energy requirement for pre-treating/pre-processing the waste; Q2 = energy required for pre-treating the sludge; Q3 = energy required for mixing the waste and sludge; Q4 = energy required for pumping the sludge from mixing tank to AD; Q5 = energy required for the AD; Q6 = energy required for the dewatering the digestate; Q7 = energy required for the transporting the dewatered digestate; Q8 = energy required during the land application of dewatered digestate; and Q = is the energy recovered from the methane. The relationships and locations of Q1 to Q8 and Q are shown in Figure 8.7.

If, “NE = positive”, the process is a source of revenue through utilisation of methane energy; if “NE = negative”, then the process is not a source of revenue. Municipal anaerobic digesters become regional renewable energy facilities by recovering the energy from the methane and utilizing it to run equipment, heat the

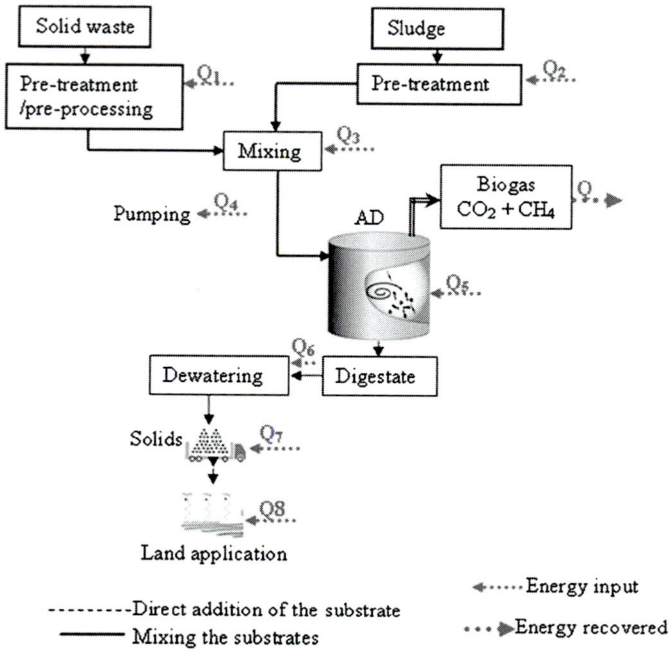

Figure 8.7. Energy requirements (Q_1 to Q_8) in and the renewable energy recovered (Q) from a co-digestion process

digester, and heating buildings. Moreover the energy input and energy recovered can be expressed as the energy ratio (E(O/I)) as given in Eq. 8.9.

$$\text{Energy ratio}(E(O/I)) = \frac{\text{Energy output}}{\text{Energy input}} \quad \text{(Eq. 8.9)}$$

If E(O/I) 1, then the process is an energy efficient process; if $E(O/I) < 1$ the process requires additional energy input. In the literature, some authors have described that AD is enhanced with biogas production increased. However, the authors have not evaluated the process in the context to the energy ratio, and thus the evaluation is not completed. Evaluation of the energy ratio for AD is necessary in order for decision makers to compare the process with other waste treatment technologies and make decisions in implementing the technology at wastewater treatment plants.

It is challenge to accurately evaluate different energy requirements (Q1 to Q8 as shown in Fig. 8.7) for the AD or co-digestion process include. Keller and Hartley (2003) identified the total energy required for pumping and mixing the sludge in an anaerobic digester to be 0.01 kW/m^3 of the reactor volume. More often than not, a heat balance is necessary to determine the heat requirements for a given installation, to size the heating system components, and evaluate the energy amounts needed for many other purposes. For example, the amount of heat necessary to raise the temperature of the incoming raw sludge to the desired level can be calculated as

$$H = Q \times Cp \times (T_2 - T_1) \qquad \text{(Eq. 8.10)}$$

where Q is the volume of sludge (m^3/day); Cp is the specific heat of sludge (4.2×103 kJ/m^3°C); T2 is the temperature of sludge in the tank (°C); and T_1 is the temperature of raw sludge entering the tank (°C). The heat requirements include the amounts needed to heat the incoming solid waste or sludge to the AD operating temperature and to compensate for the heat loss to the surrounding medium. The magnitude of heat losses from the digester is dependent upon the shape of the reactor and the type of construction used. Different materials of construction have different thermal transfer coefficients. The heat loss in the anaerobic digester is around 2–8% of the sludge heating requirements (Zupancic and Ros 2003; Bolzonella et al. 2003). Fig. 8.8 shows an example of simplified energy balance in a solid waste treatment plant with biogas generation/utilization components. In general, if the MSW management system is complicated, the energy balance can be very challenge, particularly when mass balance, carbon footprint, and life cycle assessment (LCA) are also needed to be evaluated. Most of current studies are related to evaluation of different energy requirements, the biogas production, GHG emissions, and economic feasibility of AD or co-digestion processes.

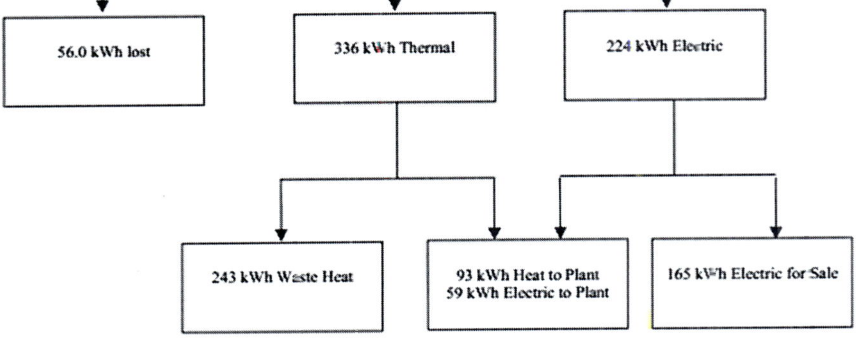

Figure 8.8. An example of simplified energy balance in a solid waste treatment plant with biogas generation/utilization components

Source: Adapted from Ostrem 2004; reproduced with permission from Columbia University

8.6.3 Biogas Production and Energy Recovery

Biogas produced during AD mainly contains a mixture of CH_4 and CO_2. A well-operated digester should contain ~65–80%CH_4 and 35–20 % CO_2 by volume with only traces of other gases (Angelidaki et al. 2006; Alvarez et al. 2010; Khalid et al. 2011). During digester upsets, the percentage of CO_2 in the off gases will increase. Biogas is an important parameter as it indicates overall performance of the process. However, the biogas production rate alone cannot be used for indicating process imbalance since its changes depends on the hydraulic and organic loading rate; the biogas yield also depends on feed composition and other conditions. Moreover, it has low sensitivity to overloading compared to other process indicators. For example, a decrease in biogas production often occurs after the process is severely inhibited or already broken down. Thus, the biogas production rate is not an effective early warning indicator (Switzenbaum et al. 1990; Moletta et al. 1994).

The fuel value of the methane gas is 31.79 $Mj/m^3\,CH_4$ (Tchobanoglous et al. 2003). Typical value of biogas production is usually estimated from the percentage of VS reduction in full scale AD (Tchobanoglous et al. 2003). Usually, Vgas ranges from 0.75 to 1.12 $m^3/kg\,VS_{destroyed}$, and the average value of V_{gas} is about 0.935 $m^3/kg\,VS_{destroyed}$. Haug (1977) outlined methane production as 0.449 $m^3\,CH_4/kg\,VS_{destroyed}$. Combined heat and power generator (CHP) are commonly used to generate energy from biogas. In parallel to power generation, a more or less high percentage of heat is developed in CHPs, depending on the power generator technology. More modern technologies like fuel cells or micro gas turbines are very seldom to be found. The total efficiency, i.e., the sum of the electrical and thermal efficiencies, is within the range 85–90% with modern CHPs (Deublein and Steinhauser 2008). Only 10–15% of the energy of the biogas is wasted. But the electrical efficiency (maximum 40%) is still very low: from 1 m^3 biogas only 2.4 KWh electric current can be produced (Deublein and Steinhauser 2008).

8.7 CASE STUDIES OF CO-DIGESTION OF DIFFERENT FEED STOCKS

There are huge numbers of case studies on co-digestion in UK, European Union, USA, and a few studies in other countries. Case studies on anaerobic co-digestion have reported a wide range of biogas yield improvements. Some of the studies are summarised in the Table 8.8.

East Bay Municipal Utilities District (EBMUD) reported that the energy production from food waste is around 730–1300 kWh/tonnes of food waste loaded; whereas treating the solids will generates 560–940 kWh/ tonnes of solids loaded (Gray et al. 2008). The biogas plant located at Bruck and Der Leitha (Austria) handles 30,000 tonnes of feedstocks per year, mainly from organic residues (food and beverage industry) but also from crops framers. In this plant, the biomethane is producing at a rate of 100 $m^3\,CH_4/h$, and the plant has two combined heat and power units and generates about 17 × 105 kWh energy

Table 8.8. Case studies of anaerobic co-digesters[a]

Country	*Type of Waste*	*Waste quantity Tonnes/d*	*Digester Capacity* (m^3)	*Temp. (°C)*	*Biogas/energy production*	*Ref.*
Denmark (Lemvig Biogas)	AM + OW	AM = 500 OW = 120	14300	53.5	10.2 million Nm^3/year	[1]
Denmark (Ribe Biogas Plant)	AM + OW	AM = 390 OW = 160	5235	53.0	5.5 million Nm^3/year	[2]
Swizerland Swiss Farmer Power Biogas plant	PM + IW + biowaste	PM = 82 IW = 41 Biowaste = 44	1350 × 2	36.0	3.1 million m^3/year	[3]
European Union (St. Martin biogas plant)	slaughterhouse waste	NA	600 × 2	55.0	4.7 MWh/d of electricity and 7 MWh/d of heat.	[4]
Austria Rohkraft biogas plant	Energy crops + PM + leachates	EC = 30.1 PM + leachates = 20	2000 &1850	39.0	4.02 million m^3/year	[5]
Sweden (Laukaa) Kalmari biogas farm	CW	2000 m^3/yr	1000	35.0	Electricity 75 MWh/year Heat 150 MWh/year Biomethane for traffic fuel 1000 MWh/year	[6]

Note: AM = animal mature; OW = organic waste; PM = pig manure; IW = industrial waste; CW = cow manure; NA = not applicable. References: [1] = http://www.iea-biogas.net/download/success_story_denmark_lemvik2013.pdf; [2] = http://www.iea-biogas.net/download/Success%20Story%20Ribe2012.pdf; [3] = http://www.iea-biogas.net/download/success_story_switzerland2012.pdf; [4] = http://www.iea-biogas.net/download/st_martin.pdf; [5] = http://www.iea-biogas.net/download/reidling_final.pdf; [6] = http://www.iea-biogas.net/success-stories.html?file=files/...kalmari2012.pdf

(http://www.iea-biogas.net/_download/success_bruck_austria2013.pdf). At Zeewolde in the Netherlands, two digesters (operated at 37°C) having the capacity of 2500 m^3 each are digesting manure, corn, grass and waste products from the food industry and producing 7500 MWh electricity as well as heat energy of 7100 MWh (http://www.iea-biogas.net/_download/success_story_zeewolde2011.pdf).

On the other hand, the National Renewable Energy Laboratory evaluated the Feasibility Study of AD of Food Waste in St. Bernard, Louisiana and concluded that AD of food waste is not economical favorable due to high initial costs for the technology, low area tipping fee, and energy prices. The other possible option suggested is co-digestion with sludge solids at WWTP (http://www.nrel.gov/docs/fy13osti/57082.pdf). In 2011, Biogen Company has designed and commissioned an anaerobic co-digester plant, having a capacity of 26,000 tonnes/yr, at Harper Adams University College, UK to treat local food waste and farm waste (400 cow dairy units). The plant generates 4000 MW electricity and 4100 MW heat every year.

8.8 ADVANTAGES AND DISADVANTAGES OF AD

8.8.1 Multiple Benefits

AD of the solid waste creates an integrated management system which reduces the environmental pollution compared with other treatment or disposal methods. The principal advantages of AD are energy recovery (from the produced methane) and the digestate with fertilizer value. But energy recovery may not be always positive! The environmental benefits, energy benefits, treatment benefits, and socio-economic benefits of AD of the waste are discussed below (Khalid et al. 2011).

Environmental Benefits. The environmental benefits associated with AD of waste are:

- Elimination of malodorous compounds: i.e., usually the organic matter present in the waste will decompose and produces ammonia, hydrogen sulphide and VFAs, which are responsible for the most odors. During the AD the VFAs are metabolized by the anaerobic bacteria to methane and carbon dioxide, which result in significant odor emissions (Khalid et al. 2011).
- Reduction of pathogens: AD result in the production of Class A solids (Lukehurst et al. 2010). The U.S.EPA has evaluated that there is 99% and 90% reduction in concentration of fecal coliform and fecal streptococci, respectively.
- Deactivation of weed seeds: AD produces the digestate with no weed seeds; therefore there is a less need of herbicides (Lukehurst et al. 2010).
- Reduction of Pollution: AD minimizes the bio-chemical oxygen demand for the waste to biodegrade (i.e., the measure of the polluting strength of a waste). On an average it can lower up to 40% of that in the digestate

(http://www.ciwm.co.uk/web/FILES/Technical/CIWM_Report_2012.pdf). It will avoid the contamination of the surface water and ground water resources.

- Production of the sanitized and nutrient rich compost: utilization of the digestate as a soil conditioner will enhance the plant growth or it can further processes into organic fertilizer or combustible fuel on the basis of the energy value. Thus AD will reduce the dependence of the inorganic fertilizers for soil enrichments by capturing and reuse of the nutrients (Lukehurst et al. 2010).
- GHG emissions: uncontrolled emissions of methane (which is 23 time more potent as a GHG than carbon dioxide) into the atmosphere from landfill, disposal of waste/sludge or storage of farm slurries will be controlled by using the AD (USEPA 2011). The captured methane under control conditions can be used for energy recovery. This will offset the utilization of fossil fuels and the GHG emissions associated. Reduced fertilizer usage will reduce the N2O. AD will result in the amount of waste trucked to distant landfill and will minimise the transportation emission (Khalid et al. 2011).

Energy Benefits of AD

- AD is a net energy producing process (Khalid et al. 2011; Mata-Alvarez al. 2011; Atandi and Rahman 2012).
- Producing sustainable biofuel to offset the usage of the fossil fuels (Fehrenbach et al. 2008; Weiland 2010; Khalid et al. 2011; Mata-Alvarez al. 2011)
- The AD with energy recovery facility can contribute to decentralized distribute power systems (Khalid et al. 2011; Mata-Alvarez al. 2011;USEPA 2011)
- Compared to raw waste/slurry, the digestate requires less energy to mix and spread (Tuomisto et al. 2012). Moreover applying the digestate will increase tillage, field capacity, reduces the soil erosion and increase the crop production.

Socio-economic Benefits

- Methane produced during AD will give direct financial returns when electricity is recovered (Weiland 2010)
- Processing the digestate as fertilizers to generate income (Lukehurst et al. 2010).
- Generating electricity from the methane can contribute the power tax credits (which are obtained from each kWh of power generated) (AEBIOM 2009).
- Methane energy reduces the dependence on energy imports (Weiland 2010; Khalid et al. 2011).
- The energy, manpower, required for moving, handling, and processing the digestate will be minimised (Al-Sedai 2008).
- Generation of sustainable energy and minimization of the operation cost (Al-Sedai 2008).

- AD of waste reduces the health hazards to humans and animals by decreasing more pathogens than the traditional treatment methods (Al-Sedai 2008)
- Construction and maintenance of the AD creates employment (Moriarty 2013).

Treatment Benefits

- Reducing the polymer requirement for sludge dewatering (Appels et al. 2008).
- AD requires less area to treat the waste compared with other treatment methods (Tchobanoglous et al. 2003).
- Solid waste disposal onto the land or landfill causes environmental pollution (e.g., soil pollution). AD reduces the volume of waste to be landfilled or for disposal; moreover it prevents the pollutions (Khalid et al. 2011).
- Energy contained in the waste is converted into methane which sustainably reduces the uncontrolled emissions from the waste degradation (Appels et al. 2008; Khalid et al. 2011).

8.8.2 Disadvantages of AD

The principal disadvantages of AD are: i) the capital cost is high because large closed digestion tanks require feeding, heating, and mixing the waste or mixture of waste and sludge; ii) microorganisms involved in AD are sensitive to small changes in the environment. Therefore, the process is susceptible to upsets; iii) monitoring of performance and strict process control are required to prevent upsets; and iv) the slow microbial conversion rates during AD processes results in high HRT (30–60 d) in the digester and larger digester volume, which are the major concerns (draw backs) of the AD process. The slow digestion process also limits the speed with which the system can adjust to changes in waste loads, temperature, and other environmental conditions.

8.9 FUTURE PERSPECTIVES

It is the authors' opinion that the future research about AD and co-digestion should be focused on the following issues:

- AD or co-digestion of the source separated organic waste with sewage sludge is not well established.
- The pre-treatment technologies to enhance the AD are mainly studied for sludge digestion. However, there is little research on the pre-treatment of organic waste for enhancing the efficiency of anaerobic co-digestion.
- Utilisation of organic acids produced during the co-digestion as carbon source for denitrification in WWTPs are not fully explored.
- Optimisation of the organic acid production from the source separated organic waste or through co-digestion has not been thoroughly evaluated.

- Pre-treatment of the lignocellulose waste and increased biogas production during AD or co-digestion needs further studies.
- Fate of the emerging contaminates during the AD or co-digestion has not been fully evaluated.
- The efficient biogas collection system and separation of the gases for the energy generation from anaerobic co-digesters is required.

8.10 CONCLUSIONS

AD technology is proven and well established technology and has experienced significant growth during last decade. AD/co-digestion technology is superior over other treatment technologies, which has been installed globally for treating the organic waste, MSW, or source separated organic waste (SSOW). In AD, the waste is converted into biogas, which is collected and used as direct energy source or converted to electricity. For example, in the developing countries they use the biogas directly as energy for cooking, heating, etc. AD with different configurations (e.g., mainly high solids, low solids one-stage, two-stage digestions) has been used for co-digesting the sludge and organic waste. However, two-stage AD is not being widely used as one-stage AD due to the operation complexity and its sensitivity to temperature variation.

The organic matter, OLR, HRT, pH, temperature alkalinity, ammonia, nutrient, and mixing are the important parameters that can affect the AD efficiency and biogas production. During the AD the organic matter is solubilised, digested and mineralised to more soluble form, providing a nutrient fertilizer. Moreover, the digestate required less energy to dewater and less volume for truck loading for disposal. Utilising digestate as fertilizer will increases the soil characteristics (water holding capacity, fertility, smooth tilling, etc.), crop production, and minimises the demand of industrial fertilizers for crop production. Moreover, AD minimise the GHG emissions by reducing the sludge production, lower volume of sludge to be dewatered and trucking, replacing the industrial fertilizer, increasing the soil properties and crop production. Increasing the soil fertility will increase easy movement of the vehicles tractors used in agriculture thus can reduces the emissions. Further the methane produced can replace the fossil fuel which can lower the GHG emissions.

Under thermophilic process the AD at high solids with well maintenance and operation can produce up to 125 cubic meters of biogas per ton of feed stock with 50–60% methane concentration. Co-digestion of organic fraction of AD has showed higher energy production. The quality of the feedstock (having the organic fraction) will influence the AD efficiency and biogas production and methane content. The collection of the source separated organic waste is the major challenge. People need to be educated regarding the separation of waste before disposal, which will minimise the energy requirements for pre-treatment of waste before AD. Out of the available technologies installed on the industrial scale Valorga has proven to be most widely used AD technology with the large number of plants and installed capacity.

It is the authors' opinion that the future research about AD and co-digestion should be focused on i) AD or co-digestion of the source separated organic waste with sewage sludge; ii) the pre-treatment technologies to enhance the AD; and iii) the efficient biogas collection system and separation of the gases for the energy generation from anaerobic co-digesters, among other issues.

8.11 ACKNOWLEDGMENT

Sincere thanks are due to the Natural Sciences and Engineering Research Council of Canada (Grant A 4984, Canada Research Chair) for their financial support. Views and opinions expressed in this article are those of the authors.

References

AEBIOM. (2009). "A biogas road map for Europe." European Biomass Association, Brussels, Belgium.

Ağdağ, O. N., and Sponza, D. T. (2007). "Co-digestion of mixed industrial sludge with municipal solid wastes in anaerobic simulated landfilling bioreactors." *J. Hazard. Mater.*, **140**(1-2), 75–85.

Ahring, B. (2003). "Perspectives for anaerobic digestion." *Biomethanation I*, B. Ahring, I. Angelidaki, E. C. Macario, H. N. Gavala, J. Hofman-Bang, A. J. L. Macario, S. J. W. H. O. Elferink, L. Raskin, A. J. M. Stams, P. Westermann, and D. Zheng, eds., Springer, Berlin, 1–30.

Al Seadi, T. (2008). *Biogas handbook*, Syddansk Universitet, Odense.

Alvarez, J. A., Otero, L., and Lema, J. M. (2010). "A methodology for optimizing feed composition for anaerobic co-digestion of agro-industrial wastes." *Bioresour. Technol.*, **101**(4), 1153–1158.

Angelidaki, I., Chen, X., Cui, J., Kaparaju, P., and Ellegaard, L. (2006). "Thermophilic anaerobic digestion of source-sorted organic fraction of household municipal solid waste: Start-up procedure for continuously stirred tank reactor." *Water Res.*, **40**(14), 2621–2628.

Angelidaki, I., Ellegaard, L., and Ahring, B. (2003). "Applications of the anaerobic digestion process." *Biomethanation II*, B. Ahring, B. K. Ahring, I. Angelidaki, J. Dolfing, L. Euegaard, H. N. Gavala, F. Haagensen, A. S. Mogensen, G. Lyberatos, P. F. Pind, J. E. Schmidt, I. V. Skiadas, and K. Stamatelatou, eds., Springer, Berlin, 1–33.

Anjum, M., Khalid, A., Mahmood, T., and Arshad, M. (2012). "Anaerobic co-digestion of municipal solid organic waste with melon residues to enhance biodegradability and biogas production." *J. Mater. Cycles Waste Manage.*, **14**(4), 388–395.

Appels, L., Baeyens, J., Degrève, J., and Dewil, R. (2008). "Principles and potential of the anaerobic digestion of waste-activated sludge." *Prog. Energy Combust. Sci.*, **34**(6), 755–781.

Aragaw, T., Andargie, M., and Gessesse, A. (2013). "Co-digestion of cattle manure with organic kitchen waste to increase biogas production using rumen fluid as inoculums." *Int. J. Phys. Sci.*, **8**(11), 443–450.

Atandi, E., and Rahman, S. (2012). "Prospect of anaerobic co-digestion of dairy manure: A review." *Environ. Technol. Rev.*, **1**(1), 127–135.

Baere, L. D., and Mattheeuws, B. (2010). "Anaerobic digestion of MSW in Europe." *BioCycle*, **51**(2), 24–26.

Blank, A., and Hoffmann, E. (2011). "Upgrading of a co-digestion plant by implementation of a hydrolysis stage." *Waste Manage. Res.*, **29**(11), 1145–1152.

Bolzonella, D., Battistoni, P., Mata-Alvarez, J., and Cecchi, F. (2003). "Anaerobic digestion of organic solid wastes: Process behavior in transient conditions." *Water Sci. Technol.*, **48**(4), 1–8.

Bond, T., Brouckaert, C., Foxon, K., and Buckley, C. (2012). "A critical review of experimental and predicted methane generation from anaerobic codigestion." *Water Sci. Technol.*, **65**(1), 183–189.

Boone, D. R., and Xun, L. (1987). "Effects of pH, temperature, and nutrients on propionate degradation by a methanogenic enrichment culture." *Appl. Environ. Microbiol.*, **53**(7), 1589–1592.

Braun, R., and Wellinger, A. (2010). "Potential of codigestion." ⟨http://www.iea-biogas.net⟩ (Jul. 1, 2013).

Bridgeman, J. (2012). "Computational fluid dynamics modeling of sewage sludge mixing in an anaerobic digester." *Adv. Eng. Software*, **44**(1), 54–62.

Carrère, H., et al. (2010). "Pretreatment methods to improve sludge anaerobic degradability: A review." *J. Hazard. Mater.*, **183**(1-3), 1–15.

Carrere, H., Rafrafi, Y., Battimelli, A., Torrijos, M., Delgenes, J. P., and Motte, C. (2012). "Improving methane production during the codigestion of waste-activated sludge and fatty wastewater: Impact of thermo-alkaline pretreatment on batch and semi-continuous processes." *Chem. Eng. J.*, **210**, 404–409.

Carucci, G., Carrasco, F., Trifoni, K., Majone, M., and Beccari, M. (2005). "Anaerobic digestion of food industry wastes: Effect of codigestion on methane yield." *J. Environ. Eng.*, 10.1061/(ASCE)0733-9372(2005)131:7(1037), 1037–1045.

Castrillon, L., Fernandez-Nava, Y., Ormaechea, P., and Maranon, E. (2011). "Optimization of biogas production from cattle manure by pre-treatment with ultrasound and co-digestion with crude glycerin." *Bioresour Technol.*, **102**(17), 7845–7849.

Castrillón, L., Fernández-Nava, Y., Ormaechea, P., and Marañón, E. (2013a). "Methane production from cattle manure supplemented with crude glycerin from the biodiesel industry in CSTR and IBR." *Bioresour. Technol.*, **127**, 312–317.

Castrillón, L., Marañón, E., Fernández-Nava, Y., Ormaechea, P., and Quiroga, G. (2013b). "Thermophilic co-digestion of cattle manure and food waste supplemented with crude glycerin in induced bed reactor (IBR)." *Bioresour. Technol.*, **136**, 73–77.

Cesaro, A., Naddeo, V., Amodio, V., and Belgiorno, V. (2012). "Enhanced biogas production from anaerobic codigestion of solid waste by sonolysis." *Ultrason. Sonochem.*, **19**(3), 596–600.

Chae, K. J., Jang, A., Yim, S. K., and Kim, I. S. (2008). "The effects of digestion temperature and temperature shock on the biogas yields from the mesophilic anaerobic digestion of swine manure." *Bioresour. Technol.*, **99**(1), 1–6.

Chartrain, M., Bhatbagar, L., and Zeihus, J. G. (1987). "Microbial ecophysiology of whey biomethanation: Comparison of carbon transformation parameters, species composition, and starter culture performance in continuous culture." *Appl. Environ. Microbiol.*, **53**(5), 1447–1156.

Chen, Y., Cheng, J. J., and Creamer, K. S. (2008). "Inhibition of anaerobic digestion process: A review." *Bioresour. Technol.*, **99**(10), 4044–4064.

CIWMB (California Integrated Waste Management Board). (2008). "Current anaerobic digestion technologies used for treatment of municipal organic solid waste." Contractor's Rep. to The Board, #IWMB-2008-011, California Environmental Protection Agency, Sacramento, CA.

Cuetos, M., Fernández, C., Gómez, X., and Morán, A. (2011). "Anaerobic co-digestion of swine manure with energy crop residues." *Biotechnol. Bioprocess Eng.*, **16**(5), 1044–1052.

Demirel, B., and Yenigün, O. (2002). "Two-phase anaerobic digestion processes: A review." *J. Chem. Technol. Biotechnol.*, **77**(7), 743–755.

Demirel, B., Göl, N., and Onay, T. (2013). "Evaluation of heavy metal content in digestate from batch anaerobic co-digestion of sunflower hulls and poultry manure." *J. Mater. Cycles Waste Manage.*, **15**(2), 242–246.

Deublein, D., and Steinhauser, A. (2008). *Biogas from waste and renewable resources*, Wiley-VCH, Weinheim.

Elbeshbishy, E., Aldin, S., Hafez, H., Nakhla, G., and Ray, M. (2011). "Impact of ultrasonication of hog manure on anaerobic digestability." *Ultrason. Sonochem.*, **18**(1), 164–171.

El-Mashad, H. M., and Zhang, R. (2007). "Co-digestion of food waste and dairy manure for biogas production." *Trans. ASABE.* **50**(5), 1815–1821.

Esposito, G., Frunzo, L., Giordano, A., Liotta, F., Panico, A., and Pirozzi, F. (2012). "Anaerobic co-digestion of organic wastes." *Rev. Environ. Sci. Biotechnol.*, **11**(4), 325–341.

Fehrenbach, H., Giegrich, J., Reinhardt, G., Sayer, U., Gretz, M., Lanje, K., Schmitz, J. (2008). "Kriterien einer nachhaltigen Bioenergienutzung im globalen Maßstab." *UBA-FB*, **206**, 41–112.

Ferrer, I., Garfí, M., Uggetti, E., Ferrer-Marti, L., Calderon, A., and Velo, E. (2011). "Biogas production in low-cost household digesters at the Peruvian Andes." *Biomass Bioenergy*, **35**(5), 1668–1674.

Forgács, G., Lundin, M., Taherzadeh, M. J., and Horváth, I. S. (2013). "Pretreatment of chicken feather waste for improved biogas production." *Appl. Biochem. Biotechnol.*, **169**(7), 1–13.

Frigon, J.-C., Roy, C., and Guiot, S. (2012). "Anaerobic co-digestion of dairy manure with mulched switchgrass for improvement of the methane yield." *Bioprocess. Biosyst. Eng.*, **35**(3), 341–349.

Fuchun, Z., Deguo, X., Xuefu, X., and Longjun, X. (2006). "Research achievements and application in anaerobic treatment of organic solid wastes—A review." *Chin. J. Geochem.*, **25**(2), 178–181.

Gautam, R., Sumit, B., and Sunil, H. (2009). "Biogas as a sustainable energy source in Nepal: Present status and future challenges." *Renew. Sustainable Energy Rev.*, **13**(1), 248–252.

Ghosh, S., and Pohland, F. G. (1974). "Kinetics of substrate assimilation and product formation in anaerobic digestion." *J. Water Pollut. Control Fed.*, **46**(4), 748–759.

Gray, D. M., Auto, P., and Peck, C. (2008). "Anaerobic digestion of food waste."U.S. EPA No. EPA-R9-WST-06, U.S. EPA, Oakland, CA.

Harper, S. R., and Pohland, F. G. (1986). "Recent development s in hydrogen management during anaerobic biological wastewater treatment." *Biotechnol. Bioeng.*, **28**(4), 585–602.

Hartmann, H., and Ahring, B. K. (2006). "Strategies for the anaerobic digestion of the organic fraction of municipal solid waste: An overview." *Water Sci. Technol.*, **53**(8), 7–22.

Haug, R. T. (1977). "Sludge processing to optimize digestibility and energy production." *J. Water Pollut. Control Fed.*, **49**(7), 1713–1721.

Hidalgo, D., Sastre, E., Gómez, M., and Nieto, P. (2012). "Evaluation of pre-treatment processes for increasing biodegradability of agro-food wastes." *Environ. Technol.*, **33**(13), 1497–1503.
Ho, L., and Ho, G. (2012). "Mitigating ammonia inhibition of thermophilic anaerobic treatment of digested piggery wastewater: Use of pH reduction, zeolite, biomass and humic acid." *Water Res.*, **46**(14), 4339–4350.
Holm-Nielsen, J. B., Al Seadi, T., and Oleskowicz-Popiel, P. (2009). "The future of anaerobic digestion and biogas utilization." *Bioresour. Technol.*, **100**(22), 5478–5484.
Hoornweg, D., and Bhada-Tata, P. (2012). "What a waste: A global review of solid waste management." Urban Development and Local Government Unit, World Bank, Washington, DC.
Hublin, A., Zokić, T., and Zelić, B. (2012). "Optimization of biogas production from co-digestion of whey and cow manure." *Biotechnol. Bioprocess Eng.*, **17**(6), 1284–1293.
IEA (International Energy Agency). (2008). "IEA bioenergy task 37 plant list 2008." ⟨http://www.ieabiogas.net/anlagelisten/Plantlist_08.pdf⟩ (Jul. 1, 2013).
IEA (International Energy Agency). (2011). "IEA Bioenergy Task 37." ⟨http://www.iea-biogas.net/country-reports.html⟩ (Jun. 12, 2015).
Kacprzak, A., Krzystek, L., and Ledakowicz, S. (2010). "Co-digestion of agricultural and industrial wastes." *Chem. Papers*, **64**(2), 127–131.
Kangle, K. M., Kore, S. V., Kore, V. S., and Kulkarni, G. S. (2012). "Recent trends in anaerobic codigestion: A review." *Universal J. Environ. Res. Technol.*, **2**(4), 210–219.
Kaparaju, P., Buendia, I., Ellegaard, L., and Angelidakia, I. (2007). "Effects of mixing on methane production during thermophilic anaerobic digestion of manure: Lab scale and pilot scale studies." *Bioresour. Technol.*, **99**(11), 4919–4928.
Kassuwi, S., Mshandete, A., and Kivaisi, A. (2012). "Anaerobic co-digestion of biological pre-treated Nile perch fish solid waste with vegetable fraction of market solid waste." *J. Agric. Biol. Sci.*, **7**(12), 1016–1031.
Kawai, M., Kishi, M., Hamersley, M. R., Nagao, N., Hermana, J., and Toda, T. (2012). "Biodegradability and methane productivity during anaerobic co-digestion of refractory leachate." *Int. Biodeterior. Biodegrad.*, **72**, 46–51.
Keller, J., and Hartley, K. (2003). "Greenhouse gas production in wastewater treatment: Process selection is the major factor". *Water Sci. Technol.*, **47**(12), 43–48.
Khalid, A., Arshad, M., Anjum, M., Mahmood, T., and Dawson, L. (2011). "The anaerobic digestion of solid organic waste." *Waste Manage.*, **31**(8), 1737–1744.
Khanal, S. K., Grewell, D., Sung, S., and Leeuwen, J. (2007). "Ultrasound applications in wastewater sludge pretreatment: A review." *Crit. Rev. Env. Sci. Technol.*, **37**(4), 277–313.
Kim, D.-H., and Oh, S.-E. (2011). "Continuous high-solids anaerobic co-digestion of organic solid wastes under mesophilic conditions." *Waste Manage.*, **31**(9-10), 1943–1948.
Kim, M., Gomec, C. Y., Ahn, Y., and Speece, R. E. (2003). "Hydrolysis and acidogenesis of particulate organic material in mesophilic and thermophilic anaerobic digestion." *Environ. Technol.*, **24**(9), 1183–1190.
Komatsu, T., Kudo, K., Inoue, Y., and Himeno, S. (2007). "Anaerobic codigestion of sewage sludge and rice straw." *Wastewater biosolids sustainability: Technical, managerial and public synergy*, GMSC, New Brunswick, Canada, 495–501.
Levis, J. W., and Barlaz, M. (2011), "Is biodegradability a desirable attribute for discarded solid waste? Perspectives from a national landfill greenhouse gas inventory model." *Environ. Sci. Technol.*, **45**(13), 5470–5476.

Levis, J. W., Barlaz, M. A., Themelis, N. J., and Ulloa, P. (2010), "Assessment of the state of food waste treatment in the United States and Canada." *Waste Manage.*, **30**(8-9), 1486–1494.

Lim, J. W., and Wang, J. Y. (2013). "Enhanced hydrolysis and methane yield by applying microaeration pretreatment to the anaerobic co-digestion of brown water and food waste." *Waste Manage.*, **33**(4), 813–819.

Lissens, G., Vandevivere, P., De Baere, L., Bley, E. M., and Verstraete, W. (2001). "Solid waste digesters: Process performance and practice for municipal solid waste digestion." *Water Sci. Technol.*, **44**(8), 91–102.

Li, Y., Park, S. Y., and Zhu, J. (2011). "Solid-state anaerobic digestion for methane production from organic waste." *Ren. Sust. Energ. Rev.*, **15**(1), 821–826.

Long, J. H., Aziz, T. N., de los Reyes III, F. L., and Ducoste, J. J. (2012). "Anaerobic co-digestion of fat, oil, and grease (FOG): A review of gas production and process limitations." *Process Saf. Environ. Prot.* **90**(3), 231–245.

Lou, X. F., Nair, J., and Ho, G. (2012). "Field performance of small scale anaerobic digesters treating food waste." *Energy Sustainable Dev.*, **16**(4), 509–514.

Lukehurst, C. T., Frost, P., and Al Seadi, T. (2010). "Utilisation of digestate from biogas plants as biofertiliser." IEA Bioenergy, Ireland.

Luste, S., Luostarinen, S., and Sillanpää, M. (2009). "Effect of pre-treatments on hydrolysis and methane production potentials of by-products from meat-processing industry." *J. Hazard. Mater.*, **164**(1), 247–255.

Madsen, M., Holm-Nielsen, J. B., and Esbensen, K. H. (2011). "Monitoring of anaerobic digestion processes: A review perspective." *Ren. Sust. Energ. Rev.*, **15**(6), 3141–3155.

Mahapatra, S., Chanakya, H. N., and Dasappa, S. (2009). "Evaluation of various energy devices for domestic lighting in India: Technology, economics and CO2 emissions." *Energy Sustain. Dev.*, **13**(4), 271–279.

Mata-Álvarez, J. (2003). *Biomethanization of the organic fraction of municipal solid wastes*, IWA Publishing, London.

Mata-Alvarez, J., Dosta, J., Macé, S., and Astals, S. (2011). "Codigestion of solid wastes: A review of its uses and perspectives including modeling." *Crit. Rev. Biotechnol.*, **31**(2), 99–111.

Mata-Alvarez, J., Mace, S., and Llabres, P. (2000). "Anaerobic digestion of organic solid wastes. An overview of research achievements and perspectives." *Bioresour. Technol.*, **74**(1), 3–16.

McMahon, K. D., Stroot, P. G., Mackie, R. I., and Raskin, L. (2001). "Anaerobic codigestion of municipal solid waste and biosolids under various mixing conditions-II: Microbial population dynamics." *Water Res.*, **35**(7), 1817–1827.

Menardo, S., Airoldi, G., and Balsari, P. (2012). "The effect of particle size and thermal pre-treatment on the methane yield of four agricultural by-products." *Bioresour. Technol.*, **104**, 708–714.

Moletta, R., Escoffier, Y., Ehlinger, F., Coudert, J. P., and Leyris, J. P. (1994). "On-line automatic control system for monitoring an anaerobic fluidized-bed reactor: Response to organic overload." *Water Sci. Technol.*, **30**(12), 11–20.

Monnet, F. (2003). "An introduction to anaerobic digestion of organic wastes." Remade Scotland, Scotland.

Monou, M., Kythreotou, N., Fatta, D., and Smith, S. (2009). "Rapid screening procedure to optimise the anaerobic codigestion of industrial biowastes and agricultural livestock wastes in Cyprus." *Waste Manage.*, **29**(2), 712–720.

Monteny, G.-J., Bannink, A., and Chadwick, D. (2006). "Greenhouse gas abatement strategies for animal husbandry." *Agric. Ecosyst. Environ.*, **112**(2), 163–170.

Montusiewicz, A., and Lebiocka, M. (2011). "Co-digestion of intermediate landfill leachate and sewage sludge as a method of leachate utilization." *Bioresour. Technol.*, **102**(3), 2563–2571.

Moriarty, K. (2013). "Feasibility study of anaerobic digestion of food waste in St. Bernard, Louisiana." National Renewable Energy Laboratory, Golden, CO.

Mosier, N., et al. (2005). "Features of promising technologies for pretreatment of lignocellulosic biomass." *Bioresour. Technol.*, **96**(6), 673–686.

Mshandete, A., Björnsson, L., Kivaisi, A. K., Rubindamayugi, M. S. T., and Mattiasson, B. (2006). "Effect of particle size on biogas yield from sisal fibre waste." *Renew. Energy*, **31**(14), 2385–2392.

Mudhoo, A. (2012). *Biogas production: Pretreatment methods in anaerobic digestion*, Wiley, Hoboken, NJ.

Muhammad Nasir, I., Mohd Ghazi, T., and Omar, R. (2012). "Production of biogas from solid organic wastes through anaerobic digestion: A review." *Appl. Microbiol. Biotechnol.*, **95**(2), 321–329.

Naik, S., Goud, V. V., Rout, P. K., and Dalai, A. K. (2010). "Production of first and second generation biofuels: A comprehensive review." *Renew. Sustainable Energy Rev.*, **14**(2), 578–597.

Nallathambi Gunaseelan, V. (1997). "Anaerobic digestion of biomass for methane production: A review." *Biomass Bioenergy*, **13**(1), 83–114.

Neyens, E., and Baeyens, J. (2003). "A review of thermal sludge pre-treatment processes to improve dewaterability." *J. Hazard. Mater.*, **98**(1), 51–67.

Nguyen, P. H. L., Kuruparan, P., and Visvanathan, C. (2007). "Anaerobic digestion of municipal solid waste as a treatment prior to landfill." *Bioresour. Technol.*, **98**(2), 380–387.

Omar, F. N., Nor'Aini, A. R., Hafid, H. S., Phang, L. Y., and Hassan, M. A. (2009). "Separation and recovery of organic acids from fermented kitchen waste by an integrated process." *Afr. J. Biotechnol.*, **8**(21), 5807–5813.

Ostrem, K. (2004). "Greening waste: Anaerobic digestion for treating the organic fraction of municipal solid wastes." M.S. thesis, Columbia Univ., New York.

Palmowski, L., and Müller, J. A. (2000). "Influence of the size reduction of organic waste on their anaerobic digestion." *Water Sci. Technol.*, **41**(3), 155–162.

Palmowski, L., and Müller, J. A. (2003). "Anaerobic degradation of organic materials-significance of the substrate surface area." *Water Sci. Technol.*, **47**(12), 231–238.

Park, S., and Li, Y. (2012). "Evaluation of methane production and macronutrient degradation in the anaerobic co-digestion of algae biomass residue and lipid waste." *Bioresour. Technol.*, **111**, 42–48.

Pastor, L., Ruiz, L., Pascual, A., and Ruiz, B. (2013). "Co-digestion of used oils and urban landfill leachates with sewage sludge and the effect on the biogas production." *Appl. Energy*, **107**, 438–445.

Pauss, A., Nyns, E. J., and Naveau, H. (1984). "Production of methane by anaerobic digestion of domestic refuse." *Anaerobic digestion and carbohydrate hydrolysis of waste*, G. L. Ferrero, M. P. Ferranti, and H. Naveau, eds., Elsevier Applied Science, London, 209–222.

Pesta, G. (2007). "Anaerobic digestion of organic residues and wastes." *Utilization of by-products and treatment of waste in the food industry*, V. Oreopoulou and W. Russ, eds., Springer, New York, 53–72.

Petersson, A., and Wellinger, A. (2009). "Biogas upgrading technologies-developments and innovations." *IEA Bioenergy Task Force*, **37**, 20.

Pilli, S., Bhunia, P., Yan, S., LeBlanc, R. J., Tyagi, R. D, and Surampalli, R. Y. (2011). "Ultrasonic pretreatment of sludge: A review." *Ultrason. Sonochem.*, **18**(1), 1–18.

Press, J. G. (2008). "Toronto moves forward with anaerobic digestion of residential SSO." *BioCycle*, **49**(9), 44.

Rapport, J. L., Zhang, R., Williams, R. B., and Jenkins, B. M. (2012). "Anaerobic digestion technologies for the treatment of municipal solid waste." *Int. J. Environ. Waste Manage.*, **9**(1-2), 100–122.

Romero, M., Romero, L., and Álvarez-Gallego, C. (2010. "Comparision of three preatments for organic matter solubilization from OFMSW." *Pulse*, **5**, 1.

Saha, M., Eskicioglu, C., and Marin, J. (2011. "Microwave, ultrasonic and chemo-mechanical pretreatments for enhancing methane potential of pulp mill wastewater treatment sludge." *Bioresour. Technol.*, **102**(17), 7815–7826.

Saifuddin, N., and Fazlili, S. A. (2009). "Effect of microwave and ultrasonic pretreatments on biogas production from anaerobic digestion of palm oil mill effleunt." *Am. J. Engg. Appl. Sci.*, **2**(1), 139–146.

Salminen, E., and Rintala, J. (2002). "Anaerobic digestion of organic solid poultry slaughterhouse waste—A review." *Bioresour. Technol.*, **83**(1), 13–26.

Sasaki, K., et al. (2010). "Bioelectrochemical system stabilizes methane fermentation from garbage slurry." *Bioresour. Technol.*, **101**(10), 3415–3422.

Shan, G.-B., Zhang, T. C., and Surampalli, R. Y. (2010). "Bioenergy from landfills." *Bioenergy and bioefuel from Biowastes and Biomass*, S. K. Khanal, R. Y. Surampalli, T. C. Zhang, B. P. Lamsal, R. D. Tyagi, and C. M. Kao, eds., ASCE, Reston, VA, 88–115.

Sheng, F., Lijuan, P., Zhixiang, Y., and Jianwei, M. (2011). "A review of different pretreatment techniques for enhancing biogas production." *Proc., Materials for Renewable Energy & Environment (ICMREE), 2011 Int. Conf.*, Institute of Electrical and Electronics Engineers (IEEE), New York.

Shimizu, T., Kudo, K., and Nasu, Y. (1993). "Anaerobic waste activated sludge digestion—A bioconversion mechanism and kinetic model." *Biotechnol. Bioeng.*, **41**(11), 1082–1091.

Siddiqui, Z., Horan, N., and Anaman, K. (2011). "Optimisation of C:N ratio for co-digested processed industrial food waste and sewage sludge using the BMP test." *Int. J. Chem. Reactor Eng.*, **9**(1), S4.

Sobotka, M., Votruba, J., Havlik, I., and Minkevich, I. G. (1983). "The mass-energy balance of anaerobic methane production." *Folia Microbiol.*, **28**(3), 195–204.

Steffen, R., Szolar, O., and Braun, R. (1998). "Feedstocks for anaerobic digestion." Institute for Agrobiotechnology, Tulln Univ. of Agricultural Sciences, Viences.

Stroot, P. G., McMahon, K. D., Mackie, R. I., and Raskin, L. (2001). "Anaerobic codigestion of municipal solid waste and biosolids under various mixing conditions-I. Digester performance." *Water Res.*, **35**(7), 1804–1816.

Sulaiman, A., et al. (2009). "The effect of mixing on methane production in a semi-commercial closed digester tank treating palm oil mill effluent." *Aust. J. Basic Appl. Sci.*, **3**(3), 1577–1583.

Sullivan, D. (2012). "Anaerobic digestion in the Northwest." *BioCycle*, **53**(3), 33.

Switzenbaum, M. S., Giraldo-Gomez, E., and Hickey, R. F. (1990). "Monitoring of the anaerobic methane fermentation process." *Enzyme Microb. Technol.*, **12**(10), 722–730.

Taherzadeh, M. J., and Karimi, K. (2008). "Pretreatment of lignocellulosic wastes to improve ethanol and biogas production: A review." *Int. J. Mol. Sci.*, **9**(9), 1621–1651.

Taricska, J. R., Long, D. A., Chen, J. P., Hung, Y. T., and Zou, S. W. (2007). "Anaerobic digestion." *Biosolids engineering and management*, Humana Press, New York, 135–137.

Tchobanoglous, G., Burton, F. L., and Stensel, H. D. (2003). *Meltcalf & Eddy, Inc.'s wastewater engineering: Treatment, disposal, and reuse*, 4th ed., McGraw-Hill, New York.

Thangamani, A., Rajakumar, S., and Ramanujam, R. A. (2010). "Anaerobic co-digestion of hazardous tannery solid waste and primary sludge: Biodegradation kinetics and metabolite analysis." *Clean Technol. Environ. Policy*, **12**(5), 517–524.

Thiele, J. H., Chartrain, M., and Zeihus, J. G. (1988). "Control of interspecies electron flow during anaerobic digestion: Role of floc formation in syntrophic methanogenesis." *Appl. Environ. Microbiol.*, **54**, 10–19.

Tumutegyereize, P., Muranga, F. I., Kawongolo, J., and Nabugoomu, F. (2011). "Optimization of biogas Production from banana peels: Effect of particle size on methane yield." *Afr. J. Biotechnol.*, **10**(79), 18243–18251.

Tuomisto, H., Hodge, I., Riordan, P., and Macdonald, D. (2012). "Comparing energy balances, greenhouse gas balances and biodiversity impacts of contrasting farming systems with alternative land uses." *Agric. Syst.*, **108**, 42–49.

Turovskly, I. S., and Mathai, P. K. (2006). *Wastewater Sludge Processing*, Wiley, New York.

Tyagi, V., and Lo, S.-L. (2011). "Application of physico-chemical pretreatment methods to enhance the sludge disintegration and subsequent anaerobic digestion: An up to date review." *Rev. Environ. Sci. Biotechnol.*, **10**(3), 215–242.

USEPA. (2010). "Energy recovery from waste." ⟨www.epa.gov/waste/nonhaz/municiapl/wte/⟩ (May 2, 2013).

USEPA. (2011). "AgSTAR digester database." ⟨http://www.epa.gov/agstar/⟩ (May 2, 2013).

Van Haandel, A. C., and van der Lubbe, J. G. M. (2012). *Handbook of biological wastewater treatment: Design and optimisation of activated sludge systems*, International Water Assn, London.

Viana, M. B., Freitas, A. V., Leitão, R. C., Pinto, G. A. S., and Santaella, S. T. (2012). "Anaerobic digestion of crude glycerol: A review." *Environ. Technol. Rev.*, **1**(1), 81–92.

Ward, A. J., Hobbs, P. J., Holliman, P. J., and Jones, D. L. (2008). "Optimisation of the anaerobic digestion of agricultural resources." *Bioresour. Technol.*, **99**(17), 7928–7940.

Weiland, P. (2010). "Biogas production: Current state and perspectives." *Appl. Microbiol. Technol.*, **85**(4), 849–860.

Wu, W. (2007). "Anaerobic co-digestion of biomass for methane production: Recent research achievements." *Optimization*, **1**, 1VS.

Yamashiro, T., et al. (2013). "Anaerobic co-digestion of dairy cow manure and high concentrated food processing waste." *J. Mater. Cycles Waste Manage.*, **15**(4), 539–547.

Yang, Y. Q., Shen, D. S., Li, N., Xu, D., Long, Y. Y., and Lu, X. Y. (2013). "Co-digestion of kitchen waste and fruit-vegetable waste by two-phase anaerobic digestion." *Environ. Sci. Pollut. Res.*, **20**(4), 2162–2171.

Yu, H., Samani, Z., Hanson, A., and Smith, G. (2002). "Energy recovery from grass using two-phase anaerobic digestion." *Waste Manage.*, **22**(1), 1–5.

Zhang, P., Zeng, G., Zhang, G., Li, Y., Zhang, B., and Fan, M. (2008). "Anaerobic co-digestion of biosolids and organic fraction of municipal solid waste by sequencing batch process." *Fuel Process. Technol.*, **89**(4), 485–489.

Zhang, R., El-Mashad, H. M., Hartman, K., Wang, F., Liu, G., Choate, C., and Gamble, P. (2007). "Characterization of food waste as feedstock for anaerobic digestion." *Bioresour. Technol.*, **98**(4), 929–935.

Zhang, T. C., and Noike, T. (1993). "Influence of retention time on reactor performance and bacterial trophic populations in anaerobic digestion processes." *Water Res.*, **28**(1), 27–36

Zhang, T. C., and Noike, T. (1991). "Comparison of one-phase and two-phase anaerobic digestion processes in characteristics of substrate degradation and bacterial population levels." *Water Sci. Technol.*, **23**(7–9), 1157–1166.

Zhang, T. C., and Noike, T. (1994). "Influence of retention time on reactor performance and bacterial trophic populations in anaerobic digestion processes." *Water Res.*, **28**(1), 27–36.

Zhou, F., Xiong, D., Xian, X., and Xu, L. (2006). "Research achievements and application in anaerobic treatment of organic solid wastes-A review." *Chin. J. Geochem.*, **25**(2), 178–181.

Zupancic, G. D., and Ros, M. (2003). "Heat and energy requirements in thermophilic anaerobic sludge digestion." *Renew. Energy*, **28**(14), 2255–2267.

CHAPTER 9

Composting for Organic Waste Management

Jonathan W. C. Wong[1]
Ammaiyappan Selvam[2]
Mukesh K. Awasthi[3]

9.1 INTRODUCTION

Organic solid waste management is a major challenge throughout the world, and a poor management can result in environmental deterioration and human health hazards. Effective management of organic waste relies on the availability and acceptability of the technology for the treatment. Every year about 1.3 billion tonnes of municipal solid wastes (MSW) are generated globally in urban setups alone and is expected to increase to 2.2 billion tonnes by 2025 (Hoornweg and Bhada-Tata 2012). About 30–50% of the MSW are putrescible; whose sustainable disposal requires appropriate technologies. Besides, significant quantities of biosolids, animal manures, agricultural wastes and other organic wastes generated also require suitable treatments. Most of the putrescible wastes in the MSW and other wastes were landfilled while only a small fraction was treated in the past. Decomposition of organic fractions in the landfills causes emission of greenhouse gases contributing to global warming, and odour causing human health hazards, and contaminates the groundwater and soil through leachate production if they are not properly managed. To prevent these environmental damages, recycling of these organic wastes is essential.

Considering the application, level of scientific expertise required and the demand for the recycled products, composts appear to be an ideal technology due to its robustness to handle a variety of organic wastes. Flexibility and the robustness make composting suitable for a broad range of situations, thus there is a composting system for nearly every type of organic waste i.e., simple systems for early stages of industrial development to relatively complex, mechanized

[1]Hong Kong Baptist Univ., Hong Kong SAR.
[2]Hong Kong Baptist Univ., Hong Kong SAR.
[3]Hong Kong Baptist Univ., Hong Kong SAR.

systems for advanced industrial development (Chatzouridis and Komilis 2012). In the developing countries, constraints related to economics, technology, and qualified personnel have narrowed the choice of treatment and composting best fits these requirements. Composting in a rural setup is relatively easy because the odour and vectors could not attract the focus due to the large area of land available and the less dense population and human interaction although the problems are unsolved. However, performing composting in urban setting requires stringent controls over the process to prevent the odour and vectors, and the need of high and effective decomposition to shorten the composting time.

9.2 PRINCIPLES OF COMPOSTING

9.2.1 Basic Considerations

Composting is a controlled microbial decomposition of organic solid waste, under aerobic condition, where microorganisms convert waste into a chemically humus-like stable product. The term controlled indicates that the process is managed or optimized to achieve the desired objectives. The main objectives of composting are to decompose organic fraction of waste to a stable condition; reduce its volume, weight and moisture content; minimize potential odor; destruct pathogen and increase potential nutrients for agricultural application. Complete stabilization is neither practical nor desirable because complete stabilization would destroy all the slowly degradable organic matter that gives compost its soil-building properties. The degree of stabilization and pathogen destruction depends on the purpose of composting and intended use of the compost product. Achieving a stable and mature composting product depends on proper management of the process through controlling the following factors:

- Nutrient balance is determined by the ratio of carbon (C) to nitrogen (N) in the compost mix, which affects the microbial growth in the composting system. C is the energy source while protein is important for cell differentiation. The optimal initial C/N ratio should range from 25 to 35 to facilitate rapid composting which is consistent with the nutrient needs of the bacteria and fungi in the compost pile.
- Moisture content of the composting mix ideally should be 55–60% throughout the active period of the composting process. As a rule of thumb, the moisture content should neither be too dry to limit microbial growth nor too wet to fill the pore space to make oxygen supply a limiting factor.
- Oxygen supply at adequate levels enables biological processes to thrive with optimum efficiency and the oxygen content in the pile should be around 12 to 15%. Aeration affects temperature, moisture, carbon dioxide (CO_2) and oxygen (O_2) content of the air in the pile, and the rate of removal of potentially toxic gasses.

- Temperature increases during composting as a result of microbial activity; and a suitable insulation and controlling the temperature between 55°C and 65°C is a prerequisite for efficient and rapid composting. A temperature higher than 70°C will be inhibitory to the microbial degradation.
- pH is often self-regulating between 6.0 and 7.5, which is an optimal range for composting bacteria. However, some feedstocks require the addition of external pH adjusting materials such as lime for rapid composting since self-regulation takes longer time and odour problem are very serious during the low pH conditions.

If these factors are balanced and properly managed, the decomposition of organic matter can be maximized and this can significantly reduce the composting time. Besides the efficient composting process can generate sufficient heat to destroy weed seeds, pathogens, and fly larvae.

A typical composting process can be divided into four distinct phases covering the active composting and curing stages mainly characterized by the temperature changes and also the types of microorganisms that dominate the at various temperatures.

1. **Initial mesophilic phase:** During this first phase, a diverse population of mesophilic bacteria and fungi proliferates, degrading primarily the readily available nutrients and thereby raising the temperature to about 45°C. At this point their activities decrease, and heat resistant bacteria become dominant.
2. **Thermophilic phase:** After a short lag period (not always discernible) there occurs a second more or less steep rise of temperature. This phase is characterized by the development of a thermophilic microbial population comprising bacteria, actinomycetes and fungi. The temperature optimum of these microorganisms is between 55°C and 65°C, their activities terminate at temperature higher than 70°C. The pH typically decreases since organic acids are produced. Additionally, pathogenic microbes and helminthes eggs are eliminated as a result of heat generated during this process. Thus, the organic compost is safer for use by farmers.
3. **Cooling phase:** The phase is characterized by a gradual temperature decline as a result of declining readily available organic substrates and a decrease in microbial population. Mesophilic microorganisms having survived the high temperature phase or invading the cooling down material from the outside succeed the thermophilic ones and extend the degradation process as far as it is intended.
4. **Compost maturation phase:** The final stage is known as stabilization or curing phase, where the temperature declines further and remains about 25–30°C or close to ambient. In this step, the process of humification of organic compost occurs resulting in increases in humic matter content and cation exchange capacity (CEC) of the compost.

9.2.2 Microorganism and their Succession during Composting

The composting mass contains a wide variety of microorganisms with different temperature preferences, such as the mesophiles and thermophiles, and the turning operation can deliver the microorganisms to various temperature conditions. Initial microbial population is highly influenced by the substrate; however, the air passing through the composting mass add a significant microbial load resulting the eventual change in the diversity as influenced by the prevailing temperature and other corresponding physicochemical conditions. For example, in source-separated household waste, only few mesophilic fungi were observed initially while after certain period, as the decomposition proceeded, numerous thermophilic bacteria and fungi were observed (Ryckeboer et al. 2003a). Similarly, initial low pH of the food wastes favours the proliferation of fungi and yeast and slow down bacterial growth.

Bacteria are small and fast decomposers responsible for initial decomposition of most easily available nutrients. A wide variety of bacteria, such as *Clostridium thermocellum, Streptomyces* spp., *Ruminococcus* spp., *Pseudomonas* spp., *Cellulomonas* spp., *Bacillus* spp., *Serratia, Proteus, Staphylococcus* spp., and *Bacillus subtilis* were reported to produce diverse group of enzymes involved in saccharification of organic substrates that are useful in the degradation of organic substrates dominated by plant biomass. Many workers have studied *Acetivibrio cellulolyticus* as it differs from most of the cellulolytic organisms in its ability to utilize only cellulose, cellobiose or salicin for growth (Perez et al. 2002; Huang et al. 2010). *Cellulomonas* sp. is important and extensively studied aerobic cellulolytic waste degrading microorganism. It is amusing to note that *Bacillus* strains are known for their ability to enter into resting states and are good producers of secondary metabolites that could contribute them further advantages over competitors under conditions of slow growth on cellulosic substrates. The aerobic, mesophilic *Sporocytophaga myxococcoides* has been illustrated as one of the most active cellulolytic bacteria, produces several carboxyl-methyl cellulases, at least two of which act as endoglucanases (Crecchio et al. 2004; Zhang et al. 2011).

Fungi tend to be present in the later phase of composting because of their preferred foods include hard-to-degrade ingredients such as waxes, complex proteins, hemicelluloses, lignin, and pectin. Fungi are less sensitive to environments with low moisture and pH than bacteria, but because most fungi are obligate aerobes (require oxygen to grow), they have a lower tolerance for low-oxygen environments than bacteria. Fungi also cannot survive above a temperature of 60°C. A diverse spectrum of cellulolytic fungi (Herrmann and Shann 1997) have been isolated and identified over the years. Cellulolytic microbes are primarily carbohydrate degraders and cellulose is the most abundant source of substrate for the growth of fungi at thermophilic phase. Besides, most cellulolytic fungi can utilize a variety of other carbohydrates in addition to cellulose. While several fungi can metabolize cellulose as an energy source, only a few strains are capable of secreting a complex of cellulase enzymes, which could have practical appliance in the enzymatic hydrolysis of cellulose.

During the initial stage of the composting process, the microbial biomass increases drastically. The changes in chemical properties of the substrates also impact the succession of the microbes. Initially, the easily available organic matter triggers the fast growing heterotrophic microbes; while the degradation of polymeric material is a slow process and their degradative products can be used by other microorganisms (Ryckeboer et al. 2003b). Quantitatively, the main components of organic matter are carbohydrates, protein, lipids and lignin. Different microorganisms participate and produce different enzymes needed for degradation of these substrates. Depending on the input substrates, nitrogen may also be available in significant quantities. At a C/N ratio from 25 to 35, microbial activity is vigorous resulting in the "active phase". In general, the highest microbial populations and enzyme activities are observed during this phase.

Heterogeneous and non-uniform composting mass tends to have confined alcove with fluctuating physicochemical conditions that was observed to impact the survival of specific microbial communities. Gram-negative bacteria and fungal population expanded with increasing temperature of up to ~55°C, but decreased at higher temperature; while increase in population once the composting mass cool to ≤55°C. During the active phase, due the high temperature, fungal activity was described to be comprehensively inhibited while Gram-positive bacteria tend to dominate this thermophilic phase; and the populations decrease when the temperature decreases below 50°C. The most persuasive factor is temperature, which itself is an outcome of the microbial activities. In addition, the O_2 level tends to affect the microbial diversity and dominance. For example, a member of the *Bacillaceae* was the major dominating population (98%) when thermophilic conditions and an optimum aeration were maintained (Wichuk and McCartney 2007).

Furthermore, the low O_2 condition highly influences the fate and dynamics of nitrifying and denitrifying microbial communities (deGuardia et al. 2008). During the initial stage of composting, presence of micro-environment devotes to the activities of anaerobic bacteria; however, when the clusters are cracked during the subsequent composting process, these sites are disclosed reducing the anaerobic condition. Similarly, not all microorganisms have similar obligation in terms of water requirement and a difference in moisture content can cause a shift in the composition of the microbial diversity. In general, high water content favours bacteria over fungi. Too high a moisture content has a negative effect on the biological activity due to the increased compaction that tends to prevent the movement of air through the matrix.

9.3 SUBSTRATES FOR COMPOSTING

Natural decomposition occurs in any pile of waste material even if the C/N ratio, moisture content, and aeration are outside the recommended limits for composting; however, the decomposition proceeds at a slow rate and generate rotting

Table 9.1. Selected physicochemical properties of some composting substrates

Parameter	*Sludge*	*Pig manure*	*Horse stable bedding waste*	*Abattoir blood meal*	*Food waste*
Moisture content (%)	73–85	72–82	62–63	58–66	55–74
pH	5.3–8.3	6.7–7.3	7.0–7.2	8.2–9.8	4.6–6.6
Total organic carbon (%)	20.7–32.2	32.9–39.4	39.7–42.3	32.0–36.1	48.2–58.8
Total nitrogen (%)	3.7–10.6	3.2–3.8	1.2–1.6	12.0–14.1	1.9–6.7
C/N ratio	3.1–6.4	9.0–11.1	25.9–33.7	2.3–2.9	7.5–17.7
Organic matter (%)	57–66	78–84	91–91	96–98	83–98

Sources: Wong 2005; Wong et al. 2009; Wong 2010, Wong et al. 2011, 2012

odors (USCC 2008). To hasten the rate of microbial activity during composting, the ingredient and conditions are to be adjusted optimally. The important ingredients of a compost mix are the substrate, and/or bulking agent. The substrate is the waste to be treated while an amendment and bulking agents are used to adjust the C/N ratio, pH, bulk density and moisture content. The bulking agents create a good physical structure enabling the air movement and maintaining sufficient free air space. Physicochemical properties of many composting substrates are not directly suitable for the composting. Substrates that are available in huge quantities include biosolids, organic fraction of the MSW mainly containing the food wastes and livestock wastes; and their composting properties relevant are presented in Table 9.1.

9.3.1 Biosolids

Composting of biosolids has been a practice for nearly a century by now with notable works from Clay Kellogg Sr., Roger T. Haug and others (Johnson 1996; Haug 1993). The huge volume of biosolids and the unacceptable option of landfill disposal have made composting an attractive alternative for biosolids treatment. The disposal of biosolids is an expensive and environmentally serious issue due to the potential pathogens and pollutants in biosolids. Composting followed by land application represents one of the most economical means for the treatment and final disposal of biosolids. Biosolids is highly heterogeneous across the regions and a generalized design is very difficult to propose if a rapid, optimal and effective composting is targeted. The heavy metal contents and the chemicals used for sludge settling and dewaterability provide unique properties for biosolids in different countries. However, some general properties can be used to draw a basic level of operational strategies. Generally, biosolids is characterized by high moisture content, neutral pH and a bulk density of ~1.0 (Lau et al. 2001), which will reduce the free air space (FAS) and the physical structure required for composting. In addition, biosolids contains high N contents, which requires a good bulking agent with high C content and good moisture absorbing capability (Wong et al. 2001).

Traditionally, most studies on sewage sludge composting have been carried out using conventional outdoors technologies such as windrows and static piles. However, odour and leachate formation are critical in the acceptance of sludge composting. In contrast, rotary drum composters are considered to produce uniform end product without any odour or leachate related problems provided adequate biofilter or other odour removal mechanisms are included. There were lot of reports on the composting of biosolids and some of the previous works clearly demonstrated the feasibility of obtaining quality compost from biosolids (Wong et al. 1997; Fang et al. 1998, 1999; Wong and Fang 2000; Chen 2012; Cukjati et al. 2012; Gould et al. 2013; Maulini-Duran et al. 2013). One critical issue is heavy metal contents of the sludge composts. Since heavy metal contents of biosolids often vary geographically (Wong and Henry 1988; Tyagi et al. 1991), the compost obtained from different sources could vary significantly and the interpretation with the compost heavy metal standards should be done carefully. Often, the labile fractions rather than the total metal concentrations tend to determine the biosolids compost toxicity upon application. Therefore, minimizing the mobility of heavy metals during biosolids composting using various additives, such as coal fly ash and lime, have received significant attention in the past (Wong et al. 1997; Ho and Qiao 1998; Fang et al. 1999; Zorpas et al. 2000; Chiang et al. 2011; Wong and Selvam 2006).

Agricultural residue, yard waste or dry plant material, shredded tires and wood chips are some of the commonly used bulking agents. Rice straw, used as bedding material in horse stables was also demonstrated to be a good bulking agent for sludge composting in a pilot-scale in-vessel composting (Wong et al. 2011). The bulking materials not just improve the physical and chemical structure of the sludge, also tend to dilute the toxic substances such as heavy metal contents of the compost product. Recently, a recyclable plastic bulking agent was reported to successfully establish the thermophilic temperature enabling reuse of the bulking materials (Zhou et al. 2014). However in such cases, the reduction in organic matter will increase the heavy metal contents jeopardizing the quality of the final compost. Therefore, caution should be exercised when selecting an ideal bulking agent.

Obtaining a farm-suitable compost from biosolids in terms of heavy metals is a challenge. Table 9.2 presents the heavy metal contents of the biosolids co-composted with horse stable straw bedding waste at two different initial C/N ratios as obtained through different mixing ratios of these substrates. Heavy metals contents indicate that the compost product is not suitable for organic farming while suitable for general agricultural purposes (Wong et al. 2011). Especially, cadmium, chromium and selenium contents are higher than the limits for organic farming. Although it is suitable for agricultural application, during continuous

Table 9.2. Heavy metals contents of the final composting products from sewage sludge-horse stable straw bedding waste co-composting at 1:1:5 (HSL) and 1:2:9 (LSL) mixing ratios

	Compost Standards			HSL	LSL
Heavy metals	*HKORC[a]*	*HKORC[b]*	*TMECC[b]/CCME[c]*	*(C/N-19)*	*(C/N-30)*
Arsenic (mg/kg)	≤13	≤10	≤13	5	5
Cadmium (mg/kg)	≤32	≤1	≤3	1	1.6
Chromium (mg/kg)	≤210	≤100	≤210	127	101
Copper (mg/kg)	≤700	≤300	≤400	129	115
Lead (mg/kg)	≤150	≤100	≤150	16	14
Mercury (mg/kg)	≤1	≤1	≤0.8	0.77	0.45
Nickel (mg/kg)	≤62	≤50	≤62	32	22
Selenium (mg/kg)	≤2	≤1.5	≤2	2	1
Zinc (mg/kg)	≤1300	≤600	≤700	449	398

Sources: Wong 2009b; Wong et al. 2011

[a]Canadian Council of Ministers of Environment (2005), Guidelines for Grade A compost quality

[b]Compost and Soil Conditioner Quality Standards for General Agricultural Use (HKORC 2005)

[c]Compost and Soil Conditioner Quality Standards for Organic Farming (HKORC 2005)

[d]TMECC (2002): Test Methods for the Examination of Composts and Composting

applications, the soil background metal concentrations should also be considered (Smith 2009) in order to avoid the plant and human health hazards.

Apart from heavy metals, the level of organic pollutants in the compost material, especially pharmaceuticals, personal care products (PCPs), endocrine disrupters (EDCs), pesticides, and PBDEs (Richardson and Ternes 2005; Oleszczuk 2007; Smith 2009) are of serious concerns currently and require adequate process control. Besides, mass-burning is evolving as a main strategy to handle the sludge in order to cope up with the ever increasing sludge quantities (Fytili and Zabaniotou 2008). However, in some developing countries, where modern technologies have not been established, composting at least provides more economic alternatives to treat biosolids, which simultaneously generates a value added product.

9.3.2 Food Waste

As per the study of the Food and Agricultural Organization, about one-third of food produced for human consumption is lost or wasted globally, which amounts to about 1.3 billion tons per year. On a per-capita basis, 95–115 kg/year of food is wasted in industrialized world than in developing countries such as Sub-Saharan Africa and South/Southeast Asia, where the per-capita waste generation is 6–11 kg/year (Gustavsson et al. 2011). However, another study by Institution of Mechanical Engineers indicates that the food loss can be up to 2 billion tonnes (Fox 2013). With an increase in population, it is important to reduce food wastage on one hand; but it is also necessary to avoid inappropriate disposal of the food waste, which poses a challenge to the environmental quality through greenhouse gas emission. Since food is intended for human consumption, the level of toxic substances and pathogens will not be a problem if these wastes are composted. Food waste has unique properties as a raw compost feedstock due to its high moisture content, low C/N ratio, high fat or oil (15–40%), low physical structure and the significant quantities of easily degradable organic matter. Food waste is highly susceptible to odor production posing challenges of its composting in the urban setups (Chang et al. 2006; Ke et al. 2010; Wong et al. 2011). The other key problems include low pH, nitrogen loss and eventual composting failure; and these factors attracted a great deal attraction in the past decades. Despite all these odds and difficulties, the quality of the composts obtained from food wastes are of high quality and suitable for organic farming.

9.3.2.1 Controlling pH during Food Waste Composting

The critical issue with composting of food waste is the initial rapid degradation of readily available organic matter resulting in the generation of significant quantities of organic acids, often referred as low molecular weight carboxylic acids (LMWCAs). Production of LMWCAs decreases the pH of composting mass and this problem is very serious with food wastes due to its high volatile solid contents (Sundberg and Johnson 2005). Bergersen et al. (2009) reported that the amount of these organic acids constitute approximately 25% of dry matter in composting at the start of the experiment. Despite the acidic conditions, the temperature increases during the initial phase of composting, reaching a stage known as "thermoacidophilic", which subsequently inhibits both the mesophilic as well as thermophilic microbes. Without sufficient microbial activity, the decomposition process is retarded which eventually leads to composting failure (Nakasaki et al. 1998; Cheung et al. 2010).

The most common approach to control acidity is the addition of alkaline materials such as lime (Nakasaki et al. 1993; Bergersen et al. 2009; Wong et al. 2009). However, high concentrations of lime can be inhibitory to the microbes. Respiration was completely suppressed during composting of food waste with 10% lime, where pH remained high (Bergersen et al. 2009). To reduce the impact of lime, alkaline coal fly ash amendment could also achieve an equivalent pH control (Figure 9.1) as well as organic decomposition (Wong et al. 2009). Besides, sodium acetate as a buffering agent during composting was also reported (Yu and Huang 2009). Other methods such as increased aeration so as to lengthen the mesophilic phase and enhance the degradation of organic acids (Smars et al. 2002), inoculation of thermophilic yeasts (Choi and Park 1998), recycling the active compost up to 24% of the initial mix as microbial inoculum (Sundberg and Jonsson 2005) and high quantity of (up to 60%) bulking agents (Sundberg et al. 2004; Aasen et al. 2006) were also reported to control the acidity problem during food waste composting with varying degree of success. Of these, the recycling of active or

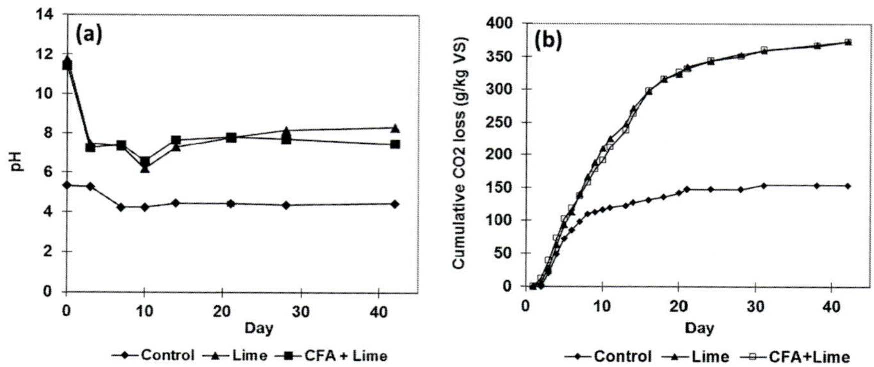

Figure 9.1. Effect of lime and coal fly ash (CFA) on controlling the pH (a) and organic decomposition (b) during food waste composting. The alkalinity of both lime and CFA (10%) + lime was equivalent to 1.88% $CaCO^3$

Source: Adapted from Wong et al. 2009

mature compost appears to be promising as they offer multiple advantages such as: adjusting the feedstock properties as the food wastes have a poor physical structure, increases the microbial population and reduces the requirement for bulking agents.

9.3.2.2 Controlling Nitrogen Loss during Food Waste Composting

The addition of alkaline materials to buffer against the initial acidity of food waste will increase the pH of the composting mass to above 11.0, which will favour for the volatilization of ammonium in the composts resulting in the release of ammonia from the composters. The loss of nitrogen, as NH_3, decreases the nutrient content of the final product and also leads to odour problem. During composting of organic wastes, total nitrogen loss could amount up to 50% of the initial nitrogen content (Wang et al. 2013). Therefore, any effort to control the pH should be considered for the conservation of N during composting. Nitrogen loss depends on feedstock property and other environmental factors, such as pH, carbon: nitrogen (C/N) ratio, particle size, temperature and aeration conditions (Bernal et al. 2009). However, considering the optimum conditions for microbial growth, there is not much space to adjust these conditions drastically to reduce N loss. Recently, struvite formation is being focused to be an effective approach to control N loss (Jeong and Kim 2001; Jeong and Hwang 2005; Ren et al. 2010; Wang et al. 2013).

Struvite (magnesium ammonium phosphate) is a mineral, which is formed when the concentrations of Mg^{2+}, NH_4^+ and PO_4^{3-} exceed the solubility product. In food waste composting, Jeong and Kim (2001) reported that addition of Mg and P equivalent to 20% molar ratio of total N led to the formation of struvite and trapped ammonia. With a cumulative ammonia production of about 33–36% of the initial total N in aerobic composting, a theoretical dose for complete conversion of ammonia into struvite crystals seemed to be about 35% of the initial N. The

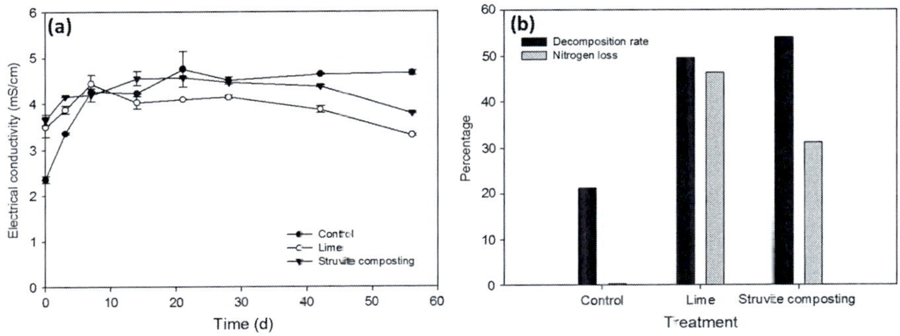

Figure 9.2. Decomposition efficiency and nitrogen loss during struvite based composting of food waste
Source: Adapted from Wang et al. 2013

addition of Mg and P salts at this level, however, caused adverse effects on the degradation of organic materials. Therefore addition of struvite salts could trap about 20% of the initial N in the compost mixture, which was suggested to be the maximum amount for nitrogen conservation during composting (Jeong and Hwang 2005). Recently, Wang et al. (2013) supplemented MgO and K_2HPO_4 for struvite formation during food waste composting with lime amendment, which significantly reduced N loss from 40.8% observed in lime amended food waste composting to 23.3%. As presented in Figure 9.2, the acidic pH reduced the organic decomposition and did not alter the equilibrium towards NH_3 while lime and struvite based composting enabled efficient organic decomposition. However, the electrical conductivity (EC) of the struvite-compost increased to around 6.14 mS/cm due to the addition of more Mg and P salts indicating that the salinity of the struvite based composts would be of a big concern for crop growth.

Struvite based composting could also significantly decrease the volatilization of NOx and VFAs during composting (Figure 9.3). The lower NOx emission monitored in the control treatment was due to its poor decomposition at low pH. Comparing with the treatment with lime amendment, a 17% reduction in NOx was observed in the struvite treatment. Meanwhile, the volatilization of VFAs was reduced by 63% as compared with the control treatment. When compost was mixed with the soil at 5% (W/W) and incubated in the dark at 26°C, the mineralization of organic matter was increased (Table 9.3). N and P released from struvite-based compost were higher than that of normal compost and soil, indicating the potential advantage of the application of struvite-based composts.

A critical problem of food waste composting is the oil content that affects the physical structure of composting mass. The oil content leads to the formation of lumps that prevent the air passage as well as resist the mixing mechanism as provided in small community-level composters. Recent studies have reported that 1.34 to 7.99% of the crude fat content of the composts was removed through the

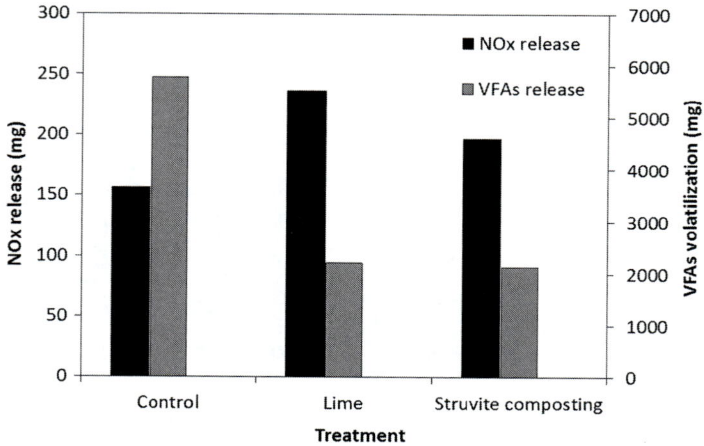

Figure 9.3. The volatilization of NOx and VFAs during food waste composting

inoculation of biofertilizers derived from food wastes with *Brevibacillus borstelensis* SH168 (Tsai et al. 2008) and 1.0 to 11.8% inoculated with *Thermoactinomyces vulgaris* A31 (Ke et al. 2010). When composting post-consumption food wastes, more than 95% of oil was degraded by an oil degrading inoculum as compared to <40% in the control reactors. Results indicate that appropriate microbial inoculum could help to overcome the oil-laden composting mass; however, more systematic research is required to develop an efficient microbial consortium.

9.3.3 Livestock Wastes

Livestock industries generate a vast amount of organic wastes and disposing them in an environmentally-appropriate way is a challenge. Replenishment with organic matter such as livestock excretions is a major approach to restore the degraded soils due to excessive application of inorganic fertilizer. However pathogens and other trace organic pollutants such as antibiotics pose great challenges in the disposal of livestock wastes. Proper composting or stabilization can eliminate the

Table 9.3. Mineralization and nutrient release of compost amended soil

Treatment	*Carbon mineralization rate (%)*	*Phosphorus release rate (%)*	*Nitrogen release rate (%)*
Soil	4.34 ± 0.08	0.24 ± 0.00	6.75 ± 1.46
Normal compost 5%	8.86 ± 0.23	0.47 ± 0.01	7.07 ± 3.55
Struvite compost 5%	13.79 ± 0.13	3.33 ± 0.20	10.93 ± 1.64

pathogens as well as reduce the concentration of antimicrobial compounds (Selvam et al. 2012a). Composting of animal manures has been traditionally in small scales at the point of production such as farms. However, when the fertilizer value of animal manures has been recognized, composting in commercial big units is gaining attention as an alternative way of recycling the manures. Compared with direct application of manure, nutrient recycling through composting will increase the cost of production. Nevertheless, removal of pathogens and weeds, reduction of volume and moisture, removal of odour, removal of potential organic contaminants, ease of storage, transport and use are the ultimate benefits favor composting (Bernal et al. 2009; Selvam et al. 2012a, 2012b)

Generally, physicochemical properties of the animal manures are not conducive for composting (refer Table 9.1) that could limit the degradation efficiency and generate potential odour. The notable properties are high moisture, low porosity, low C/N ratio, and in some cases high pH values. Thus, critical consideration is required for substrate conditioning-feedstock formulation, type and quantity of bulking agents and process control options to be adopted to ensure the production of quality composts (Lau et al. 1992; Solano et al. 2001; Michel et al. 2004). Animal manures can be from a variety of sources such as cattle, sheep, swine, horse and poultry; and each has its own specific issues to be taken care of. For example, manures from poultry and swine tend to have high concentration of ammonium nitrogen that tends to promote ammonia emission during the early stage of composting process (Petric et al. 2009; Dias et al. 2010; Fukumoto et al. 2011; Lü et al. 2013). This would lead to a loss of N and reduce the nutrient content of the compost product; therefore, suitable bulking materials must be evaluated to address this issue.

Lignocellulosic agricultural and forestry by-products are commonly used as bulking agents in composting of animal manures. The most commonly used materials are cereal straw (Barrington et al. 2002; Wang et al. 2004; Petric and Selimbašić 2008), cotton waste (Paredes et al. 1996), hay (Barrington et al. 2002) and wood chips/shavings/saw dust and leaves (Guerra-Rodríguez et al. 2001; Huang et al. 2004; Wang et al. 2004; Ahn et al. 2007). Low moisture and high organic-C contents of these materials can alleviate the manure physical and chemical properties suitable for composting. The success of the organic matter stabilization can be related to the type and quality of the bulking agent used. Addition of organic materials containing more recalcitrant carbon, reduces the degradation of the organic matter and enhances the humification and the quality of the final composts (Mahimairaja et al. 1994; Goyal et al. 2005). These recalcitrant materials facilitate the rapid formation of humic substances during composting that improve the quality of the compost, since humic acids can affect directly or indirectly the chemical, physical and biological properties of the soil (Senesi and Plaza 2007). When using biochar as a bulking material, the humification was observed to be rapid and as high as 70% organic degradation was also observed (Dias et al. 2010).

Wong (2009b) demonstrated the use of horse stable bedding waste, a waste by itself, as a bulking agent to compost with pig manure and abattoir blood meal in a

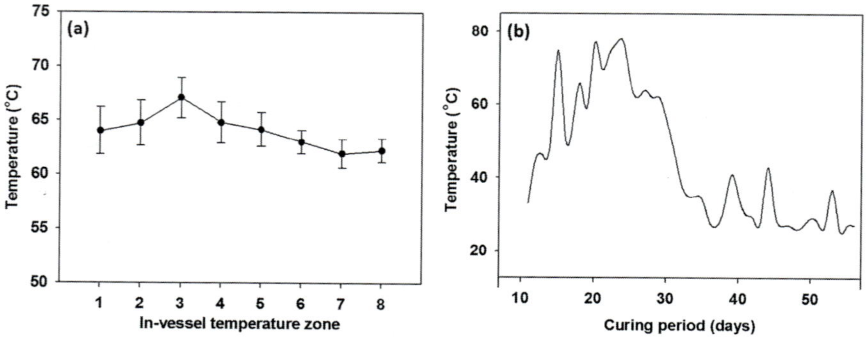

Figure 9.4. Temperature profiles of the in-vessel composter (a) and curing piles (b) during co-composting of abattoir blood meal with horse stable straw bedding waste (Modified from Wong 2009; Wong et al. 2012)

10-tpd pilot-scale in-vessel composter followed by curing in bunkers. Results from this study indicated that around 60% organic reduction was achieved during the in-vessel composting for seven days. Subsequently, the pre-mature composts were cured in negative-aerated bunkers until maturity. Figure 9.4 presents the temperature profiles of the in-vessel composter and the subsequent curing of the abattoir blood meal with horse stable bedding waste with an initial C/N ratio of 32. The temperature of the in-vessel compost ranged from 62 to 67°C while that of the curing piles was more than 65°C for about two weeks indicating an active microbial decomposition. These high temperatures ensured complete pathogen elimination. An overall organic degradation of 62% on dry weight basis was achieved and the compost maturity was achieved in about 6–7 weeks (Wong et al. 2012). A similar temperature profile ranging between 60°C and 70°C was also observed for the pig manure-horse stable bedding waster with high initial curing temperature. These trials clearly indicate that appropriate substrate mix and operation are the keys to achieve optimum composting in a commercial scale.

In contrast to typical composting, an alkaline stabilization could be a useful approach for manures. For example, 8-days alkaline stabilization of pig manure with lime at 4% and CFA at 50% completely devitalized *Salmonella* (Figure 9.5), fecal coliforms, *E.coli* and fecal *Streptococcus* in the manure-ash-lime mixtures, whereas in the control, incubation reduced the pathogen and total bacterial population by 2 ~ 3 logs. Besides, alkaline stabilization inhibited the re-growth of the pathogen during subsequent 21-days incubation (Wong and Selvam 2009). However, sufficient care must be exercised and long-term monitoring is required to reveal the complete effectiveness.

Antibiotics are used in animal husbandries for both prophylactic and therapeutic purposes. However, up to 90% of the administered antibiotics can be excreted from the animal body as non-metabolized parent compounds. These antibiotics in the animal manure become a significant source of antibiotics and

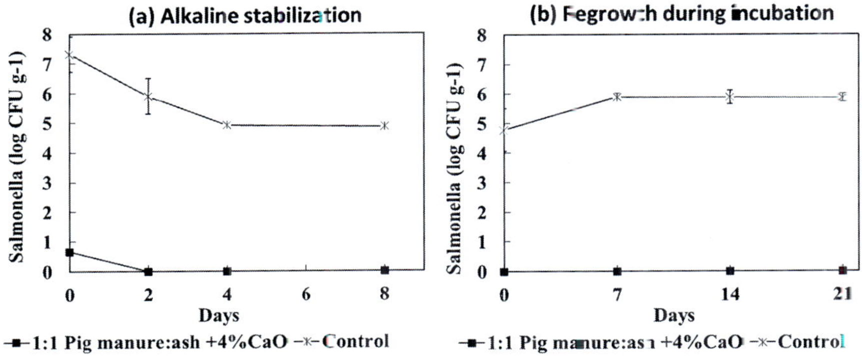

Figure 9.5. Effect of alkaline stabilization on the inactivation of Salmonella *(a) and its subsequent re-growth during greenhouse incubation (Modified from Wong and Selvam 2009)*

cause the development of antibiotic resistant microbes in the environment (Heuer et al. 2011). Composting as a potential treatment to degrade these contaminants were reported (Zhang et al. 2006; Kakimoto et al. 2007; Arikan et al. 2007; Arikan et al. 2009; Ramaswamy et al. 2010; Selvam et al. 2012a, 2013). The results of a study on the degradation of chlortetracycline (CTC), sulfadiazine (SDZ) and ciprofloxacin (CIP) during pig manure composting indicated that antibiotics at high concentrations delayed the initial decomposition and affected the nitrogen mineralization. CTC and SDZ were completely removed from the composting mass within 21 and 3 days, respectively; whereas, 17–31% of the spiked CIP remained in the compost. Therefore, composting could effectively remove the CTC and SDZ spiked at high concentrations (100 and 20 mg/kg), but the removal of fluoroquinolone antibiotics needed to be improved, indicating this antibiotic may get into the ecosystem through land application of livestock compost. Besides composting had significant influence on the antibiotic resistant gene (ARG) too. Tetracycline resistance genes (*tet*Q, *tet*W, *tet*C, *tet*G, *tet*Z and *tet*Y) and sulfonamide resistance genes (*sul*1, *sul*2, *dfr*A1 and *dfr*A7) represented 0.02–1.91% and 0.67–10.28%, respectively, of the total 16 S rDNA copies in the initial composting mass. After 28–42 days of composting, these ARGs were undetectable in the composting mass indicating that composting is a potential method for managing antibiotics in manure. Polymerase chain reaction-Denaturing gradient gel electrophoresis analysis (PCR-DGGE) of bacterial 16 S rDNA extracted from the composting mass indicate that the addition of antibiotics elicited only a transient perturbation and the bacterial diversity was restored in due course of composting.

The degradation of antibiotic was observed to be higher during the thermophilic phase, and hence the high temperature could promote effective degradation of antibiotics (Arikan et al. 2009; Selvam et al. 2013). To resolve this effect, a continuous thermophilic pig manure composting was performed (Selvam et al. 2013) and the degradation profile was almost similar to the one with the conventional composting (Selvam et al. 2012b). Therefore, mere thermophilic temperature alone could not effectively degrade the antibiotics during composting; instead a mesophilic phase preceding the thermophilic phase could enhance the degradation. However, direct involvement of such mesophilic microbes needs to be explored further.

The large abundance of livestock wastes is difficult to be effectively utilized for direct land application due to limited land area. Besides, fresh livestock wastes are also not suitable for land application not only because of its pathogens, contaminants and unstable nutrient contents but also because of the high transportation cost. Therefore, composting of livestock wastes is an alternative method to reduce the quantity of material, odor problems, pathogens and ground water pollution, and to improve the handling process.

9.3.4 Other Feed Stocks

A number of other organic wastes such as MSW, yard waste, agriculture biomass, etc., are the notable ones receiving some attention. However, the separation of wastes to a reasonable purity in case of MSW, and the availability of other plant-based wastes in reasonable quantity are the factors determining the feasibility of composting and quality of compost products to be obtained. Nevertheless, composting is a robust process and performing it in a de-centralized way may provide opportunities to recycle a variety of wastes with localized applications.

Recently, Chinese medicinal herb residues (CMHRs) caught the attention as they are produced in significant quantities in the Chinese regions. One way is that it can be used as a bulking agent for food waste composting enabling the production of compost with anti-plant pathogenic effects. It was observed that that the thermophilic phase of composting was maintained for almost three weeks that shortened the maturity period from 7–9 weeks to 5–6 weeks with the CMHRs addition. The decomposition rate of 67% was achieved with CMHRs addition while the lignin components of the herbal residues provide nucleus for the rapid development of humic substances during composting (Zhou et al. 2014). Additionally, 40% of the active compounds found in the CMHRs can be left after normal decoction and most of which have anti-pathogenic effects. After composting, the anti-pathogenic effects were still observed in the composts and exhibited inhibition rates of 16% against *Alternaria solani* and 20% against *Fusarium oxysporum*. Such a novel new substrates would add value to the compost product.

9.4 ODOUR CONTROL AND MANAGEMENT

Emission of odour is a serious concern during composting of any organic substrate. The generation of odorous compounds is the results of microbial degradation of carbohydrates, proteins and fats in the substrate. Volatile fatty acids (VFAs), amines, aromatics, inorganic sulfide, organic sulfides, terpenes and ammonia are the potentially significant odour emitted gases during composting (Table 9.4). The odorous gases are mainly emitted during the early stage of composting, especially the thermophilic stage. The factors affecting the odour

Table 9.4. Different odor-causing compounds from composting of pig manure and food waste

Odorous Compounds	*Chemical Formula*	*Molecular Weight*	*Odor Detection Threshold (mg m^{-3})*
Nitrocompounds			
Ammonia	NH_3	17	0.03–37.8
Trimethyl amine	$(CH_3)_3N$	59	0.00026–2.1
Sulfides			
Hydrogen sulfide	H_2S	34	0.0001–0.27
Dimethyl sulfide	$(CH_3)_2S$	62	0.0003–0.16
Dimethyl disulfide	$(CH_3S)_2$	94	0.0011–0.046
Mercaptans	CH_3SH	48	0.0003–38
Volatile organic compounds			
Indole	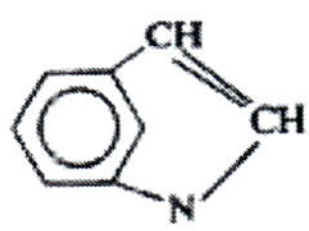	117	0.0006–0.0071
Skatole	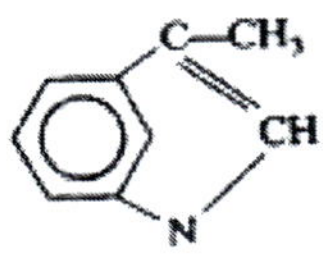	131	0.00035–0.00078
Phenol	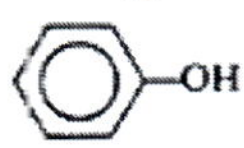	94	0.022–4
p-cresol	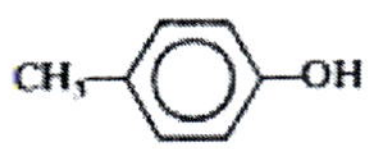	108	0.00005–0.024
Volatile fatty acids			
Acetic acid	CH_3COOH	60	0.025–10
Propionic acid	CH_3CH_2COOH	74	0.003–0.089
Butyric acid	$CH_3(CH_2)_2COOH$	88	0.0004–42
Iso-butyric	$(CH_3)_2CHCOOH$	88	0.005–0.33
n-valeric acid	C_4H_9COOH	102	0.0008–0.12
Iso-valeric	$(CH_3)_2CHCH_2COOH$	102	0.0002–0.0069
Hexanoic acid	$CH_3(CH_2)_4COOH$	116	0.02–0.52
Heptanoic acid	$CH_3(CH_2)_5COOH$	130	0.022–0.033

Sources: Van Gemert and Nettenbreijer (1977), Schaefer (1977), Spoelstra (1980) and Louhelainen et al. (2001), Font et al. (2011)

emission are the nature of the feedstock, pH, aeration, nutrient balance (C/N ratio and C/S ratio), moisture, turning, bulk density and available free air space (CIWMB 2007). Among these factors, substrates play the key role determining the odour potential in relation to other operational parameters. During the early stages of food wastes composting, significant quantities of low molecular weight carboxylic acids (LMWCAs) are produced intensively which acidify the compost mass, which results in the emission of acidic odour mainly contributed by the VFAs. Whereas, if alkaline materials are amended to control the pH, it may lead to the emission of ammonia. In contrast, animal manures such as pig manure contain high quantities of ammonia (>20% of the total N) and relatively alkaline pH that results in rapid emission of NH_3 from the composting mass. In addition, the aeration that supplies O_2 for the microbes tends to affect the generation of odourous compounds.

9.4.1 Controlling Odour Generation

The major challenge in adopting composting as the treatment technology is to develop effective means to control odour generation. The approaches for odour control include: (1) Technologies that aim at preventing odour generation such as the addition of feed additives and the proper process control of parameters during composting (Schlegelmilch et al. 2005; Wong et al. 2009; Sundberg et al. 2011, 2013); (2) Technologies that capture and destroy odors include storage covers or biofilters; and (3) Technologies that disperse or mask odors include stacks, wind breaks, site selection, perfumes, or masking agents (Schmitt et al. 1995).

Organic decomposition and odor generation problems during composting are highly related to the O_2 level within the composting system. If the O_2 content in the composting mass drops below 17%, anaerobic pockets are created, which will lead to the formation of VFAs, subsequently lowering the pH. Aerobic environment during composting can rapidly facilitate the breakdown of VFAs and produce non-offensive compounds such as water and CO_2 resulting in the reduction in the emission of VFAs (Michel and Reddy 1998; Elwell et al. 2004). Therefore, maintaining adequate aerobic conditions during composting is one of the most critical lines of defense against odours.

Besides, inoculation of suitable bacteria is shown to reduce the emission of volatile fatty acids (VFAs) from the composting mass. For example, inoculation of a thermophilic bacterial consortium reduced the emission of different VFAs (Figure 9.6). In addition, higher aeration also reduced the concentrations of VFAs in the exhaust air from the compost reactor due to enhanced microbial activity, and probably the dilution of VFA with the higher rate of air flow. Nevertheless, the influence of appropriate inoculum as well as optimal aeration have a potential in controlling odour that can enhance the adaptability of composting even in urban setups.

Ammonia is generated upon breakdown of nitrogenous compounds in the compost and is one of the main nuisances during the composting of manure. It represents 99% of the nitrogenous compounds emitted during the composting of

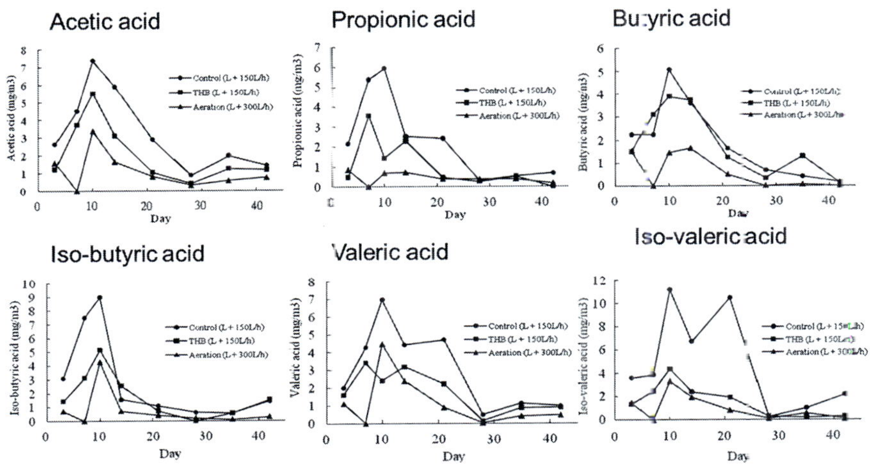

Figure 9.6. Influence of microbial inoculation and aeration on the emission of VFA during food waste composting. Treatments: Control (2.25% lime + aeration at 150° L/h), THB (thermophilic bacterial consortium + 2.25% lime + aeration at 150° L/h), and Aeration (2.25% lime + aeration at 300° L/h)

biosolids (Goldstein 2002). Ammonia is formed during the volatilization of nitrogen compounds from the compost under high pH (Seekins 1999). This is due to the shift of the ammonia-ammonium ion equilibrium to ammonia while the reverse occurs at low pH, e.g. mesophilic phase. Ammonia can also be formed by microbial ammonification in the composting process utilizing readily available nitrogen. When there is an excess of nitrogen in the compost, i.e. a low C/N ratio (<20), ammonia will be produced from the compost (Wong et al. 2012). Sikora (1999) also showed that the adjustment of C/N ratio to 25 using municipal solid waste for manure will help reduce the nitrogen loss by volatilization. Walker (1993) showed that increased process aeration led to a reduction of odorous compounds concentration but increased the total mass emission of these compounds. Use of coal fly ash was observed to reduce the NH_3 loss during food waste composting (Wong et al. 2009); however, the actual mechanism is not clear. Whether they provide adsorption sites or control the rate of ammonification needs to be investigated to exploit a cost-effect method. Similarly, zeolite, a mineral that naturally occurs on earth, can also provide adsorption sites for NH_3 released during composting (Turan and Ergun 2007).

9.5 INOCULA AND STARTER CULTURES

To enhance microbial activities for fast composting, microbial inocula or compost starters may be used, especially for commercial composters. Specific waste types, like post-consumption food waste processed in a community composter, are often promoted by microbial starters. Therefore, supplementing appropriate inocula may be necessary to activate the biodegradation of organic matter and enhance the quality of the compost.

Although the inoculation of compost with specific microbes to decompose different substrates at a broad range of environmental conditions is still contentious, several earlier researchers had attempted using bacterial (*Bacillus* sp., *Pseudomonas aeruginosa, Pseudomonas striatum*) and fungal species (*Trichoderma viride, Aspergillus niger,* Chaetomium globosum and white-rot fungi) to degrade the cellulolytic and lignolytic wastes (Crecchio et al. 2004; Yu et al. 2007; Zeng et al. 2010). Numerous microbial strains with diverse physiological ability co-exist in nature. They may compliment to each other in a mixed populations that survive under various environmental constraints and conditions (Lin et al. 2011; Wang et al. 2011). However, the choice of the microorganisms selected for the inoculation is very important that they have to be competitive enough to colonize the material and the quantity of inoculum must be sufficient. Otherwise, the competition of the native microorganism does not allow the inoculum to develop in the compost. Besides, the influence of inoculation on composting depends on the conditions of the process and on the characteristics of the raw materials (Mondini et al. 2004; Vargas-Garcia et al. 2007). The major difficulty of the use of compost inocula is usually situation specific and the inoculation strategy has to be adapted to each situation. No single microbial species is suitable for all types of wastes and reactors; and it is well established that a succession of different groups of microbes will appear along the composting time (Tsai et al. 2008; Van-Heeren 2009; Rashad et al. 2010). Alternatively, a cheap raw multi-enzyme produced through solid state fermentation (SSF) was also shown to be a good substitute for expensive conventional enzymes. Preliminary cost calculations show beneficial aspects and economic feasibility of enzyme or microbe addition to the composting process (Recktenwald et al. 2008; Liu et al. 2011).

Nowadays, commercial interests are emerging for the development of effective microbial starter cultures for efficient composting of all types of substrates and to reduce the odour. While a number of commercial microbial starter cultures are already available in the market; they are either substrate specific or effective for a narrow range of compounds. Besides regional differences in the substrate composition could also dent the effectiveness of the starter culture. Therefore, region specific development of microbial consortium is a prerequisite, especially for community level composters. Some microbial consortia with excellent characteristics have been successfully applied in industry, although the relationship among the strains in mixed cultures is relatively unknown. Awasthi et al. (2014a) inoculated the organic fraction of MSW with *Trichoderma viride, Aspergillus niger* and *A. flavus* (supplemented as spore suspensions) in windrow and observed enhanced organic degradation and accelerated the rate of composting when compared with a control without inoculation. Similarly, our investigation of using bacterial consortium having oil degrading ability during the composting of post-consumption efficiency effectively improved the oil degradation and accelerated the composting (Awasthi et al. 2014b) Therefore, a potential requirement for the development of starter inoculum exists at least for the small-scale community level composters.

9.6 TYPES OF COMPOSTING SYSTEMS

Compost systems, based on the mode of operation can be classified into three broad categories namely: windrows, in-vessel systems, and vermicomposting.

9.6.1 Windrow Systems

Windrow composting is one of the simple and less costly systems (Van-Haaren et al. 2010), in which, organic substrate is heaped into long piles called windrows and are aerated by either manual or mechanical turning (Figure 9.7). The turning frequency should satisfy the O_2 requirement of the composting mass. In between turnings, air cannot effectively diffuse into the core of the windrows, and hence some anaerobic pockets may exist in the inner parts of the windrows leading to methane formation. However, with adequate turning, the composting process can be enhanced and the methane generation can be reduced (Komilis and Ham 2004; Wong et al. 2001; Awasthi et al. 2014a). Windrow composting is more suitable for the substrates with adequate free air space. Therefore, substrates such as food waste and animal waste may not be a good option as they need more bulking agents ultimately enhancing the production cost. The original height of piles should be 2–2.5 m high depend on the substrate porosity, climate conditions, and the capacity of the equipment used to build the pile. Extra height is beneficial in the wintertime to retain heat. It may be beneficial to top off the pile with 6

Figure 9.7. Windrows operation

inches of matured compost as they prevent the surface of the pile from drying, insulates it from heat loss, dampen flies, and filters ammonia and odours generated within the pile.

9.6.2 Static Aeration Pile Systems

A variation to the windrow systems is the 'aerated static pile system'; in which the air is forced through the piles using aeration pumps. Figure 9.8 shows the variations of facilities applying the aerated static pile. The aeration involves either mechanically forcing air up (positive pressure) or pulling it down (negative pressure) through the undisturbed composting mass. No turning or agitation of the materials occurs once the pile is setup. With the aerated static pile approach, the initial feed stock mixture is piled on a base layer of wood chips, hack straw or other material facilitating the movement of air from or into the air delivery pipes. The pipe is connected to a blower, which either pulls or pushes air through the pile. The average initial height of piles is 3.0–3.5 m depending on the material porosity, weather conditions, and the reach of the equipment used to build the pile. Extra height is advantageous in the wintertime to retain heat. It may be necessary to top off the pile with 6 inches of finished compost or bulking agent. The layer of finished compost protects the surface of the pile from drying, insulates it from heat loss, discourages flies, and filters ammonia and potential odours generated within the pile. Following composting, the material is cured either in non-aerated or aerated piles. Aeration will speed up the process so that less land area is needed. Typically curing is done under very low positive aeration. The composting turning equipment is also used for curing.

In general, the aerated static pile method is best suited for granular and relatively dry feedstock materials that have a relatively uniform particle size of less than 3.5 to 5 cm in diameter. This is because large or wet materials and materials

Figure 9.8. (a) Open extended aerated static pile (EASP) showing blowers along wall. (b) EASP in a building. Aeration is provided through trenches in the floor. Construction of piles done by front-end loaders. Curing is located in an attached covered area. Davenport, IA

of diverse sizes have a tendency to clump. Clumping constricts air flow through the piles, leads to short circuits of air pumping equipment, produces anaerobic pockets, and otherwise limits the rate of decomposition. Aerated static piles are commonly used for composting wet materials (such as biosolids). Clumping is controlled by proper mixing of bulky materials that adjust porosity and moisture.

9.6.3 In-Vessel Systems

Lack of large land area, and environmental issues like odour and leachate problems associated with open window composting favour an alternative composting system, in-vessel system. In-vessel system is a high-technology method in which composting is conducted within a fully enclosed environment. All critical environmental conditions are mechanically controlled with this method, and most in-vessel systems are fully automated These systems are rarely used to compost yard trimmings because it is expensive to maintain this degree of control. More and more facilities are selecting in-vessel systems for their MSW composting program. In-vessel technologies range from relatively simple to extremely complex systems. Two broad categories of in-vessel technologies available are rotating drum (Figure 9.9a) and tank systems (Figure 9.9b).

Rotating drum systems rely on a tumbling action to continuously mix the feedstock materials. The drums typically are long cylinders, approximately 2.7 m in diameter, which are rotated slowly. Oxygen is forced into the drums through nozzles from exterior air pumping systems. The tumbling of the materials allows oxygen to be maintained at high and relatively uniform levels throughout the drum. During the in-vessel composting, the substrates are subjected to high rate composting in the in-vessel reactor for about one week and discharged. Complete stabilization of the composting material is not possible within this timeframe, therefore, further composting and curing for 1 to 2 months is necessary depending upon the properties of the substrates or the initial composting mix. The exhaust

Figure 9.9. Types of in-vessel composters: (a) Rotary drum composter, and (b) a tank composter

gas is collected and routed to a biofilter for odour and contaminant removal. Thus the environmental impacts during the high-rate composting period can be highly reduced. Studies have clearly demonstrated the feasibility and efficiency of in-vessels composters' suitability to highly populated cities (Said-Pullicino et al. 2007; Wong 2005; Selvam et al. 2011; Wong et al. 2011; Wong et al. 2012).

Rectangular tanks are one type of horizontal in-vessel system. These tanks are long vessels in which aeration is accomplished through the use of external pumps that force air through the perforated bottom of the tanks. Mixing is accomplished by mechanically passing a moving belt or paddle wheel, through the composting material. This agitates the material, breaks up clumps of particles, and maintains porosity. Composting materials are retained in the system for 6 to 28 days and then cured in windrows for 1 to 2 months.

9.6.4 Vermicomposting

Vermicomposting is another process that uses red earthworms to transform energy-rich and complex organic substances into a stabilized humus-like product called vermicompost. The most common types of earthworms used for vermicomposting are brandling worms (*Eisenia foetida*) and red-worms or red wigglers (*Lumbricus rubellus*) (Khwairakpam and Bhargava 2009; Yadav and Garg 2011). During vermicomposting, worms feed on the organic matter and produce worm castings, which tend to concentrate the nutrients in soluble forms used by the plants. Naturally occurring organisms such as bacteria also assist in the aerobic degradation of the organic matter. Vermicomposting is especially useful for biowaste since the worms consume the material quickly and there are fewer problems with odor.

The traditional open system of vermicomposting includes beds or windrows on the ground containing materials up to 45 cm deep. But such methods require large areas of land for large scale production and relatively labor intensive. There are increased interests in in-vessel vermicomposting systems, which use bins or containers often stacked in racks. However, the rack system poses operational constraints, as considerable handling and lifting machinery are required, which pose problems when adding water and additional layers of materials.

Operating mid- to large-scale systems pose temperature control challenges. These systems generate more heat from the decomposing organic materials and hold the heat longer than the smaller units. High temperature is not conducive for the worms. For this reason, vermicomposting is more appropriate when a pretreatment of thermo-composting (thermophilic phase of the normal composting) is applied (Nair et al. 2006). This pretreatment allows and completes the active phase of composting; and the curing phase is suitable for the vermicomposting. This integration can also enables adequate pathogen inactivation and good quality product. A comparison of windrow, in-vessel and vermicomposting systems are presented in Table 9.5.

Table 9.5. Comparison of different composting methods

Aerated static pile	*Windrow*	*In-vessel*	*Vermicomposting*
Highly affected by weather; can be lessened by covering, but at increased cost	Highly affected by weather; can be lessened by covering, but at increased cost	No major effect of weather but slightly affected by ambient temperature	Normally set as indoor or in shades; outdoor shallow piles are prone to precipitation
Requires minimal management once the compost pile has been established	Piles needs regular turning	Process automated in the rotary drum; turning required for curing piles	Substrate to be added in suitable intervals, monitoring piles is crucial, minimum management
Comprehensive long operating history For both small and large scale	Reliable technology for large or small scale	Comparatively short operating history; mid-to large scale, reliable and rapid	Comparatively short operating history; mostly demonstrated for small-scale systems
Large volume of bulking agent required increasing material handling at each stage	Large volume of bulking agent required increasing material handling at each stage	Generally, less amount of bulking agents and less feed stock to handle at each phase	Less amount of bulking agent; but the shallow piles increase the handling area.
Adaptable to changes in substrate and bulking agent characteristics	Adaptable to changes in substrate and bulking agent characteristics	Sensitive to changes in substrate and bulking agents	Worms are sensitive to changes in substrate and environmental conditions to remain healthy
Wide-ranging capital cost	Low capital cost	High capital costs and maintenance costs	Low capital costs and labor intensive operation

(Continued)

Table 9.5. Comparison of different composting methods (Continued)

Aerated static pile	*Windrow*	*In-vessel*	*Vermicomposting*
Equipment and key facilities include blowers and pipes, vehicles for waste handling	Equipment include windrow turners, and vehicles for waste handling	Equipment include rotary drum and process controls, aerators, sensors, biofilters and vehicles for waste handling	Equipment include vehicles for waste handling
Moderate labor requirements	Moderate labor requirements; if turners are not used labor intensive	Labor intensive	Labor intensive, even when machinery is used for adding materials in the beds
Large land area needed	Large land area needed	Small land area adequate	Large land area needed; with the rack systems requirement can be decreased
Large volumes of air to be treated for odor control	High potential for odor generation during turning; difficult to capture/contain air for treatment; leachate production	Small volume of process air that is more easily captured for treatment	No air treatment required; odour production depends on operation efficiency
Moderately dependent on mechanical equipment	minimally dependent on mechanical equipment	Highly dependent on mechanical equipment and electric power	minimally dependent on mechanical equipment
Moderate energy requirement	Very low energy requirement	Moderate to high energy requirement	Very low energy requirement

Sources: Modified from U.S. EPA 2002; Wong 2005

9.7 CENTRALIZED COMPOSTING

Centralized composting refers to mid- to large scale composting of organics collected from different areas and transported to the composting facility. A prerequisite for such composting is the availability of large quantities of substrate in reasonable purity. Therefore, substrates like biosolids from wastewater treatment facilities, livestock wastes from concentrated animal feeding operations, yard wastes and agricultural wastes. In case of MSW, separation of putrescible wastes at source enables their processing at the centralized composting facilities. Thus, those countries such as USA, Canada, and some members of the European Union, which have mandatory source separation of MSW can recycle the organic waste through such centralized composting.

In addition, material recovery facilities (MRFs) having provision for the separation of organic wastes provide an excellent opportunity to recycle them through composting; however, the quality of the compost may not be as good as the compost produced from source separated organic wastes. Centralized composting facilities typically have a design capacity of 50 to 200 tonnes per day (Reddy 2011).

The principal systems adopted for the centralized composting are in-vessel and/or windrow composting. For example, most European compost installations include an in-vessel composter to cover the intensive composting phase while cured in static aerated piles. In contrast, windrow composts are suitable for regions where land availability is plenty; however, there are some restrictions in terms of substrates to be handled. For example, composting of animal manures and food wastes will pose some operational constraints due to the emission of ammonia and acidic odour because these emissions cannot be controlled in windrow composting. In contrast in-vessel composting systems are capable of processing almost all kind of organic wastes, and provide opportunities to control the emission of odour effectively. Besides, for cities and countries, where land availability is a serious issue, in-vessel composting systems will be appropriate.

As a typical successful example from Hong Kong, Wong et al. (2010) demonstrated the composting of food wastes from five different sources using an in-vessel composter followed by curing in negatively aerated bunkers. Food wastes generated from lunch box caterers, hotel, shopping mall, in-flight catering service and wet market were collected and transported to the 4 tpd pilot plant for composting. Wastes were mixed with wood chips and lime to adjust the initial composting properties and to control pH, respectively. Subsequently the mix was composted in an in-vessel composter for 7 days followed by curing for ~35 days. The thermophilic temperature was readily established in the composters for all the food wastes except wet market waste. Representative temperature profiles for the wastes from food waste caterers and wet market are presented in Figure 9.10. Since the wet market wastes contain more vegetables, their degradation pattern was slightly different from food waste consisted of a mix of pre-and post-consumption wastes. For all the food wastes, the average operation temperatures in different zones of the composter ranged from 60.7 to 66.3°C indicating high efficiency on pathogen removal and fly control.

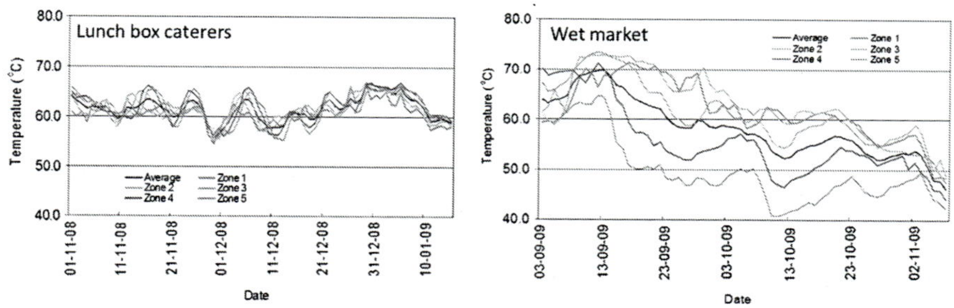

Figure 9.10. Temperature profile of the in-vessel composters treating food wastes from lunch box caterer and wet market

Matured composts were obtained after curing for about 5 weeks in the bunkers with weekly turning and native aeration in the bunkers. About 0.21 t of compost was obtained per tonne of food waste and an average weight reduction of ~56% was achieved on dry weight basis. The quality of the compost obtained from all the food wastes satisfied the composting standard for the organic farming (HKORC 2005) as presented in Table 9.6. The total NPK contents of compost samples were higher than 4% and the compost could be classified as an organic fertilizer. Meanwhile, the seed germination rates of the compost samples collected from these piles were above 80%, and therefore they will not inhibit seed germination.

The content of genetically modified organisms was <0.1% in the final product and hence the food waste compost can be used for organic farming. The environmental performance of the in-vessel composting was acceptable indicating the significance of this operation, because the facility situated in the middle of commercial centers and residential buildings. Therefore, odour is a critical issue that was efficiently controlled using a biofilter and a good housekeeping.

9.8 DECENTRALIZED COMPOSTING

The term "decentralized composting" is normally used for on-site composting, which possesses many advantages. Particularly, decentralized composting is appropriate for countries/cities where source separation of MSW is not followed and financial constraints does not allow the establishment of large scale MRFs. Besides, the organic wastes tend to have very high moisture contents ranging from 60% to >90%. Therefore transporting them to a central processing facility is demanding economically. Instead of transporting them to a faraway processing facility, if the waste is treated/recycled on-site, then the revenue required for collection and transportation can be saved. Since waste collection is the expensive process of the entire waste management system, even a small improvement in the management system would save the resources which otherwise should be met by the

Table 9.6. Properties of the compost obtained from food wastes collected from lunch box caterers

Parameter	Compost standards[a]	Food waste compost
Compost Maturity		
Ammonium (mg/kg dw)	≤700	44.7
CO_2 evolution rate (g C/kg VS/day)	≤2	0.55
C:N ratio		20.8
Seed germination rate (%)	≥80	90
Physical Properties		
pH Value	5.5–8.5	7.55
Organic matter (% dw)	>20	87.5
Moisture (%)	25–35	40.9
Nutrient Contents		
Total nitrogen (as N % dw)		2.00
Total phosphorous (as P_2O_5%dw)		0.78
Total potassium (as K_2O%dw)		1.23
Total N, P, K	≥4%dw	4.01
Heavy metals		
Arsenic (mg/kg dw)	≤10	0.31
Cadmium (mg/kg dw)	≤1	0.07
Chromium (mg/kg dw)	≤100	4.10
Copper (mg/kg dw)	≤300	9.62
Mercury (ug/kg dw)	≤1000	<2.5
Nickel (mg/kg dw)	≤50	2.79
Lead (mg/kg dw)	≤100	3.70
Selenium (mg/kg dw)	≤1.5	0.10
Zinc (mg/kg dw)	≤600	200
Pathogens		
Salmonella sp. (MPN/4 g)	≤3	0
Fecal coliforms (MPN/g)	≤1,000	<3

[a]Compost and Soil Conditioner Quality Standards for Organic Farming, 2005, Hong Kong Organic Resource Centre (HKORC)

governments. Besides, if the organic waste recycling is efficient and the products generates revenue that can be useful for the particular community. A comparison of centralized and decentralized composting systems is presented in Table 9.7.

Decentralized community on-site composting is a potential approach to reduce food waste requiring collection also resulting in significant reduction of monetary requirements. Small-scale in-vessel composting system for composting organic residues was first introduced in early 1990s (Goldstein 2002) and its potential was quickly recognized as this technology made it possible for the

Table 9.7. Comparison of centralized and decentralized composting system

Centralized composting system	*Decentralized composting system*
Involves collection and transportation of source separated wastes to a treatment facility	Source separated wastes are treated on-site that reduces the organized collection and transportation.
High capital cost for man power, machineries and vehicles; high operation cost (depends on type of composting system)	Low capital cost, public involvement reduce the operation cost
Require large land areas, electricity, regular operation and maintenance, Some odor and noise control methodology are required	Requires very small area, low maintenance costs and low level skills required, odour problems need adequate treatments and housekeeping
Less interference and involvement of the residents	Direct involvement of the public; efficient operation is a key for successful adaption of composting at the community
Operational control and routine periodic maintenance is necessary to ensure the proper functioning of this type of treatment system	Decentralized systems allow for flexibility in management and a series of processes can be combined to meet treatment goals and address environmental and public health protection requirements
Less efficient in cold areas and thus may require longer retention time	Good insulation and very efficient in cold areas and require short retention time

institutions and other establishments to perform composting with limited land availability (Papadopoulos et al. 2009; Schwalb et al. 2011). Other advantages of this technology include better process control, low labor cost, user- friendly operational demands and fewer equipment requirements. However, the key to the success is the performance of the composters and the quality of the composts obtained with these small-scale composters of 50–200 kg/d capacity (Figure 9.11). If not properly managed, they tend to produce acidic odour and the oily nature of food waste intensively acidifies the compost. Besides, the oil forms lumps of material critically affecting the composting process.

The intensive acidification leads to odour and poor composting resulting in immature compost that cannot be used immediately without further curing. Trial using a 50 kg/day reactor for vegetable waste composting demonstrated the

Figure 9.11. Small-scale community level composter

decomposition was effective in achieving ~90% and 82–85% on weight and volume basis, respectively in the presence of enhanced microorganisms (Selvam and Wong 2011). The final pH and EC of the compost products were ~8.5 and ~12 dS/m, respectively (Table 9.8). Total organic matter decreased rapidly from 96% to 84% in the first 21 days and then decreased marginally thereafter. The total nitrogen of the composting mass was nearly half of the waste indicating a significant loss of nitrogen during composting. Fecal coliforms and salmonella were not detectable after 5 weeks indicating that the effectiveness in the removal of pathogens. However, due to the very high EC, application of >1% compost decreased germination rate and growth of Chinese cabbage indicating the limitation of the compost application. The compost inoculum improved the degradation of organic matter; however, did not improve its property in terms of salinity and seed germination index. Therefore, microbial inoculum of specific

Table 9.8. Quality of food waste compost obtained from small food waste composter

Parameter	*Control*	*Enhanced microorganisms*
pH	10.1 ± 0.1[a]	10.1 ± 0.05
Electrical conductivity (dS/m)	12.0 ± 0.03	12.2 ± 0.10
Total organic matter (%)	82.7 ± 0.20	78.5 ± 0.37
Total organic carbon (%)	26.8 ± 0.16	26.6 ± 0.26
Total phosphorus (%)	1.18 ± 0.79	1.24 ± 0.11
C/N ratio	16.0 ± 1.17	16.8 ± 0.59

[a]Mean ± standard deviation (h = 3).

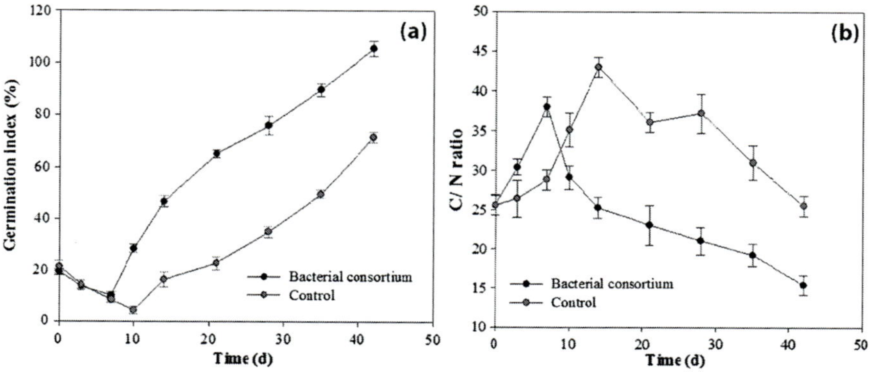

Figure 9.12. Germination index (%) (a), C/N ratio (b) during food waste composting employing bacterial consortium in in-vessel composter

functions and adopted to the local substrate conditions is an essential part of such small-scale composters.

Recently, a composting mix including alkaline substance, and microbial consortium capable of degrading oil and organic acid was developed for food waste composting (Awasthi et al. 2014b). Inoculation of this bacterial consortium increased the rate of organic degradation up to 68% and buffered the pH to neutral (7.7) range for optimum composting that reduced the odour significantly. In addition, the quality of the compost obtained was drastically improved. Mature compost with a seed germination index of >80% was obtained in about 4–5 weeks (Figure 9.12). This innovation in substrate mixing and operation made community on-site composting a possible technology in the urban setups.

Decentralized composting reduces the burden of municipality and an increased environmental awareness among residents; and citizens welcome the positive changes in their immediate environment. Proper segregation and collection of organic wastes also contributes to reducing the environmental impact at disposal sites.

9.9 CONCLUSIONS

Composting is a robust technology capable of accommodating any organic wastes, and is being practiced for many decades. Plenty of resources in terms of composting systems, scale of composting, suitability of substrate, process controls and compost quality and applications are available. Yet, when comes into practice, extending the past practice to a new substrate often not successful and requires consideration of basic factors affecting the composting performance. A classic example is the composting of food waste, which poses very critical operational constraints. If these constraints were not properly addressed, it would affect and even lead to the failure of the entire recycling programme. In addition, the heterogeneity of the organic wastes as substrate needs meticulous planning for selecting appropriate technologies, specific nutrient balance, bulking agent or co-substrate, supplementation of microbes and process controls. This brief fundamental aspects of composting and applications for different waste types, level and scale of composting, and the advantages of such systems presented here would be a guide on which new practice can be constructed and streamlined. Compost replenishes the soil with organic matter and key nutrients; however, the quality determines the market and the application. Cost effective and sustainable decentralized composting technology has many advantages but the success lie in the quality of the compost product.

References

Aasen, R., Boen, A., and Bergersen, O. (2006). "Effect of increased porosity on short organic acids, bacterial counts and respiration rate during composting of acidic organic waste." *Proc., Int. Conf. ORBIT 2006*, Weimar, Germany.

Ahn, H. K., Richard, T. L., and Choi, H. L. (2007). "Mass and thermal balance during composting of a poultry manure—Wood shavings mixture at different aeration rates." *Process Biochem.*, **42**(2), 215–223.

Arikan, O. A., Mulbry, W., Ingram, D., and Millner, P. (2009a). "Minimally managed composting of beef manure at the pilot scale: Effect of manure pile construction on pile temperature profiles and on the fate of oxy-tetracycline and chlortetracycline." *Bioresour. Technol.*, **100**(19), 4447–4453.

Arikan, O. A., Sikora, L. J., Mulbry, W. W., Khan, S. U., and Foster, G. D. (2007). "Composting rapidly reduces levels of extractable oxy-tetracycline in manure from therapeutically treated beef calves." *Bioresour. Technol.*, **98**(1), 169–176.

Awasthi, M. K., Pandey, A. K., Khan, J., Bundela, P. S., Wong J. W. C., and Selvam, A. (2014a). "Evaluation of thermophilic fungal consortium for organic municipal solid waste composting." *Bioresour. Technol.*, **168**, 214–221.

Awasthi, M. K., Selvam, A., Chan, M. T., and Wong, J. W. C. (2014b). "In-vessel composting of food waste employing indigenous microbial consortium." *Proc., 4th Int. Conf. on Solid Waste Management and Exhibition on Municipal Services, Clean Technology*, ISWMAW, Kolkata, India.

Barrington, S., Choinière, D., Trigui, M., and Knight, W. (2002). "Effect of carbon source on compost nitrogen and carbon losses." *Bioresour. Technol.*, **83**(3), 189–194.

Bergersen, O., Boena, A. S., and Sorheim, R. (2009). "Strategies to reduce short-chain organic acids and synchronously establish high-rate composting in acidic household waste." *Bioresour. Technol.*, **100**(2), 521–526.

Bernal, M. P., Alburquerque, J. A., and Moral, R. (2009). "Composting of animal manures and chemical criteria for compost maturity assessment—A review." *Bioresour. Technol.*, **100**(22), 5444–5453.

CCME (Canadian Council of the Ministers of the Environment). (2005). "Guidelines for compost quality." Ministry of Public Works and Government Services, Canada.

Chang, J., Tsai, J. J., and Wu, K. H. (2006). "Thermophilic composting of food waste." *Bioresour. Technol.*, **97**(1), 116–122.

Chatzouridis, C., and Komilis, D. (2012). "A methodology to optimally site and design municipal solid waste transfer stations using binary programming." *Resour. Conserv. Recycl.*, **60**, 89–98.

Chen, Y. (2012). "Sewage sludge aerobic composting technology research progress." *AASRI Procedia*, **1**, 339–343.

Cheung, H. N. B., Huang, G. H., and Yu, H. (2010). "Microbial-growth inhibition during composting of food waste: Effects of organic acids." *Bioresour. Technol.*, **101**(15), 5925–5934.

Chiang, K.-Y., Yoi, S.-D., Lin, H. N., and Wang, K.-S. (2011). "Stabilization of heavy metals in sewage sludge composting process." *Water Sci. Technol.*, **44**(10), 95–100

Choi, M. H., and Park, Y. H. (1998). "The influence of yeast on thermophilic composting of food waste." *Lett. Appl. Microbiol.*, **26**(3), 175–178.

CIWMB (California Integrated Waste Management Board). (2007). "Comprehensive compost odor response project." ⟨http://www.calrecycle.ca.gov/publications/Documents/Organics%5C44207001.pdf⟩ (Mar. 4, 2014).

Crecchio, C., Curci, M., Pizzigallo, M. D. R., Ricciuti, P., and Ruggiero, P. (2004). "Effect of municipal solid waste compost amendments on soil enzyme activities and bacterial genetic diversity." *Soil Biol. Biochem.*, **36**(10), 1595–1605.

Cukjati, N., Zupančič, G. D., Roš, M., and Grilc, V. (2012). "Composting of anaerobic sludge: An economically feasible element of a sustainable sewage sludge management." *J. Environ. Manage.*, **106**, 48–55.

deGuardia, A., Petiot, C., Rogeau, D., and Druilhe, C. (2008). "Influence of aeration rate on nitrogen dynamics during composting." *Waste Manage.*, **28**(3), 575–587.

Dias, B. O., Silva, C. A., Higashikawa, F. S., Roig, A., and Sánchez-Monedero, M. A. (2010). "Use of biochar as bulking agent for the composting of poultry manure: Effect on organic matter degradation and humification." *Bioresour. Technol.*, **101**(4), 1239–1246.

Elwell, D. L., Borger, D. C., Blaho, D. V., Fahrni, J. K., Keener, H. M., and Willett, L. B. (2004). "Changes in concentrations of malodorous compounds during controlled aeration composting." *Compost. Sci. Util.*, **12**(2), 102–107.

Fang, M., Wong, J. W. C., Li, G. X., and Wong, M. H. (1998). "Changes in biological parameters during co-composting of sewage sludge and coal fly ash residues." *Bioresour. Technol.*, **64**(1), 55–61.

Fang, M., Wong, J. W. C., Ma, K. K., and Wong, M. H. (1999). "Co-composting of sewage sludge and coal fly ash: Nutrient transformations." *Bioresour. Technol.*, **67**(1), 19–24.

Font, X., Artola, A., and Sanchez, A. (2011). "Detection, composition and treatment of volatile organic compounds from waste treatment plants." *Sensors*, **11**(4), 4043–4059.

Fox, T. (2013). "Global food waste not, want not." ⟨http://www.imeche.org/docs/default-source/reports/Global_Food_Report.pdf?sfvrsn=0⟩ (Apr. 4, 2014).

Fukumoto, Y., Suzuki, K., Kuroda, K., Waki, M., and Yasuda, T. (2011). "Effects of struvite formation and nitratation promotion on nitrogenous emissions such as NH_3, N_2O and NO during swine manure composting." *Bioresour. Technol.*, **102**(2), 1468–1474.

Fytili, D., and Zabaniotou, A. (2008). "Utilization of sewage sludge in EU application of old and new methods—A review." *Renewable Sustainable Energy Rev.*, **12**(1), 116–140.

Goldstein, N. (2002). "Getting to know the odor compounds." *Biocycle*, **43**, 42–44.

Gould, J. C., Rossano, M. G., Lawrence, L. M., Burk, S. V., Ennis, R. B., and Lyons, E. T. (2013). "The effects of windrow composting on the viability of Parascaris equorum eggs." *Vet. Parasitol.*, **191**(1-2), 73–80.

Goyal, S., Dhull, S. K., and Kapoor, K. K. (2005). "Chemical and biological changes during composting of different organic wastes and assessment of compost maturity." *Bioresour. Technol.*, **96**(14), 1584–1591.

Guerra-Rodríguez, E., Diaz-Raviña, M., and Vázquez, M. (2001). "Co-composting of chestnut burr and leaf litter with solid poultry manure." *Bioresour. Technol.*, **78**(1), 107–109.

Gustavsson, J., Cederberg, C., Sonesson, U., van Otterdijk, R., and Meybeck, A. (2011) *Global food losses and food waste: Extent, causes and prevention*, Food and Agriculture Organization of the United Nations, Rome, 29.

Haug, R. T. (1993). *The practical handbook of compost engineering*. Lewis Publishers, Boca Raton, FL.

Herrmann, R. F., and Shann, J. F. (1997). "Microbial community changes during the composting of municipal solid waste." *Microb. Ecol.*, **33**(1), 78–85.

Heuer, H., Schmitt, H., and Smalla, K. (2011). "Antibiotic resistance gene spread due to manure application on agricultural fields." *Curr. Opin. Microbiol.*, **14**(3), 236–243.

HKORC (Hong Kong Organic Resource Centre). (2005). "Compost and soil conditioner quality standards." ⟨http://www.hkorc-cert.org/download/COMPOST-SD-080124-A-Eng.pdf⟩ (Sep. 6, 2013).

Ho, G., and Qiao, L. (1998). "Chromium speciation in municipal solid waste: Effects of clay amendment and composting." *Water Sci. Technol.*, **38**(2), 17–23.

Hoornweg, D., and Bhada-Tata, P. (2012). "What a waste: A global review of solid waste management (urban development series knowledge papers)." Urban Development & Local Government Unit, World Bank, Washington, DC.

Huang, G. F., Wong, J. W. C., Wu, Q. T., and Nagar, B. B. (2004). "Effect of C/N on composting of pig manure with sawdust." *Waste Manage.*, **24**(8), 805–813.

Huang, D., et al. (2010). "Change of microbial population structure related to lignin degradation during lignocellulose waste composting." *Bioresour. Technol.*, **101**(11), 4062–4067.

Jeong, Y. K., and Hwang, S. J. (2005). "Optimum doses of Mg and P salts for precipitating ammonia into struvite crystals in aerobic composting." *Bioresour. Technol.*, **96**(1), 1–6.

Jeong, Y. K., and Kim, J. S. (2001). "A new method for conservation of nitrogen in aerobic composting processes." *Bioresour. Technol.*, **79**(2), 129–133.

Johnson, K. H. (1996). "Seven decades of sludge compost marketing." *The science of composting*, M. de Bertoldi, P. Sequi, B. Lemmens, and T. Papi, eds., Blackie Academic & Professional, Glasgow, 551–556.

Kakimoto, T., Osawa, T., and Funamizu, N. (2007). "Antibiotic effect of amoxicillin on the feces composting process and reactivation of bacteria by intermittent feeding of feces." *Bioresour. Technol.*, **98**(18), 3555–3560.

Ke, G. R., Lai, C. M., Liu, Y. Y., and Yang, S. S. (2010). "Inoculation of food waste with the thermo-tolerant lipolytic actinomycete Thermoactinomyces vulgaris A-31 and maturity evaluation of the compost." *Bioresour. Technol.*, **101**(19), 7424–7431.

Khwairakpam, M., and Bhargava, R. (2009). "Bioconversion of filter mud using vermicomposting employing two exotic and one local earthworm species." *Bioresour. Technol.*, **100**(23), 5846–5852.

Komilis, D., and Ham, R. (2004). "Life-cycle inventory of municipal solid waste and yard waste windrow composting in the United States." *J. Environ. Eng.*, 10.1061/(ASCE)0733-9372(2004)130:11(1390), 1390–1400.

Lau, A. K., Lo, K. V., Liao, P. H., and Yu, J. C. (1992). "Aeration experiments for swine waste composting." *Bioresour. Technol.*, **41**(2), 145–152.

Lau, S. S. S., Fang, M., and Wong, J. W. C. (2001). "Effects of composting process and fly ash amendment on phytotoxicity of sewage sludge." *Arch. Environ. Contam. Toxicol.*, **40**(2), 184–191.

Lin, H., Wang, B., Zhunag, R., Zhou, Q., and Zhao, Y. (2011). "Artificial construction and characterization of a fungal consortium that produces cellulolytic enzyme system with strong wheat straw saccharification." *Bioresour. Technol.*, **102**(22) 10569–10576.

Liu, D., et al. (2011). "Changes in biochemical and microbiological parameters during the period of rapid composting of dairy manure with rice chaff." *Bioresour. Technol.*, **102** (19), 9040–9049.

Louhelainen, K., Kangas, J., Veijanen, A., and Viilos, P. (2001). "Effect of in situ composting on reducing offensive odors and volatile organic compounds in swineries." *Am. Ind. Hyg. Assoc. J.*, **62**(2), 159–167.

Lü, D. A., Yan, B. X., Wang, L. X., Deng, Z. Q., and Zhang, Y. B. (2013). "Changes in phosphorus fractions and nitrogen forms during composting of pig manure with rice straw." *J. Integr. Agric.*, **12**(10), 1855–1864.

Mahimairaja, S., Bolan, N. S., Hedley, M. J., and Macgregor, A. N. (1994). "Losses and transformation of nitrogen during composting of poultry manure with different amendments: An incubation experiment." *Bioresour. Technol.*, **47**(3), 265–273.

Maulini-Duran, C., Artola, A., Font, X., and Sánchez, A. (2013). "A systematic study of the gaseous emissions from biosolids composting: Raw sludge versus anaerobically digested sludge." *Bioresour. Technol.*, **147**, 43–51.

Michel, F. C., Jr., Pecchia, J. A., Rigot, J., and Keener, H. M. (2004). "Mass and nutrient losses during the composting of dairy manure amended with sawdust or straw." *Compost Sci. Util.*, **12**(4), 323–334.

Michel, F. C., and Reddy, C. A. (1998). "Effect of oxygenation level on yard trimmings composting rate, odor production, and compost quality in bench-scale reactors." *Compos. Sci. Util.*, **6**(4), 6–14.

Mondini, C., Farnasier, F., and Sinicco, T. (2004). "Enzymatic activity as a parameter for the characterization of the composting process." *Soil Biol. Biochem.*, **36**(10), 1587–1594.

Nair, J., Sekiozoic, V., and Anda, M. (2006). "Effect of pre-composting on vermicomposting of kitchen waste." *Bioresour. Technol.*, **97**(16), 2091–2095.

Nakasaki, K., Akakura, N., and Atsumi, K. (1998). "Degradation patterns of organic material in batch and fed-batch composting operations." *Waste Manage. Res.*, **16**(5), 484–489.

Nakasaki, K., Yaguchi, H., Sasaki, Y., and Kubota, H. (1993). "Effects of pH control on composting of garbage." *Waste Manage. Res.*, **11**(2), 117–125.

Oleszczuk, P. (2007). "Changes of polycyclic aromatic hydrocarbons during composting of sewage sludges with chosen physico-chemical properties and PAHs content." *Chemosphere*, **67**(3), 582–591.

Papadopoulos, A. E., et al. (2009). "Performance of a new household composter during in-home testing." *Waste Manage.*, **29**(1), 204–213.

Paredes, C., Bernal, M. P., Cegarra, J., Roig, A., and Navarro, A. F. (1996). "Nitrogen transformation during the composting of different organic wastes." *Progress in nitrogen cycling studies, developments in plant and soil sciences*, Vol. **68**, Kluwer Academic Publishers, Dordrecht, 121–125.

Perez, J., Munoz-Dorado, L., Dela-Rubia, T., and Martinez, J. (2002). "Biodegradation and biological treatment of cellulose, hemicellulose and lignin: An overview." *Int. J. Microbiol.*, **5**(2), 53–63.

Petric, I, and Selimbašić, V. (2008). "Composting of poultry manure and wheat straw in a closed reactor: Optimum mixture ratio and evolution of parameters." *Biodegradation*, **19**(1), 53–63.

Petric, I., Šestan, A., and Šestan, I. (2009). "Influence of initial moisture content on the composting of poultry manure with wheat straw." *Biosyst. Eng.*, **104**(1), 125–134.

Ramaswamy, J., Prasher, S. O., Patel, R. M., Hussain, S. A., and Barrington, S. F. (2010). "The effect of composting on the degradation of a veterinary pharmaceutical." *Bioresour. Technol.*, **101**(7), 2294–2299.

Rashad, F. M., Saleh, W. D., and Moselhy, M. A. (2010). "Bioconversion of rice straw and certain agro-industrial wastes to amendments for organic farming systems. 1: Composting, quality, stability and maturity indices." *Bioresour. Technol.*, **101**(15), 5952–5960.

Recktenwald, M., Wawrzynczyk, J., Dey, E. S., and Norrlöw, O. (2008). "Enhanced efficiency of industrial-scale anaerobic digestion by the addition of glycosidic enzymes." *J. Environ. Sci. Health. Part A Toxic/Hazard. Subst. Environ. Eng.*, **43**(13), 1536–1540.

Reddy, P. J. (2011). *Municipal solid waste management: Processing–energy recovery–global examples*, BS Publication, Hyderabad, Andhra Pradesh, 449.

Ren, L. M., Schuchardt, F., Shen, Y. J., Li, G. X., and Li, C. P. (2010). "Impact of struvite crystallization on nitrogen losses during composting of pig manure and cornstalk." *Waste Manage.*, **30**(5), 885–892.

Richardson, S. D., and Ternes, T. A. (2005). "Water Analysis: Emerging contaminants and current issues." *Anal. Chem.*, **77**(12), 3807–3838.

Ryckeboer, J., Mergaert, J., Coosemans, J., Deprins, K., and Swings, J. (2003a). "Microbiological aspects of biowaste during composting in monitored compost bin." *J. Appl. Microbiol.*, **94**(1), 127–137.

Ryckeboer, J., et al. (2003b). "A survey of bacterial and fungi occurring during composting and self heating processes." *Ann. Microbiol.*, **53**(4), 349–410.

Said-Pullicino, D., Erriquens, F. G., and Gigliotti, G. (2007). "Changes in the chemical characteristic of water extractable organic matter during composting and their influence on compost stability and maturity." *Bioresour. Technol.*, **98**(9), 1822–1831.

Schaefer, J. (1977). "Sampling, characterization and analysis of malodours." *Agric. Environ.*, **3**(2-3), 121–127.

Schlegelmilch, M., Streese, J., Biedermann, W., Herold, T., and Stegmann, R. (2005). "Odour control at biowaste composting facilities." *Waste Manage.*, **25**(9), 917–927.

Schmitt, M.A., Schmidt, D. R., and Jacobson, L. D. (1995). "A manure management survey of Minnesota swine producers: Effect of farm size on manure application." *Appl. Eng. Agric.*, **12**(5), 595–599.

Schwalb, M., Rosevear, C., Chin, R., and Barrington, S. (2011). "Food waste treatment in a community center." *Waste Manage*, **31**(7), 1570–1575.

Seekins, B. (1999). "Troubleshooting the compost pile." *BioCycle*, **40**(11), 53–55.

Selvam, A., and Wong, J.W.C. (2011). "Decentralized small-scale food waste composting: Feasibility study." *Proc., IEEE Int. Conf. on Waste Recycling, Ecology and Environment*, Southwest Univ. of Science and Technology, Mianyang, Sichuan, P.R. China, 731–734.

Selvam, A., Xu, D., Zhao, Z., and Wong, J. W. C. (2012b). "Fate of tetracycline, sulfonamide and fluoroquinolone resistance genes and the changes in bacterial diversity during composting of swine manure." *Bioresour. Technol.*, **126**, 383–390.

Selvam, A., Zhao, Z. Y., Wong, J. W. C., Yu, S. M., Law, A., and Chung, P. (2011). "Co-composting of pig manure with horse stable bedding waste: Carbon and nitrogen transformation." *Proc., 2nd Int. Conf. on Solid Waste Management and Exhibition on Municipal Services, Urban Development, Public Works & Clean Technology*, Jadavpur Univ., ISWMAW, Kolkata, India, 820–824.

Selvam, A., Zhao, Z., and Wong, J. W. C. (2012a). "Composting of swine manure spiked with sulfadiazine, chlortetracycline and ciprofloxacin." *Bioresour. Technol.*, **126**, 412–417.

Selvam, A., Zhao, Z., Li, Y., Chen, Y., Leung, K. S. Y., and Wong, J. W. C. (2013). "Degradation of tetracycline and sulfadiazine during continuous thermophilic composting of pig manure and sawdust." *Environ. Technol.*, **34**(16), 2433–2441.

Senesi, N., and Plaza, C. (2007). "Role of humification processes in recycling organic wastes of various nature and sources as soil amendments." *Clean–Soil Air Water*, **35**(1), 26–41.

Sikora, L. J. (1999). "MSW compost reduces nitrogen volatilization during dairy manure composting." *Compost. Sci. Util.*, 7(4), 34–41.

Smars, S., Gustafsson, L., Beek-Friis, B., and Jonsson, H. (2002). "Improvement of the composting time for household waste during an initial low pH phase by mesophilic temperature control." *Bioresour. Technol.*, **84**(3), 237–241.

Smith, S. R. (2009). "A critical review of the bioavailability and impacts of heavy metals in municipal solid waste composts compared to sewage sludge." *Environ. Int.*, **35**(1), 142–156.

Solano, M. L., Iriarte, F., Ciria, P., and Negro, M. J. (2001). "Performance characteristics of three aeration systems in the composting of sheep manure and straw." *J. Agric. Eng. Res.*, **79**(3), 317–329.

Spoelstra, S. F. (1980). "Origin of objectionable odorous components in piggery wastes and the possibility of applying indicator components for studying odour development." *Agric. Environ.*, **5**(3), 241–260.

Sundberg, C., and Jönsson, H. (2005). "Process inhibition due to organic acids in fed-batch composting of food waste—Influence of starting culture." *Biodegradation*, **16**(3), 205–213.

Sundberg, C., et al. (2011). "Characterization of source-separated household waste intended for composting." *Bioresour. Technol.*, **102**(3), 2859–2867.

Sundberg, C., et al. (2013). "Effects of pH and microbial composition on odour in food waste composting." *Waste Manage.*, **33**(1), 204–211.

Sundberg, C., Smårs, S., and Jönsson, H. (2004). "Low pH as an inhibiting factor in the transition from mesophilic to thermophilic phase in composting." *Bioresour. Technol.*, **95**(2), 145–150.

TMECC. (2002). *Test methods for the examination of composts and composting*, W. Thompson, P. Leege, P. Millner, and M. E. Watson, eds., U.S. Composting Council, U.S. Government Printing Office, Bethesda, MD.

Tsai, C. J., Chen, M. L., Ye, A. D., Chou, M. S., Shen, S. H., and Mao, I. F. (2008). "The relationship of odor concentration and the critical components emitted from food waste composting plants." *Atmos. Environ.*, **42**(35), 8246–8251.

Turan, N. G., and Ergun, O. N. (2007). "Ammonia uptake by natural zeolite in municipal solid waste compost." *Environ. Prog.*, **26**, 149–156.

Tyagi, R. D., and Tran, F. T. (1991). "Microbial leaching of metals from digested sewage sludge in a continuous system." *Environ. Technol.*, **12**(4), 303–312.

USCC (U.S. Composting Council). (2008). "U.S. Composting Council fact sheet: Composting, global climate change and carbon trading." Ronkonkoma, NY.

U.S. EPA. (2002). "Biosolids technology fact sheet, use of composting for biosolids management." Municipal Technology Branch, Washington, DC.

Van Gemert, L. J., and Nettenbreijer, A. H. (1977). *Compilation of odour threshold values in air and water*, National Institute for Water Supply (RID), Central Institute for Nutrition and Food Research, TNO (CIVO), Zeist, Netherlands.

Van-Haaren, R. (2009). "Large scale aerobic composting of source separated organic wastes: A comparative study of environmental impacts, costs, and contextual effects." M.S. thesis, Columbia Univ., New York.

Van-Haaren, R., Themelis, N. J., and Barlaz, M. (2010). "LCA comparison of windrow composting of yard wastes with use as alternative daily cover (ADC)." *Waste Manage.*, **30** (12), 2649–2656.

Vargas-Gracia, M. C., Lopez, M. J., Suarez, F., and Moreno, J. (2007). "Effect of inoculation in composting processes: Modification in lignocellulosic fraction." *Waste Manage.*, **27**(9), 1099–1107.

Walker, J. M. (1993). "Control of composting odors." *Science and engineering of composting: Design, environmental, microbiological and utilization aspects*, H. A. J. Hoitink and H. M. Keener, eds., Renaissance Publications, Ohio, 185–218.

Wang, P., Chang, C. M., Watson, M. E., Dick, W. A., Chen, Y., and Hoitink, H. A. J. (2004). "Maturity indices for composted dairy and pig manures." *Soil Biol. Biochem.*, **36**(5), 767–776.

Wang, W., Yan, L., Cui, Z., Gao, Y., Wang, Y., and Jing, R. (2011). "Characterization of a microbial consortium capable of degrading lignocellulose." *Bioresour. Technol.*, **102**(19), 9321–9324.

Wang, X., Selvam, A., Chan, M. T., and Wong, J. W. C. (2013). "Nitrogen conservation and acidity control during food wastes composting through struvite formation." *Bioresour. Technol.*, **147**, 17–22.

Wichuk, K. M., and McCartney, D. (2007). "A review of the effectiveness of current time-temperature regulation on pathogen inactivation during composting." *J. Environ. Eng. Sci.*, **6**(5), 573–586.

Wong, J. W. C. (2005). "Organic waste generation and management study (I)—Major generators of food waste and yard waste. Organic waste generation and management study." Final Rep., Hong Kong Environmental Protection Dept., Earth Tech Consultancy, Hong Kong Baptist Univ., Hong Kong SAR, P.R. China.

Wong, J. W. C. (2009a). "Pilot plant development of biodegradable waste treatment facilities- investigation." Final Rep., Hong Kong Environmental Protection Dept., Earth Tech Consultancy, Hong Kong SAR, P.R. China.

Wong, J. W. C. (2009b). "The additional work on livestock waste/sludge composting trials at Ngau Tam Mei animal waste composting plant." Electrical and Mechanical Services Dept., Sino-Forest Applied Research Centre for Pearl River Delta Environment, Hong Kong Baptist Univ., Hong Kong SAR.

Wong, J. W. C. (2010). "Pilot plant development of biodegradable waste treatment facilities-investigation." Hong Kong Environmental Protection Dept., Earth Tech Consultancy, Hong Kong Baptist Univ., Hong Kong SAR, P.R. China.

Wong, J. W. C., and Fang, M. (2000). "Effects of lime addition on sewage sludge composting process." *Water Res.*, **34**(15), 3691–3698.

Wong, J. W. C., and Selvam, A. (2006). "Speciation of heavy metals during co-composting of sewage sludge with lime." *Chemosphere*, **63**(6), 980–986.

Wong, J. W. C., and Selvam, A. (2009). "Reduction of indicator and pathogenic microorganisms in pig manure through fly ash and lime addition during alkaline stabilization." *J. Hazard. Mater.*, **169**(1-3), 882–889.

Wong, J. W. C., et al. (2001). "Co-composting of soybean residues and leaves in Hong Kong." *Bioresour. Technol.*, **76**(2), 99–106.

Wong, J. W. C., Fang, M., Li, G. X., and Wong, M. H. (1997). "Feasibility of using coal ash residues as co-composting materials for sewage sludge." *Environ. Technol.* **18**(5), 563–568.

Wong, J. W. C., Fung, S. O., and Selvam, A. (2009). "Coal fly ash and lime addition enhances the rate and efficiency of decomposition of food waste during composting." *Bioresour. Technol.*, **100**(13), 3324–3331.

Wong, J. W. C., Selvam, A., Zhao, Z., Yu, S. M., Law, A. C. W., and Chung, P. C. P. (2011). "Influence of different mixing ratios on in-vessel co-composting of sewage sludge with horse stable straw bedding waste: Maturity and process evaluation." *Waste Manage. Res.*, **29**(11), 1164–1170.

Wong, J. W. C., Selvam, A., Zhao, Z., Yu, S. M., Law, A. C. W., and Chung, P. C. P. (2012). "In-vessel co-composting of horse stable bedding waste and blood meal at different C/N ratios: Process evaluation." *Environ. Technol.*, **33**(22), 2561–2567.

Wong, K., and Henry, G. (1988). *Bacterial leaching of heavy metals from anaerobically digested sludge*, Vol. **III**, D. L. Wise, ed., CRC Press, Boca Raton.

Yadav, A., and Garg, V.K. (2011). "Recycling of organic wastes by employing Eisenia fetida." *Bioresour. Technol.*, **102**(3), 2874–2880.

Yu, H., and Huang, G. H. (2009). "Effects of sodium acetate as a pH control amendment on the composting of food waste." *Bioresour. Technol.*, **100**(6), 2005–2011.

Yu, H., et al. (2007). "Microbial community succession and lignocelluloses degradation during agricultural waste composting." *Biodegradation*, **18**(6), 793–802.

Zeng, G., et al. (2010). "Effect of inoculation with Phanerochaete chrysosporium at various time points on enzyme activities during agricultural waste composting." *Bioresour. Technol.*, **101**(1), 222–227.

Zhang, J., et al. (2011). "Effects of physico-chemical parameters on the bacterial and fungal communities during agricultural waste composting." *Bioresour. Technol.*, **102**(3), 2950–2956.

Zhang, S. Q., Zhang, F. D., Liu, X. M., Wang, Y. J., and Zhang, J. F. (2006). "Degradation of antibiotics and passivation of heavy metals during thermophilic composting process." *China Agric. Sci.*, **39**(2), 337–343.

Zhou, H. B., et al. (2014). "Application of a recyclable plastic bulking agent for sewage sludge composting." *Bioresour. Technol.*, **152**, 329–336.

Zorpas, A. A., Constantinides, T., Vlyssides, A. G., Haralambous, I., and Loizidou, M. (2000). "Heavy metal uptake by natural zeolite and metal partitioning in sewage sludge compost." *Bioresour. Technol.*, **72**(2), 113–119.

CHAPTER 10

Biochar for Waste Management and Environmental Sustainability

Meththika Vithanage[1]
Anushka Upamali Rajapaksha[2]
Mahtab Ahmad[3]
Yoshiyuki Shinogi[4]
KyeHoon Kim[5]
Geonha Kim[6]
Yong Sik Ok[7]

10.1 INTRODUCTION

Waste management is inadequate in most cities in developing countries, as a significant portion of the population does not have access to a waste collection service and only a fraction of the generated waste is actually collected (Visvanathan et al. 2003). Although if municipalities collect the waste, final location is usually an open dumpsite. Systems for transfer, recycling, and disposal of solid waste are unsatisfactory (Schübeler et al. 1996). Problems of cover and liner shortages, lack of leachate collection and treatment facilities, inadequate compaction, poor site design, poor location selection, animals, and encroachment by garbage pickers are common at dumpsites (UNEP 2000). Landfill leachate is

[1]National Institute of Fundamental Studies, Kandy, Sri Lanka.

[2]National Institute of Fundamental Studies, Kandy, Sri Lanka.

[3]Kangwon National Univ., Chuncheon, Korea.

[4]Kyushu Univ., Japan.

[5]The Univ. of Seoul, Seoul, Korea.

[6]Hannam Univ., Daejeon, Korea.

[7]Kangwon National Univ., Chuncheon, Korea.

a common ground and surface water contamination source. However, many other environmental and social dilemmas are inherently related to open landfilling.

Many other waste management methods have been practiced in different parts of the world besides landfilling, and recycling, composting, waste to energy and incineration are included. Burning solid waste in incinerators reduces the volume of waste by 90% and weight by 75%. The heat released from the burning of solid waste in waste-to-energy incinerators can be used to heat nearby buildings or sold to generate electricity (Smargon 1999). Recycling remains the most popular environmental activity among industrialized nations. Resource recovery provides many benefits; saves energy, minimizes pollution and land disruption, reduces waste disposal costs, and increases the life of landfills by preventing waste accumulation. However, not all solid waste can be recycled except glass, plastics, paper, and metals. France, Sweden, Switzerland, and Finland recycle at least one-third of their paper expenditures (Smargon 1999). Japan, Mexico, and the Netherlands were at a 44% rate about two decades ago, which was the highest in the world (Miller 1990). However, recycling represents a very small percentage (1%) of agricultural waste, forestry, logging residues, food waste, livestock and poultry waste, sludge, invasive plants, and sea weed (Smargon 1999).

When the prevailing situation is so critical there is often a tendency to apply the "western way" of managing solid waste-related issues for a better solution. Nevertheless, a western approach used in developing countries very rapidly results in malfunctioning facilities and inefficiently managed operations (Bodelius and Rydberg 2000). Furthermore, most local municipalities in developing countries cannot afford the luxury of constructing containment landfills or incineration facilities. Thus, the prevailing conditions in these regions demand comprehensive research on solid waste and landfill management issues to improve protection of the local environment for a sustainable landfill management approach. Biochar has received recent attention as a universal solution to many environmental problems (Lehmann and Joseph 2009), At this point, biochar, a type of charcoal produced by low temperature pyrolysis of biomass, may play a vital role as a solution to never ending municipal solid waste issue.

10.2 BIOCHAR: AN EMERGING MATERIAL TO REDUCE WASTE

Biochar can be an immediate solution to reducing the global impact of bio-waste and wastewater (Lehmann and Joseph 2009). The growing cost of waste disposal is likely to make the production and application of biochar an economically viable waste management option. This section discusses biochar and its properties and uses as a potential material for waste management.

Biochar is a porous carbonaceous material obtained from thermochemical decomposition of biomass in an oxygen depleted atmosphere through a process known as pyrolysis (Shackley et al. 2012). This process is quite similar to producing charcoal, which is one of the most ancient industrial technologies developed by mankind. Biochar is being promoted for its potential to improve soil properties, crop productivity, and carbon sequestration in soil (Lehmann and Joseph 2009). Biochar is not pure carbon but consists of carbon (C), hydrogen (H), oxygen (O), nitrogen (N), sulfur (S), and ash (Duku et al. 2011a).

10.2.1 Biochar Feedstocks

Biochar is generally derived from biomass originating from plant- and animal-based waste material. Figure 10.1 summarizes possible feedstocks of biochar from different sources. Various types of biomass such as agricultural crop residues, forestry residues, invasive plants, wood waste, and organic portion of municipal solid waste, manures, and different types of sludge can be used as feedstock for biochar production.

Pyrolysis of these types of wastes produces both energy and biochar with relatively high levels of plant nutrients such as phosphorous, potassium, nitrogen, magnesium, and calcium. Hence, containment and use of nutrient-rich manures and animal products for producing biochar may also have positive environmental effects, including reduced nutrient run-off and corresponding reductions in GHG emissions such as methane and nitrous oxide (He et al. 2000). Although manure and municipal waste can be used for pyrolysis, the high risk of contamination from toxic chemicals and heavy metals may limit its use on agricultural soils (Lehmann et al. 2006). Therefore, not all waste materials are suitable for biochar production for agricultural purposes (Lehmann et al. 2006; McHenry 2009).

The abundance of biological ligno cellulosic biomass worldwide makes it a potential feedstock for biochar production (Duku et al. 2011a). The suitability of a particular biomass as a potential feedstock for biochar production depends upon

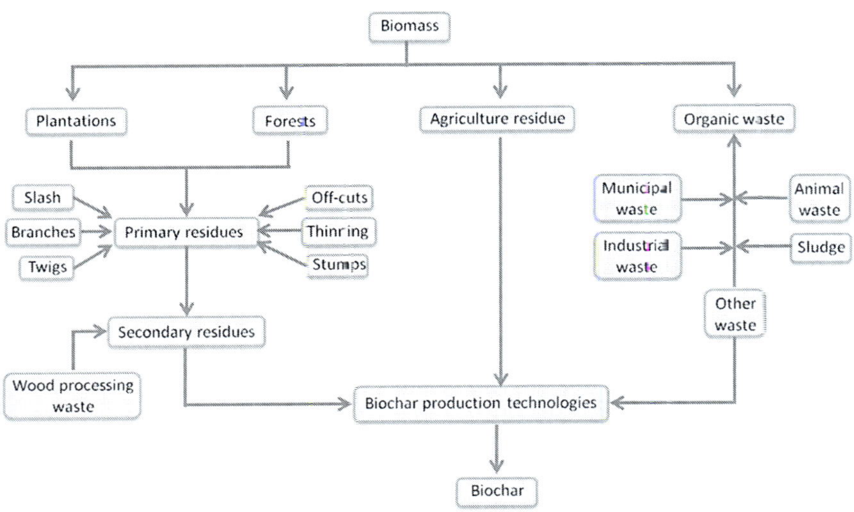

Figure 10.1. Different feedstock types for biochar production

various characteristics such as moisture content, calorific value, fixed carbon content, oxygen, hydrogen, nitrogen, volatiles, ash content, and the cellulose/lignin ratio (McHenry 2009). The most suitable biomass that produces the highest biochar yield when pyrolyzed has a high lignin concentration. Almost any organic material and some plastics can potentially be used as feedstock, including rural wastes, agro-industrial waste, sewage, animal manure, aquatic algae, and aquatic plants (Barrow 2012).

10.2.1.1 Agricultural Residue

Global annual production of agricultural residues is estimated at >500 million tons (Sánchez 2009). Much work has been conducted on agricultural waste-derived biochar for sorbing organic and heavy metal pollutants. The feedstocks used were soybean stover, rice husk, rice straw, wheat residue, and sugar beet tailing (Ahmad et al. 2012; Chen et al. 2011). Underutilized olive husks from olive oil production have been used for biochar production in Australia (El Hanandeh 2011). Yuan and Xu (2011) used nine plant residues, including non-leguminous straw from canola, wheat, corn, rice, and rice hulls as well as leguminous straw from soybean, peanut, faba bean, and mung bean to produce biochar to improve acidic soil quality. Post-harvest residues of corn, cotton, rice, sorghum and wheat produced an average 23% conversion to biochar (Jung 2011).

Other processing residues include nut shells and hulls, fruit and olive pits, and cotton gin trash. An estimate of such agricultural residues conducted in California (USA) has shown that large shells and pits total $>400{,}000\ \mathrm{Mg\,y^{-1}}$, whereas rice hulls and cotton gin trash add about $>450{,}000\ \mathrm{Mg\,y^{-1}}$ (grain and fiber processing). Many of these wastes can be used as sources for biochar production (Matteson and Jenkins 2007).

10.2.1.2 Logging and Wood Processing Residue

Removing forest biomass such as slash from logging residues and pre-commercial thinning is now becoming a viable option with the use of bio-oil. (McElligot, 2011). Logging residues including stumps, off-cuts, branches, thinning, and twigs are referred to as primary residues. According to Parikka (2004), <66% of the woody biomass volume is generally removed from the forest for further processing, whereas the remaining biomass is either left on-site, burned on-site, or utilized as wood fuel. *Jatropha* spp. plantations are being explored in India for biodiesel, alcohol, and biochar production (Hooda and Rawat 2006).

Wood processing wastes such as discarded logs, bark, sawdust, and off-cuts generated through sawmill and plywood mill processing activities are also a viable option for producing biochar (Duku et al. 2011a).

10.2.1.3 Municipal Solid Waste

It has been estimated that about 1.3 billion tons of MSW are generated globally each year, which is increasing rapidly as urbanization, mass consumption, and throw-away lifestyles become more prevalent (Hoornweg and Bhada-Tata 2012).

Most food waste is disposed in landfills. A report on the assessment of California's compost- and mulch-producing infrastructure concluded that only about 15% of food waste is diverted from landfills to conversion into bio products (Matteson and Jenkins 2007). Hence, there is a great potential in generating biochar as a byproduct from the waste to energy process.

10.2.1.4 Livestock and Poultry Waste

Manure has many applications other than its use as a fertilizer to improve crop production. It also has multiple uses as a household fuel, and biogas production, although these purposes are not being fully exploited. One estimate suggests that only 1% of global manure production is recycled as biogas (Thøy et al. 2009). Manure is beginning to play a new role as a biochar source (Cao and Harris, 2010; Uchimiya et al. 2012). Dairy manure biochar has been used for metal ion and atrazine sorption in aqueous media (Cao and Harris 2010).

Biochar produced from anaerobically digested dairy fiber is beneficial for reducing phosphorus in dairy industry (Streubel et al. 2011). Additionally, manure-based biochars are rich in exchangeable ions (Singh et al. 2010) and show high potential for metal stabilization (Uchimiya et al. 2012).

10.2.1.5 Wastewater and Sewage Sludge

Sludge- and wastewater-derived biochar plays an important role in environmental management. Although biochar produced from wastewater sludge significantly increases soil electrical conductivity and phosphorus and nitrogen contents, bioavailability of metals present in the biochar is low for food, indicating the potential of biochar for improving agricultural soil quality (Hossain et al. 2010). Sewage sludge-derived biochar could be a sustainable management option for infertile soils. Studies on evaluating the effects of sewage sludge biochar on soils have shown a reduction in mobile and plant-available heavy metals, which could be due to the high metal binding ability of sewage sludge-derived biochar (Lu et al. 2012a). The thermostability index, which is the ratio between weight losses of the exothermic reactions, decreases after incubation in soil amended with biochar compared to that with sewage sludge (Glaser et al. 2002). These results provide valuable information for utilizing carbon from waste products.

10.2.1.6 Other Wastes

Many different feedstocks have been used for biochar production such as black wattle, vineyard annual prunings, sugar cane, green tide algae, bamboo etc. (Uras et al. 2012; Chen et al. 2010; Bird et al. 2011). Algal biochars show similar characteristics to those produced from poultry litter relative to those derived from lingo cellulosic feedstocks. Hence, macro algal biochars are volumetrically less able to provide carbon sequestration benefits (Bird et al. 2011). Biochar from bamboo is as almost the same quality as wood biochar but it has a larger surface area index, greater absorption capacity and stronger alkaline content (Xu et al. 2012).

10.2.2 Biochar Production Technologies

A number of processes have been applied to produce biochar. Figure 10.2 is a flow chart of different biochar production technologies. Biochar was historically produced in traditional kilns in batch and continuous modes. Batch type kilns are still preferred in developing countries because of their simplicity (Brown 2009). The main problem associated with these traditional kilns is the emission of smoke, soot, and combustible gases into the atmosphere. There are some primitive and comprehensive ways to manufacture char with pan recently such as Moki-kiln (Japan). Torrefaction is a type of pyrolysis and is considered an upgraded technology to convert woody biomass into high energy density fuel with a high energy yield. The pyrolysis technique has been recently and efficiently utilized to convert biomass into valuable products such as bio-oil, syngas, and biochar. The advantage of the pyrolysis process over traditional kilns is that the emissions during thermochemical decomposition of biomass (combustible gases) are condensed to generate bio-oil and bio-energy. Pyrolysis can be classified into slow, intermediate, and fast pyrolysis. During slow pyrolysis, the process conditions include long vapor residence times of 10 s, temperatures of 450–650°C, increased atmospheric pressure, and very low heating rates of 0.01–2.0°C s^{-1} (Sohi et al. 2009), which results in increased cracking reactions that reduce the liquid organic yield and increase biochar yield. Slow and intermediate pyrolysis both result in higher biochar yields under low and moderate temperatures (35 and 25% respectively) than that from fast pyrolysis (about 10%). Other recent techniques of producing biochar are hydrothermal liquefaction processes and wood-gas stoves. Advances in biochar production technologies are continuing to develop and are mainly focused on cost-effectiveness.

Pyrolysis has gained much popularity because of its multiple advantages. During pyrolysis, the polymeric building blocks of the biomass, namely cellulose, hemicellulose, and lignin undergo cross-linking, depolymerization, and

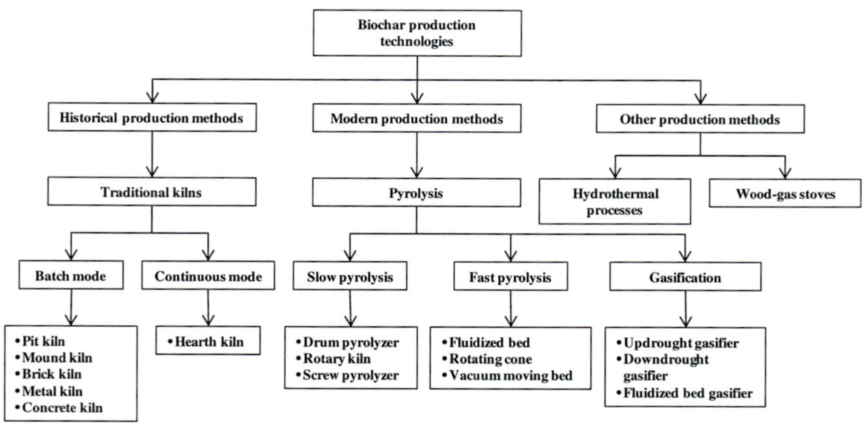

Figure 10.2. Different biochar production technologies

fragmentation at various temperatures. The primary products of biomass pyrolysis are usually referred to as condensable (tars or bio-oil) and non-condensable volatiles (gas) and char (Duku et al. 2011b).

Biochars produced from different algal species (Spirulina spp., Spyrogyra spp., and Cladophora spp.) result in 28–31% yields (Chaiwong et al 2012). The ultimate analysis of diverse feedstocks showed biochar yields of 30–38% (Downie et al. 2009). Fast pyrolysis is a good process to produce bio-oil, where a slow and intermediate pyrolysis seem to be the most appropriate to produce biochar. To sum up, pyrolysis conditions that favor high biochar yields are: (i) high lignin, ash, and nitrogen contents in the biomass, (ii) low pyrolysis temperature (<400°C), (iii) high process pressure, (iv) long vapor residence time, (v) extended vapor/solid contact, (vi) low heating rate, (vii) large biomass particle size, and (viii) optimized heat integration (Demirbas 2004).

Generally, biochar production is influenced by feedstock type and pyrolysis conditions. Massive yield of biochar could be obtained from animal litter and solid waste compared to wood biomass and crop residue (Ahmad et al. 2013). It is also recommended to adjust the pyrolysis conditions such as temperature, heating rate and atmospheric gases of pyrolysis chamber to achieve the required biochar with specific characteristics. These factors are considered to be important while evaluating the efficiency of the produced biochar.

10.2.3 General Characteristics of Biochars

Biochar can be produced from a number of wastes via pyrolysis. A temperature range of 100–900°C is generally applied to convert waste into biochar using slow to moderate pyrolysis processes. Pyrolysis temperature is the key factor influencing biochar characteristics. Table 10.1 shows the effect of pyrolysis temperature on biochar characteristics produced from a variety of feedstocks. A sharp decrease in biochar yield is observed at ≤400°C, which is related to the loss of volatile matter and non-condensable gases such as CO_2, CO, H_2, and CH_4 (Uchimiya et al. 2010). Biochar yield becomes steady at >400°C, and an increase in aromatic C and condensation reactions occur (Keiluweit et al. 2010). Biochar yield is also dependent on feedstock type. Biochars produced from animal litter and solid waste produce high yields due to their high ash content. The elemental composition of biochar is also influenced by pyrolysis temperature. Biochars produced at ≥700°C contain ~90% elemental C, which is attributed to graphitization into well-organized layers. The H and O contents of biochars decreases with increasing pyrolysis temperature due to dehydration and deoxygenation of the feedstock (Keiluweit et al. 2010). Molar H/C and O/C ratios have also been used as indicators of the aromaticity and polarity of biochars, respectively (Chen et al. 2008). A low H/C molar ratio is related to high aromaticity, whereas a low O/C molar ratio indicates low polarity of the biochars (Uchimiya et al. 2010).

The surface structure of the biochar is important for determining the sorption properties for organic and inorganic contaminants in soil and water, in addition to their impacts on moisture and nutrient retention in soil. In general, the specific

Table 10.1. Characteristics of biochars produced from wastes as affected by pyrolysis temperature

Agricultural/ solid waste	*Pyrolysis temperature ° C*	*Yield %*	*Mobile matter %*	*Fixed Matter %*	*Ash %*	*C %*	*H %*	*O %*	*N %*	*Surface area m^2/g*	*Pore volume cm^3/g*	*Reference*
Cottonseed hull	200	83.4	69.3	22.3	3.1	51.9	6.0	40.5	0.6	–	–	Uchimiya et al. (2011)
	350	36.8	34.9	52.6	5.7	77.0	4.5	15.7	1.9	4.7	–	
	500	28.9	18.6	67.0	7.9	87.5	2.8	7.6	1.5	0.0	–	
	650	25.4	13.3	70.3	8.3	91.0	1.3	5.9	1.6	34.0	–	
	800	24.2	11.4	69.5	9.2	90.0	0.6	7.0	1.9	322.0	–	
Peanut shell	300	36.9	60.5	37.0	1.2	68.3	3.9	25.9	1.9	3.1	–	Ahmad et al. (2012a)
	700	21.9	32.7	58.1	8.9	83.8	1.8	13.3	1.1	448.2	0.200	
Saw dust	450	–	40.1	57.2	1.1	72.0	3.5	24.4	0.1	–	–	Lin et al. (2012)
	550	–	13.6	82.6	2.8	85.0	1.0	13.7	0.3	–	–	
Soybean stover	300	37.0	46.34	38.7	10.4	68.8	4.3	25.0	1.9	5.6	–	Ahmad et al. (2012a)
	700	21.6	14.7	67.7	17.2	82.0	1.3	15. 5	1.3	420.3	0.190	
Wheat straw	400	34.0	–	–	9.7	65.7	4.1	–	1.1	4.8	–	Kloss et al. (2012)
	460	–	–	–	12.0	72.4	3.1	–	1.1	2.8	–	
	525	–	–	–	12.7	74.4	2.8	–	1.0	14.2	–	

Broiler litter	350	–	–	–	–	45.6	4.0	18.3	4.5	60.0	0.000	Uchimiya et al. (2010)
	700	–	–	–	–	46.0	1.4	7.4	2.8	94.0	0.018	
Feed lot	350	51.1	47.9	23.5	28.7	53.3	4.1	15.7	3.6	1.3	–	Cantrell et al. (2012)
	700	32.2	19.8	36.3	44.0	52.4	0.9	7.2	1.7	145.2	–	
Poultry litter	350	54.3	42.3	27.0	30.7	51.1	3.8	15.6	4.5	3.9	–	Cantrell et al. (2012)
	700	36.7	18.3	35.5	46.2	45.9	2.0	10.5	2.1	50.9	–	
Swine solid	350	62.3	49.8	17.7	32.5	51.5	4.9	11.1	3.5	0.9	–	Cantrell et al. (2012)
	700	36.4	13.4	33.8	52.9	44.1	0.7	4.0	2.6	4.1	–	
Tire rubber	200	93.5	–	–	15.0	74.7	6.4	3.9	–	–	–	Lian et al. (2011)
	400	59.3	–	–	15.4	77.7	3.6	3.3	–	24.2	0.080	
	600	54.5	–	–	15.6	81.3	1.7	1.4	–	51.5	0.120	
	800	43.0	–	–	10.5	86.0	0.9	2.2	–	50.0	0.110	
Turkey litter	350	58.1	42.1	23.1	34.8	49.3	3.6	15.4	4.1	2.6	–	Cantrell et al. (2012)
	700	39.9	20.8	29.2	49.9	44.8	0.9	5.8	1.9	66.7	–	

surface area of biochar increases with pyrolysis temperature due to the destruction of aliphatic alkyl and ester groups and exposure of the lignin aromatic core (Chen and Chen 2009). Development of micropores due to the thermal loss of volatiles is another contributor to the high surface area of biochars (Downie et al. 2009). However, biochars derived from animal litter and solid waste have lower specific surface areas compared to those of crop residue-derived biochars, which may be related to their lower C content and high H/C and O/C ratios, resulting in the deformation of extensive cross-linkages (Bourke et al. 2007). Changes in the surface functional groups of biochars with pyrolysis temperature have also been reported (Ahmad et al. 2012a; Chen et al. 2008). The loss of H- and O-containing functional groups with increasing pyrolysis temperature results in the high aromaticity and low polarity of biochars. Interestingly, the Fourier transform infrared spectra of biochars produced at ≥700°C is very similar to that of activated C, indicating that biochar exhibits an analogous behavior to that of activated C (Chen et al. 2008).

Biochar pH can range from 4 to 12 depending upon the feedstock type and pyrolysis conditions (Sparkes and Stoutjesdijk 2011). However, most biochars have neutral to alkaline pH that imparts their liming effect on acidic soils resulting in improved crop productivity (Verheijen et al. 2010). The pH of biochar increases with pyrolysis temperature due to the accumulation of alkali salts from the feedstock organic matrix (Shinogi and Kanri, 2003). Additionally, the removal of various acidic functional groups at high temperature causes the surfaces of biochars to become more basic (Li et al. 2002).

Uzoma et al. (2011) reported on maize productivity in soil treated with cow manure-derived biochar. They examined the effect of cow manure biochar pyrolyzed at 500 °C on maize yield, nutrient uptake, and physicochemical properties of a dry land sandy soil. Authors reported that applying cow manure biochar improves field-saturated hydraulic conductivity of the sandy soil, and net water use efficiency increases as a result. These results indicated a significant increase in pH, total C, total N, Oslen-P, exchangeable cations, and CEC. Biochar from cattle manure contains a larger content of phosphorus with citric type and, therefore, it is easy to be absorbed by crop roots and can replace phosphorus in chemical fertilizer (Shingyoji et al. 2009). Chan et al. (2008) used poultry litter biochar on radishes and showed an increase in yield from 42 to 96% relative to an unamended control.

Quality variations in poultry litter biochars pyrolyzed at different temperatures show maximal transformation of feed organic carbon (OC) into biochar recalcitrant OC at 500°C (Song and Guo 2012). Biochar from swine separated-solids, paved-feedlot manure, dairy manure, poultry litter, and turkey litter have been studied, and swine biochar shows the highest P, N, and S contents and the lowest pH and Electrical Conductivity (EC) values. Poultry litter biochar exhibits the greatest EC values. Turkey litter biochars has the highest ash content and the highest biochar mass recovery, whereas feedlot biochars demonstrate the lowest (Cantrell et al. 2012).

Overall, pyrolysis temperature is the pivotal parameter affecting biochar properties. Thus, pyrolysis conditions should be optimized according to the objectives, feedstock type, and biochar application to obtain maximum output and avoid detrimental effects.

10.3 ENVIRONMENTAL BENEFITS OF BIOCHARS

The majority of biochar research has focused on short-term impacts of applying biochar on soil nutrients, metals, gaseous, and organic contaminants dynamics. Few long-term field studies have focused on nutrient use efficiency, water use efficiency, net C sequestration, net GHG emissions, and changes in soil quality, plant, or microbial community dynamics (Ippolito et al. 2011). Promising benefits of biochar are listed in Table 10.2.

The physicochemical properties of black wattle and vineyard annual pruning biochars have been compared to sugar cane bagasse biochar, and the results demonstrated that sugar cane bagasse biochar is a promising sorbent and soil conditioner due to its high surface area, high surface acidity, and microporous structure (Uras et al. 2012).

Biochar from bagasse contains potassium; thus, it can be applied to sugarcane fields. According to Chen et al. (2010) applying bagasse based biochar enabled an increase in soil water content in sugarcane field and increased sugar content of sugarcane. Kameyama et al. (2012) reported that applying biochar from bagasse enables increased NO_3^- residence time in the crop root-zone and, therefore, provides a greater opportunity for crops to absorb NO_3^-.

An evaluation of the physiochemical properties and potential uses of macro algal (seaweed) biochar from eight species of green tide algae sourced from fresh, brackish, and marine environments indicated comparatively low carbon content, surface area, and CEC but high pH, ash, nitrogen, and extractable inorganic nutrients including P, K, Ca, and Mg (Bird et al. 2011). In comparison with woody materials, sewage sludge and animal manure require drying prior to pyrolysis and will generate finer grained, dustier materials. Sewage sludge biochar shows increasing EC with time, dissolution of phosphate and sulfuric ions, decreasing nitrogen with increasing temperature, and high phosphorous and potassium concentrations (Shinogi and Kanri 2003). Biochar contains potassium ions (Shinogi and Kanri 2003; Shinogi et al. 2003) that can be easily absorbed by crop roots and, therefore, can replace chemical fertilizer. However, the potential for converting animal manure and sewage sludge into biochar is tremendous and offers a solution to a range of problems that are associated with processing these waste materials. The benefits include a reduction in odors; elimination of pesticides, pharmaceutical, and other organic chemical contaminants; and the removal of weed seeds and undesired microorganisms such as human and plant pathogens (Navia and Crowley 2010). In particular, wooden char has been utilized

Table 10.2. Possible benefits of biochar based on the literature

Benefits	*Reference*
Agriculture sector	
Enhance plant growth	(Mao et al. 2012; Sohi et al. 2010; Sparkes and Stoutjesdijk 2011)
Raise and sustain crop yields	(Sparkes and Stoutjesdijk 2011)
Improve nutrient-poor soils, including acidic tropical humid and drier environment soils	(Gaskin et al. 2010; Sparkes and Stoutjesdijk 2011)
Compensate for greenhouse gas emissions associated with agricultural development, Store carbon in soil	(Awad et al. 2012; Awad et al. 2013; Mao et al. 2012)
Help reduce phosphate and nitrate and agrochemicals pollution of streams and groundwater	(Chen et al. 2011; Sparkes and Stoutjesdijk 2011)
Reduce the need for fertilizer/ manure/compost	(Ippolito et al. 2012; Mao et al. 2012)
Retention of plant nutrients, notably retention of N on permeable soils under rainy conditions	(Sparkes and Stoutjesdijk 2011; Chen et al. 2010)
Reduce soil acidity/raise pH	(Lehmann et al. 2011; Streubel et al. 2011)
Reduce aluminum toxicity	(Lehmann et al. 2011)
Increase cation exchange capacity	(Mao et al. 2012)
Improve moisture retention	(Mao et al. 2012)
Reduce the demand for irrigation and make cropping more secure	(Lehmann 2011; Chen et al. 2010)
Increase soil microbial biomass	(Mao et al. 2012)
Support other beneficial organism like earthworms	
Support nitrogen fixation	(Cheng et al. 2012; Ippolito et al. 2012)
Increase arbuscularmycorrhizal fungi in soil	(Ippolito et al. 2012)
Suppress methane and N	(Cheng et al. 2012)
Waste reduction	
Reduce costs of sewage and animal waste treatment and cut emissions	(Cao and Harris 2010; Chan et al. 2008)
Support biofuel production, reduce its carbon footprint and even enable it to move toward being carbon neutral	(He et al. 2000; Matteson and Jenkins 2007; Shackley et al. 2012)

Table 10.2. Possible benefits of biochar based on the literature (Continued)

Benefits	*Reference*
Act as a good material used as landfill covers due to its hydrophobicity and porosity	(Barrow 2012)
Enable production of useful materials from uncropped land making use of unused wastes	(Barrow 2012)
Pollution prevention	
Adsorptive remediation of organic contaminants in soils and water	(Ahmad et al. 2012a; Chen and Chen 2009; Lian et al. 2011)
Stabilization of toxic metals in soils	(Ahmad et al. 2012b)
Removal of metals from aqueous media	(Lu et al. 2012a; Mohan et al. 2007; Uchimiya et al. 2010)
Adsorptive remediation of inorganic contaminants in soils and water	(Lu et al. 2012a; Uchimiya et al. 2010)
Reduce plant uptake of pesticides from contaminated soils	(Kookana 2010; Yu et al. 2011)

for humidity control in Japanese residences. It is applied at the soil surface of the residence and is reported to regulate humidity in the residence.

10.4 CHALLENGES AND FUTURE PERSPECTIVES

Biochar has the potential to deliver a number of benefits to the environmental sector, but economic viability of biochar must be considered to ensure significant development within this emerging industry, particularly in the case of waste management. Production of biochar using waste material is a fascinating method to reduce biological waste material. However, producing biochar is highly dependent on a number of factors, including feedstock costs, the process itself, and the value of end-products. It may be necessary to enhance the technical aspects of biochar production to receive the maximum benefit with zero emissions. In this case, best pyrolysis technique should be chosen (Lu et al. 2012b). Hydrothermal carbonization (HTC) is a novel thermal conversion technique that could serve as an environmentally beneficial waste management/treatment process. HTC seems to be very promising for producing hydrochar with higher energy and can serve as an attractive alternative for anaerobic digestion and energy production (Lu et al. 2012b). However, research evaluating conversion of wastes via different thermal conversion techniques is still lacking, and much work is needed to better understand the environmental implications. A life cycle assessment of each of the thermal conversion processes together with a cost-benefit analysis would provide greater insight into the overall environmental impact of the system. For example, Kameyama et al. (2010) studied total CO_2 emissions during bagasse pyrolysis and concluded that if initial water content of the feeding material (bagasse) can be reduced, sequestration of carbon in soil would be increased (Kameyama et al. 2010).

Despite widespread interest in applying biochar for soil amelioration and potential mitigation of climate change, substantial uncertainties remain about the impact, capacity, and environmental sustainability of biochar production and its applications, particularly in the long term. Due to the irreversibility of applying biochar to soils, any potential risks should be thoroughly examined before widespread use of biochar is adopted.

Biochar application rates are another important factor, as some biochars may adversely affect plant growth. Additionally, not all the soils behave in a similar way after applying biochar, particularly because biochar affects soil physiochemical properties differently (Sparkes and Stoutjesdijk 2011). Hence, it is essential to develop guidelines for biochar as soil amendments with different properties under different climatic conditions and for different crops. Similarly, the ability of biochar to inhibit microbial degradation of organic molecules may allow the persistence of some adverse compounds in soils (Kookana 2010). Therefore, more in-depth knowledge would be very useful for future applications of biochar in agriculture.

Modeling may also be very effective for predicting chemical changes in complex soil mineral systems. Soil carbon and nitrogen cycle simulation modeling would be essential to understand the functional behavior of biochar and the impact on soil-based GHG emissions. It is obvious that biochar can act as an universal substance for many environmental problems, and, particularly, as a solution for reducing waste and climate change but assessments, measures of certainties, and predictions are challenges for future research.

10.5 ACKNOWLEDGMENTS

This study was supported by the National Research Foundation of Korea (NRF) (RF-2015R1A2A2A11001432). This work was carried out with the support of the "Cooperative Research Program for Agricultural Science & Technology Development (Project No. PJ010182042015)," Rural Development Administration, Republic of Korea.

References

Ahmad, M., et al. (2012). "Effects of pyrolysis temperature on soybean stover- and peanut shell-derived biochar properties and TCE adsorption in water." *Bioresour. Technol.*, **118**(1), 536–544.

Ahmad, M., et al. (2013). "Biochar as a sorbent for contaminant management in soil and water: A review." *Chemosphere*, **99**(1), 19–33.

Ahmad, M., Soo Lee, S., Yang, J. E., Ro, H. M., Han Lee, Y., and Sik Ok, Y. (2012b). "Effects of soil dilution and amendments (mussel shell, cow bone, and biochar) on Pb availability and phytotoxicity in military shooting range soil." *Ecotoxicol. Environ. Saf.*, **79**(1), 225–231.

Awad, Y. M., Blagodatskaya, E., Ok, Y. S., and Kuzyakov, Y. (2012). "Effects of polyacrylamide, biopolymer, and biochar on decomposition of soil organic matter and plant residues as determined by 14C and enzyme activities." *Eur. J. Soil Biol.*, **48**(1), 1–10.

Awad, Y. M., Blagodatskaya, E., Ok, Y. S., and Kuzyakov, Y. (2013). "Effects of polyacrylamide, biopolymer and biochar on the decomposition of 14C-labeled maize residues and on their stabilization in soil aggregates." *Eur. J. Soil Sci.*, **64**(4), 488–499.

Barrow, C. J. (2012). "Biochar: Potential for countering land degradation and for improving agriculture." *Appl. Geogr.*, **34**(1), 21–28.

Beesley, L., Moreno-Jiménez, E., and Gomez-Eyles, J. L. (2010). "Effects of biochar and greenwaste compost amendments on mobility, bioavailability and toxicity of inorganic and organic contaminants in a multi-element polluted soil." *Environ. Pollut.*, **158**(6), 2282–2287.

Bird, M. I., Wurster, C. M., de Paula Silva, P. H., Bass, A. M., and de Nys, R. (2011). "Algal biochar–production and properties." *Bioresour. Technol.*, **102**(2), 1886–1891.

Bodelius, L., Rydberg, A. (2000). "MSW management in China." *J. Waste Manage.*, **20**(1), 27–49.

Bourke, J., Manley-Harris, M., Fushimi, C., Dowaki, K., Nunoura, T., and Antal, M. J. (2007). "Do all carbonized charcoals have the same chemical structure? 2. A model of the chemical structure of carbonized charcoal." *Ind. Eng. Chem. Res.*, **46**(18), 5954–5967.

Brown, R. (2009). "Biochar production technology." *Biochar for environmental management: Science and technology*, J. Lehmann and S. Joseph, eds., Earthscan, London.

Cantrell, K. B., Hunt, P. G., Uchimiya, M., Novak, J. M., and Ro, K. S. (2012). "Impact of pyrolysis temperature and manure source on physicochemical characteristics of biochar." *Bioresour. Technol.*, **107**(1), 419–428.

Cao, X., and Harris, W. (2010). "Properties of dairy-manure-derived biochar pertinent to its potential use in remediation." *Bioresour. Technol.*, **101**(14), 5222–5228.

Chaiwong, K., Kiatsiriroat, T., Vorayos, N., and Thararax, C. (2012). "Biochar production from freshwater algae by slow pyrolysis." *Maejo Int. J. Sci. Technol.*, **6**(2), 186–195.

Chan, K. Y., Van Zwieten, L., Meszaros, I., Downie, A., and Joseph, S. (2008). "Using poultry litter biochars as soil amendments." *Soil Res.*, **46**(5), 437–444.

Chen, B., and Chen, Z. (2009). "Sorption of naphthalene and 1-naphthol by biochars of orange peels with different pyrolytic temperatures." *Chemosphere*, **76**(1), 127–133.

Chen, B., Chen, Z., and Lv, S. (2011). "A novel magnetic biochar efficiently sorbs organic pollutants and phosphate." *Bioresour. Technol.*, **102**(2), 716–723.

Chen, B., Zhou, D., and Zhu, L. (2008). "Transitional adsorption and partition of nonpolar and polar aromatic contaminants by biochars of pine needles with different pyrolytic temperatures." *Environ. Sci. Technol.*, **42**(14), 5137–5143.

Chen, Y., Shinogi, Y., and Taira, M. (2010). "Influence of biochar use on sugarcane growth, soil parameters, and groundwater quality." *Soil Res.*, **48**(7), 526–530.

Cheng, Y., Cai, Z., Chang, S., Wang, J., and Zhang, J. (2012). "Wheat straw and its biochar have contrasting effects on inorganic N retention and N2O production in a cultivated Black Chernozem." *Biol. Fertil. Soil*, **48**(8), 941–946.

Demirbas, A. (2004). "Effects of temperature and particle size on bio-char yield from pyrolysis of agricultural residues." *J. Anal. Appl. Pyrol.*, **72**(2), 243–248.

Downie, A., Crosky, A., and Munroe, P. (2009). "Physical properties of biochar." *Biochar for environmental management: Science and technology*, J. Lehmann and S. Joseph, eds., Earthscan, London.

Duku, M. H., Gu, S., and Hagan, E. B. (2011a). "Biochar production potential in Ghana—A review." *Renew. Sustainable Energy Rev.*, **15**(8), 3539–3551.

Duku, M. H., Gu, S., and Hagan, E. B. (2011b). "A comprehensive review of biomass resources and biofuels potential in Ghana." *Renew. Sustainable Energy Rev.*, **15**(1), 404–415.

El Hanandeh, A. (2011). "Trade-offs in the production and end-use of biochar and bio-oil from the solid waste generated from the olive oil industry in Australia." 19th Int. Congress on Modelling and Simulation, F. Chan, D. Marinova, and R. S. Anderssen, eds., Modelling and Simulation Society of Australia, New Zealand, 12–16.

Gaskin, J. W., et al. (2010). "Effect of peanut hull and pine chip biochar on soil nutrients, corn nutrient status, and yield." *Agron. J.*, **102**(2), 623–633.

Glaser, B., Lehmann, J., and Zech, W. (2002). *Ameliorating physical and chemical properties of highly weathered soils in the tropics with charcoal: A review*, Springer, Berlin.

He, B., Zhang, Y, Funk, T. L, Riskowski, G. L., and Yin, Y. (2000). "Thermochemical conversion of swine manure: An alternative process for waste treatment and renewable energy production." *Am. Soc. Agric. Eng.*, **43**(6), 1827–1833.

Hooda, N., and Rawat, V. R. S. (2006). "Role of bio-energy plantations for carbon-dioxide mitigation with special reference to India." *Mitigation Adaptation Strategies Global Change*, **11**(2), 437–459.

Hoornweg, D., and Bhada-Tata, P. (2012). "What a waste: Waste management around the world." World Bank, Washington, DC.

Hossain, M. K., Strezov, V., Yin Chan, K., and Nelson, P. F. (2010). "Agronomic properties of wastewater sludge biochar and bioavailability of metals in production of cherry tomato (Lycopersiconesculentum)." *Chemosphere*, **78**(9), 1167–1171.

Ippolito, J. A., Barbarick, K. A., and Elliott, H. A. (2011). "Drinking water treatment residuals: A review of recent uses." *J. Environ. Qual.*, **40**(1), 1–12.

Ippolito, J. A., Laird, D. A., and Busscher, W. J. (2012). "Environmental benefits of biochar." *J. Environ. Qual.*, **41**(4), 967–972.

Jung, W. K. (2011). "Characterization of crop residue-derived biochars produced by field scale biomass pyrolyzer." *Korean J. Soil Sci. Fertil.*, **44**(1), 1–7.

Kameyama, K., Miyamoto, T., Shiono, T., and Shinogi, Y. (2012). "Influence of sugarcane bagasse-derived biochar application on nitrate leaching in calcaric dark red soil." *J. Environ. Qual.*, **41**(4), 1131–1137.

Kameyama, K., Shinogi, Y., Miyamoto, T., and Agarie, K. (2010). "Estimation of net carbon sequestration potential with farmland application of bagasse charcoal: Life cycle inventory analysis through a pilot sugarcane bagasse carbonisation plant." *Soil Res.*, **48**(7), 586–592.

Keiluweit, M., Nico, P. S., Johnson, M. G., and Kleber, M. (2010). "Dynamic molecular structure of plant biomass-derived black carbon (Biochar)." *Environ. Sci. Technol.*, **44**(4), 1247–1253.

Khatib, I. A. (2011). "Municipal solid waste management in developing countries: Future challenges and possible opportunities." Chapter 3, *Integrated waste management—Volume 2*, S. Kumar, ed., InTech, 25–48.

Kim, K. H., Kim, J. Y., Cho, T. S., and Choi, J. W. (2012). "Influence of pyrolysis temperature on physicochemical properties of biochar obtained from the fast pyrolysis of pitch pine (Pinusrigida)." *Bioresour. Technol.*, **118**(1), 158–162.

Kloss, S. F, et al. (2012). "Characterization of slow pyrolysis biochars: Effects of feedstocks and pyrolysis temperature on biochar properties." *J. Environ. Qual.*, **41**(4), 990–1000.

Kookana, R. S. (2010). "The role of biochar in modifying the environmental fate, bioavailability, and efficacy of pesticides in soils: A review." *Soil Res.*, **48**(7), 627–637.

Lehmann, J., and Joseph, S. (2009). *Biochar for environmental management: Science and technology*, Earthscan, Sterling, VA.

Lehmann, J., Gaunt, J., and Rondon, M. (2006). "Bio-char sequestration in terrestrial ecosystems–A Review." *Mitigation Adaptation Strategies Global Change*, **11**(2), 395–419.

Lehmann, J., Rillig, M. C., Thies, J., Masiello, C. A., Hockaday, W. C., and Crowley, D. (2011). "Biochar effects on soil biota—A review." *Soil Biol. Biochem.*, **43**(9), 1812–1836.

Lian, F., Huang, F., Chen, W., Xing, B., and Zhu, L. (2011). "Sorption of apolar and polar organic contaminants by waste tire rubber and its chars in single- and bi-solute systems." *Environ. Pollut.*, **159**(4), 850–857.

Li, L., Quinlivan, P. A., and Knappe, D. R. U. (2002). "Effects of activated carbon surface chemistry and pore structure on the adsorption of organic contaminants from aqueous solution." *Carbon*, **40**(12), 2085–2100.

Lima, I. M., and Marshall, W. E. (2005). "Granular activated carbons from broiler manure: Physical, chemical and adsorptive properties." *Bioresour. Technol.*, **96**(6), 699–706.

Lin, Y., Munroe, P., Joseph, S., Henderson, R., and Ziolkowski, A. (2012). "Water extractable organic carbon in untreated and chemical treated biochars." *Chemosphere*, **87**(2), 151–157.

Lu, H., Zhang, W., Yang, Y., Huang, X., Wang, S., and Qiu, R. (2012a). "Relative distribution of Pb^{2+} sorption mechanisms by sludge-derived biochar." *Water Res.*, **46**(3), 854–862.

Lu, X., Jordan, B., and Berge, N. D. (2012b). "Thermal conversion of municipal solid waste via hydrothermal carbonization: Comparison of carbonization products to products from current waste management techniques." *Waste Manage.*, **32**(7), 1353–1365.

Mao, J. D., et al. (2012). "Abundant and stable char residues in soils: Implications for soil fertility and carbon sequestration." *Environ. Sci. Technol.*, **46**(17), 9571–9576.

Matteson, G. C., and Jenkins, B. M. (2007). "Food and processing residues in California: Resource assessment and potential for power generation." *Bioresour. Technol.*, **98**(16), 3098–3105.

McElligot, K. M. (2011). "Biochar amendments to forest soils: Effects on soil properties and tree growth." Univ. of Idaho, Moscow.

McHenry, M. P. (2009). "Agricultural bio-char production, renewable energy generation and farm carbon sequestration in Western Australia: Certainty, uncertainty and risk." *Agric. Ecosyst. Environ.*, **129**(1-3), 1–7.

Miller, G. T. J. (1990). *Living in the environment: An introduction to environmental science*, Wadsworth Publishing Company, Belmont, CA.

Mohan, D., et al. (2007). "Sorption of arsenic, cadmium, and lead by chars produced from fast pyrolysis of wood and bark during bio-oil production." *J. Colloid Interface Sci.*, **310**(1), 57–73.

Navia, R., and Crowley, D. E. (2010). "Closing the loop on organic waste management: Biochar for agricultural land application and climate change mitigation." *Waste Manage. Res.*, **28**(6), 479–480.

Parikka, M. (2004). "Global biomass fuel resources." *Biomass Bioenergy*, **27**(6), 613–620.
Riitta, P. S., and Masato, Y. (2006). "Waste generation, composition and management data." IPCC Guidelines for National Greenhouse Gas Inventories, Geneva.
Sánchez, C. (2009). "Lignocellulosic residues: Biodegradation and bioconversion by fungi." *Biotechnol. Adv.*, **27**(2), 185–194.
Schübeler, P., Wehrle, K., and Christen, J. (1996). "Urban management and infrastructure: Conceptual framework for municipal solid waste management in low-income countries." Swiss Centre for Development Cooperation in Technology and Management (SKAT), Gallen, Switzerland.
Shackley, S., et al. (2012). "Sustainable gasification-biochar systems? A case study of rice-husk gasification in Cambodia, part 1: Context, chemical properties, environmental and health and safety issues." *Energy Policy*, **42**(1), 49–58.
Shingyoji, T., Matsumaru, T., and Inubushi, K. (2009). "Residual phosphorus availability and salt influence of carbonized cattle manure and washed material on successive cultivated Komatsuna (Brassica rapa L.)." *Jpn. J. Soil Sci. Plant Nutr.*, **80**(4), 355–364.
Shinogi, Y., and Kanri, Y. (2003). "Pyrolysis of plant, animal and human waste: Physical and chemical characterization of the pyrolytic products." *Bioresour. Technol.*, **90**(3), 241–247.
Shinogi, Y., Yoshida, H., Koizumi, T., Yamaoka, M., and Saito, T. (2003). "Basic characteristics of low-temperature carbon products from waste sludge." *Adv. Environ. Res.*, **7**(3), 661–665.
Singh, B. P., Hatton, B. J., Singh, B., Cowie, A. L., and Kathuria, A. (2010). "Influence of biochars on nitrous oxide emission and nitrogen leaching from two contrasting soils." *J. Environ. Qual.*, **39**(4), 1224–1235.
Singh, B., Singh, B. P., and Cowie, A. L. (2010). "Characterisation and evaluation of biochars for their application as a soil amendment." *Soil Res.*, **48**(7), 516–525.
Smargon, J. A. (1999). "Global perspective on solid waste management." ⟨http://www.afn.org/~recycler/waste.html⟩ (Feb. 18, 2016).
Sohi, S., Loez-Capel, S., Krull, E., and Bol, R. (2009). "Biochar's roles in soil and climate change: A review of research needs." CSIRO Land and Water Science, Glen Osmond, SA, Australia, 64.
Sohi, S. P., Krull, E., Lopez-Capel, E., and Bol, R. (2010). "A review of biochar and its use and function in soil." *Advances in agronomy*, D. L. Sparkse, Academic Press, Burlington, 47–82.
Song, W., and Guo, M. (2012). "Quality variations of poultry litter biochar generated at different pyrolysis temperatures." *J. Anal. Appl. Pyrol.*, **94**(1), 138–145.
Sparkes, J., and Stoutjesdijk, P. (2011). "Biochar: Implication for agricultural productivity." ABARES Technical Rep. 11.06, Australian Bureau of Agricultural and Resource Economics and Science, Canberra.
Streubel, J. D., Collins, H. P., Garcia-Perez, M., Tarara, J., Granatstein, D., and Kruger, C. E. (2011). "Influence of biochar on soil pH, water holding capacity, nitrogen and carbon dynamics." *Soil Sci. Soc. Am. J.*, **75**(4), 1402–1413.
Thøy, K., Wenzel, H., Jensen, A. P., and Nielsen, P. (2009). "Biogas from manure represents a huge potential for reduction in global greenhouse gas emissions." IOP Conf. Series: Earth and Environmental Science, Vol. **6**, 242020, IOP Publishing, London.
Uchimiya, M., Cantrell, K. B., Hunt, P. G., Novak, J. M., and Chang, S. (2012). "Retention of heavy metals in a TypicKandiudult amended with different manure-based biochars." *J. Environ. Qual.*, **41**(4), 1138–1149.
Uchimiya, M., Lima, I. M., Thomas Klasson, K., Chang, S., Wartelle, L .H., and Rodgers, J. E. (2010). "Immobilization of heavy metal ions (CuII, CdII, NiII, and PbII)

by broiler litter-derived biochars in water and soil." *J. Agric. Food. Chem.*, **58**(9), 5538–5544.

Uchimiya, M., Wartelle, L. H., Klasson, K. T., Fortier, C. A., and Lima, I. M. (2011). "Influence of pyrolysis temperature on biochar property and function as a heavy metal sorbent in soil." *J. Agric. Food. Chem.*, **59**(6), 2501–2510.

UNEP. (2000). "International source book on environmentally sound technologies for wastewater and stormwater management." Osaka, Japan.

UNEP. (2004). "Vital waste graphics. Basel convention secretariat and produced in partnership and division of environmental conventions (DEC)." 7th Meeting of the Conf. to the Parties of the Basel Convention, Geneva, Switzerland.

Uras, Ü., Carrier, M., Hardie, A. G., and Knoetze, J. H. (2012). "Physico-chemical characterization of biochars from vacuum pyrolysis of South African agricultural wastes for application as soil amendments." *J. Anal. Appl. Pyrol.*, **98**(1), 207–213.

Uzoma, K. C., Inoue, M., Andry, H., Fujimaki, H., Zahoor, A., and Nishihara, E. (2011). "Effect of cow manure biochar on maize productivity under sandy soil condition." *Soil Use Manage.*, **27**(2), 205–212.

Verheijen, F., Jeffery, S., Bastos, A. C., van der Velde, M., and Diafas, I. (2010). "Biochar application to soils. A critical scientific review on soil properties, processes and functions." European Commission, Italy.

Visvanathan, C., Trankler, J., Basnayake, B. F. A., Chiemchaisri, C., Joseph, K., and Gonming, Z. (2003). "Landfill management in Asia- Notions about future approaches to appropriate and sustainable solutions." Proc., 9th Int. Waste Management and Landfill Symp., International Waste Working Group, Hamburg, Germany.

World Bank. (2004). "World development indicators." Washington DC.

Xu, T., Lou, L., Luo, L., Cao, R., Duan, D., and Chen, Y. (2012). "Effect of bamboo biochar on pentachlorophenol leachability and bioavailability in agricultural soil." *Sci. Total Environ.*, **414**(1), 727–731.

Yuan, J. H., and Xu, R. K. (2011). "The amelioration effects of low temperature biochar generated from nine crop residues on an acidic Ultisol." *Soil Use Manage.*, **27**(1), 110–115.

Yu, X. Y., Mu, C. L., Gu, C., Liu, C., and Liu, X. J. (2011). "Impact of woodchip biochar amendment on the sorption and dissipation of pesticide acetamiprid in agricultural soils." *Chemosphere*, **85**(8), 1284–1289.

CHAPTER 11

Innovative Technologies: Plasma Arch Gasification

Qunxing Huang[1]
Xu Cai[2]
Moussa-Mallave Alhadj-Mallah[3]
Jun Wang[4]
Feiyan Mao[5]
Xu Han[6]

11.1 INTRODUCTION

Waste treatment has become a big problem in many developing countries especially those with huge population and limited land resources, like China. In 2012, World Bank report forecasted that till 2025, more than 40% of the world increased municipal solid waste (MSW) generation comes from the countries located in East Asia and pacific region (World Bank 2012); and in China more than 180 million tons of MSW had been collected from urban cities alone (Byun et al. 2012). Despite the rapid volume reduction capability and global advances in air pollution control, mass burning continues to have a poor public image. One of the most notable problem is the potential risk to human health that might result from the emission of pollutants generated by the incineration process. Some of these pollutants have been found to cause adverse health effects. Although such effects have generally been observed with much higher ambient concentrations than those emitted from WTE plants, questions persist about the possible effects of the smaller amounts of toxic species from incineration facilities, especially when these are combined with the mix of pollutants emitted from other sources. The possible social, economic, and psychological effects associated with local residents

[1]Zhejiang Univ., Hangzhou, P.R. China.
[2]Zhejiang Univ., Hangzhou, P.R. China.
[3]Zhejiang Univ., Hangzhou, P.R. China.
[4]Zhejiang Univ., Hangzhou, P.R. China.
[5]Zhejiang Univ., Hangzhou, P.R. China.
[6]Zhejiang Univ., Hangzhou, P.R. China.

that live or workg near an incineration facility also have been topics of concern and many incineration projects have been suspended or stopped.

To alleviate such public concerns, many researchers have devoted themselves to the advanced thermal treatment technology with reduced condition: gasification (Byun et al. 2012), which is considered to be one of the most promising and environmental friendly solution. During gasification heat from partial self-combustion or external is used in an oxygen starving environment to break down hydro carbon based materials into combustible gases. It is closely related to combustion or pyrolysis and the air fuel equivalence ratio (ϕ) is defined as the actual air fuel ratio divides by the air fuel ratio for complete combustion) is the key to understand the difference between them. When ϕ is larger than 0 and less than 1.0, it is called as gasification. O_2 is strictly controlled and limited and the feedstock is not allowed to be completely burned. When the burnable materials are heated up under reduction environment, they will go through the progress of pyrolysis, producing char and tar. The char and tar are broken down into syngas, mainly composed of H_2 and CO, as the gasification process continues. The global gasification reactions is commonly formulated as follows based on the equivalent ultimate analysis (CHxOy) (Byun et al. 2012)

$$\begin{aligned} & CH_xO_y + wH_2O + mO_2 + 3.76mN_2 \\ & \quad \rightarrow aH_2 + bCO + cCO_2 + dH_2O + eCH_4 + fN_2 + gC \qquad \text{(Eq. 11.1)} \end{aligned}$$

When the temperature is known, the composition of the syngas can be deduced through Gibbs free energy minimization. However, traditional gasification technologies operate the reactor below 1,000°C (Mountouris et al. 2006) and do not use any external heat source and rely on the process itself to sustain the reactions. So traditional gasifiers are actually "partial combustors", and a substantial portion of the hydro carbon resources is combusted just to support the thermal chemical reaction. The low reaction temperatures cannot break down all the materials. As a result, in traditional gasification, many materials must be sorted out of the waste stream and landfilled or processed in other ways. The gas generated from traditional gasification process which is fed with municipal solid waste or other mixed wastes is very dirty and contains char, tars, soot and particulate matters (Mountouris et al. 2006). In additional to reducing energy converting efficiency, the tar is very difficult to remove. Other contaminants including HCl, NO_X, SO_X, et al., must be further cleaned if the gas is targeted for gas engine or other chemical utilization. Due to the low temperature, the total organic carbon (TOC) in the residue may be up to 15% of the weight of the fed material and the residue must be well treated further. (Moustakas et al. 2005). Due to these drawbacks, most traditional gasification systems cannot process mixed wastes, e.g., MWS directly from the truck (Mountouris et al. 2006).

In order to destruct all toxic components and convert combustible material into high quality gases, the temperature should be increased to high enough. Plasma arc, which is developed for stabilizing radiative wastes, provides a very promising solution. In this chapter, the developing history and fundamentals of plasma arch gasification of solid waste and its current practical application worldwide are introduced.

11.2 PLASMA FUNDAMENTALS

Word "Plasma", comes from Greek "πλαμα" which means "*to mold*", was named by Langmuir at 1928 (Winner of 1932 Nobel prize) who described his observation about the plasma as (Langmuir 1928):

> *"The ionized gas contains ions and electrons in about equal numbers so that the resultant space charge is very small. We shall use the name "plasma" to describe this region containing balanced charges of ions and electrons"*

Plasmas is considered as quasi-neutral particle systems in the form of fluid-like gaseous mixtures in which there are roughly equal numbers of positively and negatively charged particles, produced when the atoms in an agent gas is ionized when it passing through the high voltage electrical node. Although plasma jets or clouds contain electrons and ions with a large mean kinetic energy of the electrons, it is still electrically neutral and because it is fulfilled with free electrons and ions. Plasma is a good electrical conducting medium and is referred to as the fourth state of matter in many textbooks, which is distinct from the solid, liquid, and gaseous states. Generally, plasmas can be categorized into two types according to electron temperature, low-temperature plasmas (LTP) and high-temperature plasmas (HTP) (Willis et al. 2010). LTP can then be subdivided into thermal LTP and non-thermal or non-equilibrium LTP. The plasma used for waste gasification belongs to thermal low temperature plasma and HTP is only used for high energy physics or nuclear energy. See Table 11.1.

The temperature of non-thermal LTP is almost close to room temperature and the electron energy distribution is non-Maxwellian. For thermal plasma, the high temperature electrons are of high mobility and will transfer part of their

Table 11.1 Subdivision of plasma

Low-temperature plasmas (LTP)		*High-temperature plasma (HTP)*
Thermal LTP $T_e \approx T_t \approx T \approx 2 \times 10^4$ K e.g., arc plasma at normal pressure	Non-thermal LTP $T_i \approx T_e \approx 300$ K $T_i \approx T_e \approx 10^5$ K e.g., low-pressure glow discharge	$T_i \approx T_e > 10^7$ K e.g., fusion plasmas

Source: Hippler et al. 2008; reproduced with permission from John Wiley & Sons energy to the heavy particles by elastic collisions leading to an equal distribution of the energy, i.e., as an approximation, the mean heavy particle energy is the same as the electron energy and there is thermal equilibrium among all the kinds of particles allowing a single temperature to be defined for the plasma at a given position (Langmuir 1928). In the hottest regions of thermal plasmas, this mean kinetic energy is of the order of 1 eV, which corresponds to a temperature of the order of 104 K.

11.2.1 Non-Equilibrium Plasma

Non-equilibrium plasmas are plasmas with a low ionization degree, characterized by their relatively low electron or particle densities (less than 10^{20} m^{-3}). They have lower energy densities and they are very stable and occur most commonly at relative low pressures. This kind of plasmas is mainly produced in various types of electric glow discharges by direct current, radio frequency or microwave sources, and also corona discharges.

In non-equilibrium plasma, the ionization and dissociation of molecular precursors are initiated by the collisions between free electrons and gas phase species and then form a special chemical condition. Non-equilibrium plasma has gained more and more attention in the research and industrial application. For example, many researches think of cold atmospheric plasma as an innovative tool in medicine and hygiene. In material processing, low-pressure non-equilibrium plasmas are widely used in etching and deposition processes and surface modifications. To the efficient non-equilibrium plasma, the type and density of the radicals are more important than the total energy in the plasma (Hippler et al. 2008).

11.2.2 Arc Plasma

Thermal plasmas are partially or strongly ionized gases. They are typically characterized by the high-energy density and the equality of temperature of the heavy particles and electrons. Because of their much higher mobility, the energy given to the plasma is captured by the electrons and transferred to the heavy particles by elastic collision under high frequency Thermal plasmas can also be produced with direct current (DC) plasma torches or in radio frequency (RF) inductively coupled discharges (Boulos 1991; Gleizes et al. 2005). Depending on the design of the anode, the plasma torches can be classified into non-transferred arc plasma torch and transferred one (Figure 11.1–11.3). In the non-transferred arc torch, the two electrodes do not participate in the process and are only used for plasma generation. The couple of electrodes could have different structure type, such as a cylindrical bar (cathode) and a hollow cylinder (anode) or two coaxial cylinders that could operate with opposite polarities (Bonizzoni et al. 2002).

To avoid the material overheating, the electrodes should be cooled by water or gases. Some plasma torches employ magnetic fields to control the distribution of generated high temperature electrons to force the attack point of the arc on the whole electrodes surface.

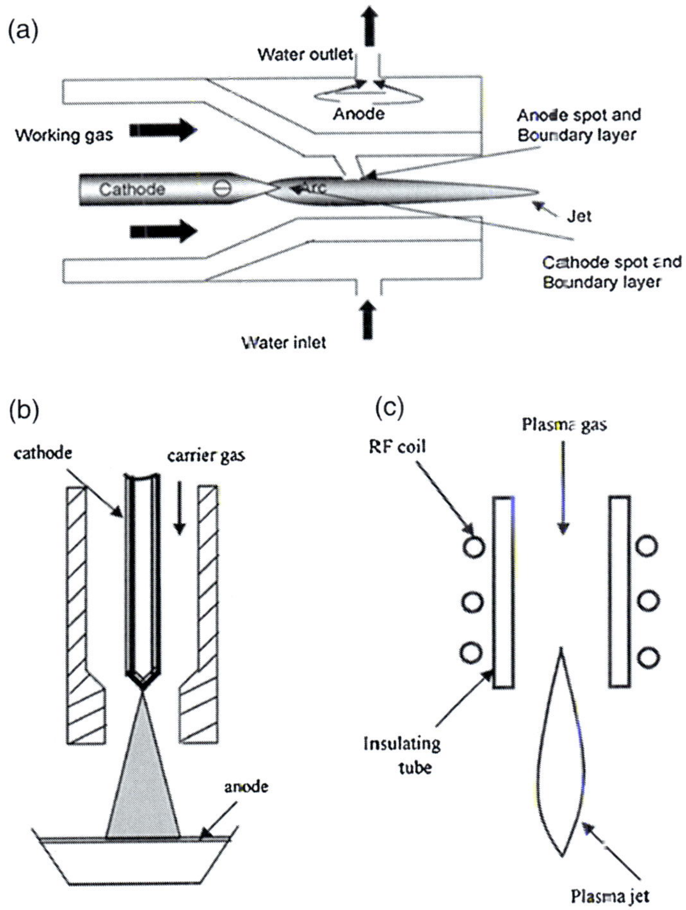

Figure 11.1. Three kinds of arc plasma torches: (a) Schematic diagram of a DC non-transferred arc plasma torch; (b) Schematic diagram of a DC transferred arc plasma torch; and (c) Schematic diagram of RF inductively coupled discharge device
Source: Gomez et al. 2009; reproduced with permission from Elsevier

In the transferred arc reactors, anode is separated from the torch and placed outside and when the plasma is generated, the arc will transfer to the extent of anode as electrically elevated thermal fluxes. Usually, the external anode is used as the metallic container of the substance to be treated. For example, when the transfer plasma is used for fly ash stabilization/vitrification, the arc starts between the cathode and the ash container heating the ash to melting temperatures and the fly ash are liquefied, subsequently vitrified reaching the complete inertisation.

The cathode can be constructed by a water cooled metal or by a "consumable through sublimation" conductor material (Bonizzoni et al. 2002). Graphite or

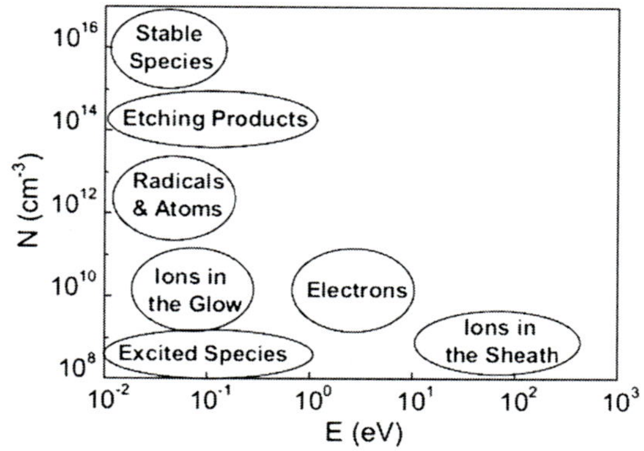

Figure 11.2. Typical concentrations and energies of different species generated in growing discharges
Source: Tanarro et al. 2011; reproduced with permission from Elsevier

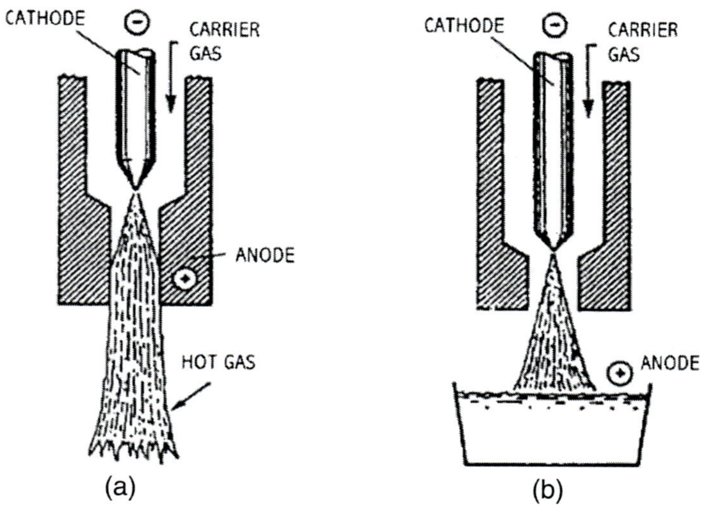

Figure 11.3. (a) Non-transferred arc plasma torch; and (b) transferred arc plasma torch
Source: Bonizzoni et al. 2002; reproduced with permission from Elsevier

other carbon material is also commonly used as the consumable cathode. The problem of consumable carbon cathode is that the products may be polluted by carbon and requires a more frequent replacement. The non-consumable electrodes, generally in copper or irons, have a longer life but need more attention to prevent erosion to avoid fluid leak. The selection of transfer or non-transfer plasma torch should be determined on the application conditions. For example, when the solid waste is pyrolyzed or the waste is in a liquid phase, non-transferred arc reactors should be considered because it can deal with dangerous and corrosive elements. In most conditions, the materials will be put in the downstream of the electrodes to prolong the life of the electrodes. The transferred arc reactors are usually used in metallurgical processes (Bonizzoni et al. 2002). Both the kinds of plasma arc reactors have high heating efficiency up to about 90%, and the transferred arc reactor is a little lower. But the electrodes of non-transferred arc reactor are consumed much faster. Similar with non-equilibrium plasma, thermal plasma can also be generated by DC, AC and radio frequency (RF) inductively coupled discharges. The plasma is formed by transmitting a high electric current between two electrodes in the presence of a gas carrier between them (Pourali 2010).

When plasma is used as a heating source for industrial applications, like melting, metal cutting and waste gasification, a plasma torch should be used in which the gas carrier will be broken down and heated to an extremely high temperature forming plasma jets. The typical temperature outside a thermal plasma jet can reach 12,000°C (Pourali 2010). The plasma is associated with ions and according to plasma definition, the degree of ionization is proportional to the number of atoms that have lost (or gained) electrons. For thermal plasmas that used for MSW gasification, it is controlled mostly by the temperature of the electric ions (Gomez et al. 2009). Because of the electrical resistivity across the system, a dramatic heat can be generated, which strips away electrons from the gas molecules resulting in an ionized gas stream. For example, if the temperature reachs 2,000°C, gas molecules will break into atomic state and when the temperature increases to 3000°C gas molecules lose electrons and become ionized. In this state, gas has a liquid-like viscosity at atmospheric pressure and the free electric charges confer relatively high electrical conductivities that can approach those of metals (Gomez et al. 2009).

When the thermal plasma is used for waste gasification, it basically provides an extremely high heating source and the simplified heat transfer mechanism involved is presented schematically in Figure 11.4 The net energy contributing to heating and melting the solid waste (Qn) is the difference between the conductive and convective energy transferred from the plasma jet to the waste particles, and the radiative energy loss from the surface of the particle to the surroundings (Boulos 1991)

$$Q_y = hs(T_\infty - T_s) - \sigma\varepsilon S(T_s^4 - T_\alpha^4) \qquad \text{(Eq. 11.2)}$$

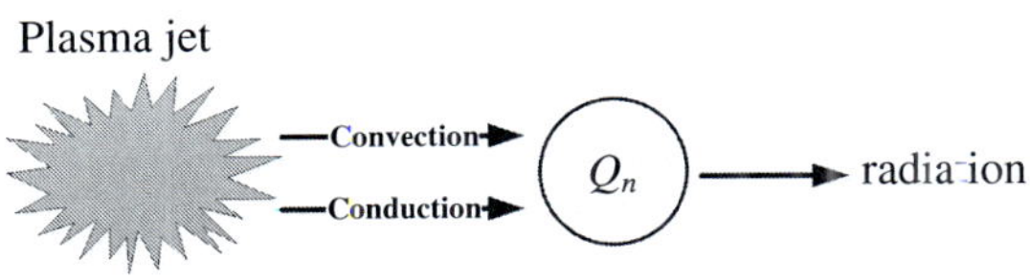

Figure 11.4. Basic heat transfer mechanism involved in plasma heating and melting of solid materials
Source: Gomez et al. 2009; reproduced with permission from Elsevier

Where, h is the heat transfer coefficient between plasma medium and solid particles, S is the surface area of the particle, T_∞ is the equivalent plasma temperature, T_s and T_α are respectively the temperature of particle surface and the reactor wall temperature, σ is the Stephan-Boltzmann constant and ε is the particle emissivity.

This equation can be used to provide a basic idea about the energy balance when the plasma is heating a solid waste particle. But this equation is not accurate enough, for example, the chemical reaction between the plasma ions with the waste is ignored and heating transfer between the waste particles and walls through conduction is also neglected. The most outstanding character of plasma gasification is the extremely high temperature which provides a flash gasification environment for MSW; and the reaction time is pretty short and the wastes can be gasified in short times and the reformation of persistent organic pollutants can be minimized because of the large temperature gradients. In addition, there is no need of oxidants in the process to produce heat source, therefore, the gas stream volume is smaller. The temperature of the plasma relates with the ionization degree, and it's described by Saha equation as follows (Janajreh et al. 2013):

$$\frac{p\alpha^2}{1-\alpha^2} = K(T)^{5/2}\exp\left(-\frac{eV_i}{RT}\right) \qquad \text{(Eq. 11.3)}$$

Where p is the gas pressure; α is ionization degree; V_i is the ionization potential of the gas; $K = \left(\frac{2\pi m_e k}{h^2}\right)^{3/2}$, h and k are Planck constant and Boltzmann constant, respectively. Another form of the equation is like,

$$\frac{\alpha^2}{1-\alpha^2}p = \frac{(2\pi m_e)^{3/2}}{h^3}(kT)^{5/2}\exp\left(-\frac{eV_i}{RT}\right) \qquad \text{(Eq. 11.4)}$$

It can be concluded that the ionization degree of gas relies on the species, temperature and pressure. Based on SI unit of pressure, the equation can be simplified to,

$$\frac{x^2}{1-x^2}p = 3.2\times10^{-2}T^{5/2}\exp\left(-\frac{11600V_i}{RT}\right) \qquad \text{(Eq. 11.5)}$$

The temperature of plasma arch can range from 10,000°C to 25,000°C depending on the gas that used in the torch and the extremely high temperature making the direct measurement impossible. Currently, the temperature is obtained from arch spectrum. Figure 11.5 shows the emission spectrum of plasmas that are generated under the atmosphere using N_2 and Ar gas carrier.

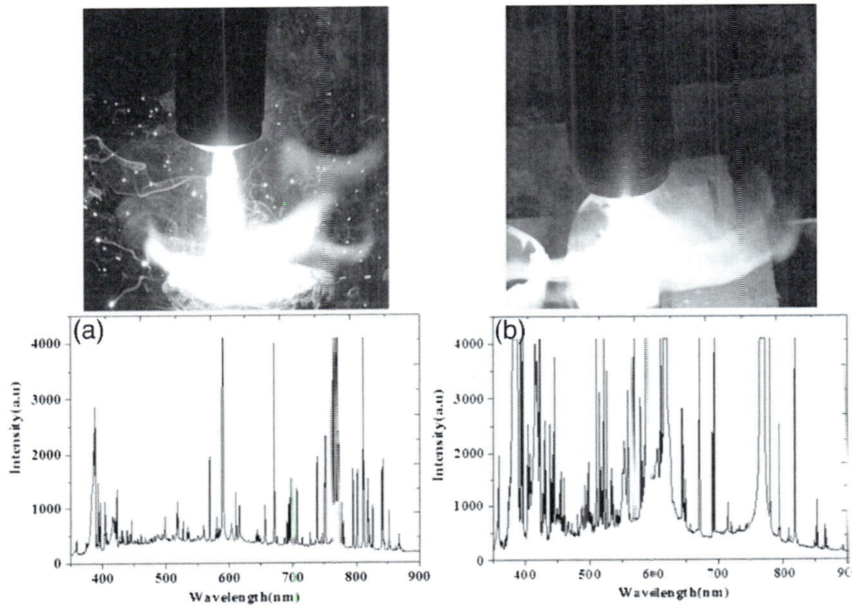

Figure 11.5. Plasma flame spectra from Zhejiang University, China, 2013. (a) Plasma arc image and associated spectrum using N2 (~25,000°C); and (b) Plasma arch spectrum and associated spectrum using Ar (~16,000°C)

The electron temperature of the plasma arc can be deduced through the following equation (Tu, 2007):

$$T = \frac{E_2 - E_1}{k\text{In}\left(\frac{I_1\lambda_1 A_2 g_2}{I_2\lambda_2 A_1 g_1}\right)} \qquad \text{(Eq. 11.6)}$$

Here, T is the electrons temperature of the plasma. k is Boltzmann constant; I_1 and I_2 are monochromatic radiative intensities at λ_1 and λ_2 respectively. A and g are energy transition probability and weight; E is the exciting energy.

11.3 PLASMA SYSTEMS FOR WASTE GASIFICATION

The concept of treating MSW with plasma arc technology was first proposed by Dr. S. L. Camacho in his patent applied at the mid of 1973 (Pourali 2010). He proposed a furnace with multiple plasma torches to continuously pyrolyze household and industrial wastes. By means of his furnace, the reacting process can produce many useful things, such as combustible gas and inert solid materials that can be used for construction purpose. His experiment also demonstrated that the gas emission could be controlled; and nothing like ashes that needs to be sent to a landfill would be produced in the process.

In 1988, Resorption Canada Limited (now renamed to Plasco Energy Group) introduced their prototype plasma gasification for treatment of MSW. Their system used a 150 kW plasma torch in a furnace to pyrolyze approximately 500 pounds of MSW per hour (Pourali 2010). In 1991 Georgia Institute of Technology (Georgia Tech) started a plasma arc research technology program and established Plasma Applications Research Facility (PARF) laboratory which was the largest university based research facility for plasma technology at that time. Through the years, the lab ran several experiments and under supervision of Dr. L. J. Circeo published numerous papers and contributed many project developments in the United States and worldwide (Pourali 2010). In 1992, Camacho was granted another patent for his invention related to mixed refuse materials being heated by a plasma torch for the purpose of converting its organic components into fuel gases, and its inorganic components into inert solid matter (Pourali 2010). He was also granted a third patent on this subject in 1996.

Also in 1990 s, two companies, Hitachi metals from Japan and Westinghouse Electric Corporation decided to jointly develop a plasma demonstration facility based on the 12 tonnes per day (TPD) test reactor located at Westinghouse Electric Corporation's Madison. In 1999, Hitachi Metals built a 24 TPD facility in Yoshi, Japan and put into operation for one year and this demonstration project was scaled up to a 220 TPD system at 2003, the Eco-Valley plasma gasification facility built at Utashinai, Japan on the island of Hokkaido. Eco-Valley is one of the first WTE plants using Plasma Gasification Technology (Willis et al. 2010). As mentioned upper, gasification is one of the thermochemical pathways that can be used to generate cleaner fuel gases and any carbonaceous fuel (fossil fuel, waste stream, hazardous waste) can be converted into syngas (H_2, CO, CH_4) by using this technology. Through plasma gasification process, the organic fraction of solid feedstock is converted into high calorific value syngas that can be used as fuel or chemical feedstock. The inorganic fraction is vitrified into a non-leachable slag that can be safely disposed of or even reused as construction material (Janajreh et al. 2013). Pyrolysis, gasification, and vitrification are the three main thermochemical processes that occur in a thermal plasma furnace when combustible solid waste is fed for treatment. Among these processes, gasification and vitrification are considered to be the most important processes for the energy recovery and volume reduction.

Compared with commonly used mass combustion methods for waste treatment, thermal plasma provides the following advantages (Byun et al. 2012): (1) the potential for high production with small furnace due to its capability to reach high temperature in short time under lower temperature conditions; (2) rapid start-up and shutdown times because of the high heat flux densities at the boundary of reactors; and (3) No or only a small amount of oxidant is required making it easier and less expensive to operate. These features make thermal plasma process to be a very attractive alternative to mass burning or other conventional methods for solid waste treatment (Byun et al. 2012) except for the large quantities of electricity required for the generation of high temperature arch. The total energy efficiency is the key obstacle restricting its commercialization. This is the reason why plasma is still not commonly used in waste processes in practice, and the main target is to study how to decrease the use of electrical energy or to enhance the efficiency of energy use.

11.3.1 Plasma Arc Torch

The plasma torches provide superheated medium to the gasifier and allow the operator to control the gasification process. The majority of plasma arc generators used in materials processing use DC as it has less noise, more stable operation with better control and lower electrode consumption (Gomez et al. 2009). The plasma torch system is capable of increasing the specific energy of the process gas to between two and ten times higher than conventional combustion equipment. As an example, the plasma torches (as shown in Figure 11.6 used at Eco Valley provided by Westinghouse Plasma Corporation) is capable of superheating the process gas to a temperature in excess of 5,500°C (Willis et al. 2010).

The working gas in the torch can also affect the plasma temperature. When nitrogen is used, as shown in the Figure 11.5(a), the plasma jets look very bright and emitting extremely high radiation and the temperature may exceed 20,000°C. When Argon is used, the temperate is approximately 10,000°C and the color of the plasma flame turns out to be purple and red.

11.3.2 Gasification Systems

Similar with traditional gasification system, a MSW plasma gasification plant consists of feeding system, plasma torches and power supplies, gasification reactor, slag handling equipment, syngas treatment system, and operation control system (Janajreh et al. 2013). The feeding system is used to prepare different types of feedstock, and when the size of the waste is two big, milling or grinding before it

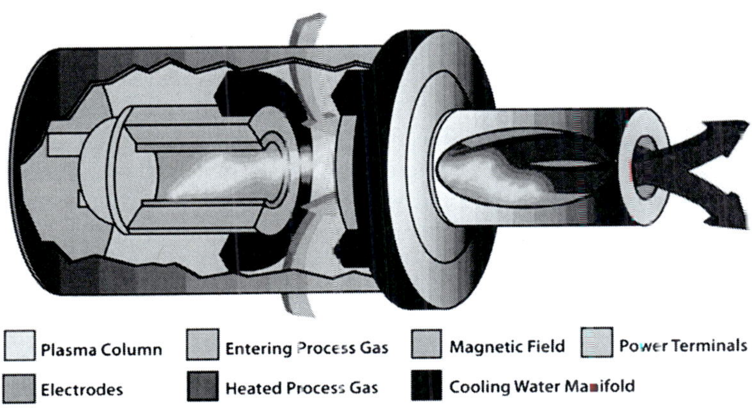

Figure 11.6. The structure of Westinghouse Plasma Corporation plasma torch
Source: Willis et al. 2010; reproduced with permission from ASME

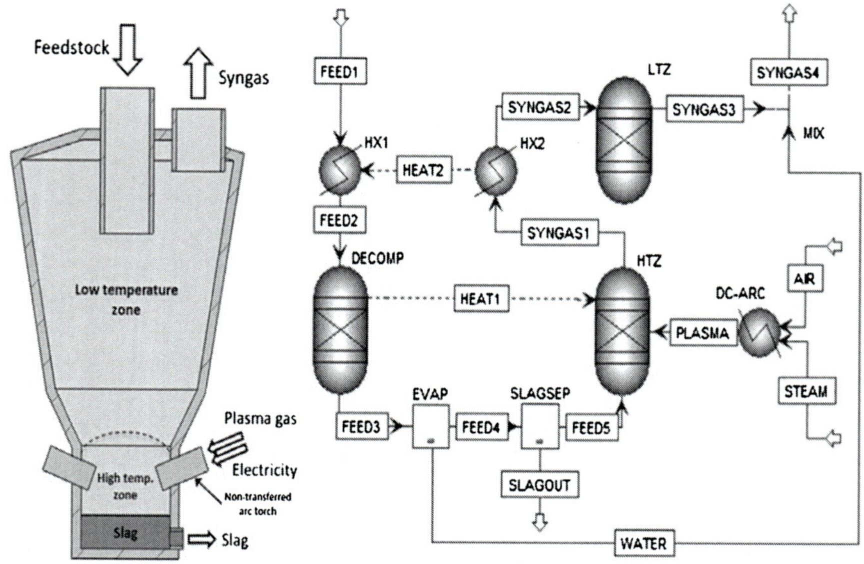

Figure 11.7. Plasma gasification reactor (left) and flow sheet of gasification system (right)
Source: Janajreh et al. 2013; reproduced with permission from Elsevier

enters the reactor is required. As shown in Figure 11.7, plasma gasification reactors consist of a reactor chamber connected to accommodate plasma torches those supply the energy necessary for the gasification. The wall of the reactor chamber is made of refractory material able to withstand the high diffused temperatures from the plasma torch. As the waste enters the reactor, it will be decomposed by the high temperature plasma and the inorganic/ash portion is reduced to a molten slag (Gleizes et al. 2005). The molten slag is collected at the bottom of the reactor where it is removed either continuously or periodically (Willis et al. 2010) and it can be poured into molds to create ingots or bricks, or quenched in water to create granules to be reused or disposed of in landfills (Fauchais and Vardelle 2000). The plasma gasification essentially results in the form of syngas and traces of pollutant gases (Janajreh et al. 2013). Therefore, the syngas is subjected to clean up and heat recovery. Once cleaned, it can be combusted in the steam or gas cycle or gas engine or used for chemical production. The whole process is monitored and controlled to ensure optimal performance of the system's temperature, feeding rates, plasma power and slag removal rate (Janajreh et al. 2013).

Figure 11.8 shows a process flow sheet of 20 tons per day demonstration plant constructed in Yblin, Israel. Pretreated MSW is fed into the reactor through the airtight chambers placed at the upper part of the reactor, wherein gasification reactions occur. Syngas produced from gasification flows into the second combustion chamber and is burnout there (Zhang et al. 2012). Then, hot flue gas from second combustion chamber is used to produce over heated steam, which drives a

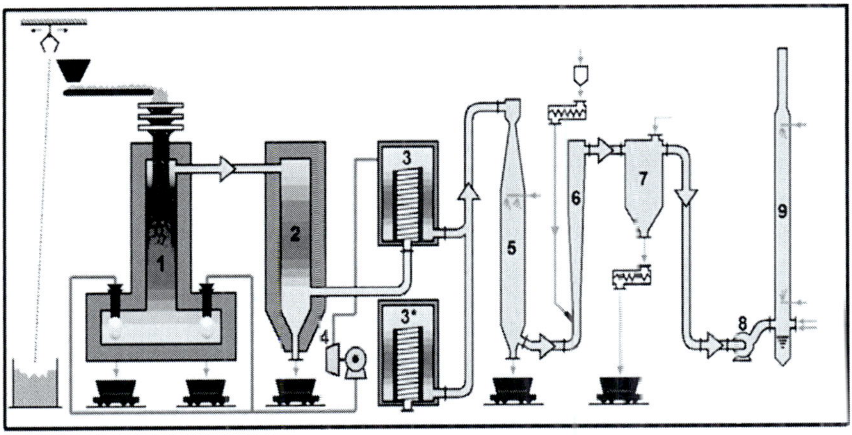

Legend: 1 - Plasma chemical reactor
2 - Afterburner
3 - Waste heat boiler
3*- Once through boiler
4 - Steam turbo generator
5 - Scrubber evaporator
6 - Reactor absorber
7 - Bag filter
8 - Exhaust fan
9 - Polishing scrubber

Figure 11.8. Illustration of the plasma gasification melting plant
Source: Zhang et al. 2012; reproduced with permission from Elsevier

steam turbine connected to an electrical generator. The generated electricity, besides providing power for the plasma torches and the rest of the system, can be sold to grid. The fly ash, SOx and other pollutantsis removed from the flue gas by APC system which usually include a scrubberand bag house filter. The solid residue from gasification is melted by the plasma jet and collected by the slag collectors.

11.4 PRACTICAL APPLICATION

The plasma gasification process has been demonstrated in many research and studies as one of the most e environmentally friendly methods and it has been considered as a viable and effective method for solid waste treatment and including hazardous waste and holds strong promise for energy recovery (Byun et al. 2012). Plasma arch gasification uses an external energy source and inert atmosphere, resulting in very little combustion of the waste material. The high temperature plasma essentially destroys all chemical bonds resulting in large amounts of radicals, electrons, ions and excited molecules in the system. These active species, and the high radiation intensity from the plasma, has the advantage to substantially increase the rates of reactions, allow reactions to proceed that are otherwise hard to achieve under normal lower temperature conditions and utilizing a small reactor size at high throughput (Janajreh et al. 2013). As a result, most of the carbon is converted to fuel gas. Plasma gasification is the closest technology available to pure gasification, i.e., it is a “true gasification” (Mountouris et al. 2008). Through plasma gasification process, the organic fraction of solid feedstock is converted into high calorific value syngas (Ducharme 2010). Due to the high temperatures, all the tars, char and dioxins are broken down. The exit gas from the reactor is cleaner; while the inorganic fraction can be vitrified into a non-leachable slag that can be safely disposed (Janajreh et al. 2013) or reused as construction material. In addition, plasma gasification technology can combine with other techniques. For example, plasma gasification in integrated gasification combined cycle (IGCC) is promoted as cleaner and more efficient energy conversion method than the conventional pulverized coal steam cycle (Gleizes et al. 2005). Furthermore, fast start-up and shutdown are amenable due to high heat flux densities at the reactor boundaries. Fast start-up and shutdown of the reactor can then be obtained without compromising refractory performance. As combustion is absent in the plasma reactor, less gas stream volume is produced compared to traditional combustion processes. Plasma torch is an independent heat source that can be easily controlled to regulate temperature within the reactor. The plasma gas can be selected to manipulate the reaction environment (Janajreh et al. 2013).

However, after more than three decades of developing and struggling in market, currently, plasma arc gasification system has still not been thoroughly proven to be commercial available comparing with traditional mass burning waste to energy system. But the situation is improving as more and more countries are tightening their pollution emission control regulations and residents are more like to accept the concept of gasification rather than incineration.

11.4.1 Overview of Commercial Plasma Facilities

There have been plasma waste processing facilities since 20th century. Here is a table showing the plasma used on MSW treatment all over the world. Currently, most practical applications of plasma gasification of waste are located in Japan where limited land space is forcing local government to find alternatives to prolong the lifetime of landfill. Hitachi metals and Alter NRG are jointly operating a 220 TPD commercial plasma arc facility in Utashinai since 2002. In Europe, several small scale demonstration plants were also reported. Table 11.2 gives a brief survey of the reported plasma arc gasification projects worldwide (UCR, 2009). As we can see, most of the plasma torches are provided by Europlasma and Westinghouse. In the following sections, several projects are introduced in more detail. Most of the introduction is directly cited from their open accessible website or brochures from technology provider.

11.4.2 Westinghouse Plasma Gasification Technology

The non-transferred plasma torch developed by Westinghouse consists of a pair of tubular water-cooled copper electrodes. The operating gas is introduced through a small slot between the electrodes (Ducharme 2010). Their torch can be used in metal melting, hazardous waste destruction, fly ash vitrification and also waste gasification. Westinghouse plasma torch has two different sizes, the Marc 3 and Marc 11 which can provide a power rating of 150 kW and 2000 kW, respectively.

Table 11.2. Survey of the reported plasma arc gasification projects

Company	*Capacity*	*Website*
AdaptiveNRG	33,000 TPY	www.adaptivenrg.com
AlterNRG (Westinghouse Plasma Technology)	n/a	www.alternrg.com
Europlasma	n/a	www.europlasma.com
Geoplasma LLC	24 TPD	www.geoplasma.com
InEnTech Chemical	22 TPD	www.inentec.com
Peat International Inc	9 TPD	www.peat.com
Plasco Energy Group	31,100 TPY	www.plasco.com
Plasma Arc Technologies	n/a	www.plasmaarctech.com
Plasma Environmental Technologies Inc.	n/a	www.plasmaenvironmental.com
Recovered Energy Inc. (Westinghouse)	3 TPD	www.recoveredenergy.com
Sun Energy Group LLC	n/a	www.sunenergygrp.com
Westinghouse Plasma/ GeoPlasma	n/a	www.westinghouse-plasma.com

TPD: Tonnes Per Day; TPY: Tonnes Per Year; TPH: Tonnes Per Hour

In the mid 1990's, Westinghouse Plasma Corp. designed and installed a pilot scale, 24 TPD MSW gasification plant in Yoshii, Japan (Figure 11.9). Lessons learned at Yoshii were applied to full scale facilities in Mihama-Mikata and Utashinai Japan, both began commercial operation in 2002 and 2003, respectively. These facilities are still under operation today The Mihama-Mikata plant processes 20 tonnes of MSW and 4 tons of sewage sludge from the towns of Mihama and Mikata; and heat generated from syngas combustion is used to dry the sewage sludge. All of the slag from the Mihama-Mikata plant is used as construction materials. The EcoValley plasma gasification facility is located in Utashinai, Japan, in the island of Hokkaido. The plant was constructed in 2002 and has been fully operational since 2003. The facility processes up to 220 TPD of MSW or up to 165 TPD of a 50/50 mixture of MSW and auto shredder residue (Willis et al. 2010).

The gasification technology provided by Westinghouse was considered to be one of the most advanced plasma gasification technology. Not only they have Japanese plants operating with their gasification vessel, they also have an operating pilot plant and several contracts worldwide. The most convincing reference plant is the EcoValley plant that encountered several issues in the past but is now operating on 100% MSW (Ducharme 2010). One interesting point in their technology is the ability of the gasification vessel to be applied to different kind of waste without modification. This is a strength that allows them to conquer different types of clients without modifying their core technology.

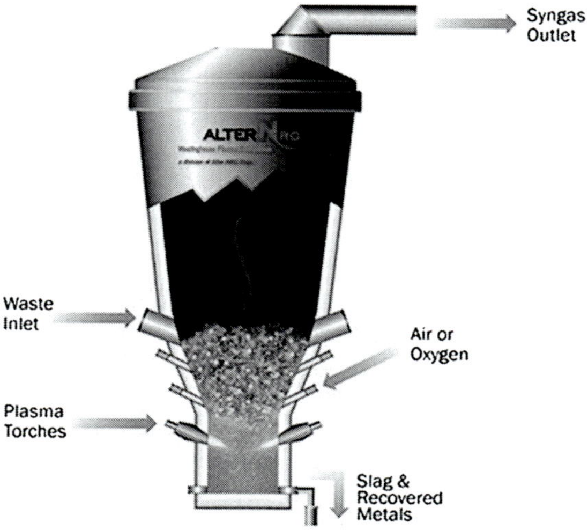

Figure 11.9. Alter NRG plasma (Westinghouse Plasma) gasifier
Source: Willis et al. 2010; reproduced with permission from ASME

In 2006, Alter NRG acquired the Westinghouse Plasma Corporation (WPC). Based on the experience gained at the two Japanese facilities, Alter NRG is proving the next generation plasma waste gasification system and one of the new generation gasifier was commissioned in 2009 in Pune, India, which was originally designed to treat 72 TPD of hazardous wastes from over 40 different industries. Contrary to the other companies, Alter NRG/ Westinghouse has put the plasma jets at the bottom of the gasifier. Up to six plasma torches are typically used to provide sufficient heat for the gasification to take place. A bed coke is created by inputing metallurgical coke (met coke) to absorb and retain the heat energy from the plasma torches and provides the environment for melting the inorganic materials into a glassy material. The coke is fed along with the MSW feedstock and is paramount for the operation of the gasifier (Ducharme 2010). As the mix of waste and coke is going down through the reactor, the waste will start gasifying whereas the coke will slowly gasify but will remain at the bottom (Ducharme 2010); and the waste will lie on top of it. The only materials that will escape the bed coke are the slag and melted metals. They will be collected at the end. Metals are separated from the slag through an ultimate quick process.

Willis et al. (2010) has make a summary of the lessons learned from Ecovalley plasma WTE facility. They pointed out that the three issues that prohibit the continuous operation in the past were: bottom internal size of the reactor, the life time of refractory wall and the carry-over of particulate particles. Except the first issue, the second and third issues caused problems and are commonly encountered in other plasma gasification system also. In order to resist the extremely high temperature, the initial refractory in the reactor of EcoValley was composed of alumina internal layer and silicon carbide outer layer. When the reactor was put into operation for several months, they realized that the high thermal conductivity of the silicon carbide caused significant heat loss and the melted slagging froze. So Hitachi finally exchanged the internal and outer layer to maintain the wall temperature above the melting point of the slag. The carry-over of particulates was mainly the result of the fine dust particles in the feedstock and inorganic species with low volatile temperature. Hitachi solved this problem by lowering the syngas temperature to 750°C resulting in lower heat recovery efficient. The energy balance provided by Hitachi shows that the plasma torches cost totally about 5% of the total heating value of the treated waste.

11.4.3 Plasco Energy Group Inc.'s Plasma Gasification Technology

Plasco Energy Group ("Plasco") was formed on 2005 with the amalgamation of Resorption Canada Limited, which was founded in 1974 to research the use of plasma for the destruction of hazardous waste. In the 1990s, Resorption Canada Limited has built a 5 TPD demonstration plant for waste gasification along with its Spanish joint-venture partner in Spain. Based on this pilot plant, a full scale system, Plasco Trail Road facility (PTR) was built at the City of Ottawa, Canada, with a capacity of 220 TPD in 2008. It is the only operating commercial-scale plasma gasification system that converts MSW into syngas that fuels GE Jenbacher engines to generate electricity.

Unlike the Westinghouse system, the plasma torch in Plasco is used for syngas refining and bottom ash melting rather than directly gasifying the waste. Before fed into the gasification chamber, the waste is pretreated to remove any materials with high recycling. When shredded MSW enters the gasification chamber, the waste is converted into a crude syngas using recycled heat. The crude syngas flows into the upgrading chamber, where thermal plasma is employed to refine the gas. The clean gas can be burned in a gas turbine or fuel the internal combustion engines that efficiently generate electricity. The solid residue from the conversion chamber is sent to a separate high temperature Carbon Recovery Vessel (CRV), where the solids are melted into a liquid slag through a plasma torch. The cooled slag pellets are sold as construction aggregate. In October 2011, Plasco passed all environmental assessments and was given approval to process up to 85 tonnes of waste at their Trail Road Facility. In December, 2011, Plasco signed a contract with the city of Ottawa to process 300 TPD wastes for the next 20 years, with initial approval for the construction of a new 300 TPD capacity facility (University of California, 2009).

Accoring to their technology brochure, on April 12, 2010, Plasco announced that they had reached an agreement to construct a demonstration facility in Beijing, China, capable of processing 200 TPD municipal wastes and they aslo were planing to construct a larger facility capable of processing 1,000 TPD upon acceptance of the proved concept. On January 20, 2011, the Salinas Valley Solid Waste Authority has announced that they would begin testing the Plasco plasma gasification system. By not feeding all waste to the plasma torches, the Plasco Gasification System uses significantly less energy than other plasma gasification systems, resulting in more potential net energy gain than other systems

(University of California, 2009). Unfortunately, public accessible data shows that the annual continuous operation time of Plasco is very short. In 2008, it totally operated for less than 1000 hours and treated 2000 tonnes of waste and there is no detailed information about the kernel issue.

11.4.4 Euro Plasma Gasification Technology

Similar to the plasma system of Plasco, Euro plasma is also combined with traditional gasification system. Two non-transfer DC plasma arch torches are used for gas refining and slag melting. Each torch consists of two tubular, coaxial, water-cooled, copper electrodes separated by a gap into which the plasma forming gas is injected (Ducharme 2010). The thermal efficiency of the DC plasma torch calculated from the energy imparted to the plasma jet divided by the electrical energy supplied to the torch system is in the order of 75–80% (A. Vardelle, University of Limoges).

The gasification system is shown in Figure 11.10. A traditional forward moving grate is employed as main gasification reactor. Along the grate moving direction, the chamber is separated into two sections. In the first section, the waste is heated to around 200–300°C and the water vapor is collected and removed away from the reactor. It can significantly improve the heating value of the waste and because the de-chlorination of PVC occurs in this temperature range, the chlorine species can also be removed away ahead of gasification. In the second section, the

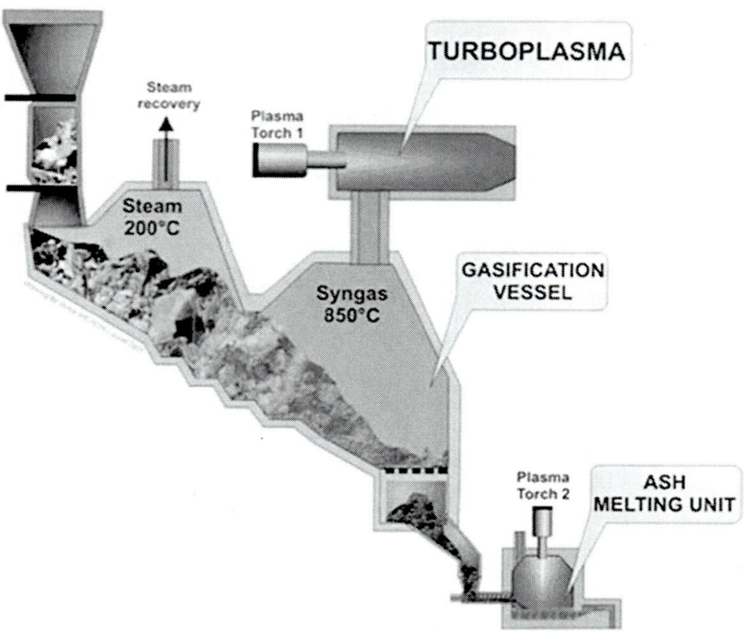

Figure 11.10. Europlasma auto-thermal gasifier
Source: Ducharme et al. 2010; reproduced with permission from ASME

dried waste is gasified and the gas produced is reformed in another chamber equipped with plasma torch. The bottom ash is conveyed to another reactor where it is melted by the second plasma torch.

The overall process uses an auto-thermal gasification process as described in Figure 11.10. In addition to traditional gasification reactor and plasma gas reforming chamber, a heat exchanger that recovers the sensible heat of the gasification gas, dust and acids scrubbing and finally gas engines to produce electricity from the syngas are the key components of the facility (Ducharme 2010). The advantage of the Europlasma gasification system is that, the plasma torch is designed only for slag melting and syngas reforming, such as thermal cracking of tar. The whole process does not rely on the full use of plasma torches (Ducharme 2010). Primary gasification of the waste is achieved by using the recycled heat of combustion. The syngas is then heated to over 1200°C with a plasma torch and prepared for electricity production in a gas turbine (Gomez et al. 2009). At this stage, all organic free radicals are destroyed. Lastly, a second plasma torch vitrifies the ash, consisting of metals and minerals (Ducharme 2010).

In 2012, the first commercial plant of Europlasma, CHO-power in Morecenx, France, was put into operation. CHO-Power is claimed to be a new concept for converting residual waste and biomass into energy, based on enhancing the well-established gasification. The plant capacity is 50,000 tons per year, corresponding to a net electrical output of 12 MW of electricity (Ducharme 2010). In a statement in Mar. 2013, the CHO Power said that the plant has produced 2235 MWh electricity delivered to the grid since its construction was completed in summer of 2012 its construction. Further information about its current status is not available.

11.4.5 InEnTec Technology

The plasma technology provided by InEnTec (which stands for Integrated Environmental Technologies, LLC) was patented since 1998 and has been used in two plants both of which are closed due to technical and financial difficulties of the plasma system. It is InEnTec that combined plasma and glass melting together to develop Plasma Enhanced Melter (PEM) which is shown in Figure 11.11).

According to the available introduction material, the InEnTec gasification system was designed to handle different types of solid waste, from MSW to hazardous waste. The overall system has three main components, a downdraft pre-gasifier, a plasma process vessel (is also called the PEM process chamber), and a thermal residence chamber (Ducharme 2010). The main unit in the system design is the downdraft gasifier, where about 80% of organic portion in the waste is converted to syngas. In the PEM process chamber, there are two power systems: (1) A Direct Current (DC) plasma arc for high temperature organic waste destruction and gasification. The plasma arc is created between graphite electrodes (Ducharme, 2010), and (2) An Alternative Current (AC) powered resistance (Joule) heating system to maintain an even temperature within the molten bath (EvTEC 2002). The plasma arc provides the heat to the process chamber, and the Joule heating system allows a good distribution of this heat within the molten bath (Ducharme 2010).

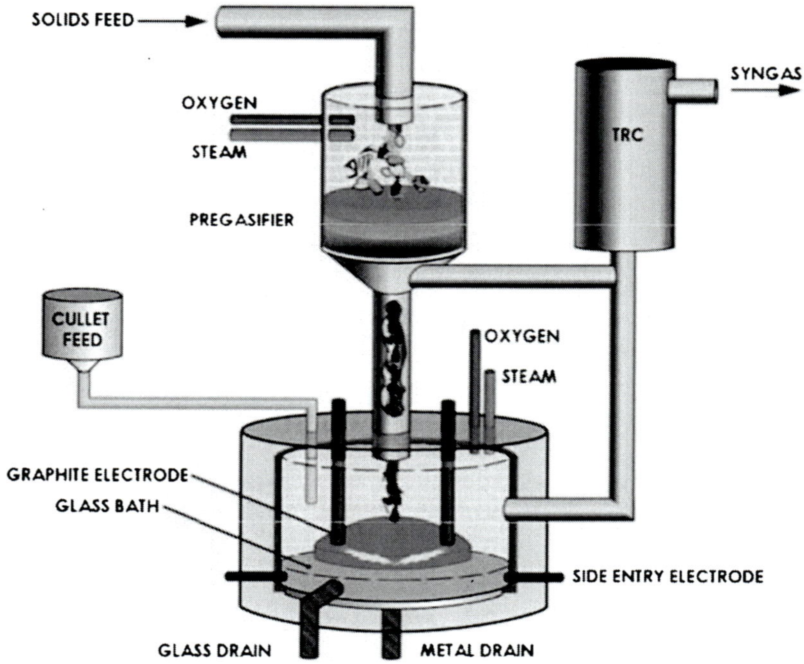

Figure 11.11. Technology principle of the PEM process
Source: Ducharme et al. 2010; reproduced with permission from ASME

The public accessible report "Environmental Technology: Verification report for the Plasma Enhanced Melter (May 2002)" on the InEnTec website described the results of pilot tests which were carried out in a 25 tons per day plants located in Richland, MA. The reports says when biomass, MSW or medical waste were treated the around 100 to 200 ppm CO and total hydrocarbon are detected in burned off-gases indicating the syngas from the gasifier was carrying relative high tar. A comparative test conducted by WTERT ground showed that except the reduced gas flow flux, the emission was about the same level as that from a mass burning WTE plants.

11.5 EMISSION FROM PLASMA PROCESSES

By converting the combustible MSW components into syngas with extremely high temperature, the pollutants generated from plasma gasification system are significantly lower than traditional mass burning system. The pollutants can be removed during the cleaning of the syngas or after the combustion of the syngas. The nonvolatile toxic heavy metals can be stabilized in the slag.

11.5.1 Emission of Dioxins

Dioxin is one of the most toxic organic species which is formed normally produced during the incomplete combustion of MSW contaminated with chlorine species including organic chlorine or salts. According the survey carried out by Chinese EPA in 2004, the incineration of MSW has contributed about 6.5% of total dioxin generation in China at 2008. When the MSW is gasified by high temperature plasma arc, the thermal reaction is carried out in reduced environment. The formation and "De-Novo Synthesis" of dioxin is inhibited by insufficient of oxygen and deactivated metal catalysts. In addition to the inhabitation of the generation, the original dioxins introduced in raw MSW can be destroyed completely by plasma arc. However, due to financial and operational problem, the only two full scale commercial plants that designed to treat municipal solid waste by plasma arc are all located in Japan, One is at Utashinai and another is at Mihama-Mikata. The syngas is directly burned after gasification. The dioxin emission data measured from 2003 to 2007 are listed in Tab. 11.3. As we can see, the toxic equivalent quotient ranges from the 0.00004 to 0.01 ng-TEQ/m3 which in much lower than Japanese emission criterion for the gasification system (Juniper 2008).

U.S.A EPA has measured the emission of the 10 TPD InEnTec plasma arc pilot plant at Washington for circuit boards treatment and they found the emission of dioxin was 0.000013 ng-TEQ/m3. Result of the third-party demonstration source test for the 110 TPD Plasco plasma arc gasification plant at Ottawa (Canada) was 0.009245 ng-TEQ/m3 in the report provided by Georgia technology research institute. From these data, we can conclude that the dioxin emission from a plasma arc gasification plant is at least one order of magnitude below the criterion defined for waste incineration plants in EU and Japan.

11.5.2 Emission of Other Gas Pollutants

When the syngas is directly burned after gasification, the flue gas may contain the similar pollutants as those from traditional waste mass burning plants. The emission of pollutant species measured by EPA and third-party respectively for InEnTec and Plasco is given in the Table 11.3 (UCR, 2009).

As expected, for mercury, HCl, and Sox, all plasma arc gasification can EU emission standard easily. For NOx, when sometimes would appear to be close to the limit because the oxygen and nitrogen contained in waste originally may react to form NOx under high temperature. But the particulate matter may be an issue at limit Plasco indicating they may be facing the same ash problem with the gasification facility at EcoVelley.

11.5.3 The Heavy Metal Leaching Character of Vitrified Slags

Solid residues, especially fly ash discharged after waste incineration usually contains high concentration of toxic metals and in many countries, fly ash is classified as hazardous waste which should be solidified and stabilized before send to landfill. The plasma gasification process can fully melt solid residue into vitrified slag which is very suitable to be recycled as aggregate material for

Table 11.3. Gas emission of two plasma arc gasification plants

	InEnTec	*Plasco*	
Location	Washington U.S.A	Ottawa, Canada	EC 2000/76
Inspector	EPA	Third-Party	Standard
Year	2000	2008-2009	
Scale (TPD)	10	110	
Waste	Circuit Boards,	MSW	
PM (mg/m^3)	3.3	12.8	30
HCL (mg/m^3)	2.7	3.1	10
NOx (mg/m^3)	162	150	280
SOx (mg/m^3)	—	26	50
Hg (mg/m^3)	0.00067	0.0002	0.05

Source: Data from UCR, 2009.

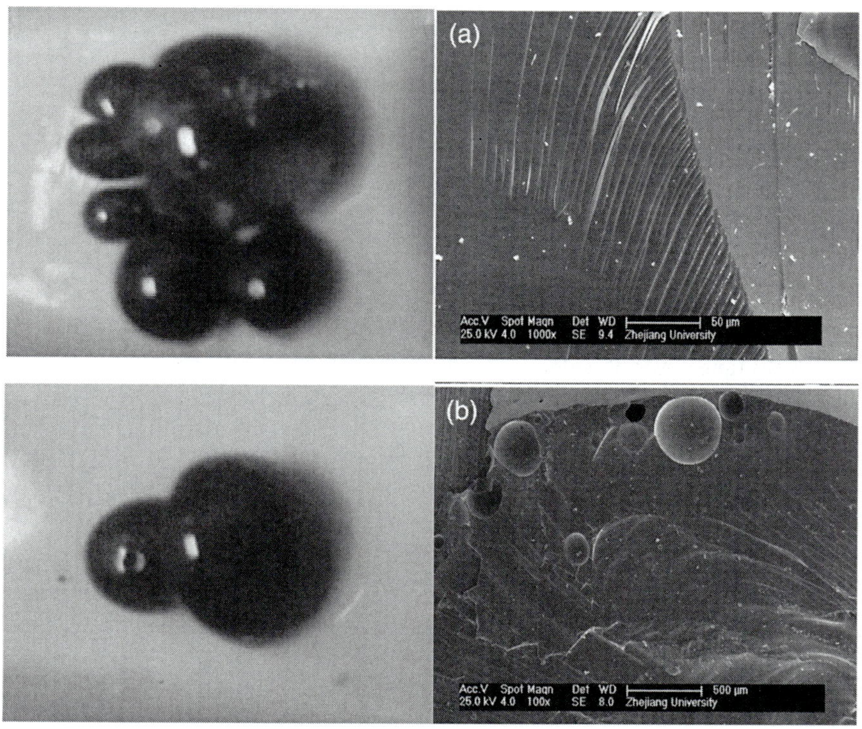

Figure 11.12. Outlook and SEM of plasma vitrified fly ash slag using (a) Ar; (b) N2

construction application. During the slagging process, the heavy metals can be encapsulated within the solid matrix to avoid leach out. Figure 11.12 show the appearance and scanning electron microscope (SEM) image of vitrified fly ash samples collected at a waste incineration plant at Hangzhou city (China 2013). As we can see, the surface of the slag becomes smooth, and glassy.

In Japan, the slag produced from Utashinai and Mihama-Mikata plasma gasification plants is re-used in the Japanese construction industry either as aggregate material for backfilling and road construction or as material for producing paving slab and roof tiles. Obviously, before any re-utilization, these slags must pass the heavy metal leaching test. Alter NRG has reported the test results for the slag sample from Mihama-Mikata. The test was carried out through TCLP (Toxicity characteristic leaching procedure) protocol according to EPA SW846 methods 1311 and 6010. The metal concentrations in the leaching liquid are orders of magnitude lower than Japanese standards.

11.6 SUMMARY

In recent studies, plasma gasification has been widely used and was considered to be very efficient and no-pollution methods. Many tests demonstrate that there is no heavy metal and dioxins generated and the emission of NOx and SOx can be controlled. Plasma arch gasification don't have inner energy source and the atmosphere is inert. Thus, the waste materials cannot combust and this will reduce the chance of generating hazardous organic compounds. The external energy source helps control the temperature of the reactor. The plasma arc can reach rather high temperature; this will promote the breaking up of all chemical bonds and form active species like radicals, electrons and so on. Both can increase the reaction rate and shorter reaction time, and then the reactor can be made in smaller scale with high throughput. Due to high temperature, the chars and complex organic compounds can be broken down and the products are very simple and environmentally friendly. Besides, plasma torches can combine with other techniques. And this can promote the efficiency of the system and the exit gas will be cleaner. There are high heat flux densities at the boundaries of the reactor, so it can start-up and shutdown rapidly.

Currently, plasma arch gasification is still under improving and experience obtained from past pilot and industrial applications has pointed out clearly that the key disadvantage of plasmas arc gasification system is the high capital investment and operational cost compared with combustion system. The expensive material used to resist the high temperature and the frequent replacement of plasma torch have made it unsuitable for the mass treatment of raw MSW. The plasma gasification systems reported to be financial profitable and available for continuous operation when fed with very high quality industrial waste or well pretreated MSW. Therefore, in the future, most attention will be paid on optimizing the integrated system with affordable investment and reduced operational cost.

References

Bonizzoni, G., and Vassallo, E. (2002). "Plasma physics and technology: Industrial applications." *Vacuum*, **64**(3–4), 327–336.

Boulos, M. I. (1991). "Thermal plasma processing." *IEEE Trans. Plasma Sci.*, **19**(6), 1078–1089.

Byun, Y., Cho, M., Hwang, S. M., and Chung, J. (2012). "Thermal plasma gasification of municipal solid waste (MSW)." *Gasification for practical applications*, Y. Yun, ed., INTECH, Rijeka, Croatia.

Ducharme, C. (2010). *Technical and economic analysis of plasma-assisted waste-to-energy processes [D]*, Columbia Univ., New York.

Ducharme, C., and Themelis, N. (2010). "Analysis of thermal plasma-assisted waste-to-energy processes." *Proc., 18th Annual North American Waste-to-Energy Conf. NAWTEC18*, ASME, New York.

EvTEC (Environmental Technology Evaluation Center). (2012). "Environmental technology verification report for the plasma enhanced melter." ASCE, Reston, VA.

Fauchais, P., and Vardelle, A. (2000). "Pending problems in thermal plasmas and actual development." *Plasma Phys. Controll. Fusion*, **42**(12B), B365–B383.

Gleizes, A., Gonzalez, J. J., and Freton, P. (2005). "Thermal plasma modeling." *J. Phys. D Appl. Phys.*, **38**(9), R153–R183.

Gomez, E., Rani, D. A., Cheeseman, C. R., Deegan, D., Wise, M., and Boccaccini, A. R. (2009). "Thermal plasma technology for the treatment of wastes: A critical review." *J. Hazard. Mater.*, **161**(2), 614–626.

Hippler, R., Kersten, H., Schmidt, M., and Schoenbach, K. H. (2008). *Low temperature plasmas: Fundamentals, technologies and techniques*, Wiley, Berlin.

Hoornweg, D., and Bhada-Tata, P. (2012). "What a waste: A global review of solid waste management." *Urban development series, World Bank*, Washington, DC.

Janajreh, I., Raza, S. S., and Valmundsson, A. S. (2013). "Plasma gasification process: Modeling, simulation and comparison with conventional air gasification." *Energy Convers. Manage.*, **65**, 801–809.

Juniper. (2008). "The alter NRG/Westinghouse Plasma gasification process." Independent Waste Technology Rep., Juniper Consultancy Services, England.

Langmuir, I. (1928). "Oscillations in ionized gases." *PNAS*, **14**(8), 627–637.

Mountouris, A., Voutsas, E., and Tassios, D. (2006). "Solid waste plasma gasification: Equilibrium model development and energy analysis." *Energy Convers. Manage.*, **47**(13–14), 1723–1737.

Mountouris, A., Voutsas, E., and Tassios, D. (2008). "Plasma gasification of sewage sludge: Process development and energy optimization." *Energy Convers. Manage.*, **49**(8), 2264–2271.

Moustakas, K., Fatta, D., Malamis, S., Haralambous, K., and Loizidoum, M. (2005). "Demonstration plasma gasification/vitrification system for effective hazardous waste treatment." *J. Hazard. Mater.*, **123**(1), 120–126.

Pourali, M. (2010). "Application of plasma gasification technology in waste to energy challenges and opportunities." *IEEE Trans. Plasma Sci.*, **1**(3), 125–130.

Tanarro, I., Herrero, V. J., Carrasco, E., and Jiménez-Redondo, M. (2011). "Cold plasma chemistry and diagnostics." *Vacuum*, **85**(12), 1120–1124.

Tu, X. (2007). "Characterization of DC plasma jets aimed at the treatment of hazardous waste." Ph.D. thesis, Zhejiang Univ., Hangzhou.

UCR. (2009). "Evaluation of emissions from thermal conversion technologies processing municipal solid waste and biomass." Univ. of California, Riverside.

University of California. (2009). "Evaluation of emissions from thermal conversion technologies processing municipal solid waste and biomass." BioEnergy Producers Association, Los Angeles, CA.

Willis, K. P., Osada, S., and Willerton, K. L. (2010). "Plasma gasification: Lessons learned at eco-valley WTE facility." *Proc., 18th Annual North American Waste-to-Energy Conf. NAWTEC18*, ASME, New York.

Zhang, Q., Dor, L., Fenigshtein, D., Yang, W., and Blasiak, W. (2012). "Gasification of municipal solid waste in the plasma gasification melting process." *Appl. Energy*, **90**(1), 106–112.

CHAPTER 12

Sustainable Hazardous Waste Management/Treatment: Framework and Adjustments to Meet Grand Challenges

Prangya R. Rout[1]
Puspendu Bhunia[2]
Anushuya Ramakrishnan[3]
Rao Y. Surampalli[4]
Tian C. Zhang[5]
Rajeshwar D. Tyagi[6]

12.1 INTRODUCTION

Hazardous waste is a waste that is dangerous or potentially harmful to our health or the environment. Hazardous wastes can be liquids, solids, gases, or sludges. They can be discarded commercial products, like cleaning fluids or pesticides, or the by-products of manufacturing processes. During the past five decades, significant quantities of hazardous wastes have been released into the environment due to poor management of these wastes or failures of treatment facilities. Ample evidences exist to reveal that improper disposal of these wastes may result in contamination of air (via fugitive dust emissions and volatilization); surface water (ground water seepage, surface runoff or overland flow); ground water (leaching/infiltration); soils (fugitive dust generation/deposition and tracking); sediments (surface runoff/overland flow seepage and leaching) and biota (due to

[1]Indian Institute of Technology, Bhubaneswar, India.
[2]Indian Institute of Technology, Bhubaneswar, India.
[3]Univ. of Nebraska-Lincoln, Lincoln, NE.
[4]Univ. of Nebraska-Lincoln, Lincoln, NE.
[5]Univ. of Nebraska-Lincoln, Lincoln, NE.
[6]INRS-ETE, Université du Quebéc, Québec, QC, Canada.

bioaccumulation and biological uptake). For example, one of the major environmental concerns arises from the contamination of groundwater by landfill leachate (a water-based solution of compounds from waste), posing significant risk to downstream surface waters and wells (Christensen et al. 1994). In order to safeguard our environment for future generations, it becomes important to regulate and manage hazardous wastes in a manner that is ecologically and socially sound, as well as economically viable and sustainable.

However, sustainable management of hazardous wastes is full of challenges mainly because of the complicity and uniqueness associated with hazardous wastes. Wastes that are released to the air as emanations naturally are regulated on site, as per legal requirements. Liquid effluents are also treated on site through common effluent treatment plants (ETPs), and legal requirements are fulfilled before release. However, the situation is more complex with hazardous wastes because total storage, disposal of or utilization of these solid wastes is not feasible within the premises on a long run. Such wastes need special collection, treatment, and recycle. More often than not, specific issues are associated with hazardous wastes from industrial units, sediment, sludge of ETPs, and other polluted habitats. The management problem is further complicated by the heterogeneous nature of the waste in terms of chemistry, bioavailability and toxicity. Moreover, the entire society is called upon to be responsible for the sustainable management of hazardous wastes. To this end, sustainable management of hazardous wastes calls for the reduction of the impacts of hazardous wastes on human health and the environment; it promotes a paradigm shift in emphasis from remedial measures to preventive measures such as reduction at source, reuse, recycling and recovery. More importantly, it is critical to recognize sustainable management of hazardous wastes as part of the wider issues of sustainable management for water, food and solid waste, climate change, and economic/social development.

Traditionally, hazardous waste management includes four major components: fundamentals (e.g., processes, fate, transport and behavior of hazardous wastes, toxicity and risk assessment, etc.), management, treatment and disposal methods, and site remediation (LaGrega et al. 2001). However, to meet the grand challenges (e.g., climate change, energy, development and health), hazardous waste management has been evolving very fast with the central theme focusing on sustainability. New information is now being generated rapidly. This chapter is meant as a starting point for further analysis, with a focus on a) introduction to the newly developed concepts about sustainable hazardous waste management and treatment, and b) linking the key elements of sustainable hazardous waste management with the current broader issues (e.g., renewable energy and climate change). In order to approach these objectives in a focused and easy-to-understand manner, we choose to use hazardous industrial wastes as a platform for our elaboration and discussion, although many other sectors may generate significant amount of hazardous wastes (e.g., biomedical waste and municipal waste). While the topic can be very broader, we limit our elaborations to the following: a) fundamentals and basic components of sustainable hazardous waste management (Section 12.2), b) technical options of sustainable hazardous waste treatment and remediation (Section 12.3), c) disposal of hazardous waste (Section 12.4), and d) adjustments to meet global challenges (Section 12.5). The chapter identifies a set of future trends and challenges for sustainable hazardous waste management/treatment with recommendations being made about research needs, which helps achieve the broader goals of sustainability.

12.2 FUNDAMENTALS AND BASIC COMPONENTS

Fundamentals of sustainable hazardous waste include a) classification and characteristics of hazardous waste sources, b) regulations governing hazardous waste management and management programs, c) the concept of "Environmental Audits," d) the concept of "Pollution prevention," e) the development of hazardous waste management systems and facilities, and f) integrated sustainable hazardous waste management and practices. Details of these fundamentals are described below.

12.2.1 Classification and Characteristics of Hazardous Waste Sources

12.2.1.1 Waste Classification and Categorization

It is important to recognize that there are varying degrees of hazards associated with different waste streams. The major group of industries that generate hazardous wastes and the categories of hazardous wastes are shown in Table 12.1. Hazardous wastes are classified into listed wastes, characteristic wastes, universal wastes and mixed wastes.

Waste Categorization Based on Waste Emission. There are three different categories of waste that contain four different lists of hazardous wastes classified by the U.S. Environmental Protection Agency (USEPA) that are located in Title 40 of the CFR at Part 261. These three categories are:

- Nonspecific Source Wastes (The F-List): Wastes from common manufacturing and industrial processes such as solvents used in cleaning and degreasing operation are classified in the F-List.
- Source-specific Wastes (The K-List): Wastes from specific industries such as petroleum refining or pesticide manufacturing are classified in the K-List. Wastewaters from treatment and production and certain sludges in these industries are examples of source-specific wastes.
- Discarded Commercial Chemical Products (The P-list and the U-list): Specific commercial products in unused form are included in this list. Some pesticides and some pharmaceutical products tend to become hazardous when discarded.

Table 12.1. Major groups of industries generating hazardous wastes

Industries	*Potentially hazardous waste generated*
Leather manufacturing	Heavy metals, organic solvent, alkaline solvents, toxic sulfides, toxic dyes, alcohols
Equipment repair	Acids, bases, organic solvents, paints, ignitable wastes
Furniture manufacturing	Organic solvents, paints, dyes, VOCs, organic chlorine compounds, alcohols
Metal processing	Heavy metals, fluorides, cyanides, acid and alkaline cleaners, solvents, pigments, abrasives plating salts, oils, phenols
Dry Cleaning	Perchloroethylene, Trichloroethane, chlorofluorocarbons
Pharmaceutical	Organic solvents and residue, heavy metals e.g. Hg and Zn
Oil, gasoline and other petroleum product	Oil, phenols and other organic compounds, heavy metals, ammonia, salts, acids and caustics
Plastic processing	Organic chlorine compounds
Printing	Heavy metals, organic solvents, organic chlorine compounds
Pesticides	Organic chlorine compounds, organic phosphate compounds
Photo processing	Silver, acids
Paints manufacturing	Heavy metals, pigments, solvents and organic residues
Textile manufacturing	Heavy metals, dyes, organic chlorine compounds, solvent

Waste Categorization Based on Waste Nature. Hazardous wastes may be categorized based on the risk that they pose. Such wastes are classified into three categories (Misra and Pandey 2005):

- High-risk wastes: They contain significant concentrations of constituents that are highly toxic, mobile, persistent and bio-accumulative (e.g. chlorinated solvents, waste from metal degreasing, cyanide waste, dioxin-based waste, PCB waste).
- Intermediate-risk wastes: They contain metal hydroxide sludges (excluding Cr^{6+} which is under the high-risk waste due to its extreme toxicity). The toxic metals of this category are relatively insoluble with low mobility.
- Low risk wastes: They include primarily high volume low hazard wastes and some putrescible wastes.

There are good economic advantages for ranking wastes according to the level of hazards they present and also to identify compatibilities and/or non-compatibilities of chemical constituents of waste streams for development of an effective waste management program. For example, some wastes containing aluminum and chemical or alkaline cleaners form more dangerous toxic or hazardous substances when mixed together. In contrast, waste containing both selenium and mercury would be less likely to create toxicity problems in comparison to waste containing mercury only. Practically, the degree of hazard posed by wastes may be dependent on several factors, such as:

- Physical form, composition and reactivity (fire and explosion) and quantities;
- Biological and ecological effects (i.e., bioavailability, toxicity and Eco toxicity);
- Mobility (i.e., transport in various environmental media, leaching potential);
- Persistence (including fate in environment, detoxification potential, secondary transformation);
- Indirect health effects that may result from pathogens, vectors, etc.; and
- Local conditions (e.g., temperature, soil types, ground water table conditions, humidity, light etc.).

12.2.1.2 Waste Characterization

Waste characterization is an essential requirement in the development of an effective industrial waste management plan (Sharma and Lewis 1994). Prior to establishing strategies for the control and treatment of waste, it becomes necessary to evaluate whether the waste contains hazardous components (Hoornweg and Thomas 1999). One of the simplest systems for the characterization of industrial waste is to review industries that have shown to possess high potential for hazardous waste production, e.g. lead acid battery, pesticides, electroplating, petroleum, tanning, etc. and to collect information regarding the nature of waste, taking the following points into consideration:

- A matrix of waste classes that identifies compatible and non-compatible wastes should be developed during the waste characterization process. Critical chemicals in each type should be identified for monitoring.
- Potentially non-compatible wastes should be identified, together with the potential consequences of such mixtures.
- In addition, compatible and even friendly wastes should be identified.

After identifying the waste, the hazardous waste may be characterized on the basis of the risk they pose. Four important characteristics that evaluate the hazardous nature of waste are corrosivity, ignitability, reactivity and toxicity. Characteristics of each of these wastes are discussed below:

- *Corrosivity:* A liquid waste that has a pH of less than or equal to 2.0 or greater than or equal to 12.5; a liquid that corrodes plain carbon steel at a rate greater than 6.35 mm per year. Examples include waste rust removers and waste battery acids.

- *Ignitability:* A waste is regarded as ignitable hazardous waste if it is a liquid with a flash point less than 140°F, an ignitable compressed gas or oxidizer, or other material that can cause fire through friction absorption of moisture or spontaneous chemical changes. The flash point of a waste is usually determined with a material safety data sheet. Common examples include used oil-based paint, adhesives, epoxy resins, naphtha, lacquer, thinner, used paint thinner, adhesives and mineral spirits
- *Reactivity:* A solid waste is classified as reactive hazardous waste if it is normally unstable, readily undergoes violent changes without detonating, reacts violently with water, forms a potentially explosive mixture with water, or generates toxic gases or fumes when mixed with water or noncorrosive materials, is incapable of detonation or explosive reaction, or is a forbidden Class A or Class B explosive. Examples of reactive wastes include, dynamite, gunpowder, sodium metal or wastes containing cyanides or sulfides.
- *Toxicity:* A waste is toxic, hazardous if (according to the Toxicity Characteristic Leaching Procedure (TCLP) it exceeds the regulatory levels for any of the eight metals, six pesticides or 25 organic chemicals listed in the appendix under toxic substances. This list includes metals like chromium, lead, mercury, silver and organic chemicals like benzene, chloroform, methyl ethyl ketone, and butadiene. TCLP is a refined form of EP toxicity test, that has been used to determine if a waste is toxic hazardous. The TCLP is more accurate and aggressive for many chemicals than the formerly used EP Toxicity test.

12.2.2 Regulations and Management Programs

Selection of suitable technology for the remediation of hazardous wastes needs to be supported by environmental regulations. Certain regulations that govern the selection and application of specific remediation technology include the resource conservation and recovery act (RCRA); the Comprehensive Environmental Response compensation and Liability Act (CERCLA) and the Superfund amendments and Hazardous and Solid Waste (Jenkins et al. 2009).

RCRA and HSWA. Regulation of hazardous waste management evolved primarily from amendments to the Solid Waste Disposal Act (SWDA), passed in 1965, to focus on municipal waste disposal practices at sites such as dumps and landfills. Before enactment of RCRA in 1976, there were few Federal laws or standards regulating the management of hazardous waste. Regulation of these materials was primarily left up to the state and local governments. The 1984 passage of the Hazardous and Solid Waste Amendments (HSWA) of RCRA established the current framework for hazardous waste management requirements.

RCRA deals with the disposal of hazardous liquid, sludge or semi-solid waste. The USEPA believes that bulk liquids, sludge and semi-solid wastes could be placed in a landfill under certain controlled conditions, such as with a secure liner

Table 12.2. Main features of Subtitle C of RCRA

No.	*Main features of Subtitle C of RCRA*
1	The Subtitle C of RCRA ascertains a system for controlling hazardous wastes from its generation to its disposal ("cradle-to-grave" approach).
2	For hazardous waste management, requirements of three categories of facilities include, generators, transporters and disposers.
3	Generators should do "cradle-to-grave" tracking of hazardous wastes. The large quantity of hazardous waste generators (1000 Kg per month) must undergo personnel training, record keeping, contingency planning and temporary storage standards
4	Transporters are subject to different necessities concerning storage-time limits and spill management
5	Disposers are subject to, obtaining permits, meeting financial assurance requirements for closure/post-closure and meet specific design and operating requirements for each hazardous waste management unit at the facility.

and a system for the collection and removal of leachate. If these set-up and measures were not in place, USEPA requires treatment by mixing the waste with materials such as soil, fly ash or cement kiln dust to stabilize or solidify the waste. RCRA, as amended by HSWA, consists of 10 Subtitles: A through J. Subtitle C is the RCRA section pertinent to hazardous waste (Jenkins et al. 2009). Main features of Subtitle C are shown in Table 12.2.

Congress in 1984 strengthened RCRA with HWSA, the only major set of amendments ever made to RCRA. HSWA required USEPA to establish new criteria for municipal solid waste landfills. These criteria mandated landfill liners, collection systems for the liquids that seep from landfills, gas ventilation systems, gas and leachate monitoring and other requirements. The number of landfills operating in the United States fell from almost 8,000 in 1988 to about 1,800 in 2006 (U.S. EPA 2007). The main achievements of HSWA are shown in Table 12.3.

CERCLA (Superfund). Established in 1980, CERCLA is focused on a single issue-managing risk at contaminated sites-that is considered to be a highly complex factor. CERCLA authorizes USEPA to locate, investigate and respond to land-based spills and other threatened or actual uncontrolled releases of hazardous substances. Hazardous substances include a wide variety of materials identified by the Clean Water Act, the Clean Air Act and other environmental statutes, but nuclear materials and petroleum are specifically excluded (Jenkins et al. 2009). Table 12.4 shows the main features of CERCLA.

The 1986 Superfund Amendments and Reauthorization Act (SARA) added a new corporate environmental income tax and authorized annual tax

Table 12.3. Main achievements of HWSA

No.	*Main achievements of HWSA*
1	Established new criteria for municipal solid waste landfills
2	Eliminated land disposal of untreated hazardous wastes
3	Extended most hazardous waste regulations to facilities generating between 100 and 1,000 kilograms of hazardous waste per month
4	Expanded EPA's authority to require cleanups at RCRA facilities that treat, store or dispose hazardous wastes
5	Responded to a growing problem of petroleum contamination and other hazardous substances leaking from underground storage tanks (UST)

Table 12.4. Main features of CERCLA

No.	*Main features of CERCLA*
1	Recognizing the sites where releases of hazardous substances had occurred or might occur, and pretense a severe risk to human health or the environment.
2	Establishing procedures and funds for two broad categories of cleanups of the sites: removal actions and remedial actions
3	Taking appropriate action in connection to emergency and time-critical removals, quick cleanup sites that pose, or have the potential to pose imminent health risk
4	Taking appropriate action to remedy those releases, though the remedial actions are lengthy and complex, but less imperative
5	Remedial sites that pose significant risks and require federal funding are placed on the National Priority List (NPL), which was initially limited to 400 (Greenstone and Justin 2006)
6	A retrospective, strict, joint and several liability was assigned among all responsible parties identified at a particular site
7	Seeking that the potential responsible parties (PRP) pay for the cleanup activities out of a special trust fund known as Superfund, or initially funded with trust monies which are later recouped through settlements with PRPs (U.S. EPA 2008a). About 70% of NPL cleanups have been financed by PRPs (Ramseur and Mark Anthony 2006)

collections of $1.7 billion per year for five years. Special tax collections expired in 1995 and, despite occasional congressional benefits to reinstate them, were not renewed. At the end of 2007, there were 1,569 sites on NPL and 66% of them had reached the final stage of remedy, construction and completion (U.S. EPA 2008a).

Table 12.5. Federal laws that regulate hazardous waste management

Federal Law	*Main Purpose*
Hazardous Materials Transportation Act (HMTA)	Outlines requirements for the transportation of Hazardous Waste
Toxic Substances Control Act (TSCA)	Regulates management of PCBs and other toxic wastes
Occupational Safety and Health Act (OSHA)	Regulates work safety, emergency response, clean up, and training requirements
Clean Air Act (CAA)	Regulates air emissions from RCRA TSDFs and incinerators
Federal Facility Compliance Act (FFCA)	Waives sovereign immunity for penalties under RCRA
Emergency Planning and Community Right-to-Know Act (EPCRA)	Outlines requirements for emergency planning, notification, response, working with the local community, and reporting
Pollution Prevention Act (PPA)	Outlines principles guiding waste management practices

Other Applicable Federal Laws and Regulations. Other legislation and regulations that impact hazardous waste management are shown in Table 12.5 (Jenkins et al. 2009). The impact of these laws varies depending on the types of waste produced as well as the media (land, water, or air) into which the wastes can be released.

Management Programs. The evolution of state and federal programs over the years has resulted in a plethora of targeted programs. State Superfund, Voluntary cleanup programs (VCP) and brown fields programs and federal national priority lists (NPL), emergency removal, RCRA Corrective Action, and UST programs have all been tailored to address different needs. The USEPA introduced the "One Cleanup Program" in 2003 to improve the coordination of cleanups that might otherwise occur under different statutes. This approach aims to generate clearer and more useful information for communities and to improve measurement of cleanup performance results (U.S. EPA 2003). To better establish the idea of "One Cleanup Program" and address reuse of remediated sites, USEPA established a "Land Revitalization Initiative" in 2004. This initiative emphasized that reuse and cleanup are mutually supportive and that anticipating property reuse should be an integral part of EPA's cleanup decisions. Activities associate with this initiative include the development of cross-program measures of land reuse outcomes, public outreach and promotion of tools that address barriers to land reuse (U.S. EPA 2008b). At the state level, similar efforts are developing.

Table 12.6. Goals of implementing an environmental audit program

No.	*Goals*
1	To identify environmental problems through environmental audits
2	Development and implementation of cost-effective potential solutions
3	Recognize and regulate risks
4	Identify the strengths and weaknesses of current environmental efforts
5	Anticipate future environmental issues
6	Assessment of the efficacy of internal environmental management programs
7	Minimize the potential for misdemeanor or legal accountabilities for damages to human health or the environment
8	Sustain the image of the corporation
9	Meet Securities and Exchange Commission disclosure requirements

12.2.3 Environmental Audits

The adverse consequences of mismanaging environmental responsibilities have prompted the use of environmental audits as a preventative tool. The concept of an environmental audit had its beginnings in the late 1970 s. In comparison with a financial audit, an environmental audit has broader objectives. In fact, environmental auditing can take many forms, each with its own set of objectives. The USEPA defines environmental auditing as a systematic, documented, periodic, and objective review by a regulated entity of facility operations and practices related to meeting environmental requirements. The goals of implementing such a program are listed in Table 12.6 (Tusa 1990).

Types of Environmental Audits. These are briefly described as follows:

- *Compliance audits:* Compliance Audits focus on regulatory compliance. Verification of compliance was the driving force behind the initial developments in environmental auditing. Today, it still remains one of the main reasons for conducting audits. Determining what specific regulatory requirements are imposed upon an operation, finding out whether the operation is in compliance, and pinpointing possible violations in time to take proactive measures are the major goals of compliance auditing.
- *Management audits:* Management audit determines whether an adequate compliance management "system" is established, implemented, and used correctly, to integrate environmental compliance into everyday operating procedures. Such an audit examines cultural, management, and operational elements to include internal policies, human resources, training programs, budgeting and planning systems, monitoring and reporting systems, and information management systems. A management audit detects potential systematic breakdowns that could manifest themselves as an environmental problem.

- *Liability definition audit.* Liability Definition Audits are typically done for prospective buyers of real estate and for proposed mergers and acquisitions. Such audits identify environmental problems that could reduce the value of a property or expose, the buyer to liability.
- *Other types of audits:* The waste contractor audit uses features of both a compliance audit and a liability definition audit to analyze commercial facilities used to store, treat and/or dispose of wastes. The risk definition audit analyzes the operations of facilities that handle hazardous materials and substances. Waste minimization audits examine waste generated by a community facility with the objective of identifying viable actions to reuse, recycle, or otherwise reduce the quantity and toxicity of each waste stream, in response to waste diversion goals or needs.

Features of Effective Environmental Auditing. A review of mature auditing programs being conducted by both the public and the private sector has identified many common features deemed to be an important part of an effective auditing program (La Grega et al. 2001):

- Explicit top management support for environmental auditing and commitment to follow-up on audit findings;
- An environmental audit team separate from and independent of the persons and activities to be audited;
- Adequate team staffing and auditor training;
- Explicit audit program objectives, scope, resources, and frequency;
- A process which collects, analyzes, interprets, and documents information sufficient to achieve audit objectives;
- A process which includes specific procedures to promptly prepare candid, clear, and appropriate written reports on audit findings, corrective actions, and schedules for implementation;
- A process which includes quality assurance procedures to ensure the accuracy and thoroughness of environmental audits.

Stages of Environmental Auditing. Regardless of the type of audit being performed, it typically would unfold in three stages as follows:

- Audit program planning (including pre-visit data collection);
- On-site activities; and
- Evaluation and reporting.

Program planning. A compliance audit program consists of many elements, including a visit to the site. Prior to the actual visit, it is essential to plan the overall program. Planning would include obtaining commitment by management, defining the objectives and requirements, and developing an information management system. It is also important to organize the audit team and make sure that the program has the resources and tools necessary to address the issues that need to be investigated. In addition, one must consider the legal measures, which may be appropriate to protect the results of the program. In summary, the auditing program must be tailored to meet the needs of the facility being audited.

On-site activities. Each site audit begins with an overview of the plant audit with facility representatives. Then the team reviews in-plant operations and environmental records and conducts a preliminary inspection and a windshield tour of the plant vicinity. It is important at this time to review the audit checklist and questionnaire, survey the plant in detail and then complete the questionnaire. The completed questionnaire brings together the most comprehensive responses to each survey question based on written research, responses of facility representatives and the results of the inspections. Once done, an exit briefing can be compiled to identify areas that must be immediately addressed by the plant manager, When the on-site audit is completed the team reconvenes to interpret the data and summarize the findings. At this stage the team compares results and identifies non-compliance areas and significant liability issues. It also assesses potential costs and develops specific recommendations.

Evaluation and reporting. The audit data are evaluated during two periods while still at the site. The first occurs in the form of immediate feedback as the auditor collects information. The second occurs as the audit team performs a comprehensive analysis upon completion of all physical inspections, interviews, and document reviews. The evaluation of data is such that the findings for the subsequent written report are developed before the audit team leaves the site.

Benefits of Environmental Auditing. New international standards for environmental auditing are now being actively promoted by public authorities and adopted by private firms. One important feature of these standards is their emphasis on managerial systems and incentives that support a wiser use of environmental resources. It is found that the insertion of environmental audits within current management systems would certainly induce a manager to care more about the environment; although this may come at the expense of less concern for other activities, we find plausible circumstances in which properly designed environmental audits overcome such a tradeoff and increase the manager's attention to both environmental and traditional tasks (Sinclair-Desgagné and Landis Gabel 1997).

The audit brings an increased attention to environmental issues, prompting facility personnel to pay more attention to these issues, and elevating the issues to the level of consciousness and concern that they deserve. With a good audit program, a facility should experience a decreasing pattern of serious discrepancies. This will provide the facility and corporate management increased confidence in the facility's control of environmental issues, including anticipating new issues as new regulations occur.

12.2.4 Pollution Prevention

Audits consist in methods used to identify environmental problems at the origin of waste production. Pollution Prevention provides practices that reduce or eliminate waste generation at the source.

Basic Concepts. Pollution prevention consists of all those activities that reduce the generation of waste. The U.S. EPA (1989) defines pollution prevention as "the use of materials, processes, or practices that reduce or eliminate the creation of pollutants or wastes at the source. It includes practices that reduce the use of hazardous materials, energy, water, or other resources and practices that protect natural resources through conservation or more efficient use." The idea underlying the promotion of pollution prevention is that it makes far more sense for a generator not to produce waste than to develop extensive treatment schemes to insure that the waste poses no threat to the quality of the environment. The USEPA suggested source reduction, recycling, treatment and disposal in a hierarchy towards the management of wastes.

The Congressional Office of Technology Assessment defines waste reduction as: "In plant practices that reduce, avoid or eliminate the generation of all wastes so as to reduce risks to health and the environment. Actions taken away from the waste generating activity, including waste recycling or treatment of wastes after they are generated are not considered waste reduction. Also, an action that merely concentrates the hazardous content of a waste to reduce waste volume or dilutes it to reduce degree of hazard is not considered waste reduction" (Herdman 1989). In this context the term waste minimization applies to any management technique or process modification that ultimately reduces the mass or toxicity of waste sent on to processing, treatment, storage, and disposal facilities. The somewhat broader term pollution prevention has an evolving definition that includes: a) managing chemicals to reduce risk; b) identifying and estimating all releases; and c) waste minimization.

Benefits of Pollution Prevention. Pollution prevention can help achieve the following (U.S. EPA 1989): improve the firm's "bottom line;" make compliance with environmental regulations easier; demonstrate a proactive commitment to genuinely pursuing a pollution prevention program. Pollution prevention is a compelling strategy for many reasons. If no pollution is generated, there are no pollutants to be managed. Thus, future problems are avoided, such as the problems, which occur when previously accepted land disposal methods are discovered to be major sources of environmental contamination. Preventing pollution before it occurs also prevents situations that not only might endanger members of the community, but workers involved in the management of pollution as well (U.S. EPA 1991). Two major incentives are described below:

Economic incentives. One of the significant benefits of pollution prevention is that it is often an economical approach. When wastes are reduced or eliminated, cost savings in materials result and more products are produced from the same starting materials. It is common practice today to think of wastes simply as resources out of place. It is often the case that a second look at the manufacturing process can identify additional ways that wastes can be reduced, sometimes at little or no cost. A frequently cited example is the 3M Company's 3 P (Pollution Prevention Pays) program. Reportedly, the following amounts of waste were eliminated due to their source reduction programs implemented in 1975

Table 12.7. Economic incentives achieved by pollution prevention

Waste type	*Eliminated*	*(U.S. Only)*
Air pollutants (tons)	121,000	(110,000)
Water pollutants (tons)	14,600	(13,500)
Sludge/solid waste (tons)	314,000	(303,000)
Wastewater (10^6 gallons)	1600	(1000)

(Table 12.7) (3M Sustainability, 1975). The processes and product changes made to achieve these reductions yielded cumulative savings already in the first year. These savings were achieved on capital costs of unneeded pollution control facilities and sales from new products from former wastes. Several changes in the economic environment justify a second look at the way a manufacturer handles waste. The most dramatic is the sharp increase in the cost of hazardous waste disposal.

Legal and regulatory incentives. A second motivation for conducting waste minimization evaluations in the United States is that they are required by law. Because reports released by industries will become public information under the SARA, this act provides another strong incentive to develop an effective source reduction program. Another legal element that forces waste reduction is the potential future liability for any hazardous wastes produced. An industry that generates a hazardous waste is responsible for that waste forever. Because these future liabilities are of unknown magnitude, many industries do not want to ship any wastes to what may become future Superfund sites.

12.2.5 Development of Hazardous Waste Management Systems

The principal objective of any hazardous waste management system is to ensure safe, efficient and economic collection, transportation, treatment and disposal of wastes. It should also ensure that the system operates satisfactorily for current as well as foreseeable future. A typical hazardous waste management system constitutes components for collection, transportation, treatment and disposal of wastes. The overview of hazardous management plan (Misra and Pandey 2005) is shown in Figure 12.1. It involves selection of suitable treatment and disposal techniques and sites, allocation of hazardous waste from generator to the sites and selection of the transportation routes (Cutter 1983).

Steps towards Effective Management of Hazardous Wastes. These include:

- Waste characteristics, including waste types, degree of hazards, chemical and physical stability, waste compatibilities, and the ability to segregate ignitable, reactive or incompatible wastes and selection of suitable treatment and disposal techniques;
- Fate and transport characteristics of chemical constituents of wastes and their projected degradation products;
- Critical media of concern (air, surface water, ground water, soils/sediments, terrestrial and aquatic biota);

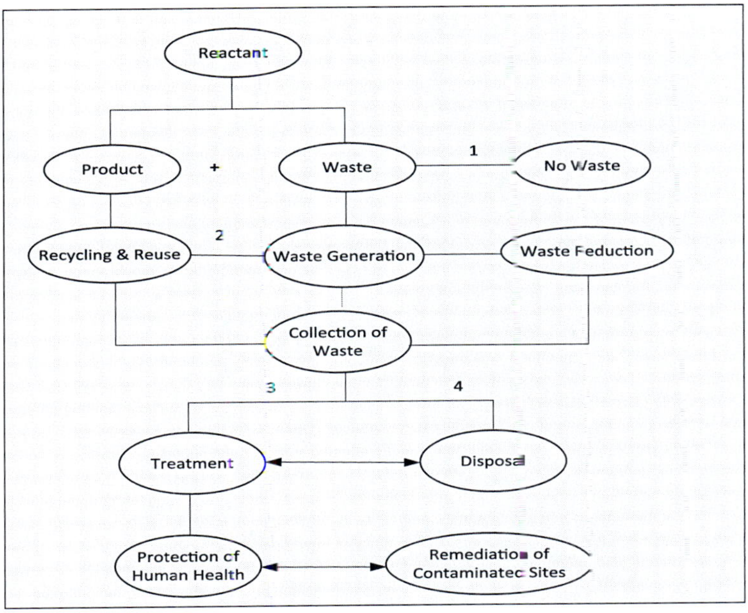

Figure 12.1. A typical hazardous waste management system

- Evaluation of potential release and exposure pathways of waste constituents and the potential for human and ecosystem exposures;
- Assessment of the environmental and health impacts of the wastes, if such wastes reach critical human and ecological receptors;
- Characterization of disposed sites, including site geology, topography, hydrogeology and meteorological conditions;
- Determination of extent of service area for proposed waste facility, i.e. waste handling from local industry Vs. from regional or national generators;
- Suitability of proposed location for waste facility based on environmental, social and economic concerns including proximity to populations, ecological systems, water resources etc.;
- Best available technology (BAT) for handling the particular wastes. In addition, there should be contingency plans and emergency procedures in the design of waste management plans; and
- Provision for effective long-term monitoring and surveillance programs including post-closure maintenance of facilities.

Planning and Design of Hazardous Waste Management Systems. An optimal plan for hazardous waste management should consider the influence of system components to one another (Misra and Pandey 2005) (Figure 12.2). Selection of treatment technology should be governed by the characteristics of

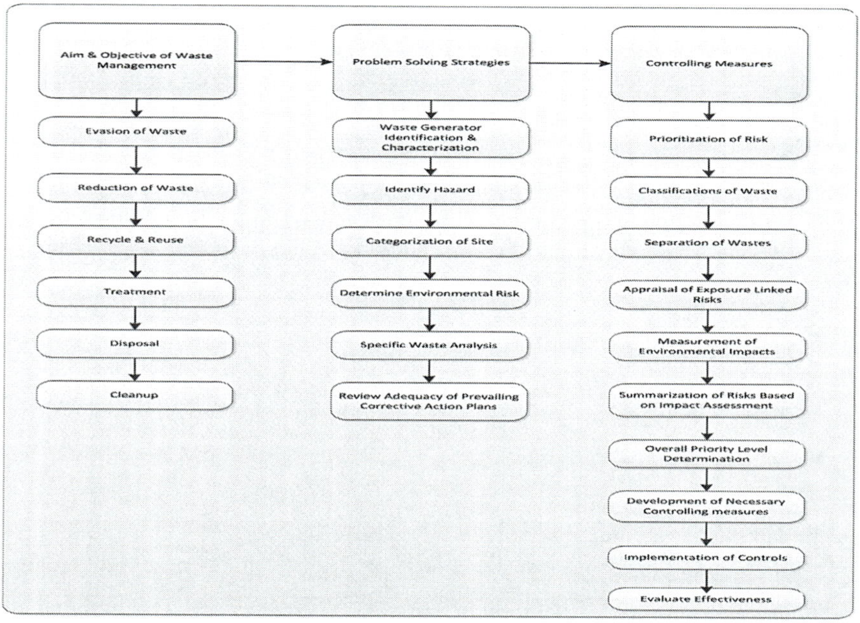

Figure 12.2. Overview of a hazardous waste management plan profile

the waste to be treated and the available capacity of the disposal facilities. For example, if the landfill capacity is limited, one has to opt for a treatment technology that generates fewer residues, e.g. incineration (which has a residue generation factor between 0.15 and 0.30). On the other hand, if a sufficient landfill capacity is available, one may opt for a treatment technology that generates more residues, but has some other advantages. For example, the chemical fixation and solidification technology has a relatively less impact on the risk to the air environment and is less capital intensive. However, its residue generation factor is relatively high (between 1.1 and 1.3) as compared to incineration technology (Nema and Modak 1998).

The capacity of waste treatment facility is regarded to remain constant over time; moreover, the capacity can be enhanced if required. However, the capacity of a disposal facility is an exhaustible resource with the shipment of hazardous waste residue to the disposal sites remaining a continuous process.

In fact, the quantity of wastes arriving to a treatment/disposal facility may even increase over a period of time due to industrial growth, unless waste minimization measures are implemented. Therefore, the difference between the available and required capacities of the disposal facilities is expected to narrow down with time. Once a disposal facility is exhausted, it has to be replaced by an alternative disposal facility. At this juncture, the decision makers may consider the following options to reallocate the wastes/waste residues (i.e., waste which was coming to the exhausted disposal facility may go to some other disposal facility which still can accommodate wastes/waste residues) to locate a new disposal facility (Nema and Modak 1998). Rehabilitation of abandoned sites and re-entry and reuse also has to be done in due course.

Out of the above mentioned options, the selection of a suitable option depends on the cost and risk associated with it. Once a site is chosen for a specific type of technology, it is difficult to change the type of technology. For example, once an incinerator is installed at a particular site, then it may be impractical to replace it by a chemical fixation/solidification technology after a few years. A change in the routes of wastes shipment and reallocating waste to other compatible treatment technologies may be a feasible option. Rising cost of land and increasing population growth with expanding city limits make the task of locating new disposal facilities more difficult and less feasible. Similarly, increased cost of transportation makes the option of transferring waste/waste residues to a treatment facility outside the region unattractive (Sharma and Lewis 1994).

As shown in Figure 12.3, the environment is both a source for the resources and a sink for discarded material. A system for sustainable solid waste management should be within the societal capacity on various issues such as financial affordability or availability of technology as well as assimilative capacity of the adjoining environment. Such a system would be pragmatic and would provide a realistic approach for improvement.

Sustainable hazardous waste management and practices are within the framework shown in Figure 12.3, as hazardous waste is a specific form of solid waste. Due to the uniqueness and complexity of issues related to hazardous wastes, sustainable hazardous waste management has been evolving slowly, from the

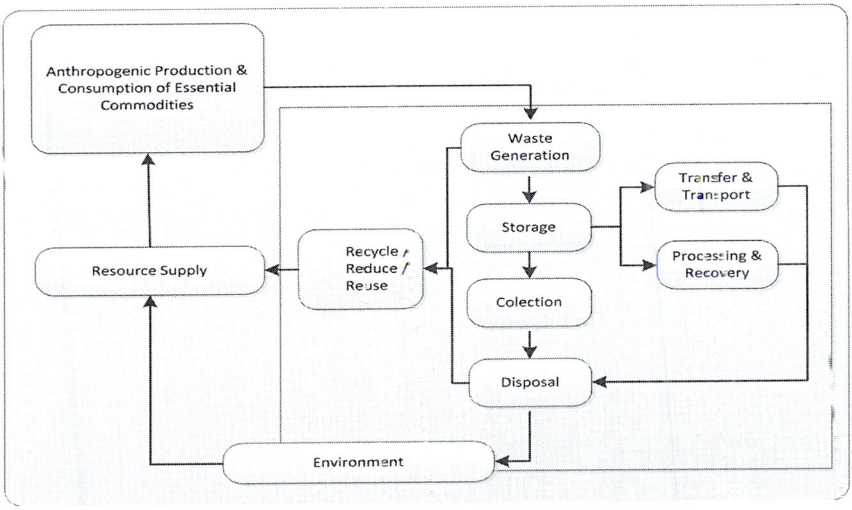

Figure 12.3. A sustainable solid waste management system

traditional approach to waste minimization (that usually is within the boundary of a manufacture entity) to newly developed systematic approach, i.e., Integrated Sustainable Waste Management (ISWM). This section is to briefly introduce these two different paradigms.

12.2.6 Integrated Sustainable Hazardous Waste Management/ Practices

12.2.6.1 Waste Minimization within a Manufacturing Entity

Towards the goal of minimizing the generation of hazardous waste at a manufacturing facility, the following steps are suggested:

- Planning and organization;
- Characterization of waste and losses;
- Development of waste minimization options;
- Technical, regulatory, and economic feasibility;
- Implementation (including training); and
- Monitoring and optimization.

The first step includes involving the management in discussion to result in a definite commitment and establishment of a clear set of goals. It would typically involve identifying individual roles and responsibilities in the organization towards the establishment of waste minimization plan. The second step is to understand the types and sources of wastes; an outcome of the aforementioned audits. Steps three and four analyze this information in a framework that includes a range of perspectives: environmental, economic, legal, political, and technical. The end result of this analysis should be a comprehensive waste minimization plan. Implementation of this plan must concern virtually all aspects of the company, including research and development, marketing, production, training, maintenance, and raw materials procurement. Finally, these efforts should be examined to ensure that they are accomplishing the objectives in an optimal way.

Waste Minimization Options. The options/stages for waste minimization are:

- *Common sense waste reduction:* Immediate reduction opportunities without having to change production processes. Examples are reducing the quantity of water used in cleaning equipment and ceasing the mixture of hazardous and non-hazardous waste.
- *Information-driven waste reduction:* Relatively easy reduction opportunities resulting from simple changes in production motivated by similar practices in other industries. Examples are changing raw materials (e.g., replacing an organic solvent with a water-based product, or installing a closed-loop recycling unit).

- *Audit-dependent waste reduction:* Waste reduction opportunities requiring capital investment and production changes with a very good economic payback.
- *Research-and-development-based waste reduction:* Waste reduction opportunities facing severe technical obstacles requiring extensive study

The first and second options (stages) can be implemented with a fairly simple audit using a checklist. The third stage requires a formal audit. A formal waste minimization audit surveys each operation that generates waste to determine how the waste is generated, what its characteristics are, how it is managed, and what the costs are. This provides baseline conditions for evaluating progress toward meeting waste minimization goals. It also provides the necessary data to prioritize waste streams and identify alternatives towards the minimization of high-priority wastes (U.S. EPA 1986).

Methods and Approaches to Waste Minimization. The simplest waste minimization approach would be the changes in housekeeping. It includes segregating hazardous waste, sweeping floors prior to washing, and training employees to be sensitive to the implications of their actions can dramatically reduce waste generation. Once these measures are complete, it would be wise to undergo production changes that ensure waste minimization. Production changes implemented by the industries often result in one or many of the following changes:

- Changes in process operating variables (water content/temperature);
- Changes in feedstock;
- Introduction of new process chemistry;
- Equipment changes; and
- Elimination of product.

In general, the following approaches are taken individually or simultaneously to minimize waste production from an entity.

Volume reduction. The right approach to waste minimization involves examination of ways to reduce the volume of hazardous waste. It can be accomplished by modifying production processes, segregation, and re-use. It should be ensured that volume reduction of waste should also be accompanied by toxicity reduction. For example, under California's Hazardous Waste and Management Review Act of 1989, "actions that merely concentrate the constituents of a hazardous waste to reduce its volume" are explicitly excluded from the definition of "source reduction."

Modifications of operations and processes. Modifications of operations and processes are different from production changes where major modifications may involve. Processes and/or operations may be modified by changing: a) material storage, b) raw materials, c) operating procedures, d) equipment and e) end product. For example, in the electroplating process, the levels of different control parameters can be changed so that the cyanide waste can be reduced significantly.

Segregation. An important caveat of source reduction is to avoid mixing wastes. Mixing a small amount of a hazardous waste with a larger amount of non-hazardous waste results in a large amount of material that must be treated as a hazardous waste. Eliminating the interaction of solid wastes with liquids, unless specified in special containers, diverting spent cooling water from waste streams, having specific task order in the house keeping operations are some measures that can substantially reduce waste volumes. Keeping wastes segregated greatly facilitates any required treatment. Proper labeling of all lines and containers will greatly increase the likelihood that plant personnel will follow any changes in practices intended to enhance segregation of wastes.

Reuse. Certain materials disposed of as hazardous waste will have potential elements in them that can be reclaimed for another application. For example, toluene is often used as cleaning agent for processes and as a thinning agent for inks in printing industry. If toluene is limited to clean one specific color ink, it may be reused to thin the same color ink. Hazardous materials that are past their expiration date may be subjected to re-assay and re-qualification, so that they may be used for production.

Toxicity reduction. A number of waste minimization techniques reduce the concentration of contaminant in a liquid or solid waste stream, without reducing the volume of wastes produced. It is often possible to reduce the toxicity substantially so that the remaining waste is no longer considered a hazardous waste.

Material substitution. Most manufacturing processes were designed with product quality and profitability as the primary design criteria. The changing economics of waste management are forcing a new look at production formulations. Many material changes have proven to be cost-effective and resulted in waste reduction. Examples are the use of non-cyanide baths in metal plating operations and the use of less toxic solvents in cleaning operations.

Recycling. It may be possible to recycle waste into another process or plant if volume or toxicity reduction becomes challenging. It is effective to recycle waste on-site to eliminate chances of mishandling the waste by shipping it off-site. Key issues in recycling include:

- Identification of recyclable materials;
- Identification of recycling opportunities;
- Market issues: Low value and low density materials are disincentives for plastic recycling, compared to the use of new raw material;
- Cost of recycling;
- Infrastructure for collection recycling materials; and
- Specifications for recycled materials.

12.2.6.2 Framework and Principles ISMN

From the systems point of view, Integrated Sustainable Waste Management (ISWM) has also been defined as the selection and application of suitable techniques, technologies and management approaches to achieve specific objectives and goals. This paradigm selects all the component systems simultaneously to allow for a rational planning and effective execution. The resulting system components operate in a way to ensure mutual compatibility of the components, thereby improving the overall performance (Shedkar 2009). Here, we introduce four basic principles and three major components of ISMN.

Four Basic Principles. The ISWM concept takes as a point of departure four basic principles: Equity, effectiveness, efficiency and sustainability.

- *Equity:* All citizens are entitled to an appropriate waste management system for environmental health reasons.
- *Effectiveness:* The waste management model applied will lead to the safe removal of all waste.
- *Efficiency:* The management of waste is carried out to maximize the benefits, minimize the costs and optimize the use of resources, taking into account, equity, effectiveness and sustainability.
- *Sustainability:* The waste management system is appropriate to the local conditions and feasible from a technical, environmental social, economic, financial, institutional and political perspective. It can maintain itself over time without exhausting the resources upon which it depends.

Equity gains priority as pollution affects the quality of life of citizens, and it results in communicable diseases spread through vectors including flies, insects, rats and air and water pollution. Moreover, polluted areas result in poor environmental conditions that encourage social unrest and anti-governmental activities. Polluted neighborhoods also affect city's economic development, increase the number of unhealthy laborers and turn away the investors.

Effectiveness of waste management pertains to the removal of all waste as planned and recovery of all materials. Effectiveness of service should ensure that the less visible parts of the city as well as the city center are maintained well.

Efficiency of service is the result of the optimal combination of labor, money, equipment, machinery and management and it needs to be balanced by reasonable cost paid by beneficiaries.

Sustainability refers to the utilization of resources to fit into the local culture, context and society. Resources include human (manpower), material (equipment) or natural resources (water, air, soil). Utilization of resources should ensure the distinction between renewable and non-renewable ones. Interaction of social and political aspects with technical and environmental should ensure that the system is sustainable, that is, it can reproduce itself without reducing the possibilities open to future generation.

ISWM Components. ISWM has three major components: (1) the stakeholders involved in waste management, (2) the (practical and technical) elements of the waste system and (3) the aspects of the local context for assessing and planning a waste management system. Descriptions of these components are as follows.

Stakeholders. Stakeholders have various interests and roles in their particular waste management, but they can cooperate for a common interest (Snel and Ali

1999). Their influence (the extent to which stakeholders are able to persuade or coerce others into making certain decisions or following certain courses of action) and importance (the extent to which the problems, needs and interests of a particular stakeholder are a priority in a project or plan) varies (Arroyo et al. 1999).

Waste system elements. All waste system elements should be looked upon as being stages in the movement, or flow, of materials from the mining stage, via processing, production and consumption stage towards final treatment and disposal. A waste management system is a combination of several stages in the management of the flow of materials within the city and the region. A waste management plan is part of an integrated materials management strategy, in which the city makes deliberate and normative decisions about how materials should flow. The waste elements then become specific tactics to deal with specific materials after they have been consumed. ISWM recognizes the high-profile elements 'collection', 'transfer' and 'disposal' or 'treatment'. It gives equal weight to the less well understood elements of 'waste minimization', 'reuse' and 'recycling and composting'. The major elements that need to be integrated to govern performance within the system are shown in Figure 12.4.

ISWM Aspects. The ISWM concept distinguishes six aspects through which the existing waste system can be assessed and with which a new or expanded system can be planned. The ISWM aspects give a municipal manager a set of tools to perceive study and balance priorities and create measures to give the desired results. The six aspects of ISWM are described below:

1. Environmental aspects focus on the effects of waste management on land, water and air; on the need for conservation of nonrenewable resources; pollution control and public health concerns.

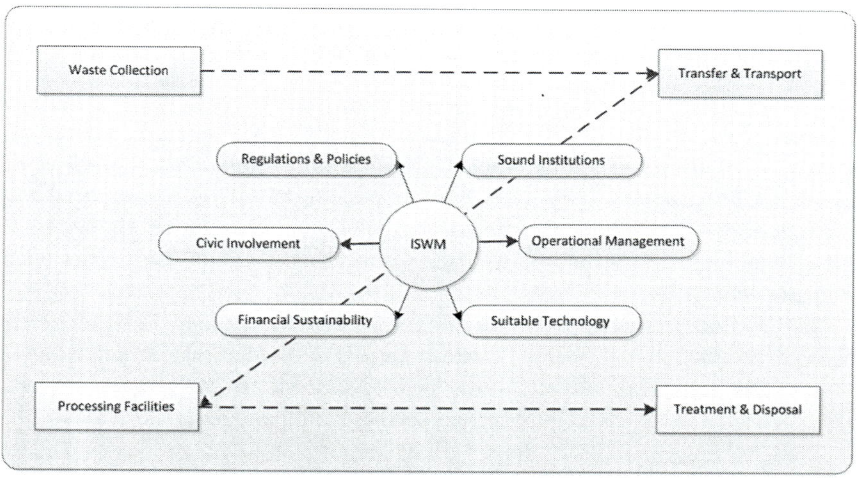

Figure 12.4. Integrated sustainable solid waste management system

2. Political/legal aspects address the boundary conditions in which the waste management system exists: setting goals and priorities; determination of roles and jurisdiction; the existing or planned legal and regulatory framework; and the basic decision making processes.
3. Institutional aspects relate to the political and social structures which control and implement waste management: the distribution of functions and responsibilities; the organizational structures, procedures and methods implicated; the available institutional capacities; and the actors such as the private sector who could become involved. Planning is often considered the principal activity in relation with institutional and organizational aspects.
4. Socio-cultural aspects include the influence of culture on waste generation and management in the household and in businesses and institutions; the community and its involvement in waste management; the relations between groups and communities, between people of various age, sex, ethnicity and the social conditions of waste workers.
5. Financial-economic aspects pertain to budgeting and cost accounting within the waste management system and in relation to the local, regional, national and international economy. Some specific issues are: privatization; cost recovery and cost reduction; the impact of environmental services on economic activities; the commodities marketplace and how the recycling infrastructures connect to it; efficiency of municipal solid waste management systems; macroeconomic dimensions of resource use and conservation; and income generation.
6. Technical and performance aspects concern the observable practical implementation and maintenance of all of the waste elements: what equipment and facilities are in use or planned; how they are designed; what they are designed to do; whether they work in practice; and how clean the city is on a consistent basis.

The above analysis indicates that sustainable hazardous waste management is of paramount importance for proper health, environmental protection and natural resource management, and sustainable development. Prevention of the generation of hazardous wastes and the rehabilitation of contaminated sites are the key elements, and both require knowledge, experienced people, facilities, financial resources and technical and scientific capacities. In the following sections, we introduce technical options for hazardous waste treatment and/or remediation as well as disposal methods. The central theme is that hazardous waste treatment at a centralized facility depends upon the types of wastes encountered. However, according to the principles of ISWN, there should be minimum facilities for storage, treatment and disposal at this facility.

It is imperative to develop a proper hazardous waste treatment and disposal plan to ensure that it is properly sited to protect people and the environment from any adverse impacts. In addition, attention should be paid towards the assessment of abandoned hazardous waste disposal sites and remediation of the affected sites to prevent further damage to human health and environmental quality. Materials included in Sections 12.3 and 12.4 are introductory but essential to understand the linkage between current practices in hazardous waste management/ treatment and the broader issues of sustainable management for water, food and solid waste, climate change, and economic/social development.

12.3 HAZARDOUS WASTE TREATMENT/REMEDIATION TECHNOLOGIES

The main objective of hazardous waste treatment is to alter the physical and/or chemical properties of the waste to make it amenable for safe disposal. Several treatment technologies exist (e.g., acid-base neutralization, incineration, chemical fixation, solidification, etc.), which may be used prior to the ultimate disposal of the hazardous wastes. These treatment technologies aim at converting the waste into a more innocuous form, immobilizing the toxic components, or reducing the quantity of the waste. They are characterized by their capacity to treat specific waste types, their waste residue generation factor, the cost and the risk associated with their application (Blackman 1996). The methods used in various types of treatment processes are shown in Table 12.8. Detailed descriptions of some processes are as follows.

12.3.1 Solidification/Stabilization Techniques

Stabilization is a process of converting a toxic waste to a physically and chemically more stable form, that is, alters hazardous waste chemically to produce a less toxic

Table 12.8. Hazardous waste treatment processes/technologies

Classification	*Process/technology option*
Physical	Sedimentation; thickening; centrifugation; clarification; foaming; floatation; electrostatic precipitation; fabric filter; wet scrubbing; activated carbon adsorption
Physico-chemical	Adsorption; reverse/forward osmosis; dialysis; crystallization; distillation, evaporation; leaching; solvent extraction; stripping, ultra/nano filtration; coagulation-flocculation
Chemical	Oxidation/reduction; chemical precipitation; neutralization; ion exchange; pyrolysis; solidification/stabilization; hydration; catalysis
Biological	Activated sludge; trickling filters; waste stabilization ponds aerobic/anaerobic sludge treatment; anaerobic lagoons

Table 12.9 Types and applications of stabilization/solidification

Types	*Application*
Microencapsulation	Particulate matter, liquids, sludge, heavy metals, radioactive wastes
Surface encapsulation	Sulfates, halides, heavy metals, radioactive materials
Vitrification	Radioactive and hazardous wastes.
Chemfix process	Industrial waste, contaminated soil, sludge, metal salts, in-situ treatment, metal salts and radioactive wastes
Self-cementation	Inorganic wastes and flue gas desulfurization
Thermoplastic stabilization	Radioactive wastes
Organic polymer stabilization	Radioactive wastes and contaminated sludge
Lime based solidification	Inorganic wastes and flue gas desulphurization
Pozzolan based	Solid organics, oxidizers, heavy metals, radioactive wastes

or less mobile form. It involves chemical interactions between waste and the binding agent. By comparison, solidification converts liquid waste, semi-solid sludge or a powder into a monolithic form or granular material that will allow relatively easy handling and transportation to landfill sites (Conner and Hoeffner 1998; Glasser 1997; Poon et al. 2004). Methods of stabilization (Lehman 1982) are listed in Table 12.9, and they vary according to the agents employed.

Some advantages of the cement based solidification (Chemfix process) technique for sludge treatment are that: (1) it is cheap, as most additive materials are cost-effective and are readily available; (2) the technique is able to withstand wide variations in chemical composition of the sludge and is pretty robust; and (3) equipment operation is simple and easily available. Some disadvantages are: (1) there is sludge-bulking due to the addition of cement and other additives, as it adds considerably to the weight and bulk of the sludge; (2) often, low strength cement-waste mixtures are vulnerable to leaching of contaminants; (3) the presence of metals including, Mn, Sn, Cu and Pb salts will increase the setting time and reduce the physical strength of the cement matrix; and (4) organic matters will hinder the settling process (Visvanathan 1996).

Among the other stabilization techniques, encapsulation is a thermal process that is considered as highly energy intensive. Previous studies however have revealed that encapsulation would provide a high degree of control over the release of unwanted hazardous wastes into the environment (Batstone et al. 1989).

Self-cementation is a solidification technique that is applicable to inorganic wastes containing large amounts of calcium sulfate or calcium sulfite. In this

Table 12.10. Performances, advantages and disadvantages of different treatment technologies for metal uptake

Treatment Types	*Heavy metals*	*Treatment Efficiency (%)*	*Advantages*	*Disadvantages*
Ion exchange	Cr(III), Ni(II), Zn(II), Cu(II), Cd(II)	100 100 100 100 100	Quick process, no sludge	Costly, specific use
Coagulation-flocculation	Zn(II)	99	Superior sludge settling and shorter settling time	Extra cost for sludge disposal
Precipitation	Ni(II)	99.3	Cost effective, simple operation	Extra cost for sludge disposal
Reverse osmosis (RO)	Cu(II)	100	High rejection rate and temperature resistant	Energy intensive, susceptible to membrane fouling
Nano-filtration	Cd(II)	99	Lower pressure than RO	Costly, prone to membrane fouling
Ultra-filtration	Cu(II)	100	Smaller space requirement	Costly, prone to membrane fouling

technique, a small quantity of waste is dried, calcined and mixed with the rest of the waste and other additives to form a hardened mass (Albino et al. 1996).

12.3.2 Physico-chemical Treatment Techniques

Table 12.10 presents an overview of various physico-chemical treatments for wastewater laden with heavy metals. The table also summarizes the advantages and disadvantages of various physico-chemical techniques for hazardous inorganic wastewaters (Kurniawana et al. 2006). Among the treatments presented in the table, some of them have been tested successfully in laboratory or even on a pilot scale. Some may even be ready for commercialization. However, no individual treatment has been found to be universally effective and applicable for heavy metal removal. As future regulations focus upon the ability of industrial users to comply with the legal requirements of the residual metal concentration in

Table 12.11. Chemical stabilization processes for hazardous waste

Process	*Waste Type*
Neutralization	Strong acids and alkalis
Molten salt combustion	Range of organic compounds
Wet Oxidation	Organic compounds, cyanide, cyanate, phenols
Reduction	Reduction of Cr(VI) to Cr(III) by ferrous sulphate
Microwave based destruction	Poychlorinated biphenyls, pesticides, etc.
Hydrolysis	Certain carbamate insecticides and organophosphorus pesticides,
Precipitation	Precipitation of cadmium as its hydroxide by sodium hydroxide
Ozonation	Phenols, cyanides and lead in air and water
Dechlorination	Polychlorinated organic compounds
Treatment with formaldehyde	Cyanide ion and complexes, chromium (VI)
Catalytic hydrogenation	Polychlorinated hydrocarbons
Catalytic Reduction	Chlorinated organic compounds present in wastewater
Combustion	Organic compounds

discharge, it would be very beneficial to investigate various treatments, which can assist industrial users to comply with the environmental law.

12.3.3 Chemical Treatment Techniques

Chemical methods include oxidation; reduction, precipitation, as well as complexation and neutralization participate in the chemical stabilization of hazardous waste. It would be wise to choose a method that is fast, inexpensive and does not generate residual secondary pollutants that need further handle prior to disposal. Table 12.11 (Kassim 2005) lists chemical stabilization processes employed for specific hazardous wastes.

12.3.4 Biological Treatment Techniques

Bioremediation technologies can be broadly classified as ex situ and in situ. Ex situ technologies are those treatments which involve the physical removal of the contaminated material for treatment process. In contrast, in situ techniques involve treatment of the contaminated material in place. Some of the examples of in situ and ex situ bioremediation are described in Table 12.12 (Boopathy 2000).

Table 12.12. Examples of in situ and ex situ bioremediation techniques

Technique	*Description*
Phyto-remediation	Treatment of environmental problem through the plants, can be ex or in situ
Bio-farming	Treatment system for contaminated soil, may be done in ex situ condition
Bio-leaching	Metal extraction through the use of living beings.
Composting	Treatment process in which contaminated material is mixed with a bulking agent
Bio-piles	An ex situ static pile composting process in which compost is formed into piles and aerated with blowers.
Bioreactors	A reactor facilitating ex situ biodegradation may be used to treat liquids/slurries
Bioventing	Treatment of contaminated soils by drawing oxygen through the soil to stimulate microbial activity, an in situ approach
Bioslurping	Bioventing along with vacuum-enhanced free-product recovery is an in situ approach
Rhizo-filtration	Filtration of water through roots to remove toxic substances.
Biofilters	Microbial stripping columns are used to treat air emissions
Bio-augmentation	Addition of microbial cultures to a contaminated medium, often used in both in situ and ex situ system
Bio-stimulation	Stimulation of indigenous microbial populations by providing necessary nutrients
Intrinsic bioremediation	Unaided bioremediation of contaminant
Natural attenuation	Essentially an in situ biological remediation which facilitate the self-cleaning ability of the nature.
Pump & treat	Pumping ground water to surface, followed by treating and re-injecting.

12.4 DISPOSAL OF HAZARDOUS WASTE

Disposal of wastes is the final process and a key issue in the overall hazardous waste management program (Millano 1996). The disposal facilities act as permanent repositories for the waste residues generated from the treatment facility. Even the most advanced treatment methods result in residues that are no longer amenable to cost-effective treatment. The options for disposal include deep well disposal, dilution and dispersal, solidification and stabilization, incineration, ocean dumping, sanitary landfill and land burial (Dawson and Mercer 1986). Economics of waste disposal will govern the amounts and types of wastes that will be moved to distant disposal sites. The choice of disposal should be based on evaluation of economics and potential pollution risks. Details of some disposal options are described as follows.

12.4.1 Incineration

Incineration is a controlled process involving the oxidative conversion of combustible solid material to harmless gases suitable for atmospheric release. It converts waste to a less bulky, less toxic, or less noxious material. The principal products of incineration are water, carbon dioxide and ash, while the products of primary concern due to their environmental effects are compounds of sulfur, nitrogen and halogens (Rabel et al. 1998). Incineration requires adequate controls in place, and their absence can lead to the release of undesired materials into the atmosphere. For example, hydrogen chloride generated during incineration of chlorinated hydrocarbons is now being removed through absorption in water. The hydrochloric acid, so formed is then either neutralized or is concentrated for reuse. This also reduces the risk of dioxin formation. The solid and liquid effluents from the secondary treatment processes will occasionally require treatment prior to ultimate disposal.

Incineration causes reduction of bulky water (e.g., sludge) to an ash residue that can be handled more easily for ultimate disposal. However, it could also lead to the formation of poisonous soluble metal oxides. This concentration in the incinerator ash has to be handled by recovery/detoxification/disposal.

12.4.2 Landfill

Common Landfill. This disposal method involves the mixing of waste with soil, evaporation and infiltration and/or shallow burial. Solid wastes will normally be incorporated in a landfill and buried. Liquids, slurries and sludge might also be incorporated into a landfill. The composition and quantity of that waste governs the potential for environmental damage by landfilled hazardous wastes. Small quantities of hazardous wastes containing cyanides and arsenic compounds and some heavy metal compounds are accepted in landfills, if they are present in drums or plastic containers.

Approved Landfill. A site which is suitable for the disposal of inert solid wastes and decomposable organic materials. The site must ensure the separation of the wastes from underlying or adjacent usable water due to leachate possibilities

Secured Landfill. Properly sited and designated landfills minimize the adverse impacts on the environment. A site suitable for disposal of all wastes, including liquid and solid hazardous wastes must ensure that there will be no discharge of these materials or their byproducts to usable ground or surface waters by leaching, percolation or any other means (Lidskog 1993). Primarily, the hazardous contents of the landfill must be isolated from the surrounding environment. Depending on the nature of the waste, various types of secure landfills such as open, control and closed are used. In open landfills, wastes such as metal cans and glass bottles are buried in open landfills. In control landfill, weakly toxic or potentially toxic wastes, such as incinerator ash or sewage sludges are buried in controlled landfills. In closed landfills, highly toxic wastes such as soils contaminated with polychlorinated biphenyls or mercury compounds are buried in closed landfills. It is protected by walls, roof and concrete roof from outer environment. Leachate generation is minimal as rainwater does not intrude.

A major problem with secured landfill is ground water contamination due to infiltration of leachate. The USEPA has proposed Double Liners and Leachate Collection and Removal System (LCRS). Selection of liner material depends on hydraulic conductivity, chemical compatibility with leachate landfill gas permeation, aging and durability characteristics, stress and strain characteristics and ease of installation. Darcy's law (1856) is usually used to calculate infiltration of three types of liners: a) soil liner; b) geomembrane liner; and c) composite liner (geomembrane liner overlying compacted clay liner) as follows:

$$\text{Flow rate (q)}(\text{m}^3/\text{s}) = \text{K}_\text{s} i \text{A}(\text{Soil liners}) \ (\text{Darcy's law})$$

$$= \text{C}_\text{B} a (2\text{gh})^{0.5}(\text{Geomembrane Liners}) \ \text{Bernoulli's orifice Eq.}$$

$$= 0.21 n^{0.9} a^{0.1} k^{0.74}(\text{composite with good contact})$$

$$= 1.15 n^{0.9} a^{0.1} k^{0.74}(\text{composite with poor contact})$$

where K_s = hydraulic conductivity of soil (m/s); i = hydraulic gradient [(h + d)/d]; h = liquid head (m); d = soil thickness (m); A = area of soil (m^2); $C_B = 0.6$ (flow coefficient); a = Area of hole (m^2); g = Acceleration due to gravity (m/s^2); n = porosity = volume of voids/total volume.

USEPA has compared theoretical leachate collection efficiencies for landfills having compacted soil bottom liners to those having composite bottom liners. Leachate collection efficiency is defined as the percentage of liquid collected and removed in Local Resource Center System (LRCS) to the total liquid entered into LRCS as leakage from top layer. It was found that composite bottom liners with a tear in geo-membrane layer can give 95% leachate collection efficiency, whereas compacted soil liners could give only 50% efficiency.

12.4.3 Other Disposal Methods into Contiguous Water Bodies

Several methods are used widely for hazardous waste disposal, such as disposal into contiguous water bodies, lagooning, subsurface injection, etc. For the method of disposal into contiguous water bodies, the solid waste usually is temporarily stored in the lagoon where it is treated to minimize environmental impact. For example, acids and bases are neutralized and suspended solids are permitted to settle. Lagoons may be employed for either temporary or permanent storage. They are frequently employed for interim storage and water treatment. Potential environmental problems may arise from lagooning related to the subsurface transport of leachate and surface water run-off. In the case of subsurface injection, the solid waste is slurried and then pumped into underground cavities. Subsurface injection of hazardous waste should only occur when leachate is fully contained.

12.5 ADJUSTMENTS TO MEET GLOBAL CHALLENGES

In many places, particularly some developing countries, hazardous waste crisis is occurring, and it is likely it will be worsen in the future due to population explosion, increased urbanization/industrialization, improved living standards, and over-exploitation of resources. The world is facing many grand challenges with a major one being energy challenge, which consists of two phases: 1) in the next 12 years (2013–2025), fossil fuels (especially oil) will be continuously used as the predominant source of energy and chemical feedstocks, where managing carbon, reducing the intense use of energy resources, and educational efforts to promote sustainability thinking will be critical; and 2) in 2025–2105, with the use of fossil fuels being phased out, the ability to carry out green technologies and having access to alternative renewable sources of fuels and feedstocks will be realized.Sustainable hazardous waste management is being adjusted to meet global challenges such as population explosion, increasing demand for food and water, resource depletion, and changing climate conditions. In this section, we will describe some of these adjustments and then introduce some examples to demonstrate the current practices and trends of sustainable hazardous waste management/treatment.

12.5.1 Adjustments to Meet Grand Challenges

As shown in Figure 12.3, a system for sustainable solid and hazardous waste management should be within the societal capacity on various issues. Such a system would be pragmatic and would provide a realistic approach for improvement, which requires people to embrace the concept that the environment is both a source for the resources and a sink for discarded material. Sustainable hazardous waste management depends on a dynamic balance among resources, energy consumption and conservation, financial affordability, availability of technology as well as assimilative capacity of the adjoining environment. Therefore, some adjustments centering on the following four categories have been or will be made to meet grand challenges.

Shifting to New Policies and Implication Directions. Climate change and the exhaustion of raw material and fossil fuel have forced governments to establish new policies. As a result, a new waste policy direction goes beyond preventing pollution and moves towards the construction of a low-carbon and sustainable resource recycling society. Managing waste for carbon sequestration, transferring waste as a recyclable resource and improving resource productivity have become the focal points in our society. Because of these changes, the policy structure has been modified from one focusing on reduction, recycling, treatment and disposal to one orienting to recycling, energy harvesting, and the improvement of treatment and disposal methods, causing a paradigm shift in hazardous waste management (UN 2012a) (Table 12.13).

One very important shift is to low-energy and low-waste product production and resource recovery. Nowadays, different programs have been established in many countries to a) convert waste into energy, b) prohibit the use of hazardous materials, and c) promote the reduction of resource consumption before production and effectively recycle previously-used resources. Restricting the use of disposable products becomes a common practice in many counties. Preventing packaging waste generation and waste charge policy are also implemented in many countries. For example, the South Korean government is devising policies to increase the percentages of recycled inflammable and organic waste to 47% (1,820,000 tons/year) and 26% (2,040,000 tons/year), respectively, by 2013, and will build 600 low-carbon green villages by 2020 (UN 2012a). In the US, the President's Executive Order 13514, Federal Leadership in Environmental, Energy, and Economic Performance, emphasizes the importance of sustainability and requires U.S. Federal agencies to meet a number of energy, water, and waste reduction targets,including 50% recycling and waste diversion by 2015 (USNR 2010). Working in close partnership with the States, Tribes, industry, environmental groups, and the public, the USEPA has been promoting the shifting from a waste management approach to materials management: a "cradle-to-cradle" approach aimed at reducing environmental impacts throughout the life cycle of products, materials, and activities. In 2009, USEPA published a report "*Sustainable Materials Management: The Road Ahead,*" in which, a path to move towards materials management was suggested, with the following focuses:

- knowing and reducing the life cycle impacts across the supply chain;
- using less material inputs (reduce, reuse, recycle);
- using less toxic and morerenewable materials; and
- considering whether services can be substituted for products.

In addition, using green remediation principles, the USEPA is taking actions to encourage thesustainable reuse of hazardous materials by considering all aspects of environmental protection.In addition to clean up, these actions can result in lower energy demand, reduced greenhousegas emissions, less water use, and other health and environmental benefits that contribute to economicand environmental sustainability. For example, EPA's Brownfields Sustainability Pilotsfund actions that further long-term environmental stewardship. More examples are described below.

Shifting to Robust Financial Support Framework and Policies. Currently, central and local governments are striving to a) share burdens and benefits of hazardous waste management equally among the taxpayers;

Table 12.13. Paradigm shift in waste management

Category	*Previous Policy Direction*	*Shift*	*New Policy Direction*
Central theme	Waste	→	Recyclable resources
Motivation	Deteriorationenvironmental condition	→	Prevention of climate change, resource exhaustion
Objective	Build fresh livelihood state	→	Establish sustainable living condition
Executable tactics	Recyclables used to be what we kept out of the garbage	→	Garbage is what we have left over after we reuse and recycle.
Key assignment	Systems for collection/transportation/ disposal/treatment; waste minimization/ reuse/recycling	→	Transformation of waste to energy; Pilot scale treatment, Evaluation of resource recyclability; Quality certification for recycled products;

b) spread widely cost allocation mechanisms for different treatment/remediation projects; c) establish the commercial market at economically viable prices for hazardous waste treatment/remediation technologies; and d) increase funding for demonstration projects. For example, one of the manners to be taken by UN to strengthen and upgrade its Environment Program is to have secure, stable, adequate and increased financial resources from the regular budget of UN and voluntary contributions to fulfill its mandate (UN 2012b). Increasing the financial contributions to sustainable hazardous waste management systems is key to achieving the goals of sustainability and development.

The USEPA's Smart Growth Program is a good example in providing funding to facilitate the sharing of best practices and act as a catalyst for implementation of ideas. The related USEPA funding includes (USNR 2010):

- The State Clean Diesel Grant Program (http://www.epa.gov/otaq/diesel/prgstate.htm) allocates funding to states to implement state grant and loan programs.
- The SmartWay Innovative Financing Program (http://www.epa.gov/otaq/smartway/transport/what-smartway/financing.htm) provides information on state incentive programs and several lenders who offer loans to help finance technologies that improve freight fuel economy.
- USEPA's Environmental Financial Tools (http://www.epa.gov/efinpage/efinfin.htm) include links to sources of financing such as the Environmental Finance Program (EFP) and EPA Programs and Offices.
- A Guidebook of Financial Tools (http://www.epa.gov/efinpage/guidebook.htm) provides an overview and analysis on how to pay for environmental programs.
- Environmental Financial Advisory Board (EFAB) Report on Innovative Finance Programs for Air Pollution Reduction (12 pp, 6 K, November 2007) provides recommendations for financing programs that reduce emissions from mobile sources.

Shifting to Universal Membership/Networks and Well-Prepared Society. To this end, major adjustments include: a) raising public awareness via education, training, and revitalizing the marketplace; b) providing transparent information about planned projects related to hazardous waste management/treatment/remediation in a timely manner to increase public understanding of the benefits and risks, and c) stimulation of collaboration at different levels and engaging major groups and other stakeholders to track and implement changes in a timely manner.

To establish a culture of resource recycling, SSSR (Save, Share, Swap, and Reuse) campaign has been promoted in different countries to bring the reuse of goods into everyday habits. Some countries establish and run permanent secondhand marketplaces in different cities and districts for exchanging or trading secondhand goods. For example, in South Korea, there are 190 local governments (city, district, and regional) opened marketplaces in January 2009 in collaboration

Table 12.14. WHO's guideline for better hazardous waste management practices

- Separating hazardous waste from non-hazardous ones
- Storing hazardous waste materials in covered areas, to avoid possible leakage of liquid components into the environment
- Marking the waste stored in each container clearly, and to make sure that the storage is done away from the working areas and residential areas.
- Making it sure that the industries and municipalities work together to develop superior engineered landfills.
- Prohibiting mobilization of chemical constituents as leachates by covering up existing open dumps with a low permeability liner.
- Pretreatment of hazardous wastes is highly desirable, to meet the stringent criteria of the volume the toxicity level prior to disposal.

with the government. The government has developed an online information system for secondhand trading by 2012 (UN 2012a). In addition it is important to disseminate and share evidence-based environmental information, and raise public awareness on critical as well as emerging hazardous waste management issues.

There are many examples regarding stimulation of collaboration at different levels. One important shift is to the thriving network of partnerships–domestic and international, governmental and non-governmental–that help to clearly define needs, formulate priority approaches to meeting those needs, discovering new and better ways ofresponding to them, and then making broad use of new tools (USNR 2010). An array of international, national, regional, state and/or localpartnerships is key to successful implementation of more comprehensive guidelines in routine practices (Table 12.14) for achieving better hazardous waste management practices (Wyes 1997).

Many programs have been created for establishing universal membership and networks. The Change Management Process (CMP) Program, for example, enables the USEPA and the states to continue the ongoing partnership with the goals of a) representing the interests of the user community, b) tracking and implementing changes in a timely manner, c) evaluating changes prior to implementation, and d) integrating with RCRA program and development activities. One significant mechanism of the CMP program is to have a) the Coordination Group (CG) as the keystone to represent varying viewpoints of the user community, b) Expert Groups that provide valuable technical insight into major issues, and act as sounding boards to help the CG formulate its recommendations, and c) the Design and Implementation Groups to handle both major system changes and system development (CMCG 2011). As another example, Sustainable Futures is a partnership among USEPA, the chemical industry, and other stakeholders. Participation in the program allows companies to quickly and cost-effectively screen chemicals for hazards/risks early in the development process so that they can more quickly commercialize environmentally preferable new chemicals and identify safer alternatives for existing chemicals.

Participants in the training sessions also included government scientists from Australia, Europe (Poland, Germany, Slovakia, and the Netherlands) and Japan, and scientists from several consulting firms (USNR 2010). As another example, to establish strong network, UN invite relevant stakeholders (e.g., the UN regional commissions, UN organizations and bodies, other relevant intergovernmental and regional organizations, international financial institutions) to achieve sustainable development, in particular in least developed countries (UN 2012b). UN will ensure the active participation of all relevant stakeholders, drawing onbestpractices and models from relevant multilateral institutions and exploring new mechanisms to promote transparency and the effective engagement of civil society. In addition, WHO worked with international agencies, organizations and industries to promote better hazardous waste management practices, and came up with the guidelines (Table 12.14).

- USEPA's Resource Conservation Challenge (RCC) is an important national effort developed by EPA to conserve natural resources. The RCC builds on the strengths of many EPA programs and partners to help prevent pollution, reduce the use of toxic chemicals, promote recycling and reuse of materials, and conserve energy. The current focus of the RCC is fourfold: 1) achieving a national 35% recycling rate for municipal solid waste; 2) fostering beneficial reuse of secondary materials; 3) reducing priority and toxic chemicals; and 4) promoting green initiatives, with an initial focus on electronics.

Shifting to Better Tools and Strong Promotion of Science-policy Interface. One biggest shift in recent years is to manage materials and products on a life-cycle basis. Life-cycle assessment (LCA) represents an important change in how we think about hazardous waste management (USNR 2010). For example, an Economic Input-Output Life Cycle Assessment (EIO-LCA) developed by Dr. Leontief classifies all production and service industries in the entire U.S. economy into 428 sectors, and then estimates impacts resulting from additional demand in each sector. The environmental impacts associated with the monetary transactions in each sector can be estimated with software developed by the faculty at Carnegie Mellon University with the following steps (Hendrickson et al. 2006): 1) estimate output changes in final demand by sector (F); 2) assess direct and indirect economic change with input-output model (X); 3) assess environmental discharges as a result of sector output changes (E); and 4) sum sector discharges to find overall discharges. The environmental effects estimated include: resource inputs (e.g., electricity consumption, fuel use, ore consumption, fertilizer use, and water consumption) and environmental outputs (e.g., toxic emissions from the Toxics Release Inventory (TRI), toxicity-weighted chemical emissions (CMU-ET), RCRA hazardous waste generation and management, ozone depletion potential, global warming potential, conventional pollutant emissions) and the social costs of pollution. Nowadays, the method has been used in many different applications to determine the environmental sustainability and economic feasibility of different systems or designs through EIO-LCA and benefit/cost analyses.

Along with LCA, many different models or methods have been developed, which allow us to act less wastefully and consider system-wide impacts in the design, marketing, reuse, recycling, and disposal of products. One of the models is the USEPA's Waste Reduction Model (WARM). WARM can help solid waste planners and organizations track and voluntarily report greenhouse gas (GHG) emissions reductions from several different waste management practices. WARM calculates and totals GHG emissions of baseline and alternative waste management practices—source reduction, recycling, combustion, composting, and landfilling. The model calculates emissions in metric tons of carbon equivalent (MTCE), metric tons of carbon dioxide equivalent (MTCO2 E), and energy units (million BTU) across a wide range of material types commonly found in municipal solid waste (U.S. EPA 2012).

In addition, LCA provides a vehicle to bring together information and assessment to support informed decision-making. For example, the USEPA conducts economic analyses as part of its decision making process, such as strategic planning or priority setting, development of voluntary actions and regulations, and the measurement of results. Economic analyses include market studies, financial feasibility studies, and industry sector studies. LCA usually is a very useful tool in these analyses. Furthermore, science-policy interface has been advanced. One example is that ToxCast™ has been launched in 2007, which provides a cost-effective approach for prioritizing the toxicity testing of large numbers of chemicals in a short period of time, and thus, allows for incorporating risk findings with more detailed economic analyses or regulatory impact analyses (RIAs) to assess benefits of regulatory actions being considered and compare these benefits to estimated costs. ToxCast™ can use data from cutting-edge high throughput screening (HTS) bioassays developed in the pharmaceutical industry to build computational models. These modes can be used to forecast the potential human toxicity of chemicals. These hazard predictions will provide USEPA with science-based information to prioritize chemicals for more detailedtoxicological evaluations, and lead to more efficient use of animal testing (USNR 2010).

12.5.2 Examples and Case Studies: Meeting Grand Challenges

Concrete Recycle and Economic Comparison. There is a huge generation of construction waste that has led to a space constraint in landfill areas. One of the best ways to deal with this problem is to promote recycling of construction waste. As concrete waste accounts for the major source of construction solid waste that contributes to 50% of the total consumption waste. Recycling concrete waste is the best option to reduce quantities of construction waste (Tam 2008). The case study presents the cost and benefit of conventional practice of dumping the construction waste to landfills and utilizing them to produce new natural materials for new concrete production and the modern concrete recycling method to recycle the construction waste as aggregate for new concrete production

Figure 12.5 shows the flow chart for conventional recycling method. In this method, the on-site construction waste is dumped to landfills and new products

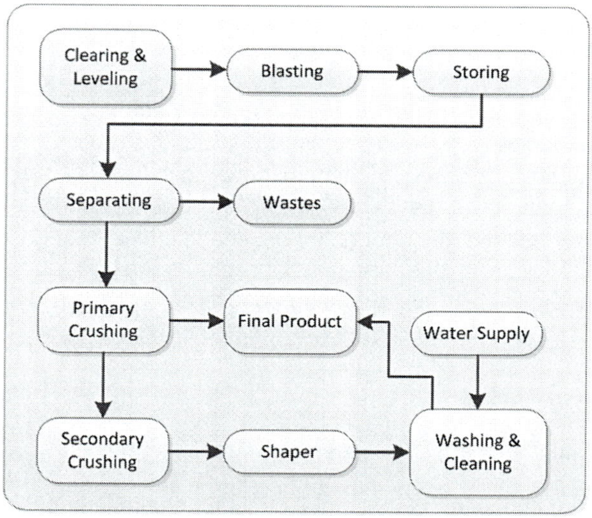

Figure 12.5. The conventional method for handling on-site construction waste

are produced from rocks supplied on the construction site for new concrete production. Figure 12.6 shows the flow chart for the modern sustainable concrete recycling method. In this method, concrete is recycled as an aggregate for new concrete production. The concrete recycling method is more beneficial than the conventional method. The recycling method receives a positive net benefit of about $30,916,000 per year while the conventional method receives a negative net benefit of about –$44,076,000 per year (Tam 2008). It is found that the concrete recycling method can result in a huge sum of saving. The benefits resulting from the recycling method can balance the cost expended for the current practice.

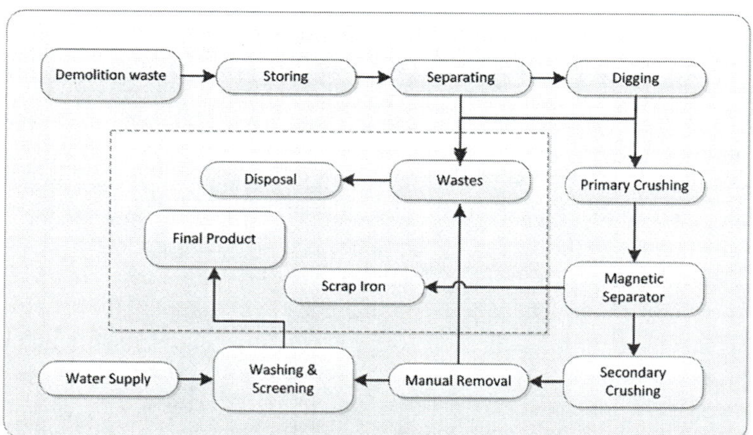

Figure 12.6. Flow chart of the sustainable concrete recycles method

Therefore, recycling concrete waste for new production is a cost effective method that also helps in protecting the environment and achieves construction sustainability.

Bioenergy from Landfills. The use of landfills currently represents the most common practice for waste disposal, and about 84–90% of municipal solid waste (MSW) is disposed of in landfills (Shan et al. 2010). In landfills, landfill gas (LFG) is produced during anaerobic decomposition of organic waste. LFG is of great concerns because it is hazardous for three key reasons: a) LFG becomes explosive when it escapes from the landfill and mixes with oxygen; b) the methane is 20 times more potent as a greenhouse gas than carbon dioxide; and c) LFG contains volatile organic compounds (VOCs) that can contribute to the formation of photochemical smog.

One sustainable solution is to collect LFG to generate energy, which not only reduces greenhouse emissions into the atmosphere but also generates revenues for operators and local governments.Currently, many different ways are usedfor energy recovery from FLG (Shan et al. 2010). In 1995, methane emissions from landfills were estimated to be 216.1 Teragrams (Tg) of CO_2 equivalents. In 1997, these emissions were 207.5 Tg of CO_2 equivalents, and by 2001 they had been reduced to 202.9 Tg of CO_2 equivalents. The USEPA estimated that, between 1995 and 2001, 253 Tg of CO_2 equivalents were used in landfill-gas-to-energy projects, while 350.3 Tg of CO_2 equivalents was flared (U.S. EPA 2004). In these projects the LFG is used as a direct fuel in industrial processes or as the fuel that runs electricity-generating equipment. LFG can be used to produce bio-methane by removing the CO_2 from the LFG. LFG can also be used to produce: a) liquid hydrogen (LH_2) by steam reforming (SMR), partial oxidation, pyrolysis, autothermal pyrolysis, and autothermal SMR followed by cryogenic separation or pressure swing adsorption (PSA) purification; b) synthetic natural gas via methanation; c) methanol via methanol synthesis; and d) synthetic hydrocarbons via Fischer-Tropsch (FT) synthesis (Muradov and Smith 2008).

However, to make LFG energy recovery projects be more successful, non-technical barriers must be overcome, such as awareness of LFG energy recovery and risk perception by potential end-users and financiers, project and energy supply economics, market access, finance, planning, permitting, licensing, and bureaucracy (He et al. 1997). Moreover, there is a need to increase the public awareness of LFG utilization as a valuable, free and renewable energy resource, make related policy, regulations or laws to ensure efficient utilization of LFG, and encourage and stimulate the operators of landfills and the potential building owners to use the energies produced by the LFG trigeneration plants off-site the landfills. In particular, the government could tie the gap between the LFG operators and the potential building owners (Shan et al. 2010).

Carbon Capture and CO_2 Beneficial Use. Global energy models suggest that with current global coal use patterns, it will not be possible to stabilize atmospheric greenhouse gases (GHG) concentrations at acceptable levels. Therefore, it is necessary to make major reductions in GHG emissions. To mitigate climate changes, we must aggressively pursue carbon capture, storage and sequestration (CCS) from fossil fuel power plants, other industries, and places where CCS technology can be used.

There are many carbon capture technologies. These technologies can be categorized as a) physical/chemical and biological technologies and b) technologies for carbon capture from concentrated point sources and mobile/distributed point- or non-point sources. In general, on-site capture is the most viable approach for large sources and initially offers the most cost-effective avenue to sequestration. For mobile/distributed sources like cars, on-board capture at affordable cost would not be feasible, but are still needed (Zhang and Surampalli 2012). Usually, EIO-LCA should be considered for processes or concepts to ensure that additional CO_2 is not produced, and extra energy not consumed, due to CO_2 reuse processes.

Efforts for CO_2 beneficial uses focus on using captured CO_2 for value-added products (e.g., chemicals, cements, or plastics) or beneficial activities so that a portion of the CO_2 capture costcan be offset. Many different pathways/reuse technologies have been explored, in which CO_2 is either being used a) as a substrate for boosting production (e.g., algae cultivation) or an additive for enhancing the processes (extraction of alumina from bauxite residue), b) as a feedstock for synthesizing stable product (e.g., for urea yielding boosting, polymer processing, liquid fuel production), c) for integration into pre-existing products (e.g., in concrete curing, carbonate mineralization), or d) as a working fluid for enhancing other activities (e.g., CO_2 enhanced oil recovery, CO_2 enhanced coal-bed methane) (Zhang and Surampalli 2012). The global CO_2 reuse market currently amounts to ~80 million tonnes/year, of which ~50 million tonnes per year are used for EOR at a price of $15–19/tonne. Potentially, the global supply of anthropogenic CO_2 is ~500 million tonnes of low-cost (<$20/tonne) and high concentration CO_2; at a much higher cost ($50–100/tonne), around 18,000 million tonnes per year could also be captured for CO_2 reuse (PB and GCCSI 2011). Advances of CO_2 reuse technologies depend on future carbon restrictions and prices and their interaction with other CCS technologies.

However, there's a huge gap between what can technically do and what we are doing. High costs, inadequate economic drivers, remaining uncertainties in the regulatory and legal frameworks for CCS deployment, and uncertainties regarding public acceptance are barriers to large-scale applications of CCS technologies in the world (CCCSRP 2010). Therefore, it is imperative to overcome the technical, regulatory, financial and social barriers.

Converting Waste into Energy and Value-Added Products. One important new concept in sustainable hazardous waste management is to convert waste into energy and/or value-added products. Nowadays, we can produce ethanol from municipal wastewater sludge, wastepaper containing cellulose in combination with lignin, and beer brewery wastewater. We also can produce biofuel or chemical products from olive mill wastewater. For example, a "white rot" basidiomycete fungus, *Pleurotussajor-caju*, was used to reduce phenolic compounds by 50–68% from olive mill wastewater. Furthermore, *P. sajor-caju* pretreated olive mill wastewater enhanced the ethanol production of 14.2 g/L after 43 h of yeast fermentation (Massadeh and Modallal 2008). Another example is the beneficial reuse of sludge generated by wastewater treatment plants for value-added products, such as construction aggregates, adsorbents, fuels, biotechnological products (e.g., biopesticides, bioherbicides, enzymes, bioplastics, and bioflocculants) and miscellaneous products like animal feed, secondary metabolites and electricity. Stricter regulations imposed on sustainable sludge management in different countries are catalyzing the re-orientation of sludge to value-addition. Here, a value-added product implies that the value of the final product should exceed the cost of processing.The emerging value-added products are at different levels of production, facing challenges in terms of process conditions, yield, scale-up or formulation (Tyagi et al. 2009).

Here, we briefly introduce *Bacillus thuringiensis* (Bt) based biopesticides production. Bt-based biopesticides are the most successful biopesticide as of today being used actively in agriculture, forestry and the public health sector. The Bt based biopesticides were first commercially used in the name of "sporeine" in 1938 in France. Conventional Bt formulations are based on traditional strains of Bt var. *kurstaki* due to strain stability.Recently, inexpensive zero-cost wastewater and/or wastewater sludge has been used to produce Bt-based biopesticides. Intensive studies have been conducted in different places such as the one at the University of Quebec (Figure 12.7).

The production process consists of the following steps: a) sludge fermentation, b) product recovery/harvesting, c) product formulation.The Bt-INRS process has shown immense potential for this process and the alternative is certainly **BENEFICIAL** with the acronym suitably defined as: **B**uffering capacity; **E**ntomotoxicity, extracellular polymeric substances and enzymes production; **N**utrient

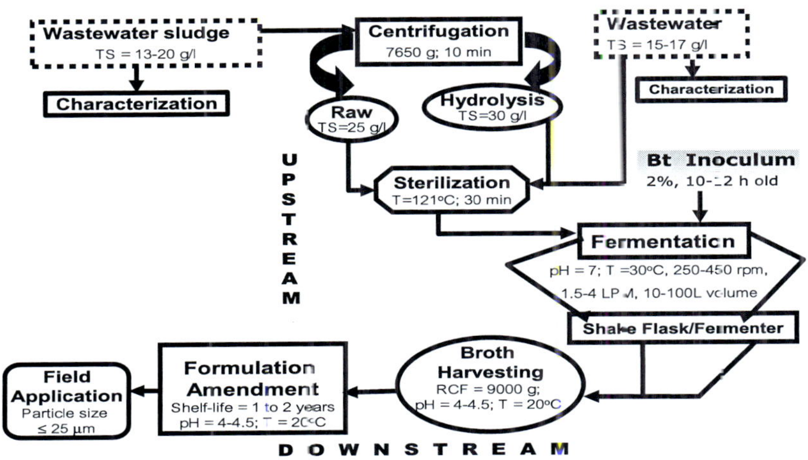

Figure 12.7. Bacillus thuringiensis *biopesticides production trainwith wastewater/wastewater sludge as raw materials (RCF = relative centrifugal force; TS = total solids; T = temperature)*

source; **E**conomical; **F**locs (aid in harvesting and UV protection); **I**nherent source of many natural features; **C**hromopheric compounds (UV protection); **I**ndicators of; **A**pplication and **L**onger shelf and field life (Brar et al. 2009).

Sustainable Pesticides Management. Pesticides management for controlling their environmental contamination is very complex and full of challenges. At least, sustainable pesticides management requires a) implement of Integrated Pest Management (IPM) and Best Management Practices (BMPs), b) understanding of landscape capacity (or vulnerability) to contamination from leaching and runoff at a watershed and/or regional scale, and c) knowledge/modeling to evaluate the effects of conservation behaviors and management practices on concentration and transport of pesticides (Zhang et al. 2009). IPM combines cultural, biological, genetic, and chemical controls to form a comprehensive program for managing pests.

Implementation of IPM can reduce pesticide use to the minimum amount which is, however, still enough to produce high quality food, thereby maximizing farmers' profits (Waskom 1995). In addition to IPM, proper pesticide application, handling and disposal of pesticides, and pesticide selection are also the essential parts of BMPs. Currently, an important part of BMPs is implemented to control the sources of pesticides and minimize their pollution. Proper storage, handling, and disposal of pesticides can greatly reduce the risk of point source contamination to water resources. Other BMPs for protecting the water resources from the pesticide contamination include conservation tillage and conservation buffers practices. However, an information gap needs to be filled of evaluating the effects of management practices (e.g., conservation behaviors, IPM and BMP implementation) on concentration and transport of pesticides. Without this information, it will be very difficult to achieve anything simply relying on monitoring programs and the Total Maximum Daily Load (TMDL) recommendations.

Watershed vulnerability to pesticide contamination is determined by a combination of physical setting (e.g., soils, topography and climate) and management practices for land (e.g., BMPs) and pesticides applications (e.g., IMP). Currently, the USDA-NRCS SSURGO data sets, with scales of 1:12,000 to 1:63,360, provide information for exploring regional processes and relationships and reflect what one might expect to find at a field scale. For example, Shea et al. (2006) used SSURGO and the USGS/USEPA National Hydrography Datasets (NHD), and pesticide property data as model inputs to assess landscape vulnerability to pesticide leaching and runoff. To include the behavioral dimension of farm conservation and environmental farming into sustainable pesticides management, some theories have been developed such as *Theory of Planned Behavior* (Ajzen 2002) and *Metaeconomic Theory* (Lynne 2002). Important behavioral areas are those related to implementation of IMP and BMPs for pesticide applications and land conservation. Willock et al. (1999) made efforts to integrate the spatial and temporal data with personality characteristics and behavioral expectations being summarized across a resident population in an area. In many cases, a survey tool is usually used to determine the practices, motivations and behaviors of producers in vulnerable areas; a statistical model is used to predict their responsiveness to change, and decision typologies were mapped. Georeference of the behavioral dimension (i.e. mapping personological typologies) is deemed as essential to effective policy implementation and control of pesticides (or other pollutant) contamination.

12.6 CONCLUSIONS

Hazardous wastes continue to pose a significant problem throughout the world today in terms of their generation, their increasing quantity and toxicity. In the meantime, government regulations to restrict their release and disposal are becoming more stringent. Environmental audits act as effective tools in identifying environmental problems at the origin of waste production. Pollution prevention methods are also effectively pursued to reduce the quantum of hazardous waste generation. Separation and destruction technologies can play a role in decontamination of waste streams. On the other hand, decontamination and discharge may not be the long term solution, since many of the conventional technologies are inefficient in managing trace contaminants. Waste stream recycle and pollution prevention may be more attractive, especially as discharge limits continue to decrease and as design criteria tend to become more stringent. As a consequence, there are more opportunities for the development of innovative technologies that are more effective. Biological waste treatment processes exhibit considerable potential due to their low cost and relatively benign nature. Microorganisms will continue to play an important role in wastewater treatment through their immense potential for immobilization and bio-accumulative properties. Their abilities to oxidize and reduce various species must be exploited in liberating and recycling metal species. Waste-recycle and process integration will become pertinent, and they often utilize existing equipment and capabilities. Though specialized unit operations can provide new and improved alternatives, it is the overall system performance that governs the effectiveness of any specific waste management plan. Process synthesis and integration will play an important role in the area of pollution prevention, but it is imperative to have an overall strategy and clear vision for each application. Generally, the sustainable waste management approach will convert wastes to assets, reduce resource consumption and production costs and provide more robust and competitive production and marketing networks. Improvements in waste management plan will enhance product recovery, promote reagent reuse, and these factors play an important role in the overall plan.

References

3M Sustainability. (1975). "Sustainability in-depth: The pollution prevention pays (3P) program." ⟨http://www.mmm.com/sustainability⟩ (Oct. 5, 2015).

Ajzen, I. (2002). "Perceived behavioral control, self-efficacy, locus of control, and the theory of planned behavior." *J. Appl. Social Psychol.*, **32**(4), 665–683.

Albino, V., Cioffi, R, Santoro, L., and Valenti, G. L. (1996). "Stabilization of residue containing heavy metals by means of matrices generating calcium trisulphoaluminate and silicate hydrates." *Waste Manage. Res.*, **14**(1), 29–41.

Arroyo, J., Rivas, F., and Lardinois, I., eds. (1999). *Solid waste management in Latin America —The role of micro- and small enterprises and cooperatives*, Urban Waste Series 5, Gouda, The Netherlands.

Batstone, R., Smith, J. E., and Wilson, D. (1989). "The safe disposal of hazardous wastes: The special needs and problems of developing countries." World Bank, Washington, DC.

Blackman, Jr., W. C., (1996). *Basic hazardous waste management*, CRC Press, Boca Raton, FL.

Boopathy, R. (2000). "Factors limiting bioremediation technologies." *Bioresour. Technol.*, **74**(1), 63–67.

Brar, S. K., Verma, M., Tyagi, R. D., Valéro, J. R., and Surampalli, R. Y. (2009). "Biopesticides-bacillus thuringiensis." *Sustainable sludge management: Production of value added products*, R. D. Tyagi, R. Y. Surampalli, S. Yan, T. C. Zhang, C. M. Kao, and B. N. Lohani, eds., ASCE, Reston, VA.

CCCSRP (California Carbon Capture and Storage Review Panel). (2010). "Findings and recommendations by the California carbon capture and storage review panel." ⟨www.climatechange.ca.gov/carbon_capture_review_panel/.climatechange.ca.gov/carbon_capture_review_panel/⟩ (Feb. 1, 2012).

Christensen, T. H., et al. (1994). "Attenuation of landfill leachate pollutants in ground water." *Crit. Rev. Env. Sci. Technol.*, **24**(2), 119–202.

CMCG (RCRAInfo Change Management Coordination Group). (2011). "RCRA info change management process." Washington, DC.

Conner, J. R., and Hoeffner, S. L. (1998). "The history of stabilisation/solidification technology." *Crit. Rev. Env. Sci. Technol.*, **28**(4), 325–396

Cutter, S. L. (1983). *Living with risk: The geography of technical hazards*, Edward, Arnold, London.

Darcy, A. (1856). *Les Fontaines publiues de la ville de Dijon*, Dalmont, Paris.

Dawson, G. W., and Mercer, B. W. (1986). *Hazardous waste management*, Wiley, New York.

Glasser, F. P. (1997). "Fundamental aspect of cement solidification and stabilization." *J. Hazard. Mater.*, **52**(2–3), 151–170.

Greenstone, M., and Justin, G. (2006). "Does hazardous waste matter? Evidence from the housing market and superfund program." MIT Dept. of Economics, Cambridge, MA.

He, C., Herman, D., Minet, R. G., and Tsotsis, T. T. (1997). "A catalytic/sorption hybrid process for landfill gas cleanup." *Ind. Eng. Chem. Res.*, **36**(10), 4100–4107.

Hendrickson, C. T., Lave, L. B., and Matthews, H. S. (2006). "Environmental life cycle assessment of goods and services." Resources for the Future, Washington, DC.

Herdman, R. C. (1989). "The congressional office of technology assessment." *JAMA.*, **262**(4), 550.

Hoornweg, D., and Thomas, L. (1999). "What a waste: Solid waste management in Asia." *Working paper 1, Urban and Local Government*, World Bank, Washington, DC.

Jenkins, R. R., Kopits, E., and Simpson, D. (2008). "Policy monitor-the evolution of solid and hazardous waste regulation in the Unites States." *Rev. Environ. Econ. Policy*, **3**(1), 104–120.

Kassim, T. A. (2005). *Environmental impact assessment of recycled wastes on surface and ground waters: Engineering modeling and sustainability*, Springer Science & Business Media, NewYork.

Kurniawana, T. A., Gilbert, Y. S., Chana, W. H. L., and Sandhya, B. (2006). "Physico-chemical treatment techniques for wastewater laden with heavy metals." *Chem. Eng. J.*, **118**(1–2), 83–98.

LaGrega, M. D., Buckingham, P. L., and Evans, J. C. (2001). *Hazardous waste management*, 2nd Ed., McGraw-Hill, New York.

Lehman, J. P. (1982). *Hazardous waste disposal*, Plenum Press, New York, 396.

Lidskog, R. (1998). "Social aspects of the siting of facilities of hazardous waste management." *Waste Manage. Res.*, **16**(5), 476–483.

Lynne, G. D. (2002). "Agricultural industrialization: A metaeconomics look at the metaphors by which we live." *Rev. Agric. Econ.*, **24**(2), 410–427.

Massadeh, M. I., and Modallal, N. (2008). "Ethanol production from olive mill wastewater (OMW) pretreated with Pleurotussajor-caju." *Energy Fuels*, **22**(1), 150–154.

Millano, E. F. (1996). "Hazardous waste: Storage, disposal, remediation and closure." *Water Environ. Res.*, **68**(4), 586–608.

Misra, V., and Pandey, S. D. (2005). "Hazardous waste, impact on health and environment for development of better waste management strategies for future in India." *Environ. Int.*, **31**(3), 417–431.

Muradov, N., and Smith, F. (2008). "Thermocatalytic conversion of landfill gas and biogas to alternative transportation fuels." *Energy Fuels*, **22**(3), 2053–2060.

Nema, A. K., and Modak, P. M. (1998). "A strategic design approach for the optimization of regional hazardous waste management systems." *Waste Manage. Res.*, **16**(3), 210–224.

PB (Parsons Brinkderhoff) and GCCSI (Global CCS Institute). (2011). "Accelerating the uptake of CCS: Industrial use of captured carbon dioxide." Docklands, Australia.

Poon, C. S., Qiao, X. C., and Lin, Z. S. (2004). "Effects of flue gas desulphurization sludge (FGD) on the pozzolanic reaction of reject fly ash blended cement pastes." *Cem. Concr. Res.*, **34**(10), 1907–1918.

Rabel, A., Spadaro, J. V., and McGavron, P. D. (1998). "Health risks of air pollution from incinerator; a perspective waste management." *Waste Manage. Res.*, **16**(4), 365–388.

Ramseur, J. L., and Mark Anthony, R. (2006). "Overview and selected issues." *Congressional Research Service Report for Congress*, Congressional Research Service (CRS), Washington, DC.

Shan, G. B., Zhang, T. C., and Surampalli, R. Y. (2010). "Bioenergy from landfills." *Bioenergy and bioefuel from biowastes and biomass*, S. K. Khanal, R. Y. Surampalli, T. C. Zhang, B. P. Lamsal, R. D. Tyagi, and C. M. Kao, eds., ASCE, Reston, VA.

Sharma, H. D., and Lewis, S. P. (1994). *Waste containment systems, waste stabilization, and landfills: Design and evaluation*, Wiley, New York.

Shea, P. J., Maribeth, M., Martin, A. R., Lynne, G. D., and Burbach, M. E. (2006). "Targeting watershed vulnerability and behaviors leading to adoption of conservation management practice." U.S. Dept. of Agriculture, Washington, DC

Shedkar, A. V. (2009). "Sustainable solid waste management: An integrated approach for Asian countries." *Waste Manage*, **29**(4), 1438–1448.

Sinclair-Desgagné, H., and Landis Gabel, H. (1997). "Environmental auditing in management systems and public policy." *J. Environ. Econ. Manage.*, **33**(3), 331–346.

Snel, M., and Ali, M. (1999). "Stakeholder analysis in local solid waste management schemes." WELL/WEDC, London/Loughborough.

Tam, V. W. Y. (2008). "Economic comparison of concrete recycling: A case study approach." *Resour. Conserv. Recycl.*, **52**(5), 821–828.

Tusa, W. (1990). "Developing an environmental audit program." *Risk Manage.*, **37**(8), 60–67.

Tyagi, R. D., Surampalli, R. Y., Yan, S., Zhang, T. C., Kao, C. M., and Lohani, B. N., eds. (2009). *Sustainable sludge management: Production of value added products*, ASCE, Reston, VA.

UN (United Nation). (2012a). "Waste management, Korean national report." ⟨www.un.org/esa/dsd/dsd_aofw_ni/ni_pdfs/NationalReports/korea/WasteManagement.Pdf⟩ (Nov. 1, 2012).

UN (United Nation). (2012b). "Modalities for the sixth high-level dialogue on financing for development."

U.S. EPA (U.S. Environmental Protection Agency). (1986). "Report to congress: Minimization of hazardous waste." EPA/530-SW-86-033A, Washington, DC.

U.S. EPA (U.S. Environmental Protection Agency). (1989). "Pollution prevention benefits manual." Rep. No. WAM-1, PPIC, Falls Church, VA.

U.S. EPA (U.S. Environmental Protection Agency). (1991). "Pollution prevention 1991: Progress on reducing industrial pollutants." Washington, DC.

U.S. EPA (U.S. Environmental Protection Agency). (2003). "One cleanup program: Working in concert to make our cleanups better." ⟨http://www/epa.gov/oswer/onecleanupprogram/docs/OCPFactSheet.pdf⟩ (May 2, 2012).

U.S. EPA (U.S. Environmental Protection Agency). (2004). "Inventory of U.S. Greenhouse gas emissions and sinks: 1990–2002." Office of Atmospheric Programs, Government Printing Office, Washington, DC.

U.S. EPA (U.S. Environmental Protection Agency). (2007). "Municipal solid waste generation, recycling and disposal in the United States: Facts and figures for 2006." ⟨www/epa.gov/oswer/non-hw/muncpl/pubs/msw06.pdf⟩ (May 2, 2012).

U.S. EPA (U.S. Environmental Protection Agency). (2008a). "Fiscal year 2007 superfund annual report." ⟨www/epa.gov/superfund/accomp/pdfs/sf_annual_report_2008.pdf⟩ (May 18, 2012).

U.S. EPA (U.S. Environmental Protection Agency). (2008b). "Land revitalization initiative." ⟨http://www.epa.gov/swerrims/landrevitalization.lrso.htm⟩ (May 28, 2012).

U.S. EPA (U.S. Environmental Protection Agency). (2012). "Waste reduction model." ⟨http://www.epa.gov/climatechange/waste/calculators/Warm_home.html⟩ (Nov. 18, 2012).

USNR (U.S. National Report). (2010). "The United States of America national report: Transport, chemicals, waste management, mining, and sustainable consumption and production." ⟨http://www.un.org/esa/dsd/dsd_aofw_ni/ni_pdfs/NationalReports/usa/Full_text.pdf⟩ (Nov. 18, 2012).

Visvanathan, C. (1996). "Hazardous waste disposal." *Resour. Conserv. Recycl.*, **16**(1–4), 201–212.

Waskom, R. M. (1995). "Best management practices for agricultural pesticide use." Colorado State-Univ., Fort Collins, CO.

Willock, J., et al. (1999). "The role of attitudes and objectives in farmer decision making: Business and environmentally-oriented behaviour in Scotland." *J. Agric. Econ.*, **50**(2), 286–303.

Wyes, H. W. (1997). "Hazardous wastes: Its impact on human health in Europe." *Toxicol. Ind. Health*, **13**(2–3), 109–119.

Zhang, T. C., and Surampalli, R. Y. (2012). "Carbon capture and storage for mitigating climate changes." *Greenhouse gas emissions and climate change: Modeling, mitigation and adapting*, R. Surampalli, T. C. Zhang, C. S. P. Ojha, R. D. Tyagi, and C. M. Kao, eds., ASCE, Reston, VA.

Zhang, T. C., Lai, K. C. K., and Surampalli, R. (2009). "Pesticides." *Contaminants of emerging environmental concern*, A. Bhandari, R. Surampalli, C. D. Adams, P. Champagne, S. K. Ong, R. D. Tyagi, and T. C. Zhang, eds., ASCE, Reston, VA.

CHAPTER 13

Biohydrometallurgical Processing of Metallic Components of E-Wastes

Marjorie Valix[1]

13.1 INTRODUCTION

The revolution in information and communication technology (ICT) has brought huge technical benefits and wealth, but has created a major global problem: the phenomenal generation of vast amounts of electronic waste, or 'e-waste' through product obsolesce. Waste electrical and electronic equipment (WEEE) including mobile phones, printers computer and their peripherals, popularly known as 'e-wastes', have become a significant threat to both the ecology and human health worldwide. The rapid advance in information technology and the relentless desire of consumers for the "state of the art" products have led to an accelerated obsolescence of electronic goods. Without the systems and the technology in place to effectively capture and manage these wastes, the accumulation of e-wastes has become a pervasive problem at a local, regional and national level. The problem with e-waste is not just its growing volume but also its toxicity, and its content of valuable resources (e.g., gold, copper) which are lost when e-waste is disposed (Barba-Gutierrez et al. 2008; Chancerel and Rotter 2009; Tuncuk et al. 2012; Oleszek et al. 2013; Zheng et al. 2013). The average 12% annual growth of e-waste is about three times that for municipal waste streams (Pariatamby and Victor 2013; Xu et al. 2013). They contain toxic components such as brominated flame retardant, lead and mercury, and intricate structures of glass, ceramic and polymeric composites that currently make it extremely difficult to extract valuable materials (Barba-Gutierrez et al. 2008; Komilis et al. 2013; Feldt et al. 2014). Millions of tonnes of these toxins are released into the air, water and soil each year from landfilling, incineration, and chemical and thermal treatment in unregulated recycling processes (Townsend 2011; Zhang et al. 2013); these technologies are expensive, achieve only partial reclamation, and when unregulated are hazardous for e-workers (Robinson 2009). In addition, e-waste continues to be shipped to

[1]Univ. of Sydney, NSW, Australia.

underdeveloped countries, which have lower environmental standards, either illegally or under the guise of international 'aid', despite the Basel Convention, an international treaty that aims to reduce the movements of hazardous waste between nations (Wei and Liu 2012; Pariatamby and Victor 2013; Sthiannopkao and Wong 2013). Unregulated and ad-hoc, low-tech methods of processing e-waste have exacerbated the environmental and health risks in these countries (Chi et al. 2011; Akormedi et al. 2013; Sthiannopkao and Wong 2013; Tue et al. 2013).

International environmental policies and the escalating issues of e-waste placed significant pressures on each country globally to develop a national position on e-waste management. The management of e-waste is currently approached via two principles: Extended Product Responsibility (EPR), which shifts the environmental costs associated with products throughout their life cycles to producers and Product Stewardship, which suggests all stakeholders, producers, sellers and consumers, have roles and responsibility for the full environmental and economic impacts of the product (Wilts et al. 2011; Nash and Bosso 2013; Wagner 2013). EPR now underpins most of the environmental policies in the OECD countries (Kiddee et al. 2013). The most stringent of these are the EU's Waste Electrical and Electrical and Electronic Equipment (WEEE) (European-Parliamentary-Council 2002b) and Restriction of Hazardous Substances (RoHS) (European-Parliamentary-Council 2002a). These directives cover both the end-of-life management and use of hazardous materials in production. These legal initiatives to promote the recycling of e-waste are an important acknowledgement of the issue, but need to be supported by technological solutions to make a substantive impact. Mandates, on their own, however cannot solve this problem without the technological solutions and markets for recycled products that will enable mandated recycling targets to be met. In the absence of technological solutions, it is well established that environmental mandates simply perpetuate other insidious problems—e.g., e-waste exportation (Wei and Liu 2012; Pariatamby and Victor 2013; Sthiannopkao and Wong 2013). The challenge is to develop sustainable recycling technologies to address the volume and complexity of this waste using cost-effective and ecologically sensitive methods (Dalrymple et al. 2007). In the light of these concerns, the search for simple, cost effective and environmentally friendly method for reclaiming value added materials and energy from e-waste is an urgent goal.

End-of-life programs by reusing and repairing end of use electronic units is challenged by credible markets (de Oliveira et al. 2012; Herat and Agamuthu 2012; Ongondo et al. 2013). Various technologies including pyrometallurgical, hydrometallurgical and bioprocessing of electronic wastes (Cui and Zhang 2008; Lee and Pandey 2012; Tuncuk et al. 2012) and hybrid technologies combining both hydrometallurgical and bioprocessing methods (Pant et al. 2012) have been used in recycling e-waste. It is apparent that despite significant work that has been achieved in these areas, these processes continue to be confronted by economic, technical and ecological issues.

Bioleaching offers a promising technology. This technology uses naturally occurring biological micro-organisms and their metabolic products in extracting valuable metals from the waste such spent catalyst and electronic waste (Brandl et al. 2008; Mishra et al. 2008; Valix 2001b). Biologically assisted degradation of wastes has tremendous potential as a recycling technology. The technology is environmentally sound with a huge potential to lower operational cost and energy requirements. The wide range of waste treated by bioleaching is a clear indication of the ease of implementation of the process (Mishra et al. 2009).

This chapter presents a review on the use of biohydrometallurgical technology in the recovery of metals from e-wastes specifically focussing on the extraction of copper. This discussion will include an analysis of bioleaching of copper from printed circuit boards with particular emphasis on the use of chemolithoautotrophs bacteria and heterotrophic fungi, the leaching chemistry, secondary reactions and kinetics of the chemical and bioleaching processes.

13.2 END-OF-LIFE PROGRAMS FOR E-WASTE

The biggest challenge in reclaiming e-waste is developing a process that can adapt to the complexity, changing compositions, toxicity and volume of e-waste. Current management of e-waste can involve reusing and/or recycling.

Reusing of end-of-use (or still working) units involves repairing, reconditioning and refurbishing. These are the lowest cost options as they allow the original resources within the unit to be retained. However these practices are challenged by the lack of credible and stable markets for refurbished products (de Oliveira et al. 2012; Herat and Agamuthu 2012; Ongondo et al. 2013).

Recycling by material reclamation of end-of-life (or uneconomical to repair) units involve: (1) physical processing (dismantling, sorting and shredding); (2) thermal processing (pyrolysis and pyrometallurgy) and (3) hydrometallurgical processing (chemical treatment) (Dalrymple et al. 2007). Because e-wastes are complex mixtures of materials, these reclamation processes are characterised by high chemical and energy demands and low recycling efficiencies (Tanskanen 2013). Current best physical recycling practices can recover only 20% of the components and chemical treatments cannot reclaim plastics. Toxic secondary products from recycling operations are unacceptable as feedstock to existing processes, e.g., mineral processing operations and their disposal in landfills results in adverse environmental problems. These issues are magnified by huge volumes of e-waste.

Although the technologies for recycling of relatively pure forms of metals including steel, aluminum, copper, zinc, lead and nickel are well established, recycling metals from e-wastes is greatly more challenging. There is often reluctance by smelters and scrap metal buyers to recycle e-waste. This is attributed to complexity, composition and the presence of toxic materials in the waste. Only a limited number of thermal processes are available for formal processing of the metallic components of e-waste worldwide. This include Aurubis smelter in Germany (Kahhat and Williams 2009), Noranda copper smelting in Quebec, Canada and Ronnskar smelter in Sweden (Cui and Zhang 2008). These pyrometallurgical processes are challenged in upgrading the final metal products because of the nature of the metallic feedstock, which in e-wastes consist of both pure metals and alloys. Metals that are in their pure forms are easily processed by melting in smelters (Reck and Graedel 2012). However the stability of the alloys and their separation behavior during melting makes the recovery of the pure metals by thermochemical methods more energy intensive and relatively difficult or impossible (Nakajima et al. 2010). Upgrading of the final metals often necessitates the subsequent use of hydrometallurgical processing (Cui and Zhang 2008). In addition smelting of electronic waste can generate dioxin arising from the thermal decomposition of the plastic components of e-wastes (Kahhat and Williams 2009). Heavy investments by thermal processes are therefore required to manage gaseous emissions resulting from e-waste re-processing.

Unfortunately an alternative route for processing the metallic components of e-waste is through informal or un-regulated processes. Often conducted in open burning sites (Tsydenova and Bengtsson 2011; Chi et al. 2011; Labunsk et al. 2013) these emit large quantities of toxic heavy metals (Hg, Zn) and polychlorinated dibenzodioxins and dibenzofurans resulting in both unprecedented and long term ecological and health problems (Cui and Zhang 2008; Kahhat and Williams 2009; Tsydenova and Bengtsson 2011).

Interest, on the other hand in using hydrometallurgical methods to process e-waste has increased because of its simplicity, the exactness of the technology and the requirement of lower capital investment and operating costs in comparison to pyrometallurgical methods (Oishi et al. 2007). Hydrometallurgical methods involve the dissolution of the metallic fractions of e-wastes in either acidic or alkali solutions, the nature of which has been driven by the recovery of precious metals (Cui and Zhang 2008). The selective recovery of pure metals is a potential advantage for these processes (Dalrymple et al. 2007). Leaching reagents that have been used include cyanide, thiourea, thiosulfate and halide solutions (Cui and Zhang 2008; Tuncuk et al. 2012; Zhang et al. 2012). Leaching is followed by conventional metal recovery methods including precipitation, solvent extraction, adsorption and ion-exchange and electrowinning (Cui and Zhang 2008; Rimaszeki et al. 2012; Robotin et al. 2012; Coman et al. 2013). Although the technical feasibility of using hydrometallurgical routes have been proven, the economics of processing and the environmental impact, in particular of using toxic reagents such as cyanide and thiourea, continue to be an issue (Cui and Zhang 2008; Tuncuk et al. 2012).

Bioleaching of electronic scrap has become a rapidly-expanding area of research in the last decade (Brandl and Wegmann 2001; Cui and Zhang 2008; Adhapure et al. 2013). The primary organism that have been used are the chemolithoautotrophs bacteria (Ilyas et al. 2010; Hong et al. 2013b) and heterotrophic fungi (Brandl and Wegmann 2001; Valix et al. 2012) via normal metabolic processes. Bioleaching has numerous environmental and economic benefits that arise from processing that avoids high temperatures, high pressures and hazardous materials. The process is further discussed in the next section.

13.3 E-WASTE CHARACTERISATION AND RESOURCE VALUATION

The composition of electronic scrap varies widely with age, origin and manufacturer and as such it is impossible to give precise average compositions (Cui and Zhang 2008). Photographs of various electronic wastes are shown in Figure 13.1. The various compositions of e-wastes quoted by various researchers are given in Table 13.1. Approximate analysis presents 60–70% of general electronic scrap as non-metals and 30–40% as metals (Brandl et al. 2008; Brandl and Wegmann 2001). Some studies have specifically targeted printed circuit boards (PCB), and have cited the same proportion of non-metals to metals (Huang et al 2009; Park and Fray 2009). There is also agreement among the studies of PCB that copper is the most abundant metal (10–30%); other base metals such as nickel, lead, tin, aluminium and iron might occupy 1–10% of the total weight, and that precious metals (gold, silver and palladium) are <1% (Ilyas et al. 2007; Huang et al. 2009; Yang et al. 2009). A summary of compositions of various e-wastes are given in Table 13.1. As a short note of interest, Cui and Zhang (2008) and Park and Fray (2009) examined the value distribution of various metals in electronic scrap. They both observe that the precious metals, while in very low concentration, have the highest price to weight ratio, followed by the base metals (of these, copper and zinc are the most prized). Despite this there is a fairly even distribution of research on recovery of all metals from electronic scrap.

PCB Copper rich waste

Figure 13.1. Photographs of electronic waste

Table 13.1. Metallic Components of E-wastes

E-waste	*Al*	*Fe*	*Cu*	*Ni*	*Pb*	*Co*	*Zn*	*Sn*	*Au*	*Ag*	*Pd*	*References*
	Weight (%)								*ppm*			
General electronic waste	23.7	—	8.0	15	2.0	—	26	23	—	—	—	Brandl and Wegmann (2001)
General electronic waste	2	8	20	2	2	—	—	—	1000	2000	50	Cui and Zhang (2008)
PCB	0.71	8.3	8.5	2	3.15	—	8.0	0.0007	12	29	—	Ilyas et al. (2007)
PCB	—	8	20	2	2	—	1	4	1000	2000	—	Huang et al. (2009)
PCB	5	5	16	1	2	—	1	3	250	1000	—	Park and Fray (2009)
PCB	2.6	0.81	23.1	0.19	2.89	0.001	1.75	1.88	14.4	217	3.60	Xiang et al. (2010)
PCB	1.4	1.2	12.6	2.4	3.1	—	5.6		14	33	—	Liang et al. (2010)
PCB	1.33	2.05	18.5	0.43	2.66	—	—	4.91	86	694	97	Yazici and Deveci (2013)

13.4 BIOLEACHING OF METALLIC COMPONENTS OF WEEE

Although the success of biomining in processing low grade Cu minerals is evident in the 20 plants commercialised in the last 30–40 years (Watling 2006), its application in recycling of waste materials such as electronic waste is still emerging. Biomining offers a viable option for processing WEEE with potential benefits including lower cost and lower energy option over conventional approaches. Its potential to re-process these secondary resources to chemically inactive residues and zero production of toxic gases represents significant environmental benefits. It's potential to offer future ecologically sustainable technique in managing e-waste is motivating the thriving research in this field.

13.4.1 Micro-Organisms

The process of bioleaching harnesses the natural abilities of micro-organisms and their metabolic products to extract metals in precisely engineered operations. Three groups of micro-organisms are currently used in leaching: chemolithoautotrophs bacteria which derive their energy from the biooxidation of inorganic compounds containing reduced S and Fe^{2+}, heterotrophic bacteria and fungi, both of which require carbon source for growth and energy (Brandl and Wegmann 2001; Brown et al. 2010; Ilyas et al. 2010; Xiang et al. 2010). These organisms affect the solubilisation of metals from metal-rich materials via a series of reactions, including oxidation, reduction, complexation by the generation of inorganic (e.g., sulphuric acid) and organic-acid leachates (e.g., citric acid). The use of bacteria from the genus *Acidithiobacillus* has dominated the key research in bioleaching of e-waste. This is attributed to their greater tolerance to heavy metals (Leduc et al. 1997; Brandl and Wegmann 2001; Wang et al. 2009). The toxicity will increase with the concentration of metals and will vary with mixtures of metals (Aston et al. 2010). Two different types of bacteria (mesophilic and thermophilic) in this genus are important in the extraction of metals. The mesophilic autotroph *Acidithiobacillus ferrooxidans*, *A. thiooxidans*, and *Leptospirillum ferrooxidans* have been used extensively in leaching e-wastes (Leduc et al. 1997; Brandl and Wegmann 2001; Wang et al. 2009; Lewis et al. 2011). *A. thiooxidans* oxidises S compounds to generate sulphuric acid and *A. ferrooxidans* oxidises ferrous iron and reduced sulphur compounds to ferric ion and sulphuric acid. Because e-wastes do not contain sulphur compounds, these are supplemented with inorganic compounds. Three different culturing media have been used, the most common is Lundgren–Silverman 9 K medium (Das et al. 1999; Choi et al. 2004; Lewis et al. 2011; Adhapure et al. 2013; Zhu et al. 2013) followed by Mackintosh and Baering medium (Lewis et al. 2011). The Silverman Lundgren 9 K medium composition (g/L) is: part A – $(NH_4)_2SO_4$ – 3.0, KCl – 0.20, K_2HPO_4 – 0.050, K_2HPO_4 – 0.050, $MgSO_4 \cdot 7H_2O$ – 0.050; part B – $FeSO_4 \cdot 7H_2O$ – 45; Mackintosh medium is composed of (g/L): $(NH_4)2SO_4$ – 0.07, KH_2PO_4 – 0.01, $MgCl_2 \cdot 6H_2O$ – 0.03, $CaCl_2 \cdot 2H_2O$ – 0.07, $FeSO_4 \cdot 7H_2O$ – 15, $MnCl_2$ – 0.03, ZnC_{l2} – 0.03, $CoCl_2$ – 0.03, H_3BO_3 – 0.05, Na_2MoO_4 – 0.01, $CuCl_2$ – 0.03; and

Baering medium is composed of (g/L): $Na_2S_2O_3 \cdot 5H_2O$ – 5, K_2HPO_4 – 0.1, $NaHCO_3$ – 0.2, NH_4Cl – 0.1.

These can be supplemented by elemental sulphur (Brandl and Wegmann 2001; Wang et al. 2009; Hong et al. 2013a). More recently the use of waste inorganic such as pyrite (FeS_2) with ability to supply inexpensive S and Fe^{2+} has also been considered as growth nutrient (Bas et al. 2013; Shibayama et al. 2013).

The role of the organisms in improving the leaching efficiencies is exemplified from comparative tests involving abiotic chemical leaching and with bacterial leaching. Leaching for example at pH 1.5 resulted in 50% Cu extraction in 12 days in comparison to 95% Cu extraction within the same period with mixed culture of acidophiles inoculated at pH 1.5 (Xiang et al. 2010). These demonstrate the contribution of bacterial leaching to chemical leaching is in improving the kinetics of metal extraction.

Moderately thermophilic bacteria including *Sulfobacilllus thermosulfidooxidans* and *Thermoplasma acidophilum* have been used in bioleaching of e-wastes (Ilyas et al. 2007; Ilyas et al. 2010). The advantage of thermophilic organisms is their ability to grow at elevated temperatures (moderate thermophiles <50°C, extreme thermophiles >80°C). The higher temperature of leaching promoted by thermophilic bacteria is an advantage as it promotes greater metal recovery and faster leaching kinetics (Das et al. 1999).

The use of consortia or pairing of organisms in leaching waste has been favoured by researchers and is beginning to receive greater interest. The mixed culture of acidophilic bacteria (MCAB) are isolated directly from natural acid mine drainage with no other subculturing to isolate individual strains (Xiang et al. 2010; Zhu et al. 2011). These consortia have proven to provide greater metal extraction efficiency when compared to individual bacteria in leaching e-wastes (Ilyas et al. 2007; Xiang et al. 2010; Zhu et al. 2011; Liang et al. 2013). Although their application in leaching electronic wastes are still in its infancy, the impact of using consortia of organisms in bioleaching of ores has been profound (Qiu et al. 2005; Hong and Valix 2014). The synergy is exemplified by the actions of consortia consisting of S and Fe oxidising bacteria (*A. ferrooxidans*, *A. thiooxidans*) and of autotrophs and heterotrophs. *A. ferrooxidans* works cooperatively with *A. thiooxidans* by leaching ferric iron, that subsequently reacts with sulfidic minerals to generate sulphuric acid and ferrous iron. Dissolved sulphide compounds, which are otherwise inaccessible are used as nutrient and oxidised by the *A. thiooxidans* to sulphuric acid. Whilst the use of both autotrophs and heterotrophs provide advantages including the reduction in the toxic effects of dissolved organic carbon during bacterial growth, improved mineral attachment, increased acid production, growth and rates of leaching (Okibe and Johnson 2004; Noel et al. 2010; Liu et al. 2011). Ilyas et al. (2007) has shown that leaching of washed PCB with metal adapted *S. thermosulfidooxidans* gave copper leaching rates of 3.9 g/L · day. Using mixed cultures of S. *Thermosulfidooxidans* and acidophilic heterotroph of washed PCB resulted in almost twice the rate of 5.83 gCu/L · day demonstrating the efficacy of these consortia. Thermophilic consortia of *Sulfobacilllus thermosulfidooxidans* with *Thermoplasma acidophilum* and *S. thermosulfidooxidans* with *Sulfobacilllus acidophilus* have leached 85% Cu in 18 days and 82% in 12 days (Ilyas et al. 2013).

For practical applications, the additional advantage offered by the use of consortia is that organisms could be extracted directly from their source without the need to isolate individual organisms. In addition, the need to address concerns with controlling microbial populations in applications that are unsterile such as e-wastes is overcome by the use of mixed consortia.

Organism adaptation as a means to increase leaching efficiency is also favoured among researchers. Adaptation is achieved by subculturing of micro-organisms in media of increasing metal concentrations (Ilyas et al. 2007; Ilyas et al. 2010). Because of the toxicity of metals and other components of e-waste to bacteria are well recognised, organisms are often adapted by subculturing in heavy metals of various concentrations prior to leaching (Brandl and Wegmann 2001; Ilyas et al. 2007; Ilyas et al. 2010; Liang et al. 2010; Lewis et al. 2011). The results are organisms with greater tolerance to heavy metals that proves to be more effective in leaching electronic wastes.

Studies involving heterotrophic fungi have included the use of *Aspergillus niger*, *Penicillium simplicissimum* and *P. chrysogenum* (Brandl and Wegmann 2001; Brown et al. 2010; Ilyas and Lee 2013). E-wastes are alkaline (Ilyas et al. 2007). The key advantages of fungi bioleaching of e-waste is therefore their ability to dissolve metals under alkaline conditions by complexing the metallic fractions with organic acid metabolites and the lower toxicity of the complexed metals (Burgstaller and Schinner 1993). The lack of carbon based nutrients in e-wastes, however, means this must be supplemented typically with glucose, sucrose and waste carbon substrate (e.g., molasses) that could contribute to higher cost of processing. In general it appears the recovery of metals achieved with fungi are typically lower than those achieved with the acidophilic bacteria (Brandl and Wegmann 2001).

In addition to base and ferrous metals, precious metals including gold, platinum and silver have also been recovered by bio-cyanidation of metallic wastes. This is typically achieved using cyanogenic bacterial strains, the most notable are *Chromobacterium violaceum*, *Pseudomonas fluorescens*, *Bacillus megaterium* (Maketon and Ogden 2009). Although HCN are known to be generated by bacteria including *Chromobacterium violaceum*, *Pseudomonas fluorescens*, *P. Aeruginosa* and fungi including *Marasmius oreades*, *Clitocybe* sp., *Polysporus* sp., only bacteria have been utilised in research and most of the HCN formation data available for these organisms are only in the form of qualitative information (Guzman et al. 2003). Because of the different requirements in leaching of base and precious metals, two stage process involving an oxidative acid leaching of base metals (Cu in particular) followed by leaching of precious metals from e-waste has been proposed (Tuncuk et al. 2012).

13.4.2 Method of Bioleaching

Studies of bioleaching of e-waste have employed predominantly batch reactors (Brown et al. 2010; Xiang et al. 2010; Zhu et al. 2011), with few continuous column studies (Ilyas et al. 2010; Ilyas et al. 2013). Batch tests are generally favoured because of the ease of use and these tests generate data required for scaled design of bioreactors.

Bioleaching in batch reactors has been conducted by (Brandl and Wegmann 2001; Wang et al. 2009; Brown et al. 2010; Zhu et al. 2011; Yazici and Deveci 2013; Hong and Valix 2014):

Direct leaching (in-situ). This is where leaching is achieved in the presence of the organism. This can be achieved by introducing the waste at the inoculation stage, often referred to as one stage or by adding the waste when metabolic acids have begun to form, referred to as two-step process.

In-direct leaching (ex-situ). This is also a two-step process whereby the metabolic acids or spent medium are harvested and used in leaching the metallic fractions in the absence of the organism.

The majority of test work has been conducted by direct leaching methods (Wang et al. 2009; Xiang et al. 2010; Zhu et al. 2013) with few indirect bacterial leaching test (Hong and Valix 2014). From an industrial perspective the use of indirect leaching is considered more appropriate as it offers greater flexibility in the optimisation of the process. Running the leaching process independently of the abiotic effects of the leaching process would provide greater flexibility. The metabolic activities of the organisms could be influenced by the toxic effect of the metals that are being dissolved and by other environmental conditions such as temperature, pH and dissolved oxygen concentration. This would allow conditions including higher pulp density, lower pH and higher temperatures to be considered in the optimisation of metal extraction. Recent studies have shown the use of two-step by the direct method can also achieve some of the advantages of the in-direct method (Liang et al. 2010; Xiang et al. 2010; Zhu et al. 2011). This aspect will be considered further in this review.

13.5 LEACHING CHEMISTRY AND SECONDARY REACTIONS

The key in applying the bioleaching technology in reclaiming the metallic components from e-waste lies in precisely understanding the complex parallel and interdependent reactions in this process. In this section we present the leaching reactions that promote the dissolution of copper. The dissolution of any metallic fractions, however, is also subject to other side reactions referred to as secondary reactions. These are reactions other than leaching that could promote or alternatively hamper the dissolution and recovery of the metallic fractions by leaching (Valix et al. 2001b; Viramontes-Gamboa et al. 2007). These reactions could include adsorption, precipitation and co-precipitation, passivation and galvanic coupling. Optimising the recovery of the metallic fractions from e-waste will require confirmation of the occurrence of these reactions and identification of the windows of conditions under which these reactions could be minimised or avoided. Currently there is only very scarce information in regards to the secondary reaction. The key study has been carried out in in-direct bioleaching of waste (Hong and Valix 2014).

13.5.1 Leaching Reactions

Copper dissolution will proceed by two stage redox reactions (see equations 1 and 2). The relative extent of each reaction is determined by pH, the nature of anions that are present, temperature, and dissolved oxygen concentration. Here reduced copper metal is first oxidised rapidly to cuprous ion (Cu^+) with an $E° = -0.52$ V followed by its subsequent slower oxidation to cupric ion (Cu^{2+}) with $E° = -0.16$ V (Mattsson and Bockris 1959; Andersen et al. 1975).

$$Cu \rightarrow Cu^+ + e^- \quad \text{(Eq. 13.1)}$$

$$Cu^+ \rightarrow Cu^{2+} + e^- \quad \text{(Eq. 13.2)}$$

The corresponding cathodic reaction to support the oxidation of copper can be achieved using various redox couple including O_2/H_2O, H_2O_2/H_2O and Fe^{3+}/Fe^{2+}. In the absence of additional oxidant reagent the O_2/H_2O redox couple with $E° = 1.23$ V achieves this role:

$$O_2 + 4H^+ + 4e^- \rightarrow 2H_2O \quad \text{(Eq. 13.3)}$$

The resulting reaction is (Nandeesh and Sheshadri 1988):

$$2Cu^+ + O_2 + 4H^+ \rightarrow 4Cu^{2+} + 2H_2O \quad \text{(Eq. 13.4)}$$

Copper ions then react with the anions of the acidic lixiviant. In sulphuric acid the copper ion will react with SO_4^{2-} ion to form a soluble salt:

$$Cu^{2+} + SO_4^{2-} \rightarrow CuSO_4 \quad \text{(Eq. 13.5)}$$

In organic acid, for example citric acid, copper will react with the citrate anion ($C_6H_5O_7^-$ or C_{it}) to form the following metal complexes:

$$Cu^{2+} + H_3C_{it} \rightarrow CuH_2C_{it}^{1+} + H^+ \quad \text{(Eq. 13.6)}$$

$$Cu^{2+} + H_2C_{it}^{1-} \rightarrow CuHC_{it}^{0} + H^+ \quad \text{(Eq. 13.7)}$$

$$Cu^{2+} + H_2C_{it}^{2-} \rightarrow CuC_{it}^{1-} + H^+ \quad \text{(Eq. 13.8)}$$

13.5.2 Secondary Reactions

The dissolution of copper and other metals from electronic waste in leaching experiments could be used to examine and assess the kinetics and mechanisms involved in the leaching process. The quantitative analysis of the leaching kinetics, however, is challenged by complexity of the dissolution mechanism introduced by secondary reactions or competing reactions. These reactions and their potential impact on metal leaching are considered below.

13.5.2.1 Adsorption

Secondary reactions are reactions other than metal leaching which can influence the dissolution or stability of the metals solution (Valix et al. 2001a). These can include the adsorption of the metal ions and metal complexes on the epoxy waste. The substrate most commonly used in printed circuit boards is a glass fibre reinforced epoxy resin with a copper foil bonded on to one or both sides. Adsorption of dissolved copper species on this substrate was considered by examining the copper metal speciation and the surface charge of the reinforced epoxy resin. This was examined for the copper-citrate complexes. The analysis for the copper ions present in the copper sulphate system would also fall under similar analysis.

Copper Speciation. The presence of ligands such as the citrate anion modifies the metal speciation and as such the sorption mechanism. Copper uptake in solution is considered to proceed by the electrostatic attraction between the dominant copper – citrate complexes and the surface of the reinforced epoxy. To consider this effect, the predominance diagram of the copper-citrate species is presented. The species that were considered for the citric acid and copper-citrate system diagrams were the aqueous citrate (L = ligand) species: L^{3-}, HL^{2-}, H_2L^-, H_3L; and the aqueous copper species: Cu^{2+}, CuL^-, $CuHL^0$, CuH_2L^+ and $Cu(OH)L^{2-}$. The distribution of citrate and copper species was calculated using the formation constants shown in Table 13.2. The acid and metal complex speciations are shown graphically in Figures 13.2 and 13.3.

Table 13.2. Formation constants for citric acid and copper citrate complexes at 298 K (L = citrate)

Reactions	*Log K*	*References*
$H^+ + L^{3-} \leftrightarrow HL^{2-}$	6.29	(Maketon and Ogden 2009)
$2H^+ + L^{3-} \leftrightarrow H_2L^-$	11.23	
$3H^+ + L^{3-} \leftrightarrow H_3L$	14.41	
$Cu^{2+} + L^{3-} \leftrightarrow CuL^-$	5.95	(Guzman et al. 2003)
$Cu^{2+} + H^+ + L^{3-} \leftrightarrow CuHL^0$	8.68	
$Cu^{2+} + 2H^+ + L^{3-} \leftrightarrow CuH_2L^+$	11.34	
$Cu^{2+} + H_2O + L^{3-} \leftrightarrow Cu(OH)L^{2-} + H^+$	2.16	

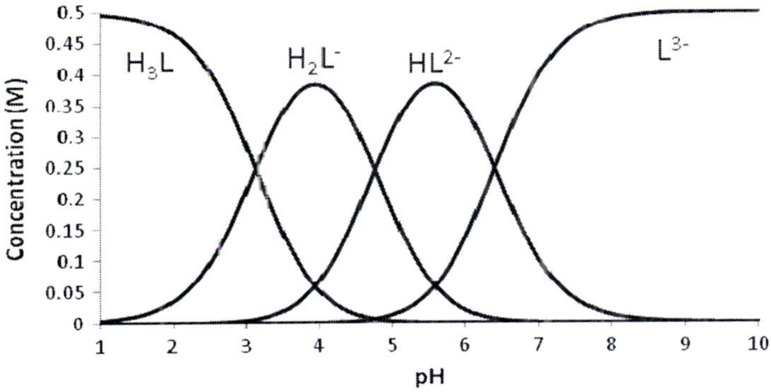

Figure 13.2. Predominance diagram for citric acid, total ligand activity = 0.5 M, T = 298 K

Source: Bucknell et al. 2010

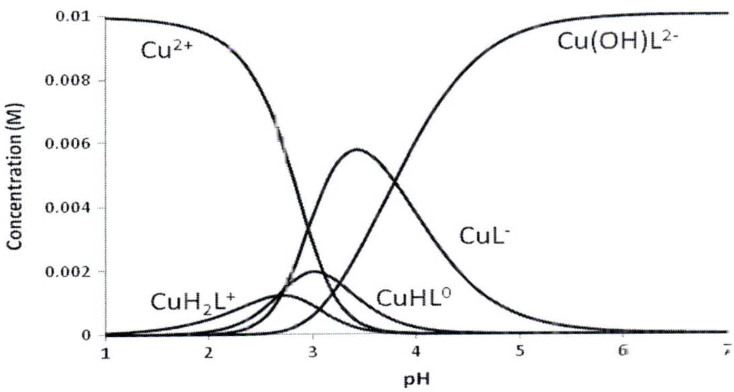

Figure 13.3. Speciation distribution of copper citrate complexes: total copper activity = 0.01 M; T = 298 K

Source: Bucknell et al. 2010

Figure 13.2 shows that below pH 2.0, the citrate ligand is highly protonated. It is therefore not surprising that the key species (see Figure 13.3) below pH 2.0 is the Cu2+ ion. The most dominant copper-citrate species under typical leaching condition at pH 2.0–3.0 is $CuH_2 L^+$. Leaching under these conditions (pH < 3.0) would predominantly generate positively charged copper species. This has an important bearing in controlling the removal of dissolve copper by adsorption.

Surface Charge of Epoxy Based Substrate. The surface charge of the reinforced epoxy was estimated by zeta potential measurement (Brown et al. 2010; Bucknell et al. 2010). This was conducted to establish the potential for electrosorption of copper species on the epoxy residue during leaching. The magnitude of the zeta potential is synonymous to the relative charge on the

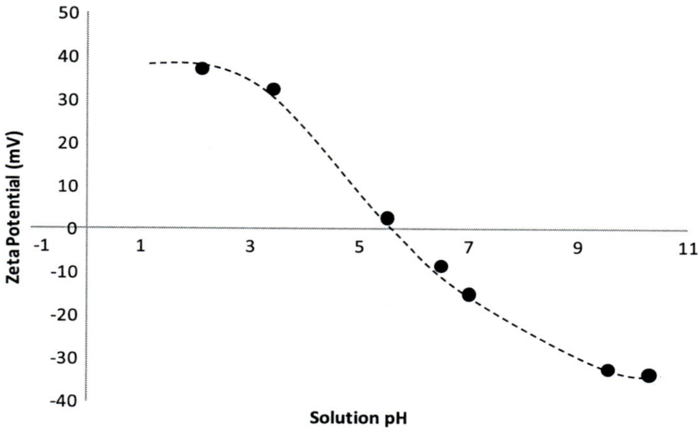

Figure 13.4. Zeta potential of reinforced epoxy residue
Sources: Brown et al. 2010; Bucknell et al. 2010

surface of the particle. When the zeta potential is positive, the net charge around the diffuse layer of the particle is generally considered positive and negative zeta potential indicates a negatively charged particle. Zeta potential will vary with pH. A zeta potential versus pH curve will be positive at low pH and lower at or negative at high pH. There is a point where the plot passes through a zero zeta potential. This point is called the isoelectric point. The zeta potential measurement for the reinforced epoxy is shown in Figure 13.4.

As shown the IEP of the epoxy is approximately pH 5.7 (Bucknell et al. 2010). Considering the copper species that are likely to be generated at pH below 5.7 are positive, leaching below this pH should avoid the adsorption of the dissolved copper species. Considering studies of various commercial fibrous materials, similar to those used in many electric and electronic waste materials, reports an average IEP of 4.3 (Bismarck et al. 2004). These results suggests that adsorption of copper-citrate in the presence of epoxy will have little effect on the recovery of copper metal. Guzman et al. (2003) has reached similar conclusion in suggesting that copper citrate adsorption on similar cationic solid, chitosan, is negligible at pH less than 3.0.

13.5.2.2 Precipitation

Precipitation the transformation of dissolved metals to less soluble solids could reduce metal recovery significantly during leaching and could partially hamper leaching kinetics by fouling metallic surfaces. Precipitation in hydrometallurgical processing could be promoted by saturation of dissolved metals in solution (Park and Fray 2009) and the presence of precipitating agents such as hydroxides, sulfides, sulfate and oxalates. Precipitation in bioleaching of e-wastes that have been reported have focused principally in hydrolysed ferric precipitates

(Wang et al. 2009; Xiang et al. 2010; Lewis et al. 2011; Zhu et al. 2011). Precipitation in autotrophic leaching of electronic wastes appears to be predominated by the formation of hydrolysed iron precipitates. The formation of the iron precipitate appears to be controlled by the concentration of ferric ion in solution, pH and the presence of monovalent ions. Ferric ions are present in substantial quantities in bacterial leaching as it is often used as Fe^{2+}/Fe^{3+} redox couple to support the oxidation and dissolution of copper:

$$Cu^{\circ} + 2Fe^{3+} \leftrightarrow Cu^{2+} + 2Fe^{2+} \qquad \text{(Eq. 13.9)}$$

In this process, the ferrous iron is regenerated by its biooxidation catalysed by the bacteria:

$$2Fe^{2+} + \frac{1}{2}O_2 + H^+ \overset{\text{bacteria}}{?} 2Fe^{3+} + H_2O \qquad \text{(Eq. 13.10)}$$

This will result in the cycling of Fe^{2+}/Fe^{3+} that promotes the dissolution of metallic copper, but will also increase the pH. The increase in pH (pH > 2.0) will result in the formation of Fe oxyhydroxides through ferric hydrolysis (Hofmann et al. 2013):

$$Fe^{3+} + H_2O \rightarrow Fe(OH)_3 + H^+ \qquad \text{(Eq. 13.11)}$$

The formation of ferric hydroxide will generate H^+ that in turn will reduce the pH. As the pH is reduced down to between pH 1.5–2.3, ferric hydroxide will react with monovalent cations (e.g., H_3O^+, Na_2^+, K^+, Rb^+, NH_4^+, Ti^+, $1/2Pb^{2+}$ and $1/2Hg^{2+}$) and sulphate ions to form Fe hydroxysulphate or jarosite precipitates (Norton et al. 1991):

$$Fe(OH)_3 + SO_4^{2-} + 3Fe^{3+} \rightarrow KFe_3(SO_4)(OH)_6 + H^+ \qquad \text{(Eq. 13.12)}$$

The concentration of other dissolved metals including Cu, Ni, Co, Zn in solution will be controlled through their sequestration by adsorption and co-precipitation in both the ferric hydroxide and jarosite precipitates (Gunsinger et al. 2006; Hofmann et al. 2013). Most researchers emphasize the need to control the leaching pH to low values (pH < 2.5) to control the formation of hydrolysed iron precipitates in bioleaching of e-wastes (Xiang et al. 2010; Lewis et al. 2011).

There are however other pathways that could lead to metal precipitation during e-waste bioleaching primary from metabolites generated from the organisms and sulphide oxidation species. Metals ions of Ca^{2+}, Sr^{2+}, Ba^{2+}, Ra^{2+}, Pb^{2+}, Ag^+, Hg^{2+} can be removed from solution by forming less soluble sulphates. Ilyas et al. (2007) has observed the precipitation of Pb as $PbSO_4$ and Sn as SnO. Our

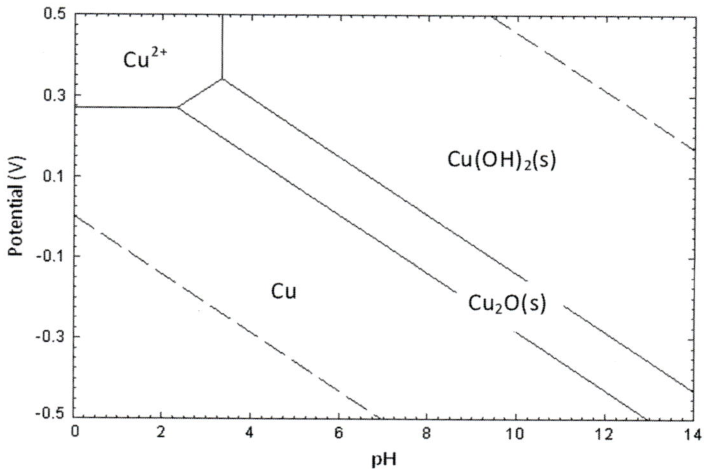

Figure 13.5. Pourbaix diagram for Cu-H2O system at temperature = 363 K and copper activity = 0.01 M
Source: Bucknell et al. 2010

recent study has revealed the passivation of copper surface by formation of CuS (Hong and Valix 2014).

The occurrence of precipitation in leaching with fungi and their metabolic products, organic acids does not appear to be as well documented (Bucknell et al. 2010). Thermodynamic occurrence of $Cu(OH)_2$ precipitation, the passivation through the formation of oxidized copper species such as Cu_2O and CuO, and the effect of organic acids were examined from constructed Pourbaix diagram of the Cu-H_2O and Cu-citrate systems (Bucknell et al. 2010). The Pourbaix diagrams were established by solving the appropriate Nernst equation using the Gibbs free energy and stability constants for formation of aqueous citrate (L = ligand) species L^{3-}, HL^{2-}, H_2L^-, H_3L; the aqueous copper species Cu^{2+}, CuL^-, $CuHL^0$, CuH_2L^+, $Cu(OH)L^{2-}$; the solid copper species Cu, CuO, Cu_2O and $Cu(OH)_2$ (Wu and Tsai 2007b). The resulting Pourbaix diagrams for the Cu-H_2O and Cu-citrate systems are shown in Figures 13.5 and 13.6 respectively. As shown as the pH approached pH > 3.0 precipitation of copper as $Cu(OH)_2$ occurred.

The presence of organic ligand, as shown in Figure 13.6, significantly modified the copper-citrate Pourbaix diagram. The stability of the citrate ligand shifted the $Cu(OH)_2$ predominance region from pH 3.5 to pH 11. The ligand therefore allows the copper-citrate complex to remain in solution at higher pH without precipitating. These results show that precipitation of $Cu(OH)_2$ is unlikely to hamper the recovery of copper from e-waste using citric acid at the typical pH of leaching (pH below 3.0).

In general organic acids form stable metal complexes that could avoid the formation of precipitates (Burgstaller and Schinner 1993). This is with the exception of oxalic acid. There is wide evidence in literature to suggest insolubility

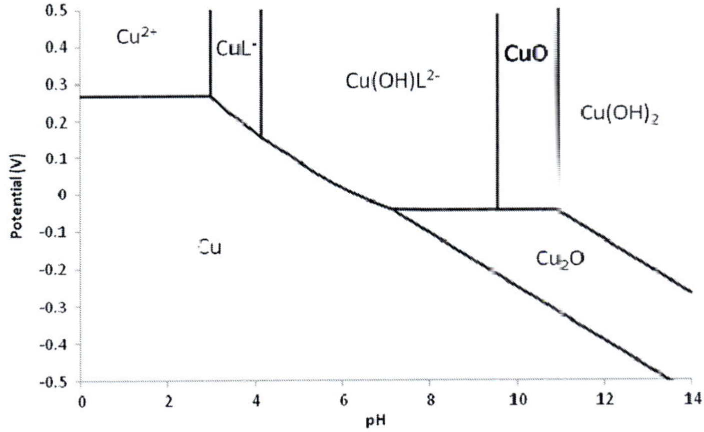

Figure 13.6. Pourbaix diagram for Cu-citrate-H2O system at 363 K: total copper activity = 0.01 M; total ligand activity = 0.5 M
Source: Bucknell et al. 2010

of some metal oxalates (Fernandes et al. 2012; Sun and Qiu 2012; Cai et al. 2013; Fernandes et al. 2013).

Oxalic acid is avoided in leaching of Co, Mn, Mg, Cu, Al, Ni, Fe, Cr and Zn because of the formation of metal oxalate precipitates (Tang and Valix 2006). Most metal oxalates are insoluble with the exception of sodium, potassium, lithium and iron (Strasser et al. 1994; Gadd 2001). Although the occurrence of oxalate precipitation in e-waste leaching has not been observed, it would be presumed that its formation during the fungal fermentation of organic acids may contribute to precipitation. Fermentation conditions should therefore be selected to minimise its formation.

13.5.2.3 Passivation

Passivation of the copper surface by cuprous and cupric ions has been inferred by on XPS analysis of the copper particles after leaching (see Figure 13.7) (Hong and Valix 2014). In this work the $Cu(2p)_{3/2}$ spectrum was fitted to four components. The component at binding energy 932.5 eV represents the Cu(I) state. The three peaks located at 935, 939.5 and 943.5 eV are due to Cu(II) species (Cano et al. 2001).

Further analysis of the surface with SEM- EDS spot analysis in Figure 13.8 showed the copper surface is covered by two specific copper species. It was inferred from the atomic ratios of the elemental analysis in Table 13.3 that the copper surface was primarily covered with Cu_2O (spectrum 1) with deposits of $CuSO_4$ precipitate (spectrum 2 and 3).

Examination of the electrochemical reactions also sheds some light into the nature of this passivation. The first stage oxidation of copper has been suggested to occur as follows (Schweickert et al. 1976):

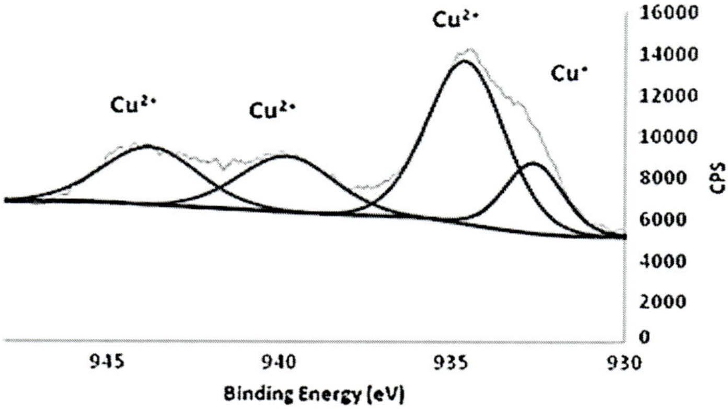

Figure 13.7. Curve-fitting results for the Cu^{2+}/Cu^{+} *XPS spectrum of Cu-rich waste-after leaching*

Source: Hong and Valix 2014; reproduced with permission from Elsevier

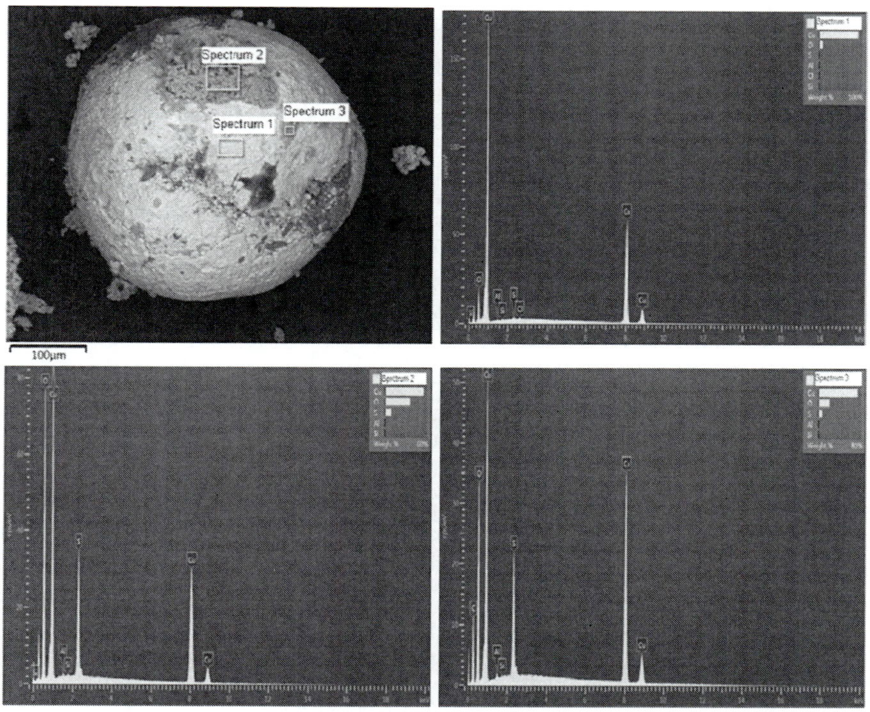

Figure 13.8. SEM micrograph showing the surface morphology of the copper particle leached in H_2SO_4*, pH 1.5 for 6 hours at 90°C and EDS analysis of spectrums 1, 2 and 3*

Source: Hong and Valix 2014; reproduced with permission from Elsevier

Table 13.3. EDS-elemental analysis of copper particle leached in H_2SO_4, pH 1.5 for 6 hours at 90°C (see Figure 13.8)

Elements	Spectrum 1 (atomic %)	Spectrum 2 (atomic %)	Spectrum 3 (atomic %)
O	26.93	66.39	48.32
Al	0.33	0.29	0.58
S	0.38	7.89	7.91
Cu	72.16	25.43	43.19

Source: Hong and Valix 2014; reproduced with permission from Elsevier

$$Cu \rightarrow Cu^{+}_{ads} + e^{-} \rightarrow Cu^{+}_{sol} + e^{-} \qquad \text{(Eq. 13.13)}$$

Because the rate determining step in the dissolution of copper is the oxidation of cuprous to cupric ions (Mattsson and Bockris 1959), this infers that cuprous ions, at least at the beginning of the leaching process, are present in large concentrations in the vicinity of the dissolving copper particles.

Cuprous ions are considered thermodnamically unstable (Chen and Dutrizac 1989) and their fate, that is they can either be displaced or disproportionate, determines the recovery of copper. The work by Liu et al. (2011) suggests that one fate would be the conversion of Cu^+ to Cu_2O. This begins with the electrochemical reaction of molecular dissolved oxygen with $E° = 0.4$ V:

$$O_2 + 2H_2O + 4e^{-} \rightarrow 4OH^{-} \qquad \text{(Eq. 13.14)}$$

This leads to the precipitation (see equation 13) and dehydration of the precipitate to cuprous oxide (see equation 14).

$$Cu^{+} + OH^{-} \rightarrow CuOH^{-} \qquad \text{(Eq. 13.15)}$$

$$2CuOH \rightarrow Cu_2O + H_2O \qquad \text{(Eq. 13.16)}$$

An alternative process for the oxidation of the cuprous ion has been postulated to occur through the peroxide route. According to (Andersen et al. 1975) the oxidation reaction in equation (5) can also proceed as follows:

$$2Cu^{+} + O_2 + 2H^{+} \rightarrow 2Cu^{2+} + H_2O_2 \qquad \text{(Eq. 13.17)}$$

The presence of H_2O_2 as suggested by the work of (Stewart and Gewirth 2007) can also lead to the formation of Cu-OH. This in turn could also induce the formation of Cu_2O as suggested by equation (10). These results suggest passivation of the copper surface during bioleaching occur principally from the formation of oxidised cuprous ions and support the SEM-EDS analysis in Figure 13.8. Cupric ions (Cu^{2+}) on the copper surface are present as deposits of

$CuSO_4$ precipitate. These may foul the surface but would not have similar passivating effects as Cu_2O. These results suggest that competitive reactions of the Cu^{2+} disproportionation to Cu^{2+} and reaction with sulphate or hydroxyl ions in the vicinity of the copper ions dictate if leaching or passivation will proceed.

Passivation of copper surface by the ferric hydrolysis has also been inferred from bacterial leaching of waste PCB (Zhu et al. 2011; Bas et al. 2013)

13.5.2.4 Galvanic Interaction

Another feature of e-waste that has been suggested to influence metal recovery from e-waste is the galvanic coupling of Cu with the other metallic components in the waste (Hong and Valix 2014). This process, which is also referred to as cementation, involves the electrochemical interaction of metals where metals with more electropositive electrode reduction potential have a tendency to be reduced, whilst those with lower standard potential are oxidised and are preferentially dissolved (Demirkiran et al. 2007). The ranking of electrode potential at 25°C among the metals in the waste studied is: Cu^+/Cu (0.52 V) > Cu^{2+}/Cu (0.34 V) > Zn^{2+}/Zn (−0.76 V) > Al^{3+}/Al (−1.67 V) > Mg^{2+}/Mg (−2.35 V). The presence therefore of metals with lower electrode potentials has the potential to limit the selective leaching of copper from the wastes. Recent studies of metal cementation reveal that in addition to standard potential various factors including solution temperature, pH, concentration, presence of dissolved oxygen, leaching reagents and time can also influence the selective removal or preferential dissolution of metals (Sulka and Jaskula 2003; Sulka and Jaskula 2004; Keles 2009). The effects of pH, temperature and time on the galvanic coupling of metals has been considered in the study of in-direct bioleaching of e-waste (Hong and Valix 2014).

Effect of Acid pH. The relative metal mobilisation of copper and other metals in sulphuric and citric acids have been shown to be affected by the acid pH or concentration in in-direct leaching (Hong and Valix 2014) (see Figure 13.9). As shown leaching in sulphuric acid (Figure 13.9a) at the higher pH (pH 1.4 and 1.7) the selective dissolution of the metals followed the ORP ranking, that is Mg > Al > Zn > Cu. Metals with the least electrode potential, for example Mg, were preferentially dissolved in comparison to Cu. However the lower pH (<1.0) appears to overcome this effect leading to the selective dissolution of Cu. In citric acid the galvanic effects were overcome at all pH considered. Figure 13.9b showed that Cu was always selectively dissolved.

Effect of Temperature. Hong and Valix (2014) have shown that increasing temperature of leaching increased the mobilisation of all the metals, in particular Cu in both sulphuric and citric acids (see Figure 13.10). Although the overall Cu mobilisation in sulphuric acid is higher (99%) than in citric (65%) for the fixed period of 24 hours of leaching at 90°C, the selectivity of Cu recovery is significantly higher in citric than in sulphuric acid. These results suggest the potential benefit of mixed consortia and in particular thermophiles in enhancing both the rates of leaching and metal recovery.

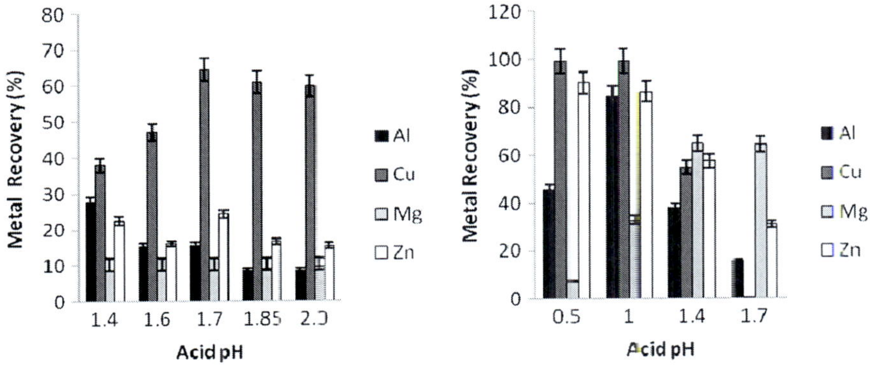

Figure 13.9. Metal leaching as a function of: a) sulphuric acid pH and b) citric acid pH (10 g/L, 90°C, and 24 hours)
Source: a) Hong and Valix 2014 and b) Valix, unpublished results; reproduced with permission from Elsevier

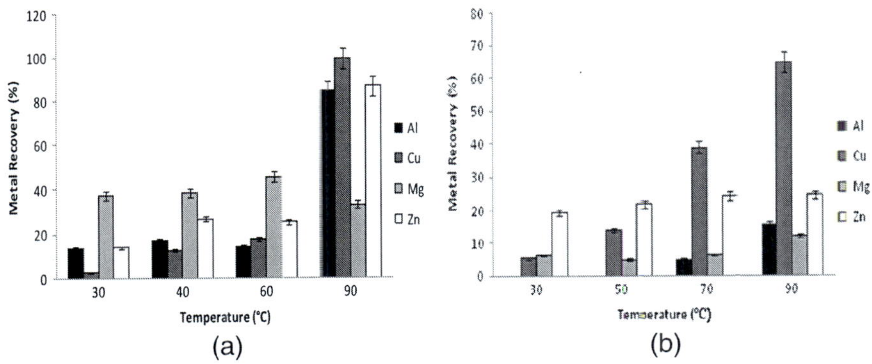

Figure 13.10. Metal leaching as a function temperature in: a) sulphuric acid (pH 1.0) and b) citric acid (pH 1.7) (10 g/L, 24 hours)
Source: a) Hong and Valix 2014 and b) Valix, unpublished results; reproduced with permission from Elsevier

Effect of Time. The period of leaching has been shown to overcome galvanic coupling in sulphuric acid (Hong and Valix 2014) (see Figure 13.11a). It would be anticipated that longer periods of leaching would leach the less electropositive metals first eventually allowing the dissolution of the remaining electronegative metals. This appears true for Al and Zn, however despite only dissolving about 40% of Mg; about 99% of Cu was dissolved in 24 hours. The effect could be observed in Figure 13.11b, where Cu is selectively dissolved at periods >10 hours. These results show the galvanic effects in leaching e-wastes do not follow conventional rules and reflects the complexity of the system.

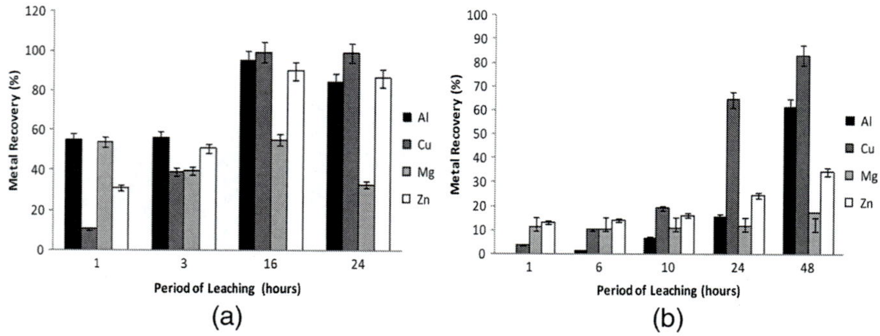

Figure 13.11. Metal leaching as a function time in: a) sulphuric acid (pH 1.0) and b) citric acid (pH 1.7) (10 g/L, 90°C)

Source: a) Hong and Valix 2014 and b) Valix, unpublished results; reproduced with permission from Elsevier

13.6 KINETICS OF COPPER EXTRACTION FROM PCB USING BACTERIA AND FUNGI

Numerous studies of the kinetics of metal leaching from electronic wastes have been conducted covering the examination of a wide range of parameters. This review examined the effect of factors including the nature of organisms, inoculation concentration, inoculation pH, type and concentration of oxidants, waste particle size and composition.

13.6.1 Organisms

Various organisms have been used for the extraction of metals from e-waste. The organisms considered are mesophilic bacteria (*T. ferrooxidans* and *T. thiooxidans*), thermophile (*S. thermosulfidooxidans*), consortia of autotrophic and heterotrophic bacteria and fungi. Comparison of the efficiency and rates of copper extractions of various organisms are summarised in Table 13.4. These results have excluded data that used oxidants (e.g., Fe^{2+}). Because most extraction data are reported in Figures rather as numerical data, the author endeavored to read the data accurately. However reported data in this review could be presumed to be approximate of the actual results.

A general comparison shows that bacteria provide greater efficiency in copper recovery and rates of copper extraction compared to fungi. This could be attributed to stronger acidity of sulphuric acid generated by the bacteria in comparison to organic acids generated by fungi. In addition, although the organic acids generated by fungi are able to stabilise metals in solution by complexation, they have limited ability to oxidise copper. Sulphuric acid is a strong oxidising agent, whereas organic acids tend to promote reduction (Valix et al. 2001).

Adaptation of organisms is shown to have a significant effect on the kinetic of leaching. This is exemplified by the greater performance of adapted mesophiles

Table 13.4. Efficiency and rate of copper extraction by various organisms

Organisms	Period of Leaching (days)	Waste Composition Cu (%)	Pulp Density (g/ml) %	Metal Extraction Efficiency (%)	Cu (g/L · days) $\times 10^3$	References
A. thiooxidans and *A. ferrooxidans*	10	8	0.5	95	38.0	
	10	8	1	98	78.4	
	10	8	5	68	272.0	
	10	8	10	23	184.0	
Penicillium simplicissimum	21	8	0.1	42	1.6	
	21	8	0.5	63	12.0	Brandl and Wegmann (2001)
	21	8	1	42	16.0	
Aspergillus niger	21	8	0.1	60	2.3	
	21	8	0.5	10	1.9	
	21	8	1	1	0.4	
A. ferrooxidans	5	30.8	0.78	38	182	Wang et al. (2009)
A. thiooxidans	5	30.8	0.78	36	173	
A. thiooxidans and *A. ferrooxidans*	5	30.8	0.78	61	293	
A. thiooxidans and *A. ferrooxidans* (metal adapted)	10	12.6	0.8	97.5	98.3	Liang et al. (2013)

(Continued)

Table 13.4. Efficiency and rate of copper extraction by various organisms (Continued)

Organisms	*Period of Leaching (days)*	*Waste Composition Cu (%)*	*Pulp Density (g/ml) %*	*Metal Extraction Efficiency (%)*	*Cu* $(g/L \cdot days) \times 10^3$	*References*
A. ferrooxidans, A. thiooxidans, and *L. ferrooxidans* (adapted)	1.4	39.44	10	99.5	27701	Lewis et al. (2011)
Mixed culture of acidophilic bacteria (MCAB)	3	64.15	0.8	7.4	126.6	Zhu et al. (2011)
Acid mine drainage (AMD) acidophilic bacteria	8	23.1	2	10	57.8	Xiang et al. (2010)
T. ferrooxidans and *T. thiooxidans* (activated sludge)	10	12.6	0.4	79	39.8	Liang et al. (2010)
S. thermosulfidooxidans	NA	8.5	10	NA	2340	Ilyas et al. (2007)
S. thermosulfidooxidans (adapted)	NA	8.5	10	NA	3900	

A. ferrooxidans, A. thiooxidans used by Liang et al. 2013 that gave copper rates of leaching of 98.2×10^{-3} g/L · day. This is slightly higher in comparison with the rate of unadapted organisms used by Brandl et al. (2001) of 78.4×10^{-3} g/L · day. This comparison was made because of closely similar composition of waste (12.6 and 8%) and pulp density (0.8 and 1%) of these studies, which are known to affect the rates of leaching. The effect of adaptation trend is also true for the thermophile *S. thermosulfidooxidans* (Ilyas et al. 2007) where adapted organisms showed 1.67 times the rate of the unadapted organism.

Among the bacteria, it is clear the consortia of metal adapted *A. ferrooxidans, A. thiooxidans*, and *L. ferrooxidans* provided the highest rate of copper extraction (Lewis et al. 2011). Wang et al. (2009) demonstrated there is no significant difference in the individual performance of *A. ferrooxidans* and *A. thiooxidans*. However combining the two organisms resulted in 1.7 times higher rate of copper extraction. In general the consortia isolated from metal laden wastes also provided greater rates of leaching. A comparison of the performance of mixed culture of acidophilic bacteria (MCAB) enriched from natural acid mine drainage gave, approximately, a rate of 126.6×10^{-3} g/L · day at a pulp density of 0.8% (Zhu et al. 2011) whilst subcultured *A. ferrooxidans, A. thiooxidans*, and *L. ferrooxidans* gave, approximately, a rate of 78.4×10^{-3} g/L · day at a pulp density of 1% (Brandl and Wegmann 2001). These results demonstrate the efficacy of mixed cultures compared to pure cultures. In addition the environment where the mixed cultures are isolated from appear to also have an effect. Cultures isolated from acid mine drainage appear to show greater performance than microorganisms isolated from activated sludge (Liang et al. 2010; Zhu et al. 2011). This could be attributed to greater exposure of organism to metallic components, which is presumed to have acted in adapting the organisms to the toxicity of heavy metals.

In general thermophile (*S. thermosulfidooxidans*) where leaching was conducted at higher temperature (45°C) showed greater performance than mesophiles. Table 13.4 show the increased in rate to be about 12.7 times that of subcultured mesophiles. Despite the higher temperature used for thermophiles, its performance is still lower than some mixed consortia used at lower temperatures 28–30°C, such as those used by Lewis et al. (2011). These results demonstrate the advantage in using thermophiles and specific consortia of organisms in the extraction of the metallic components from e-waste.

13.6.2 Method of Bioleaching

Bioleaching can be conducted by direct and indirect leaching. The following compares the effect of method of bioleaching on the rates of copper extraction.

Abiotic vs. Direct Bioleaching. The role of the live organisms in enhancing the rates of metal leaching in addition to supplying the biogenic acid has been well noted in various direct bioleaching studies by comparing abiotic to bacterial leaching (Wang et al. 2009; Yang et al. 2009; Xiang et al. 2010; Lewis et al. 2011; Zhu et al. 2011). A comparison between bacterial and abiotic leaching is shown in Table 13.5. The introduction of bacteria to the leaching process can increase the rate of copper by 200–680% in comparison to abiotic leaching. The results by Zhu

Table 13.5. The efficiency of copper extraction by direct bacterial leaching and abiotic leaching

Organisms	*Period of Leaching (days)*	*Waste Composition Cu (%)*	*S/L (g/L)*	*pH*	*Inoculum (%)*	*Cu Extraction (%)*	*Rate of Cu Extraction (g/L · day)*	*References*
AMD, acidophilic bacteria	8	23.1	20	1.5	10	78	0.45	Xiang et al. (2010)
Abiotic (H_2SO_4)						35	0.20	
Mixed culture of acidophilic bacteria (MCAB)	3	64.15	8	2	5	35	0.6	Zhu et al. (2011)
Abiotic (H_2SO_4)					10	48	0.82	
					20	62	1.06	
					30	68	1.16	
					0	10	0.17	
A. ferrooxidans, A. thiooxidans, and *L.ferrooxidans* (adapted)	1.4	39.44	100	1.9	10	99.5	27.7	Lewis et al. (2011)
Abiotic (H_2SO_4)	5					70	5.52	
A. ferrooxidans	1	25.06	15	2.0	20	49.13	1.85	Yang et al. (2009)
Abiotic (H_2SO_4)					40	58.74	2.21	
					60	84.64	3.18	
					80	98.06	3.69	
					100	98.59	3.71	
					0	25.0	0.94	

et al. (2011) showed that increasing the inoculation concentrations by addition of 5% to 30% of bacterial inoculum resulted in copper rates of leaching approximately 0.6 to 1.16 g/L · day. Whilst the results from Yang et al. (2009) showed that the addition of 20 to 100% of bacterial inoculums can raise the rate of copper extraction from, approximately, 1.85 to 3.81 g/L · day. This showed the greater concentration of bacteria increased the rates of leaching. This was attributed to the greater oxidation of ferrous ion to ferric iron that is catalysed by the bacteria (Yang et al. 2009; Zhu et al. 2011). The greater ferric concentration in turn allows greater oxidation of copper (see equation 9). The greater quantity of bacteria is able to perpetuate the generation of greater quantities of biogenic acids and the required redox reactions. It can also be presumed the greater bacterial population would also allow greater resistance to bacteriostatic effect of the waste. A comparison between the purely abiotic and the bacterial leaching carried at temperatures at and above 50°C did not indicate notable differences in copper extraction (Lewis et al. 2011). This is consistent with observations made by Rossi (1990) and Hong et al. (2013b). This suggests that above 45°C bacterial activity decreases significantly and as a consequence, ferric iron regeneration and copper solubilisation kinetics are affected. At the elevated temperature of leaching pure chemical solubilisation mechanisms predominate over the bacterial leaching.

Direct Leaching (One Step vs. Two Steps). One step direct bioleaching involves the addition of the waste at the inoculation of the organism. In this process the biological activities and leaching occur simultaneously. In the two-step direct leaching, the waste is added once the biogenic acids have formed. It has been noted in various studies that a two-step process enhances the efficiency of leaching and could sharply shorten the leaching period by nearly half in comparison to a one-step process (Liang et al. 2010; Xiang et al. 2010). These results show that that powdered PCB could restrain the growth of bacteria (Xiang et al. 2010). It has been demonstrated that the time at which the waste is added can slightly increase the metal recovery. Optimal copper recovery was achieved (93%) by adding 4 g/L of PCB at 144 hours rather than at 48 hours (86%), 96 hours (89%) and 192 hours (86%) of bacterial inoculation (Liang et al. 2010). Sequential multiple addition of 4 g/l and 6 g/L of PCB using mixed cultures of bacteria at two points of time from 44, 96 and 144 hours of bacterial inoculation revealed even higher metal recovery from 94 to 97% Cu (Liang et al. 2010). These results demonstrate the allowing the organisms to grow and metabolise enables them to cultivate greater resistance to the bacteriostatic effects of the waste.

Direct vs. Indirect Leaching. In-direct leaching or spent medium leaching involves the generation of biogenic acids under optimal conditions, followed by harvesting of the acid and their use in leaching of the metallic components of the waste. Biogenic acids will consist of various metabolic products including the biogenic mineral or organic acids, polysaccharides, proteins, enzymes and unused growth nutrients. A comparison of indirect and direct bacterial leaching by one-step and two-step methods are shown in Figures 13.12 and 13.13 respectively. As shown, in general in-direct leaching provides better rates of leaching compared to the one-step method, with the exception of the results obtained by

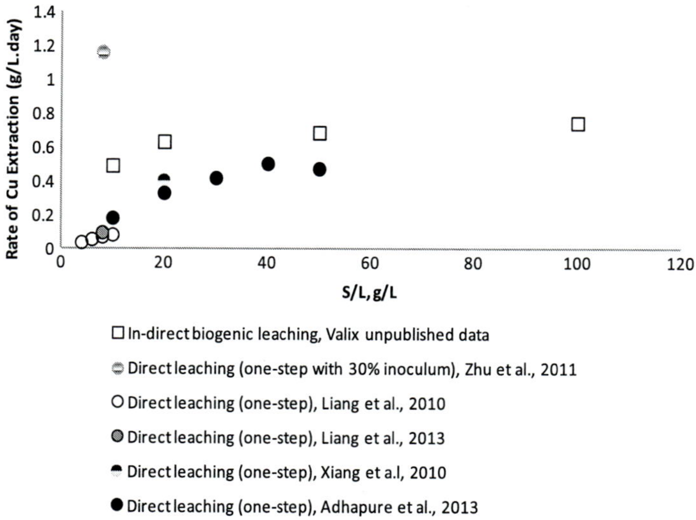

Figure 13.12. Comparing the rates of copper extraction based on direct (one-step) bacterial and in-direct bacterial leaching

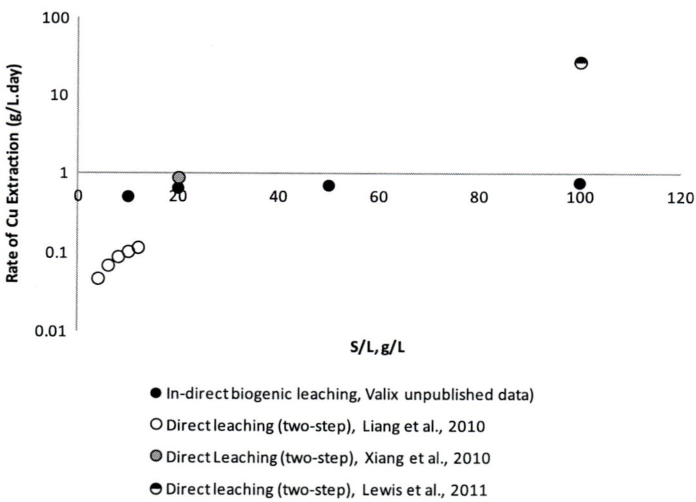

Figure 13.13. Comparing the rates of copper extraction based on direct (two-step) bacterial and in-direct bacterial leaching

Zhu et al. (2011), with comparatively higher rate of leaching. This was achieved with 30% inoculums compared to typically 10% inoculum in other studies, suggesting the concentration of bacteria in the direct leaching can supersede the advantage of in-direct leaching. Figure 13.13 shows at the higher pulp densities (>20 g/L), the direct leaching by two-step method is able to exceed the rates of leaching by in-direct method. These results demonstrate the organisms are able to enhance the rate of leaching with roles other than chemical solubilisation. It appears the presence of the organisms in the direct attack of the metals allows a more efficient metal recovery by bioleaching.

13.6.3 Effect of Other Bioleaching Factors on the Kinetics of Leaching

13.6.3.1 Pulp Density

Increasing the pulp density in bacterial leaching is known to hamper the rates of copper extraction. This effect has often been attributed to the toxic effect of the waste on the bioleaching organisms (Brandl and Wegmann 2001; Ilyas et al. 2010; Pradhan et al. 2010; Bas et al. 2013). Figure 13.14 shows a summary of direct bacterial leaching confirming the negative effect of increasing pulp density on copper extraction. Superimposed on these results is the in-direct leaching results (Valix, unpublished data). As shown it appears the pulp density affected the copper extraction in the same way regardless of the presence of the organism. That is increasing the pulp density reduced the copper extraction regardless of the presence of the organism, suggesting the effect of pulp density may include other factors other than the bacteriostatic effect of copper.

To examine the bacteriostatic effect of the waste as a function of pulp density, the rate data were replotted as a function of dissolved copper concentration in Figure 13.15. These results demonstrate the toxic effect of the waste. As shown

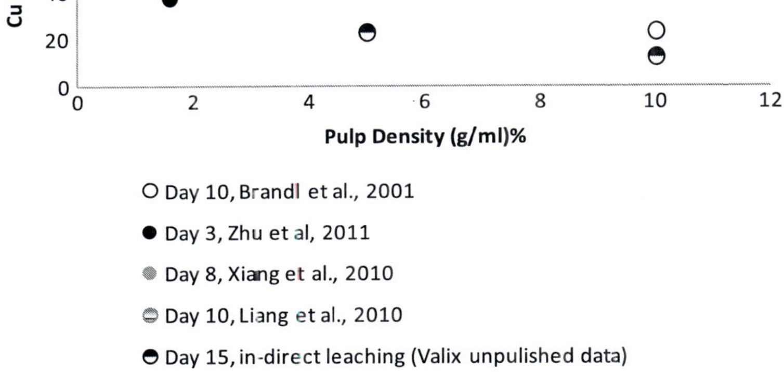

Figure 13.14. Comparing the effect of pulp density on the direct and in-direct bioleaching of copper from e-waste

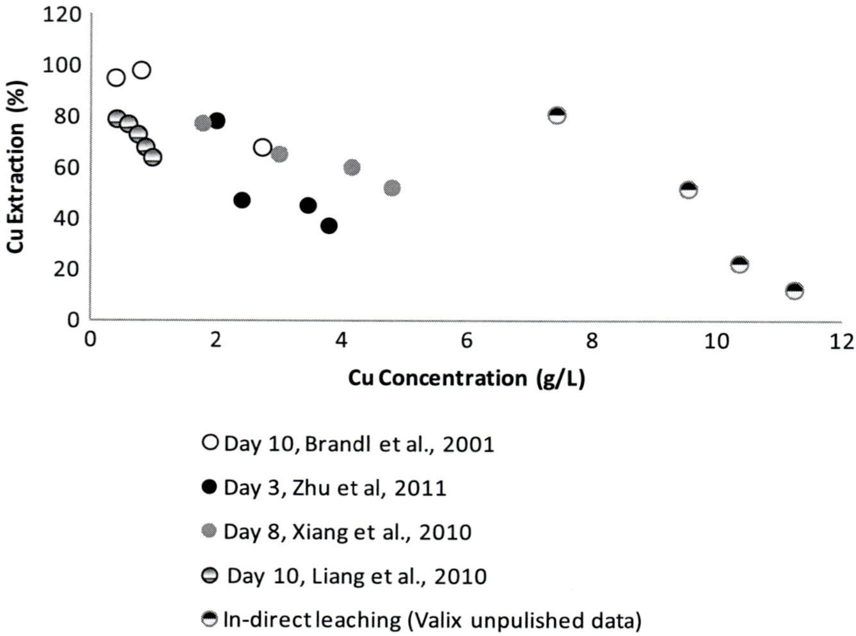

Figure 13.15. Comparing the effect of dissolved copper on the direct and in-direct bioleaching of copper from e-waste

there is a significant difference between the direct and in-direct leaching. The extent of copper extraction is significantly lower in direct leaching than in in-direct leaching when compared on the same dissolved copper extraction. These results demonstrate that increasing the dissolved copper concentration does have a toxic effect on the bacteria.

Nonetheless, the effect of pulp density trend for in-direct method where leaching is conducted in the absence of the organism suggest that in addition to the toxic effect of dissolved copper, chemical solubilisation is affected by the pulp density. The exact nature of this feature has not been established in the literature. It is envisaged that one or more of the many components of the waste may contribute to this effect.

An interesting feature of the effect of pulp density is its effect on the rates of leaching. Figures 13.12 and 13.13 show that increasing the pulp density results in higher rates of leaching. A comparison of the effect of metal concentration on the rate of copper extraction suggests that in general the rates by direct leaching gives greater rates of leaching than in-direct leaching (see Figure 13.16). This is in contrast with the effect of copper concentration on extent of metal extraction. That is the presence of the organisms is important in enhancing the rates of leaching. This is consistent with the results in Table 13.5 that show the inoculum concentration can enhance the rate of leaching. The relatively higher rates

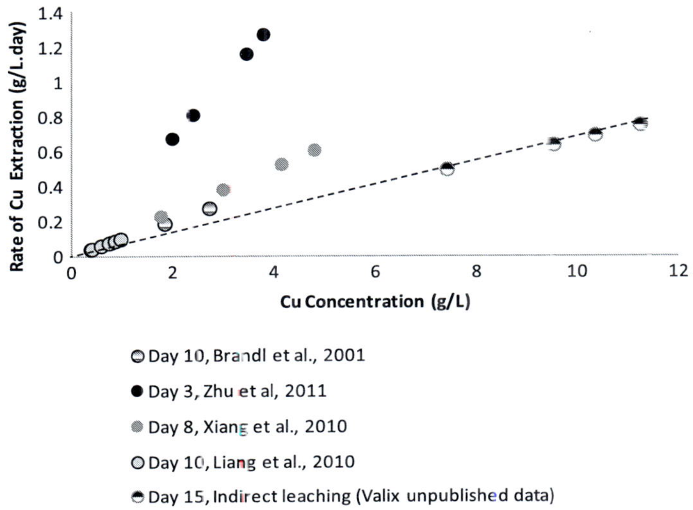

Figure 13.16. Comparing the effect of dissolved copper on the rates of copper extraction by direct bioleaching of copper from e-waste

observed by Zhu et al. (2011) compared to the other test work could not be established in this review. It is proposed that the nature of the organisms used by these researchers, which were a mixed culture of acidophile from acid mine drainage may have contributed to the greater efficiency of leaching. The heavy metal tolerance of these organism may have allowed the strain to provide greater resistance to the bacteriostatic effect of increasing pulp density.

13.6.3.2 Inoculation pH

The pH at which the organisms are inoculated and grown play an important aspect of controlling the rates of copper extraction from e-wastes (Ilyas et al. 2007; Yang et al. 2009; Xiang et al. 2010; Zhu et al. 2011; Bas et al. 2013). Electronic waste are alkaline requiring the partial neutralisation to enable control of pH during the leaching process (Ilyas et al. 2007). Washing of the waste enabled the rates of copper extraction to be increased from 0.23 g/L · day to 0.39 g/L · day (Ilyas et al. 2007).

The effect of pH of bacterial inoculation on the rates of copper extraction is compared in Figure 13.17. Although both organisms from Xiang et al. (2010) and Zhu et al. (2011) were isolated from pyrite wastes, the response of the organisms as shown differed. The AMD used by Xiang et al. (2010) showed minimal response to pH, however the AMD used by Zhu et al. (2011) demonstrated higher rates with increasing pH from 1.25 to 2.5 resulting in rates from 0.07 to 1.1 g/L · day. The data from Yang et al. (2009), were obtained using *A. ferrooxidan* only. The trend here suggests the more acidic condition results in higher rates. It appears the pH of inoculation does affect the performance of the organisms and although the

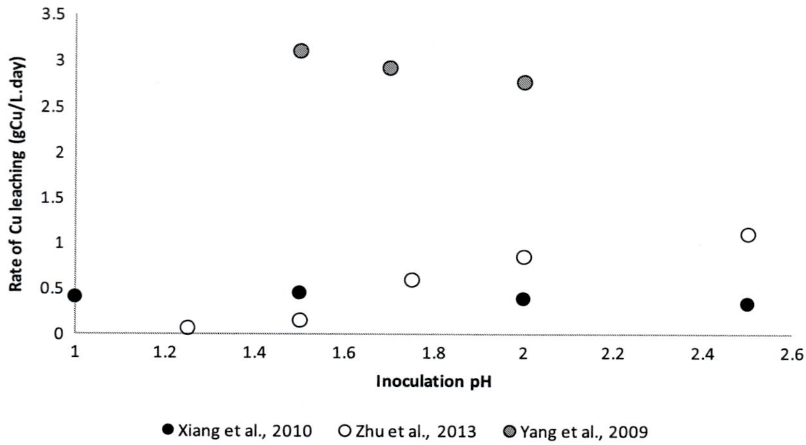

Figure 13.17. The inoculation pH effect on the rates of copper extraction by direct bioleaching of copper from e-waste

bacteria are all acidophilic, the optimal pH appears to vary with the type of organisms within pH of 1.0–2.5.

13.6.3.3 Oxidising Agents

In addition to the involvement of the O_2/H_2O redox couple (see equation 3) (Yazici and Deveci 2013), other oxidants have also been used to optimize the dissolution of copper and other metals from e-waste. Redox couples including H_2O_2/H_2O and Fe^{3+}/Fe^{2+} with their high reduction potentials of 1.78 and 0.7 V respectively makes them ideal as oxidizing agents (Zhu et al. 2011; Birloaga et al. 2013; Oh et al. 2003). Among these oxidants, only the Fe^{3+}/Fe^{2+} couple has been used in the bacterial leaching of metallic waste (Ilyas et al. 2007; Lewis et al. 2011; Zhu et al. 2011). In addition to supporting the oxidation of copper, Fe^{2+} is used a substrate to support bacterial growth and energy requirements. The addition of dissolved iron has been observed to markedly enhance the dissolution rate and extent of copper. Iron is added as soluble iron salt (e.g., $FeSO_4{\bullet}7H_2O$) and from waste pyrite mineral (FeS_2) (Zhu et al. 2011; Adhapure et al. 2013; Bas et al. 2013). A comparison of the effect of dissolved iron and pyrite on the extent and rates of copper extraction is shown in Figures 13.18 and 13.19 respectively. As shown dissolved iron in general provide greater efficiency than FeS_2 mineral. This could be attributed to lower solubility of FeS_2 and thus limiting the available iron to the bacteria. It is also apparent that an optimal quantity of Fe^{3+} is achieved from 10–15 g/L. The reduction in rates and extent of copper recovery at the higher Fe^{3+} concentration has been attributed the hydrolysis of iron and its subsequent precipitation to $Fe(OH)_3$ and jarosite precipitates (Xiang et al. 2010; Zhu et al. 2011; Bas et al. 2013). Both the precipitates can reduce metal extraction and rate of leaching by co-precipitating dissolved metals and by fouling the surface of the dissolving copper.

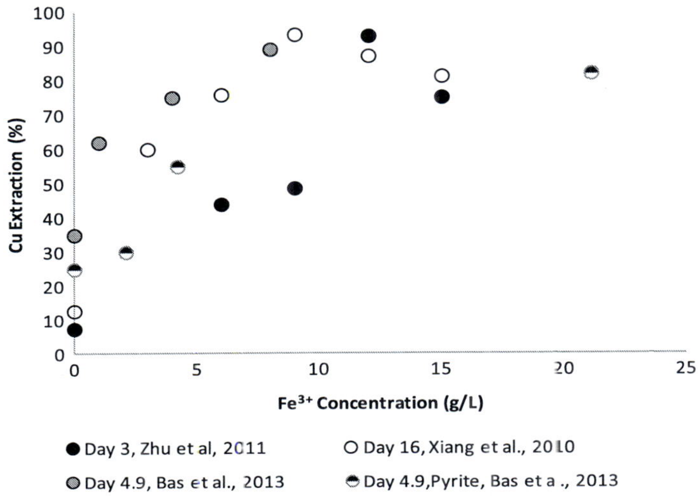

Figure 13.18. Copper extractions as a function of dissolved iron and pyrite by direct bioleaching of copper from e-waste

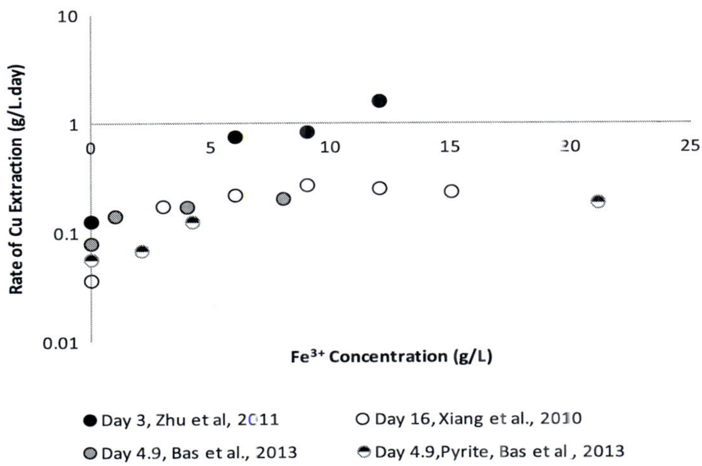

Figure 13.19. Rates of copper extraction as a function of dissolved iron and pyrite by direct bioleaching of copper from e-waste

A comparison of the effect of oxidants (H_2O_2/H_2O and Fe^{3+}/Fe^{2+}) in Table 13.6 shows the significant effect of H_2O_2/H_2O on the rates of copper extraction. The comparison is based on the use of Fe^{3+}/Fe^{2-} by direct bacterial leaching and of H_2O_2/H_2O in abiotic leaching. As shown the use of H_2O_2/H_2O increased the rate by as much as 11713%. The biocidal nature of H_2O_2 precludes

Table 13.6. Comparing the effect of oxidants on the extent and rates of copper extraction

Bacteria/Leaching Medium	*Period of Leaching (days)*	*Waste Composition Cu (%)*	*S/L (g/L)*	*Oxidant*	*Cu Extraction (%)*	*Rates of Cu Extraction (g/L · day)*	*References*
Mixed culture of acidophilic bacteria	3	64.15	8	0	7.4	0.13	Zhu et al. (2011)
				6 g/L Fe^{3+}	43.9	0.75	
				9 g/L Fe^{3+}	48.6	0.83	
				12 g/L Fe^{3+}	92.9	1.59	
				15 g/L Fe^{3+}	75.1	1.28	
Acid mine drainage (AMD) acidophilic bacteria	16	23.1	20	0	12.7	0.037	Xiang et al. (2010)
				3 g/L Fe^{3+}	60	0.17	
				6 g/L Fe^{3+}	75.8	0.22	
				9 g/L Fe^{3+}	93.3	0.27	
				12 g/L Fe^{3+}	87.1	0.25	
				15 g/L Fe^{3+}	81.2	0.23	
1 M H_2SO_4	0.125	30.57	100	10% H_2O_2	49.2	120	Birloaga et al. (2013)
2 M H_2SO_4				10% H_2O_2	61.04	149	
1 M H_2SO_4				20% H_2O_2	43.7	107	
2 M H_2SO_4				20% H_2O_2	76.12	186	

its use in bacterial leaching. However its use in in-direct leaching appears to outperform bacterial leaching.

The key issue however in using H_2O_2 is its rapid dissociation requiring a high quantity of reagent to maintain a certain reduction potential, coupled with its high cost, challenges the economics of its use. Whilst the addition of Fe^{3+} can introduce its own challenges. The first is that it can increase the amount of metal contaminant in the final leachates. The typical optimal Fe^{3+} dosage has been reported at 9–12 g/L in 8–20 g/L of waste containing approximately 23% copper (Xiang et al. 2010; Zhu et al. 2011). The presence of additional metal contaminants could introduce technical difficulties in the final metal recovery and refining by methods including solvent extraction and electrowinning (Lewis et al. 2011; Yazici and Deveci 2013; Zhang and Cheng 2007). The second issue is that Fe^{3+} can contribute to metal loss to co-precipitation with hydrolysed ferric ($Fe(OH)_3$) (Zhu et al. 2013) and with jarosite ($KFe_3(SO_4)(OH)_6$) precipitates (Xiang et al. 2010). The regeneration of the oxidant, Fe^{2+} to Fe^{3+}, could be achieved using bacteria but this reaction would consume additional acid. Although the consumption of acid by the Fe^{2+}/Fe^{3+} oxidation can be overcome using wastes such as pyrite (FeS_2), the S in pyrite provides the energy source for the mesophilic bacteria and in the process is able to generate additional sulphuric acid (Bas et al. 2013; Ilyas et al. 2013), it does not avoid the issues of ferric hydrolysis.

13.6.3.4 Particle Size and Composition of Waste

The effect of particle size on the rate of copper extraction has often been difficult to establish (see Figure 13.20). As shown, although increasing the particle size results in lower rates in general, the scatter of data between 0 to 1000 μm is wide. These show that reducing the particle size does not itself enhance the rate of leaching as would be anticipated from a mass transfer point of view. Correlation of the rate of

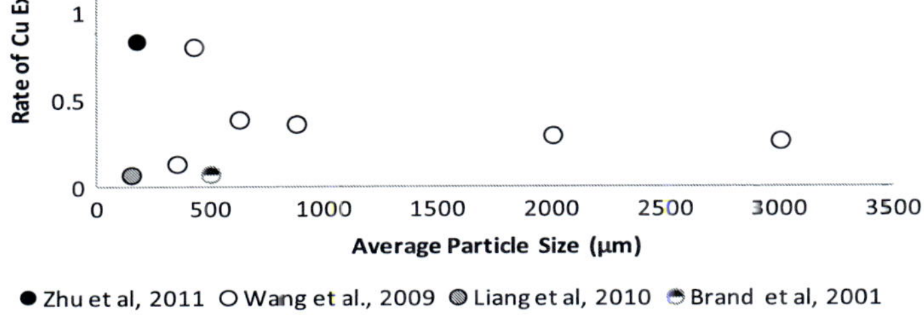

Figure 13.20. Rates of copper extraction as a function of waste particle size by bioleaching of copper from e-waste (S/L 7.8–10 g/L)

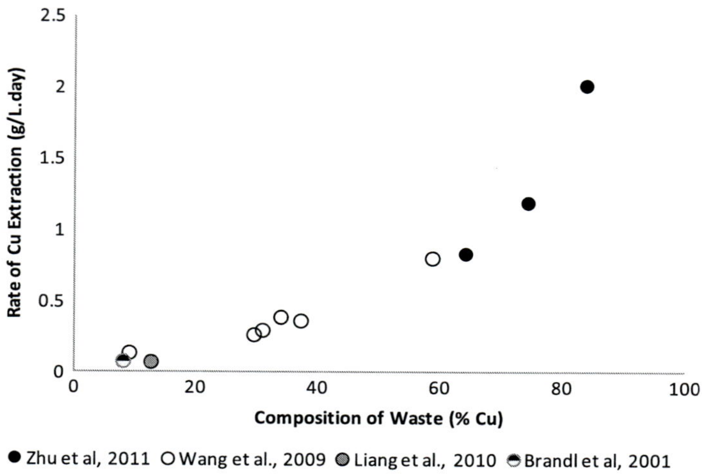

Figure 13.21. Rates of copper extraction as a function of waste composition (% Cu) by bioleaching of copper from e-waste (S/L 7.8–10 g/L)

copper extraction to the composition of the waste in terms of % Cu, as shown in Figure 13.21, shows a stronger correlation. As shown increasing the % Cu resulted in higher rates of copper extraction. This is attributed to the galvanic coupling effect of other metal that are present in the waste. As suggested metals with more electropositive electrode reduction potential have a tendency to be reduced, whilst those with lower standard potential are oxidised and are preferentially dissolved (Demirkiran et al. 2007). Thus the typical presence of lower reduction potential metals including Fe, Pb, Sn, Ni, Al and Cr can hamper the copper extraction by galvanic coupling effects. Thus the purity of the waste has a significant effect on copper extraction from the waste and poses the challenge in recycling lower grades of waste.

13.7 CONCLUSIONS

The two aspects that were examined in this study are the mechanism and the factors influencing the kinetics involved in bioleaching Cu by autotrophic and heterotrophic organisms from Cu-rich waste. From our results the following conclusions were derived.

Mechanism of bioleaching copper e-wastes. The primary reaction involved in leaching of copper proceeds by two stage redox reactions, the first involving the rapid oxidation of copper to cuprous ion (Cu^{+}) followed by its subsequent slower oxidation to cupric ion (Cu^{2+}). This is coupled with the cathodic reaction, for example of O_2 and H^{+} to form water. In the presence of complexing agents such as organic acids, copper forms series of metal-complexes determined by the pH and nature of acid. The relative extent of each reaction is determined by pH, the nature of anions that are present, temperature, and dissolved oxygen concentration.

Secondary reactions can also affect the extent and rates of reactions. The secondary reactions include adsorption, precipitation, passivation and galvanic coupling. The occurrence of these reactions can challenge the optimisation of metal extraction from e-wastes.

Examination of metal speciation at various pHs suggests that adsorption of positively charged copper ion and copper-complexes at the typical pH used for leaching (pH 1.0–3.0) are unlikely to be removed by adsorption on epoxy, which has an isoelectric point of 5.7.

Examination of metal salt solubility and predominance diagram of metal speciation suggests that precipitation by metal saturation and hydroxide precipitation are unlikely to remove dissolved metals from solution. However the presence of precipitating agents (e.g., oxalic acid) could induce precipitation. The hydrolysis of ferric ions to form $Fe(OH)_3$ and jarosites are notable in most bacterial leaching studies where Fe^{2+}/Fe^{3+} redox couple is used as oxidant. The precipitation of metals (Cu, Sn, Pb) as sulphates have also been noted. These suggest the effect of precipitation affects metal recovery in bacterial leaching and should be considered in bioleaching efforts.

Passivation has been observed to hamper the kinetics of leaching. This occurs primarily as a result of cuprous ion oxidation. It appears the competitive reactions of the Cu^+ disproportionation to Cu^{2+} and reaction with sulphate or hydroxyl ions in the vicinity of the copper ions dictate if leaching or passivation will proceed. The fouling effect of jarosite has been implied to also cause passivation of waste.

Galvanic interaction or the electrochemical interaction of metals where metals with more electropositive electrode reduction potential have a tendency to be reduced, whilst those with lower standard potential are oxidised and are preferentially dissolved. Copper, being the more electropositive among the metals prevalent in e-wastes is likely to be influence by galvanic interaction. It appears however that galvanic effects are reduced by leaching at lower pH ($pH < 1.0$), high temperature (90°C) and longer period of leaching. In general, the greater presence of lower reduction potential metals in the waste results in lower extent of copper extraction and rates of leaching. Galvanic coupling has therefore challenge the processing of lower grade e-waste.

Kinetics of leaching. The effect of various factors on the kinetics of copper extraction by bioleaching was considered in this review. These included:

Organisms: The organisms that have been considered for bioleaching of electronic waste include bacteria and fungi. The performance of bacteria in general is greater than fungi. Mesophilic bacteria are able to leach, approximately, at the rate of 0.038–27.701 g/L · day, whilst fungi leach, approximately, at the rate of 4×10^{-3}–0.016 g/L · day. Thermophiles and consortia appear to provide an advantage over individual mesophiles. Thermophiles are able to leach at 10–20 times the rate of most mesophilic bacteria. Whilst consortia can increase the rate by 2–195 times the rate of copper extraction.

The cultivation of organisms with tolerance to heavy metals is also an advantage. The tolerance development of the organism can be achieved by gradual acclimatization of the organism to increasing concentration of heavy metals. It appears isolating the cultures directly from waste environment laden with heavy metals (e.g., acid mine drainage) also have similar effect as metal adaptation of the organisms. Adaptation can increase the rates of copper extraction by up to 2.0.

Method of Bioleaching: Bioleaching can be performed directly in the presence of the organisms and without the organism by the in-direct method. The significant role of bacteria in the bioleaching of waste has been demonstrated by 200 to 680% increase in rates based on direct bacterial leaching compared to abiotic leaching.

Most bioleaching has been carried out by direct bacterial leaching either by one-step or two-step methods. The introduction of waste after the generation of biogenic acid by the two-step method has been shown to significantly enhance the rate of leaching in comparison to one-step leaching. Rates by two-step method could be 2.5 times higher than one-step method. It appears allowing the organisms to grow first can enable it to cultivate greater resistance to the bacteriostatic effect of the waste that would be introduced at a later stage.

There is currently a scarcity of in-direct bioleaching data results. In this review, unpublished data of the author on rates from in-direct leaching were compared to direct leaching. The advantage of in-direct leaching is its independence from the bacteriostatic effect of the waste. However it appears this advantage could be overcome in direct leaching by controlling the inoculum concentration. At <30% inoculum, direct one-step leaching showed lower rates compared to in-direct leaching, whilst at >30% inoculums the rates were higher. This demonstrates the bacterial concentration could overcome the bacteriostatic effect leading to enhanced rates of copper extraction. In general the direct (two-step) leaching showed higher rates of copper extraction. Both results emphasises the important role of bacteria in the extraction of metals from waste.

Pulp density: Increasing the pulp density resulted in lower extent of copper extraction, however the rates of copper extraction increased with increasing pulp density. Further examination of the effect of increasing pulp density on the extent of copper extraction revealed this could be attributed to bacteriostatic effect of copper and components that affects the chemical dissolution of the waste. Comparison of the extent of copper extraction as a function of dissolved copper resulting from various pulp densities showed the extent of extraction were higher by indirect rather than direct leaching. These demonstrate the bacteriostatic effect of the waste. However in the absence of the organisms, increasing the pulp density also resulted in the lowering of the extent of copper extraction. This suggests other factors other than bacteriostatic effect causes the reduction in extent of copper extraction.

The effect of pulp density on the rates are in direct contrast with metal extraction. It appears the direct leaching offered higher rates of copper extraction with increasing pulp density and dissolved copper extraction. The important role of the bacteria in enhancing the rates are further emphasized by the higher rates obtained with greater inoculum concentration.

Inoculation pH: The rates of copper extraction were found to be influenced by the initial inoculation pH of the leaching process. The pHs that have been considered are from 1.0–2.5. The optimal pH appears to vary with the type of organisms. A. ferrooxidan and a mixed acidophile isolated from pyrite favoured the lower pH (1.5), whilst another mixed acidophile isolated also pyrite favoured pH of 2.5.

Oxidising agents: Redox couples including H_2O_2/H_2O and Fe^{3+}/Fe^{2+} have been showed to significantly enhance the rates of copper extraction. The Fe^{3+}/Fe^{2+} couple is the only oxidizing agent used in addition to the O_2/H_2O redox couple in bacterial leaching. Iron is added either as dissolved iron salt and/or as iron sulphide mineral (FeS_2). Iron provides nutrients for bacterial growth and promotes the oxidation of copper. The addition, for example of 12 g/LFe^{3+} can increase the rate of copper extraction by as much as 12.5 times.

H_2O_2/H_2O can substantially increase the rate of copper extraction by as much as 120 times when compared with Fe^{3+}/Fe^{2+}. The bacteriostatic effect of peroxide precludes its used in bacterial leaching. However its application in in-direct leaching could lead to significant improvements of kinetics.

There are issues with the use of oxidising agents. With Fe^{3+}/Fe^{2+} it is related to increased contamination of leachates and the greater cost of secondary treatment processing. Whilst H_2O_2/H_2O is costly and is highly volatile.

Particle Size and Composition of Waste: In general larger particle sized waste leads to lower rates of leaching. However the effect of particle size, particularly at the lower size (0–1000 μm) is less well defined. It appears the reducing the particle size does not necessarily lead to higher rates of leaching as would be expected. It was determined the composition or the grade of the waste has a stronger impact on the rates of leaching. Waste with higher copper concentration gave higher rates of copper leaching, irrespective of the particle size. It appears the presence of lower reduction potential metallic contaminants causes the galvanic coupling effects that hamper the rate of leaching.

13.8 ACKNOWLEDGMENTS

This research was supported under Australian Research Council Discovery Project Scheme (DP1096342). I would also like to acknowledge Young Hong, Amy Chan, Masha Saidan, John Bucknell and Brendan Brown for performing the experiments for this study.

References

Adhapure, N. N., Waghmare, S. S., Hamde, V. S., and Deshmukh, A. M. (2013). "Metal solubilization from powdered printed circuit boards by microbial consortium from bauxite and pyrite ores." *Appl. Biochem. Microbiol.*, **49**(3), 256–262.

Akormedi, M., Asampong, E., and Fobil, J. N. (2013). "Working conditions and environmental exposures among electronic waste workers in Ghana." *Int. J. Occup. Environ. Health*, **19**(4), 278–286.

Andersen, T. N., Ghandehari, M. H., and Eyring, H. (1975). "Limitation to mixed potential concept of metal corrosion—Copper in oxygenated sulfuric-acid solutions." *J. Electrochem. Soc.*, **122**(12), 1580–1585.

Aston, J. E., Peyton, B. M., Lee, B. D., and Apel, W. A. (2010). "Effects of ferrous sulfate, inoculum history, and anionic form on lead, zinc, and copper toxicity to Acidithiobacillus caldus strain BC13." *Environ. Toxicol. Chem.*, **29**(12), 2669–2675.

Barba-Gutierrez, Y., Adenso-Diaz, B., and Hopp, M. (2008). "An analysis of some environmental consequences of European electrical and electronic waste regulation." *Resour. Conserv. Recycl.*, **52**(3), 481–495.

Bas, A. D., Deveci, H., and Yazici, E. Y. (2013). "Bioleaching of copper from low grade scrap TV circuit boards using mesophilic bacteria." *Hydrometallurgy*, **138**, 65–70.

Birloaga, I., De Michelis, I., Ferella, F., Buzatu, M., and Veglio, F. (2013). "Study on the influence of various factors in the hydrometallurgical processing of waste printed circuit boards for copper and gold recovery." *Waste Manage.*, **33**(4), 935–941.

Bismarck, A., Boccaccini, A. R., Egia-Ajuriagojeaskoa, E., Hulsenberg, D., and Leutbecher, T. (2004). "Surface characterization of glass fibers made from silicate waste: Zeta-potential and contact angle measurements." *J. Mater. Sci.*, **39**(2), 401–412.

Brandl, H. B., and Wegmann, M. (2001). "Computer-munching microbes: Metal leaching from electronic scrap by bacteria and fungi." *Hydrometallurgy*, **59**(2–3), 319–326.

Brandl, H., Lehmann, S., Faramarzi, M. A., and Martinelli, D. (2008). "Biomobilization of silver, gold, and platinum from solid waste materials by HCN-forming microorganisms." *Hydrometallurgy*, **94**(1–4), 14–17.

Brown, B., Saidan, M., and Valix, M. (2010). "Bioleaching of E-wastes." *Proc., 25th Int. Mineral Processing Congress*, Australasian Institute of Mining and Metallurgy (AusIMM), VIC, Australia.

Bucknell, J., Saidan, M., and Valix, M. (2010). "Kinetics of reclamation of e-waste by a bioleaching route." *Proc., 25th Int. Mineral Processing Congress*, Australasian Institute of Mining and Metallurgy (AusIMM), VIC, Australia.

Burgstaller, W., and Schinner, F. (1993). "Leaching of metals with fungi." *J. Biotechnol.*, **27**(2), 91–116.

Cai, L., Xiao, H. R., Huang, S. M., Li, H., and Zhou, G. T. (2013). "Solubilization of magnesium-bearing silicate minerals and the subsequent formation of glushinskite by Aspergillus niger." *Geomicrobiol. J.*, **30**(4), 302–312.

Cano, E., Torres, C. L., and Bastidas, J. M. (2001). "An XPS study of copper corrosion originated by formic acid vapour at 40% and 80% relative humidity." *Mater. Corros.*, **52**(9), 667–676.

Chancerel, P., and Rotter, V. S. (2009). "Assessing the management of small waste electrical an electronic equipment through substance flow analysis: the example of gold in Germany and the USA." *Rev. Metall./Cah. Inf. Tech.*, **106**(12), 547–553.

Chen, T. T., and Dutrizac, J. E. (1989). "A mineralogical study of the deportment and reaction of silver during copper electrorefining." *Metall. Trans. B*, **20**(3), 345–361.

Chi, X. W., Streicher-Porte, M., Wang, M. Y. L., and Reuter, M. A. (2011). "Informal electronic waste recycling: A sector review with special focus on China." *Waste Manage.*, **31**(4), 731–742.

Choi, N. S., Cho, K. S., Kim, D. S., and Kim, D. J. (2004). "Microbial recovery of copper from printed circuit boards of waste computer by Acidithiobacillus ferrooxidans." *J. Environ. Sci. Health Part A Toxic/Hazard. Subst. Environ. Eng.*, **39**(11–12), 2973–2982.

Coman, V., Robotin, B., and Ilea, P. (2013). "Nickel recovery/removal from industrial wastes: A review." *Resour. Conserv. Recycl.*, **73**, 229–238.

Cui, J. R., and Zhang, L. F. (2008). "Metallurgical recovery of metals from electronic waste: A review." *J. Hazard. Mater.*, **158**(2–3), 228–256.

Dalrymple, I., et al. (2007). "An integrated approach to electronic waste (WEEE) recycling." *Circuit World*, **33**(2), 52–58.

Das, T., Ayyappan, S., and Chaudhury, G. R. (1999). "Factors affecting bioleaching kinetics of sulfide ores using acidophilic micro-organisms." *Biometals*, **12**(1), 1–10.

Demirkiran, N., Ekmekyapar, A., Kunkul, A., and Baysar, A. (2007). "A kinetic study of copper cementation with zinc in aqueous solutions." *Int. J. Miner. Process.*, **82**(2), 80–85.

de Oliveira, C. R., Bernardes, A. M., and Gerbase, A. E. (2012). "Collection and recycling of electronic scrap: A worldwide overview and comparison with the Brazilian situation." *Waste Manage.*, **32**(8), 1592–1610.

European-Parliamentary-Council. (2002a). "Directive 2002/96/EC of the European Parliament and of the Council of 27 January 2003 of the RoHS." *Official Journal L 037, 13/02/2003 P. 0019 – 0023, European Council and Parliament*, New Europa Building, Brussels, Belgium.

European-Parliamentary-Council. (2002b). "Directive 2002/96/EC of the European Parliament and of the Council of 27 January 2003 of WEEE." *Official Journal L 037, 13/02/2003 P. 0019 – 0023, European Council and Parliament*, New Europa Building, Brussels, Belgium.

Feldt, T., et al. (2014). "High levels of PAH-metabolites in urine of e-waste recycling workers from Agbogbloshie, Ghana." *Sci. Total Environ.*, 369–376 466–467.

Fernandes, A., Afonso, J. C., and Dutra, A. J. B. (2012). "Hydrometallurgical route to recover nickel, cobalt and cadmium from spent Ni-Cd batteries." *J. Power Sourc.*, **220**, 286–291.

Fernandes, A., Afonso, J. C., and Dutra, A. J. B. (2013). "Separation of nickel(II), cobalt(II) and lanthanides from spent Ni-MH batteries by hydrochloric acid leaching, solvent extraction and precipitation." *Hydrometallurgy*, **133**, 37–43.

Gadd, G. M. (2001). "Microbial metal transformations: Minireview." *J. Microbiol.*, **39**(2), 83–88.

Gunsinger, M. R., Ptacek, C. J., Blowes, D. W., Jambor, J. L., and Moncur, M. C. (2006). "Mechanisms controlling acid neutralization and metal mobility within a Ni-rich tailings impoundment." *Appl. Geochem.*, **21**(8), 1301–1321.

Guzman, J., Saucedo, I., Revilla, J., Navarro, R., and Guibal, E. (2003). "Copper sorption by chitosan in the presence of citrate ions: Influence of metal speciation on sorption mechanism and uptake capacities." *Int. J. Biol. Macromol.*, **33**(1–3), 57–65.

Herat, S., and Agamuthu, P. (2012). "E-waste: A problem or an opportunity? Review of issues, challenges and solutions in Asian countries." *Waste Manage. Res.*, **30**(11), 1113–1129.

Hofmann, A., et al. (2013). "Interaction of Fe(III) and Al(III) during hydroxylation by forced hydrolysis: The nature of Al-Fe oxyhydroxy co-precipitates." *J. Colloid Interface Sci.*, **407**, 76–88.

Hong, Y., and Valix, M. (2014). "Bioleaching of electronic waste using acidophilic sulfur oxidising bacteria." *J. Cleaner Prod.*, **65**, 465–472.

Hong, Y., Saidan, M., and Valix, M. (2013a). "Bioleaching of electronic waste using acidophilic sulfur oxidising bacteria." *Proc., Int. Conf. on Solid Wastes*, Sino-Forest Applied Research Centre for Pearl River Delta Environment (ARCPE), Hong Kong Baptist University, Hong Kong SAR, P.R. China.

Hong, Y., Saidan, M., and Valix, M. (2013b). "Kinetics of copper bioleaching of E-waste using acidophilic bacteria." *Proc., Int. Conf. on Solid Wastes*, Sino-Forest Applied Research Centre for Pearl River Delta Environment (ARCPE), Hong Kong Baptist University, Hong Kong SAR, P.R. China.

Huang, K., Guo, J., and Xu, Z. M. (2009). "Recycling of waste printed circuit boards: A review of current technologies and treatment status in China." *J. Hazard. Mater.*, **164**(2-3), 399–408.

Ilyas, S., Anwar, M. A., Niazi, S. B., and Ghauri, M. A. (2007). "Bioleaching of metals from electronic scrap by moderately thermophilic acidophilic bacteria." *Hydrometallurgy*, **88**(1–4), 180–188.

Ilyas, S., Lee, J. C., and Chi, R. A. (2013). "Bioleaching of metals from electronic scrap and its potential for commercial exploitation." *Hydrometallurgy*, **131**, 138–143.

Ilyas, S., Ruan, C., Bhatti, H. N., Ghauri, M. A., and Anwar, M. A. (2010). "Column bioleaching of metals from electronic scrap." *Hydrometallurgy*, **101**(3–4), 135–140.

Ilyas, S., and Lee, J. C. (2013). "Fungal leaching of metals from electronic scrap." *Miner. Metall. Proc.*, **30**(3), 151–156.

Kahhat, R., and Williams, E. (2009). "Product or waste? Importation and end-of-life processing of computers in Peru." *Environ. Sci. Technol.*, **43**(15), 6010–6016.

Keles, O. (2009). "An optimization study on the cementation of silver with copper in nitrate solutions by Taguchi design." *Hydrometallurgy*, **95**(3–4), 333–336.

Kiddee, P., Naidu, R., and Wong, M. H. (2013). "Electronic waste management approaches: An overview." *Waste Manage.*, **33**(5), 1237–1250.

Komilis, D., Tataki, V., and Tsakmakis, T. (2013). "Leaching of heavy metals from personal computer components: Comparison of TCLP with a european leaching test." *J. Environ. Eng.*, 10.1061/(ASCE)EE.1943-7870.0000752, 1375–1381.

Labunsk, I., Harrad, S., Santillo, D., Johnston, P., and Brigden, K. (2013). "Levels and distribution of polybrominated diphenyl ethers in soil, sediment and dust samples collected from various electronic waste recycling sites within Guiyu town, southern China." *Environ. Sci. Process. Impacts*, **15**(2), 503–511.

Leduc, L. G., Ferroni, G. D., and Trevors, J. T. (1997). "Resistance to heavy metals in different strains of Thiobacillus ferrooxidans." *World J. Microbiol. Biotechnol.*, **13**(4), 453–455.

Lee, J. C., and Pandey, B. D. (2012). "Bio-processing of solid wastes and secondary resources for metal extraction—A review." *Waste Manage.*, **32**(1), 3–18.

Lewis, G., Gaydardzhiev, S., Bastin, D., and Bareel, P. F. (2011). "Bio hydrometallurgical recovery of metals from Fine Shredder residues." *Miner. Eng.*, **24**(11), 1166–1171.

Liang, G. B., Mo, Y. W., and Zhou, Q. F. (2010). "Novel strategies of bioleaching metals from printed circuit boards (PCBs) in mixed cultivation of two acidophiles." *Enzyme Microb. Technol.*, **47**(7), 322–326.

Liang, G., Tang, J., Liu, W., and Zhou, Q. (2013). "Optimizing mixed culture of two acidophiles to improve copper recovery from printed circuit boards (PCBs)." *J. Hazard. Mater.*, **250**, 238–245.

Liu, H. W., et al. (2011). "The co-culture of Acidithiobacillus ferrooxidans and Acidiphilium acidophilum enhances the growth, iron oxidation, and CO_2 fixation." *Arch. Microbiol.*, **193**(12), 857–866.

Maketon, W., and Ogden, K. L. (2009). "Synergistic effects of citric acid and polyethyleneimine to remove copper from aqueous solutions." *Chemosphere*, **75**(2), 206–211.

Mattsson, E., and Bockris, J. O. (1959). "Galvanostatic studies of the kinetics of deposition and dissolution in the copper + copper sulphate system." *Trans. Faraday Soc.*, **55**(9), 1586–1601.

Mishra, D., Ahn, J. G., Kim, D. J., Roy Chaudhury, G., and Ralph, D. E. (2009). "Dissolution kinetics of spent petroleum catalyst using sulfur oxidizing acidophilic microorganisms." *J. Hazard. Mater.*, **167**(1–3), 1231–1236.

Mishra, D., Kim, D. J., Ralph, D. E., Ahn, J. G., and Rhee, Y. H. (2008). "Bioleaching of metals from spent lithium ion secondary batteries using Acidithiobacillus ferrooxidans." *Waste Manage.*, **28**(2), 333–338.

Nakajima, K., Takeda, O., Miki, T., Matsubae, K., Nakamura, S., and Nagasaka, T. (2010). "Thermodynamic analysis of contamination by alloying elements in aluminum recycling." *Environ. Sci. Technol.*, **44**(14), 5594–5600.

Nandeesh, L. S., and Sheshadri, B. S. (1988). "The anisotropic nature of copper corrosion and inhibition in sulfuric-acid containing strychnine." *Corrosion Sci.*, **28**(1), 19–27.

Nash, J., and Bosso, C. (2013). "Extended producer responsibility in the United States full speed ahead?" *J. Ind. Ecol.*, **17**(2), 175–185.

Noel, N., Florian, B., and Sand, W. (2010). "AFM & EFM study on attachment of acidophilic leaching organisms." *Hydrometallurgy*, **104**(3–4), 370–375.

Norton, G. A., Richardson, R. G., Markuszewski, R., and Levine, A. D. (1991). "Precipitation of jarosite compounds as a method for removing impurities from acidic wastes from chemical coal cleaning." *Environ. Sci. Technol.*, **25**(3), 449–455.

Oh, C. J., Lee, S. O., Yang, H. S., Ha, T. J., and Kim, M. J. (2003). "Selective leaching of valuable metals from waste printed circuit boards." *J. Air Waste Manage. Assoc.*, **53**(7), 897–902.

Oishi, T., Koyama, K., Alam, S., Tanaka, M., and Lee, J. C. (2007). "Recovery of high purity copper cathode from printed circuit boards using ammoniacal sulfate or chloride solutions." *Hydrometallurgy*, **89**(1–2), 82–88.

Okibe, N., and Johnson, D. B. (2004). "Biooxidation of pyrite by defined mixed cultures of moderately thermophilic acidophiles in pH-controlled bioreactors: Significance of microbial interactions." *Biotechnol. Bioeng.*, **87**(5), 574–583.

Oleszek, S., Grabda, M., Shibata, E., and Nakamura, T. (2013). "Distribution of copper, silver and gold during thermal treatment with brominated flame retardants." *Waste Manage.*, **33**(9), 1835–1842.

Ongondo, F. O., Williams, I. D., Dietrich, J., and Carroll, C. (2013). "ICT reuse in socio-economic enterprises." *Waste Manage.*, **33**(12), 2600–2606.

Pant, D., Joshi, D., Upreti, M. K., and Kotnala, R. K. (2012). "Chemical and biological extraction of metals present in E waste: A hybrid technology." *Waste Manage.*, **32**(5), 979–990.

Pariatamby, A., and Victor, D. (2013). "Policy trends of e-waste management in Asia." *J. Mater. Cycles Waste Manage.*, **15**(4), 411–419.

Park, Y. J., and Fray, D. J. (2009). "Recovery of high purity precious metals from printed circuit boards." *J. Hazard. Mater.*, **164**(2–3), 1152–1158.

Pradhan, D., Kim, D. J., Ahn, J. G., Chaudhury, G. R., and Lee, S. W. (2010). "Kinetics and statistical behavior of metals dissolution from spent petroleum catalyst using acidophilic iron oxidizing bacteria." *J. Ind. Eng. Chem.*, **16**(5), 866–871.

Qiu, M. Q., Xiong, S. Y., Zhang, W. M., and Wang, G. X. (2005). "A comparison of bioleaching of chalcopyrite using pure culture or a mixed culture." *Miner. Eng.*, **18**(9), 987–990.

Reck, B. K., and Graedel, T. E. (2012). "Challenges in metal recycling." *Science*, **337**, 690–695.

Rimaszeki, G., Kulcsar, T., and Kekesi, T. (2012). "Application of HCl solutions for recovering the high purity metal from tin scrap by electrorefining." *Hydrometallurgy*, **125**, 55–63.

Robinson, B. H. (2009). "E-waste: An assessment of global production and environmental impacts." *Sci. Total Environ.*, **408**(2), 183–191.

Robotin, B., Coman, V., and Ilea, P. (2012). *Nickel recovery from electronic waste III. Iron nickel separation*. Studia Universitatis Babes-Bolyai Chemia, **57**(3), 81–90.

Rossi, G. (1990). *Biohydrometallurgy*, McGraw-Hill, Hamburg.

Schweickert, H., Elmiligy, A. A., Melendez, A., and Lorenz, W. J. (1976). "Mathematical treatment of galvanostatic double pulse method applied to consecutive charge-transfer reactions." *J. Electroanal. Chem.*, **68**(1), 19–30.

Shibayama, A., Tongamp, W., Altansukh, B., Haga, K., and Hosoi, A. (2013). "Electronic waste treatment. Part 1: Autoclave oxidation-leaching using pyrite waste from mine tailing." *Hydrometallurgy*, **137**, 92–100.

Stewart, K. L., and Gewirth, A. A. (2007). "Mechanism of electrochemical reduction of hydrogen peroxide on copper in acidic sulfate solutions." *Langmuir*, **23**(19), 9911–9918.

Sthiannopkao, S., and Wong, M. H. (2013). "Handling e-waste in developed and developing countries: Initiatives, practices, and consequences." *Sci. Tot. Environ.*, **463**, 1147–1153.

Strasser, H., Burgstaller, W., and Schinner, F. (1994). "High-yield production of oxalic-acid for metal leaching processes by Aspergillus niger." *FEMS Microbiol. Lett.*, **119**(3), 365–370.

Sulka, G. D., and Jaskula, M. (2003). "Study of the kinetics of silver ions cementation onto copper from sulphuric acid solution." *Hydrometallurgy*, **70**(1–3), 185–196.

Sulka, G. D., and Jaskula, M. (2004). "Study of the mechanism of silver ions cementation onto copper from acidic sulphate solutions and the morphology of the silver deposit." *Hydrometallurgy*, **72**(1–2), 93–110.

Sun, L., and Qiu, K. Q. (2012). "Organic oxalate as leachant and precipitant for the recovery of valuable metals from spent lithium-ion batteries." *Waste Manage.*, **32**(8), 1575–1582.

Tang, J. A., and Valix, M. (2006). "Leaching of low grade limonite and nontronite ores by fungi metabolic acids." *Miner. Eng.*, **19**(12), 1274–1279.

Tanskanen, P. (2013). "Management and recycling of electronic waste." *Acta Mater.*, **61**(3), 1001–1011.

Townsend, T. G. (2011). "Environmental issues and management strategies for waste electronic and electrical equipment." *J. Air Waste Manage. Assoc.*, **61**(6), 587–610.

Tsydenova, O., and Bengtsson, M. (2011). "Chemical hazards associated with treatment of waste electrical and electronic equipment." *Waste Manage.*, **31**(1), 45–58.

Tue, N. M., Takahashi, S., Subramanian, A., Sakai, S., and Tanabe, S. (2013). "Environmental contamination and human exposure to dioxin-related compounds in e-waste recycling sites of developing countries." *Environ. Sci. Processes Impacts*, **15**(7), 1326–1331.

Tuncuk, A., Stazi, V., Akcil, A., Yazici, E. Y., and Deveci, H. (2012). "Aqueous metal recovery techniques from e-scrap: Hydrometallurgy in recycling." *Miner. Eng.*, **25**(1), 28–37.

Valix, M., Saidan, M., and Brown, B. (2012). "Leaching of electronic waste using biometabolised acids." *Chin. J. Chem. Eng.*, **20**(3), 530–534.

Valix, M., Usai, F., and Malik, R. (2001a). "The electro-sorption properties of nickel on laterite gangue leached with an organic chelating acid." *Miner. Eng.*, **14**(2), 205–215.

Valix, M., Usai, F., and Malik, R. (2001b). "Fungal bio-leaching of low grade laterite ores." *Miner. Eng.*, **14**(2), 197–203.

Viramontes-Gamboa, G., Rivera-Vasquez, B. F., and Dixon, D. G (2007). "The active-passive behavior of chalcopyrite—Comparative study between electrochemical and leaching responses." *J. Electrochem. Soc.*, **154**(6), C299–C311.

Wagner, T. P. (2013). "Examining the concept of convenient collection: An application to extended producer responsibility and product stewardship frameworks." *Waste Manage.*, **33**(3), 499–507.

Wang, J. W., Bai, J. F., Xu, J. Q., and Liang, B. (2009). "Bioleaching of metals from printed wire boards by Acidithiobacillus ferrooxidans and Acidithiobacillus thiooxidans and their mixture." *J. Hazard. Mater.*, **172**(2–3), 1100–1105.

Watling, H. R. (2006). "The bioleaching of sulphide minerals with emphasis on copper sulphides—A review." *Hydrometallurgy*, **84**(1–2), 81–108.

Wei, L., and Liu, Y. S. (2012). "Present status of e-waste disposal and recycling in China." *Proc., 7th Int. Conf. on Waste Management and Technology*, Vol **16**, L. Jinhui and H. Hualong, eds., Elsevier, Beijing, 506–514.

Wilts, H., Bringezu, S., Bleischwitz, R., Lucas, R., and Wittmer, D. (2011). "Challenges of metal recycling and an international covenant as possible instrument of a globally extended producer responsibility." *Waste Manage. Res.*, **29**(9), 902–910.

Wu, Y. F., and Tsai, T. H. (2007b). "Effect of organic acids on copper chemical mechanical polishing." *Microelectron. Eng.*, **84**(12), 2790–2798.

Xiang, Y., et al. (2010). "Bioleaching of copper from waste printed circuit boards by bacterial consortium enriched from acid mine drainage." *J. Hazard. Mater.*, **184**(1–3), 812–818.

Xu, G. Y., Zhou, S. Q., Hu, H. L., and Xu, Z. C. (2013). "Generation and management: A general industrial hazardous waste survey in China." *Asian J. Chem.*, **25**(12), 6601–6608.

Yang, T., Xu, Z., Wen, J. K., and Yang, L. M. (2009). "Factors influencing bioleaching copper from waste printed circuit boards by Acidithiobacillus ferrooxidans." *Hydrometallurgy*, **97**(1–2), 29–32.

Yazici, E. Y., and Deveci, H. (2013). "Extraction of metals from waste printed circuit boards (WPCBs) in H_2SO_4-$CuSO_4$-NaCl solutions." *Hydrometallurgy*, **139**, 30–38.

Zhang, K., Wei, Y. L., and Zeng, E. Y. (2013). "A review of environmental and human exposure to persistent organic pollutants in the Pearl River Delta, South China." *Sci. Total Environ.*, **463**, 1093–1110.

Zhang, W. S., and Cheng, C. Y. (2007). "Manganese metallurgy review. Part III: Manganese control in zinc and copper electrolytes." *Hydrometallurgy*, **89**(3–4), 178–188.

Zhang, Y. H., Liu, S. L., Xie, H. H., Zeng, X. L., and Li, J. H. (2012). "Current status on leaching precious metals from waste printed circuit boards." *Proc., 7th Int. Conf. on Waste Management and Technology*, Vol. **16**, L. Jinhui and H. Hualong, eds., Elsevier Science Bv., Amsterdam, 560–568.

Zheng, G. N., Xu, X. J., Li, B., Wu, K. S., Yekeen, T. A., and Huo, X. (2013). "Association between lung function in school children and exposure to three transition metals from an e-waste recycling area." *J. Exposure Sci. Environ. Epidemiol.*, **23**(1), 67–72.

Zhu, N., et al. (2011). "Bioleaching of metal concentrates of waste printed circuit boards by mixed culture of acidophilic bacteria." *J. Hazard. Mater.*, **192**(2), 614–619.

Zhu, P., Chen, Y., Hao, Y., Fan, Z. Y., Qian, G. R., and Zhou, M. (2013). "Leaching behavior of copper powders from waste printed circuit board by electrogenerated chlorine and aeration." *Miner. Metall. Proc.*, **30**(2), 79–84.

CHAPTER 14

Management of Special Wastes: Biosolids

Tanaji T. More[1]
Sridhar Pilli[2]
Song Yan[3]
Rajeshwar D. Tyagi[4]
Rao Y. Surampalli[5]
Tian C. Zhang[6]

14.1 INTRODUCTION

Wastewater sludge (also called biosolids) is typically defined as the residue generated from wastewater treatments. According to Eurostat, sewage sludge is the accumulated settled solids separated from various types of water either moist or mixed with a liquid component as a result of natural or artificial processes. In relation to environmental matters, the definition of sludge supplied by the European Environment Agency (EEA 2005) may be useful: 1) a semi-fluid mass of sediment resulting from treatment of water, sewage and/or other wastes; and 2) a soft, soupy, or muddy bottom deposit, such as found on tideland or at the bottom of a water body. Sewage sludge consists of organic and inorganic solids removed in the primary and/or secondary clarifier or in a separate thickening process. The generated sludge is usually in the form of a liquid or semi-solid, containing 0.25–12% solids by weight, depending on the treatment operations and processes used (Metcalf and Eddy 2003). Wastewater sludge contains considerable quantities of pollutants, such as heavy and toxic metals, toxins (pesticides, and household and industrial chemicals), protein and organics (carbohydrates, fats, oils, greases, chemicals, etc.). There are many chemicals (more than 330 synthetic

[1]INRS-ETE, Université du Quebéc, Québec, QC, Canada.
[2]INRS-ETE, Université du Quebéc, Québec, QC, Canada.
[3]INRS-ETE, Université du Quebéc, Québec, QC, Canada.
[4]INRS-ETE, Université du Quebéc, Québec, QC, Canada.
[5]Univ. of Nebraska-Lincoln, Lincoln, NE.
[6]Univ. of Nebraska-Lincoln, Lincoln, NE.

chemical contaminants) in the sludge that are known or suspected toxins. Sludge also contains various disease-causing bacteria (including Escherichia coli and Salmonella), viruses (hepatitis and polio), germs and parasites (that can contaminate soils and crops) (Gaspard et al. 1997). Therefore, untreated sludge poses a risk to human health, agricultural productivity and the environment. Sludge also contains foul odours which may lead to respiratory problems, creates unpleasant environment and reduces quality of life.

In recent years, sewage sludge has become an international topic of concern over various common problems. General agreement is that the sludge production in the world is on a relentless growth curve. Large quantities of sludge are being produced annually by wastewater treatment plants (WWTPs) all over the world. Due to world-wide population growth and increasing stringent environmental regulations, the municipal sludge production is expected to increase significantly in the future (Metcalf and Eddy 2003; LeBlanc et al. 2008). The cost associated with the treatment of the large quantities of sludge can be considerable. For example, the annual cost of sludge management in the USA ($4.0–6.5 billion) is 30–50% of the annual costs of wastewater treatment (Dentel 2012). Therefore, the sludge management has become the most significant challenge in wastewater management (Metcalf and Eddy 2003), and it represents both an economic and environmental concern of considerable importance. Since last few decades, the rapid increase in sludge production, following massive investments in new WWTP installations and upgrading existing facilities, have created the need for adaptation of economically and environmentally acceptable sludge management schemes, which take into account the worldwide growing interest in reuse and recovery possibilities (LeBlanc et al. 2008).

Environmental quality requirements for sludge are becoming increasingly stringent and disposal outlets are decreasing. Economic and social pressures are demanding for low-cost solutions to sludge disposal problems. It is well known that sludge handling, treatment and disposal are complex, and there are varieties of sludge treatment processes and their combinations. Traditionally, thickening, conditioning, dewatering and drying are primarily used to remove moisture from sludge, while digestion, composting, incineration and wet oxidation are used to treat or stabilize the organic material in the sludge. In general, land application, incineration and landfilling are the most widely used sludge disposal methods. More recently, the traditional sludge disposal routes are under legislative, social and economic pressure. For example, some disposal routes (e.g., sea disposal) have been already phased out. On the other hand, recycling and reusing the sludge are the most preferred options over the traditional land application, landfilling or incineration. However, due to growing concerns over pathogens, toxic pollutants, heavy metals and emerging contaminants (ECs), the sludge handling and disposal methods are under constant scrutiny. Overall, sludge management is facing a great challenge of finding cost-effective, innovative and sustainable solutions.

This chapter discusses current understanding of sludge production, characteristics, legislation, sludge handling and disposal technologies and recent trends in sludge management. Specifically, this chapter reviews published reports on frameworks of sludge management all over the world, unifies existing data to outline the current situation of sustainable management of sludge, including sludge treatment/disposal options, methods for energy recovery and production of value-added products from sludge, and discusses future perspectives of sludge management issues and specific conditions for different countries.

14.2 SLUDGE PRODUCTION

Sludge is an inevitable and major product of the municipal wastewater treatment processes. Sludge arises mainly from the settlement of wastewater solids and from the surplus growth of the organisms present in the wastewater. Sludge generation in biological WWTPs is difficult to estimate because all suspended solids of the influent wastewater do not appear in the sludge. Therefore, very detailed data are needed or to be collected on wastewater characteristics, the treatment processes and the operating parameters to reliably predict sludge production (Koch et al. 1997). The sources of sludge in a WWTP vary according to the type of the plant and its operational method (Metcalf and Eddy 2003). The principal sources of sludge at municipal WWTPs are primary sedimentation basin (residue generated during the primary physical and/or chemical wastewater treatment) and secondary clarifier (residue generated during secondary biological wastewater treatment). Additional sludge may also come from chemical precipitation, facilities, screening, grinder and filtration devices. If the advanced (tertiary) treatments (such as for N and P removal) are used, excess sludge is produced in the tertiary clarifier.

In general, specific sludge production in wastewater treatment varies widely from 35 to 85 g dry solids per population equivalent per day (Ginestet and Camacho 2007). Gomez et al. (2010) reported that the production of dry biosolids ranges from 20 to 32.85 kg per person per year. Specific primary sludge mainly consists of settleable solids from raw wastewater and its production is 50–60 gTSS per population equivalent per day (Parker 2012). Some sludge solids are biologically metabolized to soluble or gaseous end-products in the biological treatment process or sludge digestion process. Specific production of excess sludge in WWTPs is about 0.32 ± 0.08 gTSS/gCOD removed (with digestion) and 0.27 ± 0.07 gTSS/gCOD removed (without digestion) (Ginestet and Camacho 2007; Parker 2012).

Table 14.1 shows the annual sludge production in different countries/regions. Sewage sludge produced in kg per person per year in different parts of the world in 1999 was 20.0, 18.0, 12.5, 7.5, 1.0 and 1.0, respectively in Western Europe, North America, Japan, Eastern Europe, East Asia and Asia Pacific (Oleksandr 2005). There was no recorded annual sewage sludge production for Africa, South Asia, South America and Middle East region. The high-income countries that have the most comprehensive infrastructure and wastewater treatment technologies (e.g., secondary and tertiary treatments) produce the largest masses of wastewater sludge per person (LeBlanc et al. 2008). The middle-income countries (which have

Table 14.1. The annual sludge production in different countries/regions

Region	*Country*	*10^3 Tons dry solids/yr*	*References*
North America	United States	6,515	LeBlanc et al. (2008)
	Canada	550	
South America	Brazil	372	LeBlanc et al. (2008)
Australasia	Australia & New Zealand	360	LeBlanc et al. (2008); AWA (2012)
Europian Union	EU-27	10,957	Kelessidis and Stasinakis (2012)
	EU-15	9,806	
	EU-12	1,151	
Asia	China	2,966	LeBlanc et al. (2008)
	Japan	2,000	
	Turkey	580	
	Thialand	63	
Africa	South Africa	465*	NSER-SA (2013)

*Estimated sludge production in South Africa for year 2011

less-developed wastewater treatment infrastructure) collect and treat wastewater from lower percentages of their populations, and produce far less wastewater sludge per person on a national level. Whereas, inadequate or absence of sewerage systems means no recorded sewage sludge. Looking ahead, it is certain that there will be steady increase in wastewater sludge production in many parts of the world in the years to come. In contrast, most developed countries in the world are now seeing a leveling of wastewater sludge production rate, because most of their populations have already been connected with the wastewater management infrastructure, either directly or indirectly. However, in such countries, growing population and economic considerations will be the deciding factor for sludge management practices.

14.3 SLUDGE CHARACTERISTICS

To treat and dispose the sludge effectively, it is very important to know the characteristics of the sludge. Wastewater sludge is a heterogeneous material varying in quantity and composition from one place to another and even from one day to the next in the same WWTP because of variations in the incoming wastewater composition. The characteristics of the sludge vary around the world since the properties of sludge depend upon several factors such as population density, their habits, the proportion and nature of community's industrial base, the composition of the wastewater, the age of sludge and method of wastewater treatment (U.S. EPA

2006). Wastewater is typically processed through physical (e.g., sedimentation, floatation), chemical (e.g., coagulation, flocculation) and biological treatment methods and special wastewater treatment processes (e.g., tertiary). These wastewater treatment processes are either combined or individually used in a variety of systems to meet the stringent legislation limits concerning wastewater effluent discharge from the WWTPs. Based upon these wastewater treatment processes, generated sludge can be grouped as primary, secondary or biological, mixed and tertiary wastewater sludge (Table 14.2). Typically, sludge is the dispersion of solid waste particles in water. A characteristic feature of different types of sludge is the very high water content, with only 1–5% solids on a mass basis. Table 14.3 shows a typical composition and properties of the municipal wastewater sludge.

The sludge characteristics can be broadly classified as physical, chemical and biological. Physical parameters (e.g., solids, moisture and rheology) give general information on sludge processability and handleability; chemical parameters give necessary information about the presence of nutrients and toxic/dangerous compounds; whereas biological parameters give information on microbial activity and the presence of organic matter/pathogens. According to their importance from the sludge management point of view, the components of wastewater sludge can be conveniently divided into four principal groups: water content, nutrient contents (macro and micro), organic and inorganic toxic pollutants as well as pathogens. Many nutrients and chemical constituents are important in accordance with the ultimate disposal of the processed sludge and the liquid removed from the sludge during treatment. According to Metcalf and Eddy (2003), control of pH levels, alkalinity and organic acid content are important in the process of anaerobic digestion. The energy (thermal) content of sludge is significant when thermal processes (e.g., gasification, pyrolysis, combustion, and wet oxidation) are considered. Brief descriptions about the sludge characteristics are as follows.

14.3.1 Water Content

Water is present in wastewater sludge in different forms. The most widely accepted classification of water content in wastewater sludge is the operational one which is based on the arbitrary hypotheses. The water content of sludge is generally considered to be constituted by two fractions: free water and bound water (Colin and Gazbar 1995). This classification does not characterize the binding energy of water to the solid phase. The general classified water content of the sludge is free water 70%, adsorption and capillary water 22% and intracellular water 8% (Jonassen et al. 2008). However, in the literature various nomenclatures are used for the water content present in the sludge (Table 14.4). The water content of wastewater sludge is important because it is directly linked with sludge handling/disposal methods and overall sludge management costs.

14.3.2 Nutrient Contents

Many of the nutrients present in the wastewater sludge are important when considering the final disposal of the processed sludge (e.g., land applications) and

Table 14.2. General classification of wastewater sludge

Type of sludge	*Origin*	*Solids conc. (%)*	*Specific gravity of sludge*	*Remarks*
Primary	Produced during primary treatment of sewage	5.0–8.0	1.02–1.05	Contains heavy solids, soap, oil, grease, grit, sand; overall mostly inorganic matter
Biological/2nd treatment	Produced during the secondary wastewater treatment	0.5–2.0	1.005–1.01	Contains mostly organic matter, biomass, metals and inorganic chemicals
Mixed	Mixture of primary and biological sludge	2.5–4.0	—	Mixture of primary and biological sludge
Tertiary	Produced during tertiary or advanced wastewater treatment	—	1.005–1.01	Content depends on the type of wastewater treatment and chemical used if chemical treatment is used
Digested	Produced after aerobic or anaerobic digestion of the sludge	—	1.005–1.01	—

Source: Manara and Zabaniotou (2012) and Metcalf and Eddy (2003)

Table 14.3. Typical composition and properties of wastewater sludge

Parameters (units)	Untreated primary		Digested primary		Untreated activated sludge
	Range	Typical	Range	Typical	
pH	5.0–8.0	6.0	6.5–7.5	7.0	6.5–8.0
Total dry solids (% of TS)	2.0–8.0	5.0	6.0–12.0	10.0	0.83–1.16
Volatile solids (% of TS)	60–80	65	30–60	40	59–88
Proteins (% of TS)	20–30	25	15–20	18	32–41
Nitrogen (% of TS)	1.5–4	2.5	1.6–6.0	3.0	2.4–5.0
Phosphorous (% of TS)	0.8–2.8	1.6	1.5–4.0	2.5	2.8–11.0
Grease and fats					
Ether soluble (% of TS)	6–30	—	5–20	—	
Ether extract (% of TS)	7–35	—	—	—	5–12
Potash (% of TS)	0–1	0.4	0.0–3.0	1.0	0.5–0.7
Cellulose (% of TS)	8.0–15.0	10.0	8.0–15.0	10.0	—
Iron (not as sulfide)	2.0–4.0	2.5	3.0–8.0	4.0	—
Silica SiO_2 (% of TS)	15.0–20.0	—	10.0–20.0	—	—
Alkalinity (mg/L as $CaCO_3$)	500–1500	600	2500–3500		580–1100
Organic acids (mg/L as HAc)	200–2000	500	100–600	3000	1100–1700
Energy content (kJ TS/kg)	23000–29000	25000	9000–14000	12000	19000–23000
Fecal coliform (MPN/g TS)	2,016,543	—	—	—	—

Source: Metcalf and Eddy (2003); Izrail et al. (2006); Smith et al. (2009)

Table 14.4. Classification of different types of water present in wastewater sludge

Water bonding	*Classification of the water within activated sludge*	*Remarks*
Free (un-bound)	1. **Free water** (Smollen 1986; 1990) 2. **Free water** (Kopp and Dichtl 2000; 2001) 3. **Free water** (Lee and Lee 1995) 4. **Surface condensation** (Keey 1972; Halde 1979) 5. **Free moisture:** non-associated with solid particles include void water not affected by capillary force (Tsang and Vesilind 1990; Vesilind 1994; Smith and Vesilind 1995; Vesilind and Hsu 1997)	Thermodynamically, it behaves as pure water; it can be removed by simple thickening or by the application of weak mechanical strain (Colin and Gazbar 1995)
Chemical-bound	1. **Chemically bound:** bound by powerful chemical linkage (Smollen 1986; 1990) 2. **Intracellular water** (Kopp and Dichtl 2000; 2001) 3. **Internal water** (Lee and Lee 1995) 4. **Ionic:** hydration (Keey 1972; Halde 1979) 5. **Bound (or hydration) water molecular crystallisation** (Tsang and Vesilind 1990; Vesilind 1994; Smith and Vesilind 1995; Vesilind and Hsu 1997)	Only removable by the addition of heat (Smith and Vesilind 1995); can be eliminated by thermal drying at temp. >105°C (Colin and Gazbar 1995)
Physical-bound	1. **Bound water:** adsorbed by individual solid particle (Smollen 1986; 1990) 2. **Surface water:** bound by adhesive forces (Kopp and Dichtl 2000; 2001) 3. **Surface water:** physically/chemically adsorbed (Lee and Lee 1995)	It can be reduced by sludge chemical conditioning to replace water molecules by the adsorbed conditioners; over dosage of the chemical conditioners results in an increase in the bound water content

	4. **Adsorptive:** hydrogen bonding/solvatation, physico-chemical adsorption, osmotic inhibitions; structural: dissolution in gels (Keey 1972; Halde 1979) 5. **Surface (or vicinal) water:** held on to the surface of solid particles by adsorption and adhesion (Tsang and Vesilind 1990; Vesilind 1994; Smith and Vesilind 1995; Vesilind and Hsu 1997)	
Mechanical-bound	1. **Immobilised water:** capillary, trapped or adhesive (Smollen 1986 1990) 2. **Interstitial water:** bound by capillary forces between the flocs (Kopp and Dichtl 2000; 2001) 3. **Interstitial water** (Lee and Lee 1995) 4. **Micro capillary:** capillary condensation (Keey 1972; Halde 1979) 5. **Interstitial water:** trapped in crevices and interstitial spaces of flocs and organisms (perhaps water trapped within cells) (Tsang and Vesilind 1990; Vesilind 1994; Smith and Vesilind 1995; Vesilind and Hsu 1997)	Only part of the interstitial water can bc removed by mechanical dewatering

the liquid removed from the sludge during treatment (e.g., landfill leachate, ground water contamination). Wastewater sludge contains varying concentrations of macronutrients such as nitrogen (N), phosphorus (P), potassium (K), sulphur (S), calcium (Ca), magnesium (Mg) and other micronutrients (Table 14.5). Typical major nutrients (present in sludge of some countries) such as N, P and K are presentenced in Table 14.6.

Wastewater sludge contains both organic and inorganic N. Organic N present in sludge occurs as nitrogen bound in organic molecules such as proteins. Forms of inorganic N present in sludge include nitrate (NO3-), nitrite (NO2-), ammonia (NH3) and ammonium (NH4+). The amounts of organic and inorganic N present in sludge are dependent on the waste source and wastewater and sludge treatment processes. Most of the organic N in sewage sludge is associated with the sludge solids, and thus organic N levels are not appreciably altered by sludge dewatering or drying procedures. However, water soluble inorganic forms of N and their concentrations decreases dramatically during dewatering. Phosphorus is present in wastewater sludge in organic (predominant) and inorganic (phosphate) forms. Potassium is not generally found in high concentrations in wastewater sludge

Table 14.5. Typical nutrient contents in wastewater sludge

Nutrients	*Remarks*
Macronutrients	
Nitrogen (N):	
• Organic (e.g., amino acids)	• Mostly biological origin (~60% on a dry basis); not readily available
• Inorganic (e.g., NO^{3-}, NO^{2-}, NH_3/NH_4^+)	• Soluble; readily available
Phosphorus (P)	Readily form insoluble compounds with sludge solids
Potassium (K)	Required to be recycled
Sulphur (S)	—
Calcium (Ca)	Readily form insoluble compounds with sludge solids
Magnesium (Mg)	Readily form insoluble compounds with sludge solids
Micronutrients	
Boron (B)	—
Copper	—
Iron (Fe)	Readily form insoluble compounds with sludge solids
Chlorine (Cl)	—
Manganese (Mn)	—
Molybdenum (Mo)	—
Zinc (Zn)	—

Table 14.6. Major nutrient content of sewage sludge in different countries

Country	*Total N*[a]	*Total P*[a]	*Potassium*[a]	*Ref.*[b]
USA	20.0–60.0	20.0–30.0	4.3	[1, 2]
Canada	32.5–53.7	8.7–16.9	1.0	[1, 2]
UK	38.0	22	—	[1, 2]
Spain	16.3–45.6	4.2–16.7	—	[1, 2]
China	22.6	11.5	5.6	[1, 2]
Argentina	19.3	7.2	—	[1, 2]
India	26.0–38.7	13.4–14.4	—	[1, 2]
Hong Kong	43.8–65.0	10.6–21.8	—	[1, 2]
Australia	70.4	18.0	9.7	[1]
Brazil	57.5	18.2	3.6	[1]
Italy	52.0	14.0	—	[1]
Finland	34.0	24.0	—	[1]
Turkey	16.8	6.8	4.9	[1]

[a]Units in g/kg of dry solids.
[b][1] = LeBlanc et al. (2008); and [2] = Pathak et al. (2009).

because potassium forms are soluble in wastewater and remain in the liquid phase. Of the water soluble constituents that do remain in the sludge, dewatering of sludge reduces their concentrations in sludge while air or heat drying results in increased levels of nonvolatile constituents.

14.3.3 Toxic Inorganic and Organic Pollutants

Heavy Metals. Various processes of wastewater treatment finally accumulate heavy metals in the sludge. Typical metal concentrations of the wastewater sludge are presented in Table 14.7.

Metal concentrations in sewage sludge vary widely depending on several factors such as sludge origin and sludge pretreatment processes (Sanchez-Martin et al. 2007). Industrial wastes usually contain higher levels of heavy metals than domestic wastes. Metal contents and distribution also differs between aerobically and anaerobically digested sludge (Doelsch et al. 2006). Microbial transformations during mesophilic anaerobic digestion increase the stability of organic-metal complexes and precipitate metals as low solubility sulphides. Metals in sewage sludge are distributed in the form of soluble, exchangeable, organic and precipitated. Heavy metals such as zinc (Zn), copper (Cu), nickel (Ni), cadmium (Cd), lead (Pb), mercury (Hg) and chromium (Cr) are principal elements restricting the use of sludge for agricultural purposes. Their potential accumulation in human tissues and biomagnifications through the food chain create both human health and environmental concerns. High loading of sewage sludge to soil may give rise to accumulation of heavy metals, and this increased heavy metal may lead to phytoxicity for the plants grown on soil (Sanchez-Martin et al. 2007). The mobility of trace metals, their bioavailability and related eco-toxicity to plants, depend strongly on their specific chemical forms or ways of binding. Therefore, the contents of heavy metals, pesticides and hydrocarbons need to be determined when sewage sludge is to be disposed of by land application.

Emerging Contaminants. Recent studies have revealed the presence of different ECs with wide range of concentrations in wastewater sludge (Díaz-Cruz et al. 2009; Clarke and Smith 2011; Eljarrat et al. 2013). Modern society depends on a large range of organic chemicals. Synthetic organic chemicals used in food production, pharmaceuticals, personal care products, plastics manufacturing, and other industrial processes such as flame retardants, dioxins and steroid hormones end up in wastewater and then finally in the wastewater sludge (Table 14.8).

In Europe, most of the sludge (53%) is used in agriculture directly or after composting, whereas in North America more than 50% of the sludge is disposed of by land applications. By these land and agricultural routes, ECs present in the sludge, can migrate to natural environment such as soil, groundwater and natural surface water resources. The possible threats for the environment and human health due to the agricultural use of sludge containing ECs are reported comprehensively in recent reviews (Díaz-Cruz et al. 2009; Clarke and Smith 2011; Stasinakis 2012; Eljarrat et al. 2013). Many of these chemicals are toxic, carcinogenic, and some display endocrine-disrupting properties in organisms exposed to

Table 14.7. Typical metal content in wastewater sludge and regulatory limits

		Regulatory limits, mg/kg dry solids	
Metal	*Typical concentration in wastewater sludge[a] (mg/kg dry solids)*	*USA best quality standard for general public contact users[b]*	*European directive regulatory limits[b]*
Arsenic (As)	1.1–230	41	—
Cadmium (Cd)	1.0–3.410	39	20–40
Chromium (Cr)	10.0–990,000	—	—
Cobalt (Co)	11.3–2490	—	—
Copper (Cu)	84.0–17,000	1500	1000–1750
Iron (Fe)	1000–154,000	—	—
Lead (Pb)	13.0–26,000	300	750–1200
Manganese (Mn)	32.0–9870	17	—
Mercury (Hg)	0.6–56	17	16–25
Molybdenum (Mo)	0.1–214	75	—
Nickel (Ni)	2.0–5300	420	300–400
Selenium (Se)	1.7–17.2	100	—
Tin (Sn)	2.6–329	—	—
Zinc (Zn)	101–49,000	2800	2500–4000

[a]Metcalf and Eddy (2003)
[b]LeBlanc et al. (2008)

Table 14.8. Emerging contaminants occurring in wastewater sludge

Group of ECs	*Representative compounds*	*Conc. of ECs(μg/kg of dry solids)*	*Ref.*[a]
Pharmaceuticals and Personal care products	Carbamazepine (CBZ), Diclorenac (DCF), Trimethoprim, Clarithromycin, Ciprofloxacin, Ofloxacin, Triclocarbon (TCC), Sulfapyridine, Sulfamethoxazole (SMX), Sulfadiazine, Sulfathiazole, Triclosan (TCS), Galaxolide (HHCB)	0.1–1508	[1–4]
Estrogens	E1, E2, α-E2, E3, E1-3S, E2-3S, EE2	0.64–406	[1–4]
Surfactants	Sodium dodecylbenzenesulfonate (C12LAS)	1.0–52,333,000	[1–4]
	Nonylphenol (NP)		
	Dedecyl trimethyl ammonium chloride (C12TMA)		
Phthalate acid esters	Bis(2-ethylhexyl) phthalate (DEHP)		[1–4]
Perfluorinated compounds	Perflurooctanoic acid (PFOA)	0.4–3390	[1–4]
Organotins	Bis(tributyltin)oxide (TBTO)	20.0–9000	[1–4]
	Dibutyltin (DBT)		
	Triphnyltin (TPhT)		
Brominated flame retardants	Tetrabromobisphenol A (TBBPA)	0.2–58800	[1–4]
Benzotriazoles	Benzotriazole (BTr)		[1–4]
Nanoparticles	Fullerenes		[4]
Illicit drugs	4-MBC, OMC, OC, OT	4.0–300	[4]
UV filters	AM, MA	10.0–27700	[4]
Bisphenol-A		4.0–325,000	[1]

[a] [1] = Clarke and Smith (2011); [2] = Díaz-Cruz et al. (2009); [3] = Eljarrat et al. (2013); and [4] = Stasinakis (2012)

critical concentrations over a certain period of time (Díaz-Cruz et al. 2009). The presence, the frequency of occurrence, uncertainty about the source and the fate of these ECs during conventional sludge handling and disposal processes of these contaminants pose great concern currently.

Significant amounts of organic contaminants present in the wastewater system get removed by degradation and attenuation during wastewater and sludge treatments. However, many of organic contaminants have lipophilic properties and hence retained to sewage sludge; they may be present in varying residual concentrations in dry solids, depending on the initial amounts present, their lipophilicity and the extent of destruction during wastewater and sludge treatment. During wastewater treatment most of the ECs are sorbed to some extent on wastewater solids and end up in the sludge through sedimentation and/or biological flocculation during the treatment processes (e.g., in primary and secondary clarifiers). Fate of these chemicals is often affected by solids retention time, temperature, and sludge characteristics, adaptation of biomass and bioavailability of these compounds. During sludge anaerobic digestion, nonylphenol ethoxylates (NPE), some pharmaceuticals and most phthalates are biotransformed to some extent, while contradictory results have been reported for linear alkylbenzene sulfonate (LAS), NP, synthetic musks, estrogens and DEHP. No significant removal has been observed for TCS, TCC, fluoroquinolones, carbamazepine, TBT and PFCs; the available data for these compounds is yet very limited. Moreover, they are persistent, difficult to degrade and can be bioaccumulated.

One of the major issues with ECs is the methods of analysis due to their ultra-low concentrations in sludge. Different methodologies have been developed for the analysis of ECs. However, very few studies are referred to the analysis of ECs in sewage sludge. Sample preparation procedures play a fundamental role in developing analytical methodology for complex matrices, such as sludge. In general, the instrumental determination is mainly based on the use of mass spectrometer (MS) coupled with gas chromatography or liquid chromatography (LC). LC-MS has shown advantages in terms of versatility and relatively simplified sample preparation (Eljarrat et al. 2013). Overall, concentrations of ECs in sludge range between few μg/kg to some (~30) g/kg of dry sludge solids (Eljarrat et al. 2013).

Clarke and Smith (2011) have ranked the ECs, in a decreasing order of priority, as follows: 1) perfluorinated chemicals (PFOS, PFOA) 2) polychlorinated alkanes (PCAs) and polychlorinated naphthalenes (PCNs), 3) polybrominated diphenyl ethers (PBDEs), organotins (OTs), triclosan (TCS), and triclocarban (TCC), 4) benzothiazoles, 5) antibiotics and pharmaceuticals, 6) synthetic musks, bisphenol-A, quaternary ammonium compounds (QACs), and steroids, 7) phthalate esters (PAEs) and 8) polydimethylsiloxanes (PDMSs). Perfluorinated chemicals and polychlorinated alkanes currently are reported as of great concern because these chemicals have been detected in human blood and environmental samples (Oehlmann et al. 2008; Clarke and Smith 2011). The highly water soluble chemicals are likely to increase exposures to humans through water contamination, plant accumulation and grazing animal accumulation. Continued assessments of the significance and implication of ECs in sludge are necessary to support and ensure the long-term sustainability and security of the beneficial agricultural route for future sludge management.

14.3.4 Pathogens

In wastewater sludge the primary source of pathogens is human wastes. In general, municipal sludge contains human pathogenic bacteria, protozoa, viruses and helminthes (Table 14.9), which are generally classified as primary (e.g., bacteria, viruses, protozoa, and helminthes) or secondary (e.g., fungi). A primary pathogen can invade and infect a healthy person, whereas a secondary, or opportunistic, pathogen invades and infects highly debilitated or immunosuppressed individuals (Lu et al. 2012). The pathogens of concern in sludge include bacteria: E. coli O157: H7, Salmonella, Listeria, and Helicobacter pylori; the viruses: coxsackievirus, echovirus, hepatitis A, rotavirus, and norovirus; and the parasites: Cryptosporidium, Cyclospora, Toxoplasma, Microsporidia and Giardia (Sidhu and Toze 2009). Because of the extremely large numbers of pathogens that exist in the world, it is impossible to test sludge for all types of pathogens.

Wastewater treatment processes achieve a certain degree of influent pathogen reduction, and most of the surviving pathogens end up in the sludge. Pathogens occur in the sludge from relatively large volumes of treated wastewater, and hence, they become concentrated and very infectious. The concentrations of pathogens in wastewater basically depend on the health condition of the community. During epidemics, often there is an increase in the sludge pathogens' concentration. Complete sterilization of the sludge is practically difficult to achieve and also undesired as only a small portion of them are the causes of concern. Land spread sludge can be treated to nearly eliminate pathogens. Composting, if conducted properly is a very effective way of disinfection and destruction of pathogens primarily by exposing pathogens to high temperatures (e.g., 55°C) for a prolonged period of time (Lu et al. 2012). Researchers have found that pathogens can survive in sludge for weeks, months, or even years after reduction treatment processes.

14.3.5 Odor

Degradation of organic matter including proteins, amino acids and carbohydrates leads to the formation of odorous compounds (Walker 1991). Organic and inorganic forms of sulfur, mercaptans, ammonia, amines and organic fatty acids are present in the sludge. These compounds are considered as the most offensive odor causing compounds and are released from sludge by heat, aeration and digestion.

The odors vary by the type of residual solids processed and the methods of processing. Anaerobic digestion of primary wastewater residuals produces hydrogen sulfide and other sulfur-containing gases, while alkaline stabilization of the solids volatilizes ammonia along with other volatile compounds. Composting odors can be caused by ammonia, amine, sulfur-based compounds, fatty acids, aromatics and hydrocarbons such as terpenes from the wood products used as bulking agents. Aerobically digested and air-dried sludge solids may contain little

Table 14.9. Pathogens occurring in the sludge and potential indicators

Type	*Some species/strains*	*Potential indicators*	*Ref.*[a]
Human			
Bacteria	*Salmonella* spp., *Shigella* spp., *Campylobacter* spp., Enterohaemorrhagic *E. coli* O157:H7, *Helicobacter pylori* and *Listeria monocytogenes*, *Pseudomonas aeruginosa*, *Yersinia enterolitica*, *Clostridium perfringens*, *E. coli* (enteropathogenic strains), *C. botulinum*, *Bacillus anthracis*, *Listeria monocytogenes*, *Vibrio cholera*, *Mycobacterium* spp., *Leptospira* spp., *Staphylococcus*, *Streptococcus*	Total & fecal coliforms, F-specific coliphage, faecal streptococci, Somatic coliphage, B-fragilis phage, *Enterococcus* sp., Thermotolerant coliforms, *C. perfringens*	[1,2]
Protozoa parasites	*Cryptosporidium parvum*, *Cryptosporidium hominis*, *Giardia intestinalis*, *Entamoeba histolytica*, *Giardia lambia*, *Toxoplasma gondii*, *Sarcocystis*	Giardia cysts	[1,2]
Fungi	*Aspergillus* spp., *A. fumigates*, *Phialophora richardsii*, *Geotrichum candidum*, *Trichophton* spp., *Epidermophyton* spp.	—	[1,2]
Helminths	*Ascaris*, *A. suum*, *A. lumbricoides*, *Capillaria*, *Taenia saginata*, *Taenia solium*, *Diphyllobothrium latum*, *Echinococcus gramulosus*, *Ancylostoma duodenale*, *Necator americanus*, *Hymenolepis diminuta*, *Toxocara canis*, *T. cati*, *Trichuris trichura*	—	[1,2]
Viruses	Enteric virus, Poliovirus, Coxsackievirus, Echovirus, 'New' enterovirus, Adenovirus, Reovirus, Hepatitis A-virus, Rotavirus, Astrovirus, Calicivirus, Coronavirus, Norwalk-like calicivirus, Small round viruses, Parvovirus, Adenoassociated viruses, Influenza virus, Polyomavirus, Norovirus, Enterovirus	Virus indicator, Retrovirus, Adenovirus, Somatic coliphage, F. RNA phage	[1,2]

Yeast	*Candida albicans, C. krusi, C. tropicalis, C. guillermondii, Cryptococcus neoformans, Trichosporon*	—	[2]
Prasites	*Ascaris* spp., *Taenia* spp., *Giardia lamblia, Crytosporidium parvum*	—	
Plants			
Fungi	*Chalara elegans, Cylindrocarpon destructans, Fusarium oxysporum, Phythophtora cryptogea, Plasmodiophora brassicae, Pythium ultimum, Rhizoctonia solani, Sclerotinia sclerotiorum*	—	[3]
Viruses	Beet necrotic yellow vein virus, Tobacco mosaic virus, Tobacco necrosis virus	—	[3]
Nematodes (oocysts, larva, ova)	*Ditylenchus dipsaci, Longidorus* spp., *Xiphiema* spp.	—	[3]

[a] [1] = Carrington (2001); [2] = Sidhu and Toze (2009) and [3] = Lepeuple et al. (2004)

hydrogen sulfide, but have mercaptan and dimethyl sulfide odors. Nuisance odors occur commonly at sludge processing facilities and sludge recycling locations where proper management and control are not implemented. Failure to acknowledge the potential for odors and to work to prevent odor emissions can result in complaints, shutdowns, expensive retrofits, and non-acceptance of the finished products.

Sludge processing facilities can be operated and managed to reduce odor generation and emissions. The quantity and intensity of odorous compounds can be reduced by proper sludge management. This includes several incentives such as addressing the operation and maintenance procedures to prevent anaerobic conditions, addition of oxidizing agents to prevent formation of hydrogen sulfide, selection of polymers which are resistant to breakdown at high temperatures and pH, optimizing all stabilization processes such as anaerobic digestion, aerobic digestion, or alkaline stabilization, evaluate the impacts of blending different types of solids and storage (U.S. EPA 2006).

14.4 WASTEWATER SLUDGE REGULATIONS

14.4.1 North America

The USA and Canada have continuing extensive research programmes focused on potential risks from dioxins and furans, accumulation of anti-microbial agents in sludge, pathogens reactivation and regrowth, odor sources and relevant mechanisms, etc. The public and political perceptions have ultimately guided sludge/biosolids policy and practices. These perceptions are often based upon scientific principles and risk assessments. The USEPA has established pollutant numerical limits and management practices (40 CFR Part 503) for the reuse and disposal of solids generated from the processing of municipal wastewater and septage. These regulations cover specifically land application of biosolids, surface disposal of biosolids, pathogen and vector reduction in treated biosolids and incineration. The reuse of reclaimed water and biosolids undertaken in accordance with the Best Practice Environmental Management Guidelines for the Use of Reclaimed Water (Publication 464) and the Draft Best Practice Environmental Guidelines for Biosolids Management, respectively, are exempt from EPA works approval and licensing provisions. The disposal of effluent and biosolids is, however, subject to EPA works approval and licensing. The exemption provisions are intended to encourage reuse schemes rather than disposal, by allowing them to proceed without unnecessary regulatory control. In Canada, the various steps in biosolids management including production, treatment, sale and end use/disposal are regulated under a multi-faceted regulatory system, involving federal and provincial/territorial legislation of Canada (CCME 2010). Environment Canada administers the Canadian Environmental Protection Act at the national level, whereas Canadian Food inspection Agency (CFIA) regulates the sale and import of biosolids intended for use as fertilizers or supplement. The provinces manage the wastewater treatment and sludge disposal through provincial/territorial acts and regulations.

14.4.2 European Union (EU)

The generation of waste is increasing continuously each year within the EU and therefore, since last few decades, a number of directives on waste treatment have been approved by the EU. These directives establish a legal framework for the treatment of waste within the community. The waste framework directive of 1975 regulates EU member states to manage waste by encouraging prevention and environmentally friendly disposal. The sewage sludge directive (86/278/EEC) seeks to encourage the use of sewage sludge in agriculture (CEC 1986). However, any potential harmful effect of sewage sludge use on soil, vegetation, animals and human beings must be prevented. According to the above principle, the use of untreated sludge in agriculture is prohibited, unless it is injected or incorporated in the soil. The urban wastewater treatment directive (UWWT, 91/271/EEC) requires all discharges from municipal wastewater plants serving more than 2000 population equivalent to be treated according to the sensitivity of the receiving water in specified phases that were completed by 31 December 2005 (CEC 1991). It is declared that "sludge arising from waste water treatment shall be re-used whenever appropriate." This directive also stops the disposal of sludge at surface waters (dumping from ships or discharge from pipelines or by any other means) by 31 December 1998. Accordingly, all sludge disposals must be land based in the future. This directive also set the rules for handling the hazardous waste. This directive was amended by 98/15/EC to set more stringent quality standards for wastewaters. The implementation deadline for UWWT directive was 2005 for the older EU-15 member states and 2015 or 2018 for the countries that joined the EU after 2004 (CEC 2009). The decision 2001/118/EC in chapters 19 and 20 has enlisted sludge in non-hazardous wastes (CEC 2001).

In order to better protect the environment, the member states should take measures for the waste management based on basic principles (hierarchy of priority order), i.e., prevention, minimization, recycling, reuse, energy recovery and improving final disposal and monitoring. All the measures taken by the EU in accordance with these principles must not endanger human health and/or harmful to the environment. Based on these principles, the Landfill Directive 99/31/EC (CEC 2009) prohibited landfilling of both liquid and untreated wastes and set restrictions as well as quantitative targets for bio-degradable municipal solid wastes (such as sewage sludge) that are disposed in landfills. According to these, a reduction of 50% and 65% of produced solid waste by years 2013 and 2020 should be achieved, respectively. National legislations of some of the EU member states (e.g., Germany, Austria, the Netherlands) have set very strict limits for the organic matter or total organic carbon (TOC) present in sludge, practically prohibiting sludge landfilling. Sewage sludge directive 86/278/EEC (CEC 1986) describes the beneficial sludge use on soils. It also specifies the rules for the sampling and analysis of sludge and soils, sets out record keeping requirements and limit values for concentrations of heavy metals in sewage sludge and soil. Most of the EU countries have set more stringent national requirements for heavy metals concentrations in sludge and the concentrations of heavy metals in soil as compared the EU directive. Therefore, there is a wide variation in national limit values for heavy metals among EU members. Most of the EU member states have prohibited the disposal of untreated sludge in soil (Kelessidis and Stasinakis 2012). In the case of Austria, Denmark, Finland and Poland, specific obligations for treatment have been set such as biological stabilization or chemical stabilization before reuse, while France, Sweden and Estonia allow the use of untreated sludge under certain authorized conditions. Use of sludge in forests, silviculture, mines and green areas has been prohibited in Austria, Belgium, Germany, Luxembourg, Netherlands and Poland. Very few countries have rules limiting the concentration of any organic chemicals in sewage sludge. EU standards for sludge deposition on the soils are far more stringent than that of the USEPA standards. For example, the allowed limit for the Hg and Cd concentration in the sludge deposited soils are 1.5 and 3 ppm for EU but 8 and 20 ppm for the USEPA.

14.4.3 Australasia

The regulatory framework in Australasia has traditionally focused on ensuring that the impacts of treated effluent on the receiving environment are acceptable. As a consequence, water businesses throughout the country historically tended to stockpile sludge as a means of reducing costs. However, more recently there is strong unanimity of view amongst state regulators that more sustainable management of sludge is desirable. The sustainable sludge management include recovery of the amount of residual energy, but this option can be difficult to be implemented because of community concerns and energy costs. Within the regulatory framework, both Australia and New Zealand are making strong efforts to carry beneficial recycling of sludge to land as a main management strategy. Australia is already well advanced on that way, with almost total beneficial reuse applied to sludge from major cities, where over 50% of population is residing. However, New Zealand has a long way to go, still relying heavily on landfilling, but with a lot of research activity on beneficial outlets around bigger cities. In this context, the National Biosolids Research Programme by the Commonwealth Scientific and Industrial Research Organisation (CSIRO) of Australia also should focus on assessing the effects of metals, nutrients, pathogens and trace organics.

14.4.4 Asia

In the Eastern Asia region, Japan and South Korea show a particularly major contrast. Japan is opting for the highest technological approaches to sludge management, with problems of higher energy costs, while until 2008 South Korea had practiced oceanic disposal of the sludge. In Japan, the rate of agricultural reuse is limited and is now decreasing. Composting seems to be an alternative, so it has been planned to make composting a bigger part of reuse operations as the high energy costs of thermal processes become more problematic. However, this option is subject to strict regulation for heavy metals, and also the acceptance of farmers, who prefer animal manures. In Japan, the use of incinerated ash or dewatered cake is growing each year, as an ingredient of Portland cement and as the additives in other thermal solidification processes. However, the problem associated with increasing energy costs is a real limiting factor. In 2008, South Korea had planned to ban ocean dumping, and current national law on fertilizers prohibits the use of sludge on agricultural land. A law change is, therefore, essential for future security of outlets. It has also had successful trials with sludge as feedstock for earthworms, and sludge admixtures with coagulants and plant seed (cultivating vegetation, grass, etc.) as a protective surface against erosion on sloping cut ground. It clearly appears that in both cases changes in legislation are becoming essential.

In South and South-East Asian countries the sludge management situation is highly different. For example, in India, onsite sanitation and wastewater management is highly neglected in most of urban areas (only 30% of the sewage is being treated). Fecal sludge is most commonly managed in the line of municipal solid waste (EAI 2014). At the present, sewage sludge and effluents from sewage treatment plants are frequently disposed of on agricultural lands for irrigation/manure purposes. Only recently (in 2008), Government of India has raised the budget for urban sanitation plans and began to address the management of septage. However, according to ECO-Asia (2010), lack of awareness and motivation are the main obstacle in the implementing wastewater management policies. Apart from the specific case of China, the range of population for countries in South Asian or South-East Asia is wide, and consequently, sludge management approaches are also significantly distinct. The sludge lagoons are used in Vietnam (78 million in population with only one sewage treatment plant in operation and 6 more planned), whereas the sludge almost totally landfilled on an off-shore island site in Singapore (4.3 million with all sewage being collected and treated). Malaysia plans to cover 85% of population by 2022 and to replace the use of drying beds and lagoons with centralised mechanical dewatering and disposal to landfill or land reclamation in coming years. Taiwan is following a policy of total reuse by 2010 from landfilling to a mix of composting for horticulture and co-incineration with municipal solid waste with ash reuse.

In China, the number of municipal WWTPs has more than tripled from 718 in 2005 to 2,823 in 2010 and to >5,200 in 2012 (CGTI 2011), while in the 1980s there were only few WWTPs in the entire county. The sludge discharge over the past five years has increased by 5% per year, and it is now estimated to reach 22 to 30 million tons annually (CGTI 2011). Currently, about 48% of sludge goes to fertilizer; 34% to landfills; 14% is discharged (dumped) into the environment; and the remaining 4% is incinerated or used in other ways (Zuo 2010). Annual sludge treatment expenses in China are estimated to be $200 million in 2010 and $800 million in 2015 (CGTI 2011). The awareness of sludge management is rising up in China. The 11th five-year plan for 2006–2010 put the sludge management issue on the table for the very first time (Hua 2011). Since 2009,

China has published sludge technology guidelines and treatment standards, including preferred technologies and restrictions on sludge transportation and storage. In 2009, the Circular Economy Promotion Law required county-level governments to support use by private enterprises of sludge treatment systems. In 2010, guidelines on best-available technologies were issued by the National Planning Commission (NPC). The central government has issued a sludge treatment technology guide and has announced has planned to spend $10 billion in sludge treatment over the 12th Five-Year Plan period (2011–2010), which is barely enough to meet infrastructure investment needs. National regulations are promulgating the construction of treatment plants for sludge, including drying, composting, incineration, and brick manufacture as one way to utilize the treated sludge. Nevertheless, China's sludge industry faces many challenges, ranging from lack of comprehensive national policies to insufficient funding. Moreover, sufficient and transparent information on sludge management still is not available in China.

14.4.5 South America

In Latin American countries, sludge management has been neglected or given secondary importance than the wastewater treatment. Economy is the main issue for lack of proper sludge management in the region. Majority of the sanitary investments are directed towards the sewerage and wastewater treatment considering that 20% of population is without adequate sanitation. The quite small attention given to sludge management is reflected in the local legislations that often simply adopt regulations of industrialized or more advanced countries without any attempt to adapt it to local situations. In this context, developing appropriate low cost methods to make sludge safe and suitable for reuse in agriculture is the primary need of the region. Another important step required to be taken is the development of legislation adapted to the area with the improvement of institutional capacity to guarantee its enforcement. Based on the importance of agriculture in Latin America, as well as the degree of soil erosion, land application for crop production and reclamation are considered the best management practices in the region.

14.4.6 Africa

Poor economy and inadequate legislation are the two major problems in African countries. Sanitation systems mainly include on-site systems, such as septic tanks, bucket latrines, pit latrines, etc. The management of sludge is typically limited to on-site storage, and in many cases the treatment plant consists of integrated pond systems that have never been de-sludged. Any cases, a number of encouraging initiatives for improving the management of sludge from on-site sanitation systems have been reported. The treatment of faecal sludge includes pond-based systems (Ghana, Benin and Mali), constructed wetlands (Mali), co-treatment in ponds (Burkina Faso), co-treatment with wastewater in ponds (Botswana and Tanzania) and co-treatment with wastewater in activated sludge plants (South Africa).

There is a lack of easily accessible information on sludge management practices or on relevant legislation or guidelines in nearly all the countries of this continent. Similar to South American countries, some African countries have given priority to wastewater treatment but neglected the sludge management. In most of the African countries there is inadequate or no sanitation management adopted in rural areas and cities. African countries have mostly adopted legislations of the industrialised countries. Therefore, often these countries have neglected the local situations in their legislations. South Africa is an exception for the sludge management practices. South Africa (a developing country) has relatively progressive and comprehensive environmental regulations for sewage sludge management. Recently, new South African sanitary guideline series (5 volumes) are being implemented to rectify the shortcomings of the previous sludge guideline and to encourage proper sludge management practices (Snyman and Herselman 2006). These volumes deals with the selection of appropriate sludge management options based on the characteristics and classification of the sludge and requirements for the different sludge management strategies such as agricultural use, on-site and off-site disposal, beneficial use, thermal processes and obtaining commercial products from sludge.

14.5 SUSTAINABLE MANAGEMENT OF SLUDGE

The 21st century witnesses an increasing interest in sustainable sludge management. Currently, stricter regulations imposed on sustainable sludge management in different countries are catalyzing the re-orientation of the sludge to energy recovery and value-added products. Processes which promote sustainability will become viable options for resource management if conversion into a "value-added product" or energy recovery/generation can be realized (Tyagi et al. 2009). In this section, the current situation of sustainable management of sludge is described, including sludge treatment/disposal technologies, methods for energy recovery and production of value-added products from sludge.

14.5.1 Treatment Technologies

14.5.1.1 Sludge Thickening

Sludge thickening is the process used for sludge volume reduction. Sludge volume is reduced by reducing the moisture content of the sludge, i.e., increasing the dry solids content of sludge. Thickeners are used in WWTPs, where solids must be concentrated to increase the efficiency of further treatment. Sludge thickening influences the reliability and performance, of the entire sludge treatment system. The volume reduction obtained by sludge thickening is beneficial to subsequent treatment processes, such as digestion, dewatering, drying and combustion (Metcalf and Eddy 2003). In these treatments, the sludge thickening is important because of the limitations over several factors such as the capacity of tanks and equipment, quantity of sludge conditioning chemicals, the amount of heat required by digesters and the amount of auxiliary fuel required for heat drying and/or incineration. Sludge volume reduction is also very desirable when liquid sludge is transported via trucks for direct application to land as a soil conditioner or other applications. The proper location of the thickener in a WWTP is important. If sludge is to be digested, thickening a blend of primary and waste activated sludge (WAS) is a good practice. If the sludge is to be dewatered, then it should be thickened separately and mixed immediately before dewatering. Generally, thickening is used as the pre-processing of sludge prior to sludge dewatering. Thickening is economically attractive because considerable volume reduction is achieved with even modest increases in sludge solids concentrations. Such reduction can provide significant savings in the cost of dewatering, digestion, or other downstream activities.

The three commonly used methods for sludge thickening are gravity settling, dissolved-air-floatation, and mechanical thickening such as centrifugation, gravity belt and rotary drum methods (Metcalf and Eddy 2003). Thickening by gravity is the most common concentration process in use at WWPTs due to its simple and inexpensive operation. It is the process of concentrating the solids of sludge to reduce their volume. Sedimentation is the downward settling of heavy particles to the bottom of the tank. Flotation is the upward travel of lighter and finely divided suspended particles to the top surface of the tank. It is primarily used in chemical treatment and to treat activated sludge. Air is first dissolved into the raw sludge under pressure and then released at atmospheric pressure. Separation of solids takes place by introducing fine air bubbles into the sludge. The bubbles rise, causing flotation of solids. Coagulants as alum and salts of iron are added to hasten flotation. Dissolved air flotation moves up the solids rapidly which otherwise settle slowly under gravity thickening. Flotation thickening units are useful in sewage treatment plants for handling waste-activated sludge. They have the advantage over gravity thickening tanks on offering higher solids concentrations and lower initial equipment cost due to higher hydraulic and solids loadings which can be applied. Centrifugation was shown to be an effective method to thicken a portion of the wastewater sludge before digestion. The thickened portion was then diluted with wastewater sludge to achieve the desired six percent solids content in the digester. Although centrifuges have been used widely for dewatering, they have had limited use for thickening because of their relatively high cost. They have been used for thickening WAS where space limitations or sludge characteristics make other methods unsuitable.

14.5.1.2 Sludge Stabilization

Sludge stabilization can be biological, chemical or physical. Biological stabilization includes anaerobic digestion, thermophilic aerobic digestion and sludge pasteurization. Sludge digestion is the process of decomposition of organic matter into relatively stable, thick and readily disposable end products. It is done to (i) control the potential for purification of organic matter and to reduce biochemical oxygen demand (BOD) considerably, readily dewater and easy to dry, (iii) reduce pathogens, (iv) recover useful end products such as methane, and (v) reduce volume of sludge.

Anaerobic stabilization utilizes anaerobic digestion processes to stabilize sludge. Most of anaerobic digestion processes are green technologies as they can be operated in concert with energy recovery from sludge and/or biowaste. Details of anaerobic digestion are described in Section 14.5.3.1. Aerobic digestion is the process of oxidizing and decomposing the organic part of the sludge by microorganisms under aerobic conditions. Aerobic digestion produces a stable Class B product, reduces mass/volume of the sludge and pathogenic organisms. Aerobic sludge digestion is the most widely used stabilization process in plants with average flows less than 5 MGD. In chemical stabilization, sludge is treated with high doses of chlorine or compounds of chlorine or other oxidizing agents as hydrogen peroxide and ozone, which deodorized the sludge, and microbial activity is slowed down. In lime stabilization, lime is added to raw sewage to raise its pH to 12 or higher. At this pH, sludge doesn't putrefy nor create odours nor pose a health hazard. A high lime dosage for a three-hour contact time is needed for the effective pathogen kill. Microorganisms are not destroyed but their activity is slowed, and sludge must be disposed of before the organic matter starts to putrefy again. In physical stabilization, the sludge is heated to a temperature of 140°C to 240°C in a reaction vessel at a pressure of 1720 to 2760 kN/m^2 for 15 to 40 minutes. Heat coagulates the solids, breaks down the gel structure and reduces their affinity for water. The resulting sludge is sterilized, almost deodorized and easily de-watered on vacuum filters or filter presses without the need of chemicals. It is costly and is adopted where biological method not practicable.

14.5.1.3 Sludge Conditioning

The solid particles of domestic sludge are extremely fine and hydrophilic (water-loving) and hence, dewatering is difficult. Conditioning is applied to the sludge to separate water entrapped in sludge flocs and to increase dewatering and thickening efficiency (Watanabe and Tanaka 1999). Sludge conditioning increases particle size, overcoming hydration effects and repelling electric charges between particles. It causes dispersed and colloidal fine particles in the sludge to accumulate and their bonding water molecules to release. Conditioning can be carried out by means of chemicals, physical washing, ultrasonication, freeze/thawing processes. Sludge coagulation and flocculation are the fundamental processes of the conditioning. In coagulation, the sludge particles are destabilized by decreasing magnitude of the repulsive interactions between particles, followed by flocculation in which colloidal, and finely divided suspended particles are agglomerated by gentle mixing (U.S. EPA 2006). Sludge particles are negatively charged and repel each other. Therefore, charge neutralization is an important stage in conditioning operation which reduces the repulsing effect of the sludge particles and allows particle collision and aggregation. Both inorganic and organic reagents have been used for sludge conditioning. However, use of inorganic conditioning chemicals increases not only the sludge mass by 15–30% but also reduces the fuel value of the sludge for incineration. As a result, high molecular-weight polymeric materials have been widely accepted in the flocculation of suspended particles in sludge dewatering and water treatment. Roughly 5 kg of polymer are used to condition each 1000 kg of dry solids in typical dewatering processes (Metcalf and Eddy 2003). Cost of the chemical polymers used for the conditioning operations is very high. It is estimated that 25–50 million kg of polymer ($2–4 per kg of cationic synthetic polymers with the total cost of $130 million per annum) are used in the U.S. for sludge conditioning (Subramanian et al. 2010).

In general, flocculating agents are classified into three groups: (i) inorganic flocculants, such as aluminum sulfate and polyaluminum chloride; (ii) organic synthetic flocculants, such as polyacrylamide derivatives and polyethylene amine; and (iii) naturally occurring flocculants, such as chitosan, sodium alginate and bioflocculants (Subramanian et al. 2010). The total operating cost of filter press is Euro 100–280/ton dry solids. However, the polymers/flocculants covers 50% of the total operating cost (Jonassen et al. 2008).

14.5.1.4 Sludge Dewatering

Sewage sludge generated at WWPTs characteristically has a high water and organic content. Both sludge thickening and dewatering are the methods of sludge volume reduction. However, degree of sludge volume reduction is different in both. Sludge thickeners concentrate sludge at lower than 15% sludge solids whereas the sludge dewatering operations results in final sludge solids concentration higher than 15%. Thickened sludge still behaves like liquid and can be pumped. Dewatering is the mechanical or natural removal of water from the sludge (increasing the concentration of total solids). Sludge dewatering is always required prior to treatments such as composting, incineration or landfilling. Despite the growing interest for technologies which allow reducing the specific sludge production (Vergine et al. 2007), mechanical dewatering is still a crucial step to limit the total amount of sludge to be disposed of. Currently, dewatering is mainly performed by belt filter presses or filter presses and centrifugation (Table 14.10).

Sludge resulting from WWT operations contains 0.5–15% of total solids (TS) and 50–80% of volatile solids (VS) (von Sperling and Gonçalves 2007; Uggetti et al. 2010). By sludge thickening about 8% solids concentration can be achievable before dewatering. A typical value is in the 3–5% range. Sludge dewatering can increase the solids content to around 35%, and drying (sludge drying beds, incineration) can remove additional moisture. Energy or reagents required for sludge conditioning increases with the increase in the bound water content of the sludge. The separation of water from the solids depends upon the size of the solids particles. Water separation from the sludge becomes more difficult with an increase in the number of smaller size particles.

Mechanical dewatering of biological sludge requires chemical conditioning or very seldom physical conditioning to enhance the sludge dewaterability,

Table 14.10. Total solids content of the sludge after sludge dewatering treatments

Dewatering methods	*Type of sludge*	*FinalTS (%)*	*Advantages*	*Disadvantages*	*Ref.*[a]
Sludge wetlands	Anaerobic digestion	32	Low energy requirements; reduced operating and maintenance costs; reasonable integration in the environment	Numbers of plants still are relatively less	[1,2]
	Bio-aerated filters (BAF) and raw slurry solids	20			
	Imhoff tanks	50			
	Activated sludge and WAS from settling tank	20			
	Activated sludge, extended aeration	15–28			
Belt Filter Press (BFP)	WAS	12–18	Established technology; relatively simple to operate; pressure and shearing simple, visual process	Difficult to control odor mitigation; good flocculation is necessary; often operated with high hydraulic loads or low residence times	[1,3,4]
	Anaerobic digestor (mixture)	17–23			
	Raw sludge conditioning with polyelectrolytes	18–30			
	Digested sludge conditioning with polyelectrolytes	12–25			
Centrifuge	Activated sludge	14–20	Established technology; reliable and operator friendly; need regular maintenance	Energy costs but offset by high solids; additional flow sheet unit operations vital	[1,3,4]
	Aerobic digester	8–10	Operational flexibility; easy odor control; combined thermal and drying and dewatering; high G forces and high shearing		
	Conditioning with polyelectrolytes	15–35			

(Continued)

Table 14.10. Total solids content of the sludge after sludge dewatering treatments (Continued)

Dewatering methods	*Type of sludge*	*FinalTS (%)*	*Advantages*	*Disadvantages*	*Ref.*[a]
Filter Press	Activated sludge	27–33	Established technology; high pressure without shearing	Semi–continuous but automated	[1,3,4]
	Anaerobic digester	29–35			
	conditioning with $FeCl_3$, lime or polyelectrolytes	20–45			
Vacuum Filter	Vacuum filter activated sludge	12–18	Established technology; low pressure without shearing	Low operational speed, Low cake solids contents	[1,3,4]
	Anaerobic digester (mixture)	17–23			
Rotary Disc Press	—	—	New technology; tolerates poor flocculation; pressure and shearing; simple, visual process	Low operational speed	[1]
Screw Press	—	—	New technology; high pressure and high shearing, low capital costs	Low operational speed, Low cake solids content, Prefers high solids content	[1]
Composting	Aerobic digestion and centrifuge	83	Established old method	Temperature is important factor; not practiced very extensively	[5]
	Dewatered raw sludge	44	Established old method		[6]

[a][1] = Miller et al. (2008); [2] = Uggetti et al. (2010); [3] = Goncalves et al. (2007); [4] = Metcalf and Eddy (2003); [5] = Bertran et al. 2004; and [6] = (Ruggieri et al. 2008)

promoting aggregation of small flocs due to ionic interactions and bridging phenomena (Spinosa and Vesilind 2001). The selection of the chemicals depends on the device used for the dewatering and on the nature of the sludge. Inorganic chemicals (e.g., iron, alum salts or lime) are often used with frame filter presses where high pressure of 10–15 bar is applied, while organic polymers are used with belt presses and centrifuges, where high shear stresses are present. The prediction of the sludge dewaterability is important for the selection of chemicals and the optimization of their doses. The specific resistance to filtration (SRF) and the capillary suction time (CST) are the most commonly used measures of dewaterability. The dewatered sludge generally behaves as a solid and can be trucked in most cases.

Dewatering may be carried out by using other processes such as drying beds, wetlands and thermal drying. The thermal drying includes direct drying, indirect drying, and fluidized-beds systems. Sludge drying beds are the oldest method of sludge dewatering and are still used extensively in small-to-medium sized plants. They are relatively inexpensive and provide dry sludge cake. Drying constitutes an important process for wastewater sludge management, as it can reduce the mass and the volume of the product and consequently the cost of storage, handling and transport. In recent years, many advances have been made to conventional drying bed technology, and new systems are used in medium- and large-sized plants. The drying period is 10–15 days (sand beds) and 30–40 days (for paved drying beds), and the moisture content of the cake is 60–70% and 40–50%, respectively. The mechanical dewatering is not always sufficient to satisfy new environmental regulations; and additional drying step is often needed. Artificial media (made of stainless steel wire wedge) and panels (of polyurethane) have been successfully utilized in wedge-wire drying beds. Drainage and evaporation are the two mechanisms utilized to form a sludge cake. The advantages of these systems are no clogging of the media; constant and rapid drainage; higher throughput rate than sand beds; easier removal of sludge cake; ability of drying difficult-to-dewater sludge; and ease of maintenance (U.S. EPA 2006).

The technology known as sludge treatment wetlands has been used for sludge dewatering since the late 1980s. Sludge is applied over the wetland bed in 200 to 300 mm thick layers and allowed to dry for one week to many weeks depending on the solar insolation. Water trickles down and gets collected in perforated drains while a part of the water gets evaporated due to air currents. Sludge cake formed over the beds is separated when got dried. Major advantages include its low energy requirements, reduced operating and maintenance costs, and a reasonable integration in the environment. Sludge treatment wetlands have been used in the USA and Europe for sludge dewatering and stabilization since 1980s. The largest experience comes from Denmark, where there are over 140 full-scale systems currently in operation (Nielsen 2008). Other systems are implemented in the USA, Poland, the UK, Italy, France, and Spain (Uggetti et al. 2010). Several pilot plant trials have been carried out in Palestine, Cameroon, China, and Canada.

14.5.2 Disposal Methods

During the last couple of decades, there has been a major change in the ways the sludge is disposed. Presently in Europe, North America, Japan and other developed countries, most common sludge management options are land application, landfilling and incineration (Roy et al. 2011), as shown in Table 14.11. Other alternative sludge management options are composting and anaerobic digestion. The latest trends in sludge disposal options include combustion, wet oxidation, pyrolysis, gasification and co-combustion of sewage sludge with other materials for further use as energy source.

14.5.2.1 Land Application

Land application is a commonly used and an attractive option for sewage sludge management because application of sewage sludge to land recycles N, P, other macronutrients (Ca, K and S), micronutrients (Cu and Zn) and organic matter (Metcalf and Eddy 2003). This gives sludge unique advantage as 'fertilizer' since most of the elements present in sludge are essential to plant growth. Utilizing sewage sludge on land as an agricultural fertilizer is the most common option adopted worldwide. Sludge also can be used as a forest fertilizer, as a soil conditioner for the restoration of disturbed soils, as a soil forming material for reclaiming derelict land, and for producing soil for use on green areas in the urban environment. However, land application of sewage sludge has met much opposition in recent years due to the presence of pathogens, high concentrations of heavy metals, toxic chemicals, pharmaceuticals and solvents in the sewage sludge (Roy et al. 2011). Problems of sludge on land may be environmental, agronomic, and operational.

Environmental concerns regarding land disposal are surface water and groundwater pollution, contamination of the soil and crops with toxic substances, and transmission of human and animal diseases. Environmental problems are also largely associated with emissions of odor, ammonia, and pathogens. Land application of sludge also has met social and technical obstacles. The agronomic and operational aspects are very much interrelated, and primary concerns are soil conditions and crop response in relation to accuracy and timing of applications. The relationship between farmers and their customers, the food industry and retailers is of vital importance for accepting sludge use in agriculture. The technical problems arise due to the fact that sludge is being produced all year round whereas its application on land takes place once or twice a year; consequently the sludge should be stored until its land application. Concerns over real and perceived risks from the heavy metals and organic contaminants in sludge, and caution over the addition of N and P-rich manures to land, will continue to be major factors in limiting the use of sludge. Stringent legislations have made sludge treatments necessary to reduce heavy metal contents prior to its land application. The quality of sludge throughout Europe hasimproved markedly over the last 20 years. Therefore, it is anticipated that in the future, the amount of sludge used in agriculture could be increased substantially by appropriate promotion and the

Table 14.11. Percent wastewater sludge disposed by different methods

Region	Year	Land application	Incineration	Land-filling	Compost	Others	Ref.[a]
USA	—	52	22	17	9	—	[1]
Western Canada	—	66	—	4	27	—	[1]
Japan	—	—	80 more	—	—	—	[1]
China	2010	48.5	3.5	34.5	—	13.5	[2]
South Korea	2004	9.8	11.7	1.4	—	77.1 Ocean dumping	[3]
Australia	2011	55	15 less	30	15 less	—	[4]
Finland	2005	3	97	—	—	—	[5]
Luxembourg	2008	56	11	—	33	—	[5]
Cyprus	2007	50	12	—	38	—	[5]
Portugal	2007	87	—	7	—	6	[5]
Czech Republic	2008	47	1.5	12	31.5	—	[5]
Estonia	2009	78.5	1.5	20.5	—	—	[5]
France	2008	47	19	7	27	—	[5]
U.K.	2005	69	14	5	1	9	[5]
Ireland	2007	69	—	6	—	25	[5]
Slovakia	2007	—	—	24	68	8	[5]
Spain	2006	64.5	4	15.5	—	16	[5]
Hungary	2007	57	1	29	3	10	[5]

(Continued)

Table 14.11. Percent wastewater sludge disposed by different methods (Continued)

Region	*Year*	*Land application*	*Incineration*	*Land-filling*	*Compost*	*Others*	*Ref.*[a]
Denmark	2007	59	16	6	—	19	[5]
Sweden	2009	24	1	24	33	18	[5]
Latvia	2007	42	—	—	11	47	[5]
Germany	2008	28	51	—	17.5	3.5	[5]
Italy	2005	23	3	42	21	11	[5]
Lithuania	2007	33	—	12	9	46	[5]
Austria	2008	16	38	7	22	17	[5]
Bulgaria	2009	37	—	27	—	36	[5]
Poland	2009	22	2	14	5	57	[5]
Romania	2009	—	—	74	20	6	[5]
Belgium	2008	14	53.5	—	—	32.5	[5]
Netherlands	2006	—	67	4.5	13	15.5	[5]
Slovenia	2009	1.5	61.5	19	—	18	[5]
Greece	2008	1	46	38	—	15	[5]
Malta	2009	—	—	100	—	—	[5]
EU-15	—	44	21	15	10	9	[5]
EU-12	—	16	1	28	21	35	[5]
EU-27	—	41	19	17	12	12	[5]

[a][1] = LeBlanc et al. (2008); [2] = Chen et al. (2012); [3] = Park et al. (2009); [4] = AWA (2012); and [5] = Kelessidis and Stasinakis (2012)

adsorption of quality-assured best operational practices, and by the development of sludge-based products such as compost.

14.5.2.2 Incineration

Incineration technology is well established as a method of sludge treatment and disposal throughout the world. Incineration involves the combustion of sewage sludge at high temperatures in an enclosed structure in an excess of oxygen. Incineration converts sludge into carbon dioxide and ash by rapidly oxidizing organic matter. This process reduces the sludge to a mass of ash that is less than 20 percent of its original volume. Sewage sludge incineration has been a solution for urban areas lacking disposal space and eliminates some environmental and health problems by destroying pathogens and toxic organic chemicals. The incineration technology is attractive because of its several advantages, including significant reduction of sludge volume, thermal destruction of toxic organic compounds, recovering the energy (in terms of heat) and minimization of odor (Roy et al. 2011). The incineration technology has great potential to be used effectively and in a large scale, both for volume reduction and energy recovery and will certainly play an important role in sludge management in the future with continuous developments in technology and conversion efficiencies.

Incineration processes can be distinguished as mono-incineration and co-incineration in waste plants, power plants and cement industry. Modern incineration configuration includes the use of heat produced by sludge combustion to heat water that becomes steam. Produced steam can turns a turbine to generate electricity. Therefore, this technology is also regarded as waste to energy plants rather than sludge disposal plants. Modern energy incinerators are capable of operating auto-thermally and are often able to produce surplus and usable heat and power (Hall 1995). The technology of incineration in terms of the process engineering, energy efficiency and plant compactness has experienced great improvement lately. Modern fluidized bed incinerators have become more and more attractive both in terms of capital as well as operating costs, in comparison to the conventional multiple hearth type (Mininni et al. 1997). Introducing new technologies for controlling gaseous emissions can minimize the adverse effects of incineration and reduce the corresponding cost, which gives incineration considerable advantages in the future as compared to other available disposal routes.

Due to EU legal limitations concerning landfilling and agricultural reuse as well as the ban on sea disposal, the dependence on incineration is bound to increase in the EU in the near future (Malerius and Werther 2003). Incineration of sludge is currently a relatively minor method of sludge disposal; however, it remains as the most attractive disposal method currently in Europe. The amount of sludge being incinerated has already reached 24% of the sludge produced in Denmark, 20% in France, 15% in Belgium, 14% in Germany, 25% in the USA and 55% in Japan (Lundin et al. 2004).

Important for the sludge-combustion is its heat-value, depending on the organic matter and the water content of the sludge that can be influenced by dewatering and drying systems. Important considerations in evaluating incineration methods include the composition of the sludge feed and the amount of auxiliary fuel required. Air pollution constraints and resultant equipment and treatment requirements, as well as ash disposal are also important. The effect of incineration of municipal sludge on air quality has been studied by several researchers. There are several issues (ash disposal, pollutant emissions, moisture content, and high operational and installation costs) of concern regarding the incineration technology which limits its application. Incineration produces large quantities of ash (30% of dry sludge solids) (Malerius and Werther 2003). The ash is considered as highly toxic due to its metal content. The ash produced by incinerator is generally disposed of in the landfill. Harmful emissions produced by sludge incineration are the major constrains to its widespread use.

The main problem confronting the incineration process is how to reduce the high water content of the sewage sludge causing a negative energy balance. On contrary, dried sludge as powder is difficult to store (self-ignition) and could be explosive if the atmosphere is not inert. The calorific value of sewage sludge largely depends on the type of sewage sludge and its degree of dewatering. The characteristics of the sludge vary because of the treatment processes and different compositions of the pollutants. For decreasing the moisture content, the pretreatment (e.g., dewatering and/or drying device) are needed. Other issues, such as ammonia from the sludge, may cause problems in the bottom ash by increasing the leachability of Cu. Very sophisticated systems are required for incineration. Therefore, in general, incineration is always an expensive solution, even if it is the most effective method to destroy critical organic pollutants and gives the possibility to separate and collect salts and inorganic substances in the residues of the flue gas cleaning system.

Ash produced from sewage sludge incineration has a high phosphorus content (Lundin et al. 2004). Incineration of sewage sludge with subsequent recovery of P is a relatively new technique. Mono-incineration nutrient recovery from sludge ash is more promising. In Germany, 20% of the total sewage sludge is treated in mono-incineration plants. For a long-term effective and economic thermal treatment of sewage sludge, mono-incineration plants have to work to capacity; and additional burning with refuse derived fuels has to be optimized to provide the requirements for P recovery. Leaching with acid and base is a promising method for P recovery from the ash and slag that remain after combustion before entering a set of ion exchangers (Lundin et al. 2004). Phosphorus removal, recovery and recycling would be very useful for the control over nutrient emissions from the sewage sludge. According to recent studies, heavy metals can be effectively removed from sewage sludge ashes by a thermochemical treatment (Adam et al. 2009).

14.5.2.3 Landfills or Surface Application

The most common waste management strategy is landfilling (Lou and Nair 2009). Sludge dewatering is usually required prior to landfilling for reducing volume to be transported and to reduce leachate from landfills. Solid concentration is an important factor while accepting sludge for landfilling. Landfilling has become an expensive option for several reasons such as increase in sludge production, lack of readily available disposal space, stringent legislations (requiring new landfills to be highly engineered), need to minimize odor generation from landfill, and aesthetic objections from nearby residents (Roy et al. 2011). Many modifications in conventional landfilling technique have been explored recently, such as energy recovery from landfilling gases, aerox landfills, composting of waste prior to landfilling and compost capping (Lou and Nair 2009). These modifications have positive impacts on greenhouse gas (GHG) emissions, reduction in leachate contamination and space limitations (Lou and Nair 2009). Disposal of incinerated ash in landfills has also been an option for the management of heavy metal contents of the ash. Landfilling with energy recovery from landfill gas is a modern sludge management option.

14.5.2.4 Composting

Composting is a well-established low cost technology for the degradation of organic matter and reduction of pathogens in the solid wastes (an organic fertilizer called compost) (Metcalf and Eddy 2003). Composting is a aerobic process by which different microbial communities initially degrade organic matter into simpler nutrients and in a second stage, complex organic macromolecules such as humic acids are formed. Temperature, oxygen and moisture content are the main control variables in the composting process, apart from other chemical, biochemical or microbiological properties (Gea et al. 2007). Composting followed by land application represents one of the most economical ways for the treatment and disposal of sewage sludge because it combines material recycling and sludge disposal at the same time. The presence of high levels of heavy metals often demands the improvement of the composting process to minimize the mobility of heavy metals using various additives (Wong and Selvam 2006). During composting the humification of organic matter will have a significant effect on the physicochemical form of existence of heavy metals. In this context, lime is a suitable material (less than 1% w/w) to co-compost with sewage sludge to reduce the availability of heavy metals (Wong and Selvam 2006). Co-composting processes of fats and sludge can be successfully carried out to obtain a sanitized and stabilized product. Addition of fat content up to 50% is possible although a maximum fat content of 30% is recommended to achieve a high fat content reduction (85%) and to avoid long composting periods at full-scale facilities. Fat-enriched wastes can be added to low energy content wastes to fulfil the international requirements on compost sanitation. From this point of view, co-composting strategies are of special interest for plant managers (Gea et al. 2007).

14.5.2.5 Ocean Disposal

Over a century, sewage sludge was routinely dumped into the ocean. Today, in the most part of the world, ocean or sea disposal is either banned, restricted or the least favored sludge disposal alternative from environmental pollution point of view. The concerns associated with ocean disposal of sewage sludge include the buildup of heavy metals, pathogenic organisms, decreased dissolved oxygen, increased turbidity, and adverse effects to water quality, bottom sediments and marine organisms. With ocean disposal, there is no recycle of the rich nutrients present in the sludge. The sludge disposal into the Northeast costal area of the U.S. had severely damaged the marine organisms and beaches in 1988. This followed up by the Ocean Dumping Ban Act of 1988, which was enacted to prohibit all municipal sewage sludge and industrial waste dumping into the ocean after December 31, 1991 (U.S. EPA 1995, 2006). Disposal of wastewater sludge to the surface waters by dumping from ships, by discharge from pipelines or by other means is prohibited by the EU since 1998 (CEC 1999). Despite growing concerns of marine life and the environment, sludge disposal into the ocean or sea is still practiced in many parts of the world because it is relatively simple and inexpensive way of discarding wastes. The volatile hydrocarbons evaporate into the atmosphere, while grease, oil and scum remain on the water surface and may be transported long distances by winds and water currents. The remaining sludge solids either settle to the ocean floor or are retained in dispersed forms at various depths. Many contaminants that are in the fine particles accumulate below the surface thereby causing contamination to sea organisms. In 2006, more than 4,933 ton/day of the sewage sludge was disposed into the sea by South Korea to avoid expensive sludge treatment and disposal options (Park et al. 2010). Around 362,000 m3 of sewage sludge was disposed of in 2011 in the Yellow sea (Hong et al. 2012). Yellow sea is surrounded by the contiguous China-Korea land mass. The continuous disposal of sewage sludge has disturbed the natural marine ecosystem significantly. In this case, ocean color imagers have been used for monitoring of sewage sludge disposal into the sea. Korean Geostationary Ocean Color Imager and Moderate Resolution Imaging Spectroradiometer Aqua observations revealed that algal blooms persisted in spring annually at the dump site in the Yellow Sea between 2000 and 2012 (Hong et al. 2012).

14.5.3 Energy Recovery from Sludge

There are many sludge-management options in which production of energy (heat, electricity, or biofuel) could be the most attractive for sustainable sludge management. The most important energy recovery options are described in the following sections (Table 14.12).

14.5.3.1 Anaerobic Digestion

Anaerobic digestion is a well-known microbial process carried out in the absence of oxygen. It consists of three steps, i.e., hydrolysis, acidification and biogas generation (Rulkens 2008). Anaerobic digestion technology is currently in practice at many large modern WWPTs to recover energy through methane production and also as a method to stabilize/reduce the mass of biosolids; and to convert part of the volatile compounds into biogas (Metcalf and Eddy 2003). Generally, biogas consists of 60% CH_4 and 40% CO_2 and can be applied as an energy resource either at the WWPT itself or elsewhere. Biogas production strongly depends upon the

Table 14.12. Technologies/processes for energy recovery/generation from sludge

Process & Tech.	*Form of energy*	*Current status*	*Advantages*	*Disadvantages*	*Ref.*[a]
Anaerobic digestion	Methane/ heat	Applied on large scale	Biogas produced can be applied as an energy resource either at the WWTP itself or elsewhere; it can be applied on large and medium sized WWTPs	After energy recovery of the biogas the energy content of the residual fraction is low; toxic compounds are partly removed; digested sludge needs further treatment as it still contains heavy metals, soluble phosphates, and inorganic.	[1,2]
Mono and co-incineration	Heat/ electricity	Applied on large scale	Energy recovery in the form of heat and electricity; heat recovered can be used for the sludge drying	The emissions of the pollutants in the exhaust gases; quality of the ash and disposal; expensive technology	[3–5]
Pyrolysis	Heat	Applied on large scale	Energy recovery in the form of heat and electricity; valuable gases (chemicals/ fuels) recovery such as char, oil; flexible than incineration	Complicated process; uncertainties about the real practical and financial performance	[4]
Gasification	Heat	Applied on large scale	Energy recovery in the form of heat and electricity; flexible than incineration	Complicated process; uncertainties about the real practical and financial performance	[4,6,7]
Wet oxidation	Heat	Not yet applied on large scale	Phosphates can be easily removed from the ashes	Need for high pressure piping; high pressure reactors; corrosion issues	[2,6,8]

(Continued)

Table 14.12. Technologies/processes for energy recovery/generation from sludge (Continued)

Process & Tech.	*Form of energy*	*Current status*	*Advantages*	*Disadvantages*	*Ref.*[a]
Microbial fuel cells	Electricity	Initial research phase	Promising innovative technology; direct electricity recovery from all types of wastewater and wastewater sludge; high organic matter removal	Technology is still at research stage; power recovery is still very low to be utilized; fate of toxic compounds during reactor operation is unknown	[9-12]
Biodiesel production	Bio-diesel	Initial research phase	Innovative technology; reuse of the waste	Technology is still at research stage; post treatment requirement; fate of toxic compounds during biodiesel production unknown	[2]
Bioethanol production	Bio-ethanol	Initial research phase	Innovative technology; reuse of the waste	Technology is still at research stage.	[2]
Hydrogen production	Hydrogen	Initial research phase	Innovative technology; reuse of the waste	Technology is still at research stage.	[2,5]
Acetone production	Acetone	Initial research phase	Innovative technology; reuse of the waste	Technology is still at research stage.	[2]
Butanol production	Butanol	Initial research phase	Innovative technology; reuse of the waste	Technology is still at research stage.	[2]

[a][1] = Metcalf and Eddy (2003); [2] = Rulkens (2008); [3] = Luts et al. (2000); [4] = Rulkens (2008); [5] = Wang et al. (2008); [6] = Fytili and Zabaniotou (2008); [7] = Mathieu and Dubuisson (2002); [8] = Svanstrom et al. (2004); [9] = Dentel et al. (2004); [10] = Logan et al. (2006); [11] = Ok et al. (2010); and [12] = Xiao et al. (2011)

type of sludge and also the operating conditions of the digester (Nishio and Nakashimada 2007). The gas production from a mixture of primary and secondary sludge roughly amounts to 1 m3 of biogas/kg of organic solids biodegraded (Rulkens 2008). Biogas production can be increased substantially by applying appropriate pretreatment such as hydrothermal, heating, microwave heating, ultrasonic treatment, ozone, enzymes, liquid jets, hydrolysis, wet oxidation, etc. The anaerobic processes applied in practice are mostly mesophilic. It is also possible to digest sludge thermophilically at a temperature of about 55°C. In comparison to mesophilic digestion, thermophilic treatment has some advantages, such as slightly higher biogas production, higher destruction degree of pathogens, and a larger reduction in the amount of organic solids. The sludge retention time of these processes is approximately 20 days that can be reduced. With the standard digestion technologies, only approximately 20–30% of the organic matter is mineralized (Rulkens 2008). However, there are also reports suggesting that the anaerobic digestion be the most efficient means to recover the energy from organic waste materials with a moisture content of >50%, and combustion for energy recovery was only efficient for materials with a moisture content of <50% (Matteson and Jenkins 2007). Typically, the moisture content in raw sludge ranges from 95% to 99%, and in dewatered sludge, ranges from 70 to 85%. Therefore, currently compared with other technologies used to recover energy from sludge, anaerobic digestion is still considered the most effective means of producing energy (Wang et al. 2012). With regard to the energy recovery from sludge using anaerobic digestion, the biogas production, the total energy balance, the amount of residual sludge, and the total costs have to be taken into account and analyzed.

Anaerobic digestion of sewage sludge is mainly applied at large- and medium-sized WWPTs. However, the interest in anaerobic digestion is currently growing also for small-sized WWPTs (Rulkens 2008). There are some issues with anaerobic digestion, such as toxic organic compounds are only partly removed in the anaerobic digestion process. It also contains the heavy metals, soluble phosphates and inorganics. This demands further treatment of the digested sludge (Rulkens 2008). Digested sludge holds less heat value. Therefore, incineration of such sludge is not very attractive. Moreover, a large area is required for the conventional aqueous anaerobic digestion facility because of the low solid contents (1–5%) in the sewage sludge. Therefore, currently dry anaerobic digestion technology is being evaluated to produce methane from dewatered biosolids, which could significantly reduce the area required for the digestion process (Nishio and Nakashimada 2007). Use of dry anaerobic digestion can also significantly reduce dewatering costs because undigested sludge are much easier to condition and dewater with less demand for polymer (Wang et al. 2012).

14.5.3.2 Mono- and Co-Incineration

Incineration technology has already been discussed earlier in this section. Incineration is a very popular method of energy recovery in Japan, the USA, Belgium, Demark, France, and Germany (Wang et al. 2012). Dry matter content or lower heating value and composition of biosolids are the most important factors influencing energy recovery. There is a linear negative relationship between biosolids water content and heating values. Usually, an external energy supply is essential to dry and combust dewatered biosolids. High water content makes incineration a poor source of energy as incineration itself consumes high energy (Brown 2007). Incineration performance depends upon the performance of mechanical dewatering and drying processes. Incineration may be considered as a means of waste minimization rather than energy generation (Wang et al. 2012).

Mono-incineration of dewatered sludge is expensive and limited in its capacity. Co-incineration is depending on the kind of fuel and needs an optimal mixture. Thermal refuse processing plants for domestic refuse or coal, with controlled incineration technology and highly effective flue gas cleaning systems, can normally be operated on both domestic refuse and sewage sludge from the same area dewatered to 30–35% DS, without any significant fall in capacity (Reimann 1999). Co-incineration of sewage sludge in coal-fired power plants has been applied in practice, e.g., in Germany for more than 10 years (Wang et al. 2012). To avoid the high costs of a stand-alone incineration plant for sludge and also to improve the energy recovery efficiency, possibilities have been investigated in Belgium to incinerate dried sludge in a coal-fired power plant (Luts et al. 2000). Co-incineration was suggested as a valuable and responsible sludge disposal option because it can be carried out in the existing units (existing coal combustion installations and existing exhaust gas treatment systems) with a high electrical efficiency. Also, sludge disposal can be achieved with a limited influence on emissions and residual ashes, without creating additional waste streams. Wet sludge or dewatered sludge also has been utilized as alternative fuel at cement kiln. Zabaniotou and Theofilou (2008) have recommended the co-incineration processing of wet sewage sludge blended with pet coke or meat and bone meal (as an alternative fuel). The maximal sewage sludge feed rate should not be more than 5% of the clinker production capacity of the cement plant (Fytili and Zabaniotou 2008). Zabaniotou and Theofilou (2008) used wet sludge of moisture content 65–70% as an alternative fuel in cement kiln. Co-incineration of sludge in existing power and heating plants and cement kilns represents an advantage for a low investment cost and rapid implementation (Cartmell et al. 2006; Wang et al. 2012). However, co-incineration also presents significant risks to water utilities. The energy recovery from the co-incineration of the sludge can be enhanced by improvement of the dewatering and drying processes of the sludge and by utilizing heat from the exhaust gases. In addition, the use of the polluted drying gas in the co-incineration process can contribute to a better total combustion performance (Rulkens 2008). Also, other co-incineration processes, for example, in combination with municipal waste, are applied in practice.

14.5.3.3 Pyrolysis

Pyrolysis is the decomposition of organic material by extreme heat in the absence of oxygen to produce the mixture of gaseous and liquid fuels and the solid inert residue of carbon (Wang et al. 2008). The ratio of what is produced during the decomposition process varies according to the temperature used for decomposition (Kim and Parker 2008). At elevated pressures, pyrolysis efficiently converts the organic matter in the biosolids to gases that can be treated to produce hydrocarbons or used as basic chemical or as fuel. The resulting oil is similar in characteristics to heavy crude oil. The conversion of the combustible gases of both systems into electrical power can be achieved more efficiently by pyrolysis.

Pyrolysis is applied in practice to produce oil from sewage sludge (Bridle and Pritchard 2004). In this process, dried sludge pellets are heated in the absence of oxygen in a gas generator to a temperature of approximately 450°C. The organic compounds are converted into char, oil, and non-condensable gases. Oil production can be optimized by maintaining temperatures at 400°C. A cost analysis of drying and pyrolysis for treatment of primary, waste activated and digested biosolids was conducted using the price of crude oil as a basis. The outcome revealed a net value for each feedstock at each of the process temperatures tested; it was concluded that the temperature of pyrolysis and the volatile solids content in biosolids were the major factors affecting oil and char yield. The char produced during the low- and medium-temperature pyrolysis may be used as fuel to dry biosolids to reduce external energy input (Brown 2007). As a chemically stable organic C, char can also be applied to soil to increase long-term soil C sequestration and to improve soil productivity (Lehmann 2007). Besides oil and char, some researchers focus on syngas production using microwave-induced pyrolysis of biosolids.

The first commercial biosolids pyrolysis plant was built in Western Australia (Bridle and Skrypski-Mantele 2004). However, this plant has now been discontinued, as the resultant product was found to be unsuitable for diesel engines and the economy of the system were poor (U.S. EPA 2006). Also, it is necessary to treat large amounts of waste gases. Because of the presence of toxic organic pollutants in the sewage sludge, the treatment process of the gases can be complicated. In general, the process performance of pyrolysis is much more complicated than incineration.

14.5.3.4 Gasification

Technologies which allow utilization of wastes as fuel are of great importance. Theoretically, almost all kinds of organic wastes with moisture content of 5–30% can be gasified. Gasification is a series of complex sequential chemical and thermal sub-processes. However, only specific biofuels can lead to the successful gasification (Fytili and Zabaniotou 2008). Gasification depend upon the fuel properties such as surface, size, shape as well as moisture content, volatile matter and carbon content. As for the feedstock utilized in gasification, inexpensive materials, such as sludge, would be practical. The energy content of the gas produced through gasification depends on numerous factors, such as the feed fuel, reactor type, etc. The effects of sludge on gasification are need to extensive investigation to obtain a desirable gas for electricity production. Gasification technology can be applied to convert the sewage sludge into a useable energy and to reduce the environmental problem (Mathieu and Dubuisson 2002). Few studies have been conducted on sewage gasification. Electricity is an attractive product for its relatively high value, ease of distribution and ease of adapting to local or national market specifications. In addition, the electricity market in Europe offers significant incentives, up to 0.1 EUR/kWh, for power generated from renewable resources. Engines and turbines need further development to provide necessary performance guarantees. The main attraction of using gas turbines to convert the energy in biofuel to electricity is the potentially high cycle efficiency. The selection of either gas engines or gas turbines depends mainly on size. Gas turbines are more appropriate for larger size plants, whereas gas engines are more suitable for smaller ones.

The gasification is actually energetically self-sustaining and usually in steady conditions and no energy input is necessary (Mathieu and Dubuisson 2002). During the gasification process, sewage sludge undergoes a complex physical and chemical change, starting with the drying or removal of water and then pyrolysed or thermally decomposed. In the final step, the pyrolysis products, condensable and non-condensable vapors and char undergo gasification, where they are concurrently oxidized and then reduced to permanent gases at the reduction zone. In the drying zone, sewage sludge descends into the gasifier and moisture is evaporated using the heat generated in the zones below. The rate of drying depends on the surface area of the fuel, the recirculation velocity, the relative humidity of these gases and temperature differences between the feed and hot gases as well as internal diffusivity of moisture within the fuel. Characteristically, sludge with less than 15% moisture loses all moisture in this zone (Marrero et al. 2004).

14.5.3.5 Wet Oxidation

Wet oxidation of sewage sludge belongs to the category of thermal processes. It takes place in an aqueous phase at temperatures of 150–320°C and pressures of 1–15 MPa using pure or atmospheric oxygen. The high pressure temperature is needed to prevent boiling at the temperatures required for the process (Hall 1995). During the process, the organic content of sewage sludge is thermally degraded, hydrolyzed, oxidized and converted to carbon dioxide, water and nitrogen. The whole process occurs first at sub-critical conditions (below 374°C and a pressure of 10 MPa) and then at supercritical conditions (above 374°C and a pressure of 21.8 MPa) (Fytili and Zabaniotou 2008).

Supercritical wet oxidation occurs at temperatures and pressures above the supercritical point of water (Wang et al. 2008). Supercritical water has special properties, such as a superior ability to dissolve oxygen and organic compounds. However, it has a low solubility for mineral compounds, such as sodium chloride (Marrone et al. 2004). The organic compounds get completely oxidized by with supercritical oxidation. Nitrogen, present in nitrogen-containing compounds, such as ammonia and amino acids, is converted into nitrogen gas. Also toxic organic compounds are completely oxidized. The oxidation rate at supercritical conditions is much higher than at subcritical conditions. The required retention time for the oxidation of sewage sludge in a supercritical reactor is in the order of a few seconds to 1 min (Rulkens 2008). This means that the required reactor size is relatively small. Energy recovery from this oxidation process can occur directly by heat exchange in the reactor or from the exit flow from the reactor. In comparison to sludge incineration, supercritical oxidation has the advantage that off-gas treatment is very simple and negligible. It is also not necessary to dewater the sludge prior to the oxidation process. The inorganics present in the treated sludge can easily be removed from the water phase as ash.

The wet oxidation process can be designed for above- or below-ground operation. System designs have been developed that have reduced the capital and operating costs of processing municipal sludge below that of incineration (Svanstrom et al. 2004). Supercritical wet oxidation has a high potential to solve the sludge problem in a sustainable way and also appears to be economically competitive with existing high-thermal sludge treatment systems. However, large-scale practical experience is not available yet. Accumulation of salts on any surface within the reactor can lead to scaling and down time for the system (Wang et al. 2008). Use of oxygen in the process, use of high-pressure piping, the need of high-pressure reactors, and potential corrosion problems if chlorides are present in the sludge might be bottlenecks in the acceptance and further development of this technology (Rulkens 2008). Future works on the approach to controlling scale build up and corrosion are necessary for successful applications of this technology (Marrone et al. 2004). The phosphate can be recovered as well as heavy metals can be removed from the ashes from a supercritical wet oxidation process. The solid residue has the potential for P extraction (Wang et al. 2008).

14.5.3.6 Biofuels Production from Sewage Sludge

Recently considerable research has been carried out on conversion processes that can be used to produce liquid or gaseous energy carriers from sewage sludge. Biodiesel, hydrogen and electricity production are the three major biofuels which are being investigated while other biofuels of interest are ethanol, butanol or acetone.

Biodiesel Production. Recently, municipal sludge has been utilized as a lipid feedstock for biodiesel production (Kargbo 2010). Municipal sludge contains significant concentrations of lipids derived from the direct adsorption of lipids onto the sludge. These energy-containing lipids include triglycerides, diglycerides, monoglycerides, phospholipids, and free fatty acids contained in the oils and fats. During wastewater treatment process microorganisms utilize organic and inorganic compounds in wastewater as a source of energy, carbon, and nutrients. The cell membrane of these microorganisms is a major component of sewage sludge and is composed primarily of phospholipids. It is reported that 36.8 wt% of the dry sludge is comprised of fatty acids and steroids. With the fatty acids from sludge predominantly in the range of C10 to C18, these are excellent for the production of biodiesel. Biodiesel production from sludge could be a viable alternative to land disposal that will help to solve both energy and environmental problems.

However, biodiesel production from sewage sludge poses huge challenges to overcome if commercial opportunities are to be realized. Some of these challenges are not unique to biodiesel production from waste sludge but to the biodiesel industry as a whole. The first challenge is to select the best feedstock, i.e., primary and/or secondary sludge and the pretreatments for maximum lipids extraction. The second challenge is the cost of lipid extraction from sludge, which is quite expensive and requires large volume of organic solvents. Lipid extraction efficiency and cost is greatly affected by sources and type of sludge, solvent selection, sludge to solvent ratio, extraction time and temperature and solvent recovery. Optimization of these factors is necessary for efficient lipids extraction and cost reduction. The third challenge is the selection of the biodiesel production methods. The type and amount of catalysts and the rate of transesterification need to be studied extensively to optimize the methods of biodiesel production.

Microbial Fuel Cells (MFCs) for Electricity Production. Recently, MFCs have been integrated with WAS treatment for recovery energy (Wang et al. 2012). MFCs are devices that can convert (oxidize) the chemical energy of organic and inorganic matter directly into useful electrical energy by catalytic reaction of microorganisms (Logan et al. 2006). MFCs have generated considerable interest among the academic community in recent years and MFCs research is a rapidly evolving field (Dentel et al. 2004; Ok et al. 2010). The combination of MFCs with WAS treatment can expand the application fields of MFCs and may simultaneously help in sludge management in future. According to an energy analysis conducted by Ge et al. (2013), the total energy in the MFC systems was comparable with that of anaerobic digesters. However, the electric energy from MFC had a minor contribution, and methane gas still dominated the total energy production. Currently, one of the major drawbacks of MFCs using WAS is the lower power output as compared to those using simple or pure substrates as fuels. Maximum power density of 15.75 W/m^3 has been reported in dual chambered MFCs powered by WAS, whereas it was relatively higher (253 W/m^3), for similar MFCs using glucose as substrate (Wang et al. 2012). WAS generated from WWTPs is mostly present in the insoluble and large particulate form, whereas the microbes in MFCs can only make use of soluble and easily biodegradable organic matters as their substrates, such as proteins, carbohydrates, and other small organic molecules. Therefore, providing available substrates for microbes is one of the challenges facing MFCs, which also governs the efficiency of electricity generation. Implementing sludge pretreatment prior to MFCs can enhance the performance of the MFCs (Xiao et al. 2011). Ultra-sonication, microwave and alkalization have been used to disintegrate the insoluble and large particulate materials. Other potential pretreatment methods, such as ozonation, thermal hydrolysis and advanced oxidation processes can be used.

There are several challenges of MFCs using WAS as fuel such as a higher internal resistance, low power density, relatively slow sludge degradation compared to the MFCs using glucose, acetate, and other compounds as substrate (Logan et al. 2006). Sludge contains a lot of colloidal particles and polymer substances that tend to adsorb to all types of surfaces and can block the internal structure of the MFC. The fate of toxic organic matters and inorganic compounds (e.g., heavy metals) present in the sludge during MFCs operation is not yet investigated. In such cases, the post treatment of the residual waste will be required. Besides the issues identified above, the MFCs using WAS as fuel also face some common challenges, such as the scale-up problem, low catalytic rate of microbes, biofilm growth, mass transfer, electron transfer, and so on. Most of the designs reported in the literature cannot be scaled to the level that is needed for a large WAS treatment facility, which may require hundreds of cubic meters of reactor volume. The design should be simplified and the intrinsic conversion rate should be enhanced in order to develop cost-effective and large-scale systems. The selection of high-efficient microbes, enhancement of the mass/electron transfer rate, and improvement of the biofilm growth on anode materials are also future tasks that should be addressed in order to achieve real applications of the MFC technology in WAS treatment. In the systems using proton exchange membrane (PEM) as separators, the membrane fouling issue and the transfer resistance due to the use of membranes should also be paid attention in the future (Wang et al. 2012). Overall, MFCs technologies are still largely carried out at the laboratory scale and therefore, their potential in direct production of electricity from sewage sludge needs rigorous laboratory and pilot-scale studies. With the worldwide increase in energy demands and interest in utilizing sludge as an energy source, MFCs could be a fascinating future technology for energy recovering from sludge.

Hydrogen Production. Hydrogen is universally accepted as an environmentally safe, renewable energy source as well as an ideal and clean alternative to fossil fuels. Almost all of the hydrogen produced today is by steam reforming of natural gas. For the near term, this method of production will continue to dominate. Hydrogen can be produced from sludge by applying different thermal techniques and/or biological (dark or anaerobic fermentation) processes. A few studies have been focused on the production of hydrogen from sewage sludge using dark fermentation. Hydrogen production from wastewater sludge utilizing a clostridium strain isolated from the sludge sample was reported first by Wang et al. (2003b). Later, it was observed that hydrogen production was enhanced by the application of sludge filtrate than that obtained with whole sludge (Wang et al. 2003a). Although these studies successfully established the feasibility of producing hydrogen from wastewater sludge, the hydrogen generated during the first 16–24 h of fermentation was consumed during a later stage. The pathway of hydrogen consumption, which remains unknown, is of academic interest.

Hydrogen production by thermal processes is quite promising. Researchers from the UK and Turkey have investigated the hydrogen production potential from sewage sludge, by applying downdraft gasification technique (Aznar et al. 1998; Mathieu and Dubuisson 2002). Wet sewage sludge can be assumed as one of the most common feedstock to manufacture hydrogen gas all over the world (Dogru et al. 2002). Hydrogen can also be produced by thermal gasification of biomass including sewage sludge (Fytili and Zabaniotou 2008). The processes involved in producing hydrogen from waste resemble the processes in production from fossil fuel. Under high temperatures, the waste breaks down to gas. The gas consists mainly of H_2, CO and methane (CH_4). Steam is then introduced to reform CH_4 to H_2 and CO. CO is then put through the shift process to attain a higher level of hydrogen. The by-product from this process is CO_2, but CO_2 from biomass is considered "neutral" with respect to GHG, as it does not increase the net CO2 concentration in the atmosphere. The mixed gas can also be used in fuel cells for electricity production. Compared to conventional processes for production of electric energy from waste, integrated gasification fuel cell systems are preferable. Electrical efficiency over 30% is possible for these systems (Fytili and Zabaniotou 2008). Taking into account the simple way to produce methane and the huge experience with methane production processes, it will be doubtful whether hydrogen production from sewage sludge will become much more attractive than the production of methane in the short term.

Production of Other Biofuels. To our knowledge, there are hardly any studies on the production of ethanol, butanol, or acetone from sewage sludge. This is because the production of these biofuels from sewage sludge is less attractive because of the complexity of the components separation system. It is possible to produce these biofuels from a mixture of sewage sludge and other residues of biological origins, such as food wastes or biomass. However, the presence of toxic components and inorganic compounds in the sludge and/or biowastes may interfere with the quality of the residue and further processing of this residue.

There are several studies on developing innovative biological conversion processes of biomass and improving existing biological conversion processes of biomass. The most studies are focused on to improve the technical performance and decrease the total costs of biological conversion processes, also including costs of pretreatment and post-treatment. A better overall performance is possible by improvement of the pretreatment, focused on a better accessibility of the substrate component for microbiological conversion, application of new types of microorganisms, such as hyper thermophilic microorganisms or genetically modified microorganisms, improved reactor concepts, and better process control.

14.5.4 Value-Added Products from Sludge

Composition of sewage sludge makes it a potential source for the recovery or production of value-added products. In that context, recently a large number of products are being investigated extensively at the lab- and/or pilot-scale or commercialized in a large scale. Table 14.13 summaries important technologies/processes for recovery/generation of value-added products from wastewater sludge. Brief descriptions on these value-added products are as follows.

14.5.4.1 Phosphorus Recovery

Phosphorus recovery and reuse has become an important issue in recent years since P is a limited resource (Rulkens 2004). P occurs in wastewater from a variety of sources such as urine and feces, food wastes, detergents, etc. In WWTPs with enhanced biological P removal or using alumina salts to precipitate P, most P in wastewater is transferred into the sludge, and therefore, sewage sludge represents an important secondary phosphorus source (Blöcher et al. 2012). Thus, for WWPTs currently using iron precipitation, a process change regarding P removal is required. P recovery is often combined with the removal of heavy metals from

Table 14.13. Technologies/process for value-added products generation from sludge

Value-added products	*Advantages*	*Disadvantages*	*Remarks*	*Ref.*[a]
Phosphorus	Recycling of valuable nutrients; can be used as fertilizer; economical; with metal recovery it can improve the quality of sludge if used as an energy source	Still at research phase or at pilot study phase; if used as fertilizers, the impact on health needs to be investigated.	In the near future, it is expected that interest in phosphorus recovery will increase.	[1–3]
Construction materials	Use in cement kilns is in practice; quality of the cement is in the acceptable limit; recycling of waste; economical	Application is at low loading rate; fate of certain emerging toxicants must be studied	It is being practiced at many developed countries.	[4]
Biofertilizers	Economical; environmental friendly; nontoxic; reduction in carbon foot print; sludge reuse	Heavy metals can have adverse effect on microbial populations in the soil; ecotoxicological effects on plants over long periods	Recent field applications of the biofertilizers are very positive; it is more likely for larger-scale applications.	[5–7]
Bioplastics	Environmental friendly; non toxic; biodegradable; production is simple; reduction in carbon foot print; sludge reuse	Still at research phase; low yields; recovery issues	Significant developments in production and recovery at pilot scale in near future are expected.	[5,7]
Biopesticides	Economical; environmental friendly; non-toxic; reduction in carbon foot print; sludge reuse	Still at research or at pilot study stage; sludge pathogens and heavy metals; recovery and maintenance issues	Pilot scale studies are being carried out; the future field applications are more likely possible.	[5,7]

(Continued)

Table 14.13. Technologies/process for value-added products generation from sludge (Continued)

Value-added products	*Advantages*	*Disadvantages*	*Remarks*	*Ref.*[a]
Bioflocculants	Non-toxic; environmental friendly; biodegradable; sustainable; reduction in carbon foot print; sludge reuse	Still in research phase; feasibility in practical applications is yet to investigate.	Recent studies have attracted more attention to explore the future implications of the bioflocculants.	[7,8]
Biosorbents	Environmental friendly; non-toxic; biodegradable; reduction in carbon foot print; sludge reuse	Still at research or pilot study phase	The interest in investing various aspects of developing Bioabsorbent from sludge is growing.	[5,7]
Biosurfactants	Biodegradability diversity; low production costs; *in-situ* applications; reduction in carbon foot print; sludge reuse	High recovery costs; high waste volume	Strain improvement; whole-cell immobilization; improved fermentation technology; metabolic enzymes	[5,7]
Enzymes	Reduction in carbon foot print; sludge reuse; economical	Still at research phase; no large scale enzyme production and/or extraction from the sludge that has been reported so far;	Significant research is required to implement the enzymes production process in environmental application	[9,10]
Alum or coagulants	Sludge reuse; economy; recycling chemicals	Still not widely applied	Potential future applications due to growing concerns over toxicants appearing in the sludge	[11]
Volatile fatty acids	Reduction in carbon foot print & Sludge reuse	Still at research or pilot study phase	Possible future large scale production	[12,13]

[a][1–3] = Jaffer et al. (2002); Rulkens (2004); and Blöcher et al. (2012); [4] = Lin and Weng (2001); [5] = (Metcalf and Eddy 2003; Brar et al. 2009a; Tyagi et al. 2009); [8] = More et al. (2012); [9] = Drouin et al. (2008); [10] = Subramanian et al. (2009); [11] = Jiménez et al. (2007); [12] = Zhang et al. (2009); and [13] = Gomez et al. (2009)

the sludge in order to improve the quality of the sludge and/or the use of sludge as an energy source. There are several technological developments for the recycling of phosphorus from the liquid phase, from sewage sludge or from sewage ash. There are mainly three approaches to recover phosphorus from the sludge and/or wastewater. The first approach is from the sludge by thermal, chemical or microbiological methods or a combination of these methods. The second approach is the direct from the wastewater in the water line. In this case chemical precipitation can be used for recovering phosphorus from the water phase in the anaerobic zone after the sludge has been separated. The third approach is the phosphorus recovery from the ashes of sludge incineration plants, for example by means of extraction processes (Rulkens 2004).

Recovery of phosphorus from liquid phase is based mostly through precipitation or crystallization processes (Jaffer et al. 2002). In the crystallization process, phosphorus is released from sewage sludge that originates from an enhanced biological phosphorus removal process, and precipitated as calcium phosphate or magnesium ammonium phosphate. During anaerobic digestion of enhanced biological phosphorus removal sludge, most of the phosphorus stored as polymers of phosphorus and part of the phosphorus present in the organic matter is released. The rejected liquors from anaerobically digested sludge show high phosphorus concentration which makes these streams very appropriate to recover phosphorus in a crystallization process. Other rejected liquors, as the thickener supernatant, could also be suitable for that process.

Most of the phosphorus recovery processes focus on a separation technology with limited synergies to the existing sludge treatment. Sewage sludge treatment today has to integrate resource recovery aspects, and especially phosphorus recovery. Phosphorus recovery as struvite ($MgNH_4PO_4{\bullet}6H_2O$) requires optimization of the crystallization process (Martí et al. 2010). The future investigations on the sludge line management strategy are necessary in order to improve the phosphorus recovery by struvite crystallization. Since it is expected that the problem of mineral phosphorus depletion will become more acute in the next decades, the implementation of this hybrid process will be of increasing interest. The phosphate products will mainly be used as fertilizer and thus for food production (Blöcher et al. 2012). Therefore, it is essential to assure within a phosphorus recovery process that compounds, which possibly have a negative health impact, do not enter the food chain.

14.5.4.2 Construction Materials

The sludge disposal problems can be drastically reduced by recycling sludge into building and construction materials (Lin and Weng 2001). The dried sludge has been used for cement manufacture in many parts of the world such as European countries, the USA and Japan. Sludge utilization for construction works as cement ingredient has been progressed over the last decade. About 26% of generated sludge in 2002 was utilized as cement ingredient. The sludge can be used as a raw material for construction materials, including building bricks, lightweight artificial aggregates, and cement like materials. Sewage sludge contains clay, silica, aluminum, calcium, iron and other inorganic materials. Therefore, sewage sludge can be is used as cement ingredient, which consists of the same elements.

Cement can be produced by using digested sludge mixed with limestone powder (Tay and Show 1993). The ash produced from lime stabilized and conditioned sludge (normally 0.3 to 0.5 kg CaO/kg dry sludge) has compositions closer to those of cement. Sludge also has been put to use in the manufacture of slag, pumice and even Portland cement (Onaka 2000). Sludge aggregates made from 100% sludge can be used to replace granite in concrete without any compromise in compressive strength. Cement-like materials made from sludge can also replace ordinary Portland cement for up to 20% by weight (Payá et al. 2002). Sludge-marine clay combinations containing 0, 20, 50, 80 and 100% clay contents, respectively, have also been transformed into pelletized aggregates displaying lower particle densities between 1.48 and 2.25 g/cm^3, compared to the density of granite at 2.56 g/cm^3. Good 28-day concrete compressive strength of 38.5 N/mm^2 was achieved by the 100% sludge aggregates, similar to 38.0 N/mm^2 of granite. The leachate contamination levels from the aggregates after 150 days were found acceptable when used in concrete, indicating insignificant environmental contamination (Tay et al. 2004). Ash produced from the sludge can be used for production of fine aggregates or as filler in cement and road construction, pozzolanic material by addicting water reducing agent. Sludge ash has been used in asphalt paving mixes to replace both fine aggregate and mineral filler size fractions in the mix. Sludge ash can also be vitrified to produce a frit for use as an aggregate substitute material (USDT 2012). Several studies have shown that sludge or sludge ash could be blended with clay to produce brick (Lin and Weng 2001). Oven-dried sludge and incinerated sludge ash could be used as brick making materials (Tay 1987). The dried sludge in the mixture might create uneven surfaces while burning the organic components. According to Tay (1987), the maximum amount of sludge ash required for brick making is 50% by weight. Up to 10% ash could be added to the bricks to produce the strength of equivalent normal clay bricks. The studies also showed that the bricks made from sludge ash exhibited strength higher than those incorporating dried sludge (Lin and Weng 2001). Mixing wastewater sludge with dam sediments has also been reported by many studies (Brar et al. 2009b). About 20% of the generated sludge in Japan in 2002 was utilized for construction work. In Japan, dewatered sludge is incinerated to reduce its volume and incinerator ash is used as soil stabilization additive, roadbed material, etc. The slag generated in the sludge melting process can be used as roadbed material, concrete aggregate, etc. It can also be utilized as material for tile, permeable block, etc. Incinerated sludge ash can be mixed with other material, and tile, brick, clay pipe and so on. Recently in Japan, new technology making brick only from incineration ash was developed (SBMC 2013).

14.5.4.3 Biofertilizers

The application of sludge to agricultural soils improves soil's physical and biological properties because it contains organic matter and plant nutrients (Metcalf and Eddy 2003). Wastewater sludge application to agricultural soils is an economical way of disposal. It improves the physical characteristics of the soil and increases the organic matter content and essential plant nutrients; in particular N and P. Sewage sludge contains numerous components required for microbial growth and may increase the activity of soil microorganisms, including rhizobial growth. Contaminants associated with sewage sludge may also negatively affect the survival of various soil microorganisms, when sewage sludge used as a fertilizer. Concerns about the use of sewage sludge contaminated by heavy metals have increased. Heavy metals remain in the soil over long periods and this may have ecotoxicological effects on plants and soil microorganisms. There are concerns about adverse effects on microbial processes related to nutrient cycling in these types of soils.

14.5.4.4 Biopesticides

Bacillus thuringiensis (Bt) is the most successful biopesticide. Production cost of conventional biopesticides (pesticides derived from animals, plants, bacteria and certain minerals), could be substantially reduced by using wastewater sewage sludge as a raw material (Vidyarthi et al. 2002). Wastewater sludge based bt biopesticides hold the future key to broad acceptability and marketability of biopesticides (Brar et al. 2009a). The biopesticide production using wastewater and/or wastewater sludge as a raw material has been investigated extensively at L'Institut national de la recherche scientifique-Eau, Terre et Environnement (INRS-ETE) since the 2000s. Production of biopesticides from sludges warrants intensive investigation into the development of alternative biopesticidal formulations, microencapsulations, dry powder, and granules for use in agriculture, which are currently in progress in the laboratory at INRS-ETE.

Simultaneous production of enzymes and biopesticides results in further cost reduction. The optimization of different fermentation parameters viz. temperature, pH, agitation and aeration, volume and age of inoculum, sludge suspended solids concentration, C/N ratio, and various pretreatment methods of sludge in batch culture indisputably allowed to achieve higher entomotoxicity and biopesticidal potential of the final product (Brar et al. 2009a). Recently, wastewater sewage sludge has been successfully used as a raw material for the production of Trichoderma sp.-based bioherbicides/biopesticides, which have broad spectrum activity in comparison to Bt (Verma et al. 2005).

Two principal problems associated with the use of wastewater sludge as a sole raw material are the presence of toxic heavy metals and pathogens. As far as pathogens are concerned, the sludge is sterilized prior to biopesticide and enzymes fermentation where all types of pathogens are eliminated. Moreover, as the quantity of the sludge required for the Bt pesticide production is significantly less, and the toxic effect of sludge metals is less likely to be prevalent (Brar et al. 2009a).

14.5.4.5 Biosurfactants

Many of the potential applications that have been considered for biosulfactant depend on the production costs. Much effort is still needed for process optimization at the engineering and biological levels. At present the use of biosufactants is limited by the cost of production and insufficient experience in applications. Development of cost effective substrates and effective microorganisms are the prerequisites for the field applications of biosurfactant in competition with synthetic surfactants. Moreover, stringent regulations are being mandated on industrial activities because of increasing environmental pollution and health issues. Therefore, the interest in biodegradable biosurfactants is continuously growing. The demand for natural products is increasing and therefore, high quality microbially produced biosurfactants shall replace the currently used synthetic surfactants.

14.5.4.6 Bioplastics

Production of bioplastics (polyhydroxyalkanaoates, PHAs) using WAS is very attractive to produce cost competitive product and treat the wastewater or decrease the pollution load of the wastewater. PHA production costs can significantly reduce by the use of mixed cultures and cheaper substrates as well as cheaper and non-sterile equipments (fermenters/bioreactors etc). Therefore, if the PHA content of mixed cultures can be improved, the mixed-culture process has a great potential for PHA content of mixed cultures for PHA production. The PHA production can be increased to 62% in microaerophilic-aerobic sludge process (Takabatake et al. 2002). The recent studies have reported that mixed cultures for PHA production have a high PHA storage capacity, high PHA yields, and high specific PHA productivity. The mixed cultures have revealed very high polymer yield (0.56 PHB/g substrate) and very high specific productivity (0.77 g PHB/g cell dry weight/h). The specific productivity of mixed cultures is significantly higher than the recombinant E. coli cultures. The physical properties of the PHAs produced by the mixed culture of activated sludge were comparable to PHAs produced by pure cultures. However, there are limited studies on cultivation strategies and the process control to improve the PHA content, yield and productivity in mixed cultures. There are many research challenges that need to be further studied. The major technical difficulty with PHA production processes from wastewater sludge is the optimization of low organic loading rates and lower yields (Brar et al. 2009a). However, overall, the cost of these processes will be substantially reduced due to cheaper substrates and non-sterile reactors with little process control. Furthermore, inherent culture stability and use of open fermentation eliminates the traditional bottlenecks of continuous monoseptic fermentations.

14.5.4.7 Bioflocculants

Recently there is growing interest in the production of bioflocculants using different microorganisms such as bacteria, fungi and yeasts. Bioflocculants holds promise to be an alternative to conventional chemical coagulants and polymers. Wastewater sludge has recently been used as a raw material for the production of bioflocculants in the form of extracellular polymeric substances or filamentous fungi (More et al. 2012). Bioflocculants production generally consists of three main steps: pretreatment, fermentation and harvesting. Wastewater sludge is potentially economical media as it is a rich source of carbon, nitrogen, phosphorus and other nutrients. The use of sludge will be advantageous for growth of the microorganisms isolated from the wastewater sludge which are already well adapted to it. Suspended sludge solids concentrations affects the growth rate and product formation by the bacterial strain cultured due to the fact that nutrients required for bacterial growth are mostly embedded into it (Drouin et al. 2008; Vidyarthi et al. 2002). Pretreatment of the sludge is essential for solubilising complex carbon sources to simpler ones which helps in accelerating growth and product formation by bacteria. Bioflocculants produced in wastewater sludge were shown the comparable flocculation performance in comparison with the bioflocculants produced in the synthetic medium. Kaolin flocculation activity of about 80% was obtained by bioflocculants produced in the sterilized sludge (More et al. 2012). There are certain issues, such as pretreatments, low nutrient availability, presence unwanted inorganic matter, which need to be addressed before utilizing wastewater sludge at the large scale. Bioflocculants production using wastewater sludge is in the laboratory phase.

14.5.4.8 Enzymes

Enzymatic activity and/or extraction of different enzymes from WAS have been investigated by many researchers for protease (Yu et al. 2007), dehydrogenase (Nybroe et al. 1992), catalase, peroxidase and o-diphenol oxidase (Subramanian et al. 2009), esterase and dehydrogenase (Boczar et al. 1992), α-amylase (Yu et al. 2007), glycosidase (Frolund et al. 1995), and α-glucosidase (Yu et al. 2007). Enzymes are biological catalysts (proteins) that increase the rate of the chemical reactions without undergoing any net chemical change during the reaction. Enzymes are popularly used in the detergent, food, pharmaceutical, diagnostics, and fine chemical industries. Enzymes in activated sludge have been used as indicators of specific microbial populations, measures of active biomass and specific process indicators such as phosphorus and COD removal (Subramanian et al. 2009). Similarly, the presence of redox enzyme activity in activated sludge is a direct indicator of effective treatment plant operation and sludge composition. The presence of enzymatic (esterase, dehydrogenate, α-glycosidase and alanine-aminopeptidase) activity in activated sludge confirmed the microbial abundance and their heterotrophic activity in sludge (Nybroe et al. 1992). The wastewater sludge can be used as a cost effective raw material to produce useful enzymes for industrial applications. Wastewater sludge is a complex raw material that can support growth and protease production by Bacillus licheniformis (Drouin et al. 2008).

Activated sludge contains 60–70% of the organic matter, which is mainly composed of protein, carbohydrates and lipids. The microorganisms dwelling in wastewater sludge are able to produce hydrolytic enzymes in order to degrade/convert these macromolecules (organic matter) into simpler forms which facilitate them to utilize these simpler compounds for their energy and growth. Most of the microbial enzymes produced in sludge are membrane bound intracellular and/or even extracellular enzymes that are adsorbed onto sludge bioflocs (Jung et al. 2002). Different extraction methods can be used for extracting enzymes from activated sludge. These enzymes extraction methods consist of partial or complete enzyme inactivation during the enzyme recovery process. The enzymes naturally produced in sludge have potential applications in improving the WWPT operation by efficient sludge solids degradation, hydrolysis of organic matter and biodegradation of toxic pollutants. For example, the enzymes extracted from the sludge were used for the biodegradation of plastic (Polyethylene terephthalate, PET) (Zhang et al. 2006). Large scale enzyme production and/or extraction from the sludge have not been reported so far. Therefore, further studies must be carried out to optimize the enzyme extraction process, and techno-economic studies should also be undertaken. There is growing research interest in the recovery of enzymes from activated sludge and their potential environmental applications.

14.5.4.9 Alum Recovery

The concept of using a recovery system for coagulant reuse has been explored in several studies that have mainly focused on alum recovery from the wastewater sludge (Jiménez et al. 2007). Acidification, alkalization, ion exchange and membrane separation are the methods of recovering coagulants from sludge. Combination of these methods has also reportedly been used to increase the recovery efficiency. In acidification, the sulphuric acid (5.44 g of $H_2SO_4/g\,Al^{3+}$) is applied to the sludge with insoluble aluminum hydroxide, which forms dilute liquid alum. Acidification is a high efficiency and low cost method. The aluminum recovery by acid extraction (pH 1.0–3.0) from sludge can reach 70–90%. Moreover, recovery of alum reduces the operating cost of the coagulation-flocculation method by 50–60% due to savings on chemicals.

However, acidification process is non-selective as it recovers alum along with all other substances that are soluble (e.g., manganese, zinc, and lead) under high acidic conditions. This will result in the build-up of impurity in water if it is recycled for use as coagulant. In addition, if this recovered alum is reused there will be trihalomethane formation during the chlorination stage of the water treatment. According to USEPA, Trihalomethanes are carcinogens. However, the recycled coagulant can be used in wastewater treatment. It is because of its higher ability to reduce COD, total nitrogen (TN) and total phosphorus (TP) in the clarification of sewage and wastewater from coastal landfill site in comparison with that of commercial Al2(SO4)3 and poly-aluminum chloride (PACl). Acid recovery reduces sludge volume by 40–50% and decreases its disposal cost by 28%. Recovery must be performed in sludge with 2% TS, at a pH of 2 and with a 15 minute contact time. Aluminum oxides also have been used for alum recovery from sludge under alkaline conditions. Coagulant recycling by alkaline reaction of aluminum hydroxide sludge using sodium hydroxide (NaOH) and lime $Ca(OH)_2$ at pH ranges of 11.4–11.8 and 11.2–11.6, respectively gives the yield of aluminates at 80% with NaOH and 30% with $Ca(OH)_2$. The recovered solution contains high amount of natural organic matter.

14.5.4.10 Volatile Fatty Acids

Fermenting WAS to produce volatile fatty acids (VFA) offers many benefits such as cost-effective reduction of mass of sludge requiring disposal thus the carbon footprint. Fermentation is a sequential two-step process. In the first step, disintegration and hydrolysis of complex organic material to soluble substrate occurs by biological and abiotic processes. The second step is acidification of soluble organics. The acid-forming bacteria use soluble substrates for their metabolic activities and produce VFA, lactate and ethanol. The hydrolysis and acidification depend upon pH, temperature, substrate loading, hydraulic and solids retention times (Veekan and Hamelers 1999). Among these factors, temperature is one of the most important factors when operating in a climate subjected to significant seasonal changes in temperature. Temperature can affect biochemical reactions in many ways, such as reaction rate, the reaction pathway, microorganism yields, and death rate (ADM1 2002).

The hydrolysis process is well described in a first order reaction proposed by Eastman and Ferguson (1981). Veeken and Hamelers (1999) investigated the hydrolysis rates of biowastes components at 20, 30 and 40°C. The hydrolysis rate constants had increased at higher temperatures for all the studied biowastes components. Several studies reported on the effect of temperature on VFA production from primary sludge fermentation. Perot et al (1988) observed that the best hydrolysis conditions of primary sludge were pH of 6.8, 50°C and 545 rpm. Skalsky and Daigger (1995) showed that VFA production from primary sludge at 14°C was approximately 42% less than that observed at 21°C. Zhang et al. (2009) also observed that the VFA production from primary sludge was improved with the enhancement of temperature from 10 to 35°C. However, so far no study has been reported on the effect of temperature on WAS hydrolysis rate and VFA generation.

Agitation or mixing is required to keep the organic material in suspension in order to increase the mass transfer rate (Pfeffer 1974). Stafford (1982) studied the effect of mixing on anaerobic sludge digestion and reported that for the impeller speed between 140 and 1000 rpm, no improvement in gas yields was observed. According to Gomez et al. (2009), by adopting proper agitation the performance of the fermentation system can be improved significantly. In case of sludge fermentation biomass settles under the incompletely mixed conditions (mechanical stirring at 1–2 rpm) in a thickener-fermenter. Danesh and Oleszkiewicz (1997), also had observed that the short-time and intermittent mixing of primary sludge in a clarifier-fermenter was sufficient as diffusion was thought to be responsible for most of the VFA release to the over-laying liquid.

14.6 CHALLENGES

Sludge is a very complex mixture of different chemicals, biological components and varying percentages of moisture content. Sludge management must explore the options of recovering or reuse of the valuable products from the sludge. Most of the toxic and ECs from wastewater end up in the wastewater sludge. Most of these ECs are toxicants, carcinogens and endocrine disrupting compounds. However, there is scarcity of information available of these chemicals. Therefore, extensive investigations are warranted the fate and transport of these contaminants during the conventional treatment and disposal options.

From a global point of view, it is clear that the development of correct sludge management procedures requires: i) assessing management options which are capable of maximizing the recycle/recovery benefits; ii) developing operational systems appropriate to local circumstances, including economy, geography, climate, etc.; iii) developing realistic and enforceable regulations adapted to local situations accompanied by standardized procedures and methods for sludge characterization and by guidelines for good management practices; and iv) assuring long-term services and sustainable operations. In practice, different sludge treatment options are often applied in combination with each other. For example, mechanical sludge dewatering is often applied prior to thermal processes such as land applications. Combined treatment of sludge along with other organic or inorganic wastes also can be a beneficial or sustainable option. Examples of such combinations are the co-incineration of sludge in an incinerator for municipal waste, the use of sludge for cement production and the co-digestion of sludge and another organic waste to increase biogas production. The innovations in new strategies and modifications in existing methods of sludge management will continue to evolve in the future.

14.7 SUMMARY

Sludge treatment and disposal strategies are continuously evolving with respect to the specific local, national and international environmental regulations and legislations. The fundamental interest is in developing sustainable sludge management by adopting safe disposal practices along with beneficial sludge reuse and achieving environmental and public health protection. Developed countries have already established sludge management practices according the local environmental and legislations restrictions. The integrated application of various sludge treatment steps or processes is an important task in the sludge management. Sludge management includes processes that not only have been applied in practice but also have application potential (and thus, are under development). For example, sludge incineration can be improved substantially with respect to energy recovery. Many sludge treatment steps are still in the development stage, such as gasification, pyrolysis, supercritical wet oxidation, application of protozoa and metazoa, phosphate recovery processes, and sludge pretreatment processes focused on an increase in the biogas production from sludge.

Due to very complex composition of sludge and the presence of various types of hazardous substances, sludge treatment is unlikely to become less expensive. In the future, sludge treatment must be focused more on the recovery of potential valuable products from sludge and the development of markets for these products. Product classification, management and control may bring revolution in the future sludge management. The integration of sludge treatment with other industrial activities or waste treatment processes may result in a reduction of the net costs (e.g., sludge use in cement kilns or thermal processes). The broad assessment of the various sludge treatment scenarios and options has to be based on current practice, research results, specific relevant local circumstances, general policy regarding wastewater treatment and sludge treatment, technical and economic feasibility and environmental benefits. Basic aspects of sludge management are the recovery of valuable products from sludge, the destruction of toxic pollutants and/or pathogens, and the prevention of discharge of toxic pollutants into the sewerage system. Wastewater treatment and sludge treatment have to be considered simultaneously as an integral part of the overall waste (wastewater and sludge) management. In this context, the future wastewater treatment systems may have to change significantly. The current national and international policy with respect to sludge, the expected changes in the future, and the public involvement and acceptance, especially regarding treatment location and health aspects, has to be taken into account during assessing various sludge management strategies.

References

Adam, C., Peplinski, B., Michaelis, M., Kley, G., and Simon, F. G. (2009). "Thermochemical treatment of sewage sludge ashes for phosphorus recovery." *Waste Manage.*, **29**(3), 1122–1128.

ADM1. (2002). "IWA task group for mathematical modelling of anaerobic digestion processes." Rep. No. 13, IWA Publishing, London.

AWA (Australian Water Association). (2012). "The management of biosolids in Australia." Australia.

Aznar, M. P., Caballero, M. A., Gil, J., Martín, J. A., and Corella, J. (1998). "Commercial steam reforming catalysts to improve biomass gasification with steam-oxygen mixtures. 2: Catalytic tar removal." *Ind. Eng. Chem. Res.*, **37**(7), 2668–2680.

Bertran, E., Sort, X., Soliva, M., and Trillas, I. (2004). "Composting winery waste: Sludges and grape stalks." *Bioresour. Technol.*, **95**(2), 203–208.

Blöcher, C., Niewersch, C., and Melin, T. (2012). "Phosphorus recovery from sewage sludge with a hybrid process of low pressure wet oxidation and nanofiltration." *Water Res.*, **46**(6), 2009–2019.

Boczar, B. A., William, M. B., and Robert, J. L. (1992). "Characterization of enzyme activity in activated sludge using rapid analyses for specific hydrolases." *Water Environ. Res.*, **64**(6), 792–797.

Brar, S. K., Verma, M., Tyagi, R. D., and Surampalli, R. Y. (2009a). "Value addition of wastewater sludge: Future course in sludge reutilization." *Pract. Period. Hazard. Toxic Radioact. Waste Manage.*, 10.1061/(ASCE)1090-025X(2009)13:1(59), 59–74.

Brar, S. K., Verma, M., Tyagi, R. D., Surampalli, R. Y., and Zhang, T. C. (2009b). "Emerging value added products and miscellaneous products." *Sustainable sludge management: Production of value added products*, R. D. Tyagi, R. Y. Surrampalli, S. Yan, T. Zhang, C. K. Kao, and B. N. Lohani, eds., ASCE, Reston, VA, 66–100.

Bridle, T. R., and Pritchard, D. (2004). "Energy and nutrient recovery from sewage sludge via pyrolysis." *Water Sci. Technol.*, **50**(9), 169–175.

Bridle, T. R., and Skrypski-Mantele, S. (2004). "Experience and lessons learned from sewage sludge pyrolysis in Australia." *Water Sci. Technol.*, **49**(10), 217–223.

Brown, S. (2007). "Green incineration." *BioCycle*, **48**(7), 21.

Carrington, E. G. (2001). "Evaluation of sludge treatments for Pathogen reduction." Final Rep. No 5026/1, European Commission, Medmenham, Marlow, Bucks.

Cartmell, E., et al. (2006). "BiosolidsA fuel or a waste? An integrated appraisal of five co-combustion scenarios with policy analysis." *Environ. Sci. Technol.*, **40**(3), 649–658.

CCME (Canadian Council of Ministers of the Environment). (2010). "A review of the current canadian legislative framework for wastewater biosolids." Winnipeg, MB, Canada.

CEC (Council of European Communities). (1986). "Directive on the protection of the environment, and in particular of the soil, when sewage sludge is used in agriculture (86/278/EEC)." Brussels.

CEC (Council of European Communities). (1991). "Directive concerning urban wastewater treatment (91/271/EEC)." Brussels.

CEC (Council of European Communities). (1999). "Implementation of Council Directive 91/271/EEC of 21 May 1991 concerning urban wastewater treatment, as amended by Commission Directive 98/15/EC of 27 February 1998." Brussels.

CEC (Council of European Communities). (2001). "Disposal and recycling routes for sewage sludge." .

CEC (Council of European Communities). (2009). "Fifth Commission summary on the imple Fifth Commission summary on the implementation of the urban waste water treatment directive." Brussels.

CGTI (China Greentech Initiative). (2011). "Sludge treatment in China." ⟨http://www.china-greentech.com/sites/default/files/CGTI2011-CW-WS3-Sludge.pdf⟩ (Feb. 1, 2014).

Chen, H., Yan, S. H., Ye, Z. L., Meng, H. J., and Zhu, Y. G. (2012). "Utilization of urban sewage sludge: Chinese perspectives." *Environ. Sci. Pollut. Res.*, **19**(5), 1454–1463.

Clarke, B. O., and Smith, S. R. (2011). "Review of 'emerging' organic contaminants in biosolids and assessment of international research priorities for the agricultural use of biosolids." *Environ. Int.*, **37**(1), 226–247.

Colin, F., and Gazbar, S. (1995). "Distribution of water in sludges in relation to their mechanical dewatering." *Water Res.*, **29**(8), 2000–2005.

Danesh, S., and Oleszkiewicz, J. A. (1997). "Volatile fatty acid production and uptake in biological nutrient removal systems with process separation." *Water Environ. Res.*, **69**(6), 1106–1111.

Dentel, S. (2012). "Sludges, residuals and biosolids." *Global trends and challenges*, International Water Association (IWA), London.

Dentel, S. K., Strogen, B., and Chiu, P. (2004). "Direct generation of electricity from sludges and other liquid wastes." *Water Sci. Technol.*, **50**(9), 161–168.

Díaz-Cruz, M. S., et al. (2009). "Analysis of selected emerging contaminants in sewage sludge." *Trends Anal. Chem.*, **28**(11), 1263–1275.

Doelsch, E., Deroche, B., and Van de Kerchove, V. (2006). "Impact of sewage sludge spreading on heavy metal: Speciation in tropical soils (Reunion, Indian Ocean)." *Chemosphere*, **65**(2), 286–293.

Dogru, M., Midilli, A., and Howarth, C. R. (2002). "Gasification of sewage sludge using a throated downdraft gasifier and uncertainty analysis." *Fuel Process. Technol.*, **75**(1), 55–82.

Drouin, M., Lai, C. K., Tyagi, R. D., and Surampalli, R. Y. (2008). "Bacillus licheniformis proteases as high value added products from fermentation of wastewater sludge: Pretreatment of sludge to increase the performance of the process." *Water Sci. Technol.*, **57**(3), 423–429.

Eastman, J. A., and Ferguson, J. F. (1981). "Solubilization of particulate organic carbon during the acid phase of anaerobic digestion." *J. Water Pollut. Control Fed.*, **53**, 352–366.

Eco-Asia (Environmental Cooperation-Asia). (2010). "A rapid assessment of septage management in Asia: Policies and practices in India, Indonesia, Malaysia, the Philippines, Srilanka, Thailand and Vietnam." AECOM International Development and the Swiss Institute of Aquatic Sciences and Technology, Switzerland.

EEA (European Environment Agency). (2005). "The European environment–State and outlook." Rep. No. 1/2005, Denmark.

EIA. (2014). "Sewage." ⟨http://www.eai.in/ref/ae/wte/typ/clas/sewage.html⟩ (Feb. 14, 2014).

Eljarrat, E., Díaz-Cruz, M. S., Farré, M., Alda, M. J. L., Petrović, M., and Barceló, D. (2013). "Analysis of emerging contaminants in sewage sludge." *Emerging organic contaminants in sludges*, T. Vicent, G. Caminal, E. Eljarrat, and D. Barceló, eds., Springer, Berlin, 31–71.

Frolund, B., Griebe, T., and Nielsen, P. H. (1995). "Enzymatic activity in the activated-sludge floc matrix." *Appl. Microbiol. Biotechnol.*, **43**(4), 755–761.

Fytili, D., and Zabaniotou, A. (2008). "Utilization of sewage sludge in EU application of old and new methods—A review." *Renewable Sustainable Energy Rev.*, **12**(1), 116–140.

Gaspard, P., Schwartzbrod, J., Wiart, J., Galvez, L., and Dumourtier, N. (1997). "Parasitological risks associated with the use of biosolids in agriculture: Nematode Egg survival on grass, vegetables and within soils." *Proc., Beneficial Reuse of Water and Biosolids*, Asociación para la Defensa de la Calidad de las Aguas, Water Environment Federation, Alexandria, VA.

Gea, T., Ferrer, P., Alvaro, G., Valero, F., Artola, A., and Sánchez, A. (2007). "Co-composting of sewage sludge: Fats mixtures and characteristics of the lipases involved." *Biochem. Eng. J.*, **33**(3), 275–283.

Ge, Z., Zhanga, F., Grimaucb, J., Hurstb, J., and Hea, Z. (2013). "Long-term investigation of microbial fuel cells treating primary sludge or digested sludge." *Bioresour. Technol.*, **136**, 509–514.

Ginestet, P., and Camacho, P. (2007). "Technical evaluation of sludge production and reduction." *Comparative evaluation of sludge reduction routes*, IWA Publishing, Londra, 1–15.

Gomez, A., Zubizarreta, J., Rodrigues, M., Dopazo, C., and Fueyo, N. (2010). "Potential and cost of electricity generation from human and animal waste in Spain." *Renewable Energy*, **35**(2), 498–505.

Gomez, X., Cuetos, M. J., Prieto, J. I., and Moran, A. (2009). "Bio-hydrogen production from waste fermentation: Mixing and static conditions." *Renewable Energy*, **34**(4), 970–975.

Goncalves, R. F., Luduvice, M., and von Sperling, M. (2007). *Sludge thickening and dewatering*, IWA Publishing, London.

Halde, R. E. (1979). "Sewage sludge characterization by vacuum drying." *Filtr. Sep.*, **5**, 238–242.

Hall, J. E. (1995). "Sewage sludge production, treatment and disposal in the European Union." *Water Environ. J.*, **9**(4), 335–343.

Hong, G., et al. (2012). "Surveillance of waste disposal activity at sea using satellite ocean color imagers: GOCI and MODIS." *Ocean Sci. J.*, **47**(3), 387–394.

Hua, L. (2011). "Business opportunity of municipal sewage sludge treatment in China." M.S. thesis, Lahti Univ. of Applied Sciences, Finland.

Izrail, S., Turovskiy, P., and Mathai, K. (2006). *Wastewater sludge processing*, Wiley, New York.

Jaffer, Y., Clark, T. A., Pearce, P., and Parsons, S. A. (2002). "Potential phosphorus recovery by struvite formation." *Water Res.*, **36**(7), 1834–1842.

Jiménez, B., Martinez, M., and Vaca, M. (2007). "Alum recovery and wastewater sludge stabilization with sulfuric acid." *Water Sci. Technol.*, **56**(8), 133–41.

Jonassen, O., Leonard, A., and Stenström, S. (2008). "Technical reports 1-Drying and dewatering sludge." *Annex X energy efficient drying and dewatering technologies*, Industrial Energy-Related Technologies and Systems (IETS), Liège, Belgium.

Jung, J., Xing, X.-H., and Matsumoto, K. (2002). "Recoverability of protease released from disrupted excess sludge and its potential application to enhanced hydrolysis of proteins in wastewater." *Biochem. Eng. J.*, **10**(1), 67–72.

Kargo, D. M. (2010). "Biodiesel production from municipal sewage sludge." *Energy Fuels*, **24**(5), 2791–2794.

Keey, R. B. (1972). "Introduction to industrial drying operations." *International series of monographs in chemical engineering*, Vol. **13**, Pergamon Press, Oxford, 19–232.

Kelessidis, A., and Stasinakis, A. S. (2012). "Comparative study of the methods used for treatment and final disposal of sewage sludge in European countries." *Waste Manage.*, **32**(6), 1186–1195.

Kim, Y., and Parker, W. (2008). "A technical and economic evaluation of the pyrolysis of sewage sludge for the production of bio-oil." *Bioresour. Technol.*, **99**(5), 1409–1416.

Koch, C. M., Lee, J. S., Bratby, J. R., and Barber, D. B. (1997). "A critical evaluation of procedure for estimating biosolids production." *Water residuals biosolids management: Approaching the year 2000*, Water Environment Federation, Philadelphia.

Kopp, J., and Dichtl, N. (2000). "Prediction of full-scale dewatering results by determining the water distribution of sewage sludges." *Water Sci. Technol.*, **42**(9), 141–149.

Kopp, J., and Dichtl, N. (2001). "Influence of the free water content on the dewaterability of sewage sludges." *Water Sci. Technol.*, **44**(10), 177–183.

LeBlanc, R. J., Matthews, P., and Roland, P. R. E. (2008). "Global, Atlas of Excreta, wastewater sludge, and biosolids management: Moving forward the sustainable and welcome uses of a global resource." UN-HABITAT, New Brunswick, Canada.

Lee, D. J., and Lee, S. F. (1995). "Measurement of bound water content in sludge: The use of differential scanning calorimetry (DSC)." *J. Chem. Technol. Biotechnol.*, **62**(4), 359–365.

Lehmann, J. (2007). "A handful of carbon." *Nature*, **447**(7141), 143–144.

Lepeuple, A. S., Gaval, G., Jovic, M., and de Roubin, M. R. (2004). "Literature review on levels of pathogens and their abatement in sludges, soil and treatment." Framework 6 Programme, European Commission, Brussels, Belgium. ⟨https://www.ecn.nl/docs/society/horizontal/hor6_pathogens.pdf⟩ (Mar. 4, 2016).

Lin, D., and Weng, C. (2001). "Use of sewage sludge ash as brick material." *J. Environ. Eng.*, 10.1061/(ASCE)0733-9372(2001)127:10(922), 922–927.

Logan, B. E., et al. (2006). "Microbial fuel cells: Methodology and technology." *Environ. Sci. Technol.*, **40**(17), 5181–5192.

Lou, X. F., and Nair, J. (2009). "The impact of landfilling and composting on greenhouse gas emissions-A review." *Bioresour. Technol.*, **100**(16), 3792–3798.

Lundin, M., Olofsson, M., Pettersson, G. J., and Zetterlund, H. (2004). "Environmental and economic assessment of sewage sludge handling options." *Resour. Conserv. Recycl.*, **41**(4), 255–278.

Lu, Q., He, Z. L., and Stoffella, P. J. (2012). "Land application of biosolids in the USA: A review." *Appl. Environ. Soil Sci.*, 11.

Luts, D., Devoldere, K., Laethem, B., Bartholomeeusen, W., and Ockier, P. (2000). "Co-incineration of dried sewage sludge in coal-fired power plants: A case study." *Water Sci. Technol.*, **42**(9), 259–268.

Malerius, O., and Werther, J. (2003). "Modeling the adsorption of mercury in the flue gas of sewage sludge incineration." *Chem. Eng. J.*, **96**(1–3), 197–205.

Manara, P., and Zabaniotou, A. (2012). "Towards sewage sludge based biofuels via thermochemical conversion–A review." *Renewable Sustainable Energy Rev.*, **16**(5), 2566–2582.

Marrero, T. W., McAuley, B. P., Sutterlin, W. R., Steven Morris, J., and Manahan, S. E. (2004). "Fate of heavy metals and radioactive metals in gasification of sewage sludge." *Waste Manage.*, **24**(2), 193–198.

Marrone, P. A., Hodes, M., Smith, K. A., and Tester, J. W. (2004). "Salt precipitation and scale control in supercritical water oxidation—Part B: Commercial/full-scale applications." *J. Supercrit. Fluids*, **29**(3), 289–312.

Martí, N., Pastor, L., Bouzas, A., Ferrer, J., and Seco, A. (2010). "Phosphorus recovery by struvite crystallization in WWTPs: Influence of the sludge treatment line operation." *Water Res.*, **44**(7), 2371–2379.

Mathieu, P., and Dubuisson, R. (2002). "Performance analysis of a biomass gasifier." *Energy Convers. Manage.*, **43**(9–12), 1291–1299.

Matteson, G. C., and Jenkins, B. M. (2007). "Food and processing residues in California: Resource assessment and potential for power generation." *Bioresour. Technol.*, **98**(16), 3098–3105.

Metcalf and Eddy. (2003). *Wastewater engineering—Treatment and reuse*, Tata McGraw Hill Education, New York.

Miller, S. A., et al. (2008). "Innovations in dewatering sludges." *Wastewater treatment and reuse*, Water Environment Research Foundation, Alexandria, VA.

Mininni, G., Di Bartolo Zuccarello, R., Lotito, V., Spinosa, L., and Di Pinto, A. C. (1997). "A design model of sewage sludge incineration plants with energy recovery." *Water Sci. Technol.*, **36**(11), 211–218.

More, T. T., Yan, S., Hoang, N. V., Tyagi, R. D., and Surampalli, R. Y. (2012). "Bacterial polymer production using pretreated sludge as raw material and its flocculation and dewatering potential." *Bioresour. Technol.*, **121**, 425–431.

Nielsen, S. (2008). "Sludge treatment and drying reed bed systems 20 years of experience." *Proc., Processing of Conf.: European Conf. on Sludge Management*, Univ of Liege, Liège, Belgium.

Nishio, N., and Nakashimada, Y. (2007). "Recent development of anaerobic digestion processes for energy recovery from wastes." *J. Biosci. Bioeng.*, **103**(2), 105–112.

NSER-SA. (2013). "Sustainability of Terrestrial Ecosystems. State of terrestrial ecosystem: 2. National State of the Environment Report." South Africa.

Nybroe, O., Jørgensen, P. E., and Henze, M. (1992). "Enzyme activities in waste water and activated sludge." *Water Res.*, **26**(5), 579–584.

Oehlmann, J., Oetken, M., and Schulte-Oehlmann, U. (2008). "A critical evaluation of the environmental risk assessment for plasticizers in the freshwater environment in Europe, with special emphasis on bisphenol A and endocrine disruption." *Environ. Res.*, **108**(2), 140–149.

Ok, J., Park, D., Jung, J. W., Ghodake, G., and Lee, D. S. (2010). "Feasibility study on electricity production from a sewage sludge extract using microbial fuel cell." *J. Nanoelectron. Optoelectron.*, **5**(2), 227–231.

Oleksandr, P. (2005). "Sewage sludge." ⟨http://www.worldmapper.org/⟩ (Feb. 14, 2014).

Onaka, T. (2000). "Sewage can make Portland cement: A new technology for ultimate reuse of sewage sludge." *Water Sci. Technol.*, **41**(8), 93–98.

Parker, C. (2012). "Sludge production." IWA Water Wiki, U.K.

Park, H. J., et al. (2010). "Clean bio-oil production from fast pyrolysis of sewage sludge: Effects of reaction conditions and metal oxide catalysts." *Bioresour. Technol.*, **101**(1), S83–S85.

Park, J. M., Lee, S. B., Kim, J. P., Kim, M. J., Kwon, O. S., and Jung, D. I. (2009). "Behavior of PAHs from sewage sludge incinerators in Korea." *Waste Manage.*, **29**(2), 690–695.

Pathak, A., Dastidar, M. G., and Sreekrishnan, T. R. (2009). "Bioleaching of heavy metals from sewage sludge: A review." *J. Environ. Manage.*, **90**(8), 2343–2353.

Payá, J., et al. (2002). "Advantages in the use of fly ashes in cements containing pozzolanic combustion residues: silica fume, sewage sludge ash, spent fluidized bed catalyst and rice husk ash." *J. Chem. Technol. Biotechnol.*, **77**(3), 331–335.

Perot, J. C., Sergent, M., Richard, P., and Millot, N. (1988). "The effect of pH, temperature and agitation speed on sludge anaerobic hydrolysis-acidification." *Environ. Technol. Lett.*, **9**(8), 741–752.

Pfeffer, J. T. (1974). "Reclamation of energy from organic waste." , U.S. EPA, Washington, DC.

Reimann, D. O. (1999). "Problems about sewage sludge incineration." *Workshop on Problems around Sludge*, The Directorate-General for the Environment and the Joint Research Centre of the European Commission, Italy.

Roy, M. M., Dutta, A., Corscadden, K., Havard, P., and Dickie, L. (2011). "Review of biosolids management options and co-incineration of a biosolid-derived fuel." *Waste Manage.*, **31**(11), 2228–2235.

Ruggieri, L., Gea, T., Artola, A., and Sánchez, A. (2008). "Influence of different co-substrates biochemical composition on raw sludge co-composting." *Biodegradation*, **19**(3), 403–415.

Rulkens, W. H. (2004). "Sustainable sludge management-what are the challenges for the future?" *Water Sci. Technol.*, **49**(10), 11–19.

Rulkens, W. (2008). "Sewage sludge as a biomass resource for the production of energy: Overview and assessment of the various options." *Energy Fuels*, **22**(1), 9–15.

Sanchez-Martin, M. J., Garcia-Delgado, M., Lorenzo, L. F., Rodriguez-Cruz, M. S., and Arienzo, M. (2007). "Heavy metals in sewage sludge amended soils determined by sequential extractions as a function of incubation time of soils." *Geoderma*, **142**(3–4), 262–273.

SBMC (Sewerage Business Management Centre). (2013). "Reuse of sewage sludge." ⟨http://www.sbmc.or.jp/english/050427/Reuse_of_Sewage_Sludge.htm⟩ (Feb. 1, 2014).

Sidhu, J. P. S., and Toze, S. G. (2009). "Human pathogens and their indicators in biosolids: A literature review." *Environ. Int.*, **35**(1), 187–201.

Skalsky, D. S., and Daigger, G. T. (1995). "Wastewater solids fermentation for volatile acid production and enhanced biological phosphorus removal." *Water Environ. Res.*, **67**(2), 230–237.

Smith, J. K., and Vesilind, P. A. (1995). "Dilatometric measurement of bound water in wastewater sludge." *Water Res.*, **29**(12), 2621–2626.

Smith, K. M., Fowler, G. D., Pullket, S., and Graham, N. J. D. (2009). "Sewage sludge-based adsorbents: A review of their production, properties and use in water treatment applications." *Water Res.*, **43**(10), 2569–2594.

Smollen, M. (1986). "Categories of moisture content and dewatering characteristics of biological sludges." *Proc., 4th World Filtration Congress*, Vol. **14**, IV, KVIV-Technologisch Instituut, Antwerpen, Belgium, 35–41.

Smollen, M. (1990). "Evaluation of municipal sludge drying and dewatering with respect to sludge volume reduction." *Water Sci. Technol.*, **22**(12), 153–161.

Snyman, H. G., and Herselman, J. E. (2006). "Guidelines for the utilisation and disposal of wastewater sludge. Volume 1: Selection of management options." Rep. No. TT261/06, WRC, South Africa.

Spinosa, L., and Vesilind, P. A. (2001). *Sludge into biosolids: Processing, disposal and utilization*, IWA Publication, London.

Stafford, D. A. (1982). "The effects of mixing and volatile fatty acid concentrations on anaerobic digester performance." *Biomass*, **2**(1), 43–55.

Stasinakis, A. S. (2012). "Review on the fate of emerging contaminants during sludge anaerobic digestion." *Bioresour. Technol.*, **121**, 432–440.

Subramanian, B. S., Yan, S., Tyagi, R. D., and Surampalli, R. Y. (2009). "Enzymes: Production and extraction." *Sustainable sludge management: Production of value added products*, R. D. Tyagi, R. Y. Surampalli, S. Yan, T. Zhang, C. K. Kao, and B. N. Lohani, eds., ASCE, Reston, VA, 231–261.

Subramanian, B. S., Yan, S., Tyagi, R. D., and Surampalli, R. Y. (2010). "Extracellular polymeric substances (EPS) producing bacterial strains of municipal wastewater sludge: Isolation, molecular identification, EPS characterization and performance for sludge settling and dewatering." *Water Res.*, **44**(7), 2253–2266.

Svanstrom, M., Modell, M., and Tester, J. (2004). "Direct energy recovery from primary and secondary sludges by supercritical water oxidation." *Water Sci. Technol.*, **49**(10), 201–208.

Takabatake, H., Satoh, H., Mino, T., and Matsuo, T., (2002). "PHA (polyhydroxyalkanoate) production potential of activated sludge treating wastewater." *Water Sci. Technol.*, **45**, 119–126.

Tay, J. (1987). "Bricks manufactured from sludge." *J. Environ. Eng.*, 10.1061/(ASCE)0733-9372(1987)113:2(278), 278–284.

Tay, J., and Show, K. (1993). "Manufacture of cement from sewage sludge." *J. Mater. Civ. Eng.*, 10.1061/(ASCE)0899-1561(1993)5:1(19), 19–29.

Tay, J. H., Show, K. Y., Lee, D. J., and Hong, S. Y. (2004). "Reuse of wastewater sludge with marine clay as a new resource of construction aggregates." *Water Sci. Technol.*, **50**(9), 189–196.

Tsang, K. R., and Vesilind, P. A. (1990). "Moisture distribution in sludges." *Water Sci. Technol.*, **22**(12), 135–142.

Tyagi, R. D., Surrampalli, R. Y., Yan, S., Zhang, T., Kao, C. K., and Lohani, B. N. (2009). *Sustainable sludge management—Production of value added products*, ASCE, Reston, VA.

Uggetti, E., Ferrer, I., Llorens, E., and García, J. (2010). "Sludge treatment wetlands: A review on the state of the art." *Bioresour. Technol.*, **101**(9), 2905–2912.

USDT (U.S. Department of Transportation). (2012). "User guidelines for waste and byproduct materials in pavement construction." Federal Highway Administration, Washington, DC.

U.S. EPA. (1995). "A guide to the biosolids risk assessment for the EPA Part 503 rule." Washington, DC.

U.S. EPA. (2006). "Emerging technologies for biosolids management." Washington, DC.

Veeken, A., and Hamelers, B. (1999). "Effect of temperature on hydrolysis rate of selected biowast components." *Bioresour. Technol.*, **69**(3), 249–254.

Vergine, P., et al. (2007). "Partial ozonation of activated sludge to reduce excess sludge production: Evaluation of effects on biomass activity in a full scale demonstration test." *Proc., Int. Water Association Specialist Conf.*, Greater Moncton Sewerage Commission, Canada.

Verma, M., Brar, S. K., Tyagi, R. D., Valéro, J. R., and Surampalli, R. Y. (2005). "Wastewater sludge as a potential raw material for antagonistic fungus (Trichoderma sp.): Role of pretreatment and solids concentration." *Water Res.*, **39**(15), 3587–3596.

Vesilind, P. A. (1994). "The role of water in sludge dewatering." *Water Environ. Res.*, **66**(1), 4–11.

Vesilind, P. A., and Hsu, C. C. (1997). "Limits of sludge dewaterability." *Water Sci. Technol.*, **36**(11), 87–91.

Vidyarthi, A. S., Tyagi, R. D., Valero, J. R., and Surampalli, R. Y. (2002). "Studies on the production of B. thuringiensis based biopesticides using wastewater sludge as a raw material." *Water Res.*, **36**(19), 4850–4860.

von Sperling, M., and Gonçalves, R. F. (2007). "Sludge characteristics and production." *Sludge treatment and disposal*, V. A. Cleverson, M. von Sperling, and F. Fernandes, eds., IWA Publishing, London.

Walker, J. M. (1991). "Fundamentals of odor control." *BioCycle*, **32**, 50.

Wang, C. C., Chang, C. W., Chu, C. P., Lee, D. J., Chang, B. V., and Liao, C. S. (2003b). "Producing hydrogen from wastewater sludge by Clostridium bifermentans." *J. Biotechnol.*, **102**(1), 83–92.

Wang, C. C., et al. (2003a). "Using filtrate of waste biosolids to effectively produce bio-hydrogen by anaerobic fermentation." *Water Res.*, **37**(11), 2789–2793.

Wang, H., et al. (2008). "Technological options for the management of biosolids." *Environ. Sci. Pollut. Res. Int.*, **15**(4), 308–317.

Wang, Z., Mei, X., Ma, J., and Wu, Z. (2012). "Recent advances in microbial fuel cells integrated with sludge treatment." *Chem. Eng. Technol.*, **35**(10), 1733–1743.

Watanabe, Y., and Tanaka, K. (1999). "Innovative sludge handling through pelletization/thickening." *Water Res.*, **33**(15), 3245–3252.

Wong, J. W. C., and Selvam, A. (2006). "Speciation of heavy metals during co-composting of sewage sludge with lime." *Chemosphere*, **63**(6), 980–986.

Xiao, B., Yang, F., and Liu, J. (2011). "Enhancing simultaneous electricity production and reduction of sewage sludge in two-chamber MFC by aerobic sludge digestion and sludge pretreatments." *J. Hazard. Mater.*, **189**(1–2), 444–449.

Yu, G. H., He, P. J., Shao, L. M., and Lee, D. J. (2007). "Enzyme activities in activated sludge flocs." *Appl. Microbiol. Biotechnol.*, **77**(3), 605–612.

Zabaniotou, A., and Theofilou, C. (2008). "Green energy at cement kiln in Cyprus–Use of sewage sludge as a conventional fuel substitute." *Renewable Sustainable Energy Rev.*, **12**(2), 531–541.

Zhang, J., Gong, J., Shao, G., Qin, J., and Gu, Z. (2006). "Biodegradability of diethylene glycol terephthalate and poly(ethylene terephthalate) fiber by crude enzymes extracted from activated sludge." *J. Appl. Polym. Sci.*, **100**(5), 3855–3859.

Zhang, P., Chen, Y., Haung, T. Y., and Zhou, Q. (2009). "Waste activated sludge hydrolysis and short-chain fatty acids accumulation in the presence of SDBS in semi-continuous flow reactors: Effect of solids retention time and temperature." *Chem. Eng. J.*, **148**(2–3), 348–353.

Zhang, Y., et al. (2011). "Extracellular polymeric substances enhanced mass transfer of polycyclic aromatic hydrocarbons in the two-liquid-phase system for biodegradation." *Appl. Microbiol. Biotechnol.*, **90**(3), 1063–1071.

Zuo, J. (2010). "Status and development of sewage sludge treatment and disposal in China." Tsinghua Univ., China.

CHAPTER 15

Healthcare Waste Management

Kurian Joseph[1]
Ammaiyappan Selvam[2]
Jonathan W.C. Wong[3]

15.1 INTRODUCTION

"Health Care Wastes" (HCW) or "biomedical wastes (BMW)" include any wastes generated during the diagnosis, treatment or immunization of human beings or animals at hospitals, dispensaries, laboratories, blood banks or veterinary and research institutions. It consists wholly or partly of human or animal tissue, blood or other body fluids, excretions, drugs or other pharmaceutical products, swabs or dressings, or syringes, needles or other sharp instruments and any other waste arising from medical, nursing, dental, veterinary, pharmaceutical or similar practice, investigation, treatment, care, teaching or research, or the collection of blood for transfusion which unless rendered safe may prove hazardous and may cause infection to any person coming into contact with it. Wide variety of antibiotics, cytotoxics (toxic to cells and inhibit cell growth and multiplication), corrosive chemicals and radioactive substances used in modern hospitals and health care institutions ultimately become part of HCW. About 75–90% of these wastes produced by health-care providers is non-risk or general health-care waste, comparable to domestic waste (Marinkovic et al. 2008; Chartier et al. 2014). It comes mostly from the administrative and housekeeping functions of health-care establishments and may also include waste generated during maintenance of health-care premises. Although the proportion of infectious and hazardous waste in overall medical waste is relatively small, ranging 10–25%, any improper waste management, where infectious waste is mixed with general waste, can render all the waste potentially infectious and hazardous (Chaerul et al. 2008; Cheng et al. 2009).

[1]Anna Univ., Chennai, India.
[2]Hong Kong Baptist Univ., Hong Kong SAR.
[3]Hong Kong Baptist Univ., Hong Kong SAR.

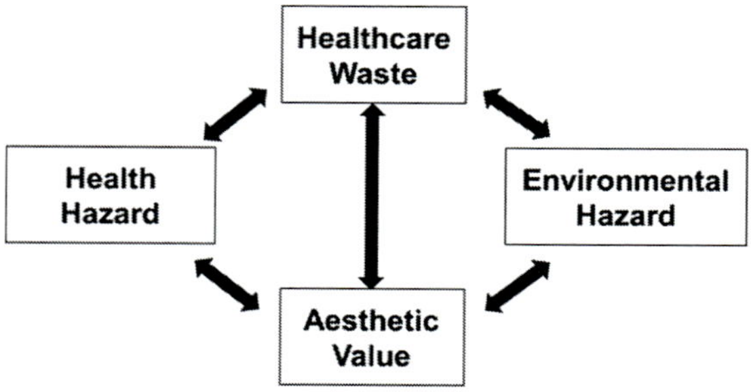

Figure 15.1. Concerns associated with health care wastes

Though health care activities are vital for our wellbeing, the wastes generated from such activities can be hazardous due to their high potential for causing infections, injuries and environmental hazards as well as aesthetic concerns, if these are not properly managed (Figure 15.1). The problem of medical waste is not of quantity but of its infectious and hazardous potential.

The most important pathways for the transmission of disease to healthcare workers are percutaneous injuries with contaminated sharps (e.g., through the skin), contamination through the fecal–oral route (e.g., salmonellosis, hepatitis A), and contamination through airborne transmission (e.g., tuberculosis, measles) (Diaz et al. 2005). The rag pickers and waste workers are often the worst affected, because of their potential for contacts with such wastes while scavenging or handling municipal solid wastes, mixed with HCW. Increased use of disposables resulting in inappropriate recycling and illegal re-use is also another critical issue.

15.2 CATEGORIES OF HEALTHCARE WASTE

Health care waste comprises general waste or municipal solid waste, e.g. packaging materials, food scraps, newspapers, bottles; and hazardous health care waste. Classification of wastes vary depending on the national/local regulations. Health Care Wastes categorized by World Health Organization (WHO) is presented in Table 15.1.

A WHO (2005) report indicates that an estimated 16 000 million injections are administered every year globally; however, not all needles and syringes are properly disposed of, creating a risk of injury and infection and opportunities for re-use. About 21 million hepatitis B virus (HBV) infections, 2 million hepatitis C virus infections and 0.26 HIV infections worldwide were linked to the injections with contaminated syringes. Ironically, many of these infections were avoidable upon safe disposal of the syringes. In developing countries, additional hazards occur from scavenging at waste disposal sites and the

Table 15.1. Classification of healthcare waste by World Health Organization

Waste category	*Descriptions and examples*
Hazardous health-care waste	
Sharps waste	Used or unused sharps (e.g. hypodermic, intravenous or other needles; auto-disable syringes; syringes with attached needles; infusion sets; scalpels; pipettes; knives; blades; broken glass)
Infectious waste	Waste suspected to contain pathogens and that poses a risk of disease transmission (see section 2.1.2) (e.g. waste contaminated with blood and other body fluids; laboratory cultures and microbiological stocks; waste including excreta and other materials that have been in contact with patients infected with highly infectious diseases in isolation wards)
Pathological waste	Human tissues, organs or fluids; body parts; fetuses; unused blood products
Pharmaceutical waste, cytotoxic waste	Pharmaceuticals that are expired or no longer needed; items contaminated by or containing pharmaceuticals Cytotoxic waste containing substances with genotoxic properties (e.g. waste containing cytostatic drugs – often used in cancer therapy; genotoxic chemicals)
Chemical waste	Waste containing chemical substances (e.g. laboratory reagents; film developer; disinfectants that are expired or no longer needed; solvents; waste with high content of heavy metals, e.g. batteries; broken thermometers and blood-pressure gauges)
Radioactive waste	Waste containing radioactive substances (e.g. unused liquids from radiotherapy or laboratory research; contaminated glassware, packages or absorbent paper; urine and excreta from patients treated or tested with unsealed radionuclides; sealed sources)
Non-hazardous or general health-care waste	Waste that does not pose any particular biological, chemical, radioactive or physical hazard

Source: Chartier et al., 2014; reproduced with permission from World Health Organization

manual sorting of hazardous waste from health-care establishments. These practices are common in many regions of the world. The waste handlers are at immediate risk of needle-stick injuries and exposure to toxic or infectious materials.

The transmission of infection and its control follows "chain of infection" with the following components in order (Chartier et al., 2014):

- *Infectious agent*: a microorganism that can cause disease
- *Reservoir*: a place where microorganisms can thrive and reproduce (e.g. in humans, animals, inanimate objects)
- *Portal of exit*: a means for a microorganism to leave the reservoir (e.g. respiratory, genitourinary and gastrointestinal tracts; skin and mucous membranes; and the placenta)
- *Mode of transmission*: how the microorganism moves from one place to another (e.g. contact, droplets, airborne)
- *Portal of entry*: an opening allowing the microorganism to invade a new host
- *Susceptible host*: a person susceptible to the disease, lacking immunity or physical resistance to prevent infection

Each component has unique significance and must be understood in order to prevent the transmission. Each link in the chain must be present and in the precise sequential order for an infection to occur. Breaking any link in the chain will prevent infection, although control measures for healthcare are most often directed at the "mode of transmission" stage in the chain of infection (Chartier et al., 2014).

As mentioned previously, about 10–25% of the overall HCW represents the infectious and hazardous waste. Of this hazardous wastes, infectious and anatomic wastes together is the predominant waste (Figure 15.2); thus most commonly need to be safeguarded against. Other potentially hazardous health care waste components (i.e. chemicals such as mercury in thermometers, pharmaceuticals such as cytotoxic drugs or materials contaminated with cytotoxics, radioactive waste and pressurized containers) comprise a very small fraction of the total waste quantity. Pathological wastes are generated during surgery/autopsy or other medical procedures and include human tissues, organ, body parts, body fluids and specimens along with their containers.

Infectious wastes may contain sufficient population of infectious (viral, bacterial or parasitic diseases) agents that are capable of causing and spreading infections among people, livestock and vectors. Such wastes include cultures and stocks of infectious agents from laboratory work, waste from infected patients in isolation wards, human tissues, anatomical waste, organs, body parts, placenta, animal waste (tissue/ cell cultures), any pathological/ surgical waste, microbiology and biotechnology waste (cultures, stocks, specimens of micro-organism, live or attenuated vaccines, etc.), cytological, pathological wastes, solid waste (swabs, bandages, mops, any item contaminated with blood or body fluids), infected syringes, needles, other sharps, glass, rubber, metal, plastic disposables and other such wastes. Therefore, mixing this infectious waste with other waste leads to potential risk of infection to various stakeholders dealing with the HCW. The potential infections caused by exposure to health-care wastes, causative organisms and transmission vehicles are presented in Table 15.2.

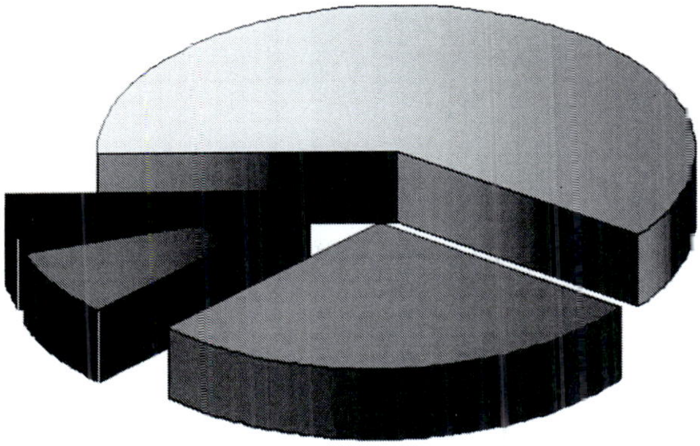

Figure 15.2. Composition of hazardous health care wastes
Source: Kühling (2014)

Table 15.2. Potential infections caused by exposure to HCW, causative organisms and transmission vehicles

Type of infection	*Examples of causativeorganisms*	*Transmission vehicles*
Gastroenteric infections	Enterobacteria, e.g. *Salmonella*, *Shigella* spp., *Vibrio cholerae*, *Clostridium difficile*, helminths	Faeces and/or vomit
Respiratory infections	*Mycobacterium tuberculosis*, measles virus, *Streptococcus pneumoniae*, severe acute respiratory syndrome (SARS)	Inhaled secretions, saliva
Ocular infection	Herpesvirus	Eye secretions
Genital infections	*Neisseria gonorrhoeae*, herpesvirus	Genital secretions
Skin infections	*Streptococcus* spp.	Pus
Anthrax	*Bacillus anthracis*	Skin secretions
Meningitis	*Neisseria meningitidis*	Cerebrospinal fluid
Acquired immunodeficiency-syndrome (AIDS)	Human immunodeficiency virus (HIV)	Blood, sexual secretions, body fluids
Haemorrhagic fevers	Junin, Lassa, Ebola and Marburg viruses	All bloody products andsecretions
Septicaemia	*Staphylococcus* spp.	Blood

(Continued)

Table 15.2. Potential infections caused by exposure to HCW, causative organisms and transmission vehicles (Continued)

Type of infection	*Examples of causativeorganisms*	*Transmission vehicles*
Bacteraemia	Coagulase-negative *Staphylococcus* spp. (including methicillian-resistant *S. aureus*), *Enterobacter, Enterococcus, Klebsiella* and *Streptococcus* spp.	Nasal secretion, skin contact
Candidaemia	*Candida albicans*	Blood
Viral hepatitis A	Hepatitis A virus	Faeces
Viral hepatitis B and C	Hepatitis B and C viruses	Blood and body fluids
Avian influenza	H5N1 virus	Blood, faeces

Source: Chartier et al. (2014); reproduced with permission from World Health Organization

15.3 WASTE GENERATION RATE

All round technological progress has led to increased availability of health related consumer goods, which have the propensity for production of increased wastes. Waste generation depends on numerous factors such as established waste management methods, type of health-care establishment, hospital specializations, proportion of reusable items employed in health care, and proportion of patients treated on a day-care basis (Prüss et al. 1999). Estimates of health care waste generation depend on the number and types of medical areas, number of beds in use in each medical area and the number of inpatients and outpatients (Cheng et al. 2009). The largest quantities of potentially infectious health care waste are produced typically in surgical, maternity, isolation and medical laboratory areas and emergency rooms. Waste production data are usually expressed as kilogram/occupied bed per day for inpatients and kilogram/appointment per day for outpatients. Total HCW generation from different countries are presented in Table 15.3.

In Hong Kong, more than 2,000 tonnes of clinical wastes are generated every year (HKEPD 2014). More stringent sorting requirements at hospitals implemented since 1992 has paid off as clinical waste requiring special handling in 2000 was about 0.13 kg/bed/day in government hospitals (Ha et al. 2000; Chung and Lo 2003). Since the general wastes are excluded, this value can be considered as the quantity of infectious and hazardous waste closely matching the value of 0.12 kg/bed/day for the year around 2007 as compiled by Emmanuel (2007). Data from 150 health care establishments in Taiwan have indicated that the average waste generation rates ranged from 2.41 to 3.26 kg/bed/day for general medical wastes, and 0.19–0.88 kg/bed/day for infectious wastes (Cheng et al. 2009). Total health care waste in a primary health care clinic may be about 0.1 kg/patient per day and that in a general hospital may be 2 kg/bed per day

Table 15.3. Comparison of hospital waste generation rates in different countries

Country	*Generation rate (kg bed^{-1} day^{-1})*	*Reference(s)*
Bangladesh	0.57–1.28	Sarkar et al. 2006; Alam et al. 2008
Brazil	3.2–4.5	Da Silva et al. 2005
China	1.8–2.2	Shen et al. 2003
Cambodia	0.098	Visvanathan 2008
Greece	1.9	Tsakona et al. 2007
Iran	2.75–4.58	Askarian et al. 2004; Masoumbeigi et al. 2008; Farzadkia et al. 2009; Taghipour and Mosaferi 2009; Bazrafshan and Mostafapoor 2011
Ethiopia	0.361–0.669	Debere et al. 2013
India	0.30–2.31	Patil and Pokhrel 2005; Visvanathan 2008
Indonesia	0.75	Visvanathan 2008
Jordan	0.83	Abdulla et al. 2008
Laos	0.84	Visvanathan 2008
Libya	1.3	Sawalem et al 2009
Mongolia	0.13–0.52	Visvanathan 2008
Norway	3.9	Bdour et al. 2007
Pakistan	1.28–3.47	Pescod and Saw 1998
Palestine	1.86–2.3	Al-Khatib and Sato 2009
Portugal	3.9	Diaz et al. 2008
Spain	4.4	Bdour et al. 2007
Taiwan	2.41–3.26	Cheng et al. 2009
Thailand	1.0	Kerdsuwan 2000
Turkey	0.63	Birpinar et al. 2009
USA	2.59–5.83	Emmanuel 2007
Vietnam	1.42	Diaz et al. 2008

(Rushbrook and Zghondi 2005). Estimates say that the total quantity of medical waste, in a city like Delhi, is less than 60 metric tonne/day as compared to a total municipal waste stream of 5000 metric tonnes. Based on occupancy ratio of the hospitals, 700 g/day waste is reported per bed in hospitals at Ahmedabad. Immunization injections in India itself is estimated to be about 210 million a year, resulting in huge quantities of immunization waste (Singh 2004). Biomedical waste generated in 30 Delhi government hospitals varied from 0.78 kg/bed to 0.30 kg bed^{-1} day^{-1}. Some other hospitals have even much lesser generation rate indicating an overall average of 0.26 kg bed^{-1} day^{-1} (Visvanathan 2008).

The Table indicates that the generation rate of hospital waste differs not only from country to country but also within countries. For example, in India, the generation rate ranged between 0.30 and 2.31 kg/bed/day. The variation in waste generation among hospitals may be attributed to a variety of reasons, such as the type of healthcare establishment, income level, welfare of patients and visitors, diversity of departments (e.g. surgical, general, pediatric), type of hospital in terms of private or public, level of instrumentation and location, hospital specialization, proportion of disposable substances used in healthcare activities and efficiency of segregation of the hazardous hospital waste from the non-hazardous hospital waste stream. It is also reported that the range of generation rate values for countries of similar income levels is probably as wide in high-income countries as in less wealthy countries (Prüss et al. 1999).

15.4 SOURCE REDUCTION AND SEGREGATION

The key to minimization and effective management of health-care waste is segregation (separation) and identification of the waste. Only 10–25% of HCW is of real concern, because of its infectious nature, indicating the importance of "Source reduction and segregation" as the key to proper medical waste management. If this waste is not properly segregated at the source of generation, the whole of the waste gets mixed up, making the whole waste infectious. Appropriate handling, treatment and disposal of waste by type reduces costs and does much to protect public health. By segregation, different categories and types of HCW are sorted and placed in different containers, enabling minimizing the amount of waste to be treated besides more efficient treatment of each category of waste.

The management of infectious waste is the responsibility of the health care institutions generating the same. The person who produces a waste item should have the knowledge to decide if it is a hazard risk or non-risk item and must have the discipline at all times to place it in the correct container. Rigorous measures for source reduction and segregation can substantially reduce the risks and ensure the safety of staff, patients, general public and the environment. Minimizing the amount of waste generated at the source itself is possible through product substitution, technology change and good operating practices. Changes in purchasing policies and product substitution can also reduce the toxicity of the waste generated. For example, mercury based thermometer can be substituted by electronic sensing devices.

Waste should be identified and segregated at the point of generation. Non-hazardous waste, such as paper and cardboard, glass, aluminum and plastic, should be collected separately and recycled. Food waste should be segregated and composted. Infectious and/ or hazardous wastes should be identified and segregated according to its category using a color-coding system. The colour coding of bins may vary according to national legislation. Typically,

- Yellow bins are used for infectious waste, like anatomical, pathological, soiled dressings contaminated with blood or other body fluids.
- Red bins are used for infected plastics like IV sets, tubing, catheters microbiological waste etc.

- Blue Glass/Stainless steel or translucent puncture proof container are recommended for metal sharps.
- Black bins are used for cytotoxic drugs, incinerator ash, chemical waste and expired medicines.
- White bins may be used for all sorts of non-infected general waste which are dry and recyclable.
- Green bins are suggested for collecting and transporting biodegradable wastes such as food waste from wards, canteens and dining halls.

Mercury is to be stored separately in sealed and impermeable containers in a secure location. Containers should be kept covered to prevent contact with the open air. Sharps and potentially infectious waste should be kept in separate containers, with well-fitting lids, either removable by hand or preferably operated by a foot pedal, in each medical area. Containers should be clearly labeled and clear signs placed on containers and bags to differentiate between general and hazardous health care waste. Practices such as having yellow bags in black bins – should be avoided, because it will increase the potential for confusion and poor segregation. A minimum of one removal from each storage point should be provided in every working shift. Thus scientific segregation practices require investments for the purchase of bins, polythene bags, hypochlorite solution, wheel barrows etc., and the recurring cost should be kept in mind while planning for waste management. It is essential that they insist on the use of clean gloves and uniforms, along with other protective clothing, face masks, or eye protection, as appropriate, by the employees to provide protection against exposure to infectious waste. It is important to keep in mind that sustaining staff motivation is threatened if segregated wastes are re-mixed during temporary storage and final treatment and disposal.

Albany Medical Center, a 500-bed research hospital in New York has recycled 7,273 tons of waste and saved the hospital $4 million in six years by recycling 43% of its total waste stream (HCWH 2001). In addition to a host of typical items it recycles, such as paper, cardboard and steel cans, the Center is also able to recycle five different types of waste chemicals into usable products through the use of a chemical distillery that can convert waste alcohol, formalin, xylene, mineral spirits and paint into pure products for use in its laboratories, reducing the waste production of hazardous chemicals from 29 tonnes to 6 tonnes and save $250,000 per year in disposal and chemical purchasing costs.

Holy Family Hospital, a 300-bed multi-specialty hospital in New Delhi, segregates its plastic waste from patient care into eight categories. Recyclers appreciate this practice as it reduces their work and they collect the different kinds of plastic, paying the hospital a good price. The hospital also reuses discarded bed sheets by cutting out the unusable portion and converting the remaining sheet into nappies to be used in its nursery, or for cloth dusters. A 600-bed multi-specialty hospital in Delhi, through an effective mercury spill management programme, has collected 1.6 kg of mercury (spilt due to breakage of thermometers and sphygmomanometers) and returned it to a thermometer manufacturing unit (Patil and Pokhrel 2005).

15.4.1 Transport

The HCW collected for its treatment and disposal at the site should be packed as per the regulations and transported to its treatment and disposal site. For this purpose, waste is collected and segregated according to its specifications and stored for further treatment. The transport of these wastes can be classified into on-site and off-site transport. It is essential that hazardous and non-hazardous wastes should always be transported separately. In addition, infectious waste should be transported separately from hazardous waste, to prevent the possible spread of infectious agents.

15.4.1.1 On-Site Transport

Health-care waste should be transported within the hospital or other facility by means of wheeled trolleys, containers, or carts that are not used for any other purpose and meet the following specifications,

- easy to load and unload;
- no sharp edges that could damage waste bags or containers during loading and unloading and
- easy to clean.

The vehicles should be cleaned and disinfected daily with an appropriate disinfectant. All waste-bag seals should be in place and intact at the end of transportation.

15.4.1.2 Off-Site Transport

The health-care waste producer is responsible for safe packaging and adequate labeling of waste to be transported off-site and for authorization of its destination. Packaging and labelling should comply with national regulations governing the transport of hazardous wastes, and with international agreements if wastes are shipped abroad for treatment.

In general, the waste should be packaged according to the recommendations, in sealed bags or containers, to prevent spilling during handling and transportation. The bags or containers should be appropriately robust for their content (puncture-proof for sharps, for example, or resistant to aggressive chemicals) and for normal conditions of handling and transportation, such as vibration or changes in temperature, humidity, or atmospheric pressure. In addition, radioactive material should be packed in containers whose surfaces can be easily decontaminated. The United Nations recommend further packing requirements for infectious substances. For infectious health-care wastes, it is recommended that packaging should be design type-tested and certified as approved for use. Health-care wastes that are known or suspected to contain pathogens likely to cause human disease should be considered as “Infectious Substances” and should comply with the packaging requirements. The packaging recommended for most health-care wastes, with a relatively low probability that infectious substances are present and which are not likely to cause human disease, is simpler.

15.4.1.3 Labelling

All waste bags or containers should be labelled with basic information on their content and on the waste producer. This information may be written directly on the bag or container or on preprinted labels, securely attached. For health-care waste, the following additional information should be marked on the label:

- Waste category
- Date of collection
- Place in hospital where produced (e.g. ward)
- Waste destination.

In case of problems involving questions of liability, full and correct labeling allows the origin of the waste to be traced. Labeling also warns operative staff and the general public of the hazardous nature of the waste. The hazards posed by container contents can be quickly identified in case of accident, enabling emergency services to take appropriate action.

15.4.1.4 Transportation Vehicles or Containers

Before transportation of the waste, dispatch documents should be completed, all arrangements should be made between consignor, carrier, and consignee, and, in case of exportation, the consignee should have confirmed with the relevant competent authorities that the waste can be legally imported and that no delays will be incurred in the delivery of the consignment to its destination.

Waste bags may be placed directly into the transportation vehicle, but it is safer to place them in further containers (e.g. cardboard boxes or wheeled, rigid, lidded plastic or galvanized bins). This has the advantage of reducing the handling of filled waste bags but results in higher disposal costs. These secondary containers should be placed close to the waste source. Any vehicle used to transport HCW should fulfill the following design criteria:

- The body of the vehicle should be of a suitable size commensurate with the design of the vehicle, with an internal body height of 2.2 metres.
- There should be a bulkhead between the driver's cabin and the vehicle body, which is designed to retain the load if the vehicle is involved in a collision.
- There should be a suitable system for securing the load during transport.
- Empty plastic bags, suitable protective clothing, cleaning equipment, tools, and disinfectant, together with special kits for dealing with liquid spills, should be carried in a separate compartment in the vehicle.
- The internal finish of the vehicle should allow it to be steam-cleaned, and the internal angles should be rounded.

- The vehicle should be marked with the name and address of the waste carrier.
- The international hazard sign should be displayed on the vehicle or container, as well as an emergency telephone number.

Vehicles or containers used for the transportation of health-care waste should not be used for the transportation of any other material. They should be kept locked at all times, except when loading and unloading. Articulated or demountable trailers temperature-controlled if required) are particularly suitable, as they can easily be left at the site of waste production. Other systems may be used, such as specially designed large containers or skips; however, open-topped skips or containers should never be used for transporting health-care waste. Where the use of a dedicated vehicle cannot be justified, a bulk container that can be lifted on to a vehicle chassis may be considered. The container may be used for storage at the health-care establishment and replaced with an empty one when collected. The finish of these bulk containers should be smooth and impervious and permit easy cleansing or disinfection. The same safety measures should apply to the collection of hazardous HCW from scattered small sources. Health-care establishments that practice minimal programmes of health-care waste management should either avoid off-site transportation of hazardous waste or at least use closed vehicles to avoid spillage.

Refrigerated containers may be used if the storage time exceeds or transportation times are long. The infectious waste should be kept cool or refrigerated at preferably in a range of 3°C to 8°C if stored for more than a week. If refrigerated storage room is not available, storage times (the time gap between generation and treatment) should not exceed 72 and 48 h for temperate climate and 48 and 24 h for warm climate in winter and summer seasons, respectively (Chartier et al., 2014).

15.4.1.5 Routing

Health-care waste should be transported by the quickest possible route, which should be planned before the journey begins. After departure from the waste production point, every effort should be made to avoid further handling. If handling cannot be avoided, it should be pre-arranged and take place in adequately designed and authorized premises. Handling requirements can be specified in the contract established between the waste producer and the carrier.

15.4.2 Risk Involved During Healthcare Waste Management

In many developing countries, HCW seldom receives due attention and often handled as part of the municipal waste stream. Physicians, healthcare staff and workers often do not realize the flipside of their negligence and the disorder it causes later. Nevertheless, awareness of the perceived and real problems of handling and disposal of HCW is now increasing. While procedures for proper HCW management are lavishly discussed, not much thought has gone into disseminating the health risk arising from poor management of HCW. A rapid survey of literature can prove this beyond argument. Mishandling of HCW can turn into a threat to public and the environment at large if the most common pathways of infection such as direct contact, contact through vectors, airborne transmission and water sources are not obstructed. Nonetheless, the risks due to improper HCW management differ between population groups. Specific groups coming into direct contact with HCW include waste pickers, waste recyclers, drug addicts (who scavenge for used needles and disposed medicines); hospital sweepers and other low-grade hospital staff. They can be affected either through direct contact with waste at work every day or through dependency on existing HCW for scavenging it to generate income for livelihoods. Healthcare workers, HCW workers and municipal waste workers are usually underprivileged and have little education thus turning to waste picking as an additional revenue generating option. This group of people should be provided with protective equipment and receive basic training. On the other hand, very low-income families, such as those who depend on waste picking, living in squatter settlements or even within a waste dump are prone to high risk. Table 15.4 presents the health risks at various stages of contact with HCW.

15.4.2.1 Risks during Handling and Transportation

In most cases, the concentration of hazardous chemicals present in clinical waste is generally too low to be considered an occupational problem or a danger to the public. Physical injuries caused by discarded sharps among clinical waste are a more significant risk as they may directly contribute to the transmission of microbial infectious agents. Clinical waste components may also pose microbiological risks by creation of infectious aerosols, by blood splashes during handling or by airborne endotoxins from microorganisms in the waste. The major risks during handling and transportation of clinical waste would be the physical injuries caused by discarded sharps and microbiological risks by creation of infectious

Table 15.4. Health risks from HCW Management

Occupational	*Recycling*	*Environmental*	*Accidental*
Nosocomial Infections	*Type of Infections*	*Environmental Pollution*	*Pressurized containers and Radioactive Isotopes*
HIV	Respiratory	Carbon Monoxide	Explosion
Hepatitis B	Gastroenteric	Hydrocarbons	Injuries
Tuberculosis	Skin	Airborne pathogens	Radiation
Tetanus	Ocular	Pathogenic organisms	Epidemics
Pneumonia	Viral		
Cholera		Toxic Chemicals	

Source: Visvanathan 2008; reproduced with permission from Asia Institute of Technology, Thailand aerosols, by blood splashes during handling or by airborne endotoxins from microorganisms in the waste.

A Clinical Waste Management Plan should be developed by the waste producers to set out the detailed requirements and procedures for proper handling of clinical waste for reference by the staff. A responsible person, who should have relevant experience and appropriate training, should be identified for coordinating the various activities relating to clinical waste management.

15.4.2.2 Risks during Waste Treatment

Health risks may be generated through the release of toxic pollutants during waste treatment. Incineration is a key process in the treatment of hazardous wastes and clinical wastes. It is often imperative that clinical waste be subjected to the high temperatures of incineration to destroy pathogens and toxic contaminants it contains. Where waste acceptance or pre-acceptance procedures are poor, and/or containment is inadequate or unproven, this assumption cannot be made. The onus is therefore on the operator to demonstrate that emissions from the plant are controlled during both commissioning and more importantly during routine operation. Clinical wastes incinerators are an acknowledged source of persistent organic pollutants (dioxin, PAH, PCB) and mercury emission (Sawyer et al. 1993; Salkin 2004; Blenkharn 2005; Lovoie et al. 2006). However, modern clinical waste incinerators operated under good practice with appropriate containment should not cause any adverse impact on the environment and public health because of the incorporation of appropriate air pollution control equipment to meet the stringent air emission standards.

The potential emissions from clinical waste incineration plants include physicochemical pollutants such as total suspended particulates (TSP), respirable suspended particulates (RSP), sulphur dioxide, nitrogen oxides, heavy metals and dioxins. Since Dioxins is the most toxic and most persistent among all potential organic contaminants in the flue gas, it is recommended to monitor its presence in the flue gas to indicate the effectiveness of the incineration process for the removal of organic contaminants.

In addition, bioaerosols containing viable microbes could also pose risk. Reception, storage and handling of clinical waste at the treatment site may cause the spreading of bioaerosols generated from the waste, and therefore pose potential risk to human health. Monitoring of key indicators including, at least total bacteria and total fungi is therefore necessary to evaluate the impact of incineration of HCW at treatment site on the ambient air quality.

15.5 TREATMENT AND DISPOSAL

Poor HCW management results in adverse effects on the environment and the public health. The most commonly noted issues in HCW management are in terms of safe disposal of infectious and sharp waste, occupation safety for workers handling the waste and illegal scavenging of salvageable materials. In 2002, the results of a WHO assessment carried out in 22 developing countries showed that around 18–64% of the healthcare facilities do not use appropriate disposal methods. Hence, effective and efficient treatment and disposal of such waste receives a lot of attention worldwide because of the potential environmental hazards and public health risks.

Treatment involves processing of wastes that changes the character of hazardous/infectious waste to render them less hazardous or non-hazardous so as to protect public health through the destruction or isolation of hazardous HCW from people, grazing animals and disease vectors. The common treatment and disposal methods utilized in the management of infectious HCWs are combustion technologies, autoclave and microwave disinfection systems, chemical disinfection, encapsulation by means of immobilizing compounds and disposal on land (UNEP 2003). In most of the developing countries infectious, pathological wastes and sharps are incinerated / open burned and the bottom ash is disposed along with the municipal waste while the flyash is disposed as hazardous waste.

Treatment of HCW should depend upon nature of the waste, volume of waste and technology (i.e. technologically viable, environmentally safe and meeting the regulatory standards at an affordable cost). The critical problem of medical waste is primarily of disinfecting the waste than destroying it With respect to risk reduction associated with microorganisms, the most important objective is to destroy them prior to release of the contaminated material into the environment. Pathogen destruction can be achieved through proper thermal, chemical, and irradiative treatment.

All the techniques for the treatment of infectious waste have both advantages and disadvantages. The choice of technique should not be based on economic characteristics but on safety characteristics that will enable reliable care of both human health and the environment. Proper management of medical waste requires the careful separation of the hospital waste stream, as a different treatment technique is expected to be suitable for each fraction (Bdour et al. 2007; Komilis et al. 2011). The final choice of treatment system should be made carefully on the basis of various factors, many of which depend on local conditions (Prüss et al. 1999). The factors to consider are:

- Disinfection efficiency
- Health and environmental considerations
- Volume and mass reduction
- Occupational health and safety considerations
- Quantity of wastes for treatment and disposal/capacity of the system
- Types of waste for treatment and disposal
- Infrastructure requirements

- Locally available treatment options and technologies
- Options available for final disposal
- Training requirements for operation of the method
- Operation and maintenance considerations
- Available space, location of the treatment site and disposal facility
- Investment and operating costs
- Public acceptability and regulatory requirements.

Shredding of health-care waste before disinfection is essential to increase the extent of contact between waste and disinfection by increasing the surface area and eliminating any enclosed spaces. It is also required to render any body parts unrecognizable and to avoid any adverse visual impact on disposal. If treatment or disposal is at off-site locations then a system of transporting the waste is required. Protective gloves, goggles and overalls should be provided to transport staff who loads wastes to reduce further the risk of infection and injury. Transport and treatment systems should aim to be large enough to handle each day's HCW production, and disposal sites should be big enough to provide a minimum of 6 months disposal capacity.

15.5.1 Incineration

Incineration is the process of burning waste material in the presence of oxygen. Airflow in excess of the stoichiometric requirement for combustion is essential for the incinerator to enhance the combustion process. The most reliable and commonly used treatment process for HCW is double –chamber incineration. The incinerator of this kind mainly comprises of a chamber, where the waste is thermally decomposed through an oxygen deficient, medium-temperature combustion process (800–900°C), producing solid ashes and gases. The chamber includes a fuel burner, used to start the process. The waste is loaded in suitable bags or containers. The gases produced in this way are burned at high temperature (900–1200°C) by a fuel burner in the post-combustion chamber, using an excess of air to minimize smoke and odours. Their ashes will contain less than 1% unburnt material, which can be disposed of in landfills.

Incinerators have significant air emission control problems if operated at low temperatures (400–500°C). The demand for excess airflow limits the achievable temperature and the insufficient temperature in the process chamber of the incinerators can produce extremely toxic products like furans and dioxins if the waste contains chlorinated substances like PVC plastics such as blood bags, urine bags, IV tubes and syringes, or any other chlorine containing material like bleached paper. In addition, waste to be incinerated should not be chemically treated with any chlorinated disinfectants. Potentially hazardous HCW components such as chemicals like mercury in thermometers, pharmaceuticals from cytotoxic drugs or materials contaminated with cytotoxics, radioactive waste and pressurized containers and gas cylinders shall not be treated by thermal methods.

A HCW incinerator may release a wide variety of pollutants including the most known dioxins and furans, heavy metals (such as lead, mercury, and cadmium), TSP, RSP, acid gases (hydrogen chloride and sulfur dioxide), carbon monoxide, and nitrogen oxides. Dioxins have been linked to cancer, immune system disorders, diabetes, birth defects, and other health effects. Suitably designed pollution control devices should be installed/retrofitted with the incinerators to achieve the prescribed emission limits. Minimum height of the stack for the incinerator should be 30 m above the ground; and only low sulfur fuels shall be used as fuel in the incinerator (MoEF 2000).

Case Study from Hong Kong Incineration Plant. In Hong Kong, most clinical waste were disposed of at landfills prior to 2011; whereas a small amount comprising human and animal tissues is incinerated at the existing hospital pathological waste incinerators and at the crematoria managed by the Food and Environmental Hygiene Department. The existing disposal method, though safe and proper, is not entirely satisfied the requirements in the longer run for Hong Kong. Therefore, the incineration facility at the Chemical Waste Treatment Centre (CWTC) was upgraded to treat the medical wastes and begun operation in 2011. Emission control measures through installation of effective pollution control devices adopted at the CWTC include: (i) the rotary kiln operates at above 1,000°C; (ii) The secondary combustion chamber operates at 1,100–1,250°C which can retain the flue gas for more than 2 seconds so as to disintegrate harmful substances such as dioxins in the flue gas; and (iii) The flue gas is cooled abruptly to below 200°C to prevent dioxins reformation.

In order to further enhance the environmental performance of CWTC, upgrade works on air pollution control system through installation of sulphur and nitrogen oxides removal facilities were also installed to meet the latest European Union (EU) air emission standards. A Selective Non-Catalytic Reduction (SNCR) process was installed at the CWTC to improve the nitrogen oxides removal efficiency. The efficiency of SNCR in removing the nitrogen oxides depends on the system design, operating conditions and catalytic performance. The normal typical nitrogen oxides reduction efficiency of SNCR systems is 30–60%, which is sufficient to satisfy the required reduction efficiency of 33%. For the removal of acid gases and sulfur oxides, the Incinerator Flue Gas Cleaning System (IFGCS) was upgraded. In the upgrade works, a new spray dryer absorber with a new atomizer, a new LAB-Loop dray reactor and a new bag house filter were installed to improve the efficiency of reagent administration and reaction to remove the pollutants in the flue gas. The control strategies for mercury and other heavy metals include (i) the injection of activated carbon at the upstream of the spray drier, and (ii) controlling the feeding of wastes by the installation of on-line mercury monitoring system to the incineration system. This should be adequately enough to prevent the potential exceedance of heavy metal emissions.

The effectiveness of the operation procedures, including handling and transportation, disposal of the clinical wastes generated from the clinical waste producers to the CWTC for treatment was monitored. In addition, effectiveness of pollution control mechanism during the incineration of the clinical waste at the

CWTC by determining levels of pollutants in stack emissions and ambient air at the site and nearby area were also determined in a study during 2011-2012 that included a baseline study, and subsequent monitoring for six months (Wong 2012). Bioaerosol levels in the ambient air samples were also monitored and assessed for possible biohazard associated with the incineration of clinical waste. The summarized data of stack emission for various physicochemical pollutants are presented in Tables 15.5 and 15.6.

The concentrations of all the key physicochemical pollutants (including particulate matter, sulphur dioxide, nitrogen oxides, mercury, toxic metal group I and group II and dioxins) under the monitoring period were found significantly lower than their respective control limits set for the CWTC by Hong Kong EPD of the HKSAR Government as well as the European Union (EU) air emission standards as stated in the Directive 2000/76/EC of the European Parliament and of the Council (2000). In most cases, the average results under the 6 months' monitoring were below or close to the values observed during the baseline monitoring. For those results which were higher than the baseline values, they were well below their control limits. Therefore, the monitored results indicated that incineration of clinical waste at the CWTC did not increase the levels of selected key physicochemical pollutants in the stack emission. Similarly, the concentrations of monitored physicochemical pollutants in the ambient air of the surrounding 11 sites were also complied with requirements of the recommended guideline values, where applicable.

The results of the fungal and bacterial populations in the bioaerosols of the ambient air are summarized in Table 15.7. During this monitoring programme, the total fungal population was within the low or at the low side of the intermediate guideline value of American Conference of Governmental Industrial Hygienists (ACGIH) (Gots et al. 2003); while the total bacterial population was below the guideline value of Hong Kong Indoor Air Quality Objectives (HKIAQO) for excellent air quality. Some of the monitoring results were higher than the baseline values, but they were still within the guideline values. Therefore, the incineration of clinical waste at the CWTC had no direct effect on both fungal and bacterial populations in the surrounding environment.

Figure 15.3 and 15.4 show the comparisons of the levels of total bacteria and total fungi in this study and others reports and research papers. To clearly indicate the differences, the Log_{10} values (lg) of the concentrations of bacteria and fungi were used to draw the figures. The total bacteria in ambient air inside the CWTC fell in the middle range reported in the literatures while that for sites outside the CWTC fell in the low range. The concentrations of total fungi in ambient air both inside and outside the CWTC are in the low range reported in the literatures.

In conclusion, the monitoring results of this assessment indicated that the incineration of clinical waste at the CWTC did not increase the pollutants level in the stack emission or pose any adverse effects on the ambient air quality of adjacent areas. This study indicates that if properly managed with appropriate pollution control mechanisms then incineration can be an effective method to treat the HCW.

Table 15.5. Range and average values of particulate matter, sulphur dioxide, and nitrogen oxides concentrations in the stack emission

	Particulate matter(mg/dscm)[c]		*Sulphur dioxide(mg/dscm)*[c]		*Nitrogen oxides(mg/dscm)*[c]	
Monitoring data	*Daily Average*	*Range of half-hourly average*	*Daily Average*	*Range of half-hourly average*	*Daily Average*	*Range of half-hourly average*
Range	0.037–5.2	0.007–10.7	0.0002–19.0	0.0–166.1	93.5–157.1	54.6–301.2
Average	3.06	1.81–4.97	5.04	0–53.3	131.4	86.8–212.5
Baseline study	4.4	3.2–5.7	19.5	3.8–85.4	124	81–183
HK[a]/EU[b] Emission standards[a]	10	30	50	200	200	400

[a]Maximum emission standards set for the CWTC by HKEPD
[b]European Parliament and of the Council 2000
[c]Concentrations of pollutants in stack emission are standardized and expressed for dry standard cubic meter (dscm) containing 11% O_2

Table 15.6. Concentrations of mercury, toxic metal group I, toxic metal group II, and dioxins in the stack emission

Monitoring data	*Mercury (mg/dscm)*[c]	*Toxic metal group I*[c,d] *(mg/dscm)*	*Toxic metal group II*[c,e] *(mg/dscm)*	*Dioxins (ng I-TEQ/dscm)*[c]
Range	<0.0005–0.0015	BDL–0.0040	0.0026–0.1929	0.007–0.035
Average	0.0012	0.0021	0.0411	0.020
Baseline data	0.0003	BDL	0.1823	0.0025
HK Emission standards[a]	0.05	0.05	0.5	0.075
EU Emission standards[b]	0.05	0.05	0.5	0.1

[a]Maximum emission standards set for the CWTC by HKEPD: Average value over the sampling period of a minimum of 30 minutes and a maximum of 8 hours
[b]Maximum emission standards of the Annex V of the European Parliament and of the Council 2000.
[c]Concentrations of pollutants in stack emission are standardized and expressed for dry standard cubic meter (dscm) containing 11% O_2
[d]Toxic metal group I includes cadmium, thallium and their compounds.
[e]Toxic metal group II includes antimony, arsenic, chromium, cobalt, copper, lead, manganese, nickel, vanadium and their compounds

Table 15.7. Comparison of bioaerosols at different types of sampling sites in baseline monitoring programme

Monitoring data	*Total fungi (CFU/m*^{3}*)*	*Total bacteria (CFU/m*^{3}*)*
Range of values for sites in CWTC	21–133	19–393
Range of values for sites adjacent to the CWTC	7–262	4–302
Baseline for all sites	131–283	26–240
Guideline value	100-1000	500-1000

Sources: Total fungi, ACGIH 1995, Gots et al. 2003; Total bacterial, HKIAQO 2003

15.5.2 Non-Incineration Methods

Non-incineration type thermal treatment methods such as autoclaves and hydroclaves are commonly used to sterilize waste, which at a later stage may be burnt in an incinerator. These techniques employ steam sterilization with direct and indirect heating methods. Microwave disinfection is another non-incineration process that is used for sterilization (disinfection) of medical waste. Subsequently, the sterilized waste is either dumped along with the municipal waste or is treated

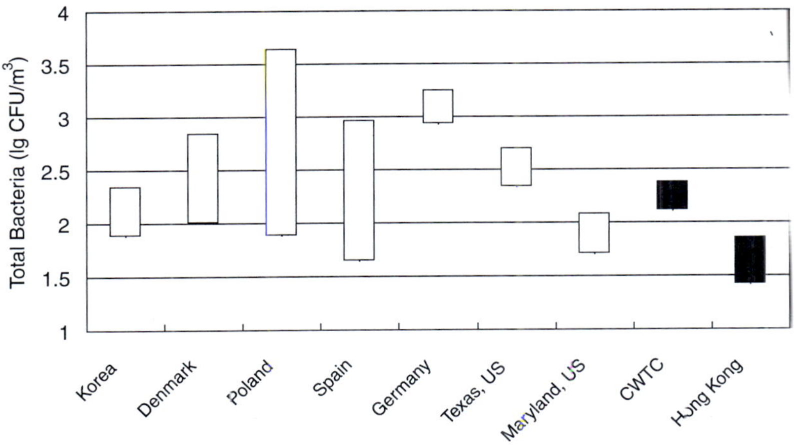

Figure 15.3. Comparison of the ranges of total bacteria observed in this study and others reports and research papers. The two columns in the right side indicate the range of total bacteria in the ambient air at sampling sites inside the CWTC (CWTC) and outside the CWTC (Hong Kong)

Sources: Jones and Cookson 1983; Nielsen et al. 1995; Pastuszka et al. 2000; Herr et al. 2003; Mota et al. 2008; Vilavert et al. 2009

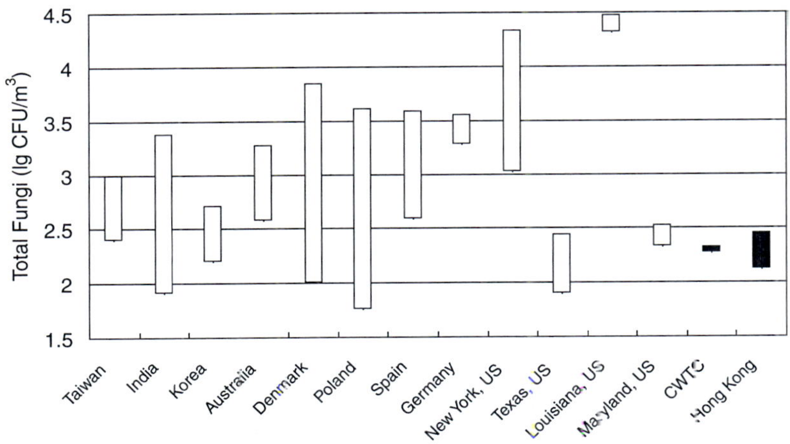

Figure 15.4. Comparison of the ranges of total fungi observed in this study and others reports and research papers. The two columns in the right side indicate the range of total fungi in the ambient air at sampling sites inside the CWTC (CWTC) and outside the CWTC (Hong Kong)

Source: Jones and Cookson 1983; Li and Kuo 1992; Nielsen et al. 1995; Pastuszka et al. 2000; Hargreaves et al. 2003; Herr et al. 2003; Adhikari et al. 2004; Solomon et al. 2006; LeBouf et al. 2008; Mota et al. 2008; Vilavert et al. 2009 by incineration. The major disadvantage of the non-incineration process is that the reduction of waste volume is not significant. In general, it is not advisable to treat large body parts, animal carcasses, or other large items in autoclaves or microwave disinfection systems as it require more time for the entire material to reach the prescribed temperatures.

15.5.2.1 Autoclave

Autoclaving is the process of steam disinfection wherein, the waste is exposed to steam at high pressure and temperature to decontaminate the infectious waste. Autoclaves use moist heat (steam) to achieve disinfection (HCWH 2004). The steam required is generated by means of boilers heated by means of conventional fuels such as furnace oil, gas, diesel, coal, or biomass. An autoclave essentially is a steel vessel hermetically sealed with a hinged door (with gaskets) and designed to resist high temperatures and pressures. Normally, a steam jacket surrounds the vessel to reduce the amount of condensation on the inside wall of the vessel and thus reduce heat loss. Disinfection in an autoclave is carried out in batches. The entire process from loading and disinfection to unloading is called a cycle. Air generally is removed from the autoclave by using a vacuum pump at the beginning of the cycle and prior to steam injection.

The HCW loaded bins or carts are introduced into the autoclave and the door is sealed. At this point, air is evacuated from the unit either by vacuum or by introducing steam into the vessel. Steam is forced into the unit until the required temperature is achieved. Steam is added as necessary in order to maintain a prescribed temperature for a given period of time. Upon reaching the necessary time - temperature requirements, the vents in the unit are opened and the steam is released through a condenser so that the water can be re-used. Once the waste is cool enough to be handled, the material is removed and taken to the disposal site or treated mechanically (size reduction or compaction or both). Size reduction essentially changes the appearance of the waste and increases the bulk density of the mixture, thereby reducing the volume that the mass would occupy 60–80%. When operating a gravity flow autoclave, medical waste shall be subjected to:

(i) a temperature of not less than 121°C and a pressure of 15 pounds per square inch (psi) for an autoclave residence time of not less than 60 minutes; or

(ii) a temperature of not less than 135°C and a pressure of 31 psi for an autoclave residence time of not less than 45 minutes; or

(iii) a temperature of not less than 149°C and a pressure of 52 psi for an autoclave residence time of not less than 30 minutes.

The autoclave should completely and consistently kill the approved biological indicator at the maximum design capacity of each autoclave unit. Biological or chemical indicators are inserted in waste loads in order to evaluate the degree of disinfection. Usually, the spores of *Bacillus stearothermophilus* available commercially in disks, strips, or in ampoules are placed in the waste as it is introduced into the autoclave and removed after the process is finished to test the effectiveness of the process. The spores are collected and incubated for about a week. Use of heat sensitive tape or other device for each container is recommended to indicate that the container has been autoclaved. The chemical indicator strip/tape changes colour when a certain temperature is reached and can be used to verify that a specific temperature has been achieved. It may be necessary to use more than one strip over the waste package at different location to ensure that the inner content of the package has been adequately autoclaved.

15.5.2.2 Microwave Disinfection

Microwave disinfection is a steam-based process, where steam is generated by microwave energy to disinfect the waste through the action of moist heat and microwave (HCWH 2004). Microwave disinfection systems use short waves in the electromagnetic spectrum within the range of the radio frequency band generated by means of magnetrons or traveling wave oscillators that convert electrical energy into microwave energy. The high frequency of the microwaves makes the molecules in the receiving body vibrate very rapidly as they attempt to align to the changing electromagnetic field. The very high level of vibration results in friction generating substantial heat. Steam produced from the moisture in the waste by the microwave energy brings about the destruction of the pathogenic organisms in the waste. Microwave systems are designed and built in a variety of sizes, ranging from a few kg per hour to more than 400 kg per hour. The units can be operated as a batch process or in a semi-continuous mode.

Microwave treatment shall not be used for cytotoxic, hazardous or radioactive wastes, contaminated animal carcasses, body parts and large metal items. The microwave should completely and consistently kill the bacteria and other pathogenic organisms that is ensured by approved biological indicator at the maximum design capacity of each microwave unit.

15.5.2.3 Plasma Pyrolysis

Plasma is the state of matter formed by removing the bound electrons from atoms. It is an electrically conducting fluid consisting of charged and neutral particles. The charged particles have high kinetic energies. When the ionized species in the plasma recombine with the stripped electrons, significant amounts of energy in the form of ultraviolet radiation are released. The particle kinetic energy takes the form of heat and can be used for decomposing chemicals. In addition, the presence of charged and excited species renders the plasma environment highly reactive, which can catalyze homogeneous and heterogeneous chemical reactions.

Plasma-arc employing carbon electrodes were first used in 1960 s as a source of intense heat. The short lifetime of the electrodes was a major impediment in the successful development of this technology. With rapid developments in plasma sources, plasma technology is being applied to destroy highly toxic compounds and to modify refractory compounds in an environment-friendly way.

Plasma pyrolysis integrates the thermo-chemical properties of plasma with the pyrolysis process (Nema and Ganeshprasad 2002). Plasma pyrolysis uses extremely high temperatures of plasma-arc in an oxygen starved environment to completely decompose waste material into simple molecules. The most likely compounds that form during plasma pyrolysis of carbonaceous matter are methane, carbon monoxide, hydrogen, carbon dioxide and water molecules. These gases are burned and produce a high temperature (around 1,200°C). In the plasma pyrolysis process, the hot gases are quenched from 500° to 70°C to avoid recombination reactions of gaseous molecules that inhibit the formation of dioxins and furans. The organic mass to gas conversion is more than 99% and it does not require segregation of chlorinated hydrocarbons. High temperature and UV radiation present in the plasma kill bacteria completely.

15.5.3 Chemical Disinfection

Chemical disinfection, used routinely in health care to kill microorganisms on medical equipment and on the floors and walls, is now being extended to the treatment of health-care waste. Chemical disinfection relies on the properties of the chemical agent to inactivate pathological organisms. The effectiveness of a certain chemical agent depends upon temperature, pH, and on the possible presence of other compounds, which can have a negative impact on the effectiveness of the chemical agent. The most resistant microorganisms to chemical treatment include bacterial spores and hydrophilic viruses.

Chemicals used to achieve chemical disinfection include hypochlorites, alcohols, acids, alkalis, phenols, halogens, detergents (including quaternary ammonium compounds) and peroxides. Common disinfectant could be 5.25% sodium hypochlorite. Since chlorine has been demonstrated to be a precursor to the formation of dioxins in combustion, non-chlorine based disinfectants such as aldehydes (such as formaldehyde and glutaraldehyde), calcium oxide, ozone etc. are being preferred. Calcium oxide (lime) is generally applied to HCWs in developing countries as lime added in sufficient quantities raises the pH to 11 or higher and such a high pH creates an environment that inhibits the survival of microorganisms. Sufficiently high concentration of the chemical agent, sufficient time in which the wastes are in contact with the chemical agent and waste with small particle sizes are essential considerations determining the effectiveness of chemical disinfection.

Powerful disinfectants are often hazardous and harmful to skin and mucous membranes. Users should therefore wear protective clothes, including gloves and protective clothes, or goggles. Disinfectants are also aggressive to certain building materials and should be handled and stored accordingly.

15.5.4 Burial in Special Pits

The key point in disposal of the waste by deep burial is to isolate the waste to prevent casual access by people and to destroy the infectious pathogens by natural decomposition. Burial of HCW in special pits is practiced in relatively small healthcare facilities located in isolated (rural) areas where the geological and hydrological conditions are suitable and the area is sparsely populated. The pit should be about 2–3 m deep and approximately 2 m diameter. If possible, the entire pit should be lined with a 30-cm layer of compacted clay or any other suitable low permeability material such that percolation of any liquids into the soil underneath is reduced to a minimum. The top portion of the pit should be slightly elevated and properly sloped to keep surface waters from entering the pit. Ideally, the pit should be covered with a simple, but sturdy removable cover. The entire disposal area should be properly fenced off in order to keep unauthorized personnel or animals from entering into the area.

15.5.5 Training

Potentially infectious and hazardous wastes, generated in healthcare facilities throughout the world, are often managed by poorly educated and untrained laborers, who perform most activities without proper guidance and insufficient protection. Facilities involved in infectious waste management activity shall provide all employees, staff, housekeeping, and janitorial personnel who are involved in the generation and/or handling of infectious waste with infectious waste management training. Some basic training that healthcare facility staff should undergo, in the context of waste management is:

- Knowledge training about different waste streams, their infectiousness, and the intensity of consequences if not correctly handled, etc.
- Skills training on procedures to be followed to prevent injuries to self and others while handling infectious HCW, emergency response in case of spills, occupational health and safety, etc.
- Attitude guidance on approaching noninfectious waste as a resource for recycling, team working with counterparts in organizing recycling schemes within the facility.

Training is intended to make the waste management plan effective and to explain what people should do. Such training shall include, at least, the following:

- Definition and identification of infectious waste being generated at the facility.
- Explanation of the requirements in relation to handling, treatment, transportation, storage, and disposal of infectious waste.
- Assignment of roles and responsibilities for implementation.

Personal responsibilities are assigned for handling waste and reinforced through training to improve hygiene and personal safety. Arrangements shall also be made to educate and raise the awareness of patients and visitors through distribution of brochures, posters and discussion with patients and visitors about the risk of health care waste, improvements being made, safe practices they can do to prevent their exposure and their role in reducing secondary infections in health care establishments.

The Himalayan Institute Hospital Trust (HIHT), a 700-bed multi-disciplinary hospital in Uttaranchal, India, for example, is reported to have successfully proven that effective planning and training can help manage the huge quantity of immunization waste generated at outreach locations (Singh 2004). The rural outreach wing of the HIHT, undertaking immunization activities since 1994 has managed waste generated from approximately 334,920 syringes over a period of eight years. At the outreach clinic, the health workers carry a bucket with yellow liner, to contain the cotton swabs generated, and a sharps container for the injection units. After administering an injection, the injection unit is dropped in the sharps container. Both the containers are brought back to the centre. Cotton swabs are incinerated and contents of sharps containers are autoclaved and shredded. The treated and shredded waste from the hospital is handed over to the waste contractor every alternate day. The contractor takes this waste to his facility and segregates the plastic from the metal. The plastic is dried and sold to recycling plants and the metal waste is deep buried at an identified site.

15.6 CENTRALIZED FACILITIES

Centralized treatment and disposal should be encouraged wherever practicable as it is more economic and relies on a high level of trust and commitment among health care establishments to work together. Central facilities allow health care establishments to share and reduce the cost of achieving highly controlled disposal of potentially infectious health care waste.

A common Bio-Medical Waste Management Treatment Facility (CBWTF) shall have all the facilities like incinerator, autoclave/ microwave/hydroclave, shredder and effluent treatment plant, etc. Facilities receiving hazardous HCW should have all applicable permits and capacity to handle specific types of HCW. Wastes from each category should be treated according to the prescribed treatment methods and technologies. When selecting a waste disposal technology, operators should consider other potential health and environmental issues that may be generated by the treatment.

The hospitals/Nursing Homes utilizing the services of this CBWTF can be charged at the rate of per bed after conducting the cost analysis. The operator of such infectious waste storage and/or treatment facility shall have and adhere to a management/operation plan which shall include a method of receiving wastes ensuring that infectious wastes are handled separately from other waste until treatment is accomplished and prevents unauthorized persons from having access to or contact with the wastes. Resources should be ensured and responsibilities (who does what and when) clearly defined. Progress should be monitored and success acknowledged (Rushbrook and Zghondi 2005). A senior manager should be responsible for safe handling and subsequent treatment and disposal of health care waste. Quality control procedures needs to be defined to provide information on achievements made, tracking of waste movements, checking of waste generation rates, incidents and accidents when handling waste and ensuring accountability of supervisors.

15.7 LEGAL AND INSTITUTIONAL FRAMEWORK

Establishment of a sustainable health care waste management system benefits from a national legal framework that regulates and organizes different elements of a health care waste management system. Five guiding principles commonly adopted in waste-related laws are the polluter pays, precautionary, proximity, duty of care, and cradle to grave control (Rushbrook and Zghondi 2005).

Realizing the hazards of improper treatment of hazardous HCW, the Government of India enacted the Bio-Medical Waste (Management and Handling) Rules (MoEF 1998). The Rule applies to all persons who generate, collect, receive, store, transport, treat, dispose, or handle bio medical waste in any form. The Rule places clear guidelines on the segregation, packaging, transportation and storage of bio-medical waste in conformance to internationally accepted standards. Standards have been prescribed to protect consumers, employees, patients and the environment. Similar regulatory frameworks exist in many countries. The law include a clear definition of hazardous HCW and its various categories. For example the Indian Legislation has classified the biomedical wastes into 10 Categories and proposed a waste treatment option for these wastes (Table 15.8).

Apart from providing detailed guidelines on the categories of bio-medical waste, the law should also recommend treatment options that are to be adhered. Key aspects covered in the Indian Rules are:

- Colour Coding and Type of Container for Disposal of Bio-Medical Wastes
- Standards for Treatment and Disposal of Bio-medical Wastes
- Standards for Incinerators (Operation and Emission)
- Standards for Waste Autoclaving (operational parameters, Spore testing, Routine Test)
- Standards for Liquid Waste
- Standards of Microwaving
- Standards for Deep Burial

Specific requirements for record-keeping and reporting, details for an inspection system to ensure enforcement of the law, and for penalties to be imposed for contravention are also included in the legal framework. The legal framework shall ensure that the HCW is segregated and clearly labelled at source. A system of colour code for containers and types of waste should be followed. It shall be mandatory that the biohazard symbol must be clearly seen on the bins and facilities handling HCW. Containers should be removed when they are three-quarters full. Waste must be transported away from the areas of generation at regular intervals or every morning. No untreated bio-medical waste shall be stored

Table 15.8. Biomedical waste classification and treatment methods as per Indian Regulatory Framework

Waste category	*Waste description*	*Waste treatment options*
Category 1	Human anatomical wastes (human tissues, organs, body parts)	Incineration/deep burial
Category 2	Animal Wastes (animal tissues, organs, body parts, carcasses, bleeding parts, fluid, blood and experimental animal used in research, waste generated by veterinary hospitals, colleges, and discharge from hospitals)	Incineration/deep burial
Category 3	Microbiology and biotechnology waste(wastes from laboratory cultures stocks or specimens of microorganisms live or attenuated vaccines, human and animal cell culture used in research and infectious agents from research and industrial laboratories, wastes from production of biologicals, toxins, dishes and devices used for transfer of cultures)	Incineration/deep burial
Category 4	Waste sharps (needles, syringes, scalpels, blades, glass, etc., that are capable of causing puncture and cuts. This includes both used and unused sharps.)	Disinfection by chemical treatment/autoclaving/ microwaving and mutilation/shredding
Category 5	Discarded medicines and cytotoxic drugs (wastes comprising outdated contaminated and discarded medicines)	Incineration/destruction and drugs in secured landfill
Category 6	Soiled waste (items contaminated with blood and body fluids including cotton, dressing, soiled plaster casts, linens, bedding, other material contaminated with blood)	Incineration/Autoclaving/ Microwaving

Table 15.8. Biomedical waste classification and treatment methods as per Indian Regulatory Framework (Continued)

Waste category	*Waste description*	*Waste treatment options*
Category 7	Solid waste (items generated from disposal items other than the waste sharps such as tubing's, catheters, intravenous sets, etc.)	Disinfection by chemical treatment/autoclaving/ microwaving and mutilation shredding
Category 8	Liquid wastes (wastes generated from laboratory washing, cleaning, housekeeping and disinfecting activities)	Disinfection by chemical treatment and discharge into drain
Category 9	Incineration ash (ash from incineration of any bio-medical waste)	Disposal in municipal landfill
Category 10	Chemical wastes (chemicals used in production of biologicals, chemicals used in disinfection, as insecticides, etc.)	Chemical treatment and discharge into drains for liquids and secured landfill

Source: MoEF (1998)

beyond a period of 48 hours. Transport of waste from areas of generation must be done only by designated staff, who are aware of the hazards of the material they handle and of protective measures to be taken. Untreated HCW shall be transported only in specially designed vehicles.

15.8 CONCLUSION

Healthcare or biomedical waste management is a basic public health concern and needs to be integrated from the beginning of a health care project planning. Essential elements include segregation at source, proper collection, safe storage, systematic transportation and environmentally sound processing and disposal. 'All waste generated in a healthcare facility is infectious' is a myth that has to be cracked and a resource respecting practice should be put into action in a HCW management system. Mixing an infectious constituent to the non-infectious constituent would render the whole stream infectious, thus pushing it to be treated with special care. The problem of HCW needs a systematic approach, with investments and training of staff in segregation, waste minimization and safe technologies, as also centralized facilities. About 75% of all HCW can be disposed through regular municipal waste methods or sent for recycling as in the case of other domestic waste. It is the other 25% that are potentially infectious and supposed to create serious health threats to healthcare workers and communities if not disposed through prescribed methods. Without the source segregation and recycling activities in place, the hospital would have to dispose the general waste along with the infectious waste resulting in unwanted disposal costs.

An understanding of the waste category and the volume is vital before one decides on the technology to be adopted. The most commonly proclaimed treatment technology for HCW is incineration, where the waste undergoes combustion under controlled conditions. Since there are several types of situations on the ground and various site-specific requirements, the processes should be done through ground level micro-planning which will also ensure involvement of a larger number of local stake-holders. It should be kept in mind that without a will by the leadership of an institution to make improvements little can be achieved by staff alone.

References

Abdulla, F., Qdais, H. A., and Rabi, A. (2008). "Site investigation on medical waste management practices in northern Jordan." *Waste Manage.*, **28**(2), 450–458.

ACGIH (American Conference of Governmental Industrial Hygienists). (1995). "Guidelines for the assessment of bioaerosols in the indoor environment." Cincinnati, OH.

Adhikari, A., Sen, M. M., Gupta-Bhattacharya, S., and Chanda, S. (2004). "Airborne viable, non-viable, and allergenic fungi in a rural agricultural area of India: A 2-year study at five outdoor sampling stations." *Sci. Total Environ.*, **326**(1-3), 123–141.

Alam, M., Sujauddin, M., Iqbal, A., and Huda, S. (2008). "Report: Healthcare waste characterization in Chittagong medical college hospital, Bangladesh." *Waste Manage. Res.*, **26**(3), 291–296.

Al-Khatib, I. A., and Sato, C. (2009). "Solid health care management status at health care centers in the West Bank-Palestinian Territory." *Waste Manage.*, **29**(8), 2398–2403.

Askarian, M., Vakili, M., and Kabir, G. (2004). "Results of a medical waste survey in private hospitals in Fars province, Iran." *Waste Manage.*, **24**(4), 347–352.

Bazrafshan, E., and Mostafapoor, K. F. (2011). "Survey of medical waste characterization and management in Iran: A case study of Sistan and Baluchestan Province." *Waste Manage. Res.*, **29**(4), 442–450.

Bdour, A., Altrabshen, B., Hadadin, N., and Al-Shareif, M. (2007). "Assessment of medical waste management practice: A case study of the northern part of Jordan." *Waste Manage.*, **27**(6), 746–759.

Birpinar, M. E., Milgili, M. S., and Erdogan, T. (2009). "Medical waste management in Turkey: A case study of Istanbul." *Waste Manage.*, **29**(1), 445–448.

Blenkharn, J. I. (2005). "Safe disposal and effective destruction of clinical waste." *J. Hosp. Infect.*, **60**(4), 295–297.

Chaerul, M., Tanaka, M., and Shekdar, A. (2008). "A system dynamics approach for hospital waste management." *Waste Manage.*, **28**(2), 442–449.

Chartier, Y., et al., eds. (2014). *Safe management of wastes from health-care activities*, 2nd Ed., World Health Organization, Geneva, 308.

Cheng, Y. W., Sung, F. C., Yang, Y., Lo, Y. H., Chung, Y. T., and Li, K. C. (2009). "Medical waste production at hospitals and associated factors." *Waste Manage.*, **29**(1), 440–444.

Chung, S. S., and Lo, C. W. H. (2003). "Evaluating sustainability in waste management: The case of construction and demolition, chemical and clinical wastes in Hong Kong." *Resour. Conserv. Recycl.*, **37**(2), 119–145.

Da Silva, C., Hoppe, A., Ravanello, M., and Mello, N. (2005). "Medical waste management in the South of Brazil." *Waste Manage.*, **25**(6), 600–605.

Debere, M. K., Gelaye, K. A., Alamdo, A. G., and Trifa, Z. M. (2013). "Assessment of the health care waste generation rates and its management system in hospitals of Addis Ababa, Ethiopia, 2011." *BMC Public Health*, **13**(1), 9.

Diaz, L. F., Eggerth, L. L., Enkhtsetseg, S., and Savage, G. M. (2008). "Characteristics of healthcare wastes." *Waste Manage.*, **28**(7), 1219–1226.

Diaz, L. F., Savage, G. M., and Eggerth, L. L. (2005). "Alternatives for the treatment and disposal of healthcare wastes in developing countries." *Waste Manage.*, **25**(6), 626–637.

Emmanuel, J. (2007). "Survey of health care waste characteristics and generation data from different Countries." United Nations Development Programme, Global Environment Facility Global Demonstration Project on Health Care Waste New York.

European Parliament and of the Council. (2000). "Directive 2000/76/EC of the European parliament and of the council." ⟨http://www.central2013.eu/fileadmin/user_upload/Downloads/Document_Centre/OP_Resources/Incineration_Directive_2000_76.pdf⟩ (Apr. 10, 2014).

Farzadkia, M., Moradi, A., and Mohammadi, S. M. (2009). "Hospital waste management status in Iran: A case study in the teaching hospitals of Iran University of Medical Sciences." *Waste Manage. Res.*, **27**(4), 384–389.

Gots, R. E., Layton, N. J., and Pirages, S. W. (2003). "Indoor health: Background levels of fungi." *AIHA J.*, **64**(4), 427–438.

Ha, D., Lei, P., Lio, K. W., and Liu, R. (2000). "Development of clinical waste control scheme in Hong Kong." *Waste Management: The Challenge for Asian Cities—Search for a Sustainable Future, Proc., ISWA Int. Symp. and Exhibition on Waste Management in Asian Cities Hong Kong*, Vol. **1**, The Hong Kong Polytechnic University, Hong Kong SAR, P.R. China, 90–95.

Hargreaves, M., Parappukkaran, S., Morawska, L., Hitchins, J., and He, C. (2003). "A pilot investigation into associations between indoor airborne fungal and non-biological particle concentrations in residential houses in Brisbane, Australia." *Sci. Tot. Environ.*, **312**(1–3), 89–101.

HCWH (Health Care without Harm). (2001). "Going green: A resource kit for pollution prevention in health care." ⟨http://noharm.org/lib/downloads/waste/Recycling_Fact_Sheet.pdf⟩ (Apr. 10, 2014).

HCWH (Health Care without Harm). (2004). "Non-incineration medical waste treatment technologies in Europe." Prague, Czech Republic.

Herr, C. E. W., zur Nieden, A., Jankofsky, M., and Stilianakis, N. I. (2003). "Effects of bioaerosol polluted outdoor air on airways of residents: A cross sectional study." *Occup. Environ. Med.*, **60**(5), 336–342.

HKEPD. (2014). "Clinical waste management." Quantity of Clinical Wastes, Hong Environmental Protection Dept., Hong Kong SAR, P.R. China.

HKIAQO. (2003). "A guide on indoor air quality certification scheme for offices and public places." ⟨http://www.iaq.gov.hk/cert/doc/CertGuide-eng.pdf⟩ (Apr. 9, 2014).

Jones, B. L., and Cookson, J. T. (1983). "Natural atmospheric microbial conditions in a typical suburban area." *Appl. Environ. Microbiol.*, **45**(3), 919–934.

Kerdsuwan, S. (2000). "Case study of using hospital waste incinerator in Thailand." *Proc., 93rd Annual Meeting and Exhibition*, Air & Waste Management Association, Salt Lake City, UT, 18–22.

Komilis, D., Katsafaros, N., and Vassilopoulos, P. (2011). "Hazardous medical waste generation in Greece: Case studies from medical facilities in Attica and from a small insular hospital." *Waste Manage. Res.*, **29**(8), 807–814.

Kühling, J. G. (2014). "Treatment of healthcare waste–myths and truth in the 21st century." *Int. J. Infect. Control*, **10**(1), 1–9.

LeBouf, R., Yesse, L., and Rossner, A. (2008). "Seasonal and diurnal variability in airborne mold from an indoor residential environment in Northern New York." *J. Air Waste Manage. Assoc.*, **58**(5), 684–692.

Li, C. S., and Kuo, Y. M. (1992). "Airborne characterization of fungi indoors and outdoors." *J. Aerosol Sci.*, **23**, 667–670.

Lovoie, J., Dunkerley, C. J., Kosatsky, T., and Dufresne, A., 2006. "Exposure to aerosolized bacteria and fungi among collectors of commercial, mixed residential, recyclable and compostable waste." *Sci. Total Environ.*, **370**(1), 23–28.

Marinkovic, N., Vitale, K., Holcer, N. J., Dzakula, A., and Pavic, T. (2008). "Management of hazardous medical waste in Croatia." *Waste Manage.*, **28**(6), 1049–1056.

Masoumbeigi, H., Karimi-zarchi, A., and Tajik, J. (2008). "Survey of solid waste situation in the specialist hospital in Tehran with emphasis on quantity of waste production." *J. Military Med.*, **9**(2), 129–138.

MoEF (Ministry of Environment & Forests). (1998). "Biomedical waste (management and handling) rules, ministry of environment and forests, government of India." ⟨http://envfor.nic.in/legis/hsm/biomed.html⟩ (Apr. 9, 2014).

MoEF (Ministry of Environment & Forests). (2000). "Environmental standards for ambient air, automobiles, fuels, industries and noise." Central Pollution Control Board, India.

Mota, L. C., Gibbs, S. G., Green, C. F., Payan, F., Tarwater, P. M., and Ortiz, M. (2008). "Characterization of seasonal indoor and outdoor bioaerosols in the arid environment of El Paso, Texas." *J. Environ. Health*, **70**(10), 48–53.

Nema, S. K., and Ganeshprasad, K. S. (2002). "Plasma pyrolysis of medical waste." *Curr. Sci.*, **83**(3), 271–278.

Nielsen, E. M., Nielsen, B. H., and Breum, N. O. (1995). "Occupational bioaerosol exposure during collection of household waste." *Ann. Agric. Environ. Med.*, **2**, 53–59.

Pastuszka, J. S., Paw, U. K. T., Lis, D. O., Wlazlo, A., and Ulfig, K. (2000). "Bacterial and fungal aerosol in indoor environment in Upper Silesia, Poland." *Atmos. Environ.*, **34**(22), 3833–3842.

Patil, G. V., and Pokhrel, K. (2005). "Biomedical solid waste management in an Indian hospital: A case study." *Waste Manage.*, **25**(6), 592–599.

Pescod, M. B., and Saw, C. B. (1998). "Hospital waste management in four cities: A synthesis report." Urban Waste Expertise Programme, WASTE, Nieuwehaven, Netherlands.

Prüss, A., Giroult, E., and Rushbrook, P. (1999). "Safe management of waste from health-care activities." World Health Organization, Geneva.

Rushbrook, P., and Zghondi, R. (2005). "Better health care waste management: An integral component of health investment." Regional Office for the Eastern Mediterranean - Regional Centre for Environmental Health Activities, World Health Organization, Amman.

Salkin, I. F. (2004). "Review of health impacts from microbiological hazards in health-care wastes." Dept. of Blood Safety and Clinical Technology and Dept. of Protection of the Human Environment, World Health Organization, Geneva.

Sarkar, M. S. K. A., Haque, M. A., and Khan, T. A. (2006). "Hospital waste management in Sylhet city." *ARPN J. Eng. Appl. Sci.*, **1**(2), 32–40.

Sawalem, M., Selic, E., and Herbell, J. (2009). "Hospital waste management in Libya: A case study." *Waste Manage.*, **29**(4), 1370–1375.

Sawyer, B., Elenbogen, G., Rao, K. C., O'Brien, P., Zenz, D. R., and Lue-Hing, C. (1993) "Bacterial aerosol emission rates from municipal wastewater aeration tanks." *Appl. Environ. Microbiol.*, **59**(10), 3183–3186.

Shen, B. H., Wang, X. C., Li, J. S., and Zhang, Y. H. (2003). "Current situation and disposing countermeasure of medical waste in Jilin province." *China Environ. Manage.*, **22**, 35–36.

Singh, R. (2004). "Disposing immunization wastes in India." Toxics Link, Delhi.

Solomon, G. M., Hjelmroos-Koski, M., Rotkin-Ellman, M., and Hammond, S. K. (2006). "Airborne mold and endotoxin concentrations in New Orleans, Louisiana, after flooding, October through November 2005." *Environ. Health Perspect.*, **114**(9), 1381–1386.

Taghipour, H., and Mosaferi, M. (2009). "Characterization of medical waste from hospitals in Tabriz, Iran." *Sci. Total Environ.*, **407**(5), 1527–1535.

Tsakona, M., Anagnostopoulou, E., and Gidarakos, E. (2007). "Medical waste management and toxicity evaluation: A case study." *Waste Manage.*, **27**(7), 912–920

UNEP (United Nations Environment Programme). (2003). "Technical guidelines on the environmentally sound management of biomedical and healthcare wastes." ⟨www.basel.int⟩ (Apr. 9, 2014).

Vilavert, L., Nadal, M., Inza, I., Figueras, M. J., and Domingo, J. L. (2009). "Baseline levels of bioaerosols and volatile organic compounds around a municipal waste incinerator prior to the construction of a mechanical-biological treatment plant." *Waste Manage.*, **29**(9), 2454–2461.

Visvanathan, C. (2008). "Healthcare waste in Asia: Intuitions & insights." ⟨http://www.eawag.ch/forschung/sandec/publikationen/swm/dl/healthcare_waste.pdf⟩ (Apr. 9, 2014).

WHO (World Health Organization). (2005). "Management of solid health-care waste at primary health-care centres: A decision-making guide." ⟨http://www.who.int/water_sanitation_health/medicalwaste/decisionmguide_rev_oct06.pdf?ua=1⟩ (Apr. 9, 2014).

Wong, J. W. C. (2012). "An independent expert assessment for the incineration of clinical waste at the chemical waste treatment centre at Tsing Yi (CWTC)." Environmental Protection Dept., Hong Kong SAR Government, Earth Tech Consultancy Ltd., Hong Kong Baptist University, Hong Kong SAR, P.R. China.

CHAPTER 16

Construction and Demolition Waste

Sadhan Kumar Ghosh[1]
Sannidhya Kumar Ghosh[2]

16.1 INTRODUCTION

The excessive consumption of natural resources has caused environmental pollution and deterioration of the landscapes. The impact is caused by the uncontrolled dumping of untreated Construction and Demolition Waste (C&DW) in illegal landfills and roadside. Recycling of demolition waste was first carried out after the Second World War in Germany, to tackle the problem of disposing large amounts of demolition waste. Demolition waste was caused by the war and simultaneously generated raw materials for reconstruction. Natural calamities, for e.g., earthquake, flood, tsunami and several man-made disasters, made C&DW treatment and disposal inevitable in many countries. Historically, reuse and recycling of a portion of C&DW have been in practice in developing countries worldwide mainly to reduce the construction cost. C&DW treatment is gaining interest in industries due to the implementation of green building standards. Considerable research on recycling process, recycling technologies and management techniques of the constituents of C&DW such as concrete, masonry & bricks, bituminous and others from construction industry have been carried out in different countries. These studies have demonstrated the possibility of using C&DW to develop new materials of construction, helping in sustainable resource utilization.

16.2 WHAT IS C&D WASTE?

Waste arising from the construction, repair, maintenance and demolition of buildings, structures and roadways is generally termed as Construction and Demolition Waste (C&DW) (Figures 16.1 and 16.2). It mostly includes brick, concrete, hardcore,

[1]Jadavpur Univ., Kolkata, India.
[2]University of Colorado Boulder, Boulder, CO, USA.

Figure 16.1. C&DW at Mumbai Metropolitan Region Development Authority, India

Figure 16.2. C&DW in India for disposal

subsoil and topsoil. It also includes quantities of timber, metal, plastics etc. C&DW includes structural and functional materials comprise of the followings:

- brick, concrete, and other masonry materials
- stone
- glass
- wall coverings
- dry wall
- framing and finishing lumber
- roofing materials
- plumbing fixtures (toilets, sinks, water heaters, pipes)
- heating equipment (furnaces, duct work)
- electrical wiring and components containing no hazardous fluids or refrigerants

- insulation
- wall-to-wall carpeting
- asphaltic substances
- metal incidental to any of the above
- weathered railroad ties and weathered utility poles

C&DW had traditionally been landfilled, frequently in the same landfills as were used to dispose of the municipal solid waste (MSW). Furthermore, it became clear that the volume of C&DW, most of which is inert, was roughly varying between 15% and 50%. Mixing of C&DW with MSW is a real issue because of the non-existence of effective system and rules for segregation of C&DW in many countries. Mixing of these wastes creates problem of treatment of both MSW and C&DW. C&DW reduces the effectiveness of composting and has negative impact on calorific value (CV) of MSW while considering waste-to-energy possibilities. On the other hand, MSW reduces the C&DW recycling effectiveness (Ghosh and Ghosh 2013). Landfill space is scarcely available. The costs of improved environmental protection involved in modern landfill engineering and management is increasing day by day. It is obvious that action to re-use or recycle C&DW would reduce the proportion going to landfill, thereby relieving the pressures on MSW disposal as well as respecting the hierarchy of waste management practices set in the country specific guidelines, framework directives on waste handling rules or environmental action programmes. Effective C&DW management will help in curbing excessive consumption of natural resources leading to sustainable natural resource conservation.

The nature of today's C&DW being demolished is directly influenced by the building techniques and materials which were in vogue when the buildings, civil engineering structures and associated infrastructure were built. The nature and volumes of demolition waste arising also reflect the solidity and flexibility (and therefore the life expectation) of the structures themselves, and the balance in the economy and investment (between housing industry and the service sector) of previous years. Considerable improvement has taken place in the building materials and structural engineering in last couple of decades. Hence the nature and volume of today's construction waste, by contrast, reflects today's building materials and activity levels. The sort of considerations which affect the nature and volume of C&DW includes the following:

1. individual family dwellings are predominantly built of blocks, bricks, reinforced concrete, metal rods & flats and wood, with wood much more widely used in several countries mostly in the colder regions;
2. the 1950s and 1960s apartment buildings in a major part of the world accommodated the flood of workers to post-war urban industrial centres were generally built of reinforced concrete with copper piping replacing lead;
3. by the 1980s, plastics (especially PVC double glazing units) were becoming widespread in pipes and window frames in all sorts of residential buildings

worldwide. Many industrial and commercial buildings erected since the 1980s have benefited from faster construction techniques based on steel frames.;

4. steel-framed structures lack the 'natural' fire resistance provided by concrete and brick, requiring much more fire proofing of beams &columns (may involving hydrocarbon-based materials) and firefighting systems (may involve chemicals which may themselves be hazardous);
5. the glass cladding which characterizes many modern corporate headquarter buildings and shopping centres, will in turn affect the composition of future C&DW arising;
6. with strong conservation laws often require developers to preserve the original buildings facades, re-building behind them in many countries;
7. older buildings are more likely to contain hazardous material, e.g., asbestos, CFCs &PCBs, because controls on these materials have been tightened over time. Controls have been imposed on the use of hazardous materials in commerce and industry, so the potential for contamination of the fabric of building structures from the products used are made in them, and from the wider environment, has decreased;
8. the trend in fixatives, fillers and coatings has moved from nails, screws, plaster, mortar and emulsion paints to organic resins and solvent-based products which, although inert or at least non-hazardous in their final form, are made up on site from components which are often flammable and/or toxic, and whose residues and containers are therefore also potentially hazardous;
9. bonding also makes full separation of waste streams, more difficult;
10. the changing balance between road construction and maintenance (as well as associated infrastructure such as bridges and tunnels) affects the nature of recyclable material arising, and the demand for other C&DW derived aggregates.

The characteristics of CD&W, its sources and possible destination are to be very carefully studied for effective disposal of the same.

16.3 CHARACTERISTICS OF C&DW, ITS SOURCES, AND DESTINATION

Where there is a limited range of options for dealing with waste, there may be little incentive to differentiate between different types of 'waste'. In modern society there are a wide variety of reasons why organizations and/or individuals may seek a more detailed understanding of the characteristics of 'waste'. The type of information required and the level of detail to a large extent will depend on the motivation of the stakeholder. The composition of CD&W may comprise of materials such as, concrete, bricks, tiles, ceramics, and gypsum based materials, wood, glass, plastic, electrical wiring, asphalt, tar and tarred products, metals (including their alloys), soil and dredged soil, insulation materials, mixed CD&W, Hazardous components namely asbestos, hazardous fluids or refrigerants etc. In order to gain a very high-level understanding it is common to differentiate between the primary sources of the waste. Cross-contamination and general mixing of materials is frequently observed on construction and demolition sites. This is of greatest concern if the mixing involves hazardous materials, e.g., asbestos and to certain heavy metals (such as lead), solvents and adhesives. Each country has its own classification of C&DW that can cover a very wide range of materials. The most obvious categories are:

1. waste arising from the total or partial demolition of buildings and/or civil infrastructure;
2. waste arising from the construction of buildings and/or civil infrastructure;
3. soil, rocks and vegetation arising from land levelling, civil works and/or general foundations; and
4. road planning's and associated materials arising from road maintenance activities.

The different types of sites which generate C&DW can arise from a range of different origins, or site types, as shown in Table 16.1.

There are five basic activities, some or all of which may occur on all of the above site types those are described in Figure 16.3. The linkages between the site

Table 16.1. Different types of sites which generate C&DW

Sites	*Types of sites/Category*
'Demolish and clear' sites	Sites with structures or infrastructure to be demolished, but on which no new construction is planned in the short term.
'Demolish, clear and build' sites	Sites with structures or infrastructure to be demolished prior to the erection of new ones.
'Renovation' sites	Sites where the interior fittings (and possibly some structural elements as well) are to be removed and replaced.
'Greenfield' building sites	Undeveloped sites on which new structures or infrastructure are to be erected.
'Road build' sites	Sites where a new road (or similar) is to be constructed on a greenfield or rubble free base.
'Road refurbishment' sites	Sites where an existing road (or similar) is to be resurfaced or substantially rebuilt

1 Remove selected materials from existing structure(s), possibly after *in situ* treatment

⇩

2 Demolish the balance of the structure(s), sort into waste streams as appropriate, and treat each waste stream on- or off-site prior to recycling or final disposal

⇩

3 Clear surrounding land surface(s) and any unwanted existing services/utility connections, broken down into two sub-activities:

⇩

3a Remove (i) hard surface coverings and (ii) any unwanted existing services and utility connections for recycling/disposal, and/or

3b Clear and dispose of unwanted surface vegetation

⇩

4 Prepare site for sale or construction, broken down into two sub-activities:

4a Prepare levels and foundations for new structures, and/or

4b Prepare to leave site clear and vacant

⇩

5 Erect new structure(s), then treat/dispose of construction waste materials

Figure 16.3. Basic activities found on demolition and construction sites

types identified in Table 16.1 and the five activities and their sub-activities in Figure 16.3 are to be established during implementation.

While handling C&DW, a number of hazardous materials may be found. Followings are some description of the common wastes in construction and demolition that may be dangerous/hazardous waste. Appropriate country specific actions are to be taken while coming across these types of waste.

- **Asbestos in CDW**: Asbestos can lodge in the lungs causing serious problems. Older types of floor tile, insulation, or other materials are Asbestos-containing materials in CDW
- **Mercury in CDW**: Thermostats can be characterized as toxic, due to mercury; they should be handled as per country specific rules in case of Mercury-containing thermostats & switches and relays within CDW. Bulbs and tubes can be characterized as toxic, due to mercury and they should be handled as per country specific rules Mercury-containing light bulbs/lamps (fluorescent bulbs).
- **Methylene chloride in CDW:** Waste methylene chloride paint: The paint stripper (unused) in CDW is the discarded commercial chemical product which may be toxic in nature. The paint sludge stripped from buildings contains solvent blend, before use, 10% or more of methylene chloride. Waste paints, waste or expired oil-based varnish, solvents, sealers, thinners, resins, roofing cement, adhesives, machinery lubricants, and caulk. Ignitable if flash point is below 140°F; State-only waste, often, due to metal content.
- **Solvents in CDW:** Waste paint thinner solvents are ignitable wastes. The solvent blend contains, before use, 10% or more of solvents such as, toluene, and MEK. The mixture also has a flash point below 140°F. Waste gun-cleaning solvent is ignitable waste. The solvent blend contains, before use,

10% or more of solvents such as, toluene, and MEK. The mixture also has a flash point below 140°F. Sludge or "bottoms" from solvent still that recycles gun cleaner or thinner are listed as ignitable waste: Still bottoms from a still where the solvent blend contains, before use, 10% or more of solvents such as, toluene, and MEK. The mixture also has a flash point below 140°F.

- **PCB in CDW:** PCB-containing light ballasts listed with PCB concentration of ≥2 parts per million
- **Dangerous/hazardous waste in CDW:** Shop towels/rags contaminated with dangerous/hazardous waste absorbents soaked with waste become dangerous/ hazardous waste. However, non-dripping rags/towels may not be dangerous/hazardous waste if they are: laundered at an appropriate facility, stored in containers away from a source of ignition and not mixed with other waste.
- **Toxic Sanding dust in CDW:** Sanding dust from older buildings sometimes characterized as toxic.
- **Aerosol cans in CDW:** Propellant is most likely ignitable, and the product may be dangerous waste in case of wastes generated from Aerosol cans.
- **Lead in CDW:** Siding or drywall with lead paint, lead flashings, lead in plumbing and pipes are usually the Lead-containing materials in CDW

Table 16.2 presents the possible destinations for C&DW which is found in most of the countries.

Table 16.2. Possible destinations for C&DW

Re-use options	1) Re-use on-site for the original intended purpose; 2) Re-use off-site for the original intended purpose;
Recycling options	3) On-site processing to recover high value saleable materials; 4) Off-site processing to recover high value saleable materials; 5) Recycling on-site for a low-value purpose (including non-essential land raising); 6) Recycling off-site for a low-value purpose (including non-essential land raising);
Incineration options	7) Off-site incineration with energy recovery; 8) Off-site incineration without energy recovery;
Landfilling options	9) Off-site land filling of segregated waste materials; 10) Off-site land filling of un-segregated waste.

16.4 C&D WASTE QUANTIFICATION MODELS

Once the characteristics and the sources are known, the amount of the generated wastes is to be quantified. Survey, past history, country specific policy and case study are the main methodologies for data collection for forecasting quantification of C&D waste. Conducting survey (sampling and weighing) at landfills is often used for determining local needs of waste management system with sufficient time and support. However, even on the local level there may be significant barriers to this method (Masudi et al. 2012). Sampling from mixed waste streams with statistical confidence is very difficult, time consuming, and expensive. Several waste quantification models with different perspective, accuracy levels and end results have been presented by researchers, namely, Yost and Halstead (1996), Franklin Associates (1998), Gheewala and Kofoworola (2009), Lu et al. (2013), Martinez-Lage et al. (2010) and Llatas (2011). Some of those are discussed here in brief for understanding.

Yost and Halstead (1996) proposed a method that combines building permits data from national statistics, which provide the number and estimated financial value, with empirical waste generation rate data. Data gathered from C&D sites are described as 'waste assessment data'. These data were utilized to produce the average waste generation rate per area, which is usually expressed in unit of amount (weight or volume) per area of activity. The waste generation rate per area is considered as 'waste factor' in the quantification model. Most studies combine the empirical waste assessment data (waste factor) obtained in the field with area of activity level extracted from national statistical database from their respective countries, which provides annual data on sector activities (number of construction or demolition permits, construction value, and area of activity level). Some of the quantification models for estimation of C&D waste generation which have been developed in different items are summarized below.

16.4.1 Models Proposed by Yost and Halstead (1996), Franklin Associates (1998)

Waste produced in a region = {Activity level of construction; demolition; or renovation in a region} X {Waste produced per activity}.

Method to use:

1. Start with total dollars of new construction, from Census Bureau, Current
2. Construction Reports.
3. Calculate sq. ft. of new construction from total dollars and $/ft^2 construction cost.
4. From empirical waste assessment, estimate lb/ft^2 of new construction.
5. Calculate total generation: Total new residential construction debris in tons/year.

Results:

1. the average waste generation rate (G_{av}) for residential construction = 4.38 lb/ft^2;
2. non-residential construction = 3.89 lb/ft^2;
3. residential demolition = 115 lb/ft^2;
4. non-residential demolition = 173 lb/ft^2; and
5. non-residential renovation = 17.67 lb/ft^2.

16.4.2 Models by Gheewala and Kofoworola (2009) for New Residential and Non-residential Projects in Thailand.

$$Q_x = A \cdot G_{av} \cdot P_x$$

where, Q_x = quantity in tons;A = area of activity in m^2; G_{av} = Average waste generation rate; P_x = percentage of waste material.

Results:

1. G_{av} for new residential = 21.38 kg/m^2; and
2. G_{av} for non-residential projects = 21.38 kg/m^2

16.4.3 Models by Lu et al. (2013) for High-rise Building Projects in Shenzhen, South China

$$WGR = \sum m_i \cdot A$$

Where WGR = Waste Generation Rate; m = the quantity of one waste material for a single bucket; and A = selected area for on-site sorting and weighing.

Results:

1. WGRs in Shenzhen = 3.275 to 8.791 kg/m^2; and
2. Timber and concrete are the largest components of waste generated.

16.4.4 Models by Martinez-Lage et al. (2010) Applied in Northwest of Spain

$$R_{build} = \sum_{counties} \{R_{Ci} + R_{Ri} + R_{Di}\} = \sum_{counties} \{C_C \cdot S_{Ci} + C_{Ri} \cdot S_{Ri} + C_D \cdot S_{Di}\}$$

where, R_{build} = C&D debris generated during a given year or horizon year (HY) distributed over counties; R_C = waste from new construction; R_R = waste from renovation; R_D = waste from demolition; S_C = total surface area for new construction; S_R = surface area for renovation; S_D = surface area for demolition; C_c = waste quantity per surface area of new construction; C_R = waste per area for renovation; and C_D = waste per area for demolition.

Results:

1. New construction work is estimated to generate 80 kg of waste/m^2, which is equivalent to 0.11 m^3 of waste/m^2 (density = 700 kg/m^3);
2. Demolition work is estimated to generate 1,350 kg of waste/m^2 and
3. Renovation work produces around 90 kg/m^2

16.4.5 Models by Llatas (2011) Dwelling Projects in Spain Involving Over 200 Building Elements in a Single Project

$$CW_B = \Sigma_j CW_{SBEj} = \Sigma_{ji} CW_{BEi} = \Sigma_{ji} CW_{Pi} + \Sigma CW_{Ri} + \Sigma_{ji} CW_{Si}$$

where, CW_B = volume of waste expected; CW_{SBEi} = volume of waste expected in the system building element 'j'; CW_{BEi} = volume of expected waste from building element 'i'; CW_{Pi} = volume of expected packaging waste element 'i'; $\sum CW_{Ri}$ = volume of remains expected from building element 'i'; CW_{Si} = volume of expected soil in building element 'i'; $CW_{py} = \Sigma_k (EWL)_{Pk} \cdot Qi \cdot F_P \cdot F_C \cdot F_I$; $CW_{Ri} = \Sigma_k (EWL)_{Rk} \cdot Qi \cdot F_R \cdot F_C \cdot F_I$; $CW_{Si} = \Sigma_k (EWL)_{Sk} \cdot Qi \cdot F_S \cdot F_C \cdot F_I$; $(EWL)_{pk}$ = Code of packaging; $(EWL)_{Rk}$ = Code of remains; $(EWL)_{Sk}$ = Code of soil; Qi = amount of building element 'i'; F_P = packaging waste factor; F_C = conversion factor; F_R = remains factor; F_S = soil factor; and F_I = increased volume factor.

Results:

1. Waste generation rate (without soil) = 0.1388 m^3/m^2;
2. Packaging waste generation = 0.0819 m^3/m^2; and
3. Soil generation = 0.2805 m^3/m^2.

With this model, chances for construction waste recovery and prevention could be increased.

Quantification model suggested by Franklin Associates (1998) is considered to be the benchmark for an accurate and comprehensive national C&D waste estimation for new construction, renovation, and demolition activities, due to strong empirical waste data obtained from rigorous waste sampling across the country. Recent quantification model employed by Martinez-Lage et al. (2010) can be regarded as the one of the models that gives accurate and comprehensive results. The model uses reasonable assumptions, taking account C&D waste from all activities, density of waste, detailed waste assessment data from all counties of the region, and supported by strong data for building permits from a number of governmental sources. It was found that concrete/bricks and timber are the main waste materials for new construction projects in Norway, with 67% and 15% respectively, while for demolition projects, 85% of waste generated is concrete, which is considered as the main waste generating activity. It was also predicted that concrete/brick waste generation will increase by four-fold and timber waste will increase by double in 2016 using Monte Carlo iterative tool by accounting building lifetime (60–90 years old). While quantification models proposed by Llatas (2011) provide a reliable and detailed approach for estimating waste generated from an individual construction project, taking account of soil remaining (wrecked) materials, packaging, and even hazardous waste to comply with stringent EU requirements. National or regional estimates provide general figures and forecasts for annual C&D waste generation. Studies found that waste generation factor differs according to the locations. The feasibility of C&DW recycling will depend on the characteristics, source and quantity of the feedstock. The recycling of C&DW will be discussed in brief in the next section with a case study.

16.5 C&DW RECYCLING

The "Sustainable building" has become a national catchphrase. In civil engineers' and architects' offices and on construction sites around the world there is increasing emphasis on reducing the environmental impacts of renovation and new construction. There are many green building rating systems and standards, such as LEED, BREEM, CASBEE, GBCI and GreenStaranc are becoming popular day by day. Recycling C&DW is one of the most important aspects of this movement. C&DW recycling is among the most visible commitments a developer can make to sustainable building. Figure 16.4 illustrates a hierarchy model of C&DW management. The change in the behavior of the consumer and the consumption pattern supported by re-design of the structures, buildings, materials, use and technology are essential for an effective C&DW management.

In providing materials to local vendors and processors, job site recycling creates employment and economic activity that help sustain local economies. On a lifecycle basis, recycling produces usable materials at much less environmental cost than materials from primary sources. That is, in addition to conserving raw

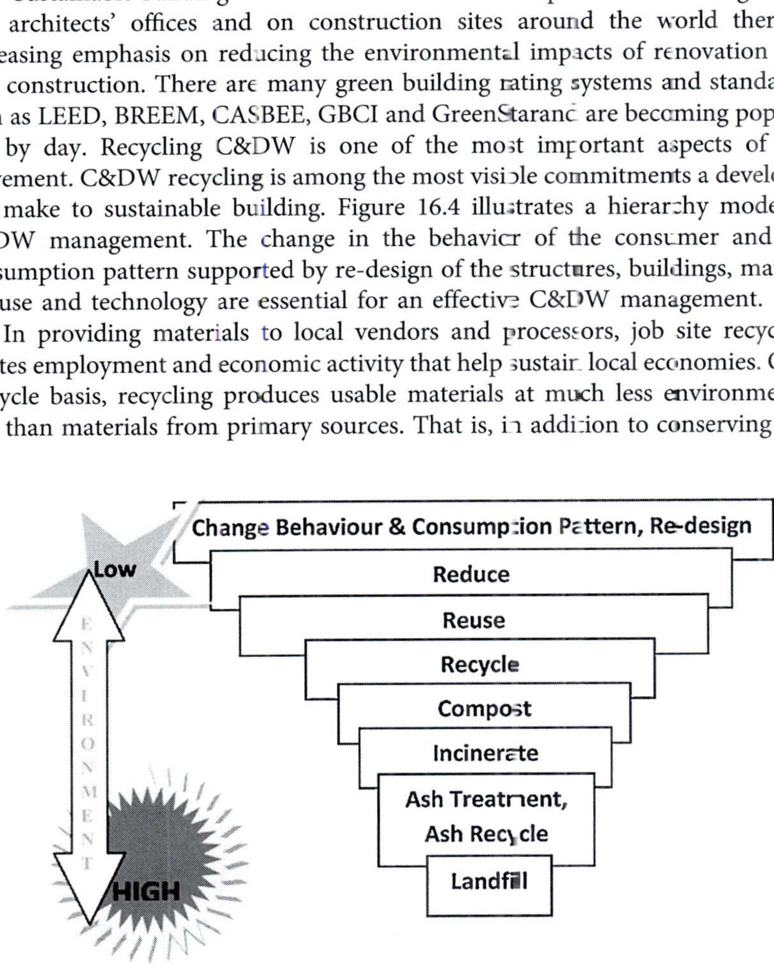

Figure 16.4. Hierarchy model of construction and demolition waste management

materials, recycling conserves energy and water, and reduces the production of greenhouse emissions and other pollutants. Almost every day, somewhere in the globe, an owner, civil engineer, architect, or contractor is proposing to recycle C&DW. On the contrary, almost every day, someone else is throwing up a barrier to recycling. Unless the proponent can address each barrier, recycling will often be abandoned before it is tried. Or worse, it will be forced on an unwilling participant who will, with or without actual malice, find a way to torpedo the effort. It is far better to address the barriers up front, with real information, than let them stand in the way, or linger and taint the whole recycling effort. ABC – asphalt, brick, and concrete – are, by weight, the dominant materials in construction and demolition waste. Any project that can capture and recycle ABC is well on the way toward an excellent recycling rate.

Angulo et al. (2010) pointed out that most of the national standards and international recommendations specify that only recycled aggregates produced from concrete wastes, with a low degree of contamination (usually less than 5% w/w) and porosity within certain limits are suitable for the production of structural concrete. Other C&DW recycled aggregates, from masonry or mixed sources, should be limited to use in non-structural or road-base applications. This is a safe approach, but carries with its certain limitations. The feasibility of C&D wastes recycling plant critically depends on the regulatory policy and support by the government, contract specifications, economics, selected technology, attitude of the developer and project management practice. Except for the impact due to improper disposal of C&D wastes in landfills that might cause a remarkable increase in environmental costs, the depletion of raw materials suitable for construction may generate renewed interest in converting C&D wastes into useful secondary materials.

Recycling C&D wastes is a very common practice in many countries. A number of C&D recycling projects have been implemented, e.g., in the USA, Canada, Europe, Australia, India, Japan, Hong Kong, Taiwan etc. The processing capacity of C&DW plants range from 100 to 1500 tons per day. Table 16.3 shows the recovery rates of common recyclable materials in different countries. Table 16.4 shows the products manufactured from recycled asphalt, brick, concrete, roofing, and others. Table 16.5 shows the technology used for recycling components of C&DW and Tables 16.6–16.8 illustrate the different types of recyclable materials in C&DW, their potential markets in general and limitations on recycling.

16.5.1 C&DW Recycling Process and Mass Balance Analysis

Figure 16.5 shows the mass balance of a C&DW recycling unit throughout the mechanical sorting process based on a continuous processing of 6 tons/h situated in Taiwan. It includes the initial screening of the waste material, sorting of recyclables and processing the recovered material for specific secondary uses. The mass balance of a full-scale construction waste sorting plant also confirmed the technical potential of recycling.

Table 16.3. Recovery rates of commonly recyclable materials (Percentages greater than 100% mean materials being recycled for more than one time)

Place	*Year*	*Paper (%)*	*Plastic (%)*	*Metals (%)*	*Glass (%)*
Hong Kong	2001	58	38	89	3
Australia	1995	51	30 [polyethylene terephthalate (PET) bottles]; 42 [high-density polyethylene bottles]	65 (aluminium (Al) can); 23 (others)	42
Japan	2000	58	14	75	78 (glass bottles)
USA	1999	42	6	35	23
Germany	1999	169	108	105 (fin plate); 87 (Al can)	88
United Kingdom	1998	38	3	43 (Al can); 35 (ferrous scrap)	22

Table 16.4. Products made from recycled asphalt, brick, concrete, roofing, and others

	Construction and demolition waste material						
Product Manufactured with Recycled C&DW Content	*Asphalt*	*Brick, concrete*	*Glass, porcelain, ceramic*	*Roofing shingles*	*Membrane roofing*	*Catch- basin grit*	*Processed contaminated soils*
Bituminous concrete mixes	★	★		★	★		
"Jack Patch" cold patch mix	★			★			
Crushed gravel substitutes		★	★	★		★	
"Reclaim" base and fill						★	★
Hot deicer	★			★			
"Top solution" manufactured soil		★					★

Table 16.5. Technology used for recycling components of C&DW

C&D materials	Recycling technology	Recycled product
Brick	· Burn to ash · Crush into aggregate	· Slime burnt ash · Filling material Hardcore
Concrete	· Crush into aggregate	· Recycled aggregate · Cement replacement (replace cement by fine portion of demolished concrete) · Protection of levee · Backfilling and Filler
Asphalt	· Cold recycling · Heat generation · Parallel drum process · Elongated drum · Microwave asphalt recycling system · Finfalt · Surface regeneration	· Recycled asphalt · Asphalt aggregate
Masonry	· Crush into aggregate · Heat to 900°C to ash	· Thermal insulating concrete · Traditional clay brick · Sodium silicate brick

(Continued)

Table 16.5. Technology used for recycling components of C&DW (Continued)

C&D materials	*Recycling technology*	*Recycled product*
Timber	• Reuse directly • Cut into aggregate • Blast furnace deoxidization Gasification or pyrolysis • Chipping • Molding by pressurizing timber chip under steam and water	• Whole timber • Furniture and kitchen utensils • Lightweight recycled aggregate • Source of energy • Chemical production • Wood-based panel • Plastic lumber • Geofibre • Insulation board
Ferrous metal	• Melt • Reuse directly	• Recycled steel scrap
Non-ferrous metal	• Melt	• Recycled metal
Glass	• Reuse directly • Grind to powder • Polishing • Crush into aggregate • Burn to ash	• Recycled window unit • Glass fibre • Filling material • Tile • Paving block • Asphalt • Recycled aggregate • Cement replacement • Man-made soil

Plastic	• Convert to powder by cryogenic milling • Clipping Crush into aggregate • Burn to ash	• Panel • Recycled plastic & Plastic lumber • Recycled aggregate • Landfill drainage • Asphalt • Man-made soil
Paper and cardboard	Purification	Recycled paper

Table 16.6. Recyclable wood in C&DW materials and potential markets

Material	*Description and sources*	*Markets*	*Limitations on recycling*
Wood, Dimensional	Two-by ends from new construction and renovation. Whole boards from renovation and demolition. Pre-fabbed walls and trusses greatly reduce waste.	Reuse markets available in most areas for whole boards. Scrap goes primarily to mulch and boiler fuel.	Generally few limitations. Some markets refuse nails and screws.
Wood, Flooring, Trim	Largest quantities from demolition and renovation. Ends and scrap from new construction and renovation.	Can be high reuse value in hardwood and softwood flooring and some trim & molding. Painted/treated wood, scrap and damaged wood goes primarily to boiler fuel.	Limited markets for painted or treated wood. All painted/treated wood should
Wood, Engineered	Plywood, OSB, glu-lam beams, etc. Significant quantities from new construction and renovation as well as demolition. Pre-fabbed walls and trusses greatly reduce waste quantities.	Some re-use value through deconstruction. Most is recycled as boiler fuel.	Generally few limitations.

Table 16.7. Recyclable C&DW materials and potential markets

Material	*Description and sources*	*Markets*	*Limitations on recycling*
Brick	Largely from demolition and renovation. Limited waste from new construction.	High-value re-use markets for some brick. More often placed in mixed aggregate markets, with concrete and block. Used in aggregate production.	Few limitations.
Concrete, formed	Largely from demolition and renovation. Limited waste from new construction.	Mixed aggregate markets with brick and block. Used in aggregate production.	Concrete w/rebar typically must be separated from brick, block, and concrete w/out rebar. Lead paints an issue where present.
Concrete block	Largely from demolition and renovation. Limited waste from new construction.	Mixed aggregate markets with brick and concrete.	Few limitations.
Asphalt pavement	Almost exclusively from parking areas. Limited waste from new construction.	Typically recycled separately from other materials. Used in production of new asphalt.	Few limitations.
Gypsum wallboard	Clean scrap from renovation and new construction. Currently no markets for demolition wallboard.	GP Gypsum and U.S. Gypsum offer recycling capabilities in N.E. A few markets grind as soil amendment.	Clean scrap from new installation only, without tape, nails, screws, corner bead.

(Continued)

Table 16.7. Recyclable C&DW materials and potential markets (Continued)

Material	*Description and sources*	*Markets*	*Limitations on recycling*
Ceiling tiles	Largely from demolition and renovation. Generally limited waste from new construction.	Armstrong Ceiling Systems accepts and recycles most ceiling tiles, when consolidated to truckload volumes (30,000 sq ft).	No mold; no asbestos or other hazmat; no vinyl, fabric, or foil faced tiles; no visible wood pulp or cardboard-like face. Armstrong takes tiles for testing prior to recycling.
Porcelain fixtures	Demolition and renovation only. Usually none from new construction.	Ground and used as aggregate or decorative chip.	Generally require removal of seats and plastic/metal fixtures.
Roofing, asphalt shingles	Large quantities from demolition and renovation. But also frequently significant scrap from new construction.	Asphalt and other paving materials	No asbestos or other hazmat. Metal (nails, flashing, etc.) typically acceptable.
Roofing, membrane	Large quantities from demolition and renovation. But also frequently significant scrap from new construction.	Asphalt and other paving materials.	No asbestos or other hazmat. Metal (flashing, dripedge, etc.) typically acceptable.
Roofing, slate	Large quantities from demolition and renovation. But also frequently significant scrap from new construction.	Often reusable. Damaged roofing ground and used as aggregate.	Few limitations.

Table 16.8. Metals, mixed debris, and textiles in C&DW materials and potential markets

Material	*Description and sources*	*Markets*	*Limitations on recycling*
Metals, ferrous	Structural and framing steel from demolition. Framing scrap from new construction and renovation. Typically very little structural steel from new construction or renovation.	Scrap markets; used in production of new steel.	Few limitations.
Metals, non-ferrous	Aluminum, copper, brass and alloys from electric, plumbing, and HVAC. Often significant scrap in new construction.	Scrap markets. Highest value if separated by metal at point of generation. Can be mixed and marketed with ferrous metals.	Few limitations.
Roofing, metal	Large quantities from demolition and renovation. But also frequently significant scrap from new construction.	Scrap markets.	Few limitations.
Carpet	Large quantities from replacement, demolition, renovation. Significant scrap from new installation.	Several manufacturers accept their own or other carpet for recycling. Other markets available but limited. Carpet is taken apart into multiple materials which are then recycled separately.	Carpet must be dry and mold free. Cost is typically very high. Recycling most often feasible with replacement installation.

(Continued)

Table 16.8. Metals, mixed debris, and textiles in C&DW materials and potential markets (Continued)

Material	*Description and sources*	*Markets*	*Limitations on recycling*
Mixed bebris	Large quantities from demolition and renovation. Small to large quantities from new construction, depending on feasibility of source separation.	Sorted mechanically and/or by hand into constituents, typically wood, metal, aggregate, and residual.	Hazmats (lead, asbestos, treated wood) may preclude recycling or increase recycling expense. Recycling rates typically less than source-separated, and costs are typically higher.

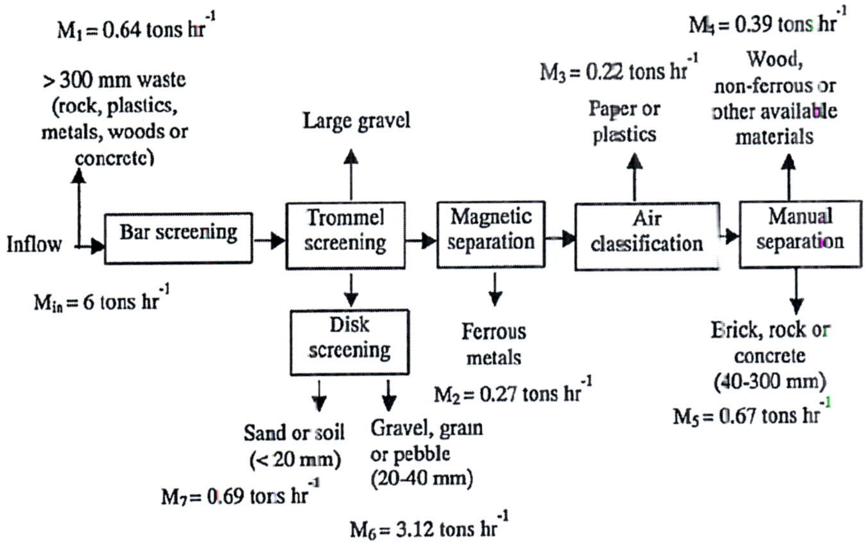

Design basis : 6 tons hr^{-1} C&D waste.

Note: 1. 8 hours per workday.
2. The sorting process is at steady state.
3. M_{in} denotes the CDW inflow.
M_1 = flow of rock, plastics, metals, woods or concrete.
M_2 = flow of ferrous metals.
M_3 = flow of paper or plastics.
M_4 = flow of wood, non-ferrous or other available materials.
M_5 = flow of brick, rock or concrete.
M_6 = flow of gravel, grain or pebble.
M_7 = flow of sand or soil.

Figure 16.5. Mass balance throughout the C&DW recycling

The mechanical sorting process illustrated in Figure 16.5 in a Taiwan facility of 6 tons/h of CD&W on one line, consisting of five operation units: 1) Bar screening, 2) Disk or trommel screening, 3) Magnetic separation, 4) Air classification, and 5) Manual separation. The bulky C&DW (<300 mm in size), e.g., rock, plastic, wood, steel, metal, or concrete, received by trucks are dumped onto the floor and sorted using a vibrating screen. Wastes, such as sand, soil, gravel, grain, or pebbles, passed through the first operation unit and sent into the trommel screen and disk screen in sequence. Ferrous metals can be extracted through overhead magnetic separator are collected for material recycling, recovery, and reuse.

The air classifier, blowing with a regular air stream of 250 m^3/ min further isolates inert materials, e.g., glass, wood, or ceramics, from the other available C&DW. At the same time, most of the fluff plastics can be blown away for additional separation. Residual waste materials, passing through the air classifier, are sent into a manual sorting unit for further separation. The output from the

Figure 16.6. A typical trommel screen for composite C&DW recycling plant

manual separation process consists mainly of wood chips and other residues, such as scrap tires etc. Figure 16.6 shows a typical trammel screen and composite C&DW recycling plant.

6 tons of C&DW processed = 3.80 tons of construction materials, M6, M7

+0.27 ton of ferrous metals, M2

+0.61 ton of combustible waste streams, M3, M4

+Disposal of other residues, M1, M5 to landfill operation efforts.

16.5.2 Jahangir Puri C&DW Recycling Plan: Case Study at Delhi, India

Considering the large scale growth of the city and the execution of infrastructural projects, the wastes generated during the demolition and construction activities are a major source of worry. To tackle this growing menace, Government decided to specifically initiate C&DW Project. The plant was established in July 2009 due to an urgent requirement of recycling huge amount of C&DW generated during the construction of villages for XIX Commonwealth Games, held in Delhi, India, from 3 to 14 October 2010.

The plant capacity is 500 tons per day of C&DW as per the concession agreement. Since the commissioning of the plant in 2009 up to 2013 end, nearly 0.70 million tons of C&DW from the city of Delhi was lifted for processing in the plant. Approximately 0.23 million tons of C&DW was lifted for processing from entire Delhi during Common Wealth Games 2010.

The objectives of the project are to maximize recovery of recyclable material, minimize waste quantity, decrease land filling, increase life of sanitary landfill sites and reduction in total costs of Municipal Waste Management. Figure 16.7 shows

Figure 16.7. The C&DW Plant at Jahangir Puri, Delhi, India

the C&DW Plant at Jahangir Puri, Delhi, India. The C&DW materials, after its processing are used as filling material (for embankments), cover material for sanitary land filling (SLF), as sub-base material in road construction and as a raw material for making of pavement tiles and new concrete cubes etc., by adding suitable admixtures.

The land lease period is of 10 years on develop, build, operate and transfer (DBOT) basis. This is the first greenfield project equipped with European Wet Processing Technology in India. Some of the products in the plant are, 1) Ready mix, 2) Kerbstones, and 3) Brick tiles on commercial level. Development of pre-fabricated facility is underway. Tiles and pavement blocks made in Indian C&DW recycling plant at Delhiare shown in Figure 16.8. The process flow chart of the C&DW Plant at Jahangirpuri, India is shown in Figure 16.9. Products manufactured from recycled asphalt, brick, concrete, roofing and others (Ghosh et al. 2011), are shown in Table 16.5.

16.5.3 Cost Effectiveness of Recycling and the Building Supply Chain

The built environment of the future is being constructed at the beginning of a new ecological era where governments are framing markets with regulation and legislation that respond to the challenges of environmental sustainability, and where industry must respond to the challenges of low-carbon economies and resource depletion. Businesses that are profiting and growing are adapting to these new challenges and responding with innovations that turn waste into valuable

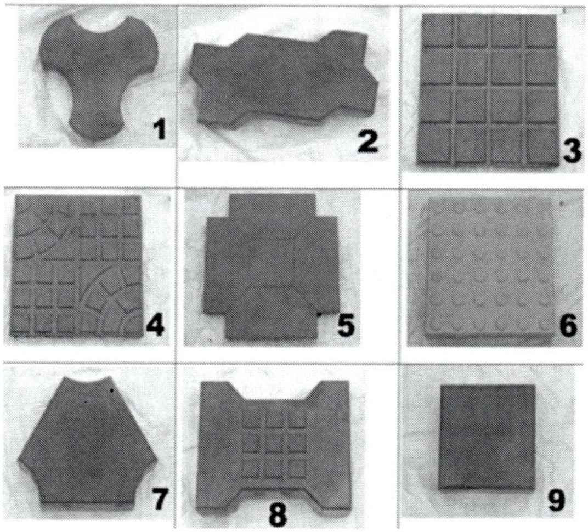

Figure 16.8. C&DW recycled products: tiles and pavement blocks

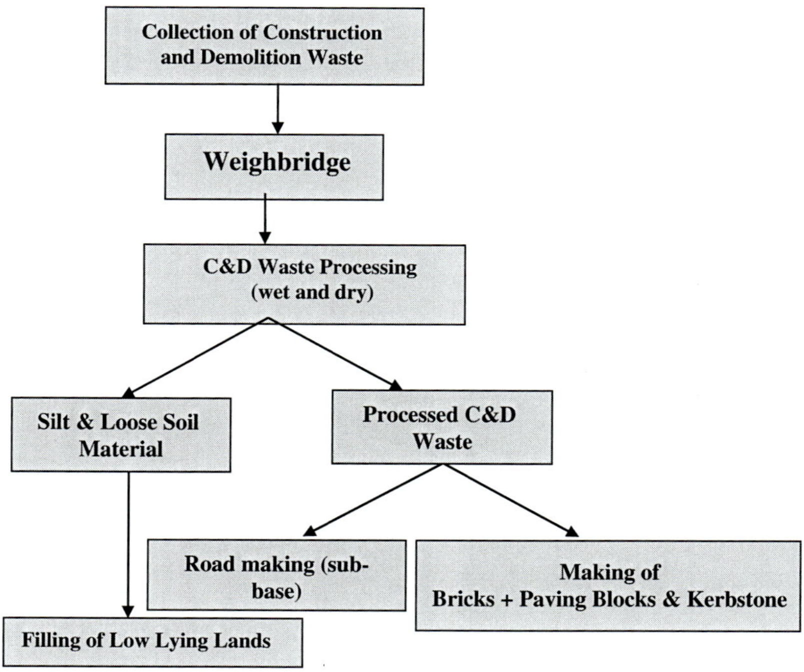

Figure 16.9. Process flow chart of the C&DW Plant at Jahangirpuri, India

resources to supply the construction industry, which has traditionally been adverse to behavioural change. There are now, more than ever, clear opportunities for business and industry to invest in activities that will create profit and improve environmental outcomes by extracting valuable resources from the C&D waste stream.

Cost of recycling vs. disposal is a critical point. If recycling costs more than disposal, then there will always be a very good reason NOT to recycle. But if recycling is cost-competitive or less expensive than disposal, than recycling should be considered as part of every job.

For each material, the total management cost, whether by disposal or recycling, has two components. The recycling cost is the cost per ton to process and recycle a material once it reaches a market (or, in the case of material that's disposed of the landfill with tipping fee). The transportation cost is the cost per ton to get a material to the market. This cost varies with the distance to market and with the quantity of material that can be hauled in a single load. The economic benefits of recycling are highest if the waste materials can be separated from each other and recycled individually. This is called "source separation". Table 16.9 shows advantages and disadvantages of Source Separation vs Commingled Recycling.

Source separation means separating different recyclable materials at the job site. That is, workers keep metals separate from wood, and wood separate from concrete, and so on, and place each material into a different container. These

Table 16.9. Advantages and disadvantages of source separation vs. commingled recycling

Recycling Method	*Advantages*	*Disadvantages*
Source separation	• Higher recycling rates • Lower recycling costs; revenues paid for some materials • Often a cleaner, safer work site	• Multiple containers on site • Workers separate materials for recycling • More complex logistics • Multiple markets; more information to manage
Commingled recycling	• Only one or two containers on site • No need for workers to separate materials for recycling • Easier logistics • One market; less information	

containers are then transported to different markets. Commingled recycling is the alternative to source separation. Commingled recycling means placing all recyclable materials into a single container, which is then transported to a processing facility, where different materials are separated by hand or by automated equipment. Source separation and commingled recycling have distinct advantages and disadvantages.

The biggest tradeoff between source separation and commingled recycling is complexity vs economics. Source separation is more complex because workers must separate waste materials before they throw them away. There are more containers on site, and there are more markets and haulers to work with and keep track of. But in most cases, source separation is economically more advantageous than commingled recycling:

1. Source separation produces materials that are ready to go directly to market; there is no need to pay a processor to sort materials.
2. Source separated materials are generally of higher quality, with fewer contaminants, so they're worth more in recycling markets.

On balance, source separation is generally preferable to commingled recycling. It costs less, and recycling rates are typically higher.

16.6 ENVIRONMENTAL CONSIDERATIONS AND LEGISLATION

Historically people had access to a local 'tip' or "dumping ground" where any unwanted materials could be dumped at little or no cost. There may have been little understanding of the potential impacts associated with different waste items, and little regulatory oversight of the dumping area. There is now, however, an increasingly detailed understanding around the different degrees of environmental, social and economic risks associated with various types of 'waste'. For example, it would generally be accepted that there is a much higher risk to human health and the natural environment associated with industrial waste from a chemical manufacturing facility than there would be associated with virgin soil excavated from a green-fields construction site. Flowing from the greater understanding of the potential impacts of waste materials, increasingly strict regulatory controls have been imposed on those organizations that manage discarded materials.

Although much C&DW is inert and the environmental impacts are therefore relatively mild, there are nevertheless adverse environmental impacts from the associated transport. Because both C&DW arising and potential locations for their re-use after processing are likely to occur in more urban settings, any impacts may well have the potential to affect more people. It should be stressed that these additional impacts do not include those associated with demolition itself, because the demolition process will occur whether or not the resultant waste materials are recycled. The breaking up, crushing, sorting and stockpiling of C&DW-derived aggregates is likely to generate

1. noise and dust;
2. air pollution (from the use of internal combustion engines);
3. potential for pollution of surface and groundwater by fuels and lubricants used in plant and machinery;
4. visual and aesthetic impacts (particularly if the processing site is a greenfield or urban site);
5. changes to natural habitats and possible destruction of historical artefacts (if the processing site is a greenfield site).

Some of these impacts are primarily environmental but others depend on the presence of people, and can more correctly be termed amenity impacts. The transport and delivery impacts associated with C&DW recycling are essentially the same as those associated with primary aggregates delivered by road, unless the C&DW can be processed and used on the original site. In general rail transport is far less likely to be a practical proposition for C&DW recycling (which moves from site to site) than it is for quarrying (which stays in one place). The above arguments concentrate on the inert fraction of C&DW which can be processed into aggregates. However, C&DW includes non-inert materials, and the environmental case for reusing or recycling these rather than sending them for final disposal to either a landfill or an incinerator is clear, particularly in the case of the more hazardous fraction. However, despite the benefits of avoiding landfilling or incineration, it should be acknowledged that processing is not an impact-free activity.

More complex C&DW processing and sorting systems, such as those found at fixed C&DW recycling centres, may well generate additional impacts. Some of these impacts (such as noise) will be similar in nature but different in intensity to those listed above. Others will go beyond those listed above, of which the most serious is likely to be potential pollution to the water environment as a result of the washing of C&DW derived aggregates to remove unwanted fractions (such as wood and plastic). Such potential impacts can be mitigated, but at a cost which will be reflected in the recycling centre's balance sheet.

Recycling is thought to be a vital part of reducing the environmental impact of the construction industry. However, the recycling process uses energy and emits CO_2 which must be balanced against any savings. For example, CO_2 is emitted when waste is transported from demolition or construction sites to recycling plants, and energy is needed (which also leads to CO_2 emissions) to run machinery that sorts and processes the waste. A C&DW plant receives mixed C&DW materials, which are then separated into waste types, such as concrete aggregates, metals, paper, plastics and many other items. Environmental impacts of different aspects of the processes exist, which may be measured either in emissions of CO_2 in tons (other greenhouse gases were included, but measured in terms of impact equivalent to tons of CO_2) and primary energy expenditure (energy contained/produced at power plants), measured as the equivalent to tons of oil.

Three forms of environmental impact may be considered: 'incorporated', 'operational' and 'transport-related'. 'Incorporated' impacts were the energy and CO_2 emissions which arose from the production of all buildings and machinery needed for the plant. Operational' impacts are mainly made up of the electricity needed to run machinery, as well as some emissions from diesel-powered engines used in the sorting process. Finally, 'transport-related' impacts are the result of transport of waste, both from the demolition or construction site to the plant, as well as from the plant to recycled materials processors or landfill sites. Studies shows that the environmental benefits in a C&DW plant can only come to be, however, if the output materials produced by the recycling facility are effectively sent out and used for the fabrication of new products, especially construction related ones. Transport to and from the plant caused the greatest environmental impact overall. The researchers suggest that energy-saving or CO_2-reduction measures targeted at the transport phase would be an effective way of reducing the impact of recycling plants, possibly through the use of electric vehicles (Kampman et al. 2006),. The researchers caution that the results are simplified because they include only CO_2 emissions and energy use. However, these are thought to be the primary environmental impacts of recycling plants.

Legislation for disposal of C&DW does not exist in many of the developing countries whereas, in most of the developed countries, the same is covered by rules and directives. Enactment of C&DW legislation and its effective implementation will be able to reduce the impact of the C&DW disposal.

16.7 INTERNATIONAL SCENARIO OF C&D WASTE MANAGEMENT

In the last two decades, significant improvement in C&D waste management has taken place worldwide. A review of the situation in some of the countries is made in the following paragraphs. C&D waste production equates to a large amount of carbon emissions that require all sectors of the economy, to a greater or lesser extent, to contribute to its reduction. The development and implementation of innovative and effective C&D waste management solutions will affect the contribution that waste management as a whole makes to climate change. The construction industry plays a leading role in improving the quality of the built environment, but its activities also impact on the wider environment in a number of ways, including waste production. The built environment consumes more natural resources than necessary and therefore generates a large amount of waste. To move forward with C&D waste agenda, a number of countries worldwide set ambitious targets and measures to monitor, support and implement ways to effectively manage C&D waste; including waste prevention, minimization, reuse, recycling, and energy recovery. Consequently, stakeholders in C&D sectors are under increasing pressure to improve their waste reduction performance. Although the idea of C&D sustainable waste management is well acknowledged and generally accepted, it is proving more difficult to implement. The IWWG set up the C&D Task Group to provide an international platform for C&D waste information exchange and dissemination; research and innovation; and practical applications for members and those engaged with the planning, procurement, design, construction, refurbishment, and demolition activities in the built environment. International societies like, International Society of Waste Association (ISWA), International Waste Working Group (IWWG), International Society of Waste Management, Air and Water(ISWMAW) and some others international agencies are developing awareness worldwide for recycling and controlled disposal of C&DW.

16.7.1 C&D Waste Management in the Asia

In India, it is estimated that the construction industry generates about 10–12 million tons of C&DW annually. There is a huge demand of aggregates in the housing and road sectors but significant gap in demand and supply exist, which can be reduced by recycling construction and demolition waste to certain specifications. Historically, a portion of C&DW items, e.g., bricks, tiles, wood, metal, etc. are re-used and recycled. Concrete and masonry constitute about 50% of the C&D waste. Projections for building material requirement by the housing sector indicate a shortage of aggregates to the extent of about 55,000 million m^3. An additional 750 million m^3 of aggregates would be required to achieve the targets of the road sector. There is also a huge demand for aggregates in the housing and road sectors, but there is a significant gap in demand and supply. C&DW recycling may bridge the gap. In the last decade, major cities like Bangalore, Chennai, Delhi, Kolkata and Mumbai have adopted considerable initiatives of C&D waste management including introduction of C&DW handling rules.

Highly-Developing Asian Island Countries (HDAIC), e.g., Japan and Hong Kong with its dense population and tightly packed skyscrapers, construction presents many challenges, not least managing waste. To tackle the issue, a series of construction waste management (CWM) policies have been introduced by the Hong Kong Government. Based on the 'polluter pays' principle, the Hong Kong government implemented a Construction Waste Disposal Charging Scheme (CWDCS) in 2006. In line with the CWDCS, a levy of HK$125 ($16) was imposed for every tonne of construction waste a contractor disposes of in landfill. However, the levy will be HK$100 ($13) per ton if the waste has first been processed at off-site sorting facilities. Further, waste will be charged at just HK$27 ($3.5) per tonne if it consists of inert materials which are accepted by public fill reception facilities. Having implemented a new construction waste management policy in Hong Kong in 2006, waste generated by the sector has fallen significantly. Solid waste arising from construction activities is a grave concern in Hong Kong. The latest figures show that in 2011 some 13,458 tons of MSW was landfilled every day, and that C&DW accounted for around a quarter of that. In addition to the environmental impacts, construction waste places tremendous pressure on valuable landfill space in the compact city. On the face of it the changes to the management of C&D wastes in Hong Kong would seem to be a major success. In 1999 the city sent on average 7890 tons of C&D waste to landfill every day - accounting for 21% of total arising, with the remainder being sent to public filling areas. An off-site Construction Waste Sorting (CWS) program was introduced and two off-site waste sorting facilities were set up. Between from commencing operations in 2006 to February 2012, the two off-site CWS facilities have handled a total of 5.11 million tons of C&D waste in Hong Kong (Lu 2013).

High economic growth in Japan during the 1970's coupled with the growing numbers of large scale public works, office buildings, and apartments meant large amounts of waste was generated by C&D works. The amount of C&D waste increased from 30 million tons in 1980 to 66.9 million tons in 1992. The amount is almost doubled in 12 years. This waste was mainly legally disposed of at landfills or illegally dumped. Historically, C&D waste management in Japan has been composed of three phases: (1) the launch of first-generation waste act in 1970's; (2) an emphasis on reuse/recycle in the 1990's; (3) the launch of a 'Recycle-based Society' in 2000. Since the 1970's, C&D waste was governed by 'Act on waste disposal and cleaning' (Waste Act 2002). Even though C&D waste was regarded as 'industrial waste' under this act, this lenient legal regime did not play an effective role in resolving the waste problems. In 2002, the government reviewed the achievement of the past efforts. The important indicators were summarized as follows:

1. The overall recycle rate of C&D waste reached 85% in 2002;
2. The recycle rate of concrete block and bitumen concrete block reached more than 90%;
3. The recycling rate of construction wood, construction dirt mud, mixed construction waste, and construction aggregate was unsatisfactory.
4. Fly-tipping remained a serious problem.

The outcome looks quite satisfying, in spite of some flaws. However, due to the fact that many old buildings would come into the stage of renovation and the ambitious goal of 100% recycle rate had been set under the scheme of recycle-based society in Japan, the new goal was moved forward. As a result, the recycle rate reaches 90% in 2005 and 95% in 2010.

In People's Republic of China, large quantity of C&DW with rapid urbanization, growth of economy and population, has placed great burden on the finite landfill space and environment since the early 1980 s. The current recycling in China is restricted to those high scrap value recycling materials like steel, while other non-profitable materials like aggregates will be sent to C&D landfills. Mixed collection, manual separation and simple landfill without collection system of leachate and odour are broadly applied in China. It is estimated that the number of C&DW generated each year in China will reach 600 million tons. "Every year, new buildings in China, total up to 2 billion square meters, use up 40 percent of the world's cement and steel, but our buildings can only stand 25 to 30 years on average", Vice-Minister of housing and urban-rural development,

People's Republic of China, said at a recent international forum on green and energy-efficient building. In China, construction waste comprises 30 to 40 percent of the total urban waste. The construction of a 10,000 m^2 building will create 500 to 600 tons of waste, while the demolition of a 10,000 m^2 old building will create 7,000 to 12,000 tons of waste. Space from building demolition in China annually constitutes about 40 percent of the total construction area. Some suggestions given for improving the C&DW management (Cui et al. 2012) are: 1) Perfecting the polluter-pays system; 2) Specialization of C&DW collection method; 3) Encouragement of new recycling technologies and mature technology imported; 4) Implementing an integrated disposal method; 5) Establishment of regulations and improvement of supervision structure; and 6) Enhancing environmental awareness.

In many other countries, efforts on formulation and establishing C&D waste management systems are taking place. Proper disposal of C&D wastes has received wide attention and again significantly due to natural calamity and man-made disasters.

16.7.2 C&D Waste Management in Australia

Information gathered by sustainability Victoria, indicates that C&DW material recovered for reprocessing in Victoria, Australia, is predominantly concrete, and that this is consistent with previous surveys for this sector as per C&DW Guide, Commonwealth of Australia (2011). In Sustainability Victoria's survey period, a total of 3.15 million tons of C&D material was recovered for reprocessing. The report highlights that the historic trend for C&D material recovery in Victoria is one of the high and strong diversions. It also indicates that historic growth in C&D recovery rates is beginning to slow, and it is expected that future growth will remain slow. This slowing in growth reflects that most large scale demolition waste is already being captured and recovered. In Australia, well defined C&D waste management system has been in practice in Victoria, Queensland, South Australia, Western Australia, Tasmania and Australian Capital Territory for long time.

16.7.3 C&D Waste Management in the Europe

In the European Union (EU), C&D activities produce the highest levels of waste, accounting for 855 million tons per year (33.3% of the total EU waste). At present, 75% of C&D waste in the EU is being landfilled, although over 80% recycling rates have been achieved in countries such as Germany and the Netherlands. In the UK, the disposal of C&D waste accounts for more than 50% of overall landfill volumes. Currently only around 30% of the materials used in construction are recycled as opposed to virgin materials, a figure that could potentially rise to 90%, helping to support sustainable construction and yielding a range of environmental and economic benefits (Waste Management World 2011). According to a new report from analysts Frost & Sullivan, the market for recycled construction materials such as PVC, gypsum, aggregates and recycled glass generated revenues of

€744.1 million in 2010, and is estimated to reach €1.3 billion by 2016. The EU directive 2008/98/EC requires that by 2020, 70% of these construction waste streams should be recovered; but current treatment facilities provide only limited scope to sort C&DW into recyclable fractions.

C&DW management in Germany is a mature and well integrated sub industry within the broader German construction market. During 2002 and 2003, Germany generated 381.3 and 366.4 million tons of waste, respectively. From these amounts, C&DW represented 63.2% in 2002 and 61.0% in 2003 (Weisleder and Nasser 2006). Despite these high numbers, only 15% of this material was disposed of in landfills, while the remaining 85% was recovered and reused in further applications or recycled. According to the report of the ArbeitsgemeinschaftKreislaufwirtschaftsträgerBau (KWTBau 2005), the total amount of C&DW generated in 2002 was 240.9 tons. From this amount, the main part corresponded to excavation material (65.9%), followed by building demolition waste (24.3%), road demolition waste (7.8%) and construction site waste (2.0%). According to the Federal Statistical Office, from the total amount of C&D waste generated in 2002, the percentage of recycling was 85.6%.

16.7.4 C&D Waste Management in the North America

In the USA, C&D activities generate over 170 million tons of waste per year, of which 48% was estimated to be recovered. In terms of waste-designed facilities' floor areas ratio, it is estimated that most of the buildings in the USA generate between 20 to 30 kg/m^2 of C&D waste. NEWMOA found that the total C&D waste generation in the Northeast USA in 2006 was approximately 12.1 million tons (NEWMOA 2009). The per-capita generation of C&D waste varies widely among states for various reasons. It includes limitations on data availability, from roughly 0.19 to 0.42 tons per person per year. C&D waste is mainly disposed in landfills. In the USA, the rate of concrete recycling is about 82%, according to the Construction Materials Recycling Association (CMRA), while the recycling rate for structural steel is around 98 percent, according to the Steel Recycling Institute. Around 70.6 million tons of urban wood waste were generated in the USA in 2010, including 48 percent from MSW and 52% from C&D activities. After current recovery efforts or wood unavailable for recycling due to chemical treatments and/or other issues, there remains an untapped resource of 17.3 million tons of urban waste wood available for recycling on an annual basis.

Late in 2008, the economy changed dramatically, reducing construction activity as well as markets for recovered materials. In Massachusetts, the Department of Environmental Protection (MA DEP) in its "Beyond 2000 Solid Waste Master Plan", committed Massachusetts to an 88% reduction in C&DW by 2010 while the C&DW materials were recycled at a rate of 66% in 2008. While this recycling rate was high, the bulk of the recycled tonnage is asphalt, brick, and concrete (ABC), which is routinely recycled (2.33 out of 2.52 million tons). When the ABC tonnage is excluded from the C&D data, the remaining material is only recycled at a rate of 14%. Considering that about 0.7 million tons of Massachusetts

C&D material is disposed of each year and an additional 0.4 million tons is managed as fines and residuals (a relatively low value use), there is considerable room for increased C&D recycling (Solid Waste Master Plan 2013). The objectives adopted were, increase the recycling rate for C&D materials excluding ABC to 50% by 2020.Based on 2007 C&D generation, this would mean reducing C&D disposal and landfill uses such as alternative daily cover and grading shaping materials by a total of 0.4 million tons annually by 2020.

It was noted that the amount of recycled and/or reused waste use in the Canadian construction industry is much lesser than many other developed nations. Canada recycles and/or reuses only 22% of the solid waste. The Leadership in Energy and Environment Design (LEED) standards are quite popular in Canada. Each rating system uses different methods and weights to analyze the sustainability in construction projects. Although many Canadian construction projects use the LEED 'ratings' to assess sustainability, it does not have a broad focus on C&D waste. The need for comprehensive and integrated waste management mechanisms, technologies, rating systems, and policies is widely visible in Canada comparing to some European countries.

16.8 CONCLUSION

Within the construction industry, there is a steady increase of recognition of the environmental and social impacts caused by extraction, construction, demolition and waste disposal process. These are accompanied by a willingness to do something about them. Increasing the amount of construction waste that is reused or recycled, is one way in which the industry is becoming more sustainable. Nevertheless, it is important to bear in mind that recycling has its own share of impacts, and there are other barriers to be overcome. Using lifecycle assessment, a number of alternative waste management strategies can be judged against a range of operational, environmental, social and economic criteria. This can enable the construction industry to make more informed waste management decisions.

As long as primary materials are displaced, the reuse of demolition waste has the lowest overall environmental and social impact, followed by a combination of reuse and recycling, with landfill as the least desirable option. The financial costs follow a similar pattern, which therefore raises the question of why recycled materials are not more widely substituted for primary materials in the construction industry. Such factors include the timing of materials supply and demand, perceived additional risks, or because the predicted financial benefits may not always be fully realized due to the market structure. Secondary materials may serve a slightly different market to primary materials, so they are not really acting as a substitute product. Alternatively, the cost of materials may only be a small part of the overall cost of a construction project and therefore, it is not worth taking even a small risk with recycled materials that have not been fully proven. However, the results are based on case studies and should be treated with caution: the best environmental option will depend on local circumstances. With the introduction of Green Building Standards, the popularity of reuse and recycle of C&D Waste are increasing.

It is concluded that the construction industry could become more sustainable by using recycling to strategically displace more primary materials, although there may be some limitations on the extent to which this occurs in practice.

References

Angulo, S. C., Carrijo, P. M., Figueiredo, A. D., Chaves, A. P., and John, V. M. (2010). "On the classification of mixed construction and demolition wastes aggregates by porosity and its impact on the mechanical performance of concrete." *Mater. Struct.*, **43**(4), 515–528.

Commonwealth of Australia. (2011). "Construction and demolition waste guide (2011)—Recycling and re-use across the supply chain." Department of Sustainability, Environment, Water, Population and Communities, Australian Government, Parkes ACT, Australia.

Cui, S., Liu, Q., Yan, J., and Dyu, X. (2012). "Key engineering materials." ⟨http://www.scientific.net⟩ (Jun. 11, 2014).

Franklin Associates. (1998). "Characterization of building-related construction and demolition debris in the United States." , U.S. Environmental Protection Agency, Municipal and Industrial Solid Waste, Division Office of Solid Waste, Washington, DC.

Gheewala, S. H., and Kofoworola, O. F. (2009). "Estimation of construction waste generation and management in Thailand." *J. Waste Manage.*, **29**(2), 731–738.

Ghosh, S. K., and Ghosh, S. K. (2013). "Review of C & D wastes supply chain management in India." *Proc., 2nd Int. Conf. on Solid Waste*, Hong Kong Baptist University, Hong Kong SAR, P.R. China.

Ghosh, S. K., Ghosh, S. K., and Aich, A. (2011). "Rebuilding C&D waste recycling efforts in India." *Waste Manage. World*, **12**(5).

Kampman, B., Bruyn, S. D., and Boer, E. D. (2006). "Cost effectiveness of CO2 mitigation in transport." *European Conf. of Ministers of Transport*, Delft, The Netherlands, 7–10.

KWtBau. (2005). "Voluntary commitments of the construction industry (ArbeitsgemeinschaftKreislaufwirtschaftsträgerBau)." Berlin, Germany.

Llatas, C. (2011). "A model for quantifying construction waste in projects according to the European waste list." *J. Waste Manage.*, **31**(6), 1261–1276.

Lu, W. S. (2013). "Construction waste—Hong Kong style." *Waste Manage. World*, **14**(4), 1056–1064.

Martinez Lage, I., Martinez Abella, F., Vazquez Herrero, C., and Perez Ordonez, J. L. (2010). "Estimation of the annual production and composition of C&D debris in Galicia, Spain." *J. Waste Manage.*, **30**(4), 636–645.

Masudi, A. F., Hassan, C. R. C., Mahmood, N. Z., Mokhtar, S. Z., and Sulaiman, N. M. (2012). "Waste quantification models for estimation of construction and demolition waste generation: A review." *Int. J. Global Environ.*, **12**(2–4), 269–281.

NEWMOA. (2009). "Construction & Demolition Waste Management in the Northeast in 2006." ⟨http://www.newmoa.org⟩ (Jun. 11, 2014).

Solid Waste Master Plan. (2013). "Pathway to zero waste, Massachusetts 2010-2020." Massachusetts Dept. of Environmental Protection, Executive Office of Energy and Environmental Affairs, Boston, MA.

Waste Act. (2002). "Act on waste disposal and cleaning, enacted on 25 Dec 1970, the amendment: 1 Dec 2002, Japan." ⟨http://law.e-gov.go.jp/htmldata/H12/H12HO110.html⟩ (Jun. 11, 2014).

Waste Management World. (2011). "Recycled materials could supply 90% of Europe's construction needs." ⟨http://www.waste-management-world.com/articles/print/volume-12/issue-5/regulars/news/recycled-materials-could-supply-90-of-europes-construction-needs.html⟩ (Jun. 11, 2014).

Weisleder, S., and Nasser, D. (2006). "Construction and demolition waste management in Germany." ZEBAU GmbH, EU-ASIA PRO ECO IIB Post-Tsunami Programme, Hamburg, Germany.

Yost, P. A., and Halstead, J. M. (1996). "A methodology for quantifying the volume of construction waste." *Waste Manage. Res.*, **14**(5), 453–461.

CHAPTER 17

Treatment and Use of Ashes from Solid Waste Processing

Kaimin Shih[1]
Kui-Hao Chuang[2]

17.1 INTRODUCTION

Municipal solid waste (MSW) is generated mainly by households. This includes garbage, organic waste, recyclable materials, and hard waste. It also contains waste that municipal activities produce through emptying litter bins, maintaining parks, and sweeping streets. The amount of municipal solid waste generated is thousands of millions of tons every year. With social, financial and environmental consequences, the growing waste generation is gaining more attention and MSW management is becoming a major issue. Thermal treatment is an effective technique for treating solid waste, since it can reduce the mass and volume of waste. The major types of thermal treatments for solid waste include pyrolysis, gasification and incineration, depending on the amount of air used in the treatment. At present, municipal solid waste incineration (MSWI), as a waste-to-energy strategy, has been considered to be an environmentally friendly solution and a common alternative to landfilling. The incineration process can usually be classified into three main stages which include incineration, energy recovery, and air pollution control. Air pollution control units remove particulates and acid gases to significantly reduce the air emissions from incinerators. However, this operation often generates additional solid residues and/or wastewater that need to be treated. The core problem associated with the MSWI residues is the fly ashes (FA) and bottom ashes (BA). Fly ash refers to the fine particulate matters still in the flue gas downstream from the boiler and recovered in electrostatic precipitators or fabric filters. The amount of fly ash produced by an MSW incinerator is in the order of 1–3% of the waste input mass on a wet basis. Bottom ash is a heterogeneous mixture of slag, containing various proportions of glass and metal residues collected at the outlet of the combustion chambers in quenching/cooling tanks.

[1]Univ. of Hong Kong, Hong Kong SAR.
[2]Central Taiwan Univ. of Science and Technology, Taiwan.

Sewage sludge is an inevitable by-product of the wastewater treatment process and its management is also one of the most critical environmental issues. The need for water supply is increasing along with improvement in the quality of life, meaning that the amount of sewage sludge is expected to rapidly increase. For example, in Europe, the total amount of sewage sludge has increased from 7.2 million tons of dry matter in 1998 to approximately 9.4 million tons in 2005 (European Environment Agency 2009). The quantity of ash derived from incinerating sewage sludge is much smaller than that of MSWI ash, which may explain the relatively limited research carried out on sewage sludge ashes (SSA). In addition, the management of sewage sludge by incineration processes is a relatively recent development and the majority of it was previously reused in agricultural activities. Approximately 20% of the dry solids of sewage sludge is inert inorganic material and forms a substantial quantity of ash from incinerating the sludge. Incineration of sewage sludge effectively reduces the volume of sludge and renders it biologically and chemically inert before landfilling. The SSA is not considered as a hazardous waste, despite generally having higher levels of metals in the ashes. The disposal of ash to landfill is not ideal and therefore the development of economically viable reuse and recycling strategies is of great interest.

The severe environmental constraints and declining energy supplies associated with fossil fuels have increased the interest in developing bio-energy technologies. The aim of increasing the percentage of biomass fuels in primary energy consumption has been defined in the White Paper of the European Union (European Commission 1991), as they provide a reasonable option and a remarkable potential to reduce greenhouse emissions (Petersen Raymer 2006; Vamvuka and Kakaras 2011). Residues from agricultural generation and processing industries are readily available in large quantities and municipal solid wastes are commonly produced through social activities. Since the disposal of biomass is no longer viable for environmental and economic reasons, the application of thermo-chemical technologies, particularly with combustion, is thought to be a promising solution for energy production. Biomass fuels normally have a high content of alkali metals, which, together with other mineral components in the ash, give rise to severe ash deposition problems in combustion. Biomass ashes generated during the thermo-chemical conversion of biomass for energy production are also recognized as a type of solid waste. The utilization of biomass ashes is still in its initial stage and needs further investigation. It is commonly assumed that biomass ashes are without toxic metals, as in the case of MWSI ash, but certain published results on biomass ash composition are very inconsistent and this uncertainty may significantly offset the overall cost-effectiveness of incinerating biomass for energy. However, the recycling of biomass ashes is also of increasing interest, because some biomass ashes, such as olive ash with valuable phosphorus and potassium nutrient elements, have great potential for agricultural and forestry applications.

This chapter provides the characteristics of incineration ashes derived from municipal solid waste, sewage sludge and waste biomass with a main focus on the chemical properties of the ashes. Furthermore, the common ash treatment methods, including separation, solidification/stabilization and thermal treatment, are also introduced. Potential strategies for the beneficial utilization of ashes, including for construction materials, adsorbents for wastewater, zeolite synthesis, and agricultural applications, are also presented in the discussion.

17.2 ASH PROPERTIES FROM SOLID WASTE PROCESSING

It is necessary to offer an unambiguous understanding of ash properties, particularly their chemical composition, before determining suitable treatment processes or beneficial utilization strategies. The chemical properties of incineration ashes can generally be categorized according to the different sources of solid waste, i.e. municipal solid waste, swages sludge, and biomass (Figure 17.1).

17.2.1 Municipal Solid Waste

The contents of the MSWI fly ash and bottom ash reported in some representative studies are shown in Tables 17.1–17.5. For oxides, SiO_2, Al_2O_3, CaO, Fe_2O_3, Na_2O, and K_2O are the common oxides found in the ash. It can be observed that the

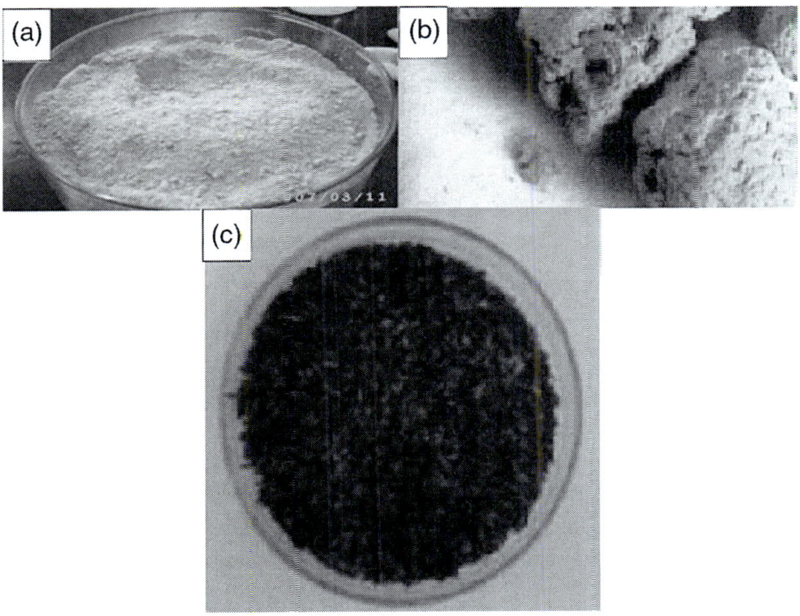

Figure 17.1. Incineration ashes from (a) municipal solid waste, (b) sewage sludge, and (c) biomass

Sources: (a) Chou et al. 2009; reproduced with permission from Elsevier. (b) Helena Lopes et al. 2003; reproduced with permission from Elsevier; (c) Imyim and Prapalimrungsi 2010; reproduced with permission from Elsevier

Table 17.1. Major chemical compositions of MSWI fly ash (wt.%)

Reference / Oxide	(Qian et al. 2008)	(Shi and Kan 2009)	(Wu et al. 2011)	(Ginés et al. 2009)	(Wang et al. 2010)	(Pan et al. 2008)	(Andreola et al. 2008)
SiO_2	5.44	24.5	13.87	6.35	32.8	13.6	18.5
Al_2O_3	3.1	7.42	3.71	3.5	8.6	0.92	7.37
CaO	42.55	23.37	29.82	43.05	28.4	45.42	37.5
Fe_2O_3	1.69	4.01	1.46	0.63	1.5	3.83	2.26
MgO	1.83	2.72	1.54	1.38	2.4	3.16	2.74
Na_2O	4.82	4	3.23	5.8	3.9	4.16	2.93
K_2O	4.31	4.6	3.47	4.59	4.2	3.85	2.03
SO_3	12.73	12.03	6.82	4.64	N/A	6.27	14.4
P_2O_5	1.62	N/A	N/A	N/A	N/A	1.72	1.56
TiO_2	0.92	N/A	N/A	N/A	1.6	3.12	1.56

Table 17.2. Major chemical compositions of MSWI bottom ash (wt.%)

Reference / Oxide	(Wei et al. 2011)	(Li et al. 2012)	(Andreola et al. 2008)	(Pan et al. 2008)	(Dabo et al. 2009)	(Filipponi et al. 2003)	(Rendek et al. 2006)
SiO_2	31.93	59.59	46.7	13.44	40.5	47.76	49.3
Al_2O_3	16.65	18.61	6.86	1.26	2.8	10.55	7.5
CaO	33.4	7.58	26.3	50.39	31.5	16.45	16.3
Fe_2O_3	5.97	5.5	4.69	8.84	10	8.61	7.6
MgO	3.33	1.32	2.22	2.26	1.9	3.67	2.6
Na_2O	2.53	1.32	4.62	12.66	3.4	3.51	6
K_2O	0.85	2.29	0.888	1.78	0.8	1.41	1.1
SO_3	0.4	0.65	2.18	1.79	0.86	N/A	N/A
P_2O_5	0.02	N/A	0.855	3.19	N/A	1.29	1.2
TiO_2	1.45	N/A	0.77	2.36	N/A	0.79	0.6

majority of components in fly ash are non-hazardous elements (Tables 17.1 and 17.2). Although the quantitative compositions of oxides may vary, high alkali metal oxides have been commonly observed. This is mainly due to the use of calcium hydroxide or sodium carbonate for acid gas removal. According to Tables 17.3 and 17.4, certain hazardous metals, e.g. Pb, Cr, Ni, Cu, Zn, Cd, and Hg, are more frequently found in MSWI ashes, with Zn and Pb usually present with the highest levels. Hazardous metal release is the major concern of the leachability problem of ashes and may strongly affect the environment, without appropriate treatment and confinement. The vaporization of metals during incineration and the mechanism

Table 17.3. Hazardous metals in MSWI fly ash (mg/kg)

Reference / *Element*	*(Wu et al. 2011)*	*(Chang et al. 2009)*	*(Hjelmar 1996)*	*(AWG 1997)*	*(NCHRP 2000)*
Ag	N/A	N/A	31–95	N/A	N/A–700
As	113.2	93	49–320	37–320	15–751
Ba	N/A	4,300	920–1,800	330–3,100	88–9,001
Cd	68.3	470	230–430	50–450	5–2,211
Co	N/A	N/A	29–69	N/A	2.3–1,671
Cr	451	863	140–530	140–1,100	21–1,901
Cu	1,122	1,300	860–1,400	600–3,200	187–2,381
Hg	5.58	N/A	0.8–7	0.7–30	0.9–73
Mn	869.2	1,600	N/A	800–1,900	171–8,500
Mo	24.08	N/A	15–49	15–150	N/A
Ni	70.5	124	95–240	60–260	10–1,970
Pb	1,607	10,900	7400–19,000	5,300–26,000	200–2,600
Sb	N/A	412	N/A	260–1,100	N/A
Se	81.94	41	6.1–31	N/A	0.48–16
Sn	N/A	N/A	1,400–1,900	N/A	N/A
Sr	N/A	433	<80–250	N/A	N/A
V	N/A	37	32–150	29–150	N/A
Zn	3269	25,800	19,000–41,000	9000–70,000	2,800–152,000

of metal adsorption on the fly ash surface are usually the main causes of observing higher hazardous metal levels in fly ash than in bottom ash. Volatile heavy metals, such as Cd, are concentrated in fly ash more or less independently of temperature, air/fuel, and chlorine/metal ratio (Sørum et al. 2003).

Gaseous heavy metal compounds condense in the boiler and on ash particles (Obernberger 1998); the lower the temperature the higher the condensed amount (Sørum et al. 2004). Semi-volatile elements, such as Cr, Cu, and Ni, remain mostly in bottom ash, although a small portion is also likely to be carried over to fly ash due to entrainment and evaporation (Zhang et al. 2008). MWSI fly ash has a much higher chloride content than MSWI bottom ash, and this could be attributable to the lime scrubber in the air pollution control system, which removes acid gas such as HCl from the emission stream. The loss on ignition of ashes is generally around 4–13% for fly ash and 3–5% for bottom ash, respectively.

17.2.2 Sewage Sludge Incineration

Major chemical compositions of sewage sludge incineration ash reported in the literature are summarized in Table 17.6. The table shows that besides

Table 17.4. Hazardous metals in MSWI bottom ash (mg/kg)

Reference / *Element*	*(Bruder-Hubscher et al. 2002)*	*(Rocca et al. 2012)*	*(Hjelmar 1996)*	*(IAWG 1997)*	*(NCHRP 2000)*
Ag	N/A	N/A	4.1–14	N/A	2–38
As	N/A	4–5	19–80	0.1–190	1.3–45
Ba	N/A	N/A	900–2,700	400–3,000	47–2,000
Cd	N/A	<0.02	1.4–40	0.3–70	0.3–61
Co	N/A	N/A	<10–40	N/A	22–706
Cr	350–420	310–340	230–600	23–3,200	13–1,400
Cu	1,290–1,610	3,430–4,270	900–4,800	190–8,200	80–10,700
Hg	N/A	N/A	<0.01–3	0.02–8	0.003–2
Mn	780–840	490–570	N/A	80–2,400	50–3,100
Mo	N/A	24–30	2.5–40	2–280	N/A
Ni	N/A	125–160	60–190	7–4,200	9–430
Pb	1,700–2,100	875–1,100	1,300–5,400	100–13,700	98–6,500
Sb	N/A	130–140	N/A	10–430	N/A
Se	N/A	N/A	0.6–8	N/A	ND–3.4
Sn	N/A	N/A	<100–1,300	N/A	N/A
Sr	N/A	N/A	170–350	N/A	N/A
V	N/A	1,6–2.8	36–90	20–120	N/A
W	N/A	N/A	<20–50	N/A	N/A
Zn	2,600–3,000	4,090–4,330	1,800–6,200	610–7,800	200–12,400

Table 17.5. Chloride content and Loss on Ignition (LOI) of MSWI fly ash and bottom ash (g/kg)

	Fly ash	*Bottom ash*
Cl	5,749–215,000	800–201,000
LOI	4.3–13.36	3.24–4.59

Sources: IAWG (1997); Lam et al. (2010); Sabbas et al. (2003).

oxygen the five major elements present in sewage sludge incineration ash were Si, Fe, Al, Ca, and P. The amount of P_2O_5 and CaO is comparable with common mineral mixtures, such as clay or the earth materials. The orange/red color of ash was probably due to the elevated levels of Fe in ash, and the ash generally showed elevated levels of Ca due to the different ash handling methods, such as partial blending with spent lime/coke from the flue gas treatment system. Other elements present at lower but noticeable levels were Mg, K, Na, Ti and S, and the trace element analysis (Table 17.7) confirms the existence of heavy

Table 17.6. Major chemical compositions of the incineration ashes of sewage sludge (wt.%)

Reference *Oxide*	*(Lin et al. 2006)*	*(Wang et al. 2012)*	*(Wang et al. 2005)*	*(Merino et al. 2007)*	*(Wang et al. 2005)*	*(Donatello et al. 2010)*	*(Garcés et al. 2008)*
SiO_2	46.27	26.71	43.6	25.4	64.95	32.2	19.2
Al_2O_3	14.12	31.74	16.6	7.64	15.21	13.5	8.9
CaO	4.80	13.08	5.61	21.05	1.93	11.9	30.6
Fe_2O_3	7.46	6.8	10.4	20	6.95	16.6	10
MgO	2.01	1.08	1.4	1.63	1.45	3.9	2.7
Na_2O	N/A	0.47	0.82	0.48	0.79	2.1	0.8
K_2O	1.46	0.69	2.34	0.78	2.52	1.8	1.4
SO_3	N/A	2.27	0.24	N/A	N/A	1.5	11.1
P_2O_5	17.75	16.71	12.1	14.2	1.51	15.1	12.3
TiO_2	0.09	0.45	N/A	0.29	N/A	1.4	1

Table 17.7. Trace metals in MSWI bottom ash (mg/kg)

Reference *Element*	*(Adam et al. 2009)*						*(Cyr et al. 2007)*
As	13.1	17.8	12.1	40	16.3	4.25	23
Ba	N/A	N/A	N/A	N/A	N/A	N/A	1430
Cd	3.73	4.71	3.8	3.6	3.03	2.23	14
Co	N/A	N/A	N/A	N/A	N/A	N/A	669
Cr	70	126	115	130	88.4	91.9	2636
Cu	947	1166	1230	1000	1267	470	2483
Hg	0.23	<0.1	<0.1	<0.1	<0.1	<0.1	N/A
Mo	11.6	22	22.3	20	79.5	4.92	N/A
Ni	45.5	60	80.2	98	39.5	42.7	621
Pb	118	201	184	264	134	113	720
Sb	N/A	N/A	N/A	N/A	N/A	N/A	73
Sn	55.6	46.6	36.1	60	55	50.4	283
Sr	N/A	N/A	N/A	N/A	N/A	N/A	623
V	N/A	N/A	N/A	N/A	N/A	N/A	63
Zn	1763	2181	1839	2170	2151	1577	N/A

metals such as zinc, chromium, and copper in the ash. Zn was found to be the richest hazardous metal in sewage sludge incineration ash, and its levels may reach up to 2,181 mg/kg. Therefore, hazardous metals in sewage sludge ash must be considered in the further planning of beneficial uses, such as for fertilizers.

17.2.3 Biomass Thermal Processing

The biomass ash composition is affected by boiler tube corrosion, biomass type, and boiler ash deposit formation and agglomeration. Table 17.8 shows the chemical composition of some typical biomass ashes reported in the literature. The ash mainly contains K, Na, Ca, Mg, Si, P, Al, Fe, and S, particularly for alkali metals (Na and K) and alkali earth metals (Mg and Ca). Biomass ash levels are usually higher in Ca, K, and P but to a lesser extent in Na. All of these are necessary plant nutrients and soil improvement agents (Steenari et al. 1999). From Table 17.8, it can be seen that rice straw, wheat straw and sunflower shells generally contain larger amounts of potassium. There is a considerable amount of SiO_2 in rice husk, rice straw, wheat straw, and bagasse, while wood chip and olive residue are rich in CaO. The Ca and Mg in the biomass may further react with the Si content and result in alkaline-earth metal silicates after the thermal processing (Xiao et al. 2011).

17.3 OVERVIEW OF CURRENT ASH MANAGEMENT

Generally, the reduction of municipal waste solid through incineration can reach 80% by weight and 90% by volume. However, currently the ash derived from incinerating municipal solid waste cannot be sustainably managed and may sometimes even result in secondary pollution. Ash management can usually be separated into three aspects: before incineration, during incineration and after incineration.

17.3.1 Waste Sorting before Incineration

Waste sorting not only recycles useful materials but also reduces the noncombustible materials to reduce the amount of ash. It can decrease the emission of pollutants in flue gas during the process of incineration and also the need to treat toxic compounds in ash.

17.3.2 Operation in Incineration Process

When feeding MSW into the furnace for incineration, the combustion efficiency is related to temperature, turbulence, and time. Therefore, the combustion must follow three principles: the temperature must be high enough (≥850°C) to ensure effective drying and combustion; the mixture of air and inflammable gases must be optimized; and the resident time for the burning must be sufficient (>2 s). It is generally necessary to perform and optimize the processing parameters first in the laboratory, then in the pilot scale, and finally during full-scale operation. The optimal operation parameters for incineration can reduce the amount of ash and prevent or minimize the formation of toxic compounds generated through incineration.

17.3.3 Recovery and Reuse of Ash after Incineration

After waste incineration, the ashes are usually disposed of in landfill or beneficially reused as material resources. Reusing the ash materials is clearly a preferable

Table 17.8. Elemental analyses of ashes derived from different waste biomasses

Reference	*Wood chip*	*Rice husk*	*Rice straw*	*Wheat straw*	*Olive residue*	*Sunflower shell*	*Bagasse*
SiO_2	9.35–31.7	73–98.61	44.7–75.38	27.96–78	10.7–45.05	1.1–29.3	50.36–61.69
Al_2O_3	3.12–14.62	0.064–1	0.07–1.94	0.77–2	2.33–6.98	0.5–2.9	11.56–15.3
CaO	19.59–41.16	0.21–1.96	1.6–10.12	2.7–8.1	10.21–32.9	15.8–16	1.9–8.81
Fe_2O_3	1.14–8.89	0.05–2	0.1–0.98	0.39–2	1.7–6.24	0.9–2.1	9.78–18.78
MgO	4–5.7	0.1–0.96	1.64–3.5	1.73–4	3–6.73	6.1–13.1	2.2–3.24
Na_2O	0.63–1.95	0.01–0.4	0.14–2.8	0.22–5.22	0.93–7.24	0.2–1.5	0.9–1.06
K_2O	2.96–10.72	0.16–6.2	15.3–38.92	7–25	9.18–27.23	35.6–45.1	0.49–19.34
SO_3	1.03–3.13	0.01–0.4	0.86–4.95	3.8–3.83	1.57–4.97	1.3–11.7	0.04
P_2O_5	4.02–13.23	0.06–1.5	0.61–1.93	1.05–4	2.36–14.7	1.3–11.7	4.05

Sources: Affandi et al. 2009; Foo and Hameed 2009; Tan and Lagerkvist 2011; Tantawy et al. 2012; Vamvuka and Kakaras 2011 management strategy, but a number of associated problems may also occur, as follows: high organic contents and leachability, high soluble salt levels (chlorides, sulphates, and possibly fluorides), and enriched hazardous metals and leachability.

17.4 ASH TREATMENT TECHNOLOGIES

The treatments of ash are intended to reduce their environmental impact and/or to facilitate their reuse potential and can be categorized into three types (IAWG 1997): (i) separation; (ii) solidification/stabilization (S/S); and (iii) thermal methods. It is common to begin the treatment with separation techniques, and then with stabilization/solidification methods or thermal treatment (Mangialardi 2003). Current processing options within these three types of treatment strategies are organized in Figure 17.2.

17.4.1 Separation Methods

Separation is commonly used for the pre-treatment during resource extraction in the mining industry. The purpose of separation for ash materials is to improve the ash quality with the aim of enhancing the utilization potential. Common separation methods include ash washing, leaching procedures, and electrochemical processes.

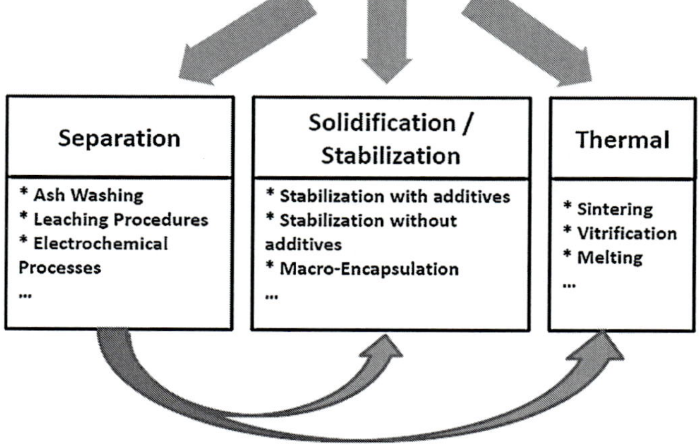

Figure 17.2. Main processing options for ash treatment

17.4.1.1 Ash Washing

The process of ash washing is to reduce the amount of heavy metals, chloride, alkali, and soluble salts in the ash by using liquid solution such as acid and water, to dissolve them from the ash phase. Alkali and heavy metals are often seen in bottom ash, and high chloride and other salt concentrations are found in fly ash. High chlorine contents, such as 11,000 and 156,000 mg/kg, have been found in MSWI fly ash and bottom ash, respectively (Toshihisa et al. 1998). Some beneficial uses of ash, such as for cement materials, require the removal of chlorides and thus the washing of ash to reduce the chloride content is needed. The chlorides in MSWI fly ash, which mainly include alkali chlorides such as NaCl, $CaCl_2$, and KCl, are readily water-extractable and can be eliminated by dissolution with water (Derie 1996). Boghetich et al. pointed out that the removal of chlorides from ash depends on temperature, granulometric properties, and solid/liquid ratio of extraction (Boghetich et al. 2005). The levels of chloride, sodium, and potassium can be significantly reduced as a consequence of washing, whereas the reduction of sulphates and most heavy metals, which are solubility controlled, may not be practically effective (IAWG 1997; Schneider et al. 1994). However, the addition of chemical additives or an appropriate control of solution pH may enhance the dissolution of heavy metals during the washing process (Ito et al. 2008; Quina 2005; Zhu et al. 2010). Although often not sufficient to achieve the objective of detoxifying the ash when used alone, the ash washing method is also often used as a pretreatment for the subsequent ash treatment method(s) to enhance the overall quality of treatment products.

17.4.1.2 Leaching Procedures

The leaching procedures are designed to extract heavy metals from ash and sometimes even further to recover them from leachants. This is essential for economically recovering metals from concentrated source solutions. The leaching of heavy metals is affected by the leachant pH, the type of extraction agent, the ratio of liquid to solid, and the leaching time (Hong et al. 2000; Mizutani et al. 1996; Shi and Kan 2009). The most commonly used leachants are acidic aqueous solution, since many metal compounds display high solubility at low pH (Brunori et al. 1999; Liu et al. 2005; van der Sloot et al. 2001). Besides the control of leachant pH, another main strategy to improve the metal leachability is the addition of chemical additives. Bio-leaching has been suggested for the removal of metals from fly ashes (Bosshard et al. 1996; Ishigaki et al. 2005; Wang et al. 2009) by using a bacteria leaching mechanism, but unfortunately the L/S (liquid to solid) ratios often need to be very high for the possible survival of microbes. Several leaching additives, such as EDTA, ammonium nitrate, ammonium chloride and a number of organic acids, were used for enhancing the metal leaching from MSWI fly ashes. Current results show that EDTA can mobilize Cu and Pb well, and NH_4NO_3 is especially effective for leaching Cu (Karlfeldt Fedje et al. 2010). In addition, hydrothermal treatments with the modification of chemical compositions of ash-containing materials may promote the reaction between acids and toxic metals (e.g. Pb, Cd, Cr) to further enhance the extractability of metals and the recycling benefit of MSWI ash (Zhang and Itoh 2006).

17.4.1.3 Electrochemical Processes

Electrochemical reactions involve the reduction/oxidation reactions that occur at the interface of electrodes and electrolytes. The driving force of the redox reactions comes from the potential difference between the cathode and anode. During electrolysis, chemical oxidations occur at the anode, while chemical reductions take place at the cathode. Therefore, oxygen and hydrogen are generated at the anode and cathode, respectively, and metals are deposited on the cathode as well. The operational parameters affecting the electrolytic recovery of metals are current density, pH, temperature, mixing conditions, and electrode distance (Quina et al. 2008). Yang and Tsai (1998) determined that current density was important to electrolytic recovery of Cu and Pb, but affected the metal recovery in different ways. The extraction of Pb usually increases with higher current density, but the recovery of Cu requires a lower level of current density.

17.4.2 Solidification/Stabilization

The solidification/stabilization process is currently the most widespread treatment method applied for MSWI ashes prior to their landfilling. The method provides physical and/or chemical bonding for hazardous constituents by using a binder or additive, and, in many cases, initiates chemical changes within the constituents themselves (Wiles 1996). The main purpose of the solidification/stabilization treatment is to reduce the leachability of heavy metals present in ashes and to allow their safer disposal in a sanitary landfill. Solidification refers to a physical process that converts a slurry mixture of ashes with chemical additives, such as Portland cement and kiln dust, into solids and thus restrains pollutant movement and leachability. Stabilization generally refers to a chemical process with actual bonding of pollutants in the ashes to minimize the toxicity and solubility of the targeted pollutants. The best S/S technique uses a combination of physical solidification and chemical (stabilization) reactions that first initiates chemical stabilization and then subsequent physical solidification. A number of authors have applied S/S methods to incineration residues and the most frequent methods can be categorized into solidification with binders (Mangialardi et al. 1999; Polettini et al. 2001; Ubbrìaco and Calabrese 1998), chemical fixation and solidification with binder (Kawakami et al. 1996; Youcai et al. 2002), chemical stabilization with chemicals (Mizutani et al. 2000; Sivula et al. 2008; Youcai et al. 2002), and other miscellaneous methods (He et al. 2006; Jianguo et al. 2009; Massardier et al. 1997). However, from the extensive studies it has been concluded that the drawback of the S/S process is that it is not effective for immobilizing soluble salts and preventing the long-term leachability of the materials. Derie suggested a preferred S/S technique should be with four treatment stages: (1) eliminating the alkali chlorides by dissolution, (2) adding phosphoric acid, (3) calcination, and (4) solidification with clinker or cement (Derie 1996). This design can destroy hazardous organics, reduce heavy metal reactivity, and solidify the toxic waste. However, it is clear that the result of S/S will produce a larger volume of waste (almost double) requiring landfill, due to the addition of the hydraulic binder. Therefore, the S/S is usually more suitable for treating those ashes with high levels of pollutants (Alba et al. 2001; Quina et al. 2008).

17.4.3 Thermal Methods

The thermal treatment of ashes is used to reduce leaching from the materials and also to reduce the volume of waste. The products can be more environmentally stable and reused for raw materials in beneficial applications. Dioxins, furans and other toxic organic compounds in ashes will usually be further destroyed in a high temperature environment (Sakai and Hiraoka 2000). The types of thermal treatment can usually be classified into: sintering, vitrification and melting (IAWG 1997).

17.4.3.1 Sintering

Sintering involves heating the residues to achieve the bonding of particles and the reconfiguration of chemical phases. Typical sintering temperatures are around 900 °C, and a denser product with a higher strength and less porosity is obtained. A certain amount of sintering will also take place in the incineration furnace when MSW is incinerated, particularly in a rotary kiln.

17.4.3.2 Vitrification

Vitrification is a process of mixing ashes with glass precursor materials (e.g. quartz sand) at high temperatures (1000–1500°C). The material will then produce chemical bonding and cool to form a single solid phase, amorphous, glassy product. This method is not widely adopted for MSWI residue due to the high cost of operation.

17.4.3.3 Melting

Melting (or fusion) is similar to vitrification, but does not require the addition of glass precursors. The product usually results in a multiple-phase composition, although the operation temperature may be close to those used in vitrification. Due to the resulting phase separation, it is possible to use further specific metal phases for recycling purposes. Eighmy and Kosson (Eighmy and Kosson 1996; IAWG 1997) suggested melting as an efficient method for stabilizing hazardous ashes, particularly for those containing Cd, Cu, Pb and Zn metals.

17.4.4 Combined Methods

Combining treatment methods may be a preferred strategy, due to the different advantages offered by different treatment methods, for treating the ashes produced from MSW, sewage sludge and biomass incinerations. For example, Derie (1996) developed a stabilization technique using a four-stage treatment:

(1) elimination of the alkali chlorides by dissolution, (2) addition of a moderate quantity of phosphoric acid, (3) calcination, and (4) solidification by Portland clinker or cement. The main advantages of this process include the decomposition of PCDD/PCDF, the significant decrease of heavy metal reactivity, the satisfactory mechanical properties of final products, and the limited increase of waste (weight increase less than 25%). Katsuura et al. (1996) developed a fly ash treatment method, the "Acid Extraction-Sulfide Stabilization (AES) Process", in which the fly ash is acid-extracted and stabilized by NaHS to change the residual heavy metal into insoluble metal sulfides. Mangialardi (Mangialardi 2003) reported a method combining water washing and S/S processes for treating fly ashes. Water washing can first remove large amounts of chloride and sulfate salts from fly ash, and promote the formation of heavy metals into hydrated phases. The treated fly ash was then incorporated into cement for further disposal. Overall, the removal of soluble salts as the first treatment step is often considered to be feasible and highly beneficial. The adoption of the S/S method as the second step can generally be an effective strategy to fix heavy metals in ashes.

17.5 BENEFICIAL USES OF ASHES

The value of ash is generally increased after ash treatments, and, therefore, the planning of beneficial uses for ash is favored over disposal in landfill. The advantages of utilizing the ash materials include the reduction of the disposal cost for ash in landfill, together with the saving of raw materials used for beneficial applications. Three main factors are often considered during the ash application: suitability for processing, technical performance and environmental impact (Ferreira et al. 2003). In the following, we review nine potential ash applications reported in the literature, which are grouped into four main categories: construction materials (cement and concrete, ceramics, road sub-base, lightweight aggregate), adsorbents for wastewater treatment (hazardous metals, dyes, and other organic compounds), synthesis of zeolites, and agricultural applications.

17.5.1 Construction Materials

17.5.1.1 Cement and Concrete

The main compositions of MSWI ash (CaO, SiO_2, Al_2O_3 and Fe_2O_3) are similar to the raw materials for cement production. Since cement is the most cost- and energy-intensive component of concrete, the unit price of concrete can be reduced if ash partially replaces the cement. Mangialardi et al. (1999) reported that ash could be useful for cement production. Wu et al. (2011) showed that MSWI fly ash was successfully used as a major cement raw material in sulphoaluminate cement (SAC) clinker, and they also pointed out that the expansive properties of SAC strongly depended on the gypsum content. The water demand for the consistency, setting time, strength, and hydration characteristics of MSWI bottom ash–cement matrix was investigated (Li et al. 2012). Experimental results show that the incorporation of MSWI bottom ash into blended cements should be kept under 30%, but the reactivity is lower than Portland cement and its addition to cement may lead to the retardation of cement hydration. Sewage sludge ash has only a limited content of SiO_2 and Al_2O_3, which are the two oxides responsible for the pozzolanic activity in cement-based materials (Cyr et al. 2007). Usually, the biomass ashes are characterized by higher contents of CaO, Al_2O_3, and SiO_2 (Rajamma et al. 2009; Wang and Baxter 2007). The effective role of biomass fly ash in the mitigation of the alkali-silica reaction in concretes was reported by Esteves et al. (2012).

Furthermore, because MSWI bottom ash, fly ash, and biomass ash contain lime instead of calcium carbonate, they can be beneficial in the reduction of CO_2 emission during the process of cement production. The current technical challenge of this application is the product quality, due to the high chloride content, and the associated corrosion issue in the cement kilns (Ampadu and Torii 2002). In addition, the further enriched heavy metals in the ash will be an environmental concern and require further investment in the treatment cost.

17.5.1.2 Ceramics

It is necessary to find another material to replace clay because of a shortage of clay for producing ceramic bricks and tiles across the world. Large quantities of SiO_2, Al_2O_3, and CaO in the ash could be recycled as raw materials for the production of glasses, glass-ceramics, and ceramics at high temperature (>1000 C). Fly ash, featuring a high content of SiO_2, Al_2O_3, and CaO, can replace part of the clay for a sustainable raw material source of ceramic products. Using fly ash in making ceramic tiles for sound product performance on the one hand and stabilization of heavy metals and decomposition of dioxins on the other was reported by Haiying et al. (2007). By adding 20% MSWI fly ash, the ceramic tile showed high compressive strength, low water absorption and low toxic heavy metal leachability with 960°C sintering. The recycling of untreated and vitrified bottom ashes into traditional ceramic tiles was investigated by Rambaldi et al. (2010). By substituting less than 10 wt.% of feldspar with vitrified bottom ash in a porcelain stoneware mix, the incipient crystallization was facilitated and the sintering process could be achieved with a 20°C decrease of maximum firing temperature. The overall result suggests that the vitrified bottom ash is more suitable as a secondary raw material in silicate-based ceramics, in particular for porcelain stoneware tiles. Siqueira et al. (2009) prepared ceramic matrix composites (CMCs) from a polysiloxane network filled with rice husk ash as a reactive filler. Besides the economic benefit of using this waste material, the CMCs obtained by pyrolysis were observed to have a low porosity and high flexural strength.

Glass-ceramics are polycrystalline materials of fine microstructures that are usually produced by controlled crystallization (devitrification) during heating (Rawlings et al. 2006). A large volume of research has been conducted on the preparation of glass-ceramics from incineration residues, such as fly ashes (Cheng 2004; Francis et al. 2002; Rincó et al. 1999; Romero et al. 2001), bottom ashes

Figure 17.3. Glass–ceramic material produced by the bottom ash of municipal solid waste
Source: Monteiro et al. (2008); reproduced with permission from Elsevier

(Aloisi et al. 2006; Appendino et al. 2004; Schabbach et al. 2012), and mixtures of bottom ash and fly ash (Barbieri et al. 2000; Qian et al. 2006). The glass-ceramic can be obtained by heating a mixture of 80% bottom ash and 20% fly ash at 950°C to obtain a durable material showing a strong chemical resistance and the potential of being a superior construction product (Vu et al. 2012). Monteiro et al. (2008) used the bottom ash produced by municipal solid waste incinerator to produce glass glass–ceramic material aiming for construction applications (Figure 17.3). Park et al. (2003) demonstrated different types of glass and glass-ceramics made by fly ash from sewage sludge incinerators. They pointed out that glass-ceramics containing large amounts of diopside (produced by 1050°C/2 h) generally showed better physical and chemical properties than their anorthite counterparts (produced by 1150°C/3 h), due to the interlocking microstructure of the diopside crystals.

Produced by vitrifying the waste ash and controlled crystallization during cooling process, glass-ceramics often have a much lower metal and pollutant release property than their original ash forms (Park and Heo 2002). These glass-ceramics are attractive building materials for beneficial uses in the construction industry and also have the potential to be tailored into different features for functional ceramic products.

17.5.1.3 Road Sub-Base

Road pavement is a stratified, multi-layered structure consisting of a surface layer (made of bitumen or asphalt), a middle layer (base course and sub-base) and the lowest layer (subgrade) (Ferreira et al. 2003). The base course lies immediately below the wearing surface of a pavement and plays an essential role in structural integrity. It consists of granular material (like stone fragments or slag) that can be stabilized with cementitious materials (cement, natural pozzolans, etc.). Between the base course and the subgrade there is the sub-base material, which supports the base course and is built with the same type of material as the base course but with an inferior quality.

The possible applications of MSWI bottom and fly ashes may include the substitution of the sand or earth material used for road sub-base. The soluble salts in MSWI fly ash can be first removed by ash washing (Colangelo et al. 2012), and the washed ash can be stabilized for sub-base materials. The specific environmental considerations arising from the use of MSWI bottom ash for pavement were studied by Forteza et al. (2004). The results concluded that the engineering properties of bottom ash are close to those of natural aggregates, and the metal contents in both the bottom ash and its leachates fulfill the environmental regulations for such beneficial use. To effectively reduce the environmental impact on the underlying soil and groundwater due to the adoption of ash materials for sub-base, it is suggested that the ash wash pre-treatment is required to first decrease the concentration of soluble pollutants (Aberg et al. 2006). However, attention should be paid to the possibility that the materials and pretreatments qualified for environmental regulations of road construction purposes may not meet the leaching standards for other types of construction materials.

17.5.1.4 Lightweight Aggregate

Lightweight aggregate (LWA) is a granular material less dense than gravel in concrete. It also has a lower compressive strength, but is an essential component in a variety of building materials, such as prefabricated structural units, lightweight thermo-acoustic insulation concrete, and lightweight structural concrete. It is especially used more in high-rise buildings, track ballast, road coating along with bituminous materials, and other geotechnical applications (González-Corrochano et al. 2009). Production of LWAs from mineral and organic wastes is considered to be a very satisfactory economic and environmental alternative. A positive effect (Cicek and Tanrıverdi 2007) on the compressive strength was observed from the addition of fly ash, sand and hydrated lime into bricks. On average, the weight of fly ash bricks is 1/3 less than that of conventional clay-fired bricks (Reidelbach 1970), enabling a truck to carry more bricks per load and reducing shipping costs. Incineration bottom ash with less than 8 mm fraction was milled, pelletized and rapidly sintered at temperatures between 1000 and 1050°C (Cheeseman et al. 2005) for LWA material, and the product density, water absorption and crushing strength are all comparable to those of commercially available LWA products. Sewage sludge ash was used to produce the foamed lightweight material (SSAFLM), and Figure 17.4 indicates that the volume shrinkage rate of SSAFLM increased as the amount of ash increase under high temperatures (Chen et al. 2006). The preparation of LWA from rice husk ash (RHA) was investigated and the results showed the ground RHA–LWA gave better performances in expansion, solubility, and disintegration properties, compared to the as-received RHA–LWA (Chindaprasirt et al. 2009). The stability of such lightweight aggregate was observed by the incorporation of 2–7% boric acid by weight of RHA in the mixture and the aggregate density could achieve 0.20–0.40 g/cm^3.

Application of incineration ashes as lightweight aggregates is not expected to pose environmental problems, since the leaching of heavy metals from ash is generally not significant, as long as the lightweight concrete is used in sheltered

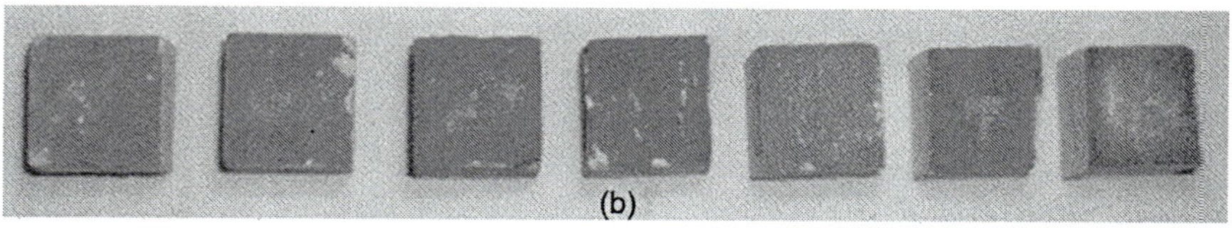

Figure 17.4. Foamed lightweight materials produced from sewage sludge ash and fired at different temperatures

Source: Chen et al. (2006); reproduced with permission from Elsevier

conditions. However, a subsequent problem may arise after the demolition of structures, where the rain water is expected to come into contact with fly ash and produce a leaching effect for metals (Ferreira et al. 2003).

17.5.2 Adsorbents for Wastewater Treatment

Adsorption is widely used for the removal of water pollutants due to its efficiency in separating a wide range of chemical compounds, together with the easy operation procedures. Although activated carbon has been one of the most efficient materials for absorbing hazardous compounds from wastewater, it is an expensive material for wastewater treatment purposes and requires a complicated activation process, when the aim is to recycle the sorbents. Therefore, the use of low-cost sorbents, such as the waste incineration ash, has been attracting substantial research and development activity.

17.5.2.1 Hazardous Metals

Heavy metals are among the most commonly seen environmental pollutants in wastewater and may pose a severe public health threat. Waste ash has been widely explored as a low-cost adsorbent for heavy metal uptake, and as early as 1975 Gangoli et al. (1975) investigated the potential for using fly ash for removing heavy metals from industrial wastewater. Both precipitation and/or adsorption are involved in the removal of hazardous metals from water. Precipitation of heavy metals may result from the presence of calcium hydroxide in changing the solution pH, while the adsorption mechanism is mainly due to the presence of silica and alumina in fly ash. The ash derived from incinerating sewage sludge can also be used for removing heavy metals from wastewater, due to its similar chemical composition to that of fly ash and blast-furnace slag. Pan et al. (2003) demonstrated that the primary mechanisms of removing copper by sewage sludge ash were electrostatic attraction, surface complexation, and cation exchange. Furthermore, the precipitation of copper hydroxide may provide additional removal efficiency, if with a higher level of sewage sludge ash dosage (30–40 g/l) and wastewater pH (>pH 6.2 in equilibrium). A feasibility study of using bagasse fly ash (BFA) as an adsorbent to remove Cd(II) and Ni(II) metal ions from aqueous solutions was conducted by Srivastava et al. (2006). The adsorption capacity of Ni(II) was found to be higher than that of Cd(II) in both binary- and single-metal solutions. The regeneration of spent sorbents is generally not recommended because of its low adsorbent cost.

17.5.2.2 Dyes

Dyes and pigments are discharged into wastewater from various industrial sources, and can cause skin irritation, allergic dermatitis, cancer, and mutations. The mixture of dried sewage sludge and sewage sludge incineration ash was studied for the removal of acid red 119 (AR119) dye as a more environmentally friendly and low-cost adsorbent (Moghaddam et al. 2012). The result showed the success of >95% dye removal for the 200 mg dye/L initial concentration at pH 6 after a contact time of 60 min. The adsorption behavior of Reactive Black 5 (RB) and Reactive Yellow 176 (RY) on coal fly ash (FA-CO), HCl-treated coal fly ash (TFA-HCl), and biomass fly ash (FA-BM) was studied (Pengthamkeerati et al. 2008), and the biomass fly ash was found to be the best adsorbent of dyes. The adsorption pattern of both RB and RY dyes on the biomass fly ash was pH-dependent with a maximum adsorption capacity at the final pH of 8.1–8.5.

17.5.2.3 Other Organic Compounds

Phenolic compounds are among the common organic pollutants in wastewater and may cause water to have an unpleasant taste and odor. Biological processes, activated carbon adsorption and solvent extraction are the most widely used methods for removing phenols and their derivatives from wastewaters. In recent years, fly ash has been showing great promise in removing phenolic compounds from wastewater. For example, bagasse fly ash was converted into a low-cost adsorbent and successfully used for the removal of phenolic compounds (Gupta et al. 1995; Gupta et al. 1998; Srivastava et al. 1995).

17.5.3 Zeolite Synthesis

The main compositions of zeolite are SiO_2 and Al_2O_3, which indicate the strong potential of using ash for fabricating zeolite materials due to their similar compositions. Bac et al. used the air-cooled ash and water-cooled ash obtained from a municipal solid waste incineration plant to synthesize zeolites (Bac et al. 2010). The samples were synthesized under different NaOH concentrations and reaction times. The tobermorite and pectolite-1A phases could be formed in most reaction conditions and the hydroxycancrinite phase can be derived by 200°C synthesis. Yang and Yang (1998) reported the synthesis of zeolite-like materials from MSW fly ash via a hydrothermal alkaline processing, similar to that for coal fly ash. Although the zeolite quality obtained was lower than that of commercial zeolites, the synthesized zeolite was completely safe to use according to the toxicity characteristic leaching procedure (TCLP) result. Lin et al. synthesized ZSM-type zeolites from biowaste gasification ash using alkaline hydrothermal and fusion treatment (Lin et al. 2003). The sorption capacities of synthesized zeolites showed their potential as low-cost sorbents for treating p-nitrophenolic wastewater, although the capabilities were not as good as the more expensive commercial

Figure 17.5. High-purity Na-X and Na-A zeolites obtained from using the fine particles of bagasse fly ash
Source: Purnomo et al. (2012); reproduced with permission from Elsevier

activated carbon. Fine particles of bagasse fly ash have been reused to prepare zeolite Na–X and Na–A with high purity (Purnomo et al. 2012). The derived octahedral-shape faujasitic zeolite (zeolite X) and cubic-shape zeolite A are showed in Figure 17.5. Besides their application in wastewater, the zeolites synthesized from the solid waste incineration ashes can also be used in gaseous pollutant treatment (Ferreira et al. 2003).

17.5.4 Agriculture Applications

Agriculture plays a major role in the global fluxes of carbon dioxide, nitrous oxide, and methane. Robertson et al. (2000) suggested the additional opportunities for lessening the potential global warming by changing current agronomic practices. Nitrogen, phosphorous and potassium are the three main elements in plant fertilizers. As MSWI fly ash contains acceptable amounts of two main nutrients, i.e. phosphorous and potassium, MSWI fly ash may be used for agricultural applications to substitute for commercial fertilizers (Ferreira et al. 2003). A study reported that plant growth in soils amended by fly ash was 1.5–2 times greater than in non-amended soils (Rosen et al. 1994). Besides its potential as a fertilizer, MSWI fly ash may also be a liming agent to reduce the acidity of soil, due to its high level of lime (calcium oxide) (Eighmy et al. 1995; Hansen et al. 2001; Rosen et al. 1994). Sewage sludge incineration ash contains a similar amount of phosphorus to commercial superphosphate. The plant weight with fertilizer made from sewage sludge incineration ash was proven to be similar to that with conventional phosphate fertilizer after 6 weeks of growth (Franz 2008). However, two main concerns need attention when using waste incineration ashes for agriculture applications, namely the potential damage to plants and animals due to the high levels of heavy metals and the potential salt stress of plants due to the salt contents in the ash. Another important aspect is that the change of soil pH may affect the mobility of metals in soils and may possibly contaminate ground and surface waters (Ferreira et al. 2003). The reutilization of biomass ashes in agriculture has been an important issue in creating nutrient cycles and saving on fertilizers. Biomass ashes can be applied in agriculture as a valuable fertilizer without concern about harmful heavy metals and other toxic substances (Schiemenz et al. 2011).

17.6 CONCLUSION

As municipal solid waste and sewage sludge often contain hazardous metals (Pb, Cd, Cr, Zn), their incineration residues (ashes) may be further enriched by metal pollutants. This implies that the direct application of such waste ashes for beneficial uses and landfill disposal are not environmentally safe and sustainable. In the cases of metals existing in soluble salt forms, a simple ash washing process is usually recommended for increasing the value of ash material and its reuse potential. This chapter highlights several applications for ashes derived from the solid waste processing and also demonstrates the relative advantages/disadvantages of different applications from the environmental/engineering point of view. It has been found that most of the environmental concerns regarding ash application are related to the leaching behavior of the final products. Therefore more mechanistic studies of material leaching processes and product performance designs will provide significant contributions in promoting feasible waste-to-resource strategies for solid waste processing ashes.

References

Aberg, A., Kumpiene, J., and Ecke, H. (2006). "Evaluation and prediction of emissions from a road built with bottom ash from municipal solid waste incineration (MSWI)." *Sci. Total Environ.*, **355**(1-3), 1–12.

Adam, C., Peplinski, B., Michaelis, M., Kley, G., and Simon, F. G. (2009). "Thermochemical treatment of sewage sludge ashes for phosphorus recovery." *Waste Manage.*, **29**(3), 1122–1128.

Affandi, S., Setyawan, H., Winardi, S., Purwanto, A., and Balgis, R. (2009). "A facile method for production of high-purity silica xerogels from bagasse ash." *Adv Powder Technol.*, **20**(5), 468–472.

Alba, N., Vázquez, E., Gassó, S., and Baldasano, J. M. (2001). "Stabilization/solidification of MSW incineration residues from facilities with different air pollution control systems. Durability of matrices versus carbonation." *Waste Manage.*, **21**(4), 313–323.

Aloisi, M., Karamanov, A., Taglieri, G., Ferrante, F., and Pelino, M. (2006). "Sintered glass ceramic composites from vitrified municipal solid waste bottom ashes." *J. Hazard. Mater.*, **137**(1), 138–143.

Ampadu, K. O., and Torii, K. (2002). "Chloride ingress and steel corrosion in cement mortars incorporating low-quality fly ashes." *Cem. Concr. Res.*, **32**(6), 893–901.

Andreola, F., et al. (2008). "Reuse of incinerator bottom and fly ashes to obtain glassy materials." *J. Hazard. Mater.*, **153**(3), 1270–1274.

Appendino, P., Ferraris, M., Matekovits, I., and Salvo, M. (2004). "Production of glass-ceramic bodies from the bottom ashes of municipal solid waste incinerators." *J. Eur. Ceram. Soc.*, **24**(5), 803–810.

Bac, B. H., Song, Y., Moon, Y., Kim, M. H., and Kang, I. M. (2010). "Effective utilization of incinerated municipal solid waste incineration ash: Zeolitic material synthesis and silica extraction." *Waste Manage. Res.*, **28**(8), 714–722.

Barbieri, L., Corradi Bonamartini, A., and Lancellotti, I. (2000). "Alkaline and alkaline-earth silicate glasses and glass-ceramics from municipal and industrial wastes." *J. Eur. Ceram. Soc.*, **20**(14-15), 2477–2483.

Boghetich, G., Liberti, L., Notarnicola, M., Palma, M., and Petruzzelli, D. (2005). "Chloride extraction for quality improvement of municipal solid waste incinerator ash for the concrete industry." *Waste Manage. Res.*, **23**(1), 57–61.

Bosshard, P. P., Bachofen, R., and Brandl, H. (1996). "Metal leaching of fly ash from municipal waste incineration by Aspergillus niger." *Environ. Sci. Technol.*, **30**(10), 3066–3070.

Bruder-Hubscher, V., Lagarde, F., Leroy, M. J. F., Coughanowr, C., and Enguehard, F. (2002). "Application of a sequential extraction procedure to study the release of elements from municipal solid waste incineration bottom ash." *Anal. Chim. Acta*, **451**(2), 285–295.

Brunori, C., Balzamo, S., and Morabito, R. (1999). "Comparison between different leaching/ extraction tests for the evaluation of metal release from fly ash." *Int. J. Environ. Anal. Chem.*, **75**(1-2), 19–31.

Chang, C. Y., Wang, C. F., Mui, D. T., Cheng, M. T., and Chiang, H. L. (2009). "Characteristics of elements in waste ashes from a solid waste incinerator in Taiwan." *J. Hazard. Mater.*, **165**(1-3), 766–773.

Cheeseman, C. R., Makinde, A., and Bethanis, S. (2005). "Properties of lightweight aggregate produced by rapid sintering of incinerator bottom ash." *Resour. Conserv. Recycl.*, **43**(2), 147–162.

Chen, C. H., Chiou, I. J., and Wang, K. S. (2006). "Sintering effect on cement bonded sewage sludge ash." *Cem. Concr. Res.*, **28**(1), 26–32.

Cheng, T. W. (2004). "Effect of additional materials on the properties of glass-ceramic produced from incinerator fly ashes." *Chemosphere*, **56**(2), 127–131.

Chindaprasirt, P., Jaturapitakkul, C., and Rattanasak, U. (2009). "Influence of fineness of rice husk ash and additives on the properties of lightweight aggregate." *Fuel*, **88**(1), 158–162.

Chou, S. Y., Lo, S. L., Hsieh, C. H., and Chen, C. L. (2009). "Sintering of MSWI fly ash by microwave energy." *J. Hazard. Mater.*, **163**(1), 357–362.

Cicek, T., and Tanrıverdi, M. (2007). "Lime based steam autoclaved fly ash bricks." *Constr. Build. Mater.*, **21**(6), 1295–1300.

Colangelo, F., Cioffi, R., Montagnaro, F., and Santoro, L. (2012). "Soluble salt removal from MSWI fly ash and its stabilization for safer disposal and recovery as road basement material." *Waste Manage.*, **32**(6), 1179–1185.

Cyr, M., Coutand, M., and Clastres, P. (2007). "Technological and environmental behavior of sewage sludge ash (SSA) in cement-based materials." *Cem. Concr. Res.*, **37**(8), 1278–1289.

Dabo, D., Badreddine, R., De Windt, L., and Drouadaine, I. (2009). "Ten-year chemical evolution of leachate and municipal solid waste incineration bottom ash used in a test road site." *J. Hazard. Mater.*, **172**(2-3), 904–913.

Derie, R. (1996). "A new way to stabilize fly ash from municipal incinerators." *Waste Manage.*, **16**(8), 711–716.

Donatello, S., Tyrer, M., and Cheeseman, C. R. (2010). "EU landfill waste acceptance criteria and EU hazardous waste directive compliance testing of incinerated sewage sludge ash." *Waste Manage.*, **30**(1), 63–71.

Eighmy, T. T., and Kosson, D. S. (1996). "U.S.A. National overview on waste management." *Waste Manage.*, **16**(5-6), 361–366.

Eighmy, T. T., et al. (1995). "Comprehensive approach toward understanding element speciation and leaching behavior in municipal solid waste incineration electrostatic precipitator ash." *Environ. Sci. Technol.*, **29**(3), 629–646.

Esteves, T. C., Rajamma, R., Soares, D., Silva, A. S., Ferreira, V. M., and Labrincha, J. A. (2012). "Use of biomass fly ash for mitigation of alkali-silica reaction of cement mortars." *Constr. Build. Mater.*, **26**(1), 687–693.

European Commission. (1991). "Council Directive of 21 May 1991 concerning urban waste water treatment (91/271/EEC)." ⟨http://eur-lex.europa.eu/LexUriServ/LexUriServ.do?uri=CONSLEG:1991L0271:20081211:EN:PDF⟩ (Nov. 12, 2003).

European Environment Agency. (2009). "Sewage sludge—A future waste problem?" ⟨http://www.eea.europa.eu/data-and-maps/indicators/generation-and-treatment-of-sewage-sludge⟩ (Dec. 1, 2010).

Ferreira, C., Ribeiro, A., and Ottosen, L. (2003). "Possible applications for municipal solid waste fly ash." *J. Hazard. Mater.*, **96**(2-3), 201–216.

Filipponi, P., Polettini, A., Pomi, R., and Sirini, P. (2003). "Physical and mechanical properties of cement-based products containing incineration bottom ash." *Waste Manage.*, **23**(2), 145–156.

Foo, K. Y., and Hameed, B. H. (2009). "Utilization of rice husk ash as novel adsorbent A judicious recycling of the colloidal agricultural waste." *Adv. Colloid Interface Sci.*, **152** (1-2), 39–47.

Forteza, R., Far, M., Seguí, C., and Cerdá, V. (2004). "Characterization of bottom ash in municipal solid waste incinerators for its use in road base." *Waste Manage.*, **24**(9), 899–909.

Francis, A. A., Rawlings, R. D., and Boccaccini, A. R. (2002). "Glass-ceramics from mixtures of coal ash and soda-lime glass by the petrurgic method." *J. Mater. Sci. Lett.*, **21**(12), 975–980.

Franz, M. (2008). "Phosphate fertilizer from sewage sludge ash (SSA)." *Waste Manage.*, **28**(10), 1809–1818.

Gamgli, N., Markey, D. C., and Thodos, G. (1975). "Removal of heavy metal ions from aqueous solutions with fly ash." *Proc., National Conf. on Complete Watereuse*, 2nd Ed., A.I. Ch.E, New York, 270–275.

Garcés, P., Pérez Carrión, M., García-Alcocel, E., Payá, J., Monzó, J., and Borrachero, M. V. (2008). "Mechanical and physical properties of cement blended with sewage sludge ash." *Waste Manage.*, **28**(12), 2495–2502.

Ginés, O., Chimenos, J. M., Vizcarro, A., Formosa, J., and Rosell, J. R. (2009). "Combined use of MSWI bottom ash and fly ash as aggregate in concrete formulation: Environmental and mechanical considerations." *J. Hazard. Mater.*, **169**(1-3), 643–650.

González-Corrochano, B., Alonso-Azcárate, J., and Rodas, M. (2009). "Characterization of lightweight aggregates manufactured from washing aggregate sludge and fly ash." *Resour. Conserv. Recycl.*, **53**(10), 571–581.

Gupta, V. K., Mohan, D., Srivastava, S. K., and Yadav, I. S. (1995). "Removal of 2, 4-dinitrophenol using bagasse fly ash- a sugar industry waste material." *Fresenius Environ. Bull.*, **4**(9), 550–557.

Gupta, V. K., Sharma, S., Yadav, I. S., and Mohan, D. (1998). "Utilization of bagasse fly ash generated in the sugar industry for the removal and recovery of phenol and p-nitrophenol from wastewater." *J. Chem. Technol. Biotechnol.*, **71**(2), 180–186.

Haiying, Z., Youcai, Z., and Jingyu, Q. (2007). "Study on use of MSWI fly ash in ceramic tile." *J. Hazard. Mater.*, **141**(1), 106–114.

Hansen, H. K., Pedersen, A. J., Ottosen, L. M., and Villumsen, A. (2001). "Speciation and mobility of cadmium in straw and wood combustion fly ash." *Chemosphere*, **45**(1), 123–128.

Helena Lopes, M., Abelha, P., Lapa, N., Oliveira, J. S., Cabrita, I., and Gulyurtlu, I. (2003). "The behaviour of ashes and heavy metals during the co-combustion of sewage sludges in a fluidised bed." *Waste Manage.*, **23**(9), 859–870.

He, P. J., Cao, Q. K., Shao, L. M., and Lee, D. J. (2006). "Aging of air pollution control residues from municipal solid waste incinerator: Role of water content on metal carbonation." *Sci. Total Environ.*, **359**(1-3), 26–37.

Hjelmar, O. (1996). "Disposal strategies for municipal solid waste incineration residues." *J. Hazard. Mater.*, **47**(1-3), 345–368.

Hong, K.-J., Tokunaga, S., and Kajiuchi, T. (2000). "Extraction of heavy metals from MSW incinerator fly ashes by chelating agents." *J. Hazard. Mater.*, **75**(1), 57–73.

Imyim, A., and Prapalimrungsi, E. (2010). "Humic acids removal from water by aminopropyl functionalized rice husk ash." *J. Hazard. Mater.*, **184**(1-3), 775–781.

International Ash Working Group (IAWG), Chandler, A. J., et al. (1997). "Municipal solid waste incinerator residues." *Studies in environmental science*, Vol. **67**, Elsevier, Amsterdam.

Ishigaki, T., Nakanishi, A., Tateda, M., Ike, M., and Fujita, M. (2005). "Bioleaching of metal from municipal waste incineration fly ash using a mixed culture of sulfur-oxidizing and iron-oxidizing bacteria." *Chemosphere*, **60**(8), 1087–1094.

Ito, R., Dodbiba, G., Fujita, T., and Ahn, J. W. (2008). "Removal of insoluble chloride from bottom ash for recycling." *Waste Manage.*, **28**(8), 1317–1323.

Jianguo, J., Maozhe, C., Yan, Z., and Xin, X. (2009). "Pb stabilization in fresh fly ash from municipal solid waste incinerator using accelerated carbonation technology." *J. Hazard. Mater.*, **161**(2-3), 1046–1051.

Karlfeldt Fedje, K., Ekberg, C., Skarnemark, G., and Steenari, B. M. (2010). "Removal of hazardous metals from MSW fly ash—An evaluation of ash leaching methods." *J. Hazard. Mater.*, **173**(1-3), 310–317.

Katsuura, H., Inoue, T., Hiraoka, M., and Sakai, S. (1996). "Full-scale plant study on fly ash treatment by the acid extraction process." *Waste Manage.*, **16**(5-6), 491–499.

Kawakami, I., Esaki, M., Tetsuyama, I., and Sumitomo, M. (1996). "Immobilization of fly ash from MSW incinerators and ash-melting furnaces." *Waste Manage.*, **16**(5-6), 483–489.

Lam, C. H. K., Ip, A. W. M., Barford, J. P., and McKay, G. (2010). "Use of incineration MSW ash: A review." *Sustainability*, **2**(7), 1943–1968.

Lin, K. S., Wang, H. P., Chang, N. B., Jou, C. J. G., and Hsiao, M. C. (2003). "Synthesis of ZSM-type zeolites from biowaste gasification ashes." *Energy Sources*, **25**(6), 565–576.

Liu, F., Liu, J., Yu, Q., Jin, Y., and Nie, Y. (2005). "Leaching characteristics of heavy metals in municipal solid waste incinerator fly ash." *J. Environ. Sci. Health., Part A*, **40**(10), 1975–1985.

Li, X. G., Lv, Y., Ma, B. G., Chen, Q. B., Yin, X. B., and Jian, S. W. (2012). "Utilization of municipal solid waste incineration bottom ash in blended cement." *J. Cleaner Prod.*, **32**, 96–100.

Mangialardi, T. (2003). "Disposal of MSWI fly ash through a combined washing-immobilisation process." *J. Hazard. Mater.*, **98**(1-3), 225–240.

Mangialardi, T., Paolini, A. E., Polettini, A., and Sirini, P. (1999). "Optimization of the solidification/stabilization process of MSW fly ash in cementitious matrices." *J. Hazard. Mater.*, **70**(1-2), 53–70.

Massardier, V., Moszkowicz, P., and Taha, M. (1997). "Fly ash stabilization-solidification using polymer-concrete double matrices." *Eur. Polym. J.*, **33**(7), 1081–1086.

Merino, I., Arévalo, L. F., and Romero, F. (2007). "Preparation and characterization of ceramic products by thermal treatment of sewage sludge ashes mixed with different additives." *Waste Manage.*, **27**(12), 1829–1844.

Mizutani, S., van der Sloot, H. A., and Sakai, S.-I. (2000). "Evaluation of treatment of gas cleaning residues from MSWI with chemical agents." *Waste Manage.*, **20**(2-3), 233–240.

Mizutani, S., Yoshida, T., Sakai, S.-I., and Takatsuki, H. (1996). "Release of metals from MSW I fly ash and availability in alkali condition." *Waste Manage.*, **16**(5-6), 537–544.

Moghaddam, S. S., Alavi Moghaddam, M. R., and Arami, M. (2012). "Response surface optimization of acid red 119 dye adsorption by mixtures of dried sewage sludge and sewage sludge ash." *Clean-Soil Air Water*, **40**(6), 652–660.

Monteiro, R. C. C., Figueiredo, C. F., Alendouro, M. S., Ferro, M. C., Davim, E. J. R., and Fernandes, M. H. V. (2008). "Characterization of MSWI bottom ashes towards utilization as glass raw material." *Waste Manage.*, **28**(7), 1119–1125.

NCHRP (National Cooperative Highway Research Program). (2000). "Waste and recycled materials use in the transportation industry." Transportation Research Board, National Research Council, Washington, DC.

Obernberger, I. (1998). *Ashes and particulate emissions from biomass combustion: Formation, characterisation, evaluation, treatment*, dbv-Verlag der Technischen Uninersitat Graz, Austria.

Pan, J. R., Huang, C., Kuo, J. J., and Lin, S. H. (2008). "Recycling MSWI bottom and fly ash as raw materials for portland cement." *Waste Manage.*, **28**(7), 1113–1118.

Pan, S. C., Lin, C. C., and Tseng, D. H. (2003). "Reusing sewage sludge ash as adsorbent for copper removal from wastewater." *Resour. Conserv. Recycl.*, **39**(1), 79–90.

Park, Y. J., and Heo, J. (2002). "Vitrification of fly ash from municipal solid waste incinerator." *J. Hazard. Mater.*, **91**(1-3), 83–93.

Park, Y. J., Moon, S. O., and Heo, J. (2003). "Crystalline phase control of glass ceramics obtained from sewage sludge fly ash." *Ceram. Int.*, **29**(2), 223–227.

Pengthamkeerati, P., Satapanajaru, T., and Singchan, O. (2008). "Sorption of reactive dye from aqueous solution on biomass fly ash." *J. Hazard. Mater.*, **153**(3), 1149–1156.

Petersen Raymer, A. K. (2006). "A comparison of avoided greenhouse gas emissions when using different kinds of wood energy." *Biomass Bioenergy*, **30**(7), 605–617.

Polettini, A., Pomi, R., Sirini, P., and Testa, F. (2001). "Properties of portland cement—Stabilised MSWI fly ashes." *J. Hazard. Mater.*, **88**(1), 123–138.

Purnomo, C. W., Salim, C., and Hinode, H. (2012). "Synthesis of pure Na-X and Na-A zeolite from bagasse fly ash." *Micropor. Mesopor. Mater.*, **162**, 6–13.

Qian, G. R., Shi, J., Cao, Y. L., Xu, Y. F., and Chui, P. C. (2008). "Properties of MSW fly ash-calcium sulfoaluminate cement matrix and stabilization/solidification on heavy metals." *J. Hazard. Mater.*, **152**(1), 196–203.

Qian, G., Song, Y., Zhang, C., Xia, Y., Zhang, H., and Chui, P. (2006). "Diopside-based glass-ceramics from MSW fly ash and bottom ash." *Waste Manage.*, **26**(12), 1462–1467

Quina, M. J. (2005). "Treatment processes and valorization of APC residues-municipal solid waste incineration." Ph.D. thesis, Univ. of Coimbra, Portugal.

Quina, M. J., Bordado, J. C., and Quinta-Ferreira, R. M. (2008). "Treatment and use of air pollution control residues from MSW incineration: An overview." *Waste Manage.*, **28** (11), 2097–2121.

Rajamma, R., Ball, R. J., Tarelho, L. A. C., Allen, G. C., Labrincha, J. A., and Ferreira, V. M. (2009). "Characterisation and use of biomass fly ash in cement-based materials." *J. Hazard. Mater.*, **172**(2-3), 1049–1060.

Rambaldi, E., Esposito, L., Andreola, F., Barbieri, L., Lancellotti, I., and Vassura, I. (2010). "The recycling of MSWI bottom ash in silicate based ceramic." *Ceram. Int.*, **36**(8), 2469–2476.

Rawlings, R., Wu, J., and Boccaccini, A. (2006). "Glass-ceramics: Their production from wastes—A review." *J. Mater. Sci.*, **41**(3), 733–761.

Reidelbach, J. A. (1970). *An industrial evaluation of fly ash bricks*, U.S. Dept. of the Interior, Washington.

Rendek, E., Ducom, G., and Germain, P. (2006). "Carbon dioxide sequestration in municipal solid waste incinerator (MSWI) bottom ash." *J. Hazard. Mater.*, **128**(1), 73–79.

Rincó, J. M., Romero, M., and Boccaccini, A. R. (1999). "Microstructural characterisation of a glass and a glass-ceramic obtained from municipal incinerator fly ash." *J. Mater. Sci.*, **34** (18), 4413–4423.

Robertson, G. P., Paul, E. A., and Harwood, R. R. (2000). "Greenhouse gases in intensive agriculture: Contributions of individual gases to the radiative forcing of the atmosphere." *Science*, **289**(5486), 1922–1925.

Rocca, S., van Zomeren, A., Costa, G., Dijkstra, J. J., Comans, R. N. J., and Lombardi, F. (2012). "Characterisation of major component leaching and buffering capacity of RDF incineration and gasification bottom ash in relation to reuse or disposal scenarios." *Waste Manage.*, **32**(4), 759–768.

Romero, M., Rincon, J. M., Rawlings, R. D., and Boccaccini, A. R. (2001). "Use of vitrified urban incinerator waste as raw material for production of sintered glass-ceramics." *Mater. Res. Bull.*, **36**(1), 383–395.

Rosen, C. J., Olson, D., and Bierman, P. M. (1994). "Swiss Chard and Alfalfa responses to soils amended with municipal solid waste incinerator ash: Growth and elemental composition." *J. Agric. Food Chem.*, **42**(6), 1361–1368.

Sabbas, T., et al. (2003). "Management of municipal solid waste incineration residues." *Waste Manage.*, **23**(1), 61–88.

Sakai, S.-I., and Hiraoka, M. (2000). "Municipal solid waste incinerator residue recycling by thermal processes." *Waste Manage.*, **20**(2-3), 249–258.

Schabbach, L. M., et al. (2012). "Post-treated incinerator bottom ash as alternative raw material for ceramic manufacturing." *J. Eur. Ceram. Soc.*, **32**(11), 2843–2852.

Schiemenz, K., Kern, J., Paulsen, H. M., Bachmann, S., and Eichler-Löbermann, B. (2011). "Phosphorus fertilizing effects of biomass ashes." *Recycling of biomass ashes*, H. Insam and B. A. Knapp, eds., Springer, Berlin, 17–31.

Schneider, J., Vehlow, J., and Vogg, H. (1994). "Improving the MSWI bottom ash quality by simple in-plant measures." *Stud. Environ. Sci.*, **60**, 605–620.

Shi, H. S., and Kan, L. L. (2009). "Leaching behavior of heavy metals from municipal solid wastes incineration (MSWI) fly ash used in concrete." *J. Hazard. Mater.*, **164**(2-3), 750–754.

Siqueira, E. J., Yoshida, I. V. P., Pardini, L. C., and Schiavon, M. A. (2009). "Preparation and characterization of ceramic composites derived from rice husk ash and polysiloxane." *Ceram. Int.*, **35**(1), 213–220.

Sivula, L., Väisänen, A., and Rintala, J. (2008). "Stabilisation of MSWI bottom ash with sulphide-rich anaerobic effluent." *Chemosphere*, **71**(1), 1–9.

Sørum, L., Frandsen, F. J., and Hustad, J. E. (2003). "On the fate of heavy metals in municipal solid waste combustion Part I: devolatilisation of heavy metals on the grate." *Fuel*, **82**(18), 2273–2283.

Sorum, L., Frandsen, F. J., and Hustad, J. E. (2004). "On the fate of heavy metals in municipal solid waste combustion. Part II. From furnace to filter." *Fuel*, **83**(11-12), 1703–1710.

Srivastava, S. K., Gupta, V. K., Johri, N., and Mohan, D. (1995). "Removal of 2, 4, 6-trinitrophenol using bagasse fly ash: A sugar industry waste material." *Indian J. Chem. Technol.*, **2**, 333–336.

Srivastava, V. C., Mall, I. D., and Mishra, I. M. (2006). "Equilibrium modelling of single and binary adsorption of cadmium and nickel onto bagasse fly ash." *Chem. Eng. J.*, **117**(1), 79–91.

Steenari, B. M., Karlsson, L. G., and Lindqvist, O. (1999). "Evaluation of the leaching characteristics of wood ash and the influence of ash agglomeration." *Biomass Bioenergy*, **16**(2), 119–136.

Tantawy, M. A., El-Roudi, A. M., and Salem, A. A. (2012). "Immobilization of Cr(VI) in bagasse ash blended cement pastes." *Constr. Build. Mater.*, **30**, 218–223.

Tan, Z., and Lagerkvist, A. (2011). "Phosphorus recovery from the biomass ash: A review." *Renewable Sustainable Energy Rev.*, **15**(8), 3588–3602.

Toshihisa, M., Kazumasa, S., and Hiroaki, T. (1998). "Utilisation of municipal solid waste incinerator residue for cement production." *J. Japan Waste Manage. Assoc.*, **51**, 557–561.

Ubbrìaco, P., and Calabrese, D. (1998). "Solidification and stabilization of cement paste containing fly ash from municipal solid waste." *Thermochim. Acta*, **321**(1-2), 143–150.

Vamvuka, D., and Kakaras, E. (2011). "Ash properties and environmental impact of various biomass and coal fuels and their blends." *Fuel Process. Technol.*, **92**(3), 570–581.

van der Sloot, H. A., Kosson, D. S., and Hjelmar, O. (2001). "Characteristics, treatment and utilization of residues from municipal waste incineration." *Waste Manage.*, **21**(8), 753–765.

Vu, D. H., Wang, K. S., Chen, J. H., Nam, B. X., and Bac, B. H. (2012). "Glass-ceramic from mixtures of bottom ash and fly ash." *Waste Manage.*, **32**(12), 2306–2314.

Wang, K. S., Chiou, I. J., Chen, C. H., and Wang, D. (2005). "Lightweight properties and pore structure of foamed material made from sewage sludge ash." *Constr. Build. Mater.*, **19**(8), 627–633.

Wang, K. S., Tseng, C. J., Chiou, I. J., and Shih, M. H. (2005). "The thermal conductivity mechanism of sewage sludge ash lightweight materials." *Cem. Concr. Res.*, **35**(4), 803–809.

Wang, L., Skjevrak, G., Hustad, J. E., and Grønli, M. G. (2012). "Sintering characteristics of sewage sludge ashes at elevated temperatures." *Fuel Process. Technol.*, **96**, 88–97.

Wang, Q., Yang, J., Wang, Q., and Wu, T. (2009). "Effects of water-washing pretreatment on bioleaching of heavy metals from municipal solid waste incinerator fly ash." *J. Hazard. Mater.*, **162**(2-3), 812–818.

Wang, Q., Yan, J. H., Chi, Y., Li, X. D., and Lu, S. Y. (2010). "Application of thermal plasma to vitrify fly ash from municipal solid waste incinerators." *Chemosphere*, **78**(5), 626–630.

Wang, S., and Baxter, L. (2007). "Comprehensive study of biomass fly ash in concrete: Strength, microscopy, kinetics and durability." *Fuel Process. Technol.*, **88**(11-12), 1165–1170.

Wei, Y., Shimaoka, T., Saffarzadeh, A., and Takahashi, F. (2011). "Mineralogical characterization of municipal solid waste incineration bottom ash with an emphasis on heavy metal-bearing phases." *J. Hazard. Mater.*, **187**(1-3), 534–543.

Wiles, C. C. (1996). "Municipal solid waste combustion ash: State-of-the-knowledge." *J. Hazard. Mater.*, **47**(1-3), 325–344.

Wu, K., Shi, H., and Guo, X. (2011). "Utilization of municipal solid waste incineration fly ash for sulfoaluminate cement clinker production." *Waste Manage.*, **31**(9-10), 2001–2008.

Xiao, R., Chen, X., Wang, F., and Yu, G. (2011). "The physicochemical properties of different biomass ash es at different ashing temperature." *Renew. Energy*, **36**(1), 244–249.

Yang, G. C. C., and Tsai, C. M. (1998). "A study on heavy metal extractability and subsequent recovery by electrolysis for a municipal incinerator fly ash." *J. Hazard. Mater.*, **58**(1-3), 103–120.

Yang, G. C. C., and Yang, T. Y. (1998). "Synthesis of zeolites from municipal incinerator fly ash." *J. Hazard. Mater.*, **62**(1), 75–89.

Youcai, Z., Lijie, S., and Guojian, L. (2002). "Chemical stabilization of MSW incinerator fly ashes." *J. Hazard. Mater.*, **95**(1-2), 47–63.

Zhang, F.-S., and Itoh, H. (2006). "Extraction of metals from municipal solid waste incinerator fly ash by hydrothermal process." *J. Hazard. Mater.*, **136**(3), 663–670.

Zhang, H., He, P. J., and Shao, L. M. (2008). "Fate of heavy metals during municipal solid waste incineration in Shanghai." *J. Hazard. Mater.*, **156**(1-3), 365–373.

Zhu, F., Takaoka, M., Oshita, K., Kitajima, Y., Inada, Y., Morisawa, S., and Tsuno, H. (2010). "Chlorides behavior in raw fly ash washing experiments." *J. Hazard. Mater.*, **178**(1-3), 547–552.

CHAPTER 18

Landfill Design and Operation

Jay N. Meegoda[1]
Hiroshan Hettiarachchi[2]
Patrick Hettiaratchi[3]

18.1 INTRODUCTION

Municipal solid waste landfills (MSWLFs) receive household waste. MSWLFs can also receive non-hazardous sludge, industrial solid waste, and construction and demolition debris. Modern landfills are well-engineered facilities that are located, designed, operated, and monitored to ensure compliance with federal regulations. Solid waste landfills must be designed to protect the environment from contaminants which may be present in the solid waste stream. The landfill siting plan prevents the siting of landfills in environmentally-sensitive areas while on-site environmental monitoring systems monitor for any sign of groundwater contamination and for landfill gas, and provides additional safeguards. In addition, many new landfills collect potentially harmful landfill gas emissions and convert the gas into energy.

This chapter provides a comprehensive but brief discussion on all aspects associated with landfill design, construction and operation. Siting, regulations and other important steps that need to happen before design stage are also presented in brief. Information on monitoring and post closure requirements is discussed at the end of the chapter. It should be noted that discussions provided on landfill design are qualitative as they do not include detailed designs. Readers are encouraged to refer to references included for detailed information on landfill designs.

18.2 TYPES OF LANDFILLS

Prior to 1950s landfills in the USA were merely dumpsites and there are still waste dumpsites in most developing nations. Contamination issues were not considered seriously due to a general belief that the leachate produced by waste is completely

[1]New Jersey Institute of Technology, Newark.
[2]United Nations Univ. (UNU-FLORES), Germany.
[3]Univ. of Calgary, Calgary, Alberta, Canada.

purified by soil and ground water. However, several studies conducted after mid 1950's proved otherwise (Vesilind et al. 2002). This led to the introduction of new and stricter regulations. Considering municipal solid waste (MSW) is less harmful compared to leachate produced by industrial wastes that included chemicals, waste in general was divided into two types: hazardous and nonhazardous (Bagchi 2004). New regulations eventually transformed nonhazardous waste dumpsites into "sanitary landfills." In the past any landfill that practiced use of a daily cover was considered to be a sanitary landfill. Today sanitary landfill refers to an engineered MSW disposal facility designed and operated to minimize environmental and health impacts. Sanitary landfilling continues to be and affordable and environmentally acceptable method of solid waste disposal (Reddy et al. 2009).

In a conventional sanitary landfill, waste is kept effectively dry by using a base liner and cover. Figure 18.1 shows a cross sectional view of a conventional sanitary landfill. Such design would prevent rain infiltration and the liquid resulting from biodegradation of waste, termed leachate, is collected at the bottom of the landfill and securely disposed. Containment and restriction of liquid infiltration retards degradation of MSW resulting in less complete and less predictable degradation that drags for years (Sharma and Reddy, 2004). Recent landfill mining activities show newsprint in a state of degradation such that news articles could be read in their entirety. Excavated waste was dated to 1949 by newsprint. Slow degradation of waste increases liability and also decreases return on investment of valuable landfill space.

In recent years, there has been a shift in philosophy of landfill design from the dry storage concept towards the bioreactor approach (Reddy et al. 2009). In the bioreactor approach, the moisture content of the MSW is increased by recirculation of leachate to enhance the biodegradation. This is a financially attractive

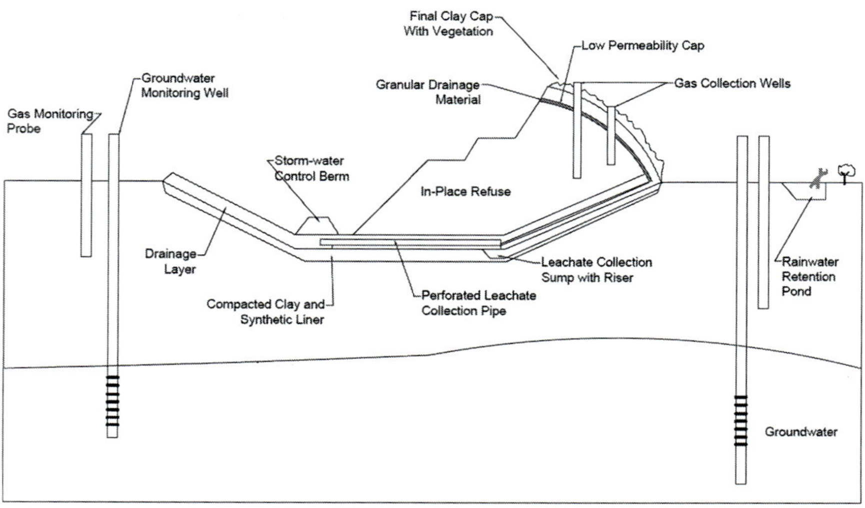

Figure 18.1. Cross section of a typical modern sanitary landfill

option as to securely dispose collected leachate would be expensive and by recirculating leachate one can eliminate the expensive treatment cost of leachate. In addition, due to high moisture content due to leachate recirculation, there is rapid degradation of waste. Hence bioreactor landfills offer a significant reduction in post-closure management time and operation cost (Reddy and Bogner 2003). SWANA (2001) defines a bioreactor landfill as:

> *"Any permitted Subtitle D landfill (see Section 20.3) or landfill cell, subject to new source performance standard/emissions guidelines, where liquid or air, in addition to leachate and landfill gas condensate, is injected in a controlled fashion into the waste mass in order to accelerate or enhance bio-stabilization of the waste."*

Bioreactor landfills can be categorized into three types: anaerobic, aerobic and hybrid. In anaerobic bioreactor landfills, biodegradation is accelerated by anaerobic microorganisms that do not use oxygen for cellular respiration. These microorganisms convert organic wastes into organic acids and ultimately into methane and carbon dioxide (Sharma and Reddy 2004). Aerobic bioreactor landfills utilize aerobic microorganisms that require oxygen for cellular respiration and produces carbon dioxide. Hybrid bioreactor landfills use a combination of above two approaches.

A "Sustainable Landfill" is a variation of the bioreactor landfill approach, which entails the operation of a waste cell in different modes to maximize resource and space recovery. One example is the Landfill Biocell pilot project run by the City of Calgary. During the first phase, the Biocell operates as an anaerobic bioreactor with leachate recirculation and gas extraction for power generation. The second phase is operated as an aerobic bioreactor and converts MSW to a compost-like product. The third phase of operation is mining to recover resources and space, allowing the empty cell to again receive waste and for the cycle to be repeated. These phases of operation are summarized in Figure 18.2. The Calgary Biocell is a full-scale pilot project that has been implemented to acquire data and demonstrate the applicability of the Biocell concept under extremely cold temperatures.

Bioreactor landfills are becoming increasingly popular as a sustainable alternative to dry landfills. New design guidelines and regulations are being introduced for bioreactor landfills, but it happens at a very slow pace. Sustainable landfills have not become a reality at a commercial level yet. Within this context, rest of the topics in this chapter is primarily focused on design, construction and maintenance of conventional MSW landfills. Bioreactor landfill approach is also briefly discussed wherever it is appropriate.

18.3 SITING AND REGULATORY REQUIREMENTS

Siting a landfill is a lengthy process and is very important that the selected site needs to be acceptable to general public. Usually the process begins with drawing a circle using a search radius keeping the MSW generator at its center. Search radius

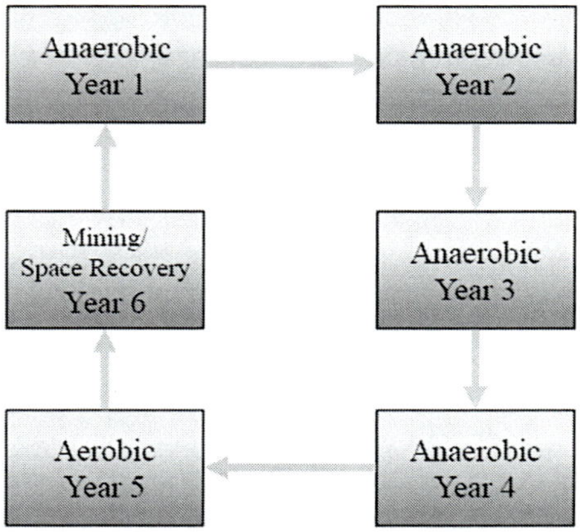

Figure 18.2. Phases of Biocell Operation with example durations
Source: Hsieh et al. 2008; © ASCE

is determined by the economics of MSW hauling. If the general public is going to be affected due to location chosen, they need to be informed immediately. Key aspects of the siting process are discussed briefly in the following paragraphs.

18.3.1 Size of the landfill

The required volume of the landfill should be calculated based on the expected future waste generation and the expected life of the landfill. Foot print area of the landfill can be estimated once the depth is finalized. Service, buildings, access roads, buffer zones and etc. must be included in estimating the general size of the landfill as a facility.

18.3.2 Traffic and Access

Landfill generated traffic (during construction and operation) can give rise to noise, vibration, exhaust emissions, dust, dirt and visual intrusion. Heavy vehicles on narrow roads may create traffic management issues including delays to other traffic, damage to roads and can be a source of complaint. The following should be evaluated as part of the site selection process:

- Distance of potential sites from waste generation areas where regard should be taken to the Proximity Principle and objectives of County Development Plans and Local Area Plans.
- Proximity to the existing national/regional road or rail network and expected vehicle movements where landfill should have good access to national or regional road routes or rail lines.
- Any required upgrading or new road infrastructure to accommodate additional traffic.
- The residential nature of potential access routes.

18.3.3 Site-specific Information

A comprehensive study should be performed to collect all pertinent information about the site and the vicinity that may affect the final decision of selection as well as construction/maintenance of the landfill. Following list includes the most important aspects that need attention:

- Soil, topographic, floodplain, land use, road, and utility maps
- Soil/geological reports
- Presence of water wells
- Aerial photographs
- Flood insurance rating
- Nearby wetlands and nature preserves
- Real estate values
- At least 15 miles away from an airport to avoid impact from birds

18.3.4 Site Hydrology

A hydrogeological report should be prepared for the selected site. To compile this report, a comprehensive site investigation program should be performed with the following key components (Sharma and Reddy 2004): soil boring, test pits, groundwater sampling and testing, and field and laboratory hydraulic conductivity.

18.3.5 Permits

Once the site selection is completed and an environmental impact study that includes traffic analysis is performed, the project details are presented to state and local authorities for approval. Obtaining a development permit is the first of many steps. In addition to submission of all documents and proof of addressing any issues encountered, the process may also include public hearings about the proposed project (Sharma and Reddy 2004). Development permit only permits the construction of a landfill. Once the construction is completed, the owner must request for an operating permit to become operational.

18.3.6 Other Regulatory Issues

Many countries have adopted statutes governing standards for solid waste management. These statutes allow regulatory bodies to impose minimum standard for landfills. Therefore, it is necessary to obtain information about all relevant authorities that have the control over different aspects related to landfill design and operation. To ease the process, Bagchi (2004) recommends construction of a flow diagram indicating regulatory requirements and the name of the authority involved in each operation to plan for the approval process. Design and maintenance of solid waste landfills are federally mandated by the Resources

Conservation and Recovery Act (RCRA), which was passed in 1976 by the US Congress. Amendments made to this act in 1984 made it stronger to address issues with hazardous waste but also specified guidelines for nonhazardous solid waste as well. MSW is covered by the Subtitle D of this legislation and hence landfills that fall under these requirements are usually called "Subtitle D landfills" (Vesilind et al. 2002). Local regulations should be taken serious as they may be stricter than the federal regulations. State and local authorities have the ability to adopt federal regulations or to develop new laws and regulations that are more stringent. Fulfilling regulatory requirements is usually a lengthy process and it may take few years. Note that as opposed to North American policy on waste management, the European Union (EU) encourages recycling and discourages landfilling. The policy of EU on waste management can be summarized as (Environmental Protection Agency of Ireland 1997):

- Prevention of waste
- Reducing the quantity of non-recoverable waste
- Recycling and reusing waste to the maximum extent for raw material and energy
- Disposing safely of any remaining wastes which cannot be recovered

18.3.7 Additional Regulatory Requirements for Bioreactor Landfills

Similar to conventional dry landfills, bioreactor landfills must also meet requirements of RCRA Subtitle D. As per the Code of Federal Regulation 40CFR part 258, introduction of leachate or other nutrients into waste is allowed (USEPA 2002). However, leachate recirculation is only allowed in MSW landfills that have composite liners with minimum of 0.6 m thick compacted clay with hydraulic conductivity equal or less than to 10^{-7} cm/s.

18.4 TYPICAL LANDFILL CONFIGURATIONS

Both above-ground and below-ground landfill configurations are commonly used. Four such configurations are given in Figure 18.3. Figure 18.3 (a) is an *area fill* (above ground) that requires no excavation. It is the best suited for flat terrains with shallow groundwater elevations. Figure 18.3 (d) is a below-ground option called *valley fill* (or *canyon fill*) used in mountainous terrains. Figure 18.3 (c) is *above and below ground fill* and Figure 18.3 (b) is a *Trench fill*, and both are combined above and below ground options. These are suitable for relatively flat areas. Trench fills are often used for small scale waste streams. Few other important aspects related to landfill configurations are briefly discussed in the following subsections. Figure 18.4 shows excavation for Calgary Biocell which is *above and below ground fill*.

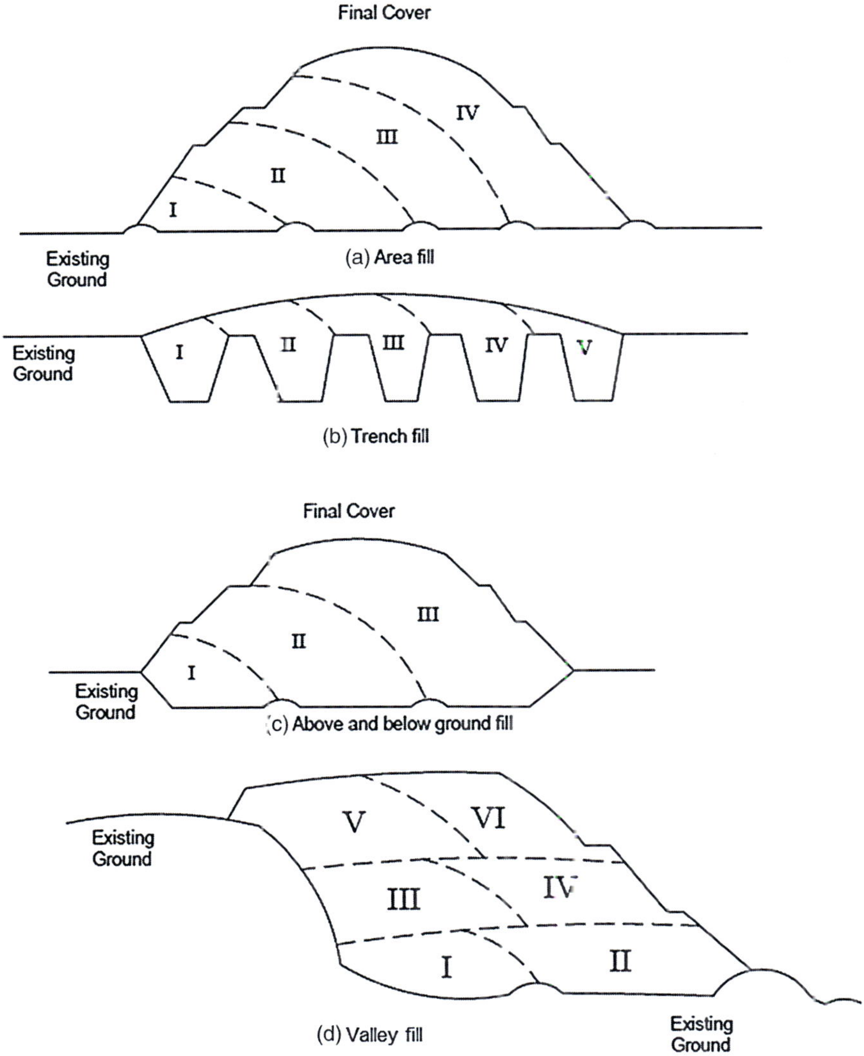

Figure 18.3. Different landfill configurations
Source: Adapted from Qian et al. (2002)

18.4.1 Cell Layout

Waste footprint area is divided into cells and usually the landfill is constructed one cell at a time to keep the initial capital cost to a minimum. This will also help to minimize the leachate generation and the costs associated with stockpiling of excavated soils. Typical cell size often varies between 2 to 3 acres (Sharma and Reddy 2004). Cell filling may take from 1 to 3 years.

Figure 18.4. Excavation of Calgary Biocell

18.4.2 Water Table, Aquifers, and Bedrock

Regulatory requirements usually dictate the distance to the water table from the sub-base. If not, sub-grade must be kept above water table. Excavating bedrock adds a drastic increase to the total cost. Keeping the base near or above bedrock is encouraged wherever possible.

18.4.3 Landfill Foundation and Slope Stability

Landfill foundation should avoid any previously mined areas or sink holes beneath. If located below water table, uplifting can become an issue as well. This is really critical during the construction phase. If there are any aquifers with artesian pressures then a sufficient gap between the sub-base and the top of the aquifer must be maintained to achieve a factor of safety much greater than one. Stability of the excavated slopes as well as waste slopes needs to be checked to prevent failures. Limit equilibrium method is often used for slope stability analysis. It is important to evaluate both intermediate and final waste slopes. Seismic stability analysis is also required in addition to static analysis, if the landfill is located in a seismically active zone.

18.4.4 Site Development Plan

A master plan is needed to identify the location of the property boundary, optimal waste footprint, cell implementation phases and locations for various facilities. The facilities needed within the property boundary may include:

- Office buildings
- Scale and scale house

- Areas for truck loading and washing
- Stockpiling areas
- Leachate holding tanks
- Access roads
- Landfill gas collection system

18.5 KEY ASPECTS OF DESIGN AND CONSTRUCTION

Modern landfills are well-engineered facilities that are located, designed, operated, and monitored to ensure compliance with federal regulations. Landfill design involves designing of physical elements of the landfill as well as the operational systems. The major design components of a landfill are illustrated in Figure 18.1. They include: sub-base, liner, leachate management system, gas management system, final cap, and stormwater management.

18.5.1 Preparation of Landfill Sub-Base

It is important to have a properly prepared sub-base as the landfill liner is constructed directly on the sub-base. If sub-base is not properly compacted, waste compaction in the first few lifts becomes challenging. In Sandy soils, the general recommendation is to compact the sub-base to 85–90% of the relative density. During construction, density is usually checked at 30 m grid points (Bagchi 2004). Nuclear density gauge may be used to measure the *in situ* density. If there is clay in the sub-base, then consolidation characteristics of the material need to be investigated as well.

18.5.2 Liner Design

The primary objective of a liner system is to prevent contamination of soil and groundwater. It also facilitates collection and removal of leachate produced by the waste. In general a liner system consists of multiple layers of clay or geo-synthetics (mostly geo-membranes) to prevent movement of any liquid between the landfill and surrounding site. Liner material selection should be based on the type of waste and method of landfill operation. Leachate must not make any adverse effect on the liner material. Traditionally liners were made out of clay. If clay is in short supply, it can be mixed with sand to make a relatively impermeable soil liner. Clay does not cut down permeability completely. However, permeability in clay liners decreases with time, possibly due to clogging of the pore spaces by the contaminant from the landfill. One advantage of clay liners is that they are not easily damaged during installation or service.

Geosynthetic liners can limit leakage more efficiently, but they are susceptible to damage during installation. If heavy compaction equipment is expected to be used on a geosynthetic liner, the thickness of the drainage blanket needs to be

Figure 18.5. Calgary Biocell HDPE liner construction

Figure 18.6. Calgary Biocell composite liner construction

increased to protect the geosynthetic layer. High density polyethylene (HDPE) geomembranes are recommended for landfill use due to their resistance to most chemicals. Figure 18.5 shows Calgary Biocell HDPE liner construction and Figure 18.6 shows Calgary Biocell composite liner construction.

Landfills are designed as single-lined or multiple-lined landfills, depending on the applicable local, state or federal regulations. Some example cross-sections of both types are shown in Figure 18.7. A single liner system only includes either a clay liner or a geo-membrane. A composite single liner consists of a geo-membrane sitting on clay (or Geo-synthetic Clay Liner i.e. GCL). This

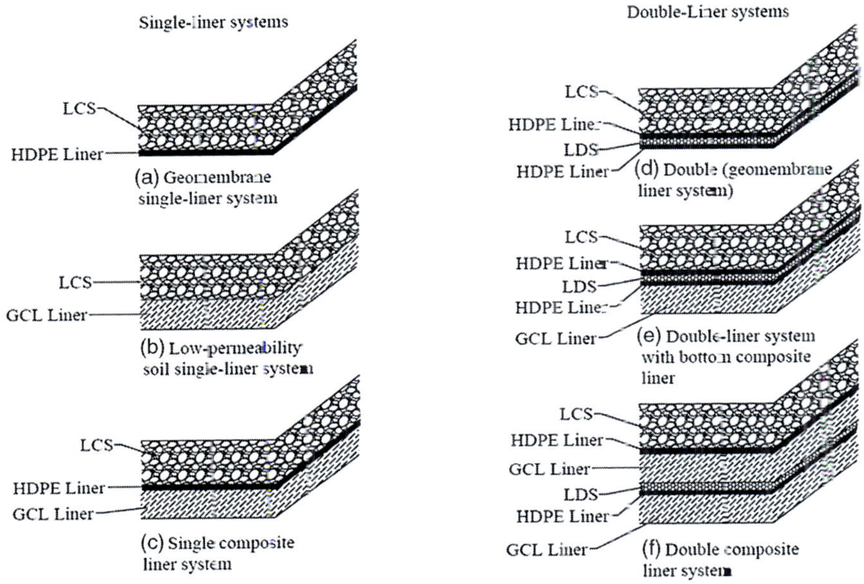

LCS=leachate collection system, GCL= geo-synthetic clay liner, LDS=leachate detection system

Figure 18.7. Example liner systems
Source: Adapted from Vesilind et al. 2002

configuration is the minimum requirement set by the RCRA for Subtitle D landfills.

Thickness of a liner always depends on the type of material selected. With geo-synthetics, the main concerns are puncture resistance during installation and the degradation due to ultraviolet rays. In general 1.5–2.0 mm thick geomembranes are used for lining landfills (Bagchi 2004). Clay liner thicknesses are governed by the construction related issues and freeze–thaw degradation. On theoretical grounds at a minimum a 15–30 cm thick clay layer is recommended. However, thicker layers are used to better handle the construction related issues.

18.5.3 Leachate Management

RCRA regulations restrict leachate head on the liner to 30 cm for Subtitle D landfills. Leachate in excess of 30 m should be removed from the landfill. Leachate can be removed by using gravity flow or by pumping. Leachate collected from the landfill may be stored on site to be treated later, or transported for treatment and disposal off site. Surface impoundments and tanks are the typical leachate storage methods. The most economical option is to transport leachate to an offsite facility for treatment and disposal (Vesilind et al. 2002). This allows the owner/operator to focus on their primary goal of managing the landfill while the leachate is handled by an expert on wastewater treatment. From the owner/operator's side, this option

Table 18.1. Summary of leachate treatment options

Treatment Option	*Removal Objective*	*Comments*
Biological		Best used on "young" leachate
Activated sludge	BOD/COD	Flexible, shock resistant, proven, minimum SRT increases with increasing Organic strength, >90% BOD removal possible
Aerated lagoons	BOD/COD	Good application to small flows, >90% BOD removal possible
Anaerobic	BOD/COD	Aerobic polishing necessary to achieve high-quality effluent
Powdered activated carbon/act. sludge	BOD/COD	>95% COD removal, >99% BOD removal
Physical/Chemical		Useful as polishing step or for treatment of "old" leachate
Coagulation/ Precipitation	Heavy metals	High removal of Fe, Zn; moderate removal of Cr, Cu, Mn; little removal of Cd, Pb, Ni
Chemical oxidation	COD	Raw leachate treatment requires high chemical dosages, better used as polishing step
Ion exchange	COD	10–70% COD removal, slight metal removal
Adsorption	BOD/COD	30–70% COD removal after biological or chemical treatment
Reverse osmosis	TDS	90–96% TDS removal

Notes: COD – Chemical oxygen demand; BOD – biological oxygen demand; TDS – Total dissolved solids

Source: Data from Vesilind et al. 2002 and King and Mureebe 1992

also eliminates the burden associated with the permitting, testing, monitoring and reporting requirements.

Leachate treatment is challenging mainly because of the irregular production rates and variable compositions. A summary of commonly used treatment methods is presented in Table 18.1. More than one method is often required to achieve the intended goal. Leachate treatment practiced at the Al Turi Landfill in New York is an example for use of multiple methods (Vesilind et al. 2002). They used polymer coagulation, flocculation, sedimentation, anaerobic biological treatment, aerobic biological treatment, and filtration in their process (King and Mureebe 1992).

18.5.4 Landfill Gas Management

Gas collection systems can be designed as active or passive. Passive systems use vents to release landfill gas into the atmosphere. They are usually used in small landfills (<40,000 m^3). Passive gas venting system may consist of a series of isolated gas vents. A schematic of a passive vent is shown in Figure 18.8. Depth of a passive vent can be as deep as 75% of the landfill depth or as shallow as just few feet below the cap. One vent per every 7,500 m^3 is typically used as spacing (Vesilind et al. 2002).

An active venting system consists of a series of deep extraction wells linked by a header pipe. Active systems use vacuum pressure to collect landfill gas through extraction wells. Passive landfill gas collection systems use vertical as well as horizontal wells. Vertical wells are more popular and a cross sectional view is given in Figure 18.9. Gas is extracted using central blowers that are sized according to the volume of the gas that needs to be transported (Vesilind et al 2002). Use of large pipes and minimizing the number of bends and valves used in them can help minimize head losses.

Spacing of wells is based on expected gas flow from the landfill. It is recommended to install them so the zones of influence overlap each other by a certain percentage. A 100% overlap may be achieved by installing them on corners of hexagons and a 60% overlap can be achieved if they are installed in a square array (Bagchi 2004). Gas extraction wells are connected to a blower through HDPE header pipes (15–20 cm diameter). Header pipes are embedded

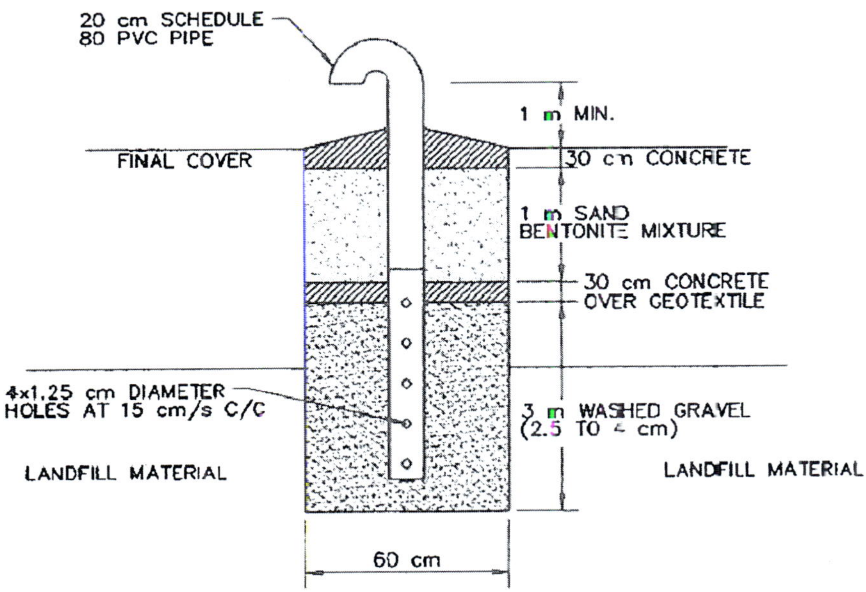

Figure 18.8. Schematic of a passive landfill gas vent
Source: Adapted from USACE 1995

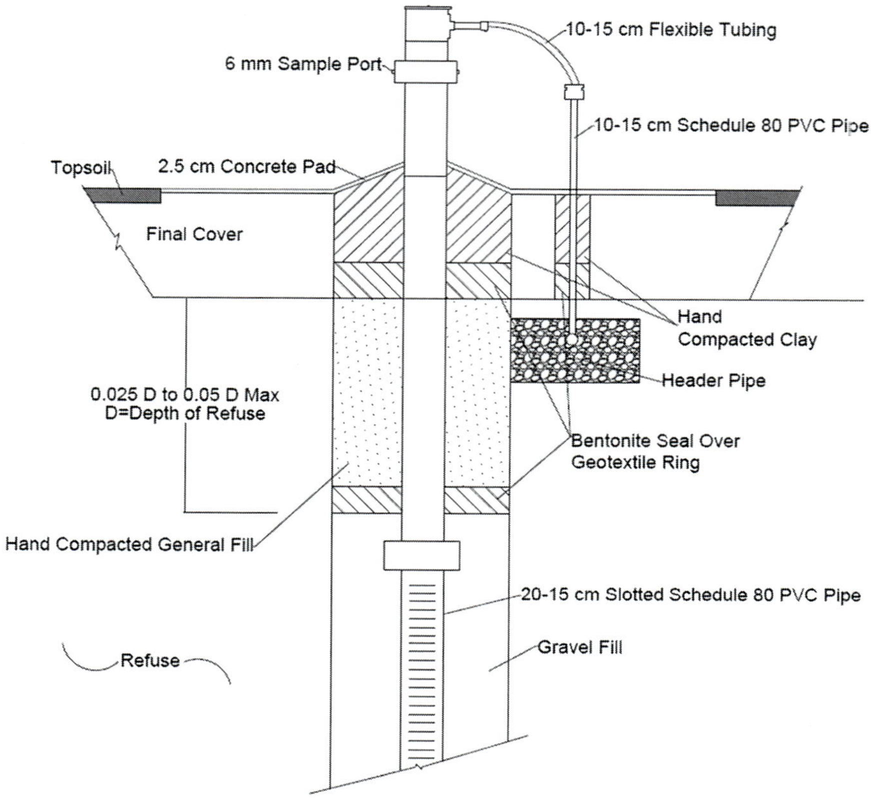

Figure 18.9. Vertical landfill gas extraction well
Source: Adapted from Bagchi 2004

in trenches filled with sand. The blower needs to be installed at a slightly higher elevation than the header pipes to facilitate dripping of condensate (Bagchi 2004). Condensation is a common occurrence due to the drop in temperature of the saturated landfill gas while moving in the header pipe. Condensate collected in the pipes may be released back into the landfill or treated similar to leachate. In colder climates one should be mindful of freezing of condensate and clogging of pipes.

18.5.5 Daily, Intermediate and Final Cover Design

Daily cover is the name given to the layer of compressed soil or earth which is laid on top of a day's deposition of waste on an operational landfill site. The cover helps prevent the interaction between the waste and the air, reducing odors and enabling a firm support upon which for vehicles to operate. While soils are the traditional materials employed in daily cover, alternative options such as "green waste", mixtures of paper sludge and tire derived aggregate (TDA) have displayed mechanical characteristics desirable for daily cover. When compared to traditional soil layers, the paper sludge paste was 2-3 times lighter, at least two orders of magnitude more impermeable, and comparable in shear strength (Ng and Lo 2007).

Intermediate covers are placed on previously active working faces of the landfill that will not be covered with waste for an extended period of time, typically from about 7 to 60 days or longer. Intermediate covers have traditionally consisted of a layer of soil, and in some circumstances included the use of an additional layer of plastic scrim or geotextile material. The reason for using intermediate covers are in part the same as those for using daily covers – controlling odors from contaminants, blowing litter, vectors and fires.

A final cover (or cap) is placed on MSW, when it reaches its design height to minimize rainwater intrusion, spread of waste, and odor. USEPA regulations require that the cover be less permeable than that of the liner (Vesilind et al. 2002). Typical side slopes are from 1 : 3–1 : 4. As illustrated in Figure 18.10, they

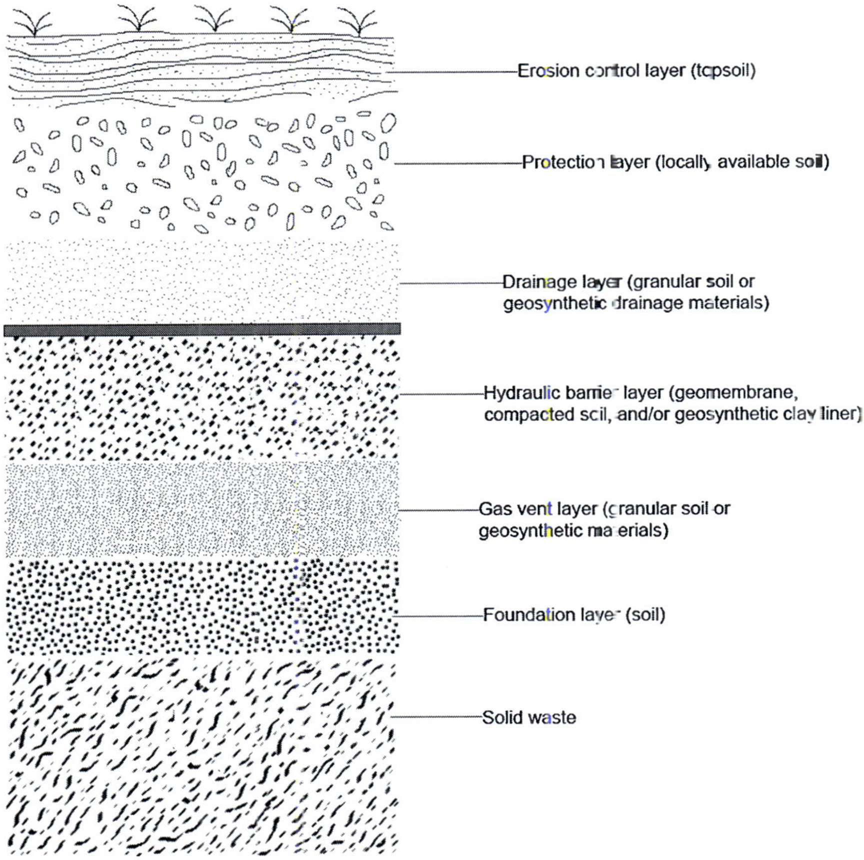

Figure 18.10. Cross-section of a typical final cover system

Source: Adapted from Qian et al. 2002

typically consists of a vegetation layer (at the top) followed by a supporting soil later, filter/drainage layer, hydraulic barrier, gas control layer, and a foundation layer.

A foundation layer of soil may be first introduced to ease the construction process. This layer is also known as the grading layer as it provides a stable foundation to build upon rest of the cover layers. A layer of geotextiles may be used below this layer. The gas control layer is then built on the foundation soil layer which consists of 15–60 mm of coarse grained material. Thickness selection should be based on the stability and the design of gas collection system.

A barrier layer is constructed on top of the gas control layer, using clay or GCL or a geo-membrane. If clay is the choice, it needs to be a minimum of 60 cm in thickness. If geo-membrane is used, minimum of 1 mm thick Low Density Polyethylene (LDPE) layer is recommended. Construction of a low permeability layer below the geo-membrane is also required to minimize infiltration in the event of a formation of a leak in the membrane.

Coarse sand may be used to construct the drainage layer (30 cm minimum thickness) on the clay barrier. Role of this layer becomes very critical if the barrier is geosynthetic. Soil-geosynthetic interface tends to lose friction when saturated. But an efficient drainage in the drainage layer can prevent that by avoiding unstable soil conditions leading to failure. A 45–90 mm layer of soil may be used as a protective layer. However, if the protective layer is sandy, having a separate drainage layer is not required. To facilitate vegetation, about 15 cm organic soil should be spread on the top of the protective layer. To encourage the growth of vegetation, necessary nutrients may be added to the top soil after consulting a horticulturist (Bagchi 2004). It should be noted that all these layers may not be needed in some cases. For example the gas control layer may be combined with the foundation layer or gas control layer may not be necessary if the waste in concern is not expected to produce large quantities of landfill gas (Qian et al. 2002).

18.5.6 Stormwater Management

Storm-water run-on and runoff is regulated by the RCRA for subtitle D landfills. Run-on and runoff control is needed to prevent stormwater getting into the active phase of the landfill to minimize leachate production. This is achieved by diverting stormwater from the active phase in the landfill. A facility must be capable of handling a peak volume from a 24 hour, 25 year storm (Vesilind et al. 2002; U.S. EPA 2005). Precipitation water falling on landfill needs to be routed to natural drainage paths. Stormwater is collected using stormwater ditches and then directed to stormwater basins before releasing to the natural environment (Figure 18.11). There can be several ditches running over a landfill depending on the estimated volume of runoff water through each section of the landfill. Principles used in open channel flows are applicable to ditch design, and therefore, the Manning equation is usually employed. One of the main difficulties associated with this process is to find an appropriate roughness values to be used in the

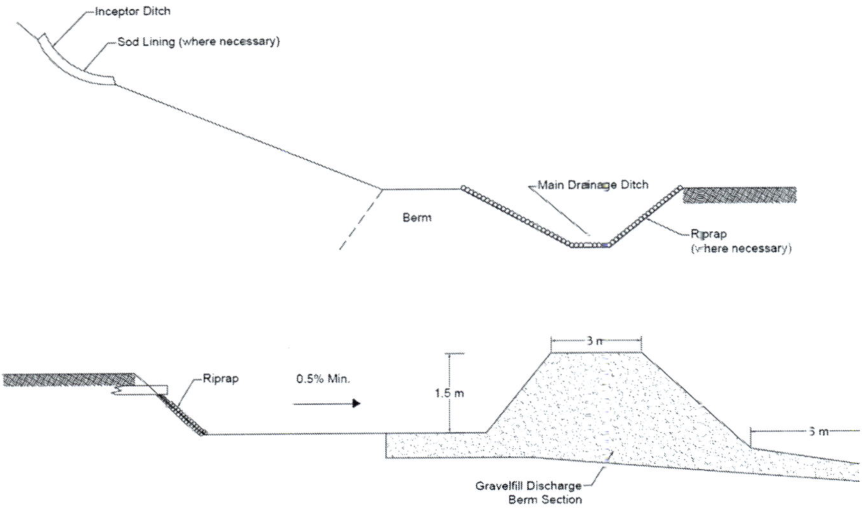

Figure 18.11. Example cross sections of a storm water ditch and a basin used in landfill designs
Source: Adapted from Bagchi 2004

Manning equation. Usually triangular or trapezoidal sections are used. If higher velocities are expected additional erosion protection measures such as rip-rap lining or erosion mats may be used (Bagchi 2004). About 10% base slopes are recommended to discourage erosion.

Sedimentation basins are used to reduce the total dissolved solids in in the surface water. Theses basins need periodic cleaning to remove collected sediments. Sedimentation is usually modeled using the Stoke's law. However, currents and particle interaction need to be taken into account (Bagchi 2004). Usually 2 : 1 length to width ratio with a depth of 1.5 m is recommended for storm-water basins in landfills. A new generation of landfill cover termed Phytocap is developed that could also address storm water issues. Phytocap is a natural soil-plant alternative to the conventional engineered landfill cover design. It requires less engineering input and has a lower cost than conventional impermeable covers as it only utilizes local recourses. It also offers the advantage of oxidizing methane to reduce landfill greenhouse emissions. This type of covers has the potential to make a significant difference in the way that developing countries are capping their waste sites.

18.5.7 Bioreactor Landfill Design

Most components of a bioreactor landfill are designed similar to that of a traditional dry landfill. Key exceptions are discussed in the next few paragraphs.

Usually liner consists of a geo-membrane placed over a clay layer. About 1.5 mm minimum thickness of geo-membranes and 1.2–1.5 m of clay layers are recommended. Some states in the US have more stringent requirements for bioreactor landfills and specify a multiple liners. Gravel is recommended for the drainage blanket and a geotextile must be placed above it.

Traditional dry landfills are maintained at 12–15% of moisture whereas in the bioreactor landfills it is expected to be near 40% (Tchobanoglous and Kreith 2002). This is achieved through leachate recirculation. Leachate can be recirculated using many different techniques including surface wetting, spraying, and horizontal as well as vertical injection. Recirculation operation should be moved from one place to another with intense pumping rates to achieve better results. On site or offsite leachate treatment may not be necessary for bioreactor landfills due to recirculation of leachate. However, for developing nations with high organic content waste, excess leachate is generated, so on site or offsite leachate treatment may be necessary. The down side of the recirculation is that it is not a complete alternative to treatment. When there is a surplus of leachate the facility must be ready to address. When there is a shortage in leachate to be recirculated, there should be contingency plan for that as well.

Due to the increased unit weight of MSW as a result of leachate recirculation and also the large deformations expected from differential settlements, structural integrity of the leachate collection system must be checked. Increased diameters are also recommended for leachate collection system. Increased MSW moisture content poses stability issues in bioreactor landfills. Tchobanoglous and Kreith (2002) suggested conducting stability analysis considering 10% increase and 30–40% increase in moisture content in addition to a regular stability checks. Internal stability as well as stability along the geomembranes on side slopes must be checked. Theses checks should also take potential leachate mounding in to account.

18.5.8 Landfill Construction

Bagchi (2004) describes two different types of construction specifications: work-type based and performance based. In the work-type, the contractor is given detailed instructions on "what" and "how" and it is more close to traditional design-bid-build delivery method. Performance based method requires to specify the final product and not the process and more similar to the emerging delivery method of design-build. Performance based specification method is often preferred as the details necessary for the work-type method is difficult to gather and there is huge uncertainly. Hiring independent quality control staff is essential to the successful implementation of the project.

18.6 LANDFILL OPERATION

Having a detailed operation plan is a must to make the day-to-day landfill operation smooth. Tchobanoglous and Kreith (2002) present a detailed account on important factors that must be considered during the operation of a landfill. Operating schedule, filling plan, equipment requirements, operating records,

Table 18.2. Factors that must be considered during the operation of a landfill

Factors	*Remarks*
Communications	Telephone for emergencies
Days and hours of operation	Usual practice is 5 to 6 days/week and 8 to 10 h/day
Employee facilities	Rest rooms and drinking water should be provided
Equipment maintenance	A covered shed should be provided for field maintenance of equipment
Litter control	Use movable fences at unloading areas; crews should pick up litter at least once per month or as required
Pest control	Implementation and enforcement of daily cover
Operation plan	With or without the co-disposal of treatment plant sludges and the recovery of gas
Operational records	Tonnage, transactions, and billing if a disposal fee is charged
Salvage	No scavenging; salvage should occur away from the unloading area; no salvage storage on site
Security	Provide locked gates and fencing, lighting of sensitive areas
Spreading and compaction	Spread and compact waste in layers less than 0.6 m thick
Unloading Area	Keep small, generally under 30 m on a side; operate unloading areas for automobiles and commercial trucks

Source: Data from Tchobanoglous and Kreith 2002

billing information, traffic control, safety and security are among the top priorities. Other factors that must be considered are listed in Table 18.2. A proper plan assures safe working environment, optimizes use of the space, and also minimizes environmental damage. Important aspects related to landfill filling and special considerations for bioreactor landfills are discussed briefly in the following paragraphs.

18.6.1 Waste Acceptance at Landfills

During landfill operations, the waste collection vehicles are weighed at a weigh-bridge on arrival and their load is inspected for wastes that do not accord with waste acceptance criteria of the landfill. Afterward, the waste collection vehicles use the existing road network on their way to the tipping face or working front where they unload their contents. After loads are deposited, compactors or bulldozers are used to spread and compact the waste on the working face. Before leaving the landfill boundaries, the waste collection vehicles pass through a wheel cleaning facility. If necessary, they return to the weighbridge in order to be weighed without their load. Through the weighing process, the daily incoming waste tonnage can be calculated and listed in databases for record keeping. In addition to trucks, some landfills may be equipped to handle railroad containers. The use of 'rail-haul' permits landfills to be located at more remote sites, without the problems associated with many truck trips. At the end of each operational day, the working face of the landfill must be covered for environmental and public health reasons. This helps to control odor, diseases, rodents, and spread of litter by wind and animals.

18.6.2 Waste Filling and Compaction

When the landfill is ready to accept waste, it is filled in lifts (see Figure 18.12). First lift is directly placed on the top of the liner. Each lift is compacted using heavy compactors to optimize usable landfill space. Types of waste, moisture content, lift thickness are the primary factors that govern the compaction. Compaction equipment selection depends on the slopes. Slopes steeper than 1:3 are better compacted by track-type tractors while shallower slopes are better handled by landfill compactors (Vesilind et al. 2002). Figure 18.13 shows the filling operation of Calgary Biocell and Figure 18.14 shows a truck unloading the collected waste. Waste in the first lift needs to be screened to remove any sharp objects so they may not damage the liner. Filling sequence must be defined during the design/permitting stage (Vesilind et al. 2002). The working face must be large enough to concurrently handle unloading waste from several trucks.

18.6.3 Bioreactor Landfill Operations

Sharma and Reddy (2004) points out that a bioreactor landfill should be treated similar to a large biological digester. The operational procedure must be aimed at enhancing the efficiency of the biodegradation. Therefore, close monitoring is a must. A daily operational plan may consider some type of waste pretreatment such as shredding and/or management of nutrients in the waste mass.

Shredding is the most common pretreatment option. Shredding allows waste to come out of plastic bags to enhance biodegradation. Separation or segregation of organic waste is another option. For an example, separation and exclusion of construction and demolition waste improve the percentage of biodegradable fraction in waste. Nutrients necessary for the biodegradation activities is generally supplied by the waste itself or through the recirculation of the leachate collected from the same landfill. The deficient nutrients may also be added to the recirculated leachate. In addition, other chemical or biological supplements may also enhance the biodegradation activities (Sharma and Reddy 2004).

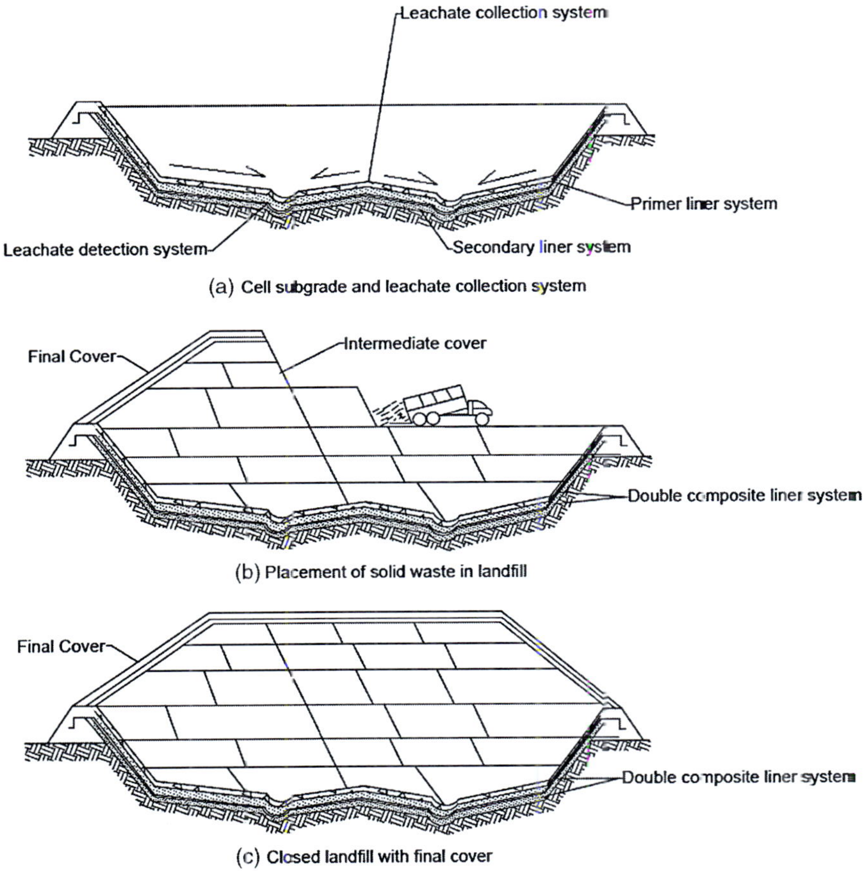

Figure 18.12. Construction sequence of a solid waste landfill
Source: Adapted from Qian et al. 2002

Heavy compaction of MSW is always encouraged in dry landfills. However, in bioreactor landfills it is somewhat questionable as heavy compaction can result in lower hydraulic conductivity and can make a negative impact on biodegradation. In the Calgary Biocell the deposited waste was neither compacted by earth moving equipment nor used daily cover. The waste in the Biocell was therefore allowed to compact by its own weight (Hettiarachchi et al. 2013). A back analysis performed by Hunte et al. (2007) revealed that during the construction of lift two of the Calgary Biocell, the unit weight of MSW in the bottom layer was approximately 5.6 kN/m^3.

The enhancement in biodegradation due to leachate recirculation may result in increased gas production even during filling. Therefore, the gas collection system may be installed during active filling, which is challenging in terms of construction and maintenance point of view.

Figure 18.13. Filling operation of Calgary Biocell

Figure 18.14. A truck unloading the collected waste

18.7 POST-CONSTRUCTION MONITORING

Monitoring is an important task during landfill construction/operation as well as after closure. Leachate, possible leakage through liner system, groundwater quality, landfill gas migration, and stability are among the most commonly monitored aspects.

18.7.1 Leachate Monitoring and Leakage Detection

Leachate monitoring includes monitoring leachate head on the liner as well as the quality and quantity of leachate produced. Monitoring leachate head on the liner is required to control the potential leakages due to high heads. Leakage is usually detected using a lysimeter installed below the crest of the liner, which is considered the point of maximum leakage potential. Location and the number of lysimeters can vary but at least more than one need to be installed (Vesilind et al. 1988). Leakage detection is performed by monitoring the unsaturated zone between the sub-base and the seasonal high water level. Two approaches are used: direct monitoring and indirect monitoring (Bagchi 2004). Direct monitoring approach involves instrumentation that collects samples. Figure 18.15 is a schematic of a lysimeter (porous cup suction) used for liquid sample collection. Indirect method uses instruments to detect water percolation.

18.7.2 Groundwater Monitoring

Groundwater quality monitoring is conducted through sampling of monitoring wells. Ideally, a few groundwater wells should be installed at different depths in the vicinity of the landfill. Additional wells are placed near or at the property boundary (Vesilind et al. 2002). A schematic of a typical monitoring well is shown in Figure 18.16. The wells should cover up-gradient as well as the down-gradient to compare. If down-gradient results vary significantly from up-gradient, then more thorough investigations are conducted.

18.7.3 Landfill Gas Migration

Quality and composition of the landfill gas in the landfill and in the nearby soil need to be monitored. Gas monitoring probes are installed to monitor the concentration of methane and other hazardous air contaminants. Bagchi (2004) points out that there is a high chance for gas migration in the soil when the ground is frozen or saturated. Schematic of a vadoze zone gas monitoring probe is shown in Figure 18.17.

18.7.4 Stability of the Final Cover

Excessive deformation due to differential settlement can fail the cover especially when synthetic covers are employed. Stability of clay covers also need to be monitored if the waste slopes are higher than usual. Settlement of cover should be often monitored using a 100 ft grid. Monitoring should be performed quarterly or biannually (Bagchi 2004).

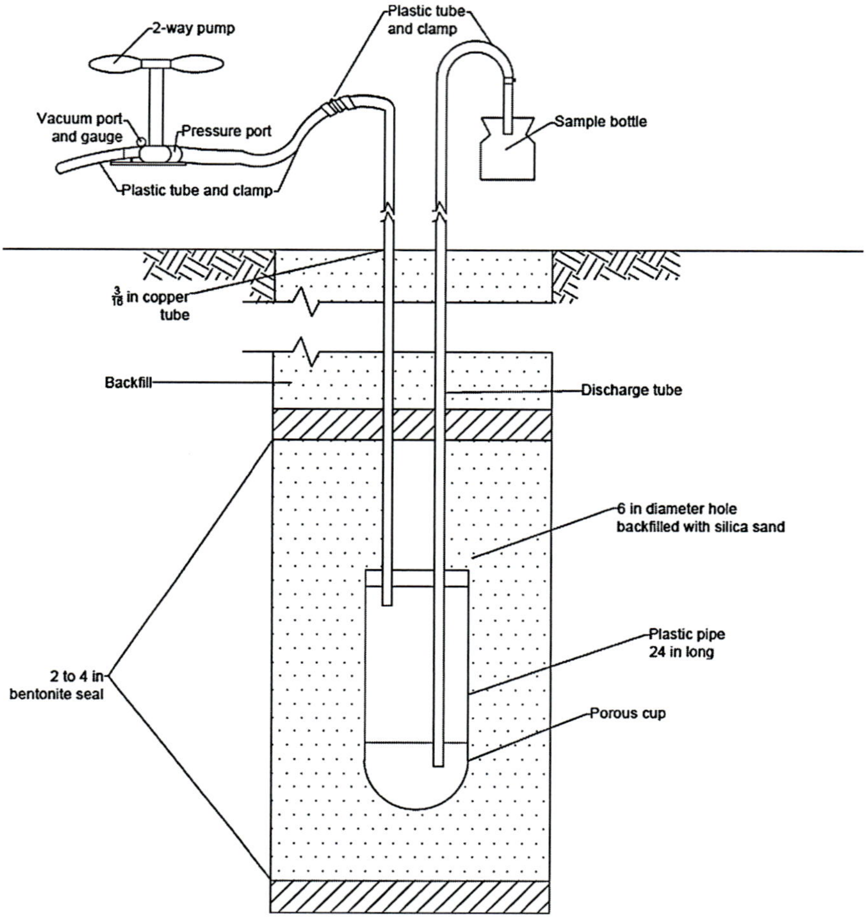

Figure 18.15. Schematic of a porous cup suction lysimeter used for liquid sample collection

Source: Adapted from Tchobanoglous and Kreith 2002

18.8 LANDFILL POST CLOSURE

Closure activities must begin within 30 days of the last receipt of waste and must be completed within 180 days. Critical technical issues that must be faced by the designer regarding the closure include the following.

- Degree and rate of post-closure settlement and stresses imposed on soil liner components
- Long-term durability and survivability of cover system
- Long-term waste decomposition and management of landfill leachate and gases

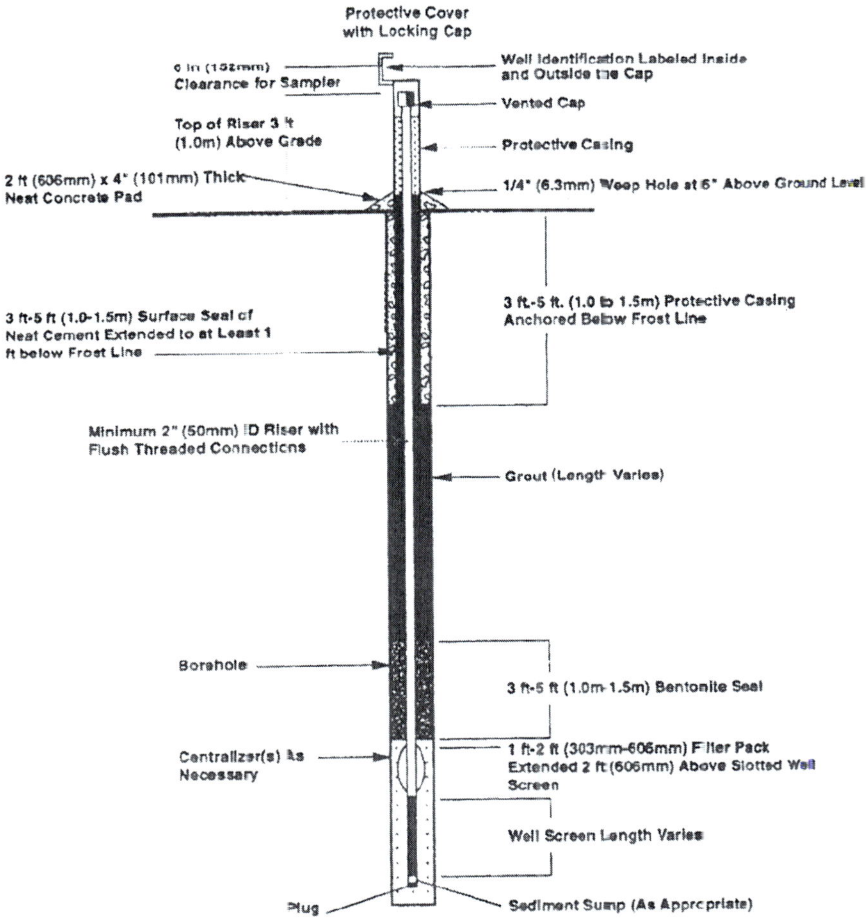

Figure 18.16. Schematic of a groundwater monitoring well
Source: U.S. EPA 1993

- Environmental performance of the combined bottom liner and final cover system.

Post-closure care requirements are focused on operating and maintaining the proper functions of four systems that prevent or monitor releases from the MSWLF unit:

- Cover system
- Leachate collection system
- Ground-water monitoring system
- Gas monitoring system

Owners or operators must comply with these environmental requirements for a period of 30 years following closure. Following completion of the post-closure

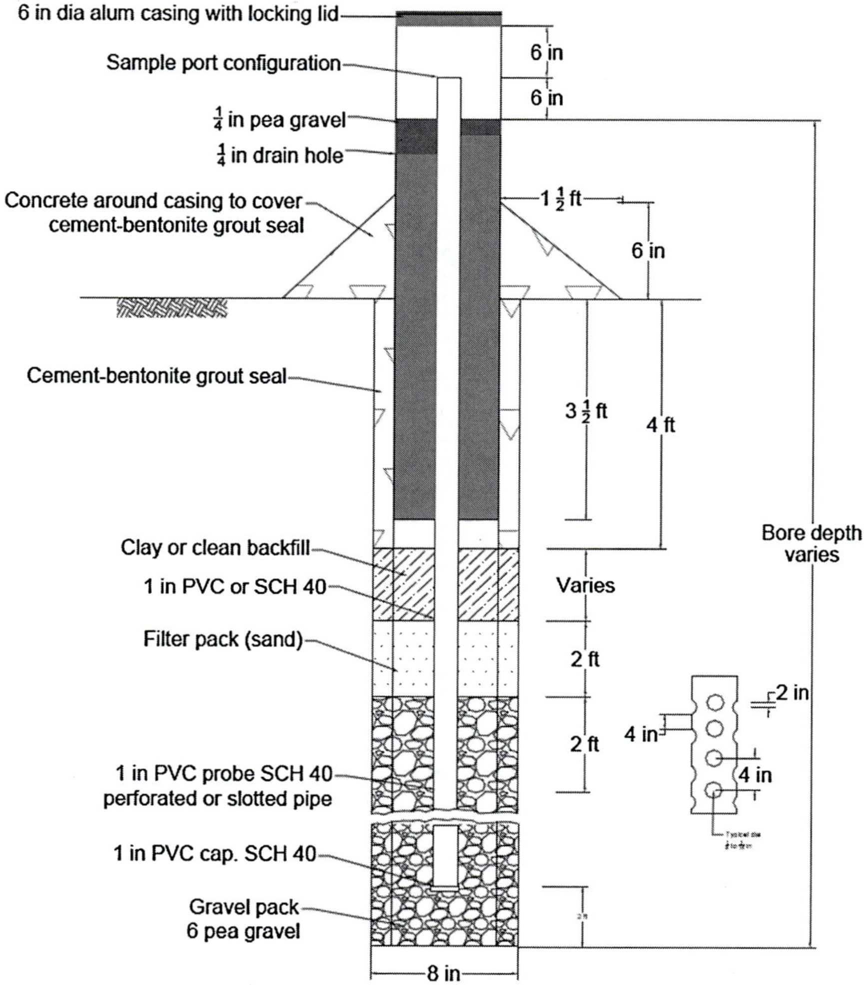

Figure 18.17. Schematic of a vadoze zone gas monitoring probe
Source: Adapted from Tchobanoglous and Kreith 2002

care period, the State Director must be notified that an independent registered professional engineer has verified and certified that post-closure care has been completed in accordance with the post closure plan and that this certification has been placed in the operating record (U.S. EPA 1993)

18.9 LANDFILL REDEVELOPMENT

The movement of brownfields redevelopment has helped invigorate existing slow process of remediation of contaminated sites. The use of risk based clean-up approaches now allowed in many states has facilitated brownfields redevelopment. Landfills are a particular subset of brownfields, particularly older landfills in industrialized areas. Older landfills that were never properly closed are true brownfields with idle land from which pollutants are often discharged. With investigation and limited remediation, this subset of brownfields-like sites presents unique opportunities for redevelopment. Redeveloping landfills is particularly challenging not only due to clean-up issues, but also the geotechnical issues of building on waste.

Most of the unregistered landfills were never properly closed. Only a handful of unregistered landfills were properly closed and received a Closure and Post Closure Plan Approval pursuant to the Amended State Solid Waste Management Act of 1975 and/or the Sanitary Landfill Facility Closure and Contingency Fund Act of 1992. The few properly closed landfills were either large private commercial landfills, sole source industrial landfills owned by major corporations, or municipal landfills. Hundreds of registered landfills were never properly closed because the owners lacked the resources to comply with regulatory closure requirements. While the Sanitary Landfill Facility Closure and Contingency Fund Act provides a revenue source through a tax on operating landfills, the State has historically not utilized these funds for closure of abandoned landfills and reserved the public funds for emergency actions, such as extinguishing a landfill fire or remediating methane migration.

A wide variety of remediation techniques can be utilized in landfill redevelopment. In the simplest case, waste can be capped in place with one foot of silty, clayey material and one and one-half feet soil cover. In the most complex case, a slurry wall/sheet pile wall can be used to contain leachate from outflow from the site and an interior leachate system can be installed. The degree of capping, containment and leachate collection depends on the underlying geology, leachate strength and site specific cap design.

18.10 CONCLUSIONS

Modern landfills are well-engineered facilities. They are sited, designed, operated, and monitored in accordance with federal regulations and local regulations. Three types of landfills are identified in literature. Conventional dry landfills are the most widely used option. Bioreactor Landfills are becoming popular as a more environmental friendly alternative to dry landfills. Sustainable landfills are the most recent addition to the list. Sustainable landfills allow resource mining and refilling.

Landfills can be regarded as a viable and abundant source of materials and energy. In the developing world, this is widely understood and one may thus often find waste pickers scavenging for usable materials. In the developing world either landfilling is discouraged or landfills are mined to recover resources. Within this context sustainable landfills may be viewed as a concept that provides a common solution to waste disposal in both developed and developing nations.

18.11 ACKNOWLEDGEMENTS

Authors would like to acknowledge the City of Calgary, Canada for allow sharing information about the Biocell landfill and also Ms. Masroor Khan of New Jersey Institute of Technology for drafting all figures.

References

Bagchi, A. (2004). *Design of landfills and integrated solid waste management*, 3rd Ed., Wiley, Hoboken, NJ.

Environmental Protection Agency of Ireland. (1997). "Landfill operational practices." Ireland.

Hettiarachchi, C. H., Hettiaratchi, J. P. A., Hunte, C., and Meegoda, J. N. (2013). "Operation of a landfill bioreactor in a cold climate: Early results and lessons learned." *J. Hazard. Toxic Radioactive Waste Manage.*, 10.1061/(ASCE)HZ.2153-5515.0000159, 307–316.

Hsieh, H. N., Hettiaratchi, J. P. A., El Haggar, S., Meegoda, J. N., and Stressel, R. I. (2008). "Towards the development of sustainable landfills." *Geo Congress*, ASCE, Reston, VA, 344–351.

King, D., and Mureebe, A. (1992). "Leachate management successfully implemented at landfill." *Water Environ. Technol.*, **4**(9), 1–18.

Ng, K. T., and Lo, I. M. (2007). "Mechanical behaviors of a synthetic paste of tire chips and paper sludge in MSW landfill daily cover applications." *Can. Geotech. J.*, **44**(8), 928–941.

Qian, X., Koerner, R. M., and Gray, D. H. (2002). *Geotechnical aspects of landfill design and construction*, Prentice Hall, Upper Saddle River, NJ.

Reddy, K. R., and Bogner, J. E. (2003). "Bioreactor landfill engineering for accelerated stabilization of municipal solid waste." *Int. e-Conf. on Modern Trends in Foundation Engineering: Geotechnical Challenges and Solutions*, Indian Institute of Technology, Madras, 22.

Reddy, K. R., Hettiarachchi, H., Gangathulasi, J., Parakalla, N., Bogner, J., and Lagier, T. (2009). "Compressibility and shear strength of landfillled municipal solid waste at Orchard Hills Landfill." *Waste Manage. Res.*, **27**(6), 578–587.

Sharma, H. D., and Reddy, K. R. (2004). *Geoenvironmental engineering: Site remediation, waste containment, and emerging waste management technologies*, Wiley, New Jersey.

SWANA (Solid Waste Association of North America). (2001). "Request for comment on bioreactor landfill definition." Silver Spring, MD.

Tchobanoglous, G., and Kreith, F. (2002). *Handbook of solid waste management*, 2nd Ed., McGraw-Hill, New York.

USACE. (1995). "Landfill off-gas collection and treatment systems." U.S. Dept. of the Army, Washington, DC.

U.S. EPA. (2002). "2002 code of federal regulation 40CFR." USGPO, Washington, DC.

U.S. EPA. (2005). "Introduction to United States environmental protection agency land disposal units (40 CFR parts 264/265, subparts K, L, M, N)." Office of Solid Waste and Emergency Response, USGPO, Washington, DC.

Vesilind, P. A., Peirce, J. J., and Weiner, R. F. (1988). *Environmental engineering*, 2nd Ed., Butterworth Publishers, Stoneham, MA.

Vesilind, P. A., Worrell, W. A., and Reinhart, D. R. (2002). *Solid waste engineering*, Brooks/Cole-Thomson Learning, Pacific Grove, CA.

CHAPTER 19

Landfill Leachate Collection and Treatment

Tian C. Zhang[1]
Rao Y. Surampalli[2]

Landfills are planned and designed to discard solid waste material by burying it between the layers of earth, in an effort to decrease human health hazards. However, landfill can create hazardous problems to the environment, such as air pollution caused by toxic gases released and groundwater pollution due to landfill leachate. Collection and treatment of landfill leachate in a sustainable way is of tremendous importance in order to prevent landfill leachate from causing serious environmental problems. This chapter focuses on landfill leachate collection and treatment for sustainable development of human society, including the principle and operation of landfill, leachate production, collection, characterization, and treatments under different operational and environmental conditions, current status of landfill leachate treatment and future directions.

19.1 LANDFILLS, LEACHATE PRODUCTION, AND COLLECTION

19.1.1 Landfills

Landfill is defined as the system designed and constructed to contain discarded waste so as to minimize releases of contaminants to the environment (LaGrega et al. 2000). Landfills are the old form of waste treatment. Before the advent of modern landfills, landfills were in the form of unregulated garbage dumps and burning sites. Modern landfills (Figure 19.1) are now the only sanctioned garbage disposal sites for most municipalities and are used in many places around the world. Newer types of landfills are bioreactor landfills; they are expected to reduce the leachate amount and the costs associated with leachate management, to increase the methane production and to reduce the amount of land required for landfills.

---

[1]Univ. of Nebraska-Lincoln, Lincoln, NE.

[2]Univ. of Nebraska-Lincoln, Lincoln, NE.

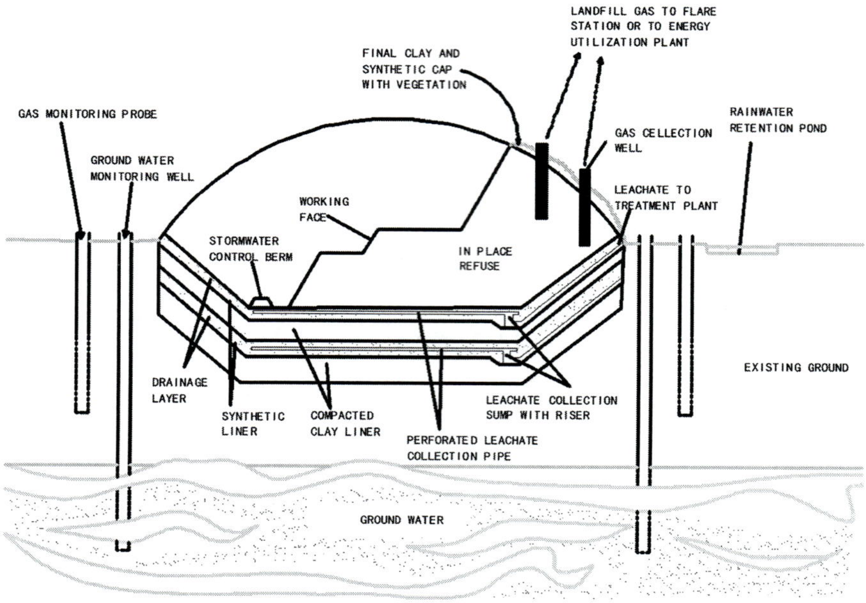

Figure 19.1. Schematic diagram of a modern landfill
Source: Adapted from Peavey 2010

In general, there are several kinds of landfills: municipal solid waste landfills (MSWLFs), Industrial waste landfills (IWLFs), construction & demolition debris landfills (CDDLFs), and hazardous waste landfills. Currently, 21 hazardous waste landfills are engaged in commercial disposal of the Resource Conservation and Recovery Act (RCRA) subtitle C hazardous waste in the U.S. (EHSO 2013), and there are 3,536 MSW landfills in the U.S. (NLR 2013). The design of these landfills is very different. However, since MSWLFs receive significant quantities of household hazardous waste, many of the design concepts and materials used for MSFLFs are the same as hazardous waste landfills (HWLFs) (LaGrega et al. 2000).

19.1.2 Leachate Production

In modern landfills, the waste is actually encapsulated and sealed off to prevent migration of pollutants and pathogens. The key parts of landfill design are i) the cover system to minimize air emissions and infiltration of precipitation and ii) the bottom liner and leachate collection system to prevent leachate from getting outside the landfill boundaries or down into groundwater. It is important to keep the landfill as dry as possible to reduce the amount of leachate. This is done by collecting surface runoff with ditches and catch basins, and the collected runoff is sent to collection ponds, treated for suspended soil particles and possible contaminants, and then pumped off-site.

It is unavoidable for a landfill to generate leachate due to i) rainwater soaking into the landfill and ii) the moisture within the trash itself. Leachate seeps downward to the base of the landfill by gravity, carrying particulate materials in suspension along with the dissolved constituents. Usually, the Hydrologic Evaluation of Landfill Performance (HELP) computer model is used to estimate leachate generation (Schroeder et al. 1994). The HELP program was developed to conduct water balance analyses of landfills, cover systems, solid waste disposal and containment facilities. The model facilitates rapid estimation of the amounts of runoff, evapotranspiration, drainage, leachate collection, and liner leakage that may be expected as a result of the operation of a wide variety of landfill designs. The model, applicable to open, partially closed, and fully closed sites, is a tool for designers to "assist in the comparison of design alternatives as judged by their water balances" (Schroeder et al. 1994).

It should be pointed out that the model a) should not be used for sizing leachate treatment facilities; b) should be used with particular caution under different weather conditions; c) is very sensitive to the allowance of dry periods and evapotranspiration, and accurately estimating evapotranspiration of the site can be a major limitation; and d) omits some conditions in open landfills and seldom considers landfills with a final cover. It is important to realize that climate changes in recent years have caused many problems for rainfall intensity and return periods, which calls for new evaluation on how to conduct water balance analyses of landfills. For example, the HELP version 3 was examined in a German climate in an extensive validation study, and a German adaptation was developed. However, for arid or semi-arid regions, or locations with frequent extreme weather conditions, the HELP model may have limitations, weak points, and errors (LLTEW 2011), and thus, requires further research.

19.1.3 Leachate Collection

Landfills, particularly HWLFs, usually have a multiple layer system to collect the highly contaminated leachate. The system consists of a filter zone (a geotextile), a primary leachate collection zone (with a leachate collection piping system), primary barrier liner (flexible membrane liner, FML), secondary leachate collection zone (with another leachate collection piping system), secondary barrier liner, and compacted clay liner (LaGrega et al. 2000).

In the past, tremendous efforts have been made on developing materials that are suitable for use in order to satisfy drainage and filter requirements. Geotextiles (also called filter fabrics) and geosynthetic drainage nets (geonets) have been developed and used to replace soil drainage layers to save the space in the filter zone. Long-term creep and clogging are major design considerations. Currently, it is unknown exactly what criteria should be used in assessing the potential for the clogging of geonets. Wastewater quality criteria for evaluation of drip irrigation clogging hazards is recommended by others (Oweis and Khera 1998).

Leachate is collected and conveyed through a network of pipes and sumps. The leachate collection pipe is sized to handle the anticipated flow with Manning equation:

$$Q = 1.486\,AR^{2/3}S^{0.5}/n \qquad \text{(Eq. 19.1)}$$

where, Q = flow rate, A = the cross-sectional area, R = hydraulic radius, S = slope, and n = Manning roughness coefficient. Assuming a circular plastic pipe (with n = 0.01) running full, the pipe diameter (D) can be calculated as:

$$D = 0.327(Q/S^{0.5})^{0.375} \qquad \text{(Eq. 19.2)}$$

The leachate is directed to a separate leachate collection pond for contaminants detection and then follow-up treatment. The treatment may occur on-site or off-site. The options for treatment include recirculating the leachate back to the landfill, treating for sanitary sewer discharge or for local surface water discharge.

Considerable research has been conducted on materials that can be used as the liner. Liners are broadly classified as i) soil liners, e.g., compacted clay liners (CCL), or soil (silt sand) mixed with bentonite; ii) geomembranes; and iii) geosynthetic clay liners (GCL) where soil (bentonite) is sandwiched between two sheets of fabric. The current dilemma for landfill liners are that "once the unit is closed, the bottom layer of the landfill will deteriorate over time and, consequently, will not prevent the transport of leachate out of the unit" (U.S. EPA 1988a, b). It is commonly recognized that even the best of the landfill liner systems being advanced would eventually fail to prevent groundwater pollution by MSW leachate (U.S. EPA 1988b). Therefore, one important direction is to develop a more sophisticated leachate detection system located between the two composite liners. When leachate is detected in that layer, action will be taken to prevent leachate from leaking out the landfill. Moreover, it would require the availability of adequate post-closure funding and responsibility to maintain the cover, operate the leachate collection system and leak detection system, and the ability to effect repairs in all systems, for as long as the wastes in the landfill can generate leachate when it is contacted by water (Lee and Jones-Lee 2009).

19.2 LEACHATE CHARACTERISTICS

19.2.1 Stabilization Processes in Landfills

Farquhar and Rovers (1973) first described the landfill phases as i) the aerobic phase (a few days), ii) the anaerobic acid phase (12–18 months for warmer climates, and up to 3 years in more temperate countries, e.g., the UK, Robinsion 2007), iii) the initial methanogenic phase (a few years), and iv) the stable methanogenic phase (a few decades). Later, additional phases are speculated as v) methane oxidation, vi) air intrusion, vii) carbon dioxide, and viii) soil air (Christense and Kjeldsen 1995); phases v to vii are not supported with field observation as most well monitored landfills are still younger than 30 years (Kjeldsen et al. 2002). The decomposition of landfill waste in these phases were described by Kjeldsen et al. (2002) that follows the general knowledge reported in the literature about aerobic and anaerobic digestion processes (Metcalf and Eddy 2013).

19.2.2 Leachate Quantity and Quality

Leachate production rates are often between 12 and 22% (mean value = 18%) of the annual precipitation in many Northern German landfills, lower than 10% for very young landfills in operation, and above 25% (up to 50%) for landfills with storage capacity (Stegmann et al. 2005). As shown in Table 19.1, landfill leachate is a complex mixture consisting of (Kjeldsen et al. 2002; Robinsion 2007):

- *Dissolved organic matter*: Quantified as Chemical Oxygen Demand (COD) or Total Organic Carbon (TOC), volatile fatty acids and more refractory compounds such as fulvic-like and humic-like compounds.
- *Inorganic macrocomponents*: Calcium (Ca^{2+}), magnesium (Mg^{2+}), sodium (Na^{+}), potassium (K^{+}), ammonium (NH_4^{+}), iron (Fe^{2+}), manganese (Mn^{2+}), sulfate (SO_4^{2-}), chloride (Cl^{-}), and hydrogen carbonate (HCO_3^{-}).
- *Heavy metals*: Cadmium (Cd^{2+}), chromium (Cr^{3+}), copper (Cu^{2+}), lead (Pb^{2+}), nickel (Ni^{2+}) and zinc (Zn^{2+}).
- *Xenobiotic organic compounds (XOCs)*: A variety of aromatic hydrocarbons, phenols, chlorinated aliphatics, pesticides and plasticizers.

Despite the complexity of leachate composition, leachate can be characterized as follows (Kjeldsen et al. 2002):

1. Landfill leachate contains four major groups of pollutants.
2. High leachate concentrations of all components exist in the early acid phase, while a more stable leachate, with lower concentrations and a low BOD/COD-ratio, exist in the long methanogenic phase. Roughly, BOD/COD $\geq$ 0.4 for acid phase; 0.2–0.4 for transient phase; and $\leq$0.2 for methanogenic phase (Kruse 1994).
3. Generally, very low concentrations of heavy metals are observed. The enhanced release of accumulated heavy metals would not take place within the time frame of thousands of years.
4. A broad range of XOCs exist in leachate, but they do not constitute a major long-term problem in most cases.
5. Ammonia often is a major pollutant in leachate and is of most concern in the long run. High concentration of ammonia affects COD removal so that pre-treatment to selectively precipitate NH_4 may be needed.

In general, it is believed that representative leachate samples are difficult to obtain because: 1) leachate composition varies significantly among landfills

Table 19.1. Leachate composition affected by landfill age and anaerobic digestion

Composition[a]	*New landfill (<2 years)*[b]	*Mature landfill (>10 years)*[b]	*Acetogenic Leachate*[c]	*Methanogenic leachate*[c]
BOD_5	2000–30000	100–200	4000–30000	<500–1000
TOC	1500–20000	80–160	3000–20000	< 10–300
COD	3000–60000	100–500	10000–50000	2000–6000
TSS	200–2000	100–400	NA[d]	NA
Organic N	10–800	80–120	NA	NA
NH_4-N	10–800	20–40	750–2000	1500–3000
NO_3-N	5–40	5–10	<1–< 1	<0.1–< 0.1
Total Phosphorus	5–100	5–10	5–20	1000–3000
Orthophosphorus	4–80	4–8	NA	NA
Alkalinity as $CaCO_3$	1000–10000	200–1000	2000–10000	10000–30000
pH	4.5–7.5	6.6–7.5	5.5–7.0	7.5–8.5
Total hardness as $CaCO_3$	300–10000	200–500	NA	NA
Ca^{2+}	200–3000	100–400	1000–2500	100–300
Mg^{2+}	50–1500	50–200	200–600	500–1500
K^+	200–1000	50–400	300–800	20–200
Na^+	200–2500	100–200	500–2000	50–300
Cl^-	200–3000	100–400	1000–2000	2000–4000
SO_4^{2-}	50–1000	20–50	200–1500	10–50
Total Fe	50–1200	20–200	100000–800000	100–500

[a] All units = mg/l except pH (no unit)
[b] Adapted from O'Leary and Tchobanoglous (2002)
[c] Adapted from Robinson (2007)
[d] Not available

Table 19.2. Typical leachate composition of landfills with different ages

Composition	*Recent (<5 yrs)*	*Medium (5–10 yrs)*	*Old (>10 yrs)*
pH	6.5	6.5–7.5	>7.5
COD (mg/l)	>10000	4000–10000	<4000
BOD_5/COD	> 0.3	0.1–0.3	< 0.1
NH_4-N	1500–3000	2000	2000
Organic compounds	80% VFA	5–30% VFA + HAs and FAs	HAs and FAs
Heavy metals	Low-medium	Low-medium	Low
Biodegradability	High	Medium	Low

Note: VFA = volatile fatty acids, HAs = humic acids, and FAs = fulvic acids
Sources: Renou et al. (2008); Robinson (2007)

depending on waste composition, waste age, landfilling technology, and location (weather conditions); and 2) sampling locations, sampling methods and sample handling routines affect the leachate composition. The spatial distribution of the leachate quality is important if leachate is released to groundwater as in the case of older landfills without the liner. Currently, no standard protocols exist for sampling, filtration and storage of leachate samples (Kjeldsen et al. 2002). However, recent studies indicate that: a) it is difficult to predict the acetogenic leachate quality at tropical sites because of the rapid transition there, and actual results are very responsive to short-term weather patterns; and b) the quality of leachates during methanogenic phases of waste decomposition at very large landfills is remarkably consistent, in spite of differences in climate, waste types irrespective of developed and developing countries. The review on different reports about leachate composition and concentrations, as presented in Table 19.1, are more or less representative and thus, can be used with caution. However, to simplify the issue caused by the variations of leachate composition, Table 19.2 is recommended.

19.3 LEACHATE TREATMENTS

19.3.1 Treatment Methods

Tables 19.3–19.6 summarize the current methods for leachate treatment. In general, options based on different mechanisms include physical/chemical and biological treatment. Those based on treatment train related to the landfill include recirculation of leachate through landfill site (e.g., by injection, by spray irrigation) and on-site treatment (e.g., any physical/chemical and/or biological treatment) and off-site treatment (e.g., discharge to a municipal water treatment facility).

In the past, the most commonly used methods are aerobic biological systems, including aerated lagoons, activated sludge systems, and sequential batch reactors

Table 19.3. Summary of landfill leachate transfer

Method	*Description*
~20% of leachate is handled this way; getting biogas from recirculation is a sustainable practice	
A. Combined treatment with domestic sewage	Is preferred for its easy maintenance, low operating costs, and leachate as nitrogen source for biological WWT; however, the organic inhibitory compounds and heavy metals may reduce treatment efficiency; also, many landfills' locations do not allow this option
B. Recirculation	One of the cheapest options available; enhances biodegradation by increasing moisture and movement of moisture; improves the leachate quality and shortens the stabilization time of landfills from several decades to 2-3 years; can generate biogas, which once collected, is a renewable energy; however, a high recirculation rate may cause saturation, ponding, acidic conditions, pipe clogging; some landfills are not designed with vertical wells

Sources: Stegmann et al. (2005); Cath et al. (2006); Renou et al. (2008); Zhang et al. (2012)

(SBRs). While these biological methods are essentially successful, they do suffer some problems specifically associated with the properties of landfill leachate. For example, the retention times are long to enhance performance. The microbes are sensitive to toxic heavy metals, loading rates, and temperature variations. It is difficult to remove NH_4-N since enough organic carbon may not be available after BOD removal in the activated sludge process. Total dissolved solids (TDS) often cannot be reduced by biological treatment processes (some even increase it); several treatment technologies often need to be used in a train of unit processes in order to meet effluent limits.

Recently, membrane technologies are used for leachate treatment, including microfiltration (MF), ultrafiltration (UF), nanofiltration (NF), reverse osmosis (RO), and forward osmosis (FO). RO and FO membranes are capable of selective separations and can achieve filtrate quality better than any biological systems. Usually, they are not subject to the loading rate, toxic metals, and temperature limitations of biological systems, and thus, represent the future direction. Renou et al. (2008) reported landfill leachate treatment distribution in France as follows:

- For 2000: biological = 62%, physicochemical = 9%, membrane = 10% and transfer + others = 21%
- For 2002: biological = 46%, physicochemical = 13%, membrane = 18% and transfer + others = 23%

Table 19.4. Summary of landfill leachate biological treatment methods

Method	*Description*
~40–60% of leachate is handled this way; with reliability, simplicity and high cost-effectiveness; used for BOD and nitrogen removal, particularly efficient for immature leachates (BOD/COD > 0.5)	
A. Aerobic treatment	Efficient for BOD removal and nitrification; heterotrophic denitrification may require addition of external carbon sources
a) Suspended growth	very common, worldwide used
1) Lagoon	long retention time (50-100 d); green technology for leachate treatment; low cost and less maintenance; N, P, and Fe removals >70%, ~60% COD removal; temperature dependence is a major limitation
2) Activated sludge processes	Extensively used, but inadequate for handling leachate due to poor sludge settleability, longer aeration time, high energy demand, excess sludge production, inhibition of microbes by high NH_3-N, which is released when pH is about 9 due to aeration
3) SBR	Flexibility to handle non-steady flow rate of leachate; COD removal >75%; ideally suited to nitrification-denitrification processes
b) Attached-growth	Attracts more studies recently; not suffer from loss of active biomass; nitrification less affected by low temperature; high nitrogen content induces inhibition
1) Trickling filters	90% nitrification of leachate is possible due to longer solids retention time (SRT); 87% BOD removal is achievable
2) Moving-bed biofilm reactor	Suspended porous polymeric carriers with attached biofilm as active biomass kept in the aeration tank; higher biomass concentration, no long sludge settling periods, less sensitive to toxic compounds; high organic (60–80%) and ammonia (85–90%) removals
B. Anaerobic treatment	Suitable for dealing with high strength organic effluents; save energy; low solids production; a slow process with low reaction rates
a) Suspended-growth	Commonly used; digester is green technology if biogas is recovered
1) Digester	Conventional anaerobic suspended-growth anaerobic digester used very often; 80-90% BOD and ~55% COD removals

(Continued)

Table 19-4. Summary of landfill leachate biological treatment methods (Continued)

Method	*Description*
2) SBR	Good performance; sequential anaerobic-aerobic operations results in better removal of COD (62%), ammonia-N (31%), and phosphorus (19%)
3) UASB	Has high treatment efficiency and a short hydraulic retention time; works at low temperature; sensitive to toxic substances
b) Attached growth	Often high rate systems
1) Anaerobic filter	A high rate system; gathers the advantages of other anaerobic systems and minimizes the disadvantages; biomass is retained as biofilms on support material; may reduce COD by 90%; high supporting material cost
2) Hybrid bed filter	A UASB at the bottom and an anaerobic filter on top; enhances solid's retention; the biomass concentration in the system increases and the performance enhanced.
3) Fluidized bed reactor	Carbon-assisted fluidized bed reactor is effective for treating old landfill leachate

Sources: Stegmann et al. (2005); Cath et al. (2006); Renou et al. (2008); Zhang et al. (2012)

Therefore, the market share of membrane processes increases rapidly, which reflects the worldwide trend. However, two problems are associated with the membrane process, particularly the RO process, that is, membrane fouling and formation of concentrated brine. Many studies have been conducted to overcome these problems.

19.3.2 Most Applied Methods, Treatment Effectiveness, and Treatment Strategy

Madu (2013) surveyed 27 different landfills concerning the treatment methods they used and found that SBR, lagoon, wetland, and RO are more popular than other methods (Figure 19.2), with SBR being the most applied one (55.5% plants use it). Compared with other technologies (particularly activated sludge processes), SBRs have several advantages, such as i) energy is optimized through control of aeration rate and duration; ii) no secondary clarifier is needed so that the cost is reduced; iii) flexibility in adjusting reaction time and tank volume to meet variable loading; and iv) small foot-print. Lagoon processes are technically and economically viable at full scale while the wetland and the reverse osmosis are very good secondary polishing processes for the biological pretreatments (Madu 2013).

Table 19.5. Summary of leachate conventional physical/chemical treatment methods

Method	*Description*
	~10% of leachate is handled this way; often used for treating a specific pollutant (e.g. stripping off ammonia) or in concert with other treatments (pre-treatment or last purification)
A. Flotation	Mainly used for reduction of suspended solids (SS), colloidal particles, and floating material; usually used with other processes
B. Coagulation-flocculation	Used as a pre-treatment before biological or membrane treatment or after GAC or PAC sorption treatment for removing loaded activated carbon; normal coagulants used in water and/or wastewater treatment plants can be used, but iron salts are better than aluminum salts; the process produces a large quantity of sludge with increased Fe and Al concentration
C. Chemical precipitation	Widely used as a pre-treatment for selectively precipitating NH_4 (e.g., adding $MgCl_2 \cdot 6H_2O$ and $Na_2HPO_4 \cdot 6H_2O$ to get magnesium ammonium phosphate (MAP) or precipitating NH_4 as struvite ($Mg:NH_4:PO_4 = 1:1:1$) under anaerobic condition
D. Chemical oxidation	A widely studied method for removal of refractory compounds in leachate; advanced oxidation processes (AOP) are the focus, including O_3, H_2O_2, H_2O_2/Fe^{2+} (Fenton's), UV, ultrasound (US), electron beam (EB), transition metal ions or photocatalyst; using AOP after biological treatment is more efficient; using AOP for direct leachate treatment would be less effective (~30% COD depletion), but may increase BOD/COD ratios (i.e., improve the biodegradability); AOP + other treatments would be better; however, high energy demand and needs to change leachate conditions (e.g., lower pH for Fenton's process to work)
E. Air stripping	The most common method for eliminating a high concentration of NH_4 (removal efficiency = 75–99%); high pH must be achieved by adjusting the leachate (causing $CaCO_3$ scaling and foams of the stripping tower if lime is used), and the off-gas needs treatment (otherwise, air pollution by NH_3)

(Continued)

Table 19-5. Summary of leachate conventional physical/chemical treatment methods (Continued)

Method	*Description*
F. Sorption	Often used to reduce the residual COD and AOX below the discharge limits; GAC or PAC is effective to sorb pollutants, much better than chemical methods, but carbon regeneration is an issue. Other materials can be used as sorbents, e.g., zeolite, resin, peat, limestone (for metal removal); biotreatment + GAC sorption (either added in aeration tank + flocculation/precipitation or in a follow-up column) is very effective

Sources: Stegmann et al. (2005); Cath et al. (2006); Renou et al. (2008); Zhang et al. (2012)

Table 19.7 indicates that for treating leachate of young landfills, there is a wide range of treatment techniques that can be used. Membrane techniques are versatile for treatment of different landfill leachates, and can save space. Some methods can be operated under the watchful eyes of less skillful personnel. Table 19.8 compares some leachate treatment methods by considering leachate characteristics and landfill ages. It seems that the strength of landfill leachate is often too high to be treated with a single treatment method (even with the RO process) to meet the stringent discharge regulations (particularly for removal of TDS, COD, and TOC) (see Table 19.9). However, using combinations of different methods can achieve treatment goals (Table 19.10). Combinations shown in Table 19.11 have been operated in Germany for several years, and their reliability has been demonstrated (Stegmann et al. 2005).

According to Renou et al. (2008), the major criteria for the selection of suitable treatment strategy include: i) consideration of the initial leachate quality; ii) the discharge limitations and requirements; iii) the treatment efficiency; iv) methods to reduce the cost of treatment residues; v) pre-treatment (e.g., biological treatment is not a good pre-treatment for RO); vi) variability to be viable over the lifetime of the landfill, and to cope with temporal changes in leachate quality and composition, and vii) cost (capital and operation/maintenance cost, e.g., cleaning cost for RO systems). Usually, treatability trials are needed to derive the treatment process, plant size, predicted emissions and the required abatement. For sustainable development, biogas capture must be promoted.

19.3.3 Cost of Landfill Leachate Treatments

Charges for trade effluent to sewer are based on the Mogden formula (IPPC 2007):

$$\text{Bill} = R + [(V + Bv) \text{ or } M] + B(Ot/Os) + S(St/Ss) \qquad \text{(Eq. 19.3)}$$

Table 19.6. Summary of leachate membrane treatment methods

Method	*Description*
~10-18% of leachate is handled this way; emerged since 1990s; separating leachate into two streams, i.e., pure permeate with a recovery about 90% and brine with a concentration factor of 10 (or ~20% of the leachate will become brine); two major drawbacks (fouling and brine production) need more research	
A. Microfiltra-tion (MF)	Used as a pre-treatment for another membrane (e.g., UF and RO) or prefiltration; mainly used for removal of SS and colloids; COD removal = ~25–35%
B. Ultrafiltration (UF)	Used as a pre-treatment for another membrane (e.g., RO) or used as a post-treatment after the biological treatment; mainly used for particle removal but may remove COD by 50%; as a new focus, UF is used in MBRs to achieve high biomass concentration and a low sludge production with excellent effluent quality, and also used in a bio-GAC + UF system for efficient removal of TOC (95-98%) and specific pollutants (97%); in other combined systems, slow growth organisms (nitrifiers) can be kept and function actively in the systems
C. Nanofiltration (NF)	Versatile for removing organic, inorganic, and microbial contaminants (e.g., 60–70% COD and 50% NH_4): using physical methods + NF can achieve 70-80% COD removal; rejecting sulfate ion, but not chloride and sodium; fouling is a major issue
D. Reverse osmosis (RO)	Being one of the most promising and efficient membrane methods; rejecting COD, heavy metals, and most of ions (98–99%) with a flux of 15 LMH at 30–60 bar pressure. The energy consumption = ~5 kWh/m^3 of permeate for a recovery of 80%, but separation of NH_4 is not sufficient unless multiple stages are used; the disc-tube-module (DT-module) developed by Pall-Exekia and Rochem is widely used; major issues: fouling and brine; fouling cleaning is expensive but can be reduced by several ways. One way is to use torsional oscillation of the membrane stack to increase turbulence near the membrane as VSEP systems; brine can be handled by recycling back to the landfills (not the best option), on-site evaporation (need land), solidification (costly), deep mine landfilling (not widely available), and incineration (maybe the best option so far)

(*Continued*)

Table 19-6. Summary of leachate membrane treatment methods (Continued)

Method	*Description*
E. Forward osmosis (FO)	Can be very effective for leachate treatment; often uses pre-treated leachate as feed, but raw leachate can also be used; the draw solution often goes to RO units for separating pure water and concentrating the draw solution for reuse; field tests result in water recoveries of 92% with high contaminant rejection (99%); fouling of both FO and RO membrane is reduced, and flux can be restored after cleaning

Sources: Stegmann et al. (2005); Cath et al. (2006); Renou et al. (2008); Zhang et al. (2012)

where, Bill = total charge rate for disposal (pence/m^3); R = unit cost for reception and conveyance (pence/m^3); V = unit cost for primary volumetric treatment (pence/m^3); Bv = additional volume charge if there is biological treatment; M = treatment and disposal where effluent goes to a sea outfall; B = unit cost for biological treatment (oxidation of settled sewage) (pence/m^3); Qt = COD of trade effluent (mg/l); Qs = COD settled sewage (mg/l); S = unit cost for sludge disposal (pence/m^3); St = solids value trade effluent (mg/l); and Ss = solids value*settled sewage (mg/l).

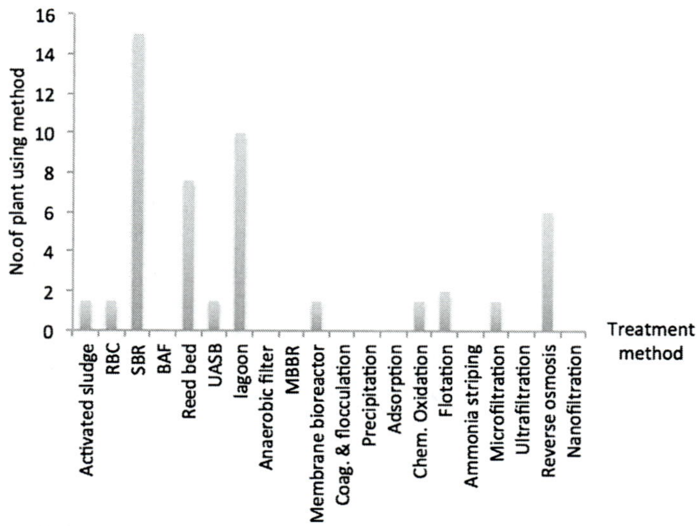

Figure 19.2. Most popular treatment methods of 27 landfills
Source: Aadapted from Madu (2013)

Table 19.7. Comparison of different treatment methods based on treatment efficiency, space utilization, and capital and O/M costs

	Leachate					
Treatment Process	*Young*	*Medium*	*Old*	*Space utilization*	*Installation & operation cost*	*Requiring less skilled personnel*
Biological						
Activated sludge	Good	Fair	Poor	Poor	Expensive	No
RBC	Good	Fair	Poor	Good	Expensive	Yes
SBR	Good	Fair	Poor	Good	Less expensive	No
Reed beds	Fair	Fair	Good	Poor	Less expensive	Yes
BAF	Good	Fair	Fair	Good	Expensive	Yes
Lagoons	Good	Fair	Poor	Poor	Expensive	Yes
UASB	Good	Fair	Fair	Good	Less expensive	Yes
AF	Good	Fair	Fair	Good	Expensive	Yes
MBBR	Good	Fair	Poor	Poor	Expensive	No
MBR	Good	Fair	Fair	Poor	Expensive	No
Physicochemical						
Coagulation and flocculation	Poor	Fair	Fair	Fair	Less expensive	No
Precipitation	Poor	Fair	Poor	Fair	Less expensive	No
Adsorption	Poor	Fair	Good	Good	Less expensive	No
Flotation	Poor	Fair	Fair	Poor	Expensive	Yes
Chemical oxidation	Poor	Fair	Fair	Good	Expensive	No
Ammonia stripping	Poor	Fair	Fair	Poor	Expensive	No
Membrane process						
Microfiltration	Poor	Poor	Poor	Good	Expensive	Yes
Ultrafiltration	Fair	Fair	Fair	Good	Expensive	Yes
Nanofiltration	Good	Good	Good	Good	Expensive	Yes
Reverse osmosis	Good	Good	Good	Good	Expensive	Yes

Source: Madu (2013)

Table 19.8. Estimated effluent concentration after single treatment[a]

Process	BOD (mg/l)			COD (mg/l)			TKN (mg/l)		
	Recent	Medium	Old	Recent	Medium	Old	Recent	Medium	Old
Transfer									
Treat with WWTP	Depending on WWTP			Depending on WWTP			Depending on WWTP		
Recirculation	150	123	40	2000	2250	1600	–	–	–
Physicochemical									
Coagulation-flocculation	–	–	–	6000	3750	2000	1575	1400	1400
Chemical precipitation	–	–	–	7000	2250	2800			–
Sorption	600	300	80	3000	1500		–	–	–
Chemical oxidation	–	–	–	7000	3000		–	–	–
Stripping	–	–	–	7000	5250	2800	–	–	
Biological									
Lagoon	600	300	40	6000	750	–	450	1300	1300
Aerobic processes	600	300	80	4000	750	1600	450	400	400
Anaerobic processes	600	300	80	4000	1500	1200	450	400	400
membrane bioreactor (MBR)	600	300	80	1500	1125	600	450	400	400
Membrane filtration									
Ultrafiltration	–	–	–	5000	3750	2000	675	600	600
Nanofiltration	600	300	80	4000	1500	800	450	400	400
Reverse osmosis	150	75	20	500	375	200	113	100	100

[a]Calculations are based on Table 19.2 and removal efficiencies in Table 22 of Renou et al. (2008). For example, for recirculation's effluent BOD of young landfills, 10000*0.3*0.05 (good for young landfills) = 150 mg/l. RO's removal efficiency for BOD, COD, and TKN are assumed to be 95%

Table 19.9. Selected discharge limits of treated leachate

Parameter	*Germany*	*France*	*Parameter*	*Germany*
COD	200	125 (for Q > 100 kg/d)	Total phosphorus	3
BOD	20	30 (for Q > 30 kg/d)	Hydrocarbons	10
TN	70	30 (for Q > 50 kg/d)	NO_2-N	2
TSS	–	35 ((for Q > 15 kg/d)	AOX	0.5

Sources: Stegmann et al. (2005), Renou et al. (2008)

The costs of leachate treatment vary roughly and depend on the volume and composition of leachate and the final disposal route. Table 19.11 shows the examples of two different landfills in the UK. The costs of leachate treatment in Germany is between 10 and 70€/m^3 leachate (Stegmann et al. 2005). Vaidya (2012) analyzed 9 landfills and their methods for leachate treatment, and summarized the findings (Table 19.12). For example, details about the MSW landfill of Hall County, GA include: a total disposal area of ~39 acres, handling ~175 tonne MSW/d, and generating leachate ~2.7 million gal/yr (=69230 gal/ac-yr). The landfill has been using RO systems + concentrate reuse since 1999 with the current capacity of ~20,000 gal/d. The leachate treatment costs are: maintenance = $40,320; operator = $16,775; power = $22,000; and total = $79,095 (thus cost/gal = $0.036). If hauling to WWTP, the hauling cost per gallon = $0.18, and thus, total cost to haul/yr = $394,200. Vaidya (2012) summarized the typical baseline below:

- Hauling + sewer/treatment = ~$55.0 to $110/1000 Gallons
- Feasible to treat leachate if typically Q ≥ 5000 GPD
- Wetlands = ~20–30% of baseline cost if land is inexpensive/available
- RO = ~40–60% of baseline
- Chemical/AOP/GAC/PAC + others = ~55–70% of baseline

19.4 SUSTAINABLE DEVELOPMENT OF LEACHATE TREATMENT TECHNOLOGY

Sustainable development of leachate treatment techniques are critical to our society as human beings are facing global challenges such as population explosion, increasing demand for food and water, resource depletion, and climate change. We have been forced to move forward towards sustainable society. From previous sections, it is understandable that developing on-site techniques for leachate treatment may make more sense than transferring leachate for remote treatment. In this section, several techniques are described as they have potential to be sustainable techniques.

Table 19.10. Effectiveness of different combinations for leachate treatment

COD (mg/l)		NH_4-N (mg/l)		AOX (mg/l)	
Influent	*Effluent*	*Influent*	*Effluent*	*Influent*	*Effluent*
1: biological + active carbon + flocculation/precipitation + neutralization					
1571	76	579	0.09	1.45	0.18
686	51	528	0.7	1.65	0.23
2: biological + UF + active carbon					
1000–12000	< 200	400–800	<10	1–2.5	0.1-0.7
3: biological + chemical oxidation (O_3 + UV)					
320–5796	30–137	125–1350	0.4–36.2	–	–
4: biological + chemical oxidation (O_3 + UV) + biological					
1200–4000	18–150	600–1900	0.1–9	1–3.8	0.04–0.18
5: biological + chemical oxidation (O_3 + UV) + biological + activated carbon					
758–1332	1–85	375–885	0.1–0.6	0.85–2.1	0.17–0.43
6: one step RO					
4124	20	577	8	–	–
1550	68	750	7	1.4	< 0.01
7: Two-step RO					
1590–2980	4–25	900–1800	4.4–8.8	1.5–1.9	0.002–0.02
8: biological + one-step RO					
446–872	5.3–27	80–396	0.03–10.1	0.4–1.4	< 0.01–0.05
9: biological + two-step RO					
1366–3010	< 2	130–854	6.3	1.09–2.24	0.045

Source: Adapted from Stegmann et al. (2005)

Table 19.11. Landfill leachate treatment costs[a]

	Capital expenditure (£)		Operational cost (£/m^3) (O/M/reagent or transport cost + discharge costs)	
Treatment method	*Landfill 1*	*Landfill 2*	*Landfill 1*	*Landfill 2*
Removal by tanker & disposal at WWTP	–	–	17.50 (15 + 2.5)	15.38 (15 + 0.38)
Air/CH_4 stripping + sewer connection & disposal costs[b]	300,000	100,000	3.10 (0.60 + 2.5)	0.98 (0.60 + 0.38)
SBR and disposal to sewer[c]	1,000,000	250,000	1.72 (0.80 + 0.92)	1.15 (0.80 + 0.35)
SBR + air floatation + CW + discharge to surface water[d]	1,500,000	400,000	1.50 (no disposal cost for discharge to surface water)[e]	

[a]Landfill 1: leachate Q = 400 m^3/d, raw COD = 6,000 mg/l and SS = 250 mg/l;. Landfill 2: Q = 60 m^3/d, COD = 150 mg/l, SS = 90 mg/l

[b]Methane stripping reduces methane concentrations sufficient to allow discharge to sewer but does not significantly reduce COD or SS

[c]Landfill 1: COD is reduced to 1500 mg/l and SS remain at 250 mg/l

[d]Landfill 2: COD is reduced to 42 mg/l and SS remain at 90 mg/l

[e]It is possible that the operational expenditure figure quoted could range from £0.75–5.50/m^3 depending on the concentration of ammonia in the leachate and how much of this requires removal

Table 19.12. Performance and cost estimation of some treatment methods of landfill leachate

Input	*New Hanover, NC*	*Confidential*	*Hall County, GA*	*Broward County, FL*
Source	MSW	MSW + C & D Waste	MSW	Class III cells
Age (yr)	22	–	13	18
Volume (M^3/d)	227	2,844	28	144
COD/removal efficiency	450/70%	5400/50%	~800	870
BOD/removal efficiency	41/80%	2160/80%	~500	630
TDS (mg/L)/ removal	4673/72%	–	~1000	3500
NH_3-N (mg/L)/removal	108/95%	1580/79%	~20	~7.6
pH	7.8/7.6	–	~7.6	7.6
Treatment method	CW	SBR to sewer	RO/on-site reuse	Study CW/RO/Haul
Net % cost (\$/M^3)	4.60	15.06	9.51	2.19-3.70/5.38/10.04
Input	*Pinellas County, FL*	*Pasco County, FL*	*Pasco County, FL*	*Pasco County, FL*
Source	MSW closed	Class III cells	MSW cells	MSW cells
Age (yr)	22-40	18	1-16	16-22
Volume (GPD)	473	227	96	40
COD/removal efficiency	–	1800	1500	800
BOD/removal efficiency	–	1200	1150	300
TDS (mg/L)	–	2400	2200	1100
NH_3-N (mg/L)	–	14	60	24
pH	–	6.5	6.9	6.8
Treatment method	Haul to WTP	Pump to WWTP	Haul to WWTP/sewer	Haul to WWTP/sewer
Net % cost (\$/M^3)	6.34	13.50	To be confirmed	13.50/0.92

Source: Adapted from Vaidya 2012

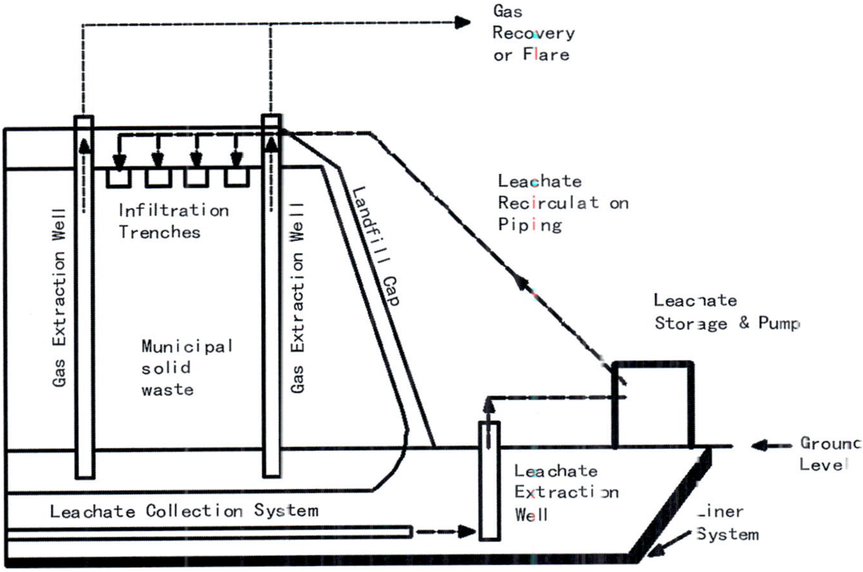

Figure 19.3. Schematic diagram of a bioreactor landfill (adapted Hughes et al. 2013)

19.4.1 Bioreactor Landfills

A bioreactor landfill (see Figure 19.3) changes the goal of landfilling from the storage of waste to the treatment of waste, making the landfill a natural bioreactor that simultaneously degrades waste material and generates landfill gas (LFG), which, is a renewable energy. Turning landfills into bioreactor landfills is not difficult in both concept and practice; it only needs to treat leachate with recirculation methods and collect/utilize LFG (Reinhart and Townsend 1998; Shan et al. 2010). The bioreactor landfill provides control and process optimization, primarily through the addition of leachate or other liquid amendments. Sufficient experience now exists to define recommended design and operating practices (Reinhart et al. 2002), and to collect and utilize LFG (Shan et al. 2010).

However, technical challenges and research needs remain related to sustainability, liquid addition, leachate hydrodynamics, leachate quality, the addition of air, and cost analysis. Moreover, the valorization and application of bioreactor landfills is affected by landfill systems, LFG collection, LFG concentration and purity, LFG online monitoring, policies, and so on. Future work on landfill technology may focus on: i) the development and optimization of large-scale bioreactor landfills, ii) LFG collection and purification, iii) improvement of methodologies for in-situ and post-closure monitoring, iv) life-cycle assessment (LCA), and v) sustainability, broader visions and non-technical barriers (Reinhart et al. 1998; Shan et al. 2010).

19.4.2 Membrane Technology for Leachate Treatment

As discussed in previous sections, among many potential techniques, membrane technology is one of the most direct, effective and feasible approach for on-site treatment of landfill leachate. However, current RO technique is ~40–60% of the baseline cost of landfills, which is still very high. However, membrane techniques for leachate treatment can be improved, particularly in the areas of membrane fouling control, membrane units design and operation, and pre-treatment techniques. A few examples are described as follows.

RO with vibration: New logics Vibratory Shear Enhanced Process (VSEP) RO system uses high crossflow and high turbulence to keep the feed liquid homogenous and evenly concentrated. It also employs torsional oscillation (50 times/second and the membrane displacement = ¾" peak to peak at the perimeter) of the membrane stack. The resulting shear is 150,000 inverse seconds, which is ten times greater than the shear in crossflow systems. In principle, conventional membrane systems could develop cakes of colloids that would grow large enough to completely clog the membrane. In VSEP, irrespective of colloidal amount, equal numbers are removed as the diffusion layer is limited in size due to the gravitational pull (G force) of the vibrating membrane. In addition, the vibration and oscillation of the membrane surface itself inhibits crystal formation. High shear significantly reduces the fouling of many materials. Thus, the VSEP system can achieve fluxes as high as 13 times that of the conventional RO system (Figure 19.4). The resistance to fouling can be enhanced with membrane selection where virtually any commercially available membrane materials such polyamide, polypropylene, polyester, polysulfone and Teflon can be used (NLR 2013).

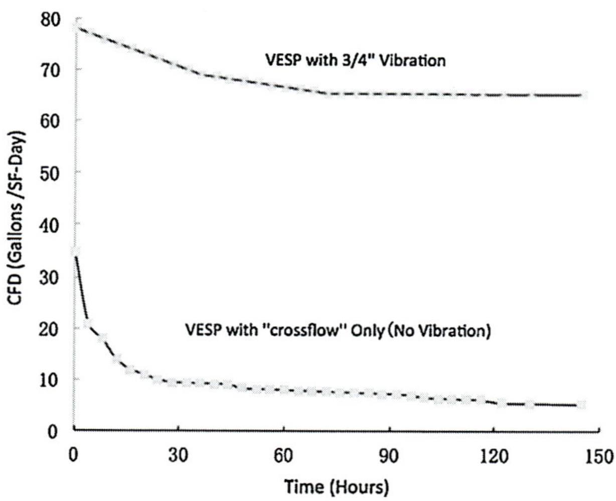

Figure 19.4. Effect of vibration on RO performance. Test conditions: 400 psi, 25°C, saturated calcium sulfate slurry

Source: Adapted from NLR (2013)

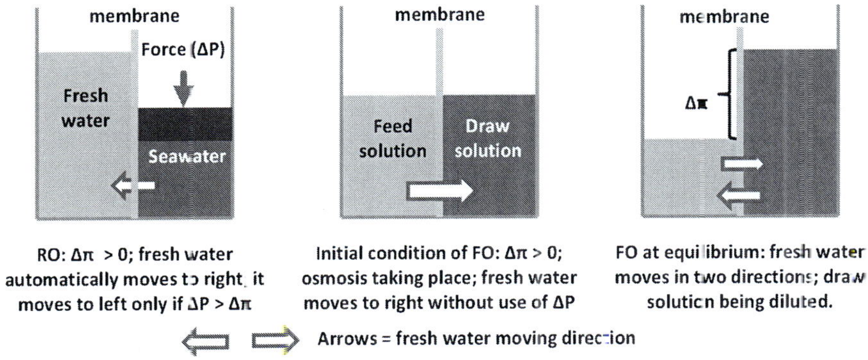

Figure 19.5. Schematic of water flows in RO and FO.
ΔP = applied hydraulic pressure. Δπ=osmotic pressure of the right chamber – that of the left chamber
Source: Zhang and Surampalli (2012); © ASCE

FO+RO systems: FO processes can be used to improve membrane techniques for leachate treatment. In the FO process, the osmotic pressure gradient is used to let freshwater flow from the feed solution chamber (fed with raw leachate) to the draw solution chamber with concentrated draw solute (Figure 19.5). Osmotic driving forces in FO can be significantly greater than hydraulic driving forces in RO, potentially leading to higher water flux rates and recoveries (Zhang et al. 2012).

Currently, there are several reported field-scale FO systems for leachate treatment. Osmotek (2003) reported the successful operation of a full-scale FO system in 1999 (Figure 19.6). The treatment plant treated over 18,500 m^3 of leachate within 9 months, with an average water recovery of 91.9% and an average RO permeate conductivity of 35 μS/cm. Most contaminants had greater than 99% rejection and final effluent concentrations were much lower than the National

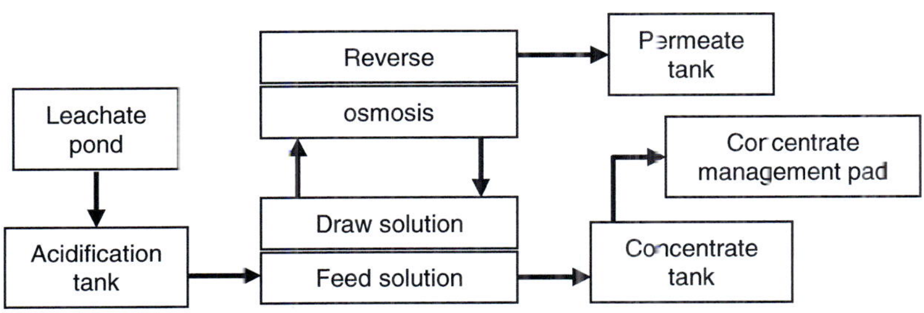

Figure 19.6. A flow diagram of a full-scale Osmotek's FO leachate treatment system. The system uses Osmotek's CTA membrane and NaCl as the draw solution with water recoveries of 94–96%. The concentrate is solidified by fly ash and Portland cement and then sent to landfill

Pollutant Discharge Elimination System's (NPDES) total maximum daily load (TMDL) levels (York et al. 1999).

Hydration Technology Innovations (HTI) developed its Closed Loop Hybrid FO Wastewater System using plate and frame cassettes of HTI's proprietary OsMem FO membrane. The system has 2,500 m^2 of FO membranes and produces a flux rate of 25,000 gal/d. A bag filter system is used to filter the leachate to 100 μm which then is used as the feed of the FO unit. The FO system is cleaned on a weekly basis. To achieve permeate discharge conductivities of less than 10 μS/cm, RO polishing is used. The RO system uses high pressure seawater spiral wound elements with a nano-membrane brine booster. HTI's system treats leachate for $0.5/gal and meets tough NPDES discharge regulations (HTI 2013).

FO systems do have additional complications, however, when compared with RO systems. In particular, the FO process does not provide high quality water for use in a single step; after the FO step, the high quality water is mixed with the osmotic agent. These are the major obstacles of FO process for desalination. To separate permeate from draw solution and concentrate the draw solution for its reuse, a RO unit usually is used. While the treatment train of pre-treatment + FO + RO looks awkward, the system has better performance than two-step RO system. This is mainly because both the FO and RO system are not easy to be fouled, and the FO system is easy to install with much less initial cost and easy to operate as the FO system is a low pressure system (Zhang et al. 2012; Zhang and Surampalli 2012).

19.4.3 Using Green Technology for Leachate Treatment

Constructed wetland, lagoon, MBR, and SBR are green technologies due to their obvious advantages compared with other conventional wastewater treatment processes. Their applications for leachate treatment should be promoted in the future. Also, some other techniques should be promoted, such as passive treatment of leachate using the granular drainage blanket of the leachate collection system, and microbial fuel cells (MFCs) that can generate electricity from landfill leachate. These novel techniques provide an economical route for leachate treatment and energy recovery.

19.5 CONCLUSIONS

This chapter presents leachate collection and treatment methods. The selection of these methods depends on regulations, cost analysis, technical feasibility and sustainability. Whenever possible, natural processes (lagoon, wetland) and sustainable methods (bioreactor landfills) should be considered first. RO, SBR, and MBR have many advantages, particularly with their small footprint. In many cases, combinations of different treatment methods are needed. Some methods can be operated under the watchful eyes of less skillful personnel. Membrane techniques are versatile for treatment of different landfill leachates, and can save space; thus, represent the future trend of leachate treatment. Bioreactor landfills for waste management will be the norm in the near future.

19.6 ABBREVIATIONS/NOMENCLATURE

BOD	Biochemical oxygen demand
CCL	Compacted clay liners
CDDLFs	Construction & demolition (C&D) debris landfills
COD	Chemical oxygen demand
CTA	Cellulose triacetate
FAs	Fulvic acids
FO	Forward osmosis
GAC	granular activated carbon
GCL	Geosynthetic clay liners
HAs	Humic acids
HELP	Hydrologic Evaluation of Landfill Performance
HTI	Hydration Technology Innovations
HWLFs	Hazardous waste landfills
IWLFs	Industrial waste landfills
LMH	L/m^2-h
LFG	Landfill gas
MBR	Membrane bioreactor
MF	Microfiltration
MSWLFs	Municipal solid waste landfills
NF	Nanofiltration
NH_4-N	Ammonium nitrogen
NPDES	The National Pollutant Discharge Elimination System
PAC	Powdered activated carbon
PRO	Pressure retarded osmosis
RCRA	The Resource Conservation and Recovery Act
RO	Reverse osmosis
SBR	Sequencing batch reactor
TDS	Total dissolved solids
TMDL	Total maximum daily load
TOC	Total organic carbon
UASB	Up-flow anaerobic sludge blanket
UF	Ultrafiltration
USEPA	United States Environmental Protection Agency
VFA	Volatile fatty acids
VSEP	Vibratory shear enhanced process
XOCs	Xenobiotic organic compounds
ΔP	Hydraulic pressure
$\Delta \pi$	Osmotic pressure difference

References

Christensen, T. H., and Kjeldsen, P. (1995). "Landfill emissions and environmental impact: An introduction." *Proc., 5th Int. Landfill Symp. SARDINIA '95*, Vol. **3**, T. H. Christensen, R. Cossu, and R. Stegmann, eds., CISA, Cagliari.

EHSO (Environment, Health and Safety Online). (2013). "Commercial hazardous waste landfills." ⟨http://www.ehso.com/cssepa/tsdflandfills.php⟩ (Apr. 1, 2013).

Farquhar, C. J., and Rovers, F. A. (1973). "Gas production during refuse decomposition." *Water Air Soil Pollut.*, **2**(4), 483–495.

HTI (Hydration Technology Innovations). (2013). "Applying HTI's forward osmosis technology to landfill leachate treatment." ⟨http://futuretechme.com/downloads/forward-osmosis⟩ (Mar. 1, 2013).

Hughes, K. L., Christy, A. F., and Heimlich, J. E. (2013). "Bioreactor landfills." ⟨http://ohioline.osu.edu/cd-fact/0139.html⟩ (Mar. 1, 2013).

IPPC (Integrated Pollution Prevention and Control). (2007). "Guidance for the treatment of landfill leachate." ⟨http://www.environment-agency.gov.uk⟩ (Mar. 1, 2013).

Kjeldsen, P., Barlaz, M. A., Rooker, A. P., Baun, A., Ledin, A., and Christensen, T. H. (2002). "Present and long-term composition of MSW landfill leachate: A review." *Crit. Rev. Environ. Sci. Technol.*, **32**(4), 297–336.

Kruse, K. (1994). "Langfristiges Emissionsgeschehen von Siedlungsabfalldeponien." Veröffentlichungen des Instituts für Stadtbauwessen, Technische Universität Braunschweig, Heft 54, Braunschweig, Germany.

LaGrega, M. D., Buckingham, P. L., and Evans, J. C. (2000). *Hazardous waste management*, 2nd Ed., McGraw-Hill, New York.

Lee, G. F., and Jones-Lee, A. (2009). "Flawed technology of subtitle D landfilling of municipal solid waste." ⟨http://www.gfredlee.com/Landfills/SubtitleDFlawedTechnPap.pdf⟩ (Mar. 1, 2013).

LLTEW (Landfill leachate treatment expert website). (2011). "Avoiding over reliance on the hydrologic evaluation of landfill performance (HELP) computer program." ⟨http://leachate.co.uk/main/⟩ (Mar. 1, 2013).

Madu, J. I. (2013). "New leachate treatment methods." ⟨http://www.chemeng.lth.se/exjobb/E486.pdf⟩ (Sep. 16, 2015).

Matcalf and Eddy. (2013). *Wastewater engineering treatment and reuse*, 5th Ed., McGraw Hill, New York.

NLR (New Logic Research). (2013). "Membrane filtration of landfill leachate: Advances in module designs enhance landfill effluent treatment." Emeryville, CA.

O'leary, P. R., and Tchobanoglous, G. (2002). "Landfiling." Chapter 14, *Handbook of soild waste management*, 2nd Ed., G. Tchobanoglous and F. Kreith, eds., McGraw-Hill, New York, 14.1–14.93.

Osmotek. (2003). *Landfill leachate treatment.* ⟨http://www.rimnetics.com/osmotek.htm⟩ (Jan. 20, 2007).

Oweis, I. S., and Khera, R. P. (1998). *Geotechnology of waste management*, 2nd Ed., PWS Publishing, Boston.

Peavey, C. (2010). "Food waste as a resource—Part 1: Landfill technology." ⟨http://www.begreenjournal.com/post/2010/08/19/⟩ (Mar. 1, 2013).

Reinhart, D. R., and Townsend, T. G. (1998). *Land bioreactor design and operation*, CRC Press, New York.

Reinhart, D. R., McCreanor, P. T., and Townsend, T. G. (2002). "The bioreactor landfill: Its status and future." *Waste Manage. Res.*, **20**(2), 172–186.

Renou, S., Givaudan, J. G., Poulain, S., Dirassouyan, F., and Moulin, P. (2008). "Landfill leachate treatment: Review and opportunity." *J. Hazard. Mater.*, **150**(3), 468–493.

Robinson, H. (2007). "The composition of leachates from very large landfills: An international review." *CWRM*, **8**(1), 19–32.

Schroeder, P. R., Dozier, T. S., Zappi, P. A., McEnroe, B. M., Sjostrom, J. W., and Peyton, R. L. (1994). "The hydrologic evaluation of landfill performance (HELP) model: Engineering documentation for version 3." EPA/600/R-94/168b, U.S. EPA, Office of Research and Development, Washington, DC.

Shan, G.-B., Zhang, T. C., and Surampalli, R. Y. (2010). "Bioenergy from landfills." Chapter 5, *Bioenergy and biofuel from biowastes and biomass*, S. K. Khanal, R. Y. Surampalli, T. C. Zhang, B. P. Lamsal, R. D. Tyagi, and C. M. Kao, eds., ASCE, Reston, VA.

Stegmann, R., Heyer, K. U., and Cossu, R. (2005). "Leachate treatment." *Proc., Sardinia 2005, 10th Int. Waste Management and Landfill Symposium*, S. Margherita di Pula, Cagliari, Italy.

U.S. EPA. (1988a). "Criteria for municipal solid waste landfills." Washington, DC.

U.S. EPA. (1988b). "Solid waste disposal facility criteria proposed rule." *Federal Register*, **53**(168), 33314–33422.

Vaidya, R. (2012). "Comparison of leachate treatment alternatives from MSW and C&D waste." ⟨http://www.swanafl.org/Resources⟩ (Mar. 1, 2013).

York, R. J., Thiel, R. S., and Beaudry, E. G. (1999). "Full-scale experience of direct osmosis concentration applied to leachate management." *Proc. 7th Int. Waste Management and Landfill Symp. (Sardinia '99)*, S. Margherita di Pula, Cagliari, Sardinia, Italy.

Zhang, T. C., and Surampalli, R. Y. (2012). "Novel membrane-separation techniques and their environmental applications." *Membrane technology and environmental applications*, T. C. Zhang, R. Y. Surampalli, S. Vigneswaran, R. D. Tyagi, S. L. Ong, and C. M. Kao, eds., ASCE, Reston, VA, 696–726.

Zhang, T. C., Surampalli, R. Y., Vigneswaran, S., Tyagi, R. D., Ong, S. L., and Kao, C. M., eds. (2012). *Membrane technology and environmental applications*, ASCE, Reston, VA

CHAPTER 20

Landfill Aftercare and Maintenance

Lee Man Chu[1]

20.1 INTRODUCTION

Landfilling, the most common option for municipal solid waste (MSW) management, requires a larger land area for waste disposal and burial than other treatment methods. A landfill with its organic putrescible waste mass and unique microbial community is regarded as a complex bioreactor from a biotechnological viewpoint. Starting from the first day of operation, site managers have to face the environmental problems of landfill leachate and gas produced, and other associated factors such as subsidence, odor and windblown litter. Some of these landfill factors have to be dealt with properly throughout the life of a landfill. When the site's fill capacity is exhausted the landfill is closed, and the resulting area, sometimes considered a byproduct of the operation, has the general features of degraded or damaged land that needs restoration work to restore or improve its value. Landfill aftercare commences when the site is closed and its surface cover installed. The need to manage leachate and gas continues after landfill closure, which together with restoration and surveillance should be integral components of the total landfill management. Old landfills that are smaller may be left vacant with no planned afteruse function. However, as the pressure on land for development intensifies in many parts of the world, completed landfills are usually not abandoned, but are converted and put to good use as parks, golf courses, conservation sites, or other amenities. Most are designated for soft enduses, as construction on closed landfills is prohibited by landfill gas hazards, attack of building materials and subsidence (Emberton and Parker 1987), although there is a trend in some cities to erect new commercial buildings on landfills, which presents novel challenges in monitoring and maintenance, particularly in landfill gas control (Stearns and Petoyan 1984; Card 1992).

[1]The Chinese Univ. of Hong Kong, Hong Kong SAR.

20.2 LANDFILLING

Landfilling is primarily an engineering topic, which takes into account site selection, design, construction, and operation. It is a solid waste disposal method that entails the burial of the waste and its isolation from the surroundings to reduce public health and environmental impacts. The earlier landfills are the attenuation type, which is no longer environmentally acceptable, due to the absence of isolation or a barrier layer to prevent the release of landfill leachate and gas, which migrate to the surroundings or deeper strata. Modern landfills belong to the containment (bioreactor) and entombment (dry storage) types in which the bottom or the entire landfill is lined with impervious membranes to contain and store the buried waste. The latter type is particularly applicable to hazardous wastes. In both types the bottom of the site is sealed with an impervious liner system composed of multiple synthetic membranes coupled with collection pipes to retain the leachate and gas produced (Figure 20.1).

For MSW landfills, operations involve the deposition of waste on working faces, and spreading and compaction by heavy machinery such as bulldozers and steel-wheel compactors. An inert layer is used to cover the waste at the end of a

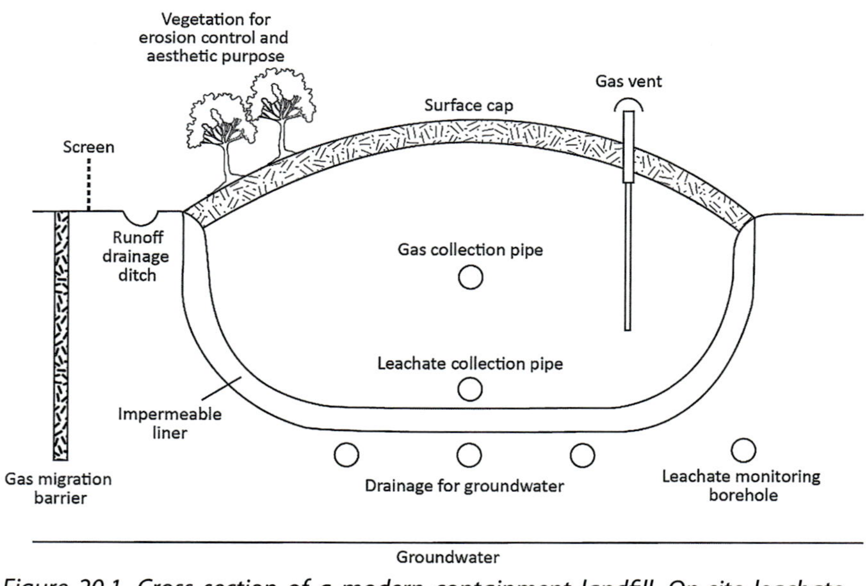

Figure 20.1. Cross section of a modern containment landfill. On-site leachate treatment facilities are not shown. In landfills with high gas production rate, landfill gas is collected to generate power for on-site usage, with surplus gas flared off for safety reason

working day, and the operational cycles repeated to form individual cells and layered lifts. During the filling process, extraction wells and pipes are installed for landfill leachate and gas, which are collected for treatment and potential utilization as landfill gas. When the fill reaches its final capacity, the site is closed to isolate the buried waste and its potentially negative effects from the landfill site and its environment. The surface is usually capped with a lower impermeable layer to prevent rainwater percolation into the fill, a middle porous layer of sandy soil to promote lateral drainage, and a superficial soil layer to allow vegetation establishment. The integrity and functioning of the final cover are essential to the aesthetic improvement and ecological development of the completed landfill in the postclosure period.

20.3 LANDFILL FACTORS

20.3.1 Landfill Leachate

Landfill leachate is a mixed (inorganic and organic) liquor produced by water that percolates through the buried MSW, carrying suspended and dissolved solids produced when the absorbent characteristic of the waste is exceeded. It is the product of a mass transfer process, and involves the hydrolysis of macromolecules and biodegradation of its organic fractions, the solubilization of salts contained in the wastes, and particulate drag (Andreottola and Cannas 1992). Landfill leachate is usually ten times higher in organic strength than municipal sewage. It has very high salinity and redox potential, and contains very high levels of chemical oxygen demand (COD), biochemical oxygen demand (BOD) and ammoniacal-nitrogen. Its quantity and composition is mainly governed by waste properties, site factors and climatic conditions. Specifically, waste composition, fill volume and depth, decomposition rate, landfill age, surface capping and percolate volume are major influencing components. Peak concentrations of COD and total solids exceeding 40,000 mg L-1 are common. High concentrations of various pollutants are observed near the onset of leaching from young landfills while the dilution and microbial utilization of organics reduce the strength of leachate in older landfills. The change in leachate composition over time follows a pattern similar to the various stages of fermentation (Figure 20.2) (Andreottola and Cannas 1992; Chu et al. 1994). If the leachate is allowed to flow out of the site, as in the case of old landfills without an impermeable lining at the bottom, its strength will be affected by attenuation in the substratum beneath the landfill and dilution by groundwater below the water table. This will pollute groundwater, which is a potential problem for sites with high water table. Owing to its strength, landfill leachate collected is processed physically, chemically or biologically in a treatment facility before discharge, or recirculated back to the landfill for waste stabilization (Christensen et al. 1992). Leachate treatment works are designed to provide service over the landfill life expectancy of decades.

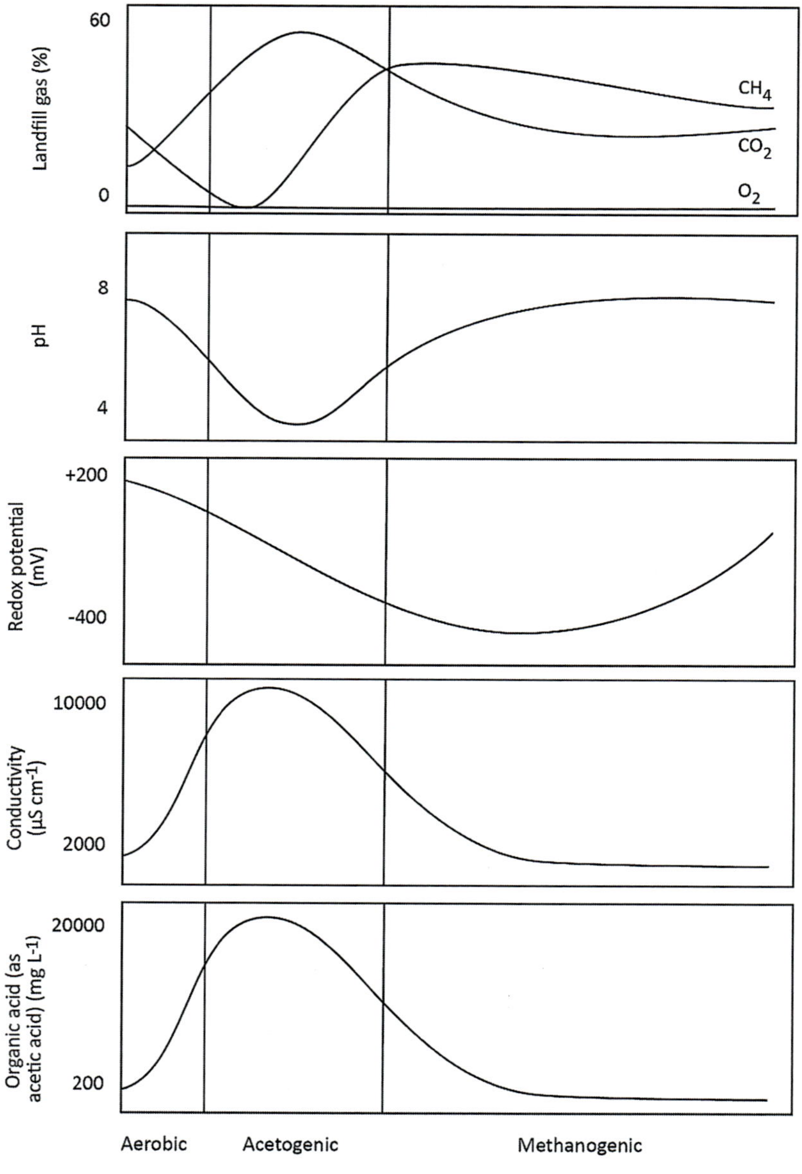

Figure 20.2. Changes in the composition of landfill gas and leachate during the course of waste degradation. The period of time for different phases can vary substantially among landfills under different climates, so no exact timescale is given

Source: Andreottola and Cannas 1992; Pfeffer 1992; Tchobanoglous et al. 1993)

20.3.2 Landfill Gas

Landfill gas produced as a result of anaerobic decomposition of MSW in the landfill bioreactor contains a high concentration of methane gas (50–60%) and carbon dioxide (40–50%), with trace levels of other gases such as nitrogen, ammonia, carbon dioxide, hydrogen sulfide, hydrogen and volatile organic compounds (<5%). These are formed during waste degradation as soon as the oxygen concentration inside the bioreactor is used up. This is not only a postclosure problem, but also an environmental nuisance once operation has started.

Gas production rate will peak after a few years, depending on the decomposition rate of the waste, which is under the influence of temperature and rainfall of the site. Methane concentration will reach a maximum of 60% in the landfill gas and become stable for some years. A landfill can produce landfill gas for more than 10–15 years, though the production rate and methane yield will decline as the site ages. Large landfill which receives waste for a prolonged period will have different phases of methane production in different areas of the site.

Landfill gas has a calorific content of 20 MJ m-3 while methane has a value of 35 MJ m-3, so it is an attractive alternative for power generation. Methane is explosive when mixed with air at around 5–15%, while at higher concentrations it will displace oxygen to produce an asphyxiation risk. Hydrogen sulfide in the landfill gas creates odor and may also cause asphyxiation. Gas migration on landfills is governed by meteorological conditions, waste factors and soil properties (Kjeldsen 1996). The landfill gas problem persists as long as its emission and impact remain high. It is usually either vented or collected for energy production.

Landfill odor is another nuisance to the general public, especially to the neighboring communities. Malodorous gaseous emission is associated with odorous compounds such as sulfides (hydrogen sulfide, dimethyl sulfide, dimethyl disulfide, and carbonyl sulfide) and mercaptans (methyl and ethyl mercaptans) produced from the biological decay of the organic fractions of MSW during operation and storage in the landfill body. Landfill odor is a sensitive issue which often becomes health concerns and frequently provokes opposition from residents nearby against landfill operation and extension. Landfill odor problem is less serious for closed landfills than operational ones, as odor is more difficult to control in active landfills, with refuse transport vehicles as another important odor source. Odor gases are present in trace amount in landfill gas after closure.

20.3.3 Subsidence

Solid waste mass contains putrescibles that will decompose and reduce in bulk volume. MSW is heterogeneous in composition, and there will be spatial difference in moisture content and compaction across the landfill. As a result, waste material is likely to break down and settle at differing rates, hence differential settlement of the filled material and surface. There is short-term deformation due to initial compression and reorientation of waste particles, and long-term settlement attributed to the decomposition of organic material (Grisolia and Napoleoni 1995; Liu et al. 2006). These create depressions and cracks over the surface of the landfill, and instability in the covering soil layer. A number of variables affect the magnitude of subsidence, including waste composition, cover thickness and the degree of compaction. Subsidence of a closed landfill is unavoidable and will lead to deformed surface, slope collapse, damages of surface and underground drains as well as vegetation decline. It also restricts or sometimes prevents the construction of high-rise buildings on closed landfills, as cracking of the soil cover will inevitably lead to damage of the surface capping system which may result in landfill gas leakage and explosion risk. Subsidence is a geotechnical issue which causes problems for afteruse development. Settlement can last for as long as 20 years and there will be continuous residual settlement which can extend to as long as 50 years (Sarsby 2000; Park and Park 2009). Accurate prediction of the long-term settlement of landfills is challenging and significant for the afteruse development of postclosure landfills (Sharma and De 2007).

20.4 POSTCLOSURE MANAGEMENT

When the landfill is closed and becomes inactive, an additional final cap is placed on the top to isolate the waste from the environment. Postclosure landfills are monitored and managed to control landfill associated factors and to develop endues of the site after closure. A plan is required which lays out the design and requirements for closure, and the long-term elements of postclosure maintenance. Landfill stability, leachate, gas, subsidence and cover system will be the focal points for management. All the surface areas within the landfill boundary at the time of closure will need rehabilitation, both ecological and environmental. Maintenance of the capping and drainage system as well as the vegetation on the cap must be adequately managed to ensure the long-term sustainability of MSW landfills after closure.

20.4.1 Leachate Management

Information on the volume and composition of the landfill leachate produced is important for the formulation of a sound leachate management strategy. It is a common practice to decrease the volume of landfill leachate requiring treatment by reducing the ingress of rainwater into the buried waste through the use of a less permeable surface cap. This is done by installing a hydraulic barrier layer comprising porous gravel layer to promote drainage, and an impervious clay layer or synthetic liner underneath the top cover soil layer to block and retain water, and prevent it from permeating into the waste (Table 20.1). While this is relatively well demonstrated to implement, it has been suggested that establishing a vegetation layer to increase evapotranspiration is also effective in removing water and minimizing its downward percolation. However, these

Table 20.1. Management of landfill leachate by volume reduction in postclosure landfills

Management option	*Installation*
Reduce runoff from surrounding	Intercepting ditches at landfill boundary
Reduce water entering fill	Impermeable surface cap
Reduce groundwater intrusion	Impermeable bottom lining
Increase runoff from fill	Impermeable surface cap and convex surface
Increase evapotranspiration	Vegetation cover on surface

two methods seem to conflict in nature, as plant establishment requires a well-drained layer for root development. The enhancement of water removal via evapotranspiration requires the selection of species with high evapotranspiration potential.

Landfill leachate that is formed is usually collected and treated prior to discharge. Pretreatment either on-site within the landfill or off-site away from the landfill is usually required because of the high strength of landfill leachate which will overload a conventional treatment works for municipal sewage. Pretreatment methods can be physical, chemical or biological. In modern landfills, pretreatment is mostly included in-house by operations such as aeration, thermal process and reverse osmosis which aim to remove ammonia, a major pollutant in leachate. Pretreated leachate is then transferred to municipal sewage works for conventional treatment.

20.4.2 Gas Control

The barrier layer prevents the upward migration of landfill gas and volatile organics, and stops it from escaping into the ambient environment. Landfill gas is conventionally dealt with by releasing into the atmosphere via passive gas flow under a concentration gradient using gravel vents and wells sunk into the fill. More commonly, landfill gas is controlled by active collection using pipes to check lateral gas migration, or recovered by extraction using pumps via a network of horizontal gas trenches and vertical wells. The gas with a high energy value can be upgraded by passing it through scrubbers to remove gaseous impurities such as carbon dioxide, water vapor and particulates for power generation. Such energy recycling practice is especially viable at large landfills. Landfill gas can be reclaimed for direct use for heating or indirect use for electricity generation. If the gas extraction rate is too low to justify power generation, the gas can be burnt at a flare stack.

Odor from landfills can be managed by maintaining a small working face during routine operation and covering the waste spread after daily operation on active landfills. Better land use planning and site selection, taking into account the location of sensitive recipients and the effects of wind and topography, could certainly help reduce odor complaints. On postclosure landfills, odor can be effectively controlled by collecting landfill gas using horizontal collectors and covering up manholes.

20.4.3 Final Cover System

An engineered capping system is placed on the final surface of the closed landfill to cover up and isolate the waste underneath, including any product produced from the anaerobic decomposition of waste, as well as preventing infiltration of water into the waste, and providing a soil layer for vegetation establishment. This superficial cover is one of the most important components of postclosure landfills, and must be carefully chosen and designed. It is typically a multilayer system constructed to minimize rainwater infiltration from the surface into the underlying waste. Most incorporate an impermeable barrier concept to achieve waste containment.

The minimum thickness of the landfill cap is determined by the number of layers in the design, construction limits, and the vegetation to be planted. Containment landfills have more (≥4) layers than attenuation sites, and include a surface soil layer, drainage layer, hydraulic/gas barrier layer, gas collection/venting layer, and an optional foundation layer (Figure 20.3) (Daniel 1995; Jesionek et al. 1995). Inclusion of a geomembrane and a geosynthetic clay liner are also common features to restrict the vertical migration of water or gas from or into the waste mass. Countries such as the United States, England, and Germany have regulatory requirements and technical specifications for the individual

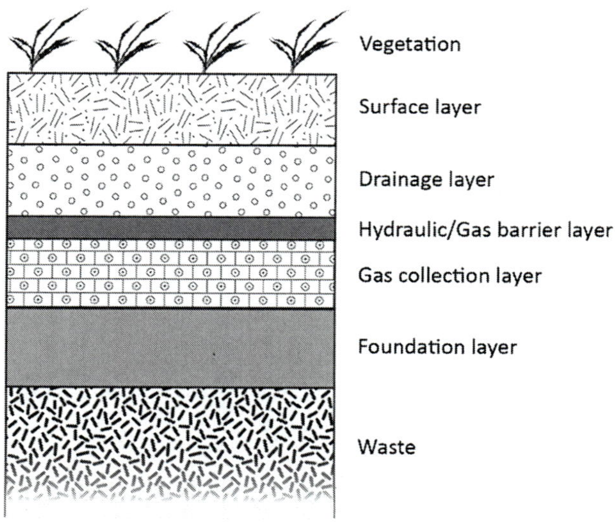

Figure 20.3. Basic components of a final cover system on postclosure landfills

Table 20.2. Key soil properties important to landfill restoration

Properties	*Parameters*
Physical	Soil texture, soil structure, soil profile and depth, bulk density and porosity
Chemical	pH, salinity, nutrient contents, contamination level
Biological	Microflora composition and populations, microbial and enzyme activities, soil microinvertebrate diversity, earthworm abundance

component layers (Koerner and Daniel 1997; O'Leary and Tchobanoglous 2002). However, their effectiveness and performance are influenced by temperature extremes, cyclic wetting and drying, differential settlement, water and/or wind erosion, slope slippage, movement of heavy machinery, plant roots, and burrowing by animals (Daniel 1995).

The topmost cover, which is usually thin, is the surface vegetative soil layer for plant establishment. It serves to provide a rooting medium for vegetation cover, promote site aesthetics, control erosion, protect the barrier layer, and meet other ecological needs. Soil plays a major role in ecological restoration, and is the soul of the landfill cap regarding site end use. The design, selection of soil type, and quality of final cover material should be geared towards postclosure landfill aftercare and maintenance. The physical, chemical, and biological properties of the soil for the final cover are the key to restoration (Table 20.2), as they affect root development, plant growth, drainage, erodibility, and cap integrity. The soil type available will mostly determine the soil quality achieved. Soils are variable according to regions, and the availability of suitable soil is a common problem. Soil present on the site, regardless of its quality, is normally removed and stockpiled for use after closure before landfilling operations begin, but its quantity may not be sufficient for covering the postclosure site, and is likely to be adversely affected during storage (Harris et al. 1989; Williamson and Johnson 1990). Care must be taken in the handling of soil during removal, storage, and respreading. Materials may need to be imported from elsewhere, though their quality is not guaranteed. Poor soil may not be amenable to good plant growth. Soil of coarse texture is not good for retaining water and will dry out quickly, while soil that is too fine will impede water movement and ventilation. To support good vegetation growth, substandard soil should be improved with fertilizer or organic compost to address its deficiencies. Natural colonization of restored landfills by earthworms has been reported (Judd and Mason 1995), though earthworm inoculation does not necessarily work for soil amelioration (Butt et al. 1999). The thickness of the top layer depends on factors such as plant types to be grown, soil material to be used and its availability, and climate (e.g., temperate versus tropical). On the other hand, soil thickness will affect the type of vegetation to be planted. A minimum of 150–400 mm is

Table 20.3. Recommended soil depth for different afteruses on postclosure landfills

Afteruse	*Subsoil minimum depth (mm)*	*Topsoil minimum depth (mm)*	*Total depth (mm)*
Low maintenance amenity sward	1,000	Not specified	1,000
Intensive pasture	700–900	150–300	1,000
Trees (synthetic barrier system present)	700–850	150–300 (preferable)	1,000
Trees (clay barrier system present	1,200–1,350	150–300 (preferable)	1,500

Source: Ireland EPA (1999)

sometimes proposed to control erosion and sustain the growth of shallow-rooted plants such as amenity grasses (Table 20.3). In most cases, 300–800 mm soil is laid to accommodate the healthy growth of plant roots for revegetation, and a thicker soil layer (≥1,000 mm) is needed if tree planting is desired for site afteruse.

In old-style landfills without barrier layer, though the anaerobic degradation of organic MSW in the landfill is less exothermic, heat is still produced from the waste mass and transferred to the surface by the movement of warm landfill gas. Consequently, the temperature of landfill cover soil is higher than the ambient; elevated temperature of 30–40°C is common and the difference could be as high as 30°C (Leone et al. 1977; Chan et al. 1991; Moffat and Houston 1991).

20.5 LANDFILL RESTORATION AND AFTERCARE

Ecological restoration is the process of assisting the recovery of ecosystems that have been disturbed or degraded through human activities. The term restoration can be broadly defined or narrowly used, whereas rehabilitation refers to partial restoration of ecosystem structure and function (Table 20.4). A closed landfill may be regarded as an ecosystem under stress that requires rehabilitation. The development of a functional ecosystem on the final cover is fundamental to the rehabilitation of landfills. Basic principles of restoration apply, which entail different approaches or their combination (Table 20.5). Planning and implementation details are important and should take the interests of the various stakeholders into account. It is equally important to understand which species/communities/ecosystems can thrive on the rehabilitated landfill, and if the intended afteruse is compatible to the objectives of restoration.

Table 20.4. Terminologies and definitions for different levels of restoration targets

Terminology	*Definition*
Restoration	Return of an ecosystem to a close approximation of its condition prior to disturbance/degradation Re-establishment of former ecosystem structure and function
Rehabilitation	Restoration/Reinstatement of part of the ecosystem structure and/or functions
Reclamation	Return of an ecosystem to a proper useful state

Table 20.5. Major approaches for restoration of degraded ecosystems

Approach	*Remarks*
Biodiversity	Based on richness and assemblages of species (including indicator species, keystone species and endangered species)
Ecosystem function	Based on biological inventories and material and energy flows between biotic and abiotic components of ecosystem
Ecosystem service	Based on the provisions of habitats, natural resources and life-support process regulation

20.5.1 Postclosure Afteruse

Landfill aftercare and maintenance are basically part of the landfill management activities. They essentially involve the environmental monitoring and routine inspection of various landfill parameters of the completed site, which usually relate to issues of public health impact, environmental pollution, ecological disturbance and landscape quality. As soon as the site reaches filling capacity and stops receiving waste, the landfill manager and contractor need to close the landfill properly. The most appropriate approaches are based on the ecological and economic assessment of the completed site. The postclosure plan and operation must encompass the proper handling and control of leachate and gas emission, their monitoring to ensure site safety, control of subsidence, choice and availability of cover soil, establishment of vegetation, and site maintenance. Maintenance and rehabilitation of landfills after closure usually aim to minimize and mitigate the environmental impacts of waste disposal and prevent threats to public safety.

Enduses include amenities, recreation, wilderness, agriculture and forestry, although commercial development is no longer forbidden. Most sites will be reclaimed to become golf courses, playing fields, parks and botanical gardens.

They may provide community services and facilities for green spaces, leisure sports, cycling, jogging and running. Agricultural rehabilitation is more demanding for soil quality, and it may be difficult to support good productivity. Afteruse options will reflect local culture and land use needs. The design of the postclosure landfill depends on the afteruse and environmental standard of the completed site, which affect the restoration scheme and cost. Initial restoration works on closed landfills generally require artificial revegetation and landscaping, irrespective of the planned soft endues. The use of vegetation management tools is common, as almost all restoration works involve the establishment and development of vegetation on the postclosure landfill (Howell et al. 2012). There is a wide maintenance spectrum for postclosure landfills with different enduses. Although low maintenance is the central theme for land rehabilitation, particularly those left for nature conservation and ecological diversity, sites reclaimed for parks, golf courses and football pitches will need more engineering and landscaping input.

20.5.2 Revegetation of Soil Cover

Revegetation of the final soil cover is essential to most soft enduses, and its success relies on the quality of the cover material, plant requirements and aftercare strategies. The functions of the established vegetation on the cover soil include hydrological and mechanical ones. Plantations on landfill are beneficial to the closed site in terms of visual appearance, erosion control and successional changes. To this end, large landfills can be completed, covered and released in progressive phases, with revegetation occurring in different periods synchronically with landfill operation.

Selection of plant species is critical to the success of revegetation, and usually depends on the enduse and conservation value of the site, its relationship with the surrounding environment, amenities and visual qualities, and the degree of management (Cairney and Hobson 1998). The final cover is usually hydroseeded with grass species in the initial phase to produce an instant plant cover for effective erosion control. Old landfills without barrier systems are not suitable to woodland establishment (Rawlinson et al. 2004), probably due to the landfill gas emission and other site constraints. The planting of trees, and shrubs to certain extent, is also controversial, due to concerns that deep-rooting, a growth feature of woody species, may penetrate the drainage and barrier layers, and damage the integrity of the cap system. However, if properly designed and worked out, the capping system can be maintained while supporting a variety of plant species on the landfill. Ordinary tree growth, reportedly, will not disrupt the landfill cap, especially on sites where thicker topsoil (1,200–1,500 mm) is laid for revegetation (Dobson and Moffat 1993; Robinson and Handel 1995; Moffat et al. 2008). The risk is acceptable even when roots are exploiting weakness in the cap fabric (Hutchings et al. 2006). A shallow final soil cover (<1,000 mm) may restrict root proliferation and rooting depth, and is likely to result in uprooting of trees, especially in strong winds.

To plant on closed landfills, key factors for considerations are soil depth, soil quality, topography, soil moisture, and the presence of landfill gas, pests and invasive species. Landfill gas ingress into the root zone is a major factor causing poor grass growth on old landfills (Trotter and Cooke 2005), though grasses tend to grow better than trees and shrubs because of their shallow rooting depths which help to avoid the reducing atmosphere in deeper soil (Lan and Wong 1994). Although the effects of landfill gas and leachate are usually mitigated in modern closed landfills, the final soil cover may still be stressful for plant establishment. Nutrient shortage, drought and mammal browsing are usual limiting factors for tree performance on landfills. Species that can survive poor soil structure and low fertility are potentially better candidates for use. Wild species have been advocated. Locally adapted and available species that are perennial in life cycle are the best choices, with nitrogen-fixing legumes being the preferred plant types as they do not rely on the input of nitrogen fertilizers and they are relatively more tolerant to landfill gas (Wong 1995; Chan et al. 1996, 1998). Standardized planting procedures have been developed for completed landfills (Gilman et al. 1985). For landfills that have no impermeable cap, plant species with high evapotranspiration (ET) potential are especially suitable, as they are capable of removing water from the soil layer. They are considered alternative covers and referred to as ET cover, which outperforms conventional landfill caps in terms of hydrological performance, especially in arid areas where a high ET rate and low precipitation work together to remove all infiltrated water (USEPA 2011). One version of ET cover is the phytocap, an engineered agronomic system that harnesses the natural transpiration process of plants to prevent water from percolating beyond the root zone in deeper substrata (Suthersan 2002; WMAA 2011). It is also termed water balance cover which is designed to maximize evapotranspiration and water storage, instead of interception by an impermeable barrier. It is suggested to be a cost-effective option for developing countries (Yuen et al. 2010). It can be monolithic, or multilithic with a capillary break for hydraulic control at the bottom, Selection of the right tree species will enhance the performance (Venkatraman and Ashwath 2009). Soil properties are crucial in governing the amount of water stored, infiltration and plant growth. Organic and inorganic amendments such as polymer gel, fly ash, compost, peat mosses and biochar may be added to improve water storage capacity by functioning as an effective sponge.

20.5.3 Ecosystem Development on Closed Landfills

Rehabilitated landfills can be considered as ecosystems manipulated to speed up secondary succession and the recovery of major ecosystem attributes (Table 20.6). In fact, there are various approaches to rehabilitating landfills (Simmons 1992): (a) natural regeneration without ecological intervention; (b) ecological intervention and subsequent natural habitat development; and (c) habitat creation and intense management. Each has its merits, shortcomings, and site-specific suitability.

Regardless of the outcome of rehabilitation, it is essential to rebuild the biological status and activity of the soil-vegetation compartments. The manipulation of ecological processes in the initial stages is generally accepted as the best available option for postclosure landfill development. Closed landfills can be categorized according to their closure period, with vegetation coverage positively correlated with site age and soil nutrient content, and negatively with soil

Table 20.6. Ecological attributes important to ecosystem restoration (The first four attributes are amenable to direct manipulation)

Attribute	*Parameter*
Composition	Species diversity, relative abundance & dominance
Structure	Biomass, trophic structure, levels of interaction
Pattern	Horizontal arrangement of system components
Heterogeneity	Complexity of composition, structure & pattern
Function	Performance of basic ecological processes
Dynamics	Temporal change, e.g. succession; resistance and resilience to disturbance

penetrability (Choi 2000). The natural succession of grassland to woodland ecosystem is slow and may take decades to complete. Accompanying these changes are faunal community changes and colonization by invertebrates such as insects and spiders, and vertebrates such as reptiles and birds (Davis 1989: Chan et al. 1997; So and Chu 2010; Rahman et al. 2011). This is important as ecological succession should not solely benefit vegetation, and postclosure landfills can compensate for habitat loss and have certain value for nature conservation (Watson and Hack 2000). Native tree species grow better than exotics commonly planted on waste-lands, and old landfills appear to favor high biodiversity (Dickinson et al. 2005). Planting natives either in the beginning or at a later stage by enrichment planting can help to increase species diversity, and could make a difference to the habitat value for wildlife conservation and other ecological services.

However, given the properties of most soils available for use as top cover, fertilizers and organic soil amendments are commonly added to improve soil deficiency in the initial stages. Care and maintenance are essential to the successful restoration of landfills. Nonetheless, besides environmental compatibility, there are no standard or consistent quantitative evaluation methodologies or criteria to assess the restoration status of postclosure landfills (Morris and Barlaz 2011; Scharff et al. 2011).

20.5.4 Landfill Aftercare

Closed landfills will undergo different management stages during which there are different level of regulatory aftercare, environmental liabilities and range of reuse options in relation to the prevailing site conditions and environmental threats (Figure 20.4). The initial period is important to future development, as planting failure is not uncommon during this establishment stage, particularly when smaller whips and transplants are used (Cairney and Hobson 1998). Trained personnel of the management team should carry out regular site inspections and soil monitoring. Information on surface integrity and settlement, slope profile, drainage systems, irrigation systems, plant growth, landfill gas wells and odor nuisance are important.

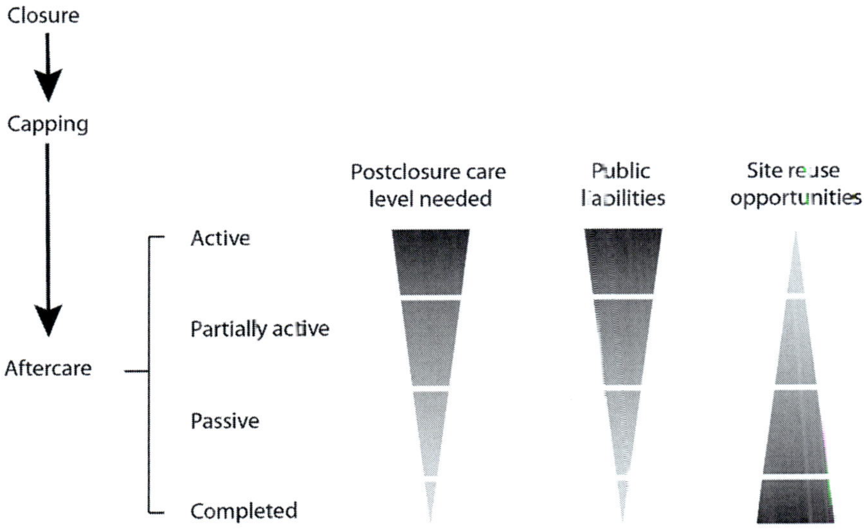

Figure 20.4. Various management stages during the postclosure period of a MSW landfill. The decreased level of regulatory aftercare and environmental liabilities, and increased range of reuse options with time is related to improved site conditions and reduced threats. The time needed to end aftercare is site-specific

In the first few years, the surface cover system has to be inspected with care, especially after heavy storms. Poor soil fertility and structure must be addressed, and there are general guidelines for planting on postclosure landfills (Dobson and Moffat 1993). The survival and health of trees planted should be evaluated and checked for site adversity. When the replacement of dead trees is necessary, the optimal annual planting time should be considered. Other tasks such as fence inspection and weed and pest control are also important. Field operators should keep good records of the inspection data and any observed defect. Surface emissions are important to general public health and safety. Borehole point sampling and testing for leachate and gas are also regular monitoring tasks (Table 20.7). The monitoring and postclosure care period is at least 30 years as required by regulation (USEPA 1993), though sometimes this period is considered as unrealistically long for reasons of liability and cost. Discussion has been raised as to when postclosure landfill can be released from aftercare and what criteria should be used (Stegmann et al. 2003; Laner et al. 2011). Some countries have spelled out the aftercare period in their formal regulations, but so far there is a paucity of procedures and criteria for ending postclosure landfill care because there is no standard guidance or definition of the assessment methods to determine whether the potential risks of a closed landfill are acceptable (Laner et al. 2012). This renders it difficult for authorities or operators to make decisions to stop, extend or shorten aftercare for compliance (Barlaz et al. 2002).

Table 20.7. Checklist for monitoring and surveillance of postclosure landfills during aftercare period

Landfill item	*Variable*	*Study and monitoring*
Site	Environmental compatibility	Characterization study, vulnerability study, risk analysis
Waste	Biological activity	Respiration test
Landfill leachate	Leachate treatability	Elution test, lysimeter test
Landfill gas	Emission potential for methane	Scenario analysis, field monitoring
Odor	Emission potential for odorous compounds	Field monitoring; field olfactometry
Subsidence	Differential settlement	Strain sensors test
Soil	Soil quality	Soil organic matter and nutrient analyses, soil microbiological analysis
Vegetation	Plant diversity and health	Vegetation analysis

The activities generated from the use of the reclaimed land from postclosure landfills must avoid damage to the integrity of the capping systems. If wilderness is the end result, the rehabilitated site should be left to natural modification with minimal human interference. On the other hand, if a sports field is the target afteruse, the site should be leveled and the turf heavily maintained with efficient irrigation, regular mowing and fertilization as well as weed control.

Despite the challenges that restoration ecologists face, the rehabilitation of postclosure landfills offers great opportunities for creating new land, developing alternative habitats, recovering ecosystem functions and repairing degraded landscapes. Certainly there are limitations, both in terms of resources and knowledge, but with the evolution of landfill closure design, technology and management, we will see more postclosure landfills transformed to practicable afteruse. It is understood that land used for waste disposal can be recycled, and provided long-term liabilities and legal issues are resolved, landfilling can remain a viable MSW management option, at least from a backend viewpoint.

20.6 LANDFILL MINING

Landfill mining which reclaims useful resources from waste materials buried in the ground (Krook et al. 2012) is another revisited alternative for closed landfills. There is growing interest on its development which looks promising, but it is at present technologically immature and economically unproven.

Closed landfills are in fact untapped reserves of valuable recyclables (Kapur and Graedel 2006; Zanetti and Godio 2006); this is particularly true in areas with poor recycling efforts. To reclaim these resources, mining of old unlined dumpsites for recyclable materials and energy theoretically sound and practically feasible, which also frees the land when excavation is done (Spencer 1990). Wastes buried are dug out and processed using conveyor belts, trommel screens and magnetic separator which segregate the wastes into discrete fractions such as soil, metals, glass, plastics and oversizes. Non-biodegradables are more accessible and recoverable, but degraded materials are of lower quality. The recovery of secondary materials will inevitably reduce the use of virgin resources and landfill footprint, and reclaim space for beneficial uses. There have been concerns regarding dust, odor, toxic gases and stormwater runoff, while the viability of the operation is affected by technology efficiency, economic feasibility as well as recoverable product quality and marketability (Baas et al. 2010; Frandegard et al. 2013). Despite a history of more than six decades, landfill mining is still immature and has yet to be developed. It is most applicable to hazardous landfill remediation, or installing liner for outdated municipal landfills. The concept and technology deserve closer examination, and the integration of landfill mining into waste management will definitely change the nature of landfills to temporary storage and enhance their role as an option of sustainable material management.

20.7 CONCLUSION

The reuse of a former landfill site can only be achieved with comprehensive afteruse planning and aftercare management. This is enhanced by increasing the reliability of environmental impact assessment exercise and landfilling technology. Prolonged postclosure care is usually needed, and the length of the aftercare period has obvious financial implications. As soon as the site is environmentally safe and functionally stable, mandatory aftercare ends and public liability is much reduced. Additionally, postclosure care is considered adequate only if developments in the site vicinity, which are important to the long-term functioning of landfill technology, are followed (Rodic 2000). Postclosure landfill aftercare is a good example of putting resilience management into practice. This provides a means of attaining sustainable landfilling which justifies the management of MSW for the local community, when a landfill ban or zero landfilling is not always feasible. With the growing pressure on resource shortages, closed landfills may have a role to play in resource recovery and sustainable material management.

References

Andreottloa, G., and Cannas, P. (1992). "Chemical and biological characteristics of landfill leachate." *Landfilling of Waste: Leachate*, T. H. Christensen, R. Cossu, and R. Stegmann, eds., Elsevier Applied Science, London, 65–88.

Baas, L., Krook, J., Eklund, M., and Svensson, N. (2010). "Industrial ecology looks at landfills from another perspective." *Reg. Dev. Dialogue*, **31**(2), 169–183.

Barlaz, M. A., Rooker, A. P., Kjeldsen, P., Gabr, M. A., and Borden, R. C. (2002). "Critical evaluation of factors required to terminate the postclosure monitoring period at solid waste landfills." *Environ. Sci. Technol.*, **36**(16), 3457–3464.

Butt, K. R., Frederickson, J., and Lowe, C. N. (1999). "Colonization, survival and spread of earthworms on a partially restored landfill site." *Pedobiologia*, **43**(6), 684–690.

Cairney, T., and Hobson, D. M. (1998). *Contaminated land. Problems and solutions*, E & Fn Spon, London.

Card, G. B. (1992). "Development on and adjacent to landfill." *Water Environ. J.*, **6**(4), 362–371.

Chan, G. Y. S., Wong, M. H., and Whitton, B. A. (1991). "Effects of landfill gas on subtropical woody plants." *Environ. Manage.*, **15**(3), 411–431.

Chan, Y. S. G., Wong, M. H., and Whitton, B. A. (1996). "Effects of landfill factors on tree cover: A field survey at 13 landfill sites in Hong Kong." *Land Contam. Reclam.*, **4**(1), 115–128.

Chan, Y. S. G., Chu, L. M., and Wong, M. H. (1997). "Influence of landfill factors on plants and soil fauna—An ecological perspective." *Environ. Pollut.*, **97**(1), 39–44.

Chan, Y. S. G., Wong, M. H., and Whitton, B. A. (1998). "Effects of landfill gas on growth and nitrogen fixation of two leguminous tress (*Acacia confusa* and *Leucaena leucocephala*)." *Water Air Soil Pollut.*, **107**(1-4), 409–421.

Choi, C. Y. (2000). "Ecological studies of the cover soils on completed landfills." M. Phil. thesis, Chinese University of Hong Kong, Hong Kong.

Christensen, T. H., Cossu, R., and Stegmann, R. (1992). *Landfilling of waste: Leachate*, Elsevier Applied Science, London.

Chu, L. M., Cheung, K. C., and Wong, M. H. (1994). "Variation in the chemical properties of landfill leachate." *Environ. Manage.*, **18**(1), 105–117.

Daniel, D. E. (1995). "Soil barrier layers versus geosynthetic barriers in landfill cover systems." *Landfill Closures- Environmental Protection and Land Recovery*, J. Dunn and U. P. Singh, eds., American Society of Civil Engineers, New York, 1–18.

Davis, B. N. K. (1989). "Habitat creation for butterflies on a landfill site." *Entomologist*, **108** (1-2), 109–122.

Dickinson, N., Rawlinson, H., Nolan, P., and Putwain, P. (2005). "Community forestry on old landfill sites." *Quar. J. Forestry*, **99**(4), 263–270.

Dobson, M. C., and Moffat, A. J. (1993). "The potential for woodland establishment on landfill sites." Dept. of the Environment, HMSO, London.

Emberton, J. R., and Parker, A. (1987). "The problems associated with building on landfill sites." *Waste Manage. Res.*, **5**(1), 473–482.

Frändegård, P., Krook, J., Svensson, N., and Eklund, M. (2013). "A novel approach for environmental evaluation of landfill mining." *J. Cleaner Prod.*, **55**(1), 24–34.

Gilman, E. F., Flower, F. B., and Leone, I. D. (1985). "Standardized procedures for planting vegetation on completed sanitary landfills." *Waste Manage. Res.*, **3**(1), 65–80.

Grisolia, M., and Napoleoni, Q. (1995). "Deformability of waste and settlements of sanitary landfills." Proc., World Congress on Waste Management, ISWA, Vienna.

Harris, J. A., Birch, P., and Short, K. C. (1989). "Changes in microbial community and physic-chemical characteristics of topsoils stockpiled during opencast mining." *Soil Use and Manage.*, **5**(4), 161–168.

Howell, E. A., Harrington, J. A., and Glass, S. B. (2012). *Introduction to restoration ecology*, Island Press, Washington.

Hutchings, T. R., Sinnett, D., Peace, A. J., and Moffat, A. J. (2006). "The effect of woodland growth on a containment landfill site in Hertfordshire, UK." *Forestry Urban Greening*, **5**(4), 169–176.

Ireland EPA. (1999). "Landfill manuals, landfill restoration and aftercare." Wexford, Ireland.

Jesionek, K. S., Dunn, R. J., and Daniel, D. E. (1995). "Evaluation of landfill final covers." Proc., Sardinia 95, 5th Int. Landfill Symp., CISA, Cagliari.

Judd, K. W., and Mason, C. F. (1995). "Earthworm populations of a restored landfill site." *Pedobiologia*, **39**(2), 116–125.

Kapur, A., and Graedel, T. E. (2006). "Copper mines above and below ground. Estimating the stocks of materials in ore, products, and disposal sites opens up new ways to recycle and reuse valuable resources." *Environ. Sci. Technol.*, **40**(10), 3135–3141.

Kjeldsen, P. (1996). "Landfill gas migration in soil." *Landfilling of Waste: Biogas*, T. H. Christensen, R. Cossu, and R. Stegmann, eds., E & Fn Spon, London, 87–132.

Koerner, R. M., and Daniel, D. E. (1997). *Final covers for solid waste landfills and abandoned dumps*, ASCE Press, Virginia.

Krook, J., Svensson, N., and Eklund, M. (2012). "Landfill mining: A critical review of two decades of research." *Waste Manage.*, **32**(3), 513–520.

Laner, D., Crest, M., Scharff, H., Morris, J. W. F., and Barlaz, M. A. (2012). "A review of approaches for the long-term management of municipal solid waste landfills." *Waste Manage.*, **32**(3), 498–512.

Laner, D., Fellner, J., and Brunner, P. H. (2011). "Environmental compatibility of closed landfills—Assessing future pollution hazards." *Waste Manage. Res.*, **29**(1), 89–98.

Lan, C., and Wong, M. H. (1994). "Environmental factors affecting growth of grasses, herbs and woody plants on a sanitary landfill." *J. Environ. Sci.*, **6**(4), 504–513.

Leone, I. A., Flower, F. B., and Gilman, E. F. (1977). "Damage to woody species by anaerobic landfill gases." *J. Arboric.*, **3**(12), 221–225.

Liu, C. N., Chen, R. H., and Chen, K. S. (2006). "Unsaturated consolidation theory for the prediction of long-term municipal solid waste landfill settlement." *Waste Manage. Res.*, 10.1177/0734242X06062579, 80–91.

Moffat, A., Foot, K., Kennedy, F., Dobson, M., and Morgan, G. (2008). "Experimental tree planting on UK containment landfill sites: Results of 10 years' monitoring." *Arboric. Urban Forestry*, **34**(3), 163–172.

Moffat, A. J., and Houston, T. J. (1991). "Tree establishment and growth at Pitsea landfill site, Essex, U.K." *Waste Manage. Res.*, **9**(1), 35–46.

Morris, J. W. F., and Barlaz, M. A. (2011). "A performance-based system for the long-term management of municipal waste landfills." *Waste Manage.*, **31**(4), 649–662.

O'Leary, P. R., and Tchobanoglous, G. (2002). "Landfilling." *Handbook of solid waste management*, 2nd Ed., G. Tchobanoglous and F. Kreith, eds., McGraw-Hill Professional, New York.

Park, H. I., and Park, B. (2009). "Prediction of MSW long-term settlement induced by mechanical and decomposition-based compressions." *Int. J. Environ. Res.*, **3**(3), 335–348.

Pfeffer, J. T. (1992). *Solid waste management engineering*, Prentice Hall, Englewood Cliffs.

Rahman, M. L., Tarrant, S., McCollin, D., and Ollerton, J. (2011). "The conservation value of restored landfill sites in the East Midlands, UK for supporting bird communities." *Biodivers. Conserv.*, **20**(9), 1879–1893.

Rawlinson, H., Dickinson, N., Nolan, P., and Putwain, P. (2004). "Woodland establishment on closed old-style landfill sites in N.W. England." *For. Ecol. Manage.*, **202**(1–3), 265–280.

Robinson, G. R., and Handel, S. N. (1995). "Woody plant roots fail to penetrate a clay-lined landfill: Management implications." *Environ. Manage.*, **19**(1), 57–64.

Rodic, L. (2000). "Reliability of landfill technology." Eburon, Delft.

Sarsby, R. W. (2000). *Environmental geotechnics*, Thomas Telford Ltd., London.

Scharff, H., van Zomeren, A., and van der Sloot, H. A. (2011). "Landfill sustainability and aftercare completion criteria." *Waste Manage. Res.*, **29**(1), 30–40.

Sharma, H. D., and De, A. (2007). "Municipal solid waste landfill settlement: Postclosure perspectives." *J. Geotech. Geoenviron. Eng.*, 10.1061/(ASCE)1090-0241(2007)133:6(619), 619–629.

Simmons, E. (1992). "Landfill site restoration for wildlife." *Waste Plann.*, **3**(6), 3–7.

So, W. Y., and Chu, L. M. (2010). "Ant assemblages on rehabilitated tropical landfills." *Biodivers. Conserv.*, **19**(13), 3685–3697.

Spencer, R. (1990). "Landfill space reuse." *BioCycle*, **31**(2), 30–33.

Stearns, R. P., and Petoyan, G. S. (1984). "Utilization of landfills as building sites." *Waste Manage. Res.*, **2**(1), 75–83.

Stegmann, R., Yeyer, K.-U., Hupe, K., and Ritzkowski, M. (2003). "Discussion of criteria for the completion of landfill aftercare." Proc., Sardinia 2003, 9th Int. Waste Management and Landfill Symp., CISA, Cagliari.

Suthersan, S. S. (2002). "Engineered vegetative landfill covers." *Natural and enhanced remediation systems*, Lewis, Boca Raton, 313–364.

Tchobanoglous, G., Theisen, H., and Vigil, S. A. (1993). *Integrated solid waste management: Engineering principles and management issues*, McGraw-Hill Book Co., Singapore.

Trotter, D. H., and Cooke, J. A. (2005). "Influence of landfill gas on the microdistribution of grass establishment through natural colonization." *Environ. Manage.*, **35**(3), 303–310.

USEPA. (1993). "Solid waste disposal facility criteria: Technical manual." EPA Rep. EPA 530-R-93-017, Office of Solid Waste and Emergency Response, Washington, DC.

USEPA. (2011). "Fact sheet on evapotranspiration cover systems for waste containment." EPA 542-F-11-001, Office of Solid Waste and Emergency Response, Washington, DC.

Venkatraman, K., and Ashwath, N. (2009). "Phytocapping: Importance of tree selection and soil thickness." *Water Air Soil Pollut. Focus*, **9**(5–6), 421–430.

Waste Management Association of Australia. (2011). "Guidelines for the assessment, design, construction and maintenance of phytocaps as final covers for landfills." Burwood, NSW, Australia.

Watson, D., and Hack, V. (2000). *Wildlife management and habitat creation on landfill sites: A manual of best practice*, Ecoscope Applied Ecologists, Muker.

Williamson, J. C., and Johnson, D. B. (1990). "Determination of the activity of soil microbial populations in stored and restored soils at opencast coal sites." *Soil Biol. Biochem.*, **22**(5), 671–675.

Wong, M. H. (1995). "Growing trees on landfills." *Environmental biotechnology: Principles and applications*, M. Moo-Young, W. A. Anderson, and A. M. Chakrabarty, eds., Kluwer Academic, Amsterdam, 63–77.

Yuen, S. T. A., Michael, R. N., Salt, M., Jaksa, M. B., and Sun, J. (2010). "Phytocapping as a cost-effective and sustainable cover option for waste disposal sites in developing countries." *Proc., ICSBE 2010*, Islamic Univ. of Indonesia, Jogjakarta, Indonesia, 145–151.

Zanetti, M., and Godio, A. (2006). "Recovery of foundry sands and iron fractions from an industrial waste landfill." *Resour. Conserv. Recycl.*, **48**(4), 396–411.

CHAPTER 21

Legal and Institutional Framework for Sustainable Solid Waste Management

Moe N. Chowdhury[1]

21.1 INTRODUCTION

Solid waste management and more particularly trash disposal was not a big issue in most parts of the world when population was scarce, lands were plenty and subsistence living the norm. But the industrial revolution that began in the late 18th century England and later spread to other countries changed forever the management of societal discards. Wealthy nations of Europe, the U.S., Japan and other industrialized nations learned earlier how to deal with rising waste volume as their manufacturing systems started churning up a variety of consumer products and their impacts on the environment became discernible. Many advancing nations of Asia, Africa and Latin America are now following the footsteps of the industrial societies and experiencing rapid rural to urban migration; their modernization of physical and economic infrastructure, rising income and consumerism is contributing to growing amounts of solid waste. The world is facing serious environmental and health related problems which can be partly blamed to the growing per capita waste generation and failure to manage it in an environmentally safe manner. Accelerated growth of megacities, each over 10 million people, is challenging waste management planners in developing nations to come up with innovative and environmentally sustainable practices for reducing, recycling and disposing of increasing throwaways. Poorer and transitional nations therefore must be prepared with effective policy resources and legal and institutional tools to cope with lurking environmental problems (Shekdar 2012).

Legal and institutional framework is a broad concept and it concerns the functions of governments, private enterprises, political jurisdictions, judicial systems, legislative bodies and regulatory agencies. More specifically legal framework incorporates laws, amendments, treaties, acts, ordinances, mandates,

[1]Jackson State Univ., Mississippi.

regulations and their enforcement mechanisms. Institutional system refers to the structure of government and its agencies, independent think tanks and private sector services. Existing financing system, technological level and practices using local engineering, and distribution of responsibilities among various levels of governments are also part of the institutional system. Together, legal and institutional elements are closely intertwined and impact upon each other. In many parts of the world, the practice of solid waste management predates the emergence of any legal infrastructure and replete with inadequate or absence of solid waste laws. The judicial system of a nation, the bodies of laws and regulations and their effects on its institutional structure define the role of all stakeholders in the processing and handling of societal discards. Solid waste statutes affect every citizen, service provider and generator. But it is only the governments at the national, provincial, regional and local level that have the statutory authority to enact laws and regulations for its citizens and economic pursuits.

This chapter scrutinizes the solid waste legal and institutional arrangement in a select group of countries across developed and underdeveloped parts of the world. The U.S. and European Union have been chosen to represent the advanced nations and the seven South Asian countries including Bangladesh, Bhutan, India, Maldives, Nepal, Pakistan and Sri Lanka are used as samples for transitional nations. A global comparative approach has been applied to unveil the nature and characteristics of solid waste laws and institutions *vis-a-vis* each country's and European Union's unique socio-economic, demographic and political conditions. Occasional attempts have been made to detect if the statutes and regulations of the U.S. and European Union or certain features serve as examples for transitional nations. The purpose of this case study is to contribute towards a methodology for critiquing the existing solid waste legal and institutional system, and make recommendations for improving the handling of discards in both developed and advancing countries.

21.2 WHY LEGAL FRAMEWORK MATTERS

A well-defined legal framework articulates the purpose for solid waste prevention, reduction, segregation, recycling, combusting and landfilling. Waste management is challenging in the densely populated parts of the world, where land disposal is competing with the needs for habitation and agricultural use. Rapidly growing cities are failing to cope with escalating waste volume, leaving behind uncollected waste on the streets and posing threats of unsanitary conditions. Leftover waste is often mixed with human and animal waste, causing flooding and breeding of insects and rodent vectors. Often waste is dumped and burned in open pits without cover, leading to pest problems and spread of diseases. Absence or a weak system of legal framework with no standards, regulations and absence of a proper enforcement mechanism will lead to health and environmental degradations (CalRecovery 2005a).

Despite private sector's increasing role, it is only the government that must lay out the ground rules for waste management. Legal framework is the result of national, regional and municipal governments establishing and implementing laws, regulations and enforcement mechanism. The role of government in crafting a legal system varies across countries. For example, the U.S. federal government plays an important role over "municipal solid waste landfills", "wastewater discharges from municipal solid waste management facilities", "landfill gas emissions from municipal solid waste landfills", "air emissions from municipal solid waste-to-energy plants", and "ash from municipal solid waste-to-energy plants". The Canadian federal government has no such regularity and enforcement system in place which is delegated to the provincial governments (Hickman 1999).

Solid waste laws and regulations are aimed at building institutions to address *a) direct services* for waste pre-treatment, collection, transportation, recycling, reduction, recovery and disposal, and *b) indirect or support services* like public education, awareness, information dissemination, technical expertise and financing (Mahar 2010). Non-government organizations (NGOs) and the private sector are playing an increasingly important role in waste management. Public awareness is becoming an important factor in the direct services. An understanding of the level of participation by the NGOs and the private sector and their role in direct services is critical. For example, informal sector participation in waste collection and disposal may be absent in certain countries but it may be in control of the recycling and recovery activities. Taking account of such anomalies will help build a more effective legal system. As economic growth and industrialization are making countries well-off, support services like education and awareness are becoming essential and integrated with the direct services. The involvement of government and non-government entities in building effective institutions is one of the primary goals of a statutory framework (Schubeler et al. 1996).

Literature on the mutual impacts of the legal system and institutions is scanty. Waste management researchers, to understand such interactions, may rely on the linkages between national laws, regulations, economic development and the writings of the political economy on the importance of legal system for economic welfare (Hobbes 1651, 1996; Ogus 2004). Adam Smith purported that the welfare of a state and its prosperity is dependent upon a good administration of justice (Smith 1775, 1980). Similarly Max Weber supports that economic development flows from a "rational" legal system (Weber 1925, 1954). A well-oiled legal framework may assist enticing investors to a solid waste industry, where upholding of any contract and agreement will have the support of a state and its judicial system. But the transaction costs and expenses for enforcing legal instruments have the potential for impeding economic development (Williamson 1985). Low cost legal procedures and institutions are therefore in order for enforcing or adjudicating disputes over contracts and agreements.

Challenges remain for a normative analysis of the Western institutions and principles and their adaptations to the developing world, where legal systems have yet to be developed in their fullest form. According to some legal scholars, the Western style capitalist and liberal democracy that comes with well articulated property rights, contractual entitlements and the independent courts may not be easily transferrable to the underdeveloped world (Tamanaha 1995; Davis and Trebilcock 2001; Trubek 1972). The U.S. legal virtuosos' push for American style legal framework for developing countries did not produce much success (Merryman and Clark 1978). The missing point in such endeavors was the failure to take into consideration the embedded laws and institutions in poorer countries and their cultural and political milieus. The newly independent Eastern European countries that were under the former Soviet Union went through the challenges of privatizing solid waste industry and the overhauling of legal frameworks. Some Western countries went through the concentration of privatization and deregulation of their industries in which the states had a much greater role. They were forced to retrofit their laws and regulations to redefine the role of the state and the private sector in solid waste industry. The predominance of the informal nature of the solid waste industry in transitional countries may make the effort of embodying components of the Western legal system more difficult. With the infusion of capital in the solid waste industry and increased efficiency with the use of machine power, Western model may become more relevant. However, the Western model that has evolved for a liberal capitalist economy may not be a match for developing nations, where state intervention and Keynesian measures in waste management are very strong (Posner 1998; Ogus 2004).

The legal and judicial infrastructure of the industrializing nations must address "company law, joint ventures and public-private partnerships; and fair and predictable tax laws" for the infusion of foreign capital in their economies (Perry 2001). Studies suggest that economic development is dependent upon the protection of property rights and the existence of a framework for rule of law that reduces the influence of powerful people in the developing countries (World Bank 1997; Beck et al. 2001). As countries move toward the path of economic advancement, they should be able to invest more in improving their legal system and consequently draw foreign capital to their solid waste industry (Posner 1998). Some East Asian countries were able to move ahead economically and improve their legal and institutional system at the same time.

The positive correlation between legal infrastructure and economic development has its parallel to the importance of laws and regulations in solid waste management. The more technical aspects of the legal framework that are generally applied to the corporate management, finance and accounting may be less relevant for countries that are in the process of modernization and not expecting in the near future of any infusion of overseas investment in their solid waste industry. However, the general principle of the legal system that establishes the "rule of law" remains paramount for all countries irrespective of their stage of economic development (Ogus 2004). A distinction must be made that the "rule of law" is different from the "rule by law" that is sometimes used by despotic governments to exert arbitrary decisions and where mediation and negotiation sit in the back burner. Rule of law applies to everyone including the government. The draft legal framework designed for solid waste industry and its regulations should be made available to the public before it is finalized, so everyone has a chance to review and comment. To be effective, rules must be enforceable, enforced, certain and unambiguous. The judicial process must be credible and its decisions must be binding to send a message that no one is above the rule of law including the elites in the position of power and those in control of the political and economic institutions (Frischtak 1997; World Bank 1992; 1997; Sherwood et al. 2009; Faundez 2000; Carothers 1998; Perry 2001).

21.3 NATURE AND CHARACTERISTICS OF LEGAL FRAMEWORK

An ill-defined legal framework without the articulation of Integrated Solid Waste Management (ISWM) may raise the costs for waste services and the potential for negative impacts from waste disposal (Ahmadi et al. 2012). According to Tchobanoglous and Kreith (2002) "ISWM can be defined as the selection and application of suitable techniques, technologies, and management programs to achieve specific waste management objectives and goals." The U.S. Environmental Protection Agency (USEPA) adopted an ISWM hierarchy that includes "source reduction, recycling, waste combustion, and landfilling". ISWM has traditionally been the domain of the local governments and municipalities in both industrialized and developing countries. Local governments look forward to the state, provincial or federal government guidance, which must define and incorporate the elements of ISWM in the laws and regulations (U.S. EPA 2013).

The evolution of a legal and institutional framework is an incremental process. It took many decades for the developed nations to take control over the management of the societal discards. The task however remains incomplete as industrialized countries are still striving to further improve their solid waste management through judicial and institutional reform. For example, the U.S. has more than 100 years of experience in passing solid waste laws and enforcing regulations, first through federal intervention and later state and local initiatives in formalizing and implementing provisions within the federal guidelines. Each year many states introduce new laws, most of which fail but some get through and become new acts. The three-tier U.S. legal system i.e. federal, state and local judicial process permits municipalities and counties to provide solid waste services with participation from the private and government entities and often joint public-private efforts. The framework for local implementation however is provided by the state and federal legislation in a way that allows innovation and flexibility on the part of the local governments. The federal government encourages creativity in the crafting and implementation of state statutes (Chowdhury 2011). The gradual changes in the existing solid waste infrastructure of the developing nations, in response to economic growth, will provide signals for any need for retrofitting of the prevailing legislative framework. Some countries have the options of starting from a blank slate where no solid waste laws currently exist. A review of the solid waste institutions may reveal the nature of services and inequity in reaching out to the low income communities. The legal framework must ensure that monopolistic behavior is reduced, competition among service providers encouraged and minimum services for all residents guaranteed.

The European colonial rule has a lot to do with the type of legal system in a vast majority of countries. Historically, there are four major global bodies of legal systems: common, civil, socialistic and Islamic laws. Because of dual or multiple colonial heritages, a mixture of common and civil laws are practiced in some countries. A few have based their jurisprudence on customary laws. Common laws are precedential meaning they are developed by judges in their court proceedings and any subsequent judgments must take into consideration prior decisions. Common laws are practiced in the United Kingdom and its former colonies in North America with the exception of the State of Louisiana in the U.S. and the Province of Quebec in Canada. They are also prevalent in many African, Asian and the Caribbean countries. Civil laws are codified in a constitution and are rooted in the Roman laws. Widespread use of the civil laws in the continental Europe helped their diffusion to Latin America, the Caribbean, Russia, Africa and the Southeast Asia (Rogowski 1996). Islamic laws can be found mostly in Muslim-majority counties of the Middle East and the North and West African nations. Socialistic system of laws has declined with the demise of the former Soviet Union and its allies, which embraced the civil laws following the end of the Cold War in the late 1980 s and early 1990 s. The multiplicity of legal systems has implications for their adaptations. It may be challenging for a developing nation without any solid waste laws and where civil laws form the core of its jurisprudence, to adopt a version of a legal system based on common laws (JuriGlobe 2014)).

21.4 TRAJECTORY OF THE US EXPERIENCE

Communities across the U.S. did not provide solid waste collection service until the mid-19th century. As the country was prospering and population increasing, geographically confined cities, in the absence of advanced transportation technology, were facing the crisis of garbage being thrown on the streets for animals to lounge upon and scavengers to recover resources. Some municipalities began to furnish refuse collection in the third quarter of the 19th century. Disposal was primarily in the form of burning and/or dumping at open sites or discharging into rivers and oceans. Limited waste collection during the pre-automobile transportation technology left mounting trash volume on the streets which caused deterioration of public health (Melosi 2005).

Waste collection practices became more frequent as the 20th century dawned on and when truck transportation made waste removal more frequent and expeditious. But the solid waste management continued to be characterized by open burning, incineration and underground burying. The geographic size of the country with a relatively smaller population combined with improved transportation to haul waste further away attributed to the continuance of such practice well into the 20th century. Less than 100 million people lived in the U.S. at the turn of the 20th century which is less than 1/3rd of the current population. As the U.S. population was rapidly increasing and industrial growth accelerating, especially after the 2nd World War, waste generation skyrocketed and its management became a major issue. The rising standard of living and the use of new consumer products, conveniently packaged in plastic wrapping, added significantly to the escalating waste volume (Small 1970) (Garrick 2004).

The United States has accumulated more than 100 years of experience in regulating rejectamenta and managing one of the highest per capita waste generations. The first such legislation, Rivers and Harbours Act of 1899 also known as the Refuse Act, was enacted for curbing indiscriminate dumping into the oceans, rivers and other water bodies (United States Code 1899). Many emerging nations have incorporated similar laws to protect their inland waters and ocean fronts. Such ban on ocean and river dumping improves not only the water quality but chances for better political relationships with neighbouring countries in case the washed-away solid waste is creating tensions.

A succession of major environmental laws, including those for societal discards, was approved in the 1960 s, 1970s and 1980 s. These enactments have accomplished safer disposal by improving upon the design and engineering of solid waste facilities. The principal legislative goals were to reduce the potential for groundwater contamination and incinerator emissions. In the past three decades however there has been an impasse in the passage of new laws and amendments to the old ones. Political gridlock revealed by the increasing partisan politics in the Washington D.C. has prevented further environmental statutes and also made it harder for changes in the prevailing laws that are needed to overhaul the regulations. Nationwide the recycling rate is only about 32% (U.S. EPA 2013). While some states in the Northeast, East, Midwest and West coast have achieved modest success in reducing waste and reaching above the national recycling rate, a majority in the South and sparsely populated West failed to put a dent on landfilling or incinerating. As a result, the total volume of solid waste landfilled or burned is continuing to rise almost every year. The recent slight reduction in the disposal of municipal solid waste is largely due to the slowing down of the U.S. economy (U.S. EPA 2013). The vast size of the country and low density of population, 80 persons per square mile, relatively inexpensive lands in the remote and rural parts of the country and lack of recycling mandates are partly to blame for the current practices of waste management. The costs for purchasing lands for new landfills are a small part of the overall expenses for burying waste under the ground. Landfilling therefore remains the preferred choice of disposal for more than 70 percent of all discarded wastes (Chowdhury 2011).

The takeaways from the U.S. experience for economically rising nations are many. First, solid waste laws and regulations are essential for reducing the occurrences of environmental catastrophes as experienced by many U.S. communities. Second, passing new laws or implementing policy may not be difficult at the early stage of economic advancement before special interest groups gain strength and throw gauntlets of obstacles opposing changes in the status quo. Third, a culture of regulatory infrastructure is imperative at national, regional and local levels as well as unwavering commitment to allocate resources for the enforcement of new laws and regulations. With the hindsight of the U.S. experience, the advancing economies will benefit in devising customized policy to suit the needs of their unique historical, cultural, and economic realities. Each country must come up with its own solution in the light of its idiosyncratic political, legal, institutional and geographical advantages and barriers. When it comes to solid waste management, one set of tools will not be justifiable for all countries. Variations and adaptations commensurate with local conditions are keys in the successful development of legal and institutional infrastructure.

21.4.1 Significance of the Major US Solid Waste Laws

Refuse handling in countries experiencing rising middle class and higher disposable income is reminiscent of the early U.S. waste management practices. Open-burning and land dumping in many U.S. communities during this era is revisiting at the outskirts of rapidly growing metropolises of Asia, Africa and Latin America. Widespread dumping of industrial waste in many American rivers as late as the 1960 s has been well documented (OHS 2013). However, the passage of the federal Clean Water Act has forced to stop such discharge into rivers. The large geographic size of the U.S., comparable to China but with less than 1/4th of its population, assisted in absorbing the environmental impacts of increased refuse dumping and caused delay in the implementation of any meaningful laws and regulations. Over 9 million square kilometres of land (3.5 million square miles) and relatively smaller population of 314 million (2013 estimate) are making it difficult for waste reduction and recycling programs to make any major inroads and become ubiquitous. Emerging economies, a majority with much higher population density and not much land to spare for landfilling, may not be able to wait too long before adopting strategies and implementing sustainable solid waste management practices.

Three sets of U.S. solid waste laws are particularly relevant for advancing countries: a) Solid Waste Disposal Act, b) Resource Recovery Act, and c) Resource Conservation and Recovery Act. In 1945, the U.S. Department of Public Health (USDPH) was put in charge of the solid waste program in the wake of the debate over what to do with the massive generation of refuse. Solid waste dialogue continued in the 1950 s and 1960 s and especially after 1970 when the USEPA was established, which created the Office of Solid Waste. Although the USDPH was not a regulatory agency, it managed to raise public conscious about the health and environmental issues of solid waste disposal and lobbied for the passage of the country's first major Solid Waste Disposal Act (SWDA) in 1965 (Hickman 2003).

The primary purposes of SWDA are to a) encourage research and development on disposal and resource conservation by reducing the amount of refuse and increasing recovery, and b) "to provide technical and financial assistance to state and local governments and interstate agencies in the planning, development and conduct of solid waste disposal programs" (U.S. Congress 1965). The SWDA provided for the first time the definitions for "Solid Waste" and "Solid Waste

Disposal". Its provisions may be beneficial in developing new policy and in the preparation of nascent legislations. The SWDA made state and local governments responsible for solid waste management and limited the federal role to passing a national legislation. Geographically large countries like Brazil, Democratic Republic of Congo, Nigeria, India and China may want to replicate the U.S. model of a three-tier administrative system to allow flexibility and stimulate creativity in the implementation of national guidelines by the states, provinces and local governments. Federal policy and guidelines allow lower tiers of government to design their own laws and regulations in the context of their regional and local peculiarities.

The SWDA was the beginning of any significant federal involvement but it needed overhaul immediately following its passage to give better directions for refuse management. The act did nothing to stop the open burning and dumping at un-engineered landfills and concerns for its amendment became imminent. The proximity of the Kenilworth Dump near Washington D.C., where trash was burned on a daily basis sending smoke to the direction of White House, U.S President's residential quarter, and Capitol Hill, where the U.S. Congress is housed, played a key role in raising awareness for remediating the pervasive problem of waste management among legislators and President Lyndon B. Johnson. Engineers at the Office of Solid Waste, United States Public Health Service (USPHS) and academics were still not sure about the long term policy for disposal and collection (USPHS 1964). However, research supported by the federal grants and approved by the SWDA continued and led to a better understanding of solid waste management issues. The SWDA was amended by the Resource Recovery Act (RRA) of 1970 that kept some of the original provisions, replacing some with the newer ones and adding more in response to the research findings (U.S. Congress 1970).

Federal enforcement and increase in regulatory authority did not arrive until the signing into law of the Resource Conservation and Recovery Act (RCRA) in 1976, amending the SWDA and RRA. The RCRA divided waste streams into hazardous (Subtitle C) and non-hazardous solid waste (Subtitle D) (Figure 21.1). The objective of the Subtitle D is to channel federal directives to state and regional solid waste plans, establish criteria for the selection of sanitary landfill sites and upgrading open dumps to sanitary landfills.

Until the passage of the RCRA, the 50 states in the union had limited role in solid waste management. However, with the federal grants since the passage of SWDA and additional funding authorized by RCRA, states began accumulating better understanding of waste collection, storage and disposal practices. States designated a single agency whose purpose would be handling solid waste issues and meeting federal mandates to qualify for receiving grants. In the early 1960 s, most states would have only a handful of solid waste professionals assigned to an agency whose responsibilities will also span to health, air, and water issues. Soon, states passed their own solid waste legislations within the federal stipulations and developed their own design criteria for sanitary landfills. States like Missouri adopted RCRA early enough to integrate it into their enactments and were able to

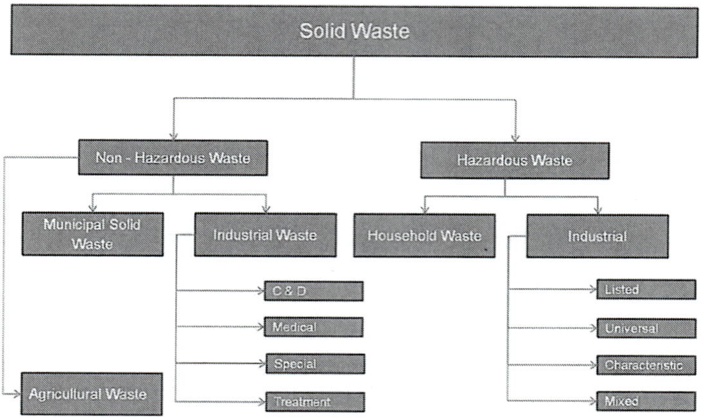

Figure 21.1. Solid Waste Tree, based on RCRA, USEPA
Source: Solid Waste Policy in the U.S. Wikipedia. Retrieved December 5, 2012

close most of the unsanitary disposal sites. By the late 1980 s and early 1990 s old dumps in communities everywhere begin to be replaced by sanitary landfills with liners, heavy clay and other barriers to protect the groundwater from contamination. This is also the time when some states on the East Coast faced landfill space crisis. Images of barges full of garbage sailing around the world to find disposal sites were etched into the memories of many solid waste mandarins. However, the solid waste industry responded to such crisis and built newly engineered sanitary landfills to meet the demand for a growing trash volume. Since then the landfill disposal costs have stabilized throughout most of the U.S. except where population density and land prices are relatively higher (U.S. Congress 1976).

U.S. legislative experience has ramifications for nations with limited exposure to developing a solid waste legal framework. First, the execution of new enactments must be predicated upon the consideration of solid waste origination from various sources like residential, commercial, industrial and agricultural activities and their proportionalities in the total generation. Second, the types of solid waste must be categorized into rubbish and garbage from households, construction and demolition debris, commercial waste from businesses and industries, industrial hazardous discards and the by-products of the mining and agricultural pursuits. Research on waste characterization will determine the relative importance of each waste type and its percentage of the total waste stream. Third, the nature of collection, storage and processing will also play a role in the new policy and laws. The prevailing disposal practice and the social, political, technological and environmental constraints related to waste handing will play a critical role in decision making. Fourth, the recovery and recycling rates in developing countries are much higher than in the U.S., because of wage differences and local demand for recyclables, which will influence specific sections of legislations dealing with their separation and processing into feedstock for industry. The high cost of labour-intensive processing is a contributing factor towards export of recyclables from the U.S. to countries where wages are much lower. Finally a healthy and open debate among solid waste professionals, industry representatives, academics and researchers should precede the concoction of legal and institutional framework.

21.4.2 Solid Waste Planning for Waste Reduction and Recycling

Landfilling is the primary method for waste disposal in the U.S. and almost 70% of all municipal discards is managed this way. But many African, Asian and Latin American countries may not be able to afford the cost of building sanitary landfills. Under many circumstances, siting of new landfills in closer proximity to human settlements will be unavoidable due to higher population density and settlement pattern. The clean air regulations, requiring the use of best available technology and expensive equipment, imposed higher cost of combustion forcing many incinerators out of business and making landfilling cheaper in most parts of the U.S. According to the USEPA, the MSW generation has been increasing every year for the last few decades (Figure 21.2). Although the recycling rate is increasing, it failed to rein in the rising tide of refuse (U.S. EPA 2013). The golden era of solid waste legislation ended with the amendment of the Resource Conservation and Recovery Act in 1984, and new legislative efforts for change in favour of waste reduction and recycling did not come after that. The private solid waste industry was very active during the passage of all major enactments and lobbied heavily to make sure that the proposed legislations favour their business practice. It has since invested heavily in building the infrastructure of sanitary landfills which is the most profitable sector of the business. The regulatory

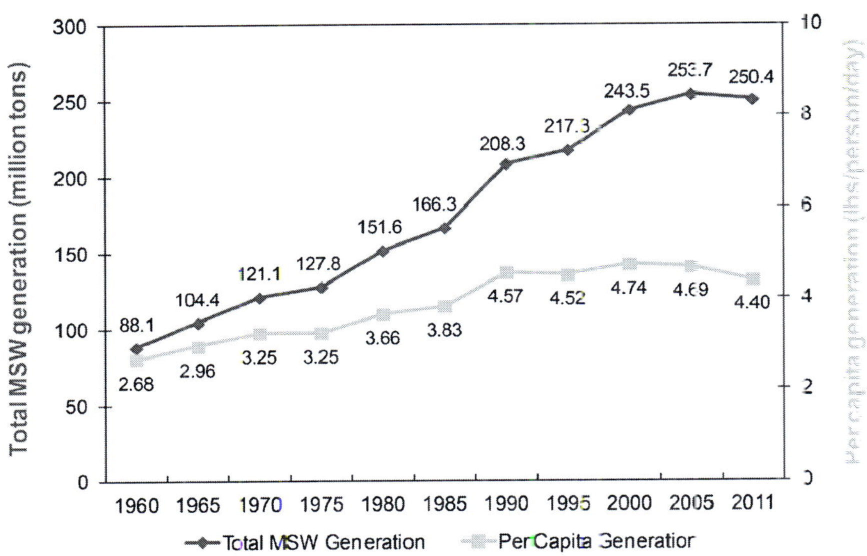

Figure 21.2. MSW Generation in the U.S. (1960–2011)
Source: U.S. EPA (2013)

environment established by the USEPA has been under attack since the 1980 s and has been accused for the rising cost of doing business (Klyza et al. 2008; Vig et al. 2010; Rosenbaum 2008). The emerging countries may have a totally different political and institutional *milieu*, where the passage of new legislation to reduce environmental externalities and encourage waste reduction and recycling may not be that difficult.

21.4.3 Possible Courses of Action

The U.S. solid waste management remains an incomplete success story when judged *vis-a-vis* achieving at least 50% or higher rates of waste reduction and recycling. The RCRA and its amendments had solved the landfill space crisis by increasing the number of disposal sites, but failed to achieve significant waste minimization. The prevailing political and economic conditions are not amenable to rectify existing laws in favour of recycling and waste reduction. For lack of a federal mandate some states were discouraged to develop state solid waste plans, but most prepared and updated them periodically. A majority of states also require local governments to create and implement their own plans. Waste reduction and recycling goals have been set by the USEPA. States have their own benchmarks for waste reduction and recycling programs set up by the municipalities and counties. But the funding opportunities to support such measures have remained insufficient in many states, adversely impacting community recycling programs. The recycling industry has struggled over the last three decades for low market demand for recyclables, but recently experienced growth despite economic recession. However, recycling processing facilities are not getting enough feedstock due to inadequate collection infrastructure for recyclables as community supported drop-off and curbside (or kerbside) recycling are not available in every municipality.

Without a broad legislative system, environmental effects from solid waste generation and disposal will be difficult to control. Despite flaws, the U.S. federal legislation remains a potential model for emulation as it provides a frame of reference for lower tier of governments; it is flexible and adaptable for sub-national governance. Expatriates working in the solid waste field may provide a readily available source for their native countries to tap into for developing and implementing a legal and institutional system. The recommendations in this chapter are not intended to copycat U.S. strategy but to suggest that some countries may benefit from its legislative experience, research and development. Japan with over 50% recycling rate and much higher population density may provide a better alternative for some advancing countries (Hershkowitz and Salerni 1987). Germany's Green Dot program is another par excellence. The legal and institutional arrangement of the industrialized countries is worth closer examination to find out what works and which policies and their components are candidates for direct adoption or adaptation. The first attempt towards formulating a long-term policy however is to start a dialogue and carry out research to increase the understanding of the existing solid waste management. A panel consisting of local experts, academics, government officials, environmental groups, industry representatives and expatriates with knowledge of their native countries and adopted homelands will provide an appropriate forum for starting a debate.

21.5 EUROPEAN UNION'S LEGISLATIVE EFFORTS

A majority of European countries, compared to the U.S., are densely populated and each has a long history of government intervention in providing public services. Western European countries, most belonging to the European Union, have implemented integrated system of waste management with prevention receiving the highest priority followed by reuse, recycling, materials recovery, energy recovery and disposal. Material recovery trumps energy recovery as a policy except in France, which cherishes both equally. France also prefers, unlike the rest of the Western Europe, a centralized system of decision making which influenced the evolution of its laws and regulations. However, a common denominator in Europe is the public acceptance of government responsibility for waste management with a trend towards increased privatization including the newly democratized Eastern Europe. A complicating factor however for legal and institutional framework is the expanding European Union (EU) Directives. Europeans are paying attention to this increasing EU role that is impinging upon the long standing legal and institutional structure. Even non-EU members are heeding to the dominant role of EU directives in anticipation of joining the community (CalRecovery 2005b).

EU consists of 27 countries, where almost half a billion people live. In 2006, a total of 3 billion tons of throwaways were produced in EU, 2 billion of which were mineral and solidified waste from mining, quarrying and construction and demolition waste. The average per capita waste generation is about 6 tons, but significant variations exist among member countries due to the methodology used in estimation. Bulgaria claims to have the highest per capita generation of 30 tons, followed by Luxembourg (20 tons), Romania (20 tons), Estonia (15 tons), Finland (14 tons) and Sweden (14 tons). Each EU household discards about half a ton of hazardous waste, composing of toxins and heavy metals, amounting to about 100 million tons (EU 2010a, b). Municipal waste generation and treatment fluctuates significantly among EU members. Denmark, Cyprus and Ireland show some of the highest per capita waste generation of over 700 kilograms while Czech Republic, Poland and Slovakia are on the other end of the spectrum producing less than 400 kg (Table 21.1). Again caution should be applied in using these per capita generations as a benchmark for policy formulation because of the methodology and definitions used for waste.

The legacy of EU waste policy and legislative framework goes back to the 1970 s, but the key features of the current strategy can be found in the Directive 2008/98/EC, which has seven chapters divided into 42 articles (Table 21.2). The first part of the directive deals with the definitions, scope and the hierarchy of

Table 21.1. Variations in municipal waste generation and treatment practices in 2009

	Municipal waste generated (kg/person)	Municipal waste treated %			
		Landfilled	Incinerated	Recycled	Composed
EU 27	522	42	20	22	17
Belgium	492	4	34	39	23
Bulgaria	468	100	0	0	0
Czech Republic	294	84	13	2	1
Denmark	801	5	53	24	17
Germany	564	1	35	46	18
Estonia	536	64	0	34	2
Ireland	786	64	0	34	2
Greece	448	84	0	14	2
Spain	588	60	10	13	17
France	541	34	36	16	14
Italy	550	46	11	11	33
Cyprus	754	87	0	13	0
Latvia	377	86	0	13	1
Lithuania	400	96	0	2	2
Luxembourg	694	25	47	0	28
Hungary	456	77	9	13	1
Malta	652	93	0	2	5
Netherlands	630	3	38	32	28
Austria	597	13	28	21	38
Poland	322	90	0	6	4
Portugal	472	63	19	8	10
Romania	379	99	0	1	0
Slovenia	441	66	0	34*	—
Slovakia	309	82	11	2	5
Finland	507	53	12	26	10
Sweden	518	4	47	37	12
United Kingdom	572	57	9	22	12

Source: Eurostat 2009; Hall (2010)

waste to set the foundation for the rest of the legislation. Technical standards for facilities and provisions for specific waste streams like discarded batteries, packaging, electrical and electronic gadgets comprise the rest of the directive for incorporation by the member countries.

The major thrust of the directive is on protecting the environment and improving public health through reducing the impact of waste. Member states

Table 21.2. Characteristics of European Union Directive 2008/98/EC

CHAPTER I: SUBJECT MATTER AND DEFINITIONS		**CHAPTER IV: PERMITS AND REGISTRATIONS**	
Article 1	Subject matter and scope	Article 23	Issue of permits
Article 2	Exclusions from the scope	Article 24	Exemptions from permit requirements
Article 3	Definitions	Article 25	Conditions for exemptions
Article 4	Waste hierarchy	Article 26	Registration
Article 5	By-products	Article 27	Minimum standards
Article 6	End-of-waste status	**CHAPTER V: PLANS AND PROGRAMS**	
Article 7	List of waste	Article 28	Waste management plans
CHAPTER II: GENERAL REQUIREMENTS		Article 29	Waste prevention programs
Article 8	Extended producer responsibility	Article 30	Evaluation and review of plans and programmes
Article 9	Prevention of waste	Article 31	Public participation
Article 10	Recovery	Article 32	Cooperation
Article 11	Re-use and recycling	Article 33	Information to be submitted to the Commission
Article 12	Disposal	**CHAPTER VI: INSPECTIONS AND RECORDS**	
Article 13	Protection of Human Health & Environment	Article 34	Inspection
Article 14	Costs	Article 35	Record keeping
CHAPTER III: WASTE MANAGEMENT		Article 36	Enforcement and penalties
Article 15	Responsibility for waste management	**CHAPTER VII: FINAL PROVISIONS**	
Article 16	Principles of self-sufficiency and proximity	Article 37	Reporting and reviewing
Article 17	Control of hazardous waste	Article 38	Interpretation and adaptation to technical progress
Article 18	Ban on the mixing of hazardous waste	Article 39	Committee procedure
Article 19	Labeling of hazardous waste	Article 40	Transportation
Article 20	Hazardous waste produced by households	Article 41	Repeal and transitional provisions
Article 21	Waste oils	Article 42	Entry into force
Article 22	Bio-waste		

Source: Data from Official Journal of the European Union, November 22, 2008 must follow a hierarchy of prevention, reuse, recycling and recovery before deciding for disposal. The directive starts with what materials are excluded from the definition of waste, then defines certain terms like "waste", "hazardous waste", "waste oils", "waste producer", "waste holder", "waste management", "prevention", "reuse", "recovery", "recycling", "disposal" and "best available technique". According to the criteria used for defining 'end-of-waste', certain categories of materials are no longer considered as waste if they have undergone recovery or recycling. The directive encourages the reuse of industrial by-products by excluding them from the definition of waste as long as they meet certain criteria. EU members must reuse or recycle 50% of household waste that includes at minimum discarded paper, metal, plastic and glass by the year 2020. Member states are also mandated to reduce non-hazardous construction and demolition waste by 70%. Following the efforts of reuse and recycling, any leftover waste must be disposed of safely and according to the standards of protecting the environment and human health (EU 2010a).

One of the major highlights of the directive is the "extended producer responsibility" (EPR), which is part of the producer pay principle, whose primary purpose is to make reuse and waste prevention work. EPR allows member states to pass laws requiring producers to take the financial responsibility for accepting and managing returned products following the end of their use. Member states may encourage design and development of products to reduce their environmental impacts. The cost of waste management must be borne by the waste producers or final users making them responsible for once their products are thrown away. EPR also provides incentives to product manufacturers to avoid generating waste and resort to recycling and recovery for unavoidable discards (Hall 2010).

Among the directive provisions for waste management are obligations to develop an integrated waste system among EU countries in order to reduce cross-boundary waste shipment both among member countries and between EU and non-EU members. Part of the objective is to achieve self-sufficiency in disposal and recovery. Member countries must develop waste management plans and carry out comprehensive analysis of all waste streams, inventorying waste recycling, recovery and disposal facilities. Needs for additional facilities must be assessed in the plans.

21.6 SOUTH ASIA

South Asian countries share a long tradition of passing legislation as many laws were enacted for protecting wildlife, canal irrigation, energy development, pesticide use, noise pollution and forests during the British colonial era in the 18th and 19th centuries. The history of laws even dates back to the Moguls beginning in the 16th century although it was limited to hunting only (Naureen 2009). The legacy of the archaic British legal system and colonial-style administration still lingers in Bangladesh, Bhutan, India, Maldives, Nepal, Pakistan and Sri Lanka and their cities even after more than six decades marking the end of the foreign rule. Rapid urban and economic growth and multiplication of populations have made the old planning, management and decision-making obsolete.

South Asian countries may give an illusion of similarity in physical and economic characteristics but a closer observation will reveal variations. Bangladesh and Maldives lead among the pack in terms of population density, but India claims to be a giant in territorial extent and total population (Table 21.3). An intersection of geography, population density and level of economic development in Bhutan, Maldives, Nepal and Sri Lanka create unique waste management challenges. Compared to their neighbors, these countries are relatively smaller in terms of number of inhabitants and areal extension. All four have set up a three tier governmental system i.e., national, provincial and local. While Nepal and Bhutan are geographically landlocked nestled just north of India, Sri Lanka and Maldives are island countries off the southern Indian Coast. A mixture of mountainous, foothills and plains in Nepal and Bhutan have limited development of transportation infrastructure, creating impediments for an efficient system of waste management. Island countries of Sri Lanka and Maldives in the Indian Ocean, with very limited landmass, require an idiosyncratic waste management strategy to suit their particularized conditions. They face the problem of finding proper disposal sites and where indiscriminate ocean dumping threatens immediate environment and tourism industry (Ravi 2007).

Relatively low level of urbanization combined with smaller per capita gross national income has so far constricted waste generation rates in South Asian countries (Table 21.3), but the recent high economic growth and rising middle class is poised to change that. Solid waste as a phrase has not entered into the jargon of urban administration in most municipalities except for the megacities that have been influenced by the donor agencies' refuse management surveys (UNEP 2005). Some of the world's largest metropolises are located in India, Pakistan and Bangladesh where waste collection and disposal is becoming ever challenging (ADB 2012a). Theoretically, higher income and more urban population are positively correlated with per capita waste generation, but the data in Table 21.3 do not support that conclusively. However, evaluation of waste generation data is a difficult task as the methodology of the survey and the definitions for waste are not constant. Urban areas and resort centers produce more waste compared to rural areas, giving wide latitude for per capita waste generation (APO 2007).

21.6.1 India

India with 1.2 billion people, the second largest population, is expected to surpass China's 1.3 billion in a few years. Its urbanization rate of 28% or 336 million people suggests the prospect for more urban dwellers and rising discards (ADB 2012a). Each of the four Indian mega-cities has more than 4 million inhabitants

Table 21.3. Socio-economic data and per capita waste generation in South Asia

South Asian Countries	Total surface area ('000 sq. km)	Population 2012		Per capita gross national income US $ (2011)	% of Urban population (2011)	Waste generation (kg/capita/day)
		Total (million)	Density			
Bangladesh[1]	144.00	151.82	1,054	780	28.4	0.41 (urban)
Bhutan[2]	38.39	0.72	19	2,130	35.6	0.53
India	3,287.26	1,230.01	374	1,420	31.2	0.3–0.6
Maldives	0.30	0.33	1,102	5,720	41.1	0.8–1.0
Nepal	148.18	26.85	182	540	17.0	0.2–0.5
Pakistan[3]	796.10	180.71	227	1,120	36.2	0.5–0.7
Sri Lanka	65.61	20.27	309	2,580	18.2	0.2–0.9

Source: Data for total surface area, population, per capita gross national income and % of urban population from Asian Development Bank. Basic 2013 Statistics. Economics and Research Department, Development Indicators and Policy Research Division. (www.adb.org) Retrieved May 25, 2013. Waste generation data for India, Maldives, Nepal and Sri Lanka from Asian Productivity Organization (APO), 2007. Solid Waste Management: Issues and Challenges in Asia. Report of the APO Survey on Solid-Waste Management 2004–2005

[1]County Analysis Paper: Bangladesh, Third Meeting of the Regional 3R Forum in Asia Technology, Singapore, October 5–7, 2011

[2]Phuntsho, Sherub; Dulal, Ichharam; Yangden, Dechen; Tenzin, Ugyen M.; Herat, Sunil; Shon, Hokyong; Vigneswaran, Saravanamthu; Studying municipal solid waste generation and composition in the urban areas of Bhutan. Waste Management & Research, 2010: 28: 545–551

[3]Haider, Irteza; Haider, Murtaza; Badami, Madhav; Household Solid Waste Generation in Urban Pakistan: A Case Study of Rawalpindi. Draft working paper, April 2013 (http://ssrn.com/abstract=2258903)

besides 400 cities with a minimum of 100,000 in every one of them. India has a parliamentary form of government and a bicameral national legislature and the country is divided into 28 states and 7 union territories.

The establishment of the Ministry of Environment in 1980, later renamed Ministry of Environment and Forests (MoEF), marked the beginning of environmental regulation in India. The first environmental legislation however did not enunciate until 1986 when Environmental Protection Act was passed, followed by the rules established in the same year. The passing of this landmark statute took place in the wake of the tragic Bhopal natural gas incidence in which several thousand Indians lost their lives in 1984. Rules concerning hazardous, biomedical wastes and batteries were respectively implemented in 1989, 1998 and 2001. In 2000 the Indian Parliament also passed the Municipal Solid Waste (Management and Handling) Rules (MSW rules) (Tables 21.4 and 21.5). Despite the passage of laws and development of rules, their compliance by the states and local governments are minimal (Ravi 2007).

The history of Indian legislative process suggests that natural disasters or mass public coercion have better chances in making meaningful reform. The solid waste rules implemented in 2000 was in response to the 1996 lawsuit against the central and state governments for an inadequate refuse management which led Supreme Court to form an investigative committee.

The MSW rules contain four schedules and apply to all municipalities, making them responsible for collection, segregation, storage, transportation, processing and disposal of wastes generated within their boundaries (Table 21.4).

In 2007 the Comptroller and Auditor General of India carried out a Performance Audit, required by the Constitution of India, on the implementation of the MSW rules. The audit investigated the MoEF and Central Pollution Control Board (CPCB), 24 states and their State Pollution Control Boards, State Forest Departments, State Urban Development Departments, municipalities in 20 states, 5 districts in 20 states and 180 hospitals in 15 states. Following are the highlights

Table 21.4. Four Schedules of Municipal Solid Waste Rules 2000

Schedule-I	Relates to implementation Schedule
Schedule-II	Specifications relating to collection, segregation, storage, transportation, processing and disposal of municipal solid waste (MSW).
Schedule-III	Specifications for land filling indicating; site selection, facilities at the site, specifications for land filling, Pollution prevention, water quality monitoring, ambient air quality monitoring, Plantation at landfill site, closure of landfill site and post care.
Schedule-VI	Indicate waste processing options including; standards for composting, treated l lactates and incinerations.

Source: Central Pollution Control Board, Ministry of Environment and Forests, Government of India

Table 21.5. Authorities and Responsibilities in Municipal Solid Waste Rules 2000

	Agencies/Authorities	*Responsibility*
1	Municipal Authorities	i. Ensuring that municipal solid wastes to be handled as per rules. ii. Seeking authorization from State Pollution Control Board (SPCB) for setting up waste processing and disposal facility including landfills. iii. Furnishing annual report. iv. Complying with Schedule I, II, III and IV of the rules
2(i)	State Government Secretary In-Charge of Department of Urban Development	Overall responsibility for the enforcement of the provisions of the rules in the metropolitan cities.
r(ii)	District Magistrates/Deputy Commissioner	Overall responsibility for the enforcement of the provisions of the rules within the territorial limits of their jurisdiction.
3	Central Pollution Control Board (CPCB)	i. Co-ordinate with State Boards and Committees with reference to implementation and review of standards and guidelines and compilation of monitoring data. ii. Prepare consolidated annual review report on management of municipal solid wastes for forwarding it to Central Government along with its recommendations before 15 December every year. iii. Laying down standards on waste processing/disposal technologies including approval of technology.
4.	State Pollution Control Board (SPCB)	i. Monitor the compliance of the standards regarding ground water, ambient air leachate quality and the compost quality including incineration standards as specified under Schedule II, III & IV. ii. Issuance of authorization to the municipal authority or an operator of a facility stipulating compliance criteria and standards. iii. Prepare and submit to the CPCB an annual report with regard to the implementation of the rules

Source: Central Pollution Control Board, Ministry of Environment and Forests, Government of India and recommendations of the audit report (Comptroller and Auditor General of India 2008):

1. Assessment of quantum of waste being generated in the country and identification of the risks to environment and health posed by waste.

   - Neither MoEF nor the states had completely assessed the quantity of various kinds of waste like MSW, bio-medical waste, hazardous waste, e-waste etc., being generated in the country.
   - MoEF was unable to make any projections about the amounts of waste that might be produced in future. Only 25% of the sampled states had made projections about the growth in waste. Adequacy of capacity to handle waste currently and in the future was assessed only by 29% of the states.
   - MoEF/CPCB had not completely assessed the risks to environment and public health posed by waste. Only 25% of the sampled states had assessed the risks to public health.

Recommendations

   - CPCB, as the nodal agency for pollution related issues should carry out, periodically, a comprehensive assessment of the amounts of waste being generated, according to the major waste types. All the states in India should be involved in this exercise so that a comprehensive database on waste is generated for aiding policy-making and intervention.
   - MoEF, with involvement of all the states, may collect data about growth of the various kinds of waste, analyse the factors contributing to its growth and the increase in waste quantities to arrive at strategies for waste management.
   - MoEF/CPCB, in conjunction with the states, may estimate the current capacity to handle all kinds of waste all over the country and ensure that additional capacity of waste infrastructure, if required, is created for safe disposal.
   - MoEF may carry out waste related pollution impact monitoring on a regular basis, to study the effects of improper disposal of waste on the environment. MoEF along with the states may also carry out regular surveillance including epidemiological surveillance of waste related impacts on public health.

2. Existence of policies and strategies for management of wastes and reflection of priority to waste reduction and waste minimization as against waste disposal.

   - Waste management efforts were not directed by a separate policy. MOEF not adopted a hierarchical approach to waste management, in the order of environmental priority. No effective strategies have been introduced to implement the 3 Rs (reduce, reuse and recycle), the current focus being only on disposal of waste. Only 8% of the sampled states had implemented the 3 Rs.

- MoEF/CPCB as well as 79% of the sampled states did not set any targets/ timelines for reduction of MSW, bio-medical waste, plastic waste, hazardous waste etc,. In the absence of clear targets/timelines, efforts made by the government to reduce waste were not measurable.MoEF had not appropriately addressed the role of informal sector in handling waste. Only 17% of the sampled states had recognized the role of rag pickers.
- MoEF and the states have not taken effective action to promote the use of recycled and environmentally friendly products. The implementation of MoEF's environment labeling programme called "ECOMARK" was tardy and was granted to only three product categories ever since the programme was introduced in 1991.

Recommendations

- MoEF may consider framing a specific policy for the management of wastes in India, incorporating the internationally accepted hierarchy for management of wastes.
- MoEF and the states may consider introducing effective strategies for the reduction and recycling of household waste like deposit refund schemes, promoting the use of jute bags rather than plastic bags, waste exchanges, etc., for reduction of waste at source.
- MoEF, in consultation with the states, should prepare an action plan for the reduction, reuse and recycling of waste with clearly defined numerical targets as well as timelines for the achievement of targets.
- MoEF should consider the introduction of Environmentally Preferred Purchases and lay down guidelines for the purchase of recycled products to promote the purchase of eco-friendly goods by the government and the agencies controlled by it.
- MoEF should include more products under the "ECOMARK" scheme and monitor adherence to environmental standards of these products. It should also prescribe standards for classifying products as environmentally friendly and carry out environmental impact studies of such products.

3. Existence of legislations specifically dealing with disposal of each kind of waste, incorporating penalty for violation.
   - Laws have not been framed for all kinds of waste, leaving the safe disposal of many kinds of waste like construction and demolition waste, agricultural waste, e-waste etc., unmonitored.

- The polluters were not being effectively held responsible for unsafe disposal, thereby creating no deterrence for non-implementation of the rules. In only 25% of the sampled states, some token action had been taken by PCBs/state governments against defaulters for illegal dumping of waste.

Recommendations

- MoEF should consider framing laws/rules for the management of all major kinds of waste like construction & demolition waste, end of life vehicles, packaging waste, mining waste, agriculture waste and e-waste being generated in the country.
- Considering the fact that the provisions of Environment Protection Act are seldom used, both at the central and the state level for punishing the polluter, there is a need to incorporate the Polluter Pays Principle (PPP) in the waste rules/legislations itself. This would act as a deterrent against open dumping of waste.

4. Allocation of clear responsibility and accountability to various agencies involved in the process of waste management.
   - There was no single body taking ownership of waste issues both at the central level and at the state level, leading to diffusion of responsibility and weak accountability.
   - Only 15% of states constituted the Solid Waste Missions for implementation of MSW rules, despite directives of CPCB in 2004-05 that all states should set up such missions.
   - There was no clear identification of bodies for monitoring of waste rules at the centre as none of the four central ministries, i.e., MoEF, Ministry of Urban Development, Ministry of Health and Family Welfare and Department of Petrochemicals took responsibility for monitoring of MSW, biomedical waste rules and plastic waste rules.
   - In the states, only 33% of the sampled states had allocated responsibility to PCBs for monitoring of MSW rules; 46% of the states had allocated responsibility for monitoring of bio-medical waste rules and only 37% of the sampled states were monitoring the implementation of the plastic waste rules.

Recommendations

- Since waste causes pollution and pollution issues are necessarily the responsibility of MoEF, the Central Government should consider appointing MoEF as the nodal body for managing all kinds of waste.
- MoEF should clearly identify, at the central level, bodies which would be responsible for the implementation of the waste management rules relating to MSW, biomedical waste and plastic waste. The states should

also clearly identify the agency responsible for implementation of the waste rules.

- Solid Waste Missions for dealing with overall issues relating to implementation of MSW rules should be set up in all the states. The government should assign clear responsibility to MoEF or any central body/ agency for monitoring.

Solid waste management in India is primarily a local affair. State government laws and local ordinances have provisions for the transportation, disposal and collection of solid waste but without much specificity concerning the "technical or organizational" details. State laws lay out the local government duties for keeping the streets clean, supplying garbage containers and carrying out waste to disposal sites. The missing elements in municipal ordinances are the responsibilities of the citizens for not littering on streets and keeping waste inside receptacles. There are no stipulations for sanitary landfills, transfer stations, waste reduction, recycling and keeping waste carriers covered during their transportation. No state monitoring system for municipalities is in place. Disposal sites accept all kinds of discards including municipal, industrial, biomedical and even sewage (Zhu et al. 2008).

21.6.2 Bangladesh and Pakistan

Besides India, South Asia has two other populous countries that remained together as one sovereign nation until 1971. According to the Asian Development Bank database for 2011, Bangladesh has a population of 142 million. The estimated population of Pakistan is 177 million (ADB 2012a). Geographically much larger Pakistan with an area of 307,374 sq miles (796,095 square kilometers without disputed areas) is more than five times the size of Bangladesh (56,977 square miles or 147,570 square kilometers). Pakistan is divided into four provinces, several federal territories and tribal areas. The local governments consisted of more than 100 districts (zillahs), over 400 sub-districts (tehsils) and several thousand union councils. Unlike Pakistan which has three tiers of administration i.e., federal, provincial and local, Bangladesh set up two spheres of government, national and local. Bangladesh is carved out into six divisions, but the principle administrative jurisdictions are its 64 districts or zillahs, 469 upazillahs or sub-districts (parishads), and 4,484 union parishads. Six city corporations include Dhaka, Chittagong, Rajshahi, Khulna, Barisal and Sylhet.

Massive increase in solid waste generation, especially in urban areas of both countries, is fueled by expanding population and rural to urban migration. A recent estimate by the Pakistan Environmental Agency reveals that the country generates approximately 20 million tons of refuse a year which is about 59,000 tons per day. Pakistan's population is also growing at a faster rate of 1.8% compared to Bangladesh's 1.3% (Hassan 2012). Populations of both countries have increased several times during the last five decades and will continue to swell in the coming years albeit at a much lower rate. Nowhere the challenges of solid waste management manifest itself more than their two megacities, Dhaka in Bangladesh (14.65 million) and Karachi in Pakistan (13.12 million) (UNDESA 2010). Stark realities of waste generation and disposal in Dhaka are unveiled in a study sponsored by the Japan International Cooperation Agency. Dhaka City Corporation website indicates a daily waste generation of 3,000 to 4,000 tons, whereas an estimated 9,500 tons/day are generated in Karachi (Pakistan Observer 2011). Dhaka's website may have used an old source that reported a daily generation of 3,500 tons in the early 1990 s when Dhaka's population was less than half of what it is now (Enaytullah 1994). This low generation total has been quoted in subsequent research without factoring into rapidly changing population figures (Islam and Shafi 2004).

Pakistan and Bangladesh have passed environmental laws and set up environmental ministry at the national level. Both countries were influenced by their participation at internationally organized environmental forums held during the last several decades. Pakistan established the Environment and Urban Affairs Division in 1974 under the Ministry of Housing and Works. The division was entrusted to formulate environmental policy and with procedures for environmental impact assessments. Pakistan Environment Protection Council and Pakistan Environment Agency were created in 1983. Pakistan Environmental Protection Act was enacted in 1997. The Water Pollution Control Ordinance, adopted in 1970, is the first major environmental law in Bangladesh (during erstwhile East Pakistan) which was repealed by the passage of the Environmental Pollution Control Ordinance in 1977. Subsequently, Environmental Conservation Act (ECA) was passed in 1995, replacing the 1977 ordinance and followed by the Environmental Conservation Rules in 1977. More recently Biological Diversity Act was enacted in 2012 and the ECA was amended in 2010 (Naureen 2009)

Bangladesh and Pakistan have enacted environmental laws, but the missing elements are the specific laws addressing solid waste management (Bhuiya 2007). Sanitary landfills are non-existent, municipal wastes continue to be dumped at non-engineered sites and open burning is a common practice. Studies funded by the donor agencies reveal the severity of the situation in Dhaka and Karachi that are running out of disposal space but neither government is serious about the consequences of indiscriminate dumping and rapidly expanding waste volume. Absence of any solid waste laws and rules simply reinforce the lack of commitment on the part of the central governments. They have lumped together the responsibility of environmental policy under ministries responsible for forestry and local government alike. Recently Pakistan has created the Ministry of Climate Change replacing the Ministry of Environment, Local Government and Rural Development. A parallel central government agency in Bangladesh is Ministry of Environment and Forests.

21.6.3 Sri Lanka

The central government of Sri Lanka oversees 9 provinces and 330 local authorities divided into 18 municipal councils, 42 urban councils and 270 rural pradeshiya sabhas or councils. The total land area is approximately 62,705 square kilometers or 24,211 square miles and the estimated population 20 million. The number of local governments changes often with the division of the existing jurisdictions into new administrative territories. Provincial governments have wider latitudes in terms of collecting taxes and fees, including for solid waste services, and responsibility for internal law and order, economic plans, education and housing. Local authorities have jurisdictions over health, utility services and roads (CLGF 2012). The Ministry of Environment is the lead agency for solid waste management planning and policy.

Sri Lanka has passed environmental laws and set up national offices for pollution control, but delayed implementing any direct measures for reducing the effects of waste disposal. The Article 27 (14) of the new constitution, adopted in 1978, reiterates that "The state shall protect, preserve, and improve the environment for the benefit of the community". The National Environment Act was passed in 1980 creating the Central Environmental Authority in 1981, and it was put in charge of directing the local governments on environmental issues. The act was subsequently amended in 1988 and 2000. Section 23 A of the act stipulated license requirements for discharging solid waste that has pollution potential. A committee was formed to coordinate environmental related activities of the various government agencies. It took almost another 19 years, following the passage of the act, to adopt the National Strategy for Solid Waste Management in 1999. Policies were also developed at the same time for forestry, wildlife conservation, land development and mining, agriculture, fisheries, plantations, coastal and marine areas, industry, tourism, energy, transportation, health and sanitation and urban development. Adoption of such measures reflects the governmental response to curtail the externalities of rapid economic growth Sri Lanka has been experiencing since the late 1970 s. The country is transitioning from a predominantly agricultural economy to a semi-industrial society (Abeysuriya 2007; NSWMSC and JICA 2008).

The Sri Lankan government has gone beyond the formulation of a national solid waste strategy and into the implementation phase by setting up an institutional framework that appears to be working. Ministry of Local Government and Provincial Councils set up the National Solid Waste Management Support Center which works closely with local authorities. The center took a three-prong strategy of a) waste minimization, b) maximum resource recovery and c) sanitary landfill. The second strategy is based upon segregation of waste, home and centralized composting and "strengthening the social capacity" (Ministry of Local Government and Provincial Councils 2014). The ministry encourages the utilization of local technology to keep cost down and make projects sustainable by using existing manpower, citizens and non-government organizations. Technical manuals have been developed by the ministry to share with the local governments. The ministry is also engaged in carrying out investigations of waste management practices to provide such information. Local authorities are also incentivized to seek technical assistance from the ministry. Newsletters are posted on the ministry website to highlight on-going and completed annual reports and solid waste projects in various provinces.

21.6.4 Maldives

Maldives consists of 1,192 islands of various sizes ranging from 0.5 km^2 to about 5km^2, grouped into 26 atolls, straddling north and south of the equator. Only 197 of these islands are currently inhabited. Maldives has a land area of only about 300km^2 or 115 square miles, but scattered over 90,000 km^2 or 35,000 square miles making it one of the most territorially dispersed countries. The total population of Maldives is only about 329,000 (ADB 2007, 2012b).

Vast distances between atolls influenced the formation of Maldives' government structure. Local government empowerment is embedded in the Decentralization Act 2010 and the Local Council Election Act 2010. The Ministry of Home Affairs oversees local governments through the Local Government Authority Department. Local governments consist of 20 atoll councils, two city councils (Male and Addu) and 66 island councils. Councils can charge fees or rents and are entrusted to create security instruments like bonds and bills to finance development efforts (CLGF 2012).

Maldives has put a very high priority on protecting its fragile coral reef environment. Growing population and increased economic activities, which includes thriving tourism industry contributed to high per capita waste generation. The most recent State of the Environment report reveals that approximately 298,000 tons of wastes were created in 2009 or 0.91 kg/person/day or 2lbs/person/day (ADB 2007). Other research has cited a much higher rate of waste generation (UNEP 2002; JICA 1999). Waste generation varies significantly among different parts of the country with capital city Male responsible for generating 0.8 to 1.0 kg/person/day to about 0.3 to 0.5 kg/person/day in other islands. The resort islands experience much higher rate of generation to the tune of 2.5kg/person/day. Municipal waste collection is absent on most islands. In Male, which has a population 100,000 or one-third of the country total, individual private collectors gather waste from door to door and commercial establishments and take them to two collection yards. Waste is transported from Male and other atolls by barges to Thilafushi Island, which accepts almost 80% of the waste generated in Maldives, where it is burned. Some recycling of iron and plastics are taking place in Thilafushi. Although the resort islands generate more waste, better management has assisted to their cleaner appearance. In a majority of inhabited islands however household wastes are indiscriminately dumped or burned in backyards or at designated sites not far from the Indian Ocean (IFC 2011; Latheefa et al. 2011).

Maldivian constitutional provision for environmental protection, under Chapter 1 Article 22, and the environmental law sets the cornerstone for solid waste legal and institutional framework (Hussain 2003). Environmental awareness is further revealed by the adoption of sustainable measures laid out in its first National Environmental Action Plan (NEAP) in 1990 (MHTE 2009).

Environmental Protection and Preservation Act was passed in 1993. It contains many environmentally related provisions including one for "waste management and trans-boundary movement of hazardous wastes" (ADB 2007). Developed in 1999, National Environmental Action Plan II continued emphasis on solid waste management.

According to NEAP III 2009, Maldivian government is very aware of corals reefs' importance for its tourism industry, which is the mainstay of its economy, and therefore has committed to protect such natural resources from the harmful effects of growing solid waste volume, climate change and rising sea level. The 7th largest concentration of coral reefs is crucial for the tourism and fishing industry. About one-third of the GDP comes from the eco-friendly tourism industry. The small size of the country, lack of available land and the topography of the islands demand that solid waste is not dumped indiscriminately. More than 80% of its land surface is less than one meter above the sea level. The high groundwater level makes it easier for run-off from waste to contaminate the underground water. The Ministry of Environment and Energy, a single dedicated agency, is entrusted to enforce environmental laws and regulations, representing a major shift when the environmental issues were previously addressed by the multipurpose Ministry of Housing, Transport and Environment (ADB 2007).

The Maldivian government approached the International Finance Corporation (IFC), a subsidiary of the World Bank, for seeking assistance in addressing the country's waste management dilemma. The IFC carried out an investigative research and recommended a strategy to mitigate the adverse impacts resulting from its tourism development, growing volume of solid waste and climate change. An integrated strategy for waste collection, transportation, storage and recycling is part of the proposed project developed by the IFC. Once completed, the project will address 60 to 70% of the waste generated in Maldives including from the capital city Male and the surrounding islands. The chosen international firm from among 15 bidders will invest $50 million to implement the waste management strategy that proposes to build a 2.7 megawatt waste-to-energy plant on the Thilafushi Island and operating the entire project for 20 years. Maldivian government will have the option to take over the project thereafter or extend the contract (Singh 2011; Dado 2012)

21.6.5 Nepal

Nepal is inhabited by about 27 million people and it occupies portions of the Himalayas, where well-known and the world's highest peak Mount Everest is located, and its foothills, spreading over a territory of 147,181 km^2 (56,827 square miles). Communities generating waste must dispose of them locally as rugged topography limits transportation options. Democracy took roots in Nepal following a tumultuous period of political unrest that led to the adoption of a new interim constitution in 2007 and constitutional monarchy that allows the king at the apex but with significantly reduced power. A Council of Ministers headed by a prime minister is the executive body. The Council oversees the Ministries,

Departments and five Development Regional Offices that are further divided into 14 zones and 75 districts. Districts are further broken into small "Ilakas" or areas. Nepalese constitution transferred most of the administrative powers to the Council of Cabinet Ministers and has a clause for impeachment of the king by the National Assembly (UNESCAP 2013).

Nepal's legal and institutional guidelines for environmental policy began almost 40 years ago in the 1970 s, but meaningful and significant solid waste legislation came much later. At first, solid waste issues were addressed tangentially as a component of the larger environmental strategy without any specific actions or plans. The National Park and Wildlife Conservation Act of 1973 marks the announcement of the first environmental legislation. Up until the mid 1980 s solid waste did not get much attention; it changed with the initiation of the Solid Waste Management and Resource Mobilization Act of 1987 for the regulation, transportation, recycling and disposal of waste (Table 21.6) The act was immediately implemented in the capital city of Kathmandu, surrounding municipalities and later to other communities. It created the Solid Waste Management and Resource Mobilization Center and spelled out its functions, duties and authorities, followed by a set of rules in 1989. A Board of Directors, consisting of a chairman appointed by the King of Nepal and representatives from several ministries and the general manager, was responsible for overseeing the center. The major provision of the act was directed at giving the center authority over controlling illegal dumping of waste including the prohibitions and penalties for indiscriminate disposal. Tuladhar (2007) reported that the center is currently not functioning and has been replaced by another agency. Subsequent environmental policy lacked the integration of the 1987 Act.

Nepalese government devised a National Conservation Strategy in 1988 to manage and implement pollution control policy and solid waste strategy. In 1993 the Nepal Environmental Policy and Action Plan was developed but without any reference to solid waste. Except for the construction of a landfill near Kathmandu, the tenth plan (2002–2007) missed the opportunity for laying out a solid waste strategy. The first breakthrough for any concrete solid waste policy came in 1996 when National Solid Waste Management Policy was established (Table 21.6). Not much was known about solid waste generation, collection or disposal in Nepal at the time of this policy initiation. The major objectives included making local governments more competent in managing their wastes, raising public awareness and participation, using appropriate technology, privatizing solid waste related services, etc. A National Council was formed for solid waste management consisting of members from the National Planning Commission, secretaries from various ministries, chiefs of various municipalities and private sector representatives from the Nepal Chambers of Commerce, Industries and Hotel Association. Tuladhar (2007) reported that the National Council failed to meet for the last five years.

The passage of the more comprehensive Solid Waste Management Act in 2011 is the most recent development. It repealed the 1987 Solid Waste Management and Resource Mobilization Act and also replaced the Solid Waste

Table 21.6. Nepal's efforts for environmental and nature conservation

1. Enactment of National Park and Wildlife Conservation Act, 1973
2. National Park and Wildlife Conservation Rules, 1974
3. The National Forestry Policy (NFP), 1976
4. Leased Forest Rules, 1978
5. Watershed Conservation Act, 1982
6. King Mahendra Trust for Nature Conservation Act, 1982
7. Soil and Watershed Conservation Regulation, 1985
8. National Conservation Strategy, 1987
9. *Solid Waste (Management and Resource Mobilization) Act, 1987*
10. *Nepal Environmental Policy and Action Plan (NEPAP), 1993*
11. Forestry Policy, 1992
12. Hydropower Development Policy, 1992
13. Industrial Policy, 1992
14. National Shelter Policy, 1996
15. Buffer Zone Management Regulation, 1996
16. *National Solid Waste Management Policy, 1996*
17. *Environment Protection Act, 1996*
18. *Environmental Protection Rules, 1997*
19. Buffer Zone Management Guidelines, 1999
20. Formulation of Sustainable Development Agenda for Nepal (SDAN), 2002
21. Nepal Biodiversity Strategy, 2002
22. Nepal Biodiversity Strategy Implementation Plan, 2006-2010
23. National Wetland Policy, 2002
24. Water Resources Strategy, 2002
25. Leasehold Forestry Policy Guidelines, 2003
26. National Action Program on Land Degradation and Desertification, 2004
27. National Agricultural Policy, 2004
28. Community Forestry Guidelines, 1996 revised in 2002 and 2005
29. Agro-biodiversity Policy, 2007
30. National Adaptation Programme of Action (NAPA), 2010
31. Climate Change Policy, 2011

Source: National Planning Commission, Government of Nepal. (2012) Nepal Status Paper: United Nations Conference on Sustainable Development 2012 (Rio + 20) Synopsis. Singhadurbar, Kathmandu, Nepal

Management and Resource Mobilization Center with Solid Waste Management Technical Assistance Center. The center's Board of Directors consists of the secretary of the Ministry of Local Development, joint secretary of the Municipality Management Division of the Ministry of Local Development, joint secretaries of the Ministry of Finance and Ministry of Environment, two solid waste experts including a woman nominated by the board and the executive director of the center. The act has a provision for a Solid Waste Management

Council to determine the policy concerning solid waste management. The council is composed of the

1. Minister of Local Development
2. Secretaries of the Ministries of Industry, Environment, Health and Population, Local Development and National Planning Commission
3. Chief of the Kathmandu Metropolitan City
4. nominated representatives of five development regions and areas affected by solid waste
5. representative of the Federation of Nepalese Chambers of Commerce and Industry
6. two solid waste experts or scientists including a woman
7. institutional representative nominated by the Council from among community organizations working in the solid waste management sector

Local entities like metropolitan city, sub-metropolitan city, municipality or the village development committee are now responsible for managing solid waste. Besides, revisions of the older provisions for production, collection, minimization and disposal of solid waste, examples of some newer stipulations can be found for transfer station, sanitary landfill site, involvement of the private sector, public-private partnership, service fees for solid waste management, pollution control measures and the management of medical waste.

21.6.6 Bhutan

The small size and relatively sparse population did not stop Bhutan from setting up a legal and institutional outline par excellence. About 720,000 people in 38,394 km^2 or 14,824 square miles make the country the least densely populated in South Asia. Landlocked Bhutan with relatively new democratic institutions, challenging relief and strong prospect for tourism has made good progress in passing solid waste related legislations and regulations (Statistical Yearbook of Bhutan 2012).

The decentralized administrative system of Bhutan is commensurate with the responsibility for waste management falling on the local governments. Poor transportation and the hilly nature of the territories have limited Bhutan's ability to provide an integrated system of waste management. Each district and village must rely on developing its own waste collection and disposal system. The country is divided into 20 districts (dzongkhags) that are administered by the district committees (dzongkhag tshogdus). The districts are further divided into 205 blocks (gewogs) and several villages make up each block. There are also 54 urban centers and satellite towns. According to the Bhutan Municipal Act of 1999, "waste impounding" and "litter control" is the prerogative of municipal corporations. For sustainable waste management to work, the impacts of solid waste laws and regulations must be felt at these local governments (Department of Urban Development and Engineering Services,

Ministry of Works and Human Settlement 2013; Statistical Yearbook of Bhutan 2012).

Despite a balanced approach between economic development planning and preservation of pristine environmental resources, pollution control is becoming difficult with impending population increase and upwardly waste volume (European Union External Action 2014). Waste collection services are available in a handful of urban areas and sanitary landfills are yet to be built. Solid waste generation and disposal has particularly become a serious problem in the two largest cities, Thimphu and Phuentsholing. Memelakha landfill, the only disposal site for the capital city of Thimphu, is overflowing (Penjor 2007). Between 1994 and 2005, MSW generation increased more than four-fold from 8 tons to almost 37 tons per day. Growing industrial, electronics and medical waste are becoming major concerns as well (NEC 2008).

Bhutan's environmental laws and regulations are driven by the philosophy of the Gross National Happiness (GNH) and not by the Gross National Product. The adoption of sustainable waste management strategy is the natural outcome when GNH is the benchmark for economic development and planning (GNH 2012; Penjor 2007). The four pillars of GNH form the basis of the nation's five year plans are:

1. promotion of equitable and sustainable socio-economic development
2. preservation and production of cultural values
3. conservation of the natural environment, and
4. good governance

The Article 5 of the Bhutan's constitution mandates that 60% of the country's total lands must be under forest cover perpetually. As of 2008, 72% of the land coverage was under forest, and it will become increasingly challenging, as population increases, to meet the minimum threshold. Bhutan boasts of remaining carbon neutral (Constitution of the Kingdom of Bhutan 2008).

A flurry of waste management legislations, regulations and policies in the last few years show commitment on the part of the Bhutanese government to manage environmental pollution. The National Environmental Protection Act of 2007 adopted the principle of 3Rs (reduce, reuse and recycle). It also created the National Environment Commission as an independent agency with the jurisdiction and power to make decisions and formulate policy as necessary. The commission is an autonomous and regulatory authority for all environment associated matters that strives to achieve sustainable development through the prudent use of natural resources; it is authorized to oversee environmental laws, policies and mandates, and also serves as the coordinating agency for inter-sectoral environmental programs. The agency works with both public and private sectors in matters related to incentives, enforcement and regulation. The commission also has the stipulation for "polluter pay" where waste producers will be responsible for managing waste by reusing, recycling or disposing of safely.

The passage of the National Environmental Protection Act was quickly followed by the adoption of the National Strategy & Action Plan: Integrated

Solid Waste Management. The plan lays out an inventory of current waste collection, disposal, processing conditions, institutional arrangements and strategies for enforcement. It suggests specific actions necessary for improving the system of waste management focusing primarily in urban areas where environmental threat is greatest. Two of the newer strategies in the plan that did not exist in the act are a) Community Mobilization Program (Program I) and Capacity Building Program (Program II). Program I is focused on public awareness and education, aiming at a cross-section of citizens of all ages ranging from school children, members of sports and youth clubs, grownups and interest groups such as industrial and hazardous waste producers. The objectives of Program II, geared primarily to municipalities and other local governments, are to improve the capacity of service providers. A Mobile Integrated Solid Waste Management Action Team consisting of waste management professionals will assist Thimphu City Corporation and develop a model program for rural municipalities, villages and more rural areas of Bhutan. However, the waste professionals in Bhutan realized that having an action Plan is not enough without some teeth for implementation and enforcement, and therefore moved to developing specific legislative mandates and regulations.

Waste Prevention and Management Act is the first waste specific environmental law passed in 2009. The act incorporated the principles of polluter pay and the three-tier strategy for waste reduction, reuse and recycling, both proposed in the National Environmental Protection Act. It further defined the responsibilities of the individual citizens, identified implementing government agencies at the national and local levels, their powers and duties. The Ministry for Works and Human Settlements and City Corporations are to oversee waste management in towns and cities (thromdes); the Ministry of Economic Affairs for wastes resulting from trade and industries; the Ministry of Health for managing medical waste; the Ministry of Education "for ensuring education and awareness in waste prevention.....”; the Ministry of Agriculture for wastes accruing from agriculture related activities. Some of the ministries have designated departments to carry out directions for certain categories of waste streams. The Department of Information Technology under the Ministry of Information and Communication has been assigned the responsibility for e-waste management.

The Waste Prevention and Management Act broadened the definition of waste to include hazardous, non-hazardous, medical and e-waste. The implementing agencies now have much wider authority for developing rules, regulations and plans. They can request the central government "to issue executive orders prohibiting or restricting the sale, distribution, or use of excessive packaging or products resulting in large volumes of waste". They can prepare waste management plans, including those for private commercial and industrial establishments, and impose restrictions on littering and waste disposal. Tipping fees and similar charges can be imposed by the implementing agencies to cover operational costs. Tax incentives for environmental services are now available and provisions are allowed for making funds available for co-financing with civil society, public-private partnerships and rewards for exemplary waste management initiatives on the part of scrap dealers, recyclers and similar entities.

The most recent addition to Bhutan's legislative repertoire is the Waste Prevention and Management Regulations adopted in 2012 for the purpose of establishing guidelines to implement the Waste Prevention and Management Act. The regulations clarified the roles of implementing agencies for an effective system of waste management. They prohibit location of sanitary landfills and dumps in forest, parks, sensitive and biologically diverse habitats, human settlements and similar environmentally protective areas. The new regulations have reclassified wastes into four categories: medical, municipal, industrial and E-waste, providing detailed directions for their management, regulations and enforcement. The implementing agencies are given prerogatives to form Sanitary Committees for coordinating with district environmental officers, "review waste management plans....", "review and assess needs for human, financial or technical support... ", "recommend educational and awareness programs....", "review and recommend waste collection and disposal service fee.." and similar functions.

21.7 CONCLUSION

Developed and transitional nations' legal and institutional frameworks are in flux irrespective of where they are on the economic development spectrum. Copious experience in creating and amending laws and regulations has provided wisdom to advanced industrialized societies. Emerging countries are facing challenges from expanding economic activities and rapidly rising waste volume is making it necessary for its safer disposal. Can transitional nations gain from the lessons of advanced nations in their legal and institutional experiment with societal throwaways? Are legislative and institutional parameters of rich countries transferable to countries that are on the path to prosperity? This chapter suggests the answer is yes but with cautions as circumstances of each country is different. There is no harm in probing the experiences of rich nations but direct transfer of legal and institution frameworks is not recommended. Certain elements of the advanced legal system may be amenable for direct adoption while others need to be adapted to suit the needs of the discrete institutional, political and economic realities.

Federal legislative efforts have stalled in the U.S., albeit many of the 50 states in the union have not disengaged themselves from improving upon their waste management through new enactments. Prevailing political conditions however have a lot to do with new legislative efforts as some states with a majority liberal inclination and environmental consciousness, for example California, Washington, Oregon and those in the New England, have made significant strides in terms of waste minimization, prevention and recycling. However, a majority of states in response to federal inaction have made little or no efforts in reducing the volume of waste. Total landfilling and incineration in the U.S. therefore is on rise almost every year with solid waste generation increasing faster than recycling and recovery amounts. The U.S. legislative experience of over 100 years however remains an untapped resource for industrializing countries to probe into and look for areas that can be readily adopted or retrofitted for incorporation in their own legal system.

European Union (EU) countries are at the forefront of waste management as evinced from their recent and ongoing legislative measures centered on waste prevention, reuse, recovery, recycling and disposal as the last resort. EU directives provide a framework for member countries to devise their own laws and regulations. However, it is too early to evaluate the effects of the EU directive, promulgated in 2008, on member countries and their statutes. Some Western European members already had their solid waste legislative infrastructure in place before the EU directives were issued. Current fiscal and economic crisis may not allow EU countries to pay much attention to the new EU mandates. But EU members especially those on the Eastern and Southern parts may not want to wait too long before implementing the directives because of the job creating opportunities of recycling and recovery and prospect for reducing environmental threat.

This chapter reviewed the solid waste legislative experience of seven South Asian countries that are very divergent from geographic and environmental viewpoints. Their common denominators are the British colonial rule and position on economic development spectrum where they are clustered at the bottom end. Island countries of Maldives and Sri Lanka enjoy a relatively higher per capita income compared to the mainland South Asian countries. India is the giant among these countries with its vast territorial extent and population. Bhutan is relatively sparse and Bangladesh has one of the world's highest population densities. Pakistan and Nepal are large relative to their population sizes.

Legislative attempts to manage solid waste vary significantly in South Asian countries. Some had their first taste of environmental legislation as early as the 1970 s, but most did not have any until more recently. International agencies are assisting in the development and passage of their new legislations or amending the old ones. However, most attempts are limited in the enactments rather than implementation and enforcement. The crown jewels of legislative efforts are Maldives, Bhutan and Sri Lanka that have gone beyond the passage of laws and into the implementation phase through regulations and amendments, but it is too early to evaluate their efforts. Nepal passed its solid waste laws in 1987 and more recently in 2011 replacing the former. Bangladesh and Pakistan are the only two countries without specific solid waste legislation although both have provisions for waste management in their comprehensive environmental statutes. The case for India suggests that mere passing environmental laws and adopting rules are not a testimony for progress in terms of sustainable solid waste management. The Comptroller and Auditor General of India in a 2007 investigation found some serious inadequacies of MSW rules and mandates.

References

Abeysuriya, T. K. (2007). "Country reports—Sri Lanka." *Solid-waste management: Issues and challenges in Asia*, Asian Productivity Organization, Tokyo.

ADB (Asian Development Bank). (2007). "Maldives: Environment assessment." Tokyo.

ADB (Asian Development Bank). (2012a). "Fact sheets for individual countries." Manila.

ADB (Asian Development Bank). (2012b). *Key indicators for Asia and the pacific: Highlights*, 43rd Ed., Manila.

Ahmadi, M., Hashim, H. S., Mohamed, A. F., and Moharamnejad, N. (2012). "The institutional and legal arrangements for stakeholder participation in urban waste management in Iran-Tehran case." *6th Int. Symp. on Advances in Science and Technology*, Khavaran Institute of Higher Education, Mashhad, Iran.

APO (Asian Productivity Organization). (2007). "Solid-waste management: Issues and challenges in Asia." Tokyo.

Beck, T., Clarke, G., Groff, A., Keefer, P., and Walsh, P. (2001). "New tools in comparative political economy: The database of political institutions." *World Bank Econ. Rev.*, **15**(1), 165–176.

Bhuiya, G. M. J. A. (2007). "National reports—Bangladesh." *Solid-waste management: Issues and challenges in Asia*, Asian Productivity Organization, Tokyo.

CalRecovery. (2005a). *Solid waste management (volume I: Solid waste management)*, International Environmental Technology Centre, Osaka.

CalRecovery. (2005b). *Solid waste management (volume II: Regional overviews and information sources)*, International Environmental Technology Centre, Osaka.

Carothers, T. (1998). "The rule of law revival." *Foreign Affairs*, **77**(2), 95–106.

Chowdhury, M. (2011). "The U.S. solid waste planning experience and insights for rapidly advancing nations." Proc., Int. Conf. of Solid Waste 2011: Moving Towards Sustainable Resource Management, Sino-Forest Applied Research Centre for Pearl River Delta Environment (ARCPE), Hong Kong Baptist University, Hong Kong SAR.

CLGF (Commonwealth Local Government Forum). (2012). "Country profile." London, U.K.

Comptroller and Auditor General of India. (2008). "Report of the comptroller and auditor general of India for the year ended March 2007: Performance audit of management of waste in India." Rep. No. PA 14, Union Government Scientific Departments, New Delhi, India.

Constitution of the Kingdom of Bhutan. (2008). Royal Government of Bhutan, Thimphu, Bhutan.

Dado, M. E. (2012). "Maldives—Maldives environmental management project: P108078—Implementation status results report: Sequence 10." ⟨http://documents.worldbank.org/curated/en/2012/12/17110988/maldives-maldives-environmental-management-project-p108078-implementation-status-results-report-sequence-10⟩ (Apr. 2, 2013).

Davis, K. E., and Trebilcock, M. J. (2001). "Legal reforms and development." *Third World Quar.*, **22**(1), 21–36.

Department of Urban Development and Engineering Services, Ministry of Works and Human Settlement. (2013). Bhutan Municipal Act of 1999, ⟨http://www.dudh.gov.bt/Towns/Towns.htm⟩ (Jan. 25, 2013).

Enaytullah, I. (1994). "A study of solid waste management for environmental improvement of Dhaka City." M.S. thesis, Bangladesh Univ. of Engineering and Technology, Dhaka, Bangladesh.

EU (European Union). (2010a). "Environmental statistics and accounts in Europe." European Commission, Luxembourg.

EU (European Union). (2010b). "Being wise with waste: The EU's approach to waste management." Publications Office of the European Union, Luxembourg.

European Union External Action. (2014). EU development cooperation with Bhutan, ⟨http://eeas.europa.eu/bhutan/index_en.htm⟩ (Jan. 7, 2014).

Eurostat. (2009). "Waste management in Europe: Framework, trends and issues." European Federation of Public Service Union (EPSU), Luxembourg.

Faundez, J. (2000). "Legal reform in developing and transition countries: Making haste slowly." *Governance, development and globalization*, M. Footer and J. J. Norton, eds., Blackstone, Oxford.

Frischtak, L. (1997). "Political mandate, institutional change and economic reform." *Good government and law: Legal and institutional reform in developing countries*, J. Faundez, ed., Macmillan, London.

Garrick, L. E. (2004). "A historical context of municipal solid waste management in the United States." *Waste Manage. Res.*, **22**(4), 306–322.

GNH (Gross National Happiness Commission), Royal Government of Bhutan. (2012). "Framework to mainstream environment, climate change and poverty (ECP) concerns into the eleventh five year plan (2013–2018)." Thimphu.

Hall, D. (2010). "Waste management in Europe: Framework, trends and issues." Public Services International Research Unit (PSIRU), Rep. European Federation of Public Service Unions (EPSU), London, U.K.

Hassan, R. T. (2012). "The cost of inefficient solid waste management." *Express Tribune*, Karachi, Pakistan.

Hershkowitz, A., and Salerni, E. (1987). *Garbage management in Japan: Leading the way*, Inform, New York.

Hickman, H. L., Jr. (1999). *Principles of integrated solid waste management*, American Academy of Environmental Engineers, Annapolis, MD.

Hickman, H. L., Jr. (2003). *American alchemy: The history of solid waste management in the United States*, Forester Press, California.

Hobbes, T. (1651). *Leviathan*, R. Tuck, ed., Cambridge University Press, Cambridge.

Hussain, D. (2008). "Functional translation of the constitution of the republic of Maldives." Ministry of Legal Reform, Information and Arts, Male, Maldives.

IFC (International Finance Corporation). (2011). "Maldives: Solid waste. Success stories, public-private partnerships." World Bank Group, Washington, DC.

Islam, N., and Shafi, S. A. (2004). "Solid waste management and the urban poor in Dhaka." *Forum on Urban Infrastructure and Public Service Delivery for the Urban Poor: Regional Focus Asia*, Comparative Urban Studies Project, Woodrow Wilson International Centre for Scholars, Washington, DC.

JICA (Japan International Cooperation Agency). (1999). "The study on solid waste management for male' city in the republic of Maldives." Tokyo.

JuriGlobe. (2014). "The World Legal System." ⟨http://www.juriglobe.ca/eng/index.php⟩ (Jan. 7, 2014).

Klyza, C. M., and Sousa, D. (2008). *American environmental policy, 1990-2006*, MIT Press, Cambridge, MA.

Latheefa, A., Shafia, A., and Shafeega, F. (2011). *State of the environment 2011*, Ministry of Environment and Energy, Male.

Mahar, R. B. (2010). "Legal framework and financing mechanisms for waste agricultural biomass (WAB) solid waste in district Sanghar, Pakistan in converting waste agricultural biomass into energy source." United Nations Environmental Programme, International Environmental Technology Centre, Osaka, Shiga.

Melosi, M. V. (2005). *Garbage in the cities: Refuse, reform and the environment*, University of Pittsburgh Press, Pittsburgh, PA.

Merryman, J. H., and Clark, D. S. (1978). *Comparative law: Western european and latin american legal systems*, Bobbs Merill, Indianapolis, IN.

MHTE (Ministry of Housing, Transport and Environment). (2009). "Third national environment action plan." Government of the Maldives, Male.

Ministry of Local Government and Provincial Councils. (2014). ⟨http://www.pclg.gov.lk/en/sub_pgs/institutions_NSWMSC.html⟩ (Jan. 7, 2014).

Naureen, M. (2009). "Development of environmental institutions and laws in Pakistan." *Pak. J. His. Cul.*, **30**(1), 93–112.

NEC (National Environment Commission). (2008). "Royal government of Bhutan. Bhutan environment outlook 2008." Thimphu.

NSWMSC (National Solid Waste Management Support Center) and Japan International Cooperation Agency. (2008). "National solid waste management: Status report 2007." Colombo.

Ogus, A. (2004). "The importance of legal infrastructure for regulation (and deregulation) in developing countries." Center on Regulation and Competition, Institute for Development Policy and Management, Univ. of Manchester, Manchester.

OHS (Ohio Historical Society). (2013). ⟨http://www.ohiohistorycentral.org/w/Cuyahoga_River_Fire?rec=1642⟩ (May 13, 2013).

Pakistan Observer. (2011). "Karachi lacks solid waste management policy." Karachi, Pakistan.

Penjour, Y. (2007). "Enhancing municipal solid waste management system with 3R Options in Thimphu, Bhutan." M.S. thesis, Asian Institute of Technology, Khlong Luang, Pathum Thani, Thailand.

Perry, A. (2001). *Legal systems as a detriment of FDI: Lessons from Sri Lanka*, Kluwer Law International, Alphen aan den Rijn, Hague.

Posner, R. A. (1998). "Creating a legal framework for economic development." *World Bank Res. Obs.*, **13**(1), 1–11.

Ravi, D. (2007). "National reports: India." *Solid-waste management: Issues and challenges in Asia*, Asian Productivity Organization, Tokyo.

Rogowski, R. (1996). *Civil law*, New York University Press, New York.

Rosenbaum, W. A. (2008). *Environmental politics and policy*, 7th Ed., CQ Press, Washington, DC.

Schübeler, P., Wehrle, K., and Christen, J. (1996). "Conceptual framework for municipal solid waste management in low-income countries." UNDP/UNCHS/World Bank, Gallen.

Shekdar, A. V. (2012). "Sustainable solid waste management: An integrated approach for asian countries." *Waste Manage.*, **29**(4), 1438–1448.

Sherwood, R. M., Shepherd, G., and De Souza, C. M. (2009). "Judicial systems and economic performance." *Q. Rev. Econ. Finance*, **34**, 101–116.

Singh, D. (2011). "Cleaning up Thilafushi: Making the Maldives carbon neutral." *Q. J. Int. Finance*, **2**.

Small, W. E. (1970). *Third pollution: The national problem of solid waste disposal*, Praeger Publishers, New York.

Smith, A. (1775). *Essays on philosophical subjects*, W. P. D. Wightman and J. C. Bryce, eds., Clarendon Press, Oxford.

Statistical Yearbook of Bhutan. (2012). *National statistics bureau*, Thimphu, Bhutan.

Tamanaha, B. Z. (1995). "The Lessons of law and development studies." *Am. J. Int. Law*, **89**(2), 470–486.

Tchobanoglous, G., and Kreith, F., eds. (2002). *Handbook of solid waste management*, 2nd Ed., McGraw-Hill, New York.

Trubeck, D. M. (1972). "Max Weber on law and the rise of capitalism." *Wisconsin Law Rev.*, 720–753.

Tuladhar, B. (2007). "County reports–Nepal." *Solid-waste management: Issues and challenges in Asia*, Asian Productivity Organization, Tokyo.

UNDESA (United Nations Department of Economic and Social Affairs, Population Division). (2010). "World urbanization prospects." New York.

UNEP (United Nations Environment Programme). (2002). "Key environmental issues: Management of solid waste and sewage." *Maldives: State of the environment*, Klong Luang, Pathumthani.

UNEP (United Nations Environment Programme). (2005). *Solid waste management (Volume II)*, Osaka.

UNESCAP (United Nations Economic and Social Commission for Asia and the Pacific). (2013). "Country reports on local government systems: Nepal." ⟨http://www.unescap.org/huset/lgstudy/new-countrypaper/Nepal/Nepal.pdf⟩ (Jun. 1, 2013).

United States Code. (1899). "Rivers and Harbours Act of 1899, Title 33, Chapter 9, Subchapter 407." Washington, DC.

U.S. Congress. (1965). "Solid Waste Disposal Act, 42 U.S.C. 3251. P.L. 89-272, 79 Stat. 992." Washington, DC.

U.S. Congress. (1970). "The Resource Recovery Act (PL 91-512, 42 U.S.C 6901 et seq.)." Washington, DC.

U.S. Congress. (1976). "The Resource Conservation and Recovery Act (PL 94-580, 42 U.S.C 6901 et seq.)." Washington, DC.

U.S. EPA. (2013). "Municipal solid waste in the United States: 2011 Facts and figures." Washington, DC.

USPHS (United States Public Health Service). (1964). "Surgeon general's advisory committee on urban health affairs. Solid waste handling in Metropolitan areas." Washington, DC.

Vig, N. J., and Kraft, M. E. (2010). *Environmental policy: New directions for the twenty-first century*, 7th Ed., CQ Press, Washington, DC.

Weber, M. (1925). *Law in economy and society*, M. Rheinstein, ed., Harvard University Press, Cambridge, MA.

Williamson, O. E. (1985). *The economic institutions of capitalism: Firms, markets, relational contracting*, Free Press, New York.

World Bank. (1992). *Governance and development*, Washington, DC.

World Bank. (1997). *World development report 1997: The state in a changing world*, Oxford University Press, New York.

Zhu, D., Asnani, P. U., Zurbrägg, A., and Mani, S. (2008). *Improving municipal solid waste management in India: A sourcebook for policy makers and practitioners*, International Bank for Reconstruction and Development/World Bank, Washington, DC.

CHAPTER 22

LCA for Decision-Making in Solid Waste Management

Sébastien Bonnet[1]
Shabbir H. Gheewala[2]

22.1 INTRODUCTION

Wastes are produced at all level of society, from households to industrial facilities at an ever increasing rate and managed at various levels from local government to private organizations. For instance, municipal solid waste (MSW) has been growing faster than the rate of urbanization. This is because waste is a by-product of human lifestyle which in the industrialized world has evolved to a consumption based society. As reported by the World Bank (2012), in over a decade from 2.9 billion urban residents generating 0.68 billion tonnes of solid waste per year, the population has grown to about 3 billion residents generating about 1.3 billion tons per year. It is predicted that following this trend of urbanization and waste generation rate, by 2025, there will be 4.3 billion residents living in urban areas contributing 2.2 billion tonnes of solid waste per year. This tremendous increase in solid waste generation will further impact the environment and humans that come in contact with it or with the resulting pollution.

Poorly managed waste leads to substantial impacts on human health, the environment and of course the economy and so to unsustainable development of societies. Therefore, the greatest challenge to both the public and private sector is cost-effective and environmentally efficient management of wastes. From a sustainability perspective, waste generation results from inefficient use of natural resources. In the life cycle context this means wastes are generated from extraction of natural resources, processing activities leading to products manufacturing, packaging, transportation, consumption and ultimate disposal of remaining fractions. Such a holistic approach of looking at waste

[1]King Mongkut's Univ. of Technology Thonburi, Thailand.
[2]King Mongkut's Univ. of Technology Thonburi, Thailand.

generation through the various stages of the supply - chain, has been looked at as a means to reduce impacts from waste generation and disposal. This approach is termed "Life Cycle Assessment" or LCA and focuses primarily in assessing the relative overall environmental impacts of a particular waste production and/or management system and providing possible options for improvement. The results from an LCA can therefore lend to the overall objective of sustainable waste management in providing basic environmental information to policy makers to identify sustainable improvement options. This means coming up with waste minimization and/or management alternatives that are environmentally friendly, cost effective and socially acceptable (Liamsanguan and Gheewala 2008a; Del Borghi et al. 2009; Weitz 2012; Menikpura et al. 2013).

LCA provides a key piece to understanding the broad environmental aspects of waste management and is to be combined with technological, economic and social information to make effective decisions. LCAs of waste management targeting decision making (including policy) focus on investigating and assessing the performance of waste management alternatives. However, past decisions on waste management have been linked to the underlying concept of the waste hierarchy as promoted in EU or US (USEPA 2012; Cleary 2009). This concept stipulates that a priority order should be respected when assessing and making recommendations for improvement of waste management. The waste hierarchy concept relies on the underlying assumption that waste prevention is favored before re-use, recycling, recovery and landfilling as the last option. It assumes a superiority of environmental performance based on this hierarchy principle which induces LCA studies and results to be designed and interpreted in line with this concept. LCAs performed on the basis of such pre-defined conditions may result in narrowing down, in particular cases, the set of strategies available for improving a particular solid waste management system and therefore related decision making. In this chapter, the concept and use of LCA for environmental performance evaluation of solid waste management, and the potential and challenges of such a holistic approach for decision making are reviewed and discussed.

22.2 POTENTIAL AND CHALLENGES OF LCA FOR SOLID WASTE MANAGEMENT AND DECISION MAKING

22.2.1 Solid Waste Generation and Environmental Impacts

22.2.1.1 Global Overview

As of 2002, the global quantity of solid waste generation amounted to 12 billion tonnes of which 11 billion tonnes was industrial wastes and 1.6 billion tonnes were MSW (Pappu et al. 2007). About 19 billion tonnes of solid wastes are expected to be generated annually by the year 2025. Solid waste streams can be characterized by their sources, by the types of wastes produced, as well as by generation rates and

Table 22.1. Waste generation per capita categorized by income level

Income level	Waste Generation Per Capita (kg/capita/day)		
	Lower Boundary	Upper Boundary	Average
High	0.7	14	2.1
Middle	0.13	5.4	0.99
Lower	0.09	4.3	0.6

Note: Adapted from World Bank (2012)

composition. Accurate information in these three areas is necessary in order to monitor and control existing waste management systems and to make regulatory, financial, and institutional decisions (World Bank 1999). Knowledge of the composition of a particular solid waste stream is crucial to the design of an appropriate treatment system. Generally, in most developed countries the highest fraction of MSW, consists of paper up to 70%, while food waste represents the highest fraction in developing countries, up to 80% (Tchobanoglous et al. 1993).

Waste generation rates are affected by socio-economic development, degree of industrialization and climate. The generation of solid waste differs from nation to nation according to economic status, population density, urban lifestyles, food habits, geographical conditions and other socio-economic and cultural factors. Generally, the greater the economic prosperity and higher the percentage of urban population, the greater the amount of solid waste produced (Al-Khatib et al. 2010; Kathiravale and Yunus 2008). An illustration of waste generation rate based on income level is shown in Table 22.1.

22.2.1.2 Environmental Impacts

As already mentioned, solid wastes are produced in proportion to the economic productivity and consumption rate of the population of a particular country. In high income countries, larger amounts of wastes are produced, particularly packaging waste, and high end technologies are used for treatment and ultimate disposal. In low income countries, a large fraction of organic waste is produced and simple treatment technologies are used with open dumping dominating as the ultimate disposal method (Shekdar 2009; Kathiravale and Yunus 2008).

MSW, a fraction of solid waste of major concern for many countries worldwide, is known to contribute to global warming and climate change. This is because of the greenhouse gases (GHGs) emitted from the collection, transportation and final disposal of MSW. Thus, for instance, methane emissions from open dumping and landfilling of MSW contribute a major share of the overall anthropogenic emission of methane, which is a foremost GHG (IPCC 2007). The leachate generated from the degradation of solid waste in landfills is characterized by a high organic load in addition to containing many toxic pollutants which can easily contaminate surface water bodies and groundwater. These also create severe problems in terms of environmental pollution (including eutrophication and acidification issues) and human health (Wanichpongpan and Gheewala 2007; Kathiravale and Yunus 2008; Liamsanguan and Gheewala 2008b). More sophisticated MSW treatment methods, such as incineration and composting, also create some environmental and health effects such as generation of toxic fly ash, flue gas containing heavy metals, dioxins and furans from incinerators and production of GHGs, odor, and leachate from inappropriate composting piles. The potential environmental degradations caused by different solid waste management methods are summarized in Table 22.2 for illustrative purposes.

Solid waste management not only results in environmental impact but also economic and social problems, which hinder the development of sustainable waste management system. To evaluate the performance of alternative MSW management systems, a life cycle approach can be used as a useful method to make appropriate strategies as to which waste management strategy to opt for (CALCAS 2011; Eriksson et al. 2005). By applying life-cycle approach, priorities can be identified more easily and policies can be targeted more effectively so that the maximum benefit can be achieved relative to the effort expended (Buttol et al. 2007). Details of the LCA tool and its application for solid waste management and decision making are provided in the next sections.

22.2.2 Concept of LCA in the Context of Solid Waste Management

22.2.2.1 Birth of the Tool

The concept of Life Cycle Assessment (LCA) goes back to the 1960 s when scientists started to be preoccupied by the rapid consumption and therefore depletion of fossil fuel resources. LCA was developed as an approach to understand the impacts associated to energy consumption. In the late 1970 s and early 1980 s, environmental concern shifted to issues of hazardous waste management. As a result, life cycle logic was incorporated into the emerging method of risk assessment, which was used with increasing frequency in the public policy community to develop environmental protection standards. When solid waste became a worldwide issue in the late 1980 s, the life cycle analysis method developed in the REPA studies again became a tool for analyzing the problem. In 1990, for example, a life cycle assessment was completed for the Council for Solid Waste Solutions, which compared the energy and environmental impacts of paper to that of plastic grocery bags (CSWS 1990). Thereafter, environmental groups around the world adopted life cycle analysis for environmental assessment of products' impacts. Initially it was limited to the public sector, but little by little the LCA concept was adopted by an increasing number of corporations and non-profit organizations as an aid to understanding the environmental impacts of their actions (Svoboda 1995). In 1999, the International Organization for Standardization (ISO) developed the standard ISO 14040 series on Environmental Management – Life Cycle Assessment.

Table 22.2. Major environmental impacts from MSW management

	Environmental damage				
Type of technology	*Damage to Water*	*Damage to Air*	*Damage to Soil*	*Damage to Landscape*	*Damage to Climate*
Landfilling	Leachate (heavy metals, synthetic org. compounds)	CO_2, CH_4, odour, noise, VOCs	Heavy metals, synthetic org. compounds	Visual effects, vermin	Worst option for GHG emissions
Incineration	Fall-out of atmospheric pollutants	SO_2, NOx, N_2O, HCl, HF, CO, CO_2, dioxins, furans, PAHs, VOCs, odour, noise	Fly ash, slag	Visual effects	GHG emissions
Composting	Leachate	CO_2, CH_4, VOCs, dust, odour, bioaerosols	Minor impact	Some visual effects	Small GHG emissions
Anaerobic digestion	Leachate	VOCs, NH_3, H_2S, odour	Minor impact	Some visual effects	Small GHG emissions
Land spreading	Bacteria, viruses, heavy metals	Bioaerosols, dust, odour	Bacteria, viruses, heavy metals, PAHs, PCBs	Vermin, insects	Small GHG emissions
Recycling	Wastewater	Dust, noise	Land filling of residues		Minor GHG emissions
Waste transportation	Spills	CO_2, SO_2, NO_x, SO_x, dust, odour, noise, spills	Spills		Significant contribution from CO_2

Source: Alvarez et al. 2000; Giusti 2009

22.2.2.2 Assessment Framework

LCA can be defined as a tool enabling to assess the environmental performance of products or services, regardless of where a particular operation is located or who has ownership of environmental pollution. LCA therefore takes a comprehensive global perspective of the system assessed. According to ISO 14040, LCA is "a technique for assessing the environmental aspects and potential impacts associated with a product, service or function by compiling an inventory of relevant inputs and outputs of a product system" (McDougall et al. 2001). It enables to evaluate the potential environmental and health impacts of those inputs and outputs and interpret the results in relation to the objectives of the study. Rebitzer et al. (2004) has defined LCA as a methodological framework for estimating and assessing the environmental impacts attributable to the life cycle of a product. Common mid-point level impact categories assessed include, climate change (global warming), stratospheric ozone depletion, eutrophication, acidification, abiotic resource depletion, toxicity and so on. Because of its standard and methodological structure, LCA can therefore be applied as a tool to characterize the environmental performance and potential impacts of various waste management alternatives.

According to the ISO 14040, the methodology for conducting LCA includes four main phases that are (1) goal and scope definition, (2) inventory analysis, (3) impact assessment and (4) interpretation (McDougall et al. 2001). The life cycle assessment framework can be represented as illustrated in Figure 22.1.

An explanation of these LCA terms and their importance in the context of waste management is provided below.

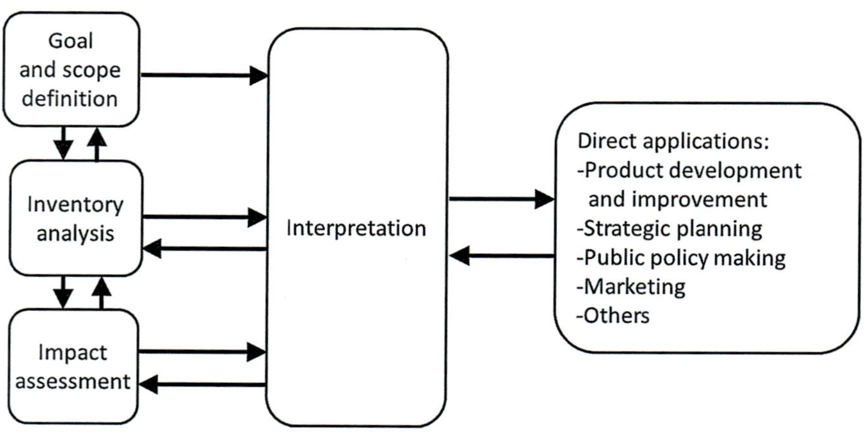

Figure 22.1. Life cycle assessment framework according to ISO 14040
Source: ISO (2006); reproduced with permission from International Standards Organization

22.2.2.3 Goal and Scope Definition

It is the first step in which the initial choices that determine the working plan of the entire LCA are made (Guinée et al. 2001). The goal of a LCA study shall clearly state the intended application, the reason for carrying out the study and who the target audience is (McDougall et al. 2001).

When the LCA is presented, the objective(s) of the study should be described and the scope identified. According to ISO 14040, the scope of the study should cover among other, the functional unit, the system boundary, geographical location and time scale. Other considerations to be included are the methodological choices and assumptions/limitations, the sources and quality of the data used, and the format of reporting (McDougall et al. 2001; Guinée et al. 2001).

With regard to waste management, LCA enables to assess the environmental impacts associated with a particular waste management system and trade-offs as compared to waste treatment alternatives. It also enables to evaluate the environmental profile of a particular waste management system but also the impacts associated to each stage and each contributing sources of the life cycle of that system. This information is valuable to identify options enabling to minimize environmental impacts from most contributing life cycle stages and therefore to optimize the overall performance of a waste management system through a series of possible alternatives.

The functional unit defined under the goal and scope of an LCA is a very crucial parameter to define as it provides the basis on which the assessment is made. The functional unit describes the primary function(s) fulfilled by the product (waste in this case) of the system studied, and indicates how much of this function is to be considered in the intended LCA study. It is used as a basis for selecting one or more alternatives to the system considered to satisfy the same function. The functional unit also enables different systems to be treated as functionally equivalent that allow reference flows to be determined for each of them (Guinée et al. 2001). In relation to solid waste management, a quantity based functional unit referring to the amount of a given composition of solid waste treated for a particular location is commonly used.

Under the scope of the study, it is also important to identify the boundary of the system studied, i.e. the life cycle stages and processes included and those not included in the assessment, the geographical boundary, and time scale under which the LCA results are valid. Defining adequately such a scope is critical as it conditions the extent of the assessment and the associated data requirement for it. LCAs for waste management with a view to provide options for policy making are usually focused on analysis of alternative scenarios or strategies (typically, all basic activities associated to solid waste management are included, i.e. waste collection and transportation, recovery and sorting of waste, treatment and ultimate disposal). Generally, also, in terms of geographical boundary, LCA of waste management are site specific from a community level to a regional level. With regard to time scale, most LCAs of waste management are based on a one year time boundary. However, for landfill much longer time scale can be considered to

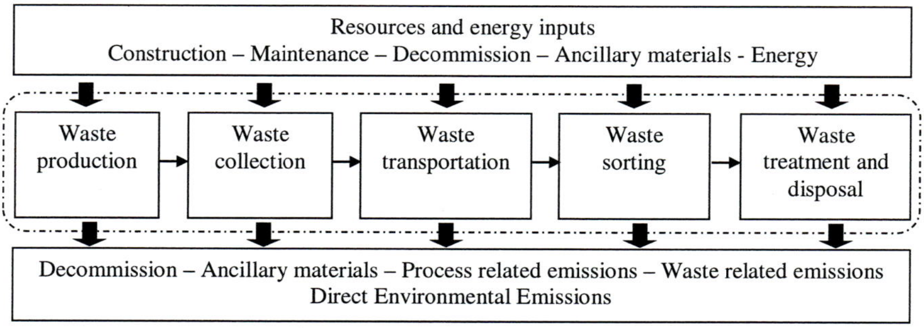

Figure 22.2. Defining the system boundary for life cycle inventory analysis

reflect the time over which emissions from the decomposition of waste may occur (Weitz 2012).

22.2.2.4 Inventory Analysis

This step involves data collection and calculation procedures to quantify the relevant inputs and outputs of a particular system. Thus, in the case of a solid waste management system, all of the raw materials and energy inputs and outputs (releases to air, water and land associated with the system) over the whole life cycle (cradle to grave) should be accounted for as illustrated in Figure 22.2. At this stage, setting the system boundary, designing the flow diagrams with unit processes, collecting data for each of these processes, performing allocation for multifunctional processes and completing the final calculations (Guinée et al. 2001) has to be done. The results of this phase generate an inventory database of all inputs out outputs of the system studied and so of the environmental burdens associated with the functional unit. These burdens can be segregated by life cycle stage, by media (air, water, land) or by specific process.

22.2.2.5 Life Cycle Impact Assessment

This step aims at a further interpretation of the life cycle inventory data and to analyze the environmental burdens associated with the material and energy flows determined in the inventory analysis. Life Cycle Impact Assessment or LCIA in short, includes the creation of impact categories, the assignment of inventory data to specific impact categories, and the modeling of the magnitudes of the inventory data within impact categories and (Bengtsson and Steen 2000). Hence Impact categories, category indicators, characterization models and equivalency factors (McDougall et al. 2001) are the important components of LCIA to measure the potential impacts on environment. LCIA may also include the valuation or weighting of impacts in order to estimate the environmental performance of a particular scenario. The results produced from a LCIA are more useful to policy-makers than is the life cycle inventory (Craighill and Powell 1996), as it provides information about environmental impacts for different categories each defined by a specific indicator. With regards to LCIA of waste management, the most common mid-point impact categories include; Global warming, Acidification, Eutrophication and Abiotic resource depletion. Toxicity impacts in the form of human toxicity, and eco-toxicity are less frequent (Weitz 2012; Cleary 2009). Also the weighting of environmental impacts to be able to make a comparative assessment across various impact categories, for instance to compare the relative importance of the impact of a waste management strategy on global warming as compared to eutrophication can be performed. However, this introduces large uncertainties in the results obtained due to the subjectivity associated with the weighting factor chosen. Although the end result may be of obvious interest for policy making, the lack of a scientific basis for the application of weighting techniques that reduce LCA results to a single index value is not recommended when intended for comparative assertions of waste strategies that are to be disclosed to the public (ISO 2006).

22.2.2.6 Interpretation

This phase of the LCA focuses on combining the findings from the inventory analysis and the impact assessment, consistent with the goal and scope in order to reach the conclusions and the recommendations. This may involve the iterative process of reviewing/revising the scope of the LCA, as well as the nature and the quality of the data collected consistent with the defined goal (McDougall et al. 2001). This part of the LCA process when assessing solid waste management systems focuses on the analysis of various strategies and the formulation of recommendations that can be linked to policy decision making if that is the intended purpose of the LCA study.

22.3 APPLICATION OF LCA FOR SOLID WASTE MANAGEMENT

22.3.1 Lessons Learnt from Review Studies

Over the past decade, there have been numerous LCA studies to analyze life cycle emissions of municipal solid waste (MSW) management systems most specific to a waste stream at local (community) or regional level and some at national level. Based on this information, reviews have been published discussing and comparing the goal, scope, methods and results obtained from such LCA studies. The purpose of these reviews were in general to highlight some of the common trends and challenges associated to SWM and the use of LCA as the evaluation tool (Cleary 2009; De Borgi et al. 2009; Weitz 2012; Barlaz and Dunn 2009; RTI international 2010). Some highlights are provided here for selected LCA review works. Weitz for instance reviewed 18 LCA papers that fulfilled the following criteria (1) use LCA methodologies and are either an LCA or provide an LCI (2) deal with end-of-life resource management process that includes at least one energy recovery process. The impact categories considered were abiotic resource depletion, acidification, energy consumption, eutrophication, eco-toxicity, global warming, human toxicity and photo-oxidant formation. The results of the study showed that recycling performs better for all the impact categories considered followed by waste-to-energy treatment and then landfilling with or without energy recovery. However, Weitz (2012) reported that the system boundary and assumptions of each of the LCA studies reviewed varied largely, which of course influence the results obtained. However, some key basic highlights could be made from this review of LCA studies, which hold for most waste management situations, as detailed below.

- Landfill is among the least preferred options for solid waste management with respect to environmental impacts which is in line with the waste hierarchy concept.
- Recycling, although the most favorable option in terms of environmental impacts, is only an appropriate option under specific conditions. There are issues of waste sorting at source, collection efficiency, material suitability, contamination with other materials, logistics, costs and others, that would lead to recycling being less performant than waste-to-energy management.
- The environmental performance of recycling depends on the type and amount of material substituted by recyclables. Recycling performs better when it is assumed that recycled materials replace an equivalent amount of virgin material.
- The difference in environmental performance between recycling and waste-to-energy narrows down when WTE substitutes electricity and heat generation using fossil fuels. Of course energy conversion efficiency and calorific content of feedstock greatly influence WTE environmental performance.

Cleary (2009) examined a total of 20 process-based life cycle assessments of MSW management systems. All of these LCAs focused on waste management systems that addressed residential or commercial (non-industrial) solid waste, including at least, plastic, paper, and organic materials. Only studies that addressed a wide array of environmental emissions and/or impacts were selected by Cleary (2009) to be part of this analysis. In general, all the studies reviewed were comparative LCAs evaluating the environmental emissions and/or performance of various types of solid waste management systems. Some compared one scenario with a defined baseline, often landfilling, while most compared a number of options to one another. Most of the reviewed LCAs included landfilling, recycling, and thermal treatment of MSW, with fewer studies modeling composting and anaerobic digestion. Basic LCA criteria were looked at including, definition of goal and scope, functional unit, geographical scale, waste management system boundary, types and sources of data, and environmental impacts. The following observations were made:

- Most waste management LCA studies are site or region specific. This is in line with the nature of the assessment which usually concerns assessing and/or improving the environmental performance of a particular waste management system.

- With regard to the definition of goal, scope functional unit and system boundary, gaps and incomplete information were noticed for most of the studies reviewed. The ISO 14044 specifies that such aspects need to be clearly defined at the onset of any LCA studies to assure a proper and well informed assessment. These gaps reveal the difficulties in complying with ISO 14044 standard and the potential issue this may lead to in terms of environmental performance assessment and so appropriate waste management decision making.
- Concerning LCIA, the main impact categories concerned include abiotic resource depletion, global warming, acidification and eutrophication. Toxicity assessment, including human toxicity and eco-toxicity are much less popular and were not part of the assessment in more than half of the studies reviewed.
- In terms of weighting of impacts, to make comparisons across environmental impact categories, most reviewed studies did not perform such assessment. In fact, ISO (2006) advises LCA practitioners to acknowledge the lack of a scientific basis for the application of weighting techniques which can also reduce LCA results to a single overall score to compare overall environmental performance of alternative scenarios.
- Sensitivity analysis is usually not included in waste management LCA studies. Sensitivity analysis is useful in assessing the effects of altering selected model or assessment parameters.
- With regard to data sources, most studies relied on emission databases rather than data from direct measurements from a particular waste treatment facility. This may in some cases lead to appropriateness issues of the data being used. This is to be clearly identified in the assumptions of the study as part of the goal and scope.
- Most of the reviewed LCA studies were found to have make use of LCA computer models to facilitate the modeling of the waste management systems and the calculation of environmental emissions and impacts. The LCA models include the US EPA Decision Support Tool, IWM-1, IWM-2, Simapro, WISARD, ORWARE, EASEWASTE and ISWM.

This review of LCAs indicates that there is still a challenge in following the general requirements and guidelines for LCA as stipulated by the ISO (2006). A major issue concerns the identifications of system boundary and restricted use of impact categories, unlike what is required by ISO (2006) for LCIAs intended to be used in comparative assertions for public disclosure. Despite these gaps and several other associated to datasets, methodologies and computer models, most of the LCA results converged to confirm the validity of the waste hierarchy principle as also observed by Weitz (2012) in his review.

22.3.2 Computer Models

The increased number of solid waste management LCAs has been accompanied with the development of a substantial number of LCA computer models

(Gentil et al. 2010). These models have been developed notably to help waste managers and policy makers assessing the environmental impacts of a specific solid waste management system without requiring them to have an in-depth knowledge of the LCA methodology. These models were also designed to enable understanding of how changes in the system studied affect environmental impacts through scenario analysis. Therefore these models were also developed to help with the process of decision making in solid waste management (Del Borghi et al. 2009). It is generally agreed that a good LCA computer model should be user-friendly, flexible to allow entry and recording of user's data, and with good and easy to understand graphical/tabular display of life cycle impact results (Winkler and Bilitewski 2007; ICF Consulting Canada 2007). A selection of ten popular models for solid waste management is presented in Table 22.3 where a summary of their respective characteristics and limitations is provided.

The use of computer models has therefore become popular for the LCA assessment of solid waste management and many published literature refer to the use of such models to facilitate the calculations of environmental impacts from solid waste management systems.

22.4 POTENTIAL OF LCA FOR SOLID WASTE MANAGEMENT DECISION MAKING

22.4.1 Contribution to the Decision Making Framework

There are many tools for assessment of environmental impacts, but as illustrated in above sections, LCA is a useful concept and tool for environmental analysis of solid waste management. It helps expanding the perspective beyond the waste management system. This is important since the environmental consequences of waste management often depend more on the impacts on surrounding systems than on the emissions from the waste management system itself (Ekvall 1999). The broad system perspective makes LCA a powerful tool for environmental comparison of different options for waste management of a specific product, a material, or a complex waste flow. However, the definition of a scientifically sound baseline is necessary for decision making to compare the magnitude of impacts relative to the baseline and determine whether or not they are acceptable. This baseline is essential for decision making to answer the question "is this MSW management system sustainable?"

Apart from environmental issues, economic and social aspects are also important decisive factors for sustainable MSW management to provide a complete picture for decision making (Aye and Widjaya 2006; Finnveden et al. 2009). As stated by Weitz (2012), key aspects of the decision making process in waste management include:

- Life cycle cost of the waste management system assessed
- Technical feasibility based on waste composition and generation rate, and technology development

Table 22.3. Characteristics of popular LCA computer models used for solid waste management

Model	*Developer*	*Software*	*Scope*	*MSW Focus*	*User-friendliness*	*Flexible*	*Limitations*	*References*
EASEWASTE	Danmarks Techniske Universitet (DTU), Denmark	Danish LCA-Database	LCI/LCIA	Yes	Poor/Fair	Yes	Lacking a recycling use sector Emphasis on household waste	Del Borghi et al. (2009); ICF Consulting Canada (2007)
GaBi4.4	PE Europe GmbH	Database	LCI/LCIA	No	Fair	Yes	For experienced LCA practitioners	ICF Consulting Canada (2007)
Integrated Waste Model (IWM)	Corporations Supporting Recycling and Environmental Plastics Industry Council	SQL Database	LCI	Yes	Poor/Fair	No	Emission factors outdated Only one scenario can be quantified at a time	ICF Consulting Canada (2007); Winkler and Bilitewski (2007)
MSW Decision Support Tool (DST)	U.S. EPA and RTI International	Database in Excel	LCI/LCIA	Yes	Poor	Yes	This software is only run at RTI, EPA and University of Carolina Application to U.S. EPA is needed if user has to use the software	ICF Consulting Canada (2007); Winkler and Bilitewski (2007)

(Continued)

Table 22.3. Characteristics of popular LCA computer models used for solid waste management (Continued)

Model	*Developer*	*Software*	*Scope*	*MSW Focus*	*User-friendliness*	*Flexible*	*Limitations*	*References*
ORWARE	Swedish Environmental Protection Agency	Mathematical software MatLab with the tool Simulink	LCI/LCIA	Yes	Poor	Yes	Specifically deals with waste, and does not encompass energy or transportation Incorporates non-established technologies Data availability, data quality/uncertainty Third party reviews of software indicate scale issues	ICF Consulting Canada (2007); Winkler and Bilitewski (2007)
SimaPro 7.2	Product Ecology Consultants B.V.	Database	LCI/LCIA	No	Good	Yes	Lack of graphical interface Sensitivity analysis was mentioned as an area of concern	ICF Consulting Canada (2007)
UMBERTO	Institute for Energy and Environmental Research Heidelberg	Borland Database Engine	LCI/LCIA	No	Excellent	Yes	Extremely broad scope	ICF Consulting Canada (2007); Winkler and Bilitewski (2007)

WARM	ICF International	Excel database	LCA for GHG emissions	Yes	Excellent	Yes	Model designed as a LCA tool for GHG emissions only	ICF Consulting Canada (2007)
WISARD	Environmental Agency of England and Wales	Database	LCI/LCIA	Yes	Good	Yes	The software has limitations in its assessment of environmental impacts	ICF Consulting Canada (2007)
WRATE	Environmental Agency, UK	Database	LCI/LCIA	Yes	Fair	Yes	Only UK technologiesLong calculation time	ICF Consulting Canada (2007)

- Environmental performance based on LCA results and environmental impact assessment
- Market opportunities for sale/recovery of energy and by-products from a waste management system
- Social acceptability and participation

All these aspects need to be integrated as part of a common sustainability assessment framework for solid waste management. LCA as one piece of this overall decision making framework has gained in acceptance as a tool for waste management planning and policy-making. It is now being used in various contexts, ranging from local planning to policy making at national and international levels.

22.4.2 Challenges and Limitations

Waste management is a complex process with a range of environmental as well as socio-economic consequences for society. The holistic nature of LCA makes it a powerful tool for environmental assess is based on a standard method, it is important to recognized its limitations. The environmental information obtain from LCAs is not complete nor absolutely objective or accurate. These issues have been addressed by Ekvall and co-workers (2007) and are summarized below to illustrate the limitation of the tool for decision making. First of all, as also indicated earlier, LCAs should follow the basic steps of ISO14040. However, while following the standard, methodological choices are still to be made to perform the studies. These are based on expert judgments from LCA practitioners. Those choices include:

- Choice of time perspective
- Assumptions made in the study
- Sources of input data
- Allocation of environmental burdens to different life cycles
- Modeling of environmental impacts

Because of the nature of these choices, the results obtained from LCA studies cannot provide answers that are truly objective. However, following the LCA standard enables to give an estimate of potential environmental impacts under the specified conditions of the study and so to provide guidance for decision making of alternative strategies that are valid under those conditions. The challenge for this case, as illustrated in section 22.3, concern the gaps still observed from many LCA studies on solid waste management in adequately defining the scope and goal, functional unit, and system boundaries of the solid waste management systems assessed. Another, challenge is related to inherent characteristics of LCA that prohibit it from giving adequate answers to certain questions. As reported by Ekvall and co-workers (2007), this includes (1) temporal and spatial limitations, traditional LCA models are static (provide results for current time) and provide impacts at a global level (not site- specific impacts). These restrict the long-term and site specific validity of the decisions made based on the results of a particular solid waste management. Also LCA models are typically linear steady-state models of physical flows. Therefore LCA results can only indicate what waste-management option contributes the least to different environmental impacts. Another most obvious limitation is that only environmental impacts are addressed in traditional LCAs. Also as reported by Gheewala (2009), addressing the long-term sustainability of a waste management system requires knowledge of the financial costs and social impacts of available waste management options. Such aspects need therefore to be included along with LCA to broaden the depth of the analysis and relevance for decision making so that the waste management strategy selected for a particular system is environmentally friendly, cost effective and socially acceptable.

References

Al-Khatib, I. A., Monou, M., Salam, F. A., Zahra, A., Shaheen, H. Q., and Kassinos, D. (2010). "Solid waste characterization, quantification and management practices in developing countries. A case study: Nablus district—Palestine." *J. Environ. Manage.*, **91**(5), 1131–1138.

Alvarez, J. M., Macè, S., and Llabrès, P. (2000). "Anaerobic digestion of organic solid wastes. An overview of research achievements and perspectives." *Bioresour. Technol.*, **74**(1), 3–16.

Aye, L., and Widjaya, E. R. (2006). "Environmental and economic analyses of waste disposal options for traditional markets in Indonesia." *Waste Manage. (Oxford)*, **26**(10), 1180–1191.

Barlaz, M., and Dunn, S. (2009). "A comparison of alternative solid waste management practices." North Carolina State Univ., Raleigh, NC.

Bengtsson, M., and Steen, B. (2000). "Weighting in LCA—Approaches and applications." *Environ. Prog.*, **19**(2), 101–109.

Buttol, P., Masoni, P., Bonoli, A., Goldoni, S., Belladonna, V., and Cavazzuti, C. (2007). "LCA of integrated MSW management systems: Case study of the Bologna district." *Waste Manage.*, **27**(8), 1059–1070.

CALCAS (Co-ordination Action for innovation in Life-Cycle Analysis for Sustainability). (2011). ⟨http://www.calcasproject.net⟩ (Jul. 1, 2013).

Cleary, J. (2009). "Life cycle assessments of municipal solid waste management systems: A comparative analysis of selected peer-reviewed literature." *Environ. Int.*, **35**(8), 1256–1266.

Craighill, A. L., and Powell, J. C. (1996). "Lifecycle assessment and economic evaluation of recycling: A case study." *Resour. Conserv. Recycl.*, **17**(2), 75–96.

CSWS (Council for Solid Waste Solutions). (1990). "Resource and environmental Profile analysis of polyethylene and unbleached paper grocery sacks." Washington, DC.

Del Borghi, A., Gallo, M., and Del Borghi, M. (2009). "A survey of life cycle approaches in waste management." *Int. J. Life Cycle Assess.*, **14**(7), 597–610.

Ekvall, T. (1999). "Key methodological issues for life cycle inventory analysis of paper recycling." *J. Cleaner Prod.*, **7**(4), 281–294.

Ekvall, T., Assefa, G., Bjoerklund, A., Eriksson, O., and Finnveden, G. (2007). "What life-cycle assessment does and does not do in assessments of waste management." *Waste Manage.*, **27**(8), 989–996.

Eriksson, O., et al. (2005). "Municipal solid waste management from a systems perspective." *J. Cleaner Prod.*, **13**(3), 241–252.

Finnveden, G., et al. (2009). "Recent developments in life cycle assessment." *J. Environ. Manage.*, **91**(1), 1–21.

Gentil, E. C., et al. (2010). "Models for waste life cycle assessment: Review of technical assumptions." *Waste Manage.*, **30**(12), 2636–2648.

Gheewala, S. H. (2009). "LCA of waste management systems - research opportunities." *Int. J. Life Cycle Assess.*, **14**(7), 589–590.

Giusti, L. (2009). "A review of waste management practices and their impact on human health." *Waste Manage.*, **29**(8), 2227–2239.

Guinée, J. B., et al. (2001). "Life cycle assessment—An operational guide to the ISO Standards." Ministry of Housing, Spatial Planning and the Environment (VROM), and Centre of Environmental Science, Leiden Univ. (CML), Netherlands.

ICF Consulting Canada. (2007). "Review of life cycle assessment (LCA) tools for municipal solid waste (MSW) management." Canada.

IPCC. (2007). "Contribution of working group III to the fourth assessment report of the intergovernmental panel on climate change." Cambridge University Press, Cambridge.

ISO (International Organization for Standardization). (2006). "Environmental management– Life cycle assessment– Requirements and guidelines." Geneva, Switzerland.

Kathiravale, S., and Yunus, M. N. (2008). "Waste to wealth." *Asian Eur. J.*, **6**(2), 359–371.

Liamsanguan, C., and Gheewala, S.H. (2008a). "LCA: A Decision support tool for environmental assessment of MSW management systems." *J. Environ. Manage.*, **87**(1), 132–138.

Liamsanguan, C., and Gheewala, S.H. (2008b). "The holistic impact of integrated solid waste management on greenhouse gas emissions in Phuket." *J. Cleaner Prod.*, **16**(17), 1865–1871.

McDougall, F. R., White, P. R., Franke, M., and Hindle, P. (2001). *Integrated solid waste management: A life cycle inventory*, 2nd Ed., Blackwell Science, Malden, MA.

Menikpura, S. N. M., Gheewala, S. H., Bonnet, S., and Chiemchaisri, C. (2013). "Evaluation of the effect of recycling on sustainability of municipal solid waste management in Thailand." *Waste Biomass Valorization*, **4**(2), 237–257.

Pappu, A., Saxena, M., and Asolekar, S. R. (2007). "Solid wastes generation in India and their recycling potential in building materials." *Build. Environ.*, **42**(6), 2311–2320.

Rebitzer, G., et al. (2004). "Life cycle assessment. 1: Framework, goal and scope definition, inventory analysis, and applications." *Environ. Int.*, **30**(5), 701–720.

RTI international. (2010). "Literature review of LCA studies for end-of-life management processes, focusing on energy recovery processes and plastics." American Chemistry Council, Washington, DC.

Shekdar, A. V. (2009). "Sustainable solid waste management: An integrated approach for Asian countries." *Waste Manage.*, **29**(4), 1438–1448.

Svoboda, S. (1995). "Note on life cycle analysis." National Pollution Prevention Center for Higher Education, Univ. of Michigan, Ann Arbor, MI.

Tchobanoglous, G., Theisen, H., and Vigil, S. A. (1993). *Integrated solid waste management: Engineering principles and management issues*, McGraw Hill, Singapore.

USEPA. (2012). "Solid waste management hierarchy." ⟨http://www.epagov/wastes/nonhaz/municipal/hierarchy.htm⟩ (Jul. 1, 2013).

Wanichpongpan, W., and Gheewala, S. H. (2007). "Life cycle assessment as a decision support tool for landfill gas-to energy projects." *J. Cleaner Prod.*, **15**(18), 1819–1826.

Weitz, K. A. (2012). "Life cycle assessment and end of life materials management." *Life cycle assessment handbook– A guide for environmentally sustainable products*, M. A. Curran, ed., Scrivener Publishing LLC, MA, 249–266.

Winkler, J., and Bilitewski, B. (2007). "Comparative evaluation of life cycle assessment models for solid waste management." *Waste Manage. (Oxford)*, **27**(3), 1021–1031.

World Bank. (1999). "What a waste: Solid waste management in Asia." ⟨http://www.worldbank.org⟩ (Jul. 1, 2013).

World Bank. (2012). "What a waste: A global review of solid waste management." ⟨http://www.worldbank.org⟩ (Jul. 1, 2013).

Index

Page numbers followed by *e*, *f*, and *t* indicate equations, figures, and tables.